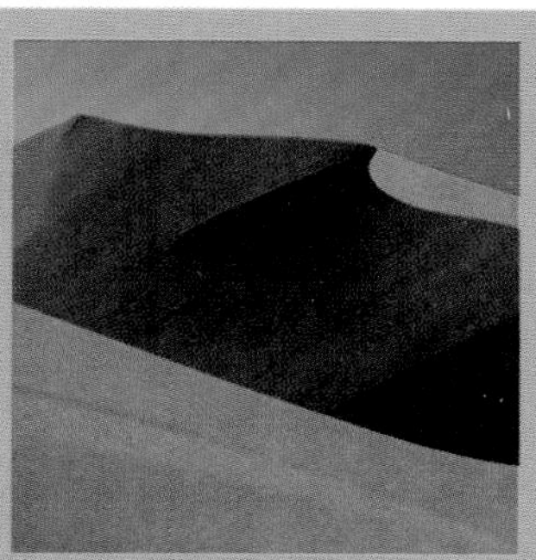

HOLT World Geography Today

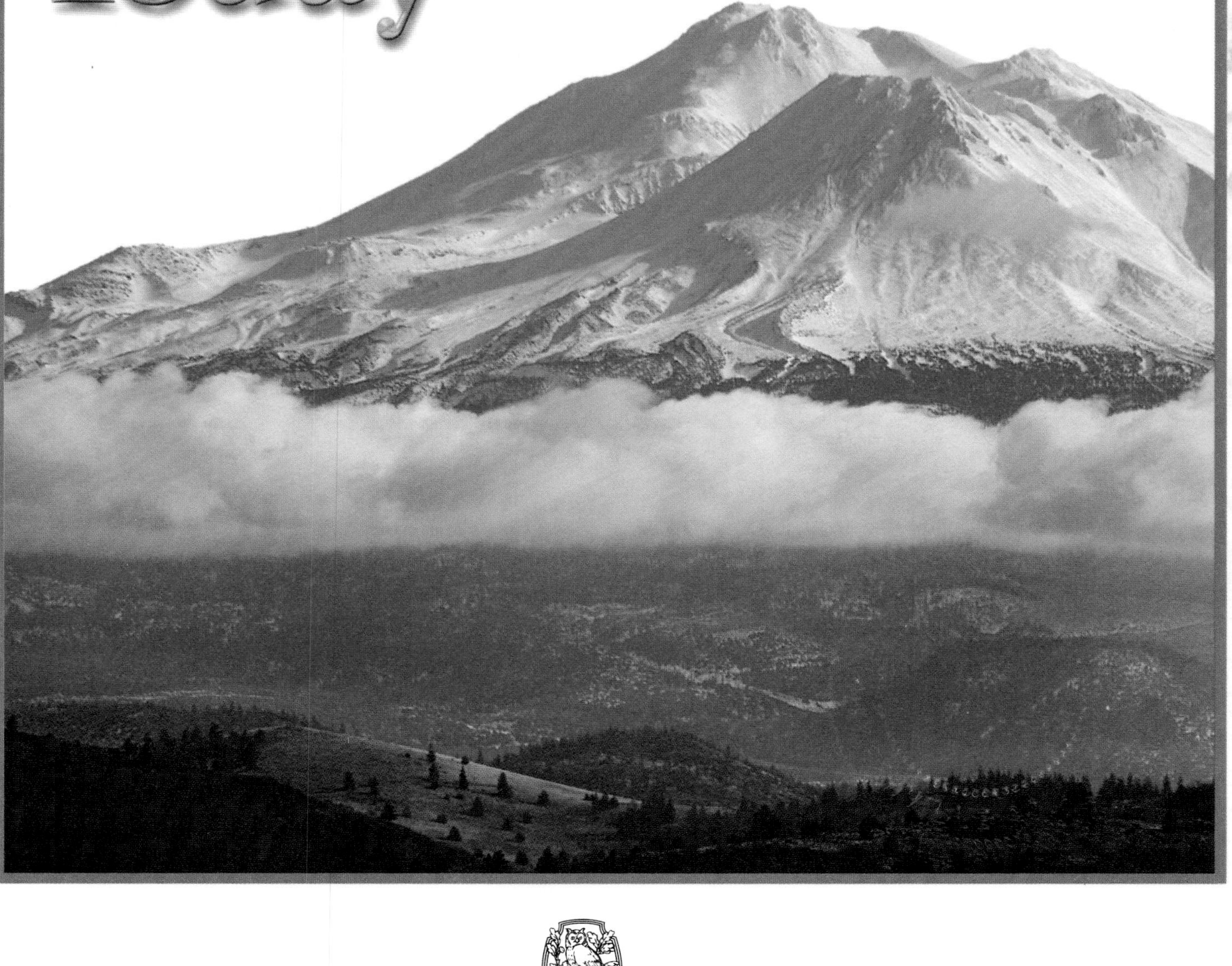

HOLT, RINEHART AND WINSTON

A Harcourt Education Company

Austin • Orlando • Chicago • New York • Toronto • London • San Diego

The Authors

Prof. Robert J. Sager is Chair of Earth Sciences at Pierce College in Lakewood, Washington. Prof. Sager received his B.S. in geology and geography and M.S. in geography from the University of Wisconsin and holds a J.D. in international law from Western State University College of Law. He is the coauthor of several geography and earth science textbooks and has written many articles and educational media programs on the geography of the Pacific. Prof. Sager has received several National Science Foundation study grants and has twice been a recipient of the University of Texas NISOD National Teaching Excellence Award. He is a founding member of the Southern California Geographic Alliance and former president of the Association of Washington Geographers.

Prof. David M. Helgren is Director of the Center for Geographic Education at San Jose State University in California, where he is also Chair of the Department of Geography. Prof. Helgren received his Ph.D. in geography from the University of Chicago. He is the coauthor of several geography textbooks and has written many articles on the geography of Africa. Awards from the National Geographic Society, the National Science Foundation, and the L. S. B. Leakey Foundation have supported his many field research projects. Prof. Helgren is a former president of the California Geographical Society and a founder of the Northern California Geographic Alliance.

While the chapter openers come from actual interviews, the young people's identities have been changed to protect their privacy.

Cover and Title Page photographs: Mount Shasta, California; Sahara Desert, Africa; and background mountains

Cover and Title Page Photo Credits: Mount Shasta, California; Sahara Desert, Africa; and background mountains, Artbase Inc.

For acknowledgments, see page 797, which is an extension of the copyright page.

Printed in the United States of America

ISBN 0-03-038036-7

2 3 4 5 6 7 8 9   032   09 08 07 06 05 04

Academic Reviewers

Robin Elisabeth Datel
Instructor in Geography
California State University, Sacramento

Dennis Dingemans
Professor of Geography
University of California, Davis

Jeffrey Gritzner
Professor of Geography
The University of Montana

W. A. Douglas Jackson
Emeritus Professor, Geography and
International Relations
Henry M. Jackson
School of International Studies
University of Washington

Robert B. Kent
Chair and Professor of Geography and
Planning
University of Akron

Kwadwo Konadu-Agyemang
Professor of Geography and Planning
University of Akron

Nancy Lewis
Professor of Geography
University of Hawaii

Garth Andrew Myers
Associate Professor of Geography and
African Studies
University of Kansas

Eric P. Perramond
Assistant Professor of Geography &
Environmental Science
Stetson University

Bill Takizawa
Professor of Geography
San Jose State University

Brent Yarnal
Professor of Geography
Pennsylvania State University

Teacher Reviewers

Melissa Counihan
Rockdale High School
Rockdale, Texas

Tom Fischer
St. Clare School
St. Louis, Missouri

William Fisher
Bryan High School
Bryan, Texas

Steve Gargo
Appleton West High School
Appleton, Wisconsin

Lisa Haydel
Evergreen Junior High School
Houma, Louisiana

Nancy Lehmann-Carssow
Instructional Specialist
Lanier High School
Austin, Texas

C. Eugene Price
Heritage Junior-Senior High School
Monroeville, Indiana

Ron Scholten
Social Studies Department Chair
Providence High School
Burbank, California

Catharine Stoner
Wentzville High School
Wentzville, Missouri

Frank Thomas, Jr.
Hays County CISD
Austin, Texas

Susan W. Walker
Beaufort County Schools
Beaufort, South Carolina

Jane Young-Leatherman
Social Studies Department Chair
Durant High School
Durant, Oklahoma

Field Test Teachers

Amber Acuña
Luther Burbank High School
San Antonio, Texas

J. Mark Buffington
Mexico High School
Mexico, Missouri

Sandra Dawson-O'Bryan
Angleton High School
Angleton, Texas

Marie Ervin
James Madison High School
Houston, Texas

Patrick Haney
Carter High School
Fort Worth, Texas

Randy Kindschuh
Wisner-Pilger High School
Wisner, Nebraska

Jim Long
Holmes High School
San Antonio, Texas

Christine A. Moore
Fort Zumwalt South High School
St. Peters, Missouri

Carol Ragsdale
Central High School
West Helena, Arkansa

Juliann Warner
Martin High School
Arlington, Texas

It's All About

YOUTH INTERVIEWS

with young people from around the world begin each chapter, making real world peer connections for your students. Plus students can read the interviews in their entirety on our Web site.

The first step to success in the social studies classroom is capturing and sustaining the interest of your students. *HOLT WORLD GEOGRAPHY TODAY* is designed to be open and friendly to all students, so that they develop an enthusiasm for learning and an appreciation for their world.

HOLT WORLD GEOGRAPHY TODAY offers
- Built-in Reading Support
- Technology with Instructional Value
- Standardized Testing and Skill Building
- The Best Teacher Management System in the Industry

RELEVANCE

CNNfyi.com™ is designed to give students in grades 6–12 access to the news about people, places, and environments around the globe while offering "real-world" articles, career and college resources, and online activities.

In-Text Features that Put Geography into Perspective

- Case Studies
- Cities & Settlements
- Connecting to Anthropology
- Connecting to Economics
- Connecting to Government
- Connecting to History
- Connecting to Literature
- Connecting to Math
- Connecting to Science
- Connecting to Technology
- Connecting to the Arts

- Focus on Citizenship
- Focus on Culture
- Focus on Geography
- Focus on History
- Focus on Science, Technology, and Society
- Geography for Life
- Our Amazing Planet
- Major World Religions
- Why it Matters

Reading for

At Holt, we don't assume that students know how or have any desire to make sense of what they're reading, and we develop our programs based on that assumption. We don't just ask students questions about content, we give them strategies to get to that content. Through design, research, and the help of experts like Dr. Judith Irvin, we make sure students' reading needs are covered with our programs.

Helping Students Make Sense of What They're Reading

An Essay by Dr. Judith Irvin, Ph.D.

Who in middle and high schools helps students become more successful at reading and writing informational text? When I ask this question of a school faculty, the Language Arts/English teachers point to the social studies and science teachers because they are the ones with this type of textbook. The social studies and science teachers point to the Language Arts/English teachers because they are the ones that "do" words.

I advocate teachers taking an active role in helping students learn how to use text structure and context to understand what they read. Through consistent and systematic instruction that includes modeling of effective reading behavior, teachers can assist students in becoming better readers while at the same time helping them learn more content material.

The strategies in this book are designed to assist students with getting started, maintaining focus with reading, and organizing information for later retrieval. They engage students in learning material, provide the vehicle for them to organize and reorganize concepts, and extend their understanding through writing.

When teachers combine the teaching of reading and the teaching of content together into meaningful, systematic, and corrected instruction, students can apply what they have learned to understanding increasingly more difficult and complex texts as they progress through the school years.

READING STRATEGIES FOR THE SOCIAL STUDIES CLASSROOM

by Dr. Judith Irvin,
Ph.D. Reading Education

Additional Reading Support

- **Graphic Organizer Activities**
- **Guided Reading Activities**
- **Main Idea Activities for English-Language Learners and Special Needs Students**
- **Audio CD Program**

MEANING

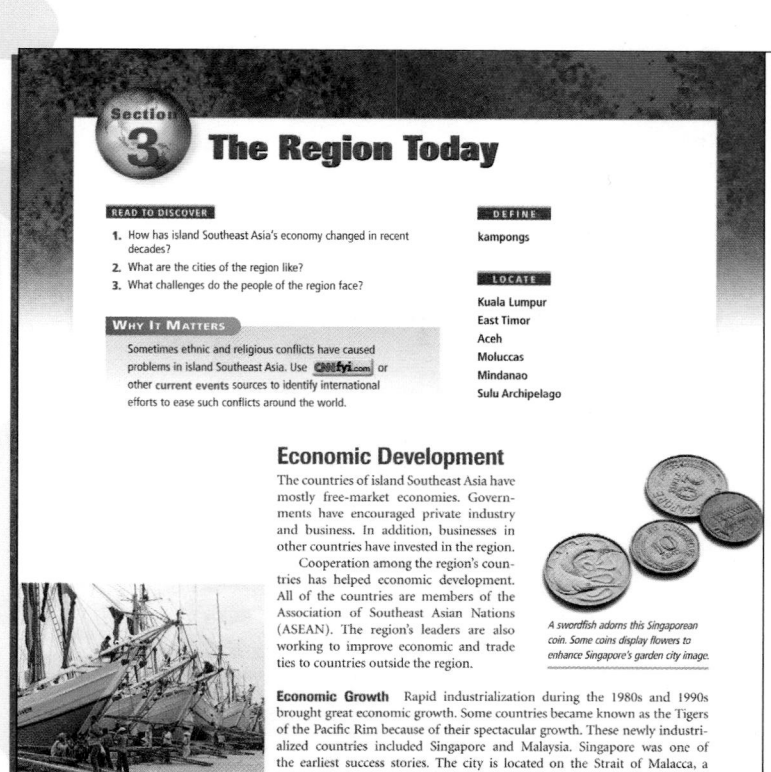

Successful Readers must have:

1 AN ENGAGING NARRATIVE

Great care is taken in selecting and presenting content in a way that students will find motivating and engaging. Features such as **Youth Interviews** help students connect their own lives to the lives and cultures of other students around the world.

2 A FORECAST OF WHAT THEY WILL LEARN

Read to Discover questions give students insight into the content they will cover in the chapter to come. In features such as **Why It Matters** students gain insight into regional issues.

3 VOCABULARY DEFINED IN CONTEXT

Important new terms are identified at the beginning of every section and are defined in context so students will develop an understanding of the contextual meaning of all terms.

4 STRATEGIES FOR UNDERSTANDING WHAT THEY READ

Through the design of the text, students are led through the content using built-in reading strategies. For example, **Reading Checks** in the text are used as a comprehension tool. The checks remind students to stop and engage with what they have read, functioning as a "Tutor in the Text."

Get Your Students

Your students love activities that get them involved with the content. That's why Holt offers active-learning resources that link directly to program content and provide a multitude of different lessons for large-group, small-group, and individual projects.

CREATIVE TEACHING STRATEGIES

These innovative teaching strategies can be utilized at various points in your lesson. The wide range of cooperative-learning activities, including learning stations and simulations, motivate your students and help them develop critical-thinking skills.

HANDS-ON GEOGRAPHY ACTIVITIES

From hands-on study of world cultures to hands-on practice with skill building, the following booklets cover it all. *Cultures of the World Activities* is a stand-alone booklet containing recipes, games, and craft activities. *Geography for Life Activities with Answer Key* contains a a problem-solving activity for each chapter, reflecting the skills and knowledge called for in the **National Geography Standards**.

GEOGRAPHY APPLICATIONS

For use in geography as well as earth science courses, here are two stand-alone booklets that organize applications with special relevance. *Environmental and Global Issues Activities* contains activities related to current environmental and global issues. *Lab Activities for Geography and Earth Science* contains laboratory activities related to physical geography and earth science.

Involved in Learning

Resources for Active Lessons

- **Block Scheduling Handbook with Team-Teaching Strategies**
- **Creative Teaching Strategies for World Geography**
- **Critical Thinking Activities**
- **Cultures of the World Activities**
- **Environmental and Global Issues Activities**
- **Geography and Cultures Visual Resources**
- **Geography for Life Activities**
- **Lab Activities for Geography and Earth Science**
- **Map Activities**
- **Regions of the World Map Transparencies**
- **World and Regional Outline Maps**
- **World History and Geography Document-Based Questions Activities**

Joining Forces

CNN® Presents

to Enrich your Classroom

CNNfyi.com

At **CNNfyi.com**, students will love exploring news stories written by experienced journalists as well as student bureau reporters complete with links to homework help and lesson plans.

CNN PRESENTS VIDEO LIBRARY

The **CNN PRESENTS** video collection tackles the issue of making content relevant to students head on. Real-world news stories enable students to see the connections between classroom curriculum and today's issues and events around the nation and the world.

CNN PRESENTS...

- **America: Yesterday and Today, Beginnings to 1914**
- **America: Yesterday and Today, 1850 to Present**
- **America: Yesterday and Today, Modern Times**
- **Geography: Yesterday and Today**
- **World Cultures: Yesterday and Today**
- **American Government**
- **Economics**
- **Texas**
- **September 11, 2001: A Turning Point in American History**

Holt is proud to team up with CNN/TURNER LEARNING® to provide you and your students with exceptional current and historical news videos and online resources that add depth and relevance to your daily instruction. This information collection takes your classroom to the far corners of the globe without students ever leaving their desks!

Your Multi-talented Classroom

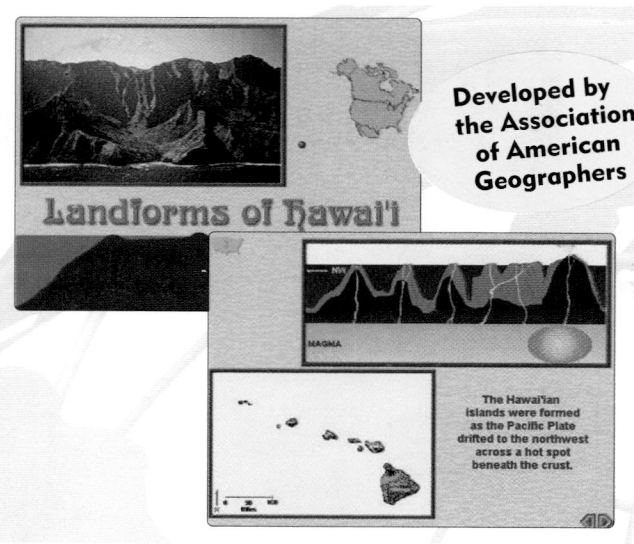

Developed by the Association of American Geographers

ACTIVITIES AND READINGS IN THE GEOGRAPHY OF THE WORLD

Integrate real geography into the topic you're studying with **Activities and Readings in the Geography of the World (ARGWorld).** This CD-ROM features world geography case studies with a multitude of activities that focus around geographical themes, population geography, economic geography, political geography, and environmental issues. Case studies will help teachers address the National Geography Standards.

HOLT RESEARCHER ONLINE: WORLD HISTORY AND CULTURES

New and online—students can access this outstanding research tool at **www.hrw.com**. A fully searchable database provides biographies, nation profiles and statistics, a glossary, and powerful graphic capabilities.

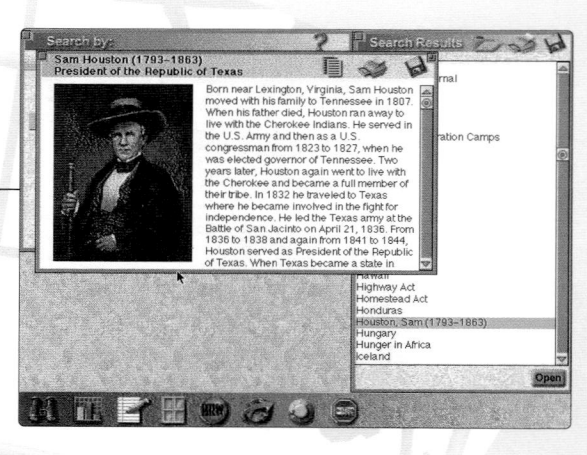

GLOBAL SKILL BUILDER CD-ROM

This CD-ROM is a comprehensive program containing interactive lessons that motivate your students to strengthen their map, graph, and computer skills. A handy **User's Guide and Teacher's Manual** provides student project sheets for each lesson along with optional suggestions for using the Internet to help complete the activity.

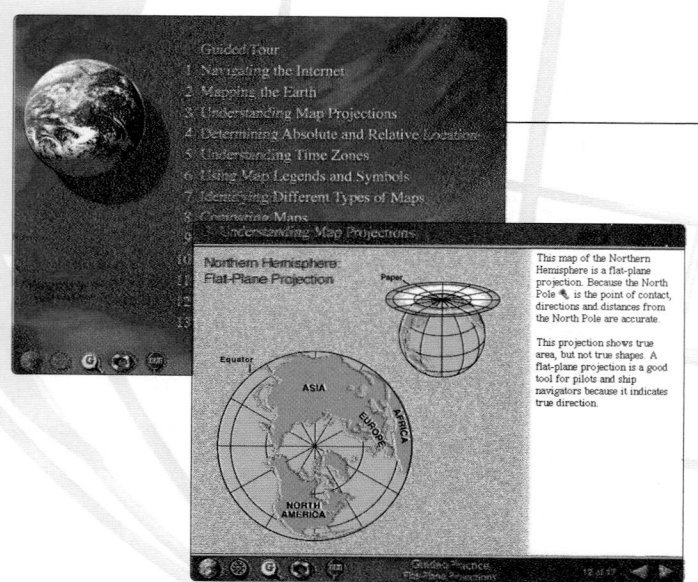

needs Multimedia Tools

THE WORLD TODAY VIDEODISC PROGRAM

This unique resource offers a stimulating outlook on world geography by showing your students the different ways geographers organize the world, and challenging them to contemplate and discuss significant world issues. Compelling video segments with in-depth content cover contemporary culture in every major world region.

HOLT WORLD GEOGRAPHY TODAY AUDIO CD PROGRAM

The **Audio CD Program** provides in-depth audio section summaries and self-check activity sheets to help those students who respond to auditory learning. Available in English and Spanish.

Other Multimedia Products

- **CNN Presents Geography: Yesterday and Today**
- **CNN Presents World Cultures: Yesterday and Today**
- **Holt Researcher Online: World History and Cultures**

M9

Technology with

go.hrw.com FOR TEACHERS

Throughout the *Annotated Teacher's Edition*, you'll find **Internet Connect** boxes that take you to specific chapter activities, links, current events, and more that correlate directly to the section you are teaching. Through **go.hrw.com** you'll find a wealth of teaching resources at your fingertips for fun, interactive lessons.

...ut materials for your students
...M with Test Generator.

Reinforcement, Review, and Assessment

ELL	Main Idea Activity 30.1
ELL	English Audio Summary 30.1
ELL	Spanish Audio Summary 30.1
REV	Section 1 Review, p. 682
A	Daily Quiz 30.1

ELL	Main Idea Activity 30.2
ELL	English Audio Summary 30.2
ELL	Spanish Audio Summary 30.2
REV	Section 2 Review, p. 689
A	Daily Quiz 30.2

ELL	Main Idea Activity 30.3
ELL	English Audio Summary 30.3
ELL	Spanish Audio Summary 30.3
REV	Section 3 Review, p. 693
A	Daily Quiz 30.3

internet connect

HRW ONLINE RESOURCES
GO TO: go.hrw.com
Then type in a keyword.

TEACHER HOME PAGE
KEYWORD: SW3 Teacher

CHAPTER INTERNET ACTIVITIES
KEYWORD: SW3 GT30
Choose an activity to:
- test knowledge with an interactive map.
- create a poster about rain forests.
- analyze tectonic forces that cause volcanoes and earthquakes.

CHAPTER ENRICHMENT LINKS
KEYWORD: SW3 CH30

CHAPTER MAPS
KEYWORD: SW3 MAPS30

ONLINE ASSESSMENT
Homework Practice
KEYWORD: SW3 HP30
Standardized Test Prep
KEYWORD: SW3 STP30
Rubrics
KEYWORD: SS Rubrics

COUNTRY INFORMATION
KEYWORD: SW3 Almanac

CONTENT UPDATES
KEYWORD: SS Content Updates

HOLT PRESENTATION MAKER
KEYWORD: SW3 PPT30

ONLINE READING SUPPORT
KEYWORD: SS Strategies

CURRENT EVENTS
KEYWORD: S3 Current Events

DIRECT LAUNCH TO CHAPTER ACTIVITIES

GUIDED ONLINE ACTIVITIES

LINKS FOR EVERY SECTION

MAPS AND CHARTS

INTERACTIVE PRACTICE AND REVIEW

RUBRICS FOR SUBJECTIVE GRADING

UP-TO-DATE INFORMATION

CLASSROOM PRESENTATION SUPPORT

PRACTICE FOR READING SUCCESS

WEB RESOURCES FOR CURRENT ISSUES

Chapter Review and Assessment

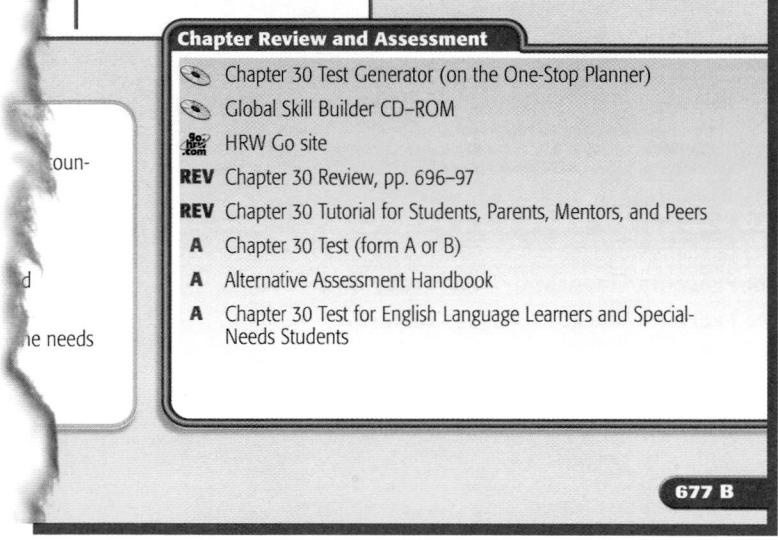

	Chapter 30 Test Generator (on the One-Stop Planner)
	Global Skill Builder CD–ROM
	HRW Go site
REV	Chapter 30 Review, pp. 696–97
REV	Chapter 30 Tutorial for Students, Parents, Mentors, and Peers
A	Chapter 30 Test (form A or B)
A	Alternative Assessment Handbook
A	Chapter 30 Test for English Language Learners and Special-Needs Students

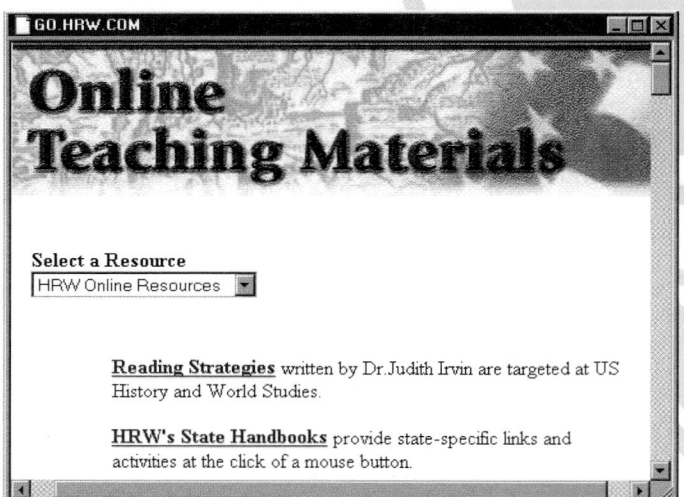

GO.HRW.COM

Online Teaching Materials

Select a Resource
HRW Online Resources

Reading Strategies written by Dr. Judith Irvin are targeted at US History and World Studies.

HRW's State Handbooks provide state-specific links and activities at the click of a mouse button.

Instructional Value

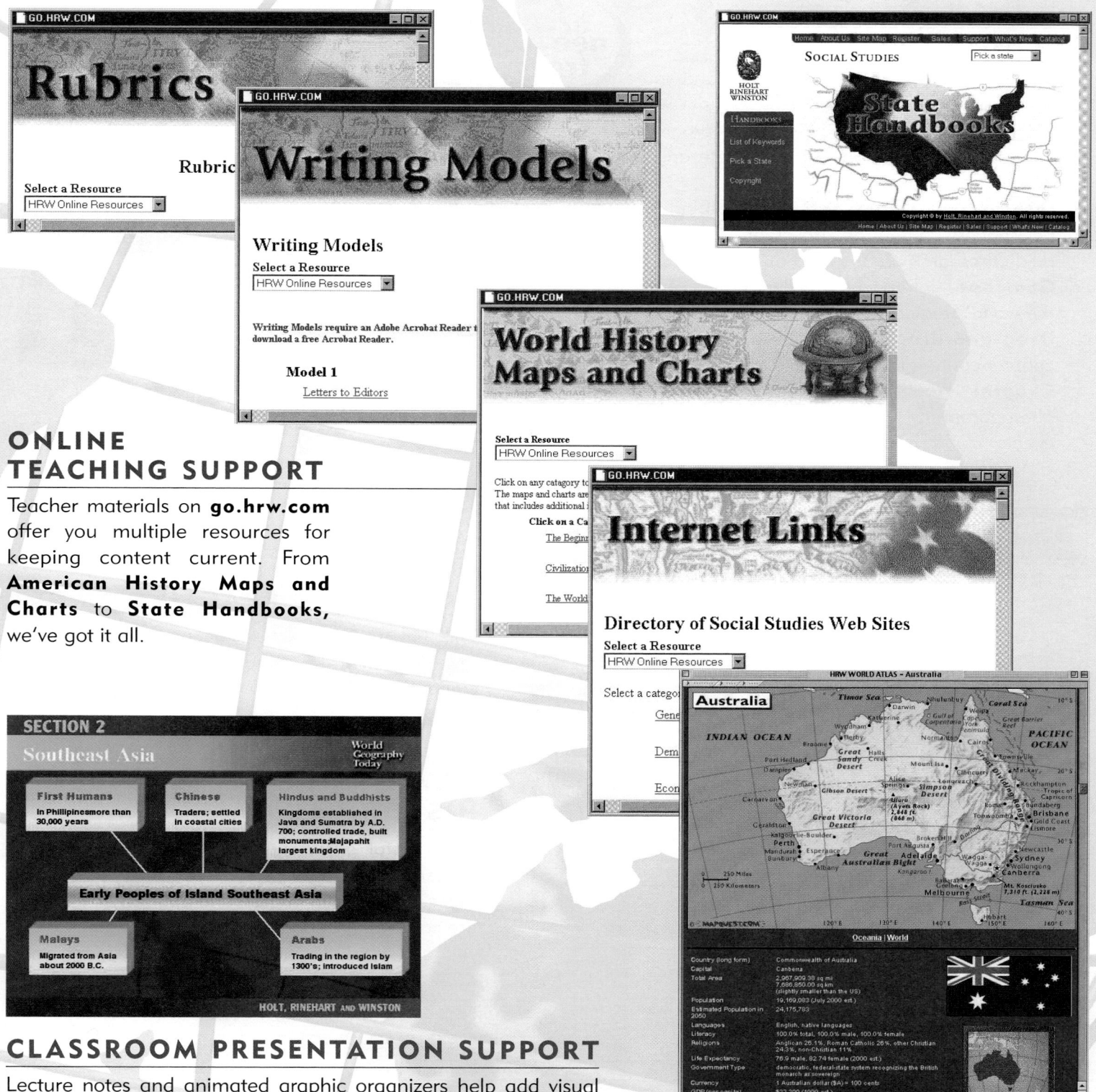

ONLINE TEACHING SUPPORT

Teacher materials on **go.hrw.com** offer you multiple resources for keeping content current. From **American History Maps and Charts** to **State Handbooks,** we've got it all.

CLASSROOM PRESENTATION SUPPORT

Lecture notes and animated graphic organizers help add visual support to your classroom presentations.

Technology that

go.hrw.com FOR STUDENTS

Your students can access interactive activities, homework help, up-to-date maps, and more when they visit **go.hrw.com** and type in the keywords they find in their text.

ONLINE WORLD TRAVEL

When you log on to **go.hrw.com**, you and your students gain passage to **GeoTreks**—a site with guided Internet activities that integrate program content, spark imaginations, and promote online research skills. You'll find:

Interactive templates for creating newspapers, postcards, travel brochures, guided research reports, and more

GeoMaps—Interactive satellite maps of the world's regions for content review

Drag-and-drop exercises to review chapter content in short, fun activities

Chapter Web Links for prescreened, age-appropriate Web sites and current events

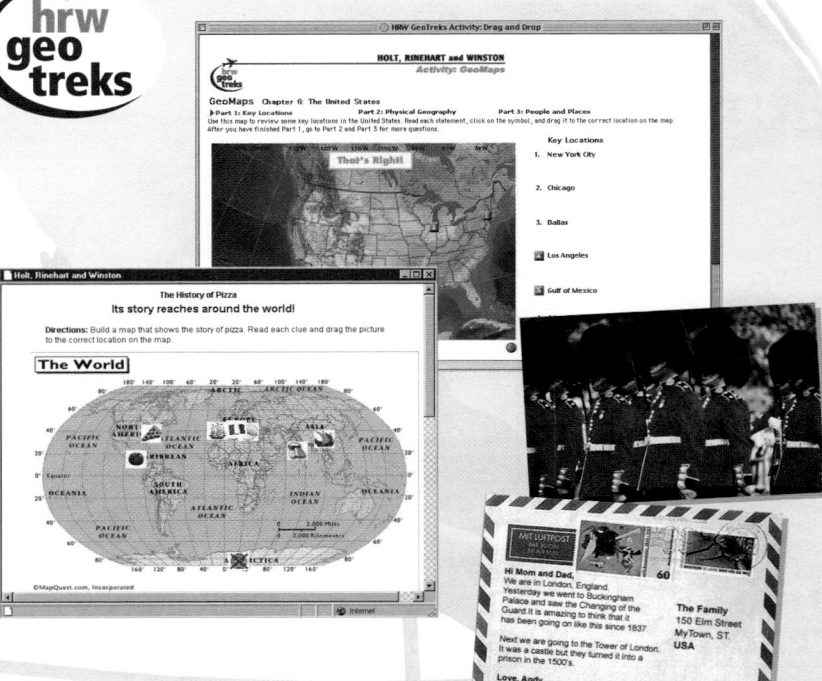

HOMEWORK PRACTICE

This helpful tool allows students to practice and review content by chapter anywhere there is a computer.

HRW ONLINE ATLAS AND HISTORICAL MAPS

The helpful online atlas contains over 300 well-rendered and clearly labeled country and state maps. Available in English and Spanish, these maps are continually updated so you can rest assured that you and your students have the latest and most accurate geographical content.

Online historical maps provide fascinating visual "snapshots" of the past. Students will relish the chance to explore medieval European trade routes, explorers' routes, ancient African kingdoms, and more.

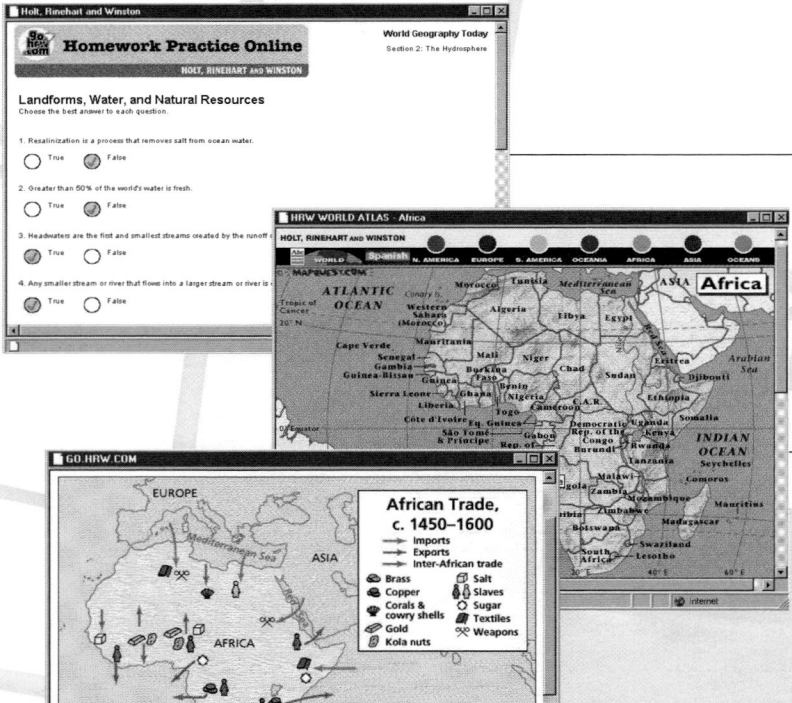

Delivers Content

New Online Textbook

You'll know what to do when you see it!
Finally, an online textbook that takes full advantage of Web technology in a way that makes sense—*HOLT WORLD GEOGRAPHY TODAY ONLINE EDITION.*

- Entire student edition online formatted to match printed text
- User-friendly navigation
- Hot links to interactive activities, practice, and assessment
- Student Notebook for online responses

Unique Teacher's

In-Text Chapter Planning

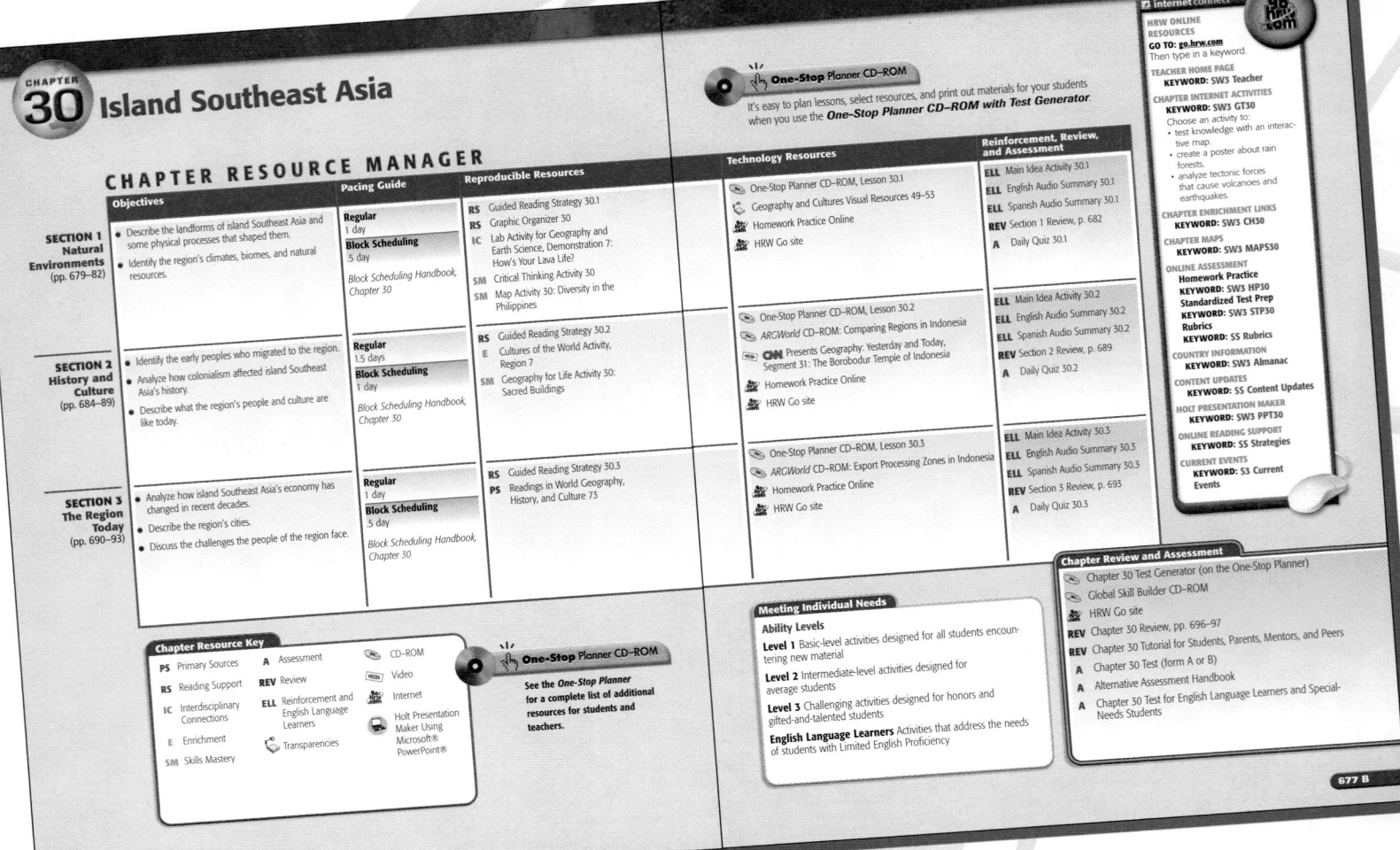

Program Resources at Your Fingertips

Reading Support
- **Graphic Organizer Activities**
- **Guided Reading Strategies**
- **Main Idea Activities for English-Language Learners and Special Needs Students**
- **Reading Strategies for the Social Studies Classroom**
- **Vocabulary Activities**

Primary Sources
- **The Complete School Atlas**
- **Readings in World Geography, History, and Culture**
- **World History and Geography Document-Based Questions Activities**

Active-Learning Support
- **Block Scheduling Handbook with Team-Teaching Strategies**
- **Creative Strategies for Teaching World Geography**
- **Cultures of the World Activities**
- **Environmental and Global Issues Activities**
- **Geography for Life Activities**
- **Lab Activities for Geography and Earth Science**

Skill-Building Support
- **Critical Thinking Activities**
- **Social Studies Skills Review**
- **Map Activities**

- **World and Regional Outline Maps**
- **Geography and Cultures Visual Resources**
- **Regions of the World Map Transparencies**
- **Regions of the World Map Posters**

Review And Assessment
- **Daily Quizzes**
- **Chapter Tutorials for Students, Parents, Mentors, and Peers**
- **Chapter and Unit Tests**
- **Chapter and Unit Tests for English-Language Learners and Special-Needs Students**
- **Alternative Assessment Handbook**

Management System

ONE-STOP PLANNER® CD-ROM WITH TEST GENERATOR

Holt brings you the most user-friendly management system in the industry with the **One-Stop Planner CD-ROM with Test Generator.** Plan and manage your lessons from this single disc containing all the teaching resources for **Holt World Geography Today,** valuable planning and assessment tools, and more.

- **Editable lesson plans**
- **Classroom Lecture Notes and Animated Graphic Organizers**
- **Easy-to-use test generator**
- **Previews of all teaching and video resources**
- **Easy printing feature**
- **Direct launch to go.hrw.com**

Everything you need is on one disc!

BLOCK SCHEDULING HANDBOOK WITH TEAM TEACHING STRATEGIES

This is more than a pacing guide—it provides daily lesson plans that suggest practical ways to cover more than one textbook section in an extended class period and ways to make inter-disciplinary connections.

Energize

Presentations That Benefit Learning

Classroom presentations and lecture notes can be accessed with ease when you use Holt's **Presentation** tool found on the *One-Stop Planner CD-ROM.* This resource helps you spice up your presentations and gives you ideas to build on. You'll find Microsoft® PowerPoint® presentations that include lecture notes and animated graphic organizers for each chapter and section of your text.

Your Classroom Presentations

OBJECTIVE-BASED LESSON CYCLE

With lively activities and presentation strategies such as **Let's Get Started, Building Vocabulary,** and **Graphic Organizers,** your step-by-step lesson cycle makes planning your lessons easy and productive.

Section 2

OBJECTIVES

1. Identify the early peoples who migrated to the region.
2. Analyze how colonialism affected island Southeast Asia's history.
3. Describe what the region's people and culture are like today.

LET'S GET STARTED

Copy the following passage onto the chalkboard: *Look at the large photo in the unit opener. What other statues in the textbook look similar? What can you conclude about culture in island Southeast Asia?* Discuss responses. *(Similar statues from China or India indicate that the region shares religious traditions with those countries.)* Then point out a photo of a mosque and tell students that more Muslims live in Indonesia than in any other country. In Section 2 students will learn more about the region's history and culture.

Building Vocabulary

Write the key terms on the chalkboard. Call on a student to read the definition of the word *heterogeneous* from a dictionary and contrast it with the meaning of **homogeneous** as it appears in the Section 2 text. Ask another student to propose a definition for **slash-and-burn** agriculture based on the term's parts. Finally, have volunteers read the definitions aloud from the glossary.

Teach Objective 1

LEVEL 1: Copy the following graphic organizer onto the chalkboard, omitting the italicized answers. Have students fill in details. ENGLISH LANGUAGE LEARNERS

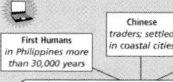

First Humans *in Philippines more than 30,000 years*

Chinese *traders; settled in coastal cities*

Hindus and Buddhists *kingdoms established in Java and Sumatra by A.D. 700; controlled trade, built monuments; Majapahit largest kingdom*

Early Peoples of Island Southeast Asia

Malays *migrated from Asia about 2000 B.C.*

Arabs *trading in the region by 1300s; introduced Islam*

LEVELS 2 AND 3: Have students use the graphic organizer in the Level 1 activity as a basis for this one. Tell students to imagine that they are among the early peoples who migrated to island Southeast Asia. Ask them to write a diary entry for the day their boat lands in the region, a day on the Borobudur construction site, time spent trading with Chinese merchants, or another episode in the region's early history. Have students use the illustrations in this chapter and the unit opener as inspiration for detail. You may want students to consult other resources also.

Close

Refer students to the population pyramids in Section 2. Call on volunteers to create one-sentence summaries of what the diagrams depict or what can be inferred from them. *(Example: In 2000, all three countries had more children under the age of five than in any other age category.)*

Review and Assess

Have students complete the **Section Review.** Then have students complete **Daily Quiz 30.2.**

Reteach

Have students complete **Main Idea Activity for English Language Learners and Special-Needs Students 30.2.** Then have each student write 10 questions based on Section 2. Ask students to exchange papers, answer the questions, and return them for checking. Discuss questions that caused difficulty. ENGLISH LANGUAGE LEARNERS

Extend

Have interested students conduct research on the Philippines' revolt against American authority (1899–1901) and the role that the revolt played in Filipinos' perceptions of national unity. BLOCK SCHEDULING

Portfolio Activity

Have students conduct research on the *wayang* shadow puppets of Java. These puppets are usually made from sheets of leather pierced with decorative holes. Have students create their own puppets and write a shadow play about events or issues in island Southeast Asia. Heavy paper can be substituted for leather. Students can mount their puppets on bamboo skewers and use brads to make the limbs flexible. The performance takes place behind a screen—perhaps an old white sheet—and in front of a light source. After students have performed their plays, place puppets and notes in portfolios.

Food Festival

During the time the Dutch occupied Indonesia they adopted a style of dining called a *rijsttafel* (RYS-tah-fuhl), which means "rice table" in Dutch. A *rijsttafel* consists of rice accompanied by many small, highly-seasoned dishes, including fruits, meats, seafoods, vegetables, and sauces. Have students locate and prepare recipes for the side dishes. Provide a big bowl of rice, and let students help themselves. Note that peanut sauce is a popular condiment in Indonesia. Some students are severely allergic to peanuts. You may want to eliminate peanut sauce or peanuts from the menu completely.

Writing

TEACHER TO TEACHER

These strategies are offered in the columns of your *Annotated Teacher's Edition* and provide you with valuable, classroom-tested ideas and activities that have been developed and successfully applied by your peers.

Side-Column Annotations that Spark Curiosity

- **Across the Curriculum**
- **Case Studies**
- **Cities & Settlements**
- **Connecting to Anthropology**
- **Connecting to Economics**
- **Connecting to Government**
- **Connecting to Literature**
- **Connecting to Math**
- **Connecting to the Arts**
- **Cultural Kaleidoscope**
- **Daily Life**
- **Essential Element**
- **Eye on Earth**
- **Focus on Citizenship**
- **Focus on Culture**
- **Focus on History**
- **Focus on Science, Technology, and Society**
- **Geography for Life**
- **Linking Past to Present**
- **Major World Religions**
- **Our Amazing Planet**
- **Why We Should Know More**
- **You Don't Say!**

Assessment for

CRITICAL THINKING
REINFORCEMENT

INTERACTIVE
PRACTICE
ACTIVITIES
FOR EACH SECTION

CHAPTER 30 Review

Section 3 Review

go.hrw.com Homework Practice Online
Keyword: SW3 HP30

Define
kampongs

Working with Sketch Maps On the map you created in Section 2, label Kuala Lumpur, East Timor, Aceh, Moluccas, Mindanao, and Sulu Archipelago. Which island region declared independence in 1999?

Reading for the Main Idea
1. *Human Systems* Why are some of the countries in island Southeast Asia known as the Tigers of the Pacific Rim?
2. *Environment and Society* What factor helped make Singapore a major trade and industrial center?

Critical Thinking
3. *Comparing and Contrasting* What are some differences between the city of Singapore and other big cities in the region? What are some reasons for these differences?
4. *Making Generalizations* Why do you suppose Indonesia's various ethnic groups each view their country differently?

Organizing What You Know
5. Copy the table below. Use it to identify economic, environmental, and political challenges the region's countries face.

	Challenges	
Economic	Environmental	Political

Island Southeast Asia • 693

Building Vocabulary

On a separate sheet of paper, explain the following terms by using them correctly in sentences.

archipelago	homogeneous
lahars	slash-and-burn agriculture
endemic species	kampongs

Locating Key Places

On a separate sheet of paper, match the letters on the map with their correct labels.

Malay Peninsula	Manila
Malay Archipelago	Jakarta
Strait of Malacca	Kuala Lumpur
Philippines	Singapore

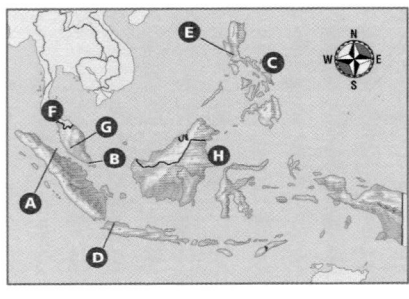

Understanding the Main Ideas

Section 1
1. *Physical Systems* Why is the land of island Southeast Asia so unstable in some areas?
2. *Places and Regions* How do the region's location and monsoons influence its climates?

Section 2
3. *Human Systems* Which ethnic group makes up the majority of Singapore's population? Do members of that ethnic group tend to live in urban or rural areas throughout island Southeast Asia?
4. *Places and Regions* What is the dominant religion of island Southeast Asia? What other religions are practiced there?

Section 3
5. *Places and Regions* What economic, environmental, and political challenges face the region today?

696 • Chapter 30

Thinking Critically

1. **Analyzing Information** How has the region's cultural diversity affected the structure and goals of educational systems there?
2. **Drawing Inferences and Conclusions** Why do you think wet-rice cultivation is the most common method of growing rice in the region?
3. **Identifying Cause and Effect** How do you think island Southeast Asia might be different today if Europeans had never colonized the region?

Using the Geographer's Tools

1. **Analyzing Population Pyramids** Study the population pyramids in Section 2. What clues might these pyramids give us about the relative standard of living in the countries?
2. **Analyzing Statistics** Review the Fast Facts and Comparing Standard of Living tables at the front of this unit. Then go to **go.hrw.com** to access the World Factbook. Use information you find from these sources to rank the countries of island Southeast Asia by standard of living. Write a paragraph explaining your ranking.
3. **Preparing Maps** Draw an outline map of island Southeast Asia and identify areas that are experiencing ethnic and religious conflict.

Writing about Geography

Review the various ways farmers in island Southeast Asia modify their physical environment to grow rice. Compare those methods with the ways farmers in dry climate regions of the United States modify their environment to grow wheat. You may want to use the Internet or consult library resources for more information. Write a short report about your findings. When you are finished with your report, proofread it to make sure you have used standard grammar, spelling, sentence structure, and punctuation.

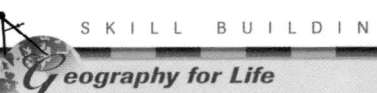

SKILL BUILDING

Geography for Life

Creating a Benefits-Cost Balance Sheet
Environment and Society As you have read, the tropical rain forests of island Southeast Asia are being cleared rapidly. What might be some advantages and disadvantages of clearing the forests for farmland, timber, or other purposes? Identify and evaluate those benefits and costs in the form of a two-column balance sheet.

Every Student

CONSISTENT STANDARDIZED TEST PREPARATION

Building Social Studies Skills

Per Capita GDP

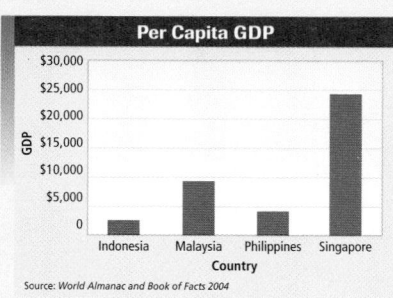

Source: World Almanac and Book of Facts 2004

Interpreting Graphs

Study the bar graph above. Then use the information in the graph to help you answer the questions that follow. Mark your answers on a separate sheet of paper.

1. Singapore's per capita GDP is
   a. less than $10,000.
   b. more than five times the per capita GDP of Malaysia.
   c. more than $20,000.
   d. the lowest of the four countries.

2. Rank the countries from first to last by the value of each country's per capita GDP.

Analyzing Secondary Sources

Read the following passage and answer the questions. Mark your answers on a separate sheet of paper.

"Farmers in the region grow rice in three ways. Wet-rice, or paddy, cultivation is the most productive and common method. Rice paddies are constructed with dikes in lowland areas or with mud terraces in hilly areas. Water flow down steep slopes is controlled, and erosion is limited. This, along with the area's warm and wet climate, allows farmers to grow more than one rice crop each year."

3. In the selection the word *terraces* refers to
   a. beautiful structures on the region's old colonial farmhouses.
   b. scenic spots along rivers in Malaysia.
   c. city gardens that have been transformed into rice farms.
   d. flat areas carved into hillsides so that rice can be grown there.

4. Why do you think being able to grow more than one rice crop a year is important in island Southeast Asia? Give specific reasons for your answer.

Alternative Assessment

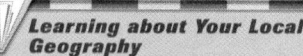

PORTFOLIO ACTIVITY

Learning about Your Local Geography

Group Project: Field Work

Plan, organize, and complete a research project with a partner about the plant and animal life of your state. Identify important geographic questions about the species you identify in the area. What environmental conditions, such as climate, are common there? What species have adapted to that environment? What effect has interaction with humans had on the local environment and the species that live there? Organize your questions and answers in a table and identify connections between your local environment and that of island Southeast Asia. Finally, prepare a short report comparing the natural environments of your state and island Southeast Asia.

Internet Activity: go.hrw.com
KEYWORD: SW3 GT30

Choose a topic about island Southeast Asia to:
* test your knowledge of island Southeast Asia with an interactive map.
* research the rain forests of island Southeast Asia and make a poster.
* analyze tectonic forces that cause volcanoes and earthquakes.

Island Southeast Asia • 697

ACCESS ONLINE RUBRICS FOR GRADING PROJECTS AND PORTFOLIO ASSIGNMENTS

WORLD HISTORY AND GEOGRAPHY DOCUMENT-BASED QUESTIONS ACTIVITIES

This resource provides a wide variety of primary sources and thought-provoking questions to help students develop intelligent, formed opinions. Important historical and geographical themes are grouped together, allowing for scaffolded instruction.

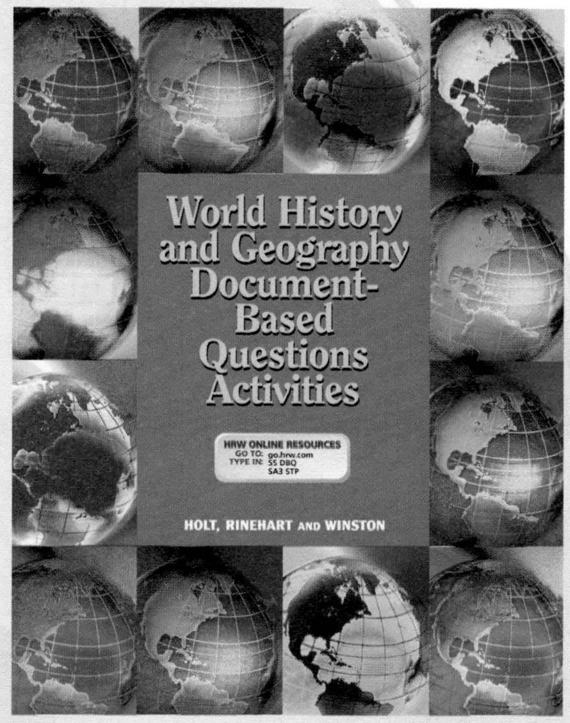

World History and Geography Document-Based Questions Activities

HRW ONLINE RESOURCES
GO TO: go.hrw.com
TYPE IN: SS DBQ
SA3 STP

HOLT, RINEHART AND WINSTON

Additional Print and Technology Assessment Resources

- **Daily Quizzes**
- **Chapter Tutorials for Students, Parents, Mentors, and Peers**
- **Chapter and Unit Tests**
- **Chapter and Unit Tests for English-Language Learners and Special-Needs Students**
- **Alternative Assessment Handbook**
- **Test Generator (located on the One-Stop Planner)**

HOLT
World Geography Today

TABLE OF CONTENTS

Inuit man and igloo in Canada

*Parliament building in
Budapest, Hungary*

Vancouver, Canada

Cornfield and farm, Minnesota

Lavender field, France

Tian Shan, Central Asia

Ancient stone heads, Turkey

Chimanimani National Park, Zimbabwe   *Gold staff, Ghana*

Gorilla in Rwanda

Rice crops in Sri Lanka

Korean fan dancers

Food vendors in Bangkok, Thailand

Aboriginal art, Australia

Geography and Your World

Interdisciplinary Activities

connecting to *Literature*

FOCUS ON

Interdisciplinary Activities

Connecting to Other Disciplines

Technology Activities

Special Features

Technology Activities

internet connect

Skill-Building Activities

Skills Handbook

Skill Building: Using the Geographer's Tools

Skill-Building Activities

Skill-Building Activities

Skill-Building Activities

Charts, Diagrams, and Graphs

How to Use Your Textbook

Use the chapter opener to preview the region you are about to study.

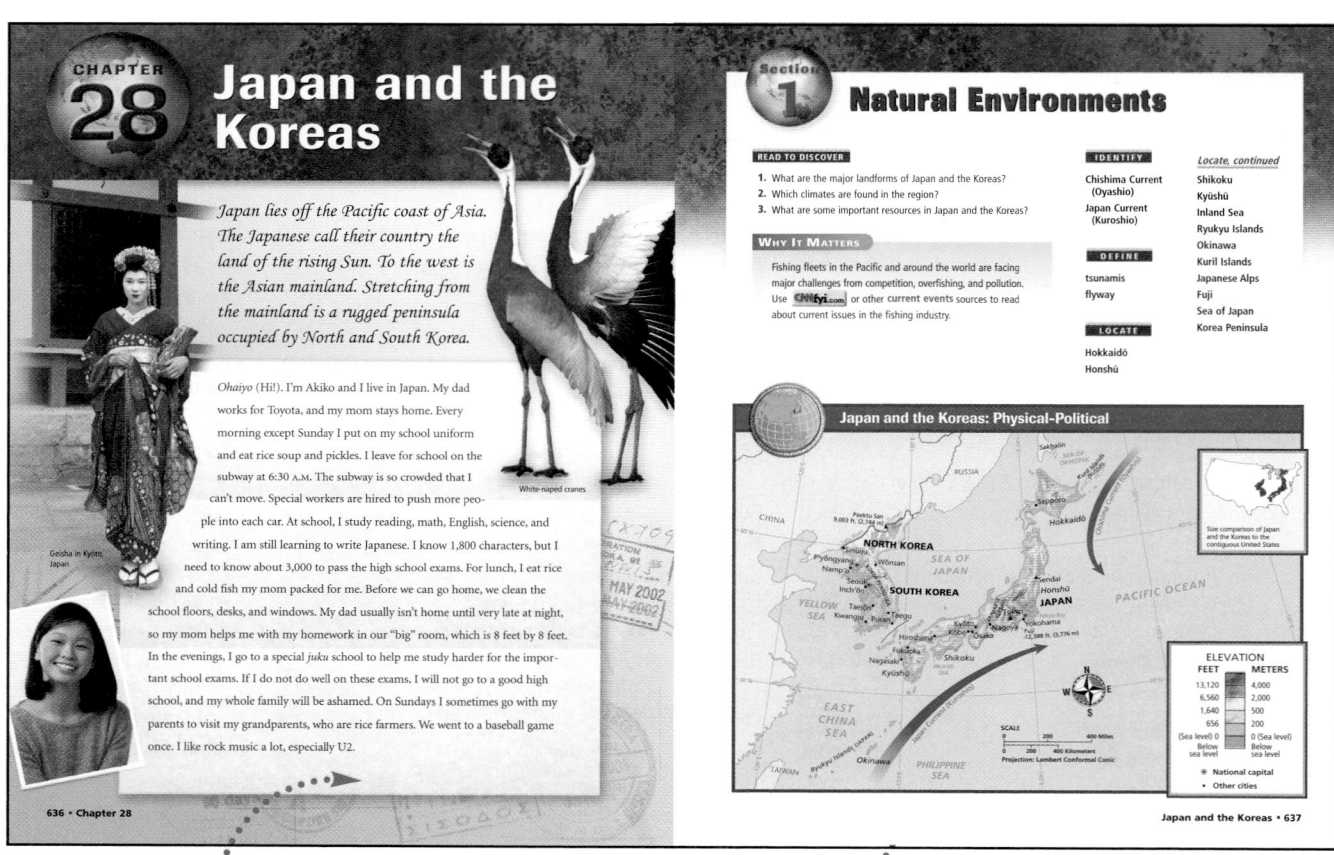

An interview

with a student begins each regional chapter. These interviews give you a glimpse of what life is like for some people in the region you are about to study.

Chapter Map

The map at the beginning of Section 1 in regional chapters shows you the countries you will read about. You can use this map to identify country names and capitals and to locate physical features. These chapter maps will also help you create sketch maps in section reviews.

Use these built-in tools to read for understanding.

Read to Discover questions begin each section of *World Geography Today*. These questions serve as your guide as you read through the section. Keep them in mind as you explore the section content.

Why It Matters is an exciting way for you to make connections between what you are reading in your geography textbook and the world around you. Explore a topic that is relevant to our lives today by using **CNNfyi.com**.

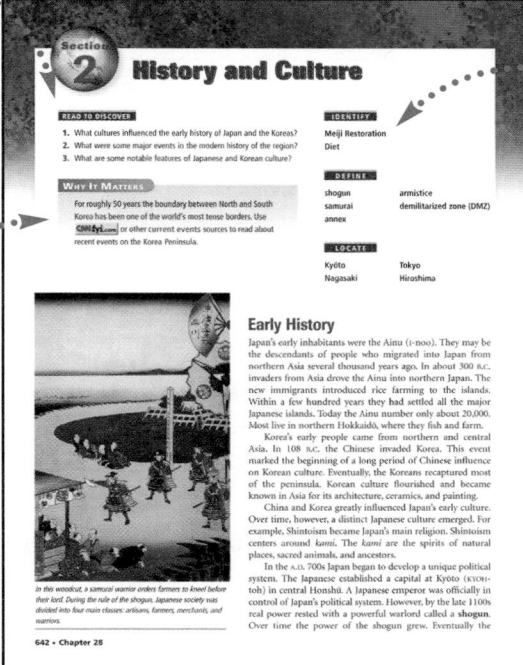

Identify and Define terms are introduced at the beginning of each section. The terms will be defined in context. They will include terms important to the study of geography and to the understanding of the region you are studying.

Interpreting the Visual Record features accompany many of the textbook's rich photographs. Pictures are one of the most important primary sources geographers can use to study our planet. These features invite you to analyze the images so that you can learn more about their content and their links to what you are studying in the section. Other captions ask you to interpret maps, graphs, and charts.

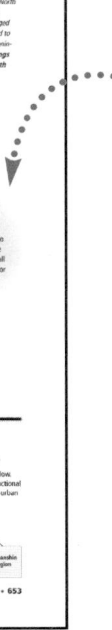

Our Amazing Planet features provide interesting facts about the region you are studying. Here you will learn about the origins of place-names and fascinating tidbits like the Earth's coldest place or largest living thing.

Reading Check questions appear often throughout the textbook to allow you to check your comprehension. As you read, pause for a moment to consider each Reading Check. If you have trouble answering the question, review the material that you just read.

Use these tools to pull together all of the information you have learned.

Reading for the Main Idea questions help review the main points you have studied in the section.

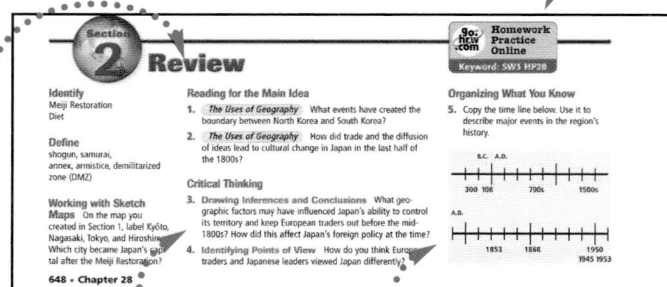

Homework Practice Online lets you log on to the go.hrw.com Web site to complete an interactive self-check of the material covered in the section.

Critical Thinking activities in section and chapter reviews allow you to explore a topic in greater depth and to build your skills.

Graphic Organizers will help you pull together important information from the section. You can complete the graphic organizer as a study tool to prepare for a test or writing assignment.

Locating Key Places activities ask you to identify and locate places you have read about in the chapter.

Using the Geographer's Tools, Geography for Life, and Building Social Studies Skills activities help you develop the skills you need to study geography and to answer standardized-test questions.

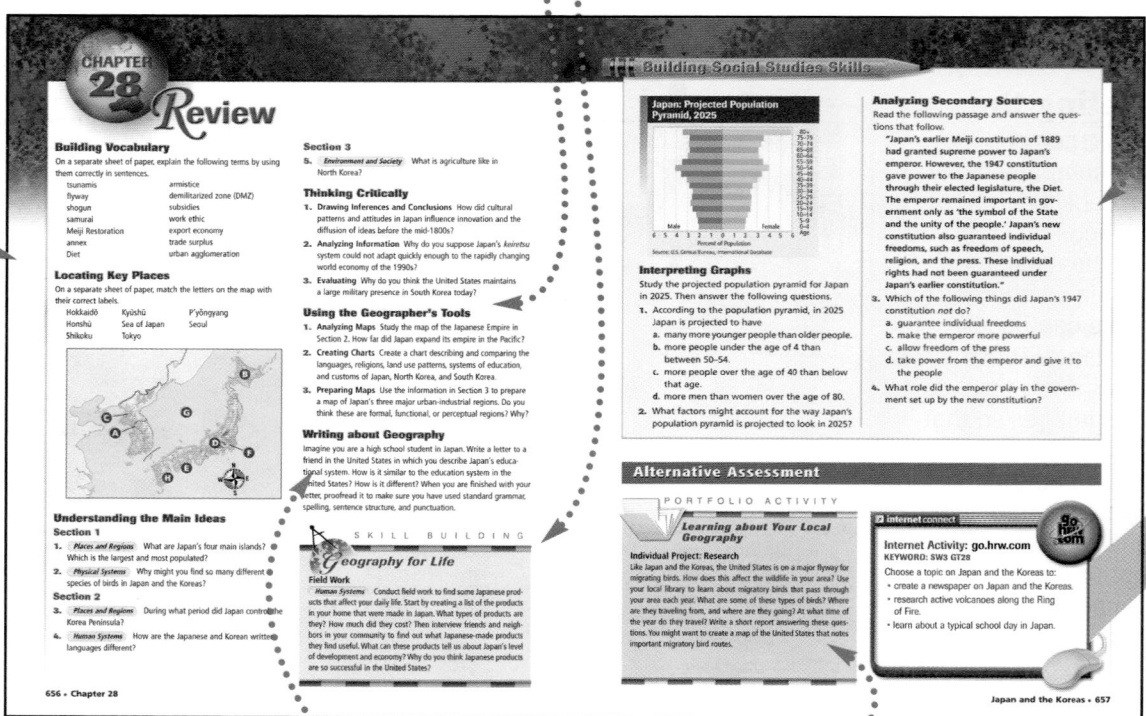

Writing about Geography activities let you practice your writing skills to explore in more detail topics you have studied in the chapter.

Portfolio Activities are exciting and creative ways to explore your local geography and to make connections to the region you are studying. Some activities ask you to work cooperatively, and others include projects to complete on your own.

Use these online tools to review and complete online activities.

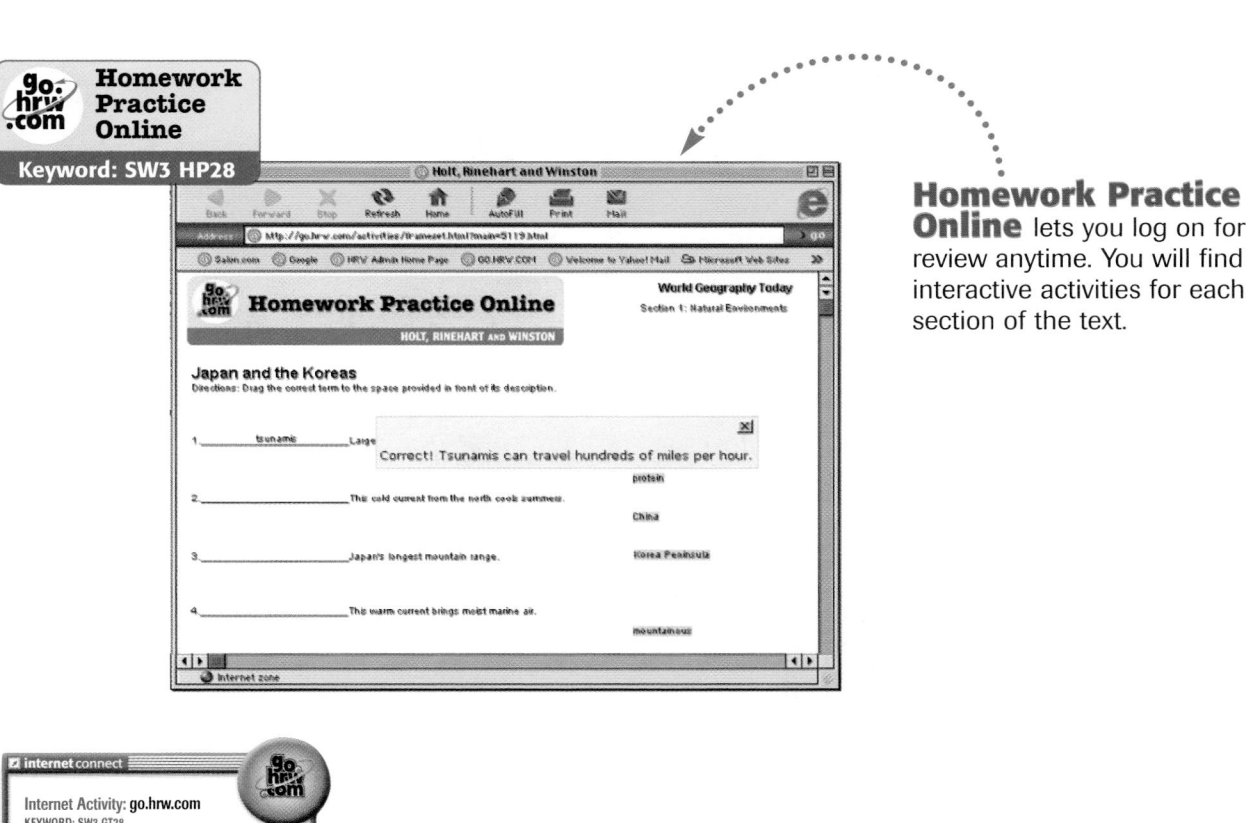

Homework Practice Online lets you log on for review anytime. You will find interactive activities for each section of the text.

internet connect

Internet Activity: **go.hrw.com**
KEYWORD: SW3 GT28

Choose a topic on Japan and the Koreas to:
* create a newspaper on Japan and the Koreas.
* research active volcanoes along the Ring of Fire.
* learn about a typical school day in Japan.

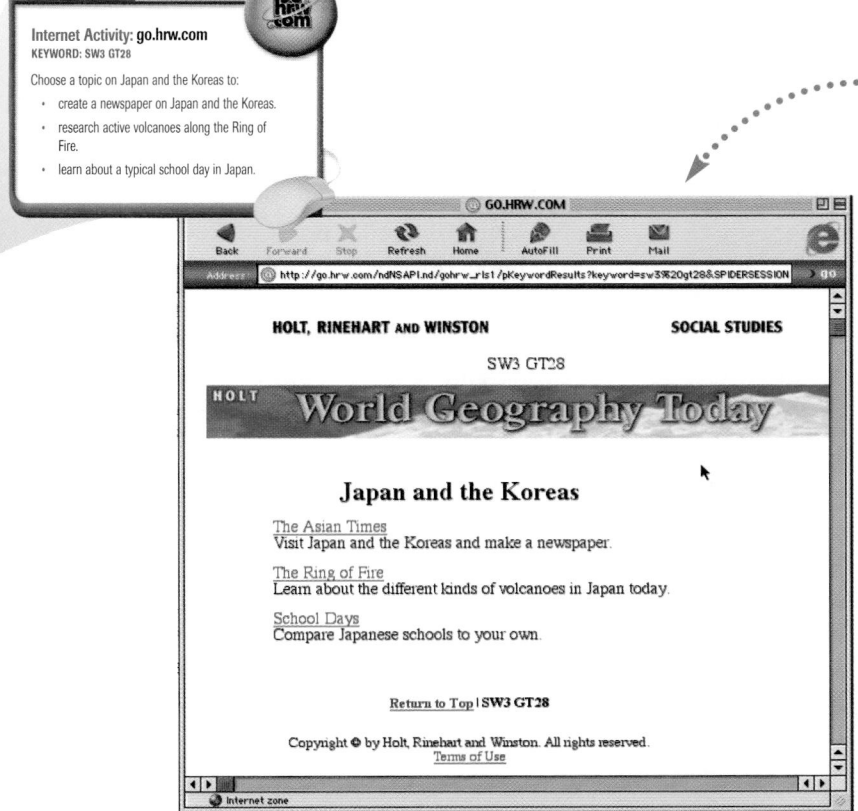

Internet Connect activities are just a part of the world of online learning experiences that await you on the go.hrw.com Web site. By exploring these online activities, you will take a journey through some of the richest world geography materials available on the World Wide Web. You can then use these resources to create real-world projects, such as brochures, databases, newspapers, reports, and even your own Web site!

Why Geography Matters

Have you ever wondered. . .

why some places are deserts while other places get so much rain? What makes certain times of the year cooler than others? Why do some rivers run dry? Maybe you live near mountains and wonder what processes created them.

Student reporters contribute to **CNNfyi.com.**

Do you know why the loss of huge forest areas in one part of the world can affect areas far away? Why does the United States have many different kinds of churches and other places of worship? Perhaps you are curious why Americans and people from other countries have such different points of view on many issues. The key to understanding questions and issues like these lies in the study of geography.

Geography and Your World

All you need to do is watch or read the news to see the importance of geography. You have probably seen news stories about the effects of floods, volcanic eruptions, and other natural events on people and places. You likely have also seen how conflict and cooperation shape the relations between peoples and countries around the world. The Why It Matters feature beginning every section of World Geography Today uses the vast resources of **CNNfyi.com** or other current events sources to examine the importance of geography. Through this feature you will be able to draw connections between what you are studying in your geography textbook and events and conditions found around the world today.

My fall semester project testing lake water.

Geography and Making Connections

When you think of the word geography, what comes to mind? Perhaps you simply picture people memorizing names of countries and capitals. Maybe you think of people studying maps to identify features like deserts, mountains, oceans, and rivers. These things are important, but the study of geography includes much more. Geography involves asking questions and solving problems. It focuses on looking at people and their ways of life as well as studying physical features like mountains, oceans, and rivers. Studying geography also means looking at why things are where they are and at the relationships between human and physical features of Earth.

The study of geography helps us make connections between what was, what is, and what may be. It helps us understand the processes that have shaped the features we observe around us, as well as the ways those features may be different tomorrow. In short, geography helps us understand the processes that have created a world that is home to more than 6 billion people and countless billions of other creatures.

Geography and You

Anyone can influence the geography of our world. For example, the actions of individuals affect local environments. Some individual actions might pollute the environment. Other actions might contribute to efforts to keep the environment clean and healthy. Various other things also influence geography. For example, governments create political divisions, such as countries and states. The borders between these divisions influence the human geography of regions by separating peoples, legal systems, and human activities.

Governments and businesses also plan and build structures like dams, railroads, and airports, which change the physical characteristics of places. As you might expect, some actions influence Earth's geography in negative ways, others in positive ways. Understanding geography helps us evaluate the consequences of our actions.

Skills Handbook and Atlas

Critical Thinking

Throughout World Geography Today, *you are asked to think critically about some of the information you are studying. Critical thinking is the reasoned judgment of information and ideas. The development of critical thinking skills is essential to effective citizenship. Such skills empower you to exercise your civic rights and responsibilities as well as learn more about the world around you. Helping you develop critical thinking skills is an important goal of* World Geography Today. *The following critical thinking skills appear in the section reviews and chapter reviews of the textbook.*

Summarizing involves briefly restating information gathered from a larger body of information. Much of the writing in this textbook is summarizing. The geographical data in this textbook has been collected from many sources. Summarizing all the qualities of a region or country involves studying a large body of cultural, economic, geological, and historical information.

Finding the main idea is the ability to identify the main point in a set of information. This textbook is designed to help you focus on the main ideas in geography. The Read to Discover questions in each chapter help you identify the main ideas in each section. To find the main idea in any piece of writing, first read the title and introduction. These two elements may point to the main ideas covered in the text. Also, formulate questions about the subject that you think might be answered in the text. Having such questions in mind will focus your reading. Pay attention to any headings or subheadings, which may provide a basic outline of the major ideas. Finally, as you read, note sentences that provide additional details from the general statements that those details support. For example, a trail of facts may lead to a conclusion that expresses the main idea.

Mud mosque, Mali

Comparing and contrasting

involve examining events, points of view, situations, or styles to identify their similarities and differences. Comparing focuses on both the similarities and the differences. Contrasting focuses only on the differences. Studying similarities and differences between people and things can give you clues about the human and physical geography of a region.

Buddhist shrine, Myanmar

Stave church, Norway

Supporting a point of view

involves identifying an issue, deciding what you think about it, and persuasively expressing your position. Your stand should be based on specific information. When taking a stand, state your position clearly and give reasons that support it.

Identifying points of view

involves noting the factors that influence the outlook of an individual or group. A person's point of view includes beliefs and attitudes that are shaped by factors such as age, gender, race, and economic status. Identifying points of view helps us examine why people see things as they do. It also reinforces the realization that people's views may change over time or with a change in circumstances.

Political protest, India

Identifying bias

is an important critical thinking skill in the study of any subject. When a point of view is highly personal or based on unreasoned judgment, it is considered biased. Sometimes, a person's actions reflect bias. At its most extreme, bias can be expressed in violent actions against members of a particular culture or group. A less obvious form of bias is a stereotype, or a generalization about a group of people. Stereotypes tend to ignore differences within groups.

Probably the hardest form of cultural bias to detect has to do with perspective, or point of view. When we use our own culture and experiences as a point of reference from which to make statements about other cultures, we are showing a form of bias called ethnocentrism.

Analyzing

is the process of breaking something down into parts and examining the relationships between those parts. For example, to understand the processes behind forest loss, you might study issues involving economic development, the overuse of resources, and pollution.

Ecuador rainforest

Cleared forest, Kenya

Evaluating

involves assessing the significance or overall importance of something. For example, you might evaluate the success of certain environmental protection laws or the effect of foreign trade on a society. You should base your evaluation on standards that others will understand and are likely to consider valid. For example, an evaluation of international relations after World War II might assess the political and economic tensions between the United States and the Soviet Union. Such an evaluation would also consider the ways those tensions affected other countries around the world.

Identifying cause and effect

is part of interpreting the relationships between geographical events. A cause is any action that leads to an event; the outcome of that action is an effect. To explain geographical developments, geographers may point out multiple causes and effects. For example, geographers studying pollution in a region might note a number of causes.

Drawing inferences and drawing conclusions

are two methods of critical thinking that require you to use evidence to explain events or information in a logical way. Inferences and conclusions are opinions, but these opinions are based on facts and reasonable deductions.

Drought in West Texas   *Dallas, Texas*

For example, suppose you know that people are moving in greater and greater numbers to cities in a particular country. You also know that poor weather has hurt farming in outlying areas while industry has been expanding in cities. You might infer from this information some of the reasons for the increased migration to cities. You could conclude that poor harvests have pushed people to leave outlying areas. You might also conclude that the possibility of finding work in new industries may be pulling people to cities.

Making generalizations and predictions

are two critical thinking skills that require you to form concise ideas from a large body of information. When you are asked to generalize, you must take into account many different pieces of information. You then form a unifying concept that can be applied to all of the pieces of information.

Many times making generalizations can help you see trends. Looking at trends can help you form a prediction. Making a prediction involves looking at trends in the past and present and making an educated guess about how these trends will affect the future.

Communications technology, rural Brazil

Writing about Geography

Writers have many different reasons for writing. In your study of geography, you might write to accomplish many different tasks. You might write a paragraph or short paper to express your own personal feelings or thoughts about a topic or event. You might also write a paper to inform your class about an event, person, place, or thing. Sometimes you may want to write in order to persuade or convince readers to agree with a certain statement or to act in a particular way.

You will find various kinds of questions at the end of each section, chapter, and unit throughout this textbook. Some questions will require in-depth answers. The following guidelines for writing will help you structure your answers so that they clearly express your thoughts.

The Writing Process

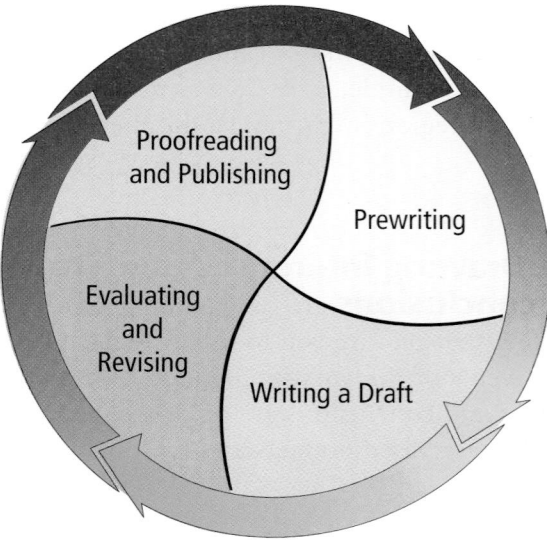

Prewriting Prewriting is the process of thinking about and planning what to write. It includes gathering and organizing information into a clear plan. Writers use the prewriting stage to identify their audience and purpose for what is to be written.

Often, writers do research to get the information they need. Research can include finding primary and secondary sources. You will read about primary and secondary sources later in this handbook.

Writing a Draft After you have gathered and arranged your information, you are ready to begin writing. Many paragraphs are structured in the following way:

- **Topic Sentence:** The topic sentence states the main idea of the paragraph. Putting the main idea into the form of a topic sentence helps keep the paragraph focused.

- **Body:** The body of a paragraph develops and supports the main idea. Writers use a variety of information, including facts, opinions, and examples, to support the main idea.

- **Conclusion:** The conclusion summarizes the writer's main points or restates the main idea.

Evaluating and Revising Read over your paragraphs and make sure you have clearly expressed what you wanted to say. Sometimes it helps to read your paragraphs aloud or to ask someone else to read it. Such methods help you identify rough or unclear sentences and passages. Revise the parts of your paragraph that are not clear or that stray from your main idea. You might want to add, cut, reorder, or replace sentences to make your paragraph as clear as possible.

Proofreading and Publishing
Before you write your final draft, read over your paragraphs and correct any errors in grammar, spelling, sentence structure, or punctuation. Common mistakes include misspelled place-names, incomplete sentences, and improper use of punctuation, such as commas. You should use a dictionary and standard grammar guides to help you proofread your work.

After you have revised and corrected your draft, neatly rewrite or type your paper. Make sure your final version is clean and free of mistakes. The appearance of your final version can affect how your audience perceives and understands your writing.

Practicing the Skill

1. What are the steps in the writing process?
2. What elements are found in many paragraphs?
3. Write a paragraph or short paper about your community for a visitor. When you have finished your draft, review it and then mark and correct any errors in grammar, spelling, sentence structure, or punctuation. At the bottom of your draft, list key resources—such as a dictionary— that you used to check and correct your work. Then write your final draft. When you are finished with your work, use pencils or pens of different colors to underline and identify the topic sentence, body, and conclusion of your paragraph.

Doing Research

Research is at the heart of geographic inquiry. To complete a research project, you may need to use resources other than this textbook. For example, you may want to research specific places or issues not discussed in this textbook. You may also want to learn more about a certain topic that you have studied in a chapter. Following the guidelines below will help you plan and execute any research project you would like to undertake.

Planning The first step in approaching a research project is planning. Planning involves deciding on a topic and finding information about that topic.

- **Decide on a Topic.** Before starting any research project, you should decide on one topic. If you are working with a group, all group members should participate in choosing a topic. Sometimes a topic will be assigned to you, but at other times you may have to choose your own. Once you have settled on a topic, make sure you can find resources to help you research it.

- **Find Information.** In order to find a particular book, you need to know how libraries organize their materials. Libraries classify their books by assigning each book a call number that tells you its location. To find the call number, look in the library's card catalog. The card catalog lists books by author, by title, and by subject. Many libraries today have computerized card catalogs. Libraries often provide instructions on how to use their computerized card catalogs. If no instructions are available, ask a library staff member for help.

Most libraries have encyclopedias, gazetteers, atlases, almanacs, and periodical indexes. Encyclopedias contain geographic, economic, and political data on individual countries, states, and cities. They also include discussions of historical events, religion, social and cultural issues, and much more. A gazetteer is a geographical dictionary that lists significant natural physical features and other places. An atlas contains maps and visual representations of geographic data. To find up-to-date facts, you can use almanacs, yearbooks, and periodical indexes. References like *The World Almanac and Book of Facts* include historical information and a variety of statistics. Periodical indexes, particularly *The Reader's Guide to Periodical Literature*, can help you locate informative articles published in magazines. *The New York Times Index* catalogs the newspaper articles published in the *New York Times.*

You may also want to find information on the World Wide Web. The World Wide Web is the part of the Internet where people put files called Web sites for other people to access. To search the World Wide Web, you must use a search engine. A search engine will provide you with a list of Web sites that contain keywords relating to your topic. Search engines also provide Web directories, which allow you to browse Web sites by subject.

● Organizing Organization is key to completing research projects of any size. If you are working with a group, every group member should have an assigned task in researching, writing, and completing your project. You and all the group members should keep track of the materials that you used to conduct your research. Then compile those sources into a bibliography and turn it in with your research project.

In addition, information collected during research should be organized in an efficient way. A common method of organizing research information is to use index cards. If you have used an outline to organize your research, you can code each index card with the appropriate main idea number and supporting detail letter from the outline. Then write the relevant information on that card. You might also use computer files in the same way. These methods will help you keep track of what information you have collected and what information you still need to gather.

Some projects will require you to conduct original research. This original research might require you to interview people, conduct surveys, collect unpublished information about your community, or draw a map of a local place.

Before you do your original research, make sure you have all the necessary background information. Also, create a pre-research plan so that you can make sure all the necessary tools, such as research sources, are available.

● Completing and Presenting Your Project Once you have completed your research project, you will need to present the information you have gathered in some fashion. Many times, you or your group will simply need to write a paper about your research. Research can also be presented in many other ways, however. For example, you could make an audio tape, a drawing, a poster board, a video, or a Web page to explain your research.

Practicing the Skill

1. What kinds of references would you need to research specific current events around the world?
2. Work with a group of four other students to plan, organize, and complete a research project on a topic of interest in your local community. For example, you might want to learn more about a particular individual or event that influenced your community's history. Other topics might include the economic features, physical features, and political features of your community.

Analyzing Primary and Secondary Sources

When conducting research, it is important to use a variety of primary and secondary sources of information. There are many sources of firsthand geographical information, including diaries, letters, editorials, and legal documents such as land titles. All of these are primary sources. Newspaper articles are also considered primary sources, although they are generally written after the fact. Other primary sources include personal memoirs and autobiographies, which people usually write late in life. Paintings and photographs of particular events, persons, places, or things make up a visual record and are also considered primary sources. Because they allow us to take a close-up look at a topic, primary sources are valuable geographic tools.

Secondary sources are descriptions or interpretations of events written after the events have occurred by persons who did not participate in the events they describe. Geography textbooks such as this one as well as biographies, encyclopedias, and other reference works are examples of secondary sources. Writers of secondary sources have the advantage of seeing what happened beyond the moment or place that is being studied. They can provide a perspective wider than that available to one person at a specific time.

How to Study Primary and Secondary Sources

1 Study the Material Carefully. Consider the nature of the material. Is it verbal or visual? Is it based on firsthand information or on the accounts of others? Note the major ideas and supporting details.

2 Consider the Audience. Ask yourself, "For whom was this message originally meant?" Whether a message was intended for the general public or for a specific private audience may have shaped its style or content.

3 Check for Bias. Watch for words or phrases that present a one-sided view of a person or situation.

4 Compare Sources. Study more than one source on a topic. Comparing sources gives you a more complete and balanced account of geographical events and their relationships to one another.

Practicing the Skill

1. What distinguishes secondary sources from primary sources?
2. What advantages do secondary sources have over primary sources?
3. Why should you consider the intended audience of a source?
4. Of the following, identify which are primary sources and which are secondary sources: a newspaper, a private journal, a biography, an editorial cartoon, a medieval tapestry, a deed to property, a snapshot of a family vacation, a magazine article about the history of Thailand, an autobiography. How might some of these sources prove to be both primary and secondary sources?

Becoming a Strategic Reader

by Dr. Judith Irvin

Everywhere you look, print is all around us. In fact, you would have a hard time stopping yourself from reading. In a normal day, you might read cereal boxes, movie posters, notes from friends, t-shirts, instructions for video games, song lyrics, catalogs, billboards, information on the Internet, magazines, the newspaper, and much, much more. Each form of print is read differently depending on your purpose for reading. You read a menu differently from poetry, and a motorcycle magazine is read differently than a letter from a friend. Good readers switch easily from one type of text to another. In fact, they probably do not even think about it, they just do it.

When you read, it is helpful to use a strategy to remember the most important ideas. You can use a strategy before you read to help connect information you already know to the new information you will encounter. Before you read, you can also predict what a text will be about by using a previewing strategy. During the reading you can use a strategy to help you focus on main ideas, and after reading you can use a strategy to help you organize what you learned so that you can remember it later. *World Geography Today* was designed to help you more easily understand the ideas you read. Important reading strategies employed in *World Geography Today* include:

A Tools to help you **preview and predict** what the text will be about

B Ways to help you **use and analyze visual information**

C Ideas to help you **organize the information** you have learned

A. Previewing and Predicting

How can I figure out what the text is about before I even start reading a section?

Previewing and **predicting** are good methods to help you understand the text. If you take the time to preview and predict before you read, the text will make more sense to you during your reading.

1 Usually, your teacher will set the purpose for reading. After reading some new information, you may be asked to write a summary, take a test, or complete some other type of activity.

"After reading about Japan, you will work with a partner to create a cultural museum exhibit describing . . ."

Previewing and Predicting

step 1 Identify your purpose for reading. Ask yourself what you will do with this information once you have finished reading.

▼

step 2 Ask yourself what is the main idea of the text and what are the key vocabulary words you need to know.

▼

step 3 Use signal words to help identify the structure of the text.

▼

step 4 Connect the information to what you already know.

2 As you preview the text, use **graphic signals** such as headings, subheadings, and boldface type to help you determine what is important in the text. Each section of *World Geography Today* opens by giving you important clues to help you preview the material.

Section 2 — History and Culture

READ TO DISCOVER

1. What cultures influenced the early history of Japan and the Koreas?
2. What were some major events in the modern history of the region?
3. What are some notable features of Japanese and Korean culture?

WHY IT MATTERS

For roughly 50 years the boundary between North and South Korea has been one of the world's most tense borders. Use **CNN fyi.com** or other current events sources to read about recent events on the Korea Peninsula.

IDENTIFY

Meiji Restoration
Diet

DEFINE

shogun   armistice
samurai   demilitarized zone (DMZ)
annex

LOCATE

Kyōto   Tokyo
Nagasaki   Hiroshima

Looking at the section's **main heading** and subheadings can give you an idea of what is to come.

Read to Discover questions give you clues as to the section's main ideas.

Identify and Define terms let you know the key vocabulary you will encounter in the section.

A. Previewing and Predicting (continued)

3 Other tools that can help you in previewing are **signal words**. These words prepare you to think in a certain way. For example, when you see words such as *similar to*, *same as*, or *different from*, you know that the text will probably compare and contrast two or more ideas. Signal words indicate how the ideas in the text relate to each other. Look at the list below for some of the most common signal words grouped by the type of text structures they include.

SIGNAL WORDS

Cause and Effect	Compare and Contrast	Description	Problem and Solution	Sequence or Chronological Order
because	different from	for instance	the question is	not long after
since	same as	for example	a solution	next
consequently	similar to	such as	one answer is	then
this led to...so	as opposed to	to illustrate		initially
if...then	instead of	in addition		before
nevertheless	although	most importantly		after
accordingly	however	another		finally
because of	compared with	furthermore		preceding
as a result of	as well as	first,		following
in order to	either...or	second ...		on (date)
may be due to	but			over the years
for this reason	on the other hand			today
not only...but	unless			when

4 Learning something new requires that you connect it in some way with something you already know. This means you have to think before you read and while you read. You may want to use a chart like this one to remind yourself of the information already familiar to you and to come up with questions you want answered in your reading. The chart will also help you organize your ideas after you have finished reading.

What I know	What I want to know	What I learned

B. Use and Analyze Visual Information

How can all the pictures, maps, graphs, and time lines with the text help me be a stronger reader?

Using visual information can help you understand and remember the information presented in *World Geography Today*. Good readers make a picture in their mind when they read. The pictures, charts, graphs, time lines, and diagrams that occur throughout *World Geography Today* are placed strategically to increase your understanding.

1 You might ask yourself questions like these:

Why did the writer include this image with the text?

What details about this image are mentioned in the text?

Analyzing Visual Information

step 1 As you preview the text, ask yourself how the visual information relates to the text.

▼

step 2 Generate questions based on the visual information.

▼

step 3 After reading the text, go back and review the visual information again.

▼

step 4 Connect the information to what you already know.

2 After you have read the text, see if you can answer your own questions.

→ Why is the train important?

→ What infrastructure is needed for this train to operate?

→ Why is the area around the rail line built up?

B. Use and Analyze Visual Information *(continued)*

2 Maps, graphs, and charts help you organize information about a place. You might ask questions like these:

> How does this map support what I have read in the text?
>
> What does the information in these pie and bar graphs add to the text discussion?

→ What is the purpose of this map?

→ What special features does the map show?

→ What do the colors, lines, and symbols on the map represent?

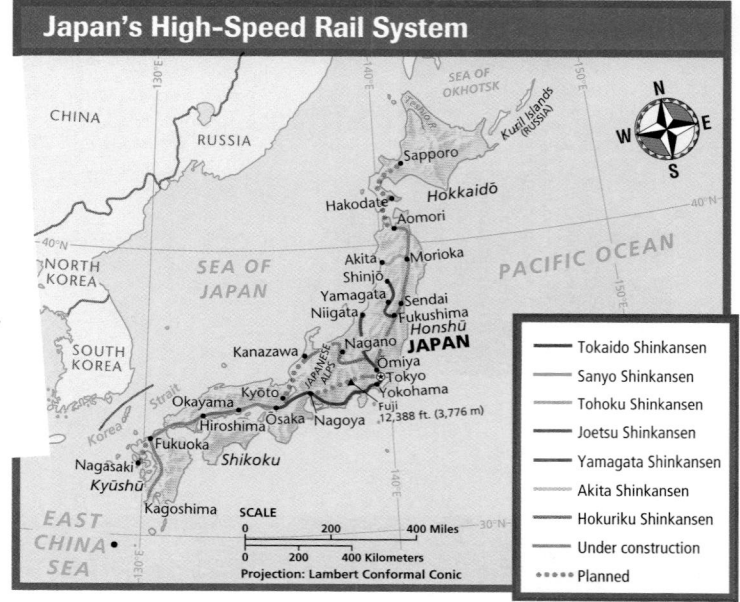

Japan's High-Speed Rail System

| Tokaido Shinkansen |
| Sanyo Shinkansen |
| Tohoku Shinkansen |
| Joetsu Shinkansen |
| Yamagata Shinkansen |
| Akita Shinkansen |
| Hokuriku Shinkansen |
| Under construction |
| Planned |

SCALE 0 200 400 Miles
0 200 400 Kilometers
Projection: Lambert Conformal Conic

→ What information is the writer trying to present with these graphs?

→ Why did the writer use a pie graph and a bar graph to organize this information?

→ Is information in the pie graph important to understanding the bar graph?

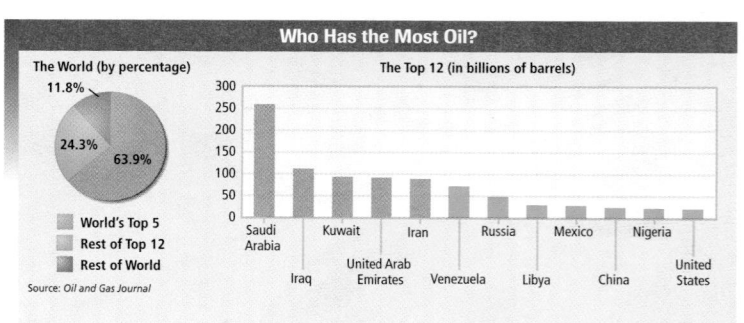

Who Has the Most Oil?

The World (by percentage)
11.8%
24.3%
63.9%

World's Top 5
Rest of Top 12
Rest of World

The Top 12 (in billions of barrels)
300
250
200
150
100
50
0

Saudi Arabia, Kuwait, Iran, Russia, Mexico, Nigeria, Iraq, United Arab Emirates, Venezuela, Libya, China, United States

Source: Oil and Gas Journal

3 After reading the text, go back and review the visual information again.

4 Connect the information to what you already know.

C. Organize Information

Once I learn new information, how do I keep it all straight so that I will remember it?

To help you remember what you have read, you need to find a way of **organizing information**. Two good ways of doing this are by using graphic organizers and concept maps. **Graphic organizers** help you understand important relationships—such as cause-and-effect, compare/contrast, sequence of events, and problem/solution—within the text. **Concept maps** provide a useful tool to help you focus on the text's main ideas and organize supporting details.

Identifying Relationships

Using graphic organizers will help you recall important ideas from the section and give you a study tool you can use to prepare for a quiz or test or to help with a writing assignment. Some of the most common types of graphic organizers are shown below.

▶ Cause and Effect

Events in history cause people to react in a certain way. Cause-and-effect patterns show the relationship between results and the ideas or events that made the results occur. You may want to represent cause-and-effect relationships as one cause leading to multiple effects,

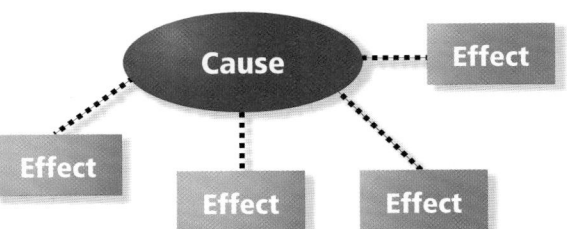

or as a chain of cause-and-effect relationships.

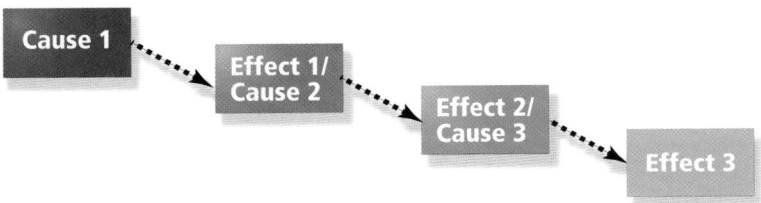

Constructing Graphic Organizers

step 1 Preview the text, looking for signal words and the main idea.

▼

step 2 Form a hypothesis as to which type of graphic organizer would work best to display the information presented.

▼

step 3 Work individually or with your classmates to create a visual representation of what you read.

▶ Comparing and Contrasting

Graphic Organizers are often useful when you are comparing or contrasting information. Compare-and-contrast diagrams point out similarities and differences between two concepts or ideas.

Characteristics   Shared Characteristics   Characteristics

▶ Sequencing

Keeping track of dates and the order in which events took place is essential to understanding the history and geography of a place. Sequence or chronological-order diagrams show events or ideas in the order in which they happened.

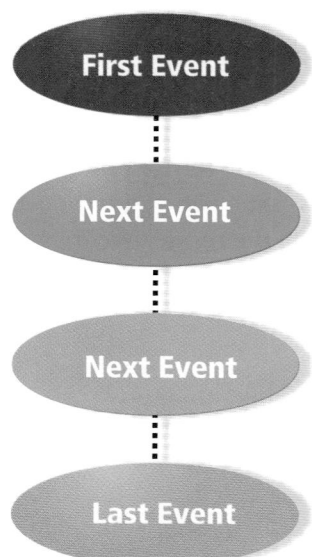

First Event

Next Event

Next Event

Last Event

▶ Problem and Solution

Problem-solution patterns identify at least one problem, offer one or more solutions to the problem, and explain or predict outcomes of the solutions.

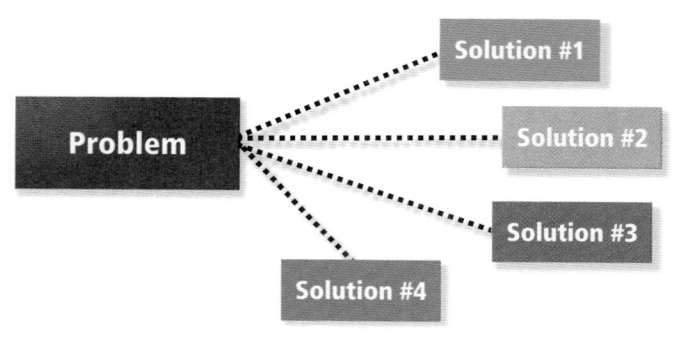

Problem

Solution #1

Solution #2

Solution #3

Solution #4

Identifying Main Ideas and Supporting Details

One special type of graphic organizer is the concept map. A concept map, sometimes called a semantic map, allows you to zero in on the most important points of the text. The map is made up of lines, boxes, circles, and/or arrows. It can be as simple or as complex as you need it to be to accurately represent the text.

Here are a few examples of concept maps you might use.

Constructing Concept Maps

step 1

Preview the text, looking at what type of structure might be appropriate to display on a concept map.

▼

step 2

Taking note of the headings, bold-faced type, and text structure, sketch a concept map you think could best illustrate the text.

▼

step 3

Using boxes, lines, arrows, circles, or any shapes you like, display the ideas of the text in the concept map.

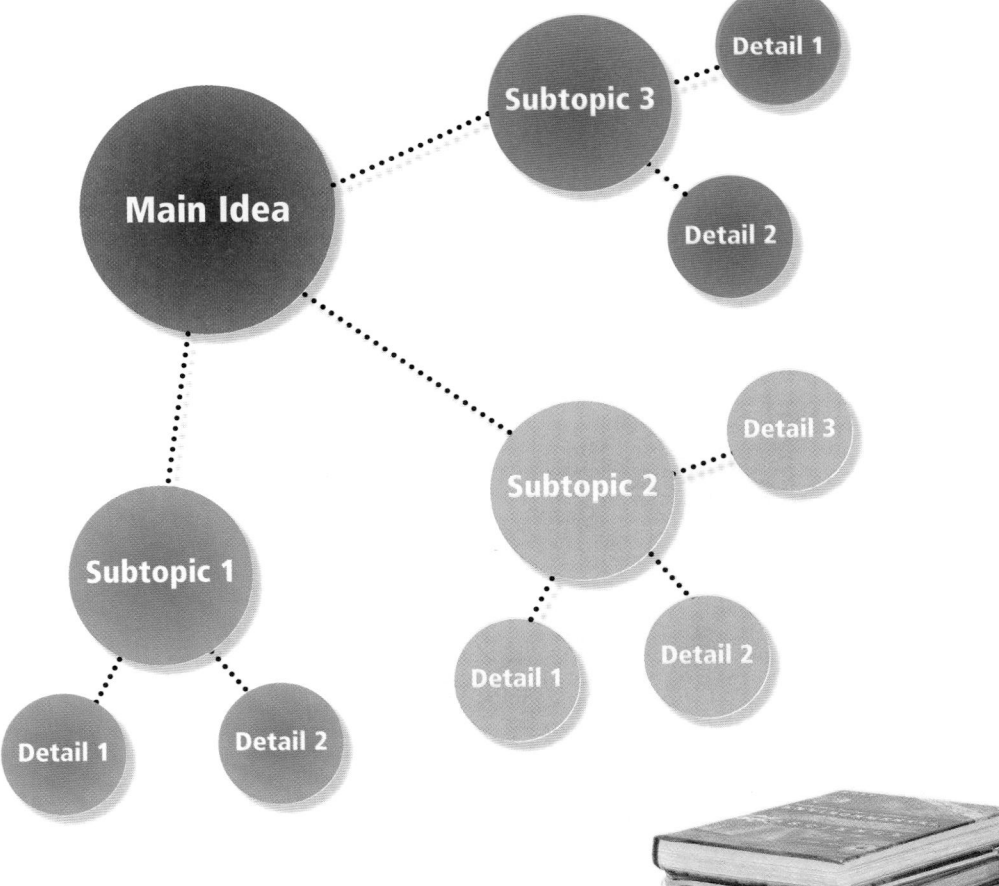

Standardized Test-Taking Strategies

A number of times throughout your school career, you may be asked to take standardized tests. These tests are designed to demonstrate the content and skills you have learned. It is important to keep in mind that in most cases the best way to prepare for the test is to pay close attention in class and take every opportunity to improve your general social studies, reading, writing, and mathematical skills.

Tips for Taking the Test

1. Be sure that you are well rested.
2. Be on time, and be sure that you have the necessary materials.
3. Listen to the teacher's instructions.
4. Read directions and questions carefully.
5. **DON'T STRESS!** Just remember what you have learned in class, and you should do well.

Practice the strategies at go.hrw.com .....

STANDARDIZED TEST PREP ONLINE
keyword: SW3 STP

Tackling Social Studies

The social studies portions of many standardized tests are designed to test your knowledge of the content and skills that you have been studying in one or more of your social studies classes. Specific objectives for the test vary, but some of the most common include the following:

1. Demonstrate an understanding of issues and events in history.
2. Demonstrate an understanding of geographic influences on historical issues and events.
3. Demonstrate an understanding of economic and social influences on historical issues and events.
4. Demonstrate an understanding of political influences on historical issues and events.
5. Use critical thinking skills to analyze social studies information.

Standardized tests usually contain multiple-choice and, sometimes, open-ended questions. The multiple-choice items will often be based on maps, tables, charts, graphs, pictures, cartoons, and/or reading passages and documents.

Tips for Answering Multiple-Choice Questions

1. If there is a written or visual piece accompanying the multiple-choice question, pay careful attention to the title, author, and date.

2. Then read through or glance over the content of the written or visual piece accompanying the question to familiarize yourself with it.

3. Next, read the multiple-choice question first for its general intent. Then reread it carefully, looking for words that give clues or can limit possible answers to the question. For example, words such as *most* or *best* tell you that there may be several correct answers to a question, but you should look for the most appropriate answer.

4. Read through the answer choices. Always read all of the possible answer choices even if the first one seems like the correct answer. There may be a better choice farther down in the list.

5. Reread the accompanying information (if any is included) carefully to determine the answer to the question. Again, note the title, author, and date of primary-source selections. The answer will rarely be stated exactly as it appears in the primary source, so you will need to use your critical thinking skills to read between the lines.

6. Think of what you already know about the time in history or person involved and use that to help limit the answer choices.

7. Finally, reread the question and selected answer to be sure that you made the best choice and that you marked it correctly on the answer sheet.

Strategies for Success

There are a variety of strategies you can prepare ahead of time to help you feel more confident about answering questions on the social studies TAKS. Here are a few suggestions:

1. Adopt an acronym—a word formed from the first letters of other words—that you will use for analyzing a document or visual piece that accompanies a question.

Helpful Acronyms

For a document, use **SOAPS**, which stands for

S	Subject
O	Overview
A	Audience
P	Purpose
S	Speaker/author

For a picture, cartoon, map, or other visual piece of information, use **OPTIC**, which stands for

O	Occasion (or time)
P	Parts (labels or details of the visual)
T	Title
I	Interrelations (how the different parts of the visual work together)
C	Conclusion (what the visual means)

2. Form visual images of maps and try to draw them from memory. The TAKS will most likely include maps showing many features, such as states, countries, continents, and oceans. Those maps may also show patterns in settlement and the size and distribution of cities. For example, in studying the United States, be able to see in your mind's eye such things as where the states and major cities are located. Know major physical features, such as the Mississippi River, the Appalachian and Rocky Mountains, the Great Plains, and the various regions of the United States, and be able to place them on a map. Such features may help you understand patterns in the distribution of population and the size of settlements.

3. When you have finished studying a geographic region or period in history, try to think of who or what might be important enough for the TAKS. You may want to keep your ideas in a notebook to refer to when it is almost time for the test.

4. The TAKS will likely test your understanding of the political, economic, and social processes that shape a region's history, culture, and geography. Questions may also ask you to

understand the impact of geographic factors on major events. For example, some may ask about the effects of migration and immigration on various societies and population change. In addition, questions may test your understanding of the ways humans interact with their environment.

5. For the skills area of the tests, practice putting major events and personalities in order in your mind. Sequencing people and events by dates can become a game you play with a friend who also has to take the test. Always ask yourself "why" this event is important.

6. Follow the tips under "Ready for Reading" below when you encounter a reading passage in social studies, but remember that what you have learned about history can help you in answering reading-comprehension questions.

Ready for Reading

The main goal of the reading sections of most standardized tests is to determine your understanding of different aspects of a piece of writing. Basically, if you can grasp the main idea and the writer's purpose and then pay attention to the details and vocabulary so that you are able to draw inferences and conclusions, you will do well on the test.

Tips for Answering Multiple-Choice Questions

1. Read the passage as if you were not taking a test.

2. Look at the big picture. Ask yourself questions like, "What is the title?", "What do the illustrations or pictures tell me?", and "What is the writer's purpose?"

3. Read the questions. This will help you know what information to look for.

4. Reread the passage, underlining information related to the questions.

5. Go back to the questions and try to answer each one in your mind before looking at the answers.

6. Read all the answer choices and eliminate the ones that are obviously incorrect.

Types of Multiple-Choice Questions

1. **Main Idea** This is the most important point of the passage. After reading the passage, locate and underline the main idea.

2. **Significant Details** You will often be asked to recall details from the passage. Read the question and underline the details as you read, but remember that the correct answers do not always match the wording of the passage precisely.

3. **Vocabulary** You will often need to define a word within the context of the passage. Read the answer choices and plug them into the sentence to see what fits best.

4. **Conclusion and Inference** There are often important ideas in the passage that the writer does not state directly. Sometimes you must consider multiple parts of the passage to answer the question. If answers refer to only one or two sentences or details in the passage, they are probably incorrect.

Tips for Answering Short-Answer Questions

1. Read the passage in its entirety, paying close attention to the main events and characters. Jot down information you think is important.

2. If you cannot answer a question, skip it and come back later.

3. Words such as *compare, contrast, interpret, discuss,* and *summarize* appear often in short-answer questions. Be sure you have a complete understanding of each of these words.

4. To help support your answer, return to the passage and skim the parts you underlined.

5. Organize your thoughts on a separate sheet of paper. Write a general statement with which to begin. This will be your topic statement.

6. When writing your answer, be precise but brief. Be sure to refer to details in the passage in your answer.

Targeting Writing

On many standardized tests you will occasionally be asked to write an essay. In order to write a concise essay, you must learn to organize your thoughts before you begin writing the actual composition. This keeps you from straying too far from the essay's topic.

Tips for Answering Composition Questions

1. Read the question carefully.

2. Decide what kind of essay you are being asked to write. Essays usually fall into one of the following types: persuasive, classificatory, compare/contrast, or "how to." To determine the type of essay, ask yourself questions like, "Am I trying to persuade my audience?", "Am I comparing or contrasting ideas?", or "Am I trying to show the reader how to do something?"

3. Pay attention to keywords, such as *compare*, *contrast*, *describe*, *advantages*, *disadvantages*, *classify*, or *speculate*. They will give you clues as to the structure that your essay should follow.

4. Organize your thoughts on a separate sheet of paper. You will want to come up with a general topic sentence that expresses your main idea. Make sure this sentence addresses the question. You should then create an outline or some type of graphic organizers to help you organize the points that support your topic sentence.

5. Write your composition using complete sentences. Also, be sure to use correct grammar, spelling, punctuation, and sentence structure.

6. Be sure to proofread your essay once you have finished writing.

Gearing Up for Math

On most standardized tests you will be asked to solve a variety of mathematical problems that draw on the skills and information you have learned in class. If math problems sometimes give you difficulty, use the tips below to help you work through the problems.

Tips for Solving Math Problems

1. Decide what is the goal of the question. Read or study the problem carefully and determine what information must be found.

2. Locate the factual information. Decide what information represents key facts—the ones you must have to solve the problem. You may also find facts you do not need to reach your solution. In some cases, you may determine that more information is needed to solve the problem. If so, ask yourself, "What assumptions can I make about this problem?" or "Do I need a formula to help solve this problem?"

3. Decide what strategies you might use to solve the problem, how you might use them, and what form your solution will be in. For example, will you need to create a graph or chart? Will you need to solve an equation? Will your answer be in words or numbers? By knowing what type of solution you should reach, you may be able to eliminate some of the choices.

4. Apply your strategy to solve the problem and compare your answer to the choices.

5. If the answer is still not clear, read the problem again. If you had to make calculations to reach your answer, use estimation to see if your answer makes sense.

ARCTIC OCEAN
80°N
BEAUFORT SEA
Victoria Island
Baffin Bay
Greenland
Bering Strait
Mackenzie River
Great Bear Lake
Baffin Island
Davis Strait
Denmark Strait
Iceland
Yukon River
60°N
Great Slave Lake
BERING SEA
GULF OF ALASKA
HUDSON BAY
Aleutian Islands
Lake Winnipeg
Vancouver Island
ROCKY MOUNTAINS
Great Lakes
St. Lawrence River
ATLANTIC OCEAN
Bay of Biscay
40°N
NORTH AMERICA
Missouri River
Colorado River
Mississippi
APPALACHIAN MTS.
Strait of Gibraltar
ATL
Rio Grande
Tropic of Cancer
Hawaiian Islands
GULF OF MEXICO
Bahamas
20°N
Greater Antilles
Niger
CARIBBEAN SEA
Lesser Antilles
PACIFIC
Isthmus of Panama
GUIANA HIGHLANDS
0°
Equator
OCEAN
Amazon River
N
W E
SOUTH AMERICA
ANDES
S
BRAZILIAN HIGHLANDS
20°S
ANDES
Tropic of Capricorn
ATLANTIC
Paraná River
40°S
OCEAN
Strait of Magellan
Falkland Islands
Tierra del Fuego
Cape Horn
60°S
160°W 140°W 120°W 100°W 80°W 60°W 40°W 20°W
Antarctic Circle
Weddell Sea

ELEVATION

FEET		METERS
13,120		4,000
6,560		2,000
1,640		500
656		200
(Sea level) 0		0 (Sea level)
Below sea level		Below sea level
	Ice cap	

SCALE

0   1,000   2,000 Miles

0   1,000   2,000 Kilometers

Projection: Mollweide

ARCTIC 80°N OCEAN
LAPTEV EAST SIBERIAN SEA
SEA
North BARENTS KARA
Cape SEA SEA
BALTIC 60°N KAMCHATKA
SEA Ob Yenisey Lena River Kolyma River SEA OF PENINSULA
EUROPE River River OKHOTSK
URAL MOUNTAINS Sakhalin
ALPS Lake Amur River
Baikal Hokkaidō
ARAL ALTAY SHAN SEA 40°N
.S. SEA Lake OF
CASPIAN SEA Balkhash GOBI JAPAN Honshū
BLACK SEA ASIA Yellow River Shikoku
MEDITERRANEAN SEA Euphrates River Shikoku
HIMALAYAS EAST Kyūshū
H A R A Tigris River Persian Gulf THAR CHINA
RED SEA ARABIAN DESERT SEA
PENINSULA Ganges River Tropic of Cancer
AFRICA Taiwan 20°N
ARABIAN Bay PACIFIC
SEA of
Bengal
SOUTH OCEAN
Sri CHINA
Lanka Strait SEA Philippine
Nile River of Islands
Congo Malacca
River MALAY
Lake PENINSULA
Lake Victoria Borneo New
Tanganyika Sumatra Guinea Solomon
Sulawesi Islands Equator 0°
INDIAN OCEAN Java (Celebes)
Mozambique Channel
Madagascar CORAL
SEA
GREAT New
KALAHARI SANDY Caledonia 20°S
DESERT DESERT Tropic of Capricorn
AUSTRALIA
GREAT
Cape of VICTORIA GREAT DIVIDING RANGE
Good Hope DESERT Darling River North
Island
TASMAN
SEA
20°E 40°E 60°E 80°E 100°E 120°E 140°E 160°E 60°S Tasmania South
Island
ANTARCTICA

Denmark Strait North
Cape
Iceland BARENTS KARA
SEA SEA
N
W E 60°N
S BALTIC SCALE
NORTH SEA 0 250 500 750 Miles
SEA
50°N British 0 250 500 750 Kilometers
Isles URAL MTS. Projection: Mollweide
Rhine Volga River
Danube EUROPE
ATLANTIC Bay of ALPS
Biscay
OCEAN BLACK SEA
40°N
MEDITERRANEAN SEA
Strait of
Gibraltar Crete

Boundaries
⊛ National capitals
• Other cities

SCALE
0   1,000   2,000 Miles
0   1,000   2,000 Kilometers
Projection: Mollweide

ARCTIC OCEAN

Greenland (DENMARK)

ICELAND

ALASKA (U.S.)

Aleutian Islands

CANADA

Vancouver   Winnipeg

NORTH AMERICA

Ottawa   Montreal
Chicago   Toronto
UNITED STATES   New York City
Washington, D.C.

ATLANTIC OCEAN

Los Angeles

Houston

Bermuda (U.K.)

Rabat
Casablanca
MOROCCO

WESTERN SAHARA (Claimed by Morocco)

MEXICO

Tropic of Cancer

Mexico City

MAURITANIA
Nouakchott

HAWAII (U.S.)

Caracas
VENEZUELA   GUYANA   SURINAME
Georgetown   FRENCH GUIANA (FRANCE)
Bogotá   Paramaribo
COLOMBIA

CAPE VERDE
Dakar   SENEGAL   Bamako
GAMBIA   BUR
GUINEA-BISSAU   GUINEA
SIERRA LEONE   COTE
LIBERIA   D'IVOIR

PACIFIC OCEAN

KIRIBATI

Equator

Quito
ECUADOR

Galápagos Islands (ECUADOR)

PERU

SOUTH AMERICA

SAMOA

American Samoa

Lima

BRAZIL
Brasília

BOLIVIA
La Paz
Sucre

TONGA

Tropic of Capricorn

PARAGUAY   Rio de Janeiro
CHILE   Asunción   São Paulo

ATLANTIC OCEAN

ARGENTINA
URUGUAY
Santiago   Buenos Aires   Montevideo

Falkland Islands (U.K.)

South Georgia Island (U.K.)

South Sandwich Islands

Antarctic Circle

N
W   E
S

FLORIDA (U.S.)

GULF OF MEXICO

Nassau

BAHAMAS

Turks and Caicos Is. (U.K.)

ATLANTIC OCEAN

Havana

CUBA

Cayman Is. (U.K.)

HAITI   DOMINICAN REPUBLIC
Port-au-Prince
JAMAICA   Santo Domingo
Kingston

Puerto Rico (U.S.)
Virgin Islands (U.S. and U.K.)

MEXICO

BELIZE
Belmopan

CARIBBEAN SEA

1

GUATEMALA   HONDURAS
Guatemala City   Tegucigalpa
San Salvador   NICARAGUA
EL SALVADOR   Managua

Netherlands Antilles (NETHERLANDS)
Aruba (NETHERLANDS)

2

Guadeloupe (FRANCE)

3

Martinique (FRANCE)

4

5   6

PACIFIC OCEAN

COSTA RICA
San José

Panama City

Port-of-Spain

7

TRINIDAD AND TOBAGO

SCALE
0   200   400 Miles
0   200   400 Kilometers
Projection: Mercator

PANAMA

COLOMBIA

VENEZUELA

GUYANA

COUNTRY	CAPITAL
1 Antigua and Barbuda	St. Johns
2 St. Kitts-Nevis	Basseterre
3 Dominica	Roseau
4 St. Lucia	Castries
5 St. Vincent and the Grenadines	Kingstown
6 Barbados	Bridgetown
7 Grenada	St. George's

ARCTIC OCEAN

RUSSIA

EUROPE

Moscow

Astana

KAZAKHSTAN

Ulaanbaatar

Harbin

MONGOLIA

GEORGIA

Almaty

ASIA

Beijing

Tianjin

NORTH KOREA

P'yŏngyang

JAPAN

Tokyo

Istanbul

ARMENIA

Baku

UZBEKISTAN

Tashkent

KYRGYZSTAN

Seoul

Pusan

Nagoya

Yokohama

Ankara

TURKEY

AZERBAIJAN

TURKMENISTAN

Ashgabat

TAJIKISTAN

CHINA

SOUTH KOREA

Osaka

Tunis

Nicosia

SYRIA

Tehran

Kabul

Islamabad

Wuhan

Shanghai

lgiers

CYPRUS

LEBANON

Damascus

Baghdad

IRAN

AFGHANISTAN

Chongqing

Beirut

Jerusalem

Amman

IRAQ

KUWAIT

Delhi

NEPAL

Kathmandu

Guangzhou

Taipei

RIA

Cairo

JORDAN

ISRAEL

SAUDI ARABIA

BAHRAIN

QATAR

OMAN

PAKISTAN

New Delhi

BHUTAN

Hong Kong

TAIWAN

Tropic of Cancer

LIBYA

EGYPT

Riyadh

UNITED ARAB EMIRATES

Karachi

INDIA

BANGLADESH

Dhaka

MYANMAR (BURMA)

LAOS

Hanoi

PACIFIC

20°N

AFRICA

OMAN

Mumbai (Bombay)

Kolkata (Calcutta)

Northern Mariana Islands (U.S.)

NIGER

CHAD

ERITREA

Khartoum

YEMEN

Asmara

Sanaa

Chennai (Madras)

Yangon (Rangoon)

THAILAND

Bangkok

VIETNAM

CAMBODIA

Manila

PHILIPPINES

Guam (U.S.)

OCEAN

Niamey

NIGERIA

Abuja

SUDAN

DJIBOUTI

Addis Ababa

Colombo

SRI LANKA

Phnom Penh

Ho Chi Minh City

PALAU

MARSHALL ISLANDS

NIN

GO

Lagos

CENTRAL AFRICAN REPUBLIC

ETHIOPIA

SOMALIA

BRUNEI

FEDERATED STATES OF MICRONESIA

CAMEROON

RIAL

INEA

UGANDA

KENYA

MALDIVES

Kuala Lumpur

MALAYSIA

TOME AND

RINCIPE

GABON

REP. OF THE CONGO

RWANDA

BURUNDI

DEMOCRATIC REP. OF THE CONGO

Nairobi

Singapore

SINGAPORE

INDONESIA

Equator

0°

NAURU

KIRIBATI

CABINDA (ANGOLA)

Kinshasa

TANZANIA

SEYCHELLES

PAPUA NEW GUINEA

Luanda

Dar es Salaam

Jakarta

Surabaya

Port Moresby

SOLOMON ISLANDS

TUVALU

ANGOLA

ZAMBIA

MALAWI

COMOROS

MOZAMBIQUE

INDIAN OCEAN

EAST TIMOR

Lusaka

Harare

NAMIBIA

ZIMBABWE

MADAGASCAR

Antananarivo

Réunion (FRANCE)

MAURITIUS

VANUATU

FIJI

Windhoek

BOTSWANA

Maputo

New Caledonia (FRANCE)

20°S

Gaborone

Pretoria

SWAZILAND

AUSTRALIA

Tropic of Capricorn

Johannesburg

LESOTHO

SOUTH AFRICA

Cape Town

Sydney

Canberra

NEW ZEALAND

Melbourne

Tasmania

Wellington

NTARCTICA

	COUNTRY	CAPITAL
1	Czech Republic	Prague
2	Slovakia	Bratislava
3	Slovenia	Ljubljana
4	Croatia	Zagreb
5	Bosnia and Herzegovina	Sarajevo
6	Macedonia	Skopje
7	Serbia and Montenegro	Belgrade
8	Lithuania	Vilnius
9	Latvia	Riga
10	Estonia	Tallinn

SCALE

0 250 500 750 Miles

0 250 500 750 Kilometers

Projection: Mollweide

ICELAND

Reykjavik

Arctic Circle

SWEDEN

FINLAND

Helsinki

NORWAY

Oslo

Stockholm

St. Petersburg

RUSSIA

NORTH SEA

DENMARK

Copenhagen

10

UNITED KINGDOM

9

Dublin

IRELAND

NETHERLANDS

Amsterdam

The Hague

London

Brussels

BELGIUM

GERMANY

Berlin

Warsaw

8

Minsk

BELARUS

Moscow

POLAND

UKRAINE

Kiev

ATLANTIC OCEAN

Paris

LUXEMBOURG

1

Bern

Vienna

AUSTRIA

Budapest

HUNGARY

2

MOLDOVA

Chişinău

FRANCE

SWITZERLAND

LIECHTENSTEIN

3

4

ROMANIA

Bucharest

MONACO

Corsica (FRANCE)

ITALY

5

7

BULGARIA

Sofia

BLACK SEA

ANDORRA

SAN MARINO

Rome

6

PORTUGAL

Madrid

VATICAN CITY

Balearic Is. (SPAIN)

Sardinia (ITALY)

Tiranë

ALBANIA

GREECE

Lisbon

SPAIN

MEDITERRANEAN

Sicily

SEA

Athens

Gibraltar (U.K.)

MALTA

Crete

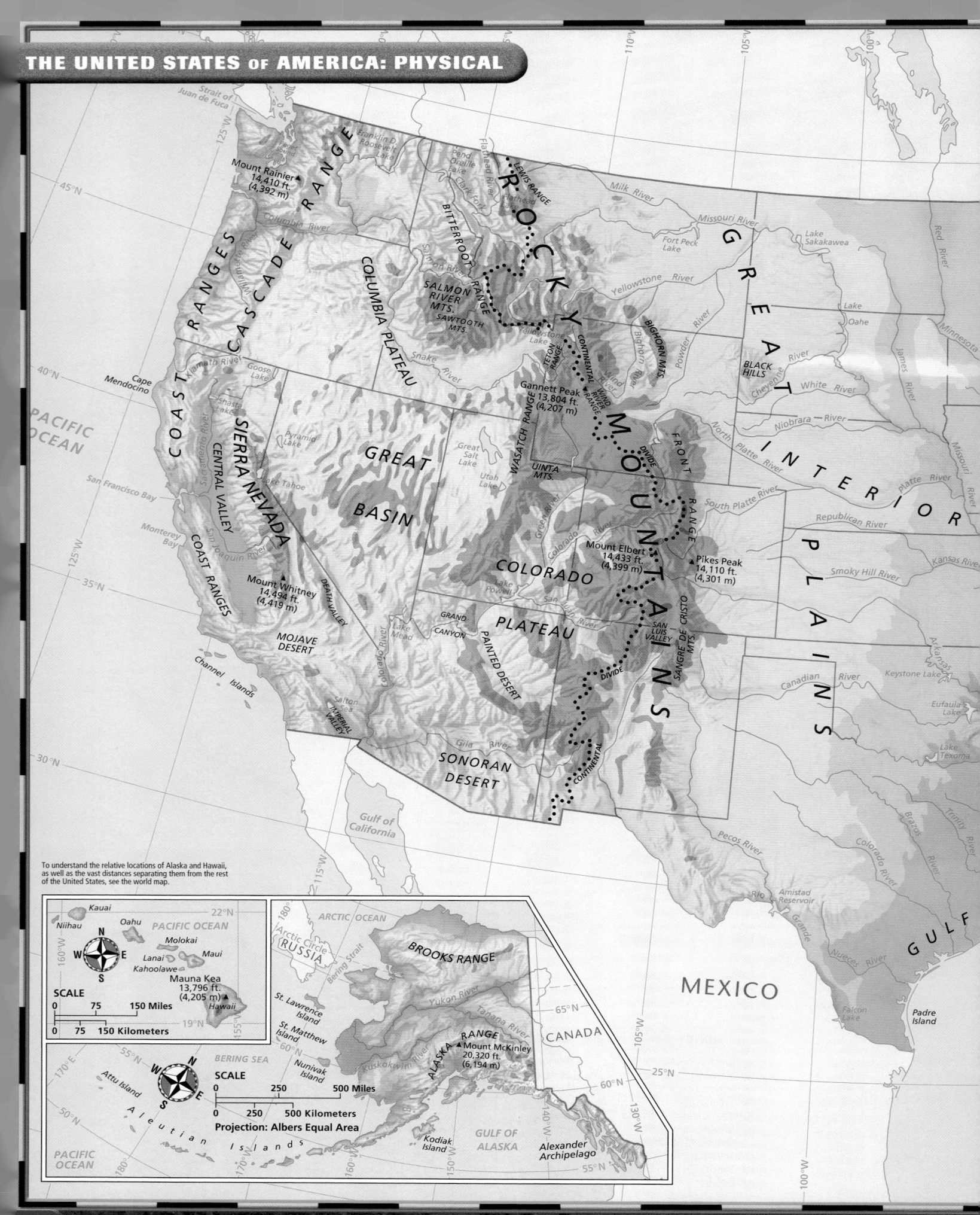

Strait of Juan de Fuca

Franklin D. Roosevelt Lake

Mount Rainier ▲
14,410 ft.
(4,392 m)

COAST RANGES

CASCADE RANGE

COLUMBIA PLATEAU

Columbia River

Pend Oreille Lake

Flathead Lake

Clark Fork

LEWIS RANGE

ROCKY

Milk River

Missouri River

Fort Peck Lake

Lake Sakakawea

GREAT

BITTERROOT RANGE

Salmon River

SALMON RIVER MTS.

SAWTOOTH MTS.

Snake River

Yellowstone Lake

Yellowstone River

CONTINENTAL

TETON RANGE

DIVIDE

BIGHORN MTS.

Bighorn River

Powder River

Lake Oahe

BLACK HILLS

INTERIOR

Red River

Minnesota River

James River

Cape Mendocino

Klamath River

Goose Lake

WIND RIVER RANGE

Gannett Peak
13,804 ft.
(4,207 m)

Wind River

North Platte River

Cheyenne River

White River

Niobrara — River

Missouri River

PACIFIC
OCEAN

San Francisco Bay

Shasta Lake

Sacramento River

Pyramid Lake

SIERRA NEVADA

CENTRAL VALLEY

GREAT

BASIN

Great Salt Lake

Utah Lake

WASATCH RANGE

UINTA MTS.

Green River

FRONT RANGE

DIVIDE

South Platte River

North Platte River

Republican River

Kansas River

Smoky Hill River

Monterey Bay

COAST RANGES

San Joaquin River

Lake Tahoe

Mount Whitney
14,494 ft.
(4,419 m)

DEATH VALLEY

Colorado River

COLORADO

PLATEAU

Lake Powell

Colorado River

Mount Elbert
14,433 ft.
(4,399 m)

Pikes Peak
14,110 ft.
(4,301 m)

M O U N T A I N S

Arkansas River

Keystone Lake

Channel Islands

MOJAVE DESERT

Lake Mead

GRAND CANYON

PAINTED DESERT

San Juan River

SAN LUIS VALLEY

SANGRE DE CRISTO MTS.

DIVIDE

Eufaula Lake

Salton Sea

IMPERIAL VALLEY

Gila River

SONORAN DESERT

CONTINENTAL

Canadian River

Lake Texoma

Gulf of California

Pecos River

Colorado River

Brazos River

Trinity River

To understand the relative locations of Alaska and Hawaii, as well as the vast distances separating them from the rest of the United States, see the world map.

Rio Grande

Amistad Reservoir

G U L F

MEXICO

Nueces River

Falcon Lake

Padre Island

PLAINS

Hawaii Inset

Kauai

Niihau

Oahu

Molokai

Lanai

Maui

Kahoolawe

PACIFIC OCEAN

22°N

19°N

Mauna Kea
13,796 ft.
(4,205 m) ▲

Hawaii

N
W E
S

SCALE
0   75   150 Miles
0   75   150 Kilometers

Alaska Inset

ARCTIC OCEAN

Arctic Circle

RUSSIA

Bering Strait

BROOKS RANGE

St. Lawrence Island

St. Matthew Island

Nunivak Island

Yukon River

Tanana River

ALASKA RANGE

Mount McKinley
20,320 ft.
(6,194 m)

65°N

60°N

CANADA

25°N

Kuskokwim River

BERING SEA

N
W E
S

SCALE
0   250   500 Miles
0   250   500 Kilometers

Projection: Albers Equal Area

Attu Island

Aleutian Islands

Kodiak Island

GULF OF ALASKA

Alexander Archipelago

PACIFIC OCEAN

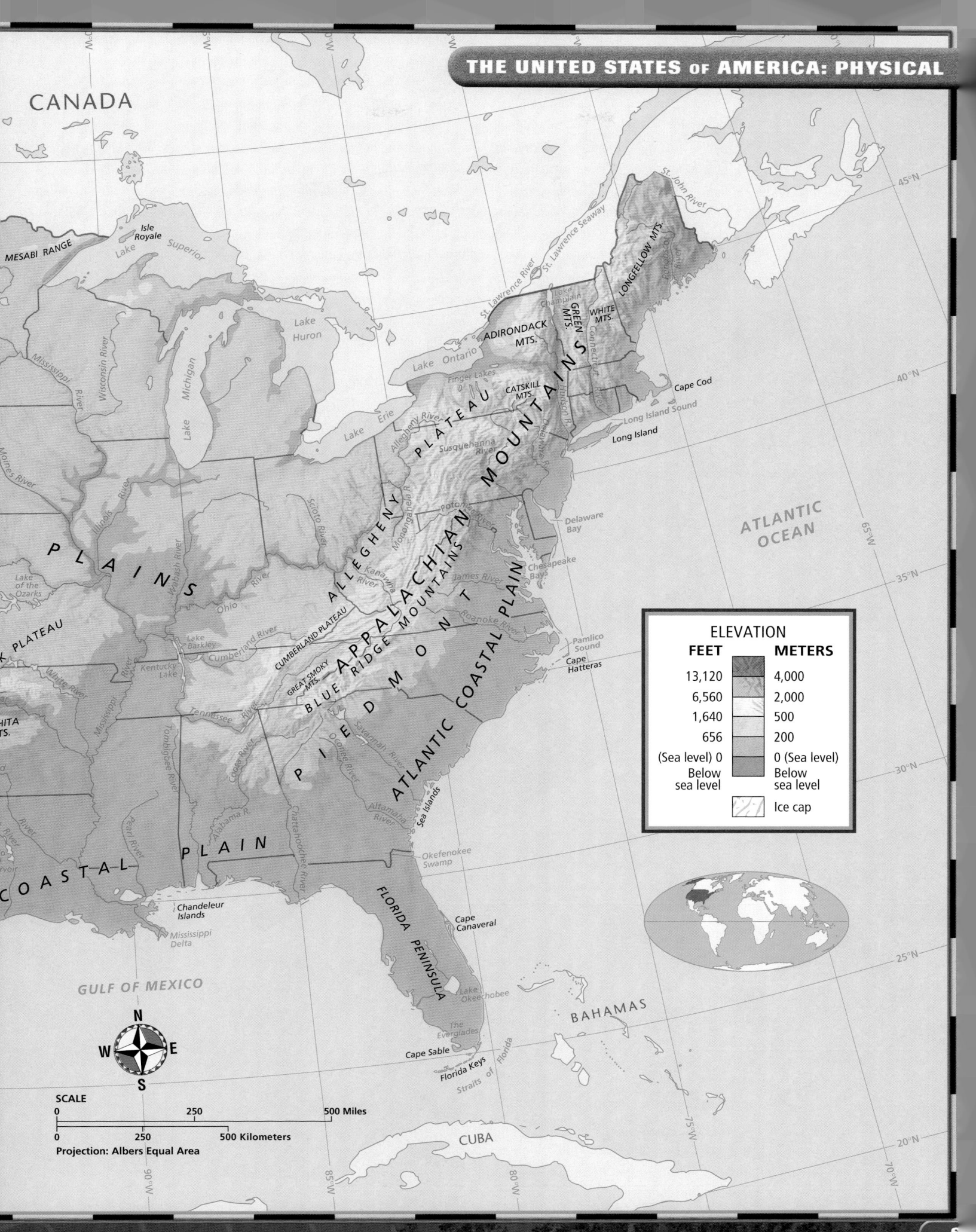

CANADA

MESABI RANGE

Isle Royale

Lake Superior

Lake Huron

Lake Michigan

Wisconsin River

Mississippi River

Des Moines River

Illinois River

Lake of the Ozarks

PLATEAU

PLAINS

Lake Ontario

Lake Erie

Finger Lakes

St. Lawrence River

St. Lawrence Seaway

Lake Champlain

ADIRONDACK MTS.

GREEN MTS.

WHITE MTS.

LONGFELLOW MTS.

St. John River

Penobscot River

Kennebec River

Connecticut River

Cape Cod

CATSKILL MTS.

Allegheny River

Susquehanna River

Hudson R.

Delaware R.

Long Island Sound

Long Island

ATLANTIC OCEAN

ALLEGHENY

APPALACHIAN MOUNTAINS

Scioto River

Ohio River

Wabash River

Kanawha River

Monongahela R.

Potomac River

James River

Roanoke River

Delaware Bay

Chesapeake Bay

OZARK PLATEAU

Lake Barkley

Kentucky Lake

Cumberland River

Tennessee River

CUMBERLAND PLATEAU

GREAT SMOKY MTS.

BLUE RIDGE MOUNTAINS

PIEDMONT

ATLANTIC COASTAL PLAIN

Pamlico Sound

Cape Hatteras

OUACHITA MTS.

White River

Mississippi River

Coosa River

Alabama River

Tombigbee River

Chattahoochee River

Flint River

Oconee River

Savannah River

Ocmulgee River

Altamaha River

Sea Islands

COASTAL PLAIN

Pearl River

Red River

Chandeleur Islands

Mississippi Delta

Okefenokee Swamp

FLORIDA PENINSULA

GULF OF MEXICO

Cape Canaveral

Lake Okeechobee

The Everglades

Cape Sable

Florida Keys

Straits of Florida

BAHAMAS

CUBA

ELEVATION

FEET		METERS
13,120		4,000
6,560		2,000
1,640		500
656		200
(Sea level) 0		0 (Sea level)
Below sea level		Below sea level
	Ice cap	

N
W E
S

SCALE

0   250   500 Miles

0   250   500 Kilometers

Projection: Albers Equal Area

45°N

40°N

35°N

30°N

25°N

20°N

65°W

70°W

75°W

80°W

85°W

90°W

To understand the relative locations of Alaska and Hawaii as well as the vast distances separating them from the rest of the United States, see the world map.

HAWAII

SCALE
0   75   150 Miles
0   75   150 Kilometers

ALASKA

SCALE
0   250   500 Miles
0   250   500 Kilometers
Projection: Albers Equal Area

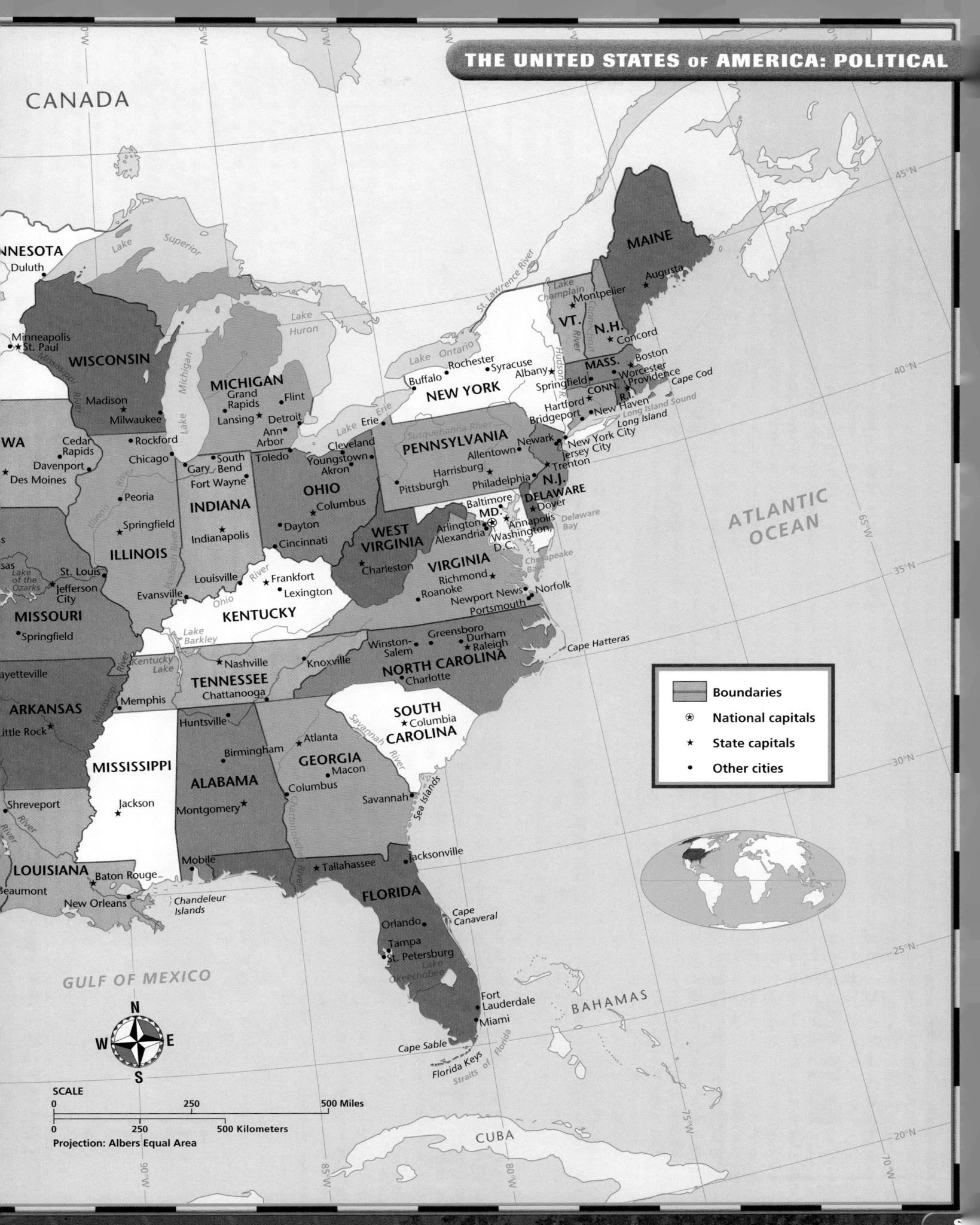

CANADA

MINNESOTA
Duluth

Minneapolis
St. Paul

WISCONSIN
Madison
Milwaukee

IOWA
Cedar Rapids
Rockford
Davenport
Des Moines

MICHIGAN
Grand Rapids
Flint
Lansing ★ Detroit
Ann Arbor

Lake Superior
Lake Huron
Lake Michigan
Lake Ontario
Lake Erie

St. Lawrence River

MAINE
Augusta ★

Lake Champlain
Montpelier ★
VT.
N.H.
Concord ★
Boston ★
Worcester
MASS.
Springfield CONN. Providence Cape Cod
Hartford ★ R.I.
Bridgeport New Haven
Long Island Sound
Long Island

Buffalo Rochester Syracuse
Albany ★
NEW YORK

Chicago
Gary South Bend
Fort Wayne
Toledo
Cleveland
Youngstown
Akron
PENNSYLVANIA
Allentown
Harrisburg ★
Pittsburgh
Philadelphia
Newark New York City
Jersey City
Trenton ★
N.J.

Susquehanna River
Hudson R.

Peoria

INDIANA
Springfield ★
Indianapolis ★
Dayton
Cincinnati

OHIO
Columbus ★

ILLINOIS

St. Louis
Jefferson City ★

Louisville
Frankfort ★
Lexington
Evansville

WEST VIRGINIA
Charleston ★

Baltimore
MD. ★
Arlington
Alexandria Washington, D.C.
Annapolis ★
Delaware Bay
DELAWARE
Dover ★

ATLANTIC OCEAN

Lake of the Ozarks

MISSOURI
Springfield ★

Wabash River
Ohio River

KENTUCKY

VIRGINIA
Richmond ★
Roanoke
Newport News Norfolk
Portsmouth

Chesapeake Bay

Lake Barkley
Kentucky Lake

Nashville ★
Knoxville
TENNESSEE
Chattanooga

Winston-Salem Greensboro Durham
Raleigh ★
NORTH CAROLINA
Charlotte

Cape Hatteras

Fayetteville

ARKANSAS
Little Rock ★

Memphis
Huntsville

Birmingham

Atlanta ★

SOUTH CAROLINA
Columbia ★

Savannah River

MISSISSIPPI
Jackson ★

ALABAMA
Montgomery ★

GEORGIA
Columbus
Macon

Savannah
Sea Islands

Shreveport

LOUISIANA
Baton Rouge ★
Beaumont
New Orleans

Mobile

Chattahoochee River

Tallahassee ★

Jacksonville

Chandeleur Islands

FLORIDA

Cape Canaveral

BAHAMAS

Orlando
Tampa
St. Petersburg
Lake Okeechobee
Fort Lauderdale
Miami

GULF OF MEXICO

Cape Sable
Florida Keys
Straits of Florida

CUBA

	Boundaries
⊛	National capitals
★	State capitals
•	Other cities

N
W   E
S

SCALE
0   250   500 Miles
0   250   500 Kilometers
Projection: Albers Equal Area

45°N
40°N
35°N
30°N
25°N
20°N

90°W
85°W
80°W
75°W
70°W

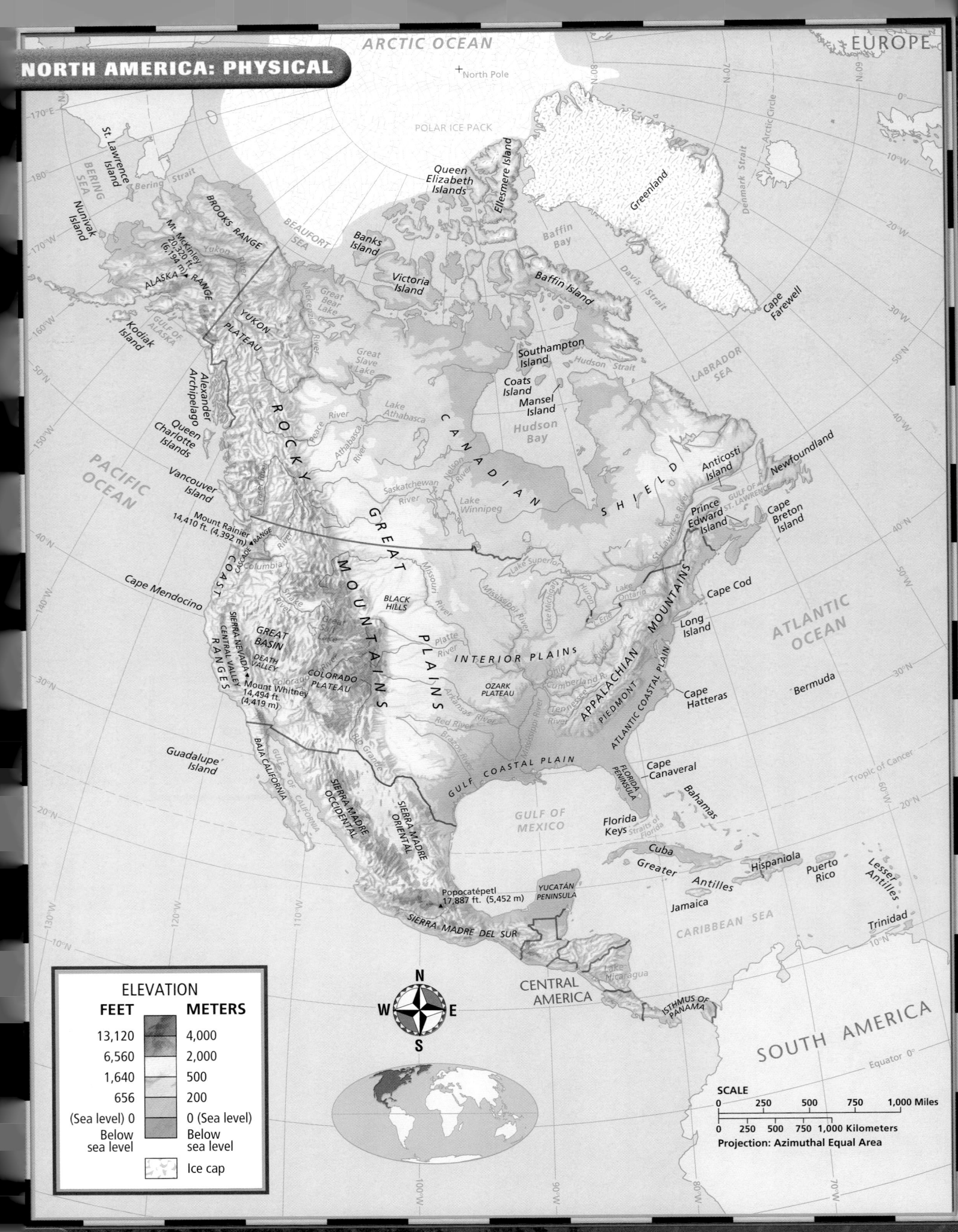

NORTH AMERICA: PHYSICAL

ARCTIC OCEAN

EUROPE

+ North Pole

POLAR ICE PACK

PACIFIC OCEAN

ATLANTIC OCEAN

Greenland

Ellesmere Island

Queen Elizabeth Islands

Banks Island

Victoria Island

Baffin Island

Baffin Bay

Denmark Strait

Cape Farewell

Davis Strait

LABRADOR SEA

Arctic Circle

BEAUFORT SEA

Mackenzie River

Great Bear Lake

Great Slave Lake

Lake Athabasca

C A N A D I A N   S H I E L D

Southampton Island

Coats Island

Mansel Island

Hudson Strait

Hudson Bay

Newfoundland

Anticosti Island

GULF OF ST. LAWRENCE

Prince Edward Island

Cape Breton Island

BERING SEA

Bering Strait

St. Lawrence Island

Nunivak Island

BROOKS RANGE

Mt. McKinley 20,320 ft. (6,194 m)

ALASKA RANGE

Yukon

YUKON PLATEAU

GULF OF ALASKA

Kodiak Island

Alexander Archipelago

Queen Charlotte Islands

Vancouver Island

Peace River

Athabasca River

Saskatchewan River

Nelson River

Lake Winnipeg

R O C K Y   M O U N T A I N S

G R E A T   P L A I N S

Missouri River

Lake Superior

St. Lawrence R.

Lake Huron

Lake Michigan

Lake Ontario

Lake Erie

Cape Cod

Long Island

A P P A L A C H I A N   M O U N T A I N S

Mount Rainier 14,410 ft. (4,392 m)

CASCADE RANGE

Columbia R.

Snake R.

Cape Mendocino

COAST RANGES

SIERRA NEVADA

CENTRAL VALLEY

GREAT BASIN

DEATH VALLEY

Mount Whitney 14,494 ft (4,419 m)

COLORADO PLATEAU

Colorado R.

Great Salt Lake

BLACK HILLS

Platte River

I N T E R I O R   P L A I N S

Arkansas River

OZARK PLATEAU

Ohio R.

Cumberland R.

Tennessee River

Mississippi River

PIEDMONT

ATLANTIC COASTAL PLAIN

Cape Hatteras

Bermuda

Red River

Rio Grande

Brazos R.

GULF COASTAL PLAIN

FLORIDA PENINSULA

Cape Canaveral

Tropic of Cancer

Guadalupe Island

BAJA CALIFORNIA

GULF OF CALIFORNIA

SIERRA MADRE OCCIDENTAL

SIERRA MADRE ORIENTAL

GULF OF MEXICO

Florida Keys

Straits of Florida

Bahamas

Cuba

Greater Antilles

Hispaniola

Puerto Rico

Lesser Antilles

Jamaica

CARIBBEAN SEA

Trinidad

Popocatépetl 17,887 ft. (5,452 m)

YUCATÁN PENINSULA

SIERRA MADRE DEL SUR

Lake Nicaragua

CENTRAL AMERICA

ISTHMUS OF PANAMA

SOUTH AMERICA

Equator 0°

N
W   E
S

ELEVATION

FEET		METERS
13,120		4,000
6,560		2,000
1,640		500
656		200
(Sea level) 0		0 (Sea level)
Below sea level		Below sea level
	Ice cap	

SCALE

| 0 | 250 | 500 | 750 | 1,000 Miles |

| 0 | 250 | 500 | 750 | 1,000 Kilometers |

Projection: Azimuthal Equal Area

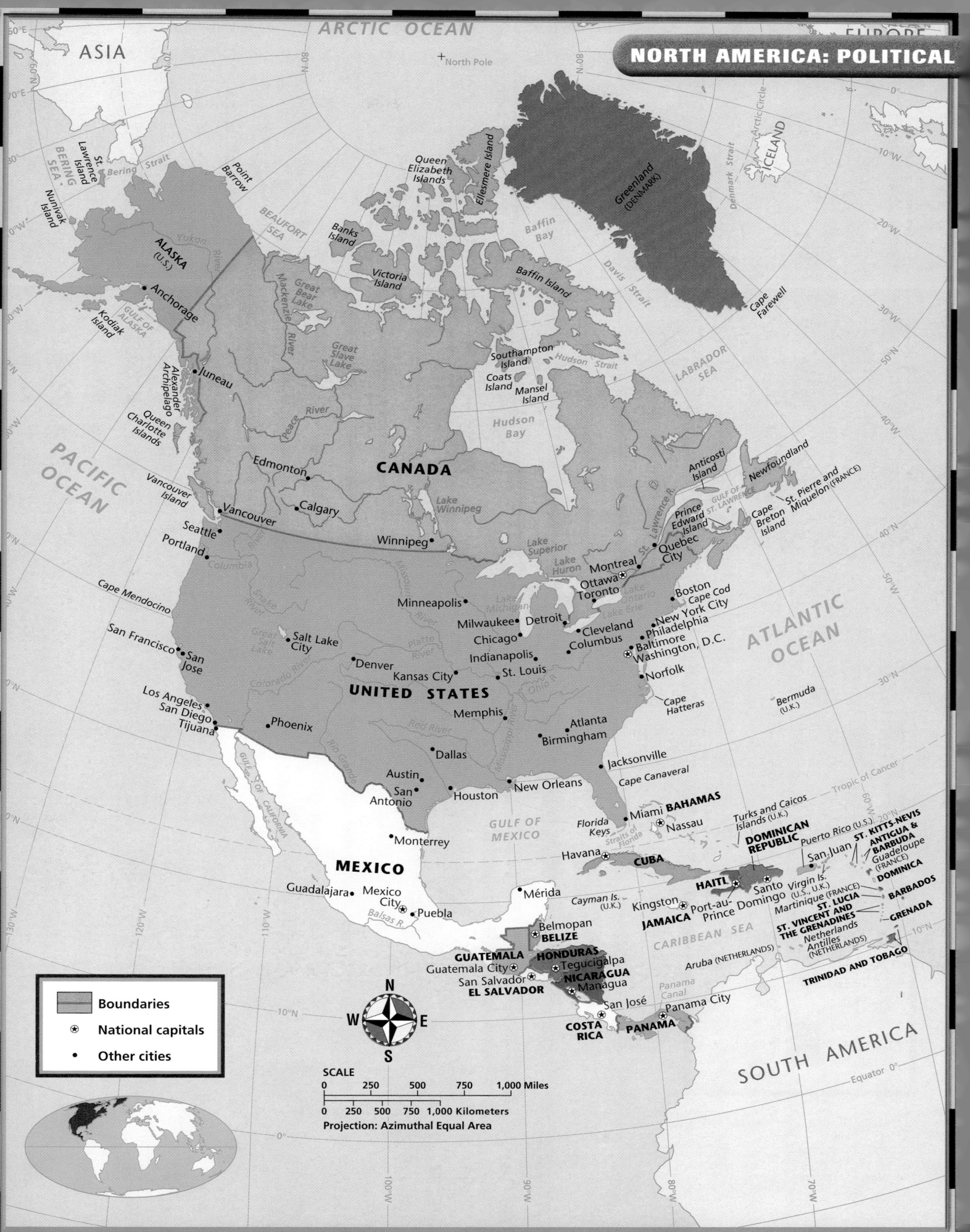

NORTH AMERICA: POLITICAL

ASIA

ARCTIC OCEAN

EUROPE

North Pole

Point Barrow

St. Lawrence Island

BERING SEA

Nunivak Island

Kodiak Island

ALASKA (U.S.)

• Anchorage

GULF OF ALASKA

Alexander Archipelago

• Juneau

Queen Charlotte Islands

PACIFIC OCEAN

Vancouver Island

Cape Mendocino

Queen Elizabeth Islands

Ellesmere Island

Banks Island

Victoria Island

BEAUFORT SEA

Baffin Island

Southampton Island

Coats Island

Mansel Island

Baffin Bay

Davis Strait

Greenland (DENMARK)

ICELAND

Denmark Strait

Cape Farewell

Hudson Strait

LABRADOR SEA

Hudson Bay

Great Bear Lake

Mackenzie River

Peace River

Great Slave Lake

CANADA

Edmonton •

• Vancouver

Seattle •

Portland •

San Francisco •

San Jose •

Los Angeles •

San Diego •

Tijuana •

• Calgary

Winnipeg •

Lake Winnipeg

Minneapolis •

Milwaukee •

Chicago •

Salt Lake City •

Great Salt Lake

Denver •

Kansas City •

Phoenix •

UNITED STATES

Columbia

Snake River

Colorado River

Platte River

Missouri River

Lake Superior

Lake Huron

Lake Michigan

Lake Erie

Lake Ontario

Detroit •

Indianapolis •

St. Louis •

Memphis •

Ohio R.

Montreal

Ottawa

Toronto

Cleveland •

Columbus •

Anticosti Island

Newfoundland

Prince Edward Island

St. Lawrence R.

GULF OF ST. LAWRENCE

Quebec City

Cape Breton Island

St. Pierre and Miquelon (FRANCE)

Boston •

Cape Cod

New York City

Philadelphia •

Baltimore •

Washington, D.C.

Norfolk •

ATLANTIC OCEAN

Cape Hatteras

Bermuda (U.K.)

Red River

Mississippi River

Dallas •

Austin •

San Antonio •

Houston •

New Orleans •

Atlanta •

Birmingham •

Jacksonville •

Cape Canaveral

Rio Grande

GULF OF CALIFORNIA

• Monterrey

MEXICO

Guadalajara •

Mexico City

Puebla •

Balsas R.

Mérida •

Belmopan

GUATEMALA

Guatemala City

San Salvador

EL SALVADOR

GULF OF MEXICO

Miami •

Florida Keys

Straits of Florida

Havana

CUBA

Cayman Is. (U.K.)

BELIZE

Kingston

JAMAICA

HONDURAS

Tegucigalpa •

NICARAGUA

Managua •

San José •

COSTA RICA

Panama Canal

Panama City •

PANAMA

BAHAMAS

Nassau

Turks and Caicos Islands (U.K.)

DOMINICAN REPUBLIC

Puerto Rico (U.S.)

San Juan •

HAITI

Port-au-Prince

Santo Domingo

Virgin Is. (U.S., U.K.)

ST. KITTS-NEVIS

ANTIGUA & BARBUDA

Guadeloupe (FRANCE)

DOMINICA

Martinique (FRANCE)

ST. LUCIA

BARBADOS

ST. VINCENT AND THE GRENADINES

GRENADA

Netherlands Antilles (NETHERLANDS)

CARIBBEAN SEA

Aruba (NETHERLANDS)

TRINIDAD AND TOBAGO

Tropic of Cancer

SOUTH AMERICA

Equator 0°

Arctic Circle

Yukon River

Legend

	Boundaries
⊛	National capitals
•	Other cities

N
W E
S

SCALE

0   250   500   750   1,000 Miles

0   250   500   750   1,000 Kilometers

Projection: Azimuthal Equal Area

SOUTH AMERICA: PHYSICAL

CENTRAL AMERICA

CARIBBEAN SEA

Panama Canal

Margarita Island

Tobago
Trinidad

Lake Maracaibo

Orinoco River Delta

LLANOS

Angel Falls

GUIANA HIGHLANDS

Devil's Island
Cape Orange

ATLANTIC OCEAN

Cauca River

Meta River

Orinoco River

Amazon River Delta

▲ Mount Tolima
18,425 ft. (5,616 m)

Malpelo Island

ANDES

Caguetá River

Rio Negro

Japurá River

AMAZON BASIN

Amazon River

Equator 0°

Galápagos Islands

0° Equator

GULF OF GUAYAQUIL

▲ Mount Chimborazo
20,561 ft. (6,267 m)

Amazon River

Marañón River

Ucayali River

Juruá River

Purus River

Madeira River

Tapajós River

Xingu River

Tocantins River

Parnaíba River

BRAZILIAN HIGHLANDS

10°S

▲ Mount Huascarán
22,205 ft. (6,768 m)

Beni River

Mamoré River

MATO GROSSO PLATEAU

São Francisco River

Araguaia River

PACIFIC OCEAN

Lake Titicaca

▲ Ancohuma Peak
20,958 ft. (6,388 m)

Lake Poopó

ATACAMA DESERT

ANDES

Pilcomayo River

GRAN CHACO

PLATEAU OF BRAZIL

Tropic of Capricorn

20°S

San Ambrosio Island

San Félix Island

Paraguay River

Salado River

Juan Fernández Islands

▲ Mount Aconcagua
22,834 ft. (6,960 m)

Salado River

Paraná River

Uruguay River

Rio de la Plata

ATLANTIC OCEAN

30°S

PAMPAS

Colorado River

SAN MATÍAS GULF

Chiloé Island

PATAGONIA

Chonos Archipelago

SAN JORGE GULF

Cape Tres Puntas

Bahía Grande

Strait of Magellan

Falkland Islands

Tierra Del Fuego

South Georgia Island

Cape Horn

ELEVATION

FEET	METERS
13,120	4,000
6,560	2,000
1,640	500
656	200
(Sea level) 0	0 (Sea level)
Below sea level	Below sea level

SCALE

0   250   500   750   1,000 Miles

0   250   500   750   1,000 Kilometers

Projection: Azimuthal Equal Area

SOUTH AMERICA: POLITICAL

CENTRAL AMERICA

CARIBBEAN SEA

N
W   E
S

ATLANTIC OCEAN

Barranquilla
Cartagena
Caracas
Lake Maracaibo
VENEZUELA
Georgetown
Paramaribo
GUYANA
Cayenne
SURINAME FRENCH GUIANA (FRANCE)
Medellín
Bogotá
COLOMBIA
Orinoco River
Cali
Malpelo Island (COLOMBIA)

Equator 0°

Quito
ECUADOR
Guayaquil
Galápagos Islands (ECUADOR)
Rio Negro
Amazon River
Belém
Amazon River

BRAZIL

Recife

Marañón River
PERU
Trujillo
Ucayali River
São Francisco River

Callao
Lima
Arequipa

Lake Titicaca
BOLIVIA
La Paz
Lake Poopó
Sucre

Brasília

Salvador

Belo Horizonte

PACIFIC OCEAN

PARAGUAY
Paraguay River
Asunción
Campinas
São Paulo
Rio de Janeiro
Curitiba

San Ambrosio Island (CHILE)
San Félix Island (CHILE)

CHILE

Uruguay River
Paraná River
Pôrto Alegre

Juan Fernández Islands (CHILE)

Córdoba
Rosario
URUGUAY
Valparaíso
Santiago
Buenos Aires
Montevideo
Rio de la Plata

ATLANTIC OCEAN

ARGENTINA

Legend
- Boundaries
- ⊛ National capitals
- ★ Other capitals
- • Other cities

Strait of Magellan
Tierra del Fuego
Falkland Islands (U.K.)
South Georgia Island (U.K.)

SCALE
0   250   500   750   1,000 Miles
0   250   500   750   1,000 Kilometers
Projection: Azimuthal Equal Area

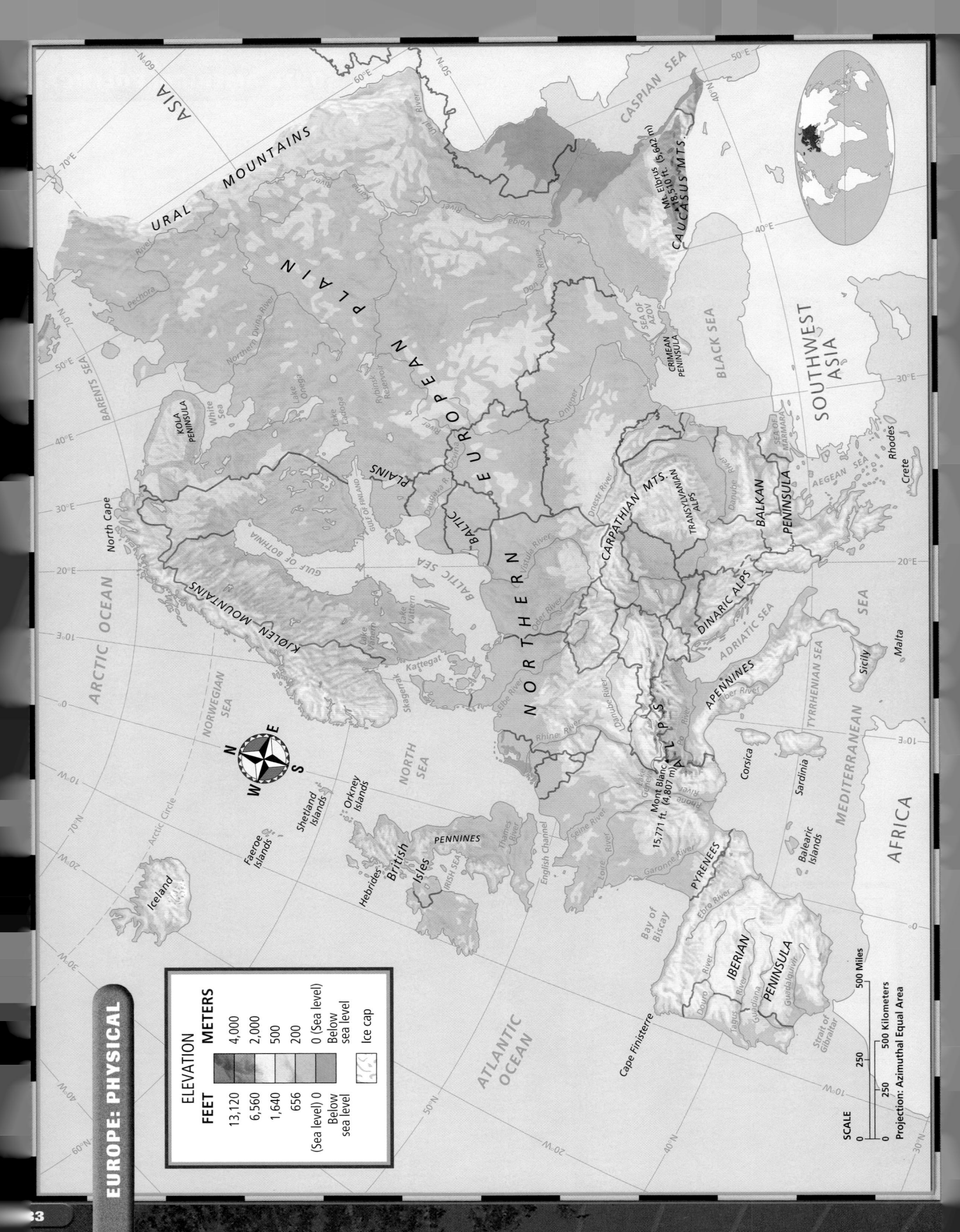

ASIA

URAL MOUNTAINS

CAUCASUS MTS.
Mt. Elbrus (5,642 m)
18,510 ft.

CASPIAN SEA

NORTHERN EUROPEAN PLAIN

BARENTS SEA

Pechora River

KOLA PENINSULA

White Sea

Lake Onega

Lake Ladoga

Rybinsk Reservoir

Northern Dvina River

Volga River

Don River

Dnieper

SEA OF AZOV

CRIMEAN PENINSULA

BLACK SEA

SOUTHWEST ASIA

ARCTIC OCEAN

North Cape

GULF OF BOTHNIA

KJØLEN MOUNTAINS

BALTIC PLAINS

GULF OF FINLAND

Daugava R.

Western Dvina River

BALTIC SEA

Vistula River

Oder River

CARPATHIAN MTS.

TRANSYLVANIAN ALPS

Dniestr River

Danube

BALKAN PENINSULA

SEA OF MARMARA

AEGEAN SEA

Rhodes

Crete

NORWEGIAN SEA

Lake Vänern

Lake Vättern

Kattegat

Skagerrak

Elbe River

DINARIC ALPS

ADRIATIC SEA

APENNINES

Tiber River

TYRRHENIAN SEA

Sicily

Malta

MEDITERRANEAN SEA

AFRICA

N E S W

NORTH SEA

Shetland Islands

Orkney Islands

Rhine River

Danube River

A L P S

Po River

Corsica

Sardinia

Balearic Islands

Faeroe Islands

British Isles

PENNINES

Thames R.

Seine River

English Channel

Loire

Geneva

15,771 ft. Mont Blanc (4,807 m)

Rhône River

Garonne River

PYRENEES

Hebrides

IRISH SEA

Iceland

Bay of Biscay

Arctic Circle

Cape Finisterre

Douro River

IBERIAN PENINSULA

Tagus River

Guadiana River

Guadalquivir

Ebro River

Strait of Gibraltar

ATLANTIC OCEAN

ELEVATION

FEET	METERS
13,120	4,000
6,560	2,000
1,640	500
656	200
(Sea level) 0	0 (Sea level)
Below sea level	Below sea level

Ice cap

SCALE
0   250   500 Miles
0   250   500 Kilometers
Projection: Azimuthal Equal Area

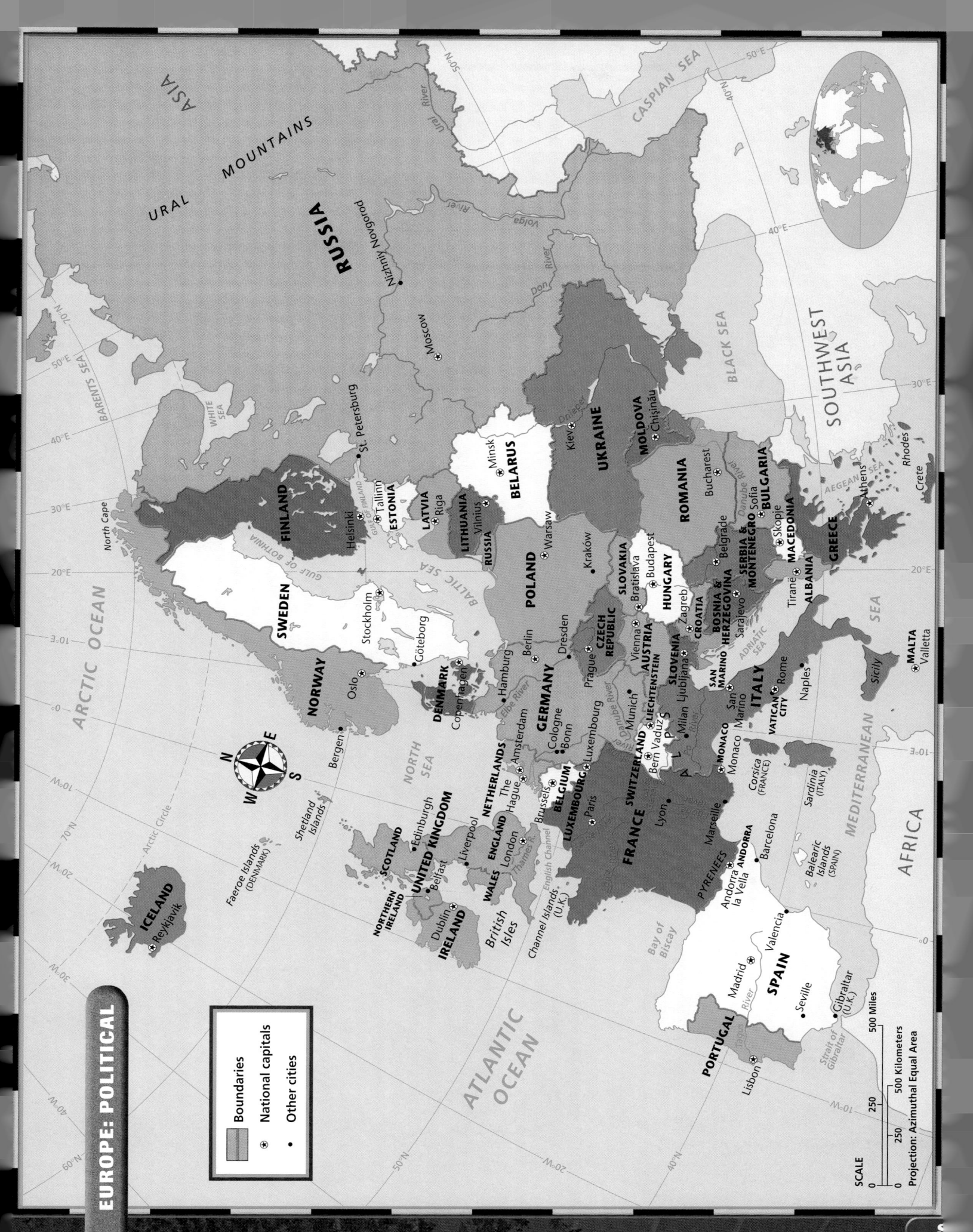

EUROPE: POLITICAL

Boundaries

⊛ National capitals

• Other cities

SCALE

0   250   500 Miles

0   250   500 Kilometers

Projection: Azimuthal Equal Area

ASIA: PHYSICAL

ELEVATION

FEET	METERS
13,120	4,000
6,560	2,000
1,640	500
656	200
(Sea level) 0	0 (Sea level)
Below sea level	Below sea level

Ice cap

SCALE
0   500   1,000 Miles
0   500   1,000 Kilometers
Projection: Two-Point Equidistant

NORTH AMERICA

EUROPE

AFRICA

AUSTRALIA

SIBERIA

Aleutian Islands
KAMCHATKA PENINSULA
BERING SEA
SEA OF OKHOTSK
CENTRAL RANGE
Kuril Islands
Sakhalin
KOLYMA MTS.
CHERSKIY RANGE
VERKHOYANSK MOUNTAINS
STANOVOY MOUNTAINS
Lena River
Aldan River
Amur River
YABLONOVYY RANGE
SHILKA RANGE
Lake Baikal
MONGOLIAN PLATEAU
GREATER KHINGAN RANGE
GOBI
Wrangel Island
New Siberian Islands
LAPTEV SEA
Lower Tunguska River
Yenisey River
Angara River
CENTRAL SIBERIAN PLATEAU
SAYAN MOUNTAINS
ALTAI SHAN
TIAN SHAN
KARA SEA
TAYMYR PENINSULA
North Land
Franz Josef Land
Novaya Zemlya
BARENTS SEA
Ob River
Irtysh River
WEST SIBERIAN PLAIN
KAZAKH UPLANDS
Lake Balkhash
TARIM BASIN
TAKLIMAKAN DESERT
KUNLUN SHAN
URAL MOUNTAINS
Ural River
Syr Dar'ya
TURAN LOWLAND
KYZYL KUM
Amu Dar'ya
KARA-KUM
HINDU KUSH
PLATEAU OF TIBET
Mount Everest 29,035 ft. (8,850 m)
HIMALAYAS
Nu Jiang
CASPIAN SEA
USTYURT PLATEAU
ARAL SEA
DASHT-E KAVIR (GREAT SALT DESERT)
GANGETIC PLAIN
THAR DESERT
Indus River
Sutlej River
Ganges River
Godavari River
DECCAN PLATEAU
EASTERN GHATS
WESTERN GHATS
CAUCASUS MTS.
Mount Ararat 16,945 ft. (5,165 m)
BLACK SEA
SEA OF AZOV
Bosporus
ANATOLIA
Cyprus
MEDITERRANEAN SEA
SINAI PENINSULA
SYRIAN DESERT
AN NAFUD
Euphrates River
Tigris River
ZAGROS MTS.
PERSIAN GULF
Strait of Hormuz
GULF OF OMAN
RUB' AL-KHALI
Socotra
ARABIAN SEA
GULF OF ADEN
RED SEA
Maldives
Lakshadweep
Sri Lanka
Bay of Bengal
Andaman Islands
Nicobar Islands
ANDAMAN SEA
INDIAN OCEAN
Arctic Circle
HONSHU
HOKKAIDO
Shikoku
Kyushu
SEA OF JAPAN
Korea Strait
YELLOW SEA
EAST CHINA SEA
Ryukyu Islands
Okinawa
Taiwan
NORTH CHINA PLAIN
CHINA
Huang (Yellow) River
HUANG OF
GREAT WALL
QIN LING SHANDI
Chang (Yangtze) River
Hong River
GULF OF TONKIN
Hainan
SOUTH CHINA SEA
Luzon
Luzon Strait
Mindanao
Philippine Islands
CELEBES SEA
Sulawesi (Celebes)
BANDA SEA
Moluccas
ARAFURA SEA
New Guinea
MAOKE MOUNTAINS
PACIFIC OCEAN
INDOCHINA PENINSULA
Mekong River
Chao Phraya River
GULF OF THAILAND
MALAY PENINSULA
Strait of Malacca
Sumatra
Borneo
JAVA SEA
Java
Bangka
Mentawai Islands
Irrawaddy River
Tropic of Cancer
Equator

180
170°W
160°E
150°E
140°E
130°E
120°E
110°E
100°E
90°E
80°E
70°E
60°E
50°E
40°E
30°E
20°E
10°E
0°
170°E
40°N
50°N
60°N
70°N
80°N
90°N

85

ASIA: POLITICAL

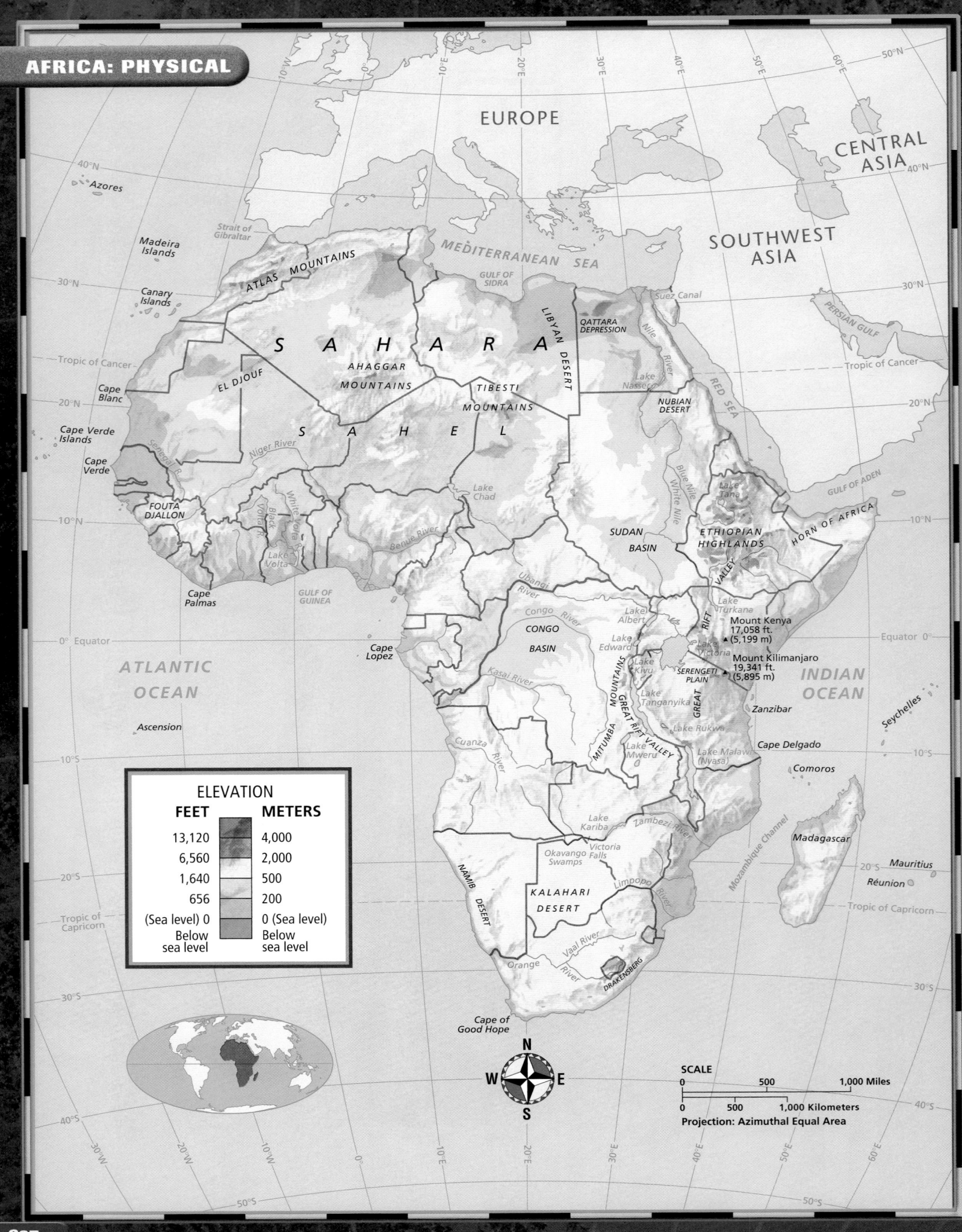

EUROPE

CENTRAL ASIA

SOUTHWEST ASIA

Azores

Madeira Islands

Strait of Gibraltar

ATLAS MOUNTAINS

Canary Islands

Cape Blanc

EL DJOUF

S A H A R A

AHAGGAR MOUNTAINS

TIBESTI MOUNTAINS

LIBYAN DESERT

QATTARA DEPRESSION

MEDITERRANEAN SEA

GULF OF SIDRA

Suez Canal

Nile River

Lake Nasser

NUBIAN DESERT

RED SEA

PERSIAN GULF

Tropic of Cancer

Cape Verde Islands

Cape Verde

FOUTA DJALLON

Senegal R.

Niger River

S A H E L

Black Volta R.

White Volta R.

Lake Volta

Cape Palmas

GULF OF GUINEA

Lake Chad

Benue River

SUDAN BASIN

Blue Nile

White Nile

Lake Tana

ETHIOPIAN HIGHLANDS

GULF OF ADEN

HORN OF AFRICA

Cape Lopez

Ubangi River

Congo River

CONGO BASIN

Kasai River

Lake Albert

Lake Edward

Lake Kivu

Lake Victoria

RIFT VALLEY

Lake Turkana

Mount Kenya 17,058 ft. (5,199 m)

Mount Kilimanjaro 19,341 ft. (5,895 m)

SERENGETI PLAIN

Zanzibar

INDIAN OCEAN

Seychelles

Equator

ATLANTIC OCEAN

Ascension

Cuanza River

MITUMBA MOUNTAINS

GREAT RIFT VALLEY

Lake Tanganyika

Lake Rukwa

Lake Mweru

Lake Malawi (Nyasa)

Cape Delgado

Comoros

Lake Kariba

Zambezi River

Mozambique Channel

Madagascar

Mauritius

Réunion

NAMIB DESERT

Okavango Swamps

Victoria Falls

Limpopo River

KALAHARI DESERT

Tropic of Capricorn

Vaal River

Orange River

DRAKENSBERG

Cape of Good Hope

ELEVATION

FEET		METERS
13,120		4,000
6,560		2,000
1,640		500
656		200
(Sea level) 0		0 (Sea level)
Below sea level		Below sea level

N W E S

SCALE

0 · 500 · 1,000 Miles

0 · 500 · 1,000 Kilometers

Projection: Azimuthal Equal Area

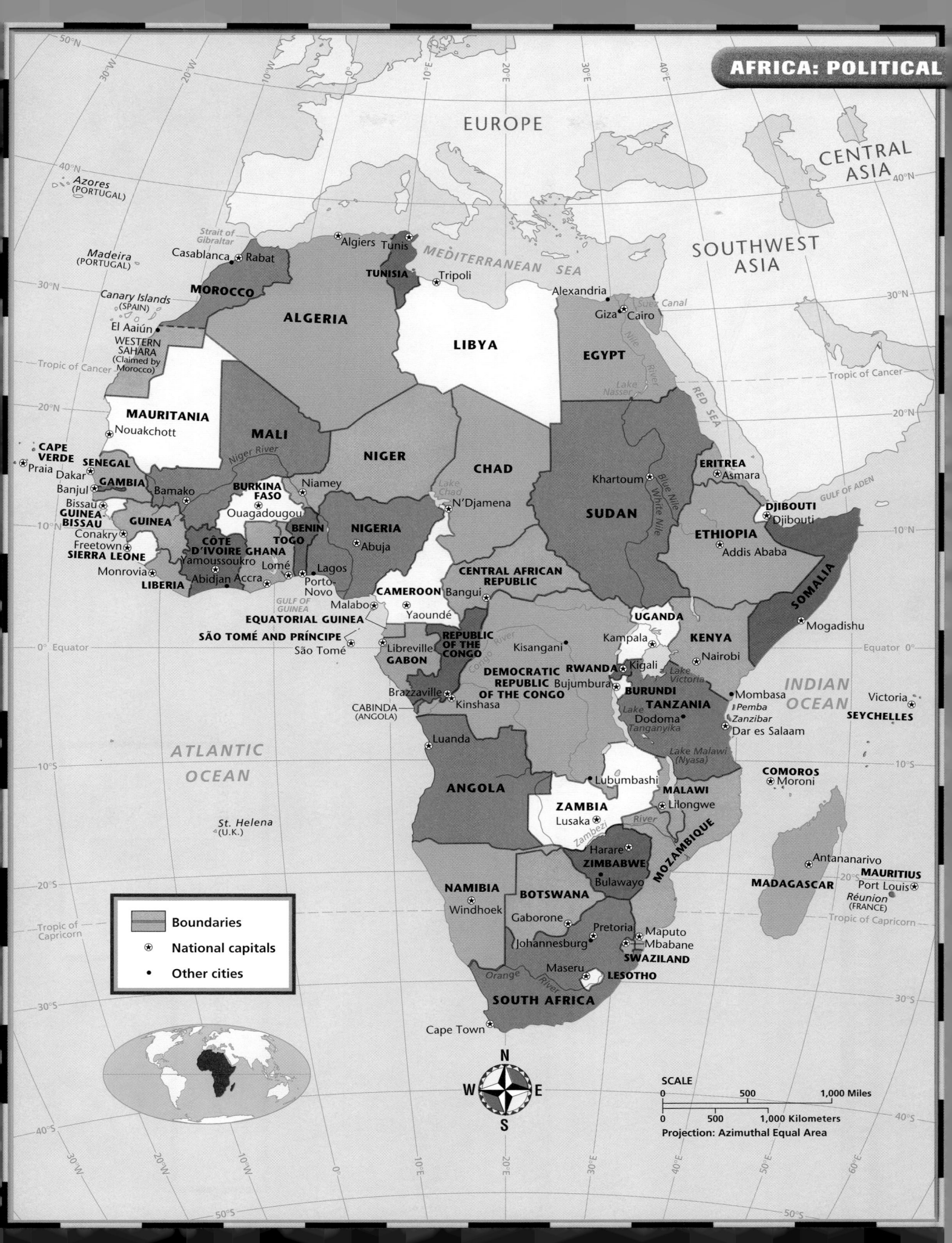

AFRICA: POLITICAL

EUROPE

CENTRAL ASIA

SOUTHWEST ASIA

MEDITERRANEAN SEA

Azores (PORTUGAL)

Madeira (PORTUGAL)

Strait of Gibraltar

Algiers • Tunis

Casablanca • Rabat

TUNISIA • Tripoli

Alexandria

MOROCCO

Canary Islands (SPAIN)

El Aaiún •

WESTERN SAHARA (Claimed by Morocco)

ALGERIA

LIBYA

EGYPT

Giza • Cairo

Suez Canal

Lake Nasser

Tropic of Cancer

MAURITANIA

⊛ Nouakchott

MALI

NIGER

CHAD

Khartoum •

RED SEA

GULF OF ADEN

ERITREA
⊛ Asmara

CAPE VERDE
⊛ Praia

Niger River

SENEGAL

Dakar ⊛

GAMBIA
Banjul ⊛ Bamako ⊛

Niamey ⊛

BURKINA FASO

Lake Chad

N'Djamena •

SUDAN

Blue Nile

White Nile

DJIBOUTI
Djibouti ⊛

Bissau ⊛

GUINEA BISSAU

GUINEA

Ouagadougou ⊛

ETHIOPIA

Addis Ababa •

Conakry ⊛

Freetown ⊛

SIERRA LEONE

CÔTE D'IVOIRE GHANA
Yamoussoukro ⊛

BENIN
TOGO
Lomé ⊛

NIGERIA
⊛ Abuja

CENTRAL AFRICAN REPUBLIC

SOMALIA

Monrovia ⊛

Abidjan • Accra ⊛

Lagos •

LIBERIA

Porto-Novo ⊛

GULF OF GUINEA

CAMEROON Bangui •

⊛ Mogadishu

Malabo ⊛

EQUATORIAL GUINEA

Yaoundé ⊛

UGANDA

Kampala ⊛

KENYA

SÃO TOMÉ AND PRÍNCIPE
São Tomé ⊛

Congo River

Kisangani •

⊛ Nairobi

INDIAN OCEAN

Libreville ⊛

REPUBLIC OF THE CONGO

RWANDA
Kigali ⊛

Lake Victoria

Victoria ⊛

GABON

DEMOCRATIC REPUBLIC OF THE CONGO

Bujumbura ⊛

BURUNDI

Mombasa •

SEYCHELLES

0° Equator

Brazzaville ⊛

TANZANIA

Pemba

Equator 0°

CABINDA (ANGOLA)

Kinshasa ⊛

Dodoma ⊛

Zanzibar •

Congo River

Lake Tanganyika

Dar es Salaam •

Luanda ⊛

Lubumbashi •

Lake Malawi (Nyasa)

COMOROS
⊛ Moroni

ATLANTIC OCEAN

ANGOLA

MALAWI

Lilongwe ⊛

ZAMBIA

Lusaka ⊛

River

St. Helena (U.K.)

Zambezi

Harare ⊛

MOZAMBIQUE

Antananarivo ⊛

MAURITIUS
Port Louis ⊛

ZIMBABWE

Bulawayo •

MADAGASCAR

Réunion (FRANCE)

NAMIBIA

BOTSWANA

Windhoek ⊛

Gaborone ⊛

Pretoria ⊛

Maputo ⊛

Tropic of Capricorn

Johannesburg •

Mbabane ⊛

SWAZILAND

Maseru ⊛

LESOTHO

Orange River

SOUTH AFRICA

Cape Town •

Legend

	Boundaries
⊛	National capitals
•	Other cities

SCALE

0   500   1,000 Miles

0   500   1,000 Kilometers

Projection: Azimuthal Equal Area

N W E S

ASIA

INDIAN OCEAN

TIMOR SEA

ARAFURA SEA

Darwin

ARNHEM LAND

KIMBERLEY PLATEAU

GREAT SANDY DESERT

GIBSON DESERT

HAMERSLEY RANGE

NORTHERN TERRITORY

WESTERN AUSTRALIA

GREAT VICTORIA DESERT

MACDONNELL RANGES

Alice Springs

AUSTRALIA

Broome

Carnarvon

North West Cape

Geraldton

Laverton

Perth
Fremantle

120°E

130°E

Great Australian Bight

GULF OF CARPENTARIA

Cape York

CAPE YORK PENINSULA

Torres Strait

GREAT

BARRIER

REEF

CORAL SEA

Cloncury

Flinders River

QUEENSLAND

GREAT DIVIDING

GREAT ARTESIAN BASIN

Lake Eyre
(52 ft. [16 m]
below sea level)

SOUTH AUSTRALIA

Port Pirie

Adelaide

Kangaroo Island

Rockhampton

Bundaberg

Brisbane
Gold Coast

RANGE

Sydney

Canberra
AUSTRALIAN
CAPITAL TERRITORY
Mount Kosciusko
7,310 ft. (2,228 m)

NEW SOUTH WALES

Lachlan River

Wagga Wagga

Murrumbidgee River

Murray River

VICTORIA

Ballarat

Melbourne
Geelong

Bass Strait

Launceston

Hobart

TASMANIA

PACIFIC OCEAN

TASMAN SEA

North Cape

Auckland
Hamilton

North Island

NEW ZEALAND

South Island

SOUTHERN ALPS

Mount Cook
12,349 ft.
(3,764 m)

Wellington

Cook Strait

Christchurch

Dunedin

Stewart Island

Tropic of Capricorn

170°E

160°E

150°E

140°E

130°E

120°E

110°E

180°

170°E

160°E

150°E

140°E

10°S

20°S

30°S

40°S

50°S

N
E
S
W

SCALE: At Equator

0   250   500 Miles

0   250   500 Kilometers

Projection: Lambert Conformal Conic

ELEVATION

FEET	METERS
13,120	4,000
6,560	2,000
1,640	500
656	200
(Sea level) 0	0 (Sea level)
Below sea level	Below sea level

⊛ National capitals
★ State/territorial capitals
• Other cities

89

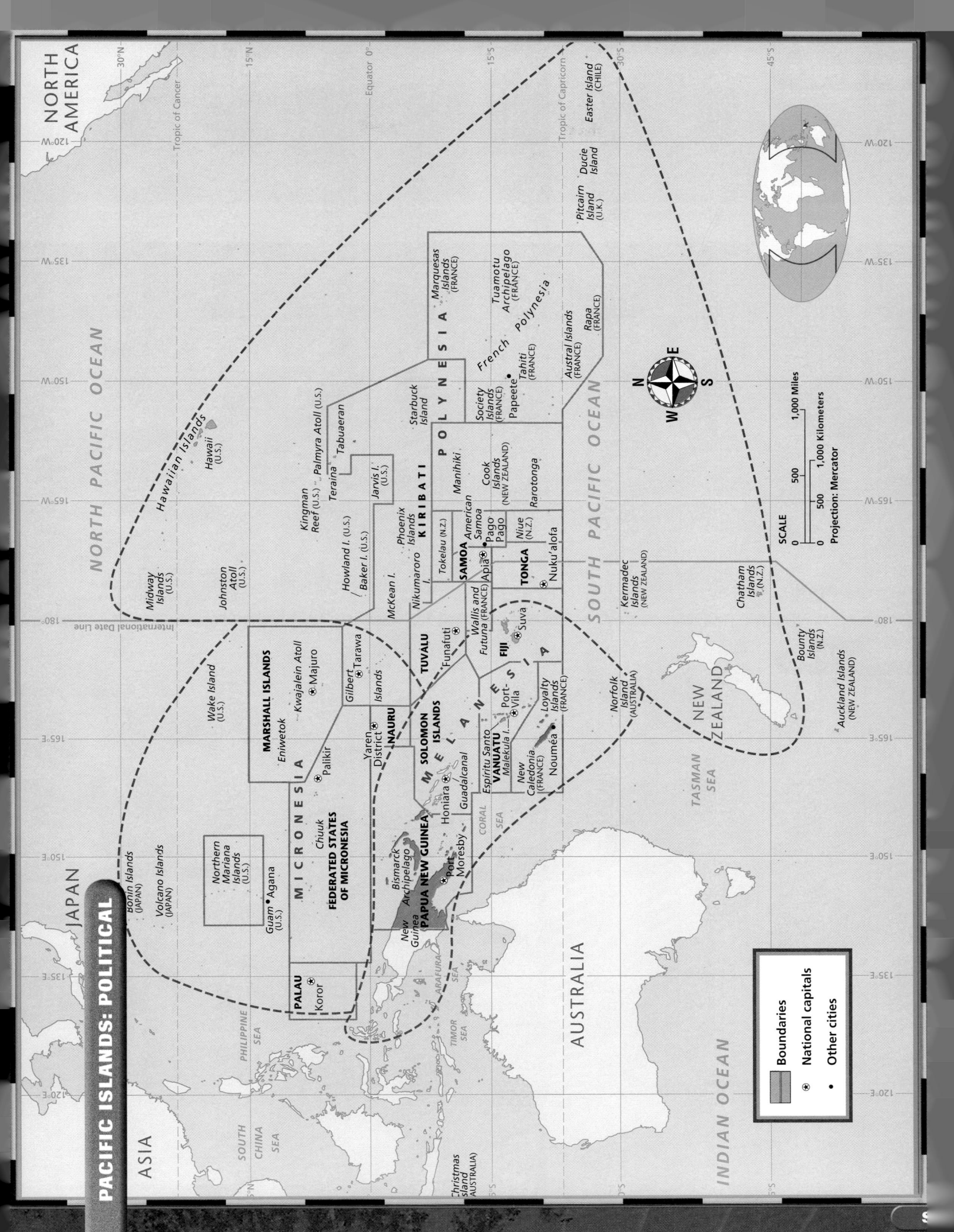

PACIFIC ISLANDS: POLITICAL

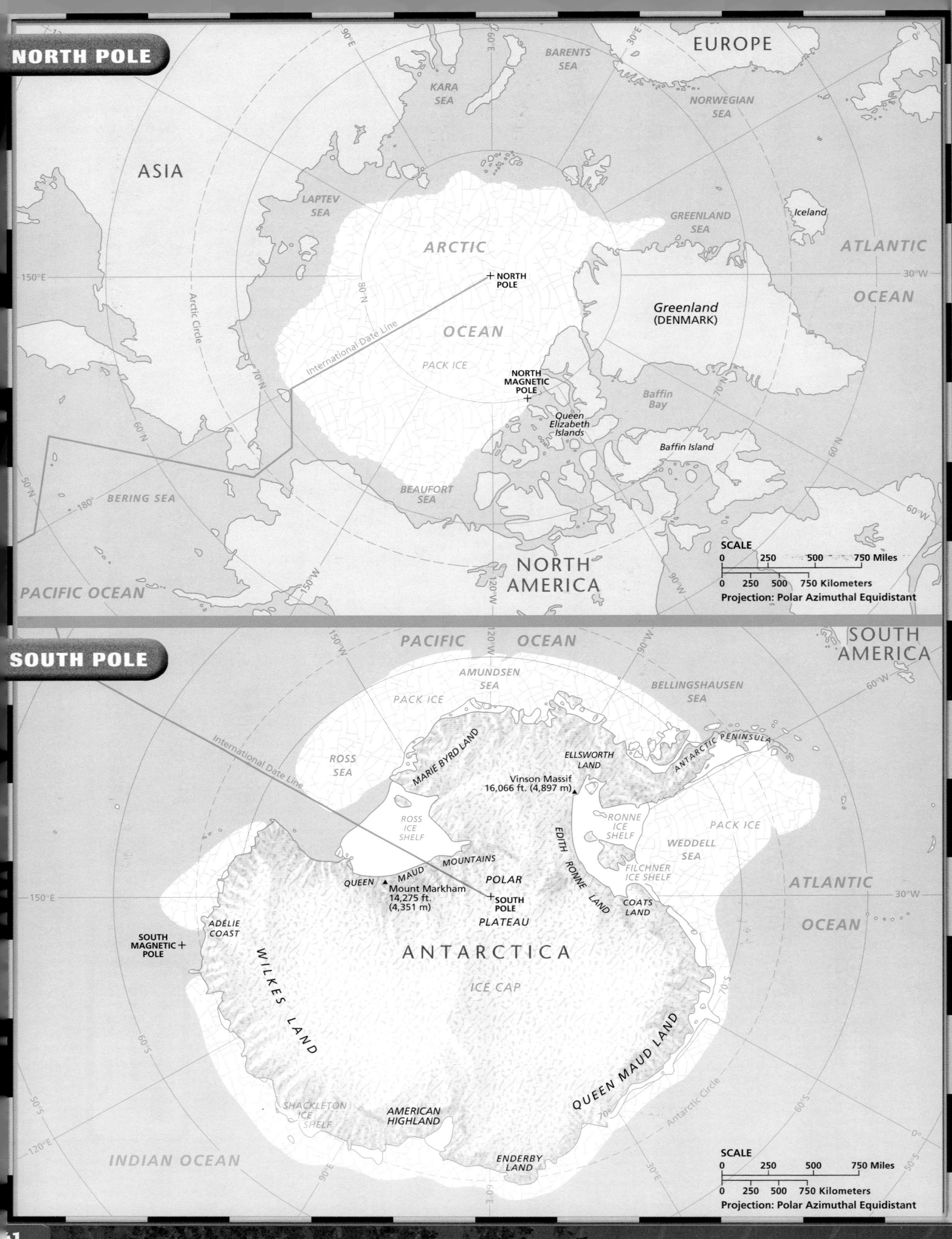

EUROPE

BARENTS
SEA

KARA
SEA

NORWEGIAN
SEA

ASIA

LAPTEV
SEA

ARCTIC

GREENLAND
SEA

Iceland

ATLANTIC

OCEAN

+ NORTH
POLE

Greenland
(DENMARK)

OCEAN

PACK ICE

NORTH
MAGNETIC
POLE
+

Baffin
Bay

Queen
Elizabeth
Islands

Baffin Island

BEAUFORT
SEA

BERING SEA

NORTH
AMERICA

PACIFIC OCEAN

International Date Line

Arctic Circle

SCALE

| 0 | 250 | 500 | 750 Miles |

| 0 | 250 | 500 | 750 Kilometers |

Projection: Polar Azimuthal Equidistant

PACIFIC   OCEAN

SOUTH
AMERICA

AMUNDSEN
SEA

PACK ICE

BELLINGSHAUSEN
SEA

ROSS
SEA

MARIE BYRD LAND

ELLSWORTH
LAND

ANTARCTIC PENINSULA

Vinson Massif
16,066 ft. (4,897 m) ▲

ROSS
ICE
SHELF

RONNE
ICE
SHELF

PACK ICE

WEDDELL
SEA

MOUNTAINS

QUEEN   MAUD

POLAR

Mount Markham
14,275 ft.
(4,351 m)

+ SOUTH
POLE

EDITH RONNE LAND

FILCHNER
ICE SHELF

ATLANTIC

COATS
LAND

PLATEAU

ADÉLIE
COAST

OCEAN

SOUTH
MAGNETIC +
POLE

WILKES   LAND

ANTARCTICA

ICE CAP

INDIAN OCEAN

SHACKLETON
ICE
SHELF

AMERICAN
HIGHLAND

QUEEN MAUD LAND

ENDERBY
LAND

Antarctic Circle

International Date Line

SCALE

| 0 | 250 | 500 | 750 Miles |

| 0 | 250 | 500 | 750 Kilometers |

Projection: Polar Azimuthal Equidistant

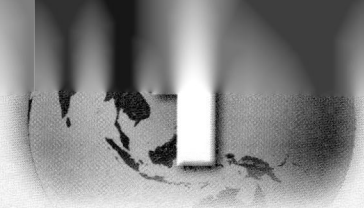

Using the Illustration

Direct students' attention to the photo below. This area has been named a World Heritage Site by the United Nations Educational, Scientific, and Cultural Organization (UNESCO). Since 1972 UNESCO has recognized locations of universal cultural or natural significance. Each unit in this book opens with a photo of one of these sites.

Unit 1 Overview

This unit introduces students to the basic tools and information geographers use to understand our world. Various factors work together to shape our planet. Solar energy and the planet's rotation create climate patterns that distinguish regions from each other. Forces deep below the surface build new landforms while ice, water, and wind gradually wear away those already in existence. The physical landscapes thus created are important factors in determining where and how people will live.

At the same time, the presence of human societies can change natural landscapes. Farmers till land to grow crops. People build villages, towns, and cities in which to live. Governments dam rivers to prevent floods and to generate electricity. Centuries of interaction between physical and human landscapes have created the world in which we live today.

UNIT 1
The Geographer's World

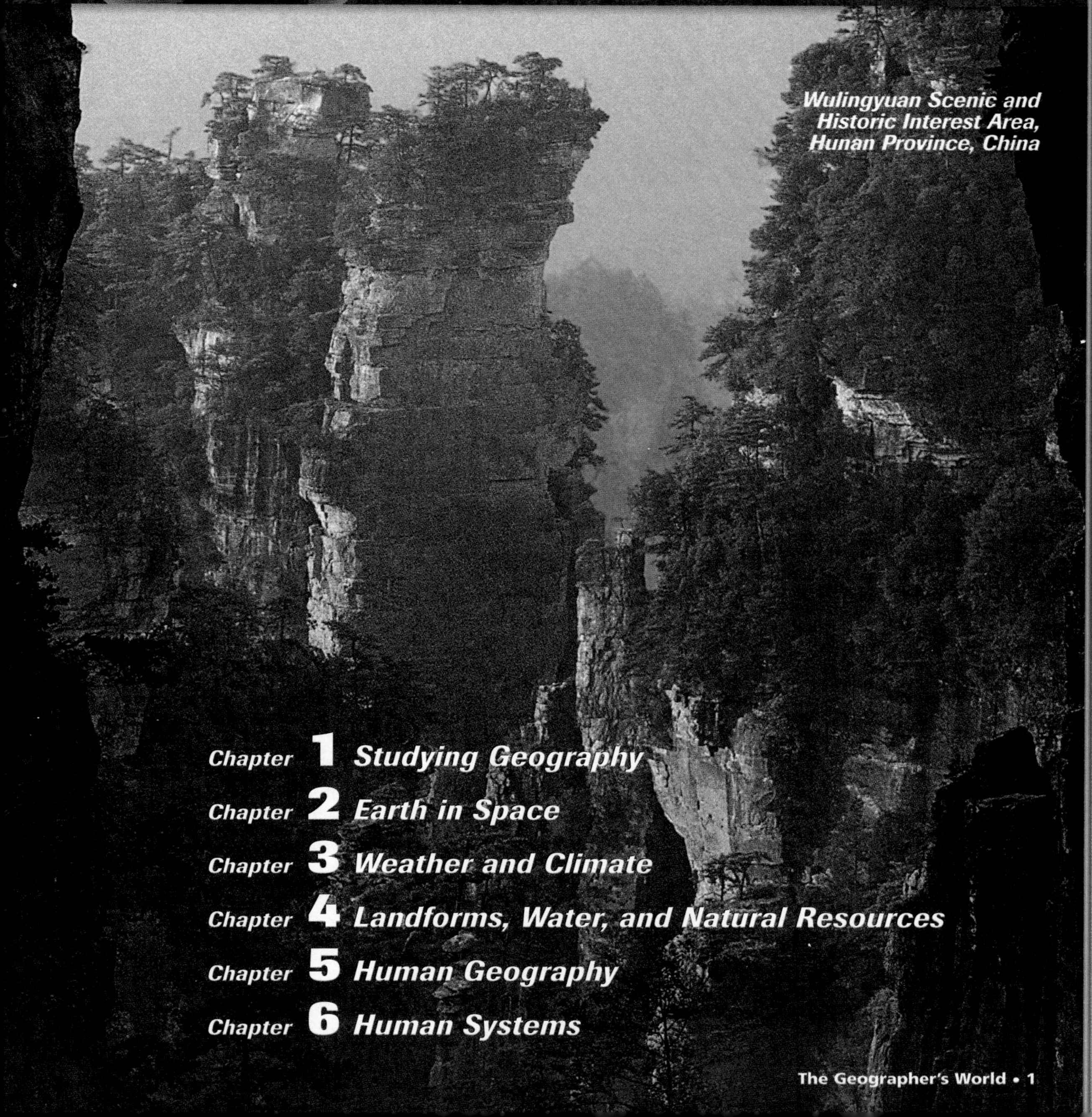

Wulingyuan Scenic and Historic Interest Area, Hunan Province, China

Unit Objectives

1. Introduce geography as a field of study.
2. Explain Earth's position in space and the forces acting on Earth's land and water.
3. Analyze the interrelationships of solar energy, climate, and vegetation.
4. Examine the physical processes that create and shape the world's landforms, water features, and resources.
5. Describe the development of cultures and the results of population expansion.
6. Investigate ways people depend on, adapt to, and modify the physical environment.
7. Use geographic information systems to examine relationships between physical and human geography.

Your Classroom Time Line

To help you create a time line for your classroom, the most important dates and time periods discussed in each unit's chapters are compiled for you. The lists begin in the sidebar on the page with the unit atlas political map. Some additional dates have been added for clarity. Note that many dates are approximate. In making the classroom time line, you may want to have students use colored markers to differentiate among political, scientific, religious, and artistic achievements.

Studying Geography

CHAPTER RESOURCE MANAGER

Objectives	Pacing Guide	Reproducible Resources
SECTION 1 **Themes and Essential Elements** (pp. 3–8) • Identify the two main branches of geography. • Explain how we use geography. • Describe some ways we can organize our world and the study of geography.	**Regular** 2 days **Block Scheduling** 1.5 days *Block Scheduling Handbook, Chapter 1*	**RS** Guided Reading Strategy 1.1 **RS** Graphic Organizer Activity 1 **PS** Readings in World Geography, History, and Culture 1
SECTION 2 **Skill Building: Using the Geographer's Tools** (pp. 9–20) • Explain how geographers and mapmakers organize our world. • Identify kinds of special maps geographers use. • Describe how geographers use climate graphs and population pyramids.	**Regular** 2 days **Block Scheduling** 1.5 days *Block Scheduling Handbook, Chapter 1*	**RS** Guided Reading Strategy 1.2 **PS** Readings in World Geography, History, and Culture 2 **IC** Lab Activity for Geography and Earth Science 2: Looking for Buried Treasure **E** Creative Strategies for Teaching World Geography, Lessons 1–3 **SM** Critical Thinking Activity 1: Climate **SM** Geography for Life Activity 1: Two Geographic Grids **SM** Map Activity 1: City Map

Chapter Resource Key

PS Primary Sources
RS Reading Support
IC Interdisciplinary Connections
E Enrichment
SM Skills Mastery

A Assessment
REV Review
ELL Reinforcement and English Language Learners

 CD–ROM
 Video
 Internet
 Holt Presentation Maker Using Microsoft® PowerPoint®
Transparencies

 One-Stop Planner CD–ROM

See the *One-Stop Planner* for a complete list of additional resources for students and teachers.

 **One-Stop** Planner CD–ROM

It's easy to plan lessons, select resources, and print out materials for your students when you use the **One-Stop Planner CD–ROM with Test Generator**.

☑ internet connect

HRW ONLINE RESOURCES

GO TO: go.hrw.com
Then type in a keyword.

TEACHER HOME PAGE
　KEYWORD: SW3 Teacher

CHAPTER INTERNET ACTIVITIES
　KEYWORD: SW3 GT1
　Choose an activity to:
　• understand GIS systems.
　• explore geographic regions and create a brochure.
　• learn to use online maps.

CHAPTER ENRICHMENT LINKS
　KEYWORD: SW3 CH1

CHAPTER MAPS
　KEYWORD: SW3 MAPS1

ONLINE ASSESSMENT
　Homework Practice
　KEYWORD: SW3 HP1
　Standardized Test Prep
　KEYWORD: SW3 STP1
　Rubrics
　KEYWORD: SS Rubrics

COUNTRY INFORMATION
　KEYWORD: SW3 Almanac

CONTENT UPDATES
　KEYWORD: SS Content Updates

HOLT PRESENTATION MAKER
　KEYWORD: SW3 PPT1

ONLINE READING SUPPORT
　KEYWORD: SS Strategies

CURRENT EVENTS
　KEYWORD: S3 Current Events

Technology Resources

- One-Stop Planner CD–ROM, Lesson 1.1
- *ARGWorld* CD–ROM
- Homework Practice Online
- HRW Go site

- One-Stop Planner CD–ROM, Lesson 1.2
- **CNN** Presents Geography: Yesterday and Today, Segment 1: Mapping Change
- Homework Practice Online
- HRW Go site

Reinforcement, Review, and Assessment

ELL Main Idea Activity 1.1
ELL English Audio Summary 1.1
ELL Spanish Audio Summary 1.1
REV Section 1 Review, p. 8
A Daily Quiz 1.1

ELL Main Idea Activity 1.2
ELL English Audio Summary 1.2
ELL Spanish Audio Summary 1.2
REV Section 2 Review, p. 20
A Daily Quiz 1.2

Meeting Individual Needs

Ability Levels

Level 1 Basic-level activities designed for all students encountering new material

Level 2 Intermediate-level activities designed for average students

Level 3 Challenging activities designed for honors and gifted-and-talented students

English Language Learners Activities that address the needs of students with Limited English Proficiency

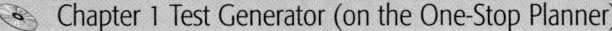

Chapter Review and Assessment

- Chapter 1 Test Generator (on the One-Stop Planner)
- Global Skill Builder CD–ROM
- HRW Go site
REV Chapter 1 Review, pp. 22–23
REV Chapter 1 Tutorial for Students, Parents, Mentors, and Peers
A Chapter 1 Test (form A or B)
A Alternative Assessment Handbook
A Chapter 1 Test for English Language Learners and Special-Needs Students

Launch into Learning

Bring a newspaper to class. Discuss with students some of the articles in the newspaper and give students a brief summary of what topics the articles cover. Be sure to select articles from the main section, world news, and local and state news. Then ask students how the articles might relate to the study of geography. (*Possible answer: Some articles may deal with international events involving many countries, how human activities have affected the environment, current political issues, or other topics.*) Next, have students note any maps printed in the newspaper. Ask students to describe the maps, analyze what the maps are designed to show, how effective they are, and how they support the text of the various articles or sections where they are found. Ask students if maps convey some types of information more effectively than words and, if so, what kinds of information maps are ideal for presenting. (*Possible answers: spatial information, patterns*) Finally, ask students to summarize how newspapers contain a wealth of geographic information.

Why We Should Know More

You may want to share with your students the following reasons for gaining a basic understanding of geography as a field of study:

▶ Knowing the fundamentals of geography will help students learn more about the world around them. A geographically informed person understands the world's countries and cultures, current events, and global issues.

▶ Geography is a growing field that includes a wide range of specializations. Some students might want to pursue careers in geography.

▶ Maps play important roles in people's daily lives, and a basic understanding of map elements and design has great practical value.

▶ Throughout this text, connections are made to the six essential elements of geography. These elements are explained fully in Chapter 1.

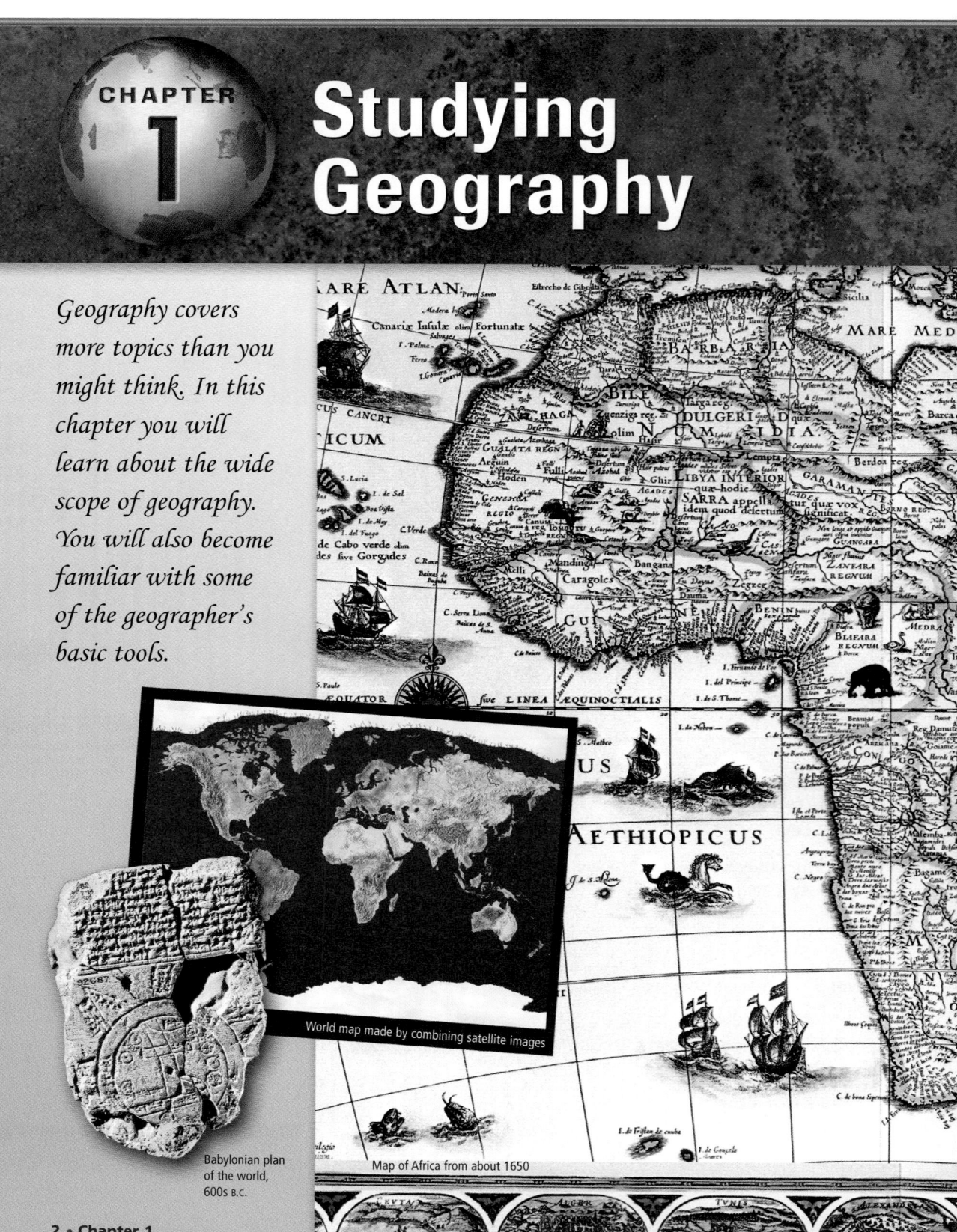

CHAPTER 1 Studying Geography

Geography covers more topics than you might think. In this chapter you will learn about the wide scope of geography. You will also become familiar with some of the geographer's basic tools.

World map made by combining satellite images

Babylonian plan of the world, 600s B.C.

Map of Africa from about 1650

Section 1 — Themes and Essential Elements

READ TO DISCOVER

1. What are the two main branches of geography?
2. How do we use geography?
3. What are some ways we can organize our world and the study of geography?

DEFINE

geography
perspective
landscapes
cartography
meteorology
region
formal region
functional region
perception
perceptual regions

WHY IT MATTERS

Geography plays a major role in many international issues. Use CNNfyi.com or other current events sources to learn about a recent international news event that involves important geographical concepts.

What Is Geography?

Many people do not really know what **geography** is. Some people think geography is just memorizing lists of countries and state capitals. Other people think it is the study of rocks. Still others think geographers just look at maps and pictures of faraway places. However, there is much more to geography!

Geography is the study of everything on Earth, from rocks and rainfall to people and places. Geographers study how the natural environment influences people, how people's activities affect Earth, and how the world is changing. To do this, geographers look at many different things, including cities, cultures, plants, and resources. Geographers focus on where these things are or where related events happen.

internet connect

GO TO: go.hrw.com
KEYWORD: SW3 CH1
FOR: Web sites about studying geography

INTERPRETING THE VISUAL RECORD

Light shines through the shelter built by this Inuit man in Nunavut, Canada. **How do you think the natural environment affects this person, and how do his activities affect Earth? How might his world be changing?**

LEVEL 1: Copy the following graphic organizer onto the chalkboard, omitting the italicized answers. Call on students to provide words and phrases to describe similarities and differences in the characteristics of the two main branches of geography. **ENGLISH LANGUAGE LEARNERS**

Human Geography
• *study of distribution and characteristics of the world's people (where people live and what they do)*
• *examines how people make and trade things that they need to survive*

Both
• *interaction of people with their environments*

Physical Geography
• *focuses on Earth's natural environments including landforms, water features, plants, animals, and other physical features*
• *studies the processes that shape the physical environment*

Across the Curriculum

► Science ◄

Perspectives from Space
Satellite images have provided additional resources for geographers using a spatial perspective in their study of Earth. In the 1960s, *TIROS 1* transmitted the first satellite photographs used to analyze Earth's atmosphere and surface. Improvements in satellite technology and imagery now allow geographers to analyze climate and vegetation patterns over time.

CRITICAL THINKING: What types of changes related to the effect of human actions on the environment might geographers observe in satellite images? *(Possible answer: Increased use of land for agricultural purposes might change natural vegetation patterns.)*

VISUAL RECORD ANSWER

Possible answers: that the women live in a warm climate, depend on the sea for their livelihood, and contribute to the family's income

VISUAL RECORD ANSWER

Possible answers: physical—questions about the area's climate, wildlife, effect of forest fires on plant life, and so on; human—questions about effect of tourism on area's economy, about the planning of new transportation routes, and so on

INTERPRETING THE VISUAL RECORD

Indonesian women sell fish at their local market. **What might the photo indicate about the physical and human landscapes in which these women live?**

Perspective—or the way a person looks at something—is an important part of learning about geography or any other subject. Geographers use a spatial perspective to study the world. That is, they look for patterns in where things are located on Earth and how they are arranged. Geographers then try to explain these patterns. They also look at a world that is shaped by **landscapes**. A landscape is the scenery of a place, including its physical, human, and cultural features. Geographers look at landscapes and try to explain what they see. For geographers the word *landscape* is almost magical. Each of us lives in a landscape. When studying world geography, we discover the amazing diversity of our world's landscapes.

Geography has two main branches—human geography and physical geography. Those who study human geography look at the distribution and characteristics of the world's people. They study where people live and work as well as their ways of life. They also look at how people make and trade things that they need to survive. The study of physical geography focuses on Earth's natural environments. These include Earth's landforms, water features, atmosphere, animals, plants, soils, and the processes that affect them. The interaction of people with their environment links human and physical geography together.

In this textbook you will use both human and physical geography to study the world. First, you will study an area's natural environment. Then you will learn about the area's human aspect and how it relates to the physical setting.

✓ **READING CHECK:** *The Uses of Geography* What are the two main branches of geography? human and physical geography

INTERPRETING THE VISUAL RECORD

Mount McKinley, the highest peak in North America, rises over Denali National Park and Preserve. Fireweed blooms in the foreground. **What questions might physical and human geographers ask about this place?**

LEVELS 2 AND 3: Have students create collages that illustrate the characteristics of the two main branches of geography. Give each student one piece of newsprint, scissors, glue, and several magazines. On one half of the paper, ask students to place words, pictures, and drawings that relate to human geography such as people, food, and modes of transportation. On the other half, have students place images and words depicting physical geography such as mountains, animals, and volcanoes. **ENGLISH LANGUAGE LEARNERS**

Teach Objectives 1–2

LEVEL 1: As a class, review the two main branches of geography *(human and physical)*. Then organize the class into groups. Have each group write a short report that demonstrates some of the ways we use human and physical geography in our everyday lives. *(Possible answers: deciding where to go and how to get there, planning cities, preparing for weather events or hazards)* Tell students to include examples as well as graphics that illustrate these uses. When the groups have finished, call on a volunteer from each group to present a report to the class. **COOPERATIVE LEARNING**

Who Uses Geography?

Now that you know what geography is, you might be wondering, who uses geography? You do. In fact, people all over the world use geography every day. We use it when we find our way to school or work and go on trips. We also use it when we watch the news on television and read about other countries. We think geographically every time we decide where to go and how to get there.

Most jobs require an understanding of geography. For example, a restaurant owner must find a good location for his or her business. Politicians need to know the geography of their districts. They must also understand the issues that are important to people there. In addition, a number of professions rely heavily on specially trained geographers.

Subfields of Geography Geography has many different subfields. One of the most well known is **cartography**—the study of maps and mapmaking. Maps are important because they help geographers study locations. Although some maps are still drawn by hand, computers have completely changed mapmaking. Computers store information from satellite images, photographs, and other sources. A cartographer then creates a map on a computer. Cartographers work for companies that publish maps, atlases, newspapers, magazines, and books. They also work for city planning agencies and other areas of government.

Another subfield of geography is **meteorology**—the study of weather. Meteorologists forecast how the weather will develop so that people know what to expect. You have probably watched these meteorologists, or weather forecasters, on local television.

Geographers at Work Many geographers work for governmental agencies. In fact, one of the largest employers of cartographers in the United States has been the United States Geological Survey (USGS). The USGS produces detailed maps of the whole country. Other agencies that hire geographers include the offices of most city, county, and state governments.

Many businesses hire geographers. Those geographers decide where to place new stores and plan shipping and trucking routes. They also help identify new markets. Geographers work in many different areas of business, such as tourism and travel and international sales.

Schools also hire geography teachers, who help people learn about the world. This knowledge is becoming more important as the different areas of the world become more closely linked. Geographic knowledge is also needed for good citizenship. Citizens who feel strongly about important geographic issues can try to influence public policies and decisions. Should we allow suburbs to be built over good farmland? Where should we put our garbage and dangerous materials? Helping citizens and governments find answers to these questions is the job of geographers.

✓ **READING CHECK:** *The Uses of Geography* What are some organizations and companies that employ geographers? government planning and other agencies, cartography companies, TV and other media, weather forecasting services, tourism, sales, schools

A technician scans a map to create a digital version.

LEVELS 2 AND 3: Organize the class into pairs. Have each pair locate a magazine article that relates to some aspect of human or physical geography. Have pairs write paragraphs explaining how geography is reflected in the articles.

Teach Objective 3

LEVEL 1: Copy the following graphic organizer onto the chalkboard, omitting the italicized answers. Call on students to provide two examples of each type of region that they list in the diagram. **ENGLISH LANGUAGE LEARNERS**

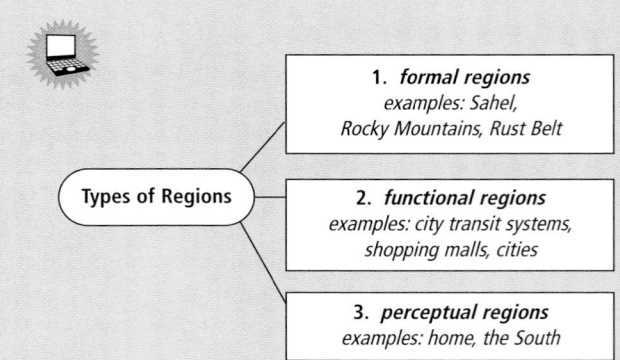

Types of Regions

1. *formal regions*
examples: Sahel,
Rocky Mountains, Rust Belt

2. *functional regions*
examples: *city transit systems,
shopping malls, cities*

3. *perceptual regions*
examples: *home, the South*

Essential Element 2

▶ **Places and Regions** ◀

Louisiana The French explorer La Salle sailed down the Mississippi River in the 1600s. He claimed the river and all land drained by it for France. This meant that all land between the Appalachian and Rocky Mountains was French territory. He named the region Louisiana in honor of King Louis XIV. Land east of the Mississippi ceased to be called Louisiana when France ceded the land to Britain in 1763. When the United States purchased Louisiana in 1803, no specific boundaries were stated. The state of Louisiana was carved out of the region and admitted to the Union in 1812.

CRITICAL THINKING: In what ways might Louisiana have been a formal region, a functional region, and a perceptual region? *(Possible answers: La Salle's Louisiana was a functional region linked by the Mississippi, the Louisiana Purchase was a perceptual region because it was purchased as a single entity, and the state of Louisiana is a formal region.)*

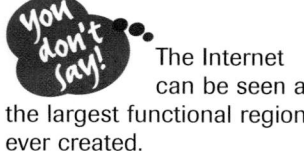 The Internet can be seen as the largest functional region ever created.

VISUAL RECORD ANSWER

Possible answers: soil type, rainfall level, employment in lumber industry

The word *geography* was first used by the ancient Greek geographer Eratosthenes (er-uh-TAHS-thuh-neez). It comes from two Greek words—*geo,* meaning "Earth," and *graphia,* which means "to describe."

γεωγραφία

INTERPRETING THE VISUAL RECORD

The Piney Woods of East Texas qualifies as a formal region, mainly because of the pine trees that grow throughout the area. **What other features might help define the Piney Woods?**

How Do We Study Geography?

An important concept in geography is the idea of a **region**. A region is an area with one or more common features that make it different from surrounding areas. Cities, states, countries, and continents are examples. Organizing Earth's surface into smaller regions makes it easier to study our complex world.

Regions are defined by their physical and human features. Physical features include the kinds of climate, river systems, soils, and vegetation you find there. Human features include the languages, religions, and trade networks of an area. Sometimes the boundaries of a region are clear. For example, the United States is a political region with clear boundaries. In other places, the boundaries are harder to set. For example, the Corn Belt is a farming region in the midwestern United States. However, the Corn Belt does not have clearly set boundaries. It stretches across a number of states. Exactly where the Corn Belt begins and ends is not clear.

Regions can be any size. Countries, deserts, and mountain ranges are examples of large regions. Smaller regions include suburbs and neighborhoods. Regions can also be divided into smaller areas called subregions. For example, the Great Plains is a subregion within North America.

✓ **READING CHECK:** *Places and Regions* What are some physical and human features that can define a region? soils, climate, vegetation, languages, trade networks, river systems, religions

Types of Regions Geographers define regions in three basic ways. The first is a **formal region**. A formal region has one or more common features that make it different from surrounding areas. An example is the Sahel in Africa. This dry region lies between the Sahara, a vast desert to the north, and wetter forested areas to the south.

Formal regions can be based on almost any feature or combination of features. Those features might include population, income levels, crops, temperature, or rainfall. Physical features might define a formal region, such as the Rocky Mountains in the western United States. Economic features also might define such a region. For example, an industrial area in the northeastern and midwestern United States is also a formal region. This region was once called the Rust Belt because so many old factories there had shut down. Today new industries have revived the region's economy.

The second type of region is a **functional region**. These are made up of different places that are linked together and function as a unit. For example, a city transit system is a functional region. It includes many different places. However, the flow of people, trains, and buses link those places together.

Many functional regions are organized around a central point. Surrounding areas are linked to this point. For example, shopping malls are centers of functional regions linked to surrounding neighborhoods. Cities are also examples of these centers. They connect to suburbs, areas in the country, and industry, which all function together.

The third type of region involves human **perception**—our awareness and understanding of the environment around us. People view regions very differently. Our views are influenced not only by what is in a region but also by what is in us. Our ways of life and experiences influence how we perceive the world. Therefore, **perceptual regions** are regions that reflect human feelings

LEVEL 2: Organize the class into six groups. Assign each group one of the six essential elements of geography. Give each group a sheet of butcher paper. Then have each group find pictures and create drawings that demonstrate an understanding of the assigned element. When the groups have finished, have them display their projects. **COOPERATIVE LEARNING**

LEVEL 3: Have students work in the same groups as in the Level 2 activity. Instruct each group to write an article that addresses the group's assigned essential element. You may want to provide a list of topics for students to choose among, or you may want to allow students to choose their own topics. Have students use Internet or library sources to conduct research on their chosen topics. Each group's article should explain how the topic it selected illustrates the appropriate essential element. When groups have completed their articles, compile them and display them by the posters created in the previous activity. **COOPERATIVE LEARNING**

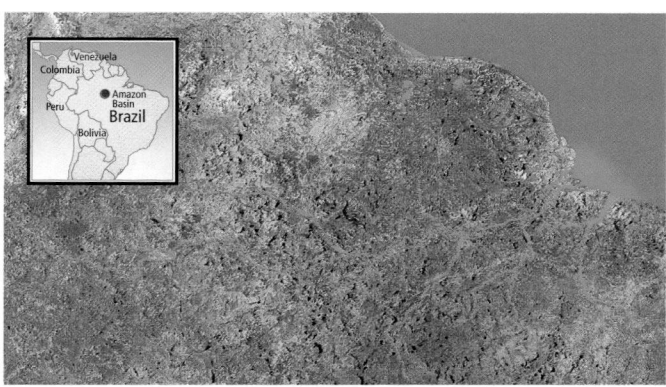

INTERPRETING THE VISUAL RECORD   *The Amazon River system in South America, at the left, is a functional region. Places within it are linked by the river's flow. For many Americans, southern California, at the right, is a perceptual region.* **What elements of southern California as a perceptual region are illustrated in this photo of the beach at Venice?**

and attitudes. For example, "back home" is a perceptual region for most people. However, it may be hard to define exactly. The U.S. Midwest may be easier to define. The South—a region sometimes called Dixie—is another example. Many people perceive these areas to be distinct regions. These areas have their own special features that make them different from anywhere else. Yet people may view—or perceive—those features in differing ways.

READING CHECK: *Places and Regions*   What are the three types of regions?
formal, functional, perceptual

The Six Essential Elements   The study of geography has long been organized according to five important themes. One theme, *location*, deals with the exact or relative spot of something on Earth. Another term for something's exact location on Earth's surface is its absolute location. Something's relative location is its position on Earth relative to other locations. *Place* includes the physical and human features of a location. *Human-environment interaction* covers the ways people and environments interrelate with and affect each other. *Movement* involves how people and things change locations and the effects of these changes. *Region* organizes Earth into geographic areas with one or more shared characteristics. Another way to look at geography is to identify essential elements in its study. These six essential elements will be used throughout this textbook. They share many properties with the five themes of geography.

1. *The World in Spatial Terms*   This element focuses on geography's spatial perspective and uses maps to study people, places, and environments.

2. *Places and Regions*   Our world has a vast number of unique places and regions. This element deals with the physical and human features of those places and how we define and perceive various regions.

3. *Physical Systems*   Physical systems shape Earth's features. Geographers study earthquakes, mountains, rivers, volcanoes, weather patterns, and so on. They also study how plants and animals relate to these systems.

Essential Element 3

▶ **Physical Systems** ◀

Great Sand Dunes
Great Sand Dunes National Monument demonstrates how physical processes shape the patterns of Earth's surface. Located in south-central Colorado, the dunes are formed by winds that blow sand across the San Luis Valley. The sand was eroded from rock by alpine glaciers during the Ice Age and by the flow of the Rio Grande. Over time, the river changed its course, leaving large deposits of sand on the valley floor. Prevailing winds carry the sand northeastward to the high Sangre de Cristo Mountains and deposit the sand as they sweep up and over the western side of the mountains.

DISCUSSION: Lead a discussion addressing the following question: What other physical processes shape Earth's surface? *(Possible answers: weathering, erosion, plate tectonics)*

VISUAL RECORD ANSWER

Possible answers: warm sunny climate, palm trees, sand, ocean, people walking and skating along the shore

INTERPRETING THE VISUAL RECORD

Workers from an environmental organization called the Green Belt Movement teach schoolchildren in Kenya, East Africa, how to care for tree seedlings. **If you were to label this photo with one of the six essential elements, which would you use? How does the essential element you chose relate to the photo?**

4. (**Human Systems**) People are central to geography. Our activities, movements, and settlements shape Earth's surface. The ways of life we follow and the things we produce and trade are part of the study of human systems. Geographers look at the causes and results of conflicts between peoples. The study of governments we set up and the features of cities and other settlements we live in are also part of this study.

5. (**Environment and Society**) Human actions, such as using oil or water, affect the environment. At the same time, Earth's physical systems—from weather to volcanoes—affect human activities. We depend on what the Earth provides to survive. The relationship between people and the environment is an important part of geography.

6. (**The Uses of Geography**) Geography helps us understand the relationships among people, places, and environments over time. Geography can help us to interpret the past and the present or to plan for the future.

✓ **READING CHECK:** (**The Uses of Geography**) What are six essential elements in the study of geography? The World in Spatial Terms, Places and Regions, Physical Systems, Human Systems, Environment and Society, The Uses of Geography

Section 1 Review

go.hrw.com Homework Practice Online
Keyword: SW3 HP1

Define geography, perspective, landscapes, cartography, meteorology, region, formal region, functional region, perception, perceptual regions

Reading for the Main Idea

1. *Places and Regions* What are examples of functional, formal, and perceptual regions?

2. *The Uses of Geography* What six essential elements are used to organize the study of geography?

Critical Thinking

3. Contrasting How is the study of human geography different from the study of physical geography?

4. Making Generalizations and Predictions What do you think a "geographical approach" to studying an issue might be?

Organizing What You Know

5. Create a graphic organizer like the one shown below. Use it to identify some of the jobs that geographers have.

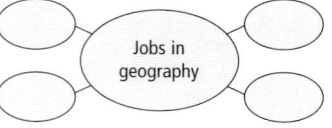

Jobs in geography

OBJECTIVES

1. **Explain how geographers and mapmakers organize our world.**

2. **Identify special kinds of maps geographers use.**

3. **Describe how geographers use climate graphs and population pyramids.**

 **LET'S GET STARTED**

Copy the following passage onto the chalkboard: *Read the definitions of grid, longitude, and latitude in this section.* Then ask students if they use grid systems in daily life. Discuss their responses. *(Students may mention that some streets and home addresses use a grid system and that many city maps use a grid system.)* Tell students they will learn more about the features and types of maps and globes in Section 2.

Building Vocabulary

Have students read the definitions of the key terms in Section 2. Then have students examine maps throughout this textbook for examples of map elements such as **meridians**. Ask students to locate **climate graphs** and **population pyramids** in other chapters and describe the information they contain.

Skill Building: Using the Geographer's Tools

READ TO DISCOVER

1. How do geographers and mapmakers organize our world?
2. What kinds of special maps do geographers use?
3. How do geographers use climate graphs and population pyramids?

WHY IT MATTERS

News reporters often use maps to identify the locations of important events. Use  or other **current events** sources to find out how maps are used in specific news stories.

DEFINE

grid	atlas
latitude	map projections
longitude	great-circle route
equator	compass rose
parallels	legend
meridians	contiguous
prime meridian	precipitation
degrees	topography
hemispheres	climate graphs
continents	population pyramids

Organizing the Globe

We begin our study of geography by looking at a globe. A globe is a scale model of Earth. It is useful for looking at the whole planet or at large areas of its land and water surface. One of the first things you will notice on the globe on this page is a pattern of lines. These lines circle the globe in east-west and north-south directions. This pattern is called a **grid**. The grid is made up of lines of **latitude** and **longitude**. Lines of latitude are drawn in an east-west direction. Lines of longitude are drawn in a north-south direction. The intersection of these imaginary lines helps us find the exact or absolute location of places.

Latitude and Longitude

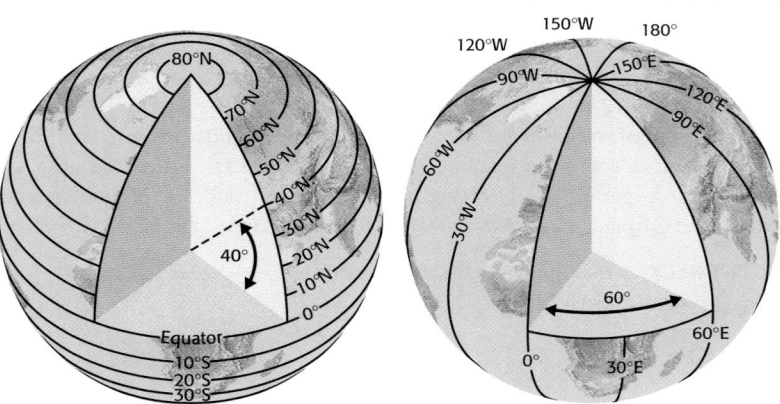

The illustration on the left shows lines of latitude. The north-south lines shown on the right are lines of longitude. Notice that lines of latitude are always the same distance apart.

Teach Objective 1

ALL LEVELS: On the chalkboard, draw a grid to represent your school grounds. Include landmarks such as key buildings, classrooms, statues, and courtyards. Number the lines on the grid in increments of 10, starting with zero. Point out that locations on the grounds can be identified by naming the intersection of horizontal and vertical lines. Point out the similarities between this grid and the latitude and longitude lines on a map. Remind students that the numerical values located where horizontal and vertical lines intersect are known as *coordinates*. Then name some school landmarks and have students indicate their coordinates.

HOMEWORK: Have each student draw a map of the contiguous United States that includes major lines of latitude and longitude. Then have students use their maps to answer these questions: What is the approximate latitude of the Florida Keys? *(24°N)* What is the approximate latitude of the U.S.-Canada border from Washington to Minnesota? *(49°N)* What is the approximate longitude of the West Coast? *(124°W)* What is the approximate longitude of the East Coast? *(75°W)* Which major U.S. city is located at about 40°N and 74°W? *(New York)*

Across the Curriculum

► **Technology** ◄

Harrison's Chronometer
For centuries, sailors could not calculate longitude at sea. This problem increased the danger of ocean voyages. To measure longitude in the open ocean required two accurate clocks—one that showed the time onboard and another that showed the time in a different location where the longitude was known, such as the home port. Each hour's difference between the two clocks would represent 15 degrees of longitude. However, the pendulum clocks that were commonly used at the time were not accurate at sea.

In 1714 England's Parliament offered £20,000 (more than $1 million in today's money) to anyone who could design a clock that would remain accurate at sea. Clockmaker John Harrison succeeded by eliminating pendulums, combining metals with different properties, and creating rust-free mechanisms. In 1762 Harrison's marine chronometer was only five seconds off after a voyage to Jamaica. The new clock and the ability to determine longitude in the open ocean helped Britain establish dominance of the seas.

MAP ANSWER

Northern and Western Hemispheres; Antarctica and Australia

NORTHERN HEMISPHERE

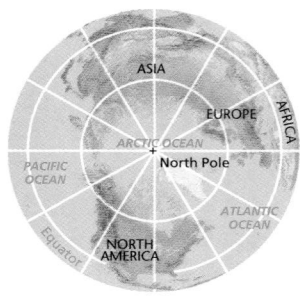

SOUTHERN HEMISPHERE

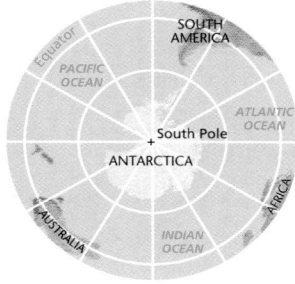

EASTERN HEMISPHERE

WESTERN HEMISPHERE

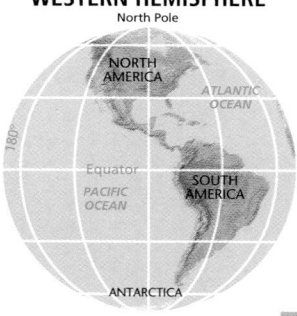

Lines of latitude measure distance north and south of the **equator**. The equator is an imaginary line that circles the globe halfway between Earth's North Pole and South Pole. Lines of latitude are also called **parallels**. This is because they are always parallel to the equator and each other. Lines of longitude are called **meridians**. They measure distance east and west of the **prime meridian**. This is an imaginary line drawn from the North Pole through Greenwich, England, to the South Pole. Parallels and meridians measure distances in **degrees**. The symbol for degrees is °. Degrees are further divided into minutes, for which the symbol is ´. There are 60 minutes in a degree.

As you can see on the globe on the previous page, parallels north of the equator are marked with an *N*. Those south of the equator are marked with an *S*. Lines of latitude range from 0°, for locations on the equator, to 90°N and 90°S, for locations at the North Pole and South Pole. Lines of longitude range from 0° on the prime meridian to 180° on a meridian in the mid–Pacific Ocean. Meridians west of the prime meridian to 180° are labeled with a *W*. Those east of the prime meridian are labeled with an *E*.

Hemispheres, Continents, and Oceans The globe's grid does more than help us locate places. Geographers also use those grid lines to organize the way we look at our world. For example, the equator divides the globe into two halves, or **hemispheres**. The half lying north of the equator is the Northern Hemisphere. The southern half is the Southern Hemisphere. The prime meridian and the 180° meridian divide the world into the Eastern Hemisphere and the Western Hemisphere. The prime meridian separates parts of Europe and Africa into two different hemispheres. To avoid this separation, some geographers divide the Eastern and Western Hemispheres in the Atlantic Ocean at 20°W. Doing so places all of Europe and Africa in the Eastern Hemisphere.

We can also organize our planet's land surface into seven large landmasses, called **continents**. There are seven continents: Africa, Antarctica, Asia, Australia, Europe, North America, and South America. Asia, the largest, is more than five times the size of Australia, the smallest. Landmasses smaller than continents and completely surrounded by water are called islands. Greenland is the world's largest island.

Geographers also organize Earth's water surface into separate areas. The largest area is the global ocean. Geographers further divide this ocean into four areas: the Atlantic Ocean, the Arctic Ocean, the Indian Ocean, and the Pacific Ocean. The Pacific is the largest ocean and the world's largest geographic feature. It is more than 12 times the size of the smallest ocean, the Arctic.

Smaller bodies of waters include seas, gulfs, and lakes. Gulfs and seas, such as the Gulf of Mexico and the Caribbean Sea, are areas of salt water that are connected to the larger oceans. Lakes are inland bodies of water. Although it is called a sea, the Caspian Sea in Asia is really the world's largest lake.

✓ SKILLS CHECK: **The World in Spatial Terms** What are some ways geographers organize our world? grids, hemispheres, continents, oceans

INTERPRETING THE MAPS *In which hemispheres is the United States located? Which continents are located entirely within the Southern Hemisphere?*

LEVEL 1: Use the map of the school grounds drawn on the chalkboard in the All Levels activity. Have students work in pairs to devise questions about the completed grid. *(Examples: At what coordinates is the library located? At what coordinates is the auditorium? Using the grid, how would you reach the main office from your classroom?)* When pairs have completed their questions, have them exchange quizzes with other pairs. Students should quiz each other with their questions.* **COOPERATIVE LEARNING**

LEVEL 2: Display a map and a globe at the front of the classroom. Remind students that a globe is a scale model of Earth and a map is a two-dimensional depiction of all or part of Earth. Call on students to come to the front of the room and point out each of the following: lines of longitude, lines of latitude, Western and Eastern Hemispheres, the equator, the prime meridian, continents, and oceans. Lead a discussion on this question: Why might cartographers have created these systems of organization?

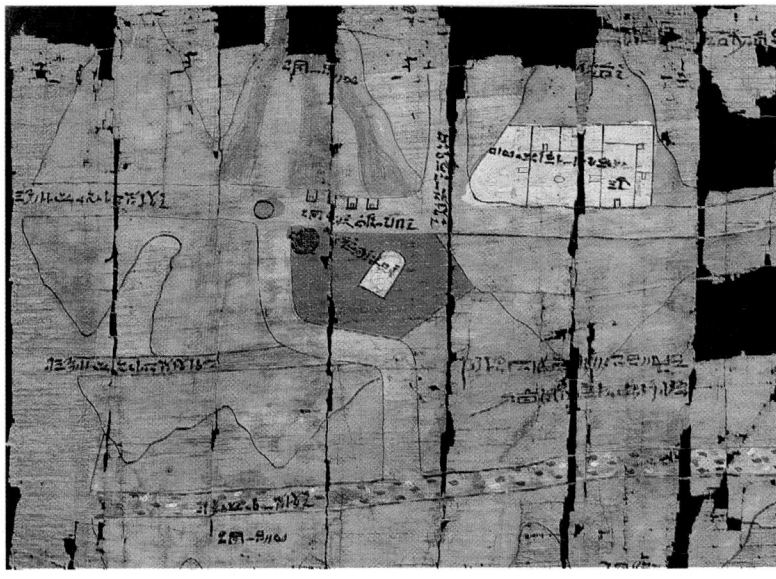

This map, which is more than 3,000 years old, shows an ancient mining and quarrying area in southeastern Egypt.

Making Maps

Globes are not the only useful visual tools for studying Earth. In organizing and identifying places in our world, geographers also use maps. Maps are flat representations of all or part of Earth's surface. A collection of maps in one book is called an **atlas**. You will find an atlas of world and regional maps at the front of this textbook.

Mapmakers have different ways of presenting our round Earth on flat maps. These different ways are called **map projections**. Because our planet is round, all flat maps—no matter their projection—have some distortion. For example, some maps do not show the true sizes of landmasses. This is particularly true at higher latitudes. For example, on some maps Greenland—which lies mostly within the Arctic Circle—might appear larger than Australia, a continent. However, those maps might be useful because they show true direction and true shapes. Other maps show size in true proportions but distort shapes. Mapmakers must choose the type of projection that is best for their purposes. The most common projections are cylindrical, conic, and flat-plane.

Map Projections Maps with cylindrical projections are designed as if a cylinder has been wrapped around the globe. The cylinder touches the globe only at the equator. The meridians are pulled apart and are parallel to each other instead of meeting at the poles. This causes landmasses near the poles to appear larger than they really are.

A Mercator map is a cylindrical projection. The Mercator map is useful for navigators because it shows true direction and shape. However, landmasses at high latitudes—such as Europe and North America—are exaggerated in size. They appear larger than they really are. Landmasses in lower latitudes may appear relatively smaller than they really are. (See the diagram.)

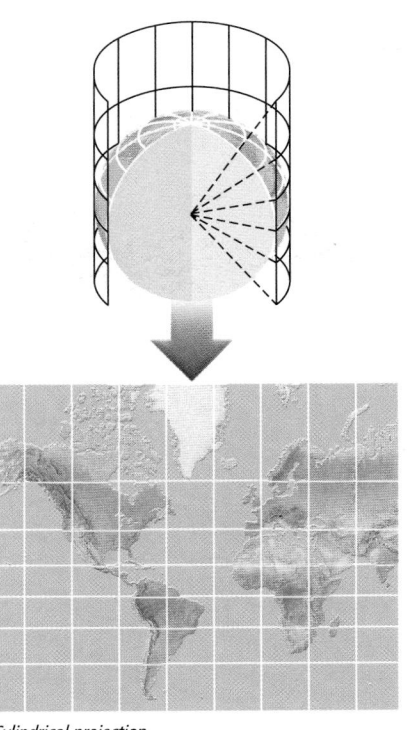

Cylindrical projection

LEVEL 3: Have each student create an illustrated book for elementary school students that explains how geographers and mapmakers organize our world. Encourage students to make the drawings and text appropriate for the younger age level. Topics may include an explanation of the purpose of globes and maps or their basic features. You may also want students to include discussions of hemispheres, continents, and oceans. When the books are complete, arrange for your class to share their books with younger students. Be sure students use geographic terminology correctly and use standard grammar, spelling, sentence structure, and punctuation.

Teach Objective 2

ALL LEVELS: As a class, review the characteristics of the three types of map projections. Display a globe at the front of the classroom and have a piece of paper large enough to encircle it at hand. Call on volunteers to demonstrate how each type of map projection is created by wrapping the paper around the globe in the appropriate fashion.

Have students locate special-purpose maps throughout this textbook. Ask them to formulate questions that can be answered by examining these maps. Challenge students to answer each other's questions.

Daily Life

A Flight from Dallas to London A passenger on a nonstop flight from Dallas, Texas, to London, England, might be surprised at the route taken if he or she looked at a Mercator world map before the trip. The flight would not follow the most direct route as it appears on the map because planes fly the shorter great-circle route to save time and fuel. On Mercator maps lines of longitude are the same distance apart at the poles and at the equator, so true distances are not measured. A plane would fly the Mercator route only if it could fly through, rather than above, Earth.

ACTIVITY: Have students use chalk to mark the most direct route between Dallas and London on a Mercator map. Then mark the most direct route on a globe. Compare the landmarks and coordinates crossed. Then have students trace a Quito, Ecuador, to Nairobi, Kenya, route. Ask them to explain why the map and globe routes are similar in this case. *(The equator is a great circle.)*

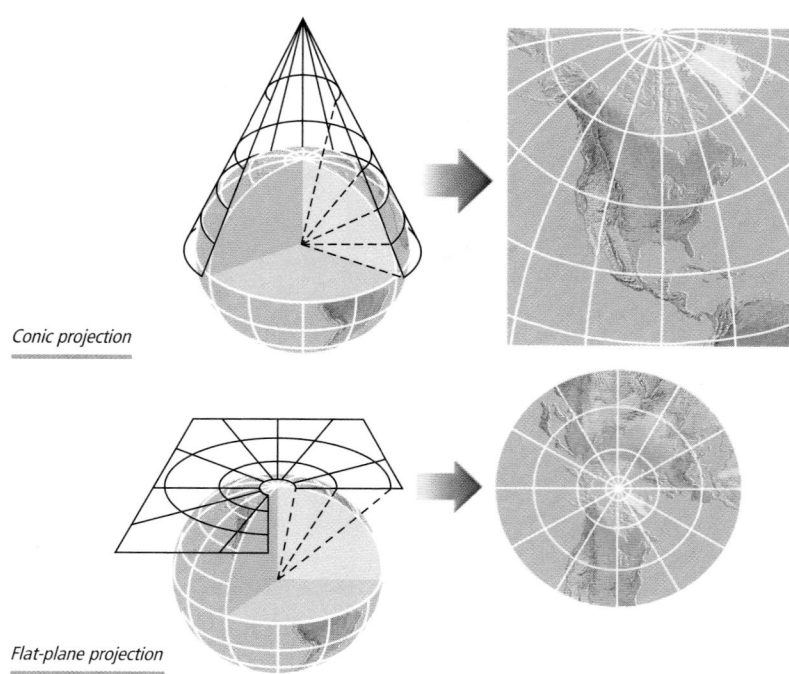

Conic projection

Flat-plane projection

Conic projections are designed as if a cone has been placed over the globe. A conic projection is most accurate along the lines of latitude where it touches the globe. It retains almost true shapes and sizes of landmasses along those locations. Conic projections are most useful for areas that have long east-west dimensions, such as the United States and Russia.

Flat-plane maps are those that appear to touch the globe at one point, such as the North Pole or the South Pole. A flat-plane projection is useful for showing true direction for airplane pilots and ship navigators. It also shows true area sizes, but it distorts shapes.

Great-Circle Route Drawing a straight line on a flat map will not show the shortest route between two places. Remember that maps represent a round world on a flat plane. The shortest route between any two places on the planet is called a **great-circle route**. (See the illustrations on this page.) Airline pilots and ship captains use great-circle routes to help them navigate. You can see

Using great-circle routes saves time and fuel for travelers. Use your atlas to find more examples of how great circle routes shorten other trips, such as those between Canada and Japan.

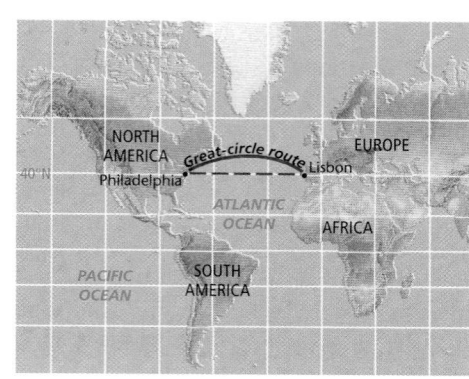

LEVEL 1: Copy the following graphic organizer onto the chalkboard, omitting the italicized answers. Call on students to provide words and phrases to describe the characteristics of the three projection types. List the information in the organizer on the chalkboard as students provide it. **ENGLISH LANGUAGE LEARNERS**

Map projection	Advantages	Disadvantages
cylindrical	*used by navigators because it shows true direction and shape*	*exaggerates landmasses at high latitudes*
conic	*accurate for areas with long east-west dimensions*	*not as accurate for areas that extend mostly north to south*
flat-plane	*used by pilots and navigators because it shows true direction, area, and sizes*	*distorts shapes*

how a great-circle route shows the shortest distance between two points by using a round globe in your classroom or library.

✓ **SKILLS CHECK:** (*The World in Spatial Terms*) What are three kinds of map projections that mapmakers use? cylindrical, conic, flat-plane

Understanding Map Elements

The study of geography involves more than simply looking at big places you see on a globe or map. It also involves looking at places at different scales—the size of an area and the level of detail that is shown. In fact, a map can show small areas, such as the floor plan of a building. It might show a neighborhood or a voting precinct. Maps can also show larger areas. For example, they might show whole cities, states, countries, continents, and oceans. You can see examples of maps at different scales throughout this textbook. For example, on this page are maps of Washington, D.C., and surrounding areas. Notice how details on each map change as the scale changes.

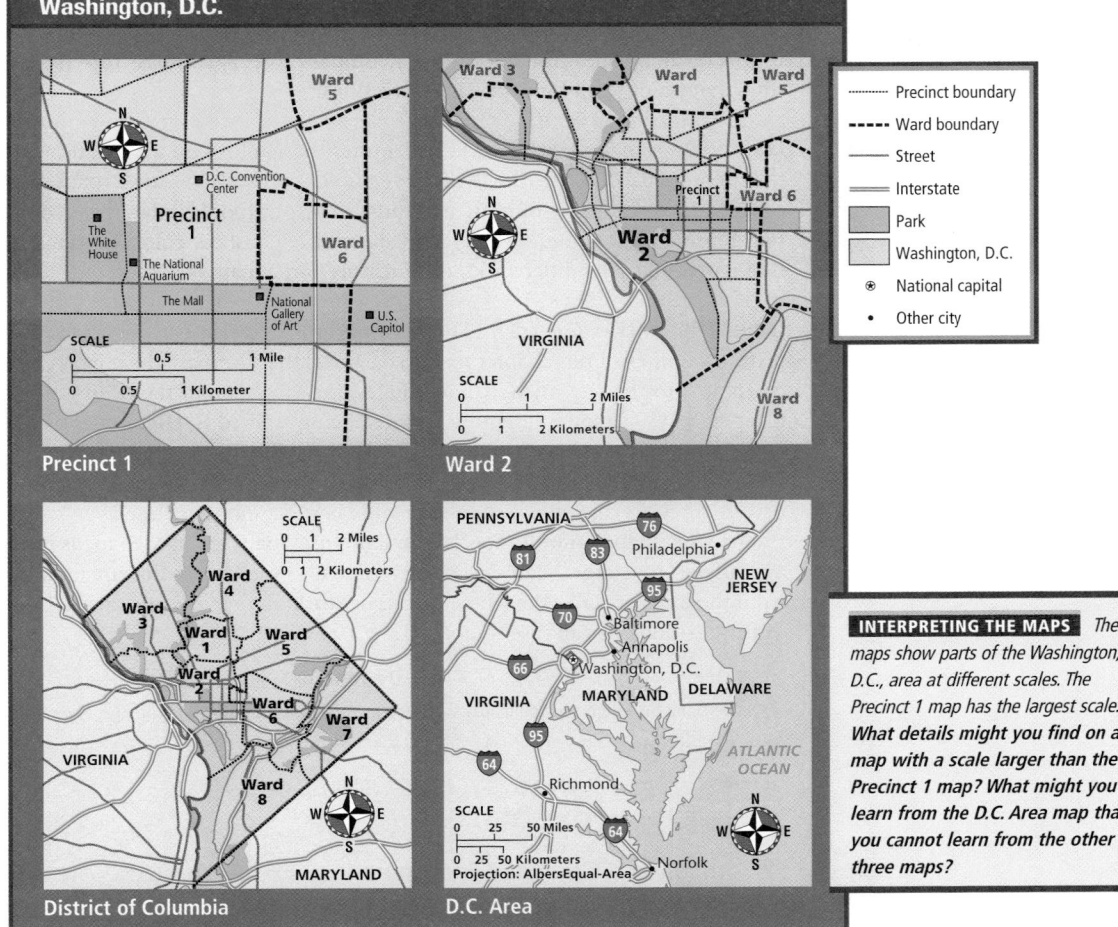

INTERPRETING THE MAPS *These maps show parts of the Washington, D.C., area at different scales. The Precinct 1 map has the largest scale.* **What details might you find on a map with a scale larger than the Precinct 1 map? What might you learn from the D.C. Area map that you cannot learn from the other three maps?**

MAP ANSWER

Possible answers: more streets, buildings and other local features; how to drive from Washington, D.C., to other cities, how far the cities are from the ocean

LEVEL 2: Refer students to the chart of map projections written on the chalkboard during the previous exercise. Then display two examples of each of the three types of map projections in front of the class. Label the projections A–F. Have students use the information in the chart to decide into which map projection category each of the displayed maps falls. When students have finished, call on volunteers to share their responses with the class. Tape the maps to the chalkboard under their correct categories as students identify them. ENGLISH LANGUAGE LEARNERS

LEVEL 3: Have students examine the atlas at the beginning of this textbook to identify the type of projection of each map in the atlas. Students should write a sentence about each map to describe the characteristics that led them to their conclusions about the projections used. Then have students draw conclusions about what types of maps were used for which purposes in this atlas. Encourage students to create a chart to organize their observations.

Essential Element 1

► The World ◄ in Spatial Terms

Alaska and Hawaii on U.S. Maps   Before 1959 maps of the United States provided easy visual comparison of the relative sizes of different states. After Alaska and Hawaii were admitted to the United States, cartographers had to find a way to add them to U.S. maps. The new states were distant from the rest of the country, which presented a problem. In addition, they were vastly different in size. Alaska is more than twice as large as Texas, while the total land area of the Hawaiian Islands is less than all but three other states. In addition, the individual islands are extremely small when compared to the whole country. The solution to these cartographic problems was to show each new state as an inset map with its own scale.

DISCUSSION: Have students look at the United States map in the atlas at the front of this book. Ask this question: What problems would be created if the same scale were used for the Alaska and Hawaii maps? What problems would result from using the same scale for all three maps?

Our Amazing Planet

People probably made maps even before written languages existed. However, the earliest known map is one discovered on a clay tablet at a place called Nuzi in what is now Iraq. The map dates to about 2500 B.C.

Maps may show different details depending on scale. However, they usually have the same basic elements. This is because maps, in some ways, are like messages sent out in code. Cartographers provide basic map elements to help us translate these codes. Thus, we can understand the information, or message, in a map. Almost all maps have several common elements. They include a distance scale, a directional indicator, and a key. The key is a guide that identifies symbols. You can see these elements on the Washington, D.C., maps.

Distance Scales   A map's distance scale helps us determine real distances between points on a map. Remember that maps of small areas can show more detail than maps of large areas. Because of this, scales on those maps can indicate short distances. Some might show just one or two miles or kilometers. Others might show smaller distances—even just hundreds of feet or meters. Maps showing large areas, such as state and country maps, must have scales that indicate longer distances. Those scales might show distances in tens or even hundreds of miles or kilometers.

Directional Indicators   A directional indicator shows which directions on a map are north, south, east, and west. Some mapmakers use a "north arrow," which points toward the North Pole. Most maps have north at the top. Maps in this textbook show direction with a **compass rose**. A compass rose has arrows that point to all four principal directions.

Legends   A map's **legend**, or key, identifies the symbols on a map and what they represent. Legends might show symbols representing cities, roads, and other features. Some legends, such as those in this textbook's atlas, show colors that represent elevation, or height. In short, legends show colors or symbols that represent many different kinds of features on a map.

Other Elements   You will find other important map elements as you use this textbook. For example, chapter maps have boxes that compare the physical size of an area to the size of the United States. (See the map at the beginning of Chapter 9.) A shape representing the area being studied is in red. It is placed over an outline of the **contiguous** United States. *Contiguous* means "connecting" or "bordering." The 48 states between Canada and Mexico are contiguous because they are all connected. Alaska and Hawaii are not included.

An inset map is another special element. Inset maps are used to focus in on a small part of a larger map. Some inset maps also show areas that are far away from the main areas on the whole map. The world map in the atlas at the front of this textbook has an inset map.

✓ SKILLS CHECK: *The World in Spatial Terms*   What special map elements help us understand the information on maps? distance scales, directional indicators, legends, size comparisons, insets

FOCUS ON HISTORY

Exploration and Changing Perceptions   For centuries, people have used maps to reflect their perceptions of the world around them. For example, early Greeks created maps of the Mediterranean world. They had limited knowledge

Climate Graphs

Description
show the average temperatures and precipitation in places

Uses
compare climate and weather patterns in different places

Population Pyramids

Description
show percentages of males and females by age group in countries' populations

Uses
help us understand population trends in countries

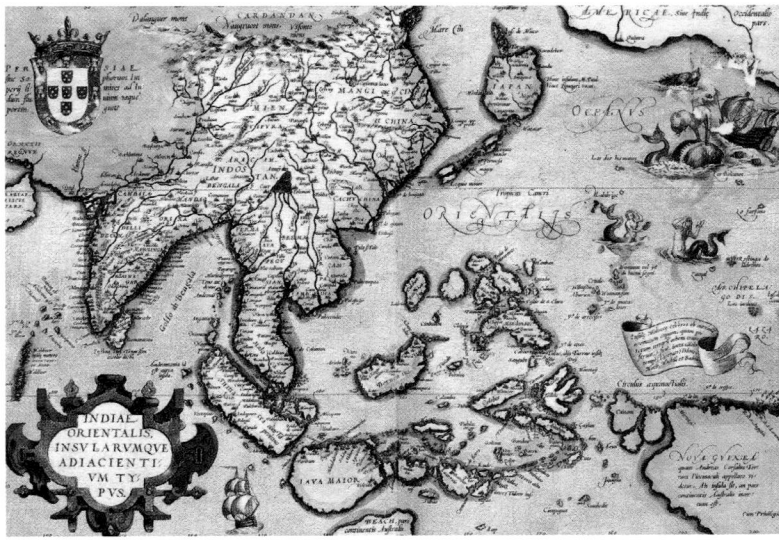

INTERPRETING THE VISUAL RECORD

This map of India and the Pacific Ocean dates from 1570. Note the enlarged detail near the bottom of the page. **How does this map show that sailors were uncertain about what dangers lay ahead for them? How do the distances and shapes shown on the map compare to those on today's maps?**

of what lay beyond that world. Some early European mapmakers even placed sea monsters at the edges of their maps. This reflected their uncertainty about what lay beyond the seas that surrounded the land areas. Early Chinese mapmakers placed China at the center of the world. They believed Chinese culture to be superior to all other cultures.

All of these limited perceptions began to change as sailors explored the outside world. The Chinese explorer Cheng Ho sailed to many places in Southeast Asia and eastern Africa in the 1400s. He took back to China his impressions of the people and places he encountered. About the same time, European explorers began sailing far from home. Soon European explorers sailed around Africa to reach Asia. However, some Europeans believed that a shorter route to Asia lay to the west—across the Atlantic Ocean.

In 1492 Christopher Columbus sailed westward from Spain. He and later explorers found two huge continents—the Americas—that lay between Europe and Asia. Knowledge of the Americas changed European perceptions of the size and features of their world. These changing perceptions drew more Europeans to the Americas and elsewhere. They also sparked a race for European colonies around the world. European countries became more powerful. However, local cultures in the Americas, Africa, and Asia were weakened or destroyed. Even parts of China—once seen by Chinese as the center of the world—fell under the influence of Europe.

 READING CHECK: *The Uses of Geography* How did European perceptions of the world change after the voyages of Christopher Columbus? What changes did this bring to societies around the world?

Knowledge of new lands and a larger world led to European colonization and the weakening or destruction of societies in the Americas, Africa, and Asia.

Using Special-Purpose Maps

Geographers use many different kinds of maps. Many maps, for example, focus on certain kinds of information about a place or region. You will find

Global Perspectives

Chinese Cartography

Ancient Chinese cartographers produced excellent maps. Today these maps provide information about Chinese knowledge in the past and Chinese perceptions of the world around them.

In the A.D. 200s one Chinese official and mapmaker produced a map of the politically organized part of China on 18 silk rolls. The official, whose name was Phei Hsiu, used a grid of east-west and north-south lines to plot the features on his map. These features included rivers, mountain ranges, cities, and the coastline. Two other detailed maps were carved in stone in A.D. 1137. One labels China and the "barbarian countries," which indicates how many Chinese perceived the areas that bordered their great empire. Both carved maps show roughly the area from the Great Wall in the north to the island of Hainan in the south and from the mountains of the Asian interior to the Chinese coast.

VISUAL RECORD ANSWER

inclusion of sea monsters on the map; India and Pacific Ocean too small, other shapes inaccurate, basic arrangement fairly accurate

Linking Past to Present

Collecting Weather Data

One of the first attempts to collect global weather data for mapmaking purposes was made by the famous German geographer Alexander von Humboldt. After traveling across Russia and Siberia in 1829, von Humboldt observed that temperatures at the same latitude varied with distance from the ocean. When he returned to St. Petersburg, von Humboldt asked the czar to set up a network of weather stations across Russia to record data. He then asked the British government to establish weather stations throughout its empire. One of his colleagues convinced other European governments to do the same. For the first time, meteorological data from around the world became available. Von Humboldt used this data to draw the first world map of average temperatures, an important step toward creating a world climate map.

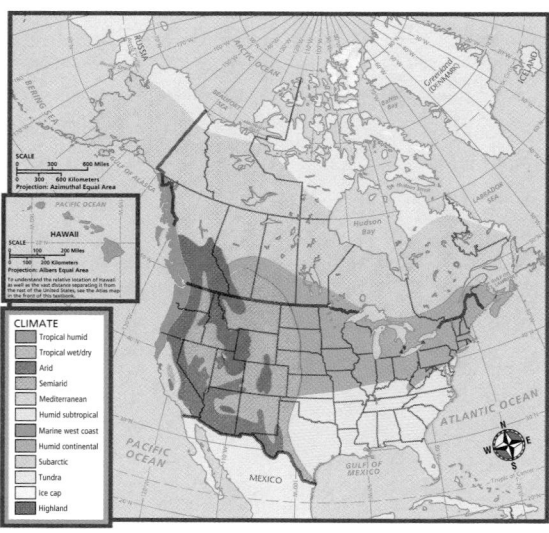

Climate Map, United States and Canada

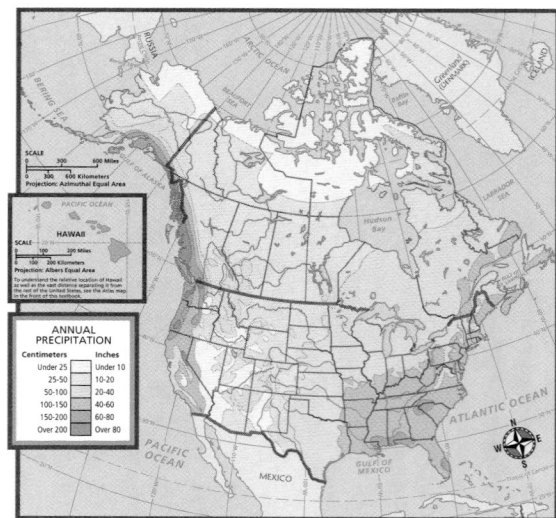

Precipitation Map, United States and Canada

these special-purpose maps throughout this textbook. Some are political and physical atlas maps. You will find such maps at the front of this textbook and at the beginning of units and chapters. Political maps show the world's borders, cities, countries, states, and other political features. Physical maps show natural features like mountains, rivers, and other bodies of water. You will read about these kinds of features in Chapter 4. You will find symbols that identify the features on physical maps in legends. Of course, studying our world means more than looking at country borders, cities, and physical features. Each unit in this textbook opens with other special-purpose maps. These maps show climate, precipitation, population, and economic features.

Climate and Precipitation Maps Mapmakers use some maps to show weather patterns and atmospheric conditions. Climate maps use color to show the various climate regions of the world. You will read about climates in Chapter 3. Colors that identify climate types are found in a legend. However, boundaries between climate regions do not indicate a sudden change in average weather conditions. Instead, those boundaries mark areas of gradual change between climates. **Precipitation** maps are paired with climate maps at the beginning of units. The word *precipitation* refers to condensed droplets of water that fall as rain, snow, sleet, or hail. These kinds of maps show the average amount of precipitation that a region gets each year. Each map's legend uses colors to identify those amounts. By using the map's legend, you can see what areas get the most or least precipitation.

Population and Economic Maps Population maps give you a snapshot of the distribution of people in a region. You will read about population features in Chapter 5. Each color on a population map represents an average number of people living within a square mile or square kilometer. Sometimes symbols identify cities with populations of a certain size. The map's legend identifies these colors and symbols.

LEVEL 3: As a class, review the characteristics and purposes of climate graphs and population pyramids. Then have students create a list of three questions that could be answered by using each of these types of graphs. *(Possible questions: Which U.S. cities have similar rainfall patterns in the summer months? What is the proportion of males to females over the age of 60 in the United States?)* When students have completed their questions, call on volunteers to share them with the class.

Teach Objectives 2–3

LEVEL 1: Lead a class discussion on the utility and limitations of various special-purpose maps, climate graphs, and population pyramids. Ask students questions such as these: What are some of the strengths and weaknesses of climate maps? *(Possible answer: They show general patterns of weather, but they lack detailed information and simplify weather patterns.)* How can population pyramids help to predict future population trends? *(Possible answer: By illustrating the composition of a country's people by age, they can suggest trends in birth and death rates for the near future.)* Encourage students to ask questions of their own.

Population Map, United States and Canada

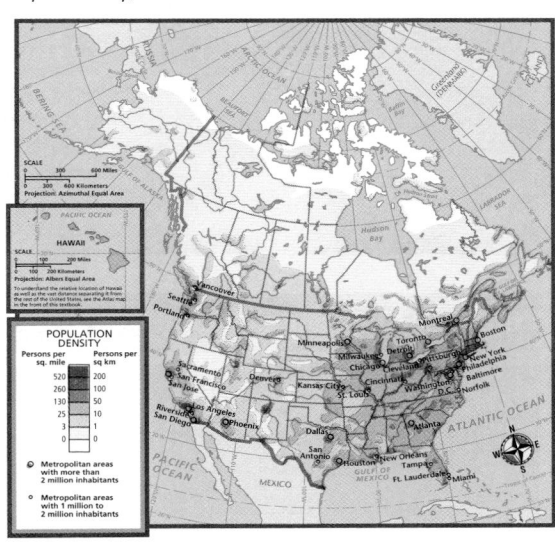

Land Use and Resources Map, United States and Canada

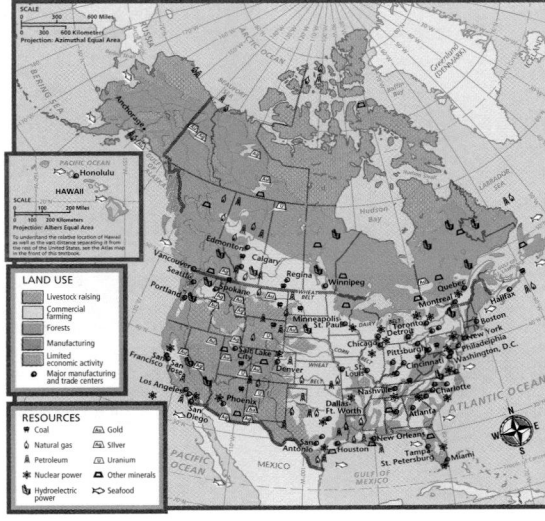

By looking at the population map of the United States in Unit 2 you can identify the most populated areas of the country. Those areas are mainly in the eastern half of the country, in the Midwest, and along the West Coast. In general, fewer people live in the interior and western states.

Economic maps show a region's important natural resources and the ways in which land is used. You will read about economic features in Chapter 6. Symbols show the location of resources, such as oil and gold. Colors show where land is used for farming or other economic activities.

✓ **SKILLS CHECK:** **The World in Spatial Terms** How are colors used in climate, precipitation, population, and economic maps? to identify important features, such as climate types, precipitation levels, population density, land use

Elevation Profiles and Topographic Maps Some maps focus on an area's land features. You can see that each physical map in this textbook uses color to show land elevations. Elevation is the height of the land above sea level. Each color represents a different elevation.

In each unit atlas, you will also find an elevation profile like the one shown on the next page. An elevation profile shows a side view of a place or area. The profile shows the physical features of Guadalcanal, an island in the South Pacific. These features lie along a line from Point A to Point B in the elevation map below it.

Vertical (bottom to top) and horizontal (left to right) distances are calculated differently on elevation profiles. The vertical distance (such as the height of Mount Makarakomburu) is exaggerated when compared to the horizontal distance between Point A and Point B. This technique is called vertical exaggeration. If the vertical scale were not exaggerated, even tall mountains would appear as small bumps on an elevation profile.

The purpose of some maps is to show just the **topography**—or elevation, layout, and shapes—of the land. A special kind of topographical map is called a contour map. Contour maps provide a way of looking at the shapes of land in an

Essential Element 1

► **The World** ◄
in Spatial Terms

Dot Maps Another way to show population distribution is to create dot maps. Cartographers first used dot maps as early as 1863. These maps have remained popular because they clearly show the spatial distributions and patterns of populations.

On most population dot maps, each dot represents a certain number of people. For example, one dot may represent 100,000 people. Dots are then placed on the map according to where people are concentrated.

ACTIVITY: Provide students with outline maps of your state and its counties and a list of each county's population from an almanac or other source. Then have students create a dot map of your state. As a class, decide how many people each dot should represent to most effectively show the state's population distribution. Finally, have students compare their maps. How are they different or similar? Are some clearer than others? Why?

📶 internet connect

GO TO: go.hrw.com
KEYWORD: SW3 CH1
FOR: Web sites about satellite pictures of Earth

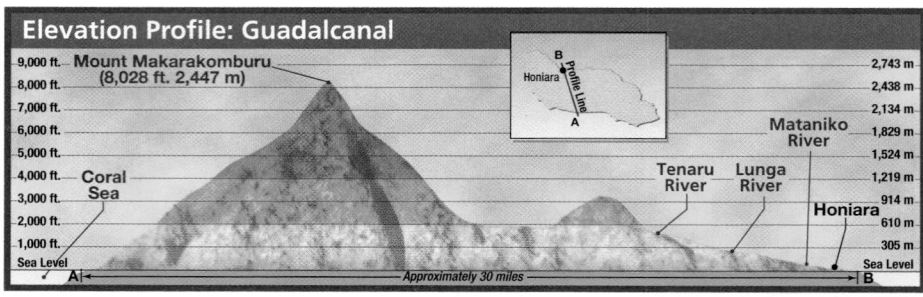

Elevation Profile: Guadalcanal

LEVEL 2: Instruct students to examine the population pyramid for Nigeria in this section, the population density map of Africa in the Unit 7 Atlas, and the statistics for Nigeria in the Unit 7 Fast Facts and Comparing Standard of Living charts. Then instruct students to write paragraphs describing Nigeria's population and to predict likely future population patterns in the country.

LEVEL 3: Organize the class into four groups labeled A, B, C, and D. Have groups A and B conduct research about population distribution and age structure in your community. Group A should then create a population distribution map of your area. Group B should create a population pyramid. Have groups C and D research the climate in your community. Group C should create a climate map of your area. Group D should create a climate graph. Have the groups display their work. Then lead a discussion in which students describe the methods they used to compile their research and discuss how geographers might use similar methods to gather information about larger areas.

Cooperative Learning

Understanding Contour Maps Organize the class into groups. Provide each group with a block of styrofoam to cut into the shape of a mountain. After the styrofoam mountains are cut, have each group draw four horizontal lines around the mountain that are evenly spaced between the mountain's base and top. Then have students label these contour lines. The base of the mountain should be labeled zero, and the four lines should be labeled 1,000 feet, 2,000 feet, 3,000 feet, and 4,000 feet. Next, have students place the entire mountain on a piece of construction paper and trace around the base. Instruct students to cut each mountain at the 1,000-feet line and discard the bottom sections. Then have each group trace its mountain again. Explain to students that the second line shows those parts of the mountain with an elevation of exactly 1,000 feet. Ask students to repeat the process until the top part of the mountain has been traced. Finally, have students label their contour maps with the corresponding elevations from the mountain.

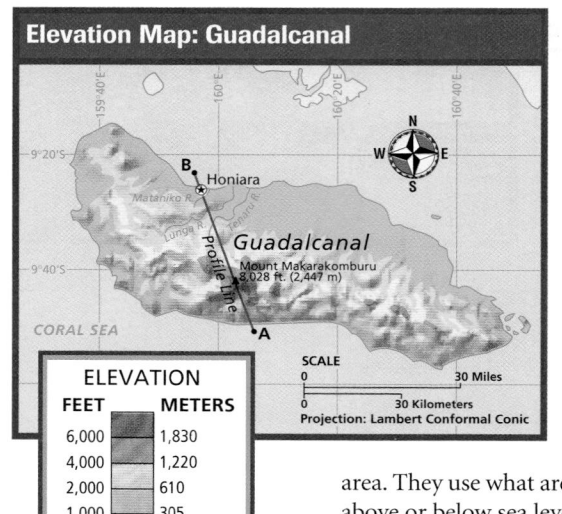

Elevation Map: Guadalcanal

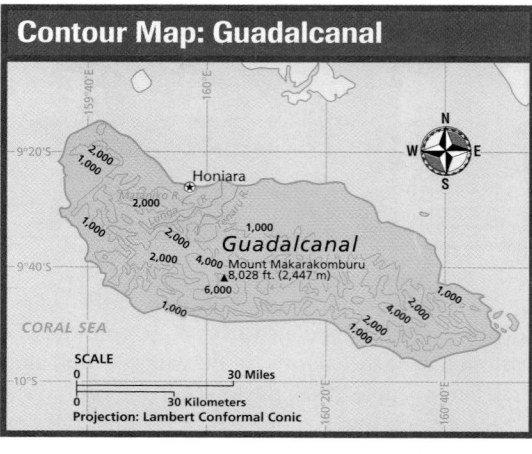

Contour Map: Guadalcanal

area. They use what are called contour lines to connect points of equal elevation above or below sea level. Elevation levels are written on the lines. The closer together the lines are, the steeper the land. For example, the land south of Mount Makarakomburu is steeper than the land on the northwestern end of the island.

✓ SKILLS CHECK: (*The World in Spatial Terms*) Where is the largest, lowest, and flattest area of Guadalcanal? northern-central coast, moving inland

Climate Graphs and Population Pyramids

Maps are not the only special tools geographers use to study the world around us. They also use many tools that you will find useful in economics, government, history, and other social studies. You can read more about bar graphs, charts, pie graphs, and tables in the Social Studies Skill-Building Handbook before Unit 1. In fact, you will find many charts, graphs, and tables in this textbook. You will also find two other common diagrams that show important geographic characteristics: **climate graphs** and **population pyramids**.

A climate graph shows the average temperatures and precipitation in a place. As you can see on the opposite page, along the left side of the climate graph is a range of average monthly temperatures. Along the right side is a range of average monthly precipitation amounts. The months of the year are labeled across the bottom. In this textbook's climate graphs, a red line shows the average monthly temperatures at the location. Green bars show average monthly precipitation amounts.

Teacher to Teacher

C. Eugene Price of Fort Wayne, Indiana, suggests the following activity to help students review the concepts taught in this section. Organize the class into six groups. Give each group one of the three Read to Discover questions at the beginning of the section. Give the groups 15 to 20 minutes to study their questions and to prepare complete answers. Next, allow each group five minutes to report its answer to the class. You may wish to ask or have other students ask additional questions or add information to further answer each question.

Using National Geography Standard 1:
The World in Spatial Terms: How to Use Maps and Other Geographic Representations, Tools, and Technologies to Acquire, Process, and Report Information from a Spatial Perspective Have students collect maps from newspapers, magazines, books, and other sources. Then have students attach descriptions to each map identifying the purpose for which it was made and what special information can be found on it. Have students write short paragraphs comparing the quality and effectiveness of the different maps.

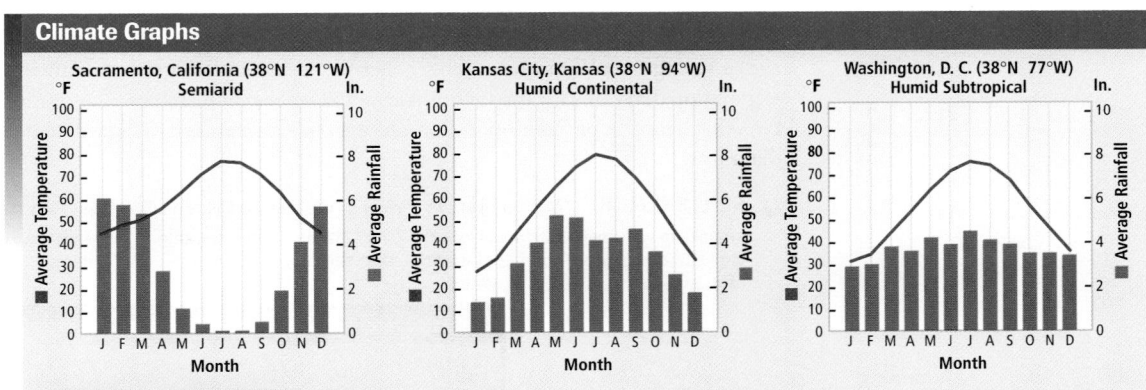

Climate Graphs

INTERPRETING THE GRAPHS *Which city usually gets at least 2 inches of rainfall every month? Which city has the driest summers? Which city has the widest difference between summer and winter temperatures?*

GRAPH ANSWER

Washington, D.C.; Sacramento; Kansas City

GRAPH ANSWER

yes, because each year the percentage of females of child-bearing age increases; 40–44, indicates that Russia will not grow rapidly, because most of these people are beyond child-bearing years

A population pyramid shows the percentages of males and females by age group in a country's population. As you can see in the pyramid on this page, these diagrams are split into two sides. Each bar on the left shows the percentage of a country's population that is male and of a certain age. The bars on the right show the same information for females. The percentages are labeled across the bottom of the diagram.

How do you think these population diagrams got their names? The base of a population pyramid shows the percentage of the youngest people in a country. The top shows the percentage of the oldest people. In many countries there are many more younger people than older people. Thus, the base of a country's diagram is often much wider than the top. The result is a diagram shaped like a pyramid.

Population pyramids help us understand population trends in countries. Countries that have large percentages of young people have populations that are growing rapidly. They have pyramids with very wide bases. Nigeria is an example of such a fast-growth country. On the other hand, in countries like

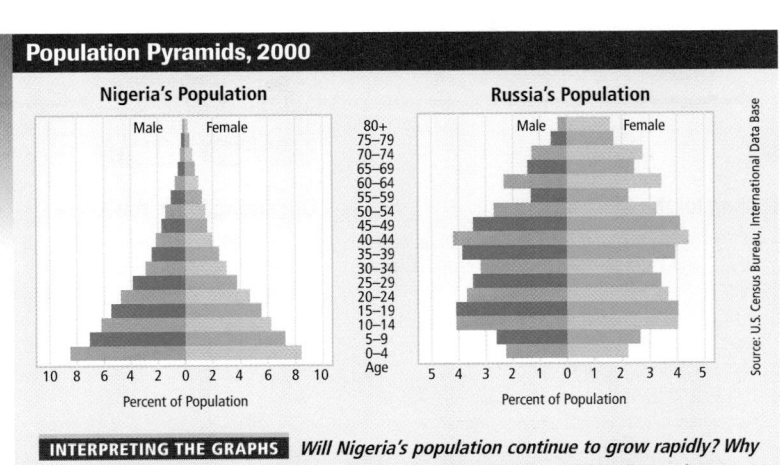

Population Pyramids, 2000

Source: U.S. Census Bureau, International Data Base

INTERPRETING THE GRAPHS *Will Nigeria's population continue to grow rapidly? Why or why not? Which Russian age group is the largest? What does this indicate about Russia's future population growth?*

Section 2 Review Answers

Define For definitions, see: grid, p. 9; latitude, p. 9; longitude, p. 9; equator, p. 10; parallels, p. 10; meridians, p. 10; prime meridian, p. 10; degrees, p. 10; hemispheres, p. 10; continents, p. 10; atlas, p. 11; map projections, p. 11; great-circle route, p. 12; compass rose, p. 14; legend, p. 14; contiguous, p. 14; precipitation, p. 16; topography, p. 17; climate graphs, p. 18; population pyramids, p. 18

Working with Sketch Maps Maps will vary, but listed places should be labeled in their approximate locations. Students may identify specific bays, seas, gulfs, or lakes.

Close

Write the following statement on the chalkboard: *A cartographer's work is never done.* Ask students why this statement might be true. *(Possible answers: Physical processes change Earth's surface. New roads and suburbs are built. Political boundaries change.)*

Review and Assess

Have students complete the **Section Review**. Then have students complete **Daily Quiz 1.2**.

Reteach

Have students complete **Main Idea Activity for English Language Learners and Special-Needs Students 1.2**. Then have students illustrate one of the section's topics. ENGLISH LANGUAGE LEARNERS

Extend

Have interested students use Internet and library resources to conduct research on the history of cartography in an area that has been occupied and mapped since antiquity. Ask students to investigate how maps of that region have evolved over time and to present visual samples of their findings to the class. BLOCK SCHEDULING

Section 2 Review Answers

Reading for the Main Idea

1. for latitude—parallels north and south of the equator; for longitude—meridians east and west of the prime meridian

2. physical, political, climate, precipitation, population, and economic features

3. because population diagrams of slow-growth countries may not be shaped like pyramids

Critical Thinking

4. climate map, precipitation map, climate graph (NGS 1)

Organizing What You Know

5. cylindrical—designed as if a cylinder has been wrapped around the globe, parallel meridians that do not meet at poles; conic—designed as if a cone has been placed over the globe, best for areas with long east-west dimensions; flat-plane—appear to touch the globe at one point, useful for showing true direction

CONNECTING TO HISTORY ANSWER

Possible answers: religions practiced, foods eaten

Russia, populations are growing much more slowly or not at all. The percentage of young people may be much smaller there than in fast-growth countries. Population diagrams for those countries actually lose their shapes as pyramids. For this reason, population pyramids are sometimes called age-structure diagrams.

✓ SKILLS CHECK: (*Human Systems*) What do you think a climate graph or population pyramid for your community would look like? Why? Possible answers: might show distinct wet and dry or warm and cool months; that there are many young or old people, affecting the shape of the pyramid

Connecting to HISTORY

Using Historical Maps

Studying geography often requires us to understand how human and physical processes have shaped a place or region over time. Geographers can use historical maps to study these processes and their effects. For example, the map here uses colors to show areas of North and South America that Europeans had settled and taken over before 1700. Students of geography can use this map to predict where various European languages are probably spoken in the Americas today.

Making Generalizations What else might a historical map like this one suggest about the human geography of North and South America today?

European Empires in the Americas, 1700

- Spanish territory
- Portuguese territory
- English territory
- Dutch territory
- French territory

SCALE
0   1,000   2,000 Miles
0 1,000 2,000 Kilometers
Projection: Miller Cylindrical

Section 2 Review

go.hrw.com
Homework Practice Online
Keyword: SW3 HP1

Define grid, latitude, longitude, equator, parallels, meridians, prime meridian, degrees, hemispheres, continents, atlas, map projections, great-circle route, compass rose, legend, contiguous, precipitation, topography, climate graphs, population pyramids

Working with Sketch Maps Sketch a map of the world. Label Earth's seven continents and the Arctic, Atlantic, Indian, and Pacific Oceans. In the margin, identify at least three bodies of water that are smaller than oceans.

Reading for the Main Idea

1. (*The World in Spatial Terms*) What do the letters *N, S, E,* and *W* mean when they accompany labels for latitude and longitude?

2. (*The World in Spatial Terms*) What regional features are found on special maps at the beginning of each unit throughout this textbook?

3. (*Human Systems*) Why are population pyramids sometimes called age-structure diagrams?

Critical Thinking

4. **Evaluating Information** What tools might geographers use to study a region's weather patterns?

Organizing What You Know

5. Copy the chart below and use it to describe cylindrical, conic, and flat-plane map projections.

Cylindrical	Conic	Flat-plane

Technology: Uses of GIS

The use of GIS technology has increased dramatically in recent years, and this growth is likely to continue in the future. As new technologies allow scientists to collect more and more spatial data, they need an efficient way to manage the information. Often, a GIS system is the best way to process, store, analyze, and display large amounts of spatial data. Ask the class how issues such as deforestation and the search for new energy resources may contribute to the growth of spatial data. *(Possible answer: Scientists may need to collect global data on forest cover to study deforestation or use enhanced satellite images to search for new energy resources.)* Point out that as advanced technologies are developed and scientists collect new spatial data, GIS systems can be an invaluable tool to help us learn about human impacts on the environment and to study the changing nature of the planet. Ask students to describe some ways that GIS systems can handle spatial data that conventional maps could not. *(Possible answer: GIS maps can be easily updated as new data is collected, and GIS software can be used to combine, statistically analyze, and highlight different data sets.)* Have each student select a topic that they think could be studied using a GIS. Then have students write short proposals outlining the kind of information they would collect and how they think using a GIS could help them study the topics they selected.

The World in Spatial Terms

Geography for Life

Geographic Information Systems

A geographic information system, or GIS, is a special kind of computer system. A GIS stores, displays, and maps locations and their features. Geographic information systems have become important tools in geography and many other fields. Among those fields are city planning, real estate, and environmental studies. Scientists also use them for a variety of tasks. For example, they can use a GIS to monitor natural hazards like volcanoes. They can also map crop growth and track the movements and locations of endangered species. Emergency workers can use a GIS to learn the shortest route to someone calling 911. Geographic information systems are also helpful in understanding the spread of disease and charting power outages. In short, they are valuable tools for many tasks. Today students can even earn a college degree in GIS technology.

How does a GIS work? The information collected for a map is called spatial data. With a GIS, different sets of this data are saved in a computer as "layers." (See the diagram.) These data layers can then be manipulated, compared, or combined. Cartographers can instantly display this data on a computer screen and get detailed information about each feature. They can also use a GIS to study features and their locations and to uncover relationships between them.

For example, imagine that you plan to open a restaurant. How would you find the best location for your new business? Using a GIS could help. With a GIS, you could find the locations of vacant buildings. You could then plot those located near major streets. You could also show the buildings' rental costs and information about the people who live nearby. That information might include how much money the people make. With this kind of information, a GIS could help you identify the best locations for your business.

Applying What You Know

1. **Summarizing** What is a GIS? What are some fields that use GIS technology?

2. **Problem Solving** What are some geographic problems and issues that could be studied using a geographic information system? What other kinds of spatial data might be useful?

Information Layering

Highways and streets

Buildings
- Locations
- Vacancies
- Rental costs
- Square footage

Population
- Zip codes
- Income levels

Property 1
604 Lost Creek Blvd.
2,500 sq. ft.
$2,000/month

Property 2
1200 Enfield Road
2,600 sq. ft.
$2,500/month

Property 3
1709 Great Hills Trail
1,500 sq. ft.
$2,000/month

Average Income
- ☐ $35,000 – $45,000
- ☐ $25,000 – $35,000
- ☐ $15,000 – $25,000
- ■ Vacant buildings

INTERPRETING THE DIAGRAM
Examine the map above and the layers at the left that would be combined in a GIS. **Which of the three locations would you choose for a new restaurant? What influenced your choice?**

Applying What You Know Answers

1. a computer system that stores, displays, and maps locations and their features; geography, city planning, real estate, environmental studies
2. Possible answers: erosion, deforestation; aerial photographs

DIAGRAM ANSWER

Possible answer: Property 2; near intersection, high-income area

This Geography for Life feature addresses National Geography Standards 1, 17, and 18.

CHAPTER 1 Review Answers

Building Vocabulary For definitions, see: geography, p. 3; cartography, p. 5; region, p. 6; formal region, p. 6; functional region, p. 6; perceptual regions, p. 6; equator, p. 10; parallels, p. 10; meridians, p. 10; prime meridian, p. 10; continents, p. 10; atlas, p. 11; contiguous, p. 14; topography, p. 17; population pyramids, p. 18

Locating Key Places
- **A.** South America
- **B.** Europe
- **C.** Africa
- **D.** North America
- **E.** Pacific Ocean
- **F.** Atlantic Ocean
- **G.** Antarctica
- **H.** Australia
- **I.** Arctic Ocean
- **J.** Asia
- **K.** Indian Ocean

CHAPTER 1 — Review and Assessment Resources

TECHNOLOGY

▶ Chapter 1 Test Generator (on the One-Stop Planner)
▶ Global Skill Builder CD–ROM
▶ HRW Go site

REINFORCEMENT, REVIEW, AND ASSESSMENT

▶ Chapter 1 Review, pp. 22–23
▶ Chapter 1 Tutorial for Students, Parents, Mentors, and Peers
▶ Chapter 1 Test (form A or B)
▶ Alternative Assessment Handbook
▶ Chapter 1 Test for English Language Learners and Special-Needs Students
▶ Unit 1 Test
▶ Unit 1 Test for English Language Learners and Special-Needs Students

Assess

Have Students complete a Chapter 1 Test.

Reteach

Organize the students into four groups. Assign two of the groups the first section of this chapter and the other two groups the second section. Have each group summarize the key points of its section in a paragraph. Call on volunteers to read the paragraphs to the class. **ENGLISH LANGUAGE LEARNERS, COOPERATIVE LEARNING**

CHAPTER 1 — Review Answers

Understanding the Main Ideas

1. human and physical

2. The World in Spatial Terms, Places and Regions, Physical Systems, Human Systems, Environment and Society, The Uses of Geography

3. formal—has one or more common features that make it different from surrounding areas; functional—made up of different places that are linked together and function as a unit; perceptual—reflect human feelings and attitudes

4. cylindrical—shows true direction and shape but distorts near the poles; conic—accurate along the lines of latitude where it touches the globe, useful for areas that have long east-west dimensions; flat-plane—useful for showing true direction for pilots and navigators, shows true area but distorts shapes

5. physical, political, climate, precipitation, population, economic

Thinking Critically

1. formal—defined by sameness; functional—defined by linkages; perceptual—defined by perceptions and feelings (NGS 5)

2. Students should note the utility of maps (The World in Spatial Terms), identifying the community as a place or a region (Places and Regions), studying its physical environment (Physical Systems), studying its human features, (Human Systems), studying the

---

CHAPTER 1 — Review

Building Vocabulary

On a separate sheet of paper, explain the following terms by using them correctly in sentences.

geography	perceptual regions	continents
cartography	equator	atlas
region	parallels	contiguous
formal region	meridians	topography
functional region	prime meridian	population pyramids

Locating Key Places

On a separate sheet of paper, match the letters on the map with their correct labels.

Africa	Europe	Atlantic Ocean
Antarctica	North America	Indian Ocean
Asia	South America	Pacific Ocean
Australia	Arctic Ocean	

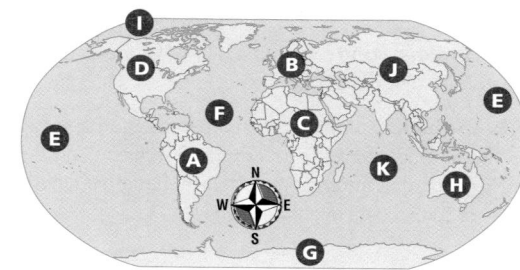

Understanding the Main Ideas

Section 1

1. *(The Uses of Geography)* What are two main branches in the study of geography?

2. *(The Uses of Geography)* What six essential elements help us organize the study of geography?

3. *(Places and Regions)* How are the three kinds of regions defined?

Section 2

4. *(The World in Spatial Terms)* What are the advantages and disadvantages of the three main map projections?

5. *(The World in Spatial Terms)* What are six kinds of special-purpose maps?

Thinking Critically

1. Analyzing Information How are formal, functional, and perceptual regions different from each other?

2. Evaluating How would the six essential elements help you organize the study of your community's geography?

3. Identifying Points of View How might different perspectives affect the way different people perceive the region around your community?

Using the Geographer's Tools

Constructing a Population Pyramid Use the following percentages to construct a population pyramid for the United States. You can use the pyramids in Section 2 as a model. When you are finished, describe the population features that the diagram shows about the United States. Do you think the U.S. population is growing rapidly or slowly? Which age groups make up the largest parts of the population?

Age Groups	Male	Female
0–4	3.5%	3.3
5–9	3.7	3.5
10–14	3.7	3.5
15–19	3.7	3.5
20–24	3.4	3.3
25–29	3.2	3.3
30–34	3.5	3.6
35–39	4	4.1
40–44	4.1	4.1
45–49	3.5	3.7
50–54	3	3.2
55–59	2.3	2.5
60–64	1.8	2
65–69	1.6	1.9
70–74	1.4	1.8
75–79	1.1	1.6
80+	1.1	2.2

Writing about Geography

Imagine that you are a geography teacher preparing for your first day of class. Prepare a short lecture describing geography and the kinds of things students will learn about during the course of their study. Include ways that students use geography in daily life and how they may use it in a future job. When you have finished your lecture, proofread it to make sure you have used standard grammar, spelling, sentence structure, and punctuation.

SKILL BUILDING

Geography for Life

Creating a Precinct Map

Environment and Society Use your library, the Internet, and other resources to locate information about voting precinct boundaries in your city or county. Then prepare a map showing those precincts and shade the precinct in which you live. Your map should also include notable human and physical features, such as major roads and rivers. In the margin of your map, note how those features may have helped shape precinct boundaries.

Different cultures use different methods for showing directions and locations. For example, long ago Polynesians developed shell maps to help them navigate in the vast Pacific Ocean. Have students work in groups to devise new ways to record spatial information about a region familiar to them. Ask each group to include a standard map of the area along with a new map that group members create. A legend should accompany each new map. **COOPERATIVE LEARNING**

Food Festival

Have students bring food items to class and use them to discuss human and physical geography. For example, beans grown in a certain area can reflect the soil and climate conditions there. In relation to culture and human geography, people use beans in many different recipes. A Midwestern cook may add beans to a dish called Cincinnati five-way chili, while a Bostonian may bake beans and add salt pork, molasses, and brown sugar.

Building Social Studies Skills

Interpreting Maps

Study the map below. Then use the information from the map to help you answer the questions that follow.

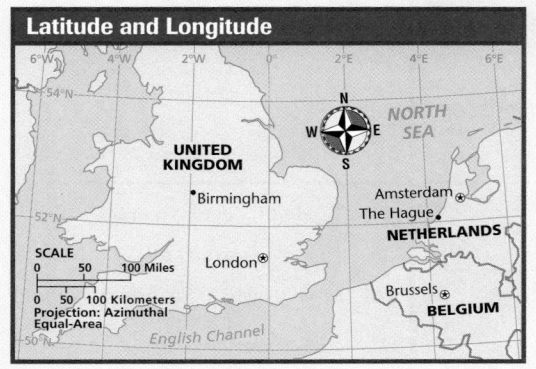

Latitude and Longitude

1. What is the approximate latitude and longitude of The Hague, capital of the Netherlands?
   a. 52°S, 8°E
   b. 52°N, 4°E
   c. 51°N, 1°W
   d. 54°N, 4°W

2. Imagine that your best friend went to Europe with her parents. Your friend decided to play a location game with you. She sent a postcard with the following message: "We flew in to a city near 52°N, 4°E. Two days later we went to a place close to 51°N, 4°E for a festival. Finally, we were near 52°N, 0° for three days to visit my mom's cousins." Write a few sentences in which you describe your best friend's trip, naming the cities she and her family visited.

Building Vocabulary

To build your vocabulary skills, answer the following questions.

3. *Perspective* is an important part of learning about geography.

   In which sentence does *perspective* have the same meaning as it does in the sentence above?
   a. This particular perspective is the east side of the building.
   b. The perspective from the lookout point was beautiful.
   c. From the child's perspective, the stranger looked 10 feet tall.
   d. Renaissance painters used new techniques to create perspective.

4. *Contiguous* means the same as
   a. bordering.
   b. distant.
   c. vertical.
   d. parallel.

interaction of humans and the environment (Environment and Society), and identifying historical events that shaped the community (The Uses of Geography). **(NGS 3)**

3. Answers will vary. Students might note differing points of view of people living in rural or urban areas. **(NGS 3)**

Using the Geographer's Tools

population growing slowly, largest groups aged 35–39 and 40–44

Writing

Students should discuss various uses of geography in daily life and how geography might be used in jobs. Use Rubric 24, Oral Presentations, to evaluate student work.

Geography for Life

Maps should accurately reflect the boundaries of voting precincts in your area. Use Rubric 20, Map Creation, to evaluate student work.

Social Studies Skills

1. b
2. traveled through The Hague, Brussels, London
3. c
4. a

PORTFOLIO ACTIVITY

Climate graphs should accurately reflect local climate data. Use Rubric 30, Research, to evaluate student work.

Alternative Assessment

PORTFOLIO ACTIVITY

Learning about Your Local Geography

Individual Project: Researching Climate Statistics

Plan, organize, and complete a research project about climate conditions in your community or a nearby big city. Use the Internet, almanacs, and other resources to find information about average monthly temperatures and precipitation in the location you choose. Then use the data you collect to create a climate graph.

internet connect

Internet Activity: go.hrw.com
KEYWORD: SW3 GT1

Choose a topic about studying geography:

Use the Internet to research geographic information systems (GIS) and how they can be applied to solve geographic and locational problems. Access the American Fact Finder mapmaker and create a map of your state that includes data on population and population density, education, economic and industrial development, and trade. When you create each map, locate your county and answer the questions posed about the data you have collected.

CHAPTER RESOURCE MANAGER

Objectives	Pacing Guide	Reproducible Resources
SECTION 1 **The Solar System** (pp. 25–28)	• Describe Earth's position in the solar system. • Explain how rotation and revolution affect Earth.	**Regular** 1 day **Block Scheduling** 1 day *Block Scheduling Handbook, Chapter 2*	**RS** Guided Reading Strategy 2.1 **SM** Geography for Life Activity 2 **IC** Lab Activity for Geography and Earth Science, Demonstration 2: Tubby Terra **PS** Readings in World Geography, History and Culture 4
SECTION 2 **Earth-Sun Relationships** (pp. 29–31)	• Explain how the angle of the Sun's rays affects the amount of solar energy received at different locations on Earth. • Distinguish between solstices and equinoxes.	**Regular** 1 day **Block Scheduling** 1 day *Block Scheduling Handbook, Chapter 2*	**RS** Guided Reading Strategy 2.2 **SM** Critical Thinking Activity 2: Marking Time
SECTION 3 **The Earth System** (pp. 34–36)	• Identify Earth's four spheres. • Analyze how Earth's environment is unique in our solar system.	**Regular** 1 day **Block Scheduling** 1 day *Block Scheduling Handbook, Chapter 2*	**RS** Guided Reading Strategy 2.3 **RS** Graphic Organizer 2 **SM** Map Activity 2: Seawater Pollution

Chapter Resource Key

PS Primary Sources

RS Reading Support

IC Interdisciplinary Connections

E Enrichment

SM Skills Mastery

A Assessment

REV Review

ELL Reinforcement and English Language Learners

  Transparencies

  CD–ROM

  Video

  Internet

  Holt Presentation Maker Using Microsoft® PowerPoint®

 One-Stop Planner CD–ROM

See the *One-Stop Planner* for a complete list of additional resources for students and teachers.

 **One-Stop** Planner CD–ROM

It's easy to plan lessons, select resources, and print out materials for your students when you use the ***One-Stop Planner CD–ROM with Test Generator***.

Technology Resources	Reinforcement, Review, and Assessment
One-Stop Planner CD–ROM, Lesson 2.1	**ELL** Main Idea Activity 2.1
CNN. Presents World Cultures: Yesterday and Today, Segment 37: Mission to Mars	**ELL** English Audio Summary 2.1
Homework Practice Online	**ELL** Spanish Audio Summary 2.1
HRW Go site	**REV** Section 1 Review, p. 28
	A Daily Quiz 2.1
One-Stop Planner CD–ROM, Lesson 2.2	**ELL** Main Idea Activity 2.2
Geography and Cultures Visual Resources 1–2	**ELL** English Audio Summary 2.2
Homework Practice Online	**ELL** Spanish Audio Summary 2.2
HRW Go site	**REV** Section 2 Review, p. 31
	A Daily Quiz 2.2
One-Stop Planner CD–ROM, Lesson 2.3	**ELL** Main Idea Activity 2.3
ARGWorld CD–ROM	**ELL** English Audio Summary 2.3
Homework Practice Online	**ELL** Spanish Audio Summary 2.3
HRW Go site	**REV** Section 3 Review, p. 36
	A Daily Quiz 2.3

internet connect

HRW ONLINE RESOURCES

GO TO: go.hrw.com
Then type in a keyword.

TEACHER HOME PAGE
　　KEYWORD: SW3 Teacher

CHAPTER INTERNET ACTIVITIES
　　KEYWORD: SW3 GT2
　　Choose an activity to:
　　• investigate the cause of seasons.
　　• explore recycling methods and create an action plan.
　　• report on the south polar regions and conditions in Antarctica.

CHAPTER ENRICHMENT LINKS
　　KEYWORD: SW3 CH2

CHAPTER MAPS
　　KEYWORD: SW3 MAPS2

ONLINE ASSESSMENT
　　Homework Practice
　　KEYWORD: SW3 HP2
　　Standardized Test Prep
　　KEYWORD: SW3 STP2
　　Rubrics
　　KEYWORD: SS Rubrics

COUNTRY INFORMATION
　　KEYWORD: SW3 Almanac

CONTENT UPDATES
　　KEYWORD: SS Content Updates

HOLT PRESENTATION MAKER
　　KEYWORD: SW3 PPT2

ONLINE READING SUPPORT
　　KEYWORD: SS Strategies

CURRENT EVENTS
　　KEYWORD: S3 Current Events

Meeting Individual Needs

Ability Levels

Level 1 Basic-level activities designed for all students encountering new material

Level 2 Intermediate-level activities designed for average students

Level 3 Challenging activities designed for honors and gifted-and-talented students

English Language Learners Activities that address the needs of students with Limited English Proficiency

Chapter Review and Assessment

	Chapter 2 Test Generator (on the One-Stop Planner)
	Global Skill Builder CD–ROM
	HRW Go site
REV	Chapter 2 Review, pp. 38–39
REV	Chapter 2 Tutorial for Students, Parents, Mentors, and Peers
A	Chapter 2 Test (form A or B)
A	Alternative Assessment Handbook
A	Chapter 2 Test for English Language Learners and Special-Needs Students

Launch into Learning

Focus students' attention on the photo of the Eagle Nebula in Section 1. Explain that the photograph was taken in 1995 from the Hubble Space Telescope and shows areas where new stars are forming. If possible, display a photo, also taken by the Hubble Space Telescope, of the Hubble Deep Field. This remarkable photo of a small patch of sky shows galaxies that may be 10 billion light-years away from Earth. That is, the light from these ancient galaxies traveled at 186,282 miles per second for 10 billion years before reaching the telescope's lens. Point out that a few years ago these advances in astronomical information were almost unimaginable. In fact, one prominent astronomer says that half of all the research papers in astrophysics ever published have appeared in the last 15 years. Tell students that in this chapter they will learn more about humanity's corner of the universe and how other objects in our solar system affect Earth.

Why We Should Know More

These are among the reasons why the subject of Earth in space should interest students:

▶ It is easier to understand physical processes here on Earth if we first understand our planet's relationship to the solar system.

▶ New discoveries about the solar system are made almost every day. We need background knowledge if we are to understand those findings.

▶ To protect our supplies of freshwater and clean air we should know more about how these precious resources fit into the Earth system.

CHAPTER 2

Earth in Space

Earth is our home in the universe. In this chapter, you will learn about Earth's position in the solar system and relation to other objects in space. You will also learn about the Earth system— the interactions of Earth's land, water, air, and life.

Yerkes Observatory, Wisconsin

LET'S GET STARTED

Copy the following instructions onto the chalkboard: *Imagine that you were born and raised in a dark cave. When you finally come out, you see the Moon and stars in the night sky. How would you explain what you see?* Discuss responses. *(Possible answer: that the objects are very close or attached to a ceiling)* Then ask students to describe relationships among Earth, the Sun, the Moon, and the stars. Tell students that in Section 1 they will learn more about how these relationships affect us on Earth.

Building Vocabulary

Write the key terms on the board in random order. Have students group the terms by using information they already know or from clues in the words themselves. *(Students are likely to group terms relating to the **solar system** and those that relate directly to Earth.)* Discuss responses. Call on volunteers to find the definitions in Section 1 or the glossary. Discuss the definitions and connections among the terms. Ask also for other meanings of **rotation** and **revolution** with which students are familiar.

Section 1

The Solar System

READ TO DISCOVER

1. What is Earth's position in the solar system?
2. How do rotation and revolution affect Earth?

WHY IT MATTERS

Today scientists are making many exciting discoveries about the universe. Use  or other **current events** sources to learn about how a recent scientific discovery has increased our knowledge of the universe.

DEFINE

solar system

planets

moons

satellite

solar energy

rotation

revolution

Space and the Universe

If you look at the sky on a clear night, you can see thousands of stars. With a telescope you can see millions more. Beyond the telescope's view are trillions more. All these stars are part of the universe. The universe is made up of all existing things, including space and Earth. Most astronomers believe that the universe is 10 to 20 billion years old. Since its birth, the universe has been expanding continuously. It is unimaginable in size.

Space is filled with large objects called stars. Most stars are grouped together in huge clusters called galaxies. Many objects that look like individual stars to the naked eye are really billions of stars in a faraway galaxy. The Milky Way is the galaxy in which we live. The part of this galaxy that we can see from Earth looks like a bright milky streak across the night sky. The Sun is a medium-size star near the edge of the Milky Way.

When you look up at the night sky, do you ever wonder if there are places with life other than Earth? Scientists wonder about this and continue to search for clues. The vast universe holds unlimited possibilities for future discovery.

✓ **READING CHECK:** *Physical Systems* What are large groups of stars such as the Milky Way called? galaxies

This image of the Eagle Nebula shows an area where new stars are formed. The pillars of gas and dust contain material that condenses and ignites, forming stars.

The Planets The Sun and the group of bodies that revolve around it are called the **solar system**. The Sun's great size attracts the other objects through gravity. Besides the Sun, the solar system includes **planets**, asteroids, and comets. Planets are major bodies that orbit a star.

There are four inner and five outer planets in the solar system. The inner planets are closest to the Sun. They are Mercury, Venus, Earth, and Mars. The inner planets are terrestrial, meaning they

LEVEL 1: Copy the following graphic organizer onto the chalkboard, omitting the italicized answers. Point out that the sizes and distances are not to scale. Call on volunteers to label the objects. **ENGLISH LANGUAGE LEARNERS**

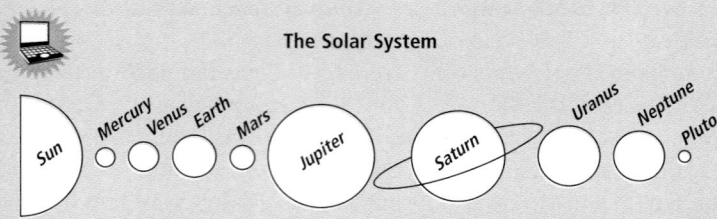

The Solar System

LEVELS 2 AND 3: Assign students to represent the Sun and each of the planets. Other students could represent moons, including Earth's, or asteroids between Mars and Jupiter. Have students wear identifying labels. Then take the class to the school's football field for a simulation. Have the student representing the Sun stand at one goal line. Then ask other students to stand the correct distance from the "Sun," according to the following chart. Later, lead a discussion on what students learned from the demonstration. *(Possible answer: The distances between outer planets are greater than between inner planets.)*

Across the Curriculum

▶ **Science** ◀

The Tunguska Event
Small rocky objects called asteroids also have their place in the solar system. They can have a direct impact on planets. When asteroids collide, they may cast off fragments of dust, some of which become meteors. If a meteor explodes just above Earth's surface, widespread damage can result. Many scientists believe such an event occurred in the Tunguska region of Siberia, Russia, on June 30, 1908. Thousands of square miles of subarctic forests were devastated by the blast. Calculations equate the power of the Tunguska explosion to 15 megatons of dynamite— 1,000 times the force of the atomic bomb released on Hiroshima, Japan, in 1945. Effects recorded after the explosion included disturbances in Earth's magnetic field and genetic mutations in certain plant and animal species.

CRITICAL THINKING: Why might scientists want to learn more about the Tunguska event? *(Having more information may help us prepare for another catastrophe of this sort.)*

DIAGRAM ANSWER

Possible answer: Tides affect marine life, recreation, shipping routes, and harbors.

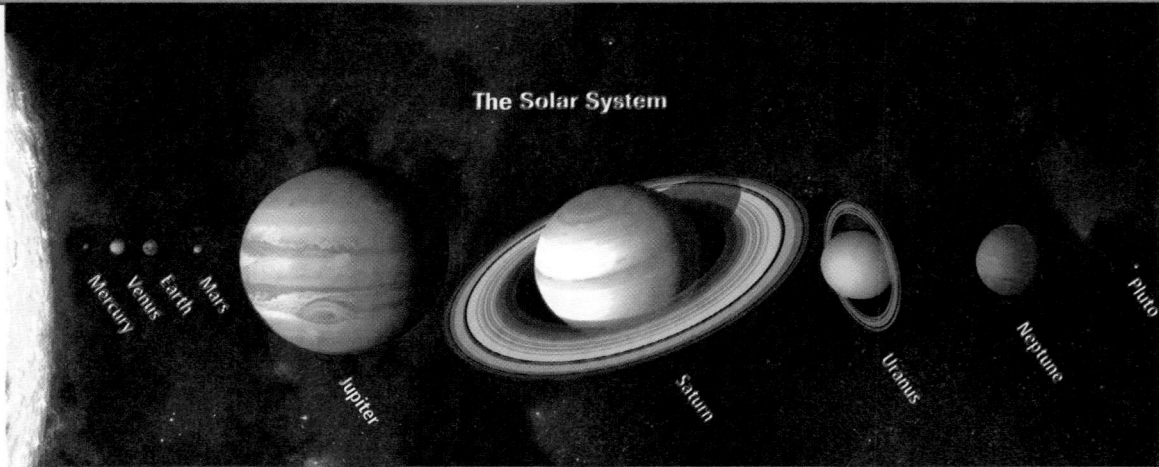

The Solar System

This diagram shows the nine planets and their relative sizes. However, the relative distance of each planet from the Sun is not accurate.

internet connect

GO TO: go.hrw.com
KEYWORD: SW3 CH2
FOR: Web sites about Earth in space

have a solid rocky surface. The outer planets are generally much larger and farther from the Sun. They are Jupiter, Saturn, Uranus, Neptune, and Pluto. The outer planets are mainly gaseous. An exception is Pluto, a small planet with a rocky core. Unlike stars, planets do not generate their own light. They are visible to us because they reflect sunlight.

Earth and the other planets look like they are perfectly round. However, this is not exactly correct. Earth's polar areas are slightly flattened. Around the equator, Earth is slightly bulged. This slight variation of a perfect sphere is called an oblate spheroid. *Oblate* means "flattened."

Moons are smaller objects that orbit a planet. A body that orbits a larger body can also be called a **satellite**. Moons are natural satellites. Earth has one moon. However, hundreds of human-made satellites also orbit Earth. Mercury and Venus are the only planets in the solar system with no moons. Saturn has at least 18.

✓ **READING CHECK:** *Physical Systems* What are the main bodies that make up the solar system? the Sun, planets, asteroids, comets, and moons

Effects of the Moon and Sun on Tides

INTERPRETING THE DIAGRAM *Tides are higher than normal when the gravitational pull of the Moon and Sun combine. These tides, called spring tides, occur twice a month during the full moon and new moon. High tides are lower during neap tides, when the pull of the Sun is at a right angle to the pull of the Moon. Neap tides occur during quarter moons.* **Why might knowing the schedule of high and low tides be important to people living in coastal areas?**

Neap tides

Pull of Sun — Earth
Pull of Moon
Moon
Orbital path of Moon
Orbital path of Earth
Pull of Sun
Moon
Pull of Moon — Earth
Sun

Spring tides

Planet	Distance from the Sun in AU*	Scaled distance in yards
Mercury	0.39	1
Venus	0.72	1.8
Earth	1	2.5
Mars	1.52	4
Jupiter	5.20	13
Saturn	9.51	24
Uranus	19.19	49
Neptune	30.06	76
Pluto	39.53	100

* AU stands for astronomical unit, the average distance between Earth and the Sun.

Teach Objective 2

ALL LEVELS: Have students imagine that Earth's rotation takes 365 days instead of 24 hours and that, therefore, the same side always faces the Sun. Ask them to speculate about whether humans could live on the planet under this condition and, if so, where. Ask what would probably happen to lakes and oceans.

HOMEWORK: Have each student select one of the planets. Then have students use the information they have learned about the relative locations of planets and other information to describe what the planet's physical environment might be like.

The Sun, Earth, and Moon   The Sun is an average star in age, brightness, and size. It is small compared to huge stars called supergiants. However, compared to Earth, the Sun is very large. The diameter of Earth is about 8,000 miles (12,900 km). The diameter of the Sun is about 865,000 miles (1,390,000 km)—more than 100 times greater. The Sun operates like a giant thermonuclear reactor, releasing enormous amounts of energy.

Earth is the third planet from the Sun and the fifth-largest of the nine. Earth's orbit around the Sun is not a perfect circle. It is elliptical, or oval-shaped. Earth's orbit averages about 93 million miles (150 million km) from the Sun.

The Moon is about 240,000 miles (385,000 km) from Earth and is about one fourth the size of Earth. The Moon orbits Earth every 29.5 days, or about once every month. On clear nights you can see its surface. The Moon has a simple geography, with a barren volcanic surface and many craters. These craters were made by the impact of meteors and comets. Unlike Earth, the Moon has no air, water, or life.

The Sun, Earth, and Moon exert gravitational forces on one another that influence physical processes on Earth. The most obvious of these are the rise and fall of ocean tides. Tides are the response of the fluid ocean surface to the gravitational pull of the Moon and Sun. (See the diagram of tides.)

✓ **READING CHECK:** *Physical Systems* What causes ocean tides?
the gravitational pull of the Moon and Sun

Earth's Rotation, Revolution, and Tilt

Most of Earth's energy comes from the Sun. This type of energy is called **solar energy** and reaches Earth as light and heat. All life on Earth depends on solar energy. Solar energy affects weather, plants, animals, and human activities. It influences the clothes we wear, the homes we live in, the foods we grow and eat, and even which sports we play. Three different relationships between Earth and the Sun control how much solar energy is received at different locations. Do you know what these are?

Rotation   Imagine that Earth has a rod running through it from the North Pole to the South Pole. This rod represents Earth's axis, and the planet spins around on it. One complete spin of Earth on its axis is one **rotation**, which takes 24 hours. Earth rotates in a west-to-east direction. We see the effects of Earth's rotation as the Sun "rising" in the east and "setting" in the west. To us, it appears that the Sun is moving across the sky. Actually, it is only Earth rotating on its axis.

Solar energy strikes only the half of Earth facing the Sun. If Earth did not rotate on its axis—creating day and night—only the half facing the Sun would receive solar energy. That side of Earth would be very hot. The half of the planet facing away from the Sun would always be dark and cold. Earth's rotation allows the entire planet's surface to receive the warming effects of daylight and the cooling effects of darkness.

Revolution   In addition to rotating on its axis, Earth revolves around the Sun. It makes one elliptical orbit, or **revolution**, every 365 ¼ days—one Earth

Between 1968 and 1972 NASA's Apollo missions landed astronauts on the Moon six times. In this photo, astronaut Harrison Schmitt studies a boulder in the Taurus-Littrow Valley. Schmitt made detailed descriptions of craters, boulders, and debris and helped bring back some 249 pounds (113 kg) of lunar material.

Unique in the solar system, Earth is the only planet with liquid water at the surface, active mountain-building processes, and life.

Eye on Earth

Bay of Fundy Tides   Some of the highest tides on Earth occur in the Bay of Fundy, located off the southeastern Canadian province of Nova Scotia. Tides there reach heights of more than 50 feet (15.2 m). The shape of the bay's shoreline helps explain why the tides are so high. Because the bay is funnel-shaped and the water cannot escape as the tide rises, the water piles up as the tide moves into the ever-narrowing bay.

Another spectacular phenomenon, the tidal bore, also occurs in the region. As incoming water rushes into the Bay of Fundy, it also travels up rivers and streams that empty into the bay. This action causes the rivers to temporarily reverse their flows. The tidal bores—large crested waves up to six feet (2 m) tall—rush upstream. A tidal bore can travel two to three times faster than the normal tidal current.

ACTIVITY: Have students create diagrams to explain the high tides and tidal bores of the Bay of Fundy.

internet connect

GO TO: go.hrw.com
KEYWORD: SW3 CH2
FOR: Web sites about tides

Close

Call on students to suggest recent movies or television shows that depict our solar system, galaxies, or stars in some way. Ask them to compare the cinematic treatment of the scientific topics with what they have learned. Lead a discussion about the accuracy of the filmed versions.

Review and Assess

Have students complete the **Section Review**. Then have students complete **Daily Quiz 2.1**.

Reteach

Have students complete **Main Idea Activity for English Language Learners and Special-Needs Students 2.1.** Then have students discuss each of the Section 1 key terms by linking it to at least one other term in the list. **ENGLISH LANGUAGE LEARNERS**

Extend

Organize interested students into groups to conduct research on the discoveries made by major figures in the history of astronomy. Then have the groups work together to write and perform a script for a conference call in which the scientists discuss their work. **COOPERATIVE LEARNING, BLOCK SCHEDULING**

Section 1 Review Answers

Define For definitions, see: solar system, p. 25; planets, p. 25; moons, p. 26; satellite, p. 26; solar energy, p. 27; rotation, p. 27; revolution, p. 27

Reading for the Main Idea
1. inner planets closer to the Sun, with solid rocky surfaces; outer planets generally larger, farther from the Sun, and gaseous in form

2. barren volcanic surface and many craters; no air, water, or life

3. rotation—Earth spins on its axis, completing one rotation every 24 hours. revolution—Earth revolves around the Sun once every 365¼ days, or year.

Critical Thinking
4. Answers will vary. **(NGS 1)**

Organizing What You Know
5. Sun—average star in size, age, and brightness; near edge of the Milky Way; diameter about 100 times greater than Earth; at center of solar system; Earth— inner terrestrial planet; oblate spheroid; one moon; elliptical orbit around the Sun; tilted on its axis 23½ degrees; Moon— diameter about one-fourth that of Earth's; orbits Earth every 29.5 days; barren volcanic surface and many craters; no air, water, or life

CONNECTING TO TECHNOLOGY ANSWER

Possible answer: has made new methods for exploration possible

Connecting to TECHNOLOGY

Astronomy

Astronomers study objects and matter beyond Earth. For most of history, people could see space only with the naked eye. However, with the help of modern technology, astronomers can now see far into space.

In the early 1600s the famous scientist Galileo used telescopes to study space. Telescopes magnify objects in space, making faint stars and galaxies visible. Telescopes have become central to astronomy. With the development of space-age technology, astronomy made even greater advances. For example, scientists sent out probes to collect new information about planets, moons, and other objects. The Hubble Space Telescope (shown below), built between 1978 and 1990, led to many exciting discoveries. The Hubble is the first powerful telescope ever placed in orbit around Earth, which gives it a much clearer view of space. Astronomers have used the Hubble to study black holes, galaxies, stars, and other distant objects.

Summarizing How has the development of modern technology advanced astronomy?

Hubble Space Telescope image of interacting spiral galaxies NGC 2207 and IC 2163

year. Each time you celebrate your birthday we have just completed another orbit around the Sun. For convenience, our calendars have 365 days in a year. To account for the one-fourth day gained each year, an extra day—February 29—is added to the calendar every four years. This year, one day longer than the previous three, is called a leap year.

Tilt If Earth's axis always pointed straight up and down in relation to the Sun, daylight hours would be the same at every location on Earth. Each day would consist of 12 hours of daylight and 12 hours of darkness. This would be true throughout the year. However, this is not the case because Earth's axis is tilted in relation to the Sun.

As Earth revolves around the Sun, its axis points toward the same spot in the sky. The North Pole points to a star known as the North Star. The position of the axis is fixed in respect to the North Star. Yet it is not fixed in relation to our Sun. Earth's orbit lies on a plane that runs from the center of the Sun to the center of the planet. Earth's axis is tilted 23½ degrees from the perpendicular, or 90 degrees, to the plane of its orbit. Thus, as Earth revolves around the Sun, the North Pole points at times toward the Sun and at times away from the Sun. (See the diagram in Section 2.) The tilt of Earth on its axis affects the amount of solar energy that different places receive during the year.

✓ **READING CHECK:** *Physical Systems* How do rotation, revolution, and tilt affect the amount of solar energy received at different locations on Earth? Rotation allows the warming effects of sunlight and cooling effects of darkness. Tilt and revolution cause hemispheres to receive varying amounts of solar energy during the year.

Section 1 Review

Define
solar system
planets
moons
satellite
solar energy
rotation
revolution

Reading for the Main Idea

1. *Physical Systems* How are inner planets and outer planets different?

2. *Physical Systems* What is the geography of the Moon like?

3. *Physical Systems* What is the difference between rotation and revolution?

Critical Thinking

4. *Drawing Inferences and Conclusions* How do you think technologies developed to study space can affect people's lives on Earth?

go. hrw .com **Homework Practice Online**
Keyword: SW3 HP2

Organizing What You Know

5. Create a chart like the one shown below. Use it to describe the Sun, Earth, and Moon.

Sun	Earth	Moon

Section 2

OBJECTIVES

1. **Explain how the angle of the Sun's rays affects the amount of solar energy received at different locations on Earth.**

2. **Distinguish between solstices and equinoxes.**

 **LET'S GET STARTED**

Copy the following questions onto the chalkboard: *What is the greatest distance north or south that you have traveled? During which season did you travel? What was the difference in the length of day and night?* Discuss responses. *(Example: Students from Florida who have been to Canada during the summer may note that the days were longer in Canada.)* Point out the diagram on this page. Discuss how students' travel experiences relate to it. Tell students they will learn more about Earth-Sun relationships in Section 2.

Building Vocabulary

Write the key terms on the chalkboard. Point out the **tropics** and **polar regions** on a wall map. Then underline *sol* in **solstice**. Explain that *sol* is the Latin word for "sun." Ask which phrases in Section 1 with the same prefix relate to the Sun *(solar system, solar energy)*. Point out the two parts of **equinox**—*equi*, which in Latin means "equal," and *nox*, which means "night." On the equinox, day and night last the same length of time. Call on volunteers to read all the definitions aloud.

Section 2

Earth-Sun Relationships

READ TO DISCOVER

1. How does the angle of the Sun's rays affect the amount of solar energy received at different locations on Earth?

2. What are solstices and equinoxes?

WHY IT MATTERS

Many festivals and celebrations around the world take place during a solstice or equinox. Use **CNNfyi.com** or other **current events** sources to learn about a celebration at one of these times.

IDENTIFY

Tropic of Capricorn   Arctic Circle
Antarctic Circle   Tropic of Cancer

DEFINE

tropics   solstice
polar regions   equinox

Solar Energy and Latitude

As you know, different places on Earth receive different amounts of solar energy. Areas near the equator receive a lot of solar energy all year. These places are generally warm. We call these warm low-latitude areas near the equator the **tropics**. Other places get very little solar energy. These areas are at high latitudes and are cold most of the time. Because these areas surround the North and South Poles, we call them the **polar regions**. The areas between the tropics and the polar regions are called the middle latitudes. The amount of solar energy reaching these areas changes greatly during the year. They may be warm or cool, depending on the time of year.

The amount of solar energy that a place receives relates to the angle at which the Sun's rays strike Earth. Direct vertical solar rays heat Earth's surface more than angled rays. This is because the amount of solar energy in a direct ray is concentrated on a smaller area. The same amount of energy in an angled ray is spread over a larger area.

When the North Pole points toward the Sun, direct rays strike the Northern Hemisphere. (See the diagram.) Thus, the Northern Hemisphere receives more concentrated solar energy, making temperatures warmer. The length of time between sunrise and sunset also grows longer. At this time, the Southern Hemisphere receives more angled rays and is cooler.

When the North Pole tilts away from the Sun, the most direct rays strike the Southern Hemisphere. Now, the Southern Hemisphere receives more solar energy, experiences longer days, and thus has warmer temperatures. At this time, the Northern Hemisphere receives less solar energy and has cooler temperatures.

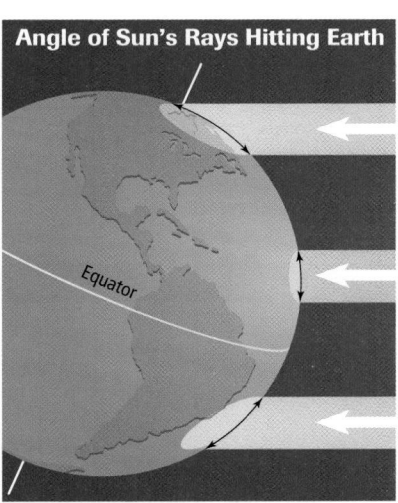

Angle of Sun's Rays Hitting Earth

Equator

This diagram shows summer in the Northern Hemisphere. At this time, Earth tilts toward the Sun, and direct rays are more concentrated in the Northern Hemisphere. Meanwhile, in the polar regions and Southern Hemisphere, solar rays are spread over a larger area.

✓ **READING CHECK:** *Physical Systems* Which latitudes receive the most solar energy throughout the year? the low latitudes, or tropics

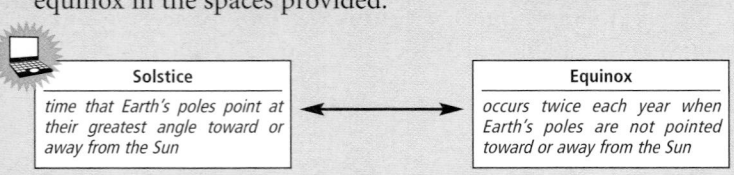
ALL LEVELS: Lead a discussion on why different places receive different amounts of sunlight and how the amount received affects ways of life. You may want to have students illustrate their observations. Then copy the following graphic organizer onto the chalkboard, omitting the italicized answers. Have students describe the concepts of solstice and equinox in the spaces provided.

Solstice
time that Earth's poles point at their greatest angle toward or away from the Sun

⟷

Equinox
occurs twice each year when Earth's poles are not pointed toward or away from the Sun

Using National Geography Standard 7:

Physical Systems: The Physical Processes that Shape the Patterns of Earth's Surface Draw a line to represent the equator around a tennis ball. Draw outlines of the continents on the ball. Then pass a knitting needle or skewer through the ball perpendicular to the line, to represent Earth's axis. Have a student hold the ball by the needle at an angle, to represent Earth's tilt. Then, in a darkened classroom, have the student walk in a circle around a small lit lamp, rotating the "Earth" on its axis while maintaining the tilt. Point out that the rotation causes day and night, while the tilt causes seasons.

Linking Past to Present

The Winter Solstice in Ancient Ireland Around the world there are many archaeological sites that display ancient peoples' knowledge of the solstices. Newgrange, an Irish tomb constructed about 5,000 years ago, is one of the most famous. Known in Gaelic as "the cave of sun," Newgrange contains a passage 62 feet (19 m) long that leads to a burial chamber. An opening at the entrance allows the rising Sun's rays to reach all the way to this chamber, but only on the winter solstice, the shortest day of the year. The interior remains illuminated for about 15 minutes. Then it is left in darkness again for another year.

The Newgrange tomb lies within a large grass-covered mound 280 feet (85 m) wide. It is one of the most popular tourist sites in Ireland. However, few visitors are allowed into the tomb during the solstice event.

ACTIVITY: Have students conduct research on Newgrange and create a simulation of how the Sun's rays reach the tomb's interior.

DIAGRAM ANSWER

December 21

Our Amazing Planet

In the 200s B.C. the Greek astronomer Aristarchus proposed that Earth rotated on its axis—creating day and night—and revolved around the Sun each year.

INTERPRETING THE DIAGRAM

*As Earth revolves around the Sun, the tilt of the poles toward and away from the Sun causes the seasons to change. **On which day is the North Pole pointed directly away from the Sun?***

The Seasons

We refer to the times of greater and lesser heat as the seasons. There are four general seasons: winter, spring, summer, and fall. Some regions, particularly the tropics, are warm year-round but have alternating wet and then dry seasons.

In each hemisphere, the Sun's energy is stronger during the summer. Daytime lasts longer. In the winter, daytime is shorter, and the Sun's energy is weaker. During spring and fall, the Sun's energy is more evenly distributed. At these times, daylight and darkness are closer to equal length. The tilt of Earth's axis causes the Northern and Southern Hemispheres to have opposite seasons at the same time of the year.

Solstices Twice during the year, Earth's poles tilt toward or away from the Sun more than at any other time. The time that Earth's poles point at their greatest angle toward or away from the Sun is called a **solstice**. Solstices occur each year about December 21 and June 21.

In the Northern Hemisphere, the December solstice has the fewest daylight hours of the year and is the first day of winter. The Southern Hemisphere on the same day has its greatest number of daylight hours, and it is the first day of summer. During the December solstice, the Sun's most direct rays strike Earth in the Southern Hemisphere along a parallel 23 1/2 degrees south of the equator. This parallel is called the **Tropic of Capricorn**. The South Pole is tilted toward the Sun and receives constant sunlight. All areas located south of the **Antarctic Circle** have 24 hours of daylight. The Antarctic Circle is the parallel 66 1/2 degrees south of the equator. Meanwhile, the area around the North Pole experiences constant darkness and is very cold. The parallel beyond which no

The Seasons

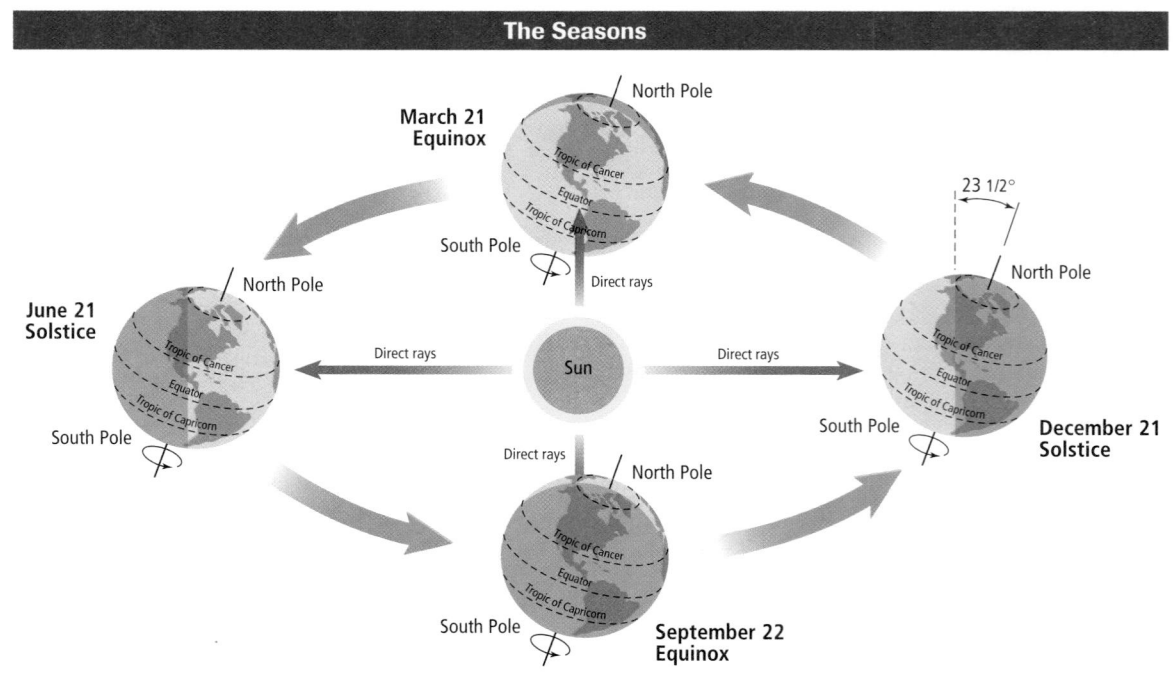

sunlight shines on this day is known as the **Arctic Circle**. It is located 66 1/2 degrees north of the equator.

During the June solstice, the Northern Hemisphere experiences the greatest number of daylight hours of the year and the first day of summer. On this day, the North Pole tilts toward the Sun. The Sun's direct rays are at their most northerly position, striking Earth at a line 231/2 degrees north of the equator. This line is called the **Tropic of Cancer**. If you traveled to Australia on the June solstice, it would be the first day of winter, which has the fewest daylight hours of the year. During the June solstice, the Sun never sets north of the Arctic Circle. Daylight lasts 24 hours. During this time, the opposite occurs south of the Antarctic Circle, where darkness lasts 24 hours.

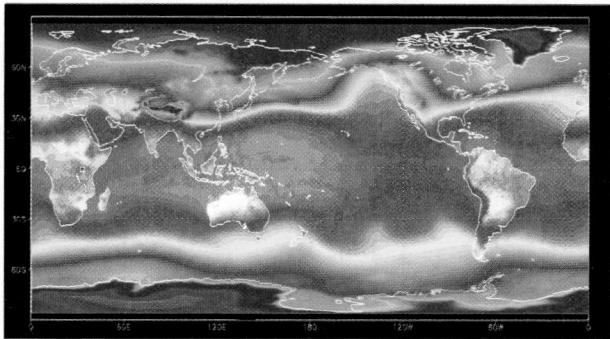

INTERPRETING THE MAP *This map shows Earth's average temperatures during December 19–25, 1999. Warmer temperatures are shown in red and yellow and colder temperatures are shown in purple and blue.* **Where was the North Pole tilted in relation to the Sun when this data was collected? How did this affect the distribution of temperatures on Earth?**

✓ **READING CHECK:** *Physical Systems* When is the first day of winter in the Northern Hemisphere? In the Southern Hemisphere? about December 21 and June 21

Equinoxes An **equinox** occurs twice each year when Earth's poles are not pointed toward or away from the Sun. *Equinox* means "equal night" in Latin. At this time, the direct rays of the Sun strike the equator, and both poles are at a 90 degree angle from the Sun. Both hemispheres receive an equal amount of sunlight—12 hours each.

Equinoxes occur on about March 21 and September 22. In the Northern Hemisphere, the March equinox marks the beginning of spring. To people living in the Southern Hemisphere, however, the March equinox marks the beginning of fall. The opposite situation occurs about September 22. Days between the solstices and equinoxes gradually become warmer or cooler, and daytime becomes longer or shorter, depending on where you live. This cycle is repeated each year, creating the four seasons.

✓ **READING CHECK:** *Physical Systems* When is the first day of spring in the Northern Hemisphere? In the Southern Hemisphere? about March 21 and September 22

Section 2 Review

Homework Practice Online
Keyword: SW3 HP2

Identify
Tropic of Capricorn
Antarctic Circle
Arctic Circle
Tropic of Cancer

Define
tropics
polar regions
solstice
equinox

Reading for the Main Idea

1. *Places and Regions* Where are the tropics and polar regions found?
2. *Places and Regions* Where does the amount of solar energy Earth receives vary most during the year?

Critical Thinking

3. **Analyzing** When does winter occur on Earth? Explain your answer.
4. **Drawing Inferences and Conclusions** How do you think the tilt of Earth on its axis might affect life in the polar regions?

Organizing What You Know

5. Create a chart like the one shown below. Use it to describe the location, amount of solar energy received, and the time of year when solar energy is received for each region.

Tropics	Middle latitudes	Polar regions

Calculating Time Differences

Refer students to the time-zone map. Provide students with a series of questions that can be answered by using the time-zone map. Students should assume that no countries are on daylight savings time. *(Examples: If it is 1 A.M. in São Paulo, Brazil, what time is it in* Lagos, Nigeria? [5 A.M.]) *If it is 9 P.M. in Sydney, Australia, what time is it in Moscow, Russia? [2 P.M.] If you live in Houston, Texas, and want to watch a British soccer match that airs at 10 A.M., London time, how early will you have to be awake? [4 A.M.]*)

Atomic Clocks Universal time, measured at Greenwich, is calculated through a series of measurements of the Earth's rotation relative to the Sun. Since 1972, however, most scientists and governments have depended on atomic clocks for precise time measurements. Factors on Earth's surface can actually change the rate of the planet's rotation slightly, so the length of a year measured in solar time might differ from the same year measured in atomic time by several seconds. Occasionally, astronomers at the Paris Observatory analyze such differences and announce how many "leap seconds" will be added to atomic measurements to keep the two systems even.

Atomic clocks work on the same principles as other clocks, but vibrating atoms take the place of pendulums or gears. The most accurate clocks, which use cesium atoms, are estimated to lose less than a second in 3 million years. Countries across the globe maintain such clocks in national laboratories. Their displays form the basis for Coordinated Universal Time, which is broadcast via radio signals all around the world.

Skill-Building Activity

Using a Time Zone Map

Because the Sun is not directly overhead every place on Earth at the same time, clocks are adjusted to reflect the difference in the Sun's position. Earth rotates on its axis once every 24 hours, so in one hour it makes 1/24 of a complete revolution. Since there are 360 degrees in a circle, we know that Earth turns 15 degrees of longitude each hour ($360° ÷ 24 = 15°$).

Earth turns in a west-to-east direction. As a result, the Sun rises first in New York, for example, and later in Los Angeles, which is farther west. If one place has the Sun directly overhead at noon, another place 15 degrees to the west will have the Sun directly overhead one hour later. The planet will have rotated 15 degrees during that hour. To account for this, we divide Earth into 24 time zones. Each time zone covers about 15 degrees of longitude. The time is an hour earlier for each 15 degrees you move westward on Earth. It is an hour later for each 15 degrees you move eastward. For example, if the Sun rises at 6:00 A.M. in New York, it is still only 3:00 A.M. in Los Angeles, three time zones to the west.

By international agreement, longitude is measured from the prime meridian, which passes through the Royal Greenwich Observatory in Greenwich (GREN-ich), England. Time is also measured from Greenwich and is called Greenwich mean time (GMT) or universal time (UT). For each time zone east of the prime meridian, clocks are set one hour ahead of GMT. For each time zone west of Greenwich, clocks are set one hour behind GMT. For example, when it is noon in London, England, it is 1:00 P.M. in Rome, Italy, one time zone to the east. At the same time, it is 7:00 A.M. in New York City, five time zones to the west.

Halfway around the planet from Greenwich is the international date line. It is a north-south line that runs through the Pacific Ocean. It generally follows the

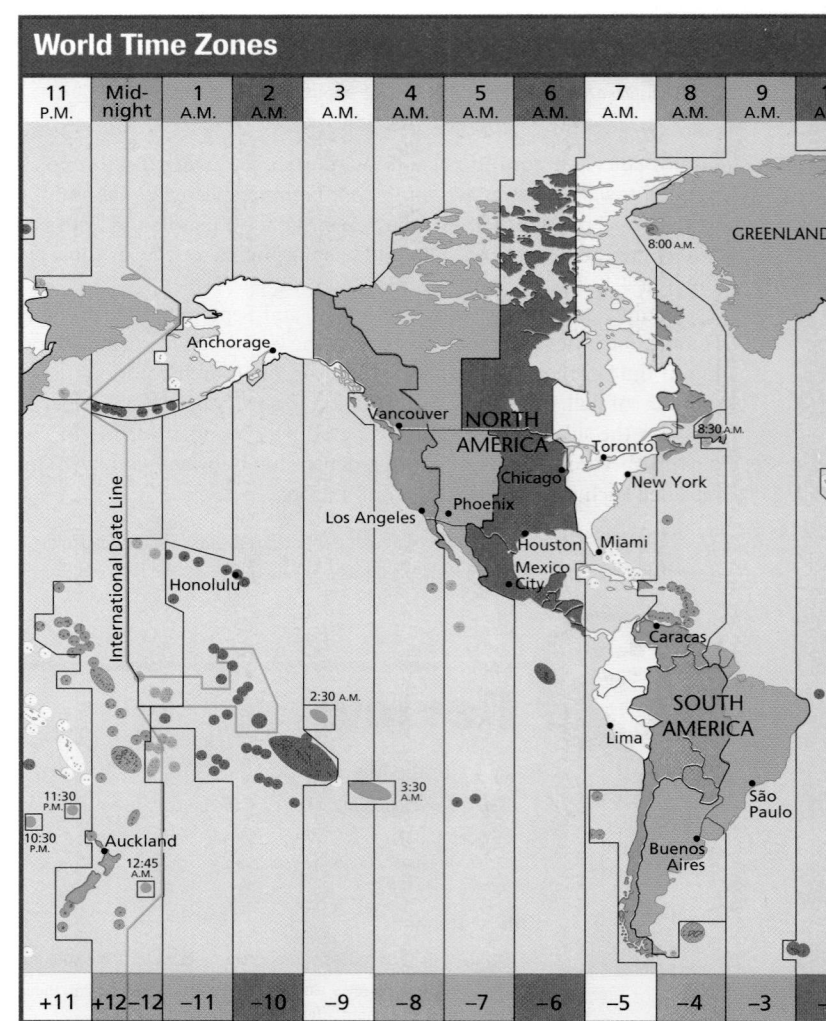

Traveling across Time Zones

Point out to students that many people suffer jet lag when traveling long distances. Because they cross many time zones, their sleep patterns are disrupted. For example, if a traveler from Los Angeles arrives in London at lunchtime, back home in California it is still early in the morning—when the traveler would most likely be asleep.

Have students determine how many time zones they would have to cross traveling west to east for the following journeys: London to Moscow, Russia *(3)*, New York to Cairo, Egypt *(7)*, Chicago to Jakarta, Indonesia *(13)*.

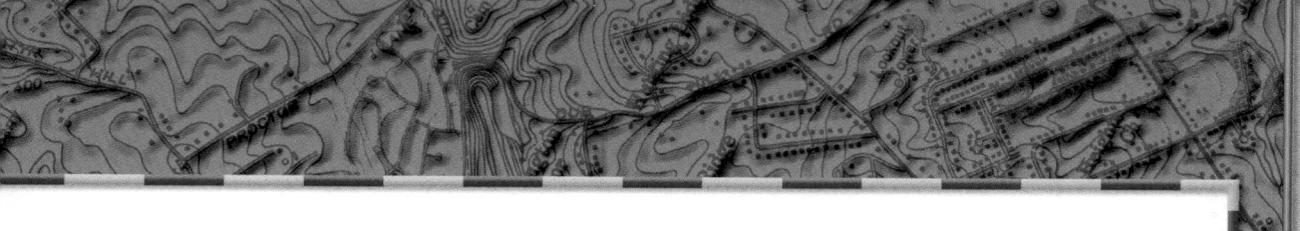

180° line of longitude. However, sometimes it leaves this line to avoid dividing island countries. When you cross the international date line, the date and day change. If you cross the date line from west to east, you gain a day. If you travel from east to west, you lose a day.

As you can see from the World Time Zones map, time zones do not follow meridians exactly. Instead,

time zone lines often follow political boundaries. For example, in Europe and Africa many time zones follow country borders. The contiguous United States has four major time zones: eastern, central, mountain, and Pacific. Alaska and Hawaii are in separate time zones to the west. Some countries make local adjustments to the time in their time zones. For example, most of the United States has daylight savings time in the summer. People in places with daylight savings time adjust their clocks to have more daylight during the evening hours.

Practicing the Skill

1. In which time zone do you live? Check your time now. What time is it in New York?

2. How many time zones does China have?

3. If it is noon in New York, what time is it in London?

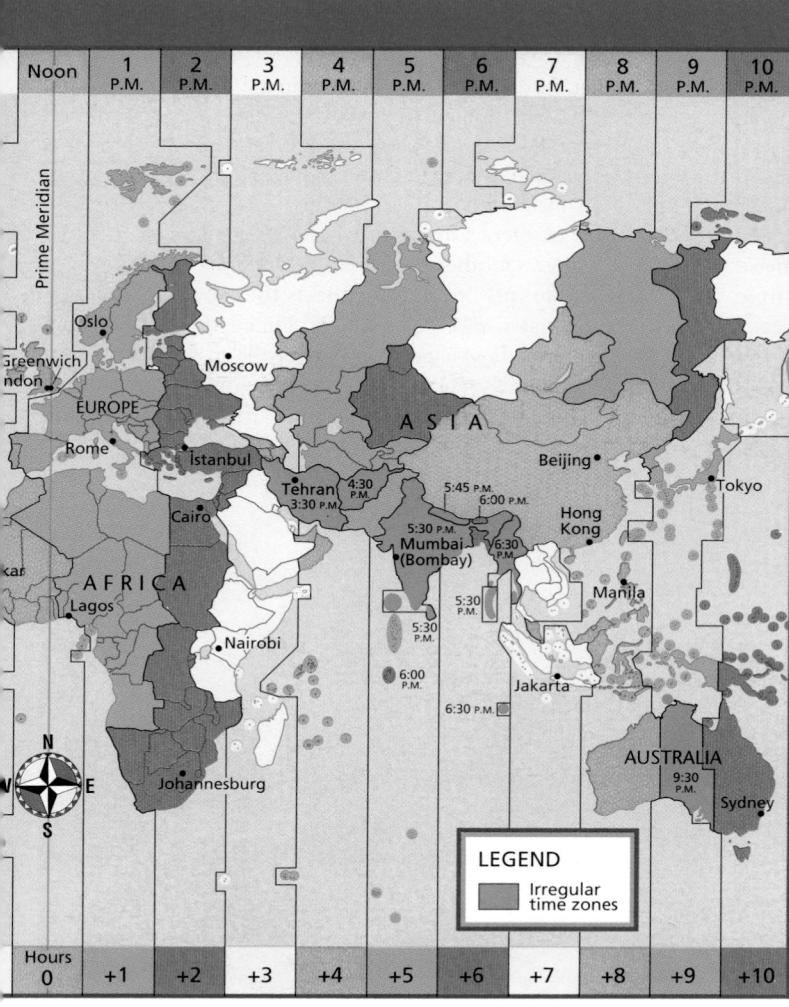

Practicing the Skill Answers

1. Answers will vary. If students are in the Eastern time zone, the time is the same as New York's. The time is one hour behind New York for each time zone moving westward.
2. one
3. 5:00 P.M.

Section 3

OBJECTIVES

1. Identify Earth's four spheres.

2. Analyze how Earth's environment is unique in our solar system.

 **LET'S GET STARTED**

Copy the following instructions onto the chalkboard: *When you see or hear the word "sphere" what shape comes to mind? Discuss responses. (Students may mention globes, round objects, Earth.)* Explain to the students that a sphere can be all of these things and more, such as a place or space within which something exists. Tell students they will learn more about the Earth system and its spheres in Section 3.

Building Vocabulary

Write the key terms on the chalkboard in random order. Have students group the terms by the suffix. *(Students are likely to group terms including "sphere".)* Ask students to suggest the meaning of each "sphere" word by using information they already know. *(Examples: Students might relate "hydro" to water and "bio" to biology.)*

Section 3 The Earth System

READ TO DISCOVER

1. What are Earth's four spheres?
2. How is Earth's environment unique in the solar system?

WHY IT MATTERS

Many global environmental issues are hotly debated today. Use **CNNfyi.com** or other **current events** sources to learn about the debate over one of those issues.

DEFINE

atmosphere
lithosphere
hydrosphere
biosphere
environment

Earth's Four Spheres

Earth is a complex planet. Its different parts interact in a vast number of ways. The scale of some Earth interactions is so small that they are hard to notice. For example, ants and termites help mix decayed plant matter and soil in some places. Other interactions, such as rainfall and flooding, affect large regions.

Geographers call all these interactions the Earth system. A system is a group of different parts that interact to form a whole. Major parts of Earth can be viewed as separate from each other. However, they interact constantly. For example, when a volcano erupts, it not only affects the mountain where it is located but also the air, water, and life around it. Some large eruptions can affect global weather patterns. People are also part of the Earth system. Our actions affect Earth in many ways. For example, people convert large areas of Earth into farmland. This affects plant and animal life, soil fertility, and water use.

Kilauea volcano, on the island of Hawaii, is one of the largest, most active volcanoes in the world. Kilauea has been erupting steadily since 1983 and has generated numerous lava flows, which have added more than 500 acres (200 hectares) of land to the island. In 1986 Kilauea began releasing large amounts of sulfur dioxide gas into the air. This gas reacts chemically with particles in the air to produce volcanic smog, or "vog," which contaminates rainwater and causes a health hazard.

Teach Objectives 1–2

LEVEL 1: Copy the following graphic organizer onto the chalkboard, omitting the italicized answers. Use it to help students describe the four parts of the Earth system. **ENGLISH LANGUAGE LEARNERS**

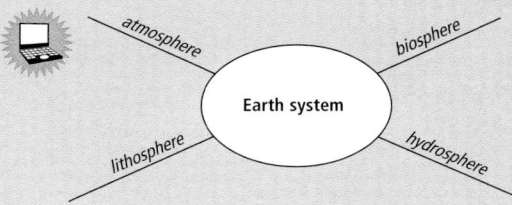

LEVELS 2 AND 3: Have students complete the Level 1 activity. Then organize students into groups to research the other planets in the solar system and some of the larger moons, such as Europa and Io. Then have students discuss whether each planet or moon might have the elements to sustain human life. **COOPERATIVE LEARNING**

Geographers divide the Earth system into four major parts. Each part is called a sphere because it occupies a shell around the planet. The **atmosphere** is the envelope of gases that surrounds Earth. It is the least dense and outermost sphere, extending from Earth's surface into space. Earth's gravity holds the atmosphere around the planet. About 78 percent of Earth's atmosphere is a gas called nitrogen, and about 21 percent is oxygen. The rest is made up of carbon dioxide, ozone, and other gases. These gases and water vapor sustain life on Earth. The atmosphere also protects the planet from the Sun's harmful radiation.

The **lithosphere** is the solid crust of the planet. This outer crust includes rocks and soil. It forms Earth's continents, islands, and ocean floors.

The **hydrosphere** is all of Earth's water. Water covers about 70 percent of Earth's surface. The hydrosphere includes water in liquid, solid, and gaseous forms. Liquid water is found in oceans, lakes, rivers, and underground. Clouds and fog are made up of liquid droplets. Solid water, or ice, is found on both land and sea. Large amounts of ice are locked in glaciers in the polar regions. Earth is the only planet in the solar system known to have large amounts of surface water. Water is essential to all living organisms.

The **biosphere** is the part of Earth that includes all life forms. It includes all plants and animals. The biosphere overlaps the other three spheres. It extends from deep ocean floors to high in the atmosphere.

Earth's four major spheres are all interconnected. Each one affects the other. For example, the hydrosphere supplies people with water, which we need to live. It is also a home for plants and animals. The hydrosphere affects the lithosphere when rain breaks up rocks and washes them away. It also constantly interacts with the atmosphere, causing clouds and rain.

✓ **READING CHECK:** *Physical Systems* How are Earth's four spheres different from each other? atmosphere: the gases around Earth; lithosphere: the solid outer crust; hydrosphere: all of Earth's water; biosphere: all life on the planet

Earth's Environment

Earth's four spheres make up the **environment**, or surroundings. The environment includes all the biological, chemical, and physical conditions that interact and affect life. Within our solar system, no other planet has an environment as complex as Earth's. Our closest neighbors, Venus and Mars, each have an atmosphere and a lithosphere. However, neither has a vast supply of liquid water. In recent years, scientists have looked far into space. They have discovered many other stars that have planets. In fact, scientists have found more than 50 planets outside our solar system. Do you think there is another planet in the universe that has the right environmental conditions for life?

Earth's atmosphere from space

Earth's lithosphere—a volcano in Hawaii

Earth's hydrosphere—waves off the Oregon coast

Earth's biosphere—a rain forest in New Zealand

internet connect

GO TO: go.hrw.com
KEYWORD: SW3 CH2
FOR: Web sites about nature preserves

Close

Display and discuss a newspaper story about a natural disaster or weather event. Ask students how the event may involve the components of the Earth system. *(Example: A thunderstorm is the result of changes in the atmosphere. Resulting floods change the hydrosphere and move the soil of the lithosphere. Plants and animals of the biosphere may be drowned or displaced.)*

Review and Assess

Have students complete the **Section Review**. Then have students complete **Daily Quiz 2.3**.

Reteach

Have students complete **Main Idea Activity for English Language Learners and Special-Needs Students 2.3**. Then have students draw each of the four spheres, labeling each accordingly. Call on volunteers to display their drawings. **ENGLISH LANGUAGE LEARNERS**

Extend

Have interested students conduct research on a specific environmental problem in your community and explain how the problem affects the four parts of the Earth system. **COOPERATIVE LEARNING, BLOCK SCHEDULING**

Section 3 Review Answers

Define For definitions, see: atmosphere, p. 35; lithosphere, p. 35; hydrosphere, p. 35; biosphere, p. 35; environment, p. 35

Reading for the Main Idea
1. in countless ways, such as a volcanic eruption, which affects land, air, water, and life around it

2. persuade government officials, make legal challenges, force votes on issues; example: building a dam

Critical Thinking
3. weathers rocks of lithosphere, provides oxygen to biosphere, evaporates water from hydrosphere (NGS 7)

4. Pollution can damage the hydrosphere and atmosphere. Eliminating wild areas can destroy plants and animals of the biosphere. (NGS 14)

Organizing What You Know
5. atmosphere—envelope of gases that surrounds Earth; least dense and outermost sphere; made up mostly of nitrogen and oxygen; protects the planet from the Sun's harmful radiation; lithosphere—solid outer crust of the planet; includes soil, land, and rock formations; hydrosphere—all of Earth's water; in liquid, solid, and gaseous forms; covers about 70 percent of Earth's surface; biosphere—includes all plant and animal life forms; environment—formed by interaction of all four spheres; includes lakes, rivers, rocks, valleys, plants, trees, clouds, wind, and so on

Glen Canyon Dam, located in northern Arizona on the Colorado River, was completed in 1964 and created Lake Powell, a large reservoir that provides water and electricity to western states. Some environmental groups argue that the dam has disrupted the river's natural ecosystems and that Lake Powell should be drained. For some, the controversy surrounding the construction of the dam marked the beginning of the modern environmental movement.

Earth's environment is the key to our survival and quality of life. Therefore, we must be aware of how our activities affect the planet. One way we can learn more about Earth's environment is by studying environmental issues.

FOCUS ON CITIZENSHIP

Environmental Issues Many people today are concerned about the effects humans have on the environment. In fact, many geographers study a variety of environmental issues. These geographers study how to solve environmental problems at local, state, national, and international levels. Environmental studies can help us learn how to solve difficult environmental problems. They can also help us learn more about the world around us.

People often have different points of view about environmental issues. These different points of view can influence public policies and affect how decisions are made. For example, suppose government officials want to build a new dam. Some citizens might support the idea. A new dam could create jobs, prevent flooding, and increase water supplies. However, other people might oppose the construction of a new dam. They might be concerned that it would harm the environment. Some fish species might be threatened if they could not swim upriver to spawn. This, in turn, could affect the fishing industry. In addition, some land would be permanently flooded to protect other areas from occasional floods.

How might this problem or similar ones be resolved? Individuals and groups such as conservationists, farmers, and others may try to influence governmental decisions. This is called lobbying. Sometimes, legal challenges are made to a planned development. Other times, a group of citizens will force a vote on the issue. This allows the public to decide. Can you think of other ways to resolve difficult environmental issues?

✓ **READING CHECK:** *Environment and Society* How can people with different points of view on environmental issues influence decision making?

Individuals or groups can lobby, make legal challenges to planned developments, or try to force a vote on the issue.

Section 3 Review

Homework Practice Online
Keyword: SW3 HP2

Define
atmosphere
lithosphere
hydrosphere
biosphere
environment

Reading for the Main Idea
1. *Physical Systems* How are Earth's four spheres connected to form a system?

2. *Environment and Society* How do different points of view on environmental issues influence public policies and decision-making processes? Give an example.

Critical Thinking
3. **Drawing Inferences and Conclusions** How does the atmosphere affect the other spheres?

4. **Making Generalizations and Predictions** How can human activities affect the Earth system?

Organizing What You Know
5. Copy the graphic organizer below. Use it to describe each of Earth's four spheres. In the center, describe ways the four spheres interrelate to influence Earth's environment.

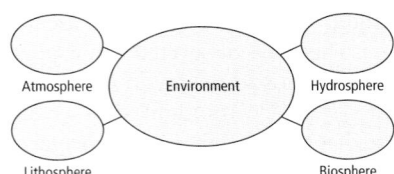

Art: Ice Ages and Cave Paintings

During the most recent ice age, the people who lived in what is now Europe had to survive in an extremely harsh environment. Some animals, such as big cats and bears, threatened the humans' safety. Other animals, such as bison, were hunted for food by humans. Early artists painted these and other animals expertly on the walls of caves. Show the class examples of some cave paintings from Lascaux or Chauvet in France, Altamira in Spain, or other famous caves. Point out how accurately the animals are portrayed with just a few lines and how some of the animals' images follow the shapes or lines of the cave wall. Have students conduct further research on the cave paintings and then create their own paintings on butcher paper. Before students start to paint, crumple the paper slightly and then smooth it out to mimic the roughness of cave walls.

Teacher to Teacher

Ron Scholten of Los Angeles, California, suggests students complete the following address information to review Earth in space and the Earth system: name, street, city, county or parish, state or province, country, continent, planet, system, galaxy, and universe.

Physical Systems

Geography for Life

Ice Ages and the Earth System

Throughout this textbook you will read about the forces and processes that have shaped the Earth system. Among the most important of these processes were Earth's ice ages. Ice ages are long periods of time during which thick ice sheets cover vast areas of land. During the ice ages, Earth's temperatures cooled for thousands or even millions of years, and ice sheets spread across the planet. During the most recent ice age, which ended about 10,000 years ago, ice sheets covered almost one third of Earth's present land area. (See the map of Pleistocene Glaciation.) Earth is currently experiencing a warm period between ice ages.

Ice ages greatly affect Earth's hydrosphere. For example, water from the oceans is frozen and locked in the expanding ice sheets. During the most recent ice age, enough water was frozen to lower global sea levels some 300 to 400 feet (90 to 120 m). As a result, more land was exposed at Earth's surface. The shapes and locations of coastlines differed greatly from how they look today. For example, the British Isles and mainland Europe were connected. North America and Asia were joined across what is now the Bering Strait. Rising ocean waters eventually covered those land bridges, however. Ocean levels rose because the warm period between ice ages causes ice to melt and ocean levels to rise. In fact, this process is partly responsible for the gradual rise in ocean levels today.

Some of the effects that ice ages have had on Earth's landscapes are easy to see. Slowly moving ice sheets carved and scraped the lithosphere, removing soil and forming holes. When the ice melted some of these holes filled with water and became lakes and swamps. The Great Lakes were formed in this way. Moving ice also wore down mountains and created great valleys.

In addition to their effects on the hydrosphere and lithosphere, ice ages also affected Earth's atmosphere and biosphere. Colder temperatures caused shifts in wind patterns. Ice-age temperature changes also caused a major redistribution of plants and

Pleistocene Glaciation

| | Ice sheet or glacier |

ATLANTIC OCEAN
PACIFIC OCEAN
PACIFIC OCEAN
INDIAN OCEAN

SCALE
0   4,000 Miles
0   4,000 Kilometers

Projection: Robinson

INTERPRETING THE MAP *The most recent ice age occurred mainly during the Pleistocene Epoch. This epoch began about 2 million years ago and ended about 10,000 years ago. Therefore, this ice age is known as the Pleistocene Ice Age, or Pleistocene Glaciation. During the Pleistocene Glaciation there were perhaps 12–16 periods of major ice advances. Between these cold periods were warmer times called interglacials.* **Which parts of the world were covered with ice during the Pleistocene Glaciation?**

animals. For example, expanding ice shifted the habitats of plant and animal species in the Northern Hemisphere from north to south. As a result, many species died off. During the last ice age, the distribution of human populations also changed. Most researchers believe that a land bridge across the Bering Strait allowed humans to move from eastern Asia to the Americas. Humans also moved across land links between areas in Europe, Asia, and Australia.

Applying What You Know

1. **Summarizing** In what ways did ice ages affect the Earth system? How can some of those effects be seen today?

2. **Analyzing Information** In what ways might a new ice age affect Earth's biosphere?

Applying What You Know Answers

1. Ocean levels fell and rose as ice expanded and melted; ice sheets created shapes on the land and land bridges; wind patterns changed; land bridges aided the movement of humans across vast areas; habitats of plant and animal species shifted, and many species died off; shapes on land, such as lakes and swamps, were affected.
2. Possible answer: It could cause plant and animal extinctions and shifts in habitat.

MAP ANSWER

Alaska, Canada, parts of the northern United States, Greenland, northern Europe, northern Asia, the southern Andes, Antarctica

This Geography for Life feature addresses National Geography Standards 1, 7, 9, and 15.

CHAPTER **2** **Review Answers**

Building Vocabulary For definitions, see: solar system, p. 25; planets, p. 25; solar energy, p. 27; rotation, p. 27; revolution, p. 27; tropics, p. 29; polar regions, p. 29; solstice, p. 30; equinox, p. 31; atmosphere, p. 35; lithosphere, p. 35; hydrosphere, p. 35; biosphere, p. 35; environment, p. 35

TECHNOLOGY

▶ Chapter 2 Test Generator (on the One-Stop Planner)
▶ Global Skill Builder CD–ROM
▶ HRW Go site

REINFORCEMENT, REVIEW, AND ASSESSMENT

▶ Chapter 2 Review, pp. 38–39
▶ Chapter 2 Tutorial for Students, Parents, Mentors, and Peers
▶ Chapter 2 Test (form A or B)
▶ Alternative Assessment Handbook

▶ Chapter 2 Test for English Language Learners and Special-Needs Students
▶ Unit 1 Test
▶ Unit 1 Test for English Language Learners and Special-Needs Students

Assess

Have students complete a Chapter 2 Test.

Reteach

Have one group of students list the planets of the solar system in order of distance from the Sun, a second group to write the definitions of solstice and equinox, and a third to write a description of each of Earth's spheres. Have the groups exchange papers and discuss any questions. **ENGLISH LANGUAGE LEARNERS, COOPERATIVE LEARNING**

CHAPTER 2 Review Answers

Locating Key Places

A. Tropic of Capricorn
B. Arctic Circle
C. Antarctic Circle
D. Tropic of Cancer

Understanding the Main Ideas

1. oblate spheroid
2. 24 hours; 365¼ days
3. about June 21 (the summer solstice); raises temperatures
4. at greatest angle toward or away from the Sun
5. All the complex interactions on our planet make up the Earth system, including inter-actions among the atmosphere, lithosphere, hydrosphere, and biosphere.

Thinking Critically

1. in the high latitudes, there-fore receives low amounts of solar energy (NGS 7)
2. Because Earth is tilted on its axis, as it revolves around the Sun each pole is pointed at times toward the Sun and at times away from the Sun. This creates times when differ-ing amounts of solar energy are received. (NGS 7)
3. No. Equatorial areas are warm year-round, but they may experience alternating wet and dry seasons. (NGS 7)

CHAPTER 2 Review

Building Vocabulary

On a separate sheet of paper, explain the following terms by using them correctly in sentences.

solar system	solstice
planets	equinox
solar energy	atmosphere
rotation	lithosphere
revolution	hydrosphere
tropics	biosphere
polar regions	environment

Locating Key Places

On a separate sheet of paper, match the letters on the map with their correct labels.

Tropic of Cancer	Arctic Circle
Antarctic Circle	Tropic of Capricorn

Understanding the Main Ideas

Section 1

1. *Physical Systems* What term describes Earth's shape?
2. *Physical Systems* How long does it take Earth to make one complete rotation on its axis? One complete revolution around the Sun?

Section 2

3. *Physical Systems* When do the Sun's most direct rays strike in the Northern Hemisphere? What effect does this have on temperatures there?
4. *Physical Systems* During the solstices, where do Earth's poles point in relation to the Sun?

Section 3

5. *Physical Systems* What is the Earth system?

Thinking Critically

1. Drawing Inferences and Conclusions Suppose all you knew about a place was that it was located at 60° north latitude. What would this tell you about how much solar energy the place receives each year?
2. Identifying Cause and Effect How do Earth's revolution and tilt cause seasons?
3. Making Generalizations and Predictions If you lived in an area near the equator, would you experience all of the seasons? Explain.

Using the Geographer's Tools

1. Analyzing Diagrams Study the diagram of the seasons in Section 2. What is today's date? Using the diagram, estimate where Earth is on its path around the Sun.
2. Interpreting Maps Study the World Time Zones map that follows Section 2. Suppose a ship left Honolulu, Hawaii, at 10:00 P.M. on December 5 and traveled to Auckland, New Zealand. If the trip lasted six hours, what date and time would the ship arrive in Auckland?
3. Preparing Diagrams Using the information in this chapter, draw a diagram of Earth showing it tilted on its axis. Draw and label major lines of latitude, such as the equator, Tropics of Cancer and Capricorn, and Arctic and Antarctic Circles. How many degrees of latitude separate the Tropics of Cancer and Capricorn?

Writing about Geography

Imagine you have been asked to help search for life beyond Earth. Write a short proposal outlining where you will search for life, what technologies you will use, and what you hope to find. When you are finished with your proposal, proofread it to make sure you have used standard grammar, spelling, sentence structure, and punctuation.

SKILL BUILDING

Geography for Life

Preparing Sketch Maps

Places and Regions Find a detailed surface map of the Moon from an atlas or other source. Use it to prepare a sketch map of the Moon's surface features and topography. When you are done, com-pare your map to a world map of Earth. What features are common on the Moon but not on Earth? How is the surface of the Moon dif-ferent from the surface of Earth?

Portfolio Activity

1. Have students conduct research on the use of solar energy in their communities. Ask them to write a report that explains why using solar energy in their area is practical or impractical.
2. Have students research the average monthly temperatures throughout the year in your region. Ask them to use the data to create a bar graph, making note of the coldest and warmest months and the solstices. Have them write a short report analyzing the data.

Food Festival

The Sun makes life on Earth possible. It can also be used to process foods. Have students make sun tea by filling a jar with water, adding tea bags, covering the jar, and then placing it in sunlight to steep. After a few hours, add ice to the tea and enjoy. You might also have students locate and bring to class sun-dried food items, such as beef jerky, raisins, apricots, cranberries, or tomatoes. Challenge the class to devise ways that people in regions where fuel is scarce could use solar energy to cook their food.

Building Social Studies Skills

Sunrise and Sunset Times for Selected 2001 Dates in Sydney, Australia		
Date	Sunrise	Sunset
March 21	6:59 A.M.	7:06 P.M.
June 21	6:58 A.M.	4:56 P.M.
September 22	5:45 A.M.	5:51 P.M.
December 21	5:43 A.M.	8:03 P.M.

Interpreting Charts

Study the chart above. Then use the information in the chart to help you answer the questions that follow.

1. On which date does summer begin in Sydney, Australia?
 a. March 21
 b. June 21
 c. September 22
 d. December 21

2. Look carefully at the times listed in the chart. How does this information relate to the tilt of Earth on its axis?

Analyzing Secondary Sources

Read the following passage and answer the questions.

"Telescopes have become central to astronomy. With the development of space-age technology, astronomy made even greater advances. For example, scientists sent out probes to collect new information about planets, moons, and other objects. The Hubble Space Telescope, built between 1978 and 1990, led to many exciting discoveries. The Hubble is the first powerful telescope ever placed in orbit around Earth, which gives it a much clearer view of space. Astronomers have used the Hubble to study black holes, galaxies, stars, and other distant objects."

3. The Hubble Space Telescope
 a. was built in the late 1800s.
 b. is located in Hawaii.
 c. is a probe sent out to study Mars.
 d. is the first powerful telescope ever placed in orbit around Earth.

4. Why might the Hubble's location give it such a clear view of space?

Alternative Assessment

PORTFOLIO ACTIVITY

Learning about Your Local Geography

Individual Project: Research
Plan, organize, and complete a research project on an environmental issue in your community. Check your local newspaper to find a current environmental issue. Then write a summary that explains what the issue is and describes different points of view about it. How might this issue be resolved? What role does public policy play in this issue? Have citizens tried to influence public policy regarding this issue? If so, how? How might a geographical perspective be used to help resolve this issue?

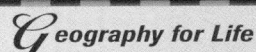

internet connect

Internet Activity: go.hrw.com
KEYWORD: SW3 GT2

Choose a topic on Earth in space to:
 • investigate the cause of seasons.
 • explore recycling methods and create an action plan.
 • report on the south polar regions and conditions in Antarctica.

Using the Geographer's Tools

1. Answers will vary.
2. 2:00 A.M. on December 6
3. 47

Writing

Answers will vary but should display an understanding of what is necessary for life. Use Rubric 42, Writing to Inform, to evaluate student work.

*G*eography for Life

Maps should show Moon's features clearly. Craters are common on the Moon, but not on Earth. The Moon lacks bodies of water and distinct continents. Use Rubric 20, Map Creation, to evaluate student work.

Social Studies Skills

1. d
2. The tilt affects the amount of sunlight that Sydney receives during the year.
3. d
4. Possible answer: because it is above the Earth's atmosphere and pollution that can create haze

PORTFOLIO ACTIVITY

Answers will vary. Use Rubric 37, Writing Assignments, to evaluate student work.

CHAPTER RESOURCE MANAGER

Objectives	Pacing Guide	Reproducible Resources
SECTION 1 **Factors Affecting Climate** (pp. 41–45) • Explain how the Sun affects Earth's atmosphere. • Tell how atmospheric pressure distributes energy around the globe. • Describe how global wind belts affect weather and climate. • Identify the ways that the oceans affect weather and climate.	**Regular** 1 day **Block Scheduling** .5 day *Block Scheduling Handbook, Chapter 3*	**RS** Guided Reading Strategy 3.1 **PS** Readings in World Geography, History, and Culture 5 **IC** Environmental and Global Issues Activity 1: That Greenhouse Effect! **IC** Lab Activity for Geography and Earth Science, Demonstrations 11 and 12 **E** Creative Strategies for Teaching World Geography, Lesson 13 **SM** Geography for Life Activity 3
SECTION 2 **Weather Factors** (pp. 46–49) • Identify the common forms of precipitation and how they are formed. • Describe how mountains and elevation affect weather and climate. • Explain what the different types of storms are and how they form.	**Regular** 1 day **Block Scheduling** .5 day *Block Scheduling Handbook, Chapter 3*	**RS** Guided Reading Strategy 3.2 **IC** Lab Activity for Geography and Earth Science, Hands–On 1 **SM** Critical Thinking Activity 3: El Niño **SM** Map Activity 3: Hurricanes
SECTION 3 **Climate and Vegetation Patterns** (pp. 50–57) • Explain how the two tropical climates differ. • Identify conditions common in dry climates. • Tell what climates are found in the middle latitudes. • Detail the characteristics of high-latitude and highland climates.	**Regular** 2 days **Block Scheduling** 1.5 days *Block Scheduling Handbook, Chapter 3*	**RS** Guided Reading Strategy 3.3 **RS** Graphic Organizer Activity 3 **PS** Readings in World Geography, History, and Culture 6 and 7

Chapter Resource Key

PS Primary Sources

RS Reading Support

IC Interdisciplinary Connections

E Enrichment

SM Skills Mastery

A Assessment

REV Review

ELL Reinforcement and English Language Learners

Transparencies

CD-ROM

Video

 Internet

Holt Presentation Maker Using Microsoft® PowerPoint®

One-Stop Planner CD–ROM

See the *One-Stop Planner* for a complete list of additional resources for students and teachers.

 **One-Stop** Planner CD–ROM

It's easy to plan lessons, select resources, and print out materials for your students when you use the **One-Stop Planner CD–ROM with Test Generator**.

Technology Resources

- One-Stop Planner CD–ROM, Lesson 3.1
- *ARGWorld* CD–ROM
- Geography and Cultures Visual Resources 3
- **CNN** Presents Geography: Yesterday and Today, Segment 3: Global Warming
- Homework Practice Online
- HRW Go site

- One-Stop Planner CD–ROM, Lesson 3.2
- Geography and Cultures Visual Resources 4
- Homework Practice Online
- HRW Go site

- One-Stop Planner CD–ROM, Lesson 3.3
- Geography and Cultures Visual Resources 5
- Homework Practice Online
- HRW Go site

Reinforcement, Review, and Assessment

ELL Main Idea Activity 3.1
ELL English Audio Summary 3.1
ELL Spanish Audio Summary 3.1
REV Section 1 Review, p. 45
A Daily Quiz 3.1

ELL Main Idea Activity 3.2
ELL English Audio Summary 3.2
ELL Spanish Audio Summary 3.2
REV Section 2 Review, p. 49
A Daily Quiz 3.2

ELL Main Idea Activity 3.3
ELL English Audio Summary 3.3
ELL Spanish Audio Summary 3.3
REV Section 3 Review, p. 57
A Daily Quiz 3.3

☑ internet connect

HRW ONLINE RESOURCES

GO TO: go.hrw.com
Then type in a keyword.

TEACHER HOME PAGE
  KEYWORD: SW3 Teacher

CHAPTER INTERNET ACTIVITIES
  KEYWORD: SW3 GT3
  Choose an activity to:
  - use satellite maps to learn about weather fronts and surface conditions.
  - create a poster on hurricanes and the technology of hurricane tracking.
  - investigate global warming.

CHAPTER ENRICHMENT LINKS
  KEYWORD: SW3 CH3

CHAPTER MAPS
  KEYWORD: SW3 MAPS3

ONLINE ASSESSMENT
  Homework Practice
  KEYWORD: SW3 HP3
  Standardized Test Prep
  KEYWORD: SW3 STP3
  Rubrics
  KEYWORD: SS Rubrics

COUNTRY INFORMATION
  KEYWORD: SW3 Almanac

CONTENT UPDATES
  KEYWORD: SS Content Updates

HOLT PRESENTATION MAKER
  KEYWORD: SW3 PPT3

ONLINE READING SUPPORT
  KEYWORD: SS Strategies

CURRENT EVENTS
  KEYWORD: S3 Current Events

Meeting Individual Needs

Ability Levels

Level 1 Basic-level activities designed for all students encountering new material

Level 2 Intermediate-level activities designed for average students

Level 3 Challenging activities designed for honors and gifted-and-talented students

English Language Learners Activities that address the needs of students with Limited English Proficiency

Chapter Review and Assessment

- Chapter 3 Test Generator (on the One-Stop Planner)
- Global Skill Builder CD–ROM
- HRW Go site
- **REV** Chapter 3 Review, pp. 60–61
- **REV** Chapter 3 Tutorial for Students, Parents, Mentors, and Peers
- **A** Chapter 3 Test (form A or B)
- **A** Alternative Assessment Handbook
- **A** Chapter 3 Test for English Language Learners and Special-Needs Students

Launch into Learning

Lead a discussion about how people who live in your area perceive its climate and weather. Does the weather allow a wide range of outdoor activities? Do residents feel that the weather often hampers their daily lives? Do residents tell jokes about the weather? For example, Texans sometimes say that television weather forecasters have only two predictions for the summer months—"sunny and hot, and dark and hot."

Others say that if you don't like the weather, just wait a minute—it will change. Then ask students what they already know about the factors that affect your area's weather. They may mention location in the far north or south, elevation, nearness to the ocean, or other factors familiar from weather reports. Tell students that they will learn about these factors and their effect on climate, weather, and vegetation in this chapter.

Why We Should Know More

There are many reasons why studying weather and climate is important. Here are some to share with your students:

▶ Ocean and wind currents can influence everything from vacation plans to our food supply.

▶ Every day, weather affects us all. We can predict it or accommodate ourselves to it better if we understand it.

▶ We can more successfully prevent damage to the environment if we comprehend the connections among climate, plants, animals, and people.

▶ Environmental changes often determine world events. We can predict events better if we recognize those cause–and–effect relationships.

CHAPTER 3 Weather and Climate

You have read about Earth in space and our planet's relationship to the Sun. In this chapter you will learn more about the Sun's energy and its effects on Earth. You will see how that energy affects atmospheric conditions and life on our planet.

Fog gathers over San Francisco Bay, California.

A tornado strikes Pampa, Texas.

Heavy snow blankets New York City.

Section 1

OBJECTIVES

1. Explain how the Sun affects Earth's atmosphere.
2. Tell how atmospheric pressure distributes energy around the globe.
3. Describe how global wind belts affect weather and climate.
4. Identify the ways that the oceans affect weather and climate.

🔊 LET'S GET STARTED

Copy the following question onto the chalkboard: *What would happen if, for one week, all the winds in the world were still?* Discuss responses. *(Possible answers: Temperature differences would become extreme, air pollution would get worse, and ocean waves would calm.)* Point out that wind is one of the factors that makes life as we know it possible on Earth. Tell students that in Section 1 they will learn more about winds and other factors that affect climate.

Building Vocabulary

Write the key terms on the chalkboard. Ask students what they have read or heard about the terms in the news media. *(In addition to* **weather**, **climate**, *and* **temperature**, *many students will know about the controversies surrounding the* **greenhouse effect** *and* **global warming**.*)* Discuss students' prior knowledge of the terms. Call on volunteers to read the definitions. Point out that although some of the terms entered common usage recently, others are old. For example, sailors were referring to the **doldrums** by the mid-1800s.

Section 1 — Factors Affecting Climate

READ TO DISCOVER

1. How does the Sun affect Earth's atmosphere?
2. How does atmospheric pressure distribute energy around the globe?
3. How do global wind belts affect weather and climate?
4. How do the oceans affect weather and climate?

WHY IT MATTERS

Has your state experienced a very hot summer or an unusually cold winter this year? Use **CNN fyi.com** or other **current events** sources to learn about the effects of weather extremes in the United States.

IDENTIFY

Gulf Stream

DEFINE

weather	cyclones
climate	prevailing winds
temperature	doldrums
greenhouse effect	front
global warming	

INTERPRETING THE VISUAL RECORD *Much of Texas suffered terrible heat and drought during 2000.* **How might harsh weather affect how people view a region? How might their views differ from those held by people who live elsewhere?**

The Sun and Latitude

The Sun plays the major role in Earth's **weather** and **climate** patterns. Weather is the condition of the atmosphere at a given time and place. Weather conditions in a geographic region over a long time are called climate. As you read earlier, solar energy heats Earth unevenly. The tilt of Earth as the planet revolves around the Sun is important. It determines which hemisphere receives the Sun's most direct rays at a given time of year. This process causes the changing seasons. Areas in the middle and high latitudes have distinct seasons. On the other hand, tropical locations in lower latitudes receive the most direct rays year-round. Thus, they are warm all year. Polar areas receive the least amount of energy from the Sun and are very cold all year.

What happens to the Sun's energy when it reaches Earth? About half is reflected back into space or absorbed by the atmosphere. Earth's surface absorbs the other half. Once absorbed, solar energy is converted into heat. The measurement of heat is called **temperature**.

Earth's atmosphere traps heat energy in a process called the **greenhouse effect**. Like the clear glass of a greenhouse, the atmosphere allows much sunlight to pass through. Earth's air then slows the rate at which the heat escapes into space. The greenhouse effect helps keep the planet warm.

Evidence seems to show that Earth has gotten warmer in recent decades. Most scientists believe that this process, called **global warming**, is caused by human activities. They point out that burning coal, natural gas, and oil adds carbon dioxide to the lower atmosphere. Because carbon dioxide absorbs heat, increased amounts

Section 1 — RESOURCES

REPRODUCIBLE

▶ Guided Reading Strategy 3.1
▶ Readings in World Geography, History, and Culture 5
▶ Environmental and Global Issues Activity 1
▶ Lab Activity for Geography and Earth Science, Demonstrations 11 and 12
▶ Creative Strategies for Teaching World Geography, Lesson 13
▶ Geography for Life Activity 3

TECHNOLOGY

▶ One-Stop Planner CD–ROM, Lesson 3.1
▶ Geography and Cultures Visual Resources 3
▶ CNN Presents Geography: Yesterday and Today, Segment 3
▶ Homework Practice Online
▶ HRW Go site

REINFORCEMENT, REVIEW, AND ASSESSMENT

▶ Main Idea Activity 3.1
▶ English Audio Summary 3.1
▶ Spanish Audio Summary 3.1
▶ Section 1 Review, p. 45
▶ Daily Quiz 3.1

VISUAL RECORD ANSWER

cause them to view the region less positively; might be more realistic

ALL LEVELS: Copy the following graphic organizer onto the chalkboard, omitting the italicized answers. Call on students to fill in the blanks to complete the sentences and identify the processes by which the Sun interacts with Earth's atmosphere. Then lead a discussion about how different views about the greenhouse effect and global warming can influence public policies and decision-making processes.
ENGLISH LANGUAGE LEARNERS

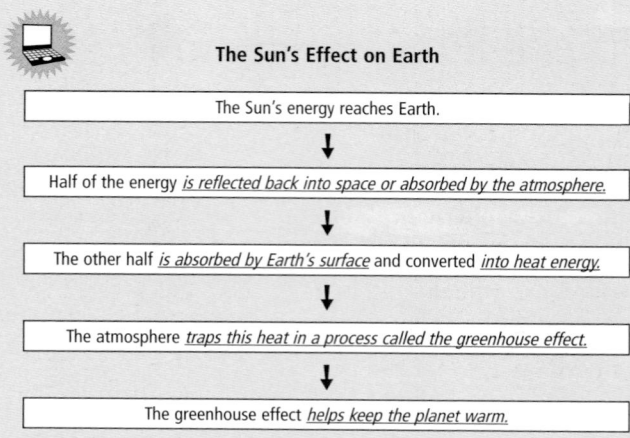

The Sun's Effect on Earth

The Sun's energy reaches Earth.

↓

Half of the energy *is reflected back into space or absorbed by the atmosphere.*

↓

The other half *is absorbed by Earth's surface* and converted *into heat energy.*

↓

The atmosphere *traps this heat in a process called the greenhouse effect.*

↓

The greenhouse effect *helps keep the planet warm.*

Across the Curriculum

► Science ◄

Global Warming on Venus
Not all scientists agree about the effects of global warming, partly because they do not know if recent increases in global temperatures are permanent or temporary. However, they can look to Venus for more information.

The surface temperature of Venus is 890°F (475°C)—too hot to sustain life as we know it. Because carbon dioxide accounts for 96 percent of the atmosphere on Venus, astronomers believe that the thick CO_2 blanket surrounding the planet traps all heat in the atmosphere. Earth's atmosphere contains only .03 percent carbon dioxide, but the level is rising. However, the connection between high CO_2 levels on Venus and high temperatures might explain why Earth's temperatures have risen 1.1°F (.6°C) during the past century.

DISCUSSION: Have the class predict what would happen if Earth's overall temperature increased a few degrees. *(Possible answers: Icebergs would melt, seas would rise, animals might relocate, many plants would die, and so on.)*

Our Amazing Planet

During the Little Ice Age of the 1500s to 1800s, the average global temperature dropped about 1°C. Ice sheets advanced over Greenland farms, and the Baltic Sea and Thames River froze more often than they do now.

---

could enhance the greenhouse effect. These scientists urge us to reduce the production of carbon dioxide and other so-called greenhouse gases.

However, some people believe that global warming is part of a natural cycle. They believe the process may not be related entirely to human activities. This difference of opinion can affect public policies on the local, state, national, and international level. For example, citizens may convince a city council to ban gasoline-powered leaf blowers within city limits.

✓ **READING CHECK:** **Physical Systems** How does Earth's atmosphere help keep the planet warm? through the greenhouse effect, slowing the loss of heat into space

Atmospheric Pressure

Earth distributes the Sun's heat through a number of processes. These processes affect climate. Some of these effects occur in the atmosphere. Although you do not notice it, air in our atmosphere has weight. The air around you, extending to the top of the atmosphere, is always pushing on you. This force is called atmospheric pressure or air pressure. If you were climbing a mountain, as you moved higher there would be less air above you pushing down. The force pushing against you—the air pressure—would drop. At very high altitudes the air is too thin for humans to breathe. This is why high-flying aircraft are sealed and pressurized.

Air pressure also changes at different places on Earth's surface. Earth's unequal heating causes most of these changes. For example, when air is heated it expands, becomes less dense, and rises. This creates a low-pressure area. The rising air usually cools as it moves higher in the atmosphere. As the air rises and cools, the water vapor it carries may form clouds. These clouds may bring rain or even storms. For this reason, low pressure usually accompanies unstable weather conditions. All centers of low pressure are called **cyclones**. They can vary in intensity. Some might take the form of slight breezes and cloud cover. Others might become powerful storms with heavy rain and high winds.

These climbers use oxygen tanks as they scale Mount Everest in Nepal. Atmospheric pressure at the top of Mount Everest—at 29,035 feet, or 8,850 meters—is about one third what it is at sea level. Since the atmospheric pressure is so low at this altitude, about two-thirds less oxygen is available in each breath. As a result, climbers must force supplemental oxygen into their lungs.

LEVEL 1: Organize the class into groups. Have each group create a storyboard explaining the process by which atmospheric pressure distributes energy around the globe. Remind students that a storyboard contains one picture per key event with a caption describing the action. Students may want to create characters to personify concepts such as cyclones. Group roles may include researcher, writer, illustrator, and presenter. When groups have completed their work, call on a volunteer from each group to present its storyboard to the class. **ENGLISH LANGUAGE LEARNERS, COOPERATIVE LEARNING**

LEVELS 2 AND 3: Have students work in pairs to create a chart that depicts how atmospheric pressure distributes energy around the globe. Students' charts should have two columns—one labeled *cause* and the other labeled *effect*. The charts should have two rows as well—one labeled *low-pressure areas* and the other labeled *high-pressure areas*. When pairs have completed their charts, have them trade charts with another pair and verify the information. (*Possible answers: low/cause—Air is warmed, expands, becomes lighter, and rises. low/effect—Unstable air results in clouds and storms. high/cause—Air is cold, dense, and heavy. high/effect—Stable air results in calm, clear weather.*) **COOPERATIVE LEARNING**

On the other hand, cold air is dense and sinks toward Earth's surface. This process creates high-pressure areas. As air sinks, it heats and dries. Centers of high pressure usually bring stable, clear, and dry weather. However, they can also bring extreme heat in summer or bitter cold in winter.

On a global scale there are four major air pressure zones. They are the equatorial low, the subtropical highs, the subpolar lows, and the polar highs. Together they carry air back and forth between the equator and the Poles and between Earth's atmosphere and its surface. How does this work? Along the equator, the direct rays of the Sun cause warm air to rise, forming the equatorial low. This rising air cools in the upper atmosphere and flows toward the Poles. At about 30° latitude the cooled air begins to sink to the surface. This sinking causes the subtropical highs in each hemisphere. At the Poles, dense cold air sinks to the surface, causing the polar highs. The cold air then flows along the surface away from the Poles. At about 60° latitude the cold air forces warmer air flowing toward the Poles higher. This rising air forms the subpolar lows. (See the diagram.)

READING CHECK: *Physical Systems* What kind of weather is usually associated with an area of low pressure? unstable weather, anywhere from slight breezes and clouds to storms with high wind and heavy rain

Global Wind Belts

Air pressure affects global wind patterns. Wind is the horizontal flow of air. Wind always flows from high to low pressure areas. For example, when air is released from a tire—an area of high pressure—it flows outward. Air will not flow into the tire unless a high-pressure hose pumps it in.

Pressure and Wind Systems

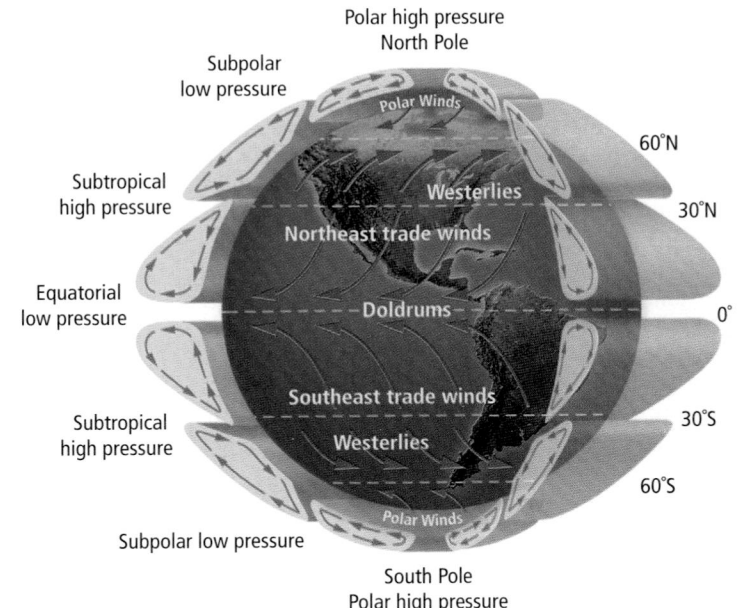

Polar high pressure
North Pole
Subpolar low pressure
Polar Winds
60°N
Subtropical high pressure
Westerlies
30°N
Northeast trade winds
Equatorial low pressure
Doldrums
0°
Southeast trade winds
Subtropical high pressure
30°S
Westerlies
60°S
Polar Winds
Subpolar low pressure
South Pole
Polar high pressure

INTERPRETING THE DIAGRAM

Earth's rotation deflects the trade winds. Otherwise, they would flow more directly north or south. **How do you think the trade winds affected exploration by European sailors?**

DIAGRAM ANSWER

aided exploration, by blowing from Europe in a southwesterly direction toward Central America

Teach Objective 3

🌐 **ALL LEVELS:** Review the ways that global winds affect weather and climate. *(Winds travel from high to low pressure areas, moving heat and cold across Earth's surface.)* Ask students to locate on a world map the following places: Nuuk (Godthåb), Greenland; New York, New York; the Galápagos Islands; Tasmania; and the outer edges of Antarctica. Have students identify the types of winds likely to be found in each of these areas *(Greenland and Antarctica—cold polar easterlies; New York City and Tasmania—westerlies; Galápagos Islands—no prevailing winds)*. Call on volunteers to share their information with the class.

Teach Objective 4

🌐 **ALL LEVELS:** Using a world map have students select one coastal location and one inland location. Then have students predict the climate in both areas and explain how the ocean and currents contribute to the climates. *(Possible answers: Coastal locations will have less variance in weather as oceans heat and cool less quickly than does land. Inland locations will have greater variance because land heats and cools quickly.)* Ask students to add a sentence explaining how ocean currents affect the relationship between tropical and polar regions. *(Possible answers: Ocean currents are responsible for moving heat between tropical and polar regions.)*

Linking Past to Present

The Jet Stream in War and Peace During the last months of World War II, Japan used the jet stream that flows across the Pacific Ocean to North America to deliver bombs. The bombs were carried by hot-air balloons. Their unique delivery system gave the weapons their name—*Fu-go*, which means "windship weapon." Some of these bombs were designed to light devastating forest fires. Although they did not start fires, hundreds of bombs landed in the United States—some as far east as Michigan.

Today the jet stream serves a peaceful purpose. Because it flows west to east, aircraft flying east "hitch a ride" on the jet stream, which saves time and fuel. In contrast, westbound aircraft avoid the jet stream. Eastbound flights that can make use of the jet stream are usually faster than westbound flights on the same routes.

ACTIVITY: Challenge students to calculate the difference that the jet stream makes in aircraft flight times. They might identify parallel flights and compare the duration of those that make use of the jet stream with those that cannot.

A warm front heading northwest from the Gulf of Mexico brings rain to the plains of southern New Mexico. Cold fronts can lower the temperature rapidly—as much as 15°F within an hour.

Winds move heat and cold across the Earth's surface. This movement helps maintain a global energy balance. Some areas of the world have winds that blow from the same direction most of the time. These winds are known as **prevailing winds**. For example, the trade winds blow from the northeast and southeast toward the equator. They flow from the subtropical highs toward the equatorial low. These winds are so named because trading ships once used them to sail across the ocean.

Not all areas of the world lie in prevailing wind belts. The zone along the equator is calm, with no prevailing winds. This area is sometimes called the **doldrums**. Because the area has little wind, sailing ships could be caught there for long periods of time.

In the middle latitudes the prevailing winds are called westerlies. These winds flow generally from the west, from the subtropical highs toward the subpolar lows. Most of the contiguous United States is located in the westerlies. These winds carry most weather patterns and storms across the United States from west to east.

In the high latitudes the winds are more variable but come mainly from the east. These areas are subject to the cold polar easterlies. These strong winds blow from Arctic and Antarctic areas into the middle latitudes. In the United States, only Alaska is far enough north to be within this polar wind belt.

A **front** occurs when two air masses of widely different temperatures or moisture levels meet. Precipitation often occurs along these fronts. The warm westerlies meet the cold polar winds between about 40° and 60° latitude. This shifting zone where the cold and warm air masses meet is called the polar front.

There are also prevailing winds in the upper atmosphere, miles above the ground. The fastest of these high-speed westerly winds are called the jet streams. Wind speeds within jet streams can reach more than 300 miles (480 km) per hour. Usually there are two or three jet streams flowing in each hemisphere. Although we do not feel them directly, the jet streams move heat and steer major weather patterns.

✓ **READING CHECK:** *Physical Systems* How is wind direction related to differences of air pressure? Wind flows from areas of higher air pressure to lower air pressure.

Oceans and Currents

Oceans also affect climate. Water heats and cools more slowly than land. Thus, land areas near oceans do not have such great temperature ranges as areas in the interior of continents. For example, Kansas City, Missouri, and San Francisco, California, are both located near the same latitude. However, Kansas City's winters are much colder and summers much warmer than those in San Francisco. This is because San Francisco lies on the Pacific coast. Kansas City is much farther from the ocean's moderating effects on temperatures.

Great rivers of seawater, called currents, are also important to climate. (See the map of world climate regions in Section 3.) Earth's winds and rotation as well as varying ocean temperatures create these ocean currents. The currents generally flow in circular paths. They move clockwise in the Northern Hemisphere and counterclockwise in the Southern Hemisphere.

Ocean currents move heat back and forth between the tropics and the polar regions. This movement helps maintain Earth's energy balance. Warm currents carry heated water from the tropics toward the cooler middle latitudes. The northward-flowing **Gulf Stream**, along the U.S. East Coast, is a good example of a warm current. Cool currents return cooled water from the middle and high latitudes toward the equator where it becomes warmer again. The southward-flowing California Current off the West Coast is a cool current. Cold ocean currents cool nearby land areas. Warm ocean currents make nearby land areas warmer. For example, consider the British Isles, which lie in high latitudes. You might expect to find cold climates there. However, a warm ocean current moderates the islands' climate.

✓ **READING CHECK:** *Physical Systems* What are the main forces that create the ocean currents? wind, Earth's rotation, varying ocean temperatures

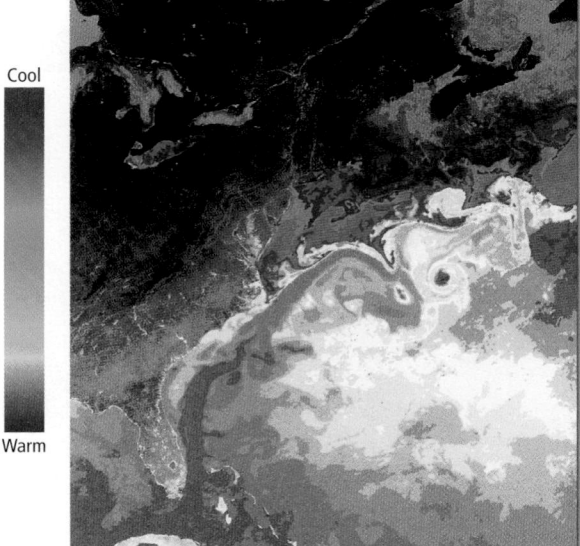

Cool

Warm

This infrared satellite image shows the Gulf Stream moving warm water from lower latitudes to higher latitudes. The dark red shape alongside Florida's east coast is the Gulf Stream.

Section **1** Review

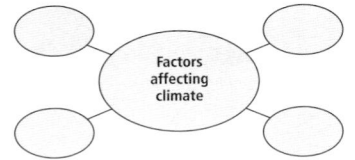

go.hrw.com **Homework Practice Online**
Keyword: SW3 HP3

Identify Gulf Stream

Define weather, climate, temperature, greenhouse effect, global warming, cyclones, prevailing winds, doldrums, front

Reading for the Main Idea

1. *Physical Systems* How do air pressure, global wind belts, and ocean currents affect Earth's energy balance?

2. *Physical Systems* How is air pressure affected when an area of Earth is heated?

3. *Places and Regions* Why are temperature ranges greater for places in the interior of continents than they are for places near oceans?

Critical Thinking

4. **Comparing** How do views on the causes of global warming differ among scientists and other people? Why is the issue important?

Organizing What You Know

5. Copy the word web shown below. Use it to identify major factors affecting climate discussed in this section.

Factors affecting climate

Weather and Climate • 45

LET'S GET STARTED

Copy the following question onto the chalkboard: *Have you ever been affected by a violent storm? If so, what kind of storm was it, and how did it affect you?* Discuss students' responses. Then ask students if they know what causes such storms. *(Using knowledge from the previous section, students may mention the Sun, winds, oceans, and rapid weather shifts over land.)* Tell students they will learn more about the causes of storms in Section 2.

Building Vocabulary

Write **evaporation**, **condensation**, and **humidity** on the chalkboard. Ask students to scan the section to find out how these words relate to one another. *(They are components in a cause-effect relationship.)* Explain that evaporation leads to water vapor in the air—humidity—which when cooled becomes condensation. Point out that the definitions for *evaporation* and *condensation* apply to liquids besides water. Have students read the definitions for the rest of the words on the list.

Section 2 Weather Factors

READ TO DISCOVER

1. What are the common forms of precipitation, and how are they formed?
2. How do mountains and elevation affect weather and climate?
3. What are the different types of storms, and how do they form?

WHY IT MATTERS

A long period of unusually low rainfall is called a drought. Use **CNNfyi.com** or other **current events** sources to learn about the effects of drought on human activities in the United States or other countries.

DEFINE

evaporation
humidity
condensation
orographic effect
rain shadow
tornadoes
hurricanes
typhoons

Precipitation

Water vapor plays an important role in many atmospheric processes. Without it, there would be no clouds, rain, or storms. The process by which water changes from a liquid to a gas is **evaporation**. Most water vapor that becomes rain is evaporated from the oceans. Some also evaporates from lakes, rivers, soils, and vegetation.

The amount of water vapor in the air is called **humidity**. The higher the temperature, the more water vapor the air can hold. When air cools, it will reach a temperature at which it cannot hold any more water vapor. At this point, **condensation** occurs. Condensation is the process by which water vapor changes from a gas into liquid droplets. Often you see condensation as clouds, dew, fog, or frost. If the condensed water droplets become large enough, they will fall as precipitation. There are four common forms of precipitation: rain, snow, sleet, and hail. Rain is, of course, a liquid. Snow is made up of generally six-sided ice crystals formed in the clouds. Sleet is rain that freezes as it falls. Hailstones are chunks of ice that form in storm clouds.

Precipitation is not evenly distributed around the world. It is generally highest in the persistent low-pressure zones. These zones are the equatorial low and the subpolar lows in the middle latitudes. Precipitation is generally lowest in the high-pressure zones. Those zones are the subtropical highs and the polar highs. Of course, precipitation also varies from season to season and from year to year in any given place.

High humidity contributes to this Virginia gardener's discomfort. Perspiration evaporates very slowly when the air has a high water vapor content.

✓ **READING CHECK:** *Physical Systems* What are the four common forms of precipitation? rain, snow, sleet, and hail

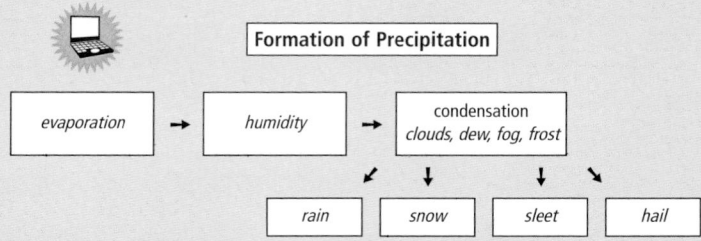

Elevation and Mountain Effects

High elevation affects weather and climate. An increase in elevation—height on Earth's surface above sea level—causes a drop in temperature. The temperature drops 3.5°F per 1,000 feet of elevation (1°C/100 m). Thus, it may be warm at the base of a mountain while the summit is covered with snow or glaciers.

Mountains influence climates through the **orographic effect**. This effect occurs when moist air pushes against a mountain. The barrier forces the air to rise. The rising air cools and condenses, forming clouds and causing precipitation. As a result, the side of the mountain facing the wind receives a great deal of moisture. This side is known as the windward side. The side of the mountain facing away from the wind is the leeward side. As the air moves down the leeward side, it warms and dries. This drier area is called a **rain shadow**. Deserts are often located in rain shadows.

INTERPRETING THE VISUAL RECORD

Pictured on the left are mountains in the western part of the Sierra Nevada, in Sequoia National Park. At the right is part of the eastern slope of the same range, near Bishop, California. **How does the vegetation of these places differ? What accounts for the difference?**

✓ **READING CHECK:** *Physical Systems* How do high elevations and mountains influence weather and climate? increasing elevation—causes drop in temperature; orographic effect—can create different levels of precipitation on the windward and leeward sides of mountains

Skill-Building Activity

Reading a Weather Map

You have probably seen weather maps like this one on television or in the newspaper. Weather maps show atmospheric conditions either as they currently exist or as they are predicted to be in the future. Most weather maps have legends that explain what the colors and symbols on the map mean. The map on this page shows two cold fronts sweeping through the United States. Low-pressure systems are at the center of storms bringing rain and snow to the Northwest and upper Midwest. Notice that temperatures behind the cold front are significantly cooler than temperatures ahead of the front.

Analyzing Maps What forms of precipitation does this map show?

Reading a Weather Map

(H) High pressure	Sunny	Thunderstorms
(L) Low pressure	Partly cloudy	Rain
Cold front	Cloudy	Snow
Warm front	Showers	Ice

Temperatures

| 20s | 30s | 40s | 50s | 60s | 70s | 80s | 90s |

ALL LEVELS: Have students draw simple illustrations that depict elevation and its relationship to the orographic effect. *(Illustrations should show that when moist air flowing from the ocean meets a mountain barrier, the air rises, cools, condenses, and produces precipitation.)* Ask students to include labels to explain how weather and climate are affected in the different areas of the illustration by both elevation and the orographic effect. **ENGLISH LANGUAGE LEARNERS**

ALL LEVELS: Have each student select a storm type and write several words or phrases to describe the storm type. *(Possible answers: tornado—twisting, spiral, destructive; hurricanes—westward, rotating, tropical; typhoons—Pacific hurricanes, sweep away beaches)* Then have students read their lists to the class and have the other students guess which storm type is being described. Once a storm has been properly identified, ask the describer to provide more detail about it.

Essential Element 1

► The World ◄ in Spatial Terms

Mapping Hurricanes
Meteorologists use computer programs to predict the paths of hurricanes. The National Hurricane Center in Miami, Florida, issues hurricane forecasts based on more than 15 different programs, or models. Often, these models accurately predict a hurricane's path.

However, prediction software can be wrong. In 2000, 15 sophisticated models predicted that Hurricane Debby would strike southern Florida. Some 15,000 people were evacuated and warned to prepare for the worst, but Debby died at sea without reaching land.

ACTIVITY: If the class is studying this material during hurricane season, have students download tracking charts from the Internet and map a storm's path. At other times of the year, have students conduct research on hurricane tracking and map the predicted and actual paths of specific storms.

VISUAL RECORD ANSWER

Possible answer: It could wash dirt from under the pavement, creating a larger hole.

Our Amazing Planet

In the United States, lightning strikes the ground an estimated 30 million times a year. In an average year lightning kills more Americans than do tornadoes or hurricanes.

INTERPRETING THE VISUAL RECORD

Hondurans look at a bridge destroyed during Hurricane Mitch in 1998. Hurricane Mitch has been labeled the deadliest Atlantic storm in two centuries. It caused massive damage throughout much of Central America and killed more than 9,000 people. **How might continued rainfall affect the place in the photo?**

Storms

Storms are sudden and violent weather events. They can cause high winds, flooding, blowing snow, lightning, and turbulent seas. As a result, they can be extremely dangerous to human life and property. They also cause major problems for transportation.

Some very low-pressure cyclones that have rising unstable air become very large storms. They usually move from west to east in the middle latitudes, pushed by the jet stream flow. They are known as middle-latitude storms or extratropical cyclones. They can be huge, up to 1,000 miles (1,600 km) in diameter. These storms can travel across an entire continent or ocean. Most middle-latitude storms occur along a polar front. They form when cold dry polar air mixes with moist warm air from the tropics.

Middle-latitude storms may produce thunderstorms and **tornadoes**. These twisting spirals of air affect fairly small areas, but they can destroy almost anything in their path. The United States experiences more tornadoes than any other country.

Storms in the tropics differ from middle-latitude storms. Tropical cyclones are usually much smaller. Because there is no cold air present, they lack fronts. Also, they mainly travel westward, pushed by the trade winds. **Hurricanes** are the most powerful and destructive tropical cyclones. These rotating storms can bring heavy rain and winds higher than 155 miles per hour (249 kph). They begin over warm tropical seas. Those that strike the United States usually form in the tropical Atlantic Ocean. They become stronger as they move westward.

Hurricanes, which are called **typhoons** in the western Pacific Ocean, can create dangerously high waves. They may also produce thunderstorms and tornadoes. High seas brought by the storms can erode and destroy beaches and coastal areas. Sometimes large stretches of beach are swept away. Flooding can threaten people, property, and plant and animal life. Fortunately, hurricanes weaken as they move inland.

✓ **READING CHECK:** *Physical Systems* What are examples of very violent middle-latitude and tropical storms? tornadoes; hurricanes

Visible light image

Radar image

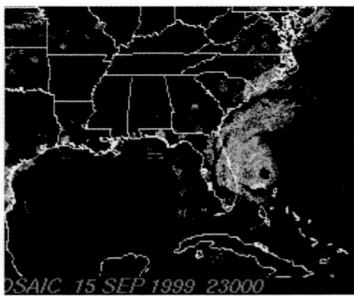

Infrared image

Hurricane Floyd, pictured above in three types of images, battered the eastern United States in mid-September 1999.

 FOCUS ON SCIENCE, TECHNOLOGY, AND SOCIETY

Weather Satellites At one time, hurricanes could strike coastal areas with little warning. For example, in 1900 a hurricane caught residents of Galveston, Texas, unprepared. The storm killed thousands of people. Today satellites orbiting Earth help weather forecasters track the development of storms and save lives.

Weather satellites carry instruments that collect information about the Earth's surface and atmosphere. For example, they carry special cameras that make use of both visible light and invisible infrared light. Infrared sensors detect heat given off by clouds, land, and water. Radar can also be used to gather atmospheric data. With the information from these instruments, scientists can predict the path of a hurricane. They can also identify areas that may flood. People who live in affected areas can then leave before conditions worsen.

As technology improves, scientists can gather data on storms earlier in their development. Two satellites now use microwaves to peer through heavy clouds. They gather information on young hurricanes while the storms are still far out at sea. One of the satellites measures the sea's roughness. The other measures the rainfall in a given area.

For centuries, people looked to the skies for information about the weather. Now, scientists can use satellites to look down on Earth from above.

✓ **READING CHECK:** *Physical Systems* How do weather satellites help people live safely in places affected by hurricanes? allow scientists and weather forecasters to track the development of hurricanes, other storms, and flooding patterns and warn residents of the threats

Homework Practice Online
Keyword: SW3 HP3

Section 2 Review

Define
evaporation
humidity
condensation
orographic effect
rain shadow
tornadoes
hurricanes
typhoons

Reading for the Main Idea

1. *Physical Systems* How is temperature change related to condensation and precipitation?

2. *Physical Systems* What determines the direction of storm movement?

Critical Thinking

3. **Contrasting** How do storms in the tropics differ from storms in the middle latitudes?

4. **Analyzing Information** In what places near the equator might you expect to see snow and ice all year? Why?

Organizing What You Know

5. Using the weather map in the chapter as a guide, create a sketch map of the United States showing imaginary weather events. Use the standard weather map symbols. Then write a short description of the weather patterns you have displayed.

OBJECTIVES

1. **Explain how the two tropical climates differ.**

2. **Identify conditions common in dry climates.**

3. **Tell what climates are found in the middle latitudes.**

4. **Detail the characteristics of high-latitude and highland climates.**

 **LET'S GET STARTED**

Copy the following question onto the chalkboard: *If you could visit any place on Earth, where would it be?* Allow students to write down their answers. When they are finished, have them share their answers with the class. Then ask students what type of weather they would expect to find in each place, and ask how they could find out. (*Possible answers: an encyclopedia, the Internet, the National Weather Service.*) Tell students they will learn more about Earth's climates in Section 3.

Building Vocabulary

Write the key terms on the chalkboard. Call on volunteers to find and read aloud the definitions in the text. Point out that **monsoon** actually has two meanings. Originally the word referred to the wind systems that bring wet and dry seasons to tropical areas. Its more common usage refers to heavy rains brought in by the winds during summer.

Section 3 RESOURCES

REPRODUCIBLE
▶ Guided Reading Strategy 3.3
▶ Graphic Organizer Activity 3
▶ Readings in World Geography, History, and Culture 6 and 7

TECHNOLOGY
▶ One-Stop Planner CD–ROM, Lesson 3.3
▶ Geography and Cultures Visual Resources 5
▶ Homework Practice Online
▶ HRW Go site

REINFORCEMENT, REVIEW, AND ASSESSMENT
▶ Main Idea Activity 3.3
▶ English Audio Summary 3.3
▶ Spanish Audio Summary 3.3
▶ Section 3 Review, p. 57
▶ Daily Quiz 3.3

Section 3 Climate and Vegetation Patterns

READ TO DISCOVER

1. How do the two tropical climates differ?
2. What conditions are common in dry climates?
3. What climates are found in the middle latitudes?
4. What characterizes high-latitude and highland climates?

WHY IT MATTERS

Scientists worry that climate changes could cause terrible problems in some parts of the world. Use **CNNfyi.com** or other **current events** sources to learn more about expected consequences of changes in global climate patterns.

DEFINE

ecosystems
monsoon
savannas
arid
deciduous forests
coniferous forests
permafrost

Tropical Climates

All of the world's climates support a variety of **ecosystems**. An ecosystem is the community of plants and animals in an area. It also includes the nonliving parts of their environment, such as soil and climate. (See Geography for Life: World Ecosystems and Biomes in the next chapter.)

Among the world's most diverse ecosystems are those found in tropical climate regions. We will begin our discussion of world climate and vegetation patterns there. You can follow our discussion by studying the map and chart of world climate regions.

Tropical Humid Climate The tropical humid climate region is found in areas close to the equator. Those regions generally have warm temperatures and plentiful rainfall all year. They never have truly cold weather. Because the equator receives the Sun's direct rays all year, the area is always warm. As a result, warm air is always rising in the tropics. This continuous rising of warm unstable air brings almost daily thunderstorms and heavy rainfall.

Western Java, in Indonesia, has a tropical humid climate. This scene is from the Bogor Botanical Gardens. Some tropical plants, like the one growing on the tree trunk, need no more moisture than what is in the warm humid air.

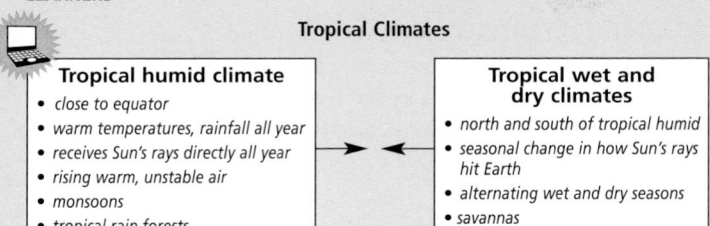

Teach Objective 1

LEVEL 1: Copy the following graphic organizer onto the chalkboard, omitting the italicized answers. Have students complete it with the characteristics of tropical humid and tropical wet and dry climates. **ENGLISH LANGUAGE LEARNERS**

LEVELS 2 AND 3: Tell students to imagine that they have just moved from a tropical humid climate to a tropical wet and dry climate. Have students use climate maps in the textbook's unit atlases to choose two specific places in these climate regions. Then have students write a diary entry in which they describe how life is different in the new climate.

Tropical Climates

Tropical humid climate
- *close to equator*
- *warm temperatures, rainfall all year*
- *receives Sun's rays directly all year*
- *rising warm, unstable air*
- *monsoons*
- *tropical rain forests*

→ ←

Tropical wet and dry climates
- *north and south of tropical humid*
- *seasonal change in how Sun's rays hit Earth*
- *alternating wet and dry seasons*
- *savannas*

Climate and Vegetation

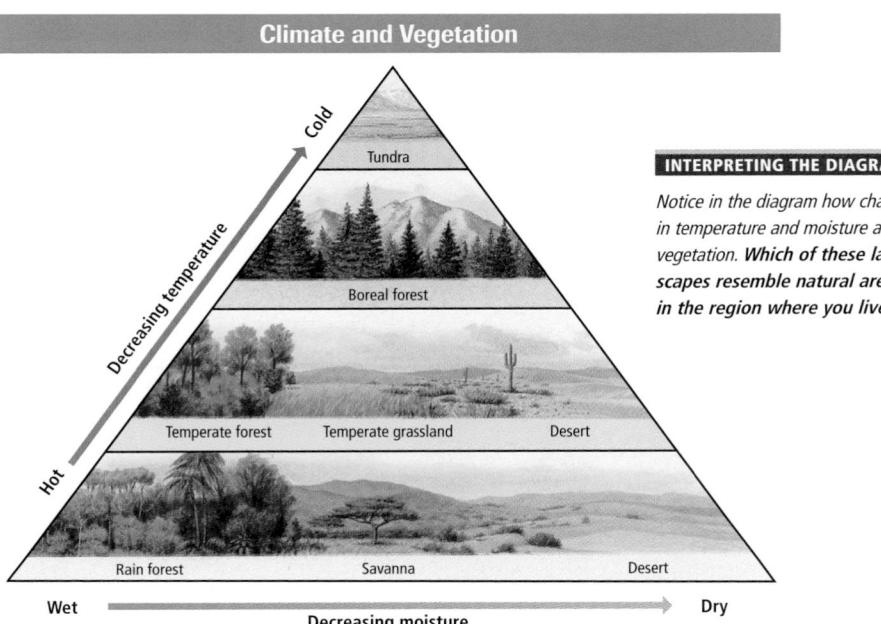

INTERPRETING THE DIAGRAM

Notice in the diagram how changes in temperature and moisture affect vegetation. **Which of these landscapes resemble natural areas in the region where you live?**

In some tropical areas, rainfall is concentrated in one wet season. India and Southeast Asia have seasons of this type. During the summer months moist air flows into these areas from the warm ocean, a high-pressure area. The air flows to the hotter land, a low-pressure area. This air flow brings heavy rains to areas along the coast and inland. During the winter, dry air flows off the cooling continent (high-pressure areas). It flows toward the warm oceans (low-pressure areas). The air flow causes dry conditions on the continent. This wind system, in which winds completely reverse direction and cause seasons of wet and dry weather, is called the **monsoon**.

Warm temperatures and heavy rainfall create ideal conditions for plant growth. Thus, dense tropical rain forests thrive in tropical humid areas. These forests are the most complex land ecosystems in the world. Thousands of kinds of plants and animals live there.

Tropical Wet and Dry Climate Just to the north and south of the tropical humid climate is the tropical wet and dry climate. It is sometimes called the tropical savanna climate. This climate results from the seasonal change in the way the Sun's rays strike areas just north and south of the equator. During summer in these areas, the Sun's rays strike most directly. As a result, temperatures rise, creating low pressure and unstable, rising air. This, in turn, leads to heavy rainfall. During the winter the Sun's direct rays move to the opposite hemisphere. High pressure replaces low pressure. The high pressure system brings stable, cool, sinking air and a dry season. This alternating pattern of wet and dry seasons supports **savannas**. Savannas are areas of tropical grasslands, scattered trees, and shrubs.

✓ **READING CHECK:** *Physical Systems* How does the tropical wet and dry climate differ from the tropical humid climate? tropical wet and dry—distinct wet and dry seasons; tropical humid climate—wet throughout the year

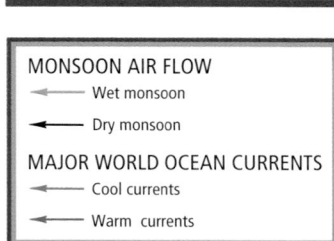 **LEVEL 1:** Have students find illustrations in this textbook that seem to show scenes of arid or semiarid climate regions. When a student locates such a photo, he or she should note the page on which it appears and cover the caption with a piece of paper. Then have the student describe orally how the scene indicates an arid or semiarid climate. Students should distinguish between the two climate types. (*Example: The Baja California photo in this section clearly shows an arid climate. Vegetation is sparse and appears to be adapted to the environment. The ground is bare and rocky.*) Call on volunteers to describe pictures on certain pages while other students examine the photos also. Discuss any features of arid and semiarid climates not apparent in photos. **ENGLISH LANGUAGE LEARNERS**

LEVELS 2 AND 3: Have students use the Level 1 activity as a basis for this one. Ask them to read the captions for the photos they have identified, locate the places on a world map, and connect information in the World's Climate Regions map and chart with the photos. Then have them use all these sources to write brief descriptions of the places they have identified.

Essential Element 5

► Environment ◄ and Society

The Olive Tree Some plants are closely associated with certain climate regions. The olive tree, for example, is common in areas with a Mediterranean climate. It is important to the culture, history, and economy of the area surrounding the Mediterranean Sea. Not only are olives and olive oil valuable food products, but the wood is strong and the olive branch is a traditional symbol of peace.

The first wild olive trees grew in what is now Turkey. Ancient societies of Southwest Asia cultivated olives as early as 3000 B.C. The Romans brought the tree to places they conquered, resulting in olives being grown throughout the Mediterranean region. During the 1700s, Spanish missionaries planted olive groves in California.

ACTIVITY: Assign an agricultural product to each student. Have each student research the origin of that crop and the path of its diffusion. Then have students use the World's Climate Regions map in this chapter to predict other places where these crops might grow well.

The World's Climate Regions

MONSOON AIR FLOW
- Wet monsoon
- Dry monsoon

MAJOR WORLD OCEAN CURRENTS
- Cool currents
- Warm currents

SCALE
0 ... 1,500 ... 3,000 Miles
0 ... 1,500 ... 3,000 Kilometers
Scale is accurate only along the equator.
Projection: Robinson

	Climate	Geographic Distribution	Major Weather Patterns	Vegetation
Tropical	TROPICAL HUMID	along equator; particularly equatorial South America, Congo Basin in Africa, Southeast Asia	warm and rainy year-round, with rain totaling anywhere from 65 to more than 450 in. (165–1,143 cm) annually; typical temperatures are 90°–95°F (32°–35°C) during the day and 65°–70°F (18°–21°C) at night	tropical rain forest
	TROPICAL WET AND DRY	between humid tropics and deserts; tropical regions of Africa, South and Central America, South and Southeast Asia, Australia	warm all year; distinct rainy and dry seasons; precipitation during the summer of at least 20 in. (51 cm); monsoon influences in some areas, such as South and Southeast Asia; summer temperatures average 90°F (32°C) during the day and 70°F (21°C) at night; typical winter temperatures are 75°–80°F (24°–27°C) during the day and 55°–60°F (13°–16°C) at night	tropical grassland with scattered trees
Dry	ARID	centered along 30° latitude; some middle-latitude deserts in interior of large continents and along western coasts; particularly Saharan Africa, Southwest Asia, central and western Australia, southwestern North America	arid; precipitation of less than 10 in. (25 cm) annually; sunny and hot in the tropics and sunny with great temperature ranges in middle latitudes; typical summer temperatures for lower-latitude deserts are 110°–115°F (43°–46°C) during the day and 60°–65°F (16°–18°C) at night, while winter temperatures average 80°F (27°C) during the day and 45°F (7°C) at night; in middle latitudes the hottest month averages 70°F (21°C)	sparse drought-resistant plants; many barren, rocky, or sandy areas
	SEMIARID	generally bordering deserts and interiors of large continents; particularly northern and southern Africa, interior western North America, central and interior Asia and Australia, southern South America	semiarid; about 10–20 in. (25–51 cm) of precipitation annually; hot summers and cooler winters with wide temperature ranges similar to desert temperatures	grassland; few trees
Middle Latitudes	MEDITERRANEAN	west coasts in middle latitudes near cool ocean currents; particularly southern Europe, part of Southwest Asia, northwestern Africa, California, southwestern Australia, central Chile, southwestern South Africa	dry sunny warm summers and mild wetter winters; precipitation averages 14–35 in. (35–90 cm) annually; typical temperatures are 75°–80°F (24°–27°C) on summer days; the average winter temperature is 50°F (10°C)	scrub woodland and grassland
	HUMID SUBTROPICAL	east coasts in middle latitudes; particularly southeastern United States, eastern Asia, central southern Europe, southeastern parts of South America, South Africa, and Australia	hot humid summers and mild humid winters; precipitation year-round; coastal areas are in the paths of hurricanes and typhoons; precipitation averages 40 in. (102 cm) annually; typical temperatures are 75°–90°F (24°–32°C) in summer and 45°–50°F (7°–10°C) in winter	mixed forest

ALL LEVELS: Have students use the World's Climate Regions map and other maps in this textbook to select a particular place in each of the following regions: one from the middle-latitude climate regions, one from the high-latitude climate regions, and one from a highland climate region. Ask students to work in pairs to create idea webs that describe how the climate of each location might affect building construction, food production, leisure activities, occupations, settlement patterns, or other aspects of daily life. Call on volunteers to draw their idea webs on the chalkboard. Continue until all the related climate regions have been examined. **ENGLISH LANGUAGE LEARNERS, COOPERATIVE LEARNING**

HOMEWORK: Have students write 10 to 20 quiz questions based on material in the map and chart of world climate regions. Challenge them to incorporate monsoon air flow patterns and ocean currents, as displayed on the map, into some of their questions. *(Example: How might the cool current off the west coast of South America affect the climate along the Pacific coast of Peru and Chile? Answer: It appears to contribute to an arid climate.)*

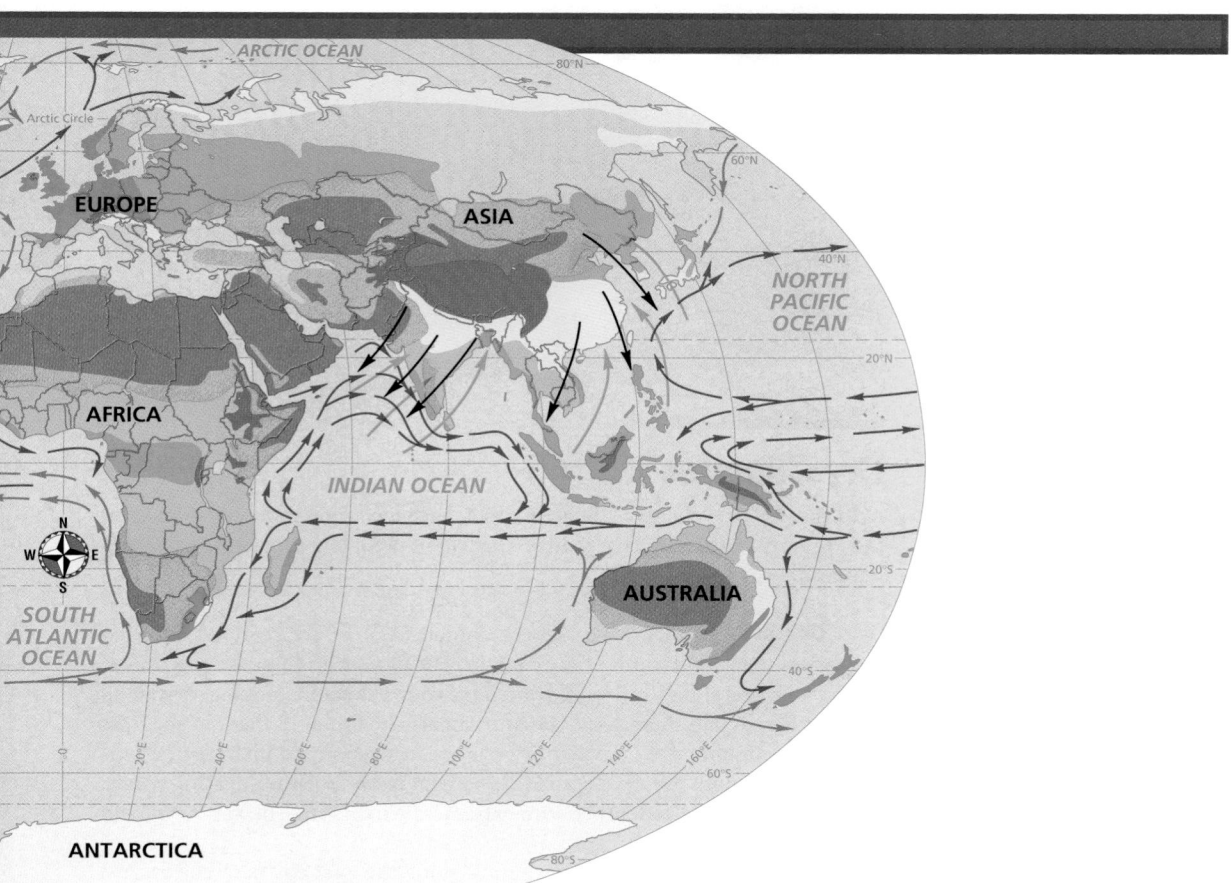

	Climate	Geographic Distribution	Major Weather Patterns	Vegetation
Middle Latitudes	**MARINE WEST COAST**	west coasts in upper-middle latitudes; particularly northwestern Europe and North America, southwestern South America, central southern South Africa, southeastern Australia, New Zealand	cloudy mild summers and cool rainy winters; strong ocean influence; precipitation averages 20–98 in. (51–250 cm) annually; westerlies bring storms, rain; average temperature in hottest month is usually between 60°F and 70°F (16°–21°C); average temperature in coolest month usually is above 32°F (0°C)	temperate evergreen forest
	HUMID CONTINENTAL	east coasts and interiors of upper-middle-latitude continents; particularly northeastern North America, northern and eastern Europe, northeastern Asia	four distinct seasons; long cold winters and short warm summers; precipitation amounts vary, usually 20–50 in. (51–127 cm) or more annually; average summer temperature is 75°F (24°C); average winter temperature is below freezing	mixed forest
High Latitudes	**SUBARCTIC**	higher latitudes of interior and east coasts of continents; particularly northern parts of North America, Europe, and Asia	extremes of temperature; long cold winters and short mild summers; low precipitation all year; precipitation averages 5–15 in. (13–38 cm) in summer; temperatures in warmest month average 60°F (16°C) but can warm to 77°F (25°C); winter temperatures average below 0°F (–18°C)	northern evergreen forest
	TUNDRA	high-latitude coasts; particularly far northern parts of North America, Europe, and Asia, Antarctic Peninsula, subantarctic islands	cold all year; very long cold winters and very short cool summers; low precipitation amounts; precipitation average is 5–15 in. (13–38 cm) annually; warmest month averages less than 50°F (10°C); coolest month averages a little below 0°F (–18°C)	moss, lichens, low shrubs; permafrost bogs in summer
	ICE CAP	polar regions; particularly Antarctica, Greenland, Arctic Basin islands	freezing cold; snow and ice year-round; precipitation averages less than 10 in. (25 cm) annually; average temperatures in warmest month do not reach higher than freezing	no vegetation
	HIGHLAND	high mountain regions, particularly western parts of North and South America, eastern parts of Asia and Africa, southern and central Europe and Asia	greatly varied temperatures and precipitation amounts over short distances as elevation changes; prevailing wind patterns can affect rainfall on windward and leeward sides of highland areas	forest to tundra vegetation, depending on elevation

Religion and Weather
Cultural and religious explanations of climate and weather have influenced people's understanding of the physical world. The Hopi's rain dance reveals that culture's attitude toward nature. By dancing and making offerings, the Hopi show respect to the kachinas, or deified ancestral spirits, who they believe control the natural world. The Hopi believe the kachinas will return the favor by sending rain. Many other cultures have ceremonies designed to affect the weather.

Drought conditions in Texas during 1999 prompted community and church leaders there to pray for rain. The mayor of one city proclaimed a day of prayer for rain. One church pastor noted that although farmers can use the most modern equipment and seed, the success of a harvest is often beyond human control.

ACTIVITY: Have students conduct research on ways that weather has been incorporated into other cultures' religious beliefs or folklore.

Teach Objectives 1–4

🌐 **ALL LEVELS:** Copy the following graphic organizer onto the chalkboard, omitting the italicized answers. Have students work in pairs to complete the organizer by filling in the correct latitudinal groupings. Ask students which climate regions are not represented on the organizer (*arid, semiarid, and highland*) and why. (*Possible answer: Although most arid and semiarid climate regions are centered at about 30 degrees north and south of the equator, they lie at other latitudes also. Highland climates can be found at any latitude.*) Then have students describe the climate regions and the types of plants that commonly live in each one. **ENGLISH LANGUAGE LEARNERS**

Climate and Latitude

high	subarctic, tundra, and ice cap
middle	*Mediterranean, humid subtropical, marine west coast, and humid continental*
low	*tropical humid and tropical wet and dry*
middle	*Mediterranean, humid subtropical, marine west coast, and humid continental*
high	*subarctic, tundra, and ice cap*

Linking Past to Present

Oases in the Sahara The Sahara is the largest tropical desert in the world, stretching from Morocco to Egypt. People have lived in the Sahara for over 7,000 years, but with high temperatures reaching above 100°F (38°C) and rainfall limited to 5 inches (13 cm) a year, life is difficult. Scattered throughout the Sahara are oases—areas where water is available for use. Oases were once used as stops along caravan routes. Since caravans averaged only 30 miles (48 km) a day, the relief provided by oases was necessary for journeys that lasted hundreds or thousands of miles.

Sea, road, and rail routes have ended reliance on caravans for transport. Also, with the nomadic lifestyle giving way to permanent settlement, oases have evolved from travel stops to irrigated communities where fruits, vegetables, and grains are grown.

A boojum tree, at left, and saguaro cactus, at right, grow in the desert of Baja California, Mexico. Both of these unusual plants are well adapted to the desert's heat and lack of moisture. The boojum drops its leaves to save water, and the saguaro stores water in its fleshy trunk. Hurricanes sometimes sweep from the Pacific Ocean across Baja California. Then desert plants burst into bloom.

☐ **internet** connect

GO TO: go.hrw.com
KEYWORD: SW3 CH3
FOR: Web sites about the Sahara

Dry Climates

Arid means "dry." All dry climate regions have low annual rainfall. However, their temperatures may vary greatly. The two types of dry climates are arid and semiarid.

Arid Climate Most arid, or desert, climate regions are centered at about 30 degrees north and south of the equator. The dryness in these areas is caused by the subtropical high-pressure zone. This zone has stable, sinking, dry air all year. Little rain falls there. Arid climates can also be found in the rain shadows of some mountain ranges. Other dry regions lie deep in the interior of continents. Such regions are dry because they are far from moisture-bearing winds. These areas experience temperature extremes. Winters may be very cold, and summers are very hot.

Small dry areas are sometimes found along the west coasts of continents. They lie where cool ocean currents create highly stable atmospheric conditions. Dry coastal deserts of this type are found along the west coasts of Australia, Mexico, South America, and southwestern Africa.

Plants and animals in these dry regions must be hardy. Most desert plants are small, and they may only sprout and flower after a rare rainfall. Soils tend to be thin and rocky.

Semiarid Climate The semiarid climate is a transition zone between the arid climate and the more humid climates. Semiarid climates receive more moisture than the deserts but less than the more humid areas. Where rainfall is heavy enough, grasses make up much of the plant life. Humans use many of these areas for growing grain. In the drier areas, plant life is more limited.

✓ **READING CHECK:** **Physical Systems** What are four factors that can create dry climates? *persistent high air pressure, a rain shadow, location in the interior of a continent, and highly stable conditions created by a cool ocean current*

Middle-Latitude Climates

Several types of climates are found in the middle latitudes. These are generally temperate climates. For the most part, they do not experience the extreme conditions found in tropical and high-latitude climates.

Mediterranean Climate   The Mediterranean climate exists mainly in two kinds of areas. One is along coastal areas of southern Europe. The other is along west coasts of continents with cool ocean currents. Mediterranean climates do not usually extend far beyond inland mountain ranges. The stable sinking air of the subtropical high-pressure zone causes long, sunny, dry summers. During the mild winter, however, cool middle-latitude storms bring rain. The natural plant life of this climate is the Mediterranean scrub woodland. It includes short trees and shrubs with a few scattered large trees.

Humid Subtropical Climate   The humid subtropical climate is much more widespread than the Mediterranean climate. It is found on the eastern side of continents where there are warm ocean currents. The moist air flowing off the warm ocean waters greatly affects this climate area. Summers are hot and humid. Winters are mild, with occasional frost and some snow. Hurricanes or typhoons can be a danger here. The natural plant life of humid subtropical and other middle-latitude regions includes temperate forests. These forests can be divided into two main types. One type is **deciduous forests**. Trees there lose their leaves during part of the year. The other type is **coniferous forests**. Trees in those forests remain green all year.

Marine West Coast Climate   The marine west coast climate is heavily influenced by oceans. This climate is generally found on the west coasts of continents in the upper middle latitudes. Temperatures in these areas are mild all year. Storms traveling across the oceans in the westerlies bring most of the rainfall. Winters are foggy, cloudy, and rainy. However, summers can be warm and sunny. This climate is widespread in northwestern Europe. Lowlands along the coast there allow cool moist ocean air to spread far inland. In some places, the rainy weather of this climate supports dense coniferous forests. These areas are called temperate rain forests.

Like many other areas in southern Europe, Greece has a Mediterranean climate. This photo was taken on the island of Corfu.

Left: A coniferous forest grows in Banff National Park in the Canadian Rocky Mountains. Right: Deciduous trees in Portland, Oregon, change color in the fall before losing their leaves.

Daily Life

San Francisco Summers
A quotation, often attributed to Mark Twain, reads "The coldest winter I ever spent was a summer in San Francisco." With year-round temperatures that average 63°F (17°C), San Francisco will never be confused with Alaska in January. Still, the Bay Area offers visitors a unique experience—thick summertime fog. San Francisco Bay moderates temperatures in the area. However, during summer mornings and afternoons, air in adjacent inland valleys warms and rises, creating a vacuum. Warm ocean air is pulled across the bay. As the wind comes into contact with the cold water, thick fog forms. Summer tourists are often caught off guard. If they have not brought warm clothing, they may shiver for the duration of their vacation.

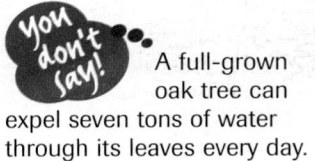   A full-grown oak tree can expel seven tons of water through its leaves every day.

HOMEWORK: Compile a list of 10 to 20 major cities that span the world's climate regions. Have students copy the list and identify the climate region within which each city is located.

Close

Point out to the class that human beings live in almost all climates of the world, including some of the most extreme. Ask students to use climate maps in the unit atlases to identify places with extreme climates. Then have them predict adaptations people have made so they could survive in these climates. (*Example: Greenland— import food, wear layers of warm clothing, live only near coast*)

INTERPRETING THE VISUAL RECORD

Tundra vegetation grows near Hudson Bay, Canada. **Why might growing low to the ground be an advantage for these plants?**

VISUAL RECORD ANSWER

Possible answers: keeps plants out of cold wind, conserves any heat absorbed by rocks during daylight hours

Humid Continental Climate The humid continental climate is found in the interiors and east coasts of upper-middle latitude continents. Invasions of both warm and cold air regularly affect these areas. The humid continental climate has the most changeable weather conditions. In fact, it experiences four distinct seasons. Because this climate type is situated along the polar front, storms bring rain throughout much of the year. Snow falls in winter. Precipitation is heavy enough to support forests.

✓ **READING CHECK:** (*Physical Systems*) In areas with a Mediterranean climate, which season brings most of the rainfall? winter

High-Latitude and Highland Climates

There are three main types of high-latitude climates. They are subarctic, tundra, and ice cap climates.

Subarctic Climate The subarctic climate is located generally above 50° north latitude. However, a warm ocean current moderates the climate in areas above this latitude in northern Europe. The subarctic climate has long cold winters. Temperatures stay well below freezing for half of the year. The short summers can be warm, however. This climate also has the greatest annual temperature ranges in the world. Although severe, the climate supports vast evergreen forests. These northern forests are also called boreal forests.

The subarctic climate region is very large. It stretches across far northern North America, Europe, and Asia. In the Southern Hemisphere there is practically no land at these latitudes.

Tundra Climate Coastal areas in high latitudes have a tundra climate. These areas also have long winters. Temperatures are above freezing only during the short summers. In some places, water and soil below the ground's surface remain frozen throughout the year. The permanently frozen soil is called **permafrost**. During summer the permafrost makes it difficult for water from melting snow to seep into the ground. As a result, swamps and bogs form on the surface.

The tundra climate takes its name from the only kinds of vegetation that can survive there. Tundra vegetation is made up of lichens, mosses, herbs, and low shrubs. Trees cannot grow there. However, during the short summer the area bursts into flower.

Review and Assess

Have students complete the **Section Review**. Then have students complete **Daily Quiz 3.3**.

Reteach

Have students complete **Main Idea Activity for English Language Learners and Special-Needs Students 3.3**. Then organize the class into four groups. Assign each group one of the following climate categories: tropical, dry, middle latitude, or high latitude and highland. Have each group draw pictures illustrating the characteristics of the climates and vegetation patterns found in the assigned general region. **ENGLISH LANGUAGE LEARNERS, COOPERATIVE LEARNING**

Extend

Have interested students conduct research on unusual plants that have developed unique strategies for living in various climates. One example is the strangler fig, which grows in rain forests of tropical humid climate regions. In the rain forest trees compete for sunlight. The strangler fig sends roots down to the ground from a seed caught in another tree. Eventually, the strangler fig's foliage shades that of the host tree, a lattice of its roots may completely surround the older tree's trunk, and the host tree dies. **BLOCK SCHEDULING**

Guanaco feed on the plains of Torres del Paine National Park, Chile. Behind them rise the Andes. This long South American mountain range contains a variety of climates.

Ice Cap Climate   Ice cap climates are found in Earth's polar regions. Those areas are always covered by huge flat masses of ice and snow. Most parts of Antarctica and Greenland are covered by ice caps. Few land plants and animals can survive in these climates. Hardy plant life may grow on exposed rocks. In contrast, the cold seas of the ice cap regions have rich marine ecosystems. Many birds, fish, and marine mammals live there. For example, seals and some whales feed on the fish of the Arctic Ocean. Polar bears feed on the seals.

Highland Climate   Highland areas can have varying climates. This is partly because temperatures change with elevation. In addition, prevailing wind patterns can affect rainfall on windward and leeward sides of highland areas.

The lowest elevations of a mountain generally have a climate and vegetation similar to that of the surrounding area. Higher up the mountain, temperatures and air pressure are lower. The cooler temperatures limit the kinds of plants that can grow. No trees can grow above a certain level called the tree line. Climate conditions at the highest elevations are similar to those of the ice cap climate. Ice and snow are always present, and plant and animal life is scarce. Some mountains in the tropics have snow all year. For example, Africa's snowcapped Kilimanjaro lies only about 200 miles (322 km) south of the equator.

✓ **READING CHECK:** *Physical Systems*   Which of the polar climates supports large forests? subarctic

Section 3 Review

go.hrw.com
Homework Practice Online
Keyword: SW3 HP3

Define
ecosystems, monsoon, savannas, arid, deciduous forests, coniferous forests, permafrost

Reading for the Main Idea

1. *Physical Systems*   Why are there wet and dry seasons in some tropical areas?

2. *Places and Regions*   What kinds of precipitation levels and temperatures would you expect in arid climates?

Critical Thinking

3. **Drawing Inferences and Conclusions**   Why do you think Mediterranean climates do not extend far inland past mountain ranges?

4. **Comparing**   How do conditions on high mountaintops within highland climate regions resemble the ice cap climate?

Organizing What You Know

5. Create a graphic organizer like the one below. Add as many rows as you need to describe the vegetation that is typical in each of Earth's climate types.

Climate type	Vegetation

Section 3 Review Answers

Define   For definitions, see: ecosystems, p. 50; monsoon, p. 51; savannas, p. 51; arid, p. 54; deciduous forests, p. 55; coniferous forests, p. 55; permafrost, p. 56

Reading for the Main Idea
1. Changes in the ways the Sun's rays strike areas north and south of the equator create the seasons. Summer heat creates a low-pressure zone, bringing unstable, wet weather. In winter, cooler weather creates a high-pressure zone, bringing stable, drier weather.

2. low rainfall levels, range of temperatures

Critical Thinking
3. the orographic effect of the mountains (NGS 7)

4. Temperatures remain very cold all the time, ice and snow are always present, and few living things can survive. (NGS 7)

Organizing What You Know
5. tropical humid—tropical rain forest; tropical wet and dry—grassland, scattered trees and shrubs; arid—sparse drought-resistant plants; semiarid—grassland, few trees; Mediterranean—scrub woodland and grassland; humid subtropical—mixed forest; marine west coast—temperate rain forest; humid continental—mixed forest; subarctic—evergreen forest; tundra—mosses, lichens, herbs, low shrubs; ice cap—few plants; highland—varies according to elevation

Setting the Scene

Have students read Case Study: The Poles. Then tell them the following information about protecting Antarctica. In 1959 an agreement called the Antarctic Treaty established Antarctica as a place for science and peace. The treaty was originally signed by 12 countries and was later agreed to by more than 12 others. The Antarctic Treaty banned military activity in the region. It also made Antarctica a nuclear-free zone and encouraged scientific research. However, the Antarctic Treaty did not cover mining rights on the continent. When Antarctica's oil and mineral riches were discovered, some countries wanted rights to this new source of wealth.

Building a Case

In 1991 the countries that had signed the original Antarctic Treaty signed a new agreement called the Madrid Protocol. It forbids most activities on Antarctica that do not have a scientific purpose. It also bans mining and drilling and sets limits on tourism. Pollution concerns are addressed specifically. In 50 years the agreement can be changed if enough of the signing countries agree. Then it becomes the responsibility of another generation to preserve Antarctica. Today's students are part of that generation. What should tomorrow's leaders do about Antarctica?

Linking Past to Present

The Northwest Passage

Long before the Panama Canal allowed ships to travel quickly from the Atlantic Ocean to the Pacific Ocean, explorers dreamed of finding a trade route called the Northwest Passage through the Arctic Ocean. However, the icy waters north of Canada proved to be impassible. Many explorers lost their lives in search of a safe route. Finally, in 1969 a commercial ship successfully crossed the Northwest Passage, but only with the aid of a strengthened hull and ice-breaking technology.

MAP ANSWER

Possible answer: because it is the area closest to other continents; Argentina, Chile, United Kingdom

Applying What You Know Answers

1. The South Pole is in the center of a large landmass—Antarctica—the coldest, driest, highest, and windiest place on Earth. The ice there is thousands of feet thick. The North Pole is in the center of the Arctic Ocean. Arctic ice is thin, loose, seasonal, and highly mobile.

2. established research stations, left trash and sewage, oil spills; causing melting of Antarctic ice sheet

This Case Study feature addresses National Geography Standards 4, 5, 7, 13, and 14.

CASE STUDY

The Poles

Environment and Society Earth's axis intersects its surface at the North and South Poles. The Poles are the same distance from the equator, and both have ice cap climates. However, the Poles also have varying geographic features. They and their surrounding regions offer unique opportunities for expanding human knowledge about our world.

A Frozen Land

The South Pole is in the heart of Antarctica—a continent buried under snow and ice. It is the coldest, driest, windiest, and most isolated continent on Earth. Antarctica also has the highest average elevation of any continent. Russian scientists recorded the world's lowest temperature there, −128.6°F (−89.2°C). Even in the summer, the average temperature in Antarctica's interior stays far below freezing. The air is so cold that it cannot hold moisture. As a result, central Antarctica gets only 2 inches (5 cm) of precipitation per year. However, the ice has been building up for millions of years!

What does the bottom of the world look like? More than 95 percent of Antarctica's surface is ice. It is Earth's deep freeze, storing more than 90 percent of the planet's ice. Ice does not cover the Transantarctic Mountains, however. At 6,500 to 13,000 feet (2,000 to 4,000 m), these mountains break through the ice that blankets most of the continent. Mountains, plateaus, and valleys lie buried under the ice throughout the rest of the continent. Much of Antarctica's rock foundation lies under the frozen surface, some of it below sea level. If all of Antarctica's ice melted, parts of the continent would become island chains. In addition, without the weight of the ice pushing it downward, the underlying rock would rise.

A Frozen Sea

The North Pole lies in the middle of the Arctic Ocean instead of deep within a continent. Asia, Europe, and North America surround the Arctic Ocean. A permanent layer of sea ice covers the ocean's center. During the warmest months of the year, when the

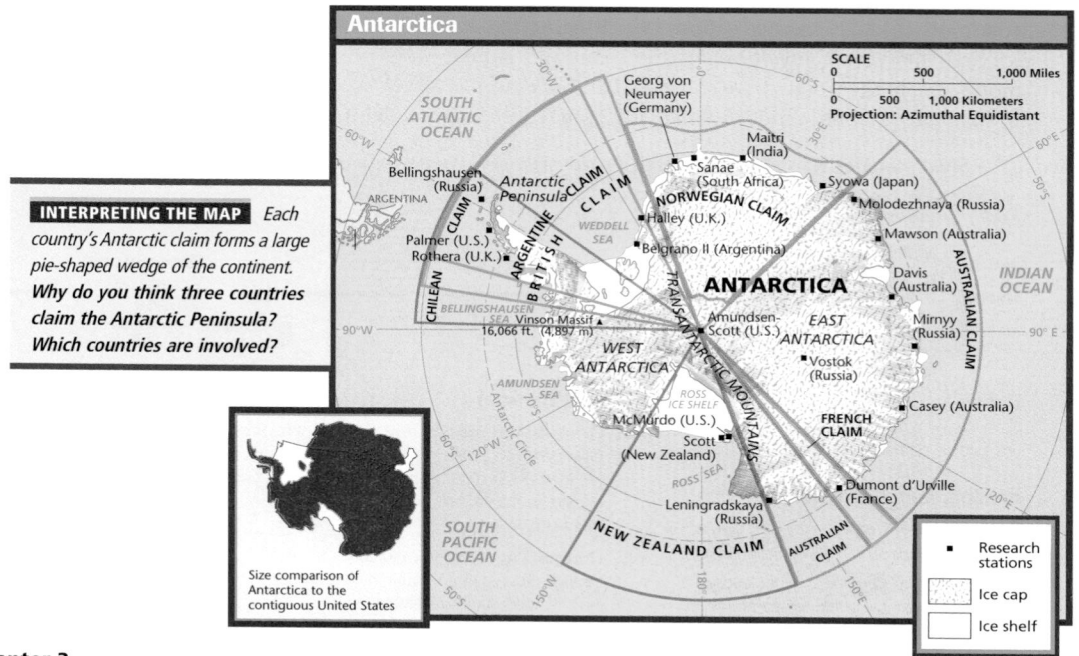

Antarctica

INTERPRETING THE MAP Each country's Antarctic claim forms a large pie-shaped wedge of the continent. **Why do you think three countries claim the Antarctic Peninsula? Which countries are involved?**

Size comparison of Antarctica to the contiguous United States

- ■ Research stations
- Ice cap
- Ice shelf

Drawing Conclusions

Organize the class into two groups. Have one group conduct research on how Antarctic ice affects climate and then draw a diagram of their findings. Ask the other group to investigate the food web based on the tiny shrimp-like creatures called krill that thrive in Antarctic waters. This group should draw the food web for display. Have both groups present their research. Then lead a discussion about how mining or drilling for oil might affect these two aspects of Antarctica. Students should conclude that these economic activities could severely alter the significant roles that Antarctica and the surrounding seas play in regulating Earth's weather and in supporting entire ecosystems.

Going Further: Thinking Critically

Then organize the class into four new groups: research scientists, members of environmental organizations, a mining company, and an international tourism company. Have each group prepare remarks for a press conference during which the members outline their goals for Antarctica, their response to the Madrid Protocol, and their vision for the continent in the year 2050. Encourage students to make sketches to show how their group envisions the Antarctic region's future physical, economic, and population geography. Hold the "press conferences," letting other students play the roles of reporters, asking questions of the presenters. **COOPERATIVE LEARNING**

temperature hovers near freezing, the edges of this frozen crust melt. When winter cold returns, so does the ice.

Compared to the stable and thick Antarctic ice cover, Arctic ice is very thin and loose. Antarctic ice is typically thousands of feet thick. In contrast, Arctic ice averages only 10 to 16 feet (3 to 5 m) in thickness. Arctic ice floats in large chunks on the ocean's surface. Currents, tides, and wind push and pull the ice. Cracks and ridges spread. Large sections of ice collide, combine, break up, and collide again.

Research at the Poles

Scientists have conducted research at both Poles for many decades. During the 1950s and 1960s American and Soviet scientists mapped the Arctic Ocean's floor from research stations on the ice. However, because Arctic ice is always changing, these stations were not permanent.

In contrast, Antarctic research stations have been more durable. Some 29 countries have sponsored research projects in Antarctica. These projects address a wide range of topics. Some scientists search the ice for meteorites that provide information about our solar system. Others have compared air trapped in ancient ice bubbles with today's air. They learned that the use of fossil fuels has raised the amount of carbon dioxide in the air to the highest levels in human history. Some researchers concentrate on how animals survive in the frigid climate of Antarctica and the waters surrounding it.

A Harsh but Fragile Place

Antarctica is not immune to damage or change. For some time, people at research stations piled up trash and sewage and pushed it into the ocean. In addition, some energy companies have hoped to extract the

INTERPRETING THE VISUAL RECORD *Bottom: The polar bear is at home in the Arctic Ocean and on land or ice. Top: At the North Pole, passengers from a Russian ship walk through all the time zones.* **Through which hemispheres could these people walk?**

continent's store of oil and minerals. Oil spills have already caused environmental damage. In 1991, 32 countries forged an agreement to protect Antarctica. The agreement forbids most activities in Antarctica that do not have a scientific purpose. It bans mining and drilling and limits tourism.

Still, Antarctica is changing. Satellite images show that the continent's ice sheet is shrinking. In 2001, scientists found that since 1992 about 8 trillion gallons of water had melted from a glacier in western Antarctica. That is enough to cover about 23 million acres of land with about 1 foot of water. Scientists believe this finding is further evidence of global warming.

Applying What You Know

1. **Summarizing** What are some of the important differences in the physical geography of the North and South Poles?

2. **Identifying Cause and Effect** How have humans modified the physical environment of Antarctica? How may global warming be affecting Antarctica?

TECHNOLOGY

- Chapter 3 Test Generator (on the One-Stop Planner)
- Global Skill Builder CD–ROM
- HRW Go site

REINFORCEMENT, REVIEW, AND ASSESSMENT

- Chapter 3 Review, pp. 60–61
- Chapter 3 Tutorial for Students, Parents, Mentors, and Peers
- Chapter 3 Test (form A or B)
- Alternative Assessment Handbook

- Chapter 3 Test for English Language Learners and Special-Needs Students
- Unit 1 Test
- Unit 1 Test for English Language Learners and Special-Needs Students

Assess

Have students complete a Chapter 3 Test.

Reteach

Organize the class into three groups, one for each section of the chapter. Have the groups discuss the material among themselves, focusing on the key concepts as set by the Read to Discover questions. Have students prepare and give oral presentations that demonstrate their understanding of the section's main ideas. **ENGLISH LANGUAGE LEARNERS, COOPERATIVE LEARNING**

CHAPTER 3 Review Answers

4. erode and damage coastal areas, bring flooding that threatens people, plants, property, and animals

5. tropical regions; subtropical high-pressure zones that bring stable, sinking, dry air all year

Thinking Critically

1. because the warm, low-pressure air rises, carrying water vapor into the atmosphere, where it cools and condenses to create precipitation (NGS 7)

2. Possible answers: short growing season and permafrost preventing tree growth (NGS 7)

3. 30° north and south latitude, because these zones have persistent high pressure and stable, dry air (NGS 7)

Using the Geographer's Tools

1. polar winds, Northern Hemisphere westerlies, northeast trade winds, southeast trade winds, Southern Hemisphere westerlies, southeast trades winds

2. See the chart for answers.

3. Summer diagram should show Sun's rays hitting directly, increasing temperature and causing low pressure and unstable, rising air and rainfall. Winter diagram should show Sun's rays at an angle, creating high pressure, stable cool air, and dry weather.

CHAPTER 3 Review

Building Vocabulary

On a separate sheet of paper, explain the following terms by using them correctly in sentences.

weather	prevailing winds
condensation	climate
doldrums	rain shadow
temperature	front
hurricanes	greenhouse effect
evaporation	monsoon
cyclones	humidity
arid	

Locating Key Places

On a separate sheet of paper, match the letters on the map of Australia with their correct labels.

tropical humid climate	tropical wet and dry climate
marine west coast climate	humid subtropical climate
semiarid climate	Mediterranean climate
arid climate	

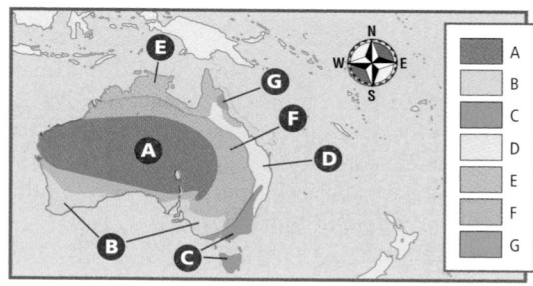

Understanding the Main Ideas

Section 1

1. *Physical Systems* How does latitude relate to climate?

2. *Physical Systems* How do atmospheric pressure zones and ocean currents affect Earth's energy balance?

Section 2

3. *Places and Regions* On a global scale, where is precipitation most common?

4. *Physical Systems* What effects can hurricanes have on local environments?

Section 3

5. *Places and Regions* Where will you find climates that are generally warm and wet all year? What creates dry weather conditions in arid regions?

Thinking Critically

1. Identifying Cause and Effect Why is precipitation heavier in the persistent low-pressure zones of the world?

2. Drawing Inferences and Conclusions What climate factors do you think keep trees from growing in tundra regions?

3. Analyzing Information Around what lines of latitude are most arid regions centered? Why?

Using the Geographer's Tools

1. Analyzing Diagrams Study the diagram of Earth's pressure and wind systems. Then identify Earth's main prevailing wind patterns.

2. Summarizing Review the map and chart of Earth's climate regions. Create a word web for each climate type. Use the webs to describe the distribution and precipitation, temperature, and wind patterns of each climate. Then describe the factors that influence climate.

3. Preparing Diagrams Create two diagrams to show how the seasonal change in the way the Sun's rays strike the areas north and south of the equator helps create the tropical wet and dry climate. Draw one diagram to show how the climate is affected in the summer. The other diagram should show how the climate is affected in the winter.

Writing about Geography

Conduct research on the latest scientific studies of global warming. You may want to concentrate on evidence from the polar regions. Pay particular attention to differences of opinion. Write a short newspaper article about your findings. When you are finished with your article, proofread it to make sure you have used standard grammar, spelling, sentence structure, and punctuation.

SKILL BUILDING

Geography for Life

Using Research Skills

Physical Systems Conduct research on three desert regions on different continents or at different latitudes. Use the climate and precipitation maps in this textbook's unit atlases to help you locate the deserts. Then write a report comparing and contrasting the three areas. Discuss the factors that create each arid climate and the range of temperatures and rainfall in each location. In addition, describe the plants and animals that live in each region.

Portfolio Activity

Have students conduct research on ways that sailors from the 1400s to 1700s used global wind currents to help them reach their destinations. Students might compare these early sea voyages to the use that Bertrand Piccard and Brian Jones made of global wind currents during their nonstop balloon trip around the world in 1999. On a world map have students use different colors to draw and label the currents and the routes taken by the early and recent adventurers. Then have them write paragraphs summarizing the results to accompany their maps.

Food Festival

Climate affects not just which plants will grow in a certain area but also entire cuisines. Have students investigate why people in warm climates are more likely to eat foods flavored with chili peppers and what recipes they enjoy. Cuisines of equatorial Africa, the Caribbean Islands, India, Mexico, South America, Spain, and Thailand all include dishes using chilies. Have students search markets for different types of chilies and try recipes with them. Caution students to use great care when handling chilies—they should not touch their eyes, mouths, or noses after handling the peppers. They should wash their hands well with soap and water or even wear rubber gloves.

Building Social Studies Skills

Interpreting Maps

Study the precipitation map below. Then use the information from the map to help you answer the questions that follow. Mark your answers on a separate sheet of paper.

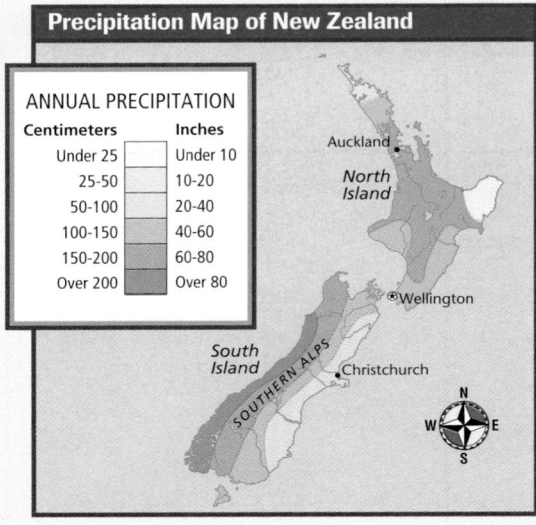

Precipitation Map of New Zealand

ANNUAL PRECIPITATION

Centimeters	Inches
Under 25	Under 10
25-50	10-20
50-100	20-40
100-150	40-60
150-200	60-80
Over 200	Over 80

1. Of the following amounts of precipitation, which is the amount that falls in Wellington, according to the precipitation map?

a. 19 cm
b. 27 cm
c. 88 cm
d. 120 cm

2. Describe the difference in precipitation levels of the western and eastern coasts of New Zealand's South Island.

Using Language

The following passage contains mistakes in grammar, punctuation, or usage. Read the passage and then answer the following questions on a separate sheet of paper.

"[1] A humid tropical climate is warm all year. [2] A humid tropical climate is rainy all year. [3] People living in this climate do not see a change from summer to winter. [4] The heat in the tropics cause almost daily rainstorms."

3. Which word group contains an error in subject-verb agreement?
a. 1
b. 2
c. 3
d. 4

4. Write a sentence that effectively combines word groups 1 and 2.

Writing

Geography for Life

Research should include information on three desert regions on different continents or at different latitudes. Students should support their opinions. Use Rubric 9, Comparing and Contrasting, to evaluate student work.

Social Studies Skills

1. d
2. The eastern side of South Island gets less precipitation than the western side because the eastern side is in a rain shadow.
3. d
4. A humid tropical climate is warm and rainy all year.

Alternative Assessment

PORTFOLIO ACTIVITY

Learning about Your Local Geography

Group Project: Field Work

Within a group, use a barometer, rain gauge, and thermometer to keep a daily record of air pressure, rainfall, and temperature at your school. Track these values over several weeks. If possible, note how these factors change with the seasons. Then conduct research on weather records for your area and compare your observations to the official records. Write several generalizations about changes or lack of change in weather patterns.

internet connect

Internet Activity: go.hrw.com
KEYWORD: SW3 GT3

Choose a topic on weather and climate to:
- use satellite maps to learn about weather conditions, fronts, and surface conditions around the world.
- create a poster to display research on hurricanes and the technology of hurricane tracking.
- investigate global warming.

PORTFOLIO ACTIVITY
Students' reports will vary, but students should use various instruments to conduct their research. Use Rubric 14, Group Activity, to evaluate student work.

Landforms, Water, and Natural Resources

CHAPTER RESOURCE MANAGER

Objectives	Pacing Guide	Reproducible Resources
SECTION 1 **Landforms** (pp. 63–69) • Identify the physical processes inside Earth that build up the land. • Describe the physical processes on Earth's surface that wear down the land. • Explain how these physical processes interact to create landforms.	**Regular** 1.5 days **Block Scheduling** 1 day *Block Scheduling Handbook, Chapter 4*	**RS** Guided Reading Strategy 4.1 **IC** Lab Activities for Geography and Earth Science, Demonstration 8 **SM** Geography for Life Activity 4
SECTION 2 **The Hydrosphere** (pp. 70–73) • Identify the forms and locations in which we find water on Earth. • Explain the causes and effects of floods.	**Regular** 1 day **Block Scheduling** .5 day *Block Scheduling Handbook, Chapter 4*	**RS** Guided Reading Strategy 4.2 **SM** Map Activity 4: Dams
SECTION 3 **Natural Resources** (pp. 74–81) • Explain why soil and forests are important resources. • Clarify concerns over water quality and air quality. • Identify some of the ways that minerals are used. • List the main energy resources and discuss how they are used.	**Regular** 1 day **Block Scheduling** .5 day *Block Scheduling Handbook, Chapter 4*	**RS** Guided Reading Strategy 4.3 **RS** Graphic Organizer Activity 4 **PS** Readings in World Geography, History, and Culture 7 and 8 **SM** Critical Thinking Activity 4: Rising Global Energy Consumption

Chapter Resource Key

PS Primary Sources **A** Assessment CD–ROM

RS Reading Support **REV** Review Video

IC Interdisciplinary Connections **ELL** Reinforcement and English Language Learners Internet

E Enrichment Transparencies **Holt Presentation Maker Using Microsoft® PowerPoint®**

SM Skills Mastery

 One-Stop Planner CD–ROM

See the *One-Stop Planner* for a complete list of additional resources for students and teachers.

 **One-Stop** Planner CD–ROM

It's easy to plan lessons, select resources, and print out materials for your students when you use the ***One-Stop Planner CD–ROM with Test Generator***.

☐ **internet** connect

HRW ONLINE RESOURCES

GO TO: go.hrw.com
Then type in a keyword.

TEACHER HOME PAGE
   KEYWORD: SW3 Teacher

CHAPTER INTERNET ACTIVITIES
   KEYWORD: SW3 GT4
   Choose an activity to:
   • create a newspaper on causes and effects of earthquakes.
   • understand the formation and uses of fossil fuels.
   • analyze the different stages of the hydrologic cycle.

CHAPTER ENRICHMENT LINKS
   KEYWORD: SW3 CH4

CHAPTER MAPS
   KEYWORD: SW3 MAPS4

ONLINE ASSESSMENT
   Homework Practice
   KEYWORD: SW3 HP4
   Standardized Test Prep
   KEYWORD: SW3 STP4
   Rubrics
   KEYWORD: SS Rubrics

COUNTRY INFORMATION
   KEYWORD: SW3 Almanac

CONTENT UPDATES
   KEYWORD: SS Content Updates

HOLT PRESENTATION MAKER
   KEYWORD: SW3 PPT4

ONLINE READING SUPPORT
   KEYWORD: SS Strategies

CURRENT EVENTS
   KEYWORD: S3 Current Events

Technology Resources

- One-Stop Planner CD–ROM, Lesson 4.1
- *ARGWorld* CD–ROM
- Homework Practice Online
- HRW Go site

- One-Stop Planner CD–ROM, Lesson 4.2
- Geography and Cultures Visual Resources 7
- **CNN** Presents Geography: Yesterday and Today, Segment 4: Farming with Saltwater
- Homework Practice Online
- HRW Go site

- One Stop Planner CD–ROM, Lesson 4.3
- Geography and Cultures Visual Resources 8
- Homework Practice Online
- HRW Go site

Reinforcement, Review, and Assessment

ELL Main Idea Activity 4.1
ELL English Audio Summary 4.1
ELL Spanish Audio Summary 4.1
REV Section 1 Review, p. 69
A Daily Quiz 4.1

ELL Main Idea Activity 4.2
ELL English Audio Summary 4.2
ELL Spanish Audio Summary 4.2
REV Section 2 Review, p. 73
A Daily Quiz 4.2

ELL Main Idea Activity 4.3
ELL English Audio Summary 4.3
ELL Spanish Audio Summary 4.3
REV Section 3 Review, p. 81
A Daily Quiz 4.3

Meeting Individual Needs

Ability Levels

Level 1 Basic-level activities designed for all students encountering new material

Level 2 Intermediate-level activities designed for average students

Level 3 Challenging activities designed for honors and gifted-and-talented students

English Language Learners Activities that address the needs of students with Limited English Proficiency

Chapter Review and Assessment

- Chapter 4 Test Generator (on the One-Stop Planner)
- Global Skill Builder CD–ROM
- HRW Go site
- **REV** Chapter 4 Review, pp. 84–85
- **REV** Chapter 4 Tutorial for Students, Parents, Mentors, and Peers
- **A** Chapter 4 Test (form A or B)
- **A** Alternative Assessment Handbook
- **A** Chapter 4 Test for English Language Learners and Special-Needs Students

Launch into Learning

Write *air, earth, fire,* and *water* on the chalkboard. Tell students that long ago people thought these four substances were elements and that everything in the world was made from them. Ask students to identify different forms of air, earth, fire, and water or different ways in which the substances can affect our lives—directly or indirectly. Write their responses under the appropriate headings. *(Possible answers: air—wind, tornadoes, ozone; earth—garden soil, landslides, mountains; fire—forest fires, gas flame, volcanoes; water—oceans, rain, rivers, tap water)* Point out that although we now know air, earth, fire, and water are not elements, understanding the characteristics of these substances helps us comprehend how landforms are built and worn down, the role that water plays in life, and the importance of natural resources. Tell students they will learn more about these characteristics in this chapter.

Why We Should Know More

You may want to emphasize the importance of studying the land, water, and other materials on which we depend by discussing these points with your students:

▶ By learning more about how landforms are created and change, we can save lives threatened by earthquakes, volcanoes, and other hazards.

▶ Our quality of life—and life itself—can depend on how well we protect Earth's air and water.

▶ We make decisions about using resources every day, even if we do not realize it. We need information in order to make good decisions.

▶ We must develop new sources of energy that will replace those, such as oil and gas, that cannot be renewed.

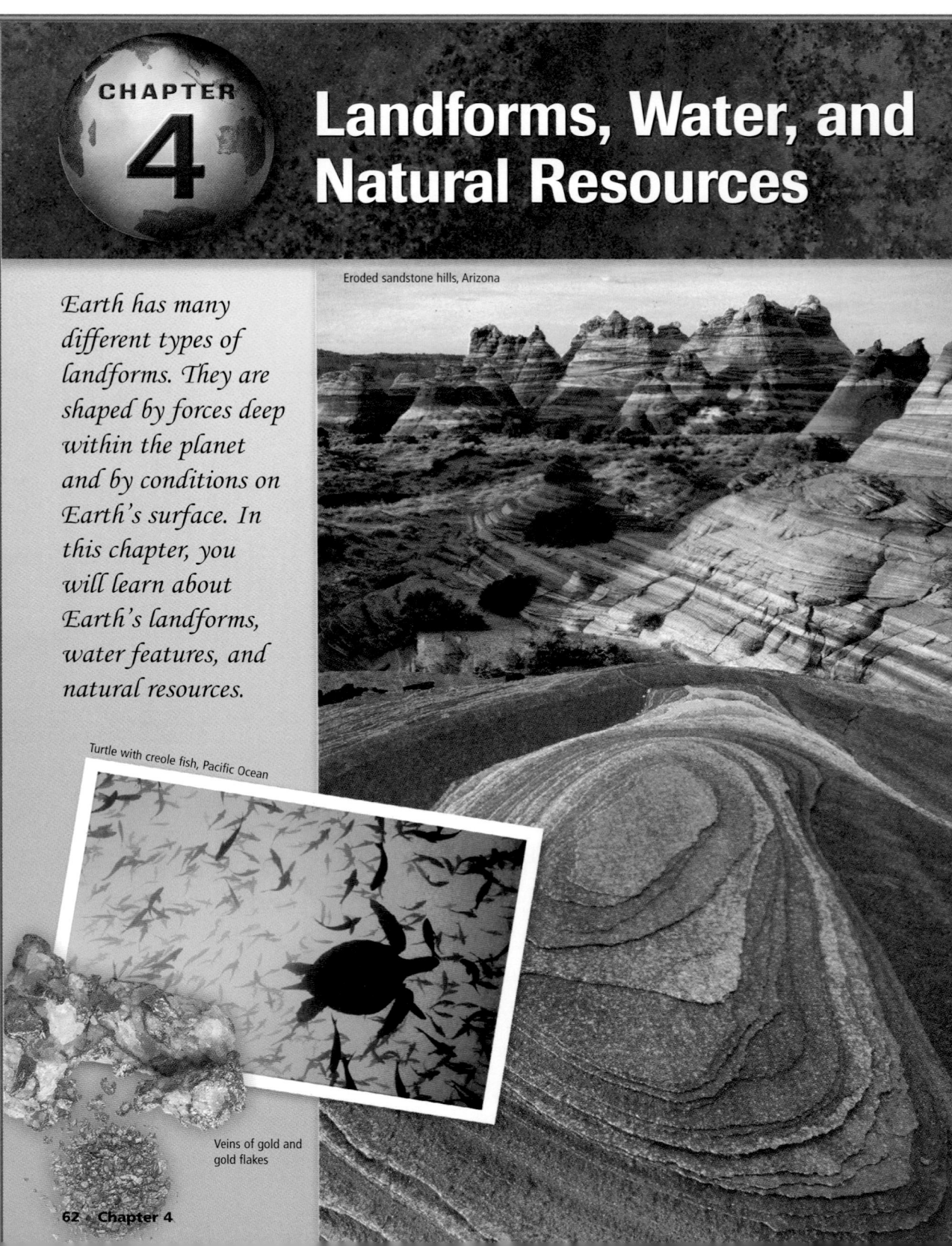

CHAPTER 4
Landforms, Water, and Natural Resources

Earth has many different types of landforms. They are shaped by forces deep within the planet and by conditions on Earth's surface. In this chapter, you will learn about Earth's landforms, water features, and natural resources.

Eroded sandstone hills, Arizona

Turtle with creole fish, Pacific Ocean

Veins of gold and gold flakes

LET'S GET STARTED

Copy the following question onto the chalkboard: *What do the phrases "solid as a rock," "mountain of strength," or "older than dirt" suggest about Earth?* Discuss responses. *(Possible answer: that it is unchangeable and permanent)* Point out that, although many processes on Earth are slow and mountains appear permanent, the planet is actually changing constantly. Some of those changes, such as volcanic eruptions, are very dramatic. Tell students they will learn more about both slow and swift changes to Earth's landforms in Section 1.

Building Vocabulary

Write the key terms on slips of paper and place the slips in a box or similar container. (Write duplicate slips for some terms so that you have enough slips for all members of the class.) Direct each student to draw a slip from the container. Have students skim the text or the glossary to find the definitions of the terms they selected. Then direct them to make sketches illustrating the terms. Pair students and have partners try to identify each other's illustrated term.

Section 1 — Landforms

READ TO DISCOVER

1. What physical processes inside Earth build up the land?
2. What physical processes on Earth's surface wear down the land?
3. How do these physical processes interact to create landforms?

WHY IT MATTERS

Earthquakes are among the natural forces that shape the land. They have changed landscapes in the United States in the distant past and in recent years. Use **CNNfyi.com** or other **current events** sources to learn about earthquakes and how they have affected different parts of the world.

DEFINE

core	**folds**
mantle	**faults**
magma	**weathering**
plate tectonics	**sediment**
continental drift	**erosion**
rift valleys	**glaciers**
abyssal plains	**plateau**
continental shelves	**alluvial fan**
trench	**delta**

Section 1 RESOURCES

REPRODUCIBLE
▶ Guided Reading Strategy 4.1
▶ Lab Activities for Geography and Earth Science, Demonstration 8
▶ Geography for Life Activity 4

TECHNOLOGY
▶ One-Stop Planner CD–ROM, Lesson 4.1
▶ Homework Practice Online
▶ HRW Go site

REINFORCEMENT, REVIEW, AND ASSESSMENT
▶ Main Idea Activity 4.1
▶ English Audio Summary 4.1
▶ Spanish Audio Summary 4.1
▶ Section 1 Review, p. 69
▶ Daily Quiz 4.1

Forces below Earth's Surface

Geology—the study of Earth's physical structures and the processes that have created them—is important to geographers. We live and build on hills, mountains, valleys, and other landforms. They can make travel easy or difficult. Landforms can also give us clues to what Earth was like in the past. Forces below Earth's surface are a key to the shaping of landforms.

Scientists have identified four important zones in Earth's interior, as you can see in the diagram. The planet's center is like a nuclear furnace, where decaying radioactive elements generate heat. At Earth's center, or **core**, both temperatures and pressures are very high. The core is divided into inner and outer layers. The inner core is solid. The outer core is mostly dense liquid metal, mainly iron and nickel. Beyond the core is the **mantle**—the zone that has most of Earth's mass. The uppermost layer is the crust. Although it is up to about 25 miles (40 km) thick, the crust is comparatively thin. Huge currents carry heat from the core through the mantle to the crust. Liquid rock within Earth is called **magma**. When this liquid rock spills out onto the surface it is called lava. Magma erupts from vents called volcanoes.

Internal Forces The theory of **plate tectonics** explains how forces within the planet create landforms. This theory views Earth's crust as divided into more than a dozen rigid, slow-moving plates. The plates can be compared to the cracked shell of a hard-boiled egg. Some plates are as large as a quarter of the planet, but others are only a few hundred miles across. The plates slowly move across the upper mantle, usually less than an inch per year. This process is called **continental drift**. Along the plate boundaries, the crust is subject to stresses that lead to melting, bending, and breaking. In the middle of

☐ internet connect

GO TO: go.hrw.com
KEYWORD: SW3 CH4
FOR: Web sites about landforms, water, and natural resources

The Interior of Earth

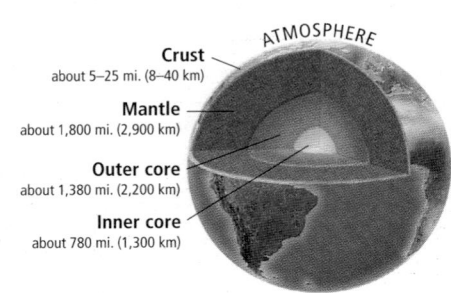

Crust about 5–25 mi. (8–40 km)
ATMOSPHERE
Mantle about 1,800 mi. (2,900 km)
Outer core about 1,380 mi. (2,200 km)
Inner core about 780 mi. (1,300 km)

INTERPRETING THE DIAGRAM *Temperatures within Earth increase with depth. In the inner core, temperatures may reach as high as 12,000°F (7,000°C).* **How far below the surface is Earth's inner core?**

DIAGRAM ANSWER

approximately 3,185 miles (5,108 km) to 3,205 miles (5,148 km)

63

LEVEL 1: Have students work in small groups to create informational pamphlets titled When Plates Collide. Tell students that their pamphlets should include descriptions of landforms that are created by the collision of tectonic plates, illustrations of the processes involved in the creation of these landforms, and examples of locations around the world where these processes are taking place. Call on groups to display and discuss their pamphlets. **ENGLISH LANGUAGE LEARNERS, COOPERATIVE LEARNING**

LEVELS 2 AND 3: Pair students and provide each pair with sheets of heavy paper. Then ask students to experiment with the paper on their desktops to mimic the formation of fold mountains and fault-block mountains. Call on volunteers to demonstrate the movements to the class. **COOPERATIVE LEARNING**

Continental Drift Alfred Wegener was not the first scientist to notice similarities in the coastlines of the continents. However, he was the first to weave seemingly unrelated facts from geology and paleontology into the theory of continental drift. Wegener noted that some ancient rock layers, glacial markings, and fossils matched on continents now separated by oceans. His theory, published in 1915, was ridiculed because he could not explain why the continents moved. Wegener's visionary ideas served as the framework for the current theory of plate tectonics.

ACTIVITY: Invite students to cut out continents from paper maps and fit them together, as Wegener did, into jigsaw-puzzle maps to illustrate continental drift.

MAP ANSWER

Pacific, Antarctic, Eurasian, African; Caribbean, Cocos, Juan de Fuca, Scotia; North American, Eurasian, South American, African

Plate Tectonics

INTERPRETING THE MAP

This map shows Earth's tectonic plates and the directions in which they are moving. **Which plates appear to be the largest? Which are the smallest? Which plates meet in the center of the Atlantic Ocean?**

NORTH AMERICAN PLATE
JUAN DE FUCA PLATE
EURASIAN PLATE
PACIFIC PLATE
CARIBBEAN PLATE
COCOS PLATE
ARABIAN PLATE
AFRICAN PLATE
PHILIPPINE PLATE
PACIFIC PLATE
Equator
NAZCA PLATE
SOUTH AMERICAN PLATE
INDO-AUSTRALIAN PLATE
SCOTIA PLATE
ANTARCTIC PLATE

SCALE
0   2000 Miles
0   2000 Kilometers
Scale is accurate only along the equator.
Projection: Robinson

—— Plate boundaries
←— Direction of plate movement

plates, however, little tectonic activity takes place. Continental areas in the middle of plates are steadily eroded. Ocean floor areas in the middle of plates are steadily buried with sediment.

Volcanoes often form long rows and signal that a plate boundary is nearby. Earthquakes—sudden shakings of Earth's crust—take place when tectonic forces cause masses of rock inside the crust to break. Earthquakes are also common near plate boundaries.

Scientists use the theory of plate tectonics to explain the long history of Earth's surface. They believe that about 200 million years ago all the modern continents were part of one supercontinent called Pangaea (pan-JEE-uh). Pangaea then broke into two smaller supercontinents called Gondwana and Laurasia. These two, in turn, broke into the modern continents during the last 100 million years. The theory of plate tectonics helps explain the "fit" between the coastlines of Africa and South America. Rock formations that match up across the boundaries provide more evidence. The theory also helps geographers understand the origins of mountains and the landforms of the ocean floors.

Continental Drift

Pangaea, a supercontinent about 200 million years ago, was surrounded by Earth's single ocean, Panthalassa. New oceanic crust formed as Pangaea broke apart and continental plates drifted away from each other.

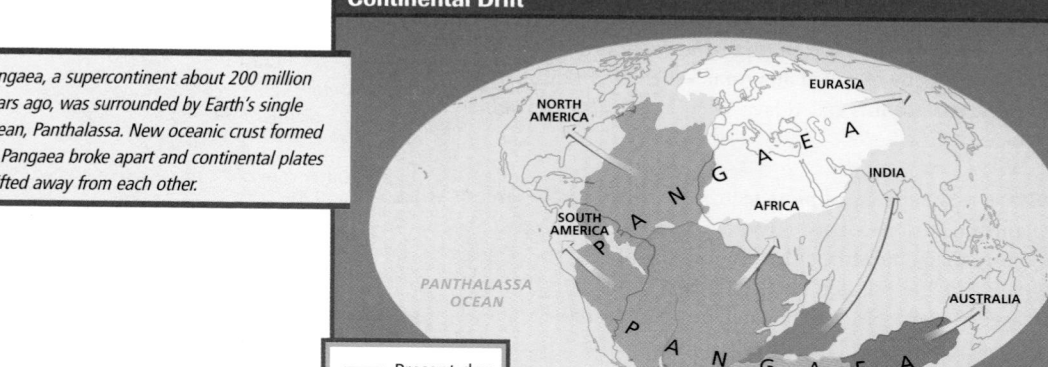

NORTH AMERICA
EURASIA
PANGAEA
INDIA
AFRICA
SOUTH AMERICA
PANTHALASSA OCEAN
AUSTRALIA
PANGAEA
ANTARCTICA

—— Present-day shoreline

LEVEL 1: Copy the following graphic organizer onto the chalkboard, omitting the italicized answers. Have students complete it by adding detail about the processes by which weathering and erosion wear down the land. **ENGLISH LANGUAGE LEARNERS**

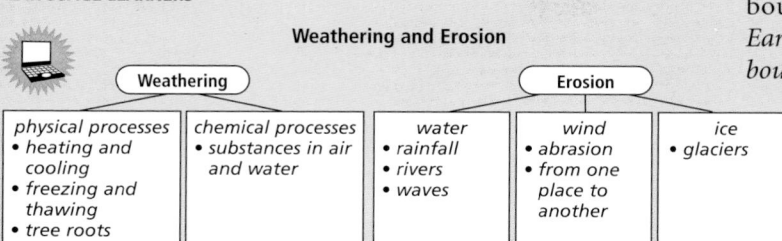

Weathering and Erosion

Weathering		Erosion		
physical processes • *heating and cooling* • *freezing and thawing* • *tree roots*	*chemical processes* • *substances in air and water*	*water* • *rainfall* • *rivers* • *waves*	*wind* • *abrasion* • *from one place to another*	*ice* • *glaciers*

HOMEWORK: Have students mark the tectonic plate boundaries on an outline map of the world. Next have students use the textbook's index to learn some of the places where volcanoes and earthquakes are common and to mark those locations on the map. Have students write a caption to summarize the relationship between tectonic plate boundaries and earthquakes and volcanoes. *(Example: Earthquake and volcanic activity is common along tectonic plate boundaries.)*

Movement at Plate Boundaries

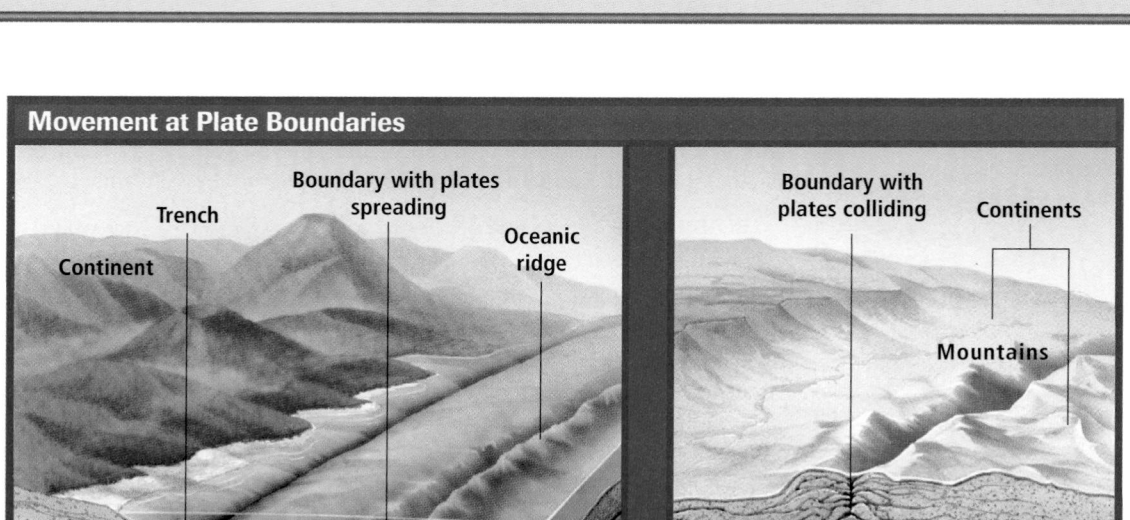

Left: Where Plate 2 pushes under Plate 1, a deep trench forms. Where Plate 2 and Plate 3 move apart, lava creates an oceanic ridge. Right: Where Plate 4 and Plate 5 collide, the crust is pushed up, forming a mountain range.

Plate Movement Three types of movements at plate boundaries are possible. First, the plates can move apart, or spread. Second, the plates can collide. Third, the plates can move laterally, slipping past each other. The movement of plates creates distinctive landforms.

Long ago in Earth's history the crust sorted itself into two layers of different kinds of rocks. The lower layer, made of heavier rock, is found on the ocean floors. Lying on top is a patchy layer of lighter rock. This layer makes up the continents. Nearly all spreading plate boundaries are found on the ocean floors. As fresh lava wells upward, this new crust pushes the plates apart. The rising heat also lifts the crust upward, building a chain of young volcanic mountains, called an oceanic ridge. Earth's oceanic ridges connect in a nearly continuous submarine mountain chain. This chain is nearly 40,000 miles (60,000 km) long and contains some of the world's highest peaks. The part of this chain that runs through the Atlantic Ocean is called the Mid-Atlantic Ridge. However, except for on a few islands, such as Iceland, these mountains are hidden beneath the waves.

A few spreading plate boundaries lie under continents. In these places, the crust stretches until it breaks, forming **rift valleys**. The biggest rift valleys are in eastern Africa.

Away from the oceanic ridges, rocks of the ocean floors gradually sink because they have no supporting heat below them. Here we find **abyssal plains**, the world's flattest and smoothest regions. In these areas a thick layer of mud, clay, and other materials has buried all features.

This map shows some of the landforms on the oceans' floors. Note how the Mid-Atlantic Ridge runs the length of the ocean and continues into the Indian Ocean.

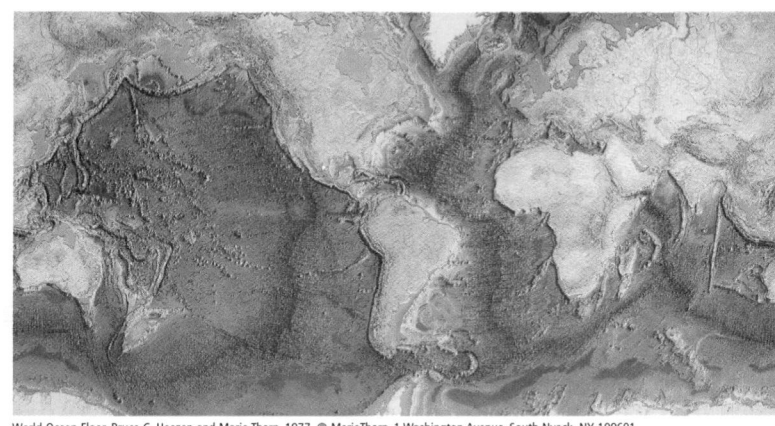

World Ocean Floor, Bruce C. Heezen and Marie Tharp, 1977, © MarieTharp, 1 Washington Avenue, South Nyack, NY 109601

Essential Element 3

▶ **Physical Systems** ◀

An Oceanic Ecosystem
In some places along the crests of oceanic ridges, hydrothermal vents called black smokers release scalding acidic water into the ocean. In 1977 a bizarre community of creatures was found thriving in this harsh environment. More than 300 species of vent creatures have been identified.

Because they can withstand higher temperatures than other organisms, bacteria are the first to colonize new vents. The bacteria use hydrogen sulfide discharged at the vents as their energy source and form white mats on the sea floor. Unique worm species, tiny lobsters, blind shrimp, and clams and mussels as big as dinner plates come to feed on the bacteria. Giant tube worms up to 4 feet (1.2 m) tall may be the most exotic of the vent creatures. Their swaying red plumes are filled with blood, which transports sulfides to bacteria living within the tube worms' spongy tissues. In return for housing, the bacteria convert the hydrogen sulfide into food for the worms, which have no mouths or digestive systems.

ACTIVITY: Have students conduct research on the geology and biology of oceanic ridges and black smokers. Then have them draw diagrams of their findings.

Eye on Earth

Tsunamis Earthquakes on the seafloor can trigger sea waves that can in turn devastate coastal areas, particularly in the Pacific Rim. In the deepest parts of the oceans, tsunamis may travel at speeds of over 450 miles per hour (720 kph), the speed of a jet airplane. However, at the surface these waves are so small they cannot be felt on board a ship.

As a tsunami enters shallow water, it slows because its speed is tied to the water's depth. Because the total energy of the tsunami remains constant, however, the water piles up. As it approaches a coast, the tsunami's crest may exceed 100 feet (30 m).

CRITICAL THINKING: How long would it take for a tsunami that began near Honolulu, Hawaii, to reach the Los Angeles, California, coast? *(The distance is about 2,500 miles [4,023 km], so if the tsunami traveled at 450 miles per hour it would arrive after about 5.5 hours.)*

Folds and Faults

Mountains formed by faults

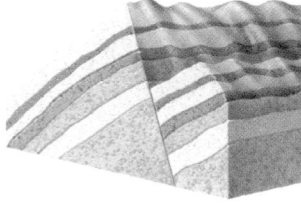

Mountains formed by folds

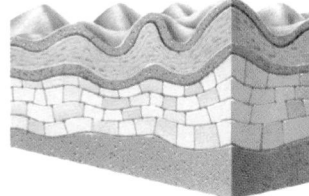

Top: Fault-block mountains form where parts of Earth's crust are pushed up along a fracture line, or fault. Bottom: Many large mountain ranges are created by folding, which occurs when two plates collide and their edges wrinkle.

The continental surface extends under the shallow ocean water around the continents. These areas are called **continental shelves**. At the edges of the continental shelves, the seafloor drops steeply down to the abyssal plains.

✓ **READING CHECK:** (*Physical Systems*) What physical process forms oceanic ridges? *rising lava that lifts the crust upward, separating plates*

When Plates Collide Colliding plate boundaries are found both on ocean floors and along continental edges. When two plates on the ocean floor collide, one slides underneath the other. This plate boundary is called a subduction zone, and the deep valley marking the plate collision is called a **trench**. The plate sliding downward generates heat as it grinds against the plate above it. This heat may produce a row of volcanoes. Some of these volcanoes rise high enough to become island chains. This process created places like the Tonga Trench and the nearby Tonga Islands in the western Pacific.

Sometimes one of the colliding plates is carrying a continent. In this case the heavier oceanic plate dives under the lighter continental plate. The squeezing of the continental plate causes volcanoes, **folds**, and **faults**. Folds are places where rocks have been compressed into bends. Faults are places where rock masses have broken apart and moved away from each other. (See the diagrams.)

It is also possible for two continental plates to collide. The result is a great long-lasting episode of mountain building with awesome folding, faulting, and volcanism. The Himalayas of Asia formed along such a plate boundary. There, scientists believe, a small continent marked by the triangle of modern India began crashing into Asia millions of years ago. The Himalayas, the mountains formed by this collision, are still growing. In fact, plate collisions have built most of the world's mountains.

Plates Moving Laterally When plates move laterally past each other, long fractures develop along the edges of both plates. The pressures along these boundaries are seldom uniform. While a little squeezing produces low mountains, a little spreading generates broad valleys. Earthquakes can be frequent in

Left: The San Andreas Fault cuts through the Carrizo Plain in California. Right: This freeway in Oakland, California, collapsed during a 1989 earthquake.

LEVEL 2: After students have completed the Level 1 activity, organize the class into pairs. Have pairs explore the school grounds or a nearby natural area to find examples of erosion. Ask them to collect rocks and to classify them according to which force appears to have eroded them. Also have students determine if and why any of the three forces of erosion are not present in their region. *(Possible answer: Glaciers are not a force of erosion in many regions.)* **COOPERATIVE LEARNING**

LEVEL 3: Have students write a news story to follow the headline "Between weathering and erosion, no surface is safe!" Students may want to dramatize the story and act it out as a 1940s-style radio play.

these areas. Again, most of these plate boundaries lie under the oceans. However, the San Andreas Fault in California is a famous example on land. Here a sliver of continental crust on the Pacific plate's edge is moving northwestward along the edge of North America. Meanwhile, the North American plate is being pushed westward by the oceanic ridge in the Atlantic Ocean. As this plate boundary occasionally skids along, earthquakes occur from Los Angeles to north of San Francisco.

✔ **READING CHECK:** (*Physical Systems*) What physical processes created the Himalayas and the San Andreas Fault in California? two plates colliding, two plates sliding past each other

Forces on Earth's Surface

While tectonic forces are building up Earth's surface, other forces are wearing it down and making it more level. It is the interaction of these two kinds of forces that shapes the landforms we see.

Weathering and Erosion   Rocks break and decay over time in a process called **weathering**. Weathering is usually slow and difficult to detect. However, even the hardest rocks will eventually wear down. Chemical processes cause some weathering. Substances in air and water react with the rock, creating new chemical compounds and slowly dissolving the rock. Weathering is also caused by physical processes that break rocks into smaller pieces. In desert areas, daytime heating and nighttime cooling can cause rocks to crack. In high mountains, repeated freezing and thawing of water inside a cracked rock can cause it to break even more. The roots of trees can pry rocks apart. Weathering breaks rock into smaller particles of gravel, sand, and mud called **sediment**.

Along with weathering, the other process changing landforms on Earth's surface is **erosion**. Erosion is the movement of surface material from one location to another. Water, wind, and ice cause erosion.

Water, Waves, and Wind   Water is the most important force of erosion. Rainfall can cause rapid erosion where few plants protect the ground. Water erosion can begin as tiny channels on hillsides. If erosion is severe, a channel may grow into a gully. Running water can even carve deep canyons, such as the Grand Canyon. Rivers usually carry water and sediment from mountains in the center of continents all the way to the ocean. Wave action is another powerful force of erosion. During a storm, waves can tear away tons of beach sand in a few hours. Waves can also change shorelines slowly over many years.

Wind is another force that causes erosion. Plants protect most land surfaces from wind action. However, in dry lands, on beaches, and in places where people or animals have destroyed the vegetation, wind can cause significant erosion. Wind works in two ways. One is abrasion—that is, it blasts particles of sand against rock. The other way is by blowing sand and dust from one place to the next. Hills of wind-deposited sand are called dunes. Sand dunes are common on beaches and in deserts. Wind also lifts dust high into the atmosphere and transports it great distances. For example, dust from the Sahara in Africa is carried across the Atlantic Ocean to the Caribbean islands!

✔ **READING CHECK:** (*Physical Systems*) How do wind and water shape the land? wear away and deposit material

INTERPRETING THE VISUAL RECORD

Erosion wears away Earth's surface at an island beach off the Florida coast (top) and at Theodore Roosevelt National Park, North Dakota (bottom). **What physical processes are causing erosion in these places?**

Global Perspectives

Windblown African Dust
Evidence in the soils of Caribbean islands indicates that for thousands of years dust particles have blown across the Atlantic Ocean from Africa. Now some biologists believe the dust contributes essential nutrients to the Amazon rain forest. However, it also transports bacteria, viruses, and fungi that kill Caribbean Sea coral. Furthermore, people who live in the southern United States sometimes have to breathe the brown haze that blows across the Atlantic.

Since the 1970s, there has been a dramatic increase in the quantity of windblown African dust. Drought conditions, the loss of plants that held dirt in place, and strong winds created by intense heat have contributed to the increase. About a billion tons of African dust fall on the Caribbean area alone every year.

ACTIVITY: Have students conduct an Internet search for newspaper articles about African dust affecting people in the United States. Ask them to describe those effects.

VISUAL RECORD ANSWER

wave action (top) and wind action (bottom)

67

Teach Objective 3

ALL LEVELS: Have students use clay to create a model of the various landforms discussed in the paragraphs under the heading Shapes on the Land. Encourage them to make their landforms look as realistic as possible. Then have students create small flaglike labels for the various landforms. Ask them to include the name and a brief description of the landform on the label. (*Example: mountain—a highland area created by volcanoes, folding, or faulting*) Call on individual students to discuss their models. Display models in the classroom.

Close

Focus students' attention on the photograph of the San Andreas Fault in this section. Then refer them to the photo of Ayers Rock (Uluru) in the chapter on Australia. Point out that the first photograph illustrates the rapid change that tectonic processes can bring to Earth's crust, while the second shows the effect of millions of years of weathering and erosion on an ancient mountain range.

Eye on Earth

Sand Dunes Some landforms do not stay in one place on Earth's surface. Sand dunes can move slowly across beaches and deserts. True dunes, as opposed to those that form around vegetation, can cover huge areas and reach great heights. The vast sand seas of the Sahara and the Arabian Desert are called ergs. Dunes in the ergs commonly reach more than 650 feet (200 m) in height. In Algeria, large networks of dunes known as *draa* may be more than 980 feet (300 m) tall. Some of the highest dunes in the world are in the United States. The Great Sand Dunes of Colorado rise to nearly 800 feet (245 m).

ACTIVITY: Have students use sand from a garden center to create small models of the main types of dunes—crescent, sword, parabolic, star, and pyramidal. They will need to use Internet or library resources for more information.

INTERPRETING THE VISUAL RECORD

Weathering is slowly wearing down this mountainside. Rock that has been shattered by frost and falls from a mountain forms what is called a talus slope. Gravity slowly pulls the rock down in a process called talus creep. **In what other ways does gravity affect erosion?**

Cyclists in Alaska descend to Black Rapids Glacier. In 1936 this glacier surged forward at the rate of 10 feet (3 m) per hour. It has slowed since then. The inset photo shows the face of a glacier. Here chunks may break off, or calve, and form icebergs.

VISUAL RECORD ANSWER

pulls liquid water and ice, both of which cause erosion, downhill

The Power of Ice Thick masses of ice—called **glaciers**—also erode rock and move sediment. Glacier ice builds up when winter snows do not melt during the following summer. Glaciers may be great ice sheets covering whole regions. Today ice sheets more than two miles (three kilometers) thick cover most of Antarctica and Greenland. As these giant domes of ice sag with the pull of gravity, the edges of the ice sheets are pushed outward.

Glaciers are also found in the valleys of high mountains. These mountain glaciers flow slowly downhill, often sliding on a thin layer of liquid water at their bases. Glaciers can level anything in their paths. They can move rocks as big as houses over long distances. They can grind rocks into sediment as fine as flour. During the ice ages, glaciers reached to what are now St. Louis in North America and Moscow in Russia.

Glaciers can be found in high mountains all around the world, even on high mountains near the equator. Flowing downhill like slow rivers of ice, mountain glaciers carve great U-shaped valleys and sharp mountain peaks. The carving action of glaciers created many of the spectacular valleys we see around the world today. Yosemite Valley in California is a good example.

✓ **READING CHECK:** **Physical Systems** What role do gravity and ice sheets play in shaping the land? Gravity pulls down ice sheets, pushing their edges outward, where they carve the landscape.

Shapes on the Land

One way to better understand landforms is to divide them into three groups. One group is built by tectonic processes. These landforms, such as mountains and some valleys, are created by volcanoes, folding, and faulting. Erosion and the depositing of sediment may be changing these landforms, rounding and smoothing their edges. Yet the basic landform is the result of tectonic processes.

A second type of landform is created by erosion. It is made of rock and has a thin layer of sediment or soil on the surface. The forces of erosion are slowly

lowering its surface. These landforms often reflect the hardness of the underlying rock. Harder rocks resist erosion and over time will stand above the surrounding land. One example of this landform is a **plateau**. A plateau is an elevated flatland that rises sharply above nearby land on at least one side. A plain—a nearly flat area—is often the final stage of a landscape wearing smooth.

A third kind of landform is formed by sediment deposited by ice, water, or wind. A sand dune in a desert is an example of this kind of landform. Another example is a floodplain. A floodplain is a landform of level ground built by sediment deposited by a river or stream.

The terrain in most regions is a jigsaw puzzle of many landforms. For example, a mountain range is formed by tectonic activity. Erosion may then form deep valleys between the mountains. The sediment eroded from the mountains may then be deposited at the mountains' bases. The result of this process can be an **alluvial fan**. This is a fan-shaped deposit of mud and gravel often found along the bases of mountains. Still later a stream may erode the sediments in the alluvial fan, carrying them all the way to a river mouth. There the sediment may move out into the ocean and sink, or the sediment may accumulate, building a **delta**. Eventually this sediment from the distant mountains could travel still farther. It might finally be deposited in an oceanic trench.

The location, shape, and size of landforms have influenced human settlement and transportation throughout history. For example, people tend to settle in flat areas where they can farm. People use rivers for water supplies and transportation. Many railroads and highways have been built along river valleys as well. People have also changed Earth's surface to suit their needs. Large machines can smooth ground for home construction, and explosives clear the way for roads. Governments build dams across rivers, turning valleys into lakes.

Landforms

Volcano · Mountains · Glacier · River · Plain · Alluvial fan · Valley · Lake · Plateau · Canyon · Isthmus · Hills · Desert · Peninsula · Delta · Floodplain · Sandy beach · Strait · Coastline · Island

INTERPRETING THE VISUAL RECORD

Many of the landforms shown here can be found together in various regions of Earth. **Which landforms can you identify where you live?**

✓ **READING CHECK:** *Physical Systems* What are two kinds of landforms created by deposits of sediment? sand dunes and floodplains

Section 1 Review

go.hrw.com Homework Practice Online
Keyword: SW3 HP4

Define
core, mantle, magma, plate tectonics, continental drift, rift valleys, abyssal plains, continental shelves, trench, folds, faults, weathering, sediment, erosion, glaciers, plateau, alluvial fan, delta

Reading for the Main Idea

1. *Physical Systems* What are some processes that shape landforms?

2. *Physical Systems* What are the three kinds of landforms?

Critical Thinking

3. **Analyzing Information** How might a plateau show the effects of both tectonic process and erosion?

4. **Drawing Inferences and Conclusions** Why do you think farming and raising livestock may lead to rapid erosion?

Organizing What You Know

5. Copy the table below. Use it to list the three types of tectonic plate boundaries, the landforms that result from each type, and an example of each type.

Type of plate boundary	Resulting landforms	Example

Section 2

OBJECTIVES

1. Identify the forms and locations in which we find water on Earth.

2. Explain the causes and effects of floods.

LET'S GET STARTED

Copy the following questions onto the chalkboard: *What does water have to do with a computer, a sheet of paper, and a new Japanese car?* Discuss responses. *(The silicon chips in computers must be washed with ultrapure water, paper is manufactured by drying a water/pulp mixture, and cars come here from Japan by ship.)* Point out that not only do our bodies need water to live but we also depend on water in many other ways. Tell students they will learn more about the importance of water and where it is found in Section 2.

Building Vocabulary

Write the key terms on the chalkboard. Ask students to locate the definitions of the terms by skimming the section. Explain that *sal* means "salt" and that the prefix *de-* indicates that something is being removed. Therefore, **desalinization** refers to removing salt from water. Then have students refer to a map of the United States. Call on volunteers to compose sentences that relate key terms to bodies of water in the United States. *(Example: The Arkansas, Missouri, and Ohio Rivers are all **tributaries** of the Mississippi.)*

The Hydrosphere

READ TO DISCOVER

1. In what forms and where do we find water on Earth?
2. What are the causes and effects of floods?

WHY IT MATTERS

Rainfall in any given area varies from year to year. Use **CNNfyi.com** or other current events sources to learn about record high and low rainfall levels in your area.

DEFINE

desalinization
hydrologic cycle
headwaters
tributary
watershed

drainage basin
estuaries
wetlands
groundwater
water table

Water on Earth

Necessary for life, water is abundant on Earth. Yet 97 percent of the world's water is in oceans and is too salty for most uses. The salt in ocean water can be removed through **desalinization**. For example, some countries in the very dry region of Southwest Asia use this process to get freshwater. However, it is expensive because of the large amount of energy needed.

Less than 3 percent of the world's water is fresh. Most of this water is frozen in the ice caps of Antarctica and Greenland. The remainder is less than 1 percent of the world's water. This water is found in clouds, lakes, and rivers and in the ground. It is only this tiny proportion of the world's water that is available for human use.

The Hydrologic Cycle   The amount of water on Earth stays much the same over time. However, its physical state is always changing, from gas (water vapor)

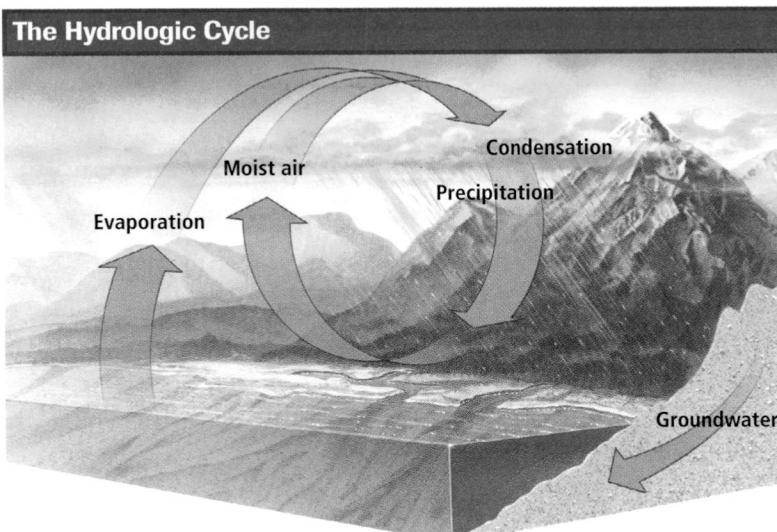

The Hydrologic Cycle

Condensation

Moist air

Precipitation

Evaporation

Groundwater

The circulation of water from one part of the hydrosphere to another is driven by solar energy. Following evaporation, water eventually returns to the surface in various forms of precipitation.

LEVEL 1: Copy the following graphic organizer onto the chalkboard, omitting the italicized answers. Have students complete the organizer by adding information on the roles that evaporation, condensation, and precipitation play in the hydrologic cycle. **ENGLISH LANGUAGE LEARNERS**

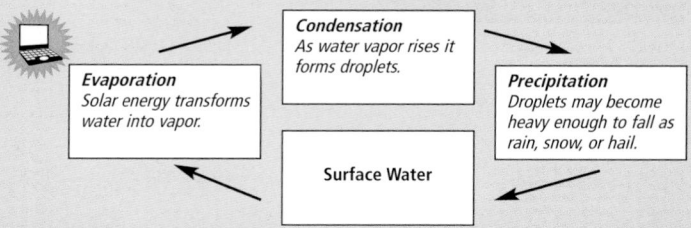

LEVELS 2 AND 3: After students have completed the Level 1 activity, organize the class into small groups. Provide a detailed map of a state, region, or country to each group. Ask students to examine the maps for as many of the water features discussed in Section 2 as they can find. Tell students to list the tributaries, wetlands, estuaries, and so on that they find and to determine what regions, if any, these features appear to define. **COOPERATIVE LEARNING**

---

to liquid to solid (ice). The movement of water through the hydrosphere is called the **hydrologic cycle**. Solar energy, winds, and gravity drive this cycle. When heated by solar energy, water may change to vapor. This process is called evaporation. Most water in the atmosphere has evaporated from the ocean. As water vapor rises with the air, it cools and forms tiny liquid droplets in a process called condensation. Joining to form clouds, these droplets may grow to become raindrops heavy enough to fall to Earth. Water falling to the surface, either as rain, snow, or hail, is called precipitation. If it falls on land, water may be stored. It can build up in plants, in a river or stream, in a lake, or below the ground. Surface water either flows to the sea in a river or evaporates again and returns to the atmosphere.

✓ **READING CHECK:** *Physical Systems*
What forces drive the hydrologic cycle? solar energy, winds, and gravity

Surface Water As precipitation falls on continents and islands, it flows down hills and mountains toward the lowlands and coasts. The first and smallest streams from this runoff are called **headwaters**. As these headwaters join, they form larger streams, and farther downstream they eventually form rivers. Any smaller stream or river that flows into a larger stream or river is called a **tributary**. In the central United States the Arkansas, Missouri, and Ohio Rivers are all major tributaries of the Mississippi River. The whole region drained by a river and its tributaries is called a **watershed** or **drainage basin**. Thus, the Mississippi River watershed has hundreds of tributaries and a watershed of more than 1.2 million square miles.

On continents, surface water may collect in lakes. Many lakes, including the Great Lakes along the U.S.-Canada border, were carved by glaciers during the ice ages. Other lakes, like the Dead Sea in Israel and Jordan, lie in valleys created by continental rifts. Some lakes lie in volcanic craters, like Crater Lake in Oregon. Lakes provide water supplies, fish, and recreation opportunities. Because water heats and cools more slowly than land, large lakes can also reduce the severity of the climates on their shores. This moderating effect allows palm trees to grow on the shore of Lake Geneva in Switzerland. In fact, this effect is common on almost any lakeshore.

Some lakes are not connected to an ocean. These lakes, like the Great Salt Lake in Utah, may collect minerals from runoff. As water continually flows in and evaporates, more and more minerals build up in the lake water. Thus, the lake becomes more and more salty.

Surface water is also found in **estuaries**. An estuary forms where a river meets an inlet, or small arm, of the sea. In this semi-enclosed body of coastal

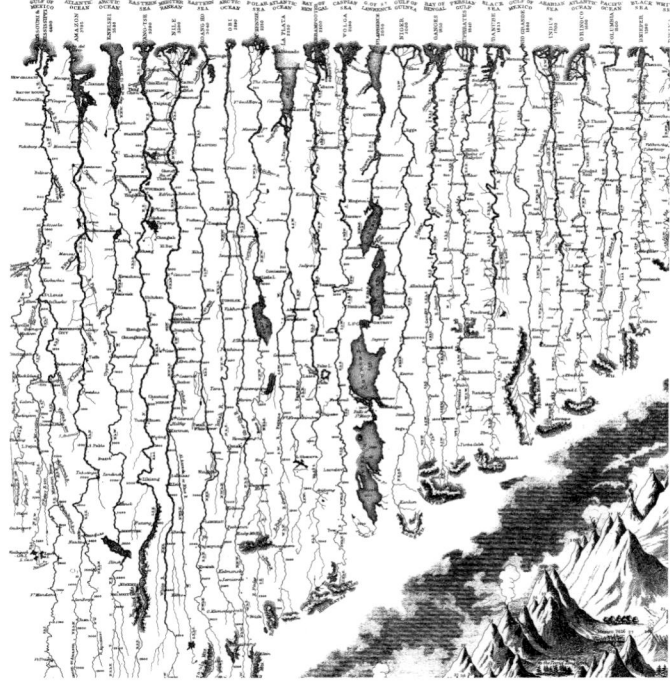

INTERPRETING THE VISUAL RECORD *This diagram showing the headwaters, tributaries, lakes, and deltas of major rivers was published in 1864. The rivers appear side by side.* **Use a map of North America to identify the lake and river system in the diagram's center. This river is the heart of what functional region?**

 The Great Lakes on the border between the United States and Canada together make up the largest area of freshwater in the world.

VISUAL RECORD ANSWER

St. Lawrence River; Great Lakes/southeastern Canada

Teach Objective 2

LEVEL 1: Call on volunteers to tell about recent flooding incidents they have experienced personally, read about in newspapers, or seen reported on television. Encourage them to discuss where and when the flooding occurred, what caused the flooding, and the extent of the damage. Then lead the class in a discussion of steps that might be taken to prevent flooding.

LEVELS 2 AND 3: Provide each student with a topographical map of your state or region. Have students shade the areas on the map they think might be prone to flooding. Call on students to explain why flooding might be a threat to the areas that they marked.

Close

Lead a discussion about water-related issues that your community may face in coming years. Those issues may include a dropping water table, threats to nearby wetlands, recurring floods, or other related topics.

Global Perspectives

Losing Groundwater

Since pumps have become available worldwide, many of the world's most populous countries have been removing groundwater at a rapid rate. India, for example, is pumping out groundwater twice as fast as it can be replenished. In Palestine's Gaza Strip, the Indian state of Gujarat, and Florida—areas where withdrawal of water for irrigation has been heavy—seawater is seeping into the groundwater.

Although there is 60 times as much water in underground aquifers as in lakes, ponds, and rivers, water tables in many areas are dropping. Most countries have no limit on how much water can be withdrawn from these irreplaceable natural resources.

DISCUSSION: Lead a discussion on this question: Should a landowner be allowed to pump groundwater even if doing so causes a neighbor's well to run dry?

VISUAL RECORD ANSWER

those adapted to true desert conditions

Our Amazing Planet

The Everglades is still one of the world's great wetlands, although it is now about one half of its original size. The marshland once covered almost 9 million acres (3.6 million hectares).

INTERPRETING THE VISUAL RECORD

The Okavango River empties into an area that receives little rainfall, forming a large swamp in what is otherwise a desert. The swamp is rich in vegetation, including reeds and water lilies. Examples of the animals that live there are storks, cranes, ducks, hippopotamuses, crocodiles, and lions. **What kinds of plants and animals might live there if the Okavango River flowed to the sea?**

water, seawater and freshwater mix. Chesapeake Bay on the Atlantic coast of the United States is a good example. Many estuaries provide good harbors, and most of the largest port cities in the United States lie on estuaries. Examples include New York, New Orleans, and Seattle. Estuaries are rich in shellfish and fish, so they must be carefully protected from pollution. Their usefulness as harbors often conflicts with their importance as shelters for animals.

Surface water is also found in **wetlands**. A wetland is any landscape that is covered with water for at least part of the year. They include bogs, coastal marshes, river bottomlands, and wooded swamps. The Everglades in Florida and the Okavango inland delta in southern Africa are two of the world's most famous wetlands. Wetlands support large populations of fish, shellfish, birds, and native plants. Many migrating birds depend on wetlands for food, water, and rest during their long journeys. Unfortunately, people have drained, paved, or filled many wetlands and used the land for farms, houses, and industrial sites. More than half the original wetlands in the United States have been destroyed. Preserving the remaining wetlands has become an important environmental issue.

Groundwater Water found below ground is called **groundwater**. When rainwater sinks into the ground, it is first stored in the soil. Plant roots reach down through the soil to take up this water. Water in the soil slowly moves downslope and deeper underground. The cracks and spaces in the rock immediately below the soil contain both air and water. However, farther down all the spaces inside the rock are filled with water. The level at which all the spaces are filled with water is called the **water table**. The level of the water table rises and falls with rainy and dry seasons.

In some areas, groundwater is being used up as the water is pumped out for both farms and cities. As a result, water tables may drop. In addition, the ground may settle or slump where it is no longer supported by water below. This process can damage buildings, roads, sewer lines, and other structures. In some places, big cracks open in Earth's surface. For example, these cracks have caused considerable damage in southern Arizona.

✔ **READING CHECK:** *Physical Systems* What are some common sources of water on and near land? rivers, lakes, wetlands, estuaries, groundwater

INTERPRETING THE VISUAL RECORD

Floods can devastate human settlements, but they can also make human settlement possible. This satellite photo of the Nile River delta in northern Egypt shows how the river turns the desert into fertile farmland. For thousands of years, the Nile flooded along its length every year, bringing rich mud to the Egyptian fields. A dam now controls the Nile's waters. **How do you think the Nile's flooding has affected the size of Egypt's population?**

Floods

Floods occur when rivers carry more water than the stream channels can hold. They usually result from heavy rains or sudden snow melts. Floods erode land and destroy vegetation. They can also drown people and livestock. People who do not drown may die from starvation after the flood has destroyed their crops. Others may die from diseases caused by polluted drinking water. Flooding is natural, but human activity can make floods worse. For example, people increase surface runoff when they clear vegetation from land and cover the ground with pavement and buildings. Flooding increases if rainwater cannot sink quickly into the soil.

The damage flooding causes is often increased because, for many reasons, people choose to live next to rivers. Floodplains are fertile farmland. Cities and industries are typically located along rivers because of the easy water transport, waterpower, and abundant water supply. For thousands of years, governments have been trying to control floods with dams and artificial channels. These measures work to some degree, but flood disasters still occur.

✓ **READING CHECK:** *Environment and Society* How do floods affect the environment and people? erode land, destroy vegetation, drown people and livestock, destroy crops, pollute water supplies

2 Review

Homework Practice Online
Keyword: SW3 HP4

Define desalinization, hydrologic cycle, headwaters, tributary, watershed, drainage basin, estuaries, wetlands, groundwater, water table

Reading for the Main Idea

1. (*Physical Systems*) How much of the water on Earth is liquid freshwater?

2. (*Physical Systems*) How and why do humans destroy wetlands?

Critical Thinking

3. **Evaluating** What might be some advantages and disadvantages of policies that limit economic development and water use in and around estuaries and wetlands?

4. **Drawing Inferences and Conclusions** How does the presence of buildings and roads make flooding worse?

Organizing What You Know

5. Create and label a pie graph showing the distribution of water on Earth.

OBJECTIVES

1. Explain why soil and forests are important resources.

2. Clarify concerns over water quality and air quality.

3. Identify some of the ways that minerals are used.

4. List the main energy resources and discuss how they are used.

 **LET'S GET STARTED**

Copy the following instructions onto the chalkboard: *Select an object in this classroom. From what materials is the object made? From where, in turn, do those materials come? (Example: zippered jacket—cotton, synthetic fibers, metal; cotton—needs soil and water to grow; synthetic fibers—from petroleum; zipper—from a metal ore)* Point out that the soil, water, petroleum, and metal ore are resources—materials that people need and value. Tell students they will learn more about the types and uses of resources in Section 3.

Building Vocabulary

Write the key terms on the chalkboard. Underline the following words and word parts: *leach-, contour, exhaustion, rotat-, de-, re-, acid, aque-, -ducts, aqui-, fossil, petro-, hydro-, geo-,* and *-thermal.* Have students work individually or in pairs to find one or more of the terms in a dictionary, relate the word or word part to the term's meaning, and explain the complete term to the class. Then call on volunteers to read any key-term definitions not covered by the discussion of the word parts.

 Section 3 # Natural Resources

READ TO DISCOVER

1. Why are soil and forests important resources?
2. What are the concerns about water quality and air quality?
3. What are some of the ways minerals are used?
4. What are the main energy resources, and how are they used?

WHY IT MATTERS

Preventing erosion of topsoil is a major concern in many countries, including the United States. Use **CNNfyi.com** or other **current events** sources to learn about soil conservation programs.

DEFINE

humus	acid rain
leaching	aqueducts
contour plowing	aquifers
soil exhaustion	fossil water
crop rotation	ore
irrigation	fossil fuels
soil salinization	petrochemicals
deforestation	hydroelectric power
reforestation	geothermal energy

Soil and Forests

Landforms and water are among Earth's most visible features. They are also essential natural resources. A resource is any physical material that makes up part of Earth and that people need and value. Some of Earth's natural resources are renewable, while others are nonrenewable. Renewable resources, such as soil and forests, are those that natural processes continuously replace. Nonrenewable resources are those that cannot be replaced naturally after they have been used.

Soil Building   Soil is natural material that includes both rocky sediments and organic matter. The sediments can come from the rocks below the ground surface. They may also be transported to the site by streams and the wind. The organic material can be decayed plant and animal matter. Soil differs from place to place, and there are hundreds of soil types.

Thick soil makes the state of Mississippi a fertile farming area. The layer of soil in the picture was probably picked up elsewhere by the wind and deposited there.

Near the surface, bacteria, insects, and worms break down the plant and animal material into a mixture called **humus**. These creatures—along with larger animals like rodents—open up spaces in the soil. The spaces allow air and water into the soil. Weathered rock material, organic matter, gases, and water must all be present in the soil to support plant life.

One important factor in the composition of soil is the rock the soil particles come from—the parent rock. Minerals in the soil vary depending on the type of parent rock. For example, eroded sandstone produces a different type of soil than does eroded granite or limestone. Climate is another major factor in soil variation. Moisture, sunlight, temperature, wind, and plants and animals in the soil all influence rock weathering. Soils vary between landforms as well. For example, sloping ground, like the side of a ridge, tends to erode steadily. As a result, the soil there is usually shallow. In contrast, sediments generally build up on valley floors, creating deep soils.

Soil develops very slowly, taking a few or even hundreds of years. It forms distinct layers, called soil horizons. (See the diagram.) Most soils have three layers, called the A, B, and C horizons. The upper layer, the A horizon, is also called the topsoil. This horizon is rich in humus and has active populations of plants and small animals. This horizon also has the most plant roots. The B horizon, sometimes called the subsoil, is the middle layer. This horizon is mostly minerals, but the roots of large plants penetrate to these depths. The C horizon is composed of weathered rock fragments of the soil's parent material. The depths of all these horizons can vary greatly from place to place. Taken together, the soil horizons at any place are called the soil profile.

Particularly important to soil profile development is the amount of rainwater moving through it. **Leaching** eventually moves the nutrients out of the reach of plants' roots. Leaching is the downward movement of minerals and humus in soils. Areas with abundant rain often have deep soils where many minerals have been leached downward. The result can be soil with low fertility even though the area's vegetation may be a rain forest. On the other hand, soils in dry areas are often shallow and leaching is limited. Soils here may contain abundant supplies of the minerals needed by plants.

Around the world, soil profiles vary according to climate and major vegetation types. The soils beneath semiarid grasslands in central Kansas are much the same as soils beneath semiarid grasslands in Ukraine. Similarly, the soils in the rainy tropics of South America resemble the soils in the rainy tropics of Africa. However, soil profiles also vary based on local environmental conditions. For example, soils next to swamps have abundant moisture and humus, no matter what the climate is like.

✓ **READING CHECK:** *Physical Systems* Why do some dry areas have very fertile soil?
limited leaching, so minerals remain near the surface

Sustaining Soil Resources Producing the world's food depends on soil. For this reason, preserving soil resources is important to all of us.

Soil erosion is a natural process. However, farming usually leads to increased erosion, removing soil faster than new soil can be formed. Maintaining the A horizon is particularly important because this is where most plant nutrients are found. Farmers can control erosion by plowing less and by using **contour plowing**. Contour plowing works across the hill, rather than up and down the hill. This slows the movement of water down the slope, reducing erosion.

Soil Horizons

Topsoil (Humus)

Subsoil

Broken Rock

Parent Rock

INTERPRETING THE DIAGRAM *The three layers of soil are the topsoil, subsoil, and broken rock.* **What physical processes may have broken the rock beneath the subsoil?**

LEVEL 1: Copy the following graphic organizer onto the chalkboard, omitting the italicized answers. Have students complete the organizer by adding information on the causes and effects of air pollution. (You may want to point out that acid rain can also damage building exteriors.) Then have students create a similar graphic organizer to depict the causes and results of issues surrounding our water supply and water quality. **ENGLISH LANGUAGE LEARNERS**

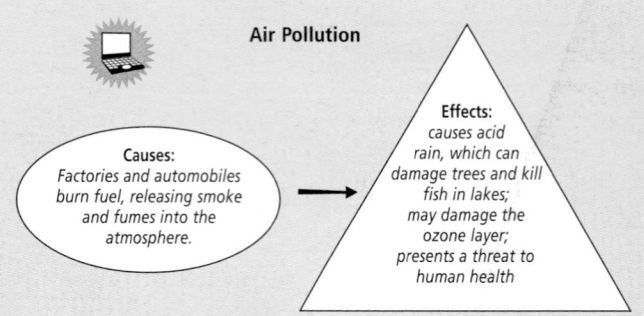

Air Pollution

Causes:
Factories and automobiles burn fuel, releasing smoke and fumes into the atmosphere.

Effects:
causes acid rain, which can damage trees and kill fish in lakes; may damage the ozone layer; presents a threat to human health

---

Linking Past to Present

Salty Soil Sank Sumer

Sumer, in what is now Iraq, was one of the world's earliest civilizations. Evidence shows that soil salinization contributed to Sumer's decline. Careless irrigation may have been partly to blame. For a time, Sumerian farmers controlled salinization by limiting irrigation and allowing farmland to rest every other year. As the growing population required more food, however, these practices ended.

At one point, the Sumerian city of Umma blocked the canals that fed water to an area also claimed by the city of Girsu. In return, Girsu's king dug a new canal. Although it brought in much-needed water, the canal caused the water table to rise and brought more salt into the soil.

Sumerian farmers gradually changed from wheat to salt-tolerant barley. Eventually, wheat was no longer cultivated in southern Mesopotamia. Overall crop yields dropped by two thirds. With its agricultural base weakened, Sumer lost political power.

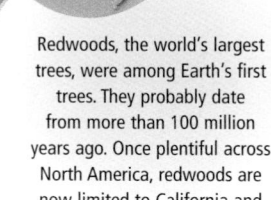

Our Amazing Planet

Redwoods, the world's largest trees, were among Earth's first trees. They probably date from more than 100 million years ago. Once plentiful across North America, redwoods are now limited to California and Oregon in a narrow area near the Pacific Ocean.

VISUAL RECORD ANSWER

major role, because the steeper the slope, the more important contour plowing would be

---

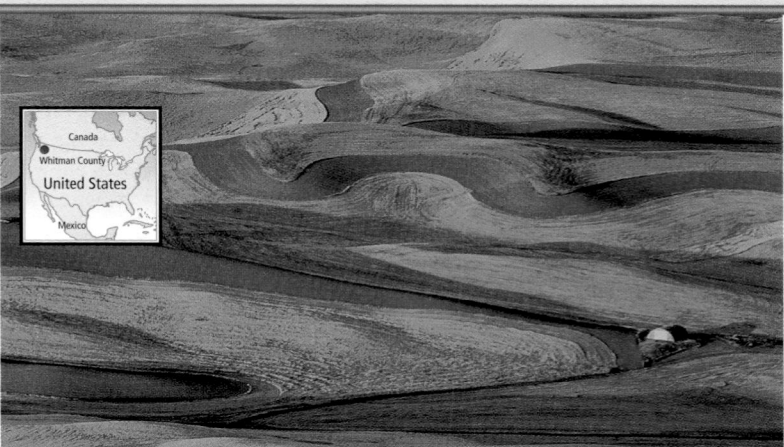

INTERPRETING THE VISUAL RECORD

Farmers are trying to preserve the soil of this field in southeastern Washington State by using contour plowing. **What role might the angle of a field's slope play in contour plowing?**

Crops draw certain nutrients out of the soil. If the same crop is planted year after year, a field may suffer nutrient loss. In extreme cases this loss can lead to **soil exhaustion**. This is a condition in which the soil becomes nearly useless for farming. The usual way to build up soil nutrients is with fertilizer. Some fertilizers are natural organic materials like manure and decayed plants. Chemical fertilizers are also used. Yields can also be maintained by **crop rotation**, or planting different crops in alternating years. Crop rotation gives a field time to replace naturally the nutrients used by each different crop.

People often look to the world's drylands to expand farming. Many dry regions can support farming if water is artificially supplied to the land—a process called **irrigation**. However, irrigation can lead to **soil salinization**, or salt buildup in the soil. Salt is destructive to nearly all crops. Small amounts of salt are often present in irrigation water. Evaporation of the water concentrates the salt on the surface of the soil. Irrigation can also cause salt to rise into the topsoil from the parent materials below. Soil salinization is a serious problem in many dry areas of the world, including the southwestern United States.

READING CHECK: *Environment and Society* How do farmers adapt to the need to avoid soil exhaustion? with fertilizers and crop rotation

Forests Forests protect soil from erosion, provide habitats for many different species, and yield useful products. In addition to wood, we get food, medicines, oils, and rubber from forests.

Deforestation is the destruction or loss of forests. The tropical rain forests of Africa, Asia, and Latin America are being cleared at a rapid rate today. Logging for wood, clearing for farmland and ranch land, and cutting wood for fuel are the main causes of deforestation. Deforestation is also a problem in North America.

Many countries are taking steps to protect their forests. For example, cutting trees has been outlawed in some areas. Many countries require that new trees be planted after an area has been cut. This replanting process is called **reforestation**. In many cases the young trees can be harvested after a few decades. Still, these new forests cannot fully replace the ecological diversity of the original old-growth forests.

READING CHECK: *Environment and Society* What policies and regulations have countries taken to prevent deforestation? making some areas off-limits to cutting, requiring reforestation

LEVEL 2: Have students design covers for a magazine titled Water Monthly. Have them include titles and one-sentence descriptions of several articles on the economic importance of maintaining a dependable supply of clean water. Call on volunteers to display and discuss their magazine covers.

LEVEL 3: Have students complete the Level 1 and Level 2 activities. Then ask them to write a brief article for one of the titles that appears on their Water Monthly covers. Provide supplemental materials for use in the classroom. In addition, have students write letters to the magazine's editor, refuting or supplementing their articles.

Teacher to Teacher

Frank Thomas Jr. of Austin, Texas, suggests the following activity to help students understand the different types of irrigation. Organize the class into several small groups and assign each group a different type of irrigation—canal, center-pivot, drip, sprinkler, and so on. Direct groups to search Internet and library resources to find information on the assigned irrigation method. Then have groups use their findings to create a visual aid that demonstrates how their irrigation method works. Point out that the visual aid might be a poster, diagram, model, or even a skit. Call on groups to present their visual aids to the class.

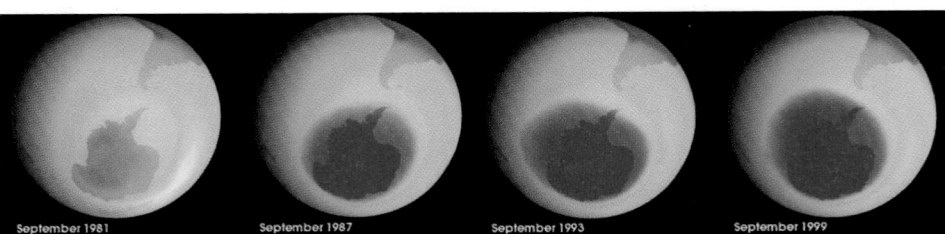

September 1981   September 1987   September 1993   September 1999

Air and Water

All living things on Earth depend on clean air and water. Fortunately, they are renewable resources. However, as the population grows and industry expands, our air and water supplies are at risk.

Air Humans, animals, and plants need the gases in air to survive. Pollution threatens our air supply. Automobiles and factories release smoke and fumes into the atmosphere. These chemicals build up in the air, particularly in large cities. The chemicals can interact with sunlight and create a mixture called smog. Some cities, such as Los Angeles and Mexico City, have terrible smog. At high enough levels, air pollution is dangerous to human health.

Some chemicals in air pollution form a mild acid when they combine with water vapor in the atmosphere. The acid can be similar in strength to vinegar. When it falls, this **acid rain** can damage trees and kill fish in lakes.

The effects of smog and acid rain are generally short-term. Air pollution also has the potential to cause long-term change in Earth's atmosphere. Certain kinds of pollution can damage a part of the atmosphere called the ozone layer. The ozone layer is important because it absorbs harmful ultraviolet radiation

These satellite images show how the Southern Hemisphere's ozone layer has been damaged. The purple area is a hole in the ozone. During September 2000 the hole extended for the first time over a populated area—the city of Punta Arenas, Chile. People there were briefly exposed to high levels of ultraviolet radiation.

Connecting to TECHNOLOGY

Irrigation

Irrigation water is delivered to farmland by many different methods. The oldest ways involve directing water into a field, usually from a small canal. This process can be as simple as flooding, in which water covers the entire surface of the field. Water can also be run into furrows—shallow trenches plowed between rows of crops. In either case, ditches are usually dug at the lower edge of the field. They allow the excess water to drain off the field. Proper drainage is very important for preventing soil salinization. If water is allowed to remain on a field and evaporate, salts are likely to concentrate in the soil.

More modern irrigation techniques use water more efficiently. Sprinkler irrigation is widely used in the United States. This method allows water flow to be controlled, from a mist to a heavy soaking. Sprinkler systems often use an automated device to travel over a field. The center-pivot system uses a long arm with sprinklers mounted on it. Anchored in the center of the field, the arm sweeps in a circle.

Drip irrigation, the most efficient method, applies water directly to the base of the plant with special "leaky" hoses. Drip irrigation is relatively expensive to set up and maintain. As a result, farmers generally use it only where water is very scarce. However, it is also used where a crop may be valuable enough to offset the high cost.

Analyzing Information How have different kinds of modern irrigation systems aided agriculture?

Sprinklers irrigate hay fields in Idaho.

77

LEVEL 1: Prepare a series of flash cards, with each card carrying the name of one of the minerals discussed in the text. Hold up each flash card in turn and call on a volunteer to describe how the mineral named on the card is used. As you work through the cards, have students write down the minerals and their uses. **ENGLISH LANGUAGE LEARNERS**

LEVEL 2: Have students use the notes that they took in the Level 1 activity as the basis for a newspaper article on using mineral resources for the Teen News section of the newspaper. Articles should be titled "Mineral Resources— What They Are and How We Use Them." Suggest that students illustrate their articles with appropriate visual materials. Call on volunteers to read, display, and discuss their articles.

Linking Past to Present

Roman Aqueducts The first Roman aqueduct was built in 312 B.C. It was 10 miles (16 km) long. During the next 500 years, 11 aqueducts brought water from springs in the hills east of Rome to the city.

Pipes were made mostly of stone and terra cotta, but some were made of wood, leather, lead, or bronze. They ran at ground level or underground to protect the water supply from enemies. About 10 percent of the aqueducts were carried on beautiful high arches made of stone, brick, and cement.

While gravity brought water to Rome from up to 59 miles (95 km) away, modern aqueducts far surpass those built by the Romans in both length and capacity. New York's three main aqueducts bring water from up to 120 miles (192 km) away, and water flows 230 miles (370 km) to Los Angeles.

ACTIVITY: Have students work in groups to investigate aqueducts and water transport systems in ancient cultures. Possibilities include Egypt, Assyria, Babylonia, Persia, India, and Greece. You may want to have students build scale models of the systems.

The ancient Romans built this aqueduct in southern Spain. Rows of arches support the water channel at the top of the aqueduct. Some Roman aqueducts are still in use. Modern aqueducts are also crucial links in water supply systems. California's aqueduct system is the largest in the world.

from the Sun, protecting living things. As pollution damages the ozone layer, more solar radiation may cause an increase in skin cancer. You read about another long-term result of air pollution in the previous chapter. Some polluting gases may contribute to global warming.

Water In many parts of the world, maintaining a dependable supply of clean water is a major challenge. Cities and farms in dry regions often require water to be stored or transferred from other areas. For centuries, people have built complex systems of dams, canals, reservoirs, and **aqueducts**—artificial channels for transporting water.

In many dry areas, groundwater is the only dependable source of water. Thousands of years ago people learned to reach groundwater by digging wells. Today wells are drilled into the ground, and pumps bring the water to the surface. Wells are drilled to the depth of rock layers where groundwater is plentiful. These layers are called **aquifers**. Groundwater must be protected from pollution. This pollution is hard to remove.

Most groundwater is fed by rain. Because it can be preserved and managed like surface water, groundwater is usually a renewable resource. However, some groundwater, particularly in desert areas, was deposited long ago when the climate was wetter. This **fossil water** is not being replenished by rain. Today fossil water is being pumped to farmlands in many desert areas. These wells will run dry someday.

✓ **READING CHECK:** **Environment and Society** How is pollution caused by humans affecting Earth's air and water? creating smog, acid rain, creating ozone problems, global warming

Our Amazing Planet

Mining is dangerous! The deepest shafts of one South African gold mine reach 13,000 feet (3,962 m) below the surface. Temperatures in the mine would reach 140°F (60°C) if huge refrigeration units did not pump cool air from the surface.

Mineral Resources

Minerals are solid substances that come out of the ground. Examples include metals, rocks, and salt. Some minerals, such as the quartz in sand, are quite common. Others are hard to find—gold, for example. Over the centuries, people have developed countless uses for minerals.

LEVEL 3: Have students complete the Level 1 and Level 2 activities. Call on volunteers to share with the class what they know about recycling programs in the community. If there are no organized programs, or students do not know of any, ask them to describe the kinds of programs they would like the community to adopt. Encourage students to take notes as the various programs are discussed. Have them use these notes to write letters to the editor of a local newspaper urging the community to begin, continue, or expand a recycling program. Select students to read their letters to the class. Use these letters as a starting point for a class discussion on the value of recycling.

Teach Objective 4

LEVEL 1: Organize students into teams. Have each team make two oversize flash cards, one with *renewable* written on it, the other with *nonrenewable* written on it. Then name an energy resource and ask each team to hold up a flash card identifying the resource as renewable or nonrenewable. Award one point for correct answers. When you have read through all the energy resources mentioned in the text, tally scores and announce the winning team. Then discuss how these energy resources are used. **ENGLISH LANGUAGE LEARNERS, COOPERATIVE LEARNING**

Uses of Minerals Minerals are used in many processes and products, including construction, jewelry, and manufacturing. Rock, such as limestone, can be cut into blocks for building. Metals, from aluminum to zinc, are shaped into a vast range of products. Our daily lives are filled with metal objects, from airplanes to soft drink cans. Other important materials are made up of many different minerals. For example, sand that contains the mineral quartz is a key ingredient in glass.

Workers extract minerals from holes in the ground called mines. Some mines, like the gold mines in South Africa, tunnel deep underground. Other mines are open pits, some of which are miles across. Long ramps allow giant trucks to move in and out of the pit. The trucks move the **ore**—the mineral-bearing rock—to a processing plant where the valued mineral is removed.

Some minerals are truly rare. For example, gemstone-quality diamonds are minerals found in just a few countries. In addition, the deposits that contain diamonds are often only a few hundred yards across.

A mine employee shows the reddish ore of the largest iron mine in the world. This open-pit mine produces some 45 million tons of iron annually, along with gold and other metals. It is located deep in the forest of northern Brazil.

Mineral Recycling Today many products are made from recycled minerals. The advantages of recycling are obvious. For one thing, the more we recycle, the slower we will use up nonrenewable resources. For another, mines often cut great scars into the landscape. Processing plants belch dust and smoke and consume large amounts of energy. In contrast, using mineral products again can save money and spares damage to the environment. Recycling also saves space in landfills.

Recycling is not new. Precious metals like gold and silver have been reused over and over again through the centuries. However, increasing environmental awareness has encouraged more reuse of mineral products. Today aluminum, copper, and steel are often recycled.

✓ **READING CHECK:** *Environment and Society* Why is recycling resources important?
nonrenewable resources used up more slowly, helps the environment

Energy Resources

Some of the most important nonrenewable resources are sources of energy. These energy resources have become increasingly useful and valuable as people develop new technologies that need them for power. Energy resources include coal, natural gas, and petroleum. These substances are called **fossil fuels** because they were formed from the remains of ancient plants and animals. Coal comes from ancient swamps and bogs. Petroleum and natural gas are the remains of tiny plants and animals that lived in seas and lakes. Over millions of years, heat and pressure converted the dead plants and animals into oil and gas.

Another energy resource is uranium, a metallic element. Uranium is radioactive and provides the energy for nuclear power plants. Nuclear power does not pollute the air, but it creates waste material that remains dangerous for thousands of years. The problem of disposing of nuclear waste has limited the use of nuclear power.

🔲 **internet** connect

GO TO: go.hrw.com
KEYWORD: SW3 CH4
FOR: Web sites about alternative energy sources

LEVEL 2: Have students work in small groups to design newspaper advertisements for companies that harness and sell renewable energy—hydroelectric power, wind power, geothermal energy, and solar energy. Tell groups that their advertisements should focus on why it is important to develop and use renewable energy resources. Then display and discuss the advertisements. **COOPERATIVE LEARNING**

LEVEL 3: Have students undertake a cost-benefit analysis of the various energy resources discussed in this section. Ask them to present their analyses in brief summaries. *(Example: Coal, a nonrenewable resource, has many uses and is relatively cheap and plentiful, but its use can cause considerable environmental damage.)* Have students share and compare their analyses.

Close

Lead a discussion about the many things—from easy to difficult—that students can do to conserve resources.

Across the Curriculum

▶ Technology ◀

Hibernia Oil Platform As petroleum supplies run short in some areas, oil companies have begun to drill in more hazardous locations. To protect drilling operations off Newfoundland in an ice-plagued region of the Atlantic Ocean called Iceberg Alley, engineers built the Hibernia offshore oil platform. It is capable of withstanding a collision with a 5-million-ton iceberg.

Although the Hibernia should survive even the largest of icebergs—the kind that comes along once in 10,000 years—no one wants to test the claim. Special crews lasso and tow away smaller icebergs if they threaten the platform.

ACTIVITY: Have students conduct further research on offshore oil drilling, the hazards it presents, and efforts to solve those problems.

MAP ANSWER

Global trade patterns develop to link resources with consumers.

Fossil Fuels Coal was the first fossil fuel to be used as a major energy source. It is a solid and is often found near the land surface. People have long used coal for heat. Coal also powered early steam engines and steel mills. Coal remains an important fuel for electric power generation. However, it has environmental disadvantages. Often it contains minerals that do not burn. This waste material either goes into the air as pollution or must be disposed of in landfills. Burning coal that contains sulfur is one of the causes of acid rain. However, coal has the advantage of being plentiful. At current rates of consumption, the world has enough coal for centuries to come.

Petroleum and natural gas come from wells drilled into deep sedimentary deposits. About 150 years ago, people began to use petroleum as fuel for lamps. The invention of the automobile later greatly expanded the market and need for petroleum, also called oil, as a fuel source. Today oil is made into gasoline, diesel and heating fuel, asphalt, and many other products. As oil has become more useful, it has become more valuable. Today its production and processing is a major industry. However, supplies are limited. Reducing the need for oil is important for our future.

Natural gas is usually odorless, but it is given an artificial smell so that leaks can be detected. Natural gas is particularly important for home and industrial heating. Both natural gas and oil are also used to generate electricity. They usually burn more cleanly than coal. Natural gas is growing in importance as a fuel. It is abundant, and supplies will certainly be available for decades to come. On the other hand, natural gas prices have been rising.

Fossil fuels are also important sources of chemicals. Coal has long been important in making dyes. Oil is the source of many important products called **petrochemicals**. Petrochemicals include the raw materials for many explosives, food additives, medicines, pesticides, and plastics. Having dependable sources of fossil fuels is a key factor for a country's economic success. Unfortunately, as

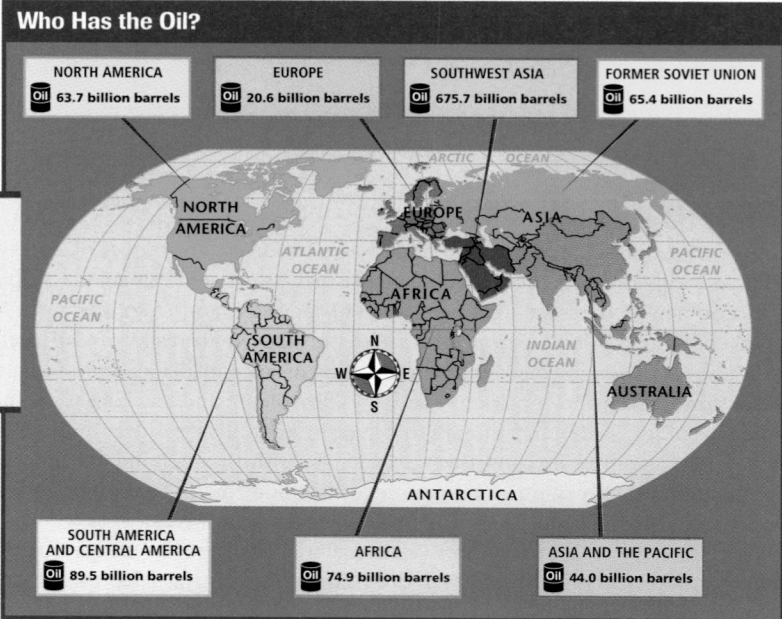

Who Has the Oil?

NORTH AMERICA	EUROPE	SOUTHWEST ASIA	FORMER SOVIET UNION
Oil 63.7 billion barrels	Oil 20.6 billion barrels	Oil 675.7 billion barrels	Oil 65.4 billion barrels

INTERPRETING THE MAP *About two thirds of all known oil deposits are located in Southwest Asia. New deposits are still being found, but Southwest Asia will probably continue to hold the largest share.* **How do you think the location of oil deposits affects global trade routes?**

SOUTH AMERICA AND CENTRAL AMERICA	AFRICA	ASIA AND THE PACIFIC
Oil 89.5 billion barrels	Oil 74.9 billion barrels	Oil 44.0 billion barrels

fossil fuels are burned, they release carbon dioxide. Carbon dioxide is one of the gases, called greenhouse gases, that keep extra heat in the atmosphere.

✓ **READING CHECK:** *Environment and Society* How are fossil fuels used?
energy resources and sources of chemicals

FOCUS ON SCIENCE, TECHNOLOGY, AND SOCIETY

Renewable Energy Resources Renewable energy sources include waterpower, wind power, heat from Earth's interior, and solar energy. **Hydroelectric power**—electricity produced by moving water—is the most widely used type. Dams hold water back and force it to flow through narrow openings. As the water is pulled through these openings by gravity, it turns generators to produce electricity. Dams produce about 10 percent of the electricity used in the United States.

Wind has been used to propel ships for thousands of years. Windmills were also invented centuries ago. They were used to pump water out of wells or grind grain. Modern versions of windmills, called wind turbines, create electricity. Although wind contains great power, it may take thousands of wind turbines to equal the electricity of a conventional power plant. This fact limits the use of wind power.

The heat of Earth's interior, called **geothermal energy**, can also be used to generate electricity. Geothermal power plants, which capture this energy, are built in places with volcanoes or hot springs.

The energy of the Sun—solar energy—can also be used to create power. Special solar panels can absorb solar energy and convert it to electricity. Also, the Sun's rays can be concentrated and used to heat water or homes. Although solar energy has great promise, the technology needed to harness it is costly and complex. Breakthroughs in the future may make solar energy a major power source.

✓ **READING CHECK:** *Environment and Society* What has been the effect of new technologies and new perceptions of fossil fuel resources?

Dams harness waterpower. Turbines and solar panels harness wind and solar power.

INTERPRETING THE VISUAL RECORD *Some renewable energy sources can also be called flow resources. These resources must be used when and where they occur. They include running water, sunlight, and wind. For example, wind power can only be effective in places that have strong steady winds. Wind turbines, like these in southern California, harness wind to produce energy.* **How might stricter regulations on the use of fossil fuels affect the use of flow resources?**

Section 3 Review

go.hrw.com
Homework Practice Online
Keyword: SW3 HP4

Define humus, leaching, contour plowing, soil exhaustion, crop rotation, irrigation, soil salinization, deforestation, reforestation, acid rain, aqueducts, aquifers, fossil water, ore, fossil fuels, petro-chemicals, hydroelectric power, geothermal energy

Reading for the Main Idea

1. *Physical Systems* From which soil horizons do plants draw nutrients?

2. *Physical Systems* What are the two causes of soil salinization?

Critical Thinking

3. **Drawing Inferences and Conclusions** Are old-growth forests a renewable or nonrenewable resource? Why?

4. **Analyzing Information** How have changing perceptions of oil's usefulness made it more valuable over time?

Organizing What You Know

5. Copy the table below. Use it to list two short-term and two long-term threats to the atmosphere.

Short-term threats	Long-term threats

Economics: The Importance of Fishing

Point out that water biomes shelter a vast range of species, from microscopic organisms to gigantic mammals and fish. In areas with access to water biomes, fishing has been an important economic activity for thousands of years. There are two basic water biomes—freshwater and saltwater, also called marine. The freshwater biome includes creeks, lakes, ponds, rivers, and wetlands. Seas, oceans, and marine wetlands make up the marine biome. These biomes can be further subdivided according to water depth.

Organize students into several groups. Have each group select a U.S. region that has easy access to a large freshwater or marine

biome. Have groups conduct research on commercial fishing in the selected region to answer these questions: Which fish or shellfish species are important to the local economy? What percentage of the region's workers are employed by the fishing industries? What other industries, such as canning or boat repair, depend on the fishing industries? How do fluctuations in other industries, such as oil production, affect fishing? How have new technologies affected fish catches and the number of people employed by fisheries? How do changes in transportation and communication affect the fisheries? Have groups present their findings in a report illustrated with maps, sketches, charts, or graphs. Call on groups to present and discuss their reports. **COOPERATIVE LEARNING, BLOCK SCHEDULING**

The Flooded Forest Point out the large area in South America designated as tropical rain forest on the world ecosystems and biomes map. This is the largest tropical rain forest in the world. The Amazon River and its many tributaries flow through it.

Biomes can expand or shrink laterally, depending on climate changes. Within the South American rain forest, however, there is also a vertical element. Twice during the year the upper course of the Amazon floods. In places, floods raise the river level 40 to 50 feet (12 to 15 m). This flooded forest is known as the *várzea*. Then pink river dolphins and relatives of crocodiles called caimans swim among the trees. Some fish eat fruit that drops from the branches.

Organize the class into two groups—one to study the plants and animals of the Amazon rain forest as it is at low water and the other to study how flooding affects the plants and animals. Then have the groups combine their findings on a mural to depict the forest's annual cycle.

This Geography for Life feature addresses National Geography Standards 7 and 8.

Physical Systems

Geography for Life

World Ecosystems and Biomes

Plant and animal communities that cover large land areas are called biomes. Areas that share the same biome have similar ecosystems. These ecosystems display complex relationships among the area's climates, vegetation, soil, and geology. Some biomes even have the same names as some climate regions.

Different biomes also can be found in the oceans. In contrast to land biomes, marine biomes depend mainly on the depth at which the plants and animals live. For now, however, we will limit our discussion to biomes on land.

What's in a Biome?

A single biome can exist in places scattered around the world. Different species that live in areas with the same biome may have much in common. We can use rain forests

as one case. The plants and animals in Africa's rain forests are different species than those of Brazil's rain forests. Monkeys of the Western Hemisphere, for example, are different from African monkeys. However, the two groups' appearance and behaviors are often similar because they have adapted to similar environments. Monkeys of both hemispheres spend much of their time in trees, because they find their main foods in trees.

These similarities are what make biomes such a useful way to classify large parts of the world. Geographers can make generalizations about factors like a biome's structure and seasonal growth patterns. Geographers can also generalize about ways people use the biome's plants and animals. In turn, this information can help people manage and conserve these environments.

World's Biomes

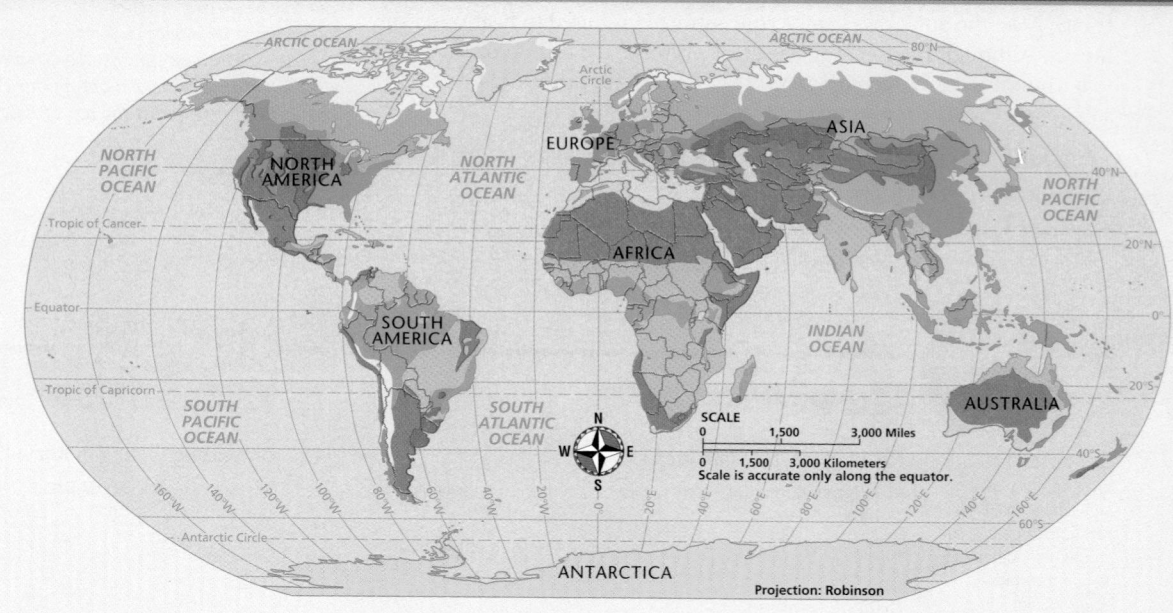

Going Further: Thinking Critically

As students may discover while working on the Interdisciplinary Study activity, removing fish from a biome affects other organisms in that biome. Often the effects are minimal, and the overall health of the biome is maintained. Sometimes, however, removing too many fish affects other species adversely—particularly those that prey on the commercially valuable fish.

Have students conduct research to identify ecosystems whose health is threatened by the removal of a single species or type of species. An example is menhaden, a herringlike fish used in animal feed. The decline of krill in Arctic and Antarctic waters is another. Students should first identify the source of the threat and then analyze the direct and indirect effects. Have students use their findings to produce public service announcements that focus on the role the animal plays in the ecosystem's health. Have students videotape their PSAs and present them to the class. **BLOCK SCHEDULING**

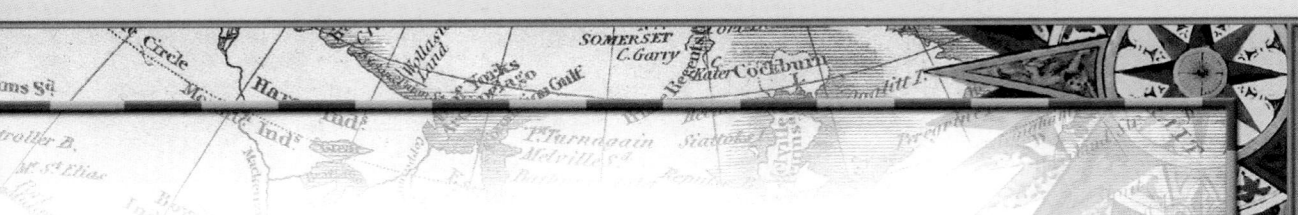

Distribution of Biomes

Latitude and the world climate zones have influenced the distribution of biomes. However, a biome results from the interaction of many factors. A sharp variation in one factor can result in a biome existing where you might not expect it. Perhaps the most dramatic examples are caused by variations in elevation. For example, the plants and animals that make up the tundra biome live in northern Alaska near the Arctic Circle. Similar life-forms can be found near the top of Kilimanjaro, a mountain on the equator in eastern Africa.

Variation in soil and geology can also determine a given place's biome type. The process of desertification provides a clear example. If animals overgraze a grassland area, wind or water can erode the topsoil. Without topsoil to support plant growth, the area can become a desert. This process can take place in a dramatically short time. On the other hand, fertile volcanic soil can support denser and more varied vegetation than thin nutrient-poor soil. Different types of fertile soil also support different plants.

For example, an isolated pocket of sandy acidic soil supports a small forest of tall pines in Bastrop County, Texas. Surrounding the so-called Lost Pines are oak and pecan trees growing in different soils. Thus, two places with the same climate but different soils have different biomes.

Applying What You Know

1. **Drawing Inferences and Conclusions** Compare the map of world biomes to the map of world climate regions in Chapter 3. What connections do you see between the two distribution patterns?

2. **Analyzing Information** Describe the climate, soils, vegetation, and representative animals of the biome in which you live. What geological features do you think have influenced plant and animal life in your region? How?

Applying What You Know Answers

1. Answers should note that climate, soil, vegetation, and typical animals are all interrelated.
2. Students might mention mountains, faults, or other geological features.

Review Answers

Building Vocabulary For definitions, see: magma, p. 63; continental drift, p. 63; abyssal plains, p. 65; sediment, p. 67; plateau, p. 69; delta, p. 69; headwaters, p. 71; watershed, p. 71; wetlands, p. 72; groundwater, p. 72; humus, p. 75; crop rotation, p. 76; irrigation, p. 76; fossil water, p. 78; fossil fuels, p. 79

Locating Key Places

A. groundwater
B. moist air
C. evaporation
D. condensation
E. precipitation

Understanding the Main Ideas

1. Plates can be moving apart, colliding, or moving laterally past each other.
2. weathering and erosion
3. Runoff carries minerals into the lake. The water continually evaporates, concentrating minerals and salts.
4. smog and acid rain
5. creates air pollution

Biome Legend

Biome	Climate	Soil	Vegetation	Typical Animals
Tropical rain forest	warm and rainy year-round	thin; poor in nutrients as a result of leaching	most diverse of all biomes; huge variety of trees, vines, and other plants	monkeys, lemurs, parrots, snakes, tree frogs, bats, pigs, small antelopes, tigers, jaguars, and leopards
Savanna	warm year-round; distinct rainy and dry seasons	generally poor in nutrients	tall grasses, some trees, and thorny shrubs	gazelles, rhinoceroses, giraffes, lions, hyenas, ostriches, crocodiles, and elephants
Semiarid and desert	dry; sunny and hot in tropical regions; wide temperature variations in middle latitudes	poor in organic matter but may have plentiful minerals	moisture-retaining plants such as cacti; shrubs and thorny trees	kangaroo rats, lizards, scorpions, snakes, birds, bats, and toads
Grassland	low rainfall; warm summers; cold winters	most fertile soils of all biomes	grasses; trees near water sources	lions; large grazing animals, including buffaloes, kangaroos and antelopes
Temperate forest	plentiful rainfall; warm summers; cold winters	very fertile; rich in organic nutrients	deciduous and evergreen trees; shrubs; herbs	deer, bears, badgers, squirrels, wolves, wild cats, red foxes, owls and many other birds
Mediterranean scrub forest	hot dry summers; cool wet winters	rocky; nutrient-poor soils	evergreen shrubs; scrubby trees; herbs	ground squirrels, deer, elk, mountain lions, coyotes, and wolves
Boreal forest	short warm summers; long cold winters	acidic soil	mosses, lichens, and coniferous trees	birds, rabbits, moose, elk, wolves, lynxes, and bears
Tundra	short cool summers; long cold winters	permafrost	mosses, lichens, sedges, and dwarf trees	rabbits, lemmings, reindeer, caribou, musk oxen, wolves, foxes, birds, and polar bears
Barren	cold year-round	thin; poor soils	few mosses and lichens in coastal areas	animals that depend on the sea including penguins, other birds, and seals

83

Review and Assessment Resources

TECHNOLOGY

▶ Chapter 4 Test Generator (on the One-Stop Planner)
▶ Global Skill Builder CD–ROM
▶ HRW Go site

REINFORCEMENT, REVIEW, AND ASSESSMENT

▶ Chapter 4 Review, pp. 84–85
▶ Chapter 4 Tutorial for Students, Parents, Mentors, and Peers
▶ Chapter 4 Test (form A or B)
▶ Alternative Assessment Handbook

▶ Chapter 4 Test for English Language Learners and Special-Needs Students
▶ Unit 1 Test
▶ Unit 1 Test for English Language Learners and Special-Needs Students

Assess

Have students complete a Chapter 4 Test.

Reteach

Organize students into five groups for these topics: landform creation, water distribution, soils and forests, air quality and water supply, and minerals and energy resources. Have each group create an idea web using the chapter material for its assigned topic. Display and review the webs. **ENGLISH LANGUAGE LEARNERS, COOPERATIVE LEARNING**

CHAPTER 4 Review Answers

Thinking Critically

1. the old mountain chain; worn down and smoothed by erosion over longer period of time (NGS 7)

2. Rain forests ordinarily have leached soil; any nutrients remaining could be depleted rapidly by planting. (NGS 14)

3. coal used to power early industries; oil used for lighting, then fuel for autos, used also for other products; scars landscape, pollutes (NGS 14)

Using the Geographer's Tools

1. Possible answer: Major river systems drain huge areas and provide transportation routes that link places in a region.

2. Diagrams should show rainwater taking nutrients below the reach of plant roots.

3. African plate—northeast; Antarctic plate—south and east; Arabian plate—northeast; Eurasian plate—southeast; Indo-Australian plate—northeast; Nazca plate—east; North American plate—northwest; Pacific plate—northwest; Philippine plate—northwest; South American plate—west

Writing

Students should list all the natural resources they use in a day and then list those resources a person living many years ago might have used.

CHAPTER 4 Review

Building Vocabulary

On a separate sheet of paper, explain the following terms by using them correctly in sentences.

magma	wetlands
continental drift	groundwater
abyssal plains	humus
sediment	crop rotation
plateau	irrigation
delta	fossil water
headwaters	fossil fuels
watershed	

Locating Key Places

On a separate sheet of paper, match the letters on the diagram of the hydrologic cycle with their correct labels.

evaporation	precipitation
condensation	groundwater
moist air	

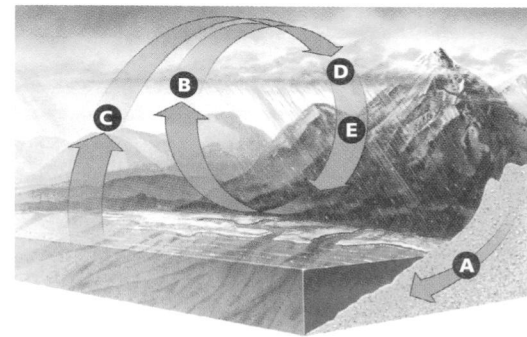

Understanding the Main Ideas

Section 1

1. _Physical Systems_ What are the three types of movements possible at plate boundaries?

2. _Physical Systems_ What are the two physical processes that wear down landforms on Earth's surface?

Section 2

3. _Physical Systems_ What process makes some inland lakes salty?

Section 3

4. _Environment and Society_ What are two short-term effects of air pollution?

5. _Environment and Society_ What is a major drawback of using coal as an energy source?

Thinking Critically

1. Analyzing Information Which would have a smoother shape, a young mountain chain or an old mountain chain? Why?

2. Drawing Inferences and Conclusions In an area where a rain forest has been cleared, why would you expect the soil to become exhausted quickly?

3. Identifying Cause and Effect What new technologies made fossil fuels more valuable over time? What effect have development and use of these resources had on the physical environment?

Using the Geographer's Tools

1. Creating Maps Review the maps of the United States in the atlas at the front of this textbook. Then sketch a map of the United States and shade the region drained by the Mississippi River system. In what ways do you think this and other river systems constitute a region?

2. Analyzing Information Sketch the soil horizons diagram. Label the A, B, and C horizons. Add arrows and symbols to the diagram to show the leaching process.

3. Preparing Charts Study the Plate Tectonics map. Create a chart listing each plate and the direction or directions of its movement, if any.

Writing about Geography

Keep a journal listing all the natural resources you use during a single day. Then create a journal from the point of view of a person living 50, 100, or 200 years in the past. Remember that technology changes the way people use resources. When you are finished with your two journals, proofread them to make sure you have used standard grammar, spelling, sentence structure, and punctuation.

SKILL BUILDING

Geography for Life

Preparing Diagrams and Models

Physical Systems Use construction paper or modeling clay to create posters or three-dimensional models of the different types of plate boundaries. Show landforms that may result, such as mountain ranges.

Portfolio Activity

To demonstrate the importance of water, have students conduct research on where the school's drinking water comes from. *(Possible sources include rivers, lakes, reservoirs, and groundwater.)* Students should also investigate how the school disposes of its wastewater. Have students use their findings to create diagrams of the school's water-supply and water-disposal systems. Place diagrams in portfolios.

Food Festival

In places where farmland is scarce, farmers are increasingly using hydroponic agriculture. Plants grown hydroponically are raised in a water-based nutrient solution. Ask students to visit a local grocery and to locate items in the produce department that have been grown hydroponically. Encourage students to compare the flavor of hydroponic products and those grown in soil. The class may also want to grow plants such as tomatoes hydroponically.

Building Social Studies Skills

Interpreting Graphs

Study the line graph below. Then use the information from the line graph to help you answer the questions that follow. Mark your answers on a separate sheet of paper.

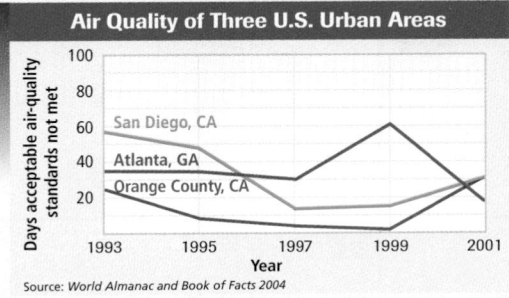

Air Quality of Three U.S. Urban Areas

San Diego, CA
Atlanta, GA
Orange County, CA

Source: World Almanac and Book of Facts 2004

1. How many days did Atlanta fail to meet acceptable air-quality standards in 1997?
   a. 31
   b. 61
   c. 14
   d. 3

2. Describe the overall pattern of air quality in Orange County and San Diego between 1993 and 2001. Then contrast those patterns with that of Atlanta during the same period.

Building Vocabulary

Answer the following questions on a separate sheet of paper.

3. *Magma* specifically means
   a. layer of Earth's interior between the crust and core.
   b. liquid rock.
   c. rock that contains minerals.
   d. Earth's solid center.

4. Many earthquakes have been caused by movement along the San Andreas Fault. In which sentence does *fault* have the same meaning as it does in the sentence above?
   a. It is partly the fault of the geologist that the exploration company has gone bankrupt.
   b. The fault in this circuit is preventing the light from coming on.
   c. The area of deforestation is a fault in an otherwise beautiful landscape.
   d. When layers of rock break and move, a fault results.

Alternative Assessment

PORTFOLIO ACTIVITY

Learning about Your Local Geography

Group Project: Field Work
As a group, learn about recycling in your community. If possible, tour your local recycling facility. Find out what materials are recycled and how they are gathered, transported, and processed. Interview workers at the recycling center or public officials to learn the costs and benefits of recycling to your community's economy and natural environment.

internet connect

Internet Activity: go.hrw.com
KEYWORD: SW3 GT4

Choose a topic about landforms, water, and natural resources to:
- create a newspaper on the causes and effects of earthquakes.
- understand the formation and uses of fossil fuels.
- analyze the different stages of the hydrologic cycle.

Journal entry will probably display increased usage of energy resources, particularly fossil fuels. Students should use standard grammar, spelling, sentence structure, and punctuation. Use Rubric 15, Journals, to evaluate student work.

Geography for Life

Students' models should reflect correct information. Use Rubric 3, Artwork, to evaluate student work.

Social Studies Skills

1. a

2. Based on the number of days in which air standards were not met, air quality in Orange County and San Diego improved considerably after 1993. Atlanta's air quality has worsened.

3. b

4. d

PORTFOLIO ACTIVITY

Students should research recycling in your community. Use Rubric 14, Group Activity, and Rubric 1, Acquiring Information, to evaluate student work.

Human Geography

CHAPTER RESOURCE MANAGER

Objectives	Pacing Guide	Reproducible Resources

**SECTION 1
Population
Geography**
(pp. 87–93)

- Explain how geographers study population.
- Identify some important trends in world population.

Regular
1 day

Block Scheduling
1 day

*Block Scheduling Handbook,
Chapter 5*

RS Guided Reading Strategy 5.1
RS Graphic Organizer Activity 5
PS Readings in World Geography,
History, and Culture 9 and 10
SM Geography for Life Activity 5

**SECTION 2
Cultural
Geography**
(pp. 94–98)

- Explain how geographers study culture.
- Analyze how cultures change over time.

Regular
1 day

Block Scheduling
1 day

*Block Scheduling Handbook,
Chapter 5*

RS Guided Reading Strategy 5.2
E   Creative Strategies for Teaching
World Geography, Lessons 4 and 5

**SECTION 3
World
Languages
and Religions**
(pp. 100–03)

- Examine the geography of the world's languages.
- Describe the three main types of religions that geographers identify.

Regular
1 day

Block Scheduling
.5 day

*Block Scheduling Handbook,
Chapter 5*

RS Guided Reading Strategy 5.3
SM Critical Thinking Activity 5: Buddhism
SM Map Activity 5: European Religions

Chapter Resource Key

PS Primary Sources	**A** Assessment	CD-ROM	
RS Reading Support	**REV** Review	Video	
IC Interdisciplinary Connections	**ELL** Reinforcement and English Language Learners	Internet	
E Enrichment		 Holt Presentation Maker Using Microsoft® PowerPoint®	
SM Skills Mastery	Transparencies		

One-Stop Planner CD–ROM

See the *One-Stop Planner*
for a complete list of additional
resources for students and
teachers.

One-Stop Planner CD–ROM

It's easy to plan lessons, select resources, and print out materials for your students when you use the **One-Stop Planner CD–ROM with Test Generator**.

internet connect

HRW ONLINE RESOURCES

GO TO: go.hrw.com
Then type in a keyword.

TEACHER HOME PAGE
KEYWORD: **SW3 Teacher**

CHAPTER INTERNET ACTIVITIES
KEYWORD: **SW3 GT5**
Choose an activity to:
- research world population growth.
- examine factors that cause the migration of refugees.
- answer demographic questions about the United States.

CHAPTER ENRICHMENT LINKS
KEYWORD: **SW3 CH5**

CHAPTER MAPS
KEYWORD: **SW3 MAPS5**

ONLINE ASSESSMENT
Homework Practice
KEYWORD: **SW3 HP5**
Standardized Test Prep
KEYWORD: **SW3 STP5**
Rubrics
KEYWORD: **SS Rubrics**

COUNTRY INFORMATION
KEYWORD: **SW3 Almanac**

CONTENT UPDATES
KEYWORD: **SS Content Updates**

HOLT PRESENTATION MAKER
KEYWORD: **SW3 PPT5**

ONLINE READING SUPPORT
KEYWORD: **SS Strategies**

CURRENT EVENTS
KEYWORD: **S3 Current Events**

Technology Resources

- One-Stop Planner CD–ROM, Lesson 5.1
- *ARGWorld* CD–ROM
- Geography and Cultures Visual Resources 10 and 11
- CNN Presents Geography: Yesterday and Today, Segment 5: World Population–Hitting Six Billion
- Homework Practice Online
- HRW Go site

- One-Stop Planner CD–ROM, Lesson 5.2
- Homework Practice Online
- HRW Go site

- One-Stop Planner CD–ROM, Lesson 5.3
- Homework Practice Online
- HRW Go site

Reinforcement, Review, and Assessment

- **ELL** Main Idea Activity 5.1
- **ELL** English Audio Summary 5.1
- **ELL** Spanish Audio Summary 5.1
- **REV** Section 1 Review, p. 93
- **A** Daily Quiz 5.1

- **ELL** Main Idea Activity 5.2
- **ELL** English Audio Summary 5.2
- **ELL** Spanish Audio Summary 5.2
- **REV** Section 2 Review, p. 98
- **A** Daily Quiz 5.2

- **ELL** Main Idea Activity 5.3
- **ELL** English Audio Summary 5.3
- **ELL** Spanish Audio Summary 5.3
- **REV** Section 3 Review, p. 103
- **A** Daily Quiz 5.3

Meeting Individual Needs

Ability Levels

Level 1 Basic-level activities designed for all students encountering new material

Level 2 Intermediate-level activities designed for average students

Level 3 Challenging activities designed for honors and gifted-and-talented students

English Language Learners Activities that address the needs of students with Limited English Proficiency

Chapter Review and Assessment

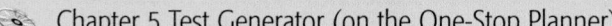

- Chapter 5 Test Generator (on the One-Stop Planner)
- Global Skill Builder CD–ROM
- HRW Go site
- **REV** Chapter 5 Review, pp. 110–11
- **REV** Chapter 5 Tutorial for Students, Parents, Mentors, and Peers
- **A** Chapter 5 Test (form A or B)
- **A** Alternative Assessment Handbook
- **A** Chapter 5 Test for English Language Learners and Special-Needs Students

Launch into Learning

Display an article or photo from a newspaper or newsmagazine that contains elements of both physical and human geography. (*Possible topics include a natural disaster, a new medicine created from a recently discovered plant, and new environmental legislation.*) Ask students to list the physical and human elements shown or implied in the story or photo. (*For a hurricane, examples of physical elements are rain, mud slides, beach erosion, and high tides; possible human elements are people being forced from their homes and a decline in business income.*) Point out to students that their interpretation of the article or photo would be incomplete if they left out either the physical or human issues. Similarly, the study of geography must include the human element. Tell students they will learn more about the human element of geography in this chapter.

Why We Should Know More

You may want to emphasize the importance of learning the basics of human geography by sharing these points with your students:

▶ Population issues affect food supplies, drive international conflicts, and pose environmental questions.

▶ Cultural differences influence world events and our daily lives. We should understand why cultures are different.

▶ Identification with a certain language or religion has been a driving force throughout history and continues to affect world events.

CHAPTER 5 Human Geography

People have shaped much of the world around us. Cities, farms, airports, and other features have been created to meet our ever-changing needs. The study of the world's people— how they live and how their activities vary from place to place—is called human geography.

The Shibuya district in Tokyo, Japan

Selling goods in Latacunga, Ecuador

Population Geography

READ TO DISCOVER

1. How do geographers study population?
2. What are some important trends in world population?

WHY IT MATTERS

People are always moving to new places. Use CNN**fyi**.com or other **current events** sources to find a news item about a group of people leaving an area for a new one.

IDENTIFY

demography
population density
birthrate
death rate
migration

emigrants
immigrants
push factors
pull factors
refugees

Studying Population

Is the population of your state growing or decreasing? Where do people choose to live? What is the average age of people in your state? These are the kinds of questions that population geographers ask. These geographers study the relationships between populations and their environments. They use maps, graphs, population pyramids, and a spatial perspective to study population patterns and trends. Population geography is closely related to **demography**— the statistical study of human populations. Statistics are information in number form. They help us learn about populations. Demographers collect statistics about populations and use them to forecast what future populations will be like. Such forecasts and demographic information can be very useful. For example, it can be used to decide where to build new schools or how to redraw political boundaries to reflect changing populations.

internet connect

GO TO: go.hrw.com
KEYWORD: SW3 CH5

FOR: Web sites about human geography

The high population density of Shanghai, China, can be seen along the city's bustling shopping streets. Geographers and demographers use statistics like population density to study the world's population.

ALL LEVELS: From your school's administration, obtain enrollment records and student demographic information for the last several years. Provide students with these figures and have them plot the data on a chart or graph. Ask students if the school's population is growing or shrinking. Then have students calculate the population densities of your school and of your classroom. Be sure that students use the same units of measurement in both calculations. Ask students to compare the two density figures. *(The classroom density will likely be higher, because it is a small area and excludes places with few "residents" like cafeterias and gymnasiums.)*

Finally, ask students to observe the school's population distribution during a lunch period or break. Have students draw a population density map of the school grounds that reflects their findings. Then ask where large numbers of students congregate and where few students are found. Call on volunteers to suggest factors that might influence this distribution. *(Possible answer: Many students may be in the cafeteria to buy lunch, and few will be in classrooms between periods.)* Lead a class discussion about how these studies of your school's population relate to similar studies of the populations of countries. **ENGLISH LANGUAGE LEARNERS, BLOCK SCHEDULING**

Essential Element 4

▶ Human Systems ◀

Density Facts About 90 percent of the world's people live on 10 percent of its land. The tiny microstate of Monaco has the highest population density of any country—more than 42,000 people per square mile. The Netherlands, however, is generally considered Europe's most densely populated country. Its density is 991 people per square mile.

Mongolia, which has only 4 people per square mile, and Australia, with 6 people per square mile, are among the least densely populated countries in the world.

CRITICAL THINKING: What factors might create low density in Mongolia and Australia and high density in very small countries? *(Mongolia and Australia have large areas of uninhabitable desert land. Very small countries are often highly urbanized.)*

CHART ANSWER

Possible answer: Blue Earth County probably has far fewer towns, and therefore fewer businesses and less ethnic diversity.

MAP ANSWER

Answers will vary, but students should support their predictions.

Population Density in Selected U.S. Counties

County, State	People per Square Mile
Loving, TX	0.10
Forest, WI	10
Blue Earth, MN	75
Honolulu, HI	1,448
San Francisco, CA	16,256
New York (Manhattan), NY	67,255

Source: *The World Almanac and Book of Facts 2004*

INTERPRETING THE CHART *Population densities in U.S. counties vary dramatically.* **Based on the population data in this chart, how would you expect the human geography of Blue Earth County, Minnesota, to be different from that of New York (Manhattan), New York?**

Population Density One important statistic geographers use is **population density**. This is the average number of people living in an area. It is usually expressed as persons per square mile or square kilometer. In this textbook you will find the population densities of the world's countries in the Fast Facts table in each unit.

Population densities around the world vary greatly. For example, Canada has just 8 persons per square mile. On the other hand, Bangladesh has 2,361 persons per square mile. Population density reflects the size of a country, the size of its population, and its environmental conditions. For example, Canada is a huge country. However, much of it is too cold to support large numbers of people. Most of Canada's 31.6 million people live along the country's southern border. Temperatures are warmer there. As a result, southern Canada is more densely populated than the rest of the country. Bangladesh is much smaller in area than Canada. Yet nearly all of Bangladesh is covered by rich farmland. Many of the country's roughly 131 million people are crowded onto that farmland. Within a country, smaller areas may have an even greater range of population densities. (See the chart of population density in selected U.S. counties.)

✓ **READING CHECK:** *Human Systems* Why do population densities vary greatly among the world's countries? reflect a country's size, population, environment

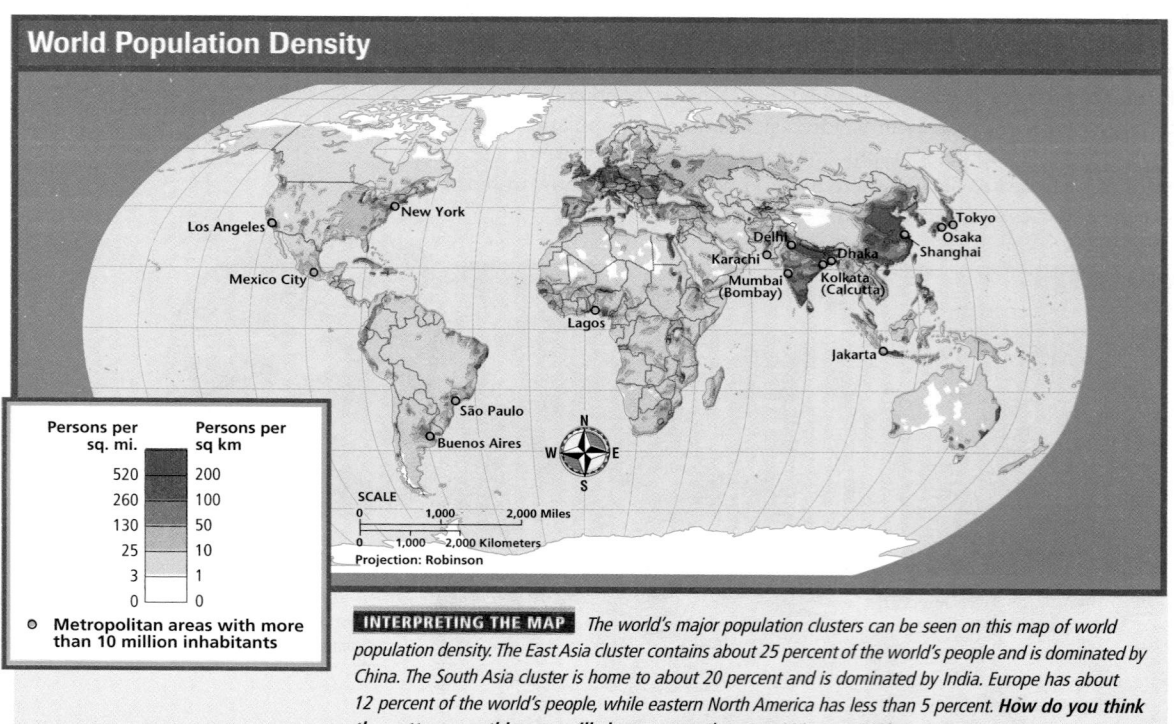

World Population Density

INTERPRETING THE MAP *The world's major population clusters can be seen on this map of world population density. The East Asia cluster contains about 25 percent of the world's people and is dominated by China. The South Asia cluster is home to about 20 percent and is dominated by India. Europe has about 12 percent of the world's people, while eastern North America has less than 5 percent.* **How do you think the patterns on this map will change over the next 100 years? Why?**

Nigeria is an example of a country with a high birthrate—there are about 40 births per 1,000 persons each year. The high birthrate, combined with a much lower death rate, has resulted in a young, rapidly growing society. About 44 percent of Nigeria's population is currently below the age of 15.

Population Distribution

People are spread unevenly across Earth. Some places are crowded with people, while others are empty. Where do most of the world's people live? Why do people live in some places and not in others? About 90 percent of the world's people live in the Northern Hemisphere. About two thirds of these people live in the middle latitudes between 20° and 60° north. Many people there live in lowland areas, particularly along fertile river valleys near the edges of continents. Four areas of the world have great clusters of population. Those areas are East Asia, South Asia, Europe, and eastern North America. There are also smaller clusters, such as the Nile Valley in Egypt and the island of Java in Indonesia. (See the World Population Density map.)

Why is the world's population so unevenly distributed? The simple reason is that people tend to live in areas that are favorable for settlement. They also tend to avoid areas that are unfavorable. Therefore, places with mild climates, fertile soils, and an adequate supply of freshwater usually have more people than areas without these features. For the same reason, polar regions, deserts, and rugged mountains usually have few people.

✔ **READING CHECK:** *Human Systems* Why is the world's population so unevenly distributed? because people tend to live in areas that are favorable for settlement and avoid areas that are not

Population Change

Geographers also use statistics to study how populations are changing. The number of people in any place is the result of three major factors. The first is how many people are born each year. This factor is expressed as the **birthrate**. The birthrate is the number of births each year for every 1,000 people living in a place. The second factor is how many people die each year. This is known as the **death rate**—the total number of deaths each year for every 1,000 people. The third factor is **migration**. Migration is the process of moving from one place to live in another. People who leave a country to live somewhere else are called **emigrants**. People who come to a new

LEVEL 1: Copy the following graphic organizer onto the chalkboard, omitting the italicized answers. Call on students to provide words and phrases to identify technological improvements that have contributed to the world's population growth. As students name advancements, ask how each contributed to population growth. Lead a class discussion about the causes and effects of the world's rapid population growth. Ask students if they think this population growth will continue. Have students support their answers. **ENGLISH LANGUAGE LEARNERS**

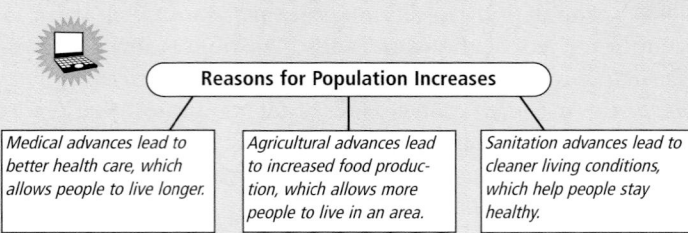

Reasons for Population Increases

| Medical advances lead to better health care, which allows people to live longer. | Agricultural advances lead to increased food production, which allows more people to live in an area. | Sanitation advances lead to cleaner living conditions, which help people stay healthy. |

Daily Life

Refugee Camps There are some 39 million refugees throughout the world. About 14 million have crossed international borders and 25 million have been displaced within their own countries. Most refugees reside in tent cities in the developing world. Southwest Asia and Africa host the largest numbers.

Sometimes these refugees must make their own tents with plastic sheeting. Water and food supplies are limited. There is no electricity. Overcrowded and unsanitary conditions can lead to the rapid spread of diseases.

Initially, many refugees are too depressed to take decisive action, but eventually they settle into activities of daily life—collecting water, clearing a small plot of land, and going to the market. Over time, some create sheltered meeting places to gather or teach their children. Others set up stalls to sell or trade items such as shoes made from old tires. Economic activities for refugees are severely limited, however.

INTERPRETING THE VISUAL RECORD

In the late 1990s unrest in Kosovo, Yugoslavia, disrupted the lives of hundreds of thousands of ethnic Albanians and forced many people to flee the region. Here, refugees in neighboring Macedonia wait to be transported to nearby transition camps. **How might such a sudden population increase strain the resources of Macedonia?**

VISUAL RECORD ANSWER

Macedonia must provide food and shelter for the refugees.

country to live are called **immigrants**. If large numbers of people are leaving or entering a country, it can have major effects on a population.

FOCUS ON GEOGRAPHY

Migration Migration is a common theme in human history. Scientists think that humans and their ancestors began migrating out of Africa between 1 and 2 million years ago, moving into Asia and Europe. They may have been looking for better hunting, or perhaps they just wanted to know what was over the horizon. The last continents discovered and settled by humans were Australia at least 40,000 years ago and, later, North and South America. To reach the Americas, people may have crossed a land bridge from Asia to Alaska that became available when sea levels fell during the last ice age. However, there is growing evidence that humans may have reached the Americas even earlier. Some may have traveled to the region in small boats along the coast.

Geographers study migration by analyzing **push factors** and **pull factors**. A push factor causes people to leave a location. A pull factor attracts people to a new location. Most people migrate for economic reasons. The push may be a lost job or lack of opportunity for promotion. The pull may be a better job or higher pay somewhere else.

Many other push and pull factors can cause people to move. For example, environmental conditions and hazards cause migration. Droughts and floods have often forced people to move. Over time, warm sunny climates have drawn people to Arizona, Florida, and Texas. People also flee political unrest and wars. People who have been forced to leave and cannot return to their homes are **refugees**. They often leave because they do not feel safe where they live or suffer discrimination. On the other hand, political or personal freedom can be a pull factor. For example, people have moved to find freedom in democratic countries like the United States and Canada. Finally, physical geography can affect the routes, flows, and destinations of migrants. For example, many early Americans migrated from the East Coast to the interior plains to farm. The physical

LEVEL 2: Organize students into pairs and have each pair write a newspaper article reporting census information that might be released in 2050. Student articles should predict the total world population in that year, speculate about migration trends, and provide support for their predictions. You may want to provide supplemental resources. Call on volunteers to share their articles with the class.
COOPERATIVE LEARNING

LEVEL 3: Assign each student a period of world history. Then have each student create an illustrated time line that notes major changes in world population figures and patterns during the assigned time period. Encourage students to create pictures or design symbols to represent periods of urbanization, migration, or other shifts in population. Combine and display the completed projects to create a single time line of changes in the world's population. Lead a class discussion about population changes that have occurred through time. **BLOCK SCHEDULING**

geography of the United States, including the locations of the Appalachian Mountains and major rivers, greatly influenced these migration patterns.

✓ READING CHECK: (*Human Systems*) What kinds of factors cause migration?

push factors such as lack of jobs, droughts, floods, political unrest, wars, persecution; pull factors such as good jobs or pay, warm climates and other favorable environmental conditions, lack of persecution, physical geography

Natural Increase   Geographers are also interested in the rate of natural population growth, or natural increase. This rate is based just on births and deaths—it does not take migration into account. You can find the rate by subtracting the birthrate from the death rate. This final number is expressed as a percentage. In the United States, the rate of natural increase is about 0.6 percent each year.

The rate of natural increase varies greatly in countries around the world. The highest rates are found in countries in Africa and Southwest Asia. Rates there are sometimes 3 percent or higher. The number of people living in those places is rising rapidly. Most countries in Central and South America and in Southeast Asia have more moderate rates of natural increase. The rates for these areas are somewhere between 1 and 3 percent. The lowest rates—less than 1 percent—are found in most European and North American countries. Australia, New Zealand, and Japan also have such low rates. Some countries, such as Italy and Russia, actually have negative growth rates. The number of people living in these countries is decreasing.

These percentages—1, 2, and 3—may sound small. However, they can add up to large population increases in a short period of time. For example, suppose the number of people living in a country grows at a rate of 3 percent. That country's population will double in only about 23 years! Geographers call the number of years needed to double a country's population its doubling time.

✓ READING CHECK: (*Human Systems*) Where are the highest rates of natural increase found in the world? The lowest? highest: Africa and Southwest Asia; lowest: Europe, North America, Australia, New Zealand, Japan

World Population Trends

The world's population has increased rapidly in the last 200 years. (See the graph of world population growth.) Earth is now home to more than 6 billion people. This number is increasing by nearly 80 million each year. That is almost 220,000 people a day! It is hardly surprising that worries about crowding and environmental problems are in the news.

In the year A.D. 1 the world's population was probably about 300 million. By 1600, it had doubled to 600 million. Then populations began to increase steadily. Farm technology and public sanitation improved, and the basics of modern medicine began. Cities grew larger. The world's population distribution was also changing as large population clusters grew in North America, South Asia, and East Asia. By about 1850, world population had doubled again to 1.2 billion. Since then, Earth's population growth has exploded. World population passed the 2 billion mark before 1930 and doubled again to 4 billion by 1975. In 2000 there were more than 6 billion people.

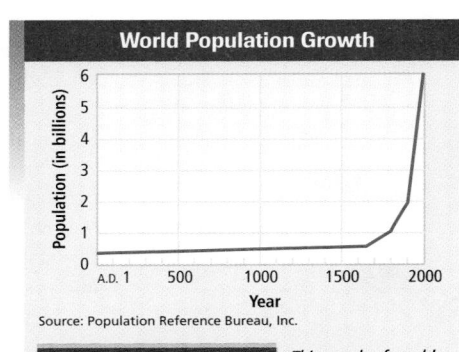

World Population Growth

Population (in billions) / Year
A.D. 1 500 1000 1500 2000

Source: Population Reference Bureau, Inc.

INTERPRETING THE GRAPH   *This graph of world population growth shows the dramatic increase in the world's population over the last several hundred years. While population growth for most of human history was slow, it exploded as improved health, nutrition, sanitation, and medical care combined to lower death rates around the world.* **What trends in past world population growth does this graph show? How has the world's population distribution changed during this time?**

Essential Element 4

► **Human Systems** ◄

The U.S. Population   Every 10 years, the U.S. Census Bureau counts the people who live in this country and gathers information about them. Following are some interesting facts about the 2000 census form:
► For the first time, there was a question about grandparents as caregivers.
► Respondents could mark more than one racial category.
► The form was printed in English, Spanish, Chinese, Tagalog, Vietnamese, and Korean. Assistance guides were available in 49 other languages.
► Census workers counted homeless people at camping sites, shelters, soup kitchens, and other places.

DISCUSSION: Lead a discussion on what these facts about the census form can tell us about changes in American society.

you don't say! There are now four times as many people on the planet as there were one century ago, in 1900.

GRAPH ANSWER

slow growth until about 1800, rapid increase since then; has become more urbanized

Aging Populations In the third stage of demographic transition, the average age of human beings rises as birthrates decline and life expectancies increase. Life expectancy at birth currently averages 66 years worldwide. It has increased by about 20 years since 1950.

Those over 60 make up the fastest growing segment of the population. This population is expected to rise from 10 percent of the present world population to 22 percent by 2050. High costs of long-term care for people over 85 and of medical services for all elders will strain social security and health care systems worldwide.

ACTIVITY: Ask students to research and create charts or bar graphs for current and projected life expectancies of the most and least populous countries of each continent.

Between 1950 and 1998 the average life expectancy in the developing world rose from 40 to 63 years, and the world's population more than doubled.

GRAPH ANSWER

Stage 2; to explain that the human, economic, medical, and urban geography of those countries changed while they experienced changes in birthrates and death rates

The Demographic Transition

per thousand / per year — Time →

Stage 1 | Stage 2 | Stage 3

Birthrate

Death Rate

INTERPRETING THE GRAPH *The demographic transition model is based on changes that occurred in the populations of many Western European countries as they industrialized. New economic and social conditions shaped these countries' populations. As medical care and health improved, the death rate fell. Eventually, the birthrate also fell as modern urban societies developed.* **During which stage in the model is population growth highest? How could you apply the demographic transition model to present geographic information about the countries of Western Europe?**

Some people worry about overpopulation. Overpopulation is a situation in which the existing number of people is too large to be supported by available resources. However, there is much debate among geographers and other scientists about how many people the world can support. Also, birthrates in many countries have dropped in recent years. These changes will affect future population growth and resource needs around the world.

✓ **READING CHECK:** *Human Systems* How many people are there in the world? How is this number changing each year? more than 6 billion; growing by about 80 million each year

The Demographic Transition The demographic transition is a model that shows how birthrates and death rates dropped in many Western countries as they developed modern economies and industries. Most of the world's richest and most technologically advanced countries have experienced a transition from high birthrates and death rates to low birthrates and death rates. In addition, many of the world's poorer countries are now in the middle of similar population changes. While many demographers think these countries will also experience a transition to lower birthrates and death rates, no one can predict exactly how this will happen.

During the first stage of the demographic transition, both birthrates and death rates are high. (See the graph of the demographic transition.) Parents have many children, but poor health conditions mean that many do not live to become adults. The infant mortality rate—the number of infants that do not survive their first year of life—is high. Since both the birthrate and death rate are high, the population neither grows nor decreases much. Instead, the total population remains relatively stable. This stage of the demographic transition is typical of countries that are mostly agricultural.

During the second stage, the death rate begins to fall. This happens partly because of improvements in medicine and health care, particularly for infants and children. As a result, the infant mortality rate drops significantly during this stage, and more children survive their early years. Better food production and distribution may also help lower the death rate. While the death rate falls, the birthrate remains high. Families grow as children and adults live longer. As a result, the total population grows. Also during this stage, economic improvements and the switch to more advanced farming technologies cause rural to urban migration and the rapid growth of cities. Near the end of the second stage, the birthrate begins to fall. As more people switch to life in a modern urban society, they marry later and have fewer children. As a result, population growth begins to slow again.

During the third stage, both birthrates and death rates are low. Thus, total population growth is low. All of the world's economically advanced countries have reached this final stage. These countries include the United States, Canada, Japan, Singapore, Australia, New Zealand, and nearly all European countries.

These countries have all experienced a demographic transition from an agricultural society with high birthrates and death rates to an urban and industrial society with low birthrates and death rates.

✓ **READING CHECK:** (**Human Systems**) How does the demographic transition explain trends in past world population growth? Populations in many societies grew as they switched from agricultural to urban and industrial societies and as birthrates and death rates fell.

Future Populations   How will the world's population change in the future? Will it continue to grow rapidly, or will lower birthrates lead to a slower rate of growth and a more stable population?

While we cannot be certain about how populations will change in the future, we can use demographic information to make population projections. Population projections are estimates of a future population's size, age, growth rate, or other characteristics based on current data. Forecasting future population growth is not easy. Population growth is tied to future birthrates, which cannot be known ahead of time. As a result, demographers often make several different projections. (See the graph.) In general, the farther into the future a projection is made, the less reliable it is.

Future world growth rates will depend on what happens in the countries that have high rates of natural increase. In some of these countries birthrates have been falling over the last 20 years. These countries seem to be moving toward the third stage of the demographic transition.

Regardless of how the world's population changes in the future, all countries will face population-related challenges. For example, caring for a growing number of children or older people will strain the resources of many countries. Maintaining public health, producing food, and protecting the environment will be important issues for populations around the world.

✓ **READING CHECK:** (**Human Systems**) Why is it so difficult to predict population growth? because future birthrates cannot be known beforehand

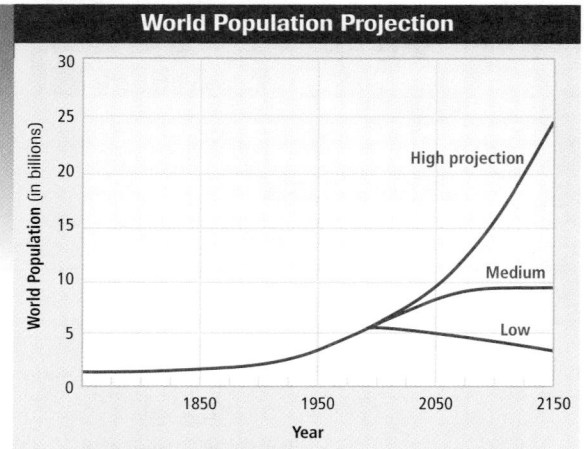

World Population Projection

Source: United Nations Development Programme

INTERPRETING THE GRAPH   *This graph shows three different population projections for the next 150 years: a low, medium, and high projection. The three projections are based on different assumptions about whether birthrates will increase or decrease in the future. **How different are these three projections? Which do you think most accurately predicts future population patterns? Why?***

Section 1 Review

go.hrw.com **Homework Practice Online**
Keyword: SW3 HP5

Define demography, population density, birthrate, death rate, migration, emigrants, immigrants, push factors, pull factors, refugees

Reading for the Main Idea

1. (**Human Systems**) What are some statistics that population geographers and demographers use to study populations?

2. (**Places and Regions**) Where are the four great clusters of world population?

Critical Thinking

3. **Making Generalizations and Predictions** How can the demographic transition model help geographers predict future population changes?

4. **Analyzing Information** What is one example of physical geography affecting the route and flow of migration?

Organizing What You Know

5. Create a chart like the one shown below. Use it to identify five push factors and five pull factors that can cause migration.

Push factors	Pull factors

LET'S GET STARTED

Copy the following instructions onto the chalkboard: *What are some organized events that occur regularly in your community?* Discuss responses. *(Possible answers: concerts, festivals, parades, sports events, swap meets)* Point out to students that the events they listed are part of the community's culture. Then ask if students know of similar events in other places. Compare those events with the students' lists to note ways in which other communities have different cultures. Tell students that in Section 2 they will learn more about culture.

Building Vocabulary

Write the key terms on the chalkboard. Underline the word *culture* in **culture traits** and **culture region** and "cultur" in **acculturation**. Discuss the relationship of the root to these terms. Then ask students to suggest definitions for the words *innovate, diffuse, global, tradition,* and *fundamental*. Call on volunteers to suggest meanings for **innovation, diffusion, globalization, traditionalism,** and **fundamentalism** based on these definitions. Have students read the definitions to check for accuracy.

Section 2 RESOURCES

REPRODUCIBLE

▶ Guided Reading Strategy 5.2
▶ Creative Strategies for Teaching World Geography, Lessons 4 and 5

TECHNOLOGY

▶ One-Stop Planner CD–ROM, Lesson 5.2
▶ Homework Practice Online
▶ HRW Go site

REINFORCEMENT, REVIEW, AND ASSESSMENT

▶ Main Idea Activity 5.2
▶ English Audio Summary 5.2
▶ Spanish Audio Summary 5.2
▶ Section 2 Review, p. 98
▶ Daily Quiz 5.2

Section 2 Cultural Geography

READ TO DISCOVER

1. How do geographers study culture?
2. How do cultures change over time?

WHY IT MATTERS

Our world's billions of people have many different customs and traditions. Use **CNNfyi.com** or other **current events** sources to learn about some customs and traditions that people have brought from other places to the United States.

DEFINE

culture	innovation
culture traits	diffusion
culture region	globalization
ethnic groups	traditionalism
acculturation	fundamentalism

Studying Culture

A key term for understanding human geography is **culture**. Culture includes all the features of a people's way of life. It is learned and passed down from parents to children through teaching, example, and imitation. Important parts of culture include a group's language, religion, architecture, clothing, economics, family life, food, and government. It also includes a people's beliefs, institutions, shared values, and technologies as well as its members' skills. What are some of the features of your culture?

Culture Traits Activities and behaviors that people often take part in are called **culture traits**. There are many kinds of culture traits. Some, such as learning to read and do math, are much the same around the world. This is true even though different cultures use different alphabets and symbols. Other culture traits vary from place to place. For example, most Americans eat with a knife, fork, and spoon. However, Chinese eat with chopsticks. Ethiopians eat

INTERPRETING THE VISUAL RECORD

The Amish are a religious group in North America with many distinct culture traits, such as simple clothing styles and the use of traditional farming techniques. These unique culture traits give Amish communities a distinctive cultural landscape. What elements of Amish culture can you see in this photograph?

LEVEL 1: Copy the graphic organizer at right onto the chalkboard. Ask students to copy the organizer and fill in each box with descriptions of their own culture. Call on volunteers to share their organizers with the class. Lead a discussion about the different cultures found in your community. **ENGLISH LANGUAGE LEARNERS**

LEVEL 2: Instruct each student to select a country or cultural group from across the world and to conduct research to create a graphic organizer like the one in the Level 1 activity. Lead a discussion about differences between each culture presented and local cultures.

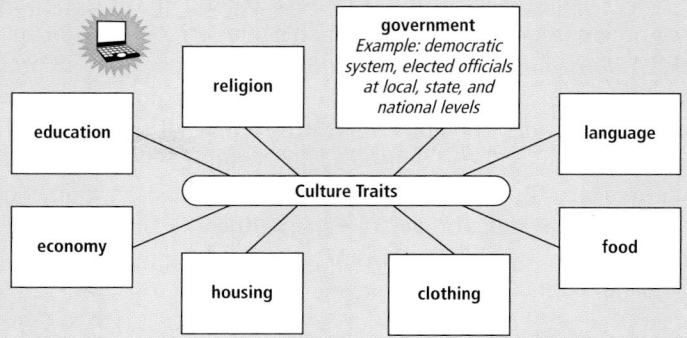

with their fingers or use bread to scoop their food. Each of these traits is considered correct where it is practiced.

Many culture traits are linked. For example, the Amish are a religious group in the United States and Canada. Traditional Amish beliefs favor a simple way of life and separation from many parts of modern society. People in Amish communities may not use automobiles, electricity, or telephones. Men are known for their simple dress, which includes broad-brimmed black hats and plain clothes. Women wear bonnets, long dresses, and shawls. Most Amish farmers do not use modern technology, such as tractors. Instead, they use horses to plow their fields. This example shows how the culture traits of religion, farming, and the use of technology can be linked.

✓ **READING CHECK:** (*Human Systems*) How are Amish cultural traits linked, and how do they separate the Amish from the rest of society? Traditional Amish beliefs affect how people use technology, dress, and farm and require a simple way of life and separation from modern society.

Culture Regions The world is made up of many different culture groups. Each group has its own way of life. This includes the way they use their land, the resources they depend on and value, and what religions they practice. It also includes settlement patterns, attitudes toward the role of women in society, forms of government, and other customs. These different ways of life generate distinctive cultural landscapes around the world.

An area in which people have many shared culture traits is called a **culture region**. An individual country may be a single culture region. For example, Japan has one dominant culture throughout the country. However, sometimes countries include many culture regions. For example, many countries in Africa include dozens of different **ethnic groups**. An ethnic group is a human population that shares a common culture or ancestry. African countries that are home to many different ethnic groups include Kenya, Nigeria, and South Africa.

Country borders sometimes divide culture regions and separate members of one ethnic group. For example, the Kurds are an ethnic and linguistic group in Asia. However, the borders of Iran, Iraq, and Turkey divide their traditional homeland. A culture region can also be made up of several countries. Australia and New Zealand make up one such culture region. The two countries share the same language and have similar traditions and systems of government.

✓ **READING CHECK:** (*Human Systems*) What traits make a place's cultures distinctive? land use, resource use, religion, settlement patterns, attitudes toward role of women in society, forms of government, other customs

Culture Change

Culture traits change through time. Sometimes, these changes are as simple as wearing new clothing styles. Ask some adults what clothing styles were popular when they were your age. Clearly, for many people clothing is more than just something that covers our bodies. It also reflects personal values, tastes, and style.

Often, changes in culture traits are even more complex than new fashions. For example, in the 1800s people in the United States walked or rode streetcars downtown to do their shopping. In the 1900s, buses and cars eventually replaced streetcars. Today modern expressways and subways take shoppers away

One important change in many cultures involves attitudes toward the role of women in society. In the 1900s women in many countries entered the workforce in large numbers, became politically active, and enjoyed more freedoms and independence. These cultural changes allowed many women to seek careers in traditionally male fields, such as steelmaking.

The Masai Despite pressure from the governments of Kenya and Tanzania, the Masai, a nomadic herding people of Africa's Great Rift Valley, retain many of their traditional customs. One of their most notable customs is the division of its society by age classes. Of these classes, the best known are the Masai warriors. Boys are initiated as junior warriors— or *morans*—at about age 14. For approximately 15 years they live in the bush, isolated from their families and the rest of the Masai. There they develop strength, courage, and endurance—traits for which Masai warriors are renowned around the world. After this period they join the ranks of the senior warriors.

The primary task of Masai warriors is to herd cattle and protect them from lions and other predators. Warriors braid their long hair and decorate their hair and bodies with ocher, an iron ore used as pigment. They also wear distinctive beads on neckbands, bracelets, anklets, earrings, and the belts that carry their short swords.

CRITICAL THINKING: Why might the Masai resist cultural changes occurring in East Africa? *(Possible answer: The Masai might fear that abandoning nomadism and the warrior system might threaten their cultural identity or even their survival.)*

LEVEL 3: Organize students into small groups and assign each group a culture region in your state or region or elsewhere in the United States. *(Examples include the Amish country in Pennsylvania, Cajun areas in Louisiana, and Chinatown in San Francisco.)* Have each group conduct research to identify cultural features that distinguish residents of their assigned culture region from the people around them. Have students create posters that illustrate some of these features. Encourage students to include maps on their posters that show the locations of the culture regions. **COOPERATIVE LEARNING**

Teach Objective 2

LEVEL 1: Show students pictures of clothing from the United States in 1776. Ask students to note differences between the styles popular then and those currently in fashion. *(Possible answers: different cuts of clothing, clothing made by hand rather than by machine)* Then have the class examine a picture of the Declaration of Independence while a volunteer reads an excerpt. Ask students how it differs from government documents of today. *(Possible answers: language and style differences, handwritten rather than typed, handwriting differences)* Lead a class discussion about cultural change and factors that inspire it. **ENGLISH LANGUAGE LEARNERS**

Daily Life

Blue Jeans Blue jeans, a popular symbol of Western culture, can be traced back to the California Gold Rush. However, they also have connections to France and Italy. In 1853 Levi Strauss arrived in San Francisco where he first made trousers for gold miners from brown tent canvas. He soon switched to a more durable fabric called *jean* after Genoa, the Italian city where it was made. Eventually "Levi's waist overalls" were made entirely of cotton denim, originally called *serge de Nimes* (duh-NEEM) for the French city where the fabric was first made.

Hollywood Westerns of the 1930s spread the image of cowboys and their blue jeans. Television, American GIs, and college students of the 1960s helped create global demand for this American icon.

DISCUSSION: Ask students for their opinions about blue jean styles and brands. What is the future of blue jeans?

internet connect

GO TO: go.hrw.com
KEYWORD: SW3 CH5
FOR: Web sites about world clothing styles

VISUAL RECORD ANSWER

Possible answer: People may or may not adopt various practices, depending on how they fit into established cultural patterns.

INTERPRETING THE VISUAL RECORD

Innovation and diffusion of ideas cause culture change in places like this area near Breves, Brazil. With satellite connections and other modern technologies, diffusion across the world is now possible in a matter of days. **How might cultural patterns in this part of Brazil influence the process of diffusion?**

from city centers to huge shopping malls in the suburbs. As transportation systems have changed, so have culture traits such as shopping patterns and personal mobility. Can you think of other examples of changing cultural traits in the United States?

Throughout history, general processes such as migration, war, and trade have caused cultures to change. These processes expose culture groups to new ways of life, including new languages, resources, and technologies. For example, when Spain colonized the Americas, Spanish settlers brought horses with them. Before Spanish settlement, American Indian cultures had not been exposed to these animals. Soon, however, they acquired horses, which became an important part of their culture. Many American Indian groups became skilled riders and used horses in hunting and in war.

The general processes of migration, war, and trade still cause culture change around the world. New culture traits are added as older ones fade from memory. Today modern communications and transportation systems have speeded culture change. Many of the world's young people seem to enjoy such fast change. They look forward to new technologies and ways of thinking.

When an individual or group adopts some of the traits of another culture, the process is called **acculturation**. Immigrants to the United States provide good examples. They often have to learn a new language and adopt a new way of life in this country. It is often difficult and stressful. However, through time, they often fully accept a new culture. When immigrant groups adopt all of the features of the main culture, it is called assimilation.

✓ **READING CHECK:** *Human Systems* How do general processes such as migration, war, and trade cause culture change? expose culture groups to new ways of life

Innovation and Diffusion Two concepts help us understand how cultures change. The first is **innovation**—new ideas that a culture accepts. People are always thinking of new ways to do things. However, only new ideas that are

useful will last. Some innovations happen just once and then spread throughout the world. For example, baseball was developed in the United States and later spread to the Caribbean and Asia. Other innovations are discovered in different places at different times. For example, ways for building boats developed independently among people all over the world.

The second important concept for understanding how cultures change is **diffusion**. Diffusion happens when an idea or innovation spreads from one person or group to another and is adopted. For example, jazz is an American form of music that took hold in New Orleans. Later, it spread to other parts of the United States and the world. Certain factors can aid or slow diffusion. For example, physical barriers like mountains and deserts can slow diffusion. Cultural similarities, such as shared languages, can aid diffusion from one group to another.

Types of Diffusion   How does diffusion happen? Culture traits can spread to new places in several ways. Sometimes, information about a new idea or innovation spreads throughout a society. This is called expansion diffusion.

Culture traits also spread when people move to new places and take their culture with them. This is called relocation diffusion. That is how English, a European language, became the main language in Australia and New Zealand. Religions often spread through this type of diffusion. For example, Judaism diffused throughout the United States with European immigrants.

Culture traits sometimes spread from places of greater size and influence to smaller places. This is called hierarchical diffusion. For example, new fashions and music styles often begin in Los Angeles or New York. These large cities are centers of the entertainment industry. From these cities, the new styles spread to other large cities. Eventually, the latest fashions reach small towns.

✓ **READING CHECK:**   (*Human Systems*)   What are some factors that might aid or slow diffusion? cultural similarities such as shared languages, physical barriers like mountains and deserts

Globalization   Today communications networks like the Internet and satellite television deliver information constantly to people around the world. These innovations are spreading culture traits more quickly than at any other time in human history. As a result, a global set of culture traits is taking hold. For example, people around the world now eat the same kinds of fast-food and

INTERPRETING THE VISUAL RECORD

The spread of global businesses is an important part of the process of globalization. For example, American fast-food restaurants have diffused to Russia and Eastern Europe, Latin America, Asia, and other world regions. This fast-food restaurant is in Moscow. **How might the spread of U.S.-based fast-food franchises such as this one be an example of cultural convergence?**

97

Close

Call on students to suggest foods that are typical of a culture in your area. (*Examples: enchiladas, dim sum, scrapple, gumbo, grits*) Ask students to speculate how the popularity of that dish spread to the region. (*Possible answers: immigration, national restaurant chains, television ads*) Point out how food is related to the main topics of the section.

Review and Assess

Have students complete the **Section Review.** Then have students complete **Daily Quiz 5.2.**

Reteach

Ask each student to create a list of five culture traits of his or her family and to determine how each became part of the family's culture. **ENGLISH LANGUAGE LEARNERS**

Extend

Have interested students conduct research on the assimilation of ethnic groups into American society. Ask them to compare assimilation experiences of immigrants of the late 1800s and early 1900s to those of the recent past. Challenge them to create hypotheses that explain any differences. **BLOCK SCHEDULING**

Section 2 Review Answers

Define For definitions, see: culture, p. 94; culture traits, p. 94; culture region, p. 95; ethnic groups, p. 95; acculturation, p. 96; innovation, p. 96; diffusion, p. 97; globalization, p. 98; traditionalism, p. 98; fundamentalism, p. 98

Reading for the Main Idea

1. all the features of a society's way of life, including language, religion, government, economics, food, clothing, architecture, family life, shared values, beliefs, institutions, and technologies, as well as its members' skills

2. Diffusion spreads cultural traits to new areas.

Critical Thinking

3. Answers will vary, but students should support their choices. (NGS 10)

4. Answers will vary, but students should support their views. (NGS 6)

Organizing What You Know

5. Expansion diffusion—Innovation gradually spreads out from its source. Relocation diffusion—Culture traits spread as people migrate. Hierarchical diffusion—Culture traits spread from places of greater influence to places of lesser influence. Examples will vary.

VISUAL RECORD ANSWER

Answers will vary, but students might mention style of clothing.

INTERPRETING THE VISUAL RECORD

Geographers are fascinated by globalization and traditionalism. Our world is full of places that are modern and innovative and places that are traditional and historic. Here, bagpipers in Edinburgh, Scotland, play traditional Scottish music. The bagpipe has become a well-known symbol of Scotland. **Besides bagpipes, what other examples of cultural traditions can you see in this photograph?**

wear the same types of jeans. Many also drive the same kinds of cars and enjoy listening to the same music. This process, in which connections around the world increase and cultures become more alike, is called **globalization.** Today it often has its roots in the United States. For example, American English, slang, popular culture, and businesses have spread to many countries and caused culture change. However, globalization also affects American culture. For example, Japanese electronics, German cars, and Italian fashions are all popular in the United States.

Globalization is an example of cultural convergence—different cultures blending together. This happens when the ideas, habits, and institutions of one culture come in contact with those of another culture. The spread of name-brand soft drinks throughout the world is an example. So is the popularity of Mexican food in the United States.

✓ **READING CHECK:** *Human Systems* What are some examples of globalization and the spread of cultural traits? spread of American English, slang, popular culture, businesses; popularity of Japanese electronics, German cars, Italian fashions

Traditionalism The opposite of globalization is **traditionalism.** Traditionalism means following longtime practices and opposing many modern technologies and ideas. In some cases, increasing religious **fundamentalism** has been a reaction to the spread of modern culture. The word *fundamentalism* can describe any movement in which people believe in strictly following certain established principles or teachings. In fact, many people believe that old ways of doing things should not be changed. They argue that old traditions tie people to their community, religion, and ancestors.

Traditionalism contributes to cultural divergence—the process of cultures becoming separate and distinct. This process happens when one group protects its culture from outside influences and another group welcomes change. Places with traditional cultures preserve the past, and their landscapes change very little. For example, ethnic celebrations and festivals keep cultures separate and distinct. These festivals celebrate a culture's unique history, identity, and way of life.

✓ **READING CHECK:** *Human Systems* Why do some favor traditionalism? Does traditionalism contribute to cultural convergence or cultural divergence? believe traditions tie people to their community, religion, ancestors; cultural divergence

Review

Homework Practice Online
Keyword: SW3 HP5

Define
culture, culture traits, culture region, ethnic groups, acculturation, innovation, diffusion, globalization, traditionalism, fundamentalism

Reading for the Main Idea

1. *Human Systems* What do geographers study when they look at a region's culture?

2. *Human Systems* How does diffusion cause culture change?

Critical Thinking

3. Supporting a Point of View What do you think are the most important parts of your culture? Why?

4. Contrasting How might globalists and traditionalists view cultures, places, and regions differently?

Organizing What You Know

5. Copy the graphic organizer below. Use it to describe the three different types of diffusion. Then give two examples of each.

Expansion diffusion	Relocation diffusion	Hierarchical diffusion

Mythology: Landforms and Culture

Just as physical features can affect a region's history, they can shape the beliefs of local peoples. Many myths and legends attempt to explain the creation of prominent or unusual landforms. For example, a popular Irish myth tells of the origin of the Giant's Causeway, a formation of about 40,000 stone columns on the coast of Northern Ireland. Scientists say the causeway's basalt columns were formed 50 to 60 million years ago when lava cooled as it reached the sea. According to one legend, however, a heroic giant named Finn MacCool built the causeway so he could walk to the island of Staffa in Scotland to fight a rival giant named Benandonner. The exhausted MacCool then returned to Ireland and fell asleep. Benandonner found the causeway and walked to Ireland to challenge MacCool. MacCool's wife, Oonagh, fearful for her husband's safety, told Benandonner that the sleeping giant he saw was her child. Benandonner saw that the sleeping figure was nearly as large as he and became afraid, because the father of a child this size would have been truly immense. He fled back to Scotland, destroying the causeway as he ran. All that remained were a few support columns.

Organize students into groups and assign each group a landform that figures in its mythology. (*Examples: the Grand Canyon, the Nile River in Egypt, Ayers Rock (Uluru) in Australia, Mount Olympus in Greece*) Have groups conduct research to find legends that describe the creation of the landform and compare their findings to scientific explanations. **COOPERATIVE LEARNING**

The Uses of Geography

Geography for Life

Geography and History

To understand our world, we must study its geography and its history at the same time. All historical events happen at a certain place. In turn, environmental conditions affect historical events.

Earth's landscapes and environmental conditions change over time. These changes in physical geography can determine the fates of entire civilizations. Most of these changes are slow and their effects very gradual. Other changes are more dramatic. For example, to the ancient Greeks, Ephesus was a key commercial city on the coast of what is now Turkey. Over time a river filled the city's harbor with sediment. When it was no longer useful as a port, Ephesus faded from importance.

Environmental changes such as climate shifts can also affect history. For example, a period of weather extremes began about A.D. 1300. In Europe heavy rains and early freezes caused crops to fail. In some places, many people starved to death. Because many survivors were poorly nourished, they were vulnerable to diseases, such as the plague. The spread of the plague, or Black Death, across Europe turned the 1300s into a century of disasters.

Physical features can also influence history and the distribution of culture groups. Consider why people have settled in certain places. You may live in the Mississippi River watershed in a city with a French name, such as St. Louis or Des Moines. Why did French explorers and settlers go there instead of farther east or west? The region's waterways give us the answer. The Great Lakes and the Mississippi River provided a route south and west from French territory in Canada. The Mississippi became a highway into the continent's heart for those who had crossed the Atlantic from France. Similarly, high mountains between Afghanistan and Pakistan seem like a huge wall in the path of travelers. However, the Khyber Pass cuts through this wall. In places this ravine is only 50 feet (15 m) wide. Yet over the centuries invading armies and migrating peoples have poured through the pass. Each group has left its stamp on the region's diverse cultures.

Just as physical features can make diffusion of peoples and ideas easier, they can also make it more difficult. For example, the Sahara forms a vast barrier across northern Africa. During the A.D. 600s, Arab armies brought Islam to the area north of the desert. However, the Sahara slowed the spread of Arab influence to the south. Other features of Africa's natural environments also limited European influences for a time there. The mosquitoes that carry malaria require a warm, rainy climate. Because the insects thrived in Africa's tropical forests, movement by Europeans into many areas was delayed. Many Africans who had lived in those regions for generations had some resistance to malaria. Europeans did not.

In turn, history affects both physical and human geography. For example, European sailors of the 1500s changed forever the human geography of the places they opened to settlement. An area's languages, religions, ethnic makeup, architecture, clothing, customs—all these would change. Humans have also changed physical geography. We have altered the land throughout our history.

Since ancient times, the Khyber Pass has been an important travel route between Afghanistan and Pakistan. Persians, Greeks, Mughals, Afghans, and other peoples have all passed through it.

Applying What You Know

1. **Summarizing** What are some ways in which geography affects history?

2. **Drawing Inferences and Conclusions** Look at a physical map of the world. How do you think certain deserts, mountain ranges, rivers, and other physical features affected migration patterns and the distribution of culture groups?

Essential Element 6

▶ **Uses of Geography** ◀

Historical Geography

Our world is a product of the past. Cities, farms, roads, and other landscape features were created in the past and have evolved through time into their present form. The study of how places have developed and changed through time is called historical geography.

Knowledge of historical geography is crucial for understanding the world today. For example, many immigrant groups have migrated to the United States. Knowing where these groups settled can help us understand the cultures of places today. The Cajun region of southern Louisiana and the Cuban community of Miami are examples of places with distinctive cultures tied to their historical geography.

Applying What You Know Answers

1. Changes in physical features and climates may change the way of life of people in an area; physical features may act as aids to diffusion or as barriers.

2. Possible answers: Australian deserts limited European settlement; Amazon River allowed travel, settlement by many culture groups in the heart of the rain forest; the deserts and mountains of Central Asia kept Chinese civilization isolated.

This Geography for Life feature addresses National Geography Standards 15 and 17.

OBJECTIVES

1. Examine the geography of the world's languages.

2. Describe the three main types of religions that geographers identify.

 **LET'S GET STARTED**

Copy the following passage onto the chalkboard: *Several British writers have said that America and Britain are separated by a common language. What do you think they meant?* Discuss responses. Point out that people who speak the same language may use different dialects with different vocabularies, pronunciations, or spellings. For example, in Great Britain an apartment is a "flat," *schedule* is pronounced "shedule," and *jail* is often spelled "gaol." Tell students they will learn more about dialects, other aspects of language, and religion in Section 3.

Building Vocabulary

Write the key terms on the chalkboard. Call on volunteers to read the definitions. Point out that **lingua franca** is Italian for "Frankish language" but came to mean "European language" when the Franks ruled much of the continent. It now refers to a language used for trade. Compare *mono-* and *poly-;* the first means "one," and the other means "many." *Theism* means "belief in a god or gods." Point out that **mosques** and **hajj** are from Arabic, although the word *mosque* has also undergone other influences.

 **Section 3 RESOURCES**

REPRODUCIBLE

▶ Guided Reading Strategy 5.3
▶ Critical Thinking Activity 5: Buddhism
▶ Map Activity 5: European Religions

TECHNOLOGY

▶ One-Stop Planner CD–ROM, Lesson 5.3
▶ Homework Practice Online
▶ HRW Go site

REINFORCEMENT, REVIEW, AND ASSESSMENT

▶ Main Idea Activity 5.3
▶ English Audio Summary 5.3
▶ Spanish Audio Summary 5.3
▶ Section 3 Review, p. 103
▶ Daily Quiz 5.3

CHART ANSWER

Possible answers: Chinese and Indian populations are growing quickly. English was spread to many parts of the world during colonial era, now spread by electronic media.

 **Section 3**

World Languages and Religions

READ TO DISCOVER

1. What is the geography of the world's languages?
2. What are the three main types of religions that geographers identify?

WHY IT MATTERS

Language and religion give character to places. They also are key causes of cultural differences. Use **CNNfyi.com** or other **current events** sources to learn about a conflict where religion is an important issue.

DEFINE

dialect
lingua franca
ethnic religions
animist religions
polytheism

universalizing religions
monotheism
missionaries
mosques
hajj

Geography of Languages

Language is one of the most important areas of study in geography. Language is important to culture because it is the main means of communication. For example, one generation passes customs and skills to the next mainly through language. Language is an important part of a culture's traditional celebrations, rituals, and ceremonies. Language also influences the routes and patterns of cultural diffusion and migration. For example, new information spreads more easily among places that have a common language. In addition, language differences can be a barrier to diffusion.

For geographers, one of the most basic facts about languages is that they have spatial characteristics. Languages are spoken in specific regions of the world and shape people's lives there. Patterns of speech also help make specific regions of the world distinctive. For example, Spanish is spoken throughout most of South America. The widespread use of Spanish in the region helps make South America distinct. However, most people in South America's largest country, Brazil, speak Portuguese. As a result, Brazil is a clear linguistic subregion within South America.

About 3,000–6,500 languages are spoken in the world today. Experts divide these languages into more than a dozen families. These families are groups of languages that experts believe have a common origin. (See the map of world language families.) Language families, in turn, are divided into language branches. For example, English is a language in the Germanic branch of the Indo-European family. Spanish and French are languages in the Romance branch of the same family. About 50 percent of the world's people speak an Indo-European language.

The language with the most speakers is Mandarin Chinese. (See the chart of principal languages of the world.) Mandarin belongs to the Sino-Tibetan language family. About 20 percent of the world's people speak a Sino-Tibetan language. Speakers of any one language might use a particular **dialect**. A dialect

Principal Languages of the World

Language	Speakers (in millions)
Mandarin	874
Hindi	366
English	341
Spanish	322
Bengali	207
Arabic	202
Portuguese	176
Russian	167
French	77
Malay-Indonesian	35

Source: The World Almanac and Book of Facts 2004

INTERPRETING THE CHART

Mandarin has by far the most "first language" speakers of any language in the world, while English is a distant third. **Why do you think English has the third-highest number of "first language" speakers?**

100

LEVELS 1 AND 2: Prepare slips of paper with the names of countries written on them. Ask each student to select a country and to locate it in an atlas. Then have students consult the world language families map in their textbooks to determine which language family or families are found in their assigned countries. Have students draw a border around their slips in the color that represents the appropriate language family on the textbook map. Then tell students to consult almanacs to discover their country's primary language and to use foreign language dictionaries or the Internet to find examples of common words in that language. (*Possible words: mother, father, hello, thank you*)

Have students write these words and their English equivalents on their slips. Attach the slips to a large wall map near the appropriate country. Lead a discussion on the wide variety of languages in the world and the distribution of language families. **ENGLISH LANGUAGE LEARNERS**

LEVEL 3: Organize the class into pairs. Have each pair choose a country in which English is the official language and conduct research to learn more about that country's dialect. (*Possible countries: Australia, Bahamas, Belize, Botswana, Canada, Fiji, Jamaica, Kenya, United Kingdom*) Then have each pair create a dialogue using examples of the country's distinctive vocabulary. Call on volunteers to share their dialogues. **COOPERATIVE LEARNING**

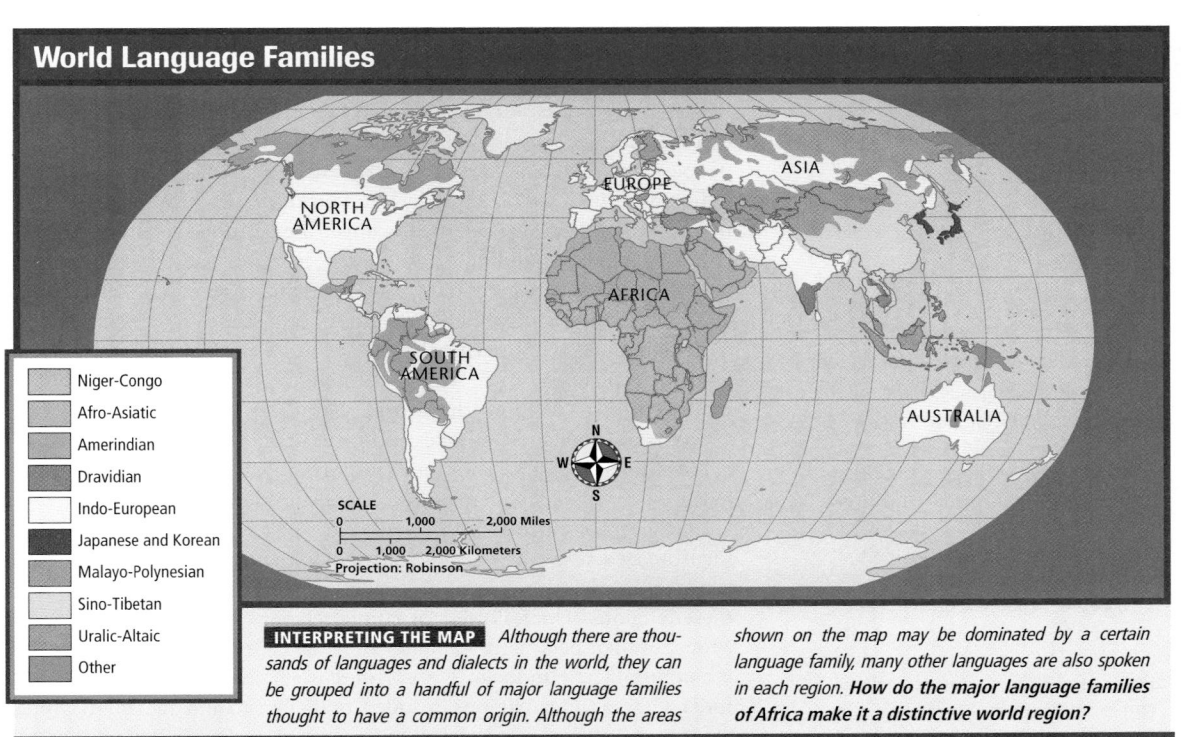

World Language Families

Legend:
- Niger-Congo
- Afro-Asiatic
- Amerindian
- Dravidian
- Indo-European
- Japanese and Korean
- Malayo-Polynesian
- Sino-Tibetan
- Uralic-Altaic
- Other

SCALE
0   1,000   2,000 Miles
0 1,000 2,000 Kilometers
Projection: Robinson

INTERPRETING THE MAP *Although there are thousands of languages and dialects in the world, they can be grouped into a handful of major language families thought to have a common origin. Although the areas shown on the map may be dominated by a certain language family, many other languages are also spoken in each region.* **How do the major language families of Africa make it a distinctive world region?**

is a regional variety of a language. For example, British English and American English are dialects of the same language. While different dialects can usually be understood by speakers of the same language, they contain distinctive words and pronunciation. These unique words and phrases can usually provide clues to where a speaker is from. Can you think of some dialects of English found in the United States?

To understand language patterns, geographers study where a language comes from and how it has spread. Languages spoken by very few people usually indicate remote populations whose people have not migrated. We can see examples of this in some mountain areas. For example, in the Caucasus Mountains many different languages are spoken by small groups of people. Some of these groups have been isolated in mountain valleys since ancient times.

English has become the most widespread language in the world. English traces its beginnings to Anglo-Saxon, which became distinct about 1,500 years ago on the island of Great Britain. By about 1500, it had developed into Modern English. The language spread rapidly when Britain began to colonize large areas of the world a few hundred years ago. Today English is the main language of globalization and the Internet. Millions of people now speak English as a second language. English words have also become part of other languages. For example, baseball is called *beisbol* in Spanish. The French refer to the weekend as *le weekend*. English has become a **lingua franca**—a language of trade and communication—for the whole world.

✔ **READING CHECK:** (**Human Systems**) How has English spread around the world? It spread several hundred years ago when England began to colonize the world.

Our Amazing Planet

Some linguists believe that 90 percent of the languages spoken today may die out in the next 100 years.

Global Perspectives

Language and the Internet English is the most widely used language on the Internet, but other languages are surviving and thriving in cyberspace. For example, one can access newspapers from 80 or so countries—not only large countries like France and Germany but also smaller ones like Qatar. There are discussion groups in more than 100 languages, including Basque, Hmong, Navaho, Swahili, and Yoruba.

Some critics charge that the Internet may help drive minority languages to extinction. Because even small language groups can send their messages worldwide, however, the Internet may actually help preserve threatened tongues. For example, immigrant groups can communicate with friends or family in their homelands—in their native languages—over the Internet.

CRITICAL THINKING: Although people in practically all the world's countries are online, what factors still limit Internet use in developing countries? (*Lower incomes, limited access to electricity, and illiteracy diminish Internet use.*)

MAP ANSWER

Much of Africa is dominated by the Niger-Congo family.

Teach Objective 2

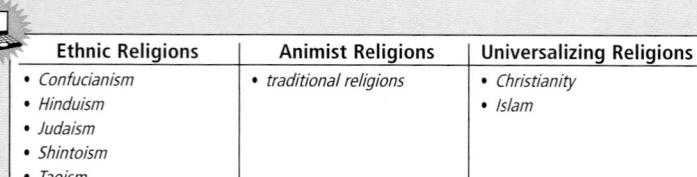

LEVEL 1: Copy the following graphic organizer onto the chalkboard, omitting the italicized answers. Call on students to provide examples of each of the three main types of religions identified by geographers. Then lead a class discussion about similarities and differences between the types. **ENGLISH LANGUAGE LEARNERS**

Ethnic Religions	Animist Religions	Universalizing Religions
• *Confucianism* • *Hinduism* • *Judaism* • *Shintoism* • *Taoism*	• *traditional religions*	• *Christianity* • *Islam*

LEVELS 2 AND 3: Organize students into small groups. Direct students' attention to the map of world religions in this section. Have each group conduct research to learn where each of the religions on the map began and how many followers each has today. Have groups use their findings to create charts. Call on volunteers to speculate how various religions have diffused from their places of origin. **COOPERATIVE LEARNING**

Daily Life

Falun Gong Falun Gong is one of the world's newest religious movements. Founded in China in 1992, it is a branch of the Qi Gong (CHEE GONG) tradition that practices slow physical movements and meditation. Many teachings of Falun Gong are based on virtues traditionally related to Buddhism, Confucianism, and Taoism. Some of its views are considered radical, however, and are not related to traditional Chinese religion. For example, Falun Gong's leader, who lives in exile in New York, believes extraterrestrials are interfering with Earth and corrupting humanity by influencing the expansion of computers and technology.

The Chinese government has persecuted followers of Falun Gong. They maintain that it is a cult which promotes superstition and harms Chinese society. The government fears Falun Gong's ability to use the Internet to organize large numbers of people for demonstrations.

CONNECTING TO TECHNOLOGY ANSWER

Different cultures are blending together with the widespread use of English on the Internet.

Connecting to TECHNOLOGY

English on the Internet

The Internet links millions of the world's computers. It developed out of efforts to make sure that U.S. authorities would be able to communicate after a nuclear war. Ever since then the Internet has been dominated by the United States and the English language. Today more than 80 percent of the Internet's home pages, or Web sites, are in English. Also, about 90 percent of Internet users are English-speakers. Many scientists and businesspeople around the world use the Internet—and therefore English—to communicate with each other.

People in non-English-speaking countries often build their Web sites in English. This is particularly helpful if they use the Web for international communication. In some countries, using English has become a status symbol. It suggests that a Web site's information may interest people around the world.

Analyzing In what way is the use of English on the Internet an example of cultural convergence?

An Internet cafe in Ecuador

Geography of Religion

Religion is another important topic in human geography. It is a key culture trait that binds many societies together and gives meaning to people's lives. Geographers are interested in religion for many reasons. Religions differ from place to place and produce culture traits that can be mapped. Religions also can greatly affect the cultural landscape. For example, religious buildings and sacred locations are usually clearly marked places where certain types of behavior are required. Religion can also have more indirect effects on the landscape. For example, the Christian practice of burying the dead requires Christian societies to use land for cemeteries. Also, religious differences are a key component of many conflicts around the world. Conflicts between Hindus and Muslims in India and Catholics and Protestants in Northern Ireland are two examples. Perhaps even more importantly, religion is an important part of many people's identity and connection to a certain place.

Geographers identify three main types of religions. **Ethnic religions** focus on one ethnic group and generally have not spread into other cultures. Old beliefs, legends, and customs of different ethnic groups shape these religions. The followers of ethnic religions do not seek to convert people to their beliefs. Instead, they practice their religion as part of their ethnic heritage. Hinduism is the largest such religion. It is centered in India. Other ethnic religions include Confucianism and Taoism in China, Shintoism in Japan, and Judaism in Israel. (See the World Religions map.)

In **animist religions**, people believe in the presence of the spirits and forces of nature. **Polytheism**, or the belief in many gods, is an essential part of most of these religions. Animist religions are also often considered ethnic religions because particular peoples practice them. They are common in many traditional societies and likely were practiced long before more modern types of religions became common.

In contrast to ethnic religions, **universalizing religions** seek followers all over the world. They hope to appeal to people of many different cultures. More than half of the world's people follow these religions. Christianity and Islam are universalizing religions. Each is based on **monotheism**—the belief in one god. These religions also have **missionaries**, who help spread the religion. New converts are accepted through symbolic rituals and initiations. The ultimate goal of universalizing religions is to spread their beliefs to the entire world. Therefore, it is not surprising that universalizing religions are the most rapidly growing religions.

Religion provides a rhythm to daily life and shapes distinctive cultural patterns. In Islamic countries, for example, people stop to pray several times a day. When praying, they face toward Mecca, the spiritual center of Islam. Houses of worship called **mosques** are common. Important mosques are beautifully designed and decorated. Large crowds gather for prayers there, particularly on Friday at noon. Many people in Islamic countries also make a special religious journey to Mecca called a **hajj**. Islam requires its followers to make this journey at least once in their lifetime.

Close

Instruct students to look through the infomation about various cultures in this book and call out the names of as many languages as they can find. Point out that only languages used by large numbers of people are discussed. In some countries there are hundreds of languages spoken.

Review and Assess

Have students complete the **Section Review**. Then have students complete **Daily Quiz 5.3**.

Reteach

Organize the class into pairs and have each pair write six sentences about the importance of languages and beliefs to human geography. Call on volunteers to share their sentences with the class. **ENGLISH LANGUAGE LEARNERS, COOPERATIVE LEARNING**

Extend

Have interested students conduct research on the historic relationships between languages and religions, such as the ties between Arabic and Islam or Latin and Christianity. Ask students to learn why these ties exist and how the spread of a language and its corresponding religion are related. **BLOCK SCHEDULING**

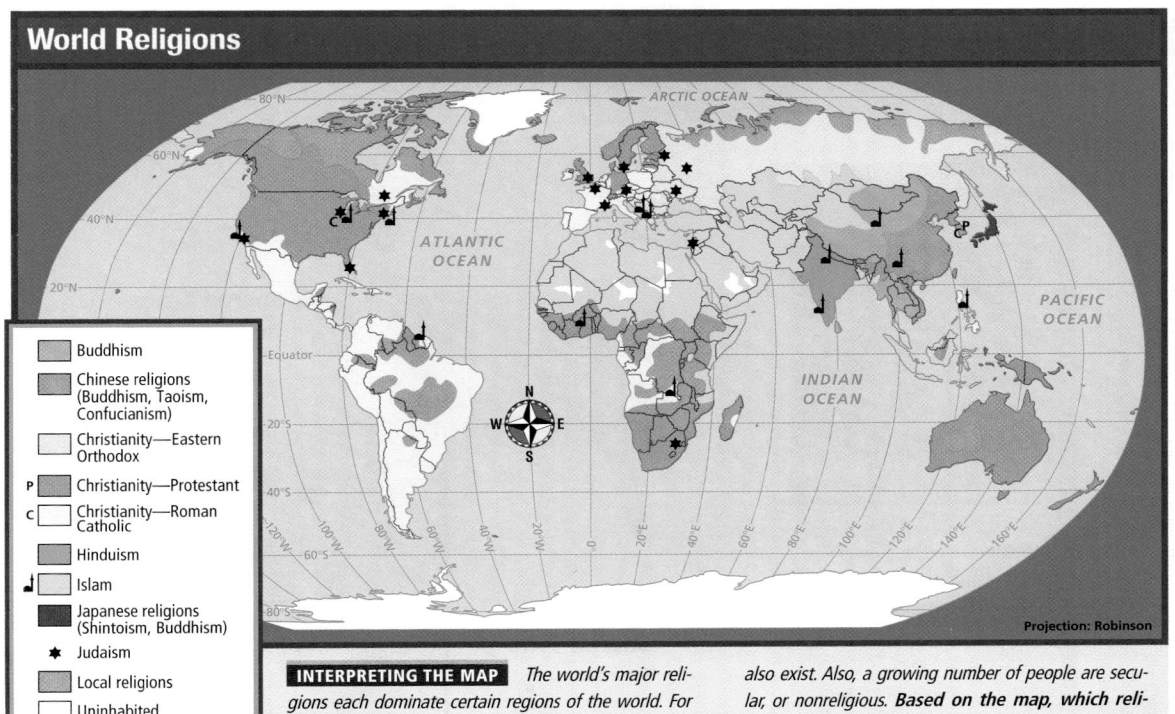

World Religions

- Buddhism
- Chinese religions (Buddhism, Taoism, Confucianism)
- Christianity—Eastern Orthodox
- P Christianity—Protestant
- C Christianity—Roman Catholic
- Hinduism
- Islam
- Japanese religions (Shintoism, Buddhism)
- ✶ Judaism
- Local religions
- Uninhabited

Projection: Robinson

INTERPRETING THE MAP *The world's major religions each dominate certain regions of the world. For example, Buddhism is most widely practiced in Asia, while Hinduism dominates India. However, within each major world region, a huge number of smaller religions* *also exist. Also, a growing number of people are secular, or nonreligious.* **Based on the map, which religions are practiced in the United States? In Africa? How widespread is Christianity? How widespread is Islam?**

While religion remains important around the world, in some places it has declined. A growing number of people in these places are secular, or nonreligious. Europe in particular has experienced declines in the number of people actively participating in religious activities, such as attending church each week. There are many possible reasons for this decline in religion. For example, the development of modern technologies and science have caused some people to question the beliefs of traditional religions.

✓ **READING CHECK:** *Human Systems* What are the three main types of religions?
ethnic religions, universalizing religions, animist religions

Section 3 Review

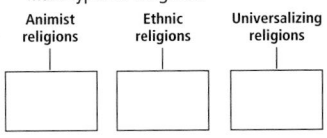

Homework Practice Online
Keyword: SW3 HP5

Define dialect, lingua franca, ethnic religions, animist religions, polytheism, universalizing religions, monotheism, missionaries, mosques, hajj

Reading for the Main Idea

1. *Human Systems* Why are communication and beliefs so important in human geography?

2. *Human Systems* About how many languages are spoken in the world today? Which language has the most speakers?

3. *Human Systems* What are some examples of ethnic religions?

Critical Thinking

4. **Analyzing** How has religion shaped the cultural landscape of your community?

Organizing What You Know

5. Copy the chart below. Use it to describe, compare, and contrast the three main types of religions.

Animist religions	Ethnic religions	Universalizing religions

Section 3 Review Answers

Define For definitions, see: dialect, p. 100, lingua franca, p. 101; ethnic religions, p. 102; animist religions, p. 102; polytheism, p. 102; universalizing religions, p. 102; monotheism, p. 102; missionaries, p. 102; mosques, p. 102; hajj, p. 102

Reading for the Main Idea
1. because language and religion shape the cultures of regions and are key causes of cultural differences

2. about 3,000–6,500; Mandarin

3. Hinduism, Confucianism, Taoism, Shintoism, Judaism

Critical Thinking
4. Answers will vary according to the community and students' perceptions.
(NGS 10)

Organizing What You Know
5. Animist religions—worship of spirits and forces of nature; Ethnic religions—focus on one ethnic group, generally have not spread into other cultures; Universalizing religions—seek followers all over the world, hope to appeal to people of many different cultures

MAP ANSWER

Christianity, Islam, Judaism; Islam, Judaism, Christianity, local religions; Christianity and Islam both widespread throughout the world

103

Judaism developed in the eastern Mediterranean region during ancient times. Over the centuries, Jews were dispersed to many other parts of the world. This dispersion is known as the diaspora (dy-AS-puh-ruh). The first major dispersion occurred in 586 B.C. and is known as the Babylonian Exile. In that year the Babylonians conquered the Kingdom of Judah and later forced many Jews into slavery in Babylon (modern Iraq). During Roman times, Jews were dispersed throughout the Mediterranean region. Later, they settled in Europe, North Africa, and Arabia. The dispersion of Jews led to the development of cultural subgroups. Jews in Mediterranean lands were known as *Sephardim*, while those in central and eastern Europe were known as *Ashkenazim*. In the late 1800s and 1900s many *Ashkenazim* migrated to North America. The Holocaust in Europe forced many Jews to flee to the United States and Palestine. The creation of Israel in 1948 encouraged further migration to that area. Have students conduct research to create a map of the Jewish diaspora. The map should use arrows to show major migrations and include other information such as dates and numbers of migrants. Finally, have students create a pie graph showing the distribution of Jews today by major countries.

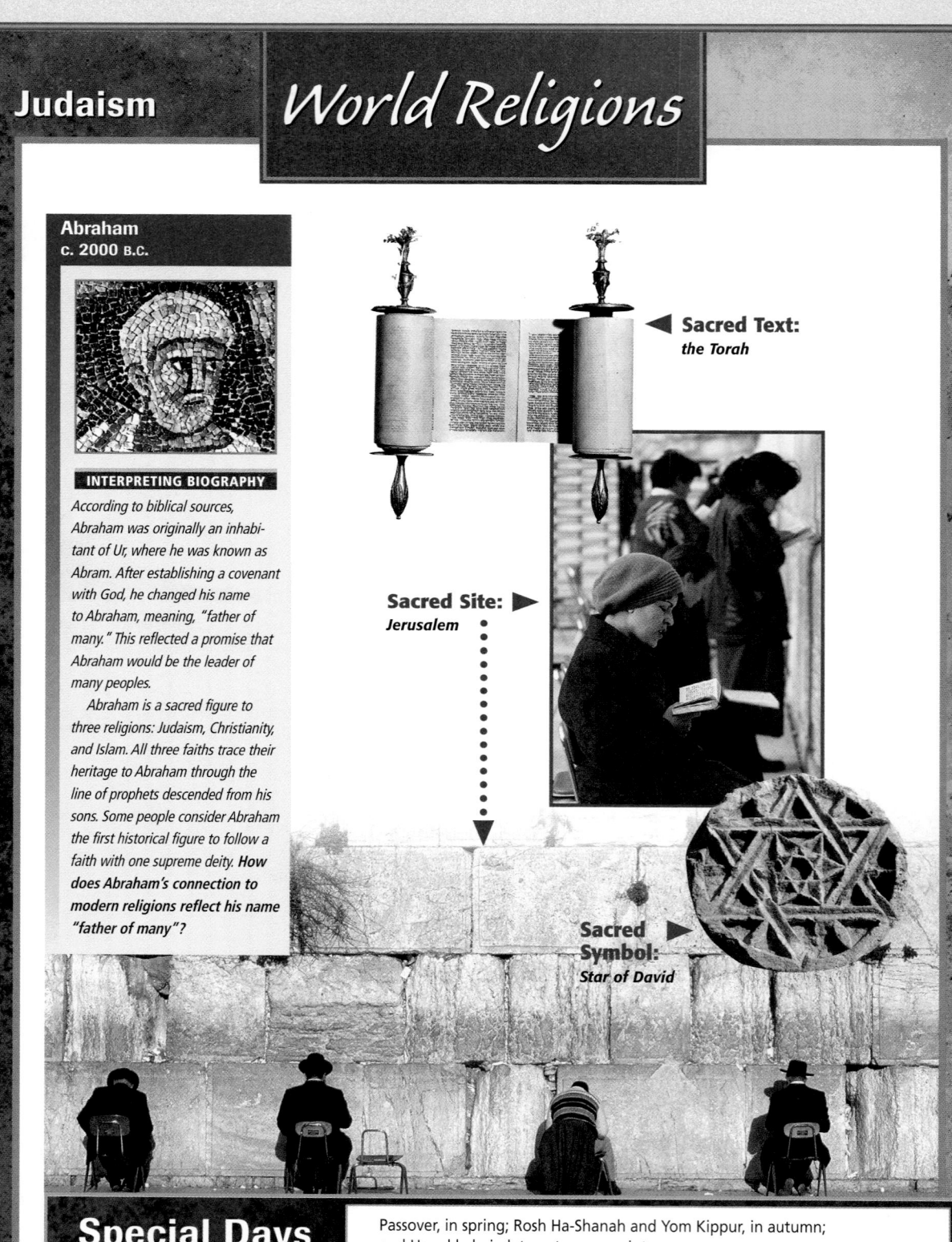

World Religions

Judaism

Abraham
c. 2000 B.C.

INTERPRETING BIOGRAPHY

According to biblical sources, Abraham was originally an inhabitant of Ur, where he was known as Abram. After establishing a covenant with God, he changed his name to Abraham, meaning, "father of many." This reflected a promise that Abraham would be the leader of many peoples.

Abraham is a sacred figure to three religions: Judaism, Christianity, and Islam. All three faiths trace their heritage to Abraham through the line of prophets descended from his sons. Some people consider Abraham the first historical figure to follow a faith with one supreme deity. **How does Abraham's connection to modern religions reflect his name "father of many"?**

◀ **Sacred Text:** *the Torah*

Sacred Site: ▶ *Jerusalem*

Sacred Symbol: ▶ *Star of David*

Special Days

Passover, in spring; Rosh Ha-Shanah and Yom Kippur, in autumn; and Hanukkah, in late autumn or winter

Studying Hinduism

Hinduism began to develop some 4,000 years ago in the Indian subcontinent. Today Hinduism is the main religion of India, is closely tied to India's history and culture, and is considered the largest ethnic religion in the world. Unlike most other major world religions, Hinduism has no founder, no clear beginning, and no central authority, hierarchy, or organization. A polytheistic religion, Hinduism includes the worship of many different gods. These gods range from local deities to pan-Indian gods or even a single high God. The range of beliefs in Hinduism is equally diverse, and there is no single belief that unites all followers of the religion. There are, however, several widely held beliefs among Hindus. For example, most believe in the sanctity of the ancient religious writings known as the Vedas and in an eternal and infinite source of reality called *brahman*. Other common beliefs include *ahimsa*—noninjury to living things—and *samsara*—a continuous cycle of rebirth based on one's previous acts. Have students conduct research on Hinduism. Some questions to consider might include: How many people in the world are followers of Hinduism? In which countries other than India is Hinduism practiced? What are some common beliefs, practices, and characterisitics of Hinduism?

Hinduism

Hindus consider it a sacred duty to bathe in the holy waters of the Ganges River. This ritual cleanses the bather's mind and spirit.

Sacred Texts:
The Vedas, Bhagavad Gita

Sacred Sites:
the Ganges River,
the city of Varanas

Festival of Holi

Sacred Creature:
the Cow

The cow is a particularly sacred animal in the Hindu faith in part because of the important role it has played in sustaining life.

Special Days

Festival of Holi, in spring; Diwali, or Deepavali (Festival of Lights) in autumn

Mahavira
c. 599–527 B.C.

INTERPRETING BIOGRAPHY

Over the years many religious leaders added to and expanded Hindu thought. One such person was Mahavira, also known as Vardhamana. He was born into a warrior clan in northeastern India. At the age of 30 he left his home and entered the forest to find spiritual fulfillment. He got rid of all his personal possessions. For more than 12 years, he wandered the countryside with nothing to his name and little contact with other people.

After he felt he had gained the answers to his questions about life, Mahavira began teaching others. He believed the key to enlightenment was to live apart from the material world as much as possible. Many early Hindus were influenced by his ideas. Eventually his beliefs became the basis of Jainism, a new religion. **How did Mahavira influence Hindus?**

Mapping the Diffusion of Buddhism

Buddhism originated some 2,500 years ago in what is now India. Buddhism developed initially as an offshoot of Hinduism based on the teachings of the Buddha. The Buddha sent monks out to preach the faith to others. During the Buddha's lifetime, the new religion spread throughout northern India. In the 200s B.C. missionaries introduced Buddhism to Sri Lanka. Missionaries and traders later introduced Buddhism to China (about 100 B.C. to A.D. 200), Korea and Japan (about A.D. 300 to 500), Southeast Asia (about A.D. 400 to 600), Tibet (about A.D. 700), and Mongolia (about A.D. 1500). As it diffused, Buddhism developed many regional forms. In China,

Korea, and Japan it merged with native belief systems such as Confucianism, Taoism, and Shintoism. In these areas, people follow a type of Buddhism known as Mahayana. In Sri Lanka and mainland Southeast Asia, a form of Buddhism known as Theravada is practiced. Buddhists in Tibet and Mongolia practice a version called Lamaism. While Buddhism diffused throughout much of Asia, it declined in its source area of India. Have students create a map showing the origin and diffusion of Buddhism in Asia. Maps should use arrows to show the religion's spread and include dates to indicate when it reached certain areas.

Buddhism

Siddhartha Gautama
563 B.C. – 483 B.C.

INTERPRETING BIOGRAPHY

Siddhartha Gautama was born the son of an Indian prince. At the age of 29, he left his palace and was shocked by the suffering he saw. As a result, he wondered about the great problems of life. "Why does suffering exist?" he asked. "What is the meaning of life and death?"

Gautama decided to spend the rest of his life seeking answers to his questions. In what is now called the Great Renunciation, he put aside all his possessions, left his family, and set out to search for truth. One day, while meditating under a bodhi tree near the town of Bodhgaya, Gautama realized the key to ending suffering. This led to the development of the Four Noble Truths and the Eightfold Path, which all Buddhists follow. After his experience under the bodhi tree, Gautama became known as the Buddha, or "enlightened one." **How did Siddhartha Gautama reject his old life?**

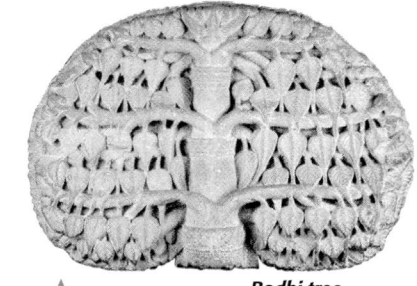

▲ **Sacred Site:**
Bodhgaya

Bodhi tree

Buddha Day festival

Sacred Objects:
Statues of the Buddha
▼

Sacred Text:
the Pali Canon

Special Days

Buddha Day, celebrated at the full moon in May

Learning about Confucius

Confucius was China's most famous and influential teacher and philosopher. His ideas and personal views on how to live a moral and proper life have influenced Chinese culture for more than 2,000 years. In addition, Confucian ideas diffused to other countries, particularly Japan, the Koreas, and Vietnam. One of the fundamental beliefs of Confucianism's followers is that all people have the ability to learn and improve themselves. Confucius himself spent many years studying the works and ideas of ancient Chinese scholars. He studied Chinese rituals, history, music, poetry, arithmetic, archery, and charioteering. When he was in his 30s, Confucius began teaching. He demanded from his students a willingness to learn. Confucius wanted to make education widely available. At first he taught a small group of students who were his followers. Later his fame spread, and more people came to listen to his teachings. Confucius lectured on topics such as the ancient classics and personal self-improvement.

Have students conduct research on the life, philosophy, teachings, and influence of Confucius. Then lead a discussion about his ideas on education and learning. Ask students to discuss what educational philosophies of Confucius are still important concepts in education today.

Confucianism

Because Confucius emphasized the importance of education and learning, his followers celebrate his birthday as Teacher's Day.

▲ **Sacred Text:**
The Analects

Sacred Symbol: ▶
Yin-yang

Sacred Site:
Confucian Temple
▼

Confucius
551 B.C. – 479 B.C.

INTERPRETING BIOGRAPHY

Westerners know K'ung Ch'iu as Confucius. He was born in the Chinese province of Lu. He spent much of his life tutoring and working in low-level government positions.

Confucius grew frustrated by other officials around him. In midlife he left his government position and traveled the countryside promoting reform. After 13 years he returned to Lu to teach about the ideas he had formed during his travels.

Confucius had little to say about gods, the meaning of death, or the idea of life after death. For this reason some people do not consider Confucianism a religion, although the goal of Confucianism is to be in "good accord with the ways of heaven." Many followers of Confucianism practice his ideas as religion. **Why do some people not consider Confucianism a religion?**

Yin-Yang In ancient Chinese thought there are two complementary forces contained in all things—yin and yang. Yin is thought of as Earth, female, dark, passive, and absorbing. It is associated with the Moon, winter, even numbers, and valleys and streams. Yin is represented by the tiger, the color orange, and a broken line. Yang is conceived of as heaven, male, light, active, and forceful. It is associated with the Sun, summer, odd numbers, and mountains. Yang is represented by the dragon, the color blue, and an unbroken line.

According to the belief, a balance of yin-yang forces is essential to universal harmony. As one of the forces increases, the other decreases. When the two forces are in balance, they are represented as the light and dark halves of a circle.

ACTIVITY: Have students draw a circle with light and dark halves representing a balance of yin-yang forces. Then have them list some ideas and symbols associated with the yin half of the circle and those associated with the yang half of the circle.

BIOGRAPHY ANSWER
Confucius had little to say about gods, the meaning of death, or the idea of life after death.

Special Days

Teacher's Day, in August or September

Christianity is the world's largest religion, both by number of followers and by area. More than 1.8 billion people are Christian—roughly one third of the world's population. Areas where Christianity is the dominant religion include most of North and South America, Europe, Russia, Australia, New Zealand, the Philippines, and many parts of sub-Saharan Africa. Despite Christianity's widespread distribution, it has long been fragmented into many denominations. These denominations have different hierarchies and practices and are concentrated in certain areas. The fragmented nature of Christianity is perhaps best illustrated in the United States, where there are hundreds of different Christian groups. Some came to the United States long ago to escape persecution in Europe. Others were founded and developed here. Some major regional patterns include: Baptist and other conservative denominations in the South; Lutherans in the upper Midwest; Mormons in Utah; and Roman Catholicism in southern Louisiana, the Southwest, and Northeast. Have students conduct research on the distribution of Christian groups in the United States and create a map showing major regional patterns. Then ask students to analyze their maps to identify historical and cultural factors that might help explain them.

Daily Life

Church Architecture

Many geographers are interested in how religion is visibly expressed on the landscape. Buildings such as churches can reflect the history, culture, beliefs, and regional geography of the people and places where they are found.

Christian churches vary greatly in architectural style, size, ornateness, building materials, and other factors. Europe's Gothic cathedrals are considered houses of God and are typically huge stone structures with vaulted ceilings, stained glass windows, flying buttresses, gargoyles, and other decorations. In contrast, many Protestant churches in the United States are viewed simply as places to assemble for worship and are therefore very simple in design and decoration.

CRITICAL THINKING: How might geographers use church architecture to learn about the historical, cultural, and physical geography of different regions? *(Possible answer: They can study and map the distribution of such factors as the age of churches, building materials used, and cultural characteristics.)*

BIOGRAPHY ANSWER

Possible answer: to teach moral lessons to ordinary people in simple terms

Christianity

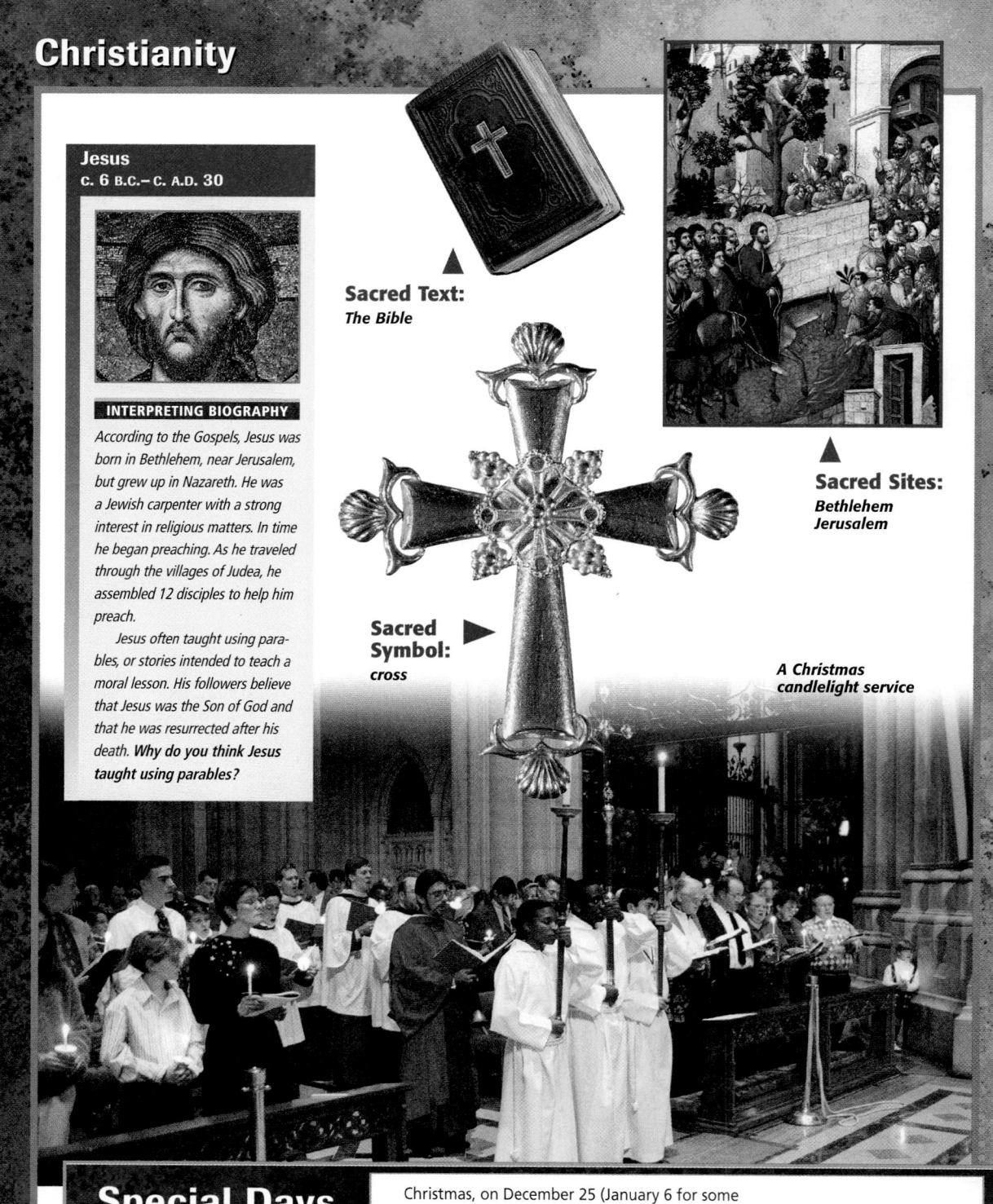

Jesus
c. 6 B.C.– c. A.D. 30

INTERPRETING BIOGRAPHY

According to the Gospels, Jesus was born in Bethlehem, near Jerusalem, but grew up in Nazareth. He was a Jewish carpenter with a strong interest in religious matters. In time he began preaching. As he traveled through the villages of Judea, he assembled 12 disciples to help him preach.

Jesus often taught using parables, or stories intended to teach a moral lesson. His followers believe that Jesus was the Son of God and that he was resurrected after his death. **Why do you think Jesus taught using parables?**

Sacred Text:
The Bible

Sacred Sites:
Bethlehem
Jerusalem

Sacred Symbol:
cross

A Christmas candlelight service

Special Days   Christmas, on December 25 (January 6 for some Orthodox churches); Easter, in the spring

For Muslims around the world, five basic acts provide the basis for a righteous way of life. These five acts are known as the five pillars of Islam. The five pillars are: 1) a profession of faith demonstrated by the statement "There is no god but Allah, and Muhammad is His messenger"; 2) daily prayer; 3) giving charity to the poor; 4) fasting from dawn to sunset during the holy month of Ramadan; and 5) making a pilgrimage to the holy city of Mecca once in a lifetime. The five pillars are important and symbolic acts of faith. For example, giving charity to the poor helps Muslims show their compassion for the less fortunate and symbolizes people's social responsibility to care for the community. Fasting during the month of Ramadan is a demanding physical exercise, as no food or water can be consumed during the daylight hours. This act of faith binds members of the Muslim community together and symbolizes in part the importance of spiritual matters over the physical demands of the body. Every evening after dark, families and friends congregate to break their fast and celebrate their faith. Have students conduct research on the five pillars of Islam. Then lead a discussion on the importance of these symbolic acts of faith in Muslim communities around the world.

Islam

Sacred Text:
Qur'an

Sacred Objects:
prayer rugs

Muslim woman praying during Ramadan

Thousands of Muslim pilgrims gather around the Ka'bah, in Mecca.

Sacred Sites:
Mecca (Makkah), Medina, Jerusalem

Muhammad
C. A.D. 570 – A.D. 632

INTERPRETING BIOGRAPHY

In Islam, Muhammad is a messenger or prophet of God. Muhammad was born in Mecca (Makkah) and orphaned at an early age. He was from a respected but poor family. They belonged to a leading tribe of caravan merchants and keepers of Abraham's shrine and pilgrimage site, the Ka'bah.

Islam prohibits the use of images for Muhammad. The symbol above, which means "Muhammad is the Prophet of God," is often used in place of his picture. **Why is a symbol used in place of Muhammad's image?**

Special Days

Fast of Ramadan, during the entire first month of the Islamic year and Id al-fitr at the end of Ramadan; Id al-Adha at the end of the hajj

---

109

CHAPTER 5 Review and Assessment Resources

TECHNOLOGY

▶ Chapter 5 Test Generator (on the One-Stop Planner)
▶ Global Skill Builder CD–ROM
▶ HRW Go site

REINFORCEMENT, REVIEW, AND ASSESSMENT

▶ Chapter 5 Review, pp. 110–11
▶ Chapter 5 Tutorial for Students, Parents, Mentors, and Peers
▶ Chapter 5 Test (form A or B)
▶ Alternative Assessment Handbook

▶ Chapter 5 Test for English Language Learners and Special-Needs Students
▶ Unit 1 Test
▶ Unit 1 Test for English Language Learners and Special-Needs Students

Assess

Have students complete a Chapter 5 Test.

Reteach

Organize the class into three groups and assign each group a section of the chapter. Have each group create a poster with captions that summarize the main ideas of their assigned section. Display the posters in your classroom. **ENGLISH LANGUAGE LEARNERS, COOPERATIVE LEARNING**

CHAPTER 5 Review Answers

Thinking Critically

1. Possible answer: Such innovations make communication faster, link distant places better, and therefore aid more rapid diffusion of cultural traits and contribute to cultural change. (NGS 10)

2. Possible answer: Other cultures, places, or regions might be viewed as unfamiliar and strange. (NGS 10)

3. Answers will vary, but students might note the presence of churches and other places of worship and their special architecture. They might also mention various religious practices commonly seen here. (NGS 10)

Using the Geographer's Tools

1. There was slow growth before 1600 and more rapid growth after. Students might say that the graph could be a good indicator of future growth, but that various factors could change trends.

2. parts of the Southwest and extreme northeast; Southwest—immigration from Mexico, northeast—French Catholic cultural influence

3. Graphs should reflect data in the chart. North America, India, and Europe have many English-speakers.

Writing

Entries should accurately describe various aspects of local culture. Use Rubric 15, Journals, to evaluate student work.

CHAPTER 5 Review

Building Vocabulary

On a separate sheet of paper, explain the following terms by using them correctly in sentences.

demography
population density
migration
culture
culture traits
ethnic groups
acculturation
innovation

diffusion
dialect
ethnic religions
animist religions
polytheism
universalizing religions
monotheism

Locating Key Places

On a separate sheet of paper, match the letters on the map with their correct labels.

Hinduism
Buddhism
Islam

Christianity—Eastern Orthodox
Christianity—Protestant
Christianity—Roman Catholic

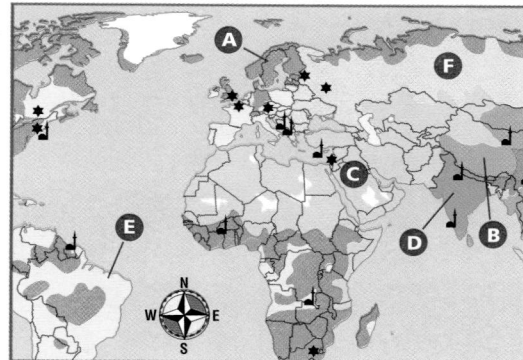

Understanding the Main Ideas

Section 1

1. *Human Systems* What do population geographers use to study population patterns and trends?

2. *Human Systems* What factors influence migration?

Section 2

3. *Human Systems* What are culture traits? What is one example of how culture traits vary from place to place?

Section 3

4. *Human Systems* Why is language fundamental to culture?

5. *Human Systems* Why is religion fundamental to culture?

Thinking Critically

1. **Making Generalizations** How do you think innovations like computers, the Internet, and other modern telecommunications affect culture change today?

2. **Identifying Points of View** How might a people's culture influence the way they view other cultures, places, or regions?

3. **Identifying Cause and Effect** In what ways do you think religion shapes daily life and cultural patterns in this country as it does in Islamic countries? What are some examples?

Using the Geographer's Tools

1. **Interpreting Graphs** Study the World Population Growth graph in Section 1. How would you describe the world's population growth before about 1600? Afterward? What might this graph tell us about future population growth?

2. **Interpreting Maps** Study the World Religions map in Section 3. Which areas of the United States does the map identify as Roman Catholic? Why do you think this is so?

3. **Drawing Graphs** Use the information in the chart of Principal Languages of the World in Section 3 to draw a bar graph of the world's major languages. Which major world regions have many English-speakers?

Writing about Geography

Imagine a typical day where you live. What activities are going on? How do these activities reflect local culture? What kinds of words and expressions do people use? What kinds of architecture, clothing, food, and music are common? Write a journal entry about the culture of your area. When you are finished with your journal entry, proofread it to make sure you have used standard grammar, spelling, sentence structure, and punctuation.

SKILL BUILDING

Geography for Life

Using Questionnaires

Human Systems Use a questionnaire to find out what other students in your class think would be a good innovation. First, design a short questionnaire that asks the questions you want the other students to answer. For example, you could ask: What innovation would you like to see developed in your lifetime? Why do you think this would be a good innovation? How would it change your life? When the questionnaires are completed, compare them. What did you learn?

Portfolio Activity

1. Have students collect or draw symbols and logos from signs, clothing, food and beverage labels, and the like that are part of American culture. Ask students how they might interpret these images if they lived in another culture. Have students write brief explanations of the symbols' meanings and place them, with the objects or pictures, in portfolios.

2. Have students conduct research on the population of your city or town since its founding. Then have students create a line graph that illustrates population changes over time. Ask students to explain any dramatic population increases or decreases. Place the graphs in student portfolios.

Food Festival

Ask students to bring foods to class that reflect their families' cultures. If most of the students are from similar backgrounds, have them concentrate on family versions of a common dish. For example, there are many ways to make pierogis, which are Polish filled dumplings. Jewish students of Eastern European heritage might compare recipes for kugel, a baked pudding of noodles or potatoes. Some students may feel that their traditions are fast food and mainstream American dishes. Encourage them to find recipes from their ancestors' heritage.

Building Social Studies Skills

Interpreting Graphs

Study the bar graph below. Then use the information to help you answer the questions that follow.

1. Which country is a source of between 500,000 and 1,000,000 refugees?
   a. Sudan
   b. Myanmar
   c. Afghanistan
   d. Palestine

2. Which regions of the world appear to have the most difficult refugee problems? Support your answer.

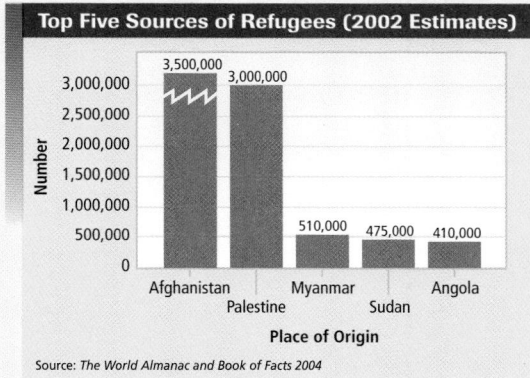

Top Five Sources of Refugees (2002 Estimates)

Afghanistan: 3,500,000
Palestine: 3,000,000
Myanmar: 510,000
Sudan: 475,000
Angola: 410,000

Number / Place of Origin

Source: *The World Almanac and Book of Facts 2004*

Analyzing Secondary Sources

Read the following passage and answer the questions on a separate sheet of paper.

> "The demographic transition is a model that shows how birthrates and death rates dropped in many Western countries as they developed modern economies and industries. Most of the world's richest and most technologically advanced countries have experienced a transition from high birthrates and death rates to low birthrates and death rates. In addition, many of the world's poorer countries are now in the middle of similar population changes."

3. Birthrates and death rates dropped in many Western countries as they developed what?
   a. widespread farming
   b. modern economies and industries
   c. resistance to diseases
   d. strong militaries

4. What do you think are examples of low birthrates and death rates? Of high birthrates and death rates?

Alternative Assessment

PORTFOLIO ACTIVITY

Learning about Your Local Geography

Group Project: Research
Plan, organize, and complete a research project about historical migration in your community. Check your local library to find information about the early settlement and history of your community. Talk to local historians to find out what they know about the settlement of your area. As a group, put together a short presentation on historical migration in your community that answers the following questions: When was your area settled? Where did the settlers come from? What brought them to your area? How did they influence the area's culture? How did physical geography affect the routes migrants took to reach your community?

internet connect

Internet Activity go.hrw.com
KEYWORD: SW3 GT5

Choose a topic on human geography to:
- research and report on world population growth.
- examine push and pull factors that cause the migration of refugees.
- answer questions on demographic calculations and data from the United States Census Bureau.

CHAPTER RESOURCE MANAGER

Objectives	Pacing Guide	Reproducible Resources
SECTION 1 **Economic Geography** (pp. 113–18)	• Identify the three main types of economic systems. • Contrast developed and developing countries.	**Regular** 1 day **Block Scheduling** 1 day *Block Scheduling Handbook, Chapter 6*	**RS** Guided Reading Strategy 6.1 **RS** Graphic Organizer Activity 6 **PS** Readings in World Geography, History, and Culture 8 **SM** Geography for Life Activity 6
SECTION 2 **Urban and Rural Geography** (pp. 119–26)	• Analyze how people have used land throughout history. • Explain how urban geography describes human settlements. • Identify some of the ways people use land in rural areas.	**Regular** 1.5 days **Block Scheduling** 1 day *Block Scheduling Handbook, Chapter 6*	**RS** Guided Reading Strategy 6.2 **IC** Environmental and Global Issues Activity 8: There's a Space for Us **SM** Map Activity 6: Urbanization
SECTION 3 **Political Geography** (pp. 128–31)	• Explain how government and geography are connected. • Identify the three main types of geographic boundaries. • Analyze how conflict and cooperation affect international relations.	**Regular** 1 day **Block Scheduling** .5 day *Block Scheduling Handbook, Chapter 6*	**RS** Guided Reading Strategy 6.3 **SM** Critical Thinking Activity 6: The Cuban Missile Crisis

Chapter Resource Key

PS Primary Sources

RS Reading Support

IC Interdisciplinary Connections

E Enrichment

SM Skills Mastery

A Assessment

REV Review

ELL Reinforcement and English Language Learners

Transparencies

CD–ROM

Video

Internet

Holt Presentation Maker Using Microsoft® PowerPoint®

 One-Stop Planner CD–ROM

See the *One-Stop Planner* for a complete list of additional resources for students and teachers.

 **One-Stop Planner CD-ROM**

It's easy to plan lessons, select resources, and print out materials for your students when you use the **One-Stop Planner CD-ROM with Test Generator**.

Technology Resources	Reinforcement, Review, and Assessment
One-Stop Planner CD-ROM, Lesson 6.1	**ELL** Main Idea Activity 6.1
ARGWorld CD-ROM	**ELL** English Audio Summary 6.1
Homework Practice Online	**ELL** Spanish Audio Summary 6.1
HRW Go site	**REV** Section 1 Review, p. 118
	A Daily Quiz 6.1
One-Stop Planner CD-ROM, Lesson 6.2	**ELL** Main Idea Activity 6.2
Geography and Cultures Visual Resources 9	**ELL** English Audio Summary 6.2
Homework Practice Online	**ELL** Spanish Audio Summary 6.2
HRW Go site	**REV** Section 2 Review, p. 126
	A Daily Quiz 6.2
One-Stop Planner CD-ROM, Lesson 6.3	**ELL** Main Idea Activity 6.3
Homework Practice Online	**ELL** English Audio Summary 6.3
HRW Go site	**ELL** Spanish Audio Summary 6.3
	REV Section 3 Review, p. 131
	A Daily Quiz 6.3

⊿ internet connect

HRW ONLINE RESOURCES

GO TO: go.hrw.com
Then type in a keyword.

TEACHER HOME PAGE
  KEYWORD: SW3 Teacher

CHAPTER INTERNET ACTIVITIES
  KEYWORD: SW3 GT6
  Choose an activity on human systems to:
  • create a data profile using online databases.
  • understand and compare economic systems.
  • learn about different systems of government.

CHAPTER ENRICHMENT LINKS
  KEYWORD: SW3 CH6

CHAPTER MAPS
  KEYWORD: SW3 MAPS6

ONLINE ASSESSMENT
  Homework Practice
  KEYWORD: SW3 HP6
  Standardized Test Prep
  KEYWORD: SW3 STP6
  Rubrics
  KEYWORD: SS Rubrics

COUNTRY INFORMATION
  KEYWORD: SW3 Almanac

CONTENT UPDATES
  KEYWORD: SS Content Updates

HOLT PRESENTATION MAKER
  KEYWORD: SW3 PPT6

ONLINE READING SUPPORT
  KEYWORD: SS Strategies

CURRENT EVENTS
  KEYWORD: S3 Current Events

Meeting Individual Needs

Ability Levels

Level 1 Basic-level activities designed for all students encountering new material

Level 2 Intermediate-level activities designed for average students

Level 3 Challenging activities designed for honors and gifted-and-talented students

English Language Learners Activities that address the needs of students with Limited English Proficiency

Chapter Review and Assessment

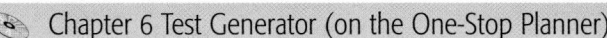
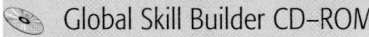

Chapter 6 Test Generator (on the One-Stop Planner)

Global Skill Builder CD-ROM

HRW Go site

REV Chapter 6 Review, pp. 134–35

REV Chapter 6 Tutorial for Students, Parents, Mentors, and Peers

A Chapter 6 Test (form A or B)

A Alternative Assessment Handbook

A Chapter 6 Test for English Language Learners and Special-Needs Students

Launch into Learning

Call on volunteers to review the definitions of human geography and physical geography. *(Human geography is the study of the distribution and characteristics of the world's people. Physical geography focuses on Earth's natural environments.)* Ask students to speculate about how human settlements and systems are related to physical environments.

(Possible answers: People use and preserve natural resources. People build cities, tend crops, and otherwise manipulate the environment. People establish political and other boundaries.) Tell students that they will learn more about how human systems depend on and relate to physical geography in this chapter.

You may want to emphasize the importance of this chapter's content by discussing these points with your students:

▶ In our shrinking world, no country's economy is entirely separate from others. A strike or crop failure in one place can affect countries on the other side of the world.

▶ We may be able to make our communities, whether they are in the city or in the countryside, more livable if we understand the basic concepts of urban and rural geography.

▶ A country's basic political system plays a major role in its foreign policy. Other countries' relationships with the United States directly or indirectly affect many aspects of our daily lives, from our food supply to national security.

CHAPTER 6 Human Systems

People affect Earth's geography. They use resources, earn a living, build communities, and compete for control of Earth's surface. All these activities shape the world around us.

Parliament building in Budapest, Hungary

The skyline of Dallas, Texas

Dogon women selling vegetables in Mali

 **LET'S GET STARTED**

Copy the following instructions onto the chalkboard: *In your notebooks, list some economic activities involved in the creation and sale of a pencil. Discuss responses. (Possible answers: wood harvested; graphite mined; rubber processed; pencils assembled, shipped, and sold; marketing and research conducted to sell and improve the product)* Point out that the creation of goods and products involves many connected activities that are carried out in different places. Tell students that they will learn more about economic systems in Section 1.

Building Vocabulary

Write the key terms on the chalkboard. Call on volunteers to read the definitions aloud. Then ask students to draw charts in their notebooks and to identify the terms that name ways to measure economic development (**gross national product**, **gross domestic product**, **literacy rate**, **developed countries**, and **developing countries**) and those that relate to economic or political systems (**market economy**, **free enterprise**, **capitalism**, **command economy**, and **communism**).

Section 1
Economic Geography

READ TO DISCOVER

1. What are the three main types of economic systems?
2. How are developed countries and developing countries different?

WHY IT MATTERS

The world's poorer countries are constantly struggling to improve their economies. Use **CNN fyi.com** or other **current events** sources to find news about a country facing challenges as it tries to develop its economy.

DEFINE

market economy
free enterprise
capitalism
command economy
communism
gross national product (GNP)
gross domestic product (GDP)
industrialization
literacy rate
developed countries
infrastructure
developing countries

 **RESOURCES**

REPRODUCIBLE
▶ Guided Reading Strategy 6.1
▶ Graphic Organizer Activity 6
▶ Readings in World Geography, History, and Culture 8
▶ Geography for Life Activity 6

TECHNOLOGY
▶ One-Stop Planner CD–ROM, Lesson 6.1
▶ Homework Practice Online
▶ HRW Go site

REINFORCEMENT, REVIEW, AND ASSESSMENT
▶ Main Idea Activity 6.1
▶ English Audio Summary 6.1
▶ Spanish Audio Summary 6.1
▶ Section 1 Review, p. 118
▶ Daily Quiz 6.1

Economic Systems

Economic geography deals with how people earn a living and use resources and with the links among economic activities. Economic geographers group money-making activities into four categories. These are primary (first order), secondary (second order), tertiary (third order), and quaternary (fourth order) activities. (See the chart.)

Primary Activities	Secondary Activities
• use natural resources directly	• use raw materials to produce or manufacture something new
• **location:** at the site of the natural resource being used	• **location:** close to the resource or close to the market for the finished product
• **examples:** wheat farming, iron mining	• **other factors affecting location:** labor, energy, and land costs
	• **examples:** processing wheat into flour, manufacturing steel

Tertiary Activities	Quaternary Activities
• provide services to people and businesses	• process and distribute information
• **location:** usually near customers	• **location:** anywhere
• **examples:** bakeries, car dealerships	• **other factors affecting location:** access to skilled workers, good transportation and communications systems, places with pleasant climates and a high quality of life
	• **examples:** plant-genetics research, automotive engineering

INTERPRETING THE CHART *The most basic economic activities, such as farming, extract resources directly from Earth and have locations that are closely tied to environmental conditions and the distribution of resources. More complex activities, such as software development, produce more sophisticated goods and have locations that are not as tied to environmental conditions.* **What factors affect the location of the different types of economic activities?**

CHART ANSWER

location of resources, markets, and customers; labor, energy, and land costs; access to skilled workers; good transportation and communications systems

113

ALL LEVELS: Call on students to suggest primary, secondary, tertiary, and quaternary economic activities that are found in your state. As students list activities, write them on the chalkboard. After the class has identified several examples of each type of activity, organize students into groups. Have each group draw a map of your state that shows the locations of each type of activity. You may want to provide background information from your state economic development agency. Call on volunteers from each group to share their maps with the class. **ENGLISH LANGUAGE LEARNERS, COOPERATIVE LEARNING**

LEVEL 1: Copy the following graphic organizer onto the chalkboard, omitting the italicized answers. Call on students to complete the organizer with words and phrases that describe the three main types of economic systems. Then lead a class discussion about the details of each type. **ENGLISH LANGUAGE LEARNERS**

Daily Life

Sweet-grass Baskets

Visitors to the Meeting Street Marketplace in Charleston, South Carolina, might observe African American artisans weaving sweet-grass baskets. Weavers use various grasses, leaves, needles, and rushes found in the nearby marshlands and swamps to create intricate baskets of different shapes and sizes. Parents pass weaving techniques along to their children, much as their ancestors did in West Africa.

Historically, people used the baskets to store household items and to barter for other goods. Field hands also used baskets to separate chaff from rice. Today collectors buy sweet-grass baskets because of the beauty and quality of their design.

CRITICAL THINKING: Which type of economic activity does the making of sweet-grass baskets represent? *(secondary activity)*

VISUAL RECORD ANSWER

Possible answer: wide range, from common goods like clothing to items for a highly specialized market

---

internet connect

GO TO: go.hrw.com
KEYWORD: SW3 CH6
FOR: Web sites about human systems

Economic Activities Economic activities that use natural resources directly are called primary activities. They include farming, fishing, forestry, herding, and mining. Primary activities provide the basic raw materials for industry. They are located at the site of the natural resource being used.

Secondary activities use raw materials to produce or manufacture something new. Examples include steelmaking, processing wheat into flour, and making lumber into plywood. Dairies take a raw material—fresh milk—and process it into products like cheese and ice cream. Often, a raw material goes through several stages before it becomes a finished product. Secondary activities are usually located close to the resource being used or close to the market for the finished product.

Tertiary activities provide services to people and businesses. For example, doctors, teachers, and dry cleaners provide personal and professional services. Store clerks, truck drivers, and restaurant staff provide retail and wholesale services. Tertiary activities are usually located near customers to serve them better. However, the use of computers and the Internet is dramatically changing the locations of many service industries. For example, more customers now shop online and buy products from Web sites.

In advanced economies many workers process and distribute information. These jobs are called quaternary economic activities. They require workers with specialized skills and knowledge. Jobs include research scientists, computer programmers, and government administrators. Quaternary activities are not tied directly to resources, environmental conditions, or access to markets. Increasingly, workers in this category can be located almost anywhere.

Economic Systems The four types of economic activities are found in different economic systems around the world. There are three main kinds of economic systems. Each uses resources and produces goods in a distinctive way. The most basic economic system is a traditional or subsistence economy. In this economy, people make goods for themselves and their families. There is

INTERPRETING THE VISUAL RECORD

Large shopping malls, such as this one in New York City's Trump Tower, are common in countries that have market economies and a free enterprise system. Shoppers here can find a wide range of stores and goods concentrated in one area. **What types of goods would you expect to find in this shopping mall?**

Economy	Motivator	Description	Location
traditional or subsistence	survival	People make goods for themselves or their families with little surplus.	Most subsistence economies are found in poor countries and rural areas.
market	profit	People freely choose what to buy and sell.	Most of the world's rich countries have market economies.
command	government regulations	The government establishes products, locations, and prices.	Communist countries have command economies.

LEVEL 2: Organize the class into groups and have each group choose a large country. Then have each group conduct research to learn about the various economic activities that take place in its assigned country. Instruct groups to identify examples of traditional, market, and command economies and to explain if and how each system operates in that country. Students should note the location of each system within the country and the products created under each system. Have each group prepare a poster that illustrates its findings. **COOPERATIVE LEARNING**

little surplus or exchange of goods. As a result, there are few markets—places to buy and sell things. Traditional economies are found mostly in the world's poorer countries, particularly in rural areas.

A **market economy** is another type of economy. In a market economy, people freely choose what to buy and sell. Market economies are guided by the system of **free enterprise**. This system lets competition among businesses determine the price of products. The businesses' need to make profits drives their decisions. Businesses supply products and set prices to meet demand. Free enterprise is the basis of **capitalism**. In a capitalist system, businesses, industries, and resources are privately owned.

Market economies are characterized by specialization. Businesses and regions make the goods that they can sell for the highest profit. For example, the midwestern United States has ideal conditions for producing corn, hogs, and soybeans. As a result, farmers there produce these goods and export them around the world. Most of the world's rich countries have market economies. These countries include Australia, Japan, the United States, and most European countries. Goods made in these countries are traded around the world.

The third type of economy is one in which the government makes the major economic decisions. In a **command economy**, the government decides what to produce, where to make it, and what price to charge. Prices are not based on the market forces of supply and demand. For example, it may cost $1.00 to produce a loaf of bread. However, the government may set the price at 25 cents so that people can easily afford it. Communist countries have command economies. **Communism** is an economic and political system in which the government owns or controls almost all the means of production. Cuba and North Korea are communist countries.

✓ **READING CHECK:** *Human Systems* What are the three main kinds of economic systems? traditional, market, and command economies

Economic Patterns, Resources, and Technology
The creation and distribution of resources affect the locations of economic activities. Resources also affect the movement of products, capital, and people. How does this happen? The need for a resource draws businesses and workers to the place where it is found. Related businesses grow nearby. Businesses must then find ways to ship their products to markets.

Consider the California Gold Rush of 1849. People from all over the world went to California to mine for gold. As a result, San Francisco grew from a small town to a city of 25,000 people in

INTERPRETING THE VISUAL RECORD

The California Gold Rush of 1849 quickly transformed the city of San Francisco from a small quiet town to a city attracting thousands of immigrants from all over the world. As people and capital flocked to the city, a wide range of businesses sprang up to serve the miners, and real estate prices soared. **How did the gold rush affect the movement of people and products to San Francisco?**

VISUAL RECORD ANSWER

Thousands of people immigrated to San Francisco, and businesses sprang up to serve the miners.

115

Ask each student to select a product or resource and to examine various ways in which people use it to satisfy their basic needs. For example, students might compare the cultivation of rice by subsistence farmers in Southeast Asia to production by commercial rice growers in the southeastern United States, or traditional weaving in Central Asia to large-scale rug making in Europe. Have students present their findings in written reports. Encourage students to illustrate their reports with maps, charts, or graphs.

Teach Objective 2

LEVEL 1: Organize students into pairs. Have each pair examine the unit Fast Facts tables in this textbook to identify developing countries and developed countries. Call on volunteers to share their lists with the rest of the class. As students name countries, ask them what information led to its classification as developing or developed. Then lead a class discussion about the factors geographers use in determining a country's level of development. **ENGLISH LANGUAGE LEARNERS, COOPERATIVE LEARNING**

Across the Curriculum

► **Economics** ◄

Texas Cattle Drives The location of resources often leads to the movement of people, but sometimes the location of people can lead to the movement of resources. For example, when in the 1850s money became scarce in Texas local demand for beef fell, and cattle prices dropped to between $5 and $10 a head. To turn a profit, cattle ranchers needed to get the cattle to markets in the North, East, and West where prices were higher. Their solution was to drive whole herds north to Kansas where railroads crossed the Plains. These cattle drives contributed to the romanticization of the cowboy way of life. They also increased ranchers' interest in obtaining grazing lands in Indian Territory.

DISCUSSION: Lead a discussion about the effects of resources and markets on the locations of economic activities.

TABLE ANSWER

Japan and the United States; India

---

just one year. The city also became an important financial and banking center. Banks and other businesses provided money to invest in the new mines. Transportation connections developed between gold-mining areas and ports.

Changes in technology, transportation, and communication also affect the location and patterns of economic activities. The development of refrigerated railroad cars, ships, and trucks is an example. The use of refrigerated transportation allowed farm goods to be shipped around the world without spoiling. As a result, the global production and trade of farm goods changed. For example, in the last 150 years, Argentina has become a major meat exporter. This would not have been possible without these changes in technology. Likewise, the development of computers and the Internet has affected the locations of many service and information industries. A growing number of people are now able to work at home and send their work around the world by e-mail.

✓ **READING CHECK:** *Human Systems* How have changes in technology, transportation, and communication affected the locations and patterns of economic activities? Refrigerated railroad cars, ships, and trucks have allowed countries like Argentina to produce and export farming goods.

Level of Development

One of the most important topics for understanding world geography is understanding the level of development in countries around the world. Development refers to steady improvements in a country's economy and in people's quality of life.

Economic progress varies greatly among different countries and also within countries. Geographers use general measures of development to analyze such progress. You will find many of these measures in the tables at the beginning of Units 2 through 10.

Measures of Development One common measure of development is **gross national product (GNP)**. GNP is the total value of goods and services that a country produces in a year. GNP includes goods and services made by businesses owned by that country's citizens but located in foreign lands. **Gross domestic product (GDP)** includes only those goods and services created within the country. GDP becomes more useful when it is divided by the number of people living in a country. This gives us the per capita GDP, which can be used to compare income levels in different countries.

Another measure of development is the level of **industrialization**. Industrialization is the process by which manufacturing based on machine power becomes widespread in an area. In industrialized countries, many people work in manufacturing, service, and information industries. In other countries, most people work in primary economic activities, particularly farming. There are a number of additional measures of development. They include the average amount of energy people use. In addition, some measures look at the size and quality of a country's transportation and communications systems. For example, countries with more telephones per person tend to be more developed than countries with fewer telephones.

Standard of Living A country's level of development, in turn, determines the standard of living of its people. Standard of living is

Literacy Rates	
Country	**Literacy Rate**
China	86%
India	60%
Japan	100%
Mexico	93%
United States	97%

Source: *World Almanac and Book of Facts 2004*

INTERPRETING THE TABLE *Literacy rates reflect a country's standard of living. Along with other data, these rates can be used to compare levels of economic development in different countries.* **Based on the information in this table, which countries do you think have the highest standards of living? Which country do you think has the lowest level of economic development?**

LEVEL 2: Have students write diary entries for high school students in a developed country. Then have them do the same for a person of the same age living in a developing country. Call on volunteers to present their entries to the class. Lead a class discussion on how the diary entries differ. Ask students what factors lead to different standards of living in developed and developing countries.

LEVEL 3: Have students consult reference books and the unit Fast Facts tables in this textbook to create a list of middle-income countries like those listed on the chart in this section. Then instruct each student to choose one of these countries and to examine its economic activity for the past several years to determine whether it seems likely to become a developed country or to remain a middle-income country. Ask students to predict the future growth of their chosen country and to provide written and graphic evidence to support their predictions.

measured by factors like amount of personal income, levels of education, and food consumption. **Literacy rate**—the percentage of people who can read and write—also reflects standard of living. Other measures of standard of living include quality of health care, technology level, and life expectancy—the average length of people's lives.

✓ **READING CHECK:** *Human Systems* How is a country's standard of living measured? amount of personal income, levels of education, food consumption, literacy rate, availability of health care, level of technology, life expectancy

Developed and Developing Countries Geographers organize the world's countries into two main groups. The richest countries are called **developed countries**. They have high levels of industrialization, and their people enjoy high standards of living. The world's developed countries include most countries in Europe, the United States, Canada, Japan, Australia, and others. Less than 25 percent of the world's people live in developed countries.

Developed countries share many features. (See the chart.) For example, each has a high per capita GDP. These countries also have high levels of education and good health care. Literacy rates are high. More literacy leads to more educated workers, who are more productive economically. Life expectancy is also high. Both birthrates and death rates are usually low. As a result, overall population growth is low.

Most people in developed countries live in cities and work in service or manufacturing industries. Few work in agriculture. The small number of farmers use advanced technology to produce large amounts of food. Developed countries also have good **infrastructure**. An infrastructure is a system of roads, ports, and other facilities needed by a modern economy. Developed countries have global market economies.

The world's poorer countries are called **developing countries** or less developed countries. These countries are less productive economically and have

Selected Countries' Statistics

		Per Capita GDP	Life Expectancy	Literacy Rate	Urban	TV Sets (per 1,000 persons)	Physicians
Developed Countries	Australia	$27,000	77, male; 83, female	100%	91%	716	1 per 395 persons
	Japan	$28,000	78, male; 84, female	100%	79%	719	1 per 525 persons
	United States	$36,300	74, male; 80, female	97%	75%	844	1 per 357 persons
Developing Countries	Afghanistan	$ 700	48, male; 46, female	36%	22%	14	1 per 9,090 persons
	Haiti	$ 1,700	50, male; 53, female	53%	36%	5	1 per 8,418 persons
	Mali	$ 860	45, male; 46, female	46%	31%	13	1 per 21,269 persons
Middle-Income Countries	Brazil	$ 7,600	67, male; 75, female	86%	82%	333	1 per 774 persons
	Mexico	$ 9,000	69, male; 76, female	92%	75%	272	1 per 812 persons
	Thailand	$ 6,900	69, male; 74, female	96%	20%	274	1 per 4,192 persons

Sources: *The World Almanac and Book of Facts 2004*; U.S. Census; *Britannica Book of the Year, 2002*

INTERPRETING THE TABLE *Demographic, economic, and social data for the world's countries varies greatly, as this table shows. Based on the information in this table, what features do developed countries have in common? What features do developing countries have in common? How might literacy rates affect per capita GDP?*

Close

Have students flip through their textbooks to find pictures of people engaged in various economic activities. Ask students to determine if they are primary, secondary, tertiary, or quaternary activities. Then ask the class if they think the pictures illustrate traditional economies or market economies. Have students justify their responses.

Review and Assess

Have students complete the **Section Review**. Then have students complete **Daily Quiz 6.1.**

Reteach

Have students complete **Main Idea Activity for English Language Learners and Special-Needs Students 6.1.** Then refer students to the list of key terms. Have each student write sentences that explain the relationship between pairs or groups of words. ENGLISH LANGUAGE LEARNERS

Extend

Have interested students conduct research to discover economic links between the United States and another country. Have them use almanacs and other sources to identify and map the flow of resources and products between countries. BLOCK SCHEDULING

Section 1 Review Answers

Define For definitions, see: market economy, p. 115; free enterprise, p. 115; capitalism, p. 115; command economy, p. 115; communism, p. 115; gross national product (GNP), p. 116; gross domestic product (GDP), p. 116; industrialization, p. 116; literacy rate, p. 117; developed countries, p. 117; infrastructure, p. 117; developing countries, p. 117

Reading for the Main Idea
1. traditional—goods produced for personal use, little surplus or exchange of goods; market—people free to choose what to buy and sell, free enterprise, specialization; command—major economic decisions made by government

2. levels of industrialization, standard of living, per capita GDP, workforce structure, trade activity

Critical Thinking
3. Answers will vary but should demonstrate how the distribution of resources affects the location of economic activities. (NGS 16)

4. Possible answer: caused them to perceive agricultural resources as more valuable (NGS 11)

Organizing What You Know
5. See the subsection on developed and developing countries for possible answers.

VISUAL RECORD ANSWER

Answers will vary.

INTERPRETING THE VISUAL RECORD *These two photos from Australia and Afghanistan show some of the differences between developed and developing countries.* **Based on these photos, how do you think daily life in these two places is different? What can you see in the photographs that might indicate the level of development in each place?**

lower standards of living. This group includes most countries in Africa, Asia, Central and South America, and the Pacific Islands. Most of the world's people live in developing countries. These countries have low per capita GDPs. (See the chart.) In general, birthrates are high, and life expectancy is low. Grade schools are often available, but few people go to high school or college. Most people farm, and many homes do not have electricity. There are also not many computers, refrigerators, or televisions per person. There are few service businesses and manufacturing industries to provide jobs. Because rural areas offer few jobs, many people move to cities.

Between the world's richest and poorest countries are what some geographers call middle-income countries. Examples include Mexico, Brazil, Thailand, and Malaysia. They have features of both developed and developing countries. Their cities may be modern, but rural areas and small towns are often poor. Many of these countries have new industries, and many people are switching from rural life to city life. In many of these countries incomes are rising quickly. They may soon join the developed countries. However, some countries, like Argentina and South Africa, seem stuck in the middle-income category.

✓ **READING CHECK:** *Places and Regions*
What are some examples of developed and developing countries? Canada, Japan, the United States, Australia, and most countries in Europe are developed countries. Most countries in Africa, Asia, the Pacific Islands, and Central and South America are developing countries.

Homework Practice Online
Keyword: SW3 HP6

Section 1 Review

Define market economy, free enterprise, capitalism, command economy, communism, gross national product (GNP), gross domestic product (GDP), industrialization, literacy rate, developed countries, infrastructure, developing countries

Reading for the Main Idea

1. *Human Systems* What are the main characteristics of traditional, market, and command economies?

2. *Places and Regions* What factors would you use to measure the level of economic development in a country?

Critical Thinking

3. *Analyzing Information* How might the discovery of valuable minerals nearby affect economic activities in your community?

4. *Analyzing Information* How do you think the development of new technologies such as refrigerated railroad cars, ships, and trucks can change people's perception of resources?

Organizing What You Know

5. Create a chart like the one shown below. Use it to list characteristics of developed and developing countries.

Developed countries	Developing countries

Section 2

OBJECTIVES

1. Analyze how people have used land throughout history.

2. Explain how urban geography describes human settlements.

3. Identify some of the ways people use land in rural areas.

 **LET'S GET STARTED**

Copy the following instructions onto the chalkboard: *Think about a city that you have lived in or visited. How does the city compare to the surrounding area? How does the downtown area differ from the city's outskirts?* Discuss student responses. Tell students that urban geographers are interested in the growth and development of cities. They ask questions similar to the ones above and try to uncover patterns that explain how cities are shaped. Tell students that they will learn more about the characteristics of the world's cities in Section 2.

Building Vocabulary

Write **subsistence agriculture** and **agribusiness** on the chalkboard and underline the prefix *agri-* in each term. Point out that it comes from the Latin word that means "field." Then have students suggest definitions for the two terms, based on what they know about subsistence economies. Then write **urbanization** and **pastoralism**. Tell students they are derived from *urban,* meaning "of the city," and *pastoral,* meaning "of shepherds." Call on students to read the definitions of all the key terms.

 Section **2**

Urban and Rural Geography

READ TO DISCOVER

1. How have people used land throughout human history?
2. How does urban geography describe human settlements?
3. What are some of the ways people use land in rural areas?

WHY IT MATTERS

U.S. cities share many of the same challenges. Use **CNNfyi.com** or other **current events** sources to learn about a U.S. city that is trying to solve some of its problems.

DEFINE

domestication
urbanization
world cities
central business district (CBD)
edge cities
subsistence agriculture
shifting cultivation
pastoralism
market-oriented agriculture
agribusiness

Using the Land

Not all of the resources that humans need can be found in one place. Usually people must travel or trade to find everything they need. How people get resources from the land has greatly affected Earth's geography.

Origins of Domesticated Plants

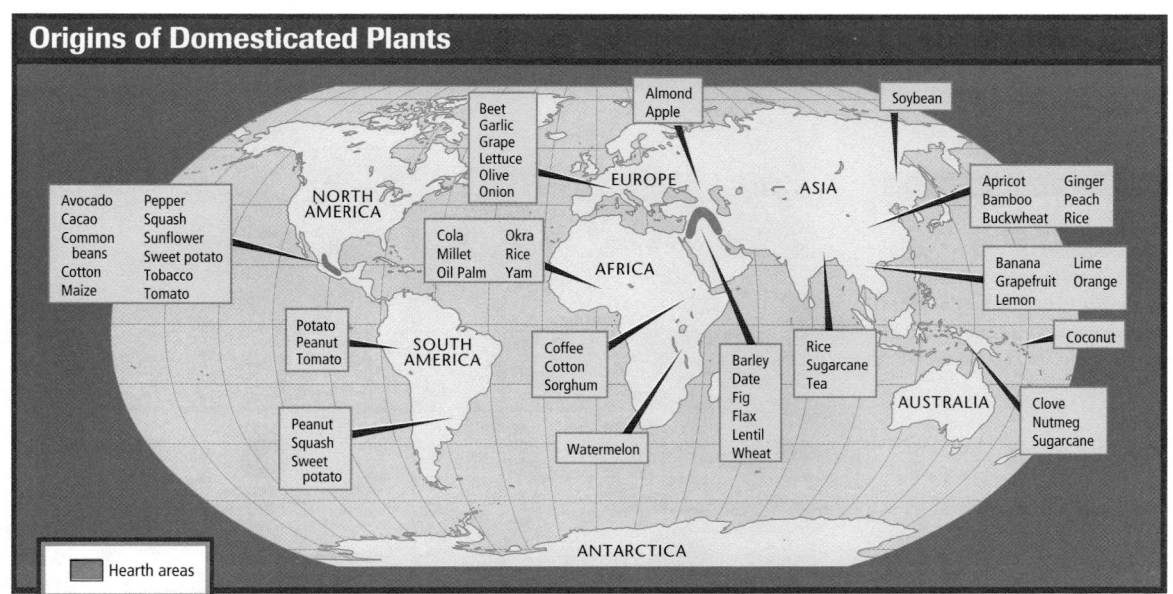

| Hearth areas |

INTERPRETING THE MAP *This map shows the probable origins of many of the world's most important domesticated plants. It is a generalized map, as the exact origins of many domesticated plants are* unknown, *though their general area of origin is. Hearth areas were particularly important centers of plant domestication.* **Which plants are native to Asia? To North America?**

Section **2** RESOURCES

REPRODUCIBLE
▶ Guided Reading Strategy 6.2
▶ Environmental and Global Issues Activity 8
▶ Map Activity 6: Urbanization

TECHNOLOGY
▶ One-Stop Planner CD–ROM, Lesson 6.2
▶ Geography and Cultures Visual Resources 9
▶ Homework Practice Online
▶ HRW Go site

REINFORCEMENT, REVIEW, AND ASSESSMENT
▶ Main Idea Activity 6.2
▶ English Audio Summary 6.2
▶ Spanish Audio Summary 6.2
▶ Section 2 Review, p. 126
▶ Daily Quiz 6.2

MAP ANSWER

Asia—soybean, apricot, bamboo, buckwheat, ginger, peach, rice, banana, grapefruit, lemon, lime, orange, sugarcane, tea; North America—avocado, cacao, common beans, cotton, maize, pepper, squash, sunflower, sweet potato, tobacco, tomato

Teach Objective 1

 LEVEL 1: Copy the following graphic organizer onto the chalkboard, omitting the italicized answers. Have students fill in the boxes to explain how people have used land throughout human history. Then ask each student to choose a step on the diagram and to illustrate it. Display illustrations in the classroom. **ENGLISH LANGUAGE LEARNERS**

Agriculture and Human History

| Hunter–gatherers move with the seasons in search of food. | → | Humans domesticate plants and animals. | → | People develop agriculture and transform their environments. | → | Agriculture provides surplus food and allows people to learn new crafts and skills. | → | Towns and cities grow as civilizations develop. | → | Trade between cities increases cultural diffusion. |

Cultural Kaleidoscope

The Penan of Borneo

One of the few surviving hunter-gatherer societies can be found deep within the Gunung Mulu National Park on the island of Borneo. The 300 to 400 Penan people live there and refuse to be resettled outside their forest, which provides all their needs. The forest supplies poles and leaves for shelter, wild game for meat, and plants for medicines. Forest vines provide twine and fiber for baskets. The Penan crush sago palm into a paste, which they then dry to make flour. This is their staple food. Fruits, wild mushrooms, and greens add variety to the Penan diet.

CRITICAL THINKING: How might logging in the forest affect the life of the Penan people? *(Possible answer: Logging would remove all things necessary for the Penan way of life and might force the Penan to change their lives completely.)*

VISUAL RECORD ANSWER

Possible answers: cats, dogs, goats, horses, rabbits, sheep, and so on

Our Amazing Planet

Only about 150 of the world's 80,000 or so edible plants have been domesticated. Of these, just 20 provide about 90 percent of the world's food supply.

Hunting and gathering was the main way of life for most of human history. Often hunter-gatherers moved their camps with the seasons. They searched for different plants and animals throughout the year. Today few such societies are left. They remain mostly in environments that are too difficult for farming. One example is the Inuit. They hunt and fish along the Arctic shores of North America and Siberia.

Agriculture began about 10,000 years ago. It radically changed the way people saw the land around them and how they used it. Agriculture appeared when hunter-gatherers learned how to grow plants and tame animals for their own use. This innovation is called **domestication**. Many scientists think domestication first began in Southwest Asia. It also developed independently in other regions. (See the map of origins of domesticated plants.)

Learning how to raise animals and grow crops were two of the most important developments in human history. Farmers transformed the world's environments. They cleared land to plant crops like wheat and barley. They learned how to plow and irrigate land. Farming produced more food, so the same land could support more people. As a result, populations increased, and people could settle permanently in one place.

The first cities appeared in Southwest Asia more than 5,000 years ago. City life became possible when there was enough food so that some people did not have to farm. Instead, some of them worked as potters or weavers. Others became merchants and traders. Still others carried out governmental or religious tasks. Towns and cities began to grow. The growth in the proportion of people living in towns and cities is called **urbanization**. With the development of cities, population densities increased and communication became easier. As trade connected early cities, cultural diffusion increased.

✓ **READING CHECK:** *The Uses of Geography* How did domestication change how people see their environment and use land? Farmers cleared land to plant crops and created a larger and more reliable food supply, which led to the development of cities.

INTERPRETING THE VISUAL RECORD

This ancient Egyptian wall painting shows domesticated cattle in Egypt. The Egyptians used cattle as draft animals for farming and as a source of meat and skins. **What are some other kinds of domesticated animals that you are familiar with?**

LEVELS 2 AND 3: Organize students into groups. Have each group propose a scenario that describes what human life might be like today if the progression from hunting and gathering to civilization had been interrupted. Instruct students to create feasible scenarios and then act out their ideas. Then have the members of each group write a paragraph to explain their conclusions. **COOPERATIVE LEARNING**

Teach Objective 2

LEVEL 1: Direct students' attention to a large wall map of your state. Call on volunteers to locate the state's largest cities. Then ask the class to suggest factors that influenced each city's location. (*Possible answers: near an*

important resource, on a trade or transportation route, in a defensible position) Have students examine the distribution of cities within the state to answer the following questions: How many large cities are there? Where are they located in relation to each other? Where are small cities located in relation to large ones? Lead a class discussion about factors that have influenced the distribution of cities within an area. Then ask students if they are familiar with any problems your state's cities face. Discuss their responses.

General Areas of City Locations

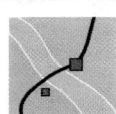 **1** River crossing

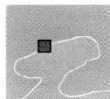 **2** Natural harbor

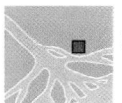 **3** Head of delta

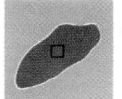 **4** Defensive hilltop site

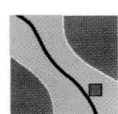 **5** Defensive site controlling a pass

Urban Geography

Cities are centers of culture, trade, government, and ideas. Urban geography—the geography of cities—is an important subject. It includes the study of city locations, sizes, land use, and urban problems.

City Growth A number of important factors have influenced the site and growth of cities. One is location near key resources. Other factors include location along transportation and trade routes and at easily defended sites. (See the diagram.) Many of the world's greatest cities grew up where two or three of these factors were present. Once cities are established, continued access to other cities and resources allows them to grow and prosper.

Many cities are found near freshwater, which is a key resource. For example, many cities lie along rivers, particularly in dry regions. Cities are also found where waterfalls or rapids kept large boats from going farther upstream. Other cities are located near important mineral resources. For example, Johannesburg in South Africa grew near huge gold deposits. Houston, Texas, became an important city when oil deposits were found nearby.

Because resources are not distributed evenly, trade routes have always been vital to human societies. As a result, cities rose along these routes. For example, London and Philadelphia grew as great river ports near the ocean. Chicago expanded due to its railway connections near the shore of Lake Michigan. Singapore, in Southeast Asia, grew along a major shipping route.

Easily defended sites protect cities from invasion and allow them to prosper. For example, Jerusalem began on a hilltop that was easier to defend than surrounding lowlands. The first settlers in Paris lived on islands in the middle

INTERPRETING THE DIAGRAM

This diagram shows the types of locations that are likely to be selected as sites for villages, towns, or cities. Topography, access to resources and transportation routes, and other factors have long affected settlement locations. **How might these locations allow cities to grow and prosper? Can you think of some examples of cities that have these types of locations?**

Linking Past to Present

City on Two Continents
Located on both sides of the Bosporus—the strait at the entrance to the Black Sea—Byzantium was founded by the Greeks about 657 B.C. The port city was a target for numerous invaders because it linked Europe and Asia.

Byzantium became a Roman city in about A.D. 200. Emperor Constantine renamed it Constantinople in A.D. 330 and built a protective wall around it. However, European Crusaders sacked the city in 1204. The Turks later took Constantinople and built many magnificent palaces, mosques, and fountains.

In 1930 the city's name was changed to İstanbul. It is a popular tourist destination and maintains its role as a major link between two continents.

ACTIVITY: Have students conduct research on trade routes through İstanbul and create product maps to illustrate their findings.

 Settlements can turn a difficult location into an advantage. Hammerfest and Honningsvaag, two Norwegian towns far north of the Arctic Circle, vie for the title of "world's northernmost town" because the distinction draws tourists.

DIAGRAM ANSWER

Answers will vary.

 **LEVEL 2:** Organize the class into small groups and provide each group with a street map of a city in your area. Have students use reference books, the Internet, and personal experience to identify and shade on the map the parts of the city and the outlying area, using the urban land use diagram in this section as an example. After students have completed their maps, have each student write a brief paragraph explaining the relationships displayed there. Call on volunteers to share their work with the class. **COOPERATIVE LEARNING**

LEVEL 3: Organize the class into small groups. Have the members of each group choose one city that they think deserves the label *world city.* Have each group create an informational brochure about its chosen city that includes general statistics about the city's population, names and descriptions of large businesses or organizations headquartered there, and other information that might be of interest to business travelers. Brochures should also mention nearby cities that are economically or socially tied to the world city. After students have completed their brochures, ask if they still consider their chosen subject a world city. Discuss their responses. **COOPERATIVE LEARNING**

Essential Element 1

▶ **The World** ◀
in Spatial Terms

Threshold and Range
Key to Christaller's Central Place Theory are two other concepts he pioneered—*threshold* and *range.* These ideas explain the relationships between cities in Christaller's model.

Threshold refers to the number of people required to sustain a particular service. Some businesses, like gas stations and grocery stores, require a relatively small population to remain profitable. These are found in even small towns. Other facilities, like hospitals and department stores, need thousands of customers to remain in operation and are found in only large cities. *Range* identifies how far people will travel to purchase a product or service. While people will not drive a long distance to buy a loaf of bread, they will to consult a cardiac specialist. Range determines how far from a central hub small cities can reasonably be established.

DISCUSSION: Lead a discussion about how threshold and range are reflected in Central Place Theory.

MODEL ANSWER

It can be used to explain why there are more small urban places than large ones.

Over the last 5,000 years, the world's population has gone from less than 1 percent urban to 45 percent urban.

of the Seine River for protection. How do you think resources, trade routes, and defensive sites influenced the location of towns and cities near you?

✓ **READING CHECK:** *Places and Regions* What factors have influenced city location and growth? location along transportation and trade routes, proximity of key resources, availability of defensive sites, continued access to other cities and resources

Patterns in the Size and Distribution of Cities The world has many villages, fewer towns, still fewer cities, and a handful of giant urban areas. The larger places are, the fewer there are. Geographers call this pattern a hierarchy of urban places. Cities and settlements are different in size because they serve different functions and purposes.

Villages have only a few hundred people. Stores in villages usually sell basic goods, such as food or farm supplies. Towns are larger than villages. They may have a few thousand people. Towns have some stores that sell goods for daily life and others that sell items needed less often. These items may include books, cars, or furniture. Towns may also have local and area government buildings, like county courthouses.

Cities, with several thousand to a few million people, have more services. Cities have large shopping centers, government offices, and many businesses. They also have hospitals and perhaps museums and universities. Some cities grow into huge urban areas called **world cities**. World cities are the most important centers of economic power and wealth. Their economies are dominated by the headquarters of global banks and businesses like advertising and insurance. London, New York, and Tokyo are all world cities. Each is the financial, business, and government center for a huge area and is important to the global economy.

FOCUS ON GEOGRAPHY

Central Place Theory In the 1930s the German geographer Walter Christaller noticed that there was a regular pattern to the locations of different-sized urban places. Why, he wondered, was this so? Christaller developed a theory to explain the patterns in the size and location of cities. We call his idea central place theory.

Christaller thought that the best arrangement of different-sized cities serving different functions would take the shape of a hexagon. At the center of the hexagon would be a first-order place. Such a place would be the area's largest city and would have the most goods and services. Evenly spread around this city would be smaller towns ranked as second-order, third-order, or fourth-order, according to their size. These smaller towns would offer fewer goods and services. However, there would be more of them. According to Christaller, this pattern represented the most efficient arrangement of urban places.

However, in the real world cities and towns do not form a perfect hexagon. This is true because the most efficient arrangement is not the only factor involved. Environmental factors, such as mountains or rivers, affect settlement location. Cultural factors, such as political boundaries or trade routes, also play an important role.

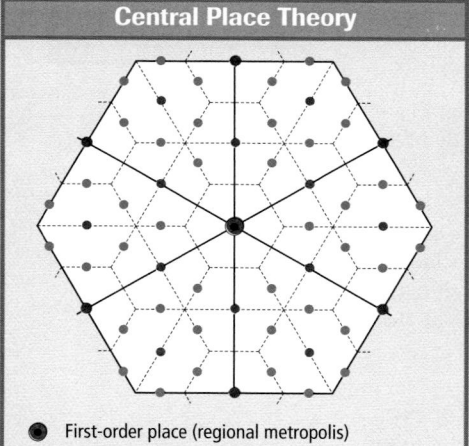

Central Place Theory

● First-order place (regional metropolis)
● Second-order place
• Third-order place
· Fourth-order place
— Main transport routes between regional metropolises

Source: *The Human Mosaic, Eighth Edition,* Terry G. Jordan

INTERPRETING THE MODEL *This model shows the distribution of different-sized urban places according to central place theory.* **How can this model be applied to present information about the locations of urban places in the real world?**

LEVEL 1: Write the following terms on the chalk-board: *subsistence agriculture, shifting cultivation, pastoralism, market-oriented agriculture,* and *agribusiness.* Ask the class what these terms have in common. *(Possible answer: They are all ways that land is used in rural areas.)* Then have students draw a picture of each type of rural land use. When students have completed their pictures, call on volunteers to explain their work to the class. Display the pictures in the classroom. **ENGLISH LANGUAGE LEARNERS**

LEVELS 2 AND 3: Organize the class into four groups and assign each group an area near your community: one to the north, one south, one east, and one west. Have each group research the land surrounding your community. Ask students to investigate how the area is used. Have them identify what kinds of agriculture, if any, are practiced in the area. Have each group create maps, charts, or graphs to illustrate their findings. **COOPERATIVE LEARNING**

Nevertheless, central place theory is useful because it does help explain certain geographic patterns. First, there are fewer large cities than small ones. Large cities are also generally farther apart. People usually find specialized goods and services in large cities but not in small towns. Finally, people in small towns often travel farther to find goods that they need less often. In short, the theory does not match the real world perfectly. However, it helps us understand how cities of different sizes are arranged and connected.

✓ **READING CHECK:** *Human Systems* How does central place theory help explain patterns in the size and distribution of cities?

explains why there are fewer large cities than small ones, why specialized goods and services are usually not found in small towns but are found in large cities, why people must travel farther to find goods that they need less frequently

Urbanization Today Both the number and proportion of people living in cities has grown dramatically over the last century. Probably in the next 15 years, more than half of the world's people will live in cities. This growth results not only from the natural increase of populations but also from people leaving rural areas for cities.

In developed countries about 75 percent of the people live in cities. However, urban growth in the world's rich countries is now slow. There are several reasons for this trend. First, the populations of most rich countries are either stable or are increasing slowly. Also, many people prefer the peace and quiet of rural areas to fast-paced cities.

Less than half of the people in developing countries live in cities. However, these areas are rapidly being urbanized. As the populations of these countries grow, more of their people come to the cities looking for jobs. In fact, many of the world's largest cities today are found in developing countries.

✓ **READING CHECK:** *Places and Regions* How do patterns of urbanization differ in developed and developing countries? *developed countries: about 75 percent but growing slowly now; less developed countries: less than half but growing quickly*

Urban Land Use Though they are all unique, cities around the world function in much the same way. They are home to businesses, government agencies, housing, industries, and religious and social groups. All these activities and groups are connected by different forms of transportation.

Most city centers are dominated by large stores, offices, and buildings. Such an area is called a **central business district (CBD)**. CBDs are transportation

INTERPRETING THE VISUAL RECORD

*The city of São Paulo, Brazil, is an example of a huge, rapidly growing city in a middle-income country. Between 1975 and 2000 the city's population grew by an incredible 77 percent. Today São Paulo has more than 17 million people and is one of the five largest cities in the world. **What do you think attracts immigrants from rural areas to cities like São Paulo?***

VISUAL RECORD ANSWER

Possible answers: jobs, excitement, cultural and educational opportunities

Traffic problems are common in large cities. Roads are clogged during rush hours, when large numbers of people travel to or from work. Traffic problems are particularly common in U.S. cities, such as Los Angeles (below), because Americans rely so heavily on cars as the main means of transportation. Cars transport few people per vehicle and require a lot of space to park.

124

Urban Land Use

■	Old CBD
■	Urban redevelopment project
■	Pedestrian mall
■	Zone of transition
■	Pre-war housing
□	Post-war housing
★	Major mall
✪	Shopping centers
H	Suburban hospitals and health care
■	Newest housing
■	Outlying office parks
■	Industrial parks
■	Agriculture

INTERPRETING THE DIAGRAM *This diagram shows a generalized pattern of land use typical in many large U.S. cities. Outlying suburbs and industries have grown as the traditional CBD declined. **Where are major malls and shopping centers located? How does the age of housing developments change as you travel away from the CBD? How might the growth of the city into new areas affect agriculture?***

hubs where roads, railroads and buses come together. Outside the CBD are factories and warehouses, which need larger amounts of land. Farther away from the city center, housing takes up most of the land. Small shopping strips with convenience stores line major streets, particularly at major road junctions. Still farther out are the suburbs, which usually have the newest houses. Some suburbs were once small towns that have been engulfed by a spreading city. Other suburbs are in areas that were once farmland. (See the Urban Land Use diagram.)

Modern cities are often ringed by one or more major highways. Along these roads, many new clusters of tall office buildings and big shopping centers have appeared. These clusters of large buildings away from the CBD are called **edge cities**. Increasingly the residents of the suburbs work and shop in edge cities. As a result, many people no longer need to visit the original CBD.

Urban Problems Most cities today face serious social, housing, transportation, or environmental problems. The key social problem in cities is often poverty. This is particularly true in developing countries. In those countries many city-dwellers lack good jobs and housing.

It should not be surprising that air, water, and land pollution are also urban problems. These problems are made worse because cities concentrate people, homes, and industries in small areas. Some countries have reduced pollution by using new technologies and public education programs. However, many cities in developing countries face serious pollution problems. Some do not have adequate sewer systems. Water supplies may not be safe to drink. Also, many poorer countries are just starting to enforce laws that control industrial pollution.

✓ **READING CHECK:** (*Human Systems*) What are some common urban problems and their causes? poverty, pollution; concentrated, growing populations, particularly in developing countries

Rural Geography

Rural landscapes are found outside cities. Agriculture is the key economic activity in most rural areas. However, people who live there also work in forestry, mining, and recreation. Rural areas may have land set aside in national parks and wildlife preserves.

Subsistence Agriculture   The kind of agriculture practiced most widely around the world is called **subsistence agriculture**. It is found in many parts of Africa, Asia, and Central and South America. In subsistence agriculture, food is produced by a family for its own needs. Anything extra—usually very little—may be sold for important supplies, like cooking oil, clothes, or fuel for a lamp. This kind of agriculture uses little if any machinery.

Subsistence agriculture takes different forms around the world. In many regions families farm the same fields for generations. In more difficult environments, such as tropical forests, **shifting cultivation** is common. In shifting cultivation, farmers clear trees or brush for planting. Sometimes they burn the debris. Then the fields are farmed for a few years, but fertility steadily decreases. The field is eventually abandoned, and then a new field is cleared. Crops such as maize (corn) in South America and manioc (cassava) in Africa are grown in this way.

Another type of subsistence agriculture is **pastoralism**—herding animals. Cattle, goats, horses, sheep, or other animals provide milk and meat for pastoralists. In addition, animal skins or hair are used for shelter and clothing. Pastoralists usually follow regular migration routes. In semiarid regions they migrate with their herds across open grasslands. Others move their animals up and down mountain slopes with the seasons. Pastoralists often trade animal products for grain or vegetables. A few animals may be sold in towns to buy other supplies. In pastoralist cultures, animals are more than just sources of food. Herds represent wealth and social prestige.

INTERPRETING THE VISUAL RECORD

Subsistence agriculture is characterized by low levels of technology. For example, this farmer in China is watering his crops without the help of modern farm machinery. **How might the technology used in this photograph be different in a developed country with advanced farming technologies?**

Connecting to ECONOMICS

The von Thünen Model

Johan von Thünen (1783–1850) was a German scholar and farmer. He noticed regular patterns of land use in farming areas of southern Germany. Based on these patterns, von Thünen developed one of the first spatial models in geography. Von Thünen's model is used to explain the location of different types of agriculture in a market economy.

Von Thünen noted that land closest to urban areas has the highest value. Land values decrease as one moves away from the city. Therefore, the most intensive forms of agriculture—those that generate the most profit on the smallest plots of land—are located closer to urban areas. Here, activities such as dairy farming and market gardening are common. Less intensive agricultural activities—those that require large amounts of land to be profitable—take place farther away from cities. In these areas, activities such as ranching and grain farming are found.

Analyzing Information   How does von Thünen's model explain the location of different types of agriculture in a market economy?

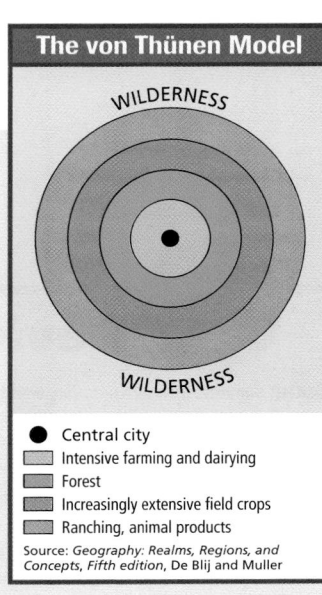

The von Thünen Model

WILDERNESS

WILDERNESS

● Central city
☐ Intensive farming and dairying
☐ Forest
☐ Increasingly extensive field crops
☐ Ranching, animal products

Source: Geography: Realms, Regions, and Concepts, Fifth edition, De Blij and Muller

125

Close

Call on students to suggest advantages and disadvantages to living in an urban environment as opposed to a rural one. As students suggest ideas, write them on the chalkboard. Then have students vote on whether they would prefer to live in an urban or rural area.

Review and Assess

Have students complete the **Section Review**. Then have students complete **Daily Quiz 6.2.**

Reteach

Have students complete **Main Idea Activity for English Language Learners and Special-Needs Students 6.2.** Then have students work in pairs to draw graphic organizers that summarize the section material. **ENGLISH LANGUAGE LEARNERS**

Extend

Have interested students conduct research on when and where various plants and animals were domesticated for use as food. Challenge them to choose a particular time period and region and to write recipes appropriate for the foods available. **BLOCK SCHEDULING**

Section 2 Review Answers

Define For definitions, see: domestication, p. 120; urbanization, p. 120; world cities, p. 122; central business district (CBD), p. 123; edge cities, p. 124; subsistence agriculture, p. 125; shifting cultivation, p. 125; pastoralism, p. 125; market-oriented agriculture, p. 126; agribusiness, p. 126

Reading for the Main Idea
1. about 10,000 years ago; Southwest Asia more than 5,000 years ago

2. The CBD houses large businesses and serves as a transportation hub. Outside it are factories and warehouses. Farther away, land is used mostly for housing and suburbs. Cities are often ringed by major highways, along which are edge cities.

Critical Thinking
3. Market-oriented agriculture uses machinery and scientific developments to produce only crops likely to be profitable. Subsistence farmers grow what they need to survive. (NGS 11)

4. Such farming would require large-scale irrigation and might strain local water resources. (NGS 14)

Organizing What You Know
5. Answers will vary.

VISUAL RECORD ANSWER

Each location will have the right amount of fertilizer applied. It will not be wasted on areas that are already fertile.

INTERPRETING THE VISUAL RECORD

The application of advanced farming technologies to increase production is a key feature of market-oriented agriculture. Here, detailed soil fertility maps, GPS technology, and computers are used to apply the correct amount of fertilizer to each part of this farmer's fields. **How might these technologies increase yields and reduce costs?**

Market-Oriented Agriculture Another type of agriculture is **market-oriented agriculture**, or commercial agriculture. Under this system, farmers grow products to sell to consumers. They specialize in growing crops that other people want to buy and that they can sell at a profit. Often, crops are first sold to companies that process and package them. Then consumers buy the final products in stores. This kind agriculture is the most common type found in developed countries.

Scientific advances have made market-oriented agriculture very productive. Those advances include new animal breeds, fertilizers, pesticides, and plant varieties. In fact, commercial farmers depend on the latest technology to be successful. Their farms and ranches are often enormous in size. Yet because of technological advances, they require fewer and fewer workers. Some commercial farms are not owned by families at all. Instead, they are parts of large corporations. Advanced commercial farming is referred to as **agribusiness**. Agribusiness is the operation of specialized commercial farms for more efficiency and profits. It is characterized by huge farms with close links to other parts of the food industry.

✓ **READING CHECK:** (**Human Systems**) What role has technology played in the development of market-oriented agriculture? increased productivity, caused farms and ranches to be larger and use fewer workers, led to agribusiness

Section 2 Review

go.hrw.com **Homework Practice Online**
Keyword: SW3 HP6

Define domestication, urbanization, world cities, central business district (CBD), edge cities, subsistence agriculture, shifting cultivation, pastoralism, market-oriented agriculture, agribusiness

Reading for the Main Idea
1. (**Human Systems**) When do many scientists think agriculture began? Where and when did cities originate?

2. (**Places and Regions**) How is land use typically arranged in urban areas?

Critical Thinking
3. *Comparing* Why do you think market-oriented agriculture is more productive than subsistence agriculture?

4. *Analyzing Information* How might large-scale commercial farming be possible in areas with arid or semiarid climates? What environmental consequences might such farming have in dry regions?

Organizing What You Know
5. Using the information in this section, draw a sketch map showing land use in a typical urban area.

History: City Plans in New Spain

From 1535 to 1821, much of North America was controlled by Spain. This territory, called New Spain, was administered by an official called a *viceroy* who lived in Mexico City. Many settlers were brought to the colony, which included Central America, Mexico, and the southwestern United States, to establish a Spanish presence in the area. Since most of these settlers had lived in Spanish cities in Europe, they also built cities in their new homeland.

Spanish cities in the New World were established according to a strict plan called the Laws of the Indies. Within each city, the pattern of streets and buildings was the same. At the city center was a plaza, or market square, around which the town was focused.

Facing the plaza was a church, because religion was very important to Spanish culture. The other three sides of the plaza were lined with government buildings so that city administrators could keep a watchful eye on the city. Streets spread outward from the plaza in a precise grid. This pattern may have been implemented to help city officials maintain order in the case of a rebellion. Soldiers could easily seal off a small area to put down an uprising.

Organize the class into groups and have each group study early settlements established by settlers in various parts of the United States. Ask them how colonists from different countries planned their communities. Then have students draw maps of some early colonies to clarify any differences. Lead a class discussion about why settlement patterns varied between countries.

Human Systems

Geography for Life

Rural Settlement Forms

Settlements are purposely grouped and organized clusters of houses and buildings. Some rural settlements have only a few buildings grouped loosely together. We call these hamlets. Other settlements may be larger, such as villages. These settlements take various forms.

Perhaps you live in or have visited a rural community. In such places, homes and other buildings may be scattered and linked by a network of roads. Wooded areas or farmlands may separate these homes and other buildings from each other. On the other hand, a rural community might grow from a cluster of farmhouses separated from agricultural fields. Hamlets like these are found in many hilly rural areas throughout Asia and Europe. Some of these hamlets have a linear form. That is, buildings are clustered along features such as roads or streams. Sometimes hamlets are groups of buildings clustered at the intersection of several roads.

Larger villages may take forms similar to hamlets. Historically, houses in villages were clustered together for defensive reasons. People could defend themselves better against outsiders by grouping together. In fact, many old villages occupy easily defended sites, such as hilltops. Round villages and walled villages are often clustered together in

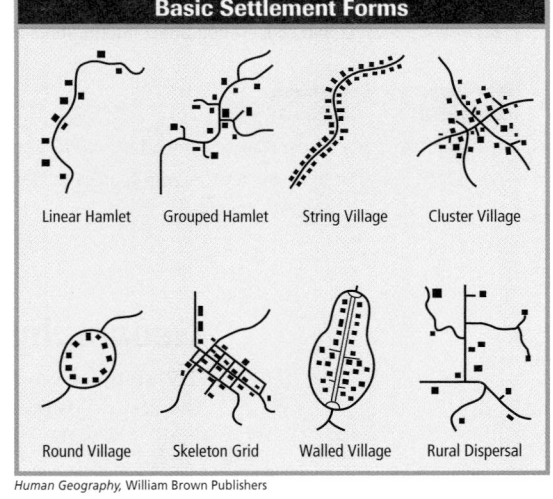

Basic Settlement Forms

Linear Hamlet Grouped Hamlet String Village Cluster Village

Round Village Skeleton Grid Walled Village Rural Dispersal

Human Geography, William Brown Publishers

this way. Many old cities around the world began behind strong defensive walls. Also, like hamlets, some villages are grouped along rivers, roads, streams, or other linear features. These places are called string villages. For example, in many lowland areas of western Europe, villages are located along waterways. Villages grouped around road intersections are called cluster villages.

Defense is not the only reason houses in a village are clustered together. Such settlement forms may also leave the best land available for farming. By occupying hilltops or rocky areas, more good farmland is available. For example, some Japanese villages are so tightly packed together that only narrow passages remain between houses. This pattern reflects the need to farm all the useful land. In the hilly parts of Europe, many villages are clustered on hillsides. This arrangement leaves the level land for farming.

A hamlet near Kathmandu, Nepal

China
Nepal Kathmandu
India Bangladesh

Applying What You Know

1. **Contrasting** How do string, or linear, settlement forms differ from cluster forms?

2. **Analyzing Information** Why do you think road intersections and rivers are common locations for hamlets and villages?

Urban Planning in the United States Until the late 1800s most American towns were laid out in a rectangular grid system. At that time, however, some critics began to complain about the monotony of "grid-iron" towns. One reaction was a "back-to-nature" movement that stressed curved roads and pretty scenery. This idea has influenced suburban planning and has resulted in a less rigid street pattern in many urban areas.

Have students examine city maps to identify areas with grid patterns and those with patterns of curved lines. Ask students in which neighborhoods it would be easier to follow directions to a particular destination and which neighborhoods might be more visually interesting. Have students discuss the advantages and disadvantages of each system.

Applying What You Know Answers

1. Linear or string forms have buildings that line rivers, roads, or other linear features. Cluster forms feature buildings grouped around connecting roads, in easily defended areas, or away from agricultural fields.

2. Possible answer: They are important trade and transportation routes.

This Geography for Life feature addresses National Geography Standards 3, 9, 10, and 12.

127

OBJECTIVES

1. **Explain how government and geography are connected.**

2. **Identify the three main types of geographic boundaries.**

3. **Analyze how conflict and cooperation affect international relations.**

 **LET'S GET STARTED**

Copy the following instructions onto the chalkboard: *Work with a partner to identify current conflicts occurring at local, national, or international levels. Identify a conflict for each level.* List responses on the chalkboard. Then ask students to suggest possible causes for the conflicts. Discuss responses. Tell students that in Section 3 they will learn about different factors that contribute to conflict and cooperation worldwide.

Building Vocabulary

Write **totalitarian governments**, **democracy**, **tariffs**, and **quotas** on the chalkboard. Tell students that totalitarian governments and democracy are nearly opposites. In the first, one person or group holds all power, while in the latter, everyone has a voice in the government. Point out that both tariffs and quotas are used to protect a country's economy. Have students read each term's definition. Then discuss how tariffs and quotas protect companies from foreign competition. Call on volunteers to read the definitions of the remaining key terms.

 **RESOURCES**

REPRODUCIBLE

▶ Guided Reading Strategy 6.3
▶ Critical Thinking Activity 6: The Cuban Missile Crisis

TECHNOLOGY

▶ One-Stop Planner CD–ROM, Lesson 6.3
▶ Homework Practice Online
▶ HRW Go site

REINFORCEMENT, REVIEW, AND ASSESSMENT

▶ Main Idea Activity 6.3
▶ English Audio Summary 6.3
▶ Spanish Audio Summary 6.3
▶ Section 3 Review, p. 131
▶ Daily Quiz 6.3

 **Section 3** # Political Geography

READ TO DISCOVER

1. How are government and geography connected?
2. What are three main types of geographic boundaries?
3. How do conflict and cooperation affect international relations?

WHY IT MATTERS

There are many border disputes around the world. Use **CNNfyi.com** or other **current events** sources to learn about a border dispute in the news.

DEFINE

natural boundaries
cultural boundaries
geometric boundaries
nationalism
totalitarian governments
democracy
tariffs
quotas

Geography and Governments

The study of government and politics is an important part of geography. In this section you will learn how control of Earth's surface is divided. We will also look at how the culture of a place influences its government.

Government and Development   There are about 200 countries in the world today. A country has an independent government, which has authority over territory within its borders. Governments are free to make their own laws and have their own leaders. Usually, governments negotiate and deal with each other in peace. We call this diplomacy. Governments interact with each other through trade agreements and international organizations. However, disputes between countries sometimes lead to conflict or war.

Good governments protect the lives and property of the people who live in a country. They also protect the freedoms and rights of their citizens. In doing so, they help ensure the conditions needed for economies to develop and for people to prosper. In many developing countries, governments are unstable. That is, they do not last long or have much authority. Corruption can also be a problem. Political leaders may use their power only to enrich themselves and their friends.

Officials sign the Treaty of Versailles in 1919, which ended World War I. International agreements and diplomacy through government representatives are basic features of the world's independent countries.

Teach Objective 1

ALL LEVELS: Call on volunteers to suggest ways in which government and geography are connected. Ask students to think of physical and human processes that can lead to the formation of national boundaries. Lead a class discussion about these processes. Then ask students to name ways in which cultural beliefs can influence or shape governmental decisions. Encourage students to provide examples of this influence in the United States or in other countries with which they are familiar. Discuss responses.

Teach Objective 2

LEVEL 1: Copy the following graphic organizer onto the chalkboard, omitting the italicized answers. Have students use it to describe the three main types of geographic boundaries. Examples will vary. **ENGLISH LANGUAGE LEARNERS**

Type of boundary	Description	Example
natural	*follow a feature of the landscape*	*U.S.-Mexico border along the Rio Grande*
cultural	*based on cultural traits such as religion or language*	*Spain-Portugal border*
geometric	*follow regular, geometric patterns such as latitude and longitude*	*U.S.-Canada border*

Cultural Beliefs and Government   The cultural beliefs of different groups affect how governments are set up and operate. These beliefs influence government decisions and public policies. For example, a people's cultural beliefs might lead to laws that force businesses to close on special religious days. Cultural beliefs also affect the way citizens see their duties and responsibilities.

To see how these cultural influences can work, consider Israel, a country in Southwest Asia. After World War II, Israel was established as a homeland for Jews from anywhere in the world. Israel's role as a Jewish homeland has guided many of its government policies. For example, any Jew who wishes to become a citizen of Israel can do so. In fact, many Jews have moved there from around the world. Still, people in Israel debate the role of religion in government and society. About 20 percent of Israeli Jews strictly follow the beliefs of Judaism. These people are called Orthodox Jews. They tend to believe that religious values should play an important role in shaping government policy. Most other Israeli Jews seek to limit the role of religion in Israel. In fact, Israeli law guarantees religious freedom for all people, Jewish or not. About 20 percent of Israeli citizens are not Jews. Most of these people are Arabs who have very different cultural beliefs and practices. They have often opposed Israel's immigration, military, and foreign policies.

✔ **READING CHECK:**   *Human Systems*   How can cultural beliefs influence citizenship practices and public policies? laws about special religious days, religious freedom, immigration, military and foreign policies

Geographic Boundaries

Three main types of boundaries separate countries from each other. Boundaries that follow a feature of the landscape are called **natural boundaries**. Mountains make good natural boundaries. They are difficult to cross and are permanent markers. Rivers, on the other hand, are often troublesome boundaries. Many rivers are important transportation routes for more than one country. In addition, river channels may move. Other natural boundaries include deserts, lakes, and oceans.

Borders that are based on culture traits, such as religion or language, are called **cultural boundaries**. For example, the border between mostly Muslim Pakistan and mostly Hindu India was established largely along religious lines. The same is true on the island of Ireland. A border divides the island between mostly Protestant Northern Ireland and the mostly Roman Catholic Republic of Ireland. Many cultural boundaries are based on language. The boundary between Portugal and Spain is an example.

Boundaries that follow regular, geometric patterns are called **geometric boundaries**. These borders are usually straight lines drawn without regard to environmental or cultural patterns. Geometric boundaries are often based on lines of latitude or longitude. For example, the border between the United States and Canada lies mostly along the 49th parallel. In colonial Africa, European countries drew many geometric boundaries that are still used today. Some of these boundaries have led to problems because they divide the territory of different ethnic groups.

✔ **READING CHECK:**   *Human Systems*   What are three different types of boundaries that geographers study? natural, cultural, and geometric

The Rio Grande is a natural boundary between Mexico and the United States.

 **LEVELS 1 AND 2:** Have students use newspapers and magazines to find examples of contemporary world conflicts. Have students plot and label the locations of the conflicts on a wall map. Then have them identify the causes of the conflicts. Use the completed map and students' findings to lead a class discussion about the role of conflict in shaping boundaries and influencing the allocation of territory.

LEVEL 3: Have each student select an organization, such as the World Trade Organization, that encourages economic cooperation between nations. Have students conduct research on their chosen organizations to learn about their roles in international politics. Have each student create a flyer that the organization might use to encourage countries to join its effort. Flyers should include information about the organization's founding, its members, its goals and objectives, and its role in shaping world affairs.

Linking Past to Present

Partition and Problems

Partition is the division of a country into two or more territorial units. Following World Wars I and II, many areas in Europe and Asia were partitioned to form new countries. These partitions were often drawn without regard for ethnic boundaries. For example, the countries of Europe's Balkan Peninsula are home to many different ethnic groups, several of which want political power. Such partitions led to the evacuation of many emigrants and refugees as people sought safe havens. Although partitions were intended to solve problems, they often created new ones. In some cases, the tensions caused by partition continue even today.

ACTIVITY: Have students conduct research on areas that were partitioned after World War I or World War II. Ask students whether partitioning solved problems in the areas they studied. Why or why not?

Section 3 Review Answers

Define For definitions, see: natural boundaries, p. 129; cultural boundaries, p. 129; geometric boundaries, p. 129; nationalism, p. 130; totalitarian governments, p. 130; democracy, p. 130; tariffs, p. 130; quotas, p. 130

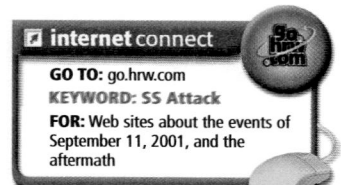

internet connect

GO TO: go.hrw.com
KEYWORD: SS Attack
FOR: Web sites about the events of September 11, 2001, and the aftermath

Conflict and Cooperation

The study of international relations is another important area of political geography. The field focuses on foreign policies and relations between the world's countries. Very few countries today are isolated. Instead, they are affected by events in both nearby and faraway places.

Political Conflicts Political conflicts, both within and among countries, are common. There are many reasons for this. Most people feel proud of their culture and country. Feeling pride and loyalty for one's country or culture group is called **nationalism**. These feelings are often expressed in special songs, symbols, and writings. Unfortunately, one group's pride can conflict with that of another. Competing feelings of nationalism have led to problems time and again. Conflicts can also result from other differing culture traits, such as religion. A modern example is Sri Lanka, where a Hindu minority has battled the Buddhist majority.

How a group of people should be governed has also been a source of conflict. One person or a small group of people governs some countries. They have full authority to make laws and decisions. In these **totalitarian governments** one person or a few people decide what is best for everyone. The people have little or no say in how their country is governed. Communist countries, such as North Korea and Cuba, have such governments. In other countries, all citizens have a voice in their government. These countries are based on **democracy**. This is a system in which the people decide who will govern. They choose their leaders by voting in free elections. Democratic governments value individual freedoms and human rights. The United States and many other countries have democratic governments.

Economic issues also lead to conflicts. For example, countries often establish **tariffs**—taxes on imports and exports. Countries also set **quotas**—limits to the amount of a product that can be imported. Tariffs and quotas usually help protect a country's industries from foreign competition. However, they can also cause trade disputes among countries.

Terrorism Some political conflicts involve the use of terrorism by one or more individuals or groups. Terrorism is the use of violence and fear as a political force. Terrorists act for many reasons. For example, some want independence for homelands that may be part of or under the control of another country. Others might work to further various political goals, such as different public or social policies. Terrorists do not usually act under the authority of a particular government. However, governments sometimes protect and even support the actions of terrorists who share similar political goals. Stopping terrorism is difficult, particularly in South America, Southwest Asia, and parts of Europe.

Even the United States has been directly affected by terrorism. One of the worst acts of terrorism in history happened on September 11, 2001. On that day, fundamentalist Muslim terrorists stunned the world by hijacking four commercial airliners. The terrorists crashed two of the planes into the twin towers of New York's World Trade Center, collapsing the buildings. The third plane hit the Pentagon, just outside Washington, D.C., and the fourth crashed in rural Pennsylvania. The attacks killed thousands of people. In the wake of these events, the U.S. government worked to enlist the cooperation of other countries in a global effort to defeat terrorism. The United States was not always able to get the level of cooperation it sought from all countries in military, political, and security matters.

International Cooperation Countries often cooperate with each other. They do this for two main reasons—political and economic benefits.

Political and military cooperation are most developed in the United Nations (UN). Nearly all of the world's countries are members of the UN. As a result, it is the most important international organization. In the UN, representatives of the world's countries can discuss international issues and voice their concerns. The UN's main goals are settling disputes between countries and trying to prevent wars. Sometimes, the UN sends peace-keeping military forces to warring regions. It also tries to solve global problems such as disease, hunger, and illiteracy.

Many countries benefit from economic cooperation and free trade. This cooperation can help countries produce goods at lower costs and reach larger markets. People can then buy those goods at lower prices. Economic cooperation can also end or reduce tariffs and quotas. For example, the World Trade Organization (WTO) works to make trade between countries fairer and easier. Most countries belong to the WTO. You will read about many other important economic organizations in this textbook.

✔ **READING CHECK:** *Human Systems* What are the main reasons countries cooperate with each other? for political and economic benefits

©2001 *The Record* (Bergen, NJ), Thomas E. Franklin, Staff Photographer

INTERPRETING THE VISUAL RECORD *The terrorist attacks on the World Trade Center and the Pentagon in 2001 killed thousands of people. Rescue workers and firefighters, such as these, faced the grim task of searching for victims.* **How might images like this one inspire people to work to defeat terrorism around the world?**

Section 3 Review

Define natural boundaries, cultural boundaries, geometric boundaries, nationalism, totalitarian governments, democracy, tariffs, quotas

Reading for the Main Idea

1. *Human Systems* How do good governments promote economic development?

2. *Places and Regions* What physical and cultural factors might influence the ways a country's boundaries are established?

3. *Human Systems* How do organizations like the United Nations and the World Trade Organization promote international cooperation?

Critical Thinking

4. **Analyzing Information** What kinds of boundaries surround your state? What kinds surround the United States?

go.hrw.com Homework Practice Online
Keyword: SW3 HP6

Organizing What You Know

5. Copy the chart below. Use it to describe the features of totalitarian and democratic forms of government. List examples of each.

Totalitarian	Democratic

Section 3 Review Answers

Reading for the Main Idea
1. by protecting people, property, rights, and freedoms

2. Natural boundaries like mountains and rivers form barriers. Cultural factors like religion and language help determine where to divide countries or other areas.

3. The UN's member countries work together to prevent war and solve world problems. The WTO promotes trade by lowering tariffs and other barriers.

Critical Thinking
4. Answers will vary. Rivers and other physical features make up some boundaries. Others are along lines of latitude and longitude. U.S. boundaries include rivers, mountain areas, lines of latitude and longitude, and oceans. (NGS 4)

Organizing What You Know
5. Totalitarian—One person or a few people make the decisions. Others have little or no say. Examples include communist countries like North Korea and Cuba. Democratic—Everyone has a voice in the government. Free elections and the protection of rights and freedoms are guaranteed. An example is the United States.

VISUAL RECORD ANSWER

Possible answer: might impress upon viewers the heroism of common people, giving them more incentive to unite with others to fight terrorism

Chemistry: Smog and Ozone

One of the most visible effects of technology on the environment is the smog that hangs in hazy brown clouds over many cities. It is created by chemicals released into the atmosphere by the burning of fossil fuels. When these fuels are burned, a chemical called nitric oxide (NO) is emitted, as are hydrocarbons, or unburned fuel. When the amount of these chemicals in the air reaches a sufficient level, sunlight triggers a chemical reaction. The nitric oxide is converted to nitrogen dioxide (NO_2), the brown gas we call smog.

In high enough quantities, nitrogen dioxide is toxic and corrosive to metal. In the atmosphere, it is capable of absorbing energy from sunlight. When this happens, the compound breaks apart. Released oxygen atoms join with oxygen present in the air (O_2) to form ozone (O_3). Like nitrogen dioxide, ozone can be harmful to buildings, plants, and people. Scientists estimate that increased ozone levels in the atmosphere cause billions of dollars in damages to crops, forests, structures, and human health in North America each year.

Organize the class into groups and have each group conduct research on pollutants that are released into the atmosphere by human activities. Have each group create a poster that illustrates the origins of these chemicals and their effects on society and the environment. **COOPERATIVE LEARNING**

Human Systems

Geography for Life

Technology and the Environment

People depend on their environment for clothing, food, fuel, and shelter. We also adapt to the environment and modify it to better suit our needs. Throughout history, major technological innovations have caused large-scale changes in the environment and in human societies.

Fire

One of the earliest examples of people using technology to change the environment is the use of fire. Early humans used fire to stay warm and to cook food. They also used fire to clear brush, improving grazing lands for the animals they hunted. Much later, people used fire to clear land for farming or raising livestock. By using fire, people changed the landscape over wide areas.

The Steam Engine

In the 1700s steam power became the main power source for industry. As well as powering factory machines, steam engines were used to power ships and trains. Steam-powered vehicles were very successful. They made travel faster, safer, and more reliable. As a result, people changed the way they thought about long-distance travel and trade. Soon railroad lines were built throughout Europe and North America, and then in countries around the world. Railroads helped unify huge countries like the United States, Canada, Russia, and India. Then steamships linked countries, and global trade increased.

As the use of steam engines increased, so did the demand for coal, their main fuel. As a result, steam engines indirectly led humans to modify the environment as they dug large coal mines around the world. In addition, burning coal caused air pollution, which damaged the environment. Industrial cities of the 1800s, such as London and Birmingham, had terrible smog.

Gasoline and Diesel Power

In the 1900s gasoline engines and then diesel engines began to replace coal-powered steam engines. Gasoline and diesel fuel were particularly suitable for cars and trucks. The rapid development of automobiles led people to modify the environment again with roads and highways. Webs of paved roads soon connected cities, towns, and rural areas. Today the United States alone has more than 2.3 million miles (3.7 million km) of paved roads.

In developed countries, gasoline and diesel machinery replaced human and animal power in agriculture. Patterns of food production and distribution around the world shifted as a result. Also, fewer farmers were needed to plow, plant, and harvest. These changes allowed more and more people to move from rural areas to the cities. This trend continues today as use of machines for agriculture is becoming more widespread in developing countries.

Electricity

With electricity, power could be transmitted over long distances. In addition, factories could be built farther from their power sources. Cities and roads could be lit more easily and safely. New electric machines also made many household jobs easier and faster, giving people more free time. In agriculture, electric motors powered farm machinery, water pumps, and machines used to process crops. Even fewer farmers were needed to feed the rest of the population. For example, in 1900 it took about four

The development of gasoline as a fuel for cars has led to environmental changes such as the expansion of roads and an increase in air pollution.

Going Further: Thinking Critically

Scientists have known since the 1800s that cities experience higher temperatures than nearby rural areas. Concrete and asphalt trap solar energy during the day and release it into the air after sunset, so temperatures do not decrease as much as they would in rural areas. Heat can also be trapped in the spaces between tall buildings. In addition, buildings and vehicles generate heat. Many cities also lack trees and plants that can provide shade and cool the air. These factors create phenomena called urban heat islands.

Many scientists now believe that these islands can affect more than just local temperatures. High atmospheric temperatures can contribute to the production of ozone above cities. These unusual temperature patterns can also alter local weather conditions. In some cases, the low pressure caused by hot air prevents rainstorms from passing over cities. Instead, the rain is diverted to surrounding communities. In other cases, the rising air collides with cooler air in the atmosphere and actually creates rain or clouds within the city. In heavily urbanized areas, like the northeastern United States, urban heat can change weather patterns for an entire region.

Organize the class into small groups and have each group conduct further research on urban heat islands. Have them examine cities in your state to determine if the phenomenon has been detected there. Encourage students to contact state or local weather researchers for additional information. Have each group prepare a report about the effects of urban heat on regional weather patterns. **COOPERATIVE LEARNING**

INTERPRETING THE VISUAL RECORD *This image of Earth's lights at night shows how people have used electricity to transform the environment. The brightest areas are generally the most urbanized, but not necessarily the most densely populated.* **Which areas appear the brightest? Which areas are dark? What might this image indicate about the level of development in these regions?**

farmers to feed 10 people. In the United States today, one farmer produces enough food to feed nearly 130 people.

The use of electric power has also caused direct environmental change. Coal and natural gas are among the resources used to generate electricity. Mining and drilling for these resources has altered, and even damaged, landscapes. Dams can also generate power. Building a dam affects the upstream environment by creating a lake where none had been. Downstream, flooding is reduced.

Environmental Change

As these examples show, technology allows people to alter their environment in important ways. The rate of technological developments has greatly increased over the last several hundred years. Can you think of other technological innovations that have allowed societies to change their environment?

This dam on the Columbia River in Oregon shows how humans modify the environment to generate electricity.

Applying What You Know

1. **Evaluating** How important are major technological innovations, such as fire, steam power, diesel, and electricity, that have been used to modify the environment?

2. **Comparing** How does environmental change caused by the development of gasoline power and diesel power compare to change caused by the steam engine?

CHAPTER 6 Review Answers

Building Vocabulary For definitions, see: market economy, p. 115; command economy, p. 115; developed countries, p. 117; infrastructure, p. 117; developing countries, p. 117; domestication, p. 120; urbanization, p. 120; central business district (CBD), p. 123; subsistence agriculture, p. 125; market-oriented agriculture, p. 126; natural boundaries, p. 129; cultural boundaries, p. 129; geometric boundaries, p. 129; nationalism, p. 130; democracy, p. 130

Locating Key Places
A. central city
B. intensive farming and dairying
C. forest
D. field crops
E. ranching, animal products
F. wilderness

Understanding the Main Ideas

1. primary; quaternary
2. GNP, GDP, per capita GDP, industrialization, energy consumption per person, levels of education, caloric intake per person, availability of health care, literacy rate, life expectancy
3. People began to work at other tasks besides farming. Communication became easier. Cultural diffusion increased.
4. produce food for their own use

Review and Assessment Resources

TECHNOLOGY

▶ Chapter 6 Test Generator (on the One-Stop Planner)
▶ Global Skill Builder CD–ROM
▶ HRW Go site

REINFORCEMENT, REVIEW, AND ASSESSMENT

▶ Chapter 6 Review, pp. 134–35
▶ Chapter 6 Tutorial for Students, Parents, Mentors, and Peers
▶ Chapter 6 Test (form A or B)
▶ Alternative Assessment Handbook

▶ Chapter 6 Test for English Language Learners and Special-Needs Students
▶ Unit 1 Test
▶ Unit 1 Test for English Language Learners and Special-Needs Students

Assess

Have students complete a Chapter 6 Test.

Reteach

Assign each of three groups a section of this chapter. Ask each group to write summarizing sentences about its assigned section. As volunteers read their sentences aloud, write them on the chalkboard to create a chapter summary. Then have students ask each other questions about the summary. **ENGLISH LANGUAGE LEARNERS**

CHAPTER 6 Review Answers

5. diplomacy, trade agreements, participation in international organizations, conflicts and war, alliances

Thinking Critically

1. Answers will vary. Students should demonstrate that they understand the various kinds of data used. (NGS 4)

2. Possible answer: They have transformed many areas into farmland and have built many cities. (NGS 14)

3. Answers will vary. Students might note the role of religion in community life or the use of bilingual educational programs in schools. Others might point out special cultural organizations that work with local governments. (NGS 10)

Using the Geographer's Tools

1. Possible answer: New forms of transportation might allow new resources to be exploited.

2. Possible answers: Numbers of physicians per person may relate to life expectancy. Literacy rate and per capita GDP might be related. Number of TV sets per person may relate to urban population.

3. Maps and answers will vary.

CHAPTER 6 Review

Building Vocabulary

On a separate sheet of paper, explain the following terms by using them correctly in sentences.

market economy
command economy
developed countries
infrastructure
developing countries
domestication
urbanization
central business district (CBD)

subsistence agriculture
market-oriented agriculture
natural boundaries
cultural boundaries
geometric boundaries
nationalism
democracy

Locating Key Places

On a separate sheet of paper, match the letters on the diagram with their correct labels.

intensive farming and dairying
forest
wilderness

field crops
ranching, animal products
central city

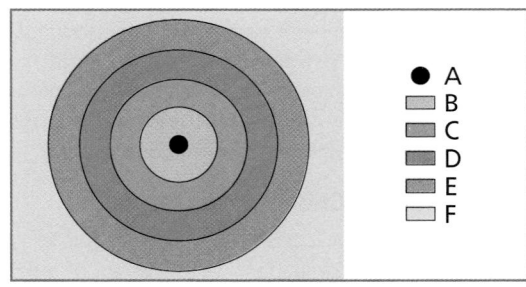

- ● A
- ▨ B
- ▨ C
- ▨ D
- ▨ E
- ▨ F

Understanding the Main Ideas
Section 1

1. (*Human Systems*) Which type of economic activity is located at the site of the resource being exploited? Which activity can be located almost anywhere?

2. (*Human Systems*) What are some measures of development?

Section 2

3. (*The Uses of Geography*) How did the development of early cities affect people's daily lives?

4. (*Human Systems*) How do people satisfy their basic needs with subsistence agriculture?

Section 3

5. (*Human Systems*) What are some of the ways countries interact with each other?

Thinking Critically

1. Evaluating Information What factors do you think are most important in determining the level of economic development and standard of living of a country? Why?

2. Summarizing How have humans modified the physical environment since the development of agriculture?

3. Analyzing How have cultural beliefs influenced public policies and citizenship practices in your community?

Using the Geographer's Tools

1. Interpreting Diagrams Study the chart of economic activities in Section 1. How might changes in transportation affect the location of different economic activities?

2. Interpreting Charts Study the chart of developed and developing countries in Section 1. How might some of the different categories of information shown in the chart be related?

3. Preparing Sketch Maps Use the student atlas at the beginning of this book and the information in Section 3 to find a country that has both a natural boundary and a geometric boundary. Draw a sketch map of the country, showing what these boundaries are based on. Why do you think these boundaries were selected?

Writing about Geography

Write a short article that summarizes how subsistence agriculture and market-oriented agriculture are different. What are the main differences between these two forms of agriculture? In general, how is technology used in each system? Where in the world is each type common? When you are finished with your article, proofread it to make sure you have used standard grammar, spelling, sentence structure, and punctuation.

S K I L L   B U I L D I N G

Geography for Life

Gathering Field Data

(*Places and Regions*) Use a notepad and pencil to gather field data about urban land use in your area. Start in the downtown or CBD of your closest urban area. Take notes about what types of buildings you see, how high they are, and how they are used. Then walk in one direction away from the downtown area, taking more notes along the way. When you are done, try to answer the following questions: How is land used differently as one travels away from the downtown? Why do you think this is?

134

Portfolio Activity

Organize the class into five groups. Assign each group one of the following roles: citizens of a developed country, government officials from a developed country, citizens from a developing country, government officials from a developing country, or delegates from an international cooperative organization. Have each group conduct research to prepare for a panel discussion on how much developed countries should help expand the economies of developing countries. Conduct the discussion and then have students write summaries of the major arguments made by panel members to place in their portfolios. **COOPERATIVE LEARNING**

Food Festival

In most developing countries, people depend upon one or two staple crops for most of their meals. These meals often consist of a starchy grain or root and a protein source like beans or other legumes. Although most people depend on crops native to their area, some staples have been spread throughout the world. Among the most common are beans, cassava, corn, peas, rice, wheat, and other grains. Have students determine which of these staple crops are available in your area. Ask them to find examples of various ways people prepare these foods and to bring samples to class.

Building Social Studies Skills

Interpreting Graphs

Study the line graph below. Then answer the questions that follow.

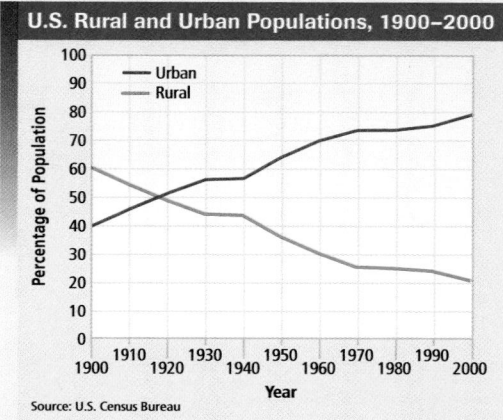

U.S. Rural and Urban Populations, 1900–2000

Source: U.S. Census Bureau

1. In 1990, how many more Americans were living in an urban environment than were living in a rural environment?
   a. 30 percent
   b. 40 percent
   c. 50 percent
   d. cannot be determined

2. During which three decades did the most dramatic increase in urban population occur?

Analyzing Secondary Sources

Read the following passage and answer the questions.

"The first cities appeared in Southwest Asia more than 5,000 years ago. City life became possible when there was enough food so that some people did not have to farm. Instead, some of them worked as potters or weavers. Others became merchants and traders. Still others carried out governmental or religious tasks. Towns and cities began to grow. The growth in the proportion of people living in towns and cities is called urbanization."

3. When did city life become possible?
   a. after people became potters and weavers
   b. more than 15,000 years ago
   c. when there was enough food that some people did not have to farm
   d. after trade connected early cities

4. Explain how the development of cities affected people's occupations.

Writing

Articles should accurately describe each type of agriculture. Use Rubric 9, Comparing and Contrasting, to evaluate student work.

Geography for Life

Answers will vary, but students should accurately report urban land use in your community. Use Rubric 1, Acquiring Information, to evaluate student work.

Social Studies Skills

1. c
2. 1940s, 1950s, 1960s
3. c
4. Possible answer: Because city dwellers did not have to farm, they were able to take up other occupations.

Alternative Assessment

PORTFOLIO ACTIVITY

Learning about Your Local Geography

Group Project: Research

Plan, organize, and complete a research project on the economic activities in your community. Assign one person each to research one of four main kinds of economic activities that are found in your community. Check your local library or chamber of commerce to find information. Try to learn about the different jobs that people do and which activities employ the most people in your community. Then create a pie graph that estimates the percentages of people working in primary, secondary, tertiary, and quaternary activities. What did you learn from your research?

internet connect

Internet Activity: go.hrw.com
KEYWORD: SW3 GT6

Access the Internet through the HRW Go site to create a data profile of one country from each of the six populated continents. Your data profile should include information from the following categories: people, economy, environment, technology, and trade. Use information from the World Bank to create your data profile.

PORTFOLIO ACTIVITY

Graphs will vary depending on economic activities in your community. Use Rubric 7, Charts, and Rubric 14, Group Activity, to evaluate student work.

Point out to students that one of the most useful features of geographic information systems is the ability to zoom in to view small areas of a map or zoom out to view large areas. Ask students to recall what they learned about map scale earlier in this unit. Then ask why GIS users might want to view one map at several different scales. *(Possible answer: Different scales allow the viewer to obtain different types of information. A map that shows a small area will show fine details that are not represented on larger maps, like the locations of specific landmarks. A map that shows a large area will show patterns of physical features that often are not visible when viewing a small area.)*

To demonstrate to students the usefulness of this feature, have them locate an online map service that allows users to zoom in or out. Have them examine a map that shows your community at a medium scale. Ask students to describe what they can see on the map. Then direct students to zoom in on a particular area of the community and describe how the map's content changes. *(Students should note that more precise details and labels become visible as they zoom further.)* Have students return to the original view of the map so they can zoom out. Ask how the map changes when it shows a larger area. *(Students should note that many details are lost, but the community's location within your state or region becomes clearer.)* Lead a discussion about reasons people might have for using maps of various scale in their daily lives.

Geography
Skill-Building Workshop

WORKSHOP

Using Geographic Information Systems

As you read in Chapter 1, a geographic information system is a special computer system. A GIS stores, displays, and maps locations and their features. Geographers can use GIS software to examine relationships among our planet's physical and human features. The software helps researchers find patterns that may help solve complex problems. For example, geographers can use a GIS to identify areas that are favorable for growing crops. Map layers may show areas with ample water, rich soils, and favorable terrain for farming. Once a GIS merges those layers, good locations for farming become easier to identify. A GIS can also be helpful in urban planning. A planner might use a GIS to help plan an urban park or locate areas ideal for business, industry, or residential use. All of the information that a researcher needs to do such work can be translated into map layers. Combining those map layers gives the researcher a more complete image of the characteristics of the area under study. With the development of powerful personal computers, GIS technology has become more accessible. Insurance companies, farmers, local governments, health departments, and schools now use GIS technology to solve problems and to display spatial information.

Developing the Skill In this workshop you will practice making a simple GIS plan with graph paper and transparency overlays. However, a real GIS plan is a complex process. Such a plan requires special software and sets of data. The software is the tool that tells the computer what to do. The software must have a database program to manage the information, a spreadsheet program to process the data, and a programming language to write functions for the data. It must also have the ability to zoom in and out and to pan across the computer screen. For example, a researcher might choose to zoom in, or enlarge, a small area of a map. Doing so will let a researcher see more

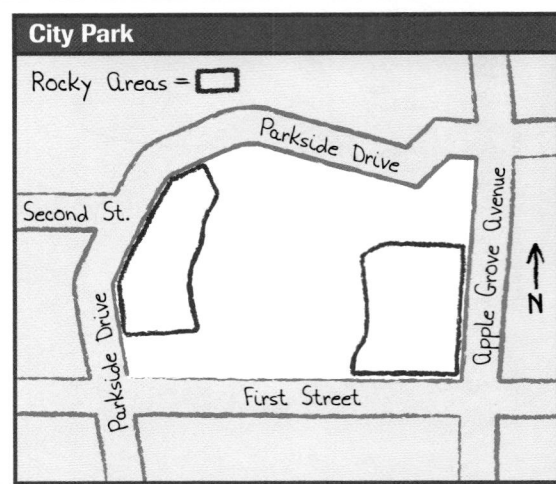

detail. At other times a researcher might want to zoom out, or view a larger area in less detail. Doing so will help the researcher see how a small place fits into surrounding areas.

The geographic data is the fuel for the whole process. For example, census data might tell a researcher the average personal income in different areas of a city. Data can also be gathered from field work. Consider how a park planner might use GIS to create or redesign a park. The planner will want to observe the terrain and other physical features of areas throughout the park. For example, that planner will want to identify areas with rocky terrain. (See the illustration.) He or she will also want to identify areas with existing vegetation, easy access to neighboring streets, and broad open spaces. That and other information can be displayed in map layers. By putting those layers together, the planner can decide which areas are best suited for gardens, parking, picnics, recreation, and other uses.

The most important part of GIS technology is the geographer or researcher using it. Because there is a risk of using too much data, the wrong data, or different scales of data, the geographer must be careful. For example, when using a GIS, it is very important that all the data provided is at the same scale and projection. In other words, each map layer should show the same area at the same scale.

Going Further: Thinking Critically

Tell students that many people besides geographers use geographic information systems. Have students examine the GIS they created in the exercise below. Then ask how the following groups of people might all use the information in the GIS for various purposes: city planners, landscapers, park rangers, and visitors. *(Possible answers: City planners might use GIS technology to identify the best location for a new park. Landscapers could use it to identify factors such as water resources and contours that might affect the planting and growth of flowers, grasses, and trees. Park rangers could use the GIS to study different areas of the park to decide which activities should be allowed in each area. Visitors might use information from the GIS to* decide which areas of the park they would most like to see.) Have each group choose one of these groups or another group that could benefit from the GIS and write a brief proposal explaining how the GIS could be used. *(Possible answer: Students might argue that a GIS could be installed in a park visitors' center and configured so that tourists could press a button to highlight areas near water, picnic areas, and so forth.)*

Practicing the Skill

Now it is your turn to practice using a GIS. Work with a group of other students to design a new park in your community or redesign an existing park.

The tools you will need include graph paper, five transparency overlays, and transparency markers of different colors and width. You will also need a map of an existing park or another map of your community that you can use to find the best spot for a new park.

Create a base map for your park, using a black marker to show only park boundaries. You can then photocopy that base map onto your transparency overlays. Each overlay will become a layer in a detailed map of your park site.

Working with your group, study the park site you have chosen. Each group member should have a different task. Each will research the park site and create a map overlay based on his or her observations. Your group should create four map overlays in all:

• The first map layer should use one color to show the location of water sources, such as creeks, ponds, rivers, or streams.
• The second map layer should use one color to show areas with rocky terrain.

• The third map layer should use colors to mark areas with thick tree growth, scattered trees, or mostly grass.
• The fourth map layer should use one color to show contour lines. In this project, the contour lines do not have to be completely accurate. Simply use lines to identify flat, gently sloping, or steeply sloping areas. You might want to review the discussion of contour lines in Chapter 1. Because you could have many contour lines, you might want to analyze this map layer separately from the others.

When the map layers are completed, use an overhead projector to place the layers on top of each other. By doing this, the group should be able to identify areas that are best suited for gardens, picnics, parking, hiking trails, and other uses. In some cases you may need to use only one or two layers to identify areas suited for certain functions.

Then work with group members to develop a final "park plan." That plan should include a map of your park and a written summary. The summary should describe why different functions were assigned to specific areas within the park. Gather all of your materials into a formal plan folder and make a presentation to your class.

The Geography of Terrorism

OBJECTIVES

1. Explain what terrorism is and describe different kinds of terrorist tactics.

2. Describe terrorist activities that have occurred throughout the world.

3. Analyze the future of terrorism, as well as the international efforts to combat terrorism.

Bombing in Buenos Aires, Argentina

The top image on this page is from the car bombing of the Israeli embassy in Buenos Aires, Argentina. In the attack, which was traced to Hezbollah, 29 people were killed, and 242 were injured. Hezbollah claimed that the bombing was in retaliation for an Israeli attack that killed a leader of Hezbollah and his family.

Have students contrast this image with the bottom photo, the destruction of the World Trade Center in New York City. You might wish to remind students that on September 11, 2001, one single day, more people were killed than in any other single terrorist attack on any nation.

The Geography of Terrorism

On September 11, 2001, just 19 men, 15 of them from Saudi Arabia, used simple weapons—small knives and razor blades—to hijack four airplanes. These men then used the planes to kill themselves and almost 3,000 innocent people, most of whom died at the World Trade Center in New York City. This horrible act set off a chain of events that resulted in two wars, thousands of additional deaths, and the destruction of billions of dollars worth of property.

How could so few, ill-equipped men bring about so much death and destruction? What cause could have been so important that they were willing to give up their lives

Above: Rescuers respond after a terrorist bombing of the Israeli embassy in Buenos Aires, Argentina. Below: The World Trade Center, September 11, 2001

to kill people they did not know? Why did these men target the buildings they targeted? What has been the effect of their actions on the way people live their lives?

To answer these questions is to take a closer look at terrorism, which has existed throughout history and occurs today at alarming rates in many regions of the world. Because it impacts the way people live around the globe, terrorism is an obvious topic of study for geographers.

ALL LEVELS: Before students enter the classroom, write the following definitions of terrorism on the chalkboard:

Terrorism is "premeditated (planned) politically motivated violence against noncombatant targets by subnational groups, usually with the goal of influencing people's opinions." Terrorism is "the calculated use of violence or the threat of violence to inculcate (instill) fear; intended to coerce (force) or to intimidate governments or societies in the pursuit of goals that are generally political, religious, or ideological."

Have students copy the two definitions. Ask them to write their own definitions, and have volunteers share these definitions with the class.

Guide the class in a discussion of what terrorism is, how it differs from civil insurrection, and reasons that groups resort to terror as a means of achieving their goals.

What Is Terrorism?

Terrorism is not an easy term to define. Different governments, international agencies, and political groups define it in very different ways. Even the U.S. State Department, FBI, and the Department of Defense rely on different definitions.

To complicate matters, whether or not an act is considered terrorism can depend on a person's perspective—that is, which side he or she is on. Members of rebel groups using violence to overthrow what they believe is an unjust government would probably describe themselves as revolutionaries or guerrillas. Those opposed to their cause, however, would call them terrorists.

For example, in 1987 a group called the Contras killed over 6,000 civilians in its attempts to overthrow the government of Nicaragua. Government officials in Nicaragua referred to the Contras as terrorists. U.S. President Ronald Reagan, who supported the Contras, called them "freedom fighters."

While definitions and perspectives may differ, most people today would agree that terrorism is criminal activity involving the use of violence, or the threat of violence, to promote a particular cause. Terrorists' targets are usually innocent civilians, and the terrorists' motives are usually political, although some experts

The terrorist group Hamas claimed responsibility for the suicide bombing of this bus in Jerusalem.

believe that more and more terrorist acts are religiously motivated. In many cases the political and religious motives overlap.

Acts of terrorism are usually, but not always, carried out by individuals or groups not affiliated with a government. In fact, throughout history, terrorism has been the weapon of the weak—those who lack political and military power. As we know, however, these "weak" terrorists have the power to cause enormous suffering.

Terrorist Weapons and Tactics

How do individuals or small groups go about creating terror? In ancient times they used daggers, swords, and vials of poison. Today they still use knives and

poisons, but they also use powerful bombs, automatic weapons, and grenade launchers.

Among the most used terrorist methods are bombings, hijackings, the taking of hostages, and assassinations. Commitment and resourcefulness are also important. In creating so much damage with such simple weapons the 9/11 terrorists demonstrated the effectiveness of dedication to a cause and careful planning.

Bombings Unfortunately terrorists have found that the materials and know-how needed to make a bomb are easy to acquire. Once

U.S. President Ronald Reagan supported the Contras in Nicaragua (at left), whom he considered to be "freedom fighters." The government of Nicaragua saw the Contras as terrorists.

Linking Past to Present

Nicaragua and the Contras

Following the overthrow of the Somoza government in Nicaragua, the Contras, a counterrevolutionary group, was formed to oppose Nicaragua's Sandinista government. The first Contra groups organized in Honduras and then in Costa Rica. President Ronald Reagan believed that the Sandinistas were establishing a Communist regime in Nicaragua and gave the CIA authority to support the Contras. Following a legal suit that Nicaragua filed in the World Court against the United States, direct U.S. military aid for the Contras was stopped.

In 1988 successful mediation efforts by Central American governments led to a ceasefire between the Nicaraguan government and the Contras. In 1990, internationally monitored elections were held, and they were won by an anti-Sandinista coalition. Free elections were held in 1996 and 2001; in both of these elections, the Sandinistas were defeated.

CRITICAL THINKING: Why were the Contras willing to agree to mediation and internationally supervised elections?
(By agreeing to mediation, the Contras achieved a peaceful resolution to the political problems in Nicaragua.)

ALL LEVELS: In the 1960s and 1970s, airplane hijackings were a frequently used terrorist tactic. According to the U.S. Department of Transportation, between 1968 and 1972, 364 airplanes were hijacked worldwide. While not all of these hijackings can be considered directly related to international terrorism, many were.

One tactic that terrorists may choose in the future is information warfare. It is a growing threat, and one that has raised concern in both economic and military fields.

According to the U.S. Department of Defense, information warfare "involves actions taken to achieve information superiority by affecting adversary information, information-based processes, information systems, and computer-based networks while defending one's own information, information-based processes, information systems, and computer-based networks." The U.S. Department of Defense has expressed serious concerns that organized terrorists could use information warfare techniques to disrupt military operations by harming command and control systems and other key military and battlefield components.

Terrorism and the Media

Terrorists rely on the media to spread the word of their actions. While it is the job of the media to broadcast reports about terrorism and terrorist attacks around the world, the coverage enables terrorist groups to instill even more fear and possibly gain support for their activities. This tactic, however, does not always work.

Immediately following the events of 9/11, global condemnation of the tragedy was tremendous. Around the world, people reacted immediately and with sympathy. The headline of the French newspaper *Le Monde* proclaimed in response to that day, "We are all Americans!"

Remind students that partly as a result of the international outrage and media coverage of the tragic events of 9/11, the United States had little difficulty recruiting international support for its efforts to remove the Taliban from power in Afghanistan. Troops from 26 nations were involved in the war effort, and these coalition forces moved quickly and forcefully to remove the Taliban from power in Afghanistan, root out al Qaeda leaders, and destroy their training bases.

You might wish to create an ongoing bulletin board of media coverage of terrorist events around the world and discuss these events with students as they occur.

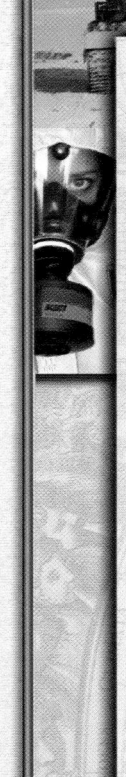

the bombs have been assembled, they can be strapped to a suicide bomber's body, loaded into cars or trucks and driven to the target site, or simply hidden along the side of a road.

Terrorists have been able to make powerful bombs. For example, an October 2002 nightclub bombing killed over 200 people on the Indonesian island of Bali. In 1983, Islamic Jihad terrorists detonated a bomb inside a U.S. Marine compound in Lebanon, killing almost 300 people.

Hostages and Hijackings

Taking hostages and hijacking airplanes are ways in which just a few terrorists can impact thousands, or even millions, of lives. At times, terrorists attempt to trade hostages for the release of other terrorists who have been captured and imprisoned. At other times, they attempt to trade hostages for safe passage to a desired location or for a ransom payment.

At the 1972 summer Olympics in Munich, Germany, a Palestinian terrorist group took several Israeli athletes hostage. As a worldwide TV audience watched the crisis unfold, the terrorists achieved their primary goal: publicity for their cause.

Cells A tactic that has helped terrorists avoid capture is their ability to maintain secrecy by working in cells. A cell is a very small group of terrorists who do not know the names or locations of other terrorists in the same organization. A network of such cells, in which only certain leaders know how to communicate with other cells, makes it difficult for counterterrorism personnel to identify and track down terrorists.

Media Coverage In today's world, terrorists plan and execute their acts to create maximum fear and to ensure widespread media exposure. Sometimes they target government or military installations, but often they attack civilians engaged in common, everyday activities, such as riding a bus or eating in a restaurant.

Terrorists often target symbols of the government, culture, or religion that they oppose. The 9/11 terrorists chose the World Trade

In the 1950s and 1960s, the National Liberation Front waged a violent campaign against the French for Algerian independence. The group is credited with developing the use of cells within terrorist organizations to protect the identity of members.

Center as a target because it symbolized the powerful U.S. economy that, in the terrorists' view, was damaging their religion and culture. The 9/11 terrorists also knew that the attack would generate massive fear and publicity.

The Beginning of Modern Terrorism

Experts debate the beginning of modern terrorism. Forced to pick a date, many would pick July 22, 1968, when three terrorists belonging to a sub-group of the

Two nightclubs on the beach in Bali, Indonesia, were bombed in October 2002. Rescuers helped search for survivors, but more than 200 people were killed in the blasts.

Information warfare may also take the form of providing propaganda to an enemy in an attempt to persuade enemy forces to give up.

Guide students in a discussion of the threat of information warfare. You might wish to have interested students research and report on information warfare and compare it to traditional terrorist tactics.

Have the students research and report on the Oklahoma City National Memorial that was built to honor the victims of the 1995 bombing of the Murrah Federal Building. Students might wish to compare and contrast this memorial to the one at the World Trade Center site in New York City.

Palestine Liberation Organization (PLO) hijacked an Israeli commercial jet.

Although this was not the first hijacking of an airplane, it was the first that had a purpose beyond the terrorists' rerouting the plane to take them to a chosen destination. These hijackers had several political reasons for the hijacking, including trading their hostages for fellow terrorists who were in Israeli jails. Publicity was also a goal.

PLO terrorist marches in funeral for suicide bomber.

Through such terrorist acts, and others involving bombings and assassinations, the PLO became the model and the trainer for many of today's international terrorists. The PLO trained terrorists for other organizations at its training camps in Jordan and Lebanon. It also made the raising of money to fund its activities a priority, another first for modern terrorist organizations. From the late 1960s to the 1980s, the PLO laid the foundation for much of today's terrorism.

International and Domestic Terrorism

The development of highly sophisticated communication and other technology has led to an increase in international terrorism, which differs from domestic terrorism.

Newspaper readers in India learn of the capture of Saddam Hussein, who practiced state terrorism in Iraq for nearly 25 years before being toppled from power. Although terrorists usually crave publicity, these are not the type of headlines they have in mind.

International terror occurs across national boundaries or involves people from multiple countries. Al Qaeda, the terrorist group that committed the 9/11 atrocities, commits violent acts all over the world, and therefore is an international terrorist group.

Terrorists who operate wholly within the boundaries of a particular country and target only citizens of that country are domestic terrorists. The Ku Klux Klan, for example, is a domestic terrorist group that promotes racism in the United States.

State or State-Sponsored Terrorism

A variation of the types of terrorism described up to this point is a particular kind known as state or state-sponsored terrorism.

We use the term *state terrorism* when discussing governments that rule through the use of fear. Leading up to and following World War II, Adolph Hitler in Germany and Joseph Stalin in the Soviet Union created terrorist states to obtain and hold onto political power.

Other, more recent, examples include the governments of Argentina (1976–1983), Cambodia (1976–1979), and Iraq (1979–2003), which used torture, assassination, and mass killings to terrorize their citizens.

Other countries stop short of actually committing acts of terror, but they do sponsor terrorism. This means that they provide sanctuary and protection for terrorists. They also may supply money and other resources to help the terrorists commit their violent acts.

Countries such as Afghanistan, Iran, Sudan, and Syria have supported terrorist groups in recent years. Doing so enabled these weaker countries to secretly attack their much stronger enemies with less risk of retaliation than would be the case in an open declaration of war.

Think About It

1. How would you define terrorism? Do you believe that civil wars should be classified as examples of terrorism?

2. Should the media give the amount of coverage they do to terrorist groups? Or does such coverage encourage groups to commit further acts of terror?

History

State and Domestic Terrorism

State terrorism has a long history. Examples from recent times include the following: Stalin and the NKVD in the Soviet Union were responsible for the deaths of as many as 25–30 million people; Hitler and the Gestapo in Nazi Germany killed between 6–10 million people; Mao Zedong in Communist China, 5 million; Pol Pot and the Khmer Rouge in Cambodia, 1 million; Idi Amin, Uganda, 100,000; Augusto Pinochet in Chile, 3,000.

Domestic terrorism is that which is perpetrated within the nation by citizens of the nation. The 1995 bombing of the Alfred P. Murrah Federal Building in Oklahoma City, Oklahoma, by Timothy McVeigh and Terry Nichols is considered domestic terrorism. Ask students to cite other incidents of domestic terrorism in the United States and around the world.

Answers to Think About It

1. Student answers will vary but should include that violence and destruction are used. Civil wars are sometimes classified as terrorism.
2. Students may argue that the job of the media is to provide unbiased coverage of events. Media coverage can hurt terrorists, but it provides them significant worldwide attention.

Jihad

Jihad is an Arabic word that means either "combat" or "striving." Many consider it a fundamental principle of Islam, and jihad is sometimes referred to as the "sixth pillar of Islam." Most manuals on Islamic law include chapters on jihad, and waging war on infidels is considered jihad. Most Islamic scholars agree on the basic definition, that only a war that has an ultimate religious purpose can be termed as jihad.

A number of modern acts of terrorism have been declared to be an expression of jihad. For some militants within the Islamic sphere, one who commits suicide while fighting oppression is considered to be a holy martyr. Many Muslims, however, do not agree with this view and consider killing a civilian as immoral and unjust.

Much has been written about Islam and jihad. You might wish to have interested students research the topic and report their findings to the class.

Classifying Terrorist Groups

Before students read the information about Global Terrorism Today, have them make a 3-column chart. The first column should be labeled *economics;* the second column, *religion;* and the third column, *nationalism.* As students read the information on these pages, have them classify each group mentioned in the text and complete their charts. Have students share their charts with the class and discuss why they chose to classify the groups as they did.

SMALL GROUP ACTIVITY:

Divide the class into six groups, one for each of the geographic regions listed on these pages. Ask each group to research recent terrorist activity in their assigned region, as well as efforts to combat terrorism. Have each group present its findings to the entire class. Presentations should include visuals, including maps and charts, that illustrate the information presented in the report.

Global Terrorism Today

Economics, religion, and nationalism motivate terrorism around the globe today. Here are some examples.

United States   Prior to the events of September 11, 2001, Americans were more concerned with domestic terrorism than international terrorism. The most devastating single act of domestic terrorism was the bombing of a federal office building in Oklahoma City in 1995, which resulted in the deaths of 168 people. The key terrorist involved in the bombing, Timothy McVeigh, was apparently motivated by anti-government, anti-gun-control beliefs that he shared with others in militia and white supremacist groups.

Middle and South America
Most of the terrorist activity in Latin America has been economically motivated. Terrorists have attempted to portray their activity as a justified means of fighting against poverty and what they believe has been an uneven and unfair distribution of wealth. Another major motive for terrorism is the economic and political frustration of American Indians, whose ancestors built great civilizations in Latin America centuries ago. Today, these American Indians are among the poorest inhabitants of the region and often lack any real political power.

In recent years Colombia has experienced extreme violence, as members of the Revolutionary Armed Forces of Colombia and the National Liberation Army of Colombia have assassinated government and military leaders and destroyed symbols of what they see as their exploitation by the upper class and foreign businesses. The use of drug trafficking to provide funding for these terrorist activities has made matters worse. In next-door Peru, the Sendero Luminoso, or Shining Path, group has also been waging a long, violent campaign against the government.

Europe   A variety of causes have motivated terrorists in modern-day Europe. Perceived discrimination against ethnic and religious minorities is foremost among them. Such concerns have been cited for decades by the Irish Republican Army (IRA) as justification for the bombings and assassinations they have committed in their fight for Northern Ireland's independence from Britain. Another terrorist group, ETA, uses violent attacks in its campaign to force Spain to allow the Basque region of the country to secede.

Russia   One of the most unusual terrorist groups is known as the Black Widows. It is made up of female suicide bombers who have lost husbands, brothers, or fathers in the efforts of largely Muslim Chechnya to gain independence from Russia. In October 2002, the Black Widows were among a group of 41 terrorists who took

Investigators examine the remains of a car bombing in Bogotá, Colombia, in October 2003. The bombing was one of a series that occurred just prior to national elections.

In April 1995, Timothy McVeigh blew up the Alfred P. Murrah Federal Building in Oklahoma City, Oklahoma, killing 168 people. Following the bombing, a police car in Oklahoma reflected the feelings of many Americans.

Weapons of Mass Destruction

The Nuclear Non-Proliferation Treaty (NPT), signed in 1968 by 188 nations, restricted the possession of nuclear weapons to the United States, the United Kingdom, France, Russia, and the People's Republic of China. These five nations possessed nuclear weapons when the treaty was adopted. These five countries are also the five permanent members of the United Nations Security Council, and they have agreed not to transfer nuclear weapons technology to other nations. Pakistan, India, and Israel also possess nuclear weapons. In addition, North Korea allegedly announced to U.S. diplomats that it possessed several nuclear weapons.

In 1988 Iraq used chemical warfare on Iraqi Kurds, killing 200,000 in that year alone. Following Iraq's defeat in the Persian Gulf War, a UN coalition forced Iraq to agree to an inspections and monitoring regime to ensure that the country had dismantled its WMD program. Arms inspections continued until 1998. In 2002, the UN resumed its arms inspection program in Iraq, but no WMDs were found. Nor were WMDs found in Iraq by the U.S. Iraq Survey Group following the ouster of Saddam.

Have students research and report on efforts to control nuclear proliferation and have them prepare maps showing the countries that are suspected of either possessing or developing nuclear weapons or other WMDs.

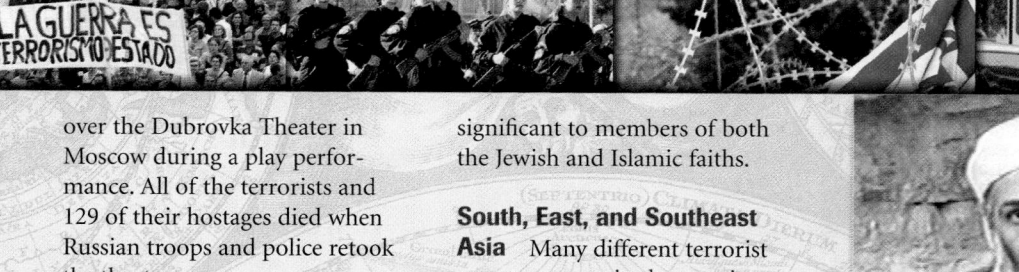

over the Dubrovka Theater in Moscow during a play performance. All of the terrorists and 129 of their hostages died when Russian troops and police retook the theater.

Southwest Asia The countries in Southwest Asia are home to more terrorist groups than the countries of any other region. At the heart of terrorist activity in the region is the conflict between the state of Israel and the Palestinians, who feel they have been unfairly expelled from their homeland. As you read earlier, terrorist activities carried out by the PLO served as the foundation of much of today's terrorism around the world.

Terrorist groups based in the region include Hamas, Fatah, the Palestinian Islamic Jihad, and Hezbollah. While these groups fight for territory and Palestinian statehood, their conflict with Israel is also based on deep religious beliefs and the importance of controlling religious sites that are significant to members of both the Jewish and Islamic faiths.

South, East, and Southeast Asia Many different terrorist groups operate in these regions. The Liberation Tigers of Tamil Eelam (LTTE) group in Sri Lanka seeks independence from Sri Lanka for the Tamil ethnic group. LTTE members have used extensive suicide bombings to further their cause.

The Aum Shinrikyo is a religious cult based in Japan whose goal is to take control of Japan first and then the world. Members of the group released deadly sarin gas in the Tokyo subway in 1995, killing 12 people and injuring over 5,000. The group has members in many countries around the world.

The largest terrorist group in the Philippines is the Moro Islamic Liberation Front (MILF). It is an Islamic fundamentalist group with reported ties to Osama bin Laden. Its aim is to establish a separate Islamic state in the Philippines.

Al Qaeda leader Osama bin Laden

Al Qaeda and the War on Terrorism

The international terrorist organization al Qaeda and its leader, Osama bin Laden, became household words after the attacks of September 11, 2001. No prior terrorist act had ever been as destructive or as widely publicized. How did al Qaeda become in just a few years the most dangerous terrorist group in the world? Many factors contributed to the rise of the organization, but most significant was its founder, Osama bin Laden.

Osama bin Laden was born into a wealthy family in Saudi Arabia around 1957. Over the years he had developed a fanatical hatred of the United States. As a strict Muslim fundamentalist, he saw America as morally corrupt and in violation of many Islamic laws. He was outraged that U.S.

Members of Japan's Self-Defense Forces respond to reports of a suspicious odor at a subway station near Tokyo in April 1995. Security personnel were on high alert following the subway gas attack carried out by Aum Shinrikyo the previous month.

History of Al Qaeda

Al Qaeda is an Arabic word meaning "the base." The al Qaeda terrorist organization was founded in the late 1980s by Osama bin Laden and others. Al Qaeda brought together Arabs who had fought against the Soviet Union in the war in Afghanistan. Its mission was to oppose non-Islamic governments through the use of force and violence. One of al Qaeda's principal goals was to drive U.S. armed forces out of Saudi Arabia and Somalia. Al Qaeda had ties to other terrorist organizations, including Hezbollah and the al Jihad in Egypt. Al Qaeda, which found a home in Afghanistan when it was ruled by the Taliban, was a well-financed, well-organized organization with cells throughout the Middle East, Africa, and Europe.

Eurasia

Have students examine the map and chart on this page carefully. Guide them in a review of the attacks and casualties that have occurred in each of the regions shown on the map and chart. Ask students to identify which has had the most attacks. *(Latin America)*. Ask students to suggest possible reasons why Latin America has had almost twice as many terrorist attacks as any other region. Then ask students to identify the region that has had the most deaths due to terrorist attacks. *(Africa)*. Again, ask students to suggest reasons why.

Some students may not be familiar with the term *Eurasia*. If this is the case, you might wish to have students read the Unit 5 opening on pages 370–371 and have them examine the maps in the Unit 5 Atlas, pages 372–377 of the text.

Point out to students that because of cultural differences between Europe and Asia, they have been considered separate continents. Some earth scientists, however, consider Eurasia to be a continent in and of itself. The U.S. Department of State considers that the following nations comprise Eurasia: Armenia, Azerbaijan, Belarus, Georgia, Kazakhstan, Kyrgyzstan, Moldova, Russia, Tajikistan, Turkmenistan, Ukraine, and Uzbekistan.

You might ask interested students to research and report on data for the current year. Compare and contrast this data with the information in the chart. Is the number of terrorist attacks increasing or decreasing? Which region or which countries have been subject to the most terrorist activity?

Guantánamo Bay, Cuba

In 1898, during the Spanish-American War, U.S. Marines established camp at the harbor known as Guantánamo Bay. The United States acquired the land from Spain at the end of the war. In 1903 President Theodore Roosevelt was able to obtain a permanent lease on the land from Cuba; Franklin Delano Roosevelt ordered that the base be expanded in 1939. It has served as a maintenance port for U.S. ships and as a facility to house Cuban and Haitian refugees intercepted at sea while trying to reach the United States. The base was also used as a prison site for suspected al Qaeda and Taliban members.

Answer to Think About It

Student answers will vary, but some students may argue that when people are able to earn a decent living, they are less likely to rebel.

U.S. soldiers escort a terrorist suspect to his prison cell at Guantánamo Bay, Cuba.

troops operated military bases in Saudi Arabia, home of the sacred shrine of Mecca. He was also infuriated by American support for Israel. His answer to what he saw as these intolerable U.S. actions was to declare a holy war, or jihad, against the United States and its supporters.

After September 11, 2001, U.S. President George W. Bush declared a war on terrorism. The United States assembled an international coalition of forces and attacked Afghanistan, the location of bin Laden's terrorist training camps. Many members of al Qaeda were captured and sent to a prison at the American naval base at Guantánamo Bay, Cuba.

As of early 2004, however, Osama bin Laden was still directing terrorist actions from a hideout, probably in the mountains of Afghanistan along the border with Pakistan. Al Qaeda remained a potent force with cells in more than 30 countries.

Think About It

If the governments of Middle and South America were able to resolve the economic problems that affect their nations, would the terrorist groups operating in those nations disband? Explain your answer.

The Future of International Terrorism

In the early years of the twenty-first century, terrorism affects the lives of millions of people around the globe. Some are affected much more than others. Citizens in Israel and Colombia, for example, live with the real and constant fear that they, or their loved ones, could be in the wrong place at the wrong time, becoming innocent victims of a suicide bomber or an assassin's stray bullet.

People in other parts of the world are safer, but still experience various degrees of inconvenience, restricted lifestyles, or a loss of certain civil liberties as a result of terrorism. Government buildings are barricaded and closely guarded, limiting access to government resources. Flights are cancelled if government agents obtain evidence of possible attacks. Delays

International Terrorist Attacks and Casualties, 1998–2002

Attacks	
Africa	167
Asia	387
Eurasia	90
Latin America	676
Middle East	135
North America	6
Western Europe	189

Casualties	
Africa	5,828
Asia	4,161
Eurasia	738
Latin America	283
Middle East	1,462
North America	4,091
Western Europe	451

North America
Latin America
Africa
Western Europe
Middle East
Eurasia
Asia

U.S. Department of State

As students read about ways to combat international terrorism, you might wish to share some of the remarks made by UN Secretary-General Kofi Annan at a September 2003 conference, "Fighting Terrorism for Humanity: A Conference on the Roots of Evil."

"Terrorism is a global threat, and it can *never* be justified. No end can give anyone the right to kill innocent civilians. On the contrary, the use of terrorism to pursue any cause — even a worthy one—can only defile that cause, and thereby damage it…

If we are to defeat terrorism, it is our duty, and indeed our interest,…to carefully examine what works, and what does not, in fighting it…

We also delude ourselves if we think that military force alone can defeat terrorism…

To fight terrorism, we must not only fight terrorists. We have to win hearts and minds. And we can only do all this effectively if we work together through multilateral institutions.…And we will not hand terrorists a victory, but a stinging rejection, both of their methods and their world view."

Organize a class debate based on Kofi Annan's statement: "We also delude ourselves if we think military force alone can defeat terrorism.…We have to win hearts and minds."

---

at airports become longer and more frequent.

Dirty Bombs Looking ahead, many counterterrorism experts are concerned about terrorists using dirty bombs, which are a combination of conventional explosives and some type of radioactive material. Although not as deadly as nuclear weapons, dirty bombs are considered easier for terrorists to produce. If assembled and detonated, however, dirty bombs could disperse deadly radioactive dust into the atmosphere over a large city, causing untold deaths and illnesses.

Weapons of Mass Destruction
Many worry that terror attacks will become even more deadly in the years to come. Of special concern are weapons of mass destruction (WMDs). The United States invaded Iraq in March 2003 primarily because Saddam Hussein was thought to have stockpiles of WMDs. Although no WMDs were found in Iraq, other concerns have surfaced. For example, the world learned in February 2004 that Pakistan's leading nuclear scientist, Abdul Qadeer Khan, controlled a network that sold plans, expertise, and even components for making nuclear weapons to countries such as Iran, Libya, and North Korea.

A Positive Outlook While some paint a bleak picture for the future, others point to signs that the world community can solve its terrorism problems. Leaders such as

U2's Bono (center) joins hands with John Hume (left) and David Trimble (right) at a rally in support of the "Agreement," the historic peace accord worked out between opposing factions in Northern Ireland.

Abdul Qadeer Khan

David Trimble and John Hume, who shared the 1998 Nobel Peace Prize, helped negotiate peace accords that have all but eliminated the terrorist activity that has gone on for decades in Northern Ireland and England. In early 2004, Libyan leader Mu'ammar Gadhafi, after years of engaging in state-sponsored terrorism, announced to the world that he would begin dismantling his nuclear weapons program. In addition, in February 2004 a U.S. counterterrorism official estimated that two-thirds of al Qaeda's leaders had been killed or captured since 9/11.

Of course it will take all of the nations of the world, working together, to bring an end to the violence and destruction caused by international terrorism. Because of the tragic events of 9/11, more resources, and more international cooperation than ever before have been directed toward ending terrorism around the world.

Think About It
What steps can the international community take to stop the development of WMDs? the spread of terrorism?

These drawings for the new tower, surrounding buildings, and memorial to be built at the World Trade Center site in New York City symbolize the world community's determination to defeat terrorism.

According to the U.S. Nuclear Regulatory Commission, the conventional explosive used in a dirty bomb would result in more immediate deaths than the radioactive material that would be released. Certain radioactive materials could, however, contaminate several city blocks.

A second type of dirty bomb might involve a radioactive source placed in a public place, where people passing by might get a significant dose of radiation.

While the NRC does not diminish the dangers of a dirty bomb, they do refer it as a weapon of mass disruption, not a weapon of mass destruction.

Student answers will vary but may include the necessity to share information, to share military resources, to organize international monitoring groups, and to provide financial aid to other countries to help track groups that might be purchasing nuclear weapons or technology and supplies to produce weapons of mass distruction.

Using the Illustration

Direct students' attention to the photo of the Statue of Liberty. Ask volunteers to describe what they think about or feel when they see the statue. *(Possible answers: national pride, freedom, democracy, welcome, immigration)* For thousands of immigrants to the United States the Statue has been a symbol of liberty and the chance for a new life.

The statue "Liberty Enlightening the World," as it is formally known, was conceived by French journalist and politician Édouard-René Lefebvre de Laboulaye as a symbol of friendship between the United States and France. Funding for its construction was provided by the people of France. Frédéric-Auguste Bartholdi was chosen to design and build the statue. It was at his suggestion that it was placed on an island in New York Harbor, where so many immigrants caught their first glimpse of the United States. The Statue of Liberty was unveiled on October 28, 1886.

Ask students to identify other structures or sites that have special meaning to the people of the United States or Canada. Ask them what these places symbolize to the countries' people. Then ask what they might represent to people in other parts of the world.

Unit Objectives

1. Describe the major land-forms, bodies of water, climates, and resources found in the United States and Canada.

2. Examine the historical and cultural geography of the United States and Canada.

3. Analyze the changes that are occurring in the economies of the United States and Canada.

4. Learn how to use mental maps to organize information about places.

5. Understand how models are used to analyze or interpret geographic data.

UNIT 2

The United States and Canada

*The Statue of Liberty
Liberty Island, New York*

Chapter **7** *Natural Environments of North America*

Chapter **8** *The United States*

Chapter **9** *Canada*

CONNECTING TO *Literature*

"LOO-WIT"* *by Wendy Rose*

Wendy Rose
(1948–) is a poet and an artist. Rose's poems reflect her Hopi heritage and focus on environmental themes. Before she began writing, she made up songs to express herself. Rose notes that "it has always been a tradition of my people [the Hopi of Arizona] to celebrate everything in life with song. For all occasions, both happy and sad, I sang my feelings."

The way they do
this old woman
no longer cares
what we think
but spits her black tobacco
any which way
stretching full length
from her bumpy bed.
Finally up
she sprinkles ash on the snow,
cold and rocky buttes
that promise nothing
but winter is going at last.
Centuries of cedar
have bound her to earth,
huckleberry ropes
lay prickly about her neck. . . .
Around her
machinery growls,
snarls and ploughs
great patches of her skin.
She crouches
in the north,
the source
of her trembling–
dawn appearing
with the shudder
of her slopes.

Blackberries unravel,
stones dislodge;
it's not as if
they weren't warned.

She was sleeping
but she heard the boot scrape,
the creaking floor;
felt the pull of the blanket
from her thin shoulder.
With one free hand
she finds her weapons
and raises them high;
clearing the twigs from her throat
she sings, she sings,
shaking the sky like a blanket
   about her
Loo-wit sings and sings and sings!

---

* Loo-Wit *is a name that the Cowlitz tribe gave to Mount St. Helens, a volcano in Washington State. Mount St. Helens erupted violently in 1980 after being inactive for more than 120 years.*

Analyzing the Primary Source

1. **Supporting a Point of View**   Why is this image of an elderly woman awakening from her sleep appropriate for a volcano such as Mount St. Helens?

2. **Analyzing**   How does Loo-Wit warn of the impending eruption? Which of Loo-Wit's actions represents the eruption itself?

A Closer Connection

Ash from Mount Saint Helens   On the morning of May 18, 1980, Mount Saint Helens shot a blast of gas and ash—the "black tobacco" of Wendy Rose's poem—more than 15 miles (24 km) into the air in less than 15 minutes. Blasts of air from the side of the mountain also pushed the ash cloud outward for more than 12 miles (19 km) from its peak. Prevailing winds picked up some 520 million tons (472 million metric tons) of ash and carried it eastward across the United States. This dust cloud was thick enough to cause complete darkness in Spokane, Washington, more than 250 miles (402 km) away. Some ash from the eruption of Mount Saint Helens was carried as far east as central Montana, a distance of more than 500 miles (804 km).

Analyzing the Primary Source
Answers

1. Possible answer: Just as a person who has been asleep begins the day sleepily and then becomes more active, the volcano's eruption began with trembling and became more violent.

2. trembling, shuddering; spitting tobacco, raising weapons, clearing throat, singing

139

In this unit, students will learn about the tremendous geographical differences within the United States and Canada and what the two countries have in common.

The landscape of the United States is immensely varied, from high mountains to flat coastal plains. There are also many climates represented. The diverse population reflects the many cultures that have played a role in the country's history. Although agriculture in the United States is highly productive, few people live on farms. Most of the population lives in cities and suburbs. Workers in the United States are part of the most technologically advanced economy in the world. The country's wealth is based on its many natural resources, free enterprise economy, and democratic government.

Many landforms extend from the United States into Canada, the second-largest country in the world. Our northern neighbor has a colder climate and a much smaller population. Most Canadians also enjoy a high standard of living. France and Britain influenced Canada's culture. Asians form another important immigrant group. In addition, the native Inuit have recently gained greater control over their traditional homelands.

Your Classroom Time Line

These are the major dates and time periods for this unit. Have students enter them on the time line you created earlier. You may want to watch for these dates as students progress through the unit.

c.* 12,000 B.C. The ancestors of Native Americans cross into North America from Asia.

c. 1000 The building of cliff dwellings begins in what is now Mesa Verde National Park.

c. A.D. 1000–1350 Vikings explore and settle the eastern shores of Canada.

c. 1300 The cliff dwellings of Mesa Verde are abandoned.

c. 1500 Spanish explorers reach the American mainland. English and French explorers soon follow.

1500s–1700s Europeans establish colonies in North America.

1530s Jacques Cartier explores the St. Lawrence River area.

1607 Jamestown, the first permanent English settlement in America, is founded.

*c. stands for *circa* and means "about."

POLITICAL MAP ANSWERS

1. California, Arizona, New Mexico, Texas
2. Manitoba, Ontario, Quebec

CRITICAL THINKING ANSWER

3. States in the east are smaller. The colonies from which eastern states were created were relatively small. In addition, many western states contain large areas that have harsh climates and sparse populations, so few people live there.

UNIT 2 ATLAS

The World in Spatial Terms
The United States and Canada:
Political

1. **Places and Regions** Which U.S. states share a border with Mexico?
2. **Environment and Society** Compare this map to the physical map. In which provinces is the Canadian Shield found?

Critical Thinking

3. **Comparing** How do the size of states in the eastern United States compare with those in the west? Why do you think this is?

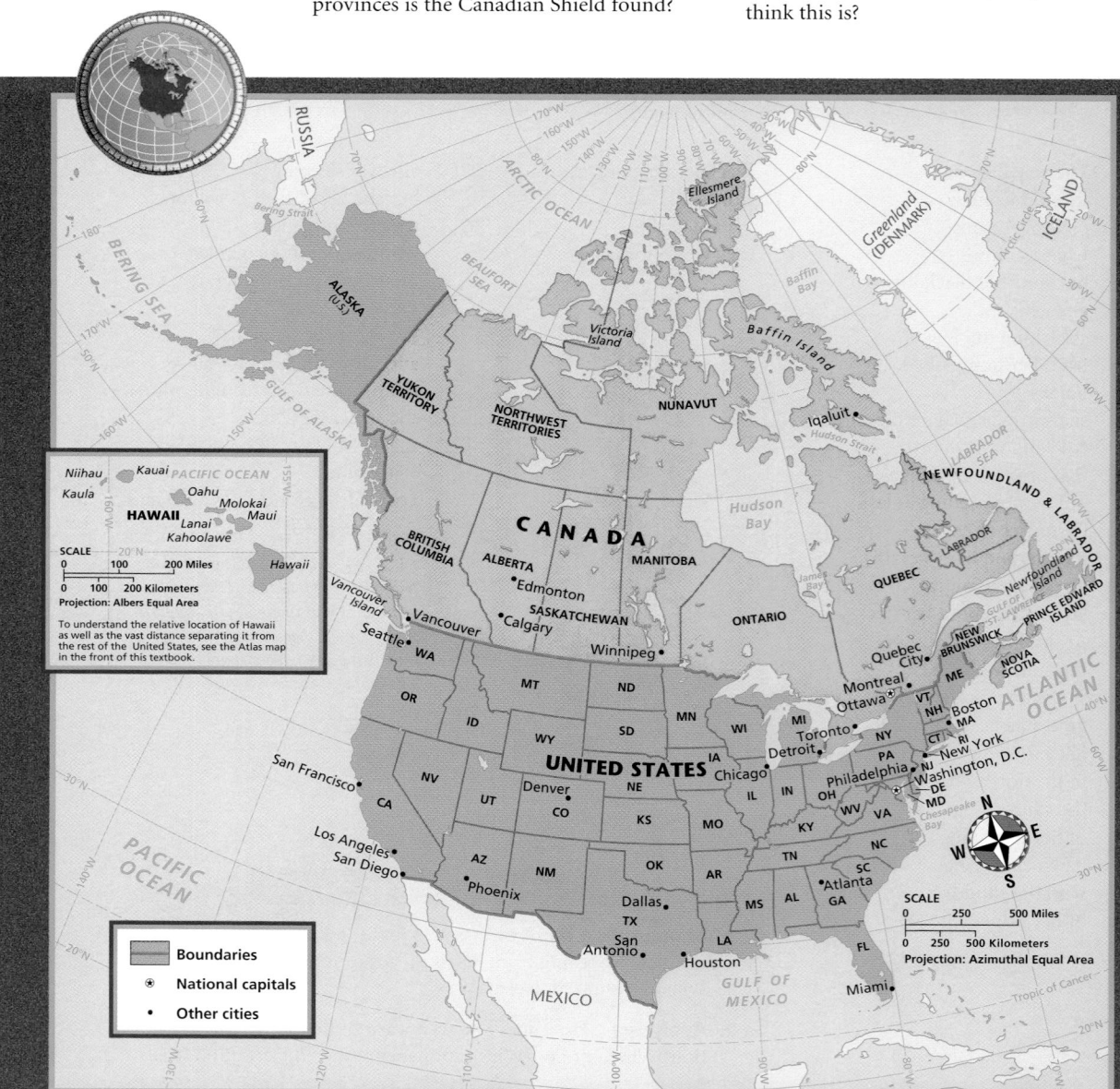

Boundaries
⊛ National capitals
• Other cities

HAWAII inset: Niihau, Kauai, Kaula, Oahu, Molokai, Lanai, Maui, Kahoolawe, Hawaii
PACIFIC OCEAN
SCALE: 0 100 200 Miles / 0 100 200 Kilometers
Projection: Albers Equal Area

To understand the relative location of Hawaii as well as the vast distance separating it from the rest of the United States, see the Atlas map in the front of this textbook.

SCALE: 0 250 500 Miles / 0 250 500 Kilometers
Projection: Azimuthal Equal Area

Using the Political Map

Have students use the **political map** on the opposite page to locate their home state. Ask them to compare it to the **physical map** to describe the state's landforms, rivers, lakes, and sea-coasts. Call on volunteers to describe those features in a few sentences. (*Example: Michigan lies on two peninsulas, which are surrounded by Lakes Superior, Michigan, and Huron. The land is fairly flat, except for hills in the northern peninsula's southeast region.*)

Ask students to name the capitals of the United States (*Washington, D.C.*) and Canada (*Ottawa*).

Using the Physical Map

Focus students' attention on the **physical map** on this page. Have each student write a sentence that describes a physical feature in terms of its relative location. (*Examples: The Labrador Peninsula is in eastern Canada between Hudson Bay and the Labrador Sea. The Rocky Mountains are in western North America. The Great Plains are east of the Rocky Mountains.*) Call on volunteers to read their sentences. Continue until all the major physical features have been located.

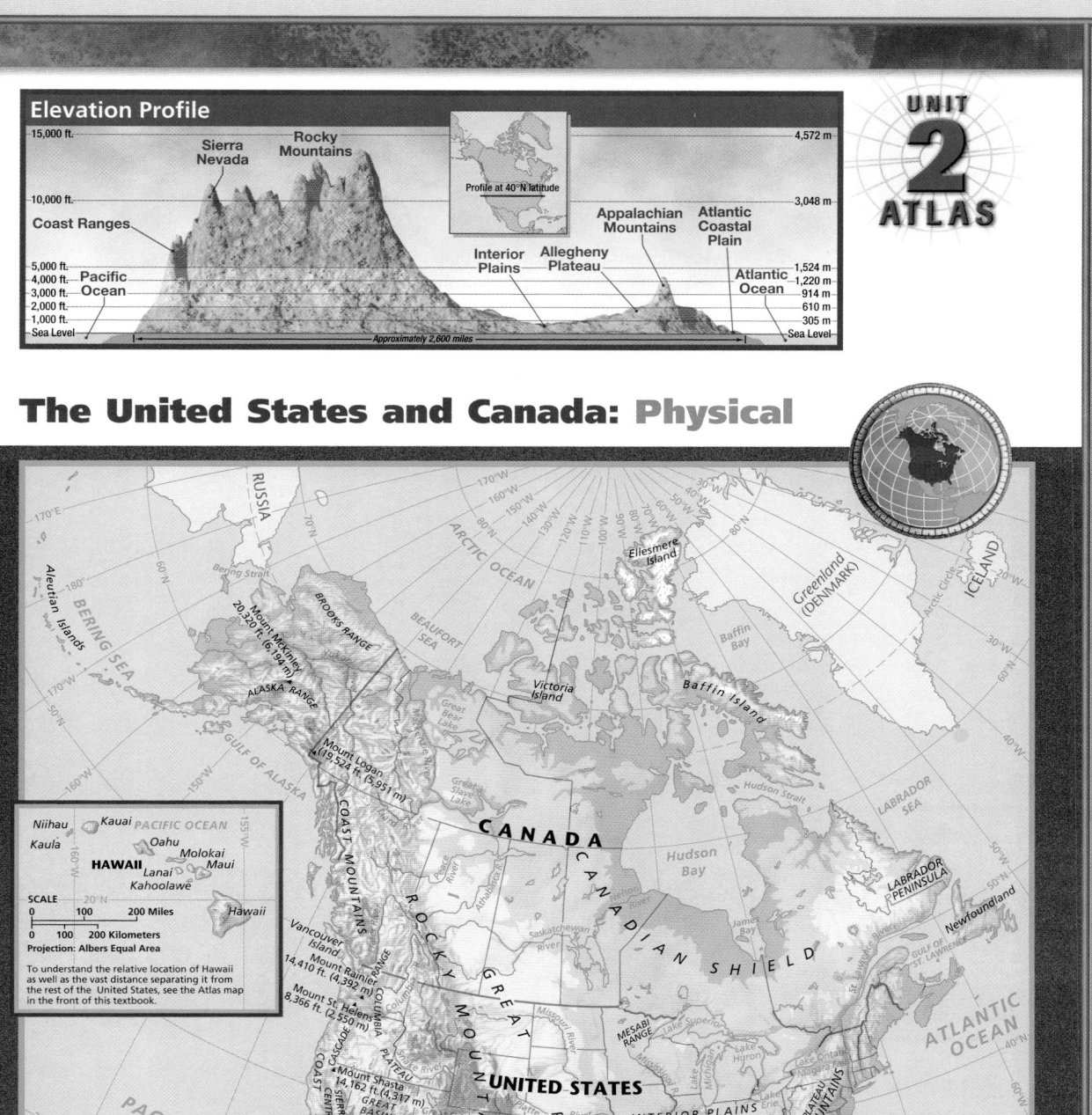

Elevation Profile

The United States and Canada: Physical

UNIT
2
ATLAS

Your Classroom Time Line, (continued)

1608 The first permanent French settlement in America, which will become Quebec City, is founded.

1619 The first enslaved Africans are brought to the American colonies.

1620s Dutch settlers establish New Amsterdam, later to become New York.

1713 The British take over Nova Scotia.

1756–63 The Seven Years' War, known in North America as the French and Indian War, is fought. After the war, the English take control of all of Canada.

1764 St. Louis is founded by the French.

1776 The 13 American colonies declare independence from Great Britain.

1781 Los Angeles is founded by the Spanish.

1790 Construction of Washington, D.C., begins.

1800s Americans and Canadians begin to move westward.

1800s Immigration to Canada increases.

1803 The Louisiana Purchase nearly doubles the size of the United States.

1825 The Erie Canal is completed, allowing easy access to the Great Lakes region from the American Northeast.

1830s The northeastern United States industrializes.

1841 Dallas, Texas, is founded.

1846 With the acquisition of the Oregon Territory, the northern boundary of the United States reaches the Pacific Ocean.

1848 The discovery of gold in California triggers further westward expansion.

1861–65 The U.S. Civil War is fought between the North and the South.

141

Direct students' attention to the **climate map** of the United States and Canada on this page. Ask students which climate type covers most of Canada's territory *(subarctic)* and which climate regions of the United States are not found in Canada *(tropical humid, tropical wet and dry, arid, Mediterranean, humid subtropical)*.

Have students compare this map to the **population map** in this atlas. Call on volunteers to identify major cities and the climate region in which they are located. *(Examples: Dallas—humid subtropical; Los Angeles—Mediterranean; Phoenix—arid; Toronto—humid continental; Vancouver—marine west coast)* Then ask students which climate regions have generally low population densities *(tundra, subarctic, arid, semiarid, highland)*. Ask students to explain why these regions might have lower populations. *(Extreme temperatures can make life uncomfortable. A lack of precipitation makes it difficult to grow food. Rugged terrain makes travel difficult.)*

Your Classroom Time Line, (continued)

1865 Vancouver is founded.

1867 The British Parliament creates the Dominion of Canada.

1867 The United States purchases Alaska from Russia.

1869 The Transcontinental Railroad, which connects the eastern United States with the West, is completed.

1870s Manitoba, British Columbia, and Prince Edward Island become provinces.

1872 Congress establishes Yellowstone National Park.

1885 The transcontinental Canadian Pacific Railroad is completed.

1898 Hawaii is annexed by the United States.

1905 Alberta and Saskatchewan become provinces.

1913 The completion of an aqueduct into Los Angeles allows the city to grow rapidly.

1914 Canada enters World War I.

1917 The United States enters World War I.

1918 World War I ends.

1930s Drought hits the southern Plains, creating the Dust Bowl.

1930s The United States suffers through the Great Depression.

1939 Canada enters World War II.

CLIMATE MAP ANSWERS

1. Rocky Mountains, Sierra Nevada, Cascade Range

2. Colder climates are found at higher latitudes.

CRITICAL THINKING ANSWER

3. Most Canadians live in southern Canada, where climates are warmer.

UNIT 2 ATLAS

The United States and Canada:
Climate

1. *Physical Systems* Compare this map to the physical map. Which major mountain ranges appear to influence the distribution of highland climate areas?

2. *Physical Systems* How do you think latitude influences the distribution of climates in Canada?

Critical Thinking

3. Analyzing Compare this map to the population map. How do you think climate has influenced the distribution of Canada's population?

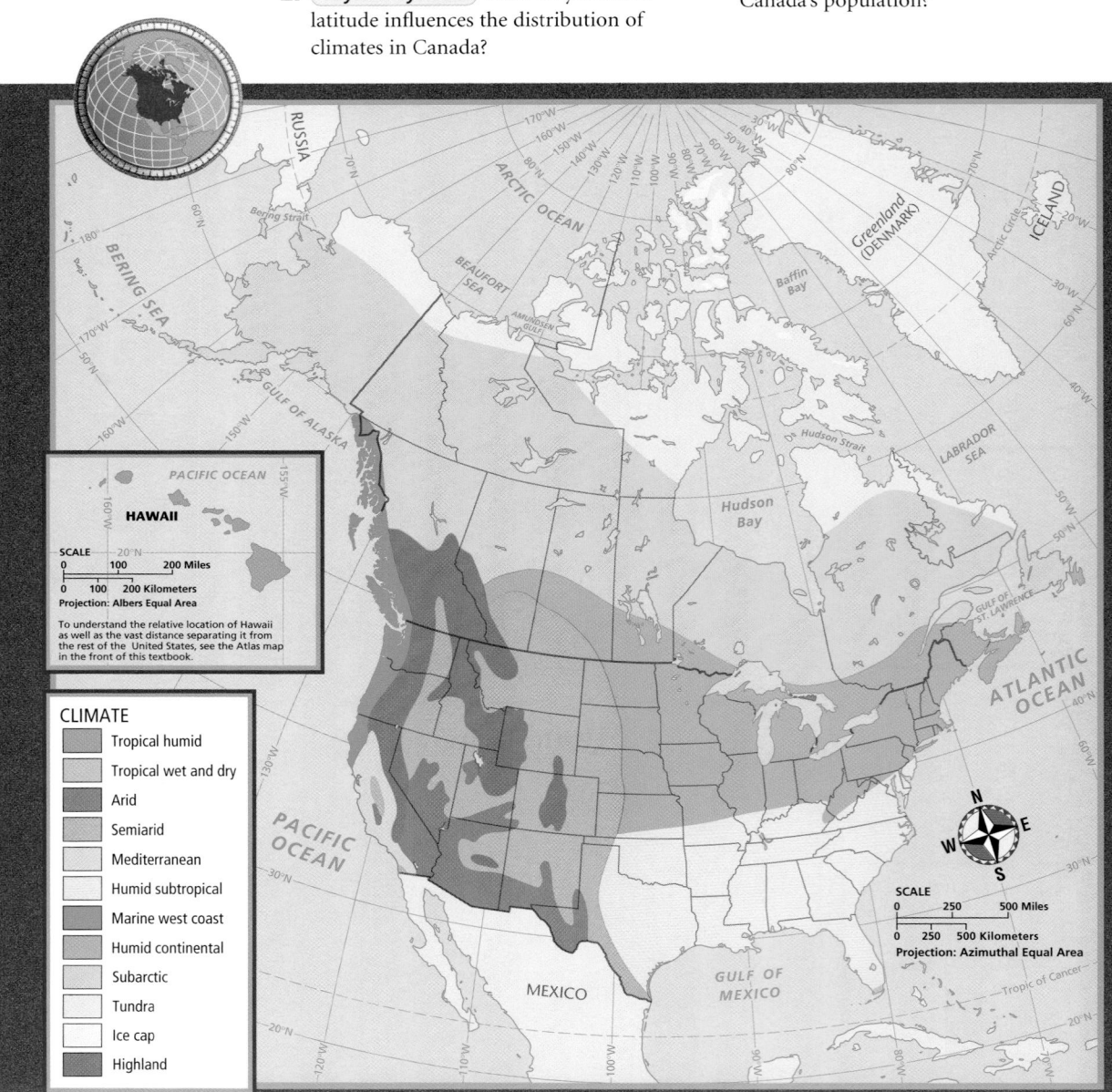

CLIMATE
- Tropical humid
- Tropical wet and dry
- Arid
- Semiarid
- Mediterranean
- Humid subtropical
- Marine west coast
- Humid continental
- Subarctic
- Tundra
- Ice cap
- Highland

HAWAII

SCALE
0   100   200 Miles
0   100   200 Kilometers
Projection: Albers Equal Area

To understand the relative location of Hawaii as well as the vast distance separating it from the rest of the United States, see the Atlas map in the front of this textbook.

SCALE
0   250   500 Miles
0   250   500 Kilometers
Projection: Azimuthal Equal Area

Focus students' attention on the **precipitation map**. Ask students to identify the regions of the United States and Canada that receive the most rainfall *(Hawaii; coastal Northwest United States and western Canada; parts of the southern United States).* Ask students to use the **climate map** to identify the climate regions that include these areas *(tropical humid; marine west coast; humid subtropical).* Then have students use the **physical map** and their knowledge of climate patterns to suggest why these areas receive more precipitation than other parts of the continent. *(Possible answers: Coastal locations receive moisture from neighboring bodies of water. Mountain ranges in the Pacific Northwest block moisture from moving inland, so the coastal sides of the mountains receive more rain.)*

The United States and Canada:
Precipitation

UNIT 2 ATLAS

1. (**Physical Systems**) Compare this map to the climate map. Which two climate regions receive the most precipitation?

2. (**Places and Regions**) Which region in the United States receives the lowest amount of precipitation?

Critical Thinking

3. **Making Generalizations** Compare this map to the climate map. How much precipitation falls in tundra climates? Why do you think these high-latitude climates have such low precipitation?

Your Classroom Time Line, (continued)

1941 Japan bombs Pearl Harbor, Hawaii. The United States enters World War II.

1945 World War II ends.

1949 Newfoundland and Labrador become a province.

1959 Alaska and Hawaii become states.

1959 The St. Lawrence Seaway is completed, linking the Great Lakes to the Atlantic Ocean.

1960s Migration to the American South increases.

1990s Congress passes laws to protect the environment.

1991 The Soviet Union collapses, marking the end of the Cold War.

1992 The United States, Canada, and Mexico sign the North American Free Trade Agreement (NAFTA), to take effect in 1994.

1995 The people of Quebec decide in a close vote to remain a part of Canada.

1999 Nunavut is created as a homeland for the Inuit.

2001 Terrorists hijack American airliners and crash them into the World Trade Center in New York and the Pentagon in Washington.

2004 NASA successfully lands two exploration rovers, *Spirit* and *Opportunity*, on Mars.

ANNUAL PRECIPITATION

Centimeters	Inches
Under 25	Under 10
25–50	10–20
50–100	20–40
100–150	40–60
150–200	60–80
Over 200	Over 80

PRECIPITATION MAP ANSWERS

1. marine west coast, humid subtropical
2. the Southwest

CRITICAL THINKING ANSWER

3. less than 10 inches; because cold air holds less water

143

While students examine the **population map** on this page, ask them to locate cities they have lived in, visited, or heard about from movies, television, or sports events. You might ask them to tell what they know about those cities.

Ask students which country has the largest area where practically no one lives (*Canada*) and which of the U.S. states has a similarly uninhabited area (*Alaska*). Have students compare this map to the **political map** to determine which of the U.S. states appear to have the lowest overall population densities (*Alaska and Nevada*). Ask also which region within the United States appears to have the highest overall population density (*northeast coast*).

POPULATION MAP ANSWERS

1. The United States is much more densely populated and has more big cities.
2. Toronto, Montreal, Vancouver

CRITICAL THINKING ANSWER

3. Early settlers traveled to the United States by sea. In addition, coastal cities have increased opportunities for trade and, therefore, for growth.

UNIT 2 ATLAS

The United States and Canada:
Population

1. (**Human Systems**) How are the population patterns of Canada and the United States different?
2. (**Places and Regions**) What are Canada's largest cities?

Critical Thinking

3. **Analyzing** Why do you think so many large U.S. cities are located on the coast?

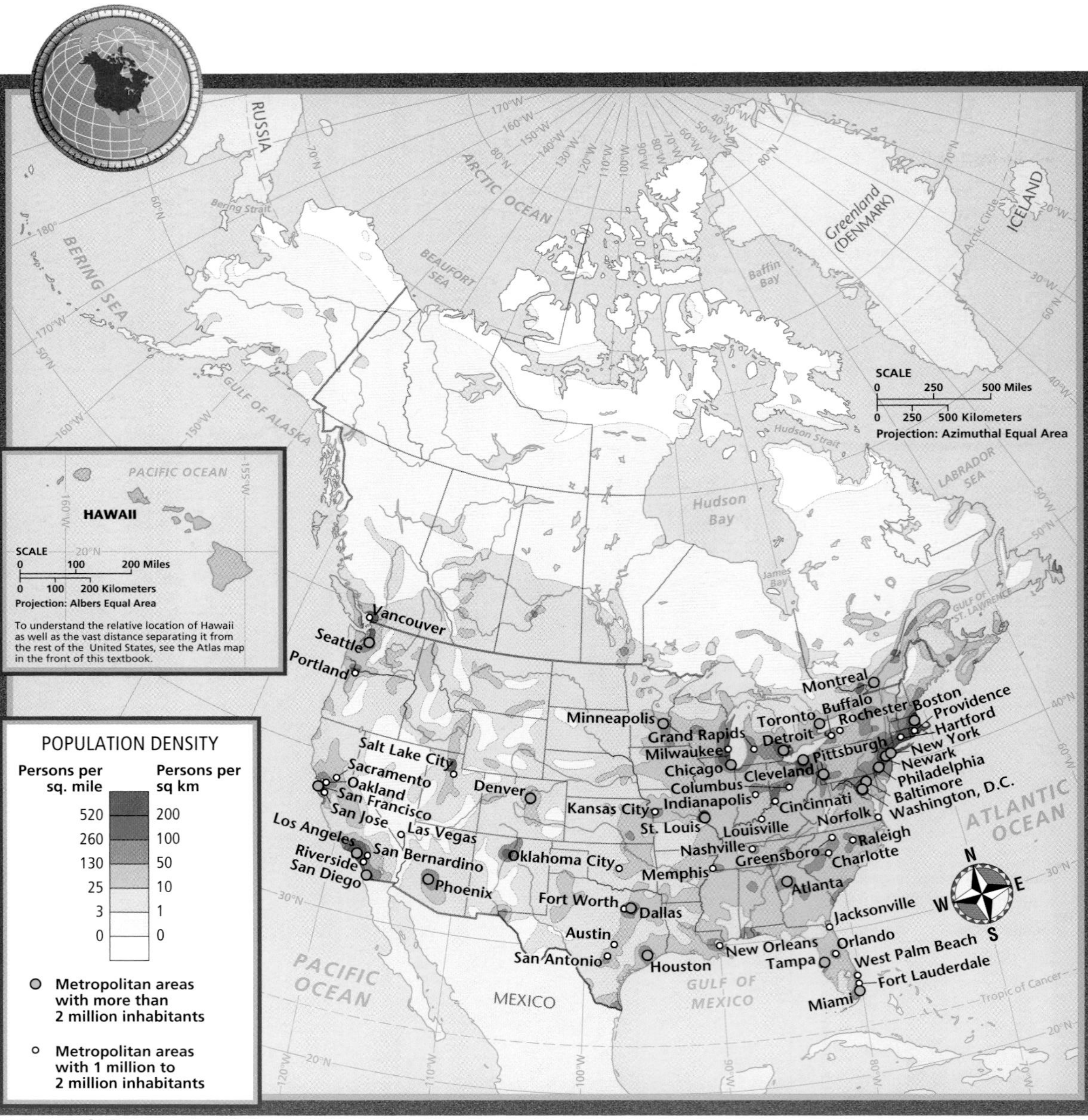

POPULATION DENSITY

Persons per sq. mile	Persons per sq km
520	200
260	100
130	50
25	10
3	1
0	0

○ Metropolitan areas with more than 2 million inhabitants

○ Metropolitan areas with 1 million to 2 million inhabitants

HAWAII

SCALE
0   100   200 Miles
0   100   200 Kilometers
Projection: Albers Equal Area

To understand the relative location of Hawaii as well as the vast distance separating it from the rest of the United States, see the Atlas map in the front of this textbook.

SCALE
0   250   500 Miles
0   250   500 Kilometers
Projection: Azimuthal Equal Area

Have students locate the state in which they live on the **land use and resources map**. Ask them to identify the economic activities and resources that are found in their state. Then ask students to use the map to identify economic patterns in the rest of the United States and Canada. Have each student write a sentence to summarize one of these patterns. Have students use other unit maps to help them write their sentences. *(Examples: There are more nuclear power plants in the eastern United States than in the western part of the country. Large parts of Canada have limited economic activity. Coal mining is common in the Appalachian Mountains.)*

The United States and Canada:
Land Use and Resources

1. (*Environment and Society*) Compare this map to the climate map. In which climates is timber production found?
2. (*Places and Regions*) Where in North America is dairy production an important economic activity?

Critical Thinking

3. **Analyzing** Compare this map to the physical map. In what physical environments is commercial farming found? Why do you think this is?

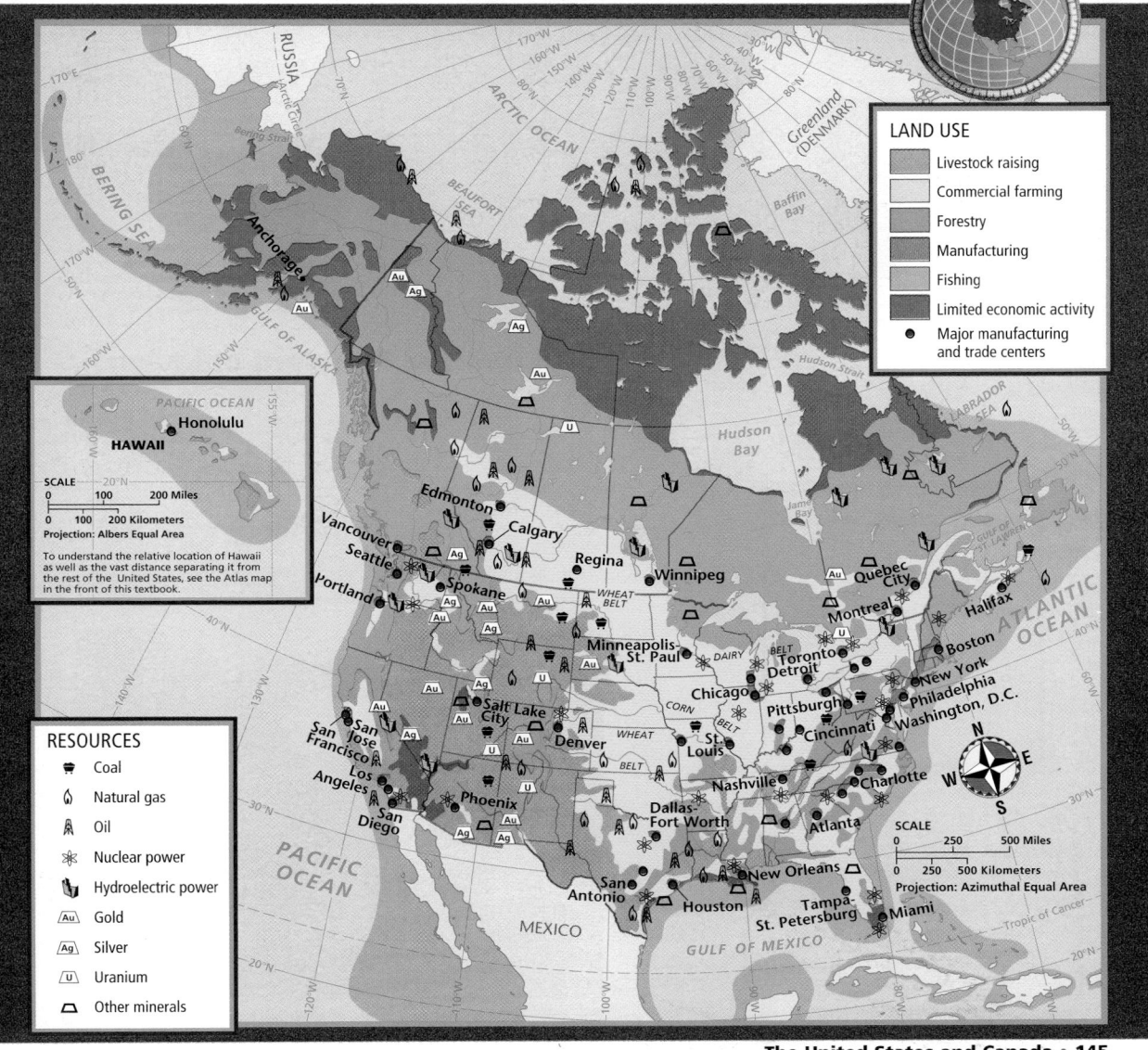

LAND USE

- Livestock raising
- Commercial farming
- Forestry
- Manufacturing
- Fishing
- Limited economic activity
- ● Major manufacturing and trade centers

RESOURCES

- Coal
- Natural gas
- Oil
- Nuclear power
- Hydroelectric power
- Au Gold
- Ag Silver
- U Uranium
- Other minerals

HAWAII

SCALE
0   100   200 Miles
0   100   200 Kilometers
Projection: Albers Equal Area

To understand the relative location of Hawaii as well as the vast distance separating it from the rest of the United States, see the Atlas map in the front of this textbook.

SCALE
0   250   500 Miles
0   250   500 Kilometers
Projection: Azimuthal Equal Area

Fast Facts Activity

Call students' attention to the area figures given for the United States and Canada. Ask students to compare the two countries in terms of size. *(Though Canada is larger, the two countries are fairly similar in size.)* Then have students focus on the population figures for the two countries. Ask them to calculate how many times larger the U.S. population is than Canada's. *(Divide the U.S. population total by Canada's population total. The result is about 9.3, which can be rounded up to 9.5.)* Ask students how their findings are reflected in the two countries' population densities. *(Because the United States has many more people in about the same area, its population density is much higher than Canada's.)*

Call on students to suggest reasons for the differences between the populations of the United States and Canada. You may wish to have them consult the climate and population maps in the unit atlas for reference. *(Most of Canada is covered by subarctic and tundra environments. Few people live in these areas. On the whole, the United States has milder climates.)*

Fast Facts Activity

Copy the following chart onto the chalkboard, omitting the italicized figures. Have students complete the chart by calculating the percentage of each country's land area occupied by the listed economic activities. Then have students use the completed chart, the Fast Facts tables, and the unit atlas to answer

Historical Geography

The War of 1812 Although the United States and Canada share the world's longest unguarded border, the exact location of that border has not always been fixed. In fact, it took a war to firmly establish the boundary between the two countries.

In the early 1800s, Americans and British Canadians came into conflict over possible uses for territory west of the Great Lakes. Tension over this issue was a major factor in the outbreak of the War of 1812. The two countries fought for more than two years without a clear winner. However, the Treaty of Ghent that ended the war called for the creation of a boundary commission to confirm the border between the two countries. This commission upheld a boundary set in 1783 that ran from the St. Lawrence River and the Great Lakes to the northwest corner of the Lake of the Woods. In 1818, the boundary was extended along the 49° N latitude line to the Rocky Mountains. The Oregon Territory, which lay west of the mountains, was controlled jointly by the United States and Great Britain until 1846.

DISCUSSION: Lead a discussion about how conflict and cooperation shaped the territories controlled by each country.

Time Line: The United States and Canada

UNIT 2 ATLAS

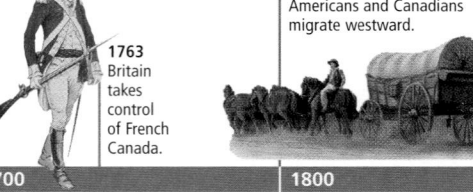

1500s–1700s Europeans establish colonies in North America.

1763 Britain takes control of French Canada.

1800s Americans and Canadians migrate westward.

1867 The British government creates the self-governing Dominion of Canada.

1914–18 Canada fights in World War I. The United States enters the war in 1917.

2001 On September 11, terrorists attacked the World Trade Center and the Pentagon.

1500 **1700** **1800** **1900** **2000**

1776 The 13 American colonies declare independence.

1848 Gold is discovered in California.

1869 The first transcontinental railroad is completed in the United States.

1939–45 Canada fights in World War II. The United States joins the war in 1941.

1991 The Cold War between the United States and the Soviet Union ends.

1861–65 The Civil War takes place in the United States.

Comparing Standard of Living

COUNTRY	LIFE EXPECTANCY (in years)	INFANT MORTALITY (per 1,000 live births)	LITERACY RATE	DAILY CALORIC INTAKE (per person)
Canada	76, male 83, female	5	97%	3,093
United States	74, male 80, female	7	97%	3,603

Sources: Central Intelligence Agency, *2001 World Factbook; Britannica Book of the Year, 2000*

The United States and Canada

Comparing Sizes

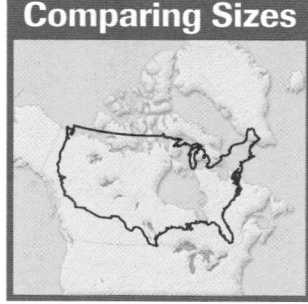

internet connect

GO TO: go.hrw.com
KEYWORD: SW3 Almanac
FOR: Additional information and reference sources

Rainbow Bridge, Lake Powell, Utah

the following questions: In which country is a greater portion of the land suitable for farming? *(United States)* Why do you think this is so? *(Much of Canada is covered with ice for most of the year. In addition, the rocky Canadian Shield is not arable.)* About how much more of Canada is forested than the United States? *(More than 2 million square miles, or 21 percent of its area)* How might the difference in forest cover be explained? *(More areas of the United States that were once covered in forests have been cleared to allow farming or to accommodate its larger population.)* Continue asking similar questions, or challenge students to think of questions of their own.

LAND USE	UNITED STATES	CANADA
Arable land	671,892 sq. mi. 1,740,202 sq km	178,011 sq. mi. 461,049 sq km
% arable	18%	5%
Permanent pastures	884,069 sq. mi. 2,289,740 sq km	106,807 sq. mi. 276,629 sq km
% pasture	24%	3%
Forests and woodlands	1,060,882 sq. mi. 2,747,688 sq km	1,922,517 sq. mi. 4,979,324 sq km
% forest	29%	50%

Source: Central Intelligence Agency, *The World Factbook 2001*

Fast Facts: The United States and Canada

UNIT **2** ATLAS

FLAG	COUNTRY Capital	POPULATION (in millions) POP. DENSITY	AREA	PER CAPITA GDP (in US $)	WORKFORCE STRUCTURE (largest categories)	ELECTRICITY CONSUMPTION (kilowatt hours per person)	TELEPHONE LINES (per person)
	Canada Ottawa	31.5 — 9/sq. mi.	3,851,806 sq. mi. 9,976,132 sq km	$ 29,400	74% services 15% manufacturing	16,008 kWh	0.63
	United States Washington, D.C.	294.0 — 83/sq. mi.	3,717,810 sq. mi. 9,629,084 sq km	$ 37,600	31% manage., prof. 29% tech., sales, admin.	12,250 kWh	0.65

Sources: Central Intelligence Agency, *World Factbook 2003; The World Almanac and Book of Facts, 2004*
The CIA calculates per capita GDP in terms of purchasing power parity. This formula equalizes the purchasing power of each country's currency.

The Grand Canyon National Park from the South Rim.

UNIT **2** Assessment Resources

▶ **Unit 2 Test**

▶ **Unit 2 Test for English Language Learners and Special-Needs Students**

▣ **internet** connect

go. hrw. com

GO TO: go.hrw.com
KEYWORD: SW3 U2
FOR: Web sites about country statistics

Highlights of Country Statistics
Links to online country statistics for the United States and Canada include:
• *CIA World Factbook*
• Library of Congress Country Studies
• Flags of the World

CHAPTER 7

Natural Environments of North America

CHAPTER RESOURCE MANAGER

Objectives	Pacing Guide	Reproducible Resources
SECTION 1 **Physical Features** (pp. 149–53)	• Identify the major landform regions of the United States and Canada. • Locate the rivers and lakes found in the region.	**Regular** 1 day **Block Scheduling** .5 day *Block Scheduling Handbook, Chapter 7*	**RS** Guided Reading Strategy 7.1 **PS** Readings in World Geography, History, and Culture 12 and 13 **SM** Geography for Life Activity 7: Landform Postcard from National Parks
SECTION 2 **Climates and Biomes** (pp. 154–57)	• Identify the climate types found in the United States and Canada. • Name the major biomes of the region and describe where they are found.	**Regular** 1 day **Block Scheduling** .5 day *Block Scheduling Handbook, Chapter 7*	**RS** Guided Reading Strategy 7.2 **SM** Critical Thinking Activity 7: Gray Wolves
SECTION 3 **Natural Resources** (pp. 158–60)	• Describe the farming, forest, and water resources of the United States and Canada. • Analyze the region's supply of energy and mineral resources.	**Regular** 1 day **Block Scheduling** .5 day *Block Scheduling Handbook, Chapter 7*	**RS** Guided Reading Strategy 7.3 **RS** Graphic Organizer Activity 7 **SM** Map Activity 7: Water Resources of North America

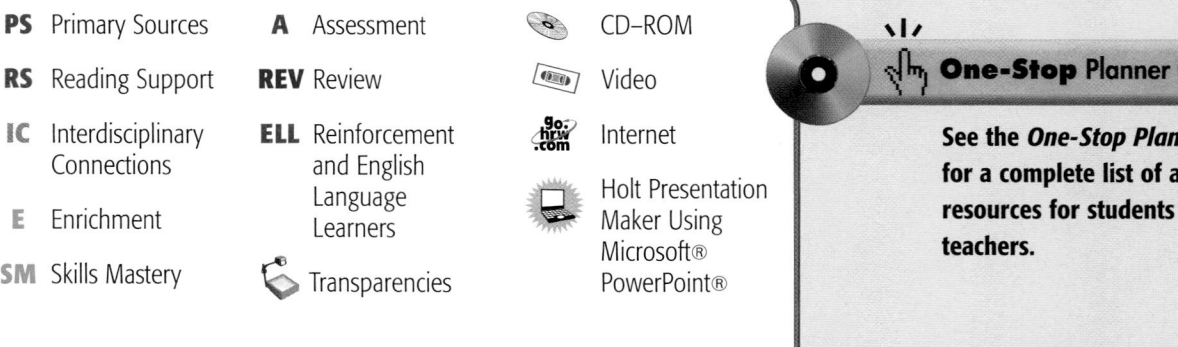

Chapter Resource Key

PS Primary Sources   **A** Assessment   CD-ROM

RS Reading Support   **REV** Review   Video

IC Interdisciplinary Connections   **ELL** Reinforcement and English Language Learners   Internet

E Enrichment   Holt Presentation Maker Using Microsoft® PowerPoint®

SM Skills Mastery   Transparencies

One-Stop Planner CD–ROM

See the *One-Stop Planner* for a complete list of additional resources for students and teachers.

 **One-Stop** Planner CD–ROM

It's easy to plan lessons, select resources, and print out materials for your students when you use the **One-Stop Planner CD–ROM with Test Generator**.

Technology Resources	**Reinforcement, Review, and Assessment**
One-Stop Planner CD–ROM, Lesson 7.1	**ELL** Main Idea Activity 7.1
Geography and Cultures Visual Resources 12	**ELL** English Audio Summary 7.1
CNN. Presents Geography: Yesterday and Today, Segment 2: Birth of an Island	**ELL** Spanish Audio Summary 7.1
Homework Practice Online	**REV** Section 1 Review, p. 153
HRW Go site	**A** Daily Quiz 7.1
One-Stop Planner CD–ROM, Lesson 7.2	**ELL** Main Idea Activity 7.2
Geography and Cultures Visual Resources 14	**ELL** English Audio Summary 7.2
Homework Practice Online	**ELL** Spanish Audio Summary 7.2
HRW Go site	**REV** Section 2 Review, p. 157
	A Daily Quiz 7.2
One-Stop Planner CD–ROM, Lesson 7.3	**ELL** Main Idea Activity 7.3
ARGWorld CD–ROM	**ELL** English Audio Summary 7.3
Geography and Cultures Visual Resources 16	**ELL** Spanish Audio Summary 7.3
Homework Practice Online	**REV** Section 3 Review, p. 160
HRW Go site	**A** Daily Quiz 7.3

☑ internet connect

HRW ONLINE RESOURCES

GO TO: go.hrw.com
Then type in a keyword.

TEACHER HOME PAGE
    KEYWORD: **SW3 Teacher**

CHAPTER INTERNET ACTIVITIES
    KEYWORD: **SW3 GT7**
    Choose an activity to:
    • examine growth and development along the Great Lakes and St. Lawrence River.
    • create a brochure on Hawaii's environment.
    • create a newspaper on the Gulf Coast of Texas.

CHAPTER ENRICHMENT LINKS
    KEYWORD: **SW3 CH7**

CHAPTER MAPS
    KEYWORD: **SW3 MAPS7**

ONLINE ASSESSMENT
    Homework Practice
    KEYWORD: **SW3 HP7**
    Standardized Test Prep
    KEYWORD: **SW3 STP7**
    Rubrics
    KEYWORD: **SS Rubrics**

COUNTRY INFORMATION
    KEYWORD: **SW3 Almanac**

CONTENT UPDATES
    KEYWORD: **SS Content Updates**

HOLT PRESENTATION MAKER
    KEYWORD: **SW3 PPT7**

ONLINE READING SUPPORT
    KEYWORD: **SS Strategies**

CURRENT EVENTS
    KEYWORD: **S3 Current Events**

Meeting Individual Needs

Ability Levels

Level 1 Basic-level activities designed for all students encountering new material

Level 2 Intermediate-level activities designed for average students

Level 3 Challenging activities designed for honors and gifted-and-talented students

English Language Learners Activities that address the needs of students with Limited English Proficiency

Chapter Review and Assessment

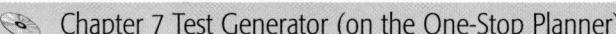 

Chapter 7 Test Generator (on the One-Stop Planner)

Global Skill Builder CD–ROM

HRW Go site

REV Chapter 7 Review, pp. 162–63

REV Chapter 7 Tutorial for Students, Parents, Mentors, and Peers

A Chapter 7 Test (form A or B)

A Alternative Assessment Handbook

A Chapter 7 Test for English Language Learners and Special-Needs Students

Launch into Learning

Call on a volunteer to sing or recite the first verse of "America the Beautiful." As the student sings or recites, write key phrases like *amber waves of grain, purple mountain majesties,* and *fruited plain* on the chalkboard. Ask students what these phrases mean and what they tell us about the physical geography of the United States. *(Possible answer: The United States has fertile farmlands and high mountains.)* Tell students that they will learn more about the physical features of the United States and Canada in this chapter.

Using the Physical-Political Map

Have students examine the map on the opposite page. Ask them to identify the major landforms and water features of the United States and Canada. Then ask them to identify the physical region in which your community is located. Call on volunteers to name places across the country they have visited and the physical regions in which these locations lie.

Why We Should Know More

You might point out these reasons for learning about the physical geography of the United States and Canada:

▶ Understanding the history of the United States and Canada requires a thorough grasp of their physical geography.

▶ Although they may occur in a distant state or province, events such as earthquakes and hurricanes affect people throughout North America. To predict and prepare for such events requires knowledge of landforms and climates.

▶ Both countries have benefited from a wealth of natural resources. For continued prosperity, we need to conserve and to use those resources wisely.

▶ Learning about the geography of the United States and Canada might help students decide where they would like to visit or live in the future.

CHAPTER 7

Natural Environments of North America

Historians think the Norse were the first Europeans to visit North America. The "Groenlendinga Saga" tells of these explorations, including Leif Eriksson's voyage to Vinland—believed to be what is now Newfoundland.

"They went ashore and looked about them. The weather was fine. There was dew on the grass, and the first thing they did was to get some of it on their hands and put it to their lips, and to them it seemed the sweetest thing they had ever tasted. Then they went back to their ship and sailed into the sound that lay between the island and the headland jutting out to the north.

. . . But they were so impatient to land that they could not bear to wait for the rising tide to float the ship; they ran ashore to a place where a river flowed out of a lake. . . . Then they decided to winter there, and built some large houses.

There was no lack of salmon in the river or the lake, bigger salmon than they had ever seen. The country seemed to them so kind that no winter fodder would be needed for livestock: there was never any frost all winter and the grass hardly withered at all.

In this country, night and day were of more even length than in either Greenland or Iceland: on the shortest day of the year, the sun was already up by 9 A.M. and did not set until after 3 P.M."

Prickly pear cactus in Arizona

Irises in Newfoundland, Canada

Black bear in Alaska

LET'S GET STARTED

Copy the following question and instructions onto the chalkboard: *Have you ever visited or seen pictures of America's national parks? Describe the physical features you saw.* Discuss student responses. Point out that the spectacular landscapes of national parks in both the United States and Canada draw millions of tourists from around the world each year. Tell students that in Section 1 they will learn about the variety of physical features found across the United States and Canada.

Building Vocabulary

Write the key terms on the chalkboard and call on volunteers to read the definitions aloud. Ask students to examine the chapter map or a wall map to locate examples of **barrier islands**, a **piedmont**, a **fall line**, and **basins**. Ask students how the phrase **hot spot** relates to its definition. *(Possible answer: Molten magma is located near the surface.)* Ask students which state was created above such a spot *(Hawaii).*

Physical Features

READ TO DISCOVER

1. What are the major landform regions in the United States and Canada?
2. What rivers and lakes are found in the region?

WHY IT MATTERS

Tourists from all over the world are drawn to the beautiful natural landscapes of the United States and Canada. Use **CNNfyi.com** or other **current events** sources to learn how important tourism is to the region's economy.

IDENTIFY

Continental Divide

DEFINE

barrier islands
piedmont
fall line
basins
hot spot

LOCATE

**Piedmont
Appalachian Mountains**

Locate, continued

**Rocky Mountains
Mississippi River
Great Lakes
Great Plains
Canadian Shield
Hudson Bay
Cascade Range
Sierra Nevada
Coast Ranges
St. Lawrence River**

Section 1 RESOURCES

REPRODUCIBLE

▶ Guided Reading Strategy 7.1
▶ Readings in World Geography, History, and Culture 12 and 13
▶ Geography for Life Activity 7: Landform Postcard from National Parks

TECHNOLOGY

▶ One-Stop Planner CD–ROM, Lesson 7.1
▶ Geography and Cultures Visual Resources 12
▶ CNN Presents Geography: Yesterday and Today, Segment 2: Birth of an Island
▶ Homework Practice Online
▶ HRW Go site

REINFORCEMENT, REVIEW, AND ASSESSMENT

▶ Main Idea Activity 7.1
▶ English Audio Summary 7.1
▶ Spanish Audio Summary 7.1
▶ Section 1 Review, p. 153
▶ Daily Quiz 7.1

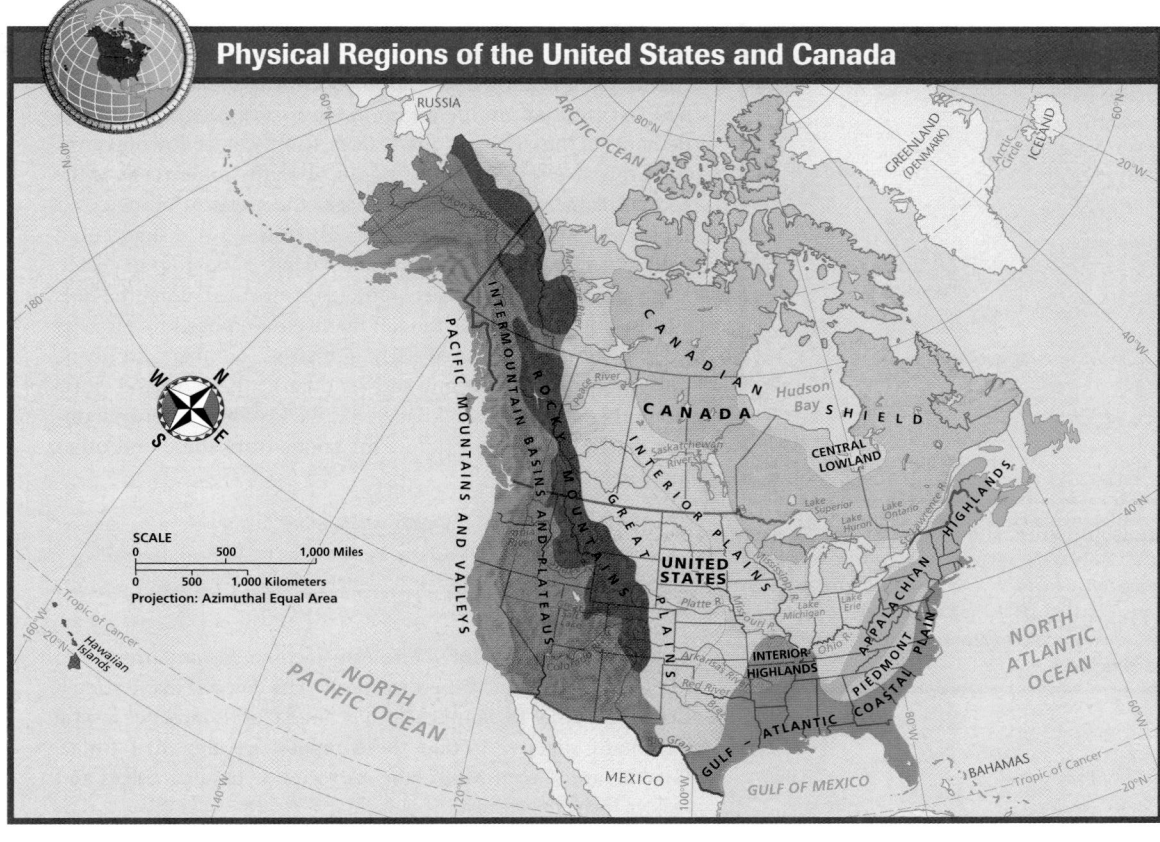

Physical Regions of the United States and Canada

SCALE
0   500   1,000 Miles
0   500   1,000 Kilometers
Projection: Azimuthal Equal Area

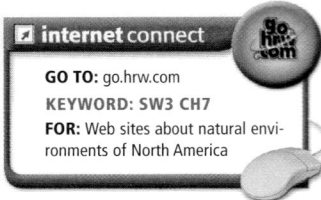 **ALL LEVELS:** Copy the following graphic organizer onto the chalkboard, omitting the italicized answers. Have students identify and describe the landform regions in the United States and Canada. Then ask students if these are formal, functional, or perceptual regions *(formal)*. **ENGLISH LANGUAGE LEARNERS**

Landform Regions of North America

Region	Location	Description
Gulf-Atlantic Coastal Plain	along the Atlantic Ocean and Gulf of Mexico	long coastal plain, begins at sea level and gradually rises
Piedmont	Alabama to New Jersey	upland region at the foot of the Appalachians
Appalachian Highlands	Alabama to southeastern Canada	several ranges, valleys, and ridges; low eroded mountains

Region	Location	Description
Interior Plains	between Appalachians and Rocky Mountains	rolling hills, many rivers and lakes, productive soils
Interior Highlands	in Missouri, Arkansas, and Oklahoma	old eroded highlands
Great Plains	south-central Canada to Texas and Mexico	high plains, subregion of interior plains
Canadian Shield	Arctic Ocean to Atlantic coast	ancient rock, heavily glaciated, little soil
Rocky Mountains	New Mexico to Canada	several high and rugged ranges
Intermountain Basins and Plateaus	between Rockies and coastal mountain ranges	high plateaus, deep canyons, isolated mountain ranges, desert basins
Pacific Mountains and Valleys	along Pacific coast	two mountain ranges separated by a series of valleys

Essential Element 3

► Physical Systems ◄

Cape Cod One of the most recognizable landforms of North America's east coast is Cape Cod in Massachusetts. It extends 65 miles (104.6 km) into the Atlantic Ocean before curving back toward the mainland. The cape's surface is generally flat and spotted with many lakes. The native vegetation consists mainly of grasses, pines, and oaks, while cranberries are grown commercially. Fish and whales inhabit offshore waters.

The unique shape of Cape Cod is the result of many physical processes. Glaciers that covered what is now Massachusetts more than 11,000 years ago carved huge indentations into the coast, leaving dozens of rocky peninsulas and capes. Strong winds, waves, and ocean currents gradually shaped the gravel and sand of Cape Cod into its distinctive hook shape.

CRITICAL THINKING: How might the physical processes that shaped places along the Gulf of Mexico be similar to those that created Cape Cod? How might they differ? *(Wind and wave action helped to create many Gulf Coast features. Glaciers did not reach the South.)*

MAP ANSWER

ports, transportation hubs to inland areas, industries such as textiles that use water power

Landforms

The United States and Canada make up about 80 percent of the continent of North America. These two countries have some of the world's most spectacular scenery. Landforms range from vast plains to high mountains, plateaus, and volcanic islands. Most major landform regions stretch from north to south across both countries. (See the chapter map.) The landforms of the eastern half of the United States and Canada are older than those of the western half. Eastern mountains have been eroded, and rolling hills and flatlands cover most of the region. In contrast, the west has a younger landscape. There you will find steep mountains, active volcanoes, deep canyons, and high plateaus.

A long coastal plain stretches along the Atlantic Ocean and Gulf of Mexico from New England to Mexico. This low plain lies close to sea level and rises gradually inland. It is narrowest along the northern Atlantic coast but widens south of New York. Along some parts of the coastal plain, **barrier islands** have formed. Ocean waves and currents create these long narrow islands by depositing sand in shallow water.

Inland from the coastal plain lies the Piedmont, an upland region. A **piedmont** is an area at or near the foot of a mountain region. The Piedmont stretches from New Jersey to Alabama.

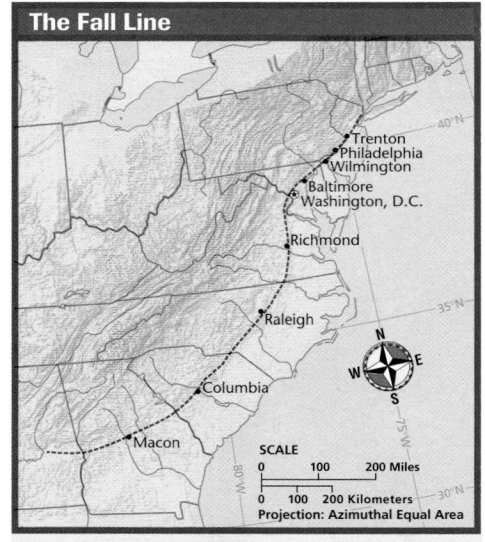

The Fall Line

SCALE
0   100   200 Miles
0   100   200 Kilometers
Projection: Azimuthal Equal Area

INTERPRETING THE MAP *Most cities on the fall line, including the ones shown on this map, mark the head of navigation on rivers. The head of navigation is the point that most ships are not able to sail past.* **What economic activities do you think these cities were established to provide?**

 FOCUS ON GEOGRAPHY

The Fall Line The boundary between the Piedmont and the coastal plain is known as a **fall line**. This natural boundary has had an important influence on the historical geography of settlements in the eastern United States. River waters flowing down from the hard rock of the Piedmont reach the softer rocks of the coastal plain along the fall line. Here these waters plunge over rapids and waterfalls. Early settlers noted that small ships could easily reach the fall line from the ocean but could not sail past it. Partly as a result, many early settlements formed along the line. The tumbling waters of the fall line were also used to turn the waterwheels that powered early industries. Lumber and textile mills were two important early industries here. Inland ports along the fall line, like Philadelphia, Pennsylvania, became important transportation points for goods from these and other industries.

✓ **READING CHECK:** *Places and Regions* Why were many settlements in the eastern United States founded along the fall line? because small ships could easily reach the fall line from the ocean

The East and Interior The Appalachian Mountains rise to the west and north of the Piedmont. The Appalachians stretch from Alabama to southeastern Canada and include several mountain ranges. Among these ranges are the Blue Ridge, Catskill, and Green Mountains. A series of parallel ridges and valleys form the eastern Appalachians.

Using National Geography Standard 4:
Places and Regions: The Physical and Human Characteristics of Places
Organize the class into five groups. Assign each group one of these coastal regions: states along the Gulf of Mexico, northern Atlantic states, southern Atlantic states, California, or the Pacific Northwest states. Have students conduct research on issues that involve living or building in their assigned areas. Suggest that they consult sources such as beach conservation organizations, insurance companies, realtors, and tourist agencies. Have each group create a chart of the advantages and disadvantages of settling along its assigned coast. **COOPERATIVE LEARNING**

HOMEWORK: Have each student draw an elevation profile of the United States or Canada along a particular line of latitude and label major landforms, rivers, and lakes on his or her profile.

Teach Objective 2

LEVEL 1: Provide students with outline maps of North America that include the continent's major rivers and lakes. Have students use their textbooks or supplemental maps to label the bodies of water. Then have each student draw a line on his or her map to represent the Continental Divide. Lead a discussion about how river systems can be used to divide a territory into regions. **ENGLISH LANGUAGE LEARNERS**

The collision of eastern North America with Africa more than 300 million years ago created the Appalachians. Erosion has since lowered and smoothed the peaks of these mountains. The highest peaks rise to just above 6,000 feet (1,829 m).

Between the Appalachians and the Rocky Mountains lie the vast interior plains. The Mississippi River and its many tributaries drain most of this region. Glaciers covered the northern interior plains, north of the Ohio and Missouri Rivers, during the last ice age. Today you will find thousands of lakes there, including the Great Lakes. This area has rolling hills, many river systems, and productive soils. The interior plains partly surround a highland region in Missouri, Arkansas, and Oklahoma. Like the Appalachians, these interior highlands are a region of old eroded uplands. They include the Ozark Plateau. Farther west are the Great Plains, a subregion of the interior plains. The Great Plains stretch from south-central Canada into Texas and Mexico and reach to the eastern edge of the Rocky Mountains. Elevations along this edge of the Great Plains reach more than 5,000 feet (1,524 m) above sea level.

North of the interior plains lies the Canadian Shield. This arc of ancient rocks covers nearly half of Canada. The Canadian Shield is centered on Hudson Bay. It stretches from the Arctic Ocean eastward to the Atlantic coast. This area has been thoroughly scraped by glaciers. This process left a rough rocky landscape with little soil for productive farmland.

The West The Rocky Mountains stretch from New Mexico to Canada. Many of the highest peaks reach more than 14,000 feet (4,267 m). The Rocky Mountains, or Rockies, are not a single range but several ranges. High plains and valleys separate these ranges. West of the Rockies lie the Cascade Range and the Sierra Nevada, two major mountain ranges located near the Pacific coast. The area between these ranges and the Rockies is called the intermountain region.

High plateaus with deep canyons, isolated mountain ranges, and desert **basins** make up most of the intermountain region. A basin is a lower area of land, generally surrounded by mountains. The Great Basin makes up a large area of the intermountain region in the United States. Most rivers there never reach the ocean. The Colorado River, which flows southward to the Gulf of California, is an exception. Farther west, at the edge of the Great Basin, is California's Death Valley. The lowest point in North America—at 282 feet (86 m) below sea level—is found there.

The Pacific coast region is made up of two major mountain systems and a series of valleys between these mountains. The Sierra Nevada and Cascade Range, or simply the Cascades, lie on the eastern edges of the Pacific coast region. The Sierra Nevada runs along California's eastern border. North of the Sierra Nevada, in northern California, Oregon, and Washington,

Our Amazing Planet

The Canadian Shield contains some of the oldest rocks in the world. Some rocks there were formed at least 3.8 billion years ago.

The Teton Range in Wyoming is one of the youngest mountain ranges of the Rockies. A combination of tectonic activity and ice age glaciation has created a spectacular landscape of rugged peaks and deep lakes in this area.

Daily Life

The Appalachian Trail
The Appalachian National Scenic Trail is the longest hiking trail in the world. It stretches more than 2,000 miles (3,218 km) along the crest of the Appalachian Mountains from Springer Mountain, Georgia, to Mount Katahdin, Maine. The original trail was established in 1937. An extension called the International Appalachian Trail was opened in 2001 to continue the path to Cape Gaspé in Quebec where the mountains plunge into the Gulf of St. Lawrence.

Volunteers cleared the first part of the trail in the 1920s, and volunteers continue to maintain shelters and campsites for hikers and to protect the environment along the trail. No other area administered by the U.S. National Parks Service was created and is managed solely by volunteers.

ACTIVITY: Have interested students conduct research on the trail and prepare maps that illustrate the topography, climate changes, vegetation, and animal life found along its route.

internet connect

GO TO: go.hrw.com
KEYWORD: SW3 CH7
FOR: Web sites about the Appalachians

Connecting to THE ARTS

Ansel Adams

Ansel Adams (1902–84) is famous for his black-and-white photographs of America's beautiful natural landscapes. Growing up in California, Adams visited the Sierra Nevada and Yosemite National Park. He began photographing these places and many others as an adult. Adams published many books of his photographs of America's rugged mountains and national parks.

Ansel Adams not only loved the natural landscapes he photographed but also tried to help protect them. From 1936 to 1973, Adams was director of the Sierra Club, a California-based conservation group. His work inspired the conservation of America's natural wonders. Through his beautiful photographs, Ansel Adams became one of the most well-known photographers of the 1900s.

Summarizing How were Ansel Adams's life and career tied to America's natural environments?

are the Cascades. A series of high volcanoes is found in this range. These volcanoes include Mount Rainier, Mount Hood, Mount Shasta, and Mount St. Helens. Along the Pacific Ocean the rugged Coast Ranges stretch from California to Canada. Between the Coast Ranges in the west and the Sierra Nevada and Cascades to the east are three fertile valleys. These are the Puget Sound lowland in Washington, the Willamette River valley in Oregon, and the Central Valley in California.

The rugged western United States is part of the Ring of Fire. The Ring of Fire is a tectonically active region around the edges of the Pacific. It has many active volcanoes and earthquake faults. On the eastern edge of this ring, the North American plate collides with the Pacific plate. All of the continental United States except some parts of California lies on the North American plate. Some areas of coastal California lie on the Pacific plate and are separated from the North American plate along the San Andreas Fault. Major earthquakes occur periodically along this fault.

The two westernmost U.S. states, Alaska and Hawaii, are also geologically active. The Hawaiian Islands are the tops of submerged volcanoes that rise from the ocean floor. They formed over a **hot spot**—a place where magma wells up to the surface from Earth's mantle. Alaska's southern coast is in a subduction zone, and powerful earthquakes sometimes strike there. The volcanic Aleutian (uh-LOO-shun) Islands extend into the Pacific from Alaska. North America's highest peak, Mount McKinley in the Alaska Range, reaches 20,320 feet (6,194 m) in elevation. Except for the Brooks Range, northern and interior Alaska have flat and hilly landscapes.

✓ **READING CHECK:** *Physical Systems* How did glaciers affect the landscape of the Canadian Shield? They scraped it down, leaving a rocky landscape with little productive farmland.

Yosemite National Park in California

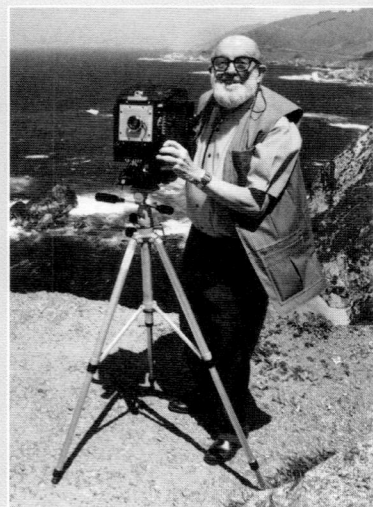

Ansel Adams photographing the California coast

Bodies of Water

The crest of the Rockies marks the **Continental Divide**. This crest divides North America's major river systems into those flowing eastward and those flowing westward. To the east the Mississippi, Missouri, and Ohio Rivers make up the continent's major river system. This system drains most of the interior plains of the United States. It also provides an important network of waterways for trade and transportation. From its delta in southern Louisiana, the Mississippi deposits huge amounts of sediment into the Gulf of Mexico.

The second major river system in the interior plains is the St. Lawrence system. The St. Lawrence connects the Great Lakes to the Atlantic Ocean and drains most of southeastern Canada. In northwestern Canada, the Mackenzie River system also drains the interior plains and part of the Canadian Shield. Three large northern lakes— Lake Athabasca, Great Slave Lake, and Great Bear Lake—drain into the Mackenzie. Water from these lakes flows northward into the Arctic Ocean.

Long rivers like the Colorado, Columbia, Fraser, and Yukon flow west out of the Rockies. The Columbia and Fraser rivers flow into the Pacific Ocean. The Colorado flows southwestward into the Gulf of California. These rivers are important water sources for many people in the western United States and Canada. They are also used to produce hydroelectricity. The Yukon River flows across Alaska to the Bering Sea.

The United States and Canada have many lakes. In fact, North America has more large lakes than any other continent. Continental ice sheets created most of these lakes. During the ice ages, the ice sheets widened and deepened existing basins. Then as the last ice age ended and the ice melted, the basins filled with water. This process formed the Great Lakes and most of Canada's large lakes. In fact, much of the Canadian Shield is a waterlogged landscape covered with lakes and wetlands.

✓ READING CHECK: *Physical Systems* How did glaciers create lakes in this region?

Continental ice sheets widened and deepened basins that filled with water as the ice melted.

INTERPRETING THE VISUAL RECORD

This image taken from the space shuttle Challenger *shows the delta of the Mississippi River. The Mississippi deposits some 220 million tons of sediment into the Gulf of Mexico each year, forming what geographers call a "bird's foot delta." The shape of the delta is caused by the compaction and sinking of sediment.* **How do you think this delta changes through time?**

Review

Homework Practice Online
Keyword: SW3 HP7

Identify Continental Divide

Define barrier islands, piedmont, fall line, basins, hot spot

Working with Sketch Maps On a map of the United States and Canada that you draw or that your teacher provides, label the Piedmont, Appalachian Mountains, Rocky Mountains, Mississippi River, Great Lakes, Great Plains, Canadian Shield, Hudson Bay, Cascade Range, Sierra Nevada, Coast Ranges, and St. Lawrence River. Where are the interior and coastal plains?

Reading for the Main Idea

1. *Places and Regions* How are the landforms of the eastern United States and Canada different from those of the western United States and Canada?

2. *Places and Regions* Where is the Great Basin located?

3. *Physical Systems* Which river drains most of the interior plains?

Critical Thinking

4. Drawing Inferences and Conclusions How do you think tectonic activity affects life in the western United States?

Organizing What You Know

5. Copy the graphic organizer below. Use it to describe the landforms and bodies of water in the interior plains, Canadian Shield, and Pacific coast region. What physical processes shaped these regions?

Interior plains	Canadian Shield	Pacific coast region

153

 **LET'S GET STARTED**

Copy the following instructions onto the chalkboard: *Write down the names of five cities or towns across the United States and Canada. What kinds of plants or animals do you think might be native to those locations? Why?* Discuss students responses. Point out that the United States and Canada include a wide variety of climates, plants, and animals. Tell students that they will learn more about the countries' climates and biomes in Section 2.

Building Vocabulary

Write the terms **natural hazards** and **lichens** on the chalkboard. Call on volunteers to find and read the definitions aloud. Ask students to describe any natural hazards that they may have experienced firsthand. (*Possible answers: floods, hurricanes, tornadoes*) Point out that lichens are able to live where many plants cannot, like on solid rock.

Section 2 Climates and Biomes

READ TO DISCOVER

1. Which climate types are found in the United States and Canada?
2. What are the major biomes of the region, and where are they found?

WHY IT MATTERS

Natural hazards in the United States and Canada destroy homes, ruin crops, and even take lives. Use CNNfyi.com or other current events sources to learn about how destructive natural hazards can be in North America.

DEFINE

natural hazards
lichens

LOCATE

Gulf of Mexico
Hawaiian Islands

Climates

The United States and Canada have a great variety of climates. (See the unit climate map.) For example, every climate type except an ice cap climate can be found in the United States. However, you will find an ice cap climate on some Arctic islands of far northern Canada.

Four major factors influence the distribution of climates in the United States. These factors are a middle-latitude location, prevailing winds, ocean currents, and high mountain ranges. Due to its more northerly location, Canada has mostly colder climates.

Use the unit climate map as we look at the distribution of climates across North America. The very southern tip of Florida has a tropical wet and dry climate. However, most of the southeastern quarter of the United States has a humid subtropical climate. This region stretches from the Atlantic coast to about 100° west longitude in western Texas and Oklahoma. Summers are hot

INTERPRETING THE VISUAL RECORD

Average winter temperatures in Toronto, Canada, are just below freezing. Although Lake Ontario does not completely freeze, the city's harbor is generally iced over from December to April. **How might cycles of freezing and thawing affect this area's environment?**

LEVEL 1: Copy the following graphic organizer onto the chalkboard, omitting the italicized answers. Have students identify the locations of each climate type in the United States and Canada. **ENGLISH LANGUAGE LEARNERS**

LEVELS 2 AND 3: Have each student choose one of the climates found in North America. Then have each student conduct research to find statistics such as annual rainfall and average temperatures for a location that has the chosen climate. Have students mark their findings on one large map. Use the map to lead a class discussion about factors that influence the locations of North America's climates.

Climate Types of North America

tropical wet and dry	*very tip of Florida, western Hawaii*
humid subtropical	*southeast United States*
humid continental	*northeastern United States and southeastern Canada*
semiarid	*Great Plains, western mountains*
arid	*areas east of Sierra Nevada and Cascades*
marine west coast	*southern Alaska through northern California*
Mediterranean	*southern and central California*
tropical humid	*eastern Hawaii*
tundra	*across northern Alaska to Newfoundland and Quebec*
subarctic	*northern Canada and Alaska*

and humid, and winters are mild. Rainfall is distributed fairly evenly throughout the year, and thunderstorms are common. The warm waters of the Gulf of Mexico and the Gulf Stream influence this climate region. The Gulf Stream is a current in the Atlantic Ocean that moves warm tropical water northward along eastern North America.

The northeastern United States and parts of southern and southeastern Canada have a humid continental climate. This climate region stretches westward from the Atlantic coast to about 100° west longitude in Kansas. The region has four distinct seasons, including a warm humid summer and a cold snowy winter. The nearness of the Great Lakes and Atlantic Ocean moderates temperatures slightly and is a source of increased precipitation.

West of 100° west longitude is the semiarid climate of the Great Plains. This climate supports vast grasslands and scattered trees. Cold air masses from the north meet warm moist air masses from the south over the Great Plains. Where these air masses come into contact with each other, violent storms can erupt. These storms can produce **natural hazards** like floods, hail, lightning, and tornadoes. Natural hazards are events in the physical environment that can destroy human life and property.

Because of its mountainous terrain, the intermountain area in the west has a variety of climates. Mountains block prevailing westerly winds that flow over the Pacific. This rain-shadow effect creates arid and semiarid climates on the leeward sides of mountains. For example, areas east of the Sierra Nevada and Cascades are dry. The Rockies have a highland climate. Temperatures and precipitation in the Rockies depend on elevation and local geography.

The Pacific coast region has two main climates, marine west coast and Mediterranean. The mild marine west coast climate dominates the coast from southeastern Alaska to northern California. These areas have cool wet winters and mild sunny summers. A Mediterranean climate is found in parts of southern and central California. This climate is known for its mild winters and long, sunny, dry summers.

Hawaii lies completely within the tropics. Because the Hawaiian Islands fall in the easterly trade wind belt, they are wetter on the windward eastern sides. These eastern areas have a tropical humid climate. Leeward, western slopes have a drier tropical wet and dry climate.

Far northern North America has a tundra climate. This extremely cold area stretches from northern Alaska across to Quebec and Newfoundland. Permafrost underlies much of the area. To the south, a subarctic climate is found. This large subarctic climate region covers most of Canada and Alaska.

✓ **READING CHECK:** *Physical Systems* What main factors influence the distribution of climate types in the United States and Canada? *middle-latitude location, prevailing winds, ocean currents, high mountain ranges, Canada's northern location*

INTERPRETING THE VISUAL RECORD

The United States is hit by more tornadoes each year than any other country. Most occur between April and June in the central United States in an area known as Tornado Alley. **How do you think hazardous environmental conditions such as tornadoes affect the natural environment?**

The world's largest living organism is a fungus in Oregon. This fungus stretches 3.5 miles (5.6 km) across and covers an area as big as 1,665 football fields. It is about 2,400 years old, lives underground, and spreads slowly from tree to tree.

Across the Curriculum

▶ **Science** ◀

Tornado Alley Tornado Alley is a name given to a stretch of the interior plains between Texas and Iowa. Though tornadoes can strike anywhere in the United States, most of them form within this area. These storms can be incredibly destructive. For example, a series of tornadoes in 1999 killed 46 people and destroyed thousands of buildings in Oklahoma and Kansas.

The storms of Tornado Alley are formed when moisture from the Gulf of Mexico is carried by prevailing winds to the plains. The point where this moist air meets dry air from the Rockies is called a dryline. As the dryline is forced eastward by strong winds from the southwest, the Gulf air is forced upward. This rising air sometimes twists into a supercell, or rotating thunderstorm. Most tornadoes develop from these storms.

ACTIVITY: Have interested students conduct research on the methods that meteorologists and researchers use to predict and study severe weather. Lead a discussion on the importance of storm predictions to residents of Tornado Alley.

VISUAL RECORD ANSWER

destroy plants and animals, erode land

LEVELS 1 AND 2: Before class, write the following titles on pieces of paper: temperate forest, semiarid and desert, grassland, boreal forest, tundra, tropical, savanna, and tropical rainforest. Then organize the class into seven groups. Give each group one of the prepared sheets and have group members record on it one fact about the listed biome. Then have each group pass its paper to another group and repeat the process. Continue until each group has commented on each biome. Then have volunteers from each group read the fact sheets aloud. Lead a class discussion about factors that influence the distribution of plants and animals across North America. Have students examine the photographs in this section and the climate and precipitation maps in the unit atlas to draw conclusions about connections between the region's climates and biomes. **ENGLISH LANGUAGE LEARNERS, COOPERATIVE LEARNING**

LEVEL 3: Have students imagine they are environmental activists who are representing either the United States or Canada at a United Nations conference about the world's biomes. They are to give presentations about how human activity has affected the region's biomes. Each student's presentation should provide a description of the current state of a chosen biome and measures that should be taken to preserve the area for future generations.

Eye on Earth

Olympic National Park

Olympic National Park occupies much of Washington's Olympic Peninsula. A temperate rain forest, the park gathers moisture from the area's abundant dew, mist, fog, rain, sleet, and snow. The western slopes of snowcapped Mount Olympus at the center of the park receive 240 inches (609.6 cm) of precipitation each year.

Parts of Olympic National Park contain more biomass per acre than any other place in the world. Douglas firs, spruce, and cedars are surrounded by mosses, lichens, ferns, huckleberries, and mosses. The park's wildlife includes Roosevelt elk, mountain goats, deer, cougars, and black bears. More than 100 species of birds, including spotted owls, tundra and trumpeter swans, and geese also live in the park.

ACTIVITY: Have students compare and contrast temperate and tropical rain forests. Ask them to note differences in location, climate, soil, vegetation, animal life, and other features of the forests.

Coast redwoods are the tallest trees in the world and can grow as high as 385 feet (117 m). Coast redwood forests are found only in a narrow belt along the Pacific Ocean in northern California and southern Oregon, where coastal fog is common. Water droplets from the fog collect on the trees' leaves and then drip down to the ground. This helps provide enough water for the huge trees to survive the dry summers.

Cacti like this cholla (CHOY-yuh) are a common sight in the Sonoran Desert of the southwestern United States. During infrequent rainstorms, wide root systems collect water, which is stored in the plant's stem. The absence of leaves and a waxy coating on the stem help prevent water loss through evaporation.

Plants and Animals

Climate patterns greatly influence North America's plant and animal life. In general, forests dominate humid areas, while grasslands or scrub cover more arid regions. However, human settlement has greatly altered the distribution of plants and animals in the region. For example, people have converted many forests and grasslands to farmland. This human activity has caused major disruptions in the natural vegetation and animal life.

The southeastern United States and much of the U.S. and Canadian west have a temperate forest biome. Different types of forests are found within these large areas. For example, mixed forests of hickory, oak, and walnut are common along the coastal plain in the southeastern United States. Deer, opossum, and raccoon are among the animals that live there. Temperate forests along the Pacific coast differ from those in the southeastern United States. Redwood trees form North America's densest and tallest forests along the coast of northern California. To the north in Oregon and Washington, Douglas fir trees become more common. Even farther north, in coastal southeastern Alaska, Sitka spruce trees are widespread. Wildlife along the Pacific coast includes black bears, eagles, hawks, and salmon.

Close

Have volunteers point to random places on a wall map of North America. Ask the class to describe the climate and biome found in each location. Then ask students how they think these factors would affect the daily lives of the people living there.

Review and Assess

Have students complete the **Section Review**. Then have students complete **Daily Quiz 7.2.**

Reteach

Have students complete **Main Idea Activity for English Language Learners and Special-Needs Students 7.2.** Have students work in pairs to write brief descriptions of each climate and biome of North America. **ENGLISH LANGUAGE LEARNERS**

Extend

Have interested students conduct research about Mount Washington in New Hampshire, which may have the worst weather of any place in the United States. Ask students to examine the mountain's fluctuating temperatures and ferocious winds and to write stories that illustrate the danger its weather can pose to climbers and tourists. **BLOCK SCHEDULING**

Much of the southwestern United States has a semiarid and desert biome. Creosote and mesquite bushes as well as many species of cacti cover open areas there. Coyotes, hawks, jackrabbits, and snakes live in this biome. A grassland biome stretches across the interior of North America. These grasslands, or prairies, once supported huge herds of bison. However, American settlers hunted the bison nearly to extinction. Over time farmers have also plowed under almost all of the original prairie. Today farmers use these grassland areas mainly for growing grains.

About half of Canada and Alaska have a boreal forest biome. This vast northern forest is one of the largest in the world. The main trees there are the spruce, fir, and pine. Great herds of caribou live in the forest during the winter. Deer, elk, moose, and wolves also inhabit the area. North of these vast forests is a treeless arctic tundra. Beneath the tundra surface is a layer of permafrost that can be up to 1,500 feet (460 m) deep. Tundra plants include grasses, small shrubs, mosses, and **lichens**—small plants that consist of algae and fungi.

Two smaller areas in the United States have tropical biomes. The tip of southern Florida has a savanna biome. The land there is swampy and covered with tall grasses. Palm trees and pine forests thrive in Florida. Hawaii has a tropical rainforest biome. Seeds carried by birds, ocean currents, and winds sprouted and grew in Hawaii's rich volcanic soils. As a result, a unique collection of plants and animals developed there. The remote location of the islands in the Pacific Ocean also influenced Hawaii's biogeography. Human activities have greatly altered Hawaii's natural ecosystems, however.

INTERPRETING THE VISUAL RECORD

Heavy rainfall and fertile volcanic soils help create a lush tropical environment in the Hawaiian Islands. More than 150 types of ferns and nearly 1,000 types of flowers grow there. Many of these plants do not grow anywhere else in the world. **What geographic features do you think helped create Hawaii's unique vegetation patterns?**

✔ **READING CHECK:** *Places and Regions* What soil conditions will you find in the tundra biome of far northern Alaska and Canada? Which plants live there? thick layer of permafrost; grasses, small shrubs, mosses, lichens

Review

Homework Practice Online
Keyword: SW3 HP7

Define
natural hazards
lichens

Working with Sketch Maps On the map you created in Section 1, label the Gulf of Mexico and the Hawaiian Islands. Which climate region do the warm waters of the Gulf of Mexico and the Gulf Stream influence?

Reading for the Main Idea

1. *Physical Systems* Why do the Great Plains have such violent weather?

2. *Places and Regions* How do climate patterns influence the distribution of vegetation in North America?

3. *Places and Regions* Why are the eastern sides of the Hawaiian Islands wetter than the western sides?

Critical Thinking

4. **Making Generalizations and Predictions** How might the tropical climates of Hawaii and southern Florida be important to agriculture in the United States?

Organizing What You Know

5. Draw a sketch map of North America. Use it to show the major biomes of the United States and Canada.

Review Answers

Define For definitions, see: natural hazards, p. 155; lichens, p. 157

Working with Sketch Maps Maps will vary, but listed places should be labeled in their approximate locations. They affect the humid subtropical climate of the southeastern United States.

Reading for the Main Idea
1. Cold air masses from the north meet warm, moist air masses from the south and create storms.

2. In general, forests dominate humid areas, while more arid regions are covered by grasslands or scrub.

3. because they are in the easterly trade wind belt

Critical Thinking
4. Possible answer: They might allow the growth of tropical crops that cannot be raised elsewhere. (NGS 4)

Organizing What You Know
5. Maps will vary but should accurately identify the major biomes of North America.

VISUAL RECORD ANSWER

plentiful rainfall, fertile volcanic soils, geographic isolation

Section 3

OBJECTIVES

1. Describe the farming, forest, and water resources of the United States and Canada.

2. Analyze the region's supply of energy and mineral resources.

LET'S GET STARTED

Copy the following question and instructions onto the chalkboard: *What kind of images come to mind when you think of a farm? Draw a picture or write a few sentences that describe these images.* Discuss student responses. Point out that the United States and Canada have some of the most productive farmland in the world. Both small traditional farms and huge commercial operations are found in each country. Tell students that they will learn more about the abundance of resources found in the United States and Canada in Section 3.

Building Vocabulary

Write the terms **alluvial soils** and **newsprint** on the chalkboard. Ask students to recall the meaning of the word *alluvial* from earlier chapters *(deposited by water)*. Point out that alluvial soils are generally extremely fertile. Bring blank sheets of newsprint or a newspaper to class and have students compare the quality of the paper with other types.

Section 3 RESOURCES

REPRODUCIBLE

▶ Guided Reading Strategy 7.3
▶ Graphic Organizer Activity 7
▶ Map Activity 7: Water Resources of North America

TECHNOLOGY

▶ One-Stop Planner CD–ROM, Lesson 7.3
▶ Geography and Cultures Visual Resources 16
▶ Homework Practice Online
▶ HRW Go site

REINFORCEMENT, REVIEW, AND ASSESSMENT

▶ Main Idea Activity 7.3
▶ English Audio Summary 7.3
▶ Spanish Audio Summary 7.3
▶ Section 3 Review, p. 160
▶ Daily Quiz 7.3

VISUAL RECORD ANSWER

Possible answer: Technological advancements allow for larger farms, which replace natural vegetation.

Section 3 Natural Resources

READ TO DISCOVER

1. What farming, forest, and water resources are found in the United States and Canada?
2. How rich is the region in energy and mineral resources?

WHY IT MATTERS

The many natural resources of the United States and Canada have helped make North America an economic powerhouse. Use **CNNfyi.com** or other **current events** sources to learn about the resources that make this region so strong economically.

DEFINE

alluvial soils
newsprint

LOCATE

Central Valley
Imperial Valley
Rio Grande valley
Colorado River
Grand Banks

Farming, Forests, and Water Resources

Abundant natural resources have helped make the United States and Canada very rich countries. The United States is an industrial giant and the world's leading agricultural producer. The country's diverse resources and strong economy help support a high standard of living. Canada also has a high standard of living, but its economy is much smaller. However, Canada is an important producer and exporter of key natural resources.

In both the United States and Canada, only about 3 percent of the population farms. However, both countries easily feed their people and have large food surpluses to export. Much of this success is a result of North America's large area and its wide variety of climates and soils. American and Canadian farmers in the Great Plains grow huge amounts of corn, soybeans, and wheat. They also raise cattle, hogs, and other livestock. In some places, fertile **alluvial soils**—soils deposited by streams or rivers—are particularly productive. These rich soils are found in the Mississippi Valley, California's Central and Imperial Valleys, and in the Rio Grande valley in Texas. With irrigation, these excellent soils support a wide range of fruits and vegetables.

Forests are another important natural resource. Both the United States and Canada are leading producers and exporters of forest products. Canada's forests provide lumber and pulp for paper. Countries like Japan and the United States look to Canada for lumber, **newsprint**, and pulpwood. Newsprint is an inexpensive paper used mainly for newspapers. About one third of the United States is forested. The country's major commercial forests are located in the southeastern states and the Pacific Northwest. Early logging cleared large forest areas in the Northeast and the Midwest. Today much of the logging there takes place on private tree farms and in national forests.

INTERPRETING THE VISUAL RECORD

Much of the United States has productive farmland, such as this area in Oregon. **How do you think the use of agricultural technology has changed the American landscape?**

ALL LEVELS: Copy the following graphic organizer onto the chalkboard, omitting the italicized answers. Call on students to provide words and phrases to describe the full range of resources found in North America. Ask students how the locations of these resources affect the locations of economic activities in the United States and Canada. Then ask how improvements in technology and transportation have affected these locations. Have students consider local, state, or federal regulations that limit or otherwise influence the use of mineral or energy resources. Ask how these policies affect the environments and economies of the United States and Canada.

ENGLISH LANGUAGE LEARNERS

Agricultural resources
- *large area*
- *good climates for farming*
- *fertile soils*
- *provide enough food to feed population and leave a surplus*

Forest resources
- *large forests and tree farms*
- *produce wood for lumber, newsprint, pulpwood, and other forest products*
- *leading producers and exporters*

Water resources
- *for irrigation and hydroelectricity production*
- *rich fisheries along coasts*

Resources of North America

Energy resources
- *huge coal reserves*
- *major oil producer, but still must import*
- *rich in natural gas*

Mineral resources
- *nickel, zinc, uranium, lead, copper, gold, and silver in Canada*
- *iron, copper, lead, zinc, gold, silver, and other minerals in United States*

As you have learned, North America has plentiful water resources. Water has been important to the economic development of both the United States and Canada. Many rivers, such as the Colorado, Fraser, and Tennessee, are used for irrigation and hydroelectricity. In fact, Canada and the United States are the world's two largest producers of hydroelectricity. North America's coastal waters also provide marine resources. Canada's fisheries are among the world's richest. However, overfishing and pollution have taken their toll, reducing fish catches in some areas. The Grand Banks area near Newfoundland is one of the most famous fishing areas in the world. Fisheries along the Atlantic coast are home to cod, haddock, lobster, and swordfish. Salmon is the main commercial fish on the Pacific coast. Fishers catch shrimp and shellfish in the Gulf of Mexico.

✓ **READING CHECK:** *Places and Regions* What factors make the Central and Imperial Valleys and the Rio Grande valley productive farming regions? *fertile alluvial soils, irrigation*

Energy and Minerals

The United States and Canada are rich in energy and mineral resources. (See the map of resources of the United States and Canada.) The United States has about 25 percent of the world's coal reserves and is a major coal exporter. Most coal is mined in the Appalachians, Rockies, and interior plains. In Canada, coal is mined in Nova Scotia and in the western provinces of Saskatchewan, Alberta,

A fishing crew brings in their catch from waters off the New England coast. New England's waters have long been a major fishing area in the United States.

Essential Element 5

► **Environment ◄ and Society**

Lobster Fishing In the 1940s and 1950s, the demand for lobsters from New England and the Maritime Provinces grew tremendously. By the early 1970s, scientists feared that overfishing had reduced the Atlantic's lobster populations to dangerously low levels. They campaigned for conservation laws to protect the crustaceans.

Since then, lobster populations have increased, and catches have nearly doubled from their previous levels. Limits on the size and number of catches are partially responsible for the increase, but researchers believe other factors have also played a role. A slight rise in water temperatures has made the lobsters more active and allowed them to grow faster. In addition, the tons of bait that fishers dump overboard to lure lobsters into their traps provide the creatures with the food they need to grow.

CRITICAL THINKING: How has human activity affected lobster populations in the Atlantic? *(Overfishing reduced the number of lobsters in the ocean, but legal restrictions on fishing and human-supplied food allowed the numbers to grow again.)*

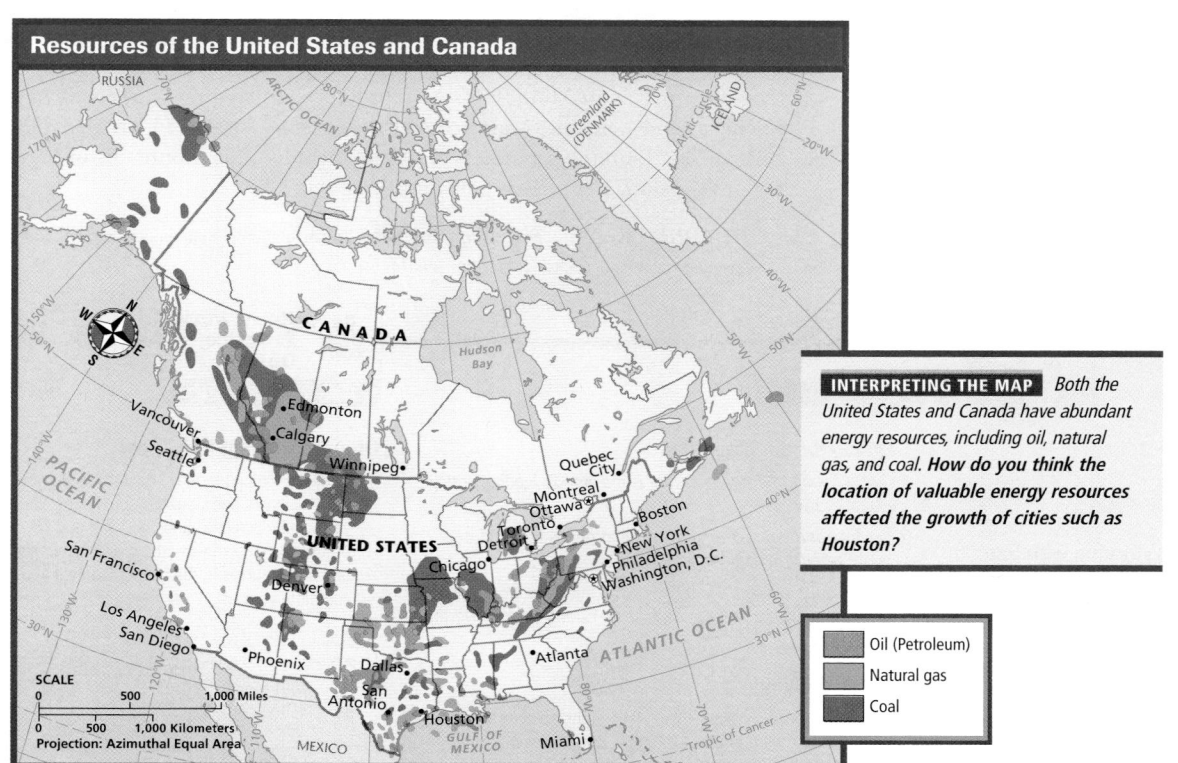

Resources of the United States and Canada

INTERPRETING THE MAP *Both the United States and Canada have abundant energy resources, including oil, natural gas, and coal.* **How do you think the location of valuable energy resources affected the growth of cities such as Houston?**

Oil (Petroleum)

Natural gas

Coal

MAP ANSWER

encouraged their growth and industrial development

Close

Ask students to identify industries or products for which your state or region is noted. As items are named, call on volunteers to list some of the mineral or energy resources that are involved in their creation. Then ask students if these resources are readily available in your area or if they must be imported.

Reteach

Have students complete **Main Idea Activity for English Language Learners and Special-Needs Students 7.3.** Then have students work in pairs to create outlines of this section. **ENGLISH LANGUAGE LEARNERS**

Review and Assess

Have students complete the **Section Review.** Then have students complete **Daily Quiz 7.3.**

Extend

Have interested students conduct research on life in mining communities of the United States and Canada. Ask students to imagine they are miners and to write journal entries describing their daily lives. They may wish to discuss the changes in mining and their communities that have been caused by new technologies. **BLOCK SCHEDULING**

Section 3 Review Answers

Define For definitions, see: alluvial soils, p. 158; newsprint, p. 158

Working with Sketch Maps Maps will vary, but listed places should be labeled in their approximate locations. The Grand Banks is famous for its fishing.

Reading for the Main Idea
1. the United States
2. lumber, pulp, newsprint
3. overfishing, pollution

Critical Thinking
4. Possible answer: The rugged environment limits access to natural resources. (NGS 16)

Organizing What You Know
5. Charts will vary but should accurately reflect the information presented on the map in this section. Coal deposits are heaviest in the Appalachians, the Midwest, the Rockies, the Prairie Provinces, British Columbia, and Alaska. Natural gas is found in California, the Appalachians, the southern Plains, and Alberta. Oil is concentrated in Alaska, California, Louisiana, Oklahoma, Texas, the Midwest, and the Prairie Provinces.

Oil wells dot the coastline of Huntington Beach, south of Los Angeles. California has been a leading producer and refiner of oil since the late 1800s, and petroleum is still the state's leading mineral product.

and British Columbia. These coal deposits are generally very thick and are located in unpopulated areas.

The United States is a major oil producer but uses much more oil than it produces. In fact, the United States has to import more than one half of the oil it needs. Most U.S. oil is produced on the Gulf Coast of Texas and Louisiana and in California and Alaska. These same areas are rich in natural gas, which is often found with oil. About 65 percent of Canada's oil and about 80 percent of its natural gas come from Alberta. Oil and gas deposits have been discovered off Canada's eastern and Arctic coasts as well.

The United States and Canada have a wide range of valuable mineral resources. The rocky Canadian Shield, once considered a wasteland, has many mineral deposits. Canada is a leading source of the world's nickel, zinc, and uranium. It is also a major producer of lead, copper, gold, and silver. Northern Canada even has diamond deposits. In the United States, valuable minerals are mined in the Appalachians, Rockies, and western mountain ranges. Iron has long been mined in Minnesota's Mesabi Range. Today iron is also mined in Michigan and Alabama. Major copper deposits are located in Arizona. Lead and zinc are found in a number of places, including the mountains of Idaho and in Missouri. Nevada has gold and silver deposits.

✓ **READING CHECK:** *Places and Regions* Which landform region in northern Canada has many mineral deposits? Canadian Shield

---

Section 3 Review

Homework Practice Online
Keyword: SW3 HP7

Define
alluvial soils
newsprint

Working with Sketch Maps On the map you created in Section 2, label the Central Valley, Imperial Valley, Rio Grande valley, Colorado River, and Grand Banks. What ocean area near Canada is one of the most famous fishing grounds in the world?

Reading for the Main Idea
1. *Places and Regions* What is the world's leading agricultural country?
2. *Environment and Society* What goods are produced with resources from Canada's forests?
3. *Environment and Society* What human activities can influence the size of catches in some fishing areas in North America?

Critical Thinking
4. **Making Generalizations** Why might modern technology be particularly important to mining operations in the Canadian Shield?

Organizing What You Know
5. Copy and complete the chart below, identifying U.S. states and Canadian provinces where you would expect to find coal, gas, and oil production. You can use the map of coal, gas, and oil resources in the chapter to help you complete your lists.

State or province	Coal production	Gas production	Oil production

Biology: Ecosystems of the Wetlands

The plants and animals that live in wetlands form ecosystems that are often unlike anything else in the area. Most wetlands are home to a diverse population that needs a great deal of water to survive. To help students understand the structure and diversity of these ecosystems, organize a field trip to a local wetland or a similar environment. Even the smallest creek or pond can support a miniature system that parallels those found in larger areas.

Have students keep lists of plant and animal species that thrive in the environment they study. First ask students to identify the plants and animals that live in or near the wetland. Then have students collect water samples and examine them for signs of living creatures such as tadpoles or small fish. Then have students study drops of the water under microscopes to identify smaller organisms. Provide science or reference books to help students identify what they see.

As a class, create a diagram of the wetland ecosystem's food web. Then lead a class discussion about the species that live in the wetland and how they differ from other species in nearby areas. Ask students how the plants and animals in the wetland depend on the presence of water.

If a field trip is not possible, try to obtain a sample of water from a wetland area from a local conservation agency or through the cooperation of another class located near a suitable area. If necessary, students can conduct research on wetlands in books or on the Internet. **BLOCK SCHEDULING**

Environment and Society

Geography for Life

Wetlands in the United States

Wetlands are areas that are covered with water for at least part of the year. Many different kinds of wetlands exist. They include marshes, swamps, estuaries, and coastal areas affected by ocean tides. Wetlands in the United States range from the Arctic bogs of northern Alaska to The Everglades of Florida. In fact, wetlands are found in all 50 U.S. states.

Wetlands are important natural resources for many different reasons. They provide habitat for fish and other marine life, birds, mammals, and a great variety of plants. For example, wetlands are a source of food for migratory birds. In addition, wetlands provide important water resources. They hold back and then slowly release floodwaters and snowmelt. They are a source of groundwater and act as filters that cleanse water of pollutants. Coastal wetlands can prevent erosion and protect coastal areas from powerful waves and storms. In addition, wetlands provide recreation for many people. Boaters, birdwatchers, fishers, kayakers, photographers, and tourists all enjoy visiting wetland areas.

Despite their value as natural resources, most people viewed wetlands as bug-infested wastelands for much of American history. As a result, they did not think that conserving wetlands was particularly important. In fact, about half of the natural wetlands in the United States have been destroyed. Most were drained, filled, or paved over to make room for farmland or expanding urban areas. Canal construction for irrigation and flood control and the development of coastal highways have also taken their toll. California alone has lost more than 90 percent of its original wetlands. Most of this loss is the result of agricultural and urban growth.

Beginning in the 1970s and 1980s, however, American attitudes toward wetlands changed. Many people began to realize the value of protecting these diverse and useful environments. In 1986 the U.S. Congress passed the Emergency Wetlands Resources Act. This law required the U.S. Fish and Wildlife Service to study the country's remaining wetlands. The Fish and Wildlife Service reports its findings to Congress every 10 years.

According to the most recent Fish and Wildlife report, the United States has made much progress in slowing the loss of wetlands. From 1986 to 1997 the rate of wetland loss fell by 80 percent from the previous decade. Much of this decline was the result of new wetland policies and programs. These policies helped reduce the drainage and filling in of wetlands. They also helped restore wetlands. The country has not yet achieved the goal of reducing the rate of wetland loss to zero. However, Americans are now protecting and restoring more of these valuable environments.

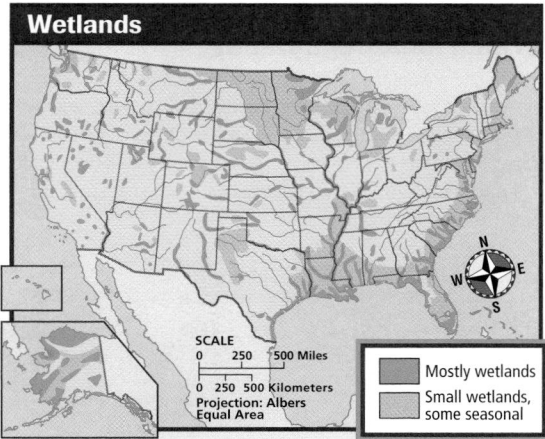

Wetlands

SCALE
0   250   500 Miles

0 250 500 Kilometers
Projection: Albers
Equal Area

- Mostly wetlands
- Small wetlands, some seasonal

INTERPRETING THE MAP *Although wetlands are found in all 50 states, their distribution is uneven. **Which regions appear to have the highest concentration of wetlands? Why do you think that is?***

Applying What You Know

1. **Summarizing** Why are wetlands such important natural resources?

2. **Analyzing Information** What human activities reduced the amount of wetlands in the United States over time? What effects have conservation efforts had on the wetlands since the 1970s and 1980s? What long-term geographic and economic advantages do you think might come from these policies?

Applying What You Know Answers

1. They are habitats for marine life and wildlife, water resources, protection against erosion, and recreational destinations.
2. The rate of wetlands loss caused by agricultural expansion and urban growth has slowed, and some areas have been restored. Plants and animals that rely on wetlands will be helped, and wetlands might be used to attract tourism.

MAP ANSWER

southeastern coast, Alaska; rivers, rainfall, coastal plains, lack of development

This Geography for Life feature addresses National Geography Standards 4, 6, 14, and 16.

CHAPTER 7 Review Answers

Building Vocabulary For definitions, see: barrier islands, p. 150; piedmont, p. 150; fall line, p. 150; basins, p. 151; hot spot, p. 152; Continental Divide, p. 153; natural hazards, p. 155; lichens, p. 157; alluvial soils, p. 158; newsprint, p. 158

Locating Key Places

A. St. Lawrence River
B. Piedmont
C. Canadian Shield
D. Mississippi River
E. Great Plains
F. Rocky Mountains
G. Great Lakes

161

TECHNOLOGY

▶ Chapter 7 Test Generator
(on the One-Stop Planner)
▶ Global Skill Builder CD–ROM
▶ HRW Go site

REINFORCEMENT, REVIEW, AND ASSESSMENT

▶ Chapter 7 Review, pp. 162–63
▶ Chapter 7 Tutorial for Students, Parents, Mentors, and Peers
▶ Chapter 7 Test (form A or B)
▶ Alternative Assessment Handbook

▶ Chapter 7 Test for English Language Learners and Special-Needs Students
▶ Unit 2 Test
▶ Unit 2 Test for English Language Learners and Special-Needs Students

Assess

Have students complete a Chapter 7 Test.

Reteach

Organize the students into small groups. Have each group create a storyboard that explains the key points of one section of this chapter. Then have groups present their storyboards to the class. **ENGLISH LANGUAGE LEARNERS, COOPERATIVE LEARNING**

CHAPTER 7 Review

Building Vocabulary

On a separate sheet of paper, explain the following terms by using them correctly in sentences.

barrier islands	hot spot	alluvial soils
piedmont	Continental Divide	newsprint
fall line	natural hazards	
basins	lichens	

Locating Key Places

On a separate sheet of paper, match the letters on the map with their correct labels.

Piedmont	Great Lakes	Canadian Shield
Rocky Mountains	Great Plains	St. Lawrence River
Mississippi River		

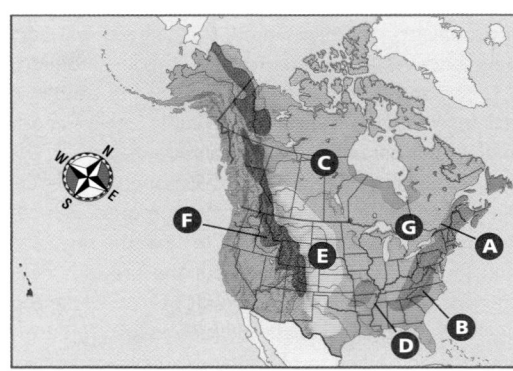

Understanding the Main Ideas

Section 1

1. *Physical Systems* What physical process forms barrier islands? Where are they found in North America?

2. *Places and Regions* What evidence of tectonic forces will you find in western areas of the United States and Canada?

Section 2

3. *Physical Systems* What can happen when different air masses come in contact with one another over the Great Plains?

4. *Places and Regions* What major factors influence climates in the United States? How does nearness to the Great Lakes and the Atlantic Ocean influence the humid continental climate region of the northeastern United States?

Section 3

5. *Places and Regions* About 25 percent of the world's reserves of which energy resource are found in the United States?

Thinking Critically

1. Drawing Inferences and Conclusions Which mountain system do you think is older, the Appalachians or the Rockies? How might the physical geographic features of each range provide clues to their relative age?

2. Analyzing Information How have people changed the natural environment of the United States and Canada over time?

3. Drawing Inferences and Conclusions How do you think the locations of lakes and rivers in the United States and Canada affected the locations of settlements? Why?

Using the Geographer's Tools

1. Analyzing Maps Look at the map of coal, gas, and oil resources in Section 3. Which states and provinces appear to have the richest energy resources?

2. Interpreting Climate Graphs Study the climate graphs for Sacramento and Washington, D.C., in Chapter 1. Which city appears to have a wetter climate? Note that Sacramento is located in California's Central Valley. How might this location account for the difference in relative precipitation amounts for the two cities?

3. Preparing Tables Look at the map of the physical regions of the United States and Canada in Section 1. Create a table that lists these geographic regions and describes the climates and biomes of each.

Writing about Geography

Choose one area of the United States and one area of Canada. Imagine that you went to both areas on a vacation. Write a letter to a friend comparing how people in each place depend on the natural resources there. When you are finished with your letter, proofread it to make sure you have used standard grammar, spelling, sentence structure, and punctuation.

SKILL BUILDING

Geography for Life

Creating Diagrams

Places and Regions Research the watershed, or drainage area, of the Mississippi River. Draw a diagram showing the Mississippi River and its main tributaries. Add other large rivers that flow into the main tributaries. Think of a way to visually represent how each tributary adds volume to the body of water it enters. Add other details, such as the length of the rivers or the areas they drain.

Portfolio Activity

Have students examine local, state, national, or international media to identify environmental issues in the United States and Canada. Organize the class into groups. Have each group conduct research on a particular issue and create a triptych that illustrates various points of view regarding the issue and how the issue can affect the natural environments of the United States or Canada. Call on volunteers from each group to report their findings to the class. Photograph triptychs for students' portfolios.

Food Festival

The natural environments of the United States and Canada make certain ingredients more readily available in some areas than others. Have students work in groups to investigate traditional foods from various regions of the two countries. Examples from the United States include chili (Southwest), grits (Southeast), and scrapple (Northeast). Canadian examples include fiddlehead ferns (Maritime Provinces) and Saskatoon pie (Prairie Provinces). Ask each group to research a region and prepare a representative dish for the class. Also, instruct students to determine whether their dishes are related to the regions' climate and vegetation.

Building Social Studies Skills

Average Precipitation in Selected U.S. Cities

City	Average Annual Precipitation (in inches)
Albuquerque, NM	6.39
Boston, MA	41.07
Chicago, IL	33.92
Houston, TX	59.71
Miami, FL	63.29
North Little Rock, AR	47.42
Reno, NV	7.08

Sources: *World Almanac and Book of Facts 2004*, National Climatic Data Center

Interpreting Tables

Study the table above. Then use the information to help you answer the questions that follow. Mark your answers on a separate sheet of paper.

1. Which city has the highest average annual precipitation?
   a. Albuquerque, NM
   b. Miami, FL
   c. North Little Rock, AR
   d. Boston, MA

2. Which cities are probably located in an arid or semiarid climate region? How do you know that?

Building Vocabulary

To build your vocabulary skills, answer the following questions. Mark your answers on a separate sheet of paper.

3. After crossing the mountains, the pioneers entered a *basin*. In which of the following sentences does *basin* have the same meaning as it does in the sentence above?
   a. Have you seen the sailing ships anchored in the basin?
   b. The settlers had transported a large basin from Virginia to California.
   c. Larry's mother washed the stained shirt in a basin.
   d. Only one of the rivers in that basin reaches the ocean.

4. Earthquakes and volcanic eruptions are deadly *natural hazards.*
   Natural hazards are
   a. the result of warfare between countries.
   b. destructive phenomena in the physical environment.
   c. dangerous obstacles along major roads and highways.
   d. human activities that alter the natural environment.

Alternative Assessment

PORTFOLIO ACTIVITY

Learning about Your Local Geography

Group Project: Research

Plan, organize, and complete a research project about water resources. As a group, determine how your local area acquires, processes, and uses freshwater. Begin planning your project by brainstorming what you know about local water resources. How much rainfall does your area receive? What does the local watershed consist of? Who uses the most water? Where do they get that water? Plan how to find the answers to these and other questions. Divide the research and other tasks among group members. When you have completed your research, communicate your results through a report. Use maps, charts, or other visual aids to support your conclusions.

🗗 internet connect

Internet Activity: go.hrw.com
KEYWORD: SW3 GT7

Access the Internet through the HRW Go site to examine the factors that have influenced city growth and economic development along the Great Lakes and St. Lawrence River. Then create a diagram or 3-D model that explains the role of transportation routes, types of economic activities, and resources of the area. Present your diagram or model to the class.

Using the Geographer's Tools

1. Answers will vary.
2. Washington, D.C., is wetter. The Coast Ranges block moist air from reaching Sacramento.
3. Tables will vary.

Writing

Student letters should demonstrate a knowledge of how people in the United States and Canada use their country's resources. Use Rubric 25, Personal Letters, to evaluate student work.

𝒢eography for Life

Student diagrams should include all major tributaries of the Mississippi and the rivers that feed into them. Use Rubric 30, Research, to evaluate student work.

Social Studies Skills

1. b
2. Reno and Albuquerque are located in an arid region. Their annual precipitation amounts are very low.
3. d
4. b

PORTFOLIO ACTIVITY

Student reports will vary based on the community but should accurately reflect the sources and uses of water in your area. Use Rubric 14, Group Activity, to evaluate student work.

163

Natural Environments of North America • 163

CHAPTER RESOURCE MANAGER

Objectives	Pacing Guide	Reproducible Resources
SECTION 1 **History and Culture** (pp. 165–71)	• Identify some important events in the history of the United States. • Describe some unique elements of American culture.	**Regular** 1.5 days **Block Scheduling** 1 day *Block Scheduling Handbook, Chapter 8*	**RS** Guided Reading Strategy 8.1 **E** Cultures of the World Activity: Region 1 **SM** Geography for Life Activity 8: Mark Twain's *Life on the Mississippi* **SM** Map Activity 8: Western Settlement of the United States
SECTION 2 **Regions of the United States** (pp. 173–79)	• Describe the economy of the Northeast. • Explain why the Midwest is an important farming area. • Analyze how the geography of the South is changing. • Explore how environmental conditions have influenced the history of the West.	**Regular** 1.5 days **Block Scheduling** 1 day *Block Scheduling Handbook, Chapter 8*	**RS** Guided Reading Strategy 8.2 **PS** Readings in World Geography, History, and Culture 12
SECTION 3 **Geographic Issues** (pp. 180–83)	• Identify important environmental issues in the United States. • Describe the natural hazards that affect the lives of Americans. • Examine how cities and population patterns in the United States are changing. • Analyze how the U.S. economy is tied to other countries around the world.	**Regular** 1 day **Block Scheduling** .5 day *Block Scheduling Handbook, Chapter 8*	**RS** Guided Reading Strategy 8.3 **RS** Graphic Organizer Activity 8 **SM** Critical Thinking Activity 8: The United States in a Global Economy

Chapter Resource Key

PS Primary Sources	**A** Assessment	CD-ROM
RS Reading Support	**REV** Review	Video
IC Interdisciplinary Connections	**ELL** Reinforcement and English Language Learners	Internet
E Enrichment		Holt Presentation Maker Using Microsoft® PowerPoint®
SM Skills Mastery	Transparencies	

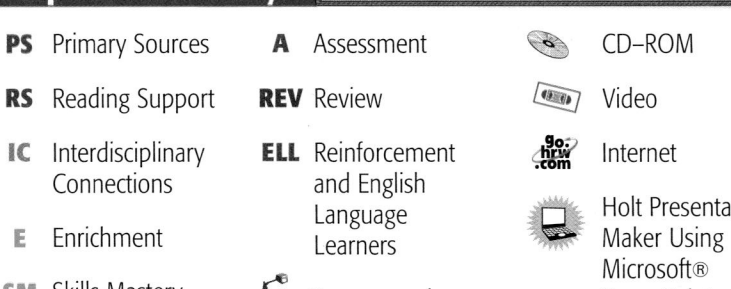

One-Stop Planner CD–ROM

See the *One-Stop Planner* for a complete list of additional resources for students and teachers.

 **One-Stop** Planner CD–ROM

It's easy to plan lessons, select resources, and print out materials for your students when you use the **One-Stop Planner CD–ROM with Test Generator**.

internet connect

HRW ONLINE RESOURCES

GO TO: go.hrw.com
Then type in a keyword.

TEACHER HOME PAGE
 KEYWORD: SW3 Teacher

CHAPTER INTERNET ACTIVITIES
 KEYWORD: SW3 GT8
 Choose an activity to:
 • test knowledge of United States geography.
 • learn about American holidays.
 • create a brochure on national parks.

CHAPTER ENRICHMENT LINKS
 KEYWORD: SW3 CH8

CHAPTER MAPS
 KEYWORD: SW3 MAPS8

ONLINE ASSESSMENT
 Homework Practice
 KEYWORD: SW3 HP8
 Standardized Test Prep
 KEYWORD: SW3 STP8
 Rubrics
 KEYWORD: SS Rubrics

COUNTRY INFORMATION
 KEYWORD: SW3 Almanac

CONTENT UPDATES
 KEYWORD: SS Content Updates

HOLT PRESENTATION MAKER
 KEYWORD: SW3 PPT8

ONLINE READING SUPPORT
 KEYWORD: SS Strategies

CURRENT EVENTS
 KEYWORD: S3 Current Events

Technology Resources

- One-Stop Planner CD–ROM, Lesson 8.1
- *ARGWorld* CD–ROM
- **CNN** Presents World Cultures: Yesterday and Today, Segment 27: The Great Depression
- Geography and Cultures Visual Resources 12–17
- Homework Practice Online
- HRW Go site

- One-Stop Planner CD–ROM, Lesson 8.2
- **CNN** Presents World Cultures: Yesterday and Today, Segment 20: Riverboats Return
- Homework Practice Online
- HRW Go site

- One-Stop Planner CD–ROM, Lesson 8.3
- Homework Practice Online
- HRW Go site

Reinforcement, Review, and Assessment

ELL Main Idea Activity 8.1
ELL English Audio Summary 8.1
ELL Spanish Audio Summary 8.1
REV Section 1 Review, p. 171
A Daily Quiz 8.1

ELL Main Idea Activity 8.2
ELL English Audio Summary 8.2
ELL Spanish Audio Summary 8.2
REV Section 2 Review, p. 179
A Daily Quiz 8.2

ELL Main Idea Activity 8.3
ELL English Audio Summary 8.3
ELL Spanish Audio Summary 8.3
REV Section 3 Review, p. 183
A Daily Quiz 8.3

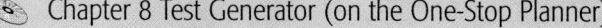

Meeting Individual Needs

Ability Levels

Level 1 Basic-level activities designed for all students encountering new material

Level 2 Intermediate-level activities designed for average students

Level 3 Challenging activities designed for honors and gifted-and-talented students

English Language Learners Activities that address the needs of students with Limited English Proficiency

Chapter Review and Assessment

- Chapter 8 Test Generator (on the One-Stop Planner)
- Global Skill Builder CD–ROM
- HRW Go site
- **REV** Chapter 8 Review, pp. 186–87
- **REV** Chapter 8 Tutorial for Students, Parents, Mentors, and Peers
- **A** Chapter 8 Test (form A or B)
- **A** Alternative Assessment Handbook
- **A** Chapter 8 Test for English Language Learners and Special-Needs Students

Launch into Learning

Create a chart titled *The United States* on the chalkboard with *History, Culture,* and *Economy* as column headings and these phrases as row headings: *Something I think I know, Something I know I know,* and *Something I know I don't know.* Have students copy the chart and complete it. Then lead a discussion about which aspects of the United States are and are not familiar to students. Have volunteers read their *Something I think I know* entries aloud and ask other students if they can confirm the statements. Do the same with the entries that describe what students do not know. Tell students they will learn more about the United States, perhaps including information for their charts, in this chapter.

Using the Physical-Political Map

Have students examine the map on the opposite page. Ask volunteers to name the principal landforms and the main bodies of water. Have students locate the areas where they live and name other areas where they have visited. Then have students compare the physical and political features of the areas mentioned.

Why We Should Know More

You might point out to your students these reasons for learning about the human geography of the United States:

▶ We can be more effective citizens if we understand the history and culture of our country.

▶ Ours is an extremely complex, mobile, and pluralistic society. To succeed today one must have a clear grasp of how our society works.

▶ Learning about the ethnic and cultural diversity of the United States can help students understand peoples and cultures from other countries.

▶ The United States faces various political, environmental, and social issues that affect our lives and require careful study.

CHAPTER
8

The United States

The United States is home to more than 280 million people, roughly 5 percent of the world's population. Most of these people live in the 48 contiguous states between Canada and Mexico. The rest live in Alaska and Hawaii.

Hopi kachina from the southwestern United States

In this textbook you will learn about the diverse regions that make up our world. You will read about the physical features that can be found in each region. You will also learn about the cultural, economic, political, and social features that make each region distinct.

In addition, this textbook will introduce you to people from the world's many countries. Most of these individuals are students like you. They will tell you about their daily lives, activities, and cultures. Through their stories, you will learn something about different ways of life around the world.

We will begin our study of the world by looking at our own country, the United States. In this chapter you will study the history, culture, regions, and important geographic issues of the United States today. You can begin your study of the United States by looking at yourself and the students around you. What languages do you speak? Do you practice a religious faith? Do you live in a big city or in a small town? What are some values and beliefs that you think are important? Do you have a favorite sport? What do you like to eat? The answers to these and many other questions will help you understand what the geography of the United States is like.

Trumpet player in New Orleans

Section 1

OBJECTIVES

1. Identify some important events in the history of the United States.

2. Describe some unique elements of American culture.

LET'S GET STARTED

Copy the following instructions onto the chalkboard: *In your notebooks, list three events that you think are the most important in American history.* Discuss student responses. As students name events, write them on the chalkboard. *(You may wish to keep a copy of this list to review at the end of the lesson cycle.)* At the end of the discussion, have the class vote on the three most important events mentioned. Tell students that in Section 1 they will learn more about the history and culture of the United States.

Building Vocabulary

Write the terms **colonies**, **plantations**, and **bilingual** on the chalkboard. Call on volunteers to suggest definitions for *colonies* and *plantations* based on their previous knowledge of the words. Then have a student locate and read their definitions aloud from the glossary or Section 1. Discuss any differences between these definitions and the ones students suggested. Point out that *lingua*, the root of *bilingual*, is the Latin word for "language." Ask students what they think *bilingual* means. Have a volunteer read the definition aloud.

Section 1
History and Culture

READ TO DISCOVER

1. What are some important events in the history of the United States?

2. What are some unique elements of American culture?

WHY IT MATTERS

American popular culture is constantly spreading to other parts of the world, where it is often a major influence on other cultures. Use  or other **current events** sources to find an article about the influence of American culture on faraway places.

DEFINE

colonies
plantations
bilingual

LOCATE

Alaska
Florida
California
Mississippi River
Texas

Model T

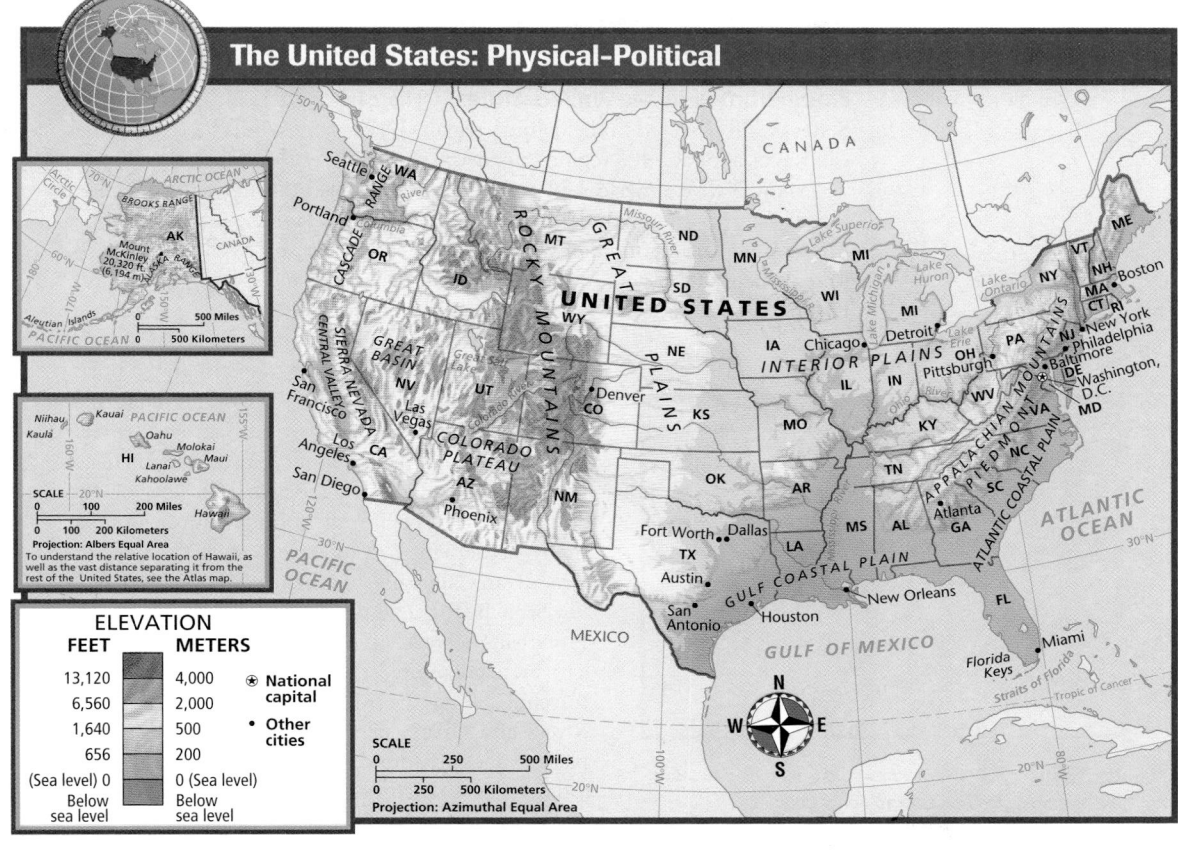

The United States: Physical-Political

Section 1 RESOURCES

REPRODUCIBLE

▶ Guided Reading Strategy 8.1
▶ Cultures of the World Activity: Region 1
▶ Geography for Life Activity 8: Mark Twain's *Life on the Mississippi*
▶ Map Activity 8: Western Settlement of the United States

TECHNOLOGY

▶ One-Stop Planner CD–ROM, Lesson 8.1
▶ CNN Presents World Cultures: Yesterday and Today, Segment 27: The Great Depression
▶ Geography and Cultures Visual Resources 12–17
▶ Homework Practice Online
▶ HRW Go site

REINFORCEMENT, REVIEW, AND ASSESSMENT

▶ Main Idea Activity 8.1
▶ English Audio Summary 8.1
▶ Spanish Audio Summary 8.1
▶ Section 1 Review, p. 171
▶ Daily Quiz 8.1

ELEVATION

FEET	METERS	
13,120	4,000	⊛ National capital
6,560	2,000	
1,640	500	• Other cities
656	200	
(Sea level) 0	0 (Sea level)	
Below sea level	Below sea level	

165

Teach Objective 1

ALL LEVELS: Copy the following graphic organizer onto the chalkboard, omitting the italicized answers. Have students complete the time line with important events from American history. **ENGLISH LANGUAGE LEARNERS**

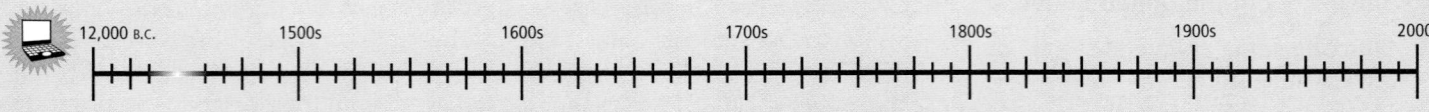

| 12,000 B.C. | 1500s | 1600s | 1700s | 1800s | 1900s | 2000s |

about 12,000 B.C.—Ancestors of American Indians arrive from Asia.

Early 1500s—The Spanish explore the Americas.

1600s—The British begin to set up colonies in America.
1619—The first enslaved Africans arrive in the British colonies.
1600s and 1700s—Spanish colonists migrate north from Mexico into the American Southwest.

1776—The American war for independence begins.

1800s—Migrants move west.
1830—American settlers reach Texas.
1830s—The northeastern United States begins to industrialize.
1848—Gold is discovered in California.
1861–65—Civil War
1869—The transcontinental railroad is completed.

1920—More Americans live in cities than in rural areas.
1917–18—U.S. involvement in World War I
1941–45—U.S. involvement in World War II
1945–early 1990s—Cold War

Across the Curriculum

► Science ◄

Kennewick Man A human skeleton discovered in 1996 is causing archaeologists to question current thinking about what life was like in America thousands of years ago. A skeleton was found in Lake Wallula on the Columbia River near Kennewick, Washington. Scientific tests have revealed that the so-called Kennewick Man is more than 9,000 years old. What puzzles scientists is that the remains bear little resemblance to any known ethnic group. Kennewick Man does not appear to have come from Asia, like the early Americans are presumed to have done, nor does he seem European.

The presence of a non-Asian people in Washington contradicts what most scientists believe about the settlement of America. Some archaeologists now think that humans must have crossed into America much earlier than they had thought, perhaps more than 20,000 years ago. They still believe that the first Americans came from what is now Mongolia but that they left Asia before developing Asian genetic characteristics.

MAP ANSWER

cities—Richmond; people—New York City, Charleston, New Orleans, St. Louis, Dubuque, San Francisco, San Diego, San Antonio, Albuquerque

166

The cliff dwellings of Mesa Verde National Park in Colorado were built between A.D. 1000 and 1300 by the Anasazi, the ancestors of the Pueblo Indians. The Anasazi grew crops on the mesa above these cliffs and in river valleys below. The cliff dwellings were abandoned about 700 years ago, possibly after severe droughts.

History

The ancestors of today's American Indians first settled North America at least 14,000 years ago. These early settlers probably came across an Ice-Age land bridge that linked Asia and Alaska. However, many researchers today think humans may have reached North America much earlier, possibly traveling by sea along the Alaskan coast. Eventually, these early settlers spread throughout the Americas. Over time many different cultures and languages emerged.

Spanish explorers reached the continent about 500 years ago. They claimed parts of what is now the United States from Florida to California. The Spanish were soon followed by the English and French. English colonists settled along the east coast of North America. The French explored the major river systems and Great Lakes in the interior of the continent.

FOCUS ON HISTORY

Place-Names in the United States The early settlement of the United States is still reflected in the country's many place-names. In fact, place-names can tell us a lot about a region's historical geography.

Many Native American place-names are still used throughout the United States. These names include the *Appalachians,* named for the Apalachee tribe. *Mississippi* is an Algonquian word that means "great river." A number of U.S. states, such as Alaska, Arizona, Kansas, Nebraska, and North and South Dakota, have names that originate from Indian words.

Spanish explorers arrived in the early 1500s. In the 1600s and 1700s Spanish settlers migrated north from Mexico into Texas, New Mexico, and California. They set up missions, towns, and forts such as San Antonio, El Paso, Albuquerque, Santa Fe, San Diego, Los Angeles, and San Francisco. Because of the Southwest's dry environment, the Spanish founded many settlements along rivers such as the Rio Grande. In California, they settled along the coast.

French fur trappers were among the first Europeans to reach the Great Lakes and Midwest. The French explored the interior of the country by traveling along the Mississippi and St. Lawrence Rivers and Great Lakes. As a result, many French place-names are found along these physical features today. These include New Orleans, St. Louis, Dubuque, and Detroit.

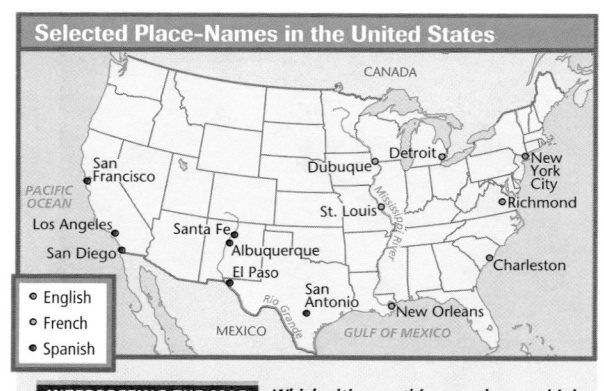

Selected Place-Names in the United States

INTERPRETING THE MAP *Which cities on this map do you think were named after European cities or to honor people?*

Many place-names in the east reflect that region's history as part of the British Empire until 1776. New York, Richmond, and Charleston are among the many English place-names there.

✓ **READING CHECK:** *Environment and Society* What physical features and environmental conditions influenced early French and Spanish migration patterns and the distribution of European settlers?

the dry environment of the Southwest, the St. Lawrence and Mississippi Rivers and Great Lakes

Colonial History The British became the major influence on the early history of the United States. Beginning in the early 1600s they set up 13 **colonies** along North America's east coast. A colony is a territory controlled by people from a foreign land. Other early settlers included Dutch and Germans. In 1619 Europeans began bringing enslaved Africans to the colonies.

Overland travel was difficult in the early colonies. For a long time, water transportation was the colonists' main link to the outside world. In fact, nearly all the early colonial settlements were ports located on natural harbors or navigable rivers. New settlers migrated by sea to the growing coastal towns and inland trading posts on rivers.

The American colonies developed regional economies. In the southern colonies, the climate and soils were ideal for growing tobacco and cotton. Soils were less fertile and farming less productive in the northern colonies. The colonies in the north became centers for trade, shipbuilding, and fishing. Forests were an important resource at a time when ships were made from lumber. Eventually the colonists grew increasingly unhappy with British control over their economies. In 1776 they began a successful rebellion against colonial rule, which led to independence for the United States.

✓ **READING CHECK:** *Environment and Society* How did physical geography affect migration patterns to colonial America? *New settlers migrated by sea to growing coastal towns because water transportation was the main link that colonists had to the outside world.*

Independence and Westward Expansion After independence, the United States set up a federal system of government. Under the U.S. federal system, power is divided between local, state, and national governments. Underlying all levels of government is the idea that ultimate power rests with the people. This idea and the U.S. model of democratic government have both diffused throughout the world.

During the 1800s many Americans and immigrants migrated westward in search of more and better farmland. By 1830, settlers had crossed the

internet connect

GO TO: go.hrw.com
KEYWORD: SW3 CH8
FOR: Web sites about the United States

internet connect

GO TO: go.hrw.com
KEYWORD: SW3 CH8
FOR: Web sites about westward expansion

Albert Bierstadt, *Emigrants Crossing the Plains*, 1867, oil on canvas, A011.IT: The National Cowboy Hall of Fame and Western Heritage Center, Oklahoma City, OK.

INTERPRETING THE VISUAL RECORD

This painting by the American artist Albert Bierstadt (1830–1902) shows a group of pioneers heading west across the Great Plains. In the 1800s some Americans thought the United States was destined to expand across the continent, an idea known as manifest destiny. **What does Bierstadt's painting suggest about American perceptions of the West during the 1800s?**

VISUAL RECORD ANSWER

Possible answers: rugged, grand, majestic, unspoiled, inviting

167

LEVEL 3: Direct students' attention to the map of Territorial Expansion of the United States. Organize the class into pairs and assign each pair one of the territories that was added to the United States after 1783. Then have each pair conduct research into the acquisition of its assigned territory and prepare a brief report on it. Call on volunteers to share their findings with the rest of the class. **COOPERATIVE LEARNING**

HOMEWORK: Have students conduct research on life in colonial America. Then have them write journal entries about a typical day in Plymouth, Santa Fe, New Orleans, or another colonial settlement. Encourage students to mention language, religion, and other cultural elements discussed in this section in their entries. Call on volunteers to share their work with the class.

Linking Past to Present

The National Road Even before the purchase of the Louisiana Territory from France, many Americans, including President Thomas Jefferson, sought to tie the eastern and western parts of the United States together. Plans were laid for a National Road to the west. Construction on the road began in 1811 but was delayed by the War of 1812. After seven years, the first section, called the Cumberland Road, opened between Cumberland, Maryland, and Wheeling, now in West Virginia. Gradually, the road was extended all the way through Ohio and Illinois to St. Louis, Missouri. Many migrants used the road to travel into the country's new territory.

The building of the National Road was the precursor to the federal highway system of today. In fact, U.S. Highway 40 follows the same path as the much older road.

CRITICAL THINKING: How might the National Road have inspired people to move westward? *(easier access across the Appalachians and into western territory)*

MAP ANSWER

through purchases and annexations; the Louisiana Purchase

VISUAL RECORD ANSWER

little settlement when seen as Great American Desert; attracted settlement when seen as Edenic; emigration during Dust Bowl

Territorial Expansion of the United States

INTERPRETING THE MAP

Within 100 years of independence, the borders of the United States had expanded to the Pacific Ocean. New land was added through negotiations, purchases, and wars. **How did the United States gain most of its territory after 1783? Which was the largest single addition?**

INTERPRETING THE VISUAL RECORD

Once considered part of the Great American Desert, the southern Great Plains attracted many settlers in the 1800s. Some even described the area as a Garden of Eden. However, in the 1930s the area was hit by a severe drought made worse by poor farming practices. Strong winds carried away topsoil and caused huge dust storms. As a result, more than 350,000 people left the area, which became known as the Dust Bowl. **How did changing perceptions of the Plains lead to shifts in settlement patterns?**

Mississippi River and settled as far south as Texas. Pioneers were soon settling on the Pacific coast. Many arrived after gold was discovered in California in 1848. However, few people settled in the deserts and mountains of the western United States or in the plains between the Mississippi River and Rocky Mountains. They called this grassland region the Great American Desert and believed it was too dry to support farming.

The boundaries of the United States shifted as settlement spread westward. By the mid-1800s the country stretched from the Atlantic to the Pacific coast. (See the map.) The government sold land cheaply or gave it away to encourage people to settle new areas.

As pioneers moved westward, they began to have bitter conflicts with American Indians. Many American Indians did not own land like the descendants of Europeans did. Instead, some Indians considered land a shared resource rather than someone's personal property. As settlers occupied and divided up land, they pushed American Indians farther west and onto reservations. Many died from warfare or from diseases carried by settlers.

Economic Development By 1830 the northeastern United States was industrializing. Industries and railroads spread. In the South, however, the economy was based on export crops like tobacco and cotton. Farmers grew these crops on **plantations**—large farms that produce one major crop. Many southern plantations used the labor of enslaved Africans. Economic differences between the North and South, and the South's insistence on maintaining slavery, eventually led to the Civil War. This war lasted from 1861 to 1865 and ended with the defeat of the southern, or Confederate, states. The federal government then moved to end slavery throughout the country.

After the Civil War, improvements in technology encouraged rapid westward migration. The transcontinental railroad was completed in 1869. This railroad made it much easier to move goods and people across the country.

Teach Objective 2

🌐 **ALL LEVELS:** Use the students in your classroom to demonstrate the cultural diversity of the United States. Have all students write down on separate small sheets of paper their ethnic heritages, the languages they or their parents speak, and their religions. Ask students not to write their names on their papers. Collect the papers. Then organize the class into three groups. Give one group all of the papers that list ethnic heritages, the second group those with languages, and the third group the ones with religions. Have each group create a graph that illustrates the information they receive. Call on volunteers from each group to share their graphs with the class. **ENGLISH LANGUAGE LEARNERS, COOPERATIVE LEARNING**

🌐 **LEVEL 1:** Direct students' attention to the population map of the United States and Cananda in the unit atlas. Lead a class discussion based on the following questions: Which half of the United States is more densely populated, the east or the west? *(the east)* Which specific areas appear to be the most densely settled? *(the Northeast, the Great Lakes area, southern California)* Which areas are the least densely populated? *(the Great Plains, Rocky Mountains, and desert Southwest)* How do these settlement patterns reflect the country's history? *(Possible Answers: Early settlements were located in the Northeast and later settlers moved westward. Immigrants have tended to move to the Northeast or large coastal cities.)* **ENGLISH LANGUAGE LEARNERS**

Railroads also allowed major cities to develop far from navigable waterways. With new agricultural machinery, farms could produce more food using fewer people than ever before. Irrigation and better plows allowed farmers to grow crops in the Great American Desert. As a result, people's perceptions of the region changed, and they began to settle the area.

The development of industry attracted more people to the country's growing cities. Some came from rural areas. However, many were immigrants, mostly from Europe. Many European immigrants settled in the industrial cities of the Northeast. By 1920 more Americans lived in cities than in rural areas.

The 1900s In the 1900s the United States experienced major social, economic, and technological changes. The country fought in World War I in 1917 and 1918 and suffered through the Great Depression in the 1930s. U.S. forces also fought in World War II from 1941 to 1945. Since then, the United States has been one of the richest and most powerful countries in the world.

After World War II, the United States and the Soviet Union became rivals in the Cold War. Both countries built huge military forces and developed nuclear weapons. The two countries never formally went to war against each other. However, they supported different sides in small wars around the world. Since the collapse of the Soviet Union in 1991, the United States and Russia have had friendlier relations.

✔ **READING CHECK:** *Human Systems* Why did many Americans and immigrants move westward during the 1800s? the availability of farmland

Culture

Because of its long history of immigration, American culture includes traditions, foods, and beliefs from all over the world. In fact, the United States is one of the world's most culturally diverse countries.

People and Languages The diversity of the American people is perhaps their major characteristic. More than 99 percent of Americans are either immigrants or the descendants of immigrants. American Indians make up less than 1 percent of the population. Today immigrants from all over the world continue to arrive in the country.

Most Americans are of European descent. This group includes people whose ancestors came from Britain, Germany, France, and other European countries. About 12 percent of Americans trace their origins to Africa. Slightly more people identify themselves as Hispanic. Hispanics and Asians are the most rapidly growing parts of the population.

Since colonial times, English has been the main language of the United States. However, the United States has no official language. The second most widely spoken language in the country is Spanish. Spanish is particularly common along the U.S.-Mexico border. Many Spanish speakers are also **bilingual**, which means they are able to speak two languages. Hundreds of other languages are spoken in the United States, particularly in large cities.

These schoolchildren in Florida illustrate the cultural and ethnic diversity of the United States. Immigration has long shaped the human geography of the country.

LEVEL 2: Organize students into small groups and have each group select an artistic form or figure to examine. For example, one group might study jazz while another looks at the works of Mark Twain. Each group should conduct research to find examples of cultural elements that influenced the development of the chosen topic. Students may also wish to examine how their subjects have influenced others around the world. Have each group create a poster that illustrates its findings. **COOPERATIVE LEARNING**

LEVEL 3: Have students perform research on the changing nature of immigration to the United States. Assign each student a period of American history and have him or her examine immigration trends for that period. Students should note the countries from which most immigrants arrived, their reasons for settling in the United States, and their destinations within this country. Then have students share their findings with the rest of the class. You may wish to have students present their reports in chronological order. **BLOCK SCHEDULING**

Cultural Kaleidoscope

Islam in the United States

Christianity is the predominant religion in the United States. However, other religious groups are growing in size and influence. For example, by 2001 Islam had surpassed Judaism to become the religion with the second-largest number of members.

The first Muslims who arrived in America were brought as slaves from western Africa. Immigrants from Africa and Asia have increased the number of Muslims living in this country to almost 6 million. Also, there is a growing population of American-born Muslims—children born into Muslim families and converts to the religion.

DISCUSSION: Lead a discussion on the contributions of various religious groups to American culture.

Diverse foods and cooking styles in the United States reflect regional and cultural differences. For example, the Cajuns of southern Louisiana are known for their spicy and flavorful dishes. Here, a restaurant owner in New Orleans shows off a dish of boiled crawfish, a local specialty.

MAP ANSWER

Point out to students that the center is defined as the place where an imaginary, flat, weightless, and rigid map of the United States would balance perfectly if all residents were of equal weight. Expansion of U.S. territory and the growth of western and southern cities accounts for the changes. The distribution will likely continue to expand west and south.

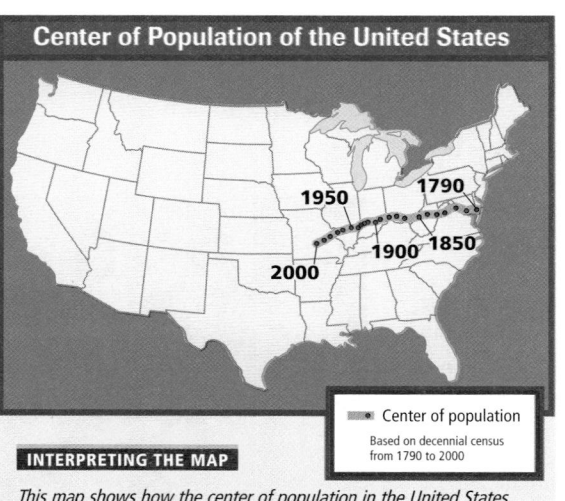

Center of Population of the United States

1790
1850
1900
1950
2000

■ Center of population
Based on decennial census
from 1790 to 2000

INTERPRETING THE MAP

*This map shows how the center of population in the United States has changed throughout the country's history. **What factors led to the changes in population distribution shown here? How do you think the distribution of the U.S. population will change in the future?***

Religion The cultural diversity of the United States can also be seen in its many religions. There are more than 1,200 religious groups in the country. Immigrants introduced many of these religions. Other religions were founded in the United States.

Christianity is the major religion in the country. European settlers brought it to North America. More than half of all Americans are Protestant Christians. Major Protestant groups include Baptists, Lutherans, and Methodists. About 25 percent of Americans are Roman Catholic Christians. Many American Catholics have Spanish, Italian, or Irish ancestors. Some 6 million Jewish Americans live in the United States—more than live in Israel. In recent years, immigration has increased the numbers of people who practice Buddhism, Hinduism, and Islam.

Settlement and Land Use The population of the United States is concentrated in the Northeast. (See the unit population map.) This pattern reflects the history of the country's settlement. Early European settlement was concentrated in the Northeast, and the first large U.S. cities were located there. Later, settlers moved westward.

Although the Northeast is still the most concentrated area of settlement, the country's population has been moving to the South and West. (See the map.) This shift in settlement reflects the decline of the country's old industrial region, once known as the Rust Belt. At the same time, warmer areas in the South and West, known as the Sun Belt, have attracted many people.

Settlement patterns also reflect land use in the country. For example, the most densely populated regions include the urban areas of the Northeast, Midwest, and Pacific coast. Less densely populated are the rich farmlands and ranch areas of the Midwest and West. The most sparsely populated areas are desert and mountain regions of the West.

✔ **READING CHECK:** *Human Systems* How has America's long history of immigration shaped its culture? ethnically diverse population, presence of many religions, use of English and other languages

Education Education has been an important factor in the development of the U.S. economy. It has also helped create a high standard of living. Basic education is free and is required of all citizens. Nearly all children complete elementary school, and more than 80 percent graduate from high school. Many go on to study at colleges and universities.

The United States has one of the largest and best systems of higher education in the world. More than 3,000 colleges and universities are located in the country. Many, such as Harvard and the University of California at Berkeley, are world leaders in research and teaching. In fact, tens of thousands of students from all over the world come to study at universities in the United States.

✔ **READING CHECK:** *Places and Regions* How does the U.S. education system make the country a distinctive region? important factor in development of the U.S. economy, excellent system of higher education

The Arts, Customs, and Traditions Americans have many different traditions and customs. In the arts, American writers, artists, musicians, filmmakers, and sculptors are internationally famous. The United States helped pioneer the development of motion pictures and still dominates the industry. American movies are shown all over the world. In architecture, the United States was the first country to build skyscrapers. This style of architecture has now diffused throughout the world. Well-known American writers and poets include Edgar Allan Poe, Mark Twain, Emily Dickinson, Ernest Hemingway, Maya Angelou, and many others.

The many ethnic and cultural groups that came to the United States brought their own musical styles. For example, Africans brought the rhythms of West African music. Europeans brought instruments and harmonies from their native lands. As many African Americans migrated to cities in the early 1900s, the musical traditions of Africans and Europeans blended together to form jazz. Jazz later diffused from cities such as New Orleans to the rest of the country. In fact, jazz is now popular around the world.

Other musical styles that originated in the United States include blues, country, rock 'n' roll, and rap. These styles have spread around the world through recordings, radio, and television. American music is a major influence on globalization and the diffusion of American culture.

Widely celebrated holidays in the United States include Christmas and Easter, which reflect America's Christian religious heritage. Many Americans celebrate other religious holidays. Americans also mark Independence Day, July 4, with fireworks and picnics. In addition to national celebrations, many towns and communities celebrate local historical events and personalities.

People in the United States enjoy watching and playing baseball, basketball, football, golf, and other sports. American athletes compete in international sporting events such as the Olympics. In fact, popular American sports such as baseball and basketball have diffused to other parts of the world and contributed to the spread of American culture.

Louis Armstrong (1901–71) was one of the most popular and influential jazz musicians of all time. Nicknamed Satchmo, he was a gifted trumpet player, singer, and bandleader. Armstrong's brilliant performances changed the way jazz was played and sung.

✓ **READING CHECK:** *Human Systems* How did the distinctive cultural patterns of the United States lead to the innovation and diffusion of jazz? The contact of different cultural groups led to the creation of jazz, which later diffused around the country and world.

Review

Homework Practice Online
Keyword: SW3 HP8

Define colonies, plantations, bilingual

Working with Sketch Maps On a map of the United States that you draw or that your teacher provides, label Alaska, Florida, California, Mississippi River, and Texas. Which state are the ancestors of Native Americans believed to have crossed into from Asia?

Reading for the Main Idea

1. *The Uses of Geography* How did migration over an Ice-Age land bridge affect the history and culture of the United States?

2. *Human Systems* How did American Indians and descendants of Europeans view land ownership differently?

3. *Environment and Society* How did irrigation and better plows change settlers' perceptions of the Great American Desert between the Mississippi River and the Rocky Mountains?

Critical Thinking

4. **Comparing** How do you think patterns of land use might lead to cultural differences between urban areas and rural farming areas?

Organizing What You Know

5. Copy the graphic organizer below. Use it to describe important cultural features that make the United States distinctive.

People and languages	
Religion	
Settlement and land use	
Education	
The arts, customs, and traditions	

Technology: Teleworking

Advances in technology and transportation have allowed business-people to travel quickly to distant locations for meetings and other functions. However, these same advances have also made it possible for people to conduct business without ever meeting face to face. For some, telephone calls, faxes, and e-mails have largely taken the place of face-to-face meetings. The popularity of home computers has also given many employees a chance to work from home. This practice is known as teleworking or telecommuting. Employees remain at home and send work material to an office through a modem and telephone line. Many companies employ teleworkers because it allows them to hire people who do not live near the company's offices. Other teleworkers may live near an office but feel more productive working at home.

Organize the class into groups and have each group create a chart listing some advantages and disadvantages of teleworking. Students should note that employers and employees might have different points of view on this matter. Encourage groups to consider both perspectives as they compile their charts. Ask students how increased teleworking might contribute to the perception that the world is getting smaller. *(Possible answer: People can work together even though they live thousands of miles apart.)*

▶ **Math** ◀

Travel Speeds New York and San Francisco lie approximately 2,500 miles (4,000 km) apart. Have students calculate the average approximate speed of the train, propeller plane, and jet airplane. Remind students that the formula to calculate speed is the distance traveled (in miles) divided by the time spent in travel (in hours). For example, a jet airplane takes five to six hours to make the trip. Dividing 2,500 miles by five to six hours yields an average speed of 417 to 500 miles per hour.

Applying What You Know Answers

1. easier to move people, products, and information to distant places
2. Physical features such as mountains might not be seen as such difficult barriers to movement. Connections between widely separated places can increase, contributing to economic and cultural diffusion and change.

DIAGRAM ANSWER

change from ocean routes to overland and air routes; increase cultural contexts

This Geography for Life feature addresses National Geography Standards 6, 9, and 10.

The World in Spatial Terms

Geography for Life

A Small World after All?

Have you ever heard the expression, "the world is getting smaller"? Have you wondered what this expression means? As technology and transportation improve, it becomes easier and quicker to move people, products, and information to distant places. As a result, these places begin to seem closer because they become more accessible. However, these places are the same distance apart as they were before.

Many geographers are interested in the relationship between changes in technology, movement, and perception. They use the term *time-space convergence* to refer to the increasing nearness of places that happens as transportation and communication technologies improve. These improvements can steadily reduce the amount of time it takes to travel from place to place.

For example, imagine you lived in the mid-1800s and wanted to travel from New York to San Francisco. How long do you think it would take? What route would you follow? How difficult would the journey be? Most importantly, how would you perceive the journey? As you can see from the diagram, it took three to five months to travel from New York to San Francisco in 1849. Today, however, it is just a five or six hour flight. What was once a long, difficult ocean voyage is now a comfortable flight complete with dinner and a movie. In fact, some people now travel from coast to coast for weekend vacations or one-day business meetings.

Other modern technologies are also tying distant places closer together. For example, faxes, the Internet, and e-mail all help the free flow of information between some of the most isolated places. The same technologies also encourage the diffusion of culture traits and globalization. Together with modern travel, these technologies are helping make the world seem smaller and smaller.

Travel Times from New York to San Francisco

jet airplane
propeller airplane
New York
railroad
San Francisco
ship

Year	Mode of travel	Time
1849	ship	3–5 months
1869	railroad	1–2 weeks
1945	propeller plane	12 hours
2000	jet airplane	5–6 hours

INTERPRETING THE DIAGRAM *As this diagram shows, changes in technology have dramatically reduced travel times between New York and San Francisco over time.* **How have new forms of transportation altered travel routes? How do you think these changes in transportation technology affect the country's cultural geography?**

Applying What You Know

1. **Summarizing** How can improvements in technology and transportation make the world seem smaller?

2. **Drawing Inferences and Conclusions** How do you think improvements in technology and transportation can change people's perceptions of geographic features? How might these changes lead to changes in human societies?

Section 2

O B J E C T I V E S

1. **Describe the economy of the Northeast.**

2. **Explain why the Midwest is an important farming area.**

3. **Analyze how the geography of the South is changing.**

4. **Explore how environmental conditions have influenced the history of the West.**

LET'S GET STARTED

Copy the following question onto the chalkboard: *What name would you give to the region of the United States in which we live?* Discuss student responses. Point out that several different answers may be correct since the country can be divided in several ways. *(For example, the Sun Belt includes large areas of what is generally considered the South.)* Tell students that they will learn more about the regions of the United States in Section 2.

Building Vocabulary

Write the terms **metropolitan area** and **smog** on the chalkboard. Have a student locate and read their definitions from the text or glossary. Point out that *metropolis* is another word for a large city. A *metropolitan area* includes the smaller cities surrounding the large one. Then have students note that the word *smog* was created by combining *smoke* and *fog*. Call on a volunteer to read the definitions for the rest of the key terms.

Section 2 — Regions of the United States

READ TO DISCOVER

1. What is the economy of the Northeast like?
2. Why is the Midwest such an important farming area?
3. How is the geography of the South changing?
4. How have environmental conditions influenced the history of the West?

WHY IT MATTERS

The U.S. Census Bureau divides the United States into regions. Use **CNNfyi.com** or other **current events** sources to learn about how the Census Bureau organizes information about the regions of the United States.

IDENTIFY

Megalopolis
Corn Belt
Dairy Belt
Wheat Belt
Silicon Valley

DEFINE

textiles
metropolitan area
arable
smog

LOCATE

Washington, D.C.
Boston
New York
Chicago
Detroit
St. Louis
Dallas
Houston
Miami
Los Angeles
Seattle
San Francisco

Section 2 RESOURCES

REPRODUCIBLE
▶ Guided Reading Strategy 8.2
▶ Readings in World Geography, History, and Culture 12

TECHNOLOGY
▶ One-Stop Planner CD–ROM, Lesson 8.2
▶ CNN Presents World Cultures: Yesterday and Today, Segment 20: Riverboats Return
▶ Homework Practice Online
▶ HRW Go site

REINFORCEMENT, REVIEW, AND ASSESSMENT
▶ Main Idea Activity 8.2
▶ English Audio Summary 8.2
▶ Spanish Audio Summary 8.2
▶ Section 2 Review, p. 179
▶ Daily Quiz 8.2

The Northeast

The Northeast is the smallest and most densely populated region in the United States. It includes Maine, New Hampshire, Vermont, Massachusetts, Rhode Island, Connecticut, New York, New Jersey, Pennsylvania, West Virginia, Maryland, Delaware, and Washington, D.C. *D.C.* is the abbreviation for "District of Columbia." The Northeast is home to about a fifth of the country's population. Most of that population is concentrated in an urban corridor known as **Megalopolis**. A megalopolis is a group of cities that have grown into one large, built-up area. Boston, New York, Philadelphia, Baltimore, Washington, D.C., and their suburbs form the Megalopolis. A web of highway, rail, and air routes links these urban areas.

The Northeast is the political and financial center and most industrialized region of the United States. In fact, the country's first industries developed there. These early industries used running water from rivers to power machinery and produce **textiles**, or cloth products. A wide variety of manufacturing

INTERPRETING THE VISUAL RECORD

New York is located in the heart of Megalopolis. The term Megalopolis *was used in the 1960s by a French geographer to describe the huge urban area stretching from Boston to Washington, D.C. Today it is used to describe other massive urban areas around the world.* **What other large urban area would you describe as a megalopolis?**

VISUAL RECORD ANSWER

Possible answers: Tokyo-Yokohama, the Los Angeles area, the San Francisco Bay Area, Mexico City

173

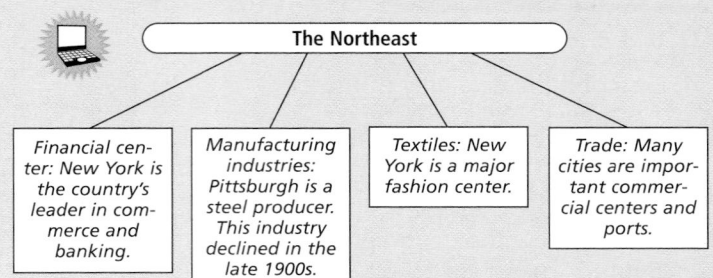

The Northeast

| *Financial center: New York is the country's leader in commerce and banking.* | *Manufacturing industries: Pittsburgh is a steel producer. This industry declined in the late 1900s.* | *Textiles: New York is a major fashion center.* | *Trade: Many cities are important commercial centers and ports.* |

Daily Life

Fishing in New England
The fishing industry has always been important in New England. Towns such as Gloucester and New Bedford, Massachusetts, have long been involved with fishing, and halibut, flounder, and cod were regularly brought to shore. Today, however, fishers face many challenges.

Changes in technology in the 1900s brought great changes to the fishing industry. Some of these changes threaten the industry's future. Fishers have overexploited the seas, so fewer fish remain. More-powerful ships allowed crews to travel greater distances in search of fish and to increase their catches. Very fine nets meant that many baby fish were caught. Some argue that there were simply too many people fishing. Diminishing fish stocks are of great concern to marine biologists and conservationists across the globe. Recent bans on fishing have increased the number of fish in certain areas, such as Georges Bank, Massachusetts. Constant diligence is required, however, to protect the future of New England's fisheries.

CRITICAL THINKING: How have improvements in technology affected fish populations near New England? (*Larger catches have led to reduced populations of fish.*)

Many farmers in the Northeast grow specialty crops that are shipped overnight to the region's large urban markets. In this photo, cranberries are being grown in southern New Jersey.

industries developed later. They developed in places favorable to certain industries, such as steelmaking. For example, consider Pittsburgh, Pennsylvania. Pittsburgh is located at the junction of two rivers and is near rich iron ore and coal deposits. This location helped make the city a major early center of the steel industry.

Good transportation connections were very important to industrial growth in the Northeast. The early growth of the region's cities was made possible by their good port sites, which allowed the movement of people and products. The completion of the Erie Canal in the 1820s provided access to the Great Lakes and the interior United States. Today a dense network of roads, railroads, and air routes crisscrosses the region and connects it to the rest of the country.

In the late 1900s manufacturing in the Northeast declined. Cheaper labor in other places forced many factories to close. Because of its declining older industries, the region became known as the Rust Belt. Meanwhile, economic growth and warm climates attracted people and industries to the South and West. The Northeast is still a major industrial area and has attracted newer industries. However, the region has lost the economic dominance it once enjoyed.

✓ **READING CHECK:** *The Uses of Geography* What features helped Pittsburgh grow as a major steel center? location near major iron ore and coal deposits and at the junction of two rivers

Cultures The Northeast is home to many different cultures. Until the mid-1960s most immigrants came from Europe. As a result, some cities in the region have large Greek, Irish, and Italian neighborhoods. In recent decades most immigrants have come from Latin America and Asia. These immigrants bring different cultures and ways of life with them, reshaping the region's cultural geography. For example, many Northeastern cities now host parades celebrating the Chinese New Year. Other neighborhoods are alive with the sounds of Latin American music and people speaking Spanish.

Cities The New York City area is the largest **metropolitan area** in the United States. A metropolitan area is a city and its surrounding built-up areas. More than 20 million people live in the New York area. Dutch colonists settled at the site of New York in the 1620s and called it New Amsterdam. The English renamed the city New York later after acquiring it from the Dutch. New York's excellent natural harbor at the mouth of the Hudson River makes it an ideal location for a port and trading center. Today New York is America's leading center of commerce, banking, advertising, fashion, and media.

Other cities in the Megalopolis have long been important commercial centers. Boston and Philadelphia date back to colonial times. Baltimore, on the western shore of Chesapeake Bay, was founded on the fall line. Baltimore became a major port because of its rail connections to interior coal mines, steel mills, and farming areas. Washington, D.C., is unique among the country's cities because it was planned and built to serve as the U.S. capital. Construction of the city began in the late 1700s.

Teach Objective 2

 LEVEL 1: Provide each student with an outline map of the United States. Then have students use the descriptions in the text and library resources to color and label the Corn Belt and Dairy Belt on their maps. Lead a discussion about various factors that make the Midwest a productive agricultural region. **ENGLISH LANGUAGE LEARNERS**

 LEVELS 2 AND 3: Have students conduct research to find examples of other agricultural products grown in the Midwest and add them to the maps they created for the Level 1 activity.

Teach Objective 3

 LEVEL 1: Copy the following graphic organizer onto the chalkboard, omitting the italicized answers. Have each student complete the organizer with words or phrases to describe the South before the Civil War and the South of today. **ENGLISH LANGUAGE LEARNERS**

The South

Before the Civil War
• *mostly rural and agricultural*
• *cotton, rice, and tobacco plantations worked by slaves*
• *lacked industries and railroads*

The South Today
• *high-tech and aerospace industries, automobile manufacturing, banking*
• *tourism*
• *immigration from Caribbean, Mexico, and other areas in Latin America*

The Great Lakes and St. Lawrence Seaway

Moving through a Canal Lock

INTERPRETING THE DIAGRAM *A system of locks along the St. Lawrence Seaway connects the Great Lakes to the Atlantic Ocean. Locks raise and lower ships so they can sail between areas with different water levels.* **Why and how have the United States and Canada altered the natural environment by building the St. Lawrence Seaway? How was technology used?**

Government employment is the largest source of income today, followed by tourism.

✓ **READING CHECK:** *Places and Regions* How is the character of the Northeast related to its cultural characteristics? A history of immigration has created a culturally diverse population, particularly in urban areas.

The Midwest

The Midwest is this country's major farming region and a leading producer of industrial goods. The region includes Ohio, Indiana, Illinois, Michigan, Wisconsin, Minnesota, Iowa, and Missouri. The combined population of these states is slightly smaller than that of the Northeast. The Great Lakes and Mississippi River link the Midwest's major cities to each other and to other regions. (See the Great Lakes and St. Lawrence Seaway map.) Railroads and highways are also major transportation connections.

Many settlers moved into the Midwest as transportation routes from the East Coast developed during the 1800s. Excellent farmland and growing cities attracted many of these settlers. In addition, between 1915 and 1930 hundreds of thousands of African Americans migrated to the region from the South. They came to find work in cities like Chicago and Detroit.

Agriculture The Midwest is one of the most productive farming regions in the world. Most of the region's land is **arable**, or fit for growing crops. In fact, some areas specialize in certain crops. One such region is the **Corn Belt**, which stretches from Nebraska to Ohio. Within the Corn Belt, Illinois and Iowa are the country's leading corn-producing states. Most of this corn is used to feed livestock such as beef cattle and hogs. The United States is also the world's major exporter of corn. Soybeans are another important Corn Belt crop. Soybeans are used to make margarine, vegetable oil, and bean curd (tofu).

The **Dairy Belt** is located north of the Corn Belt, where summers are cooler and soils are rockier and less fertile. The Dairy Belt includes Wisconsin and most of Minnesota and Michigan. Wisconsin—"America's Dairyland"—produces more butter and cheese than any other state. These dairy products are sold in the Midwest's cities and are shipped by refrigerated trucks to the rest of the country.

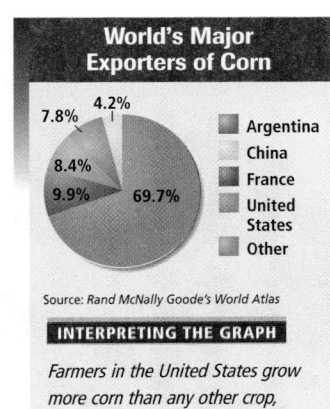

World's Major Exporters of Corn

7.8%
4.2%
8.4%
9.9%
69.7%

■ Argentina
□ China
■ France
■ United States
■ Other

Source: Rand McNally Goode's World Atlas

INTERPRETING THE GRAPH

Farmers in the United States grow more corn than any other crop, producing about 40 percent of the world's total and nearly 70 percent of world exports. **Besides the United States, what other countries are major exporters of corn?**

Essential Element 5

► **Environment** ◄
and Society

Soybeans The Corn Belt could easily be called the Soy Belt instead. The production of soybeans has been an economic boon to American farmers. It has also helped to improve soil quality. The plant was introduced into the region at the beginning of the 1800s. A visitor to China smuggled a few soybean seeds out of that country, where the plant had been domesticated for about 5,000 years, and brought them to the United States.

Farmers often rotate soybeans with corn, a process that reduces insect infestations and disease. In addition, soybeans, which are legumes, fix nitrogen in the soil. Corn needs nitrogen-rich soil to thrive. The beans themselves are consumed by humans and livestock and are made into numerous products. The United States is a leading soybean exporter. Much of the country's crop is shipped to China.

DIAGRAM ANSWER

To allow shipping and global trade, the two countries altered the flow of the St. Lawrence River by building a series of locks.

GRAPH ANSWER

France, Argentina, China

LEVEL 2: Have each student select a city or region in the South and examine how it has changed since the Civil War. Tell students to analyze changes in population, new industries that have developed, and patterns of migration and immigration to their chosen location. Ask students to present their findings in illustrated time lines.

LEVEL 3: Organize students into groups and have each group design a series of maps showing the changing geography of the South. Direct each group to draw a map for 1850, 1900, 1950, and the current year. Each map should show major cities and economic activities in the region. Then have students write brief reports that explain the changes seen in their maps. Call on volunteers to share their work with the class. COOPERATIVE LEARNING, BLOCK SCHEDULING

Cultural Kaleidoscope

The Gullah Dialect The coastal islands off South Carolina and Georgia offer a unique look into America's past. Some descendants of slaves from western Africa who live on the Sea Islands still speak an English-based dialect known as Gullah. It includes nearly 6,000 African words. Most of those words are used as personal names, such as *Anyika,* which means "She is beautiful." Other words that have passed from Gullah into English include *goober* (peanut), *juke* (as in *jukebox),* and *voodoo* (witch-craft). The crafts, music, and food of the Sea Islands also show the influence of African culture.

DISCUSSION: Ask students to make a list of questions they would like to ask Gullah speakers. Then challenge them to find the answers by conducting research on the Gullah dialect or by finding ways to contact Sea Island residents.

The Gateway Arch stands next to the Mississippi River in St. Louis. The arch is 630 feet (192 m) high and symbolizes the historical importance of St. Louis as the Gateway to the West.

Cities Chicago is the largest city in the Midwest and the third-largest U.S. metropolitan area. Chicago's location in the interior of North America has made it a major transportation center. The city has a port on Lake Michigan and is the country's most important railroad hub. Chicago's largest airport, O'Hare, is one of the busiest in the world. The city is also an important cultural center and is home to excellent universities and museums.

Other large cities located along the shores of the Great Lakes are Cleveland, Detroit, Milwaukee, and Toledo. Their locations gave the cities access to coal from the Appalachians and iron ore from upper Michigan, northern Wisconsin, and Minnesota. Each is a major manufacturing center. For example, Detroit has been the center for automobile manufacturing in the United States.

The Twin Cities of Minneapolis and St. Paul are located on the upper Mississippi River. They are major distribution centers for the agricultural products of the upper Midwest. St. Louis, Missouri, is located near the area where the Mississippi, Missouri, and Illinois Rivers flow together. The city began as a French fur-trading post in 1764. In the 1800s St. Louis was the center for pioneers heading west and was a major river port known as the Gateway to the West.

✓ **READING CHECK:** (*Places and Regions*) How did access to resources help the development of cities like Cleveland, Detroit, Milwaukee, and Toledo? Their locations on the Great Lakes gave them access to coal and iron ore needed in their manufacturing industries.

The South

The South stretches in a great arc from Virginia to Texas. The region includes Virginia, North Carolina, South Carolina, Georgia, Florida, Alabama, Mississippi, Tennessee, Kentucky, Arkansas, Louisiana, and Texas. These states are home to a little more than 30 percent of the country's population, more than any other region.

Historically, the South was mainly rural and agricultural. The majority of the population lived on farms. Enslaved Africans worked on large cotton, rice, and tobacco plantations. The Civil War wrecked the South's economy. The South became the poorest region in the country. Its economy lacked the industries and railroads that were so important to the development of the Northeast and Midwest. As a result, many southerners migrated to northern cities in search of factory jobs.

The South has attracted new industries in recent decades. Since the 1960s many people have migrated back to the region. In addition, new immigrants have moved to the South from the Caribbean, Mexico, and other areas in Latin America. In parts of southern Florida, for example, Hispanics make up a majority of the population. Spanish is more widely spoken than English in some places.

Stately oak trees and large plantation homes are common in the South. This plantation house north of Baton Rouge, Louisiana, has columns and balconies typical of plantation architecture.

LEVEL 1: Organize students into small groups and have each group create a poster or collage that illustrates the natural environments of the West. Provide students with art supplies and old magazines that include pictures of the region's landscapes. Then have each group write a series of sentences describing how environmental factors have shaped the West's history. Call on volunteers to share their posters and sentences with the class. **ENGLISH LANGUAGE LEARNERS, COOPERATIVE LEARNING**

LEVELS 2 AND 3: Have each student select a western state and write an essay relating its settlement and economic development to its landscape and climate. Have students create maps or charts to accompany their essays. Call on volunteers to share their work with the rest of the class. **BLOCK SCHEDULING**

Economy Primary industries based on local raw materials are important in the South. For example, the region is a major source of lumber. In turn, forestry supports paper, pulp, and furniture industries. Farm products, particularly cotton and tobacco, are still important. Cotton provides raw material for a large textile industry. The textile industry is concentrated in the Piedmont of Georgia, the Carolinas, and Virginia. Southern states, particularly Texas and Louisiana, are major producers of mineral and energy resources.

In recent years, many new industries have developed in the South. These new industries take advantage of the region's lower wages, cheap land, and favorable laws and regulations. For example, high-tech and aerospace industries have been growing. Foreign automobile makers have also built factories here. The headquarters of several large banks are located in North Carolina. That state has also become an important center for the biotechnology industry.

The South's warm climate has made the region a popular tourist destination. For example, Orlando and other places in Florida rely heavily on tourism. The South attracts retirees from colder parts of the country. Some cities, such as Tampa, have become popular places for retired people to live.

Cities The Dallas–Fort Worth metropolitan area is the largest in the region. Dallas was founded in 1841 as a railroad shipping stop for cattle and cotton. Fort Worth was an early cattle-marketing center and is sometimes considered part of the West. Today it is a center for oil, grain, and aerospace industries. Houston and surrounding cities make up the second-largest metropolitan area in the South. Atlanta, Miami, and New Orleans are also transportation and commercial centers.

✓ **READING CHECK:** *Human Systems* What has attracted new industries to the South in recent years? cheap land, favorable laws and regulations, lower wages

The West

The West is the largest and most sparsely populated region of the United States. About a quarter of the country's population lives in this huge area. The interior West includes the Great Plains, Rocky Mountains, and an intermountain region west of the Rockies. States there include North Dakota, South Dakota, Nebraska, Kansas, Oklahoma, New Mexico, Colorado, Wyoming, Montana, Idaho, Utah, Nevada, and Arizona. Farther west are the Pacific states: California, Oregon, Washington, Alaska, and Hawaii.

Environmental conditions have played an important role in the history of the West. Much of the area consists of dry plateaus, deserts, and high mountains.

Warm tropical weather and sandy beaches draw millions of people to Florida. Tourism is an important part of the state's economy.

The San Juan Mountains, a range in the Rockies, rise dramatically in southern Colorado. The West has long been known for its spectacular scenery and natural landscapes.

internet connect

GO TO: go.hrw.com
KEYWORD: SW3 CH8
FOR: Web sites about national monuments

Linking Past to Present

The Comstock Lode In the late 1850s prospectors discovered gold near Virginia City in western Nevada. However, frustrated miners complained that the small deposits of gold they found had to be dug out of a sticky blue-gray mud that clung to picks and shovels. This troublesome mud was soon identified as the richest deposit of silver ore ever found in the United States. It became known as the Comstock Lode. Miners became millionaires almost overnight, and Virginia City exploded into a thriving frontier town known for its wild society.

In 1864 President Lincoln admitted Nevada to the Union even though it did not yet have the population necessary to qualify for statehood. He needed the silver and gold mined from the Comstock Lode to finance northern efforts in the Civil War.

Silver mining was the most important economic activity in Nevada for many years, but the metal began to decline in value in the 1870s. The Comstock Lode, stripped of its gold and silver, was abandoned by 1900. Virginia City remains, but tourism has become its primary industry.

INTERPRETING THE VISUAL RECORD

In 1872 the U.S. Congress established Yellowstone National Park. Yellowstone was the world's first national park. Since then, the national park concept has spread around the world. **How might the diffusion of the national park concept have caused cultural change in the United States and elsewhere?**

Our Amazing Planet

The oldest known living tree is a bristlecone pine in Nevada that is thought to be about 4,900 years old.

VISUAL RECORD ANSWER

increased people's environmental awareness and led to protection of natural environments around the world

Early pioneers found the region difficult to travel across, inhospitable to live in, and almost impossible to farm. However, the opening of the transcontinental railroad in 1869 made travel much easier. As a result, many settlers moved into the region. In the 1900s aqueducts and irrigation systems opened up even more areas to settlement and farming.

The Interior West Historically, raising livestock has been a major economic activity in the interior West. In many areas, raising livestock is combined with wheat farming. The **Wheat Belt** stretches across the Dakotas, Montana, Nebraska, Kansas, Oklahoma, Colorado, and Texas. Irrigation water from the Ogallala Aquifer allows farmers to grow wheat and other crops. Overuse of this water, however, has lowered the water table of the aquifer. As a result, some farmers have begun to use more-efficient irrigation methods to try to preserve valuable groundwater.

Mining is a key economic activity in the Rocky Mountains. Early prospectors struck large veins of gold and silver there. Today Arizona, New Mexico, and Utah are leading copper-producing states. Nevada is the leading gold-mining state. Lead and many other ores are also found in the interior West. Tourism is also important. Attractions include ski resorts like Aspen and Vail in Colorado and Taos in New Mexico. Each year millions of people visit the region's stunning national parks. These parks include Glacier in Montana, Grand Canyon in Arizona, and Grand Teton and Yellowstone in Wyoming.

✓ **READING CHECK:** *Environment and Society* How did environmental conditions influence past migrations and settlement patterns in the West? Early pioneers found the dry rugged region difficult to travel across, inhospitable to live in, and almost impossible to farm.

The Pacific States Most people in the West live in the Pacific states. California is home to some 34 million people, more than any other state. Today the Pacific Coast ranks second only to Megalopolis in economic importance.

Before World War II the economy of the Pacific states was based mostly on farming, forestry, and the film industry in Los Angeles. The growth of military bases in the region during World War II boosted the economy. This growth also drew migrants to the region. After the war, dams were built on the Columbia River to provide cheap hydroelectricity for aluminum smelters and other growing industries. Aircraft manufacturing in Seattle became Washington's largest industry.

In much of California, the warm Mediterranean climate allows a year-round growing season. However, rainfall is rare in the summer, which means crops must be irrigated. Building aqueducts to carry water from the mountains of northern California and the Colorado River to central and southern California created a boom in farming. Today agriculture uses about 80 percent of California's water supply. City residents must compete with farmers for scarce water resources.

In the late 1900s the development of computer technologies brought new industries to the Pacific states. **Silicon Valley**, located south of San Francisco, became the country's leading center of computer technology. Many software companies are located in the San Francisco Bay and Seattle areas.

Ask students to summarize the economic characteristics of each region of the United States. Have students look at the pictures in the section and describe any economic activities they see. (*Possible answers: agriculture, tourism*)

Have students complete **Main Idea Activity for English Language Learners and Special-Needs Students 8.2.** Have students work in groups to outline the text material about one region of the United States. Call on volunteers to share their outlines with the class. **ENGLISH LANGUAGE LEARNERS**

| Review and Assess |

Have students complete the **Section Review**. Then have students complete **Daily Quiz 8.2.**

| Extend |

Have interested students conduct research into economic ties among the regions of the United States. For example, students might examine how raw materials produced in one region are used in another or transport routes among cities. Have students present their findings in poster form. **BLOCK SCHEDULING**

The economies of Alaska and Hawaii depend heavily on their states' locations, natural resources, and scenery. The United States bought Alaska from Russia in 1867. It became a state in 1959. Alaska is the country's largest and least densely populated state. Its economy was initially based on fishing. However, after the discovery of large North Slope oil deposits, oil became Alaska's most valuable natural resource. Hawaii also became a state in 1959. Because of its strategic Pacific location, Hawaii is home to many military facilities. The state's tropical climate and fertile soils are used to grow crops like pineapples and sugarcane. Tourism is also a major industry.

Cities Los Angeles is the largest metropolitan area in the West and the second-largest in the country. The city began as a Spanish settlement in 1781. Railroad connections boosted the town's economy in the late 1800s. However, lack of water made further growth difficult. In 1913 a new aqueduct brought water from the slopes of the Sierra Nevada nearly 250 miles (400 km) away. As a result, the city grew rapidly. Major industries today include entertainment, oil refining, chemicals, and manufacturing. Because most of its growth has occurred during the automobile era, Los Angeles is a vast sprawling city. Automobile and factory exhaust creates **smog**, which often hangs over the city. Smog results from chemical reactions involving sunlight and pollutants from automobiles and industries.

Other big cities in the region include San Francisco, Seattle, San Diego, Phoenix, and Denver. San Francisco is located on an excellent deepwater harbor. It has long attracted immigrants and is one of the most diverse cities in the country. Seattle is an important port for trade with Asia. San Diego is also a port and is home to the most important naval base on the west coast.

INTERPRETING THE VISUAL RECORD

Located in northeastern Alaska, the Arctic National Wildlife Refuge is one of the most pristine ecosystems on Earth. However, there is growing pressure to open the area to oil exploration and production. Some people worry that oil production might cause environmental damage. Supporters of this development believe the environment can be protected at the same time. **How do you think oil production might change this natural environment? Which point of view on oil production here do you agree with? Why?**

✓ **READING CHECK:** *Environment and Society* How have southern Californians modified their physical environment to make farming productive? built aqueducts to carry water from the mountains of northern California and Colorado River

---

Section 2 Review

go.hrw.com Homework Practice Online
Keyword: SW3 HP8

Identify Megalopolis, Corn Belt, Dairy Belt, Wheat Belt, Silicon Valley

Define textiles, metropolitan area, arable, smog

Working with Sketch Maps
On the map you created in Section 1, label Washington, D.C., Boston, New York, Chicago, Detroit, St. Louis, Dallas, Houston, Miami, Los Angeles, Seattle, and San Francisco. Which cities make up Megalopolis? Which city is the largest metropolitan area in the West?

Reading for the Main Idea

1. *Places and Regions* How has urban growth and industrialization in the Northeast been tied to the region's good transportation connections?

2. *Environment and Society* What makes the Midwest such an important agricultural region?

3. *Places and Regions* How have new technologies affected the economies of Silicon Valley and the San Francisco Bay area in recent decades?

Critical Thinking

4. **Comparing** How do you think life for people in Megalopolis might be similar to life in rural parts of the United States? How do you think cultural or social attitudes might differ between people who live in such different places?

Organizing What You Know

5. Copy the chart below. Use it to describe some of the major political, economic, social, and cultural characteristics of the Northeast, Midwest, South, and West.

Northeast	South
Midwest	West

OBJECTIVES

1. Identify important environmental issues in the United States.

2. Describe the natural hazards that affect the lives of Americans.

3. Examine how cities and population patterns in the United States are changing.

4. Analyze how the U.S. economy is tied to other countries around the world.

RESOURCES

REPRODUCIBLE

▶ Guided Reading Strategy 8.3
▶ Graphic Organizer Activity 8
▶ Critical Thinking Activity 8: The United States in a Global Economy

TECHNOLOGY

▶ One-Stop Planner CD–ROM, Lesson 8.3
▶ Homework Practice Online
▶ HRW Go site

REINFORCEMENT, REVIEW, AND ASSESSMENT

▶ Main Idea Activity 8.3
▶ English Audio Summary 8.3
▶ Spanish Audio Summary 8.3
▶ Section 3 Review, p. 183
▶ Daily Quiz 8.3

VISUAL RECORD ANSWER

encourage water conservation, natural landscaping

LET'S GET STARTED

Copy the following instructions onto the chalkboard: *Think of any natural disasters you remember that have struck the United States. Write a few words about them in your notebook.* Discuss student responses. *(Possible responses: tornadoes in the plains or Midwest, hurricanes along the Gulf or East Coasts, earthquakes in the West, blizzards in the Northeast, and so on)* Tell students that they will learn more about natural hazards and other geographic issues that affect the United States in Section 3.

Building Vocabulary

Write **gentrification**, **superpower**, and **trade deficit** on the chalkboard. Point out that *gentry* refers to people of the upper class. *Gentrification* involves wealthy citizens renovating homes in less affluent areas. Ask students what *super-* means *(more, over, higher)*. Ask how this relates to the meaning of *superpower*. *(Superpowers are more powerful than other countries.)* *Deficit* comes from the Latin word *deficere*, which means "to be lacking." A country with a trade deficit lacks sufficient exports to match its imports.

Section 3 — Geographic Issues

READ TO DISCOVER

1. What are some important environmental issues in the United States?
2. What natural hazards affect the lives of Americans?
3. How are cities and population patterns in the United States changing?
4. How is the U.S. economy tied to other countries around the world?

WHY IT MATTERS

Environmental issues in the United States are often in the news and are the subject of much debate. Use CNNfyi.com or other current events sources to learn about environmental issues that are important today.

IDENTIFY

North American Free Trade Agreement (NAFTA)

DEFINE

gentrification   trade deficit
superpower

LOCATE

Gulf of Mexico   Colorado River
Columbia River

INTERPRETING THE VISUAL RECORD

Demand for water is high in parts of the western United States. However, state regulations often limit the amount of water available for personal use. **How do you think regulations on water usage shape the geographic and economic character of western states?**

Environmental Issues

The United States consumes more energy than any other country. As a result, the country produces huge amounts of waste, automobile exhaust, and other pollutants. A major challenge has been finding ways to reduce pollution and protect the environment.

Population growth and economic development have contributed to this challenge. For example, power plants and factories that burn coal and oil cause acid rain. This pollution kills trees and contaminates rivers and lakes. The problem has been particularly serious in parts of upstate New York and New England. Older factories in the Ohio River Valley produce pollution that reaches these areas. Laws restricting the emissions of pollutants reduced levels of acid rain in the 1990s. However, the problem has not been resolved.

The use of fertilizers has created problems in some places. For example, the Mississippi River carries fertilizers from farms to the Gulf of Mexico. These chemicals promote the growth of algae. Then bacteria that consume the algae use oxygen that marine life needs to live. This has created a "dead zone" in the waters off the Louisiana coast. Fertilizers used in Florida's sugarcane fields have also killed plants and wildlife in the Everglades.

In the West, dams on the Columbia and Snake Rivers produce hydroelectricity. However, they also block the path of migrating salmon. Dams are one of the main causes for a dramatic decline in the salmon population. A growing population and rapid economic development have also strained water resources in the West. Competition for the limited water resources of the Colorado River has been intense. Growing populations in both Arizona and California want the water. So much water is now drawn from the Colorado River that almost none reaches the Gulf of California.

✔ **READING CHECK:** *Environment and Society* What are four examples of environmental challenges in the United States? What factors have contributed to these challenges? acid rain, overuse of fertilizers, dams, limited water resources; population growth, economic development

Teach Objective 1

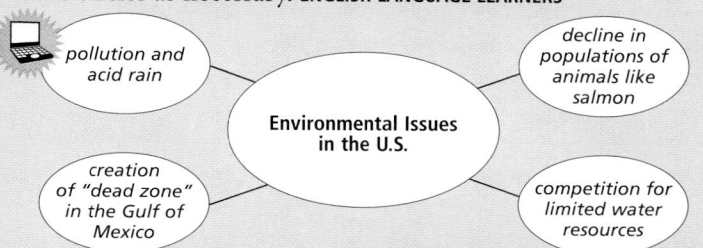

ALL LEVELS: Copy the following graphic organizer onto the chalkboard, omitting the italicized answers. Have students copy the organizer and complete it with some of the environmental issues that the United States faces. Add more circles as necessary. ENGLISH LANGUAGE LEARNERS

pollution and acid rain

decline in populations of animals like salmon

Environmental Issues in the U.S.

creation of "dead zone" in the Gulf of Mexico

competition for limited water resources

Teach Objective 2

ALL LEVELS: Organize the class into groups and assign each group one of the following natural hazards that affects parts of the United States: avalanches, blizzards, earthquakes, floods, hurricanes, tornadoes, tsunamis, or volcanic eruptions. Have each group create a poster that indicates which parts of the country are threatened by its assigned hazard. The posters should also describe the damage each hazard can cause to the environment or to society and any steps people have taken to prevent this damage. Have groups share their posters with the class. ENGLISH LANGUAGE LEARNERS, COOPERATIVE LEARNING

Natural Hazards

Natural hazards present other challenges in regions around the country. For example, some large cities on the west coast are vulnerable to earthquakes. In fact, powerful earthquakes in recent years have caused serious damage in Los Angeles, San Francisco, and Seattle. Scientists predict that more large quakes will strike in the future. In areas where earthquakes are common, developers must follow strict building codes designed to limit earthquake damage. Many use building techniques and materials that can limit damage even to large structures.

Flooding threatens many parts of the United States. In some places, dams and levees have helped control the flow of rivers and have reduced flooding. However, heavy rains and snowmelt still cause major floods. Some engineers think that by controlling rivers too much, the potential for dangerous floods actually increases.

Tornadoes are threats in the Midwest and South, particularly in spring and summer. Hurricanes threaten areas along the east coast and Gulf of Mexico. Strong hurricane winds damage buildings and vegetation and reshape barrier islands. However, the greatest danger posed by hurricanes is flooding. High seas whipped up by a hurricane's winds can flood lowlying coastal areas and cause heavy damage. The barrier islands of Texas, Florida, and the Carolinas are particularly at risk. Governments enforce strict building codes in many areas. Seawalls and other devices also help limit damage. However, coastal buildings are still damaged and destroyed.

✔ READING CHECK: *Environment and Society* What are some natural hazards that threaten parts of the United States? earthquakes, flooding, tornadoes, hurricanes

Cities and Population

The residents of U.S. cities have seen major changes over the last 50 years. Many people have moved from inner cities to the suburbs. As a result, the populations of some large cities have dropped significantly, particularly in the Northeast. Businesses also began moving to suburban shopping malls and edge cities. Because of these shifts in population, many cities are unable to collect as much money from taxes. In fact, some cities find it difficult to provide basic services to poor and minority residents who remain.

In some urban areas wealthier people have been moving back to inner cities. They are often young professionals who buy run-down houses and restore them, a process known as **gentrification**. However, this process has its opponents. These opponents argue that gentrification increases property values and property taxes. These factors in turn push out low-income residents.

Immigration and changing population patterns have also greatly affected the United States. Historically, most immigrants to the country came from Europe and settled in the Northeast. However, most immigrants today come from Asia and Latin America. These immigrants settle in many different places. As a result, the ethnic and cultural composition of the country has changed significantly. This is the case particularly in the West and in parts of the South.

INTERPRETING THE VISUAL RECORD

*People in Oakland, California, work to make a house safer in the event of an earthquake. Some engineering practices designed to make buildings more resistant to earthquakes include bolting homes to their foundations, strengthening supporting walls, and installing gas shut-off valves. **How might these measures help houses resist earthquake damage?***

San Francisco's Chinatown is the largest Chinese community outside of Asia. Most Chinese there have maintained their language and customs.

Linking Past to Present

The San Francisco Earthquake of 1906
Among the many natural hazards that Americans face, earthquakes are one of the most devastating. A particularly costly quake struck the city of San Francisco in April 1906. People as far away as Oregon and Nevada felt tremors. The quake ruptured gas lines and ignited terrible fires that swept through the city for four days. Together, the earthquake and fires may have caused more than 3,000 deaths and $500 million in damages. This figure is in 1906 dollars and would be much higher today.

Largely to avoid another tragedy like the 1906 event, California residents now work to make their homes and buildings earthquake resistant. Along with strengthened walls and braces, gas shut-off valves are installed to prevent fires from starting should gas lines be ruptured.

🖳 internet connect

GO TO: go.hrw.com
KEYWORD: SW3 CH8
FOR: Web sites about natural hazards in the United States

VISUAL RECORD ANSWER

keep houses from collapsing, prevent fires

ALL LEVELS: Organize students into groups and have each group create a segment for a news program called Our Changing Cities. Assign each group a topic for its segment and have groups research its effects on American cities. Possible topics include suburban development, gentrification, and immigration. You may wish to have students keep their reports general in nature, or you may have them focus their research on a particular city in your area. When groups have completed their research, have them perform their segments for the class. **COOPERATIVE LEARNING**

ALL LEVELS: Assign each student a country with which the United States has economic ties. Have students use resources like the *CIA World Factbook* or the *World Almanac and Book of Facts* to create a list of products the United States imports from and exports to their assigned countries. Then have students use colored string to connect their countries to the United States on a large wall map. Have them attach their lists of products to these strings. Lead a class discussion about the types of goods the United States trades with countries around the world. Then discuss why other countries want to maintain trade relations with the United States.

Essential Element 4

▶ **Human Systems** ◀

Melting Pot or Salad Bowl? The phrases *melting pot* and *salad bowl* are often used to describe the cultural diversity of the United States. In a melting pot scenario, various cultures combine to form a single new culture, much like chocolate, butter, and sugar melt together to form hot fudge sauce. The salad bowl idea refers to cultures remaining separate within a country, like lettuce and tomatoes remain separate in a salad.

ACTIVITY: Is the United States a melting pot or a salad bowl? Have students write short essays defending one of these two positions. Then stage a debate between supporters of the two positions and discuss the arguments used by each side.

CONNECTING TO GOVERNMENT ANSWER

Possible answer: Most states that gained many electoral votes in the 1900s are located along coastal and border areas, the destinations of many immigrants. As immigration continues, those states might see their populations and relative political power increase further.

MAP ANSWER

The two parties appear generally even in strength, with each winning in certain regions of the country.

Connecting to GOVERNMENT

Voting Patterns and the Distribution of Political Power

In the U.S. system of government, the states play an important role. This is particularly true in presidential elections. When a state's voters cast ballots for president in November, they are really choosing members of the electoral college, or electors. The number of electors from each state is equal to its members in the U.S. House of Representatives and the U.S. Senate. To win the election, a presidential candidate must win at least 270 of the possible 538 electoral votes.

The size of a state's population determines how many seats that state gets in the U.S. House. As a result, states with the largest populations have the most seats. It follows, then, that they also have the most electoral votes. Therefore, changes in the population of each state affect the distribution of political power in the country. In fact, political power in the United States continues to shift as the population of the country moves westward and southward.

Look at the maps from the 1920 and 2000 presidential elections. These maps show the number of electoral votes each state had in each election. In 1920 several Northeast and Midwest states had the most electoral votes. This was because they had the largest populations. However, by 2000 California, Texas, and Florida each had more electoral votes than any other state but New York.

Analyzing What role do you think immigration might play in the distribution of political power in this country? How might these maps support your answer?

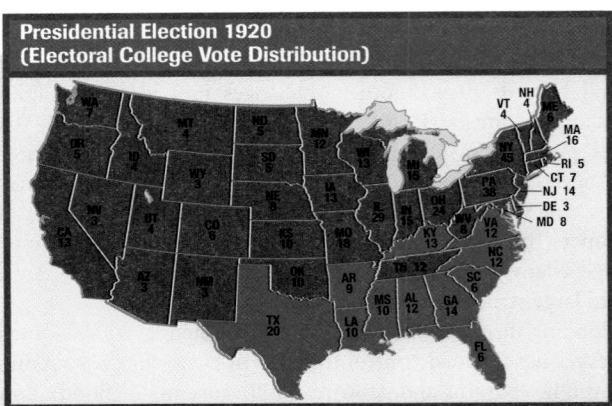

Presidential Election 1920
(Electoral College Vote Distribution)

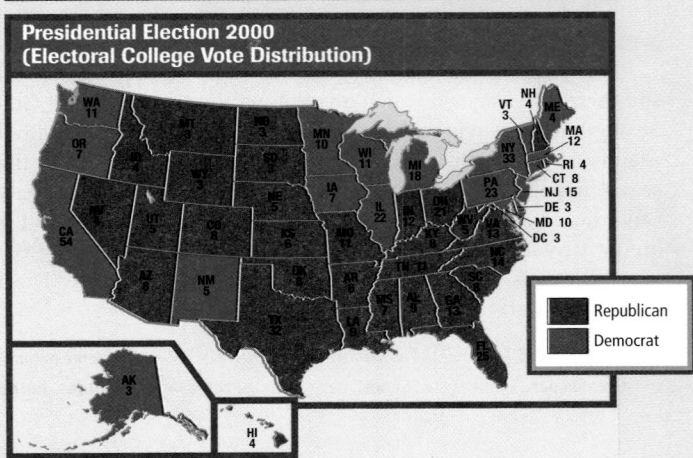

Presidential Election 2000
(Electoral College Vote Distribution)

■ Republican
■ Democrat

INTERPRETING THE MAPS *These maps show how voting patterns in the United States have changed over time. Democrats once dominated elections in the South. Republicans did better in the Northeast and West. Those voting patterns have changed somewhat. Shifts in migration patterns, political attitudes, and regional economics have all played roles in this change.* **What does the 2000 map indicate about the relative strengths of the two major U.S. political parties today?**

Changing population patterns have been a source of conflict in some places. Some people argue that the United States has historically been an English-speaking country, and they want it to remain so. They oppose bilingual education and think that new immigrants should be required to learn English. Also, some people think high levels of immigration keep wages down and make it harder for people to find jobs. Other people point to the long history of immigration to the United States. They say that immigration has strengthened the country and is needed to fill jobs.

✓ **READING CHECK:** *Human Systems* How has the character of major U.S. cities been changing over the last 50 years? People and businesses have been moving to the suburbs, lowering city populations and tax bases.

The Port of Houston is one of the busiest in the United States. Billions of dollars worth of goods pass through it each year. These goods include automobiles, chemicals, grain, machinery, and oil.

Economy and Trade

When the Soviet Union collapsed in 1991 the United States became the world's only **superpower**. A superpower is a huge powerful country. The United States has the world's largest economy and is also the most powerful country politically and militarily.

However, the U.S. economy also relies on global trade. For example, the United States imports many raw materials and manufactured goods. In most years, the country has a **trade deficit**. A country has a trade deficit when the value of its exports is less than the value of its imports. In the past the government supported American industries by imposing tariffs on many imported goods. In recent years, however, the U.S. government has moved to lower barriers to free trade. For example, in 1992 the United States, Canada, and Mexico signed the **North American Free Trade Agreement (NAFTA)**. NAFTA eliminates many tariffs on products flowing between these countries. Since NAFTA took effect in 1994, trade between the three countries has boomed. Supporters argue that free trade will help American companies sell more products and create more jobs. Opponents say free trade allows American companies to move factories to countries with lower wages and business costs, causing unemployment in the United States.

Our Amazing Planet

The United States contains roughly 5 percent of the world's population but consumes 27 percent of the world's energy production.

✓ **READING CHECK:** *Human Systems* How have shifts in U.S. government policies affected trade practices? In the past the government supported tariffs to protect U.S. industries. In recent years it has worked to lower barriers to free trade. NAFTA is one example.

Section 3 Review

go.hrw.com Homework Practice Online
Keyword: SW3 HP8

Identify North American Free Trade Agreement (NAFTA)

Define gentrification, superpower, trade deficit

Working with Sketch Maps On the map you created in Section 2, label the Gulf of Mexico, Columbia River, and Colorado River. Which river is a major water source for California and Arizona?

Reading for the Main Idea

1. *Environment and Society* How has the use of fertilizers affected the environment along the Louisiana coast?

2. *Physical Systems* How do natural hazards such as hurricanes affect the environment?

Critical Thinking

3. **Drawing Inferences and Conclusions** How might this country's history of immigration influence immigration policies today?

4. **Identifying Points of View** How might people worried by the negative effects of gentrification in inner cities influence local public policies? In what ways might gentrification help people in inner-city neighborhoods?

Organizing What You Know

5. Copy the chart below. Use it to explain how technological innovations have helped Americans adapt to natural hazards.

Natural hazards	Innovations
Earthquakes	
Floods	
Hurricanes	

183

Have students read Case Study: Creating Congressional Voting Districts. Ask students to explain why the government redraws the boundaries of congressional districts after each census. *(The census reveals changes in population numbers and patterns that have occurred in the previous 10 years. Districts are redrawn to ensure that each includes approximately the same number of people.)* Point out that many issues must be taken into account as new district boundaries are drawn.

Building a Case

Have students create a fictional county or city and organize it into six voting districts of approximately equal populations. Students should also create a profile of each district including such data as average age and ethnic makeup of its residents. Then organize the class into two groups to re-create the county 10 years later. Have one group imagine that the population has increased significantly while the other imagines it has decreased. Ask how the population has shifted within the county. Which areas might have gained residents, and which might have lost residents? Have each group decide if immigration has been a factor in the population change. If so, which areas of the county or city are most affected?

Cooperative Learning

Voting Rights Remind students that Americans 18 years of age and older are eligible to vote in local, state, and national elections. This was not always the case. In the country's early days, only white males who owned property could vote. As the country expanded, so did voting rights. Organize the class into groups. Have each group trace the expansion of voting rights to discover when various social groups, including women and minorities, won this right. Discuss student findings.

Applying What You Know Answers

1. They use census data to determine each state's representation and to draw lines so that each district has the same number of residents and minority voting strength is not diluted. The courts step in to resolve disputes.

2. Different groups likely worked together to create the district so that African Americans could elect an African American representative. Opponents and the Supreme Court challenged the district because they believed it was an example of unfair gerrymandering.

This Case Study feature addresses National Geography Standards 3, 9, 13, and 18.

CASE STUDY

Creating Congressional Voting Districts

Human Systems Every 10 years the U.S. Census Bureau counts the population of the United States. The government then uses this data to create new election boundaries for the 435 U.S. congressional districts and state legislative districts. This process is called redistricting. Guiding the redistricting process is the "one person, one vote" principle. This principle requires that each congressmember or state legislator represent roughly the same number of people. As a result, states with large populations receive more members in the U.S. House of Representatives than states with fewer people. Also, more densely populated regions of a state get more representatives.

Gerrymandering

Even guided by the "one person, one vote" principle, redistricting officials still have many decisions to make. In fact, the redistricting process is a lesson in how the forces of conflict and cooperation play out in the U.S. political system. Members of rival political parties often work to draw district boundaries that are most favorable to their own party. Even moving district boundaries slightly can shift the political balance of power within the affected districts. Sometimes the process of redistricting creates some messy district maps. Boundaries might jut out here or there, creating districts with odd shapes. These odd shapes may be the result of a redistricting technique called gerrymandering.

Gerrymandering gets its name partly from that of Elbridge Gerry, a governor of Massachusetts in the early 1800s. In 1812 Gerry's administration enacted a law that divided his state into new senatorial districts. The new boundaries placed most of the supporters of Gerry's opponents into a few districts. This process helped ensure victory for members of Gerry's party in the remaining districts. It also created some rather oddly shaped districts. One district looked something like a salamander—or, as the governor's critics wryly noted, a "gerrymander." People now use the term *gerrymander* to describe the process of drawing district boundaries that unfairly favor a political party or group.

THE GERRY-MANDER.

The term gerrymander *originated from this political cartoon. The cartoonist drew salamander features on a map of Boston-area voting districts to highlight their odd shapes. Since the districts' boundaries had been drawn by Governor Elbridge Gerry, the cartoon became known as the* Gerry-mander.

One Person, One Vote

Gerrymandering has not been the only way to unfairly favor a political party or group. During the first half of the 1900s, many states did not redraw electoral boundaries after each census. The boundaries did not move, but the people within them did. For example, many rural Americans migrated to cities, but the number of representatives from rural areas stayed the same. As a result, rural areas had relatively more political power than urban areas.

Some of the most extreme examples of political inequality between rural and urban areas occurred in the South. In 1960 just one quarter of Alabama residents elected a majority of the state's legislators. As a result, rural districts had more power than urban districts. For example, lawmakers in some urban districts of Alabama represented about 100,000 people each. In contrast, some rural legislators represented as few as 7,000 people. Tennessee's voting districts were also unfairly drawn. Populations in state senatorial

Drawing Conclusions

Have the groups redraw the voting districts in their counties or cities to reflect population changes. The group whose population increased should add one district, and the other group should remove one. Remind groups that their new districts should have an even distribution of population. Encourage group members to consider the characteristics of the districts' populations as they work. After the groups have redrawn their maps, call on a volunteer from the first group to explain its redistricting plan. Encourage the second group to respond. Then have the second group explain its plan. Ask students to describe any challenges they encountered in drawing their new boundaries. Ask how these might be similar to problems faced by state governments drawing new congressional districts.

Going Further: Thinking Critically

Organize students into small groups and have each group locate maps of your state's congressional districts before and after the most recent census. Have each group study the maps to answer the following questions:

- Did your state's representation in Congress change? How?
- How do the new boundaries of voting districts within the state reflect recent population trends?
- What factors led to the shaping of the current boundaries?
- Do you agree with the organization of political boundaries in the state?

districts ranged from 131,971 to 25,190. The most crowded representative district had 42,298 people. The smallest had only 2,340.

Over time, people in urban areas challenged the way these districts were created. They argued that the U.S. Constitution guaranteed equal representation in government—one person, one vote. Eventually, a series of U.S. Supreme Court cases forced states to redraw their electoral maps. Now all districts must contain about the same number of people.

Redistricting Today

Other laws and court cases in the last half of the 1900s further shaped redistricting processes. Officials must now take care not to create districts that dilute the voting strength of racial and ethnic minority groups. These officials use census data and sophisticated software and mapping programs, such as a geographic information system (GIS). Some states created districts to ensure the election of more minority candidates. For example, after the 1990 census, Georgia officials stretched the 11th Congressional District along a narrow strip. Snaking across 260 miles (418 km) of the state, the new district linked African American communities in Atlanta and Savannah. As the planners had intended, voters in the new district elected an African American congressmember in 1994.

Critics call such redistricting practices gerrymandering. Some also argue that grouping many minority voters into a few districts dilutes their strength in

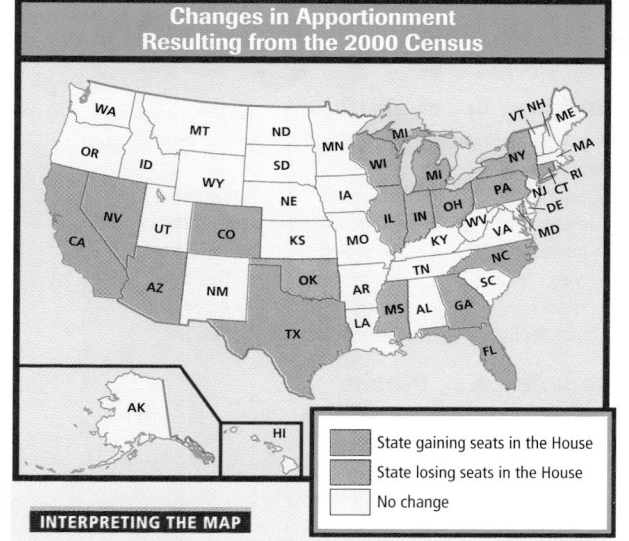

Changes in Apportionment Resulting from the 2000 Census

▨	State gaining seats in the House
▨	State losing seats in the House
☐	No change

INTERPRETING THE MAP

*Congressional districts are redrawn after each census to reflect changes in the distribution of the U.S. population. **Based on the map, which regions of the United States do you think are gaining population? Which are losing population?***

other districts. In addition, the U.S. Supreme Court later ruled that districts cannot be based mainly on racial and ethnic characteristics. Courts still sometimes step in to resolve disputes because electoral maps can be very controversial.

As you can see, the redistricting process has become quite complicated. Still, despite the challenges, the new maps must be drawn. With Census Bureau data in hand, geographers and demographers play a role in congressional redistricting. By working with government officials and citizen groups, they help redraw our country's political geography every 10 years.

Georgia's 11th District, 1992

INTERPRETING THE MAP

*In 1995 the U.S. Supreme Court declared Georgia's 11th District unconstitutional because it concluded that race was the driving force behind its boundaries. **How do you think this conflict reshaped Georgia's political geography?***

Applying What You Know

1. **Summarizing** How do officials create congressional voting districts?

2. **Analyzing** How do you think the forces of conflict and cooperation influenced the creation of Georgia's 11th Congressional District?

CHAPTER 8 Review Answers

Building Vocabulary For definitions, see: colonies, p. 167; plantations, p. 168; bilingual, p. 169; Megalopolis, p. 173; textiles, p. 173; metropolitan area, p. 174; arable, p. 175; Corn Belt, p. 175; Dairy Belt, p. 175; Silicon Valley, p. 178; smog, p. 179; gentrification, p. 181; superpower, p. 183; trade deficit, p. 183; North American Free Trade Agreement (NAFTA), p. 183

Locating Key Places

A. Washington, D.C.
B. Chicago
C. Seattle
D. Alaska
E. Los Angeles
F. Miami
G. New York

Understanding the Main Ideas

1. People once perceived the area as too dry for farming. Irrigation and changes in technology that led people to believe the area was suitable for farming allowed it to be settled.

2. Immigrants from around the world have brought many different religions and cultural changes to the United States.

Review and Assessment Resources

TECHNOLOGY
► Chapter 8 Test Generator (on the One-Stop Planner)
► Global Skill Builder CD–ROM
► HRW Go site

REINFORCEMENT, REVIEW, AND ASSESSMENT
► Chapter 8 Review, pp. 186–87
► Chapter 8 Tutorial for Students, Parents, Mentors, and Peers
► Chapter 8 Test (form A or B)
► Alternative Assessment Handbook

► Chapter 8 Test for English Language Learners and Special-Needs Students
► Unit 2 Test
► Unit 2 Test for English Language Learners and Special-Needs Students

Assess
Have students complete a Chapter 8 Test.

Reteach
Organize the class into small groups and assign each group one of this chapter's sections. Have each group examine the photographs in its assigned section and write a sentence explaining how each picture relates to the section's content. Call on volunteers to share their work with the class.
ENGLISH LANGUAGE LEARNERS

CHAPTER
8

Review Answers

3. Corn Belt—Midwest states; Dairy Belt—Wisconsin, Minnesota, Michigan; Wheat Belt—the Dakotas, Montana, Nebraska, Kansas, Oklahoma, Colorado, Texas

4. Both states have been in competition for water from the river. So much water is drawn from the river that very little reaches the ocean.

5. 1991; collapse of the Soviet Union

Thinking Critically

1. Possible answers: It is a functional region because the cities of megalopolises are connected by transportation routes and economic activities. It could be a perceptual region if ways of life and attitudes are similar within the region. (NGS 5)

2. Possible answer: It has major cities with diverse populations, major industries, the country's government, and fashion and media companies. (NGS 4)

3. Answers will vary. Students should support their points of view. (NGS 9)

Using the Geographer's Tools

1. The French settled along the Mississippi River and the Great Lakes, while the Spanish settled in the Southwest and the English colonized the northeast.

2. the 1800s

CHAPTER 8 Review

Building Vocabulary

On a separate sheet of paper, explain the following terms by using them correctly in sentences.

colonies	arable	superpower
plantations	Corn Belt	trade deficit
bilingual	Dairy Belt	North American
Megalopolis	Silicon Valley	Free Trade
textiles	smog	Agreement
metropolitan area	gentrification	(NAFTA)

Locating Key Places

On a separate sheet of paper, match the letters on the map with their correct labels.

Alaska	Miami
Washington, D.C.	Los Angeles
New York	Seattle
Chicago	

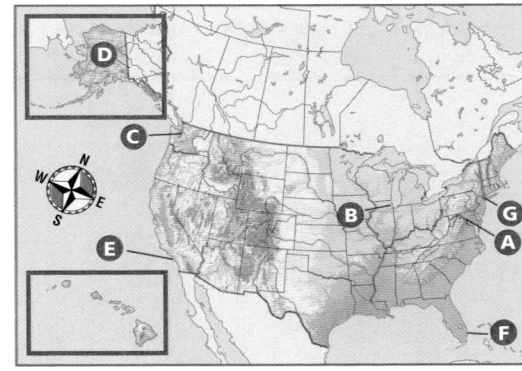

Understanding the Main Ideas

Section 1

1. *Environment and Society* How did changing perceptions of the Great American Desert lead to changes in society?

2. *Human Systems* How has immigration to the United States influenced the diffusion of religions and cultural change?

Section 2

3. *Environment and Society* How has competition for water resources from the Colorado River affected California and Arizona? How has the river itself been affected?

4. *Places and Regions* Which states are part of the Corn Belt, the Dairy Belt, and the Wheat Belt?

Section 3

5. *Places and Regions* When did the United States become the world's only superpower? Why?

Thinking Critically

1. **Making Generalizations** Do you think a megalopolis is a formal, functional, or perceptual region? Why?

2. **Analyzing** How could the Northeast be described as an important economic, political, and cultural center?

3. **Identifying Points of View** What restrictions, if any, do you think should be placed on immigration to the United States? Why? Why might immigration be such a controversial issue?

Using the Geographer's Tools

1. **Analyzing Maps** Look at the Section 1 map showing placenames in the United States. How does the map reflect French, Spanish, and English settlement patterns?

2. **Analyzing Maps** Look at the Section 1 map showing U.S. expansion. During which century did the United States acquire most of its territory?

3. **Preparing Maps** Map the locations of different types of economic activities in your state. You can use the unit land use and resources map, the chapter, and other resources.

Writing about Geography

Imagine that you are writing a script for a documentary about how life in Los Angeles has changed over time. Use the information in this chapter and in other sources to write a short introduction to your script. Be sure to answer the following questions: When was Los Angeles founded, and by whom? What resources have been important in the city's history and economic development? How big is the Los Angeles metropolitan area today? What challenges does the city face?

SKILL BUILDING

Geography for Life

Analyzing Primary Sources

Human Systems Use your local library to find firsthand accounts about the exploration, settlement, and westward expansion of the United States. What can these accounts teach us about American history and geography? What types of information do they contain? What types of information are they lacking? When might researchers want to study firsthand accounts?

Portfolio Activity

Literature that uses local color describes or incorporates the characteristics—such as speech patterns, specific locations, or foods—of a particular place. Geographers sometimes refer to such descriptions as "landscape in literature." Have each student write a short story using local color about his or her own area or another within the United States. Have students underline words or phrases that identify characteristics of the chosen locations. Call on volunteers to read their stories to the class. Place copies of the stories in student portfolios.

Food Festival

Changes in American eating habits over the years reflect the changes in American society as a whole. Have students interview their parents and older adults to find out how food fashions have changed during their lifetimes. Examples of foods that have gained or lost popularity over time include fondue, crepes, and TV dinners. Invite students to bring samples of these items to share with the class. Challenge students to identify connections between the foods and other developments in society. For example, convenience foods became more popular when women entered the workplace in larger numbers.

Building Social Studies Skills

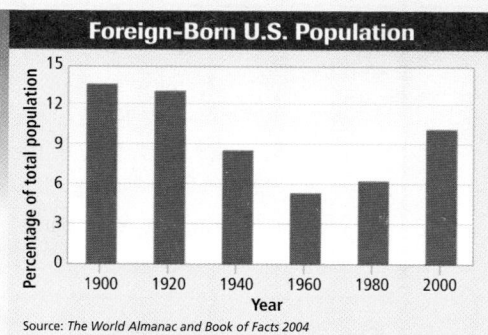

Foreign-Born U.S. Population

Source: *The World Almanac and Book of Facts 2004*

Interpreting Graphs

Study the bar graph above. Then use the information from the graph to help you answer the questions that follow.

1. In which year was the percentage of the U.S. population that was foreign-born highest?
   a. 1900
   b. 1920
   c. 1940
   d. 2000

2. Did immigration appear to increase or decrease in the middle part of the 1900s? How can you tell?

Analyzing Primary Sources

Read what historian William Cronon has to say about the Midwest. Then answer the questions that follow. Mark your answers on a separate sheet of paper.

> "I came to know and care for a landscape that few who are not midwesterners ever call beautiful. Travelers, whether in the air or on the ground, usually see the Middle West less as a destination than as a place to pass through. Only after a long while does one appreciate that the very plainness of the countryside is its beauty. . . . When people speak . . . of the American heartland, this is one of the places they mean."

3. According to Cronon, what is the source of the Midwest's beauty?
   a. its airports
   b. its travelers
   c. its plainness
   d. its cities

4. Why do you suppose Cronon thinks that travelers see the Midwest as simply a place to pass through?

Alternative Assessment

PORTFOLIO ACTIVITY

Learning about Your Local Geography

Individual Project: Research

People, places, and environments are connected and interdependent. For example, resources may attract people to a new place, and their activities may change it. Study your local town or community to learn about the relationship between its people and environment. What was the area like before it was settled? What is it like now? How have settlement and population growth affected the environment? How has economic development affected it?

internet connect

Internet Activity: go.hrw.com
KEYWORD: SW3 GT8

Choose a topic on the United States to:
- take the GeoMap challenge to test your knowledge of U.S. geography.
- learn how people from different cultures celebrate holidays in the United States.
- create a brochure on national parks of the United States.

---

3. Maps will vary but should show major economic activities found in the state.

Writing

Scripts should report accurate information about the founding of Los Angeles and should address changes in the city over time. Use Rubric 42, Writing to Inform, to evaluate student work.

Geography for Life

Students should note that primary sources can provide insights into motivations and feelings that other sources do not. However, these sources are subjective. Use Rubric 1, Acquiring Information, to evaluate student work.

Social Studies Skills

1. a
2. Possible answer: Immigration decreased. The percentage of the U.S. population that was foreign-born dropped below 6 percent in 1960.
3. c
4. Possible answer: The plainness of the region does not attract visitors to the Midwest itself.

PORTFOLIO ACTIVITY

Answers will vary based on the history of your community. Use Rubric 30, Research, to evaluate student work.

CHAPTER RESOURCE MANAGER

	Objectives	Pacing Guide	Reproducible Resources
SECTION 1 **History and Culture** (pp. 189–92)	• Identify which European countries played a role in Canada's early history. • Describe some important features of Canadian culture.	**Regular** 1 day **Block Scheduling** .5 day *Block Scheduling Handbook, Chapter 9*	**RS** Guided Reading Strategy 9.1 **E** Cultures of the World Activity: Region 1 **SM** Map Activity 9: Canada's Languages
SECTION 2 **Canada Today** (pp. 193–97)	• Identify which resources and activities drive Canada's economy. • Explain which factors and processes have influenced the growth of Canada's cities. • Analyze how Canada is organized and governed.	**Regular** 1 day **Block Scheduling** .5 day *Block Scheduling Handbook, Chapter 9*	**RS** Guided Reading Strategy 9.2 **PS** Readings in World Geography, History, and Culture 13 and 14 **SM** Geography for Life Activity 9: The Geography of Canadian Agriculture **SM** Critical Thinking Activity 9: Banff National Park
SECTION 3 **Geographic Issues** (pp. 199–201)	• Analyze how the United States influences Canada today. • Explore how geographic factors have affected Canada's national unity.	**Regular** 1 day **Block Scheduling** .5 day *Block Scheduling Handbook, Chapter 9*	**RS** Guided Reading Strategy 9.3 **RS** Graphic Organizer Activity 9

Chapter Resource Key

PS	Primary Sources	**A**	Assessment		CD–ROM
RS	Reading Support	**REV**	Review		Video
IC	Interdisciplinary Connections	**ELL**	Reinforcement and English Language Learners		Internet
E	Enrichment		Transparencies		Holt Presentation Maker Using Microsoft® PowerPoint®
SM	Skills Mastery				

 One-Stop Planner CD–ROM

See the *One-Stop Planner* for a complete list of additional resources for students and teachers.

One-Stop Planner CD–ROM

It's easy to plan lessons, select resources, and print out materials for your students when you use the **One-Stop Planner CD–ROM with Test Generator**.

Technology Resources	Reinforcement, Review, and Assessment
One-Stop Planner CD–ROM, Lesson 9.1	**ELL** Main Idea Activity 9.1
ARGWorld CD–ROM	**ELL** English Audio Summary 9.1
Homework Practice Online	**ELL** Spanish Audio Summary 9.1
HRW Go site	**REV** Section 1 Review, p. 192
	A Daily Quiz 9.1
One-Stop Planner CD–ROM, Lesson 9.2	**ELL** Main Idea Activity 9.2
Homework Practice Online	**ELL** English Audio Summary 9.2
HRW Go site	**ELL** Spanish Audio Summary 9.2
	REV Section 2 Review, p. 197
	A Daily Quiz 9.2
One-Stop Planner CD–ROM, Lesson 9.3	**ELL** Main Idea Activity 9.3
CNN Presents Geography: Yesterday and Today, Segment 8: Tradition and Change—The James Bay Project	**ELL** English Audio Summary 9.3
Homework Practice Online	**ELL** Spanish Audio Summary 9.3
HRW Go site	**REV** Section 3 Review, p. 201
	A Daily Quiz 9.3

internet connect

HRW ONLINE RESOURCES

GO TO: go.hrw.com
Then type in a keyword.

TEACHER HOME PAGE
  KEYWORD: SW3 Teacher

CHAPTER INTERNET ACTIVITIES
  KEYWORD: SW3 GT9
  Choose an activity to:
  • take the Geomap challenge!
  • research modern Quebec and create a newspaper.
  • learn about writers, artists, and musicians of Canada.

CHAPTER ENRICHMENT LINKS
  KEYWORD: SW3 CH9

CHAPTER MAPS
  KEYWORD: SW3 MAPS9

ONLINE ASSESSMENT
  Homework Practice
  KEYWORD: SW3 HP9
  Standardized Test Prep
  KEYWORD: SW3 STP9
  Rubrics
  KEYWORD: SS Rubrics

COUNTRY INFORMATION
  KEYWORD: SW3 Almanac

CONTENT UPDATES
  KEYWORD: SS Content Updates

HOLT PRESENTATION MAKER
  KEYWORD: SW3 PPT9

ONLINE READING SUPPORT
  KEYWORD: SS Strategies

CURRENT EVENTS
  KEYWORD: S3 Current Events

Meeting Individual Needs

Ability Levels

Level 1 Basic-level activities designed for all students encountering new material

Level 2 Intermediate-level activities designed for average students

Level 3 Challenging activities designed for honors and gifted-and-talented students

English Language Learners Activities that address the needs of students with Limited English Proficiency

Chapter Review and Assessment

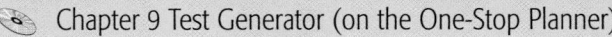

Chapter 9 Test Generator (on the One-Stop Planner)

Global Skill Builder CD–ROM

HRW Go site

REV Chapter 9 Review, pp. 202–03

REV Chapter 9 Tutorial for Students, Parents, Mentors, and Peers

A Chapter 9 Test (form A or B)

A Alternative Assessment Handbook

A Chapter 9 Test for English Language Learners and Special-Needs Students

Launch into Learning

Ask students what images or impressions of Canada and Canadians they have received from movies and television. (*Students may mention ice hockey, snow, lumberjacks, Royal Canadian Mounted Police, or other images.*) Ask which images they think are true, which are false, and which are exaggerations. Discuss with students where they think the false and exaggerated impressions may have originated. Tell students they will learn more about the reality of Canada in this chapter.

Using the Physical-Political Map

Have students examine the map on the opposite page. Call on volunteers to name the bodies of water that surround Canada (*Arctic Ocean, Atlantic Ocean, Baffin Bay, Labrador Sea, Pacific Ocean*) and to describe the land boundary that separates Canada from the United States (*straight line for most of its length*). Ask students to compare the countries' areas and to find the largest and smallest of Canada's provinces and territories. Ask how their areas compare to the size of states in the United States. (*Examples: Nunavut is larger than Alaska. Prince Edward Island is slightly larger than Rhode Island.*)

Why We Should Know More

You may want to emphasize the importance of understanding Canada's geography by sharing these points with your students:

▶ Canada and the United States share the longest unfortified border in the world and are close allies.

▶ Canada and the United States are each other's leading trading partners. As a result, the economy of each country is deeply affected by the other.

▶ Canada and the United States share a common language, a similar history, and many cultural traditions.

▶ Canada is a vast and beautiful country that offers spectacular scenery, fascinating historical sites, and multicultural entertainment to the visitor.

▶ Although Canada is a neighbor and close ally of the United States, many Americans know little about Canada's landscapes and cultures.

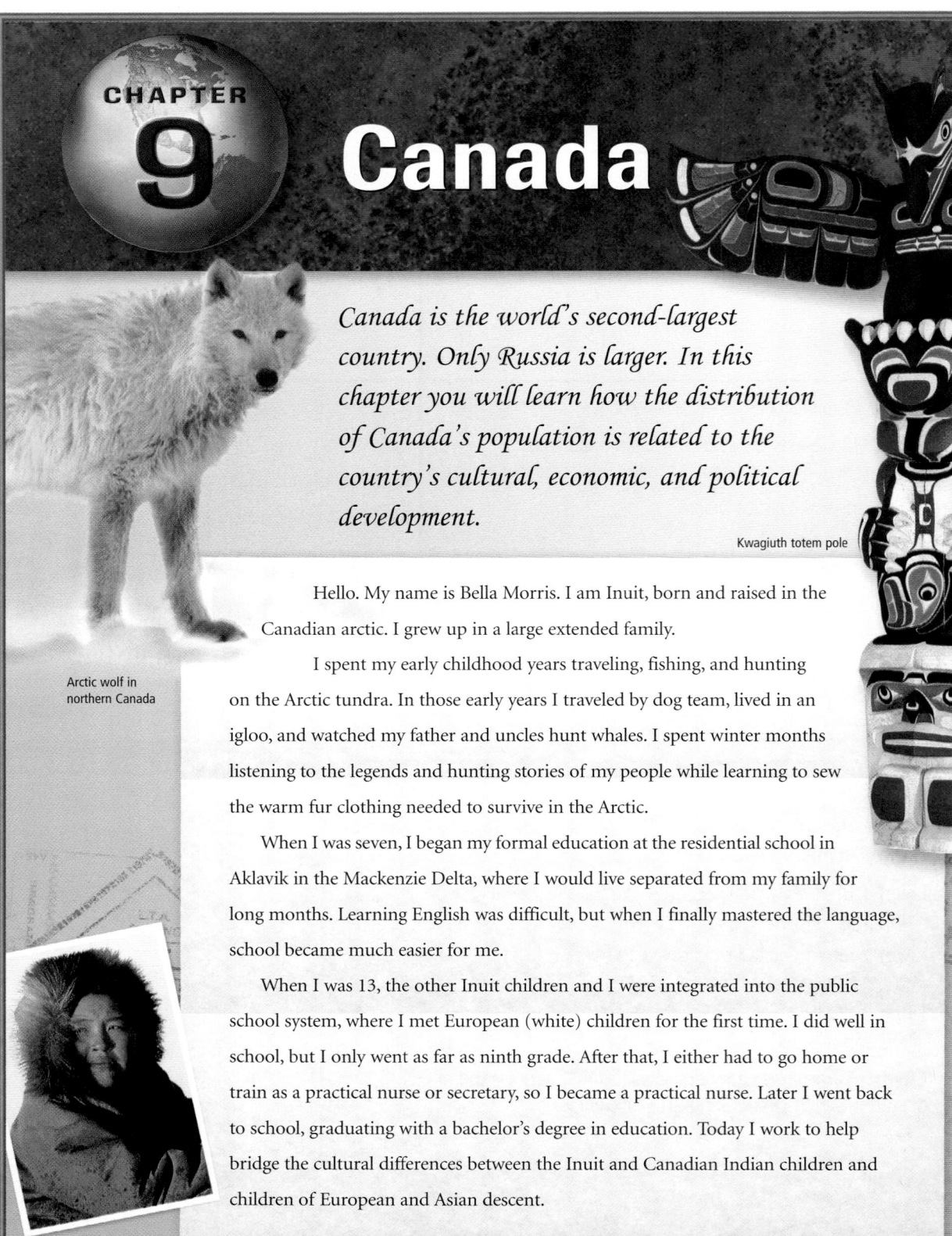

CHAPTER
9

Canada

Kwagiuth totem pole

Canada is the world's second-largest country. Only Russia is larger. In this chapter you will learn how the distribution of Canada's population is related to the country's cultural, economic, and political development.

Arctic wolf in northern Canada

Hello. My name is Bella Morris. I am Inuit, born and raised in the Canadian arctic. I grew up in a large extended family.

I spent my early childhood years traveling, fishing, and hunting on the Arctic tundra. In those early years I traveled by dog team, lived in an igloo, and watched my father and uncles hunt whales. I spent winter months listening to the legends and hunting stories of my people while learning to sew the warm fur clothing needed to survive in the Arctic.

When I was seven, I began my formal education at the residential school in Aklavik in the Mackenzie Delta, where I would live separated from my family for long months. Learning English was difficult, but when I finally mastered the language, school became much easier for me.

When I was 13, the other Inuit children and I were integrated into the public school system, where I met European (white) children for the first time. I did well in school, but I only went as far as ninth grade. After that, I either had to go home or train as a practical nurse or secretary, so I became a practical nurse. Later I went back to school, graduating with a bachelor's degree in education. Today I work to help bridge the cultural differences between the Inuit and Canadian Indian children and children of European and Asian descent.

History and Culture

READ TO DISCOVER

1. Which European countries played a role in Canada's early history?
2. What are some important features of Canadian culture?

WHY IT MATTERS

Canada and the United States share the longest unguarded border in the world. Use **CNN fyi.com** or other **current events** sources to learn about the relationship between the two countries.

DEFINE

provinces
hinterland

LOCATE

St. Lawrence River	Whitehorse
Montreal	Yellowknife
Quebec City	Windsor
Toronto	Saguenay River
Ottawa	Vancouver
Ottawa River	

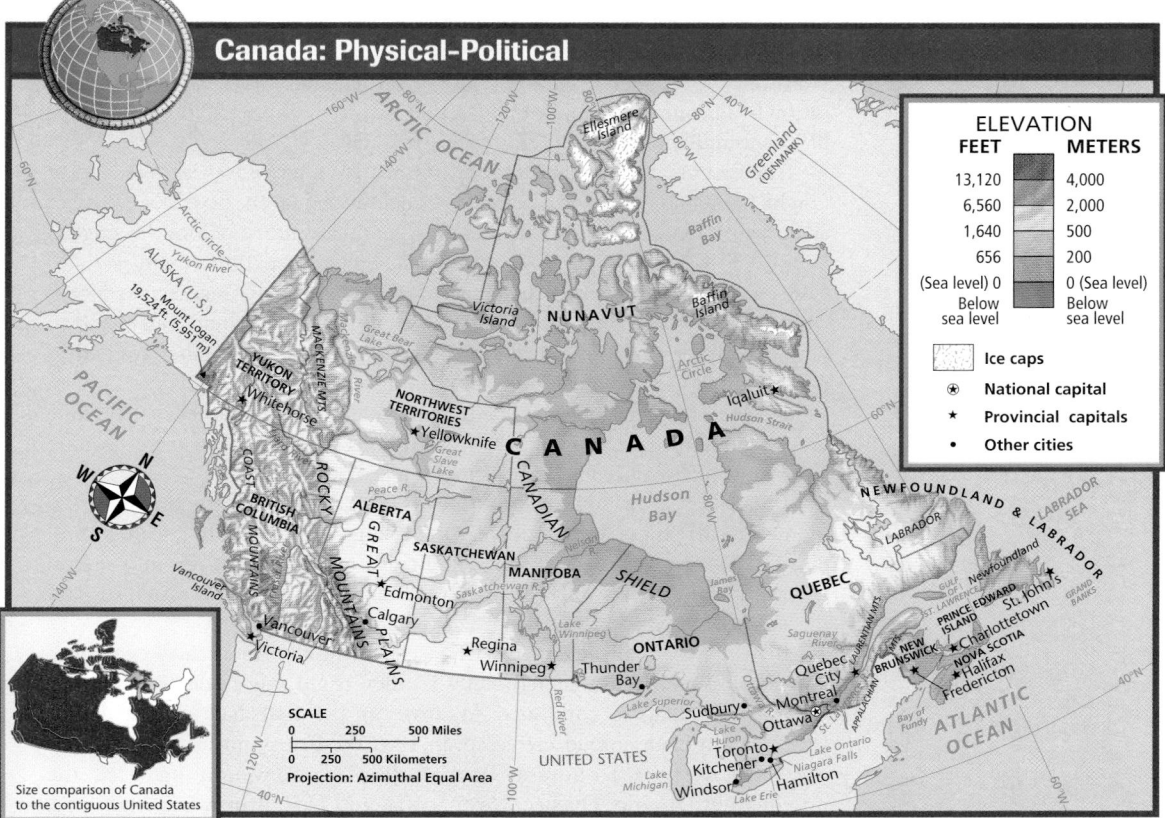

Canada: Physical-Political

ELEVATION

FEET		METERS
13,120		4,000
6,560		2,000
1,640		500
656		200
(Sea level) 0		0 (Sea level)
Below sea level		Below sea level

Ice caps
⊛ National capital
★ Provincial capitals
• Other cities

Size comparison of Canada to the contiguous United States

SCALE
0   250   500 Miles
0   250   500 Kilometers
Projection: Azimuthal Equal Area

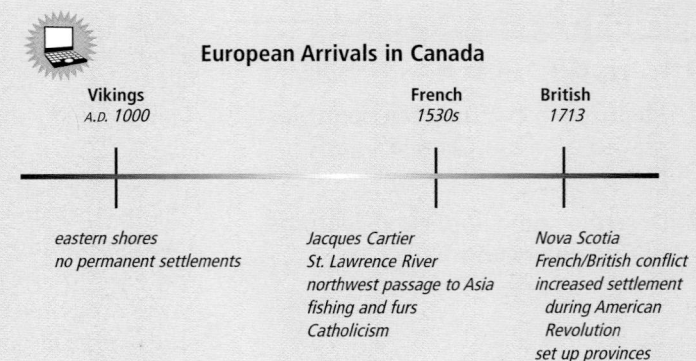

ALL LEVELS: Copy the following graphic organizer onto the chalkboard, omitting the italicized answers. Pair students and have pairs label the time line with the appropriate dates of European arrivals in Canada. Have the students use the space below the name of each group to list words or phrases that describe its settlement in Canada. Then lead a class discussion about where in Canada Europeans went and why. Ask students what elements of modern Canadian culture reflect the influences of each group on the country's history.
ENGLISH LANGUAGE LEARNERS, COOPERATIVE LEARNING

European Arrivals in Canada

Vikings A.D. 1000	French 1530s	British 1713
eastern shores *no permanent settlements*	*Jacques Cartier* *St. Lawrence River* *northwest passage to Asia* *fishing and furs* *Catholicism*	*Nova Scotia* *French/British conflict* *increased settlement during American Revolution* *set up provinces*

Cultural Kaleidoscope

Acadia Acadia, or Acadie in French, was a French colony during the 1600s in and around what is now Nova Scotia. The Acadians lived peacefully and traded with the indigenous Micmac, Malecite, and Passamaquoddy Indians. They also drained marshes along the Bay of Fundy for farming.

The British took control of Acadia in 1713 and renamed it Nova Scotia. British settlers then came to the area and founded Halifax in 1749. Tensions arose between the British and the Acadians. The Protestant British mistrusted the Catholic Acadians, so the former French colonists were deported and moved south. By 1800 nearly 3,000 Acadians had settled in Louisiana, where they became known as Cajuns (derived from *Acadian*). They relied on fishing and farming in their new communities and developed a distinct cultural identity apart from French Canadians.

DISCUSSION: Lead a discussion about environmental and cultural influences that shaped Acadian culture in the French colony. Ask what new elements might have influenced Acadian culture in Louisiana.

VISUAL RECORD ANSWER

clothing, weapons, walled fort; forests, animals

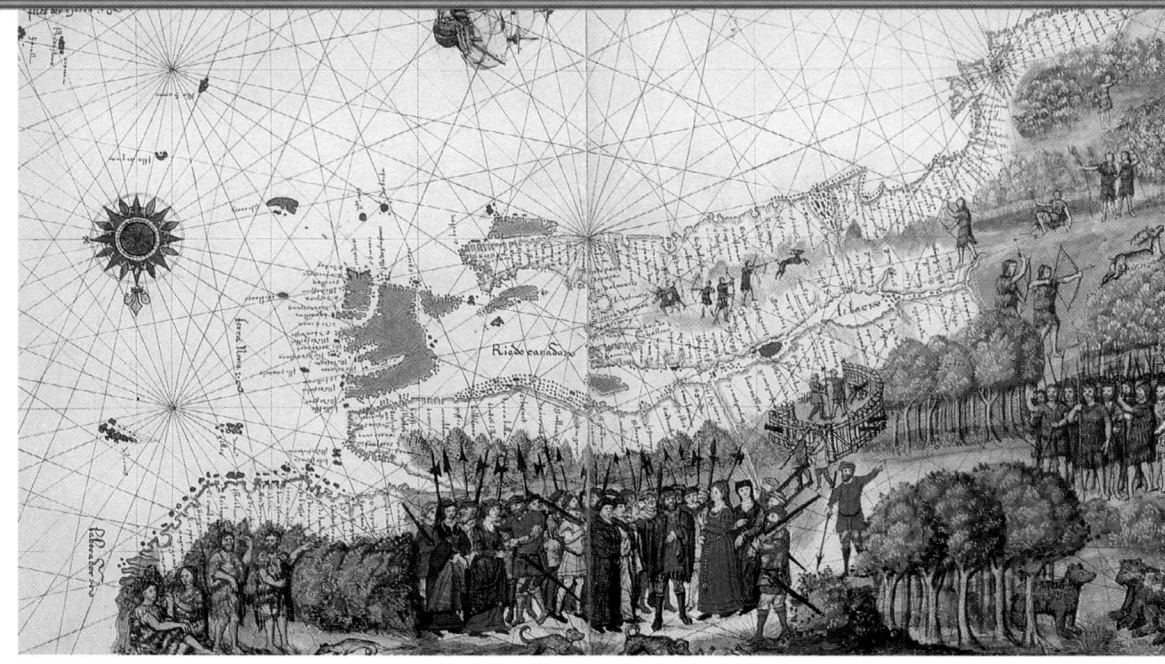

INTERPRETING THE VISUAL RECORD

This historical map shows the French explorer Jacques Cartier arriving in Canada in the early 1540s with a group of French colonists. Cartier's attempts to establish a permanent colony near what is now Quebec City failed, and he was forced to return to France. **What European and Canadian Indian cultural features can you see in this illustration? What features of the region's natural environment are visible?**

In the 1960s archaeologists discovered the first clear evidence of Viking settlement in North America. At L'Anse aux Meadows on the northern tip of the island of Newfoundland, they found the remains of eight buildings used as Viking workshops and numerous artifacts, including nails, a brass ring, and a bronze cloak pin.

History

As in the United States, Native American societies were once found across Canada. The first Europeans to sail to Canada's eastern shores were Viking adventurers. They visited between A.D. 1000 and as late as the mid-1300s. However, the Vikings left no permanent settlements. More extensive exploration by Europeans began in 1497. In that year John Cabot explored the coasts of Newfoundland and other islands for the English.

The first great European explorer of Canada's interior was Jacques Cartier (zhahk kahr-TYAY) of France. In the 1530s he traveled up the St. Lawrence River as far as present-day Montreal. This was nearly a century before the English established colonies in New England. The French had three main goals in Canada. First, they wanted to find a northwest water passage across North America to Asia. Second, they wanted to exploit nearby fishing waters and to develop a trade for animal furs from North America. Third, they wanted to convert Canadian Indians to Roman Catholicism.

By 1608 the French established a permanent settlement at what became Quebec City on the St. Lawrence River. Soon, French settlers were farming along the St. Lawrence and in nearby Nova Scotia to the east.

In 1713 Great Britain took over Nova Scotia. Eventually, the British forced many French settlers there to leave. After a long war, Britain had won control of all of French Canada by 1763. The British organized Canada into several governmental districts called **provinces**. Today Canada has 10 provinces and three special territories.

British settlement in Canada increased during the American Revolution. Many colonists left the United States so they could stay under British rule. Canada's population continued to grow in the first half of the 1800s. Immigration from abroad increased. In 1867 the British government created the self-governing Dominion of Canada. The dominion included the provinces of Ontario, Quebec, Nova Scotia, and New Brunswick. Manitoba,

Teach Objective 2

ALL LEVELS: Write the names of the following provinces and regions on the chalkboard: *Atlantic Provinces, British Columbia, northern Canada, Ontario, Prairie Provinces,* and *Quebec.* Organize students into groups and assign each group one of these areas. Have students use the text to compile information about the people of their assigned regions and to organize their findings into charts. Charts should include the names of cities and settlements, information about the history of settlement in the area, and elements of each region's culture that might set it apart from other regions of Canada. Ask each group to share its findings with the class. **COOPERATIVE LEARNING**

Using National Geography Standard 9:
Human Systems: The Characteristics, Distribution, and Migration of Human Populations on Earth's Surface Have students examine migration to and within Canada in different time periods. Ask them to plot on a map the earliest routes from Britain and France, the westward push with railroads, and expansion north following the discovery of valuable metal and mineral deposits. Students may need to consult additional resources. Then have them examine the shift to Asia as the main source of immigrants, which occurred during the 1980s. Have students record recent migration trends on their maps.

British Columbia, and Prince Edward Island joined them in the 1870s. Alberta and Saskatchewan did not become provinces until 1905. Newfoundland and Labrador became part of Canada in 1949.

✓ **READING CHECK:** **Human Systems** What European countries most influenced Canada's development? *France and Great Britain*

Culture

More than 31 million people live in Canada. French and British culture have remained strong there, along with many influences from the United States. In addition, immigration has brought other Europeans and people from the Caribbean, Asia, and Africa. Canada's government encourages each group to hold on to its culture. As a result, Canada is a multicultural country.

People, Languages, and Religion About one fourth of all Canadians live in the province of Quebec. Quebec City is the provincial capital, and Montreal is Quebec's largest city. The province is the center of French-Canadian culture. In fact, more than 90 percent of Canadians who speak French as their first language live there. Most people in Quebec are Roman Catholic, which is the largest religion in Canada. (See the graph of religions in Canada.)

French Canadians in Quebec call themselves Quebecois (kay-buh-KWAH). They have worked to maintain cultural independence from the rest of Canada. This effort has influenced public policies and laws there. For example, official papers of the provincial government are written only in French. Also, signs on businesses and along Quebec's roads are in French as well. However, a minority in Quebec do not share the province's dominant French culture.

To the west, Ontario reflects British heritage much like Quebec symbolizes the French. British customs are still widespread in Toronto, Ontario's capital. French is seldom heard or seen on signs in the city. On the other hand, Ottawa (AH-tuh-wuh), which is in Ontario and is Canada's capital, is bilingual. English and French are commonly spoken in the city, which lies just across the Ottawa River from Quebec.

Many immigrants from the British Isles and southern and Eastern Europe settled provinces in the east and west. The residents of northern Canada include people who have left Canada's cities in the south. Many Inuit (once called Eskimos) and Canadian Indians also live there. Settlements such as Whitehorse and Yellowknife are classic frontier towns.

✓ **READING CHECK:** **Places and Regions** In what way is Quebec culturally different from the rest of Canada? *It has a strong French-Canadian culture. French is widely spoken, and most people are Roman Catholic.*

Settlement and Land Use The St. Lawrence lowlands of southern Quebec and Ontario make up the most densely settled part of Canada. They are also Canada's most economically developed areas. There you will find a chain of cities that extends from Quebec City to Windsor, Ontario. This chain includes the cities of Montreal, Toronto, and Ottawa. Together all of these cities lead the country in wealth, industry, commerce, politics, and influence. This area forms the heartland of Canada. For this reason, we call Ontario and Quebec the Heartland Provinces.

🖥 internet connect

GO TO: go.hrw.com
KEYWORD: SW3 CH9
FOR: Web sites about Canada

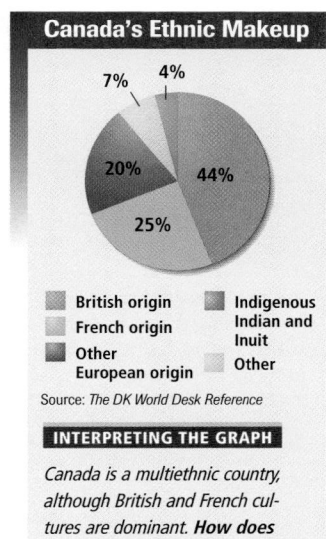

Canada's Ethnic Makeup

7% — 4%
20%
25%
44%

- British origin
- French origin
- Other European origin
- Indigenous Indian and Inuit
- Other

Source: *The DK World Desk Reference*

INTERPRETING THE GRAPH

*Canada is a multiethnic country, although British and French cultures are dominant. **How does Canada's ethnic makeup reflect its history?***

Religions in Canada

18%
42%
40%

- Protestant
- Roman Catholic
- Other

Source: Central Intelligence Agency, *The World Factbook 2001*

INTERPRETING THE GRAPH

*Canada's two main religions are Roman Catholicism and Protestantism. **In which part of Canada do you think Catholicism is dominant?***

Essential Element 4

► **Human Systems** ◄

Italians in British Columbia In the 1800s many Italian men were drawn to British Columbia by the discovery of gold in the Cariboo Mountains. More Italian laborers came later to help build the Canadian Pacific Railway. Men usually arrived first; their families came later after the men made money.

Life in British Columbia was tough for new immigrants, who often did not get the jobs they expected. They faced crowded living conditions and struggled to obtain enough food. Italian women were isolated from mainstream society and thus had difficulty learning English. Groups of Italian immigrants lived in the same areas because they shared a common language, religion, and social support network. Despite facing prejudice at first, Italians have become integrated into Canadian society. Many Canadians now enjoy the foods, games, and other cultural elements Italian immigrants brought to Canada.

DISCUSSION: Lead a discussion on how Italian immigration to British Columbia influenced Canada's culture.

GRAPH ANSWER

reflects French and British settlement

GRAPH ANSWER

Quebec

Close

Lead a discussion on the French and British settlement of Canada. Ask students to discuss the major cultural effects of these European colonizations on Canadian culture and to compare them to what they know about British and French settlement in America.

Reteach

Have the students complete the **Main Idea Activity for English Language Learners and Special-Needs Students 9.1**. Then use a map of Canada to lead a discussion about the locations and cultural identities of early European settlers in Canada. **ENGLISH LANGUAGE LEARNERS**

Review and Assess

Have students complete the **Section Review**. Then have students complete **Daily Quiz 9.1**.

Extend

Have interested students conduct research on the Canadian Indians of Nova Scotia and their relations with the French and British. Have students examine their military affiliations or lack of participation in the French/British conflict and their situations after 1867. **BLOCK SCHEDULING**

Section 1 Review Answers

Define For definitions, see: provinces, p. 190; hinterland, p. 192

Working with Sketch Maps Maps will vary, but listed places should be labeled in their approximate locations. Early settlers followed the St. Lawrence River. Many French Canadians live along this river today.

Reading for the Main Idea

1. Official documents and public signs are in French.

2. for farming; use of farm machinery encouraged by flat prairies, creating large farms that require few workers

Critical Thinking

3. Both are urban, densely populated, and highly developed in the south. French language and French-Canadian culture dominate Quebec, while English language and British customs dominate Ontario. (NGS 10)

4. northern Canada because it is more isolated (NGS 5)

Organizing What You Know

5. Atlantic—Newfoundland and Labrador, New Brunswick, Nova Scotia, Prince Edward Island; Heartland—Ontario, Quebec; Prairie—Alberta, Manitoba, Saskatchewan; Pacific Coast—British Columbia; Canadian North—Nunavut, Northwest Territories, Yukon Territory; Answers will vary but should demonstrate an understanding of the three kinds of regions.

Canada's core area includes its largest cities in the St. Lawrence lowlands, such as Montreal (top). The Prairie Provinces, which include Manitoba (bottom), are known for their productive market-oriented agriculture.

Most people in the Heartland Provinces live in the south and east. In Quebec, for example, the only densely settled areas are in the St. Lawrence, Saguenay (sa-guh-NAY), and Ottawa River valleys. The forests and rocky uplands of the vast interior are nearly empty. Only isolated government centers, trading stations, and mining districts there have many people.

The Atlantic Provinces of the east are thinly populated. These coastal provinces are Newfoundland and Labrador, New Brunswick, Nova Scotia, and Prince Edward Island. They form Canada's eastern **hinterland**. A hinterland is a region that lies far away from major population centers. Less than one twelfth of all Canadians live in the Atlantic Provinces.

In the west, settlement of most of Manitoba, Saskatchewan, and Alberta followed completion of the country's transcontinental railroad in 1885. Thousands of immigrant farmers rode the rails to their new homes in what are called the Prairie Provinces. The southern parts of these provinces were covered with prairie grasslands when European settlers arrived.

Today large farms stretch across the Prairie Provinces. However, few people are needed to work them because the level land encourages the large-scale use of farm machinery. As a result, Canada's fertile prairies remain only thinly settled. Still fewer people live farther north. The grasslands there give way to forests. Farther west is the Pacific Coast province of British Columbia. Vancouver, a major seaport, is the largest city there and the third-largest in Canada.

Forests, tundra, and rocky plains cover Canada's vast frigid north. Most of the land is underlaid by permafrost. Isolated towns and villages are scattered throughout this huge region of wilderness. It has only a few, usually gravel, highways.

✓ **READING CHECK:** *Human Systems* What area of Canada is most densely settled? southern areas of Ontario and Quebec

Section 1 Review

go.hrw.com Homework Practice Online
Keyword: SW3 HP9

Define provinces, hinterland

Working with Sketch Maps On a map of Canada that you draw or that your teacher provides, label the St. Lawrence River, Montreal, Quebec City, Toronto, Ottawa, Ottawa River, Whitehorse, Yellowknife, Windsor, Saguenay River, and Vancouver. What river did early French settlers follow into Canada? How did this route influence the distribution of ethnic groups in Canada today?

Reading for the Main Idea

1. *Human Systems* How is Quebec's French culture reflected in the province's public policies? Give two examples.

2. *Human Systems* How is the fertile land of the Prairie Provinces used? How have technology and physical geography combined to affect the population there?

Critical Thinking

3. **Comparing** What characteristics do the provinces of Quebec and Ontario have in common? In what ways are they different?

4. **Making Generalizations and Predictions** Compare the geographies of northern Canada and southern Canada. In which region do you think people would generally need to be more self-sufficient? Explain your answer.

Organizing What You Know

5. Copy the chart below. Use it to list the provinces in each of Canada's regions. You can use the chapter map to help you complete the chart. Do you think these are perceptual, formal, or functional regions? Why?

Atlantic Provinces		
Heartland Provinces		
Prairie Provinces		
Pacific Coast		
Canadian North		

OBJECTIVES

1. **Identify** which resources and activities drive Canada's economy.

2. **Explain** which factors and processes have influenced the growth of Canada's cities.

3. **Analyze** how Canada is organized and governed.

LET'S GET STARTED

Copy the following instructions onto the chalkboard: *Ontario and Quebec are called Canada's heartland. Refer to the chapter map and describe how the geography of this area might have contributed to its settlement and industrial development.* Discuss responses. *(Possible responses: Early settlement was tied to a waterway passage into the interior. Sea passage via the Great Lakes and St. Lawrence River encouraged industrial development.)* Tell students that they will learn more about Canada's economy in Section 2.

Building Vocabulary

Write **parliament** and **consensus** on the chalkboard. Call on volunteers to read the definitions from the glossary or text. Then lead a discussion about similarities and differences between the two systems. Ask how the system of consensus used in the Nunavut Territory might be applied to other places or situations. Ask students to name some advantages and disadvantages of consensus decision-making.

Section 2

Canada Today

READ TO DISCOVER

1. What resources and activities drive Canada's economy?
2. What factors and processes have influenced the growth of Canada's cities?
3. How is Canada organized and governed?

WHY IT MATTERS

The United States imports more goods from Canada than from any other country. Use **CNNfyi.com** or other **current events** sources to learn more about U.S.-Canadian trade.

DEFINE

parliament
consensus

LOCATE

Laurentian Mountains	Edmonton
Sudbury	Winnipeg
Thunder Bay	Iqaluit
Vancouver Island	Baffin Island
Calgary	

Section 2 RESOURCES

REPRODUCIBLE
- ▶ Guided Reading Strategy 9.2
- ▶ Readings in World Geography, History, and Culture 13 and 14
- ▶ Geography for Life Activity 9: The Geography of Canadian Agriculture
- ▶ Critical Thinking Activity 9: Banff National Park

TECHNOLOGY
- ▶ One-Stop Planner CD–ROM, Lesson 9.2
- ▶ Homework Practice Online
- ▶ HRW Go site

REINFORCEMENT, REVIEW, AND ASSESSMENT
- ▶ Main Idea Activity 9.2
- ▶ English Audio Summary 9.2
- ▶ Spanish Audio Summary 9.2
- ▶ Section 2 Review, p. 197
- ▶ Daily Quiz 9.2

Economic Development

Canada today is a developed country with a market economy and high standard of living. Its most important trade partner is the United States. Both countries have good transportation systems and similar business practices. As a result, Canadian and American firms can easily do business together.

Over the last century Canada has shifted away from an agricultural economy. Today its economy is based mainly on manufacturing and service industries. Mining has also long been a major activity. In fact, no other country exports more minerals and metals than Canada. Agriculture remains important. Canada is a major exporter of farm goods, producing more than its small population needs. Where do you think those farm goods are produced? Where would you expect to find large manufacturing centers? Next we will look at how the economic geography varies across the country's regions.

INTERPRETING THE VISUAL RECORD

While Canada has a strong, modern economy based on manufacturing and services, economic development is still a challenge in some areas. For example, many small towns in eastern Canada, such as Prospect, Nova Scotia, have historically depended on primary activities such as fishing. These areas lag behind Canada's most economically developed areas. **How would you describe the physical geography of this region?**

VISUAL RECORD ANSWER

coastal, rugged, rocky

193

ALL LEVELS: Copy the following graphic organizer onto the chalkboard, omitting the italicized answers. Have students complete the table with descriptions of the resources and economic activities of each region of Canada. Then lead a class discussion about how each region's resources affect its economic activities.

HOMEWORK: Provide students with outline maps of Canada and have them organize the country into new regions based on factors such as soils, climates, vegetation, river systems, land use, trade networks, languages, or religions. Students should include keys with their maps to explain any colors or symbols they have used.

```
                        Canada's Economy
```

Atlantic Provinces	Quebec and Ontario	Prairie Provinces	British Columbia	Canadian North
Resources: *fish, forests, minerals, oil* **Economic activities:** *fishing, mining, some industries*	**Resources:** *farmland, minerals* **Economic activities:** *service industries, manufacturing, farming, mining*	**Resources:** *fertile soils, potash deposits, oil* **Economic activities:** *wheat farming, mining, oil drilling*	**Resources:** *forests, fish, minerals* **Economic activities:** *manufacture of forest products, fishing, mining, trade*	**Resources:** *mineral, diamond, and fossil fuel deposits, freshwater* **Economic activities:** *mining, construction, military posts, tourism*

Newfoundland and Labrador The physical geography of the province of Newfoundland and Labrador has profoundly shaped the region's culture. The island of Newfoundland is part of the Appalachian region and consists of a large plateau. The Canadian Shield covers most of the Labrador Peninsula. The surfaces of both Newfoundland and Labrador are spotted with infertile soils, peat bogs, and rock outcroppings.

Subarctic and tundra climates cover most of Newfoundland and Labrador. Short summers and long cold winters with heavy snows are common. The few crops that grow in these cold areas include potatoes, turnips, broccoli, beets, and Chinese cabbage. Wild blueberries are the only agricultural export. Barren areas cover most of Newfoundland, broken only by reindeer moss and lichens. The ocean waters and bays along the coast provide fish—the region's best economic resource.

DISCUSSION: Lead a discussion on how the physical geography provides or limits economic opportunity where you live.

Our Amazing Planet

Canada's Bay of Fundy has the highest tides in the world. They can be as high as 70 feet (21 m). These tides bring water from the North Atlantic Ocean into the narrow bay. The bore, or leading wave of the incoming water, can roar like a big truck as the tides rush in.

INTERPRETING THE VISUAL RECORD

Completed in 1959, the St. Lawrence Seaway connects the Great Lakes to the Atlantic Ocean through a combination of artificial and natural waterways. It allows ocean-going ships to reach ports as far as Lake Superior and is important economically for both Canada and the United States. **How do you think the St. Lawrence Seaway affects the locations and patterns of economic activities in Canada?**

The Atlantic Provinces Life in this part of Canada is challenging. It has been the country's poorest region, with the lowest wages and the highest unemployment rate. In addition, long cold winters and thin rocky soils make farming difficult. Small farms that produce a variety of crops are found there. Crops do a little better in the milder climate of Nova Scotia.

Other economic activities have long depended on the resources of the sea and forests. However, the easy-to-reach old-growth forests are mostly gone. In addition, the Grand Banks area off Newfoundland—once one of the world's great commercial fishing grounds—has been overfished. As catches declined, unemployment increased. Today the government limits the number of fishing boats in the area. However, fishing is still important to the region's economy. With careful management, fish stocks may increase in the future.

The other natural resources of the Atlantic Provinces could help economic development. The mainland part of Newfoundland and Labrador has important mineral deposits. Also, oil has been found offshore. Yet mining these resources provides few jobs, and other industries have been hard to develop. Why? The region's population is too small to provide a good home market. The major population and market centers in Quebec and Ontario are far away. With high unemployment, migration to the wealthier cities of Ontario and western Canada has been common. However, the region has finally made important progress in attracting new businesses in recent years.

✔ **READING CHECK:** *Human Systems* Why has industry been hard to develop in the Atlantic Provinces? The home market is small, and the provinces are far from Canada's major market centers.

Quebec and Ontario As you read in Section 1, Quebec and Ontario make up Canada's heartland. Montreal is the industrial and financial center of Quebec. It also serves as a major port on the St. Lawrence Seaway. This is true even though much of the river is frozen for four months each year. Quebec City, located where the shores of the St. Lawrence River pinch together, is also a major port. Service industries are more important than manufacturing there. To the north, ski resorts dot the Laurentian (law-REN-chuhnz) Mountains.

Farming takes place in both southern Ontario and Quebec. However, manufacturing is the most important economic activity in Ontario. Southeastern Ontario is the chief manufacturing district of Canada. Toronto, Hamilton, Kitchener, Windsor, and other, smaller cities are located in this region. Factories there supply many of the needs of a modern industrial society and growing urban population.

Outside this part of Ontario, most of the province's cities are isolated. Some are service centers for remote mining districts. Sudbury, north of Lake Huron, developed around one of the world's largest deposits of nickel. Other

LEVEL 1: In advance, prepare slips of paper with the names of Canadian cities from the chapter and unit maps. Place the slips in a hat and have each student draw one. Then have students imagine that they are exchange students living in the cities whose names they drew. Ask each student to write a letter home to his or her family describing the city and its history. You may want to have students use library or other resources to learn more about their cities' histories and development. Students might also wish to illustrate their letters.

LEVELS 2 AND 3: Organize students into pairs. Have each pair use the text to gather information on urban centers in four areas of Canada: Ontario, Quebec, British Columbia, and the Prairie Provinces. Ask them to discuss geographic and cultural factors that make these cities in each of these areas successful and habitable. *(Possible answers: Planners in Montreal have adapted their city to the cold weather by building tunnels. Vancouver is an important port because it has a major ice-free harbor. Winnipeg's location on the Canadian Pacific Railway makes it an important shipping center.)* Ask volunteers to share their ideas with the class. Lead a class discussion on how the location, resources, or environmental characteristics of various cities have contributed to urban development in Canada. **COOPERATIVE LEARNING**

---

cities have grown up at transportation junctions. The city of Thunder Bay on northwestern Lake Superior serves as the major port for wheat from the Prairie Provinces.

The Prairie Provinces Wheat is a major crop in the Prairie Provinces, and farmers there export most of it. Changes in the global wheat market and uncertain weather conditions can cause problems for these farmers. Once, as in the United States, the government guaranteed prices. This meant that farmers could count on making a profit from the sale of their wheat. However, the government is reducing aid, and the risks for individual farmers have increased. The results are ever-larger farms that use more modern technology and machinery. Those farms can grow more food with fewer workers.

Saskatchewan's economy is mostly agricultural. However, the province has other industries as well. For example, the province has the world's largest deposits of potash. Potash mining provides an important raw material that is used in fertilizers. Alberta's income is based mostly on fossil fuels, particularly oil. Rich oil fields are found there and in western Saskatchewan. The Rocky Mountains of southwestern Alberta are also a valuable natural resource. Their spectacular scenery attracts tourists from around the world.

✓ **READING CHECK:**   *Places and Regions*   How have technology and reduced government aid changed farming patterns in the Prairie Provinces? encouraged growth of ever-larger farms with fewer workers

British Columbia The province of British Columbia, or "BC," stretches inland from the Pacific Coast. It is a land of mountains, plateaus, and fertile river valleys. British Columbia is rich in natural resources. Like the Pacific Northwest of the United States, much of the land is covered with forests of fir, spruce, and cedar trees. Income from forest products is substantial. In addition, salmon fishing and mining are important. Farmers use British Columbia's limited farmland mostly for growing fruits and vegetables as well as for dairying. Because of its location on Canada's west coast, BC trades with countries around the Pacific Rim. Japanese companies are important buyers of the province's forest products and minerals.

The Canadian North In the last 30 years, technology has helped make northern Canada less remote. Airplanes and satellite communications have tied the region more closely to the rest of the country. Now, in spite of its severe climate, the north has important promise for Canada's future. It is one of the modern world's great frontiers. Rich deposits of metals, diamonds, and fossil fuels have been discovered. In addition, supplies of freshwater there are huge. Years ago the indigenous peoples of this region lived by hunting and gathering. Some Inuit today still make a living this way. However, many now work for mining and construction companies, on military posts, or in the tourism industry.

✓ **READING CHECK:**   *Physical Systems*   What effects have new technologies and discoveries of natural resources had on northern Canada? They link it more tightly to the rest of Canada and lead to development and economic growth.

Inuit Igloos

An igloo is a traditional Inuit hunting shelter built from blocks of snow. The dome-shaped igloo remains comfortable in northern Canada's howling winter winds and subzero temperatures. How is that possible? The secret lies in turning the snow house into an ice house. The Inuit do this by heating the igloo's interior so that its inside walls begin to melt. The walls absorb the water until the snow blocks are soaked through. Then the heating is stopped, and cold outside air is allowed inside the igloo. The freezing air fuses the blocks and creates an airtight structure of solid ice.

Because ice insulates, the igloo traps warm air. Extreme cold is kept out. In fact, the temperature of the interior can be kept at nearly 55°F (about 13°C) without threatening the structure. A little water may run down the walls, but it freezes again.

Drawing Conclusions Building igloos from snow is one way humans have depended on and adapted to their environment. What traditional building materials might people in forested or warm treeless areas use?

► **Environment** ◄
and Society

The Inuit The word *Inuit*, meaning "people," was officially adopted in 1977 as a more appropriate term than *Eskimo* for the indigenous peoples of northern Canada. The Inuit migrated across the Bering Strait to North America long ago and have retained a distinctive culture.

Traditional Inuit beliefs are closely tied to the environment; every object and living being is believed to have a spirit. The extended family is their most important social unit. Men build houses, hunt, and fish, and women cook, prepare animal skins, and make clothing. Though many Inuit maintain traditional ways, they also live as part of modern society. Sculpture and printmaking are their primary economic mainstays. They have formed political organizations to represent their interests, which include the rights to land and self-government, the right to hunt whales, and the preservation of the environment.

DISCUSSION: Discuss with students why the issues of land rights and the environment are important to the Inuit.

wood, leaves, other tree parts, mud, grass, shrubs

Teacher to Teacher

William Fisher of Bryan, Texas, suggests the following activity to help students understand the unique characteristics of the political units that make up Canada. Organize the class into groups and assign each group one of the provinces or territories of Canada. Each group should conduct research into its assigned province or territory and should produce a poster describing its population, cities, economic activities, climates, wildlife, culture, and history. Encourage students to find pictures and maps with which to illustrate their posters. Have each group present its poster to the class. **COOPERATIVE LEARNING**

Teach Objective 3

ALL LEVELS: Organize students into groups. Have each group make a list of important characteristics of both the Canadian system of national government and the particularities of the Nunavut Territory. Then lead a class discussion about the organization and government of Canada. Ask students about possible advantages and disadvantages of each governmental system and organization. Try to elicit a variety of opinions. (*Possible answer: Consensus might work for a small population, such as in the Nunavut Territory, but might not work as well in Ontario or Quebec.*) **COOPERATIVE LEARNING**

Linking Past to Present

Ottawa Between 1826 and 1832 the British government built a canal between the Ottawa River and Lake Ontario. On it they founded a village named Bytown. The settlement developed into a lumber town, and the Chaudière Falls on the Ottawa River were harnessed to run lumber mills.

In 1855 Bytown was renamed Ottawa. In 1857 it became the capital of the Province of Canada. The city's location between Quebec and Ontario symbolized unity between the two regions. Parliament buildings were constructed and the federal civil-service economy grew.

ACTIVITY: Instruct students to conduct research on another capital city and to analyze the cultural and environmental factors that led to its being chosen as a capital. Then have them compare the city to Ottawa.

Toronto's location on the northern shore of Lake Ontario provides access to Atlantic shipping and to major industrial centers in the United States.

Urban Environments

Canada's cities are generally well managed, clean, and safe. Toronto is Canada's largest city. It has a metropolitan population approaching 5 million. The city is also home to Canada's largest stock exchange, major banks, and insurance companies. Many other large Canadian companies are also located there. In addition, people there can visit great museums and other cultural institutions. Recent immigrants from Eastern Europe, the Caribbean, and China help make Toronto a multicultural city.

Founded on an island in the St. Lawrence River, Montreal has a population of about 3.5 million. It is Canada's second-largest city. Underground passageways and overhead glass tunnels connect many buildings in the city center. These structures protect people from the city's cold winter weather. Montreal's residents are proud of their subway, the Métro. It is patterned after the subway system in Paris.

Victoria, the capital of British Columbia, is located at the southeastern tip of Vancouver Island. The city is the home port for a large fishing fleet. Its old English charm also attracts many tourists every year. Nearby Vancouver, on the British Columbia mainland, is western Canada's most populous city. (See Cities & Settlements: Vancouver.) It has Canada's major ice-free harbor and is Canada's main Pacific port. The city has also become a major center for movie and television productions.

Alberta has two rapidly growing cities, Calgary and Edmonton. Each is an important oil and agricultural center. Glass office towers stand out as striking structures on the Canadian prairie. However, Winnipeg, the capital of Manitoba, is the Prairie Provinces' chief city. All east-west rail traffic passes through Winnipeg. This makes the city an important collection and shipping point for the region's products.

✔ **READING CHECK:** *Places and Regions* What factors have been important to the growth of major cities in the Prairie Provinces? location near oil resources, agricultural land, along an important rail route

Government and Politics

Canada's ties to Great Britain have remained close. Britain's monarch is also Canada's monarch. Also like Britain, Canada is a democracy. It is governed by a prime minister and an elected **parliament**, or legislature. A minister, or premier, also heads each province's parliament. Provincial governments can levy taxes and set policies on issues such as education and civil rights. Canada has three northern territories spread across the Canadian Arctic and sub-Arctic: Yukon Territory, Northwest Territories, and Nunavut (NOO-na-voot). While they do not live in provinces, residents in the territories still have considerable control over local issues.

Close

Lead a discussion on regional Canadian culture. Introduce topics like the economic status of the Inuit and the multiculturalism of cities such as Toronto or Vancouver. Compare these topics to similar themes in the United States, such as the social or political situation of minority populations in U.S. cities and the economic status of American Indians in the United States.

Review and Assess

Have students complete the **Section Review**. Then have students complete **Daily Quiz 9.2**.

Reteach

Have students complete the **Main Idea Activity for English Language Learners and Special-Needs Students 9.2**. Have students work in groups to identify the major cultural groups in Canada's history and to mark locations influenced by these groups on a map of Canada. ENGLISH LANGUAGE LEARNERS, COOPERATIVE LEARNING

Extend

The Canadian Pacific Railway was begun in the late 1800s to connect Canada's east and west coasts. Have interested students conduct research on the railroad to learn how it affected the cultural geography of Canada. BLOCK SCHEDULING

FOCUS ON GOVERNMENT

Nunavut Nunavut is Canada's newest territory. Canada's government created Nunavut out of the Northwest Territories in 1999. They created it to give the Inuit of the region a self-governing homeland. In fact, Nunavut's name is an Inuit word meaning "Our Land." This land covers about one fifth of Canada. However, it has less than 13 miles (21 km) of highways. Looked at another way, Nunavut is three times the size of Texas but has fewer than 30,000 people. Its people live in just 28 widely scattered communities.

To govern, Nunavut's leaders blend tradition with technology. Nunavut's 19-member elected assembly meets in Iqaluit (ee-KAH-loo-it), on southeastern Baffin Island. With some 4,200 people, it is by far the territory's largest town. Unlike most Canadian legislatures, where one political party is in control, Nunavut's assembly makes laws by **consensus**. Consensus means "general agreement." It is the traditional Inuit way of making decisions.

Nunavut's leaders also plan to use the Internet and e-mail to bring government to the widely scattered residents. These leaders plan to open community centers where people who do not have computers can access government offices. In addition, a number of government agencies are based in towns outside the capital. Scattering government agencies allows Nunavut's people to have better access to government jobs and services. Those jobs are considered important in Nunavut. This is because nearly as many people in Nunavut still hunt and fish for a living as work for wages.

✔ READING CHECK: *Environment and Society*
In what ways do Nunavut's leaders blend tradition and technology to govern their territory?
govern by consensus; use Internet, e-mail to bring government to region's residents

INTERPRETING THE VISUAL RECORD

These children travel by boat to visit relatives in Nunavut, Canada's newest territory. The daily lives of many Inuit in northern Canada blend traditional activities with modern ones. **How do you think diffusion of technology and changing trade patterns might cause cultural change among these Inuit children?**

Section 2 Review

Define parliament, consensus

Working with Sketch Maps On the map that you created in Section 1, label Laurentian Mountains, Sudbury, Thunder Bay, Vancouver Island, Calgary, Edmonton, Winnipeg, Iqaluit, and Baffin Island. Where in Canada would you find ski and tourist resorts?

Reading for the Main Idea

1. *Physical Systems* What natural resources are important to Canada's economy?

2. *Places and Regions* What influenced the growth of important cities in Ontario and Quebec? Give some examples.

Critical Thinking

3. **Evaluating** What geographic and economic effects do you think government limits on fishing in the Grand Banks might have over time?

4. **Making Generalizations** Why did Canada create the territory of Nunavut? Why do you suppose the Inuit wanted this territory?

Organizing What You Know

5. Copy the chart. Use it to identify physical and human factors that define Canada's regions. You may want to review what you read about Canadian culture in Section 1. Note factors such as soils, climate, vegetation, language, trade, river systems, religion, economic activities, and natural resources.

Atlantic Provinces	
Ontario and Quebec	
Prairie Provinces	
British Columbia	
Canadian North	

go. hrw .com Homework Practice Online
Keyword: SW3 HP9

Section 2 Review Answers

Define For definitions, see: parliament, p. 196; consensus, p. 197

Working with Sketch Maps Maps will vary, but listed places should be labeled in their approximate locations. These resorts are in the Laurentian and Rocky Mountains.

Reading for the Main Idea
1. farmland, forests, fish, minerals, fossil fuels, metals

2. Montreal, Quebec City: on St. Lawrence Seaway; Toronto, Hamilton, Kitchener, Windsor: manufacturing centers; Sudbury: near nickel mines; Thunder Bay: port

Critical Thinking
3. Fish stocks might increase. (NGS 14)

4. to give Inuit a self-governing homeland; to reflect their historical and cultural distinction (NGS 12)

Organizing What You Know
5. Atlantic Provinces—fishing; thin soils; little industry; sea and forest resources; Ontario and Quebec—economic and political heartland; St. Lawrence Seaway; French and English heritages; mining; Prairie Provinces—rich soils; farming; oil, minerals; mountains; British Columbia—mountains; Pacific trade links; Canadian North—cold, remote; Inuit culture; mineral and energy resources

VISUAL RECORD ANSWER

wear Western clothes, use technology like motorboats

Sociology: Migration Decisions

Residential preferences reflect both the needs of people and the cultural and environmental characteristics of places. Ask students what factors they think are important in deciding where to live.

Have students imagine themselves in four stages of life: a young child, a college student or young professional, a married person with young children, and an older retired person. Ask students to think about what kinds of activities, services, organizations, resources, and environments are important to them at these different stages. They should also think about their needs in terms of environment. For example, an elderly person may not want to drive and may value a neighborhood in which he or she can walk to nearby stores. Married couples with young children will likely value good school districts. Call on volunteers to provide examples of the residential preferences of people they know from different age groups. For example, students may note where their grandparents live and why, or where older siblings in college now live.

Have each student use a decision-making process to choose the best place to live at a certain stage of life. Ask each student to select an age and a life stage and to imagine that he or she can move to any city or location. Students should make lists of the factors that are most desirable for their chosen life stages and then gather information about locations that best meet their criteria. Instruct each student to create a list of three options and to choose the most suitable location for his or her needs. Finally, have students share the reasoning behind their decisions with the class.

Global Perspectives

Immigrant Residential Preferences Many factors influence where immigrants to Canada decide to live. Many new immigrants migrate within Canada during their first few years in the country. Their residential preferences are affected by the information available to them before they move and by their economic prospects.

Immigrants to Canada come from Asia, Europe, the Americas, Africa, and the Pacific region. Their country of origin is a key factor in where people choose to live. Toronto, Montreal, and Vancouver have large communities that consist mostly of one ethnic group. As a result, many immigrants settle in these cities because they can join communities that share their culture. Economic opportunities also draw immigrants to Canada's large cities.

DISCUSSION: Lead a discussion on why most immigrants go to urban rather than rural areas.

Applying What You Know Answers

1. culture and experience
2. Possible answer: to the West Coast

Places and Regions

Geography for Life

Canadian Residential Preferences

Imagine that you could live anywhere in the United States or Canada. Where would you choose to live? Why? Would you pick a place near your home or a place far away? Are there some areas where you would not want to live? Why?

Many geographers are interested in how people perceive the world around them. These perceptions include people's thoughts about distant places. The knowledge and images that we have of different places are all a part of our mental maps—and all people have different mental maps. For example, think for a minute about the city of Montreal. What images and ideas come to your mind? Where did you get these ideas and images? Now suppose you had a friend who grew up in Montreal. How do you think his or her mental map would be different from yours? This example illustrates an important idea: people's culture and experience influence their opinions and perceptions of the world and its places.

In the 1970s and 1980s a cultural geographer named Herbert A. Whitney studied where a group of Canadians would most like to live. They could choose from places and areas in the United States and Canada. He surveyed people from all over Canada and asked them to indicate on a map which areas they liked and disliked. Most people surveyed were young, single college students. Whitney specifically chose this group because they tend to be the most mobile segment of society. He hoped that by analyzing their residential preferences, he could learn about possible future migration patterns.

Whitney's study produced some very interesting results. It showed that people in different parts of Canada had different preferences about where they wanted to live. For example, people in Quebec favored parts of the East Coast and southern United States more than people from Ontario and British Columbia. People in British Columbia had a strong preference for the western coasts of both countries.

However, people throughout Canada also shared some common perceptions. For example, most Canadians had a strong dislike of northern Canada as a place to live. In contrast, they had a strong like for the west. Also, Canadians generally had a strong like for areas near their homes and a dislike for distant areas. This last result might be expected. People usually have more knowledge of nearby areas and feel more comfortable there. For many people, distant places can seem different, strange, or even scary.

Applying What You Know

1. **Summarizing** What affects people's mental maps of distant places?

2. **Making Generalizations and Predictions** After studying the maps, where would you expect Canadians to migrate in the future?

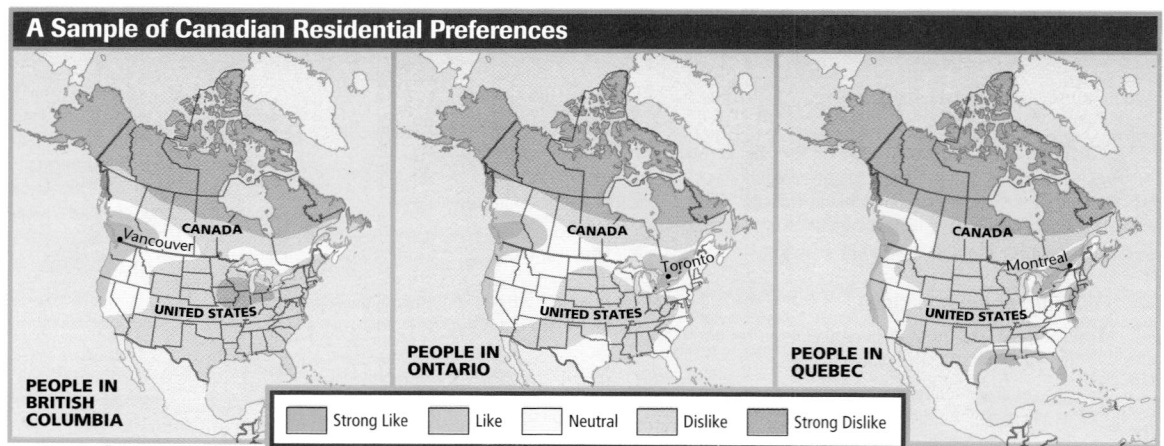

A Sample of Canadian Residential Preferences

PEOPLE IN BRITISH COLUMBIA

PEOPLE IN ONTARIO

PEOPLE IN QUEBEC

Strong Like | Like | Neutral | Dislike | Strong Dislike

LET'S GET STARTED

Copy the following question onto the chalkboard: *How might increased trade between the United States and Canada influence the diffusion of ideas and cultural values?* Discuss responses. *(Possible response: One country's media, fashion, or culture may promote a demand for them in the other country.)* Tell students that in Section 3 they will learn more about how Canada's proximity to the United States affects its culture.

Building Vocabulary

Write **regionalism** and **separatism** on the chalkboard. Ask a volunteer to read the definitions from the text or glossary. Ask students if they think there is regionalism in the United States. *(Answers will vary, but students may refer to perceptual regions such as the South or Midwest.)* Then discuss with students what factors might contribute to regionalism or separatism *(cultural, historical, and economic differences)*.

Geographic Issues

READ TO DISCOVER

1. How does the United States influence Canada today?
2. How have geographic factors affected Canada's national unity?

DEFINE

regionalism
separatism

WHY IT MATTERS

If Canada split apart, the event could have important effects on the United States. Use **CNNfyi.com** or other **current events** sources to learn about the debate over Canada's political future.

Surveyors from the International Boundary Commission mark a section of the U.S.-Canada border.

Our Northern Neighbor

The United States has always had much in common with its northern neighbor. English is the main language in the United States and most of Canada. The two countries also share similar histories as former European colonies. Many immigrants have made their new homes in these countries. In addition to such cultural and historical ties, the two countries have strong economic connections. Many companies based in the United States are giants in the Canadian economy. In addition, the U.S. and Canadian economies tend to rise and fall together. This relationship has become stronger since the signing of the North American Free Trade Agreement (NAFTA). NAFTA has resulted in increased trade and cooperation between the two countries.

In addition, Canada's cities in many ways seem more connected to U.S. cities than to each other. For example, Vancouver has common interests with Seattle. Farther east, Toronto has connections to many cities in the U.S. Midwest. Montreal has many business links to Boston.

The potential for cultural domination by the United States is a great concern for many Canadians. Canada's population is only about 31 million, and the great majority of Canadians live near the U.S. border. For this reason, the exchange of culture traits between the two countries can hardly be expected to be equal. With nine times the population of Canada, the United States often seems to overwhelm Canadians with its mass culture. For example, Canadians hear cultural, political, and economic information about the United States nearly every day. Yet many Americans seldom think about Canada.

Still, Canadians have a strong sense of political independence from the United States. Most Canadians would not want their

RESOURCES

REPRODUCIBLE
▶ Guided Reading Strategy 9.3
▶ Graphic Organizer Activity 9

TECHNOLOGY
▶ One-Stop Planner CD–ROM, Lesson 9.3
▶ CNN Presents Geography: Yesterday and Today, Segment 8: Tradition and Change—The James Bay Project
▶ Homework Practice Online
▶ HRW Go site

REINFORCEMENT, REVIEW, AND ASSESSMENT
▶ Main Idea Activity 9.3
▶ English Audio Summary 9.3
▶ Spanish Audio Summary 9.3
▶ Section 3 Review, p. 201
▶ Daily Quiz 9.3

Teach Objective 1

 ALL LEVELS: Copy the following graphic organizer onto the chalkboard, omitting the italicized answers. Have students complete the organizer with phrases that describe how Canada's relationships with the United States have led to economic, cultural, and social changes there.

U.S. Influences on Canada's Culture

Economy	Population	Culture	Cities
economies closely tied together; many American companies in Canada	*U.S. population is much larger; most Canadians' homes near U.S. border*	*American mass media heard and seen in Canada nearly every day*	*Canadian and U.S. cities linked, increases cultural exchange*

Teach Objective 2

 ALL LEVELS: Organize students into five groups. Assign four of the groups to represent the following provinces: Alberta, British Columbia, Ontario, and Quebec. The fifth group should represent a panel of ministers from the Canadian Parliament. Have the first four groups discuss reasons for staying united with Canada or reasons that they might want independence. Each group should then present an action plan to the panel of ministers, who will discuss each case and propose a solution. Then lead the class in an open discussion on the issues surrounding the geographical, cultural, and political unity of Canada. **COOPERATIVE LEARNING**

 ## Section 3 Review Answers

Define For definitions, see: regionalism, p. 200; separatism, p. 200

Reading for the Main Idea
1. shared language, histories, immigration history, economic connections; Canadian cities often closer and more closely connected to American cities than to each other

2. It would physically divide Canada and might cause it to dissolve if other provinces were encouraged to seek independence.

3. British Columbia is physically separated from the rest of Canada by high mountains and is far from the country's centers of population and government.

Critical Thinking
4. Answers will vary. Supporters may point to Quebec's cultural uniqueness and a desire to compromise with Quebecois separatists. Opponents may suggest that this policy might have actually strengthened and encouraged separatism. **(NGS 10)**

Organizing What You Know
5. Organizers will vary but should include NAFTA and other trade links, American companies in Canada, and regional connections between American and Canadian cities.

VISUAL RECORD ANSWER

architecture, Catholic churches

200

country to join their neighbor to the south. However, Canadians today are debating the nature and unity of their country. This debate could help decide the future relations between Canada's provinces and the United States.

✓ **READING CHECK:** *Human Systems* How has NAFTA affected the economy of Canada? It has tied the rise and fall of Canada's economy more closely to that of the United States.

Regionalism and Separatism

The nature of the debate over a united Canada is linked to the country's physical and cultural geography. Canadians often show considerable **regionalism** when considering their country's important issues. Regionalism refers to the feeling of strong political and emotional loyalty to one's own region.

Canada's physical geography and isolated settlements help keep regionalism alive. For example, many Albertans believe their province shares too much of its income from the oil industry with the national government. British Columbians often feel separate from the rest of Canada, which lies far to the east across the Rocky Mountains and prairies. Many Canadians outside the Heartland Provinces believe the interests of Ontario and Quebec dominate in Ottawa, the national capital.

Even more critical to Canada's future than regionalism is **separatism**. This is the belief that certain parts of a country should be independent. Separatist feelings are strongest in Quebec. The separatist movement there has grown over the last 30–40 years, which has greatly affected Canada's government. Quebecois culture is specially protected under Canadian laws. In fact, civil law in Quebec is based on the French legal system rather than the English. Also, French is one of Canada's official languages, along with English. According to law, the children of French Canadians born outside of Quebec can choose to

INTERPRETING THE VISUAL RECORD

The landscape of Quebec City reflects the dominance of French culture in the region. **What cultural features can you see in this photograph that help make French Canada a distinct region from the rest of the country?**

Close

Lead a discussion on the tension between feeling a tie to a specific region or place and the need to belong to a larger country. Ask students what such tensions in Canada are based on (*geographic or cultural differences*) and how governments try to overcome regionalism and bring people together (*national language, education system*).

Review and Assess

Have students complete the **Section Review**. Then have students complete **Daily Quiz 9.3**.

Reteach

Have students complete the **Main Idea Activity for English Language Learners and Special-Needs Students 9.3**. Then call on students to identify the significance of each of the following terms to Canada today: *NAFTA, cultural domination, regionalism, Quebec, national unity.* **ENGLISH LANGUAGE LEARNERS**

Extend

Have interested students compare maps that show voting patterns in the 1995 election in which the people of Quebec chose to remain a part of Canada to language maps of the province. Ask students what these maps suggest about the distribution of political power in Quebec. **BLOCK SCHEDULING**

have their children educated in French. Special immigration powers help Quebec to attract French-speaking newcomers. No other province has such status. Still, many French Canadians want Quebec to be an independent country. This issue has again and again threatened to break up Canada. In 1995, people in Quebec narrowly voted down a proposal that would have made their province an independent country. In fact, about 60 percent of the province's French-speaking citizens voted for it. People who pushed the proposal promise to bring the issue before voters again.

People in Ontario and Quebec have been arguing over language and culture for more than two centuries. Quebec even has what some people call "language police." These government officials make sure that signs are always in French. If English is also used, they make sure the French lettering is most prominent. Even Montreal has changed in recent years. The city once was an island of British-Canadian culture in Quebec. However, some corporations have left because they were worried about what an independent Quebec would mean for their businesses. Many have moved west to cities like Toronto and Calgary. Thousands of English-speaking residents have also left. Now Quebecois fill most important jobs in banking, education, insurance, and manufacturing.

The debate over Quebec has created the worst unity crisis in Canada's history. An independent Quebec would separate the Atlantic Provinces from the rest of Canada. Some Canadians worry that British Columbia, which has strong ties with countries around the Pacific, might also leave the union. Other provinces might demand their independence as well.

Still, Canadians seem to feel that these disputes are more like family squabbles than a national crisis. Canada remains peaceful because most Canadians believe everyone should have a chance to explain and debate their views. Most still support a united Canada rather than the idea of several small countries. Canadians continue to find strength in their diversity.

✓ **READING CHECK:** *The World in Spatial Terms* How might the independence of Quebec affect the political geography of Canada? The eastern provinces would be separated from the rest of Canada.

Street signs in French stand in front of a Catholic church in old Montreal. In Quebec both language and religion—two of the most important features of culture—are different from the rest of Canada and provide the basis for regionalism and separatism.

CHAPTER 9 Review Answers

Building Vocabulary For definitions, see: provinces, p. 190; hinterland, p. 192; parliament, p. 196; consensus, p. 197; regionalism, p. 200; separatism, p. 200

Locating Key Places

A. Baffin Island
B. Montreal
C. Winnipeg
D. Toronto
E. Vancouver
F. Ottawa
G. Vancouver Island
H. Quebec City

Understanding the Main Ideas

1. Inuit and Native American, French, British, other Europeans, people from Caribbean, Asia, Africa

2. major cities, lead country in wealth, industry, commerce, politics

3. based more on manufacturing and service industries today

4. The government recognizes Britain's monarch and follows the British parliamentary system. Nunavut was created as a homeland for the Inuit.

5. English, similar histories, trade connections, strong U.S. cultural influences

Section 3 Review

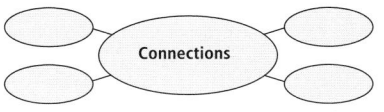

go.hrw.com **Homework Practice Online** Keyword: SW3 HP9

Define
regionalism, separatism

Reading for the Main Idea

1. *Human Systems* Why is Canada's culture so closely connected to that of the United States?

2. *Places and Regions* Why would an independent Quebec be a challenge for Canada?

3. *Places and Regions* For what reasons might regionalism in British Columbia be stronger than in most of the other provinces?

Critical Thinking

4. *Supporting a Point of View* Do you think that it has been a good policy to give Quebec such a special status in Canada? Explain your answer.

Organizing What You Know

5. Create a diagram like the one below and use it to identify the ways in which Canada and the United States are economically connected.

Connections

TECHNOLOGY

▶ Chapter 9 Test Generator (on the One-Stop Planner)
▶ Global Skill Builder CD–ROM
▶ HRW Go site

REINFORCEMENT, REVIEW, AND ASSESSMENT

▶ Chapter 9 Review, pp. 202–03
▶ Chapter 9 Tutorial for Students, Parents, Mentors, and Peers
▶ Chapter 9 Test (form A or B)
▶ Alternative Assessment Handbook
▶ Chapter 9 Test for English Language Learners and Special-Needs Students
▶ Unit 2 Test
▶ Unit 2 Test for English Language Learners and Special-Needs Students

Assess

Have students complete a Chapter 9 Test.

Reteach

Organize students into three groups and assign one section to each group. Have each group discuss its assigned section material and prepare a review for the class. Then have each group conduct the review, calling on volunteers from other groups to answer questions about the section material. **COOPERATIVE LEARNING**

CHAPTER 9 Review Answers

Thinking Critically

1. Answers will vary but should note the regionalism of culture groups like the French Canadians and Inuit. Students might suggest that vast areas, sparse settlement, and great distances between population centers promote regionalism. **(NGS 5)**

2. Answers will vary, but students should recognize that northern Canada has plentiful resources but few people, and its infrastructure is insufficient to support manufacturing and access to markets. **(NGS 11)**

3. Answers will vary. Students might say that French Canadians see their country as dominated by English speakers. English Canadians might view their country as a collection of cultures. The Inuit might see Canada as dominated by descendants of conquering peoples. **(NGS 2)**

Using the Geographer's Tools

1. British

2. Graphs will vary but should reflect the information in the tables.

3. Student maps will vary but should show an understanding of cultural, economic, and physical features of Canada.

CHAPTER 9 Review

Building Vocabulary

On a separate sheet of paper, explain the following terms by using them correctly in sentences.

provinces	parliament	regionalism
hinterland	consensus	separatism

Locating Key Places

On a separate sheet of paper, match the letters on the map with their correct labels.

Montreal	Ottawa	Winnipeg
Quebec City	Vancouver	Baffin Island
Toronto	Vancouver Island	

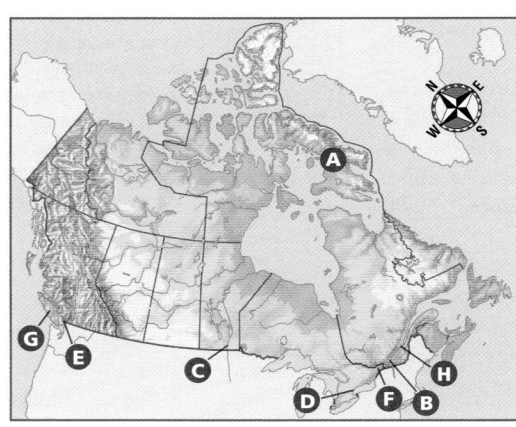

Understanding the Main Ideas

Section 1

1. *Human Systems* What major groups have settled in Canada over time?

2. *Places and Regions* What characteristics make Ontario and Quebec Canada's heartland?

Section 2

3. *Places and Regions* How has Canada's economy changed over the last century?

4. *Human Systems* How is Canada's heritage reflected in its system of government? Why was Nunavut created?

Section 3

5. *Human Systems* What are some important cultural, historical, and economic ties between the United States and Canada?

Thinking Critically

1. **Identifying Cause and Effect** How does Canada's cultural and physical geography contribute to regionalism?

2. **Drawing Inferences** Do you think major manufacturing centers will develop in northern Canada? Explain.

3. **Making Predictions** How might Canadians of French, English, and Inuit descent view their country differently?

Using the Geographer's Tools

1. **Analyzing Pie Graphs** Review the pie graph showing ethnic groups in Canada in Section 1. What are the origins of the largest ethnic group?

2. **Creating Pie Graphs** Use the unit Fast Facts table to create two pie graphs. One should show how the land area of North America is distributed between the United States and Canada, and the other how the continent's population is distributed. How does the population density of the United States compare to Canada?

3. **Creating Maps** Draw a map of what North America might look like if Canada were to dissolve. Indicate possible national boundaries, names of new countries, and likely capitals. What geographic and cultural features would link these countries?

Writing about Geography

Review the discussions of economic development and standard of living in Chapter 6. Then analyze information from chapters in this unit and from the unit Fast Facts table to compare the level of development and standard of living in Canada and the United States. Write a short report that identifies the factors you have analyzed and the conclusions you have reached.

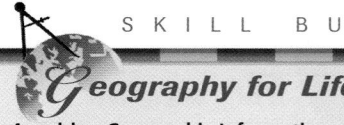

SKILL BUILDING

Geography for Life

Acquiring Geographic Information

Places and Regions Review the unit map of economic activities in Canada and the United States. Choose a Canadian province and investigate economic activities there. For example, what are the major items that the province manufactures? What crops or livestock are raised? Write a one-page summary of what you have learned. When you are finished with your summary, proofread it to make sure that you have used standard grammar, spelling, sentence structure, and punctuation.

Portfolio Activity

Have students conduct research on Asian immigration to Canada's west coast. Ask them to imagine they are second- or third-generation Asian residents of Vancouver. Have each student write an imaginary history of his or her family. Histories should note when the family arrived in Canada, why they came, and what the situation was like for them upon arrival and after several years. Place stories in students' portfolios.

Food Festival

Wild rice is not rice at all. It is a grass seed that grows in northern lakes and rivers, including many in Canada. For generations, the Ojibwa have harvested the grain the traditional way, paddling canoes among the plants and coaxing the ripe grain to fall into the boat. Now wild rice harvesting is an important industry for the Ojibwa, particularly in Saskatchewan. In the late 1990s the province's lakes produced about 2.6 million pounds (1.17 million kg) of wild rice each year. Ask students to locate recipes for wild rice and to prepare them for class. The grain can be used in many ways, such as in casseroles, griddle cakes, meat loaf, salads, soup, stuffings, or separately as a side dish.

Building Social Studies Skills

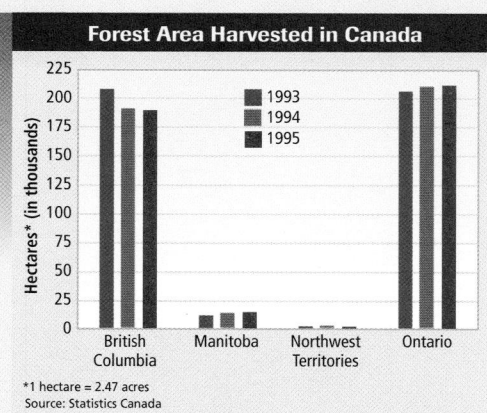

Forest Area Harvested in Canada

*1 hectare = 2.47 acres
Source: Statistics Canada

Interpreting Graphs

Use the graph above to answer the questions that follow.

1. Which province had the most forest area harvested in 1995?
   a. British Columbia
   b. Manitoba
   c. Northwest Territories
   d. Ontario

2. Why do you think the amount of forest area harvested in Manitoba and Northwest Territories is so small?

Analyzing Primary Sources

Writer Ian Darragh visited Prince Edward Island in the late 1990s and noted the Confederation Bridge, a new bridge to the mainland. Read his description, then answer the questions that follow.

> "The eight-mile-long ribbon of concrete, which opened in June 1997, connects Prince Edward Island, Canada's smallest province, to the mainland for the first time in 5,000 years. . . . While people debate how the bridge may alter the environment of Northumberland Strait—from the swirl of the currents to the spring breakup of ice—it is already changing the ebb and flow of life on the island itself, a place long defined by close-knit communities and a slow-paced way of life."

3. What features of the physical environment does the author suggest might be affected by the new bridge?
   a. a slow-paced way of life
   b. ribbons of concrete
   c. water currents and the spring breakup of ice
   d. the ebb and flow of life itself

4. In what ways does the author suggest that the bridge might affect the "ebb and flow of life on the island itself"?

Alternative Assessment

PORTFOLIO ACTIVITY

Learning about Your Local Geography

Group Project: Research
Plan, organize, and complete a research project with a partner about the cultural background of your community. From what countries or areas did early settlers in your community originate? What ethnic backgrounds do current residents have, and what languages are commonly spoken there? Working with your partner, create historical and current population maps of your community that illustrate your findings.

☑ internet connect

Internet Activity: go.hrw.com
KEYWORD: SW3 GT9

Choose a topic on Canada to:
- take the GeoMap challenge to test your knowledge of Canada's geography!
- research modern Quebec and create a newspaper.
- learn about writers, artists, and musicians of Canada.

Writing

Students should accurately compare development and standards of living in the United States and Canada. Use Rubric 9, Comparing and Contrasting, to evaluate student work.

*G*eography for Life

Summaries should include sufficient detail to represent the full economic spectrum of the area. Use Rubric 37, Writing Assignments, to evaluate student work.

Social Studies Skills

1. d
2. Possible answer: The environment, particularly the climate, may be unsuitable for timber to be easily harvested in these provinces.
3. c
4. Possible answer: It might affect how close-knit the community is and the community's slow-paced way of life.

PORTFOLIO ACTIVITY

Students projects will vary based on the area selected. Maps should accurately present the information gathered through student research. Use Rubric 20, Map Creation, to evaluate student work.

Identifying Cultural Bias

Cultural bias is the influence that a person's own culture has on his or her view toward other cultures and peoples. It can be expressed in many ways. Sometimes a person's actions reflect a cultural bias. Such actions might try to change another person's way of life. They may also be more extreme, including even violent actions against members of a particular culture group. The September 11 attacks were the result of terrorists' cultural bias against Americans.

Cultural bias is more often expressed in words than in actions. A writer's own cultural experiences sometimes lead him or her to make judgments about other cultures he or she is describing. This type of cultural bias is often unintentional and can be mistaken as simple observations of a culture group's way of life. Words like *good, bad, normal, strange, righteous,* and *evil* are often signs that a writer or speaker is expressing culturally biased judgments.

Some forms of cultural bias are less obvious. Among these are stereotypes and ethnocentrism. Using stereotypes shows that one is ignoring differences within groups and instead treating everyone in a group as if they were alike. For example, immediately after the attack on the World Trade Center, some Americans assumed that all Muslims were like the fundamentalist terrorists. Political and religious leaders were quick to refute this stereotype. Ethnocentrism involves a person using his or her own culture and experiences as the only point of reference from which to make statements or judgments about others. An ethnocentric statement suggests that the speaker's culture is superior to the one to which it is compared.

Linking Past to Present

Al Qaeda Osama bin Laden, the man accused of sponsoring the 2001 attacks, is the leader of the fundamentalist Muslim terrorist organization al Qaeda. The group was created in the late 1980s by conservative Muslims who had helped to overthrow Afghanistan's communist government. Thousands of fighters, who were called *mujahideen,* responded to calls from Islamic leaders to oust the Soviets and restore the country's Muslim leadership. Many of the *mujahideen,* unhappy with less conservative governments in the region, also wanted to spread their fundamentalist beliefs. Al Qaeda, funded by Osama bin Laden, was created to train these fighters.

Bin Laden and his followers subscribe to a conservative branch of Islam formed in the 1700s called Wahhabism. Wahhabis are opposed to changes in society (*bida* in Arabic) that they believe deviate from the teachings of the Qur'an. In the past they protested technological inventions such as telephones and televisions and social changes like educating women. They are also dedicated to spreading their beliefs and are noted for the ferocity with which they will fight for their cause. For Wahhabi extremists, Afghanistan under the Taliban represented an ideal society, one governed by the strictest laws of Islam.

CITIES & SETTLEMENTS

New York

Human Systems New York has long been a beacon for people all over the world. Ever since the city's founding, waves of immigrants entering New York have helped shape the culture of the city as well as that of the United States. Today about 80 percent of all New Yorkers are of African, Irish, Italian, Jewish, or Puerto Rican descent. Many other ethnic groups and nationalities are also represented there. These others include immigrants from Africa, the Caribbean, China, Russia, other European countries, and Colombia as well as other Middle and South American countries. Scattered throughout the city are neighborhoods like Manhattan's Chinatown and Little Italy, which reflect unique cultural landscapes. In each neighborhood, visitors will find places of worship, restaurants and other businesses, and services that cater to resident ethnic groups.

A World City

New York's global ties have also helped make it a world financial capital. The city is home to major banks, insurance companies, and other financial institutions. More than 200 international banks from every major country in the world have offices in New York. In fact, of the 25 largest foreign branches of international banks in the United States in 2001, 20 had offices in the city. In addition, the New York Stock Exchange has attracted many of the world's largest firms that buy and sell stocks in companies. The Exchange is located on Wall Street, a place that has come to symbolize the country's financial power.

New York is also a major cultural center. It has many famous museums, galleries, theaters, and performance halls. Large media companies have offices in the city. Major American television networks are based there, and many movies and television shows seen around the world have been set in New York. The concentration of media in New York has helped make the city's concrete canyons and towering skyscrapers famous. In fact, for many people the concept of a

These photos show the dramatic change in New York's skyline after the destruction of the twin towers of the World Trade Center on September 11, 2001. Today the Empire State Building is once again New York's tallest building.

"city" has been shaped by New York's striking skyline. In the same way that the Eiffel Tower has become a recognizable symbol of France and Paris, gleaming skyscrapers, many of which are found in New York, have become symbols of the United States.

September 11, 2001

For nearly three decades the twin towers of the World Trade Center dominated New York's skyline. These modern towers each soared higher than 1,360 feet. They joined the Statue of Liberty and the Empire State Building as among New York's most recognizable landmarks. It is perhaps the World Trade Center's status as a great symbol of modern capitalism and globalization that made it a target of attack in 2001. Fundamentalist Muslim terrorists flew two hijacked

Going Further: Thinking Critically

Have students examine newspapers, magazines, television broadcasts, Internet sites, and other current events sources that describe the September 11, 2001 attacks and the events that followed. Direct students to use library or Internet resources to find descriptions of the attacks from English-language newspapers from various countries to examine many perspectives of the event. Call on volunteers to describe how the accounts differ. Ask students how each article reflects the author's cultural bias.

Then ask students to find examples of each of these four types of cultural bias in the articles they have located: actions that show cultural bias, culturally biased judgments, stereotypes, and ethnocentrism. For each example that students provide, ask if the bias appears intentional or if the perpetrator of the bias seems unaware of his or her own leanings. Ask students to suggest reasons why writers may intentionally include cultural bias in their works. *(Possible answers: to appeal to readers' emotions, to sway readers' opinions about an event or subject, or to inspire cultural pride)* Then lead a class discussion about how cultural bias can cloud people's views of the world and its people.

planes into the towers on September 11 of that year. Other hijacked planes crashed into the Pentagon outside Washington, D.C., and in rural Pennsylvania.

The destruction of the World Trade Center left thousands dead. New York mayor Rudolph Giuliani called the death toll "more than we can bear." Nonetheless, he vowed, "We're going to rebuild and rebuild stronger."

Looking Forward

New Yorkers and other Americans pulled together to rebuild, inspiring people around the world with their dignity and courage. People in some countries saw the acts of the terrorists as an attack on all of the world. Some declared with respect and unity that they too were Americans.

In New York, residents pondered their city's future. How had the attacks changed their city? One change was the loss of a sense of security many New Yorkers and other Americans had felt before the attacks. Americans had long felt protected from such horrible attacks by the vast ocean distances separating this country from the world's trouble spots. Commentators observed that the terrorist attacks changed life as New Yorkers and other Americans had known it. This realization refocused the efforts of Americans to defeat terrorism on a global scale.

Another change was a renewed sense of respect and admiration among New Yorkers for their city's fire and police departments. Hundreds of firefighters

New York police officers and firefighters worked to help survivors after the attack on the World Trade Center. Cities like New York must employ large numbers of police officers, firefighters, and other emergency personnel.

and police officers died when the World Trade Center towers collapsed. They had bravely rushed into the buildings in an attempt to rescue trapped victims.

Perhaps the most obvious change after the attack was New York's altered skyline. Gone were the gleaming towers of the World Trade Center, which had soared high above the city's other skyscrapers. Some people wanted the city quickly to rebuild the towers to replace much-needed office space. Many people felt that the site should include proper memorials to the thousands of victims of the terrorist attack. More symbolically, New Yorkers had to decide what any new complex would say about their city. An international competition led to the selection of a design by Daniel Libeskind. Libeskind's design features a 1,776-foot tower that houses a multi-story garden. The new design offers a memorial as well as symbolizes New York's stature as a center of world culture and commerce.

New York and other cities have places like parks and city squares for recreation and public gatherings. New York's Washington Square became a site for memorials to victims of the September 11, 2001, terrorist attacks on the World Trade Center and the Pentagon.

Applying What You Know

1. **Summarizing** What are some economic, social, and cultural characteristics of New York?

2. **Making Generalizations** Why do you think some people outside the United States said they thought of themselves symbolically as Americans after the terrorist attacks on September 11, 2001?

Applying What You Know Answers

1. New York houses a diverse population and attracts more immigrants from around the world each year. Visitors to the city can find places of worship, restaurants, and other businesses and services that cater to various ethnic groups. The city is also home to major financial and media companies, international banks, and cultural institutions.

2. Possible answer: The destruction was so immense and horrible that many people elsewhere were united in shock and saw the danger that terrorists posed to people everywhere. Many saw the attack on New York as a symbolic attack on the entire free world. By calling themselves Americans, these people were expressing their respect and sympathy for citizens of the United States.

This Cities & Settlements feature addresses National Geography Standards 6, 10, 12, and 13.

Workshop 1
Going Further: Thinking Critically

Prepare two sets of words on slips of paper—at least one from each set for each student. One set of words should consist of active verbs, such as *arrive, climb, descend, dig, drive, fall, fly, swim*, and *walk*. The other set should consist of nouns relating to places, such as *beach, castle, cave, church, cliff, desert, downtown, highway, island, mountain, neighborhood, ocean*, and *overpass*. Have students draw at least one word of each type from a hat. Then have students use their chosen words at least once to write travel or adventure stories set in imaginary or familiar places. You may want to require that the stories mention a certain number of places. Tell students that when they finish they should be able to draw a map of the setting through which the characters travel. When all the students have completed their stories, have them exchange papers to draw maps of each other's story settings. Then have students return the stories to their writers, along with the mental maps. Ask the writers to compare the maps to how they had envisioned the settings. Lead a discussion about how mental maps can vary, depending on one's experiences.

1. Students' maps should include important places in the school, such as the cafeteria and gymnasium, along with their classrooms. Routes between classrooms should be clearly marked and labeled.

2. Students' maps should include places such as major streets, stores, churches, and parks. The route to school should be clearly marked and labeled.

3. Students' maps should recognizably depict your state. Places, features, and routes will vary according to students' experiences.

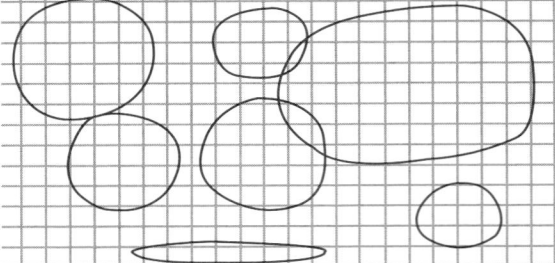

Geography
Skill-Building Workshop

WORKSHOP 1

Using Mental Maps

Mental maps are maps that represent the mental image a person has of an area. They help us make sense of the world around us by organizing information we have about places. Those places range from our own homes to the whole world.

You will not find mental maps in a textbook. Instead, mental maps are images we have in our minds. These images include knowledge of features and spatial relationships in an area. They also include our perceptions and attitudes about that particular area. As such, mental maps of even the same places vary from person to person.

Each of us uses a variety of mental maps. You can probably think of many places about which you have a mental map. You likely have a mental map of the floor plan of your home. You probably also have a mental map of the neighborhood in which you live. You know where the homes of friends, certain streets, houses, schools, and places of business and worship are located. You can use your mental map to plan routes to each of those places. You can also use a mental map of routes you take when you travel outside your community.

Developing the Skill   We can also use mental maps to organize how we think about our state, country, and world. For example, you have read that there are seven continents on Earth. Without reviewing a map of the globe, you might be able to sketch at least the locations of those continents on a map of the world. You might be able to do this even if you are not sure of the shape of each continent.

The sample sketch map on this page uses circles that show the general locations of the continents. Now close your eyes and visualize what you already know about these continents. As you think about the continents, consider the following questions:

- Can you place the name of each continent on your mental map?

- What do you know about the shapes and relative sizes of each continent?

- Which continents lie in the northern half of the globe?

- Which continents are located south of the equator?

- How are the continents connected to or separated from each other?

- What oceans lie between the continents?

After you have thought about the continents, sketch their shapes and label them and the oceans on your own map of the world. Then compare your sketch map to the world map in the atlas at the front of this textbook. How does your sketch map differ from the atlas map? How is it similar? How did your perceptions of the world guide the way you sketched your map?

Practicing the Skill

1. Sketch the floor plan of your school from memory. Locate and label important places, including classrooms you use each day. Identify the fastest or easiest routes between your classrooms.

2. Sketch a map of your neighborhood from memory. On your map locate important places, landmarks, and the route you take to school.

3. Sketch a map of your state from memory. Include important cities, rivers, travel routes, and places you have visited or would like to see in the future.

Workshop 2
Going Further: Thinking Critically

Organize the class into groups and provide each group with a large map of your state or region. Have students examine the cities shown on their maps to determine whether they are distributed according to the central place theory model. First ask students to identify the largest city on their maps, the one that offers the widest range of goods and services to its residents. Direct them to call this city the first-order place and to mark it with a large colored dot. Then ask students to identify those cities that are somewhat smaller than the first-order city but still large enough to offer a wide variety of goods and services. Tell students to label these as second-order places and to mark them with a different color. Then have students identify and mark third- and fourth-order cities on their maps. When students have finished marking their maps, have them study the patterns that they see in the distribution of cities. Ask them to identify similarities between the arrangement of cities in your state and in the central place theory model. *(Students may notice similarities to the central place theory model, such as a fairly even distribution of large cities throughout the state, but they should note that your state does not follow the model exactly.)* Lead a class discussion on geographers' uses of models. Point out that models represent the ideal arrangement of geographic phenomena and seldom show what actually exists in the world. Remind students that they are useful, however, in expressing new idea or theories.

WORKSHOP 2

Using Geographic Models

When geographers present ideas, they sometimes find it helpful to use models. In this context, *model* does not mean a small replica of something, like a model airplane. Rather, the models geographers use are more like the plans for different types of airplanes. You might look at these plans to construct an actual airplane. On the other hand, you might just study the plans to better understand how airplanes work.

Developing the Skill This textbook includes a number of models. Two of them are shown on this page and were discussed in Chapter 6. One model represents an idea called central place theory. It shows the ideal distribution of urban places of different sizes and the main transport routes that link these places. The model does not show how urban places are really distributed. Instead, it represents what one geographer thought was the best arrangement of such places, with each place serving a special function.

The second model represents patterns of land use in Germany observed by a geographer in the 1800s. Each circular band represents a different kind of land use. The distribution of land uses is related to land value and to distance from the central city. In short, this model is used to show which types of land use are most common nearest a city center and which are most common farther away.

The two models on this page have some common features. They both use color, although color may not be necessary if a model is not very complex. Both also have legends. The legends are similar to those you might find on a map. Each legend identifies symbols and colors that represent information in the model.

The von Thünen Model

WILDERNESS

WILDERNESS

- ● Central city
- Intensive farming and dairying
- Forest
- Increasingly extensive field crops
- Ranching, animal products

Source: *Geography: Realms, Regions, and Concepts*, Fifth Edition, De Blij and Muller

Practicing the Skill

1. Create a model for the ideal floor plan of a school. Your model should show how you think different areas of an ideal school might be arranged. For example, where should the cafeteria, classrooms, gymnasium, main offices, parking areas, and restrooms be located in relation to each other? Include a title, legend, any necessary labels, and a descriptive caption for your model.

2. Work with a group to study the layout of major cities around the world. Then work together to create a model that could be used to plan future cities. Your group should consider issues such as where to locate commercial and industrial areas, parks and recreational areas, and residential areas. Also consider the transportation network that should connect your model city's various areas. Draw your model on a posterboard. Then present and explain it to your class.

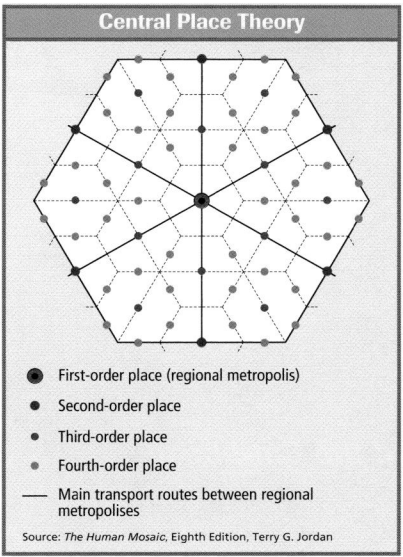

Central Place Theory

- ● First-order place (regional metropolis)
- ● Second-order place
- ● Third-order place
- ● Fourth-order place
- — Main transport routes between regional metropolises

Source: *The Human Mosaic*, Eighth Edition, Terry G. Jordan

Using the Illustration

Focus students' attention on the photograph of Chichén Itzá on Mexico's Yucatán Peninsula, once a center of Maya culture. Point out that the serpent statue in the foreground is part of the city's Temple of the Warriors. The statue may have represented Kukulcán, the plumed serpent god who, according to legend, was also a Maya ruler. The Castillo in the background is topped by a temple dedicated to him.

Chichén Itzá has a long and varied history. It was probably founded by the Maya in the early A.D. 500s, but by 700 it had been abandoned. It was reoccupied in the 900s by people who spoke the Maya language but who had been strongly influenced by the Toltec of central Mexico. The structures built by these later residents—including the Temple of the Warriors and the Castillo—reflects the influence of the Toltec, particularly in its columns and sculptures. The serpent, for example, is a popular motif in Toltec art. Chichén Itzá was abandoned for the last time around 1200.

Ask students how abandoned cities like Chichén Itzá can contribute to our understanding of a region's cultures. *(Possible answer: They can tell us about life in past societies, which can help us to understand the development of modern cultures.)* Ask them what the structures seen here suggest about Maya culture. *(Possible answer: Religion was important in Maya society.)*

Unit Objectives

1. Describe the landforms, bodies of water, climates, and resources of Middle and South America.

2. Analyze the influence of ancient American civilizations, European colonialism, and recent events on societies in Middle and South America.

3. Identify the political, economic, social, cultural, and environmental challenges facing the region's countries.

4. Learn how to draw sketch maps and to use them as geographic tools.

5. Understand how cartograms are constructed to present geographic information.

UNIT 3
Middle and South America

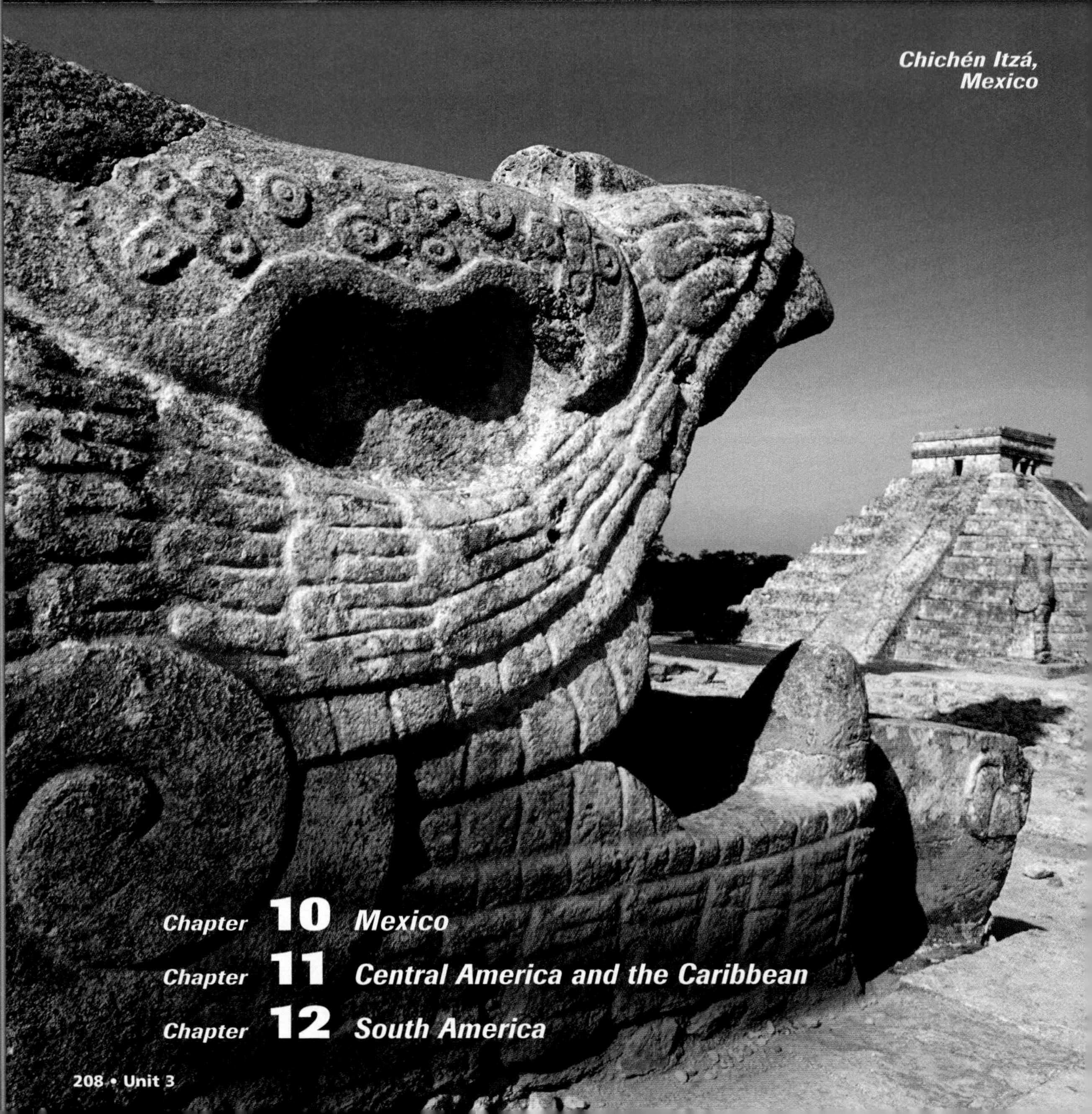

Chichén Itzá, Mexico

Chapter **10** *Mexico*

Chapter **11** *Central America and the Caribbean*

Chapter **12** *South America*

CONNECTING TO
Literature

THE TEMPLE OF THE SUN
from The Incas: The Royal Commentaries of the Inca
by Garcilaso de la Vega

Garcilaso de la Vega

(1539–1616)—also known as El Inca—was the son of a Spanish conquistador and an Inca princess. Interested in Inca history, he spoke to many people in South America about the events they witnessed and took notes about what they said. He also learned Inca oral histories. In 1560 he went to Spain, where he eventually wrote his histories of Peru.

The Temple of the Sun was located on the site that today is occupied by the Church of San Dominique, and its walls, which are made of highly polished stone, still exist. . . .

The four walls were hung with plaques of gold, from top to bottom, and a likeness of the Sun topped the high altar. This likeness was made of a gold plaque. . . . There was no other idol in this temple, nor in any other, for the Sun was the only god of the Incas, whatever people may say. . . .

On either side of this Sun, were kept the numerous mummies of former Inca kings, which were so well preserved that they seemed to be alive. They were seated on their golden thrones resting on plaques of this same metal, and they looked directly at the visitor. Alone among them, Huaina Capac's body had assumed a peculiar pose, facing the Sun, as though from childhood, he had been its favorite son who deserved to be adored for his unusual virtues. . . .

The temple was decorated with five fountains. . . . Their pipes were of solid gold and their stone pillars were covered with either gold or silver, for the sacrifices were washed in these waters. I remember the last of these fountains which was used to water the garden of the convent that the Spaniards established on this sacred ground. One day it stopped working, to the great despair of the Indians who, not knowing where the water came from, were unable to repair it; and the garden dried up, in spite of their desire and their efforts to save it. This only shows how quickly the Indians lost their traditions, since, in the space of forty-two years, there was not one left who could say from whence came the waters that circulated throughout the temple of their god the Sun.

Analyzing the Primary Source

1. **Identifying Points of View** How does Garcilaso de la Vega portray the Inca?

2. **Evaluating Sources** Did Garcilaso de la Vega witness everything he wrote about in this passage? Give reasons for your answer.

A Closer Connection

Inca Construction In his description of the Temple of the Sun, Garcilaso de la Vega notes that its original stone walls still stand. Some 400 years later, archeologists and travelers still marvel at these walls. Using no mortar, Inca builders cut huge blocks of stone and fit them together so well that no gaps have formed between them. The walls themselves tilt slightly inward to support each other. This construction prevented the temple's collapse in the earthquake-prone area. In fact, the Temple of the Sun and other Inca structures have survived quakes that destroyed more recent buildings.

Analyzing the Primary Source

Answers

1. Possible answer: He presents them as masters of traditional skills (temple building, mummifying) who are not very attached to their traditions.

2. Students will probably say that the author did witness what he describes, because his descriptions are vivid and include many details. In fact, however, much of the ornamentation the writer describes in the temple had been removed by Spanish troops by the time of his birth. He may have spoken to older people who described the temple as it had appeared.

In this unit, students will learn about the physical geography and cultures of Middle and South America.

Much of the region is within the tropics, and tropical climates dominate. Mountainous areas have highland climates. Vegetation ranges from the dense rain forest of the Amazon River basin to the hardy scrub of deserts in Mexico and along South America's Pacific Ocean coast. Although subsistence and commercial agriculture are widespread, large areas are undeveloped and almost uninhabited. Cities are concentrated along coasts and in highland areas. Industrial growth draws people to the rapidly expanding cities.

The region's cultures reflect its history of colonization by Spain, Portugal, and other European countries. In most of the region's countries people of mixed heritage are in the majority, with European and indigenous peoples in the minority.

The countries of Middle and South America face many challenges, such as improving standards of living, increasing personal freedoms, and protecting fragile ecosystems.

Your Classroom Time Line

These are the major dates and time periods for this unit. Have students enter them on the time line you created earlier. You may want to watch for these dates as students progress through the unit.

c.* 1500 B.C. Small farming villages are established in Mesoamerica.

c. A.D. 600–900 The Maya civilization flourishes in what is now southern Mexico and northern Central America.

Mid–1300s The Aztec establish Tenochtitlán.

1492 Christopher Columbus lands in the southern Bahamas.

1500s Spain and Portugal establish colonies in the Americas.

1521 Spanish conquistadores under Hernán Cortés conquer the Aztec.

1535 The Inca Empire ends.

1600s The English, French, Dutch, and Danish establish Caribbean colonies.

1800s Freed slaves begin to buy land on Caribbean islands.

* c. stands for *circa* and means "about."

POLITICAL MAP ANSWERS

1. Bolivia
2. Mexico; Brazil

CRITICAL THINKING ANSWER

3. long and narrow; could make it hard to control its territory or expensive and difficult to build infrastructure

UNIT 3 ATLAS

The World in Spatial Terms
Middle and South America:
Political

1. **Places and Regions** Which South American country has two capitals?
2. **Places and Regions** What is the largest country in Middle America? In South America?

Critical Thinking

3. **Making Generalizations** How would you describe the shape of Chile? What problems might this shape create for Chile's government?

SCALE
0   500   1,000 Miles
0   500   1,000 Kilometers

Projection: Azimuthal Equal Area

Legend:
Boundaries
⊛ National capitals
• Other cities

Using the Political Map

Focus students' attention on the capital cities on the **political map** on the opposite page. Ask students to identify capital cities that are located on or near a seacoast. (*Examples: Havana, Panama City, Caracas, Lima, Buenos Aires, Paramaribo*) Then ask why so many capitals are located on seacoasts. (*Possible answer: for trade and communication with other regions*) Ask which capitals are located closer to the countries' centers (*Mexico City, Bogotá, Quito, Santiago, Brasília*). Discuss the advantages these more centrally located capitals may have. (*Possible answer: A more centrally located capital is more accessible to all parts of the country.*)

Using the Physical Map

Direct students' attention to the **physical map.** Ask them to identify the major landform regions of South America (*Amazon Basin, Andes, highland and plateau area of Brazil*). Point out that the Andes are concentrated along the western edge of the region and are part of the same system as the Rocky Mountains.

Call on volunteers to identify the major rivers of South America (*Amazon, Orinoco, Paraguay, Paraná, Uruguay, São Francisco*). Point out that Middle America has few major rivers.

Your Classroom Time Line, (continued)

1800s Asian workers introduce Hinduism and Islam to the Caribbean Islands.

1804 Haiti gains independence from France.

1810–1821 Mexico struggles for independence from Spain.

1820s Many countries of Central and South America gain their independence from Spain.

1846 The Mexican War begins.

1848 The Mexican War ends. The United States takes control of Texas and California.

1865 Dominican Republic breaks away from Spanish control.

Late 1800s American and European investments boost Mexico's economy.

1870s Many Mexicans begin to move north into the United States.

1898 The United States takes Cuba and Puerto Rico from Spain after the Spanish-American War.

1902 Cuba gains its independence.

1903 Panama becomes independent.

1910 The Mexican Revolution begins.

1914 The United States completes the Panama Canal.

1920 The Mexican Revolution ends.

1930s Mexican migration to the United States slows during the Great Depression.

1959 Fidel Castro seizes power in Cuba and establishes a Communist government.

1960s The United States implements a ban on trade with and travel to Cuba.

1968 Arenal Volcano in Costa Rica erupts.

1970s Civil wars flare up in El Salvador and Nicaragua.

Middle and South America: Physical

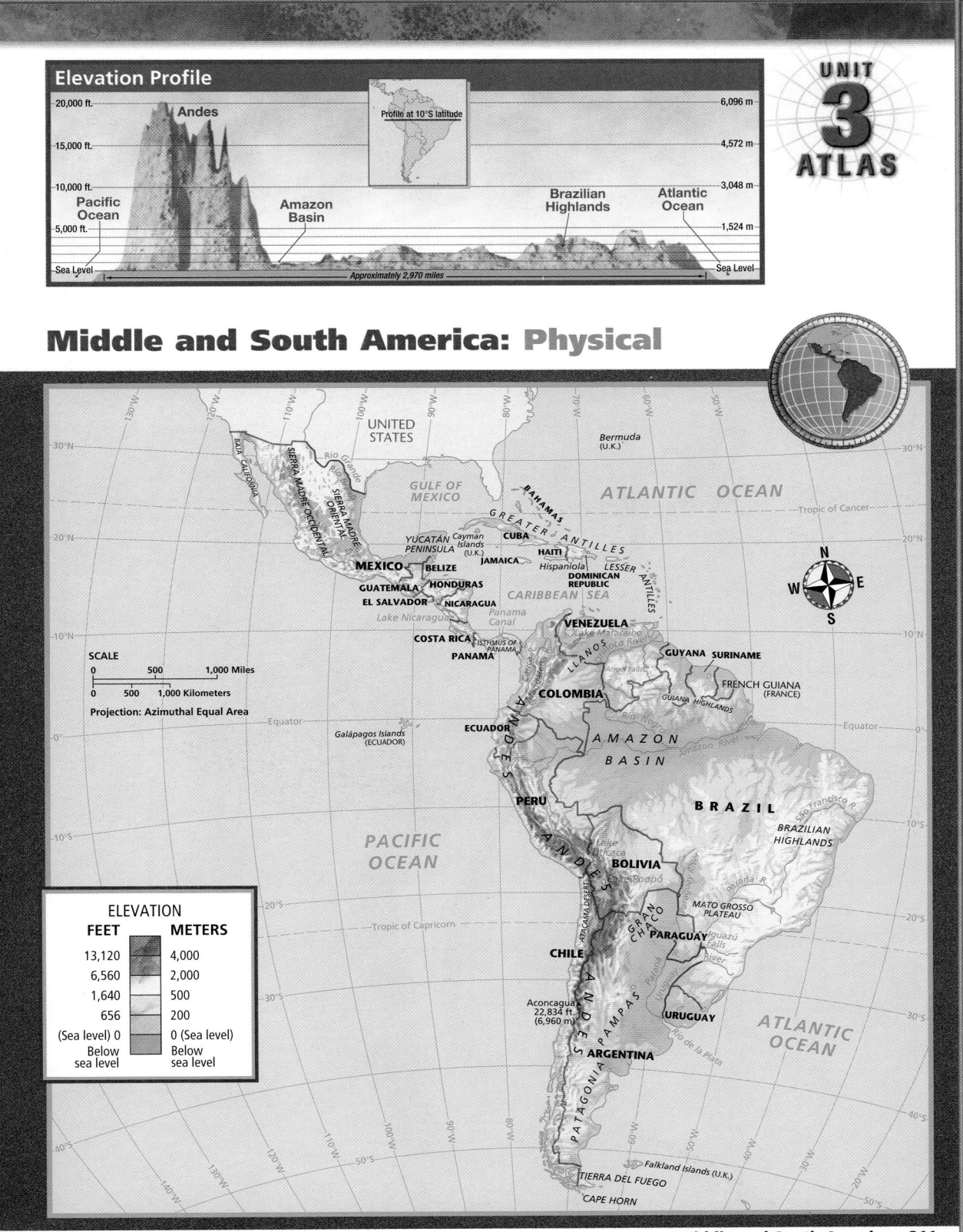

Elevation Profile

Andes — 20,000 ft. / 6,096 m
Profile at 10°S latitude
15,000 ft. / 4,572 m
10,000 ft. / 3,048 m
Pacific Ocean — Amazon Basin — Brazilian Highlands — Atlantic Ocean
5,000 ft. / 1,524 m
Sea Level — Approximately 2,970 miles — Sea Level

ELEVATION

FEET	METERS
13,120	4,000
6,560	2,000
1,640	500
656	200
(Sea level) 0	0 (Sea level)
Below sea level	Below sea level

SCALE
0   500   1,000 Miles
0   500   1,000 Kilometers
Projection: Azimuthal Equal Area

Using the Climate Map

Direct students' attention to the **climate map** of Middle and South America on this page. Point out the wide variety of climate types in the region. Ask students to compare this map to the **physical map** of the region. Have them write sentences to describe the relative locations of the main climate regions of South America. *(Examples: The highland climate runs the length of the Andes. The Amazon Basin has a tropical humid climate.)* Ask students to study the map legend and then choose the climate that is not found in the region *(subarctic).*

Your Classroom Time Line,
(continued)

1981 Belize gains independence.

Mid–1980s Costa Rica promotes ecotourism to boost its economy.

1985 Mexico City is struck by a major earthquake.

Late 1980s Mexico City officials restrict the use of cars to combat environmental problems.

1990 Immigration Act passes, increasing the number of legal immigrants from Mexico to the United States.

1990s Civil wars in Central America end.

1990s Chiapas rebels call for redistribution of wealth in Mexico.

1992 Mexico signs NAFTA with the United States and Canada, to take effect in 1994.

1995–97 Soufrière Volcano erupts, causing many people to flee Montserrat.

1999 The United States gives up control of the Panama Canal to Panama.

2001 Two major earthquakes kill hundreds of people in El Salvador.

CLIMATE MAP ANSWERS

1. They create a large highland climate region and a rain shadow in Argentina.
2. humid subtropical, tropical wet and dry

CRITICAL THINKING ANSWER

3. The cold water cools air temperatures, which prevents the air from carrying moisture to the region.

212

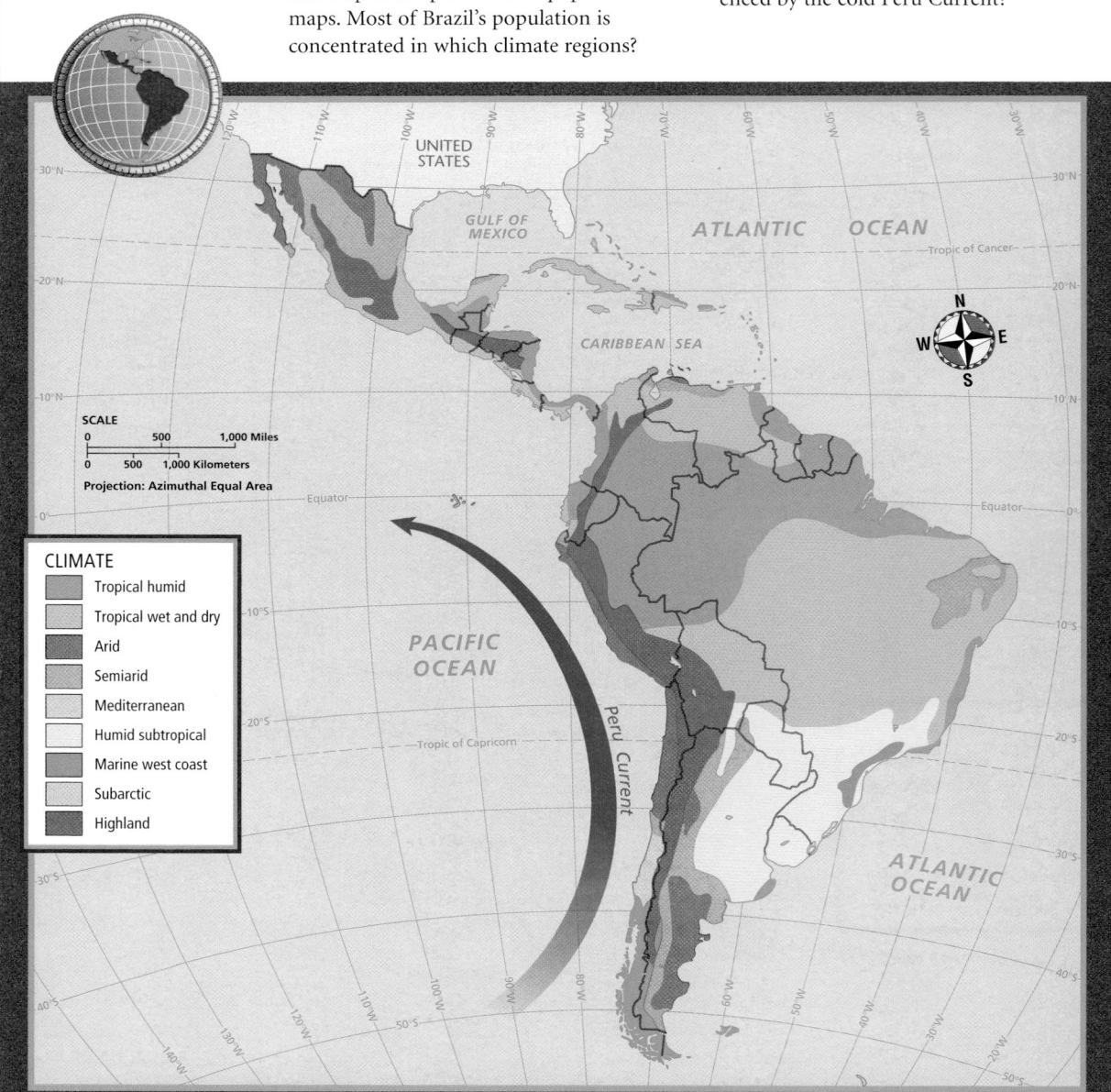

UNIT 3 ATLAS

Middle and South America:
Climate

1. (*Physical Systems*) Compare this map to the physical map. How do the Andes appear to affect climates?

2. (*Environment and Society*) Compare this map to the political and population maps. Most of Brazil's population is concentrated in which climate regions?

Critical Thinking

3. **Making Generalizations** Much of South America's west coast has an arid climate. How might this climate be influenced by the cold Peru Current?

CLIMATE
- Tropical humid
- Tropical wet and dry
- Arid
- Semiarid
- Mediterranean
- Humid subtropical
- Marine west coast
- Subarctic
- Highland

SCALE
0   500   1,000 Miles
0   500   1,000 Kilometers
Projection: Azimuthal Equal Area

Middle and South America:
Precipitation

UNIT 3 ATLAS

1. **Places and Regions** Compare this map to the physical map. Which regions receive the most precipitation?
2. **Environment and Society** Compare this map to the land use map. How might precipitation patterns affect the location of hydroelectric power sites?

Critical Thinking

3. **Drawing Conclusions** Compare this map to the physical map. Which region of Argentina lies in a rain shadow? Based on the precipitation map, from which direction do you think the prevailing winds come?

ANNUAL PRECIPITATION

Centimeters	Inches
Under 25	Under 10
25–50	10–20
50–100	20–40
100–150	40–60
150–200	60–80
Over 200	Over 80

SCALE
0   500   1,000 Miles
0   500   1,000 Kilometers
Projection: Azimuthal Equal Area

213

Using the Population Map

Have students examine the **population map** on this page. You might have them describe the general pattern of population distribution in South America. *(Possible answers: Most people live near the coast or in highland regions. The interior is very sparsely populated.)* Then ask students to suggest reasons why the population is distributed in this way. *(Possible answers: The coasts are more accessible than the interior. The highlands have a more temperate climate than lower elevations.)*

POPULATION MAP ANSWERS

1. Most areas have 0 to 3 people per square mile.
2. the east coast

CRITICAL THINKING ANSWER

3. Baja California and parts of northern Mexico are sparsely populated. These areas have arid climates.

UNIT
3
ATLAS

Middle and South America:
Population

1. *Environment and Society* Compare this map to the physical map. How densely populated is the Amazon Basin?

2. *Places and Regions* Which region of South America is the most densely populated?

Critical Thinking

3. **Making Generalizations** Compare this map to the physical and climate maps. Which regions of Mexico are the least densely populated? How might climate affect population density there?

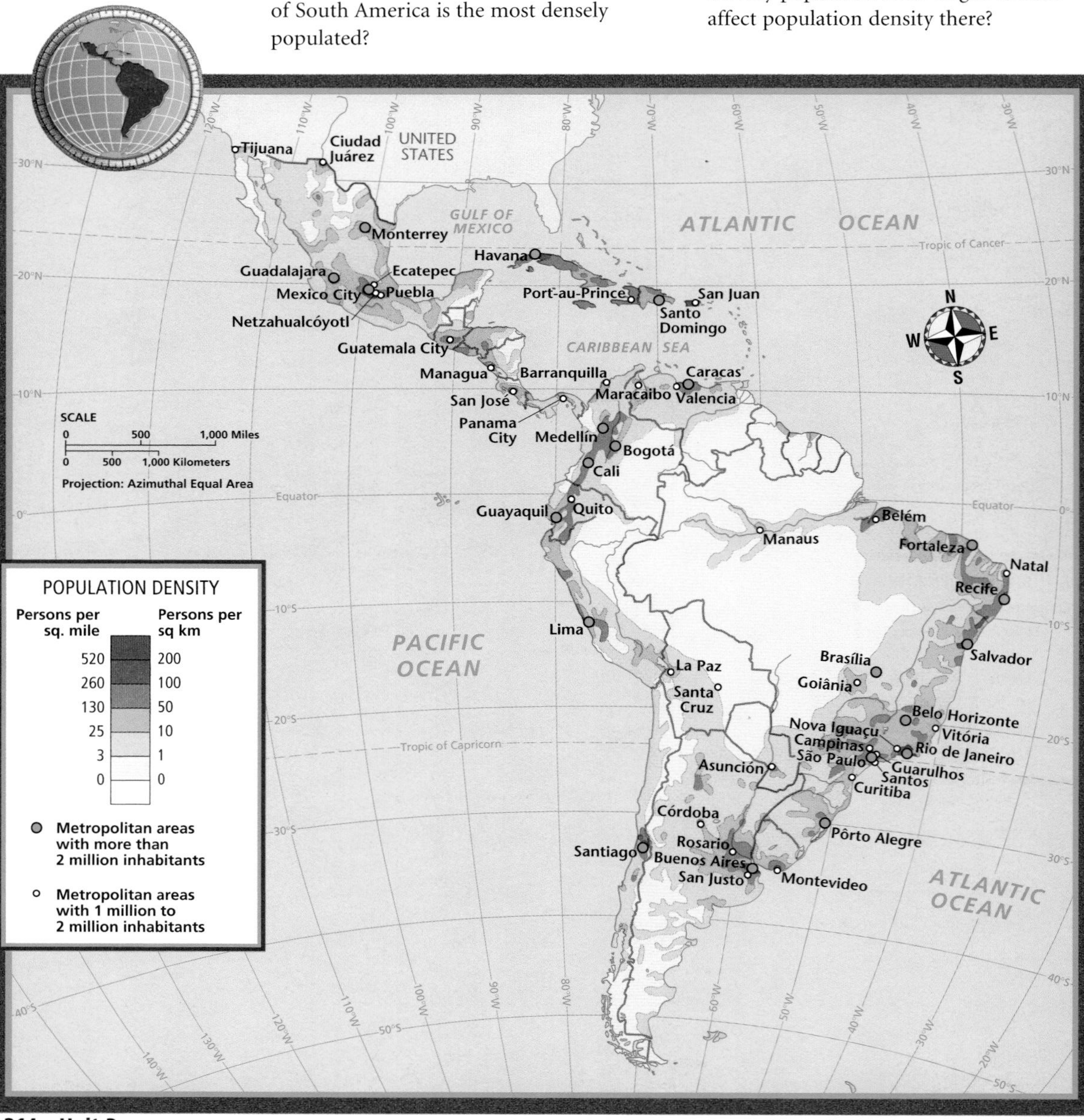

POPULATION DENSITY

Persons per sq. mile		Persons per sq km
520		200
260		100
130		50
25		10
3		1
0		0

● Metropolitan areas with more than 2 million inhabitants

○ Metropolitan areas with 1 million to 2 million inhabitants

SCALE
0   500   1,000 Miles
0   500   1,000 Kilometers
Projection: Azimuthal Equal Area

Map labels: Tijuana, Ciudad Juárez, UNITED STATES, GULF OF MEXICO, Monterrey, Havana, ATLANTIC OCEAN, Tropic of Cancer, Guadalajara, Ecatepec, Mexico City, Puebla, Port-au-Prince, San Juan, Santo Domingo, Netzahualcóyotl, Guatemala City, CARIBBEAN SEA, Managua, Barranquilla, Caracas, San José, Maracaibo, Valencia, Panama City, Medellín, Bogotá, Cali, Equator, Guayaquil, Quito, Belém, Manaus, Fortaleza, Natal, Recife, Lima, PACIFIC OCEAN, Brasília, Salvador, La Paz, Goiânia, Santa Cruz, Belo Horizonte, Vitória, Nova Iguaçu, Campinas, Rio de Janeiro, Asunción, São Paulo, Guarulhos, Santos, Curitiba, Córdoba, Pôrto Alegre, Tropic of Capricorn, Rosario, Santiago, Buenos Aires, San Justo, Montevideo, ATLANTIC OCEAN

214

Using the Land Use and Resources Map

Have students examine the **land use and resources map** on this page. Call on volunteers to name the countries that have oil or gas resources *(Mexico, Venezuela, Colombia, Ecuador, Bolivia, Chile, Argentina, Brazil).* Ask students in which parts of South America hydroelectric power has been developed *(on rivers in the northeast and southeast).* How might the availability of oil, gas, or hydroelectric power affect the industrial development of individual countries? *(Possible answer: Because factories use large amounts of power, industrial development could probably increase if these energy resources were available.)*

Middle and South America:
Land Use and Resources

UNIT
3
ATLAS

1. **Environment and Society** Where in Middle and South America would plantation agriculture likely be found?
2. **Places and Regions** Compare this map to the political map. In which countries is oil production an important activity?

Critical Thinking

3. **Analyzing** Compare this map to the climate map. Which climate areas have limited economic activity? What are some economic activities that are found in these areas?

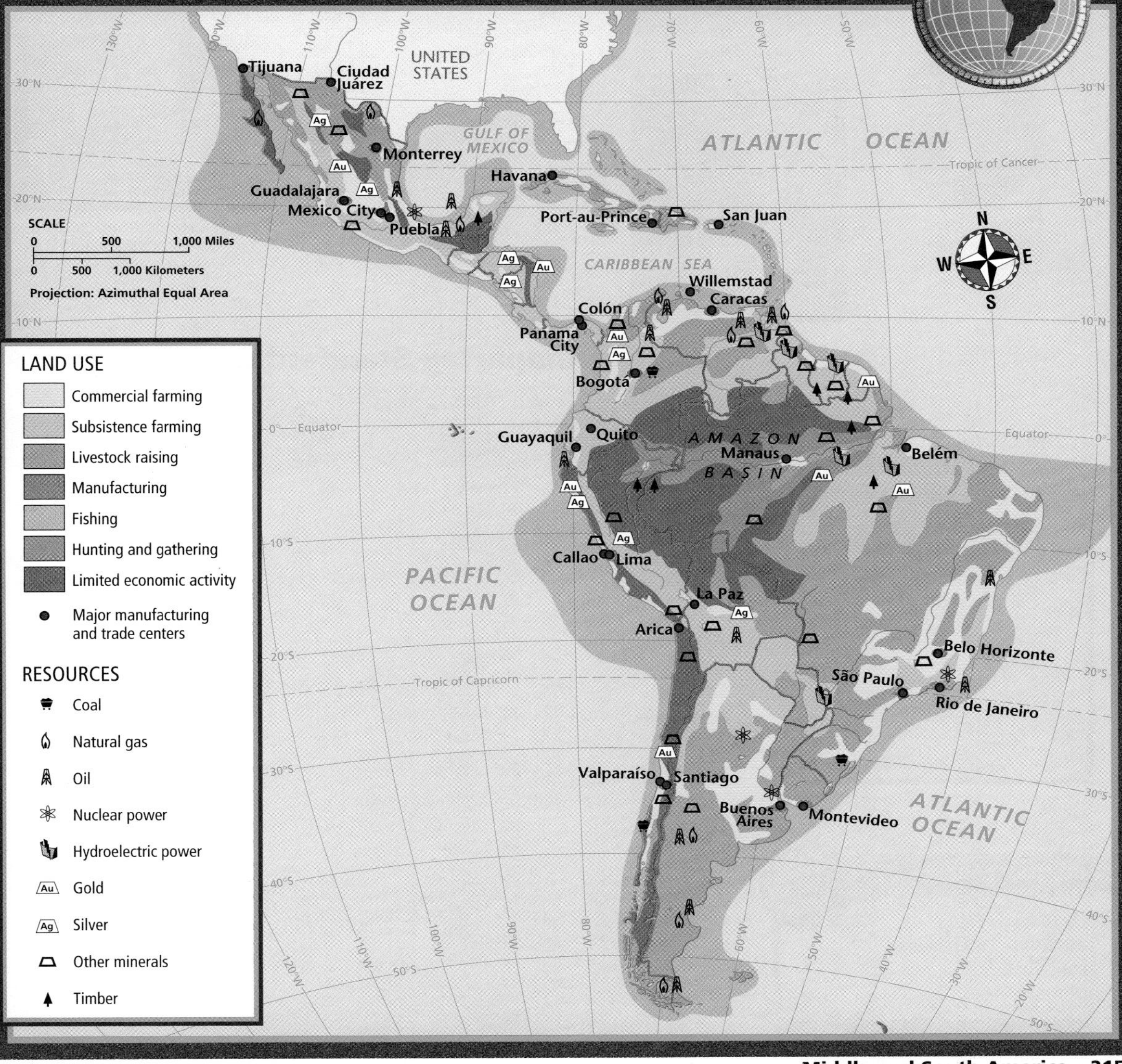

Map

UNITED STATES

Tijuana
Ciudad Juárez
Monterrey
GULF OF MEXICO
ATLANTIC OCEAN
Tropic of Cancer
Guadalajara
Mexico City
Puebla
Havana
Port-au-Prince
San Juan
CARIBBEAN SEA
Willemstad
Caracas
Colón
Panama City
Bogotá
Guayaquil
Quito
AMAZON BASIN
Manaus
Belém
Callao
Lima
La Paz
Arica
Belo Horizonte
São Paulo
Rio de Janeiro
Valparaíso
Santiago
Buenos Aires
Montevideo
PACIFIC OCEAN
ATLANTIC OCEAN
Equator
Tropic of Capricorn

SCALE
0 — 500 — 1,000 Miles
0 — 500 — 1,000 Kilometers
Projection: Azimuthal Equal Area

LAND USE
- Commercial farming
- Subsistence farming
- Livestock raising
- Manufacturing
- Fishing
- Hunting and gathering
- Limited economic activity
- ● Major manufacturing and trade centers

RESOURCES
- Coal
- Natural gas
- Oil
- Nuclear power
- Hydroelectric power
- Au Gold
- Ag Silver
- Other minerals
- Timber

215

Copy the following table onto the chalkboard. Not all of the region's countries are listed because figures are not available for some. Have students compare the information listed in the chart with that listed in the Comparing Standard of Living table. Ask students what connections they can see among literacy rates, poverty, and birthrates. Point out that high birthrates are generally associated with low education and income while low birthrates are associated with high education and high income. Ask students if the information in these charts supports this statement. Call on volunteers to explain the reasoning behind the generalization. Ask class members if they agree with the statement or not. (*Of the countries listed on the Comparing Standard of Living table, Haiti has the lowest literacy rate, the highest birthrate, and the greatest percentage of its population living in poverty. Nicaragua, which also has a relatively low literacy rate, has a high birthrate and widespread poverty. Countries with high literacy rates, like Costa Rica and Argentina, have lower birthrates and fewer people living in poverty. Other countries on the chart would also seem to support the statement, as countries with high birthrates tend to have large numbers of people living in poverty.*)

Historical Geography

Treaty of Tordesillas After word of Christopher Columbus's discoveries reached Spain, King Ferdinand and Queen Isabella wanted reassurance that the lands he had discovered would remain in Spanish hands. In response to their requests, Pope Alexander VI established a line about 320 miles (515 km) west of the Cape Verde Islands. He declared that all lands discovered west of the line would belong to Spain. Everything found to the east would belong to Portugal.

King John II of Portugal was not satisfied with this arrangement and called for a meeting with the Spanish monarchs at Tordesillas in northwestern Spain. The treaty they signed there moved the line of demarcation, or division, farther west to 1,185 miles (1907 km) west of Cape Verde. The new line lay at 46°30' west longitude.

CRITICAL THINKING: Have students examine a world map to answer this question: Why was the king of Portugal unhappy with the original line drawn by the pope? (*It gave control of all of the Americas to Spain. After the line was moved, Portugal discovered that it had received control of Brazil.*)

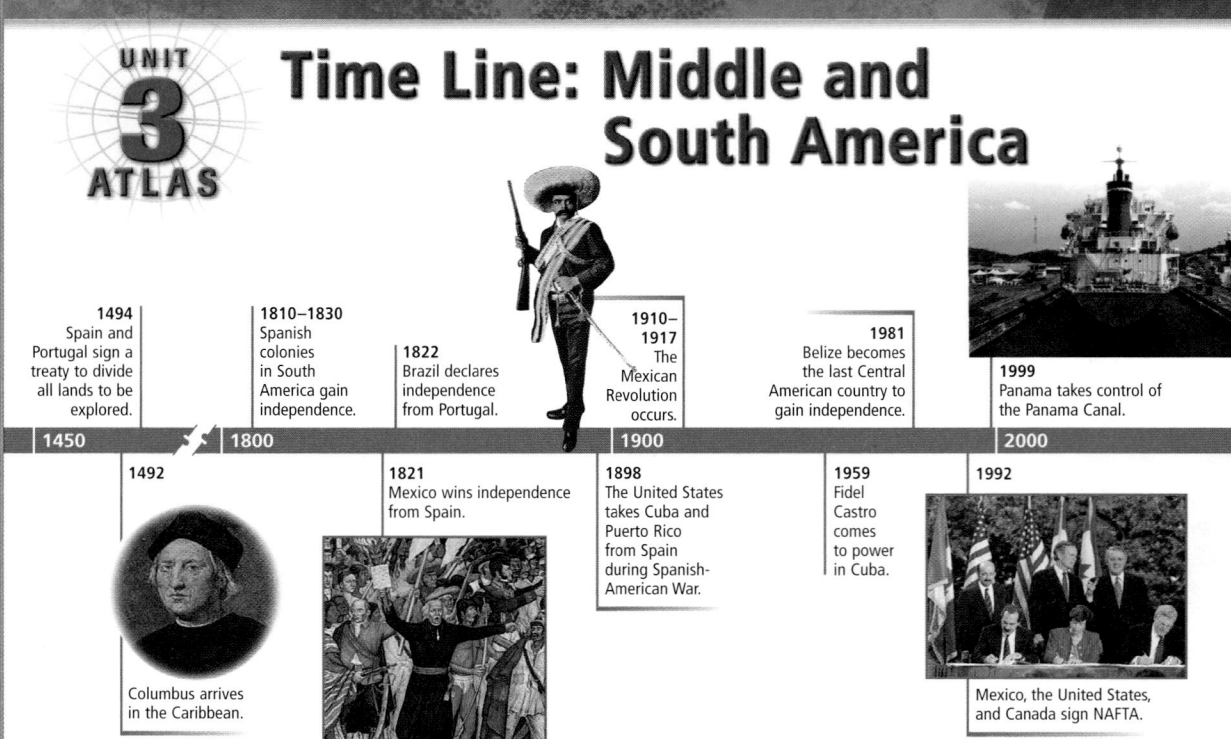

UNIT 3 ATLAS

Time Line: Middle and South America

1494
Spain and Portugal sign a treaty to divide all lands to be explored.

1810–1830
Spanish colonies in South America gain independence.

1822
Brazil declares independence from Portugal.

1910–1917
The Mexican Revolution occurs.

1981
Belize becomes the last Central American country to gain independence.

1999
Panama takes control of the Panama Canal.

| 1450 | 1800 | 1900 | 2000 |

1492
Columbus arrives in the Caribbean.

1821
Mexico wins independence from Spain.

1898
The United States takes Cuba and Puerto Rico from Spain during Spanish-American War.

1959
Fidel Castro comes to power in Cuba.

1992
Mexico, the United States, and Canada sign NAFTA.

The United States and Middle and South America

Comparing Sizes

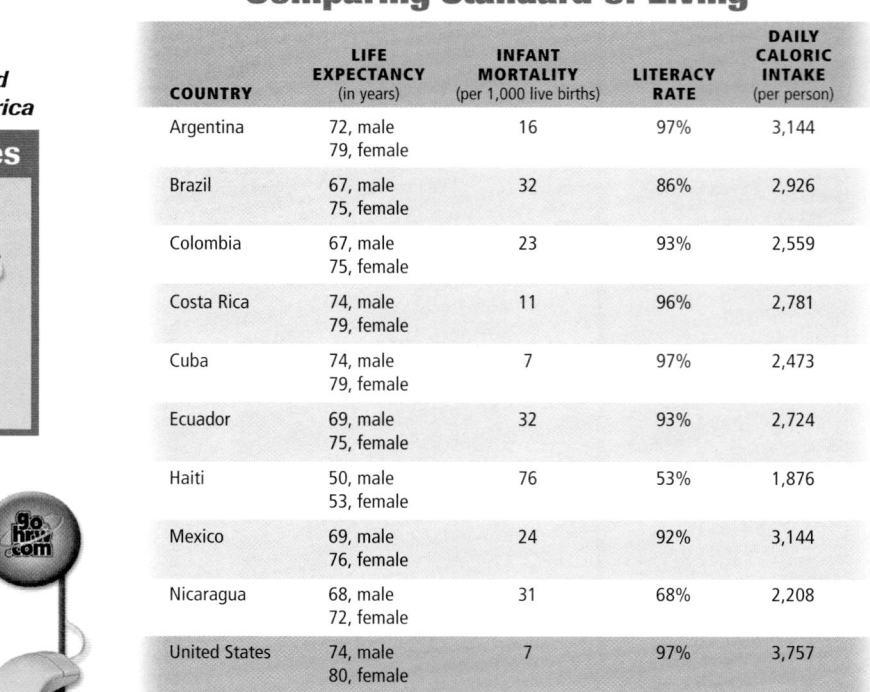

GO TO: go.hrw.com
KEYWORD: SW3 Almanac
FOR: Additional information and reference sources

Comparing Standard of Living

COUNTRY	LIFE EXPECTANCY (in years)	INFANT MORTALITY (per 1,000 live births)	LITERACY RATE	DAILY CALORIC INTAKE (per person)
Argentina	72, male 79, female	16	97%	3,144
Brazil	67, male 75, female	32	86%	2,926
Colombia	67, male 75, female	23	93%	2,559
Costa Rica	74, male 79, female	11	96%	2,781
Cuba	74, male 79, female	7	97%	2,473
Ecuador	69, male 75, female	32	93%	2,724
Haiti	50, male 53, female	76	53%	1,876
Mexico	69, male 76, female	24	92%	3,144
Nicaragua	68, male 72, female	31	68%	2,208
United States	74, male 80, female	7	97%	3,757

Sources: *World Almanac and Book of Facts 2004*; *Britannica Book of the Year, 2002*

Country	Birthrate (births per 1,000 population)	Percent of Population below Poverty Line
Argentina	18.4	37%
Belize	31.7	33%
Bolivia	27.3	70%
Brazil	18.5	17%
Chile	16.8	22%
Colombia	22.4	55%
Costa Rica	20.3	21%
Dominican Republic	24.8	25%
Ecuador	26.0	50%
El Salvador	28.7	48%
Guatemala	34.6	60%
Haiti	31.7	80%
Honduras	31.9	53%
Jamaica	18.1	34%
Mexico	22.8	27%
Nicaragua	27.6	50%
Panama	19.1	37%
Paraguay	30.9	36%
Peru	23.9	49%
Trinidad and Tobago	13.7	21%
Venezuela	20.7	67%
United States	14.2	13%

Fast Facts: Middle and South America

FLAG	COUNTRY / Capital	POPULATION (in millions) / POP. DENSITY	AREA	PER CAPITA GDP (in US $)	WORKFORCE STRUCTURE (largest categories)	ELECTRICITY CONSUMPTION (kilowatt hours per person)	TELEPHONE LINES (per person)
	Antigua and Barbuda / Saint John's	0.07 / 400/sq.mi.	171 sq. mi. / 443 sq km	$ 11,000	82% comm., services / 11% agriculture	1,442 kWh	0.55
	Argentina / Buenos Aires	38.4 / 36/sq. mi.	1,068,301 sq. mi. / 2,766,887 sq km	$ 10,200	24% services / 21% manufacturing	2,397 kWh	0.21
	Bahamas / Nassau	0.3 / 406/sq. mi.	5,382 sq. mi. / 13,939 sq km	$ 17,000	50% tourism / 40% other services	4,621 kWh	0.40
	Barbados / Bridgetown	0.3 / 1,626/sq. mi.	166 sq. mi. / 430 sq km	$ 14,500	75% services / 15% industry	2,687 kWh	0.48
	Belize / Belmopan	0.3 / 29/sq. mi.	8,867 sq. mi. / 22,965 sq km	$ 4,900	55% services / 27% agriculture	725 kWh	0.12
	Bolivia / La Paz, Sucre	8.8 / 21/sq. mi.	424,164 sq. mi. / 1,098,580 sq km	$ 2,500	39% agriculture / 14% public admin., services	413 kWh	0.06
	Brazil / Brasília	178.5 / 55/sq. mi.	3,286,486 sq. mi. / 8,511,960 sq km	$ 7,600	53% services / 23% agriculture	1,882 kWh	0.22
	Chile / Santiago	15.8 / 55/sq. mi.	292,260 sq. mi. / 756,950 sq km	$ 10,000	59% services / 27% industry	2,539 kWh	0.22
	Colombia / Bogotá	44.2 / 110/sq. mi.	439,735 sq. mi. / 1,138,908 sq km	$ 6,500	46% services / 30% agriculture	900 kWh	0.18
	Costa Rica / San José	4.2 / 213/sq. mi.	19,730 sq. mi. / 51,100 sq km	$ 8,500	58% services / 22% industry	1,464 kWh	0.25
	Cuba / Havana	11.3 / 264/sq. mi.	42,803 sq. mi. / 110,859 sq km	$ 2,300	51% services / 25% industry	1,184 kWh	0.05
	Dominica / Roseau	0.07 / 240/sq. mi.	291 sq. mi. / 754 sq km	$ 5,400	40% agriculture / 32% ind., commerce	962 kWh	0.36
	Dominican Republic / Santo Domingo	8.7 / 468/sq. mi.	18,815 sq. mi. / 48,731 sq km	$ 6,100	59% services, gov. / 24% industry	977 kWh	0.11
	Ecuador / Quito	13.0 / 122/sq. mi.	109,483 sq. mi. / 283,560 sq km	$ 3,100	45% services / 30% agriculture	5,380 kWh	0.11
	El Salvador / San Salvador	6.5 / 814/sq. mi.	8,124 sq. mi. / 21,041 sq km	$ 4,700	55% services / 30% agriculture	580 kWh	0.10

Sources: Central Intelligence Agency, *World Factbook 2003; The World Almanac and Book of Facts, 2004*
The CIA calculates per capita GDP in terms of purchasing power parity. This formula equalizes the purchasing power of each country's currency.

Historical Geography

The War of the Pacific

In 1879 Chilean forces invaded Bolivia to begin the War of the Pacific. Chile wanted to gain control of valuable deposits of minerals, particularly copper and sodium nitrate, in the Atacama Desert. Peru sided with Bolivia in the conflict largely because its mining industry was closely tied to Bolivia's.

By 1883 Chile had defeated both of its opponents. As a result of its defeat, Bolivia had to give up the province of Antofagasta on the Pacific Ocean. Bolivia was thus rendered a landlocked country. Peru also lost territory to its southern rival. Chile acquired Tarapacá, formerly part of southern Peru, and gained control of the provinces Tacna and Arica until 1929. In that year, Tacna returned to Peru, but Arica remained part of Chile.

DISCUSSION: Lead a class discussion about how conflict and cooperation have shaped political boundaries in Pacific South America. Ask students how they think the loss of its Pacific territory affected Bolivia's economy.

Duplicate the following chart onto handouts, omitting the italicized figures. Have students use the data in the chart and information from the unit Fast Facts tables to determine what percentage of each country's highways are paved. *(Divide the mileage of paved highways by the total mileage of highways in the country.)* Note that not all of the region's countries are listed. Ask students which countries have paved most of their roads *(Barbados, Uruguay, Jamaica).* Call on volunteers to suggest reasons for these percentages. *(Possible answers: Barbados and Jamaica depend on tourism for revenue, so they want to make their roads easy for tourists to travel. Uruguay is a relatively wealthy country that can afford to pave nearly all of its highways.)*

Then ask students to calculate the ratio of each country's population to its highway mileage. *(Divide the population by the total highways, giving the number of people in each country per mile of highway.)* Ask students which countries appear to have the fewest roads available to their populations *(Haiti, Guatemala, Dominican Republic).* Challenge students to suggest reasons why the countries of Middle and South America may not have as extensive a transportation network as some other parts of the world. Encourage them to consult the Fast Facts tables and unit atlas for reference. *(Possible answers: Many countries in the region are poor. Many people may prefer other means of transportation to roads, like river travel or airplanes. Some areas, like the deserts of northern Mexico, do not attract enough residents or visitors to require roads.)*

Historical Geography

Pancho Villa During the Mexican Revolution of 1910–20, a guerilla leader named Francisco "Pancho" Villa fought against dictators Porfirio Díaz and Victoriano Huerta. After Huerta resigned in 1914, Villa came into conflict with his former partner, Venustiano Carranza, and fled into the mountains and deserts of northern Mexico.

Villa had spent much of his childhood in these same mountains and was intimately familiar with them. He became a bandit, escaping capture by hiding out in the rough terrain. In March of 1916 he attacked the town of Columbus, New Mexico. President Woodrow Wilson of the United States sent forces under General John Pershing into Mexico to capture the raider. Pershing's cavalry, trucks, and airplanes scoured the area, but Villa's knowledge of the countryside allowed him to elude capture.

CRITICAL THINKING:

What are some ways in which Pancho Villa and his troops might have used the physical environment to escape their pursuers? *(Possible answer: They knew good hiding places in northern Mexico's deserts and mountains. They also knew where to find water and which of the region's plants were edible, so they could survive in the wilderness.)*

UNIT 3 ATLAS
Fast Facts: Middle and South America

FLAG	COUNTRY Capital	POPULATION (in millions) POP. DENSITY	AREA	PER CAPITA GDP (in US $)	WORKFORCE STRUCTURE (largest categories)	ELECTRICITY CONSUMPTION (kilowatt hours per person)	TELEPHONE LINES (per person)
	Grenada Saint George's	0.09 / 682/sq. mi.	133 sq. mi. 344 sq km	$ 5,000	62% services 24% agriculture	1,442 kWh	0.38
	Guatemala Guatemala City	12.3 / 295/sq. mi.	42,043 sq. mi. 108,891 sq km	$ 3,700	50% agriculture 35% services	450 kWh	0.06
	Guyana Georgetown	0.8 / 10/sq. mi.	83,000 sq. mi. 214,969 sq km	$ 4,000	39% agric., forest, fish. 24% mining, manuf., const.	1,036 kWh	0.18
	Haiti Port-au-Prince	8.3 / 782/sq. mi.	10,714 sq. mi. 27,749 sq km	$ 1,700	66% agriculture 25% services	65 kWh	0.02
	Honduras Tegucigalpa	6.9 / 161/sq. mi.	43,278 sq. mi. 112,090 sq km	$ 2,600	45% services 34% agriculture	551 kWh	0.05
	Jamaica Kingston	2.7 / 634/sq. mi.	4,244 sq. mi. 10,992 sq km	$ 3,900	60% services 21% agriculture	2,200 kWh	0.20
	Mexico Mexico City	103.5 / 139/sq. mi.	761,605 sq. mi. 1,972,548 sq km	$ 9,000	56% services 24% industry	1,805 kWh	0.14
	Nicaragua Managua	5.5 / 118/sq. mi.	49,998 sq. mi. 129,494 sq km	$ 2,500	43% services 42% agriculture	437 kWh	0.03
	Panama Panama City	3.1 / 106/sq. mi.	30,193 sq. mi. 78,200 sq km	$ 6,000	61% services 21% agriculture	1,180 kWh	0.12
	Paraguay Asunción	5.9 / 38/sq. mi.	157,047 sq. mi. 406,750 sq km	$ 4,200	45% agriculture	449 kWh	0.05

Sea lion on the shore, Puerto Egas, Galápagos Islands

Country	Total Highways	Total Paved Highways	Percent Paved	People per Mile
Antigua and Barbuda	722 mi.	238 mi.	33.0	96.9
Argentina	133,569 mi.	39,403 mi.	29.5	277.0
Bahamas	1670 mi.	959 mi.	57.4	179.7
Barbados	992 mi.	978 mi.	98.6	302.4
Belize	1781 mi.	303 mi.	17.0	168.5
Bolivia	30,628 mi.	1,550 mi.	5.1	271.0
Brazil	1,227,600 mi.	114,167 mi.	9.3	142.1
Chile	49,476 mi.	6,827 mi.	13.8	309.2
Colombia	68,200 mi.	16,120 mi.	23.6	590.9
Costa Rica	23,109 mi.	4,853 mi.	21.0	164.4
Cuba	37,732 mi.	18,488 mi.	49.0	296.8
Dominican Republic	7,812 mi.	3,859 mi.	49.4	1,100.9
Ecuador	26,782 mi.	5,062 mi.	18.9	492.9
El Salvador	6,218 mi.	1.231 mi.	19.8	997.1
Grenada	645 mi.	396 mi.	61.3	139.6
Guatemala	8,591 mi.	2,709 mi.	31.5	1,513.3
Guyana	4,941 mi.	366 mi.	7.4	141.7
Haiti	2,579 mi.	627 mi.	24.3	2,714.0
Honduras	9,548 mi.	1,938 mi.	20.3	670.3
Jamaica	11,780 mi.	8,328 mi.	70.7	229.2
Mexico	200,866 mi.	59,657 mi.	29.7	507.3
Nicaragua	10,157 mi.	1,127 mi.	11.1	482.4
Panama	7,187 mi.	2,529 mi.	35.2	389.6
Paraguay	16,059 mi.	1,902 mi.	11.8	354.9
Peru	45,198 mi.	5,394 mi.	11.9	597.4
St. Kitts-Nevis	198 mi.	84 mi.	42.5	201.6
St. Lucia	750 mi.	39 mi.	5.2	213.3
Trinidad and Tobago	5,158 mi.	2,636 mi.	51.1	232.6
Uruguay	5,569 mi.	5,013 mi.	90.0	610.5
Venezuela	59,616 mi.	20,031 mi.	33.6	400.9
United States	3,949,419 mi.	3,554,477 mi.	90.0	71.3

Source: Central Intelligence Agency, *The World Factbook 2001*

FLAG	COUNTRY / Capital	POPULATION (in millions) / POP. DENSITY	AREA	PER CAPITA GDP (in US $)	WORKFORCE STRUCTURE (largest categories)	ELECTRICITY CONSUMPTION (kilowatt hours per person)	TELEPHONE LINES (per person)
	Peru / Lima	27.2 / 55/sq. mi.	496,226 sq. mi. / 1,285,219 sq km	$ 4,800	33% agriculture 28% services	705 kWh	0.07
	St. Kitts-Nevis / Basseterre	0.04 / 385/sq. mi.	101 sq. mi. / 262 sq km	$ 8,800	54% services 25% manuf. ind., commerce	2,391 kWh	0.60
	St. Lucia / Castries	0.15 / 632/sq. mi.	238 sq. mi. / 616 sq km	$ 5,400	54% services 25% ind., comm., manuf.	750 kWh	0.30
	St. Vincent and the Grenadines / Kingstown	0.12 / 917/sq. mi.	150 sq. mi. / 388 sq km	$ 2,900	57% services 26% agriculture	717 kWh	0.22
	Suriname / Paramaribo	0.4 / 7/sq. mi.	63,039 sq. mi. / 163,270 sq km	$ 3,500	40% services 13% trade	4,179 kWh	0.18
	Trinidad and Tobago / Port-of-Spain	1.3 / 658/sq. mi.	1,980 sq. mi. / 5,128 sq km	$ 9,500	64% services 14% manuf., mining, quarrying	3,794 kWh	0.25
	Uruguay / Montevideo	3.4 / 51/sq. mi.	68,039 sq. mi. / 176,220 sq km	$ 7,800	70% services 16% industry	1,801 kWh	0.28
	Venezuela / Caracas	25.7 / 75/sq. mi.	352,144 sq. mi. / 912,049 sq km	$ 5,500	64% services 23% industry	3,170 kWh	0.11
	United States / Washington, D.C.	294.0 / 83/sq. mi.	3,717,810 sq.mi. / 9,629,084 sq km	$ 37,600	31% manage., prof. 29% tech., sales, admin.	12,250 kWh	0.65

Machu Picchu, Peru

CHAPTER RESOURCE MANAGER

Objectives	Pacing Guide	Reproducible Resources
SECTION 1 **Natural Environments** (pp. 221–23)	• Describe the main landforms of Mexico. • Identify Mexico's climates, biomes, and natural resources.	**Regular** 1 day **Block Scheduling** .5 day *Block Scheduling Handbook, Chapter 10*	**RS** Guided Reading Strategy 10.1 **RS** Graphic Organizer Activity 10
SECTION 2 **History and Culture** (pp. 224–27)	• Describe the cultures of Mexico before the Spanish arrived. • Examine how Spanish control changed Mexico. • Trace Mexico's history since independence.	**Regular** 1 day **Block Scheduling** .5 day *Block Scheduling Handbook, Chapter 10*	**RS** Guided Reading Strategy 10.2 **E** Cultures of the World Activity: Region 2 **SM** Map Activity 10: Early Mexican Civilization
SECTION 3 **Mexico Today** (pp. 229–33)	• Describe the economic and cultural regions of Mexico. • Identify the challenges that face Mexico.	**Regular** 1 day **Block Scheduling** .5 day *Block Scheduling Handbook, Chapter 10*	**RS** Guided Reading Strategy 10.3 **PS** Readings in World Geography, History, and Culture 15 and 16 **SM** Critical Thinking Activity 10: Mexican Immigrants **SM** Geography for Life Activity 10: Mexican Tourism

Chapter Resource Key

PS Primary Sources   **A** Assessment   CD–ROM

RS Reading Support   **REV** Review   Video

IC Interdisciplinary Connections   **ELL** Reinforcement and English Language Learners    Internet

E Enrichment   Transparencies   Holt Presentation Maker Using Microsoft® PowerPoint®

SM Skills Mastery

One-Stop Planner CD–ROM

See the *One-Stop Planner* for a complete list of additional resources for students and teachers.

One-Stop Planner CD–ROM

It's easy to plan lessons, select resources, and print out materials for your students when you use the **One-Stop Planner CD–ROM with Test Generator**.

internet connect

HRW ONLINE RESOURCES

GO TO: go.hrw.com
Then type in a keyword.

TEACHER HOME PAGE
KEYWORD: **SW3 Teacher**

CHAPTER INTERNET ACTIVITIES
KEYWORD: **SW3 GT10**
Choose a topic on Mexico to:
- take the Geomap challenge to test your knowledge of Mexico's geography.
- create a brochure on Mexican holidays and culture.
- learn about the physical features and climate of Mexico's coastlines.

CHAPTER ENRICHMENT LINKS
KEYWORD: **SW3 CH10**

CHAPTER MAPS
KEYWORD: **SW3 MAPS10**

ONLINE ASSESSMENT
Homework Practice
KEYWORD: **SW3 HP10**
Standardized Test Prep
KEYWORD: **SW3 STP10**
Rubrics
KEYWORD: **SS Rubrics**

COUNTRY INFORMATION
KEYWORD: **SW3 Almanac**

CONTENT UPDATES
KEYWORD: **SS Content Updates**

HOLT PRESENTATION MAKER
KEYWORD: **SW3 PPT10**

ONLINE READING SUPPORT
KEYWORD: **SS Strategies**

CURRENT EVENTS
KEYWORD: **S3 Current Events**

Technology Resources

- One-Stop Planner CD–ROM, Lesson 10.1
- Geography and Cultures Visual Resources 18–22
- Homework Practice Online
- HRW Go site

- One-Stop Planner CD–ROM, Lesson 10.2
- Geography and Cultures Visual Resources 23
- CNN Presents World Cultures: Yesterday and Today, Segment 34: Mexico's Jewish Community
- Homework Practice Online
- HRW Go site

- One-Stop Planner CD–ROM, Lesson 10.3
- ARGWorld CD–ROM
- CNN Presents Geography: Yesterday and Today, Segment 4: Farming with Saltwater
- CNN Presents Geography: Yesterday and Today, Segment 9: Baseball, Mexican Style
- Homework Practice Online
- HRW Go site

Reinforcement, Review, and Assessment

- **ELL** Main Idea Activity 10.1
- **ELL** English Audio Summary 10.1
- **ELL** Spanish Audio Summary 10.1
- **REV** Section 1 Review, p. 223
- **A** Daily Quiz 10.1

- **ELL** Main Idea Activity 10.2
- **ELL** English Audio Summary 10.2
- **ELL** Spanish Audio Summary 10.2
- **REV** Section 2 Review, p. 227
- **A** Daily Quiz 10.2

- **ELL** Main Idea Activity 10.3
- **ELL** English Audio Summary 10.3
- **ELL** Spanish Audio Summary 10.3
- **REV** Section 3 Review, p. 233
- **A** Daily Quiz 10.3

Meeting Individual Needs

Ability Levels

Level 1 Basic-level activities designed for all students encountering new material

Level 2 Intermediate-level activities designed for average students

Level 3 Challenging activities designed for honors and gifted-and-talented students

English Language Learners Activities that address the needs of students with Limited English Proficiency

Chapter Review and Assessment

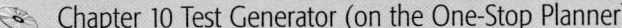

- Chapter 10 Test Generator (on the One-Stop Planner)
- Global Skill Builder CD–ROM
- HRW Go site
- **REV** Chapter 10 Review, pp. 236–37
- **REV** Chapter 10 Tutorial for Students, Parents, Mentors, and Peers
- **A** Chapter 10 Test (form A or B)
- **A** Alternative Assessment Handbook
- **A** Chapter 10 Test for English Language Learners and Special-Needs Students

Launch into Learning

Have students examine the Mexican flag pictured in the unit Fast Facts table. In the center, an eagle with a snake in its beak rests on a cactus. This symbol, Mexico's national coat of arms, comes from the legends of the Aztec, one of the country's indigenous peoples. Explain that the design on a country's flag may reveal what the people of that country value. Ask students what the Mexican flag's design may tell us. *(Possible answer: The Mexican people are proud of their history.)* Tell students that in this chapter they will learn more about Mexico's land and people.

Using the Physical-Political Map

Have students examine the map on the opposite page. Ask them to describe the relative location of Mexico, the Mexican Plateau, the Yucatán Peninsula, the Isthmus of Tehuantepec, the Río Bravo, the Gulf of California, and Baja California. Explain that *baja* means "lower."

Why We Should Know More

Consider using the following points to explain the importance of learning about Mexico:

- Mexico is one of only two countries that have land borders with the United States.
- More immigrants come to the United States from Mexico than from any other country.
- Many products, particularly vegetables and fruits, come from Mexico. In fact, nearly 90 percent of Mexico's exports come to the United States.
- Mexico is the second-largest importer of goods produced in the United States.
- Millions of Americans of many different backgrounds enjoy the foods, music, festivals, and other traditions of Mexico.

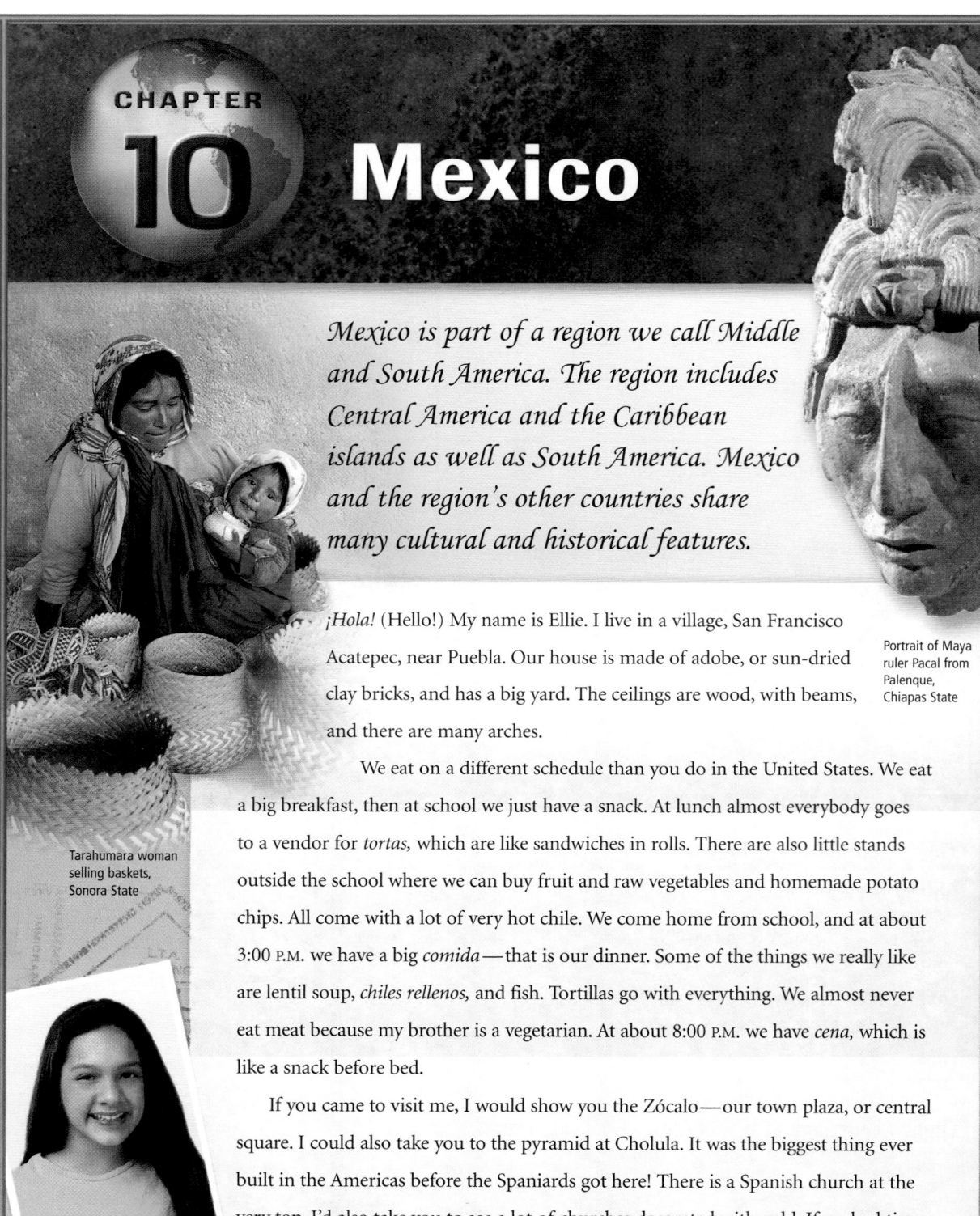

CHAPTER 10 Mexico

Mexico is part of a region we call Middle and South America. The region includes Central America and the Caribbean islands as well as South America. Mexico and the region's other countries share many cultural and historical features.

¡Hola! (Hello!) My name is Ellie. I live in a village, San Francisco Acatepec, near Puebla. Our house is made of adobe, or sun-dried clay bricks, and has a big yard. The ceilings are wood, with beams, and there are many arches.

Portrait of Maya ruler Pacal from Palenque, Chiapas State

We eat on a different schedule than you do in the United States. We eat a big breakfast, then at school we just have a snack. At lunch almost everybody goes to a vendor for *tortas,* which are like sandwiches in rolls. There are also little stands outside the school where we can buy fruit and raw vegetables and homemade potato chips. All come with a lot of very hot chile. We come home from school, and at about 3:00 P.M. we have a big *comida*—that is our dinner. Some of the things we really like are lentil soup, *chiles rellenos,* and fish. Tortillas go with everything. We almost never eat meat because my brother is a vegetarian. At about 8:00 P.M. we have *cena,* which is like a snack before bed.

Tarahumara woman selling baskets, Sonora State

If you came to visit me, I would show you the Zócalo—our town plaza, or central square. I could also take you to the pyramid at Cholula. It was the biggest thing ever built in the Americas before the Spaniards got here! There is a Spanish church at the very top. I'd also take you to see a lot of churches decorated with gold. If we had time, we'd go to the beach in Veracruz, a really pretty town about three hours away.

 **LET'S GET STARTED**

Copy the following instructions onto the chalkboard: *Look at the chapter map. What kind of climate do you think the Mexican Plateau has? Why do you think so? Discuss responses. (Possible answers: It is fairly dry because the mountains along Mexico's eastern and western edges probably create rain shadows. The absence of rivers suggests that there is little surface water and, presumably, little precipitation.) Tell students that they will learn more about Mexico's natural environments in Section 1.*

Building Vocabulary

Write **isthmus** and **sinkholes** on the chalkboard. Ask students to locate the Isthmus of Tehuantepec on the chapter map. Then have a student read the definition of *isthmus*. Call on volunteers to locate other isthmuses on a world map. *(Possible answers: Isthmus of Panama, Isthmus of Suez in Egypt, Isthmus of Kra in Thailand)* Call on a volunteer to locate and read the definition of *sinkholes*. Then ask if students know of any sinkholes in your state or region. *(Florida, Kentucky, and Missouri are some states where sinkholes are common.)*

Section 1
Natural Environments

READ TO DISCOVER

1. What are the main landforms of Mexico?
2. What climates, biomes, and natural resources does Mexico have?

WHY IT MATTERS

Mexico shares a long border with the United States. As more industries have built factories along this border, many people have raised concerns about environmental damage. Use **CNNfyi.com** or other **current events** sources to learn about pollution along the border.

DEFINE

isthmus
sinkholes

LOCATE

Mexican Plateau
Sierra Madre
  Oriental
Sierra Madre
  Occidental
Sierra Madre del Sur
Valley of Mexico

Mexico City
Orizaba
Isthmus of
  Tehuantepec
Yucatán Peninsula
Baja California

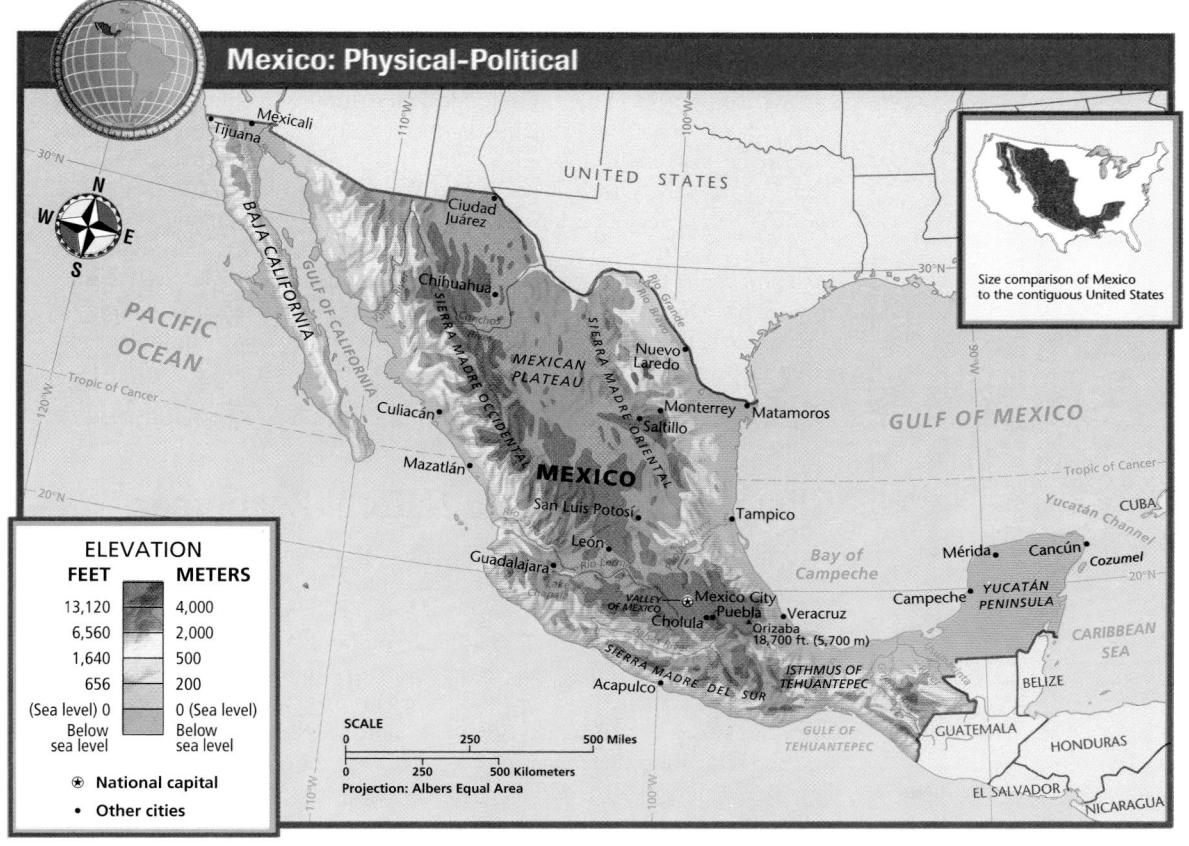

Mexico: Physical-Political

Size comparison of Mexico to the contiguous United States

ELEVATION

FEET	METERS
13,120	4,000
6,560	2,000
1,640	500
656	200
(Sea level) 0	0 (Sea level)
Below sea level	Below sea level

⊛ National capital
• Other cities

SCALE
0   250   500 Miles
0   250   500 Kilometers
Projection: Albers Equal Area

ALL LEVELS: Copy the following graphic organizer onto the chalkboard, omitting the italicized answers. Have students complete the organizer with words or phrases that describe the landforms, climates, vegetation, and resources found in each region of Mexico. Then ask students what type of region these areas represent (*formal regions*). Lead a class discussion about the effects of Mexico's landforms on the distribution of climates and vegetation across the country. **ENGLISH LANGUAGE LEARNERS**

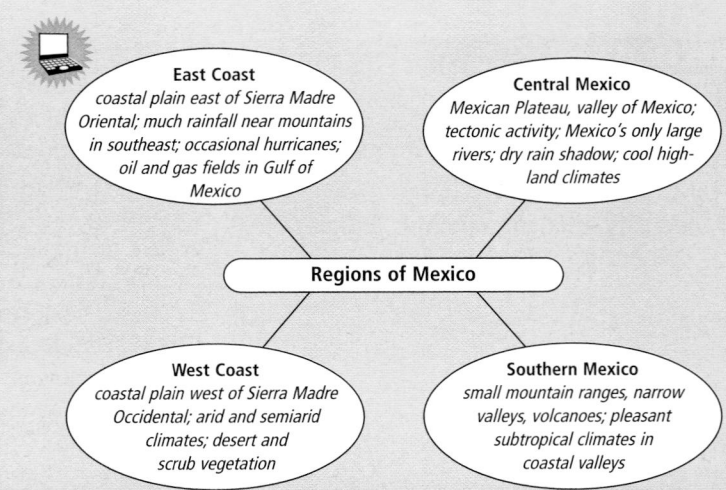

East Coast
coastal plain east of Sierra Madre Oriental; much rainfall near mountains in southeast; occasional hurricanes; oil and gas fields in Gulf of Mexico

Central Mexico
Mexican Plateau, valley of Mexico; tectonic activity; Mexico's only large rivers; dry rain shadow; cool highland climates

Regions of Mexico

West Coast
coastal plain west of Sierra Madre Occidental; arid and semiarid climates; desert and scrub vegetation

Southern Mexico
small mountain ranges, narrow valleys, volcanoes; pleasant subtropical climates in coastal valleys

Cueva de Villa Luz A cave in the state of Tabasco, in southern Mexico, is providing scientists with clues about how life functions on parts of the ocean floor and perhaps even on Mars. The Cueva de Villa Luz, or Cave of the Lighted City, is home to a unique ecosystem. Bacteria, fish, and other creatures live in an environment that receives no sunlight at all. Scientists believe that a study of this ecosystem may allow them to better understand life in other places that never see the sun, like ridges on the seafloor.

The bacteria at the heart of the cave's ecosystem feeds on sulfur in the rocky walls. The bacteria is visible as deposits of white slime—called "snot-tites" by some scientists—that drip from the walls and ceiling. This slime is extremely acidic because the bacteria secrete sulfuric acid, the same chemical used in car batteries. Few organisms on Earth can survive in so harsh an environment. Some scientists believe, however, that places like the Cueva de Villa Luz might be found on Mars and could sustain life under that planet's sulfurous surface.

Popocatépetl, an active volcano, rises behind an ancient pyramid at Cholula and the church at its summit. More than 20 million people live within 50 miles of the volcano.

internet connect

GO TO: go.hrw.com
KEYWORD: SW3 CH10
FOR: Web sites about Mexico

Our Amazing Planet

Every fall, thousands of monarch butterflies migrate to the mountains of Mexico. Some of the butterflies travel more than 1,800 miles (2,900 km) to their winter home.

Landforms

Mexico is a large country, almost three times the size of Texas. Most of the country is made up of a rugged central plateau, called the Mexican Plateau. In places this plateau is as high as 9,000 feet (2,700 m). Three great mountain ranges border the Mexican Plateau. The Sierra Madre Oriental is in the east and the Sierra Madre Occidental in the west. Along the southern Pacific coast is the Sierra Madre del Sur. Coastal plains separate the mountain ranges from the sea. The eastern coastal plain is generally wider than the plain along the Pacific coast.

At the southern end of the Mexican Plateau lies the Valley of Mexico, where Mexico City is located. The floor of this broad valley is about 7,500 feet (2,280 m) above sea level. Until they get used to the altitude, visitors to the city may feel a shortness of breath and lack of energy due to the thin air. The mountains southeast of Mexico City include great volcanoes. The highest, Orizaba, soars to 18,700 feet (5,700 m). Because the area is tectonically active, earthquakes are also common.

In southern Mexico, the landforms become more complex, with many small mountain ranges, narrow valleys, and volcanoes. The rugged terrain makes overland travel difficult. Many villages are connected only by single-lane roads. Similar landscapes continue southward into Central America.

Mexico narrows in the south to form an **isthmus**. An isthmus is a narrow strip of land connecting two larger land areas. The Pacific Ocean and Gulf of Mexico lie just about 150 miles (240 km) apart at Mexico's Isthmus of Tehuantepec (tay-WAHN-tah-pek). The Yucatán (yoo-kah-TAHN) Peninsula, Mexico's flattest region, is located in southeastern Mexico. Limestone lies beneath the Yucatán's surface. Water tends to drain through this limestone, rather than flow across the surface. As a result, the Yucatán has few rivers even though the area has a humid climate. Erosion has created many caves and **sinkholes** in this area. A sinkhole is a steep-sided depression that forms when the roof of a cave collapses.

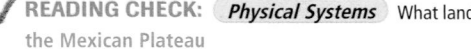

READING CHECK: *Physical Systems* What landform makes up most of central Mexico?
the Mexican Plateau

Climate, Biomes, and Natural Resources

Mexico's climate varies by region. Vegetation ranges from desert plants in the north to tropical forests in the south. The country also has a variety of natural resources. Mexico stretches through both the subtropical and tropical latitudes. Partly as a result of this location, the country has a range of tropical and dry climates. Three factors help explain Mexico's climates. First is a regional high-pressure system known as the Pacific subtropical high pressure cell. The dry weather this system creates dominates northwestern Mexico. The result is arid and semiarid climates and scrub vegetation in the north and west, particularly Baja California. The influence of this high-pressure system extends east and south as well. It limits the amount of rainfall over about two thirds of

Close

Point out on a map the major landforms and climate regions of Mexico. Ask students how the country's natural environments might affect how people live in each region.

Review and Assess

Have students complete the **Section Review**. Then have students complete **Daily Quiz 10.1.**

Reteach

Have students complete **Main idea Activity for English Language Learners and Special-Needs Students 10.1.** Have students work in pairs to write questions for a quiz game about Mexico's physical geography. Then have pairs take turns quizzing each other. **ENGLISH LANGUAGE LEARNERS, COOPERATIVE LEARNING**

Extend

Have interested students conduct research on the cenotes of the Yucatán. These water-filled sinkholes were once sacred to the Maya and now attract divers from around the world. Have students create travel brochures about the cenotes that explain their creation, history, and popularity. **BLOCK SCHEDULING**

Mexico's land area. Therefore, dry grasslands and brush cover much of northern Mexico's plains.

The northeast tradewinds are the second factor affecting Mexico's climates. These winds bring humid air from the Gulf of Mexico and the Caribbean Sea. As a result, rains sweep into Mexico mostly from the east and southeast, particularly during the summer. In fact, the forested plains of southeastern Mexico have a tropical humid climate. The easterly winds also steer hurricanes toward Mexico's east coast. Hurricanes also strike the Pacific coast but are less common there.

Finally, climates vary dramatically with elevation. Mexico's highest levels of rainfall occur where the humid trade winds rise against the mountains of the southeast. This also leads to dry rain-shadow climates on the western sides of the mountains. Many people live in the mild environments of the mountain valleys. The valleys along Mexico's southern coast also have pleasant subtropical climates. The high elevations on the Mexican Plateau can have cool highland climates. During the winter, cold polar air sometimes flows southward across the Mexican Plateau. Snow can fall in the Sierra Madre Occidental. This cold air brings freezing temperatures as far south as Mexico City.

Mexico does not have many major rivers. Its largest rivers drain areas of central Mexico. These rivers provide the country with hydroelectric power and water for irrigation.

Mexico has many mineral resources. Centuries ago silver was the country's most valuable mineral product. Mexico remains the world's leading silver producer. The country also produces many other metals, including gold, iron, lead, and mercury.

Today petroleum is Mexico's most valuable natural resource. Mexico's great oil and natural gas fields lie along the Gulf of Mexico. Most of Mexico's oil is exported to the United States.

✓ **READING CHECK:** *Physical Systems* What three factors influence Mexico's climates? What is Mexico's most important natural resource? subtropical high pressure, northeast tradewinds, elevation; oil

INTERPRETING THE VISUAL RECORD

Cuatro Ciénegas ("Four Marshes") is a unique area of sparkling water in the Chihuahua Desert of northern Mexico. These spring-fed pools shelter many species found nowhere else. Many other desert-dwelling species depend on the water. The horned lizard (inset photo) lives in the region's dry areas. **Which mountain range do you think appears in the photo?**

Section 1 Review

Define
isthmus
sinkhole

Working with Sketch Maps
On a map of Mexico that you draw or that your teacher provides, label the Mexican Plateau, Sierra Madre Oriental, Sierra Madre Occidental, Sierra Madre del Sur, Valley of Mexico, Mexico City, Orizaba, Isthmus of Tehuantepec, Yucatán Peninsula, and Baja California. Where are most of Mexico's oil deposits located?

Reading for the Main Idea

1. *Physical Systems* What three mountain ranges border the Mexican Plateau?

2. *Physical Systems* Why does the Yucatán Peninsula have sinkholes?

Critical Thinking

3. **Analyzing Information** Why are the western sides of Mexico's mountains drier than the eastern sides?

4. **Drawing Inferences and Conclusions** What geographical factors do you think might benefit industrial growth in Mexico?

go.hrw.com Homework Practice Online
Keyword: SW3 HP10

Organizing What You Know

5. Copy the following graphic organizer. Use it to describe Mexico's physical geography.

Landforms	Climate	Resources

LET'S GET STARTED

Copy the following passage onto the chalkboard: *Mexico and the United States both began as European colonies. Based on what you already know, what are some ways in which the countries are different today? What might be the causes of these differences? Discuss responses. (Possible answers: differences in the goals of the original settlers, differences in relations with Europe, differences in independence movements)* Tell students that in Section 2 they will learn more about Mexico's development during its colonial period and since its independence.

Building Vocabulary

Write the key terms on the chalkboard and call on volunteers to locate and read their definitions. Underline **conquistadores**, **haciendas**, **plaza**, and **mestizos**. Point out that all of these words come from Spanish. Ask students to think of English words that resemble *conquistadores* and that have related meanings. *(Possible answers: conquest, conquer)* Then tell students that *haciendas, plazas,* and *mestizos* are found in many regions of the world besides Mexico. They are common in many areas once controlled by Spain.

VISUAL RECORD ANSWER

Possible answer: made farming relatively easy, so people could pursue other interests, such as art or astronomy

Section 2 — History and Culture

READ TO DISCOVER

1. What were the cultures of Mexico like before the Spanish arrived?
2. How did Spanish control change Mexico?
3. What has Mexico's history been like since independence?

WHY IT MATTERS

For most of the 1900s, a single political party ruled Mexico. Recently Mexico's government has become more democratic. Use CNNfyi.com or other **current events** sources to find out about changes in Mexican politics.

DEFINE

conquistadores   mestizos
haciendas   dictator
plaza

LOCATE

Acapulco
Cancún
Mazatlán

Early Mexico

Mexico's early peoples belonged to many cultures, each with its own language. Some people were hunter-gatherers. Others were farmers. Their main crops included beans, corn, peppers, and squash. Farmers grew these crops together in the same plots. The bean plants climbed the corn stalks, while the squash and peppers grew between the corn plants. This combination maintained soil fertility and also provided a complete diet. This style of cultivation, called milpa, is still widespread today. In addition, corn tortillas and beans flavored with peppers are still bases of Mexican cooking.

Some of the American Indian peoples of what are now Mexico and Central America created highly complex and accomplished civilizations. These peoples included the Maya, Olmec, Toltec, and Zapotec. Many had large city centers with splendid avenues, plazas, and pyramids. The last of these civilizations was the Aztec. Skilled in warfare, the Aztec built an empire in what is now

INTERPRETING THE VISUAL RECORD

Palenque (puh-LENG-kay), in Chiapas State, was one of the main Maya cities. The building shown is Palenque's palace complex, which includes a four-story observatory. By watching the stars and planets, the Maya were able to devise a complex and accurate calendar. **Examine the unit and chapter maps. How might the natural environment of the Chiapas area have affected Maya culture?**

Teach Objective 1

ALL LEVELS: Organize the class into groups and have each group create a banner for a parade celebrating Mexico's early cultures. Each group may choose to represent either a particular culture, such as the Aztec or the Toltec, or traits shared by several groups, such as city layouts, common foods, or milpa agriculture. Provide groups with encyclopedias and other reference materials so that they can learn more about their chosen subjects. Call on volunteers from each group to explain the finished banners to the class. Display the banners around the classroom. **ENGLISH LANGUAGE LEARNERS, COOPERATIVE LEARNING**

Teach Objective 2

ALL LEVELS: Copy the following graphic organizer onto the chalkboard, omitting the italicized answers. Have students complete it to describe changes in Mexico during the colonial period. Ask how Spaniards' perceptions of Mexico led to these changes (*saw Mexico as a source of wealth to be colonized and exploited*). **ENGLISH LANGUAGE LEARNERS**

Disease	→	*Many indigenous people were killed by European diseases.*
Farming	→	*Small communal ejidos were replaced by large haciendas.*
Government	→	*Aztec Empire was replaced by a Spanish colonial government.*
Population	→	*Mestizos became largest element of population.*
Religion	→	*Most people were converted to Roman Catholicism.*

central and southern Mexico. Their splendid capital city, Tenochtitlán (tay-nawch-tee-TLAHN), occupied an island in a lake in the Valley of Mexico. When the Spaniards arrived, this city was one of the largest in the world. (See Cities & Settlements: Mexico City.)

✓ **READING CHECK:** *Places and Regions* What were the main crops that farmers grew in ancient Mexico? corn, beans, squash, peppers

The Colonial Period

In 1519 a band of Spanish adventurers landed on the eastern coast of Mexico. As they traveled inland, these **conquistadores** (kahn-kees-tuh-DAWR-ez), or conquerors, formed crucial military alliances with peoples who resented the Aztec. The Spaniards also had muskets and horses, which were unknown in the Americas at that time.

New diseases, such as smallpox, arrived along with the Spaniards. The American Indians had no resistance to these European diseases, which spread rapidly through their population, killing many. The high death rate weakened the Aztec Empire. This weakness, along with the Spaniards' other advantages, helped the small Spanish band capture the Aztec capital. With the city's fall, the Aztec Empire ended. The conquerors called their colony Nueva España, or New Spain. Colonists would build Mexico City on the ruins of Tenochtitlán.

Desire for gold and silver had been a major motive for Spain to colonize the Americas. The Spaniards expanded the existing mining operations. Gradually, agriculture became an important part of the colonial economy as well. The Indians had mostly owned and worked the land in groups. Lands they worked in common were called *ejidos* (e-HEE-thos). The Spanish organized these lands into **haciendas** (hah-see-EN-duhs). Haciendas were large estates usually owned by wealthy families but worked by many peasants, usually Indians.

Roman Catholic missionaries tried to convert the American Indians to Christianity. They established frontier outposts called missions. Towns often grew up around these churches. The open space, or **plaza**, in front of the church might become a center for the community market. The plaza is a common feature in towns throughout middle and South America, Spain, and other parts of southern Europe.

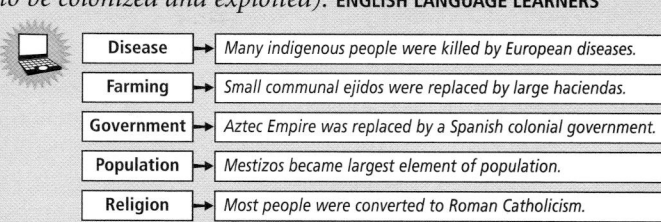

Our Amazing Planet

Before the Spanish conquest, the people of Mexico enriched their diets with protein-rich grasshoppers, larvae, locusts, and worms. Now these foods are popular in some Mexican restaurants.

INTERPRETING THE VISUAL RECORD

This painting from 1579 shows the initial meeting in 1519 of Hernán Cortés, a Spanish conquistador, and Moctezuma, the Aztec ruler. Cortés took Moctezuma prisoner and conquered the Aztec lands. **How can you tell from the picture that the relationship between the two peoples had changed over time?**

Across the Curriculum

▶ **Literature** ◀

Accounts of the Conquest
After the 1519 conquest of Tenochtitlán, many people in Europe were eager to read about Cortés and his deeds. In 1552, Francisco López de Gómara, Cortés's personal chaplain, published an account of the conquest in a long work called *Hispania Victrix*, or *Spain the Conqueror*. Gómara's account, however, is heavily biased and contains so many historical inaccuracies that it was banned in Spain from 1553 to 1727.

Unhappy with Gómara's work, Bernal Díaz del Castillo, a companion of Cortés, wrote the *Historia verdadera de la conquista de la Nueva España*, or *True History of the Conquest of New Spain*. Unlike Gómara, Díaz had actually been in Mexico during the expedition. His chronicle, published after his death, is both more accurate and more engaging than his predecessor's and has been more popular through time.

ACTIVITY: Have students read descriptions of a single event from both chronicles. Then ask students to judge which account seems to be a more reliable information source.

VISUAL RECORD ANSWER

Moctezuma is welcoming Cortés with gifts, so the relationship had probably started on better terms than it ended.

HOMEWORK: Ask students to imagine that they are Mexican Indians who have recently returned to Mexico during the time of Spanish exploration and colonization. Tell them that they have been away for 10 years and were not present for the arrival of the Spanish. Have students write accounts of how Mexico has changed during their absence.

Teach Objective 3

LEVELS 1 AND 2: Organize the class into three groups and assign each group one of the following eras of Mexican history: the struggle for independence, the Mexican Revolution, or modern history. Have each group create an illustrated time line for its assigned era that describes significant events and cultural changes that occurred during that period. When the groups have finished, combine their work into a single time line that can be displayed in the classroom. **ENGLISH LANGUAGE LEARNERS, COOPERATIVE LEARNING.**

LEVELS 3: Have students conduct research to learn about important leaders in Mexico during and since the struggle for independence. Then have each student write a eulogy for one of these historical figures. Tell students that their eulogies should emphasize the honored person's achievements and highlight changes that he or she brought about in Mexican culture. Call on volunteers to share their eulogies with the class. **BLOCK SCHEDULING**

Linking Past to Present

The Treaty of Guadalupe Hidalgo In 1845 the United States annexed Texas, which had declared its independence from Mexico nine years earlier. This upset the Mexican government, which had never recognized Texas independence. At the same time, the United States sent an ambassador to Mexico to purchase California and New Mexico. Mexico's leaders refused to negotiate with him, and the two countries soon went to war.

The Mexican War was ended in 1848 with the signing of the Treaty of Guadalupe Hidalgo. Under its terms, all Mexican territory north of the Rio Grande was ceded to the United States. The United States also acquired the territories of California and New Mexico. Mexico received $15 million in exchange for its lost land.

CRITICAL THINKING: How did conflict and cooperation shape the modern boundaries of Mexico? *(Possible answer: Both fighting and purchase shaped its border with the United States.)*

You don't say! The Mexican holiday Cinco de Mayo commemorates the defeat of French troops at Puebla on May 5, 1862.

CONNECTING TO ANTHROPOLOGY ANSWER

strengthen community, family, historical, religious ties

Oaxacan families decorate graves on the Day of the Dead. On this holiday, which combines Christian and pre-Christian traditions, Mexicans remember their dead and celebrate the continuity of life.

Connecting to
ANTHROPOLOGY

Rites of Passage

Societies hold ceremonies, such as weddings, to mark the times when its members move from one stage of life to another. Anthropologists call these ceremonies rites of passage. In Mexico, the *quinceañera* (keen-se-ahn-YE-rah) celebrates a girl's fifteenth birthday. It also marks the end of her childhood. A *quinceañera* typically begins with a Catholic Mass and ends with a big fiesta. The event can be similar to a wedding in terms of preparation, special clothes, food, and costs. *Quinceañeras* may date back to an old Aztec ceremony that stressed to young women the importance of following society's rules.

Making Generalizations and Predictions How might *quinceañeras* contribute to a society's unity?

Today most Mexicans are Roman Catholic. However, over time Christianity in Mexico has changed. In many cases pre-Christian beliefs and holidays combined with Christian beliefs. These distinctively Mexican traditions continue into the present.

Throughout Mexico's history the Spanish and American Indian cultures mixed. Early in the colonial period, most colonists were men, and marriage with American Indian women was common. Today the majority of Mexicans are **mestizos**, people of mixed European and Indian ancestry.

✓ **READING CHECK:** *Human Systems* What factors helped the Spanish conquer the Aztec? muskets, horses, alliances with enemies of the Aztec, disease

Mexico since Independence

The recent history of Mexico has been marked by rapid, sometimes violent, political change. Economic and cultural change there has been just as dramatic.

Independence and Revolution In 1810, Mexicans began to revolt against Spanish rule. Fighting continued until 1821, when Mexico finally won its independence. Although Spanish administrators returned to Spain, little really changed in Mexico. A few powerful families still controlled the economy and the government. In 1848 Mexico lost its northern territory from Texas to California following a war with the United States. Still, American and European investments during the late 1800s did lead to economic growth for Mexico. New mines were developed, and railroads were built. Modern industries grew in cities. Plantation agriculture expanded along the eastern coast. These plantations were large estates farmed by workers who lived on the property.

However, while a few Mexicans became rich, most remained poor. This economic inequality led to the Mexican Revolution, which lasted from 1910 to 1920. Following the revolution, the new government took the outward form of a democracy. The reality was much different. The president, who ruled much like a **dictator**, held almost all of the power. A dictator is a leader who rules with almost absolute authority. Mexico's government became much more involved in the national economy. It even began to directly control some industries. As a result, many foreign-owned businesses, particularly oil companies, were forced out of Mexico.

One result of the revolution was land reform. Large haciendas were broken up and given to peasant villages according to the old *ejido* system. Over time about half of Mexico's farmland became *ejidos*. While politically popular, the new land system was not very successful. The land given out was often poor and the plots small in size. Few farmers had the money to buy fertilizers and modern machinery. Many discouraged farmers moved to the cities. In 1992, *ejido* farmers won the right to sell their land, and many did. Some farmland has been combined into larger commercial farms. Some land near towns has become suburban housing tracts.

✓ **READING CHECK:** *Human Systems* What kind of government resulted from the Mexican Revolution? The government had the form of a democracy, but the president ruled almost like a dictator.

Close

Ask students to recall the principal crops grown by the early inhabitants of Mexico (*beans, corn, peppers, squash*). Then ask them to think of how these same ingredients are used in Mexican cooking today.

Review and Assess

Have students complete the **Section Review.** Then have students complete **Daily Quiz 10.2.**

Reteach

Have students complete **Main idea Activity for English Language Learners and Special-Needs Students 10.2.** Then organize the class into four groups and have each group create a poster about one of the time periods discussed in this section.
ENGLISH LANGUAGE LEARNERS, COOPERATIVE LEARNING

Extend

Have interested students conduct research about the ancient Maya. Ask students how modern perceptions of this people have changed as archaeologists and historians have deciphered their language and written records. BLOCK SCHEDULING

Modern Mexico Since about 1990, Mexico has again opened its economy to foreign businesses. Mexico's factories turn out nearly all the products a large country needs. The country, once largely rural, is increasingly urban. In fact, three quarters of Mexicans now live in towns and cities. Many work in industrial and service jobs.

In 1992 Mexico joined Canada and the United States in signing the North American Free Trade Agreement, or NAFTA. This agreement lowered trade barriers between the three countries. Since the door to markets in the north opened wider, manufacturing in Mexico has expanded even more.

Tourism has become increasingly important to Mexico. Resort cities on the east and west coasts draw tourists from around the world. Popular spots include Acapulco, Cancún, and Mazatlán. Beach attractions, grand hotels, crafts, and restaurants are central to these cities' economies. Mexico City, old colonial towns, and dramatic scenery draw visitors also to the country's interior.

Daily life in Mexico is changing rapidly. Like their American neighbors, most Mexicans watch television and shop in modern stores. Traffic jams clog the cities during rush hours. Mexican family life is also changing. Families tend to be smaller now, with only two or three children. More and more women are working outside the home. Many more Mexicans are graduating from universities. Protestant churches are attracting new members, which challenges the traditional role of the Roman Catholic Church. Mexican politics have become more democratic than ever before. Even Mexican Indians, who for a long time had little influence on the government, are becoming participants in politics. While not fully economically developed, Mexico is a powerful partner in North American affairs.

✓ **READING CHECK:** *Places and Regions* How has the balance between urban and rural settlement changed in Mexico? Mexico was once mostly rural but is now mostly urban.

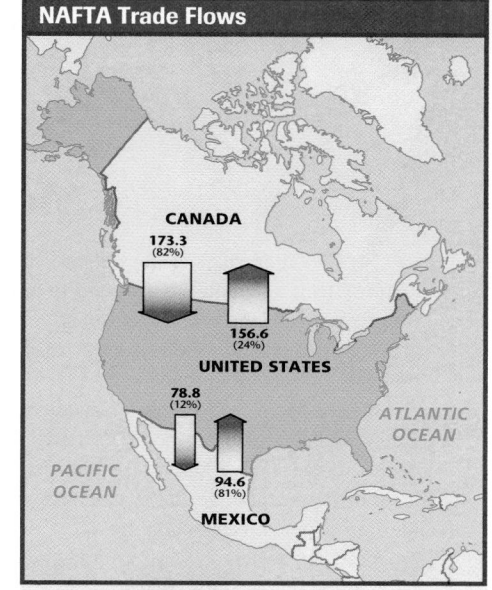

NAFTA Trade Flows

CANADA
173.3
(82%)

156.6
(24%)

UNITED STATES

78.8
(12%)

ATLANTIC OCEAN

PACIFIC OCEAN

94.6
(81%)

MEXICO

INTERPRETING THE MAP *This map shows trade among the United States and its NAFTA partners in billions of U.S. dollars. Numbers in parentheses show the percentage of each country's total exports. Trade among all three countries has greatly increased since NAFTA went into effect in 1994.* **Which country does more business with the United States? Which country sends the smaller percentage of its exports to the United States?**

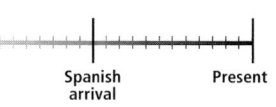

Section 2 Review

Define conquistadores, haciendas, plaza, mestizos, dictator

Working with Sketch Maps On the map you created for Section 1, label Acapulco, Cancún, and Mazatlán. Which city is on the Caribbean coast?

Reading for the Main Idea

1. *The Uses of Geography* What powerful empire ruled much of what is now Mexico at the time of Spanish arrival?

2. *Human Systems* What were the two most important economic activities of colonial Mexico?

3. *Human Systems* What agreement has lowered trade barriers between Mexico and the rest of North America? When was it signed?

Critical Thinking

4. Analyzing Information In what ways did Spanish colonization shape Mexican culture? How did precolonial beliefs shape the practice of Christianity in Mexico?

go.hrw.com Homework Practice Online
Keyword: SW3 HP10

Organizing What You Know

5. Create a time line of Mexican history. Use it to list key events from the time of Spanish arrival to the present.

Spanish arrival ———————— Present

Section 2 Review Answers

Define For definitions, see: conquistadores, p. 225; haciendas, p. 225; plaza, p. 225; mestizos, p. 226; dictator, p. 226

Working with Sketch Maps Maps will vary, but listed places should be labeled in their approximate locations. Cancún is on the Caribbean.

Reading for the Main Idea
1. the Aztec Empire
2. mining and agriculture
3. the North American Free Trade Agreement (NAFTA); 1992

Critical Thinking
4. Most Mexicans are now mestizo and Roman Catholic. Under the Spanish they moved from the *ejido* land system to the hacienda system. Many pre-Christian beliefs and holidays have become mixed with Christian beliefs. (NGS 10)

Organizing What You Know
5. 1519—conquest of Aztec; 1810–21—struggle for independence; 1848—loss of California, Texas, and other territory to United States; 1910–20—Mexican Revolution; 1992—signing of NAFTA

MAP ANSWER

Canada; Mexico

Mathematics: Immigration Statistics

Copy the following table onto the chalkboard. It shows the number of immigrants admitted to the United States from Mexico in 1998. Ask students if more men or women are moving from Mexico to the United States *(women)*, what percentage of immigrants are male *(42.86 percent)*, which age range contributes the most immigrants to the United States *(20–29)*, which contributes the fewest *(above 80)*, and what percentage of immigrants are below the age of 40 *(77.35 percent)*. Why do you think this is the case? *(People in this range are most likely to be seeking work.)* Then organize the class into groups and assign each group a country. Have each group conduct research to learn about migration patterns from its assigned country to the United States. Have each group compile its findings into a table.

Mexican Immigrants to the United States

Age	Total	Male	Female
Under 10	12,494	6,464	6,012
10–19	33,324	17,195	16,095
20–29	35,169	17,291	17,803
30–39	20,787	6,046	14,699
40–49	9,658	2,222	7,430
50–59	9,300	2,558	6,736
60–69	7,463	2,999	4,460
70–79	2,676	1,284	1,391
80 or above or unknown	704	340	363
All ages	131,575	56,399	74,989

Source: *Statistical Yearbook of the Immigration and Naturalization Service,* 1998

Human Systems

Geography for Life

Mexican Migration

Migration changes populations, cultures, and economic systems. Migrants are attracted to new areas by pull factors, such as jobs and better opportunities. They leave areas because of push factors, like war, changing economies, droughts, or poor living conditions.

Mexico has a long history of migration to the United States. The idea of going north for better opportunities is deeply rooted among Mexican youths, particularly in rural areas of west-central Mexico. In the 1870s large numbers of Mexican migrants went north to work. In the early 1900s American landowners actively recruited Mexicans to work on farms in the United States. However, the Great Depression in the 1930s temporarily slowed migration. Later, an agreement between the U.S. and Mexican governments encouraged further migration of temporary farmworkers from Mexico from 1942 to 1964. The Immigration Act of 1990 greatly increased the number of immigrants allowed to enter the United States. As a result immigration, both legal and illegal, increased.

More people emigrate from Mexico than from any other country. Some 7 million people now living in the United States were born in Mexico. The major cause of migration between the two countries is the difference in wages. Simply put, Mexican workers can earn more money in the United States

The Nogales border crossing connects Arizona and Sonora.

than they can in Mexico. Many travel to the United States temporarily to work and send money to relatives in Mexico.

This migration has created a unique cultural landscape along the U.S.-Mexico border. Many Mexican Americans live along the border in Texas, New Mexico, Arizona, and California. Many border towns are linked economically, and border crossings between the two countries are among the busiest in the world. This benefits the economies of both countries. The border region has developed a unique culture that blends American and Mexican ways of life. In fact, many geographers recognize the U.S.-Mexico border area as a distinct region. This region is sometimes called the borderlands.

However, it is estimated that about 2 million Mexicans are living in the United States illegally. In recent years, the United States has stepped up its border patrols and tried to block illegal immigration. Many people caught by patrols and returned to Mexico will try to cross the border again. Many Mexican migrants suffer mistreatment from "guides" called coyotes whom they pay to lead them across the border. Some illegal immigrants try to enter the United States in remote desert areas. Unprepared for the harsh conditions, some have died.

INTERPRETING THE VISUAL RECORD *Mexicans wade across the shallow Rio Grande in the El Paso, Texas, area. Here the river has often cut new channels, sometimes changing the country in control of hundreds of acres.* **From what you can see in the photo, how have Mexico and the United States improved their ability to control their territory?**

Applying What You Know

1. **Summarizing** How has migration from Mexico to the United States shaped the distribution of culture groups today?

2. **Drawing Inferences and Conclusions** What are some of the political, economic, social, and environmental factors that contribute to Mexican migration to the United States?

 **LET'S GET STARTED**

Copy the following question onto the chalkboard: *What are some factors that distinguish one region of a country from another?* Discuss responses. *(Possible answers: cultural, economic, historical, physical, or political factors)* Challenge students to speculate about factors that have created different types of regions within Mexico. *(Example: Northern Mexico's connection to the United States makes it a functional region.)* Tell students that in Section 3 they will learn more about life in the various regions of Mexico today.

Building Vocabulary

Write **maquiladoras** on the chalkboard. Call on a volunteer to read its definition aloud. Point out that it is derived from the Spanish word *maquila*, which refers to the portion of grain given to millers by farmers as payment for their services. A *maquiladora* was the place where such payment was made. Ask students how this root relates to the current meaning of the term. *(Possible answer: In these factories, workers are paid for their services—the making of goods.)*

Section 3

Mexico Today

READ TO DISCOVER

1. What are the economic and cultural regions of Mexico?
2. What challenges face Mexico?

WHY IT MATTERS

Economic connections with Mexico are very important to the United States. Use **CNNfyi.com** or other **current events** sources to learn about recent economic developments in Mexico and how they affect this country.

DEFINE

cash crops
maquiladoras

LOCATE

Guadalajara   Monterrey
Campeche   Tijuana
Tampico   Ciudad Juárez
Veracruz

Mexico's Regions

Mexico is divided into 31 states and the capital district. Different parts of the country exhibit great geographical, economic, and cultural diversity. It is useful to divide the country into four regions for study.

Greater Mexico City   Greater Mexico City is the cultural, economic, and political center of Mexico. This huge metropolis includes many smaller cities and may hold a fourth of Mexico's entire population. (See the graph on the next page.) It also generates much of the country's GDP.

Mexico City has monumental government buildings. It is also home to the country's largest university and greatest museums and theaters. The

This cat figure from the state of Oaxaca is an example of Mexican folk art.

The States of Mexico

1. AGUASCALIENTES
2. QUERÉTARO
3. TLAXCALA
4. MÉXICO
5. DISTRITO FEDERAL (Federal District)
6. MORELOS

INTERPRETING THE MAP

The official name of Mexico is Estados Unidos Mexicanos, or United Mexican States. Compare this map to the physical map. What features seem to form natural borders for Sinaloa and Veracruz? Why might the smaller states be in south-central Mexico? What area has a political status similar to that of Washington, D.C.?

LEVEL 1: Provide students with outline maps of Mexico and have them mark the locations of Mexico's major cities. Then have them sketch the boundaries of the economic and cultural regions of Mexico. Finally, have students create a set of symbols to indicate the economic activities found in each region. Encourage them to use different colors to identify primary, secondary, tertiary, and quaternary activities. Then lead a class discussion about factors that influence the locations of Mexico's cities and economic activities and how these factors help to divide the country into regions. Ask students how economic factors have influenced the cultures of each region. **ENGLISH LANGUAGE LEARNERS**

LEVELS 2 AND 3: Organize the class into four groups and assign each group one of the regions of Mexico discussed in this section. Have students conduct research about their assigned regions to learn more about their physical and cultural landscapes. Then have each group write a script for a documentary film about its assigned region. Each group's film should describe both the region's physical features and the cultures that have developed there. The scripts should also describe how these environments and cultures have influenced the innovation of new ideas or technologies and the diffusion of these ideas into other regions. **COOPERATIVE LEARNING**

Guadalajara Guadalajara has often been called the "most Mexican" city in Mexico. Its buildings more closely resemble traditional Mexican styles than do those of Mexico City. In addition, many features that visitors think of as central to Mexican culture developed in or near Guadalajara.

For example, the *jarabe tapatío,* or Mexican hat dance, considered by many people the country's national dance, was created there. In this dance, men throw their sombreros—wide-brimmed hats also originally from Guadalajara—to the ground for their partners to dance around. Guadalajara and its state, Jalisco, are also believed to be the home of mariachi, a musical style popular with Mexican street bands. Other cultural contributions of Guadalajara and Jalisco include various Mexican dishes, tequila, and *charreada,* or Mexican rodeo.

ACTIVITY: Have students trace the origins and diffusion of other elements of traditional Mexican culture.

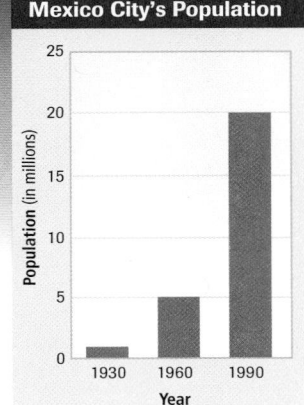

Mexico City's Population

Population (in millions) vs. *Year*

Year	Population
1930	~1
1960	~5
1990	~20

Source: *Latin America and the Caribbean,* Blouet and Blouet, eds.

INTERPRETING THE GRAPH

*Mexico City has experienced phenomenal growth. **Approximately how many times larger was the 1990 population than the 1930 population? What do you predict for Mexico City's future growth?***

headquarters of leading industries and banks, as well as the Mexican stock exchange, are located there. Millions of tourists come to Mexico City to see Aztec ruins and grand old colonial buildings. Major industries include cement, chemicals, construction, plastics, textiles, and tourism.

Mexico City's economic activity also draws many new residents. However, there are too few jobs. As a result, some people live in huge settlements of shacks built from waste wood and sheetmetal. Many people live without electricity, sewers, or a safe water supply.

Alongside this poverty, great wealth also exists in Mexico City. Exclusive boutiques and world-class restaurants attract tourists and upper-class residents. Districts with grand homes and luxury apartments extend for miles. At the city's center is the Zona Rosa, a stylish urban area.

Mexico City suffers from terrible air pollution, however. The city is located in a broad valley ringed by mountains. The mountains trap the pollution of thousands of factories and millions of cars. The government is trying to clear the air by reducing these emissions.

✓ **READING CHECK:** *Places and Regions* What are some of Mexico City's leading industries? construction, chemicals, plastics, cement, textiles, and tourism

Central Mexico Central Mexico stretches northwest of Mexico City and across the Mexican Plateau. Many cities there began as colonial mining or ranching centers. Mexico's second-largest city, Guadalajara, is located in this area. Guadalajara contains many buildings from the Spanish colonial period. Small towns with city squares and colonial churches are common.

Fertile valleys dot central Mexico. This was once colonial Mexico's great grain-producing region. Agriculture there is a mix of small family farms and medium-sized commercial farms. These farms grow a number of **cash crops**. Cash crops are crops grown for sale in a market. Trucks bringing fruits and vegetables from the area to the United States often jam border crossings. In recent years central Mexico has also been attracting new factories.

20 times larger; will continue to grow

Construction of Guadalajara's immense cathedral began when the city was founded in 1542. It was not finished until the early 1700s. After being severely damaged by earthquakes, the old towers were replaced by the yellow-tiled spires that have become the city's symbol.

ALL LEVELS: Copy the following graphic organizer onto the chalkboard, omitting the italicized answers. Ask students to provide words or phrases that describe the challenges Mexico faces. Lead a class discussion about how these issues are interrelated. Then ask students to create hypotheses about how these challenges will affect Mexico's future economic development. **ENGLISH LANGUAGE LEARNERS**

Challenges for the Future

Resolving Economic Inequality
- *poverty*
- *wealth in the hands of a few rich people*
- *few opportunities for Mexican Indians*
- *loss of skilled laborers through migration*

Reducing Crime
- *result of widespread poverty*
- *main route for drug smuggling into United States*
- *government corruption*

Improving Infrastructure
- *lack of clean water and modern sewers*
- *many roads and railways out of date*
- *difficult to move goods to market*

Gulf Lowlands and Southern Mexico

Throughout much of Mexican history, the Gulf lowlands between the cities of Campeche and Tampico were lightly settled. People living there used the region's hot humid tropical forests and savannas for grazing or growing sugarcane. Now large forest areas have been cleared for commercial farming and ranching. The region also includes Veracruz, an important seaport and communications center.

Rich deposits of oil and natural gas have long been key to this region's economy. The area is booming because of these deposits. With the development of the oil industry have come oil refineries, pipelines, petrochemical complexes, port facilities, and fertilizer plants.

Southern Mexico includes the mountainous areas south of Mexico City and the plains of the Yucatán Peninsula. This is Mexico's poorest region. It has few cities and little industry. In addition, transportation and telephone service are poorly developed. Schools are inadequate. More and more migrants head north hoping for a better life, either in Mexico City or in the United States.

Southern Mexico is also Mexico's most traditional region. Village life there has changed little over the last hundred years. Subsistence agriculture is common, and handicrafts provide much of the cash income. Mexican Indians make up about half of this area's population. Many speak Indian languages. For example, many of Yucatán's rural people speak Mayan.

✓ **READING CHECK:** **Human Systems** In what ways is southern Mexico the country's most traditional region? village life changed little, subsistence agricultural common, handicrafts provide much of income, many Mexican Indians who speak Indian languages

Northern Mexico

The large dry region of northern Mexico has become one of the most prosperous parts of the country. Much of the region's infrastructure is new and modern. The roads are good, and the telephones work well.

Monterrey is the great industrial city of the north. However, many other cities and towns have also industrialized and grown rapidly. Northern Mexico's factories and commercial farms draw migrant workers from all over Mexico. Cattle ranching, mining, and tourism are also important to northern Mexico's economy.

In the 1990s armed rebels from Chiapas, in southern Mexico, called for a redistribution of wealth. In this photo, rebellion supporters carry an image of Emiliano Zapata, who supported land reform during the Mexican Revolution of 1910 to 1920.

INTERPRETING THE VISUAL RECORD

Pinacate National Park is among the many scenic attractions in northern Mexico. The region contains both bustling cities and sparsely populated desert areas. **From what you see in the photo, what are some of the physical processes that have affected northwestern Mexico in the past?**

Nahuatl Many people in the Gulf Lowlands region of southern Mexico still speak Nahuatl, the language of the Aztec. Some 1.5 million people use the language in their daily lives, mostly in the states of Guerrero, Hidalgo, Puebla, and Veracruz.

Nahuatl belongs to the Uto-Aztecan branch of the Amerindian language family. It is related to many languages native to the American West, including Comanche, Hopi, and Ute. At the time of the Spanish Conquest of Mexico, written Nahuatl consisted of complex pictograms, but these characters were later replaced by the Latin alphabet. Spanish scholars and priests were therefore able to preserve a great deal of Nahuatl literature and poetry. Many words have made their way from Nahuatl into Spanish and from there into English. For example, the Nahuatl word *tomatl* became the Spanish *tomate*, which evolved into the English *tomato*. Some other words that were derived from Nahuatl are *avocado (ahuacatl), chocolate (xocolatl), coyote (cóyotl),* and *ocelot (ocelotl).*

VISUAL RECORD ANSWER

volcanoes, as evidenced by the craters, and erosion

Signs in the small border town of Nuevo Progreso, Tamaulipas, advertise various services. Many tourists take day trips from South Padre Island or Brownsville, Texas, to Nuevo Progreso.

FOCUS ON GEOGRAPHY

Border Towns The part of Mexico along the border definitely "faces north" in more than one sense. Many businesses have links with the United States. In fact, American companies own many of the special factories, called **maquiladoras** (mah-kee-lah-DOHR-ahs), that lie along the Mexican side of the border. These factories employ hundreds of thousands of Mexicans. Workers there assemble products for export, mostly to the United States. These products range from auto parts to toys. In addition, irrigated farms provide fruits and vegetables to markets in the United States and Europe during winter.

The border region also has many cultural links to the United States. American music, television, and other forms of entertainment are popular. The Spanish spoken in this part of Mexico is full of English words. Pairs of towns that straddle the border are beginning to function more like single cities. For example, Tijuana is increasingly linked to San Diego, California. Every day thousands of people move back and forth between cities like Ciudad Juárez and El Paso, Texas.

✓ **READING CHECK:** *Places and Regions* What links with the United States are found along Mexico's border region? business links such as maquiladoras, farm products; cultural links; closer border city ties

INTERPRETING THE MAP *Pairs of cities and towns lie along the U.S.-Mexico border.* **How do some of the cities' names reflect their locations?**

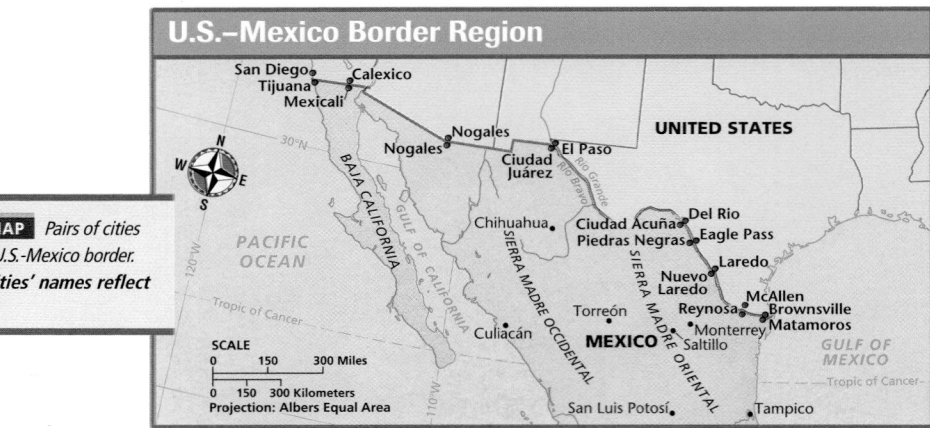

U.S.–Mexico Border Region

Close

Have students provide examples of how humans interact with the physical environment in each of Mexico's regions.

Review and Assess

Have students complete the **Section Review**. Then have students complete **Daily Quiz 10.3**.

Reteach

Have students complete **Main idea Activity for English Language Learners and Special-Needs Students 10.3**. Draw a large map of Mexico on butcher paper and cut it into the country's four regions. Have students work in groups to write phrases on each piece that describe the region. Reassemble and discuss the map. **ENGLISH LANGUAGE LEARNERS, COOPERATIVE LEARNING**

Extend

The Institutional Revolutionary Party (PRI) was the most powerful political party in Mexico for about 70 years. In 2000 presidential power shifted to the National Action Party (PAN). Have interested students conduct research on recent political developments in Mexico and how they may affect the country. **BLOCK SCHEDULING**

Challenges for the Future

Mexico is changing rapidly. Mexican politics are becoming more democratic. To stimulate growth, the country's leaders are reducing government controls on the economy. NAFTA's supporters believe that the treaty will boost economic growth. The Mexican middle class continues to grow. However, problems remain.

Geographers see several interrelated challenges in Mexico. One such challenge is economic inequality. Many Mexicans are poor, and much of Mexico's wealth lies in the hands of a few rich people. In general, Mexican Indians have fewer economic opportunities than other citizens. Reducing poverty might help create greater political stability as well as improve the country's economy. It might also slow migration out of Mexico. Migration to the United States, both legal and illegal, is common. Many Mexicans have achieved success in the United States. Some send money to their families back in Mexico. However, migration also takes skilled workers away from Mexico's economy.

Another challenge is reducing crime, much of which results from widespread poverty. Mexico is a main route for smuggling drugs into the United States. Profits from the drug trade have even tempted some government officials, including police, to break the law. Curbing crime will be key for further economic and social progress.

Finally, improving the country's poor infrastructure presents another challenge. Many Mexican communities do not have clean water supplies or modern sewers. This can cause health problems. Many of Mexico's roads and railways are worn and out-of-date. Mexicans cannot sell their products if their goods cannot get to market. All these are problems that further economic progress may help solve.

✓ **READING CHECK:** *Human Systems* What are some important challenges facing Mexico? economic inequality, crime, and poor infrastructure

A young girl looks out over the Matamoros city dump. Several families routinely search the garbage for items they can sell or use. Hazardous wastes are often dumped there illegally. Garbage that is burned in the open sends thick smoke over the area. **How might conditions at this dump affect the water supply?**

Section 3 Review

go.hrw.com **Homework Practice Online**
Keyword: SW3 HP10

Define cash crops, maquiladoras

Working with Sketch Maps
On the map you created for Section 2, label Guadalajara, Campeche, Tampico, Veracruz, Monterrey, Tijuana, and Ciudad Juárez. Which cities are located at or near sea level? Which cities are located in mountainous areas or on the Mexican Plateau?

Reading for the Main Idea

1. *Environment and Society* What geographical features make Mexico City's pollution problems worse?

2. *Human Systems* Why is migration to the United States a disadvantage for the Mexican economy?

Critical Thinking

3. *Making Generalizations* In what ways is Mexico both a rich country and a poor country?

4. *Drawing Inferences* How might economic progress help increase political stability in Mexico?

Organizing What You Know

5. Copy the graphic organizer shown below. Use it to list the four economic and cultural regions of Mexico and their major economic activities.

Region	Economic activities
Greater Mexico City	
Central Mexico	
Gulf lowlands and southern Mexico	
Northern Mexico	

Section 3 Review Answers

Define For definitions, see: cash crops, p. 230; maquiladoras, p. 232

Working with Sketch Maps Maps will vary, but listed places should be labeled in their approximate locations. Campeche, Tampico, and Tijuana are at or near sea level. Guadalajara, Monterrey, and Ciudad Juárez are in mountainous areas or on the Mexican Plateau.

Reading for the Main Idea
1. Surrounding mountains trap air pollution.

2. Migration draws skilled workers away from Mexico.

Critical Thinking
3. great wealth in some modern areas, poverty and inequality elsewhere (NGS 11)

4. Possible answer: People with money usually work for stability because they have more at stake than poor people. (NGS 11)

Organizing What You Know
5. Greater Mexico City—construction, chemicals, plastics, cement, textiles, tourism; Central Mexico—farming, manufacturing; Gulf Lowlands and Southern Mexico: oil and natural gas, farming, ranching; Northern Mexico—manufacturing, farming, ranching, mining, tourism

VISUAL RECORD ANSWER

Runoff could cause pollution.

Cities & Settlements

Relating Geography and Soils

Have students read Cities & Settlements: Mexico City and review the many environmental problems the city's residents face daily (*air pollution, a lowered water table, buildings sinking*). Point out that just as the region's physical landscape makes air pollution worse by trapping pollutants in a basin, the makeup of the region's soil layers plays a major role in the sinking, or subsidence, of Mexico City's land and buildings. The city is built on volcanic deposits and permeable soil sandwiched between layers of clay. The clay shrinks as the city's aquifer is depleted. Subsidence affects the city in many ways. Buildings lean or crack, subway tracks buckle, and water pipes that were once underground poke high into the air.

CITIES & SETTLEMENTS

Mexico City

Environment and Society

Mexico City is one of the largest cities in the world and one of the oldest cities in the Americas. The city's rich history and tremendous growth have made it Mexico's cultural, economic, and political center as well. However, Mexico City also has serious environmental problems which the city's people are working to solve.

An Indian and European City

In the mid-1300s the Aztec founded a town called Tenochtitlán in central Mexico. According to legend, their god advised them to build where they saw an eagle eating a snake while perched on a cactus. This legend is now represented on Mexico's flag. When the Spanish arrived in 1519, Tenochtitlán was home to perhaps 300,000 people. It was larger than any city in Europe at that time. In 1521 the Spanish defeated the Aztec and destroyed their city. On its ruins a new Spanish settlement rose—what became known as the Ciudad de México, or Mexico City.

Mosaics decorate the library of the National Autonomous University of Mexico.

Today Mexico City is a lively modern city that is proud of its history and culture. One can visit more than 100 museums there. Modern skyscrapers tower over Aztec ruins and Spanish colonial buildings. Few other cities offer such cultural variety.

The historical center of Mexico City is its main square, the Zócalo, where the center of Tenochtitlán once stood. Today it is one of the world's largest public squares. Bordering the Zócalo are some of Mexico City's oldest and grandest buildings. The Metropolitan Cathedral dates to the mid-1500s and is the largest church in Latin America. In the 1700s and 1800s, the block-long National Palace was home to Mexico's rulers. Today it houses government offices. Inside, huge murals by Diego Rivera, one of Mexico's famous artists, trace Mexican history from Aztec times to the early 1900s. On the square's northeast corner are the remains of the Templo Mayor, the great Aztec ceremonial pyramid. Southwest of the Zócalo, Chapultepec Park is one of the world's largest urban parks. The area was once used by Aztec emperors.

Pockets of old historic buildings are scattered throughout Mexico City. Many mark former villages that have been absorbed by the city's rapid expansion. The modern city completely surrounds these old towns. A well-known modern area includes the National University in southern Mexico City. Many of its buildings are covered with murals. The main library is decorated with tile mosaics showing Mexico's history and scientific achievements.

Cars zoom around a traffic circle on Reforma Avenue. The monument honors those who fought for Mexico's independence from Spain.

Going Further: Thinking Critically

Demonstrate Mexico City's subsidence with an experiment. You may want to have students assist with the demonstration. First, place a deep heavy cardboard box on two or three bricks in a pan or tub that is several inches wider than the box. Line the box with thick plastic film. Next, place a layer of vermiculite, available from a garden shop, into the box. Pour enough water into the box to fully saturate the vermiculite. This layer represents the aquifer that lies beneath Mexico City. Place another sheet of plastic over the vermiculite to represent the less permeable clay that overlies the aquifer. Add a thin layer of regular garden soil or gravel on top of the "clay." Then drill a hole with an ice pick through the box wall near the bottom. Be sure the hole reaches all the way through the plastic liner into the vermiculite layer. Then put several bricks on the soil. These stand for the buildings of Mexico City. As the water drains out of the hole, just as pumping drains water from the aquifer, the brick "buildings" will sink. Ask students to infer what other problems might arise from the situation. (*Water may puddle around the brick "buildings." This relates to the problem Mexico City has with rainwater and waste, which must be pumped away, particularly from the city's center. Or, the bricks may lean, which also mimics what happens to some Mexico City structures that have become unstable.*)

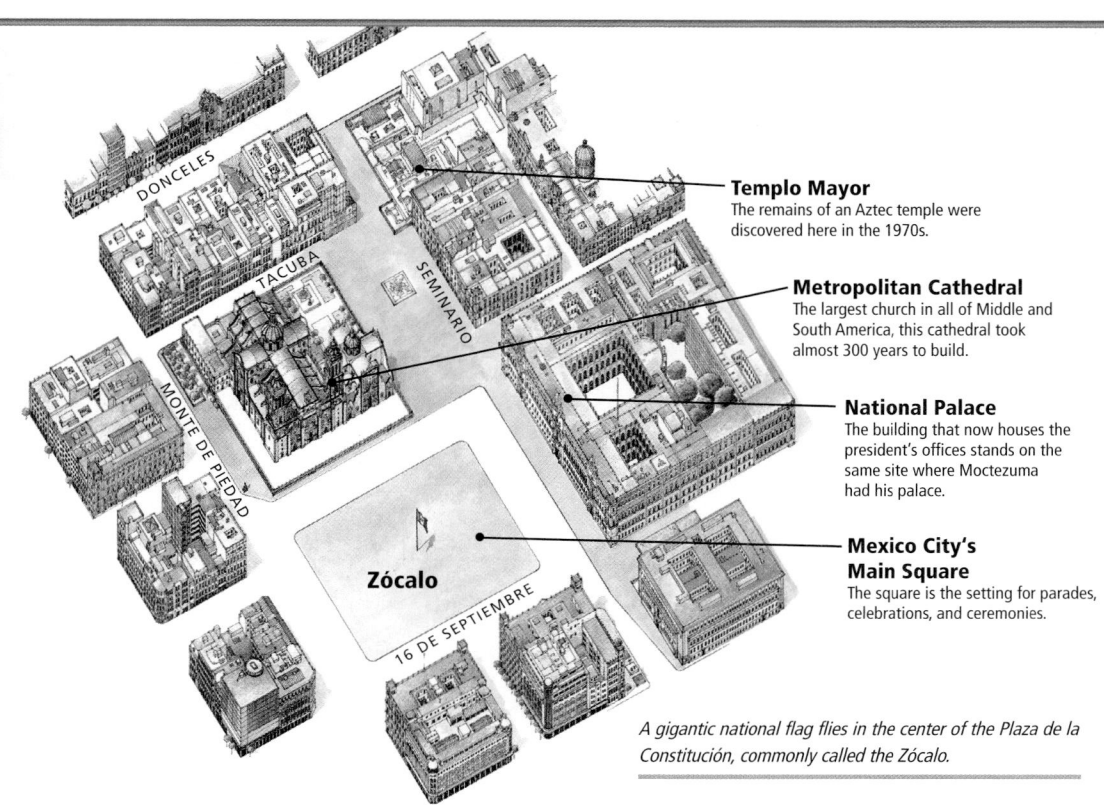

Templo Mayor
The remains of an Aztec temple were discovered here in the 1970s.

Metropolitan Cathedral
The largest church in all of Middle and South America, this cathedral took almost 300 years to build.

National Palace
The building that now houses the president's offices stands on the same site where Moctezuma had his palace.

Mexico City's Main Square
The square is the setting for parades, celebrations, and ceremonies.

A gigantic national flag flies in the center of the Plaza de la Constitución, commonly called the Zócalo.

Environmental Problems and Solutions

Some 18 to 25 million people live in the Mexico City metropolitan area today. By most accounts, Mexico City is now the second-largest urban area in the world, after Tokyo.

Mexico City's huge growing population has caused serious environmental problems including air pollution. Some 4.5 million motor vehicles and numerous factories release chemicals into the air. The resulting smog can cause breathing problems, sting eyes, and burn throats. As you read earlier, the city's location in a basin makes the situation worse. In addition, the city's industries and people use huge amounts of water. Heavy pumping from wells has lowered the water table under the city. This has caused the ground to settle and buildings in many areas to sink. Because the city lies on a dry lake bed the soils underneath are soft and unstable. In some parts of downtown, buildings have sunk as much as 29 feet (9 m) since 1900. The ground beneath the Metropolitan Cathedral has shifted so much that the building has cracked down the middle.

In 1985 a major earthquake damaged much of Mexico City. This experience raised awareness of the city's pressing environmental needs. As a result, work began in 1991 to save the Metropolitan Cathedral. Private companies have spent $300 million to repair other old buildings. Many have been turned into much-needed apartments.

Efforts were also launched to improve the city's air quality. Many factories were forced to close or cut production. In the late 1980s city officials developed a system to restrict the use of cars. Now cars with certain license plate numbers can only be driven on specific days. Since 1991 all new cars must have anti-pollution devices. Other laws ban nearly half of the city's older cars from the streets on "emergency days," when pollution levels are high. These measures seem to be working. In 1999 Mexico City experienced only 5 such days, compared to 95 of them in 1994.

Applying What You Know

1. **Summarizing** How has Mexico City changed since it was founded in the mid-1300s?

2. **Comparing** How do environmental changes in Mexico City compare to changes in large U.S. cities you have read about, such as Los Angeles?

CHAPTER 10 Review and Assessment Resources

TECHNOLOGY

▶ Chapter 10 Test Generator (on the One-Stop Planner)
▶ Global Skill Builder CD–ROM
▶ HRW Go site

REINFORCEMENT, REVIEW, AND ASSESSMENT

▶ Chapter 10 Review, pp. 236–37
▶ Chapter 10 Tutorial for Students, Parents, Mentors, and Peers
▶ Chapter 10 Test (form A or B)
▶ Alternative Assessment Handbook

▶ Chapter 10 Test for English Language Learners and Special-Needs Students
▶ Unit 3 Test
▶ Unit 3 Test for English Language Learners and Special-Needs Students

Assess

Have students complete a Chapter 10 Test.

Reteach

Organize the class into five groups and assign each group one of the following topics: vegetation, wildlife, mining and petroleum resources, cultures, or economic activities. Have each group design symbols for its topic and use them to label a large map of Mexico. Discuss the completed map. **ENGLISH LANGUAGE LEARNERS, COOPERATIVE LEARNING**

CHAPTER 10 Review Answers

Thinking Critically

1. functional because it is distinguished by the tourist industry; perceptual because it is distinguished by images of a vacation culture, and it has no formal boundaries (NGS 5)

2. Mexico City shares Washington's role as the country's political center, New York's as a financial center, and New York and Los Angeles's as cultural centers. As large cities, they probably share problems like traffic and pollution. (NGS 4)

3. Possible answer: Spanish, Roman Catholicism more common in areas settled by immigrants (NGS 9)

Using the Geographer's Tools

1. Possible answers: It might be as high as 60–100 million. The Valley of Mexico might be approaching the physical limit to its population. Pollution and crowding could discourage migration there. Conditions elsewhere might attract migrants who would have moved to Mexico City in the past.

2. Maps should accurately reflect the locations of Mexico's states and cities.

3. Answers will vary.

CHAPTER 10 Review

Building Vocabulary

On a separate sheet of paper, explain the following terms by using them correctly in sentences.

isthmus   haciendas   dictator
sinkholes   plaza   cash crops
conquistadores   mestizos   maquiladoras

Locating Key Places

On a separate sheet of paper, match the letters on the map with their correct labels.

Sierra Madre Oriental   Baja California
Sierra Madre Occidental   Acapulco
Mexico City   Monterrey
Yucatán Peninsula   Ciudad Juárez

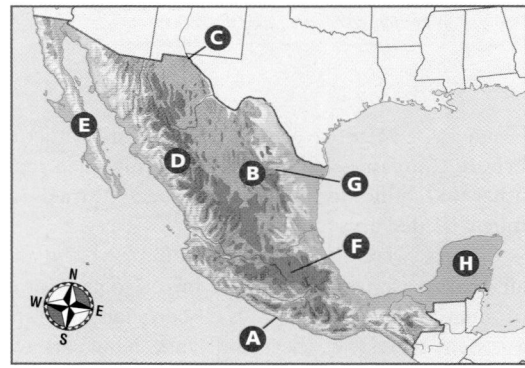

Understanding the Main Ideas

Section 1

1. (**Physical Systems**) Why does northwestern Mexico have dry weather?

Section 2

2. (**Human Systems**) What peoples ruled Mexico before the arrival of Spaniards?

3. (**Human Systems**) Why is Roman Catholicism the most common religion in Mexico?

Section 3

4. (**Places and Regions**) In what ways is Greater Mexico City important to the entire country?

5. (**Human Systems**) How have Mexican politics changed in recent years?

Thinking Critically

1. **Drawing Inferences and Conclusions** Review what you read about the resort cities on the east and west coasts of Mexico. How might "resort Mexico" be considered a functional region? How might it be a perceptual region?

2. **Comparing** How is Mexico City's status in Mexico similar to that of New York, Los Angeles, and Washington, D.C., in the United States? What problems might these cities share?

3. **Making Generalizations and Predictions** How might migration from Mexico shape cultural change in the United States?

Using the Geographer's Tools

1. **Analyzing Graphs** Look at the graph of Mexico City's population. If growth continued at the same pace, what do you think the city's population might be in 2020? What factors might affect the future growth of Greater Mexico City?

2. **Preparing Maps** Draw a sketch map of Mexico and its individual states. Then use the chapter map to locate and label the main cities. Use maps you find in library resources to check if you placed the cities correctly within the states. Revise your map if necessary.

3. **Preparing Maps** Create a map of the region near the border of Mexico and the United States. Use this textbook and other sources to identify Mexican and U.S. towns that lie across the border from each other. How might the economic and cultural features of Mexican border towns be different from cities farther south? What links would these border towns have with U.S. cities that other Mexican cities probably do not?

Writing about Geography

Write the text of an imaginary interview with a Mexican farmer who grows one or more cash crops. Include the farmer's viewpoint on the importance of trade with the United States and Canada. How does lifting trade barriers affect this farmer?

SKILL BUILDING

Geography for Life

Illustrating Geographic Information

(**Human Systems**) Create a poster or sketch map illustrating the four regions of Mexico. Use pictures and captions to describe the main cultural, economic, and natural features of these regions. Include information about the relationships among these features.

Mosaics decorate the walls of many modern buildings in Mexico. A famous example by Juan O'Gorman beautifies the library of the Universidad Nacional Autónoma de México in Mexico City's outskirts. Instruct students to locate illustrations of Mexican mosaics and use them as inspiration to create their own mosaics depicting aspects of Mexican culture, using small pieces of colored paper. Ask each student to write a paragraph explaining his or her mosaic and any symbols it includes. Place mosaics and paragraphs in student portfolios.

Food Festival

Salsa is a Mexican condiment that is now more popular than ketchup in the United States. It is based on tomatoes, onions, and chilies but may also contain cilantro, garlic, lime juice, tomatillos, or other ingredients. Salsas may be cooked or fresh, red or green, smooth or chunky. Have students prepare samples of different types of salsa. Establish prizes for the best salsa of each type. To scoop it up supply plenty of tortilla chips. Since some students are not used to spicy foods, urge them to use caution in preparing and tasting salsas. Provide cold water to soothe burning tongues.

Building Social Studies Skills

Interpreting Graphs

Use the information from the bar graph below to answer the questions that follow.

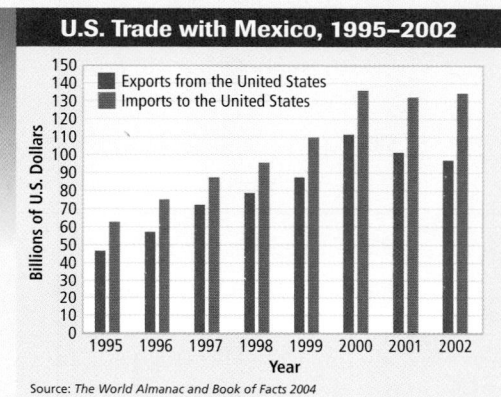

U.S. Trade with Mexico, 1995–2002

- Exports from the United States
- Imports to the United States

Billions of U.S. Dollars

Year

Source: *The World Almanac and Book of Facts 2004*

1. U.S. imports from Mexico increased every year except which one?
   a. 1998   c. 2000
   b. 1999   d. 2001

2. The North American Free Trade Agreement (NAFTA) took effect in 1994. Before this agreement, U.S. exports to Mexico were greater than imports. What general changes in U.S. exports and imports can you see since then?

Analyzing Primary Sources

Read Michael Parfit's comments about Mexico. Then answer the questions that follow on a separate sheet of paper.

"Rough by reputation, exotic even in name, the Sierra Madre Occidental (or Mother Mountains of the West) remains one of the last untrammeled [untamed] wildernesses in Mexico. The northern portion of the 800-mile-long range is famed for its immense, mile-deep gorges—Mexico's answer to the Grand Canyon. Such daunting topography has proved a barrier to settlement for centuries—and a godsend to renegades from Geronimo to Pancho Villa."

3. What makes part of the Sierra Madre Occidental "Mexico's answer to the Grand Canyon"?
   a. its 800-mile-long length
   b. mile-deep gorges
   c. its tall mountains
   d. a desert climate

4. Why do you think these mountains are both a barrier to settlement and a good place for people trying to hide from authorities?

Alternative Assessment

PORTFOLIO ACTIVITY

Learning about Your Local Geography

Individual Project: Field Work

You have read about efforts to improve air quality in Mexico City. Conduct research to learn about air pollution in your area. Contact your local air quality board, television station weather department, or a local environmental group. Request information on the air quality of your area. If possible, collect data going back several years. Graph the data. What patterns do you see over time?

internet connect

Internet Activity: go.hrw.com
KEYWORD: SW3 GT10

Choose a topic on Mexico to:
- take the GeoMap challenge to test your knowledge of geography!
- create a brochure on Mexican holidays and culture.
- learn about the physical features and climate of Mexico's coastlines.

Writing

Student interviews should note the benefits to Mexican farmers of increased trade with the United States and Canada. Use Rubric 41, Writing to Express, to evaluate student work.

*G*eography for Life

Student maps should identify and describe the regions of Mexico. Captions should explain the relationships among each region's cultural, economic, and natural features. Use Rubic 20, Map Creation, to evaluate student work.

Social Studies Skills

1. d

2. Possible answer: Except for 1995, annual growth in both U.S. exports and imports has increased rapidly. However, the United States now imports more from Mexico than it exports to that country.

3. b

4. Possible answer: The rugged landscape probably makes transportation and communication difficult. As a result, settlement would also be difficult. On the other hand, the difficult terrain would make it difficult to track people trying to hide from authorities.

PORTFOLIO ACTIVITY

Student graphs will vary according to pollution in your area. Use Rubric 7, Charts, to evaluate student work.

Central America and the Caribbean

CHAPTER RESOURCE MANAGER

Objectives	Pacing Guide	Reproducible Resources
SECTION 1 **Natural Environments** (pp. 239–42)	• Identify the physical processes that have shaped the landforms of Central America and the Caribbean. • Describe the region's climate. • Analyze the natural resources and environmental hazards found in the region.	**Regular** 1 day **Block Scheduling** .5 day *Block Scheduling Handbook, Chapter 11*	**RS** Guided Reading Strategy 11.1 **PS** Readings in World Geography, History, and Culture 19
SECTION 2 **Central America** (pp. 244–48)	• Explain how Central America's history shapes the region today. • Describe the economic, political, and social conditions in Central America.	**Regular** 1 day **Block Scheduling** .5 day *Block Scheduling Handbook, Chapter 11*	**RS** Guided Reading Strategy 11.2 **RS** Graphic Organizer Activity 11 **PS** Readings in World Geography, History, and Culture 17 **E** Cultures of the World Activity: Region 2 **SM** Critical Thinking Activity 11: The Panama Canal
SECTION 3 **The Caribbean** (pp. 249–53)	• Describe some important events in the history of the Caribbean. • Analyze the cultural and population patterns found in the region. • Discuss the economic activities of the region.	**Regular** 1 day **Block Scheduling** .5 day *Block Scheduling Handbook, Chapter 11*	**RS** Guided Reading Strategy 11.3 **PS** Readings in World Geography, History, and Culture 18 **E** Cultures of the World Activity: Region 2 **SM** Geography for Life Activity 11: The Geography of Caribbean Music **SM** Map Activity 11: Colonies and Independent States

Chapter Resource Key

PS Primary Sources

RS Reading Support

IC Interdisciplinary Connections

E Enrichment

SM Skills Mastery

A Assessment

REV Review

ELL Reinforcement and English Language Learners

 Transparencies

 CD-ROM

 Video

 Internet

 Holt Presentation Maker Using Microsoft® PowerPoint®

 One-Stop Planner CD–ROM

See the *One-Stop Planner* for a complete list of additional resources for students and teachers.

One-Stop Planner CD–ROM

It's easy to plan lessons, select resources, and print out materials for your students when you use the **_One-Stop Planner CD–ROM with Test Generator_**.

internet connect

HRW ONLINE RESOURCES

GO TO: go.hrw.com
Then type in a keyword.

TEACHER HOME PAGE
   KEYWORD: SW3 Teacher

CHAPTER INTERNET ACTIVITIES
   KEYWORD: SW3 GT11
   Choose an activity to:
   • create a postcard of an ecotour.
   • create a poster of the Panama Canal.
   • take the GeoMap challenge to test your knowledge of Central American and Caribbean geography!

CHAPTER ENRICHMENT LINKS
   KEYWORD: SW3 CH11

CHAPTER MAPS
   KEYWORD: SW3 MAPS11

ONLINE ASSESSMENT
   Homework Practice
   KEYWORD: SW3 HP11
   Standardized Test Prep
   KEYWORD: SW3 STP11
   Rubrics
   KEYWORD: SS Rubrics

COUNTRY INFORMATION
   KEYWORD: SW3 Almanac

CONTENT UPDATES
   KEYWORD: SS Content Updates

HOLT PRESENTATION MAKER
   KEYWORD: SW3 PPT11

ONLINE READING SUPPORT
   KEYWORD: SS Strategies

CURRENT EVENTS
   KEYWORD: S3 Current Events

Technology Resources	Reinforcement, Review, and Assessment
One-Stop Planner CD–ROM, Lesson 11.1	**ELL** Main Idea Activity 11.1
Geography and Cultures Visual Resources 18–22	**ELL** English Audio Summary 11.1
Homework Practice Online	**ELL** Spanish Audio Summary 11.1
HRW Go site	**REV** Section 1 Review, p. 242
	A Daily Quiz 11.1
One-Stop Planner CD–ROM, Lesson 11.2	**ELL** Main Idea Activity 11.2
CNN Presents Geography: Yesterday and Today, Segment 10: Preserving Heritage	**ELL** English Audio Summary 11.2
CNN Presents World Cultures: Yesterday and Today, Segment 15: Worlds Meet in the Americas	**ELL** Spanish Audio Summary 11.2
Homework Practice Online	**REV** Section 2 Review, p. 248
HRW Go site	**A** Daily Quiz 11.2
One-Stop Planner CD–ROM, Lesson 11.3	**ELL** Main Idea Activity 11.3
ARGWorld CD–ROM	**ELL** English Audio Summary 11.3
CNN Presents World Cultures: Yesterday and Today, Segment 35: The History of Haiti	**ELL** Spanish Audio Summary 11.3
Homework Practice Online	**REV** Section 3 Review, p. 253
HRW Go site	**A** Daily Quiz 11.3

Chapter Review and Assessment

- Chapter 11 Test Generator (on the One-Stop Planner)
- Global Skill Builder CD–ROM
- HRW Go site
- **REV** Chapter 11 Review, pp. 254–55
- **REV** Chapter 11 Tutorial for Students, Parents, Mentors, and Peers
- **A** Chapter 11 Test (form A or B)
- **A** Alternative Assessment Handbook
- **A** Chapter 11 Test for English Language Learners and Special-Needs Students

Meeting Individual Needs

Ability Levels

Level 1 Basic-level activities designed for all students encountering new material

Level 2 Intermediate-level activities designed for average students

Level 3 Challenging activities designed for honors and gifted-and-talented students

English Language Learners Activities that address the needs of students with Limited English Proficiency

Central America and the Caribbean

Launch into Learning

Ask students if anyone in the class has experienced a hurricane or seen footage of one on television. Call on volunteers to describe this type of storm. Tell students that hurricanes strike Central America and the Caribbean nearly every year, causing major damage. For example, Hurricane Mitch in 1998 caused severe flooding. Entire towns were swept away, thousands of people died, and more than $8 billion of damage was done to property. Tell students that they will learn more about the environments of Central America and the Caribbean in this chapter.

Using the Physical-Political Map

Have students examine the map on the opposite page. Discuss the meaning of the term *isthmus* (*a narrow piece of land that connects two larger landmasses like a bridge*). Point out that Central America is an isthmus. Ask students what two large landmasses Central America connects (*North and South America*). Have students identify the region's main bodies of water, islands, and other physical features. Point out the strategic location of the Panama Canal.

Why We Should Know More

You might want to reinforce interest in Central America and the Caribbean region by pointing out the following facts:

▶ Many Central American and Caribbean immigrants have come to the United States to escape poverty and political problems.

▶ The United States has a long history of military and economic involvement in the region.

▶ Some forms of music students might enjoy, such as reggae and salsa, originated in the Caribbean region.

▶ Historic sites, mild weather, and sunny beaches attract many American tourists to Central America and the Caribbean.

CHAPTER 11 Central America and the Caribbean

Central America and the Caribbean stretch across a large area of land and water between North and South America. Countries there share similar histories and face similar challenges today.

Mola design from Panama

Carnival in Trinidad

Good morning. I am Aldayne, and I live in Kingston, Jamaica. My house is gray—the door is green and the windows are blue. It is all on one floor, with a yard where we grow roses.

I am a student at Kingley Preparation School, about five or six minutes from my house. I get up at 6:00 A.M. and have a peanut butter sandwich and tea—or sometimes I just have an apple and tea at school. When school is over, I go downtown to my mother's shop—she is a dressmaker. I hang out until 6:00 P.M. when we can go home together. I do my homework and sometimes do errands for her like buying buttons. By the time I get home, all my homework is done. While my mother makes dinner, I clean my shoes, get my clothes together, and get ready for the next day. My favorite dinners are rice and peas with spring chicken, or crispy fried chicken with chips (french fries).

We have three school terms a year starting in September. We go to school for three months and then have one month off. My favorite vacation month is summer when I can go to the country with my father. I go to the river, catch fish, and have cook-outs. When I grow up, I want to be a pilot.

Section 1

OBJECTIVES

1. Identify the physical processes that have shaped the landforms of Central America and the Caribbean.

2. Describe the region's climate.

3. Analyze the natural resources and environmental hazards found in the region.

 **LET'S GET STARTED**

Copy the following questions onto the chalkboard: *What are the largest countries of Central America? Which two major island groups are found in the Caribbean? How do they differ?* Have students use the chapter map to answer the questions. Discuss student responses. *(Guatemala, Honduras, Nicaragua; Greater Antilles, Lesser Antilles; larger islands in the Greater Antilles)* Tell students that in Section 1 they will learn more about the natural environments of Central America and the Caribbean.

Building Vocabulary

Write **mangrove** and **bauxite** on the chalkboard. Call on volunteers to locate their definitions in the text and to read them aloud. Ask students where and why they would expect to find mangroves in the United States *(Gulf Coast because of warm subtropical environment, Florida, Hawaii).* Then ask students to name some common products made from aluminum, which comes from bauxite *(airplanes, building materials, consumer goods such as cans).*

Section 1

Natural Environments

READ TO DISCOVER

1. What physical processes have shaped the landforms of Central America and the Caribbean?

2. What is the climate like in Central America and the Caribbean?

3. What natural resources and environmental hazards are common in the region?

WHY IT MATTERS

Central America and the Caribbean experience dangerous natural hazards. Use CNNfyi.com or other **current events** sources to learn about the effects of a recent earthquake, hurricane, or volcanic eruption in the region.

DEFINE

mangrove

bauxite

LOCATE

West Indies	Jamaica
Greater Antilles	Puerto Rico
Lesser Antilles	Bahamas
Cuba	Martinique
Hispaniola	Cayman Islands

Section 1 RESOURCES

REPRODUCIBLE

▶ Guided Reading Strategy 11.1

▶ Readings in World Geography, History, and Culture 19

TECHNOLOGY

▶ One-Stop Planner CD–ROM, Lesson 11.1

▶ Geography and Cultures Visual Resources 18–22

▶ Homework Practice Online

▶ HRW Go site

REINFORCEMENT, REVIEW, AND ASSESSMENT

▶ Main Idea Activity 11.1

▶ English Audio Summary 11.1

▶ Spanish Audio Summary 11.1

▶ Section 1 Review, p. 242

▶ Daily Quiz 11.1

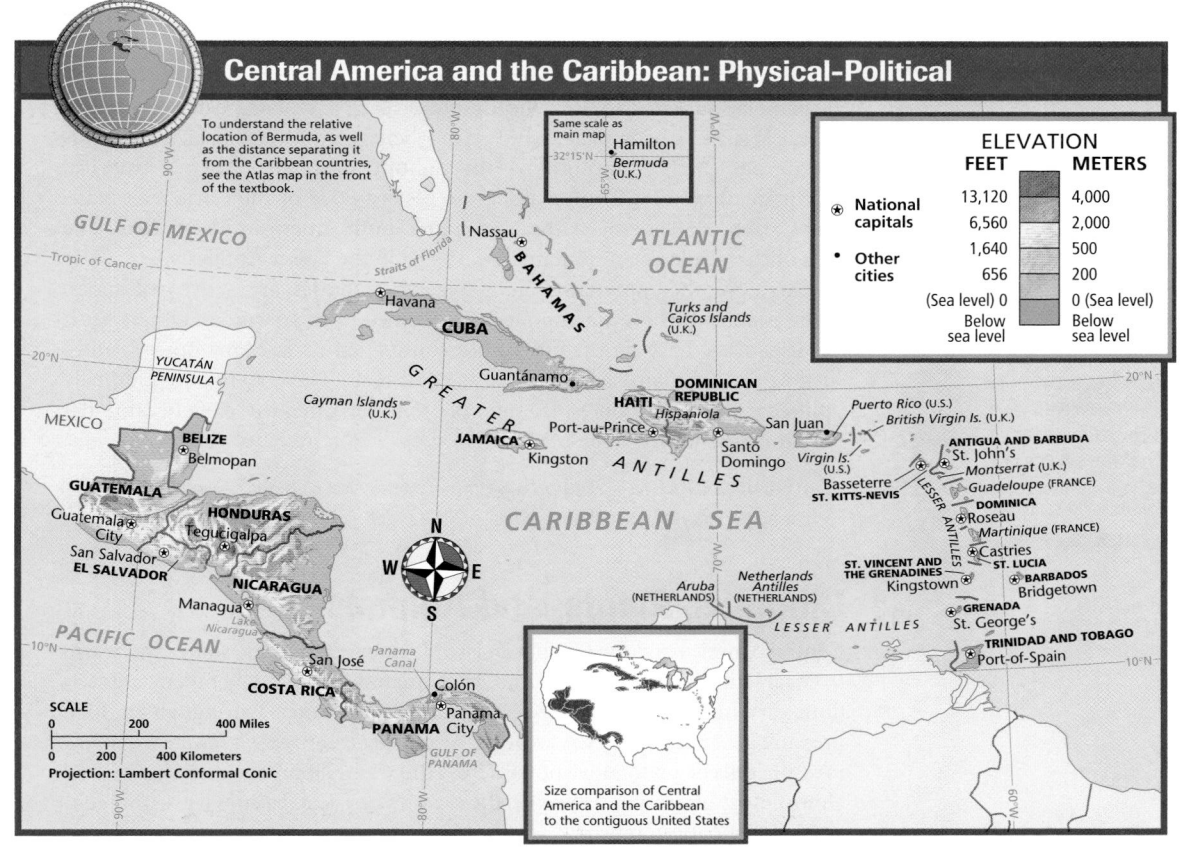

Central America and the Caribbean: Physical-Political

To understand the relative location of Bermuda, as well as the distance separating it from the Caribbean countries, see the Atlas map in the front of the textbook.

ELEVATION

	FEET	METERS
National capitals	13,120	4,000
	6,560	2,000
Other cities	1,640	500
	656	200
(Sea level) 0	0 (Sea level)
Below sea level	Below sea level

Size comparison of Central America and the Caribbean to the contiguous United States

SCALE
0   200   400 Miles
0   200   400 Kilometers
Projection: Lambert Conformal Conic

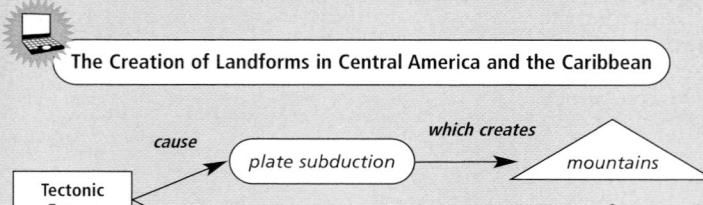

LEVEL 1: Copy the following graphic organizer onto the chalkboard, omitting the italicized answers. Have students complete the chart. **COOPERATIVE LEARNING**

LEVELS 2 AND 3: Have students use atlases to compare the physical geography of the Bahamas and Dominica *(Bahamas—many low islands, no rivers; Dominica—one mountainous island, many rivers)*. Then have students develop hypotheses about what physical processes formed the different islands. Have students use encyclopedias to check the accuracy of their theories. *(The Bahamas were formed from coral remains and limestone. Dominica was formed by volcanic processes.)*

The Creation of Landforms in Central America and the Caribbean

cause → plate subduction → *which creates* → mountains

Tectonic Forces

volcanic eruptions → new islands

Eye on Earth

The Belize Atolls Not all the landforms of Central America and the Caribbean were created by tectonic activity. Just off the coast of Belize are three oval, island-like bodies called atolls. An atoll is a coral formation, usually ring- or horseshoe-shaped, that can be large enough for human habitation. Enclosed within the atoll is a shallow lagoon. The Belize atolls began forming some 70 million years ago. They are home to more than 60 species of stony coral and 200 species of fish. The clear Caribbean waters near Belize make the atolls a favorite destination for divers.

Atolls are generally found only in the Pacific and Indian Oceans. Only four exist in the Western Hemisphere. The Belize atolls, the barrier reef to their west, and the vegetation native to the atolls and reef constitute a rich and fragile ecosystem.

INTERPRETING THE VISUAL RECORD

The beautiful twin peaks near Soufrière, St. Lucia, rise above the island's lush rain forests. **What physical processes do you think created these peaks?**

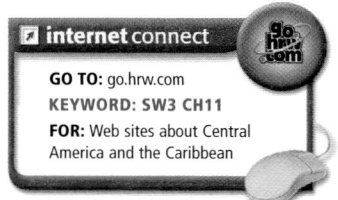

internet connect

GO TO: go.hrw.com
KEYWORD: SW3 CH11
FOR: Web sites about Central America and the Caribbean

Landforms

Central America is an isthmus that links North and South America. The Pacific Ocean lies to the west. To the east is the Caribbean Sea and a group of islands called the West Indies. These islands extend in an arc from just south of Florida to Venezuela. The major island groups of the West Indies are the Greater Antilles and the Lesser Antilles.

The Greater Antilles include the large islands of Cuba, Hispaniola, Jamaica, and Puerto Rico. Northeast of Cuba are some 700 islands that make up the Bahamas. They lie entirely in the Atlantic Ocean. Hispaniola is divided between the countries of Haiti and the Dominican Republic. The Lesser Antilles include more than 20 small island countries and territories. Central America has seven countries. Beginning in the north, these countries are Belize, Guatemala, Honduras, El Salvador, Nicaragua, Costa Rica, and Panama.

Narrow coastal plains are found in much of Central America and the Caribbean. Rugged hills or mountains lie in the interior. The rugged terrain makes travel and communication difficult. The region has few long rivers.

Tectonic processes have shaped the region. Central America, Jamaica, Hispaniola, and Puerto Rico lie on the Caribbean plate. West of Central America is the Cocos plate, which dives beneath the Caribbean plate. This action has created mountains throughout most of Central America. Mountains have also formed along the eastern edge of the Caribbean plate. The Caribbean plate meets the North and South American plates there. The tops of these volcanic mountains are islands in the Lesser Antilles.

Some of the mountains along plate boundaries are active volcanoes. Some islands, such as Martinique, were formed from volcanoes. Others, such as Hispaniola, have mountains of old continental rock. Many other islands, such as the Cayman Islands, began as coral reefs. These reefs were gradually uplifted by the collision of tectonic plates. Over time, the reefs became flat limestone islands.

✔ **READING CHECK:** **Physical Systems** What physical process is responsible for the creation of many islands in the Lesser Antilles? the meeting of two tectonic plates

Climates, Plants, and Animals

Central America and the Caribbean islands extend across the sunny and warm tropical latitudes. Tropical wet and dry climates are typical. Temperatures seldom vary more than 10°F between summer and winter. During winter, high pressure generally brings dry weather. A summer rainy season results when low pressure cells begin to move north across the region. Rain can then be expected almost every afternoon. However, the region's physical features cause this general climate pattern to vary.

Teach Objective 2

ALL LEVELS: Have students look at the chapter map and speculate on where different climates in the region are found. Have them note the latitudes of the region and major landforms that might influence the distribution of climates. Then have students examine the climate map in the unit atlas to find out exactly how climates are distributed in Central America and the Caribbean.

Teach Objective 3

ALL LEVELS: Ask students to identify the main natural hazards that threaten Caribbean islands and list them on the chalkboard (*earthquakes, hurricanes, volcanic eruptions*). Then ask students what natural resources are found in the region. List them on the chalkboard as well (*physical beauty and warm sunny climates, fertile soils, fishing grounds, mineral resources*). Finally, ask students to suggest how natural hazards might affect the environments and resources of Central America and the Caribbean. (*All the hazards can damage or destroy beautiful areas, which can disrupt tourism. Natural hazards can also damage crops and soils and destroy infrastructure needed for many economic activities.*)

In Central America, the climate zones follow the terrain. The Caribbean coast gets the full effect of moist trade winds. This results in a tropical humid climate with frequent rain, even in winter. Dense forests are common. The mountainous interior has a cooler highland climate. The eastern side of the mountains gets heavy rain as moist air rises and cools. However, mountain valleys in the west often lie in a rain shadow. These valleys and the western side of the mountains, along the Pacific coast, have drier climates. Scattered trees, shrubs, and tropical grasslands are found here. However, much of this vegetation has been cleared for plantations and ranches.

Throughout the Caribbean, elevation greatly affects climate. Islands with mountains and volcanoes often have heavy rain on their windward side. Rain shadows make other island locations somewhat drier. Lower islands do not produce any orographic effect. The rain that does fall quickly sinks into the limestone bedrock. As a result, the low islands have limited water resources. Bare ground and even cacti are common.

Thickets of **mangrove** trees dominate muddy tropical coastlines in Central America and all around the world. Mangroves are unusual because their roots grow in salt water. Shrimp and dozens of kinds of fish live among these roots. Marine life also thrives in shallow warm water and reefs off the Caribbean coast of Central America and the West Indies.

READING CHECK: **Physical Systems** How do mountains affect climates in the region? Mountains create rain shadows, with wet windward sides and drier leeward sides. Highlands are also cooler.

Natural Resources and Environmental Hazards

The region's natural resources include its physical beauty. However, this beautiful environment is also known for its dangerous environmental hazards.

Natural Resources One of the region's greatest natural resources is its warm and sunny climate. That climate attracts millions of tourists from the United States, Canada, and Europe. Fertile soils, particularly in highland valleys, are also important natural resources. Rich fishing grounds lie along the coasts. They provide fish, lobsters, shrimp, and other seafood to coastal and island populations.

Mineral resources are found in Central America and in the continental rocks of the Greater Antilles. Small gold fields have been discovered throughout Central America. Jamaica has major deposits of **bauxite**, the ore from which aluminum is made. In addition, Cuba and the Dominican Republic produce nickel. Trinidad has oil.

Climate Graph for San Salvador

San Salvador, El Salvador (14°N 89°W)
Tropical Wet and Dry Climate

INTERPRETING THE GRAPH *San Salvador is located at about 2,150 feet (655 m) above sea level, which helps moderate its temperatures. Heavy rains known as temporales fall from May to October.* **About how much rain does San Salvador receive during these months?**

Blue Hole, a circular limestone sink-hole, lies off the coast of Belize. Many divers come to explore this interesting formation.

Global Perspectives

Beach Erosion Caribbean beaches are vital to the region's economic and environmental health. Sandy shores protect areas farther inland from wave action and provide habitats for plants and animals. These beaches are essential elements in the region's profitable tourist industry. However, natural processes sometimes erode the beaches rapidly. Human actions can also cause beach damage.

ACTIVITY: Have students conduct research on various ways people have tried to prevent beach erosion. Which have been successful over the long term? Which have not stopped erosion? Which have actually speeded up the erosion process? Some students may want to conduct further research to determine which methods, if any, may slow beach erosion in specific areas of Central America or the Caribbean.

You don't say! One of the loudest of all animals, the Puerto Rican *coquí* frog can belt out at an amazing 100 decibels (db). (Noise is classified as physically painful above 120 db.) Just 1 inch (2.54 cm) long, the frog is disappearing from its rainforest home.

GRAPH ANSWER

about 64 inches

Close

Tell students that many beaches in the Caribbean are famous for having fine white sand. Point out to them that this sand is actually composed of eroded particles from coral reefs. Ask students to suggest other features of the region that attract tourists. *(Possible answers: sunshine, clear waters, warm temperatures)*

Review and Assess

Have students complete the **Section Review**. Then have students complete **Daily Quiz 11.1.**

Reteach

Have students complete **Main Idea Activity for English Language Learners and Special-Needs Students 11.1.** Then have students identify and describe major physical factors that constitute the region such as climates and vegetation. **ENGLISH LANGUAGE LEARNERS**

Extend

The Maya, one of the region's early civilizations, believed that rain gods called *Chacs* poured rain on Earth by emptying gourds and pots full of water. Have interested students find out more about how Maya beliefs were related to the natural environment. **BLOCK SCHEDULING**

Section 1 Review Answers

Define For definitions, see: mangrove, p. 241; bauxite, p. 241

Working with Sketch Maps
Maps will vary, but listed places should be labeled in their approximate locations. Central America, Jamaica, Hispaniola, and Puerto Rico lie on the Caribbean plate.

Reading for the Main Idea
1. tropical humid, highland, tropical wet and dry; latitude, elevation, mountain barriers

2. Martinique formed from volcanoes; Cayman Islands formed from uplifted coral reefs.

Critical Thinking
3. Answers will vary, but students should recognize that climate features can produce economic benefits such as attracting tourists. (NGS 4)

4. the Greater Antilles because most islands are farther from plate boundaries than Central America or the Lesser Antilles (NGS 7)

Organizing What You Know
5. Landforms—coastal plains, coral islands, rugged hills and mountains, volcanic islands; Climates—highland, tropical humid, tropical wet and dry; Resources—bauxite, fertile soils, fishing grounds, gold, nickel, oil, physical beauty and warm climates

On July 29, 1968, Costa Rica's Arenal Volcano erupted, killing more than 70 people and destroying the village of Pueblo Nuevo. Today the volcano regularly produces ash and lava, and tourists often watch its spectacular nighttime eruptions.

In 1835 a powerful volcanic eruption in Nicaragua spread ash as far as Mexico City, more than 850 miles (1,400 km) away.

Environmental Hazards Central America and the Caribbean have some of the most beautiful natural landscapes in the world. However, deadly natural hazards exist in the region as well. Earthquakes and volcanic eruptions often occur in Central America and in the Lesser Antilles. These earthquakes and eruptions have caused terrible destruction from time to time. For example, in 2001 two earthquakes just a month apart killed hundreds of people in El Salvador.

Hurricanes are another major hazard in the region. These tropical storms occur mostly in late summer. They bring destructive winds, heavy rains, and flooding. Damage has increased as populations have grown. This is true in the highlands, where farmers have cleared some slopes of forests. The heavy rains cause severe erosion and even mud slides in such areas. These mud slides can destroy homes, roads, and bridges. Crops and farm animals are also lost. In addition, it is difficult for aid to reach people needing food, shelter, and medical help.

✓ **READING CHECK:** *Environment and Society* Why does the clearing of forests make hurricanes more destructive? Their heavy rains cause severe erosion.

Section 1 Review

Homework Practice Online
Keyword: SW3 HP11

Define mangrove, bauxite

Working with Sketch Maps On a map of Central America and the Caribbean that you draw or that your teacher provides, label the West Indies, Greater Antilles, Lesser Antilles, Cuba, Hispaniola, Jamaica, Puerto Rico, Bahamas, Martinique, and Cayman Islands. What parts of the region lie on the Caribbean Plate?

Reading for the Main Idea

1. *Physical Systems* What climate types are found in Central America? What are some factors that influence these climates?

2. *Places and Regions* How and why does the physical geography of Martinique differ from that of the Cayman Islands?

Critical Thinking

3. *Supporting a Point of View* Do you think that the region's climate should be considered a natural resource? Why or why not?

4. *Making Generalizations and Predictions* Where are volcanic eruptions and earthquakes *least* likely to occur: Central America, the Greater Antilles, or the Lesser Antilles? Why?

Organizing What You Know

5. Copy the chart below and use it to identify important landforms, climates, and natural resources in the region.

Landforms	Climates	Resources

Science: Predicting Volcanic Eruptions

Since ancient times, people have been fascinated by volcanoes. In the 1800s this curiosity was formalized into a field of study known as volcanology, a branch of geology. Volcanologists study the formation, distribution, classification, and eruption of volcanoes. One of the primary goals of their study is to find some way to predict volcanic eruptions.

Millions of people live in or near sites of volcanic activity, so the potential for an eruption to affect large numbers of people is great. A warning system that could accurately predict future eruptions could save many lives.

Scientists have learned how to use tools like laser range finders, satellites, seismographs, and spectrometers to predict the likelihood of eruptions. These methods are very helpful, but they are also expensive. Some active volcanoes have not erupted in hundreds or even thousands of years but could still erupt in the future. It is not possible for volcanologists to monitor every site. Furthermore, some conditions that seem to indicate an impending eruption do not actually result in any volcanic activity. Encourage students to conduct research on recent advances in volcano-eruption prediction and debate whether scientists will ever be able to make accurate predictions.

Environment and Society

Geography for Life

Montserrat's Soufrière Volcano

You may have seen dramatic movies about volcanoes and other natural disasters that destroy entire cities. In many places around the world, such destruction is not a fictional threat. Take, for example, Montserrat (mawnt-ser-RAHT), a small island in the Lesser Antilles. Montserrat is a possession of Great Britain. On the southern part of the island sits a volcano known as Soufrière (soo-free-ER). For most of Montserrat's history, this volcano has been inactive. Islanders lived a quiet life, and tourism and agriculture provided the basis for the economy.

In 1995, however, the Soufrière Volcano began a series of major eruptions. The eruptions threw superheated clouds of ash, rock, and steam into the air. Debris rained down on the island, and piles of ash collected in many places. Lava and other material raced down the volcano at nearly 100 miles (160 km) an hour, mowing down trees.

As you might expect, the spectacular eruptions completely disrupted life on the island. Thousands of people had to flee their homes and move to the northern part of

The eruption of the Soufrière Volcano buried Montserrat's capital, Plymouth, in mud and ash.

the island. In the evenings, they listened to radio updates on the volcano's daily activity. Hotels, restaurants, and other businesses closed their doors. The capital of Plymouth, located near the volcano, was abandoned and was later buried in mud and ash.

The government set up an exclusion zone on the southern half of the island. Authorities limited access to this zone to short trips. However, eventually some people stayed in the exclusion zone for longer time periods. That was a mistake. In 1997 an eruption killed at least 19 people. Eventually, some two thirds of the population fled the island. Most migrated to other Caribbean islands, Britain, or Canada.

Montserrat

ATLANTIC OCEAN

CARIBBEAN SEA

Cudjoe Head
St. John's
St. Peter's
Northern Zone
Old Towne
Salem
Harris
Daytime Entry Zone
Cork Hill
Dyer's
Exclusion Zone
Long Ground
SOUFRIÈRE VOLCANO
Plymouth
St. Patrick's
Maritime Exclusion Zone

SCALE
0   2   4 Miles
0   2   4 Kilometers
Projection: Mercator

INTERPRETING THE MAP The eruption of the Soufrière Volcano forced the evacuation of more than half of Montserrat. **Why do you think the eruption of the volcano disrupted life on so much of the island?**

Applying What You Know

1. **Summarizing** How have the eruptions of the Soufrière Volcano affected the physical and human geography of Montserrat?

2. **Problem Solving** What, if anything, do you think should be done to make Montserrat safe for its inhabitants?

243

LET'S GET STARTED

Copy the following statement and questions onto the chalkboard: *Central America is home to many different ethnic groups. What are some of these groups? How have they contributed to Central American culture?* Discuss responses. *(Possible answers: mestizos, mulattoes, Indians, Africans; introduced foods, arts, languages, religions, and other culture traits to the region)* Tell students that they will learn more about the cultural and ethnic diversity of Central America in Section 2.

Building Vocabulary

Write the key terms on the chalkboard and have volunteers read the definitions aloud from the text or glossary. Then ask students to explain how the terms **cacao**, **ecotourism**, and **indigenous** might relate to the United States. *(Possible answers: Chocolate, which comes from cacao, is popular in the United States. Ecotourism is common in scenic areas of the United States. Native Americans were indigenous here.)*

Section 2 Central America

READ TO DISCOVER

1. How does Central America's history continue to shape the region today?
2. What economic, political, and social conditions exist in the region?

WHY IT MATTERS

Much of the food that we eat and clothing that we wear comes from Central America. Use **CNNfyi.com** or other **current events** sources to learn about goods imported to the United States from Central America.

DEFINE

indigenous
mulattoes
cacao
ecotourism

LOCATE

Panama Canal
Colón
Panama City

History and Culture

Together, the seven countries of Central America are only about three fourths the size of Texas. However, the combined population of these small countries is almost double that of Texas. Their colonial history continues to shape the culture of these countries today.

Colonization and Independence Nearly all of Central America's native, or **indigenous**, peoples were farmers when Spanish explorers arrived in the early

The Maya civilization included parts of present-day Guatemala, Belize, and Honduras. Tikal, in northern Guatemala, was a major Maya settlement that flourished between about A.D. 600 and 900.

Teach Objective 1

LEVEL 1: Copy the following graphic organizer onto the chalkboard, omitting the italicized answers. Have students complete the chart, using the text and pictures in this section for reference. **ENGLISH LANGUAGE LEARNERS**

LEVELS 2 AND 3: Ask students whether they think modern society in Central America was shaped more by indigenous peoples or Spanish settlers. Have them back their arguments with facts from the chapter.

Indigenous peoples
* *mostly farmers*
* *many enslaved or killed by disease*
* *Maya controlled parts of Guatemala, Belize, Honduras.*

Spanish colonists
* *introduced Spanish-style architecture and town planning, Catholicism, Spanish language*
* *mestizos a major ethnic group*
* *created unequal distribution of wealth*

1500s. Over time, the Central American Indian population declined dramatically. One of the causes of this decline was the spread of European diseases. In addition, the Spaniards enslaved many Indians.

Early Spanish settlement spread from the Pacific coast into the highlands. Settlers built towns around central plazas and Roman Catholic churches. Climates along the Pacific were compatible with Spanish-style agriculture, and large estates developed. A few rich families owned the land, and most of the workers had no land rights. The Spaniards mostly ignored the Caribbean coast. British Honduras (what is now Belize) became the only non-Spanish colony in Central America. Eventually, Europeans brought enslaved Africans from the Caribbean islands to the area.

Independence came to Spanish Central America in the 1820s. Little changed, however. Spanish officials left, but wealthy families continued to run the countries and their economies. Foreign companies, mainly from the United States and Great Britain, built railroads to cross the isthmus. Coffee plantations were founded along with railroads and became very important to the region. Bananas also became an important commercial crop. Large American firms controlled the banana business.

In the early 1900s the United States built the Panama Canal across central Panama. This canal has been an important economic resource. It allows ships to move from the Atlantic Ocean and Caribbean Sea to the Pacific Ocean. The United States controlled the canal until turning it over to Panama in 1999.

✓ **READING CHECK:** *Human Systems* Which European country had the greatest influence on Central America's development? Spain

People, Languages, and Religion The legacy of the colonial past continues in Central America. Wealth is still concentrated in the hands of a small number of families. The Roman Catholic Church remains important. Spanish is the official language of all the region's countries except Belize, where English is spoken. Many Central American Indian languages are also spoken.

The majority of Central Americans are mestizos. Some are **mulattoes**, people with both African and European ancestors. Small groups of Asians and Africans also live in the region. These groups are largely descended from laborers brought to the region to work on the plantations.

Most of the region's Central American Indians live in Guatemala. That country's population is almost evenly split between mestizos and Indians. The other countries have much smaller Indian populations. The populations of El Salvador, Honduras, Nicaragua, and Panama are overwhelmingly mestizo. Indians live on the Caribbean sides of Nicaragua and Panama. People of African descent are a significant group in Belize and Panama.

Costa Rica's people are mostly of Spanish descent. Few Central American Indians lived in the area when European colonists first arrived. As a result, the colonists did not have local labor to develop large estates. Instead, small family farms developed in this country.

INTERPRETING THE VISUAL RECORD

Coffee is one of Guatemala's main export crops. It is grown mostly on large plantations in cool highland areas. **To which countries do you think Guatemala exports coffee?**

Religions in Costa Rica

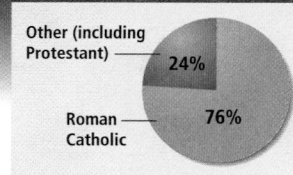

Other (including Protestant) — 24%

Roman Catholic — 76%

Source: *The DK World Desk Reference*

INTERPRETING THE GRAPH

Roman Catholicism is Costa Rica's main religion. **How might the importance of Catholicism in Costa Rica be visible in the country's landscapes?**

Cooperative Learning

Maya Calendars The Maya of Central America used two calendars—a ceremonial calendar of 260 days and a solar year calendar of 365 days. Together, these calendars described a recurring cycle of 18,980 days. By using their complex calendar, the Maya could predict astronomical events such as eclipses. The calendar also guided the observance of rituals.

ACTIVITY: Organize the class into groups to conduct research on aspects of the Maya calendar, such as the naming and numbering of dates, the time-measurement system called the Long Count (which extends back to 3114 B.C.), and how the calendar was depicted visually. Then have students locate and use a conversion site on the Internet to change the current day's date and their birthdays into Maya designation.

internet connect

GO TO: go.hrw.com
KEYWORD: SW3 CH11
FOR: Web sites about the Maya

VISUAL RECORD ANSWER

United States, European countries, Japan, Canada

GRAPH ANSWER

Possible answers: churches, cemeteries, place-names

245

LEVEL 1: Compile or have students compile a selection of newspaper or magazine articles concerning economic, political, and social conditions in Central America. Then lead a discussion relating the topics of these articles to the material in the text.

LEVEL 2: Have students use the articles gathered for the Level 1 activity and the text to write brief news reports about economic, political, or social conditions in an area of Central America. Call on volunteers to present their reports to the class.

LEVEL 3: Have students write letters to the editors of the publications from which articles were gathered for the Level 1 activity. Letters should respond to the articles in some way, either to the information contained in an article or to the way in which the information was presented. You may wish to provide students with examples of well-written letters from other publications on which they can model their own letters. Call on volunteers to share their completed letters with the class. Some students may even wish to mail their letters. Remind students to use standard grammar, spelling, sentence structure, and punctuation in their letters.

Cultural Kaleidoscope

Cacao Beans   When he returned to Spain, explorer Hernán Cortés brought with him cacao beans from Central America. He also brought a recipe for a new drink called *xocoatl*—or chocolate—made from the beans. Served with honey and sugar, it greatly impressed the members of the Spanish court. In fact, they were so impressed that they kept this wonderful bean a secret for 80 years. For years, Spanish monks were the only people in Europe who knew how to prepare chocolate. However, this was not a secret that could easily be kept under wraps. Today there are thousands of different recipes for chocolate.

ACTIVITY: Have students put together a chocolate recipe book. Encourage them to find chocolate recipes from the countries studied in this chapter. Then have students study how chocolate is used in European countries. How did the diffusion of chocolate to Europe affect the cuisines and economies of selected European countries?

VISUAL RECORD ANSWER

market economy

Our Amazing Planet

The Belize Barrier Reef is the largest coral reef in the Western Hemisphere, extending for more than 180 miles (290 km) along the country's coast. The reef is home to manatees, crocodiles, sea turtles, and more than 500 species of fish.

INTERPRETING THE VISUAL RECORD

Subsistence agriculture is common in the highlands of El Salvador. **Based on the photo, what type of economic system do you think El Salvador has?**

Belize was the last to gain independence in 1981. People of African descent who speak English live along the coast. The inland forests include both Indians and Spanish-speaking settlers from Mexico and Guatemala.

✔ **READING CHECK:**   *Human Systems*   What ethnic group has shaped the social character of Central America and makes up a majority of its population today? mestizos

Economic, Political, and Social Development

Central America continues to depend heavily on the export of coffee and bananas. Sugar, cotton, and **cacao** are also important commercial crops. Cacao is a type of tree from which we get cocoa beans. Those beans are then used to produce chocolate.

The influence of American and other foreign companies on these and other industries remains strong. In most countries, wealthy families have long had ties to foreign companies and to their own country's army. Such ties have often enabled these families to run their countries for their own benefit. However, population growth has strained this arrangement. As the people's need for land and demand for more political power have increased, unrest and violence have arisen. Many Central Americans have immigrated to the United States to escape these economic and political problems.

Reform and Development   The need for land reform has been an important problem in the region. In El Salvador, for example, rich landowners control most of the land. They raise cash crops on large profitable estates. However, most Salvadorans are poor subsistence farmers. They survive by raising corn and beans on small farms. Similar inequalities have troubled

Nicaragua. As a result, both countries suffered through long periods of violence and civil war, particularly in the 1970s and 1980s. With these countries now at peace, their governments have a chance to build fairer societies. In fact, El Salvador and Nicaragua have both made important economic progress in recent years.

Guatemala, which has also suffered through long periods of violence and unrest, has attempted land reform. Market-oriented agriculture exists along the country's Pacific coast. Meanwhile, more than half of all Guatemalans live and farm in isolated highland villages. In nearby market towns, they sell what few goods they do not consume. Migrants to the country's northern lowland plains have found some useful land there. However, these migrants are burning forest areas, including national parkland, to clear land for farming.

To some, Panama seems like three different countries in one. In the east, toward South America, is a densely forested region with few people. In the middle lies a more prosperous area surrounding the Panama Canal. The cities at each end of the canal—Colón on the Caribbean and Panama City on the Pacific—are major industrial centers. The country's western areas are more rural, and both small farms and large plantations are common.

Coffee is particularly important to the economies of Honduras, Guatemala, Panama, and Costa Rica. Rugged mountains and valleys dominate Honduras. The rough terrain makes transportation and large-scale farming difficult. As a result, economists often regard Honduras as Central America's least-developed country. Costa Rica, on the other hand, has the highest standard of living in the region. Costa Rica's tradition of democracy, education, and political stability has recently attracted investments by foreign computer companies. Computer chips are now a major export. Costa Rica has also been a leader in developing tourism.

✓ **READING CHECK:** (*Human Systems*) Which country has the highest standard of living in the region? What are some reasons for this? Costa Rica; tradition of education, democracy, political stability, foreign investment

FOCUS ON ECONOMICS

Ecotourism in Costa Rica One key to Costa Rica's successful tourism industry is the country's natural beauty. Would you pay to sit on a beautiful white-sand beach, where the rain forest extends nearly to the water's edge? Would you pay to walk in a misty mountain forest, surrounded by exotic birds? These are just two of many opportunities to get truly close to nature in Costa Rica.

INTERPRETING THE VISUAL RECORD

The Panama Canal, competed in 1914, is one of the most strategic artificial waterways in the world. **How do you think the Panama Canal caused changes in world trade patterns?**

A hiker in Costa Rica

Review Answers

Define For definitions, see: indigenous, p. 244; mulattoes, p. 245; cacao, p. 246; ecotourism, p. 248

Working with Sketch Maps
Maps will vary, but listed places should be labeled in their approximate locations. It allows shipping to move from the Atlantic Ocean to the Pacific Ocean.

Reading for the Main Idea
1. Spanish language and architecture, Catholicism, land ownership patterns, ethnic origins of many people

2. All three have market-oriented agriculture on large estates and poorer, smaller subsistence farms.

Critical Thinking
3. The canal draws shipping and related activities, which provide jobs. (NGS 11)

4. Answers will vary but should mention the unequal distribution of land and the social and political unrest that has resulted. (NGS 10)

Organizing What You Know
5. Belize—African descent, Indian; Costa Rica—mostly Spanish descent; El Salvador—mostly mestizo; Guatemala—Indian, mestizo; Honduras—mostly mestizo; Nicaragua—mostly mestizo, small Indian population; Panama—mostly mestizo, small Indian population, Africans

VISUAL RECORD ANSWER

It can encourage countries to preserve the natural environment for economic benefits.

248

INTERPRETING THE VISUAL RECORD

Visitors to Monteverde cloud forest in Costa Rica can explore the area's incredible diversity of plant and animal life by walking across suspension bridges located high in the rain-forest canopy. **How can ecotourism help promote economic development and environmental conservation at the same time?**

In the mid-1980s, Costa Rica found a way to make money and protect its natural environment at the same time. Its solution was **ecotourism**. This type of tourism focuses on guided travel through natural areas and on outdoor activities. It allows visitors to observe wildlife and learn about the environment. Today protected public or private nature preserves make up nearly 25 percent of Costa Rica's area. For a fee, people can tour these preserves with a naturalist who is an expert on the area's animals, plants, and physical geography.

Ecotourism has provided tremendous economic benefits to Costa Rica. Hotels and other businesses have sprung up around the preserves, providing jobs for the local people. The cost of tours also provides income for the economy. Today tourism is Costa Rica's leading industry. About two thirds of the country's vacationers are ecotourists.

✓ READING CHECK: **Environment and Society** How is Costa Rica's natural beauty a valuable resource? What policies have helped the country protect this resource? draws tourists; setting up public and private nature preserves

 Review

Homework Practice Online
Keyword: SW3 HP11

Define indigenous, mulattoes, cacao, ecotourism

Working with Sketch Maps On the map you created in Section 1, label the Panama Canal, Colón, and Panama City. Why has the Panama Canal been an important economic resource?

Reading for the Main Idea
1. **Human Systems** What evidence of Spanish colonization remains in Central America today?

2. **Environment and Society** How is the practice of agriculture similar in El Salvador, Guatemala, and Panama?

Critical Thinking
3. **Identifying Cause and Effect** Why do you think the region surrounding the Panama Canal is generally prosperous and industrial?

4. **Drawing Inferences and Conclusions** How do patterns of land use in Central America affect social, economic, and political conditions in the region?

Organizing What You Know
5. Copy the chart below. Use it to describe and compare the ethnic makeup of the Central American countries.

Belize	
Costa Rica	
El Salvador	
Guatemala	
Honduras	
Nicaragua	
Panama	

🎙️ *LET'S GET STARTED*

Copy the following question onto the chalkboard: *Which countries control island territories in the Caribbean? Use the chapter map to find out.* Discuss responses *(France, Netherlands, United Kingdom, United States)*. Point out that many independent Caribbean islands once belonged to these countries. Ask which cultural elements might reflect their influence in the region. *(Possible answers: clothing, food, place names, language, religion)* Tell students they will learn more about the history and culture of the Caribbean in Section 3.

Building Vocabulary

Write the key terms on the chalkboard and call on a volunteer to read the definitions aloud from the text or glossary. Point out to students that the terms **creole** and **voodoo** reflect the blending of cultures in the Caribbean. A **commonwealth** is politically tied to another country, like Puerto Rico is to the United States.

The Caribbean

READ TO DISCOVER

1. What are some important events in the history of the Caribbean?
2. What cultural and population patterns are found in the region?
3. What activities support the economies of the Caribbean countries?

WHY IT MATTERS

The Caribbean is a popular tourist destination for Americans. In addition, many people from the Caribbean have immigrated to the United States. Use **CNNfyi.com** or other **current events** sources to learn about emigration from the Caribbean or about tourism opportunities there.

IDENTIFY

Santeria
Caricom

DEFINE

commonwealth
creole
voodoo

LOCATE

Santo Domingo
Havana
San Juan

Section 3 RESOURCES

REPRODUCIBLE

▶ Guided Reading Strategy 11.3
▶ Readings in World Geography, History, and Culture 18
▶ Cultures of the World Activity: Region 2
▶ Geography for Life Activity 11: The Geography of Caribbean Music
▶ Map Activity 11: Colonies and Independent States

TECHNOLOGY

▶ One-Stop Planner CD–ROM, Lesson 11.3
▶ CNN Presents World Cultures: Yesterday and Today, Segment 35: The History of Haiti
▶ Homework Practice Online
▶ HRW Go site

REINFORCEMENT, REVIEW, AND ASSESSMENT

▶ Main Idea Activity 11.3
▶ English Audio Summary 11.3
▶ Spanish Audio Summary 11.3
▶ Section 3 Review, p. 253
▶ Daily Quiz 11.3

History and Culture

Like Central America, much of the Caribbean's modern identity is linked to the period of European colonization. European settlers and the peoples they brought to the region shaped the cultures of the islands.

Colonization and Independence In 1492 Christopher Columbus landed on an island in the southern Bahamas. However, he thought that he had reached islands off the coast of Asia that Europeans called the Indies. Columbus was wrong, but the term West Indies stuck. People still use the term to describe the islands of the Caribbean region.

Brimstone Hill Fortress on the island of St. Kitts takes its name from the volcanic stone, known as brimstone, that was used to build its massive walls. The fortress was used by both British and French forces during the colonial period.

Teach Objective 1

🌐 **LEVEL 1:** Copy the following list of peoples who settled in the Caribbean onto the chalkboard: enslaved Africans, Spanish, indigenous peoples, and other Europeans. Have students copy the list and rearrange it in order of the groups' arrivals in the region (*indigenous peoples, Spanish, other Europeans, enslaved Africans*). Then ask students to list some cultural traits brought to the islands by each group. (*Possible answers: indigenous peoples—language; Spanish— many crops and farm animals, language, religion; other Europeans—languages, religions, plantation system, sugar; enslaved Africans—languages, religions*)

🌐 **LEVEL 2:** Organize the class into groups. Assign each group one of the islands or island groups in the Caribbean. Have each group research its assigned topic and prepare a brief history of the island or islands, including such information as European settlements and dates of independence. Compile student findings into a time line of key dates in Caribbean history. **COOPERATIVE LEARNING**

🌐 **LEVEL 3:** Lead a discussion comparing the early histories of the Caribbean and United States. Ask students to note ways in which the histories are similar. (*Possible answers: settled by Europeans, slavery, struggle for independence*) Then explore the differences that exist between the two.

European colonists established plantations throughout the Caribbean, and export-oriented agriculture is still common on many islands. In Grenada, nutmeg and ginger are two important export crops.

The Spanish monarchs sponsored the voyage of Columbus, and the first wave of Europeans in the Caribbean was Spanish. They fanned out across the Greater Antilles looking for gold. The settlers brought bananas, citrus fruits, rice, sugarcane, and farm animals to the region.

When the Spanish found little gold, they lost interest in many of the small islands. However, the British, Dutch, and French competed for the islands. They reaped fabulous wealth from the sugar plantations they developed. Spain also developed sugarcane plantations in Cuba. To get workers for their farms, plantation owners turned to Africa. Over time, Europeans brought millions of enslaved Africans to the West Indies.

When slavery ended in the 1800s, former slaves could buy land if it was available. Some remained plantation workers. Others left the region, immigrating to Central America or the United States. Landowners then brought in laborers, mostly from South and East Asia, to work on the plantations.

Haiti won independence from France in 1804. The Dominican Republic gained independence from Spain by the mid-1800s. The United States took Cuba and Puerto Rico from Spain during the Spanish-American War in 1898. However, Cuba then became independent in 1902. Other Caribbean countries did not gain independence until the last half of the 1900s. Puerto Rico remains a **commonwealth** of the United States. A commonwealth is a self-governing territory associated with another country. Puerto Ricans are U.S. citizens. However, Puerto Rico has no voting representation in the U.S. Congress.

Indian, African, and European cultural traditions have combined to shape life throughout the Caribbean. In the Dominican Republic, for example, a blend of both Spanish and African cultures is found.

People, Languages, and Religions The Caribbean's population is largely descended from the Europeans and Africans who arrived during the colonial period. Most of

LEVELS 1 AND 2: Copy the following graphic organizer onto the chalkboard, omitting the italicized answers. Have students use information in the text and other sources to complete the chart. **ENGLISH LANGUAGE LEARNERS**

Cultural Influence in the Caribbean	
American influence	*Cuba, Puerto Rico, Virgin Islands*
British influence	*Antigua and Barbuda, Bahamas, Barbados, Bermuda, Dominica, Grenada, Jamaica, St. Kitts-Nevis, St. Vincent and the Grenadines, Trinidad and Tobago, Virgin Islands*
Dutch influence	*Netherlands Antilles*
French influence	*Grenada, Guadeloupe, Haiti, Martinique, St. Lucia*
Spanish influence	*Cuba, Dominican Republic, Haiti, Jamaica, Puerto Rico*

LEVEL 3: Organize the class into five groups. Assign each group one of the European spheres of influence listed in the previous activity. (You may wish to size the groups proportionally to the number of countries assigned to each sphere.) Have each group conduct research on the islands influenced by its assigned country to find elements of cultural influence present in the islands today. Then have each group share its findings with the rest of the class. **COOPERATIVE LEARNING**

the region's people are mulatto or otherwise of African descent. Nearly 40 percent of Cubans are of European descent. Haiti and Jamaica have the largest African populations in the region. Many descendants of Asian plantation workers live in Trinidad and Tobago. For example, more than a third of that country's people are East Indian. Small East Indian populations are also found in most other Caribbean countries. Jamaica and several smaller islands in the Lesser Antilles have small populations of Lebanese and Chinese as well. Most of the few Caribbean Indians who remain live in Dominica, St. Vincent and the Grenadines, Aruba, and the Netherlands Antilles.

The official language in each Caribbean country or territory depends on which European country colonized or controls it today. In most places, the main language is Spanish, English, French, or Dutch. Spanish is spoken in Cuba, the Dominican Republic, and Puerto Rico. English is also an official language in Puerto Rico. French and **creole** are official languages in Haiti. Creole is a blend of European, African, or Caribbean Indian languages. People on many other Caribbean islands often speak a creole language. In Aruba and the Netherlands Antilles, many people speak Papiamento. This creole language combines elements of Spanish, Dutch, and Portuguese.

Most people living on the Spanish- and French-speaking islands are Roman Catholic. Protestants make up the majority on the Dutch- and English-speaking islands. African traditions strongly influence religion in some places. In Haiti, for example, **voodoo** is important. Voodoo is a Haitian version of traditional African religious beliefs that are blended with elements of Christianity. Followers believe that good and bad spirits play an important role in daily life. Like voodoo, **Santeria** blends African traditions and Christian beliefs. It began in Cuba and has spread to other islands. In the 1800s Asian workers brought Hinduism and Islam to the islands.

Settlement and Land Use   Since 1960 the Caribbean's total population has more than doubled. About 36 million people lived in the region in 2000. About 70 percent of the population lives in Cuba, Haiti, and the Dominican Republic. Santo Domingo, the Dominican Republic's capital, is the region's largest city. More than 3.5 million people live there. Havana, Cuba's capital, is home to more than 2.2 million people.

Population growth has created difficult problems for the region. The Caribbean islands have only one tenth the land area of Central America. Unemployment and underemployment are high. This has led to immigration to the United States, Canada, and Europe. Large Caribbean communities can

Connecting to THE ARTS

Caribbean Music

Caribbean culture has contributed much to American music and dance. Calypso, a folk music from Trinidad, has ties to the music of Caribbean slaves of the early 1800s. It first became popular in the United States in the 1950s. More recently, reggae, salsa, and merengue have attracted a large American following. Reggae from Jamaica arrived in the 1970s. Salsa has its roots in the African rhythms and Spanish lyrics of Cuba and Puerto Rico. Merengue came to the United States from the Dominican Republic. It has long been part of rural folk music and dance traditions in that island country. Today people across the United States can find places to dance salsa and merengue.

Drawing Inferences and Conclusions   What other part of the world has influenced the Caribbean music that has become popular in the United States?

Steel drums, which originated in Trinidad, are a common element in Caribbean music. Steel drums are made from metal shipping drums and are played with rubber-tipped hammers.

Essential Element 4

▶ **Human Systems** ◀

The United States in the Caribbean   Puerto Rico has been a U.S. commonwealth since 1952. As such, its residents are citizens of the United States and can vote in presidential elections. Puerto Ricans are represented in Congress, although their commissioner does not vote. They also elect a local governor to head the island's government. Puerto Ricans do not pay federal income tax, but they do pay customs taxes. Periodically, Puerto Rico holds elections to determine its political future. In an election in 1998, the people chose to remain a commonwealth, though a large minority wanted full statehood. Some Puerto Ricans advocate complete independence from the United States.

The U.S. Virgin Islands are a U.S. territory. Residents there elect a local governor, but he or she must answer to the Office of Insular Affairs, a division of the federal Department of the Interior. Like Puerto Rico, the Virgin Islands are represented in Congress by a nonvoting commissioner. Residents do not vote in presidential elections.

CONNECTING TO THE ARTS ANSWER

Africa, Spain

251

Susan Walker of Beaufort, South Carolina, suggests the following activity to help students understand the cultures of Central America and the Caribbean. Organize students into groups and assign each group a musical style that originated in this region and has become popular in the United States. Musical styles that could be researched include calypso, merengue, reggae, and salsa. Have each group prepare a report on the musical style it researches. Encourage students to present their reports to the class and to include examples of music or dance in their presentations. **COOPERATIVE LEARNING**

Teach Objective 3

ALL LEVELS: Organize students into groups. Have students pretend that they are residents of fictional Caribbean islands. First ask each group to create a profile of its island based on the characteristics of real islands in the region. Then have group members discuss what sort of economic activities they should promote on the island, keeping in mind the resources available in the area. (*Possible answers: tourism, agriculture, high-tech industry*) Have students explain and justify their decisions. Lead a discussion comparing the economies of students' fictional islands with those of real islands in the Caribbean. **COOPERATIVE LEARNING**

Section 3 Review Answers

Identify For identifications, see: Santeria, p. 251; Caricom, p. 252

Define For definitions, see: commonwealth, p. 250; creole, p. 251; voodoo, p. 251

Working with Sketch Maps
Maps will vary, but listed places should be labeled in their approximate locations. Santo Domingo is the largest.

Reading for the Main Idea
1. population growth, unemployment

2. Soviet Union was a major buyer of Cuba's sugar. As a result, Cuba broadened its economy.

Critical Thinking
3. Answers will vary but should show an understanding of the benefits and drawbacks of tourism. (NGS 11)

4. Answers will vary but should note the influence of the colonial heritage, imported labor force, and plantation economy. (NGS 10)

Organizing What You Know
5. Caribbean—Dutch, French, U.S. territories; Protestant Christians; mix of Christian and African beliefs; variety of musical forms; Central America—largely Spanish-speaking; mostly mestizo; Shared—plantations; Catholicism in Spanish-speaking areas; people of African descent, mulattoes in some countries

VISUAL RECORD ANSWER

Most physical labor is performed by machines.

Since the 1960s the United States has enforced strict trade restrictions with Cuba because it opposes Cuba's Communist government. As a result, most American cars in the country date from the 1950s.

be found in those places. For example, today more Puerto Ricans live in New York City than in San Juan, Puerto Rico's capital.

This population pressure has increased urbanization. People have had to move to towns and cities in search of new opportunities. Today more than half the region's people live in towns and cities rather than farming communities. Partly because of this, most islands import much of their food.

✓ **READING CHECK:** *Places and Regions* What geographic factors have contributed to emigration from the Caribbean? small land area, population growth

Economic Development

Except for Cuba, most of the Caribbean countries and territories have market economies. Businesses and farms are privately owned, and their owners decide what and how much to produce. Cuba has a command economy. Its Communist government makes all the decisions about production.

Agriculture and Industry Despite efforts to encourage manufacturing, the region's economy has remained largely agricultural. Sugar, bananas, cacao, citrus fruits, and spices are the region's main exports. Still, overall farm production and employment continue to decline. To promote industry and trade, countries in the region have formed an economic union called **Caricom**, the Caribbean Community and Common Market. Difficulties, however, persist. Haiti, for example, remains one of the world's poorest nations. Its failure to solve its severe economic problems has also contributed to political instability. In 2004 armed rebels forced the country's president to resign.

Cuba, the largest Caribbean island, is not a member of Caricom. In 1959 Fidel Castro came to power and set up a Communist dictatorship. After Castro's 1959 revolution the United States banned trade with Cuba and restricted travel to the island. Only in 2000 did the U.S. government ease its ban on trade to allow Cuba to buy U.S. agricultural products. For years, Cuba produced sugar on government-owned farms to sell to the Soviet Union. However, that country's collapse has led Cuba to focus on the manufacture of farm machinery, steel, cement, clothing, food products, and consumer goods.

Cuba, Jamaica, Puerto Rico, and several countries in the Lesser Antilles have also developed important mining industries. Puerto Rico has the region's

INTERPRETING THE VISUAL RECORD

About three quarters of the land in Barbados is arable, and most of it is used to grow sugarcane. However, in recent years, the government has encouraged farmers to grow a wider range of crops. **Based on the photo, how important do you think modern technology is for growing sugarcane in Barbados?**

most industrialized economy. This is the result of Puerto Rico's easy access to the American market, low taxes, and special training programs for workers. In addition, wages in Puerto Rico are generally lower than in the United States. This means companies can produce goods less expensively there. Still, unemployment is high when compared to the United States.

Tourism   Many island leaders see tourism as the great hope for future economic growth. However, tourism also has its problems. On small islands, golf courses, resorts, and condominiums take land that could be used for farming or industry. Jobs in tourist industries are mostly seasonal, and pay is low. Furthermore, free-spending tourists often raise the cost of living on the islands. Plus, companies from developed countries build and operate most of the tourist facilities. As you might expect, then, most of the profits also go to these foreign companies. Still, tourism does bring needed income and reduces unemployment. Both of those benefits are important for island economies.

INTERPRETING THE VISUAL RECORD

Tourism is a huge industry in island countries such as the Bahamas, where it supplies about 60 percent of the country's GDP and employs about 40 percent of the labor force. **What problems might such a dependence on tourism cause?**

✔ **READING CHECK:** *Environment and Society*   What forms the basis of the economy in most of the Caribbean? agriculture

3 Review

Homework Practice Online
Keyword: SW3 HP11

Identify Santeria, Caricom

Define commonwealth, creole, voodoo

Working with Sketch Maps
On the map you created in Section 2, label the region's countries and Santo Domingo, Havana, and San Juan. Which city is the largest?

Reading for the Main Idea

1. *Human Systems*   What factors help to explain the high rate of emigration from some Caribbean islands?

2. *Human Systems*   How and why has the collapse of the Soviet Union affected the economy of Cuba?

Critical Thinking

3. Drawing Inferences   Do you agree that emphasizing tourism is a good approach to economic growth in the Caribbean countries? Why or why not?

4. Summarizing   How is the Caribbean's history reflected in its culture today?

Organizing What You Know

5. Copy the diagram. Use it to describe and then compare important cultural features of both the Caribbean islands and the countries of Central America. Where the circles overlap, identify cultural features that the regions share.

CHAPTER 11 Review and Assessment Resources

TECHNOLOGY
▶ Chapter 11 Test Generator (on the One-Stop Planner)
▶ Global Skill Builder CD–ROM
▶ HRW Go site

REINFORCEMENT, REVIEW, AND ASSESSMENT
▶ Chapter 11 Review, pp. 254–55
▶ Chapter 11 Tutorial for Students, Parents, Mentors, and Peers
▶ Chapter 11 Test (form A or B)
▶ Alternative Assessment Handbook

▶ Chapter 11 Test for English Language Learners and Special-Needs Students
▶ Unit 3 Test
▶ Unit 3 Test for English Language Learners and Special-Needs Students

Assess
Have students complete a Chapter 11 Test.

Reteach
Have students design a new national currency for a country from Central America or the Caribbean. Students should use their knowledge of the country to sketch designs for four denominations of bills and four denominations of coins. Then display the students' designs in the classroom. **ENGLISH LANGUAGE LEARNERS**

CHAPTER 11 Review Answers

Thinking Critically
1. Answers will vary but may include similarities such as tropical climates, mountains or rugged hills, coastal plains, European colonial legacy, economic challenges, population pressures, and limited land resources.

2. Answers should show that students understand the relationship between large-scale commercial agriculture and the need for transportation systems to reach markets. (NGS 16)

3. Answers will vary, but students should recognize that Central American and Caribbean Indian ethnicity, languages, religions, and other cultural influences would be stronger in the region. (NGS 10)

Using the Geographer's Tools

1. located on coastal plains along the Pacific coast of Central America and on the islands of the Greater and Lesser Antilles

2. yes; wet from May to October, dry from November to April, tropical wet and dry

3. Graphs should accurately represent the information in the unit table.

CHAPTER 11 Review

Building Vocabulary
On a separate sheet of paper, explain the following terms by using them correctly in sentences.

mangrove	cacao	voodoo
bauxite	ecotourism	Santeria
indigenous	commonwealth	Caricom
mulattoes	creole	

Locating Key Places
On a separate sheet of paper, match the letters on the map with their correct labels.

Greater Antilles	Puerto Rico	Santo Domingo
Lesser Antilles	Bahamas	Havana
Cuba	Panama Canal	

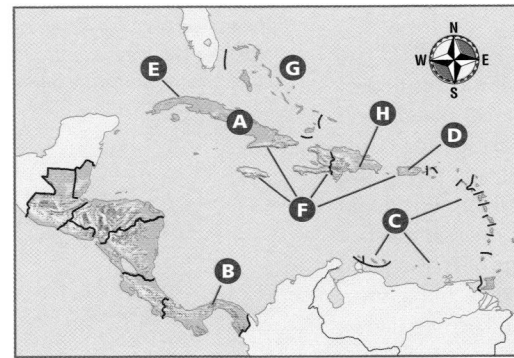

Understanding the Main Ideas
Section 1
1. *Physical Systems* What physical process has created the mountains of Central America? How do those mountains affect climates in Central America and the Caribbean islands?

Section 2
2. *Environment and Society* How are economics and politics linked in Central America?

3. *Human Systems* What economic activities are important in Central America? How is the economy of the region changing?

Section 3
4. *Places and Regions* How does Cuba differ economically from other Caribbean countries?

5. *Human Systems* What non-Caribbean countries have had the most influence in the region? Why?

Thinking Critically
1. Comparing What similarities exist between the physical and human geography of Central America and the Caribbean?

2. Identifying Cause and Effect Why would the construction of railroads in Central America have encouraged the spread of plantations and market-oriented agriculture?

3. Making Generalizations and Predictions What might cultures and economies in Central America and the Caribbean be like today if Europeans and North Americans had never become involved in the region?

Using the Geographer's Tools
1. Analyzing Maps Study the chapter physical-political map and the unit map showing land use and resources. What general conclusions can you draw about the location of plantation agriculture in the region?

2. Analyzing Graphs Review the climate graph for San Salvador in Section 1. Does the city have distinct wet and dry seasons? When do these occur? In which climate type would you expect to find this city?

3. Creating Bar Graphs Using the unit Fast Facts table, create bar graphs comparing the per capita GDPs of the countries in Central America and the Caribbean.

Writing about Geography
Describe and compare the patterns of language and religion that distinguish Central America and the Caribbean as separate cultural regions. Write a short essay about how the two regions differ in these areas of culture. When you are finished with your essay, proofread it to make sure that you have used standard grammar, spelling, sentence structure, and punctuation.

SKILL BUILDING

Geography for Life

Organizing Geographic Information
Places and Regions Imagine that you are an official in charge of promoting tourism in a Central American or Caribbean country of your choice. Use the textbook, library, and Internet resources to collect information and materials for a tourist brochure. Then use maps, pictures, or other graphics and text information to identify physical and human geographical features that will attract tourists to your country.

Building Social Studies Skills

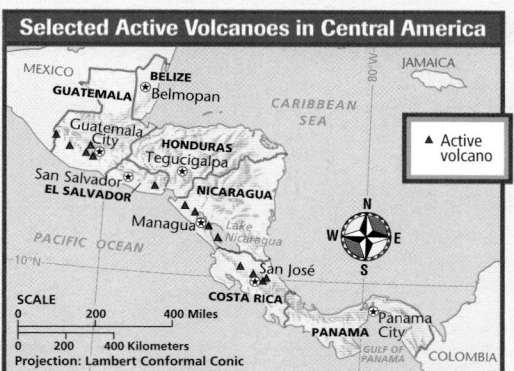

Selected Active Volcanoes in Central America

Interpreting Maps

Study the map above. Then answer the questions that follow.

1. Based on the map, which country does not have active volcanoes?
   a. Nicaragua
   b. Guatemala
   c. Costa Rica
   d. Belize

2. Which Central American countries would you expect to be most affected by earthquakes? Why?

Using Language

The following passage contains mistakes in grammar, punctuation, or usage. Read the passage and then answer the following questions. Mark your answers on a separate sheet of paper.

(1) European colonists established large plantations in Central America. (2) They grew crops like tobacco and sugarcane. (3) They forced the Central American Indians to work on plantations. (4) In gold mines elsewhere in the Americas some Indians were sent to work. (5) In addition many Africans were brought to the region as slaves.

3. In which sentence is a comma missing?
   a. 1
   b. 3
   c. 4
   d. 5

4. In which sentence does a prepositional phrase need to be moved so that it is closer to the verb it modifies?
   a. 1
   b. 2
   c. 4
   d. 5

5. Combine sentences 1 and 2 correctly.

Writing

Student essays should accurately describe the patterns of language and religion in Central America and the Caribbean. Use Rubric 9, Comparing and Contrasting, to evaluate student work.

Geography for Life

Students' brochures should reflect the results of independent research. Graphics should support written text. Use Rubric 2, Advertisements, to evaluate student work.

Social Studies Skills

1. d
2. Costa Rica, El Salvador, Guatemala, Nicaragua; These four countries have the most volcanoes and are probably most affected by earthquakes.
3. d
4. c
5. European colonists established large plantations in Central America and grew crops like tobacco and sugarcane.

Alternative Assessment

PORTFOLIO ACTIVITY

Learning about Your Local Geography

Group Project: Field Work

Plan, organize, and complete a research project about products from Central America and the Caribbean that are available in your community. Divide your community into zones. Have a group member survey stores in each zone to determine which products from Central America and the Caribbean—such as clothing, electronics, and food items—are sold there. Present your group's results on a sketch map of your community, showing the items that your group found in each location. Use the unit map showing land use and resources as a model for how to present your group's findings on your community sketch map.

✓ internet connect

Internet Activity: go.hrw.com
KEYWORD: SW3 GT11

Choose a topic on Central America and the Caribbean to:
• create a postcard of an ecotour.
• visit the Panama Canal and learn about its history.
• take the GeoMap challenge to test your knowledge of Central American and Caribbean geography!

PORTFOLIO ACTIVITY

Maps should accurately reflect the appearance of the community and the results of student research. Use Rubric 19, Map Creation, to evaluate student work.

CHAPTER RESOURCE MANAGER

Objectives	Pacing Guide	Reproducible Resources
SECTION 1 **Natural Environments** (pp. 257–61) • Identify the major landforms and rivers of South America. • Describe the climates, plants, and animals found in South America. • Tell what natural resources can be found on the continent.	**Regular** 1 day **Block Scheduling** .5 day *Block Scheduling Handbook, Chapter 12*	**RS** Guided Reading Strategy 12.1
SECTION 2 **History and Culture** (pp. 263–67) • Identify important events in the early history of South America. • Explain how the colonial era and independence affected South America. • Describe some important features of South America's cultures.	**Regular** 1 day **Block Scheduling** .5 day *Block Scheduling Handbook, Chapter 12*	**RS** Guided Reading Strategy 12.2 **E** Cultures of the World Activity: Region 2 **SM** Critical Thinking Activity 12: The Incas and Machu Picchu
SECTION 3 **South America Today** (pp. 268–71) • Explain what the economy of South America is like today. • Describe the cities of South America. • Identify the issues and challenges facing the people of South America.	**Regular** 1 day **Block Scheduling** .5 day *Block Scheduling Handbook, Chapter 12*	**RS** Guided Reading Strategy 12.3 **RS** Graphic Organizer Activity 12 **PS** Readings in World Geography, History, and Culture 20–27 **E** Creative Strategies for Teaching World Geography, Lesson 8 **SM** Geography for Life Activity 12: Latin American Urbanization **SM** Map Activity 12: Drug Traffic

Chapter Resource Key

PS	Primary Sources	**A**	Assessment	CD–ROM
RS	Reading Support	**REV**	Review	Video
IC	Interdisciplinary Connections	**ELL**	Reinforcement and English Language Learners	Internet
E	Enrichment		Transparencies	Holt Presentation Maker Using Microsoft® PowerPoint®
SM	Skills Mastery			

 One-Stop Planner CD–ROM

See the *One-Stop Planner* for a complete list of additional resources for students and teachers.

One-Stop Planner CD–ROM

It's easy to plan lessons, select resources, and print out materials for your students when you use the **One-Stop Planner CD–ROM with Test Generator**.

Technology Resources

- One-Stop Planner CD–ROM, Lesson 12.1
- *ARGWorld* CD–ROM
- Geography and Cultures Visual Resources 18–22
- Homework Practice Online
- HRW Go site

- One-Stop Planner CD–ROM, Lesson 12.2
- CNN. Presents World Cultures: Yesterday and Today, Segment 13: The Inca's Frozen Past
- Homework Practice Online
- HRW Go site

- One-Stop Planner CD–ROM, Lesson 12.3
- CNN. Presents Geography: Yesterday and Today, Segment 12: Destination Brazil
- Homework Practice Online
- HRW Go site

Reinforcement, Review, and Assessment

- **ELL** Main Idea Activity 12.1
- **ELL** English Audio Summary 12.1
- **ELL** Spanish Audio Summary 12.1
- **REV** Section 1 Review, p. 261
- **A** Daily Quiz 12.1

- **ELL** Main Idea Activity 12.2
- **ELL** English Audio Summary 12.2
- **ELL** Spanish Audio Summary 12.2
- **REV** Section 2 Review, p. 267
- **A** Daily Quiz 12.2

- **ELL** Main Idea Activity 12.3
- **ELL** English Audio Summary 12.3
- **ELL** Spanish Audio Summary 12.3
- **REV** Section 3 Review, p. 271
- **A** Daily Quiz 12.3

internet connect

HRW ONLINE RESOURCES

GO TO: go.hrw.com
Then type in a keyword.

TEACHER HOME PAGE
KEYWORD: **SW3 Teacher**

CHAPTER INTERNET ACTIVITIES
KEYWORD: **SW3 GT12**
Choose a topic on South America to:
- trek through the Guiana Highlands and create a poster.
- create a brochure on Machu Picchu.
- learn about threats to the Amazon Basin.

CHAPTER ENRICHMENT LINKS
KEYWORD: **SW3 CH12**

CHAPTER MAPS
KEYWORD: **SW3 MAPS12**

ONLINE ASSESSMENT
Homework Practice
KEYWORD: **SW3 HP12**
Standardized Test Prep
KEYWORD: **SW3 STP12**
Rubrics
KEYWORD: **SS Rubrics**

COUNTRY INFORMATION
KEYWORD: **SW3 Almanac**

CONTENT UPDATES
KEYWORD: **SS Content Updates**

HOLT PRESENTATION MAKER
KEYWORD: **SW3 PPT12**

ONLINE READING SUPPORT
KEYWORD: **SS Strategies**

CURRENT EVENTS
KEYWORD: **S3 Current Events**

Meeting Individual Needs

Ability Levels

Level 1 Basic-level activities designed for all students encountering new material

Level 2 Intermediate-level activities designed for average students

Level 3 Challenging activities designed for honors and gifted-and-talented students

English Language Learners Activities that address the needs of students with Limited English Proficiency

Chapter Review and Assessment

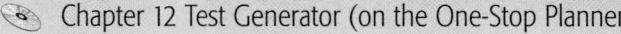

- Chapter 12 Test Generator (on the One-Stop Planner)
- Global Skill Builder CD–ROM
- HRW Go site
- **REV** Chapter 12 Review, pp. 274–75
- **REV** Chapter 12 Tutorial for Students, Parents, Mentors, and Peers
- **A** Chapter 12 Test (form A or B)
- **A** Alternative Assessment Handbook
- **A** Chapter 12 Test for English Language Learners and Special-Needs Students

Launch into Learning

Ask students to imagine that they are planning a new capital city in South America. Ask them if the capital should be near the country's densely populated coast or in the thinly settled interior. Discuss responses. Then tell students that in 1960 Brazil moved its capital to Brasília, a new city far inland. Ask in what shape students might draw the new city's limits. After a discussion, tell the class that according to tradition, the layout of the Inca capital of Cuzco resembled a mountain lion. Tell students that in Chapter 12 they will learn more about the cities and peoples of South America.

Using the Physical-Political Map

Have students examine the map on the opposite page. Call on volunteers to choose a South American country and to describe its physical features. Continue selecting countries and eliciting observations from students until all of the countries in South America have been described. Then call on students to compare the physical geography of different countries in South America. For example, how do Chile's landforms compare to those of Guyana? *(Possible answer: Chile has high mountains and many islands, while lowlands dominate Guyana's landscape, and no islands are visible.)*

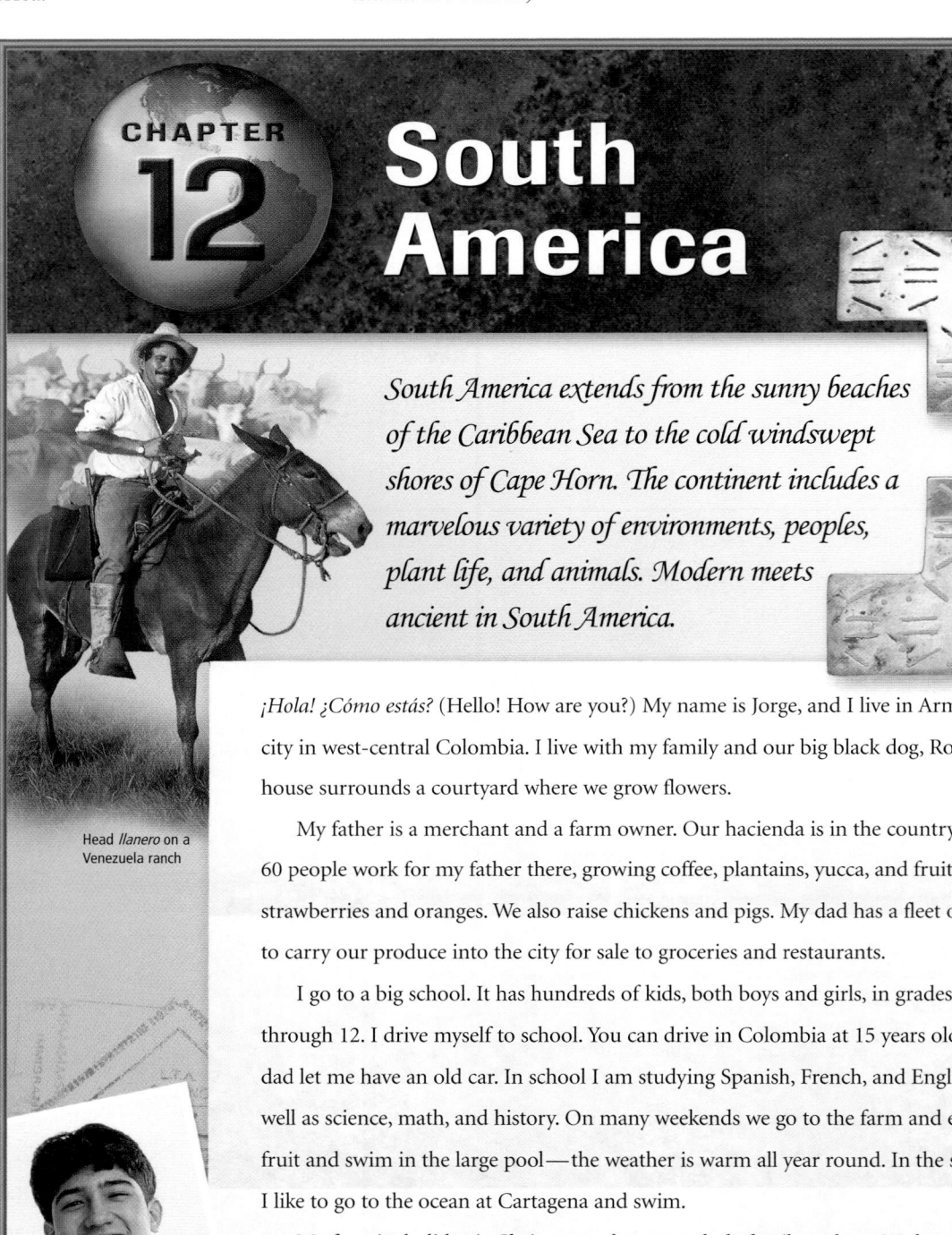

Why We Should Know More

You might want to share with your students the following reasons for learning about South America:

▶ The Pacific Ocean phenomenon known as El Niño affects the weather and economy of the United States.

▶ The United States and South America share some elements of history, such as periods of colonization and immigration, and some cultural traditions.

▶ The destruction of South America's rain forests has raised concern in the United States and may have global consequences.

▶ Peru, Bolivia, and Colombia are major sources of cocaine, which remains a problem in the United States.

▶ Copper prices in Chile, oil production in Venezuela, and other aspects of South America's economy affect the U.S. economy.

CHAPTER

12 South America

South America extends from the sunny beaches of the Caribbean Sea to the cold windswept shores of Cape Horn. The continent includes a marvelous variety of environments, peoples, plant life, and animals. Modern meets ancient in South America.

Gold chest ornament from Colombia

Head *llanero* on a Venezuela ranch

¡Hola! ¿Cómo estás? (Hello! How are you?) My name is Jorge, and I live in Armenia, a city in west-central Colombia. I live with my family and our big black dog, Rocca. Our house surrounds a courtyard where we grow flowers.

My father is a merchant and a farm owner. Our hacienda is in the country. About 60 people work for my father there, growing coffee, plantains, yucca, and fruits like strawberries and oranges. We also raise chickens and pigs. My dad has a fleet of trucks to carry our produce into the city for sale to groceries and restaurants.

I go to a big school. It has hundreds of kids, both boys and girls, in grades 1 through 12. I drive myself to school. You can drive in Colombia at 15 years old, and my dad let me have an old car. In school I am studying Spanish, French, and English, as well as science, math, and history. On many weekends we go to the farm and eat fresh fruit and swim in the large pool—the weather is warm all year round. In the summer, I like to go to the ocean at Cartagena and swim.

My favorite holiday is Christmas, when our whole family gathers. We have a big barbecue—a whole stuffed pig cooked in a large pit. One year, though, we had an earthquake after New Year's that lasted for 40 seconds. It was strange, because the earthquake moved like a wave. Some areas like the downtown were completely destroyed, while other areas, like my neighborhood, were not much affected.

Section 1

OBJECTIVES

1. Identify the major landforms and rivers of South America.

2. Describe the climates, plants, and animals found in South America.

3. Tell what natural resources can be found on the continent.

 **LET'S GET STARTED**

Copy the following question onto the chalkboard: *What images come to mind as you think of the natural environments of South America?* As students respond, list their ideas on the chalkboard. *(Possible responses include high mountains, tropical rain forests, great rivers, and plants used for medicines.)* Tell students that in Section 1 they will learn more about the landforms, rivers, climates, plants, animals, and resources of South America.

Building Vocabulary

Write the key terms on the chalkboard. Call on volunteers to read the definitions aloud. Ask students what types of plants might be found above the **tree line** (shrubs, bushes). Then ask students what might need to happen before **tar sands** become important energy sources. *(Possible answers: oil reserves depleted, technology for extraction)* Finally, ask students how the environments of *tepuís* may have influenced the types of plants found there. *(Possible answer: Inaccessibility led to the evolution of unique plants.)*

Natural Environments

READ TO DISCOVER

1. What are the major landforms and rivers of South America?
2. What climates, plants, and animals are found in South America?
3. What natural resources does the continent have?

WHY IT MATTERS

New businesses are developing products that use resources from South America's rain forests without harming the environment. Use **CNNfyi.com** or other **current events** sources to learn about these businesses and the products they sell.

IDENTIFY

Llanos
Gran Chaco
Pampas
El Niño
La Niña

DEFINE

tepuís
tree line
tar sands

LOCATE

Andes
Altiplano
Lake Titicaca
Guiana Highlands
Brazilian Highlands
Amazon River
Patagonia
Tierra del Fuego
Orinoco River
Paraná River
Río de la Plata
Atacama Desert
Lake Maracaibo

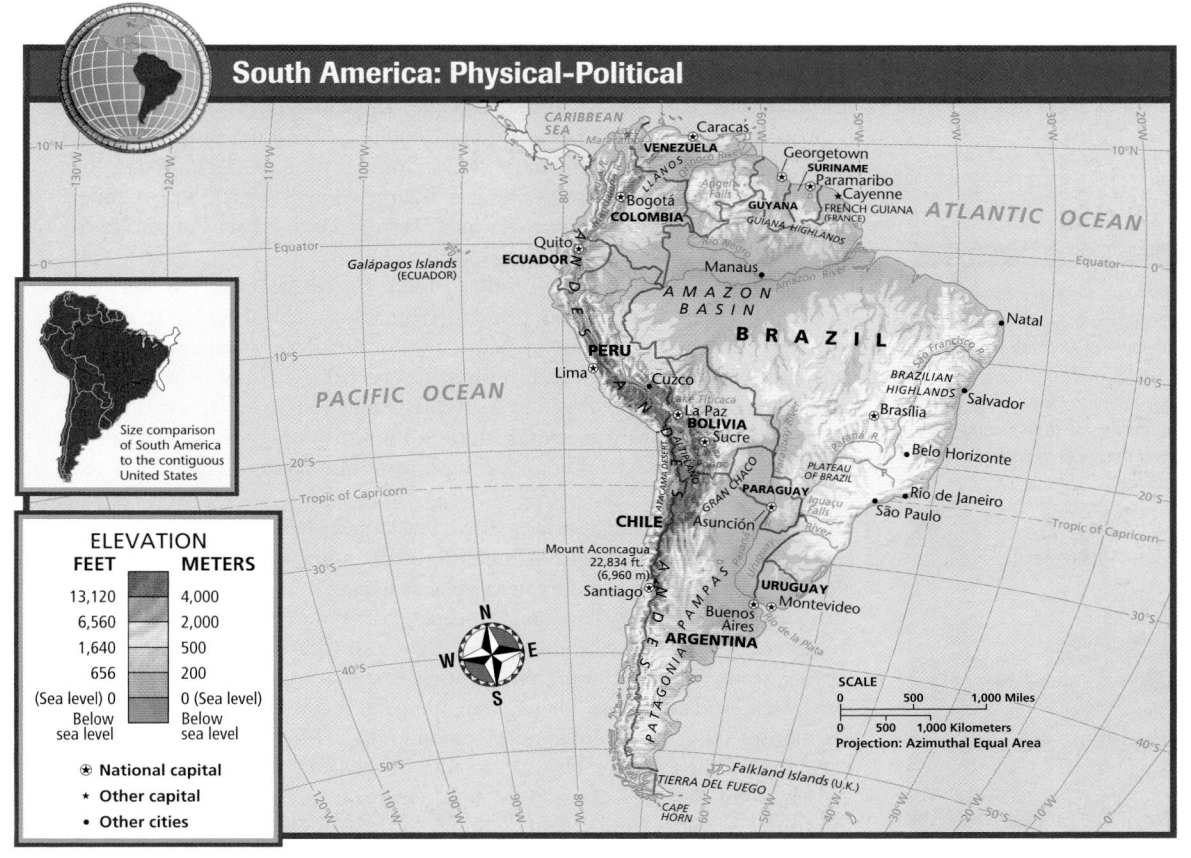

South America: Physical-Political

ELEVATION

FEET	METERS
13,120	4,000
6,560	2,000
1,640	500
656	200
(Sea level) 0	0 (Sea level)
Below sea level	Below sea level

⊛ National capital
★ Other capital
• Other cities

Size comparison of South America to the contiguous United States

SCALE
0   500   1,000 Miles
0   500   1,000 Kilometers
Projection: Azimuthal Equal Area

LEVEL 1: Copy the following graphic organizer onto the chalkboard, omitting the italicized answers. Call on students to provide words and phrases to identify and describe the landforms and rivers of South America. Then pair students and ask each pair to complete the organizer. When students have finished, call on volunteers to share their answers with the class. **ENGLISH LANGUAGE LEARNERS, COOPERATIVE LEARNING**

Landforms and Rivers
Andes
Altiplano
Guiana Highlands, tepuís
Brazilian Highlands
Amazon River basin, Llanos,
Gran Chaco, Pampas
Amazon, Orinoco, Paraná Rivers

Climates
tropical humid in Amazon Basin
highland in Andes
tropical wet and dry
Mediterranean in central Chile
marine west coase
semiarid and arid in Patagonia,
Atacama

South America

Biomes
tropical rain forest
dry forest
savannas
deserts

Natural Resources
soil, rubber, timber
hydroelectricity production
nuts and plants for medicine
gold, silver, copper, iron ore,
bauxite, oil, emeralds

Eye on Earth

Source of the Amazon

In 2000, geographers used Global Positioning System (GPS) technology to find what they believe is the source of the Amazon River. A small snow-fed stream on the upper slopes of Nevado Mismi in southern Peru marks the farthest point from which water flows continuously to the Amazon's mouth. The stream itself feeds into the Apurímac River, which then joins the Urubamba before it reaches the Amazon. However, locating the river's elusive source was easier than determining its exact length. The indistinct mouth of the Amazon, cluttered with many islands, makes exact measurement difficult. Geographers disagree on the question of where the river ends.

CRITICAL THINKING: Why might the Amazon's source be so difficult to find? *(Possible answer: It has many tributaries that form in remote areas.)*

internet connect

GO TO: go.hrw.com
KEYWORD: SW3 CH12
FOR: Web sites about South America

INTERPRETING THE VISUAL RECORD

The Altiplano is a high plain area between two ranges of the Andes. Few trees grow there. The southern Altiplano gets little rainfall. Llamas, which are related to camels, are native to the Altiplano. **What adaptations do you think llamas have developed that help them live in this environment?**

VISUAL RECORD ANSWER

Possible answers: appear to have thick fur; probably have other adaptations so they are able to live in thin atmosphere, on little water, and in severe cold

---

Landforms and Rivers

South America includes 12 countries and an overseas department of France—French Guiana. Brazil is the largest country. Colombia, Venezuela, Guyana, Suriname, and French Guiana lie along the continent's northern coasts. In the west, Ecuador, Peru, and Chile border the Pacific Ocean. Bolivia and Paraguay are cut off from the sea. Argentina and Uruguay lie along the Atlantic. The continent has bone-dry deserts, tropical rain forests, snowcapped mountains, high plateaus, and vast fertile grasslands.

Landforms South America's great mountain range, the Andes, extends along the continent's Pacific coast. The mountains stretch from the northern edge of the continent south to its tip. Mount Aconcagua, the highest peak in the range, rises to 22,834 feet (6,960 m). The collision of the Nazca and South American plates created the Andes. Continued tectonic activity causes volcanic eruptions and earthquakes.

In Peru and Bolivia the Andes divide into two great ranges. Between them lies an elevated plain known as the Altiplano. The name means "High Plateau" in Spanish. The Altiplano lies at about 12,000 feet (3,658 m) above sea level. More than 25 of the Altiplano's rivers drain into Lake Titicaca on the Peru-Bolivia border. This freshwater lake is large enough—3,200 square miles (8,288 sq km)—for small ships to navigate. Farther south is Lake Poopó, which is salty. Few people live along Poopó's banks.

In contrast to the more recently formed rugged Andes are the ancient eroded highlands of eastern South America. The Guiana Highlands rise in southern Venezuela. They stretch across part of northern Brazil and Guyana, Suriname, and French Guiana. Erosion there has left a chain of high plateaus, called *tepuís* (tay-PWEEZ), edged by high cliffs. Angel Falls in Venezuela tumbles 3,212 feet (979 m) from a *tepuí*, making it the world's highest waterfall. The Brazilian Highlands extend inland along Brazil's southeastern coast. They too are ancient and eroded.

Plains cover much of South America. The largest plain is the Amazon River basin, which occupies about 2 million square miles (5.3 million sq km). Northeastern Colombia and western Venezuela have a large plains area called the **Llanos** (YAH-nohs). This name means "Plains" in Spanish. Between the Andes and the Brazilian Highlands lies the **Gran Chaco** (grahn CHAH-koh). *Chaco* means "hunting land," and hunting is still a popular pastime in this mostly semiarid landscape. These plains are so flat that water sometimes stands for months after the summer rainy season ends. Farther south are the wide grasslands of the **Pampas**. The eastern edge of the Pampas is Argentina's most densely populated area. Erosion by both wind and water have carried fertile soil from the Andes to the Pampas. South of the Pampas, semiarid Patagonia

Angel Falls drops from the top of a tepuí. *The falls begin high above the clouds.*

LEVEL 2: Have students complete the Level 1 activity. Then ask students to imagine that they are Spaniards exploring South America for the first time in the early 1500s. Ask each student to write a journal entry about his or her explorations to describe the landforms, rivers, climates, plants, animals, and natural resources of the continent. Encourage students to use the key terms from the beginning of the section to describe some of the phenomena that they encounter.

LEVEL 3: Have students refine or revise their journal entries to include information that relates to the elevation zones as shown in the diagram. For example, if a student wrote primarily about exploring the Atacama Desert, ask him or her to "travel" farther inland and to proceed through the various elevation zones, describing the changes evident at the different levels.

stretches all the way to Tierra del Fuego. Tierra del Fuego is an island divided between Argentina and Chile.

Rivers   Only small rivers and streams flow west into the Pacific Ocean. However, three great river systems—the Amazon, Orinoco, and Paraná—drain the eastern part of the continent. The Amazon River is 4,000 miles (6,436 km) long. It is the world's largest river in volume, and no other river drains a larger area. So much water flows from the Amazon into the Atlantic Ocean that freshwater dilutes seawater more than 100 miles (161 km) from shore. Ocean-going ships can navigate up the Amazon for nearly 2,300 miles (3,700 km)—all the way to Iquitos (ee-KEE-tohs), Peru! The Orinoco River, which drains the western Guiana Highlands and the Llanos, also empties into the Atlantic Ocean. Several rivers together drain another large area far to the south. The longest of these rivers is the Paraná. The Paraná River flows into the Río de la Plata estuary between Argentina and Uruguay. The Paraná drains an area that includes the eastern slopes of the Andes and the highlands of eastern Brazil.

✓ **READING CHECK:**   *Physical Systems*   What qualities make the Amazon River unique?
largest river by volume, drains the largest area in the world

Climates, Plants, and Animals

Because South America extends across more than 60 degrees of latitude, the continent has a variety of climate regions. The Amazon River basin is the world's largest tropical humid climate region. It also has the largest tropical rain forest in the world. More than 150 inches (380 cm) of rain fall every year. Anacondas, bats, jaguars, monkeys, and countless other species live in the forest. Along the western edge of the basin, rain forests yield to the highland climates of the Andes. Environments of the central and northern Andes region can be divided into five zones, according to elevation. These zones range from hot and humid lands near sea level to frozen peaks high above the **tree line.**

Our Amazing Planet

South America's giant anacondas, or water boas, are the longest snakes in the Western Hemisphere. They are also the heaviest snakes in the world. Anacondas can be as thick as a telephone pole, measure more than 30 feet (10 m) long, and weigh up to 300 pounds (135 kg).

The Capybara   What the turkey is to Thanksgiving feasts in the United States the capybara is to Easter feasts in Venezuela. The capybara, also known as the water hog, is considered by some people to be a fish because it is an excellent swimmer and spends much time in the water. This belief allows people in Venezuela to eat capybaras during Lent, a time when many Roman Catholics do not eat meat.

The world's largest rodent, the capybara has webbed feet, can grow as long as 4 feet (1.2 m), and can weigh as much as 100 pounds (45 kg). Capybara hides are used to make gloves, and its bristles are made into brushes.

ACTIVITY: Have students conduct research on animals unique to South America and have them create brochures that present their results.

Elevation Zones in the Andes

Zone	Elevation	Description
Tierra helada	Above 16,000 feet (4,877 m)	Permanently covered with snow
Paramo	10,000 to 16,000 feet (3,048 to 4,877 m)	Potatoes, grasslands and hardy shrubs, grazing
Tree Line		
Tierra fria	6,000 to 10,000 feet (1,829 to 3,048 m)	Potatoes, wheat, oats, barley, beans, corn, rye
Tierra templada	3,000 to 6,000 feet (914 to 1,829 m)	Coffee, corn, wheat, cotton, potatoes, sugarcane, tobacco
Tierra caliente	Sea Level to 3,000 feet (914 m)	Bananas, cacao, rice, sugarcane
Sea Level		

INTERPRETING THE DIAGRAM   *Environments in the Andes change with elevation. Five different elevation zones are commonly recognized. In which zone is the Altiplano located?*

DIAGRAM ANSWER

Because it is about 12,000 feet above sea level, the Altiplano is located in the paramo.

HOMEWORK: Have students refer to the journal entries completed in the Level 2 or 3 exercise. Ask them to create illustrations for their journals. Drawings may include diagrams explaining physical processes such as El Niño, maps showing exploration routes, or depictions of plant and animal species seen in the areas explored. **ENGLISH LANGUAGE LEARNERS**

Teacher to Teacher

C. Eugene Price of Fort Wayne, Indiana, suggests the following activity to help students learn about the natural environments of South America. Organize the class into groups. Have each group write one essay question or two short-answer questions covering the information in this section. Then have the first group ask the rest of the class to answer one of its questions. Allow five to seven minutes for the other groups to prepare their answers. Have each group read aloud its answer to the question. Then repeat the process. Have the second group ask a question, then the third, and so on. **COOPERATIVE LEARNING**

Linking Past to Present

The Galápagos Islands

Ecuador's Galápagos Islands form an archipelago west of the country in the Pacific Ocean. The islands contain wildlife found nowhere else in the world. Many of these species were studied by Charles Darwin, who visited the islands and used evidence he gathered there to support his theory of natural selection.

Although the islands are now a wildlife sanctuary, some people worry that the Galápagos archipelago is threatened by its exploitation for tourism and jobs. For example, in 2001 a tanker sank while carrying oil to tourist boats on the islands. This incident highlighted the environmental threat created by the need to supply the islands' increasing population.

DISCUSSION: Lead a discussion that addresses the following questions. Do countries have a responsibility to balance economic growth with environmental protection? Why or why not? If so, to what extent?

VISUAL RECORD ANSWER

Birds and bats eat the fruit of rainforest trees and drop the seeds far from the parent trees.

260

INTERPRETING THE VISUAL RECORD

Dense rainforest vegetation grows along the Aguarico River in Ecuador. The toucan in the small photo is a rainforest resident. Birds and bats assist the survival of tropical rain forests. **How might birds and bats play a role in reforestation?**

A tree line is the line of elevation above which trees do not grow. (See the diagram.) Animals unique to South America have adapted to the Andes' harsh conditions. They include llamas, the related animals alpacas and vicuñas, and the Andean condor, a large vulture with a 10-foot (3 m) wingspan.

Many areas of South America have tropical wet and dry climates. They have wet summers and dry winters. The natural vegetation includes either dry forest or savannas, where a mixture of trees and grasses cover the plains. Southern South America has a variety of middle-latitude climates. Chile's central valley has a Mediterranean climate, with winter rains and summer droughts. Moist westerlies influence southern Chile. That area has a marine west coast climate. In southern Argentina the Andes create a rain shadow. As a result, Patagonia has semiarid and arid climates. Relatively few animals live in this area.

The driest region of South America is the Atacama Desert of northern Chile and southern Peru. A high pressure system and cool ocean currents bring dry weather to this area throughout the year. Although rain is extremely rare, fog and low clouds are common. They form when the cold Peru Current chills warmer air above the Pacific Ocean's surface. Cloud cover keeps air near the ground from being warmed by the Sun. This coastal area is one of the cloudiest—and driest—places on Earth. In fact, the area receives almost no sunshine for about six months of the year. Some people who live along this coast increase their water supply by "trapping" fog. Near the seashore, they set up plastic nets on which fog droplets condense. In this way a village can collect several thousand gallons of clean water per day.

About once or twice a decade, the dry Pacific coast is affected by an ocean and weather pattern called **El Niño** (ehl-NEEN-yoh). During an El Niño event, the eastern Pacific Ocean is warmer and the climate much wetter than normal. This pattern can alternate with **La Niña**, when Pacific waters are colder than normal. (See Geography for Life: El Niño.)

✓ **READING CHECK:** *Physical Systems* How do mountains and elevation affect the climates of South America? How does the Peru Current affect weather patterns in Chile? create five climate regions in the Andes and rain shadow in southern area; chills warmer air above Pacific surface, creating fog and clouds in Atacama Desert

Natural Resources

South America has rich mineral deposits, fertile soils, and climates suitable for growing a range of crops. Many rivers, particularly in the Paraná and Amazon Basins, have been dammed to generate electricity and store water for irrigation. The rain forests provide rubber and timber. Many nuts and plants used for medicines come from the Amazon rain forest. Scientists hope other useful plants will be discovered there.

The mineral wealth that attracted Spaniards and Portuguese to the region centuries ago is still being developed. New gold and silver deposits have been found in Brazil and Colombia. Chile is the world's largest producer and exporter of copper. Brazil has enormous reserves of iron ore and bauxite—the main aluminum ore. Colombia has long been famous for its emeralds.

Several South American countries have petroleum deposits as well. The largest oil reserves are in Venezuela. The vast oil deposits surrounding Lake Maracaibo (mah-rah-KY-boh) have made Venezuela a leading oil-exporting country. In addition, oil deposits have been developed in Colombia and the upper Amazon Basin of Peru and Ecuador. More recent oil discoveries have been made off the coasts of Argentina, Brazil, and Chile.

Venezuela will have oil resources for years to come in the form of **tar sands**. Tar sands are rock or sand layers that contain oil. However, because the oil has to be cooked out of the rocks, production is expensive. Nonetheless, tar sands may become more important once the more easily pumped oil is gone.

✓ **READING CHECK:** *Physical Systems* What are the main mineral and energy resources of South America? gold, silver, copper, iron ore, bauxite, emeralds, hydroelectricity, oil, tar sands

Chuquicamata mine, in the Atacama Desert, is the largest open-pit copper mine in the world. The ore sample shown displays copper's distinctive blue-green color.

Review

Homework Practice Online
Keyword: SW3 HP12

Identify Llanos, Gran Chaco, Pampas, El Niño, La Niña

Define *tepuís,* tree line, tar sands

Working with Sketch Maps
On a map of South America that you draw or that your teacher provides, label the Andes, Altiplano, Lake Titicaca, Guiana Highlands, Brazilian Highlands, Amazon River, Llanos, Gran Chaco, Pampas, Patagonia, Tierra del Fuego, Orinoco River, Paraná River, Atacama Desert, and Lake Maracaibo. In the margin of your map, explain why coastal Chile is both dry and cloudy.

Reading for the Main Idea
1. *Physical Systems* What is a basic difference between the Andes and the other highland regions of South America?
2. *Places and Regions* What are South America's three largest rivers, and what areas do they drain?

Critical Thinking
3. *Drawing Inferences* Look at the chapter map. To what physical conditions would you have to adjust if you moved from Rio de Janeiro to La Paz?

4. *Making Generalizations and Predictions* See the feature on the next page. How might El Niño and La Niña affect societies of the dry Pacific coast over time?

Organizing What You Know
5. Create a chart like the one below. Use it to identify the climates, vegetation, and animal life of the three landform regions listed.

	Climates	Vegetation	Animals
Amazon River basin			
Plains areas			
Andes			

Economics: The Cost of El Niño

El Niño can have far-reaching effects on the environments, economies, and infrastructures of South American countries. The economic costs of events caused by El Niño can be devastating. For example, the 1997–98 El Niño cost approximately $20 billion. Peru's fish-exporting business, which normally brought in $1 billion a year, collapsed. Chile's fishing industry was also hit hard, with reduced catches leading to the closing of many fish-processing plants. Drought in northeastern Brazil cost an estimated $4 billion, and floods in Argentina cost some $3 billion. The floods in Argentina did much of their damage to the country's farming output,

destroying about 40 percent of the cotton harvest, 30 percent of the rice crop, and 50 percent of the tobacco crop. Ecuador suffered damage to 19 large bridges and about 1,550 miles (2,500 km) of roads. The damage to roads and transportation systems caused major disruptions in travel and trade as people, goods, and resources failed to reach their destinations.

Lead a discussion with students on the possible ramifications of El Niño events on the U.S. economy. What might be some short-term and long-term economic effects of El Niño? *(Possible answers: destroyed crops and buildings, trade disruptions, drought recovery)* What might governments do to try to lessen damage caused by El Niño? *(Possible answers: build warning systems and levees, strengthen bridges, and so on)*

Across the Curriculum

► Science ◄

La Niña An El Niño is often followed by La Niña, or "the girl." Atmospheric pressure over Australia drops, and warm water pushes westward across the Pacific. The jet stream splits and weakens, allowing more hurricanes to move west across the Atlantic Ocean. The weather patterns produced are the reverse of those created by El Niño. In 2000 southern Europe experienced droughts, western Europe had floods, and parts of southern Africa experienced record flooding. Heavy rains caused flooding in Ecuador and Bolivia, and droughts in Uruguay caused crop failures.

CRITICAL THINKING: How might scientific investigation of these weather phenomena help countries? *(could help people prepare for floods and droughts)*

Applying What You Know Answers

1. warm water—flows east, blocks upwelling of cool nutrient-rich water; marine life—moves south; jet stream—stays over Canada; upper-level tropical winds altered; droughts, heavy rains, floods possible
2. because some people might benefit, while others might suffer

This Geography for Life feature addresses National Geography Standards 7, 15, and 18.

Physical Systems

Geography for Life

El Niño

Long ago, fishers noticed that once or twice a decade the normally cool waters off Peru's coast became warmer near Christmastime. Referring to the baby Jesus, they called this warming trend El Niño. *El Niño* means "The Boy" in Spanish. El Niño shows how a physical process that happens on one side of the planet can affect environments thousands of miles away.

Usually, the contrast in temperatures across the oceans helps create winds. Trade winds generally maintain a balance between warm western Pacific water and cool eastern Pacific water. Along the eastern Pacific, particularly off Ecuador and Peru, strong trade winds blow warm surface water westward. This allows colder water to rise and bring up nutrients from the depths, attracting fish.

When atmospheric pressure rises north of Australia, the winds calm. With the drop in wind, El Niño begins off the coast of South America. As easterly trade winds decrease, warm water in the western Pacific flows eastward. The warm water layer flows over cooler, nutrient-rich water. This blocks the normal upwelling of the cool water along North and South America. As a result, sea life suffers from a lack of nutrients. Fish that usually thrive off Ecuador and Peru head south in search of cooler waters

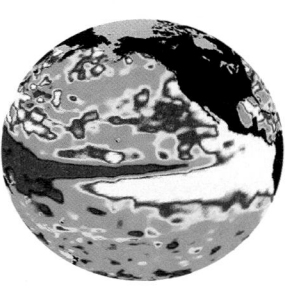

This 1997 satellite image shows an El Niño developing in the eastern Pacific. The warmer water is in white.

and more food. Chilean fishers then catch more sardines. Meanwhile, in North America, the polar jet stream stays farther north over Canada. Less cold air moves into the upper United States, and the northern states enjoy a mild winter. During the 1997–98 El Niño, for example, people in northern states saved billions of dollars in heating costs. In addition, an El Niño event alters upper-level tropical wind patterns. States along the Atlantic Ocean and Gulf of Mexico benefit because fewer hurricanes strike land.

However, El Niño can also cause terrible destruction. Because oceans affect weather patterns, severe weather changes may accompany the change in ocean temperature. Droughts can occur in the Pacific Islands. Meanwhile, heavy rains may flood areas of coastal North and South America that are usually bone-dry.

Applying What You Know

1. **Summarizing** How does an El Niño event affect the environment?

2. **Drawing Conclusions** Why might people in areas affected by El Niño view it differently?

Normal and El Niño Ocean Conditions

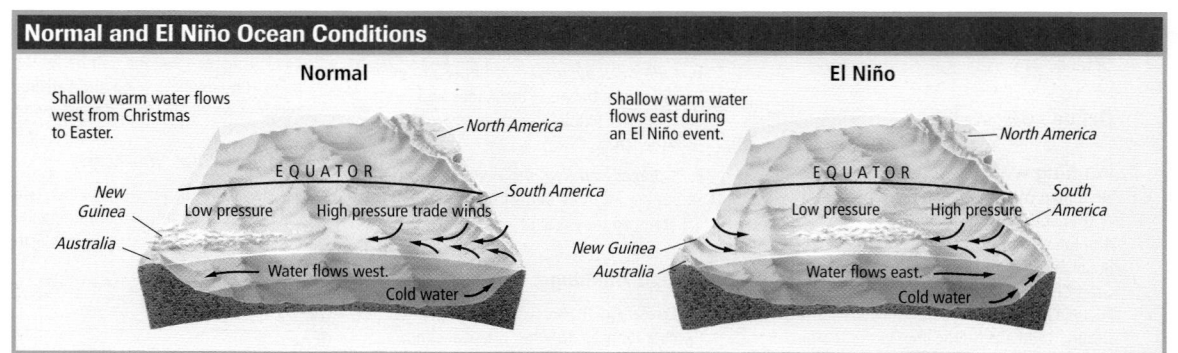

Section 2

OBJECTIVES

1. Identify important events in the early history of South America.
2. Explain how the colonial era and independence affected South America.
3. Describe some important features of South America's cultures.

LET'S GET STARTED

Copy the following passage onto the chalkboard: *All modern civilizations use wheeled vehicles and written languages. Yet the Inca civilization, which ruled a huge empire in western South America, had neither. Judging from the continent's physical geography, what methods might the Inca have developed to communicate across this rugged land?* Discuss responses. Tell students they will learn more about the early and recent history of South America in Section 2.

Building Vocabulary

Write the key terms **coup** and **latifundia** on the chalkboard. Point out that the word *coup* comes from a French word meaning "blow" or "stroke." It is a shortened form of *coup d'état*, which means "stroke of state." The term *latifundia* is comprised of the Latin words *latus* ("broad") and *fundus* ("estate"). Have students read the definitions for these terms aloud. Then have students look up the terms **buffer state** and **manioc** in the section and read their definitions aloud.

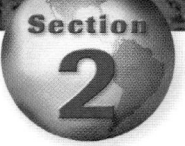

Section 2 History and Culture

READ TO DISCOVER

1. What were some important events in the early history of South America?
2. How did the colonial era and independence affect South America?
3. What are some important features of South America's cultures?

IDENTIFY

Chibcha
Inca

DEFINE

latifundia
buffer state
coup
manioc

LOCATE

Cuzco
Buenos Aires
Rio de Janeiro
Bogotá
La Paz
Quito

WHY IT MATTERS

Some South American countries have experienced many changes in government since they became independent. Use CNNfyi.com or other **current events** sources to learn how political changes in one country can affect people elsewhere.

Early History

Most researchers believe people first entered South America from the north more than 12,000 years ago. They eventually inhabited even the continent's harshest environments. The first settlers were hunter-gatherers. Farming began in the region more than 5,000 years ago.

For several thousand years before Europeans arrived, kingdoms rose and fell in western South America. In the Colombian Andes, for example, the **Chibcha** ruled and developed gold-working skills. The **Inca**, however, founded South America's greatest early civilization. At its height the Inca Empire stretched from what is now Ecuador to central Chile. The Inca built paved roads and suspension bridges to connect their empire from the Pacific coast to the Amazon lowlands. Fine examples of Inca stone construction can still be seen in Cuzco, Peru, the Inca capital. Also common and still being used for farming are Inca terraced fields braced by stone walls.

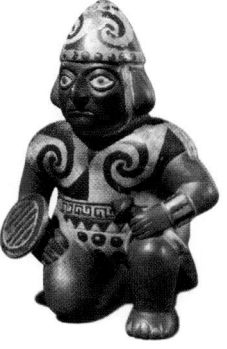

The Mochica people who made this ceramic warrior lived in northern coastal Peru from about A.D. 100 to 700.

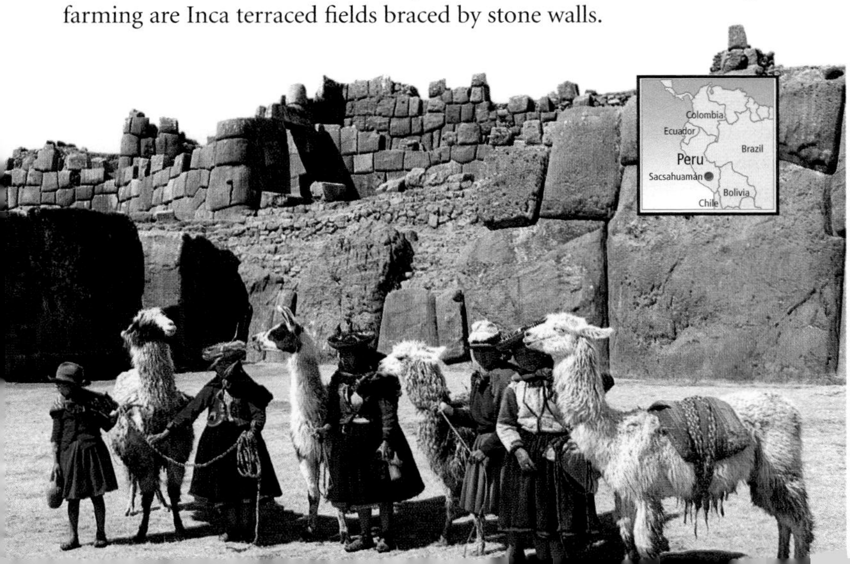

INTERPRETING THE VISUAL RECORD

Local women pose with their llamas before a wall at Sacsahuamán, an Inca site near Cuzco. Some of the stones weigh hundreds of tons. They were moved to the site without wheeled vehicles and are fitted closely together without mortar. ***What function might these walls have served?***

Section 2 RESOURCES

REPRODUCIBLE

- Guided Reading Strategy 12.2
- Cultures of the World Activity: Region 2
- Critical Thinking Activity 12: The Incas and Machu Picchu

TECHNOLOGY

- One-Stop Planner CD–ROM, Lesson 12.2
- CNN Presents World Cultures: Yesterday and Today, Segment 13: The Inca's Frozen Past
- Homework Practice Online
- HRW Go site

REINFORCEMENT, REVIEW, AND ASSESSMENT

- Main Idea Activity 12.2
- English Audio Summary 12.2
- Spanish Audio Summary 12.2
- Section 2 Review, p. 267
- Daily Quiz 12.2

VISUAL RECORD ANSWER

protection

LEVEL 1: Organize the class into groups and ask them to imagine that they are editing a book about South America's history. Have each group write a brief summary that discusses the main events in the continent's history. (*Summaries should mention such important events as Spanish and Portuguese settlement and struggles for independence.*) When groups have completed their projects, call on volunteers to read the summaries to the class. **ENGLISH LANGUAGE LEARNERS, COOPERATIVE LEARNING**

LEVELS 2 AND 3: To help students understand South America's history, have them work in groups to create mini-scrapbooks about the region's past. Students should use text and illustrations to present information about the significant events in the history of South America. (*Scrapbooks should include information about ancient civilizations as well as European colonial settlements and dates of independence.*) Provide students with paper and markers. The pages of the book can be bound using brads, staples, or string. When students are finished, have them share their scrapbooks with the class. **COOPERATIVE LEARNING**

Cooperative Learning

Regional Differences in South America Most of South America was explored and colonized by Europeans. Over time, distinct cultures emerged in each country.

Organize the class into groups. Assign each group a region of South America. Then have groups use almanacs or other reference materials to create posters comparing the dates and origins of European settlement, dates of independence, ethnic groups, languages, chief religions, major cities, and types of government in the countries of the assigned regions. Display and discuss the posters.

You don't say! When Amerigo Vespucci saw island villages built on stilts in the water off the coast of South America, he nicknamed the place Venezuela, or "little Venice."

INTERPRETING THE VISUAL RECORD

Following the arrival of Columbus in the Americas, Europeans brought many plants and animals with them. In turn, they introduced American species to Europe. This process is called the Columbian Exchange. In the painting below, workers pick coffee beans on a South American plantation. Coffee was probably first grown in eastern Africa. **How has this exchange of plants and animals affected the regions of contact?**

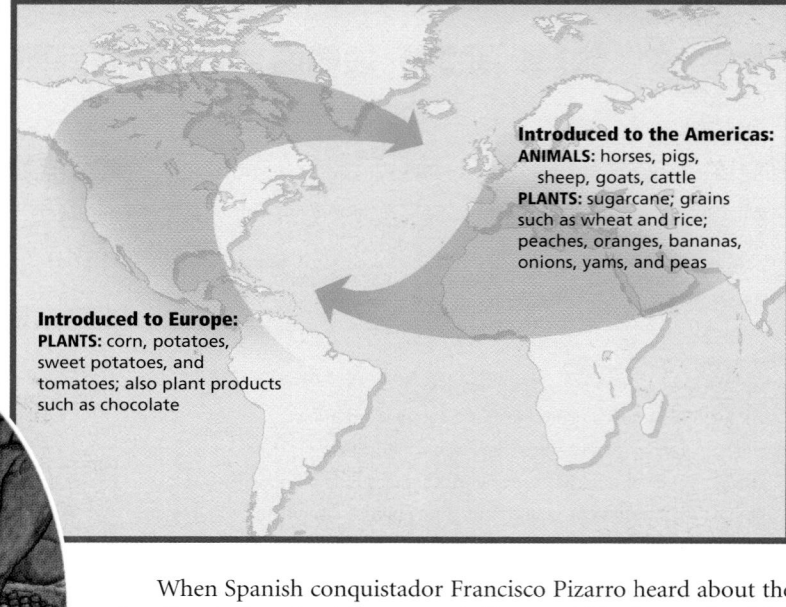

The Columbian Exchange

Introduced to the Americas:
ANIMALS: horses, pigs, sheep, goats, cattle
PLANTS: sugarcane; grains such as wheat and rice; peaches, oranges, bananas, onions, yams, and peas

Introduced to Europe:
PLANTS: corn, potatoes, sweet potatoes, and tomatoes; also plant products such as chocolate

When Spanish conquistador Francisco Pizarro heard about the Inca Empire he set out to conquer it. Unrest within the empire aided Pizarro's conquest in the 1530s. The Spanish looted and destroyed Inca buildings. They then built churches, government buildings, and Spanish-style plazas on the ruins. The Spanish also built a new capital city, Lima, near the coast.

✓ **READING CHECK:** *Human Systems* How are Inca and early Spanish influences still evident in modern South America? Inca stonework, terraced fields braced by stone wall, Spanish plazas, and new city of Lima

Spanish Settlement The Spanish focused their efforts at conquest on the western part of the continent. A 1494 treaty had divided South America between Spain and Portugal. Spain got the right to lands to the west of the treaty line and Portugal the lands to the east. The Spanish also focused on the west because the Inca, rich with gold and silver, lived there. The area was agriculturally productive and could provide the Spanish with a ready source of labor. In Spanish society owning land was a source of prestige, and the Spanish colonists soon took over South American Indian lands. They established a system of landed estates like they had known in Spain and forced the Indian peoples to work the land.

The colonists also introduced animals and agricultural products they had known in Europe. For example, they brought cattle, horses, sheep, sugarcane, and wheat. Over time, the colonists took American products like beans, chilies, corn, potatoes, and squash to Europe, Africa, and Asia.

Europeans also carried new diseases to South America. As in Mexico and Central America, these diseases killed millions of South American Indians. Europeans killed many others in the battles of conquest. Indians who labored in mines and on ranches and plantations often died from overwork. As a result, Indian populations fell sharply during the colonial period. Only a fraction of the original Indian population remained by the late 1500s.

Teach Objective 2

ALL LEVELS: Copy the following graphic organizer onto the chalkboard, omitting the italicized answers. Call on students to provide words and phrases to explain how the colonial era and independence affected South America.
ENGLISH LANGUAGE LEARNERS

Causes and Effects in South America's History

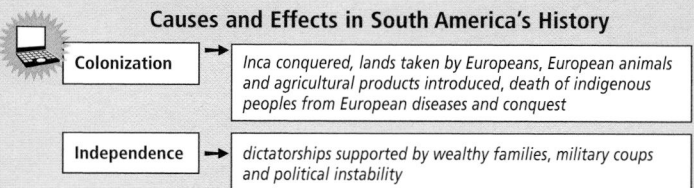

| Colonization | → | *Inca conquered, lands taken by Europeans, European animals and agricultural products introduced, death of indigenous peoples from European diseases and conquest* |

| Independence | → | *dictatorships supported by wealthy families, military coups and political instability* |

Teach Objectives 1–2

ALL LEVELS: Organize the class into groups. Have each group refer to this section and the previous activities to create a poster that compares the history of South America to that of the United States. Tell students to include such topics as indigenous peoples, European exploration, colonization, and independence. Instruct groups to illustrate their posters. For example, they may want to use time lines or graphic organizers to make their comparisons. When students have completed their posters, display them in the classroom.
COOPERATIVE LEARNING

Once the Spanish were established in Peru, their influence spread across the Altiplano into what is now Bolivia. There they expanded the Inca silver and gold mines. The Spanish colonized central Chile in the 1540s. Later they spread southeastward, herding cattle in what became Paraguay and Argentina. After a while, large estates called **latifundia** (lah-ti-FOOHN-dee-uh) spread across the Pampas. A South American Indian people known as the Guaraní already lived in the fertile lands of eastern Paraguay. The Spanish used the labor and food production skills of the Guaraní to further expand their settlement in the region.

Portuguese Settlement Portuguese settlement began in the 1530s along the eastern coast of what is now Brazil. Portuguese nobles received royal land grants to set up large plantations. Their first important crop was brazilwood, which can be used to produce red and purple dye. However, sugarcane soon became the key crop. After the decline of the South American Indian population, colonists brought in enslaved Africans to work their estates. Cities in northeastern Brazil, like Natal and Salvador, remain from the sugar and slavery era. Meanwhile, the Portuguese spread cattle ranching inland. Their expansion southward and inland in the 1600s led to major mineral discoveries. This movement eventually led to the growth of cities like São Paulo and Rio de Janeiro.

✓ **READING CHECK:** *Human Systems* How did the distribution of resources affect the location and patterns of movement of Spanish and Portuguese colonists? The availability of land, indigenous labor, and valued resources drew colonists to areas of the continent.

Colonial Era and Independence

The Spanish colonies of South America gained their independence between 1810 and 1830. Wars in Europe had weakened Spain. Spanish authorities fled their colonies after a period of unrest and minor military battles. The independence of the United States had inspired colonial leaders. There the ideals of political freedom and cooperation between the colonies had led to the formation of a single large country. However, in South America several different countries formed after the Spanish left. The wealthy elites who ruled each country were oriented more toward Europe than toward each other. In addition, these countries tended to be isolated from each other. This was because the continent's size and rugged terrain made communication difficult.

Borders of the new countries mostly followed the divisions created during the colonial period. However, the frontier between Argentina and Brazil was not controlled well by either country. As a result, the new state of Uruguay was able to form. Uruguay is an example of a **buffer state**—a small country between two larger, more powerful countries.

Brazil followed a different path to independence. European wars forced the Portuguese royal family to flee to Rio de Janeiro, where they arrived in 1808. Later, after the king's return to Portugal, Brazil declared its independence in 1822.

The British and Dutch only recently granted their small colonies in the Guianas independence. British Guiana became

Capoeira, a popular Brazilian folk dance, combines dancing, fighting, acrobatics, and music. It was possibly first developed as a martial art in the 1600s by escaped slaves who wanted to defend their freedom.

INTERPRETING THE VISUAL RECORD

Grand buildings rise on Independence Plaza in Montevideo, the capital of Uruguay. Spaniards founded the city in 1726 to balance Portuguese influence in the area. **What can you infer about Montevideo's climate from the photo?**

Essential Element 4

► Human Systems ◄
Bolivia's Two Capitals
Bolivia is unusual among the world's countries in that it has two capitals—La Paz and Sucre. This situation arose in 1898 when an effort to move the capital from Sucre to La Paz resulted in a civil war. The problem was resolved through a compromise: La Paz became the administrative capital and seat of the executive and legislative branches of the government, while Sucre remained the constitutional capital and seat of Bolivia's judiciary. La Paz, the country's largest city, is located in west-central Bolivia, near Lake Titicaca and Peru. Sucre lies on the eastern side of the Andes in southern Bolivia.

ACTIVITY: Have students analyze the unit maps and discuss possible reasons why some Bolivians may have wanted to move their capital from Sucre to La Paz. *(Possible answer: La Paz is located near Lake Titicaca, which might expand economic opportunities.)*

VISUAL RECORD ANSWER

humid tropical or subtropical climate, indicated by the palm trees

LEVEL 1: Provide students with art supplies for creating collages. Then have them refer to the information in this section and make collages that depict various aspects of South America's cultures. Remind students to include images and words that show South America's peoples, languages, religions, and other cultural traditions. Call on volunteers to present their collages to the class. **ENGLISH LANGUAGE LEARNERS**

LEVELS 2 AND 3: Organize the class into 11 groups. Assign each group one of the South American countries, but do not reveal the groups' assignments to the entire class. Have the students in each group work together to list phrases and images they would include in a book covering the culture of the assigned country. *(For example, a book on Peru might include both Roman Catholic icons and Inca artifacts.)* You may want to provide supplemental materials for use in the classroom. Call on volunteers from each group to read the lists and ask the class to identify the country being described. **COOPERATIVE LEARNING**

Suriname's Population

Suriname is slightly larger than the U.S. state of Georgia and is home to more than 430,000 people. The country's population is ethnically diverse.

Hindustanis, whose ancestors came to the area from northern India in the late 1800s, make up about 37 percent of the population. Other large groups include Creoles, Javanese, and Maroons, who are the descendants of African slaves. Smaller groups of South American Indians, Chinese, and descendants of Europeans also live in Suriname. The country's languages are as diverse as its population, and religions include Hinduism, Islam, Christianity, and indigenous beliefs.

ACTIVITY: Have students conduct research into the effect independence had on the population of Suriname and report their findings to the class.

CONNECTING TO HISTORY ANSWER

Possible answer: The theory seems logical because putting a great deal of energy into ensuring water supplies, without which people could not survive, is a logical practice for a culture to pursue.

Connecting to HISTORY

The Nazca Lines

One intriguing ancient site in South America is in the desert of southern Peru. There some 900 gigantic geometric and animal shapes are etched into the desert floor. Examples include a 1,000-foot (305 m) pelican and a 360-foot (110 m) monkey. One trapezoid covers 160,000 square yards (134,400 sq m). For the most part, these shapes can be seen clearly only from the air. The Nazca people, who lived in the area before the Inca, made the lines by moving surface stones aside to expose the lighter soil beneath. Although there are many theories, no one really knows why the Nazca Lines were drawn. One theory suggests that the lines and shapes were in some way related to water. Some researchers think they may have marked underground water sources. Many anthropologists, however, think the lines may have formed paths. They believe the Nazca walked these paths as part of rituals meant to ensure that there would be enough water for them and their lands.

Supporting a Point of View Do you think the anthropologists' theory about the Nazca Lines may be a logical one? Why or why not?

Guyana in 1966, and Dutch Guiana became Suriname in 1975. French Guiana remains a part of France.

In the end, the independence movements in South America did little to improve people's lives. Revolutions often changed the governments in the new countries. However, these revolutions only tended to replace one group of powerful families with another. Leadership was usually by one man, a dictator, who ruled with the support of wealthy colonial families. Meanwhile, life for most poor farmers, plantation workers, and city dwellers changed little. Moreover, there were few opportunities for improvement. Sometimes a group would take power by force. Such a change in government is called a **coup** (koo). Coups have been common throughout South American history. For example, Bolivia has experienced revolutions or military coups nearly 200 times since becoming independent in 1825.

✓ **READING CHECK:** *Human Systems* Why has daily life not improved much over the years for many South Americans? Power was often in the hands of ruling families or dictators.

Culture

South American Indians, Europeans, Africans, and Asians have all played a part in the peopling of South America. Each culture has left its stamp on the continent's countries.

People and Languages Today the countries of South America vary widely in their ethnic makeup. For example, 97 percent of Argentines are of European ancestry, compared to only 7 percent of Ecuadorians. Bolivia's population has the largest percentage of South American Indians—55 percent. In other countries people of mixed ancestry called mestizos are in the majority. For example, 95 percent of Paraguayans are mestizos. Asians have also immigrated to the continent. In Guyana the descendants of workers from India make up about half of the population. Many Japanese have immigrated to Brazil and Peru.

The main European language spoken in each country reflects that country's colonial history. Therefore, people in most South American countries speak Spanish, but most Brazilians speak Portuguese. English, Dutch, and French are official languages in Guyana, Suriname, and French Guiana, respectively. In the Andes region, from 10 to 13 million people speak the Inca language, Quechua. Most Paraguayans speak Spanish and Guaraní. On Uruguay's border with Brazil, a mix of Portuguese and Spanish called Portunol is widely spoken.

Settlement Patterns To a large extent, densely populated areas of South America hug the coasts and reach only a few hundred miles inland. Many major cities—such as Buenos Aires, Lima, and Rio de Janeiro—are seaports. Some cities, such as Bogotá, La Paz, and Quito, lie in high Andean valleys. Much of the South American interior is thinly populated. Large areas, particularly in the

Amazon Basin, the Andes, the Guianas, and southern Argentina, have few people.

Religion and Traditions The Spanish and Portuguese colonists were Roman Catholics. Therefore, the majority of South Americans today are also Roman Catholic. South Asians and Indonesians in the Guianas have added Hindu temples and Islamic mosques to the landscape. Many indigenous peoples, such as those who live deep in the rain forest, follow their traditional religions.

Although South America is changing rapidly, traditional ways of life can still be found. Some rainforest peoples have had little contact with the outside world. They raise bananas, **manioc**, yams, and other crops. Manioc is a tropical plant with starchy roots. Some hunt with bows and arrows or blowguns and darts. As the Amazon Basin is developed, these people's lives will change. Also, although most ranchers use modern methods, some ranch hands still live much like the cowboys of the old American West. *Llaneros* work on the Venezuelan Llanos. Argentine cowhands herd cattle and horses on the Pampas. Some wear the traditional clothing that Argentina's gauchos, or cowboys, wore in the 1700s and 1800s. The gauchos also live on in Argentine literature and popular culture. For example, the poem *The Gaucho Martín Fierro* celebrates the independent life of the Argentine gaucho.

✔ **READING CHECK:** *Places and Regions* What are two ways that Spanish and Portuguese heritage are expressed in South American culture? language and religion

Wearing a traditional derby hat, a Bolivian Indian woman sells folk medicines and good luck charms in a La Paz market. The Quechua and Aymara, Bolivia's two main ethnic groups, often combine local religious beliefs with Roman Catholicism.

Section 2 Review

Homework Practice Online Keyword: SW3 HP12

Identify Chibcha, Inca

Define latifundia, buffer state, coup, manioc

Working with Sketch Maps On the map you created in Section 1, label the countries of South America, Cuzco, Buenos Aires, Rio de Janeiro, Bogotá, La Paz, and Quito. Then shade in the areas that had been included in the Inca Empire.

Reading for the Main Idea

1. (*Human Systems*) What happened to the indigenous population following the arrival of Europeans in South America?

2. (*Places and Regions*) Where are the densely populated areas of South America? Which areas are thinly populated?

Critical Thinking

3. *Identifying Cause and Effect* How do you think the exchange of food products between the Old and New Worlds changed life in both places?

4. *Analyzing Information* What do the *llaneros* and rainforest peoples have in common?

Organizing What You Know

5. Create a time line like the one below. Use it to identify important periods and events in the history of South America.

1494 ———|———|———|——— Today

 **LET'S GET STARTED**

Copy the following question onto the chalkboard: *Judging from the photos in this section, what can you infer about South America's economy and challenges?* Allow students time to share their responses. *(Possible answers: Ranching and mining are important, and deforestation is a problem.)* Tell students they will learn more about the economy, cities, and issues of South America in Section 3.

Building Vocabulary

Write the term *minifundia* on the chalkboard. Call on a volunteer to offer an explanation for the term, based on his or her knowledge of the term *latifundia* from the previous section. *(The student should conclude that since* mini *means "small," this term means "small estates.")* Then have students read the definitions of the remaining terms aloud from the text or glossary.

Section 3 — South America Today

READ TO DISCOVER

1. What is the economy of South America like today?
2. What are South American cities like?
3. What issues and challenges face the people of South America?

WHY IT MATTERS

Every day more people move from the South American countryside to cities that are already crowded. Use **CNNfyi.com** or other **current events** sources to learn about changes in settlement patterns in South America and around the world.

IDENTIFY

Mercosur

DEFINE

minifundia
favelas
landlocked
terrorism

LOCATE

Manaus
Santiago
Lima

The Economy

Some South Americans enjoy a high standard of living, and the middle class is growing. However, all the continent's countries are considered developing or middle-income countries. Argentina, Brazil, Chile, Uruguay, and Venezuela have the strongest economies. See the unit's Fast Facts table for each country's per capita GDP. All of the countries have market economies.

Agriculture Agriculture in South America ranges from subsistence farming to huge commercial farms and ranches. When the large estates from the colonial period were broken up, small farms called **minifundia** (mi-ni-FOOHN-dee-uh), were created. Often these *minifundia* have poor land and are too small to be profitable. In most countries a few wealthy people own much of the best land. Inequality in land ownership is the basis of much poverty and unrest in South America.

Ranch workers herd horses on a flooded Argentine plain. From horseback the cowhands manage huge herds of cattle. Cattle are such a large factor in Argentina's economy that some residents eat beef at every meal. The rider pictured offers yerba maté, a tealike drink often served in a hollow gourd.

LEVEL 1: Copy the following graphic organizer onto the chalkboard, omitting the italicized answers. Call on students to provide words and phrases to describe South America's economy and cities and to identify the issues facing South America today. Then lead a discussion about how these facts are related. *(Example: Soil exhaustion in rural areas may stimulate migration to the cities.)* **ENGLISH LANGUAGE LEARNERS**

Economy	Cities	Issues
• *developing or middle-income countries*	• *large parts of each country's population living in big cities*	• *poverty*
• *agriculture—subsistence and commercial farming*	• *rural-urban migration*	• *Amazon rain forests being cleared for farming and ranching*
• *wide range of industrial products*	• *urban poor living in slums that surround the city*	• *soil exhaustion and over-grazing threatening the land*
• *Mercosur important free trade group*	• *crime and lack of services in these areas*	• *border disputes, terrorism, and political corruption*

Market-oriented farming is most highly developed in two regions—Chile's central valley and the area bridging southern Brazil and northern Argentina. Brazil produces more coffee than any other country in the world. Colombia is the second-largest coffee producer. Colombia's newest industry is selling cut flowers. These flowers are flown every night to markets around the world. Farmers in Chile's central valley grow fresh fruits and vegetables during the South American summer. This farm produce sells well in the United States during our winter. Argentina specializes in producing and exporting wheat and beef. South American countries export a wide range of agricultural products, from cacao beans to potatoes to sunflower seeds.

FOCUS ON ECONOMICS

Amazon Basin Development Agriculture and mining have played important roles in the development of the Amazon Basin. Consider the history of Manaus, a major inland port of more than 1 million people. Manaus lies on the Río Negro about 1,000 miles (1,600 km) from the Atlantic Ocean. It began as a mission in 1669 and remained relatively isolated until the late 1800s. Then the demand for rubber—used for waterproofing and tires—soared. Brazil's rain forest produced large amounts of rubber. Manaus grew rich on the profits from harvesting and shipping the precious material. Great buildings, including a cathedral and an opera house, date from this period. When sources for cheaper rubber developed in Asia, Manaus declined in importance. By 1920 the boom was over.

However, other resources from the Amazon rain forests are being exploited today. For example, the hardwood trees themselves are valuable, and the land is in demand for ranching. In addition, oil exploration is expanding into the Amazon Basin. Bauxite, copper, gold, iron ore, manganese, and tin draw in miners. Manaus is again bustling, exporting the region's riches from its river port.

Development in the Amazon Basin has had a downside, however. About 17,000 square miles (44,000 sq km) of rain forest are cleared every year. This deforestation threatens the region's unique plant and animal life. Development also threatens the ways of life of Amazonian Indians who have long lived in the forested basin.

✓ **READING CHECK:** *Environment and Society* How have the creation and distribution of resources affected development in the Amazon Basin? Rubber and other forest products have drawn in people and investment for exploiting those resources and have contributed to the growth of Manaus as a shipping point for the region's products.

Industry Most South American factories produce food items, consumer goods, or building materials for local markets. Larger countries also produce cars, trucks, and jet airplanes. Workers must assemble many export products, such as clothing and small appliances, by hand. Sometimes these industries import all the parts and raw materials needed for manufacturing, which limits profits. On the other hand, these industries provide jobs and training. This experience can lead to better jobs in more advanced industries.

Cooperation among the countries could lead to economic progress. For example, countries in the southern part of the continent have formed a trade

Prospectors dig for gold in the mud of Brazil's rain forest.

Economic development of the Amazon Basin is a cause of deforestation.

internet connect

GO TO: go.hrw.com
KEYWORD: SW3 CH12
FOR: Web sites about South America's exports

LEVELS 2 AND 3: Have students imagine that they are commerce officials or environmental consultants working in South America. Ask students to use the information from the previous exercise to write proposals that describe some of the challenges facing the region and provide ideas for solutions.

Using National Geography Standard 3:
The World in Spatial Terms: How to Analyze the Spatial Organization of People, Places, and Environments on Earth's Surface Have students analyze the spatial relationships of the cut flower export industry in Colombia, Ecuador, or Venezuela. Students should use library or other resources to answer questions such as: What factors make the industry profitable, although most consumers are far from the product's source? How do the labor costs and climate of the production area differ from those in the cities where the flowers are sent? What transportation routes are used? How has the Internet affected the industry?

Essential Element 4

▶ **Places and Regions** ◀

Emigration from Caraguatay, Paraguay
Paraguayans must obtain a visa to visit the United States. This process is not generally difficult, unless the applicant is from Caraguatay, east of Asunción. Applications from this town are scrutinized and usually denied. Most of these applications are denied because approximately 4,500 people, more than half of Caraguatay's population, moved illegally to the United States in the 1980s and 1990s. They sent back to their families money they earned as maids, gardeners, and construction workers. A high unemployment rate and a poor economic outlook in Caraguatay caused so many adolescents to leave the region that the Paraguayan government became alarmed and began offering job training and scholarships to the area's young people.

ACTIVITY: Instruct students to conduct research on Paraguay and to report on factors that might encourage emigration.

VISUAL RECORD ANSWER

The costumes resemble clothing of European colonists.

INTERPRETING THE VISUAL RECORD

Rio de Janeiro is famous worldwide for its pre-Lenten Carnaval. The days-long celebration enhances Rio's image as an exciting city. **How might the dancers' costumes reflect Brazil's history?**

organization called **Mercosur**. In Spanish, *Mercosur* stands for Southern Common Market. The purpose of Mercosur is to expand trade, improve transportation, and reduce tariffs among member countries. The full members are Argentina, Brazil, Paraguay, and Uruguay. The Andean countries have similar goals. However, each country's own interests often receive the most political support.

✓ **READING CHECK:** *Human Systems* What is one problem that many export industries face? They import needed parts and raw materials, which limits profits.

Urban Environments

In most of South America's countries the leading cities are huge in comparison to the other cities. Large parts of each country's population live in these big cities. For example, one third of Chile's people now live in or around Santiago. Nearly one third of Peru's people live in Lima. These cities grow as people move there from the countryside looking for jobs.

Both push and pull factors are at work in the migration process. Rural poverty and limited good land push people away from their small farms. The prospect of a better job and more exciting life pulls them toward the cities. However, life in the city is often just as hard as it was in the village. Rural migrants have few of the skills needed for the modern workplace. As a result, few find good jobs. Urban poor people often live in the large slums that surround major cities. These areas have their own names in each country. In Venezuela they are called *ranchos,* in Chile *callampas,* which means "mushrooms." In Brazil they are called **favelas** (fah-VE-lahs). They are home to some 25 percent of Rio de Janeiro's people. In recent years the government has worked hard to improve living conditions in the favelas. However, crime and lack of basic services, such as sanitation and schools, still trouble the people who live there.

✓ **READING CHECK:** *Human Systems* How do the leading cities of South America compare to the region's other cities? huge in comparison

Issues and Challenges

Overall, South American governments have become more democratic. Challenges remain, however. Many South Americans, both urban and rural, are poor. In some countries, such as Bolivia, high birthrates make development harder. A growing population, the need for resources, and concern for the environment create tensions throughout the region.

South America's environmental issues concern many people around the world. Of major importance are the great rain forests of the Amazon River basin. These forests have an incredible range of

Another environmental issue is protecting the unique animals of the Galápagos Islands, a small cluster of Pacific islands that belong to Ecuador. The marine iguana pictured belongs to a species found nowhere else. Oil spills and irresponsible tourism threaten the animals.

plant and animal species and hundreds of unique local ecosystems. They are also a vital source of oxygen for the whole planet. However, much of the Amazon forest may disappear within the next 100 years. As you have read, large parts of the forest are being cleared for farms and ranches. Other businesses harvest the forests' fine woods. Major mineral deposits attract prospectors and developers to the fragile forests. Other environmentally sensitive areas of the continent are also under threat.

Soil exhaustion is another environmental problem that threatens South America's future. This loss of soil nutrients has reduced the usefulness of large areas. In Brazil, for example, growing coffee—one of the country's major exports—reduces soil fertility. In other areas, overgrazing has stripped the land of plant life.

Political issues also cause conflict. Many South American countries have been involved in border disputes, often over areas with valuable resources. Following a war during the 1880s, Bolivia and Peru lost lands, and the mineral industries there, to Chile. This war left Bolivia **landlocked**. A landlocked country has no border on the ocean. Some of these border disputes still simmer. For example, Ecuador and Peru still dispute parts of their common border, as do Venezuela and Guyana.

Violence threatens the daily lives of many South Americans. **Terrorism**, or the use of fear and violence as a political force, is common. This is true today particularly in Colombia. Armed groups there often scare people away from voting places and control large parts of the country. Much of the violence is a result of the drug trade. Drug dealers have used illegal profits to support private armies and to buy off or assassinate judges and politicians. Violence has also recently troubled Bolivia and Peru.

✓ **READING CHECK:** *Environment and Society* Why is the disappearance of the rain forest a global concern? **It has a great range of plant and animal species, unique ecosystems, and an important source of oxygen.**

Derricks tap the oil of Lake Maracaibo, Venezuela. Before democratic reforms were approved, corrupt government officials skimmed off much of the oil wealth for themselves. Overdependence on oil still threatens Venezuela's stability. When oil prices fall, the whole economy suffers.

Section 3 Review

Homework Practice Online

Keyword: SW3 HP12

Identify
Mercosur

Define
minifundia, favelas, landlocked, terrorism

Working with Sketch Maps
On the map you created in Section 2, label Manaus, Santiago, and Lima. Then shade in the areas where commercial farming is most developed.

Reading for the Main Idea

1. *Places and Regions* What are some agricultural specialties of Brazil, Colombia, Chile, and Argentina?

2. *Human Systems* How does Mercosur try to improve trade in South America?

Critical Thinking

3. **Identifying Cause and Effect** What push and pull factors are causing South American cities to grow? What is the result?

4. **Comparing** What challenges do many of South America's largest cities share?

Organizing What You Know

5. Create a graphic organizer like the one below. Use it to describe the main environmental and political issues facing South America. Also, describe the basic challenges facing the region as a whole.

Basic:	
Environmental:	Political:
1.	1.
2.	2.

CASE STUDY

Urban Planning in Brasília

Places and Regions Urban geographers are interested in how cities function. They are also concerned about the quality of life in cities. Working with elected officials, urban planners try to design cities that meet residents' needs. They try to plan for future urban growth and figure out how to provide resources to growing cities. Food, water, electricity, transportation networks, and waste disposal are just a few of urban residents' needs.

Brasília, Brazil's capital, illustrates some of the issues that urban planners face. Brazil constructed its new capital in the 1950s and 1960s. Designed by architects and urban planners, the city is located deep in Brazil's interior. Beginning in 1956, workers built an airstrip and flew in heavy construction machinery. They began laying out streets and pouring building foundations. By 1960 the main buildings were completed and the federal government began to move to the new capital. Today Brasília has a population of about 1.8 million.

The urban planners who designed Brasília wanted to create a new urban environment that would lead to a better society. Segregation was one of the main problems the city's planners hoped to avoid. Segregation is the separation of different economic and social groups into distinct neighborhoods. To prevent this, city planners designed many groups of identical six-story apartment buildings. They organized the buildings into giant "superblocks." Each had its own schools, shopping areas, and parks. The goal was to have everyone live in the same type of environment. If this happened, the planners thought the new residents would develop a more integrated and equal society. Differences in income and social status would be less likely to lead to social problems. Even Brasília's address system reflects this idea. For example, think about addresses along Rodeo Drive in Los Angeles or Park Avenue in New York City. Such addresses imply high social status or wealth. Instead of proper names, numbers and letters identify streets and buildings in Brasília. For example, a typical address is SQS 106-F-504. Such addresses offer no clues about the social or economic position of people who live there.

Despite the efforts of Brasília's planners, the city is segregated. The poor live mostly on the outer edges of the city. This happened because of the new city's isolated location. More than 100,000 workers relocated to the region between 1957 and 1960. Known as *candangos*, these workers lived in wooden shacks on the edge of the huge construction site. When the workers were no longer needed, officials declared the wooden shacks slums and ordered them torn down.

However, it soon became clear that the workers would stay. The planners had not anticipated the housing needs of these people. As a result, officials changed the original plan of the city to provide housing and services to *candangos* and their families. They created satellite settlements on the edges of the central

Four large statues known as the Four Evangelists stand at the entrance to Brasília's Metropolitan Cathedral. A bell tower is at the right.

Drawing Conclusions

Lead a discussion about changes in your community or a nearby city that students have observed. Ask what role the area's physical geography may have played in those changes. *(Answers will vary according to the community's circumstances. Students may note that the area's flat terrain has allowed urban sprawl to occur, that forest fires have slowed new construction on the city's outskirts, that depletion of a natural resource has caused a decline in industry and subsequent growth, or other causes and results of the place's physical geography.)*

Going Further: Thinking Critically

Organize the class into groups. Have each group conduct research on a major U.S. city, concentrating on the original plans for streets and neighborhoods and the changes that the layout underwent over the years. So that a full range of city planning histories are covered, be sure one group examines Washington, D.C., which was a planned city, and another studies San Francisco, California, which grew haphazardly. First, have students investigate early plans, if any, for the subject city. Ask students also to address how physical geography, events, or other factors affected the city's growth. In a panel discussion, have the groups compare the cities they have studied and draw conclusions about forces that affect city planning. **COOPERATIVE LEARNING**

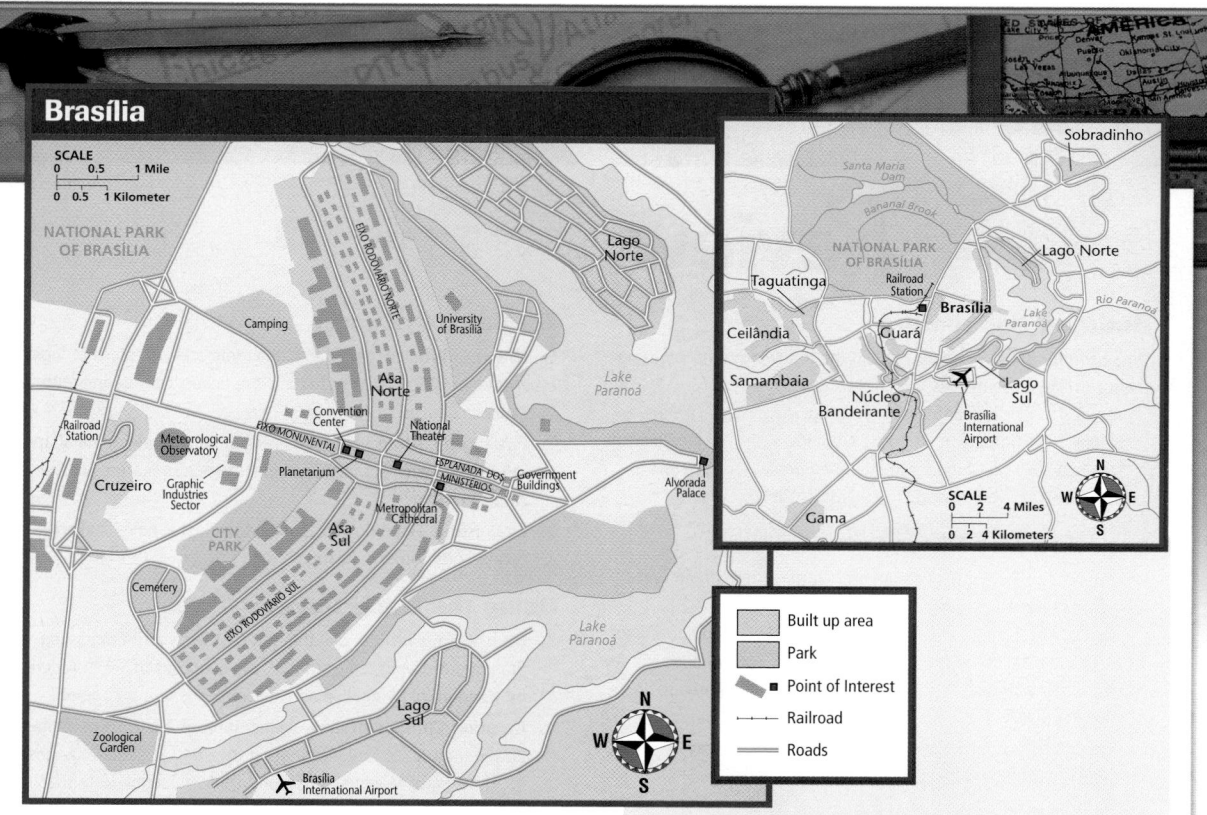

Brasília

zone. These settlements have grown dramatically as rural Brazilians continue to move to Brasília in search of better jobs and pay. By the late 1980s about 75 percent of the urban area's population lived in the satellite towns.

These towns were not as strictly planned and organized as the central zone. Instead of the same apartment buildings, single-family houses are much more common. Some of Brasília's wealthy citizens have moved away from the central zone's strict building codes. On the city's outskirts, they are free to build big homes that display their wealth and social status. The satellite towns are known for the iron fences that surround homes to deter criminals. As you can see, Brasília has not developed as the integrated city originally planned. Instead, sprawling edge settlements surround the city's heart.

Applying What You Know

1. **Summarizing** How is Brasília different than planners had hoped?

2. **Comparing** How are political, economic, and social conditions interrelated in Brasília? How does this compare to other cities you have studied?

INTERPRETING THE MAP *The large map shows Brasília itself. The small one shows the city's surroundings. A neighborhood of superblocks appears in the photo.* **How does Brasília's layout reflect its history? What are some of the satellite towns that appear on the small map?**

TECHNOLOGY

▶ Chapter 12 Test Generator (on the One-Stop Planner)
▶ Global Skill Builder CD–ROM
▶ HRW Go site

REINFORCEMENT, REVIEW, AND ASSESSMENT

▶ Chapter 12 Review, pp. 274–75
▶ Chapter 12 Tutorial for Students, Parents, Mentors, and Peers
▶ Chapter 12 Test (form A or B)
▶ Alternative Assessment Handbook

▶ Chapter 12 Test for English Language Learners and Special-Needs Students
▶ Unit 3 Test
▶ Unit 3 Test for English Language Learners and Special-Needs Students

Assess

Have students complete a Chapter 12 Test.

Reteach

Have students imagine that they are foreign exchange students living in South America. Have each student create an illustrated letter that he or she would send home after traveling through and living in the region. Encourage students to use the Read to Discover questions of each section as guides to the main ideas. ENGLISH LANGUAGE LEARNERS

CHAPTER 12 Review Answers

Understanding the Main Ideas

1. See the diagram in Section 1 for answers.

2. rubber, timber, nuts, medicinal plants

3. Chibcha—gold-working; Inca—stonework, roads, bridges, terracing; Nazca—large geometric and animal shapes

4. Spanish—established in Peru, then south along Pacific coast, across the Altiplano and Andes; Portuguese—inland from coast of Brazil

5. a growing population, the need for resources, concern for the environment

Thinking Critically

1. because Venezuela has huge oil and tar sands reserves along with water resources for farming and hydroelectricity (NGS 8)

2. fueled urban growth, such as in Manaus; attracted people looking for work, economic opportunities; species and traditional ways of life threatened by deforestation (NGS 16)

3. Answers will vary, but students should support their answers. (NGS 14)

Using the Geographer's Tools

1. Paragraphs and rankings should accurately reflect the information presented in the unit tables.

2. Student responses should reflect the information available through the Chapter Enrichment links at keyword SW3 CH12.

CHAPTER 12 Review

Building Vocabulary

On a separate sheet of paper, explain the following terms by using them correctly in sentences.

tepuís	tar sands	coup	favelas
Llanos	Inca	manioc	landlocked
tree line	latifundia	*minifundia*	terrorism
El Niño	buffer state	Mercosur	

Locating Key Places

On a separate sheet of paper, match the letters on the map with their correct labels.

Andes	Patagonia	Rio de Janeiro
Altiplano	Cuzco	Manaus
Guiana Highlands	Buenos Aires	Santiago
Amazon River		

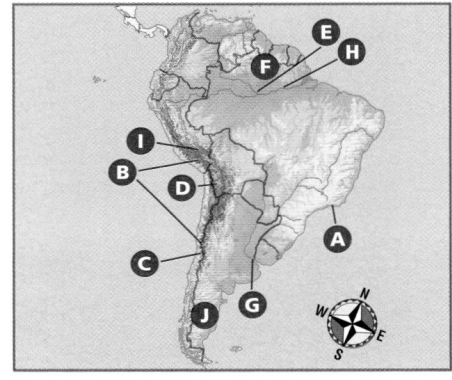

Understanding the Main Ideas

Section 1

1. (The World in Spatial Terms) What are the five climate zones of the Andes, and what vegetation is found in each?

2. (Environment and Society) What are some products that come from the rain forest?

Section 2

3. (Human Systems) What were some accomplishments of early South American peoples?

4. (Human Systems) How did Spanish and Portuguese settlement spread?

Section 3

5. (Environment and Society) What are three factors that create tensions in South America?

Thinking Critically

1. **Drawing Inferences** Why might some analysts say that logic would indicate that Venezuela is the richest country in South America instead of Argentina?

2. **Finding the Main Idea** How is economic development shaping the urban, population, and environmental features of the Amazon Basin?

3. **Evaluating** What do you think is the most serious challenge facing South Americans today? Why?

Using the Geographer's Tools

1. **Interpreting Tables** Use the unit Fast Facts and Comparing Standard of Living tables to rank the countries of South America by standard of living. Explain your rankings in a paragraph.

2. **Creating Climate Graphs** Search through the HRW Website on the Internet to find information for creating climate graphs for Lima, Peru; Manaus, Brazil; and Quito, Ecuador. Note that these three places lie near the same latitudes. In which climate regions are these places located? What are the major climate features of each place? What factors do you think influence the climate in these places?

3. **Creating Maps** Create a map of South America that shows which countries were settled by the Spanish, Portuguese, British, Dutch, and French during the colonial era.

Writing about Geography

Imagine that you are a subsistence farmer in Bolivia. Your cousin works on a large commercial farm in Chile's central valley. She has urged you to take an entry-level job there in farm management. Write a diary entry in which you reflect on your current life and how your life might change if you join her in Chile. When you are finished with your diary entry, proofread it to make sure you have used standard grammar, spelling, sentence structure, and punctuation.

SKILL BUILDING

Geography for Life

Comparing Maps

(Environment and Society) Search the Internet and library resources for aerial photographs and land use, resource, road, and vegetation maps of Brazil from different time periods. Use the photographs and maps to draw conclusions about the effects of mining, road construction, and timber harvesting on Brazil's rain forest.

Portfolio Activity

During colonial times, many South Americans perceived the rain forest as an inhospitable environment. Partly as a result, most people settled along the coast. In more recent times, many have considered the rain forest a vast resource to be exploited. Now many South Americans view the rain forest as a biological resource that should be protected and studied. These views have affected societal attitudes toward conservation. Have students write short essays in which they describe how people's changing perceptions of the rain forest have led to changes in society. Place essays in portfolios.

Food Festival

The Inca domesticated quinoa, a native Andean plant that yields nutritious seeds. It is still an important staple in South American cooking. Quinoa can be purchased in the United States at health food stores, some supermarkets, and from mail-order companies. Before using quinoa grains, the cook must remove the bitter coating by washing thoroughly in five changes of fresh cold water. The grain can then be steamed or boiled. Quinoa can be used in bread, casseroles, meat dishes, salads, soups, or even as a breakfast cereal. Have students locate recipes using quinoa and bring samples to class.

Building Social Studies Skills

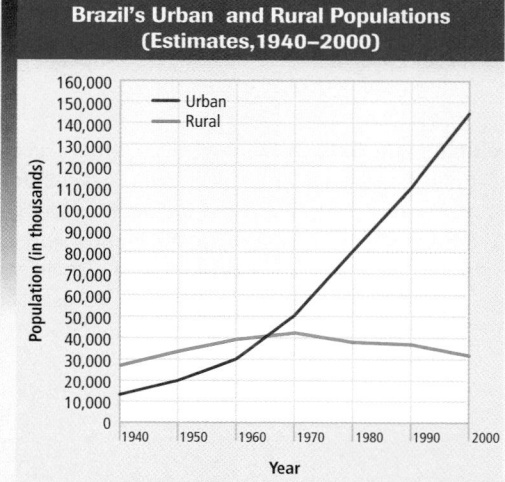

Brazil's Urban and Rural Populations (Estimates, 1940–2000)

Source: *The World Almanac and Book of Facts 2004*

Interpreting Graphs

Use the line graph above to answer the questions that follow.

1. About when were Brazil's urban and rural populations closest in number?
   a. 1960
   b. 1964
   c. 1968
   d. 1970

2. From 1940 to 2000, the rate at which Brazil's urban population increased was very different from the rate at which its rural population decreased. Describe this difference.

Using Language

The following passage contains mistakes in grammar, punctuation, or usage. Read the passage and then answer the following questions.

> "(1) Colombia is the most populous country of northern South America. (2) The national capital is Bogotá a city located high in the eastern Andes. (3) Most Colombians live in the fertile valleys and basins among the mountain ranges. (4) Rivers, such as the Cauca and Magdalena, flowing down from the Andes to the Caribbean and helping to connect settlements between the mountains and the coast."

3. In which sentence is a comma missing?
   a. 2
   b. 3
   c. 4
   d. none of them

4. Rewrite sentence 4 so that it is not a fragment.

Alternative Assessment

PORTFOLIO ACTIVITY

Learning about Your Local Geography

Group Project: Research

Indigenous people living in the Amazon rain forest are often in conflict with the loggers, miners, and ranchers who want to develop their land and its resources for profit. Plan, organize, and complete a research project on issues that have arisen in your state or region between longtime residents and business interests. Determine the relative success of conflict and cooperation in resolving those issues. Develop a hypothesis to describe what you find. Write a brief report and place it in your portfolio.

internet connect

Internet Activity: go.hrw.com
KEYWORD: SW3 GT12

Choose a topic about South America to:
- trek through the Guiana Highlands and create a poster.
- create a brochure on Machu Picchu.
- learn about threats to the Amazon Basin.

3. Student maps should accurately reflect the patterns of colonization of South America's countries.

Writing

Student diaries should reflect understanding of the cultures and people of Bolivia and Chile. Use Rubric 15, Journals, to evaluate student work.

Geography for Life

Students should address the environmental changes in the Brazilian rain forest over the time period spanned by their research. Use Rubric 1, Acquiring Information, to evaluate student work.

Social Studies Skills

1. b
2. Brazil's urban population increased rapidly, while its rural population increased slowly and then decreased slowly.
3. a
4. Possible answer: Rivers, such as the Cauca and Magdalena, flow down from the Andes to the Caribbean, helping to connect settlements between the mountains and the coast.

PORTFOLIO ACTIVITY

Student projects will vary. Issues should be valid concerns that have arisen in students' communities or states. Use Rubric 37, Writing Assignments, to evaluate student work.

Workshop 1
Going Further: Thinking Critically

Have each student draw a sketch map of your state that identifies rivers and other physical features as well as large cities. Then organize the class into groups. Have the members of each group compare their maps to find differences among them. For example, some students may have included more cities on their maps than others, or students may have drawn their maps with different levels of detail or precision. Ask students to speculate about why their maps differ. Ask them to think about questions like the following: Have students who included more detail in the maps lived in the area longer than students whose maps had less detail? Are students whose maps include more physical features interested in outdoor activities, like camping or hiking, that might have influenced their perceptions of the state's natural environment? Do students who labeled many cities on their maps regularly read or watch news reports or weather forecasts in which the locations of cities are marked? Have volunteers from each group share some of their conclusions with the class. Then lead a class discussion about how individual perceptions can affect the information included on maps.

PRACTICING THE SKILL

1. Students' maps will vary, but the countries of South America and their capitals should be labeled in their approximate locations. Rivers included on the maps should be drawn along their proper courses.
2. Students' maps should be recognizable as your county. Towns, roads, rivers, and other features should be drawn and labeled in their correct locations.

Geography
Skill-Building Workshop

WORKSHOP 1

Using Sketch Maps

As you have read, geographers use many kinds of maps to study our world. Sketch maps are simple kinds of maps. To create the simplest sketch map, you need only a pencil and a piece of paper. Depending on how complex you make the map, you might also use other tools. For example, you might use a compass, color pencils, or a ruler or other instrument with a straight edge.

Sketch maps are useful tools for communicating information in a visual way. They can show physical features, such as the locations of mountains, rivers, deserts, and oceans. They can also show human features, such as countries, cities, roads, and economic activities. Sketch maps can also show geographic distributions, such as natural resources, climate regions, vegetation patterns, and population density. Finally, sketch maps can show geographic relationships. For example, sketch maps might show relationships between climate and vegetation growth or between the locations of cities and nearby resources.

Developing the Skill One key to creating a useful sketch map is to keep it simple. Look at the sketch map of Brazil on this page. This map includes a variety of places and features, such as cities, highlands, and the Amazon River. If this map had too many labels, you might have trouble reading it. You will be asked to create sketch maps in section reviews throughout this textbook. If you are building a map from an activity in a previous section, make sure you have room in your map for new labels. If you do not have enough room on your existing map, create a new sketch map. The following guidelines will help you create useful sketch maps on your own:

- You can create sketch maps from a model, such as another map, or from a mental map.

- You can use a pencil to trace a model map's outline onto a thin sheet of paper. As an alternative, you might create your own freehand sketch map.

- Do not crowd your sketch map. Identify places that are important for your purpose. Leave out places that are not relevant or required.

- Assign symbols to represent certain kinds of places. For example, small dots or circles might represent cities. Shaded areas or special marks might show highland areas. Create a legend that identifies each symbol you use.

- Make sure the information in your map is clear. You might want to point out geographic features, distributions, and relationships in a caption.

Practicing the Skill

1. Study the political and physical maps of South America at the beginning of this unit. Then create a sketch map of the continent, labeling each country, each national capital, and at least three major rivers.
2. Locate a map of your county. Then create a sketch map that shows important towns, roads, and rivers there.

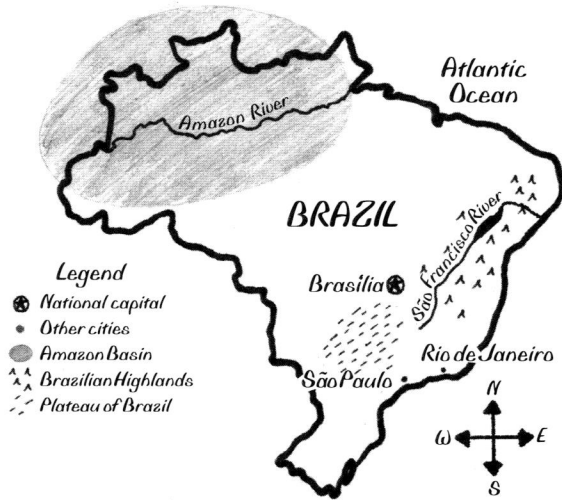

Workshop 2
Going Further: Thinking Critically

Direct students' attention to the cartogram on this page that illustrates the populations of South American countries in 2000. Then have students compare this cartogram to the ones they created in the activity below showing each country's GDP. Ask students to explain any relationships they see between countries' populations and GDPs. *(Possible answer: Students may notice that countries with large populations, like Brazil, often have large GDPs. Countries with small populations, like Suriname, tend to have smaller GDPs.)*

Ask students if the relationship between size and GDP holds true for all countries in South America. *(No, it does not. Argentina, for example, has a smaller population than Colombia but has a GDP* almost twice as high. Similarly, both Bolivia and Paraguay have larger populations than Uruguay, but Uruguay has a higher GDP than either.) Call on students to explain whether the connections they may have seen between the two cartograms are valid. *(Possible answer: They are not good connections, because many factors other than population can affect a country's GDP.)* Have students name some factors besides population that can shape a GDP. *(Possible answers: resources available in a country, technological advancement, educational facilities, and so on)*

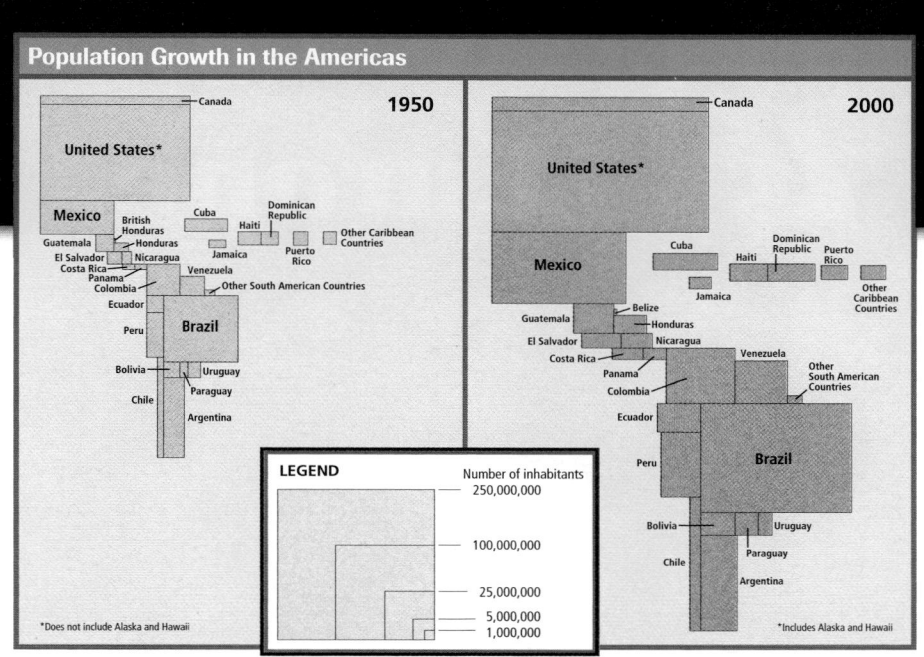

Population Growth in the Americas

1950

2000

LEGEND — Number of inhabitants

250,000,000
100,000,000
25,000,000
5,000,000
1,000,000

*Does not include Alaska and Hawaii

*Includes Alaska and Hawaii

Students' cartograms should accurately reflect the size of each country's GDP in relation to other countries in South America. Countries with strong economies, like Brazil, Argentina, and Colombia, should appear much larger than those with smaller economies, like Suriname and Guyana.

WORKSHOP 2

Using Cartograms

We are accustomed to looking at maps on which countries and states have been drawn in proportion to their geographic area. Yet it is possible to create maps that have been drawn so that sizes are proportional to something other than geographic area. Such maps are called cartograms. Actually, a cartogram is a cross between a map and a graph. Cartograms let geographers display statistical information visually. The size of countries or states can be created in proportion to many measures, such as population or GDP. For example, a cartogram showing all of the U.S. states in proportion to their population would look quite different from a map showing states by geographic area.

Developing the Skill Using a cartogram is much like reading a map. First, read the title and legend to identify the kind of information presented on the cartogram. The cartogram on this page compares the populations of countries in North and South America in 1950 and 2000. The legend shows how the size of a country changes depending on how many people live in that country. The largest countries on these cartograms are the United States, Brazil, and Mexico. Therefore, you know that those three countries are also the most populous countries in the Americas.

Practicing the Skill

You can make cartograms by hand or by using special computer software. Follow the instructions below to make a cartogram of South America by hand. You will need the following materials: color pencils or markers, scissors, an almanac or other database, graph paper, and a map of South America.

1. Find the total GDP for each country in South America.
2. Determine a scale for sizing each country by GDP. For example, you might use one square unit of area per $10 billion or $100 billion.
3. Cut out one piece of graph or grid paper for each South American country. Each piece should be cut in proportion to a country's GDP. For example, a country with a GDP of $10 billion should be shown as 10 times smaller than a country with a GDP of $100 billion.
4. Secure the pieces on paper or poster board. Each shape should be cut and pasted to reflect a country's approximate location in South America. You might color each piece so that it is easier to distinguish between countries.
5. Create a legend and title.
6. Write a short caption describing your cartogram and summarizing what it shows. Then present your cartogram to the rest of the class.

Using the Illustration

Focus students' attention on the picture on this page of Notre-Dame de Paris, one of Europe's most famous churches. Tell students that it rises above the Île de la Cité, a small island in the Seine in the heart of Paris on which the city was originally built. Ask students why the city's founders might have chosen this location for the city (*defense*).

The construction of Notre Dame began in the 1100s at the direction of Maurice de Sully, the Bishop of Paris. He wanted to build a spectacular new church that would embody the most advanced architectural knowledge of his day, an elaborate style now known as Gothic.

The cathedral's exterior illustrates several features common to Gothic architecture. Tall towers flank the main entrance, which features elaborately carved doors and statues. Graceful flying buttresses provide support to the walls, and large windows admit sunlight. Earlier cathedrals were built with thick walls and few windows to prevent the walls from collapsing. Compared to these structures, the interior of Notre Dame was light and airy. Ask students to identify other famous structures from around Europe and to compare elements of their architectural styles to that of Notre Dame.

Unit Objectives

1. Describe the major landforms, bodies of water, climates, and resources of Europe.

2. Explain the historical development of European countries and the influence of European cultures on other parts of the world.

3. Identify different cultural groups in Europe and understand how they interact with one another.

4. Discuss Europe's influence in international affairs.

5. Create and interpret graphic organizers to understand relationships between concepts and ideas.

6. Build and examine databases to compile and analyze statistical data.

UNIT 4 Europe

Notre Dame Cathedral, Paris

278

CONNECTING TO Literature

From "Pericles' Funeral Oration" in THUCYDIDES' HISTORY OF THE PELOPONNESIAN WAR

translated by Benjamin Jowett

Thucydides' *History of the Peloponnesian War* is still read for the insights it provides into the conduct of war. In this excerpt Pericles is speaking at the funeral of those Athenians who have died in the war.

Our form of government does not enter into rivalry with the institutions of others. We do not copy our neighbors, but are an example to them. It is true that we are called a democracy, for the administration is in the hands of the many and not of the few. But while the law secures equal justice to all alike in their private disputes, the claim of excellence is also recognized; and when a citizen is in any way distinguished, he is preferred to the public service, not as a matter of privilege, but as the reward of merit. Neither is poverty a bar, but a man may benefit his country whatever be the obscurity of his condition. There is no exclusiveness in our public life, and in our private intercourse we are not suspicious of one another, nor angry with our neighbor if he does what he likes; we do not put on sour looks at him which, though harmless, are not pleasant. While we are thus unconstrained in our private intercourse, a spirit of reverence pervades our public acts. . . .

. . . I would have you day by day fix your eyes upon the greatness of Athens, until you become filled with the love of her; and when you are impressed by the spectacle of her glory, reflect that this empire has been acquired by men who knew their duty and had the courage to do it, who in the hour of conflict had the fear of dishonor always present to them, and who, if ever they failed in an enterprise, would not allow their virtues to be lost to their country, but freely gave their lives to her as the fairest offering which they could present at her feast. The sacrifice which they collectively made was individually repaid to them; for they received again each one for himself a praise which grows not old. . . . Make them your examples.

A Closer Connection

Pericles and Athenian Democracy Pericles, the Athenian statesman who delivered the speech recorded by Thucydides, was one of the most powerful and respected men in all of ancient Greece. Under his leadership, Athens grew from a powerful city into the head of a great empire. Yet for all his power, Pericles was no king. The government he controlled was one of the world's first democracies.

In Pericles' Athens, the legislature included all male citizens, though women and slaves were allowed no political power. Political offices were open to all citizens, but all officials were held responsible to the general assembly. There was no separation of powers in this government. The assembly that served as the city's legislature also had executive and judicial powers.

DISCUSSION: Lead a discussion about how Athenian democracy was both similar to and different from modern American democracy.

Analyzing the Primary Source

Answers

1. He appeals to their love and dedication to their city.

2. Possible answers: Power is in the hands of the masses, not a few people. All people are considered equal before the law.

In this unit students will learn about the landscapes and cultures of one of the world's most densely settled continents.

A wide range of climates, from semiarid to icecap, provides a clue to the many ways that Europeans use the land. The farms, in general, are so productive that the majority of the population can live in cities and towns and work in manufacturing and service jobs.

Europe's past is dynamic and turbulent. Countless leaders have gone into battle to control its resources. Explorers, conquerors, and colonists spread out from Europe to influence every corner of the world. The continent was at the center of two world wars. Conflicts between religious and ethnic groups continue to trouble post-war Europe. Recently created republics struggle to find their place among the other countries.

Over the centuries, Europeans have created art, architecture, music, literature, and drama valued by people around the world. The region's international influence has grown with many of the countries joining the European Union.

Your Classroom Time Line

These are the major dates and time periods for this unit. Have students enter them on the time line you created earlier. You may want to watch for these dates as students progress through the unit.

c.* 12,000 B.C. Early peoples create cave paintings at what is now Altamira, Spain.

c. 3,100 B.C. Construction on Stonehenge begins in England.

c. 2000 B.C. A complex civilization thrives on Crete.

c. 700s B.C. City-states organize on the Greek mainland.

c. 750 B.C. Rome is established by the Latins.

c. 500 B.C. The Romans establish a republic in central Italy.

400s B.C. The Parthenon is built.

200s B.C.–100s A.D. The power of Rome expands to control large areas of Europe and the Mediterranean world.

A.D. 79 Mount Vesuvius erupts and destroys three Roman towns.

A.D. 80 The Roman Colosseum is completed.

*c. stands for *circa* which means "about."

POLITICAL MAP ANSWERS

1. Iceland, United Kingdom, Ireland, Malta
2. Norway and Sweden

CRITICAL THINKING ANSWER

3. the Pyrenees between France and Spain; the Alps between Italy, France, Switzerland, Austria, and Germany

UNIT 4 ATLAS

The World in Spatial Terms

Europe: Political

1. **Places and Regions** Which European countries are island countries?
2. **Places and Regions** Compare this map to the physical map. Which two countries occupy the Scandinavian Peninsula?

Critical Thinking

3. **Analyzing Information** Compare this map to the physical map. Where in Europe do political boundaries coincide with physical barriers?

SCALE
0   250   500 Miles
0   250 500 Kilometers
Projection: Polyconic

SCALE
0   250   500 Miles
0   250   500 Kilometers
Projection: Azimuthal Equal Area

Boundaries
National capitals
Other cities

Using the Political Map

Direct students' attention to the **political map** on the opposite page. Ask them to identify the largest country in Europe (*France*). Then ask them to name some countries that are so small that their area is not apparent on the map (*Andorra, Liechtenstein, Malta, Monaco, San Marino, Vatican City*). Ask which country would be the largest if all territory under its control were included in its area. (*Denmark would the largest because it controls Greenland.*) Point out that most of the large islands in the Mediterranean Sea are not independent countries. Call on volunteers to identify the countries to which they belong (*Balearic Islands, Spain; Corsica, France; Sardinia and Sicily, Italy; Crete, Greece*).

Using the Physical Map

Focus students' attention on the **physical map** on this page. Have them identify the continent's major mountain ranges (*Alps, Apennines, Carpathians, Dinaric Alps, Kjølen Mountains, Pyrenees*). Ask them to identify the highest point in Europe (*Mont Blanc, on France's border with Italy and Switzerland*). Then ask them to identify the part of Europe that lies below sea level (*Atlantic coastline of the Netherlands*). Ask students to name a challenge the people of the Netherlands have had to solve to settle in this area (*how to keep the North Sea from flooding their country*).

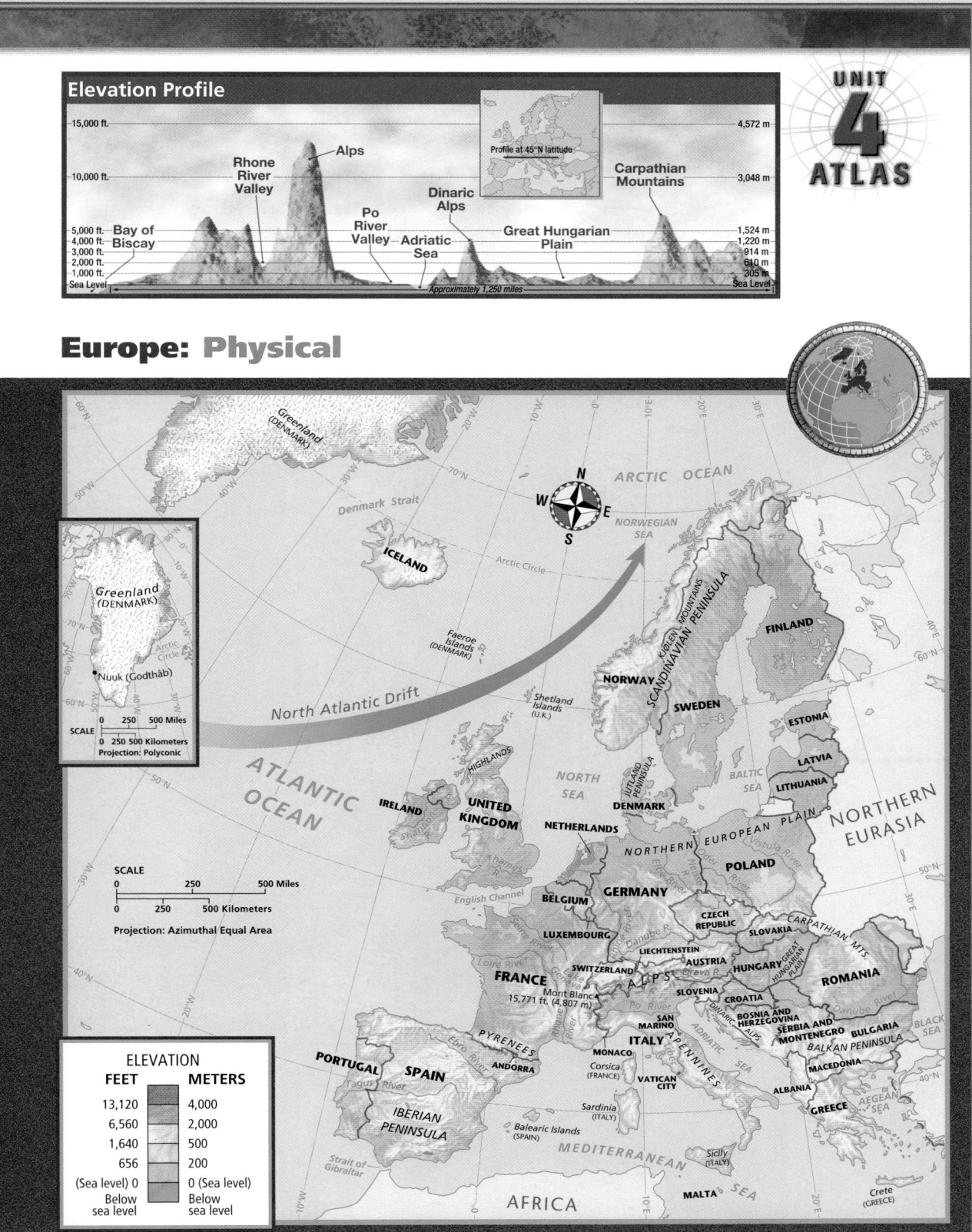

Europe: Physical

Your Classroom Time Line, (continued)

300s San Marino is founded.

370s Huns invade eastern Europe.

400s Angles and Saxons migrate to the British Isles.

476 Rome falls to invaders.

700s Charlemagne unites several German kingdoms.

700s Moors invade the Iberian Peninsula.

754 The Papal States are created in central Italy.

900s Vikings settle parts of Greenland's coast.

1066 William of Normandy conquers England.

1200s The Hanseatic League is formed.

1300s The Renaissance begins in Italy.

1347 The Black Death reaches Europe.

Late 1300s Ottomans conquer the Balkan Peninsula.

1400s The Vikings abandon Greenland.

1492 The Moors are forced out of Granada, their last stronghold in Spain.

Early 1500s The Reformation begins in Germany.

1500s–1700s European countries establish worldwide empires and global trade networks.

1500s Spain takes control of most of Middle and South America.

1540s Portugal establishes a sea trade route to Japan.

1707 England, Scotland, and Wales unite to form Great Britain.

1700s Great Britain develops coal- and iron-mining industries. The Industrial Revolution begins.

Using the Climate Map

Focus students' attention on the **climate map** on this page. Ask students to name the three largest climate regions in Europe (*marine west coast, Mediterranean, humid continental*). Have students compare this map to the **land use and resources map**. Ask students what economic activities are common in these climate regions (*commercial farming, livestock raising, manufacturing, nomadic herding, forestry, fishing*). Ask students how Europe's climates have shaped these economic activities. (*Possible answers: Climates are well-suited to supporting agriculture and herding industries. They allow the growth of forests, which has led to the creation of forestry industries. The climate is also favorable for human settlement. The cities that have been established in these regions are home to manufacturing and other industries.*)

Your Classroom Time Line, (continued)

1700s Scandinavians resettle Greenland.

1768 France annexes Corsica.

1776 The United States declares independence from Great Britain.

1789 The French Revolution begins.

1795 Protestants in Northern Ireland form the Orange Society.

Early 1800s British engineers build the world's first railroad.

1800s Wars and internal strife shake Spain's power.

1801 Ireland unites with Great Britain to form the United Kingdom.

1829 Greece gains its independence from the Ottoman Turks.

1840s Ireland suffers through the potato famine.

1860 Construction of London's underground railway system begins.

1861 Italian states unify into a single country.

1867 The Austro-Hungarian Empire is formed.

CLIMATE MAP ANSWERS

1. Mediterranean, humid subtropical, marine west coast, humid continental

2. arid, tropical wet and dry, humid tropical; location prevents these climates

CRITICAL THINKING ANSWER

3. Western Europe has mainly a marine west coast climate, while eastern Europe has a humid continental climate. Western Europe's coastal location gives it a marine influence, while eastern Europe's interior position gives it a continental influence.

Europe:
Climate

1. *Physical Systems* Which climates cover large parts of Europe?

2. *Physical Systems* Which climate types are not found in Europe? Why do you think this is so?

Critical Thinking

3. Comparing How is the climate of western Europe different from that of eastern Europe? How might the positions of these two areas on the continent of Europe influence these differences?

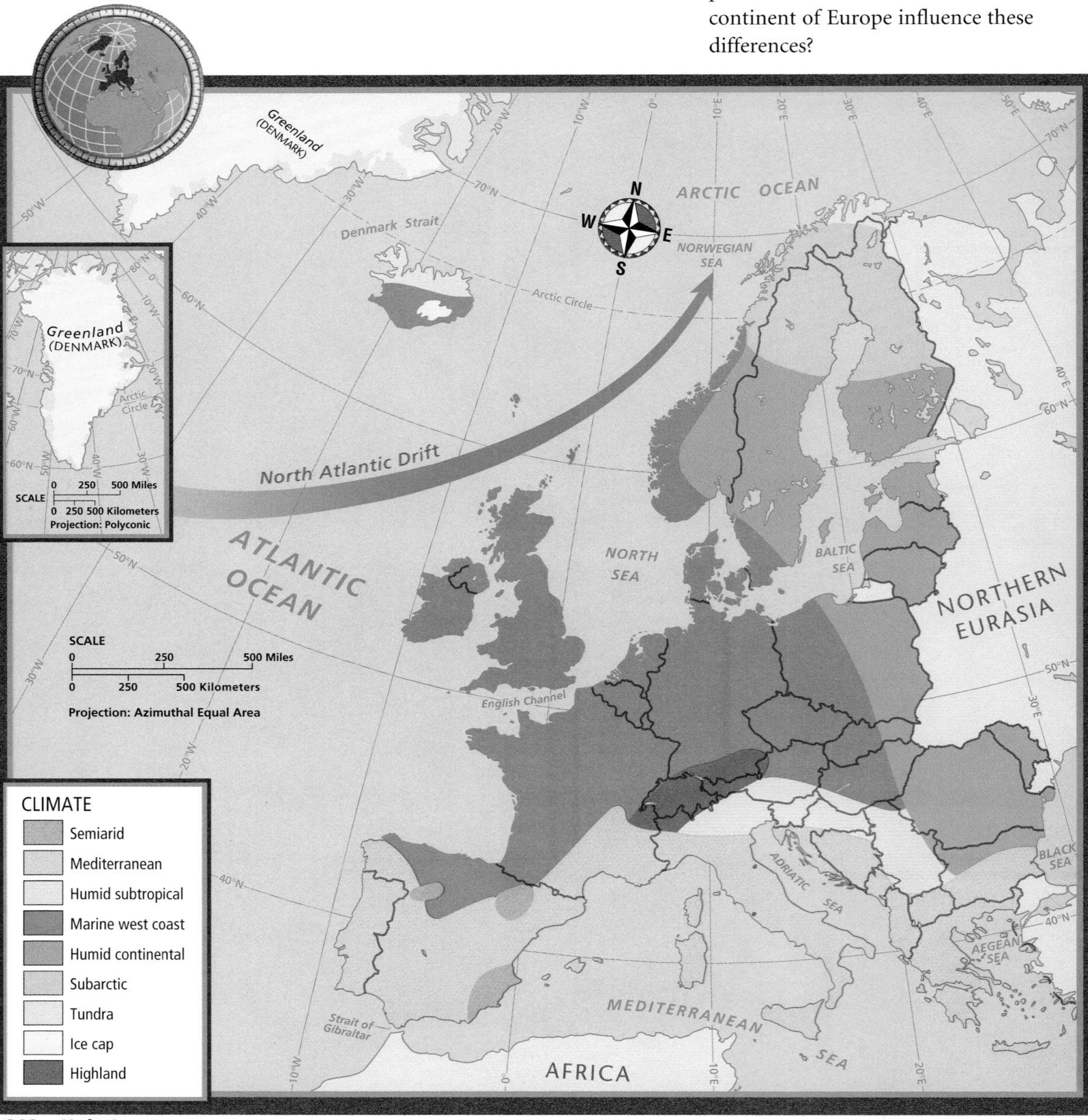

UNIT 4 ATLAS

Greenland (DENMARK)

ARCTIC OCEAN

NORWEGIAN SEA

Denmark Strait

Arctic Circle

North Atlantic Drift

ATLANTIC OCEAN

NORTH SEA

BALTIC SEA

NORTHERN EURASIA

English Channel

ADRIATIC SEA

BLACK SEA

AEGEAN SEA

Strait of Gibraltar

MEDITERRANEAN SEA

AFRICA

Greenland (DENMARK)

Arctic Circle

SCALE
0   250   500 Miles
0 250 500 Kilometers
Projection: Polyconic

SCALE
0   250   500 Miles
0   250   500 Kilometers
Projection: Azimuthal Equal Area

CLIMATE

- Semiarid
- Mediterranean
- Humid subtropical
- Marine west coast
- Humid continental
- Subarctic
- Tundra
- Ice cap
- Highland

Using the Precipitation Map

Direct students' attention to the **precipitation map** on this page. Ask them to identify the areas of Europe that receive the most precipitation each year (*northern Spain, northwestern Great Britain, the western shores of the Balkan Peninsula, northwestern Scandinavia*). Then ask them to identify the driest parts of the continent (*central and southern Spain, the northeastern Balkans, northern Scandinavia*). Have students compare this to the **physical map.** Ask them how these variations in precipitation relate to Europe's physical landscape. (*Places that receive more precipitation are found west of mountains. Dry areas are east of mountains. The continent's mountain ranges block moist air from reaching these places.*)

Europe:
Precipitation

1. (*Physical Systems*) Compare this map to the climate map. How much precipitation falls in Spain's semiarid regions?
2. (*Places and Regions*) Compare this map to the political map. How much precipitation do most areas in France and Germany receive?

Critical Thinking

3. **Analyzing Information** Which areas receive the most precipitation? Why do you think this is so?

UNIT 4 ATLAS

Your Classroom Time Line,
(continued)

Late 1800s–Early 1900s Eastern Balkans regain their independence from the Ottomans.

1871 Germany is unified under Prussian rule.

1873 The cities of Buda and Pest join to form Budapest.

1887–89 The Eiffel Tower is built.

1890–1914 Germany prospers as an industrial and military power.

1912 Albania gains its independence from Yugoslavia.

1914–18 World War I is fought.

1917 Communists take control of Russia.

1918 Poland gains its independence.

1920s Benito Mussolini takes control of Italy.

1921 Republic of Ireland gains its independence from the United Kingdom.

1929 Vatican City becomes an independent country.

1933 The Nazi Party begins its rise to power in Germany.

1936–39 The Spanish Civil War is fought. Francisco Franco takes control of the country.

1939 Germany invades Poland to begin World War II.

1943 60,000 Jews die in the Warsaw Ghetto Uprising.

ANNUAL PRECIPITATION

Centimeters	Inches
Under 25	Under 10
25–50	10–20
50–100	20–40
100–150	40–60
150–200	60–80
Over 200	Over 80

PRECIPITATION MAP ANSWERS

1. 10 to 20 inches
2. 20 to 40 inches

CRITICAL THINKING ANSWER

3. western coastal areas, such as northwestern Spain, western Balkans, western Ireland, western Norway; probably because prevailing westerlies bring moist ocean air inland

283

Using the Population Map

Have students examine the **population map** on this page. Point out that many European cities are located on or near rivers. Have students compare this map to the **physical map** to identify some of these cities and the rivers on which they were built. Students may need to use the chapter maps to confirm details. *(Possible answers: London is on the Thames. Paris is on the Seine. Rome is on the Tiber. Rotterdam is on the Rhine. Budapest is on the Danube.)*

Tell students that Europe is the most densely populated continent in the world. Ask them to identify the countries with the largest areas of high population density *(United Kingdom, Netherlands, Belgium, Germany, Italy)*. Then ask them which countries appears to have the lowest populations *(Iceland, Norway, Sweden, Finland, Latvia, Lithuania, Estonia)*. Ask students why they think these countries are less densely settled. *(Possible answers: harsher climates, remote locations)*

Your Classroom Time Line,
(continued)

1943 Mussolini is overthrown in Italy.

1944 Allies invade Normandy.

1945 Germany is defeated.

1955 Austria regains its independence.

1956 The Soviet Union invades Hungary.

1957 Belgium, the Netherlands, Luxembourg, France, Italy, and West Germany create the European Economic Community (EEC), later called the European Community (EC).

1961 The Berlin Wall is built.

1967 Military leaders seize control of Greece.

1968 The Soviet Union invades Czechoslovakia.

1974 Democracy is re-established in Greece.

1975 Francisco Franco dies. Spain becomes a democracy.

1980 Josip Broz Tito, the dictator of Yugoslavia, dies.

1981 Greece joins the EC.

1986 Spain and Portugal join the EC.

1990 Germany is reunited.

1991 Slovenia and Croatia declare their independence from Yugoslavia.

1991 Poland gains its independence from the Soviet Union.

1992 Albania breaks away from communist control.

POPULATION MAP ANSWERS

1. Northern European Plain

2. Iceland

CRITICAL THINKING ANSWER

3. It has many lowland areas with mild climates.

284

UNIT 4 ATLAS

Europe: Population

1. *Environment and Society* Compare this map to the physical map. What major lowland region in Europe has a high population density?

2. *Places and Regions* Which country in Europe has the lowest average population density?

Critical Thinking

3. **Drawing Inferences** Compare this map to the physical and climate maps. Why do you think Europe's average population density is so high?

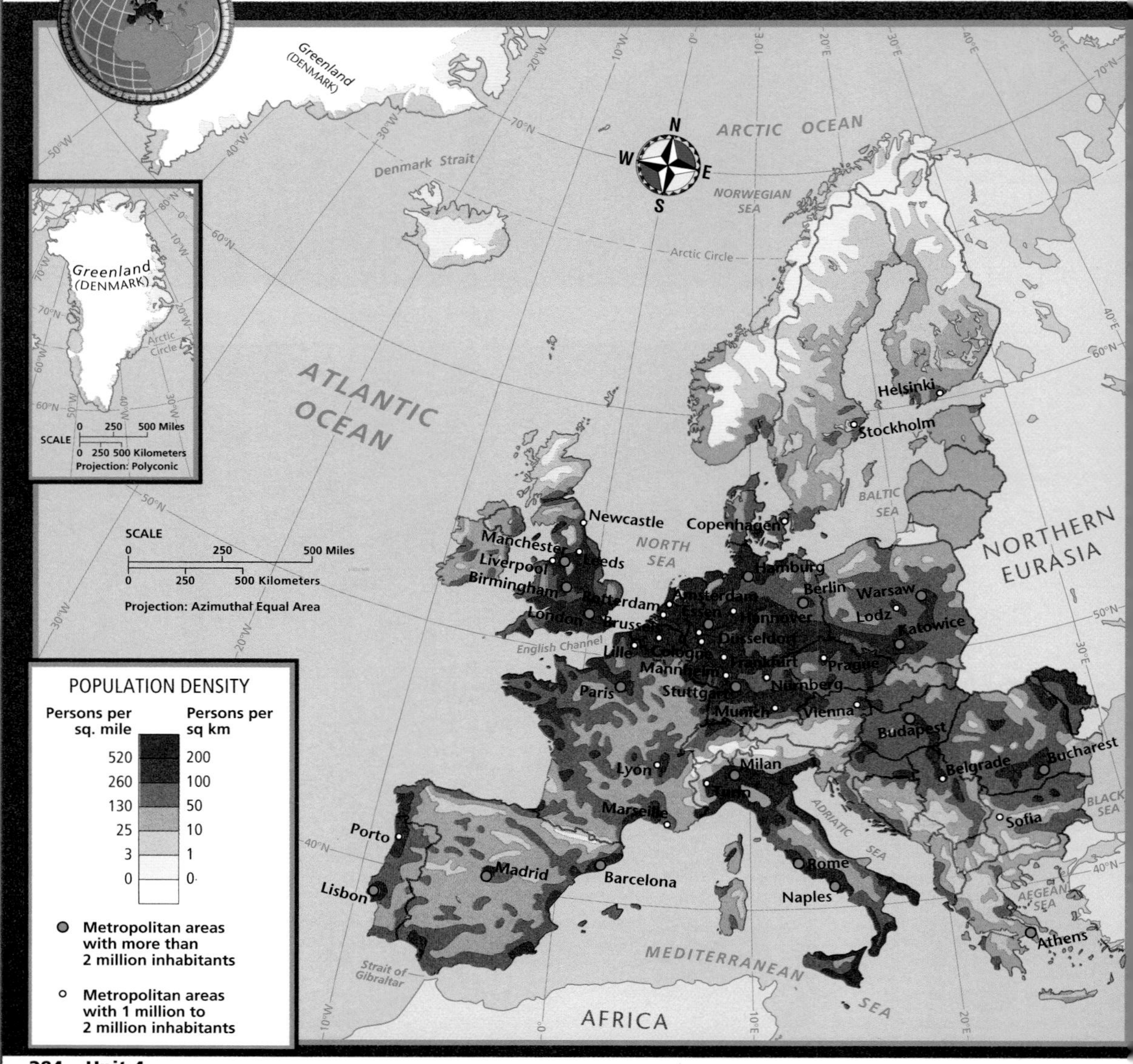

POPULATION DENSITY

Persons per sq. mile	Persons per sq km
520	200
260	100
130	50
25	10
3	1
0	0

● Metropolitan areas with more than 2 million inhabitants

○ Metropolitan areas with 1 million to 2 million inhabitants

Using the Land Use and Resources Map

Focus students' attention on the **land use and resources map** on this page. Ask students to identify countries that do not have major manufacturing and trade centers *(Iceland, Estonia, Latvia, Lithuania, Serbia and Montenegro, Albania, Macedonia, Croatia, Bosnia and Herzegovina)*. Have them identify other economic activities that are important to these countries instead *(commercial farming, fishing, livestock raising, some mineral production)*. Then ask students which countries appear to have the largest number of major manufacturing and trade centers *(Germany and Italy)*.

UNIT 4 ATLAS

Europe:
Land Use and Resources

1. *Places and Regions* What are the main economic resources of the North Sea?
2. *Human Systems* What is the most widespread type of land use in Europe?

Critical Thinking

3. **Drawing Inferences** Is subsistence agriculture found in Europe? What does this indicate about Europe's level of economic development?

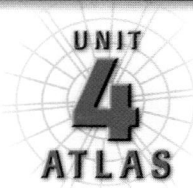

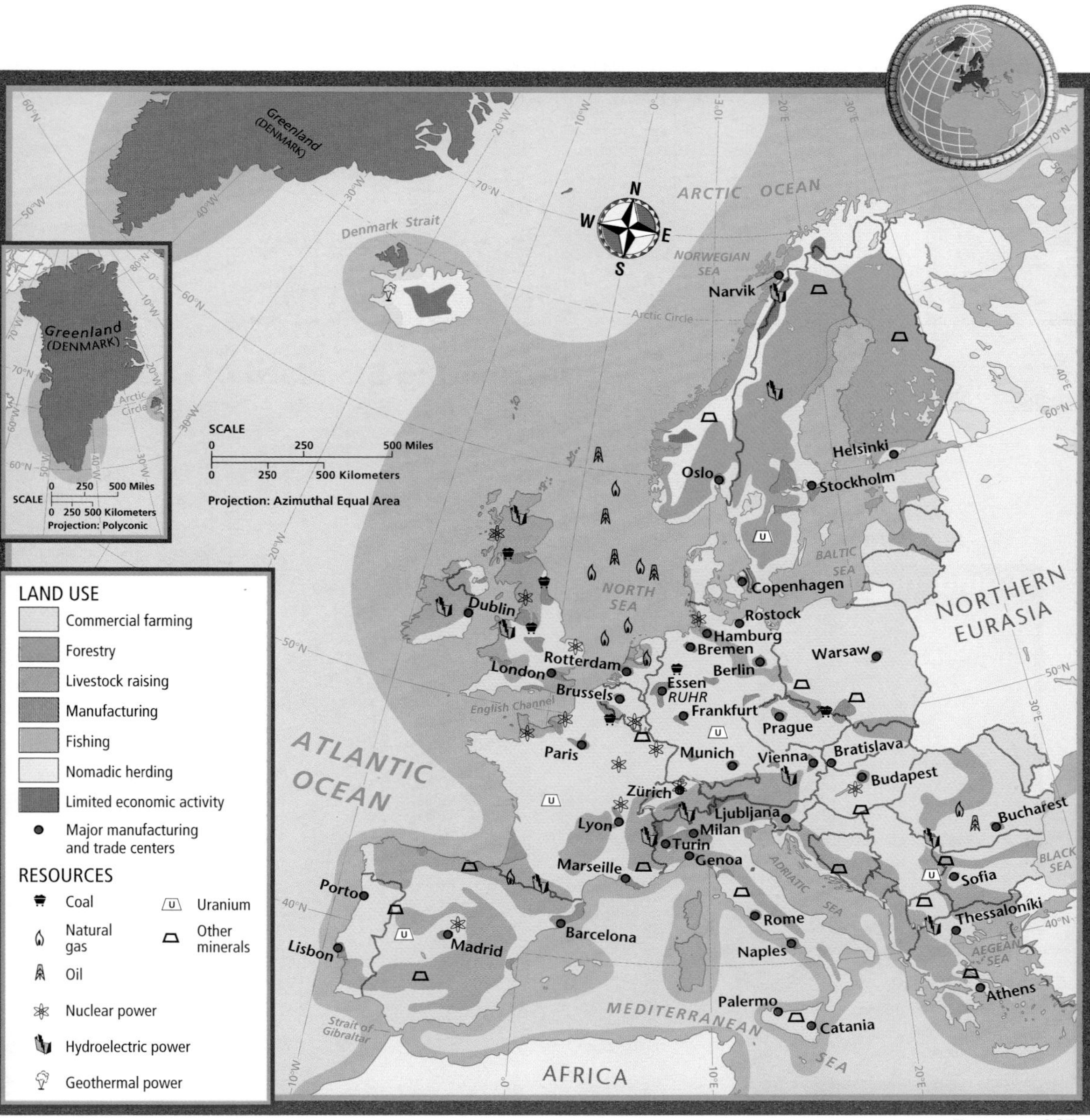

Greenland (DENMARK)
ARCTIC OCEAN
Denmark Strait
NORWEGIAN SEA
Arctic Circle
Narvik
Helsinki
Oslo
Stockholm
BALTIC SEA
NORTH SEA
Copenhagen
Dublin
Rostock
Hamburg
Bremen
Berlin
Warsaw
NORTHERN EURASIA
London
Rotterdam
Essen
RUHR
Brussels
English Channel
Frankfurt
Prague
Paris
Munich
Vienna
Bratislava
Budapest
ATLANTIC OCEAN
Zürich
Lyon
Ljubljana
Milan
Turin
Genoa
Bucharest
BLACK SEA
Marseille
ADRIATIC SEA
Sofia
Porto
Thessaloníki
Lisbon
Madrid
Barcelona
Rome
Naples
AEGEAN SEA
Athens
Strait of Gibraltar
MEDITERRANEAN SEA
Palermo
Catania
AFRICA
SCALE
0 250 500 Miles
0 250 500 Kilometers
Projection: Azimuthal Equal Area
Greenland (DENMARK)
Arctic Circle
0 250 500 Miles
SCALE
0 250 500 Kilometers
Projection: Polyconic
LAND USE
Commercial farming
Forestry
Livestock raising
Manufacturing
Fishing
Nomadic herding
Limited economic activity
Major manufacturing and trade centers
RESOURCES
Coal
Uranium
Natural gas
Other minerals
Oil
Nuclear power
Hydroelectric power
Geothermal power

Your Classroom Time Line, (continued)

1992 Bosnia declares its independence from Yugoslavia, beginning a long civil war.

1993 Czechoslovakia peacefully separates into the Czech Republic and Slovakia.

1993 The EC becomes the European Union, or EU.

1995 The Dayton Accord ends the war in Bosnia, giving Croatia and Bosnia their independence.

1997 Violence escalates in Kosovo between Albanians and Serbs.

1997 Poland establishes a free market economy.

1998 Northern Ireland, the United Kingdom, and the Republic of Ireland create a shared government in Ireland.

1999 NATO and the United States intervene to end fighting in Kosovo.

2002 The euro is adopted as the official currency of many countries.

2004 The European Union expands to 25 countries.

LAND USE AND RESOURCES MAP ANSWERS

1. oil, gas, fish
2. commercial farming

CRITICAL THINKING ANSWER

3. no; highly developed

Copy the following chart onto the chalkboard, omitting the italicized figures. Ask students to use the population and electricity consumption figures in the Fast Facts table to calculate the annual electricity consumption of each country in Europe. *(Multiply the average consumption per person by the total population to obtain the country's total consumption.)* Note that some countries do not appear on the chart. Then ask students to examine the information in the chart. Ask them which countries in Europe do not produce enough electricity to supply their populations' needs *(Albania, Bosnia and Herzegovina, Croatia, Finland, Italy, Latvia, Luxembourg,* *Moldova, Netherlands).* Have students use the information in the unit atlas and prior knowledge about these countries to suggest reasons that they may need to import power. *(Possible answers: Albania, Bosnia and Herzegovina, Croatia, and Moldova are involved in regional conflicts that might prohibit local power production. The people of Finland consume more electricity than many Europeans because their country has a harsh climate. Luxembourg and the Netherlands are small but densely populated countries whose resources cannot produce enough electricity to supply their populations.)*

Historical Geography

Córdoba Córdoba's location in southern Spain has been a major factor in the city's history. Probably founded by Carthaginians from northern Africa, Córdoba was conquered by the Romans in the 100s B.C. Under Roman rule, the city grew and prospered, but it was invaded by Visigoths in the A.D. 500s and fell into decline. The Moors who conquered Spain in the 700s destroyed much of the city but later rebuilt it and made it the capital of all Moorish Spain. Under the Moors, Córdoba flourished and became a thriving center of Islamic culture. The city's artists and thinkers were renowned throughout Europe.

In the 1200s Christian forces from Spain reclaimed the city from the Moors and made it a military base. In the 1800s Córdoba was sacked by French troops, and during the Spanish Civil War it was occupied by the forces of General Francisco Franco.

DISCUSSION: Lead a class discussion on the various groups that conquered Córdoba and their contributions to Spanish and European culture.

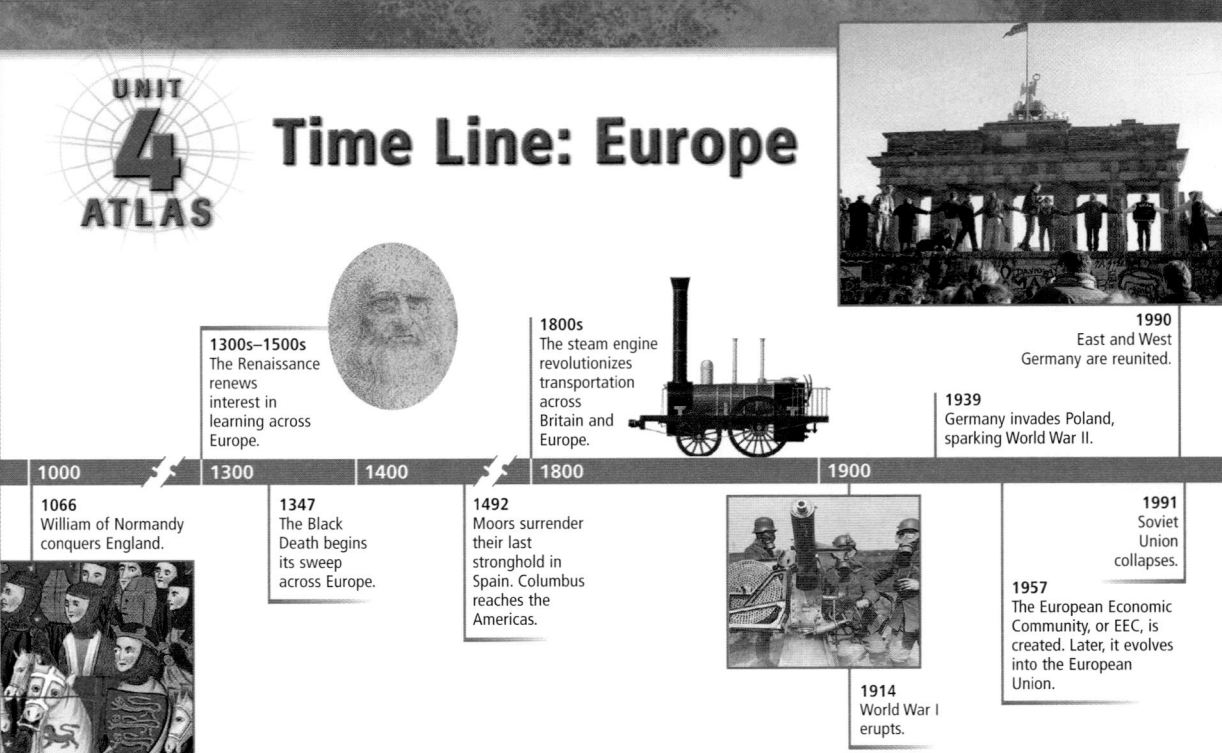

Time Line: Europe

UNIT 4 ATLAS

1300s–1500s The Renaissance renews interest in learning across Europe.

1800s The steam engine revolutionizes transportation across Britain and Europe.

1990 East and West Germany are reunited.

1939 Germany invades Poland, sparking World War II.

| 1000 | 1300 | 1400 | 1800 | 1900 |

1066 William of Normandy conquers England.

1347 The Black Death begins its sweep across Europe.

1492 Moors surrender their last stronghold in Spain. Columbus reaches the Americas.

1914 World War I erupts.

1957 The European Economic Community, or EEC, is created. Later, it evolves into the European Union.

1991 Soviet Union collapses.

The United States and Europe

Comparing Sizes

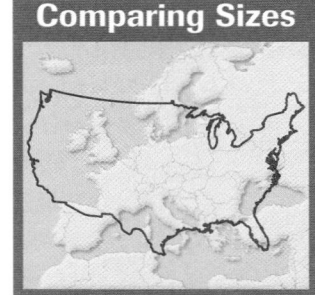

GO TO: go.hrw.com
KEYWORD: SW3 Almanac
FOR: Additional information and reference sources

Comparing Standard of Living

COUNTRY	LIFE EXPECTANCY (in years)	INFANT MORTALITY (per 1,000 live births)	LITERACY RATE	DAILY CALORIC INTAKE (per person)
Albania	70, male 75, female	37	87%	2,976
France	76, male 83, female	4	99%	3,541
Germany	76, male 82, female	4	100%	3,402
Greece	76, male 82, female	6	98%	3,630
Ireland	75, male 80, female	5	98%	3,622
Moldova	61, male 70, female	42	99%	2,763
Poland	70, male 78, female	9	100%	3,351
Serbia and Montenegro	71, male 77, female	17	98%	2,963
Spain	76, male 83, female	5	98%	3,348
United Kingdom	75, male 81, female	6	99%	3,257
United States	74, male 80, female	7	97%	3,757

Sources: *World Almanac and Book of Facts, 2004; Britannica Book of the Year, 2002*

Country	Electricity Production (in billions of kilowatt hours)	Electricity Consumption (in billions of kilowatt hours)
Albania	5.29	5.96
Austria	58.75	54.74
Belgium	74.28	78.04
Bosnia and Herzegovina	9.98	8.19
Bulgaria	41.38	32.53
Croatia	12.12	14.18
Czech Republic	70.04	55.41
Denmark	35.47	32.63
Finland	71.20	76.08
France	520.15	414.99
Germany	544.83	506.96
Greece	49.79	48.91
Hungary	34.39	35.23
Ireland	23.53	21.87
Italy	258.84	289.01
Latvia	4.37	6.03

Lithuania	14.62	8.57
Luxembourg	0.46	6.70
Macedonia	6.47	6.24
Moldova	3.39	3.24
Netherlands	88.32	99.11
Norway	120.10	114.46
Poland	134.96	118.85
Portugal	44.32	41.63
Romania	50.86	46.03
Serbia and Montenegro	31.71	32.29
Slovakia	30.29	24.40
Slovenia	13.69	13.94
Spain	222.55	210.60
Sweden	152.91	135.26
Switzerland	68.68	53.66
United Kingdom	360.93	346.37
United States	3,719.49	3,601.50

Source: Central Intelligence Agency, *The World Factbook 2004*

Fast Facts: Europe

UNIT 4 ATLAS

FLAG	COUNTRY / Capital	POPULATION (in millions) / POP. DENSITY	AREA	PER CAPITA GDP (in US $)	WORKFORCE STRUCTURE (largest categories)	ELECTRICITY CONSUMPTION (kilowatt hours per person)	TELEPHONE LINES (per person)
	Albania Tiranë	3.2 / 299/sq. mi.	11,100 sq. mi. / 28,749 sq km	$ 4,500	50% industry, services 50% agriculture	1,863 kWh	0.07
	Andorra Andorra la Vella	0.07 / 397/sq. mi.	181 sq. mi. / 469 sq km	$ 19,000	78% services 21% industry	Not available	0.51
	Austria Vienna	8.1 / 254/sq. mi.	32,378 sq. mi. / 83,859 sq km	$ 27,700	67% services 29% industry, crafts	6,758 kWh	0.47
	Belgium Brussels	10.3 / 884/sq. mi.	11,780 sq. mi. / 30,510 sq km	$ 29,000	73% services 25% industry	7,577 kWh	0.50
	Bosnia and Herzegovina Sarajevo	4.2 / 211/sq. mi.	19,741 sq. mi. / 51,129 sq km	$ 1,900	51% manufacturing 15% services	1,950 kWh	0.12
	Bulgaria Sofia	7.9 / 185/sq. mi.	42,823 sq. mi. / 110,911 sq km	$ 6,600	43% services 31% industry	4,118 kWh	0.37
	Croatia Zagreb	4.4 / 203/sq. mi.	21,831 sq. mi. / 56,542 sq km	$ 8,800	Not available	3,223 kWh	0.42
	Czech Republic Prague	10.2 / 337/sq. mi.	30,450 sq. mi. / 78,865 sq km	$ 15,300	60% services 35% industry	5,432 kWh	0.38
	Denmark Copenhagen	5.4 / 328/sq. mi.	16,639 sq. mi. / 43,095 sq km	$ 29,000	79% services 17% industry	6,042 kWh	0.70
	Estonia Tallinn	1.3 / 76/sq. mi.	17,462 sq. mi. / 45,226 sq km	$ 10,900	69% services 20% industry	4,680 kWh	0.36
	Finland Helsinki	5.2 / 44/sq. mi.	130,128 sq. mi. / 337,030 sq km	$ 26,200	32% public services 22% industry	14,630 kWh	0.55
	France Paris	60.1 / 285/sq. mi.	211,209 sq. mi. / 547,029 sq km	$ 25,700	71% services 25% industry	6,905 kWh	0.56
	Germany Berlin	82.5 / 610/sq. mi.	137,846 sq. mi. / 357,020 sq km	$ 26,600	64% services 33% industry	6,145 kWh	0.65
	Greece Athens	11.0 / 217/sq. mi.	50,942 sq. mi. / 131,939 sq km	$ 19,000	60% services 20% industry	4,446 kWh	0.51
	Hungary Budapest	9.9 / 277/sq. mi.	35,919 sq. mi. / 93,030 sq km	$ 13,300	65% services 27% industry	3,559 kWh	0.37
	Iceland Reykjavík	0.3 / 7/sq. mi.	39,768 sq. mi / 102,999 sq km	$ 25,000	60% services 13% manufacturing	25,314 kWh	0.66
	Ireland Dublin	4.0 / 149/sq. mi.	27,135 sq. mi. / 70,279 sq km	$ 30,500	63% services 29% industry	5,468 kWh	0.47

Sources: Central Intelligence Agency, *World Factbook 2003*; *The World Almanac and Book of Facts, 2004*
The CIA calculates per capita GDP in terms of purchasing power parity. This formula equalizes the purchasing power of each country's currency.

Island Security The locations of human settlements are often determined by the ease with which they can be defended. Throughout history, islands have been settled partly because the water surrounding them makes invasion difficult. The British Isles, for example, have not been successfully invaded since 1066, when Normans under William the Conqueror crossed the English Channel from France and defeated King Harold II. Since then all attempts to conquer the British Isles have failed. In 1588 the 130 ships of the Spanish Armada were ordered to invade England. Their attempt was unsuccessful, however, as severe weather and the English navy destroyed most of the fleet. In the modern era, both Napoléon Bonaparte and Adolf Hitler tried to conquer the island, but neither met with any success.

CRITICAL THINKING:
Why are island locations easier to defend than mainland areas?
(Possible answers: Invading armies must be transported by ship and cannot bring their full force into battle at once. Supplies must also be transported across the water.)

Fast Facts Activity

Copy the following chart onto the chalkboard, omitting the italicized figures. Have students examine each country's population figures to calculate the ratio of its population to the number of passenger vehicles found there. *(Divide the population by the number of vehicles. For example, Austria's 8.1 million inhabitants own 3.78 million vehicles, which represents one vehicle for every 2.17 people.)* Note that some countries do not appear on the chart. Ask how their results for European countries compare to those for the United States. *(Most European countries have a higher "persons per vehicle" number than the United States.)* Ask students to identify the countries that have the fewest vehicles to serve their populations (Moldova, Serbia and Montenegro, Romania, Latvia). Call on students to explain these findings, using the Fast Facts table for reference. *(These countries have some of the lowest per capita GDPs in Europe. Many of their residents cannot afford to own vehicles.)* Direct students' attention to the last column of the chart, which shows the total length of railroad lines in each country. Ask how Europe's railroad service compares to that of the United States. *(Europe appears to have more railroads than the United States.)* Remind students that the United States is actually larger than all of Europe. Ask how these railroad figures might help to explain Europe's lower vehicle to person ratio. *(Possible answer: Since Europe has a more developed rail system, more people might travel by train than by car.)*

Historical Geography

The Roma Originally from northern India, the nomadic Roma, or Gypsies, had moved into southeastern Europe by the 1300s. Many Roma still live in this area, but different groups have scattered across Europe. The Kalderash—the largest group—of Central Europe and the Balkans are skilled artisans. The Gitanos of Spain and southern France and the Manush of France and Germany are renowned for their music and showmanship. Regardless of their locations, however, most Roma retain some similar characteristics. They speak Romany, an Indo-European language related to those of northwest India. They also tend to practice trades that do not require permanent settlements. Many work as metalworkers, performers, traders, or mechanics. Roma women are also famous as fortunetellers.

For centuries, Europeans have been suspicious of wandering Roma and have frequently banned them from cities or countries. Hundreds of thousands were killed by the Nazis during World War II. Between 2 and 7 million Roma live in the world today. Most live in Europe, though many have given up their nomadic lifestyles.

Fast Facts: Europe

FLAG	COUNTRY / Capital	POPULATION (in millions) / POP. DENSITY	AREA	PER CAPITA GDP (in US $)	WORKFORCE STRUCTURE (largest categories)	ELECTRICITY CONSUMPTION (kilowatt hours per person)	TELEPHONE LINES (per person)
	Italy Rome	57.4 / 506/sq. mi.	116,305 sq. mi. / 301,229 sq km	$ 25,000	63% services 32% industry	5,035 kWh	0.48
	Latvia Riga	2.3 / 93/sq. mi.	24,938 sq. mi. / 64,589 sq km	$ 8,300	60% services 25% industry	2,621 kWh	0.30
	Liechtenstein Vaduz	0.03 / 533/sq. mi.	62 sq. mi. / 161 sq km	$ 25,000	51% services 47% industry	Not available	0.61
	Lithuania Vilnius	3.4 / 137/sq. mi.	25,174 sq. mi. / 65,200 sq km	$ 8,400	50% services 30% industry	2,521 kWh	0.27
	Luxembourg Luxembourg	0.5 / 454/sq. mi	998 sq. mi. / 2,585 sq km	$ 44,000	90% services 8% industry	13,400 kWh	0.77
	Macedonia Skopje	2.1 / 207/sq. mi.	9,781 sq. mi. / 25,333 sq km	$ 5,000	23% agriculture 18% mining, manuf.	2,973 kWh	0.26
	Malta Valletta	0.4 / 3,179/sq. mi.	122 sq. mi. / 316 sq km	$ 17,000	71% services 24% industry	4,173 kWh	0.53
	Moldova Chişinău	4.3 / 328/sq. mi.	13,067 sq.mi. / 33,843 sq km	$ 2,500	40% agriculture 14% industry	754 kWh	0.15
	Monaco Monaco	0.03 / 40,000/sq. mi.	0.75 sq. mi. / 1.94 sq km	$ 27,000	22% manufacturing 22% public admin.	Not available	0.97
	Netherlands Amsterdam, The Hague	16.1 / 1,232/sq. mi.	16,033 sq. mi. / 41,525 sq km	$ 26,900	73% services 23% industry	6,156 kWh	0.62
	Norway Oslo	4.5 / 38/sq. mi.	125,182 sq. mi. / 324,220 sq km	$ 31,800	74% services 22% industry	25,436 kWh	0.73
	Poland Warsaw	38.6 / 328/sq. mi.	120,728 sq. mi. / 312,684 sq km	$ 9,500	50% services 28% agriculture	3,079 kWh	0.30
	Portugal Lisbon	10.1 / 284/sq. mi.	35,672 sq. mi. / 92,390 sq km	$ 18,000	60% services 30% industry	4,122 kWh	0.43
	Romania Bucharest	22.3 / 251/sq. mi.	91,699 sq. mi. / 237,499 sq km	$ 7,400	40% agriculture 35% services	2,064 kWh	0.18
	San Marino San Marino	0.03 / 1,214/sq. mi.	24 sq. mi. / 62 sq km	$ 34,600	57% services 42% industry	Not available	0.66
	Serbia and Montenegro Belgrade	10.5 / 267/sq. mi.	39,518 sq. mi. / 102,351 sq km	$ 2,370	41% industry 35% services	3,075 kWh	0.24
	Slovakia Bratislava	5.4 / 287/sq. mi.	18,859 sq. mi. / 48,845 sq km	$ 12,200	46% services 29% industry	4,519 kWh	0.26

Country	Passenger Vehicles	Persons Per Vehicle	Railroads (in miles)
Andorra	35,358	1.98	0
Austria	4,100,000	1.98	3,779.0
Belgium	4,680,000	2.20	2,130.9
Bulgaria	1,990,000	3.96	2,662.3
Croatia	1,120,000	3.93	1,423.5
Czech Republic	3,720,000	2.74	5,855.3
Denmark	1,820,000	2.97	1,772.6
Estonia	464,000	2.80	631.2
Finland	2,130,000	2.44	3,636.3
France	27,480,000	2.19	19,802.2
Germany	42,840,000	1.93	25,312.1
Greece	2,930,000	3.75	1,579.8
Hungary	2,360,000	4.19	4,715.7
Iceland	158,900	1.88	0
Ireland	1,280,000	3.13	1,207.1
Italy	31,950,000	1.80	12,024.3
Latvia	556,800	4.13	1,495.4
Luxembourg	253,400	1.97	169.9
Malta	211,300	1.89	0
Moldova	232,300	18.53	823.4
Monaco	17,000	1.76	1.1
Netherlands	6,120,000	2.63	1,698.2
Norway	1,850,000	2.43	2,487.4
Poland	9,280,000	4.16	14,520.4
Portugal	4,930,000	2.05	1,767.0
Romania	2,980,000	7.48	7,058.7
San Marino	24,825	1.21	0
Serbia and Montenegro	1,000,000	10.7	2,538.9
Slovenia	868,300	2.30	744.6
Spain	16,850,000	2.44	8,649.0
Sweden	4,000,000	2.23	7,949.0
Switzerland	3,550,000	2.03	2,785.0
United Kingdom	24,590,000	2.41	10,464.4
United States	211,940,000	1.39	139,965.0

Sources: Central Intelligence Agency, *The World Factbook 2003; The World Almanac and Book of Facts 2004*

Flag	Country / Capital	Population (in millions) / Pop. Density	Area	Per Capita GDP (in US $)	Workforce Structure (largest categories)	Electricity Consumption (kilowatt hours per person)	Telephone Lines (per person)
	Slovenia / Ljubljana	2 / 254/sq. mi.	7,827 sq. mi. / 20,272 sq km	$ 18,000	38% manufacturing 15% trade	6,971 kWh	0.41
	Spain / Madrid	41.1 / 213/sq. mi.	194,897 sq. mi. / 504,781 sq km	$ 20,700	64% services 29% manuf., mining, const.	5,124 kWh	0.46
	Sweden / Stockholm	8.9 / 56/sq. mi.	173,732 sq. mi. / 449,964 sq km	$ 25,400	74% services 24% industry	15,198 kWh	0.73
	Switzerland / Bern	7.2 / 467/sq. mi.	15,942 sq. mi. / 41,290 sq km	$ 31,700	69% services 26% industry	7,453 kWh	0.74
	United Kingdom / London	59.3 / 635/sq. mi.	94,525 sq. mi. / 244,819 sq km	$ 25,300	74% services 25% industry	5,841 kWh	0.60
	Vatican City / Vatican City	0.0009 / 5,235/sq. mi.	0.17 sq. mi. / 0.44 sq km	Not available	Not available	Not available	Not available
	United States / Washington, D.C.	294.0 / 83/sq. mi.	3,717,810 sq. mi. / 9,629,084 sq km	$ 37,600	31% manage., prof. 29% tech., sales, admin.	12,250 kWh	0.65

The French Alps

UNIT 4 Assessment Resources

▶ **Unit 4 Test**

▶ **Unit 4 Test for English Language Learners and Special-Needs Students**

internet connect

GO TO: go.hrw.com
KEYWORD: SW3 U4
FOR: Web sites about country statistics

Highlights of Country Statistics
Links to online country statistics for Europe include:
- CIA World Factbook
- Library of Congress Country Studies
- Flags of the World

CHAPTER RESOURCE MANAGER

Objectives	Pacing Guide	Reproducible Resources
SECTION 1 **Physical Features** (pp. 291–94)	• Describe Europe's major landform regions. • Identify the major rivers and bodies of water found in Europe.	**Regular** 1 day **Block Scheduling** .5 day *Block Scheduling Handbook,* *Chapter 13*	**RS** Guided Reading Strategy 13.1 **RS** Graphic Organizer Activity 13 **SM** Critical Thinking Activity 13: The Rhine River
SECTION 2 **Climates and Biomes** (pp. 296–98)	• Analyze how ocean currents affect Europe's climates. • Identify the biomes found in Europe.	**Regular** .5 day **Block Scheduling** .5 day *Block Scheduling Handbook,* *Chapter 13*	**RS** Guided Reading Strategy 13.2 **SM** Geography for Life Activity 13: Evaluating European Forests
SECTION 3 **Natural Resources** (pp. 299–301)	• Locate Europe's forest, soil, and fishery resources. • Identify the energy and mineral resources of Europe.	**Regular** .5 day **Block Scheduling** .5 day *Block Scheduling Handbook,* *Chapter 13*	**RS** Guided Reading Strategy 13.3 **SM** Map Activity 13: European Mineral and Energy Resources

Chapter Resource Key

PS Primary Sources	**A** Assessment	CD–ROM
RS Reading Support	**REV** Review	Video
IC Interdisciplinary Connections	**ELL** Reinforcement and English Language Learners	Internet
E Enrichment	Transparencies	Holt Presentation Maker Using Microsoft® PowerPoint®
SM Skills Mastery		

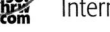

One-Stop Planner CD–ROM

See the *One-Stop Planner*
for a complete list of additional
resources for students and
teachers.

 **One-Stop** Planner CD–ROM

It's easy to plan lessons, select resources, and print out materials for your students when you use the ***One-Stop Planner CD–ROM with Test Generator***.

 **internet** connect

HRW ONLINE RESOURCES

GO TO: go.hrw.com
Then type in a keyword.

TEACHER HOME PAGE
　KEYWORD: SW3 Teacher

CHAPTER INTERNET ACTIVITIES
　KEYWORD: SW3 GT13
　Choose an activity to:
　• create a brochure on the land and rivers of Europe.
　• explore the engineering and technology that keep the Netherlands above sea level.
　• create a poster of fjords along the Scandinavian coast.

CHAPTER ENRICHMENT LINKS
　KEYWORD: SW3 CH13

CHAPTER MAPS
　KEYWORD: SW3 MAPS13

ONLINE ASSESSMENT
　Homework Practice
　KEYWORD: SW3 HP13
　Standardized Test Prep
　KEYWORD: SW3 STP13
　Rubrics
　KEYWORD: SS Rubrics

COUNTRY INFORMATION
　KEYWORD: SW3 Almanac

CONTENT UPDATES
　KEYWORD: SS Content Updates

HOLT PRESENTATION MAKER
　KEYWORD: SW3 PPT13

ONLINE READING SUPPORT
　KEYWORD: SS Strategies

CURRENT EVENTS
　KEYWORD: S3 Current Events

Technology Resources

- One-Stop Planner CD–ROM, Lesson 13.1
- *ARGWorld* CD–ROM
- Geography and Cultures Visual Resources 24
- Homework Practice Online
- HRW Go site

- One-Stop Planner CD–ROM, Lesson 13.2
- Geography and Cultures Visual Resources 26
- Homework Practice Online
- HRW Go site

- One-Stop Planner CD–ROM, Lesson 13.3
- Geography and Cultures Visual Resources 28
- Homework Practice Online
- HRW Go site

Reinforcement, Review, and Assessment

ELL Main Idea Activity 13.1
ELL English Audio Summary 13.1
ELL Spanish Audio Summary 13.1
REV Section 1 Review, p. 294
A　Daily Quiz 13.1

ELL Main Idea Activity 13.2
ELL English Audio Summary 13.2
ELL Spanish Audio Summary 13.2
REV Section 2 Review, p. 298
A　Daily Quiz 13.2

ELL Main Idea Activity 13.3
ELL English Audio Summary 13.3
ELL Spanish Audio Summary 13.3
REV Section 3 Review, p. 301
A　Daily Quiz 13.3

Meeting Individual Needs

Ability Levels

Level 1 Basic-level activities designed for all students encountering new material

Level 2 Intermediate-level activities designed for average students

Level 3 Challenging activities designed for honors and gifted-and-talented students

English Language Learners Activities that address the needs of students with Limited English Proficiency

Chapter Review and Assessment

- Chapter 13 Test Generator (on the One-Stop Planner)
- Global Skill Builder CD–ROM
- HRW Go site
- **REV** Chapter 13 Review, pp. 302–03
- **REV** Chapter 13 Tutorial for Students, Parents, Mentors, and Peers
- **A**　Chapter 13 Test (form A or B)
- **A**　Alternative Assessment Handbook
- **A**　Chapter 13 Test for English Language Learners and Special-Needs Students

Launch into Learning

Ask students to identify some outdoor sports at which European athletes have excelled at the Olympic Games or other international competitions. *(Possible answers: bicycling, golf, ice skating, mountain climbing, sailing, skiing, soccer, tobogganing)* Ask students what these sports might suggest about Europe's natural environments. Discuss their responses. *(Possible answers: Europe has a wide variety of landforms and climates. Mountain climbing and tobogganing require rugged terrain. Golf and sailing are popular in mild climates, but ice skating needs winters cold enough to freeze lakes.)* Tell students they will learn more about Europe's physical geography in this chapter.

Using the Physical-Political Map

Direct students' attention to the map on the opposite page. Point out that nearly any location in Europe lies within 500 miles of an ocean or sea. Ask students to name the major bodies of water that surround Europe *(Arctic Ocean, Norwegian Sea, Baltic Sea, North Sea, Irish Sea, Atlantic Ocean, Mediterranean Sea, Adriatic Sea, Aegean Sea, Black Sea)*. Point out that even inland locations have access to the sea through Europe's many rivers. Ask students to identify some of the continent's major river systems *(Danube, Elbe, Guadalquivir, Loire, Oder, Po, Rhine, Seine, Tagus, Thames, Vistula)*.

Why We Should Know More

These are some reasons why studying Europe's natural environments should interest students:

- ▶ Europe's configuration as a peninsula of peninsulas has dramatically affected its economic and political history.
- ▶ The spectacular scenery of Europe's mountains, beaches, fjords, rivers, and other physical features attract tourists from around the world.
- ▶ Europeans have learned a great deal about preventing pollution and repairing environmental damage—information they can share with other countries.
- ▶ Europe's waterways and other natural resources help drive strong economies that are closely connected to the United States economy.

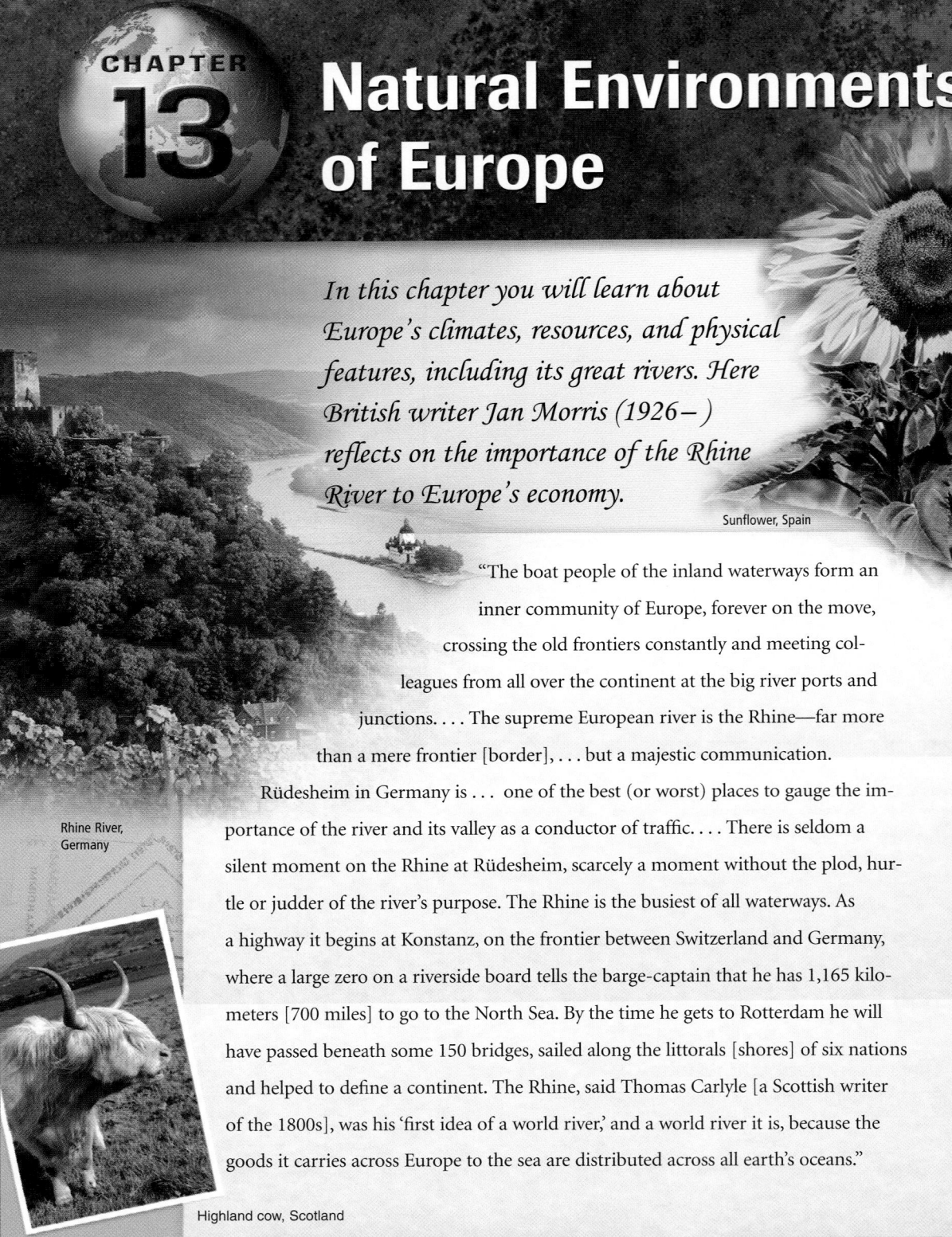

CHAPTER
13

Natural Environments of Europe

In this chapter you will learn about Europe's climates, resources, and physical features, including its great rivers. Here British writer Jan Morris (1926–) reflects on the importance of the Rhine River to Europe's economy.

Sunflower, Spain

Rhine River, Germany

Highland cow, Scotland

"The boat people of the inland waterways form an inner community of Europe, forever on the move, crossing the old frontiers constantly and meeting colleagues from all over the continent at the big river ports and junctions. . . . The supreme European river is the Rhine—far more than a mere frontier [border], . . . but a majestic communication. Rüdesheim in Germany is . . . one of the best (or worst) places to gauge the importance of the river and its valley as a conductor of traffic. . . . There is seldom a silent moment on the Rhine at Rüdesheim, scarcely a moment without the plod, hurtle or judder of the river's purpose. The Rhine is the busiest of all waterways. As a highway it begins at Konstanz, on the frontier between Switzerland and Germany, where a large zero on a riverside board tells the barge-captain that he has 1,165 kilometers [700 miles] to go to the North Sea. By the time he gets to Rotterdam he will have passed beneath some 150 bridges, sailed along the littorals [shores] of six nations and helped to define a continent. The Rhine, said Thomas Carlyle [a Scottish writer of the 1800s], was his 'first idea of a world river,' and a world river it is, because the goods it carries across Europe to the sea are distributed across all earth's oceans."

Section 1

OBJECTIVES

1. Describe Europe's major landform regions.

2. Identify the major rivers and bodies of water found in Europe.

 **LET'S GET STARTED**

Copy the following instructions onto the chalkboard: *What are some physical features in Europe that you would like to see or visit? Write your answers in your notebook.* Call on volunteers to share their answers with the class. As students name places or features, write them on the chalkboard. Use this list to discuss students' perceptions of the continent's physical geography. Tell students that they will learn more about the landforms and water features of Europe in Section 1.

Building Vocabulary

Write **polders** and **dikes** on the chalkboard. Ask students if they have witnessed or read descriptions of flood preparations. Discuss how families might protect their property from floodwaters. (*Possible answers: sand bags, walls around property, levees*) Tell students that some Europeans have taken similar measures to protect and enlarge their countries. They build dikes to block the sea and reclaim the land inside by draining water to create polders. Have volunteers locate and read the definitions for the other terms.

Physical Features

READ TO DISCOVER

1. What are Europe's major landform regions?
2. What are the region's main rivers and bodies of water?

WHY IT MATTERS

Europe's Rhine and Danube Rivers suffer from major pollution problems. Use **CNNfyi.com** or other **current events** sources to learn about river pollution and what can be done to clean up polluted rivers.

DEFINE

fjords
polders
dikes
navigable

LOCATE

Ural Mountains
Mediterranean Sea
Scandinavian Peninsula
Iberian Peninsula
Italian Peninsula

Locate, continued

Balkan Peninsula
Northern European Plain
Alps
Carpathian Mountains
Pyrenees
Black Sea
Bosporus
North Sea
Rhine River
Danube River

RESOURCES

REPRODUCIBLE
► Guided Reading Strategy 13.1
► Graphic Organizer Activity 13
► Critical Thinking Activity 13: The Rhine River

TECHNOLOGY
► One-Stop Planner CD–ROM, Lesson 13.1
► Geography and Cultures Visual Resources 24
► Homework Practice Online
► HRW Go site

REINFORCEMENT, REVIEW, AND ASSESSMENT
► Main Idea Activity 13.1
► English Audio Summary 13.1
► Spanish Audio Summary 13.1
► Section 1 Review, p. 294
► Daily Quiz 13.1

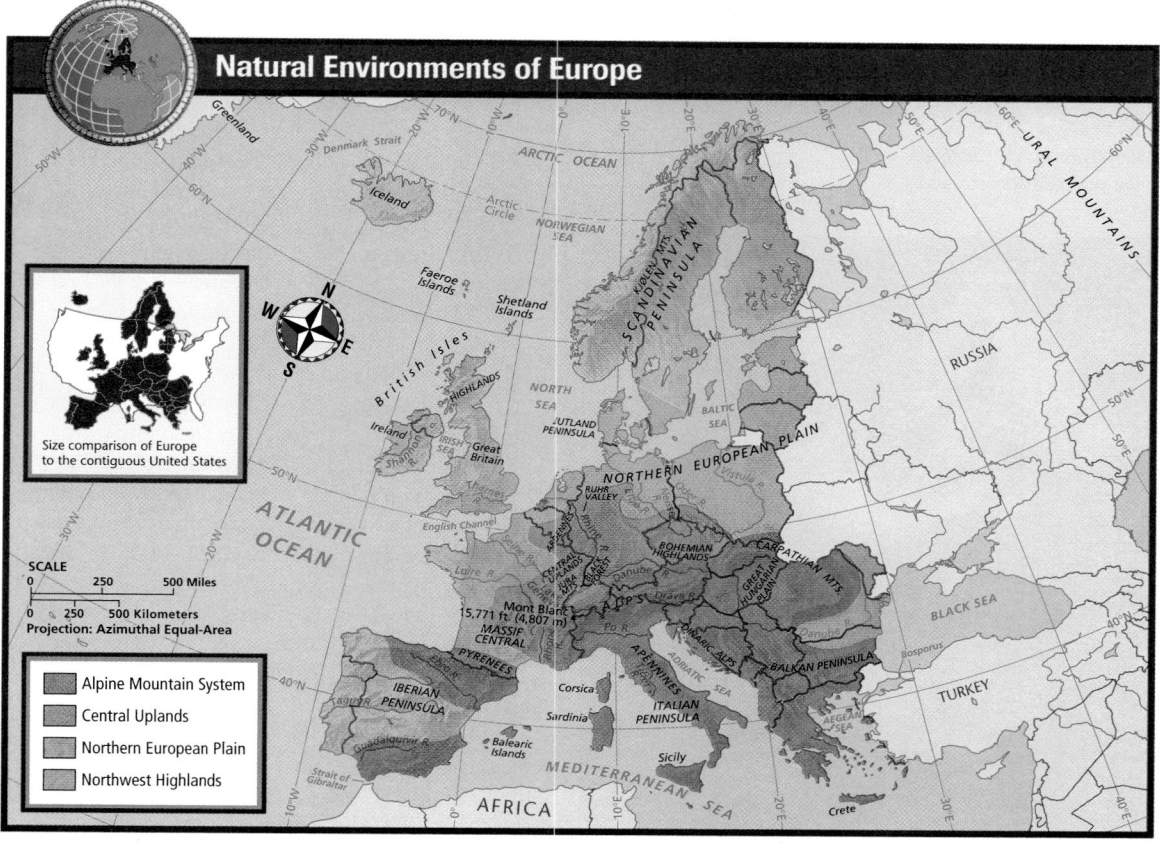

Natural Environments of Europe

Size comparison of Europe to the contiguous United States

SCALE
0   250   500 Miles
0   250   500 Kilometers
Projection: Azimuthal Equal-Area

Alpine Mountain System
Central Uplands
Northern European Plain
Northwest Highlands

ALL LEVELS: Organize the class into four groups and assign one of Europe's landform regions to each group. Then have each group use old magazines, art supplies, and other materials to create a poster about its assigned region. Each poster should include a map of the region, pictures, and written descriptions of the region's landforms and water features. Students should be creative in finding photos or preparing drawings to depict each region's physical environment and the processes that have shaped it. Ask groups to present their posters to the class. **ENGLISH LANGUAGE LEARNERS, COOPERATIVE LEARNING**

Teacher to Teacher

Steve Gargo of Appleton, Wisconsin, suggests the following activity to help students understand the effects of physical environments on human society. Organize the class into four groups and have each group conduct research on the earliest civilizations that developed in one of Europe's landform regions. Ask students to note how these cultures adapted to their surroundings and how they modified natural environments to better serve their needs. Have volunteers present each group's findings to the class. Then lead a class discussion about how differences in physical environments led to differences in society. **COOPERATIVE LEARNING, BLOCK SCHEDULING**

Essential Element 5

► Environment ◄
and Society

Climbing the Alps

Mountain climbers from around the world are drawn to Europe's Alps. Among their favorite peaks is the famous Matterhorn. For generations of mountaineers and climbers, the Matterhorn has represented a challenge to their skills. This mountain on the border between Switzerland and Italy rises 14,691 feet (4,478 m), and is one of the most frequently scaled peaks in the world. The first successful expedition to the summit was undertaken in 1865. Until then, the Matterhorn's steep ridges and glaciers had halted all attempts.

Two other Alpine peaks are particularly popular with climbers. Mont Blanc in France is the range's highest peak at 15,771 feet (4,807 m). The Eiger in southern Switzerland is lower than other mountains at 13,025 feet (3,970 m), but it is one of the world's most treacherous climbs.

internet connect

GO TO: go.hrw.com
KEYWORD: SW3 CH13
FOR: Web sites about the Alps

VISUAL RECORD ANSWER

glacial action; its steep sides and jagged edges

Our Amazing Planet

In 1783 a volcano in Iceland erupted continuously for 10 months, devastating large areas of the island. Poisonous gases released during the eruption killed about 75 percent of Iceland's livestock. Many people died from the resulting famine.

INTERPRETING THE VISUAL RECORD

The Matterhorn, on the Switzerland-Italy border, rises 14,691 feet (4,478 m).
What physical process do you think shaped this peak? What makes you think so?

Landforms

Europe stretches from the Atlantic Ocean to the Ural Mountains and from the Arctic Ocean to the Mediterranean Sea. In this unit we will study the part of Europe that lies generally west of Russia, Belarus, and Ukraine. Compared to the other continents, Europe is small. However, within Europe's small area is a complex variety of landforms, islands, and peninsulas. Major islands include Great Britain, Ireland, and Iceland. Major peninsulas include the Scandinavian, Iberian, Italian, and Balkan Peninsulas.

Europe can be divided into four major landform regions. These regions are the Northwest Highlands, the Northern European Plain, the Central Uplands, and the Alpine mountain system. (See the map.) The Northwest Highlands is an ancient eroded region of rugged hills and low mountains. In the north it includes the hills of Ireland and England, the Scottish Highlands, and the mountains of Scandinavia. Northwestern France and some of the Iberian Peninsula are also part of the Northwest Highlands. During the last ice age, glaciers scoured the landscapes of Scandinavia and much of the British Isles. Glaciers also carved **fjords** (fee-AWRDZ) along Norway's coast. Fjords are narrow deep inlets of the sea set between high rocky cliffs. When the ice melted, the retreating glaciers left behind thin soils and thousands of lakes.

To the south lies the Northern European Plain. This broad coastal plain stretches from France's Atlantic coast all the way to the Urals. Most of the plain is less than 500 feet (152 m) above sea level. Many rivers flow across it before reaching the ocean. As a result, river towns and port cities have developed there. For example, large cities like Paris and Berlin are located in this region. In fact, its many rivers, short distances, and smooth terrain have long made human contact relatively easy. These features have allowed culture groups to travel, trade, and migrate throughout the region. Today the Northern European Plain is Europe's most important farming and industrial area. As you might expect, it is also densely populated.

The third major landform region is the Central Uplands. This is an area of hills and small plateaus, with forested slopes and fertile valleys. It includes the Massif Central (ma-SEEF sahn-TRAHL) of France and the Jura Mountains on the French-Swiss border. The region stretches northeastward across southern Germany to the Bohemian Highlands. The Central Uplands are an old eroded region. As a result, the low mountains and hills in the region are often rounded. Many of Europe's productive coal fields lie in the Central Uplands. A number of industrial towns and cities grew near coal deposits.

The last and youngest region is the Alpine mountain system, which includes the Alps, Europe's major mountain range. The Alps stretch from France's Mediterranean coast to the Balkan Peninsula. Many peaks reach heights of more than 14,000 feet (4,268 m). Because of their high elevations, the Alps have large snowfields and

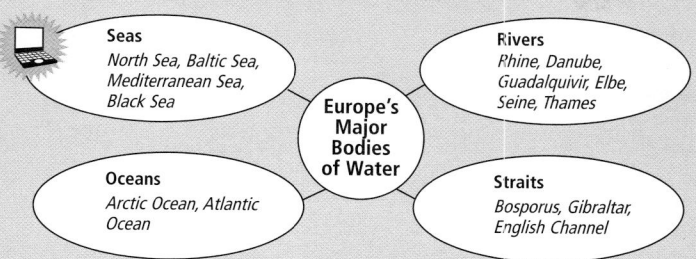 **ALL LEVELS:** Copy the following graphic organizer onto the chalkboard, omitting the italicized answers. Call on volunteers to complete the web with information from the text and chapter map. **ENGLISH LANGUAGE LEARNERS**

Seas
North Sea, Baltic Sea, Mediterranean Sea, Black Sea

Rivers
Rhine, Danube, Guadalquivir, Elbe, Seine, Thames

Europe's Major Bodies of Water

Oceans
Arctic Ocean, Atlantic Ocean

Straits
Bosporus, Gibraltar, English Channel

Using National Geography Standard 13:
Environment and Society: How Physical Systems Affect Human Systems Ask students to consider the effects that Iceland's volcanic activity, harsh climates, and short growing season have had on settlement and development patterns throughout the country's history. Have them conduct research on how Icelanders have turned some aspects of the country's unusual physical geography, such as hot springs, to their advantage. Ask students to report their findings in written reports. Finally, point out the price Icelanders pay—literally, in the form of high prices on imported goods—for living in a place that presents so many obstacles to human settlement.

glaciers. Avalanches are fairly common in winter. Although the Alps are high mountains, historically they have not been a serious barrier to human interaction. People have crossed the Alps through mountain passes for thousands of years to trade and travel. Other ranges in the Alpine system include the Carpathian (kahr-PAY-thee-uhn) Mountains in Eastern Europe and the Apennines (A-puh-nynz) in Italy. The Pyrenees (PIR-uh-neez) of France and Spain are also part of this system.

Beginning about 65 million years ago, tectonic processes formed the mountains of the Alpine system. At that time, the African plate began pushing against the Eurasian plate. This caused the mountains to rise. Tectonic activity continues today. A subduction zone off the coasts of southern Italy and Greece still creates powerful earthquakes and volcanoes. Because it lies on a tectonic plate boundary, Iceland also experiences volcanic eruptions and earthquakes.

✔ **READING CHECK:** *Physical Systems* How have physical processes affected the shapes of mountains and hills in the Central Uplands? A long period of erosion has rounded them.

 **internet connect**

GO TO: go.hrw.com
KEYWORD: SW3 CH13
FOR: Web sites about the natural environments of Europe

Connecting to
TECHNOLOGY

Polders
The Dutch have long used technology to shape their natural environment. For hundreds of years, they have been "creating" land by reclaiming it from the sea. Lands reclaimed from the sea are called **polders**.

To create polders, the Dutch built earthen walls called **dikes** along the shoreline. Then they used windmills to pump out the seawater behind the dikes. The Dutch used the drained lands for farming or for housing. By using polders to grow crops and raise livestock, the Dutch greatly increased the amount of available farmland. In fact, the Netherlands is an exporter of agricultural goods. The Dutch export products such as flowers, grains, potatoes, and sugar beets, particularly to other European countries.

Today more than 25 percent of the Netherlands lies below sea level. A national system of dams, dikes, and floodgates holds back the sea, and water is constantly pumped out. This system ranks as one of the wonders of the modern world. The largest dike, 19 miles (31 km) long and 100 yards (91 m) thick, closes off a large inlet. Completed in 1932, this dike allowed for the creation of four huge polders. Farms and cities have sprung up on these lands.

Comparing In what other areas of the world, or against what environmental hazards, might Dutch techniques for creating polders be useful?

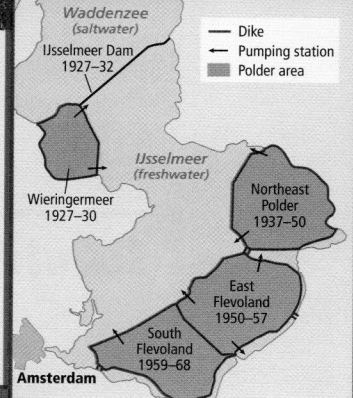

Waddenzee (saltwater)
— Dike
← Pumping station
■ Polder area

IJsselmeer Dam 1927–32
IJsselmeer (freshwater)
Wieringermeer 1927–30
Northeast Polder 1937–50
East Flevoland 1950–57
South Flevoland 1959–68
Amsterdam

Polders around the World
Although the Dutch pioneered the building of polders, the technique has since spread to other countries. For example, Estonia, Latvia, Lithuania, and Poland have used similar methods to reclaim fertile land. Estonia's polders only date to the 1970s and 1980s, but some polders on Lithuania's Nemanus River were created nearly 100 years ago. In Poland, some 770 square miles (1,994 sq. km) of farmland and pasture have been recovered.

In Asia too, countries are looking to the sea for new land. The Vietnamese city of Haiphong, for example, has built dikes and reclaimed land along the Gulf of Tonkin.

ACTIVITY: Have students conduct research to locate other areas where land could be reclaimed from the sea. Have them draw maps to reflect their findings.

You don't say! The world's deepest cave is Reseau Jean-Bernard in the Alps of eastern France. It descends more than 5,000 feet (1,524 m) below Earth's surface.

CONNECTING TO
TECHNOLOGY ANSWER

Students might note that low-lying areas prone to flooding might benefit from these techniques.

Creating a Polder

Year 1
Dikes are built around the area to be reclaimed. Pumps and canals drain the water.

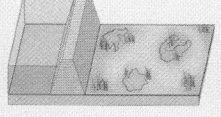

Years 2–3
The water level falls. Seeds blow into the area and salt-tolerant plants grow.

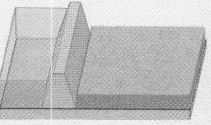

Years 4–6
Reeds are planted over a net of woven twigs. The reeds draw up more water.

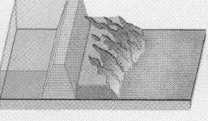

Year 7
The reeds are burned. Heavy plows turn their roots and the ash into the soil.

The land is ready for crops. Within 15 years the polder looks like it has been farmed forever.

Close

Call on volunteers to name which of Europe's landform regions they would most like to inhabit and why. After all volunteers have expressed their opinions, have the class vote on the best environment in Europe in which to live.

Reteach

Have students complete **Main Idea Activity for English Language Learners and Special-Needs Students 13.1**. Then have students create flash cards of the section's main ideas. Have students use them to study with partners in class. **ENGLISH LANGUAGE LEARNERS**

Review and Assess

Have students complete the **Section Review**. Then have students complete **Daily Quiz 13.1**.

Extend

Have interested students conduct research on Europe's peat bogs, which absorb carbon dioxide, thereby reducing levels of the gas in the atmosphere. Peat can also be used as a fuel. Have students learn more about the creation and uses of peat and share their findings with the class. **BLOCK SCHEDULING**

Section 1 Review Answers

Define For definitions, see: fjords, p. 292; polders, p. 293; dikes, p. 293; navigable, p. 294

Working with Sketch Maps Maps will vary, but listed places should be labeled in their approximate locations. The Rhine follows that course.

Reading for the Main Idea
1. Glaciers scoured Scandinavia and the British Isles, leaving fjords, lakes, and thin soils.

2. The low Northern European Plain has facilitated human contact, as have Europe's many navigable rivers and low coastlines.

Critical Thinking
3. Possible answers: The flat Northern European Plain and major rivers like the Rhine and Danube were probably the easiest migration routes. The most difficult routes would have been highland areas like the Alps, except through mountain passes. (NGS 9)

4. Possible answers: on the Northern European Plain; along navigable rivers; at seaports near the mouths of navigable rivers (NGS 12)

Organizing What You Know
5. Northwest Highlands— rugged hills and low mountains, glaciated, many lakes, thin soils; Northern European Plain—broad coastal plain, many rivers; Central Uplands— hills and small plateaus, old and eroded, coal fields, fertile valleys; Alpine mountain system—high mountains, glaciers, tectonic activity

In Romania, an artificial canal connects the Danube River to the Black Sea. The canal allows ships to avoid the marshy Danube Delta and shortens voyages by many miles. Locks, shown above, lift ships going upstream to the river's level and lower ships going downstream to sea level. By using rivers and canals, ships can now travel all the way from the Black Sea to the North Sea.

Water

Europe is nearly surrounded by water. To the south lies the Mediterranean Sea. It is connected to the Black Sea by the narrow Bosporus (BAHS-puh-ruhs). Geographers consider the Bosporus a boundary between Europe and Asia. The Arctic Ocean, North Sea, and Baltic Sea wash the shores of northern Europe. The shallow North Sea has long been important for trade and fishing. The smaller Baltic Sea freezes over during the winter months. To the west of Europe lies the North Atlantic Ocean. For centuries, European explorers, fishers, and merchants have traveled the waters of the Atlantic.

Europe's long, irregular coastline has hundreds of good natural harbors. These harbors are generally located near the mouths of **navigable** rivers, making Europe ideally situated for trade by sea. A navigable river is one that is deep enough and wide enough for shipping. Canals connect many rivers in Europe. For example, France's Canal du Midi lets boats and barges travel between the Atlantic Ocean and the Mediterranean Sea. Many interior towns and cities across Europe have access to the sea through canals and rivers.

The Rhine and Danube stand out among Europe's most important rivers. Many cities and industrial areas line their banks, and barges carry goods along their courses. The Rhine rises in the Swiss Alps. It then flows northwestward through Germany and the Netherlands before entering the North Sea. The Danube begins in the uplands of southern Germany. It flows eastward through nine countries in central and eastern Europe. It empties into the Black Sea. Unfortunately, large amounts of pollution enter the ocean from these and other rivers. Cleaning up and controlling pollution in Europe's rivers is a major environmental challenge.

✓ **READING CHECK:** *Environment and Society* How do Europe's interior towns and cities have access to the sea? *by using navigable rivers and canals*

Section 1 Review

go. hrw .com
Homework Practice Online
Keyword: SW3 HP13

Define fjords, polders, dikes, navigable

Working with Sketch Maps
On a map of Europe that you draw or that your teacher provides, label the Ural Mountains, Mediterranean Sea, Scandinavian Peninsula, Iberian Peninsula, Italian Peninsula, Balkan Peninsula, Northern European Plain, Alps, Carpathian Mountains, Pyrenees, Black Sea, Bosporus, North Sea, Rhine River, and Danube River. Which river rises in the Swiss Alps, flows through the Netherlands, and empties into the North Sea?

Reading for the Main Idea

1. *Physical Systems* How did continental ice sheets shape the landscapes of the Northwest Highlands?

2. *Environment and Society* How has Europe's natural environment made human contact relatively easy?

Critical Thinking

3. *Making Generalizations* What are some physical features that probably shaped migration routes in Europe? How do you think they did so?

4. *Analyzing Information* Considering what you know about Europe's natural environments, where would you expect to find many of its largest and most important cities and settlements?

Organizing What You Know

5. Copy the chart shown below. Use it to describe Europe's four major landform regions: the Northwest Highlands, Northern European Plain, Central Uplands, and the Alpine mountain system.

Northwest Highlands	
Northern European Plain	
Central Uplands	
Alpine mountain system	

Literature: Whirlpools in Adventure Tales

The influence of the sea on European society is reflected in its literature. Many stories tell of the danger posed to ships by wild seas, represented in many stories in the form of whirlpools. In reality, whirlpools are formed when strong tidal currents meet underwater obstructions that break up their flows. Some, like the Maelstrom and Saltstraumen in Norway or Corryvreckan in Scotland, are very powerful and dangerous to small craft. They are not, however, the swirling menaces of myths and stories.

In the *Odyssey*, Odysseus encounters the monstrous Charybdis, who alternately drinks in and spits forth seawater to destroy ships, and whom scholars now think personifies a whirlpool in the Straits of Messina near Sicily. Centuries later, American author Edgar Allan Poe wrote "A Descent into the Maelstrom" in which he tells of a small craft caught in the dark swirls of the Norwegian whirlpool. The boat is dragged toward the ocean floor, but the tale's narrator escapes when the whirlpool suddenly dissipates. In *20,000 Leagues Under the Sea* by French author Jules Verne, the submarine *Nautilus* is caught in the Maelstrom. The narrator escapes, but he does not know if the ship survived the ordeal.

Have students write adventure stories based on elements of Europe's physical geography. For example, students might write about climbers caught in an avalanche in the Alps or tourists lost on the English moors. Encourage students to read excerpts from works by Poe, Verne, or other authors for inspiration. Call on volunteers to share their stories with the class. **BLOCK SCHEDULING**

Environment and Society

Geography for Life

A Peninsula of Peninsulas

Look at a map of Europe. You will notice that the continent is actually a large peninsula made up of many smaller peninsulas. You might also notice Europe's jagged outline. What do you think created these features? How might they have affected the region's history?

Much of Europe's present-day coastline has taken shape since the end of the last ice age about 10,000 years ago. Since

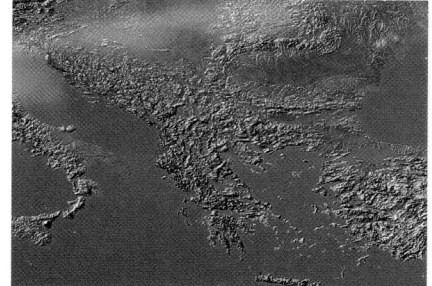

This computer-enhanced satellite image shows the Balkan and Italian Peninsulas clearly.

that time, sea levels have risen, flooding lowlands and changing Europe's shoreline. For example, in northern Europe the Baltic Sea formed from the melting Scandinavian ice sheet. The North Sea and Irish Sea also took their present form after the last ice age. Rising sea levels flooded the mouths of many rivers. This process formed estuaries that are now deep ocean ports.

Geographers have long noticed the remarkable influence of the sea on Europe. The sea greatly affects climate and rainfall patterns. Warm ocean currents bring mild temperatures and rainfall to much of western and northern Europe. In some places, these effects are felt far inland.

Europe's peninsular geography and rugged coastline have also influenced its history. Harbors along rocky shores have long offered protection for ships. Since ancient times, the joining of land and water has provided opportunities for exploration, fishing, sea trade, and political and military power. Early peoples like the Phoenicians, Greeks, and Vikings sailed and explored Europe's intricate coastline. In fact, Europe has long been a place of contact between peoples and cultures. From the Italian Peninsula, which juts into the Mediterranean Sea, the Romans ruled a vast empire. Later European culture groups turned to the sea for global colonial and economic power. Spanish and Portuguese explorers sailed around the world in the 1500s, setting up trading posts and colonies. The British, Dutch, French, and other Europeans followed. For example, in the 1700s and 1800s Great Britain used the seas to become the world's dominant colonial and sea power.

Applying What You Know

1. **Summarizing** How has Europe's peninsular geography influenced its history?

2. **Making Generalizations** How do you think Europe's peninsular geography has affected the locations of its cities and settlements?

Párga, Greece, overlooks a picturesque harbor. Like that of many towns on Europe's peninsulas, Párga's harborside location has played a major role in its history and economy. Tourists also enjoy the town's setting.

Eye on Earth

Heimaey's Harbor Like many coastal European towns, Vestmannaeyjar, on the Icelandic island of Heimaey, depends upon its harbor for communication and survival. So when a volcanic eruption threatened to destroy the harbor in 1973, the people fought to protect it.

The eruption began in late January and lasted for about five months. Residents quickly evacuated, and no one was killed in the eruption. However, lava flows destroyed many buildings and seemed likely to close off the harbor, thereby destroying the town's fishing industry. Pumps were constructed to spray the lava with icy seawater to halt its advance. This plan worked, and the town's harbor was saved. The tongue of cooled and solidified lava flow that could have destroyed the harbor now acts as a breakwater, protecting it from storms.

Applying What You Know Answers

1. Harbors provided protection for ships. The mixing of land and sea has encouraged trade, fishing, and exploration. Spain, Portugal, and other countries used the sea to establish colonies around the world.
2. Possible answer: Many cities grew as ports along Europe's coast.

This Geography for Life feature addresses National Geography Standards 4, 7, 9, 10, and 15.

295

LET'S GET STARTED

Copy the following questions onto the chalkboard: *Judging from what you have learned about Mediterranean climates, what would you expect the climate to be like in Rome, Italy?* Discuss responses. *(Possible answers: warm, sunny)* Have students examine a map to learn which U.S. cities lie at about the same latitude as Rome. *(Possible answers: Boston, Chicago, Detroit)* Point out that a variety of factors work together to create a milder climate in Rome than in these American cities. Tell students they will learn more about Europe's climates in Section 2.

Building Vocabulary

Write **North Atlantic Drift** on the chalkboard and underline the word *drift*. Point out that the word is sometimes used as a synonym for *current*. Have students consult maps of ocean currents to find an example of another current whose name includes this term. *(Possible answer: West Wind Drift)*

Section 2 Climates and Biomes

READ TO DISCOVER

1. How do ocean currents affect the distribution of Europe's climates?
2. Which biomes are found in this region?

WHY IT MATTERS

Warm ocean currents moderate Europe's climate. Use **CNNfyi.com** or other **current events** sources to learn about how the circulation of ocean waters can affect climate and weather.

IDENTIFY

North Atlantic Drift

LOCATE

British Isles

Climates

Europe has three major climate types: marine west coast, humid continental, and Mediterranean. (See the unit climate map.) The marine west coast climate is found throughout most of northern and western Europe. This climate region includes southern Iceland and the British Isles. It also stretches across northern continental Europe from northern Spain into Poland and Slovakia. Frequent Atlantic storms bring clouds and rain. Rainfall averages between 20 and 80 inches (51 and 203 cm) a year. (See the unit precipitation map.) Snow and frosts can occur in winter. Temperatures are mostly mild, and cloudy, drizzly, and foggy days are common.

INTERPRETING THE VISUAL RECORD

Ireland is one of the European countries that have a marine west coast climate. This hillside is in a part of Ireland where rainfall is very heavy—more than 60 inches (150 cm) per year. **What is the connection between Ireland's climate and its nickname—the Emerald Isle?**

Teach Objective 1

ALL LEVELS: Direct students' attention to the climate maps in this book. Ask a volunteer to identify some climates found worldwide at 50° north latitude *(highland, humid continental, marine west coast, semiarid, subarctic)*. Ask students how Europe's climates at that latitude compare to those found in North America and Asia. *(Europe's marine west coast climate is milder than the humid continental and subarctic climates of North America and Asia.)* Call on students to explain the difference. *(The North Atlantic Drift warms the air above it, which is carried over Europe by prevailing winds. This creates moderate temperatures and precipitation in northern Europe.)* **ENGLISH LANGUAGE LEARNERS**

Teach Objective 2

ALL LEVELS: Copy the following graphic organizer onto the chalkboard, omitting the italicized answers. Have students complete the chart to identify the locations, plants, and animals of Europe's biomes. **ENGLISH LANGUAGE LEARNERS**

Temperate forest	Mediterranean scrub forest	Boreal forest	Tundra
• *covers most of Europe* • *ash, beech, maple, oak trees* • *badgers, deer, many birds*	• *southern Europe* • *small trees, shrubs, drought-resistant plants* • *wild boars, wild sheep*	• *northern and central Europe* • *spruce, pine, fir trees*	• *far northern Europe* • *treeless, frozen ground most of the year* • *migratory birds, reindeer, foxes*

Areas from interior Norway and Sweden south to the Black Sea have a humid continental climate. This climate has four distinct seasons, including a cold snowy winter and mild to cool humid summer. Winters are severe in parts of this climate region. Periodic summer droughts affect Hungary and Romania.

High mountains, particularly the Alps, separate these first two climates from Europe's third major climate region to the south. Most of southern Europe has a Mediterranean climate. This region gets between 10 and 30 inches (25 and 76 cm) of rainfall a year. Most rainfall comes during the mild winter. Occasional North Atlantic storms pushed by the westerly winds bring rain at that time. Long, dry, and sunny summers are typical.

Smaller climate regions are found in other parts of Europe. For example, a subarctic climate stretches across northern Norway, Sweden, and Finland. The northernmost parts of these countries, along with interior and northern Iceland, have a tundra climate. A small humid subtropical climate region is located south and southeast of the Alps. This area stretches from Italy's Po Valley into the Balkans. In parts of Spain, high mountains block moist ocean air from reaching farther inland. A semiarid climate is found there.

Compared to world regions of similar latitude, much of Europe enjoys mild climates. Winter temperatures are particularly mild for such high latitudes. These mild temperatures are caused by the moderating influence of the **North Atlantic Drift**.

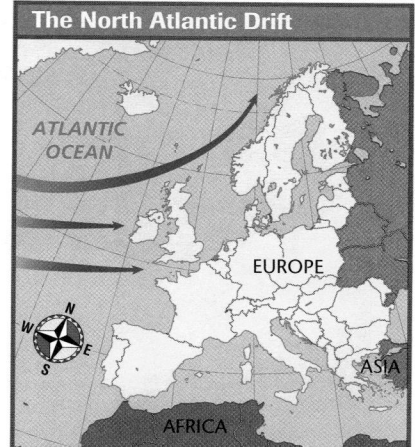

The North Atlantic Drift

INTERPRETING THE MAP *Northern Europe's temperatures are relatively mild, thanks to the North Atlantic Drift.* **What are some of the coastal countries affected by the current? Use the unit atlas to find the large island west of Iceland that is far from this ocean current and, as a result, is much colder.**

![globe icon] **FOCUS ON GEOGRAPHY**

The North Atlantic Drift The North Atlantic Drift is a warm ocean current. It originates off the coast of North America and is fed by the warm tropical waters of the Gulf Stream. (See the map.) The North Atlantic Drift warms the air above it. Then prevailing westerly winds carry this warm moist air across much of northwestern Europe. The winds bring mild temperatures and rain. These conditions allow farmers to grow crops as far north as Sweden and Iceland. Also, the warm waters keep seaports in Norway and at Murmansk, Russia, free of ice.

✓ **READING CHECK:** (*Physical Systems*) How do the North Atlantic Drift and prevailing westerly winds affect Europe's climates?

Together they bring warm air and mild temperatures to much of western and northern Europe.

Plants and Animals

Human activities have affected Europe's plants and wildlife severely. For thousands of years, people there have hunted animals and cleared forests for timber and farmland. The growth of towns, cities, and roads has also changed the natural environment. Some waterways have been polluted. As a result, many species have become extinct from loss of habitat. Some other creatures, such as bears, lynx,

White storks build their nests on a rooftop in Spain.

Cooperative Learning

Europe's Endangered Species Because of its high population density and intensive development, Europe has few natural habitats or preserves in which to protect plants and animals from extinction.

Organize the class into groups and assign each group a species that is endangered in Europe. Then have each group conduct research to create a map that shows the species' original distributions in Europe. Ask the groups to identify climates, soils, or other factors that explain this distribution. Then have each group create two transparency overlays for their maps—one that shows human activity in the area and one that illustrates the current distribution of the endangered species. Have students use their maps and overlays like the layers of a geographic information system to evaluate the role of society in the species' endangerment. Ask students to suggest ways humans might work to prevent the species' total disappearance. Students may wish to conduct further research to learn what measures have already been taken in Europe.

MAP ANSWER

Ireland, United Kingdom, Norway, Iceland; Greenland

297

INTERPRETING THE VISUAL RECORD *A member of the Sami people of northern Scandinavia works with his reindeer herd. Reindeer are known as caribou in North America. Their wide hooves allow them to walk more easily on snow.* **What food sources do you think the tundra environment provides for the reindeer during the winter?**

wolves, and wild horses, survive mainly in areas where they are protected. Despite these changes, however, Europe can be divided into four major biomes. These biomes include temperate forest, Mediterranean scrub forest, boreal forest, and tundra.

Most of Europe lies within a temperate forest biome. Trees such as ash, beech, maple, and oak are common. Badgers, deer, and a variety of birds live in this environment. Today fields and towns occupy much of the land. Only remnants of the dense forests that once covered much of the landscape remain in this region.

You will find a Mediterranean scrub forest biome in some drier areas in southern Europe. Small trees, shrubs, and drought-resistant plants are typical of the region. Animals such as wild boars and wild sheep still roam remote Mediterranean mountain areas.

Large parts of northern and central Europe lie within the boreal forest biome. These northern forests make up most of Europe's remaining woodlands. Finland, Norway, and Sweden all have large evergreen forests. Trees here, such as pine, spruce, and fir, provide most of Europe's timber for building and papermaking. However, logging and other human activities have greatly reduced the area's animal life.

Far northern Europe has a tundra biome. This biome includes much of Iceland and northern Scandinavia. The land in this cold treeless area is frozen most of the year. During the short Arctic summer, the tundra thaws, and many swamps and marshes form. Millions of migratory birds visit during the summer. Reindeer and foxes are among the tundra's mammals.

✓ READING CHECK: *Physical Systems* In which biome would you find trees such as ash, beech, maple, and oak? temperate forest biome

Section 2 Review

go. hrw .com **Homework Practice Online**
Keyword: SW3 HP13

Identify
North Atlantic Drift

Working with Sketch Maps On the map you created in Section 1, label the British Isles. Which climate dominates the British Isles?

Reading for the Main Idea
1. *Places and Regions* Where in Europe is a marine west coast climate found?

2. *Physical Systems* How do the Alps affect the distribution of climates in Europe?

3. *Environment and Society* What human activities have affected Europe's plants and wildlife?

Critical Thinking
4. **Drawing Inferences and Conclusions** What advantages might forestry industries in Norway, Sweden, and Finland have over forestry operations in other European countries?

Organizing What You Know
5. Create web diagrams like the one below to describe the climates of Europe. Use as many circles as you need to provide important details about each climate.

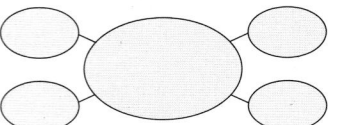

Section 3

OBJECTIVES

1. Locate Europe's forest, soil, and fishery resources.
2. Identify the energy and mineral resources of Europe.

Natural Resources

READ TO DISCOVER

1. Where are Europe's forest, soil, and fishery resources located?
2. What energy and mineral resources are found in this region, and where are they located?

DEFINE

loess

LOCATE

Po Valley
Guadalquivir River

WHY IT MATTERS

Overfishing and pollution have affected fishing industries in Europe and around the world. Use or other **current events** sources to learn about the challenges facing the world's fishing industries.

Forests, Soils, and Fisheries

Most of Europe's original forests were cleared centuries ago. For example, clearing or overgrazing in ancient times destroyed nearly all of the Mediterranean area's original oak woodlands. Only a scrub-plant community covering the hillsides remains. Air pollution and acid rain have destroyed many more trees throughout the continent. Large areas of timber-producing forests exist only in limited areas, such as Sweden and Finland. As a result, most European countries must now import lumber. However, most European countries also have reforestation and forest protection programs.

Europeans have made good agricultural use of their soils. In fact, more than half of Europe's land area is used for farming. Some of the best soils are found in the Northern European Plain. Farmers grow a variety of grains there and raise cattle and hogs. Some of these soils developed from **loess**—fine-grained, windblown soil that is very fertile. Such soils can keep their fertility for many years. In southern Europe, alluvial soils are particularly productive. River valleys, such as Italy's Po (POH) Valley and Spain's Guadalquivir (gwah-dahl-kee-VEER) River valley, are major farming centers. With the help of irrigation, these soils produce a wide range of crops. Farmers grow lemons, oranges, and many different vegetables. Farmers in southern Europe also raise goats, hogs, and sheep.

Autumn tints grapevines growing in southwestern Germany. Partly because they can grow in many different soils, grapes are grown in several European countries.

Teach Objective 1

 **ALL LEVELS:** Provide students with outline maps of Europe. Direct students to use their textbooks to create resource maps of the continent. Have students use different colors to shade areas used for forestry, commercial farming, and fishing.

HOMEWORK: Have students create charts that describe Europe's forestry, farming, and fishing industries. The charts should also note the role of technology in these industries and the challenges the industries now face.

Teach Objective 2

ALL LEVELS: Copy the following graphic organizer onto the chalkboard, omitting the italicized answers. Have students use the text and the unit atlas to identify resources found in each of Europe's physical regions. **ENGLISH LANGUAGE LEARNERS**

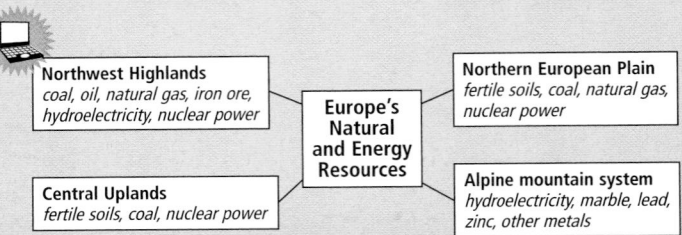

Northwest Highlands
coal, oil, natural gas, iron ore, hydroelectricity, nuclear power

Europe's Natural and Energy Resources

Northern European Plain
fertile soils, coal, natural gas, nuclear power

Central Uplands
fertile soils, coal, nuclear power

Alpine mountain system
hydroelectricity, marble, lead, zinc, other metals

Lavender, which is used in perfumes and soaps, grows in a region of southern France called Provence. Fertile soil and a temperate climate make farming possible throughout the country. Each region has its specialties. France's agricultural bounty makes it one of the world's largest exporters of farm products.

Europe produces many crops, such as large amounts of grapes, olives, potatoes, and wheat. Efficient methods and modern technology have made Europe's crop yields among the highest in the world. Farmers use chemical fertilizers to enrich the soil. They also rotate crops to maintain fertility. Modern machinery is used in planting and harvesting. However, some areas lag behind in farm production. This is the case in Europe's formerly communist countries. Farming technology is often outdated there.

Throughout history, fishing has been an important part of the European economy. Fishing villages dot Europe's coasts, and fishing boats can be found in all waters bordering the continent. Europe's best fisheries are located in the North Atlantic and Arctic Oceans and in the North Sea. Coastal waters, particularly where the warm North Atlantic Drift mixes with cold polar waters, are excellent fishing grounds. Iceland, Norway, Spain, and Denmark are major fishing countries. However, overfishing and coastal pollution threaten the future of the fishing industry in the Mediterranean and North Atlantic.

✔ **READING CHECK:** *Places and Regions* Where are some of the best farming soils in Europe found? **Northern European Plain**

Energy and Minerals

To meet its current industrial and energy needs, Europe must rely heavily on mineral imports. Europe's technologically advanced economies lack sufficient supplies of critical natural resources, such as oil, iron, and other metals.

Europe does have large deposits of coal, however. Some countries, such as Germany, Britain, and Poland, have mined coal for hundreds of years. (See the unit land use and resources map.) In fact, Germany's Ruhr coal field is one of the world's largest. During Europe's industrial era, coal and iron were essential to the creation of the steel and manufacturing industries. However, in the 1900s oil replaced coal as the most important energy source. Europe became increasingly dependent on imported oil. Today oil and gas from Southwest Asia, Russia, and Africa power the economies of most European countries.

Close

Ask students to consider how the resources found in one of Europe's countries have shaped that country's economy. Call on students to provide specific examples. (*Possible answers: France's fertile soils have allowed it to become famous for agricultural products like grapes. Great Britain's coal allowed it to become an early industrial power.*)

Review and Assess

Have the students complete the **Section Review**. Then have students complete **Daily Quiz 13.3**.

Reteach

Have students complete **Main Idea Activity for English Language Learners and Special-Needs Students 13.3**. Then give each student an outline map of Europe. Within each country's borders, have students list the natural resources found there. **ENGLISH LANGUAGE LEARNERS**

Extend

Have each student select a famous building or monument in Europe and conduct research to learn about the materials used in its construction and whether these materials reflect the minerals found in the area. Have students present their findings to the class. **BLOCK SCHEDULING**

INTERPRETING THE VISUAL RECORD

Workers cut marble from a quarry near Carrara, Italy. For centuries, sculptors have favored the white stone. **How might changes in technology have affected the marble industry of Carrara?**

Europe's main oil and natural gas deposits lie beneath the North Sea. These deposits were discovered in the early 1960s. They have greatly benefited the economies of Norway and Britain. Both countries are now energy exporters. The Netherlands also produces and exports natural gas.

Hydroelectricity is another important energy source. It is produced in mountainous Norway, Sweden, and Switzerland. France produces ocean tidal power and solar power. Iceland uses geothermal energy to heat homes and generate electricity. Nuclear power also supplies energy, particularly in France, Belgium, Bulgaria, and Sweden. However, many Europeans are worried about the long-term safety of nuclear power.

Other mineral resources in Europe include iron ore, uranium, lead, and zinc. Sweden and France both have large deposits of iron ore in upland regions. France also has sizable uranium deposits. Lead, zinc, and other metals are found in Spain and southern Europe. Marble, a stone used for building and sculpting, has long been mined in parts of southern Europe, such as in Carrara, Italy.

Our Amazing Planet

The world's largest deposits of amber are found along the shores of the Baltic Sea. Some deposits of this fossilized tree resin date back to perhaps 60 million years ago. The preserved bodies of ancient insects have been found in some deposits. The yellowish translucent amber is usually made into jewelry.

✔ **READING CHECK:** *Places and Regions* Where are Europe's main oil and natural gas fields? *beneath the North Sea*

Section 3 Review

go.hrw.com
Homework Practice Online
Keyword: SW3 HP13

Define loess

Working with Sketch Maps On the map you created in Section 2, label the Po Valley and Guadalquivir River. Which of these areas is a major agricultural region in Spain?

Reading for the Main Idea

1. *Places and Regions* What types of soils are found in river valleys such as Italy's Po Valley? How useful are these soils for agriculture?

2. *Physical Systems* How does modern farming technology affect European agriculture?

3. *The Uses of the Geography* How did the growing importance of oil as an energy source in the 1900s affect Europe?

Critical Thinking

4. **Problem Solving** How might Europeans lessen their dependence on imported oil?

Organizing What You Know

5. Copy the chart shown below. Use it to list Europe's main energy resources and where they are found. Add as many rows as you need.

Energy source	Location

301

TECHNOLOGY

▶ Chapter 13 Test Generator (on the One-Stop Planner)
▶ Global Skill Builder CD–ROM
▶ HRW Go site

REINFORCEMENT, REVIEW, AND ASSESSMENT

▶ Chapter 13 Review, pp. 302–03
▶ Chapter 13 Tutorial for Students, Parents, Mentors, and Peers
▶ Chapter 13 Test (form A or B)
▶ Alternative Assessment Handbook

▶ Chapter 13 Test for English Language Learners and Special-Needs Students
▶ Unit 4 Test
▶ Unit 4 Test for English Language Learners and Special-Needs Students

Assess

Have students complete a Chapter 13 Test.

Reteach

Have students work in small groups to prepare sketch maps of Europe that identify landform systems, water features, and climate zones. Then have students mark the locations of significant mineral deposits across the continent. Discuss the completed maps. **ENGLISH LANGUAGE LEARNERS**

CHAPTER 13 Review Answers

Thinking Critically

1. Possible answer: People who live there might be threatened by floods if the dikes break. (NGS 15)

2. The North Atlantic Drift and westerlies combine to moderate Europe's climates. (NGS 7)

3. Possible answer: Both are lowland regions important for farming, but the Great Plains have more severe climates. (NGS 4)

Using the Geographer's Tools

1. parts of the Netherlands

2. Student graphs should demonstrate these regional climate patterns: London—mild temperatures all year, rainfall constant all year; Rome—warm summers, mild winters, seasonal rainfall in the winter; Berlin—cold winters, mild summers, fairly even rainfall distribution.

3. Possible answer: The mountains would block the mild air from reaching much of interior Europe.

Writing

Student reports should demonstrate a clear understanding of the causes and effects of tectonic activity. The reports should mention areas with heavy tectonic action, including Iceland and southern Italy and Greece. Use Rubric 42, Writing to Inform, to evaluate student work.

CHAPTER 13 Review

Building Vocabulary

On a separate sheet of paper, explain the following terms by using them correctly in sentences.

fjords	navigable
polders	North Atlantic Drift
dikes	loess

Locating Key Places

On a separate sheet of paper, match the letters on the map with their correct labels.

Ural Mountains	Northern European Plain
Mediterranean Sea	Alps
Scandinavian Peninsula	Pyrenees
Iberian Peninsula	North Sea
Italian Peninsula	Rhine River

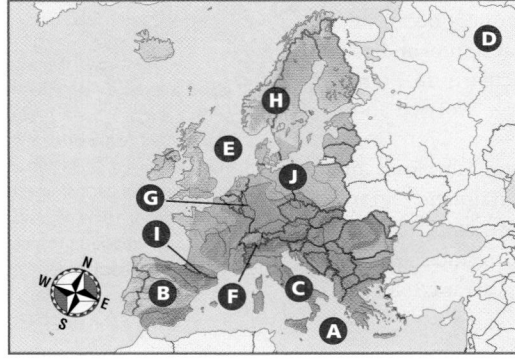

Understanding the Main Ideas

Section 1

1. (**Physical Systems**) How have tectonic processes shaped the physical environment of Europe?

2. (**Places and Regions**) Where can you find glacially scoured landscapes in Europe? What are these landscapes like?

Section 2

3. (**Places and Regions**) What are Europe's three major climate types?

4. (**Places and Regions**) Which biome do large parts of northern and central Europe have? How have people affected it?

Section 3

5. (**Environment and Society**) Where are Europe's best fisheries located?

Thinking Critically

1. **Supporting a Point of View** What potential dangers do you see with creating polders that are below sea level?

2. **Summarizing** Why does Europe have such mild climates compared to other world regions of similar latitude?

3. **Comparing** How does the environment of the Northern European Plain compare with the environment of the Great Plains in North America?

Using the Geographer's Tools

1. **Analyzing Maps** Study the unit physical and political maps. What area of Europe is generally below sea level?

2. **Creating Climate Graphs** Go to the HRW Web site on the Internet to find statistics you need to create climate graphs. Then create climate graphs for London, Rome, and Berlin. Describe the climate patterns you see in your graphs.

3. **Preparing Diagrams** Use the information in Section 2 to prepare a diagram showing how the westerlies and North Atlantic Drift bring mild temperatures to northwestern Europe. You might want to use arrows to show the general direction of wind flow. How would this pattern be different if a high mountain range bordered the coast of western and northern Europe?

Writing about Geography

Write a report about tectonic activity in Europe. Explain why, how, and where tectonic activity occurs in Europe. How does this tectonic activity affect the continent? When you are finished with your report, proofread it to make sure you have used standard grammar, spelling, sentence structure, and punctuation.

SKILL BUILDING

Geography for Life

Using Research Skills

(**Environment and Society**) Plan a research project to study the decline of commercial fishing in Europe. First, determine a goal for your research or a question that you want answered. For example, you might want to know how the amount of fish caught in Europe has changed over the last 20 years. Then find information that relates to your question, such as statistics or articles about the fishing industry. When you have gathered enough information, write a short report that answers your research question and suggests reasons for the decline of commercial fishing in the region. When you are done, you might want to suggest other possible areas for future research.

Portfolio Activity

Avalanches are a constant threat in the Alpine mountain system. Have students conduct library and Internet research to learn about Alpine avalanche control measures. Then have them construct scale models to demonstrate some common control systems. Students might use cotton batting "snow" and craft sticks to build their models on bases of crumpled newspaper. Place photos of the models and written descriptions of the projects in student portfolios.

Food Festival

Europe's temperate climate and rich soil contribute to the region's agricultural wealth. In the past, however, many Europeans ate whatever they could find to survive. Also, before refrigeration, they preserved food in many ways. Some of the resulting dishes remained popular long after a variety of fresh foods became widely available. For example, many Scandinavians enjoy *lutefisk*, which is dried cod soaked in a water-lye solution. Black pudding, which contains pig's blood and oatmeal, is available in Irish markets. Have students conduct research on these and other European foods that many Americans might consider unappetizing and, if possible, bring samples to class.

Building Social Studies Skills

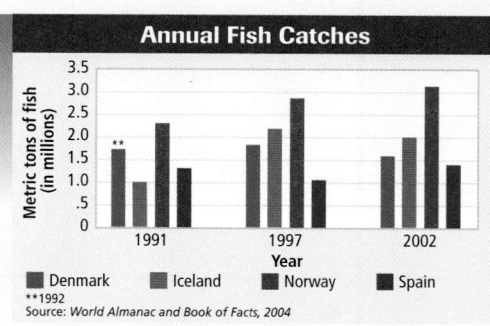

Annual Fish Catches

Metric tons of fish (in millions)

■ Denmark ■ Iceland ■ Norway ■ Spain
**1992
Source: *World Almanac and Book of Facts, 2004*

Interpreting Graphs

Study the graph above. Then use information from the graph to help you answer the questions that follow.

1. From 1991 to 1997 and from 1997 to 2002 which country had the most dramatic increase in its fish catch?
   a. Denmark
   b. Iceland
   c. Norway
   d. Spain

2. A comparison of the 1991 and 1997 fish catches of the four countries shows that only one country had a clear decrease. Which country was it? Support your answer.

Interpreting Secondary Sources

Read the following passage and answer the questions that follow. Mark your answers on a separate sheet of paper.

"To create polders, the Dutch built earthen walls called dikes along the shoreline. Then they used windmills to pump out the seawater behind the dikes. The Dutch used the drained lands for farming or for housing. By using polders to grow crops and raise livestock, the Dutch greatly increased the amount of available farmland. In fact, the Netherlands is an exporter of agricultural goods. The Dutch export products such as flowers, grains, potatoes, and sugar beets, particularly to other European countries."

3. One step that is part of the process of creating polders is
   a. flooding valleys by opening river dams.
   b. building dikes along the shoreline.
   c. using windmills to generate electricity.
   d. building homes that stand above local water levels.

4. How have the polders helped the Netherlands become an exporter of agricultural goods?

Alternative Assessment

PORTFOLIO ACTIVITY

Learning about Your Local Geography

Group Project: Field Work

Plan, organize, and complete a research project with a partner that compares plant and animal life in your area with those in Europe. First, study the plants and animals in your area by doing field work. Work together to observe your area's wildlife. You may want to make drawings of what you see. Then use a library to find information about the plants and animals that you observed. What kind of biome do you live in? What plants and animals are common to that biome? Finally, compare your biome to the biomes of Europe. Does the same biome exist in Europe? If so, where? How are the plants and animals of your area similar to and different from those in Europe?

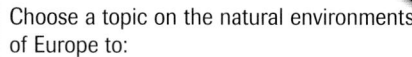

Internet Activity: go.hrw.com
KEYWORD: SW3 GT13

Choose a topic on the natural environments of Europe to:
- describe the environmental impact of oil drilling in the North Sea.
- understand how technological innovations affect the maintenance of polders.
- create a poster about fjords along the Norwegian coast.

CHAPTER RESOURCE MANAGER

Objectives	Pacing Guide	Reproducible Resources
SECTION 1 **The British Isles** (pp. 305–09) ● Discuss how history has affected the culture of the British Isles. ● Explain why the cultures of Ireland and the United Kingdom are so similar. ● Describe how the British economy has changed over the last 200 years. ● Identify the issue that has caused tension in Northern Ireland.	**Regular** 1 day **Block Scheduling** .5 day *Block Scheduling Handbook, Chapter 14*	**RS** Guided Reading Strategy 14.1 **PS** Readings in World Geography, History, and Culture 34 and 35 **E** Cultures of the World Activity: Region 3 **SM** Critical Thinking Activity 14: *From* The Speeches, May 19, 1940: Winston Churchill
SECTION 2 **France** (pp. 311–14) ● Describe French culture. ● Identify some of the main industries in France. ● Consider the challenges that France faces today.	**Regular** 1 day **Block Scheduling** .5 day *Block Scheduling Handbook, Chapter 14*	**RS** Guided Reading Strategy 14.2 **PS** Readings in World Geography, History, and Culture 31 **SM** Map Activity 14: France's Colonial Empire
SECTION 3 **The Benelux** **Countries** (pp. 315–17) ● Identify the historical ties that the Benelux countries share. ● Describe the cities and economies of the Benelux countries.	**Regular** .5 day **Block Scheduling** .5 day *Block Scheduling Handbook, Chapter 14*	**RS** Guided Reading Strategy 14.3 **RS** Graphic Organizer Activity 14 **E** Creative Strategies for Teaching World Geography, Lesson 10
SECTION 4 **Scandinavia** (pp. 318–21) ● Note how the cultures of Scandinavia are similar to and different from each other. ● Identify the industries upon which the economy of the region relies. ● Locate the areas in Scandinavia where most people live.	**Regular** .5 day **Block Scheduling** .5 day *Block Scheduling Handbook, Chapter 14*	**RS** Guided Reading Strategy 14.4 **PS** Readings in World Geography, History, and Culture 36 **SM** Geography for Life Activity 14: Linking Denmark

Chapter Resource Key

PS Primary Sources	**A** Assessment	CD-ROM
RS Reading Support	**REV** Review	Video
IC Interdisciplinary Connections	**ELL** Reinforcement and English Language Learners	Internet
E Enrichment		Holt Presentation Maker Using Microsoft® PowerPoint®
SM Skills Mastery	Transparencies	

 One-Stop Planner CD–ROM

See the *One-Stop Planner* for a complete list of additional resources for students and teachers.

One-Stop Planner CD–ROM

It's easy to plan lessons, select resources, and print out materials for your students when you use the **One-Stop Planner CD–ROM with Test Generator**.

Technology Resources

- One-Stop Planner CD–ROM, Lesson 14.1
- *ARGWorld* CD–ROM
- CNN. Presents World Cultures: Yesterday and Today, Segment 36: The Walls That Divide
- Homework Practice Online
- HRW Go site

- One-Stop Planner CD–ROM, Lesson 14.2
- Geography and Cultures Visual Resources 30
- Homework Practice Online
- HRW Go site

- One-Stop Planner CD–ROM, Lesson 14.3
- CNN. Presents World Cultures: Yesterday and Today, Segment 30: NATO at 50
- Homework Practice Online
- HRW Go site

- One-Stop Planner CD–ROM, Lesson 14.4
- CNN. Presents Geography: Yesterday and Today, Segment 14: The Fishing Life
- Homework Practice Online
- HRW Go site

Reinforcement, Review, and Assessment

ELL	Main Idea Activity 14.1
ELL	English Audio Summary 14.1
ELL	Spanish Audio Summary 14.1
REV	Section 1 Review, p. 309
A	Daily Quiz 14.1

ELL	Main Idea Activity 14.2
ELL	English Audio Summary 14.2
ELL	Spanish Audio Summary 14.2
REV	Section 2 Review, p. 314
A	Daily Quiz 14.2

ELL	Main Idea Activity 14.3
ELL	English Audio Summary 14.3
ELL	Spanish Audio Summary 14.3
REV	Section 3 Review, p. 317
A	Daily Quiz 14.3

ELL	Main Idea Activity 14.4
ELL	English Audio Summary 14.4
ELL	Spanish Audio Summary 14.4
REV	Section 3 Review, p. 321
A	Daily Quiz 14.4

☑ internet connect

HRW ONLINE RESOURCES

GO TO: go.hrw.com
Then type in a keyword.

TEACHER HOME PAGE
KEYWORD: SW3 Teacher

CHAPTER INTERNET ACTIVITIES
KEYWORD: SW3 GT14
Choose an activity on northern and western Europe to:
- learn the history of skiing in Norway.
- compare and contrast major cities in the region.
- research daily life in the region.

CHAPTER ENRICHMENT LINKS
KEYWORD: SW3 CH14

CHAPTER MAPS
KEYWORD: SW3 MAPS14

ONLINE ASSESSMENT
Homework Practice
KEYWORD: SW3 HP14
Standardized Test Prep
KEYWORD: SW3 STP14
Rubrics
KEYWORD: SS Rubrics

COUNTRY INFORMATION
KEYWORD: SW3 Almanac

CONTENT UPDATES
KEYWORD: SS Content Updates

HOLT PRESENTATION MAKER
KEYWORD: SW3 PPT14

ONLINE READING SUPPORT
KEYWORD: SS Strategies

CURRENT EVENTS
KEYWORD: S3 Current Events

Meeting Individual Needs

Ability Levels

Level 1 Basic-level activities designed for all students encountering new material

Level 2 Intermediate-level activities designed for average students

Level 3 Challenging activities designed for honors and gifted-and-talented students

English Language Learners Activities that address the needs of students with Limited English Proficiency

Chapter Review and Assessment

- Chapter 14 Test Generator (on the One-Stop Planner)
- Global Skill Builder CD–ROM
- HRW Go site
- **REV** Chapter 14 Review, pp. 324–25
- **REV** Chapter 14 Tutorial for Students, Parents, Mentors, and Peers
- **A** Chapter 14 Test (form A or B)
- **A** Alternative Assessment Handbook
- **A** Chapter 14 Test for English Language Learners and Special-Needs Students

Northern and Western Europe

CHAPTER 14

Launch into Learning

List the countries of northern and western Europe on the chalkboard. Then ask students questions about the countries. *(Examples: Which country claimed that "the sun never set" on its empire? What is the Eiffel Tower and where is it located? Which country do you associate with tulips and windmills? From which countries did the Viking warriors sail?)* Students will probably know the answers to most of these questions. Encourage students to compose other questions similar to yours. Tell students that in this chapter they will learn more about the countries and cultures of northern and western Europe.

Using the Physical-Political Map

Have students examine the map on the opposite page. Ask them to identify the region's island countries *(Iceland, Ireland, the United Kingdom)* and those that occupy peninsulas *(Denmark, Finland, Norway, Sweden)*. Ask students what is unusual about the Netherlands' land elevation. *(Much of the country is below sea level.)* Point out France's location between two seas.

Why We Should Know More

You may wish to point out to students the following reasons why we should study northern and western Europe:

▶ The United States has strong historical and economic ties to several of the countries of northern and western Europe.

▶ Belgium, Denmark, France, Iceland, the Netherlands, Norway, and the United Kingdom are members of the North Atlantic Treaty Organization (NATO). They are allies of the United States.

▶ Millions of Americans trace their ancestry to northern and western Europe.

▶ American culture has been greatly influenced by northern and western Europe. The language, law, commerce, government, literature, art, holidays, and food found in the United States bear the imprint of the region.

CHAPTER 14 Northern and Western Europe

The countries of northern and western Europe have had a tremendous influence throughout the world. Great Britain and France were among the greatest colonial powers. They spread their languages, educational, and political systems worldwide.

Gathering tulips in the Netherlands

Walrus-ivory chess pieces from Scotland

Hi. My name is Lars, and I live in Tromsø, in far northern Norway. I live in a new house in the middle of town. We are so close to the sea that we eat fish almost every day. For breakfast I have a slice of bread with a sweet goat cheese, smoked salmon, shrimp, or cod liver spread. I take two more of these sandwiches for lunch at school. For dinner we have salmon or cod or another fish about four or five times a week. The other days we like to eat pizza and hamburgers.

In the winter everyone skis to school along trails that look like snowy streets. We live above the Arctic Circle, and it is completely dark in the winter. Because of this, the trails must have streetlights. The Sun does not shine at all in Tromsø from the end of November to January 20 or so. On January 20, when the Sun appears again for just a few minutes, we have a big celebration for Sun Day.

In the summer, the Sun never sets, so there is no nighttime at all. This is my favorite time of year. We have a huge bonfire and a cookout on June 21. But it still can be very cold. Last year the temperature in the summer was usually about 6° or 7°C (about 43° or 44°F).

304

OBJECTIVES

1. Discuss how history has affected the culture of the British Isles.

2. Explain why the cultures of Ireland and the United Kingdom are so similar.

3. Describe how the British economy has changed over the last 200 years.

4. Identify the issue that has caused tension in Northern Ireland.

LET'S GET STARTED

Copy the following "formula" and question on to the chalkboard: *Celts + Romans + Angles + Saxons + Vikings + Normans + Africans + South Asians = British people.* *What does this "formula" mean?* Call on volunteers to offer answers. (*It refers to some of the many peoples that have migrated to the British Isles over time.*) Discuss responses. Then add that the British Isles have been a "melting pot" for centuries and that the population continues to become more diverse. Tell students that they will learn more about these islands in Section 1.

Building Vocabulary

Write the term **constitutional monarchy** on the chalkboard. Ask students to guess what it means based upon words they already know. For example, a monarchy is a state ruled by a king or queen, and a constitution usually establishes the branches of a country's government, so Britain's constitutional monarchy features both a monarch and a parliament. Ask a volunteer to locate the definition in the text and to read it aloud. Then have someone locate and read the definitions of the other key terms.

Section 1

The British Isles

READ TO DISCOVER

1. How has history affected the culture of the British Isles?
2. Why are the cultures of Ireland and the United Kingdom similar?
3. How has the British economy changed over the last 200 years?
4. What issue has caused tension in Northern Ireland?

WHY IT MATTERS

Since the 1960s many British musicians have helped to shape popular music in the United States. Use **CNNfyi.com** or other **current events** sources to learn about how modern British music has influenced world culture.

DEFINE

sequent occupance
famine
constitutional monarchy
nationalized

LOCATE

British Isles
England
Wales
Scotland
Northern Ireland

Locate, continued

Dublin
London
Thames River
Glasgow
Edinburgh
Birmingham
Belfast

Section 1 RESOURCES

REPRODUCIBLE

► Guided Reading Strategy 14.1
► Readings in World Geography, History, and Culture 34 and 35
► Cultures of the World Activity: Region 3
► Critical Thinking Activity 14: *From* The Speeches, May 19, 1940: Winston Churchill

TECHNOLOGY

► One-Stop Planner CD–ROM, Lesson 14.1
► CNN Presents World Cultures: Yesterday and Today, Segment 36: The Walls That Divide
► Homework Practice Online
► HRW Go site

REINFORCEMENT, REVIEW, AND ASSESSMENT

► Main Idea Activity 14.1
► English Audio Summary 14.1
► Spanish Audio Summary 14.1
► Section 1 Review, p. 309
► Daily Quiz 14.1

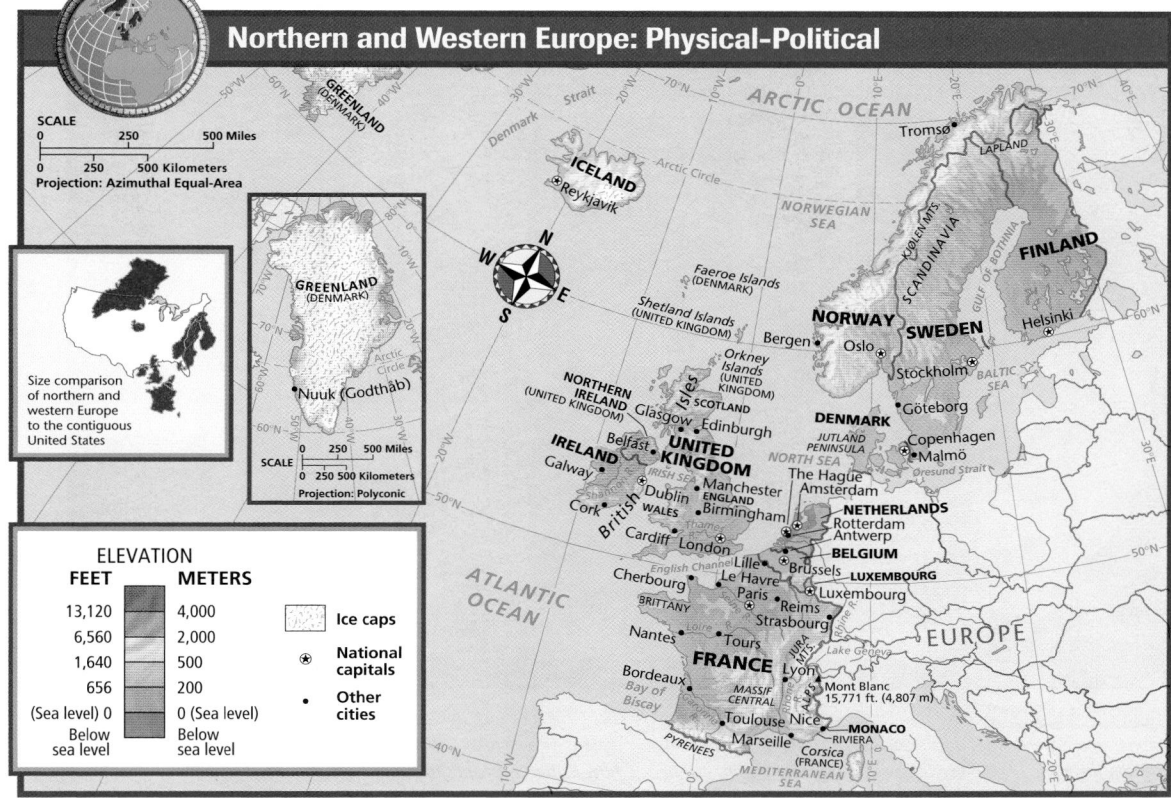

Northern and Western Europe: Physical-Political

SCALE
0   250   500 Miles
0   250   500 Kilometers
Projection: Azimuthal Equal-Area

Size comparison of northern and western Europe to the contiguous United States

GREENLAND (DENMARK)

Nuuk (Godthåb)

SCALE
0   250   500 Miles
0   250 500 Kilometers
Projection: Polyconic

ELEVATION

FEET	METERS	
13,120	4,000	Ice caps
6,560	2,000	
1,640	500	⊛ National capitals
656	200	
(Sea level) 0	0 (Sea level)	• Other cities
Below sea level	Below sea level	

Teach Objective 1

🌐 **ALL LEVELS:** Organize students into six groups and assign each group one of the following time periods: 800–1000, 1000–1200, 1200–1400, 1400–1600, 1600–1800, or 1800–2000. Have groups use library resources to create illustrated time lines of British history for their assigned periods. Suggest that groups construct their time lines on sheets of butcher paper or poster board. When groups have completed the task, have them display their time lines in chronological order around the classroom. Then lead a discussion on how the various events, developments, and individuals mentioned on the time lines have influenced British culture. **ENGLISH LANGUAGE LEARNERS, COOPERATIVE LEARNERS**

Using National Geography Standard 17:

The Uses of Geography: How to Apply Geography to Interpret the Past Have students compare old and new maps of Great Britain to examine various changes. You may want to locate in advance books that include historical maps. Encourage students to compare maps from various time periods. Point out that distances and shapes on the oldest maps may be somewhat distorted. Have students examine the maps to find such information as changes in place-names, towns that no longer exist, changes in the sizes of settlements, rivers that have changed course, or ancient roads that became major highways.

Linking Past to Present

The Legend of King Arthur The legend of King Arthur is known worldwide. He is usually portrayed as a glamorous monarch performing great deeds while surrounded by equally beautiful people. But who was this romantic figure?

Early historians described Arthur as a mighty Welsh warrior. A Roman Catholic bishop used these accounts to write a fictitious "history" of Arthur as the King of Britain and conqueror of western Europe. Later, German and French poets integrated Celtic legends of Galahad, Lancelot, Merlin, Excalibur, and the Holy Grail into the tale. Behind the legends however, there may have been a tough fighter who tried to protect Roman Britain from invading Saxons in the early A.D. 500s.

Today the Arthurian legend is the subject of numerous books, movies, operas, paintings, poems, and other works. Advertisements and brand names evoke the great king and his court. Our very language contains phrases based on King Arthur. For example, we may search for a Holy Grail of wealth or success.

VISUAL RECORD ANSWER

mark on landscape left by cultural group that is gone

MAP ANSWER

Its island geography influenced its development as a naval power.

INTERPRETING THE VISUAL RECORD

Stonehenge, an ancient complex of massive stone circles in England, was built beginning about 3100 B.C. Modern archaeologists are unsure exactly why Stonehenge was built. **How is Stonehenge an example of sequent occupance?**

History

The British Isles are made up of two independent countries—the Republic of Ireland and the United Kingdom of Great Britain and Northern Ireland. The Republic of Ireland occupies all but the northern part of the island of Ireland. The United Kingdom is often referred to as just Great Britain or even Britain. It includes four political regions—England, Wales, Scotland, and Northern Ireland. Great Britain can be divided into two physical regions, lowland Britain and highland Britain. Most of England, except the far north, is part of lowland Britain. Northern England, Wales, and Scotland make up highland Britain.

Lowland Britain has developed a complex cultural geography. About 5,000 years ago, the earliest settlers left their mark with monuments like Stonehenge. Later the Celts—the ancestors of the Scots, Welsh, and Irish—occupied the island. Then the Romans came and built fortified towns. Later, Angles and Saxons, two Germanic tribes, came and drove the Celtic peoples to highland Britain. Vikings from Scandinavia raided the coastal areas and also built settlements. In 1066 William of Normandy conquered England. Normandy is now part of France. Each of these peoples left an imprint on lowland Britain. This process of settlement by successive groups of people, each group creating a distinctive cultural landscape, is called **sequent occupance**.

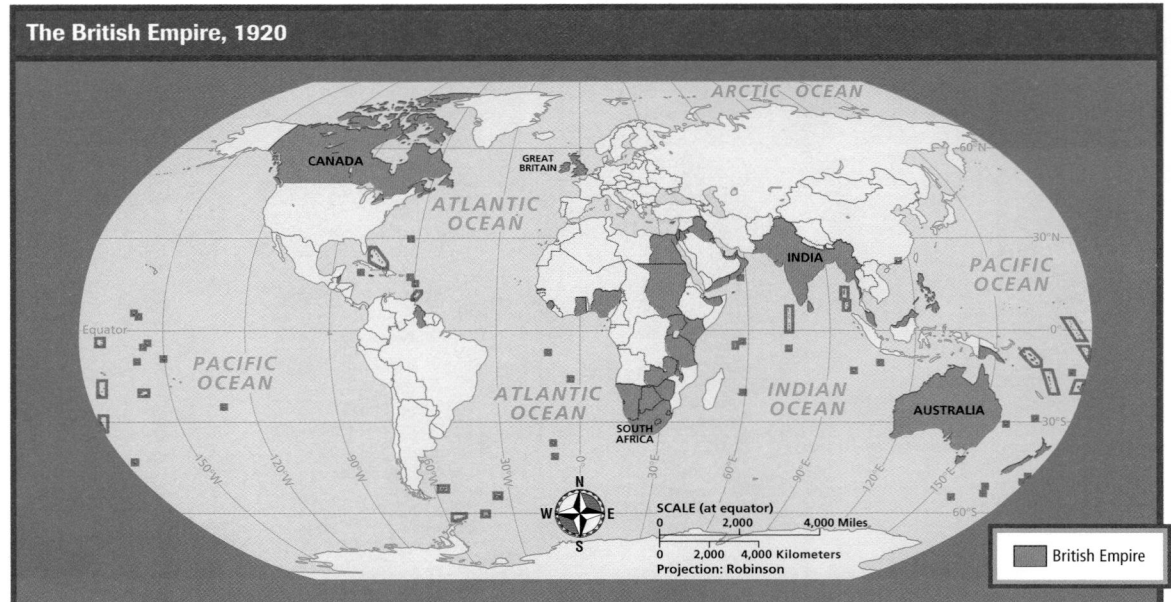

The British Empire, 1920

British Empire

INTERPRETING THE MAP
The British Empire was a worldwide system of territories and dependencies administered by the British government. Areas within the empire had different levels of self-government, and by 1920, colonies such as Canada, South Africa, and Australia largely managed their own affairs. **What geographic factors might help explain how Britain was able to control such a large empire?**

ALL LEVELS: Copy the following graphic organizer onto the chalkboard, omitting the italicized answers. Have students complete it with information about similarities and differences in the cultures of Ireland and Great Britain.
ENGLISH LANGUAGE LEARNERS

HOMEWORK: Provide each student with an outline map of the British Isles. Have students use the text and other sources to create maps of the islands' cultures. Students should use colors, shading, and symbols to identify important cultural locations and regions in Britain and Ireland.

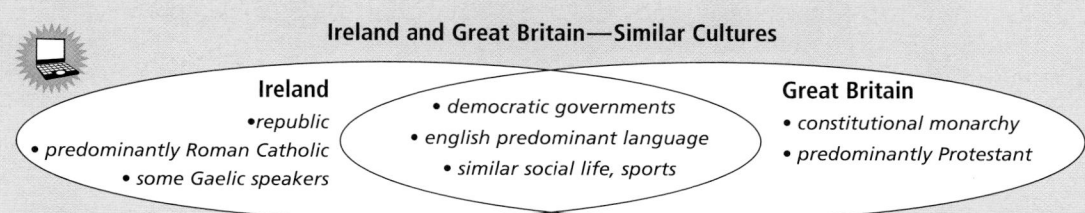

Ireland and Great Britain—Similar Cultures

Ireland
• *republic*
• *predominantly Roman Catholic*
• *some Gaelic speakers*

• *democratic governments*
• *english predominant language*
• *similar social life, sports*

Great Britain
• *constitutional monarchy*
• *predominantly Protestant*

The British Empire In the 1600s and 1700s, British explorers and settlers founded colonies around the world. By 1801 England had brought Ireland, Scotland, and Wales into the United Kingdom. The surrounding ocean helped protect this kingdom. The British also built a powerful navy to take further advantage of the sea. During the 1800s more than one fourth of the world's land was ruled by the British Empire. The empire's colonies provided raw materials for British industries. The colonies also served as markets for finished goods. The empire spread the English language, Christianity, British law, sports, and other British customs around the globe.

The colonies that became the United States were part of the British Empire when they declared independence in 1776. Over time, most parts of the empire gained independence. Most former colonies became members of the Commonwealth of Nations. They still meet to discuss economic, business, and scientific matters of common concern.

Ireland did not win independence from the British until 1921. Before independence, life had long been hard for many Irish. Then in the mid-1800s Ireland suffered from a potato **famine**. A famine is a widespread shortage of food that may lead to severe hunger and starvation. About 1 million Irish died when the potato crop failed for several years in a row. The famine, poverty, and a lack of economic opportunities led many Irish to immigrate to other countries. Many migrated to the United States. (See the map.)

READING CHECK: *Environment and Society* Why did many people emigrate from Ireland during the mid-1800s? a potato famine, poverty, lack of economic opportunities

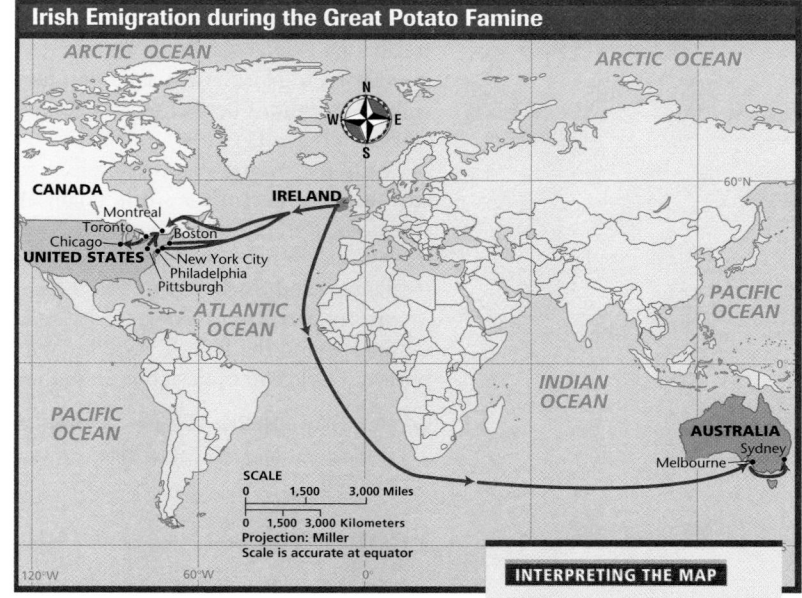

Irish Emigration during the Great Potato Famine

SCALE
0   1,500   3,000 Miles
0   1,500 3,000 Kilometers
Projection: Miller
Scale is accurate at equator

INTERPRETING THE MAP

Ireland's Great Potato Famine dramatically changed the country's population pattern as more than 1 million Irish immigrated to North America, Britain, and Australia. **What geographic factors, both physical and cultural, might explain the migration patterns shown on the map?**

Culture

Because of their shared history, Ireland and Great Britain share many cultural features. Social life is often centered around local eateries. Sports such as soccer, rugby, and cricket are popular. In addition, English is the main language of both countries. However, a small number of Irish also speak Irish Gaelic, and some Scots speak Scottish Gaelic.

The countries also differ in important ways. Both Ireland and Britain are democracies. However, their governments are organized differently. Ireland is a republic, and the president is the head of state. Britain is a **constitutional monarchy**. That is, a king or queen is the head of state, but a parliament led by a prime minister serves as the lawmaking branch of government.

Essential Element 5

► **Environment** ◄
and Society

Relying on a Resource
The "Irish," or white, potato actually originated in South America. Spanish explorers brought it to Europe in the 1500s. By the end of the 1700s, it was a major European crop. The Irish came to depend on the high nutritional properties of the potato for good health. The potato was also vital to the Irish economy. As a result, when a plant disease called blight ruined Ireland's potato crop, starvation and economic crisis were almost inevitable. The disaster made clear the danger of depending on the potato or any other single crop for many nutritional needs.

DISCUSSION: Which crops are crucial to the United States? What would we do if those crops were destroyed by disease or weather disasters? You may want to extend the discussion to Americans' dependence on fossil fuels or the importance of maintaining diverse genetic stocks of crop plants.

MAP ANSWER

Oceans served as migration routes, and migrants chose areas with cultural connections to Britain.

internet connect

GO TO: go.hrw.com
KEYWORD: SW3 CH14
FOR: Web sites about northern and western Europe

ALL LEVELS: Organize students into groups. Ask each group to create a visual essay—a series of illustrations showing how the British economy has changed over the last 200 years. Point out that the visual materials they might use include charts, graphs, maps, diagrams, sketches, photographs, clippings from newspapers and magazines, and images from the Internet. Instruct groups to include an explanatory paragraph with each visual. Have groups display, compare, and discuss their visual essays. **ENGLISH LANGUAGE LEARNERS, COOPERATIVE LEARNING**

ALL LEVELS: Ask students to assume the roles of newscasters on a radio program. Tell them that they have been asked to make three to five minute broadcasts on "the troubles" in Northern Ireland. Have students write scripts or production notes for their broadcasts. Provide reference materials and current events publications for students to use as they write. Suggest to students that their presentations cover such topics as the origin of the troubles, differing views of Protestants and Catholics, and hopes for the future. Call on volunteers to read their scripts to the rest of the class.

Essential Element 4

▶ **Human Systems** ◀

Processes, Patterns, and Functions of Human Settlement The story of England's changing fortunes can be read in the ruins of some 3,000 villages. Geographers often study reasons for the growth of settlements, but they might also analyze why towns decline or disappear.

Many different factors caused the disappearance of English villages. War depopulated many. For example, after their initial battle in 1066, the Normans marched north, killing thousands of people and destroying hundreds of villages. Disease, in particular the Black Death, wiped out many towns. During the 1500s, some landlords forced tenants from fields and villages to make room for sheep. Soil exhaustion and coastal erosion made other places uninhabitable. The rising water of reservoirs also claimed some settlements. Towns dependent on dying industries such as coal-mining lost public funds, and residents moved to new towns. The old ones were eventually bulldozed.

INTERPRETING THE VISUAL RECORD

This painting from the late 1700s shows a steam engine being used to dig a coal mine and represents the beginnings of the Industrial Revolution in England. **How are people in this painting using technology to modify the physical environment?**

VISUAL RECORD ANSWER

extracting coal, burning coal and creating air pollution

Another important difference is religion. The vast majority of people in Ireland are Roman Catholic. However, Protestants make up a majority of the population in Northern Ireland and the rest of the United Kingdom. As you will read, tensions between Catholics and Protestants have led to violence in Northern Ireland.

Dublin is Ireland's capital and its most populous city. London is the largest city, the cultural center, a world financial center, and the capital of the United Kingdom. Both cities are home to government buildings and famous landmarks. Visitors to London can see historic buildings like the Houses of Parliament and Buckingham Palace. For centuries, London's location on the Thames (TEMZ) River made that city an ideal port for trade between continental Europe and the British Isles. Today Heathrow Airport, one of the world's busiest travel centers, ties London to thousands of cities around the globe.

✓ **READING CHECK:** *Human Systems* What are some cultural features that Britain and Ireland share? What is one major difference? language, popular sports, democratic government; religion

Economy

The Industrial Revolution began in Britain. By the 1700s the country had developed coal and iron mining and a large labor force. Britain also built a good transportation network that used rivers and canals. By the early 1800s the British had built the world's first railroads. Later in the century, London built the first subway system. (See Geography for Life: The London Underground and Mass Transit.) All of these features aided industrial development.

Industrial Rise and Decline Britain's early industries included iron and steel, shipbuilding, and textiles. Innovations such as spinning machines and steam power revolutionized how fabrics were produced. Wool and cotton from British colonies as well as from the United States supplied the textile industry. Industrial growth spread from London and central England to southern Wales, Scotland, and Northern Ireland. Cities like Glasgow and Edinburgh became industrial centers. Trade of raw and finished products between Britain and its colonies further aided development.

Throughout much of the 1700s and 1800s, Britain dominated global trade. (See Case Study: Global Trade.) By 1900, however, the British had lost their dominance to foreign competition. By the mid-1900s Britain's coal mines and traditional industries were in rapid decline. British industries suffered because many were inefficient and because their products were not in demand. In the years after World War II, the United Kingdom **nationalized** many industries to try to stop the decline. Nationalized industries are those that are owned and operated by the government. They are protected from domestic competition.

Changing Fortunes Today Britain has returned most industry to private ownership, and the British economy is strong. Many early industrial cities, such as Glasgow and Birmingham, have benefited from urban renewal. They have also attracted

Close

Encourage students to look through books, encyclopedias, and magazine articles about the British Isles. Have them select a visual image that they think best represents each country. Call on volunteers to share and explain their selections.

Review and Assess

Have students complete the **Section Review**. Then have students complete **Daily Quiz 14.1**.

Reteach

Have students complete **Main Idea Activity for English Language Learners and Special-Needs Students 14.1**. Write the headings History, Culture, Economy, and Challenges on the chalkboard and have students supply terms and phrases for each heading. **ENGLISH LANGUAGE LEARNERS**

Extend

Point out that the United States and Great Britain Isles have been called "two countries separated by a common language." Have interested students conduct research to create an American-British Dictionary containing examples of different usage and vocabulary. **BLOCK SCHEDULING**

high-tech industries. Although the coal industry has declined, oil and gas wells in the North Sea have helped the economy. In addition, much of Britain's labor force works in service industries rather than manufacturing. Tourism is also an important industry.

The Irish economy traditionally was based on farming. However, Ireland now has one of Europe's most rapidly developing economies. Low taxes and a well-educated workforce have attracted foreign companies. They use the country as a door to other European markets. The main industries are now banking, computers, electronics, and food processing. Immigrants from other countries are moving to Ireland to get jobs. This situation is quite different from the past, when Irish emigrated from their poor country.

✓ **READING CHECK:** *Human Systems* How has Ireland's economy changed over time?
went from a largely agricultural economy to one that is more industrial

Issues and Challenges

One of the greatest challenges facing the people of the British Isles is violence in Northern Ireland. The Irish call the problems there "the troubles." Most of the people in Northern Ireland are descendents of Protestant English and Scottish settlers. A large minority are Irish Catholic. Many Catholics believe that union with the Republic of Ireland would protect them from discrimination in employment, housing, and government. However, Protestants want to stay part of the mostly Protestant United Kingdom.

The disagreement between the two groups has led to violence, particularly in the city of Belfast. British troops have tried to keep the peace. However, terrorist groups from both sides have killed thousands of people. A 1998 agreement created a shared government between Northern Ireland, the United Kingdom, and the Republic of Ireland. However, the future of this arrangement is not clear.

✓ **READING CHECK:** *Human Systems* How do Roman Catholics and Protestants in Northern Ireland view their region differently? Catholics want Northern Ireland to be a part of the Republic of Ireland. Protestants want Northern Ireland to remain part of the United Kingdom.

Orangemen parade in Portadown, Northern Ireland. The Orangemen are members of the Orange Society, a Protestant organization formed in 1795 to try to maintain Protestant control in Northern Ireland. In recent years, Orangemen parades in the region have led to street violence between Protestants and Catholics.

Section 1 Review

Homework Practice Online
Keyword: SW3 HP14

Define
sequent occupance, famine, constitutional monarchy, nationalized

Working with Sketch Maps
On a map of northern and western Europe that you draw or that your teacher provides, label the United Kingdom, Republic of Ireland, British Isles, England, Wales, Scotland, Northern Ireland, Dublin, London, Thames River, Glasgow, Edinburgh, Birmingham, and Belfast. In the margin of your map, identify the capital of the United Kingdom.

Reading for the Main Idea
1. *Human Systems* How has London's location affected its growth?
2. *Human Systems* How have cultural differences led to the division of Ireland?
3. *Human Systems* How did Great Britain's history as a naval power contribute to the diffusion of cultural traits? How did innovation in Britain spur the Industrial Revolution?

Critical Thinking
4. Analyzing Information What geographical factor has influenced Britain's power to control territory around the world? How do you think that factor has influenced Britain's role in foreign affairs?

Organizing What You Know
5. Create a chart like the one below. Use it to list differences between Britain's old industrial economy and its modern economy. Write a paragraph describing how Britain's economy has changed over the past 200 years.

Britain's old industrial economy	Britain's modern economy

Section 1 Review Answers

Define For definitions, see: sequent occupance, p. 306; famine, p. 307; constitutional monarchy, p. 307; nationalized, p. 308

Working with Sketch Maps
Maps will vary, but listed places should be labeled in their approximate locations. London is the capital.

Reading for the Main Idea
1. Its location along transportation routes such as the Thames River helped London become a major urban center.

2. Protestants want Northern Ireland to remain part of the mostly Protestant United Kingdom rather than a united Catholic Ireland.

3. The empire spread English language, Christianity, British law, sports, and other British customs around the globe. Spinning machines and steam power revolutionized how fabric could be produced, which contributed to industrial development.

Critical Thinking
4. Britain is an island country protected from invasion by the sea. It developed a powerful navy, which helped the country control overseas territories. Britain's strong navy made the country an important international power. (NGS 15)

Organizing What You Know
5. Answers will vary, but students should point out that the British economy once relied on traditional industries and now focuses on high-tech and service industries.

Science: Mass Transit and the Environment

Many residents of U.S. and European cities rely on automobiles for their primary means of transportation, largely because automobiles offer people great flexibility and freedom of movement. However, public transportation advocates encourage people to use mass transit systems because they cause less damage to the environment.

Automobiles typically carry just one or two passengers at a time. Buses, on the other hand, can carry more than 50 people and use less fuel per person to do it. Less fuel means fewer exhaust fumes to pollute the air, and clean air is a major concern in many cities.

Subways and trains can transport even more people than buses, and recent developments in transportation technology have created cleaner methods of propulsion. Many cities have built electric-powered light rail systems. These systems may carry as many as 14,000 people per hour.

Organize students into groups and assign each group one method of transportation such as buses, trains, subways, light rail, automobiles, or bicycles. Have each group study the effectiveness of its method of transportation and its effects on the environment. Then have each group present its conclusions to the rest of the class. Encourage listeners to ask the presenting group questions about the advantages and disadvantages of its transport system.

Across the Curriculum

► Technology ◄

The Pneumatic Subway
Seven years after the London Underground opened, an American inventor, Alfred Ely Beach, designed and built an experimental subway in New York City. Beach's subway worked like a giant straw. A powerful fan at the end of a long tunnel used air pressure to push and pull a single subway car. This pneumatic subway polluted far less than London's coal-burning steam engines.

The pneumatic subway, however, never got past the experimental stage. Not long after Beach's subway debuted, the more versatile electric train was introduced. Since then it has been the standard for almost all subway designs.

Applying What You Know Answers

1. fewer traffic jams, less air pollution, less land devoted to parking
2. Possible answer: large enough that it has major traffic and parking problems and a large tax base; buses, light rail

VISUAL RECORD ANSWER

Many passengers can fit in Underground trains.

This Geography for Life feature addresses National Geography Standards 3, 4, and 12.

The World in Spatial Terms

Geography for Life

The London Underground and Mass Transit

Greater London is home to about 7 million people. The city is also home to the world's first underground rail system. Plans for the Underground were part of an improvement plan of the mid-1800s. Construction began in 1860. Workers dug trenches along streets and built brick walls to support their sides. Then these trenches were covered with brick arches, and the roads were restored above them. London's clay soils made construction of the underground railways easier. The clay was easy to excavate and provided the raw material to make bricks for tunnel walls.

In 1863 the first subway line opened, using steam locomotives. These trains burned coal, producing unpleasant fumes. However, despite the pollution, the Underground was successful from the very beginning. During its first year, the line carried 9.5 million passengers! In 1890, electric trains began to replace steam engines.

Expansion of the Underground, which Londoners call the tube, continued over the years. Improved tunneling techniques were developed after World War I. These techniques made possible a rapid expansion of the underground network. As the system spread out from central London, large areas of rural land became prime locations for new housing developments. As a result, the expansion of the Underground contributed to the development of London's modern suburbs. Now 253 miles (408 km) of track connect 275 stations. Each year, passengers log more than 920 million journeys on the Underground.

London's transit system served as an example for others. Why are such mass-transit systems important? Cities originally developed as places where people and resources were located close together. As cities grew and places became widely separated, improved transportation networks became more important.

However, planning for transportation needs has not always kept up with urban growth. Traffic jams on major highways and roads are common. Heavy traffic increases air pollution. Parking presents other problems. Building parking lots and garages takes up valuable land. Mass-transit systems, including buses, subways, and surface trains, can help solve these problems. Cities with efficient mass-transit systems usually run more smoothly than those in which people must depend mainly on cars. Mass-transit networks help make many of the world's big cities more livable. Examples include Mexico City, Moscow, New York, and Paris.

Applying What You Know

1. **Summarizing** How has the development of mass-transit systems made big cities more livable?

2. **Drawing Inferences and Conclusions** How big do you think cities have to be to build and operate underground rail systems? What types of mass transit would you expect to find in smaller cities?

INTERPRETING THE VISUAL RECORD *London's mass-transit system includes the Underground and a citywide system of buses, which provide connections to the many Underground stations. **Why do you think London's mass-transit system is so effective at moving large numbers of people?***

🔊 *LET'S GET STARTED*

Copy the following question onto the chalkboard: *What are three things that come to mind when you think of France?* Have students write down their responses. *(Possible answers: the Eiffel Tower, fashion, food, wine)* Then ask students where they might have received their impressions of France. *(Possible responses: movies, television, magazines, news)* Ask students which subjects might not be covered by these sources. *(Possible responses: history, daily life, economy)* Tell students that in Section 2 they will learn more about France.

Building Vocabulary

Write the key term **primate city** on the chalkboard. Ask students what they already know about the word *primate.* Students will probably mention that humans and apes are biologically classified as primates. The word's root, *primus,* is from Latin and means "first." Call on a volunteer to read the term's definition *(the largest and most important city in a country).* Relate the meaning to other words based on *primus,* such as *primary, prime,* or *primal.*

Section 2
France

READ TO DISCOVER

1. What is French culture like?
2. What are some of the main industries in France?
3. What challenges does France face today?

WHY IT MATTERS

Many Europeans have been concerned about the spread of mad cow disease. Recently, the disease was found in French livestock. Use **CNNfyi.com** or other **current events** sources to learn about this and other dangerous livestock diseases.

DEFINE

primate city

LOCATE

Paris	Marseille
Seine River	Alps
Lyon	French Riviera
Lille	Corsica

Section 2 RESOURCES

REPRODUCIBLE
► Guided Reading Strategy 14.2
► Readings in World Geography, History, and Culture 31
► Map Activity 14: France's Colonial Empire

TECHNOLOGY
► One-Stop Planner CD–ROM, Lesson 14.2
► Geography and Cultures Visual Resources 30
► Homework Practice Online
► HRW Go site

REINFORCEMENT, REVIEW, AND ASSESSMENT
► Main Idea Activity 14.2
► English Audio Summary 14.2
► Spanish Audio Summary 14.2
► Section 2 Review, p. 314
► Daily Quiz 14.2

History and Culture

France is one of Europe's largest and most influential countries. Like the United Kingdom, France's culture shows the imprint of successive waves of migrants. Some of the peoples that have shaped French culture include the Gauls, Romans, Franks, and Vikings.

French Society   France has a strong cultural identity unified by language and religion. Although some people also speak regional dialects and languages, most speak French. About 90 percent of France's population is Roman

Our Amazing Planet

The mistral, a powerful wind that blows from the Alps across southern France, can reach speeds of up to 100 miles per hour (161 kmh).

France has produced many world-famous artists, including Pierre-Auguste Renoir (1841–1919). Renoir's Le Moulin de la Galette, *a scene showing life in Paris, is considered a masterpiece of impressionism. Impressionism is a style of painting that developed mainly in France in the late 1800s. It attempts to show what one's first impression of a scene is.*

Teach Objective 1

ALL LEVELS: Have students create postcards depicting various aspects of the history and culture of France. Direct students to draw an illustration on one side of each postcard. On the other side, have them write a note to a family member or friend describing the historical or cultural feature pictured in the illustration. Call on students to display and discuss their postcards. **ENGLISH LANGUAGE LEARNERS**

Teach Objective 2

ALL LEVELS: Have students write the word *FRANCE* vertically on a piece of paper. Then have students write one phrase about French industry that begins with each letter in the word. Copy this example onto the chalkboard:

Fashion design, a major industry
Roaming tourists visiting monuments
Agricultural products, particularly wine and cheese
Now heavy industries in decline
Center of high-tech industry
Economy on the rise

ENGLISH LANGUAGE LEARNERS

Cooperative Learning

The Eiffel Tower The Eiffel Tower is recognized as a revolutionary model for modern skyscrapers. Alexandre-Gustave Eiffel's open iron design was chosen from more than 100 plans in a competition. In a city of heavy brick and stone public buildings, the lightweight metal tower resembled nothing else. Eiffel used the experience he gained from his work on the Statue of Liberty to erect his 984-foot tower in Paris. For 41 years, the Eiffel Tower was the tallest structure in the world.

Organize the class into groups. Let each group select a historical building in Paris and have students research and report on it, emphasizing the role that technological change played in the building's construction.

CONNECTING TO TECHNOLOGY ANSWER

canals popular again as a way to move goods; more barge traffic

The Eiffel Tower is a landmark and symbol of Paris. It was built from 1887 to 1889 to celebrate the 100-year anniversary of the French Revolution. At 984 feet (300 m), the Eiffel Tower was the tallest human-made structure in the world until the completion of New York City's Chrysler Building in 1930.

Catholic. The French government spends money to promote French culture and language.

France has had a long and friendly relationship with the United States. However, today some French people worry about the influence of American culture in Europe. They see the spread of American fast food and media, such as movies, as a threat to their own culture. Some French dislike the fact that English words are creeping into French. They think the United States is responsible because English is becoming the global language of business and technology. To counter this trend, a 366-year-old government agency guards the French language from foreign influences. For example, it has declared that e-mail must be called *courrier électronique* (KOOH-ree-ay ay-lek-trohn-EEK). Some French are also concerned that American corporations are buying a growing number of French businesses.

Cities Paris is the capital and **primate city** of France. A primate city is one that ranks first and dominates a country in terms of population and economy. Paris is also one of Europe's largest and most important cities. About 11 million people live in the metropolitan area. The city was founded more than 2,000 years ago on an island in the middle of the Seine (SAYN) River. Today Paris is France's center for banking, business, communications, education, government, and transportation. The city is also a center for fashion, French culture, and tourism. Important regional cities include Lyon (LYOHN), Lille (LEEL), and the Mediterranean seaport of Marseille (mar-SAY).

✓ **READING CHECK:** **Human Systems** How has France reacted to the influences of American culture? Some see American culture as a threat to their own. The government promotes French culture, such as the French language.

Connecting to TECHNOLOGY

French Waterways

In the 1800s canals and rivers were a popular and economical way to move goods. Later the waterways fell from favor with the introduction of railroads and long-distance trucking. However, today there is a new interest in using waterways for commerce. France is working to increase trade on its rivers and canals. It has Europe's longest system of canals deep enough to move commercial barges. By moving more goods by water, the government hopes to reduce traffic on its roads.

To accomplish this goal, the government lifted many old regulations. River channels have also been deepened, and locks have been removed. These changes have helped speed up travel on the waterways. As a result, more businesses are shipping their goods by water.

Why is such shipping important? Barges are more than twice as energy-efficient as trains and more than five times as efficient as trucks. The increased energy efficiency is good for the economy as well as for the environment. In addition, boats are not as noisy as trains or trucks. This means that they can operate at night in urban areas without bothering people. Boats also provide a safer way to carry dangerous goods.

Analyzing How has the French perception of their water resources changed? What has been the result of this changed perception?

A canal bridge over the Loire River

🌐 **LEVEL 1:** Copy the following graphic organizer onto the chalkboard, omitting the italicized answers. Have students complete the organizer by adding information on the major issues and challenges facing France today. Then lead a discussion about each of the issues students suggest. **ENGLISH LANGUAGE LEARNERS**

💻 **Issues and Challenges Facing France Today**

| *government's influence over the economy* | *integrating increasing numbers of new immigrants into French society* | *relations with former colonies, problems with overseas departments* |

Teacher to Teacher

William Fisher of Bryan, Texas, suggests the following activity to help students understand France's productive economy. Tell students that most American workers work 40 hours a week. In France, most people work 35 hours a week but are among the most productive workers in the world. Organize students into groups to brainstorm positive and negative consequences of 35- and 40-hour workweeks. Each group should choose to support either the 35-hour or 40-hour week and write a brief defense of its position. Stage a classroom debate between supporters of the two sides. **COOPERATIVE LEARNING**

INTERPRETING THE VISUAL RECORD

A hiker views mountain scenery near Chamonix in the French Alps. France's physical and cultural landscapes attract tourists from around the world and help make tourism one of the country's leading industries. **What type of physical feature is shown in this photo? What physical processes created it?**

Economy

France has a highly diversified, developed economy. Its workers are some of the most productive in the world. This is true even though they have the shortest workweek—35 hours—and some of the longest vacations—one month—of any workers in the industrial countries.

The French are famous for fashion design. They also produce perfumes, cosmetics, jewelry, glassware, and furniture. Tourism is also important. Millions of people visit Paris, ski in the French Alps, and enjoy the famous Mediterranean coast known as the French Riviera.

Farming remains an important part of the economy. France is second only to the United States in agricultural exports. The French produce high-quality food products and a great variety of agricultural produce. Farmers provide wheat, sugar beets, olives, grapes, and dairy products. France is the world's leading wine producer in both variety and export income.

France's early industries were centered in the northeast near large deposits of coal and iron ore. Those heavy industries are now in decline. However, the country's high-tech industries are developing rapidly. The south is a growing center for aviation, communications industries, and space technology.

✓ **READING CHECK:** *Environment and Society* Where were early French industries centered? What natural resources were found there? in the northeast near the large deposits of coal and iron ore

Issues and Challenges

One of the major issues facing France today is the government's powerful influence over the economy. The French economy is both highly taxed and highly regulated, and many industries are government-owned. Many argue that these controls hurt innovation and creativity. They also make it hard for private businesses to grow. This situation is changing as state-owned businesses and industries are turned over to private owners. This process is called privatization. However, as privatization occurs, many government workers are faced with unemployment.

France's Exports

machinery and transportation equipment; chemicals; iron and steel products; agricultural products; textiles and clothing

Source: Central Intelligence Agency, *The World Factbook 2000*

France's Labor Force

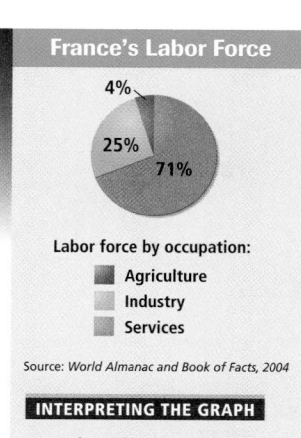

4%

25%

71%

Labor force by occupation:
- Agriculture
- Industry
- Services

Source: World Almanac and Book of Facts, 2004

INTERPRETING THE GRAPH

How does the distribution of France's labor force by occupation relate to France's exports?

Linking Past to Present

French Wines from American Vines France is known for the high quality of its wines, but years ago the industry almost died. In the mid-1800s an insect called *phylloxera* was accidentally brought from America to France. The *phylloxera* began attacking grapevine roots throughout Europe. French vineyards were devastated.

Thomas Volney Munson, a plant researcher working in Denison, Texas, saved the day. He sent rootstock of certain wild American grape species resistant to *phylloxera* to France. Grafted, or attached, to European vines, the American roots produced the "perfect" grapevine—one with pest-resistant roots and excellent fruit. The grafting process is still used today to control *phylloxera*.

VISUAL RECORD ANSWER

a glacier; snow and ice accumulation in high elevations

GRAPH ANSWER

A relatively small percentage of France's population works in industry and agriculture, but these industries produce most of France's exports.

Corsica is known for its beautiful Mediterranean landscape, which features rugged mountains and dramatic coastal cliffs. However, economically the island is less developed than much of France, which has led many Corsicans to migrate to the mainland for jobs.

Since the early 1900s many Algerians and Moroccans have been migrating to France in search of jobs. In fact, North Africans now form the largest immigrant group in the country. Immigrants have also come from former French Indochina—Vietnam, Cambodia, and Laos—and from former French colonies in West Africa. Many immigrants live in poorer sections of the major French cities. France's many immigrant communities have helped create distinctive urban landscapes. Many neighborhoods where immigrants live feature non-European restaurants and shops. Bringing these immigrants into French society is another important challenge.

After the British, the French maintained the second-largest colonial empire in the world. France has tried to maintain ties with some of those former colonies, particularly in Africa. The French also have overseas territories that are departments of France. In other words, they are considered part of France. One such department is French Guiana, in South America. Other French territories are mainly islands in the Caribbean, the South Pacific, and the Indian Ocean. They include Guadeloupe, Tahiti, and Réunion. Recent independence movements on the South Pacific island of New Caledonia have led to violence. Violence has also occurred on the large and rugged Mediterranean island of Corsica. Corsica has been part of France since 1768. However, many Corsicans consider themselves culturally distinct from France and want their island to become independent.

✓ **READING CHECK:** *Environment and Society* Why have many Moroccans, Algerians, and other groups of people been immigrating to France? They hope to find jobs in the growing economy.

 **Review**

Define primate city

Working with Sketch Maps On the map you created in Section 1, label France, Paris, Seine River, Lyon, Lille, Marseille, Alps, French Riviera, and Corsica. What is the primate city of France?

Reading for the Main Idea
1. *Human Systems* How is France's culture important to its economy?

2. *Human Systems* How are public policies and decision making influenced by French cultural beliefs?

3. *Human Systems* Where does France still have overseas possessions?

Critical Thinking
4. Making Generalizations Why do you think some French workers at government-owned companies might face losing their jobs as their companies are turned over to private ownership?

Organizing What You Know
5. Construct a word web like the one below. Use it to identify sources of American influence on French culture.

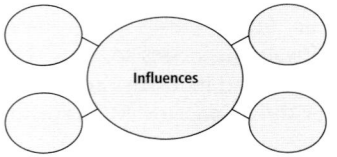

Influences

 **LET'S GET STARTED**

Copy the following instructions onto the chalkboard: *Find Benelux on the chapter map. Where in Europe is it located?* Once students have discovered that it does not appear on the map, explain that *Benelux* is not a place, but an acronym for the countries of *Be*lgium, the *Ne*therlands, and *Lux*embourg. Have students locate these countries on the map. Then tell them that they will learn more about the history and cultures of the Benelux countries in Section 3.

Building Vocabulary

Write the term **cosmopolitan** on the chalkboard. Point out that the term derives from two Greek words—*cosmos,* which means "world" or "universe," and *polis,* which means "city." So *cosmopolitan* literally means "a city of the world." Then mention that such cities as New York and London are often described as cosmopolitan. Ask students to suggest characteristics of cosmopolitan cities. Then call on a volunteer to locate and read the definition aloud from the text or glossary.

Section 3

The Benelux Countries

READ TO DISCOVER

1. What historical ties do the Benelux countries share?
2. What are the cities and economies of the Benelux countries like?

WHY IT MATTERS

In 1999 a new currency called the euro was introduced in Europe. Use **CNNfyi.com** or other **current events** sources to learn about how and why the euro was created.

IDENTIFY

European Union

DEFINE

cosmopolitan

LOCATE

North Sea
Flanders
Wallonia
Brussels
Antwerp
Amsterdam
Rotterdam
The Hague

Section 3 RESOURCES

REPRODUCIBLE
▶ Guided Reading Strategy 14.3
▶ Graphic Organizer Activity 14
▶ Creative Strategies for Teaching World Geography, Lesson 10

TECHNOLOGY
▶ One-Stop Planner CD–ROM, Lesson 14.3
▶ CNN Presents World Cultures: Yesterday and Today, Segment 30: NATO at 50
▶ Homework Practice Online
▶ HRW Go site

REINFORCEMENT, REVIEW, AND ASSESSMENT
▶ Main Idea Activity 14.3
▶ English Audio Summary 14.3
▶ Spanish Audio Summary 14.3
▶ Section 3 Review, p. 317
▶ Daily Quiz 14.3

History and Culture

Belgium, the Netherlands, and Luxembourg make up the Benelux countries. For many years, Belgium and Luxembourg were part of the Netherlands. Because of their position between three powerful countries—France, Germany, and the United Kingdom—all three of the Benelux countries have been fought over by foreign powers. After World War II, the Benelux countries established a political and economic union. Their early economic association planted the seed that eventually led to the creation of the **European Union**.

Vianden Castle in Luxembourg was built between the A.D. 1000s and 1300s.

 **FOCUS ON GOVERNMENT**

The European Union The European Union, or EU, is an organization of countries interested in increasing economic and political cooperation between its members. The EU was established on November 1, 1993. However, its origins date back to the 1950s.

   In the past, many European leaders tried to unite the continent politically and economically. They failed because they used force rather than cooperation to bring different countries together. After World War II, some European countries began forming alliances based on mutual aid rather than military strength. These alliances tried to tie members more closely together economically and politically. One way of doing this was to eliminate trade barriers among members of the same alliance.

   Belgium, the Netherlands, Luxembourg, France, Italy, and West Germany formed such an organization in 1957. They called it the European Economic Community, or EEC. Later the name was shortened to just the EC. Over time, the EC grew, joined with other organizations, and became the EU.

ALL LEVELS: Organize students into several groups. Have group members imagine that they are editors of a geography magazine written for high school students. Direct each group to draw up an outline for an article on historical and cultural ties among the Benelux countries. Each group's outline should include a brief summary of the article's topic, a list of the major points that will be covered in the article, and illustration ideas. Call on volunteers from each group to present the outlines to the class. **ENGLISH LANGUAGE LEARNERS, COOPERATIVE LEARNERS**

Teach Objective 2

ALL LEVELS: Copy the following graphic organizer onto the chalkboard, omitting the italicized answers. Have students complete it. **ENGLISH LANGUAGE LEARNERS**

The Benelux Countries

Major Cities
Belgium: *Brussels—headquarters of EU and NATO*
Netherlands: *Amsterdam—capital; Rotterdam—busy seaport; The Hague—Dutch Parliament, International Court of Justice*
Luxembourg: *none listed*

Economy
Belgium: *industry, agriculture, international business, diamond cutting, carpets, chocolate*
Netherlands: *agriculture (dairy and flowers), diversified activities, many exports, natural gas*
Luxembourg: *steel, international banking*

Members of the European Union, 2004

European Union members 2003

Countries that joined the EU in 2004 (Cyprus not shown)

SCALE
0   250   500 Miles
0   250   500 Kilometers
Projection: Azimuthal Equal Area

INTERPRETING THE MAP *As of May 2004, 25 countries belonged to the European Union. The 10 countries that joined in 2004 are shown in light green. How has the EU affected earlier boundaries and political divisions within Europe?*

The EU has increased cooperation among members in the areas of trade, lawmaking, and social issues. The EU also introduced a common currency, the euro. Most, but not all, EU countries adopted the euro in 2002.

In February 2002 the EU held a constitutional convention to address issues such as common defense and taxes and how an enlarged EU will work. In December 2002 the EU voted to admit 10 new members—mostly from Eastern Europe—and these 10 countries became members on May 1, 2004. The EU must decide whether, and how, to expand membership to other countries—such as Turkey—that want to join.

READING CHECK: *Human Systems* What role does the EU play in Europe today? promotes cooperation in the areas of trade, lawmaking, and social issues

Land Reclamation The name *Netherlands* means "low lands." Large areas of this country are below sea level. In fact, early in the country's history, much of the land in the Netherlands was coastal marshes and wetlands. However, people in the region have long worked to reclaim land from the sea. Today farms, towns, and industrial centers are located on polders below sea level. The dike and polder system has been very successful. However, the Dutch—the people of the Netherlands—worry about floods during severe North Sea storms. Rising global sea levels might also become a problem.

Language Dutch is spoken in the Netherlands and in northern Belgium. The dialect of Dutch spoken in northern Belgium is also called Flemish. About 60 percent of Belgians speak Dutch. Many Belgians also speak French. In fact, the country is divided into two cultural regions. The northern coastal region is known as Flanders. The French-speaking Belgians in the southern portion of Belgium are known as Walloons, and the region is known as Wallonia. The people in Belgium generally view themselves as either Flemish or Walloon rather than Belgian. Luxembourg, to the south, has three official languages: German, French, and Luxembourgian. Luxembourgian is a language related to German and Dutch.

READING CHECK: *Human Systems* What ties exist among the Benelux countries? political and economic union after World War II, Belgium and Luxembourg once part of the Netherlands, part of EU today

Urban and Economic Environments

The headquarters for the EU is in Brussels, the capital of Belgium. The city's central location in Europe and good transportation connections make it an ideal headquarters for the EU. Brussels is also the headquarters for the North Atlantic Treaty Organization (NATO) and many international corporations. As you might expect, Brussels is one of Europe's most **cosmopolitan** cities. A cosmopolitan city is one that is characterized by many foreign influences. The

Close

Ask students to imagine they work for an advertising agency and have been asked to compose slogans that summarize the major characteristics of each of the Benelux countries. Call on volunteers to suggest possible slogans to the class.

Review and Assess

Have students complete the **Section Review**. Then have students complete **Daily Quiz 14.3.**

Reteach

Have students complete **Main Idea Activity for English Language Learners and Special-Needs Students 14.3.** Then have students develop a list of three important things about each of the Benelux countries. **ENGLISH LANGUAGE LEARNERS**

Extend

Have interested students conduct research on the "Tulip Mania" that hit the Netherlands in the 1600s. Many Dutch people speculated on tulip prices—buying bulbs to resell them later at a much higher price. At the height of the craze, bulbs sold for incredible sums. Have students present their findings in brief written reports. **BLOCK SCHEDULING**

cosmopolitan city is one that is characterized by many foreign influences. The port of Antwerp is Belgium's second-most-important city. The Belgian economy is based on industry, agriculture, and services for international business. The country is also known for diamond cutting, quality carpets, and chocolate.

The Dutch economy is known for agriculture, particularly dairy products and flowers. For example, Dutch cheese and tulips are world famous. The Dutch economy today is very diversified, and exports are important. The economy is also one of the best performers in the EU. Natural gas deposits are found in the coastal and offshore region of the North Sea. However, the Netherlands is very dependent on imported oil.

The most urbanized and industrialized area in the Netherlands is known as the Randstad, or "Ring City." Here you will find the largest cities—including Amsterdam, the capital, and Rotterdam, one of the world's busiest seaports. These cities are strung together in a crescent shape. The Dutch parliament and International Court of Justice are located in The Hague. The Dutch population is well educated, productive, and supported by expensive government social programs.

Luxembourg is a forested and hilly country between Belgium, France, and Germany. It is the smallest member of the European Union, but it has the highest per capita GDP in the world—$34,200. Luxembourg has long been a steel producer. However, today international banking is most important. The small country has a constitutional monarchy.

INTERPRETING THE VISUAL RECORD

The Netherlands is an important exporter of flowers and is famous for its brightly colored tulip fields. Tulips are one of the most popular garden flowers in the world, and almost 4,000 varieties have been developed. **Based on the photo, what environmental factors might make the Netherlands ideally suited to using modern technology in agriculture?**

✓ **READING CHECK:** *Environment and Society* Why is Brussels an ideal location for the headquarters of many European businesses? centrally located in Europe, good transportation connections

Review

go.hrw.com Homework Practice Online
Keyword: SW3 HP14

Identify European Union

Define cosmopolitan

Working with Sketch Maps On the map you created in Section 2, label the three Benelux countries, North Sea, Flanders, Wallonia, Brussels, Antwerp, Amsterdam, Rotterdam, and The Hague. Where is the headquarters of the EU?

Reading for the Main Idea

1. *The World in Spatial Terms* How has the location of the Benelux countries influenced their history?

2. (*Human Systems*) What large European organization promotes cooperation among members in the areas of trade, lawmaking, and social issues? What are two important issues debated by members today?

Critical Thinking

3. **Making Generalizations** Why would rising sea levels be a concern in the Netherlands?

4. **Comparing and Contrasting** How do you think the Randstad in the Netherlands compares to the megalopolis of the northeastern United States? What similar political, economic, social, and environmental features would you expect to find in the two regions?

Organizing What You Know

5. Copy the chart. Use it to list the languages spoken in each of the Benelux countries.

Countries	Languages
Belgium	
The Netherlands	
Luxembourg	

Section 3 Review Answers

Identify For identification, see: European Union, p. 315, 316

Define For definition, see: cosmopolitan, p. 316

Working with Sketch Maps Maps will vary, but listed places should be labeled in their approximate locations. It is in Brussels.

Reading for the Main Idea
1. Because they are located between France, Germany, and the United Kingdom, the countries have been fought over by foreign powers.

2. EU; how much control governments should give to EU policy makers, whether to expand membership

Critical Thinking
3. Large areas of the Netherlands are below sea level and, if sea levels rise, the dike and polder system will be threatened. (NGS 4)

4. Answers will vary. Students might note concentration of political power, industrialization, cultural and social opportunities, and environmental problems like traffic, crowding, and pollution. (NGS 12)

Organizing What You Know
5. Belgium—Dutch (Flemish), French; Netherlands—Dutch; Luxembourg—German, French, Luxembourgian

VISUAL RECORD ANSWER
level land

OBJECTIVES

1. Note how the cultures of Scandinavia are similar to and different from each other.

2. Identify the industries on which the economy of the region relies.

3. Locate the areas in Scandinavia where most people live.

LET'S GET STARTED

Copy the following passage onto the chalkboard: *Greenland is not very green. It is mostly covered by ice and rock. Why do you think then that the Vikings who discovered it called the island "green land"?* Discuss responses. (Possible answer: wanted people to perceive the island as suitable for settlement) Point out that people's perceptions of a place can lead to changes in society, such as the movement of Viking settlers to Greenland in the 900s. Tell students that in Section 4 they will learn more about Greenland and the countries of northern Europe.

Building Vocabulary

Write the term **uninhabitable** on the chalkboard. Point out that the prefix *un-* means "not." Add that the base *inhabit* means "to live in," and that the suffix *-able* means "capable of." *Uninhabitable*, then, describes a place in which people are not capable of living. Ask students to suggest factors that may render a place uninhabitable. Call on volunteers to identify places in Scandinavia that they think may not support human life. Then have volunteers locate and read the definitions of the other key terms.

Section 4 RESOURCES

REPRODUCIBLE

▶ Guided Reading Strategy 14.4

▶ Readings in World Geography, History, and Culture 36

▶ Geography for Life Activity 14: Linking Denmark

TECHNOLOGY

▶ One-Stop Planner CD–ROM, Lesson 14.4

▶ CNN Presents Geography: Yesterday and Today, Segment 14: The Fishing Life

▶ Homework Practice Online

▶ HRW Go site

REINFORCEMENT, REVIEW, AND ASSESSMENT

▶ Main Idea Activity 14.4

▶ English Audio Summary 14.4

▶ Spanish Audio Summary 14.4

▶ Section 3 Review, p. 321

▶ Daily Quiz 14.4

Section 4 Scandinavia

READ TO DISCOVER

1. How are the cultures of Scandinavia similar to and different from each other?

2. What does the economy of this region rely on?

3. In what areas do most people in Scandinavia live?

WHY IT MATTERS

Many houses in Iceland use geothermal heat as an energy source. Use **CNN fyi.com** or other **current events** sources to learn about other alternative energy sources.

DEFINE

uninhabitable
geysers
socialism

LOCATE

Lapland	Stockholm
Copenhagen	Greenland
Oslo	Faeroe Islands
Helsinki	Reykjavik

Our Amazing Planet

Norway is sometimes called the Land of the Midnight Sun because the Sun does not set in the northern parts of the country for about one month every summer.

History and Culture

Five countries make up Scandinavia—Norway, Sweden, Denmark, Finland, and Iceland. These countries are the northernmost countries in Europe. In the past the region was known for the fierce Viking sailors and warriors who raided the shores of Europe. However, today the countries of Scandinavia are known for their modern economies and high standards of living.

Scandinavians share many cultural traits. For example, almost all Scandinavians are Protestant Lutheran. Except for Finnish, Scandinavian languages are closely related. Finnish belongs to the same language family as Hungarian and Estonian. However, speakers of Danish, Swedish, and Norwegian can generally understand one another. Also, all of the Scandinavian countries have democratic governments. These cultural similarities help make Scandinavia a clear cultural region.

VISUAL RECORD ANSWER

Their ships could travel long distances, which helped the Vikings conquer other lands.

INTERPRETING THE VISUAL RECORD

Viking longships, like this one from Norway, featured many technological innovations. They were lighter, faster, easier to sail, and more durable than other ships of their time. **How might Viking improvements in ship design have allowed the diffusion of Viking culture?**

Teach Objective 1

ALL LEVELS: Copy the following graphic organizer onto the chalkboard, omitting the italicized answers. Have students complete the organizer by adding information on the cultural characteristics that the Scandinavian countries share. **ENGLISH LANGUAGE LEARNERS**

mostly Lutheran Protestants — **Scandinavia** — *healthy, well-educated populations*

languages closely related (except Finnish) — *long life spans and low birthrates*

democratic systems of government — *mostly urban populations*

Teach Objective 2

ALL LEVELS: Organize students into groups and have each group develop several activity sheets dealing with the economies of the Scandinavian countries. Possible activities include crosswords, acrostics, multiple-choice questions, fill-in-the-blank questions, matching questions, and graphic organizers. After groups have finished writing the activities, have them exchange sheets and complete the activities they receive. **ENGLISH LANGUAGE LEARNERS, COOPERATIVE LEARNERS**

Lapland stretches across northern Norway, Sweden, and Finland. This region is mainly tundra and is populated by the Lapps—or Sami, as they call themselves. The Sami probably originated in central Asia. Their economy was traditionally based on reindeer herding, but today most Sami earn a living from tourism. Despite this, many Sami have maintained some of their traditional culture.

Settlement Patterns   Most Scandinavians live in the southern parts of their countries, where climates are warmer. For example, most Norwegians live in coastal plains areas or along narrow fjords. More than half of the population lives along the southeastern coast of the country. Most of the ports on that coast remain free of ice all year.

Scandinavian countries have healthy and well-educated populations with long life spans and low birthrates. In general, their populations are growing slowly and are heavily urban. About 85 percent of Sweden's population is found in urban areas. More than one fourth of Danes and Norwegians live in or around their respective capital cities of Copenhagen and Oslo. Most Finns live near Helsinki, the country's capital and leading seaport. Stockholm is Sweden's capital and largest city.

✔ **READING CHECK:**   *Human Systems*   Which country's language is not related to the languages of the other Scandinavian countries? Finland

Greenland and Iceland   Greenland is not really green. About 85 percent of it is covered by a thick ice cap. The icy interior of the island is **uninhabitable**. An uninhabitable region is one that cannot support human life and settlements. Only Greenland's rocky coastline is fit for human habitation. Most Greenlanders live along the southwestern coast. In the 900s, the Vikings founded settlements on Greenland's coast. Early Viking settlers tried to attract others by describing the island as a place with plenty of vegetation and a mild climate. This description drew new settlers to Greenland. However, these early settlements died out in the 1400s. Scandinavian settlers did not colonize the island again until the 1700s. Today Greenland is a self-governing territory of Denmark. Denmark also governs the Faeroe (FAHR-oh) Islands in the North Atlantic Ocean.

INTERPRETING THE VISUAL RECORD

*The Scandinavian countries all have a primate city. For example, Copenhagen is Denmark's capital and cultural, artistic, and economic center. **What are the primate cities of the other Scandinavian countries?***

INTERPRETING THE VISUAL RECORD

*Greenland's capital, Nuuk, is located on the island's western coast, where the warm West Greenland Current helps moderate temperatures. **How do you think this ocean current has influenced settlement patterns in Greenland?***

Across the Curriculum

► **Science** ◄

Seasonal Affective Disorder   Latitude may be to blame for the high rate of depression among Scandinavians. Doctors have identified a form of depression called seasonal affective disorder (SAD) that strikes people who do not get enough light. Because Scandinavians live at high latitudes, they see little sunlight during the winter. For example, Icelanders can expect only four hours of sunlight in December, and by 2:00 P.M. the light is already weak. The darkness can cause those who suffer from SAD to overeat, sleep too much, and withdraw from other people.

ACTIVITY: Have students collect data from world almanacs to construct line graphs showing sunrise and sunset times for cities along the 30°, 50°, and 60°N parallels. As a class, discuss the amount of sunlight each area receives. Have students identify three cities where residents might suffer from SAD.

VISUAL RECORD ANSWER

Reykjavik, Oslo, Stockholm, Helsinki

VISUAL RECORD ANSWER

encouraged settlements on the western coast

LEVEL 1: Organize students into several groups. Have each group design a web page titled Life in Scandinavia that will illustrate settlement patterns in the Scandinavian countries. Instruct groups to decide how to present their information—country by country or regionally, for example. Then have each group write out plans for the format of its site, including descriptions of maps and pictures it will include and written versions of the page's text.
ENGLISH LANGUAGE LEARNERS, COOPERATIVE LEARNERS

LEVELS 2 AND 3: Have students work in the same groups as in the Level 1 activity to implement the Web page designs they have developed. Have groups use magazines, newspapers, and other materials to design and lay out mock web pages that follow the plans they established. Remind groups that their pages should be a combination of visuals—maps, sketches, diagrams, and magazine and newspaper clippings—and written information. Call on groups to display and discuss their finished Web pages.
COOPERATIVE LEARNERS

Daily Life

Danish Design Denmark has long been a leader in designing mass-produced objects for the home. Basic to Danish design's popularity is the prominence of function—an object must first do its job well. Artistry and good craftsmanship add to the wide appeal of Danish design.

Danish porcelain factories have created beautiful dishes since the 1700s. One pattern has been in production since 1775. Silver flatware and lamps are also popular. Danish design, however, has probably made its biggest mark in the furniture trade. The Danish modern furniture style dates from after World War II. It is sleek, simple, and elegant in design.

CRITICAL THINKING: The other Scandinavian countries are also known for high-quality home decor industries. What conditions might make these countries logical leaders of these industries?

You don't say! Many Danes celebrate the American Independence Day each July 4. Some celebrations last for several days and nights.

Volcanic activity in Iceland generates many geysers. Some can shoot water as high as 1,640 feet (500 m).

In Iceland, which is greener than its name implies, all the people live along the narrow coastal plains. That is because the island's interior of ice-covered lava rock is also uninhabitable. Most Icelanders live in or near the capital Reykjavik (RAY-kyah-veek). The country is a member of NATO but has not shown an interest in joining the EU.

Iceland has tremendous geothermal energy because of its location on a mid-ocean ridge where volcanic activity is common. Underground water rises and steams as **geysers** in many locations on the island. A geyser is a hot spring that shoots water into the air. The hot water is used to heat homes and vegetable greenhouses. The island also has hydropower potential. In the future, Iceland may be able to export geothermal energy and hydroelectricity to Europe across an underwater cable.

✓ **READING CHECK:** *Environment and Society* Where will you find the human populations of Greenland and Iceland? coastal areas

Economy

The Scandinavian countries all have high standards of living. For example, Denmark is one of the EU's most prosperous countries and has one of the highest per capita GDPs in Europe. High-tech industries and export-oriented economies maintain the high standards of living.

Economic Development Finland has been transformed from an exporter of natural resources to a manufacturing country. Finland produces and exports high-tech goods. Its products include advanced telecommunications equipment, cellular phones, and computer software. The Swedes produce a variety of high-tech and high-value products. These goods include automobiles, cellular phones, aircraft, and industrial robots.

In addition to manufacturing, commercial agriculture—particularly the dairy and meatpacking industries—is important to the Danish economy. The paper- and wood-products industries are well developed in Sweden, where forests cover half the country.

Fishing is also important in Scandinavia, particularly in Iceland and Norway. Norway has a large commercial fishing fleet. However, it is offshore oil and gas from the North Sea that makes Norway a rich country. Most of Norway's oil profits are invested for the future. The country also has hydroelectric plants that produce a surplus of electricity, which is exported.

Fishing has long been important to Norway's economy. In fact, Norway has one of the largest commercial fishing industries in the world and exports more fish than any other European country.

Close

Lead a discussion on how a Viking might react if he or she visited Scandinavia today. Which aspects of the region might be familiar? Which would be unfamiliar?

Review and Assess

Have students complete the **Section Review**. Then have students complete **Daily Quiz 14.4**.

Reteach

Have students complete **Main Idea Activity for English Language Learners and Special-Needs Students 14.4**. Organize students into groups and assign each group a topic on Scandinavia. Have each group use the text to list important information about its assigned topic. ENGLISH LANGUAGE LEARNERS

Extend

Have interested students conduct research on the history or culture of the Sami and share their findings with the rest of the class. BLOCK SCHEDULING

Economic Change During much of the last half of the 1900s, Sweden's economy was a mix of capitalism and **socialism**. Socialism is an economic system in which the government owns and controls the means of producing goods. Most of Sweden's industries remained privately owned. However, the government controlled some businesses, and it levied high taxes. These high taxes still pay for a large system of government welfare and services. For example, the government pays for almost all the educational, medical, and childcare needs of its citizens. For example, all residents of Sweden are covered by national health insurance. Compared to many other countries, health conditions in Sweden are very good. The government also pays for programs to help parents raise their children. For example, parents can share up to one year of paid time off from work before their child reaches the age of eight. They also receive tax-free payments to help pay for the costs of raising children.

By the late 1990s about 60 percent of Swedes relied on the government for work or welfare payments. The Danes also have a well-developed welfare system. High taxes in Scandinavia pay for environmental protection and support sports and the arts.

Many Swedish economists blame a costly welfare system for the economic problems the country has been experiencing since 1991. In recent years Sweden's government has tried to lessen its influence over the economy. Still, many Swedes do not like the idea of cutting back the welfare system that they have. As a result, the government has preserved many of its expensive social programs.

INTERPRETING THE VISUAL RECORD *The Øresund bridge, which connects Copenhagen, Denmark, with Malmö, Sweden, opened in 2000. The massive bridge cost more than $2 billion and provides road and rail connections between the Scandinavian Peninsula and the rest of Europe.* **How might this bridge affect the locations and patterns of economic activities in Scandinavia?**

✓ **READING CHECK:** *Human Systems* Why are taxes in Scandinavian countries like Sweden so high? expensive government welfare system, including education, medical care, child care; also environmental protection and culture

Section 4 Review

Homework Practice Online
Keyword: SW3 HP14

Define
uninhabitable, geysers, socialism

Working with Sketch Maps On the map you created in Section 3, label the Scandinavian countries, Lapland, Copenhagen, Oslo, Helsinki, Stockholm, Greenland, Faeroe Islands, and Reykjavik. Which large island is part of Denmark?

Reading for the Main Idea

1. *Environment and Society* How did early Viking settlers try to shape perceptions of Greenland to draw others there? How accurate were their descriptions of the island?

2. *Human Systems* What are the standards of living and the economies of Scandinavian countries like?

3. *Human Systems* What is the major source of Norway's wealth?

Critical Thinking

4. Making Generalizations What might be one advantage and one disadvantage of the social welfare system found in many Scandinavian countries?

Organizing What You Know

5. Create a chart like the one below. Use it to list important economic activities and energy resources in the Scandinavian countries.

Countries	Economic activities and energy resources

Section 4 Review Answers

Define For definitions, see: uninhabitable, p. 319; geysers, p. 320; socialism, p. 321

Working with Sketch Maps Maps will vary, but listed places should be labeled in their approximate locations. Greenland is part of Denmark.

Reading for the Main Idea
1. said island had plenty of vegetation, a mild climate; not accurate, because much of interior covered by ice and rock

2. high standards of living; prosperous with high-tech industries, export-oriented

3. oil and gas in the North Sea

Critical Thinking
4. Possible answers: advantage—citizens assured that their basic needs will be taken care of; disadvantage—high taxes, possible negative effects on economy (NGS 10)

Organizing What You Know
5. Denmark—manufacturing, commercial agriculture (dairy and meat packing); Finland—high-tech manufacturing; Iceland—fishing, geothermal energy; Norway—fishing, oil and gas, hydroelectricity; Sweden—high-tech and automotive manufacturing, paper and wood products

VISUAL RECORD ANSWER

increase trade, encourage the development of export-oriented industries in Sweden

Setting the Scene

Have students read Case Study: Global Trade. Remind them that the roots of modern global trade go back some 500 years. Inform students, however, that many ancient cultures also engaged in long-distance trade. Such long-distance trading networks influenced the diffusion of languages, belief systems, ideas, technology, and resources. Ask students to consider why people in ancient times would engage in long-distance trade. *(Possible answers: to obtain desired goods that are unavailable locally, to learn new technologies from other cultures)* Then ask them to speculate on which ancient culture groups they are familiar with may have traded for goods from far away.

Building a Case

Have students use library and Internet resources to conduct research on ancient long-distance trading networks. Some possibilities include amber trade routes in Europe, the frankincense and myrrh trade from Southwest Asia, Phoenician trade in the Mediterranean world, and pre-Columbian trade in the Americas for items such as jade, obsidian, and quetzal feathers. Ask students to examine how long-distance trade influenced diffusion patterns and affected different regions. Also, have them analyze how the distribution and perception of resources by different cultures affected the patterns of movement of products and people.

Global Perspectives

International Space Station   A symbol of globalization and international cooperation is the International Space Station (ISS). This massive project involves materials, expertise, and investment from 16 countries, including Belgium, Denmark, France, the Netherlands, Norway, Sweden, and the United Kingdom. The ISS will employ some 100,000 people. Participating countries will all contribute toward the station's $60 billion price.

ACTIVITY: Have students use NASA's Web site to track the progress of the ISS.

MAP ANSWER

European colonialism connected the world through global trade, which changed cultural and economic patterns worldwide.

Applying What You Know Answers

1. Trade networks have linked distant parts of the world, so consumers can now buy products from around the world, and companies do business in distant places.
2. Possible answers: Innovations in different places could have created demand for new products. Countries might have formed different trade relationships.

This Case Study feature addresses National Geography Standards 3, 11, 13, and 16.

CASE STUDY

Global Trade

Human Systems   Have you ever thought about how far away some of the things you buy originate? Many of the products we buy are transported great distances before they reach the cash register. This flow of goods is part of a global system of trade. For example, raw materials like cotton, iron, oil, or wood are often harvested or mined far away from the factories that transform them into finished goods. Products like blue jeans, cars, coffee tables, and toys may then travel even greater distances from the factory to the store.

The roots of modern global trade go back some 500 years. At that time, long-distance trading networks began to develop as European countries explored and colonized the Americas and Asia. These countries were aided by improvements in ship design and navigation equipment. By the 1540s Portugal had established a chain of trading posts all the way to Japan. Portuguese ships brought valuable spices from Asia to buyers in Europe. At the same time, Spain set up its own trading networks. Spain mined gold and silver in Mexico, Bolivia, and Peru. These valuable minerals were then shipped back to Spain to enrich the royal treasury. Other ships brought tobacco from North America and chocolate from Mexico.

In the 1600s and 1700s Dutch, English, and French traders competed with Spanish and Portuguese merchants to control global trade. This competition led to a rapid growth of new colonies and trading patterns. Europeans explored distant parts of the world and discovered new foods and drinks. These new, exotic goods began to pour into Europe. For example, coffee, originally from Ethiopia, was introduced to Europe through Yemen. The first coffeehouse in London was set up in 1652. By the late 1600s coffeehouses were becoming popular in Boston, New York, and Philadelphia.

Other goods also appeared on European markets. In 1669 an English trading company made its first shipment of Chinese tea to London from Java. Dutch traders brought spices such as cloves and nutmeg from the East Indies. Later, they also traded cinnamon, coffee, jewels, and pepper. They also participated in the slave trade. In 1624 the Dutch set up a port on Manhattan and developed a fur-trading business with American Indians.

Over time, European merchants and government officials developed new economic systems to meet consumer demands for all of these new products. Specifically, they created the plantation system. This system produced large amounts of agricultural goods in the tropical and subtropical regions of the globe. The plantation system typically relies on a large labor force to raise a single crop on a large tract of land. After the crop is harvested, it is exported to distant markets. Large regions of Asia, Africa, and the Americas developed into colonial plantations. The plantations produced coffee, cotton, spices, sugarcane, tea, and many other goods.

Rotterdam is one of the busiest ports in the world, and its history is closely tied to the development of global trade. This port near the mouth of the Rhine River became internationally important in the 1600s when Dutch traders began importing goods to Europe from the East Indies. Today Rotterdam handles a huge range of goods, including oil and petroleum products, grains, and many other goods.

Drawing Conclusions

When students have completed their research, organize them into pairs. Have pairs compare what they learned about ancient trading networks and work together to draw some conclusions about the characteristics and importance of long-distance trade in different cultures. How were the trading patterns similar and different? How important was transportation technology in long-distance trade? How far did people and goods travel? Have each pair write a brief report that compares and contrasts the trading networks its members studied. Be sure students use standard grammar, spelling, sentence structure, and punctuation in their reports.

Going Further: Thinking Critically

Lead a class discussion on long-distance trade in ancient times and global trade today. Ask students the following questions:

- How is global trade different from long-distance trade in the past?
- How has technology changed, and how has this affected global trade?
- How have communications systems such as satellite television and the Internet affected global trade?
- How do you think trade in ancient times set the stage for the development of global trade today?

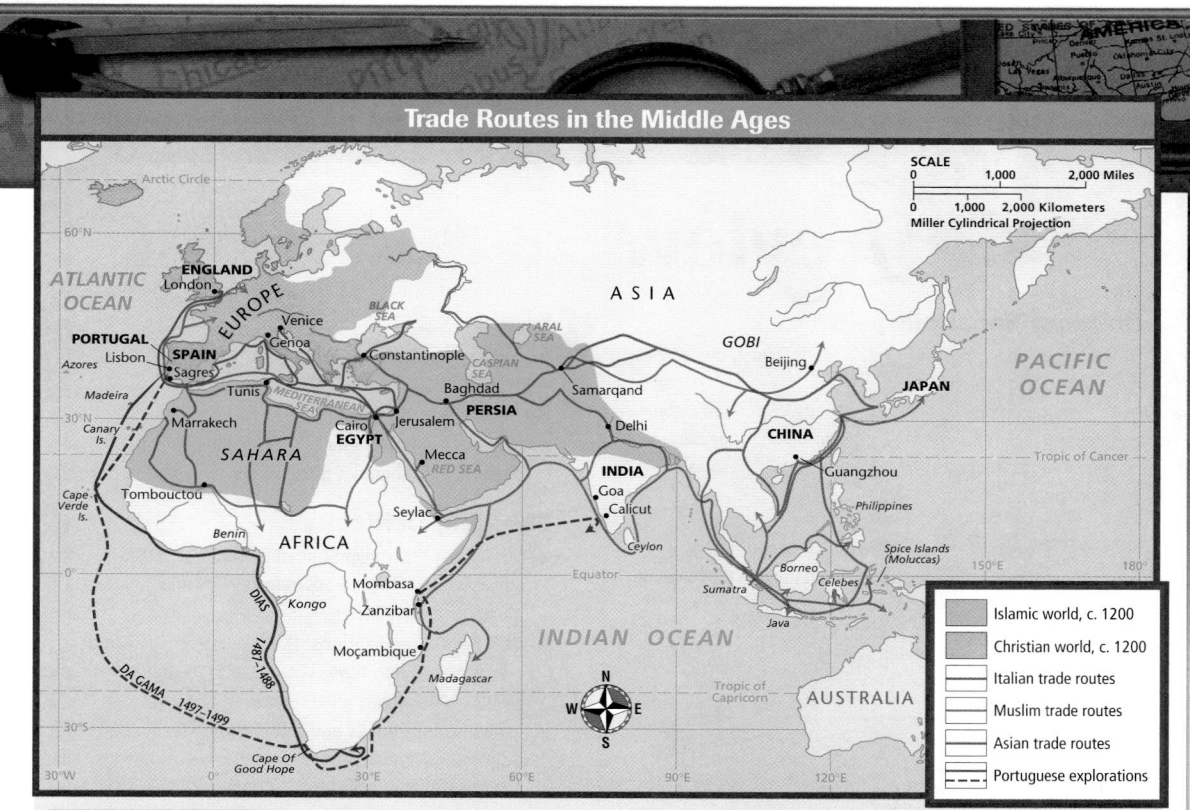

Trade Routes in the Middle Ages

SCALE
0 1,000 2,000 Miles
0 1,000 2,000 Kilometers
Miller Cylindrical Projection

- Islamic world, c. 1200
- Christian world, c. 1200
- Italian trade routes
- Muslim trade routes
- Asian trade routes
- Portuguese explorations

INTERPRETING THE MAP *In the Middle Ages, extensive trade routes had already been developed in Europe, Africa, and Asia. However, Portuguese and then Spanish explorations beginning in the late 1400s* altered these patterns. **How did global trade patterns change by the 1700s? Why did they change, and what were the implications of these changes?**

Building Vocabulary For definitions, see: sequent occupance, p. 306; famine, p. 307; constitutional monarchy, p. 307; nationalized, p. 308; primate city, p. 312; European Union, p. 315; cosmopolitan, p. 316; uninhabitable, p. 319; geysers, p. 320; socialism, p. 321

Locating Key Places

A. Seine River
B. London
C. Copenhagen
D. Northern Ireland
E. Greenland
F. Luxembourg
G. Brussels

Understanding the Main Ideas

1. Naval power helped the empire expand. It is tied to the former colonies by the Commonwealth of Nations.

2. less agricultural, one of Europe's most rapidly developing economies; drawing immigrants from other countries

3. largest city, economic, political, and cultural center

4. Flanders and Wallonia

5. People tend to live along the coasts and in the southern portions of their countries. Most Scandinavians live in cities.

In the late 1700s the Industrial Revolution had a major influence on global trade. The development of steam engines created a new way to transport goods across large distances—the railroad. A strong navy and control of the seas were no longer the only ways to develop global trading links. The vast resources of interior Asia, Africa, and the Americas became more accessible and began to appear on the global market. Also, the iron and steel needed to build railroads were suddenly in great demand. Germany's Ruhr Valley had a good supply of iron and steel and soon became a major manufacturing area. The invention of the cotton gin in 1793 also greatly affected global trade. Much of the southern United States developed a plantation economy based on cotton. By 1861 the United States grew more than 80 percent of the world's cotton. More than half of the cotton was shipped to Manchester, England, where it was made into cloth.

Today major trading routes crisscross the entire globe. Many common items that we use every day have done more traveling than most people do in a lifetime. A car may be made of German steel, Saudi Arabian plastic, and British glass. New technologies constantly reshape global trade patterns. For example, computers and electronic trading now allow people to shop on the Internet. They can order things easily from around the world. Global trade has also changed the way many companies do business. Companies can buy resources from distant places, locate factories in many different countries, and sell their products around the world. As a result, global trade is now a major force behind globalization. Expanding trade networks allow the same products to become familiar all over the world.

Applying What You Know

1. **Summarizing** How have global trade patterns changed since the 1500s? How have these patterns affected life around the world?

2. **Making Generalizations** Suppose that European countries had not developed huge colonial empires. What other factors may have influenced global trade patterns over time?

Review and Assessment Resources

TECHNOLOGY
▶ Chapter 14 Test Generator (on the One-Stop Planner)
▶ Global Skill Builder CD–ROM
▶ HRW Go Web site

REINFORCEMENT, REVIEW, AND ASSESSMENT
▶ Chapter 14 Review, pp. 324–25
▶ Chapter 14 Tutorial for Students, Parents, Mentors, and Peers
▶ Chapter 14 Test (form A or B)
▶ Alternative Assessment Handbook

▶ Chapter 14 Test for English Language Learners and Special-Needs Students
▶ Unit 4 Test
▶ Unit 4 Test for English Language Learners and Special-Needs Students

Assess
Have students complete a Chapter 14 Test.

Reteach
Organize the class into four groups representing the British Isles, France, the Benelux countries, and Scandinavia. Have each group compile a list of 10 important or noteworthy facts about its assigned area to share with the class. ENGLISH LANGUAGE LEARNERS, COOPERATIVE LEARNERS

CHAPTER 14 Review Answers

Thinking Critically

1. Catholics might believe that union would protect them from discrimination in employment, housing, and government. Protestants might fear that they would face disadvantages in a mostly Roman Catholic country. **(NGS 10)**

2. Students might say that industrial development made Britain more powerful than other countries. Over time, British culture and customs spread throughout this empire, strongly influencing other cultures. **(NGS 13)**

3. Students may say that skiing could have been an important mode of transportation in snowy and icy areas of Scandinavia. **(NGS 10)**

Using the Geographer's Tools

1. all populated continents because the British Empire had possessions on all of them

2. Students might say Eastern Europe and countries that are already surrounded by the EU, such as Switzerland.

3. Netherlands

CHAPTER 14 Review

Building Vocabulary

On a separate sheet of paper, explain the following terms by using them correctly in sentences.

sequent occupance	primate city	uninhabitable
famine	European Union	geysers
constitutional monarchy	cosmopolitan	socialism
nationalized		

Locating Key Places

On a separate sheet of paper, match the letters on the map with their correct labels.

Northern Ireland	Luxembourg	Copenhagen
London	Brussels	Greenland
Seine River		

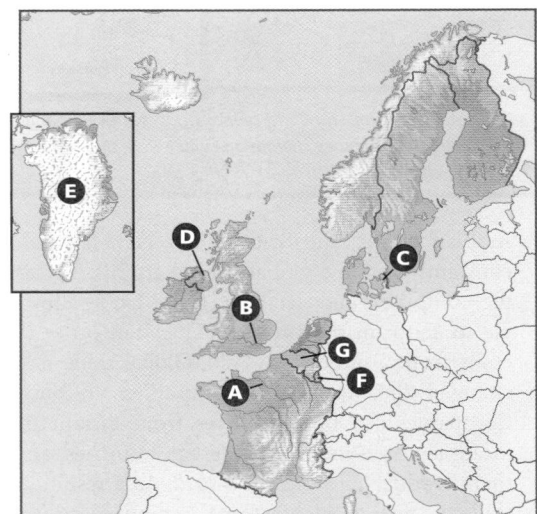

Understanding the Main Ideas
Section 1

1. (*Human Systems*) How did Britain control a vast empire? How is it tied to former colonies today?

2. (*Human Systems*) How has Ireland's economy changed in recent years? How have these changes influenced migration?

Section 2

3. (*Places and Regions*) What features make Paris the primate city of France?

Section 3

4. (*Places and Regions*) What are Belgium's two language regions?

Section 4

5. (*Environment and Society*) How is the population of Scandinavian countries distributed with regard to cities?

Thinking Critically

1. Analyzing Why might Catholics in Northern Ireland want union with the Republic of Ireland? Why might Protestants there be against it?

2. Making Generalizations How do you suppose Britain's industrial economy may have contributed to the diffusion of British culture and customs around the world?

3. Drawing Inferences and Conclusions Many sports began as activities essential to everyday life. How do you think skiing has been essential to daily life in Scandinavia?

Using the Geographer's Tools

1. Analyzing Maps Review the map of the British Empire in Section 1. Then list the continents on which you would expect to find people who speak English today. Explain.

2. Analyzing Maps Review the map of the European Union in Section 3. Which countries do you think the EU might expand to in the future? Why?

3. Creating Bar Graphs Use statistics from the unit Fast Facts table to construct a bar graph comparing the population density of countries in northern and western Europe. Which country is the most densely populated?

Writing about Geography

How is the European Union similar to and different from the United States? Do you think the EU countries will unite? Write a short report explaining your point of view.

SKILL BUILDING

Geography for Life

Observing the Weather

(*Places and Regions*) How is your local climate influenced by the wind? Set up a wind sock and record the direction and force of the wind for a week. Also, record temperature and precipitation. Graph your information and explain the information shown.

Portfolio Activity

Have students select one or more countries in northern and western Europe to which they will plan vacations. Have them plan itineraries, noting where they would like to visit and what they would like to do while they are in each location. Provide students with outline maps of the region on which to plot routes for their trips. Encourage students to share their itineraries and maps with the class. Place the completed itineraries and maps in students' portfolios.

Food Festival

Afternoon tea is a longstanding British tradition. For the upper classes it was a light meal to stave off hunger before a big formal dinner. For the working classes it was a substantial early supper. Contrary to common usage, "high tea" was the workers' meal and "afternoon tea" was the high society version. For afternoon tea in your classroom, have students bring small cucumber sandwiches, cakes, crumpets (small, thick pancakes), and scones, which are like American biscuits and are served with jam and heavy cream. Serve the food with hot tea. Encourage students to try their tea the British way—with milk and sugar.

Building Social Studies Skills

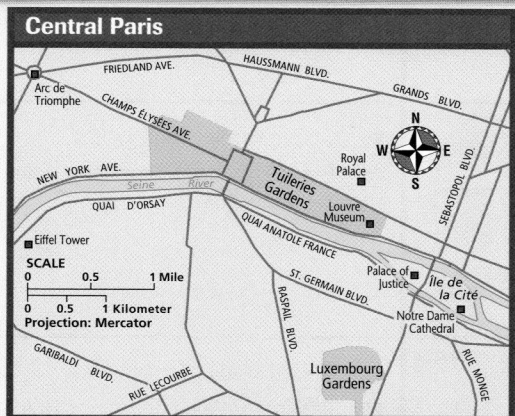

Central Paris

Interpreting Tourist Maps

Study the tourist map above. Then answer the questions that follow.

1. Which point of interest is located south of the Seine?
   a. Notre Dame Cathedral
   b. Eiffel Tower
   c. Arc de Triomphe
   d. Louvre Museum

2. Suppose you wanted to take a walking tour along the Seine. What nearby sites of interest would you be able to visit along the tour?

Analyzing Primary Sources

Read the following description of London by Simon Worrall and then answer the questions.

"The whole world lives in London. Walk down Oxford Street and you will see Indians and Colombians, Bangladeshis and Ethiopians, Pakistanis and Russians, Melanesians and Malaysians. Fifty nationalities with communities of more than 5,000 make their home in the city, and on any given day 300 languages are spoken. It is estimated that by 2010 the population will be almost 30 percent ethnic minorities, the majority born in the U.K. [United Kingdom]."

3. According to the author, in 2010 the population of London will
   a. be made up of mostly immigrants.
   b. have very few ethnic minorities.
   c. be decreasing as immigration slows.
   d. be almost 30 percent ethnic minorities.

4. What point is the author trying to make when he says "the whole world lives in London"? What details does he provide to support this point?

Alternative Assessment

PORTFOLIO ACTIVITY

Learning about Your Local Geography

Group Project: Research
In Sweden, government programs take care of the educational and medical needs of Swedish citizens. In your own area, who is responsible for paying the costs of people's educational and medical needs? With your group, research how each set of needs is addressed. Present your information in a chart.

internet connect

Internet Activity: go.hrw.com
KEYWORD: SW3 GT14

Choose a topic on northern and western Europe to:
- learn the history of skiing in Norway.
- compare and contrast major cities in the region.
- research daily life in the region.

Writing

Student reports should show an understanding of the structure of the EU. Speculations about Europe's future should be reasonable and well-supported. Use Rubric 41, Writing to Express, to evaluate student work.

Geography for Life

Student graphs should accurately display the results of their investigations. Explanations should relate the information presented in the graph to local climate patterns. Use Rubric 30, Research, to evaluate student work.

Social Studies Skills

1. b
2. Eiffel Tower, Louvre, Notre Dame, Tuileries Gardens, Champs Elysées
3. d
4. that a variety of ethnic groups live in London; names a variety of immigrant groups, describes ethnic and linguistic diversity

PORTFOLIO ACTIVITY

Charts should clearly reflect the results of student research. Use Rubric 30, Research, to evaluate student work.

Central Europe

CHAPTER RESOURCE MANAGER

	Objectives	Pacing Guide	Reproducible Resources
SECTION 1 **Germany** (pp. 327–31)	• Identify some key events in the history of Germany. • Describe some features of German culture. • Examine the German economy. • Evaluate the issues and challenges faced by Germany.	**Regular** 1 day **Block Scheduling** 1 day *Block Scheduling Handbook,* *Chapter 15*	**RS** Guided Reading Strategy 15.1 **RS** Graphic Organizer Activity 15 **PS** Readings in World Geography, History, and Culture 32 **E** Cultures of the World Activity: Region 3 **SM** Geography for Life Activity 15: Symbols of a Reunited Berlin **SM** Map Activity 15: Territorial Changes in Germany
SECTION 2 **The Alpine** **Countries** (pp. 335–37)	• Describe some important features of Austria's history, culture, and economy. • Analyze the political, cultural, and economic features of Switzerland.	**Regular** .5 day **Block Scheduling** .5 day *Block Scheduling Handbook,* *Chapter 15*	**RS** Guided Reading Strategy 15.2 **PS** Readings in World Geography, History, and Culture 33
SECTION 3 **Poland and** **the Baltics** (pp. 338–41)	• Trace the history of Poland and the Baltic countries. • Describe the urban environments and economy of Poland. • Analyze influences that have shaped the Baltic countries.	**Regular** .5 day **Block Scheduling** .5 day *Block Scheduling Handbook,* *Chapter 15*	**RS** Guided Reading Strategy 15.3 **PS** Readings in World Geography, History, and Culture 37 **SM** Critical Thinking Activity 15: Surviving the Cut
SECTION 4 **The Czech** **Republic,** **Slovakia, and** **Hungary** (pp. 342–45)	• Identify similarities and differences in the histories of the Czech Republic, Slovakia, and Hungary. • Describe the Czech Republic and Slovakia. • Explain how the fall of communism has affected Hungary.	**Regular** .5 day **Block Scheduling** .5 day *Block Scheduling Handbook,* *Chapter 15*	**RS** Guided Reading Strategy 15.4 **PS** Readings in World Geography, History, and Culture 38

Chapter Resource Key

PS Primary Sources

RS Reading Support

IC Interdisciplinary
Connections

E Enrichment

SM Skills Mastery

A Assessment

REV Review

ELL Reinforcement
and English
Language
Learners

 Transparencies

 CD–ROM

 Video

 Internet

 Holt Presentation
Maker Using
Microsoft®
PowerPoint®

 One-Stop Planner CD–ROM

See the *One-Stop Planner*
for a complete list of additional
resources for students and
teachers.

One-Stop Planner CD-ROM

It's easy to plan lessons, select resources, and print out materials for your students when you use the **One-Stop Planner CD-ROM with Test Generator**.

internet connect

HRW ONLINE RESOURCES

GO TO: go.hrw.com
Then type in a keyword.

TEACHER HOME PAGE
KEYWORD: **SW3 Teacher**

CHAPTER INTERNET ACTIVITIES
KEYWORD: **SW3 GT15**
Choose a topic on Central Europe to:
- research the change in global trade patterns of three central European countries.
- create a biography of Charlemagne.
- create a poster of Switzerland.

CHAPTER ENRICHMENT LINKS
KEYWORD: **SW3 CH15**

CHAPTER MAPS
KEYWORD: **SW3 MAPS15**

ONLINE ASSESSMENT
Homework Practice
KEYWORD: **SW3 HP15**
Standardized Test Prep
KEYWORD: **SW3 STP15**
Rubrics
KEYWORD: **SS Rubrics**

COUNTRY INFORMATION
KEYWORD: **SW3 Almanac**

CONTENT UPDATES
KEYWORD: **SS Content Updates**

HOLT PRESENTATION MAKER
KEYWORD: **SW3 PPT15**

ONLINE READING SUPPORT
KEYWORD: **SS Strategies**

CURRENT EVENTS
KEYWORD: **S3 Current Events**

Technology Resources

- One-Stop Planner CD-ROM, Lesson 15.1
- CNN. Presents Geography: Yesterday and Today, Segment 17: Castles for Sale
- Homework Practice Online
- HRW Go site

- One-Stop Planner CD-ROM, Lesson 15.2
- Homework Practice Online
- HRW Go site

- One-Stop Planner CD-ROM, Lesson 15.3
- Homework Practice Online
- HRW Go site

- One-Stop Planner CD-ROM, Lesson 15.4
- *ARGWorld* CD-ROM
- Homework Practice Online
- HRW Go site

Reinforcement, Review, and Assessment

ELL	Main Idea Activity 15.1
ELL	English Audio Summary 15.1
ELL	Spanish Audio Summary 15.1
REV	Section 1 Review, p. 331
A	Daily Quiz 15.1

ELL	Main Idea Activity 15.2
ELL	English Audio Summary 15.2
ELL	Spanish Audio Summary 15.2
REV	Section 2 Review, p. 337
A	Daily Quiz 15.2

ELL	Main Idea Activity 15.3
ELL	English Audio Summary 15.3
ELL	Spanish Audio Summary 15.3
REV	Section 3 Review, p. 341
A	Daily Quiz 15.3

ELL	Main Idea Activity 15.4
ELL	English Audio Summary 15.4
ELL	Spanish Audio Summary 15.4
REV	Section 3 Review, p. 345
A	Daily Quiz 15.4

Meeting Individual Needs

Ability Levels

Level 1 Basic-level activities designed for all students encountering new material

Level 2 Intermediate-level activities designed for average students

Level 3 Challenging activities designed for honors and gifted-and-talented students

English Language Learners Activities that address the needs of students with Limited English Proficiency

Chapter Review and Assessment

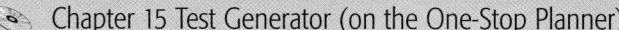

- Chapter 15 Test Generator (on the One-Stop Planner)
- Global Skill Builder CD-ROM
- HRW Go site
- REV Chapter 15 Review, pp. 346–47
- REV Chapter 15 Tutorial for Students, Parents, Mentors, and Peers
- A Chapter 15 Test (form A or B)
- A Alternative Assessment Handbook
- A Chapter 15 Test for English Language Learners and Special-Needs Students

Launch into Learning

Ask students if they are familiar with the work of scientists Nicolaus Copernicus (*realized planets revolve around the sun*), Gregor Mendel (*father of genetics*), or Albert Einstein (*introduced the theory of relativity*). Point out that all of these thinkers were from Central Europe. Their work opened new pathways for later scientists, eventually leading to space travel, genetic engineering, and countless inventions like motorcycles, microwaves, and medications. Tell students that they will learn more about the people and history of Central Europe in this chapter.

Using the Physical-Political Map

Have students examine the map on the opposite page. Ask them to name countries that fit into the following categories: countries on the Baltic Sea (*Estonia, Germany, Latvia, Lithuania, Poland*), countries with elevations over 6,560 feet (1,999 m) (*Austria, Liechtenstein, Slovakia, Switzerland*), and landlocked countries (*Austria, Czech Republic, Liechtenstein, Slovakia, Hungary, Switzerland*). Then have students list the countries in this region through which the Danube River flows (*Austria, Germany, Hungary*) and those for which it forms part of the border (*Slovakia and Hungary*).

Why We Should Know More

There are many reasons why American students should know more about the countries of Central Europe. Here are a few of them:

► Although Germany was our enemy in both world wars, it is now a major ally. The United States has military bases in Germany.

► Many Americans trace their ancestry to the region.

► The region has been home to some of the world's greatest composers, writers, artists, scientists, and philosophers.

► Millions of tourists enjoy the region's natural beauty, cuisine, historic sites, and sports.

CHAPTER 15 Central Europe

Central Europe includes some of the most industrialized and richest countries in the world. Other countries in the region are slowly recovering from decades of Communist rule.

Crown, Holy Roman Empire

Shepherd with alpenhorn, Switzerland

Grüss dich! (Hello!) My name is Lizzi (LEE-zee), and I live in the village of Deutenhausen. Deutenhausen is near Munich in Bavaria, in southern Germany.

I am in the eighth grade at the gymnasium (high school). I live on a farm with my three older sisters, my parents, and my grandmother. My parents are farmers and also own a restaurant. In the summer, I make sure the cows have enough water. I also help my parents in the restaurant by chopping vegetables for the salads. When I grow up, I hope to become a doctor and work in an emergency room.

In the morning I drink warm milk fresh from our cows and eat fresh bread baked in the restaurant. Our kitchen is a huge room where everyone hangs out. At about 7:30 A.M. each day, I take the bus to school in Weilheim, which is about 2 miles (3 km) away. My favorite subject is art. I also study German, geography, Earth science, English, and Latin. Next year I will start classical Greek.

When school is over at 12:30 P.M., I go home to have lunch with my grandmother. Afterwards, I play with my friends outdoors, even though it often rains. We splash in the creek, race our bikes, and climb up to the church steeple to hear the bells ring.

Section 1

OBJECTIVES

1. Identify some key events in the history of Germany.

2. Describe some features of German culture.

3. Examine the German economy.

4. Evaluate the issues and challenges faced by Germany.

📻 LET'S GET STARTED

Copy the following passage onto the chalkboard: *Imagine that the American Civil War had resulted in the creation of two separate countries. Imagine that the two Americas were separated for 40 years before they were reunited. What is one problem that the country might face after reunification?* Discuss responses. Explain that after World War II Germany was divided into two countries that have since been reunited. Tell students that in Section 1 they will learn more about Germany.

Building Vocabulary

Write the terms **alliance** and **balance of power** on the chalkboard. Call on volunteers to locate and read the terms' definitions from the text or glossary. Tell students that countries generally form alliances to prevent wars by maintaining the balance of power in a region. Call on a volunteer to explain how this system works. *(Small or less powerful countries form alliances to prevent attack by a more powerful common enemy. Their goal is to gather enough military power to match that of their opponent.)*

Section

1 Germany

READ TO DISCOVER

1. What are some key events in the history of Germany?
2. What are some features of German culture?
3. What is Germany's economy like?
4. What issues and challenges does Germany face today?

WHY IT MATTERS

The German government once invited foreign workers to the country in order to solve a labor shortage. Now some Germans are taking jobs in other countries. Use **CNNfyi.com** or other **current events** sources to learn about labor shortages and how countries deal with them.

DEFINE

alliances
balance of power

LOCATE

Berlin
Bavaria
Ruhr Valley

Ivory carving from Dresden, Germany

Section **1** **RESOURCES**

REPRODUCIBLE

► Guided Reading Strategies 15.1
► Graphic Organizer Activity 15
► Readings in World Geography, History, and Culture 32
► Cultures of the World Activity: Region 3
► Geography for Life Activity 15: Symbols of a Reunited Berlin
► Map Activity 15: Territorial Changes in Germany

TECHNOLOGY

► One-Stop Planner CD–ROM, Lesson 15.1
► CNN Presents Geography: Yesterday and Today, Segment 17: Castles for Sale
► Homework Practice Online
► HRW Go site

REINFORCEMENT, REVIEW, AND ASSESSMENT

► Main Idea Activity 15.1
► English Audio Summary 15.1
► Spanish Audio Summary 15.1
► Section 1 Review, p. 331
► Daily Quiz 15.1

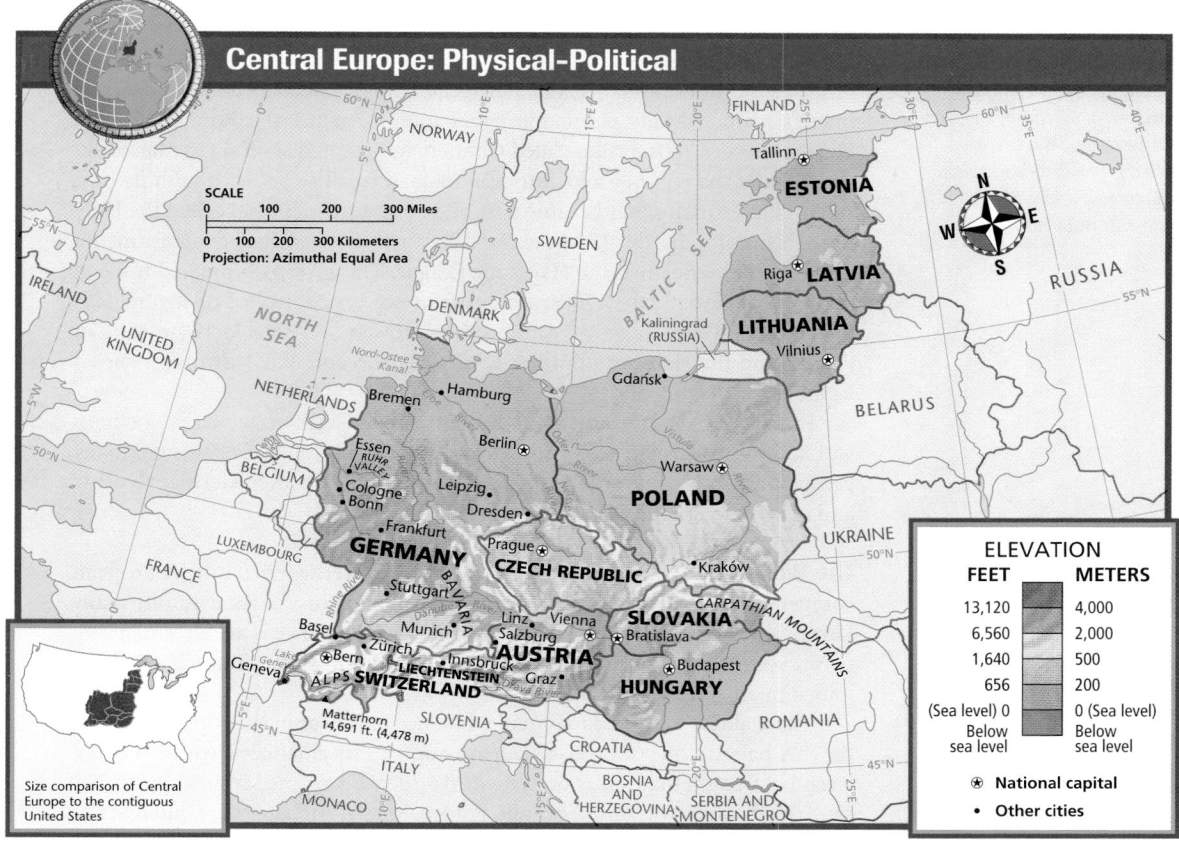

Central Europe: Physical–Political

SCALE
0   100   200   300 Miles
0   100   200   300 Kilometers
Projection: Azimuthal Equal Area

ELEVATION

FEET		METERS
13,120		4,000
6,560		2,000
1,640		500
656		200
(Sea level) 0		0 (Sea level)
Below sea level		Below sea level

⊛ National capital
• Other cities

Size comparison of Central Europe to the contiguous United States

Teach Objective 1

ALL LEVELS: Copy the graphic organizer below onto the chalkboard, omitting the italicized answers. Have students complete the chart with events from German history. Then lead a discussion describing how Germany's borders have changed over time. **COOPERATIVE LEARNING**

LEVELS 2 AND 3: Have each student create a series of maps illustrating how Germany's borders have changed under different governments. **ENGLISH LANGUAGE LEARNERS**

HOMEWORK: Have students create posters to illustrate key events in German history. Posters should be accompanied by paragraphs explaining the significance of the featured events.

Germany's Road to Unification

Germanic tribes fight against the Roman Empire. → Charlemagne begins German unification. → Holy Roman Empire and Hanseatic League control German lands. → Prussia unites northern and southern Germany. → Germany is divided after World War II. → Germany reunites after the fall of communism.

Across the Curriculum

► Art ◄

Looted Treasures During World War II, many great works of art disappeared from European museums and the homes of Jewish families. Some were destroyed by bombs or falling buildings, but many others were carried off by soldiers. However, many pieces that art historians and museum directors had presumed destroyed have begun to resurface. For example, the director of the Hermitage Museum in Saint Petersburg, Russia, announced in 1994 that he had discovered 2,700 German paintings and artifacts hidden in the museum's basement since 1945. Art experts believe that thousands of similar pieces are stashed in museums around the world.

ACTIVITY: Have students conduct further research on art theft during World War II and plot the journeys of specific paintings or artifacts on a world map.

King Louis II of Bavaria had this castle, Neuschwanstein, built in the mid-1800s. It is a fanciful version of a medieval German castle. Louis spent his family fortune and part of the country's treasury to pay for the castle. It is now a major source of income for Bavaria's tourism industry. Every year, approximately 1 million tourists visit Neuschwanstein.

History

From its location in the heart of Europe, Germany has helped shape the continent's history. Many Germanic tribes fought against the Roman Empire. During the A.D. 700s a ruler called Charlemagne united several German kingdoms. Later the region broke into hundreds of small states, each with its own ruler. The German states became part of a loose confederation called the Holy Roman Empire. By the 1300s, about 100 northern German towns formed a trading group known as the Hanseatic League. This group dominated trade in the Baltic region. By the 1700s, a number of powers controlled or strongly influenced the German states. Among these powers were the German state of Prussia and the Habsburg Empire, which later became the Austro-Hungarian Empire.

Prussia led the movement to create a single German country. Northern and southern German states united in 1871. From 1890 to 1914, Germany prospered and became a great industrial and military power. Germany's army and navy were among the strongest in Europe.

The World Wars The rapid rise of German power worried other European countries, particularly France, Great Britain, and Russia. As a result, many European countries formed military **alliances**. An alliance is an agreement between countries to support one another against enemies. Countries that are joined in an alliance are called allies.

These alliances helped maintain a **balance of power** in the region for some time. A balance of power exists when countries or alliances have such equal levels of strength that war is prevented. World War I erupted in 1914 partly because the balance was upset. Britain, France, Russia, and later the United States

internet connect

GO TO: go.hrw.com
KEYWORD: SW3 CH15
FOR: Web sites about German history

internet connect

GO TO: go.hrw.com
KEYWORD: SW3 CH15
FOR: Web sites about Central Europe

Teach Objective 2

LEVEL 1: Organize students into groups and have each group create a list of German culture traits. Call on volunteers to share items from each group's list with the class. As culture traits are named, ask the class whether they help to unify or divide eastern and western Germany. **COOPERATIVE LEARNING**

LEVELS 2 AND 3: Have each student create a design for a Web page that highlights Germany's major contributions to world culture. Remind students that their pages should include headlines, graphics, and links. Ask volunteers to share their work with the class.

Teach Objective 3

LEVEL 1: Ask students to list some German-made products with which they are familiar. Write their responses on the chalkboard. Use this list to lead a discussion about the major activities of the German economy.

LEVELS 2 AND 3: Pair students and have each pair write three newspaper headlines that describe how reunification and the changes in technology, transportation, and communication that followed it have affected Germany's economy. Then have each student write a brief article or create an editorial cartoon to accompany one of his or her group's headlines. **COOPERATIVE LEARNING**

joined forces against Germany. Germany was allied with the Austro-Hungarian Empire, Bulgaria, and the Ottoman Empire. After World War I ended in 1918, Germany had to accept harsh peace terms imposed by the victors. Germany's economy also collapsed in the 1920s. Food shortages, high inflation, and high unemployment caused severe hardships.

Germany's economic and political problems helped bring the Nazi Party to power in 1933. Adolf Hitler was the Nazi leader. Under Hitler, Germany rebuilt its military and allied itself with Italy and Japan. In 1939 Germany invaded Poland, sparking World War II. Fighting soon involved most of the European continent and later much of the world. The United States, Britain, the Soviet Union, and other allies finally defeated Germany in 1945. Some 50 million people had lost their lives. Germany and much of Europe lay in ruins.

Division and Reunification The Allied victors of World War II divided Germany. Soviet troops occupied eastern Germany as well as most of Eastern Europe. British, French, and U.S. troops occupied western Germany. Over time, two countries emerged from this division, East Germany and West Germany. Communist governments ruled East Germany as they did other Eastern European countries. West Germany became a democracy. West Germany also rebuilt rapidly with U.S. aid and soon became a global economic power. However, the economy of East Germany lagged. In 1990, following the collapse of communism, East and West Germany reunited.

✓ **READING CHECK:** *The Uses of Geography* How did World Wars I and II affect Germany? Germany was forced to accept peace and suffered an economic collapse after World War I. After World War II, a devastated Germany was split into two countries.

Culture

Today Germany has a democratic system of government. Berlin is the capital. (See Cities & Settlements: Berlin.) The country is divided into 16 states, or *Länder,* which vary in size and population. Bavaria, in the south, is the largest German state in area.

From 1961 to 1989, Berlin was split into a communist East and a capitalist West by a heavily guarded concrete wall. During this time, about 5,000 people escaped over, under, or through the wall into West Berlin.

A traditional carnival celebration winds through the streets of Mainz. The city serves as the capital of Rhineland-Palatinate, one of the Länder of southwestern Germany. Mainz is famous as the home of Johannes Gutenberg, the inventor of movable type printing.

Teacher to Teacher

Susan Walker of Beaufort, South Carolina, suggests the following activity to examine Germany as a world economic power. Have each student create a series of graphs related to Germany's economy. Possible topics for these graphs include the composition of the German labor force, Germany's major exports and imports, its major industries, and so forth. Provide students with current almanacs, encyclopedias, or Internet access so that they can locate the most recent information. Display students' graphs in the classroom.

Teach Objective 4

ALL LEVELS: Have students prepare segments for a television broadcast entitled Focus on Germany. Organize the class into groups. Assign each group an issue or a challenge that Germany faces today. Have students conduct research on their assigned issues and challenges and on solutions that have been proposed to resolve them. Then have each group write a script in which it supports one of the proposed solutions and discusses its probable effects on Germany's future. Call on groups to perform their broadcasts for the class.
COOPERATIVE LEARNING

Linking Past to Present

German Automobiles

Many Americans are familiar with German automobile companies. However, few may realize that two different German inventors developed the idea of the car. Carl Benz and Gottlieb Daimler, working separately, each designed motor-driven vehicles in the 1880s. Though the two men never met, the companies they founded eventually merged to form Daimler-Benz, the company that makes Mercedes-Benz automobiles.

Today, the automobile industry is a major contributor to Germany's economy. One out of every seven German workers works in an automobile-related industry, whether it be in manufacturing, service, sales, or raw material production. Millions of German cars—made by companies like BMW, Daimler-Benz, and Volkswagen—are sold to drivers each year.

ACTIVITY: Have students research other German-made products. Encourage students to learn about the histories and companies that produce the items they select.

VISUAL RECORD ANSWER

made them highly marketable

GRAPH ANSWER

services; far more people employed in services than in developing countries

INTERPRETING THE VISUAL RECORD

*Flowers are grown as a cash crop near western German cities. **How may strict environmental laws have affected the marketability of Germany's crops?***

Germany's Labor Force

2.8%
33.4%
63.8%

Labor force by occupation:
- Agriculture
- Industry
- Services

Source: Central Intelligence Agency, *The World Factbook 2001*

INTERPRETING THE GRAPH

What part of Germany's economy has the largest number of employees? How does Germany's labor-force distribution compare to that of most developing countries?

German is the dominant language, although it has several regional dialects. Novels, plays, and poetry written in German have enriched world literature. German composers and artists have also created great works. About a third of Germans are Roman Catholic, and a larger number are Protestant. Southern and western areas are more Catholic than northern and eastern regions. Many Germans do not attend religious services of any kind. German food features pork, sausages, veal, and cheeses. Rich pastries are popular desserts.

In recent years concern over Germany's environment has grown. Many people worry about the effects of air and water pollution and acid rain. "Green parties" are well established in government. These parties have helped pass laws to protect the country's natural environment. As a result, Germany now has some of the strictest environmental laws in the world. Interest in the environment also goes beyond legislation. For example, many Germans spend their vacations hiking, camping, or volunteering with environmental organizations.

✓ **READING CHECK:** *Environment and Society* How is concern for the environment part of Germany's political culture? Green parties are well established, and many Germans volunteer in environmental organizations.

Economy

Germany is an economic powerhouse. In fact, the country's GDP is the fourth-largest in the world, behind the United States, Japan, and China. However, Germany's per capita GDP is much higher than that of China, which has a far larger population.

Germany is one of the most prominent members of the European Union (EU). Most of Germany's trade is with other EU members. In addition, German has become a widely used language for business in Central Europe. German businesspeople are major investors in other Central European countries. The United States and Japan are other major trading partners.

The German economy is diverse. Businesses manufacture machinery, automobiles, electronics, and medical equipment. Chemicals, steel, and high-tech computer equipment are also important products.

Coal, iron ore, and other minerals helped make Germany an industrial power. In fact, the Ruhr Valley in western Germany is a major industrial center. Industries there developed around huge coal deposits. Today the Ruhr Valley is an almost continuous belt of cities and industries.

Almost half of the country's land is available for agriculture. German agriculture is efficient. Thus, farmers make up less than 3 percent of the population. Grains, potatoes, and sugar beets are major crops.

Nuclear power has provided about a third of Germany's electricity. However, in 2001 the government decided to gradually close all the nuclear power plants. Germany imports almost all of its oil.

✓ **READING CHECK:** *Places and Regions* How productive is the German economy? fourth-largest in the world, produces many industrial products

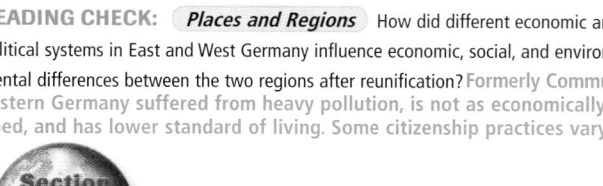

Close

Tell students that the Brothers Grimm, Jacob and Wilhelm, collected hundreds of traditional German folktales and published them. The stories in their collection are now some of the most famous in the world. These include the stories of *Sleeping Beauty, Rumpelstiltskin, Hansel and Gretel,* and *Snow White.* Bring samples of these or other German stories to share with the class.

Review and Assess

Have students complete the **Section Review**. Then have students complete **Daily Quiz 15.1**.

Reteach

Have students complete **Main Idea Activity for English Language Learners and Special-Needs Students 15.1.** Then have students work in pairs to design new state seals for Germany that reflect the country's history and culture. **ENGLISH LANGUAGE LEARNERS, COOPERATIVE LEARNING**

Extend

Have interested students conduct research on German immigration to the United States and the contributions of German Americans to our country's history, culture, or economy. Encourage students to share their findings with the rest of the class. **BLOCK SCHEDULING**

Issues and Challenges

One important issue in Germany today is the country's changing population. (See Geography for Life: Germany's Aging Population.) Low birthrates, longer life expectancies, and a large number of immigrants are all changing Germany's population. The largest group of immigrants are from Turkey. They have migrated to Germany to work in its growing industries. Most live clustered together in the neighborhoods of big cities. Their Islamic religion, Turkish language, and distinct culture add to their isolation in Germany. In some cases, prejudice and violence against Turks and other groups have been a problem. Many other immigrants to Germany are ethnic Germans from the former Soviet Union.

Since 1990 Germany has tried to bring the standard of living in the east up to that of the west. This effort has been difficult and costly. Many inefficient factories in the east were closed down. As a result, unemployment soared. With unemployment at almost 20 percent in eastern Germany, some Germans are becoming migrant workers in other European countries. Eastern Germany also suffered from heavy pollution during the Communist era, and the cleanup of the environment is just beginning.

Some easterners, or *Ossis*, feel that they are treated as second-class citizens. They resent that westerners, or *Wessis*, have a higher standard of living. Some *Ossis* also miss the lower costs of living and guaranteed jobs and housing that they had under communism. Even citizenship practices sometimes differ between *Ossis* and *Wessis*, a result of Germany's long period of division. For example, May Day, or May 1, was a major holiday in East Germany during the Communist era. It was a day to honor workers. Today some *Ossis* still organize parades to celebrate May Day. *Ossis* do this even though it is not an official holiday in the reunified Germany.

INTERPRETING THE VISUAL RECORD *More than 2 million people of Turkish descent live in Germany today. Some are victims of persecution and discrimination, even though many were born in Germany. Only recently have new laws offered immigrant Turks full German citizenship.* **How have cultural beliefs shaped the political opportunities available to German Turks?**

✓ **READING CHECK:** *Places and Regions* How did different economic and political systems in East and West Germany influence economic, social, and environmental differences between the two regions after reunification? *Formerly Communist eastern Germany suffered from heavy pollution, is not as economically developed, and has lower standard of living. Some citizenship practices vary.*

Review

Homework Practice Online
Keyword: SW3 HP15

Define alliances, balance of power

Working with Sketch Maps On a map of Central Europe that you draw or that your teacher provides, label Germany, Berlin, Bavaria, and the Ruhr Valley. In the margin of your map, identify the capital of Germany.

Reading for the Main Idea

1. *Human Systems* How was Germany divided after World War II?

2. *Human Systems* What role does Germany play in European economies?

3. *Environment and Society* What is the Ruhr Valley? Around what natural resource did its industries and cities grow?

Critical Thinking

4. *Comparing* In what ways might people in eastern Germany have mixed views about the effects of reunification on their lives?

Organizing What You Know

5. Copy the time line below. Use it to identify important periods and events in Germany's history after 1871.

1871 |—————————————| Today

Section 1 Review Answers

Define For definitions, see: alliances, p. 328; balance of power, p. 328

Working with Sketch Maps Maps will vary, but listed places should be labeled in their approximate locations. Berlin is the capital.

Reading for the Main Idea
1. The Soviets took control of eastern Germany, while the Allies occupied western Germany. Berlin was also divided.

2. Germany is a trade partner with other countries of the EU. German businesspeople are major investors in other Central European countries.

3. a major industrial and urban region of western Germany; coal

Critical Thinking
4. Western Germany's stronger economy can help rebuild eastern Germany's economy, but some *Ossis* might resent the *Wessis'* higher standard of living. They might also miss some features of communism, such as lower costs of living and guaranteed jobs and housing. (NGS 11)

Organizing What You Know
5. 1871—unification; 1890–1914—economic prosperity; 1914–1918—World War I; 1920s—economic collapse; 1933—rise of Nazi party; 1939–1945—World War II; 1990—reunification.

VISUAL RECORD ANSWER

contributed to their being viewed as outsiders, denied citizenship

Have students read Cities & Settlements: Berlin. Ask them to identify some causes behind Berlin's original rise to greatness (*capital of Brandenburg and Prussia; commercial center; capital of Germany after unification*). Then have students suggest some effects of the city's early prestige (*popular among brilliant artists, teachers, writers, political thinkers, composers; attracted immigrants from across Europe; flourished as a cultural center through the 1920s*).

Berlin's fortunes turned in the 1930s, however. Ask students to identify the primary cause behind this change in fortune (*rise of the Nazis*). Call on students to identify the immediate consequences of this change (*bombing of Berlin during World War II, division among several victorious countries after war, building of the Berlin Wall*). Then ask students to suggest more lingering effects of the city's downturn (*disputes between residents of former East and West Berlin over property or jobs, resentment between residents of various parts of the city, lingering cultural differences*).

In the 1990s new building programs were begun around the city. Ask students why they think many Germans are determined to rebuild their capital as a new and unified city. (*Possible answers: Berlin serves as a symbol of Germany both to Germans and to the rest of the world, so prosperity in Berlin suggests prosperity in the rest of the country. Many want to rebuild the city into the European cultural center it once was or simply to erase visible reminders of World War II.*) Lead a class discussion about the possible effects of recent renovation efforts in Berlin.

Essential Element 4

► Human Systems ◄

Checkpoint Charlie

When the Berlin Wall was built in 1961, only a few border crossings were left open. The most famous of these was named Checkpoint C, but it was generally referred to as Checkpoint Charlie. Over time, it became the main crossing point between the two parts of the city. By the time the wall came down, thousands of East Germans had tried to enter West Berlin, often through Checkpoint Charlie. Some tried simply to elude the checkpoint guards, but others sought more unusual passage. One man even reportedly smuggled his girlfriend through the checkpoint by fitting her inside the seat of his car. Many similarly amazing stories of daring escapes are told at the Checkpoint Charlie Museum in Berlin.

This Cities & Settlements feature addresses National Geography Standards 4, 9, 12, and 13.

CITIES & SETTLEMENTS

Berlin

Human Systems   For much of its long history, Berlin seemed to be on the road to greatness. In the mid-1900s, though, a series of events intervened to block its path. Today, barely a half century after the city's near-destruction and later division, Berlin is once again a vibrant city. It still does not enjoy the status of London, Paris, or Rome. However, Berlin at last seems ready to rejoin the ranks of the great European cities.

A History of Triumph and Tragedy

Berlin was founded in the early 1200s as a trading village. Although it became the capital of a small independent state called Brandenburg, the town grew slowly. In 1670 its population was only 12,000. By then, Brandenburg had merged with a neighboring state to form the powerful kingdom of Prussia. In the early 1700s Prussia's king made Berlin his capital. The king and later rulers turned the town into a great city. By 1750 Berlin had become a thriving commercial center and home to some 100,000 people. In the late 1800s Prussia united the region's other states to form Germany. Berlin became the new country's capital. By 1880 the city's population had reached 1.3 million.

While it was Prussia's capital, Berlin became a center for the arts, education, and literature. Among the city's residents in the 1800s were political philosopher Karl Marx and the composer Felix Mendelssohn. Over time French, Jewish, Polish, and Russian immigrants flocked to the city. Despite political and economic problems following Germany's defeat in World War I, Berlin continued to flower. The city remained a cultural center during the 1920s. In the 1930s, however, Adolf Hitler came to power and led Germany into World War II. Bombs and invading armies then destroyed much of Berlin.

After the war Britain, France, the United States, and the Soviet Union divided Berlin and the rest of Germany into zones. The British, French, and American zones in Berlin soon merged into what was called West Berlin. The Soviet zone became East Berlin. All of Berlin was located deep within Soviet-controlled East Germany. In 1961 the Communists built a high concrete wall around West Berlin to keep East Germans from fleeing to the west. The wall did not come down until 1989, as both East Germany and the Soviet Union began to collapse. In 1990 Berlin, like all of Germany, was officially reunited.

A Future of Cautious Hope

Most of the Berlin Wall is gone now. Small sections still stand as a monument to Berliners' struggle for unity and freedom. In some ways, however, the city remains divided. For example, when the Communists controlled East Berlin they seized houses owned by West Berliners and turned them into public housing. After reunification, the former owners began to reclaim their property. Many East Berliners have been forced to give up homes where they lived for years. In addition, many of East Berlin's inefficient government-run businesses closed. Some 275,000 easterners lost their jobs. The east has seen a great deal of commercial construction since reunification. However, East Berliners

In 1989 the Brandenburg Gate was opened to traffic for the first time in 28 years. The gate had been blocked by the Berlin Wall.

Going Further: Thinking Critically

Rebuilding efforts in Berlin have attracted the attention of designers and architects from around the world. They are particularly interested in what they call New Berlin architecture, a design system that combines old or historic elements with new and modern designs. Berlin's urban planners have achieved great artistic success with this kind of blending. For example, the Reichstag, formerly the home of Germany's legislature, was one of Berlin's most beautiful buildings before World War II. However, during the war it was burned by the Nazis, bombed by the Allies, and stormed by Russian forces. Some renovation of the building was begun in the 1970s, but it was not until reunification that a serious effort was made to rebuild the Reichstag. The original building was cleaned and restored but with a new twist. Designers added a huge glass dome to give the building a modern touch.

Similar efforts have been undertaken around the city. Old mixes with new around Berlin's famous squares. Dramatic changes are occurring in the formerly empty fields where the Berlin Wall stood. Potsdamer Platz, pictured on this page, is an example. Homes, shops, and entertainment venues are housed in a glitzy, modern structure built on the former site of the wall in downtown Berlin.

Have students conduct research on Potsdamer Platz or on other areas of recent construction in Berlin. Ask them to identify signs of the blending of historic and modern elements in these developments. Encourage students to bring pictures of their subjects to share with the class. Discuss some possible effects of this sort of construction on both local and global perceptions of the city.

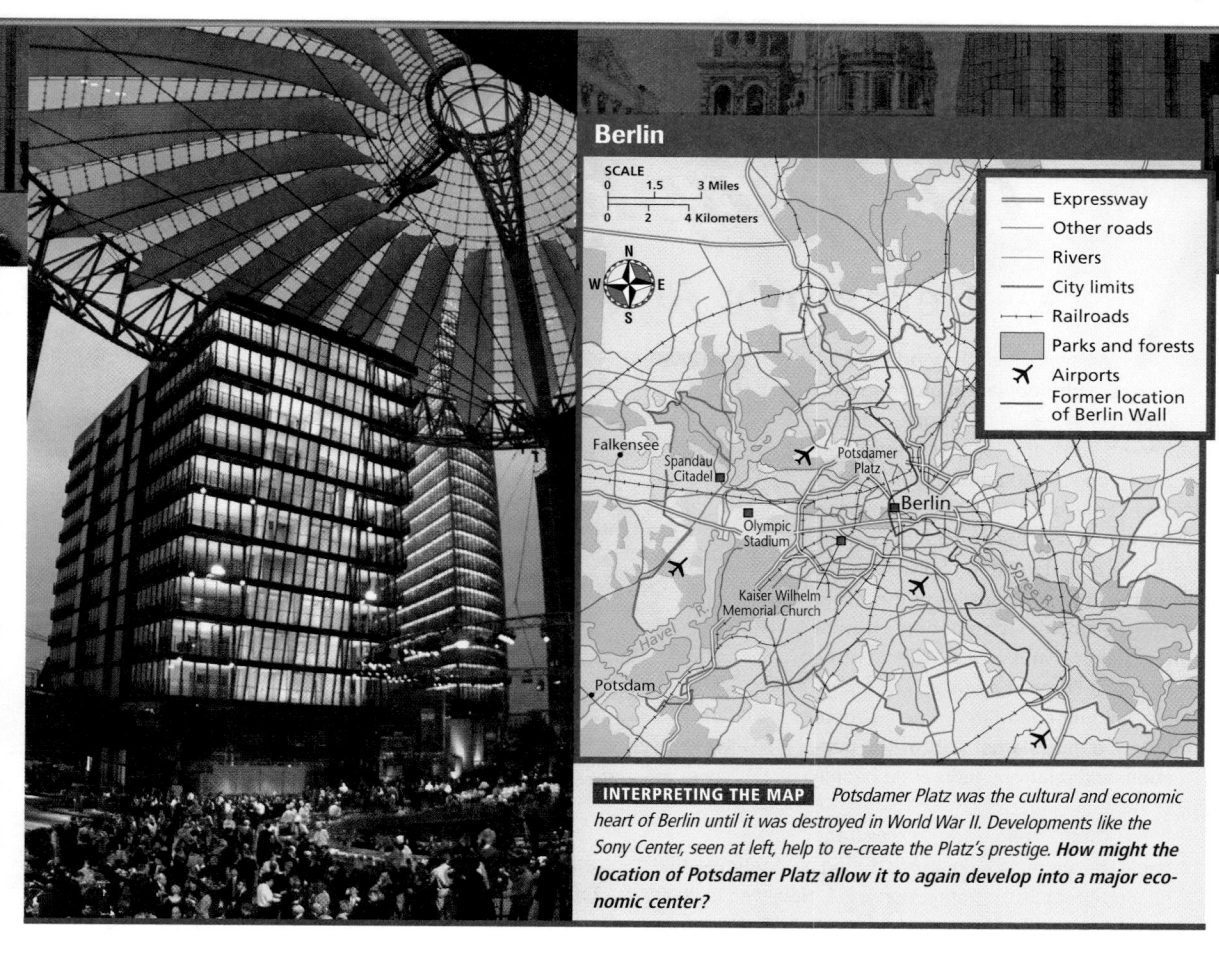

INTERPRETING THE MAP *Potsdamer Platz was the cultural and economic heart of Berlin until it was destroyed in World War II. Developments like the Sony Center, seen at left, help to re-create the Platz's prestige.* **How might the location of Potsdamer Platz allow it to again develop into a major economic center?**

MAP ANSWER

It is located in the heart of the city near the crossroads of several major highways that run through the city. The Platz is also convenient to two railroads and an airport.

resent the fact that westerners have many of the jobs created by Berlin's recent growth. For their part, West Berliners resent the attention, money, and development focused on East Berlin. Berliners have a phrase to describe the tension—*die Mauer im Kopf,* or "the wall in the head." It refers to the psychological and emotional barrier that still lingers in the city.

Yet if concrete remains a reminder of Berlin's recent past, it is also a symbol of the city's future. At one point in the late 1990s, some 1,200 construction cranes dotted the Berlin skyline. In 1991 Potsdamer Platz was a huge vacant field just east of the wall in downtown Berlin. Today it is packed with workers, shoppers, and entertainment-seekers. This gleaming new city center houses offices for international corporations, apartments, and general office buildings. Eastern Berlin needs more construction. More than 100,000 new businesses were started in Berlin during the 1990s. In addition, thousands of new emigrants from Eastern Europe and the former Soviet Union have flooded the city. In the first five years after reunification, Berlin's population jumped from 3 million to 3.5 million.

Today western Berlin is energetic and prosperous. Meanwhile, the eastern half of the city is shedding its Communist past and looking toward the future. Life in the former East Berlin is no longer dreary. Many of its beautiful old buildings have been restored as art galleries and cafés. Residents dress in the latest styles. Cultural divisions between east and west continue to lessen. The golden era of the 1920s is returning to the city. The "new" Berlin is again one of the liveliest cities in the world.

Applying What You Know

1. **Summarizing** Why are there still some divisions among Berliners even though the Berlin Wall has been torn down?

2. **Contrasting** What were the urban environments of East and West Berlin like before the wall came down? What are they like today?

Applying What You Know Answers

1. East Berliners resent competition from West Berliners for new and much-needed jobs that have been created on the city's east side. West Berliners resent the money and effort being focused on the redevelopment of East Berlin.

2. Possible answer: The west was a thriving, bustling city, while the Communist east was much less vibrant. Today the culture and economy of the east are reviving, and the two sections of the city are much more similar in their environments.

Mathematics: Calculating Dependency Ratios

Demographers are interested in how changes in a country's population affect its economic systems. One factor that they calculate in their research is a country's dependency ratio. This figure represents the number of people, both elderly citizens and children, who must be supported by every 100 people of working age, or between the ages of 15 and 64. To calculate a country's dependency ratio, one uses the formula:

$$\frac{total\ dependent\ population}{total\ working\ population} \times 100 = dependency\ ratio$$

For example, in 2001 Germany had 12,925,322 people aged 14 or younger and 13,793,279 older than 64. When added together, these numbers reveal a total of 26,718,601 Germans dependent upon others for their well-being. Germans of working age numbered 56,310,935. One can then apply the dependency ratio formula:

$$\frac{26,718,601}{56,310,935} \times 100 = 47.45$$

So for every 100 people of working age in Germany in 2001, there were about 47 people who depended upon them for support.

Have students obtain population figures from the *CIA World Factbook* or other sources to calculate dependency ratios for countries in Europe and around the world. Then lead a class discussion about how ratios from developing countries compare to those from developed countries.

Global Perspectives

Aging and Government
Germany is not the only country dealing with an aging population. In Singapore, for example, life expectancy has reached more than 80 years, so the government there has begun to address issues related to caring for the elderly. Singapore has created a Tribunal for the Maintenance of Parents. Aging parents can bring their children before the tribunal if they do not feel their needs are being met.

DISCUSSION: Lead a class discussion about how governments might address the needs of elderly citizens.

GRAPH ANSWER

about 7 percent; fewer young people, slow population growth

Applying What You Know Answers

1. The pyramids show the population becoming older, with a wide "bulge" moving upward between 2000 and 2025. The 2050 pyramid likely will show the bulge at or near the top, indicating an even older population. Slower or negative population growth might be expected.
2. Answers will vary.

This Geography for Life feature addresses National Geography Standards 1, 4, and 9.

Geography for Life

Germany's Aging Population

While many of the world's countries face high population growth rates, Germany—and most of Europe—faces a different problem. The UN estimates that Europe's population will decline from 713 million in 2025 to 658 million in 2050. In contrast, the population in most of Africa may double in about 30 years. In fact, of the 45 countries in the world with the lowest population growth rates, 30 are in Europe. This statistic means that Europe's population is aging. Thus, middle-aged and older people make up a much larger part of the population, and relatively few young people live there.

The total fertility rate—the average number of children a woman has during her lifetime—that is needed to replace a population is 2.1. However, in 2000, Germany's total fertility rate was about 1.38. Therefore, fewer children are born in Germany than are needed to replace the population. This means that young people are making up a smaller percentage of Germany's population. At this rate the population will gradually decline.

Germany's population is also aging because people are living longer. Good health care helps people survive illnesses and accidents. Now 16 percent of the German population is aged 65 or older. This figure will probably rise to more than 23 percent by 2025. Such a trend will increase demand for health care and government pensions. Pensions are regular payments to retired people. As more and more older people retire, young people will carry the increasing burden of supporting the retirees. How will Germany accommodate this change?

Germany could increase its birthrate or its worker productivity. It could decrease pension benefits or increase immigration. Immigrants are part of German society. When the economy was booming in the 1950s, the country needed workers. People from Turkey and other countries responded to the German government's invitation to become "guest workers." When this campaign ended in 1973, most Germans assumed the workers would return to their homelands. Yet many Turks stayed. Nearly 2 million Turks lived in Germany by the mid-1990s.

Since 1990, ethnic Germans and others from the former Soviet Union and Eastern Europe have also moved to Germany. Their return could help slow the aging of Germany's population. However, many Germans worry about their country's ability to absorb so many immigrants. Furthermore, the unemployment rate is currently high—as much as 20 percent in some areas. As a result, the government now discourages ethnic Germans from returning.

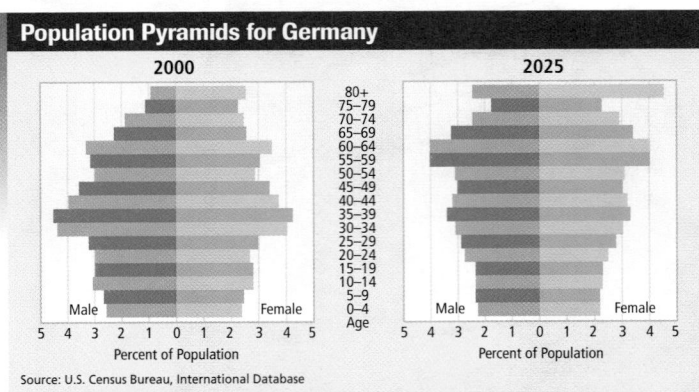

Population Pyramids for Germany

Source: U.S. Census Bureau, International Database

INTERPRETING THE GRAPH *In 2025 what percentage of Germany's population will be 80 and older? What trend do you see at the bottom of the 2025 pyramid?*

Applying What You Know

1. **Analyzing** How is the continued aging of Germany's population shown on the population pyramids? What do you think Germany's population pyramid will look like in 2050? What do you predict for Germany's future population growth?

2. **Evaluating** Compare Germany's situation to Oman's. Oman's population growth rate is 3.46 percent. Thus, Oman has a young population that will double in about 20 years. What might be the consequences of Oman's young population? Create a chart listing the advantages and disadvantages of Germany's and Oman's population structure.

Section 2

OBJECTIVES

1. **Describe some important features of Austria's history, culture, and economy.**

2. **Analyze the political, cultural, and economic features of Switzerland.**

 *LET'S GET STARTED*

Copy the following instructions on the chalkboard: *Draw or list things that come to mind when you think of Austria or Switzerland.* Discuss student responses. (*Possible answers: chocolate, cheese, mountains, châteaus, cows, traditional Alpine clothing*) Point out that many elements of Switzerland's and Austria's cultures reflect the Alpine region's landscapes. Tell students that they will learn more about these countries, their landscapes, and their cultures in Section 2.

Building Vocabulary

Write the terms **cantons** and **confederation** on the chalkboard. Call on volunteers to locate and read the terms' definitions. Tell students that the Swiss cantons formed a confederation to address national issues but that each canton maintains its own local government. Remind students that the United States once tried a similar system but abandoned it when the Constitution was written. Then have another volunteer read the definitions of the remaining terms.

The Alpine Countries

READ TO DISCOVER

1. What are some important features of Austria's history, culture, and economy?

2. What are the political, cultural, and economic features of Switzerland?

WHY IT MATTERS

The headquarters of the Red Cross and Red Crescent (the name Red Crescent is used in Muslim countries) is located in Geneva, Switzerland. Use **CNNfyi.com** or other **current events** sources to learn more about this international humanitarian organization.

DEFINE

confederation
cantons
neutral
multilingual

LOCATE

Vienna	Basel
Danube River	Bern
Geneva	Zürich

RESOURCES

REPRODUCIBLE
► Guided Reading Strategy 15.2
► Readings in World Geography, History, and Culture 33

TECHNOLOGY
► One-Stop Planner CD–ROM, Lesson 15.2
► Homework Practice Online
► HRW Go site

REINFORCEMENT, REVIEW, AND ASSESSMENT
► Main Idea Activity 15.2
► English Audio Summary 15.2
► Spanish Audio Summary 15.2
► Section 2 Review, p. 337
► Daily Quiz 15.2

Austria

Austria and Switzerland are both located in the Alps, the most mountainous region of Central Europe. Germanic culture has deeply influenced these two Alpine countries. The areas of both countries were settled by Germanic tribes after the fall of the Roman Empire. Like Germany, Austria became part of the old Holy Roman Empire. From the 1400s onward the Holy Roman emperor was a member of the Habsburgs, a powerful family of German nobles. Many different ethnic groups lived within the empire. Each had its own language, local government, and legal system. The Holy Roman Empire was united only by its allegiance to the emperor and for the defense of the Roman Catholic Church.

The Austrian Empire, under Habsburg control, eventually replaced the Holy Roman Empire. At the height of their power, the Habsburgs ruled Spain, the Netherlands, Germany, Italy, and parts of eastern Europe. The Austrian Empire was a major power in Europe in the 1800s. In 1867 the Austrians agreed to share political power with the Hungarians. The Austrian Empire became the Austro-Hungarian Empire. That empire collapsed at the end of World War I. Hungary and other parts of the empire won independence. In addition, the Habsburgs lost power in Austria, which then became a democratic republic. Germany took over Austria shortly before World War II, uniting the two countries. After the war, the Allies occupied Austria. The country became independent again in 1955.

Today Austria is a country about the same size as South Carolina. Nearly all Austrians speak German, and more than

Austria's Habsburg emperors lived in elegant palaces like Schloss Belvedere in Vienna, seen here. The schloss, which means "palace," now houses a museum and a botanical garden.

ALL LEVELS: Copy the following graphic organizer onto the chalkboard, omitting the italicized answers. Have students complete the diagram to compare the histories, cultures, governments, and economies of Austria and Switzerland. Ask students how the countries' divergent histories have led to differences in their political organizations. (*Unlike Austria, Switzerland has never ruled a powerful empire and thus was able to remain politically neutral in major European conflicts in which Austria became entangled.*) Then ask students how the countries' cultures and economies are similar. (*Both have diverse economies that depend largely on manufacturing and tourism. Both also house many foreign operations.*)

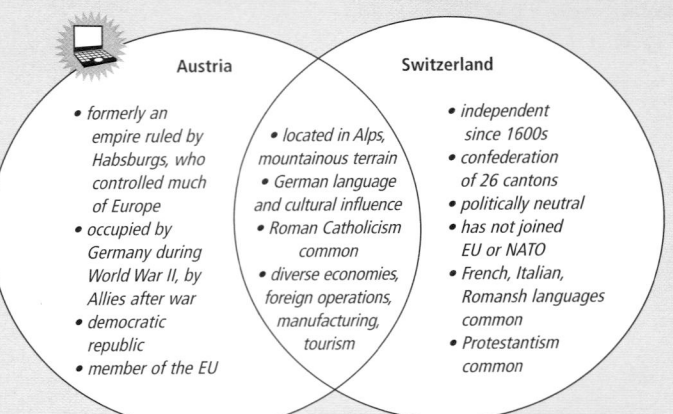

Austria
- *formerly an empire ruled by Habsburgs, who controlled much of Europe*
- *occupied by Germany during World War II, by Allies after war*
- *democratic republic*
- *member of the EU*

(overlap)
- *located in Alps, mountainous terrain*
- *German language and cultural influence*
- *Roman Catholicism common*
- *diverse economies, foreign operations, manufacturing, tourism*

Switzerland
- *independent since 1600s*
- *confederation of 26 cantons*
- *politically neutral*
- *has not joined EU or NATO*
- *French, Italian, Romansh languages common*
- *Protestantism common*

Protestant Geneva
In the mid-1500s John Calvin, one of the key figures of the Protestant Reformation, established a religious government in Geneva, Switzerland. The city became a Protestant stronghold. Legend has it that Geneva's Protestant leaders owed their success to a pot of soup.

In 1602 the Catholic Duke of Savoy in France sent an army to drive the Protestants out of Geneva. French soldiers tried to slip quietly over the city walls, but an alert cook heard them and emptied a cauldron of hot soup onto their heads. She then inspired her fellow townspeople to fight while the Genevan army readied itself for battle. The city was saved and the Protestant leaders remained in power. These same leaders later urged the city's goldsmiths to stop making jewelry, because wearing jewelry had been forbidden. The jewelers then focused their talents on what is now one of Switzerland's most famous industries—watchmaking.

internet connect

GO TO: go.hrw.com
KEYWORD: SW3 CH15
FOR: Web sites about Swiss culture

MAP ANSWER

to avoid internal tensions over language divisions

The Austrian government encourages the country's farmers to maintain traditional rural customs. Agriculture itself is not a major factor in the Austrian economy, but the preservation of rural ways helps draw millions of tourists each year.

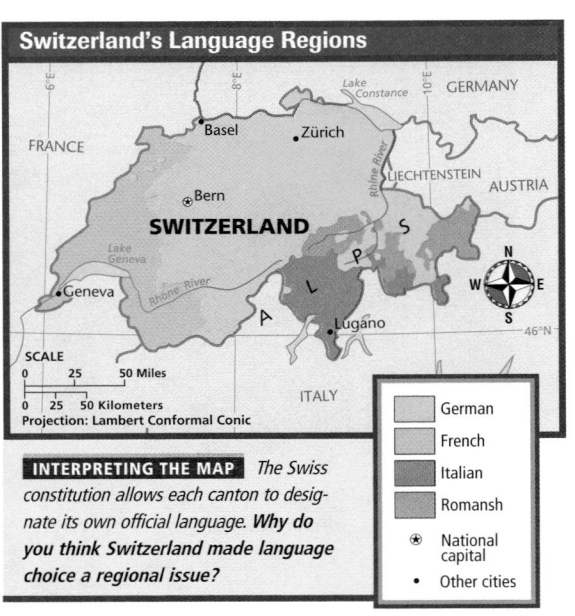

Switzerland's Language Regions

SCALE
0   25   50 Miles
0   25   50 Kilometers
Projection: Lambert Conformal Conic

German
French
Italian
Romansh
⊛ National capital
• Other cities

INTERPRETING THE MAP The Swiss constitution allows each canton to designate its own official language. **Why do you think Switzerland made language choice a regional issue?**

75 percent are Roman Catholic. Vienna, Austria's capital and largest city, is located on the banks of the Danube River. Vienna was the political and cultural capital of Central Europe during Habsburg rule. Historic palaces, churches, and performance halls beautify the city. In addition, many great artists and composers once lived there. Among the most famous were Wolfgang Amadeus Mozart and Ludwig van Beethoven.

Austria is a member of the EU and has a diverse economy. Austrian industries include steel, machinery, and chemicals. Forestry and hydropower are also important. Austria is famous for its high-quality wood, glass, textile, and ceramic handicrafts. The scenic Alps and the country's cultural attractions draw many tourists. Austrian ski resorts are world-famous.

After World War II, Austria kept up trade relations with many countries in Eastern Europe. These ties strengthened after the Soviet Union collapsed in the early 1990s. Today many American and Western European companies base their Eastern European operations in Austria.

✓ **READING CHECK:** (**Human Systems**) Why is Vienna an important city culturally and historically? political and cultural capital of Central Europe and of the Austro-Hungarian Empire, now Austria's capital and largest city

Switzerland

The history of Austria and Switzerland began to diverge in the late 1200s. At that time Swiss states began to form alliances to protect themselves against invading Austrian armies. Switzerland became independent of Habsburg rule in the 1600s. Today Switzerland is a **confederation**, or a group of states joined together for a common purpose. The country is made up of 26 **cantons**, or states. Each canton has self-government for all issues not reserved for the federal government. The federal government controls national policies, such as defense, international relations, and social programs.

Since Switzerland was formed, it has generally been a **neutral** country. A neutral country is one that does not take sides in international conflicts or alliances. In fact, Switzerland has not been involved in any recent wars. To preserve their neutrality, the Swiss have even resisted joining international organizations. For example, Switzerland is not a member of the EU. The country does participate in international affairs, however. In fact, many world and regional organizations have their headquarters in the Swiss city of Geneva. Geneva also hosts many international conferences.

Switzerland has four major languages: German, French, Italian, and Romansh. (See the map of Switzerland's language regions.) However, many Swiss are **multilingual**, meaning they speak several languages. English is rapidly becoming the country's fifth language. About 46 percent of the Swiss are Roman Catholic, and 40 percent are Protestant. As you can see, Switzerland is a culturally diverse country. However, it is one of the world's most stable countries.

Close

Tell students that the Swiss banking system is famous for its discretion. Banks will not disclose the names of depositors. Millions of dollars worth of gold and other precious items deposited into Swiss bank accounts for safekeeping during World War II still sit unclaimed in Swiss bank vaults.

Review and Assess

Have students complete the **Section Review**. Then have students complete **Daily Quiz 15.2**.

Reteach

Have students complete **Main Idea Activity for English Language Learners and Special-Needs Students 15.2**. Have students write facts about the Alpine countries on index cards taped to the wall in the shape of a mountain. Have students read their facts aloud. **ENGLISH LANGUAGE LEARNERS**

Extend

Many of the world's greatest classical musicians, including Beethoven, Brahms, Mozart, Schubert, and Strauss, lived in Vienna. Have each interested student conduct research on one of these composers, bring samples of his work to class, and present a report about his life and music. **BLOCK SCHEDULING**

INTERPRETING THE VISUAL RECORD

The Swiss Alps are a favorite destination for people from around the world. Tourists are drawn to mountain resorts like this one near Bern. **How can you tell from the photo that Alpine architecture has been adapted to the area's climate?**

Switzerland's largest cities are Basel, Bern, Geneva, and Zürich. Zürich, the largest, is a leading world-banking center. Basel is a transportation center on the Rhine River in northwestern Switzerland. Bern, the capital, is centrally located between the country's German-speaking and French-speaking populations.

Switzerland has one of the world's highest standards of living. Immigrant workers make up about one fourth of the population and are vital to Swiss business. International banking and insurance are important segments of the economy. Switzerland also produces chemicals, pharmaceuticals, watches, and some farm goods. Most of the country's farm output comes from dairy products. Swiss cheese and chocolate are world-famous. This scenic mountain country also attracts crowds of tourists. Timber production once played a role in the economy. However, air pollution has damaged the woodlands. In fact, pollution has harmed more than 35 percent of the country's forests. The Swiss government now limits tree cutting.

✓ **READING CHECK:** *Human Systems* How has Switzerland's traditional foreign policy shaped the country? Neutrality has kept Switzerland out of wars and international organizations and has made it an ideal location for international organizations' headquarters.

Section 2 Review

go.hrw.com
Homework Practice Online
Keyword: SW3 HP15

Define confederation, cantons, neutral, multilingual

Working with Sketch Maps On the map that you created in Section 1, label Austria, Switzerland, Vienna, the Danube River, Geneva, Basel, Bern, and Zürich. Which city is a leading world-banking center?

Reading for the Main Idea

1. *Places and Regions* What are some of Austria's main economic products and industries?

2. *Environment and Society* What are Switzerland's largest cities? In which language region is the capital located?

3. *Environment and Society* How has air pollution affected Switzerland's forests?

Critical Thinking

4. **Drawing Inferences and Conclusions** How do you think Switzerland's system of government has helped it maintain unity despite the country's cultural divisions?

Organizing What You Know

5. Create a chart like the one shown below. Use it to compare the cultures and economies of Austria and Switzerland.

	Austria	Switzerland
Culture		
Economy		

Section 2 Review Answers

Define For definitions, see: confederation, p. 336; cantons, p. 336; neutral, p. 336; multilingual, p. 336

Working with Sketch Maps Maps will vary, but listed places should be labeled in their approximate locations. Zürich is a leading world-banking center.

Reading for the Main Idea
1. steel, machinery, chemicals, forestry, tourism, hydropower, handicrafts

2. Basel, Bern, Geneva, Zürich; between the French- and German-speaking regions

3. damaged forests and hurt the timber industry

Critical Thinking
4. Possible answer: Switzerland's system of self-governing cantons has allowed differing cultural groups to have local control over their own affairs, thereby preventing nationwide conflicts over cultural matters. (NGS 10)

Organizing What You Know
5. Austria: culture—German language, mostly Roman Catholic; economy—diverse economy, industry, forestry, handicrafts, tourism; Switzerland: culture—four languages, both Catholic and Protestant, economy—banking, insurance, some manufacturing, dairy products, tourism

---

VISUAL RECORD ANSWER

Its steep roof and deep gables prevent snow buildup.

OBJECTIVES

1. **Trace the history of Poland and the Baltic countries.**
2. **Describe the urban environments and economy of Poland.**
3. **Analyze influences that have shaped the Baltic countries.**

 **LET'S GET STARTED**

Copy the following questions onto the chalkboard: *Have you ever witnessed or read about a political demonstration? What were the protesters' goals? Were they successful in achieving those goals?* Discuss responses. Tell students that political protests in Poland, Estonia, Latvia, and Lithuania were largely responsible for ending communist rule in those countries and contributed to the collapse of the Soviet Union. Tell students that in Section 3 they will learn more about the history and cultures of these countries.

Building Vocabulary

Write the key terms on the chalkboard and call on a volunteer to read the definitions from the text or glossary. Explain that the prefix *ex-* in **exclave** means "outside" or "away from." Ask students how this prefix relates to the term's meaning. (*An exclave is part of a country located outside or away from the main body of the country.*) Then tell students that the word **ghetto** is derived from the name of an island near Venice, Italy, where the city's Jews were at one time required to live.

 **RESOURCES**

REPRODUCIBLE

▶ Guided Reading Strategy 15.3
▶ Readings in World Geography, History, and Culture 37
▶ Critical Thinking Activity 15: Surviving the Cut

TECHNOLOGY

▶ One-Stop Planner CD–ROM, Lesson 15.3
▶ Homework Practice Online
▶ HRW Go site

REINFORCEMENT, REVIEW, AND ASSESSMENT

▶ Main Idea Activity 15.3
▶ English Audio Summary 15.3
▶ Spanish Audio Summary 15.3
▶ Section 3 Review, p. 341
▶ Daily Quiz 15.3

 **Section 3**

Poland and the Baltics

READ TO DISCOVER

1. What is the history of Poland and the Baltic countries?
2. What are the urban environments and economy of Poland like today?
3. What influences have shaped culture in the Baltic countries?

WHY IT MATTERS

In 1999 Poland became one of three former Soviet-bloc states to join NATO. Use **CNNfyi.com** or other **current events** sources to learn more about NATO's relationship with former communist countries in Europe.

DEFINE

exclave
ghetto

LOCATE

Kaliningrad
Warsaw
Vistula River
Kraków
Gdańsk

In 1386 the monarchs of Poland and Lithuania united their countries. This union, the Commonwealth of Two Nations, became a powerful force in European affairs. After about 1550, however, costly wars and poor leadership weakened the Commonwealth. Russia took it over in the mid-1700s. This Polish helmet dates from about 1640.

History

Poland gets its name from a Slavic people who moved into the area long ago. Their name, *Polanie*, came from a Slavic word meaning "plain" or "field." In fact, Poland's landscape is filled with plains and rolling hills.

Poland is the largest of the European countries that once made up what was called the Soviet bloc. These countries were allied with the Soviet Union from shortly after World War II until the early 1990s. Poland had also been under Russian control during part of the 1700s and 1800s. During that period, Austria and Prussia (and later Germany) also occupied areas that make up Poland today. Poland became independent after World War I. During and after World War II, however, the Soviet Union occupied the country. A Communist government then ruled the country for more than 40 years.

The Baltic countries also gained independence from Russia after World War I. These countries are Estonia, Latvia, and Lithuania. They stretch northward from Poland to the Gulf of Finland. The Soviet Union took over these countries during World War II. However, they regained their independence in 1991.

While most people have embraced the move toward democracy and capitalism, some people in the Baltics and Poland favor a return to Communist rule. Russian cultural influences linger in the Baltic countries. In fact, some of the countries have large Russian minority populations. Also, Russia still controls Kaliningrad (kuh-LEE-nin-grat). The city, along with its surrounding territory, is an **exclave**. An exclave is an area separated from the rest of a country by the territory of other countries.

✔ **READING CHECK:** *Human Systems* What role have foreign countries played in the history of Poland and the Baltic countries? Foreign countries, particularly Russia and the Soviet Union, have occupied or ruled all or parts of the countries during large stretches of their history.

 **ALL LEVELS:** Copy the following graphic organizer onto the chalkboard, omitting the italicized answers, and have students complete it.

Poland and the Baltics

| 1700 | 1900 | 1950 | 1990 | 2000 |

1700s to 1800s
Poland occupied by Russia, Prussia, and Austria; Baltics occupied by Russia.

1910s
Poland and Baltics become independent after World War I.

1940s
The Soviet Union occupies Poland.

1980s
Poland breaks away from communism.

1991
Baltic countries become independent.

Using National Geography Standard 5:
Places and Regions: That People Create Regions to Interpret Earth's Complexity Poland's boundaries have varied dramatically over the centuries. Poland has been large and powerful at times, weak and small at others. In fact, Poland was completely eliminated as a country three times, its territory distributed among Russia, Prussia, and Austria. Before class, obtain maps of Poland at several different times in history. Organize the class into groups and assign one map to each group. Have each group conduct research on why Poland's boundaries appear as they do on its assigned map. Ask groups to present their findings to the class in chronological order.

FOCUS ON HISTORY

Kaliningrad The Russian exclave of Kaliningrad is slightly larger than the state of Connecticut. The rest of Russia lies more than 200 miles (322 km) away, across Lithuania and Belarus. Lithuania borders the city on the north and east. Poland lies to the south.

Founded by German knights in the 1200s, the city and surrounding area were once part of Germany. The Germans called the city Königsberg (KOOH-niks-berk), or "King's City." Königsberg eventually became the seat of the Prussian government. During World War II, it became a staging area for German attacks against the Soviet Union. When the Soviets defeated Germany, they took control of the area.

The Soviets wanted to remove as much of Königsberg's German culture as possible. They renamed the area Kaliningrad after a Soviet leader. They also destroyed historical sites. Finally, the Soviets forced the German residents to leave. Some went to Germany, while the Soviets sent others to prison camps. Ethnic Russians and other Slavs then moved into the abandoned homes. Kaliningrad became a key base for the Soviet navy. The city is still Russia's only Baltic port that is free of ice all year.

After the collapse of the Soviet Union, Kaliningrad began to build a market economy. A skilled workforce and the port's access to richer European markets may contribute to that goal. In addition, both Russian and foreign companies doing business in Kaliningrad receive special tax breaks that increase profits. Still, some Russian leaders value the city's status as a military base and want to isolate it from foreign influences. Their efforts might slow Kaliningrad's economic growth. In short, the city's future may be determined as much by outsiders as by the people who live there.

✓ **READING CHECK:** *Human Systems* How did Kaliningrad become a Russian city?

The Russians took over the German city after defeating Germany in World War II.

Poland

Nearly all of Poland's people are ethnic Polish and speak Polish, a Slavic language. In addition, the population is overwhelmingly Roman Catholic.

Warsaw is Poland's capital and its transportation hub. The city lies along the Vistula River. Evidence of early settlement on the city's site dates back more than 1,000 years. Warsaw became the capital of the kingdom of Poland in the late 1500s. In the early 1900s Warsaw had the largest urban Jewish population in the world. When the Germans took over Warsaw during their invasion of Poland, they forced the Jews to live in a **ghetto**. A ghetto is a section of a city where a minority group is forced to live. The Jews defied the Germans in the Warsaw Ghetto Uprising of 1943. More than 60,000 Jews died as a result. World War II devastated

Kaliningrad

SCALE
0 — 50 — 100 Miles
0 — 50 — 100 Kilometers
Projection: Conic Equidistant

Baltic Sea

LITHUANIA

Kaliningrad

KALININGRAD (RUSSIA)

POLAND

To Russia

Roads
Railroad

INTERPRETING THE MAP *Kaliningrad is Russia's only port on the Baltic Sea that can be used all year.* **Why would Kaliningrad be of strategic military importance to Russia?**

The Solidarity Party, a labor union that became a political party, led Poland's struggle to break away from communism. Poland was the first country to leave the Soviet bloc. Here an old Soviet monument bears red paint thrown by anticommunist protesters.

Cooperative Learning

Auschwitz Oświęcim, Poland, called Auschwitz in German, was the site of the largest Nazi concentration camp during World War II. More than 4 million people died in the camp complex. Most of the Auschwitz victims were Jews, but thousands of Roma (Gypsies), Poles, and Soviet prisoners of war were also killed.

Organize the class into groups to research the history of Oświęcim since World War II. Encourage students to note the effects of the war on the city's culture.

You don't say! The Wieliczka Salt Mine in Poland contains more than 186 miles (300 km) of tunnels carved entirely out of halite, or rock salt. The mine contains statues, altars, and even a chapel carved from salt.

📶 **internet** connect

GO TO: go.hrw.com
KEYWORD: SW3 CH15
FOR: Web sites about Polish landmarks

MAP ANSWER

provides sea link for railroads and roads into Russia, allows supplies to reach Russia, permits troops and equipment to travel by sea during all seasons

Teach Objective 2

LEVEL 1: Have students create postcards that illustrate economic activities or urban environments in Poland. On the back of each postcard, students should write notes to friends sharing some interesting facts about Poland's cities or economy. Display the postcards around the classroom.

LEVELS 2 AND 3: Have students imagine that the Polish government has chosen them to spearhead a movement to strengthen Poland's economy. Have students work in groups to conduct research on areas of the country's economy that need further development and to make suggestions as to how economic reforms should be implemented.

Teach Objective 3

ALL LEVELS: Organize the class into three groups. Assign each group one of the Baltic countries. Tell groups that they are responsible for representing their assigned countries at an international cultures festival to be held in their community. Direct the groups to identify elements of the Baltic cultures such as language, religion, land use, education, and customs that might be of interest to local residents. Have each group arrange a presentation that describes its assigned country's culture and identifies various influences that have shaped the culture throughout history. Encourage students to include pictures, recordings, or samples in their presentations.

Eye on Earth

Upper Silesia Upper Silesia, a region in southern Poland, is one of the most densely populated areas in Europe. It is also one of the most polluted areas in the world. Thousands of acres of farmland have been ruined by industrial waste, air pollution, and acid rain. Fuel leakage and sewage from former military bases have added to the problem. This pollution has taken its toll on the people of Upper Silesia. Infant mortality rates in the region are the highest in Europe, and the average life expectancy of Silesians is four years lower than that of other Poles. Furthermore, cancer, respiratory diseases, and lead poisoning are much more common in Upper Silesia than any other region in Poland.

ACTIVITY: Have students conduct research on environmental cleanup programs in Central Europe. Have students use their findings to create reports or maps about the state of environmental efforts in the region.

GRAPH ANSWER

Since the beginning of the new market economy, Poland's economy has diversified, resulting in the availability of more consumer products.

Poles play a game of street chess. The people of Poland are proud of their country's diverse cultural heritage. Poland has strong traditions of art, literature, and music. Since the 1950s, film has also become a prominent means of expression in Poland. Polish movie directors have achieved worldwide fame for their work.

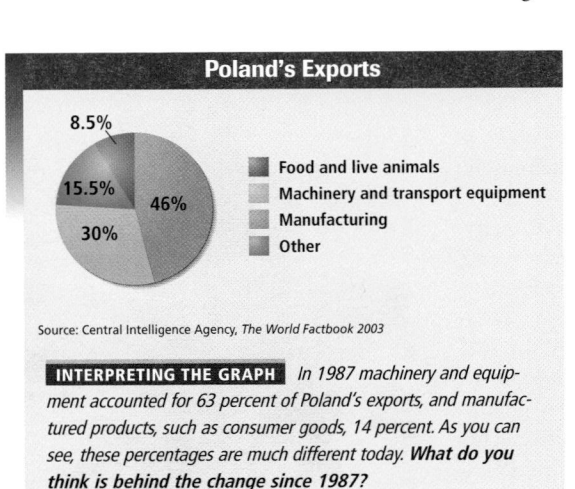

Poland's Exports

- 46% — Machinery and transport equipment
- 30% — Manufacturing
- 15.5% — Food and live animals
- 8.5% — Other

Source: Central Intelligence Agency, *The World Factbook 2003*

INTERPRETING THE GRAPH *In 1987 machinery and equipment accounted for 63 percent of Poland's exports, and manufactured products, such as consumer goods, 14 percent. As you can see, these percentages are much different today.* **What do you think is behind the change since 1987?**

the city. After the war, the Poles rebuilt Warsaw. Today more than 2.2 million people live there.

Farther south along the Vistula lies Kraków (KRAH-kow). The beautiful medieval city has a university, monuments, and museums. Poland's main seaport is Gdańsk (guh-DAHNSK), on the Baltic coast. Gdańsk has been a shipbuilding city since the 1500s.

Poland's economy has made progress since the end of the Communist era. Many successful Polish companies have emerged. In addition, Poland has attracted foreign investment. Auto and glass manufacturing has grown. Still, the traditional coal and steel industries are lagging. Poland's economic future took an upturn in 1997 when the country adopted a new constitution. The constitution committed the country to a free-market economy and to turning over many government-controlled companies to private ownership. In 2004, Poland joined the EU.

Much of Poland's farming activity takes place in productive soils created by thick deposits of loess. Cereals, potatoes, and sugar beets are the main crops. Still, farmers have suffered as the country has moved from communism to capitalism. Many do not have work. Others have moved to the cities to look for jobs.

✓ **READING CHECK:** *Human Systems* How has the operation of Poland's economy changed since the end of the Communist era? building a market economy, attracting foreign investment, Polish companies experiencing growth

The Baltic Countries

During the Middle Ages, two groups of people lived in what are now the Baltic countries. The Balts occupied modern Latvia and Lithuania. Finns from Scandinavia made up the other group. They settled in Estonia. Lithuania remained an independent country for many years, but Latvia and Estonia did not. First Vikings and then German knights called the Teutonic Order invaded and conquered these countries. The Teutonic Knights brought a strong German element into Baltic society. They also helped spread Christianity to the Latvians and the Estonians. Lithuania did not become Christian until later.

Historically, the Baltic Sea was one of the busiest trade routes in northern Europe. People who met in Baltic ports exchanged both goods and information. The influence of this meeting of cultures can still be seen in the Baltic countries. For example, the Estonian language is related to Finnish. Also, like the Finns, almost all Estonians are Lutheran. Latvia has ties to Sweden, a result of a long history of trade between the two countries. Lithuania, on the other hand, is closer culturally to Poland and the Roman Catholic Church. Most Lithuanian folk festivals are tied to church holidays. Folk music is an important part of these festivals. Many of the instruments used are similar to Polish musical instruments. One example is the *cymbaly,* a type of percussion instrument. Also, Russian minorities in each of the Baltic countries keep Russian cultural traditions alive.

Estonia, Latvia, and Lithuania share some challenges. They are trying to rebuild their economies after years of Soviet rule. Their citizens are also cleaning up environmental pollution from the Soviet era. Small populations and limited natural resources make trade essential to all three countries.

✔ **READING CHECK:** (*Human Systems*) Why is trade essential to the Baltic countries?
They have small populations and limited natural resources.

INTERPRETING THE VISUAL RECORD

Riga, the capital and largest city of Latvia, is a port city on the Baltic Sea. The city's architecture and society reflect the influence of many different cultures. Germany, Poland, Russia, and Sweden have all helped to shape Latvian society. **How has the city's location influenced Riga's cultural development?**

Section 3 Review

go. hrw .com
Homework Practice Online
Keyword: SW3 HP15

Define exclave, ghetto

Working with Sketch Maps On the map that you created in Section 2, label Poland, Estonia, Latvia, Lithuania, Kaliningrad, Warsaw, the Vistula River, Kraków, and Gdańsk. In which city did Jews living in the ghetto rise up against the Germans?

Reading for the Main Idea

1. (*Human Systems*) What countries once controlled areas that are now part of Poland?

2. (*Human Systems*) How are Poland's three main cities related to the country's politics, sea trade, and history?

3. (*Environment and Society*) What are two challenges facing the Baltic countries?

Critical Thinking

4. **Making Generalizations** Why might many people in Poland and the Baltic countries want closer ties to Western Europe than to Russia?

Organizing What You Know

5. Draw a map illustrating cultural and historical ties between the countries discussed in this section and other European countries. Note some specific cultural connections, such as language and religion.

OBJECTIVES

1. Identify similarities and differences in the histories of the Czech Republic, Slovakia, and Hungary.

2. Describe the Czech Republic and Slovakia.

3. Explain how the fall of communism has affected Hungary.

LET'S GET STARTED

Copy the following question onto the chalkboard: *What does a country need to do or have to attract tourists?* Discuss student responses. Tell students that the Czech Republic, and to a lesser extent Slovakia and Hungary, became popular tourist destinations during the 1990s. Before that time, few people traveled to the region because the governments were communist and tourist facilities were underdeveloped. Tell students that they will learn more about the region's history and economics in Section 4.

Building Vocabulary

Write the term **complementary region** on the chalkboard. Ask students to identify other instances in which they have seen the term *complementary* used. *(Possible answers: in cooking, where complementary flavors taste good together; in geometry, where complementary angles together measure 90 degrees)* Point out that complementary suggests two or more things that work well together. In a complementary region, two areas' activities or strengths benefit each other.

Section 4

The Czech Republic, Slovakia, and Hungary

READ TO DISCOVER

1. What are some similarities and differences in the histories of the Czech Republic, Slovakia, and Hungary?

2. What are the Czech Republic and Slovakia like today?

3. How has the fall of communism affected Hungary?

WHY IT MATTERS

Czechoslovakia split peacefully into two countries in 1993. Use **CNNfyi.com** or other **current events** sources to learn about other countries that have split apart in recent years.

DEFINE

complementary region

LOCATE

Prague
Bratislava
Budapest

Wood carvings from Hungary

History

The Czech Republic, Slovakia, and Hungary lie south of Germany and Poland. Slavic peoples have long lived in Slovakia and the Czech Republic. About 90 percent of Hungary's people belong to a non-Slavic ethnic group—the Magyars.

The Czech Republic, Slovakia, and Hungary were once part of the Austro-Hungarian Empire. They gained independence after World War I. The Czech Republic and Slovakia formed one country, Czechoslovakia. The union of the two parts of Czechoslovakia had economic advantages. Czech lands had mineral resources and industries. Slovakia was mostly agricultural. Together they formed a **complementary region**. The combining of two areas with different activities or strengths, each of which benefits the other, forms a complementary region. At the time, Czechoslovakia was one of the world's 10 most industrialized countries.

Germany occupied Czechoslovakia during World War II. The Soviet Union then occupied Czechoslovakia and Hungary at the end of World War II and set up Communist governments there. The Soviets invaded Hungary in 1956 and Czechoslovakia in 1968 to keep control of both countries. Soviet control finally ended in the early 1990s. In 1993 the Czechs and Slovaks decided to separate peacefully into two countries.

INTERPRETING THE VISUAL RECORD

Medieval castles and ruins are scattered across Slovakia's landscape. The fortress shown below dates back to the 1200s. **Why were castles like this one built on hilltops?**

✓ **READING CHECK:** (*Human Systems*) In what way did the Czech Republic and Slovakia form a complementary region? The Czech region provided mineral resources and industry, while the Slovak region provided agricultural products.

VISUAL RECORD ANSWER

easier to defend

LEVEL 1: Have students create a list of themes in the history of the countries discussed in this section. (*Possible themes: empire, occupation, American influence, migration, and so on*) Copy the list onto one end of the chalkboard. On the other end, write the names of the three countries described in this section. Call on a volunteer to draw a line from a country's name to an item on the list that applies to its history and explain the connection. Repeat the activity until each country's history has been discussed. Then lead a discussion comparing and contrasting the histories of the Czech Republic, Slovakia, and Hungary and how these different histories are reflected in the countries today. **COOPERATIVE LEARNING**

LEVELS 2 AND 3: Organize the class into three groups and assign each group one of the countries described in this section. Then have each group conduct research to learn about its assigned country's economic and political changes since the collapse of communism. Ask each group to prepare a brief report describing its country's efforts to break away from the Soviet Union, the process by which this break was achieved, and the effects of the break on the country's politics and economy. Lead a class discussion about differences in the countries' experiences. **COOPERATIVE LEARNING**

Austro-Hungarian Empire (before World War I)

Hungary and Czechoslovakia (after World War I)

Hungary, Slovakia, and the Czech Republic (1993)

INTERPRETING THE MAP *Political boundaries in Central Europe have changed significantly in the last century. For example, Czechoslovakia was created after World War I from part of the defeated Austro-Hungarian Empire. In 1993 the Czech Republic and Slovakia separated from each other.*

This division into two countries was not a violent event. **What cultural factors do you think led to the creation of these states as separate countries?**

The Czech Republic and Slovakia

The Czech Republic is made up of the regions of Bohemia and Moravia. About 40 percent of Czechs are Roman Catholic. About the same percentage are not religious. The country's capital and largest city is Prague (PRAHG). Prague is located on seven hills along the Vltava (VUHL-tuh-vuh) River. This historic city was founded more than 1,100 years ago. It is a cultural, university, and tourist center with a rich architectural heritage. The urban region is also the center of the country's major industries.

American culture has increasingly influenced Prague in recent years. With the end of the Communist era, trade and cultural links with the United States grew. A community of American businesspeople and students also began to grow over time. American English-language schools have popped up across the city. In addition, Prague cinemas show the latest releases from Hollywood. Even American fast food is easy to find in the central city.

The Czech Republic is a hilly region with good supplies of coal, iron ore, and uranium. The Czechs became well known for the production of fine steel and glass products. Farms grow mainly cereals and sugar beets. The country's economy had one of the strongest and most stable economies of the former Soviet-bloc countries. However, political and economic problems in the late 1990s slowed progress. Part of the problem was that the government still had too much influence over the economy. Still, the Czech Republic has attracted foreign investment and tourism. The country continues to move toward a market economy. In 1999 it joined NATO, and in 2004 it joined the EU.

Slovakia is the poorer eastern half of the two former regions of Czechoslovakia. The move to a capitalist system there has been hard. Unemployment is high. In joining the EU in 2004, Slovakia hoped to strengthen its economy. The capital and largest

Prague, the capital of the Czech Republic, has become a popular destination for tourists. Historical attractions like the Charles Bridge on the Vltava River are gathering spots for visitors and residents alike.

▶ **Human Systems** ◀

Converting to Capitalism
The economies of both the Czech Republic and Slovakia suffered when the Soviet Union collapsed. The countries were left with outdated industries and environmental problems. Under Soviet rule, Czechoslovakia had been dependent upon the sale of military products to the Soviet army. When the Soviet Union dissolved, so did the market for Czechoslovakian products.

Some newly capitalist Czechs sought to adapt their existing products to new markets. One factory in the city of Nový Jičín had produced battle tanks. After Communist rule ended, workers converted these tanks into remote-controlled fire engines. These machines can go into a fire that is too hot or too dangerous for human firefighters.

ACTIVITY: Ask students to brainstorm or conduct research to find examples of economic or cultural changes that could result from changes in an economic system.

MAP ANSWER

different ethnic groups, languages, cultures, economies

343

Teach Objective 2

 **LEVELS 1 AND 2:** Have each student divide a sheet of paper into two columns with the labels *The Czech Republic* and *Slovakia.* Prepare a series of statements that describe the two countries. As you read each statement aloud, have students write it below the appropriate heading. Then lead a discussion about differences and similarities between the two countries. **ENGLISH LANGUAGE LEARNERS**

LEVEL 3: Have students write paragraphs speculating why relations between the Czech Republic and Slovakia have remained peaceful when situations in other formerly-communist countries have turned violent.

Teach Objective 3

ALL LEVELS: Copy the following graphic organizer onto the chalkboard, omitting the italicized answers. Have students complete it. Then lead a discussion about how students think Hungary's government and economy might develop in the future.

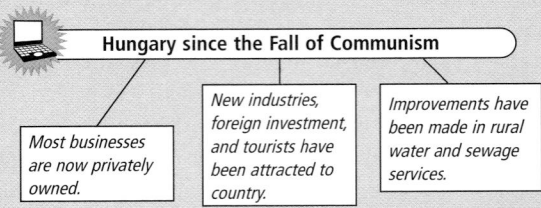

Hungary since the Fall of Communism

Most businesses are now privately owned.

New industries, foreign investment, and tourists have been attracted to country.

Improvements have been made in rural water and sewage services.

Section 4 Review Answers

Define For definition, see: complementary region, p. 342

Working with Sketch Maps Maps will vary, but listed places should be labeled in their approximate locations. Prague, Bratislava, Budapest are the capitals, respectively.

Reading for the Main Idea
1. Czech Republic, Slovakia—Slavic; Hungary—Magyar

2. increased trade with United States and exposure to its language and popular culture; growing community of American businesspeople and students

3. some private ownership of businesses allowed; one of the region's strongest economies

Critical Thinking
4. good market for previously unavailable American consumer products (NGS 11)

Organizing What You Know
5. Prague—cultural and educational center, tourist destination, American influence, industrial; Bratislava—trade center, educational and cultural center; Budapest—transportation center, growing suburbs

CONNECTING TO HISTORY ANSWER

perceptions of what constitutes Central Europe have changed over time; strong functional ties to West, countries share cultural characteristics

MAP ANSWER

transportation hub

Connecting to HISTORY

A Shifting Region

Geographically, the countries discussed in this chapter lie in the middle of Europe, which stretches eastward to the Ural Mountains. However, throughout history, people's perceptions about just what makes up Central Europe have changed.

For example, in the early 1900s Germany was considered part of Central Europe. Then Germany was divided into East Germany and West Germany from the end of World War II until 1990. Communist East Germany, Poland, Czechoslovakia, Hungary, Romania, and Bulgaria were tied to the Soviet Union. These so-called Soviet-bloc countries made up a region known as Eastern Europe. (See the map.)

Today Germany is reunited. Poland, Hungary, and other former Soviet-bloc countries have strong ties to the West. Once again, many people consider these countries part of Central Europe. In fact, in 2004 the seven former Soviet-bloc countries in Central Europe joined the EU. To the east, the former Soviet Union has split into several countries. Perhaps these countries will build stronger ties with the West. Then the boundaries of Central Europe may march farther east.

Making Generalizations In what ways is Central Europe a perceptual region? In what ways might it be a functional or formal region?

Eastern Europe

Legend:
- Central Europe
- Soviet-bloc Eastern Europe, 1988
- ⊛ National capital

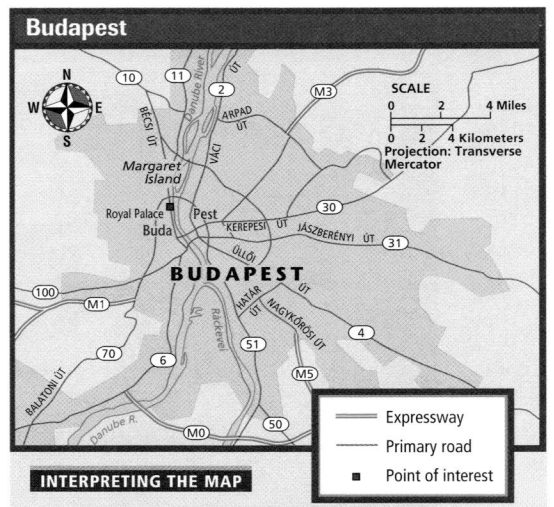

Budapest

Legend:
- Expressway
- Primary road
- ■ Point of interest

INTERPRETING THE MAP

Budapest, the capital of Hungary, benefits from international commerce. Many foreign investors have chosen the city as a base from which to ship their products around the world. **Why do you think Budapest is a suitable location for these enterprises?**

city is Bratislava. It is located on the Danube River and on a major railway junction. This location makes the city ideally located for the trade of both goods and ideas. Although Bratislava lies in the far west of Slovakia, it is the country's educational and cultural center. Institutions from Bratislava stage ballets, concerts, operas, and plays all across the country. Slovaks also have a strong folk culture, which is still evident in Slovakia's art and music.

✓ **READING CHECK:** *Environment and Society* What factor has made Bratislava central to Slovakia's culture and economy? located on important transportation routes, namely the Danube River and a major railway junction

Hungary

The Hungarians speak Magyar, rather than a Slavic language. Magyar is related to Finnish and has its origins in Central Asia. About two thirds of Hungary's population is Roman Catholic.

Hungarians have a rich history of folk and music traditions. Hungarian music is heavily influenced by Roma, or Gypsy, rhythms. To Hungarians, the delivery of a song is as important as the melody and lyrics. Therefore, many Hungarian musicians are gifted actors and dancers as well.

Budapest is Hungary's major city and capital. It is made up of Buda and Pest, which lie across the Danube from each other. The two communities joined in 1873 to form Budapest. Almost one fourth of the country's population lives in or near the capital. Most businesses and industries are also located there. Its location along highways, rail lines, and the Danube help make the city the national transportation center.

Migration from rural areas to Budapest slowed in the 1990s. The slowdown came as rural areas experienced improvements in water, sewage, and other services. Also, many people have been moving out of Budapest to live in the suburbs. The suburbs have fewer problems with crowded housing and pollution. Because of these factors, Budapest's population declined a little in the 1990s.

Hungary is located on a broad agricultural plain in the central Danube Basin. Farming still plays a major role in the country's economy. Potatoes, sugar beets, and wheat are the main crops. Farmers also raise livestock, particularly cattle and hogs. Almost half of Hungary's population lives in small farming villages and towns.

The Communist government began to allow some private ownership of businesses in the 1960s. Over time, private businesses helped the economy grow. Still, after the end of Communist rule in the early 1990s, the rapid move to a market economy was hard. Today, however, most of the country's businesses are privately owned. In recent years Hungary has attracted new industries, foreign investment, and tourists. In fact, it has one of the strongest economies in the region today. Hungary joined NATO in 1999 and the EU in 2004.

Our Amazing Planet

The origins of goulash, a traditional Hungarian stew based on meat, onions, and paprika, date from the A.D. 800s. Then the dish was a cooked mixture that was dried in the sunshine so it could be packed in bags made from sheep's stomachs.

The Great Hungarian Plain accounts for approximately half of Hungary's area. Most of Hungary's farms are located on this fertile plain.

✓ **READING CHECK:** *Environment and Society* What factors affected the growth of Budapest in the 1990s? Population declined as rural services improved and fewer people migrated to the city. Also, many city residents moved to suburbs.

Section 4 Review

Homework Practice Online
Keyword: SW3 HP15

Define complementary region

Working with Sketch Maps On the map that you created in Section 3, label the Czech Republic, Slovakia, Hungary, Prague, Bratislava, and Budapest. What are the capital cities of the Czech Republic, Slovakia, and Hungary?

Reading for the Main Idea

1. *Human Systems* What are the main ethnic backgrounds of the people in the Czech Republic, Slovakia, and Hungary?

2. *Human Systems* What American cultural influences grew in Prague after the end of the Communist era?

3. *Human Systems* What was Hungary's economy like in the 1960s? How strong is it today?

Critical Thinking

4. *Making Generalizations* Why do you think American and other Western companies might want to expand into former Communist-ruled countries like Hungary and the Czech Republic?

Organizing What You Know

5. Create a chart like the one below. Then use it to compare political, economic, and cultural features of Prague, Bratislava, and Budapest.

Prague	Bratislava	Budapest

TECHNOLOGY

▶ Chapter 15 Test Generator (on the One-Stop Planner)
▶ Global Skill Builder CD–ROM
▶ HRW Go site

REINFORCEMENT, REVIEW, AND ASSESSMENT

▶ Chapter 15 Review, pp. 346–47
▶ Chapter 15 Tutorial for Students, Parents, Mentors, and Peers
▶ Chapter 15 Test (form A or B)
▶ Alternative Assessment Handbook

▶ Chapter 15 Test for English Language Learners and Special-Needs Students
▶ Unit 4 Test
▶ Unit 4 Test for English Language Learners and Special-Needs Students

Assess

Have students complete a Chapter 15 Test.

Reteach

Organize the class into groups to represent residents of Central Europe and inquisitive tourists. Have the tourists wander between the groups of locals asking them questions about their countries' histories, cultures, and economies. Have groups change roles until all students have been members of each group. **ENGLISH LANGUAGE LEARNERS**

CHAPTER 15 Review Answers

Thinking Critically

1. Western Germans might see continued development as too expensive and argue that the government has done enough. Eastern Germans might see differences in standards of living and argue that the government has not done enough to improve life in the east. (NGS 6)

2. Many capitals and major cities appear to have grown along rivers like the Danube and the Vistula. These cities' locations probably made them early transportation centers and helped facilitate trade. (NGS 12)

3. Answers will vary. Students should note specific factors, such as Hungary's recent economic progress or the location of other countries nearer to Western markets. (NGS 11)

Using the Geographer's Tools

1. French, German, Italian, Romansh languages; German-speaking; Bern, Basel, Zürich

2. Answers will vary, but students should note economic, social, demographic, and political factors that might account for their rankings.

3. Poland—Austria, Prussia/Germany, Russia; Czech Republic and Slovakia—Austro-Hungarian Empire; Hungary—Austro-Hungarian Empire; Estonia, Latvia—Russia; Lithuania—Russia

---

CHAPTER 15 Review

Building Vocabulary

On a separate sheet of paper, explain the following terms by using them correctly in sentences.

alliances	cantons	exclave
balance of power	neutral	ghetto
confederation	multilingual	complementary region

Locating Key Places

On a separate sheet of paper, match the letters on the map with their correct labels.

Berlin	Vienna	Geneva	Prague
Ruhr Valley	Danube River	Warsaw	Budapest

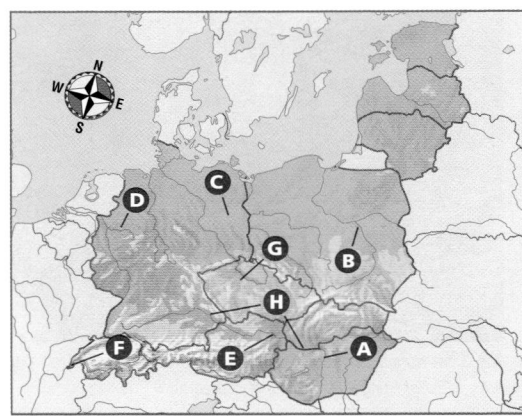

Understanding the Main Ideas

Section 1

1. (*Human Systems*) How did World Wars I and II shape the history of Germany?

2. (*Human Systems*) What problems has eastern Germany faced since reunification?

Section 2

3. (*Human Systems*) Why have many American and Western European companies chosen Austria as a base for their European operations?

Section 3

4. (*Environment and Society*) Why is loess important to Poland's economy?

Section 4

5. (*Human Systems*) How does Hungary's language differ from the languages of neighboring countries?

Thinking Critically

1. Analyzing How might differing viewpoints affect public policies regarding economic development in eastern Germany?

2. Analyzing What role have rivers played in the early development of Central Europe's major cities?

3. Comparing and Contrasting Which of Central Europe's former communist countries, other than the former East Germany, do you think may offer the brightest future for its people? Why?

Using the Geographer's Tools

1. Analyzing Maps Review the map of Swiss language regions. What are the major languages spoken in the country? In which region are three of the country's largest cities located? Which cities are these?

2. Analyzing Tables Use the unit Fast Facts and Comparing Standard of Living tables to rank countries in Central Europe by economic development and standard of living. How may the political histories of the countries have influenced their development and standard of living?

3. Creating Maps Research historical information about Central Europe since 1900. Then create a political map of the region, using colors or other tools to identify former communist countries that once were part of the Austro-Hungarian Empire, Germany, or Russia. Note that the territory of some countries was controlled by more than one of those three powers.

Writing about Geography

People in the former communist countries of Europe are eager to learn about ways of life in other countries. Write a letter to a real or imaginary teenager in one of those European countries. Note similarities and differences you might expect to see between life in this country and in your pen pal's country.

SKILL BUILDING

Geography for Life

Documenting Environmental Change

(*Environment and Society*) Environmental pollution is a legacy of the Communist era in Central Europe. What caused this pollution? Which areas are particularly polluted, and why? How might the pollution be cleaned up? Investigate these questions and write a script for a brief documentary film presenting your findings. Include a script, illustrations, maps, and other visual materials.

Portfolio Activity

1. Folk dancing is still popular in many Central European countries. Have a group of students research the steps to popular Central European folk dances and perform parts of the dances for the class. Take photographs of the performances to place in student portfolios.

2. Most nations display their flags as symbols of national pride. Have each student choose a country discussed in the chapter and design a new flag to represent it. Have students present their flags to the class and explain their designs. Place the designs in student portfolios.

Food Festival

Marzipan is a pliable paste made of ground almonds and sugar. At Christmas, German candy stores offer marzipan candies shaped like animals, fruits, sausages, vegetables, and many other things. Special favorites are marzipan potatoes and little pink pigs with chocolate coins in their mouths. For a winter holiday party, have students mold appropriate objects from mock marzipan: Use 8 ounces softened cream cheese for every 2 pounds sifted powdered sugar. Add almond extract to taste. Knead by hand. Work food color into small batches.

Building Social Studies Skills

Interpreting Graphs

Study the pie graphs below. Then use the information from the graphs to help you answer the questions that follow. Mark your answers on a separate sheet of paper.

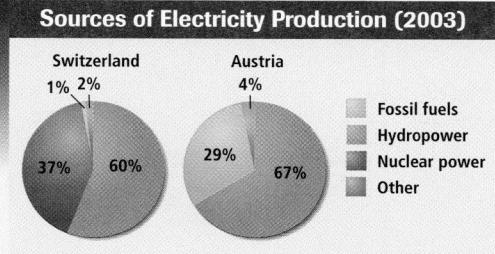

Sources of Electricity Production (2003)

Switzerland
1% 2%
37% 60%

Austria
4%
29% 67%

- Fossil fuels
- Hydropower
- Nuclear power
- Other

Source: Central Intelligence Agency, *The World Factbook 2003*

1. More than 90 percent of Switzerland's electricity is produced by
   a. hydropower alone.
   b. fossil fuels and nuclear power.
   c. nuclear power alone.
   d. hydropower and nuclear power.

2. What is the single largest source of electricity in both Austria and Switzerland? What clues might this answer give you about the physical geography of the two countries?

Building Vocabulary

To build your vocabulary skills, answer the following questions. Mark your answers on a separate sheet of paper.

3. Few countries have remained *neutral* for long periods of time. In which sentence does *neutral* have the same meaning as it does in the sentence above?
   a. The solution that resulted from the experiment was neutral.
   b. The dining area should be painted in a neutral color.
   c. The counselor has remained neutral throughout the dispute.
   d. The car rolled down the hill because it was left in neutral.

4. *Alliance* means the same as
   a. a largely self-governing state within a country.
   b. a group of states with opposing views on world issues.
   c. an agreement between countries for support against enemies.
   d. an agreement between countries to establish new borders for overseas colonies.

Writing

Student reports should demonstrate an understanding of life in Central Europe and in the United States. Use Rubric 9, Comparing and Contrasting, to evaluate student work.

Geography for Life

Students' work should accurately display the results of their research. Suggested solutions for cleaning up pollution should be reasonable and viable. Use Rubric 30, Research, to evaluate student work.

Social Studies Skills

1. d

2. Hydroelectricity is the largest electricity source. Both have rivers with large flows of water that can be harnessed for producing electricity.

3. c

4. c

Alternative Assessment

PORTFOLIO ACTIVITY

Learning about Your Local Geography

Individual Project: Research

Throughout history many peoples, empires, rulers, invasions, and wars have affected the countries of Central Europe. Use library and Internet resources to research the history of your state. Note which countries or peoples have controlled the area over time. Explain how your state's history may have shaped cultural features there today. Write a short report about your findings. Then proofread it to make sure you have used standard grammar, spelling, sentence structure, and punctuation.

internet connect

Internet Activity: go.hrw.com
KEYWORD: SW3 GT15

Access the Internet through the HRW Go site to research the change in global trade patterns of three Central European countries. Then create a database that shows those changes. Use the Holt Grapher to represent your information in graph form. Finally, write a hypothesis to explain the changes that occurred in each country's trade and note the implications of these changes.

PORTFOLIO ACTIVITY

Student projects will vary based on the community but should accurately reflect the history and cultures of your state. Use Rubric 42, Writing to Inform, to evaluate student work.

CHAPTER RESOURCE MANAGER

Objectives	Pacing Guide	Reproducible Resources
SECTION 1 **The Iberian Peninsula** (pp. 349–53)	• Analyze how past events have affected Spain. • Compare and contrast Portugal and Spain.	**Regular** 1 day **Block Scheduling** .5 day *Block Scheduling Handbook,* *Chapter 16*	**RS** Guided Reading Strategy 16.1 **PS** Readings in World Geography, History, and Culture 30 **E** Cultures of the World Activity: Region 3 **SM** Geography for Life Activity 16: Migration to Southern Europe **SM** Map Activity 16: Spain's Autonomous Communities
SECTION 2 **The Italian Peninsula** (pp. 355–59)	• Analyze how Italy's history has affected its culture. • Describe what Italy is like today.	**Regular** 1 day **Block Scheduling** .5 day *Block Scheduling Handbook,* *Chapter 16*	**RS** Guided Reading Strategy 16.2 **PS** Readings in World Geography, History, and Culture 29 **E** Cultures of the World Activity: Region 3
SECTION 3 **Greece and the Balkan Peninsula** (pp. 360–65)	• Analyze how Greece developed into a modern country. • Explain why the western Balkans are politically unstable. • Describe the changes that are occurring in the eastern Balkans.	**Regular** 1 day **Block Scheduling** .5 day *Block Scheduling Handbook,* *Chapter 16*	**RS** Guided Reading Strategy 16.3 **PS** Readings in World Geography, History, and Culture 28 **E** Cultures of the World Activity: Region 3 **SM** Critical Thinking Activity 16: Brokering Bosnia

Chapter Resource Key

PS Primary Sources	**A** Assessment	CD–ROM
RS Reading Support	**REV** Review	Video
IC Interdisciplinary Connections	**ELL** Reinforcement and English Language Learners	Internet
E Enrichment	Transparencies	Holt Presentation Maker Using Microsoft® PowerPoint®
SM Skills Mastery		

 One-Stop Planner CD–ROM

See the *One-Stop Planner* for a complete list of additional resources for students and teachers.

One-Stop Planner CD–ROM

It's easy to plan lessons, select resources, and print out materials for your students when you use the **One-Stop Planner CD–ROM with Test Generator**.

Technology Resources

- One-Stop Planner CD–ROM, Lesson 16.1
- *ARGWorld* CD–ROM
- Homework Practice Online
- HRW Go site

- One-Stop Planner CD–ROM, Lesson 16.2
- Geography and Cultures Visual Resources Activity 58
- CNN Presents World Cultures: Yesterday and Today, Segment 7: In the Shadow of Mt. Vesuvius
- Homework Practice Online
- HRW Go site

- One-Stop Planner CD–ROM, Lesson 16.3
- CNN Presents World Cultures: Yesterday and Today, Segment 5: Crete: Past and Present
- Homework Practice Online
- HRW Go site

Reinforcement, Review, and Assessment

ELL Main Idea Activity 16.1
ELL English Audio Summary 16.1
ELL Spanish Audio Summary 16.1
REV Section 1 Review, p. 353
A Daily Quiz 16.1

ELL Main Idea Activity 16.2
ELL English Audio Summary 16.2
ELL Spanish Audio Summary 16.2
REV Section 2 Review, p. 359
A Daily Quiz 16.2

ELL Main Idea Activity 16.3
ELL English Audio Summary 16.3
ELL Spanish Audio Summary 16.3
REV Section 3 Review, p. 365
A Daily Quiz 16.3

internet connect

HRW ONLINE RESOURCES

GO TO: go.hrw.com
Then type in a keyword.

TEACHER HOME PAGE
KEYWORD: SW3 Teacher

CHAPTER INTERNET ACTIVITIES
KEYWORD: SW3 GT16
Choose a topic on southern Europe and the Balkans to:
• create a postcard of the Mediterranean.
• research conflict in the Balkans.
• understand the diffusion of pizza.

CHAPTER ENRICHMENT LINKS
KEYWORD: SW3 CH16

CHAPTER MAPS
KEYWORD: SW3 MAPS16

ONLINE ASSESSMENT
Homework Practice
KEYWORD: SW3 HP16
Standardized Test Prep
KEYWORD: SW3 STP16
Rubrics
KEYWORD: SS Rubrics

COUNTRY INFORMATION
KEYWORD: SW3 Almanac

CONTENT UPDATES
KEYWORD: SS Content Updates

HOLT PRESENTATION MAKER
KEYWORD: SW3 PPT16

ONLINE READING SUPPORT
KEYWORD: SS Strategies

CURRENT EVENTS
KEYWORD: S3 Current Events

Meeting Individual Needs

Ability Levels

Level 1 Basic-level activities designed for all students encountering new material

Level 2 Intermediate-level activities designed for average students

Level 3 Challenging activities designed for honors and gifted-and-talented students

English Language Learners Activities that address the needs of students with Limited English Proficiency

Chapter Review and Assessment

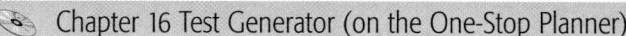

- Chapter 16 Test Generator (on the One-Stop Planner)
- Global Skill Builder CD–ROM
- HRW Go site
REV Chapter 16 Review, pp. 366–67
REV Chapter 16 Tutorial for Students, Parents, Mentors, and Peers
A Chapter 16 Test (form A or B)
A Alternative Assessment Handbook
A Chapter 16 Test for English Language Learners and Special-Needs Students

Southern Europe and the Balkans

Launch into Learning

Ask students to explain the significance of the words *democracy* and *republic* to the United States. *(Both describe the type of government upon which our country is based. A democracy is a form of government in which all citizens can participate. In a republic, the people elect leaders to represent them in the government.)* Ask students where they think Americans got these ideas. Tell them that in the 400s B.C. citizens of the Greek city-state of Athens voted on their government's decisions. Romans began to elect government representatives at about the same time. Tell students that they will learn more about these cultures, what came after them, and the nearby countries of Spain and Portugal in this chapter.

Using the Physical-Political Map

Have students examine the map on the opposite page. Remind them that Europe is often called a peninsula of peninsulas. Point out the Iberian, Italian, and Balkan Peninsulas and call on students to name the bodies of water around them. Ask students how the region's physical geography might have influenced its history and economic activity. *(Possible answer: Access to the sea encouraged the development of trade and travel.)*

Why We Should Know More

These are some reasons that studying southern Europe might interest students:

▶ Many ideas that are basic to the Western world's way of life originated with the ancient Greeks and Romans.

▶ Italian, Spanish, and Portuguese sailors opened the Americas to European exploration and settlement.

▶ Ethnic, religious, and political conflicts in southeastern Europe have involved armed forces from many countries, including the United States.

▶ Artists, musicians, writers, scientists, and scholars from southern Europe have made significant contributions to the world's cultural heritage.

CHAPTER
16

Southern Europe and the Balkans

Gondolier in Venice, Italy

Southern Europe is made up of three large peninsulas. Most of the countries there share a similar physical geography. However, the cultures of the region's countries are very different.

Red-figure pelike (wine container) from ancient Greece

Aupa! (Basque for "What's up?") My name is Nagore Perez España, and I am 17 years old. I live in Bilbao, which is the biggest city in the Basque Country in northern Spain. My parents and I live on the seventh floor of an apartment block that is 14 stories tall. Our balcony overlooks a mountain and a road. Our apartment block is near other apartment blocks just like ours, and there is a big square in between them. We have always played there since we were kids.

I attend the Elorrieta Institute, the local public high school, which is about a 15-minute walk from my apartment. I am in the equivalent of your senior year of high school. I have not decided what to do for a career, but I'm thinking of doing something related to the tourism industry. In Spain there are good possibilities of finding jobs in tourism.

I go skiing with my parents about five times a year and sometimes with my friends as well. In the summer, or when the weather is good, we go climbing mountains and also to the beach. On Saturdays I sometimes eat lunch out with my parents. We prefer the local dishes rather than hamburgers, pizzas, or hot dogs. Other people eat those things for an occasional change of pace. On Sundays I always eat lunch with my parents at my grandmother's. She cooks such good food!

Section 1

OBJECTIVES

1. Analyze how past events have affected Spain.
2. Compare and contrast Portugal and Spain.

LET'S GET STARTED

Copy the following instructions onto the chalkboard: *Look at the chapter map. How would you describe Spain's location? How do you think this location has affected the country's past?* Discuss responses. *(Students may note Spain's proximity to both North Africa and Central Europe as well as its connection by sea to other Mediterranean countries.)* Point out that many peoples have left their marks on Spain. Tell students they will learn more about how Spain's location influenced its culture and history in Section 1.

Building Vocabulary

Write the key terms on the chalkboard and call on volunteers to read the definitions from the text or glossary. Ask them to think of synonyms for **autonomy**. *(Possible answers: self-government, local authority)* Point out that the Greek word *auto* means "self." Have students think of other words that use the same root. *(Possible answers: automobile, automation)* Ask students to use the illustration in Section 1 to describe how **cork** is produced. Discuss its various uses.

Section 1
The Iberian Peninsula

READ TO DISCOVER

1. How have past events affected Spain?
2. How is Portugal both similar to and different from Spain?

WHY IT MATTERS

Spain's tourist industry is one of the largest in the world. Use **CNNfyi.com** or other **current events** sources to learn about historical and cultural sites in Spain or neighboring Portugal.

DEFINE

autonomy
cork

LOCATE

Madrid	Bay of Biscay
Balearic Islands	Lisbon
Strait of Gibraltar	Porto
Barcelona	

Section 1 RESOURCES

REPRODUCIBLE
► Guided Reading Strategy 16.1
► Readings in World Geography, History, and Culture 30
► Cultures of the World Activity: Region 3
► Geography for Life Activity 16: Migration to Southern Europe
► Map Activity 16: Spain's Autonomous Communities

TECHNOLOGY
► One-Stop Planner CD–ROM, Lesson 16.1
► Homework Practice Online
► HRW Go site

REINFORCEMENT, REVIEW, AND ASSESSMENT
► Main Idea Activity 16.1
► English Audio Summary 16.1
► Spanish Audio Summary 16.1
► Section 1 Review, p. 353
► Daily Quiz 16.1

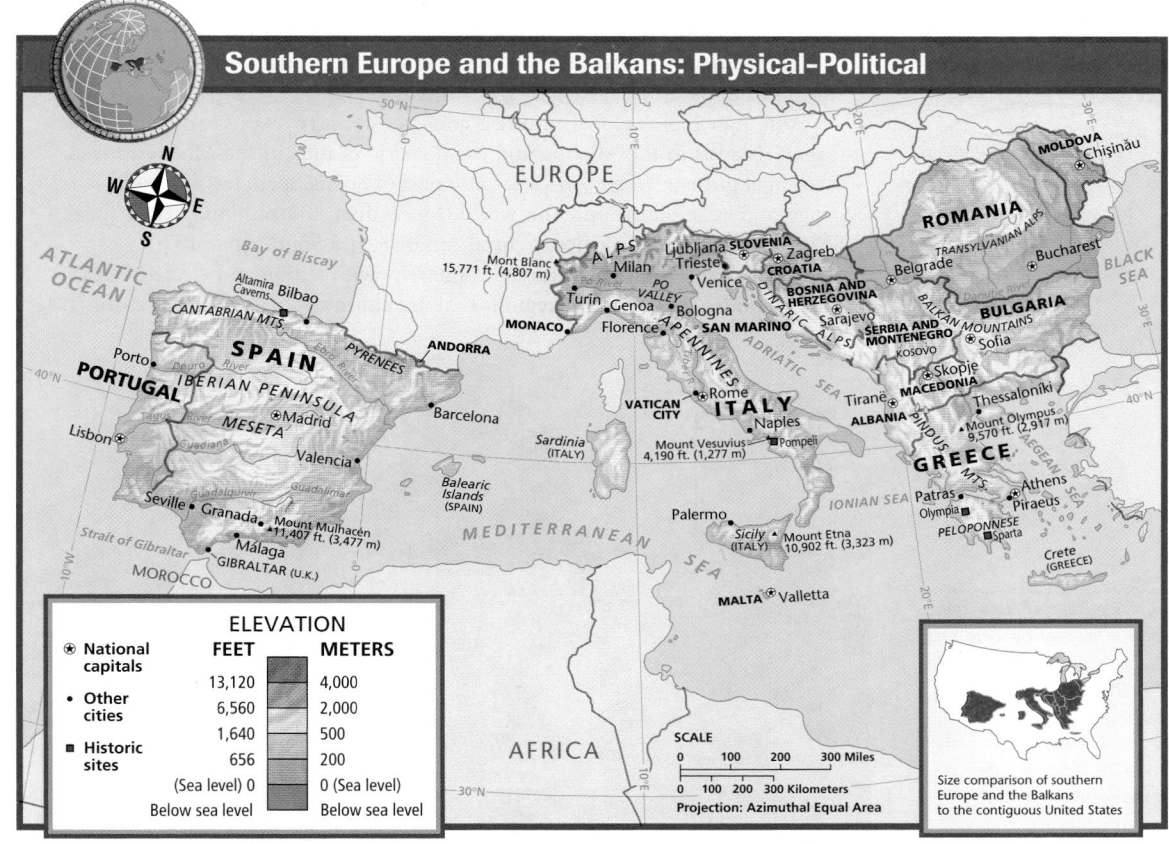

Southern Europe and the Balkans: Physical-Political

ELEVATION

	FEET	METERS
⊛ National capitals	13,120	4,000
• Other cities	6,560	2,000
	1,640	500
■ Historic sites	656	200
	(Sea level) 0	0 (Sea level)
	Below sea level	Below sea level

SCALE
0 100 200 300 Miles
0 100 200 300 Kilometers
Projection: Azimuthal Equal Area

Size comparison of southern Europe and the Balkans to the contiguous United States

ALL LEVELS: Organize the class into several groups and assign each group a period of Spain's history. *(Possible periods include Roman domination, Moorish occupation, the colonial era, and Spain's transition to democracy.)* Then have each group conduct research to create a time capsule that includes elements of the country's culture during its assigned period. Students should locate or draw pictures of typical culture elements like clothing, art, or architecture and write a list of significant accomplishments from their periods. They may wish to find examples of literature from their periods and include these in their capsules as well. Also ask students to determine which cities were important in their time periods and to write descriptions or draw maps of them. When the groups have completed all elements of their time capsules, seal the materials in boxes or envelopes. Then have each group exchange its capsule with one made by another group. Ask students to imagine that they are historians who have just uncovered these time capsules. It is their job to examine the contents to determine which elements of earlier cultures have been carried over into modern Spanish society. Have each group write a script for an interview in which the news media asks members of the group about their conclusions. ENGLISH LANGUAGE LEARNERS, COOPERATIVE LEARNING

Linking Past to Present

Religion and Colonialism
World religion patterns can sometimes provide clues to the cultural history of specific places. For example, the link between religion and colonialism is evident in the Philippines. When these islands were a Spanish colony, members of religious orders called friars acted as local administrators. This "friarocracy" administered public health and education, collected taxes, and supervised local police. The friars also worked to spread Roman Catholicism to the people. The friars' efforts were generally successful. To this day, Roman Catholicism remains a widely practiced religion in the Philippines.

ACTIVITY: Instruct students to conduct research on former Spanish colonies and to describe the effects of Spanish rule on their cultures.

internet connect

GO TO: go.hrw.com
KEYWORD: SW3 CH16
FOR: Web sites about southern Europe and the Balkans

Our Amazing Planet

Some 14,000 years ago, people painted bison, a deer, and horses on a cave ceiling near what is now Altamira, Spain. Today the paintings are treasured as some of the world's greatest artworks.

Spain

Spain and Portugal share the Iberian Peninsula, or Iberia. Spain, southern Europe's largest country in area, covers about 85 percent of Iberia. Much of Spain is surrounded by water. At one time, this feature helped make Spain a great seafaring country.

History and Government The early history of Spain is similar to that of most of the Mediterranean region. Various Mediterranean peoples, including the Romans, have ruled the area. In the A.D. 700s an Arabic people called the Moors invaded Iberia from North Africa. The Moors brought the Islamic religion, new irrigation techniques, and new crops to Iberia. They also built universities and brought many crafts and trades. However, Christian rulers eventually forced the Moors out of Spain. In 1492 the Moors surrendered their last stronghold there—Granada in southern Spain.

During the 1500s Spain used its strong navy to build a worldwide empire. At its peak, the Spanish Empire included most of Central and South America. Spain also ruled what is now the southwestern United States. In addition, the empire included small colonies in Africa and islands in the Caribbean and Pacific. However, Spain had lost almost all its overseas empire by the end of the 1800s. During the 1800s the country was also shaken by wars. Spaniards who wanted a monarchy fought those who wanted a more democratic government. The struggle for power continued in the 1900s and led to a terrible civil war in 1936. The democratic forces lost the war. A dictator, Francisco Franco, then ruled Spain from 1939 to 1975. After Franco's death Spain quickly made a transition to a democratic system of government.

Although Spain's global empire is gone, the country retains control over the Canary Islands in the Atlantic and the Balearic Islands in the Mediterranean. Two small ports in North Africa are also part of Spain. Spain has spread its language and religion around the world. More than 400 million people speak Spanish. Many live in Mexico, Central America, and South America.

Today Spain is a constitutional monarchy with a king and elected legislature. Spain has a number of regions that are culturally or historically distinct.

The Rock of Gibraltar towers over the Mediterranean Sea near the southern tip of the Iberian Peninsula. Captured by British forces from Spain in the early 1700s, Gibraltar is now a British colony and important naval base. It is strategically located near the entrance to the Mediterranean from the Atlantic Ocean.

INTERPRETING THE VISUAL RECORD

Thousands of festivals take place across Spain each year and are important events in local towns and communities. Many festivals in Barcelona feature gegants, *or "giants." These large figures are actually men on stilts underneath elaborate costumes. The* gegants *often represent biblical or historical figures.* **Why might festivals like this one be important to a culture's identity?**

In the past, some of these regions were independent kingdoms. After Franco's death in 1975, some wished to be independent again. To prevent this, Spain's central government gave the country's 17 regions different levels of **autonomy**, or self-government. These regional governments make decisions about health and social programs, urban planning, education, and other local issues. The central government still controls policies for the whole country, such as foreign relations and national defense.

✓ **READING CHECK:** *Human Systems* How did Spain spread the Spanish language and Roman Catholicism around the world? by building a large empire

People and Culture
Nearly all people in Spain today are Roman Catholic. About 75 percent speak Castilian Spanish. Castilian is a dialect from the Castile region around Madrid, the country's capital. Spain's other languages include Basque and Catalan, which are both spoken in regions near the Pyrenees.

Spain's villages and cities have many open spaces for people to meet. The plaza is a common feature in Spanish towns. A plaza is a square surrounded by public buildings, such as a church, a marketplace, and government offices. These squares serve as social gathering places, particularly on warm evenings and weekends. Plazas are also found throughout Central and South America and much of southern Europe.

Moorish influences can still be seen in Spain. For example, Arabic architectural styles are common in many towns, particularly in southern Spain. These styles include horseshoe-shaped arches and geometric decorations. Many natural features and settlements with Moorish place-names still dot the Spanish landscape. (See Geography for Life: Arabic Place-Names in Iberia.)

INTERPRETING THE VISUAL RECORD
In mostly Roman Catholic Spain the church has long played an important role in the educational system. For example, many students attend Catholic schools. However, in recent decades the government has increased its control over Spain's private religious schools. **How does this classroom compare to yours? Do you think parochial schools are as common in your community?**

351

ALL LEVELS: Copy the following graphic organizer onto the chalkboard, omitting the italicized answers. Have students complete it with details about Spain and Portugal. Then discuss points of similarity and difference between Spain and Portugal. **ENGLISH LANGUAGE LEARNERS**

Comparison of Spain and Portugal

	Early history	Language	Religion	Colonial territories	Recent history	Economy
Spain	*Romans, Moors*	*Spanish (Latin root), Basque, Catalan*	*Catholicism*	*Americas, Pacific, Africa*	*Democratic transition, European Union, immigration, independence movements*	*textiles, autos, tourism, agricultural products*
Portugal	*Romans, Moors*	*Portuguese (Latin root)*	*Catholicism*	*Brazil, Angola, Mozambique, Timor*	*European Union, democracy, immigration*	*cork, tourism, wine*

Cultural Kaleidoscope

Basques in America In the 1800s, many Basque settlers came to America and settled in California, Nevada, Oregon, Wyoming, and other western states. They were attracted by freely available land and the discovery of gold in California.

Basque Americans celebrate their cultural roots at festivals where they play traditional sports like *pelota* (similar to jai alai), weight-carrying, and woodchopping. They also dance, compete in bread-baking contests, and attend Roman Catholic masses performed in the Basque language.

You don't say! Black truffles, which are actually fungi that grow under the roots of oak and hazel trees in the Pyrenees, are among the world's most expensive delicacies. Female pigs are used to sniff out the prized fungi.

CHART ANSWER

Both countries are developed, but Spain's development may be more advanced because it appears to export a wider range of goods.

Major Exports of Spain and Portugal

Spain	Portugal
machinery, motor vehicles, foodstuffs, other consumer goods	clothing and footwear, machinery, chemicals, cork and paper products, hides

Source: Central Intelligence Agency, *The World Factbook 2003*

INTERPRETING THE CHART *Spain's most important exports are motor vehicles and a wide range of foodstuffs, such as fresh fruits, olive oil, vegetables, and nuts. Portugal's leading exports include clothing, footwear, and cork.* **What does the information in this chart suggest about the level of development in Spain and Portugal?**

Economy Spain manufactures a variety of products. These goods include textiles and clothing, footwear, ships, and automobiles. The country is also a member of the European Union (EU). Tourism is an important part of Spain's economy. Warm sunny weather and beautiful scenery attract tourists to areas like the Costa del Sol in southern Spain. Many people also visit the Pyrenees and the Balearic (ba-lee-AR-ik) Islands in the Mediterranean Sea. Famous historical and cultural sites are found throughout the country. Although tourism is a major source of income, it has also caused problems. For example, tourism has brought more traffic, pollution, and overbuilding to scenic areas.

Agriculture also plays a major role in Spain's economy. The country is a leading producer of olive oil and wine. Farmers also grow many crops that the Moors brought to the region, such as citrus fruits. The area around Valencia is particularly famous for producing oranges. Other crops, such as corn, potatoes, and tomatoes, were first imported from Spain's American colonies. Spain shipped these crops to many places in its empire. Now crops that were originally American are common in Europe and other parts of the world.

Issues and Challenges Spain's economy has been growing rapidly over the last several decades. This growth further improved when the country joined the EC in 1986. Continued economic development is one of Spain's goals. However, the country still has one of Western Europe's highest unemployment rates.

Immigration has also become an issue. Morocco lies across the Strait of Gibraltar just about 8 miles (13 km) from Spain. Many North Africans cross the strait to find work in Spain and other European countries. They usually move to the big cities, such as Barcelona, Spain's main port. However, many of these immigrants do not find jobs.

Spain also faces political challenges. As you read earlier, Spain's regions have a certain amount of autonomy. Several of those regions used to have active independence movements. However, since the regions received more autonomy, some of those movements have quieted. However, the Basques still work to make their region a separate country. Catalonia borders France and the Mediterranean Sea. The Basque Country lies along the Bay of Biscay and the Pyrenees in northern Spain. The Basques have ancient origins. Their ancestors were among the earliest settlers of Europe. The ancient Basque language seems to be unrelated to any other European language. Some Basques have turned to violence in an effort to win independence from Spain. One group, known as ETA, has killed many government officials and others. Non-Basques in the region, as well as many Basques, oppose ETA and their violent struggle for Basque independence.

✓ **READING CHECK:** **Human Systems** How has Spain addressed the desire for independence in some of its regions? grants 17 regions some form of autonomy

Portugal

Portugal lies in the western part of Iberia and faces the Atlantic. Both Portugal and Spain have many cultural similarities.

History and Culture Much of Portugal's history closely mirrors Spain's. Portugal, too, was under the control of Rome and then the Moors. Like Spanish, the Portuguese language developed from Latin, the language of the Romans. Portuguese also contains many words from Arabic. These words became part of the language when Moors ruled the region. After the Moors were driven from Iberia, the Portuguese built a powerful colonial empire. That empire included parts of Africa, Asia, and South America. Portugal's former colonies include Angola, Brazil, Mozambique, and part of Timor, an island in Southeast Asia. The Atlantic islands of Madeira and the Azores are all that remain of the old empire.

Economy Like Spain, after losing its empire Portugal entered a period of economic decline and limited personal freedoms. Now Portugal's government is democratic, and its economy, helped by membership in the EU, is growing. The country's new freeways, car factories, and high-speed trains reflect this growth. However, even the moderate increase in prosperity has drawn immigrants to Portugal, particularly from North Africa. They crowd into Portugal's cities, which include Lisbon and Porto. Lisbon is the country's largest city and capital. Porto, the second-largest city, is a major seaport in the north.

As in Spain, tourism is important to Portugal. The country is also the world's leading **cork** producer. Cork is a bark that is stripped from the trunks of cork oaks. These trees grow in southern Europe and North Africa. Some types of insulation and flooring are made from cork. However, most cork is used to seal wine and other bottles. Portugal, along with other Mediterranean countries, is a major producer and exporter of wine.

The Cork Industry

The cork oak tree grows to about 65 feet (20 m).

Workers strip cork away from the tree.

Bark is removed from the tree but will grow back over the years.

Cork products

Much of Portugal is covered with cork oak trees, which provide the basis for the country's cork industry. Some trees can provide cork for more than 100 years.

✓ **READING CHECK:** *Human Systems* How are the histories of Spain and Portugal similar?
both ruled by Romans, Moors; built and lost large empires; periods of undemocratic governments, building stronger economies, tourism important

Section 1 Review

Homework Practice Online
Keyword: SW3 HP16

Define
autonomy, cork

Working with Sketch Maps
On a map of southern Europe and the Balkans that you draw or that your teacher provides, label Spain, Portugal, Madrid, Balearic Islands, Strait of Gibraltar, Barcelona, Bay of Biscay, Lisbon, and Porto. Use the description in the text to shade in the Basque Country.

Reading for the Main Idea

1. *Places and Regions* How did Roman and Moorish rule in Spain and Portugal influence the countries' modern cultural landscapes?

2. *Places and Regions* How has Spain influenced the diffusion of foods between the Americas and Europe?

3. *Human Systems* How is immigration affecting urbanization in Spain and Portugal?

Critical Thinking

4. *Analyzing Information* Why has crossing the Strait of Gibraltar been an important migration route?

Organizing What You Know

5. Copy the chart below. Use it to show how economic development, urbanization, and environmental change are shaping Spain.

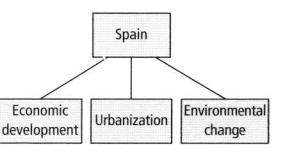

Spain

| Economic development | Urbanization | Environmental change |

History: Tāriq and the Moors

Gibraltar, like many other locations across the Iberian Peninsula, bears a name of Arabic origin. In that language it is called Jebel-al-Tāriq, or "Mountain of Tāriq," named after the general who first led the Moors into Iberia. Within 10 years of his arrival in the early A.D. 700s, Tāriq ibn Ziyād had conquered most of the peninsula.

Spain at that time was ruled by a people known as the Visigoths. When Witiza, their king, died in 710, a group of nobles elected a duke named Roderick to be the new king. This greatly upset the sons of Witiza who thought that the crown should have passed to them. Unable to defeat Roderick on their own, Witiza's sons appealed to the Muslim armies in North Africa for help. Tāriq, who was at that time serving as governor of Tangier, led his army into Gibraltar. There he met and defeated Roderick's forces.

Tāriq, however, was not content with this single victory. He immediately marched toward Toledo, which was then Spain's capital. Within a year he had conquered both that city and Córdoba. His army was soon joined by another under Tāriq's former commander, Mūsa ibn Nuṣayr. By 714, the two generals had conquered more than two thirds of the Iberian Peninsula.

Organize the class into groups and have each group identify another place in southern Europe that was named after a person. Have each group create a resumé or curriculum vitae for the person for whom its place was named, outlining the person's major accomplishments. **COOPERATIVE LEARNING**

Across the Curriculum

▶ **Art** ◀

The Alhambra The Alhambra, the palace-citadel of the Moorish kings in Granada, was built during the 1200s and 1300s. Its name comes from the Arabic word *al-hamrā*, meaning "the red house." This refers to the color of the mountain on which the palace sits and of the bricks of its walls. The Alhambra houses several court complexes, royal residences, a bath, and a mosque. Its courtyards are beautiful examples of Moorish architecture, with arches, columns, domes, fountains, and light-reflecting water basins. One area, the Lion Court, may represent the Islamic paradise as described in poetry.

MAP ANSWER

Possible answers: first areas Moors reached, homes of largest Moorish populations

Applying What You Know Answers

1. provide clues about ethnic groups, invasions, and physical conditions in the past
2. Concentration along southern coast indicates arrival from the south. Few blank areas indicate that conquest was incomplete only in small pockets and northern Spain.

This Geography for Life feature addresses National Geography Standards 4, 10, and 17.

Human Systems

Geography for Life

Arabic Place-Names in Iberia

Settlements and the patterns they form on Earth's surface can provide a historical record to modern researchers. One of the patterns that tells a story is the arrangement of place-names, also known as toponyms. Place-names can provide clues about certain ethnic groups that live in a region or about invasions that have happened there. They may also indicate other information, such as the physical conditions at the time the name was first used.

Many place-names consist of two parts—generic and specific. The generic part refers to features found in many places, like mountains, rivers, or valleys. The specific part refers to something unique to a certain place. For example, consider Battle Creek, a city in Michigan. *Battle* refers to a specific event that happened in 1824. *Creek* is the generic term and refers to a common physical feature.

The study of place-names can also help us learn about places where different groups of people lived before recorded history. For example, some European place-names date back hundreds or even thousands of years. In many places, few written records remain from these eras.

The Iberian Peninsula provides a good illustration of how place-names can be studied to learn about the past. Moors from North Africa invaded the region in the A.D. 700s and ruled for about seven centuries. They brought the Arabic language with them to Spain and Portugal. Although the Moors were eventually driven out by speakers of Romance languages, their language survives in many Iberian place-names. An example is the prefix *guada-*, which derives from the Arabic word *wadi*, meaning "river" or "stream." The prefix *guada-* appears in the names of some major Iberian rivers. *Guadalquivir*, from the Arab name *Wadi al-Kabir*, means "the great river" because *al-Kabir* means "the most great." Similarly, *Guadalimar* comes from *Wadi al-Ahmar* and indicates a red river. Note that these terms also

Arabic Place-Names in Iberia

Each dot = one Arabic place-name

INTERPRETING THE MAP *Arabic place-names are found throughout Iberia but are most common in coastal areas of southern and eastern Spain and southern Portugal.* **What might be some reasons that Arabic place-names are most densely concentrated in these areas?**

contain generic (*guada-*) and specific (*-quivir*, *-imar*) parts. Other Arabic place-names in Iberia include *Madrid*, which comes from the Arabic name *Medshrid* and relates to the abundant wood supply in the region that the Moors used for building.

The many Arabic names in Spain and Portugal indicate the Moors' cultural imprint on the region. In what other ways do you think the Moors influenced Spanish and Portuguese cultures?

Applying What You Know

1. **Summarizing** How do place-names help geographers learn about the past?

2. **Analyzing Maps** Examine the map. What might the map indicate about the direction from which the Moors invaded? What does it tell you about the extent of their empire?

OBJECTIVES

1. **Analyze how Italy's history has affected its culture.**
2. **Describe what Italy is like today.**

LET'S GET STARTED

Copy the following question onto the chalkboard: *What images come to mind when you think of Rome or the Roman Empire?* Discuss responses. *(Possible answers: gladiators, chariot races, the persecution of Christians, and so on)* Tell students that although Rome used military force to build and maintain its empire, it also made lasting contributions to architecture, city planning, engineering, language, law, literature, and other fields. Tell students that they will learn more about both ancient Rome and modern Italy in Section 2.

Building Vocabulary

Write **microstates** on the chalkboard. Circle the prefix *micro-*. Ask students what this prefix means *(small or tiny)*. Then ask them to identify other words with this prefix. *(Possible answers: microchip, microorganism, microscope)* Then ask students what the word *microstates* might mean. Call on a volunteer to locate and read the term's definition.

Section 2

The Italian Peninsula

READ TO DISCOVER

1. How has Italy's history affected its culture?
2. What is Italy like today?

WHY IT MATTERS

Italian Americans are a large ethnic group in the United States. Use CNNfyi.com or other **current events** sources to learn about the history and culture of Italian American communities in this country.

IDENTIFY

Renaissance

DEFINE

microstates

LOCATE

Sicily
Sardinia
Alps
Rome

Locate, continued

Florence
Genoa
Venice
Milan
Turin
Po River
Bologna
Trieste
Naples

Section 2 RESOURCES

REPRODUCIBLE
▶ Guided Reading Strategy 16.2
▶ Readings in World Geography, History, and Culture 29
▶ Cultures of the World Activity: Region 3

TECHNOLOGY
▶ One-Stop Planner CD–ROM, Lesson 16.2
▶ Geography and Cultures Visual Resources Activity 58
▶ CNN Presents World Cultures: Yesterday and Today, Segment 7: In the Shadow of Mt. Vesuvius
▶ Homework Practice Online
▶ HRW Go site

REINFORCEMENT, REVIEW, AND ASSESSMENT
▶ Main Idea Activity 16.2
▶ English Audio Summary 16.2
▶ Spanish Audio Summary 16.2
▶ Section 2 Review, p. 359
▶ Daily Quiz 16.2

History and Culture

Italy occupies the boot-shaped peninsula that stretches southward from the middle of Europe into the Mediterranean Sea. Italy also includes two large islands, Sicily and Sardinia. To the north, Italy is separated from the rest of Europe by the Alps. Despite this barrier, Italy has influenced European culture for more than 2,000 years. Italians have created some of the world's most beloved architecture, literature, music, painting, and sculpture.

History The Etruscans created one of the earliest civilizations to occupy the Italian Peninsula. Later, the Romans set up a republic in central Italy about 500 B.C. Over time the Romans built a huge empire. This empire stretched across much of Europe, North Africa, and Southwest Asia. The city of Rome lay

Our Amazing Planet

In A.D. 79 Mount Vesuvius near present-day Naples erupted and completely buried three Roman towns in volcanic ash. In Pompeii, the largest of the towns, archaeologists have found the remains of more than 2,000 victims.

INTERPRETING THE VISUAL RECORD

The Colosseum in Rome is just one example of the many engineering marvels built by the Romans. Officially dedicated in A.D. 80, the Colosseum was used to stage battles between soldiers, slaves, and wild animals and could hold some 50,000 spectators. **How do you think the Colosseum influenced the architecture of modern sporting arenas today?**

VISUAL RECORD ANSWER

Many modern arenas have been influenced by the seating arrangement, shape, and exterior design of the Colosseum.

LEVEL 1: Copy the following graphic organizers onto the chalkboard side by side, omitting the italicized answers. Have the class fill in details about each of the topics on the organizers. Then call on volunteers to complete the organizers on the chalkboard. Have students compare the information in the two diagrams. Ask them to identify the effects of various historical periods on Italy's culture and to name some elements of Italian culture that have been influenced by the country's history. **ENGLISH LANGUAGE LEARNERS**

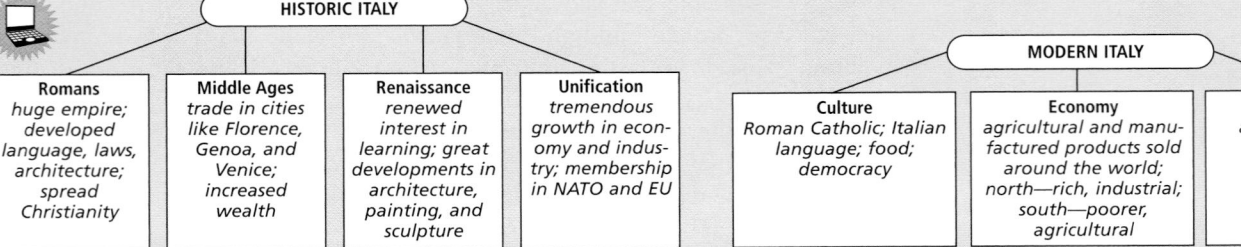

HISTORIC ITALY			
Romans *huge empire; developed language, laws, architecture; spread Christianity*	**Middle Ages** *trade in cities like Florence, Genoa, and Venice; increased wealth*	**Renaissance** *renewed interest in learning; great developments in architecture, painting, and sculpture*	**Unification** *tremendous growth in economy and industry; membership in NATO and EU*

MODERN ITALY		
Culture *Roman Catholic; Italian language; food; democracy*	**Economy** *agricultural and manufactured products sold around the world; north—rich, industrial; south—poorer, agricultural*	**Current issues** *aging population; pollution*

Essential Element 5

► Environment ◄ and Society

Venice In the Middle Ages Venice was a dynamic maritime republic and trading center. The city was built on 118 islands in a lagoon located between the mouths of the Po and Piave Rivers. Bridges connect the major islands, but Venice relies primarily on its waterways for the transportation of people and goods.

Venice faces problems caused by its unusual location. The city is sinking into the sea because its underground aquifers have been drained. Floods caused by high tides and storm winds have damaged many old buildings. Several waterways are overgrown with algae and scum. The Venetian people, the Italian government, and international organizations are working together to protect the city.

CONNECTING TO HISTORY ANSWER

Modern forms of travel allow people to spread diseases unknowingly to places far away.

MAP ANSWER

carried by traders from infected areas to previously uninfected areas; killed millions, changed social and economic structures

---

Connecting to HISTORY

The Plague

A terrible plague, which Europeans came to call the Black Death, swept across Europe in the mid-1300s. Trading ships brought it from Asia to Italian and other southern European ports in 1347. Infected fleas living on rats spread one form, bubonic plague. A second form, pneumonic plague, could also be spread by infected people. Together the two forms spread across almost all of Europe. (See the map.)

A few areas were spared, but the plague killed as many as 30 million people. That total would have been about a third of Europe's population. The many deaths disrupted people's ways of life. Workers, in short supply as so many died, demanded higher wages. Tensions between upper and lower classes increased. In addition, the power and influence of the Roman Catholic Church declined. Some people—terrified by the spreading death—turned to other beliefs and practices for comfort and protection.

Drawing Inferences and Conclusions How might modern technologies allow disease to spread even more rapidly today?

INTERPRETING THE MAP *How did trade routes influence the diffusion of the plague? What effects did the plague have on the regions into which it spread?*

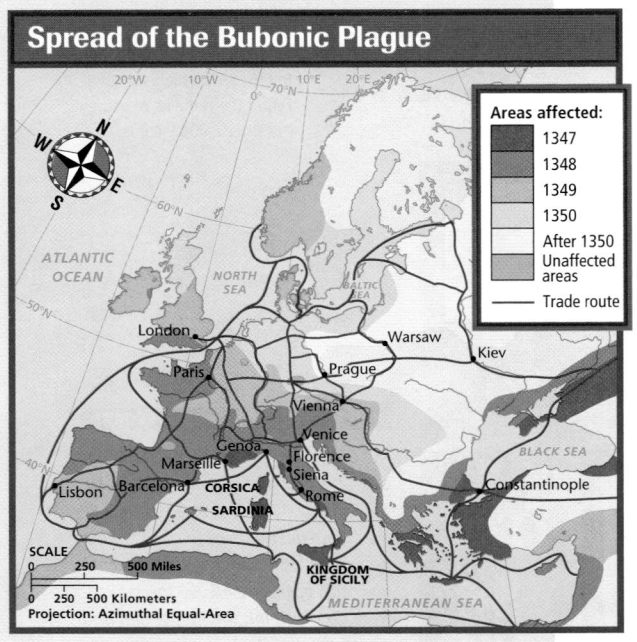

Spread of the Bubonic Plague

Areas affected:
- 1347
- 1348
- 1349
- 1350
- After 1350
- Unaffected areas
- Trade route

---

at its center. It was one of the first cities to have more than 1 million people. Roman culture, including language, laws, architecture, and urban lifestyles, diffused throughout the empire. The Romans also helped spread Christianity throughout Europe. The Western Roman Empire collapsed in the A.D. 400s. Italy remained a mix of separate states and cities for long afterward.

The influences of Roman law, literature, and language can still be seen in many European countries. For example, all the Romance languages—including Catalan, French, Italian, Portuguese, Romanian, and Spanish—are derived from Rome's Latin language. Rome is also the headquarters of the Roman Catholic Church. The seat of the church is at Vatican City, an independent country in the heart of Rome.

During the Middle Ages many Italian cities grew rich from trade. These great cities included Florence, Genoa, and Venice. During this time, northern Italy was one of the wealthiest and most culturally advanced regions in Europe. Trade not only brought wealth to Italy but also new ideas. Almost 1,000 years after the fall of the Roman Empire, Italy became the center of the **Renaissance** [re-nuh-SAHNS]. *Renaissance* is a French word meaning "rebirth." It describes the renewed interest in learning that spread throughout Europe from the 1300s to the 1500s. This time was particularly important for the development of architecture, painting, and sculpture. Some of the world's most famous artists, such as Leonardo da Vinci, Michelangelo, and Raphael, worked during this time. Today millions of people visit Florence, Rome, and Venice to see the great art and architecture of the Renaissance.

Italy did not become a united country until 1861. It fought on the side of the Allies in World War I. In the early 1920s a dictator named Benito Mussolini took control of Italy's government. He formed an alliance with Germany, and the two countries were allies during World War II. However, Mussolini was finally overthrown in 1943, and Italy was later controlled by Allied forces. Since World War II, Italy's economy and industries have grown tremendously. Today the country is a member of NATO and the EU.

✓ **READING CHECK:** *Human Systems* What role did Italy play in the Renaissance, and how does this affect its economy today? center of Renaissance; draws millions of tourists

FOCUS ON GOVERNMENT

Europe's Microstates On a mountain slope in west-central Italy may be Europe's oldest country. The people of San Marino trace their country's history back to the A.D. 300s. At that time a small group of Christians came to the area. There they found a safe place to live and practice their religion. Mountains and wise choices of allies have isolated and protected the country since then.

Today San Marino is surrounded by Italy. It is one of the world's smallest countries—just 23 square miles (60 sq km) in area. In Europe just Monaco in southern France and Vatican City in Rome are smaller. These countries are so small that we call them **microstates**. A number of other tiny countries are also found in Europe. Andorra lies in the Pyrenees between Spain and France. Liechtenstein (LIKT-uhn-shtyn) is located in the Alps between Austria and Switzerland. Malta is a small island country between Italy and Africa.

These countries have survived for a variety of reasons, such as physical isolation and international treaties. They have different kinds of governments, although all but Vatican City have elected legislatures. Vatican City is all that is left of the old Papal States. The Papal States occupied much of central Italy from A.D. 754 to 1870. The pope heads Vatican City's government as well as the Roman Catholic Church. Church officials elect the pope. He then rules with absolute authority for the rest of his life.

Tourism and trade are important to the economies of most of

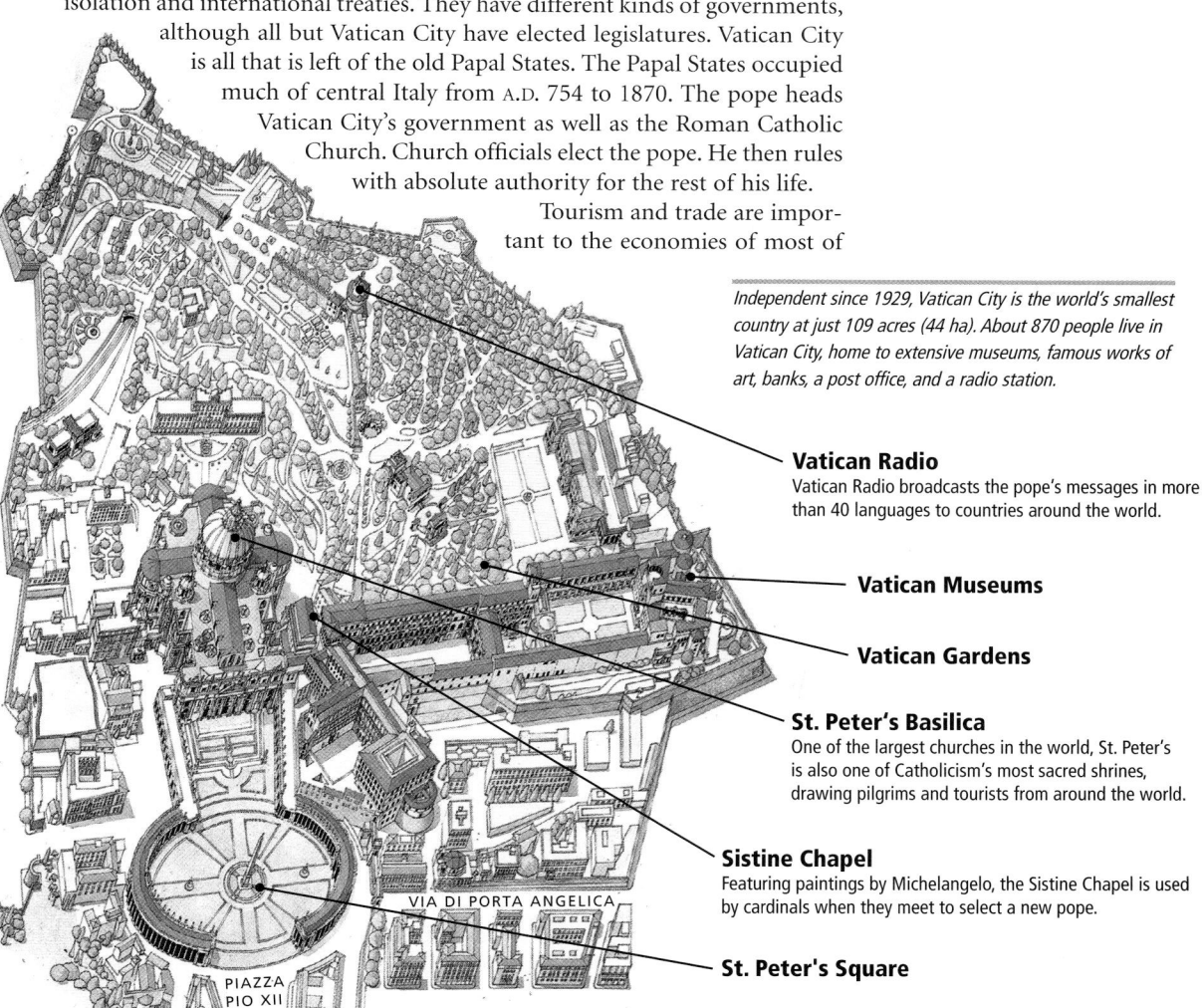

Independent since 1929, Vatican City is the world's smallest country at just 109 acres (44 ha). About 870 people live in Vatican City, home to extensive museums, famous works of art, banks, a post office, and a radio station.

Vatican Radio
Vatican Radio broadcasts the pope's messages in more than 40 languages to countries around the world.

Vatican Museums

Vatican Gardens

St. Peter's Basilica
One of the largest churches in the world, St. Peter's is also one of Catholicism's most sacred shrines, drawing pilgrims and tourists from around the world.

Sistine Chapel
Featuring paintings by Michelangelo, the Sistine Chapel is used by cardinals when they meet to select a new pope.

VIA DI PORTA ANGELICA

St. Peter's Square

PIAZZA PIO XII

LEVELS 1 AND 2: Organize students into groups of three. Have each group write a brief article for a magazine entitled Italy Today. Articles should focus on topics such as Italy's economy, its cities, or the differences between northern and southern Italy. **COOPERATIVE LEARNING**

LEVEL 3: Have each student choose one of the following topics on Italy: culture and customs, important places, industrial economy, agricultural economy, and environmental problems. Then have each student create a poster on the assigned topic, using both words and graphics. Also, have each student write a brief summary of the information shown on his or her poster. Call on volunteers to present their posters and summaries to the class. Display posters around the classroom.

Cooperative Learning

Italian Cuisine Although many people think all Italian cooking is similar, it varies greatly from region to region. Italy's diverse geography makes each region famous for particular agricultural products.

The wide plains of the north are suitable for cattle and grain production, so beef and pasta are important ingredients in this region's cuisine. In the mountainous south, sheep herding and olive cultivation are common, so people use olive oil and cheeses made from sheep's milk in their cooking. Venice's history as a trade crossroads resulted in a cuisine that includes many foreign spices, such as cloves and cinnamon.

Organize students into groups and assign each group a region of Italy. Have the groups conduct research on the cuisine of their regions. Ask them how the cuisine reflects the region's geography. Have groups contribute recipes that would comprise a typical meal in each region. You may wish to have students bring samples to class for an Italian feast.

MAP ANSWER

Milan, Rome, Turin

the microstates. Low taxes have attracted foreign citizens and businesses. Malta is Europe's only island microstate. It has limited freshwater and other natural resources. However, many tourists visit the country, which has some of the oldest stone temples in the world.

✓ **READING CHECK:** (**Human Systems**) How have Europe's microstates managed to survive until modern times? isolation and treaties

People and Culture Most people in Italy are Roman Catholic and speak Italian. Some people in the northern part of the country also speak French, German, or Slovene.

Much of modern Italian culture was first developed during the Renaissance. Italian food may be the most famous part of this culture. Delicious sauces, pastas, sausages, and pastries can be found cooking in almost every Italian home. Many of those foods come from recipes first used by Italian chefs in the 1400s. More modern Italian foods, such as pizza, have also become popular around the world.

Italian daily life is similar to that in other Mediterranean countries. The main meal is in the middle of the day. Afterwards some people rest a while before returning to work. Italians often spend evenings with their friends and family, eating and discussing the latest news.

Central Italy is the country's political and cultural center. There you will find Rome, Italy's capital and largest city. The city has spread out along the banks of the Tiber River. Ruins of ancient Roman buildings still stand there. Among the most famous are the Colosseum and the Forum.

The endurance of Rome's historical buildings stands in contrast to the endurance of its many governments. In fact, Italy has had more than 50 governments since the end of World War II. Some have lasted just a few months. They have been so unstable partly because many political parties are represented in the country's parliament. Since none has a majority, parties must join to form what are called coalition governments. These temporary alliances usually do not last long. Still, Italy has a strong democracy, and its people have many freedoms. These advantages have helped the country make great economic progress in the last half century.

✓ **READING CHECK:** (**Human Systems**) Why has Italy had many different governments since World War II? No political party has a majority in the parliament, so a number of parties must form unstable coalition governments.

Italy Today

Italy is a modern developed country. Its GDP is similar to that of France and Great Britain. Italy produces agricultural and manufactured products that are sold around the world. The most famous of these products are probably Italian automobiles, designer clothes, and fine food.

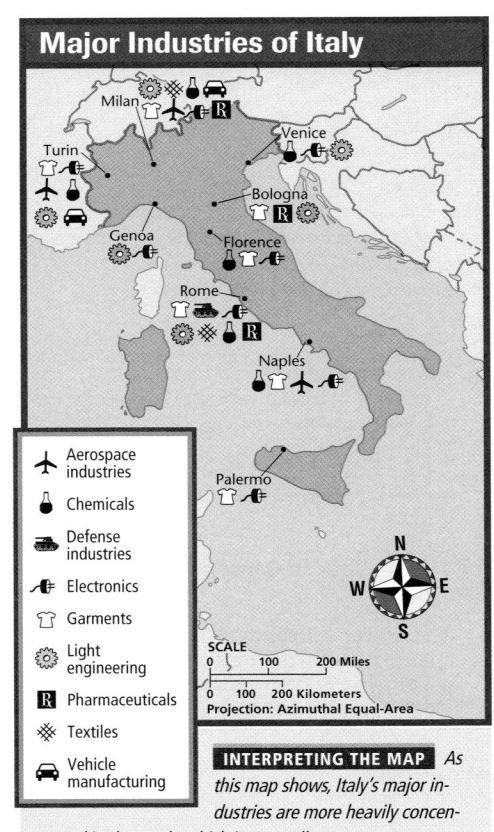

Major Industries of Italy

- ✈ Aerospace industries
- 🛢 Chemicals
- 🚙 Defense industries
- ⚡ Electronics
- 👕 Garments
- ⚙ Light engineering
- Ⓡ Pharmaceuticals
- ✳ Textiles
- 🚗 Vehicle manufacturing

SCALE
0 · · · 100 · · · 200 Miles
0 · · · 100 · · 200 Kilometers
Projection: Azimuthal Equal-Area

INTERPRETING THE MAP *As this map shows, Italy's major industries are more heavily concentrated in the north, which is generally more prosperous and developed than the south.* **Based on the map, which cities seem to be Italy's most important industrial centers?**

Close

Ask students to list the major differences between northern and southern Italy. Then ask students which region they would prefer to visit and why.

Review and Assess

Have students complete the **Section Review**. Then have students complete **Daily Quiz 16.2**.

Reteach

Have students complete the **Main Idea Activity for English Language Learners and Special-Needs Students 16.2**. Have pairs of students create graphic organizers about Italian history and culture. ENGLISH LANGUAGE LEARNERS, COOPERATIVE LEARNING

Extend

Have interested students conduct research on the works of famous Renaissance artists like Leonardo da Vinci, Michelangelo, and Raphael. Then ask students to produce collages that tell about the lives and works of the artists. BLOCK SCHEDULING

North and South Italy has two main economic regions. The north is rich and industrial. The south is poorer and more agricultural. The south is known as the Mezzogiorno (MET-soh-gee-OR-noh), which means "midday" and refers to the area's bright sunshine. The dividing line between the two regions lies just south of Rome.

In the north are the large industrial cities of Milan, Genoa, and Turin and the rich farmlands of the Po River valley. Fertile soils make the Po Valley the "breadbasket" of Italy. To the east and south of the valley are other rapidly developing cities. These cities include Bologna (boh-LOH-nyah), Florence, Trieste (tree-ES-tay), and Venice.

Southern Italy, Sicily, and Sardinia are drier and poorer. They have high poverty and unemployment rates. The south produces farm products such as olives, citrus fruits, and grapes. However, it lags far behind the north in developing a modern economy. Soil erosion and deforestation have long troubled the area. The Italian government and the EU give aid to this region. Still, the economy there has not advanced significantly. Naples is southern Italy's largest city.

Italy's Population, 2000

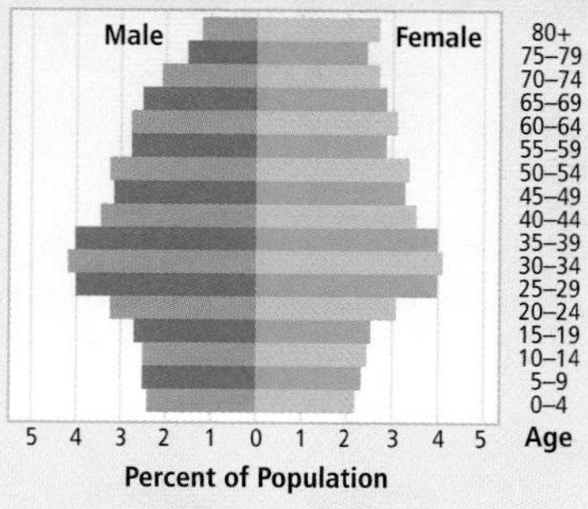

Male
Female
80+
75–79
70–74
65–69
60–64
55–59
50–54
45–49
40–44
35–39
30–34
25–29
20–24
15–19
10–14
5–9
0–4
5 4 3 2 1 0 1 2 3 4 5
Age
Percent of Population

Source: U.S. Census Bureau, International Data Base

INTERPRETING THE GRAPH *In 2000 Italy's rate of natural increase was –0.1 percent, and the country's population was expected to shrink in the coming years.* ***Based on the graph, in about how many years will Italy face the challenge of supporting large numbers of retirees?***

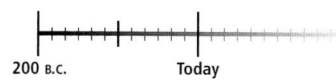

Issues and Challenges Developing southern Italy's economy remains a challenge. Another issue is the country's aging population. Italy has one of the lowest birthrates in the world. As a result, Italy's population is becoming older. This means that there are fewer young workers to replace older workers as they retire. Most of Italy's population growth is from immigration.

Pollution threatens not only Italy's future but also its past. Heavy traffic, smog, and wear and tear have damaged many historical monuments. Neglect has also taken its toll on some of Italy's cultural sites. The government has recognized the problem and is taking steps to protect and preserve important monuments. For example, Rome now limits the number of trucks that are allowed to drive through its historic center.

✓ **READING CHECK:** *Places and Regions* What are three important challenges facing Italy? economic development, an aging population, pollution that threatens historical monuments

Section 2 Review

go.hrw.com **Homework Practice Online**
Keyword: SW3 HP16

Identify
Renaissance

Define
microstates

Working with Sketch Maps On the map that you created in Section 1, label Italy, Sicily, Sardinia, Alps, Rome, Florence, Genoa, Venice, Milan, Turin, Po River, Bologna, Trieste, and Naples. Where is the seat of the Roman Catholic Church?

Reading for the Main Idea

1. *Human Systems* What great empire once ruled the Mediterranean from Italy? How is its influence seen in European culture today?
2. *Environment and Society* What importance does the Po River valley have to Italy?

Critical Thinking

3. **Comparing** How are northern Italy and southern Italy different economically?
4. **Identifying Cause and Effect** How might trade among Italian cities during the Middle Ages have spurred the Renaissance?

Organizing What You Know

5. Create a time line like the one shown below. On your time line, list important years, periods, and events in the history of Italy.

200 B.C.
Today

Section 2 Review Answers

Identify For identification, see: Renaissance, p. 356

Define For definition, see: microstates, p. 357

Working with Sketch Maps Maps will vary, but listed places should be labeled in their approximate locations. Vatican City is the seat of the Roman Catholic Church.

Reading for the Main Idea

1. Roman Empire; legal systems, religion, and Romance languages
2. Italy's "breadbasket" because of its rich fertile farmlands

Critical Thinking

3. northern—more industrial, rich farmlands; southern—poorer, higher unemployment, less industrial, agriculture limited by environmental problems such as soil erosion (NGS 4)
4. Traders carried crafts, other products, and cultural ideas beyond northern Italy. (NGS 17)

Organizing What You Know

5. 500 B.C.—Roman republic founded; A.D. 400s—collapse of the Western Roman Empire; 1300s–1500s—Renaissance; 1861—unification of Italy; 1920s—rise of Mussolini; 1943—Mussolini overthrown

GRAPH ANSWER

about 30 years

OBJECTIVES

1. Analyze how Greece developed into a modern country.

2. Explain why the western Balkans are politically unstable.

3. Describe the changes that are occurring in the eastern Balkans.

 **LET'S GET STARTED**

Copy the following instructions onto the chalkboard: *List as many different ethnic groups that live in the United States as you can. What kinds of challenges may come from many different groups of people living in the same country?* Discuss responses. Point out that fighting between ethnic groups has disrupted the lives of millions in the Balkan Peninsula. Tell students that they will learn more about this volatile region in Section 3.

Building Vocabulary

Write **enclaves** on the chalkboard. Then call on a volunteer to recall or locate a definition for the word *exclave*. Remind students that an exclave is part of a larger country, even though it is separated from the rest of that country. Then have a student locate the definition of *enclaves* in this chapter. Point out that the surrounded area is an independent entity, distinct from the region surrounding it. Unlike exclaves, enclaves are not parts of any larger units. Ethnic enclaves are groups who live surrounded by other ethnic groups.

Section 3

Greece and the Balkan Peninsula

READ TO DISCOVER

1. How did Greece develop into a modern country?
2. Why are the western Balkans politically unstable?
3. What changes are occurring in the eastern Balkans?

DEFINE

city-states
enclave

LOCATE

Crete   Belgrade
Mount Olympus   Sarajevo
Athens   Bucharest
Kosovo

WHY IT MATTERS

The Balkan Peninsula is one of Europe's most volatile regions. In recent decades it has been the scene of much ethnic conflict. Use **CNNfyi.com** or other current events sources to learn about recent efforts toward peace in the region.

Greece

At the southern tip of the Balkan Peninsula lies Greece. The country is made up of many peninsulas, islands, and rugged mountains. The largest Greek island is Crete. The highest peak in Greece is Mount Olympus. It has an elevation of 9,570 feet (2,917 m).

History   Greece was the site of one of Europe's earliest and most advanced civilizations. Civilization there can be traced back more than 2,500 years. Long ago many Greeks lived in a number of powerful **city-states**. A city-state is a self-governing city and its surrounding area. Each city was independent from the others. The people of those cities made great contributions to the arts, government, philosophy, science, and sports. Those contributions have influenced much of Europe and other places. For example, the ancient Greeks developed early systems of democratic government.

INTERPRETING THE VISUAL RECORD

The Parthenon in Athens is one of the finest examples of Greek architecture. Built in the 400s B.C. as a temple to the Greek goddess Athena, the building was later used as a Christian church and as a mosque. **What features of the Parthenon's architecture do you think might have influenced the design of important buildings in this century?**

ALL LEVELS: Copy the following graphic organizer onto the chalkboard, omitting the italicized answers. Call on students to complete the diagram with information about how each aspect of Greek culture has changed through time. Then lead a class discussion about Greece's transition from a conglomeration of city-states to a foreign possession to an independent country to a democracy. **ENGLISH LANGUAGE LEARNERS**

```
                    ┌──────────┐
                    │  Greece  │
                    └──────────┘
         ┌──────────────┼──────────────┐
┌────────────────┐ ┌────────────────┐ ┌────────────────┐
│    Economy     │ │Society and Culture│ │  Government    │
│ once poor and  │ │great contributions│ │birthplace of democ-│
│ agricultural, now│ │to the arts, government,│ │racy, military rule│
│ modernizing and│ │philosophy, science,│ │after 1967, democ-│
│ industrializing│ │and sports; strong│ │racy since 1974 │
│                │ │foreign influences;│ │                │
│                │ │increased opportuni-│ │                │
│                │ │ties for women  │ │                │
└────────────────┘ └────────────────┘ └────────────────┘
```

Over time, Greece fell under the control of outside invaders. The Persians invaded Greece but were defeated. Later, Romans and Ottoman Turks dominated Greece. It was not until 1829 that Greece became an independent country. After World War II the country slipped into a bloody civil war. A series of elected governments followed, but military leaders took over the government in 1967. In 1974 the country returned to democratic government.

People and Culture   Greece's long history of foreign rule influenced its culture. These influences are particularly evident in Greek cooking. Common foods include Turkish dishes such as baklava, a honey-based pastry. However, the Greeks kept their language. In addition, about 98 percent of the population is Greek Orthodox Christian.

Many social changes are taking place in Greece today. Once poor and agricultural, Greece is becoming an industrialized country. In fact, Greece's economy is becoming more like those of northern and central Europe. With an increase in wealth, education and opportunities for women have also increased. Unlike those of earlier generations, modern Greek women seek education and jobs outside the home. In fact, women hold important positions in the country's government, private industries, and universities.

Economy, Issues, and Challenges   Greece has made much economic progress since joining the EC in 1981. However, the country remains relatively poor by European standards. Still, many illegal immigrants enter Greece to fill low-paying jobs. Most are from Albania.

Greece's lack of population growth is much like the situation in Italy and Spain. The country is also urbanizing rapidly. Today about 30 percent of Greeks live in the capital and largest city, Athens. That city's growth has created terrible smog, traffic, and pollution. This smog and other pollution threatens the health of Greeks as well as the ancient monuments that draw tourists. As a result, the government of Athens has passed regulations designed to ease pollution problems.

✓ **READING CHECK:**   *Human Systems*   How has economic change in recent years affected women in Greece? *more educational, political, and economic opportunities*

The Western Balkans

Albania and what was once Yugoslavia make up the mountainous western Balkans. Bosnia and Herzegovina (also called just Bosnia), Croatia, Macedonia, and Slovenia are former Yugoslav republics. Serbia and Montenegro remain united and now use the name *Serbia and Montenegro*.

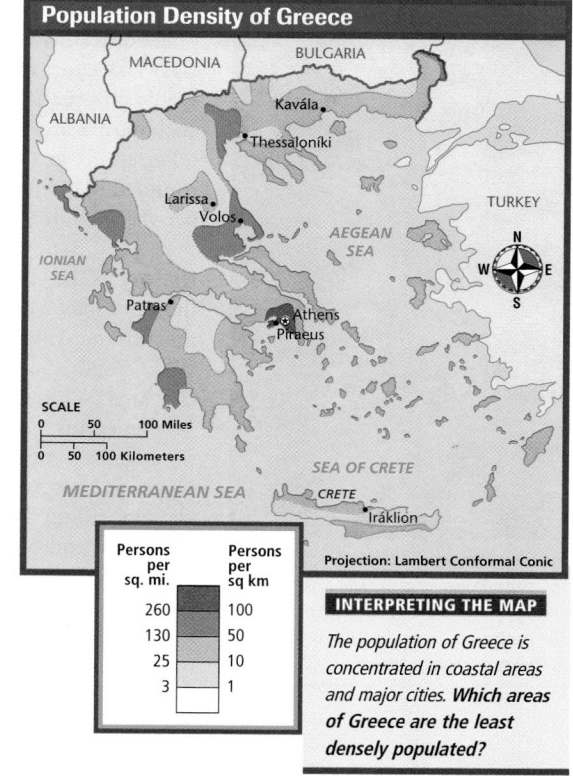

Population Density of Greece

SCALE
0   50   100 Miles
0 50 100 Kilometers

Projection: Lambert Conformal Conic

Persons per sq. mi.	Persons per sq km
260	100
130	50
25	10
3	1

INTERPRETING THE MAP

The population of Greece is concentrated in coastal areas and major cities. **Which areas of Greece are the least densely populated?**

internet connect

GO TO: go.hrw.com
KEYWORD: SW3 CH16
FOR: Web sites about the Olympic Games

MAP ANSWER

mountainous interior

LEVEL 1: Assign or have each student select a country of the western Balkans region. Be sure that each country is selected by at least one student. Instruct students to write a few sentences about what has contributed to political instability in their chosen countries. Then instruct students to compare the information on each country and to note similarities and differences in the political situation in each country. Lead a class discussion about the role of conflict in regional relations. **ENGLISH LANGUAGE LEARNERS**

LEVELS 2 AND 3: Organize the class into four groups and assign each group one of these ethnic identities: Bosnian Muslim, Albanian, Serbian, or Croat. Have groups gather information from the text and other sources about their assigned ethnic identities. Then tell them that the United Nations is sponsoring a meeting during which each group will present its point of view on the regional conflict. Ask each group to think about a possible solution to the conflict and to decide whether there should be one state or many states in the region. Have groups present their arguments and then open the floor for free discussion on relevant issues. **COOPERATIVE LEARNING**

Using Illustrations

The Fall of Belgrade
Direct students' attention to the painting on this page. Tell them that it depicts the fall of Belgrade to Süleyman I, who conquered much of southeastern Europe and greatly strengthened the Ottoman Empire. Point out that this painting is from the *Hunername* manuscript written in the late 1500s by Turkish writer Lokman ibn Seyyid.

Ask students to identify the people depicted in the scene. Do they appear to be Turks or city defenders? On what do students base their opinions? Ask whom they think the seated figure in the foreground may represent. Then ask if the Turks' strategy can be discerned from this picture. Does the illustration depict an eyewitness view of the battle? How might Lokman's perceptions have shaped the picture's content or style?

ACTIVITY: Have interested students conduct research to learn more about the fall of Belgrade. Ask them how this picture compares to written accounts of the battle.

MAP ANSWER

The war caused ethnic groups to group together to control areas; the line reflects these areas of control.

The Ottomans invaded the Balkan Peninsula beginning in the late 1300s and eventually controlled much of the region. This illustration shows the fall of Belgrade to Ottoman forces in 1521.

This area has one of the most diverse human populations in Europe. There one finds a complicated mix of languages and religions. Many ethnic **enclaves** have formed there. An enclave is a territorial or cultural unit, such as an ethnic group, that is surrounded by a different territory or culture. So, ethnic enclaves are areas where one group of people is surrounded by one or more other ethnic groups. For example, Bosnians and Albanians are Muslims surrounded by Eastern Orthodox Christian and Roman Catholic peoples. Those peoples include Serbs, Croats, and Macedonians.

History The Ottoman Turks, who were Muslim, invaded southern Europe in the late 1300s. They conquered most of the Balkan Peninsula. By World War I, most of the Balkans had won independence from the Ottomans. After World War I, Yugoslavia was formed to unite the various Slavic peoples who lived in the area. These peoples included the Bosnians, Croats, Macedonians, Montenegrins, Serbs, and Slovenes. A monarchy governed and held the country together. Albania, which had claimed independence in 1912, remained separate. However, many Albanian people had moved to neighboring countries, including the Kosovo region of Yugoslavia.

At the end of World War II, the Soviet Union occupied all of the Balkan countries but Greece. The occupied countries became communist. Yugoslavia's government was led by the dictator Tito. He kept the country united until his death in 1980. Ten years later communism crumbled in Yugoslavia and the rest of Eastern Europe. Yugoslavia then began to split apart. Slovenia and Croatia declared independence in 1991 and fighting soon broke out among Slovenes, Croats, and Serbs.

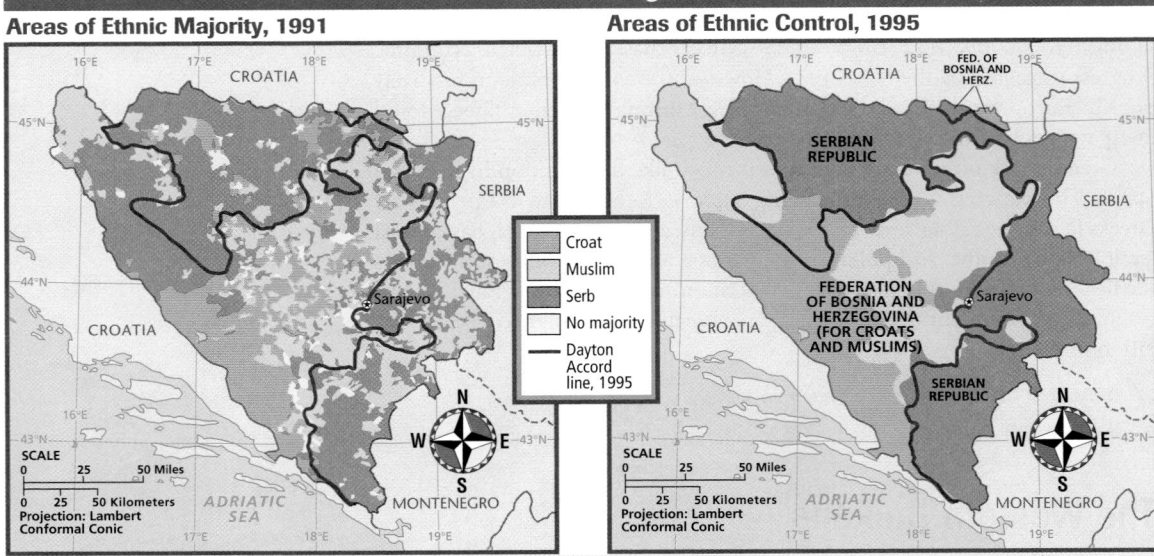

Bosnia and Herzegovina

Areas of Ethnic Majority, 1991

Areas of Ethnic Control, 1995

Legend:
- Croat
- Muslim
- Serb
- No majority
- Dayton Accord line, 1995

INTERPRETING THE MAP *Before the war in Bosnia the region's major ethnic groups were intermixed (left). After the war, however, the country's ethnic map had changed dramatically (right). In 1995 the Dayton Accord established a line between a Serbian-controlled area and* an area controlled by Croats and Muslims. **How has conflict influenced the control of different regions in the country? How might these areas of control have influenced the creation of the Dayton Accord line?**

The city of Mostar was heavily damaged during the war in Bosnia. Before the war, the city's population included many Croats, Muslims, and Serbs. However, today there are no Serbs in the city, and the Croat and Muslim populations are geographically and politically divided.

The fighting was brief in Slovenia, which borders Austria, Hungary, and Italy. However, fighting in Croatia slid into bloody civil war. Serbia sent weapons and supplies to the Serb rebels in Croatia. When Bosnia declared independence in 1992, civil war broke out there too. Bosnian Muslims, Serbs, and Croats all fought to control the country or parts of it.

Serbs in Croatia and Bosnia refused to accept separation from the rest of Yugoslavia. They did not want to live in countries controlled by people of other ethnic groups. The government in Serbia supported them. However, when the terrible fighting ended in late 1995, Croatia and Bosnia had won independence. Keeping the peace has not been easy.

In 1997 periodic fighting between ethnic Albanians and Serbs in Kosovo got much worse. Tensions between the mostly Muslim Albanians and Christian Orthodox Serbs stretch back centuries. The United States and its NATO allies intervened to stop the fighting in 1999. Kosovo's future is still uncertain.

During the war in Kosovo, many ethnic Albanians fled to Albania. These refugees were forced to leave for their own safety. The resulting population increase further strained the limited resources of Albania, Europe's poorest country. Today Albania continues to adjust from a long period of isolation under its old Communist government. That government lost power after elections in 1992.

People and Culture   As you can see, culture in the western Balkans is a complex mix of religions, ethnic backgrounds, and political views. For example, the Serbs and Macedonians are Slavic. They practice an Orthodox Christian religion. The Croatians are also Slavic, but most are Roman Catholic. Most Albanians are Muslim. All of these peoples wish to preserve their particular heritage.

Belgrade is the largest city in the region. It and Sarajevo (sar-uh-YAY-voh)—the once beautiful capital of Bosnia—have been heavily damaged by war. Many cities and towns across the region must be rebuilt. Also, many people who fled Croatia and Bosnia during the fighting want to return to their old homes. However, it is not clear whether this will be possible.

LEVELS 1 AND 2: Ask students to imagine that they are former residents of one of the eastern Balkan countries who moved to the United States immediately after the collapse of Communist control in the early 1990s. Then tell them to imagine that they are returning to their homelands to visit relatives. Have students write letters to their families in the United States explaining how things have changed in the Balkan region since their departure. Encourage students to conduct outside research to find specific examples of cultural or economic changes. Call on volunteers to share their letters with the class. **ENGLISH LANGUAGE LEARNERS**

LEVEL 3: Organize the class into three groups and assign each group one of the countries of the eastern Balkans. Then have each group assume the role of advisers from the United Nations who have been sent to the assigned country to study its recent economic development. Have the groups conduct research to learn about economic activities and industries that have been implemented in their assigned countries since the decline of communism. Then have each group prepare a series of maps and charts that illustrates its findings. Ask the groups which economic activities they believe are in need of further development. Call on volunteers from each group to share their findings with the class. **COOPERATIVE LEARNING**

Section 3 Review Answers

Define For definitions, see: city-states, p. 360; enclave, p. 362

Working with Sketch Maps Maps will vary, but listed places should be labeled in their approximate locations. Athens is the capital of Greece.

Reading for the Main Idea
1. early form of democratic government

2. Republics are dominated by certain ethnic groups and have split from the former country of Yugoslavia.

3. command to market economies

Critical Thinking
4. Possible answer: Serbs did not want to be a minority in the new countries. Croats and Bosnians may have wanted to make decisions for themselves without interference from other republics in the former Yugoslavia. (NGS 13)

Organizing What You Know
5. end of communism; independence in Slovenia and Croatia; fighting among Slovenes, Croats, and Serbs; independence in Bosnia; fighting among Albanians and Serbs in Kosovo; intervention by United States and NATO to stop the fighting

Economy, Issues, and Challenges Slovenia has rapidly built up trade with the EU. It has also attracted foreign businesses and tourists. Coal, oil, minerals, and an electronics industry support its growing economy. The country also promotes tourism, which it hopes to develop further. However, other countries in the area have had major problems. War, corruption, and a lack of modernized industries have left them with weak economies.

Continuing unrest makes this area's economic future uncertain. The United States and other NATO countries have peacekeeping troops there. Recent government changes in Serbia and Montenegro may also help. Many observers hope the new government will focus on solving economic problems and calming ethnic tensions.

✓ **READING CHECK:** *Human Systems* What characterizes the cultural geography of the western Balkans? complex mix of ethnic groups, religions, and political views

The Eastern Balkans

Bulgaria, Romania, and Moldova lie in the eastern Balkans. Those countries have not experienced the ethnic fighting seen in the western Balkans. Therefore, they are more politically stable than their neighbors. However, as in other formerly communist countries of Eastern Europe, economies are underdeveloped, and standards of living are lower than in other regions of Europe.

History and Culture The countries of the eastern Balkans have been controlled by the Roman, Byzantine, and Ottoman Empires. The region gained independence from the weakened Ottoman Empire during the late 1800s and early 1900s. After World War II the area came under the control of the Soviet Union. Communist control ended in the early 1990s.

Bulgarians are a Slavic people and share some culture traits with other Slavs. Like Russians, for example, Bulgarians use the Cyrillic alphabet. Most people in Bulgaria follow the Bulgarian Orthodox religion. In Romania about 7 percent of the population is ethnic Hungarian, but the majority is ethnic Romanian. They speak a language derived from Latin. Most Romanians follow the Romanian Orthodox religion. Most of Moldova's people are closely related to Romanians and are Eastern Orthodox Christians.

Many people in Romania live in rural areas and suffer from a lack of access to modern technology. In fact, nearly half of the country's population is rural. In general, farmers work their fields without the help of modern farm machinery. Also, there is a lack of many consumer goods, which means that people rely on homemade goods for some of their needs.

Close

Ask students to list some obstacles to future peace and prosperity in the Balkan region. Then have the class debate which issues present the greatest challenges to the region's people.

Review and Assess

Have students complete the **Section Review**. Then have students complete **Daily Quiz 16.3**. COOPERATIVE LEARNING

Reteach

Have students complete the **Main Idea Activity for English Language Learners and Special-Needs Students 16.3**. Then have students work in groups to create illustrated maps that depict the culture, history, and economy of the region. Display and discuss the maps. ENGLISH LANGUAGE LEARNERS

Extend

Have interested students conduct research on the Danube River to learn about its cultural importance in this region. Have students find literature, art, and music describing or relating to the river and bring examples to class. BLOCK SCHEDULING

CHAPTER
16 Review Answers

Building Vocabulary For definitions, see: autonomy, p. 351; cork, p. 353; Renaissance, p. 356; microstates, p. 357; city-states, p. 360; enclaves, p. 362

Locating Key Places

A. Kosovo
B. Athens
C. Rome
D. Strait of Gibraltar
E. Sicily
F. Balearic Islands
G. Bucharest
H. Madrid

Understanding the Main Ideas

1. introduced Spanish language, Roman Catholicism, and foods to Americas
2. both ruled by Romans, Moors; built and lost large empires; building stronger economies
3. northern Italy; rich farmland, more large cities
4. increased smog and other forms of pollution
5. conflict among various ethnic groups there

Economy, Issues, and Challenges The collapse of communism brought great changes to the eastern Balkans. Moldova, Bulgaria, and Romania are now working to adjust to democracy and a free-market economy. Moldova is a small densely populated country with rich soils and a mild climate. The struggling economy is based on producing fruit, grains, and wine. About one fourth of Bulgaria's labor force works in agriculture. However, the government has made efforts to attract modern industries. Romania is trying to expand an economy based on agriculture, oil, coal, and low-technology industries. Bucharest is the country's capital and major industrial center.

Most people in the eastern Balkans have living standards well below those of other Europeans. Many people of these countries lack health care and safe water, particularly in the villages. A lack of housing in urban areas is also a problem. Many people are leaving to find work in other countries. As a result, Bulgaria's population is decreasing.

INTERPRETING THE VISUAL RECORD

The huge Danube River delta in Romania has many strips of land called grinduri *that farmers use to grow crops. Reeds that grow in the region's shallow waters are also used to make paper and fibers for textiles.* **What challenges might farmers in this region face?**

✓ **READING CHECK:** *Human Systems* How have the economies and politics of the eastern Balkans changed in recent years? *movement from communist to free-market economies and democracy*

Section 3 Review

Homework Practice Online
Keyword: SW3 HP16

Define
city-states
enclave

Working with Sketch Maps On the map that you created in Section 2, label Greece, the Balkan countries, Crete, Mount Olympus, Athens, Kosovo, Belgrade, Sarajevo, and Bucharest. In the margin of your map, identify the capital of Greece.

Reading for the Main Idea

1. *Human Systems* How did ancient Greece contribute to modern government?

2. *Places and Regions* How has cultural conflict shaped the political boundaries of the western Balkans?

3. *Human Systems* How are the economies of the eastern Balkan countries changing?

Critical Thinking

4. *Identifying Points of View* Why do you think ethnic Serbs living in Croatia and Bosnia and Herzegovina opposed independence from Yugoslavia? Why do you think Croats and Bosnians may have wanted independence even at the risk of war?

Organizing What You Know

5. Copy the flowchart below and use it to trace events in Yugoslavia (now Serbia and Montenegro) since the death of Tito in 1980.

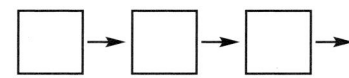

VISUAL RECORD ANSWER

flooding

TECHNOLOGY

▶ Chapter 16 Test Generator (on the One-Stop Planner)
▶ Global Skill Builder CD–ROM
▶ HRW Go site

REINFORCEMENT, REVIEW, AND ASSESSMENT

▶ Chapter 16 Review, pp. 366–67
▶ Chapter 16 Tutorial for Students, Parents, Mentors, and Peers
▶ Chapter 16 Test (form A or B)
▶ Alternative Assessment Handbook

▶ Chapter 16 Test for English Language Learners and Special-Needs Students
▶ Unit 4 Test
▶ Unit 4 Test for English Language Learners and Special-Needs Students

Assess

Have students complete a Chapter 16 Test.

Reteach

Organize the class into three groups and assign one of the chapter's sections to each group. Have each group create an outline of the major concepts in the assigned section. Then instruct groups to exchange papers and add detail to the outlines they receive. Rotate papers again. Call on students to read parts of the expanded outlines. **ENGLISH LANGUAGE LEARNERS**

CHAPTER 16 Review Answers

Thinking Critically

1. They have their own languages, history, and culture. They might view the rest of Spain as a foreign country. **(NGS 6)**

2. Northern Italy's location eases trade and the exchange of ideas, which spurs economic growth. **(NGS 11)**

3. history of Communist rule, ethnic conflict and warfare, pollution and environmental problems **(NGS 4)**

Using the Geographer's Tools

1. The diagram is not in the form of a pyramid. The base is narrow, which indicates that the proportion of young people is small.

2. Iberian Peninsula—Spain, Portugal; Italian Peninsula—Italy, San Marino, Vatican City; Balkan Peninsula—Albania, Bosnia and Herzegovina, Bulgaria, Croatia, Greece, Macedonia, Moldova, Romania, Slovenia, Yugoslavia (now Serbia and Montenegro)

3. Graphs should accurately reflect the information in the table.

Writing

Student reports should describe cultural patterns in Canada and Spain. Comparisons between countries should be logical and accurate. Use Rubric 9, Comparing and Contrasting, to evaluate student work.

CHAPTER 16 Review

Building Vocabulary

On a separate sheet of paper, explain the following terms by using them correctly in sentences.

autonomy
cork
Renaissance

microstates
city-states
enclave

Locating Key Places

On a separate sheet of paper, match the letters on the map with their correct labels.

Madrid
Balearic Islands
Strait of Gibraltar
Rome

Sicily
Kosovo
Bucharest
Athens

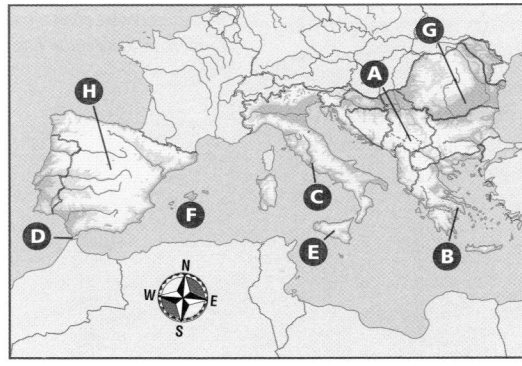

Understanding the Main Ideas

Section 1

1. (*Human Systems*) How have events in Spain's history influenced other areas of the world?

2. (*Human Systems*) How does Portugal's history mirror Spain's?

Section 2

3. (*Places and Regions*) Which of Italy's two major regions is the richest and most industrialized? Why?

Section 3

4. (*Human Systems*) How has urbanization contributed to environmental problems in Greece?

5. (*Human Systems*) What factors have shaped the boundaries of Yugoslavia (now Serbia and Montenegro)?

Thinking Critically

1. **Making Generalizations** Why might some people in the Basque Country and Catalonia see themselves more as Basque and Catalan than Spanish? How might this affect their view of the rest of Spain?

2. **Drawing Inferences and Conclusions** How might northern Italy's location near central Europe have affected its economic growth?

3. **Analyzing Information** What factors have limited economic development and improvement in the standard of living in the Balkan countries?

Using the Geographer's Tools

1. **Analyzing a Population Pyramid** Study the population pyramid for Italy. What features tell you that Italy's population is decreasing?

2. **Creating Maps** Use the unit atlas to create a map of Europe. Identify which countries make up the Iberian, Italian, and Balkan Peninsulas.

3. **Creating Bar Graphs** Use statistics from the unit Fast Facts table to construct a bar graph comparing per capita GDP in the countries of southern Europe and the Balkans.

Writing about Geography

Compare patterns of culture in Spain and Canada. How has each country been affected by cultural differences among its people? How have those differences affected political issues, boundaries, public policies, points of view, and events in each? You might want to do further research by using secondary sources of information like magazines. Write a short report about your findings. When you are finished with your report, proofread it to make sure you have used standard grammar, spelling, sentence structure, and punctuation.

SKILL BUILDING

Geography for Life

Creating an Elevation Profile

Places and Regions Create an elevation profile of Italy. The profile should show the elevation of features along a line from west to east across the peninsula, just north of Rome. Identify important points along the line. You can use the chapter map as well as atlas maps from your classroom and library to prepare your diagram.

Have students conduct research on one of the region's cities, like Madrid, Rome, Athens, Belgrade, or Sarajevo. Then have students create scrapbooks about the city that include information about its history and daily life, including how life has changed in recent years. The scrapbooks should include pictures, drawings, poetry, time lines, short stories, or other projects about the lives of city residents. Place scrapbooks in student portfolios.

Food Festival

Dolmas, also called dolmades or *dolmathes*, are stuffed grape leaves and a Greek specialty. The leaves are sold in jars. Dip 1 lb. grape leaves in boiling water, rinse in cold water, and wipe dry. Mix 1 lb. finely chopped onion with ¼ c. olive oil. Mix in 1 c. uncooked rice, 7 fl. oz. hot water, 1 bunch dill (chopped), and 1 bunch mint (chopped). Boil about five minutes. Wrap a tablespoonful of the mixture in a grape leaf. Repeat with remaining mixture and leaves. Place dolmas in a pan, with space around them. Add ¼ c. olive oil, juice of one lemon, and 15 fl. oz. water. Simmer covered at low heat for 30 minutes, until most of the water is absorbed and the rice is tender. Serve cold with lemon slices.

Building Social Studies Skills

Interpreting Graphs

Study the graph below. Then answer the questions that follow.

1. Which country is projected to lose the most people by 2050?
   a. Italy
   b. Spain
   c. Greece
   d. Portugal
2. Describe the projected general population trend in the region between now and 2050.

Building Vocabulary

To build your vocabulary skills, answer the following questions.

3. *Autonomy* means
   a. economic self-sufficiency.
   b. a large cooperative group.
   c. self-government.
   d. a territory or area of influence.
4. An *enclave* is
   a. a very small country.
   b. an extreme shortage of food.
   c. a legal restriction against the movement of freight.
   d. an area completely surrounded by another region.

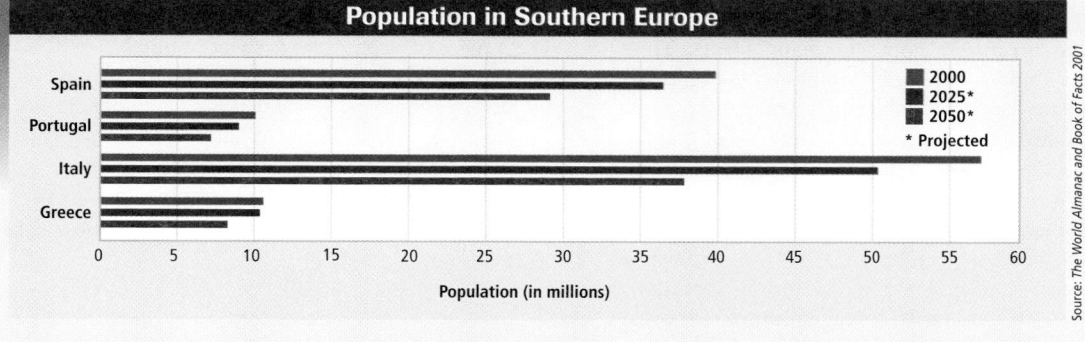

Population in Southern Europe

Legend: ■ 2000 ■ 2025* ■ 2050* * Projected

Population (in millions): 0, 5, 10, 15, 20, 25, 30, 35, 40, 45, 50, 55, 60

Source: The World Almanac and Book of Facts 2001

Geography for Life

Elevation profiles should accurately reflect Italy's physical landscape. Use Rubric 7, Charts, to evaluate student work.

Social Studies Skills

1. a
2. Possible answer: All countries are projected to have smaller populations by 2050.
3. c
4. d

Alternative Assessment

PORTFOLIO ACTIVITY

Learning about Your Local Geography

Group Project: Field Work

With a group, plan, organize, and complete a research project about features in your community that might be developed into popular tourist attractions. Identify scenic locations, historical sites, cultural events, and geographic features—such as an appropriate climate—that might draw tourists. Prepare a questionnaire to survey area residents about the attractions in your community. Present your findings in a tourism-guide brochure for your community.

⤷ internet connect

Internet Activity: go.hrw.com
KEYWORD: SW3 GT16

Choose a topic on southern Europe and the Balkans to:

- create a postcard of the islands and peninsulas of the Mediterranean.
- research conflict in the Balkans.
- understand how geographic factors and global trade influenced the development of pizza as a common food.

PORTFOLIO ACTIVITY

Students' projects will vary based on the area selected. Questionnaires should include specific, answerable questions. Use Rubric 14, Group Activity, to evaluate student work.

Workshop 1
Going Further: Thinking Critically

Point out to students that there are many types of graphic organizers, but that different types are not interchangeable. Different styles of graphic organizers are best suited to specific purposes. For example, if students want to visualize the relationships among events or ideas, idea webs are most appropriate. If the aim is to arrange a series of events in order, a time line or sequential step diagram is better. Charts or bulleted lists are best used for categorizing or comparing information, as are Venn diagrams. Other graphic organizers are useful for drawing conclusions or making generalizations. The type of organizer students create should depend on the information they are trying to present.

Organize the class into groups and assign each group a different type of graphic organizer. Then have each group prepare an organizer of its assigned type using information about life in your community. Have students compare their finished organizers to see how the content of each varies. For example, one group may have created a time line that describes the community's history, while another may have created a chart that lists important economic activities there. Lead a class discussion about how the information contained in each graphic organizer is suited to the style in which it is organized.

PRACTICING THE SKILL

Students' graphic organizers should illustrate relationships among students' perceptions of Spain's physical environments, population, and economy. Connections between ideas should be clear and logical.

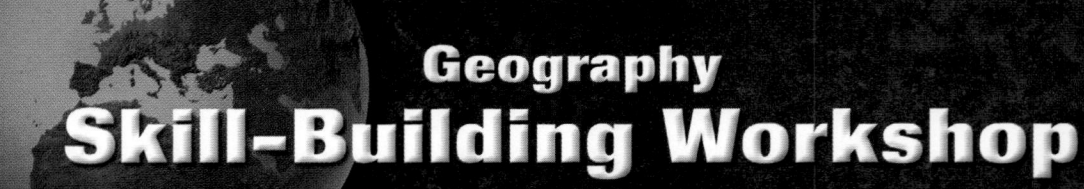

Geography
Skill-Building Workshop

WORKSHOP 1

Using Graphic Organizers

Some people say a picture is worth a thousand words. A graphic organizer creates a picture of information. It allows you to "see" patterns and relationships. We use graphic organizers every day. For example, calendars and daily planners are types of graphic organizers. The section reviews throughout this textbook include a question that asks you to use a graphic to organize what you have learned.

Graphic organizers come in many forms. Time lines, for example, identify important events in historical sequence. Charts and tables help you gather, compare, and analyze information. Another kind of graphic organizer is an idea web, which you will learn about in this workshop.

Graphic organizers can be used for many purposes. Some organizers help you brainstorm ideas, problems, and solutions. Others allow you to identify connections or tell stories. Still others help you communicate complex ideas and develop ideas for writing.

A major topic or concept is located at the web center. Supporting details radiate outward from the center. The example provides a main topic in the center: forest loss in northern Europe. The subtopics, or details, are provided in bubbles that grow from the main topic. Just as when you brainstorm and one thought leads to another, each detail can have its own bubbles. The details become more and more specific as the web grows. The web can grow until you run out of either ideas or space.

Practicing the Skill

Use a piece of paper, a large sheet of butcher paper, a chalkboard, or a dry-erase board to make an idea web. Suppose you must write a paper on the geography of Spain. What do you know about Spain's geography? Consider topics like the physical environment, demographics, and trade. Build your idea web individually or in a small group.

Idea Web

Developing the Skill Idea webs, also called word webs, are very useful graphic organizers. These organizers are maps that show how different categories of information relate to each other. They provide a structure for ideas and facts. They can also help you brainstorm and organize your writing tasks.

WORKSHOP 2

Using Databases and Analyzing Statistics

Statistics you can use in your study of geography are often organized into databases. Databases condense and organize information. They can help you answer important geographic questions. In addition, databases can help you understand geographic relationships.

A telephone book is a database. The books you keep on a shelf at home make up a database of some of the things you have read. Other examples of databases include statistical tables (like the one shown on the next page), almanacs, and CD–ROMs. Even maps are databases. Maps organize and show information about a variety of features. For example, a precipitation map shows average annual precipitation amounts for the country or region under study.

Workshop 2
Going Further: Thinking Critically

Have the class work together to plan a database that will contain information about enrollment in each grade level of your school. Call on volunteers to suggest fields that they think should be included in the database. *(Possible answers: total enrollment in each grade level, male and female enrollment for each grade level, average age of students in each grade level, average number of students per class in each grade level, and so on)* Ask students to think of ways they could collect the information to complete their database. *(Possible answers: ask school administrators for school totals, ask teachers for class statistics, physically count students)* Then ask which of these methods will yield the most reliable statistics *(asking the administrators,*

because they keep official school records). If possible, obtain enrollment figures from the administration and have the class compile them with a database software program.

Ask students to speculate about how school enrollment statistics might be used by school and district officials. *(Possible answers: in scheduling classes or hiring teachers, purchasing supplies, planning school activities, and so on)* Have students use the information in their database to ask each other questions about your school. Possible questions include: In which grade are the most students enrolled? Are there more male or female students in the school? What is the ratio of teachers to students in the school? Lead a class discussion about some uses of statistics in the study of geography.

Comparing Statistics: The Former Yugoslav Republics

Country	Population (in millions)	Per Capita GDP (in U.S. $)	Electricity Consumption (kilowatt hours per person)	Telephone Lines (per person)
Bosnia and Herzegovina	4.2	$1,900	1,950 kWh	0.12
Croatia	4.4	$8,800	3,223 kWh	0.42
Macedonia	2.1	$5,000	2,973 kWh	0.26
Serbia and Montenegro	10.5	$2,370	3,075 kWh	0.24
Slovenia	2	$18,000	6,971 kWh	0.41

Source: Central Intelligence Agency, *World Factbook 2003; The World Almanac and Book of Facts 2004*

As you work with databases, you will probably note that statistics can vary from one source to another. For example, the U.S. Central Intelligence Agency's World Factbook and some other database might differ in their population totals or economic figures for a country. The sources may have used different formulas for calculating data, such as economic statistics. In addition, they may have collected the statistics at different times in a year. One source might offer more recent information than others. As a result, it is often a good idea to check two or three sources for information you need.

Developing the Skill A number of key terms are important for learning how to create and use databases for organizing statistics and other information. A *file* stores a database on a computer. A short file name for the example on this page might be Statistics. *Fields* are names for categories of information in a database. The example above has five fields: Country, Population, Per Capita GDP, Electricity Consumption, and Telephone Lines. *Records* group fields that are all related to the database's topic or idea. All of the fields in the example are a record of statistics about the former Yugoslav republics.

To create a database on a computer, you need a database management system. Such a system is part of special software developed for record keeping. It provides instructions on how to organize and name your fields. Following are some key points to remember when creating and using any database:

- Read the headings and labels to determine the kinds of information included in the database. Are the figures provided in metric or in standard

measurement? Do the numbers represent a fraction of a larger number? If you were looking at economic data for various European countries, one source may use GNP and another GDP. It would be important to understand the differences between such data and how those differences affect the research you are conducting.

- Be sure to identify the original source for the data. When was the information collected? How was the information collected?

- Always look up any unfamiliar terms so you are sure to understand exactly what the data represents.

Practicing the Skill

Use a variety of historical and statistical databases to create a profile of population change in Europe. Collect data on Austria, Germany, Italy, and other European countries with negative or slow population growth rates. You will need to create your own databases to organize your information. You can use a software database program or create a database by hand. For example, you might create a table that tracks changes in population from 1800 to what is projected in 2050 for the countries you choose to study. As an alternative, you could create maps that use color to show population growth rates over time. For example, you might shade countries with the slowest growth rates red and the highest growth rates blue. Include a file name and identify fields you use.

Using the Illustration

Direct students' attention to the picture of St. Basil's Cathedral, which rises over Red Square in Moscow near the Kremlin. The cathedral was commissioned by Czar Ivan IV in the 1500s to celebrate his defeat over the Tatar people.

At the time of its construction, St. Basil's Cathedral was the tallest building in Moscow.

Because Ivan wanted it to be a lasting memorial of his triumph, he ordered the church built out of brick, a relatively rare building material in Russia at that time. Most buildings were made of wood or stone. For centuries, architects have marveled at the church's construction, because the complex structure was built without any written plans.

St. Basil's Cathedral is famous around the world for its grand towers topped with colorful onion-shaped domes. Each of the domes that grace the cathedral bears different colors, designs, and decorations. The shape of these domes is traditional in Russian architecture. Ask students why this shape may be popular in the region. (*Its rounded sides do not collect snow and strain the towers with excess weight.*) Call on volunteers to suggest other architectural design elements from around the world that perform similar functions.

Unit Objectives

1. Describe the physical features of Russia and northern Eurasia and link the region's natural environments with its economic development.

2. Examine the progression of the region's history, from early periods through communism to the breakup of the Soviet Union.

3. Analyze the relationships among the region's ethnic and culture groups.

4. Use charts and graphs to analyze relationships between and changes in geographic information.

5. Compile time lines to trace the chronology of events.

UNIT 5 Russia and Northern Eurasia

St. Basil's Cathedral, Moscow

Chapter **17** *Russia, Ukraine, and Belarus*

Chapter **18** *Central Asia*

introduces the physical geography of the world's largest country and its neighbors to the west, along with the historical, economic, and political issues that face the people who live there. The Caucasus countries—Armenia, Azerbaijan, and Georgia—are discussed in a feature.

discusses the past, present, and future of the Central Asian countries—Kazakhstan, Kyrgyzstan, Tajikistan, Turkmenistan, and Uzbekistan.

CONNECTING TO Literature

THE WORLD I LEFT BEHIND

by Luba Brezhneva

Luba Brezhneva (1943–), the niece of Soviet leader Leonid Brezhnev, was born in Russia and lived there until 1990. In *The World I Left Behind* she tells the story of her life in the former Soviet Union. Here Breshneva writes about when she and her husband, Mischa, moved from Moscow to live in the rural village of Korovino. She also provides some insights into the culture of rural Russia.

The bath, with its steam and heat, was a true haven for body and soul.... Yakov Maksimovich [Brezhneva's father] and his son, Ilya, would whip each other with hot birch twigs (to open the pores) and raise volleys of steam by sprinkling herbal broth on the stove. Climbing in turn to the top bunk, where it was hottest—"so the heat will enter the bones"—they stayed in the steam room till they were woozy and as red as lobsters. Then they would run to the pond—'to cast off the heat.' And then they would begin again. Finally came an unhurried conversation as they lay on benches in the dressing room. "You've gotten mighty thin, Ilya," Yakov once said as he examined his son's slim body. "The city is eating you up."

On Saturdays, while the men heated the stove, brought water, and bathed, the women would be in the house hastily finishing their domestic chores, baking pies, and cleaning. The Brezhnev men used the bathhouse before the women....

Saturdays also meant scrubbing the wooden floor of the house, a time-consuming ritual that began in the morning. It would first be swept, then doused with hot water and ashes. While still warm, it would be scraped with a knife until the wood glowed a tawny yellow brown. Then it would be rinsed. Finally patchwork rugs would be laid down. The furniture was washed in the same way, since it was made of unpainted wood.

Illness in the family also meant it was time to heat the bath. The patient was steamed in the bath and whipped with birch twigs, then smeared with goat fat and forced to drink a hot broth of bitter herbs. Finally he or she was wrapped in an old sheepskin coat and laid out on the stove to sleep, assured that "you'll be fit as a fiddle by morning."

Analyzing the Primary Source

1. **Analyzing Information** What kind of climate do you think is typical in Korovino? What clues do you find in the passage about how much life has changed in the village over time?

2. **Summarizing** According to Brezhneva, what is life like for women in rural Russia?

Analyzing the Primary Source

Answers

1. The climate is probably very cold. Brezhneva implies that life in the village has changed very little over time.

2. Possible answer: Women are responsible for the cooking and cleaning and must wait upon the men of the village.

In this unit, students will learn about the land and people of Russia and northern Eurasia—a vast region that is undergoing many political and economic changes.

Huge open plains and deep evergreen forests dominate the landscapes of Russia, Ukraine, and Belarus. The region's population is concentrated in the western plains, where the best farmland and the largest cities are located. Harsh climates keep Siberia and Russia's Far East thinly populated. Development of the region's rich resources may increase settlement there.

For most of the country's history, Russia's leaders have limited its citizens' personal freedoms. During the more than 70 years of Communist rule, the Soviet Union became a superpower. Since the collapse of the Soviet Union, Russia has struggled with rising crime, corruption, and unemployment.

The countries of northern Eurasia were also part of the Soviet Union. They are now independent republics. Their economies are emerging, with varying degrees of success. Distinctive religious and ethnic groups live in the area.

Your Classroom Time Line

These are the major dates and time periods for this unit. Have students enter them on the time line you created earlier. You may want to watch for these dates as students progress through the unit.

300s B.C. Alexander the Great invades Central Asia.

100s B.C. Chinese trade and military expeditions begin moving into Central Asia.

A.D. 600s Turkic-speaking peoples establish kingdoms in Central Asia.

700s Muslims invade Central Asia.

751 Arab armies defeat the Chinese.

800s Kiev is established and becomes an important trade center.

1100s The Eastern Orthodox Church gains prominence in Kiev.

1184–1212 Georgia experiences a Golden Age under Queen Tamara.

1218 Genghis Khan leads the Mongols through Central Asia.

1240 Mongols destroy Kiev.

1326 The seat of the Russian Orthodox Church moves to Moscow.

POLITICAL MAP ANSWERS

1. Kyrgyzstan, Tajikistan, Georgia, Armenia
2. Kazakhstan, Uzbekistan, Turkmenistan

CRITICAL THINKING ANSWER

3. They all end in -stan. Stan seems to mean "country," and the prefix of each country name refers to the major ethnic group that lives there.

UNIT 5 ATLAS

The World in Spatial Terms

Russia and Northern Eurasia: Political

1. **Places and Regions** Compare this map to the physical map. Which countries in the region are largely mountainous?

2. **Places and Regions** Compare this map to the climate map. Which countries in the region have large areas with an arid climate?

Critical Thinking

3. **Making Generalizations** What do the names Kazakhstan, Kyrgyzstan, Tajikistan, Turkmenistan, and Uzbekistan all have in common? How do you think the names of these countries might relate to the ethnic groups that live there?

Using the Political Map

Focus students' attention on the **political map** of Russia and northern Eurasia on the opposite page. Point out that all of the countries represented in color on the map used to be part of the Soviet Union, but that the Soviet Union broke up in 1991. Have students compare this map to the **physical** and **land use and resources maps** to predict how the breakup of the Soviet Union may have affected Russia's economy. *(Possible answers: lost mineral and energy resources, reduced access to Caspian and Black Seas)*

Using the Physical Map

Ask students to examine the **physical map**. Remind students that Russia is the largest country in the world. Ask them to estimate the distance between its eastern and western borders *(about 6,000 miles [9,650 km])*. Note that the Northern European Plain is a physical feature of western Russia and that the eastern portion of the country is called Siberia. Ask students what physical feature appears to separate European Russia from Siberia *(Ural Mountains)*.

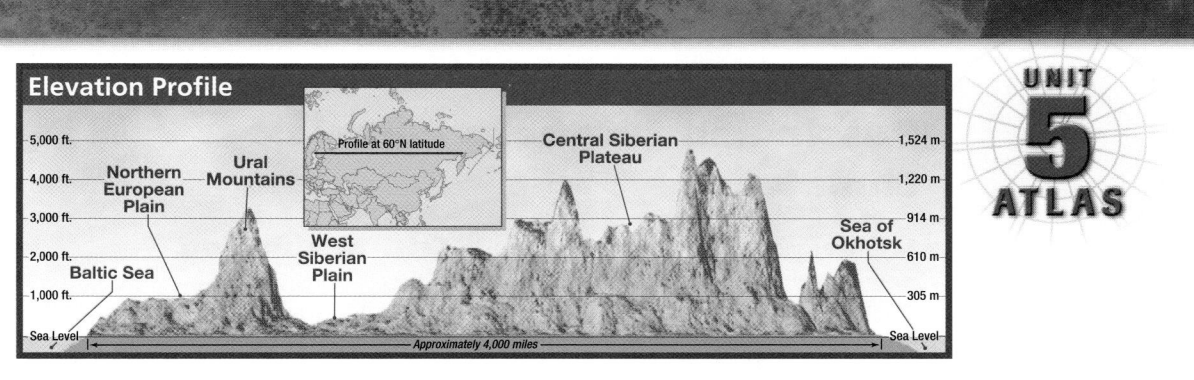

Elevation Profile

Profile at 60°N latitude

Northern European Plain — Baltic Sea — Ural Mountains — West Siberian Plain — Central Siberian Plateau — Sea of Okhotsk

5,000 ft. — 1,524 m
4,000 ft. — 1,220 m
3,000 ft. — 914 m
2,000 ft. — 610 m
1,000 ft. — 305 m
Sea Level — Sea Level

Approximately 4,000 miles

Russia and Northern Eurasia: Physical

(Map of Russia and Northern Eurasia showing physical features, countries, and labels including Greenland (Denmark), Alaska (U.S.), Atlantic Ocean, Arctic Ocean, North Pole, Bering Strait, Bering Sea, Pacific Ocean, North Sea, Barents Sea, New Siberian Islands, North Land, Novaya Zemlya, Taymyr Peninsula, Kolyma Mts., Kamchatka Peninsula, Europe, Russia, Belarus, Northern European Plain, West Siberian Plain, Siberia, Central Siberian Plateau, Cherskiy Range, Sea of Okhotsk, Sakhalin Island, Kuril Islands, Ukraine, Crimea, Donets Basin (Donbas), Black Sea, Caucasus Mts., Mt. Elbrus 18,510 ft. (5,642 m), Georgia, Armenia, Azerbaijan, Caspian Sea, Aral Sea, Kazakhstan, Lake Balkhash, Uzbekistan, Turkmenistan, Kyrgyzstan, Tajikistan, Tian Shan, Communism Peak 24,590 ft. (7,495 m), Pamirs, Altay Shan, Kuznetsk Basin, Sayan Mts., Yablonovyy Range, Stanovoy Mts., Southwest Asia, East Asia)

N W E S

SCALE
0 — 500 — 1000 Miles
0 — 500 — 1000 Kilometers
Projection: Two-Point Equidistant

ELEVATION

FEET	METERS
13,120	4,000
6,560	2,000
1,640	500
656	200
(Sea level) 0	0 (Sea level)
Below sea level	Below sea level

UNIT 5 ATLAS

Your Classroom Time Line, (continued)

1370 Timur establishes the Mongol Empire.

1400s Moscow becomes Russia's capital.

Late 1400s Europeans discover a water route to East Asia, bypassing Central Asian trade routes.

Late 1400s Ivan III of Moscow defeats the Mongols.

1547 Ivan IV crowns himself Czar of Russia.

1637 Explorers reach the Pacific Coast of Russia.

1682–1725 Peter the Great rules Russia.

1703 St. Petersburg is founded.

1712–1918 St. Petersburg serves as Russia's capital.

Late 1700s Catherine the Great rules Russia.

1800s The Russian Empire expands into Central Asia.

1812 Napoléon unsuccessfully invades Russia.

1860s Russian serfs are granted their freedom.

Late 1800s Russia begins to industrialize.

1891 Construction of the Trans-Siberian Railroad begins.

1910s Resistance to Russian rule rises in Central Asia.

1914–17 Russia fights in World War I and suffers heavy losses.

1917 During the Russian Revolution the Bolsheviks under Vladimir Lenin take control of the country.

1924 Lenin dies. Leningrad is renamed in his honor.

1924–53 Joseph Stalin rules the Soviet Union.

1949 The Soviet Union begins testing nuclear weapons in northeastern Kazakhstan.

373

Have students examine the **climate map** of Russia and northern Eurasia on this page. Ask them to name the climates of the largest areas of Russia (*subarctic and tundra*). Have students compare this map to the **population map**. Ask them which climate regions are the most densely populated (*humid continental, semiarid, Mediterranean*). Then ask students which climate regions are home to the fewest concentrations of people (*tundra, subarctic, arid*). Call on volunteers to explain this distribution. (*Possible answer: Harsh climates like tundra and desert are difficult places to live or work. Milder climates, like humid continental and Mediterranean climates, allow a wider range of agriculture and other economic activities.*)

Your Classroom Time Line,
(continued)

1957 The Soviet Union launches Sputnik.

1980s Moderate changes are introduced into Soviet politics and economics.

1989 The Baikal-Amur Mainline across Siberia is completed.

1991 The Soviet Union collapses. The Central Asian republics declare their independence.

1997 Kazakhstan moves its capital from Almaty to Astana to be closer to Europe and Russia.

1999 Rebels kidnap tourists in Uzbekistan.

2000 Vladimir Putin becomes president of Russia.

2003 Conflict continues between Chechen rebel groups and the Russian government.

CLIMATE MAP ANSWERS

1. tundra, subarctic, humid continental, semiarid
2. humid continental, semiarid, Mediterranean, highland, humid subtropical

CRITICAL THINKING ANSWER

3. Possible answer: Russia is located in high latitudes and has many interior areas that do not receive the moderating influence of oceans.

374

UNIT 5 ATLAS

Russia and Northern Eurasia:
Climate

1. (*Places and Regions*) Compare this map to the political map. What are Russia's major climates?

2. (*The World in Spatial Terms*) Compare this map to the physical map. What climate types are found near the Black Sea?

Critical Thinking

3. **Making Generalizations** Why do you think cold climates cover so much of Russia?

CLIMATE

- Arid
- Semiarid
- Mediterranean
- Humid subtropical
- Humid continental
- Subarctic
- Tundra
- Highland

Focus students' attention on the **precipitation map** on this page. Ask them to identify the driest areas in Russia and northern Eurasia (*areas around the Caspian and Aral Seas, northern Siberia*). Call on volunteers to name factors that keep these areas from receiving much precipitation. (*Possible answer: Mountains block moist air from reaching the region.*) Have

students compare this map to the **population** and **land use and resources maps**. Ask them how the lack of precipitation has affected settlement and economic activity in these areas. (*Few people live in these areas. Those that do are mostly nomadic herders.*)

Russia and Northern Eurasia:
Precipitation

UNIT
5
ATLAS

1. (*Physical Systems*) How are precipitation patterns different near the Black Sea and Caspian Sea?

2. (*Environment and Society*) Compare this map to the political and population maps. How are population density and precipitation patterns in Russia related?

Critical Thinking

3. **Making Generalizations** Why might the eastern part of the region receive little precipitation?

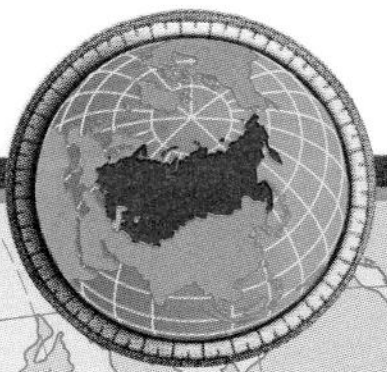

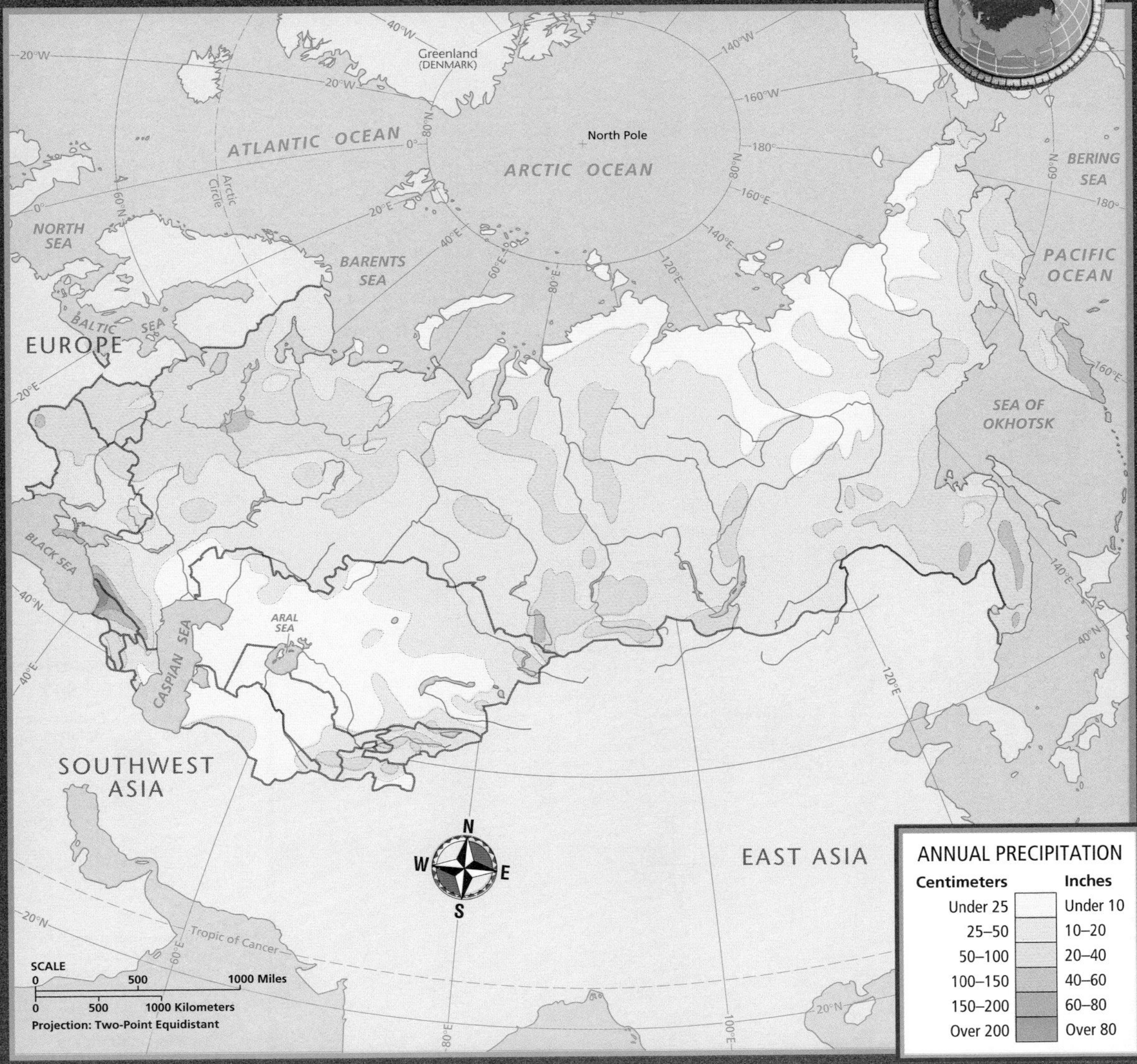

ATLANTIC OCEAN
Greenland (DENMARK)
North Pole
ARCTIC OCEAN
BERING SEA
NORTH SEA
BARENTS SEA
PACIFIC OCEAN
BALTIC SEA
EUROPE
SEA OF OKHOTSK
BLACK SEA
ARAL SEA
CASPIAN SEA
SOUTHWEST ASIA
EAST ASIA
Arctic Circle
Tropic of Cancer

ANNUAL PRECIPITATION

Centimeters	Inches
Under 25	Under 10
25–50	10–20
50–100	20–40
100–150	40–60
150–200	60–80
Over 200	Over 80

SCALE
0   500   1000 Miles
0   500   1000 Kilometers
Projection: Two-Point Equidistant

375

Using the Population Map

Direct students' attention to the **population map** on this page. Ask each student to write a sentence that summarizes the population patterns shown here. *(Possible answer: Most people live in the western part of the region; the eastern part is sparsely* *populated.)* Call on volunteers to read their statements aloud. Then ask students to suggest reasons for the patterns of population density. They may wish to consult other maps in this atlas for assistance.

POPULATION MAP ANSWERS

1. St. Petersburg, Moscow, Nizhniy Novgorod
2. humid continental, semiarid

CRITICAL THINKING ANSWER

3. Areas near rivers and lakes have higher population densities than other areas.

UNIT 5 ATLAS

Russia and Northern Eurasia:
Population

1. (*Places and Regions*) Which Russian metropolitan areas have more than 2 million inhabitants?

2. (*Environment and Society*) Compare this map to the climate map. Which climate types do the most densely populated areas in the region have?

Critical Thinking

3. **Analyzing Information** Compare this map to the physical map. How are rivers and lakes in Siberia related to population density?

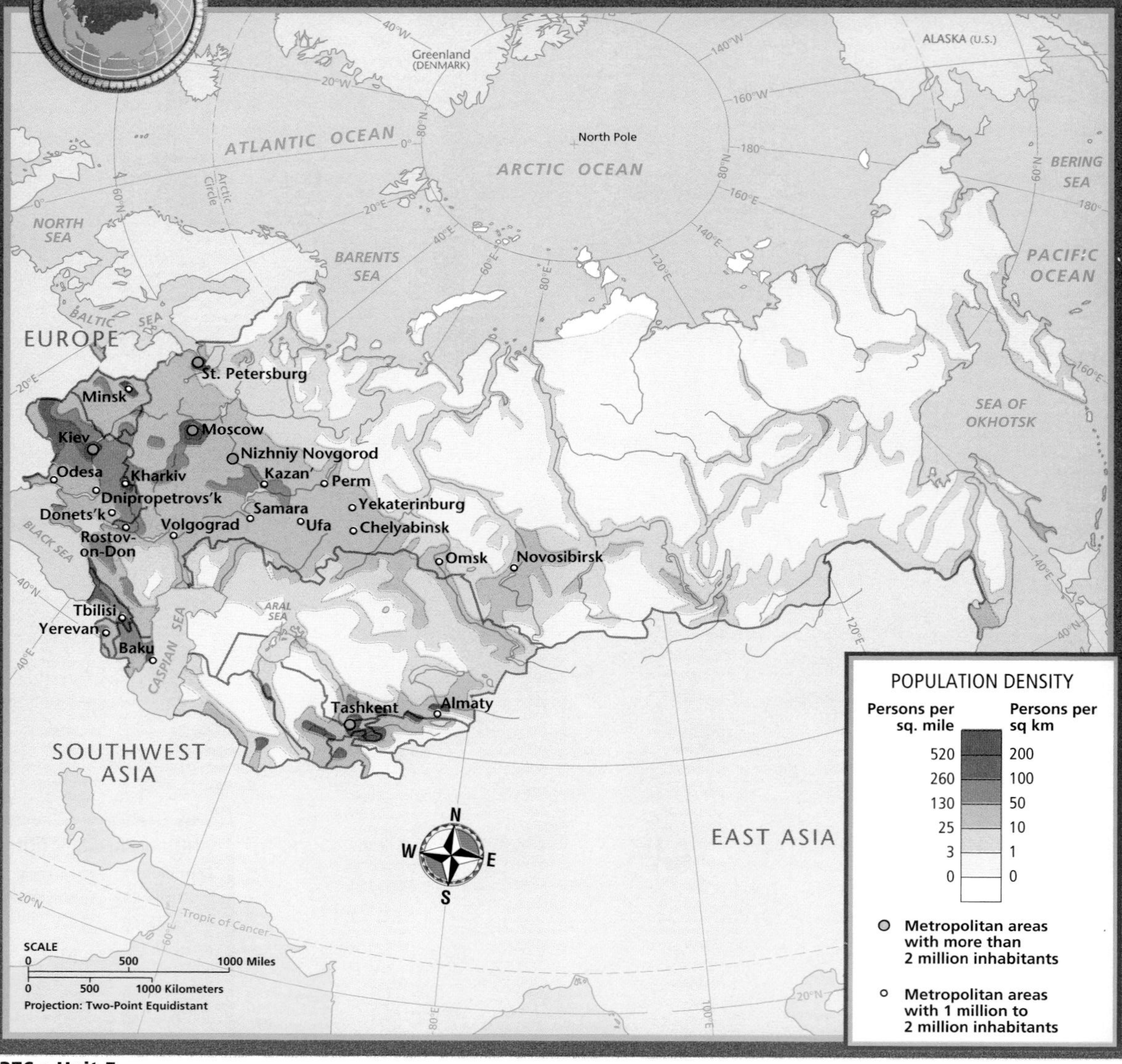

POPULATION DENSITY

Persons per sq. mile	Persons per sq km
520	200
260	100
130	50
25	10
3	1
0	0

⊚ Metropolitan areas with more than 2 million inhabitants

○ Metropolitan areas with 1 million to 2 million inhabitants

SCALE
0 — 500 — 1000 Miles
0 — 500 — 1000 Kilometers
Projection: Two-Point Equidistant

Using the Land Use and Resources Map

Call students' attention to the **land use and resources map** on this page. Have them compare it to the **population map**. Ask them which area is rich in natural resources but is sparsely populated *(Siberia)*. Challenge students to use information from the maps in this atlas to suggest why Russia may have trouble making use of Siberia's resources. *(Possible answers: vast distances from population centers, difficulties in transporting goods across the region, harsh weather)*

Russia and Northern Eurasia:
Land Use and Resources

UNIT
5
ATLAS

1. ⬭ *Environment and Society* ⬭ Compare this map to the climate map. What is the main economic activity in areas with tundra and arid climates?

2. ⬭ *Environment and Society* ⬭ How have Russians adapted Siberian rivers to produce energy?

Critical Thinking

3. **Analyzing Information** Compare this map to the climate map. How do you think farmers have adapted to environmental conditions in arid areas? How can you tell?

LAND USE AND RESOURCES MAP ANSWERS

1. nomadic herding
2. built hydroelectric projects

CRITICAL THINKING ANSWER

3. have used irrigation, particularly near rivers; commercial farming mostly along rivers in arid areas

---

LAND USE

- Commercial farming
- Subsistence farming
- Forestry
- Livestock raising
- Manufacturing
- Fishing
- Nomadic herding
- Limited economic activity
- ● Major manufacturing and trade centers

RESOURCES

- 🪨 Coal
- ⬦ Natural gas
- Oil
- ❄ Nuclear power
- Hydroelectric power
- Au Gold
- Ag Silver
- D Diamonds
- Other minerals

SCALE
0   500   1000 Miles
0   500   1000 Kilometers
Projection: Two-Point Equidistant

Copy the following chart onto the chalkboard. Explain that the unemployment rate shows the percentage of people who could work but are unable to find jobs. An unemployment rate of 4 percent means that 4 percent of the population of working age—that is, between the ages of 16 and 64—do not have jobs. When the Soviet Union broke up, many countries in this region did not have enough resources to sustain their populations independently. Have students use the land and resources map in the unit atlas to make connections between resources and unemployment rates. Point out that many countries in the area also have high underemployment rates. Underemployed workers can only find jobs that require less skill than they have or pay less money than their skills are generally worth.

COUNTRY	UNEMPLOYMENT RATE
Armenia	20 percent
Azerbaijan	16 percent
Belarus	2.1 percent (many underemployed)
Georgia	17 percent
Kazakhstan	13.7 percent
Kyrgyzstan	7.2 percent
Russia	7.9 percent (many underemployed)
Tajikistan	40 percent
Turkmenistan	Not available
Ukraine	3.8 percent (many underemployed)
Uzbekistan	10 percent (many underemployed)
United States	5.8 percent

Source: Central Intelligence Agency, *The World Factbook 2003*

Historical Geography

Russia and Invaders

Although Russia has been invaded many times during its long history, brutal weather and the Russian people have repelled each group of invaders. In 1707, for example, the army of Sweden's King Charles XII swept into Russia. By 1709 they had retreated. Bitter cold and starvation had killed many soldiers. In 1812 Napoléon Bonaparte of France led 500,000 men into Russia. Only 10,000 survived the battles, cold, disease, and hunger they found there. Similarly, when Nazi forces raided the Soviet Union in World War II, cold was a major factor in their defeat.

CRITICAL THINKING:

How do Russia's size, climate, and landforms make a successful invasion of the country difficult? *(Russia's immense size necessitates long supply lines that are difficult to maintain. The open plains of the steppe offer little shelter from the harsh weather.)*

UNIT 5 ATLAS

Time Line: Russia and Northern Eurasia

A.D. 600s
Turkic-speaking peoples establish kingdoms in Central Asia.

800s
The city of Kiev becomes an important center for trade between the Mediterranean and Baltic Sea areas.

1547
Ivan IV crowns himself czar of all Russia.

late 1700s
Catherine the Great rules Russia.

1917
The Bolsheviks overthrow the Russian government in what becomes known as the Russian Revolution.

1991
The Soviet Union collapses.

| 330 B.C. | A.D. 600 | 700 | 800 | 1200 | 1300 | 1400 | 1500 | 1600 | 1700 | 1800 | 1900 |

330 B.C.
Alexander the Great invades Central Asia.

700s
Arabic speakers invade Central Asia, bringing Islam with them.

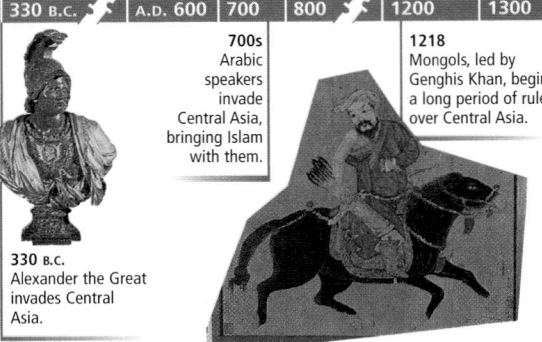

1218
Mongols, led by Genghis Khan, begin a long period of rule over Central Asia.

1682–1725
Peter the Great takes over lands along the Baltic Sea and expands Russian control into what are now Belarus and Ukraine.

1800s
Russians spread into the Caucasus and Central Asia.

1941
Germany invades the Soviet Union during World War II.

The United States and Russia and Northern Eurasia

Comparing Sizes

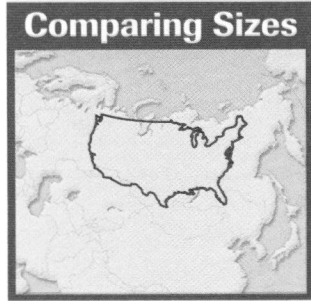

internet connect

GO TO: go.hrw.com
KEYWORD: SW3 Almanac
FOR: Additional information and reference sources

Comparing Standard of Living

COUNTRY	LIFE EXPECTANCY (in years)	INFANT MORTALITY (per 1,000 live births)	LITERACY RATE	DAILY CALORIC INTAKE (per person)
Armenia	62, male 71, female	41	99%	2,356
Azerbaijan	59, male 68, female	82	100%	2,191
Georgia	61, male 68, female	51	99%	2,252
Kazakhstan	58, male 69, female	59	98%	2,517
Kyrgyzstan	60, male 68, female	75	97%	2,535
Russia	63, male 73, female	20	100%	2,835
Tajikistan	61, male 68, female	113	99%	2,176
Turkmenistan	58, male 64, female	73	100%	2,684
Ukraine	61, male 72, female	21	100%	2,878
Uzbekistan	61, male 68, female	72	99%	2,564
United States	74, male 80, female	7	97%	3,757

Sources: *World Almanac and Book of Facts 2004; Britannica Book of the Year, 2002*

Add another column to the chart you drew for the previous activity and add the urban population figures at right to it. Ask students to name the countries of northern Eurasia with the fewest urban residents (*Kyrgyzstan, Tajikistan, Turkmenistan, Uzbekistan*). Have students consult the maps in the unit atlas to suggest reasons for this. (*Possible answer: These countries are located in arid areas, and most people there are involved in nomadic herding. They have never developed large cities.*) Ask students to note the relationships between unemployment rates and urban populations. (*Generally, unemployment is highest in countries with generally urban populations.*)

COUNTRY	URBAN POPULATION
Armenia	67 percent
Azerbaijan	52 percent
Belarus	70 percent
Georgia	57 percent
Kazakhstan	56 percent
Kyrgyzstan	34 percent
Russia	73 percent
Tajikistan	28 percent
Turkmenistan	45 percent
Ukraine	68 percent
Uzbekistan	37 percent
United States	75 percent

Source: *World Almanac and Book of Facts 2004*

Fast Facts: Russia and Northern Eurasia

UNIT **5** ATLAS

FLAG	COUNTRY / Capital	POPULATION (in millions) / POP. DENSITY	AREA	PER CAPITA GDP (in US $)	WORKFORCE STRUCTURE (largest categories)	ELECTRICITY CONSUMPTION (kilowatt hours per person)	TELEPHONE LINES (per person)
	Armenia Yerevan	3.1 / 266/sq. mi.	11,506 sq. mi. 29,800 sq km	$ 3,800	45% agriculture 30% services	1,890 kWh	0.17
	Azerbaijan Baku	8.4 / 250/sq. mi.	33,436 sq. mi. 86,599 sq km	$ 3,500	52% services 41% agric., forestry	1,989 kWh	0.12
	Belarus Minsk	9.9 / 123/sq. mi.	80,155 sq. mi. 207,600 sq km	$ 8,200	41% services 40% industry, const.	2,697 kWh	0.30
	Georgia Tbilisi	5.1 / 190/sq. mi.	26,911 sq. mi. 69,699 sq km	$ 3,100	40% services 40% agriculture	1,485 kWh	0.13
	Kazakhstan Astana	15.4 / 15/sq. mi.	1,049,155 sq. mi. 2,717,299 sq km	$ 6,300	50% services 30% industry	3,134 kWh	0.13
	Kyrgyzstan Bishkek	5.1 / 67/sq. mi.	76,641 sq. mi. 198,499 sq km	$ 2,800	55% agriculture 30% services	2,036 kWh	0.08
	Russia Moscow	143.2 / 22/sq. mi.	6,592,767 sq. mi. 17,075,188 sq km	$ 9,300	65% services 23% industry	5,396 kWh	0.25
	Tajikistan Dushanbe	6.2 / 113/sq. mi.	55,251 sq. mi. 143,099 sq km	$ 1,250	67% agriculture 25% services	2,325 kWh	0.04
	Turkmenistan Ashgabat	4.9 / 26/sq. mi.	188,456 sq. mi. 488,099 sq km	$ 5,500	48% agriculture 37% services	1,748 kWh	0.08
	Ukraine Kiev	48.5 / 208/sq. mi.	233,090 sq. mi. 603,700 sq km	$ 4,500	44% services 32% industry	3,141 kWh	0.22
	Uzbekistan Tashkent	26.1 / 151/sq. mi.	172,742 sq. mi. 447,400 sq km	$ 2,500	44% agriculture 36% services	1,804 kWh	0.06
	United States Washington, D.C.	294.0 / 83/sq. mi.	3,717,810 sq. mi. 9,629,084 sq km	$ 37,600	31% manage., prof. 29% tech., sales, admin.	12,250 kWh	0.65

Sources: Central Intelligence Agency, *World Factbook 2003; The World Almanac and Book of Facts, 2004*
The CIA calculates per capita GDP in terms of purchasing power parity. This formula equalizes the purchasing power of each country's currency.

Traditional wedding, Tashkent, Uzbekistan

Russia and Northern Eurasia

Russia, Ukraine, and Belarus

CHAPTER RESOURCE MANAGER

Objectives	Pacing Guide	Reproducible Resources
SECTION 1 **Natural Environments** (pp. 381–85)	• Identify the landforms and rivers found in Russia, Ukraine, and Belarus. • Explain the factors that influence the region's climates and vegetation. • Describe the region's natural resources.	**Regular** 1 day **Block Scheduling** .5 day *Block Scheduling Handbook, Chapter 17*	**RS** Guided Reading Strategy 17.1 **E** Creative Strategies for Teaching World Geography, Lesson 12 **SM** Geography for Life Activity 17: Waterways of Russia
SECTION 2 **History and Culture** (pp. 386–91)	• Identify some major events in the growth of the Russian Empire. • Analyze how the Soviet Union developed and explain what life was like for its citizens. • Describe some features of the region's culture.	**Regular** 1 day **Block Scheduling** .5 day *Block Scheduling Handbook, Chapter 17*	**RS** Guided Reading Strategy 17.2 **PS** Readings in World Geography, History, and Culture 40, 43, and 44 **E** Cultures of the World Activity, Region 4 **SM** Critical Thinking Activity 17: The Cold War
SECTION 3 **The Region Today** (pp. 393–97)	• Explain how the economies of areas within Russia, Ukraine, and Belarus have developed. • Identify the challenges faced by the region.	**Regular** 1 day **Block Scheduling** .5 day *Block Scheduling Handbook, Chapter 17*	**RS** Guided Reading Strategy 17.3 **RS** Graphic Organizer Activity 17 **PS** Readings in World Geography, History, and Culture 41 and 42 **E** Cultures of the World Activity: Region 4 **SM** Map Activity 17: Russia's Environmental Problems

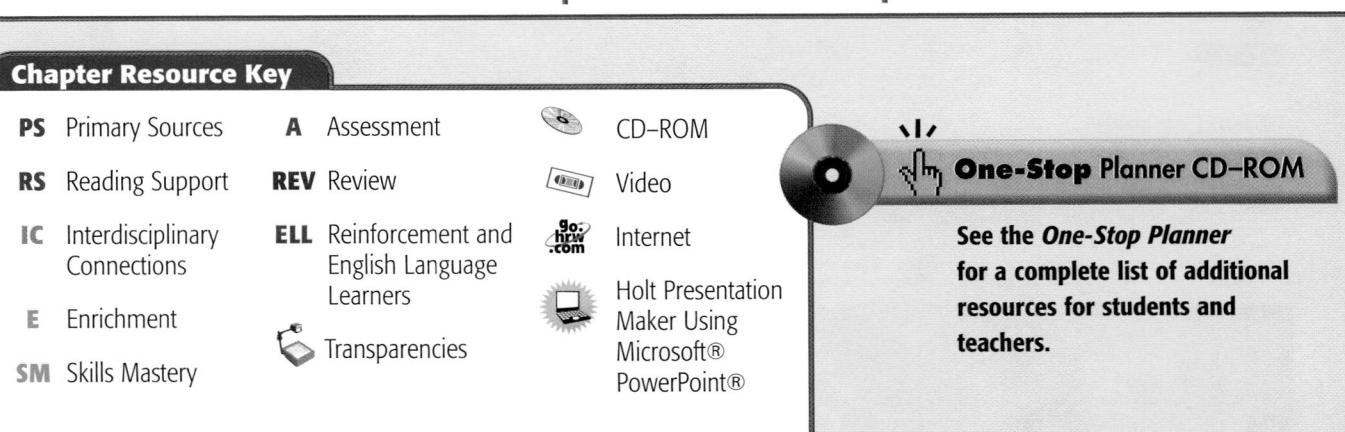

Chapter Resource Key

PS Primary Sources

RS Reading Support

IC Interdisciplinary Connections

E Enrichment

SM Skills Mastery

A Assessment

REV Review

ELL Reinforcement and English Language Learners

Transparencies

CD-ROM

Video

Internet

Holt Presentation Maker Using Microsoft® PowerPoint®

One-Stop Planner CD–ROM

See the *One-Stop Planner* for a complete list of additional resources for students and teachers.

☀ One-Stop Planner CD–ROM

It's easy to plan lessons, select resources, and print out materials for your students when you use the **One-Stop Planner CD–ROM with Test Generator**.

Technology Resources

- One-Stop Planner CD–ROM, Lesson 17.1
- CNN. Presents Geography: Yesterday and Today, Segment 18: Lake Baykal
- Geography and Cultures Visual Resources 31–35
- Homework Practice Online
- HRW Go site

- One-Stop Planner CD–ROM, Lesson 17.2
- Geography and Cultures Visual Resources 36
- Homework Practice Online
- HRW Go site

- One-Stop Planner CD–ROM, Lesson 17.3
- *ARGWorld* CD–ROM
- Homework Practice Online
- HRW Go site

Reinforcement, Review, and Assessment

- **ELL** Main Idea Activity 17.1
- **ELL** English Audio Summary 17.1
- **ELL** Spanish Audio Summary 17.1
- **REV** Section 1 Review, p. 385
- **A** Daily Quiz 17.1

- **ELL** Main Idea Activity 17.2
- **ELL** English Audio Summary 17.2
- **ELL** Spanish Audio Summary 17.2
- **REV** Section 2 Review, p. 391
- **A** Daily Quiz 17.2

- **ELL** Main Idea Activity 17.3
- **ELL** English Audio Summary 17.3
- **ELL** Spanish Audio Summary 17.3
- **REV** Section 3 Review, p. 397
- **A** Daily Quiz 17.3

⊡ internet connect

HRW ONLINE RESOURCES
GO TO: go.hrw.com
Then type in a keyword.

TEACHER HOME PAGE
KEYWORD: SW3 Teacher

CHAPTER INTERNET ACTIVITIES
KEYWORD: SW3 GT17
Choose a topic on Russia, Ukraine, or Belarus to:
- research the rise and fall of the Soviet Union.
- create a poster of a journey through the Caucasus.
- learn the causes and effects of the Chernobyl disaster.

CHAPTER ENRICHMENT LINKS
KEYWORD: SW3 CH17

CHAPTER MAPS
KEYWORD: SW3 MAPS17

ONLINE ASSESSMENT
Homework Practice
KEYWORD: SW3 HP17
Standardized Test Prep
KEYWORD: SW3 STP17
Rubrics
KEYWORD: SS Rubrics

COUNTRY INFORMATION
KEYWORD: SW3 Almanac

CONTENT UPDATES
KEYWORD: SS Content Updates

HOLT PRESENTATION MAKER
KEYWORD: SW3 PPT17

ONLINE READING SUPPORT
KEYWORD: SS Strategies

CURRENT EVENTS
KEYWORD: S3 Current Events

Meeting Individual Needs

Ability Levels

Level 1 Basic-level activities designed for all students encountering new material

Level 2 Intermediate-level activities designed for average students

Level 3 Challenging activities designed for honors and gifted-and-talented students

English Language Learners Activities that address the needs of students with Limited English Proficiency

Chapter Review and Assessment

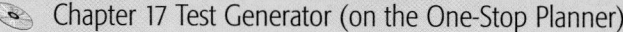

- Chapter 17 Test Generator (on the One-Stop Planner)
- Global Skill Builder CD–ROM
- HRW Go site
- **REV** Chapter 17 Review, pp. 400–01
- **REV** Chapter 17 Tutorial for Students, Parents, Mentors, and Peers
- **A** Chapter 17 Test (form A or B)
- **A** Alternative Assessment Handbook
- **A** Chapter 17 Test for English Language Learners and Special-Needs Students

Launch into Learning

Tell students that Russia occupies more land area than any other country in the world. With Ukraine and Belarus the region is even more vast. Have students use the mileage scale on the map on the opposite page to estimate distances between major cities or between physical features. *(Example: Kiev and Moscow are about 480 miles [772 km] apart.)* Then have students compare some of the mileage totals with distances on a United States map. Tell students they will learn more about how great distances affect the countries in this chapter.

Using the Physical-Political Map

Have students use the map on the opposite page to describe the general physical characteristics of Russia, Ukraine, and Belarus. *(Possible answers: mountains, flat lowlands, rivers, lakes, coastlines to the north and east)* Then ask students to suggest which areas might be most densely populated and which might be most sparsely populated. Have them support their responses by referring to the Unit 5 maps. *(The eastern part of the country is most useful for farming, hence the most densely populated. The central and western parts are least useful for farming, hence less populated.)*

Why We Should Know More

You may wish to highlight these reasons for learning more about Russia, Ukraine, and Belarus:

► In area, Russia is the largest country on Earth. It also has one of the largest populations.

► For many years, the Soviet Union and the United States were enemies. Now economic and cultural connections between the United States and Russia are increasing.

► As a nuclear power, Russia could pose a threat to U.S. national security.

► Russia continues to struggle with economic, environmental, and political problems that can affect people around the world.

► After being part of the Soviet empire, Ukraine and Belarus are rebuilding their economic and political systems. Americans may now invest in these countries.

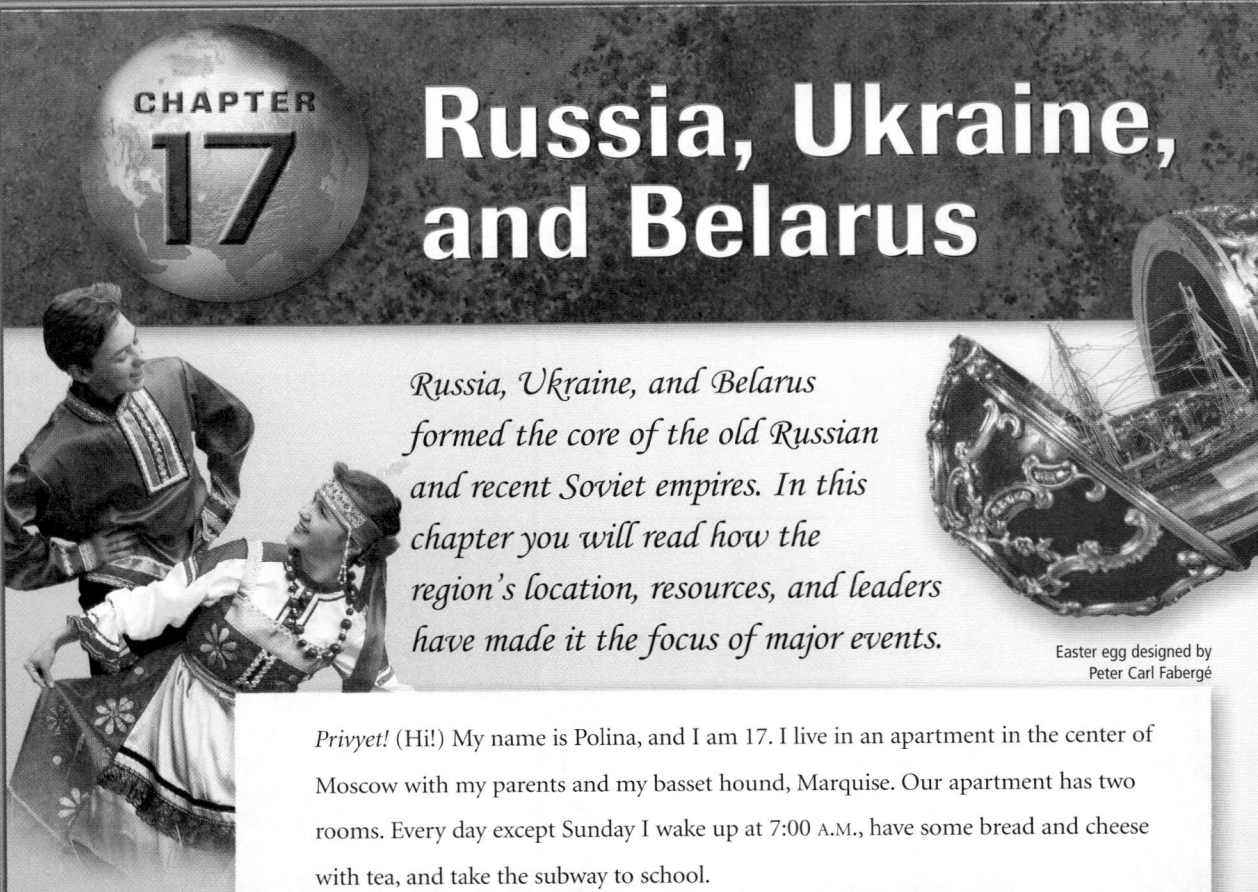

CHAPTER 17 — Russia, Ukraine, and Belarus

Russia, Ukraine, and Belarus formed the core of the old Russian and recent Soviet empires. In this chapter you will read how the region's location, resources, and leaders have made it the focus of major events.

Russian folk dancers

Easter egg designed by Peter Carl Fabergé

Privyet! (Hi!) My name is Polina, and I am 17. I live in an apartment in the center of Moscow with my parents and my basset hound, Marquise. Our apartment has two rooms. Every day except Sunday I wake up at 7:00 A.M., have some bread and cheese with tea, and take the subway to school.

I attend State School 637 and will graduate this spring. A few years ago we had to choose whether to study science or humanities. I chose humanities. My favorite subjects are history, literature, and English—my history teacher is great! During the day we have five or six classes with a 15-minute break between each one. During the breaks, I often eat a *pirozhki*, a small meat pie, at the school snack bar. At 2:00 P.M. I go home for lunch (meat, potatoes, and a salad of cooked vegetables and mayonnaise) and a nap. When I wake up, I go out with my friends to a park. On Sundays my friends and I cheer our favorite soccer team, Lokomotiv. This year, we took the train with other fans to matches in the Russian cities of St. Petersburg and Yaroslavl' and in Belarus.

In July I will take the entrance exams for university. I hope to win a place in law school, but if my exam results aren't good enough, I will go to night school. The best students get a free place in a state university just on the exam results, but other students have to pay tuition.

 **LET'S GET STARTED**

Copy the following instructions onto the chalkboard: *Look at the physical-political map at the beginning of Section 1. Choose a place on the map and write a sentence or two to describe what you think that place's natural environment might be like.* Discuss responses. Tell students that in Section 1 they will learn more about the physical geography of the region that includes Russia, Ukraine, and Belarus.

Building Vocabulary

Write the key terms on the chalkboard. Call on a volunteer to suggest a meaning for the word **icebreakers** based on its components. The word first appeared in print in 1875. Ask what this date might indicate about technology and global trade at that time. (*Possible answer: that technological innovation stimulated global trade*) Point out that **taiga** is a Russian word but that this type of forest can also be found in other countries that have cold winters and short growing seasons. Call on students to read the definitions.

Natural Environments

READ TO DISCOVER

1. What landforms and rivers are found in Russia, Ukraine, and Belarus?

2. What factors influence the region's climates and vegetation?

3. What natural resources does the region have?

WHY IT MATTERS

Russia has large coal deposits, but burning coal causes air pollution. Use **CNN fyi.com** or other **current events** sources to learn about industries that use coal and efforts to clean up the pollution they cause.

IDENTIFY

Eurasia

DEFINE

icebreakers
taiga

LOCATE

Baltic Sea
Black Sea
Ural Mountains

Locate, continued

Caucasus Mountains
Caspian Sea
Northern European Plain
Crimean Peninsula
Volga River
West Siberian Plain
Central Siberian Plateau
Kamchatka Peninsula
Lake Baikal
Murmansk
Sakhalin Island

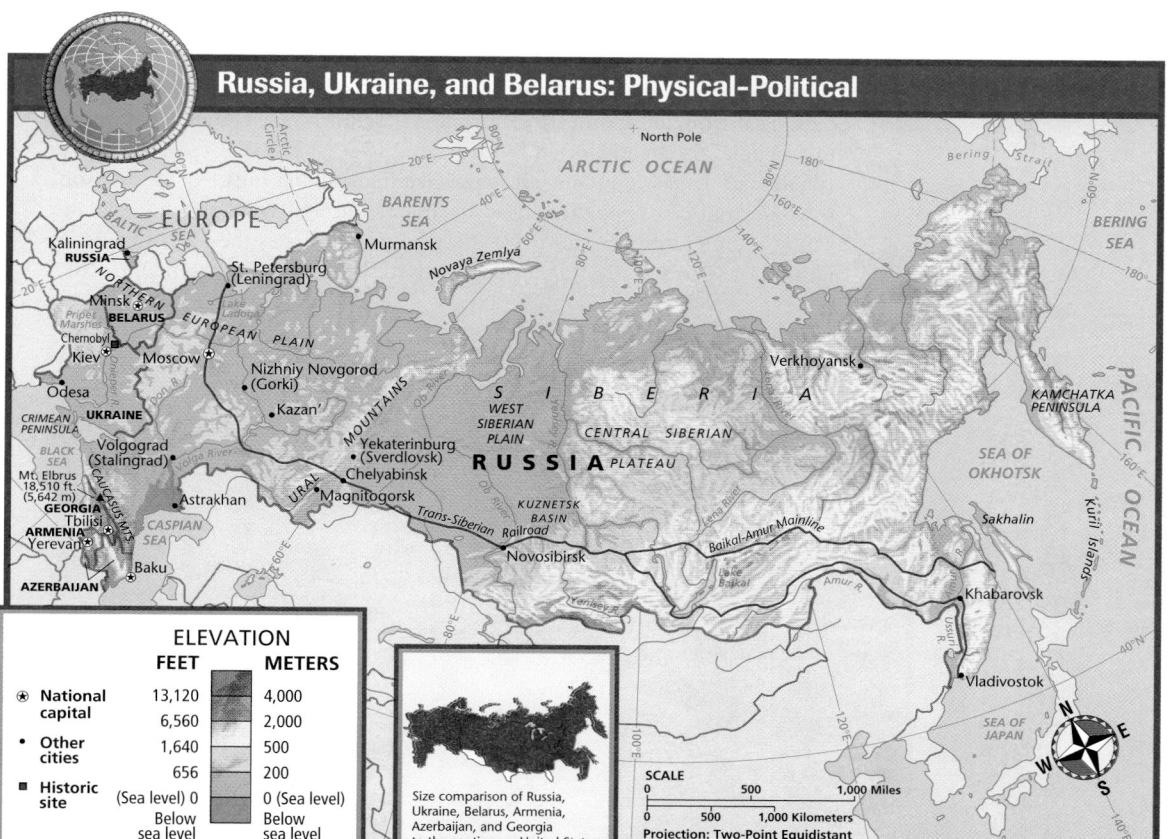

Russia, Ukraine, and Belarus: Physical-Political

North Pole

ARCTIC OCEAN

BARENTS SEA

BERING SEA

EUROPE

Kaliningrad
RUSSIA

Murmansk

Novaya Zemlya

St. Petersburg (Leningrad)

Minsk
BELARUS

Pripet Marshes

Chernobyl

Kiev

Moscow

Nizhniy Novgorod (Gorki)

Odesa

Kazan'

CRIMEAN PENINSULA
UKRAINE

Volgograd (Stalingrad)

BLACK SEA

Mt. Elbrus 18,510 ft (5,642 m)

GEORGIA

Tbilisi

ARMENIA
Yerevan

Baku

AZERBAIJAN

Yekaterinburg (Sverdlovsk)
Chelyabinsk
Magnitogorsk

Astrakhan

CASPIAN SEA

WEST SIBERIAN PLAIN

RUSSIA

CENTRAL SIBERIAN PLATEAU

S I B E R I A

Verkhoyansk

KUZNETSK BASIN

Trans-Siberian Railroad

Novosibirsk

Baikal-Amur Mainline

Khabarovsk

Vladivostok

KAMCHATKA PENINSULA

SEA OF OKHOTSK

Sakhalin

Kuril Islands

PACIFIC OCEAN

SEA OF JAPAN

BERING SEA

Bering Strait

ELEVATION

	FEET	METERS
National capital	13,120	4,000
	6,560	2,000
Other cities	1,640	500
	656	200
Historic site	(Sea level) 0	0 (Sea level)
	Below sea level	Below sea level

SCALE
0   500   1,000 Miles
0   500   1,000 Kilometers
Projection: Two-Point Equidistant

Size comparison of Russia, Ukraine, Belarus, Armenia, Azerbaijan, and Georgia to the contiguous United States

Teach Objectives 1–2

LEVEL 1: Have students work in pairs to create charts that describe the climates and vegetation of the main landform regions of Russia, Ukraine, and Belarus. Encourage students to use the Unit 5 maps and the chapter map. Then have students write phrases to identify and describe the region's main south-flowing and north-flowing rivers. Call on volunteers to write their charts and phrases on the chalkboard.
ENGLISH LANGUAGE LEARNERS, COOPERATIVE LEARNING

LEVELS 2 AND 3: Have students write an outline for a documentary film script about the physical geography of Russia, Ukraine, and Belarus. Outlines should mention the various landforms (*Ural, Caucasus, and Carpathian Mountains; Crimean Peninsula; West Siberian Plain; Central Siberian Plateau; volcanoes of the Kamchatka Peninsula*), rivers and other water features (*Dnieper, Don, Volga, Ob, Yenisey, Lena Rivers; Pripet Marshes; Lake Baikal*), climates (*humid continental, subarctic, tundra*), and vegetation types (*tundra, taiga, deciduous-coniferous forest, and steppe*). Ask students to write brief descriptions of some scenes from their films.

Across the Curriculum

► Science ◄

Ural Mountains An international team of scientists has discovered that the crustal plate under the Ural Mountains differs greatly from that under most mountain systems. These scientists used seismic waves to create geologic cross sections of the crustal plate. Like many of the world's mountain ranges, the Urals were created by the collision of two plates. In most cases, this has eventually led to a thinning of the crustal plate. Below the Urals, however, the crust was thickened by the impact and has not shrunk since the time of the collision—250 to 300 million years ago. This thickened crustal plate is what keeps Asia and Europe united as a single landmass.

ACTIVITY: Have students conduct research on the geological uses of artificial seismic waves in specific places and report their findings to the class.

internet connect

GO TO: go.hrw.com
KEYWORD: SW3 CH17
FOR: Web sites about Russia, Ukraine, and Belarus

Landforms and Rivers

Together Russia, Ukraine, and Belarus cover about 12 percent of the world's land area. Russia alone extends more than 6,000 miles (9,600 km) from east to west. The huge country stretches across **Eurasia** from the Baltic Sea and Black Sea to the Pacific Ocean. *Eurasia* is the name given to Europe and Asia when they are considered one landmass or continent. Russia is the world's largest country in area. No other country shares borders with more countries. Much of northern Russia lies above the Arctic Circle.

The Ural Mountains divide the region. Areas west of the Urals—including Ukraine and Belarus—are part of Europe. Those to the east lie in Asia. The part of Russia that is east of the Urals is known as Siberia. The region's remaining three countries—Armenia, Azerbaijan, and Georgia—are in the Caucasus Mountains. These high mountains lie between the Black Sea and the Caspian Sea. The highest point in Europe is on Russia's southern border with Georgia in the Caucasus Mountains. There Mount Elbrus soars to 18,510 feet (5,642 m). An active tectonic zone, the Caucasus region suffers from severe earthquakes.

Ice-age glaciers and long-term erosion shaped the broad plains that are the region's major landforms. Much of the European area shares the Northern European Plain with countries farther west. Thus, the European areas have low elevations. In fact, Belarus has no point over 1,135 feet (346 m) above sea level. Southern Belarus and northwestern Ukraine contain the Pripet Marshes. These marshes make up the largest swamp in Europe. Ukraine's highest point is located where the Carpathian Mountains cross the country's western borders. Peaks on the Crimean Peninsula in southeastern Ukraine, a popular tourist area, are slightly lower.

Russia's Ural Mountains are more like high rolling hills. For this reason, road and rail crossings there need no major tunnels. West of the Urals, the gently rolling terrain of the Volga River basin dominates the heart of Russia. East of the Urals is the thinly populated West Siberian Plain. The Ob River

Plains like the Ukrainian farmland pictured here spread for vast distances across Eurasia.

internet connect

GO TO: go.hrw.com
KEYWORD: SW3 CH17
FOR: Web sites about the Ural Mountains

LEVEL 1: Organize the class into groups and provide them with unlined index cards. Have each group use the Unit 5 maps and Section 1 to create several picture post-cards showing the main natural resources of Russia, Ukraine, and Belarus. On one side of the card, students should illustrate the resource. On the other side, ask students to describe where the resource can be found, its use, and other details. Then call on volunteers to draw a large sketch map of the region on butcher paper. Ask each group to read and display its postcards and then affix them to the sketch map in a place where that resource is plentiful. *(Examples: wood—Siberian taiga forests and forests west of the Ural Mountains; gold and diamonds—*

Siberia; oil and natural gas—Caspian Sea, between the Ural Mountains and the Volga River, Ob River basin; geothermal power—Kamchatka; coal mines—Ukraine) **ENGLISH LANGUAGE LEARNERS, COOPERATIVE LEARNING**

LEVELS 2 AND 3: Have students complete the Level 1 activity. Then ask them to speculate how the location of resources in the region may determine local economic activities.

---

creates a huge swamp area there. In the Russian Far East, beyond the Central Siberian Plateau, are high snowy ranges. Among these are the active volcanoes of the Kamchatka (kuhm-CHAHT-kuh) Peninsula.

The Dnieper, Don, and Volga are three of the largest south-flowing rivers in the region. These important shipping channels also supply water for hydro-electric projects and cities. The major Siberian rivers, such as the Ob, Yenisey, and Lena, flow northward to the Arctic Ocean from mountains in the south. One of the Yenisey's tributaries, the Angara River, flows through southern Siberia from Lake Baikal. Sometimes called the Jewel of Siberia for its beauty, Lake Baikal is the deepest lake in the world. It holds about one fifth of the world's freshwater!

✓ **READING CHECK:** *Physical Systems* What factors shaped the region's main landform type? Glaciers and erosion shaped the region's plains.

Climates and Vegetation

Russians sometimes joke that winter lasts for 12 months and then summer begins. As you can see on a map, much of the region is in the same latitudes as northern Canada and Alaska. The weather can be harsh. However, the region offers a wealth of resources to those who can brave the elements.

Climates Much of the country lies in the humid continental, subarctic, and tundra climate regions. During the year's five coldest months, rivers and canals throughout the region freeze. In these cold climates a polluted icy fog often hangs over cities during winter. Created by fumes and smoke from cities, this fog is trapped over the cities by the cold air. In the region's northern areas permafrost is widespread and deep. When the permafrost's surface layer melts in summer, buildings tilt, highways buckle, and railroad tracks slip sideways.

Harsh conditions prevail in the area's eastern two thirds. Any ocean winds that might bring moisture and moderate temperatures cannot reach far inland. As a result, parts of the interior are very dry. Siberia's severe winters often bring temperatures below −40°F (−40°C). At one of the coldest places outside of Antarctica, Verkhoyansk in Siberia, the thermometer has reached −90°F (−68°C).

The region's European third has the mildest climates. In addition, the soils there are better for agriculture and human settlement. Moisture from the Atlantic Ocean far to the west brings winter snow and summer rain. In the Russian Far East, coastal areas receive rain-bearing winds from the Pacific Ocean.

The cold climate and small amount of warm coastline reduce Russia's access to the sea. The Arctic Ocean can freeze all

Our Amazing Planet

Few creatures can live more than about 400 feet (120 m) below the Black Sea's surface. Too little oxygen and too much hydrogen sulfide create an environment that is poisonous to most life forms, including most bacteria.

Although in much of Siberia snowfall is relatively light, the cold temperatures ensure that the snow stays on the ground for months. The village of Ust'-Anzas, in southern Siberia, lies under a blanket of snow. The sign on the building tells travelers that inside they can buy tickets on Aeroflot, Russia's national airline.

Siberian Rivers Northern Russia's many rivers flow into the Arctic Ocean. They provide easy water transportation from far inland to the ocean, but they are blocked by ice for long periods every year. The thawing of these rivers creates an interesting phenomenon, since they thaw first near their southern sources. The resulting flow of water floods outside the frozen rivers' banks in the north and creates huge swamps. One of the largest is the Vasyuganye Swamp located at the confluence of the Ob and Irtysh Rivers. It covers some 19,000 square miles (49,000 sq km).

CRITICAL THINKING: Why might Siberia's rivers be important to Russia? *(Possible answers: The severe climate has made road-building difficult, and the rivers are the only north-south transportation routes in the area.)*

You don't say! The wels catfish of Russia can weigh as much as 600 pounds and grow to be 15 feet (5m) long.

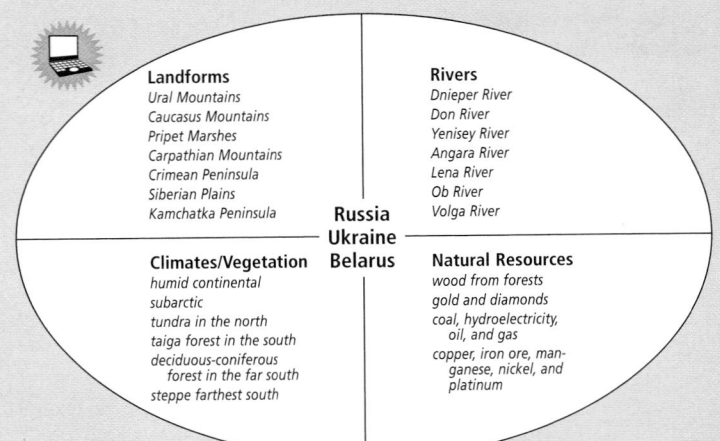

ALL LEVELS: Copy the graphic organizer at right onto the chalkboard, omitting the italicized answers. Call on students to complete it with the region's landforms, rivers, climate and vegetation types, and resources. **ENGLISH LANGUAGE LEARNERS**

LEVELS 2 AND 3: Have students complete the Level 1 activity. Then ask students to imagine they are attending a summer camp in Russia. Have students choose a place for the camp and then write a letter home. They should describe the land, weather, vegetation, and outdoor recreational activities.

Landforms
Ural Mountains
Caucasus Mountains
Pripet Marshes
Carpathian Mountains
Crimean Peninsula
Siberian Plains
Kamchatka Peninsula

Rivers
Dnieper River
Don River
Yenisey River
Angara River
Lena River
Ob River
Volga River

**Russia
Ukraine
Belarus**

Climates/Vegetation
humid continental
subarctic
tundra in the north
taiga forest in the south
deciduous-coniferous
  forest in the far south
steppe farthest south

Natural Resources
wood from forests
gold and diamonds
coal, hydroelectricity,
  oil, and gas
copper, iron ore, man-
  ganese, nickel, and
  platinum

Essential Element 2

▶ **Places and Regions** ◀

Magadan's Gold Eastern Siberia contains rich gold reserves, but connecting the resource to markets has been difficult.

Gold mining, begun in the 1830s, grew in the 1930s when Joseph Stalin developed Magadan as a transportation center on the Sea of Okhotsk. Millions of political prisoners were sent there to mine gold and to build a road connecting the sea with the Kolyma River gold fields. Many died from overwork, disease, and exposure to the harsh climate. Magadan was closed to foreigners until the collapse of the Soviet regime. Today foreign companies bid for mining licenses from the Russian government, and production continues to grow.

INTERPRETING THE VISUAL RECORD

The taiga forms a wide band across northern Russia from its western borders to its Pacific coast far to the east. It provides a wealth of forest resources. **How is taiga vegetation different from forests farther south?**

VISUAL RECORD ANSWER

Taiga contains coniferous trees, while mixed deciduous-coniferous forests grow farther south.

---

the way south to Russia's northern shores. Ship and barge traffic there requires using **icebreakers**. These are ships that can break up ice in frozen waterways. However, warm waters of the North Atlantic Drift reach around northern Norway to northwestern Russia. There you will find Murmansk, Russia's only large ice-free Arctic port.

Vegetation Differences in climate cause plant life to vary from north to south. Tundra vegetation grows along the northern coast. Low shrubs, mosses, and wildflowers are common there.

To the south is the **taiga**, a forest of mainly evergreen trees that covers half of Russia. Fir, larch, pine, and spruce are common. Farming is limited there because of the short growing season, acidic soils, and permafrost.

Farther south, in Belarus and in European Russia, you will find mixed deciduous-coniferous forest. This type of forest also grows along the coast of the Sea of Japan in the Russian Far East.

Still farther south is the drier grassland known as the steppe. Rich soil called *chernozem* (Russian for "black earth") has built up on the steppe. The grassland, long used for grazing, was plowed under by the 1800s. It has become one of the world's major grain-producing areas. In the past, people of the steppe fleeing invaders found safety in the taiga farther north. These landscapes are often featured in Russian literature.

✓ **READING CHECK:** *Places and Regions* Which vegetation area allows grain production on a large scale and why? steppe, because it has rich soil and a moderate climate

Natural Resources

Russia's forest, energy, and mineral resources are among the richest in the world. Yet much of this wealth was wasted because the government pushed production over conservation. Some of the remaining resources are in remote areas or are of low quality.

The taiga provides wood for building products and paper pulp. Steady logging west of the Ural Mountains has cleared many areas. However, in Siberia the taiga can provide forest resources for a long time to come. Eastern Siberia also has gold and diamond mines.

Coal, hydroelectricity, natural gas, and oil are the region's main energy resources. Huge oil reserves in the Caspian Sea area are being tapped by all the countries around the sea. Oil and gas fields between the Volga River and the Ural Mountains have been crucial to the region's development. They helped the Volga River basin become Russia's industrial heartland. Large reserves east of the Urals in the Ob River basin now supply

most of Russia's oil and gas. The world's largest network of pipelines carries fuel from that area to Moscow, St. Petersburg, and for export to Europe. Sakhalin (sah-kah-LEEN) Island and the Kamchatka Peninsula also have energy resources. Russia's first geothermal power station is in Kamchatka. Geothermal water is put to other uses too, such as heating greenhouses and fish farms.

Russia and Ukraine have many large coal mines. Those coal reserves could last for centuries. The region is also rich in metals, such as copper, gold, iron ore, manganese, nickel, and platinum.

✔ **READING CHECK:** *Environment and Society* Why is it hard for Russia to profit from some of its natural resources? Some are in remote areas or of poor quality; government policies wasted resource wealth.

INTERPRETING THE VISUAL RECORD

Limited access to the ocean restricts the region's fishing industry. Here fishers net sturgeon for their eggs where the Volga River flows into the Caspian Sea. Served fresh as caviar, the sturgeon's eggs are an expensive delicacy. Sturgeon are now threatened with extinction. **How might the Caspian Sea's location affect the price of caviar?**

 **Homework Practice Online** Keyword: SW3 HP17

Section 1 Review

Identify Eurasia

Define icebreakers, taiga

Working with Sketch Maps On a map of Russia, Ukraine, Belarus, and the Caucasus that you draw or that your teacher provides, label the Baltic Sea, Black Sea, Ural Mountains, Caucasus Mountains, Caspian Sea, Northern European Plain, Crimean Peninsula, Volga River, West Siberian Plain, Central Siberian Plateau, Kamchatka Peninsula, Lake Baikal, and Sakhalin Island. In the margin, identify Russia's only ice-free port.

Reading for the Main Idea

1. *Physical Systems* Why are earthquakes common in the Caucasus Mountains?

2. *Environment and Society* What are three of the European region's major south-flowing rivers? What are three functions that they serve?

3. *Places and Regions* Why is western Russia wetter than most of Siberia?

Critical Thinking

4. *Environment and Society* Why might developing Siberia's resources be difficult?

Organizing What You Know

5. Create a chart like the one shown below. Use it to describe the tundra, taiga, and steppe regions. Refer to this unit's climate and precipitation maps for more information.

	Climate	Soil conditions	Vegetation
Tundra			
Taiga			
Steppe			

OBJECTIVES

1. **Identify some major events in the growth of the Russian Empire.**

2. **Analyze how the Soviet Union developed and explain what life was like for its citizens.**

3. **Describe some features of the region's culture.**

Section 2 History and Culture

READ TO DISCOVER

1. What are some major events in the growth of the Russian Empire?
2. How did the Soviet Union develop, and what was life like for its citizens?
3. What are some features of the region's culture?

WHY IT MATTERS

Since the fall of the Soviet Union, tensions between ethnic Russians and other groups have grown. Use **CNNfyi.com** or other **current events** sources to learn about ethnic conflict in Russia.

IDENTIFY

Slavs	Cossacks
Rus	Bolsheviks

DEFINE

czar	autarky
serfs	gulag
abdicate	shatter belt
soviets	

LOCATE

Kiev	St. Petersburg
Moscow	Amur River
Sea of Okhotsk	Minsk

The Russian Empire

The roots of the Russian Empire lie in the grassy plains of the south. For thousands of years, people moved across the steppe, usually east to west. They came from what are now Mongolia, China, and the Central Asian republics. Bringing their herds with them, these peoples were often fleeing droughts and wars. Each wave of newcomers brought new ways of life to the region. The main people to settle in what are now Russia, Ukraine, and Belarus were the **Slavs**.

In the A.D. 800s the city of Kiev became an important center for trade between the Mediterranean and Baltic Sea areas. Among Kiev's early leaders were Scandinavian traders called **Rus** (ROOS). The name *Russia* comes from this word, which also referred to Slavic peoples in the region.

Merchants also traveled into the forests farther north. Over time these merchants founded new towns. Some of the towns that were located on high banks where rivers joined grew into cities. Moscow is an example. A prince ordered workers to dig ditches and build dirt walls on the site of an older settlement. The workers topped the dirt walls with a wooden wall. This fort became a large compound called the Kremlin. Its walls would eventually shelter Russia's government buildings, churches, and palaces.

INTERPRETING THE VISUAL RECORD *This intricately detailed gold necklace was made by the Scythians—one of the early peoples who moved across the Eurasian steppe. They flourished from the 700s to the 300s B.C. The Scythians were known for their skill in warfare and on horseback.* **Why do you think metalworking and jewelry making were valued art forms for a people such as the Scythians?**

Teach Objective 1

LEVEL 1: Copy the following graphic organizer onto the chalkboard, omitting the italicized answers. Call on students to provide words and phrases to identify and sequence some major events in the growth of the Russian Empire. Have students illustrate the time line.

LEVELS 2 AND 3: Have students write letters to a newspaper evaluating the actions of one of Russia's czars. *(Examples: praising Ivan IV for expanding the empire's boundaries or criticizing Peter the Great for moving the capital away from Moscow)* Letters should express a definite opinion and speculate about the effects of the czar's actions.

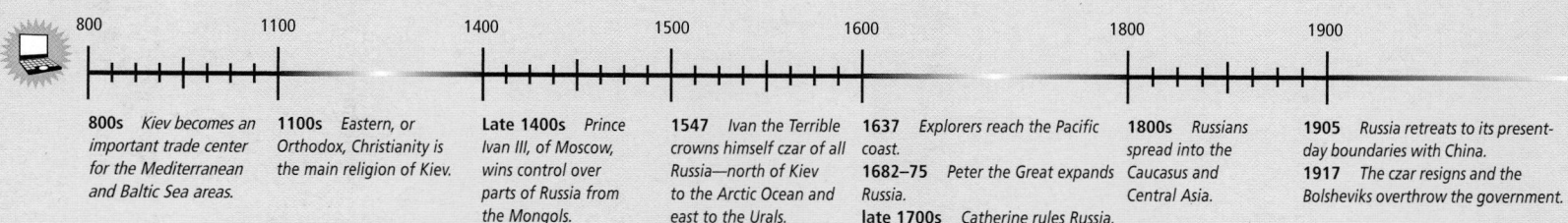

800	1100	1400	1500	1600	1800	1900

800s *Kiev becomes an important trade center for the Mediterranean and Baltic Sea areas.*

1100s *Eastern, or Orthodox, Christianity is the main religion of Kiev.*

Late 1400s *Prince Ivan III, of Moscow, wins control over parts of Russia from the Mongols.*

1547 *Ivan the Terrible crowns himself czar of all Russia—north of Kiev to the Arctic Ocean and east to the Urals.*

1637 *Explorers reach the Pacific coast.*
1682–75 *Peter the Great expands Russia.*
late 1700s *Catherine rules Russia.*

1800s *Russians spread into the Caucasus and Central Asia.*

1905 *Russia retreats to its present-day boundaries with China.*
1917 *The czar resigns and the Bolsheviks overthrow the government.*

Over time Christianity increasingly influenced the region. By the 1100s Eastern, or Orthodox, Christianity had become the main religion of Kiev. In 1240, Mongol invaders from Central Asia destroyed Kiev. They made the region the western outpost of their growing empire. For the common people, though, life went on much as it had before.

Conquest and Expansion   While the Mongols remained in power, several states emerged. The strongest was Muscovy, north of Kiev. Its chief city was Moscow. In the late 1400s Ivan III, the prince of Moscow, won control over parts of Russia from the Mongols.

In 1547 Ivan IV, who became known as Ivan the Terrible, crowned himself **czar** (ZAHR) of all Russia. The word *czar,* or *tsar,* comes from the Latin word *caesar* and means "emperor." Under Ivan IV the Russian Empire stretched from north of Kiev to the Arctic Ocean and east to the Urals.

Gradually Russian fur trappers, hunters, and pioneers migrated eastward into Siberia. By 1637, explorers reached the Pacific coast at the Sea of Okhotsk (uh-KAWTSK). **Cossacks,** a hardy people from the southern steppe frontiers, played an important role in the eastward expansion.

Russia gained some European territory under Czar Peter the Great, who ruled from 1682 to 1725. He took over lands along the Baltic Sea. He also expanded Russian control in what are now Belarus and Ukraine. Peter had St. Petersburg built for his capital. (See Cities & Settlements: St. Petersburg.) Catherine the Great ruled Russia during the late 1700s. She took the northern side of the Black Sea and encouraged settlers to move to the Volga region. This expansion brought many non-Russian peoples within the Russian Empire.

In the 1800s Russians spread into the Caucasus and Central Asia. Much of the population there was Muslim. For a brief time, Russia controlled what is now Alaska. There was even a Russian fort and farming settlement in California. By 1860 Russia had taken much of the Amur (ah-MOOR) River region that had been claimed by China. After Russia lost a war with Japan (1904–05), the country retreated to its current borders with China and North Korea.

End of an Empire   Russia started to industrialize by the late 1800s, but it remained largely a country of poor peasant farmers. These farmers, called **serfs,** worked for a lord. Serfs were bound to the land, which means they could not leave the lord's land permanently without his permission. The serfs were freed in the 1860s, but rural poverty did not end. Soon life got worse for many Russians. Poor harvests led to food shortages. There was also an economic depression. By the start of World War I in 1914, the foundations of Russian society were on shaky ground. Russia suffered huge losses in the war, and social and economic problems worsened. Finally, the czar was forced to **abdicate,** or resign, in early 1917. A republic was set up but had little success. In the fall of 1917, a small group called the **Bolsheviks** overthrew the government, an event known as the Russian Revolution. The czar and his family were killed.

✔ **READING CHECK:** *The Uses of Geography*   What are some factors that led to the fall of the czar? poverty, food shortages, losses in World War I

In Russian, Ivan IV was called Grozny, which means "Awe Inspiring." He was indeed terrible at times, as he brutally suppressed the noble class—or boyars—and lashed out at enemies. Yet during Ivan's reign, the empire expanded, and printing was introduced to Russia.

INTERPRETING THE VISUAL RECORD

Vladimir Lenin gives a speech during the Russian Revolution. Lenin believed in what he called "the dictatorship of the proletariat." Proletariat means "working class." **How would Lenin's plan for government contrast with what you have already learned about dictators or dictatorships?**

VISUAL RECORD ANSWER

authoritarian rule by an entire class of people, instead of by one person

ALL LEVELS: Organize the class into four groups to compose and perform short plays about the Soviet Union. Have two groups work on the development of the Soviet Union and the other two work on the daily lives of Soviet citizens. First, all groups should determine and record which major events or points they want to cover. Then have them compose their plots, select actors for specific characters, write dialogue and narration, create simple props, and practice their lines. Encourage students to include speaking and nonspeaking roles in their plays.

Before the groups perform their plays, have them write their designated main events or points on the chalkboard. Stage the two plays about the development of the Soviet Union first. Discuss the thoroughness with which the groups covered their main events and how they presented different aspects or interpretations of the same events. Then stage the plays about life in the Soviet Union and lead a discussion about the same questions. **ENGLISH LANGUAGE LEARNERS, COOPERATIVE LEARNING**

Stalin and Nationalism
Although he was from the Caucasus republic of Georgia, Joseph Stalin was a proponent of Russian nationalism. He sought to modernize the Soviet Union by emphasizing Russian history, culture, language, and national heroes over those of non-Russian republics, including his native Georgia. He strengthened his control by refusing to allow non-Russians to hold important offices in the republics or to serve as diplomats. Russians also controlled most of the Soviet Union's industry. To prevent other ethnic groups from gaining power, Stalin divided them by redrawing boundaries between and within republics. His divisions kept non-Russian groups from representing an overwhelming majority in any single republic.

CRITICAL THINKING: How might Stalin's actions affect stability in republics today? *(Possible answer: Ethnic groups want to unite and to redraw boundaries but have met with resistance from other groups who share their territory, resulting in violence.)*

Items of American popular culture are imported and available, when they would have been rare during the Soviet era.

These workers at a collective farm share a meal. Collective farms, or kolkhozy *in Russian, were made up of many small holdings grouped into a single unit for joint operation under government supervision. Peasants were forced to join* kolkhozy.

INTERPRETING THE VISUAL RECORD

A child and her mother walk across Red Square in Moscow. **What clues in the photograph may indicate that regional trade patterns have changed since the Soviet era?**

The Soviet Union

The Bolsheviks, led by Vladimir Lenin, wanted to remake Russia using the ideas of German philosopher Karl Marx. Marx thought that the people of the working classes were victims of capitalism. Like Marx, Lenin thought the solution was communism. Under communism, the workers were to elect governing local bodies called **soviets** to pass laws and make decisions. The Russian Empire was renamed the Union of Soviet Socialist Republics (USSR), also known as the Soviet Union. The Soviet Union eventually included 15 republics, each based largely on ethnic territories.

Life in the Soviet Union   The Soviet Union soon became a one-party, totalitarian state led at first by Lenin. After Lenin's death in 1924, Joseph Stalin took power. Stalin's brutal rule lasted until 1953. Both Lenin and Stalin tried to promote a single Soviet culture. They had names of cities and streets changed to honor communism's heroes. In addition, because it was the language of the political leadership, Russian spread to non-Russian ethnic groups.

Soviet economic planners set up a command economy. They also followed a policy of **autarky** (AW-tahr-kee). Under this system a country tries to produce all the goods that it needs. Trade with capitalist nations was very limited. Without competition, however, efficiency and product quality often fell. Production of consumer goods and services lagged far behind that of the United States and Western Europe.

The government ran large state farms, but agriculture faced constant problems. Food production was often low on the state farms. Millions of peasants died of starvation or in prison during the forced change to the new farming methods. Small private plots, which families worked in their spare time, produced about one fourth of the country's food.

Personal freedoms were strictly limited. People who disagreed with Communist leaders could be jailed. Under Stalin, millions were sent to terrible labor camps. Many of those camps were in the far north, both east and west of the Urals. This network of labor camps was called the **gulag**. Soviet leaders also tried to stop religious worship. They believed that religion would lessen people's loyalty to the state. Many Christian, Jewish, and Muslim houses of worship were closed or destroyed.

Yet the Soviet government did have some successes in education and health care. For example, by the 1980s some 90 percent of the people could read and write. Many people, including women, became doctors. In fact, basic health care was free and widely available. Most able workers had jobs.

A New Beginning   Finally, the government allowed some economic and political changes in the 1980s. However, the Soviet Union began to fall apart in 1990 and collapsed at the end of 1991. Each of the 15 former Soviet republics became independent. The new countries kept the same boundaries as the old republics even when they divided ethnic groups.

Life changed quickly for the people of the former Soviet Union. Today citizens can finally choose among candidates in elections. News from around the

LEVEL 1: Organize the class into five groups and assign one of the following topics to each group: people and languages; the peoples of the Caucasus republics; settlement; religion and education; and food, traditions, and customs. Have each group create a scrapbook of information that describes and analyzes the assigned topic. Ask each group to choose which detail within the scrapbook the members feel is the most interesting and to highlight it with a special page.
ENGLISH LANGUAGE LEARNERS, COOPERATIVE LEARNING

LEVELS 2 AND 3: Have students complete the Level 1 activity. Then ask students to create two lists—one of traditional ways of life that are being maintained by cultures of Russia, Ukraine, or Belarus, and of cultural changes that are occurring within these countries. Lead a discussion about which trend seems to be more common in the region—cultural change or the maintenance of traditional ways of life. When students have decided which trend they feel is more common, challenge them to formulate theories about reasons for such a tendency.

world now flows more freely. Religious freedoms have also expanded. In addition, communism is being replaced by capitalism. Shoppers can buy new consumer products. American fast food companies have opened restaurants there. In the new market economies, many businesses that had been owned by the government are in private hands. In many ways, however, the rapid change has caused severe hardships. You will read more about these problems in Section 3.

READING CHECK: *Human Systems* What are some ways that life has changed since the Soviet Union collapsed? freer elections, more information, expanded religious rights, more goods to buy, businesses in private hands

Among the peoples experiencing change are the Khanty of the Ob River basin. They are trying to protect their land from damage done by the oil industry.

Culture

Russia, Ukraine, and Belarus share a strong sense of cultural identity. There are many similarities in language, religion, and customs. However, there is great cultural diversity within Russia. In fact, Russia has at least 60 different ethnic groups.

People and Languages Language is an important source of national identity in the region. At least 85 percent of Russians are Slavs and speak Slavic languages. The region's Slavic languages are written in the Cyrillic alphabet, which was developed from an ancient Greek script. More than 95 percent of Ukrainians and about 98 percent of Belarusians are also Slavic. In fact, the great majority of Eurasia's more than 300 million Slavs live in these three countries.

As the Russian Empire grew, it pulled in many non-Slavic peoples. During the Soviet era, the lands where these non-Slavic peoples lived became special republics within the country. Russia has 21 of these republics today. Members of non-Slavic groups there speak different languages. Many books have been published in these languages. However, in most of the republics, Russian speakers are in the majority. In some of the republics, the non-Slavic languages are disappearing.

FOCUS ON HISTORY

The Peoples of the Caucasus Republics Some of Russia's ethnic republics are located in the Caucasus region in the south. Also in the Caucasus are the former Soviet republics of Armenia, Azerbaijan, and Georgia. Those three republics are independent countries today. The ethnic republics and countries in the Caucasus lie in a band of land that separates the Black and Caspian Seas. This area is made up of the high rugged Caucasus Mountains. The region's different ethnic groups developed within the hundreds of small isolated valleys in this mountain range.

The Caucasus is also what geographers call a **shatter belt**. It is a zone of frequent boundary changes and conflicts. Often shatter belts are located between major powers. Throughout history, peoples from the south—Turks, Persians, Arabs—and the north—Russians, Mongols, Tatars—have fought over the Caucasus. Ethnic tensions still trouble the region.

Republics of the Russian Federation and Ethnic Composition

Republic	Ethnic Composition
Adygea	Russian
Alania	Ossetian
Bashkortostan	Russian
Buryatia	Russian
Chechnya	Chechen
Chuvashia	Chuvash
Dagestan	Avar
Gorno-Altay	Russian
Ingushetia	Ingush
Kabardino-Balkaria	Kabard
Kalmykia	Kalmyk
Karachay-Cherkessia	Russian
Karelia	Russian
Khakassia	Russian
Komi	Russian
Mari El	Russian
Mordvinia	Russian
Sakha	Russian
Tatarstan	Tatar
Tyva	Tyva
Udmurtia	Russian

Source: Centre for Russian Studies

INTERPRETING THE CHART

The chart lists the ethnic group that makes up the majority in each of the republics of the Russian Federation. Note that the Russian Federation is the country's formal name.

Daily Life

Learning the Alphabet in Tatarstan Tatarstan, which straddles the Volga River, lies in the center of Russia. Most of its people are descendants of the Mongols.

For about 1,000 years the Tatar language was written in the Arabic alphabet. In 1927 reformers and the Bolsheviks changed to using the Latin alphabet. Then in 1939 Stalin forced the Tatars to switch to the Cyrillic alphabet, in which Russian is written. Today the local language is again written with Latin letters. In 2001 all first-graders in Tatar schools were required to learn their language using Latin letters. Some Russian officials feel that using the Latin alphabet is an attempt to reduce the power Moscow has over the Tatar people.

DISCUSSION: Might the alphabet change cause problems for Tatarstan? Should a country require one national alphabet? Why or why not?

internet connect

GO TO: go.hrw.com
KEYWORD: SW3 CH17
FOR: Web sites about the Tatar people

Teacher to Teacher

Steve Gargo of Appleton, Wisconsin, suggests the following activity to help students understand settlement patterns in the region. Provide each student with a map of Russia, Ukraine, and Belarus. Have students draw lines connecting St. Petersburg, Russia, Lake Baikal, and Odesa, Ukraine, to form a triangle. Point out that a large percentage of the region's population lives within that triangle. Have students use the Unit 5 atlas and the chapter text to explain why relatively few people live beyond these lines.

(Examples: north—poor, acidic soils and short growing season; south—not enough rainfall, high mountains) Why is more of the triangle in European Russia than in Asian Russia? (Possible answer: receives more moisture, allows more farming)

HOMEWORK: Have each student create a poster depicting aspects of Russian culture suggested by the subsections. Ask students to use a combination of text, graphics, pictures, and captions on their posters.

Cultural Kaleidoscope

Baku The place that is now Baku, Azerbaijan, was once a target for conquest by Persians, Greeks, and Romans. Visitors today can see evidence of later periods. Walls surround the old town, where buildings reflect Islamic and medieval cultures. In the late 1800s international oil barons, who sought to control Baku's rich oil reserves, built a boomtown there. Buildings from this era include grand mansions and museums. A new town was built in the 1900s when the Soviet Union controlled the country.

GRAPH ANSWER

World War I, the Russian Revolution, the Civil War, and World War II killed more men because they were in the armed forces. Decreasing numbers of young people indicate fewer people are entering childbearing years.

Section 2 Review Answers

Identify For identifications, see: Slavs, p. 386; Rus, p. 386; Cossacks, p. 387; Bolsheviks, p. 387

Define For definitions, see: czar, p. 387; serfs, p. 387; abdicate, p. 387; soviets, p. 388; autarky, p. 388; gulag, p. 388; shatter belt, p. 389

Georgia reached a golden age during the reign (1184–1212) of Queen Tamara. When she rallied her troops before going into battle, the soldiers cheered their "king" Tamara. At the time, there was no word for *queen* in the Georgian language.

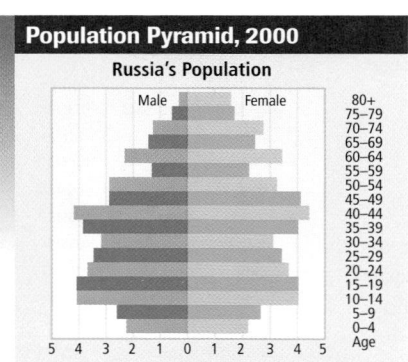

Population Pyramid, 2000
Russia's Population

Male / Female

80+ / 75–79 / 70–74 / 65–69 / 60–64 / 55–59 / 50–54 / 45–49 / 40–44 / 35–39 / 30–34 / 25–29 / 20–24 / 15–19 / 10–14 / 5–9 / 0–4 Age

5 4 3 2 1 0 1 2 3 4 5
Percent of Population

Source: U.S. Census Bureau, International Database

INTERPRETING THE GRAPH *From what you know about European and Russian history, what events might account for the larger female populations in the older age ranges? How can you tell that Russia's population is declining?*

The region's physical geography and history have shaped its cultural geography. For example, many different languages are spoken in the Caucasus today. The three main languages are very different from each other. In Azerbaijan a Turkic dialect is most common. Except for some words borrowed from Persian, Armenian seems to be unrelated to any other living language. Georgian is one of few members of the South Caucasian language family. Some people believe the language might be related to northern Spain's Basque tongue, the origins of which are mysterious.

Religions here are as diverse as languages. The Armenian Christian Church is very old. A majority of Georgians belong to an Eastern Orthodox Church that is independent from the Russian Orthodox Church. Most Azerbaijanis and some people in Russia's southern republics, like Chechnya, are Muslim. Near the northeast edge of the Caucasus are Mongolian Kalmyks, whose faith is similar to Tibetan Buddhism.

✓ **READING CHECK:** *Environment and Society* How have the physical geography and history of the Caucasus affected the region's cultural diversity?

various peoples separated by mountains and valleys; in the path of invasions because of location between seas

Settlement Just 25 percent of Russia lies in Europe. However, 80 percent of its population lives there. Russians east of the Urals are concentrated in a southern corridor of transportation routes, warmer weather, and steppe environments.

Russia, Ukraine, and Belarus all have many large cities. More than two thirds of the population lives in cities. More than 9 million people live in and around Moscow, the region's most populous urban area. More than 5 million live in St. Petersburg to the northwest. Kiev, Ukraine's capital, and Minsk, the capital of Belarus, are also among Europe's larger cities.

All three countries are losing population. Many people have emigrated. Also, the death rate is higher than the birthrate. Poor health-related behaviors, including heavy smoking and alcohol abuse, are some key reasons for this trend. In addition, the collapse of the Soviet Union plunged many people into poverty. The old health care system also fell apart, cutting people off from medical care. No other part of the world has seen such population losses.

Religion and Education Even after years of Communist rule, almost every city and village in the region has a prominent Christian church. The main religion is Eastern Orthodox Christianity. Church architecture often features an onion-shaped dome. Only the parts of Belarus and Ukraine bordering Roman Catholic Poland and Slovakia have many Catholic churches. Protestant churches are rare but increasing due to recent missionary activity. Also increasing is the number of Islamic mosques in the larger cities and in the Volga and Caucasus areas. Muslim minorities are common there.

Russia, Ukraine, and Belarus have inherited from the Soviet Union an emphasis on education and on scientific and technical training. They stress not just engineering but also the arts, humanities, and foreign languages. The best schools are in the national capitals of Kiev, Minsk, and Moscow.

Food, Traditions, and Customs   Food reveals the influences of cold climates. Over much of the north, hardy grains grown are barley, oats, rye, and wheat. Dark rye bread and barley soup are common foods. Small buckwheat pancakes called blini are served with sour cream. Cold-weather vegetables like beets, cabbage, and potatoes go into borscht. Borscht is a traditional soup that sometimes has meat.

Belarusians, Russians, and Ukrainians often drink tea. This preference comes from centuries of ties to tea-growing areas nearby in the Caucasus and Central Asia. Fruit juices and spices are often added.

Environmental differences between the forested north and the open steppes of the south are seen in rural architecture. The people of the forested north used wood to fashion their cottages, churches, and other buildings. Elaborate wood carving decorates the front of homes and other buildings. On the steppe, where there is much less wood, people often built sod homes. Roofs and walls were blocks of grassy turf. Sometimes people dug the buildings partly into the ground. This helped keep buildings cooler in summer and warmer in winter. However, today most city people of all regions live in large apartment houses.

Outside of the big cities are country cottages, called dachas (DAH-chuhs). Members of the urban middle class spend weekends and holidays in these homes. Keeping a dacha was also a way to escape being spied upon during the Soviet era.

Dachas range from mere sheds to quaint cottages, like this one, and ornate palaces. They offer relaxation, relief from city pollution, and a place to raise vegetables. Much of Russia's food is grown on dacha land.

✓ **READING CHECK:** *Human Systems*   What are some traditions that have survived changes in government? religious preferences, emphasis on education, food and drink choices, popularity of dachas

---

Section 2 Review

go.hrw.com
Homework Practice Online
Keyword: SW3 HP17

Identify
Slavs, Rus, Cossacks, Bolsheviks

Define czar, serfs, abdicate, soviets, autarky, gulag, shatter belt

Working with Sketch Maps   On the map you created in Section 1, label Russia, Ukraine, Belarus, Kiev, Moscow, Sea of Okhotsk, St. Petersburg, Amur River, and Minsk. In the margin of your map, name the two U.S. states where Russia held land during the 1800s.

Reading for the Main Idea

1. *The World in Spatial Terms*   What lands were added to the Russian Empire under Ivan IV, Peter the Great, and Catherine the Great?

2. *Human Systems*   What subjects do the region's educational systems emphasize? What other subjects do students learn?

Critical Thinking

3. **Analyzing**   Why does Russia have many non-Slavic peoples? What factors account for the cultural diversity of the Caucasus?

4. **Making Generalizations and Predictions** Using data and graphics from the chapter and the unit population map, describe the population characteristics of modern Russia. Do you think Russia's population will grow or decline in the near future? Why?

Organizing What You Know

5. Create a graphic organizer like the one shown below. Use it to provide information about the major religions of the region.

Eastern Orthodoxy	Roman Catholicism	Protestantism	Islam

Art: Depictions of Napoléon

Throughout history, military and political leaders have called upon artists to immortalize them by recreating their images in stone or on canvas. Napoléon, who was like many rulers in this regard, did not want to be remembered for the defeat represented in Minard's map. He preferred the image created by numerous painters and sculptors. Napoléon even named an official "First Painter," Jacques-Louis David (DAH-VEED) (1748–1825), to chronicle his reign. David's series of paintings, even those created after the emperor's defeat in Russia, portray him as a conquering hero. Similar works by other artists convey the same impression.

Show students or have them locate at least two of David's paintings or other depictions of Napoléon I from the early 1800s. If possible, have students select one work completed before the Russian campaign, and one after. Tell students to write an analysis comparing the depictions to the image of Napoléon conveyed by Minard's map. Have them address the following questions: How is Napoléon depicted in each piece? What images are evoked by each work? Has the presentation changed from the earlier work to the later one? Judging from what you know about Napoléon, how accurately would you say each work portrays him? You may want to have students examine portraits of other figures from Russian history for clues to how they wanted to be known and were known. Portraits, including cartoons, of Peter the Great are particularly interesting.

Across the Curriculum

▶ Music ◀

The 1812 Overture Every Independence Day Americans from coast to coast listen to Pyotr Ilich Tchaikovsky's *1812 Overture.* The piece was actually written as an anthem of Russian pride. Tchaikovsky composed the piece for the consecration of a Moscow church built as a memorial to the 1812 defeat of Napoléon's army. The overture begins with themes from a Russian hymn. At the composition's heart is a depiction of a battle between the Russian and French armies. Tchaikovsky used passages from the Russian national anthem, "God Save the Czar," and the national anthem of France, "La Marseillaise," to stage the "battle." The overture's jubilant conclusion, accented with bells and thundering blasts from actual cannons, celebrates the French defeat.

Applying What You Know Answers

1. army's size at different times, route the army followed, rivers crossed, major towns or cities, temperatures
2. Possible answers: auxiliary troops, soldiers protecting the army's path into Russia, minor skirmishes, part of the army that was delayed and did not make it all the way to Moscow

This Geography for Life feature addresses National Geography Standards 1, 13, 15, and 17.

The Uses of Geography

Geography for Life

Mapping Napoléon's Russian Disaster

Maps, diagrams, and graphs are widely used to display geographic information. The geographer must choose the best way to represent the relevant information. No single map can show everything. However, certain kinds of maps can show a remarkable amount.

Consider the approach that a French engineer took in creating a famous historical map. In 1861 Charles Joseph Minard illustrated French emperor Napoléon's 1812 invasion of Russia. On the eve of the invasion, Napoléon dominated much of Europe. However, his campaign in Russia was a disaster.

Minard's design tells a sad tale of death and misery. He used a shaded band to illustrate the changing size of Napoléon's army. Look at the left edge of the map, which shows the Polish-Russian border near the Neman River. Minard drew a thick band representing the 422,000 soldiers who swept into Russia across the river. He narrowed the band's width to show how battle losses gradually shrank the army's size as it marched eastward. When he reached Moscow, Napoléon led just 100,000 troops. After the people of Moscow burned the city, the army turned around in October to return to France. Minard drew a darker, lower band to illustrate the retreat. A temperature scale across the bottom of the map is tied to the darker band. As you can see, bitter cold weather took a terrible toll on the army. Many soldiers froze or starved to death. After crossing the Berezina River, the army straggled back to Poland. Only about 10,000 soldiers survived the complete journey.

This illustration shows the army's size, its location on certain dates, its route, and temperatures the soldiers faced. Displaying a series of events that occurred over a vast space and several months is not easy. We might call Minard a storyteller as well as an engineer and cartographer.

Applying What You Know

1. **Summarizing** What features and information did Minard show on his map?

2. **Drawing Inferences and Conclusions** What do you think is shown by the extra arms that branch off of the main bands?

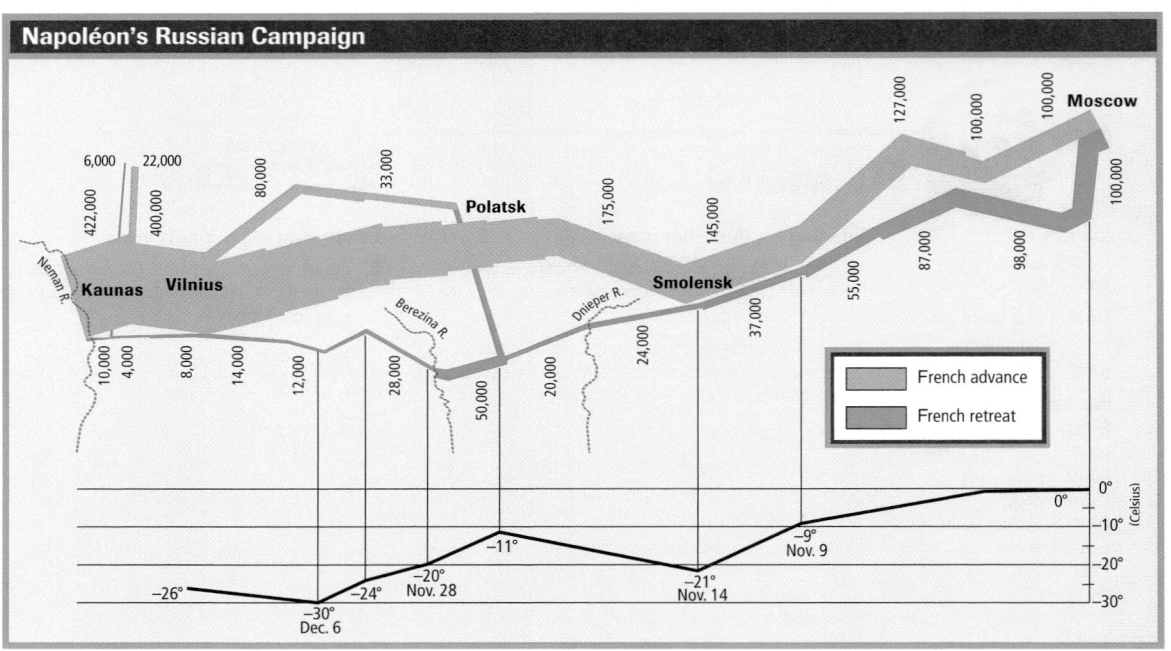

Napoléon's Russian Campaign

OBJECTIVES

1. **Explain how the economies of areas within Russia, Ukraine, and Belarus have developed.**

2. **Identify challenges faced by the region.**

Copy the following passage onto the chalkboard: *You have learned that Russia and its neighbors have undergone great changes since the collapse of the Soviet Union. Do you think that the region's people are better or worse off than they were 20 years ago? Discuss responses. (Answers will vary.)* Point out that the region's residents also hold a wide range of opinions on the subject. Tell students that they will learn more about the many issues facing the region in Section 3.

Building Vocabulary

Write the terms **light industry** and **heavy industry** on the chalkboard. Ask students to speculate what the adjective before *industry* in each of these terms does to the meaning of the term. (Light *and* heavy *may reflect the weight of the product created by the industry.)* Then have students locate and read aloud the definition of **smelters**. Call on a volunteer to describe the relationship between the terms *smelters* and *heavy industry. (Smelters process metal ores that are primarily used in heavy industry.)*

The Region Today

READ TO DISCOVER

1. How have the economies of areas within the region developed?
2. What challenges does the region face?

WHY IT MATTERS

Any major economic and political changes in the world's largest country are of worldwide interest. Use **CNNfyi.com** or other **current events** sources to learn about the latest developments in Russia.

DEFINE

light industry
heavy industry
smelters

LOCATE

Vladivostok   Kuril Islands
Khabarovsk   Chernobyl

RESOURCES

REPRODUCIBLE
- Guided Reading Strategy 17.3
- Graphic Organizer Activity 17
- Readings in World Geography, History, and Culture 41 and 42
- Cultures of the World Activity: Region 4
- Map Activity 17: Russia's Environmental Problems

TECHNOLOGY
- One-Stop Planner CD–ROM, Lesson 17.3
- Homework Practice Online
- HRW Go site

REINFORCEMENT, REVIEW, AND ASSESSMENT
- Main Idea Activity 17.3
- English Audio Summary 17.3
- Spanish Audio Summary 17.3
- Section 3 Review, p. 397
- Daily Quiz 17.3

Economic Development

Belarus, Russia, and Ukraine are changing their economies to compete in new markets. The countries are working to develop **light industry**. Light industry focuses on the production of consumer goods, such as clothing or housewares. **Heavy industry**, which usually involves manufacturing based on metals, is becoming less important. Cities are becoming more like those in richer countries. New shopping centers, stores, and sidewalk stalls are opening. Paint and better maintenance brighten old apartment houses. Single-family houses, even some luxury homes, are being built.

The Moscow Region Moscow, with its huge Kremlin, has symbolized Russia for centuries. The city became the home of the Russian Orthodox Church in the 1300s and Russia's capital in the 1400s. Most Russians have looked to Moscow as their country's heart and soul. This was true even while St. Petersburg was the capital from 1712 to 1918.

Today greater Moscow is Russia's most important economic region. It is the national center of communications, culture, education, finance, politics, and transportation. More than 70 institutions of higher learning are there. As a result, Moscow's economic advantages are many. Roads, rails, and air routes link the capital to all points in Russia. The city's location also gives its businesses access to raw materials and labor.

The economic region around Moscow stretches for many miles in all directions. Millions of Russians live and work within the area's network of transportation routes and job sites. Among the transportation links is the world's busiest subway. The area also has electrified railroads and a major beltway.

INTERPRETING THE VISUAL RECORD

Two women relax near their salon at GUM, a Moscow shopping mall of more than 150 stores that receives some 300,000 visitors per day. GUM stands for Gosudarstvenny Universalny Magazine, or "State Department Store." **How does GUM compare to your community's shopping centers?**

VISUAL RECORD ANSWER

Possible answers: much larger, more shoppers

LEVEL 1 AND 2: Copy the following graphic organizer onto the chalkboard, omitting the italicized answers. Have students complete it with the major points related to the economic development of areas within the region. Then point out that all the resources and industries were once within the same country but are now distributed among several countries. Ask students what the implications of this situation might be. *(Example: What if Ukrainian manufacturers refused to sell diesel tractor engines to industries building new factories in Siberia? How might those industries be affected?)*
ENGLISH LANGUAGE LEARNERS

Economic Development

St. Petersburg Region
Westernized, good transportation, trade with European cities, draws tourists and high-tech industries

Moscow Region
economic center, transportation hub, access to raw materials, millions of Russian workers

Ukraine and Kiev
centrally located; rich agricultural, energy, human resources; agriculture; heavy industry; metalworking

Volga and Urals
heavy industry, abundant hydroelectricity, refineries and petro-chemicals, mineral resources and smelters

Siberia
Trans-Siberian Railroad; lumber, mining, and oil; small labor force

The Russian Far East
factories, forest and mineral resources, naval bases, commercial fishing

Belarus and Minsk
few resources, educated labor force, wood products industries, outdated plants

Global Perspectives

Vladivostok At the end of the long Trans-Siberian Railway the traveler arrives in Vladivostok. The city, whose name means "Rule the East," was founded in 1860 as a military outpost. Twelve years later, the main Russian naval base was moved to this Pacific Ocean port.

Vladivostok played an important role in World War I. It was the chief entry point for supplies and equipment sent by the United States. During the Soviet era the Pacific Fleet was headquartered in Vladivostok. To protect military secrets, the city was closed to foreigners. Not until Soviet power started to falter in 1990 was Vladivostok reopened to foreign shipping or visitors. Now it is a vital port for Russia's trade with China, Japan, and other countries.

The city now receives news coverage that would have been unthinkable a few years ago. For example, in 1997 a Norwegian news source reported on a radioactive waste leak in the Vladivostok area.

Ballet dancers perform Pyotr Tchaikovsky's Swan Lake *at the Mariinsky Theater in St. Petersburg. Tchaikovsky was one of Russia's many great composers.*

The St. Petersburg Region Moscow reflects Russia's old values and traditions. In contrast, St. Petersburg represents the country's desire for Western ideas and practices. Located on the Gulf of Finland, it has been called the Venice of the North for its many canals. St. Petersburg has good transportation facilities. The city's location also eases trade and transportation links with other European cities. Major products include chemicals, machinery, ships, and textiles. Many cultural attractions and universities draw tourists and high-tech industries.

The Volga and Urals Regions Heavy industry lines Russia's Volga River and the Ural Mountains. Hydroelectricity is abundant there. Dams that produce power have also turned the Volga into a chain of lakes. Refineries and petrochemical plants process oil and gas. Russia's largest car and truck factories are in the area.

Nearly every important mineral except coal and oil has been discovered in the Urals. These resources laid the base for industrial development. Copper and iron **smelters**, factories that process metal ores, are still important.

Siberia For centuries, Russians saw Siberia as a frontier treasure chest of furs, gold, and lumber. However, opening this cold harsh region has been difficult. Now Siberian settlement, farming, and industry mostly follow the Trans-Siberian Railroad. The building of the railway started in 1891. It eventually connected Moscow to Vladivostok (vla-duh-vuh-STAHK) on the Sea of Japan. At about 5,800 miles (9,330 km), it is the longest single rail line in the world. Workers completed a more direct railway, called the Baikal-Amur Mainline (BAM), across eastern Siberia in 1989. Permafrost and other difficult conditions made building these lines a great feat.

Lumbering, mining, and oil production are Siberia's most important industries. Because wages are higher in Siberia, some Russians move there to work. Still, large areas of Siberia have few people or none at all.

INTERPRETING THE VISUAL RECORD

Workers lay a pipeline that will transport natural gas westward from Siberia. **How do you think these workers adapt to Siberia's environment in order to do their jobs?**

LEVELS 2 AND 3: Have students complete the Level 1 activity. Then ask students to imagine that they have the opportunity to start new businesses in Russia. Ask them to offer ideas for products or services that Russians may need or want that are not now readily available. Select a few ideas, and discuss the conditions that must exist for these products or services to be created, marketed, and sold at a profit. You may want to have students write brief plans for establishing their proposed businesses.

Teach Objective 2

ALL LEVELS: Organize the class into small groups. Have each group prepare a leaflet concerning one of the economic or environmental issues facing the former Soviet Union today. On one side of the leaflet, ask each group to provide information about the specific nature of the issue, including the circumstances that led to it. On the other side, have each group suggest solutions to resolve the issue along with the advantages and disadvantages of each solution offered. Encourage students to include maps or other graphics to help explain the problem areas more clearly. Have groups share their leaflets with the class. **COOPERATIVE LEARNING**

The Russian Far East Russia has a long coastline on the Pacific Ocean. There, in the Russian Far East, much land remains heavily forested. Summer weather is mild enough for farming in the Amur River valley. Khabarovsk (kuh-BAHR-uhfsk), the main inland city, has factories that process forest and mineral resources. Vladivostok is a naval base and the chief seaport and fishing center.

Sakhalin Island, with its oil and mineral resources, lies off the eastern coast of Siberia in the Sea of Okhotsk. The Kuril (KYOOHR-eel) Islands, which are important for commercial fishing, are farther east. Russia took the islands from Japan at the end of World War II. Japan claims that four of them should be returned. If the two countries settle the issue, Japan may invest more in the Russian Far East and Siberia.

Ukraine and Kiev Kiev is Ukraine's capital. Sheltered by high bluffs in the Dnieper River valley, it is an attractive city. About 10 percent of Ukraine's population lives there. The city also has a large share of the country's economic activity. Like Moscow, Kiev is centrally located in a region rich with agricultural, energy, industrial, and human resources. Kiev's winning soccer team, Dynamo, is an important symbol for the city.

Wheat, sunflowers (for cooking oil), and sugar beets are common crops in Ukraine. The country exports a wide variety of fruits, vegetables, and animal products. Ukraine's heavy industry is based on coal, iron, manganese, and other metals. These resources led to concentration of metalworking in the Donets Basin and along the Dnieper River. Ukraine's moderate climate, access to expanding markets, and resources may help it attract new investment over time.

INTERPRETING THE VISUAL RECORD *Kiev is one of the oldest cities in Europe.* **What characteristic of the region's housing patterns is visible in the photo?**

Connecting to
ECONOMICS

Trading on the Russian stock market

The Russian Stock Market

Stock markets allow businesses to grow by using other people's money. In turn, when investors buy shares in businesses they get the chance to make a profit. The New York Stock Exchange began operating in 1792. In contrast, Russia's stock market organized in 1994. For the first time, Russian citizens could buy shares in businesses that had previously been run by the government. More than 70 percent of the Russian economy was in private hands by 1995. By 1997 investment in Russian stocks by both banks and individuals was booming. However, a year later overestimation of businesses' worth, scandals, and swindles caused stock prices to fall. Many investors' profits were wiped out. Buying stock on the Russian exchange is still risky. Most stocks are cheap, but buyers can easily lose their money. On the other hand, those willing to do their homework and take risks can reap big rewards.

Making Generalizations and Predictions How would building a stable stock market contribute to Russia's efforts to build a strong market economy?

Global Perspectives

Belarus and the United States

Of all the countries that were once part of the Soviet Union, none has retained closer relations with Russia than Belarus. Reforms in Belarus have stalled, and human rights abuses are common. The government sometimes maintains surveillance of foreign visitors, and those who try to photograph what the government sees as militarily sensitive sites may get into trouble with authorities. Belarussian officials have often subjected tourists from the United States to random searches and bribery attempts.

Many observers felt that distrust of U.S. citizens was demonstrated in September, 1995, when two American hot-air balloon pilots who had entered Belarussian air space were shot down and killed. The U.S. State Department and several other parties maintained that the balloonists had permission to fly over Belarus.

VISUAL RECORD ANSWER

She probably has a much lower opinion of the market economy than someone who can profit from it.

Belarus and Minsk Belarus has few mineral resources and generally poor soil. As a result, the country has relied on its educated labor force to build its economy. The remaining forests support wood products industries. Peat is still used as a fuel, even though burning it causes air pollution. Minsk, the capital, has many of the country's industries. Its outdated motor vehicle and consumer-goods plants are left over from the Soviet era.

✓ **READING CHECK:** *Places and Regions* What economic advantages do some of these areas have? good transportation routes, plentiful resources, educated work force

Issues and Challenges

Belarus, Russia, and Ukraine face serious challenges as they move from command to market economies and democracy. Holding free elections was an early and fairly easy step. Much harder is creating the social and economic structures that support peace and prosperity.

Political and Economic Challenges Tension between supporters and opponents of reform and among ethnic groups has grown. Unemployment and crime have increased. The gap between rich and poor is also growing. Public health care has declined. Many older, unemployed, and ill people find that the safety net the old Soviet government provided is gone. Still, Russians have experienced relatively peaceful changes in government after free elections.

Placing business in private hands has had mixed results. A few people have become rich, but some did so through unfair means. Many of the newly rich do not pay their taxes. Some have turned to crime to protect their wealth and power. In addition, members of the new middle class do not feel secure. Many of them fear that the government may again take over homes and businesses.

Many economists argue that several features of the region's economies need reform. For example, factories and transportation systems need to be repaired and modernized. Corrupt officials and managers should be replaced. Also, more businesses must switch to making better goods that people around the world really want to buy. Rules that limit movement of people, money, and goods should be changed. Courts that should be able to force payment of debts, but cannot, need to be strengthened.

Geographical Challenges The Soviet Union was committed to developing local economies in remote places. This policy is less important today. People are moving from their homes in Siberia and other distant areas back to the European heartland. Some observers fear that whole industrial and mining districts will be emptied.

The Soviet history of environmental pollution created another serious challenge. In its rush to make the country an economic power, the Soviet

INTERPRETING THE VISUAL RECORD

Homeless poor people have created this tent city near Red Square in Moscow. **How might this woman's views on Russia's market economy compare to the opinions of a trader on Russia's stock market?**

Close

Ask students to speculate about what political and economic changes are still to come in Russia, Ukraine, and Belarus, based on what they have learned. Call on volunteers to state and explain their predictions. Ask students whether they think the United States should play an active role in assisting the region's people in stabilizing their governments and economies, and, if so, what it should be.

Review and Assess

Have students complete the **Section Review**. Then have students complete **Daily Quiz 17.3**.

Reteach

Have students complete **Main Idea Activity for English Language Learners and Special-Needs Students 17.3**. Then have each student write two multiple-choice questions for each of the Read to Discover question topics and exchange them with a classmate. Discuss sample questions. **ENGLISH LANGUAGE LEARNERS**

Extend

Have interested students conduct research on how changes in the region have affected minority groups there or in the Caucasus countries. **BLOCK SCHEDULING**

Smelters in the Murmansk area have contributed to high pollution levels. According to reports, acid rain has killed all forests within a 12 mile (20 km) radius of Monchegorsk, the town pictured.

government paid little attention to environmental issues. As a result, huge areas are ruined by pollution. Today the region's governments have little money to repair damage or require environmental safeguards. Therefore, these problems will remain for some time.

Perhaps the worst example of environmental damage in the former Soviet Union is in Ukraine. In 1986 a disastrous accident happened at the nuclear power plant at Chernobyl, north of Kiev. The Soviet government tried to cover up the story but failed. Radiation from explosions and fires contaminated millions of acres of forest and farmland. It spread as far away as Sweden and France. People cannot return to the immediate area for many years to come.

Finding solutions to these environmental problems and other challenges will be difficult. However, they are not impossible to overcome. The future of the countries that once belonged to the Soviet Union is not necessarily a prisoner to the past.

✔ **READING CHECK:** *Human Systems* What political, economic, social, and environmental challenges do people in the region face today? unstable politics, unemployment, crime, loss of social safety net, economic development, severe pollution

Section 3 Review

Homework Practice Online
Keyword: SW3 HP17

Define light industry, heavy industry, smelters

Working with Sketch Maps On the map you created in Section 2, label Vladivostok, Khabarovsk, Kuril Islands, Donets Basin, and Chernobyl. Circle the area at the center of a dispute with Japan.

Reading for the Main Idea

1. *Human Systems* What basic change in emphasis is occurring in Russian industry?

2. *Human Systems* Why might Ukraine attract new investment?

Critical Thinking

3. **Identifying Cause and Effect** How did Soviet economic policies affect the region's environment? What are some examples?

4. **Drawing Inferences and Conclusions** Why might Russia's unstable political system delay economic progress?

Organizing What You Know

5. Copy the graphic organizer below. Use it to identify factors that could fuel each subregion's economic growth. One is started for you.

Moscow Region	
St. Petersburg Region	good transportation
Volga and Urals Region	
Siberia	
Russian Far East	
Ukraine and Kiev	
Belarus and Minsk	

Section 3 Review Answers

Define For definitions, see: light industry, p. 393; heavy industry, p. 393; smelters, p. 394

Working with Sketch Maps Maps will vary, but listed places should be labeled in their approximate locations. The Kuril Islands are at the center of the dispute.

Reading for the Main Idea
1. from heavy industry to light industry

2. moderate climate, access to expanding markets, abundant resources

Critical Thinking
3. pushed economic development, ignored environmental protections, heavy pollution left over; Chernobyl nuclear power plant disaster of 1986 (NGS 14)

4. Possible answer: Investors are afraid they would lose their money if the government changes for the worse. (NGS 10)

Organizing What You Know
5. Moscow—center of communications, culture, education, finance, politics, transportation; centrally located; large population; St. Petersburg—good transportation facilities, close to other European cities; Volga and Urals—hydroelectricity, minerals; Siberia—railroads, resources; Russian Far East—forests, petroleum, minerals; location near other Asian countries; Ukraine and Kiev—fairly warm climate, access to expanding markets, resources; Belarus and Minsk—educated labor force

Comparing Maps

Have students read Cities & Settlements: St. Petersburg. Remind students that the city has gone by several names, reflecting the various eras of its existence, and that its location is the reason for its existence. Ask students to name some of the advantages and disadvantages that the location offers. (advantage—relatively easy access to the rest of Europe; disadvantages—buildings sinking because of all the water, seasonal flooding) Tell students that the city's marshy setting also fostered the spread of disease during its construction. So many workers died during this time that the city was said to rest on a swamp of human bones.

Have students conduct Internet and library research to locate information on St. Petersburg's setting and history. Then have them create a set of maps, all at the same scale, on heavy clear plastic sheets to display the main features of the city's natural and built environments. One map should show the rivers, original swamps, and other features of physical geography. Other maps should show the original city layout, the city in the early 1900s, the devastation of World War II, and the St. Petersburg of today. Then have students stack the maps to form one multi-layer map. Ask students to use the maps as if they comprise a geographic information system (GIS) and to write questions about St. Petersburg that such a system might help answer. (Examples: How have parts of the city changed in function? What areas damaged in the war have been rebuilt?)

Essential Element 5

► **Environment ◄ and Society**

The Siege of Leningrad
When German forces attacked Leningrad, now called St. Petersburg, in 1941, the citizens worked tirelessly to build antitank fortifications around the city. The Germans placed Leningrad under siege when they were unable to conquer the city. During the siege, which lasted until 1944, trucks and sleds crossed frozen Lake Ladoga in the winter to bring in supplies. In summer, barges managed occasionally to cross the lake with some needed goods. Every available plot of ground was used for vegetable gardens to provide some food for the starving people. Despite these efforts, at least 650,000 Leningraders died during the siege.

ACTIVITY: Have students study the location of St. Petersburg and speculate how its location both helped and hurt inhabitants during the siege.

Applying What You Know Answers

1. The city is built on more than 40 islands and is located quite far north for a major city.
2. Russia's Communist government invested much money in the city's arts and culture, stimulating the economy.

This Cities & Settlements feature addresses National Geography Standards 4, 12, 13, and 15.

CITIES & SETTLEMENTS

St. Petersburg

Places and Regions It has been called Russia's Window on the West and the Venice of the North. It has appeared on maps as St. Petersburg, Petrograd, and Leningrad. However, by any name, St. Petersburg is a beautiful and important city. In fact, in the 1990s the city was officially recognized as the cultural capital of Russia. It was also the political capital for a long period of Russia's history. Now St. Petersburg is regaining its reputation as one of the world's great historical cities.

A City Born of War

Russia was an isolated and poorly developed country for much of its early history. In the late 1600s the Russian czar Peter I wanted a seaport through which trade and the latest European ideas could enter Russia. However, Sweden controlled the Baltic Sea to the west, while the Turks controlled the Black Sea to the south. In 1703 the Russians drove the Swedes from the Baltic's eastern shore. There, where the Neva River empties into the sea, the czar—Peter the Great—founded a new city. He modeled his city, which carries his name, after London and Amsterdam. He hired French and Italian architects to design it. Peter himself laid the foundation stones for the city's fortress on May 27 of 1703. This is the city's official founding date.

St. Petersburg's success was ensured in 1712 when it replaced Moscow as Russia's capital. Peter ordered the country's nobles to move to the new city. Many built grand homes there. Over the next 200 years St. Petersburg developed into Russia's chief port and industrial center. It also became a center for art, literature, and music. The culture that developed

More than 200 pounds of gold cover the dome of St. Isaac's Cathedral. Like many other St. Petersburg monuments, it is now a museum.

there was both European and Russian.

In 1914 the city's name was changed to Petrograd—the Russian form of *St. Petersburg*. In 1917 the Russian Revolution broke out, and Petrograd was a center of revolutionary activity. The new Communist rulers then moved the capital back to Moscow. In 1924, after the Soviet leader Lenin died, the city was renamed Leningrad. It was not called St. Petersburg again until 1991.

World War II caused heavy damage in Leningrad. German troops surrounded the city for 872 days and shelled it constantly. However, Leningrad never surrendered and the siege was finally broken. When the war ended, people began restoring the city's old buildings to their original splendor. This expensive and painstaking work continues today, more than half a century later.

Environmental Challenges

Besides the destruction of war, St. Petersburg has had to deal with a difficult physical environment. The city is built on more than 40 islands. These islands are created by the Neva River delta and by smaller rivers that flow into the Neva near its mouth. Because of the many water channels that course through St. Petersburg, special construction methods have been required to keep buildings from sinking. St. Isaac's Cathedral, for example, rests on 10,000 upright tree trunks driven into the ground in the early 1800s.

St. Petersburg's location exposes it to threats from the sea. In fall and early winter, storms and strong winds move across the Baltic Sea from the west. These storms drive seawater upstream at the mouth of the

Going Further: Thinking Critically

Moscow is a vast metropolis. Travelers who come to Moscow by air fly into one of five airports. One of them serves countries outside of Russia and the former Soviet Union. The other four serve travelers to former Soviet Union countries and other Russian cities. This situation can lead to confusion and delays. For example, if you fly to Moscow from New York but want to catch a connecting flight to Vladivostok, you must find your way to another airport to complete your journey.

Tell students to imagine that they are Moscow city planners who must propose a location for a new multipurpose airport that will serve all air passengers. First, have students conduct preliminary research on contemporary Moscow, concentrating on its physical geography and the layout of streets and roads. Then have students compile a set of maps like those they created for St. Petersburg. Ask students to use the maps like a geographic information system to determine the best location for a new Moscow airport.

One of the 1,057 rooms and halls of the Hermitage

The Winter Palace's grand facade

Interior of the palace's winter garden as it appeared in 1840

The Winter Palace was built during the reign of Peter's daughter Elizabeth I in the mid-1700s. It was home to the country's rulers until the Russian Revolution in 1917. Today the palace is part of the State Hermitage Museum, which houses the art amassed by Russia's rulers.

Neva River. Since the downtown area is just a few feet above sea level, flooding often occurs. In fact, St. Petersburg has suffered more than 270 major floods during its 300-year history.

St. Petersburg, which is home to some 5 million people, is located quite far north for a major city. Winters are long and cold, with daytime highs averaging about 23°F (–5°C). Because the city is so far north, winter daylight hours are short—about six hours each day. For about three weeks in June and July, the sky does not get completely dark. From about 11:00 P.M. until 3:00 A.M. the evening twilight merges into dawn. These are St. Petersburg's famous "white nights." Many celebrations and cultural events take place during this time of the year. For example, the Stars of the White Nights Festival brings music lovers from all over the world to St. Petersburg.

Russia's Cultural Capital

Compared to other places in Russia, St. Petersburg fared well during the more than 70 years of Communist rule. The government poured large sums of money into arts and culture. Under government sponsorship, the city's Kirov Ballet became one of the world's great dance companies. Government funds were also used to rebuild historic palaces and to maintain the city's many spectacular museums. Despite the Communists' limits on free expression, a secret community of artists and musicians developed and thrived.

Today music and art are alive all over St. Petersburg. The city's best-known attraction, the State Hermitage Museum, is home to one of the world's greatest art collections. Moscow may be Russia's political capital. However, the people of St. Petersburg are determined that their city will be the capital of Russian history and culture.

Applying What You Know

1. **Summarizing** How does St. Petersburg differ from most other major world cities?

2. **Analyzing Information** How has St. Petersburg's history as a cultural and political capital influenced its economy?

TECHNOLOGY

▶ Chapter 17 Test Generator (on the One-Stop Planner)
▶ Global Skill Builder CD–ROM
▶ HRW Go site

REINFORCEMENT, REVIEW, AND ASSESSMENT

▶ Chapter 17 Review, pp. 400–01
▶ Chapter 17 Tutorial for Students, Parents, Mentors, and Peers
▶ Chapter 17 Test (form A or B)
▶ Alternative Assessment Handbook

▶ Chapter 17 Test for English Language Learners and Special-Needs Students
▶ Unit 5 Test
▶ Unit 5 Test for English Language Learners and Special-Needs Students

Assess

Have students complete a Chapter 17 Test.

Reteach

Organize the class into three groups and assign each group a section from the chapter. Have each group create a scene for a play titled *Russia—Then, Now, and Forever.* Scenes should relate to the assigned section's main points. Have students perform their scenes. **ENGLISH LANGUAGE LEARNERS, COOPERATIVE LEARNING**

CHAPTER 17 Review Answers

Thinking Critically

1. The harsh climate limits population distribution, but natural resources attract migration. Government policies pushed settlement in certain areas, particularly near natural resources, and settlement is concentrated along Trans-Siberian Railroad corridor. **(NGS 12)**

2. Possible answers: because when more money is at stake people will take bigger risks; more opportunity for corruption; during Soviet era, people more afraid of being caught by the government **(NGS 10)**

3. market economy—still developing, new stock market, more consumer choice and access to quality goods, businesses privatized; command economy—policy of autarky, government-run businesses, no competition, poor products, lagging production of consumer goods **(NGS 9)**

Using the Geographer's Tools

1. men; probably not, because the trend depends to a great extent on poverty and health care conditions, personal health decisions, and future governmental actions

2. Tables will vary, but should include essential political, economic, social, and cultural characteristics of the region.

3. Maps will vary, but listed regions and appropriate products should be labeled in their approximate locations.

CHAPTER 17 Review

Building Vocabulary

On a separate sheet of paper, explain the following terms by using them correctly in sentences.

Eurasia	czar	Bolsheviks	gulag
icebreakers	serfs	soviets	shatter belt
taiga	abdicate	autarky	light industry
Slavs			

Locating Key Places

On a separate sheet of paper, match the letters on the map with their correct labels.

Ural Mountains
Caucasus Mountains
Crimean Peninsula
Volga River
Central Siberian Plateau
Kiev
Moscow
St. Petersburg
Vladivostok

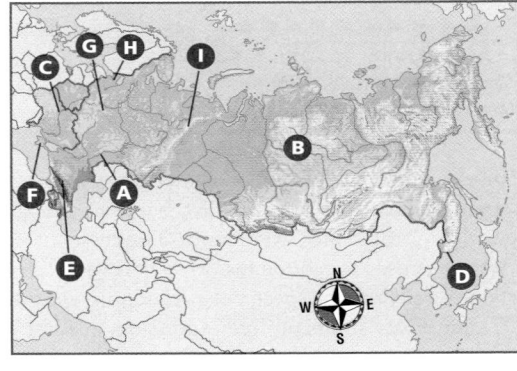

Understanding the Main Ideas

Section 1

1. *Physical Systems* What are the main physical characteristics of the two huge areas west and east of the Ural Mountains?

2. *Physical Systems* What are some resources that Russia has in large quantities?

Section 2

3. *The Uses of Geography* Across what physical region did early migrants come to Russia and its neighbors? How did those newcomers shape the region's culture?

4. *Human Systems* Where do most of the region's people live?

Section 3

5. *Environment and Society* What environmental problems remain from the Soviet era?

Thinking Critically

1. **Analyzing Information** Review the unit population map. What geographical, political, and economic factors might explain the population distribution in Siberia and the Russian Far East?

2. **Drawing Inferences and Conclusions** Why do you think crime is a big problem for Russia today, even though it was not during the Soviet era?

3. **Contrasting** How does Russia's market economy differ from the command economy of the Soviet Union?

Using the Geographer's Tools

1. **Analyzing Graphs** Study the population pyramids for Russia. Have the greatest changes been for men or women? Can you predict a trend in age distribution for the future? Why or why not?

2. **Creating Tables** Design a table that identifies important political, economic, social, and cultural characteristics of the major regions discussed in Section 3.

3. **Preparing Maps** Draw an outline map of Russia, Ukraine, and Belarus. Use three different colors to mark the extent of the tundra, taiga, and steppe regions. Create a symbol for a major product of each climate region and place it correctly.

Writing about Geography

As you have read, most people in the region live in large apartment houses. Make a list of questions you would ask people there about life in their city. How do you suppose life in such a city might be different from life in your community? What factors might account for these differences? Write a paragraph about your answers to these questions. When you are finished with your paragraph, proofread it to make sure you have used standard grammar, spelling, sentence structure, and punctuation.

SKILL BUILDING

Geography for Life

Evaluating Primary and Secondary Sources

Human Systems Ethnic conflicts have caused problems in the Caucasus region. Use the Internet and other resources to find articles from newspapers, journals, and other sources about those issues and events in the region. Compare the articles for background information provided, amount of detail, and bias. Create a chart to show your findings.

Portfolio Activity

Have students create a "family tree" of important rulers and leaders of Russia and the Soviet Union. Some of the most significant are Rurik, Ivan I, Peter the Great, Catherine the Great, Nicholas II, Vladimir Lenin, Joseph Stalin, Nikita Khrushchev, Mikhail Gorbachev, Boris Yeltsin, and Vladimir Putin. Students should provide basic information on each person's life and summarize how he or she affected Russian or Soviet history and culture. Place family trees in portfolios.

Food Festival

Here is a basic recipe for borscht, a traditional Russian soup. In a large pot, sauté a medium chopped onion in 2 tbs. butter. Stir in 1½ pounds sliced raw red beets, ¼ c. red wine vinegar, 1 tsp. sugar, 2 chopped fresh tomatoes, 1 tsp. salt, and some black pepper. Pour in ½ c. beef stock, cover, and simmer one hour. Pour in 5 c. more of beef stock and ½ pound shredded cabbage. Bring to a boil. Add ¼ pound cubed ham, 1 pound cooked sliced beef, ½ c. chopped parsley, and a bay leaf. Simmer for 30 minutes. Garnish with sour cream.

Building Social Studies Skills

Interpreting Graphs

Study the pie graph below. Then use the information from the graph to help you answer the questions that follow.

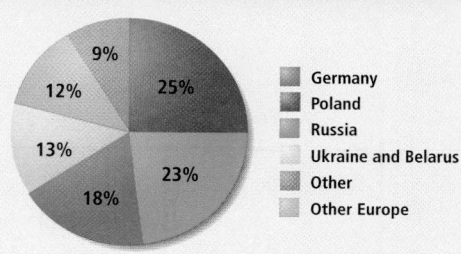

World's Major Rye Producers (1994–1996)

25% — Germany
23% — Poland
18% — Russia
13% — Ukraine and Belarus
12% — Other
9% — Other Europe

Source: *Goode's World Atlas,* 20th Edition

1. Which two countries together accounted for nearly half the world's rye production from 1994 through 1996?
   a. Germany and Belarus
   b. Russia and Belarus
   c. Poland and Russia
   d. Poland and Germany

2. In which hemisphere and on what continent is most of the world's rye produced?

Using Language

The following passage contains mistakes in grammar, punctuation, or usage. Read the passage and then answer the following questions.

> "[1] Agriculture is important to Ukraines economy. [2] The country is the world's largest producer of sugar beets. [3] Its food-processing industry makes sugar, from the sugar beets. [4] Farmers also grow fruits, potatoes, vegetables, and wheat. [5] Grain is maked into flour for baked goods and pasta."

3. Which sentence contains an error in the punctuation of a possessive form of a noun?
   a. 1
   b. 2
   c. 3
   d. All are correct.

4. Which sentence contains an error in the use of commas?
   a. 3
   b. 4
   c. 5
   d. All are correct.

5. Rewrite sentence 5 to correct an error in verb tense.

Writing

Students should prepare questions to gather information about a variety of aspects of life in a Russian, Ukrainian, or Belarussian city. Paragraphs should include comparisons between the city chosen and the student's home community. Both questions and paragraphs should use standard grammar, spelling, sentence structure, and punctuation. Use Rubric 9, Comparing and Contrasting, to evaluate student work.

Geography for Life

Student charts should accurately reflect the information presented in their source materials. Use Rubric 16, Judging Information, to evaluate student work.

Social Studies Skills

1. c
2. Northern Hemisphere, Europe
3. a
4. a
5. Grain is made into flour for baked goods and pasta.

Alternative Assessment

PORTFOLIO ACTIVITY

Learning about Your Local Geography

Group Project: Research
To build the railways across Siberia, Russian engineers had to deal with vast distances and harsh climate conditions. Highway and railway construction was also difficult in some parts of the United States. Plan, organize, and complete a group research project about how the physical geography of your state and local area affected highway or rail construction. Which bodies of water had to be crossed? Did mountains or deserts present problems? How did climate affect the workers? Prepare a short illustrated report about the answers you find to these geographic questions.

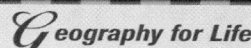
internet connect

Internet Activity: go.hrw.com
KEYWORD: SW3 GT17

Choose a topic about Russia, Ukraine, or Belarus to:
- research the rise and fall of the Soviet Union.
- create a poster of a journey through the Caucasus.
- learn the causes and results of the Chernobyl disaster.

PORTFOLIO ACTIVITY

Students' findings will vary according to the region in which they live. Use Rubric 6, Cause and Effect, to evaluate student work.

401

CHAPTER RESOURCE MANAGER

	Objectives	Pacing Guide	Reproducible Resources
SECTION 1 **Natural Environments** (pp. 403–06)	• Identify the major landforms and rivers of Central Asia. • Describe the climates, biomes, and natural resources of Central Asia.	**Regular** .5 day **Block Scheduling** .5 day *Block Scheduling Handbook, Chapter 18*	**RS** Guided Reading Strategy 18.1 **SM** Map Activity 18: Fergana Valley
SECTION 2 **History and Culture** (pp. 408–11)	• Explain how various cultures and invaders have affected Central Asia's history. • Identify some features of Central Asian cultures.	**Regular** .5 day **Block Scheduling** .5 day *Block Scheduling Handbook, Chapter 18*	**RS** Guided Reading Strategy 18.2 **RS** Graphic Organizer Activity 18 **PS** Readings in World Geography, History, and Culture 46 **E** Cultures of the World Activity: Region 4
SECTION 3 **The Region Today** (pp. 412–15)	• Identify changes in Central Asia's economy over time. • Describe Central Asia's cities. • Examine the issues Central Asia must face to improve its economy.	**Regular** 1 day **Block Scheduling** .5 day *Block Scheduling Handbook, Chapter 18*	**RS** Guided Reading Strategy 18.3 **PS** Readings in World Geography, History, and Culture 45 **SM** Critical Thinking Activity 18: Getting Ahead **SM** Geography for Life Activity 18: The Fragmented Fergana Valley

Chapter Resource Key

PS Primary Sources

RS Reading Support

IC Interdisciplinary Connections

E Enrichment

SM Skills Mastery

A Assessment

REV Review

ELL Reinforcement and English Language Learners

Transparencies

CD–ROM

Video

Internet

Holt Presentation Maker Using Microsoft® PowerPoint®

One-Stop Planner CD–ROM

See the *One-Stop Planner* for a complete list of additional resources for students and teachers.

 **One-Stop Planner CD–ROM**

It's easy to plan lessons, select resources, and print out materials for your students when you use the **One-Stop Planner CD–ROM with Test Generator**.

⊿ internet connect

HRW ONLINE RESOURCES

GO TO: go.hrw.com
Then type in a keyword.

TEACHER HOME PAGE
 KEYWORD: SW3 Teacher

CHAPTER INTERNET ACTIVITIES
 KEYWORD: SW3 GT18
 Choose a topic on central Asia to:
 • create a poster that analyzes the causes and effects of increasing technology in agriculture and its consequences for the Aral Sea.
 • research the Silk Road and its history in global trade.
 • follow nomads in Kazakhstan.

CHAPTER ENRICHMENT LINKS
 KEYWORD: SW3 CH18

CHAPTER MAPS
 KEYWORD: SW3 MAPS18

ONLINE ASSESSMENT
 Homework Practice
 KEYWORD: SW3 HP18
 Standardized Test Prep
 KEYWORD: SW3 STP18
 Rubrics
 KEYWORD: SS Rubrics

COUNTRY INFORMATION
 KEYWORD: SW3 Almanac

CONTENT UPDATES
 KEYWORD: SS Content Updates

HOLT PRESENTATION MAKER
 KEYWORD: SW3 PPT18

ONLINE READING SUPPORT
 KEYWORD: SS Strategies

CURRENT EVENTS
 KEYWORD: S3 Current Events

Technology Resources

- One-Stop Planner CD–ROM, Lesson 18.1
- *ARGWorld* CD–ROM
- Geography and Cultures Visual Resources 31–35
- Homework Practice Online
- HRW Go site

- One-Stop Planner CD–ROM, Lesson 18.2
- Geography and Cultures Visual Resources 36
- CNN. Presents World Cultures: Yesterday and Today, Segment 8: The Silk Road
- Homework Practice Online
- HRW Go site

- One-Stop Planner CD–ROM, Lesson 18.3
- CNN. Presents Geography: Yesterday and Today, Segment 20: Dateline Kazakhstan
- Homework Practice Online
- HRW Go site

Reinforcement, Review, and Assessment

- **ELL** Main Idea Activity 18.1
- **ELL** English Audio Summary 18.1
- **ELL** Spanish Audio Summary 18.1
- **REV** Section 1 Review, p. 406
- **A** Daily Quiz 18.1

- **ELL** Main Idea Activity 18.2
- **ELL** English Audio Summary 18.2
- **ELL** Spanish Audio Summary 18.2
- **REV** Section 2 Review, p. 411
- **A** Daily Quiz 18.2

- **ELL** Main Idea Activity 18.3
- **ELL** English Audio Summary 18.3
- **ELL** Spanish Audio Summary 18.3
- **REV** Section 3 Review, p. 415
- **A** Daily Quiz 18.3

Meeting Individual Needs

Ability Levels

Level 1 Basic-level activities designed for all students encountering new material

Level 2 Intermediate-level activities designed for average students

Level 3 Challenging activities designed for honors and gifted-and-talented students

English Language Learners Activities that address the needs of students with Limited English Proficiency

Chapter Review and Assessment

- Chapter 18 Test Generator (on the One-Stop Planner)
- Global Skill Builder CD–ROM
- HRW Go site
- **REV** Chapter 18 Review, pp. 418–19
- **REV** Chapter 18 Tutorial for Students, Parents, Mentors, and Peers
- **A** Chapter 18 Test (form A or B)
- **A** Alternative Assessment Handbook
- **A** Chapter 18 Test for English Language Learners and Special-Needs Students

Launch into Learning

Read to the class the following description by traveler Marco Polo: "For twelve days the course is along this elevated plain, which is named Pamer [Pamir Plateau]. . . . So great is the height of the mountains, that no birds are to be seen near their summits; and however extraordinary it may be thought, it was affirmed, that from the keenness of the air, fires when lighted do not give the same heat as in lower situations, nor produce the same effect in cooking food." Ask students if and how they have experienced the same phenomenon. (*Possible answers: while camping in mountains or adjusting baking times at high elevations*) Tell students they will learn more about how physical geography affects life in Central Asia in this chapter.

Using the Physical-Political Map

Have students examine the map on the opposite page. Call on students to name the nations that border the region (*Russia, China, Afghanistan, Iran*). Ask students why Russia might want the region's countries to remain stable and at peace. (*Possible answer: to maintain a buffer zone between itself and other countries*)

Why We Should Know More

You may want to remind students that even a region as remote as Central Asia connects with us in several ways:

▶ The region has large oil and gas deposits. The United States may want to buy petroleum products from the countries there.

▶ People in the region are struggling with the shift from communism to more open systems of government.

▶ Damage from Soviet nuclear tests in the area may last far into the future.

▶ Uzbekistan aided the U.S. military in its attacks against the Taliban in Afghanistan.

▶ The region is opening to tourism. It offers some of the most spectacular scenery in the world.

CHAPTER 18 Central Asia

Until recently Kazakhstan, Kyrgyzstan, Tajikistan, Turkmenistan, and Uzbekistan were unfamiliar to many people in other countries. Now these lands are being discovered by investors and tourists and are changing rapidly after decades of Soviet rule.

Silver and gold wolf's head from a battle flag, Kazakhstan

Man riding a donkey, Uzbekistan

Salam! (Hi!) My name is Leila. I live in Turkmenistan. Here we go to school six days a week and study 18 different subjects. There is no choice of courses. We just switched from writing with the Cyrillic alphabet to writing with the Latin alphabet.

After school we go home and have lunch. Then my sister and I divide the chores. My brothers have outdoor chores, like tending our fruit and vegetable garden or taking the sheep out to graze along the canal. The sheep are kept in a shed in back. Many of our neighbors keep cows. We grow cabbage, spinach, carrots, beets, turnips, parsley, dill, pomegranates, and grapes.

It takes a long time to cook dinner because everything is made from scratch. The oven is outside in the courtyard. For fuel we use the woody stems of cotton plants. The smoke has a wonderful smell. To keep the mosquitoes away, we scatter the seeds of desert plants on a hot dish to create smoke. We wash our hands before dinner and sit on the floor around low tables to eat with our hands, three or four people to one dish.

Every year in October, each student from seventh grade through university goes to the country to pick cotton, because hand-picked cotton is more valuable. We work for eight hours every day and have to pick at least 50 kilograms (110 lbs) a day. We get paid about 10 cents per kilo. This is the only paid job students can have.

 **LET'S GET STARTED**

Copy the following instructions onto the chalkboard: *Examine the physical-political map of Central Asia. What elements of the region's physical geography do you think might make life there difficult? Discuss student responses. (Possible answers: no access to ocean for trade or transportation, few water resources, vast distances between settlements, high mountains)* Tell students they will learn more about the region's physical features and climates in Section 1.

Building Vocabulary

Write the term **zinc** on the chalkboard. Call on a volunteer to read the definition from the text or glossary. Ask students if they are familiar with ways the metal is used from prior knowledge or other classes. *(Possible uses: essential nutrient for plants and animals, dietary supplement, protective coating for iron and steel, mixed with other metals to make alloys like brass or bronze, batteries, building construction)*

Natural Environments

READ TO DISCOVER

1. What are the major landforms and rivers of Central Asia?
2. What climates, biomes, and natural resources does the region have?

WHY IT MATTERS

Some of the largest oil reserves in the world are in the Caspian Sea region of Central Asia. Use **CNN fyi.com** or other **current events** sources to learn how the oil industry affects governments, economies, and people there.

DEFINE

zinc

LOCATE

Altay Shan	Irtysh River
Tian Shan	Lake Balkhash
Pamirs	Issyk-Kul
Kopet-Dag	Kara-Kum
Caspian Sea	Kyzyl Kum
Aral Sea	
Amu Dar'ya	
Syr Dar'ya	

Section 1 RESOURCES

REPRODUCIBLE
► Guided Reading Strategy 18.1
► Map Activity 18: Fergana Valley

TECHNOLOGY
► One-Stop Planner CD–ROM, Lesson 18.1
► Geography and Cultures Visual Resources 31–35
► Homework Practice Online
► HRW Go site

REINFORCEMENT, REVIEW, AND ASSESSMENT
► Main Idea Activity 18.1
► English Audio Summary 18.1
► Spanish Audio Summary 18.1
► Section 1 Review, p. 406
► Daily Quiz 18.1

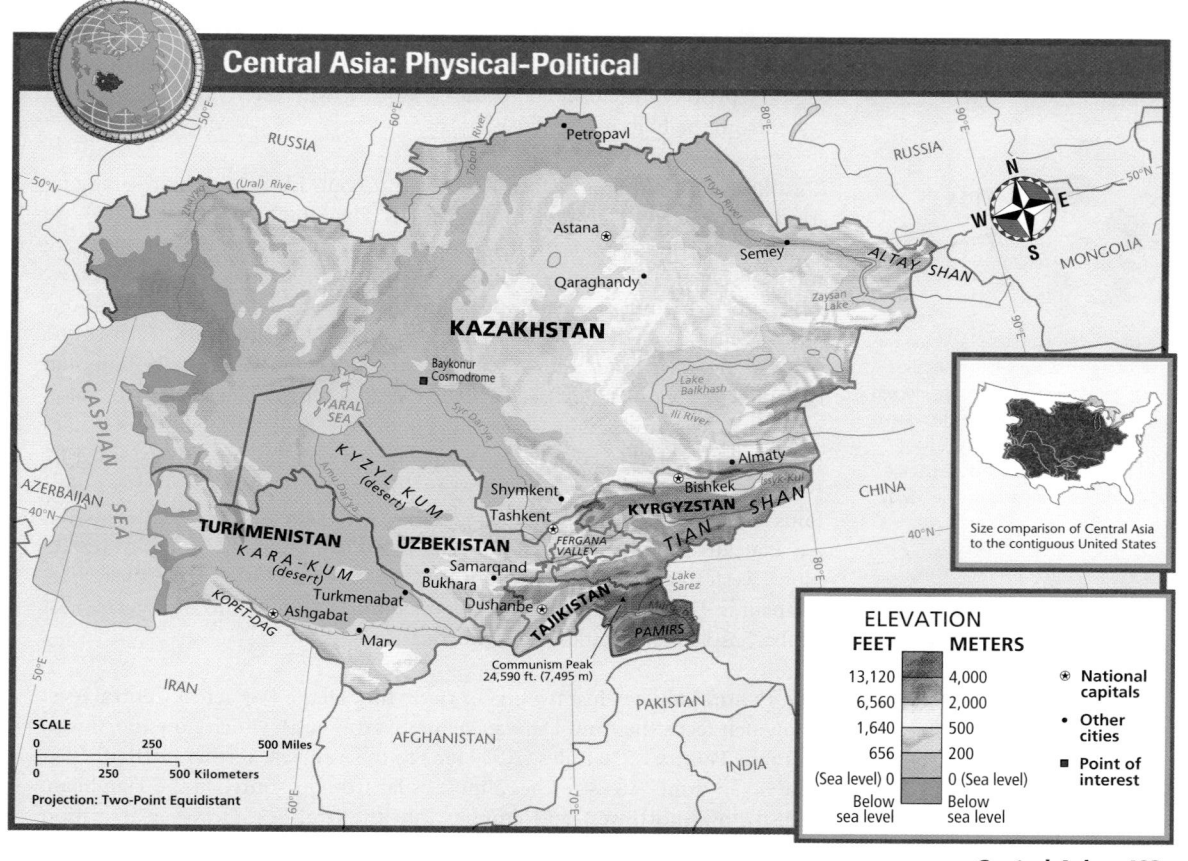

Central Asia: Physical–Political

Size comparison of Central Asia to the contiguous United States

ELEVATION

FEET	METERS
13,120	4,000
6,560	2,000
1,640	500
656	200
(Sea level) 0	0 (Sea level)
Below sea level	Below sea level

⊛ National capitals
• Other cities
■ Point of interest

SCALE
0 — 250 — 500 Miles
0 — 250 — 500 Kilometers
Projection: Two-Point Equidistant

ALL LEVELS: Provide students with blank outline maps of Central Asia. Have students locate and label the region's major physical features. After they have completed their maps, direct each student to write three interesting statements about the region's landforms and three statements about its rivers and lakes. Ask volunteers to share their statements with the class. **ENGLISH LANGUAGE LEARNERS**

LEVELS 2 AND 3: Have each student choose one of the landforms or bodies of water in this region and compare it to a similar feature studied in a previous chapter. Students should communicate their comparisons in written reports. Reports should address the physical characteristics of both features, the processes that created them, and the ways in which humans have modified or adapted them.

Eye on Earth

Problems in the Pamirs

Lake Sarez in eastern Tajikistan was created in 1911 when a massive earthquake caused part of a mountain to collapse, blocking the Murgab River. Today the lake stretches about 37 miles (60 km) behind the dam created by the landslide. Downstream, waters of the Murgab flow into the Panj River, which becomes the Amu Dar'ya.

If another earthquake were to cause the dam to collapse, a huge wall of water would destroy entire villages and threaten the lives of thousands of people. The dam's isolation in the Pamirs and the poverty of Central Asian countries make it difficult to find a solution to this potentially disastrous problem.

DISCUSSION: Ask students to consider how wealthy countries could help Central Asian and other poor countries meet the challenges posed by problems such as Lake Sarez.

The Tian Shan range contains steep ridges and deep valleys. Glaciers like this one fill many of the valleys.

Lake Sarez in eastern Tajikistan was created when a large landslide triggered by an earthquake blocked the Murgab River. If another quake breaks the natural dam, floodwaters could reach as far as the Aral Sea, hundreds of miles away.

internet connect

GO TO: go.hrw.com
KEYWORD: SW3 CH18
FOR: Web sites about Central Asia

Landforms, Rivers, and Lakes

The five countries of Central Asia are all landlocked. Semiarid grasslands are found in the north. To the east, plateaus rise above barren deserts. Most of the region's people are clustered in the southeast, where rivers bring precious water from the high mountains.

Landforms Central Asia is a land of great contrasts in elevation—ranging from below sea level to lofty mountain peaks. The Altay Shan (al-TY SHAHN) rise in the far northeast. *Shan* means "mountain" in Chinese. In the southeast are the Tian Shan (TYEN SHAHN) and Pamirs (puh-MIRZ) ranges. Tectonic forces, which push the Indian subcontinent into the rest of Asia, created these mountains. Tectonic activity sometimes causes disastrous earthquakes. Each of the five countries has mountains more than 10,000 feet (3,050 m) high. Tajikistan contains the region's tallest peak at 24,590 feet (7,495 m). Glaciers are common in Central Asia's mountains. Tajikistan's massive Fedchenko Glacier is 44 miles (71 km) long. Along Central Asia's southwestern rim, the Kopet-Dag (koh-PET-DAHG) form a rugged boundary between Turkmenistan and Iran. These mountains are lower and drier than the region's eastern ranges.

Plateaus and plains stretch north and west from the mountains. At the region's western edge lies the Caspian Sea, which is the world's largest lake. The Caspian is 92 feet (28 m) below sea level and has no outlet to the ocean. East of the Caspian is the landlocked Aral Sea.

Rivers and Lakes Just two major rivers flow all the way across Central Asia. Snowmelt feeds the Amu Dar'ya (AH-moo DAHR-yuh), which flows northwest from the Pamirs 1,578 miles (2,539 km) to the Aral Sea. Farther north, the Syr Dar'ya (sir duhr-YAH) stretches almost as far from its source in the Tian Shan. It also flows northwest and empties into the Aral Sea. However, irrigation

LEVEL 1: Copy the following graphic organizer onto the chalkboard, omitting the italicized answers. Have students work in pairs to complete the chart. ENGLISH LANGUAGE LEARNERS, COOPERATIVE LEARNING

LEVELS 2 AND 3: Have students complete the Level 1 activity. Then organize the class into groups. Have each group create a visual presentation about the climates, vegetation, and animals of Central Asia. Presentations could take the form of posters, collages, dioramas, computer presentations, or other forms. Encourage students to be creative. ENGLISH LANGUAGE LEARNERS, COOPERATIVE LEARNING

Physical Geography of Central Asia	
Climates	*arid, semiarid, Mediterranean, highland*
Plants	*deciduous forests, walnut trees, evergreen trees, grasses, shrubs, saxaul tree*
Animals	*deer, pheasants, wild boar, snow leopards, antelope, wildcats, wolves, domesticated camels, goats, sheep*
Resources	*coal, oil, natural gas, copper, lead, nickel, zinc, gold*

drains much of the water from these rivers. For example, a canal flows from the Amu Dar'ya across most of southern Turkmenistan. By the time the rivers reach the Aral Sea they are little more than trickles. (See Geography for Life: The Shrinking Aral Sea.) The Syr Dar'ya and its tributaries provide water for the densely populated Fergana Valley.

The Irtysh (ir-TUHSH) River of eastern Kazakhstan flows northward into Russia. There it joins the Ob River, which drains into the Arctic Ocean. The Irtysh also provides water for crops, reservoirs, hydropower stations, and industrial cities.

Central Asia has some interesting lakes. Shallow Lake Balkhash has freshwater where the Ili River and other streams enter it. However, the lake is salty at its eastern end. Issyk-Kul (is-sik-KUHL) never freezes over even though it lies about a mile above sea level in the Tian Shan. The lake's warm water moderates the area's otherwise cold climate. (See Focus on Geography: Issyk-Kul and Tourism.)

✔ **READING CHECK:** *Physical Systems* What is the region's physical environment like? What forces created Central Asia's high mountains? mountainous in south and southeast, plateaus and plains north and west of the mountains, three major rivers; tectonic forces

Climates, Biomes, and Natural Resources

Most places in Central Asia have harsh climates. A location in the heart of Asia means the region is far removed from oceanic influences. Also, high mountains form a barrier to warm moist winds from the Indian Ocean and create a rain shadow. As a result, precipitation totals are low, and there are extreme seasonal temperature ranges. Summer temperatures can rise to 115°F (46°C). Winter lows have dropped to −36°F (−38°C).

Climates As you can see on the unit climate map, Central Asia has mostly semiarid and arid climates. The region has two large deserts—the Kara-Kum ("Black Sand") and Kyzyl Kum ("Red Sand"). One of the world's greatest tracts of drifting sand dunes is found in the Kara-Kum. Stony ground is more typical of the Kyzyl Kum.

Southern Turkmenistan has a small area with a Mediterranean climate. With its sunny skies and rugged surrounding scenery, the city of Ashgabat was a center of Soviet filmmaking. Some areas in the foothills of the Tian Shan also have mild weather. For example, Almaty, Kazakhstan, gets about 23 inches (58 cm) of rain each year. Spring and fall are pleasant there.

Plant and Animal Life The region's highest peaks are too cold, dry, and windy for vegetation. However,

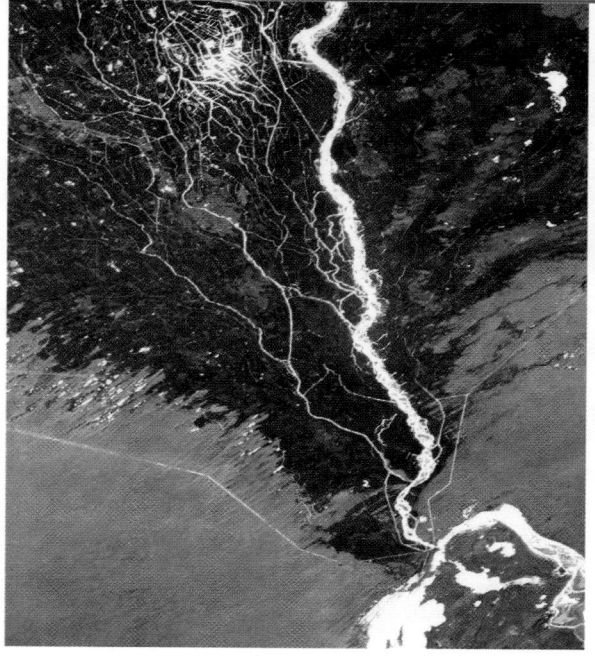

INTERPRETING THE VISUAL RECORD *What are the straight narrow lines in this aerial view of the Amu Dar'ya? How might government planners use this photo?*

INTERPRETING THE VISUAL RECORD

A Bactrian camel carries a rider across this desert in Central Asia. How did the domestication of the camel allow people to adapt to desert life?

Prjewalski's Horse Rides Again Prjewalski's horse, which once roamed the steppe of Central Asia but has been extinct in the wild since the 1960s, has returned to its native Uzbekistan. Until recently, only a few hundred of these horses—the last undomesticated horse species on the planet—remained in zoos.

Captive breeding programs increased the population, and five horses were released into the wild in 1992. They faced long odds, because these horses, which had depended on human help, were faced with the harsh Kyzyl Kum. However, the horses did begin breeding. Though their first years in the wild were rough, the horses adjusted to their environment and the population grew. By 1999, 15 of the horses were living wild in their Kyzyl Kum reserve. Similar efforts in Mongolia, also begun in 1992, may help to restore the steppe's herds of Prjewalski's horses.

VISUAL RECORD ANSWER

irrigation canals; find the canals that drain the most water from the river, determine if some canals could be closed, and so on

VISUAL RECORD ANSWER

provided transportation between oases, allowed people to carry less water

Close

Tell students that the mountains in Kyrgyzstan, particularly the Tian Shan—also known as the Celestial Mountains—are considered by some to be as beautiful as the mountains in Nepal or Switzerland. Invite students to share memories or impressions of mountains they have climbed or visited.

Review and Assess

Have students complete the **Section Review**. Then have students complete **Daily Quiz 18.1.**

Reteach

Have students complete **Main Idea Activity for English Language Learners and Special-Needs Students 18.1.** Then have students draw storyboards for documentaries about the physical landscapes of Central Asia. ENGLISH LANGUAGE LEARNERS

Extend

Tell students about the Torugart Pass, an overland route from Bishkek, Kyrgyzstan, through the Tian Shan to China. The pass was closed to international travelers until recently. Have students conduct research on trade routes through Central Asia to China and trace them on a map. BLOCK SCHEDULING

Section 1 Review Answers

Define For definition, see: zinc, p. 406

Working with Sketch Maps
Maps will vary, but listed places should be labeled in their approximate locations. The Caspian Sea is the world's largest lake.

Reading for the Main Idea
1. northeast, southeast, southwest; block winds from Indian Ocean and create a rain shadow

2. mostly arid and semiarid, but Mediterranean in southern Turkmenistan; southern Turkmenistan, foothills of the Tian Shan

Critical Thinking
3. Possible answer: High-density population may shift from the south to the west as more people go there to work in the oil fields. (NGS 12)

4. all countries landlocked, no major rivers that flow to oceans, all of which limit the ability to transport goods in and out of the region (NGS 15)

Organizing What You Know
5. Kazakhstan—Altay Shan, Aral Sea, Lake Balkhash, Syr Dar'ya, Ili River; Kyrgyzstan—Tian Shan, Issyk-Kul; Tajikistan—Pamirs, Lake Sarez; Turkmenistan—Kopet-Dag, Kara-Kum, Caspian Sea, Amu Dar'ya; Uzbekistan—Fergana Valley, Kyzyl Kum, Aral Sea, Amu Dar'ya

The number of snow leopards is decreasing quickly. Corruption and the lack of effective government controls make protecting the big cats very difficult.

deciduous forests grow at middle elevations. In fact, Central Asia is known for its many walnut trees. Evergreen trees grow at higher elevations. Many alluvial fans, foothills, and river deltas support grasses and shrubs. Animals that live in the mountains include deer, pheasants, and wild boar. One of the world's most beautiful and endangered big cats, the snow leopard, also finds shelter in the high southeastern ranges.

Like many arid places, southern Central Asia has glorious vegetation in the spring. Desert shrubs and grasses bloom briefly before the searing summer heat begins. A tree unique to Central Asia, the saxaul, is one of the few large plants found in the desert. The tree's dense, heavy wood burns like charcoal. Desert peoples have used it as firewood for thousands of years. Desert wildlife includes antelope, wildcats, and wolves. Domesticated camels, goats, and sheep also graze on desert grasses.

✓ **READING CHECK:** *Physical Systems* How is the mountain vegetation of Central Asia different from its desert vegetation? mountain—deciduous forests in middle elevations, coniferous trees higher up; desert—shrubs and grasses, saxaul tree

Natural Resources In this arid region, water is the most precious resource. Although rainfall is low, snowmelt from the eastern ranges flows into rivers. Large dams along these rivers generate hydroelectricity.

Energy and mineral resources abound in Central Asia. Coal deposits are common on Kazakhstan's eastern plateau. The richest oil fields are in the west. The area around the Caspian Sea is a particularly important source of oil. Huge oil reserves there will last for many years. (See Case Study: Pipelines or Pipe Dreams?) Large gas reserves lie below Turkmenistan's desert basin.

Kazakhstan has most of the region's mines, and it exports copper, iron, lead, and nickel. Another export is **zinc**, an element important in metal processing and other industries. Uzbekistan is a major gold producer.

✓ **READING CHECK:** *Physical Systems* Where are the region's main energy resources found? western oil fields, hydroelectricity in east, coal in eastern Kazakhstan

Section 1 Review

go. hrw .com **Homework Practice Online** Keyword: SW3 HP18

Define zinc

Working with Sketch Maps On a map of Central Asia that you draw or that your teacher provides, label the Altay Shan, Tian Shan, Pamirs, Kopet-Dag, Caspian Sea, Aral Sea, Amu Dar'ya, Syr Dar'ya, Irtysh River, Lake Balkhash, Issyk-Kul, Kara-Kum, and Kyzyl Kum. Which body of water is the world's largest lake?

Reading for the Main Idea

1. *Physical Systems* Where are the region's mountain ranges? How do they affect precipitation patterns in Central Asia?

2. *Physical Systems* What climates are found in the region? Where would you find a mild climate with reliable precipitation?

Critical Thinking

3. **Making Generalizations and Predictions** How might the location of resources affect Central Asia's population density in the future?

4. **Drawing Inferences and Conclusions** How might Central Asia's physical geography limit foreign trade for the region's countries?

Organizing What You Know

5. Create a chart like the one below. Use Section 1 and the physical-political map to complete it. Add boxes for the region's remaining countries.

Country	Landforms	Bodies of water
Kazakhstan		

Government: The Politics of Water

Political decisions like the decision to irrigate the cotton fields of Uzbekistan have contributed to the disappearance of the Aral Sea. Similar decisions in other regions have created environmental consequences and sometimes even political conflicts between the countries involved. Some experts envision a future full of wars between countries over the control of water or within countries torn between economic growth and environmental concerns. Others do not believe such conflicts will occur, because they see water as a renewable resource and therefore not as scarce as some other resources.

One area currently faced with water disputes is the Nile River Valley. Egypt, economically and militarily the most powerful country in the region, is concerned because the river's flow is greatly diminished by irrigation upstream. Another example involves the United States and Canada. The United States wants to increase the use of water from the Great Lakes to irrigate cornfields in the Midwest. Canada, on the other hand, is worried how this will affect its fisheries in the Gulf of St. Lawrence.

Organize the class into groups and have students investigate regions where countries are in conflict over water allocation. Have groups present their findings in oral reports supported by visual aids. After all groups have completed their presentations, lead a class discussion about the importance of water resources in international political relationships. **COOPERATIVE LEARNING**

Environment and Society

Geography for Life

The Shrinking Aral Sea

People often change the environment in the process of using natural resources. Sometimes when humans alter nature, they bring prosperity to some areas while creating crises in others.

Perhaps nowhere else is this process more visible than at the Aral Sea. The Aral Sea, which is really a salt lake, lies between Kazakhstan and Uzbekistan. The sea has shrunk by about 60 percent since 1960. It once covered about 26,000 square miles (68,000 sq km) and was the world's fourth-largest lake. Today the sea covers only about 11,000 square miles (28,000 sq km). The former port city of Mŭynoq now lies 30 miles (48 km) from the Aral Sea's shore. The sea's level has dropped 48 feet (16 m). As a result, the sea has become saltier, and few fish can survive in it. Moreover, some 10,000 square miles (26,000 sq km) of former seafloor have become a desert of sand and salt.

What happened? Cotton farming is largely responsible for the sea's drying. The Soviet government wanted to establish a profitable cotton industry. To water the cotton fields, the Soviets built a network of irrigation canals. This irrigation system still diverts large amounts of water from the Syr Dar'ya and Amu Dar'ya, the only rivers that flow into the Aral Sea. Very little river water reaches the shrinking sea. In fact, more water evaporates from the sea than flows into it. In addition, the sea and the surrounding land are polluted with agricultural chemicals. Wind spreads the chemicals, along with the sand and salt, ruining cropland and damaging the health of area residents.

Experts from many countries are trying to find ways to repair environmental damage in the Aral Sea region. Little hope remains for the sea itself. The Aral Sea will probably soon be just a cluster of small salty lakes.

Applying What You Know

1. **Evaluating** What geographic and economic impact have past and present water policies had on the Aral Sea and the surrounding area?

2. **Making Generalizations and Predictions** Use the map, graph, and text to infer how the shrinking of the Aral Sea will affect the surrounding area. How might settlement patterns, population distribution, economic conditions, and political conditions in the area be affected?

Global Perspectives

Shrinking Lakes The Aral Sea is not the only body of water that is disappearing. Around the world, lakes are shrinking because of human activities, particularly irrigation. Africa's Lake Chad, for example, once covered about 9,650 square miles (25,000 sq km). Irrigation and other uses have drained the lake to about 839 square miles (2,173 sq km). Severe droughts since the 1960s have kept the lake from refilling.

Applying What You Know Answers

1. The Aral Sea has shrunk and become saltier and more polluted. Chemicals from the seabed ruin cropland, limit income, and threaten people's health.
2. Possible answers: Settlement will likely shift away from the Aral Sea area, population will decline, economic conditions will worsen, and political unrest may increase.

GRAPH ANSWER

As surface area decreases, so does water level. Shallower bodies of water may evaporate more rapidly than deep ones.

This Geography for Life feature addresses National Geography Standards 4, 14, 16, and 18.

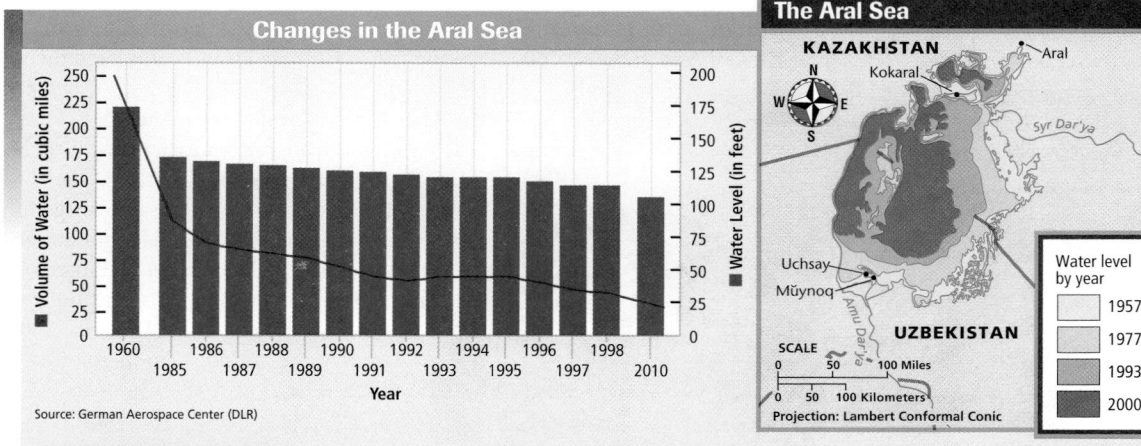

Changes in the Aral Sea

Source: German Aerospace Center (DLR)

The Aral Sea

KAZAKHSTAN
Kokaral
Aral
Syr Dar'ya
Uchsay
Mŭynoq
UZBEKISTAN
Amu Dar'ya

SCALE
0   50   100 Miles
0   50   100 Kilometers
Projection: Lambert Conformal Conic

Water level by year
1957
1977
1993
2000

INTERPRETING THE GRAPH Both this graph and the map beside it demonstrate the drying of the Aral Sea. *What relationship can you see between the surface area of the sea and its depth? How* *might changes in water level affect the rate at which the sea's volume decreases?*

LET'S GET STARTED

Copy the following instructions onto the chalkboard: *Write three sentences about what life would be like if you moved regularly from place to place. Use your knowledge of Native Americans of the Great Plains, migrant farmers, or members of a traveling tour group to describe life on the move.* Discuss responses. Tell students that Central Asia has many people who move regularly, and they live in unique ways. Tell students that in Section 2 they will learn more about the history and people of Central Asia.

Building Vocabulary

Write the terms **caravans**, **nomads**, and **yurts** on the chalkboard. Call on a volunteer to read the definitions. Then tell students that the word *nomad* comes from the Greek language, the word *caravan* is Persian in origin, and *yurt* was originally a Turkish word. Point out that the Greeks, Persians, and Turks were major cultural influences in Central Asia. Then have volunteers locate and read the definitions for the other key terms.

Section 2 RESOURCES

REPRODUCIBLE

▶ Guided Reading Strategy 18.2
▶ Graphic Organizer Activity 18
▶ Readings in World Geography, History, and Culture 46
▶ Cultures of the World Activity: Region 4

TECHNOLOGY

▶ One-Stop Planner CD–ROM, Lesson 18.2
▶ Geography and Cultures Visual Resources 36
▶ CNN Presents World Cultures: Yesterday and Today, Segment 8: The Silk Road
▶ Homework Practice Online
▶ HRW Go site

REINFORCEMENT, REVIEW, AND ASSESSMENT

▶ Main Idea Activity 18.2
▶ English Audio Summary 18.2
▶ Spanish Audio Summary 18.2
▶ Section 2 Review, p. 411
▶ Daily Quiz 18.2

Section 2 — History and Culture

A warrior's armor from Kazakhstan

READ TO DISCOVER

1. How have various cultures and invaders affected the region's history?

2. What are some features of Central Asian cultures?

WHY IT MATTERS

Central Asians are rediscovering their cultural heritage after the Soviet government suppressed that heritage for more than 70 years. Use **CNN fyi.com** or other **current events** sources to learn about Central Asian heroes, holidays, legends, and literature.

DEFINE

caravans
monoculture
nomads
transhumance
yurts

LOCATE

Samarqand
Fergana Valley

History

Humans have lived in Central Asia for thousands of years. In fact, many migrant peoples and invaders have left their imprint on the region's history and cultures. For a while, parts of the area belonged to the Persian Empire. Alexander the Great brought Greek influences when he invaded in the 300s B.C. Merchants pursuing profits instead of conquest also came through Central Asia. Great **caravans** passed through as they brought silk and other luxury goods from China along routes called the Silk Road. A caravan is a group of people traveling together for protection. Other traders would eventually sell these products in Europe and the Mediterranean region.

Armies and Empires Turkic-speaking peoples established kingdoms in Central Asia in the A.D. 600s. Not long after, Chinese armies conquered the region. Arabic speakers brought the Islamic faith when they invaded in the 700s. In 751 the Arab armies defeated the Chinese.

Beginning in 1218 Mongols from farther east, led by Genghis Khan, began a 200-year rule of Central Asia. The Mongols destroyed many cities and irrigation systems. A Turkic-speaking Mongol named Timur (tee-MOOR), also known as Tamerlane, rose to power in the 1300s and built an empire. It lasted from 1370 to 1405. Timur was a ruthless conqueror, but he supported the arts, literature, and science. He ordered the building of beautiful gardens, mosques, and palaces at his capital, Samarqand. After Timur's death, his empire broke up into city-states. In the late 1400s Europeans began

Built in Samarqand in the mid-1600s, the Tilla-Kari served both as a mosque and as a madrasa, or Islamic school. The complex's exterior walls are decorated with geometric designs worked in ceramic tiles, like the detail in the small photo. Some of Samarqand's historic buildings are decorated with gold and marble.

Teach Objective 1

🌐 **ALL LEVELS:** Copy the following graphic organizer onto the chalkboard, omitting the italicized answers. Have students complete the chart by noting how various groups have influenced the cultures of Central Asia. **ENGLISH LANGUAGE LEARNERS**

💻

Influences on Central Asian Cultures

Turkic speakers	→	*brought Turkic languages*
Arabs	→	*introduced Islam*
Mongols under Genghis Khan	→	*destroyed cities and irrigation systems*
Mongols under Timur	→	*supported arts and literature; built mosques, gardens, and palaces in Samarqand*
Russians	→	*irrigated and farmed deserts; built railroads*
Soviets	→	*expanded cotton production; redrew political boundaries; movement of Russians into region*

HOMEWORK: Have students write essays about the periods in Central Asia's history in which they would most have liked to live. Tell students they must support their decisions with historical information from the chapter or other sources. You may want to have students examine how membership within a certain group (example: as a Russian farmer) would affect their choices of historical periods. Have volunteers share their essays with the class.

to sail to East Asia, avoiding the Silk Road. As a result, trade through Central Asia declined, and the region became isolated.

Russian and Soviet Rule Under the czars, Russia conquered and colonized Central Asia in the mid-1800s. The conquest began in the north and moved south. Following the soldiers were settlers who irrigated and farmed the desert. The Russians also built railroads throughout the region. Railroads helped the Russians create a stronger military presence between India, which was under British control, and Russia. The railroads also offered better access to the region's resources. Expanded cotton and oil production followed.

Resistance to Russian rule grew during the 1910s. However, the Soviet government, which took power in Russia after the Russian Revolution in 1917, eventually crushed the resistance. In an effort to weaken that resistance, the Soviets drew new political boundaries that separated people and resources. The Central Asian countries were called republics, but they were fully under Soviet control. Russian migration to the region increased, bringing in several million people.

The Soviets built huge irrigation projects, which helped the Soviet Union become a leading cotton producer. In short, the government built a cotton **monoculture**—the cultivation of a single crop—in the region. People who had moved with their herds were forced to settle on large government-owned farms. During World War II, factories and more Russians moved east into Central Asia.

As the Soviet Union collapsed in 1991, the Central Asian republics declared their independence. Troublesome international boundaries remained from the Soviet era. For example, the fertile Fergana Valley had been divided among three countries. Those countries had to share the main rivers and irrigation canals. Today just southwest of the Fergana Valley, Uzbekistan enclaves lie in Kyrgyzstan. Enclaves of Tajikistan lie in Kyrgyzstan and Uzbekistan. The complex boundaries are now difficult to control. As a result, the governments cannot stop smugglers who ship illegal drugs through the area. The problem is most severe for Uzbekistan's government.

✓ **READING CHECK:** *Human Systems* What are some ways that Russian and Soviet rule changed Central Asia? built railroads and irrigation projects, expanded cotton production and oil exports, created a cotton monoculture, forced herders to become farmers, changed boundaries

Culture

The five Central Asian countries lie in a region sometimes called Turkistan. The suffix *-stan* means "place" or "land" in Turkish. Turkic languages and ethnic groups have long dominated the area. As a result, many people once called the whole region Turkistan. Turkic heritage is still strong in the region today.

Irrigated farming was the traditional way of life in the region's southern areas. Herding was traditional in the north. A large segment of the north's population was made up of **nomads**. Nomads are people who move often from place to place. The region's nomads moved herds from mountain pastures in the summer to lowland pastures in the winter. This practice is called **transhumance**.

A teenager skates past a statue of Vladimir Lenin in Bishkek. The statue is left over from Kyrgyzstan's Soviet era. Lenin was a leader of the Russian Revolution.

Essential Element 4

▶ **Human Systems** ◀

The Silk Road Not truly a single road but rather a series of trade routes that linked people along a 4,000-mile stretch, the Silk Road passed through Kyrgyzstan, Tajikistan, Uzbekistan, and Turkmenistan as it ran from China to the Mediterranean. Bukhara and Samarqand were great trading cities on the route.

Trade along the Silk Road developed gradually over time and continued from the 100s B.C. to the A.D. 1400s. Obviously, silk was traded along the route, but so were many other products. Paper, spices, ceramics, gunpowder, bronze, and jade went west from China in exchange for gold, ivory, horses, cotton, and frankincense.

Traffic on the route is once again increasing. Roads and rail lines that parallel the ancient route across Central Asia allow trade with China.

ACTIVITY: Organize the class into groups to create maps of the Silk Road. Have groups research and label the sources of various products traded along the route.

📡 **internet** connect

GO TO: go.hrw.com
KEYWORD: SW3 CH18
FOR: Web sites about the Silk Road

Teach Objective 2

LEVEL 1: Have students work in small groups to create collages of Central Asia's cultures. Encourage students to use words, pictures, graphs, and other elements in their collages. **ENGLISH LANGUAGE LEARNERS, COOPERATIVE LEARNING**

LEVELS 2 AND 3: Have students complete the Level 1 activity and then create designs for Central Asian carpets or other textiles typical of the region that include scenes or images from the region's cultures. Some students may want to conduct further research and to produce sample textiles.

Teacher to Teacher

Lisa Haydel of Houma, Louisiana, suggests the following activity to help students understand the importance of cultural heritage to a region's tourism industry. Organize the class into groups and have each group assume the role of a tourism development board for one of Central Asia's countries. Have the groups conduct research into their chosen countries to identify areas that could be developed as tourist attractions or points of cultural interest. Students may also wish to research festivals or events that might attract visitors to their countries. Have each group prepare an illustrated proposal of areas for development as tourist centers. **COOPERATIVE LEARNING**

Daily Life

Islam in Central Asia

Within the past decade, many Central Asians have worked to restore the practice of Islam in their region. However, these efforts have been met with a certain amount of governmental resistance. In order to limit the power of Islamic political movements, some governments in the region have imposed restrictions on some Muslims. For example, some Islamic-based political groups have been banned. Even the wearing of long beards, which can symbolize adherence to fundamentalist Islamic beliefs, has been outlawed in some areas.

Kazakhstan, Kyrgyzstan, and Turkmenistan may be less vulnerable to disruptive fundamentalist religious movements. Russian cultural influences are more widespread in these three countries than in Uzbekistan and Tajikistan. In addition to being geographically close to Iran and Afghanistan—countries where Islamic political movements have been powerful—Uzbekistan and Tajikistan have well-established Islamic institutions.

VISUAL RECORD ANSWER

A traditional housing style is used for new businesses.

GRAPH ANSWER

Soviet republics and independent states shaped around ethnicity

INTERPRETING THE VISUAL RECORD

Historically, most people of Central Asia have been farmers or nomadic herders like the Tajik man on the left. Many, however, have begun to settle in permanent communities and to seek new occupations. On the right, hungry Kyrgyz travelers can choose from several roadside restaurants housed in yurts. **How does this picture suggest a blending of traditional and modern cultures?**

Unique homes, called **yurts**, made moving with the herds possible. A yurt is a movable round house of wool felt mats. The mats are placed over a wood frame. Today these homes remain a symbol of the region's nomadic heritage. Even people in cities may put up yurts for weddings and funerals.

People, Languages, and Religion Almost two thirds of Central Asians speak a Turkic language. Many people identify themselves as Turkic. In fact, about 30 percent of the world's 125 million speakers of Turkic languages live in Central Asia. Four major Turkic languages are represented in the region. However, Tajikistan's language belongs to a different group and is related to the main language of Iran.

Russian is the main language and ethnic identity for a sizable minority of Central Asians. In fact, Russian is still an official language in some of the countries there. However, Russian language and culture in general are losing status. For example, the Latin alphabet is replacing the Cyrillic alphabet, in which Russian is written. In addition, millions of people have returned to Russia since 1991.

Traders and conquerors brought many religions to Central Asia. Islam is the main religion. Most of the region's Christians are ethnic Russians and belong to the Russian Orthodox Church. During the Soviet era, the government tried to get rid of all religions. It closed more than 35,000 mosques, churches, and Islamic schools. Many of these buildings were abandoned or destroyed over time. Since 1991 the remaining buildings have reopened. They are now powerful symbols of ethnic pride. The beautiful mosques and Islamic schools in Samarqand and other cities are major tourist attractions. Exquisite turquoise tile mosaics decorate many of them.

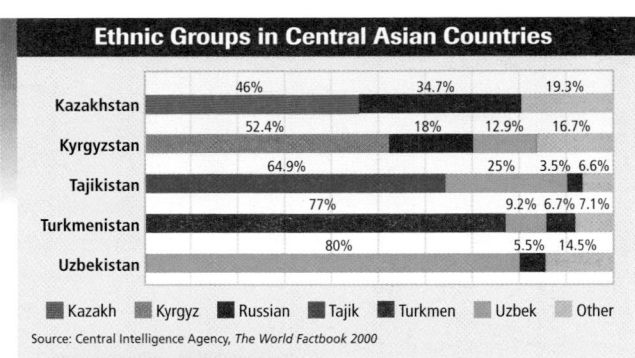

Ethnic Groups in Central Asian Countries

Country			
Kazakhstan	46%	34.7%	19.3%
Kyrgyzstan	52.4%	18% 12.9%	16.7%
Tajikistan	64.9%	25%	3.5% 6.6%
Turkmenistan	77%	9.2% 6.7%	7.1%
Uzbekistan	80%	5.5%	14.5%

■ Kazakh ■ Kyrgyz ■ Russian ■ Tajik ■ Turkmen ■ Uzbek ■ Other

Source: Central Intelligence Agency, *The World Factbook 2000*

INTERPRETING THE GRAPH *Except for the Tajiks, who are more connected to the settled culture of Iran, the region's other major ethnic groups are all Turkic peoples.* **What role do you think the distribution of ethnic groups played in shaping the borders of Central Asia?**

Close

Tell students that Genghis Khan is a hero to many Central Asians, while in Western sources he is often portrayed as a cruel and ruthless conqueror. Lead a discussion about how one's culture can affect perceptions of historical figures. Ask students to suggest American heroes who might be portrayed differently in other countries.

Review and Assess

Have students complete the **Section Review**. Then have students complete **Daily Quiz 18.2.**

Reteach

Have students complete **Main Idea Activity for English Language Learners and Special-Needs Students 18.2.** Then have them develop postage stamps that represent aspects of the history or culture of Central Asia. Students should explain their stamps. ENGLISH LANGUAGE LEARNERS

Extend

Have interested students use craft sticks, felt, and cloth to build and furnish model yurts. Point out that to be useful as movable structures, yurts must be constructed so they can be folded and loaded onto a wagon or horse. Challenge students to design their yurts so they can be collapsed and rebuilt. BLOCK SCHEDULING

Traditions and Education Textiles are among Central Asia's best-known art forms. Felt yurts, distinctive wool hats, and silk fabrics are examples. Silk textiles are produced in both homes and factories. Mulberry trees growing along roads, fields, and canals provide the leaves on which silkworms feed.

Sheep, goats, and other animals grow long hair that is excellent for weaving into carpets. Central Asian carpets, often made by Turkmen nomads, are famous. Red colors and geometric designs are common on most of the carpets. Factory-made copies from Europe and declining Silk Road trade caused production of carpets to fall. Now the region's traditional weaving crafts are being revived because the tourist trade is growing.

Like other cultural crossroads, Central Asia has varied food traditions. Noodles probably came from China. The region imports tea from China today. Tea is very popular, and teahouses are a common sight in many areas. From the nomad culture came meat dishes and dairy products, some made with soured milk. Grilled meats such as lamb are featured in many meals. The farming way of life provides bread and rice dishes. Fruits and vegetables are plentiful. Apples, apricots, lemons, pumpkins, and watermelons are among the many fruits grown locally.

Communism limited personal freedoms, but the Soviet system did stress education and health care. State-run schools replaced education in Islamic schools. The results of widely available public services have continued in the post-Soviet period. Literacy rates are well above world averages. In addition, life expectancy in most Central Asian countries is longer than the world average, particularly for women.

Distinctive headgear is part of traditional dress in Central Asia. The sheepskin hats of the Turkmen (top) contrast with the embroidered felt caps worn by Uzbeks (bottom).

✓ **READING CHECK:** *Human Systems* In what ways did education in Central Asia change during the Soviet era? The Soviets replaced Islamic schools with state-run schools.

Section 2 Review

go.hrw.com **Homework Practice Online**
Keyword: SW3 HP18

Define caravans, monoculture, nomads, transhumance, yurts

Working with Sketch Maps On the map you created for Section 1, label the countries of Central Asia, Samarqand, and the Fergana Valley. Which three countries have complex boundaries in the Fergana Valley region?

Reading for the Main Idea
1. *Human Systems* Which invaders destroyed Central Asian cities and irrigation systems? Which conqueror beautified Samarqand?

2. *Places and Regions* What are some cultural features that make Central Asia distinctive?

Critical Thinking
3. **Drawing Inferences and Conclusions** How might the region's cultural geography be different today if the Arab armies had not defeated the Chinese in A.D. 751?

4. **Analyzing Information** Why might Turkistan be considered a perceptual region?

Organizing What You Know
5. Create a graphic organizer like the one below. Use it to identify the many invasions of Central Asia. Add more boxes if necessary. Then use more arrows or other symbols to describe connections between the groups.

☐ → ☐ → ☐ → ☐ → ☐

OBJECTIVES

1. **Identify changes in Central Asia's economy over time.**

2. **Describe Central Asia's cities.**

3. **Examine the issues Central Asia must face to improve its economy.**

LET'S GET STARTED

Copy the following question onto the chalkboard: *What types of economic challenges might face a country with no access to the ocean? Write down your answers.* Discuss responses. Point out that Central Asia is rich in resources, but the costs of transportation often prohibit the region's countries from fully developing their production. Tell students that they will learn more about the economies of Central Asia and the challenges they face in Section 3.

Building Vocabulary

Write the term **dryland farming** on the chalkboard. Call on a volunteer to locate and read the term's definition from the glossary or text. Ask students why they think this farming method was named *dryland*. (*Possible answer: No water is added directly to the ground through irrigation. Moisture for the crops comes from rainfall.*)

RESOURCES

REPRODUCIBLE

▶ Guided Reading Strategy 18.3

▶ Readings in World Geography, History, and Culture 45

▶ Critical Thinking Activity 18: Getting Ahead

▶ Geography for Life Activity 18: The Fragmented Fergana Valley

TECHNOLOGY

▶ One-Stop Planner CD–ROM, Lesson 18.3

▶ CNN Presents Geography: Yesterday and Today, Segment 20: Dateline Kazakhstan

▶ Homework Practice Online

▶ HRW Go site

REINFORCEMENT, REVIEW, AND ASSESSMENT

▶ Main Idea Activity 18.3

▶ English Audio Summary 18.3

▶ Spanish Audio Summary 18.3

▶ Section 3 Review, p. 415

▶ Daily Quiz 18.3

The Region Today

READ TO DISCOVER

1. How has the economy of Central Asia changed over time?
2. What are the region's cities like?
3. What issues must Central Asia face to improve its economy?

WHY IT MATTERS

Many people in Kazakhstan suffer from ailments blamed on nuclear testing during the Soviet era. Use **CNNfyi.com** or other **current events** sources to learn about these and other health problems in Central Asia.

DEFINE

dryland farming

LOCATE

Bukhara

Tashkent

Bishkek

Almaty

Astana

Economic Changes

Agriculture remains important to Central Asia. Traditional herders raise camels, cattle, goats, horses, and sheep. Crops include cotton, fruits, barley, rice, tobacco, and vegetables. Uzbekistan is the world's third-largest exporter of cotton. To retain the soil's nutrients, farmers are now adding other crops to the region's cotton monoculture. Kyrgyzstan's fertile soils in the Tian Shan foothills allow a mix of **dryland farming** and irrigated crops. Dryland farming relies on rainfall instead of irrigation. Turkmenistan depends on the world's longest irrigation channel to water grain and cotton fields in the country's southern areas.

Mining and industry offer a chance for future wealth. Huge reserves of oil, gas, and minerals await development. However, having to use outdated equipment slows industrial growth. Corruption, poor transportation links, and a lack of cash for investment also hurt development. Moreover, many skilled Russian workers and managers are leaving the region.

The need for more foreign trade is another factor limiting development. Russia is not the reliable customer that the Soviet Union once was. Other markets are far away. Even products for local markets must often cross international borders.

✓ **READING CHECK:** *Places and Regions* What factors limit the Central Asian countries' ability to develop their mines and industries? outdated equipment, lack of cash for investment, poor transportation links, many skilled Russians leaving, no assured markets

Urban Environments

Central Asia has relatively few big cities because, throughout its history, most of the people have been nomads or farmers. As a result, few of the cities have many

INTERPRETING THE VISUAL RECORD

Workers pick cotton in a field outside Bukhara, Uzbekistan. During the harvest season, students and government workers from across the country are brought in to help with the picking. **Why might farmers be forced to rely on student labor to harvest the cotton crop?**

VISUAL RECORD ANSWER

not enough field laborers or limited funds for mechanizing the industry

Teach Objective 1

ALL LEVELS: Copy the following graphic organizer onto the chalkboard, omitting the italicized answers. Have students complete it to describe the economic activities of the Central Asian countries. **ENGLISH LANGUAGE LEARNERS**

```
            Economic Activities in Central Asia

  Agriculture            Mining and Industry        Developing Industries
  crops—barley, cotton,  huge reserves of minerals  oil and gas production
    fruits, rice,        outdated equipment, cor-   tourism
    tobacco, and         ruption, poor transporta-
    vegetables           tion, lack of funds for
  irrigated and dryland  investment
    farming              limited foreign market
  herding—camels, goats,
    horses, and sheep
```

Teach Objective 2

ALL LEVELS: Have each student choose a city in Central Asia. Have students use reference books, travel guides, and the Internet to gather information about daily life in their chosen cities. Then have them report their findings to the rest of the class as if they are guides leading tourists through the city.

Connecting to TECHNOLOGY

Baykonur Cosmodrome

The space race between the Soviet Union and the United States began in 1957. In that year the Soviets launched the first artificial satellite *Sputnik* into space. *Sputnik* blasted off from the Baykonur Cosmodrome in what is now Kazakhstan. The world's first manned orbital flight and the flight of the first woman in space also began at Baykonur. It is still Russia's most important location for launching space vehicles. However, the noise and activity have not completely changed the surrounding environment. In fact, wild camels and horses roam the site!

For decades, the Soviet Union tried to keep Baykonur's location, and even its existence, a secret. Today a detailed map of the complex is available on the Internet.

Analyzing Maps Study the chapter physical-political map. Why do you think the Soviet Union established the Baykonur Cosmodrome in that location?

In 1963 Valentina Tereshkova became the first woman in space. The letters CCCP on her helmet stand for "Union of Soviet Socialist Republics" in Russian.

old features. Those that lie along ancient trade routes, such as Bukhara and Samarqand, still have colorful markets and blue-tiled mosques. Tashkent, the region's largest city, has an Old Town of mud-brick houses and narrow streets. However, most of Tashkent is a Soviet-era city of plain apartment buildings, factories, and offices. Smaller cities, such as Bishkek, also offer few cultural or historical sites. They do serve as starting points for tourists exploring the mountains and other scenic areas. As tourists discover the Central Asian countries, some new banks, restaurants, and shops are opening in the cities.

In 1997 Kazakhstan moved its capital from Almaty to Astana. Government leaders wanted to rebuild the city into a glorious capital, but funds are not available. On the other hand, the new location is closer to Russia and Europe. This new location might strengthen links between Kazakhstan and Russia and help Kazakhstan keep its Russian population.

READING CHECK: *Human Systems* What caused the old cities of Central Asia to grow? What evidence remains from these earlier times? location along ancient trade routes; colorful markets, blue-tiled mosques, mud-brick houses, narrow streets

Issues and Challenges

The Central Asian countries must overcome tremendous challenges to ensure economic and social progress. Oil and gas deposits may bring in plenty of cash someday, but all the countries still face an uncertain future. Many of the people are poor and have few opportunities to improve their lives.

The region's location creates a basic problem. As you have read, the countries are landlocked. As a result, they are cut off from major

The bazaar in Andizhan, Uzbekistan, recalls the lively trade of the Fergana Valley city's history as a stop on the Silk Road. However, unemployment in the Fergana Valley is high. In some areas, up to half of the young men do not have jobs.

Daily Life

A Museum for Peace
Over the centuries numerous bloody battles have been fought in Central Asia, but times have changed. Now a museum dedicated to peace draws tourists to Samarqand.

The International Museum of Peace and Solidarity was built in 1986. It sponsors educational activities and international projects. On display are some 20,000 peace-promoting items from more than 100 countries, including banners, books, CDs, paintings, posters, stickers, T-shirts, and videos. The museum sponsors the Children's Peace and Disarmament Festival, during which children trade in military toys for books, craft kits, paints, and similar playthings.

CONNECTING TO TECHNOLOGY ANSWER

Possible answers: near a river but far from densely populated areas, so damage is limited in case of an accident; easier to limit access and guard perimeter in desert area

Teach Objective 3

ALL LEVELS: Organize the class into small groups. Have each group create a news report for a television program called World Issues Today. Each group should investigate a challenge facing Central Asia, write a script that explains it, and perform its broadcast segment to the class. Groups may wish to conduct further research to supplement the material in the text. **COOPERATIVE LEARNING**

Using National Geography Standard 14: Environment and Society: How Humans Modify the Physical Environment Have students conduct research on nuclear testing in an area of Kazakhstan called the Polygon. The tests, performed during the Cold War, are blamed for devastating health problems among area citizens. With the information they find, have students work in pairs to write mock interviews between news reporters and people involved in or affected by the tests. These might include Soviet military officers, nuclear scientists, farm families, doctors, Kazakh government officials, international environmental organizations, or local environmental activists. Have pairs perform their interviews for the class.

Global Perspectives

Landlocked Countries
Recent studies suggest that landlocked countries face significant disadvantages. According to these reports, landlocked locations account for annual economic growth rates that are about 1.5 percent slower than those of countries with access to the sea.

Landlocked countries must import and export goods through other countries. They also lack marine resources.

Oil discoveries in Central Asia have not been fully exploited because exporting the oil presents difficulties. Pipelines from Central Asian oil wells to seaports would have to cross other countries, where they might be vulnerable to attack or control by other governments.

DISCUSSION: Lead a discussion about agreements Central Asian countries might make with their neighbors to alleviate the disadvantages their landlocked locations present. You may want students to investigate how other landlocked countries, such as Bolivia, Ethiopia, and Switzerland, have faced the issue.

You don't say! Creatures that live in the Kara-Kum must endure temperatures above 122°F (50°C). Among these hardy animals are monitor lizards that can grow to more than 5 feet (1.5 m) long.

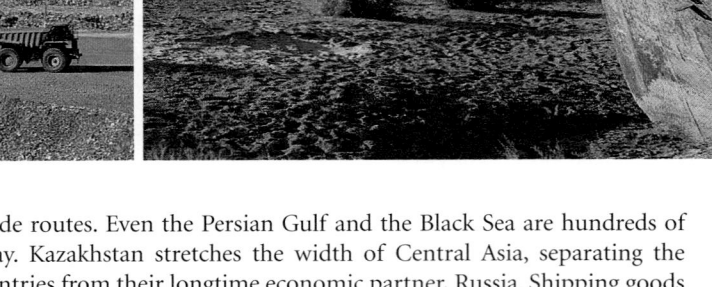

Different types of environmental challenges face Central Asia. Left: A slag heap contrasts with the snowy slope behind it. Slag is a waste product created when metal ores are refined. It is sometimes used to build roads or to make fertilizer. However, much of Central Asia's slag is contaminated with radiation or poisonous substances. Right: Stranded ships provide testimony of the Aral Sea's receding shoreline.

Our Amazing Planet

About 90 percent of the world's caviar—salted fish eggs—comes from sturgeon in the Caspian Sea. Sturgeon can grow up to 14 feet (4 m) long, weigh 1,300 pounds (590 kg), and carry 200 pounds (91 kg) of eggs worth up to $250,000 in the United States.

global trade routes. Even the Persian Gulf and the Black Sea are hundreds of miles away. Kazakhstan stretches the width of Central Asia, separating the other countries from their longtime economic partner, Russia. Shipping goods great distances for sale to other countries adds costs. Because they cost more than products from nearby countries, these goods find few buyers.

The need for water in this dry land is at the heart of many problems. The region's major rivers cross international borders. Water sources are in Tajikistan and Kyrgyzstan, while many water consumers are in Uzbekistan and Turkmenistan. Upstream users want to dam the rivers for hydroelectricity. Downstream users want the water for irrigation. Water has also been wasted. For example, Turkmenistan's canal lacks a concrete lining, so leaks are common. These leaks create swamps, high water tables, and salty soils.

Political Problems Changing from command to market economies has not been easy. The change to private enterprise is most advanced in Kazakhstan and Kyrgyzstan. It is slower in Uzbekistan and Turkmenistan. Corruption and the lack of democracy are major obstacles to economic growth. Tajikistan is recovering from years of civil war and still depends heavily on foreign aid from Russia, Uzbekistan, and international agencies.

Ethnic conflict also threatens the region. Various groups have committed violent acts. For example, in 1989 riots erupted among Uzbeks and others in the Fergana Valley. In 1999 an Islamic rebel group kidnapped foreign travelers. The region remains tense.

Environmental Issues Soviet agricultural, industrial, and military practices damaged Central Asia's land and water. For example, testing of biological weapons contaminated the Aral Sea region. In addition, beginning in 1949, the Soviet Union tested hundreds of nuclear bombs in an isolated area of northeastern Kazakhstan. Officials did not tell residents of the dangers from radiation. In fact, the testing exposed about 1.5 million people to radiation. Now birth defects, cancer, and other ailments plague the area's people.

Overuse of chemicals to increase crop yields has made some farmlands useless. In addition, places where uranium was mined and processed are now toxic. Money to clean up these sites is lacking. Many of the Russian experts and workers who know how to do the job are leaving the area.

✓ **READING CHECK:** *Environment and Society* How did various Soviet practices affect the region's environment? Testing of biological and nuclear weapons contaminated land, water, and people. Overuse of chemicals to increase crop yields made some farmlands useless. Uranium mining polluted some areas.

FOCUS ON GEOGRAPHY

Issyk-Kul and Tourism Although some areas in Central Asia have been polluted, others are clean and beautiful. For example, Kyrgyzstan's Issyk-Kul is an environmental and cultural treasure.

A gorgeous mountain setting surrounds the big lake, which covers 2,355 square miles (6,099 sq km). The lake is quite deep, at 2,303 feet (702 m), and it has a surface elevation of 5,279 feet (1,609 m). The lake never freezes, partly because of underwater hot springs. In fact, *Issyk-Kul* means "Hot Lake." Many streams tumble down to Issyk-Kul and its beaches. The clear blue water is slightly salty but pleasant for swimming. Fish are abundant. Because the lake's warmth moderates the climate, the surrounding area is a productive agricultural region. Apples are a local specialty.

Early civilizations and later conquerors seem to have valued the lake region. They left burial mounds, carvings in rock faces, and other artifacts. Some of these artifacts are up to 2,500 years old.

The Soviets built more than 100 resorts, spas, and similar establishments along the lake's shores. Most visitors were government officials because the Soviet navy tested torpedoes and other weapons in the area. Today tourists can rent converted navy boats for short trips. Local tourism prospers, as Kyrgyzstan's major cities are not far away. Hotels and travel facilities are springing up at the lake's east end. Many tour organizers now arrange hikes to the nearby mountains. A preserve the size of Switzerland has been proposed to protect Issyk-Kul's unique qualities.

✓ **READING CHECK:** *Environment and Society* How has the development of tourism affected the character of Issyk-Kul? Resorts, spas, hotels, and other facilities have been built on the lake.

INTERPRETING THE VISUAL RECORD *A rider explores a rocky area of Issyk-Kul's shoreline. The area's temperate climate provides a habitat for a wide range of wildlife. Some 40 mammal species and 200 bird species live there.* **How do you think Issyk-Kul's wildlife compares to animal populations farther from the lake?**

Review

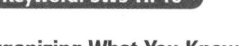

Homework Practice Online
Keyword: SW3 HP18

Define
dryland farming

Working with Sketch Maps
On the map you created for Section 2, label Bukhara, Tashkent, Bishkek, Almaty, and Astana. Where is the world's longest irrigation channel?

Reading for the Main Idea

1. *Environment and Society* How are farmers in Uzbekistan improving the soil?

2. *Places and Regions* What advantages does the new location for Kazakhstan's capital present? What problem remains?

Critical Thinking

3. **Solving Problems** How might international cooperation help the region's countries overcome the obstacles to trade that are posed by their location?

4. **Comparing Points of View** How do different countries in the region want to use water from rivers that cross their borders? How do you think these differing points of view make it difficult for countries in the region to develop policies regarding water use?

Organizing What You Know

5. Create an idea web to describe the geographic, political, and environmental challenges facing Central Asia.

Setting the Scene

Have students read Case Study: Pipelines or Pipe Dreams? Ask students to suggest ways in which the proposed pipelines could affect life in the areas they would cross. *(Possible answers: more jobs for pipeline builders and maintenance workers, increased revenues for the governments, risk of oil spills and environmental damage, possible military conflicts over control of the pipeline)* Point out that these results could have wide-ranging indirect effects.

Building a Case

If the Caspian Sea oil deposits are as vast as they seem to be, the money that might pour into the region could set big changes into motion. Compare the region's situation to Saudi Arabia's. Before the 1930s, most Saudis lived by herding animals, farming, or fishing. There were few good roads, and health care was poor. Then vast oil fields were discovered. Soon, income from oil exports brought the Saudi government money for airports, apartments, communication systems, hospitals, roads, and schools. Citizens had more money to spend and began to travel abroad. More students attended universities. The government invested in irrigation projects, which allowed farmers to raise more food. In addition, Saudi Arabia's wealth brought the country political power.

CASE STUDY

Pipelines or Pipe Dreams?

Environment and Society Azerbaijan, Kazakhstan, and Turkmenistan sit astride some of the largest oil reserves in the world. Countries inside and out of the region are working to unlock the potential of this rich natural resource. However, they have faced a number of geographic and political hurdles.

Estimates of Caspian oil reserves vary widely, from 40 billion barrels to 200 billion barrels. Even the lower amount is more than the proven oil reserves in the United States and the North Sea combined. The higher amount could be double that of Kuwait's proven oil reserves of some 100 billion barrels. Kuwait is one of Southwest Asia's richest oil-producing countries. The Caspian region might also hold huge amounts of natural gas, another valuable energy resource.

Outside powers have long tried to control the oil along the Caspian. For example, German armies tried but failed to overrun the oil fields of the nearby Caucasus during World War II. Since the collapse of the Soviet Union, oil exploration in the region has expanded. In addition, the contest to further develop the oil resources there has heated up.

An employee walks near an oil-pumping station south of Turkmenbashy, Turkmenistan. Oil fields in this area produce some 7,500 barrels of oil per day.

The main problem in developing these fields has not been how to drill for the oil. The problem has more to do with getting the oil to global markets. Remember that the Caspian Sea is really a lake and has no natural outlet to the world's oceans. The only way to get the oil out of the region is to use overland routes to seaports in other countries. Oil companies and interested countries have proposed a variety of routes, including pipeline routes, that could transport the oil to these ports. Routes through Iran, Russia, and across Georgia and Turkey have been among the proposals. Even pipelines to China and Pakistan have been suggested, but they probably would be very expensive to build. However, the main factor is not cost but who will control the pipeline, and thus the oil.

On the surface, one would think that choosing a route should not pose a problem. Simply build a pipeline across the shortest distance to the nearest port, right? Things are rarely that easy, though, particularly in this part of the world. Many proposed routes cross through the Caucasus region, a shatter belt. As you

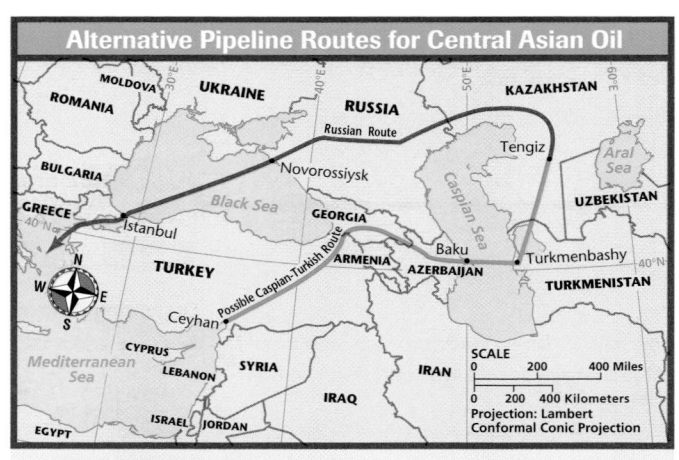
Alternative Pipeline Routes for Central Asian Oil

INTERPRETING THE MAP *Turkey is concerned about proposed pipeline routes. Because it lies between Central Asia and the Mediterranean Sea, Turkey will be involved with oil transport along either of the routes shown.* **How would the choice of each port affect the movement of capital, goods, and people through Turkey? Which route does the Turkish government favor?**

Drawing Conclusions

Have students draw idea webs to describe the changes that the discovery of oil could have on a country or region. Encourage them to follow the changes several steps away from the initial influx of money. *(Example: An increase in money for schools could result in new ideas about the role of women in society and eventual changes in those roles.)* Call on a volunteer to summarize the activity. *(Possible answer: The discovery of resources or the development of new sources of wealth may have both short-term and long-term effects on the region where those events take place.)*

Going Further: Thinking Critically

The discovery of precious metals can also have significant repercussions for a region. Finding diamonds, gold, or silver has affected many countries dramatically. Provide students with a map of global mineral resources and other research materials. Have students pursue the answers to the following questions:

- Which countries have large deposits of diamonds, gold, or silver?
- When were these resources discovered?
- What role do these resources play in the countries' economies?
- How else has discovery of these resources affected the countries?
- What are the short-term and long-term effects of these discoveries on the countries?

Tankers carry about 1.7 million barrels of oil through the Bosporus every day. Turkey's government fears a major accident in the strait and favors an alternate oil-transport route.

have read, a shatter belt is a zone of frequent boundary changes and conflicts. Often these changes are caused by the area's position between major powers. In fact, Iran, Russia, and Turkey have long competed for influence in the Caucasus and Central Asia. Some Russians see it as an area over which they have had, and should still have, influence.

The struggle for influence in the region has been particularly important in the debate over pipelines. Many of the companies that want to develop the Caspian oil fields are American or European. For political reasons, these companies and Western governments do not want to build oil pipelines across Iran or Russia. On the other hand, Russian leaders worry that Western countries—particularly the United States—already have too much influence in the region. They think moving the oil through Russian pipelines would limit those outside influences.

These competing fears and desires have led to competing pipeline plans. For example, Russia wanted oil companies to use a new pipeline across southern Russia. The pipeline could transport the oil to tankers at a Russian seaport on the Black Sea. (See the map.) The Russian plan faced opposition, however. Critics worried the Russians might use the pipeline to influence the region's affairs. They also worried about the route tankers would have to take from the Russian port. Tankers on that route would pass through the narrow Bosporus, the strait that connects the Black Sea to the Mediterranean Sea. This fact worried Turkey, which straddles the Bosporus. The Turks pointed out that the amount of maritime traffic through the narrow strait would more than double. This situation could create long traffic jams on the sea lane. Safety was also a concern because millions of people live on or near the Bosporus. An explosion or oil spill could be devastating in this densely populated area.

Turkey, its NATO ally the United States, and some oil companies campaigned for another route. Under their plan, tankers would ship the oil across the Caspian Sea to Baku in Azerbaijan. A new pipeline would then carry the oil across Georgia to a Turkish port on the Mediterranean Sea. (See the map.) However, this plan also presented challenges. One was the cost—it would be the most expensive route. Another challenge was that the pipeline would go through a territory that has experienced ethnic and religious conflict in recent years. War or other political violence could stop the movement of oil along the route. Most important, Russia, which still holds the most power in the region and is able to influence Georgia, opposed the route.

For Azerbaijan, Kazakhstan, and Turkmenistan to market their product, compromises will have to be made on many levels with the Russians, the oil companies, and regional neighbors. Economic and political cooperation among the parties involved will be necessary for the Central Asian republics to benefit from their oil resources.

Applying What You Know

1. **Summarizing** What are two of the pipeline routes that have been proposed for transporting Central Asian oil? How would each of these routes affect the ability of neighboring countries to influence Central Asian affairs?

2. **Analyzing Information** What are the drawbacks associated with each of the two pipeline routes discussed at length here?

TECHNOLOGY

▶ Chapter 18 Test Generator (on the One-Stop Planner)
▶ Global Skill Builder CD–ROM
▶ HRW Go site

REINFORCEMENT, REVIEW, AND ASSESSMENT

▶ Chapter 18 Review, pp. 418–19
▶ Chapter 18 Tutorial for Students, Parents, Mentors, and Peers
▶ Chapter 18 Test (form A or B)
▶ Alternative Assessment Handbook

▶ Chapter 18 Test for English Language Learners and Special-Needs Students
▶ Unit 5 Test
▶ Unit 5 Test for English Language Learners and Special-Needs Students

Assess

Have Students complete a Chapter 18 Test.

Reteach

Have students work in pairs to create three graphic organizers that cover the chapter information. Then have pairs exchange blank organizers and complete those they receive. Have students return the organizers to be checked for accuracy. **ENGLISH LANGUAGE LEARNERS, COOPERATIVE LEARNING**

CHAPTER 18 Review Answers

2. Possible answers: might see Russians and Russian culture as threats to their own cultural development and political independence or might see Russians as vital to their countries' economic development (NGS 11, 13)

3. revival of Islam, weaving carpets and other traditional textiles, using yurts, eating traditional foods (NGS 10)

Using the Geographer's Tools

1. Rankings may vary, but Russia will top most lists. Students may say that countries that have made more progress toward democracy have had more rapid development.

2. Students will find that in some countries the percentage of ethnic Russians has declined because many Russians are leaving Central Asia.

3. Larger percentage of workers in services indicates higher GDP.

Writing

Students should discuss the loss of status for Russians in Central Asia and their having to change alphabets. They may also mention possibilities for new opportunities in Central Asia as Kazakhstan develops, weighing the challenges that both Russia and Central Asia face, or other topics. Use Rubric 41, Writing to Express, to evaluate student work.

CHAPTER 18 Review

Building Vocabulary

On a separate sheet of paper, explain the following terms by using them correctly in sentences.

zinc   nomads   yurts
caravans   transhumance   dryland farming
monoculture

Locating Key Places

On a separate sheet of paper, match the letters on the map with their correct labels.

Tian Shan   Syr Dar'ya   Samarqand
Aral Sea   Lake Balkhash   Tashkent
Amu Dar'ya   Kyzyl Kum   Astana

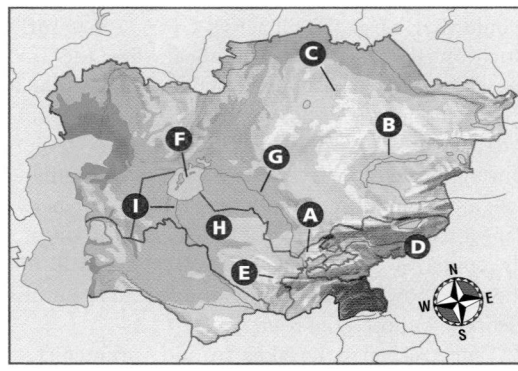

Understanding the Main Ideas

Section 1

1. (Human Systems) Why is herding common in the region's northern areas and farming more common in the south?

Section 2

2. (The World in Spatial Terms) How did boundary changes during the Soviet era affect the control that Central Asian governments have over their territory?

3. (Human Systems) What are the main languages and religions in Central Asia? Which language is losing status?

Section 3

4. (Environment and Society) What resources of Central Asia offer the possibility of wealth in the future?

5. (Places and Regions) How does Issyk-Kul contrast with some other areas in Central Asia?

Thinking Critically

1. Problem Solving If you were in charge of developing tourism in Central Asia, how would you go about doing your job?

2. Identifying Points of View How do you think Central Asia's indigenous peoples might view Russian culture and ethnic Russians who live in their countries today?

3. Analyzing Information In what ways can traditional ways of life still be seen in the region?

Using the Geographer's Tools

1. Using Statistics Use the unit Fast Facts and Comparing Standard of Living tables to rank the countries of Russia and Northern Eurasia by economic development and standard of living. What political features might help or hurt the region's development?

2. Using Databases Examine the graph of ethnic groups in Section 2. Then use the Internet or library resources to find databases that provide similar figures for a year or years before 1991. In what ways have the percentages changed? What do you suppose accounts for the changes?

3. Creating Graphs Use the unit atlas to create bar graphs of the workforce structure and per capita GDPs of the Central Asian countries. How does the workforce structure of these countries seem to affect wealth?

Writing about Geography

Imagine that you are a Russian teen living in Kazakhstan. Your parents are thinking about moving the family back to Russia. Write a diary entry in which you discuss reasons why you want to stay in Kazakhstan or why you want to leave. In your entry, analyze the economic and social effects on your family of each possibility. When you are finished with your entry, proofread it to make sure you have used standard grammar, spelling, sentence structure, and punctuation.

SKILL BUILDING

Geography for Life

Analyzing Statistics

(Environment and Society) Review the statistical and other data in Geography for Life: The Shrinking Aral Sea. How have physical processes and human activity affected settlement patterns and population distribution in the Aral Sea region? How have economic conditions suffered? What political changes might occur? How has the area's resource base lost value?

Portfolio Activity

Have interested students conduct research on Samarqand, Uzbekistan, one of the oldest cities in Central Asia. It was conquered by Alexander the Great and Genghis Khan, among others. In the late 1300s, Samarqand became the economic and cultural hub of Central Asia. It declined, however, and from the 1720s to the 1770s the city was abandoned. Samarqand has beautiful mosques and other structures, including the tomb of Timur. Ask students to compare the economies of ancient and modern Samarqand and speculate on the city's future prosperity in oral presentations. Place written copies or outlines of presentations in portfolios.

Food Festival

For special occasions, Central Asian herders serve a roasted lamb. In a Kazakh home, the honored guest may be asked to request permission from the lamb to eat the meal. A more common dish than roasted lamb throughout Central Asian countries is a savory rice dish called *palov, plov, polov,* or *pilaff.* With this dish, a large amount of rice stretches a small amount of meat to feed the whole family. Carrots, chickpeas, garlic, onions, raisins, and turnips are common additions. Have students locate recipes for the dish under its many spellings and prepare different versions for the class. Or, have students agree on a single version and contribute ingredients for its preparation.

Building Social Studies Skills

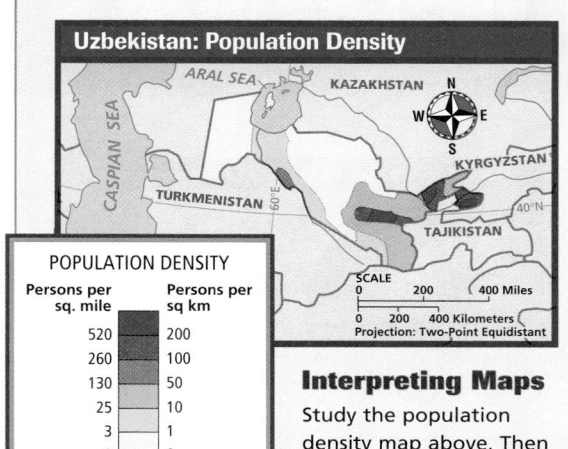

Uzbekistan: Population Density

POPULATION DENSITY

Persons per sq. mile	Persons per sq km
520	200
260	100
130	50
25	10
3	1
0	0

Interpreting Maps

Study the population density map above. Then use the information from the map to help you answer the questions that follow. Mark your answers on a separate sheet of paper.

1. Uzbekistan's population
   a. is least dense in the west and in central areas.
   b. is most dense in the far west.
   c. is evenly scattered across the country.
   d. is so small that few areas are densely populated.

2. Where would you expect to find Uzbekistan's largest cities? Why?

Using Language

The following passage contains mistakes in capitalization, grammar, and punctuation. Read the passage. Then answer the following questions. Mark your answers on a separate sheet of paper.

(1) About a third of Kazakhstan's people are Ethnic Russians. (2) Kazakh and Russian—both official languages. (3) Many ethnic Kazakh's grow up speaking Russian at home and they have to learn Kazakh in school today.

3. Rewrite line 2 to make it a complete sentence.

4. Which sentence contains incorrect capitalization?
   a. 1
   b. 2
   c. 3
   d. None of them

5. One sentence contains an error in the use of an apostrophe. It also has a missing comma. Rewrite the sentence correctly.

Geography for Life

Possible answers: Settlement near the Aral Sea has been disrupted, and more people will probably leave the area. Sources of income have diminished with the shrinking of the sea. Increasing poverty could lead to political unrest. The region's resource base—farming and industries based on the Aral Sea—has lost value. Use Rubric 6, Cause and Effect, to evaluate student work.

Social Studies Skills

1. a
2. in the east and northeast; because population densities are highest there
3. Kazakh and Russian are both official languages.
4. a
5. Many ethnic Kazakhs grow up speaking Russian at home, and they have to learn Kazakh in school today.

Alternative Assessment

PORTFOLIO ACTIVITY

Learning about Your Local Geography

Group Project: Research

You have read about the importance of the Baykonur Cosmodrome in Kazakhstan. Plan, organize, and complete a group research project about a major scientific facility in your state or area. Contact officials there to create a list of some of the projects being completed at the facility. Then have group members research the national or international significance of the projects. Use the information you find to create a brochure about the facility.

internet connect

Internet Activity: go.hrw.com
KEYWORD: SW3 GT18

Choose a topic on Central Asia to:
- create a poster that analyzes the increasing use of technology in agriculture and its consequences for the Aral Sea.
- research the Silk Road and its history in global trade.
- follow nomads in Kazakhstan.

PORTFOLIO ACTIVITY

Students' brochures should highlight achievements of the scientific facility and examine those achievements in terms of their national or international significance. Use Rubric 14, Group Activity, to evaluate student work.

Workshop 1
Going Further: Thinking Critically

Organize the class into groups and assign each group a numerical statistic of interest to your community, such as the unemployment rate or the number of families given assistance by local charities. Have students conduct research to obtain figures for their assigned topics in each of the last 10 years. Students may wish to consult locally published reference works or the Internet, or they may prefer to contact local government bureaus or community organizations to obtain these figures. Have each group compile the figures it collected into a table. Then have each group use the information in its table to create a bar graph or line graph that shows the same information. Many popular spreadsheet programs will convert data into graph form automatically. If these programs are not available, have students draw graphs themselves. Have groups examine their tables and their graphs to decide which best demonstrates changes in their data over time *(graphs)*. Tell the class that numerical tables are best when one needs to provide specific numbers or details. Graphs are better suited for illustrating the relationships or patterns found within the numbers.

PRACTICING THE SKILL

1. Students' graphs should accurately reflect the sizes of the populations of these countries. Russia's wedge should be much larger than the other two combined, and the Ukraine's should be significantly larger than that of Belarus.
2. Students' graphs should accurately reflect changes in Russia's population, which gradually declined through the 1990s into 2000.
3. Students' graphs should reflect differences in the GDPs of the countries of northern Eurasia, with Russia and Belarus having the highest and Tajikistan the lowest.

Geography
Skill-Building Workshop

WORKSHOP 1

Using Graphs

Graphs paint a clear picture of numerical data and the relationships among the data. Bar and line graphs are common graphs. Bar graphs can show how the value of a certain item changes over time, or they can compare the values of different items. For example, the bar graph on this page shows the changing water level in the Aral Sea over time. A line graph can also show changes over time. The example shows changes in the volume of water in the Aral Sea over time.

A third type of graph, a pie graph, is particularly useful when considering parts of a whole. Such a graph looks like a pie cut into slices. The complete pie, or circle, represents the whole, and each wedge represents a share of the whole. These shares are expressed in percentages. The pie graph on this page shows the world's largest rye producers. The wedges are drawn in proportion to each country's share of world rye production.

Developing the Skill Creating graphs is relatively easy. First, determine which kind of graph is best for showing the information you want to communicate. Second, give the graph a descriptive title. Third, identify the values, or numbers, you want to show.

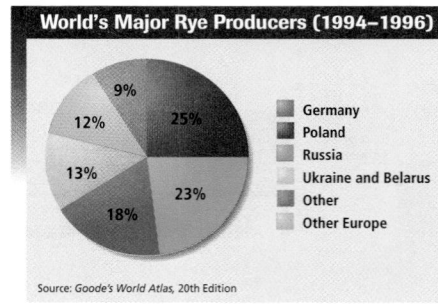

World's Major Rye Producers (1994–1996)

- 25%
- 9%
- 12%
- 13%
- 18%
- 23%

Germany
Poland
Russia
Ukraine and Belarus
Other
Other Europe

Source: *Goode's World Atlas*, 20th Edition

Bar graphs and line graphs include a side axis and a bottom axis. As you can see in the example below, values are labeled along the side axis. The years studied for volume and water level are placed along the bottom axis. The bars are then drawn so that their tops are aligned with the appropriate value along the side axis. The line graph identifies visible or invisible points that are also aligned to values along the side axis. You can also label values at the top of each bar or at dots along a line graph.

Pie graphs include a circle that represents the whole value of something. To create an effective pie graph, determine how many wedges will clearly fit into the circle. You can also include labels showing the percentage that each wedge represents. A legend, like the one above, can match colors to wedges they represent—in this case, particular countries that grow rye.

Changes in the Aral Sea

Volume of Water (in cubic miles) — left axis: 0, 25, 50, 75, 100, 125, 150, 175, 200, 225, 250

Water Level (in feet) — right axis: 0, 25, 50, 75, 100, 125, 150, 175, 200

Year: 1960, 1985, 1986, 1987, 1988, 1989, 1990, 1991, 1992, 1993, 1994, 1995, 1996, 1997, 1998, 2010

Source: German Aerospace Center (DLR)

Practicing the Skill

1. Create a pie graph showing the combined populations of Belarus, Russia, and Ukraine. Each wedge in the graph should represent one country.
2. Create a line graph showing the population of Russia from 1992 to the present.
3. Create a bar graph that compares the per capita GDP of each country you have studied throughout this unit.

Workshop 2
Going Further: Thinking Critically

Remind students that time lines, like other graphic organizers, do not include all possible information about a particular subject. The creators of time lines have specific interests or subjects with which they are concerned, so the events that are listed on one time line may not appear on another. Gather several time lines from a variety of sources that deal with the same location during the same time period. Possible sources for time lines include almanacs, encyclopedias, art history texts, military history books, government texts, world history books, and so on. Have students examine the various time lines to discover differences among them. For example, a time line from an art history book will probably include the beginning and ending dates of various artistic or musical movements while a military or political history time line might not mention these at all.

Organize the class into groups and have each group compare a series of time lines about the same time period to see what connections can be made among the events noted on them. *(Possible answers: Students might note connections like artistic developments during times of peace or prosperity or technological advancements during periods of war.)* Call on volunteers from each group to share their findings with the class. Ask how each group's conclusions compare to those of the rest of the class. Lead a discussion about how the dates noted on a time line can often reveal the creator's intentions or purpose for writing.

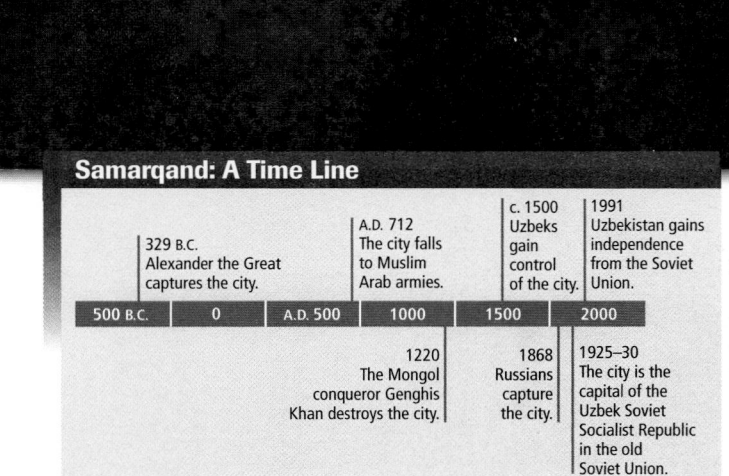

Samarqand: A Time Line

WORKSHOP 2

Using Time Lines

Time lines are visual tools that trace a chronology of events over a period of time. Time lines have a variety of uses. They can show family histories, town histories, and dates of important inventions. Time lines can also serve as records of important events in the history of a place or region. For example, the time line above shows important events in the history of Samarqand, an ancient city in Uzbekistan. Some time lines, such as those at the beginning of each regional unit in this textbook, are horizontal. However, time lines may also be vertical. The key point to remember is that a time line must progress from one point in time to another. Dates along a time line, such as the one on this page, identify key points and events.

Developing the Skill A variety of resources can help you create a time line. Newspapers, almanacs, the Internet, and encyclopedias are great resources for identifying dates and events. For example, those resources could help you create a time line of events in your state's history. You can also use biographical sources to create simple birth-to-death time lines for important historical figures. Newspapers allow you to follow events of local, national, and world news stories as they have unfolded. From this information you can create a time line for both recent and past events.

Time lines are usually read from left to right, but they can be drawn in any direction. Some time lines, for example, may read from top to bottom or bottom to top. In any case, be sure to label the events with their dates so the reader can follow the time line you create. Your labels might also include brief descriptions and illustrations. Also, space the dates and their events in proportion to the amount of real time between them. For example, if you identify important events in 1950, 1960, and 2000, you should place more space between 1960 and 2000. The extra space represents the longer elapsed time between those dates.

Remember that the years B.C. are counted backward to 1 B.C. In other words, an event that happened in 50 B.C. occurred more recently than an event in 100 B.C. Years A.D. are counted from 1 forward. If a particular date is approximated, you may place the abbreviation *c.* before it. That abbreviation stands for the Latin word *circa,* meaning "approximately." You might also include beginning and ending dates in your descriptions of events that stretch over a period of two or more years. For example, the time line on this page shows the years in which the city of Samarqand served as Uzbekistan's capital. Other examples include wars, important eras, and events that occurred repeatedly over a series of years.

Practicing the Skill

1. Study the time line above. Which dates are B.C.? Which dates are A.D.?
2. Create a time line of your life. Begin with your birth and include important events along the way. If you like, predict important events that may happen later in your life. You might include the year you graduate from high school, the year you graduate from college, and the year you hope to have reached a certain career goal.
3. Create a time line for your home state. What are the most important historical events in your state's history? You can illustrate your time line with sketches and pictures.

Using the Illustration

Focus students' attention on the photograph of the Blue Mosque in İstanbul, Turkey. Point out that the mosque was built in the early 1600s for the Ottoman Sultan Ahmed I. Much of the mosque's exterior is covered with the glazed blue tiles that have earned the structure its common name. Beneath its vast dome are both the mosque itself and a huge courtyard and pool. Six graceful minarets rise above the entire complex. Surrounding the mosque are several other structures that were also built for Sultan Ahmed, including his residence, his tomb, a *madrasa* or Islamic college, a primary school, a hospital, a soup kitchen, and several shops.

Ask students which religion's worshipers construct mosques *(Islam)*. Point out that although İstanbul is primarily a Muslim city, this has not always been the case. Islam was introduced to the city when it was conquered by the Ottoman Empire in the 1450s. Before the Ottomans, İstanbul—then known as Constantinople—was a Christian city. The interaction between religious groups is a common theme in Southwest Asia. Remind students that Judaism, Christianity, and Islam all developed in this region. Lead a discussion about how the interaction of different religions has shaped the history of Southwest Asia.

Unit Objectives

1. Describe how the physical geography and economic geography of Southwest Asia affect ways of life there.

2. Trace the history and development of religion in the region, and analyze the role that religion has played in events in Southwest Asia.

3. Identify the political, social, and environmental challenges facing the countries of Southwest Asia.

4. Understand the significance of the region's strategic location and natural resources.

5. Examine aerial photographs to study the features of Earth's surface.

6. Interpret transportation maps to learn about patterns of movement across countries and regions.

UNIT 6 Southwest Asia

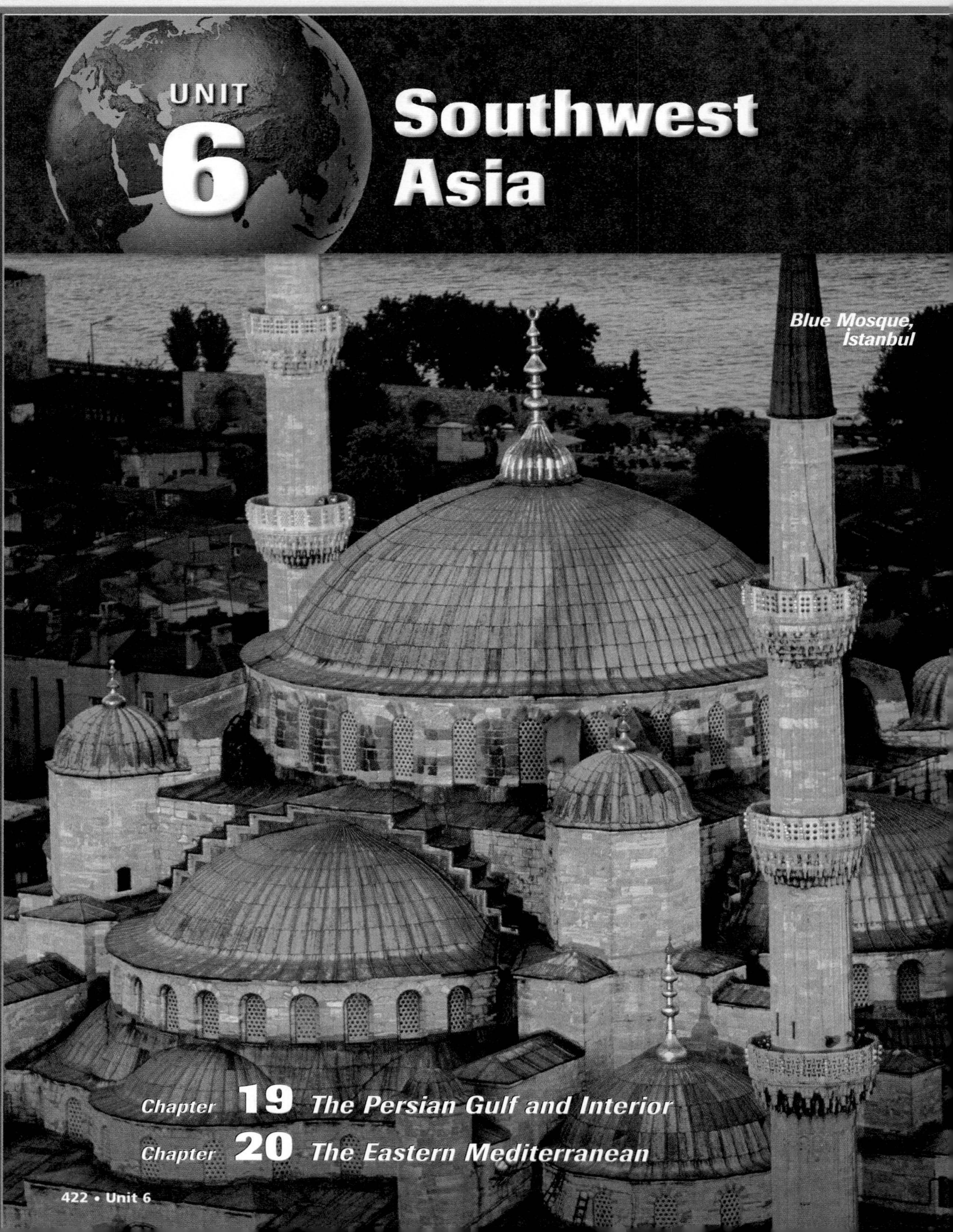

Blue Mosque, İstanbul

422

describes the landforms, climates, economies, resources, and histories of Afghanistan, Iran, Iraq, Saudi Arabia, Yemen, and the small Persian Gulf countries. The chapter concentrates on the religious heritage of the region and the role of oil in the economy.

explores the connections among natural resources, climate, and human activity in Cyprus, Israel, Jordan, Lebanon, Syria, and Turkey.

CONNECTING TO
Literature

FROM THE
EPIC OF GILGAMESH *translated by N. K. Sandars*

Gilgamesh is the hero of this ancient story that was popular all over Southwest Asia. In this passage Utnapishtim (oot-nuh-peesh-tuhm), whom the gods have given everlasting life, tells Gilgamesh about surviving a great flood.

In those days . . . the people multiplied, the world bellowed like a wild bull, and the great god was aroused by the clamor.[1] Enlil (en-LIL) heard the clamor and he said to the gods in council, "The uproar of mankind is intolerable[2] and sleep is no longer possible by reason of the babel."[3] So the gods agreed to exterminate[4] mankind. Enlil did this, but Ea (AY-uh) because of his oath warned me in a dream. . . . "Tear down your house, I say, and build a boat. . . . Then take up into the boat the seed of all living creatures."

Utnapishtim does as he is told. He builds a boat, fills it with supplies, his family, and animals. Then terrible rains come and flood Earth.

When the seventh day dawned the storm from the south subsided, the sea grew calm, the flood was stilled; I looked at the face of the world and there was silence, all mankind was turned to clay. . . . I opened a hatch and the light fell on my face. Then I bowed low, I sat down and I wept . . . for on every side was the waste of water. I looked

for land in vain, but fourteen leagues distant there appeared a mountain, and there the boat grounded; on the mountain of Nisir the boat held fast. . . . When the seventh day dawned I loosed a dove and let her go. She flew away, but finding no resting-place she returned. . . . I loosed a raven, she saw that the waters had retreated, she ate, she flew around, she cawed, and she did not come back. Then I threw everything open to the four winds. I made a sacrifice and poured out a libation.[5]

---
[1]**clamor:** *noise*
[2]**intolerable:** *not bearable*
[3]**babel:** *confusing noise*
[4]**exterminate:** *kill off*
[5]**libation:** *poured as an offering to a god*

Analyzing the Primary Source

1. **Summarizing** Why did the god bring the flood?

2. **Analyzing** Why does Utnapishtim cry?

Analyzing the Primary Source

Answers

1. He was angry because mankind had begun to disturb the peace of the gods.

2. Possible answer: He was overcome by the destruction of the entire world he had known.

In this unit, students will learn about the geography, ancient history, and rapid modernization of the region often referred to as the Middle East.

Some of the world's first civilizations developed in the Fertile Crescent, in what is now Iraq. Wave after wave of conquerors have invaded the region. In the A.D. 600s Islam spread from the Arabian Peninsula across much of the area. Tensions between major Islamic groups date back to the religion's early years and continue today.

Two other major faiths—Judaism and Christianity—originated in Southwest Asia. Over the centuries there have been many conflicts among followers of the three religions.

Arid and semiarid climates dominate Southwest Asia. Farming is possible only in limited areas, so nomadic herding is common. Overall population density is low.

The discovery of immense oil reserves in the early 1900s changed the region's economy dramatically. Some countries, particularly those along the Persian Gulf, have become very wealthy. Oil wealth gives political power to several Southwest Asian countries. However, many people still live in poverty.

Your Classroom Time Line

These are the major dates and time periods for this unit. Have students enter them on the time line you created earlier. You may want to watch for these dates as students progress through the unit.

c.* 8000 B.C. Permanent settlements are established in the eastern Mediterranean.

c. 3000 B.C. The Sumerians build the world's first known cities in Mesopotamia.

c. 2600 B.C. The roots of Judaism are seen in Southwest Asia.

c. 2350 B.C. The Akkadians conquer Sumerian cities to create the region's first empire.

c. 1000 B.C. The Hebrews establish a kingdom between the Jordan River and the Mediterranean.

c. 1000 B.C. King David establishes Jerusalem as the Israelite capital.

586 B.C. The Babylonians capture Jerusalem.

c. 550 B.C. The Persian Empire is established. The Persians soon conquer Mesopotamia.

*c. stands for *circa* and means "about."

POLITICAL MAP ANSWERS

1. Afghanistan
2. Saudi Arabia

CRITICAL THINKING ANSWER

3. Possible answer: The heart of Iraq is a lowland, and the country does not have many major mountain barriers, except in a few areas in northern Iraq.

424

UNIT 6 ATLAS

The World in Spatial Terms
Southwest Asia
Political

1. **Places and Regions** Which country in the region is landlocked?
2. **Places and Regions** Which country in the region has coastlines on both the Red Sea and Persian Gulf?

Critical Thinking

3. **Analyzing Information** Compare this map to the physical map. What geographic factors do you think allow Iraq to control its territory?

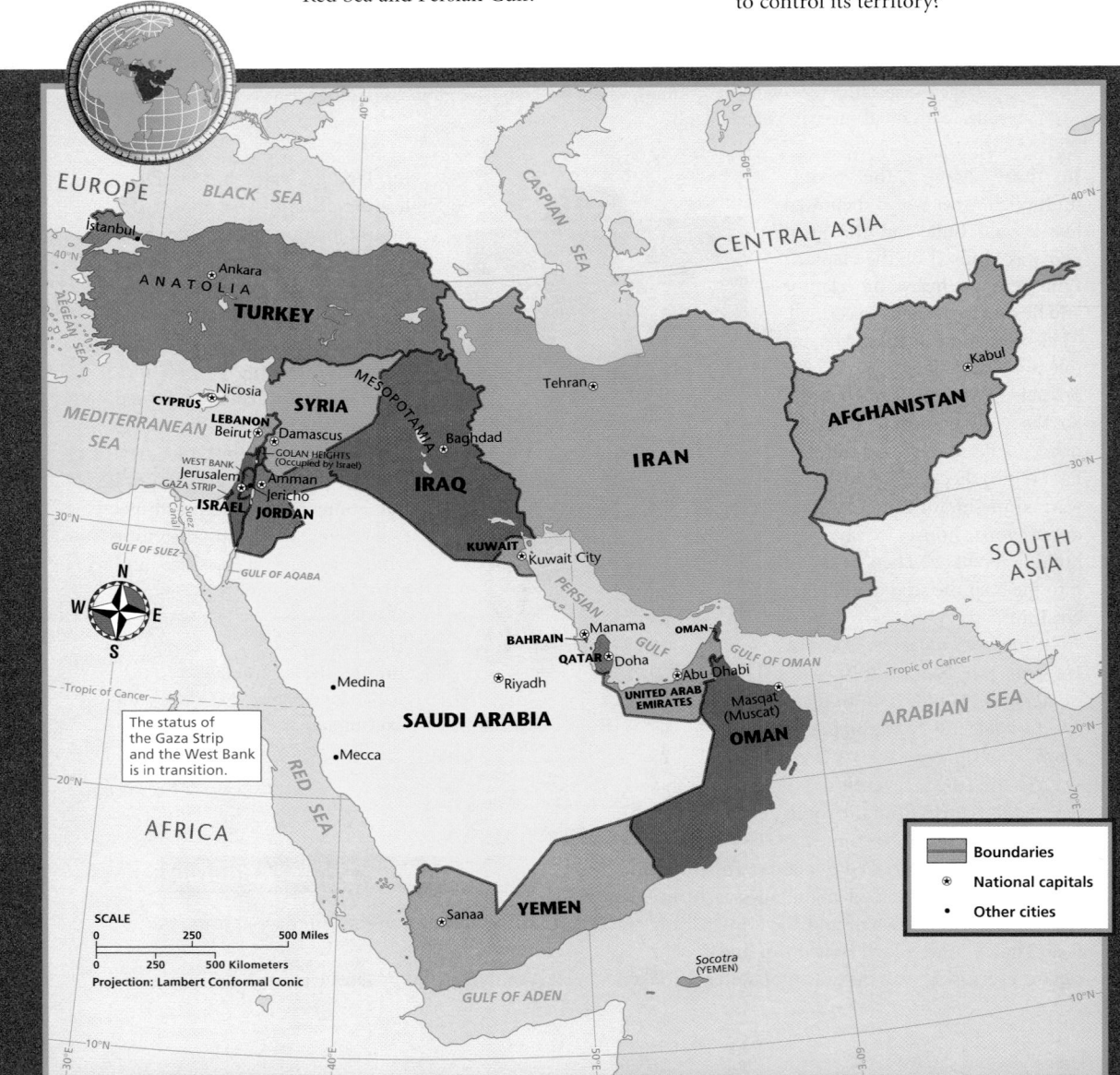

Using the Political Map

Direct students' attention to the **political map** of Southwest Asia on the previous page. Ask which three continents meet in this region *(Europe, Asia, Africa)*. Then ask how this location might have affected the region's history *(invasions, trade, and ideas from many directions)*. Which small countries border the Persian Gulf? *(Bahrain, Kuwait, Oman, Qatar, United Arab Emirates)* Which small countries border the Mediterranean Sea? *(Israel, Syria, Lebanon)* Why might all of these countries want to keep access to the sea? *(for trade, travel, communication with other countries)*

Using the Physical Map

Focus students' attention on the **physical map** on this page. Most of Southwest Asia lacks certain physical features that are important to other places the class has studied. What are these features? *(major rivers and lakes)* Call on a volunteer to identify where in Southwest Asia the lack of water seems to be most severe *(Arabian Peninsula)*. Ask students to suggest what residents of the region might do to increase their access to water. *(Possible answers: build aqueducts or pipelines to carry water from areas where it is available to dry areas, construct desalinization plants to remove salt from seawater)*

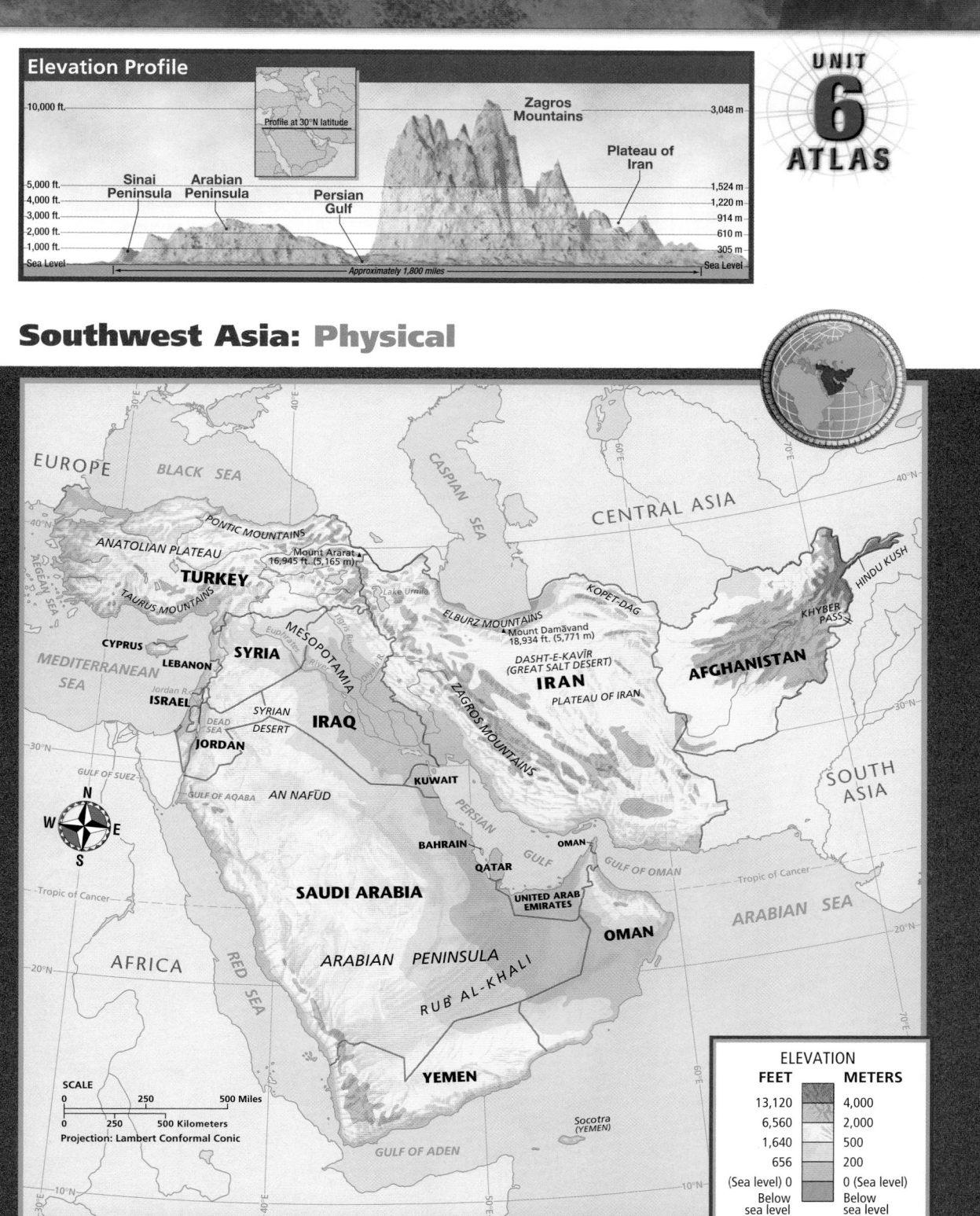

Elevation Profile

Profile at 30°N latitude

- 10,000 ft. — 3,048 m
- 5,000 ft. — 1,524 m
- 4,000 ft. — 1,220 m
- 3,000 ft. — 914 m
- 2,000 ft. — 610 m
- 1,000 ft. — 305 m
- Sea Level — Sea Level

Sinai Peninsula • Arabian Peninsula • Persian Gulf • Zagros Mountains • Plateau of Iran

Approximately 1,800 miles

UNIT **6** ATLAS

Southwest Asia: Physical

SCALE
0   250   500 Miles
0   250   500 Kilometers
Projection: Lambert Conformal Conic

ELEVATION

FEET	METERS
13,120	4,000
6,560	2,000
1,640	500
656	200
(Sea level) 0	0 (Sea level)
Below sea level	Below sea level

Your Classroom Time Line, (continued)

200 B.C.–A.D. 106 The Romans rule the eastern Mediterranean.

A.D. 100s Christianity begins to spread from the eastern Mediterranean throughout the Roman Empire.

300s Christianity is made the official religion of the Roman Empire.

c. 335 The Church of the Holy Sepulchre is completed in Jerusalem.

400s The Western Roman Empire crumbles. The Eastern Roman Empire, or Byzantine Empire, continues to flourish.

c. 610 Muhammad begins to preach the religion of Islam.

600s Arabs sweep north out of the desert, capturing much of northern Africa and southwestern Asia, including Jerusalem.

Late 600s Muslims in Jerusalem build the Dome of the Rock.

1077 Turkic Muslims capture Jerusalem.

1095–1200 European Crusaders try to reclaim Jerusalem and other cities in the region.

1200s Mongol invaders from Asia conquer what are now Afghanistan, Iraq, and Iran.

1300s The Muslim Ottoman Turks establish an empire that includes much of Southwest Asia.

1453 The Ottomans capture Constantinople and make it their capital.

1500s Mesopotamia and the western Arabian Peninsula are added to the Ottoman Empire.

1500s The Safavid Dynasty is established in Persia. Literature, architecture, and the arts flourish.

1500s The Turks conquer Cyprus, formerly a Greek possession.

1800s The British and Russian Empires attempt to extend authority over Iran and Afghanistan.

Using the Climate Map

Focus students' attention on the **climate map** on this page. Ask them to list the region's climate types in the order of the amount of area each one appears to cover *(arid, semiarid, Mediterranean, highland, humid subtropical)*. Have students compare this map to the **political map** to decide which countries they would live in or visit based solely on climate.

Ask students how they think the region's resources and climates affect farming in the region. Tell them to compare this map to the **land use and resources map** to learn where in the region agriculture is a common economic activity *(Turkey, along the Mediterranean coast, along the Tigris and Euphrates Rivers in Iraq)*. Ask what activities are common in arid parts of the region *(oil and natural gas production, mining, nomadic herding)*.

Your Classroom Time Line,
(continued)

Late 1800s The Zionist movement begins.

1878 The British take control of Cyprus.

Early 1900s British and French establish mandates in Southwest Asia.

1923 Atatürk establishes the Republic of Turkey.

1932 The Kingdoms of Iraq and Saudi Arabia are founded.

1930s–40s Jewish migration to Palestine increases as a result of World War II and the Holocaust.

1940s Syria, Lebanon, and Jordan gain independence.

1948 The State of Israel is created. Jerusalem is divided between Israel and Jordan.

1948–82 Israel and its Arab neighbors fight several wars.

1960s–70s Bahrain, Cyprus, Kuwait, Qatar, the United Arab Emirates, and Yemen gain independence from Britain.

1960 Fighting between Greek and Turkish Cypriots breaks out.

1964 The United Nations sends peace-keeping troops to Cyprus.

1964 The Palestine Liberation Organization (PLO) is established.

1967 Israel fights the Six-Day War with its neighbors. The Golan Heights and the West Bank become Israeli territory, and Jerusalem is reunited.

CLIMATE MAP ANSWERS

1. arid climate
2. Possible answer: probably dry with little vegetation in many areas

CRITICAL THINKING ANSWER

3. interior Turkey, interior Iran, Arabian Peninsula, northeastern Afghanistan

UNIT 6 ATLAS

Southwest Asia:
Climate

1. **Physical Systems** What is the most widespread climate type in Southwest Asia?

2. **Physical Systems** Compare this map to the physical map. What do you think the environment in most of the Arabian Peninsula is like?

Critical Thinking

3. **Making Generalizations** Compare this map to the physical map. What are some areas where mountain barriers might be a factor affecting climate?

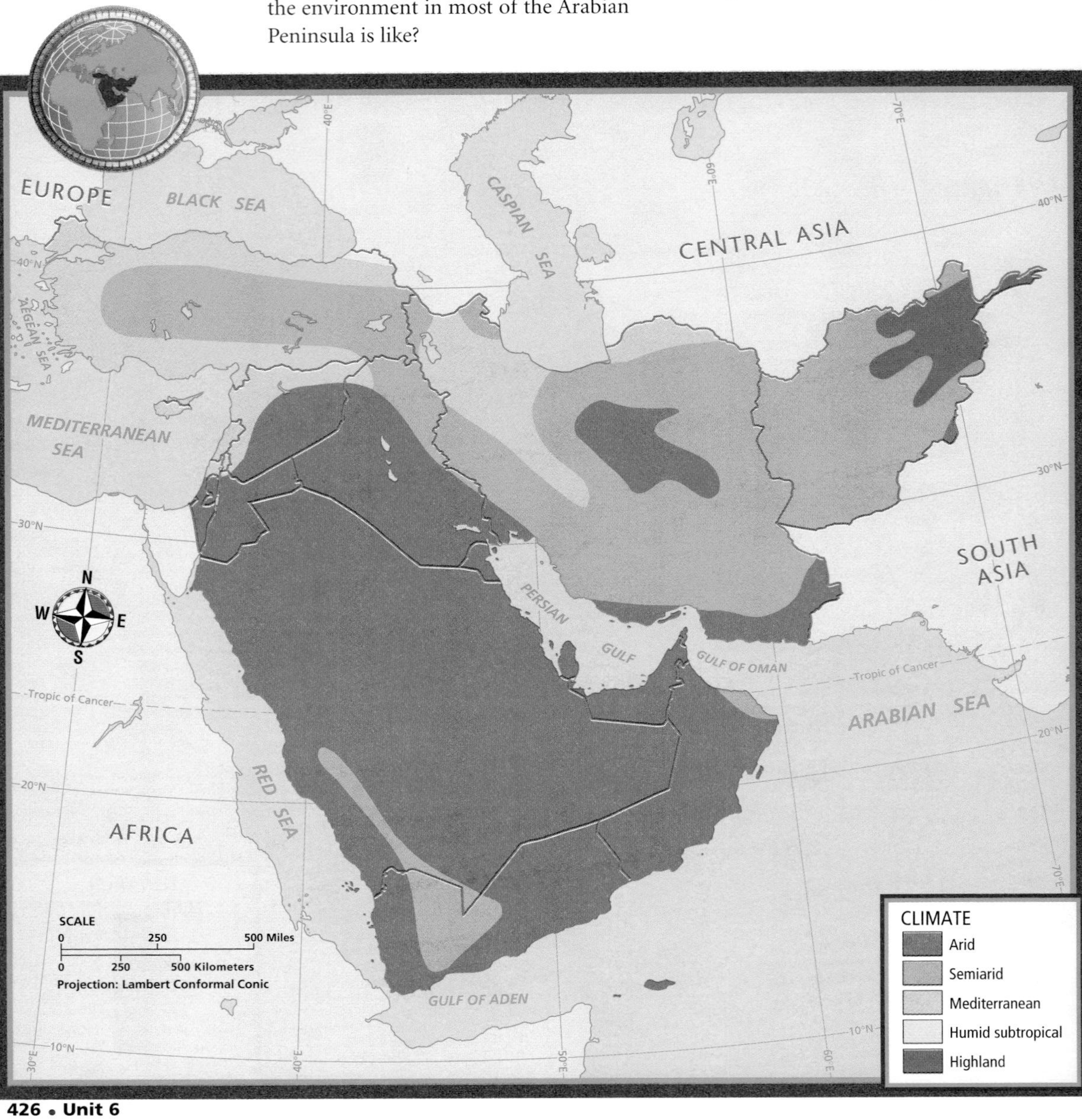

EUROPE

BLACK SEA

CASPIAN SEA

CENTRAL ASIA

AEGEAN SEA

MEDITERRANEAN SEA

40°N

30°N

20°N

Tropic of Cancer

RED SEA

AFRICA

PERSIAN GULF

GULF OF OMAN

Tropic of Cancer

SOUTH ASIA

ARABIAN SEA

GULF OF ADEN

SCALE
0   250   500 Miles
0   250   500 Kilometers
Projection: Lambert Conformal Conic

CLIMATE
- Arid
- Semiarid
- Mediterranean
- Humid subtropical
- Highland

Using the Precipitation Map

Focus students' attention on the **precipitation map** on this page. Ask students to make generalizations about the region's annual precipitation. *(Possible answers: Most of the region receives very little rain. Coastal areas along the Mediterranean, Black Sea, and Caspian Sea receive more rain than the region's interior.)* Have students compare this map to the **political map** to identify the country in the region that has the highest annual rainfall amount *(Turkey)*. Then ask students to identify countries with large areas that receive less than 10 inches (25 cm) of precipitation each year *(Afghanistan, Iran, Iraq, Jordan, Oman, Bahrain, Qatar, Saudi Arabia, Syria, United Arab Emirates, Yemen, Kuwait).*

Southwest Asia:
Precipitation

UNIT
6
ATLAS

1. (*Places and Regions*) How much precipitation do most areas on the Arabian Peninsula receive?
2. (*Places and Regions*) Which country on the Arabian Peninsula receives the most precipitation?

Critical Thinking

3. **Drawing Conclusions** Compare this map to the physical map. Based on the distribution of precipitation in the region, which area do you think is known as the Fertile Crescent?

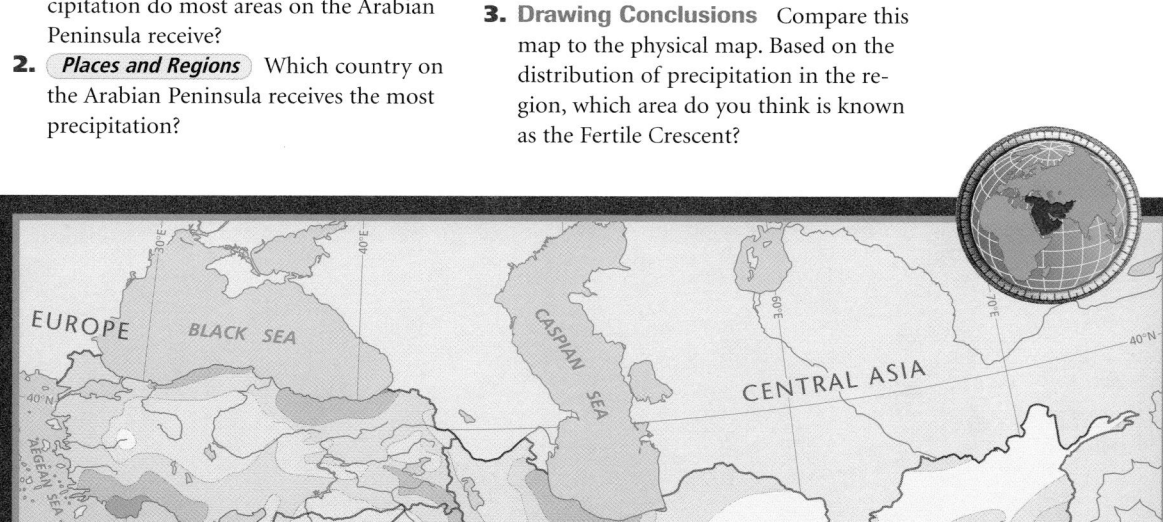

Your Classroom Time Line, (continued)

1974 Turkey invades Cyprus.

1975–1990 A civil war is fought in Lebanon.

1979 Saddam Hussein becomes president of Iraq.

1979 The Soviet Union invades Afghanistan.

1979 The Shah is overthrown in Iran.

1980 Iraq invades Iran.

1983 Turkish Cypriots declare independence for the northern part of the island, which they call the Turkish Republic of Northern Cyprus.

1989 Soviet troops withdraw from Afghanistan.

1990s The Taliban rises to power in Afghanistan.

1990s Israel and the PLO begin peace talks.

1990 Iraq invades Kuwait, triggering the Persian Gulf War.

2000 Israeli and Palestinian peace talks break down and violence erupts.

2001 The Taliban destroys several ancient Buddhist statues in Afghanistan.

2001 The United States begins military operations in Afghanistan as a response to the terrorist attacks of September 11.

2003 United States and allied forces remove Saddam Hussein from power in Iraq.

PRECIPITATION MAP ANSWERS

1. less than 10 inches
2. Yemen

CRITICAL THINKING ANSWER

3. Possible answer: the crescent-shaped area that includes Israel, Lebanon, southeastern Turkey, and northwestern Iran

ANNUAL PRECIPITATION

Centimeters	Inches
Under 25	Under 10
25–50	10–20
50–100	20–40
100–150	40–60
150–200	60–80
Over 200	Over 80

Call students' attention to the **population map** on this page. Ask students to compare it to the **political map** to identify the country in Southwest Asia that is most densely populated (*Israel*). Then have students name countries that are large in area but sparsely populated (*Saudi Arabia, Jordan, Syria, Iraq, Iran, Afghanistan, Yemen, Oman*). Ask students to identify cities in the region that are home to more than 2 million people (*Kabul, Afghanistan; İstanbul, İzmir, and Ankara, Turkey; Baghdad, Iraq; Isfahan, Mashhad, and Tehran, Iran; Beirut,* *Lebanon; Riyadh, Saudi Arabia; and Aleppo and Damascus, Syria*). Ask students to look at the **political map** again. What do Kabul, Ankara, Baghdad, Tehran, Beirut, and Riyadh have in common? (*All are national capitals.*)

Ask students to identify the most densely populated parts of the Arabian Peninsula (*western Yemen and northeastern Oman*). Ask: What might account for higher populations in these areas? (*They are located along important water trade routes and have generally milder or wetter climates.*)

POPULATION MAP ANSWERS

1. Israel
2. Its location on shipping lanes between the Black Sea and Mediterranean Sea has probably helped it grow.

CRITICAL THINKING ANSWER

3. Possible answer: Climates are probably milder along the coast and coastal areas are often lowlands. Also, they are located along important water trade routes.

UNIT
6
ATLAS

Southwest Asia:
Population

1. (**Environment and Society**) Which country in the region appears to have the highest overall population density?
2. (**Places and Regions**) Compare this map to the physical and land use maps. How do you think İstanbul's location has influenced the city's growth?

Critical Thinking

3. **Analyzing Information** Compare this map to the physical map. Why do you think many coastal areas have higher population densities than areas in the interior?

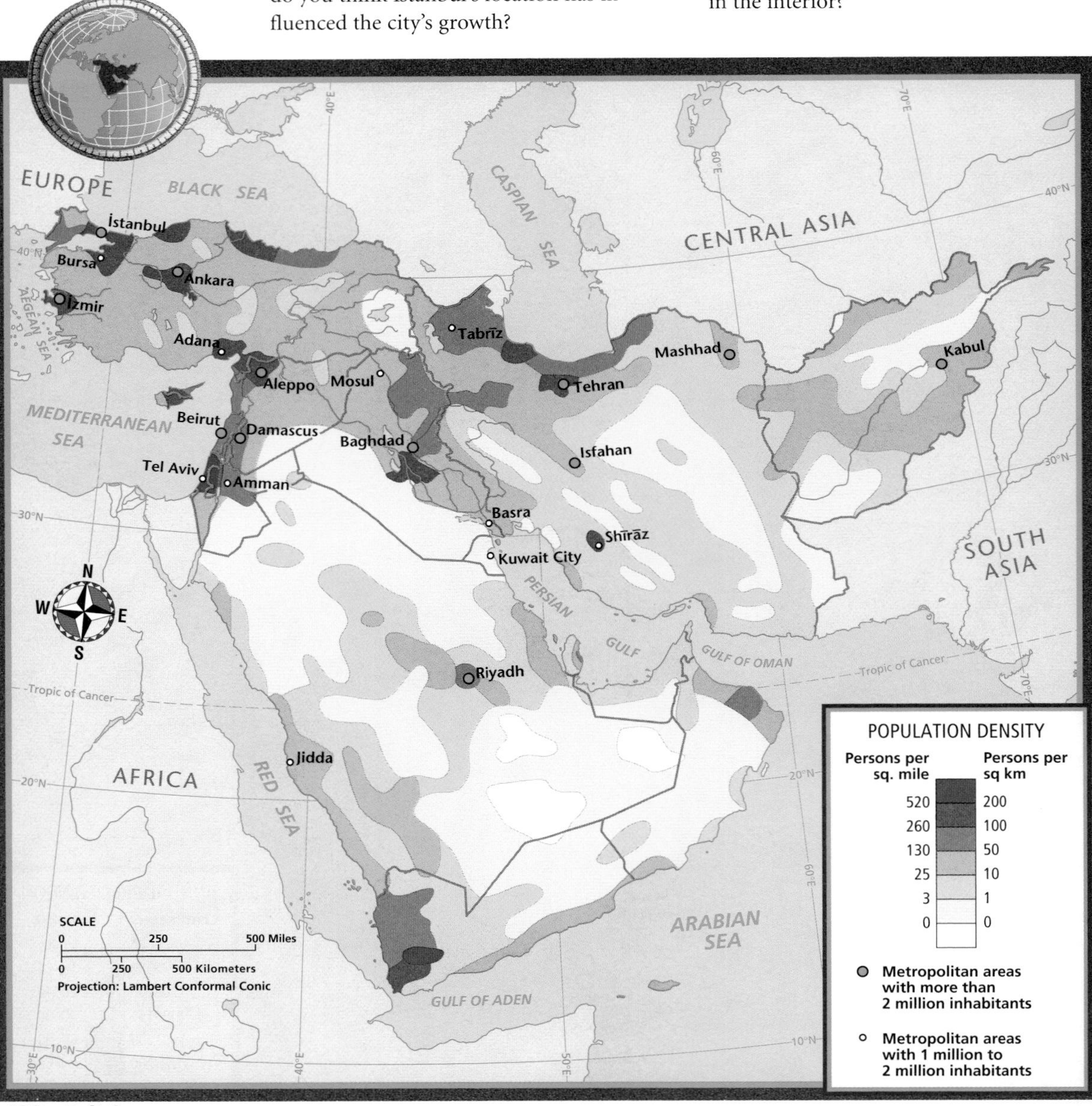

POPULATION DENSITY

Persons per sq. mile	Persons per sq km
520	200
260	100
130	50
25	10
3	1
0	0

● Metropolitan areas with more than 2 million inhabitants

○ Metropolitan areas with 1 million to 2 million inhabitants

SCALE
0 — 250 — 500 Miles
0 — 250 — 500 Kilometers
Projection: Lambert Conformal Conic

Focus students' attention on the **land use and resources map** on this page. Ask them to identify the major resources found along the Persian Gulf (*natural gas and oil*). Also have them identify other areas in the region that produce oil and gas (*northern Iraq, southeastern Turkey, Yemen*). Point out that Southwest Asia is the world's leading producer of petroleum. Ask students to examine this map to determine how oil and gas are transported through this region and to other parts of the world. (*Oil is shipped overland through pipelines to coastal cities. From there it is shipped out through the Persian Gulf, Arabian Sea, Red Sea and Mediterranean Sea.*)

Have students compare this map to the **physical map**. Ask how Southwest Asia's mineral resources appear to be related to its physical landscapes. (*Most mineral resources seem to be located in mountainous areas, like the Zagros, Pontic, Elburz, and Taurus Ranges.*)

Southwest Asia:
Land Use and Resources

UNIT
6
ATLAS

1. (*Places and Regions*) Where do oil and natural gas seem to be the most concentrated in Southwest Asia?

2. (*Environment and Society*) Compare this map to the population map. How are land use and population distribution in Israel related?

Critical Thinking

3. **Analyzing Information**   Compare this map to the climate map. How do you think farmers have adapted to arid and semiarid climates in the region? What clues to your answer do you find in the maps?

LAND USE AND RESOURCES MAP ANSWERS

1. around the Persian Gulf
2. Northern Israel is more densely populated and is dominated by commercial agriculture, while southern Israel is less densely populated and is used for nomadic herding.

CRITICAL THINKING ANSWER

3. Farming is limited mostly to areas with at least a semiarid climate, in areas along and near rivers, and in isolated desert areas, presumably where oases or other water sources are found.

RESOURCES

- Major manufacturing and trade centers
- Oil pipelines
- Shipping lanes
- Coal
- Natural gas
- Oil
- Hydroelectric power
- Au Gold
- Other minerals

LAND USE
- Commercial farming
- Subsistence farming
- Forestry
- Livestock raising
- Fishing
- Nomadic herding
- Limited economic activity

SCALE
0   250   500 Miles
0 250 500 Kilometers
Projection: Lambert Conformal Conic

EUROPE

BLACK SEA
Istanbul
Ankara
Izmir
AEGEAN SEA
MEDITERRANEAN SEA
Haifa
Tel Aviv
Damascus
Baghdad
Abādān
Jubail
Yanbu
Au
Jidda
AFRICA
RED SEA
Aden
GULF OF ADEN

CASPIAN SEA
CENTRAL ASIA
Au
Kabul
Tehran
Isfahan
SOUTH ASIA
Bandar Abbās
PERSIAN GULF
GULF OF OMAN
Abu Dhabi
Tropic of Cancer
ARABIAN SEA

Have students examine the Comparing Standard of Living table on this page. Ask them which country appears to have the highest standard of living of those listed here *(Israel)*. Call on students to explain this conclusion *(high life expectancy and literacy, low infant mortality rate)*. Then ask students which country has the lowest standard of living *(Afghanistan)*. What evidence on the chart suggests this? *(lower life expectancy and literacy rates, higher infant mortality)* Ask students to find evidence in the Fast Facts table that supports the conclusions they have drawn. *(Most of Afghanistan's population is involved in agriculture, and the country has a low per capita GDP. Telephones are rare, and the people do not use much electricity,* which suggests that they do not own many appliances or technological devices. In Israel more people are involved in services or manufacturing, and the per capita GDP is much higher. Telephones are much more common, and people use more electricity.) Have students examine the Fast Facts table to find other countries in the region that appear to have high standards of living *(Bahrain, Kuwait, Qatar, Saudi Arabia, United Arab Emirates, and the Greek part of Cyprus)*. Ask students why residents of these countries may enjoy higher standards of living than people in other countries. *(Possible answer: Most of these countries have large reserves of oil, the export of which brings money into the country.)*

The Levant The Crusades that began in the late 1000s resulted in more than battles and the taking of cities. Both during and after these wars, Europeans began to trade with the peoples of Southwest Asia for goods like currants, oils, wines, silk, and cotton. Most of their trading took place in an area known as the Levant, centered around the cities of Tyre and Sidon in what is now Lebanon.

The Levant area had been an important trade region during the time of the Phoenicians more than 2,000 years earlier. Intensive trade with western Europe did not begin, however, until Venetian merchants arrived in the Levant, seeking the source of goods brought back to Europe by Crusaders. In 1592 the Levant Company, a group of British traders, obtained a monopoly on trade with the Levant. The company remained the main source of Levantine goods in Europe until the 1800s.

CRITICAL THINKING:
Why were European traders interested in trading with the Levant? *(They could obtain goods there that were not available in Europe, and the Mediterranean provided an easy transport route.)*

UNIT 6 ATLAS

Time Line: Southwest Asia

c. 1000 B.C.
Hebrews establish a kingdom between the Jordan River and the Mediterranean Sea.

A.D. 400s
The Roman Empire and its control of the region crumble.

1453
Muslim Ottomans take control of previously Christian Constantinople.

2001
Efforts to establish a lasting Arab-Israeli peace continue.

1948
Israel becomes an independent country.

3500 B.C.	1000 B.C.	A.D. 400	1200	1400	1900

3500 B.C.
Sumerians live in Mesopotamia.

550 B.C.
The Persian Empire controls the region.

1200s
Mongols invade the region.

c. 570
Muhammad, the Prophet of Islam, is born.

1918
The Ottoman Empire is defeated in World War I and collapses four years later.

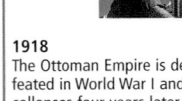

Comparing Standard of Living

COUNTRY	LIFE EXPECTANCY (in years)	INFANT MORTALITY (per 1,000 live births)	LITERACY RATE	DAILY CALORIC INTAKE (per person)
Afghanistan	48, male 46, female	143	36%	1,716
Bahrain	71, male 76, female	19	89%	Not available
Iran	68, male 71, female	44	79%	2,822
Iraq	67, male 69, female	55	40%	2,419
Israel	77, male 81, female	7	95%	3,466
Kuwait	76, male 78, female	11	84%	3,059
Saudi Arabia	67, male 71, female	48	79%	2,888
Syria	68, male 71, female	32	77%	3,378
Turkey	69, male 74, female	44	87%	3,554
Yemen	59, male 63, female	65	38%	2,087
United States	74, male 80, female	7	97%	3,757

The United States and Southwest Asia

Comparing Sizes

internet connect

GO TO: go.hrw.com
KEYWORD: SW3 Almanac
FOR: Additional information and reference sources

Sources: *World Almanac and Book of Facts, 2004; Britannica Book of the Year, 2002*

Fast Facts Activity

Copy the following chart onto the chalkboard or onto hand-outs. Ask students which countries appear to have the highest birthrates in the region (*Afghanistan, Iraq, Oman, Saudi Arabia, Syria, Yemen*). What evidence on the chart supports this conclusion? (*high percentages of population below age 14*) Call on volunteers to explain how this relates to population growth. (*Possible answer: Countries with high birthrates usually have high population growth rates as well.*) Ask students which countries' populations appear to be growing most slowly (*Cyprus, Israel*). What evidence suggests this? (*older populations, suggesting lower birthrates*)

COUNTRY	POPULATION BELOW AGED 14 (PERCENT)	POPULATION AGED 15 TO 64 (PERCENT)	POPULATION ABOVE AGE 65 (PERCENT)
Afghanistan	42 percent	55 percent	3 percent
Bahrain	29 percent	68 percent	3 percent
Cyprus	22 percent	67 percent	11 percent
Iran	32 percent	63 percent	5 percent
Iraq	41 percent	56 percent	3 percent
Israel	27 percent	63 percent	10 percent
Jordan	37 percent	60 percent	3 percent
Kuwait	28 percent	69 percent	3 percent
Lebanon	27 percent	66 percent	7 percent
Oman	42 percent	56 percent	2 percent
Qatar	25 percent	72 percent	3 percent
Saudi Arabia	42 percent	55 percent	3 percent
Syria	39 percent	58 percent	3 percent
Turkey	28 percent	66 percent	6 percent
United Arab Emirates	28 percent	69 percent	3 percent
Yemen	47 percent	50 percent	3 percent
United States	21 percent	66 percent	13 percent

Source: *The World Almanac and Book of Facts, 2004*

Fast Facts: Southwest Asia

UNIT
6
ATLAS

FLAG	COUNTRY / Capital	POPULATION (in millions) / POP. DENSITY	AREA	PER CAPITA GDP (in US $)	WORKFORCE STRUCTURE (largest categories)	ELECTRICITY CONSUMPTION (kilowatt hours per person)	TELEPHONE LINES (per person)
	Afghanistan Kabul	23.9 / 96/sq. mi.	250,001 sq. mi. / 647,500 sq km	$ 700	80% agriculture 10% industry 10% services	21 kWh	0.001
	Bahrain Manama	0.7 / 3,029/sq. mi.	257 sq. mi. / 666 sq km	$ 14,000	79% ind. comm., serv. 20% government	8,037 kWh	0.24
	Cyprus Nicosia	0.8 / 225/sq. mi.	3,571 sq. mi. / 9,249 sq km	$ 15,000 (Gr.) $ 6,000 (Tu.)	73% services (Gr.) 56% services (Tu.)	Not available	0.53
	Iran Tehran	68.9 / 109/sq. mi.	636,296 sq. mi. / 1,647,999 sq km	$ 7,000	45% services 30% agriculture	1,682 kWh	0.19
	Iraq Baghdad	25.2 / 150/sq. mi.	168,754 sq. mi. / 437,071 sq km	$ 2,400	52% services 12% agriculture	1,330 kWh	0.07
	Israel Jerusalem	6.4 / 820/sq. mi.	8,019 sq. mi. / 20,769 sq km	$ 19,000	31% public service 20% manufacturing	5,879 kWh	0.48
	Jordan Amman	5.5 / 155/sq. mi.	35,637 sq. mi. / 92,299 sq km	$ 4,300	83% services 13% industry	1,253 kWh	0.12
	Kuwait Kuwait City	2.5 / 366/sq. mi.	6,880 sq. mi. / 17,819 sq km	$ 15,000	50% gov., social svs. 40% services	11,618 kWh	0.19
	Lebanon Beirut	3.7 / 925/sq. mi.	4,015 sq. mi. / 10,399 sq km	$ 5,400	62% services 31% industry	2,037 kWh	0.19
	Oman Musqat	2.9 / 35/sq. mi.	82,031 sq. mi. / 212,459 sq km	$ 8,300	24% public admin. 15% construction	3,025 kWh	0.08
	Qatar Doha	0.6 / 144/sq. mi.	4,416 sq. mi. / 11,437 sq km	$ 21,500	51% services 22% construction	14,125 kWh	0.29
	Saudi Arabia Riyadh	24.2 / 29/sq. mi.	756,984 sq. mi. / 1,960,580 sq km	$ 10,500	63% services 25% industry	4,699 kWh	0.13
	Syria Damascus	17.8 / 250/sq. mi.	71,498 sq. mi. / 185,179 sq km	$ 3,500	40% agriculture 40% services	1,215 kWh	0.10
	Turkey Ankara	71.3 / 240/sq. mi.	301,383 sq. mi. / 780,578 sq km	$ 7,000	40% agriculture 38% services	1,579 kWh	0.27
	United Arab Emirates Abu Dhabi	3.0 / 93/sq. mi.	32,000 sq. mi. / 82,880 sq km	$ 22,000	78% services 15% industry	11,720 kWh	0.37
	Yemen Sanaa	20.0 / 98/sq. mi.	203,850 sq. mi. / 527,969 sq km	$ 840	71% agriculture 13% trade	140 kWh	0.02
	United States Washington, D.C.	294.0 / 83/sq. mi.	3,717,810 sq. mi. / 9,629,084 sq km	$ 37,600	31% manage., prof. 29% tech., sales, admin.	12,250 kWh	0.65

Sources: Central Intelligence Agency, *World Factbook 2003*; *The World Almanac and Book of Facts, 2004*
The CIA calculates per capita GDP in terms of purchasing power parity. This formula equalizes the purchasing power of each country's currency.

The Persian Gulf and Interior

CHAPTER RESOURCE MANAGER

Objectives	Pacing Guide	Reproducible Resources
SECTION 1 **Natural Environments** (pp. 433–36)	• Identify the landforms and rivers of the Persian Gulf area and interior Southwest Asia. • Explain the effect the region's geography has on its climates and biomes. • Identify the region's natural resources.	**Regular** .5 day **Block Scheduling** .5 day *Block Scheduling Handbook, Chapter 19*	**RS** Guided Reading Strategy 19.1 **PS** Readings in World Geography, History, and Culture 47 **IC** Environmental and Global Issues Activity 5: Operation Oil-Spill Cleanup
SECTION 2 **History and Culture** (pp. 437–40)	• Describe how peoples, empires, and Islam have affected the history of the Persian Gulf area and interior Southwest Asia. • Identify the major features of the region's cultures.	**Regular** .5 day **Block Scheduling** .5 day *Block Scheduling Handbook, Chapter 19*	**RS** Guided Reading Strategy 19.2 **PS** Readings in World Geography, History, and Culture 48 **E** Creative Strategies for Teaching World Geography, Lessons 14 and 15 **E** Cultures of the World Activity: Region 5 **SM** Critical Thinking Activity 19: The Fertile Crescent **SM** Map Activity 19: Persian Gulf Wars
SECTION 3 **The Region Today** (pp. 441–44)	• Identify the activities on which the region's economies depend. • Describe the region's cities. • Explain some important issues in the region today.	**Regular** 1 day **Block Scheduling** .5 day *Block Scheduling Handbook, Chapter 19*	**RS** Guided Reading Strategy 19.3 **RS** Graphic Organizer Activity 19 **PS** Readings in World Geography, History, and Culture 49–51 **SM** Geography for Life Activity 19: Speaking Graphically about Persian Gulf Oil

Chapter Resource Key

PS Primary Sources	**A** Assessment	CD–ROM
RS Reading Support	**REV** Review	Video
IC Interdisciplinary Connections	**ELL** Reinforcement and English Language Learners	Internet
E Enrichment	Transparencies	Holt Presentation Maker Using Microsoft® PowerPoint®
SM Skills Mastery		

 One-Stop Planner CD–ROM

See the *One-Stop Planner* for a complete list of additional resources for students and teachers.

 **One-Stop** Planner CD–ROM

It's easy to plan lessons, select resources, and print out materials for your students when you use the **One-Stop Planner CD–ROM with Test Generator**.

Technology Resources

- One-Stop Planner CD–ROM, Lesson 19.1
- Geography and Cultures Visual Resources 37–41
- Homework Practice Online
- HRW Go site

- One-Stop Planner CD–ROM, Lesson 19.2
- **CNN** Presents World Cultures: Yesterday and Today, Segment 9: Pilgrimage to Mecca
- Geography and Cultures Visual Resources 42
- Homework Practice Online
- HRW Go site

- One-Stop Planner CD–ROM, Lesson 19.3
- *ARGWorld* CD–ROM
- **CNN** Presents Geography: Yesterday and Today, Segment 22: Kuwait—Restoring the Environment
- Homework Practice Online
- HRW Go site

Reinforcement, Review, and Assessment

- **ELL** Main Idea Activity 19.1
- **ELL** English Audio Summary 19.1
- **ELL** Spanish Audio Summary 19.1
- **REV** Section 1 Review, p. 436
- **A** Daily Quiz 19.1

- **ELL** Main Idea Activity 19.2
- **ELL** English Audio Summary 19.2
- **ELL** Spanish Audio Summary 19.2
- **REV** Section 2 Review, p. 440
- **A** Daily Quiz 19.2

- **ELL** Main Idea Activity 19.3
- **ELL** English Audio Summary 19.3
- **ELL** Spanish Audio Summary 19.3
- **REV** Section 3 Review, p. 444
- **A** Daily Quiz 19.3

 **internet** connect

HRW ONLINE RESOURCES

GO TO: go.hrw.com
Then type in a keyword.

TEACHER HOME PAGE
KEYWORD: SW3 Teacher

CHAPTER INTERNET ACTIVITIES
KEYWORD: SW3 GT19
Choose a topic on the Persian Gulf and interior Southwest Asia to:
- report on Islamic culture.
- research oil productivity and transfer your information into a graph or chart.
- understand the technology of desalinization.

CHAPTER ENRICHMENT LINKS
KEYWORD: SW3 CH19

CHAPTER MAPS
KEYWORD: SW3 MAPS19

ONLINE ASSESSMENT
Homework Practice
KEYWORD: SW3 HP19
Standardized Test Prep
KEYWORD: SW3 STP19
Rubrics
KEYWORD: SS Rubrics

COUNTRY INFORMATION
KEYWORD: SW3 Almanac

CONTENT UPDATES
KEYWORD: SS Content Updates

HOLT PRESENTATION MAKER
KEYWORD: SW3 PPT19

ONLINE READING SUPPORT
KEYWORD: SS Strategies

CURRENT EVENTS
KEYWORD: S3 Current Events

Meeting Individual Needs

Ability Levels

Level 1 Basic-level activities designed for all students encountering new material

Level 2 Intermediate-level activities designed for average students

Level 3 Challenging activities designed for honors and gifted-and-talented students

English Language Learners Activities that address the needs of students with Limited English Proficiency

Chapter Review and Assessment

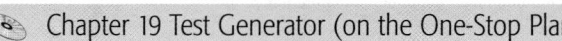

- Chapter 19 Test Generator (on the One-Stop Planner)
- Global Skill Builder CD–ROM
- HRW Go site
- **REV** Chapter 19 Review, pp. 446–47
- **REV** Chapter 19 Tutorial for Students, Parents, Mentors, and Peers
- **A** Chapter 19 Test (form A or B)
- **A** Alternative Assessment Handbook
- **A** Chapter 19 Test for English Language Learners and Special-Needs Students

Launch into Learning

Ask students if they are familiar with stories about Aladdin, Ali Baba, Sinbad the Sailor, or other tales from *The Thousand and One Nights*. These stories were compiled over many centuries and contain influences from many countries. Many of the stories are set in Baghdad, now the capital of Iraq. Lead a discussion about how Iraq is portrayed in works based on *The Thousand and One Nights* and how Iraq has been portrayed in news broadcasts. Point out that the old stories romanticize Iraq's past, and newscasts may emphasize political troubles. Tell students they will learn more about Iraq and the other nations around the Persian Gulf in this chapter.

Using the Physical-Political Map

Have students examine the map on the opposite page. Call on a volunteer to name the major bodies of water in the region. *(Possible answers: Persian Gulf, Gulf of Oman, Caspian Sea, Tigris and Euphrates Rivers)* Point out the Strait of Hormuz and the Bab al-Mandab. Ask why these waterways might be economically significant *(major trade routes)*. Then ask how the countries that border these bodies of water might be able to use their locations to their advantage. *(Possible answer: control trade, particularly oil transport)*

Why We Should Know More

Tell students that there are many reasons why we should know more about the countries of the Persian Gulf and interior Southwest Asia. Here are some of those reasons:

▶ The world's first known urban civilization, Sumer, developed in what is now Iraq. Later, the Persian Empire ruled the area. These civilizations left behind great works of art, architecture, and literature.

▶ This region is the cradle of Islam and home to many of its followers. Islam is the fastest-growing religion in the United States.

▶ This region contains most of the world's known oil reserves. Oil prices and availability affect many aspects of Americans' lives, from jobs to vacation plans.

▶ Political and military relationships between the United States and Afghanistan, Iran, and Iraq have been troubled and, at times, violent.

CHAPTER 19
The Persian Gulf and Interior

The human geography of the Persian Gulf and interior Southwest Asia is changing as the modern world mixes with the old. In this chapter you will read about changes affecting the region's cultures and economic development.

Illustrated manuscript from ancient Persia

Date harvest in Bahrain

Greetings from Iran! My name is Mitra. I live with my family on the top floor of a house in Tehran, the capital. The roof of our house is flat, and in nice weather we sit up there and look out at the mountains. In the summer we sleep there too.

My favorite holiday is No Ruz, our New Year's festival, in the spring. We get two weeks off from school. We go see all our relatives, who give us clean new money from the bank. We visit the older people first, like my mother's family in Tehran, and then they visit us. The second week we may go to Isfahan, where my father is from, and see my grandmother. Lots of sweets and pastries are served at these visits. At dinner we may have special New Year's dishes. We also set up a table with a cloth with seven food items on it that begin with the *s* sound in our language. This spread is called the cloth of seven dishes. The foods on the table symbolize rebirth, health, happiness, prosperity, joy, patience, and beauty. For example, *serkeh* is vinegar and represents age and patience. You might compare this with your putting up a Christmas tree.

My mother's family is from the north, near the Caspian Sea. During our vacation, we go to her family's summer house. I like to go to the beach and swim. There is one beach for boys and another for girls. I also love to go to the Tehran bazaars with my mother. Food shops, fruit markets, gold shops—they are all mixed together. The shop owners pull out the best things to show us. We bargain for everything.

Section 1

OBJECTIVES

1. Identify the landforms and rivers of the Persian Gulf area and interior Southwest Asia.

2. Explain the effect the region's geography has on its climates and biomes.

3. Identify the region's natural resources.

 **LET'S GET STARTED**

Copy the following instructions onto the chalkboard: *Based on books you have read or images you have seen in movies or television, make a list of words that might relate to or describe this region's physical features.* Discuss responses. *(Possible answers: deserts, mountains, sand dunes, palm trees, rocky, rugged landscape)* Point out that humans have lived in the harsh environments of this area for thousands of years. Tell students they will learn more about the landforms, rivers, climates, biomes, and resources of Southwest Asia in Section 1.

Building Vocabulary

Write the terms **exotic rivers** and **oasis** on the chalkboard. Call on a volunteer to locate and read the terms' definitions in the text or glossary. Then ask students to identify the common bond between these terms. *(They are both sources of water that can be found in deserts.)* Call on students to use each term in a sentence.

Natural Environments

1. What landforms and rivers can be found in the Persian Gulf area and the interior of Southwest Asia?

2. How does the region's physical geography affect its climates and biomes?

3. What natural resources does the region have?

Huge oil deposits lie in the Persian Gulf region. How the countries there manage this resource affects the U.S. economy. Use **CNNfyi.com** or other **current events** sources to learn about the relationships the United States has with oil-rich countries in this region.

DEFINE

exotic rivers
oasis

LOCATE

Persian Gulf
Arabian Peninsula
Red Sea
Gulf of Aden
Arabian Sea
Mesopotamia

Locate, continued

Tigris River
Euphrates River
Shatt al Arab
Zagros Mountains
Elburz Mountains
Kopet-Dag
Hindu Kush
Rub' al-Khali
An Nafūd
Caspian Sea

Section 1 RESOURCES

REPRODUCIBLE

▶ Guided Reading Strategy 19.1
▶ Readings in World Geography, History, and Culture 47
▶ Environmental and Global Issues Activity 5: Operation Oil-Spill Cleanup

TECHNOLOGY

▶ One-Stop Planner CD–ROM, Lesson 19.1
▶ Geography and Cultures Visual Resources 37–41
▶ Homework Practice Online
▶ HRW Go site

REINFORCEMENT, REVIEW, AND ASSESSMENT

▶ Main Idea Activity 19.1
▶ English Audio Summary 19.1
▶ Spanish Audio Summary 19.1
▶ Section 1 Review, p. 436
▶ Daily Quiz 19.1

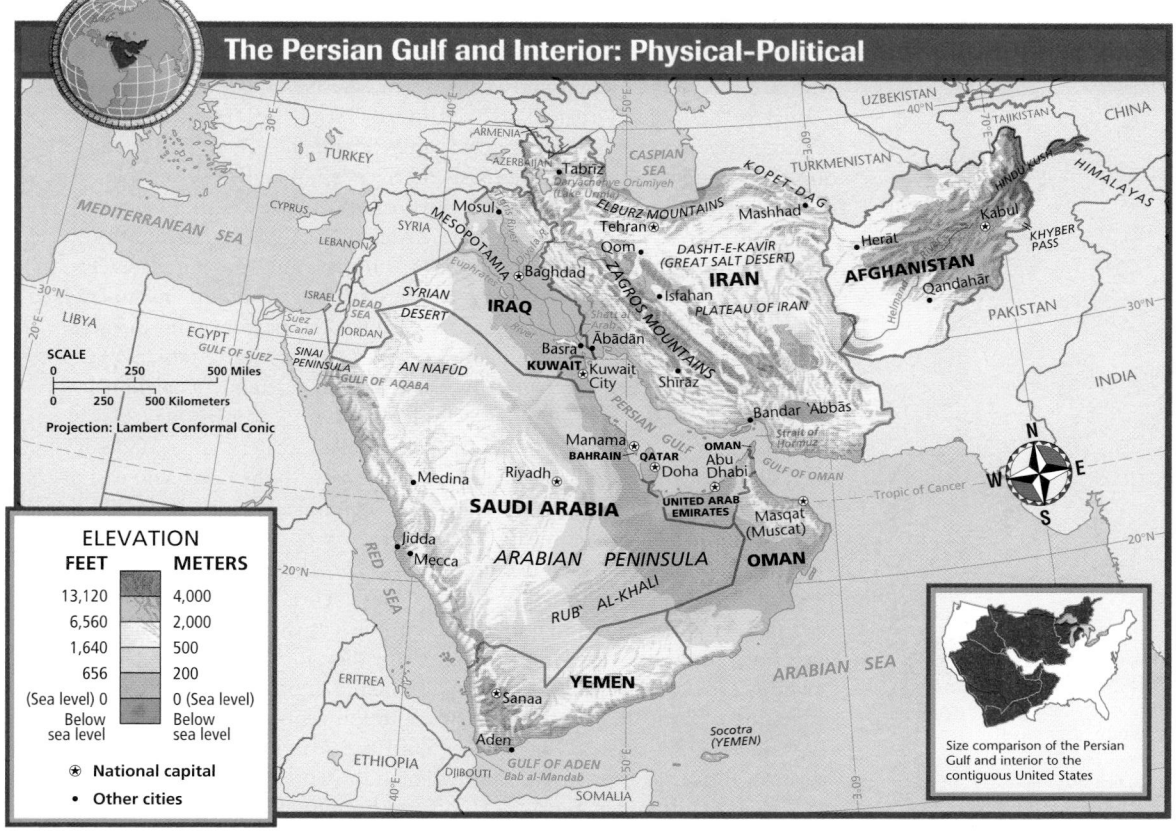

The Persian Gulf and Interior: Physical-Political

ELEVATION

FEET	METERS
13,120	4,000
6,560	2,000
1,640	500
656	200
(Sea level) 0	0 (Sea level)
Below sea level	Below sea level

⊛ National capital
• Other cities

SCALE
0   250   500 Miles
0   250   500 Kilometers
Projection: Lambert Conformal Conic

Size comparison of the Persian Gulf and interior to the contiguous United States

Teach Objective 1

ALL LEVELS: Provide each student with an outline map of the Persian Gulf area and interior Southwest Asia. Have students label the region's main landforms and bodies of water. Then have each student write two sentences about the role of tectonic forces in shaping these landforms. Ask volunteers to identify the physical features on a wall map and to read their sentences. You may then want students to conduct further research on some of the physical features to provide details on their creation, location, and features. Reports should also discuss the effects of the landforms on people. **ENGLISH LANGUAGE LEARNERS**

Teach Objective 2

ALL LEVELS: Copy the graphic organizer at right onto the chalkboard, omitting the italicized answers. Direct students to complete it. Ask volunteers to share their answers. Use the chart to introduce a discussion about the climates, plants, and animals of this region. Emphasize that the factors listed on the chart have made most of this region very hot and dry. **ENGLISH LANGUAGE LEARNERS**

Global Perspectives

The Shatt al Arab The Shatt al Arab ("Stream of the Arabs") forms part of the Iran-Iraq border. It has long been at the center of regional conflicts.

In 1847 the Ottoman Empire pressured Persia (now called Iran) to cede power over the waterway. Over the decades the Persians chafed under Iraq's dominance of the Shatt. However, in 1975 Iran and Iraq agreed to share control of the waterway. Five years later, Iraq took advantage of instability in Iran to claim both shores of the Shatt and invaded Iran, igniting a 10-year war. Many ships were sunk in the waterway, making it unnavigable. Silt and unexploded ordnance further hampered efforts to reopen it. With Iraq's invasion of Kuwait in 1990, the Shatt again closed to ships. It remains a source of conflict.

ACTIVITY: Have students investigate how changing perceptions of the Shatt al Arab have affected regional events and, by extension, regional societies.

VISUAL RECORD ANSWER

form barriers between climates, river systems, societies, languages, religions

INTERPRETING THE VISUAL RECORD

Eastern Afghanistan lies in the shadow of the rugged Hindu Kush mountain range. Crossing these mountains is difficult, as only a few major passes cut through the range. **How do landforms like the Hindu Kush divide a continent into regions?**

Landforms and Rivers

The region formed by the Persian Gulf and interior Southwest Asia includes Saudi Arabia and the smaller countries of the Arabian Peninsula. These countries are Bahrain, Kuwait, Oman, Qatar, the United Arab Emirates, and Yemen. This region is also often referred to as the Middle East. The Arabian Peninsula lies between the Red Sea and the Persian Gulf. (See the chapter map.) To the south are the Gulf of Aden and the Arabian Sea. Beyond these two bodies of water is the Indian Ocean. North and northeast of the Arabian Peninsula are three large countries that stretch farther inland into Asia. Two of these countries—Iraq and Iran—have coasts on the Persian Gulf. The third country, landlocked Afghanistan, lies to the northeast.

Tectonic forces have shaped the physical features of this region. Southwest Asia sits near the intersection of the African, Eurasian, and Arabian plates. The collision of these plates has created a mixture of rugged mountains, upland plateaus, and valleys. Plate movement has also created narrow gulfs and seas, which are bordered by coastal plains. As the African and Arabian plates move apart, the Red Sea is becoming wider. The region's frequent earthquakes are reminders of Earth's continuing tectonic activity.

Mountains stretch along the Arabian Peninsula's western edge. Wide dry plains slope down toward the Persian Gulf in the east. To the north and east of the Arabian Peninsula is a region called Mesopotamia, which lies mostly in Iraq. In Greek *Mesopotamia* means "between the rivers." In fact, Mesopotamia is a wide plain through which two great rivers flow. These rivers are the Tigris (TY-gruhs) and the Euphrates (yooh-FRAY-teez). They are **exotic rivers**, or rivers that begin in humid regions and then flow across dry areas. Before the Tigris and Euphrates reach the Persian Gulf, they join in a single channel known as the Shatt al Arab. East of Mesopotamia are the Zagros (ZA-gruhs) Mountains of Iran. In northwestern Iran lie the Elburz Mountains, where Iran's highest peak reaches 18,605 feet (5,671 m). Another range, the Kopet-Dag, rises in northeastern Iran. The rest of Iran is mainly made up of high plateaus. Afghanistan has the lofty Hindu Kush mountain range. Some peaks

internet connect

GO TO: go.hrw.com
KEYWORD: SW3 CH19
FOR: Web sites about the Persian Gulf and interior

Factors Influencing Southwest Asia's Climate

High Pressure	Orographic Effect	Winds	Elevation
• dry climates • lack of water • clear skies • plants adapted to dry conditions	• humidity at mountain peaks • rainfall on mountains • trees in mountain regions	• westerlies • bring winter rains • cyclonic storms	• cooler temperatures in highlands • resorts in mountains • skiing in Iran

Teach Objective 3

ALL LEVELS: Have students divide a sheet of notebook paper in half. Have them label one side of the sheet "Water" and the other "Oil." Ask students to list each resource's characteristics as discussed in Section 1. Details to note include the resource's availability, distribution, and uses. Discuss lists and the importance of both resources to the region. Finally, ask students to contrast this region's resources with those of other regions they have studied. (*Possible answer: Forests, metals, hydropower, fish, and soils seem to be of less significance in this region.*)

there rise higher than 24,000 feet (7,300 m). The Hindu Kush range is the western extension of the world's highest mountains, the Himalayas.

✓ **READING CHECK:** *Physical Systems* What physical process is shaping the Red Sea? the separation of the African and Arabian plates

Climates, Plants, and Animals

Hot and dry climates dominate the region. Rains come mostly during the winter when the westerly winds of the middle latitudes bring occasional cyclonic storms. The southern interior is a mostly uninhabited desert called the Rub' al-Khali (ROOB ahl-KAH-lee), or "Empty Quarter." Farther north lies the An Nafūd (ahn nah-FOOD), a desert of reddish sand. At its widest point, this desert stretches to 140 miles (225 km).

The region's mountains provide water to the valleys below. An orographic effect produces these more humid climates. The region's wettest climate is in Iran in a narrow zone along the southern shore of the Caspian Sea. Here winds blow southward over the water and pick up moisture. As the air rises along the Elburz Mountains, rain falls. The Zagros Mountains and the mountains of Yemen also have more humid climates.

The lowlands of Saudi Arabia along the Persian Gulf are among the hottest places in the world. Because of subtropical high pressure with clear skies and little shade, the summer daytime temperatures often climb above 114°F (46°C). Summer rains are almost unknown here, but nearness to the sea keeps the humidity high. However, inland areas are very dry. These areas experience rapid cooling at night because the air holds such a small amount of moisture. In fact, after sunset desert temperatures can drop some 30 degrees in just a few hours. Elevation also influences temperatures, so the region's highlands are much cooler than lowlands in general. Both Iran and Saudi Arabia have mountain resorts where people can escape the summer heat. Winter skiing is popular in Iran's Elburz Mountains.

Shrubs and grasses cover the region's wide dry plains. Trees are common only in mountain regions and the usually dry streambeds. The highest plains are grasslands. In the driest areas the ground is bare rock and sand. In some places the soil is so salty that no plants can grow. Nearly all the region's plants have adapted to survive long periods without rain. Roots either grow deep or spread out to capture as much water as possible. Many plants have developed leaves or stems that allow them to store moisture.

Hunting by humans and competition from domestic animals have made life difficult for the region's larger wild animals. Gazelles and wild goats were common a few centuries ago. Hyenas, leopards, lions, and tigers as well as herds of wild camels and donkeys once roamed the region. Most large wild mammals are now rare or restricted to a few game reserves. Today all the camels and donkeys are domesticated. Reptiles, including lizards and several poisonous snakes, are still common.

✓ **READING CHECK:** *Physical Systems* What kinds of climates dominate the region? What produces humid climates in the region? hot and dry; orographic effect

The Rub' al-Khali is the world's largest region of active sand dunes—dunes that migrate over time. The English name for this region is the Great Sandy Desert.

One of the driest areas in the world, the Rub' al-Khali receives an average of less than 4 inches (10 cm) of rainfall each year. Few water sources are found in this desert, but oil reserves have been discovered beneath its sands.

Across the Curriculum

▶ **Technology** ◀

Locating a Lost City The Rub' al Khali has not always been empty. Several lost cities are rumored to lie buried beneath the sand. Greek and Roman historians, including Herodotus and Pliny the Elder, described fabulously wealthy desert cities, rich in valuable trade goods like frankincense. In the ancient world, frankincense was used for fragrance, medicine, and embalming.

One of these fabled cities, Ubar, called the "Atlantis of the Sands," was recently rediscovered through the application of modern technology. At the request of archaeologists, NASA used the space shuttle and satellites to focus remote sensing equipment on the Rub' al Khali. The images received showed a network of ancient roads that converged in the desert. Archaeologists immediately began excavations at this site. There they uncovered a huge fortress containing Greek, Roman, and Syrian pottery—remnants of trade between Ubar and the Mediterranean world.

🖥 **internet** connect

GO TO: go.hrw.com
KEYWORD: SW3 CH19
FOR: Web sites about the deserts of Southwest Asia

INTERPRETING THE VISUAL RECORD *An Omani man collects water from a local* falaj, *or aqueduct. These channels are dug to carry water from desert springs to farms and villages. Many of the* falaj *systems in use today were built more than 1,000 years ago.* **How do you think irrigation systems like this one have changed the landscape of dry areas in Oman?**

Natural Resources

The region's two most important natural resources are oil and water. Oil is plentiful, but water is not. The Tigris and Euphrates Rivers are the main sources of water in Iraq. Canals lead away from these rivers, bringing precious water to the surrounding dry lands. In the high plateaus and mountains of northern Iran, farmers depend on rain for agriculture. Farmers in most other places must irrigate their fields.

Surface water is rare in the desert areas. It can be found only at an **oasis**, an area where a spring bubbles to the surface. People have made many of these springs into productive wells. Wells may also reach water below a dry river bed. Deep wells may tap into fossil water, groundwater that is not being replaced by rainfall. Fossil water is not a renewable resource. Desalinization of seawater provides another source of freshwater. In general, however, only wealthy countries get freshwater this way. This is because the process uses large amounts of power and is expensive. Saudi Arabia produces more desalinized water than any other country.

Oil is the region's most valuable natural resource. The oil reserves along the Persian Gulf are the largest in the world. Iraq, Oman, and Yemen also have important oil deposits. However, the region's countries have few other resources for developing industry. Only Iran has the potential to develop an industrial economy from its own reserves of metallic ores.

✓ **READING CHECK:** *Places and Regions* How can technology expand the region's supply of freshwater? The desalinization process can make freshwater from seawater.

Section 1 Review

Homework Practice Online
Keyword: SW3 HP19

Define
exotic rivers
oasis

Working with Sketch Maps
On a map of Southwest Asia that you draw or that your teacher provides, label the Persian Gulf, Arabian Peninsula, Red Sea, Gulf of Aden, Arabian Sea, Mesopotamia, Tigris River, Euphrates River, Shatt al Arab, Zagros Mountains, Elburz Mountains, Kopet-Dag, Hindu Kush, Rub' al-Khali, An Nafūd, and Caspian Sea. What do the names *Rub' al-Khali* and *Mesopotamia* mean?

Reading for the Main Idea
1. *Physical Systems* What are four physical features that influence what plants can grow in different places in the region?

2. *Environment and Society* What are the region's two most precious natural resources?

Critical Thinking
3. **Making Generalizations and Predictions** Which landforms in the region seem most favorable for human settlement? Which seem the least favorable? Why?

4. **Identifying Cause and Effect** Why do parts of Saudi Arabia have high temperature variations?

Organizing What You Know
5. Copy the chart below. Use it to list major physical features that can be found in the region's largest countries. Use the information in Section 1 and the chapter map.

Feature	Afghanistan	Iran	Iraq	Saudi Arabia
Coastal plains				
Interior plains				
Major rivers				
Plateaus				
Mountains				

Section 2

OBJECTIVES

1. Describe how peoples, empires, and Islam have affected the history of the Persian Gulf area and interior Southwest Asia.

2. Identify the major features of the region's cultures.

LET'S GET STARTED

Copy the following passage onto the chalkboard: *Some of the world's earliest civilizations were established in the Fertile Crescent in Southwest Asia. Using your chapter map, try to identify where this area was located. Why do you think it was called the Fertile Crescent?* Discuss student responses. Tell students they will learn more about the history and culture of Southwest Asia in Section 2.

Building Vocabulary

Write the words **dynasty** and **imams** on the chalkboard. Call on a volunteer to locate and read their definitions. Point out that a dynasty can be any family or group that maintains great power, wealth, or position for several generations. The power does not have to be political. Ask students to give examples. *(Example: a multigenerational family of actors)* Point out that *imam* is an Arabic word.

History and Culture

READ TO DISCOVER

1. How have peoples, empires, and Islam affected the history of the Persian Gulf area and interior Southwest Asia?

2. What are the major features of the region's cultures?

WHY IT MATTERS

Many Americans trace their heritage to Southwest Asia. Use **CNNfyi.com** or other **current events** sources to learn about recent immigrants from Southwest Asia and their cultures.

IDENTIFY

Sunni

Shi'ism

DEFINE

dynasty

imams

LOCATE

Mecca

Medina

From Empires to Independence

The world's first civilizations developed in the area known as the Fertile Crescent. This arc of productive land extends northward from the Persian Gulf and through the plains of the Tigris and Euphrates Rivers. It continues to Asia Minor and the Mediterranean coast. Asia Minor is the Asian part of what is now Turkey. Many of the plants and animals found on farms throughout the world today may have been first domesticated in the Fertile Crescent region. By about 3000 B.C. a people called the Sumerians built the world's first known cities in southern Mesopotamia. These cities depended on nearby irrigated fields of wheat and barley. City merchants traded goods from throughout the region.

The ruins of Persepolis reflect the former glory of the Persian Empire, which once stretched from Egypt to Afghanistan. King Darius I built Persepolis in about 500 B.C. as the capital of his empire.

Teach Objectives 1–2

🌐 **ALL LEVELS:** Have pairs of students create lists of peoples and empires that have affected Southwest Asia's history and note each group's contributions to the region. Then copy the following graphic organizer onto the chalkboard, omitting the italicized answers. Have students complete it with characteristics of each of the listed groups. Then ask students to combine the information on their lists with the graphic organizer to determine if and how the various groups have contributed recognizable traits to the region's contemporary cultures. (*Example: Conquest by Muslims brought Arabic, still the dominant language.*) **COOPERATIVE LEARNING**

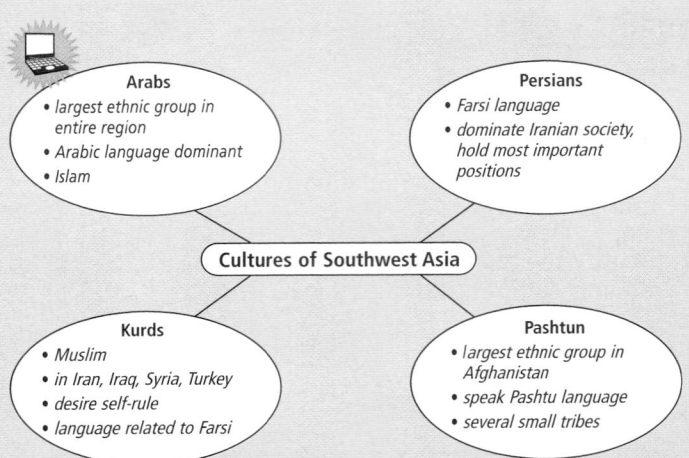

Arabs
- *largest ethnic group in entire region*
- *Arabic language dominant*
- *Islam*

Persians
- *Farsi language*
- *dominate Iranian society, hold most important positions*

Cultures of Southwest Asia

Kurds
- *Muslim*
- *in Iran, Iraq, Syria, Turkey*
- *desire self-rule*
- *language related to Farsi*

Pashtun
- *largest ethnic group in Afghanistan*
- *speak Pashtu language*
- *several small tribes*

Linking Past to Present

The World's First Money
Trade within and between the cities of Mesopotamia may have inspired the creation of a system of currency. Evidence on clay tablets found in what is now Iraq suggests that the people of Mesopotamia carried cash as early as 2500 B.C., though some scholars think the monetary system may have developed even earlier.

Silver seems to have been the standard on which the Mesopotamian economy was based. Wealthy citizens carried coils of the metal, called *har,* of predetermined weights and values. Others carried bags and jars filled with silver bits. These could be weighed by merchants as purchases were made. Mesopotamians who could not afford silver probably used copper, tin, lead, and barley when they went shopping.

You don't say! Archaeologists think the world's first wheeled vehicle was built in Erech, a Sumerian city that may date to about 5000 B.C.

CONNECTING TO MATH ANSWER

Possible answers: by making mathematical calculations and keeping business records easier

Connecting to MATH

Arabic Numerals

The symbols that most of the modern world uses to write numbers are known as Arabic numerals. Just 10 symbols are needed to write a number of any size. These symbols—0 through 9—probably evolved from letters in the Arabic alphabet. The mathematical system itself, in which a symbol's value depends on its place in the number, actually began in India. Arab traders brought the system back to Southwest Asia and later spread its use into Europe. The Arabic system proved much simpler than the old Roman system it gradually replaced. For example, CCCXXXIII in Roman numerals became 333 using Arabic numerals. This simplification of the number system aided the growth of science and commerce. Muslim scholars made important advances in art, astronomy, literature, medicine, and mathematics.

Identifying Cause and Effect How do you think Arabic numerals may have aided the growth of science and commerce?

The rich resources of Mesopotamia attracted invaders again and again. Akkadians conquered the Sumerian cities in about 2350 B.C. and created the region's first real empire. That empire extended to the Zagros Mountains and the Mediterranean coast. The islands in the Persian Gulf became major trading centers. Eventually, these trade centers linked the merchants of Mesopotamia and India.

Over time cities grew and declined in the region. Again and again invaders from the west, east, and north battled the peoples of the plains and set up new empires. About 550 B.C. an empire developed in Persia, where Iran is today. The Persians conquered both Mesopotamia and Asia Minor. Theirs was one of the largest, richest, and most powerful empires in world history. Later, the Greeks and then the Romans controlled much of the region for a time.

The Rise of Islam The prophet Muhammad, who lived in the region from about A.D. 570 to 632, established Islam. Islam is one of the world's most widely practiced religions. Muhammad was born in Mecca, a city in the western part of the Arabian Peninsula. At about the age of 40, a religious experience changed his life. He reported that a messenger of God, the angel Gabriel, told him to preach the word of God. The word *God* in Arabic is *Allah.* Muhammad spread Allah's message to his followers, called Muslims. Muslims are people who practice Islam. A holy book called the Qur'an (Koran) contains what Muslims believe to be Allah's message to Muhammad.

Muhammad established a Muslim community centered at Medina. By the time Muhammad died, Islam had spread to most of the Arabian Peninsula. After his death, Arab armies carried Islam as far west as Morocco and Spain in a little more than a century. Over time Islam spread to Central Asia, India, and Southeast Asia along land and sea trade routes. Muslims in all these places now face Mecca, Islam's holiest city, when they pray.

Gaining Independence Empires continued to rise and fall in Southwest Asia until modern times. In the 1200s the Mongols swept out of Central Asia to conquer what are now Afghanistan, Iraq, and Iran. Rulers called the Safavids (sah-FAH-weedz) came to power in Iran in the early 1500s. As their power expanded eastward, the Safavids seized Afghanistan from the Muslim rulers of India. Historians consider the rule of Safavids, which lasted more than 200 years, a golden age of Persian culture. Literature, architecture, and the arts flourished. Persian carpets, ceramics, and textiles became renowned. The Safavid **dynasty**, or line of hereditary rulers, ended in the mid-1700s. The expanding British and Russian Empires tried to control Iran and Afghanistan in the 1800s. During the 1900s both became independent countries.

The history of the western part of the region followed much the same pattern. In the 1500s the Ottoman Turks conquered Mesopotamia and the east and west coasts of the Arabian Peninsula. They held much of this area until the early 1900s, when the British took over. Iraq and Saudi Arabia emerged as independent countries in 1932. Kuwait, Bahrain, Qatar, the United Arab Emirates, and Yemen did not become independent from Britain until the 1960s and 1970s.

✓ **READING CHECK:** *Human Systems* What are some of the peoples and empires that have controlled parts of the region? Sumerians, Akkadians, Persians, Greeks, Romans, Mongols, Ottoman Turks, British

Culture

Islam is the unifying cultural feature of the region. However, the region's long history of changing empires and migrations has resulted in the presence of many ethnic groups today.

People and Languages Most people in the Persian Gulf and interior Southwest Asia are Arabs, and Arabic is the dominant language. The spread of Islam encouraged the spread of the Arabic language over time. To read the Qur'an, Muslims had to learn Arabic, which also became the common language of scholarship and trade. Trade routes have long connected the distant parts of the Islamic world. Eventually, all Arabic speakers became known as Arabs. Today Arabic place-names are found in Spain and Morocco, Central Asia, and India. This diffusion of Arabic words is a result of migration, trade, and the spread of Islam.

Like Saudi Arabia, Iraq is an Arab country. In addition, there are more than a million Arabs in southern Iran. However, non-Arab ethnic groups are also numerous in the region. For example, the Kurds, who live in the borderlands of Iran, Iraq, Syria, and Turkey, are Muslims but not Arabs. The Kurds have never had their own country. Their desire for self-rule is a source of political unrest in some countries, including Iraq.

Cultural diversity is even more complex in Iran and Afghanistan. Most of Iran's people are Persians who speak Farsi. The Kurds, Baloch, Bakhtiari, and Hazara also speak languages related to Farsi. However, Persians dominate Iran and hold most of the important positions in Iranian society. A number of other ethnic groups speak Turkic languages. Turkmen communities are found in northeastern Iran. In the northwest are the Azeri people, Iran's largest group after the Persians. The Qashqai people of the southern Zagros Mountains also speak a Turkic language.

In Afghanistan the Pashtun make up the largest ethnic group. The name *Pashtun* really refers to a number of tribes that speak the Pashtu language. They are closely related to the Tajiks to the north as well as to small tribes in eastern Iran. A number of other ethnic groups also live in Afghanistan. Yet people's loyalties often rest more with their clan and family than with their ethnic group.

✓ **READING CHECK:** *Human Systems*
What processes aided the diffusion of Arabic culture? migration, trade, and the spread of Islam

Horse races in Iran between different nomadic groups are popular events. Iran is home to many nomadic peoples, particularly in regions near the Zagros Mountains.

GO TO: go.hrw.com
KEYWORD: SW3 CH19
FOR: Web sites about Southwest Asia's ethnic groups

Each year some 2 million Muslims from around the world travel to Mecca, Islam's holiest city. All Muslims are expected to make this sacred journey, called the hajj, at least once during their lifetime. The pilgrims gather around the Ka'bah, a sacred shrine, for several days of prayer. Historically the experience of the hajj has served as a common bond among all Muslims, including followers of Sunni Islam and Shi'ism. However, overcrowding, political disputes, and religious differences have also led to tense confrontations and violence inside the holy city.

FOCUS ON CULTURE

Religion and Society Islam has split into two main branches and many different groups over the centuries. The two main branches are **Sunni** Islam and **Shi'ism.** Their differences center around who can become religious leaders, called **imams,** in Muslim society. Sunni groups choose their imams, who serve mainly as leaders of prayer. The Shia—those who practice Shi'ism—allow only descendants of the prophet Muhammad's family to become imams. The Shia also rely on imams to interpret the Qur'an and other sacred texts containing teachings that govern personal conduct.

Today about 90 percent of Muslims are Sunnis, and 10 percent are Shia. Sunnis are found everywhere Islam has spread. Because they make up the majority of Muslims, Sunnis are sometimes called orthodox Muslims. Shi'ism is concentrated in Iran, southern Iraq, Yemen, Syria, and Lebanon. Shia imams are particularly important in Iran. In that country they have considerable political as well as religious authority.

Both Sunnis and Shia share many of the same basic beliefs and practices of Islam. For example, members of both groups are expected to make a religious pilgrimage to Mecca. However, throughout history there have also been conflicts between the two groups. Some conflicts have been caused by the persecution of one group by the other.

✓ **READING CHECK:** *Human Systems* How does the role of a Sunni imam differ from that of a Shia imam?

Sunni—mainly prayer leaders; Shia—interpret the Qur'an and other sacred texts, exercise political power in Iran

Review

Homework Practice Online
Keyword: SW3 HP19

Identify Sunni, Shi'ism

Define dynasty, imams

Working with Sketch Maps On the map you created in Section 1, label the countries of the Persian Gulf and interior Southwest Asia, Mecca, and Medina. Then use the description in Section 2 to shade in the Fertile Crescent.

Reading for the Main Idea
1. *Human Systems* On what did the early cities of the Fertile Crescent depend for their growth? In what way is farming today connected with the early history of the Fertile Crescent region?

2. *Human Systems* Why will you find the Arabic language, Arabic place-names and Islam in places outside of Southwest Asia today?

3. *Places and Regions* Which ethnic group is most widely spread throughout the region? Which groups dominate Iran and Afghanistan?

Critical Thinking
4. *Making Generalizations* What major cultural feature is common throughout the region? What other cultural features could you use to further divide the Persian Gulf and interior Southwest Asia into more regions?

Organizing What You Know
5. Create a diagram like the one shown below. Use it to identify the various peoples who conquered or controlled the region. Add more boxes as needed.

LET'S GET STARTED

Write the following question on the board: *What would life be like if the United States was cut off from its major oil supplier?* Discuss student reactions. *(Possible answers: much higher gas prices, less travel, lower speed limits, and so on)* Point out that the countries of Southwest Asia play an important global economic role because they are the world's leading oil producers. Tell students they will learn more about the importance of oil to the region and life there today in Section 3.

Building Vocabulary

Write the words **ayatollahs** and **theocracy** on the chalkboard. Tell students that *ayatollah* is from the Persian and Arabic languages. Point out that *theocracy* is derived from two Greek roots—*theo-* meaning "god," and *-cracy* meaning "power" or "rule." Ask students what they think the combined word might mean. Then have a volunteer read both of the terms' definitions from the text or glossary. Ask students how the terms might be related to Southwest Asia.

The Region Today

READ TO DISCOVER

1. On what activities do the region's economies depend?
2. What are the region's cities like?
3. What are some important issues in the region today?

WHY IT MATTERS

U.S. military forces were involved in the Persian Gulf War in 1991 and remain in the region today. Use **CNN fyi.com** or other **current events** sources to learn about recent military actions in the Persian Gulf region.

IDENTIFY

Bedouins   OPEC

DEFINE

ayatollahs   theocracy

LOCATE

Tehran   Riyadh
Baghdad   Strait of Hormuz
Kabul

Economic Development

Oil and gas production is central to the economies of the countries along the Persian Gulf. Saudi Arabia alone produces some 8.25 million barrels of oil a day. Oil wealth has helped modernize the economies of the region. However, many people continue to follow traditional rural ways of life and culture. Arid climates and a rugged landscape make farming difficult. In addition, farmers find fertile soils mainly in the river valleys, on high plateaus, and at a few oases. Partly as a result of this, all the region's countries must import food. Many of the more humid areas have been overgrazed, leading to soil erosion. People have cut down most of the mountain forests that once existed. Some countries in the region are now trying to conserve soil and native plants.

Most farmers practice subsistence agriculture, producing only enough to support their families. They sell any surplus at local markets. Barley and wheat are the most common grains grown in the area. Farmers may also raise livestock—mostly sheep, goats, and some cattle. Because Islam forbids eating pork, pigs are not raised. Commercial farms are typically found near the large cities. They provide fruits and vegetables for city markets. Where irrigation is available, farmers grow citrus fruits, dates, grapes, nuts, and olives for sale. Specialized commercial farms, such as modern dairies and chicken farms, are becoming more common in the oil-rich countries. Saudi Arabia even grows flowers for export to Europe.

Nomadic herders, some of whom are known as **Bedouins**, live in outlying dry lands. They tend to move their camels, goats, or sheep in regular routes as the seasons change. These herders trade their animals, animal products, and handicrafts in towns. More and more Bedouins are leaving this traditional way of life and taking jobs in cities and on the new modern farms.

INTERPRETING THE VISUAL RECORD

Although Bedouins make up only a small percentage of Iraq's population, they move with their herds across much of the country. However, recent laws have restricted the lands that are available to Bedouins for grazing. **How do you think these laws are changing Bedouin culture?**

The Persian Gulf and Interior • 441

RESOURCES

REPRODUCIBLE

▶ Guided Reading Strategy 19.3
▶ Graphic Organizer Activity 19
▶ Readings in World Geography, History, and Culture 49–51
▶ Geography for Life Activity 19: Speaking Graphically about Persian Gulf Oil

TECHNOLOGY

▶ One-Stop Planner CD–ROM, Lesson 19.3
▶ CNN Presents Geography: Yesterday and Today, Segment 22: Kuwait—Restoring the Environment
▶ Homework Practice Online
▶ HRW Go site

REINFORCEMENT, REVIEW, AND ASSESSMENT

▶ Main Idea Activity 19.3
▶ English Audio Summary 19.3
▶ Spanish Audio Summary 19.3
▶ Section 3 Review, p. 444
▶ Daily Quiz 19.3

VISUAL RECORD ANSWER

fewer nomads, more settled in villages, seeking new economic activities, perhaps increased poverty

441

ALL LEVELS: Copy the graphic organizer at right onto the chalkboard, omitting the italicized answers. Have students complete it. Then ask students to write brief paragraphs to describe the region's economy as a whole.

Teach Objective 2

ALL LEVELS: Have groups of students create posters, using words and pictures, advertising movies about daily life to be filmed in one of the region's large cities. (*Example: "Back to Baghdad."*) Lead a discussion about how the cities are similar and different. **ENGLISH LANGUAGE LEARNERS, COOPERATIVE LEARNING**

Oil Production	Agriculture	Nomadic herding	Traditional crafts	Manufacturing
• *8.25 million barrels a day* • *economies modernized by oil wealth* • *oil-related manufacturing*	• *mostly subsistence* • *barley, wheat grown* • *livestock— sheep, goats, cattle* • *farms in river valleys* • *commercial farming near cities; citrus fruits, dates, grapes, olives with irrigation* • *import food*	• *tend herds of camels, goats, and sheep* • *make handicrafts* • *increased settling in towns due to lack of grazing lands*	• *wool rugs* • *use local materials and traditional designs*	• *building materials* • *food products* • *oil refining, chemical manufacturing* • *household supplies*

Daily Life

Cell Phones in Southwest Asia Cellular telephones are very popular in Southwest Asia. In Kuwait, at least one of every 15 people owns a cell phone. In the United Arab Emirates, the percentage is even higher. That country has a cell phone for every 2.5 residents. These phones serve as status symbols. They are also convenient, as fixed-line networks are run-down, poorly developed, and expensive.

The widespread use of cellular phones is also due to the conservatism of Southwest Asia. For example, coeducational gatherings are discouraged among young people in Kuwait. Instead, young men and young women sometimes gather in large but separate groups— often in the same restaurant or coffeehouse—and call each other on cell phones.

DISCUSSION: Lead a discussion about how new technologies might affect traditional cultures in Southwest Asia.

Baghdad is Iraq's primate city and is home to almost one third of its population. It was also one of the leading cities of ancient Mesopotamia, a cradle of early civilization. Although it has been the site of military conflicts in recent years, Baghdad remains a city of contrasts, featuring ancient mosques, bazaars, cafes, high-rise apartments, luxury hotels, and traffic congestion.

In most countries people proudly maintain traditional crafts. Artisans use local materials in products sold around the world. The region's beautiful and valuable wool rugs have been famous for centuries. Rugs are still made by hand, using traditional designs. Each community has its own distinctive designs.

Modern manufacturing in the region is limited. It focuses on building materials, food products, and household supplies. The region imports most of its electrical appliances as well as cars and trucks. The only large industries involve oil refining and related chemical manufacturing. However, these industries are highly automated and provide few jobs.

✓ **READING CHECK:** *Environment and Society* How do some of the region's people maintain their traditional ways of life? practice subsistence farming, herding, traditional crafts

Urban Environments

Among the largest cities in the region are the national capitals. These cities include Tehran, Iran; Baghdad, Iraq; Kabul, Afghanistan; and Riyadh, Saudi Arabia. As elsewhere in the world's developing regions, many people have migrated to cities from rural farms and villages. They go there looking for jobs. These people often build housing on the fringes of cities. Sometimes they bring farm animals, which then live in urban neighborhoods.

Many of the region's cities are ancient. In the older sections, life goes on today much as it did centuries ago. Old-style buildings are still in use, and many are one or two stories tall. The narrow twisting streets reflect the fact that they predate the use of cars and trucks. The stalls in a central marketplace, called a bazaar, are covered to protect them from sunlight and rain. Merchants display piles of cloth, household utensils, rugs, and spices. Craft items used to be clustered in separate sections of the bazaar. Now buyers find goods of many types, including imports, mixed together throughout the market. A neighborhood mosque is usually nearby.

In contrast, the newer sections of cities have modern buildings and air-conditioned shopping malls. The avenues are wide and clogged with cars. People live in high-rise apartment buildings. With fast-food outlets and gas stations, some of these neighborhoods look like those in the West.

Governments, Issues, and Challenges

The region's politics and concerns for the future center around three basic themes. One theme is the use of oil wealth. Another is the desire of some to preserve the authority of traditional leaders. A third theme involves the role of Islam in a modernizing world. Individual countries emphasize these themes differently.

Oil Wealth and Power Saudi Arabia's oil wealth gives the country a special position in world affairs. Saudi Arabia's huge reserves make it the world's

ALL LEVELS: In advance, prepare the names of the region's major leaders. Then review the three basic issues facing the region with the class (use of oil wealth, desire to preserve authority of traditional leaders, role of Islam in a modernizing world). Assign one of these issues to each of three groups. Then have each group write a letter to one of the leaders as if the students were citizens of the leader's country, expressing opinions about how the leader should address these issues in his or her country. Point out that, in reality, the leader may not be receptive to suggestions from the public. Have a representative from each group read its letter. **COOPERATIVE LEARNING**

Using National Geography Standard 12:
Human Systems: Processes, Patterns, and Functions of Human Settlement Have students analyze how the discovery of oil in the 1930s transformed Riyadh, Saudi Arabia, from a provincial town to a glittering, cosmopolitan city with more than a million residents. Possible sources of information include journals from individuals living during these changes, stories about Saudi Arabia in *National Geographic*, old tour guides, and historical information on the Internet. Ask students to create posters highlighting the city's technological, architectural, educational, and recreational innovations over this time period.

largest oil exporter. It is a key member of the Organization of Petroleum Exporting Countries, or **OPEC**, which influences oil prices by controlling supply. By reducing or increasing oil production, OPEC can affect the economies of many countries. Saudi Arabia's role as caretaker of Islam's holiest city, Mecca, adds to its influence. Muslims travel to this city from all over the world. A monarchy whose power is absolute rules from the capital, Riyadh. Defending Arab traditions helps the monarchy maintain control. Economic success has also limited opposition to the Saudi monarchy within the country.

In contrast to Saudi Arabia, Iran's politics have been unstable in recent decades. Rebellions by minority ethnic groups have been frequent. In 1979 a revolution toppled Iran's monarchy. A government dominated by **ayatollahs** came to power in Tehran. Ayatollahs are religious leaders of the highest authority among Shia Muslims. Today Iran is a **theocracy**, a country governed by religious law. Many religious leaders view Western ideas as a threat to public morality. As a result, some have tried to isolate the country from Western influences. However, many Iranians are seeking more personal freedoms.

Many countries pay attention to Iran's political situation. The huge tankers that ship oil from the Persian Gulf must pass through the Strait of Hormuz, which narrows to 35 miles (56 km) off Iran's southern coast. Iran could cut off a large portion of the world's oil supply by blocking the strait.

Until 2003 Iraq was ruled by a dictator, Saddam Hussein, who used the country's oil revenues to build a large military. Under Saddam, Iraq invaded its neighbors—Iran in 1980 and Kuwait in 1990—seeking to gain control of oil-rich regions. Iraq's invasion of Kuwait threatened the stability of the region. A group of countries led by the United States drove back Iraqi forces in 1991 in a conflict known as the Persian Gulf War. To prevent Iraq from launching another war, the United Nations ordered Saddam to stop producing weapons of mass destruction. In 2003, after Iraq's continued resistance to inspections and violations of UN sanctions, the United States led an invasion of the country. Iraq's forces were defeated and Saddam Hussein was eventually captured.

✓ **READING CHECK:** *Places and Regions* What geographic factors have affected Iraq's international relations in recent decades? Access to oil and historical claims led Iraq to invade Iran and Kuwait.

The al-Ghawar field in Saudi Arabia is the biggest oil field in the world. It is more than 150 miles (240 km) long and contains some 82 billion barrels of oil.

Global Perspectives

Oil Embargo In 1973 the Arab nations of OPEC announced that they would not sell oil to countries, including the United States, that had supported Israel in the Arab-Israeli war. This embargo caused American oil supplies to drop dramatically, and prices rose accordingly. Drivers waited in long lines at filling stations, many of which ran out of gas. Although the embargo itself lasted less than a year, its effects were long-lasting. Americans reduced oil consumption and increased domestic oil production. These steps reduced, but did not eliminate, American dependence on foreign oil.

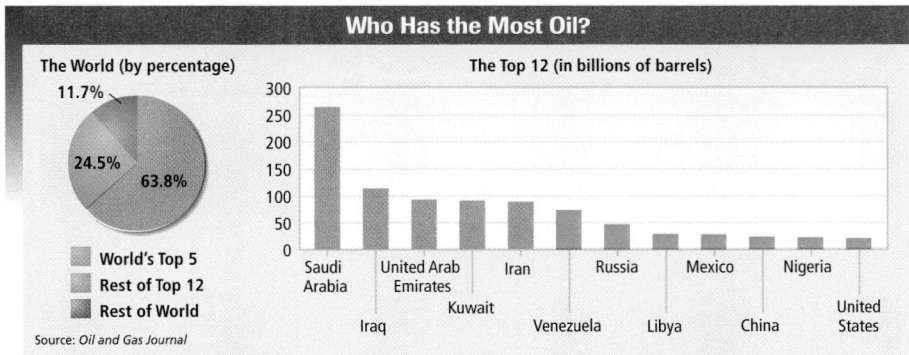

Who Has the Most Oil?

The World (by percentage)

11.7%
24.5%
63.8%

- World's Top 5
- Rest of Top 12
- Rest of World

Source: *Oil and Gas Journal*

The Top 12 (in billions of barrels)

Saudi Arabia, United Arab Emirates, Iran, Russia, Mexico, Nigeria, Iraq, Kuwait, Venezuela, Libya, China, United States

INTERPRETING THE GRAPH *Nearly two thirds of the world's known oil reserves are controlled by just five countries along the Persian Gulf. In what ways do you think the world's developed countries are affected by political developments and military conflicts in the Persian Gulf region?*

GRAPH ANSWER

can cause reduced supplies of oil and higher prices for developed countries that rely on imports

443

Close

Refer students to the subsection titled Urban Environments. Ask students to imagine that they are exchange students living in one of the region's large cities. Lead a discussion about whether they would prefer to live in the older or newer section of the city. Ask students to supply reasons for their preferences.

Review and Assess

Have students complete the **Section Review**. Then have students complete **Daily Quiz 19.3**.

Reteach

Have students complete **Main Idea Activity for English Language Learners and Special-Needs Students 19.3**. Then have students work in groups to summarize parts of Section 3 and share the summaries with the class. **ENGLISH LANGUAGE LEARNERS, COOPERATIVE LEARNING**

Extend

Have students learn more about the rigors of nomadic life by creating a plan for a family to live as nomads in the United States. Ask students to plan how the family will make its living, how the children will be educated, what provisions the family will need, and so on. **BLOCK SCHEDULING**

Section 3 Review Answers

Identify For identifications, see: Bedouins, p. 441; OPEC, p. 443

Define For definitions, see: ayatollahs, p. 443; theocracy, p. 443

Working with Sketch Maps
Maps will vary, but listed places should be labeled in their approximate locations. Many oil tankers pass through this strait.

Reading for the Main Idea
1. cash crops generally grown only in irrigated areas, farming difficult elsewhere, mainly subsistence farmers, nomadic herding in drier areas

2. the presence of bazaars, neighborhood mosques, and farm animals

Critical Thinking
3. Possible answers: water—essential to any life or economic activity; oil—basis of region's wealth and best hope for achieving development (NGS 16)

4. denied women access to education, banned them from the workforce; more opportunity in other countries even though Islam encourages traditional roles of wife and mother (NGS 10)

Organizing What You Know
5. Afghanistan—was governed by Taliban, new constitution adopted in 2004; Iran—monarchy overthrown in 1979, replaced by theocracy run by ayatollahs; Iraq—2003, Saddam Hussein removed from power; Saudi Arabia—conservative, wealthy monarchy

In January 2004, Afghanistan's president Hamid Karzai addressed the loya jirga, or Grand Assembly, after the assembly approved a new draft constitution for the country. The new constitution paved the way for democratic elections to be held in 2004.

Islam, Society, and Change In Afghanistan, ethnic and political rivalries have long plagued the country. A group called the Taliban came to power in the 1990s. They were driven by an extreme version of Sunni Islam. The Taliban established strict laws governing Afghans' lives.

After the September 11 terrorist attacks, U.S. officials focused on the aid the terrorists received from the Taliban government. U.S. and allied forces attacked terrorist camps and Taliban military targets. The Taliban regime soon collapsed. A *loya jirga*, a traditional council, with representatives from all ethnic groups was held in June 2002. It elected Hamid Karzai as the country's transitional president. In 2004, the *loya jirga* adopted a new constitution for Afghanistan. The constitution established a presidential system of government with a parliament.

Freed from the Taliban, Afghans experienced new liberties. Women were able to attend school and work outside their homes.

In other countries in the region, women have more educational and economic opportunities. Still, Islamic traditions encourage women to value roles as wives and mothers. Partly as a result, population growth rates are often high. Large families are common. Education for all segments of the population, including women, has increased. However, many school systems are inadequate, and many well-educated people cannot find jobs.

Oil, the influence of foreign cultures, and communications technology are changing the region. Over time these changes will be greater than any development since the spread of Islam. For example, much of the region's oil is available for export. Modern society runs on oil. Homes, power stations, and transportation depend on oil or oil products. Oil is also a basic raw material for products like fertilizers, industrial chemicals, and plastics. The importance of oil in global trade has given the oil-rich countries much economic and political power.

✓ **READING CHECK:** *Human Systems* What three themes are reflected in the region's politics and concerns for the future? oil wealth, preserving the authority of traditional leaders, the role of Islam in a modernizing world

Section 3 Review

go.hrw.com
Homework Practice Online
Keyword: SW3 HP19

Identify Bedouins, OPEC

Define ayatollahs, theocracy

Working with Sketch Maps On the map you created in Section 2, label Tehran, Baghdad, Kabul, Riyadh, and the Strait of Hormuz. Why is the Strait of Hormuz important to the global oil trade?

Reading for the Main Idea
1. *Environment and Society* In what ways does the water supply shape rural economic activity in the region?

2. *Human Systems* What are some ways in which traditional cultures and ways of life have been retained in the region's cities?

Critical Thinking
3. **Supporting a Point of View** Which resource do you think is more important to the future of Southwest Asia—oil or water? Explain your answer.

4. **Comparing** How did the Taliban limit economic and educational opportunities for Afghan women? How do opportunities for women there compare to opportunities in other countries in the region?

Organizing What You Know
5. Copy the chart below and use it to describe the governments and recent political histories of Afghanistan, Iran, Iraq, and Saudi Arabia.

Afghanistan	
Iran	
Iraq	
Saudi Arabia	

Anthropology: Domestication in History

Domestication, the process of adapting plants and animals to provide an advantage to humans, was one of the first steps toward civilization. Archaeologists and anthropologists believe that many species common in the world today were first domesticated in Mesopotamia. Evidence suggests that wheat and barley were cultivated as early as 9000 B.C. Dogs and sheep were probably introduced into human society at about the same time. By 7000 B.C., pigs and goats had been tamed. However, it was not until the 200s B.C. in Persia that one of the most important animals in the history of this region, the camel, was domesticated.

Camels were vital to inhabitants of this region for many reasons. Most importantly, they could be ridden across the desert. These animals are able to go five to seven days without food and require little water. Their broad feet keep them from sinking into fine sands. Domestic camels provided food in the form of milk and meat. Camels have also been a central feature in desert society. For example, a Bedouin bride would traditionally depart for her husband's home riding on a camel.

Have students conduct research on other animals or plants that have been domesticated in Southwest Asia. Students should identify the period in which the species was tamed or cultivated, its uses, and any special significance it holds in the region's culture. As students complete their research, display their findings on a class time line or map.

Places and Regions

Geography for Life

The Ecological Trilogy

The relationships between living things and their environments are important to geographers. For example, geographers study the ways pollution affects ecosystems. They also study how the overuse of natural resources like rain forests and water affect the organisms that depend on them. The study of how living things interact with and depend on each other and the environment is called ecology. Geographers are also interested in human ecology—the ways human beings interact with and depend on the environment and each other. Southwest Asia is just one region in which geographers have studied human ecology.

In the 1960s a geographer developed a model he called the ecological trilogy. This model described how the three main ways of life in Southwest Asia depended on each other. Those three ways are life as a villager, as a pastoral nomad, and as a town- or city-dweller.

At the base of the trilogy were the peasant farmers who lived in the region's small villages. They served the city people and nomads by growing basic food crops like wheat and barley. Although they may not have wanted to do so, the villagers also provided the city with soldiers, tax money, and workers. On the other hand, cities and towns offered villages technological improvements, educational opportunities, and other services.

Nomads also had an important role in the trilogy. Nomads supplied villagers with animal products like cheese, meat, milk, and wool. They also provided desert herbs and medicines. City people supplied nomads with cooking utensils and factory-made clothing. However, city people and nomads interacted less than villagers and nomads.

New patterns of living in Southwest Asia have changed the relationships within the trilogy. In recent decades, government doctors, teachers, improved roads, cellular phones, and television have come to villages and nomads. Many villagers, particularly young people, move from rural areas to the city. When they arrive, the villagers change the culture of the city. Changing ways of life for nomads are also affecting this trilogy. As nomads settle in more permanent communities, the number of people who follow their herds declines. As the cities grow, they expand outward to engulf villages.

Although changes have occurred, each of the three parts of the ecological trilogy still supports the others. In addition, no matter where they may live now, individuals still identify themselves as city people, nomads, or villagers.

Applying What You Know

1. **Summarizing** Which part of the trilogy serves as a base for the model? Why?

2. **Supporting a Point of View** How might the ecological trilogy of Southwest Asia differ from a similar model for the United States?

The three parts of the ecological trilogy are shown in the scenes below. On the left, a farmer tends crops outside his village. In the center, nomadic herders take their flock to market. On the right, people live and work in a modern urban environment in Tehran.

Review and Assessment Resources

TECHNOLOGY

► Chapter 19 Test Generator
   (on the One-Stop Planner)
► Global Skill Builder CD–ROM
► HRW Go site

REINFORCEMENT, REVIEW, AND ASSESSMENT

► Chapter 19 Review, pp. 446–47
► Chapter 19 Tutorial for Students, Parents, Mentors, and Peers
► Chapter 19 Test (form A or B)
► Alternative Assessment Handbook

► Chapter 19 Test for English Language Learners and Special-Needs Students
► Unit 6 Test
► Unit 6 Test for English Language Learners and Special-Needs Students

Assess

Have students complete a Chapter 19 Test.

Reteach

Have students create an outline to help future classes learn about the region. Organize the class into groups and assign each group a small section of the chapter to outline. Tell students to pick important or interesting facts or information for inclusion and to note why these items are important. Review the finished outlines with the class.
ENGLISH LANGUAGE LEARNERS, COOPERATIVE LEARNING

CHAPTER 19 Review Answers

Understanding the Main Ideas

1. roots deep or spread, moisture stored in leaves or stems

2. available water

3. Iran—Farsi; Iraq—Arabic; Afghanistan—Pashto; Saudi Arabia—Arabic

4. specialized commercial farms becoming more common; many people giving up old subsistence farming and herding life, moving to cities, or working on commercial farms

5. key member of OPEC and thus can influence oil supplies and prices; also the caretaker of holy city of Mecca

Thinking Critically

1. Possible answers: expand desalinization facilities, but not practical in poorer countries or landlocked Afghanistan; limit use of fossil water; expand use of qanats, other methods for bringing mountain water to dry areas; limited fossil water supplies protected and future economic development aided with these policies (NGS 16)

2. Instability could disrupt oil transport, causing problems for industrial economies of developed countries. (NGS 11)

3. Islam is important throughout the region, but Afghanistan's Taliban applied a very strict interpretation of Islamic teaching compared to other governments. Islam also plays an important role in

---

CHAPTER 19 Review

Building Vocabulary

On a separate sheet of paper, explain the following terms by using them correctly in sentences.

exotic rivers	Shi'ism	OPEC
oasis	imams	ayatollahs
dynasty	Bedouins	theocracy
Sunni		

Locating Key Places

On a separate sheet of paper, match the letters on the map with their correct labels.

Persian Gulf	Shatt al Arab	Mecca
Mesopotamia	Elburz Mountains	Baghdad
Tigris River	Rub' al-Khali	Riyadh

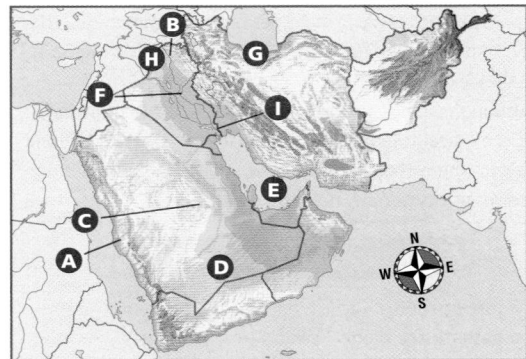

Understanding the Main Ideas

Section 1

1. *(Physical Systems)* In what ways has the plant life of the region adapted to the conditions there?

Section 2

2. *(Environment and Society)* Why has Mesopotamia been such an attractive target for invasion throughout history?

3. *(Places and Regions)* What is the major language spoken in Iran? In Iraq, Afghanistan, and Saudi Arabia?

Section 3

4. *(Human Systems)* What changes are taking place in the traditional rural economy and culture of the region?

5. *(Human Systems)* What are two important factors accounting for Saudi Arabia's influence in world affairs today?

Thinking Critically

1. **Evaluating** What policies would you recommend to governments working to protect water resources in the region? Why would those policies not be practical for some countries? What geographic and economic effects might these policies have?

2. **Drawing Inferences and Conclusions** Why might stability in the Persian Gulf region be important to developed countries?

3. **Comparing and Contrasting** How do views about Islamic teachings and the role of Islam in society vary among governments in the region? How do cultural beliefs influence public policies regarding women in Afghanistan?

Using the Geographer's Tools

1. **Analyzing Maps** Review the unit maps to study the climates and land use and resources of Iraq. In which climate region do you see large areas of commercial farming? What is the probable source of water for commercial farming there?

2. **Analyzing Statistics** Use the unit Fast Facts table to rank the countries of the region by GDP, from highest to lowest. Then prepare a second list, ranking them by literacy rate, from highest to lowest. Compare the two rankings. What does this comparison reveal about the relationship between educational level and a country's level of economic development?

3. **Creating a Population Pyramid** Use the HRW Web site to find statistics about Iran's population. Use those statistics to create a population pyramid that describes the population characteristics of Iran. Then predict future growth trends there.

Writing about Geography

Imagine that you are part of a Bedouin family that is about to move to a city in Saudi Arabia. Write a short journal entry describing how your way of life is about to change. After you have completed your journal entry, proofread it to make sure that you have used standard grammar, spelling, sentence structure, and punctuation.

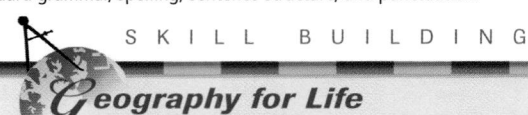

SKILL BUILDING

Geography for Life

Organizing Geographic Information

(Human Systems) Use Internet or library resources to identify the largest sources of oil imports into the United States. Also investigate how the amount of imported oil compares to the total produced from wells in the United States. Then plan, design, and create a graph that shows your findings.

Have students conduct research on the conventions of Islamic art. One such convention is a general prohibition against the depiction of animals or people. Then have students create a "name poem" of an important person, place, or term associated with the Persian Gulf area or interior Southwest Asia and decorate it in an Islamic style, with plant forms, geometric patterns, or calligraphy. A name poem arranges the letters of the name vertically. For each letter, the student provides a phrase that pertains to the poem's subject and begins with the letter.

Food Festival

Yogurt, or *maast* in Farsi, is a common ingredient in Iranian cooking. Many students like yogurt, but they probably prefer a sweet fruity version. Challenge students to make plain yogurt at home, or prepare it in the classroom in a microwave oven. To do so, heat a quart of milk until it starts to boil. Allow the milk to cool until it is still very warm to the touch. Add one tbs. of live-culture yogurt and mix lightly. Cover, wrap in towels, and keep in a warm place so that the mixture stays at about 110°F (43°C) for several hours. Then chill well. You may want to have students locate Iranian recipes using yogurt and prepare samples for the class.

Iran's government, a theocracy. The Taliban limited educational and economic opportunities for women, but this is changing. (NGS 10)

Building Social Studies Skills

Major Earthquakes in Iran (1978–2003)

Date	Location	Estimated number of deaths
Sept. 16, 1978	Northeastern Iran	15,000
June 11, 1981	Southern Iran	3,000
July 28, 1981	Southern Iran	1,500
June 20, 1990	Western Iran	40,000+
Feb. 28, 1997	Northwestern Iran	1,000+
May 10, 1997	Northern Iran	1,560
Dec. 26, 2003	Southeastern Iran	30,000+

Source: *World Almanac and Book of Facts 2003; New York Times*

Interpreting Charts

Study the chart above. Then answer the questions that follow.

1. Which region of Iran suffered the most earthquakes during the period?
 a. northern Iran
 b. southern Iran
 c. southwestern Iran
 d. southeastern Iran

2. On what dates did Iran suffer its three deadliest earthquakes during this period? List the earthquakes from deadliest to least deadly.

Building Vocabulary

To build your vocabulary skills, answer the following questions. Mark your answers on a separate sheet of paper.

3. *Theocracy* means the same as
 a. a country governed by religious law.
 b. a population that shares a common cultural background.
 c. the policy of gaining control over territory outside a country.
 d. a movement that stresses the strict following of basic traditional principles.

4. *Fossil water* means the same as
 a. water surrounded by the bones of ancient dinosaurs.
 b. Persian Gulf water processed into freshwater.
 c. groundwater not being replaced by rainfall.
 d. lake water supplied by streams and old rivers.

Using the Geographer's Tools

1. arid climate; irrigation from Tigris and Euphrates Rivers

2. Higher literacy rates and higher per capita GDPs are generally directly related.

3. should show that Iran has many young people, which suggests rapid population growth in the future

Writing

Entries should reflect an understanding of Bedouin lifestyles and of the region's cities. Use Rubric 15, Journals, to evaluate student work.

Geography for Life

Graphs should accurately display the gathered information. Use Rubric 7, Charts, to evaluate student work.

Social Studies Skills

1. b

2. June 20, 1990; December 26, 2003; September 16, 1978

3. a

4. c

PORTFOLIO ACTIVITY

Projects will vary but should accurately depict elements of the local economy. Use Rubric 14, Group Activity, to evaluate student work.

Alternative Assessment

PORTFOLIO ACTIVITY

Learning about Your Local Geography

Group Project: Field Work

The handmade rugs of Southwest Asia are famous throughout the world. Plan, organize, and complete a research project with a partner to discover what craft industries are important in your community. Conduct a survey to determine what craft items are produced there. Where are these items sold? Is your community widely known for any crafts? Present your findings in a report. Include illustrations of some of the community's craft products. When you have completed your report, proofread it to make sure that you have used standard grammar, spelling, sentence structure, and punctuation.

internet connect

Internet Activity: go.hrw.com
KEYWORD: SW3 GT19

Choose a topic on the Persian Gulf and interior to:
 • report on Islamic culture.
 • research oil productivity and transfer your information into a graph or chart.
 • understand the technology of desalinization.

CHAPTER RESOURCE MANAGER

Objectives	Pacing Guide	Reproducible Resources
SECTION 1 **Natural Environments** (pp. 449–51) • Describe the landforms and rivers found in the eastern Mediterranean region. • Identify the region's climates, biomes, and natural resources.	**Regular** 1 day **Block Scheduling** .5 day *Block Scheduling Handbook, Chapter 20*	**RS** Guided Reading Strategy 20.1
SECTION 2 **History and Culture** (pp. 454–58) • Identify the ways that various peoples and empires have influenced the eastern Mediterranean. • Analyze how the modern state of Israel developed. • Describe the region's peoples and cultures.	**Regular** 1 day **Block Scheduling** .5 day *Block Scheduling Handbook, Chapter 20*	**RS** Guided Reading Strategy 20.2 **RS** Graphic Organizer Activity 20 **PS** Readings in World Geography, History, and Culture 52 **E** Cultures of the World Activity: Region 5
SECTION 3 **The Region Today** (pp. 460–63) • Describe the activities on which the eastern Mediterranean economies rely. • Identify characteristics of the region's cities. • Analyze the challenges faced by the region's people.	**Regular** 1 day **Block Scheduling** .5 day *Block Scheduling Handbook, Chapter 20*	**RS** Guided Reading Strategy 20.3 **SM** Critical Thinking Activity 20: Oil in the Middle East **SM** Geography for Life Activity 20: Site and Situation Shape Byzantium/Constantinople/İstanbul **SM** Map Activity 20: Jerusalem **PS** Readings in World Geography, History, and Culture 50–52

Chapter Resource Key

PS	Primary Sources	**A**	Assessment	💿	CD-ROM
RS	Reading Support	**REV**	Review	📼	Video
IC	Interdisciplinary Connections	**ELL**	Reinforcement and English Language Learners	Internet	
E	Enrichment		Transparencies		Holt Presentation Maker Using Microsoft® PowerPoint®
SM	Skills Mastery				

One-Stop Planner CD–ROM

See the *One-Stop Planner* for a complete list of additional resources for students and teachers.

One-Stop Planner CD–ROM

It's easy to plan lessons, select resources, and print out materials for your students when you use the **One-Stop Planner CD–ROM with Test Generator**.

Technology Resources

- One-Stop Planner CD–ROM, Lesson 20.1
- Geography and Cultures Visual Resources 37–41
- Homework Practice Online
- HRW Go site

- One-Stop Planner CD–ROM, Lesson 20.2
- Geography and Cultures Visual Resources 42
- Homework Practice Online
- HRW Go site

- One-Stop Planner CD–ROM, Lesson 20.3
- *ARGWorld* CD–ROM
- CNN. Presents Geography: Yesterday and Today, Segment 21: Jordan's Water Crisis
- Homework Practice Online
- HRW Go site

Reinforcement, Review, and Assessment

- **ELL** Main Idea Activity 20.1
- **ELL** English Audio Summary 20.1
- **ELL** Spanish Audio Summary 20.1
- **REV** Section 1 Review, p. 451
- **A** Daily Quiz 20.1

- **ELL** Main Idea Activity 20.2
- **ELL** English Audio Summary 20.2
- **ELL** Spanish Audio Summary 20.2
- **REV** Section 2 Review, p. 458
- **A** Daily Quiz 20.2

- **ELL** Main Idea Activity 20.3
- **ELL** English Audio Summary 20.3
- **ELL** Spanish Audio Summary 20.3
- **REV** Section 3 Review, p. 463
- **A** Daily Quiz 20.3

✓ internet connect

HRW ONLINE RESOURCES

GO TO: go.hrw.com
Then type in a keyword.

TEACHER HOME PAGE
    KEYWORD: SW3 Teacher

CHAPTER INTERNET ACTIVITIES
    KEYWORD: SW3 GT20
    Choose an activity to:
    • research Jerusalem and create a newspaper.
    • send a postcard from the Dead Sea.
    • understand the sources of conflict and efforts to relieve it in Cyprus.

CHAPTER ENRICHMENT LINKS
    KEYWORD: SW3 CH20

CHAPTER MAPS
    KEYWORD: SW3 MAPS20

ONLINE ASSESSMENT
    Homework Practice
    KEYWORD: SW3 HP20
    Standardized Test Prep
    KEYWORD: SW3 STP20
    Rubrics
    KEYWORD: SS Rubrics

COUNTRY INFORMATION
    KEYWORD: SW3 Almanac

CONTENT UPDATES
    KEYWORD: SS Content Updates

HOLT PRESENTATION MAKER
    KEYWORD: SW3 PPT20

ONLINE READING SUPPORT
    KEYWORD: SS Strategies

CURRENT EVENTS
    KEYWORD: S3 Current Events

Meeting Individual Needs

Ability Levels

Level 1 Basic-level activities designed for all students encountering new material

Level 2 Intermediate-level activities designed for average students

Level 3 Challenging activities designed for honors and gifted-and-talented students

English Language Learners Activities that address the needs of students with Limited English Proficiency

Chapter Review and Assessment

- Chapter 20 Test Generator (on the One-Stop Planner)
- Global Skill Builder CD–ROM
- HRW Go site
- **REV** Chapter 20 Review, pp. 466–67
- **REV** Chapter 20 Tutorial for Students, Parents, Mentors, and Peers
- **A** Chapter 20 Test (form A or B)
- **A** Alternative Assessment Handbook
- **A** Chapter 20 Test for English Language Learners and Special-Needs Students

Launch into Learning

On a wall map, point out that the eastern Mediterranean region is located where three continents meet, at one end of a major sea, and at the entrance to an inland sea. Ask students to speculate how the location may have affected the region's cultural, economic, and political history. *(Possible answers: Cultural influences could flow easily to and from the region; the region is well situated for trade with many peoples; and the region was in the path of conquerors or ideas.)* Point out that, in addition, Turkey has endured many disastrous earthquakes. All these factors add up to a land with a rich but volatile history,

a mosaic of cultures, and an uncertain future. Tell students they will learn more about these topics in this chapter.

Using the Physical-Political Map

Have students examine the map on the opposite page. Call attention to the shapes of the countries and the landforms they occupy. Point out that there are few natural boundaries. Ask students what they already know about boundary changes in the region during the 1900s that resulted from geopolitical events. *(Possible answers: changes after the world wars, those related to Arab-Israeli conflicts)*

Why We Should Know More

You might point out to students that there are many reasons why we should know more about the countries of the eastern Mediterranean, such as these:

▶ The eastern Mediterranean region is a geographical, economic, and cultural crossroads. The region has made numerous contributions to world civilization. It has also been the scene of many conflicts over the centuries.

▶ Judaism and Christianity began in the region. Both faiths have deeply affected American culture.

▶ The United States has supported Israel since its creation. This policy has contributed to anti-American feelings and actions among some Arabs.

▶ Turkey is a close ally of the United States and has been a member of NATO since 1952.

CHAPTER 20 The Eastern Mediterranean

Complex relationships among geography, history, and religion are typical in the eastern Mediterranean region. All of the region's countries have been involved in conflicts both long ago and in recent years.

Pitcher from Syria

Bedouin woman from Israel

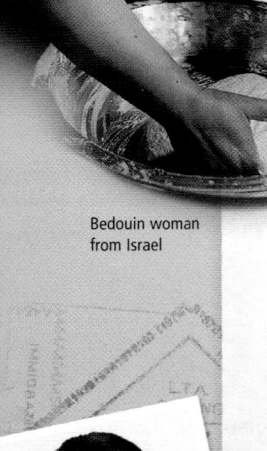

Türkiye'den selamlar! (Greetings from Turkey!) My name is Adalet, and I am in the tenth grade at a private school in Ankara, the capital of Turkey. I live in a high-rise apartment outside the city with my mom and dad. We have a view of the city, the distant mountains, and, of course, the parking lot. In the summers, I like to stay with my grandparents in their summer house on the Aegean Sea, near the ancient Greek and Roman ruins at Ephesus. I go to the beach there with my friends and stay until sundown. Back at the house, my grandpa and grandma love to cook delicious Turkish food like *kofte,* a spicy ground lamb dish.

On school days I get up at 8:00 A.M., put on my school uniform, and eat cornflakes or bread and cheese. The parents of boys and girls in my neighborhood have hired a bus to take us to school. It is a big school that goes from first grade through high school with about 800 students per grade. I am studying biology, physics, algebra, geometry, history, Turkish, and English. The science and math classes (and the English class, of course) are taught in English, the others in Turkish. At lunch my friends and I go to a little food stand nearby and eat hot dogs or grilled lamb on a skewer with bread, tomatoes, onions, and peppers. To drink, we have *ayran,* a drink made with yogurt.

Section 1

OBJECTIVES

1. Describe the landforms and rivers found in the eastern Mediterranean region.

2. Identify the region's climates, biomes, and natural resources.

LET'S GET STARTED

Copy the following questions onto the chalkboard: *Have you ever experienced an earthquake? If so, what happened? What did you do? If not, how do you think you would react if an earthquake struck?* Discuss responses. Tell students that the eastern Mediterranean region's physical geography has been shaped partly by earthquakes and that they will learn more about the region's natural environments in Section 1.

Building Vocabulary

Write **potash** and **magnesium** on the chalkboard and have students find the definitions in the text or glossary. Point out that the word *potash* literally means "pot ash" because the mineral was once prepared from wood ash boiled in pots. The word *magnesium* probably comes from Magnesia, a region in what is now Greece, where deposits of a related mineral were known many centuries ago.

Natural Environments

READ TO DISCOVER

1. What landforms and rivers are found in the eastern Mediterranean region?
2. What climates, biomes, and natural resources does the region have?

WHY IT MATTERS

In 2000 a new reservoir threatened to submerge important historical sites in Turkey. Use or other **current events** sources to learn about the many ways that dams and reservoirs affect places around the world.

DEFINE

potash
magnesium

LOCATE

Anatolia
Dardanelles
Bosporus
Sea of Marmara
Pontic Mountains
Taurus Mountains

Jordan River
Dead Sea
Tigris River
Euphrates River
Syrian Desert
Negev

Section 1 RESOURCES

REPRODUCIBLE
▶ Guided Reading Strategy 20.1

TECHNOLOGY
▶ One-Stop Planner CD–ROM, Lesson 20.1
▶ Geography and Cultures Visual Resources 37–41
▶ Homework Practice Online
▶ HRW Go site

REINFORCEMENT, REVIEW, AND ASSESSMENT
▶ Main Idea Activity 20.1
▶ English Audio Summary 20.1
▶ Spanish Audio Summary 20.1
▶ Section 1 Review, p. 451
▶ Daily Quiz 20.1

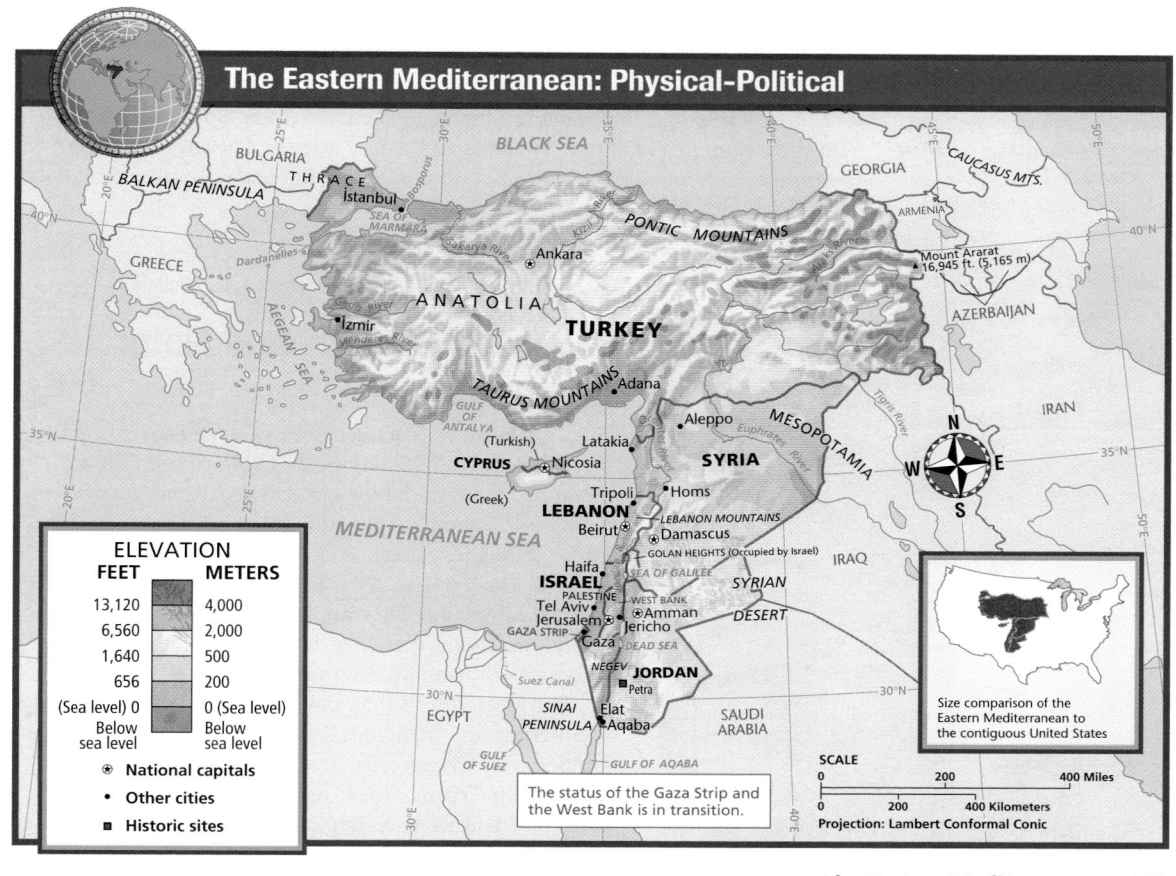

The Eastern Mediterranean: Physical-Political

ELEVATION

FEET	METERS
13,120	4,000
6,560	2,000
1,640	500
656	200
(Sea level) 0	0 (Sea level)
Below sea level	Below sea level

⊛ National capitals
• Other cities
■ Historic sites

Size comparison of the Eastern Mediterranean to the contiguous United States

The status of the Gaza Strip and the West Bank is in transition.

SCALE
0   200   400 Miles
0   200   400 Kilometers
Projection: Lambert Conformal Conic

Teach Objective 1

ALL LEVELS: Provide students with outline maps of the eastern Mediterranean on which you have drawn lines connecting these places: İstanbul to Nicosia, Ankara to Tel Aviv, Mount Ararat to Damascus, and Gaza to the easternmost point of Jordan's border with Iraq. Organize the class into four groups and assign one of the lines to each group. Have students identify the landforms and rivers they would encounter if they were to travel along the assigned line. Ask a representative of each group to share the information with the class. You may want to have students write brief travelogues of their "journeys." **ENGLISH LANGUAGE LEARNERS, COOPERATIVE LEARNING**

Teach Objective 2

ALL LEVELS: Copy the following graphic organizer onto the chalkboard, omitting the italicized answers. Have students complete it. Then have students write paragraphs to compare and contrast the region's natural environment to their own. **ENGLISH LANGUAGE LEARNERS**

Natural Environments of the Eastern Mediterranean		
Climates	**Biomes**	**Resources**
• *arid* • *semiarid* • *Mediterranean*	• *forests, particularly in Cyprus* • *diverse in semiarid areas* • *hardy species in desert areas*	• *Turkey—coal, copper, iron ore* • *Dead Sea—potash, magnesium*

Eye on Earth

Negev Wildlife The Negev is an arid climate area in the eastern Mediterranean region. By the 1960s many native species that roamed the Negev during biblical times had become endangered or could no longer be found there. Conservationists launched programs to protect and reintroduce wildlife.

Today, three fourths of the Negev has been designated as nature preserves. Visitors can see foxes, gazelles, hyenas, ostriches, wolves, and zebras. The onager, a type of wild donkey, and the oryx, a type of antelope, have also returned to the Negev area. With more prey available, the leopard again prowls the landscape. Because it is at the top of the area's food chain, the leopard's return shows that the full range of the Negev's wildlife is being restored.

You don't say! Because its hump contains fat, a camel's energy reserves allow it to travel long distances without food or water. When it does drink, a camel can gulp about 20 gallons (76 liters) within 10 minutes, an amount that would kill any other mammal.

VISUAL RECORD ANSWER

Possible answer: cause severe problems for tourist industry as the water evaporates and recedes farther away from shoreline hotels

internet connect

GO TO: go.hrw.com
KEYWORD: SW3 CH20
FOR: For Web sites about the eastern Mediterranean

INTERPRETING THE VISUAL RECORD

Top: A tourist floats in the Dead Sea. Bottom: The Dead Sea lies in a desert valley. The shoreline has receded nearly a mile in the last 40 years. Only 10 percent of the Jordan River's water flows into the sea. Growing populations and drought conditions place heavy demands on the Jordan. **How might a shrinking Dead Sea affect the area's economy?**

Landforms and Rivers

The eastern Mediterranean region is part of an area often called the Middle East. It consists of six countries. Israel, Jordan, Lebanon, Syria, and Turkey are on the mainland. Cyprus is an island in the Mediterranean Sea.

The region lies on two continents. A small part of Turkey is on Europe's Balkan Peninsula. Most of Turkey is in Asia, an area known as Anatolia (a-nuh-TOH-lee-uh). Three narrow connected bodies of water—the Dardanelles (dahrd-uhn-ELZ), the Bosporus, and the Sea of Marmara (MAHR-muh-ruh)—separate Europe from Asia.

European Turkey has plains and hills. A narrow coastal plain rims the western edge of the Asian part of Turkey. Two mountain systems run from east to west across Anatolia. They include the Pontic Mountains in the north and the Taurus Mountains in the south. Between them lies the Anatolian Plateau, which has many peaks and valleys of its own. Turkey's geology includes many faults and folds. Along with continued mountain building, these geological features also cause devastating earthquakes.

The coastal plain continues south from Turkey along the coasts of Syria, Lebanon, and Israel. Farther inland one finds plateaus, hills, and valleys. A rift valley extends northward from Africa into Syria. Hills rise on both sides of the rift. Between the rift's ridges the Jordan River flows south into the Dead Sea. This unusual sea lies 1,312 feet (400 m) below sea level. Its shore is the lowest land on Earth's surface. The Dead Sea was once part of the Mediterranean. Today it no longer has an outlet to the ocean. The sea is so salty that all swimmers can easily float in it.

The Tigris and Euphrates are the region's major rivers. Both rivers begin in the mountains of eastern Anatolia and empty into the Persian Gulf. Along with the Jordan River, the Tigris and Euphrates are important sources of irrigation water.

✓ **READING CHECK:** **Physical Systems** Why does Turkey have many earthquakes? many faults and folds and continued mountain building

Climates, Biomes, and Natural Resources

Arid, semiarid, and Mediterranean climates cover nearly all of the eastern Mediterranean. Distance from the sea, elevation, and rain shadows affect rainfall and temperatures. For example, like the nearby Greek islands, Turkey's Mediterranean coast has a mild and sunny climate. However, eastern Turkey and the Anatolian interior lie inland. They are also at higher elevations. Therefore, they experience bitterly cold winters, and heavy snowstorms are common.

Evergreen forests once covered much of the eastern Mediterranean's highlands. Lebanon, in particular, was famous for its great cedar trees. Areas of forested land still exist in Cyprus, where cedar, cypress, and pine are common. Forests have largely disappeared elsewhere,

Forests of magnificent cedar trees were once common in Lebanon. In fact, the cedar is pictured on Lebanon's flag. However, people began cutting down the big trees long ago. Now few remnants of the great forests remain, but reforestation efforts are underway.

however. They are victims of centuries of farming, herding, shipbuilding, and firewood collecting that began more than 2,000 years ago.

In semiarid areas like Anatolia, plant and animal life can be diverse. Farther south, the Syrian Desert covers much of both Jordan and Syria. Another desert, the Negev (NE-gev), lies in southern Israel. In desert biome areas, plant and animal life is scarce. Yet some hardy species, such as jackals, lizards, and snakes, are able to survive. People also live in the desert. Small groups of nomads move their flocks of sheep and goats with the seasons.

Valuable mineral and other natural resources can be found throughout the region. Turkey has supplies of coal, copper, and iron ore. Some oil and natural gas deposits are distributed across the region. The Dead Sea also provides Israel and Jordan with certain minerals, including **potash** and **magnesium**. Potash is used to process wool and to make fertilizers, glass, and soft soaps. Magnesium is a light metal that is valuable in certain industries, such as aerospace.

Our Amazing Planet

In the Cappadocia region of central Turkey, volcanic rock has eroded into an unusual landscape of columns, cones, pillars, and towers. People carved into the rock to create shelters. Some of the caves are now homes, stores, and churches.

✓ **READING CHECK:** *Places and Regions* Why do eastern Turkey and the Anatolian interior have cold winters? lie at higher elevations inland, cut off from Mediterranean's warming effects

Section 1 Review

Homework Practice Online
Keyword: SW3 HP20

Define
potash
magnesium

Working with Sketch Maps On a map of the eastern Mediterranean that you draw or that your teacher provides, label Anatolia, Dardanelles, Bosporus, Sea of Marmara, Pontic Mountains, Taurus Mountains, Jordan River, Dead Sea, Tigris River, Euphrates River, Syrian Desert, and the Negev. What parts of the region would you expect to be densely populated?

Reading for the Main Idea

1. *Places and Regions* What are the main landforms of Turkey? Which two major rivers begin in Turkey?

2. *Physical Systems* What has happened to the region's evergreen forests?

3. *Physical Systems* What factors affect rainfall and temperatures in the region?

Critical Thinking

4. **Analyzing Information** Why might the Dead Sea be considered a natural resource?

Organizing What You Know

5. Create a chart that lists the countries in the region and the different landforms found in each. Refer back to Section 1 and the chapter's physical-political map.

Country	Landforms

Setting the Scene

Have students read Case Study: Palestine—Setting Boundaries. Then point out that throughout history, conflicts over borders have been extremely contentious. In fact, many wars have been fought over which country, ethnic group, or individual would control territory. There are many examples. More than once Poland's borders have actually been erased. That is, Poland as a separate country disappeared from the map when other countries conquered it or divided it among themselves. Germany and France have repeatedly fought over a strip of land called Alsace-Lorraine. The United States has had border disputes with both Canada and Mexico.

Building a Case

Organize the class into five groups. Have each group design a new continent—labeled with mountains, plains, rivers, deserts, climate regions, and other physical features—on a large sheet of paper. Maps should then pass to another group, which will add and label rural and urban settlements of various ethnic and religious affiliations and socioeconomic conditions. After passing their maps again, the next group should add political boundaries to create individual countries. Have groups pass the maps again and write down two or three events that affect certain areas, such as an invasion, drought, or discovery of gold. Finally, groups should pass along the maps once more. Instruct the groups to predict and to write down how the events might create border conflicts.

GO TO: go.hrw.com
KEYWORD: SW3 CH20
FOR: Web sites about Israel's history

This Case Study feature addresses National Geography Standards 6, 12, and 18.

CASE STUDY

Palestine–Setting Boundaries

Human Systems   Perhaps one of the most persistent themes in human history has been the struggle over territory. Today the world is divided into almost 200 countries. As you read in Chapter 6, a variety of boundaries separate these countries. Some are natural boundaries, such as mountains and rivers. Others are cultural boundaries. The border between mostly Hindu India and mostly Muslim Pakistan is an example of a cultural boundary. Finally, some boundaries are geometric, such as lines of latitude.

Sometimes the competing desires of two or more peoples to control a piece of land have been settled peacefully. However, such disputes have also led to war. The reasons for this are many. They are often cultural, reflecting the strong ties a particular people have to a place. They can also be economic, reflecting the desire for the natural resources found in an area. Both of these factors have played a part in the long Arab-Israeli conflict.

Conflict and Cooperation

Palestine is an old Greek name for the eastern edge of the Mediterranean. It includes the modern country of Israel, the West Bank, and the Gaza Strip. For Jews and Arabs there alike, Palestine is the land of their ancestors. It is also home to many historical and cultural sites held sacred by millions of people. Many of the

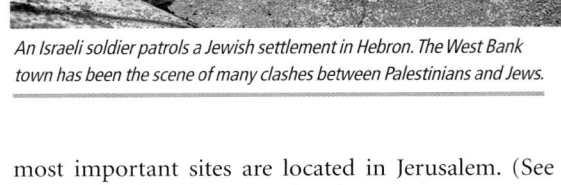

An Israeli soldier patrols a Jewish settlement in Hebron. The West Bank town has been the scene of many clashes between Palestinians and Jews.

most important sites are located in Jerusalem. (See Cities & Settlements: Jerusalem.)

Hundreds of thousands of Arabs left Palestine after Israel declared independence in 1948. Most settled in the West Bank, Gaza Strip, and neighboring Arab countries such as Jordan. Threatened by surrounding Arab armies, Israel captured the West Bank and Gaza Strip during the so-called Six-Day War in 1967. Many Israelis thought that having control over these territories was vital to their country's security. Arabs demanded that the Israelis withdraw. The dispute over the territories grew more heated and violent over time.

In the 1990s Israel's government and Arab Palestinian leaders tried to end their long dispute through negotiation. They discussed many issues. Among them was how to draw a boundary between Israel and any new Palestinian state in the occupied territories. The two sides made significant progress. In fact, over time Israel returned parts of the West Bank and Gaza Strip to Palestinian control. However, the talks between Israelis and Palestinians bogged down in 2000. Later that year, violence again erupted, taking many lives. The violence continues to this day.

A Palestinian woman pumps water from an aquifer. Control of water resources is a primary source of conflict in the West Bank.

Drawing Conclusions

Lead a discussion about the factors—direct and indirect—that created border conflicts in students' continents. *(Possible answers: expanding populations wanting more land, people flooding into a neighboring country to escape natural disasters, domestic political pressures to enlarge territory, ethnic groups wanting autonomy, citizens of one country wanting access to resources in another, and so on)* Include in the discussion the role the continent's natural environment played in the development of conflicts.

Going Further: Thinking Critically

Have groups pass their maps along once more. Ask each group to propose solutions for the border conflicts identified on the map it receives. Encourage students to develop compromises to which the conflicting parties might agree. **COOPERATIVE LEARNING**

Aquifers

West Bank

SCALE
0   25 Miles
0   25 Kilometers

SCALE
0   25   50 Miles
0   25   50 Kilometers
Projection: Transverse Cylindrical

Legend:
- Israel
- Major aquifers
- Israeli-occupied land
- Palestinian controlled
- Current and planned Israeli settlements
- Divided control
- ⊛ National capitals
- • Other cities

INTERPRETING THE MAPS *The debate over new boundaries between Israel and a possible new Palestinian country involves a number of issues. Control of water resources in the dry region is one of them. Aquifers extend across the region. New boundaries will determine who controls the water stored in these aquifers. The West Bank would become part of the new country of Palestine. The detailed map of the West Bank shows current and planned Israeli settlements as well as areas of Palestinian control.* **Why do you think establishing stable borders around Israeli settlements might be difficult? How might Israelis and Palestinians compromise over control of the region's aquifers?**

Working toward Compromise

Many issues divided Israelis and Palestinians when the talks bogged down in 2000. Three primary issues involved Jewish settlements in the occupied territories, the status of Jerusalem, and access to water.

The settlement issue is a result of Israel's government allowing Jewish settlers to set up communities in the occupied territories. Today these settlements are scattered between Palestinian-controlled lands in the West Bank and Gaza Strip. The two sides cannot agree on what to do with those settlements.

Another issue is control of Jerusalem. The city had been divided between Israel and Jordan from 1948 to 1967. It was reunited under Israeli rule during the 1967 war. Jerusalem is a difficult issue because Jewish, Muslim, and Christian religious sites are found there. In addition, both Israelis and Palestinians want the city as their capital.

Finally, water is an issue. This dry region has limited water supplies. Aquifers are a crucial water source. One large aquifer straddles the dividing line between Israel and the West Bank. Mountain rainfall,

mostly from the West Bank, supplies much of the aquifer's water. However, much of that stored water lies on the Israeli side of the aquifer. Agreement over access to the aquifer and the use of its water has been difficult to reach. In addition, Israel wanted to control land along the Jordan River, another important water source. This land would also act as a buffer zone between the West Bank and Jordan, an Arab country.

To the north the Sea of Galilee presented a similar problem. The Sea of Galilee lies along the old border between Israel and Syria. However, in the 1967 war Israel captured a Syrian area that included part of the freshwater sea's shore. In addition, springs in the area replenish the sea. Syria has demanded that Israel withdraw from this land before agreeing to a final peace treaty. However, Israelis fear that doing so would limit their supply of water from the sea.

These issues have long divided Israelis and their Arab neighbors. Until they are resolved, new boundaries dividing their countries will remain in question.

Applying What You Know

1. **Summarizing** What are some of the issues that make agreement over new boundaries between Israel and a Palestinian state difficult to reach?

2. **Comparing** Why might both Israelis and Palestinians want Jerusalem as their capital? What kind of compromise do you think might settle this issue?

OBJECTIVES

1. Identify the ways that various peoples and empires have influenced the eastern Mediterranean.

2. Analyze how the modern state of Israel developed.

3. Describe the region's peoples and cultures.

 **LET'S GET STARTED**

Copy the following passage onto the chalkboard: *The following words display evidence of cultural exchange between the eastern Mediterranean and the West: attar, damask, falafel, shalom, shish kebab. Which of them do you know?* Discuss responses. Call on volunteers to read the definitions and etymologies of the words—all of which come from languages of the eastern Mediterranean—from a dictionary. Tell students they will learn more about this cultural exchange in Section 2.

Building Vocabulary

Write **sultans** and **mandates** on the chalkboard and have volunteers read the definitions aloud from the text or glossary. Explain that the word *sultan* is derived from the Arabic word that means "authority" and that it was used to describe the leaders of many Islamic countries, not just the Ottoman Empire. Point out also that a mandate was originally a contract between two people, in which free services were performed in order to excuse debts. Ask students to describe other ways they have heard these words used.

VISUAL RECORD ANSWER

located at the crossroads of Asia and Europe

Section 2 History and Culture

READ TO DISCOVER

1. How have various peoples and empires influenced the eastern Mediterranean?
2. How did the modern state of Israel develop?
3. What are the peoples and cultures of the region like?

WHY IT MATTERS

Many people go to Israel to visit places important to Christianity, Islam, or Judaism. Use **CNNfyi.com** or other **current events** sources to learn about those sites.

IDENTIFY

Zionism   Kurds
Holocaust

DEFINE

sultans
mandates

LOCATE

İstanbul   Jerusalem
Ankara   Nicosia

The Rise and Fall of Empires

Some of history's earliest civilizations developed in the eastern Mediterranean region. Farming supported the growth of permanent settlements as far back as 8000 B.C. Over time, the Egyptians, the Hittites of Asia Minor, the Persians, and others ruled all or parts of the region. In about 1000 B.C. a people called the Hebrews set up a kingdom between the Jordan River and the Mediterranean Sea. That area has been known as Palestine and Israel at different times. The Hebrews practiced Judaism, which is the dominant religion in Israel today.

The Roman and Byzantine Empires The Romans conquered the eastern Mediterranean between 200 B.C. and A.D. 106. During this time, most Jews were exiled from Palestine because they resisted Roman authority. In the first century A.D., followers of Christianity began to spread their beliefs from the eastern Mediterranean throughout the Roman Empire. By the late 300s Christianity was the empire's official religion.

Our Amazing Planet

The Phoenicians were an early seafaring people from what is now Lebanon. Although their homeland was small, they traded as far as Spain and may even have sailed around Africa.

INTERPRETING THE VISUAL RECORD

These gigantic heads at Nemrud Dagh, now in eastern Turkey, are more than 2,000 years old. They lie in what was the kingdom of Commagene, where Greek, Persian, and Roman influences mixed. **How do you think physical geography contributed to the combination of cultural influences at Commagene?**

Teach Objective 1

ALL LEVELS: Copy the following graphic organizer onto the chalkboard, omitting the italicized answers. Have students complete it. Then have students design a flag for one of the eras. Flags should include representations of the era's main characteristics. **ENGLISH LANGUAGE LEARNERS**

Peoples and Empires in the Eastern Mediterranean

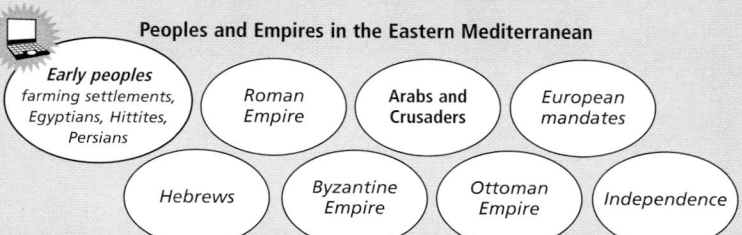

Teach Objective 2

ALL LEVELS: Compile several articles—either current or historical—from newspapers and magazines about events related to the statehood of Israel or the Arab-Israeli conflict. Pair students and give each pair two copies of one of the articles. Have each student place his or her article's events in a sequence, along with related events described in the text. Then lead a discussion about any bias students may detect in the articles. **COOPERATIVE LEARNING**

HOMEWORK: Have students write and illustrate children's books to explain how the modern state of Israel developed.

During the A.D. 400s the Western Roman Empire crumbled. The eastern part of the empire survived, however, and came to be known as the Byzantine (BIZ-uhn-teen) Empire. Christianity also became divided. The Eastern Orthodox Church broke away from the Roman Catholic Church. Constantinople—now İstanbul—was both the center of the Orthodox Church and the capital of the Byzantine Empire. It became a city of great beauty, power, and wealth. The Byzantine Empire ruled most of the eastern Mediterranean as well as areas far beyond the region. However, other peoples soon invaded, shrinking the empire.

The Arabs and Islam In the A.D. 600s Arab Muslim armies swept north out of the Arabian desert. They rapidly established an empire in southwestern Asia and northern Africa. The Byzantines lost much territory as a result, including the area that is now Israel.

Turkic Muslims captured Jerusalem from the Arabs in 1077. When Turks threatened Constantinople, the pope called for Christians to go to war against the Muslims. This series of wars was known as the Crusades. Between 1095 and the late 1200s, crusader armies from all over Europe invaded the region. For a while, the Crusaders held Jerusalem and some cities in Syria. However, Muslim armies later forced them out.

The Ottoman Empire In the 1300s the Ottoman Turks established another Muslim empire in the region. The Ottoman rulers were called **sultans**. They took Constantinople in 1453 and made it the Ottoman capital. By the 1600s the Ottoman Empire included most of Southwest Asia. The empire also stretched into parts of eastern Europe and most of North Africa. However, political struggles, corruption, and rivalries with other countries slowly weakened the empire. Over time ethnic minorities within the empire's borders also began to push for independence.

During World War I the Ottoman Empire fought on the losing side. After the war, a general named Mustafa Kemal (later known as Atatürk) took over the government. He created the Republic of Turkey and established its capital at Ankara. Beyond Turkey, the former Ottoman territories became **mandates** of Great Britain and France. These mandates were territories placed under another country's control. They were to become independent eventually. The British and French mandates included Iraq, Syria, Lebanon, and Palestine. After World War II these mandates gained independence. However, as you will see, a major dispute arose in Palestine.

Ottoman sultans lived in the Topkapi Palace from when it was built in the 1400s until the 1800s. It is a huge complex of courtyards, gates, and rooms. The Topkapi Palace Museum is now one of İstanbul's main tourist attractions.

Across the Curriculum

▶ **Technology** ◀

Greek Fire The Byzantine Empire was an extremely powerful naval force, partly because it possessed a secret weapon known as Greek fire. This deadly substance was a viscous petroleum-based mixture that burned spontaneously—even in water. It could be pumped from a tube or catapulted onto an enemy ship's deck. The material's preparation was such a closely guarded secret that to this day historians do not know the exact composition of Greek fire.

The Byzantine Empire probably would not have lasted for 1,000 years without Greek fire. After the Turkish conquest of Constantinople in 1453, however, it seems to have disappeared from use.

CRITICAL THINKING: How is Constantinople's location related to the importance of naval superiority for the Byzantine Empire? *(Constantinople is located at the entrance to major seas and waterways.)*

READING CHECK: *Human Systems* How have different empires shaped the eastern Mediterranean region? *shaped religious belief and divisions, created conflicts, changed political status*

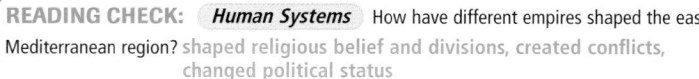

LEVEL 1: Copy the following graphic organizer onto the chalkboard, omitting the italicized answers. Have students complete and discuss it. **ENGLISH LANGUAGE LEARNERS**

Culture Traits of the Eastern Mediterranean		
	Main Languages	**Main Religions**
Cyprus	*Greek, Turkish*	*Christianity, Islam*
Israel	*Hebrew, Arabic*	*Judaism, Islam*
Jordan	*Arabic*	*Islam*
Lebanon	*Arabic*	*Christianity, Islam*
Syria	*Arabic*	*Islam*
Turkey	*Turkish*	*Islam*

LEVEL 2: Have students complete the Level 1 activity. Then have them sketch maps of the eastern Mediterranean, draw the political boundaries, and record where the main languages are spoken and the main religions are practiced.

LEVEL 3: Ask students to imagine that they are citizens of one of the region's countries and that they are members of an Internet chat room that includes teenagers from all the region's countries. Have students compose a series of chat room e-mail messages in which the teens describe and discuss their various cultures.

Palestine and Modern Israel

In the late 1800s European Jews began a movement called **Zionism**. Zionism called for Jews to set up their own country or homeland in Palestine, which was then under Ottoman rule. That land had traditionally been important to Jewish culture, history, and religion. After World War I thousands of Jews moved to the area, which became a British mandate in 1920. Later, many Jews who were fleeing persecution in Europe immigrated to Palestine. In fact, during World War II, Germany's Nazis murdered millions of Jews in what became known as the **Holocaust**. As a result, Jewish immigration to Palestine increased during the 1930s and 1940s. Arabs already living in Palestine soon felt threatened and became angry at the growing Jewish presence there. Today Jews make up about 80 percent of Israel's population. Most of the rest are Arab.

In 1947 the United Nations voted to divide Palestine into Jewish and Arab states. When the British withdrew from Palestine the next year, the Jewish leadership declared itself the independent state of Israel. Arab armies from Egypt, Iraq, Jordan (then called Transjordan), Lebanon, and Syria then invaded Israel. Israel pushed the Arab forces back and won more land. Many Palestinian Arabs fled to Jordan, Lebanon, and other Arab countries. At the same time, Jordan and Egypt took over the remaining Arab portions of Palestine. More wars between Arabs and Israelis occurred in 1956, 1967, 1973, and 1982. During the 1967 war, Israel gained Arab land west of the Jordan River, called the West Bank. It also took the Gaza Strip, a small piece of land on the Mediterranean coast. The Palestine Liberation Organization (PLO) became a force in the region around this time. Groups that wanted to establish an independent Palestinian state formed the PLO in 1964 to coordinate their efforts. Violent conflict between Palestinians and Israelis has continued despite efforts toward peace.

Jewish settlement in the occupied lands has slowed those efforts. Thousands of Jews have moved into the West Bank and Gaza Strip. Jewish settlement has also been an issue in the Golan Heights. Israel took that hilly region from Syria in the 1967 war. Many Israelis do not want to give up these areas. They think the occupied lands are important to the security of Israel. From the Golan, Israeli troops can guard the Jordan River, a crucial water

INTERPRETING THE VISUAL RECORD

Israeli soldiers patrol the West Bank. **From the photo, how can you tell that some of the West Bank's residents have adopted aspects of Western culture, while others maintain traditional ways?**

source. The Golan also blocks Syrian access to northern Israel. In addition, the West Bank separates Israel from Jordan and Arab countries to the east. Resolving the problem of the occupied lands remains an important foreign policy issue. (For more information on these issues, see Case Study: Palestine—Setting Boundaries.)

✔ **READING CHECK:** *Human Systems* Why did many Jews immigrate to Palestine during the 1930s and 1940s? persecution and murder of Jews in Europe

Culture

Language in this region is often an important part of an ethnic group's identity. The many languages spoken there reflect the area's many cultural influences. Arabic, Hebrew, and Turkish are the most common languages. Britain and France introduced English and French when they ruled the area in the 1900s. Many Jews immigrated to Israel from the former Soviet Union in the 1990s, bringing the Russian language with them. The **Kurds**, who live mainly in southeastern Turkey and neighboring countries, make up Turkey's largest minority. They have their own Kurdish language.

FOCUS ON CULTURE

Reviving a Language Hebrew was spoken in the region more than 2,000 years ago. Hebrew is the language of the Jewish Bible, or Torah. In addition, it has always been the language of Jewish prayer and religious ceremonies. Scholars of Jewish law and literature also used Hebrew. The language was not used by most Jews in daily life, however. The rebirth of Hebrew began among Jewish immigrants to what is now Israel. They developed a more modern form of the language. When Israel became independent in 1948, Hebrew became an official language. New immigrants to Israel start on the road to citizenship by learning Hebrew. Arabic is the official language for Israel's Arab minority.

✔ **READING CHECK:** *Human Systems* How do Israel's two official languages reflect the region's distinctive history?

Hebrew—official language because Israel created as homeland for Jews in 1948; Arabic—official language for Arabs, now in minority

Religion Most people in the eastern Mediterranean are Muslim, Jewish, or Christian. All of the faiths are monotheistic, or based on the belief in one God. Judaism was the first of the three to develop. Its roots date back to about 2600 B.C. Today most Jews in the eastern Mediterranean live in Israel.

Christianity developed out of Judaism and spread during the Roman era. Today there are significant Christian minorities in most countries of the region, particularly in Lebanon and Syria. Muslims form the majority in all countries of the eastern Mediterranean except Israel.

INTERPRETING THE VISUAL RECORD

Seeing Hebrew used in public daily life would have amazed Jews a hundred years ago. For centuries the language was used only in religious and scholarly contexts. **What evidence of cultural convergence do you see on this street sign?**

In İstanbul, this Muslim mystic, or dervish, seeks spiritual perfection by whirling.

VISUAL RECORD ANSWER

advertisement for popular American ice cream

Close

Tell students that Turkey has become a major international tourist destination. Ask students to suggest some cultural attractions that would draw tourists to Turkey.

Review and Assess

Have students complete the **Section Review**. Then have students complete **Daily Quiz 20.2**.

Reteach

Have students complete **Main Idea Activity for English Language Learners and Special-Needs Students 20.2**. Then have the class produce a short play about the eastern Mediterranean's history and cultures. Appoint writers, prop designers, actors, and a director. **ENGLISH LANGUAGE LEARNERS**

Extend

Have interested students conduct research on East Aramaic, Armenian, and Circassian—languages spoken by three eastern Mediterranean minority groups—and the people who speak them. **BLOCK SCHEDULING**

Section 2 Review Answers

Identify For identifications, see: Zionism, p. 456; Holocaust, p. 456; Kurds, p. 457

Define For definitions, see: sultans, p. 455; mandates, p. 455

Working with Sketch Maps
Maps will vary, but listed places should be labeled in their approximate locations. İstanbul is in both Europe and Asia.

Reading for the Main Idea
1. the Arabic language and literature

2. because Jews have immigrated there from all over the world and many Arabs still live there

Critical Thinking
3. Possible answers: fear it will damage Israel's security; might guarantee peace and security in return (NGS 13)

4. The Arabs conquered most of the area, bringing the Arabic language with them. (NGS 10)

Organizing What You Know
5. Students' time lines will vary but should include several important dates and events mentioned in the section.

These nomadic herders are spending the winter in the mountains of Lebanon. The nomadic way of life is becoming less common. Many nomads are moving to towns to look for work or settling on government-owned land.

Settlement People in the eastern Mediterranean often live in communities of similar cultural backgrounds. In fact, most of the major cities have different sections that were historically occupied by particular ethnic or religious groups. For example, Jerusalem was divided into Armenian, Christian, Jewish, and Muslim quarters. (See Cities & Settlements: Jerusalem.)

A division has occurred in modern times on Cyprus. A line runs all the way across the island. Greek Cypriots live south of the line, while Turkish Cypriots live north of it. (See Geography for Life: Cyprus—A Divided Island.)

Traditions and Customs Members of the same ethnic group may follow different religions. For example, Christian Arabs share the Arabic language and literature of Islamic Arabs. Yet they maintain their separate religious identity. In contrast, members of the same religion may belong to various ethnic groups. For example, the region's Muslims—whether Arabs, Kurds, or Turks—share many traditions. Turks and Kurds, while Islamic, are distinct from Arabs in their language and culture.

Cultural divisions also separate Turkey's urban and rural populations. Middle-class Turks tend to share the lifestyle and attitudes of middle-class Europeans. However, most rural Turks have more traditional views, such as those concerning the role of women. They prefer that women work as homemakers.

Jewish religious law influences Israel's traditions and customs. For example, because Saturday is the weekly holy day in Judaism, most Israeli businesses are closed on that day. Although most Israelis practice Judaism, Israel is also multiethnic because Jews have emigrated there from all over the world. Many of these immigrants have come from Germany, Russia, and other European countries. Others are from North Africa and Southwest Asia. Israel also has a large number of Jews from Ethiopia.

✓ **READING CHECK:** **Human Systems** How does religion shape cultural patterns in the region? *affects group identity, business, immigration*

Section 2 Review

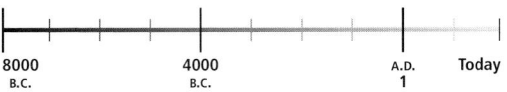
Homework Practice Online
Keyword: SW3 HP20

Identify
Zionism
Holocaust
Kurds

Define
sultans
mandates

Working with Sketch Maps
On the map you created in Section 1, label the countries of the eastern Mediterranean and İstanbul, Ankara, Jerusalem, and Nicosia. Which of these four cities is in both Europe and Asia?

Reading for the Main Idea
1. **Human Systems** What are some cultural features that Christian Arabs share with Muslim Arabs?

2. **Human Systems** Why is Israel's population so multiethnic?

Critical Thinking
3. **Drawing Inferences** Why do many Israelis oppose giving up the occupied lands and their Jewish settlements there? What might Palestinians and other Arabs offer in return for those lands?

4. **Identifying Cause and Effect** Why do many people in the eastern Mediterranean speak Arabic?

Organizing What You Know
5. Create a time line like the one shown below. Enlarge the section between A.D. 1 and Today. On your time line, list important years, periods, and events in the history of the eastern Mediterranean.

8000 B.C.	4000 B.C.	A.D. 1	Today

Environmental Science: Mining and Refining

The Greeks were first attracted to Cyprus because of the abundance of copper in the island's mountains. In pre-Roman times, Cyprus was the world's leading producer of copper ore. In fact, the word *copper* is derived from an early form of the island's name. Modern geologists have explored the ancient copper mines to learn more about how ore was processed in pre-Roman times. The scientists discovered that after the ore was mined, it was fired in wood-burning furnaces to extract the pure copper metal. It is estimated that in ancient times the island produced a total of more than 200,000 tons of copper metal. To produce this much refined metal from the ore using wood-burning furnaces, it would take about 1,200,000,000 cubic meters of pine—perhaps the most common type of wood on Cyprus at the time. To produce this much wood, some 150,000 square kilometers of forest had to be cleared. Because the total surface area of Cyprus is only 3,591 square miles (9,300 sq km), either the forests of Cyprus were regrown and then reharvested 16 times or massive amounts of wood had to be imported to support the island's copper industry. Either way, this represents a large-scale environmental disruption. Today, some mining methods, such as strip-mining, continue to damage ecosystems in some countries. Have students conduct research on modern mining and refining techniques, along with possible ways for reducing their impact on the environment.

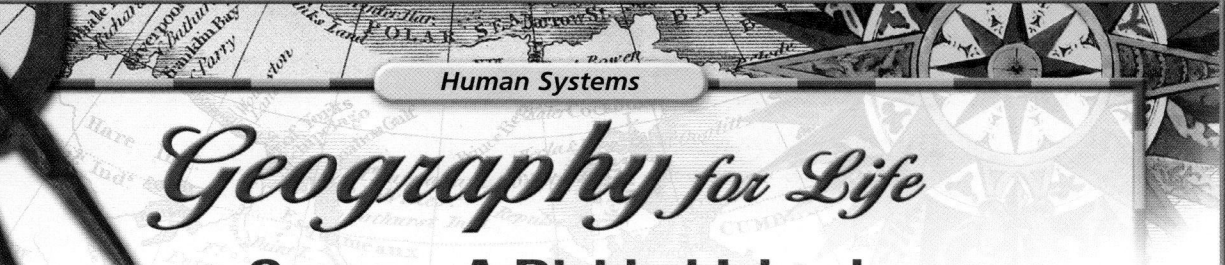

Human Systems

Geography for Life

Cyprus—A Divided Island

Cyprus lies in the Mediterranean Sea near Turkey. About 78 percent of its 763,000 people speak Greek and share a Greek heritage. Turkish Cypriots make up 18 percent of the population. Conflict between these groups marks the island's modern history.

Greeks have lived on Cyprus for at least 3,000 years. Turks later took over the island in the 1500s. Cyprus then came under British control in 1878. In 1960 Cyprus became an independent republic. Not long afterward, Turkish and Greek Cypriots clashed over sharing power in the new government. In addition, the Turkish Cypriots feared that the Greek Cypriots would join Cyprus to Greece. Fighting soon broke out. In 1964 the United Nations sent a peacekeeping force to Cyprus. In 1974 Turkey invaded and took control of the northern 40 percent of the island. Refugees—both Greek and Turkish—fled to the south. In 1983 Turkish Cypriots declared the northern territory an independent country—the Turkish Republic of Northern Cyprus (TRNC). No other country besides Turkey recognizes the TRNC. All other governments recognize Cyprus as a single country with a Greek Cypriot government.

Cyprus remains a divided island. A buffer zone called the Green Line separates Turkish Cyprus from Greek Cyprus. Nicosia, the capital of both Greek and Turkish Cyprus, lies along this line in the island's center.

Thousands of Turkish troops occupy northern Cyprus. Most countries refuse to trade with that region. As a result, the north must depend on aid from Turkey. However, the south has received help from Great Britain, Greece, the United States, and the United Nations. This aid has made construction of new businesses, housing, port facilities, and roads possible. Funds from other countries also allowed southern Cyprus to expand tourist facilities.

In 2004, the Greek state of Cyprus joined the European Union. However, residents of the TRNC do not have access to the privileges of EU membership. As a result, pro-EU parties in the TRNC have been pushing for a settlement that would unite the island.

Applying What You Know

1. **Summarizing** What are two reasons why Greek Cypriots might feel that the Turks should give up their claims to the island?

2. **Identifying Points of View** How do Greek Cypriot and Turkish Cypriot views of the island differ?

Global Perspectives

Earthquake Aid Although Greece and Turkey have clashed militarily, they have also helped each other in times of crisis. On August 17, 1999, an earthquake struck northwestern Turkey, killing more than 17,000 people. Greece promptly sent three cargo airplanes loaded with humanitarian aid. A few weeks later, on September 7, Athens was hit by an earthquake. Although it was still reeling from the August tragedy, Turkey immediately sent aid to Greece.

VISUAL RECORD ANSWER

major city in the center of the island, on a river, on the Green Line

Applying What You Know Answers

1. Greeks have lived there for at least 3,000 years, and Greek Cypriots make up 78 percent of the population.
2. Greeks—as one country under the Greek Cypriot government, with Turkish occupiers in the north; Turks—as the Turkish Republic of Northern Cyprus and a southern Cyprus under a Greek government

This Geography for Life feature addresses National Geography Standards 4, 6, 10, and 13.

Cyprus

MEDITERRANEAN SEA

Rizokarpaso

Kyrenia

Occupied by Turkey

Morphou Bay
Morphou
Khrysokhou Bay
Polis
Nicosia
Pedieos River
Lefkoniko
Famagusta Bay
Famagusta

Kyperounda
Larnaca Bay
Larnaca

Kophinou

Paphos
Episkopi
Episkopi Bay
Limassol
Akrotiri Bay
Zyyi

SCALE
0   15   30 Miles
0   15   30 Kilometers
Projection: Lambert Conformal Conic

INTERPRETING THE VISUAL RECORD *In the photo, a member of the Cyprus National Guard patrols the Green Line in Nicosia. The blue and white paint echoes the colors of the Greek flag.* **Why might both the Turkish and Greek Cypriots want Nicosia for their capital?**

ΑΠΑΓΟΡΕΥΕΤΑΙ Η ΦΩΤΟΓΡΑΦΗΣΗ

ΑΛΤ
ΠΡΟΣ
ΤΟΥΡΚΟΚΡΑΤΟΥΜΕΝΗ ΠΕΡΙΟΧΗ

OBJECTIVES

1. Describe the activities on which the eastern Mediterranean economies rely.

2. Identify characteristics of the region's cities.

3. Analyze the challenges faced by the region's people.

 **LET'S GET STARTED**

Copy the following passage onto the chalkboard: *For centuries, the eastern Mediterranean region has been famous for fine handcrafted products. What are some of the products you associate with the region?* Discuss responses. *(Possible answers: carpets, sandals, metalwork, fabric dyes)* Tell students they will learn more about the area's products and other aspects of modern life there in Section 3.

Building Vocabulary

Write the key terms on the chalkboard and call on a volunteer to read the definitions from the text or glossary. Tell students that the stalls in typical **souks** tend to be grouped by the category of product for sale. For example, all the carpet merchants will be found in the same area. Point out that in many countries there are conflicts between groups that want a **secular** government and those that want religion to play a larger role. You many want to lead a discussion about news stories related to these issues.

 **RESOURCES**

REPRODUCIBLE

▶ Guided Reading Strategy 20.3

▶ Readings in World Geography, History, and Culture 50–52

▶ Critical Thinking Activity 20: Oil in the Middle East

▶ Geography for Life Activity 20: Site and Situation Shape Byzantium/Constantinople/ İstanbul

▶ Map Activity 20: Jerusalem

TECHNOLOGY

▶ One-Stop Planner CD–ROM, Lesson 20.3

▶ CNN Presents Geography: Yesterday and Today, Segment 21: Jordan's Water Crisis

▶ Homework Practice Online

▶ HRW Go site

REINFORCEMENT, REVIEW, AND ASSESSMENT

▶ Main Idea Activity 20.3

▶ English Audio Summary 20.3

▶ Spanish Audio Summary 20.3

▶ Section 3 Review, p. 463

▶ Daily Quiz 20.3

 **Section 3**

The Region Today

READ TO DISCOVER

1. On what activities do the eastern Mediterranean economies rely?
2. What are the cities of the region like?
3. What challenges do the people of the region face?

WHY IT MATTERS

Recent developments have led to important changes in the West Bank and Gaza Strip. Use **CNNfyi.com** or other **current events** sources to learn about the latest news from those areas.

DEFINE

souks
secular

LOCATE

Tel Aviv
Haifa
Damascus
Beirut

Economic Development

Many factors have slowed the economic development of the eastern Mediterranean. For example, the land itself has caused problems for Turkey. Earthquakes there have interrupted its economic growth. Political problems are more widespread. Hostile relations between Israel and its neighbors have prevented the creation of normal economic links throughout the region. Also, hundreds of thousands of Palestinians live as refugees in other countries. Many live in Jordan and Lebanon. The resulting population increase has strained those countries' resources. At the same time, Israel has encouraged and absorbed waves of Jewish immigrants from other parts of the world. Lebanon and Cyprus have suffered devastating civil wars.

The economies of Syria and Jordan are underdeveloped and suffer from high unemployment. Lack of resources, a weak educational system, and an outdated technological base add to their problems. Jordan receives foreign aid from the United States and oil-producing Arab countries like Saudi Arabia. Israel also trades heavily with the United States and receives U.S. aid.

Agriculture   The eastern Mediterranean economies rely heavily on agriculture. In many places farming requires irrigation. Most irrigation systems used in the region are simply small canals that allow water to run directly to the fields. Some of these systems use fossil water. Large-scale irrigation depends on the major rivers. Turkey is building a network of dams and irrigation canals on the Tigris and Euphrates Rivers. Neighboring countries Syria and Iraq resent these projects, however. This is because the projects reduce the amount of water downstream that is available to them.

Farmlands in Turkey's Mediterranean coastal valleys produce most of that country's crops. Livestock raising is common in Anatolia. Despite a lack of water, agriculture is productive in Israel as well. Careful irrigation makes growing fruits and vegetables possible in semiarid parts of the country.

A Turkish carpet seller displays his wares. Turkey's national government plays a major role in basic industries such as mining, but the textile and clothing industries are almost entirely in private hands.

ALL LEVELS: Copy the following graphic organizer onto the chalkboard, omitting the italicized answers. Have students complete it with details about each country's economy. Ask students on what type of economic activity much of the region relies *(agriculture)* and on what practice that activity often depends *(irrigation)*. Have students identify the two strongest economies in the region *(Israel and Turkey)*.

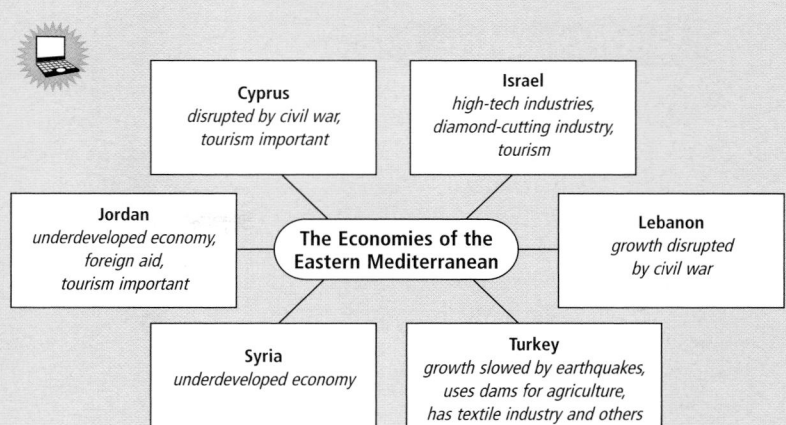

Cyprus
disrupted by civil war, tourism important

Israel
high-tech industries, diamond-cutting industry, tourism

Jordan
underdeveloped economy, foreign aid, tourism important

The Economies of the Eastern Mediterranean

Lebanon
growth disrupted by civil war

Syria
underdeveloped economy

Turkey
growth slowed by earthquakes, uses dams for agriculture, has textile industry and others

Industry   Israel is the most technologically advanced country in the eastern Mediterranean. The country has built much of its economy on high-tech industries. Diamond cutting is another major Israeli industry. Most diamond-cutting factories can be found in the Tel Aviv area. Polished diamonds are the country's leading export. Israel also has important chemical industries centered in the port city of Haifa. Turkey's industry is the second-most developed in the region. Textiles are Turkey's leading industry. The country's industrial sector includes modern urban factories and small-scale industries in rural areas.

Tourism   Tourism is a major industry for the eastern Mediterranean. In Cyprus, tourists flock to beaches along the island's southern coast. Tourists come to Turkey to see Greek and Roman ruins, Ottoman palaces, and busy carpet markets. Many who travel to Israel want to visit religious sites, such as those in Bethlehem and Jerusalem, or float in the salty Dead Sea. However, the region's continuing unrest makes the tourist industry fragile. Israel's economy can handle a drop in tourism because it is well diversified. Other countries like Jordan, however, suffer when tourism slows.

✓ **READING CHECK:**   ***Human Systems***   How do the economies of Israel and Turkey differ from those of Jordan and Syria? *more industrialized than Jordan and Syria*

Urban Environments

The urban population of the eastern Mediterranean is growing rapidly. The region's high birth rates and migration from rural areas have fueled this growth. Many cities are crowded. Housing tends to be small and cramped, and traffic congestion and smog get worse every year.

Many of the cities in the eastern Mediterranean are ancient. In fact, Damascus, Syria, is probably the oldest city in the world. These old cities' centers usually consist of twisting narrow streets where traditional craftspeople sell their wares. Open-air markets called **souks** are also common. In many of

Israeli workers cut and polish diamonds. Israel's diamond industry is one of the largest in the world and accounts for about 25 percent of the country's export earnings.

The Gaza Strip is a small, crowded piece of coastal land that has practically no resources. More than a million Palestinians live there.

Teach Objective 2

 ALL LEVELS: Display a large map of your community or a nearby city with which students are familiar. Ask how the city's plan compares to traditional cities of the eastern Mediterranean. *(Possible answers: reverse of eastern Mediterranean pattern, because government offices more likely in old part of city and shopping areas near the city's edge; streets wider in American city)* Then lead a discussion about features of specific eastern Mediterranean cities.

Teach Objective 3

 ALL LEVELS: Ask students to imagine that they represent one of the eastern Mediterranean countries

at a regional conference titled "United to Face the Future's Challenges." Have students work in groups to prepare banners, position papers, or speeches for their country's delegation. **COOPERATIVE LEARNING**

Teacher to Teacher

Catherine Stoner of Wentzville, Missouri, suggests the following activity to help students learn more about the eastern Mediterranean. Have students conduct research on the hydro-electric dams being built in Turkey and mark them on a map. Then have students write paragraphs about why the dams are controversial.

Precious Water Freshwater is not abundant in the eastern Mediterranean. For example, in 2000 Jordan had about 3,603 cubic feet (102 cu m) of renewable freshwater per person per year. In comparison, the United States had 312,158 cubic feet (8,838 cu m) for each citizen! The only country in the eastern Mediterranean with a freshwater surplus is Lebanon. Other countries struggle with drought, water pollution, and inefficient water management. For example, Syria's main source of water, the Euphrates River, is polluted with human waste and harmful chemicals. Also, some of the river's water is held upstream behind Turkish dams.

The lack of clean water in some areas and the abundance of water in others has created a contrast between the water "haves" and "have nots."

CONNECTING TO HISTORY ANSWER

application by the government of the techniques to other areas and to the country's water supply

462

Connecting to
HISTORY

Petra

Petra, a desert city built about 2,000 years ago by an Arab people, is Jordan's main tourist attraction. It features huge buildings cut directly into the cliffs. How could this city of 30,000 people thrive in the desert? Protecting the water supply was crucial. Petra's builders installed canals, terraces, and hundreds of underground storage tanks to collect and store the area's scant rainfall. Dams in the nearby hills caught the water from flash floods after rare downpours. This water conservation system is the hidden marvel of Petra.

When the caravans were routed away from Petra, the city began to decline. Earthquakes did further damage. Neglecting the water storage system, which filled with sand, probably sealed the city's fate.

Problem Solving How might further study of Petra's water system affect public policy in Jordan?

Top: A water channel at Petra Bottom: Petra's Royal Tombs

these markets, shops that sell the same items are clustered together. For example, vendors on Jerusalem's Christian Quarter Road specialize in religious souvenirs and Palestinian textiles. Typically, the newer parts of the city surround the old center. Most service-oriented business and government offices are in the newer areas.

War has damaged some cities, including Nicosia, Cyprus, and Beirut, Lebanon. In its better days Beirut had been called the Paris of the Middle East for its culture, glamour, and scenery. In recent years, Lebanon's government has worked hard to rebuild Beirut's center, making it a thriving and modern commercial zone. People there hope this rebuilding will help the country's economy recover from years of warfare.

✓ **READING CHECK:** **Human Systems** How are the centers of eastern Mediterranean cities often different from the areas that surround them? new modern areas built around old centers, which have narrow twisting streets where traditional craftspeople sell their wares

Issues and Challenges

The countries of the eastern Mediterranean face many issues and challenges. The most pressing of these are political ones. However, social and environmental problems also threaten stability.

One long-term challenge is the status of the West Bank and Gaza Strip. In the early 1990s the Israeli government began talks with the PLO. They agreed that the Palestinians would get control of parts of the West Bank and most of the Gaza Strip. However, the process has not been completed. Demonstrations, riots, and violence continue, and the future of the peace process remains uncertain.

Other ethnic and religious conflicts trouble other parts of the region. In Turkey, Armenian and Kurd minorities complain of unfair treatment by the government. Turkey's Kurds identify strongly with Kurds living in Iran, Iraq, and Syria. Some Kurds have fought for independence from Turkey. Thousands of people have been killed as a result

Close

Tell students to imagine that they are executives for an international clothing manufacturer. Their company is considering opening a factory in the eastern Mediterranean, and it is their job to select the country. Remind them to consider factors such as available resources and security.

Review and Assess

Have students complete the **Section Review**. Then have students complete **Daily Quiz 20.3**.

Reteach

Have students complete **Main Idea Activity for English Language Learners and Special-Needs Students 20.3.** Then have students work in groups to create board games based on facts about the economies, cities, and challenges of the eastern Mediterranean region. **ENGLISH LANGUAGE LEARNERS, COOPERATIVE LEARNING**

Extend

Have interested students conduct research on political developments in Jordan following the death of King Hussein in 1999. **BLOCK SCHEDULING**

INTERPRETING THE VISUAL RECORD

Students debate issues related to religious law at a Jerusalem yeshiva—a school for the study of Jewish traditions. The role of religion in public life is a source of conflict between Jews and Arabs and among Jews. For example, many ultra-Orthodox Jews refuse to celebrate Israel's independence day because they feel the country is too secular. **What are some aspects of religion in Israel that make the country distinctive?**

of this struggle. In addition, Islamic fundamentalists in Turkey have objected to the country's **secular**, or nonreligious, political system. That criticism has turned violent at times. On the other side of the issue, the Turkish government has been criticized for limiting the religious freedom of Muslims.

As you read earlier, United Nations troops have tried to keep the peace between ethnic Greeks and ethnic Turks in Cyprus. In Lebanon a civil war among several different religious militias broke out in 1975. Christians belonging to various factions fought several Muslim groups. Fighting there lasted for 15 years. Jordan is divided between "original" Jordanians and Palestinian refugees. These Palestinians now make up a slight majority of Jordan's population. Many still live in refugee camps, where health care and other social services are poor.

The eastern Mediterranean's main environmental concern is a lack of water. Droughts, a growing population, and pollution all threaten the limited freshwater of the region. Also, overgrazing has damaged semiarid grasslands in parts of Jordan and Syria.

✓ **READING CHECK:** *Human Systems* Why might some Kurds want Turkey to change its boundaries? to create their own independent country

---

Section 3 Review

go.hrw.com Homework Practice Online Keyword: SW3 HP20

Define souks, secular

Working with Sketch Maps On the map you created in Section 2, label Tel Aviv, Haifa, Damascus, and Beirut. Which city is rebuilding after a civil war?

Reading for the Main Idea

1. *Environment and Society* Why do some other countries resent Turkey's proposed dam and irrigation system?

2. *Human Systems* What factors have fueled the rapid growth of cities in the region?

3. *Environment and Society* What problems threaten the region's scarce freshwater supplies?

Critical Thinking

4. Analyzing Information In what ways does the relationship between Israel and the Palestinians affect the rest of the region?

Organizing What You Know

5. Create a word web in which you describe Israel's agriculture, cities, industry, and the challenges it faces.

Section 3 Review Answers

Define For definitions, see: souks, p. 461; secular, p. 463

Working with Sketch Maps Maps will vary, but listed places should be labeled in their approximate locations. Beirut, Lebanon, is rebuilding.

Reading for the Main Idea

1. because it reduces the amount of water available downstream in Syria and Iraq

2. high birth rates, migration

3. drought, growing population, pollution

Critical Thinking

4. reduces tourism; adds to problems of violence and refugees; creates political problems in countries with many refugees, particularly Jordan (NGS 13)

Organizing What You Know

5. agriculture—productive, careful irrigation, fruits and vegetables; cities—Jerusalem, Tel Aviv, Haifa; industry—most-technologically advanced country, diamond cutting, polished diamonds, chemicals; challenges—conflict with Palestinians, lack of water

VISUAL RECORD ANSWER

Israel was established as a homeland for Jews in the midst of an Islamic area. Now Jews are in the majority. Some citizenship practices of Israel are affected by cultural beliefs.

463

The Eastern Mediterranean • 463

Solving Problems

Have students read Cities & Settlements: Jerusalem. Then relate the following information to the class. In 1947 the United Nations recommended that Jerusalem be an international city. Neither Israel nor Jordan (then called Transjordan) agreed. When the State of Israel was created in 1948, Transjordan captured the Old City, and Jerusalem was divided between Israel and Jordan. During the Six-Day War of 1967, Israeli troops took control of the Old City. Israel's government declared Jerusalem to be unified once again.

The United Nations passed a resolution disapproving of the action. International controversy arose again in 1980, when the Knesset, Israel's parliament, declared that Jerusalem was the capital of Israel, a designation many countries refused to recognize. In fact, many of these countries moved their embassies out of Jerusalem to Tel Aviv or its twin city, Jaffa. The city's status remains a point of contention.

CITIES & SETTLEMENTS

Jerusalem

Human Systems Jerusalem is one of the world's oldest cities and is deeply sacred for three major world religions: Christianity, Judaism, and Islam. Christians honor the city as the place where Jesus preached, died, and rose again. The city has long been a center of culture and faith for Jews. Muslims believe that from the site Muhammad departed on a spiritual journey through the skies.

During Jerusalem's 5,000 year history, many groups have fought to control the city. In fact, it has been captured and occupied more than 36 times over the centuries. Even today, control of Jerusalem is central to disputes between the Israelis and Palestinians.

An Ancient and Sacred City

About 1000 B.C. King David ruled the Israelites from the city of Hebron. However, the king wanted a more centrally located capital from which he could further unite the tribes of Israel. This desire led him to seize the fortress of Zion from the Jebusites and found a new city, Jerusalem, there. When the king moved the sacred Ark of the Covenant to Jerusalem, the city became the religious center of Israel. David's son, King Solomon, expanded the city and built the Temple Mount, into which the Ark was placed.

In 586 B.C. the Babylonians captured Jerusalem and destroyed Solomon's temple. Although it was later rebuilt, the Romans destroyed it again in A.D. 70. All that remains of the temple today is a high wall on the western edge of the Temple Mount. This Western Wall—sometimes called the Wailing Wall—is the holiest place in Judaism. Each day people go to the wall to pray. They place slips of paper with prayers

A Jewish man prays at Jerusalem's Western Wall, also called the Wailing Wall.

written on them between the narrow cracks in the wall.

The Church of the Holy Sepulchre, the most sacred shrine of Christianity, is also located in Jerusalem. Completed about A.D. 335 by the Roman emperor Constantine, it occupies the site where Christians believe Jesus was crucified, buried, and resurrected. Thousands of Christian pilgrims visit the church each year and retrace "the Way of the Cross" through Jerusalem's narrow streets.

Arab Muslims captured Jerusalem in the early A.D. 600s. In the late 600s they built a mosque on the Temple Mount to house the rock from which they believe Muhammad ascended to the heavens. Known as the Dome of the Rock, it is one of Islam's holiest sites. Some Jews believe the rock may also be the foundation stone on which the Ark of the Covenant rested in its ancient temple. Some even want to rebuild the temple. However, that could not be done without destroying the Dome of the Rock and other sacred Islamic sites on the Temple Mount. Conflicts like this over space sacred to different religions are common in Jerusalem.

The Modern City

Jerusalem is located on hills overlooking the Jordan River valley. In the heart of Jerusalem lies the Old City, the most ancient part of the city. Walls surround the Old City, which is separated into four quarters. These are the Armenian Quarter, Christian Quarter, Jewish Quarter, and Muslim Quarter. Each quarter contains religious buildings and houses for the members of its community.

In the 1860s the first settlements were built outside the walls of the Old City. Today more than

Going Further: Thinking Critically

First, have students define the problem related to Jerusalem's official status. *(Possible answer: Israel's control of Jerusalem, a city sacred to both Jews and Muslims, is a sticking point in the search for peace in the eastern Mediterranean region.)* Then organize the class into two groups. Have one group conduct research on plans for Jerusalem's future that have been proposed by Palestinians or other Arab leaders. Ask the other group to examine the Israeli government's position or other proposals regarding the city's status made by Jewish Israelis. Students should use the Internet or library materials. Have the students try to answer the following questions about the proposals: What are the proposals' main components? How would the proposals affect daily life in Jerusalem for both Arabs and Jews? How might the proposals be received by the opposing side? Are there factors the plans do not take into account? If so, what are those factors? How might those factors affect the problem's ultimate solution? You may want to let students brainstorm their own solutions.

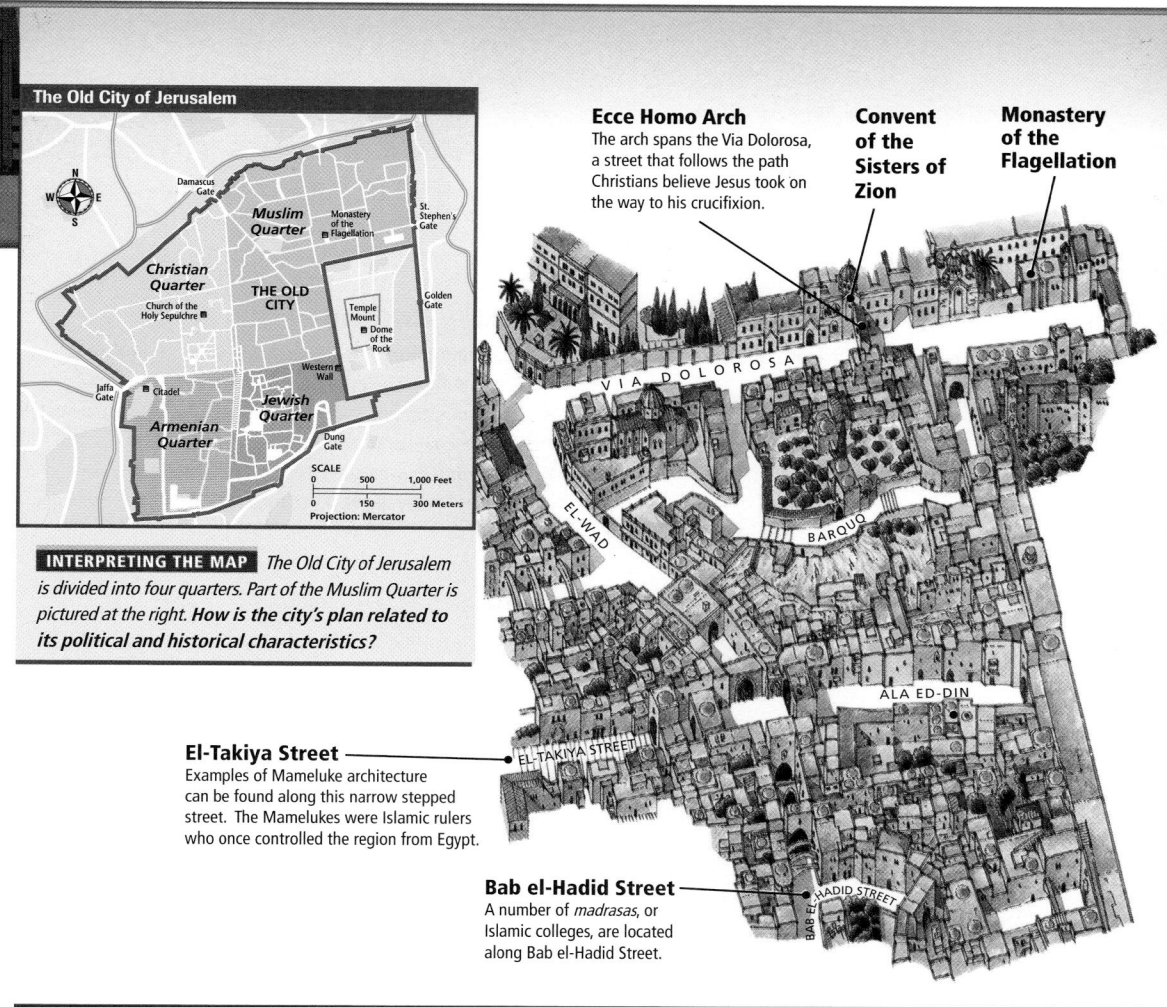

The Old City of Jerusalem

Ecce Homo Arch
The arch spans the Via Dolorosa, a street that follows the path Christians believe Jesus took on the way to his crucifixion.

Convent of the Sisters of Zion

Monastery of the Flagellation

INTERPRETING THE MAP *The Old City of Jerusalem is divided into four quarters. Part of the Muslim Quarter is pictured at the right.* **How is the city's plan related to its political and historical characteristics?**

El-Takiya Street
Examples of Mameluke architecture can be found along this narrow stepped street. The Mamelukes were Islamic rulers who once controlled the region from Egypt.

Bab el-Hadid Street
A number of *madrasas,* or Islamic colleges, are located along Bab el-Hadid Street.

600,000 people live in Jerusalem, most outside its ancient walls. Many of Jerusalem's newer neighborhoods lie to the west and north. Commercial and government centers are also located west of the Old City.

About one third of Jerusalem's residents are Palestinian Arabs. Most live in East Jerusalem. The remaining two thirds of the population are mostly Jews, and many live in West Jerusalem. Neighborhood divisions throughout the city are often based on religious practices and preferences. For example, Arab East Jerusalem includes both Muslim and Christian neighborhoods. In West Jerusalem some communities are organized according to how strictly the residents practice Judaism.

When the modern state of Israel was established in 1948, Jerusalem was divided into Israeli and Jordanian sectors. Israel captured the entire city during the Six-Day War in 1967. Since then, Israel's government has controlled all of Jerusalem, including mostly Arab East Jerusalem. This has become a major political issue as Israelis and Palestinians struggle to make peace. Both sides consider Jerusalem a sacred city and want to control it. So far, negotiators have not been able to resolve this difficult issue.

Applying What You Know

1. **Summarizing** Why have Jews and Muslims sought control of Jerusalem for so long?

2. **Comparing and Contrasting** Recall what you learned about Mecca in Chapter 19. What are some similarities and differences in the religious importance of Mecca and Jerusalem? What other places have you studied that are also important religious centers?

CHAPTER 20 Review and Assessment Resources

TECHNOLOGY
- Chapter 20 Test Generator (on the One-Stop Planner)
- Global Skill Builder CD–ROM
- HRW Go site

REINFORCEMENT, REVIEW, AND ASSESSMENT
- Chapter 20 Review, pp. 466–67
- Chapter 20 Tutorial for Students, Parents, Mentors, and Peers
- Chapter 20 Test (form A or B)
- Alternative Assessment Handbook
- Chapter 20 Test for English Language Learners and Special-Needs Students
- Unit 6 Test
- Unit 6 Test for English Language Learners and Special-Needs Students

CHAPTER 20 Review Answers

Thinking Critically

1. Although Hebrew has long been important to Jewish religious life, modern Hebrew developed among Israeli immigrants in modern times. Zionism played an important role in generating this immigration. (NGS 13)

2. It is so salty that fish cannot live in it. (NGS 7)

3. Possible answer: Israeli and other Arabs are more likely to support Palestinian desires for an independent homeland and an end to Jewish settlements in occupied lands. Many Jews see the settlements as needed for Israel's security. They may also fear a Palestinian state. Israel's government must balance the views of Israel's Jewish majority with those of its Arab minority. Arab and other countries can also put pressure on Israel to slow or stop Jewish settlement in the West Bank and Gaza Strip and to allow a Palestinian state. (NGS 13)

Using the Geographer's Tools

1. Possible answers include Gaza and the West Bank. Separation of these areas would make them difficult to defend and economic links between the regions difficult to maintain.

2. Israel, Cyprus; high GDP per capita in each country; other factors—literacy and life expectancy rates, work force structure, energy consumption, telephone lines

CHAPTER 20 Review

Building Vocabulary
On a separate sheet of paper, explain the following terms by using them correctly in sentences.

potash	mandates	Kurds
magnesium	Zionism	souks
sultans	Holocaust	secular

Locating Key Places
On a separate sheet of paper, match the letters on the map with their correct labels.

Dead Sea	İstanbul	Beirut
Euphrates River	Jerusalem	

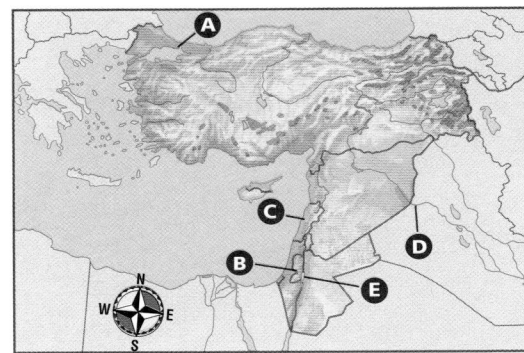

Understanding the Main Ideas
Section 1
1. (*Physical Systems*) What physical feature extends north from Africa to create the Jordan River valley and the Dead Sea?

2. (*Places and Regions*) How has the region's vegetation changed over time?

Section 2
3. (*Human Systems*) Why did many Jews immigrate to the region that later became Israel?

4. (*Places and Regions*) What are the three main religions of the eastern Mediterranean? What major feature do these religions share?

Section 3
5. (*Places and Regions*) What two countries are recovering from many years of ethnic and religious conflict?

Thinking Critically

1. **Analyzing Information** How are Zionism and the revival of the Hebrew language linked?

2. **Drawing Inferences and Conclusions** How do you think the Dead Sea got its name?

3. **Analyzing** Why do many Jewish Israelis and Arabs view political issues in the region differently? How might these differing views affect public policies like Jewish settlement in the West Bank and Gaza Strip?

Using the Geographer's Tools

1. **Analyzing Maps** Review the chapter physical-political map. What areas could become part of a new Palestinian state? What problems might the geography of this new state pose for a Palestinian government?

2. **Analyzing Statistics** Review the unit Comparing Standards of Living and Fast Facts tables. Which country or countries appear to be the most developed and have the highest standard of living? What statistics have led you to this conclusion?

3. **Preparing Graphs** Go to go.hrw.com on the HRW Web site to find the information you will need to prepare a climate graph for one of the region's cities. Write a caption describing the city's climate and the factors affecting its climate.

Writing about Geography

Review the information about the conflict between Israelis and Palestinians. Compare the Israeli and Palestinian points of view. Then find out how other countries view the conflict. Write a short report about your findings. You might want to create diagrams illustrating some of the key points in the conflict.

SKILL BUILDING
Geography for Life

Analyzing Geographic Information
Environment and Society Use Internet or library resources to investigate several of the eastern Mediterranean's non-European languages. Then answer the following questions: How similar are these languages to each other? How and when were they brought into the region? Where are they spoken today? Then draw a map showing where the languages are spoken and compare your map to the chapter physical-political map. Draw inferences about what geographic and political factors can create a language boundary. Write a paragraph describing your findings.

Portfolio Activity

1. Have students conduct research on and create models of different irrigation systems used in the eastern Mediterranean. Ask them also to suggest possible innovations. Take photographs of the models for placing in student portfolios.

2. Have students choose a foreign language spoken in the eastern Mediterranean. They should then compile small phrase books or tape cassettes with terms that a geographer could use while traveling through that language region. Have students place their phrase books or the tape transcripts in their portfolios.

Food Festival

Turkish delight (in Turkish *rahat loukoum*, meaning "rest for the throat") is a popular candy. In a small bowl, soften 3 tbs. unflavored gelatin in 1 c. cold water. Add 2 c. granulated sugar to ⅓ c. boiling water. Bring to boil. Stir in softened gelatin. Simmer 20 minutes. Add 3 to 4 tbs. lemon or orange flavoring and a few drops food color to match the flavor. Pour into lightly greased 8-inch-square pan. Chill. When firm, invert onto cutting board. Cut into 1-inch cubes. Roll in powdered sugar. Store tightly covered at room temperature. Other recipes for Turkish delight include cornstarch, pistachio nuts or almonds, or rosewater, a flavoring common in this region.

Building Social Studies Skills

Interpreting Charts

Study the chart below. Then use the information from the chart to help you answer the questions that follow. Mark your answers on a separate sheet of paper.

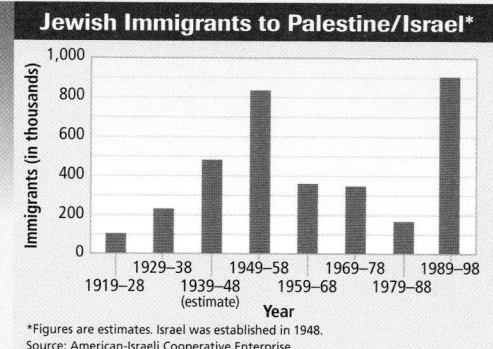

Jewish Immigrants to Palestine/Israel*

*Figures are estimates. Israel was established in 1948.
Source: American-Israeli Cooperative Enterprise

1. After Israel gained independence in 1948, the fewest number of Jewish immigrants to the area arrived between
   a. 1919 and 1928.
   b. 1929 and 1938.
   c. 1969 and 1978.
   d. 1979 and 1988.

2. What general pattern do you see in immigration to Palestine and Israel over time?

Interpreting Secondary Sources

Read the following passage and answer the questions that follow.

> "In the 1300s the Ottoman Turks established another Muslim empire in the region. They took Constantinople in 1453 and made it the Ottoman capital. By the 1600s the Ottoman Empire included most of Southwest Asia. The empire also stretched into parts of eastern Europe and most of North Africa. However, political problems and rivalries with other countries slowly weakened the empire. Over time ethnic minorities within the empire's borders also began to push for independence."

3. What event happened in 1453?
   a. The Republic of Turkey was created.
   b. The Ottoman Turks created an empire.
   c. The Ottoman Empire was defeated.
   d. The Ottomans captured Constantinople and made it their empire's capital.

4. What are three reasons that the Ottoman Empire weakened?

Alternative Assessment

PORTFOLIO ACTIVITY

Learning about Your Local Geography

Group Project: Research
You have read about how cultural beliefs can influence public policies and decisions. Plan, organize, and complete a group research project about the various ethnic or cultural groups that live in your area. You might use library or Internet resources to begin your research. Then interview community government leaders to learn how cultural beliefs have influenced local policies. Finally, prepare a short report discussing your findings.

internet connect

Internet Activity: go.hrw.com
KEYWORD: SW3 GT20

Choose a topic on the eastern Mediterranean to:
- research ancient and modern Jerusalem and create a newspaper.
- send a postcard from the Dead Sea and learn about its cultural value and its value as a natural resource.
- understand the sources of conflict and efforts to relieve it in Cyprus.

---

3. Answers will vary.

Writing
Students should discuss the conflict and support their points of view with factual information from their research. Use Rubric 9, Comparing and Contrasting, to evaluate student work.

Geography for Life
Answers will vary but should include information on how the languages are related to each other, how and when the languages were brought into the region, and where the languages are spoken today. The map should show where the languages are spoken. Use Rubric 30, Research, to evaluate student work.

Social Studies Skills
1. d
2. varied pattern of immigration
3. d
4. political problems, rivalries with other countries, ethnic minorities pushing for independence

PORTFOLIO ACTIVITY
Background research should display knowledge of the community's diversity. Interview questions and reports should display understanding of local issues. Use Rubric 14, Group Activity, and Rubric 20, Map Creation, to evaluate student work.

Workshop 1
Going Further: Thinking Critically

Before class, obtain transportation maps for several countries similar to the one shown on this page. Have students examine the maps to learn about the transportation networks that exist in each of the countries represented. Then lead a class discussion about information that can be inferred from transportation maps. Ask: What might a transportation map tell the reader about a country's level of development? *(Possible answer: Countries with well-developed transportation networks, especially those with extensive air traffic networks, have generally attained high levels of development. The absence of roads or railroads within a country can be a* sign that a country is not economically developed.) What might transportation routes tell geographers about the locations of economic activities within a country? *(Possible answer: It can often be assumed that areas served by many transportation routes are economic centers as well.)* How is the importance of tourism to a country's economy reflected in its transportation system? *(Possible answer: Countries that depend on tourism generally have extensive transportation networks, including airports, to facilitate travel within the country.)* How might transportation maps reveal areas of wilderness or sparse settlement within a country? *(These areas will not be served by many transportation routes.)*

1. railways, highways, airport
2. Antalya, Bodrum, Giresun, İskenderun, İstanbul, Mersin
3. Students' maps should accurately identify important transportation features found in your state.

Geography
Skill-Building Workshop

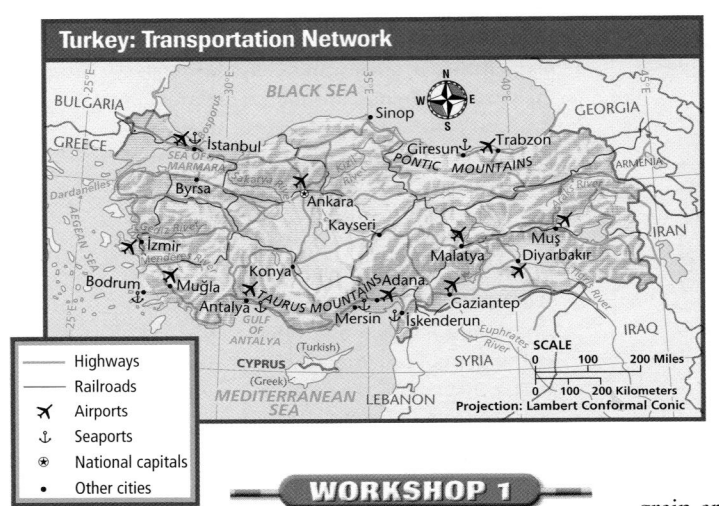

Turkey: Transportation Network

Legend:
— Highways
— Railroads
✈ Airports
⚓ Seaports
◉ National capitals
• Other cities

SCALE
0   100   200 Miles
0 100 200 Kilometers
Projection: Lambert Conformal Conic

WORKSHOP 1

Using Transportation Maps

Some maps help us do more than study physical features or locations. They can also identify important transportation routes. Road maps, for example, include roads and highways with numbers and symbols that identify each route. Some transportation maps, such as the example on this page, are more complex. Such maps include a web of railways, major roads and highways, and important seaports and airports. All of those features link different places.

Developing the Skill Transportation maps are not difficult to understand. A map legend shows the kinds of transportation routes featured on the map. Common symbols include those for major roads, railways, airports, and seaports. Sometimes highways and railways are labeled on these kinds of maps. Distances along roads and railways may also be marked in miles, kilometers, or both. If not, a distance scale helps you estimate the distance between places shown on the map. Mapmakers may also show other cities or well-known cultural features to serve as landmarks on the map.

The map on this page shows Turkey's transportation network. Turkey and the rest of Southwest Asia have long been located along major transportation and trade routes. For example, the ancient Silk Road once linked the region to eastern Asia. Today tankers and cargo ships carry oil and other goods along major shipping lanes around the region. The Bosporus, a narrow strait that separates European and Asian Turkey, is one of the region's major shipping lanes. In fact, Turkey has many seaports, some active since ancient times. Some of the most important seaports, including the city of İstanbul, are shown on the map.

Turkey also has many major airports. As you can see, the country's land transportation network includes railways, roads, and highways. Coal, other minerals, and grain are transported along Turkish railways for hundreds of miles to ports or markets. A complex road network also links towns throughout the country. Increasing automobile and truck traffic led to the construction of a bridge across the Bosporus, which was completed in 1973. This bridge, the Trans-European Motorway, has become important for trade between Europe and Southwest Asia.

Practicing the Skill

1. What transportation features link Ankara to the rest of Turkey?
2. Identify major Turkish seaports shown on the map.
3. Prepare a transportation map of your state. Identify important railways, roads, highways, airports, and seaports.

WORKSHOP 2

Analyzing Aerial Photographs

Aerial photographs are important tools geographers use to view Earth's surface. Satellites and other spacecraft orbiting the planet provide many aerial images.

Workshop 2
Going Further: Thinking Critically

Point out to students that there are many different types of aerial photographs and that each type has its own uses. Find examples of aerial photographs of the same area of the world that were created using each of the following methods: regular photography, false-color infrared photography, radar imaging, and lidar imaging. *(Aerial photographs are available online at the NASA and U.S. Geologic Survey Web sites and many other sites.)* Have students study these images to determine what information is contained in each. Then have each student write a brief paragraph explaining how each type of aerial photograph might be of use to geographers. *(Possible answers: Standard photographs taken from the air allow geographers to view phenomena from a different perspective than they can from the ground because they show subjects in normal colors. They are often hazy because atmospheric conditions prevent clear pictures. False-color photographs are not distorted by the atmosphere, so they allow geographers to study many surface features, like vegetation and bodies of water. They do not, however, show true colors so they are not as easy to interpret. Radar and lidar images are particularly useful for studying landforms and terrain because they show even small changes in elevation. Unlike infrared images, they are not blocked by cloud cover. Lidar images are often detailed enough to appear three-dimensional.)*

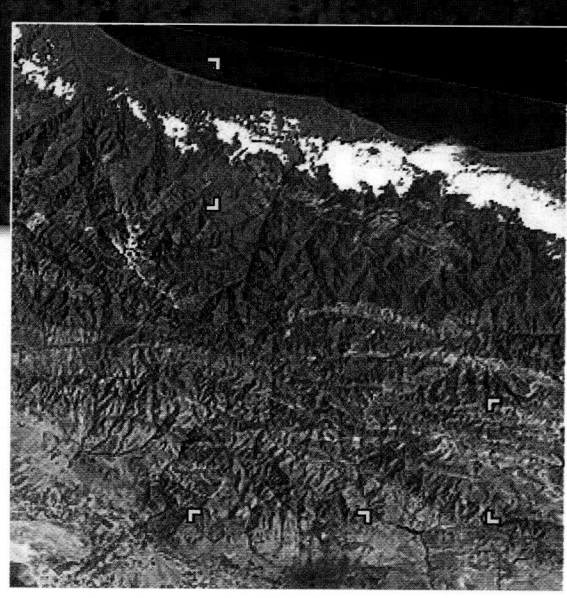

They use a variety of sensors to produce these images. These sensors include microwave detectors and scanners that measure and record electromagnetic radiation. Satellites send the data they gather to receiving stations on the ground. The data can then be converted into an image resembling a photograph.

The technology used in satellite imagery has links to World War II. Military researchers developed a special kind of color film during that war. It is known as false-color infrared film. The film captures the electromagnetic radiation that objects reflect. When this film is used, objects do not appear in their natural colors. The leaves of plants reflect very well. The healthier the vegetation is, the brighter the reflection, which appears in red on the film. Human structures like buildings, roads, and parking lots appear blue. Water appears black, unless it contains sediment or a lot of plant life.

Aerial photographs have been important to researchers. For example, geologists can use false-color aerial photographs to locate faults and other physical features on Earth's surface. The water that collects in such features encourages vegetation growth. This growth may be thicker than what is found in the surrounding area. As a result, areas along these physical features may show up in red, the color in which vegetation appears. In addition, the film clearly shows differences between vegetation and human features. As a result, it is useful in urban mapping.

Developing the Skill The following checklist can help you analyze satellite and aerial photographs. It will help you identify features in such images:

✓ Look at the shape of objects. Does the shape appear natural or human-made?

✓ Study the image's tone, or the lightness or darkness of its objects. Tone creates contrast and helps you distinguish between features.

✓ Analyze color, which fades with increasing distance. Colors also identify features.

✓ Try to locate a feature in the area that has a standard size, such as a football field. Compare other features to the one you have selected.

✓ Identify patterns, such as the repeated forms of sand dunes. Note that shadows can hide a feature or show the profile of its shape.

✓ Note the site and situation of a particular image. What is the relationship of the object to surrounding features? Look for links.

✓ Review the context of the image. Get as much information as possible about when and how the image was created.

Practicing the Skill

Use the checklist discussed above to analyze the aerial photograph on this page. The photograph shows an area of the Elburz Mountains in northern Iran. The city of Tehran (not shown) lies to the south, and the Caspian Sea is to the north. Create and complete a chart for your checklist. Then answer the following questions:

1. What do you suppose the white objects in the photograph represent?
2. What effect do the mountains appear to have on weather and climate patterns in the area?
3. Where is the thickest vegetation in the region? What might account for these vegetation patterns?

Using the Illustration

Direct students' attention to the photograph of the Great Sphinx. Point out that it is located in Giza, Egypt, near many of the country's largest pyramids, and is 241 feet (73.5 m) long. Archaeologists and Egyptologists are not sure why or when the sphinx was built, but many believe that it was intended as a guardian to protect the souls of deceased kings. Most scholars think it was built in about 2500 B.C. Unlike the pyramids near which it stands, the Great Sphinx was carved from a single piece of rock.

Ask students to describe how the physical environment appears to have affected the Great Sphinx since it was built. *(Possible answers: It appears to have been eroded, probably by wind and blown sand. The dry climate has probably protected the Sphinx from water erosion.)* Ask students to speculate about human activities that might have added to the damage done by physical processes. *(Possible answers: Structures could be damaged in wars or technological accidents. Pollution could harm the materials of which it is composed. Tourists who are not familiar with preservation techniques could worsen damage.)*

Unit Objectives

1. Describe the various landforms, bodies of water, climates, and natural resources of Africa and relate resource distribution to the economic development of the continent.

2. Identify ways that historical events have affected modern African societies.

3. Compare the cultures and ways of life of peoples in different regions of Africa.

4. Explain the economic, political, social, and environmental issues facing African countries.

5. Use questionnaires and field interviews to gather geographic information.

UNIT 7 Africa

The Great Sphinx and pyramids, Giza, Egypt

CONNECTING TO *Literature*

"AFRICAN SONG" *by Richard Rive*

Richard Rive

(1930–89) was a high school English teacher in South Africa before becoming a writer. His short stories and novels focus on the lives of black Africans under apartheid, a system of racial segregation that ended only recently. The selection below is from his short story "African Song." The story is about a young man named Muti, who attends a meeting to protest apartheid. It reflects the pride that black Africans take in their ancient roots.

And then everyone was standing and Muti watched fascinatedly as the people sang; but still he sat because he had no pass. And what Muti knew must happen was happening because the blue uniforms were coming nearer. . . . And still the people sang. . . .

And as they [the protesters] sang there was a deep calm.

And this is what they sang.
Nkosi Sikelel' Afrika . . . . .

which means God bless Africa. God bless the sun-scorched Karoo and the green Valley of a Thousand Hills. . . . God bless this Africa of heat and cold, and laughter and tears, and deep joy and bitter sorrow, God bless this Africa of blue skies and brown veld [the open grazing land], and black and white and love and hatred, and friend and enemy. . . .

God bless this Africa, this Africa which is part of us. God protect this Africa. God have mercy upon Africa.

And still they sang.

Maluphakonyisw' Upshondo Lwayo. . . . .

which is lift up our descendents. And Muti thought of himself and wondered if he were a better man than his father, and his father's father and the many before him. For he felt like the Great Bird that flies higher and higher till it is a brother to the sun and can see the land even before the White man came. . . .

But Muti did not understand. Where were the cities and the towns and the villages? And the buildings and the shops? And where were the ones who lived in the cities and the towns? And the White ones and the Black ones? . . . And when he searched even further for his own people, he found them at last, and then his heart burst with pride. For he saw proud warriors with plumes of ostrich feathers, and shaking armlets which clicked as they raised the hands. And these warriors were huge ones and proud, and lifted high the legs and stamped upon the earth so that the ground shook. For they danced the dance of the young men and it was a vigorous dance and required much strength. And they were fearsome to behold.

Analyzing the Primary Source

1. **Summarizing** What image does Rive create of Africa's physical geography?

2. **Finding the Main Idea** What message does Rive's writing convey about Africa's origins and the roles of black and white Africans in the continent's history and development?

A Closer Connection

"Nkosi sikelel' Afrika"
The passages sung at the meeting described by Richard Rive are the first lines of the anthem "Nkosi sikelel' Afrika," a song that was popular with opponents of South Africa's apartheid system. In 1925 it was named the official anthem of the African National Congress, an organization dedicated to ending apartheid. Singing it was considered an act of defiance. The song's popularity did not end, however, when apartheid was abolished. A shortened version was incorporated into the new national anthem of South Africa written in 1996. The original also serves as the national anthem of Tanzania, and it is popular in Namibia, Zambia, and Zimbabwe. Though it was written in Xhosa, the song has been translated into many languages, including Afrikaans, English, Sesotho, and Zulu.

Analyzing the Primary Source

Answers

1. It is a place of great natural beauty, variety, and contrasts—from sunscorched deserts to green hills and valleys and brown veld.

2. Possible answer: While both races have contributed to its urban and economic growth and development, its origins and heritage are rural, agricultural, and black African.

471

In this unit, students will learn that Africa's physical and human geography are extremely diverse. The continent contains the world's largest desert—the Sahara—and the longest river—the Nile. There are snowcapped mountains that border vast plains. In the tropics are dense rain forests.

Africa's human mosaic is equally varied. Hundreds of different ethnic groups speak numerous languages. Archaeological digs in East Africa have unearthed remains indicating that some of the earliest humans lived there. Wealthy, sophisticated kingdoms developed in many regions. Later, European merchants arrived, and the international slave trade began. The slave trade left a tragic legacy. European colonizers were next to arrive. Although they brought many innovations, Europeans often ignored the ancient political and cultural divisions among the people. The negative effects of colonialism are still being felt.

Today, Africa faces the future with advantages and disadvantages. Desertification and high birth rates are widespread problems. Millions of Africans live in poverty. However, the continent has rich undeveloped resources. In addition, more African governments are responding to their citizens' needs.

Your Classroom Time Line

These are the major dates and time periods for this unit. Have students enter them on the time line you created earlier. You may want to watch for these dates as students progress through the unit.

c.* 3000 B.C. The Egyptian civilization begins to flourish along the Nile River and its delta.

c. 2500 B.C. The great pyramids at Giza are built for Egyptian pharaohs.

1570–1085 B.C. Egypt expands to include what are now Libya, Israel, and Syria.

c. 800 B.C. The Phoenicians establish Carthage as a trading colony in what is now Tunisia.

332 B.C. Alexander the Great establishes the Greek city of Alexandria in Egypt.

146 B.C. The Roman Empire destroys Carthage and expands into North Africa.

A.D. 100s Bantu-speaking peoples begin to migrate into southern Africa.

350 The Kush kingdom in East Africa falls to Aksum.

*c. stands for *circa* and means "about."

UNIT 7 ATLAS

The World in Spatial Terms

Africa:
Political

1. **Places and Regions** Where are the capitals of Algeria, Tunisia, and Libya located?

2. **The World in Spatial Terms** How many landlocked countries are there in Africa? How many island countries are there?

Critical Thinking

3. **Analyzing Information** Compare this map to the physical map. Why do you think Namibia's northeastern boundary is shaped the way it is?

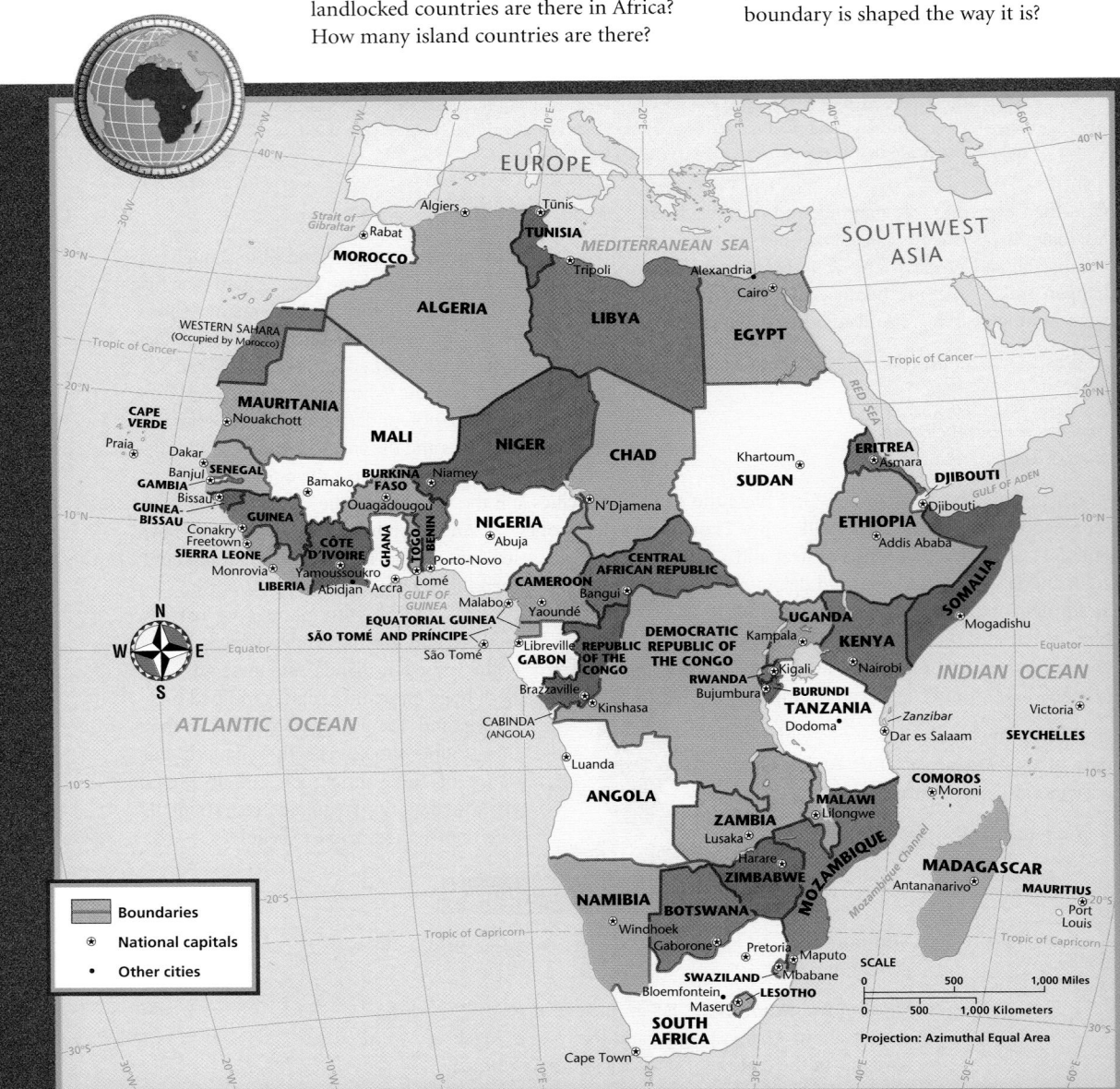

POLITICAL MAP ANSWERS

1. on the Mediterranean Sea coast
2. 15; 7

CRITICAL THINKING ANSWER

3. Possible answer: to provide access to the Zambezi River

Using the Political Map

Focus students' attention on the **political map** of Africa. Call on volunteers to name all the mainland countries that have a seacoast and others to name the landlocked countries they border. Ask students to locate the country that is divided into a small part and a large one (*Angola*). Have students compare this map to the **physical**, **climate**, and **population maps**. Ask students which boundaries appear to be unrelated to physical features (*straight lines in northern and southern Africa*). Then ask students to identify one reason that the boundaries may have been drawn in this way. (*Possible answer: They are in dry, sparsely populated areas where there are no rivers or major physical features to serve as natural boundaries.*)

Using the Physical Map

As students examine the **physical map** on this page, tell them that Africa is the second-largest continent. Then ask students what bodies of water surround Africa (*Mediterranean Sea, Atlantic Ocean, Gulf of Guinea, Indian Ocean, Gulf of Aden, Red Sea*). Africa has some great rivers and deserts. Ask students to name some of each (*deserts—Sahara, Kalahari, Namib; rivers—Nile, Congo, Niger, Zambezi, Orange*). Point out that across most of Africa the major landforms are basins and plateaus. Call on a volunteer to identify the region's major basin (*Congo Basin*). Then ask students where the region's largest plateau is located (*southern Africa*).

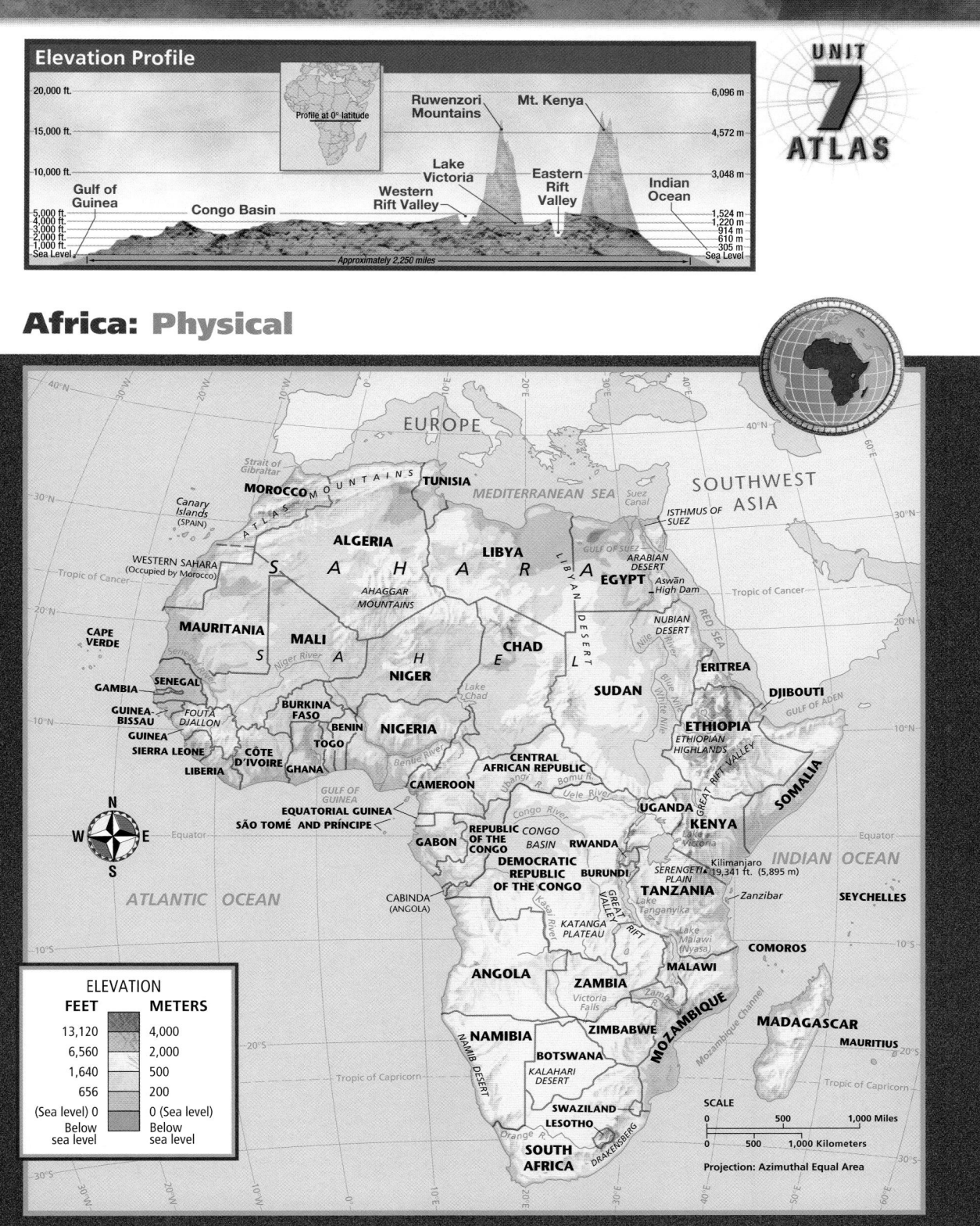

Africa: Physical

UNIT 7 ATLAS

Elevation Profile

Profile at 0° latitude

20,000 ft. — 6,096 m
15,000 ft. — 4,572 m
10,000 ft. — 3,048 m

Ruwenzori Mountains · Mt. Kenya

Gulf of Guinea · Congo Basin · Western Rift Valley · Lake Victoria · Eastern Rift Valley · Indian Ocean

5,000 ft.
4,000 ft.
3,000 ft.
2,000 ft.
1,000 ft.
Sea Level

1,524 m
1,220 m
914 m
610 m
305 m
Sea Level

Approximately 2,250 miles

ELEVATION

FEET		METERS
13,120		4,000
6,560		2,000
1,640		500
656		200
(Sea level) 0		0 (Sea level)
Below sea level		Below sea level

Your Classroom Time Line, (continued)

400s The Roman Empire collapses. The Vandals move south from Germany through Spain and into Africa.

c. 500s Arab traders sail along Africa's eastern coast.

500s The Byzantine Empire conquers much of North Africa.

600s Arab armies conquer most of North Africa, taking Morocco in the early 700s.

711 Arab armies cross into the Iberian Peninsula.

800s The Kingdom of Ghana rises to power in West Africa.

1200s The Kingdom of Mali gains power in West Africa.

1483 Portuguese sailors reach the coasts of southern Africa.

1500s The Ottomans invade North Africa.

1500s The Portuguese establish their first settlements in East Africa.

1500s–1800s Europeans trade with African kingdoms for slaves to work in their colonies.

1600s–1700s Southeast Asians are brought to southern Africa as slaves.

1652 Dutch settlers establish a colony in the Cape of Good Hope.

Early 1800s Great Britain takes over the Cape of Good Hope area.

1800s Missionaries begin to spread Christianity throughout East Africa.

1800s European explorers track the Nile to its source.

1800s The Sotho people establish a kingdom in what is now Lesotho.

1830s France takes over Algeria, Tunisia, and part of Morocco.

1848 Europeans reach the Great Rift Valley in East Africa.

473

Using the Climate Map

Focus students' attention on the **climate map** on this page, and have them identify the climate region of Africa that most resembles the climate in your area. Have students compare the latitude of their location with that of the corresponding African climate. Then ask students to write statements relating climate and latitude. *(Possible answers: Places in low latitudes generally have a hot climate. Those in middle latitudes have a temperate climate.)*

Then have students compare this map to the **physical map**. Ask them which countries have a highland climate region although they are on the equator or within 10 degrees of it *(Democratic Republic of the Congo, Uganda, Rwanda, Burundi, Tanzania, Kenya, Ethiopia)*. Then ask students how elevation affects climate. *(Higher elevations create milder climates, even at very low latitudes.)*

Your Classroom Time Line, (continued)

Mid–1800s Trade in enslaved Africans gradually slows and stops.

Mid–1800s American and European explorers begin to explore Africa's interior.

Late 1800s The discovery of diamonds and gold draws people from all over the world to southern Africa.

1884 European powers meet in Berlin to settle colonial disputes. Most of Africa is eventually divided into European colonies.

1884–1976 Spain rules Western Sahara.

1899–1902 British soldiers and the Boers of South Africa clash in the Boer War.

1910 The Union of South Africa is granted independence from Great Britain.

1912 The African National Congress (ANC) is founded in South Africa.

1922 Egypt gains limited independence from Great Britain.

1948 The apartheid system is established in South Africa.

1950s–60s Many East African countries gain their independence.

1957 Ghana becomes independent.

CLIMATE MAP ANSWERS

1. Ethiopia
2. Morocco, Algeria, Tunisia, Libya, South Africa

CRITICAL THINKING ANSWER

3. tropical humid, tropical wet and dry, semiarid, arid, possibly Mediterranean climates; tropical humid, tropical wet and dry, semiarid, possibly arid, possibly marine west coast

UNIT 7 ATLAS

Africa:
Climate

1. **Places and Regions** Compare this map to the physical map. Which country has a large area of highland climate?

2. **Places and Regions** Compare this map to the political map. Which countries have a Mediterranean climate?

Critical Thinking

3. **Comparing** If you traveled north from the equator, which climate regions would you pass through? Which would you pass through if you traveled south from the equator?

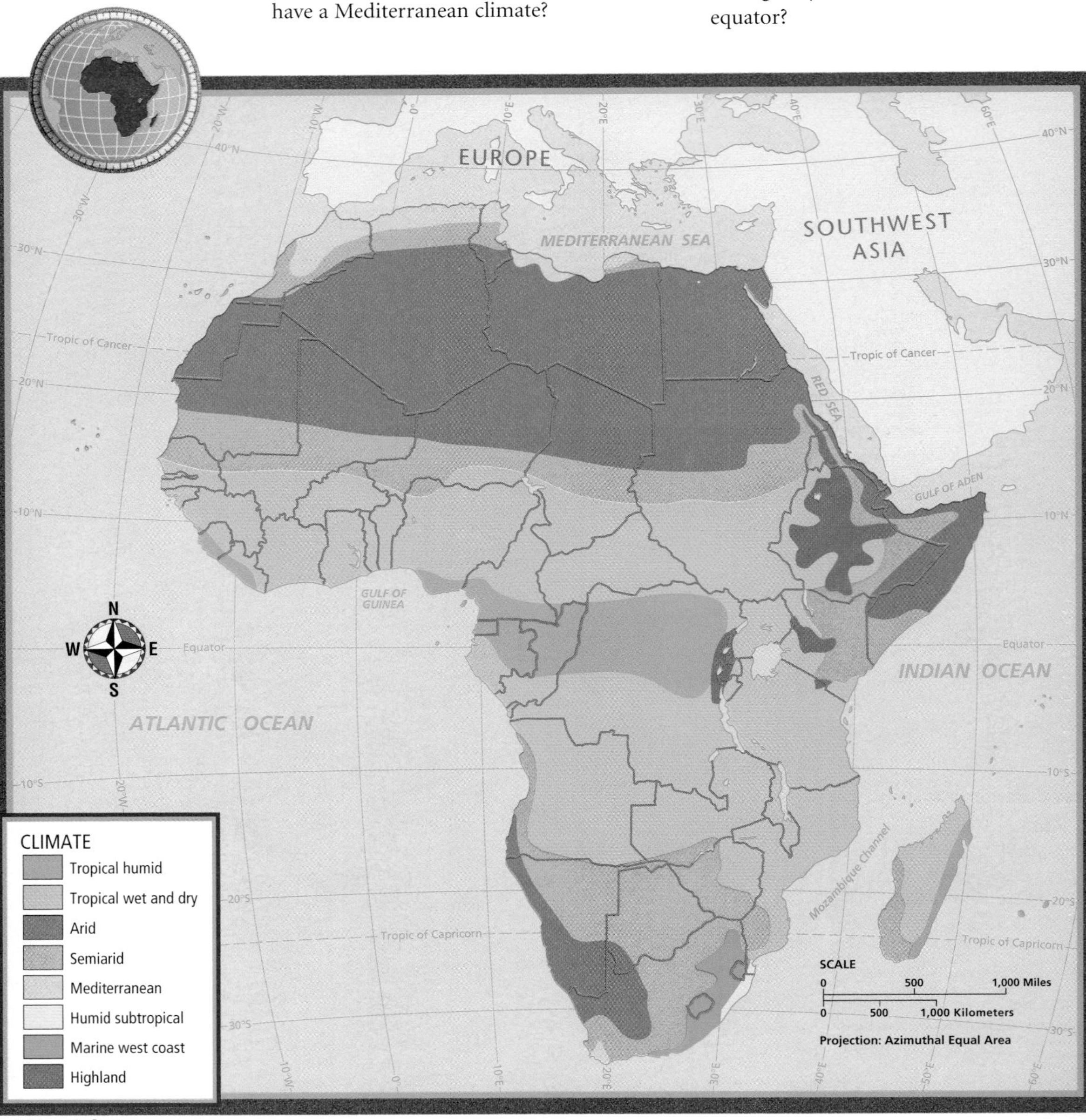

EUROPE

MEDITERRANEAN SEA

SOUTHWEST ASIA

Tropic of Cancer

Tropic of Cancer

RED SEA

GULF OF ADEN

GULF OF GUINEA

Equator

INDIAN OCEAN

ATLANTIC OCEAN

Mozambique Channel

Tropic of Capricorn

Tropic of Capricorn

CLIMATE
- Tropical humid
- Tropical wet and dry
- Arid
- Semiarid
- Mediterranean
- Humid subtropical
- Marine west coast
- Highland

SCALE
0   500   1,000 Miles
0   500   1,000 Kilometers

Projection: Azimuthal Equal Area

Using the Precipitation Map

Call students' attention to the **precipitation map** on this page. Ask them which parts of Africa receive the most rain each year *(interior Congo Basin, west coasts along Atlantic Ocean and Gulf of Guinea, east coast of Madagascar)*. Ask students what type of vegetation they would expect to find in these areas, given their precipitation and latitude *(tropical rain forests)*. Then ask students to identify the driest parts of Africa *(Sahara,*

Namib Desert, parts of Somalia, Kenya, and Ethiopia). Call on volunteers to summarize the patterns they see in Africa's annual precipitation. *(Possible answers: The central parts of the continent receive more rain than either the northern or the southern parts. Within Central Africa, western areas tend to be wetter than places in the east.)*

Africa:
Precipitation

1. **Physical Systems** Compare this map to the physical map. Which mountain range in northern Africa receives the most precipitation?
2. **Physical Systems** Between which latitudes do the highest amounts of precipitation fall?

Critical Thinking

3. **Analyzing Information** Compare this map to the climate and population maps. How do you think people have adapted to living in arid and semiarid regions?

UNIT
7
ATLAS

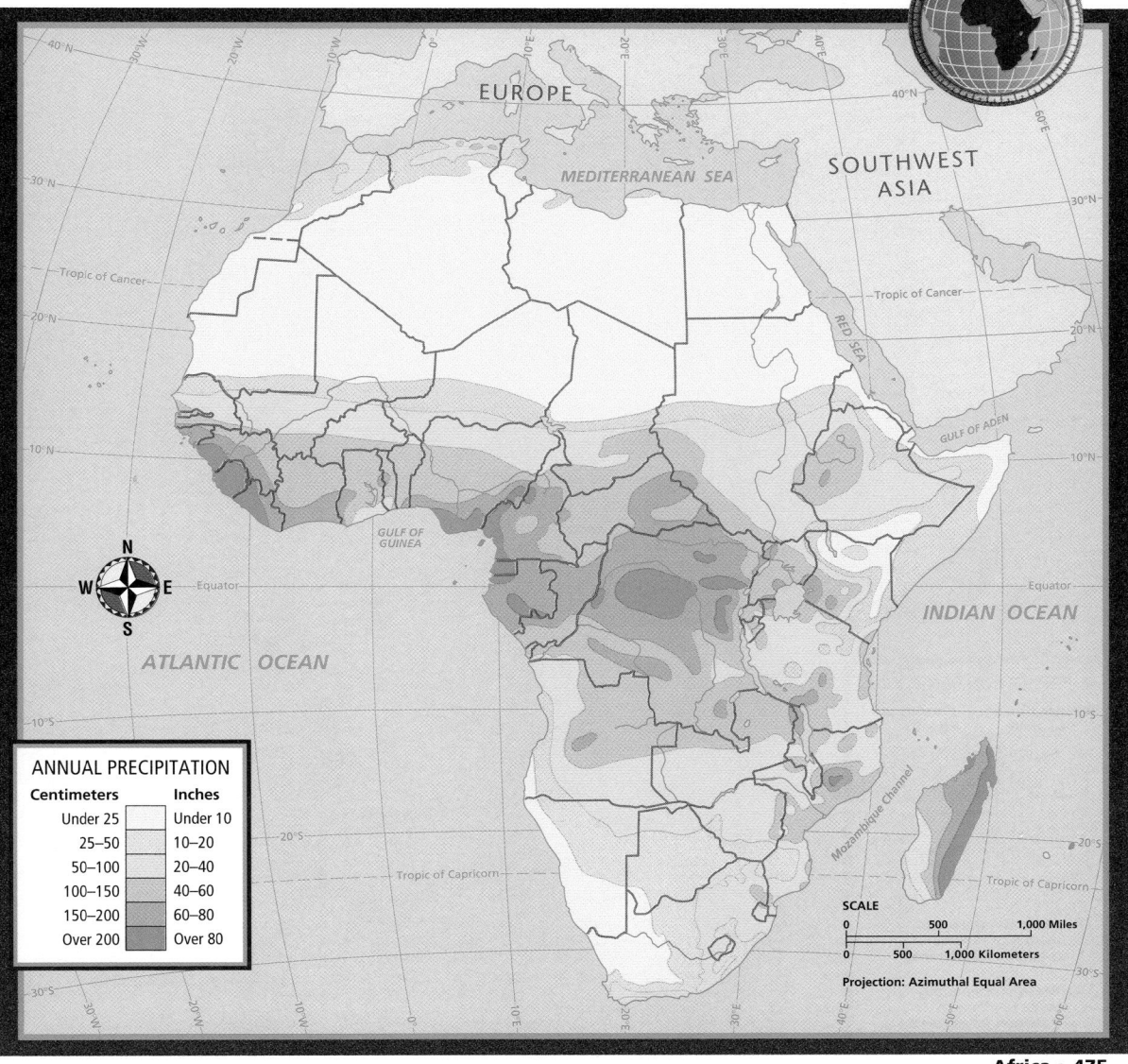

ANNUAL PRECIPITATION

Centimeters	Inches
Under 25	Under 10
25–50	10–20
50–100	20–40
100–150	40–60
150–200	60–80
Over 200	Over 80

SCALE
0   500   1,000 Miles
0   500   1,000 Kilometers
Projection: Azimuthal Equal Area

Your Classroom Time Line, (continued)

1960s–70s Many southern African nations gain their independence.

1960 Nigeria becomes independent.

1960 Construction of Egypt's Aswān High Dam begins.

1962 Nelson Mandela is sentenced to life in prison.

1960s–70s Cities in West and Central Africa experience rapid growth.

1975 Communists take control of Angola.

1976 The last European colonies in West and Central Africa are granted independence.

1977 Djibouti becomes independent.

1979 Mauritania breaks away from Western Sahara.

1980 Southern Rhodesia breaks away from the northern part of the country and is renamed Zimbabwe. The north becomes Zambia.

1984 Severe drought strikes Ethiopia.

1990s Severe drought strikes Somalia.

1990s South Africa moves away from its apartheid system.

1990 Nelson Mandela is released from prison.

PRECIPITATION MAP ANSWERS

1. the Atlas Mountains
2. between about 10 degrees north and south latitude

CRITICAL THINKING ANSWER

3. Population is mostly limited to areas where water is available, such as along the Nile River and near oases.

475

Using the Population Map

Have students examine the **population map** of Africa on this page. Ask them to state relative locations of the areas with the highest population densities. List student responses on the chalkboard. (*Possible answers: near the Gulf of Guinea, in the far northeast, near Lake Victoria in East Africa, in the far southeast*) Then ask students to compare this map to the **political map**. Ask volunteers to identify countries that correspond to the areas of high population density and to write the countries' names next to the relative location descriptions on the chalkboard.

Then ask students to identify the cities in Africa that have a population of more than 2 million (*Addis Ababa, Abidjan,*

Alexandria, Algiers, Cairo, Cape Town, Casablanca, Dakar, Dar es Salaam, Durban, Giza, Johannesburg, Khartoum, Kinshasa, Lagos, Luanda, Maputo, Nairobi). Ask students to speculate why Africa has relatively few large cities. You may wish to have students consult the **land use and resources map** for reference. (*Possible answers: Most people in Africa appear to be engaged in subsistence agriculture. Many countries are developing and have not yet reached a sufficient level of economic development to maintain cities.*)

Your Classroom Time Line, (continued)

1991 A cease-fire is imposed in Western Sahara.

1992 Political violence erupts in Algeria.

1993 Eritrea gains independence from Ethiopia.

1994 Nelson Mandela is elected president of South Africa.

1994 Violence breaks out between the Hutu and the Tutsi in Rwanda.

1999 Nigeria adopts a new democratic constitution.

1999 Flooding in Mozambique displaces hundreds of thousands of people.

2001 The Nobel Peace Prize is awarded to the United Nations and its Secretary-General, Kofi Annan of Ghana.

2003 The Nobel Prize for Literature is awarded to John Maxwell Coetzee of South Africa.

POPULATION MAP ANSWERS

1. Egypt; the Nile River
2. They have arid climates.

CRITICAL THINKING ANSWER

3. Possible answer: mostly rural because many countries have relatively even population densities and not many large cities

UNIT 7 ATLAS

Africa: Population

1. (**Places and Regions**) Compare this map to the physical map. Which country in North Africa has an area with more than 520 persons per square mile (200 per sq km)? Along what physical feature are these people concentrated?

2. (**Places and Regions**) Compare this map to the climate map. Why do you think much of North Africa and parts of southern Africa have such low population densities?

Critical Thinking

3. **Making Generalizations** Based on the map, do you think most African countries have largely rural or largely urban populations? Why?

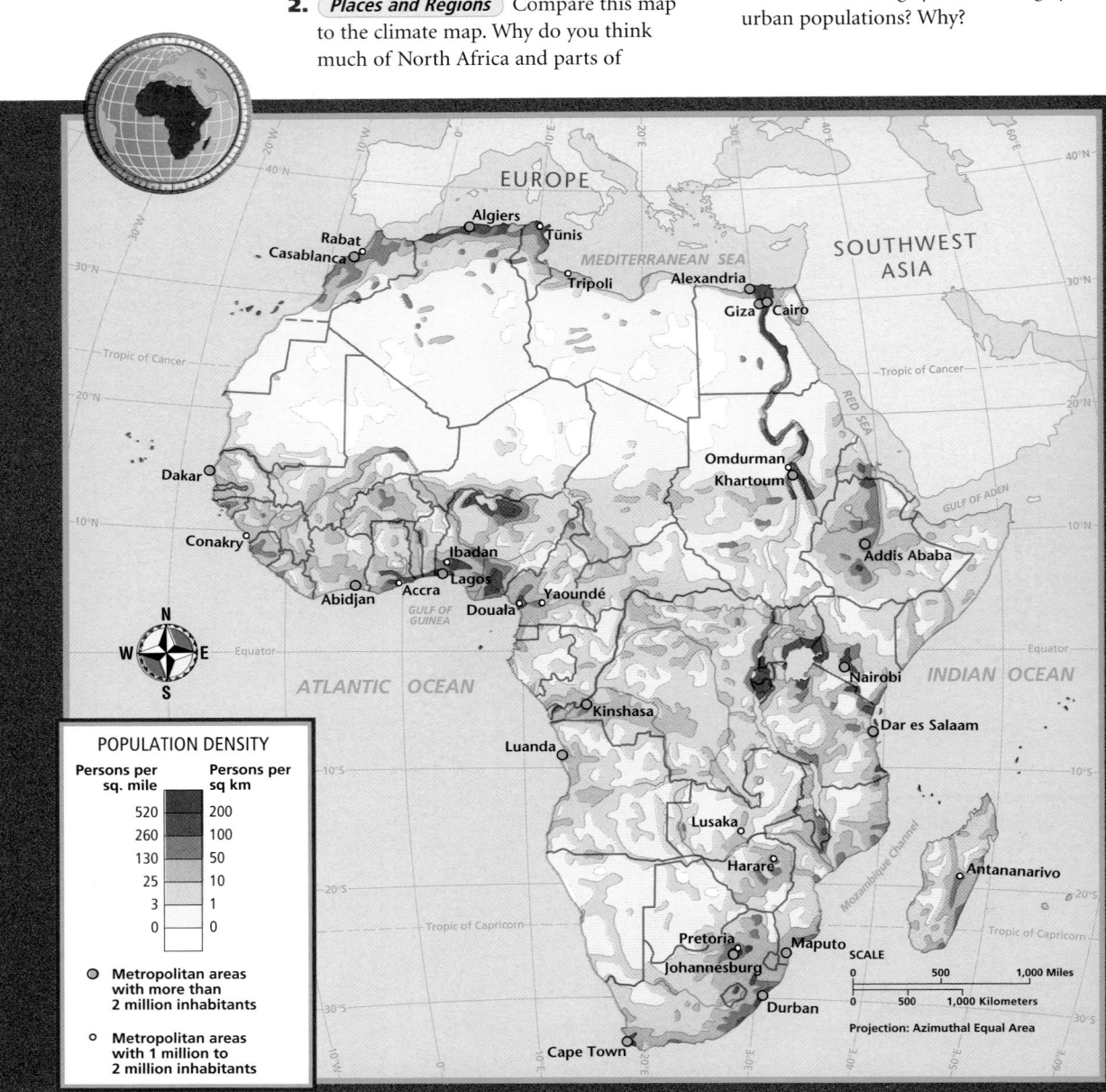

POPULATION DENSITY

Persons per sq. mile	Persons per sq km
520	200
260	100
130	50
25	10
3	1
0	0

○ Metropolitan areas with more than 2 million inhabitants

○ Metropolitan areas with 1 million to 2 million inhabitants

Focus students' attention on the **land use and resources map** on this page. Call on volunteers to make general statements about the distribution of resources. (*Examples: Resources are not distributed evenly. Most of the oil and gas industry is in northern and western Africa. Some countries have very few resources.*)

Have students compare this map to the **political map**. Ask them to identify countries with no manufacturing centers and few resources (*Mali, Chad, Djibouti, Somalia*). Ask what eco-nomic activities are located in these countries instead (*nomadic herding, subsistence farming, small amounts of commercial farm-ing, and livestock raising*). Then ask whether commercial agri-culture or subsistence agriculture is more prevalent in Africa (*subsistence*). Ask students to draw conclusions about eco-nomic development and standards of living in much of Africa, based on the information seen in this map. (*Possible answer: Most African countries appear to have developing economies, so it is likely that their standards of living are low.*)

Africa:
Land Use and Resources

UNIT 7 ATLAS

1. (*Environment and Society*) Compare this map to the climate map. How do farmers appear to have adapted to environmental conditions in arid and semiarid areas?

2. (*Human Systems*) What are some countries that appear to have mostly traditional economies?

Critical Thinking

3. **Drawing Inferences** How widespread is subsistence farming in Africa? What might this indicate about Africa's level of development?

LAND USE AND RESOURCES MAP ANSWERS

1. irrigation, particularly along rivers like the Nile River
2. Students should identify countries where subsistence farming and herding are par-ticularly common.

CRITICAL THINKING ANSWER

3. Subsistence farming is very widespread. Many countries probably have low levels of development.

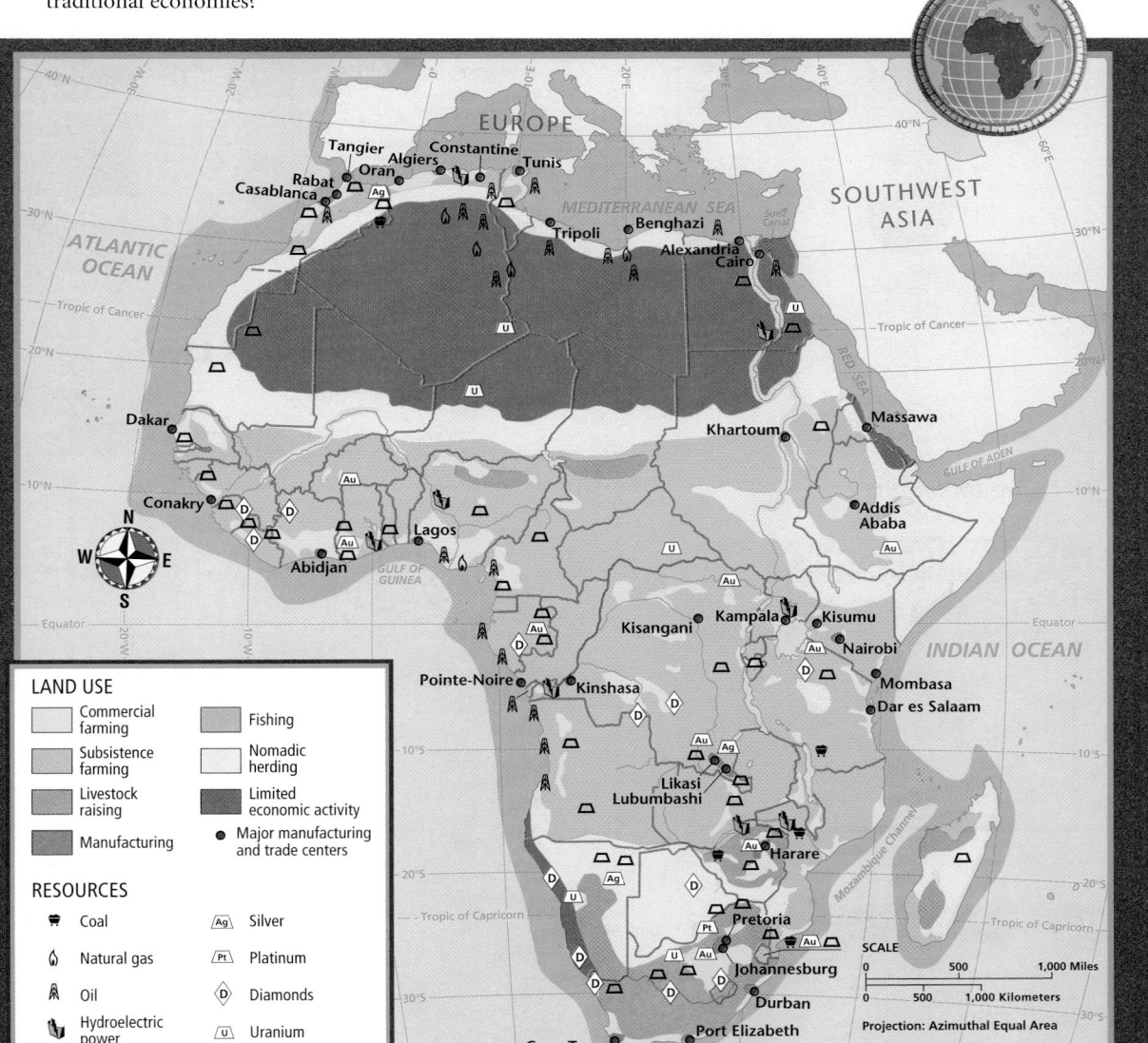

LAND USE

- Commercial farming
- Subsistence farming
- Livestock raising
- Manufacturing
- Fishing
- Nomadic herding
- Limited economic activity
- ● Major manufacturing and trade centers

RESOURCES

- 🪨 Coal
- ◊ Natural gas
- 🛢 Oil
- ⚡ Hydroelectric power
- Au Gold
- Ag Silver
- Pt Platinum
- ◇ Diamonds
- U Uranium
- ⬡ Other minerals

SCALE
0   500   1,000 Miles
0   500   1,000 Kilometers
Projection: Azimuthal Equal Area

477

Create a bar graph on the chalkboard, listing figures in increments of 5,000 (starting at 0) through 50,000 along the y-axis (vertical axis). Tell students that these figures represent the number of people per doctor in each African country. Draw a bar on the graph for the United States, which has about 408 people per doctor. The bar for the United States should be very low and barely visible on the graph. Point out that this is because the United States has a much higher ratio of doctors to people than most countries in Africa. Call on students to come to the chalkboard and create a label and bar for each of the countries listed on the following chart. Continue until all countries, or a representative sample, have been placed on the

graph. Ask students what the information on the graph suggests about health care in Africa. *(Possible answer: There are not enough trained doctors to make health care available to much of Africa's population.)*

Point out that for countries not listed on the chart, information about the number of doctors was not available. Ask students to explain why they think this data is not available for many African countries. *(Possible answers: Many of these countries are experiencing economic, political, or social tensions, and the governments are not stable enough to gather the information. Africa's physical geography—rivers, jungles, mountains, deserts—may also make data collection difficult.)*

Historical Geography

Egypt and the Sahara

Until about 6,000 years ago, much of northern Africa was covered by lakes and grasslands. Scientists believe that a combination of climate and vegetation shifts led to their replacement by a harsh desert. Some people believe that it was the growth of the Sahara that led to the establishment of Egyptian civilization.

Archaeologists have discovered settlements near the Egypt-Sudan border that were built before the Sahara became a desert. Evidence suggests that the inhabitants of these villages had developed agriculture and domesticated various animals long before similar activities were begun in the Nile Valley. These villagers also cut huge blocks of stone that some scholars believe might be the forerunners of Egypt's pyramids. Archaeologists believe that increasingly harsh conditions drove these people into the Nile Valley, where within a few centuries an advanced civilization had developed.

DISCUSSION: Lead a discussion about the possible effects of climate or vegetation changes on societies.

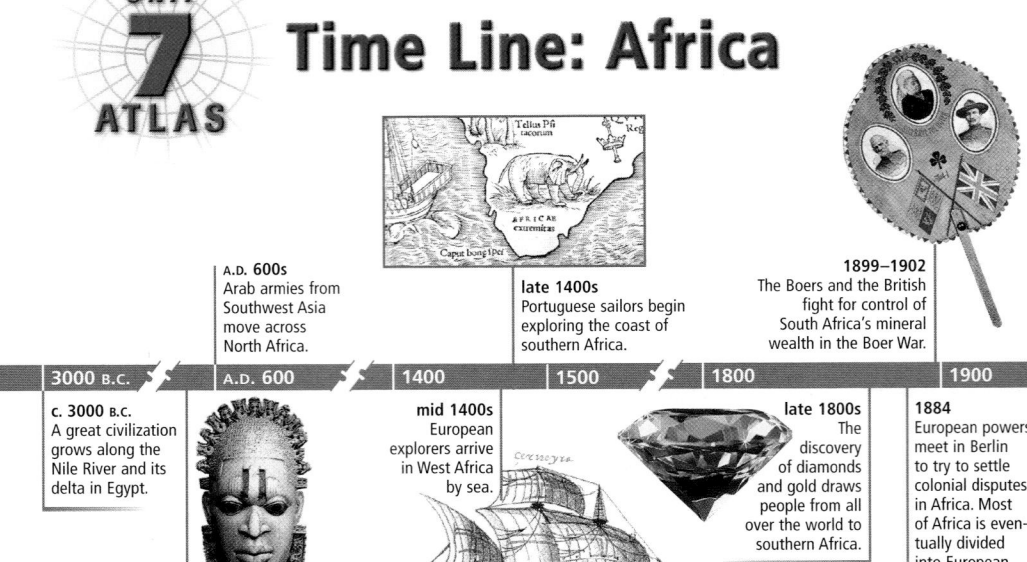

UNIT 7 ATLAS
Time Line: Africa

c. 3000 B.C. A great civilization grows along the Nile River and its delta in Egypt.

A.D. 200s Great Zimbabwe flourishes.

A.D. 600s Arab armies from Southwest Asia move across North Africa.

mid 1400s European explorers arrive in West Africa by sea.

late 1400s Portuguese sailors begin exploring the coast of southern Africa.

late 1800s The discovery of diamonds and gold draws people from all over the world to southern Africa.

1884 European powers meet in Berlin to try to settle colonial disputes in Africa. Most of Africa is eventually divided into European colonies.

1899–1902 The Boers and the British fight for control of South Africa's mineral wealth in the Boer War.

1950s and 1960s Most African countries gain their independence.

1994 Nelson Mandela is elected South Africa's first black president.

Timeline: 3000 B.C. | A.D. 600 | 1400 | 1500 | 1800 | 1900

The United States and Africa
Comparing Sizes

☑ internet connect

GO TO: go.hrw.com
KEYWORD: SW3 Almanac
FOR: Additional information and reference sources

Comparing Standard of Living

COUNTRY	LIFE EXPECTANCY (in years)	INFANT MORTALITY (per 1,000 live births)	LITERACY RATE	DAILY CALORIC INTAKE (per person)
Congo, Democratic Republic of the	47, male 51, female	97	84%	1,701
Egypt	68, male 73, female	35	58%	3,282
Ethiopia	40, male 42, female	103	43%	1,805
Kenya	45, male 45, female	63	85%	1,968
Libya	74, male 78, female	27	83%	3,267
Morocco	68, male 72, female	45	52%	3,165
Nigeria	51, male 51, female	71	68%	2,882
Rwanda	39, male 40, female	103	70%	2,035
South Africa	47, male 47, female	61	87%	2,909
Zimbabwe	40, male 38, female	67	91%	2,153
United States	74, male 80, female	7	97%	3,757

Sources: *World Almanac and Book of Facts, 2004; Britannica Book of the Year, 2002*

COUNTRY	PEOPLE PER DOCTOR				
Algeria	1,205	Ethiopia	25,000	Namibia	4,348
Benin	16,667	Gabon	5,263	Niger	33,333
Burundi	16,667	Gambia	50,000	Nigeria	4,762
Cameroon	14,286	Ghana	25,000	São Tomé and Príncipe	3,125
Cape Verde	3,448	Guinea	6,667	Senegal	14,286
Central African Republic	16,667	Guinea-Bissau	5,556	Seychelles	962
Chad	50,000	Kenya	6,667	South Africa	1,695
Comoros	10,000	Lesotho	20,000	Sudan	10,000
Congo, Republic of the	3,704	Madagascar	4,167	Tanzania	25,000
Djibouti	5,000	Malawi	50,000	Togo	16,667
Egypt	495	Mali	25,000	Tunisia	1,493
Equatorial Guinea	4,762	Mauritania	9,091	Uganda	25,000
Eritrea	50,000	Mauritius	1,176	Zimbabwe	7,143
		Morocco	2,941	United States	408

Source: *United Nations Development Programme: Health Profile*

Fast Facts: Africa

UNIT 7 ATLAS

FLAG	COUNTRY / Capital	POPULATION (in millions) / POP. DENSITY	AREA	PER CAPITA GDP (in US $)	WORKFORCE STRUCTURE (largest categories)	ELECTRICITY CONSUMPTION (kilowatt hours per person)	TELEPHONE LINES (per person)
	Algeria / Algiers	31.8 / 35/sq. mi.	919,594 sq. mi. / 2,381,738 sq km	$ 5,300	29% government / 25% agriculture	720 kWh	0.06
	Angola / Luanda	13.6 / 28/sq. mi.	481,353 sq. mi. / 1,246,699 sq km	$ 1,600	85% agriculture / 15% industry, serv.	99 kWh	0.006
	Benin / Porto-Novo	6.7 / 158/sq. mi.	43,483 sq. mi. / 112,620 sq km	$ 1,070	55% agriculture / 21% trade	94 kWh	0.009
	Botswana / Gaborone	1.8 / 8/sq. mi.	231,804 sq. mi. / 600,370 sq km	$ 9,500	24% services / 22% agriculture	876 kWh	0.08
	Burkina Faso / Ouagadougou	13.0 / 123/sq. mi.	105,869 sq. mi. / 274,199 sq km	$ 1,080	90% agriculture	20 kWh	0.004
	Burundi / Bujumbura	6.8 / 689/sq. mi.	10,745 sq. mi. / 27,829 sq km	$ 600	93% agriculture / 4% government	26 kWh	0.003
	Cameroon / Yaoundé	16.0 / 88/sq. mi.	183,568 sq. mi. / 475,439 sq km	$ 1,700	70% agriculture / 13% ind., commerce	210 kWh	0.006
	Cape Verde / Praia	0.5 / 298/sq. mi.	1,557 sq. mi. / 4,033 sq km	$ 1,400	25% agriculture / 19% construction	84 kWh	0.15
	Central African Republic / Bangui	3.9 / 16/sq. mi.	240,535 sq. mi. / 622,983 sq km	$ 1,300	74% agriculture / 8% trade	26 kWh	0.002
	Chad / N'Djamena	8.6 / 18/sq. mi.	495,755 sq. mi. / 1,284,000 sq km	$ 1,100	85% agriculture	10 kWh	0.001
	Comoros / Moroni	0.8 / 917/sq. mi.	838 sq. mi. / 2,170 sq km	$ 720	80% agriculture	26 kWh	0.01
	Congo, Democ. Republic of the / Kinshasa	52.8 / 60/sq. mi.	905,567 sq. mi. / 2,345,408 sq km	$ 610	65% agriculture / 19% services	73 kWh	0.0004
	Congo, Republic of the / Brazzaville	3.7 / 28/sq. mi.	132,047 sq. mi. / 342,000 sq km	$ 900	52% agriculture, forestry, fishing	170 kWh	0.006
	Côte d'Ivoire / Yamoussoukro	16.6 / 135/sq. mi.	124,502 sq. mi. / 322,469 sq km	$ 1,500	51% agriculture / 12% manuf., mining	179 kWh	0.02
	Djibouti / Djibouti	0.7 / 83/sq. mi.	8,880 sq. mi. / 22,999 sq km	$ 1,300	75% agriculture / 14% services	238 kWh	0.01
	Egypt / Cairo	71.9 / 187/sq. mi.	386,662 sq. mi. / 1,001,450 sq km	$ 3,900	49% services / 29% agriculture	973 kWh	0.09
	Equatorial Guinea / Malabo	0.5 / 46/sq. mi.	10,831 sq. mi. / 28,052 sq km	$ 2,700	58% agriculture / 8% services	44 kWh	0.02

Sources: Central Intelligence Agency, *World Factbook 2003; The World Almanac and Book of Facts, 2004*
The CIA calculates per capita GDP in terms of purchasing power parity. This formula equalizes the purchasing power of each country's currency.

Historical Geography

The Dyula Long before Arabs or Europeans reached West Africa, the kingdoms of that region had established trade with cultures north of the Sahara. The Arabs who conquered North Africa wanted to continue to trade for West African gold, but the kings of that region refused to allow traders to contact gold producers directly. Instead, they insisted that all trade be conducted through intermediary groups. Among these groups were the Dyula, a Mande-speaking people of West Africa who had long been renowned as traders. Their primary role in this trade was to carry gold from the mines of West Africa to trading stations at the southern edge of the Sahara. There the gold was exchanged for other goods provided by Arab and Berber merchants.

The Dyula began to abandon their trading way of life in the 1500s. Many settled in cities or towns in what are now Côte d'Ivoire, Burkina Faso, Mali, and Ghana.

CRITICAL THINKING:
Why did the kings of West Africa refuse to trade with gold miners directly? *(The kings wanted to establish firm control over all trade routes through their countries.)*

479

Focus students' attention on the Fast Facts tables below. As a class, examine the information presented for the Democratic Republic of the Congo. Have students look at the country's population density to determine whether the country is crowded. (*Possible answer: probably not, because its population density of 60 people per square mile is lower than the United States's density of 83 per square mile*) Point out that the majority of the country's population is engaged in agriculture. Have students look at the figure listed on the chart for the country's per capita GDP and predict whether most people practice commercial agriculture or subsistence agriculture. (*Possible*

answer: probably subsistence, since the per capita GDP is so low) Then ask students to speculate about the country's standard of living based on its residents' electricity consumption and telephone ownership. (*Possible answer: probably have a low standard of living, with few appliances or modern conveniences*) Ask if the information in the Comparing Standard of Living table supports this conclusion. (*Yes, it does. The country has a low life expectancy and a high infant mortality rate. These are generally signs of a low standard of living.*)

Historical Geography

The Great Trek   In the early 1800s, the British took over the former Dutch settlement of Cape Colony on the southern tip of Africa. Unhappy with British rule, between 12,000 and 14,000 Afrikaners, or Dutch colonists in Africa, left the colony and moved inland. Their migration is known as the Great Trek. Many Afrikaners regard the event as the beginning of their existence as a distinct people.

The migrants, or Voortrekkers, crossed the Orange River in the 1830s and fought a series of battles against the native peoples of the region. After winning recognition from the British in South Africa, they established two states, the Orange Free State and the Transvaal. Both states later came into conflict with Britain again, and in 1899 war broke out. When the so-called Boer War ended in 1902, the Orange Free State and the Transvaal were once again brought under British control. When South Africa became independent in 1910, they became two of its provinces.

ACTIVITY: Have students conduct research to learn more about the Great Trek and to write journal entries that describe the thoughts and lives of the Voortrekkers.

Fast Facts: Africa

FLAG	COUNTRY Capital	POPULATION (in millions) POP. DENSITY	AREA	PER CAPITA GDP (in US $)	WORKFORCE STRUCTURE (largest categories)	ELECTRICITY CONSUMPTION (kilowatt hours per person)	TELEPHONE LINES (per person)
	Eritrea Asmara	4.1 88/sq. mi.	46,842 sq. mi. 121,320 sq km	$ 740	80% agriculture 20% ind., services	50 kWh	0.009
	Ethiopia Addis Ababa	70.7 163/sq. mi.	435,186 sq. mi. 1,127,127 sq km	$ 750	80% agriculture 12% govt., services	23 kWh	0.005
	Gabon Libreville	1.3 13/sq. mi.	103,347 sq. mi. 267,668 sq km	$ 5,700	60% agriculture 25% services, gov.	559 kWh	0.03
	Gambia Banjul	1.4 369/sq. mi.	4,363 sq. mi. 11,300 sq km	$ 1,800	75% agriculture 19% ind., comm., serv.	56 kWh	0.02
	Ghana Accra	20.9 236/sq. mi.	92,456 sq. mi. 239,460 sq km	$ 2,100	60% agriculture 25% services	422 kWh	0.01
	Guinea Conakry	8.5 89/sq. mi.	94,926 sq. mi. 245,857 sq km	$ 2,000	80% agriculture 20% ind., services	87 kWh	0.003
	Guinea-Bissau Bissau	1.5 138/sq. mi.	13,946 sq. mi. 36,120 sq km	$ 800	82% agriculture	34 kWh	0.008
	Kenya Nairobi	32.0 146/sq. mi.	224,962 sq. mi. 582,649 sq km	$ 1,020	75–80% agriculture	124 kWh	0.01
	Lesotho Maseru	1.8 154/sq. mi.	11,720 sq. mi. 30,355 sq km	$ 2,700	86% subsistence agriculture	22 kWh	0.02
	Liberia Monrovia	3.4 91/sq. mi.	43,000 sq. mi. 111,369 sq km	$ 1,100	70% agriculture 22% services	129 kWh	0.002
	Libya Tripoli	5.6 8/sq. mi.	679,362 sq. mi. 1,759,540 sq km	$ 7,600	54% services 29% industry	3,381 kWh	0.11
	Madagascar Antananarivo	17.4 78/sq. mi.	226,657 sq. mi. 587,039 sq km	$ 760	76% agriculture 12% manufacturing	44 kWh	0.003
	Malawi Lilongwe	12.1 333/sq. mi.	45,745 sq. mi. 118,479 sq km	$ 670	86% agriculture	59 kWh	0.006
	Mali Bamako	13.0 28/sq. mi.	478,766 sq. mi. 1,239,998 sq km	$ 860	80% agriculture, fishing	34 kWh	0.004
	Mauritania Nouakchott	2.9 7/sq. mi.	397,955 sq. mi. 1,030,699 sq km	$ 1,900	50% agriculture 40% services	51 kWh	0.01
	Mauritius Port Louis	1.2 1,710/sq. mi.	788 sq. mi. 2,041 sq km	$ 11,000	36% const., industry 24% services	998 kWh	0.27
	Morocco Rabat	30.6 177/sq. mi.	172,414 sq. mi. 446,550 sq km	$ 3,900	50% agriculture 35% services	478 kWh	0.04

Fast Facts Activity

Organize the class into pairs or have them work individually to examine other African countries in this manner. Assign each student or pair of students three countries from various parts of the continent. Have students study the figures in the chart to determine their assigned countries' level of development and standard of living. Students may wish to consult additional resources, such as the *CIA World Factbook* or the *World Almanac* for additional information. Then have students present their conclusions about each of their countries to the class. Have students keep their summaries in their notebooks so they can review them after they have studied the countries of Africa in more depth.

FLAG	COUNTRY / Capital	POPULATION (in millions) / POP. DENSITY	AREA	PER CAPITA GDP (in US $)	WORKFORCE STRUCTURE (largest categories)	ELECTRICITY CONSUMPTION (kilowatt hours per person)	TELEPHONE LINES (per person)
	Mozambique Maputo	18.9 / 62 sq. mi.	309,495 sq. mi. / 801,588 sq km	$ 1,000	81% agriculture 13% services	74 kWh	0.005
	Namibia Windhoek	2.0 / 6/sq. mi.	318,695 sq. mi. / 825,416 sq km	$ 6,900	47% agriculture 33% services	304 kWh	0.06
	Niger Niamey	12.0 / 24/sq. mi.	489,191 sq. mi. / 1,266,999 sq km	$ 830	90% agriculture 6% ind., commerce	27 kWh	0.002
	Nigeria Abuja	124.0 / 353/sq. mi.	356,669 sq. mi. / 923,768 sq km	$ 875	70% agriculture 20% services	117 kWh	0.006
	Rwanda Kigali	8.4 / 871/sq. mi.	10,169 sq. mi. / 26,338 sq km	$ 1,200	90% agriculture	17 kWh	0.003
	São Tomé and Príncipe São Tomé	0.2 / 434/sq. mi.	386 sq. mi. / 1,000 sq km	$ 1,200	38% agriculture 23% services	98 kWh	0.03
	Senegal Dakar	10.1 / 136/sq. mi.	75,749 sq. mi. / 196,189 sq km	$ 1,500	70% agriculture	140 kWh	0.02
	Seychelles Victoria	0.08 / 457/sq. mi.	176 sq. mi. / 456 sq km	$ 7,800	71% services 19% industry	1,860 kWh	0.27
	Sierra Leone Freetown	5.0 / 180/sq. mi.	27,699 sq. mi. / 71,740 sq km	$ 580	62% agriculture 20% services	47 kWh	0.005
	Somalia Mogadishu	9.9 / 41/sq. mi.	246,201 sq. mi. / 637,658 sq km	$ 550	71% nomadic agric. 29% industry, serv.	23 kWh	0.002
	South Africa Pretoria	45.0 / 96/sq. mi.	471,010 sq. mi. / 1,219,910 sq km	$ 10,000	45% services 30% agriculture	4,024 kWh	0.11
	Sudan Khartoum	33.6 / 37/sq. mi.	967,498 sq. mi. / 2,505,808 sq km	$ 1,420	80% agriculture 10% ind., commerce	66 kWh	0.02
	Swaziland Mbabane	1.1 / 162/sq. mi.	6,704 sq. mi. / 17,363 sq km	$ 4,400	70% private sector 30% public sector	894 kWh	0.03
	Tanzania Dar es Salaam	37.0 / 108/sq. mi.	364,900 sq. mi. / 945,087 sq km	$ 630	80% agriculture 20% ind., services	74 kWh	0.004
	Togo Lomé	4.9 / 234/sq. mi.	21,925 sq. mi. / 56,785 sq km	$ 1,500	65% agriculture 30% services	125 kWh	0.01
	Tunisia Tunis	9.8 / 164/sq. mi.	63,170 sq. mi. / 163,610 sq km	$ 6,500	55% services 23% industry	991 kWh	0.12
	Uganda Kampala	25.8 / 335/sq. mi.	91,135 sq. mi. / 236,039 sq km	$ 1,260	82% agriculture 13% services	63 kWh	0.002
	Zambia Lusaka	10.8 / 38/sq. mi.	290,586 sq. mi. / 752,614 sq km	$ 890	85% agriculture 9% services	505 kWh	0.008
	Zimbabwe Harare	12.9 / 86/sq. mi.	150,804 sq. mi. / 390,581 sq km	$ 2,400	66% agriculture 24% services	761 kWh	0.02
	United States Washington, D.C.	294.0 / 83/sq. mi.	3,717,810 sq. mi. / 9,629,084 sq km	$ 37,600	31% manage., prof. 29% tech., sales, admin.	12,250 kWh	0.65

UNIT 7 Assessment Resources

▶ **Unit 7 Test**

▶ **Unit 7 Test for English Language Learners and Special-Needs Students**

internet connect

GO TO: go.hrw.com
KEYWORD: SW3 U7
FOR: Web sites about country statistics

Highlights of Country Statistics
Links to online country statistics for Africa include:
- *CIA World Factbook*
- Library of Congress Country Studies
- Flags of the World

CHAPTER RESOURCE MANAGER

Objectives	Pacing Guide	Reproducible Resources
SECTION 1 **Natural Environments** (pp. 483–86)	• Identify the landform regions found in North Africa. • Analyze the factors that influence the region's climate. • Describe the region's natural resources.	**Regular** 1 day **Block Scheduling** .5 day *Block Scheduling Handbook, Chapter 21*	**RS** Guided Reading Strategy 21.1 **RS** Graphic Organizer Activity 21 **E** Creative Strategies for Teaching World Geography, Lesson 16 **SM** Map Activity 21: Landforms of Northwest Africa
SECTION 2 **History and Culture** (pp. 487–91)	• Identify the peoples who have settled in and ruled North Africa. • Describe the people and culture of the region today.	**Regular** 1 day **Block Scheduling** .5 day *Block Scheduling Handbook, Chapter 21*	**RS** Guided Reading Strategy 21.2 **SM** Critical Thinking Activity 21: The Nile Valley **SM** Geography for Life Activity 21: People, Agriculture, and Water in North Africa
SECTION 3 **The Region Today** (pp. 492–94)	• Describe the economies and cities of North Africa. • Identify the challenges faced by people of North Africa.	**Regular** 1 day **Block Scheduling** .5 day *Block Scheduling Handbook, Chapter 21*	**RS** Guided Reading Strategy 21.3 **PS** Readings in World Geography, History, and Culture 53 and 54 **E** Cultures of the World Activity: Region 6

Chapter Resource Key

PS Primary Sources

RS Reading Support

IC Interdisciplinary Connections

E Enrichment

SM Skills Mastery

A Assessment

REV Review

ELL Reinforcement and English Language Learners

Transparencies

 CD–ROM

 Video

 Internet

 Holt Presentation Maker Using Microsoft® PowerPoint®

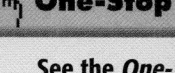 **One-Stop** Planner CD–ROM

See the *One-Stop Planner* for a complete list of additional resources for students and teachers.

 **One-Stop** Planner CD–ROM

It's easy to plan lessons, select resources, and print out materials for your students when you use the ***One-Stop Planner CD–ROM with Test Generator***.

 **internet** connect

HRW ONLINE RESOURCES

GO TO: go.hrw.com
Then type in a keyword.

TEACHER HOME PAGE
   KEYWORD: SW3 Teacher

CHAPTER INTERNET ACTIVITIES
   KEYWORD: SW3 GT21
   Choose an activity to:
   • learn about ergs, regs, and other desert features.
   • create a brochure on the pyramids of ancient Egypt.
   • take the GeoMap challenge and test your knowledge of North Africa's geography!

CHAPTER ENRICHMENT LINKS
   KEYWORD: SW3 CH21

CHAPTER MAPS
   KEYWORD: SW3 MAPS21

ONLINE ASSESSMENT
   Homework Practice
   KEYWORD: SW3 HP21
   Standardized Test Prep
   KEYWORD: SW3 STP21
   Rubrics
   KEYWORD: SS Rubrics

COUNTRY INFORMATION
   KEYWORD: SW3 Almanac

CONTENT UPDATES
   KEYWORD: SS Content Updates

HOLT PRESENTATION MAKER
   KEYWORD: SW3 PPT21

ONLINE READING SUPPORT
   KEYWORD: SS Strategies

CURRENT EVENTS
   KEYWORD: S3 Current Events

Technology Resources	**Reinforcement, Review, and Assessment**
One-Stop Planner CD–ROM, Lesson 21.1	**ELL** Main Idea Activity 21.1
Geography and Cultures Visual Resources 43–47	**ELL** English Audio Summary 21.1
Homework Practice Online	**ELL** Spanish Audio Summary 21.1
HRW Go site	**REV** Section 1 Review, p. 486
	A Daily Quiz 21.1
One-Stop Planner CD–ROM, Lesson 21.2	**ELL** Main Idea Activity 21.2
CNN Presents World Cultures: Yesterday and Today, Segment 2: Restoring the Sphinx	**ELL** English Audio Summary 21.2
Homework Practice Online	**ELL** Spanish Audio Summary 21.2
HRW Go site	**REV** Section 2 Review, p. 491
	A Daily Quiz 21.2
One-Stop Planner CD–ROM, Lesson 21.3	**ELL** Main Idea Activity 21.3
ARGWorld CD–ROM	**ELL** English Audio Summary 21.3
CNN Presents Geography: Yesterday and Today, Segment 23: Cairo—Selling the Suburbs	**ELL** Spanish Audio Summary 21.3
Homework Practice Online	**REV** Section 3 Review, p. 494
HRW Go site	**A** Daily Quiz 21.3

Meeting Individual Needs

Ability Levels

Level 1 Basic-level activities designed for all students encountering new material

Level 2 Intermediate-level activities designed for average students

Level 3 Challenging activities designed for honors and gifted-and-talented students

English Language Learners Activities that address the needs of students with Limited English Proficiency

Chapter Review and Assessment

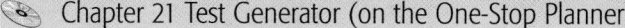

Chapter 21 Test Generator (on the One-Stop Planner)

Global Skill Builder CD–ROM

HRW Go site

REV Chapter 21 Review, pp. 496–97

REV Chapter 21 Tutorial for Students, Parents, Mentors, and Peers

A Chapter 21 Test (form A or B)

A Alternative Assessment Handbook

A Chapter 21 Test for English Language Learners and Special-Needs Students

Launch into Learning

In about 430 B.C., the Greek historian Herodotus wrote about a place in North Africa "with people dwelling round it who all of them build their houses with blocks of salt. No rain falls in these parts of Libya; if it were otherwise, the walls of these houses could not stand. . . . Beyond the ridge, southwards, in the direction of the interior, the country is a desert, with no springs, no beasts, no rain, no wood, and altogether destitute of moisture." Ask students to examine the maps in the unit atlas to determine if they think Herodotus described the area accurately. *(Possible answer: The description is* *accurate because the Sahara is extremely dry.)* Tell students they will learn more about North Africa's physical and human geography in this chapter.

Using the Physical-Political Map

Have students examine the map on the opposite page. Ask them to identify the region's main bodies of water *(Atlantic Ocean, Mediterranean Sea, Red Sea, Nile River, Lake Nasser).* Then ask students why they think there is only one river shown on the map. *(Possible answer: The Nile is the only major river because the region is mostly desert.)*

Why We Should Know More

You might want to emphasize these reasons for learning more about North Africa:

▶ The United States has good relations with Egypt, Morocco, and Tunisia.

▶ Events in the oil-producing region of North Africa can affect petroleum prices around the world.

▶ North African countries have produced beautiful forms of art, architecture, textiles, music, and literature that many Americans enjoy.

▶ The ancient Egyptians created some of history's most remarkable monuments, works of art, and technological achievements.

CHAPTER
21
North Africa

North Africa includes Morocco, Algeria, Tunisia, Libya, Egypt, and Western Sahara, which is occupied by Morocco. The culture and politics of the region are closely tied to Southwest Asia and Europe.

Pendant from Algeria

Moroccan water seller

Ahlan! (Hi!) My name is Shaimaa, and I am 18. I live with my mother and my little sister in an apartment about one hour from downtown Cairo. I am in my third year of high school.

Every day but Friday, I get up at 7:00 A.M., drink a glass of milk, and then meet my friends. We travel to school together on the metro (mass transit). My school is an all-girls school. I am in the humanities and social sciences track, so I am studying philosophy, psychology, Arabic, English, and history, my favorite. We all have religious education in school also—I study Islam with the other Muslim girls while the Christian girls meet with their religious teacher. At about 3:00 P.M., I get home from school, eat a big lunch of chicken and vegetables, and sleep for a couple of hours. When I wake up, I have a lot of homework—about six hours' worth. At 10:00 P.M., we have another small meal of cheese, yogurt, or beans before bedtime.

On Fridays I go to movies with my girlfriends and walk along the Nile River, or I stay home and listen to music. Sometimes I take a taxi to the beach with my family. For big holidays, we go to the mosque very early in the morning. Then we go to my grandmother's house with my 12 aunts and uncles and 16 cousins! After Ramadan (an Islamic holy month), we put on our new clothes and go around to all my relatives' houses, where we get presents of new money.

 **LET'S GET STARTED**

Copy the following questions onto the chalkboard: *How do you change your daily routine on hot summer days? How would you adjust to summers in North Africa, where temperatures can rise above 130°F?* Discuss responses. *(Students may discuss wearing light clothing, staying indoors during the hottest part of the day, or other strategies.)* Tell students that they will learn more about the natural environments of North Africa in this section.

Building Vocabulary

Write the key terms on the chalkboard and call on volunteers to read the definitions aloud. Have a student describe the difference between an **erg** and a **reg** by using the photo in this section. Then explain how **depressions** can be low points in a country's economy, times during which people feel low, or low areas on Earth's surface—the definition that applies in this section. Tell students that the word **wadis** comes from an Arabic word.

Natural Environments

READ TO DISCOVER

1. What landform regions are found in North Africa?
2. What factors influence the region's climates?
3. What natural resources does the region have?

WHY IT MATTERS

North Africa is part of the world's largest desert region. Use CNNfyi.com or other **current events** sources to learn about the cultures, landscapes, and ways of life in very dry places.

DEFINE

erg
reg
depressions
wadis

LOCATE

Atlantic Ocean   Sahara
Red Sea   Qattara Depression
Mediterranean Sea   Nile River
Atlas Mountains   Strait of Gibraltar

Section 1 RESOURCES

REPRODUCIBLE

▶ Guided Reading Strategy 21.1
▶ Graphic Organizer Activity 21
▶ Creative Strategies for Teaching World Geography, Lesson 16
▶ Map Activity 21: Landforms of Northwest Africa

TECHNOLOGY

▶ One-Stop Planner CD–ROM, Lesson 21.1
▶ Geography and Cultures Visual Resources 43–47
▶ Homework Practice Online
▶ HRW Go site

REINFORCEMENT, REVIEW, AND ASSESSMENT

▶ Main Idea Activity 21.1
▶ English Audio Summary 21.1
▶ Spanish Audio Summary 21.1
▶ Section 1 Review, p. 486
▶ Daily Quiz 21.1

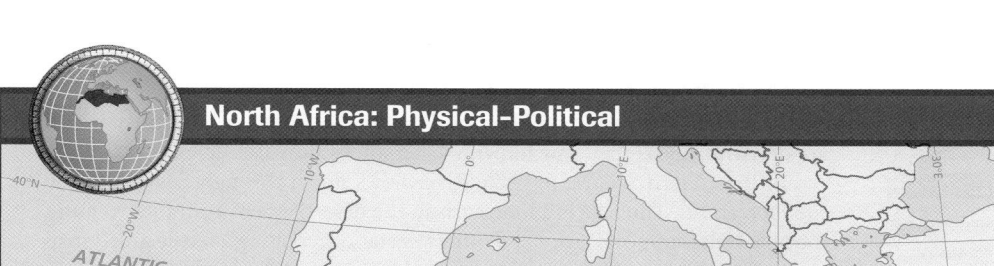

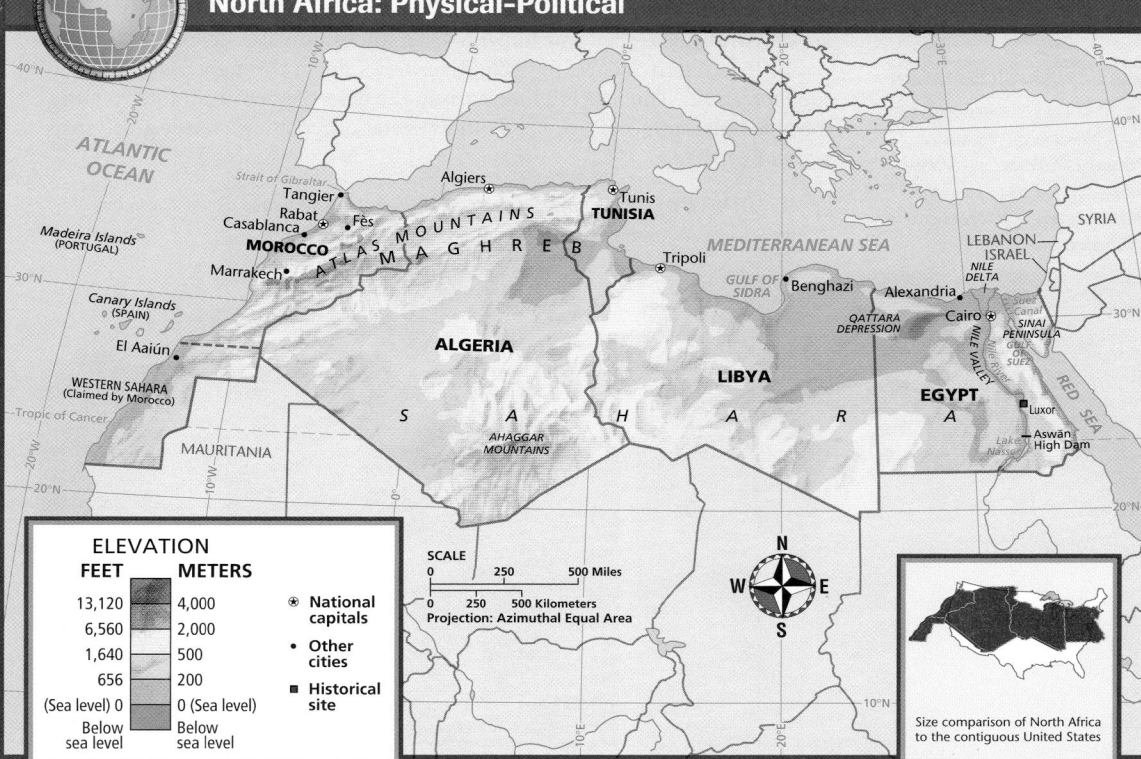

North Africa: Physical-Political

ATLANTIC OCEAN

Strait of Gibraltar

Madeira Islands (PORTUGAL)

Tangier • Algiers
Rabat • Tunis
Casablanca • Fès **TUNISIA**
MOROCCO ATLAS MOUNTAINS
Marrakech • M A G H R E B
Tripoli

Canary Islands (SPAIN)

El Aaiún

WESTERN SAHARA (Claimed by Morocco)

Tropic of Cancer

MAURITANIA

ALGERIA

S A H A R A

AHAGGAR MOUNTAINS

MEDITERRANEAN SEA

GULF OF SIDRA • Benghazi • Alexandria
QATTARA DEPRESSION • Cairo
LIBYA NILE DELTA
SINAI PENINSULA
GULF OF SUEZ

EGYPT
NILE VALLEY
• Luxor
Lake Nasser • Aswān High Dam

SYRIA
LEBANON ISRAEL

RED SEA

ELEVATION

FEET	METERS
13,120	4,000
6,560	2,000
1,640	500
656	200
(Sea level) 0	0 (Sea level)
Below sea level	Below sea level

⊛ National capitals
• Other cities
■ Historical site

SCALE
0   250   500 Miles
0   250   500 Kilometers
Projection: Azimuthal Equal Area

N W E S

Size comparison of North Africa to the contiguous United States

Teach Objective 1

ALL LEVELS: Draw an outline map of North Africa on the chalkboard. Then call on volunteers to sketch in the following features: Atlas Mountains, coastal plains, Sahara, Qattara Depression, and Nile River. **ENGLISH LANGUAGE LEARNERS**

LEVELS 2 AND 3: Discuss the role of physical processes in the creation of landforms in North Africa. Ask students what physical process created particular landscape features, like the Atlas Mountains or ergs and regs (*uplift caused by the collision of North Africa and Europe; wind*).

Teach Objective 2

ALL LEVELS: Copy the following graphic organizer onto the chalkboard, omitting the italicized answers. Have students complete it. **ENGLISH LANGUAGE LEARNERS**

Factors Affecting North African Climates

Causes	**Effects**
• Atlas Mountains • *rain shadow* • *harmattan* • *subtropical high-pressure system*	• *rain shadow* • *sandstorms* • *arid climate*

internet connect

GO TO: go.hrw.com
KEYWORD: SW3 CH21
FOR: Web sites about
Egyptian mummies

internet connect

GO TO: go.hrw.com
KEYWORD: SW3 CH21
FOR: Web sites about
North Africa

INTERPRETING THE VISUAL RECORD

The Sahara is the world's largest desert. It includes a variety of landforms, such as giant sand seas called ergs (left) and extensive gravel-covered plains known as regs (right). **What physical forces create these distinctive landforms?**

Landforms

North Africa stretches from the Atlantic Ocean to the Red Sea. On the north it is bordered by the Mediterranean Sea. Coastal plains are the main landforms where North Africa meets the Mediterranean Sea and the Atlantic Ocean. From Morocco to Tunisia, the coastal plains quickly give way to the Atlas Mountains. These mountains run parallel to the Atlantic and Mediterranean coasts from northern Tunisia to the Atlantic Ocean. The Sahara is the world's largest desert. It lies south of the Atlas Mountains and coastal plains. The Sahara extends across all of North Africa. This vast desert acts as a natural barrier between North Africa and the rest of the African continent.

Over thousands of years, the Sahara has undergone cycles of wet and dry periods. At times there has been enough rain to support grasslands and abundant animal life. However, most of the Sahara today is a barren expanse of rock and sand. The great desert covers about 3.5 million square miles (9 million sq km) of land—roughly the size of the entire United States. Because there are few plants, wind and rain can erode the land easily. As a result, bare rock surfaces are common. The basins below these rocky ridges fill with eroded sediment. Some basins are covered with high, shifting sand dunes that create a sea of sand, called an **erg**. In other areas, wind blows the sand and dust away, leaving a gravel-covered plain. This type of eroded landform is called a **reg**.

The Sahara also has large low areas called **depressions**. In Egypt the Qattara Depression, 440 feet (134 m) below sea level, is a wilderness of quicksand and salt marsh. Other depressions have large dry lakebeds where water briefly collects during rare rainstorms. Rainwater also carves out **wadis**, which are dry streambeds that only fill with water after rain falls.

In the eastern Sahara, the Nile River flows north through Egypt into the Mediterranean. The Nile is a long oasis in the desert. Water from the river and the Nile Delta supports crops and other vegetation, creating a fertile green strip across Egypt.

✓ **READING CHECK:** *Physical Systems* What has characterized the climatic history of the Sahara? alternating cycles of wet and dry periods

LEVELS 1 AND 2: Copy the following graphic organizer onto the chalkboard, omitting the italicized answers. Have students complete the chart. **ENGLISH LANGUAGE LEARNERS**

LEVEL 3: Have students complete the Levels 1 and 2 activity. Then organize students into five groups, representing Algeria, Egypt, Libya, Morocco, and Tunisia. Have each group write a draft of a political campaign speech addressing the best ways to protect its country's natural resources. **COOPERATIVE LEARNING**

Natural Resources of North Africa	
Energy and minerals	*oil, natural gas, iron ore, lead, phosphates, zinc*
Ocean resources	*fisheries, particularly for sardines*
Major crops	*grapes, olives, dates, grains, vegetables, cotton, rice*

The thorny argan tree is found mainly in Morocco. Goats climb these trees to eat the olive-like fruit that they produce.

Climates, Plants, and Animals

Most of North Africa's vegetation and wildlife are restricted to the areas with a Mediterranean climate. In the Sahara, plants and animals must be very hardy to survive. Temperatures there can climb above 130°F (54°C) in the summer. They can drop to below freezing in the winter. As in other desert areas, the daily range of temperatures is also great. Daytime temperatures are very hot, but at night, temperatures cool dramatically.

Climates   North Africa's Mediterranean climate areas are found along the coast. Warm dry summers and mild rainy winters are common there. Some areas with a semiarid climate lie between coastal areas and the Sahara.

An arid climate covers most of North Africa. A subtropical high-pressure system keeps the region very dry. The system creates a wide band of dry lands, of which the Sahara is a part, across Asia and Africa. A rain shadow caused by the Atlas Mountains also contributes to the region's dry climate. A hot dry wind called the harmattan (har-muh-TAN) often sweeps southward across the Sahara. In the Saharan ergs, the winds can cause violent sandstorms that block out sunlight for days. Sometimes hot desert winds blow Saharan dust across the Atlantic Ocean to the Caribbean and the southern United States.

Plants and Animals   Vegetation in the Sahara is limited to grasses and short shrubs. Few trees can grow in such dry regions. Plants and animals are concentrated around oases, where there are sources of water. Date palms are found around oases. Insects, snails, and small reptiles are common, along with a few larger animals such as gazelles and hyenas.

Morocco, in particular, has a great diversity of plant and animal life. The country is close to Europe, separated by just 8 miles (13 km) across the Strait of Gibraltar. As a result, many European as well as African species of plants and animals are found in Morocco. Also, migratory birds pass through Morocco every year as they fly between Europe and Africa. The country also has forests in the mountains.

Our Amazing Planet

The highest temperature ever recorded on Earth was 136°F (58°C) on the northern edge of the Sahara in Libya.

Eye on Earth

Ichkeul National Park
Ichkeul National Park is 20 miles (32 km) southwest of the Mediterranean coast in northern Tunisia. The park is an important sanctuary for water birds, particularly during the winter when Lake Ichkeul and the surrounding wetlands are temporary homes for more than 200,000 migratory birds. Various species of ducks, including the wigeon, teal, mallard, and pintail, as well as geese, storks, and pink flamingoes make their winter homes at Lake Ichkeul.

However, the recent construction of two dams has reduced Lake Ichkeul's freshwater supply. The result has been an increase in the lake's salinity. This change has caused a decline in many plant species on which the birds depend for food. Some birds have died, and others have had to go elsewhere on their annual migrations.

ACTIVITY: Have interested students create idea webs to show how the construction of the dams has affected Lake Ichkeul.

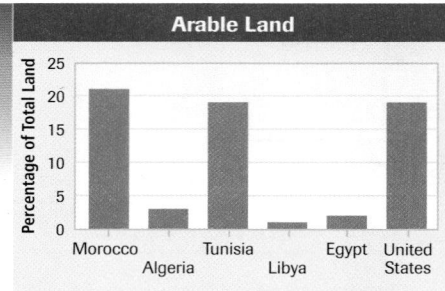

Arable Land

Percentage of Total Land

Source: Central Intelligence Agency, *The World Factbook 2000*

INTERPRETING THE GRAPH *Morocco and Tunisia have the highest percentage of arable land in North Africa, while Algeria, Libya, and Egypt have a far lower percentage.* **Based on the graph, what can you infer about environmental conditions in the different countries of North Africa?**

Egypt also has abundant bird life. More than 300 species of birds are found in the Nile Valley and Nile Delta. These include the egret, flamingo, golden oriole, heron, pelican, and stork. The bird life in Egypt benefits from the great variety of plant life along the Nile River. However, outside of the Nile Valley and Delta, Egypt is almost all desert, with limited plant and animal life. This is true of most of Algeria and Libya as well. Tunisia's milder climates support forests, grasslands, and many kinds of animals.

✓ **READING CHECK:** *Places and Regions* What two factors create the Sahara's dry climate? subtropical high pressure, rain shadow

Natural Resources

Oil and natural gas are North Africa's most valuable natural resources. Oil is found in every country in North Africa, but Libya has the largest reserves. Other important resources include iron ore, lead, phosphates, and zinc.

There are rich fishing grounds off Morocco's Atlantic coast. The main catch is sardines. Fishers there catch more than 350,000 tons of sardines every year. In fact, Morocco is the world's largest exporter of this type of fish.

Although much of North Africa is desert, rain or irrigation makes farming possible in areas with good soil. The region produces many crops, including grapes, olives, dates, grains, and vegetables. River water irrigates fields all along the Nile and in the Nile Delta. These farmlands have helped make Egypt an important cotton producer. Rice is also an important crop in the Nile Valley. Farmers in the desert oases of Algeria, Libya, and Egypt grow crops such as date palms.

✓ **READING CHECK:** *Environment and Society* Where do farmers in Egypt get water for growing crops? the Nile

Section 1 Review

Homework Practice Online
Keyword: SW3 HP21

Define erg, reg, depressions, wadis

Working with Sketch Maps
On a map of North Africa that you draw or that your teacher provides, label the countries of the region and Atlantic Ocean, Red Sea, Mediterranean Sea, Atlas Mountains, Sahara, Qattara Depression, Nile River, and Strait of Gibraltar. In the margin of your map, identify the shortest water crossing between Africa and Europe.

Reading for the Main Idea

1. **Physical Systems** How is an erg formed? How is a reg formed?

2. **Places and Regions** Where are the region's wettest climates located?

3. **Physical Systems** What is one reason that Morocco's plant and animal life is so diverse?

Critical Thinking

4. **Analyzing Information** In what way might the Mediterranean Sea be considered an important natural resource in the region?

Organizing What You Know

5. Create a chart like the one shown below. Use the chart to list and describe the region's climates, plants and animals, and natural resources.

Climates	Plants and animals	Natural resources

Section 2

OBJECTIVES

1. **Identify the peoples who have settled in and ruled North Africa.**

2. **Describe the people and culture of the region today.**

LET'S GET STARTED

Copy the following questions onto the chalkboard: *Who built the famous pyramids of the Nile Valley? How might these monuments be important today?* Discuss responses. *(Possible answers: the ancient Egyptians; might be important for tourism or as symbols of Egypt's cultural heritage)* Tell students that they will learn more about ancient and modern Egypt in Section 2.

Building Vocabulary

Write the key terms on the chalkboard and have volunteers read the definitions aloud from the text or glossary. Point out that many rivers and streams carry **silt**. If possible, locate and show examples of **hieroglyphs**—the pictures and symbols used in ancient Egyptian writing. Tell students that the word **pharaohs** comes from an Egyptian word meaning "great house." It originally referred to the royal palace in Egypt but later developed into a generic name for all ancient Egyptian kings.

Section 2 · History and Culture

READ TO DISCOVER

1. What peoples have settled in and ruled North Africa?
2. What are the people and culture of the region like today?

WHY IT MATTERS

Archaeological discoveries in Egypt and other parts of North Africa tell us much about early human history. Use **CNNfyi.com** or other **current events** sources to learn about recent archaeological research in the region.

IDENTIFY
Berbers

DEFINE
silt
pharaohs
hieroglyphs

LOCATE
Alexandria
Cairo
Marrakech
Maghreb
Suez Canal
Sinai Peninsula
Casablanca
Fès

Section 2 RESOURCES

REPRODUCIBLE
► Guided Reading Strategy 21.2
► Critical Thinking Activity 21: The Nile Valley
► Geography for Life Activity 21: People, Agriculture, and Water in North Africa

TECHNOLOGY
► One-Stop Planner CD–ROM, Lesson 21.2
► CNN Presents World Cultures: Yesterday and Today, Segment 2: Restoring the Sphinx
► Homework Practice Online
► HRW Go site

REINFORCEMENT, REVIEW, AND ASSESSMENT
► Main Idea Activity 21.2
► English Audio Summary 21.2
► Spanish Audio Summary 21.2
► Section 2 Review, p. 491
► Daily Quiz 21.2

History

The first people in North Africa were hunter-gatherers. They lived in areas where the climates were best. By 4000 B.C., much of North Africa had become desert. The human population became concentrated along the Mediterranean coast, desert oases, wadis, and the Nile River. Every year, the Nile would flood. These floods spread **silt**, which is fertile finely ground soil, over the river's banks.

Early Peoples Beginning about 3000 B.C., a great civilization grew along the Nile River and its delta in Egypt. A series of kingdoms arose, ruled by monarchs called **pharaohs**. These rulers were considered gods and had complete power over the Egyptian people. The Egyptians built great pyramids and other monuments that still stand today. They developed a writing system that used pictures and symbols called **hieroglyphs**. Egyptian astronomers created a 365-day calendar. They also learned to predict the annual floods of the Nile. Every year, when Sirius—the brightest star in the sky—appeared above the horizon at sunrise, the flood would soon follow. Because the Egyptians depended on the Nile's floods for farming, this information was key to their survival. From 1570 B.C. to 1085 B.C., Egypt expanded its power into the area that is now Syria, Israel, and Libya.

Later, as Egyptian power weakened, foreigners began to control much of North Africa. Those foreigners included the Phoenicians, Greeks, and Romans. The

This ancient Egyptian illustration from the Book of the Dead shows Horus, the falcon-god, introducing an Egyptian to the presence of Osiris, god of the underworld. The Book of the Dead was a collection of texts, spells, and formulas placed in tombs to help the dead in the afterlife. It contains many outstanding examples of ancient Egyptian art.

Linking Past to Present

The Rosetta Stone
French general Napoléon Bonaparte led an unsuccessful military campaign into Egypt from 1798 to 1801. He had planned to disrupt British trade in the region and to establish a French colony.

During the three-year expedition, one of Napoléon's soldiers discovered a stone inscribed with the ancient Egyptian writing known as hieroglyphics. Alongside the heiroglyphs was a translation of the same text in Greek. The stone was found in the town of Rosetta, about 35 miles (56 km) northeast of Alexandria, so it became known as the Rosetta Stone.

The discovery of the Rosetta Stone was the key that unlocked the mysteries of Egyptian hieroglyphics. Archeologists used the stone as a sort of dictionary to translate the ancient system of Egyptian writing. After the French surrender of Egypt in 1801, the stone passed into British hands. It is now in the British Museum in London.

In Greek mythology, the giant Atlas held the world on his shoulders somewhere near the western end of North Africa. The Atlas Mountains are named for him. In the 1500s, collections of maps or charts came to be called atlases because mapmakers included images of Atlas holding Earth in them.

Phoenicians were sailors and traders from what is now Lebanon. They set up many Mediterranean trading colonies such as Carthage, which was founded in about 800 B.C. in modern-day Tunisia. Alexander the Great, at the head of a Greek army, founded the city of Alexandria in Egypt in 332 B.C. The Roman Empire became a great power in North Africa after it destroyed Carthage in 146 B.C. After the Roman Empire crumbled in the A.D. 400s, a Germanic tribe called the Vandals moved south through Spain into Africa. They set up a kingdom in what is now Libya. In the A.D. 500s the Byzantine Empire, which had been the Eastern Roman Empire, recaptured most of North Africa.

The Arabs and Islam   Byzantine rule over North Africa was short-lived. In the 600s Arab armies from Southwest Asia swept across North Africa. They reached Africa's Atlantic coast and conquered Morocco by the early 700s. Arab armies also crossed into Iberia in the year 711. Most people in North Africa became Muslims. In addition, Arabic became the main language of the area.

Under Arab rule, Cairo and other North African cities became great centers of Islamic culture and education. Cities like Marrakech in Morocco became centers of trade between central and western Africa, Europe, and Arabia. Such cities grew rich trading gold, ivory, and spices as well as slaves. However, in the 1500s outsiders again invaded North Africa. The Ottomans—Muslims based in what is now Turkey—took control first of Egypt and then of Libya, Tunisia, and Algeria.

✓ **READING CHECK:**   *Human Systems*   What religion and language did the Arabs bring to North Africa? Islam, Arabic

Colonialism   The Ottoman Empire ruled much of North Africa until the late 1800s. Earlier in that century, western European powers had begun to take over parts of the region. Beginning in the 1830s, France moved to control the Maghreb (West), including Tunisia, Algeria, and part of Morocco. Spain took

INTERPRETING THE VISUAL RECORD

French colonists in Algeria found that the region was perfect for vineyards, and Algerian wine became a major export under French rule. However, it is mainly produced for export because most Algerians are Muslim and are forbidden to drink alcohol. How did cultural patterns influence the diffusion of crops to Algeria and affect the country's cultural landscapes?

VISUAL RECORD ANSWER

grapes and wine-making introduced by the French, which created landscapes with vineyards

LEVELS 1 AND 2: Read to students the following excerpt from Percy B. Shelley's "Ozymandias":

> . . . *Two vast and trunkless legs of stone*
> *Stand in the desert. Near them on the sand,*
> *Half sunk, a shatter'd visage lies, [with] frown*
> *And wrinkled lip and sneer of cold command . . .*
> *And on the pedestal these words appear:*
> *"My name is Ozymandias, king of kings:*
> *Look on my works, ye mighty, and despair!"*
> *Nothing beside remains: round the decay . . .*
> *The lone and level sands stretch far away.*

Then ask students how the poem might reflect North Africa's history. (*Possible answer: Many proud empires have come and gone in North Africa.*) Have students name the empires to which this statue might have belonged. (*Possible answers: Egyptian, Carthaginian, Roman, Byzantine, Ottoman*) Lead a discussion about the great empires of North African history.

LEVEL 3: Have students write an epic-style poem describing key events in North Africa's history. Students might want to concentrate on just one country. They might also want to write their poems from the perspective of inanimate objects, such as monuments, that witnessed the changes taking place over time.

---

control of northern Morocco. Thousands of French, Spaniards, and Italians settled in North Africa in the decades that followed. More than 100 years later, the large number of French living in Algeria would complicate that country's struggle for independence.

In 1882 Great Britain took over Egypt. Britain wanted control of the Suez Canal, which connects the Mediterranean with the Red Sea. The canal was an important trade link between Europe and Britain's colony of India. Italy completed the European conquest of North Africa by taking Libya from the Ottoman Empire in 1912.

Independence North Africans resented European rule. They did not have the same rights as the European settlers. Over time North Africans worked to win their independence. In 1922 Egypt gained limited independence from Great Britain. However, independence efforts across the region became stronger after World War II ended in 1945.

In 1952 a group of Egyptian military officers led a revolution that brought complete independence from Britain. France granted Tunisia and Morocco their independence in 1956. Yet the French fought a bloody war to hold on to Algeria. When Algeria finally won independence in 1962, most of the French population left.

Libya became an independent kingdom in 1951. In 1969, military officers led by Mu'ammar Gadhafi overthrew the monarchy. Gadhafi declared the country a socialist republic and adopted anti-Western policies.

✓ **READING CHECK:** *Human Systems* How did independence affect Algeria's human geography? Most French left Algeria.

Culture

The countries of North Africa share a similar history and Muslim culture. Still, there is a great deal of variation among them.

People and Languages Nearly all of the people of North Africa consider themselves Arab or Arab-Berber. The **Berbers** are a cultural group that lived in North Africa long before waves of Arab armies crossed the continent. Also, small groups of desert nomads called Bedouins live along the Sinai Peninsula in Egypt.

Arabic is the official language of every country in North Africa. However, the people of each country speak their own version of Arabic. In some rural areas Berber dialects are also common. Because of the influence of colonization, many people also speak European languages. French is still widely used in Algeria, Morocco, and Tunisia. Italian is spoken in Libya, and English is used in Egypt.

Settlement and Land Use Most North Africans live along the Mediterranean coast or in the foothills of the Atlas Mountains. An exception to this pattern is found in Egypt. About 99 percent of that country's 68 million people live in the Nile Valley and Delta. Together, those populated areas make up only 3 percent of Egypt's land. Cairo, Egypt's

---

Naguib Mahfouz

In 1988 Egyptian writer Naguib Mahfouz became the first Arab to receive the Nobel Prize for Literature. Mahfouz was born in Cairo in 1911. He has written some 40 novels and short story collections as well as many plays and screenplays.

Mahfouz's first books were set in ancient Egypt. Later works dealt with contemporary subjects. In *The Cairo Trilogy*, for example, the story of a large middle-class Egyptian family connects to colonial and modern political history.

Mahfouz has often written about social issues, such as the status of women and political prisoners. *Children of the Alley* features characters that symbolize Adam, Moses, and other religious figures. Some people criticized the book, saying it was disrespectful to Islam. Nevertheless, Mahfouz is one of the region's most respected writers.

Evaluating If you were to write a novel about North Africa, which parts of its history or culture would you use as background? Why would you choose those topics?

Daily Life

Communal Dining Customs On special occasions, communal dining is common in the North African countries of Morocco, Tunisia, and Algeria. One such special occasion is the *diffa*, a lavish feast held for weddings, births, or anniversaries of local saints. A variety of foods—breads, salads, savory pastries, meats, vegetables, and fruits—is served on one big platter that is placed in the middle of the banquet table. Everyone eats from the same plate.

During their traditional communal meals, Moroccans avoid using Western utensils such as forks or spoons. Instead, they use a piece of bread to scoop up bite-size portions of food from the platter.

ACTIVITY: Have students conduct research on specific North African dishes and prepare a menu for a *diffa* feast.

CONNECTING TO THE ARTS ANSWER

Students might note a history of invasions by outsiders, the development of Arab and Islamic cultures, or other topics. Reasons for choices will also vary.

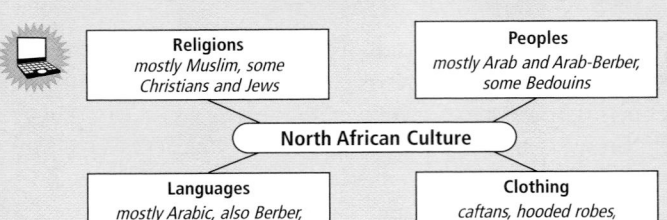

ALL LEVELS: Copy the following graphic organizer onto the chalkboard, omitting the italicized answers. Then have students complete the chart and lead a discussion about North African culture. **ENGLISH LANGUAGE LEARNERS**

Religions		Peoples
mostly Muslim, some Christians and Jews		*mostly Arab and Arab-Berber, some Bedouins*

North African Culture

Languages		Clothing
mostly Arabic, also Berber, French, Italian, English		*caftans, hooded robes, gallibiya*

HOMEWORK: Have students use the chapter map to compare the physical sizes of countries in North Africa to the physical sizes of countries in southern Europe. For example, how does Libya's area compare to that of Spain or Italy? *(Libya is much larger.)* Have students use the unit Fast Facts table to rank North Africa's countries in order from largest to smallest. Then have them compare the sizes and populations of North African countries to the sizes and populations of Spain and Italy.

Across the Curriculum

► Art ◄

The Art of *Mehndi* *Mehndi* is the ancient art of using henna paste to paint parts of the body with elaborate designs. Arabic *mehndi* is generally applied to the hands and feet. The paste is made from grinding henna leaves into a fine powder and mixing it in hot water. Different dye colors are made by adding indigo, tea, coffee, or cloves. The designs look like a tattoo but are not permanent. They can last up to four weeks.

According to some historians, the tradition of *mehndi* originated in North African and Southwest Asian cultures dating back nearly 4,000 years. One of the earliest records of henna use comes from ancient Egypt, where it is known to have been used to paint the fingers and toes of deceased pharaohs before their mummification. In recent years, the art of *mehndi* has become increasingly popular in Western culture for creating trendy temporary tattoos.

VISUAL RECORD ANSWER

high walls, narrow streets, shops

---

capital, lies in the Nile Delta. With a population of more than 10 million, Cairo is the largest urban area in North Africa. Egypt also has the largest population of any Arab country.

Urban overcrowding is a problem across all of North Africa. People from the countryside are pouring into cities in hopes of finding work and a better life. Casablanca, the largest city in Morocco, must absorb about 30,000 new migrants every year.

✓ **READING CHECK:** **Human Systems** In what areas are most of the population of North Africa found? along Mediterranean coast and in Atlas foothills

FOCUS ON CULTURE

The Medina Most old Arab cities in North Africa developed within the protective walls of a Casbah, or fort. As the population of the city grew, the buildings within the city's walls were built higher and closer together. Space was limited, and streets were as narrow as possible, often twisting at odd angles. The high walls and narrow streets also created shade that kept people and buildings cool in the hot climate.

When colonial governments took over North Africa, they built European-style cities around the old Arab city, or medina. However, people did not abandon the medina for the wide boulevards and spacious air-conditioned buildings of newer areas. Many medinas in North Africa remain lively places where people live and go for social interaction, shopping, and prayer. One of the most famous medinas in the world is in Fès, Morocco. There, tens of thousands of people crowd into a square mile of densely packed buildings.

✓ **READING CHECK:** **Human Systems** What are the medinas of old Arab cities of the region like? high walls; narrow twisting streets; lively places with social interaction, shopping, prayer

---

INTERPRETING THE VISUAL RECORD

The medina in Tunis was built during the A.D. 600s and is the cultural and historical focus of the city. During the colonial era, the French built a new area around the medina known as the ville nouvelle, *or "new city."* **What architectural features can you see in the photo that are typically found in medinas?**

Close

Ask students to pretend that they are about to host cultural exchange students from North Africa. Ask them what they could do to make the students feel more at home. *(Possible answers: learn some Arabic phrases, locate the local mosque, prepare some North African foods)*

Review and Assess

Have students complete the **Section Review**. Then have students complete **Daily Quiz 21.2**.

Reteach

Have students complete **Main Idea Activity for English Language Learners and Special-Needs Students 21.2.** Then review the photos in Section 2. Ask students to write new captions for the photos. ENGLISH LANGUAGE LEARNERS

Extend

The Berbers are the descendents of the pre-Arab inhabitants of North Africa. Most Berbers are farmers, but some are nomads. Have interested students learn more about the Berbers and how their ways of life are adapted to desert conditions. Students might present their research as interviews with Berber teenagers. BLOCK SCHEDULING

Religion Most North Africans are Muslim, except for very small Christian and Jewish minorities. Islam plays a major role in North African life. For example, the five daily prayers punctuate life and mark the time for appointments. In addition, Fridays are special days when Muslims meet in mosques for prayer. In many cities across North Africa, businesses close on Thursday and Friday before opening again on Saturday. Businesses also close early for religious holidays, such as *Id al-Adha.* During this holiday, Muslim families sacrifice a sheep in honor of the willingness of Abraham to sacrifice his own son to prove his devotion to Allah. Islamic holidays are celebrated according to a lunar calendar. As a result, holidays shift over the years, being celebrated earlier each year according to the Western calendar.

Traditions and Customs Many North Africans wear traditional clothing. While there are many regional variations, in general North African clothing is long and loose. Such styles are ideal for the region's hot climates. Men and women often wear caftans and hooded robes made with a variety of fabrics. In Egypt the caftan and the *gallibiya,* or pants and a long shirtlike garment, are popular. Many women dress according to Muslim tradition. Their clothing covers all of the body except the face and hands.

When people greet each other in the street, they often shake hands and then touch their hand to their heart. If they are family or very close friends, they will kiss each other on the cheek. The number and pattern of the kisses vary from country to country.

Celebrations such as marriages are very important to North Africans because the family is central in Arab culture. Weddings can last for several days. Except for the last day of the wedding, the women's and men's celebrations are held separately.

✓ **READING CHECK:** *Human Systems* How and why are traditional clothes of different North African countries generally similar? long and loose for the region's hot climate

INTERPRETING THE VISUAL RECORD

Cafés are popular places for men to socialize throughout North Africa. There, they play chess or dominoes, talk, or simply watch people passing by on the street. Most women socialize only at home. **How are the social customs of cafés in North Africa different from those of cafés in this country?**

Section 2 Review

Section 2 Review

Homework Practice Online
Keyword: SW3 HP21

Identify Berbers

Define silt, pharaohs, hieroglyphs

Working with Sketch Maps On the map you created in Section 1, label Alexandria, Cairo, Marrakech, Maghreb, Suez Canal, Sinai Peninsula, Casablanca, and Fès. Which city was founded by Alexander the Great?

Reading for the Main Idea

1. (*Human Systems*) Who were some early peoples that ruled over areas of North Africa?

2. (*Human Systems*) Why was control of the Suez Canal important to the British Empire?

3. (*Places and Regions*) What is the main ethnic group, language, and religion in the region?

Critical Thinking

4. **Identifying Cause and Effect** Why do many people in Algeria, Morocco, and Tunisia speak French as a second language today?

Organizing What You Know

5. Create a time line like the one shown below. On your time line, list important years, periods, and events in the history of North Africa.

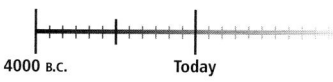

4000 B.C.   Today

Section 2 Review Answers

Identify For identification, see: Berbers, p. 489

Define For definitions, see: silt, p. 487; pharaohs, p. 487; hieroglyphs, p. 487

Working with Sketch Maps Maps will vary, but listed places should be labeled in their approximate locations. Alexandria was founded by Alexander the Great.

Reading for the Main Idea
1. Egyptians, Phoenicians, Greeks, Romans, Vandals, Byzantines, Ottomans, Europeans

2. important trade link between Europe and Great Britain's colony of India

3. Arab or Arab-Berber, Arabic, Islam

Critical Thinking
4. because these countries were once ruled by France (NGS 10)

Organizing What You Know
5. Time lines will vary but should indicate events and periods in the correct years.

VISUAL RECORD ANSWER

Only men socialize in cafes in North Africa.

 **LET'S GET STARTED**

Copy the following question onto the chalkboard: *From what you learned about North Africa's natural resources in Section 1, what economic activities do you think might be important in the region?* Discuss student responses. *(Possible answers: oil production, fishing and fish-processing in Morocco, farming in the Nile Valley)* Tell students that they will learn more about the economic activities of the North African countries in Section 3.

Building Vocabulary

Write the key terms on the chalkboard and call on a volunteer to read the definitions from the text or glossary. Ask students how much money they might save in taxes if they lived in a **free port**. Ask students if buying something like an automobile in a free port would be worth the trouble of traveling a long distance to the port. Then point out that the word **fellahin** comes from an Arabic word meaning "to cultivate." Ask students where they think most fellahin might live *(in the Nile Valley)*.

RESOURCES

REPRODUCIBLE

▶ Guided Reading Strategy 21.3
▶ Readings in World Geography, History, and Culture 53 and 54
▶ Cultures of the World Activity: Region 6

TECHNOLOGY

▶ One-Stop Planner CD–ROM, Lesson 21.3
▶ CNN Presents Geography: Yesterday and Today, Segment 23: Cairo—Selling the Suburbs
▶ Homework Practice Online
▶ HRW Go site

REINFORCEMENT, REVIEW, AND ASSESSMENT

▶ Main Idea Activity 21.3
▶ English Audio Summary 21.3
▶ Spanish Audio Summary 21.3
▶ Section 3 Review, p. 494
▶ Daily Quiz 21.3

Section 3 — The Region Today

READ TO DISCOVER

1. What are economies and cities of North Africa like?
2. What challenges do the people there face?

WHY IT MATTERS

The politics of North Africa today are strongly influenced by Islamic fundamentalism. Use **CNNfyi.com** or other **current events** sources to learn about how this movement affects countries in other parts of the world.

DEFINE

free port
fellahin

LOCATE

Tangier	Tunis
Luxor	Algiers
Tripoli	Aswān High Dam

Economic and Urban Environments

The countries of North Africa face issues typical of developing countries. These issues include controlling government spending, keeping inflation down, and reducing restrictions on businesses and foreign trade. Morocco has tried to address the last of these issues by giving the city of Tangier status as a **free port**. A free port is one where almost no taxes are placed on the goods unloaded there from other places.

Morocco and the other North African countries have taken other steps to strengthen economic ties with Europe. For example, in the late 1990s Tunisia entered an association agreement with the European Union. This agreement aims to improve trade between Tunisia and EU countries.

Economic Activities Oil and natural gas production form the backbone of the Libyan and Algerian economies. Egypt and Tunisia also have significant oil industries. When oil prices are high, these countries benefit. When prices are low, their economies weaken. As a result, they are trying to diversify their economies so they do not rely so much on oil and gas.

Agriculture is also a very important part of this region's economy, despite the dry climates. In Egypt millions of **fellahin**, or peasant farmers, work the fertile land along the Nile Valley. The fellahin make up about 40 percent of the Egyptian work force. The only country in North Africa that does not have a strong farming sector is Libya, which must import about 75 percent of its food.

INTERPRETING THE VISUAL RECORD

Economic development and rapid population growth affect life in large North African cities such as Cairo. The city's famous pyramids, visible in the background, are increasingly hemmed in by the sprawl of Africa's largest city. **Based on the photo, what environmental and urban problems do you think residents of Cairo face?**

VISUAL RECORD ANSWER

Possible answers: air pollution, crowding, housing shortages

Teach Objective 1

ALL LEVELS: Tell students to imagine that they are attending a Cairo city council meeting. Organize the class into four groups representing the following constituencies: an environmental agency, a tourist board, a city planning agency, and a chamber of commerce. Have each group form a plan to deal with issues in Cairo that affect its agency's goals. When the groups' plans are finished, have the groups present their plans as if they were appearing before the city council. Then compare and contrast the plans as a class.
COOPERATIVE LEARNING

Teach Objective 2

ALL LEVELS: Copy the following graphic organizer onto the chalkboard, omitting the italicized answers. Have students complete it. Then ask students to list some possible solutions to the problems in the chart. **ENGLISH LANGUAGE LEARNERS**

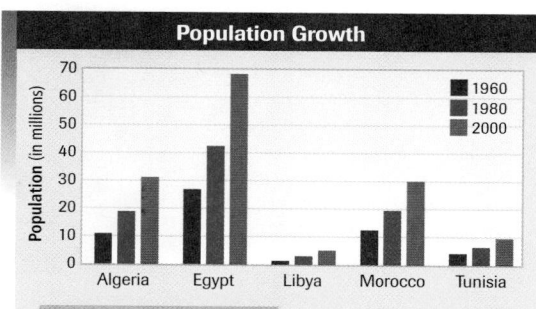

Issues in North Africa

Political	Environmental
political unrest; clashes between government and religious groups; civil war in Algeria; clashes between Libyan government and United States	*desertification; pollution from oil refining; polluted water supplies; fertilizer pollution in the Nile; Nile Delta shrinking*

Another important economic activity in North Africa is tourism. It is particularly important in Egypt, Morocco, and Tunisia. All three of these countries, and Algeria as well, have worked hard to attract tourism. However, in Algeria tourism almost entirely collapsed after political violence began in 1992. Tourism in Egypt also suffered a blow in 1997 when terrorists killed a group of foreign tourists at the ancient ruins of Luxor. Since then, tourism has slowly improved in Egypt but remains low in Algeria.

As you can see, the region has rich resources and productive industries. However, there still are not enough jobs. Rapid population growth makes this problem worse. In Libya and Algeria unemployment is as high as 30 percent. Many skilled and educated North Africans leave the region. They often find better jobs in Europe or in oil-rich Arab countries in Southwest Asia.

Population Growth

INTERPRETING THE GRAPH *The countries of North Africa have all experienced rapid population growth since the mid-1900s, largely because of falling death rates. In fact, death rates in the region are lower than in many developed countries because of the high percentage of young people.* **How do you think the populations of these countries will change in the future? Why?**

Urban Environments Among the largest cities in North Africa are Casablanca, in Morocco, and the capitals of the region's countries. Those capitals include Cairo, Egypt; Tripoli, Libya; Tunis, Tunisia; and Algiers, Algeria. These cities have a mix of modern and traditional buildings. Many large cities in North Africa are becoming more crowded as the region's population grows. At the same time, large numbers of people continue to migrate from the countryside to the cities. Thus, these large cities have grown outward with a ring of crowded slums surrounding the older core.

In cities such as Cairo, there is not enough housing. People crowd into slums, even setting up tents on rooftops or in rowboats along the Nile. Communities have even developed in cemeteries, where people convert tombs into bedrooms and kitchens.

✓ **READING CHECK:** *Human Systems* How has rapid population growth been a problem for the region's economies and cities? not enough jobs, increasingly crowded cities

Issues and Challenges

The countries of North Africa share many of the same issues and challenges. Among these challenges are poverty and political unrest. The region also has some of the same political and social problems that face its Arab neighbors in Southwest Asia.

Political Issues Islam has long been an important factor in North African politics. Today many Islamic fundamentalists believe that government should be based strictly on the laws of Islam. For the most part, the region's governments want to limit the role of Islam. Sometimes this leads to violent clashes. For example, in Algeria in 1992, an Islamic party seemed set to win an election and take power. To keep this from happening, the government canceled the election and suspended parliament. Violence broke out, and the country was soon involved in a civil war. Since then thousands have died despite some progress in efforts to end the fighting.

Essential Element 4

▶ **Human Systems** ◀

Cairo's City of the Dead
A lack of sufficient housing has prompted many homeless Egyptians to seek shelter in ancient cemeteries beyond Cairo's eastern city limits. More than 500,000 people have taken up residence throughout the graveyards. Tombs and mausoleums look like miniature houses in this area known as the City of the Dead. Although this unusual housing situation has not been officially recognized by municipal leaders, the city of Cairo now provides electricity and water to those living in the cemeteries.

DISCUSSION: Lead a discussion addressing the following questions: Should those living in the City of the Dead be allowed to take up permanent residence there? Why or why not? Should city officials provide more than utility services? If you think they should, what do you propose?

🔲 **internet** connect

GO TO: go.hrw.com
KEYWORD: SW3 CH21
FOR: Web sites about Cairo today

GRAPH ANSWER

continue to grow because of the high percentage of young people

Close

Tell students that 150 years ago, some 5 million people lived in the Nile Delta area, and there were 5 million acres of arable land in the region. Today more than 60 million people live there, and there are 7 million acres of arable land. Discuss with students the economic and environmental implications of these figures.

Review and Assess

Have students complete the **Section Review.** Then have students complete **Daily Quiz 21.3.**

Reteach

Have students complete **Main Idea Activity for English Language Learners and Special-Needs Students 21.3.** Then organize students into five groups and have each group create a newsmagazine cover featuring one of the countries of North Africa. The cover should include an illustration and titles of feature stories. **COOPERATIVE LEARNING, ENGLISH LANGUAGE LEARNERS**

Extend

Have interested students conduct research on the Aswān High Dam, including its environmental impact on the Nile Valley and Nile Delta. Students should prepare brief reports, including maps of the dam's location. **BLOCK SCHEDULING**

Section 3 Review Answers

Define For definitions, see: free port, p. 492; fellahin, p. 492

Working with Sketch Maps
Maps will vary, but listed places should be labeled in their approximate locations. The Aswān High Dam is Egypt's source of hydroelectric power.

Reading for the Main Idea
1. shifting market prices, pollution

2. historical sites

3. overcrowding, poverty

Critical Thinking
4. People must crowd into the small amount of fertile land in coastal areas and the Nile Valley. (NGS 12)

Organizing What You Know
5. Positive effects should include a reliable water supply for farming and hydroelectric power. Negative effects should include the lack of soil renewal from flooding, the damage to the fishing industry from fertilizer pollution, the shrinking of the Nile Delta, and coastal erosion.

In the 1950s, when the planned construction of the Aswān High Dam threatened to flood the ancient Egyptian temples of Abu Simbel, the United Nations and Egyptian government sponsored a project to save them. Between 1963 and 1968 a team of engineers disassembled the temples and moved them to a new location that was more than 200 feet (60 m) higher, thus saving them from destruction.

The United States has been involved in North Africa in several ways and gives large amounts of aid to Egypt. In Libya, Mu'ammar Gadhafi has seen himself as a defender of Arab causes and has pursued some anti-American policies. In 2004, however, Gadhafi agreed to open Libya to arms inspectors, paving the way to improved relations with the United States.

✓ **READING CHECK:** (**Human Systems**) What religion plays an important role in the region's politics? Why? Islam; Islamic parties want governments based on Islam.

Environmental Challenges Environmental issues in North Africa include desertification, pollution from oil refining, and polluted water supplies. In Egypt the environmental health of the Nile is a major concern. Construction of the Aswān High Dam across the upper Nile was begun in 1960. Once completed, the dam became a major source of hydroelectric power. Water stored behind the dam is used for crops year-round. This water also lets farmers open up new land for farming. These have been important benefits.

However, the dam has also stopped the annual flooding of the Nile. Before the dam was built, fields along the river were renewed every year by new deposits of silt. Now Egyptian farmers must buy fertilizer to keep the soil productive. The fertilizers have polluted the Nile and crippled the Egyptian fishing industry. Also, without the silt, the Nile Delta is slowly shrinking, and the Mediterranean coast is suffering from severe erosion.

✓ **READING CHECK:** (**Environment and Society**) How has the heavy use of fertilizers affected the Nile? pollution, threatens fishing industry

The Nile River is the world's longest river. It begins in eastern Africa and flows northward for 4,160 miles (6,693 km) to the Mediterranean Sea.

 ## Section 3 Review

Define free port, fellahin

Working with Sketch Maps On the map you created in Section 2, label Tangier, Luxor, Tripoli, Tunis, Algiers, and Aswān High Dam. What is Egypt's source of hydroelectric power?

Reading for the Main Idea
1. (**Human Systems**) What problems face the petroleum industry in North Africa?

2. (**Human Systems**) What helps make tourism an important industry in North Africa?

3. (**Human Systems**) What are some issues facing cities in North Africa?

Critical Thinking
4. Analyzing Information How might the physical geography of the region contribute to the overcrowding of its cities?

Organizing What You Know
5. Create a word web like the one below. Use it to describe the positive and negative effects of building the Aswān High Dam for Egypt.

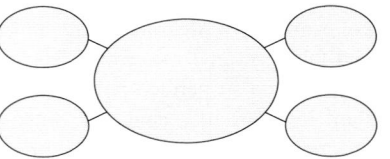

Sociology: Changing Roles of Women

The roles of women in society vary greatly around the world. Such factors as cultural traditions, religion, level of technology, economic systems, and political and legal institutions can influence the roles and responsibilities that women possess. For the Saharawis, the conflict over Western Sahara has greatly influenced women's roles. After the conflict began, the Moroccan military led a campaign to force the native Saharawis out of the territory. More than 170,000 Saharawis fled to Algeria. Because many Saharawi men stayed behind to fight the Moroccan forces, about 80 percent of the refugees were women and children. As a result, female refugees began to take over many roles typically reserved for men in their culture. They became administrators, politicians, and business owners. Education became increasingly important to these women, and literacy rates improved. Before the conflict began, only a handful of wealthy Saharawi women had the advantage of an education.

Have students conduct research on the changing roles of women in other societies around the world, such as Japan, India, and the United States. Ask them to analyze how the roles of women in each society have changed over the last 50 years, what factors have influenced these changes, and what economic opportunities have become available to women.

Human Systems

Geography for Life

Western Sahara

The phrase *Western Sahara* refers to more than a part of Africa's great desert. It is also a political unit—the only one in Africa that is not independent.

Western Sahara lies on Africa's northwest Atlantic coast. It shares borders with Algeria, Mauritania, and Morocco. Western Sahara was a Spanish colony from 1884 to 1976. When Spain withdrew from the area, no one was left as the clear authority. Three groups wanted to control Western Sahara. Both Morocco and Mauritania claimed the area. However, the Saharawis, the region's indigenous nomadic people, wanted self-rule. The Polisario—a nationalist organization—represented the Saharawis and opposed occupation by the other countries. Morocco moved some of its citizens into the western part of the disputed land. In return, the Polisario proclaimed independence. Following a change within its own government, Mauritania pulled out of Western Sahara in 1979.

During the 1980s Morocco built a sand wall along the border between Western Sahara and Mauritania. It

Two Saharawi leaders arrive for a meeting near Tindouf, Algeria, headquarters of the Polisario Front and home to thousands of refugees from Western Sahara.

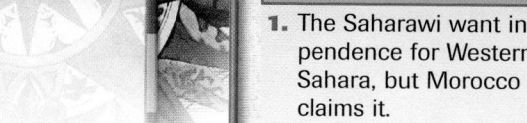

stretches from Algeria all the way to the Atlantic Ocean. (See the map.) Morocco wanted the wall to keep Polisario troops from entering the western portion of the territory and southern Morocco. In 1988 Morocco and the Polisario accepted a United Nations peace plan, which was followed in 1991 by a cease-fire. In the year of the cease-fire, the UN also established a settlement. As part of the plan, the UN was to poll the people of Western Sahara for their views. They would be asked if they wanted official union with Morocco or independence. However, it was difficult to compile a list of eligible voters. In addition, the Moroccans continued to fortify the sand wall they had constructed. They also denied UN monitors easy travel through occupied Western Sahara. While governments quarrel over their land, many Saharawis wait in refugee camps in Algeria.

Morocco and the Polisario still disagree. The main dispute involves identifying who should be allowed to vote, which further delays the election. Morocco takes advantage of the delays by moving even more people to the region. Morocco also continues to exploit Western Sahara's mineral resources. Meanwhile, the UN tries to maintain the cease-fire and avoid armed conflict.

Western Sahara

- 🏭 Phosphate mine
- 🐟 Rich fishing
- ⋯⋯ Moroccan fortified wall

Canary Islands (SPAIN)
MOROCCO
ALGERIA
El Aaiún (Laayoune)
Al Mahbas
Smara
Boukra
Tfaritiy
ATLANTIC OCEAN
Boujdour
Galtat Zemmour
Ad Dakhla
MAURITANIA
Awsard
Bir Gandouz
Techla
Aghwinit

SCALE
0   100   200 Miles
0   100   200 Kilometers
Projection: Lambert Conformal Conic

15°W   10°W   25°N

INTERPRETING THE MAP *The Moroccan defensive wall built in the 1980s stretches across much of Western Sahara, which Morocco claims.* **Based on the map, about how long is the wall? Where is it in relation to the settlements in the region?**

Applying What You Know

1. **Summarizing** Why is there a conflict between the Saharawi people and Morocco?

2. **Analyzing Maps** Examine the map. How does the distribution of resources affect which part of Western Sahara Morocco most wants to control? How is Morocco exercising this control?

Applying What You Know Answers

1. The Saharawi want independence for Western Sahara, but Morocco claims it.
2. Phosphates and fisheries are in the west, the area that Morocco most wants to control. Morocco has built a wall to enclose the area.

MAP ANSWER

about 600 miles (965 km); just to the east

This Geography for Life feature addresses National Geography Standards 10, 13, and 16.

CHAPTER 21 Review Answers

Building Vocabulary For definitions, see: erg, p. 484; reg, p. 484; depressions, p. 484; wadis, p. 484; silt, p. 487; pharaohs, p. 487; hieroglyphs, p. 487; free port, p. 492; fellahin, p. 492

Locating Key Places

A. Atlas Mountains
B. Strait of Gibraltar
C. Nile River
D. Sinai Peninsula
E. Fès
F. Suez Canal
G. Aswān High Dam

Understanding the Main Ideas

1. winds—shape dunes, clear sand from regs; rainwater—carves out wadis, fills depressions; Nile River—creates fertile valley and delta

TECHNOLOGY

▶ Chapter 21 Test Generator (on the One-Stop Planner)
▶ Global Skill Builder CD–ROM
▶ HRW Go site

REINFORCEMENT, REVIEW, AND ASSESSMENT

▶ Chapter 21 Review, pp. 496–97
▶ Chapter 21 Tutorial for Students, Parents, Mentors, and Peers
▶ Chapter 21 Test (form A or B)
▶ Alternative Assessment Handbook

▶ Chapter 21 Test for English Language Learners and Special-Needs Students
▶ Unit 7 Test
▶ Unit 7 Test for English Language Learners and Special-Needs Students

CHAPTER 21 Review Answers

2. water, good soils there; desert conditions elsewhere

3. Arabs; A.D. 600s

4. Islam; Christianity and Judaism

5. high unemployment; rapidly increasing population

Thinking Critically

1. shortened trade routes from Europe to Asia; aided trade between Europe and Asian countries (NGS 11)

2. Debate over the role of Islam has led to political violence, hurting economic development. Islam plays a major role in daily life and influences social practices and customs such as clothing styles, celebrations, and literature. (NGS 10)

3. Possible answers: Reservoir would disappear; natural cycle of flooding and silt deposition would return; agricultural production might decrease; hydropower production would decrease. (NGS 14)

Using the Geographer's Tools

1. Answers will vary. Students may note narrow streets and market areas of medinas.

2. Students should explain why they have ranked the countries as they have and note the factors they thought were most important.

496

CHAPTER 21 Review

Building Vocabulary

On a separate sheet of paper, explain the following terms by using them correctly in sentences.

erg
reg
depressions

wadis
silt
pharaohs

hieroglyphs
free port
fellahin

Locating Key Places

On a separate sheet of paper, match the letters on the map with their correct labels.

Atlas Mountains
Nile River
Strait of Gibraltar
Suez Canal

Sinai Peninsula
Fès
Aswān High Dam

Understanding the Main Ideas
Section 1

1. (*Physical Systems*) How have wind and water shaped important physical features of the Sahara?

2. (*Places and Regions*) Why are plants and animals in Egypt limited mostly to areas along the Nile Valley and Delta?

Section 2

3. (*Human Systems*) Which ethnic group makes up the majority of North Africa's population? When did this ethnic group first move into North Africa?

4. (*Places and Regions*) What is the main religion of North Africa? What are two minority religions there?

Section 3

5. (*Places and Regions*) Why have many people in the region left to find work elsewhere? What has made the problem worse?

Thinking Critically

1. Drawing Inferences Look at a world map. In what ways do you think the Suez Canal has influenced trade routes and patterns since it was built? Why might changes in these trade patterns have been important to European countries?

2. Analyzing In what ways does Islam influence the political, economic, social, and cultural characteristics of the region?

3. Drawing Inferences and Conclusions What do you think would be the effects of removing the Aswān High Dam?

Using the Geographer's Tools

1. Analyzing Photographs Look at the photograph of the medina in Section 2. How does it compare to the layout of your community's oldest area? What activities do you find in each?

2. Analyzing Statistics Review the Fast Facts and Comparing Standard of Living tables at the beginning of the unit. Then rank the levels of economic development and standard of living of the region's countries, from highest to lowest. Write a paragraph explaining your rankings. Note which statistics you believe were most important in your ranking.

3. Preparing Maps Create a map of the region that shows important agricultural areas. Shade the countries that have oil. Finally, label the Aswān High Dam and identify its importance as an energy resource.

Writing about Geography

Review what you read in Unit 6 about the Tigris and Euphrates Rivers in Southwest Asia. Then write a report comparing the importance of major rivers to the history and economies of Southwest Asia and North Africa. What role did the rivers play in the development of early civilizations in each region? Why? When you are finished with your report, proofread it to make sure you have used standard grammar, spelling, sentence structure, and punctuation.

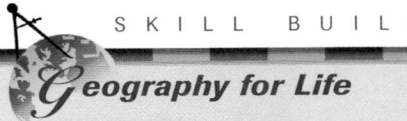

S K I L L   B U I L D I N G

Geography for Life

Analyzing Geographic Information

(*Environment and Society*) Find some recipes from North Africa and determine whether the dishes are similar to Southwest Asian, European, other African cuisine, or a mixture. Then compare the ingredients with typical crops and farm animals from the region. Which products are used most often? Write a paragraph describing your findings.

Have students design a costume that reflects the cultural traditions of North Africa. Tell students to look at the types of clothes people are wearing in the photographs in this chapter. Encourage them to conduct further research in library sources or the Internet. Then have them use these clothing styles as a guide to design their own fashions. Students should use colored pencils or markers to make sketches of their fashion designs. Have students place their designs in their portfolios.

Couscous is a staple food in North Africa. Cooks in each country prepare it differently. In Morocco, saffron is added. Algerians add tomatoes, while Tunisians top couscous with a hot pepper sauce called *harissa*. Making couscous from scratch is very time-consuming, so many cooks use a packaged product. Have students search local markets for unprocessed or packaged couscous and prepare it in one of several ways. Some might make couscous as a dessert with raisins, sugar, and cinnamon. Others might cook it with lamb or chicken, vegetables, and chickpeas.

Building Social Studies Skills

Interpreting Satellite Images

Study this satellite image of the Nile Delta in Egypt. Cultivated areas are shown in red. Then use the information from the image to help you answer the questions that follow.

1. According to this satellite image, the Nile Delta is
   a. not cultivated.
   b. only partly cultivated.
   c. highly cultivated.
   d. mostly desert.

2. Which areas in the satellite image would you expect to be densely populated? Why?

Building Vocabulary

To build your vocabulary skills, answer the following questions. Mark your answers on a separate sheet of paper.

3. *Silt* is
   a. fertile finely ground soil.
   b. a fine fabric.
   c. a sea of sand.
   d. a type of veil.

4. The caravan of traders experienced many difficulties while crossing a *depression*.
   In which sentence does depression have the same meaning as it does in the sentence above?
   a. A tropical depression is beginning to form in the gulf.
   b. A salt marsh is located at the southern edge of the depression.
   c. The depression in the cotton market took the merchants by surprise.
   d. By the third day of the sandstorm, a slight depression had set in among the tourists.

3. Maps should note banks along the Nile, the Nile Delta, coastal areas, and much of Morocco. All countries there have oil. The Aswān High Dam produces hydroelectricity.

Writing

Students should note that early civilizations grew in rich farming areas along the Tigris, Euphrates, and Nile Rivers. Use Rubric 9, Comparing and Contrasting, to evaluate student work.

Geography for Life

Answers will vary, but valid comparisons should be based on recipes and ingredients selected. Use Rubric 9, Comparing and Contrasting, to evaluate student work.

Social Studies Skills

1. c

2. The Nile Delta and Nile River valley are probably densely populated because they are cultivated areas. There is probably little or no access to water in desert areas, making it difficult for many people to live there.

3. a

4. b

Alternative Assessment

PORTFOLIO ACTIVITY

Learning about Your Local Geography

Group Project: Research

Is there an abundant supply of drinking water where you live? Does your community have to carefully protect its water resources? Work with a partner to research the location of the water resources available to your community. Find out about the quality of the water and what you can do as an individual to protect your water resources from pollution. Interview officials from your city government about any water management plans they have for the future. Create a map that illustrates some of the information you gathered. Finally, prepare a short report comparing the water resources of your state and those of North Africa.

◪ internet connect

Internet Activity: go.hrw.com
KEYWORD: SW3 GT21

Choose a topic on North Africa to:
- learn about ergs, regs, and other desert features and send a postcard from the Sahara.
- create a brochure on the design, construction, and function of the pyramids of ancient Egypt.
- take the GeoMap challenge and test your knowledge of North Africa's geography!

PORTFOLIO ACTIVITY

Students' findings will vary according to the region in which they live. Use Rubric 1, Acquiring Information, and Rubric 14, Group Activity, to evaluate student work.

CHAPTER RESOURCE MANAGER

Objectives	Pacing Guide	Reproducible Resources
SECTION 1 **Natural Environments** (pp. 499–502)	• Identify West and Central Africa's main landforms and rivers. • Describe the climates and biomes found in the region. • Name some of the region's important resources.	**Regular** 1 day **Block Scheduling** .5 day *Block Scheduling Handbook, Chapter 22*	**RS** Guided Reading Strategy 22.1 **PS** Readings in World Geography, History, and Culture 56 **E** Creative Strategies for Teaching World Geography, Lesson 16
SECTION 2 **History and Culture** (pp. 503–07)	• Identify the main eras in the history of West and Central Africa. • Describe some features of the region's cultures.	**Regular** 1 day **Block Scheduling** .5 day *Block Scheduling Handbook, Chapter 22*	**RS** Guided Reading Strategy 22.2 **RS** Graphic Organizer Activity 22 **PS** Readings in World Geography, History, and Culture 55, 57, and 63 **E** Cultures of the World Activity: Region 6 **SM** Critical Thinking Activity 22: Imperialism in Africa **SM** Map Activity 22
SECTION 3 **The Region Today** (pp. 509–11)	• Analyze West and Central Africa's economic development. • Identify the major challenges the countries face.	**Regular** 1 day **Block Scheduling** .5 day *Block Scheduling Handbook, Chapter 22*	**RS** Guided Reading Strategy 22.3 **PS** Readings in World Geography, History, and Culture 62 **SM** Geography for Life Activity 22: Nigeria's Changing Air Transport Network

Chapter Resource Key

PS Primary Sources	**A** Assessment	CD–ROM	**One-Stop** Planner CD–ROM
RS Reading Support	**REV** Review	Video	**See the *One-Stop Planner* for a complete list of additional resources for students and teachers.**
IC Interdisciplinary Connections	**ELL** Reinforcement and English Language Learners	Internet	
E Enrichment	Transparencies	Holt Presentation Maker Using Microsoft® PowerPoint®	
SM Skills Mastery			

 **One-Stop** Planner CD–ROM

It's easy to plan lessons, select resources, and print out materials for your students when you use the **One-Stop Planner CD–ROM with Test Generator**.

internet connect

HRW ONLINE RESOURCES

GO TO: go.hrw.com
Then type in a keyword.

TEACHER HOME PAGE
    KEYWORD: SW3 Teacher

CHAPTER INTERNET ACTIVITIES
    KEYWORD: SW3 GT22
    Choose a topic on West and Central Africa to:
- report on conflict, development, and migration in the Congo Basin.
- learn about Central African textiles.
- research decolonization in West and Central Africa.

CHAPTER ENRICHMENT LINKS
    KEYWORD: SW3 CH22

CHAPTER MAPS
    KEYWORD: SW3 MAPS22

ONLINE ASSESSMENT
    Homework Practice
    KEYWORD: SW3 HP22
    Standardized Test Prep
    KEYWORD: SW3 STP22
    Rubrics
    KEYWORD: SS Rubrics

COUNTRY INFORMATION
    KEYWORD: SW3 Almanac

CONTENT UPDATES
    KEYWORD: SS Content Updates

HOLT PRESENTATION MAKER
    KEYWORD: SW3 PPT22

ONLINE READING SUPPORT
    KEYWORD: SS Strategies

CURRENT EVENTS
    KEYWORD: S3 Current Events

Technology Resources

- One-Stop Planner CD–ROM, Lesson 22.1
- Geography and Cultures Visual Resources 43–47
- Homework Practice Online
- HRW Go site

- One-Stop Planner CD–ROM, Lesson 22.2
- Geography and Cultures Visual Resources 48
- Homework Practice Online
- HRW Go site

- One-Stop Planner CD–ROM, Lesson 22.3
- *ARGWorld* CD–ROM
- CNN. Presents Geography: Yesterday and Today, Segment 24: The Women of Nigeria
- Homework Practice Online
- HRW Go site

Reinforcement, Review, and Assessment

- **ELL** Main Idea Activity 22.1
- **ELL** English Audio Summary 22.1
- **ELL** Spanish Audio Summary 22.1
- **REV** Section 1 Review, p. 502
- **A** Daily Quiz 22.1

- **ELL** Main Idea Activity 22.2
- **ELL** English Audio Summary 22.2
- **ELL** Spanish Audio Summary 22.2
- **REV** Section 2 Review, p. 507
- **A** Daily Quiz 22.2

- **ELL** Main Idea Activity 22.3
- **ELL** English Audio Summary 22.3
- **ELL** Spanish Audio Summary 22.3
- **REV** Section 3 Review, p. 511
- **A** Daily Quiz 22.3

Meeting Individual Needs

Ability Levels

Level 1 Basic-level activities designed for all students encountering new material

Level 2 Intermediate-level activities designed for average students

Level 3 Challenging activities designed for honors and gifted-and-talented students

English Language Learners Activities that address the needs of students with Limited English Proficiency

Chapter Review and Assessment

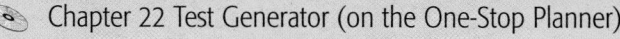

- Chapter 22 Test Generator (on the One-Stop Planner)
- Global Skill Builder CD–ROM
- HRW Go site
- **REV** Chapter 22 Review, pp. 514–15
- **REV** Chapter 22 Tutorial for Students, Parents, Mentors, and Peers
- **A** Chapter 22 Test (form A or B)
- **A** Alternative Assessment Handbook
- **A** Chapter 22 Test for English Language Learners and Special-Needs Students

Launch into Learning

Bring a jar of peanut butter to class. Ask students to name other uses for peanuts. *(Possible answers: candy, cereals, snacks, cooking oil)* Tell students that although peanuts are native to the Western Hemisphere, they were taken across the Atlantic Ocean by Europeans and were brought to the United States with enslaved Africans. Point out that peanuts are connected both to West and Central Africa's past, through the slave trade, and to its present, as a major crop and food source. Tell students they will learn more about West and Central Africa's crops, culture, food, and history in this chapter.

Using the Physical-Political Map

Have students examine the map on the opposite page. Point out that the region includes a vast area and many types of terrain, from sand desert to rain forest and mountains. Call on a volunteer to locate the region's two basins *(Chad Basin and Congo Basin)* and identify the main rivers *(Benue, Congo, Kasai, Niger, Ubangi)*. Have one student name the coastal countries and another the landlocked countries.

Why We Should Know More

You may wish to tell students that there are many reasons for us to know more about West and Central Africa:

▶ The region has a fascinating history. International relations are eased when we know the history of other regions.
▶ Many Americans are descended from Africans from this region.
▶ Americans established Liberia, on the Atlantic coast, and some freed African American slaves settled there.
▶ Political upheavals in West and Central Africa can affect U.S. foreign policy.
▶ Products such as peanuts, coffee, and chocolate are exported from the region to the United States.

CHAPTER
22
West and Central Africa

More than 315 million people live in the 24 countries of West and Central Africa. The area stretches from the deserts of Mauritania to the tropical rain forests around the Congo River.

Mask from Côte d'Ivoire

Woman carrying millet in Senegal

Ina kwanna! (Good morning!) My name is Ousseina. I have an identical twin sister named Hassana. We live in Niamey, the capital of Niger, with our grandmother, parents, and relatives. Every morning my sister and I eat breakfast with my grandmother. My favorite breakfast is millet porridge (a boiled grain) with yogurt and sugar.

When my sister and I were very small, we went to school to learn the Qur'an and the five daily prayers. We learned how to read and write in Arabic. We had to sit neatly in rows on beautiful mats and repeat verses and prayers after our religious teacher, the Malam. Now we go to a different school. In school we study literature, science, geography, history, and math. Some of our friends speak Hausa like us, but others speak a different language of Niger. School lasts from 8:00 A.M. to noon, with a 30-minute break for a snack. If we are hungry we buy mangoes, sugarcane, guavas, or bananas from street vendors. When we are thirsty, we drink baobab fruit juice or lemon and ginger juice. After school we help our grandmother pound spices for dinner, sweep the room, and wash the dishes. Sometimes after we do our homework we go to a friend's house to watch TV.

Section 1

OBJECTIVES

1. Identify West and Central Africa's main landforms and rivers.

2. Describe the climates and biomes found in the region.

3. Name some of the region's important resources.

 *LET'S GET STARTED*

Copy the following instructions onto the chalkboard: *List two ways that deserts might spread into nondesert areas.* Discuss responses. (*Possible answers: drought, over-grazing of grasslands, cutting trees, destroying plant life, changing climate*) Explain that the deserts of North Africa are gradually expanding into West Africa. Tell students that in Section 1 they will learn more about this process and other aspects of the natural environments of West and Central Africa.

Building Vocabulary

Write **desertification** on the chalkboard and underline the root *desert*. Ask students if they can identify the two suffixes that were added to the root to create the key term (*-ify and -ation*). Point out that the first suffix means "make" or "form into," and the second indicates a process. Call on a volunteer to combine meanings to form a single definition for *desertification*. Then have another student read the definition from the text or glossary.

Section 1 — Natural Environments

READ TO DISCOVER

1. What are West and Central Africa's main landforms and rivers?
2. Which climates and biomes are found here?
3. What are some of the region's important resources?

WHY IT MATTERS

Deserts are expanding in semiarid environments in West and Central Africa and around the world. Use CNNfyi.com or other **current events** sources to learn about the growth of desert areas.

IDENTIFY

Sahel

DEFINE

desertification

LOCATE

El Djouf
Niger River
Lake Chad
Congo Basin
Congo River

internet connect

GO TO: go.hrw.com
KEYWORD: SW3 CH22
FOR: Web sites about West and Central Africa

RESOURCES

REPRODUCIBLE

▶ Guided Reading Strategy 22.1
▶ Readings in World Geography, History, and Culture 56
▶ Creative Strategies for Teaching World Geography, Lesson 16

TECHNOLOGY

▶ One-Stop Planner CD–ROM, Lesson 22.1
▶ Geography and Cultures Visual Resources 43–47
▶ Homework Practice Online
▶ HRW Go site

REINFORCEMENT, REVIEW, AND ASSESSMENT

▶ Main Idea Activity 22.1
▶ English Audio Summary 22.1
▶ Spanish Audio Summary 22.1
▶ Section 1 Review, p. 502
▶ Daily Quiz 22.1

West and Central Africa: Physical-Political

Size comparison of West and Central Africa to the contiguous United States

ALGERIA · NORTH AFRICA · SAHARA · EL DJOUF · MAURITANIA · Nouakchott · CAPE VERDE · Praia · Dakar · SENEGAL · Banjul · GAMBIA · Bissau · GUINEA-BISSAU · GUINEA · FOUTA DJALLON · Conakry · Freetown · SIERRA LEONE · Monrovia · LIBERIA · Tombouctou (Timbuktu) · MALI · Gao · Niger River · Bamako · Ouagadougou · BURKINA FASO · CÔTE D'IVOIRE · GHANA · Yamoussoukro · Accra · TOGO · BENIN · Lomé · Abidjan · Porto-Novo · GULF OF GUINEA · NIGER · Niamey · Lake Chad · Kano · NIGERIA · Abuja · Lagos · Benue R. · Mount Cameroon 13,353 ft. (4,070 m) · Douala · Malabo · EQUATORIAL GUINEA · CAMEROON · Yaoundé · SÃO TOMÉ AND PRÍNCIPE · São Tomé · TIBESTI MOUNTAINS · CHAD · N'Djamena · SUDAN · CENTRAL AFRICAN REPUBLIC · Bangui · Uele R. · Lake Albert · Kisangani · Libreville · GABON · REPUBLIC OF THE CONGO · Congo River · CONGO BASIN · DEMOCRATIC REPUBLIC OF THE CONGO · Brazzaville · Pointe-Noire · Kinshasa · Kasai River · Lake Edward · Lake Kivu · Lake Tanganyika · MITUMBA MOUNTAINS · GREAT RIFT VALLEY · KATANGA PLATEAU · ANGOLA · ZAMBIA · ATLANTIC OCEAN

ELEVATION

FEET	METERS
13,120	4,000
6,560	2,000
1,640	500
656	200
(Sea level) 0	0 (Sea level)
Below sea level	Below sea level

⊛ National capital
• Other cities

SCALE
0 — 250 — 500 Miles
0 — 250 — 500 Kilometers
Projection: Azimuthal Equal Area

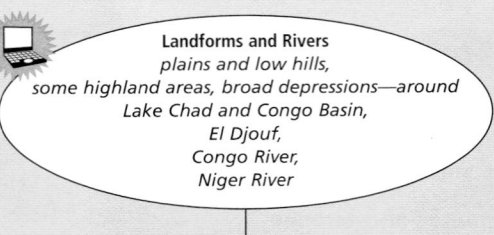

LEVEL 1: Copy the following graphic organizer onto the chalkboard, omitting the italicized answers. Call on students to provide words and phrases to identify the region's landforms, rivers, climates, vegetation, and resources. Lead a class discussion about how these aspects of the natural environment are related. **ENGLISH LANGUAGE LEARNERS**

Landforms and Rivers
*plains and low hills,
some highland areas, broad depressions—around
Lake Chad and Congo Basin,
El Djouf,
Congo River,
Niger River*

West and Central Africa

Climates and Vegetation
*zonal climate; mostly tropical;
warm throughout the year;
arid/semiarid to north with some trees, shrubs, grasses;
tropical wet and dry to south with open
grasslands, shrubs, and small trees*

Natural Resources
*tropical timber, good soils, and many minerals;
oil, particularly in Nigeria;
minerals in the Democratic Republic of the Congo;
cacao, coffee, coconuts, and peanuts
in West Africa*

Essential Element 2

► Places and Regions ◄

Upper Volta Places are often named after local physical features. Burkina Faso, a West African country slightly larger than Colorado, was previously called Upper Volta after the Volta River. The Volta's upper branches—the Black Volta and the White Volta—both originate within the country. The river itself was named for its physical features. Because its course twists across the countryside, Portuguese explorers called the river *Volta* after their word for "turn."

Upper Volta gained its independence from France in 1960. French remains the official language, but much of the population is ethnically Mossi. In August 1983 a successful coup led to the renaming of the country. Burkina Faso, the Mossi name adopted in 1984, translates roughly as "Country of Honest Men."

DISCUSSION: Lead a discussion about place names in your state. For which physical features are places named?

The tiny country of Gambia lies in an area confined to the navigable river valley of the Gambia River. Gambia extends inland along the river for about 200 miles (322 km) and is only about 15 miles (24 km) wide.

VISUAL RECORD ANSWER

Settlements probably developed where there was easy access to the ocean.

INTERPRETING THE VISUAL RECORD

Thickets of coastal mangroves, such as these in the Gambia River delta, made it very difficult for early European explorers to reach land in parts of West and Central Africa. Often these explorers had to anchor their ships offshore and use small boats to reach the mainland. **How might the physical environment of this region have influenced the locations of coastal settlements?**

Landforms and Rivers

Plains and low hills make up most of the landscapes of West and Central Africa. There are also a few highland areas and broad depressions. (See the chapter map.) In the west is El Djouf (JOOF), a desert region in eastern Mauritania and western Mali near the Niger (NY-juhr) River. Lake Chad, on the border of Cameroon, Chad, Nigeria, and Niger, lies in the middle of a depression. To the south is the Congo Basin, a huge, wet tropical lowland in Central Africa. The Congo River drains this region.

The Congo and the Niger are two of Africa's major rivers. As you can see from the map, both rivers follow unusual courses. The Congo's waters flow northward from Zambia, in southern Africa, for hundreds of miles. The river changes course in the northern Democratic Republic of the Congo. From there it flows generally westward and then toward the southwest before entering the Atlantic Ocean. The Niger River's headwaters are not far from West Africa's Atlantic coast. However, the river flows northeast across Guinea and Mali and then southeast through Niger and Nigeria before entering the Gulf of Guinea. Geographers think that the upper courses of these great rivers date back millions of years. At that time, Africa was part of the supercontinent Gondwana. Both rivers probably flowed into large inland lakes. As Gondwana broke up, the lakes drained, and each river cut a channel to the sea.

A low coastal plain runs along the Atlantic shoreline of West and Central Africa. Most of the coastline is straight, with few natural harbors. Large sandbars and mangrove trees line the coasts of Ghana, Nigeria, and Senegal.

✓ **READING CHECK:** *Places and Regions* What is the physical environment of this region like? plains, low hills, large rivers

Climates, Plants, and Animals

All of West and Central Africa lies within the tropics. Therefore, most areas are warm throughout the year. There are no major mountain ranges to break up the region's climate pattern. As a result, climate regions form bands that run east to west across the region. Geographers describe this pattern of climates as zonal. (See the unit climate map.)

Arid Environments   In the north, areas farthest from the equator have an arid climate. Here, the Sahara extends into northern Niger, Mali, and Mauritania. Along the southern edge of the Sahara is a region of semiarid climate called the **Sahel** (sah-HEL). The vegetation in this area includes scattered trees, shrubs, and grasses. The Sahel extends from Senegal and Mauritania in the west to Sudan in eastern Africa.

FOCUS ON GEOGRAPHY

The Sahel   The Sahel only receives about 4 to 8 inches (10 to 20 cm) of rainfall each year. This small amount of rain may vary greatly from year to year. In some years rain is plentiful. In other years there may be almost no rain at all in some areas. When there are several years of below average rainfall in a row, major droughts occur. In the past long droughts in the Sahel have caused widespread famines.

Most people in the Sahel are subsistence farmers. They grow crops like peanuts and grains or raise cattle and goats. In recent years the number of people and livestock has risen. The increase has put pressure on the natural environment, particularly in times of drought. People cut down trees for firewood and clear land for crops. Livestock eat grasses, leaving the soil exposed. When it rains, valuable soil is washed away. This erosion leaves the land barren and makes farming even more difficult.

The combination of droughts and a growing population in the Sahel have caused **desertification**. This means desert conditions have spread into semiarid or marginal areas. In fact, the Sahara is slowly expanding southward.

✓ **READING CHECK:**   *Physical Systems*   How have droughts and population growth affected the environment in the Sahel? They have contributed to desertification.

INTERPRETING THE VISUAL RECORD

Despite difficult environmental conditions, farmers in Africa's Sahel region attempt to grow crops where they can. Here in Mali, farmers plant their crops under shade trees to protect them from the Sun. However, rainfall is undependable, and the region often suffers from devastating droughts. A long period of drought in the late 1960s and 1970s killed some 100,000 people. **How might environmental conditions in the Sahel influence migration and population patterns?**

Tropical Environments   South of the Sahel is a zone of tropical wet and dry climate. Northeast winds from the Sahara bring hot, dry, dusty conditions in winter months. Summer winds blow in the opposite direction—from the ocean—and bring rain. Open grasslands with shrubs and small trees are common. Elephants, giraffes, zebras, and other large animals once roamed freely. However, growing human populations and the conversion of grassland into farmland have led to a serious decline in wildlife populations.

The climate zone closest to the equator is tropical humid. Rain falls year-round here, and temperatures rarely drop below 65°F (18°C). Most of Africa's dense tropical rain forests are in this part of Central Africa. Tall trees form a complete canopy in some areas. Canopies are formed by the uppermost layer of the trees, where the limbs spread out. Leaves then block sunlight from reaching plant life on the ground below. This results in areas of open ground underneath the trees. The rain forests are home to a huge number of birds and insects. Monkeys, chimpanzees, and endangered gorillas also live there.

Section 1 Review Answers

Identify For identification, see: Sahel, p. 501

Define For definition, see: desertification, p. 501

Working with Sketch Maps
Maps will vary, but listed places should be labeled in their approximate locations. The Niger River flows through Nigeria.

Reading for the Main Idea
1. Geographers think these rivers flowed inland when Africa was part of Gondwana. As Gondwana broke up, the rivers cut channels to the sea.

2. tropical location, no major mountain ranges to break up climate pattern, seasonal winds that influence tropical wet and dry climates

3. tropical timber, good soils, oil, copper, diamonds, cobalt

Critical Thinking
4. Possible answers: reduce population growth or overgrazing in the Sahel (NGS 14)

Organizing What You Know
5. Arid environments—Sahara, Sahel; scattered trees, shrubs, and grasses; Tropical environments—south of Sahel; tropical wet and dry, tropical humid climates; seasonal winds; open grasslands, shrubs, and small trees; elephants, giraffes, zebras, birds, insects, monkeys, chimpanzees, gorillas

VISUAL RECORD ANSWER

provide money for development, damage the environment

INTERPRETING THE VISUAL RECORD

Many of Africa's tropical rain forests are being cut down, which leads to the extinction of plants and animals. Rapid deforestation is a serious problem in Nigeria, where some 12 percent of the country is forested. Nigeria's large and rapidly growing population strains its forest resources as people clear trees for farmland, fuelwood, and timber bound for export. **How might deforestation in Nigeria lead to economic development and environmental change?**

tropical areas of South America; brought by Europeans during colonial era; important in Côte d'Ivoire, Ghana, Nigeria, Cameroon

Natural Resources

West and Central Africa have a wide variety of natural resources. Tropical timber, good soils for farming, and many minerals are found here. (See the unit Land Use and Resources map.) Some countries are world leaders in the production and export of certain farm or mining products.

The most valuable energy resource in the region is oil. Nigeria is Africa's largest oil producer. Smaller but significant oil reserves are found in other countries, such as Gabon and Cameroon. The Democratic Republic of the Congo is rich in minerals, including copper, diamonds, and cobalt. However, political problems and poor transportation systems have kept these resources from being fully developed.

West Africa is the world's major source of cacao, or cocoa beans. The tree that bears cacao is native to tropical areas of South America. Europeans brought it to West and Central Africa during the Colonial era. Cocoa beans are used to make chocolate. Côte d'Ivoire (koht-dee-VWAHR) is the world's leading producer. Ghana, Nigeria, and Cameroon are also important growers. Coffee, coconuts, and peanuts are also among the region's main exports. These crops grow well in the region's tropical environments.

✓ **READING CHECK:** *Places and Regions* From what world region did cacao come? How did it get to Africa and what role does it play in the region's economy?

Section 1 Review

Homework Practice Online
Keyword: SW3 HP22

Identify Sahel

Define desertification

Working with Sketch Maps
On a map of West and Central Africa that you draw or that your teacher provides, label the countries of the region, El Djouf, Niger River, Lake Chad, Congo Basin, and Congo River. Which river flows through Nigeria before entering the Gulf of Guinea?

Reading for the Main Idea

1. *The Uses of Geography* How do geographers explain the irregular course of the Congo and Niger Rivers?

2. *Physical Systems* What factors affect the distribution of climates in this region?

3. *Places and Regions* What are some major resources in West and Central Africa?

Critical Thinking

4. **Problem Solving** What could be done to try to halt desertification in this region?

Organizing What You Know

5. Copy the graphic organizer below. Use it to compare and contrast the arid and tropical environments found in this region.

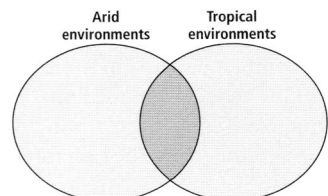
Arid environments   Tropical environments

Section 2

OBJECTIVES

1. **Identify** the main eras in the history of West and Central Africa.

2. **Describe** some features of the region's cultures.

🔊 LET'S GET STARTED

Copy the following question onto the chalkboard: *What are some aspects of our culture that reflect the history of colonization of the United States?* Discuss student responses. *(Possible answers: various foods, languages, and religions found across the United States, effects on American Indian cultures)* Tell students they will learn more about the effects of colonization on the cultures of West and Central Africa in Section 2.

Building Vocabulary

Write **staple**, **millet**, and **sorghum** on the chalkboard. Ask students to locate the terms' definitions in the section or glossary and to write a sentence that explains their relationship. *(Possible answer: Millet and sorghum are common staple crops in West and Central Africa.)*

Section 2
History and Culture

READ TO DISCOVER

1. What are the main eras in the history of West and Central Africa?
2. What are some features of this region's cultures?

WHY IT MATTERS

Many plantation crops grown in West and Central Africa during colonial times, such as cacao, are still important today. Use **CNNfyi.com** or other **current events** sources to learn about important crops in this region.

DEFINE

staple
millet
sorghum

LOCATE

Tombouctou

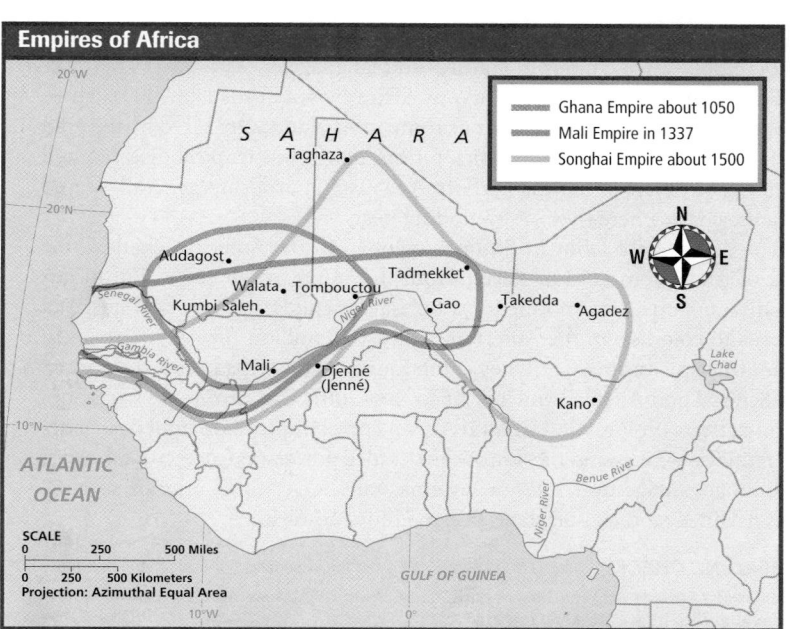

History

Great kingdoms once ruled large areas of West and Central Africa. One of the earliest was Ghana in the A.D. 800s. (See the map.) Ghana was a trading state. It exported products like gold and cloth to North Africa. The kingdom also participated in the slave trade. Traders brought to Ghana a variety of products,

Empires of Africa

Ghana Empire about 1050
Mali Empire in 1337
Songhai Empire about 1500

SAHARA
Taghaza
20°W
20°N
Audagost
Walata Tombouctou Tadmekket
Kumbi Saleh Gao Takedda
Agadez
Senegal River
Niger River
Gambia River
Mali Djenné (Jenné)
Kano
Lake Chad
10°N
ATLANTIC OCEAN
Niger River
Benue River

SCALE
0 250 500 Miles
0 250 500 Kilometers
Projection: Azimuthal Equal Area
GULF OF GUINEA
10°W 0°

INTERPRETING THE MAP *Ghana, Mali, and Songhai were all great trading empires in West Africa that grew rich from trade between tropical areas to the south and arid regions to the north.* **Which areas of West Africa did all three empires control?**

Beautiful gold objects, such as this staff from Ghana decorated with an alligator, have long attracted traders to West and Central Africa.

RESOURCES

REPRODUCIBLE
▶ Guided Reading Strategy 22.2
▶ Graphic Organizer Activity 22
▶ Readings in World Geography, History, and Culture 55, 57, and 63
▶ Cultures of the World Activity: Region 6
▶ Critical Thinking Activity 22
▶ Map Activity 22

TECHNOLOGY
▶ One-Stop Planner CD–ROM, Lesson 22.2
▶ Geography and Cultures Visual Resources 48
▶ Homework Practice Online
▶ HRW Go site

REINFORCEMENT, REVIEW, AND ASSESSMENT
▶ Main Idea Activity 22.2
▶ English Audio Summary 22.2
▶ Spanish Audio Summary 22.2
▶ Section 2 Review, p. 507
▶ Daily Quiz 22.2

LEVEL 1: Copy the following graphic organizer onto the chalkboard, omitting the italicized answers. Call on students to provide dates and phrases to identify important events and eras in the history of West and Central Africa.

ENGLISH LANGUAGE LEARNERS

Early History:
Great kingdoms, including Ghana and Mali, ruled the region. Tombouctou served as a trade and education center.

→

European Contact:
Explorers reached area by late 1400s. Slave trade of 1500s to 1800s brought at least 10 million Africans to Americas.

→

Colonial Era:
European countries wanted African resources. Many Africans began to work for wages.

→

Postcolonial Era:
All of West and Central Africa was independent by 1976. Some countries enjoy modern technology but also face many challenges.

Linking Past to Present

Mary Kingsley's Explorations Mary Henrietta Kingsley (1862–1900), a British traveler who journeyed to Central Africa in 1893 and 1894, was the first European to enter parts of Gabon. Kingsley traveled up the Ogooué River to study the Fang people, who were rumored to engage in cannibalism. She also collected specimens of beetles and fish for the British Museum. Although Kingsley died in 1900, her writings about African cultures have influenced generations of scientists.

In 1998 an international team of scientists followed in Kingsley's footsteps. Inspired by her collection of fish, they identified and cataloged numerous fish species in the Ogooué River. Fish are a substantial part of the Gabonese diet, but must be imported. The team members hoped their efforts would help to conserve native fish species and increase their populations.

ACTIVITY: Have students map Kingsley's travels.

You don't say! Early European explorers who visited what is now Ghana in the 1400s named the area the Gold Coast after local residents offered them gifts of the precious metal.

VISUAL RECORD ANSWER

African culture diffused to the Americas.

INTERPRETING THE VISUAL RECORD

The slave trade greatly affected the human geography of West and Central Africa. It altered settlement patterns, economic systems, and cultural landscapes, and caused widespread human suffering. Slaves were often held captive in coastal forts and prisons in places such as Ghana (left) and Senegal (right) before they were taken to the Americas on ships. **How do you think the slave trade in West and Central Africa shaped the distribution of culture groups around the world?**

such as salt. Islam also spread to the region along trade routes. Over time, rulers of many West African kingdoms became Muslim.

Many empires rose and fell here over the centuries. For example, Mali replaced Ghana as the most powerful kingdom during the 1200s. The city of Tombouctou (tohn-book-TOO) (also called Timbuktu) became an important center of trade and education during this time.

To the south lived the people of the tropical rain forests. Forest peoples traded less with distant lands. This was partly because the rain forest provided them with many different resources. Also, the dense vegetation of the forest made it difficult to travel long distances. Forest peoples lived in small groups and developed a wide range of cultures and languages.

European explorers arrived in West Africa by sea in the late 1400s. These Europeans were searching for a water route to Asia and were lured by the gold trade. They found an area with thick forests, dangerous tropical diseases, and few navigable rivers or natural harbors. As a result, Europeans generally stayed along the coast, where they set up trading posts.

The demand for labor in Europe's colonies in the Americas changed the focus of trade from gold to slaves. From the 1500s to the 1800s, Europeans traded with some African kingdoms for slaves. The Europeans sold enslaved Africans to colonists in the Americas. At least 10 million Africans were taken to the Americas in this way. They came mostly from areas between what are now Senegal and Angola, which is far to the south. The slave trade had wide-ranging effects on West and Central Africa. For example, it disrupted societies and families. Also, guns the Europeans traded for slaves gave coastal forest states an advantage over interior savanna states. Over time, interior savanna states declined, and coastal states became more powerful.

✓ **READING CHECK:** *Human Systems* Why did forest peoples not trade more with distant lands? Forests provided them with many resources. Dense forests also made travel across long distances difficult.

The Colonial Era By the mid-1800s the slave trade was coming to an end. Europe was industrializing, and countries there wanted minerals and tropical farm products they could not produce at home. The climates and soils of West

LEVELS 2 AND 3: Organize the class into groups and have each group write a short play based on events in the history of West and Central Africa. Encourage groups to conduct research to find additional material for their plays, such as for appropriate costumes, props, or background music. Have groups perform their plays for the class.
COOPERATIVE LEARNING

and Central Africa were good for growing many of these products, such as cocoa, peanuts, and rubber. As a result, many European countries sought political control over African territories.

During the colonial period many West and Central Africans quit subsistence farming and started working for wages. Some worked on plantations. Others moved to cities. Unlike most precolonial settlements, these new colonial cities were located along the coast. These locations offered better connections to Europe and the Americas.

The Postcolonial Era   Africa's colonial era lasted less than 100 years. In 1957 Ghana became independent. Other countries soon followed. By 1976 all African countries in this region were independent.

Although the colonial era was relatively short, it had a major effect on West and Central Africa. Before the colonial era most people in the region were subsistence farmers. Afterward, many worked in the new commercial economy. Local economies had been based on trading gold, salt, and ivory. They now depended on the export of minerals and farm products. Modern medicine and infrastructure improved many people's lives, particularly in cities. However, people also faced new and difficult problems. For example, many earned low wages or were unemployed. Also, rival ethnic groups had to share power in newly independent countries. This caused serious political rivalries.

✓ **READING CHECK:**   *The Uses of Geography*   How did the colonial era affect the economies of countries in West and Central Africa? Many economies changed from subsistence to commercial.

Connecting to
HISTORY

The Colonial Scramble

In 1884, European powers met in Berlin to settle colonial disputes in Africa. The conference set up rules designed to govern how outside powers took control of African lands. In a short time European countries had divided most of Africa among themselves. The French, British, and Belgians took most of West and Central Africa. Portugal, Spain, and Germany also claimed colonies in the region. (See the map.)

The effects of the European scramble for colonies can still be seen in Africa. Borders there today are largely those drawn by Europeans. When creating these borders, the colonial Europeans ignored Africa's existing states, ethnic groups, and natural environments. Some colonies included many different peoples. These peoples spoke many different languages and dialects. New borders also blocked traditional migration and trade routes. These borders created political problems, particularly when African colonies won independence. They also left a number of African countries landlocked. These problems remain a difficult challenge facing Africans today.

Summarizing How did the scramble for European colonies shape the borders and countries of modern Africa?

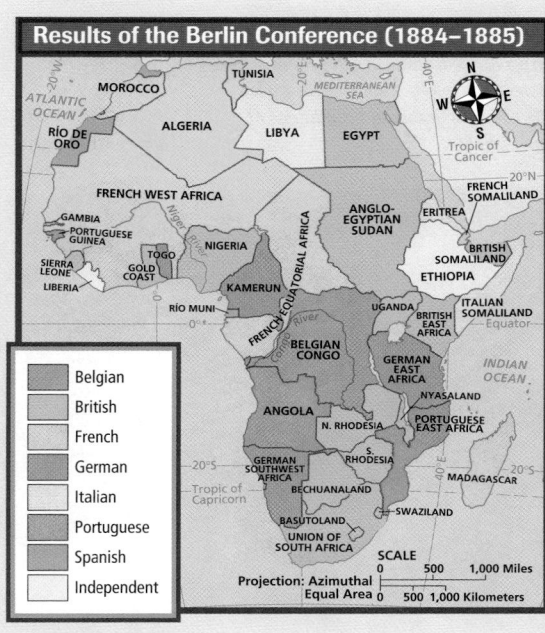

Results of the Berlin Conference (1884–1885)

Belgian
British
French
German
Italian
Portuguese
Spanish
Independent

Projection: Azimuthal Equal Area

SCALE
0   500   1,000 Miles
0   500 1,000 Kilometers

CONNECTING TO
HISTORY ANSWER

African countries mostly inherited the old colonial borders and diverse populations that existed within those borders.

ALL LEVELS: Organize the class into small groups and have each group select a culture group of West or Central Africa. Have students conduct research to learn more about their chosen peoples. Then have each group design a series of commemorative stamps that depict various elements of West and Central African culture. The stamps might illustrate culture traits related to architecture, art, clothing, food, language, religion, or traditions. Have volunteers from each group explain the stamps to the class. **COOPERATIVE LEARNING**

Using National Geography Standard 7:
Physical Systems: The Physical Processes that Shape the Patterns of Earth's Surface To help students understand desertification, demonstrate the effect of wind on overgrazed soil. Get two squares of turf from a garden center. On one, clip the grass down to the roots, to imitate how goats eat, and remove the clippings. Allow both squares to dry thoroughly. You may want to pound each square lightly with a hammer to imitate animals walking over the ground. Then turn a fan on both samples. Have students observe how the dirt blows away from the overgrazed turf square. Lead a discussion on the effects of soil erosion on West African culture.

Across the Curriculum

▶ **Art** ◀

Mbari **Houses** Traditional religious practices among some Igbo people of southeastern Nigeria have shaped a local art form. Artists called *mbari* design mud-walled houses, adorned with ceramic plates and vibrant colors, in which they place clay sculptures shaped like humans. Some *mbari* houses contain more than 100 of these figures, which often represent local deities. *Mbari* artists believe their work should reflect cultural change, so some figures now wear Western clothing and wristwatches as they talk on telephones or listen to radios.

Mbari houses are built as petitions to local spirits in times of hardship or as thanks to the spirits for gifts they have given. The houses are built in secret. Large walls are erected around a chosen site so that only people chosen to help build it can see a work in progress. Finished *mbari* houses are never repaired. They are allowed to disintegrate slowly into the earth from which they were made.

Nigeria has more than 250 ethnic groups, each with its own language. Linguists think that most Nigerian languages have been spoken in about the same places for thousands of years.

INTERPRETING THE VISUAL RECORD

*Most people in West and Central Africa live in rural areas and practice subsistence agriculture. Nomadic lifestyles are found in dry northern environments such as the Sahel. **Why might a nomadic lifestyle be common in the arid conditions of the Sahel?***

VISUAL RECORD ANSWER

Resources such as water and vegetation are scarce.

Culture

West and Central African societies are very diverse. Like the rest of Africa, peoples here reflect three major cultural influences. Those influences are traditional African cultures, Islam, and European culture. These influences have been called Africa's triple heritage.

People, Languages, and Settlement Most of the languages spoken here belong to the Niger-Congo language family. Within this family there are hundreds of different languages. Arabic is also spoken in northern areas. During the colonial era, English and French became lingua francas in much of Africa. These languages are still widely used in the region today.

Most West and Central Africans live in rural areas or small villages and rely mainly on farming. However, the populations growing the fastest are those in the cities. In most countries, the largest city is the capital. Except in landlocked countries, capitals are generally located on the coast. Most were set up as ports and government centers during the colonial era.

Religion and Education Islam is the main religion in the Sahel. However, many Christians live to the south, between the Sahel and the Atlantic coast. Many people also practice traditional African religions. These people believe that spirits—particularly the spirits of their ancestors—play an important part in their lives. People seek advice and help from spirits in times of trouble or sickness. Ancestral spirits are also honored in ceremonies.

Literacy rates are generally low throughout West and Central Africa. Only a small percentage of people in most countries finish high school. Very few have the chance to go to college. The main obstacle to education is poverty. In many poor families, parents need their children to work. Older children often take care of their younger brothers and sisters so their parents can work. At home, parents and elders provide what education they can for their children. The children learn about family and group traditions and about growing crops or raising animals. Many children do not have the opportunity to learn skills like reading and writing.

West and Central Africans have a rich tradition of dance. Here in Cameroon, Juju dancers dressed as forest animals perform a dance about the destruction of the forest.

Food, Traditions, and Customs Most people in the region produce their own food. Some grow only a few **staple** crops. Staple crops are the main food crops of a region. Such crops in West and Central Africa include cassava, corn, and yams. **Millet** and **sorghum** are also staple crops here. These two grains are resistant to drought, growing well in dry areas like the Sahel. In the cities many people eat foods introduced from other countries. For example, bread has become an important staple in some places. However, the wheat needed to make bread does not grow easily here. As a result, wheat and other introduced food crops must be imported.

Customs and traditions differ among the region's many ethnic groups. For example, the lifestyles of Muslims in the north differ greatly from those of rain forest peoples in the south. Generally, societies are based on extended families. Families consist of several households, including the head of the family, his wife or wives, their unmarried children, and their married daughters and their husbands and children. Members of the extended family work together to support the household and take care of older people and young children.

✔ **READING CHECK:** *Places and Regions* How is Africa's triple heritage reflected in the culture of the region? Arabic, Niger-Congo, and colonial languages are spoken. Islam, Christianity, and traditional African religions are all practiced.

Review

Homework Practice Online
Keyword: SW3 HP22

Define staple, millet, sorghum

Working with Sketch Maps On the map you created in Section 1, label Tombouctou. On which river is Tombouctou located?

Reading for the Main Idea

1. *Human Systems* What did kingdoms in West and Central Africa trade with people in North Africa?

2. *Places and Regions* Where will you find most capital cities in the region's countries that are not landlocked? Why?

3. *Human Systems* What are the most common religions practiced here?

Critical Thinking

4. **Drawing Inferences and Conclusions** In what ways do you think the physical environment influenced cultural patterns and the distribution of culture groups in West and Central Africa?

Organizing What You Know

5. Copy the time line below. Use it to list major events and periods in West and Central Africa's early history, colonial era, and postcolonial era.

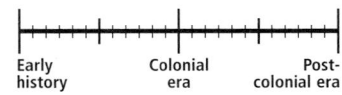

Early history — Colonial era — Post-colonial era

Government: Nigeria's Political Parties

For much of its history Nigeria has been ruled by military dictators. Under these rulers, only political parties that secured government approval could nominate candidates for public office. In the transition to a democratic government, however, the party system changed. All previous political parties were disbanded. Suddenly any Nigerian citizen could join or establish a political party.

In the elections of 1999, nine political parties fielded candidates for seats in the National Assembly, but three dominated the election. None of the parties expressed clear platforms, and there were only slight ideological differences among them. Their support therefore depended on other factors. The conservative-centrist All People's

Party (APP), which won about 20 percent of the Assembly seats, was supported mainly by members of the Yoruba ethnic group. Support for the other parties, however, spanned ethnic divisions. The centrist People's Democratic Party (PDP), which won more than 50 percent of the Assembly, attracted voters who had strongly opposed military rule. On the other hand, the Alliance for Democracy (AD), a progressive-centrist party, was popular among supporters of the last dictator. It won 20 percent of the contested seats.

Have students conduct research to learn more about the formation of the first political parties in the United States—Thomas Jefferson's Republicans and Alexander Hamilton's Federalists. Ask students to compare people's motives for supporting these parties to those seen in Nigeria's recent election. Then have each student create a chart summarizing his or her conclusions.

► Literature ◄

Chinua Achebe Born in an Igbo village in 1930, Chinua Achebe is one of Africa's most famous authors. While growing up, Achebe was introduced to European culture and Christianity, but he preferred the traditional Igbo values of his family. This regional pride is an inspiration for his writing.

Achebe's first and most widely read novel, *Things Fall Apart,* tells the story of Okonkwo, an Igbo clansman who is trying to understand his own culture when the British arrive. The novel is a poignant description of the effects of European colonization on Igbo society.

ACTIVITY: Have students conduct research to learn how the works of various Nigerian writers, artists, or musicians reflect Nigeria's diverse ethnic heritage.

Applying What You Know Answers

1. ethnic and religious conflicts among its more than 250 ethnic groups
2. Students might say the country will split along ethnic and religious lines because the peoples of Nigeria have many different cultures and beliefs. They should note that democracy allows all voices to be heard.

This Geography for Life feature addresses National Geography Standards 4, 10, and 13.

Human Systems

Geography for Life

Nigeria's Ethnic Diversity

Some countries, such as France and Japan, have a dominant culture and a clear national identity. In contrast, a national identity is hard to find in Nigeria, Africa's most populous country. Nigeria was granted full independence from Great Britain in 1960. After enduring years of harsh and corrupt military rule, Nigerians adopted a new constitution in 1999. That same year, an elected civilian government came to power. Although Nigeria's government is becoming more democratic, ethnic conflicts are still a major challenge.

To understand the country's politics, we must study Nigeria's diverse people and cultures. Nigeria has 36 states and more than 250 ethnic groups. Each group has unique customs, languages, and traditions. A strong Muslim presence dominates the north. The Hausa and Fulani together are the largest Muslim groups. Other major ethnic groups of the north include the Kanuri. Christian influence is strong in the south. The Yoruba people dominate the southwest. About half of them are Christian, but half are Muslim. The mainly Roman Catholic Igbo (Ibo) are the largest ethnic group in the southeast. The Ibibio and other peoples also live there in large numbers.

Religions in Nigeria

- 10%
- 40%
- 50%

Christian
Indigenous beliefs
Muslim

Source: Central Intelligence Agency, *The World Factbook 2003*

Some ethnic groups in the country want regional independence and a larger share of local wealth. Yet Nigeria's leaders fear that granting these requests would divide the country. Religious conflict has also appeared. For example, some northern states want to apply Islamic law to criminal offenses. Islamic law, called sharia, has long been used as the basis of family law in northern Nigeria. Sharia bans alcohol and allows severe punishments, such as cutting off a hand, for certain crimes. While officials say sharia would apply only to Muslims, many Christians are concerned.

Since military rule ended, hundreds of Nigerians have died in ethnic and religious fighting, particularly in the

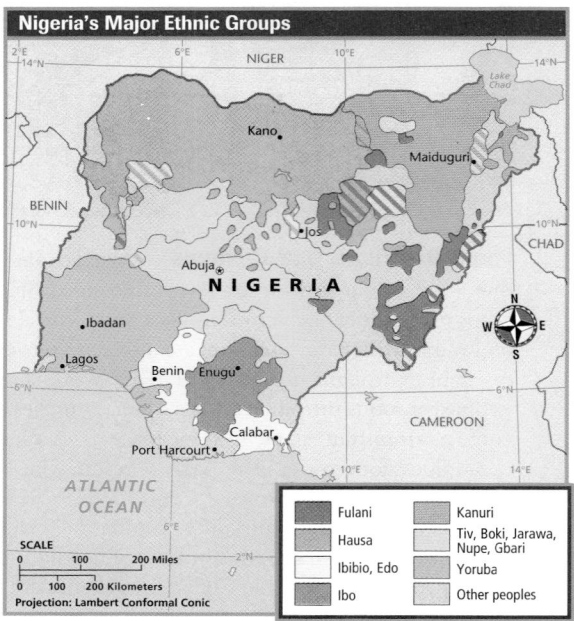

Nigeria's Major Ethnic Groups

Fulani	Kanuri
Hausa	Tiv, Boki, Jarawa, Nupe, Gbari
Ibibio, Edo	Yoruba
Ibo	Other peoples

SCALE
0   100   200 Miles
0   100   200 Kilometers
Projection: Lambert Conformal Conic

cities. Many more people have fled the violence by returning to their ethnic or religious homelands.

The previous military government might have stepped in to end the conflicts. Some observers say the military could use the ethnic clashes as an excuse to retake power. Others say that the army is now too weak and that most Nigerians would fight to defend democracy. Because no easy solution to the conflicts exists, Nigerians are finding they must learn to live with their differences.

Applying What You Know

1. **Summarizing** Why is Nigeria unstable even though its military government has been replaced by a democratically elected government?

2. **Making Predictions** Do you think Nigeria will eventually split into several countries? Why or why not? How might democracy help Nigeria's diverse peoples live with their differences?

Section 3

OBJECTIVES

1. Analyze West and Central Africa's economic development.
2. Identify the major challenges the countries face.

 **LET'S GET STARTED**

Copy the following instructions onto the chalkboard: *Look at the photograph of Lagos, Nigeria, in Section 3. Judging from this picture, what kinds of challenges do you think the people of West and Central Africa face?* Discuss responses. *(Possible answers: crowding, poverty)* Point out that the countries of this region share many of the same challenges. Tell students they will learn more about these challenges and other aspects of life in the region in Section 3.

Building Vocabulary

Write **dual economies** on the chalkboard and underline *dual*. Ask students what the word means *(two)*. Tell students that in countries with dual economies, there are two distinct economic sectors. Some people engage in subsistence activities like small-scale farming. Others practice market-driven activities like trade or plantation agriculture. Though the activities take place in the same area, they remain unrelated. People involved in one sector are not directly affected by activities or changes in the other.

Section 3 The Region Today

READ TO DISCOVER

1. How economically developed are West and Central African countries?
2. What major challenges do the countries face today?

WHY IT MATTERS

Most countries in West and Central Africa have very high population growth rates. Use CNNfyi.com or other **current events** sources to learn about the rapidly growing populations of these African countries.

DEFINE

dual economies

LOCATE

Lagos
Kinshasa
Abidjan
Accra
Douala

Level of Development

West and Central Africa is a region of developing countries. On average, people here earn less and live shorter lives than people in other parts of the world. They also have lower levels of education. (See the unit Fast Facts table.) Some countries are better off than others. For example, Gabon is one of the richest countries in Africa because of its oil reserves. In contrast, landlocked Mali is among the poorest countries in the world. It lies in the Sahel region and has few resources.

The countries of West and Central Africa have **dual economies**. In a dual economy, some goods are produced for export to wealthy countries. Meanwhile, another part of the economy produces goods and services for local people. For example, cash crops like rubber and cocoa or minerals like diamonds and bauxite might be produced and exported. On the other hand, subsistence farmers produce food for their own use. Street vendors and local markets sell clothing, food, and services to passersby.

Agriculture In the Sahel grasslands, farmers raise cattle and goats and move their herds in search of grazing lands. In tropical rain forest areas, cassava, millet, and yams are all staples. Farmers there have long planted several different crops in a single field. This kind of farming works well in tropical environments. If one crop is damaged by disease, other crops provide enough food for people to survive. Because different crops mature at different times, farmers do not have to harvest all their crops at once or keep food in storage.

The development of market economies under colonialism affected traditional farming. Plantations and ranches made it hard for herders to move their animals around to different grazing lands. Thus, herders have been forced to stay in one place. This leads to overgrazing and soil erosion. In turn, herders cannot keep as much livestock as in the past. As a result, they either have to migrate to more fertile areas or to the cities.

In a dual economy, goods like rubber are produced for export (above), while other goods are produced for local people (below).

✓ **READING CHECK:** **Human Systems** How has the development of market economies affected agriculture in the region? Plantations and ranches have made it difficult for herders to move around.

Teach Objective 1

 ALL LEVELS: Copy the following graphic organizer onto the chalkboard, omitting the italicized answers. Call on students to provide words and phrases to explain the extent to which West and Central African countries are economically developed. **ENGLISH LANGUAGE LEARNERS**

Teach Objective 2

ALL LEVELS: Organize the class into groups and have each group identify a challenge facing the people of West and Central Africa. Then have each group write a script for a public service announcement intended to inform people in the United States about the extent of the problem and the challenges it poses to people in the region. Encourage groups to conduct further research to learn more about their chosen issues. Call on volunteer groups to present their public service announcements to the class. **COOPERATIVE LEARNING**

```
                Economic Development of West and Central Africa

   Subsistence Activities                    Market Activities
 • Sahel grasslands used for herding    • primary goods exported
 • many crops planted in tropical       • heavy dependence on a few main
   rain forest                            exports
 • many crops planted in one field      • small amount of manufacturing
                                        • plantation agriculture, ranching
```

Cooperative Learning

Urban Growth Many cities in West and Central Africa are growing rapidly. For example, Nouakchott, Mauritania, grew from a small village of about 4,000 in 1957 to a city of more than half a million people today. It was a haven for refugees during Saharan droughts of the 1970s when thousands of people moved to urban areas.

Organize students into pairs. Have each pair choose a national capital or other large city from West or Central Africa. Have students use almanacs and other reference books and the Internet to track the city's growth since the mid-1900s. Have each pair create a graph to illustrate this growth. Ask each pair to investigate the factors that have affected the city's population. Then have students create population pyramids to illustrate their chosen cities' current population characteristics. Ask students what their pyramids suggest about future population growth in the cities. Then ask students to identify problems that have arisen in their cities because of the region's rapid population growth and to propose solutions to some of these problems.

GRAPH ANSWER

55.5 percent

DIAGRAM ANSWER

secondary activity

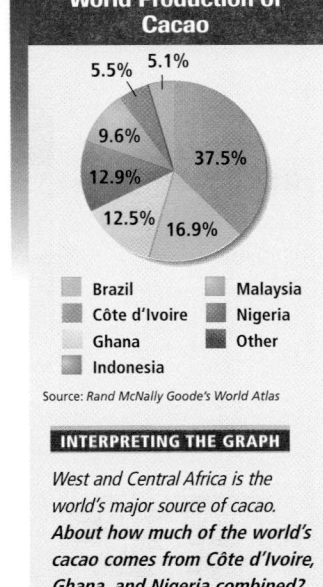

World Production of Cacao

- 37.5% Brazil
- 5.1%
- 5.5%
- 9.6%
- 12.9%
- 12.5%
- 16.9%

Brazil	Malaysia
Côte d'Ivoire	Nigeria
Ghana	Other
Indonesia	

Source: Rand McNally Goode's World Atlas

INTERPRETING THE GRAPH

West and Central Africa is the world's major source of cacao. **About how much of the world's cacao comes from Côte d'Ivoire, Ghana, and Nigeria combined?**

INTERPRETING THE DIAGRAM

Cacao beans are processed into liquid paste, which is then used to make cocoa powder, chocolate, and other products. **The production of chocolate is what type of economic activity?**

Economic Activities and Global Trade Most of the region's countries export primary rather than secondary goods. For example, Côte d'Ivoire exports cocoa beans, but the manufacture of chocolate often takes place in the developed countries. Similarly, Guinea exports bauxite, but the manufacture of aluminum takes place elsewhere.

Many countries in the region depend heavily on only a few main exports. This practice has two major disadvantages. First, it makes economies vulnerable to changes in the price of their main exports. For example, about 90 percent of Nigeria's export earnings come from selling oil. When the price of oil drops, the Nigerian economy suffers. Likewise, if the price of cocoa falls, the economy of Côte d'Ivoire can be hurt. Second, the export of primary goods is less profitable than the export of manufactured goods. For example, it is more profitable to sell peanuts processed into peanut butter and oil than to sell raw peanuts. Manufacturing does take place in some areas. However, most West and Central African countries do not have adequate facilities to process their primary products. As a result, they miss out on much of the wealth their raw materials create.

Cities In the early 1960s there were very few large cities in West and Central Africa. Only a handful of cities had populations greater than 300,000. After independence, however, cities grew very rapidly. For example, Lagos, Nigeria, grew from 760,000 people in 1960 to 4.5 million by 1980. Today some 13 million people live there. In the Democratic Republic of the Congo, Kinshasa, which had a population of 450,000 in 1960, has about 5 million people today. Other large cities include Abidjan in Côte d'Ivoire, Accra in Ghana, and Douala in Cameroon.

This rapid urban population growth has caused housing shortages. Many people live in crowded shantytowns without electricity or running water. In sharp contrast to the poor shantytowns are more prosperous downtown areas. These areas often look similar to the downtown areas of European cities. They

Cacao: From Field to Consumer

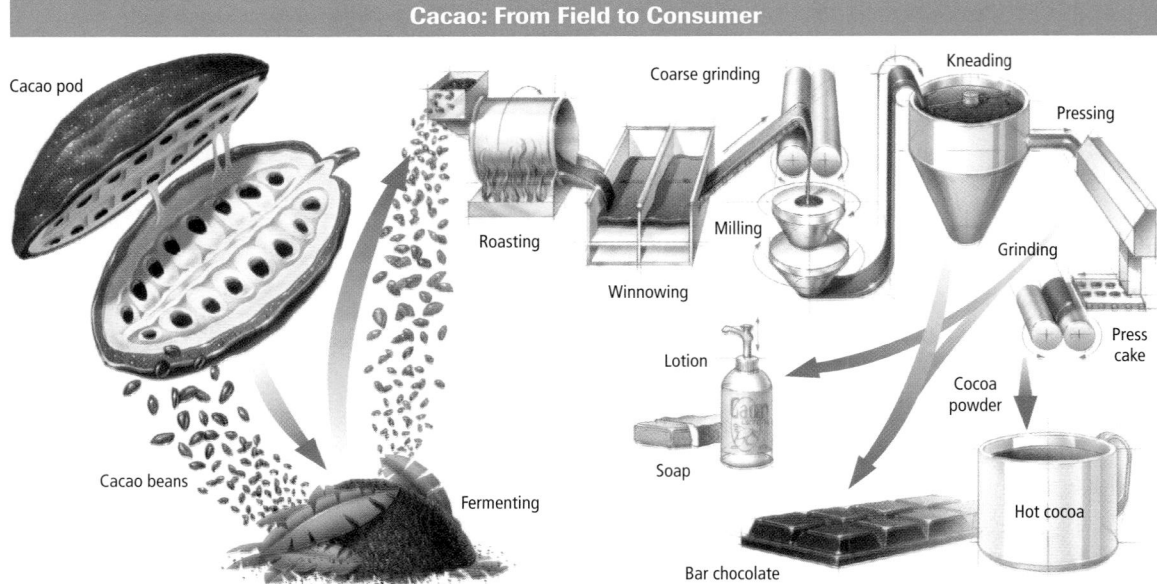

Cacao pod

Cacao beans

Fermenting

Roasting

Winnowing

Coarse grinding

Milling

Kneading

Pressing

Grinding

Press cake

Lotion

Soap

Cocoa powder

Bar chocolate

Hot cocoa

have busy roads full of cars and buses. Tall buildings dominate the central city. Neon signs and billboards advertise international products like soft drinks and electronics.

✓ **READING CHECK:** *Places and Regions* What types of goods are West and Central Africa's most important exports? **primary goods**

Issues and Challenges

The countries of West and Central Africa face many challenges. Economic development is probably the most important challenge. Issues like population growth, health care, political problems, and protecting the environment all affect development.

The region's population is growing rapidly. This rapid growth has caused many problems. Since independence, agricultural production has not kept pace with population growth. In some places, food production has even declined. Food shortages and malnutrition have become more common.

Recently, many countries have suffered wars and conflicts. In the past 10 years alone, civil wars have been fought in the Democratic Republic of the Congo, Liberia, and Sierra Leone. Thousands of people have been killed in these civil wars.

Destruction of the natural environment is a serious problem. Lumber companies harvest tropical rain forests for timber. Grasslands are cleared for farming. Such clearing has led to the extinction of some plants and animals. In the Sahel, desertification has degraded the land. As a result, people have migrated southward in search of farmland and food.

One of the most serious problems facing the region is disease. Malaria has long been a problem. In addition, HIV—the virus that causes AIDS—has spread rapidly. There is no cure for HIV infection, and treatments are very expensive. Poor, malnourished, and poorly educated people are particularly vulnerable to the disease.

✓ **READING CHECK:** *Places and Regions* What challenges does the region face? **improving countries' economies, rapid population growth, civil wars, destruction of the natural environment, disease**

With about 13 million people, Lagos, Nigeria, is one of the 15 largest cities in the world. As in many cities in developing countries, the population of Lagos is growing rapidly.

Section 3 Review

Define dual economies

Working with Sketch Maps
On the map you created in Section 2, label Lagos, Kinshasa, Abidjan, Accra, and Douala. Which city has grown to about 13 million people today?

Reading for the Main Idea

1. *Places and Regions* What level of development do all the countries share?

2. *Human Systems* What disadvantages do West and Central Africa's countries face when they depend on just a few exports?

Critical Thinking

3. **Supporting a Point of View** Who do you think benefits the most from a dual economy?

4. **Analyzing** How is economic development in West and Central Africa tied to population growth, health care, political problems, and environmental change? How might rapid urbanization make some of these problems worse?

Organizing What You Know

5. Copy the chart below. Use it to list some major challenges the countries of West and Central Africa face. In addition, suggest some possible solutions to these challenges.

Challenges	Possible solutions

go.
hrw
.com
Homework Practice Online
Keyword: SW3 HP22

Making Decisions

Have students read Cities & Settlements: The Tuareg: A Nomadic Way of Life. Reread this sentence to the class: "How hard would it be to give up a way of life that your family has treasured for generations?" Point out that some Tuareg are having to make decisions about their future. Ask students to put themselves in the place of a Tuareg family beset by drought and desertification and to decide whether the family should give up the nomadic way of life. You may want to use the Unit 10 Geography Skill-Building Workshop 2,

Making Decisions, for guidance. First, have students gather information from the feature on the issues facing the Tuareg family (*scarce water supplies, reduced pasturage, government restrictions on movement*). Then instruct students to identify the family's options, to list the advantages and disadvantages of each option, and to predict the consequences of the decision. Finally, the family would make the decision and follow through on it. Have the class vote on whether the family should settle in one place. Since the students are not in a position to maintain or abandon nomadism, ask them to list some of the steps the family might take to follow through on the decision.

Daily Life

Djenné's Architecture
Decreased rainfall in the Sahara has made life difficult for both nomads and city dwellers. For example, the residents of Djenné, Mali, risk losing their homes in a prolonged drought. Structures there are built of sun-dried mud bricks. This construction method makes the city's buildings quite fragile. If they are not regularly covered by plaster, the bricks can be destroyed by heavy seasonal rains. During the severe droughts of the 1970s and 1980s, many houses were abandoned. When rain finally fell, buildings that had not been replastered literally melted away.

Applying What You Know Answers

1. Many Tuareg are settling in towns and accepting government education. Some governments are limiting movement across the borders, and environmental changes are making it harder to graze their animals.
2. Possible answer: Their lifestyle is critical to their culture, identity, and history.

This Cities & Settlements feature addresses National Geography Standards 4, 6, 9, 10, and 12.

CITIES & SETTLEMENTS

The Tuareg: A Nomadic Way of Life

Human Systems How hard would it be to give up a way of life that your family has treasured for generations? This is the question facing many Tuareg (TWAH-reg), a nomadic people of North and West Africa. For more than 1,000 years they have raised camels, goats, and sheep in the Sahara. However, climate and politics are threatening to end the Tuareg's traditional way of life.

The "Blue Men" of the Sahara
As many as 1 million Tuareg live in Africa. Countries with Tuareg populations include Algeria, Burkina Faso, Libya, Mali, and Niger. Over the centuries, many have settled in towns in the Sahel. However, many others still live in the desert. These Tuareg graze their herds on sparse desert plants. When the plants in one area are gone, they move to a new area. Generally, the Tuareg spend only about two weeks in one place. "My father was a nomad, his father was a nomad, I am a nomad," one herder explains. "This is the life that we know. We like it." Fiercely independent, the Tuareg call their ancient way of life *adima,* meaning "far from town."

The Tuareg's independent spirit can also be seen in the arrangement of their desert camps. Tuareg travel together in small groups of relatives and friends. In camp, however, group members live apart. Each family's tent is several hundred yards away from the others. Visiting is common, but Tuareg generally do not share food or care for another family's livestock. Their diet consists mainly of fruits and grains people get through trade or from other Tuareg who farm in oases. Goat milk and cheese provide protein. On special occasions, a sheep or goat is slaughtered to provide meat.

Tents are made from goat skins that are sewn together and stretched over a rectangular frame. Inside,

Tuareg men wear blue veils in the presence of women and strangers, although this practice is less common in cities.

family members sleep on carpets or mats. A sheepskin blanket provides warmth on chilly desert nights. Family tents belongs to Tuareg women. When a woman marries, she receives a tent made by her female relatives.

Tuareg men wear cloth veils wrapped around the face and head. Because these veils are traditionally dyed blue, Tuareg are sometimes called the Blue Men. The veils help protect against windblown desert dust. Long robes are also practical in the desert. They keep sweat from evaporating too quickly, which helps protect against dehydration.

Class divisions are important in Tuareg society. A family's position is passed down from father to son. Many Tuareg oasis dwellers are members of the servant classes. Their role is to provide food and other items needed by the upper-class herders of the desert.

These Tuareg boys are making toy camels. Camels have long been an important part of the Tuareg's nomadic way of life.

Going Further: Thinking Critically

Have students review previous chapters in this book to find other peoples who face difficult decisions. Examples might include citizens of Quebec voting on secession from Canada or Russians supporting further democratic reforms. Ask students to examine one or more of these decisions in the same way they examined the Tuareg's situation. You may want to have students conduct further research on the issues in question.

The campsites of Tuareg nomads have tents and animals such as donkeys and chickens.

Living as a nomad is a sign of high status among the Tuareg and is their preferred way of life.

A Changing Way of Life

In recent years it has become harder for the Tuareg to maintain their nomadic lifestyle. Some governments in the region have limited movement across their borders because of warfare and unrest. As a result, some Tuareg cannot travel to much-needed grazing lands.

Environmental changes have also disrupted the Tuareg's way of life. For example, rainfall in the Sahara has decreased in recent decades. Since the 1960s two major droughts have dried up already scarce water resources and reduced grazing areas. The lack of water has nearly wiped out the herds of many Tuareg. Some nomads have been forced to settle near towns. They survive by gardening and selling crafts and camel rides to tourists. However, for many Tuareg the lure of their traditional life never dies. "Each time I earn a little money I buy a goat or a sheep," says one former nomad. "I save up so that I can have enough animals to return to the desert."

Still, others fear that the old ways are gone forever. This fear has weakened the Tuareg's resistance to government education for their children. For generations the Tuareg viewed schools as a government program designed to limit their movement. Now, however, many believe that education will free their children from a way of life they fear has no future. Some nomadic groups even let government teachers travel with them in the desert. Schools have also been set up at water sources in the desert. Even so, many Tuareg children still refuse to accept that they will not live as their parents and grandparents did. "I like taking care of camels," one 15-year-old insists. "I don't know the world. The world is where I am."

Applying What You Know

1. **Summarizing** How is the Tuareg's traditional nomadic life changing? What is causing these changes?

2. **Analyzing Information** Why do you think some Tuareg are determined to maintain their traditional way of life?

CHAPTER 22 Review Answers

Building Vocabulary For definitions, see: desertification, p. 501; staple, p. 507; millet, p. 507; sorghum, p. 507; dual economies, p. 509

Locating Key Places

A. Congo Basin
B. Niger River
C. Congo River
D. Tombouctou
E. Abidjan
F. El Djouf
G. Kinshasa
H. Lagos

Understanding the Main Ideas

1. the tropics; mostly warm throughout the year
2. semiarid; scattered trees, shrubs, and grasses; highly variable rainfall of only 4 to 8 inches (10 to 20 cm) each year; droughts; desertification
3. along trade routes
4. in the north
5. It has caused housing shortages and forced people to live in crowded shantytowns.

TECHNOLOGY

▶ Chapter 22 Test Generator (on the One-Stop Planner)
▶ Global Skill Builder CD–ROM
▶ HRW Go site

REINFORCEMENT, REVIEW, AND ASSESSMENT

▶ Chapter 22 Review, pp. 514–15
▶ Chapter 22 Tutorial for Students, Parents, Mentors, and Peers
▶ Chapter 22 Test (form A or B)
▶ Alternative Assessment Handbook

▶ Chapter 22 Test for English Language Learners and Special-Needs Students
▶ Unit 7 Test
▶ Unit 7 Test for English Language Learners and Special-Needs Students

Assess

Have students complete a Chapter 22 Test.

Reteach

Ask students to draw a large wall map of the region and to include the boundaries between countries. Have them label countries, rivers, basins, and other major physical features. Inside the outline of each country, have students list major events from its history, resources located there, and challenges that its people face. **ENGLISH LANGUAGE LEARNERS**

CHAPTER 22 Review Answers

Thinking Critically

1. Possible answers: Both rivers flow inland into large basins before turning and flowing out to the ocean. They may have used maps or aerial photographs. (NGS 1)

2. Possible answers: limits access to outside markets, resources harder to develop and get to markets, major cities along rivers or places with favorable climates or soils instead of along coasts; might be more politically isolated, have fewer outside allies (NGS 13)

3. left independent countries with very diverse populations and culture groups; brought European cultural features to the region, including languages and Christianity; accelerated move away from subsistence agriculture and local trading to commercial economies; growth of cities; left political problems, such as ethnic rivalries within countries (NGS 10)

Using the Geographer's Tools

1. Some countries in the region are named after these early empires although the territories of the countries and empires do not necessarily overlap.

2. Possible answers: requires expensive machinery; final product produced in locations that are closer to major markets

3. Côte d'Ivoire, Ghana, and Nigeria are the leading cacao producers. They lie along the Atlantic coast of West Africa.

CHAPTER 22 Review

Building Vocabulary

On a separate sheet of paper, explain the following terms by using them correctly in sentences.

desertification   sorghum
staple   dual economies
millet

Locating Key Places

On a separate sheet of paper, match the letters on the map with their correct labels.

El Djouf   Tombouctou
Niger River   Lagos
Congo Basin   Kinshasa
Congo River   Abidjan

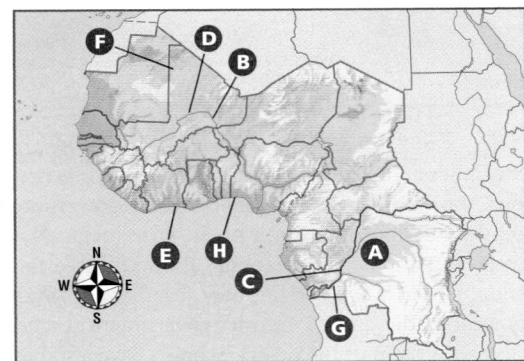

Understanding the Main Ideas

Section 1

1. *Places and Regions* In which latitudes are the countries of West and Central Africa located? How does this affect the region's climates?

2. *Physical Systems* What is the environment of the Sahel like?

Section 2

3. *The Uses of Geography* How did Islam spread into West and Central Africa?

4. *Places and Regions* In which part of West and Central Africa is Arabic spoken?

Section 3

5. *Places and Regions* How has rapid population growth affected cities in West and Central Africa?

Thinking Critically

1. Supporting a Point of View What evidence do you think geographers may have used to support their theory that the Congo and Niger Rivers once flowed into large inland lakes? What tools might geographers use to study this theory?

2. Making Generalizations How might being landlocked affect economic development, settlement, and population distribution in some countries in the region? How might it influence political conditions in a country?

3. Analyzing Information How did the colonial history of West and Central Africa affect cultural, economic, and political characteristics of the countries there?

Using the Geographer's Tools

1. Analyzing Maps Look at the map of Empires of Africa in Section 2. How are the names of these empires related to countries in the region today?

2. Interpreting Diagrams Analyze the cacao diagram in Section 3. Why do you think the manufacture of chocolate from cacao takes place in countries outside this region?

3. Preparing Tables Use the pie graph of world cacao production in Section 3 to create a table that ranks the leading producers of cacao in Africa from highest to lowest. In which part of Africa are these countries found?

Writing about Geography

Imagine that you are a researcher studying how farmers in the Sahel might adapt best to the difficult climate conditions. Compare techniques that farmers use in other semiarid regions of the world. Which might be useful in the Sahel? Why? Write a short report for a research journal explaining your suggestions. When you are finished with your report, proofread it to make sure you have used standard grammar, spelling, sentence structure, and punctuation.

SKILL BUILDING

Geography for Life

Analyzing Data

Places and Regions Use the unit Comparing Standard of Living and Fast Facts table to study the life expectancies, literacy rates, and per capita GDPs of countries in West and Central Africa. What patterns do you see? Is there a wide range of figures from country to country? What do you think are reasons for this?

Portfolio Activity

1. Have students plan a festival to highlight the musical styles of West and Central Africa, including *juju*, *makossa*, and *soukous*. Ask them to conduct research on the region's musicians and instruments. Students may want to construct their own instruments. Examples might include shakers, the *balafon*—a type of xylophone—and drums such as the *djembé*, the *doundoun*, the *kpanlogo*, or the *tama*. Place photographs or recordings of these instruments in portfolios.
2. Have students examine pictures of various types of clothing worn in West and Central Africa. Have them create an illustrated brochure describing the clothing, its origins, and the people who might wear it.

Food Festival

Plantains, which are a type of banana, are popular in the tropical regions of West Africa. This recipe for hot plantain chips is from Ghana. Slice 4 peeled firm plantains into ½-inch slices and sprinkle with lemon juice. Stir so that all surfaces are covered with lemon juice. In a separate bowl, combine 4 tsp. ground ginger and 4 tsp. cayenne pepper. Heat ¼ inch of vegetable oil in a heavy skillet until hot. Roll plantain pieces in the spice mixture to coat. Fry in skillet until crisp and golden. Remove with a slotted spoon, drain on paper towels, and sprinkle with salt. Allow to cool slightly but serve hot. The ginger and cayenne make these very spicy.

Building Social Studies Skills

Interpreting Maps

Use the transportation map below to answer the questions that follow.

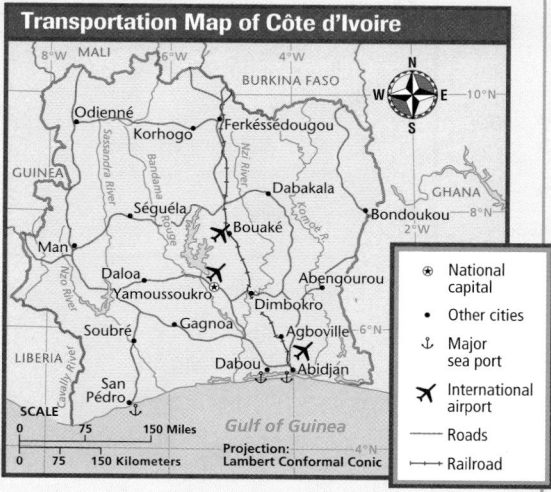

Transportation Map of Côte d'Ivoire

SCALE
0   75   150 Miles
0   75   150 Kilometers
Projection: Lambert Conformal Conic

Legend:
- ⊛ National capital
- • Other cities
- ⚓ Major sea port
- ✈ International airport
- — Roads
- ⊢⊢ Railroad

1. According to the map, what transportation facilities does the country's capital have?
   a. airport, railroad, roads
   b. seaport, railroad, roads
   c. airport and seaport only
   d. road and airport only

2. Which city do you think may be the country's busiest transportation center? Why?

Building Vocabulary

To build your vocabulary skills, answer the following questions.

3. *Millet* is
   a. a type of hammer used to build homes in the region.
   b. an army unit stationed only in capitals of the region.
   c. a drought-resistant grain grown in the Sahel.
   d. a state in Nigeria.

4. Some people in the region grow only a few *staple* crops, such as yams or corn. In which sentence does *staple* have the same meaning as it does in the sentence above?
   a. The staple of this cotton is not fine enough.
   b. Milk is a staple in the production of chocolate.
   c. Please staple together the reports on crops grown in Central Africa.
   d. The starch from cassava, a staple in West Africa, is used to make bread.

Alternative Assessment

PORTFOLIO ACTIVITY

Learning about Your Local Geography

Individual Project: Research
Research imported food crops in your area. First, look at what major food crops are produced in your area. Then list common foods that must be imported from other places. A trip to a local grocery store may help you figure out what these imports are. Often, the produce sections in grocery stores give the origin of imported foods. Once you have your list of imported foods, research where these items are produced. Do they come from other regions, states, or countries? Are some foods imported from great distances? Why do you think these foods are imported and not grown locally?

🖉 internet connect

Internet Activity: go.hrw.com
KEYWORD: SW3 GT22

Access the Internet through the HRW Go site to research the impact of political and ethnic conflict, economic development, and human migration on the Congo Basin. Then write a report in which you describe how the above factors have affected the people as well as the ecosystem in the region. Use standard grammar, spelling, sentence structure, and punctuation.

Writing

Students should discuss farming techniques in the Sahel and in other semiarid regions of the world. Suggestions for improvement should be reasonable and applicable to the Sahel. Use Rubric 42, Writing to Inform, to evaluate student work.

𝒢eography for Life

Student conclusions should be consistent with the information in the unit tables. Use Rubric 12, Drawing Conclusions, to evaluate student work.

Social Studies Skills

1. d
2. Possible answer: Abidjan, because it is a major seaport, lies along roads and a railroad, and has an international airport
3. c
4. d

PORTFOLIO ACTIVITY

Students' findings will vary according to the foods available in your region. Use Rubric 30, Research, to evaluate student work.

CHAPTER RESOURCE MANAGER

Objectives	Pacing Guide	Reproducible Resources
SECTION 1 **Natural Environments** (pp. 517–20) • Identify the landforms, rivers, and lakes of East Africa and the physical processes that shaped them. • Examine the distribution of climates, biomes, and natural resources in East Africa.	**Regular** 1 day **Block Scheduing** .5 day *Block Scheduling Handbook, Chapter 23*	**RS** Guided Reading Strategy 23.1 **RS** Graphic Organizer Activity 23 **PS** Readings in World Geography, History, and Culture 59
SECTION 2 **History and Culture** (pp. 521–24) • Identify some important developments in East Africa's early history. • Explain how European exploration and colonization affected the region. • Describe the peoples and cultures of East Africa.	**Regular** 1 day **Block Scheduling** .5 day *Block Scheduling Handbook, Chapter 23*	**RS** Guided Reading Strategy 23.2 **PS** Readings in World Geography, History, and Culture 60 **E** Cultures of the World Activity: Region 6 **E** Creative Strategies for Teaching World Geography, Lesson 17 **SM** Map Activity 23: Kenya
SECTION 3 **The Region Today** (pp. 525–28) • Explain the roles that agriculture, industry, trade, and tourism play in the economies of East Africa. • Describe the region's cities. • Identify some issues and challenges East Africans face.	**Regular** 1 day **Block Scheduling** .5 day *Block Scheduling Handbook, Chapter 23*	**RS** Guided Reading Strategy 23.3 **PS** Readings in World Geography, History, and Culture 58 **SM** Critical Thinking Activity 23: AIDS **SM** Geography for Life Activity 23: Struggles over African National Parks

Chapter Resource Key

PS Primary Sources **A** Assessment CD-ROM

RS Reading Support **REV** Review Video

IC Interdisciplinary Connections **ELL** Reinforcement and English Language Learners Internet

E Enrichment  Transparencies Holt Presentation Maker Using Microsoft® PowerPoint®

SM Skills Mastery

 One-Stop Planner CD–ROM

See the *One-Stop Planner* for a complete list of additional resources for students and teachers.

One-Stop Planner CD–ROM

It's easy to plan lessons, select resources, and print out materials for your students when you use the **One-Stop Planner CD–ROM with Test Generator**.

internet connect

HRW ONLINE RESOURCES

GO TO: go.hrw.com
Then type in a keyword.

TEACHER HOME PAGE
> **KEYWORD: SW3 Teacher**

CHAPTER INTERNET ACTIVITIES
> **KEYWORD: SW3 GT23**
> Choose a topic on East Africa to:
> • research United Nations efforts to reduce ethnic conflict in the region.
> • learn about the region's many ethnic groups.
> • create a brochure on the climate and wildlife of Kilimanjaro.

CHAPTER ENRICHMENT LINKS
> **KEYWORD: SW3 CH23**

CHAPTER MAPS
> **KEYWORD: SW3 MAPS23**

ONLINE ASSESSMENT
> **Homework Practice**
> **KEYWORD: SW3 HP23**
> **Standardized Test Prep**
> **KEYWORD: SW3 STP23**
> **Rubrics**
> **KEYWORD: SS Rubrics**

COUNTRY INFORMATION
> **KEYWORD: SW3 Almanac**

CONTENT UPDATES
> **KEYWORD: SS Content Updates**

HOLT PRESENTATION MAKER
> **KEYWORD: SW3 PPT23**

ONLINE READING SUPPORT
> **KEYWORD: SS Strategies**

CURRENT EVENTS
> **KEYWORD: S3 Current Events**

Technology Resources

- One-Stop Planner CD–ROM, Lesson 23.1
- Geography and Cultures Visual Resources 43–47
- **CNN.** Presents Geography: Yesterday and Today, Segment 25: The Nairobi National Park
- Homework Practice Online
- HRW Go site

- One-Stop Planner CD–ROM, Lesson 23.2
- **CNN.** Presents World Cultures: Yesterday and Today, Segment 32: A Kenyan Reggae Festival
- Homework Practice Online
- HRW Go site

- One-Stop Planner CD–ROM, Lesson 23.3
- Homework Practice Online
- HRW Go site
- *ARGWorld* CD–ROM

Reinforcement, Review, and Assessment

- **ELL** Main Idea Activity 23.1
- **ELL** English Audio Summary 23.1
- **ELL** Spanish Audio Summary 23.1
- **REV** Section 1 Review, p. 520
- **A** Daily Quiz 23.1

- **ELL** Main Idea Activity 23.2
- **ELL** English Audio Summary 23.2
- **ELL** Spanish Audio Summary 23.2
- **REV** Section 2 Review, p. 524
- **A** Daily Quiz 23.2

- **ELL** Main Idea Activity 23.3
- **ELL** English Audio Summary 23.3
- **ELL** Spanish Audio Summary 23.3
- **REV** Section 3 Review, p. 528
- **A** Daily Quiz 23.3

Meeting Individual Needs

Ability Levels

Level 1 Basic-level activities designed for all students encountering new material

Level 2 Intermediate-level activities designed for average students

Level 3 Challenging activities designed for honors and gifted-and-talented students

English Language Learners Activities that address the needs of students with Limited English Proficiency

Chapter Review and Assessment

- Chapter 23 Test Generator (on the One-Stop Planner)
- Global Skill Builder CD–ROM
- HRW Go site
- **REV** Chapter 23 Review, pp. 530–31
- **REV** Chapter 23 Tutorial for Students, Parents, Mentors, and Peers
- **A** Chapter 23 Test (form A or B)
- **A** Alternative Assessment Handbook
- **A** Chapter 23 Test for English Language Learners and Special-Needs Students

Launch into Learning

Have students imagine that they are going on safari in Africa. Ask them to imagine what kinds of animals they might see. *(Possible answers: elephants, lions, zebras, wildebeests, hippopotamuses, rhinoceroses)* What kinds of scenery might they see? *(Possible answers: grasslands, mountains, unusual trees, dry areas, villages with mud-walled houses and grass roofs)* Point out that this is the view many people have of Africa, but it applies most particularly to East Africa. Tell students that they will learn more about East Africa in this chapter.

Using the Physical-Political Map

Have students examine the map on the opposite page to find the Great Rift Valley. Ask students how they think the valley may have been formed *(tectonic activity along fault lines)*. What clues suggest this origin? *(Possible answer: long narrow shape of the valley, location parallel to the coast)* Point out the highland areas along the Great Rift Valley, both in central Ethiopia and farther south near Kilimanjaro. Ask students what effect these mountains may have on the region's climate. *(Possible answers: cooler temperatures, increased rainfall in east)*

Why We Should Know More

Point out to students these reasons why they should know more about East Africa:

► Many scientists believe the human species originated in East Africa. Some of the most important fossil remains of humans come from the Olduvai Gorge.

► Events in the region have influenced the foreign policy of the United States. Our country has sent both armed forces and economic aid to the region.

► The region contains wildlife not found elsewhere. Preserving the wildlife and other features of the region's natural environment is important to many people around the world.

Masai people, Kenya

CHAPTER 23 East Africa

East Africa includes Africa's largest country, Sudan, and small ones—Burundi, Rwanda, and Uganda. Kenya and Tanzania lie on the Indian Ocean. Ethiopia, Eritrea, Somalia, and Djibouti occupy the Horn of Africa.

Pyramids at Meroë, Sudan

Indemin adderu! (Good morning.) My name is Tsiyon. I live in Addis Ababa, the capital of Ethiopia, but far from the city center. I live with my parents and my older brothers. My parents sleep in the main house, while my brothers and I sleep in a smaller house near the kitchen, which is in a separate building. My father works for the game department doing research on the wild animals of our country.

In the mornings I ride to school with my oldest brother Undemageni, who works in a garage. I attend Freyhewat Number Two Junior Secondary School, where I study science, English, math, sports, and Amharic, Ethiopia's official language. Someday I want to be a doctor.

At noon, I eat a lunch of *injera* (a spongy sourdough pancake) and stew. Most days it is a meat stew, but on Wednesdays and Fridays, when Christians here fast, it is only vegetables. When I get home, after I help my mother sweep the house, I do my homework. On Saturdays, I go to church with my mother.

My favorite holidays are Christmas and New Year's. On our New Year's Day we have new clothes and a big feast. Because here in Ethiopia we use an old calendar system, our December 25 is around your January 3, and our new year begins on your September 11.

 **LET'S GET STARTED**

Copy the following instructions onto the chalkboard: *Use the chapter map to locate the source of the Nile. Where does this great river begin?* Tell students that the ancient Egyptians had probably traced the Blue Nile to its source at Lake Tana but no one found the White Nile's source until the 1870s. Ask students to suggest reasons it took so long. *(Possible answer: Mountains and deserts might have made exploration difficult.)* Tell students that in Section 1 they will learn more about the landforms and water features of East Africa.

Building Vocabulary

Write **tsetse fly** on the chalkboard. Help students pronounce the word *tsetse* (SET-see). Call on a volunteer to locate the term in the glossary or text and read the definition aloud. Tell students that the fly's name is derived from a Tswana word, *tsètsè*, meaning "fly." Afrikaans speakers adopted the name into their language, and from there it was introduced into English.

Section 1

Natural Environments

READ TO DISCOVER

1. What landforms, rivers, and lakes are found in East Africa, and what physical processes have shaped the land?

2. Why does East Africa have a variety of climates and biomes, and on what natural resources does the region depend?

WHY IT MATTERS

Illegal hunting, called poaching, and habitat destruction threaten the wildlife of East Africa. Use **CNNfyi.com** or other **current events** sources to learn about efforts to protect these magnificent animals.

DEFINE

tsetse fly

LOCATE

Great Rift Valley

Lake Malawi

Lake Tanganyika

Lake Victoria

Serengeti Plain

Nile River

Blue Nile

White Nile

Sudd

Section 1 RESOURCES

REPRODUCIBLE

► Guided Reading Strategy 23.1
► Graphic Organizer Activity 23
► Readings in World Geography, History, and Culture 59

TECHNOLOGY

► One-Stop Planner CD–ROM, Lesson 23.1
► Geography and Cultures Visual Resources 43–47
► CNN Presents Geography: Yesterday and Today, Segment 25: The Nairobi National Park
► Homework Practice Online
► HRW Go site

REINFORCEMENT, REVIEW, AND ASSESSMENT

► Main Idea Activity 23.1
► English Audio Summary 23.1
► Spanish Audio Summary 23.1
► Section 1 Review, p. 520
► Daily Quiz 23.1

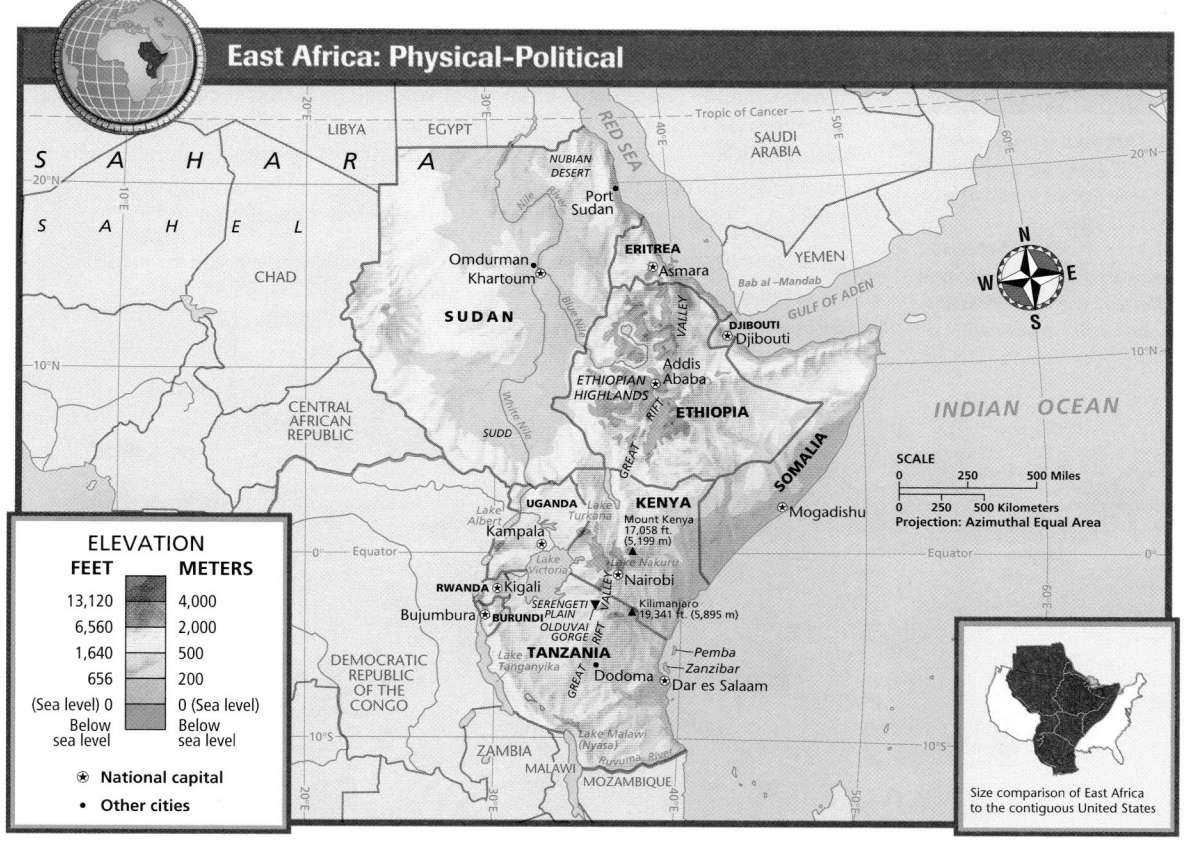

East Africa: Physical-Political

ELEVATION

FEET	METERS
13,120	4,000
6,560	2,000
1,640	500
656	200
(Sea level) 0	0 (Sea level)
Below sea level	Below sea level

⊛ National capital

• Other cities

Size comparison of East Africa to the contiguous United States

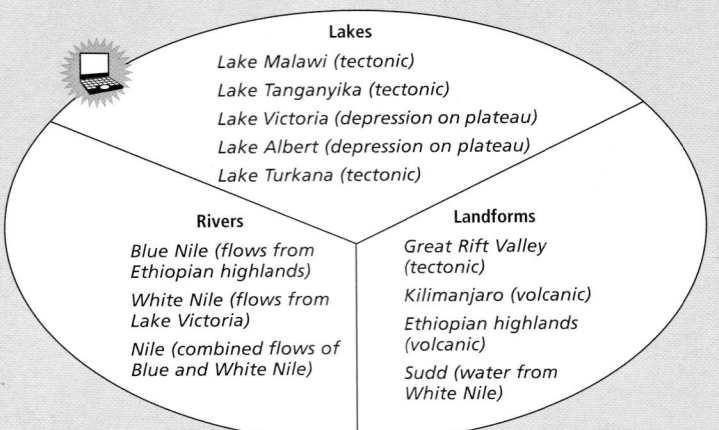

LEVELS 1 AND 2: Copy the graphic organizer at right onto the chalkboard, omitting the italicized answers. Call on students to identify the landforms, rivers, and lakes found in East Africa. As students name features, ask which physical processes created them. **ENGLISH LANGUAGE LEARNERS**

LEVEL 3: Organize students into groups. Have each group conduct research to create an elevation profile of the Eastern or Western Rift Valley. Ask students to identify on their profiles features that were created by tectonic activity. Display profiles in the classroom. **COOPERATIVE LEARNERS**

Lakes
Lake Malawi (tectonic)
Lake Tanganyika (tectonic)
Lake Victoria (depression on plateau)
Lake Albert (depression on plateau)
Lake Turkana (tectonic)

Rivers
Blue Nile (flows from Ethiopian highlands)
White Nile (flows from Lake Victoria)
Nile (combined flows of Blue and White Nile)

Landforms
Great Rift Valley (tectonic)
Kilimanjaro (volcanic)
Ethiopian highlands (volcanic)
Sudd (water from White Nile)

INTERPRETING THE VISUAL RECORD

Europeans first reached the Great Rift Valley in 1848. Until then, some Europeans did not believe that there could be snow so close to the equator, even on a mountain as tall as Kilimanjaro, which is pictured below. **How might this discovery have affected Europeans' ideas about Africa's landscapes in general?**

Landforms and Water

Tectonic processes have played an important role in shaping the physical landscape of East Africa. Forces beneath Earth's surface have lifted the region, cracking it apart. This resulted in the formation of two rift valleys. Mountains and plateaus lie along these rifts.

FOCUS ON GEOGRAPHY

The Rifts The Great Rift Valley is a series of geological faults. These faults run from the Jordan Valley in Southwest Asia all the way to Mozambique in southern Africa. In East Africa the rift system is referred to as the Western and Eastern Rift Valleys. The Western Rift Valley begins in the south near Lake Malawi. It continues northward through the valleys of Lake Tanganyika and three smaller lakes. The Western Rift then disappears in southern Sudan. Nearby rainy highlands feed this valley's great lakes. Most of the rift system's lakes are very deep. In fact, some of the lake floors are below sea level. In contrast, the continent's largest lake, Lake Victoria, is shallow. Lake Victoria fills a depression on the high plateau between the Western and Eastern Rifts.

The Eastern Rift Valley begins in Mozambique and continues northward across the Serengeti Plain. In Kenya it passes through Lake Turkana before crossing Ethiopia. At its northern end the Eastern Rift stretches to the floors of the Red Sea and the Gulf of Aden and into Southwest Asia.

Volcanoes have erupted within and near both rifts. The highlands of Ethiopia are therefore made of layers of volcanic rock. Together, the rifts and volcanoes of East Africa produce spectacular scenery. Kilimanjaro, near the Tanzania-Kenya border, is the most famous of the rift's volcanoes. Even though Kilimanjaro is near the equator, it is so high that snow always caps its twin peaks. At 19,341 feet (5,895 meters), it is the highest mountain in Africa.

✓ **READING CHECK:** **Physical Systems** What physical processes created the rift valleys of East Africa? Tectonic processes caused the land to lift and crack.

Teach Objective 2

ALL LEVELS: Provide each student with an outline map of East Africa. Have students identify the region's climate zones and biomes and shade them in different colors. Then have students label locations where the region's natural resources can be found. Remind students to create keys to accompany their maps. Call on volunteers to present their finished work to the class. Then lead a class discussion on factors that influence the locations of East Africa's climates and biomes. **ENGLISH LANGUAGE LEARNERS**

Using National Geography Standard 18:
The Uses of Geography: How to Apply Geography to Interpret the Present and Plan for the Future Tell students to imagine that they live in a village at the edge of Tanzania's Serengeti Plain. Some government officials want to include the village and its environs in a nature preserve and prohibit traditional hunting practices. Others want to build a cement factory there. Have students work to decide how the villagers can make a living while preserving their environment. Hold a mock town meeting to discuss possibilities and plans. Ask individual students to act as spokespersons for the various interests.

East Africa's major river, the Nile, flows northward through Sudan. Ancient Egypt depended on the Nile, but the Egyptians did not know where the river began. During the 1800s European explorers searched for the Nile's source. They found that the Nile's headwaters are in two different areas. The Blue Nile, which begins in the highlands of northern Ethiopia, provides most of the Nile's water. The waters that form the White Nile drain from Lake Victoria and through Lake Albert. Farther north, the White Nile almost ends in wetlands called the Sudd, in southern Sudan. The area's high temperatures cause about half of the White Nile's water to evaporate. The Blue Nile and the White Nile join in northern Sudan.

Climates, Biomes, and Natural Resources

Latitude and variations in elevation influence the climates of East Africa. Along the equator, distinct wet and dry seasons alternate. As a result, the vegetation on the high plains is a mixture of savannas and forests. Mountain slopes receive heavy rainfall, and the region's forests grow there. The highlands of Kenya and Uganda have a pleasant springlike climate year-round. British colonists found the climate comfortable and settled in these highlands.

Farther from the equator, to the north and south, seasonal droughts are more common. In areas where the dry season is long, trees are shorter and better adapted to water shortages. The mild moist Ethiopian plateau stands high above the deserts and semiarid areas of Sudan, Somalia, and northern Kenya. In these dry areas the vegetation is limited to thorny shrubs and tough grasses. Northern Sudan reaches into the Sahara. The Nile River forms oases through the desert's bare rocks and shifting sand.

Weather is often unpredictable in East Africa. Droughts have affected southern Sudan and the Horn of Africa several times in recent decades. As the grass dies, cattle also die, and people who depend on their livestock begin to starve. People and their animals then migrate to areas that still have some vegetation. This results in overgrazing and desertification. Too much rain also causes problems. After unusually heavy rains, locust populations may increase. Then these big grasshoppers swarm, devouring the plant life in their path. Again, people and animals go hungry. These natural events sometimes play a role in the region's social and political conflicts.

Animals You may have seen television programs about the wildlife of East Africa's Serengeti Plain. Oddly, the giraffes, lions, wildebeests, zebras, and other animals owe their survival partly to a pest—the **tsetse fly**. Tsetse flies carry a human disease called sleeping sickness. While many native animals are immune, the tsetse fly can spread a deadly disease to livestock. As a result, the area has few farmers or herders, leaving large wild animal populations undisturbed. Many areas with the best views, most wild animals, and fewest people have become national parks. By using modern pesticides, the savannas

Baobab (BOW-bab) trees are one of the few types of trees that grow on the African savanna. They can grow as large as 30 feet (9 m) in diameter and as high as 60 feet (18 m). The trunks store water for the tree during droughts. People sometimes hollow out the trunks to use as temporary shelters.

519

Close

Ask students to describe the relationships between landforms and climate types found in East Africa. Then call on volunteers to describe how tectonic forces act to create the region's landforms and bodies of water. Ask students to name other places in the world where tectonic forces have created landforms. (*Possible answers: Iceland, Italy, Mexico, places along the Pacific Rim, and other areas*)

Review and Assess

Have students complete the **Section Review**. Then have students complete **Daily Quiz 23.1**.

Reteach

Have students complete **Main Idea Activity for English Language Learners and Special-Needs Students 23.1**. Then ask students to write brief descriptions of the landforms, climates, and resources they might see on journeys into East Africa along particular lines of latitude. **ENGLISH LANGUAGE LEARNERS**

Extend

Have interested students conduct research on the animals of the Serengeti. Have students construct diagrams to illustrate the region's complex food chain. **BLOCK SCHEDULING**

Section 1 Review Answers

Define For definition, see: tsetse fly, p. 519

Working with Sketch Maps
Maps will vary, but listed places should be labeled in their approximate locations. The White Nile and the Blue Nile join to form the Nile.

Reading for the Main Idea
1. rift valleys, mountains, plateaus, volcanoes
2. latitude and elevation

Critical Thinking
3. Drought kills grass and causes livestock to starve, forcing humans and remaining livestock to migrate to new areas. The process leads to overgrazing and desertification. (NGS 7)
4. attraction for tourists (NGS 16)

Organizing What You Know
5. highland—forests; semiarid and arid—thorny shrubs, tough grasses; tropical wet and dry—savanna, forests

VISUAL RECORD ANSWER

Possible answers: benefits—improved economy, industrial development, foreign trade opportunities; problems—environmental damage from unregulated mining and production

520

Our Amazing Planet

In 1954 a swarm of locusts covering an area of about 77 square miles (200 sq km) invaded Kenya. Scientists estimated that there were 10 billion locusts in the swarm.

INTERPRETING THE VISUAL RECORD

A miner searches for tsavorite, a type of green garnet found, so far, only in Kenya and Tanzania. This rare gemstone was discovered in 1968. **How might the discovery of new resources create both benefits and problems for a country?**

might be made safe for grazing and farming. However, doing so may also create pressure to open the parks to people who want to graze their herds there.

Natural Resources In contrast to some other regions in Africa, East Africa in general is not rich in energy or mineral resources. Sudan began producing oil only recently. Tanzania produces gems, including sapphires and diamonds. Small gold deposits can be found along the rifts. The region's soils are also not very productive. Soils in the dry lands often have too much salt or lime to be fertile. In humid areas, the soil may be too sticky or hard to work easily. Fertile soil and a humid climate are found only in the highlands above the rifts. This rich soil helps explain why the small countries of Rwanda and Burundi can support dense populations. (See the unit population map.)

East Africa's scenery can be considered a valuable natural resource. Sparkling clean beaches stretch along the Indian Ocean. Resorts in the highlands offer both wildlife viewing and hiking through the forests. The savannas' great parks offer tourists remarkable encounters with wildlife. Expanding tourism is an economic goal of many of the region's countries.

✓ **READING CHECK:** *Places and Regions* How does the climate of Ethiopia's highlands compare to the climate of northern Sudan? Where is heavy rainfall most common? Ethiopia—cool, moist; northern Sudan—arid; mountain slopes

Section 1 Review

Homework Practice Online
Keyword: SW3 HP23

Define
tsetse fly

Working with Sketch Maps
On a map of East Africa that you draw or that your teacher provides, label the Great Rift Valley, Lake Malawi, Lake Tanganyika, Lake Victoria, Serengeti Plain, Nile River, Blue Nile, White Nile, and the Sudd. Which rivers join to form the Nile River?

Reading for the Main Idea
1. *Physical Systems* What landforms have tectonic forces created in East Africa?
2. *Physical Systems* What factors influence climates in the region?

Critical Thinking
3. *Analyzing Information* How does drought affect the region's environment?
4. *Analyzing Information* How is the region's scenery an important economic resource?

Organizing What You Know
5. Create a chart of the main climate regions of East Africa and the vegetation found in those regions. Use information from the unit climate map and Section 1 to complete your chart.

Climate regions	Vegetation

LET'S GET STARTED

Copy the following passage onto the chalkboard: "*Karibu! Hujambo?*" *is a common greeting in East Africa. What do you think it might mean?* Discuss responses. (*The greeting is in the Swahili language. It means Welcome! How are you?*) Tell students they will learn more about the history and culture of East Africa in Section 2.

Building Vocabulary

Write **ivory** and **sisal** on the chalkboard. Call on a volunteer to locate and read the terms' definitions from the text or the glossary. Point out that ivory can be obtained from animals besides the elephant. Both hippopotamus and walrus tusks are sources. Tell students that sisal is named for a town in Mexico's Yucatán from which the fiber is exported.

Section 2 History and Culture

READ TO DISCOVER

1. What were some important developments in East Africa's early history?
2. How did European exploration and colonization affect the region?
3. What are the peoples and cultures of East Africa like today?

WHY IT MATTERS

The native languages of East Africa changed through contact with Arab traders. Use **CNNfyi.com** or other **current events** sources to learn how trade is changing other languages.

IDENTIFY

Swahili

DEFINE

ivory
sisal

LOCATE

Olduvai Gorge
Nairobi
Kampala

Section 2 RESOURCES

REPRODUCIBLE
▶ Guided Reading Strategy 23.2
▶ Readings in World Geography, History, and Culture 60
▶ Cultures of the World Activity: Region 6
▶ Creative Strategies for Teaching World Geography, Lesson 17
▶ Map Activity 23: Kenya

TECHNOLOGY
▶ One-Stop Planner CD–ROM, Lesson 23.2
▶ CNN Presents World Cultures: Yesterday and Today, Segment 32: A Kenyan Reggae Festival
▶ Homework Practice Online
▶ HRW Go site

REINFORCEMENT, REVIEW, AND ASSESSMENT
▶ Main Idea Activity 23.2
▶ English Audio Summary 23.2
▶ Spanish Audio Summary 23.2
▶ Section 2 Review, p. 524
▶ Daily Quiz 23.2

Early History

The world's oldest archaeological artifacts have been found in East Africa. The evidence they provide tells us much about how the human race developed. Olduvai Gorge in Tanzania is one of the most important archaeological sites in the world. Some of the earliest human remains have been found there.

First Civilizations   Little evidence remains of East Africa's earliest cultures. Instead of keeping written records, most African peoples kept oral histories. These oral histories were stories of events and families that people memorized and passed from one generation to the next. Unfortunately, much of this information has been lost over time.

This beadwork cap was made by Masai artisans of East Africa.

Ethiopia's ancient history is unique in the region. Ethiopians had adopted Christianity several centuries before European missionaries arrived. This church at Lalībela dates from more than 800 years ago. It was carved from solid rock entirely below ground level.

521

ALL LEVELS: Copy the following graphic organizer onto the chalkboard, omitting the italicized answers. Call on students to provide phrases to identify important peoples in East Africa's early history. Then ask students to examine some of the factors that led these peoples to move into and settle the area. *(Possible answers: Nile River, access to both African interior and ports for trade)* Ask students which of the groups listed they think made the most lasting contributions to the cultures of East Africa. Call on volunteers to explain their answers. *(Possible answer: The Arabs made the most lasting contributions. Their influence is still seen in the region's architecture, language, and religion.)* ENGLISH LANGUAGE LEARNERS

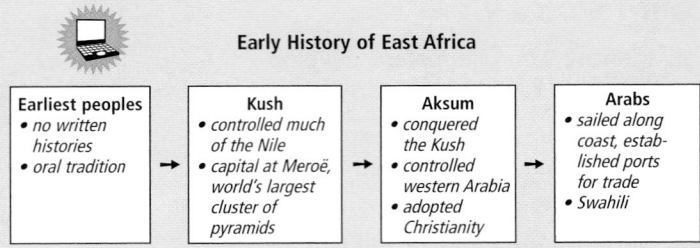

Early History of East Africa

Earliest peoples		Kush		Aksum		Arabs
• *no written histories* • *oral tradition*	→	• *controlled much of the Nile* • *capital at Meroë, world's largest cluster of pyramids*	→	• *conquered the Kush* • *controlled western Arabia* • *adopted Christianity*	→	• *sailed along coast, established ports for trade* • *Swahili*

Linking Past to Present

Zanzibar Like most of East Africa, Zanzibar Island has been home to a long series of culture groups. First settled by two peoples from the African mainland, the island attracted Persian traders by about A.D. 1100. The Portuguese took over in 1503 but were driven out some 200 years later by invading Arabs. In the 1820s British, German, Indian, and American traders established outposts on the island. Zanzibar became a British protectorate in 1890, but in 1964 it declared its independence. Shortly thereafter it united with Tanganyika to form the United Republic of Tanzania.

Zanzibar's architecture reflects this colorful past. Arab palaces and mosques stand near European cathedrals. British colonial mansions share streets with abandoned Persian bath houses.

You don't say! During World War I, several small naval battles between British and German ships occurred on Lake Tanganyika.

VISUAL RECORD ANSWER

availability of gold, ivory and slaves; access to India, Southwest Asia, East Asia, and Europe via Indian Ocean

VISUAL RECORD ANSWER

Christian priest conducting ceremony

INTERPRETING THE VISUAL RECORD

In the 1800s Arabs built the fort shown on the island of Lamu. They wanted to defend the island against other merchants who hoped to control the area's ports and trade. **Why do you think so many countries were interested in controlling trade through East Africa?**

INTERPRETING THE VISUAL RECORD

Ceremonies like this Christian wedding in Kenya often blend elements of traditional cultures and of European influences. **What evidence of European influence can you see in the photo?**

We do know about a few early civilizations in the region, however. For example, one of the wealthiest and most powerful kingdoms was Kush. Kush controlled much of the middle Nile River valley, in what is now Sudan, and briefly ruled Egypt. Like the Egyptians, the people of Kush built pyramids. The world's largest cluster of pyramids is at Meroë (MER-oh-wee), the capital of Kush. By A.D. 350, Kush had been conquered by nearby Aksum (AHK-soom). Aksum, in the highlands of what is now Ethiopia, was a city of traders and merchants. For a time Aksum controlled western Arabia. In the A.D. 300s Aksum's kings adopted Christianity.

Arab Connections Arab traders began sailing southward along Africa's Indian Ocean coast about 1,500 years ago. They established ports for trading the gold and **ivory** that Africans brought from the interior. Ivory is the cream-colored material that makes up the tusk of an elephant. It is used in making jewelry and handicrafts. At the ports these goods were traded for products from as far away as India. Slave trading also took place there. The **Swahili** (swah-HEE-lee) language developed during this period. Swahili's grammar comes from original languages of the African coast. Over time, many Arabic words were added to the language. Swahili is now a common language of the region. It is spoken as far west as the Congo.

✓ **READING CHECK:** *Human Systems* How did the spread of Arab influences affect language in East Africa? added many Arabic words to original African languages, influenced development of Swahili

European Influence

In the 1500s the Portuguese built the first European forts on the coast of East Africa. However, disease and rough terrain made travel inland difficult. As a result, for about 300 years European contact with East Africans took place mostly on the coast.

In the mid-1800s, European and American explorers, missionaries, and traders began to move inland. Sir Henry Morton Stanley, a journalist, explored the region during the 1870s. After traveling inland from the coast he followed the Western Rift Valley. In time, Stanley crossed the entire continent. Many other explorers fanned out across the region in search of precious minerals and ivory.

During the late 1800s European powers scrambled to claim territory in Africa. They drew colonial boundaries without giving thought to Africa's human or physical geography. Some colonial boundaries divided ethnic groups or grouped traditional enemies. Some even limited access to water.

The Europeans quickly established colonies, although Ethiopia remained largely free of colonial rule. Soon products from African mines and plantations supplied European economies. Cash crops included coffee, cotton, tea, and **sisal** (SY-suhl). Sisal is a strong durable plant fiber used to make rope and

Teach Objective 2

ALL LEVELS: Organize the class into groups. Have each group design a mural illustrating the influence of Europeans on East African culture. Student murals should also explain the motivations that brought European explorers and colonists to the region. Call on volunteers from each group to explain their murals to the class. **ENGLISH LANGUAGE LEARNERS, COOPERATIVE LEARNING**

HOMEWORK: Have students imagine that they are European colonists living on an East African plantation. Have them write journal entries about their experiences in the trade between Africa and Europe. These entries should also note the effects of the European presence on local cultures of the region.

Teach Objective 3

ALL LEVELS: Have each student imagine that he or she is an East African teenager writing to a pen pal in the United States. Have students write letters describing their adopted cultures as they exist today. Letters should discuss such cultural features as daily life, economic activities, food, language, or religion.

INTERPRETING THE VISUAL RECORD

Women celebrate after Eritrea's independence is declared in 1993. Eritrea's constitution praises women as influential and heroic in the country's struggle for freedom. **What might this statement imply about social and economic opportunities for women in Eritrea?**

twine. However, during the colonial era most East Africans still practiced traditional subsistence agriculture.

The Europeans built cities, hospitals, ports, roads, and schools in the areas where there were useful natural resources for export. European settlers began to farm the region's fertile highlands. Most of East Africa's modern capital cities sprang up during this time. For example, both Nairobi and Kampala served as early railway stations. The colonizers also provided a small number of Africans with a European education. Many of these Africans later led independence movements. Most countries of East Africa gained their independence during the 1950s and 1960s. In 1977 Djibouti became the last colony to win independence. Eritrea became a separate country when it broke away from Ethiopia in 1993.

READING CHECK: *Human Systems* What were some pull factors that drew Europeans to East Africa? the possibility of reaping profits from exploiting the region's resources

Culture

East Africa includes several hundred ethnic groups. Their cultures have given the world a rich heritage of architecture, art, folk tales, and music. The groups can be organized into three categories according to language. The Nilotic peoples live in the Nile River area on the plains of Sudan. Most Nilotic peoples are herders. Several Nilotic peoples migrated southward into the highlands a few centuries ago. Among these were the Masai and Tutsi. The second group is made up of Cushitic-speakers. Their lands run from the Red Sea coast through the Horn of Africa. This group includes the Amhara of the Ethiopian highlands and the Somali of the coast. The third group, Bantu-speakers, live farther south. They include the Kikuyu of Kenya and the Hutu of Rwanda. Bantus also moved into southern Africa.

People who follow Arab traditions live mostly along the Indian Ocean coast. Africans of South Asian descent also live in the region. During the colonial period, their ancestors came to work as merchants and craftspeople.

Our Amazing Planet

The Dinka of Sudan are among the world's tallest people. Manute Bol, a Dinka almost 7 feet, 7 inches tall, played in the National Basketball Association. Many Dinka women are also more than six feet tall.

Essential Element 2

▶ **Places and Regions** ◀

Djibouti Strategically located on the Bab al Mandab, the Republic of Djibouti was a French territory until 1977. The French were primarily interested in Djibouti as a trading center, a role it has maintained since independence. The capital city, also called Djibouti, is a free-trade port and is connected by rail to Addis Ababa, the capital of Ethiopia. An international airport provides links with the rest of the world. Djibouti's communications equipment is some of the best in Africa.

Sandwiched between two larger and stronger neigbors—Ethiopia and Somalia—Djibouti has observed a strict rule of neutrality in regional conflicts. In fact, it has maintained friendly relations with both Ethiopia and Somalia and hosted peace talks between leaders of the warring countries. Djibouti has also maintained a firm alliance with France.

DISCUSSION: Lead a class discussion about the factors that have shaped Djibouti's cultural, economic, and political character.

VISUAL RECORD ANSWER

more likely that social and economic opportunities will be available to them

523

Close

Have students name key events in the history of East Africa. Remind students that the region's history led to many of the aspects of its culture today. Ask students to identify some aspects of East African culture that reflect its history. *(Possible answers: architecture, language, religion)*

Reteach

Have students complete **Main Idea Activity for English Language Learners and Special-Needs Students 23.2**. Have students work in groups to outline the material in this section. Then have each student draw a picture to illustrate one point on the group's outline. ENGLISH LANGUAGE LEARNERS, COOPERATIVE LEARNING

Review and Assess

Have students complete the **Section Review**. Then have students complete **Daily Quiz 23.2**.

Extend

Have interested students conduct research about the history of Christianity in Ethiopia. Encourage students to focus their research on the rock churches of Lalībela, like the one pictured in this section. BLOCK SCHEDULING

Section 2 Review Answers

Identify For identification, see: Swahili, p. 522

Define For definitions, see: ivory, p. 522; sisal, p. 522

Working with Sketch Maps Maps will vary, but listed places should be labeled in their approximate locations. Evidence of some of the world's oldest humanlike remains has been found there.

Reading for the Main Idea
1. where there were useful resources for export; railway stations

2. Nilotic, Cushitic, Bantu; influenced development of Swahili

Critical Thinking
3. Europeans built cities, hospitals, ports, roads, and schools but also took over land and resources and drew boundaries without regard to human or physical geography. (NGS 12)

4. New foods might change or replace traditional foods or farming practices. Some new foods might provide less balanced and nutritious diets. (NGS 10)

Organizing What You Know
5. A.D. 350—Kush is conquered; 500—Arabs are trading along coast; 1500s—Portuguese build forts in East Africa; mid-1800s—Europeans move inland; late-1800s—Europeans claim colonies; 1950s and 1960s—most colonies gain independence; 1977—last colony, Djibouti, is granted independence; 1993—Eritrea gains independence.

524

Religions of East Africa

Country	Christian	Muslim	Traditional Beliefs	Other
Burundi	60%	1%	39%	
Djibouti		94%		6%
Eritrea	45%	45%		10%
Ethiopia	37%	43%	17%	3%
Kenya	66%	6%	26%	2%
Rwanda	45%		50%	5%
Somalia		100%		
Sudan	5%	70%	20%	5%
Tanzania*	26%	31%	42%	1%
Uganda	66%	16%	18%	

*except Zanzibar, which is 99% Muslim

Source: *The DK Geography of the World*

An Ethiopian woman makes injera. *Diners scoop up spicy vegetable or meat stews laid on top of the* injera *with pieces of the bread.*

Religion and family traditions are important aspects of daily life for East Africans. (See the chart.) Religions vary both within and among ethnic groups. However, most of the cultures honor ancestors. Many people believe the spirits of ancestors are strong forces in daily life.

Traditional religions are animist. Followers of animist religions believe the natural world contains spirits—in animals, mountains, trees, and waters. Many Africans also combine ancient forms of worship with later religions. Christianity came to Ethiopia more than 1,500 years ago. Missionaries during the 1800s and 1900s also spread Christianity. Arabs brought Islam to the region several centuries ago as well. New mosques throughout the region indicate Islam's continued growth. Rural communities are often dominated by one religion or the other. However, the ceremonies and holidays of many religions help make life in the cities exciting.

In the past, boiled sorghum was a main food in much of East Africa. Sorghum is a grain that can withstand drought. Sometimes the boiled grain was mixed with roasted beef or lamb. Other basic foods were sour milk and animal blood. Roots, berries, and game added to the diet. These were called bush foods because they were gathered in the wild. However, many new foods began appearing during the colonial era. Cornmeal, potatoes, rice, and wheat bread are now common. American fast food is widely available in the cities. Your favorite soft drink might be for sale in even the smallest village. Ethiopia and Eritrea have a unique food tradition based on a grain called teff. A large flat rubbery bread called *injera* is made from teff flour. *Injera* is served as large platter-shaped loaves.

✓ **READING CHECK:** *Human Systems* What are the main religions of East Africa? animism, Christianity, Islam

Section 2 Review

Identify
Swahili

Define
ivory, sisal

Working with Sketch Maps
On the map you created in Section 1, label the countries of East Africa, the Olduvai Gorge, Nairobi, and Kampala. What is the significance of the Olduvai Gorge?

Reading for the Main Idea
1. *Places and Regions* Where did Europeans build cities? What factor caused Nairobi and Kampala to grow?

2. *Human Systems* What are the three main language groups of East Africa? What influence did Arabic-speakers have on language in the region?

Critical Thinking
3. *Analyzing Information* Why might it be said that the colonial period created both benefits and problems for East Africa?

4. *Drawing Inferences and Conclusions* How might the availability of new foods influence health and culture in the region?

Organizing What You Know
5. Create a time line like the one shown below. Use it to identify and describe important events and periods in East Africa's history.

go.hrw.com **Homework Practice Online** Keyword: SW3 HP23

 **LET'S GET STARTED**

Copy the following instructions onto the chalkboard: *Look at the photographs in Section 3 and read the captions. Write down one observation about life in East Africa that you can make based on what you see.* Discuss observations. *(Possible answers: Agriculture in the region does not use much high technology. Some countries have harnessed their rivers to produce electricity. Poverty and disease are common problems in East Africa.)* Tell students they will learn more about the cities and economies of East Africa in Section 3.

Building Vocabulary

Write **genocide** on the chalkboard. Point out that it is derived from the Greek word *genos,* meaning "race" or "family." The suffix *-cide* stems from the Latin word *cīdium,* which means "a killing." *Genocide* then refers to the intentional killing of an entire people. Point out that victims of genocide are usually members of a particular ethnic, racial, or political group. Call on volunteers to locate and read this section's key terms in the glossary or text.

The Region Today

READ TO DISCOVER

1. What roles do agriculture, industry, trade, and tourism play in the economies of East Africa?
2. What are the region's cities like?
3. What issues and challenges do East Africans face?

WHY IT MATTERS

East Africa remains a politically unstable region. Use **CNNfyi.com** or other **current events** sources to learn about the most recent efforts to bring lasting peace and prosperity to East Africa.

DEFINE

gum arabic
genocide

LOCATE

Zanzibar   Addis Ababa
Dar es Salaam   Khartoum
Mombasa   Omdurman
Djibouti

Economic Development

Farming, trade, and industry all contribute to East Africa's economies. Traditional economies based on small-scale subsistence agriculture are common in rural areas. Manufacturing and global trade play minor roles.

Some of the region's poorest people earn a small income by gathering wild plant products. In Ethiopia this includes picking coffee beans from wild coffee trees. In Sudan people gather **gum arabic**, the sap of acacia trees. Gum arabic is a sticky substance that binds the ingredients of many candies and medicines.

Agriculture   Farming and herding form the basis of East Africa's economies. Humid highlands, such as those in Ethiopia, have many small subsistence farms. Cattle, goats, and sheep graze the dry lowland plains and plateaus. Farming there is possible only on irrigated land near oases.

In the region's cultures, women are often the primary farmers. Men take care of the livestock. Depending on the climate, the important food crops are beans, corn, rice, sorghum, and wheat. Farmers grow an increasing number of cash crops. These include coffee, cotton, sugarcane, and tea. Areas along the Indian Ocean coast produce their own distinctive crops, such as cloves and coconuts. Zanzibar, an island off Tanzania's coast, is a major producer of cloves.

A few large commercial farms and plantations can be found in East Africa. These farms have modern technology like tractors and trucks as well as modern seeds and fertilizers. Although few in number, these commercial farms produce crops for export and food for the region's cities.

✓ **READING CHECK:**   *Environment and Society*   What are some of the region's agricultural products? gum arabic, beans, corn, rice, sorghum, wheat, coffee, cotton, sugarcane, tea, cloves, coconut

Kenyan coffee growers prepare beans for weighing. A majority of the Kenyan farmers who grow coffee work small plots of land. Many of these growers work together in cooperative societies to process and sell their crops.

Teach Objective 1

🌍 **ALL LEVELS:** Organize the class into four groups and assign each group one of the following aspects of East Africa's economies: agriculture, industry, trade, or tourism. Have each group choose images for a slide show about the assigned topic. Images selected should describe the economic activity, explain its significance to the region, and address both large-scale commercial operations and small-scale subsistence activities. Have each group share its ideas with the class. Then lead a discussion about factors that influence the location and development of economic activities in East Africa. Ask how improvements in technology or transportation could affect economic development there. **ENGLISH LANGUAGE LEARNERS, COOPERATIVE LEARNING**

Teach Objective 2

🌍 **ALL LEVELS:** Copy the following graphic organizer onto the chalkboard, omitting the italicized answers. Call on students to describe the region's largest cities. Ask students how these cities might differ from your community. **ENGLISH LANGUAGE LEARNERS**

City	Description
Addis Ababa, Ethiopia	*largest city and capital of Ethiopia, headquarters of regional organizations*
Nairobi, Kenya	*region's most important commercial center*
Dar es Salaam, Tanzania	*vital seaport, transportation hub*
Khartoum and Omdurman, Sudan	*largest cities in Sudan, face each other across Nile*

Daily Life

The Safari Many tourists to East Africa refer to their travels as a "safari." Historically, however, the term refers to large parties of hunters who went to the region in search of big game. They were accompanied by dozens of gun bearers and trackers and were led by professional hunters. Many species were driven nearly to extinction by these excursions. National parks and preserves have since been established to protect East Africa's wildlife. Among the most famous is Serengeti National Park in Tanzania, home to millions of wildebeest and gazelles, as well as lions, elephants, rhinoceroses, and huge herds of zebras. Most safaris today consist of tourists armed only with cameras.

ACTIVITY: Have students read accounts of safaris by famous hunters like Isak Dinesen and Ernest Hemingway. Compare the perceptions of East Africa presented in these writings with attitudes toward the area today.

internet connect

GO TO: go.hrw.com
KEYWORD: SW3 CH23
FOR: Web sites about tourism in East Africa

VISUAL RECORD ANSWER

White Nile

INTERPRETING THE VISUAL RECORD

Hydroelectric power provided by Owen Falls Dam has enabled Jinja to become Uganda's main industrial center. More than 99 percent of Uganda's electricity comes from hydropower. Several other East African countries get more than 90 percent of their electricity from hydropower. **What river does the dam control? Use the small inset map and the physical-political map to find out.**

Industry and Economic Change All of the countries in the region have developing economies. Raw materials make up most of the exports. In contrast, East Africa imports many manufactured goods. The region's major ports include Dar es Salaam (dahr-es-sah-LAHM), Tanzania; Mombasa, Kenya; and the city of Djibouti (ji-BOO-tee). Manufacturing in the region centers around basic consumer goods, processed food, and building materials.

Over time the region's countries have had a mix of command and market economies. Beginning in the 1970s, for example, Ethiopia's leaders worked to create a command economy. The central government took over all of the country's agricultural land and urban rental property. It also took over banks, insurance companies, and many other businesses. However, the economy suffered throughout this period. To make matters worse, in 1984 a terrible drought struck. Famine and starvation became widespread. In the 1990s a new government began economic reforms. Ethiopia is slowly developing a market-oriented economy. Still, it remains one of the world's poorest countries.

Kenya has the highest per capita GDP in the region. However, progress has been slow, partly because of the government's mismanagement of the economy. Rapid population growth has also put pressure on the economy. (See Geography for Life: Population Growth in Kenya.)

Tourism has great potential for economic growth in the region. Fascinating animals, cool highlands, snowcapped mountains, clean beaches, and cultural events all draw tourists from developed countries. Resorts, restaurants, safari lodges, and taxi companies provide many jobs. Tourists also buy traditional arts and crafts as souvenirs. Political violence has hurt the tourist industry in some countries, however. Therefore, maintaining stability in the region is important to the industry. In addition, many people argue that only by preserving the environment will the countries of East Africa continue to draw a steady stream of visitors.

✓ **READING CHECK:** *Human Systems* How did Ethiopia's government try to create a command economy beginning in the 1970s? took over agricultural land, urban rental property, banks and insurance companies, and many other businesses

ALL LEVELS: Organize the class into several groups and assign each group a country or area of East Africa. Have students use reference books, current events sources, and the Internet to conduct research on challenges or conflicts facing their assigned countries. Then have each group prepare a script for a short documentary film about issues facing the country's people. In their scripts, students should address the roots of the issue, its effects on people's daily lives, and solutions that have been proposed to resolve the issue. Have groups share their findings with the class. Then lead a class discussion about the physical and human factors that have shaped East Africa's character. **COOPERATIVE LEARNING**

Teacher to Teacher

Nancy Lehmann-Carssow of Austin, Texas, suggests the following activity to show how population pressure and political unrest affect the natural environment. The plight of endangered mountain gorillas in Rwanda was first voiced by conservationist Dian Fossey. Since her death in 1985, movements to protect the gorillas have encountered major setbacks. Have students map the gorillas' habitats and note events in recent Rwandan history that have affected these locations. Finally, have students compose graphs depicting both Rwanda's human population and its gorilla population over several years. Ask students what conclusions they can draw from their graphs. **BLOCK SCHEDULING**

Urban Environments

Addis Ababa (AHD-dis AH-bah-bah), Ethiopia's capital, is the region's largest city. Regional organizations, such as the Organization of African Unity, have their headquarters there. A railroad through Djibouti connects Addis Ababa to the Indian Ocean coast. Nairobi, Kenya, is the most important commercial center in East Africa. The city's manufactured goods range from cement to soap. A national park—where gazelles, lions, zebras, and other animals live—lies within Nairobi's city limits. Dar es Salaam is a transportation hub. Tanzania's government is scheduled to move inland from that coastal city to Dodoma by 2005. Khartoum, which is Sudan's capital, and Omdurman (ahm-duhr-MAN) are the largest cities in Sudan. They lie across the Nile from each other.

East Africa's rapidly growing cities have superhighways and glittering skyscrapers. However, run-down buildings and slums often surround these symbols of economic progress. The number of people moving to the cities from the countryside is larger than the number of available jobs. Many of the newcomers have high hopes but possess only farming skills and little education. The unemployed and those with temporary jobs live in the slums that ring many cities. Providing better housing is a pressing issue for governments. However, providing good government under these conditions is difficult too. Political unrest sometimes is a problem. High crime rates contribute to that unrest.

✓ **READING CHECK:** *Human Systems* Why are the region's cities growing rapidly? people moving from their farms to the cities to find work

Issues and Challenges

Populations in East Africa have risen dramatically during the past 30 years. A rapidly growing population underlies many of the problems that face the region's governments and people. Overpopulation contributes to widespread poverty. In many places there is not enough food to go around. Trying to grow food where the land cannot support farming can hurt the environment. Health services and educational opportunities are also spread thin. (See Geography for Life: Population Growth in Kenya.)

Ethnic conflicts, between countries and between groups within countries, have also presented problems. Often central to these conflicts are struggles over land and fair distribution of government aid and jobs. For example, in Ethiopia the Amhara have long been the dominant ethnic group. However, the Oromo and the Tigre are now demanding what they believe is their share of power and influence. Similarly, the Baganda, whose traditional lands include areas along Lake Victoria, dominate Uganda. Complaints from the peoples of northern Uganda have led to repeated unrest. For decades, Sudan has been torn by war between the Islamic government and the animist and Christian peoples of the south. The government has tried to force Islam and Arabic ways of life on the people of southern Sudan, who want more autonomy.

Ethnic hatreds have even led to **genocide**. Genocide is the intentional destruction of a people. The worst case happened in Rwanda in 1994.

Connecting to SCIENCE

Endangered Cheetahs

Among the many animals that attract tourists to East Africa is the cheetah. However, loss of habitat and illegal hunting threaten these graceful cats. They are also threatened by their own genes. Because of their reduced numbers, Africa's cheetahs have been able to mate only with their close relatives. Today cheetahs are inbred. As a result, the cheetahs in some populations share nearly all the same genes. This lack of genetic diversity leaves cheetahs vulnerable to disease and early death. Breeding programs now aim to restore the health of the species by mixing genetic material of captive and wild cheetah populations.

Problem Solving How might the tourist industry help save the cheetah?

Global Perspectives

The Somaliland Republic
In the late 1960s conflict arose between clans in the new nation of Somalia. Fighting broke out in the late 1970s, and in 1991 the northwestern part of the country declared its independence from the rest of Somalia.

Calling itself the Somaliland Republic, the breakaway territory established its own government at Hargeisa. Though fighting between clans continued for a few years, order in the territory has been reestablished. The government of Somaliland has created an army, its own currency, a flag, and a national anthem. However, it has not been recognized as an independent state by other countries. The people of Somaliland have begun to develop their economy and have received financial assistance from the EU since the late 1990s.

CRITICAL THINKING: Why might countries be hesitant to recognize Somaliland's independence? *(Possible answer: might not want to alienate Somalia's government, might want to discourage fragmentation of existing countries)*

CONNECTING TO SCIENCE ANSWER

Possible answers: impose a special tax on tourists earmarked for cheetah programs, sponsor "adopt a cheetah" programs

Close

Have each student write three or four short statements that describe one of the countries of East Africa. On the chalkboard, write the names of East Africa's countries. Call on volunteers to write one of their statements on the board under the correct country heading. Continue calling on students until they have created a comprehensive profile of East Africa.

Review and Assess

Have students complete the **Section Review.** Then have students complete **Daily Quiz 23.3.**

Reteach

Have students complete **Main Idea Activity for English Language Learners and Special-Needs Students 23.3.** Then have each student create an idea web that shows the relationships among various aspects of East African society. ENGLISH LANGUAGE LEARNERS

Extend

Have interested students examine the roles that strategic waterways, such as the Nile or the Bab al Mandab, have had on the economic development of East Africa. Students might wish to compare these waterways to other strategic courses, like the Strait of Gibraltar or the Bosporus. BLOCK SCHEDULING

Section 3 Review Answers

Define For definitions, see: gum arabic, p. 525; genocide, p. 527

Working with Sketch Maps Maps will vary, but listed places should be labeled in their approximate locations. Addis Ababa has the largest population.

Reading for the Main Idea
1. farming and herding; by gathering products from wild plants

2. modern areas with skyscrapers and superhighways surrounded by slums; high unemployment, poor housing, political unrest, high crime

Critical Thinking
3. developing; widespread poverty, overpopulation, lack of health care and educational opportunities, high unemployment (NGS 9)

4. increases competition for limited resources (NGS 13)

Organizing What You Know
5. Maps will vary but should accurately identify areas of conflict.

Residents of a Nairobi slum carry firewood to their homes. East Africa is a poor region. Just under half of Kenya's population lives in poverty.

Diseases such as AIDS, cholera, and malaria kill thousands of East Africans each year. Healthcare workers like this Rwandan woman are working to help the sick, but with limited resources.

There the more numerous Hutu tried to wipe out the Tutsi. Armed bands killed hundreds of thousands of people.

Violence has also been a problem in Somalia. There, however, ethnic conflict has not been the issue. In Somalia most of the people are ethnic Somali and Muslim. The country has often had no central government of any kind. Instead, different clans have fought over grazing rights and the control of port cities such as Mogadishu. In addition, Somalia has tried to annex nearby parts of Ethiopia by military invasions. Relations with Ethiopia remain uneasy.

As you can see, the leaders of the East African countries face many challenges. These challenges include building stable countries in spite of ethnic fighting. Promoting economic progress, protecting the environment, and providing for the health and educational needs of growing populations are also important.

✓ **READING CHECK:** *Human Systems* How have political and economic tensions contributed to problems in the region? Tensions between ethnic groups, often over land and fair distribution of government aid, have sometimes led to fighting.

Homework Practice Online
Keyword: SW3 HP23

Section 3 Review

Define
gum arabic, genocide

Working with Sketch Maps
On the map of East Africa that you created in Section 2, label Zanzibar, Dar es Salaam, Mombasa, Djibouti, Addis Ababa, Khartoum, and Omdurman. Which of East Africa's cities has the most people?

Reading for the Main Idea
1. *Human Systems* What are the most common economic activities in East Africa? How do some of the poorest people survive?

2. *Places and Regions* What are some features that many of the region's cities have in common?

Critical Thinking
3. Categorizing Are the countries of East Africa developed or developing? On which factors do you base your answer?

4. Analyzing Information How do you think overpopulation contributes to political unrest?

Organizing What You Know
5. Create a sketch map on which you identify the places where some conflicts have occurred in recent decades in the region. Use one color or symbol for ethnic conflicts and another for conflicts not related to ethnic identity.

History: Malthusian Population Theory

The study of population growth and its effects on society is not a new discipline. In the late 1700s a British economist named Thomas Malthus argued that human population growth represents the greatest obstacle to the improvement of society. People, he wrote, always multiply faster than their food supplies increase. Malthus believed that disasters like disease, famine, and war would therefore strike overgrown human societies. Only strict limits on population growth, he wrote, could save humanity from misery and poverty.

Malthus proposed that human populations would double every 25 years unless environmental or social factors prevented it. Since he believed that this population growth remained constant, Malthus thought one could calculate the world's population at any given point in the future by using a simple mathematical calculation.

Have students examine the graph and text in this feature. Ask if they think Kenya's recent history supports Malthusian theory. *(Possible answer: It does support the theory, but Malthus's estimated growth rate was low. In 1950 Kenya's population was about 6 million. Fifty years later it had passed 30 million. This surpasses Malthus's predictions. Increased population growth led to widespread poverty and social problems. Growth also slowed after an official policy on population was implemented.)* Have interested students investigate how agricultural and medical advances have affected the Malthusian model.

The Uses of Geography

Geography for Life

Population Growth in Kenya

If the current growth rate continues, the world's population will grow by almost a billion people every 12 years! Countries where big families bring high status and ensure care for elders later in life have the highest growth rates. In addition, farming families often want many children so they can work in the fields. All these factors apply to Kenya. In 1988 Kenya had a population growth rate of 4.2 percent—the highest any country has ever recorded. Kenya's growth rate soared because more children were born, fewer children died, and modern medicine kept more people alive longer.

After Kenya won independence in 1963, the country's economy expanded quickly. However, by the 1980s the economy could no longer keep pace with the population. Too many resources went just to supply the needs of the huge population. Little was left over for development or investment.

In addition, Kenya's population explosion created conflict between people and their environment. Arable land, clean water, firewood—all are in short supply. About 75 percent of Kenyans live on 10 percent of the land. Some farmers have moved onto land not suitable for farming. Efforts to farm these lands have led to environmental damage and the destruction of wildlife habitat. Meanwhile, conservationists argue that protecting wildlife serves a long-term goal of supporting the tourist industry. Kenyans who live in urban areas have other troubles. In the cities they compete for a limited number of jobs.

Kenya's government has tried to address these problems. For example, Kenya was the first African country south of the Sahara to adopt an official policy on population. Government officials realized that only by controlling population growth would the country prosper.

Fortunately, in recent decades the average family size in Kenya has dropped, from 8 children to 5.3 children. Women's healthcare has also improved. Few women have access to family planning services, however. Women also have few economic or educational opportunities. In countries where women have access to education they soon get the chance to work outside the home. In those countries the birthrate drops, and standards of living rise. The long-term solution to Kenya's high birthrate may lie in increasing opportunities for Kenyan women.

Applying What You Know

1. **Summarizing** How does a high population growth rate limit Kenya's progress?

2. **Making Predictions** Develop and defend a hypothesis about how many people might live in Kenya 100 years from now. What factors should you take into account? What are some factors that might make the figure much higher or lower?

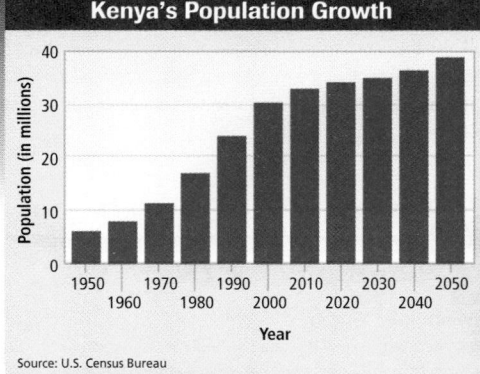

Kenya's Population Growth

Population (in millions) vs. Year (1950, 1960, 1970, 1980, 1990, 2000, 2010, 2020, 2030, 2040, 2050)

Source: U.S. Census Bureau

INTERPRETING THE GRAPH *Between 1970 and 1980 the number of Kenyan women who received an education, like those at the right, more than tripled.* **How does the graph reflect a connection between education and slowing population growth?**

TECHNOLOGY
▶ Chapter 23 Test Generator
(on the One-Stop Planner)
▶ Global Skill Builder CD–ROM
▶ HRW Go site

REINFORCEMENT, REVIEW, AND ASSESSMENT
▶ Chapter 23 Review, pp. 530–31
▶ Chapter 23 Tutorial for Students, Parents, Mentors, and Peers
▶ Chapter 23 Test (form A or B)
▶ Alternative Assessment Handbook

▶ Chapter 23 Test for English Language Learners and Special-Needs Students
▶ Unit 7 Test
▶ Unit 7 Test for English Language Learners and Special-Needs Students

Assess
Have students complete a Chapter 23 Test.

Reteach
Organize the class into three groups. Have each group discuss the material presented in one section of this chapter and prepare an oral presentation about it. Call on volunteers to deliver their material to the class.
ENGLISH LANGUAGE LEARNERS

CHAPTER 23 Review Answers

Understanding the Main Ideas

1. Mountains and highlands have lower temperatures than rift valleys.

2. development of Swahili, introduction of Islam

3. Ancient beliefs combined with Christianity; new foods are now part of the diet.

4. farming and herding

5. Animals and natural environments attract tourists. Resorts, restaurants, and other services provide jobs.

Thinking Critically

1. Mountains and rifts may have slowed or diverted east-west migration to the north or south. Mild highland climates and areas near lakes or rivers probably attracted more settlement than more arid regions. Tsetse flies and other insects kept some areas lightly settled. (NGS 15)

2. Answers will vary. (NGS 4)

3. Overpopulation contributes to urban growth, widespread poverty, food shortages, and environmental problems when farmers grow food on land that really cannot support farming. (NGS 9)

Using the Geographer's Tools

1. The mild climate of highland areas and access to rivers and lakes allow denser populations.

CHAPTER 23 Review

Building Vocabulary
On a separate sheet of paper, explain the following terms by using them correctly in sentences.

tsetse fly	sisal
ivory	gum arabic
Swahili	genocide

Locating Key Places
On a separate sheet of paper, match the letters on the map with their correct labels.

Great Rift Valley	Blue Nile	Dar es Salaam
Lake Victoria	White Nile	Addis Ababa
Serengeti Plain	Nairobi	

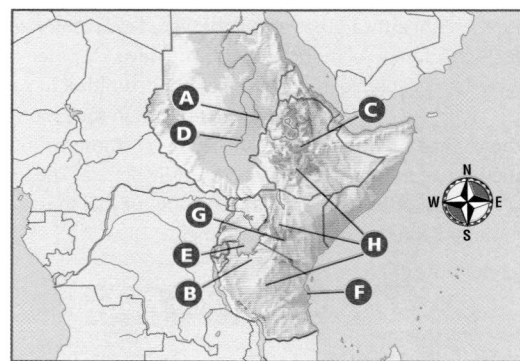

Understanding the Main Ideas
Section 1
1. *Physical Systems* How does elevation affect climate in East Africa?

Section 2
2. *Places and Regions* What are two ways that Arab traders influenced East African languages and religion?

3. *Human Systems* How have other cultures changed the traditional religions and diet of East Africa?

Section 3
4. *Human Systems* What activities form the basis of the region's economy?

5. *Places and Regions* Why does tourism hold great economic potential for the region?

Thinking Critically
1. **Drawing Inferences and Conclusions** How may East Africa's natural environments have affected the routes, flows, and destinations of African and European migration and settlement?

2. **Comparing** As you have read, humid highland areas in the region have many farms. Dry plains are used for grazing livestock. Where do you find farming and livestock grazing in the United States and other countries you have studied?

3. **Analyzing Information** How is rapid population growth shaping economic, urban, and environmental change in the region?

Using the Geographer's Tools
1. **Using Maps and Tables** Review the unit maps and the population density figures in the unit Fast Facts table. How does physical geography affect population density in East Africa?

2. **Creating Pie Charts** Convert the information in the Section 2 religions chart into pie charts. Note that you will need to make a design adjustment for Tanzania.

3. **Preparing Maps** Create an outline map of East Africa and surrounding regions. Then after reviewing information in Chapter 4, shade areas on the African and Arabian tectonic plates and note the direction of their movement. Finally, note landforms and other features created by tectonic activity.

Writing about Geography
Some farmers and herders want to clear the tsetse flies from the animal parks and use the land for agriculture. Others want to keep the parks as they are. Write a dialogue in which representatives of both groups express their points of view. Then proofread your dialogue to make sure you have used standard grammar, spelling, sentence structure, and punctuation.

SKILL BUILDING
Geography for Life
Drawing a Bar Graph
Human Systems Use the Internet or library resources to find the average number of people per doctor in each East African country and in the United States. Display your findings in a bar graph. Let the vertical axis represent the number of people. Place the countries along the horizontal axis at the bottom. Write a paragraph in which you draw conclusions about health care and the standard of living in East Africa.

1. Have students create dioramas of East African ecosystems, such as those found in the Sudd, the Serengeti, Lake Nakuru, or Kilimanjaro. Have students depict the physical features, vegetation, animal life, and other factors that make the places unique. Photograph dioramas for student portfolios.
2. Have students conduct research on the ancient kingdoms of Ethiopia and Sudan. Instruct them to create posters that depict what structures remain from those cultures and what they may have looked like when they were first built.

Food Festival

Teff, the grain from which *injera* is made, is not widely available. Make an adapted version by combining 1 c. buckwheat pancake mix, 1 c. biscuit mix, 1 egg, 1 tbs. oil, and from 1½ c. to 2 c. water. Transfer the batter to a pitcher. Heat a large skillet or griddle until medium hot. Brush ½ tsp. oil onto the pan. Pour the batter onto the pan in a thin stream, starting from the outside and moving in circles to the center. As soon as bubbles form all over the pancake, remove it from the heat and place in an oven at 325° for 1 minute. Make four more pancakes and overlap them on a large tray.

Building Social Studies Skills

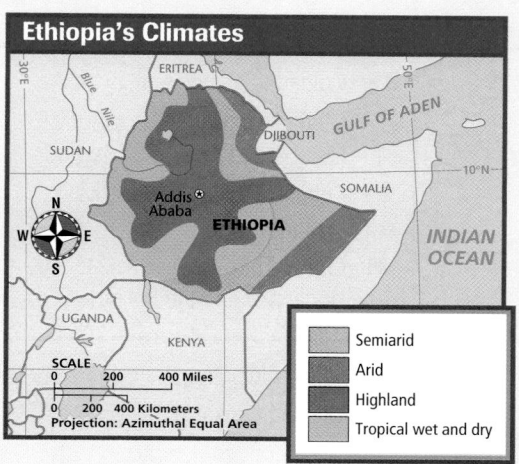

Ethiopia's Climates

Semiarid
Arid
Highland
Tropical wet and dry

SCALE
0   200   400 Miles
0   200   400 Kilometers
Projection: Azimuthal Equal Area

Interpreting Maps

Use the information from the map above to answer the questions that follow.

1. In Ethiopia, the Blue Nile passes through which climates?
   a. semiarid; highland
   b. arid; semiarid
   c. arid; tropical wet and dry
   d. tropical wet and dry; highland
2. Which area of Ethiopia would you expect to be the least densely populated? Why?

Analyzing Primary Sources

Read the following description of Ethiopia's Blue Nile, written by Virginia Morell during an expedition. Then answer the questions.

"With the river roaring through its canyon a good half mile below us, we trekked past clusters of round, thatch-roofed homes and fields of teff [a grain] edged with low stone walls and clumps of daisies. On both sides of the gorge the land rose in broad-shouldered, terraced mountains, each flat bit of land quilted with a patchwork of fields that shimmered green and gold in the sun. In many fields small groups of men, women, and children [were] weeding each row by hand."

3. What kinds of homes did the author see along the gorge?
   a. homes made from teff
   b. homes made from low stone walls
   c. thatched-roof homes
   d. homes with terraces
4. What features of the local physical geography does the passage describe?

Alternative Assessment

PORTFOLIO ACTIVITY

Learning about Your Local Geography

Group Project: Interviewing

As you have read, much of Africa's precolonial history was passed down orally. Plan, organize, and complete a group research project on oral history in your community. Interview longtime residents about how they recall a certain local weather event, such as a drought or tornado. Create a list of questions that you will ask each person. Include questions about how the person experienced the event and what may have caused it. Also ask how the event affected the interviewee's feelings for or opinions about his or her community. Videotape the interviews and show them to the class. As a group, discuss how the interviews are similar or different.

internet connect

Internet Activity: go.hrw.com
KEYWORD: SW3 GT23

Choose a topic on East Africa to:
- research United Nations efforts to reduce ethnic conflict in the region.
- learn about the region's many ethnic groups.
- create a brochure on the climate and wildlife of Kilimanjaro.

2. Pie charts should accurately reflect the information presented.
3. rift valleys, Red Sea volcanoes, created by tectonic activity

Writing

Dialogues should present both sides of the debate and demonstrate understanding of the subject. Use Rubric 41, Writing to Express, to evaluate student work.

Geography for Life

Graphs and paragraphs should accurately reflect the information gathered. Use Rubrics 7, Charts, and 12, Drawing Conclusions, to evaluate student work.

Social Studies Skills

1. d
2. arid regions along northeastern and southeastern borders; most inhospitable climate
3. c
4. river running through its canyon, gorge that contains the river bordered by broad-shouldered terraced mountains, and terraces quilted with a patchwork of fields

PORTFOLIO ACTIVITY

Student projects will vary. Questions should improve the interviewer's understanding of the subject chosen. Use Rubric 1, Acquiring Information, to evaluate student work.

531

CHAPTER RESOURCE MANAGER

Objectives	Pacing Guide	Reproducible Resources
SECTION 1 **Natural Environments** (pp. 533–35)	• Identify the major landforms and rivers of southern Africa. • Describe the climates, biomes, and natural resources of the region.	**Regular** .5 day **Block Scheduling** .5 day *Block Scheduling Handbook, Chapter 24*	**RS** Guided Reading Strategy 24.1 **PS** Readings in World Geography, History, and Culture 64 **E** Creative Strategies for Teaching World Geography, Lesson 16
SECTION 2 **History and Culture** (pp. 537–41)	• Identify some important events in the history of southern Africa. • Describe the region's cultures.	**Regular** 1 day **Block Scheduling** .5 day *Block Scheduling Handbook, Chapter 24*	**RS** Guided Reading Strategy 24.2 **PS** Readings in World Geography, History, and Culture 65 and 66 **E** Cultures of the World Activity: Region 6 **SM** Critical Thinking Activity 24: Understanding the Bushmen
SECTION 3 **The Region Today** (pp. 542–45)	• Identify the main economic activities of southern Africa. • Describe the region's cities. • Explain the challenges that face the people of southern Africa.	**Regular** .5 day **Block Scheduling** .5 day *Block Scheduling Handbook, Chapter 24*	**RS** Guided Reading Strategy 24.3 **RS** Graphic Organizer Activity 24 **SM** Geography for Life Activity 24: The Geography of Childhood Deaths in Zimbabwe **SM** Map Activity 24: White Rule in South Africa

Chapter Resource Key

PS Primary Sources

RS Reading Support

IC Interdisciplinary Connections

E Enrichment

SM Skills Mastery

A Assessment

REV Review

ELL Reinforcement and English Language Learners

Transparencies

CD-ROM

Video

Internet

Holt Presentation Maker Using Microsoft® PowerPoint®

 One-Stop Planner CD–ROM

See the *One-Stop Planner* for a complete list of additional resources for students and teachers.

  **One-Stop** Planner CD–ROM

It's easy to plan lessons, select resources, and print out materials for your students when you use the **One-Stop Planner CD–ROM with Test Generator**.

internet connect

HRW ONLINE RESOURCES

GO TO: go.hrw.com
Then type in a keyword.

TEACHER HOME PAGE
 KEYWORD: SW3 Teacher

CHAPTER INTERNET ACTIVITIES
 KEYWORD: SW3 GT24
 Choose a topic on southern Africa to:
 • learn about the Namib Desert.
 • investigate the history and legacy of apartheid.
 • create a brochure of the animal and plant life on a South African safari.

CHAPTER ENRICHMENT LINKS
 KEYWORD: SW3 CH24

CHAPTER MAPS
 KEYWORD: SW3 MAPS24

ONLINE ASSESSMENT
 Homework Practice
 KEYWORD: SW3 HP24
 Standardized Test Prep
 KEYWORD: SW3 STP24
 Rubrics
 KEYWORD: SS Rubrics

COUNTRY INFORMATION
 KEYWORD: SW3 Almanac

CONTENT UPDATES
 KEYWORD: SS Content Updates

HOLT PRESENTATION MAKER
 KEYWORD: SW3 PPT24

ONLINE READING SUPPORT
 KEYWORD: SS Strategies

CURRENT EVENTS
 KEYWORD: S3 Current Events

Technology Resources

- One-Stop Planner CD–ROM, Lesson 24.1
- **CNN** Presents Geography: Yesterday and Today, Segment 27: Mining in South Africa the Traditional Way
- Geography and Cultures Visual Resources 43–47
- Homework Practice Online
- HRW Go site

- One-Stop Planner CD–ROM, Lesson 24.2
- Homework Practice Online
- HRW Go site

- One-Stop Planner CD–ROM, Lesson 24.3
- *ARGWorld* CD–ROM
- **CNN** Presents World Cultures: Yesterday and Today, Segment 24: The Farms of Zimbabwe
- Homework Practice Online
- HRW Go site

Reinforcement, Review, and Assessment

- **ELL** Main Idea Activity 24.1
- **ELL** English Audio Summary 24.1
- **ELL** Spanish Audio Summary 24.1
- **REV** Section 1 Review, p. 535
- **A** Daily Quiz 24.1

- **ELL** Main Idea Activity 24.2
- **ELL** English Audio Summary 24.2
- **ELL** Spanish Audio Summary 24.2
- **REV** Section 2 Review, p. 541
- **A** Daily Quiz 24.2

- **ELL** Main Idea Activity 24.3
- **ELL** English Audio Summary 24.3
- **ELL** Spanish Audio Summary 24.3
- **REV** Section 3 Review, p. 545
- **A** Daily Quiz 24.3

Meeting Individual Needs

Ability Levels

Level 1 Basic-level activities designed for all students encountering new material

Level 2 Intermediate-level activities designed for average students

Level 3 Challenging activities designed for honors and gifted-and-talented students

English Language Learners Activities that address the needs of students with Limited English Proficiency

Chapter Review and Assessment

- Chapter 24 Test Generator (on the One-Stop Planner)
- Global Skill Builder CD–ROM
- HRW Go site
- **REV** Chapter 24 Review, pp. 548–49
- **REV** Chapter 24 Tutorial for Students, Parents, Mentors, and Peers
- **A** Chapter 24 Test (form A or B)
- **A** Alternative Assessment Handbook
- **A** Chapter 24 Test for English Language Learners and Special-Needs Students

Launch into Learning

Write the following words and phrases on the chalkboard: *gold, diamonds, elephants, baboons, crocodiles, sailing ships, waterfalls, mountains, cattle, nomadic herders, modern cities, mosques, crosses,* and *coffee.* Ask students what these items have in common. *(They are all found in the region of southern Africa.)* Tell students that they will learn more about how these items relate to southern Africa in this chapter.

Using the Physical-Political Map

Have students examine the map on the opposite page. Ask them to identify the general direction of the land's slope *(highest in the southeast, slopes down towards the northwest).* Then have students use the map to list the region's landforms. *(Possible answers: islands, capes, plains, escarpments, mountains)* Point out that southern Africa is what we call the region. South Africa is a country within that region. Some people in the region refer to South Africa as "the Republic."

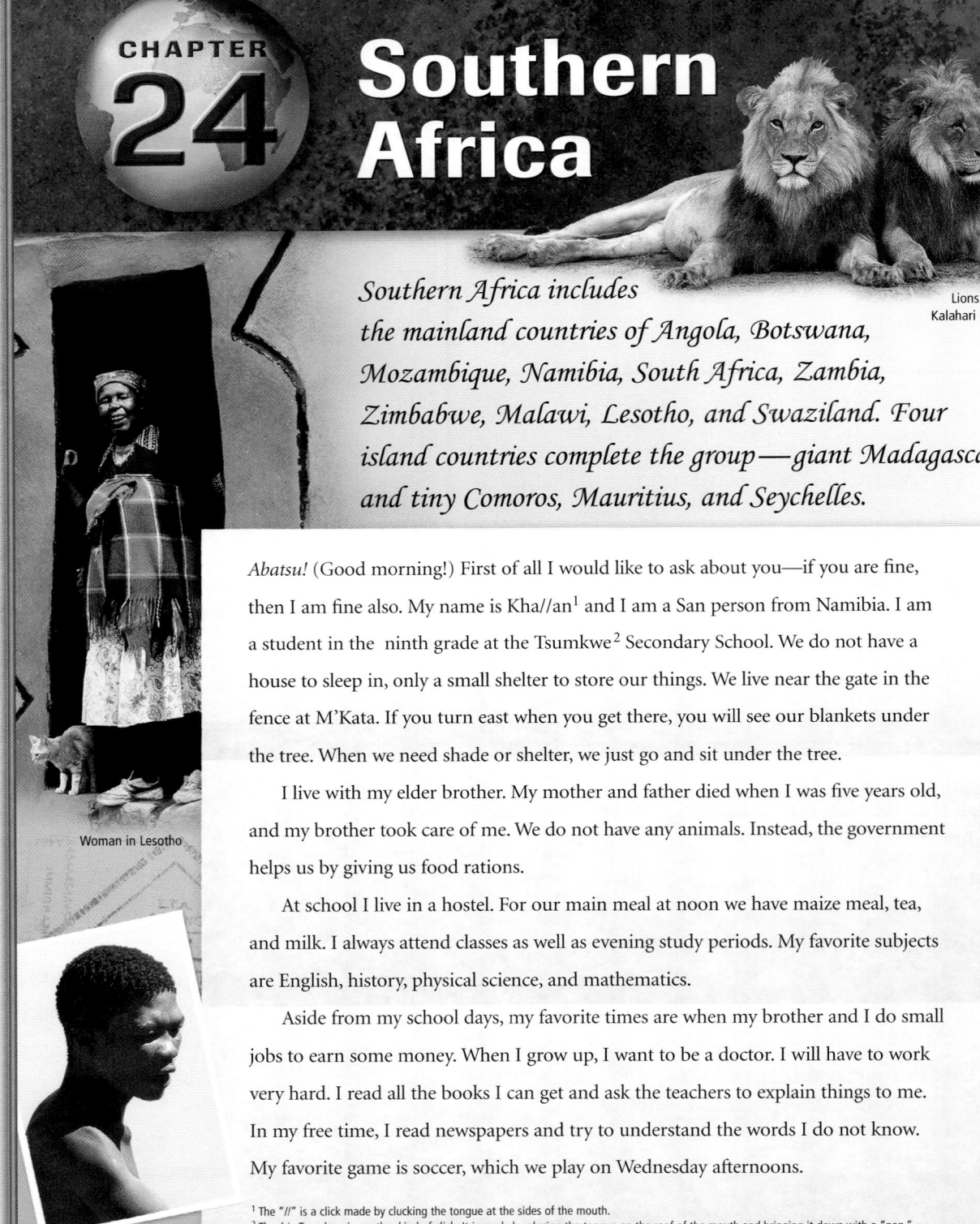

Why We Should Know More

You may wish to point out that there are many reasons that we should know more about the countries of southern Africa, including the following:

▶ The region is rich in minerals—many of them imported by the United States.

▶ Many Americans have ancestors from the region.

▶ Some of the countries of southern Africa are struggling to improve relationships among different racial and ethnic groups. Their efforts can have wide-ranging results.

▶ Many Americans travel to the region to visit its wild animal parks.

▶ Art from the region is unique and appreciated throughout the world.

Woman in Lesotho

CHAPTER 24 Southern Africa

Lions in the Kalahari Desert

Southern Africa includes the mainland countries of Angola, Botswana, Mozambique, Namibia, South Africa, Zambia, Zimbabwe, Malawi, Lesotho, and Swaziland. Four island countries complete the group—giant Madagascar and tiny Comoros, Mauritius, and Seychelles.

Abatsu! (Good morning!) First of all I would like to ask about you—if you are fine, then I am fine also. My name is Kha//an[1] and I am a San person from Namibia. I am a student in the ninth grade at the Tsumkwe[2] Secondary School. We do not have a house to sleep in, only a small shelter to store our things. We live near the gate in the fence at M'Kata. If you turn east when you get there, you will see our blankets under the tree. When we need shade or shelter, we just go and sit under the tree.

I live with my elder brother. My mother and father died when I was five years old, and my brother took care of me. We do not have any animals. Instead, the government helps us by giving us food rations.

At school I live in a hostel. For our main meal at noon we have maize meal, tea, and milk. I always attend classes as well as evening study periods. My favorite subjects are English, history, physical science, and mathematics.

Aside from my school days, my favorite times are when my brother and I do small jobs to earn some money. When I grow up, I want to be a doctor. I will have to work very hard. I read all the books I can get and ask the teachers to explain things to me. In my free time, I read newspapers and try to understand the words I do not know. My favorite game is soccer, which we play on Wednesday afternoons.

[1] The "//" is a click made by clucking the tongue at the sides of the mouth.
[2] The *k* in Tsumkwe is another kind of click. It is made by placing the tongue on the roof of the mouth and bringing it down with a "pop."

Section 1

OBJECTIVES

1. Identify the major landforms and rivers of southern Africa.

2. Describe the climates, biomes, and natural resources of the region.

 **LET'S GET STARTED**

Copy the following question onto the chalkboard: *What kinds of animals usually live in a desert? List some examples in your notebook.* Discuss student responses. *(Many students will mention camels or small animals.)* Point out that the Namib Desert of southern Africa is home to huge African elephants. Ask students how this region probably compares to areas where elephants usually live *(drier, less shade, less vegetation).* Tell students that in Section 1 they will learn more about the landforms and biomes of southern Africa.

Building Vocabulary

Write the terms **escarpment**, **biodiversity**, and **veld** on the board. Call on students to provide meanings for the prefix *bio- "life"* and the word *diverse "varied".* Then call on another volunteer to suggest a meaning for the word *biodiversity.* Have a student locate and read the term's definition from the text. Point out that *escarpment* and *veld* both refer to the region's physical features. Have another student read the definitions of these terms.

Section **1**

Natural Environments

READ TO DISCOVER

1. What are the main landforms and rivers of southern Africa?

2. What climates, biomes, and natural resources are found in the region?

DEFINE

escarpment
biodiversity
veld

WHY IT MATTERS

The countries of southern Africa struggle to balance wildlife conservation with people's need for land and resources. Use **CNNfyi**.com or other **current events** sources to learn about wildlife conservation in the region.

LOCATE

Drakensberg
Orange River
Limpopo River

Zambezi River
Namib Desert
Kalahari Desert
Okavango Swamps

Section **1** **RESOURCES**

REPRODUCIBLE

▶ Guided Reading Strategy 24.1
▶ Readings in World Geography, History, and Culture 64
▶ Creative Strategies for Teaching World Geography, Lesson 16

TECHNOLOGY

▶ One-Stop Planner CD–ROM, Lesson 24.1
▶ CNN Presents Geography: Yesterday and Today, Segment 27: Mining in South Africa the Traditional Way
▶ Geography and Cultures Visual Resources 43–47
▶ Homework Practice Online
▶ HRW Go site

REINFORCEMENT, REVIEW, AND ASSESSMENT

▶ Main Idea Activity 24.1
▶ English Audio Summary 24.1
▶ Spanish Audio Summary 24.1
▶ Section 1 Review, p. 535
▶ Daily Quiz 24.1

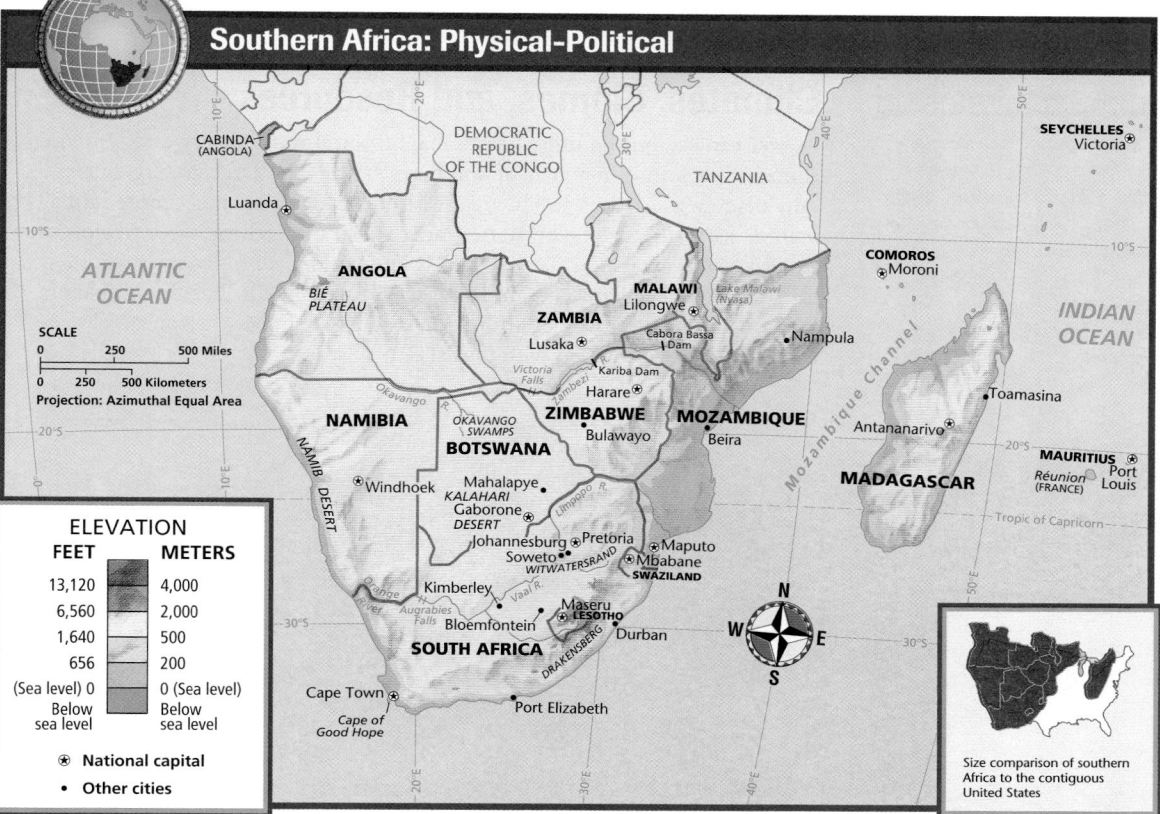

Southern Africa: Physical-Political

ATLANTIC OCEAN

CABINDA (ANGOLA)

Luanda

ANGOLA

BIÉ PLATEAU

DEMOCRATIC REPUBLIC OF THE CONGO

TANZANIA

SEYCHELLES
Victoria

COMOROS
Moroni

MALAWI
Lilongwe

Lake Malawi (Nyasa)

ZAMBIA
Lusaka

Cabora Bassa Dam

Nampula

INDIAN OCEAN

SCALE
0   250   500 Miles
0   250   500 Kilometers
Projection: Azimuthal Equal Area

Victoria Falls

Kariba Dam

Harare

Zambezi

NAMIBIA

OKAVANGO SWAMPS

ZIMBABWE
Bulawayo

MOZAMBIQUE
Beira

Mozambique Channel

Toamasina

Antananarivo

Okavango

BOTSWANA

MADAGASCAR

MAURITIUS
Réunion (FRANCE)
Port Louis

NAMIB DESERT

Windhoek

Mahalapye
KALAHARI DESERT
Gaborone

Limpopo R.

Maputo

Tropic of Capricorn

ELEVATION

FEET	METERS
13,120	4,000
6,560	2,000
1,640	500
656	200
(Sea level) 0	0 (Sea level)
Below sea level	Below sea level

Johannesburg Pretoria
Soweto   WITWATERSRAND
Mbabane
SWAZILAND

Kimberley

Vaal R.

Maseru
LESOTHO

Durban

N
W   E
S

Orange R.

Augrabies Falls

Bloemfontein

SOUTH AFRICA

DRAKENSBERG

Cape Town

Port Elizabeth

Cape of Good Hope

⊛ National capital
• Other cities

Size comparison of southern Africa to the contiguous United States

Teach Objectives 1–2

ALL LEVELS: Provide each student with an outline map of the region. First, have students label major landforms on their maps. Then have them use colors and shading to identify the region's climates and biomes. ENGLISH LANGUAGE LEARNERS

Teach Objective 2

ALL LEVELS: Copy the following graphic organizer onto the chalkboard, omitting the italicized answers. Ask students to fill in the resources of southern Africa and to note where each resource is found. ENGLISH LANGUAGE LEARNERS

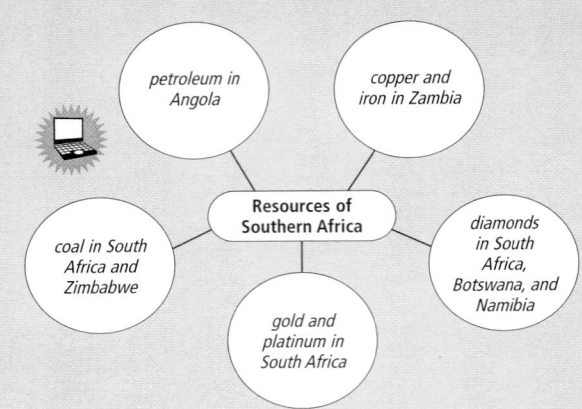

Resources of Southern Africa
- petroleum in Angola
- copper and iron in Zambia
- diamonds in South Africa, Botswana, and Namibia
- gold and platinum in South Africa
- coal in South Africa and Zimbabwe

VISUAL RECORD ANSWER

They are very rugged.

Our Amazing Planet

One of the world's great waterfalls is located on the Orange River. At flood time, Augrabies (oh-KRAH-bees) Falls is several miles wide, and 19 separate waterfalls flow into a ravine 11 miles (18 km) long.

INTERPRETING THE VISUAL RECORD

The Drakensberg is the main mountain range of southern Africa. The name Drakensberg *is Dutch and means "Dragon Mountains." The Zulu name for the range is* Kwathlamba, *which means "Piled-Up Rocks" or "Barrier of Pointed Spears."* **What does the Zulu name for these mountains suggest about their physical geography?**

Landforms and Water

Along southern Africa's coastline is a narrow coastal plain. This plain is less than 100 miles (160 km) wide in most places. Farther inland, a high plateau reaches more than 4,000 feet (1,220 m) above sea level. Most of the region lies on this plateau. Between the coastal plain and the plateau is an **escarpment**, a steep face at the edge of a plateau or other raised area. In South Africa the Drakensberg (DRAH-kuhnz-buhrk) range forms part of the escarpment. These mountains include a peak 11,425 feet (3,482 m) high.

Several major rivers flow across southern Africa. The Orange River starts in the Drakensberg and flows west across South Africa. It is the only major river in the region that drains into the Atlantic Ocean. Dams on the Orange River produce hydroelectricity. They also allow irrigation and economic development in central South Africa. The Limpopo, which is sometimes called the Crocodile River, drains into the Indian Ocean. The Zambezi River is another major source of hydroelectric power. It is more than a mile wide above Victoria Falls, where it drops 355 feet (108 m).

As southern Africa's rivers flow from the high interior plateau to the low-lying coastal plain, they form waterfalls and rapids. In addition, sandbars partially block the mouths of the Limpopo and Zambezi Rivers. As a result, large ships cannot sail upriver to the interior.

✓ **READING CHECK:** *Places and Regions* What are the three main landform regions of southern Africa? What are three major rivers in the region? a narrow coastal plain, an inland plateau, an escarpment; Orange, Limpopo, Zambezi

Climates, Biomes, and Resources

As you can see on the unit climate map, tropical wet and dry and semiarid climates are found in much of southern Africa. The Drakensberg causes a rain-shadow effect. As a result, areas east of the escarpment get more rainfall than areas to the west. The wettest area is the tropical rainforest region of eastern Madagascar. This large island is known for its **biodiversity**, or many different types of plants and animals. (See Geography for Life: Biodiversity in Madagascar.) Cape Town, South Africa, has a pleasant Mediterranean climate that helps make it popular with tourists.

Because a cold current flows off the Atlantic coast of southern Africa, the evaporation rate there is low. The dry air influences the Namib (NAH-mib) Desert, which has some of the world's highest sand dunes. Some parts of the Namib get as little as 0.5 inches (13 mm) of rain per year. Plants get water from dew and fog rather than rain. Beetles, lizards, and snakes live in the Namib, but relatively few mammals can survive there.

Moving inland, rainfall gradually increases as the Namib gives way to the Kalahari Desert. Because it is not as dry as the Namib, plant life in the Kalahari ranges from grasses to palm trees. Many kinds of animals, including antelope and elephants, live there. The Okavango Swamps of northern Botswana are particularly rich in plant and animal life. A river empties into a maze of channels and shallow lakes within the area.

Along its eastern edge, the Kalahari merges into the grassland or **veld** (VELT) of South Africa. The grassy, high plains in central South Africa north and west of Lesotho make up what is called the highveld. The middleveld, a savanna region of short trees and bushes, is found at lower elevations. Still another of South Africa's regions is the lowveld, which includes dry tall-tree savannas as well as forests at the base of the Drakensberg.

Southern Africa has many valuable energy and mineral resources. For example, Angola has petroleum reserves. South Africa and Zimbabwe (zim-BAH-bway) have enormous coal deposits. Most of the region's electricity is generated by hydropower or by burning coal. Gold and platinum are among South Africa's many metal ore deposits. South Africa, Botswana, and Namibia have productive diamond mines. Zambia's rich copper deposits make it a major producer of that metal. In fact, miners have worked Zambia's iron and copper deposits for more than 1,500 years.

✔ **READING CHECK:** *Physical Systems* How does the Drakensberg affect climate patterns in southern Africa? creates rain shadow to the west

INTERPRETING THE VISUAL RECORD *Chimanimani National Park in eastern Zimbabwe is located in a savanna region.* **What can you see in this photo that shows this is a savanna biome?**

internet connect

GO TO: go.hrw.com
KEYWORD: SW3 CH24
FOR: Web sites about southern Africa

Section 1 Review

go.hrw.com **Homework Practice Online**
Keyword: SW3 HP24

Define
escarpment
biodiversity
veld

Working with Sketch Maps
On a map of southern Africa that you draw or that your teacher provides, label the countries of the region and the Drakensberg, Orange River, Limpopo River, Zambezi River, Namib Desert, Kalahari Desert, and Okavango Swamps. What famous feature makes the Zambezi River distinctive?

Reading for the Main Idea
1. *Places and Regions* Which landform region covers the largest area of southern Africa?
2. *Physical Systems* How does the cold ocean current off the Atlantic coast influence southern Africa's climates?

Critical Thinking
3. *Finding the Main Idea* What are the two deserts in this region, and how are their environments different?
4. *Drawing Inferences and Conclusions* Do you think southern Africa's rivers are vital links between the region's interior and global trade? Why or why not?

Organizing What You Know
5. Create a graphic organizer like the one below. Use it to describe the landforms, climates, and resources of southern Africa. Use the unit atlas and the information in Section 1.

Landforms	Climates	Resources

Medicine: Medicinal Plants

Many medicines that are prescribed by doctors include or are based on natural compounds found in plants. Many of the plants from which these compounds are derived live in rainforest or jungle environments like Madagascar. Some of the medicines based on tropical plants could possibly save the lives of people with serious illnesses, like leukemia. One such plant is the rosy periwinkle, which grows only in Madagascar.

Scientists began to study this bright pink flower after residents of the island reported its use in fighting diabetes. Laboratory tests failed to indicate any success in treating that disease, but they did show that the compounds made from the periwinkle were effective

in reducing the number of white blood cells. As a result of this testing, doctors discovered two drugs called vinblastine and vincristine, which have proven effective in combating leukemia, a type of cancer characterized by an increase in white-blood-cell counts. The flower has also been used to treat Hodgkin's disease, another type of cancer. Unfortunately, the rosy periwinkle is one of the species whose habitat is threatened by deforestation in Madagascar.

Have students investigate other uses for tropical plant species. Ask students to identify products made from the plants and the uses to which these products can be put. Have students prepare oral reports to present their findings to the class. Encourage students to prepare visual aids to accompany their reports.

Masoala National Park

Madagascar's government has taken steps to protect what remains of the country's native vegetation. One such step is convincing farmers that they will benefit from increased tourism.

This ecotourism is the central feature of Madagascar's plan to preserve the island's environment. In 1997 the government established Masoala National Park on the island's northeastern coast. It covers approximately 840 square miles (2,175 sq km) and is home to several endangered species. Forests surrounding the park serve as a buffer zone. Some people who live in this area now make a living raising butterflies for zoos around the world.

MAP ANSWER

encourage policies that prevent deforestation and maintain animal habitats

Applying What You Know Answers

1. isolation from African mainland, wide range of environments
2. Possible answers: reduce population-growth rate, create nonpolluting industries so people have alternatives to farming, develop tourism

This Geography for Life feature addresses National Geography Standards 6, 8, 14, and 18.

Physical Systems

Geography for Life
Biodiversity in Madagascar

You may have heard people call various places a "Garden of Eden." The term often refers to a place of great beauty and wonder. If wonder is the defining feature, then Madagascar might be something of a Garden of Eden.

This big island off the southeast coast of Africa is home to a wide range of unique plants and animals. Some 80 percent of the country's species live nowhere else on Earth. Because Madagascar has been an island for millions of years, these species have developed in isolation. In addition, Madagascar has a wide range of environments. The east coast has a tropical rain forest. Farther west are grasslands and scattered trees. A desert lies at the island's southern tip. Madagascar's isolation and its many environments contribute to its remarkable biodiversity.

Madagascar has more than 70 species of songbirds and parrots. There are about 800 butterfly species. More than 10,000 varieties of plants exist, from 1,000 orchid species to 6 species of baobab trees. Scientists often discover previously unknown plants on the island.

Ring-tailed lemur

Geographers can use Madagascar as a laboratory for studying the ways that species interact. For example, consider the small monkeylike mammals called lemurs. Lemurs live in trees and are active mainly at night. They are native to Madagascar and are also found on the nearby islands of Comoros. Lemurs of different species may be as small as mice or as big as medium-sized dogs. They eat mainly leaves and fruit. Therefore, lemurs depend on the forests. In an interesting twist, the forests also depend on the lemurs. Why is this so? When lemurs eat fruit from the trees, they also eat the seeds. Because the seeds do not digest, the lemurs spread them far from the parent tree in their droppings. This is the main way that many trees spread into new areas.

Today, however, Madagascar's natural environments are seriously threatened. Loss of habitat from deforestation is the biggest problem. Only some 10 percent of Madagascar's forests remain. Much of the damage comes from slash-and-burn farming. As the population grows, people clear more land for crops. Many conservation organizations now call Madagascar their top priority for preserving biodiversity. The country's government is working closely with these groups. Officials encourage citizens to take pride in the island's biodiversity. Ecotourism, better farming practices, and protected national parks may offer hope for Madagascar's environment.

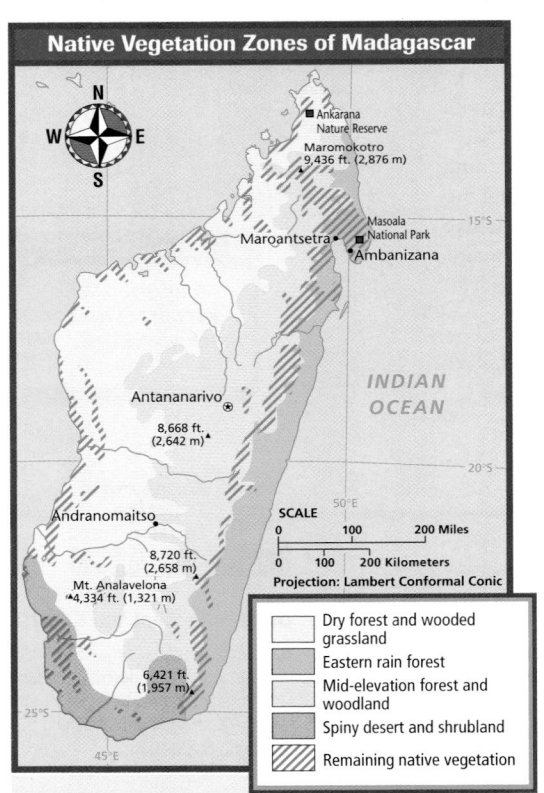

Native Vegetation Zones of Madagascar

Ankarana Nature Reserve
Maromokotro 9,436 ft. (2,876 m)

Masoala National Park
Maroantsetra • ■ Ambanizana

15°S

Antananarivo ⊛

8,668 ft. (2,642 m)

INDIAN OCEAN

20°S

Andranomaitso

8,720 ft. (2,658 m)
Mt. Analavelona 4,334 ft. (1,321 m)

50°E

SCALE
0   100   200 Miles
0 100 200 Kilometers
Projection: Lambert Conformal Conic

6,421 ft. (1,957 m)

25°S

45°E

☐ Dry forest and wooded grassland
☐ Eastern rain forest
☐ Mid-elevation forest and woodland
☐ Spiny desert and shrubland
▨ Remaining native vegetation

INTERPRETING THE MAP *Madagascar's government has passed laws to protect the country's native plants and animals. How do you think the efforts of conservation groups can influence the government's environmental policies?*

Applying What You Know

1. **Summarizing** Why does Madagascar have so many different plants and animals?

2. **Drawing Inferences and Conclusions** How might Madagascar preserve its biodiversity?

Section 2

OBJECTIVES

1. Identify some important events in the history of southern Africa.

2. Describe the region's cultures.

🔊 *LET'S GET STARTED*

Copy the following question onto the chalkboard: *How might the physical geography of southern Africa have affected its history?* Allow time for students to respond in writing. Discuss student responses. Point out that the abundant natural resources and the strategic location of southern Africa have attracted foreigners. Tell students that they will learn more about the history and culture of southern Africa in Section 2.

Building Vocabulary

Write the terms **apartheid** and **sanctions** on the chalkboard. Have volunteers locate and read their definitions from the text or glossary. Point out that *apartheid* is an Afrikaans word meaning "apartness." Locate or have students find uses of the word *sanctions* in current events magazines or newspaper headlines. Discuss the meaning of the word in the context of these examples.

Section 2 History and Culture

READ TO DISCOVER

1. What are some important events in the history of southern Africa?
2. What are the region's cultures like?

WHY IT MATTERS

South Africa's past racial policies caused years of international isolation for the country's government. Use **CNNfyi.com** or other **current events** sources to learn about South Africa's international relations today.

IDENTIFY

Afrikaners
African National Congress

DEFINE

apartheid
sanctions

LOCATE

Great Zimbabwe
Cape of Good Hope
Luanda
Maputo
Cape Town
Durban
Kimberley
Johannesburg
Pretoria
Harare

Section 2 RESOURCES

REPRODUCIBLE

▶ Guided Reading Strategy 24.2
▶ Readings in World Geography, History, and Culture 65 and 66
▶ Cultures of the World Activity: Region 6
▶ Critical Thinking Activity 24: Understanding the Bushmen

TECHNOLOGY

▶ One-Stop Planner CD–ROM, Lesson 24.2
▶ Homework Practice Online
▶ HRW Go site

REINFORCEMENT, REVIEW, AND ASSESSMENT

▶ Main Idea Activity 24.2
▶ English Audio Summary 24.2
▶ Spanish Audio Summary 24.2
▶ Section 2 Review, p. 541
▶ Daily Quiz 24.2

History

The first inhabitants of southern Africa were hunter-gatherers and animal herders. Bantu-speaking peoples, the ancestors of most modern southern Africans, began migrating to the region around A.D. 100. Most early Bantu peoples were farmers who raised crops such as beans and sorghum. They also herded cattle, goats, and sheep. Unlike the people who already lived in the region, the Bantu knew how to make iron tools.

The Bantu peoples established several powerful kingdoms. Great Zimbabwe was the center of a wealthy farming and cattle-raising community. Farther south, the Sotho established a kingdom in the 1800s in what is now Lesotho. The realm of the Zulu lay east of the escarpment.

Portuguese sailors began exploring the southern African coast in the late 1400s. They reached the coast of what is now Angola in 1483. Soon Portuguese traders and merchants, eager to trade for spices in Asia, ventured farther. To get to Asia, they had to sail around the southern tip of Africa and then cross the Indian Ocean. The journey was long and difficult. Ships had to stop along the way for supplies. Therefore, the Portuguese set up small bases along the southern African coast. However, dense vegetation, the threat of disease, and the lack of navigable rivers discouraged them from moving very far inland.

✓ **READING CHECK:** *Environment and Society* What factors discouraged the movement of early Portuguese settlers into the interior of southern Africa? dense vegetation, threat of disease, lack of navigable rivers

A San man drinks water from a hollowed ostrich egg. The San, who live in the Kalahari Desert in Botswana, Namibia, and South Africa, were historically nomadic hunter-gatherers. However, most have now settled in villages.

Teach Objective 1

LEVEL 1: Copy the following graphic organizer onto the chalkboard, omitting the italicized answers. Have students fill in the time line with events from southern Africa's history.

LEVEL 2: Organize the class into four groups and assign each group one of the following periods of southern Africa's history: early history, colonial period, independence, or modern history. Have each group design a mural for its period. Combine the groups' designs into one large mural and display it in the classroom. Discuss the finished project. **ENGLISH LANGUAGE LEARNERS, COOPERATIVE LEARNING**

1450	1650	1800	1890	1900	1910	1950	1960	1970	1980	1990	2000

late 1400s *Portuguese arrive.*
1652 *Dutch arrive.*
1800s *British take over Cape area.*

1800s *Sotho kingdom flourishes.*
late 1800s *Diamonds and gold are discovered.*
1899–1902 *Boer War is fought.*

1910 *South Africa gains independence.*
1912 *ANC is established.*
1948 *Apartheid is established.*

1960s *Most British colonies gain independence.*
1970s *Portuguese colonies gain independence.*

1975 *Angola becomes communist.*
1980 *Zimbabwe and Zambia gain independence.*

1990 *Mandela is released from prison.*
1994 *Mandela is elected president of South Africa.*

Linking Past to Present

Shaka Zulu Shaka (c. 1787–1828), the powerful founder of the Zulu Empire, was a legendary figure even in his own time. An imposing warrior, Shaka gained his authority to rule mainly through the fear he inspired. He engaged the small Zulu clan—one of hundreds in southern Africa—in a series of wars that resulted in the Zulu becoming the dominant group in the region.

Shaka Zulu has not been forgotten. He has become a symbol of Zulu strength and greatness. He also plays a role in popular culture. For example, the singing group Ladysmith Black Mambazo released an album under the name *Shaka Zulu*. Tourists can visit Shakaland, a cultural village in South Africa, to learn about Zulu traditions.

internet connect

GO TO: go.hrw.com
KEYWORD: SW3 CH24
FOR: Web sites about Zulu culture

VISUAL RECORD ANSWER

The houses display a Dutch architectural style found in Europe.

538

INTERPRETING THE VISUAL RECORD

Tulbagh, northeast of Cape Town in South Africa, was founded in 1699 by Dutch farmers. **How does the architecture in this photo reflect European influence?**

Boer soldiers pose with their weapons during the Boer War. During the war, well-armed Boer soldiers held off a much larger British army for nearly three years.

The Colonial Period In 1652 the Dutch set up a small settlement at the Cape of Good Hope. The Mediterranean climate at the Cape made it a good place to farm. Other Europeans, including French and Germans, joined the Dutch farmers. These European settlers were known as Boers (BOHRZ), which means "farmers" in Dutch. They thought of Africa as home and called themselves **Afrikaners** (a-fri-KAH-nuhrz). Over time, they developed a new language, called Afrikaans (a-fri-KAHNS). It combined elements of Dutch with African words. Speakers of German, French, and English also influenced the development of the language. Afrikaans even included words learned from Asians who had been brought to the Cape as slaves and laborers.

In the early 1800s, Great Britain took over the Cape area. In response, many Afrikaners moved inland. They wanted to be free from British rule and govern themselves. After fierce battles with Bantu-speaking peoples, the Afrikaners set up two independent republics in the interior.

During the late 1800s, the discovery of diamonds and gold drew people to southern Africa from all over the world. Both the Boers and the British wanted to control the region's mineral wealth. This led to the Boer War of 1899 to 1902. In the end, all of what is now South Africa came under British control. Britain then granted the Union of South Africa independence in 1910.

Other European countries claimed different parts of the region. The Portuguese kept control of Angola and Mozambique (moh-zahm-BEEK). The territory of South-West Africa (now known as Namibia) was a German colony. After Germany's defeat in World War I, South Africa took control of Namibia. Britain took Bechuanaland (now Botswana), Basutoland (Lesotho), Swaziland, and Northern and Southern Rhodesia (now Zambia and Zimbabwe).

✓ **READING CHECK:** **Human Systems** Why did the Afrikaners move into the interior of southern Africa? *to get away from British rule and maintain independence*

Independence Most of southern Africa remained under colonial rule until the late 1900s. During the 1960s many independence movements grew. Britain granted independence to most of its remaining colonies. The Portuguese colonies—Angola and Mozambique—did not win independence until the 1970s. Africans in Rhodesia had also been fighting to end white-minority rule. Finally, Southern Rhodesia won independence in 1980. The country changed its name to Zimbabwe. Northern Rhodesia became Zambia.

LEVEL 3: Have students complete the Level 1 and Level 2 activities. Then have students conduct research into the types of sanctions used against South Africa during the period of apartheid. Also have them examine protests staged within the country. Ask students why these sanctions and protests might have persuaded the government to change its policies. Have students present their conclusions in written reports. **BLOCK SCHEDULING**

Teacher to Teacher

C. Eugene Price of Fort Wayne, Indiana, suggests the following game to help students review the history of southern Africa. Organize the class into two teams. Have a member of the first team sit in a chair at the front of the classroom. Ask that student a question drawn from Section 2. If the student answers correctly, he or she gets a chance to answer another question. If the student does not answer correctly, have a student from the other team answer the question and take over the chair. Continue until all students have had a chance to answer questions.

Conflicts continued long after independence in some areas. In Mozambique, a civil war between the Communist government and rebels lasted until the 1990s. In 1975 a Communist government took power in Angola. Rebels supported by the United States fought against the government, which was backed by the Soviet Union and Cuba.

Although the other countries in the region had cast off rule by white minorities, Afrikaners continued to control South Africa. Whites held political power, owned most of the land, and controlled the economy. However, black Africans made up most of the population. Since 1948, black South Africans had been denied political rights under a system of laws called **apartheid** (uh-PAHR-tayt). The term means "separateness." These laws forced black South Africans to live in separate areas and use separate facilities from whites. People of mixed race, called Coloureds, and Asian immigrants made up other South African ethnic groups. Their status was slightly higher than that of black South Africans.

Ending Apartheid   Apartheid made South Africa an outcast in the world community. Eventually, some countries began to place economic **sanctions** on South Africa. Sanctions are penalties intended to force a country to change its policies. For example, a group of countries may refuse to provide economic aid to another country. Political organizations inside South Africa also pushed for an end to apartheid. The best known of these groups was the **African National Congress** (ANC), founded in 1912.

In 1990 the South African government finally began to change its policies. In that year it released the leader of the ANC, Nelson Mandela, from prison. The government began to get rid of the apartheid system. In 1994 South Africa held its first elections open to all citizens. Nelson Mandela was then elected South Africa's first black president.

FOCUS ON HISTORY

The Legacy of Apartheid   With the election of Nelson Mandela, South Africa entered a new era. According to new laws, all citizens must be treated equally. However, the country still faces serious challenges. Providing economic opportunities, education, and health care to the entire population has proven difficult. Poverty remains a problem, and on average, white South Africans are still much wealthier than black South Africans. In addition, crime has increased in the cities. One reason is that young people who had worked against apartheid are now frustrated with the slow pace of change. Poverty combined with frustration has led to violence. Finally, divisions among black ethnic groups have caused new problems. The long-term stability of the country may depend on whether the government can improve people's standard of living.

✔ **READING CHECK:**   (*Human Systems*)   What challenges has the end of apartheid left for South Africa? need to provide more economic opportunities, education, health care; mending divisions between black ethnic groups; long-term stability

Queen Nzinga is one of Angola's heroes. During the 1600s she led fighters into battle against Portuguese slave traders who were pushing farther into the interior.

Nelson Mandela was imprisoned by South Africa's government from 1962 to 1990 for his antiapartheid activities. After his release, Mandela assisted with South Africa's transition to a democratic system of government. He won the Nobel Peace Prize in 1993 and became the country's first black president in 1994.

Essential Element 5

► **Environment** ◄
and Society

Robben Island   For the first 18 years of his imprisonment, Nelson Mandela was held in a maximum security prison on Robben Island off the coast of South Africa. From 1961 to 1991 the island was used by the South African government to isolate both political prisoners and other prisoners from society. This was not the first time the island housed convicted criminals. Both the Dutch and English colonies placed prisoners on the island. The Portuguese may also have left prisoners at Robben Island in the 1500s, although this has not been proven.

In addition to its prison facilities, Robben Island has a small village and a bird sanctuary. Since the prison closed in 1991, South Africa has worked to make the island a tourist attraction. Now, several tours a day bring visitors interested in the island's history and wildlife.

CRITICAL THINKING: Why might governments build prisons on islands? *(Possible answers: Escape is more difficult. Prisoners are cut off from society.)*

ALL LEVELS: Organize students into four groups and assign each group an element of southern African culture. Have each group write a script for a segment on a television program called Life in Southern Africa. Instruct students to write dialogue for their segments and suggest images that they would choose to accompany this text. For example, a group focusing on settlement patterns may wish to show various cities around southern Africa. Encourage students to be creative. **COOPERATIVE LEARNING, BLOCK SCHEDULING**

Using National Geography Standard 10:

Human Systems: The Characteristics, Distribution, and Complexity of Earth's Cultural Mosaics Have students examine how South Africa's two largest black ethnic groups—the Xhosa and the Zulu—have affected the country's politics. Both the ANC and Inkatha Freedom Party have Zulu support. The Xhosa are more associated with the United Democratic Movement. Have each student investigate one of these ethnic groups and then write a position paper on why the group supports a particular party and how its members hope to improve life for South Africans. Also, discuss how the two groups seem to differ.

Daily Life

Celebrating Diversity in Seychelles Throughout their history, the people of Seychelles have been influenced by the presence of various groups in their islands. Africans, Arabs, and Europeans have contributed to the country's cultural heritage. Elements from each of these cultures have been woven together to form a creole, or blended, culture in the islands. Each October, the people of the Seychelles come together to celebrate their diverse heritage in the Festival Kreol.

The festival is dedicated to preserving and promoting creole cultures around the world. Events are held throughout the country, although the main events all happen on the largest island, Mahé. Creole food, fashion, games, poetry, music, and dance are highlighted in the weeklong celebration. Creole artists from around the world are invited to participate.

ACTIVITY: Ask students to conduct research on other creole festivals and to bring elements of those festivals to class to share.

CHART ANSWER

Communication difficulties may inhibit cultural diffusion.

Swazi men dance in a traditional harvest festival. Music and dance help bring a sense of unity to Swazi communities.

Culture

Southern Africa's cultural mix reflects its history. Over time, many peoples have settled in the region. Some aspects of their traditions have remained distinct. In other ways they have combined in a process of acculturation. Today African traditions are strongest in rural areas and small towns. In the cities, many people have adopted American and European customs.

People, Languages, and Religion Bantu languages of the Niger-Congo family are widely spoken in southern Africa. Two of these Bantu groups are Sotho and Nguni. Sotho speakers, including Tswana and Basuto peoples, live mainly in the interior. Nguni speakers, including the Zulu and Xhosa peoples, live closer to the coasts. The white population, concentrated in South Africa, speaks mainly English and Afrikaans. Smaller white populations live in Zimbabwe and Namibia and speak mostly English.

Because of the diversity of languages in each country, governments often rely on European languages. For example, English is commonly used in Zambia and Zimbabwe. Portuguese is the official language in Angola and Mozambique.

Before the arrival of Europeans, most southern Africans practiced traditional religions. These religions still have many followers. These followers may believe that ancestors and the spirits of the dead have divine powers. Diseases and misfortune may be explained as the work of spirits.

Europeans brought Christianity to southern Africa. Roman Catholicism is common in areas that the Portuguese colonized. The Dutch Reformed Church is the largest Christian denomination in South Africa. Today millions of Africans belong to Christian churches. Many others belong to churches that blend the teachings of Christianity with traditional African religious practices. Islam is also practiced. Some Muslims are descended from enslaved people who came from Southeast Asia in the 1600s and 1700s. South Asians brought Hinduism to southern Africa.

Language Families of Africa

	Afro-Asiatic	Niger-Congo	Khoisan	Nilo-Saharan	Malayo-Polynesian
English	**Arabic**	**Swahili**	**Nama**	**Kanuri**	**Malagasy**
mother	umm	mama mzazi	//gûs	yâ	reny
child	walad	mtoto	/gôaï	táda	zaza
head	rā`s	kichwa	tanas	kǝlâ	loha
water	mā`	maji	/gami	njî	rano
tree	šajarah	mti	heis	kǝská	hazo
house	bayt	nyumba	omi	fáto	trano
red	aḥmar	-ekundu	/awa	cimê	mena
eat	akala	la	‡û	búkin	homana
go/walk	ḏahaba/mašā	enda	!gû	lengîn	mamindra

INTERPRETING THE CHART *Like the rest of Africa, southern Africa has an incredible diversity of languages grouped into several main language families. For example, San is a Khoisan language, while Zulu is a Bantu language of the Niger-Congo family.* **How might language differences affect cultural diffusion in the region?**

Women buy and sell food in a marketplace in Moroni, the capital of Comoros. Some 33 percent of the country's population lives in urban areas.

Settlement and Land Use The wetter eastern part of the region has long been more densely populated than the drier western part. Some small rural villages have a traditional settlement pattern. In the center of the village is a pen, called a kraal (KRAWL), where cattle are kept at night. Around the pen, villagers build small houses with wooden poles, clay, and grass roofs.

Europeans founded most of the region's cities. The Portuguese established Luanda in Angola and Maputo in Mozambique. These cities were important seaports and administrative centers. The Dutch established Cape Town in what is now South Africa. Another South African city, Durban, was a major British port. As Europeans moved into the interior, they also set up mining towns and administrative centers. For example, South Africa's Kimberley and Johannesburg started as mining camps. Pretoria, South Africa, and Harare, Zimbabwe, began as government centers. Today many southern African cities are large. However, most southern Africans still live in small villages.

✓ **READING CHECK:** *Human Systems* How did colonization influence the development of religion in the region? Europeans brought Christianity, some African churches blend Christianity and African beliefs, Hinduism brought from South Asia

Connecting to ANTHROPOLOGY

Marriage Customs

Anthropologists often study how marriages are arranged. Some societies in South Africa allow polygamy. Polygamy is a practice in which a person may have more than one spouse. However, under the old apartheid government, additional wives and their children had no official status as their husband's legal family. As a result, they also had no legal protection.

Under the black-majority government, laws have changed, and women benefit. South Africa's 1998 constitution recognizes traditional marriages as legal. Now, if a man wants to divorce any of his wives he must take the case to court. Before the laws changed, the husbands' families could end a marriage, which put women and children at risk financially.

Summarizing How have cultural beliefs influenced public policy in South Africa?

A Zulu wedding in South Africa

Section 2 Review

Identify Afrikaners, African National Congress

Define apartheid, sanctions

Working with Sketch Maps On the map you created in Section 1, label Great Zimbabwe, Cape of Good Hope, Luanda, Maputo, Cape Town, Durban, Kimberley, Johannesburg, Pretoria, and Harare. Which of these places provided the name for a modern country?

Reading for the Main Idea
1. *Places and Regions* Which three European countries had colonies in southern Africa?
2. *Places and Regions* How were the cities of southern Africa established?

Critical Thinking
3. **Drawing Inferences and Conclusions** Why are European languages still used in several countries of southern Africa?

go.hrw.com **Homework Practice Online** Keyword: SW3 HP24

4. **Identifying Points of View** Why do you think nonwhite South Africans were so unwilling to accept the system of apartheid? Why do you think many whites wanted the system to continue? How did public policies change in the 1990s?

Organizing What You Know
5. Create an idea web in which you describe, compare, and contrast the languages, religions, land-use practices, and customs of southern Africa.

OBJECTIVES

1. **Identify the main economic activities of southern Africa.**

2. **Describe the region's cities.**

3. **Explain the challenges that face the people of southern Africa.**

LET'S GET STARTED

Copy the following questions onto the chalkboard: *Would you like to visit southern Africa? Why or why not?* Discuss student responses. Tell students that some countries of southern Africa, like South Africa and the Seychelles, are popular tourist destinations. Others are not visited as frequently. Ask students why they think this might be the case. Tell students that they will learn more about the economies and cities of southern Africa in Section 3.

Building Vocabulary

Write the term **informal sector** on the chalkboard. Call on a volunteer to locate and read the definition from the text or glossary. Ask students how this meaning relates to other uses of the word *informal*. (*Possible answer: Workers do not follow a set system of official rules.*) Call on volunteers to suggest some activities that may be part of an economy's informal sector.

Section 3 The Region Today

READ TO DISCOVER

1. What are the main economic activities in southern Africa?
2. What are the region's cities like?
3. What challenges face the people of southern Africa?

WHY IT MATTERS

Many musicians in southern Africa combine traditional African musical styles with American jazz and rock 'n' roll. Use **CNNfyi.com** or other **current events** sources to learn about contemporary southern African music.

DEFINE

informal sector

LOCATE

Cabinda
Soweto

Vanilla is a major export of Madagascar and Comoros. Here, a worker in Madagascar prepares vanilla beans for processing. First the beans are dried and crushed. Then alcohol is used to draw out the vanilla flavor.

Southern African Economies

All the countries in southern Africa are classified as developing countries. However, South Africa is sometimes considered a middle-income country. It has the most developed economy in Africa. South Africa's market economy includes agriculture, manufacturing, and mining industries. The country's economy is much larger than all of the other economies in the region combined. In contrast, Mozambique is one of the poorest countries in the world. Its mainly traditional economy relies on farming.

Agriculture Farming, whether market-oriented or subsistence, is the most common economic activity in southern Africa. Most farmers practice subsistence agriculture. They depend on their crops and livestock for their survival. Farmers may also sell some of their produce in local markets. In some parts of the region, market-oriented agriculture is important. These farms are generally large and rely on modern machinery. Commercial farmers either sell their products in cities or export them. In Zimbabwe, commercial farms produce the country's most important export crop, tobacco. Angola grows coffee, while Madagascar exports vanilla. South Africans grow corn, fruits, and wheat on modern, mechanized farms.

✓ **READING CHECK:** *Human Systems* How does market-oriented agriculture differ from traditional subsistence agriculture in southern Africa? Market-oriented farms are larger, use modern machinery, and produce goods mainly for export or for cities.

ALL LEVELS: Copy the following graphic organizer onto the chalkboard, omitting the italicized answers. Have students complete the chart with words or phrases that indicate the products and locations of each activity. **ENGLISH LANGUAGE LEARNERS**

HOMEWORK: Have students use the material in the text and the graphic organizer to design an informational pamphlet about southern Africa's economies. Pamphlets should provide information about the economic activities found in the region and pictures that supplement this information. Encourage students to be creative in their designs. Display the pamphlets in the classroom.

Southern Africa's Economies

Mineral exports	**Tourism**	**Informal Sector**
• *gold in South Africa* • *oil in Angola* • *diamonds in Botswana* • *copper in Zambia*	• *game parks in Botswana, Namibia, South Africa, and Zimbabwe* • *tropical islands of Comoros, Mauritius, and Seychelles*	• *foods and fruits, souvenirs* • *services such as car repair* • *found in the region's large cities*

Business and Industry Minerals and oil are increasingly important to the economies of southern Africa. The largest mineral exporter in the region is South Africa. It produces more gold than any other country. Some mines are as deep as 13,000 feet (3,962 m). The country also exports many other metals. Angola pumps oil from deposits located offshore and in the small exclave of Cabinda. An exclave is a part of a country that is separated by the territory of another country. Cabinda lies on the Atlantic coast north of the Congo River. It is separated from the rest of Angola by the Democratic Republic of the Congo.

Botswana once had a small market economy that relied heavily on beef exports. Then diamonds were discovered in the late 1960s. Today Botswana is one of the world's largest producers of diamonds. The country's economy is also one of the fastest growing in Africa.

Being dependent on exports of a few primary products such as minerals can be risky. For example, copper is Zambia's most important export mineral. However, the price of copper on the world market goes up and down. When copper prices fall, Zambia's entire economy is hurt. On the other hand, South Africa has a more diversified economy. It exports a variety of agricultural, industrial, and mineral products. As a result, the country is less affected by world price changes of certain products.

Visitors to southern African cities often notice the many small businesses operating on street corners and empty lots. Some even operate in bus stations. Women sell fruit and cooked food from roadside stands. Small children sell souvenirs to tourists. Men use portable equipment to fix cars along the roadside. These businesses are part of the **informal sector** of the region's economy. This sector is made up of people who do not work for formal businesses. These people may not have set hours, employment benefits, or even contracts. In addition, their income is not taxed. The informal sector includes self-employed people and small family-owned businesses. People in such businesses usually work long hours, and their incomes are generally low. This is the only way that many poor people, particularly those in cities, can make a living.

Parts of southern Africa have become popular tourist destinations. On the mainland, wildlife attracts thousands of visitors each year. Tourists travel to the wild-game parks of Botswana, Namibia, South Africa, and Zimbabwe to see African animals in their natural habitats. The tropical island countries of Comoros, Mauritius, and Seychelles attract visitors to their beaches and coral reefs. Partly as a result, the average income of the people of Mauritius is now among the highest in the region.

✓ **READING CHECK:** **Human Systems** In what ways do small businesses characterize the economies of the region? *Many poor people rely on such businesses to make a living.*

Southern Africa's Major Imports (in selected countries)

Angola	machinery and electrical equipment, vehicles and spare parts, medicines, food, textiles, military goods
Comoros	rice and other foodstuffs, consumer goods, petroleum products, cement, transport equipment
Mozambique	machinery and equipment, mineral products, chemicals, metals, foodstuffs, textiles
Namibia	foodstuffs, petroleum products and fuel, machinery and equipment, chemicals
South Africa	machinery, foodstuffs and equipment, chemicals, petroleum products, scientific instruments
Swaziland	motor vehicles, machinery, transport equipment, foodstuffs, petroleum products, chemicals
Zambia	machinery, transportation equipment, fuels, petroleum products, electricity, fertilizer, foodstuffs, clothing

Source: Central Intelligence Agency, *The World Factbook 2003*

INTERPRETING THE CHART *Southern Africa is dependent on imports to meet many of its basic needs.* **What does this trade pattern suggest about the economic development of these countries?**

The clear waters of the Seychelles attract divers from around the world. Tourism accounts for more than half of the country's GDP.

Radios for Southern Africa Many people in southern Africa cannot afford telephones, televisions, or electricity. However, a radio invented in Britain and manufactured in South Africa may allow these people a chance to receive information on everything from health to weather to sports to elections. Operated by windup springs, these radios require no electricity. The manufacturer is working to cut costs so that it might sell the radios for a price accessible to many people in developing countries. Many radios are also donated to groups within southern Africa. The company that makes the radios sponsors research to develop self-powered versions of other items lacking in the developing countries of southern Africa.

DISCUSSION: Lead a class discussion about the importance of mass media in American society. Ask students how their lives might be different without access to radios, telephones, televisions, or the Internet.

CHART ANSWER

They are developing countries with low levels of industrialization.

ALL LEVELS: Organize the class into small groups and assign each group a city from southern Africa. Have students look through books and magazines and on the Internet to find pictures of their assigned cities. Then have students use these images to write brief reports about what they think life is like in those cities. Encourage groups to share their reports with the class. **COOPERATIVE LEARNING**

Teach Objective 3

LEVEL 1: Pair students and have each pair develop an agenda for a conference about the future of southern Africa. The agendas should address challenges facing the region as well as opportunities for the future. **ENGLISH LANGUAGE LEARNERS, COOPERATIVE LEARNING**

LEVELS 2 AND 3: Have each student draft a campaign speech for a presidential candidate in one of the countries of southern Africa. Speeches should address challenges facing the country, possible solutions to those challenges, and the country's prospects for the future. Have volunteers read their speeches aloud to the class.

Linking Past to Present

Women's Rights in Zimbabwe
Zimbabwe's 1980 constitution prohibited discrimination against any citizen, regardless of gender. In 1982 a federal law was passed stating that, despite a southern African custom, women over 18 years old were no longer to be considered minors before the law. However, a 1999 decision passed by Zimbabwe's Supreme Court overturned this ruling. The court ruled that laws cannot supersede custom, and therefore women, regardless of their true age, should be treated as teenagers in all legal matters.

This decision stemmed from a suit filed by a Zimbabwean widow against her brother-in-law who evicted her from the house she had inherited from her father. The widow sued, believing that Zimbabwe's laws guaranteed her right to live there. The 5-to-0 decision against her spurred protests around the world.

internet connect

GO TO: go.hrw.com
KEYWORD: SW3 CH24
FOR: Web sites about southern African women

MAP ANSWER
increased trade between countries, easier movement of products to market

---

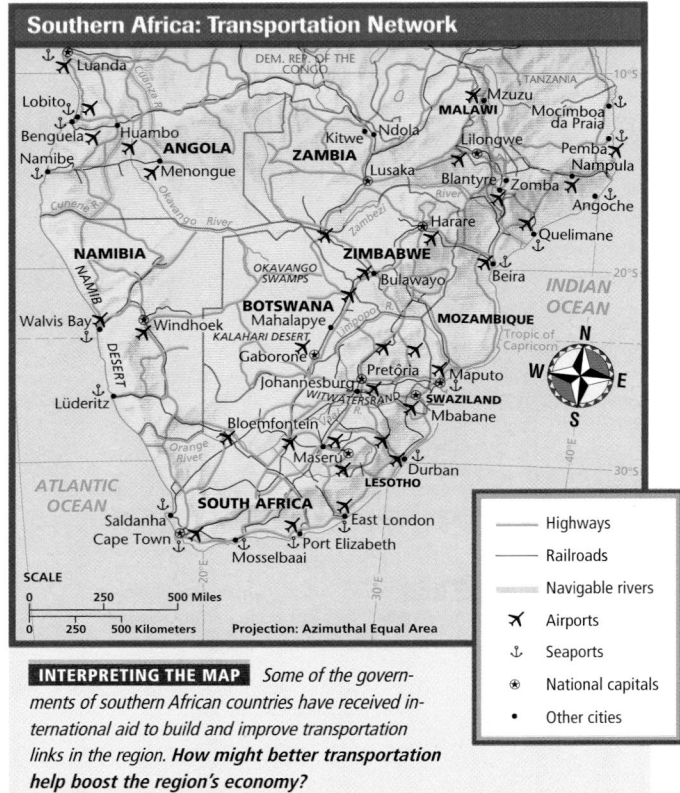

Southern Africa: Transportation Network

— Highways
— Railroads
— Navigable rivers
✈ Airports
⚓ Seaports
⊛ National capitals
• Other cities

SCALE
0   250   500 Miles
0   250   500 Kilometers
Projection: Azimuthal Equal Area

INTERPRETING THE MAP *Some of the governments of southern African countries have received international aid to build and improve transportation links in the region.* **How might better transportation help boost the region's economy?**

The township of Soweto is home to people from South Africa's many different indigenous groups.

Urban Environments

Southern African cities are places of great contrasts. In the suburbs, wealthy businesspeople, foreigners, and government officials may live in large, comfortable houses. Downtown areas have tall buildings and well-stocked stores. In the poorer areas, many people are crowded into small homes. Others live in shanties—rough houses made of scrap wood, metal sheeting, or mud bricks. Although these shantytowns may look chaotic to outsiders, many are well-organized communities. In some shantytowns, elected councils run schools. They also set rules and present residents' concerns to the government.

Greater Johannesburg is the largest urban area in southern Africa. It is home to some 4 million people and is in the center of South Africa's industrial heartland.

During the apartheid era, South Africa created residential areas for nonwhite workers. These settlements, called townships, kept the nonwhites separate from the whites. Soweto, which lies a few miles from Johannesburg, is the largest of these settlements. Soweto's name comes from the first letters of the name *South Western Townships.* It covers some 40 square miles (100 sq km). Soweto has a few grand homes. However, most people live in small houses, shacks, and apartments built as dormitories for migrant workers. Soweto was at the forefront of the struggle against apartheid in South Africa. Many of its residents struggled against the government for the equality of all South Africans.

✓ **READING CHECK:** (*Human Systems*) How did access to nearby Johannesburg contribute to the creation and growth of Soweto? It was created as a residential area for nonwhite workers in Johannesburg.

Close

Call on students to name some of the challenges facing southern Africa. Write these issues on the chalkboard as students name them. Then take a classroom poll to see which issue students think is the most critical. Discuss actions that have been taken to address that issue.

Review and Assess

Have the students complete the **Section Review**. Then have students complete **Daily Quiz 24.3**.

Reteach

Have students complete the **Main Idea Activity for English Language Learners and Special-Needs Students 24.3**. Then have students answer each of the Read to Discover questions from the section opener. Encourage students to use the terms from the Define and Locate lists in their answers. **ENGLISH LANGUAGE LEARNERS**

Extend

Have interested students conduct research on the development of South Africa's mining industry. Encourage students to note how that industry affected African and European migration and to share their findings with the class. **BLOCK SCHEDULING**

Challenges

Poverty is the most serious problem facing southern Africa. Many people cannot afford to eat a balanced diet. As a result, they are more likely to get sick. Young children are particularly vulnerable. Moreover, many of southern Africa's poor are unemployed.

The region's cities are growing rapidly, mainly because people migrate from rural areas looking for work. The region's high birthrates also contribute to the rapid growth of cities and lead to a lack of suitable housing. In addition to housing shortages, the crowded cities face serious environmental problems. Smog from cars' exhaust fumes and smoke from coal and wood fires add to urban pollution problems.

Other environmental threats include the droughts and floods that often strike the region. In 1999, for example, major floods in Mozambique left some 200,000 people stranded. Even worse flooding in 2000 displaced more than 1 million people and devastated the country's economy. In addition, Madagascar's rain forests are being cut down, and soil erosion is increasing. (See Geography for Life: Biodiversity in Madagascar.)

Disease is also a major problem in southern Africa. In some countries, more than a quarter of the population is infected with HIV, the virus that causes AIDS. For example, in Botswana some 36 percent of adults between the ages of 15 and 49 are infected with HIV. Because of HIV/AIDS, the average life expectancy of southern Africans is falling. Southern African governments are trying to educate their people about the disease in the hope of slowing its spread. (See Case Study: The Geography of Disease.) The region's governments have also called on Western drug companies to make expensive medicines more affordable for southern Africans.

✓ **READING CHECK:** *Human Systems* What are the main challenges facing southern Africa? How are they interrelated?

poverty, poor diet, unemployment, overcrowded and polluted cities, droughts and floods, environmental damage, disease; poverty can lead to poor diets, unemployment can cause poverty, overcrowded cities can compound pollution

INTERPRETING THE VISUAL RECORD *The populations of the region's cities are growing rapidly. For example, Antananarivo, Madagascar, grew from a population of about 800,000 in 1990 to 1.5 million in 2000.* **How might increased crowding compound the region's existing problems?**

Section 3 Review

Define informal sector

Working with Sketch Maps On the map you created in Section 2, label Cabinda and Soweto. Why were townships like Soweto created?

Reading for the Main Idea

1. *Places and Regions* What makes South Africa the most economically developed country in the region?

2. *Human Systems* What are some of the main cash crops grown on commercial farms in southern Africa?

Critical Thinking

3. **Finding the Main Idea** Why might governments in the region want to pursue policies that encourage the development of a wide range of resources?

4. **Analyzing Information** How might the countries of southern Africa protect and develop their tourist industries?

Organizing What You Know

5. Copy the following graphic organizer. Use it to identify the causes of some challenges facing countries in southern Africa.

Challenge	Cause
Rapid urban growth	
Poor nutrition	
Pollution	
Environmental damage	

Section 3 Review Answers

Define For definition, see: informal sector, p. 543

Working with Sketch Maps Maps will vary, but listed places should be labeled in their approximate locations. Townships were created to separate whites from nonwhites.

Reading for the Main Idea
1. diverse economy, including agriculture, manufacturing, mining

2. tobacco, wheat, corn, fruits, coffee, and vanilla beans

Critical Thinking
3. to protect against problems caused by overdependence on just one or a few resources; limit damage from price fluctuations for individual resources (NGS 16)

4. Possible answers: work to ensure political stability, protect the wildlife and natural environment (NGS 14)

Organizing What You Know
5. Rapid urban growth—immigration from rural areas; Poor nutrition—poverty; Pollution—crowded, growing cities; Environmental damage—deforestation, soil erosion

VISUAL RECORD ANSWER

higher unemployment, poverty, malnutrition, increased spread of disease

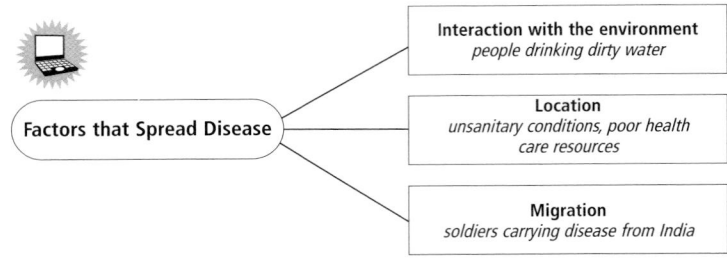
CASE STUDY

The Geography of Disease

Environment and Society   One constant throughout human history has been the threat of disease. Many societies have suffered from epidemics—the outbreak and rapid spread of a disease. Medical geographers and other health experts track the diffusion of such epidemics.

A number of factors influence the diffusion of disease. One factor is the interaction of humans with their environment. For example, consider the creation of Lake Volta in the early 1960s. The West African country of Ghana created the lake by building a huge dam across the Volta River. This dam produces hydroelectricity, and the lake provides irrigation water for crops. However, many people who settled along the lakeshore became infected with a deadly parasite. Experts tracked the source of the parasite to snails that lived along the shore.

Location is another important factor in the spread of disease. Certain climates, for example, may encourage the growth of infectious microbes and the insects that carry them. The unsanitary living conditions often found in developing countries also make it easier for diseases to spread. Worse still, poor countries often do not have the health-care resources to fight epidemics. Another factor in the spread of disease is migration. War refugees and other migrants can carry infections with them.

Medical geographers see some of these factors at work in Africa today. Two of the most difficult health threats there are malaria and Acquired Immune Deficiency Syndrome, or AIDS.

Malaria

Malaria is not a new disease. Researchers have uncovered records of ancient infections as far back as the 400s B.C. In fact, some researchers believe a malaria epidemic may have played a role in the fall of the Roman Empire. As early as the late 1400s, Europeans may have unknowingly carried the disease to the Americas.

Location and climate play important roles in the spread of malaria. Much of Africa lies in the tropics. Mosquitoes and other insects thrive in the humid tropical and subtropical climates there. One kind of mosquito that lives there carries the parasite that causes malaria. These mosquitoes are also found in

Malaria around the World, 2000

SCALE
0   1,500   3,000 Miles

0   1,500 3,000 Kilometers
Projection: Robinson
Scale is accurate at equator

Malaria-free areas

Malaria-transmission areas

Areas where malaria has been largely eliminated

INTERPRETING THE MAP

*More cases of malaria are reported each year in Africa than anywhere else in the world. Yet, as this map shows, malaria is virtually unknown in the northern and far southern parts of the continent. **What does this map suggest about the habitat of the mosquito that spreads malaria?***

Drawing Conclusions

Tell students that Dr. Snow's identification of the source of London's cholera outbreak hinged on a map. He plotted the locations of cholera deaths for several days on a city map. Ask students what pattern they think his map showed. *(The deaths centered around the contaminated pump. Few deaths occurred away from it.)* How might similar techniques be used today? *(Medical geographers and doctors can still use maps to identify general causes of disease.)* As an example, direct student attention to the malaria map on the previous page. Discuss how medical geographers could use it to draw conclusions about the disease's cause.

Going Further: Thinking Critically

Have students contact public health departments for information on allergies in the United States. Then have them answer these questions:

- Where are allergy cases common? Can any connection be made between the cases and natural features, such as types of vegetation?
- Do people in areas where air pollution is a problem suffer more from allergies? *(Students should provide evidence to support their conclusions.)*

Then ask students what actions should be taken next if mapping allergy sufferers and allergens—substances that cause allergic reactions—reveals clear connections. *(Students might provide small- or large-scale solutions to allergy outbreaks.)*

other tropical and subtropical regions around the world. (See the map.)

Experts believe there are 300 to 500 million new malaria cases worldwide each year, mostly in Africa. The annual death toll from the disease in Africa alone may be about 1 million. In fact, as many as 30 percent of all hospital admissions in Africa may be for malaria. Patients who survive can develop immunity to the disease. In the meantime, however, workers lose time at their jobs, and children miss school, often for a week or more. The death of a working family member can cause even more financial problems.

The economic costs of the disease are great. Some experts point to the benefits that might have resulted if malaria had been eliminated 35 years ago. They say that the total GDP of Africa south of the Sahara might be more than 30 percent greater than it is today.

AIDS

The movement of people has been an important factor in the diffusion of HIV, the virus that causes AIDS. HIV may infect more than 40 million people around the world today. The vast majority of those infections have occurred in Africa. (See the graph.)

Researchers still debate the origin of AIDS. However, among the first identified cases of the disease were those around the lake district of the Great Rift Valley in eastern Africa. Many researchers believe HIV probably first spread from there along trade and transportation routes. Over time, modern means of transportation, particularly air travel, helped spread HIV around the world.

AIDS has hit eastern and southern Africa particularly hard. In South Africa, for example, one in five adults is infected with HIV. In Botswana the figure is more than one in three. As a result, life expectancy has fallen dramatically. As with malaria, the economic costs of AIDS are staggering. These costs could multiply with the rapid spread of the disease in other regions, particularly Asia.

Finding Solutions

The costs of treating malaria, AIDS, and other deadly diseases are often more than developing countries can pay. In addition, some diseases, such as AIDS, have no cure. Sometimes drugs may extend life, but they are very expensive. Some experts say that richer countries should play a bigger role in providing resources for the battle against these diseases. Education and prevention programs also help.

One success story is the battle against smallpox. As recently as 1967, smallpox was a dreaded disease that caused as many as 2 million deaths each year. However, world health officials started a massive prevention program. This program included widespread vaccination and other efforts. The last natural case of smallpox occurred in Somalia in 1977. In coming years medical geographers and other experts hope to add other diseases, including malaria and AIDS, to the list of success stories.

HIV/AIDS around the World

Percentage of world HIV/AIDS cases

4.7% 2.3%
8.3%
13.2%
71.5%

- Africa south of the Sahara
- Middle and South America
- North America
- South and Southeast Asia
- Other

Source: UNAIDS/World Health Organization, 2003

> ## Applying What You Know

1. **Problem Solving** What are some factors that contribute to the diffusion of disease today? How do you think international efforts might help slow or stop the spread of diseases in developing regions such as Africa?

2. **Analyzing** How have AIDS and malaria affected the economic development of Africa south of the Sahara? What demographic effects has AIDS had on southern Africa?

CHAPTER 24 Review Answers

Building Vocabulary For definitions, see: escarpment, p. 534; biodiversity, p. 534; veld, p. 535; Afrikaners, p. 538; apartheid, p. 539; sanctions, p. 539; African National Congress, p. 539; informal sector, p. 543

Locating Key Places

A. Cabinda
B. Cape of Good Hope
C. Orange River
D. Luanda
E. Harare
F. Johannesburg
G. Zambezi River
H. Drakensberg
I. Namib Desert

Understanding the Main Ideas

1. cold water current off the west coast, rain-shadow effect caused by the Drakensberg
2. Dutch, French, and German settlers, then British; clashed, which led to the Boer War
3. set of laws in South Africa that denied political rights to nonwhite South Africans and made races live in separate areas; ended in 1990s
4. beaches and reefs; wild animals in their natural habitats
5. spread of HIV/AIDS; likely will slow growth

TECHNOLOGY

► Chapter 24 Test Generator
  (on the One-Stop Planner)
► Global Skill Builder CD–ROM
► HRW Go site

REINFORCEMENT, REVIEW, AND ASSESSMENT

► Chapter 24 Review, pp. 548–49
► Chapter 24 Tutorial for Students, Parents, Mentors, and Peers
► Chapter 24 Test (form A or B)
► Alternative Assessment Handbook

► Chapter 24 Test for English Language Learners and Special-Needs Students
► Unit 7 Test
► Unit 7 Test for English Language Learners and Special-Needs Students

Assess

Have students complete a Chapter 24 Test.

Reteach

Draw an outline map of the region on butcher paper. Organize students into groups and assign each group one of the region's countries. Have groups draw or label each country's physical features, economic activities, and cultural aspects on the map. Display and discuss the map. **ENGLISH LANGUAGE LEARNERS, COOPERATIVE LEARNING**

CHAPTER 24 Review Answers

Thinking Critically

1. Possible answers: the need for resupply ports, a wish to exploit the natural resources and agricultural products of the region, the use of Africans as laborers and consumers of European products, and the prestige value of colonies (NGS 8)

2. Answers will vary but should mention problems caused by overcrowding, pollution, and migration from rural to urban areas. Political instability contributes to economic problems. Lack of economic choices can push people into economic activities that harm the environment. (NGS 4)

3. Possible answers: The informal sector can provide income and independence in countries where women and minorities might face economic discrimination. Also, the businesses typical of the informal sector can be particularly helpful in societies where women traditionally have not worked outside the home. (NGS 10)

Using the Geographer's Tools

1. Botswana; highways and railroads

2. Maps will vary but should accurately identify the regions listed.

3. Wetter climate regions—like the eastern coastal areas—generally have more commercial farming. Drier climate regions have nomadic herding, livestock raising, or little economic activity.

CHAPTER 24 Review

Building Vocabulary

On a separate sheet of paper, explain the following terms by using them correctly in sentences.

escarpment	apartheid
biodiversity	sanctions
veld	African National Congress
Afrikaners	informal sector

Locating Key Places

On a separate sheet of paper, match the letters on the map with their correct labels.

Drakensberg	Namib Desert	Johannesburg
Orange River	Cape of Good Hope	Harare
Zambezi River	Luanda	Cabinda

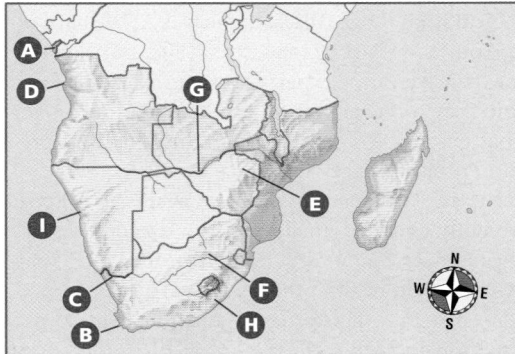

Understanding the Main Ideas

Section 1

1. *Physical Systems* What are two reasons the western part of southern Africa is drier than eastern parts?

Section 2

2. *Places and Regions* Which Europeans settled in what is now South Africa? How did these peoples interact with each other?

3. *Human Systems* What was the apartheid system? When did it end?

Section 3

4. *Environment and Society* What are some features that attract tourists to southern Africa?

5. *Human Systems* Why has life expectancy in southern Africa dropped in recent years? How do you think such changes might affect future population growth in the region?

Thinking Critically

1. Analyzing Why do you think European countries were so intent on colonizing all of southern Africa?

2. Comparing How do you think life in large cities of southern Africa compares to life in large cities in other developing countries? What political, economic, social, and environmental factors might account for similarities and differences?

3. Comparing In what world regions and kinds of societies do you think informal economic activities provide the most opportunities for women and religious minorities?

Using the Geographer's Tools

1. Analyzing Maps Examine the map of transportation routes in Section 3. Which country seems to have the largest area not served by major transportation links? How do the region's land-locked countries seem to get their goods to seaports?

2. Creating Maps Use the unit atlas to create a map of Africa. Label each country. Then use a different color to shade the regions of North Africa, West and Central Africa, East Africa, and southern Africa.

3. Creating Maps Use the climate and land use maps to create a series of maps on tissue paper or other transparent material. One map may show arid and semiarid climate regions, while the other shows wetter climate regions. Then create maps for each type of land use common in the region. Using your maps as if they were layers in a geographic information system, make generalizations about the location of climate regions and particular economic activities.

Writing about Geography

Imagine that you have moved from a small subsistence farm to a big city to find work. Will you work in industry or the informal sector? If you find work, how will your standard of living change? Write a letter to your family in which you compare your old way of life with life in the city.

SKILL BUILDING

Geography for Life

Summarizing Geographic Information
Human Systems Create an idea web in which you analyze South Africa's political, economic, social, and cultural characteristics. Note historical factors that are important in your analysis.

Tell students to imagine that they are legislators in a southern African country preparing to meet with the country's president to discuss an issue facing southern Africa. Instruct each student to select which country and issue he or she would like to represent and prepare an outline of the presentation he or she would give. Suggest that students prepare visual aids to accompany their presentations. Provide research materials for students to gather more information about their chosen topics. Place the outlines and visual aids in student portfolios.

Food Festival

To South Africans "mealie meal" is cornmeal and "green mealies" is raw corn. Green mealie bread can be baked or the batter can be dropped by spoonfuls into hot fat. The dough can also be refrigerated, sliced, and fried. To make an Americanized version of the bread, combine 2 c. biscuit mix, 1 c. creamed corn, and 1 tbs. sugar. Add one beaten egg and ½ c. whole milk. Stir just until combined. Grease a 9-inch square baking pan. Place the mealie dough in the pan and coat with melted butter. Bake at 400°F for 20 minutes.

Building Social Studies Skills

Angola: Population Density

ATLANTIC OCEAN

Luanda

DEM. REP. OF THE CONGO

ZAMBIA

NAMIBIA

BOTSWANA

POPULATION DENSITY

Persons per sq. mile	Persons per sq km
260	100
130	50
25	10
3	1
0	0

● Metropolitan areas with more than 1 million inhabitants

Interpreting Maps

Study the population density map above. Then answer the questions that follow.

1. Based on the map, which of the following statements is accurate?
   a. Luanda is located in one of Angola's least densely populated regions.
   b. Luanda is located in an area with between 25 and 130 people per square mile.
   c. Land along the coast is uninhabited.
   d. Angola has no cities with more than 1 million people.

2. Which areas of Angola are most densely populated?

Using Language

The following passage contains mistakes in grammar, punctuation, and usage. Read the passage and then answer the following questions. Mark your answers on a separate sheet of paper.

   (1) Botswana is a large, landlocked and semiarid country. (2) Thanks to mineral resources and stable political conditions. (3) Botswana is one of Africa's success stories. (4) Cattle ranching and mining of copper and diamonds is the principal economic activities.

3. Which sentence contains an error in subject-verb agreement?
   a. 1
   b. 2
   c. 3
   d. 4

4. Combine word groups 2 and 3 to form one correct sentence.

Alternative Assessment

PORTFOLIO ACTIVITY

Learning about Your Local Geography

Individual Project: Research
You have read how mineral resources contribute to southern African economies. What mineral resources are mined in or near your area? Conduct research to learn where each resource is found, how it is processed, and how it is used. Create maps or graphic organizers to organize and present your findings.

internet connect

Internet Activity: go.hrw.com
KEYWORD: SW3 GT24

Choose a topic on southern Africa to:
• learn about the Namib Desert.
• investigate the history and legacy of apartheid.
• create a brochure of the animal and plant life on a South African safari.

Writing

Student letters should demonstrate an understanding of differences in economic systems. Use Rubric 25, Personal Letters, to evaluate student work.

Geography for Life

Idea webs will vary but should accurately reflect the political, economic, social, and cultural characteristics of South Africa. Use Rubric 13, Graphic Organizers, to evaluate student work.

Social Studies Skills

1. b
2. Possible answer: Angola is most densely populated in central and coastal areas, such as around Luanda. A small area on the southern border is also among the country's most densely populated areas.
3. d
4. Sample answer: Thanks to mineral resources and stable political conditions, Botswana is one of Africa's success stories.

PORTFOLIO ACTIVITY

Student projects will vary based on the minerals available in your community. Maps or graphic organizers should accurately reflect student findings. Use Rubrics 20, Map Creation, and 30, Research, to evaluate student work.

Going Further: Thinking Critically

Tell students that bias is a common problem in writing questionnaires and conducting interviews. Biased or leading questions can affect the results of a questioner's research, making the results of the research unusable. In many cases, the people who create questionnaires do not intentionally write leading questions, but their choice of words or phrases may result in biased questions anyway. For this reason researchers must be very careful in preparing questionnaires or other information-gathering forms.

Have students collect questionnaires from various organizations in your community. Political, environmental, and social groups often prepare them to gather opinions about local issues. If possible, have students gather questionnaires from organizations that represent opposing sides of a single issue. Then have students examine the questionnaires for examples of bias. Ask students if the questions are written well or if they lead the reader to favor a particular response. Remind students that questions should be also answerable by everyone, not just by people with a particular background. If students determine that a particular questionnaire is biased, ask them how it could be rewritten to eliminate the problem. Then lead a discussion about the types of geographic information that might be obtained through questionnaires.

Geography
Skill-Building Workshop

WORKSHOP

Using Questionnaires and Field Interviews

A common instrument for gathering research data is the questionnaire. Questionnaires are documents that ask individuals to provide certain information about themselves. They have a variety of uses. For example, they can provide feedback from customers, identify special needs people in an area might have, or collect statistical information for a census.

Questionnaires can be sent thousands of miles away to people the researcher may never even see. However, researchers can also conduct field interviews, in which they talk to people face-to-face. A field researcher might want to know, for example, how people in an area use local natural resources and how government policies have affected their use of those resources. Questionnaires would not let the researcher probe for additional details he or she might need to know.

Developing the Skill Questionnaires and field interviews can be used to answer geographic questions and infer geographic relationships. For example, the U.S. Census Bureau uses questionnaires to track population growth in the United States. Answers from the questionnaires also help researchers learn how population is distributed across the country. They also reveal age, race, income, and other characteristics of a population. On this page you can see part of a questionnaire that the Census Bureau used in 2000.

Questionnaires can ask many types of questions. Some census questions ask respondents to write answers, such as names. Other questions ask respondents to check boxes next to the answers given on the form. Still other questions might match numbers or letters to answers. For example, if there are four possible answers, the answers may be coded 1, 2, 3, and 4. Answers coded in this way can help researchers create databases from the information they gather.

Guidelines for Questionnaires Focus on the information you need to learn. Then carefully choose the population you want to survey. For example, you might want to survey people in certain age groups or in certain areas to learn about living conditions in a country. Usually you can survey only a certain percentage of your target population. Often researchers will develop scientific random samples of a population. However, if that is not possible, try to question a sample that is representative of the population. You can use the following checklist to create a questionnaire:

✓ Decide what information you want to gather. Make sure your questions are designed to help you reach a specific objective.

2000 U.S. Census Questionnaire (excerpt)

4. What is this person's age and what is this person's date of birth? *Print numbers in boxes.*

Age on April 1, 2000   Month   Day   Year of Birth

→ **NOTE: Please answer BOTH Questions 5 and 6.**

5. Is this person Spanish/Hispanic/Latino? *Mark ☒ the "No" box if not Spanish/Hispanic/Latino.*

☐ **No,** not Spanish/Hispanic/Latino   ☐ Yes, Puerto Rican
☐ Yes, Mexican, Mexican Am., Chicano   ☐ Yes, Cuban
☐ Yes, other Spanish/Hispanic/Latino—*Print group.*

6. What is this person's race? *Mark ☒ one or more races to indicate what this person considers himself/herself to be.*

☐ White
☐ Black, African Am., or Negro
☐ American Indian or Alaska Native—*Print name of enrolled or principal tribe.*

☐ Asian Indian   ☐ Japanese   ☐ Native Hawaiian
☐ Chinese   ☐ Korean   ☐ Guamanian or Chamorro
☐ Filipino   ☐ Vietnamese   ☐ Samoan
☐ Other Asian—*Print race.*   ☐ Other Pacific Islander—*Print race.*

Going Further: Thinking Critically

Ask students to think about interviews they have seen conducted on television. Ask them whether interviewers appear to follow a set list of questions *(yes, in most cases)*. Then ask whether the interview appeared to ask only questions from this list *(probably not, because most interviewers ask follow-up questions in response to the answers they receive)*. Point out that these secondary questions, which frequently are not written in advance, often gather the most useful information. Though he or she may not know exactly what questions to ask in advance, an interviewer should be prepared to respond to the subject's replies.

Have each student write a list of questions that he or she might ask in an interview with a local public official about an issue that affects the community. Then have students write possible subjects for secondary questions that they might pursue, depending on the official's answers. Students should prepare secondary question topics for several possible responses to their primary questions. Call on volunteers to suggest what kind of information they would hope to learn with each line of questioning.

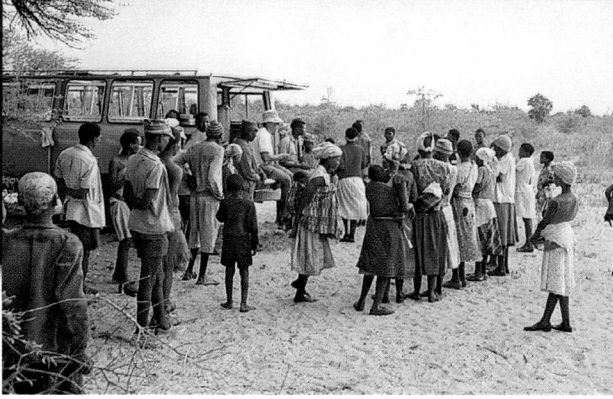

✓ Your questions must be very clear. Use a conversational style in writing your questions, avoiding technical language. Write out the question, and then put yourself in the position of the respondent. Is the question clear? Would you know how to respond to it?

✓ Ask only those questions that will provide the information you need.

✓ Structure the survey so that the questions follow a logical order.

✓ Whenever possible, use questions that require respondents to choose a specific answer. Examples include multiple-choice questions, yes-or-no questions, and questions that ask respondents to rank items in order. Respondents will better understand the purpose of your questions, and the time it takes to complete the questionnaire will be shortened.

✓ Avoid bias in writing questions. Make sure respondents of different backgrounds can answer your questions. Also, do not ask leading questions. That is, the questions should not encourage people to answer in a certain way.

✓ Test the questionnaire on a small group of people. Ask each person if he or she easily understands and can answer the questions.

✓ In order to organize and analyze the responses, you may need to code the questions. Assign a value—a number or letter—to each question's response. Then use the codes to create a database of questions and answers.

Guidelines for Field Interviews Before you begin conducting field interviews, you need to do some background research. Study your topic and the people or place you are researching. Then follow these guidelines to conduct a professional field interview:

✓ Plan a general outline for the interview and script some questions. Break the general topic into more specific parts. Then design focused questions that are not overly broad.

✓ Use a variety of questions, but avoid yes-or-no questions and questions that can be answered with a single word. Unlike with a questionnaire, you want people to explain their answers.

✓ Set up the interview in advance and send the list of questions you will ask. Doing so will help people better prepare for the interview and provide more-detailed answers.

✓ Ask for permission to record the interview.

✓ Keep questions brief and clear.

✓ Follow your plan and have a copy of the questions for your respondents in case they do not have their own copy.

✓ Be a good listener and try to determine the meaning of the responses. Ask follow-up questions that draw out additional details about previous answers.

✓ After the interview, submit a copy of the interview to the respondents. Ask respondents to check that their answers have been properly recorded. As a courtesy, you might send your final report to your respondents for their review.

Practicing the Skill

Work with a group of three or four partners to research either the ways people in your community try to conserve natural resources at home or attitudes toward local traffic issues and transportation networks. Use a questionnaire and field interviews to gather information about your topic. Create questions with other group members, then assign certain members to manage the distribution of questionnaires and conduct field interviews. Other members should organize the questions and answers into databases. Finally, analyze the information you gather and develop a report for your class.

551

Using the Illustration

Direct students' attention to the photograph of the Taj Mahal on this page. Ask students what they think of when they see images of the Taj Mahal. Discuss their responses. Tell students that it was built as a mausoleum on the banks of the Yamuna River in Agra, India, at one time the capital of the Mughal Empire. Shāh Jahān, the fifth Mughal emperor, commissioned the structure as a monument to his wife, Mumtāz Mahal, who had died in childbirth. The white marble tomb and part of the surrounding garden are seen here. The Taj Mahal complex also includes a sandstone gateway, a mosque, and a *jawab,* or guest house. Construction on the mausoleum began about 1632 and required 22 years to complete.

Ask students if they see any architectural similarities between the Taj Mahal and other structures they have seen pictured in this book. (*Students might note the similarities of the Taj Mahal's arches, domes, and minarets to other mosques they have seen.*) Point out that the complex represents a mixture of Indian, Persian, and Arabic building styles. Workers were brought to India from as far away as Persia, the Ottoman Empire, and Europe during its construction. Materials were also imported from around the world. The Taj Mahal incorporates Indian marble, Central Asian jade and crystal, Tibetan turquoise, Burmese amber, Afghan lapis lazuli, Egyptian chrysolite, and Indian Ocean coral, shells, and mother-of-pearl.

Unit Objectives

1. Describe how landforms, climates, and water resources affect the way people live in South Asia.

2. Identify the natural resources of South Asia and link them to the region's economic development.

3. Trace the history of South Asia and identify the interactions through time of the region's culture groups.

4. Analyze the social and environmental challenges facing the countries of South Asia.

5. Search the Internet to find geographic information.

6. Use a problem-solving process to identify and resolve geographic issues and challenges.

UNIT 8 South Asia

Taj Mahal, Agra, India

Chapter 25 India

Chapter 26 The Indian Perimeter

explores the geography and cultural diversity of the more than one billion people who live in the largest country in South Asia. It also describes the serious economic and environmental challenges facing India.

profiles the six countries that surround India: Pakistan, Nepal, Bhutan, Bangladesh, Sri Lanka, and the Maldives.

CONNECTING TO Literature

"HUNDRED QUESTIONS" from the Mahabharata

translated by R. K. Narayan

The "Hundred Questions" comes from Book 2 of the *Mahabharata*, an important Hindu religious text. In its present form, the *Mahabharata* is thought to have been composed sometime between the 300s B.C. and the A.D. 300s. Central to the *Mahabharata* is the Hindu concept of dharma or sacred duty. Performing these duties contributes to the universal order. In this excerpt a yaksha (forest divinity or nature spirit) challenges Yudhistira to answer questions. Yudhistira's brothers failed to heed the yaksha's challenge and died.

Yudhistira said humbly, "What god are you to have vanquished these invincible brothers of mine, gifted and endowed with inordinate strength and courage?" . . .

At this request he saw an immense figure materializing beside the lake, towering over the surroundings. "I am a yaksha. . . . If you wish to live, don't drink this water before you answer my questions." . . .

To . . . questions on renunciation, Yudhistira gave the answers: "Pride, if renounced, makes one agreeable; anger, if renounced, brings no regret; desire, if renounced, will make one rich; avarice, if renounced, brings one happiness. True tranquility is of the heart. . . . Mercy may be defined as wishing happiness to all creatures. . . . Ignorance is not knowing one's duties. . . . Wickedness consists in speaking ill of others." . . .

There were a hundred or more questions in all. Finally, the yaksha said, "Answer four more questions, and you may find your brothers—at least one of them—revived. . . . Who is really happy?"

"One who has scanty means but is free from debt; he is truly a happy man."

"What is the greatest wonder?"

"Day after day and hour after hour, people die, and corpses are carried along, yet the onlookers never realize that they are also to die one day, but think they will live for ever. This is the greatest wonder of the world."

"What is the Path?"

"The Path is what the great ones have trod. When one looks for it, one will not find it by study of scriptures or arguments, which are contradictory and conflicting." . . .

The yaksha said, "You have indeed pleased me with your humility and the judiciousness of your answers. Now let all your brothers rise up and join you."

Analyzing the Primary Source

1. **Summarizing** What does the yaksha tell Yudhistira he must do?

2. **Drawing Inferences** What do you think is meant by the Path? What features of other religions with which you are familiar might be similar to the Path?

In this unit, students will learn about the geographic and cultural contrasts of South Asia, one of the most densely populated regions on Earth.

The Indian Subcontinent is separated from the rest of Asia by the highest mountains in the world. The Himalayas drop to the fertile Gangetic Plain. Plateau and desert lands lie to the south and west. Climate patterns are greatly influenced by monsoon winds that bring wet and dry seasons to much of the region.

India dominates the subcontinent. With more than one billion people, it is the world's most populous democracy. It is a land of stunning contrasts—from nuclear technology and a strong computer industry to subsistence farming and homeless city-dwellers. Ancient customs sometimes clash with modern trends. Conflicts between India and its neighbors over territory increase political tensions.

The surrounding countries also face the difficult tasks of feeding their growing populations and developing their economies while protecting the environment.

Your Classroom Time Line

These are the major dates and time periods for this unit. Have students enter them on the time line you created earlier. You may want to watch for these dates as students progress through the unit.

c.* 2500 B.C.–1500 B.C. The Harappan civilization flourishes in the Indus Valley.

c. 1500 B.C. Aryans move into northern India from Central Asia.

c. 528 B.C. Siddhartha Gautama begins teaching the set of beliefs now called Buddhism.

c. 320 B.C. The Mauryan Empire is founded in India.

A.D. 320—c. 500 India flourishes under Gupta rulers.

800s The Bhote begin to migrate into Bhutan.

c. 1000 Turkic armies begin raiding Pakistan and northwestern India. Islam is introduced to the region.

Early 1200s A Muslim kingdom is established at Delhi.

1398 Timur conquers Delhi.

Late 1490s Europeans arrive in India and establish trade.

1520s Bābur invades India and founds the Mughal Empire.

1707 Mughal Emperor Aurangzeb dies, leaving the empire greatly weakened.

*c. stands for *circa* and means "about."

POLITICAL MAP ANSWERS

1. India
2. Bangladesh

CRITICAL THINKING

3. eastern Pakistan

UNIT 8 ATLAS

The World in Spatial Terms

South Asia:
Political

1. **Places and Regions** Which country is by far the largest in South Asia?
2. **Places and Regions** Compare this map to the physical map. Which country is almost surrounded by India and dominated by the Ganges Delta?

Critical Thinking

3. **Analyzing Information** Compare this map to the physical map. Which part of Pakistan would you expect to have the highest population density?

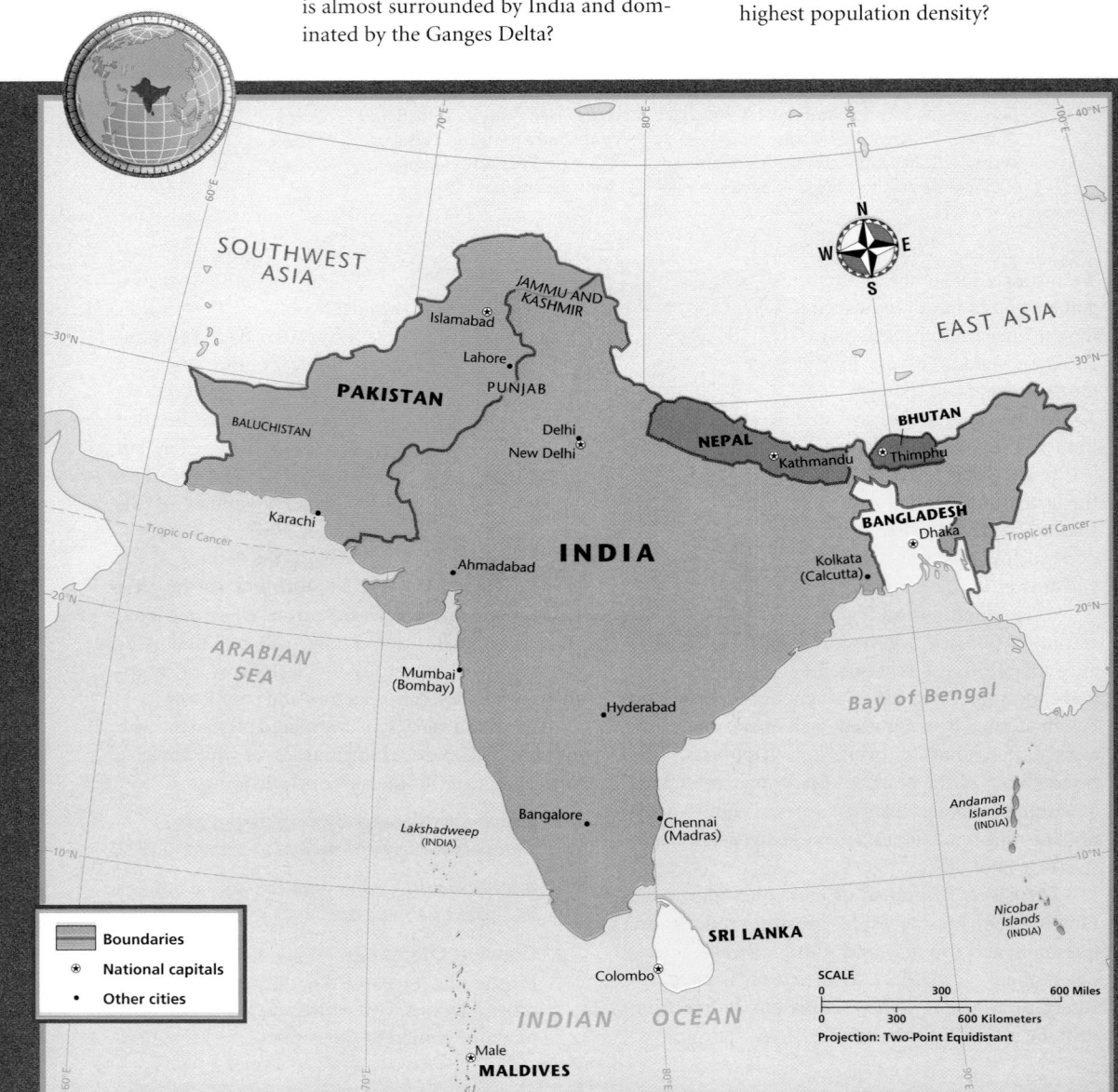

Boundaries
⊛ National capitals
• Other cities

Using the Political Map

Have students examine the **political map** on the previous page. Ask what is unusual about the eastern boundary of India. (*Students should note that a small part of the country connects with the rest of the country by a narrow strip of land.*) Call on a volunteer to suggest why the part of the regions known as Jammu and Kashmir is in Pakistan and part is in India. (*The region is claimed by both countries.*)

Using the Physical Map

Have students examine the **physical map** on this page to suggest India's three major landform regions (*the Himalayas, the Gangetic Plain, and the Deccan Plateau*). Then call on students to describe the courses of the three main rivers that flow from the Himalayas—the Indus, the Ganges, and the Brahmaputra.

Call on a volunteer to identify the two mountain ranges that are extensions of the Himalayas (*Hindu Kush, Karakoram Range*).

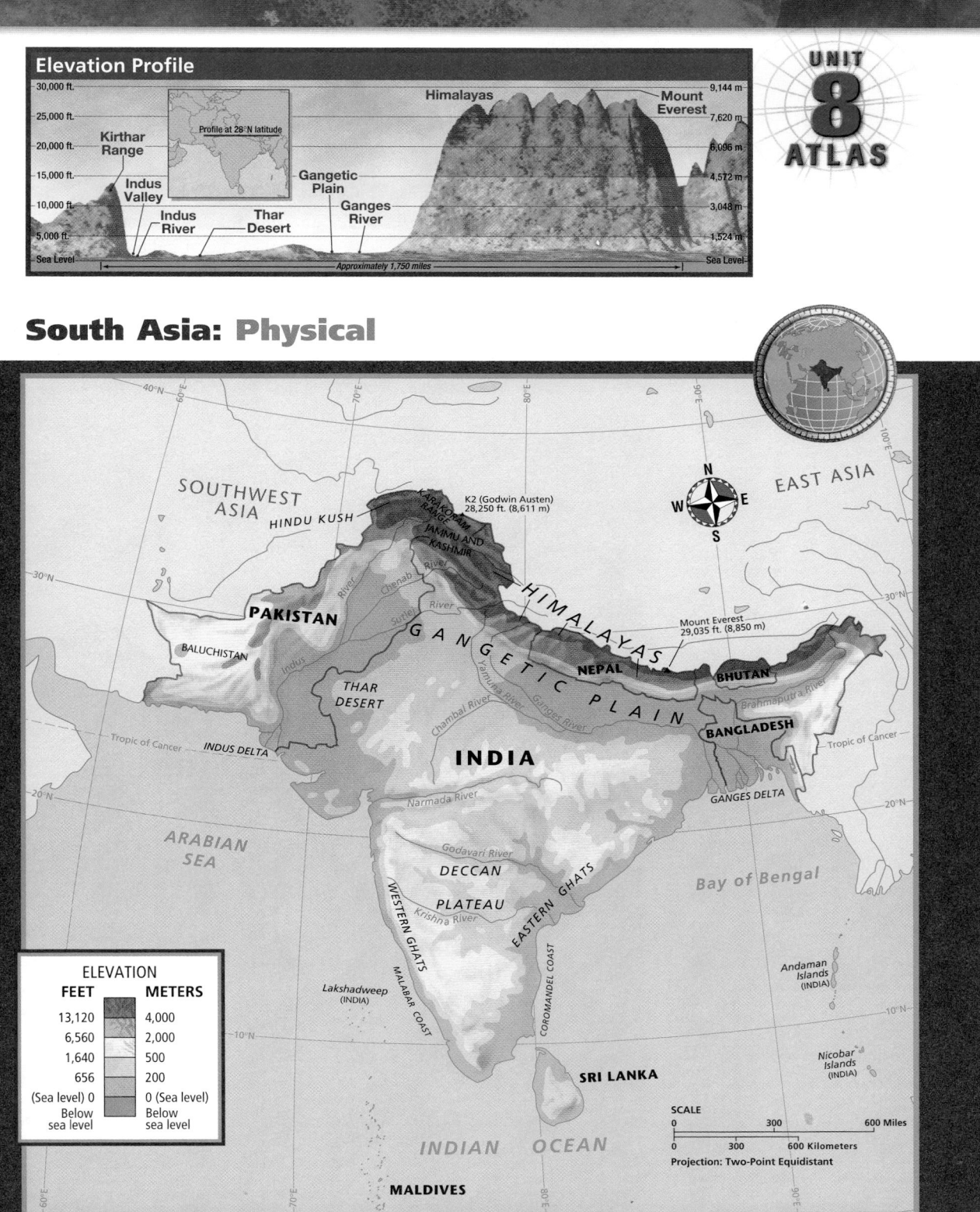

Elevation Profile

South Asia: Physical

UNIT 8 ATLAS

Your Classroom Time Line, *(continued)*

Mid–1700s British forces defeat the French and begin a period of British rule in India.

1802 Sri Lanka becomes a British colony.

Mid–1800s Tamil workers from India migrate to Sri Lanka.

1857 Indian sepoys rebel against the British.

1858 The British impose direct rule on India.

1885 Indian nationalists organize the Indian National Congress.

Early 1900s Indian demands for independence increase.

1920 Mohandas Gandhi begins a nonviolent independence movement in India.

1943 Famine strikes eastern India.

1947 The partition of India separates Pakistan from India.

1948 Mohandas Gandhi is assassinated.

1948 Sri Lanka becomes independent.

1949 Bhutan becomes independent.

1950s–60s Bangalore develops into a major scientific research center.

1951 A constitutional monarchy is established in Nepal.

1953 Norkey Tenzing and Sir Edmund Hillary reach the summit of Mount Everest.

1960s Bhutan's first cities are established.

1965 Maldives becomes independent.

1967–1978 The Green Revolution introduces new farming techniques to India.

1971 East Pakistan breaks from West Pakistan to form independent Bangladesh.

555

Using the Climate Map

Have students examine the **climate map** on this page and compare it to the **physical map** of South Asia. Ask the students the following questions: What is the dominant climate of Pakistan? *(arid)* Which country in this region has mostly humid tropical and humid subtropical climates? *(Bangladesh)* Which island country has tropical savanna and humid tropical climates? *(Sri Lanka)* What low mountain range appears to catch much of the rain from the wet monsoon? *(Western Ghats)* Which countries are mostly covered by highland climates? *(Bhutan, Nepal)* Which mountain range creates this climate? *(Himalayas)*

Your Classroom Time Line,
(continued)

1972 India and Pakistan divide Kashmir between the two countries.

Late 1970s Tensions between Hindu Tamil and Buddhist Sinhalese escalate in Sri Lanka.

Early 1980s The Grameen Bank is established in Bangladesh to provide small business loans.

1980s Sikhs in India call for the creation of an independent country, Khalistan, in Punjab.

1980s Many Indian software engineers move to the United States.

1989 Muslim Kashmiris clamor to make all of Kashmir part of Pakistan.

2000 India's population exceeds 1 billion.

CLIMATE MAP ANSWERS

1. the Deccan Plateau
2. highland climate; humid subtropical climate

CRITICAL THINKING

3. Possible answer: It probably brings large amounts of rainfall to the country and may cause flooding.

UNIT 8 ATLAS

South Asia:
Climate

1. *Places and Regions* Compare this map to the physical map. Which landform in southern India is dominated by a semi-arid climate?

2. *Places and Regions* Which climate type dominates Nepal and Bangladesh?

Critical Thinking

3. **Analyzing Information** Compare this map to the political and precipitation maps. How do you think the wet monsoon airflow affects life in Bangladesh?

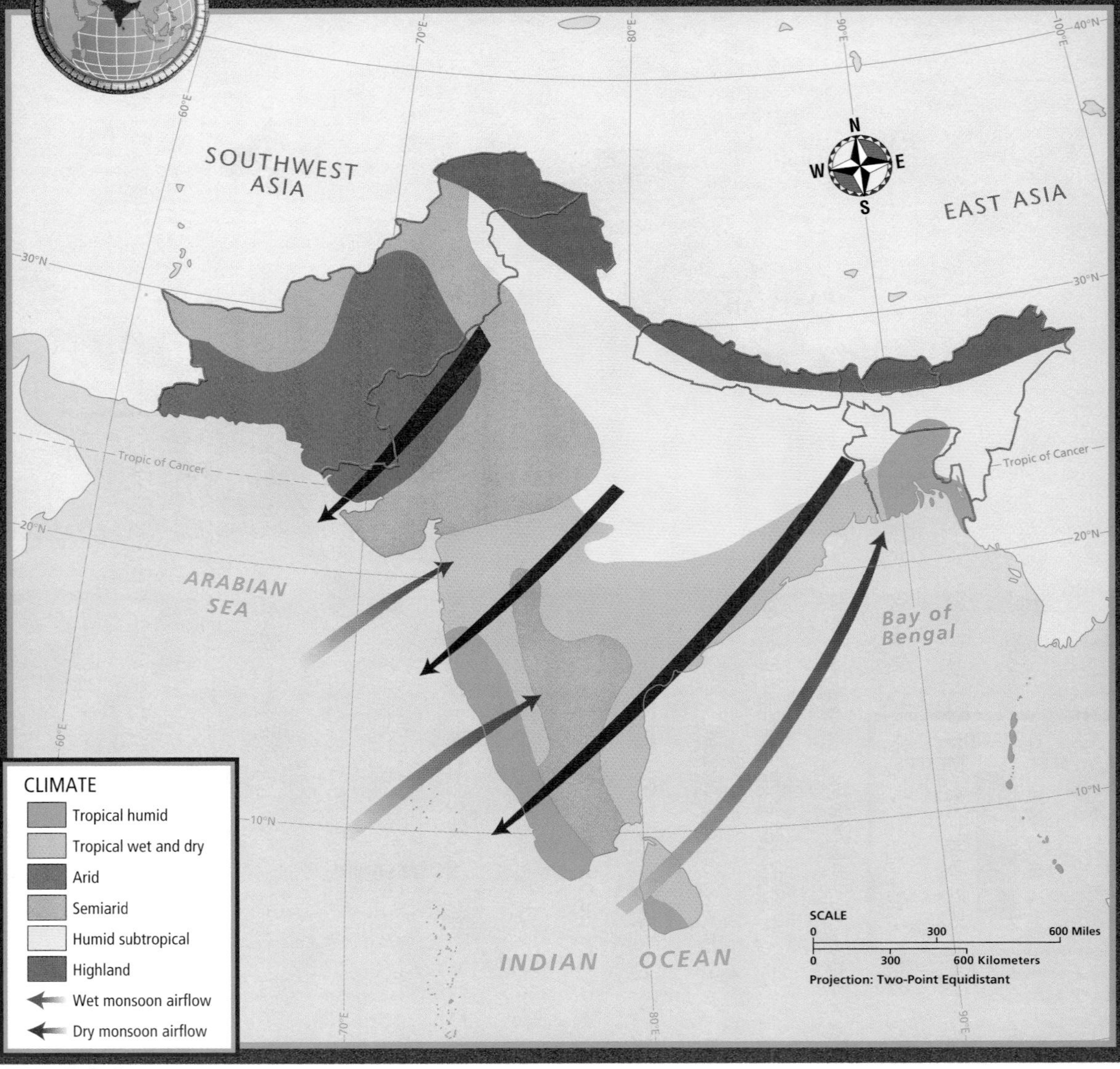

SOUTHWEST ASIA

EAST ASIA

Tropic of Cancer

ARABIAN SEA

Bay of Bengal

Tropic of Cancer

INDIAN OCEAN

CLIMATE
- Tropical humid
- Tropical wet and dry
- Arid
- Semiarid
- Humid subtropical
- Highland
- ← Wet monsoon airflow
- ← Dry monsoon airflow

SCALE
0   300   600 Miles
0   300   600 Kilometers
Projection: Two-Point Equidistant

556

Focus students' attention on the **precipitation map** on this page. Ask them to identify the areas of South Asia that receive the most rain each year (*southwestern India, Bangladesh*). Have students compare this map to the **physical** and **climate maps**. Then ask students to identify factors that lead to heavy precipitation in these areas. *(Possible answers: Mountain ranges create orographic effect; wet monsoons carry moisture into the area from the Indian Ocean and the Bay of Bengal.)* Then ask students to identify the driest country in the region *(Pakistan)*.

South Asia:
Precipitation

UNIT
8
ATLAS

1. (*Physical Systems*) Compare this map to the physical and political maps. Which mountains in southern India cause a rain-shadow effect?

2. (*Environment and Society*) Compare this map to the climate and political maps. How do you think the wet monsoon airflow affects the amount of precipitation in Bangladesh?

Critical Thinking

3. **Analyzing Information** Compare this map to the physical map. How do you think the Himalayas affect the distribution of precipitation in South Asia?

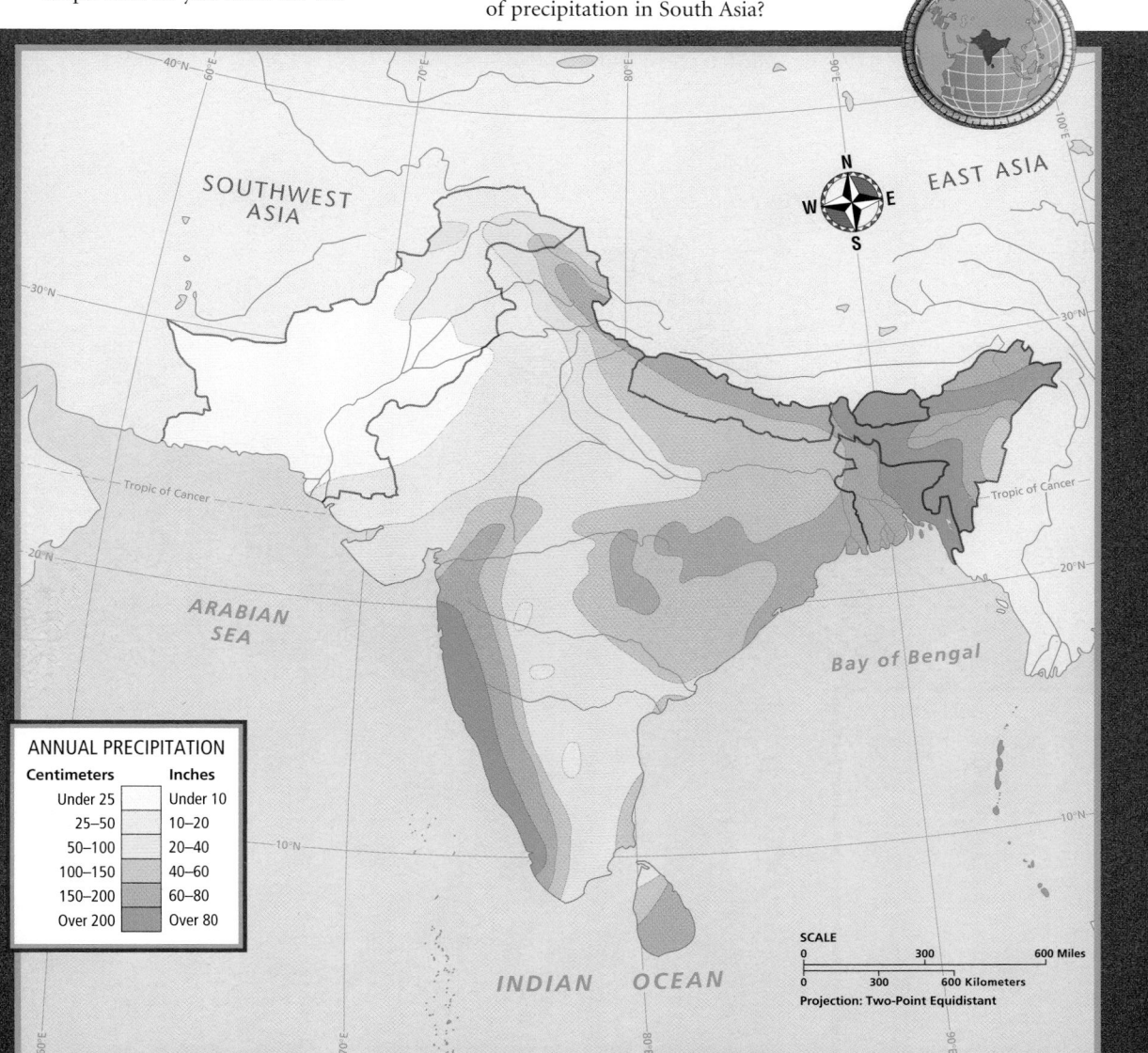

ANNUAL PRECIPITATION

Centimeters	Inches
Under 25	Under 10
25–50	10–20
50–100	20–40
100–150	40–60
150–200	60–80
Over 200	Over 80

SCALE
0   300   600 Miles
0   300   600 Kilometers
Projection: Two-Point Equidistant

557

Using the Population Map

Call students' attention to the **population map** on this page. Ask each student to write a general statement about population patterns in South Asia. *(Possible answer: South Asia is a densely populated region with many large cities.)*

Have students compare this map to the **physical map.** Call on a volunteer to identify the country with more than 520 people per square mile throughout most of its area *(Bangladesh).* Ask what desert area has more than 260 residents per square mile *(a narrow strip of land in southern Pakistan).* What physical feature makes this dense population possible? *(the Indus River)*

POPULATION MAP ANSWERS

1. Ganges River and Brahmaputra River
2. Bangladesh

CRITICAL THINKING

3. Many people live in regions with milder climates and plains, such as in river plains and in coastal areas, which provide better conditions for commercial agriculture. Population is less dense in arid and rugged highland areas.

UNIT
8
ATLAS

South Asia:
Population

1. **Environment and Society** Compare this map to the physical map. Which two major rivers flow through densely populated northern areas?
2. **Places and Regions** Which country in the region has the highest overall population density?

Critical Thinking

3. **Analyzing Information** Compare this map to the physical, climate, and land use maps. What do population density patterns suggest about the ways South Asians have adapted to their physical environments?

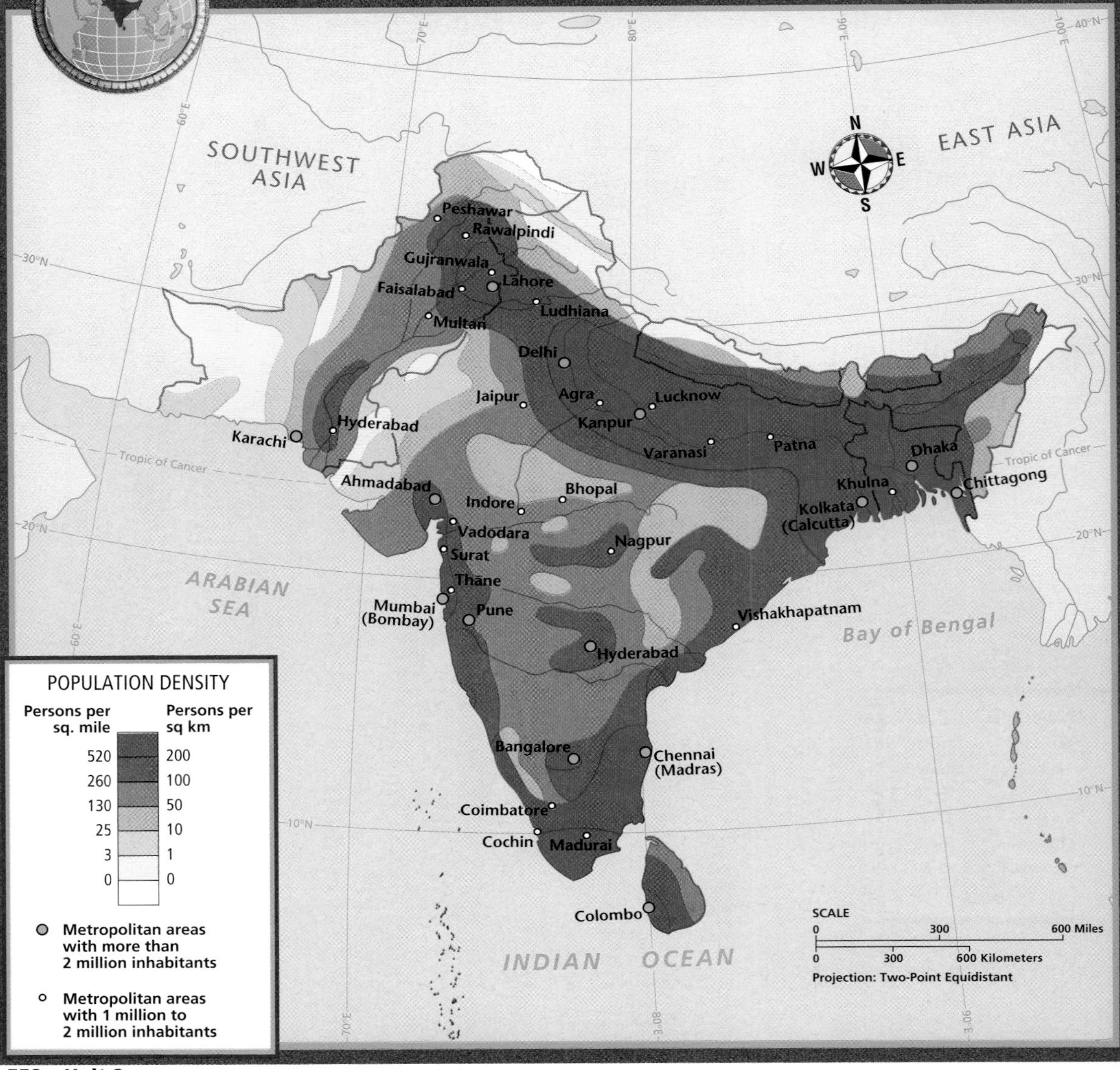

SOUTHWEST ASIA

EAST ASIA

Peshawar
Rawalpindi
Gujranwala
Faisalabad
Lahore
Multan
Ludhiana
Delhi
Jaipur
Agra
Lucknow
Kanpur
Karachi
Hyderabad
Varanasi
Patna
Dhaka
Ahmadabad
Indore
Bhopal
Khulna
Chittagong
Kolkata (Calcutta)
Vadodara
Surat
Nagpur
Thane
Mumbai (Bombay)
Pune
Vishakhapatnam
Hyderabad
Bangalore
Chennai (Madras)
Coimbatore
Cochin
Madurai
Colombo

Tropic of Cancer

ARABIAN SEA

Bay of Bengal

INDIAN OCEAN

POPULATION DENSITY

Persons per sq. mile	Persons per sq km
520	200
260	100
130	50
25	10
3	1
0	0

● Metropolitan areas with more than 2 million inhabitants

○ Metropolitan areas with 1 million to 2 million inhabitants

SCALE
0 — 300 — 600 Miles
0 — 300 — 600 Kilometers
Projection: Two-Point Equidistant

Using the Land Use and Resources Map

Focus students' attention on the **land use and resources map** on this page. Then ask students to answer the following questions: Which types of land use are most widespread in the region? *(commercial farming, subsistence farming)* Where is India's largest manufacturing and trade region? *(on the Arabian Sea)* Which cities are in that region? *(Ahmadabad, Vadodara, Mumbai, Pune)*

Have students compare this map to the **physical map** of South Asia. Call on a volunteer to identify where in India timber is harvested and which river may be used to transport the timber to manufacturing centers *(northern edge of the Deccan Plateau; Narmada River)*.

UNIT 8 ATLAS

South Asia:
Land Use and Resources

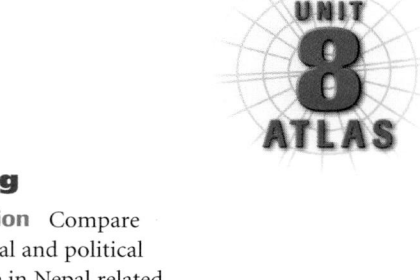

1. *Environment and Society* Compare this map to the political map. What is the dominant type of land use in coastal India?

2. *Places and Regions* Which part of India seems to have the highest concentration of minerals?

Critical Thinking

3. **Analyzing Information** Compare this map to the physical and political maps. How is land use in Nepal related to elevation?

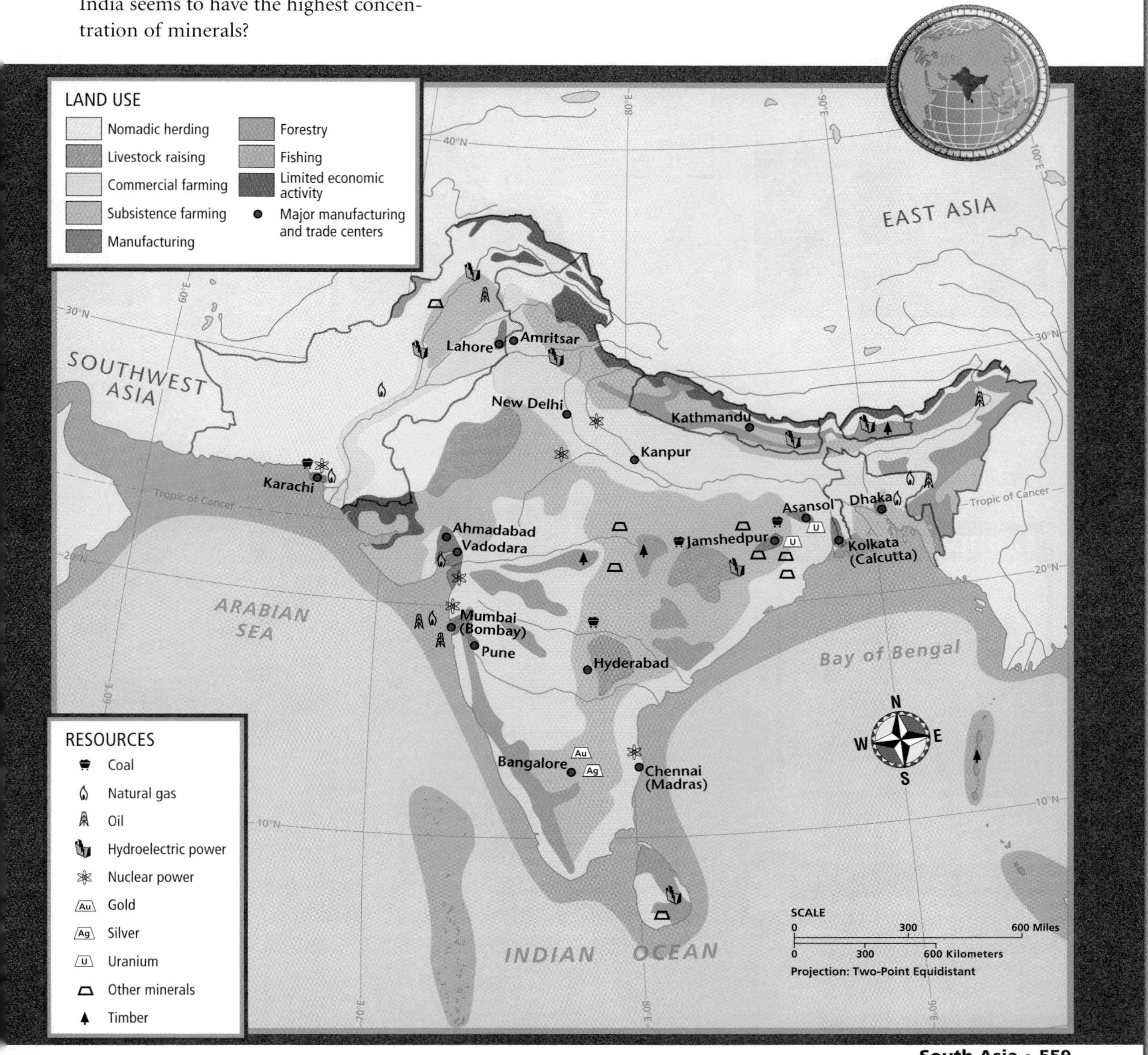

LAND USE
Nomadic herding
Livestock raising
Commercial farming
Subsistence farming
Manufacturing
Forestry
Fishing
Limited economic activity
Major manufacturing and trade centers
RESOURCES
Coal
Natural gas
Oil
Hydroelectric power
Nuclear power
Au Gold
Ag Silver
U Uranium
Other minerals
Timber
EAST ASIA
SOUTHWEST ASIA
ARABIAN SEA
Bay of Bengal
INDIAN OCEAN
Lahore
Amritsar
New Delhi
Kathmandu
Kanpur
Karachi
Dhaka
Asansol
Ahmadabad
Vadodara
Jamshedpur
Kolkata (Calcutta)
Mumbai (Bombay)
Pune
Hyderabad
Bangalore
Chennai (Madras)
Tropic of Cancer
40°N
30°N
20°N
10°N
60°E
70°E
80°E
90°E
100°E
N
S
E
W
SCALE
0 300 600 Miles
0 300 600 Kilometers
Projection: Two-Point Equidistant

LAND USE AND RESOURCES MAP ANSWERS

1. commercial farming

2. northeastern India

CRITICAL THINKING

3. The lowest areas are forested, higher areas are used for subsistence agriculture, and the highest areas are used for nomadic herding or have limited economic activity.

Provide students with a copy of the following table. Call on a volunteer to review the definition of a trade deficit *(a situation in which the value of a country's exports is less than the value of the country's imports)*. Have students examine the import and export values listed for each country to determine which countries of South Asia experience trade deficits *(all of them)*. Ask students which two countries in the region import more than twice as much as they export *(Maldives, Nepal)*. Call on volunteers to suggest effects a trade deficit might have on a country's economy.

COUNTRY	IMPORTS	EXPORTS
Bangladesh	$8.5 billion	$6.2 billion
Bhutan	$196 million	$154 million
India	$53.8 billion	$44.5 billion
Maldives	$395 million	$110 million
Nepal	$1.6 billion	$720 million
Pakistan	$11.1 billion	$9.8 billion
Sri Lanka	$5.4 billion	$4.6 billion
United States	$1.165 trillion	$687 billion

Source: Central Intelligence Agency, *The World Factbook 2003*

Historical Geography

The Sarasvati River The archaeologists who first discovered ruins of India's ancient culture named it after the nearby town of Harappa. They assumed—as have archaeologists for nearly a century—that the civilization had developed on the banks of the Indus River. Some archaeologists now believe, however, that the culture was established on the banks of another river, the Sarasvati, which has since disappeared.

Ancient Indian texts like the *Rig Veda* describe a great river called Sarasvati that sustained a great society along its banks. The writers of the *Mahabharata* noted that the river had already begun to dry. The ancient river's location remained a mystery until recently, when satellite images revealed a dry riverbed in northwestern India. This river ran right through the middle of the Harappan civilization ruins. Dozens of ancient sites line its dry banks, which leads archaeologists to believe that it was the basis for India's earliest cultures. Few sites are actually found on the Indus River itself; most lie far to the east.

ACTIVITY: Have students conduct research to find maps or aerial photographs of the Sarasvati riverbed. Ask if the images they find appear to support archaeologists' conclusions.

Time Line: South Asia

UNIT 8 ATLAS

2500 B.C. | **400 B.C.** | **1600 A.D.** | **1700** | **1800** | **1900**

c. 2500 B.C.
Indus Valley civilization flourishes.

400s B.C.
Siddhārtha Gautama founds Buddhism.

1600 A.D.
The British establish the East India Company.

c. 1630–50
A Mughal ruler builds the Taj Mahal as a tomb for his favorite wife.

1858
The British government takes control of India.

1920
Mohandas K. Gandhi begins an independence movement of nonviolent disobedience in India.

1971
Bangladesh becomes independent.

1947
India wins independence from Great Britain and is partitioned.

The United States and South Asia

Comparing Sizes

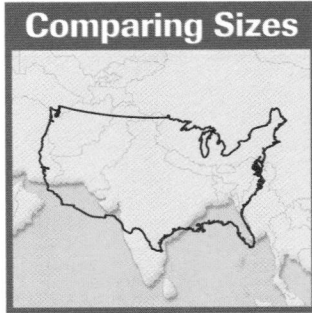

internet connect

GO TO: go.hrw.com
KEYWORD: SW3 Almanac
FOR: Additional information and reference sources

Comparing Standard of Living

COUNTRY	LIFE EXPECTANCY (in years)	INFANT MORTALITY (per 1,000 live births)	LITERACY RATE	DAILY CALORIC INTAKE (per person)
Bangladesh	62, male 61, female	66	43%	2,050
Bhutan	54, male 53, female	105	42%	Not available
India	63, male 65, female	60	60%	2,466
Maldives	62, male 65, female	60	97%	2,451
Nepal	59, male 59, female	71	45%	2,170
Pakistan	61, male 63, female	77	48%	2,447
Sri Lanka	70, male 75, female	15	92%	2,314
United States	74, male 80, female	7	97%	3,757

Sources: *World Almanac and Book of Facts, 2004; Britannica Book of the Year, 2002*

Copy the following chart onto the chalkboard. Have students examine the percentage of each South Asian country's land that is suitable for farming. Then have students examine the maps in the unit atlas. Ask them to explain relationships among each country's landforms, climates, precipitation, and arable land. Tell them to give specific examples to support their reasoning.

Then have students examine the workforce structure figures listed in the Fast Facts table. Ask them to speculate about the importance of agriculture in the region's economies. (*It is the most common economic activity in nearly all of the region's countries, practiced by most of the population.*) Ask students to

suggest why so many people in the region work in agriculture. (*Possible answer: The region has a tremendous population to feed.*)

COUNTRY	ARABLE LAND (PERCENTAGE)
Bangladesh	61 percent
Bhutan	3 percent
India	54 percent
Maldives	3 percent
Nepal	20 percent
Pakistan	28 percent
Sri Lanka	14 percent
United States	19 percent

Source: Central Intelligence Agency, *The World Factbook 2003*

Fast Facts: South Asia

UNIT **8** ATLAS

FLAG	COUNTRY / Capital	POPULATION (in millions) / POP. DENSITY	AREA	PER CAPITA GDP (in US $)	WORKFORCE STRUCTURE (largest categories)	ELECTRICITY CONSUMPTION (kilowatt hours per person)	TELEPHONE LINES (per person)
	Bangladesh Dhaka	146.7 / 2,838/sq. mi.	55,599 sq. mi. 144,001 sq km	$ 1,700	63% agriculture 26% services	97 kWh	0.005
	Bhutan Thimphu	2.3 / 124/sq. mi.	18,147 sq. mi. 47,001 sq km	$ 1,300	93% agriculture 5% services	168 kWh	0.009
	India New Delhi	1,065 / 928/sq. mi.	1,269,345 sq. mi. 3,287,588 sq km	$ 2,540	60% agriculture 23% services	467 kWh	0.04
	Maldives Male	0.3 / 2,745/sq. mi.	116 sq. mi. 300 sq km	$ 3,900	60% services 22% agriculture	342 kWh	0.09
	Nepal Kathmandu	25.2 / 476/sq. mi.	54,363 sq. mi. 140,800 sq km	$ 1,400	81% agriculture 16% services	70 kWh	0.01
	Pakistan Islamabad	153.6 / 511/sq. mi.	310,403 sq. mi. 803,940 sq km	$ 2,100	44% agriculture 39% services	405 kWh	0.02
	Sri Lanka Colombo	19.1 / 763/sq. mi.	25,332 sq. mi. 65,610 sq km	$ 3,700	45% services 38% agriculture	310 kWh	0.05
	United States Washington, D.C.	294.0 / 83/sq. mi.	3,717,810 sq. mi. 9,629,084 sq km	$ 37,600	31% manage., prof. 29% tech., sales, admin.	12,250 kWh	0.65

Sources: Central Intelligence Agency, *World Factbook 2003*; *The World Almanac and Book of Facts, 2004*
The CIA calculates per capita GDP in terms of purchasing power parity. This formula equalizes the purchasing power of each country's currency.

Bengal tiger

CHAPTER 25 India

CHAPTER RESOURCE MANAGER

Objectives	Pacing Guide	Reproducible Resources
SECTION 1 **Natural Environments** (pp. 563–66)	• Describe the major landform regions and rivers of India. • Identify the climate types and resources of India.	**Regular** 1 day **Block Scheduling** .5 day *Block Scheduling Handbook, Chapter 25*	**RS** Guided Reading Strategy 25.1 **SM** Map Activity 25: Indian Earthquakes
SECTION 2 **History and Culture** (pp. 568–73)	• Trace the major events and empires of India's early history. • Analyze how European contact affected India. • Identify the religions practiced in India. • Explain some other features of India's culture.	**Regular** 1.5 days **Block Scheduling** 1 day *Block Scheduling Handbook, Chapter 25*	**RS** Guided Reading Strategy 25.2 **RS** Graphic Organizer Activity 25 **E** Creative Strategies for Teaching World Geography, Lessons 20 and 21 **E** Cultures of the World Activity: Region 8 **SM** Critical Thinking Activity 25: The Caste System **SM** Geography for Life Activity 25
SECTION 3 **India Today** (pp. 574–79)	• Identify the main features of India's economy. • Compare life in India's villages to life in its cities. • Analyze the challenges that India faces today.	**Regular** 1 day **Block Scheduling** .5 day *Block Scheduling Handbook, Chapter 25*	**RS** Guided Reading Strategy 25.3 **PS** Readings in World Geography, History, and Culture 75 and 76

Chapter Resource Key

PS Primary Sources	**A** Assessment	CD-ROM	
RS Reading Support	**REV** Review	Video	
IC Interdisciplinary Connections	**ELL** Reinforcement and English Language Learners	 Internet	
E Enrichment	Transparencies	Holt Presentation Maker Using Microsoft® PowerPoint®	
SM Skills Mastery			

One-Stop Planner CD-ROM

See the *One-Stop Planner* for a complete list of additional resources for students and teachers.

One-Stop Planner CD-ROM

It's easy to plan lessons, select resources, and print out materials for your students when you use the **One-Stop Planner CD-ROM with Test Generator**.

Technology Resources

- One-Stop Planner CD-ROM, Lesson 25.1
- Geography and Cultures Visual Resources 55–59
- Homework Practice Online
- HRW Go site

- One-Stop Planner CD-ROM, Lesson 25.2
- Geography and Cultures Visual Resource 60
- **CNN** Presents Geography: Yesterday and Today, Segment 30: Indian Pop Music
- Homework Practice Online
- HRW Go site

- One-Stop Planner CD-ROM, Lesson 25.3
- *ARGWorld* CD-ROM
- Homework Practice Online
- HRW Go site

Reinforcement, Review, and Assessment

ELL Main Idea Activity 25.1
ELL English Audio Summary 25.1
ELL Spanish Audio Summary 25.1
REV Section 1 Review, p. 566
A Daily Quiz 25.1

ELL Main Idea Activity 25.2
ELL English Audio Summary 25.2
ELL Spanish Audio Summary 25.2
REV Section 2 Review, p. 573
A Daily Quiz 25.2

ELL Main Idea Activity 25.3
ELL English Audio Summary 25.3
ELL Spanish Audio Summary 25.3
REV Section 3 Review, p. 579
A Daily Quiz 25.3

internet connect

HRW ONLINE RESOURCES

GO TO: go.hrw.com
Then type in a keyword.

TEACHER HOME PAGE
KEYWORD: SW3 Teacher

CHAPTER INTERNET ACTIVITIES
KEYWORD: SW3 GT25

Choose an activity to:
- explore the regions of India.
- take the GeoMap Challenge.
- learn about economic and social issues in India today.

CHAPTER ENRICHMENT LINKS
KEYWORD: SW3 CH25

CHAPTER MAPS
KEYWORD: SW3 MAPS25

ONLINE ASSESSMENT
Homework Practice
KEYWORD: SW3 HP25
Standardized Test Prep
KEYWORD: SW3 STP25
Rubrics
KEYWORD: SS Rubrics

COUNTRY INFORMATION
KEYWORD: SW3 Almanac

CONTENT UPDATES
KEYWORD: SS Content Updates

HOLT PRESENTATION MAKER
KEYWORD: SW3 PPT25

ONLINE READING SUPPORT
KEYWORD: SS Strategies

CURRENT EVENTS
KEYWORD: S3 Current Events

Meeting Individual Needs

Ability Levels

Level 1 Basic-level activities designed for all students encountering new material

Level 2 Intermediate-level activities designed for average students

Level 3 Challenging activities designed for honors and gifted-and-talented students

English Language Learners Activities that address the needs of students with Limited English Proficiency

Chapter Review and Assessment

- Chapter 25 Test Generator (on the One-Stop Planner)
- Global Skill Builder CD-ROM
- HRW Go site
- **REV** Chapter 25 Review, pp. 582–83
- **REV** Chapter 25 Tutorial for Students, Parents, Mentors, and Peers
- **A** Chapter 25 Test (form A or B)
- **A** Alternative Assessment Handbook
- **A** Chapter 25 Test for English Language Learners and Special-Needs Students

Launch into Learning

Ask students to suggest images, places, or scenes that they associate with India. Discuss responses. *(Students may mention cobras, crowded cities, elephants, Mother Teresa, mountains, the Taj Mahal, tigers, or other images.)* Point out that although traditional images of India may be accurate, they are incomplete because India is undergoing rapid changes. For example, India has a thriving software industry and makes more movies than any other country. Tell students that in this chapter they will learn that India is a stunning mix of ancient and modern, urban and rural, rich and poor.

Using the Physical-Political Map

Have students examine the map on the opposite page. Call on students to describe the locations of major landforms. Ask students where they think population is most heavily concentrated and why. *(Possible answers: on the Gangetic Plain and along rivers, because the land there can be farmed)*

Why We Should Know More

There are many reasons why students in the United States should know more about India. Here are a few of them:

▶ India will probably surpass China to become the world's most populous country within the next generation.

▶ India, like the United States, is one of the world's nuclear powers.

▶ It is essential to the future health of the global economy that India, one of the world's poorest countries, continues to make economic progress.

▶ Many people of Indian heritage now live and work in the United States.

CHAPTER 25 India

About one out of every six people in the world lives in India, which is home to more than a billion people. This huge population includes people from many different ethnic and religious groups.

Royal portrait from Rajasthan

Indian woman sifting wheat in Gujarat

Hi! My name is Rojo, and I am 16. I live in Vaduthala, a small town outside the city of Cochin in the state of Kerala, southern India. My street is near a lake that is surrounded by coconut and banana trees.

When I'm relaxing at home, I like to read Indian entertainment magazines and watch television. My favorite program is about Hanuman, the monkey-god. The living room is my favorite place in the house because the ceiling fan keeps the mosquitoes away and keeps it cool. When my mom makes fish curry, she usually goes outside to the cooking terrace. That way, the whole house doesn't smell like fish. She grinds coconut there too.

I am a senior in high school. I wear a school uniform with navy blue pants and a grey shirt. Students in other schools have different uniforms, so it is very colorful in the morning when we all walk to school. We don't like walking in the summer monsoon season because we get wet. My favorite subject is math. I am studying a computer language based on math. At school we speak English. At home I speak Malayalam, the most common language of Kerala.

On Sundays we go to the Catholic church in the next town. Then we come back to the house and have a big lunch, including chicken curry, fish curry, sweet bread, eggs, and a vegetable dish. At Christmas they decorate the church with colored lights.

Section 1

OBJECTIVES

1. Describe the major landform regions and rivers of India.
2. Identify the climate types and resources of India.

 **LET'S GET STARTED**

Copy the following instructions onto the chalkboard: *Find the physical map in the unit atlas and the plate tectonics map in Chapter 4. Describe tectonic activity in the Himalayas in a sentence.* Discuss responses. *(Possible answer: The Indo-Australian Plate is pushing northward against the Eurasian Plate.)* Tell students that this mountain-building process continues, at about .2 inches (5 mm) per year, and that they will learn more about India's physical geography in Section 1.

Building Vocabulary

Write the term *subcontinent* on the chalkboard. Review the words *continent* and *subregion*. Ask how a subregion differs from a region. *(Possible answer: A subregion is similar to a region but is a smaller area within a region.)* Point out that a subcontinent differs from a continent in the same way—a subcontinent is also a very large land mass but is smaller than a continent.

Natural Environments

READ TO DISCOVER

1. What are the major landform regions and rivers of India?
2. Which climate types and resources does India have?

WHY IT MATTERS

Summer monsoon rains in India can cause floods that ruin water supplies and wash away crops, homes, and livestock. Use **CNN fyi.com** or other **current events** sources to learn about the dangers posed by floods.

DEFINE

subcontinent

LOCATE

Himalayas	Narmada River
Ganges River	Godavari River
Gangetic Plain	Krishna River
Deccan Plateau	Brahmaputra River
Western Ghats	Thar Desert
Eastern Ghats	

Section 1 RESOURCES

REPRODUCIBLE

► Guided Reading Strategy 25.1
► Map Activity 25: Indian Earthquakes

TECHNOLOGY

► One-Stop Planner CD–ROM, Lesson 25.1
► Geography and Cultures Visual Resources 55–69
► Homework Practice Online
► HRW Go site

REINFORCEMENT, REVIEW, AND ASSESSMENT

► Main Idea Activity 25.1
► English Audio Summary 25.1
► Spanish Audio Summary 25.1
► Section 1 Review, p. 566
► Daily Quiz 25.1

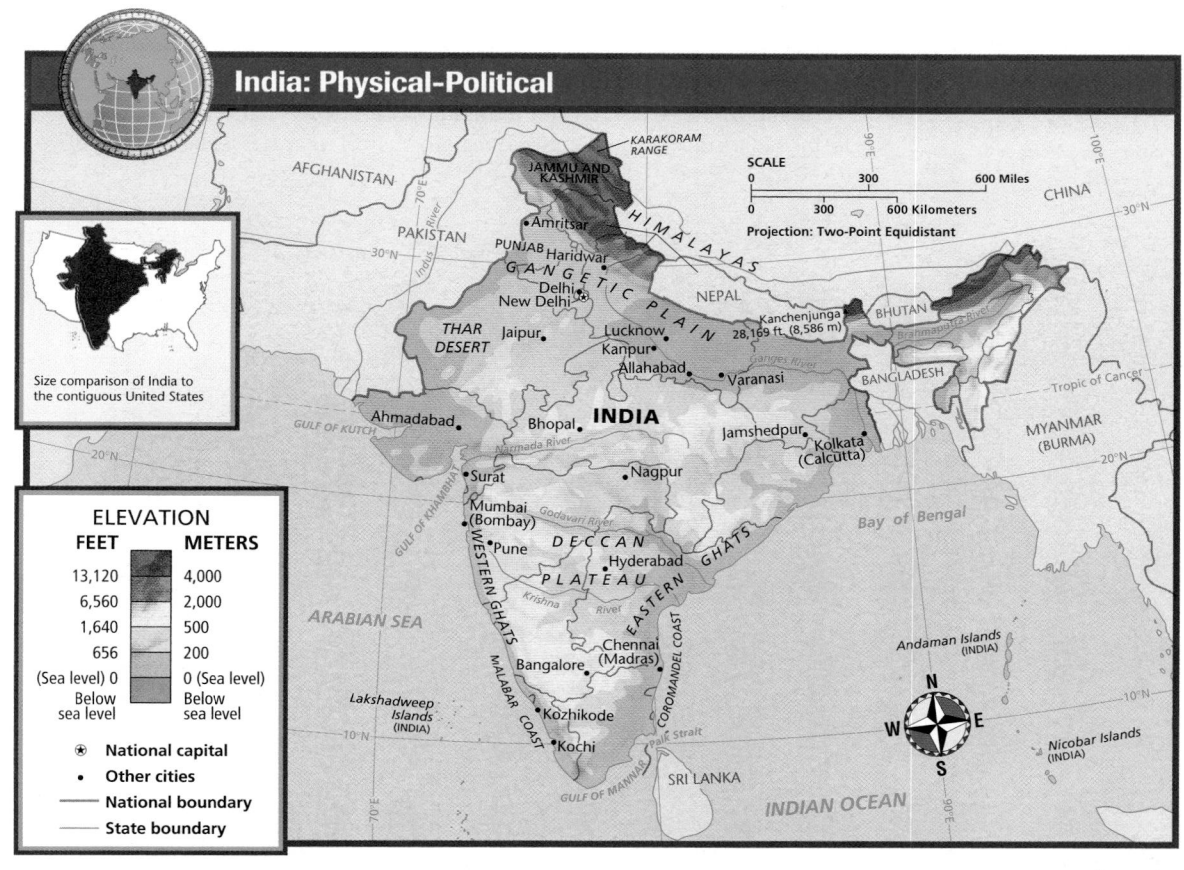

India: Physical-Political

Size comparison of India to the contiguous United States

ELEVATION

FEET	METERS
13,120	4,000
6,560	2,000
1,640	500
656	200
(Sea level) 0	0 (Sea level)
Below sea level	Below sea level

⊛ National capital
• Other cities
— National boundary
— State boundary

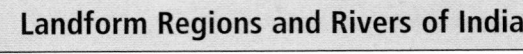

ALL LEVELS: Copy the following graphic organizer onto the chalkboard, omitting the italicized answers. Point out that it is roughly in the shape of India. Have students identify the three landform regions, starting with the most northern region in the top section. Next, ask students to list characteristics of each landform region. Last, have students add major rivers or sources of rivers to each landform region. Ask volunteers to share their charts with the class. **ENGLISH LANGUAGE LEARNERS**

Landform Regions and Rivers of India

Himalayas–mountainous, forests, farms, Kanchenjunga, Kashmir Valley, source of Ganges and Brahmaputra Rivers

Gangetic Plain–lowland, northeastern India, farming, rich alluvial soil, Ganges River

Deccan Plateau–old and eroded; volcanic; farming with irrigation; Eastern and Western Ghats; Narmada, Godavari, and Krishna Rivers

---

Essential Element 6

► **Environment** ◄
and Society

Kanchenjunga The world's third-highest peak is on the border between Nepal and the Indian state of Sikkim. Although its summit is a few hundred feet lower than Mount Everest's, Kanchenjunga is much harder to climb. In fact, mountaineers have scaled Kanchenjunga from the Indian side only four times. The mountain is scoured by winds regularly clocked at more than 100 miles per hour (161 kph). All the routes up the mountain are extremely difficult. Avalanches and rockfalls are constant dangers. Fatal accidents have increased Kanchenjunga's dangerous reputation.

In June 2000, Sikkimese authorities banned all expeditions to Kanchenjunga, known locally as Khanchen-Dzonga, for religious reasons. Sikkim's Buddhists believe Kanchenjunga is a guardian deity and therefore sacred. The local government's earlier willingness to allow expeditions in exchange for $20,000 had outraged social and religious organizations.

DISCUSSION: Lead a discussion about whether or not Sikkim's government should allow expeditions to Kanchenjunga.

internet connect

GO TO: go.hrw.com
KEYWORD: SW3 CH25
FOR: Web sites about India

The Himalayas form a barrier between the alluvial plains of northern India and the Plateau of Tibet to the north. They are the highest mountains in the world, with more than 110 peaks rising to elevations of 24,000 feet (7,325 m) or more above sea level. The Himalayas affect weather patterns in India by blocking cold air from moving south during winter and by forcing summer monsoon rains to release most of their precipitation before crossing the range to the north.

Physical Features

The world's highest mountains separate India and its neighbors from the rest of Asia. Together these countries form the Indian **subcontinent**. A subcontinent is a very large landmass that is smaller than a continent. In addition to this large landmass, India also includes two island territories located in the Indian Ocean.

Tectonic forces have shaped the Indian subcontinent. The subcontinent was once part of the supercontinent of Gondwana but then broke away. It slowly drifted northward. About 50 million years ago it pushed into Asia, forcing up a mountain system called the Himalayas (hi-muh-LAY-uhz). This mountain-building process still goes on today, causing severe earthquakes.

Landforms The Himalayas are one of India's main landform regions. They contain Kanchenjunga (kuhn-chun-JUHN-guh), which is the third-highest mountain in the world and India's highest point. At lower elevations there are farms, forests, and steep gorges. The Kashmir Valley, between two arms of the mountains, is famous for its beauty. The great Ganges (GAN-jeez) River begins high in the Himalayas' melting snow and glaciers.

South of the Himalayas is the Gangetic (gan-JE-tik) Plain. This low flat region stretches about 1,500 miles (2,415 km) across northeastern India. Its rich soil was brought down from the mountains by the Ganges River, forming the world's largest alluvial plain. In some places this alluvial layer is more than 25,000 feet (7,620 m) thick. For thousands of years, people have settled on the Gangetic Plain to farm.

South of the Gangetic Plain the land rises to the Deccan, a peninsula that forms India's third major landform region. Most of its area is a plateau. Part of

🌐 **ALL LEVELS:** Draw students' attention to the climate map and land use and resources map in the unit atlas. Have students list India's six climate regions and, for each region, describe the natural and energy resources found there. *(tropical humid—forests, fish; tropical wet and dry—coal, other minerals, nuclear power; arid—coal, natural gas, nuclear power; semiarid—gold, silver; humid subtropical—nuclear power; highland—forests, hydroelectric power)* Point out that much of India's land is used for either subsistence or commercial agriculture. Then lead a discussion about which of the resources on their lists are directly related to the climate region in which they are found. *(Possible answers: forests, hydroelectric power)*

HOMEWORK: Give each student an outline map of India. Ask students to use the unit atlas and Section 1 to map the sources and directions of the summer and winter monsoons. Direct students to use different colors for the different seasons. On the back of the map, have students write a paragraph describing the consequences of weak, heavy, or late summer monsoons on India.

---

the Deccan Plateau is very old and eroded. The newer part is made up of lava layers. Where the volcanic rocks have weathered and irrigation is possible, there are fertile grain fields. Two low mountain ranges, the Western Ghats (GAWTS) and the Eastern Ghats, form the plateau's edges. Narrow plains lie on both the Arabian Sea and Bay of Bengal coasts.

Great Rivers Several important rivers cut across the Deccan Plateau. The Narmada (nuhr-MUH-duh) River flows to the west. Hundreds of dams are being built on the Narmada. Although the reservoirs created by the dams store much-needed water, the projects have also displaced thousands of people. The Godavari (go-DAH-vuh-ree) and Krishna (KRISH-nuh) Rivers flow to the east. An irrigation and canal system linking the deltas of these two rivers has made the region a productive rice-growing area.

Like the Ganges, the Brahmaputra River begins high in the Himalayas. It flows eastward through Tibet and then turns to the south, flowing through eastern India. In Bangladesh the Brahmaputra joins the Ganges to form a huge delta.

✔ **READING CHECK:** *Places and Regions* In what landform regions are India's major rivers located? Ganges and Brahmaputra begin in the Himalayas; Ganges flows across Gangetic Plain; Narmada, Godavari, and Krishna cut across Deccan

INTERPRETING THE VISUAL RECORD *Which of India's three main landform regions can you see in this satellite image?*

Climates and Natural Resources

As you can see from the Unit 8 climate map, India has six climate types. They range from a highland climate in the Himalayas to a tropical humid climate along the southwest coast.

The Thar (TAHR) Desert, also known as the Great Indian Desert, lies in the northwest. It extends into neighboring Pakistan. The prevailing dry monsoon winds and a high-pressure zone keep the region arid and hot. Compared to the densely populated Gangetic Plain, the Thar Desert has few residents.

India's plant life tends to reflect rainfall amounts. Vegetation ranges from scrubby trees in the Thar Desert to evergreen forests in the western Himalayas. The country's animal life is varied. Crocodiles, deer, elephants, mongooses, monkeys, tigers, more than 1,200 bird species, and nearly 400 snake species live in India.

✔ **READING CHECK:** *Places and Regions* What six climate types are found in India? highland, tropical humid, tropical wet and dry, humid subtropical, semiarid, arid

The Monsoons India's monsoons strongly influence the country's climates. The wet summer monsoon usually begins about June. A large low-pressure

Our Amazing Planet

The Indian cobra has a toxic bite. Cobras kill several thousand people each year, partly because they often visit houses at night to catch rats.

Eye on Earth

Sloth Bears In India, the tiger is the "Lord of the Jungle." One of the few animals that will dispute the big cat's rule is the sloth bear. These shaggy, long-clawed bears sometimes attack tigers and humans. Most of these bears do not consider humans prey, but as encounters between sloth bears and people become more frequent, attacks become more frequent.

Insects are the sloth bear's main food. The bear breaks into a nest, blows into it, and then sucks up the ants, termites, and larvae. Because its nostrils close, the bear does not inhale the insects. The sound of a sloth bear feeding can be heard 200 yards (183 m) away. A bear may eat up to 10,000 termites at one time.

Sloth bear habitat is disappearing because of human needs for land and firewood. Poachers kill the bears for their gallbladders, which are valued in Asia as medicine. Also, traveling entertainers take cubs from their dens and train them to "dance" for paying audiences.

VISUAL RECORD ANSWER

Gangetic Plain, Deccan, part of Himalayas

📶 **internet** connect

GO TO: go.hrw.com
KEYWORD: SW3 CH25
FOR: Web sites about Indian wildlife

565

Close

What is the highest rainfall total your community has received in one year? Ask students how life in your community would be affected if it received the same amount of rain as Cherrapunji, India, which holds the world record—almost 87 feet (26.5 m)!

Review and Assess

Have students complete the **Section Review**. Then have students complete **Daily Quiz 25.1.**

Reteach

Have students complete **Main Idea Activity for English Language Learners and Special-Needs Students 25.1.** Then refer students to the physical-political map. Have them estimate where the 36°N parallel lies. Call on students to describe the landforms, rivers, climates, and resources at that latitude. Then repeat the exercise for other parallels. **ENGLISH LANGUAGE LEARNERS**

Extend

Have interested students conduct research and report on the various factors that affect the monsoons. These factors include jetstreams and the temperature of Pacific waters off the coast of South America. **BLOCK SCHEDULING**

Section 1 Review Answers

Define For definition, see: subcontinent, p. 564

Working with Sketch Maps
Maps will vary, but listed places should be labeled in their approximate locations. The world's largest alluvial plain is the Gangetic Plain.

Reading for the Main Idea
1. tectonic forces; soil brought down from the mountains by the Ganges River

2. rainfall amounts; Thar Desert; western Himalayas

3. coal, iron ore, bauxite, petroleum, uranium

Critical Thinking
4. Monsoon rains make farming possible, but monsoons can also cause disastrous floods. Also, if monsoons are late or rains stop early, crops may die and people may go hungry. (NGS 15)

Organizing What You Know
5. Himalayas—Ganges, Brahmaputra Rivers; highland; Gangetic Plain—Ganges River; humid subtropical; Deccan Plateau—Narmada, Godavari, Krishna Rivers; semiarid and tropical wet and dry

VISUAL RECORD ANSWER

Possible answer: cause mudslides, crop damage, erosion

566

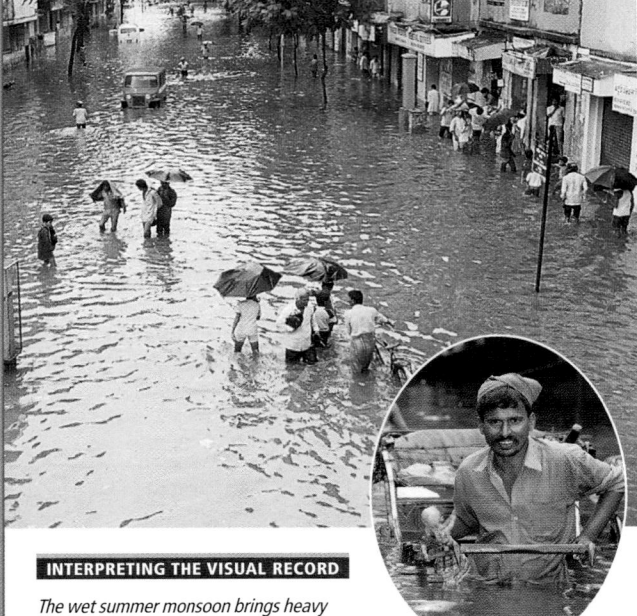

INTERPRETING THE VISUAL RECORD

The wet summer monsoon brings heavy rainfall and flooding to cities throughout India, including Mumbai (Bombay) (left) and Kolkata (Calcutta) (right). While Mumbai's average annual rainfall is some 71 inches (180 cm), about 24 inches (61 cm) fall in the month of July. **How do you think heavy monsoon rains and floods affect the environment?**

---

area over interior Asia pulls moist air inland from the Indian Ocean. As the moist air flows inland, it brings rain. The heaviest rains fall where the monsoon meets the Western Ghats and the foothills of the Himalayas. Some areas receive more than 400 inches (1,016 cm) of rain per year.

The winter monsoon lasts from November through March. During those months cold dry winds blow from Asia's interior into India. As the winds drop down from the mountains they warm and become even drier. As a result, much of India has a warm dry winter.

Unusually heavy monsoon rains can cause terrible floods. If the monsoon is late or the rains stop early, crops may die and people may go hungry. Sometimes, within the same season one part of India can suffer drought while another has floods.

✓ **READING CHECK:** *Physical Systems* What causes India's monsoons? A low-pressure area over interior Asia causes summer monsoons, and cold dry winds from Asia's interior cause the winter monsoon.

Natural Resources India's soils and rivers are its most important natural resources. About 56 percent of the country is arable, in contrast to only 19 percent in the United States.

India also has many mineral resources, including the world's fourth-largest coal reserves. There are large deposits of iron ore and bauxite, an aluminum ore, much of which is exported. India has some petroleum deposits but must import oil to meet its needs. The country's uranium mines support several nuclear power plants. Rivers supply hydroelectric power.

✓ **READING CHECK:** *Environment and Society* Why are India's soils and rivers called its most valuable natural resources? make 56 percent of the land suitable for farming

---

Section 1 Review

**go.hrw.com
Homework Practice Online
Keyword: SW3 HP25**

Define
subcontinent

Working with Sketch Maps On a map of India that you draw or that your teacher provides, label the Himalayas, Ganges River, Gangetic Plain, Deccan Plateau, Western Ghats, Eastern Ghats, Narmada River, Godavari River, Krishna River, Brahmaputra River, and Thar Desert. In the margin of your map, identify the world's largest alluvial plain.

Reading for the Main Idea
1. *Physical Systems* What physical process created the Himalayas? What physical process created the Gangetic Plain?

2. *Places and Regions* The distribution of India's plant life reflects what feature of the country's climate? Where will you find scrubby trees? Evergreen forests?

3. *Environment and Society* What are some of India's important mineral resources?

Critical Thinking
4. **Drawing Inferences and Conclusions** Why might Indian farmers look forward to the summer monsoon with both hope and fear?

Organizing What You Know
5. Create a chart like the one below. Use it to name the rivers and climates in India's three main landform regions. Use the unit maps to help you.

Landform region	Rivers	Climates

Literature: Classical Indian Literature

Have students read Geography for Life: The Holy Ganges. One reason the Ganges figures so prominently in Indian beliefs is that the Gangetic Plain was the home of early Indian cultures. Great works of literature have come down to us from some of these civilizations. Reading excerpts from these works may alter students' perceptions of early India and of traditions, such as veneration of the Ganges, that live on in modern India.

Organize the class into four groups, and assign one of the following works of literature to each group: The *Mahabharata*, probably the longest poem ever composed and an epic about two rival families; the *Bhagavad-Gita*, part of the previous work and a dialogue between a man and a disguised god; the *Panchatantra*, a collection of fables; and the *Ramayana*, an epic about Rama, who is an incarnation of the god Vishnu, and his wife Sita. Have each group conduct research on the work's historical background and then locate and read selections from the assigned work. Ask students to concentrate on connections, if any, that they see between ancient India as portrayed in the literature and modern India. Also encourage them to reflect on how the literature has affected their perceptions of Indian culture. Call on each group to report its findings to the class. Ask them to read aloud from the work as part of the reports. **COOPERATIVE LEARNING**

Places and Regions

Geography for Life

The Holy Ganges

The Ganges River begins high in the Himalayas as melting glacial ice. It then flows more than 1,500 miles (2,505 km) southeastward across India's northern plains. Millions of people depend on the Ganges. They use the river as a source of drinking water and fish. Farmers use it to irrigate their crops. The Ganges also provides the country with a trade and transportation route.

To many people, the Ganges is much more than a river. It is Hinduism's holy river. The Ganges is part of India's folklore, history, and mythology. Indians call it the goddess Ganga. Every day you can find people bathing in the river's water, drinking it, and standing in it while they pray. Hindus believe that just touching the water can wash away their sins. People with various ailments come from near and far seeking the healing powers of the water. Perhaps more than 1 million people enter the Ganges somewhere along its course each day. The cities of Allahabad, Haridwar, and Varanasi are considered particularly sacred bathing sites. Temples line the banks of the Ganges in these cities. Wide stone staircases called ghats lead down to the water.

Garbage and pollution are a common sight in the waters of the Ganges today.

Tradition says that the Ganges is pure and that nothing can pollute its water. Yet huge amounts of sewage flow into the river, making it one of the world's most polluted waterways. Waste from dozens of cities and thousands of villages is dumped into the river. Runoff from farms carries soil full of farm chemicals into it. Hundreds of factories also dump industrial waste into the Ganges. People even come to die in the river. As a result of these pollutants, many who enter the Ganges get sick.

In spite of these dangers, devout Hindus readily drink and bathe in the Ganges' waters. The government has proposed programs to reduce the river's pollution. While some Indians support a cleanup, others insist that the Ganges remains pure. Because the government lacks wide public support for its programs, it has done little to clean up the Ganges.

More than 1 million pilgrims visit Varanasi, India, each year to bathe in the holy Ganges.

Applying What You Know

1. **Summarizing** Why is the Ganges River holy to Hindus? How might non-Hindus perceive it?

2. **Supporting a Point of View** Construct an argument for or against the need to clean up the Ganges River from the perspective of either a traditional Hindu or an environmentalist.

Hindu Festivals Every 12 years the town of Allahabad in northern India celebrates the Maha Kumbh Mela or "the grand pitcher" festival. The name comes from the belief that all the forces of creation are collected in one vessel.

The Yamuna and Ganges Rivers converge at Allahabad and provide the setting for the gathering. During the 42-day celebration millions of people bathe in the Ganges. In 1989 this celebration was recorded in *The Guinness Book of World Records* as the "largest ever gathering of human beings for a single purpose." That year approximately 15 million people took part in the festival.

CRITICAL THINKING: How might participating in such an event affect people's beliefs? *(Possible answers: reaffirm beliefs, make people feel they are part of something significant and true)*

Applying What You Know Answers

1. Hindus believe that just touching the water can wash away their sins. Non-Hindus may perceive it as just a polluted river.

2. Answers will vary depending on the point of view selected to discuss.

This Geography for Life feature addresses National Geography Standards 4, 6, 14, and 16.

567

OBJECTIVES

1. Trace the major events and empires of India's early history.

2. Analyze how European contact affected India.

3. Identify the religions practiced in India.

4. Explain some other features of India's culture.

 **LET'S GET STARTED**

Copy these questions onto the chalkboard: *Which part of India do you think attracted the most invaders and why? How might European colonists have reacted to India's climate?* Discuss responses. *(Possible answers: Gangetic Plain, because of its fertility; Europeans probably dismayed by India's heat and heavy rain)* Tell students they will learn more about the many groups that invaded India and the cultures they created in Section 2.

Building Vocabulary

Tell students the story behind the word **boycott**. Charles C. Boycott was the agent for an absentee landlord in Ireland in 1880. When tenants demanded their rent be reduced, Boycott refused to comply. In return, the tenants made sure Boycott and his family suffered. The farmhands and household servants left. No one would wait on the Boycotts in the shops or deliver their mail. Boycott's name has the same meaning—of refusing to buy—in several languages. Have students read the definitions of the remaining terms.

Section 2 History and Culture

READ TO DISCOVER

1. What were the major events and empires of India's early history?
2. How did European contact affect India?
3. What religions are practiced in India?
4. What are some other features of India's culture?

WHY IT MATTERS

Hinduism is the world's oldest major religion. Use **CNN fyi.com** or other **current events** sources to learn about important Hindu beliefs and practices.

IDENTIFY

Sanskrit   Sikhism
Dalits   Jainism

DEFINE

pantheon   reincarnation
sepoys   dharma
boycott   karma
partition   caste

LOCATE

Delhi

Early Indian Civilizations

The first highly developed civilization in the Indian subcontinent developed in the Indus River valley. Named after a modern town near the site of one of its ancient cities, this culture is known as the Harappan civilization. It was centered in what is now Pakistan but also extended into India. The Harappan way of life was based on farming and trade. For India's history we look first at people who came through the western mountain passes to invade the fertile north.

The Aryans By about 1500 B.C. a warlike seminomadic people were moving into northern India from central Asia. They called themselves Aryans. The Aryans spread their influence gradually throughout northern India. In turn, Aryan culture changed as it mixed with cultures already in the area. The Dravidians, farming peoples of the interior, were slowly pushed south. They later developed advanced kingdoms in the Deccan.

The Aryans brought an Indo-European language to the subcontinent. They spoke an early form of **Sanskrit**. Hindi, the major language of modern India, developed from Sanskrit, which is still used for some religious ceremonies. The name *Himalaya* is an example of a Sanskrit word. It means "Home of Snows."

Aryan religion included many of the basic ideas that became part of Hinduism, such as a large **pantheon**. A pantheon is all the gods of a religion. The Aryans also introduced a strict system of social classes. These and other concepts developed into Hinduism, India's main religion today.

During the Aryan period, many kingdoms rose in the Gangetic Plain. These kingdoms were supported by farming the plain's rich soils. Also during the Aryan period Siddhārtha Gautama, who later became known as the Buddha, taught the concepts of Buddhism.

The Ellora Caves in western India were carved from cliffs between about 200 B.C. and A.D. 1000. They served as temples and monasteries for followers of three of India's religions: Buddhism, Hinduism, and Jainism.

LEVEL 1: Copy the following time line onto the chalkboard, omitting the italicized answers. Have students place events and their dates on the time line in the proper sequence. **ENGLISH LANGUAGE LEARNERS**

LEVELS 2 AND 3: Have students work in pairs to conduct research on a major empire or event in India's history. Provide additional resources for the students to use in the classroom. Have students include in their reports a description of the empire or event and its significance to Indian history and culture. **COOPERATIVE LEARNING**

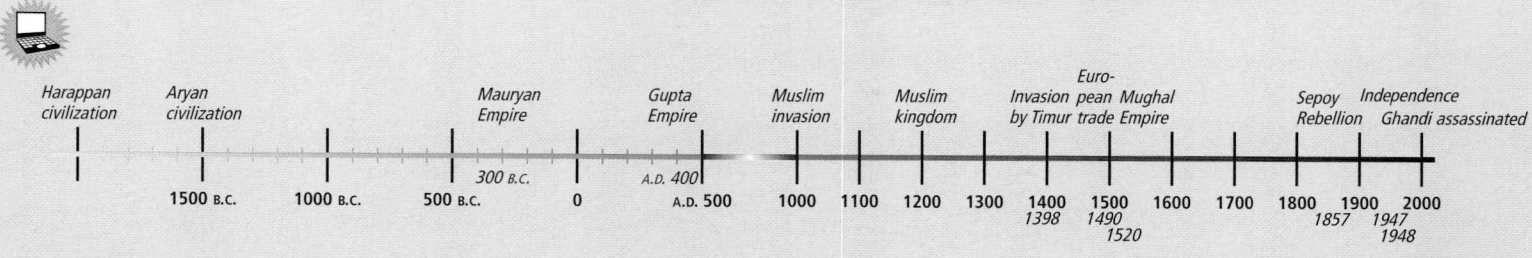

| Harappan civilization | Aryan civilization | | Mauryan Empire | Gupta Empire | Muslim invasion | Muslim kingdom | Invasion by Timur | Euro-pean trade | Mughal Empire | Sepoy Rebellion | Independence Ghandi assassinated |

1500 B.C. 1000 B.C. 500 B.C. 0 *300 B.C.* A.D. *400* A.D. 500 1000 1100 1200 1300 1400 1500 1600 1700 1800 1900 2000
1398 1490 1857 1947
1520 1948

Islamic Empires About A.D. 1000, Muslim armies began attacking northwestern India. In the early 1200s a Muslim kingdom was founded at Delhi. In 1398 Timur, a conqueror from the interior of Asia, invaded India and sacked Delhi.

A descendant of Timur and Genghis Khan nicknamed Bābur—"the Tiger"—invaded India in the 1520s. He took over most of northern India and founded the Mughal (MOO-guhl) Empire. Bābur was not only a brilliant general but also a gifted poet. After an unstable period Bābur's grandson Akbar reunited the Mughal Empire and expanded it into central India. Akbar was an effective ruler as well as a successful conqueror. He allowed the peoples he conquered to practice their own religions. Although the ruling Mughals were Muslims, most of the region's people continued to practice Hinduism.

The Gangetic Plain's fertile land and large population helped the Mughal Empire grow rich. During this period Shāh Jahān built the world-famous Taj Mahal. Many other grand monuments remain from the Mughal Empire.

Religious tolerance ended with the rule of Shāh Jahān's son Aurangzeb. He placed heavy taxes on Hindus and destroyed many temples. When Aurangzeb died in 1707, he left a weakened empire. Europeans took the opportunity to expand their influence.

READING CHECK: *Places and Regions* How did invasions alter India's early cultures? *Aryans introduced a new language, religious ideas, and social classes. Muslims introduced Islam.*

European Influence

Europeans arrived in India in the late 1490s to trade, expand their empires, and spread Christianity. The Portuguese came first, followed by the Dutch, French, and British. Powerful trading companies, such as Great Britain's East India Company, built successful businesses in the region. Indian cotton became particularly important as Britain's growing textile industry needed tons of the raw fiber. Soon Britain and France became rivals in India.

British Rule In the 1700s the British defeated the French. By the mid-1800s Britain controlled about half of the subcontinent. The East India Company controlled India for the British government.

However, foreign rule angered many Indians. In 1857 a rebellion broke out among the **sepoys**, the Indian troops under the command of

Connecting to HISTORY

Early Empires

Almost all of India was first united by the Mauryan (MOW-ree-uhn) Empire. It was founded in about 320 B.C. Aśoka, the most famous Mauryan emperor, gave up war after he saw the suffering he had caused. He adopted Buddhism and furthered its spread across India. Aśoka's thoughts and accounts of his deeds were carved on rocks and pillars placed throughout the empire.

India entered a golden age under the Gupta Empire. From A.D. 320 to about 500, architecture, art, literature, mathematics, and medicine all flourished. Gupta universities drew students from as far away as Java, in southeastern Asia, to study philosophy and other subjects. Art, learning, and trade continued to prosper after the Gupta period came to an end.

Analyzing Information

In what ways did the Mauryan and Gupta Empires contribute to the early development of India?

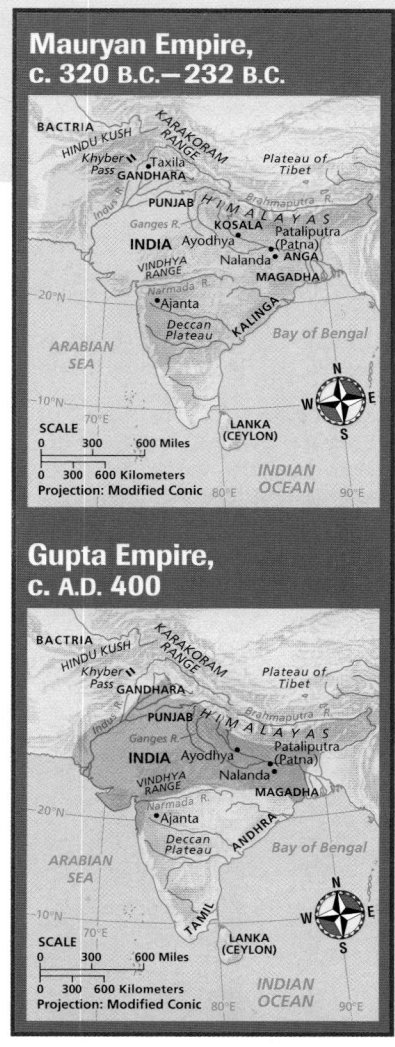

Mauryan Empire, c. 320 B.C.—232 B.C.

Gupta Empire, c. A.D. 400

Cultural Kaleidoscope

Portuguese Goa Goa has a unique character among the Indian states. The Portuguese conquered the settlement there in 1510, and a few decades later Saint Francis Xavier began missionary work in Goa. Today Roman Catholicism still has a strong presence in the state.

Several magnificent churches attest to Goa's Catholic heritage. Every 10 years, pilgrims from around the world converge on the Basilica of Bom Jesus, which contains the silver casket of Saint Francis. The Cathedral of Saint Catherine, built in the Renaissance style, is the largest Christian church in Asia. Saint Cajetan Church was modeled on Saint Peter's Basilica in Rome. Other churches display baroque, Gothic, and Tuscan styles.

ACTIVITY: Have students conduct research on Goa's cultural sites and plan an itinerary that includes the state's highlights.

CONNECTING TO HISTORY ANSWER

The Mauryan Empire adopted Buddhism and furthered its spread across India. The Gupta Empire advanced architecture, art, literature, mathematics, and medicine.

LEVEL 1: Have students work in pairs to create lists of the effects that European contact had on India. *(Possible answers: increased trade; cotton shipped to British industries; control by the East India Company; revolt among sepoys; direct rule by British government; British goods imported to India; construction of roads, railways, and ports; English language; systems of education, law, and government; unequal treatment; independence movement; independence; partition; violence as a result of partition)* Ask students to share their lists with the class. **COOPERATIVE LEARNING**

LEVELS 2 AND 3: Have students use the lists they created for the Level 1 activity as a basis for this one. Ask students to write a brief essay in which they discuss which of the effects are clearly positive, which clearly negative, and which may be seen as both positive and negative. You may want to stage a debate between students who perceive issues differently.

Linking Past to Present

Indira Gandhi Indira Gandhi was one of the most important political figures of her time. The only child of Jawaharlal Nehru, independent India's first prime minister, she showed political aptitude as early as age 12, when she established a youth organization to help with the National Congress movement. After her father became prime minister in 1947, she accompanied him on many trips to other countries. Indira Gandhi served as prime minister from 1966 to 1977 and from 1980 to 1984. Although her later terms were marred by power struggles and accusations of corruption, Gandhi guided India through a war with Pakistan and promoted social welfare and international diplomacy.

Today among the institutions named for Indira Gandhi are a conservation monitoring center, a national center for the arts, a development institute, and Delhi's international airport.

Mahatma Gandhi led India's independence movement and became internationally famous for his doctrine of nonviolent protest and civil disobedience. Gandhi also worked to unite India's Hindus and Muslims, end discrimination, and promote education and economic development for India's poor.

The Gateway of India in Mumbai (Bombay) was built to commemorate the 1911 visit of King George V and Queen Mary. On February 28, 1948, the last British troops to leave the newly independent country of India set sail from the Gateway of India.

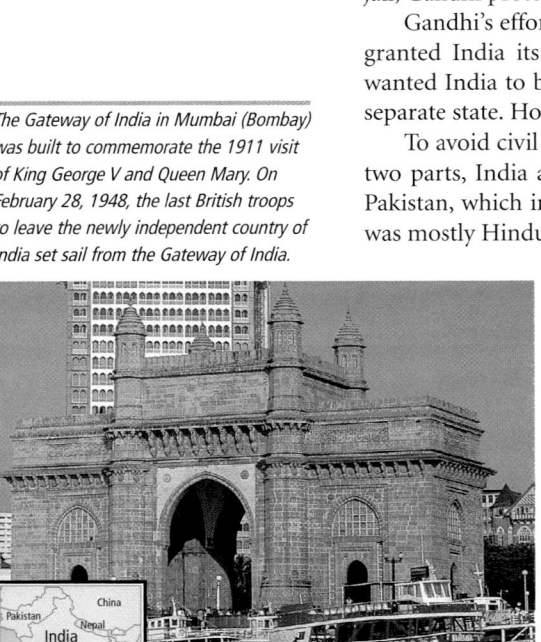

internet connect

GO TO: go.hrw.com
KEYWORD: SW3 CH25
FOR: Web sites about India's historic leaders

British officers. The sepoys killed the officers and their families. When the revolt spread across northern India, British troops rushed to the conflict. Both sides reported vicious acts of cruelty. In the end, the British crushed the revolt. The mutiny convinced the British government to end the East India Company's role and to rule India directly beginning in 1858.

India, which at that time included present-day Pakistan, Bangladesh, and Sri Lanka, became a British colony. It was called the "jewel in the crown" of the British Empire. Indian products, such as cotton, jewels, and tea, flowed to Britain. India was also a market for merchandise from British factories. To ensure an efficient flow of goods, the British organized the construction of railroads, roads, and ports. In addition, they brought the English language and English systems of education, law, and government.

Independence The British did not treat Indians as equals, and many Indians wanted independence. Others simply wanted fair treatment. In 1885, educated middle-class Indians organized the Indian National Congress. At first, the Congress asked only for more rights, such as a larger share of government jobs. The British refused this request.

Demands for independence increased in the early 1900s. A young lawyer named Mohandas K. Gandhi led the independence movement. His followers called him the Mahatma, or "Great Soul." Gandhi believed that nonviolent noncooperation was the best way to bring about change and achieve independence. Gandhi led peaceful protest marches and urged Indians to **boycott**, or refuse to buy, British goods, particularly cloth. When he was thrown into jail, Gandhi protested by going on hunger strikes.

Gandhi's efforts were effective. After World War II the British government granted India its independence. Britain and the Indian National Congress wanted India to become one country. However, India's Muslims demanded a separate state. Hostility between Hindus and Muslims grew.

To avoid civil war, in 1947 the British government divided the colony into two parts, India and Pakistan. This division is called the **partition** of India. Pakistan, which included what is now Bangladesh, was mostly Muslim. India was mostly Hindu. The region of Kashmir was divided between the two countries. However, the new boundaries left large numbers of Hindus in Pakistan and many Muslims in India. Panic broke out as some 16 million people fled to the country where their religion held a majority. Perhaps as many as 1 million people died in riots and massacres. Even Gandhi fell prey to the violence. In 1948 he was shot and killed by a Hindu extremist. Religious tension is still a pressing issue for India.

Today India is the world's most populous democracy. A large percentage of the people vote. There are 28 states, each with its own legislature, and seven territories. India's government is based on British models. It includes a multiparty parliament and a prime minister.

✓ **READING CHECK:** *Human Systems* Why did the British want control of India, and how did they lose control? They wanted trade goods and markets; conflict eventually led Britain to grant India's independence.

LEVEL 1: Have students create posters about either Hinduism or India's other major religions, describing each religion's main features, such as the number of followers in India, beliefs, places of worship, deities, symbols, and so on. Encourage students to use words, pictures, and diagrams on their posters. Display and discuss the posters. **ENGLISH LANGUAGE LEARNERS**

LEVEL 2: Have each student write a short story about an Indian teenager that includes details about the teen's religious life. For example, a story might describe how a Hindu teen's membership in a certain caste affects his or her behavior. Call on volunteers to read their stories to the class.

LEVEL 3: Have students expand their short stories to include material on whether diffusion of traits from other cultures has affected the teen's religious life.

Religion

According to its constitution, India's government is officially secular. That is, it does not support any one religion, and all faiths are equal before the law. At the same time, religion is a powerful force in Indian society. A large majority of the people take part in religious activity on a daily basis. Some Indian political parties are based on religion.

Hinduism About 80 percent of India's people are Hindu. Hinduism is an ancient religion and the largest ethnic religion in the world. There are many gods and goddesses in Hinduism. Although its pantheon is large, Hinduism teaches that all gods and all living beings are part of a single spirit.

Three Hindu beliefs, **reincarnation**, **dharma**, and **karma**, are closely related. Followers of Hinduism believe that the soul is reborn again and again in different forms. This process is called reincarnation. The importance of doing one's duty according to one's station in life is called dharma. Karma is the positive or negative force caused by a person's actions. Hindus believe that people who fulfill their dharma earn good karma and may be reborn as persons of higher status. Those with bad karma may be reborn with lower status or as animals or insects.

The **caste** system is another practice central to Hinduism. A caste is a group of people who are born into a certain position in society. The Aryans developed this system thousands of years ago. It assigned people to one of four major classes, or castes, according to their occupations. The Brahmins were the highest caste. These priests and intellectuals were the only people who could read and write. Below the Brahmins were the Kshatriyas (kuh-SHA-tree-uhz), or warriors. Next came the Vaisyas (VYSH-yuhz), the traders and merchants, and then the Sudras, the farmers and laborers. There are now thousands of subcastes.

Below these four classes were the **Dalits**, which means "the oppressed." They were not viewed as part of any caste. They held jobs that higher classes saw as unclean, such as removing dead animals from the street. Dalits were forbidden from having contact with Indians of any caste.

Under the caste system, a person is born into a caste and cannot move into another. Some Indians believe that taking a job of another caste or marrying into another caste can be punished by rebirth at a lower social level. In this way, belief in reincarnation, dharma, and karma helps maintain the caste system.

The Indian constitution abolished the caste system. It declared that the poor treatment of Dalits was illegal. Still, the caste system remains a part of Hindu life, particularly in villages. Today some Dalits are educated and have good jobs. Most live in poverty, however. Violence against the Dalits still occurs.

The statue of Bahubali in the state of Karnataka is an important pilgrimage site for followers of Jainism, one of India's indigenous religions. Every 12 to 14 years a "head anointing ceremony" is performed on the statue as milk, yogurt, butter, saffron, gold coins, sandalwood, and other items are poured over its head. Jain pilgrims believe these objects acquire a powerful spiritual energy from the statue.

571

ALL LEVELS: Organize the class into four groups—one group each for language, clothing, food, and festivals. Have each group present an oral report on its topic. Ask students to include a demonstration in the reports. For example, the clothing group might demonstrate different ways to wrap a sari. **ENGLISH LANGUAGE LEARNERS, COOPERATIVE LEARNING**

HOMEWORK: Have students imagine that they are members of India's parliament and must vote on a proposal to make Hindi and English the country's only official languages. Ask students to express their views on the issue in a brief essay.

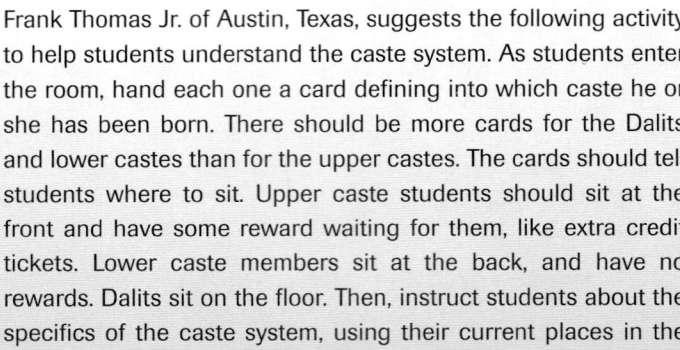

Teacher to Teacher

Frank Thomas Jr. of Austin, Texas, suggests the following activity to help students understand the caste system. As students enter the room, hand each one a card defining into which caste he or she has been born. There should be more cards for the Dalits and lower castes than for the upper castes. The cards should tell students where to sit. Upper caste students should sit at the front and have some reward waiting for them, like extra credit tickets. Lower caste members sit at the back, and have no rewards. Dalits sit on the floor. Then, instruct students about the specifics of the caste system, using their current places in the room as examples when appropriate.

Daily Life

Jainism Some 5 million Indians follow Jainism. The religion's name comes from the term *Jina,* which means "victor" or "conqueror." Believers strive to conquer their dependence on the world and the self, thus freeing themselves from contamination by the material world. Right belief, right knowledge, and right action lead to this liberation of the soul.

Jains may be monks or lay people. The life of a Jain monk is very austere. Some have no possessions at all—not even clothes. Others own a robe and a begging bowl. They may have a broom to sweep insects from the path and a facecloth to prevent breathing in tiny bugs. Jains' dedication to *ahimsa,* which means "nonviolence," prohibits harming any living creatures—even insects. Lay believers also live according to a strict code.

You don't say! Devout Sikhs follow several religious rules, including letting their hair grow. The long hair is caught in a knot at the top of the head and covered with a turban, which is usually made up of more than 16 feet (5 m) of cotton cloth.

VISUAL RECORD ANSWER

Possible answer: Clothing is one of the factors that set one region apart from those surrounding it.

Islam and Other Religions Islam is the largest minority religion in India, although estimates of its followers vary. Some experts think that India has the second-largest Muslim population in the world—second only to Indonesia. Perhaps 11 to 14 percent of the population is Muslim. Most Muslims live in northern India. Their presence there reflects northern India's history as the heart of former Muslim empires.

Christianity arrived in India long before the Portuguese landed there. Accounts from the A.D. 500s describe well-established Christian churches along the west coast. Today about 20 million Indians are Christian. Although Buddhism first developed in India and flourished before the Mughal period, its importance faded under later rulers. Less than 1 percent of the population is Buddhist today.

Two other religions that began in India are **Sikhism** and **Jainism**. Sikhism combines the Muslim belief in one God with the Hindu beliefs of reincarnation and karma. Because it preaches the equality of all people, Sikhism rejects the caste system. Jainism emphasizes a strict moral code based on preserving life.

 READING CHECK: (*Human Systems*) How does the role of religion in India's constitution contrast with its role in Indian society? The constitution declares government secular, but religion plays major role in Indian society.

India's Culture

Ways of life vary widely among India's people. The population includes many different ethnic groups. The majority of the population—72 percent—are descended from the Aryans. About one quarter of the people are Dravidians, most of whom live in the south. Smaller groups account for the rest. India's more than 1,000 languages and dialects reflect its many ethnic groups. More than 20 of these languages have at least 1 million speakers. The national language and the main language for about 30 percent of the people is Hindi. English is the most widely used language in commerce and politics. India is unique among the world's countries because it gives another 15 languages official status. Bengali, Marathi, Tamil, Telugu, and Urdu are examples. Sanskrit is also an official language.

Clothing India's tropical climates demand cool clothing. Many Indian women wear traditional clothing styles. The sari is a rectangle of cloth, usually about 18 feet (5.5 m) long, that is wrapped around the body.

INTERPRETING THE VISUAL RECORD

Saris are made from cotton, silk, or synthetic fabrics and often come in bright colors. Dressy versions may be richly embroidered in gold thread or trimmed with sequins and can be wrapped and draped in many ways. Women can express differences in status, age, occupation, region, or religion by how they wear their saris. **How might saris help make India a distinctive region?**

A tight-fitting blouse called a *choli* is worn underneath. Also popular are pajama-like pants worn under a tunic.

In the cities, Indian men wear typical Western clothing more often than women. However, a villager may be more comfortable in a garment such as the *lungi.* Like the sari, a *lungi* is a length of fabric wrapped around the waist. Loose-fitting pants are also popular for men. Some Indian men, particularly Sikhs, wear turbans. Sikhs wear turbans as a public symbol of their religious beliefs.

Food Many Americans are familiar with a packaged Indian spice mixture called curry powder. It flavors a gravy-based dish that is served over rice. However, real Indian curry powder does not come from a jar but is ground fresh every day. Cooks are proud of the unique mixtures they create from up to 20 herbs, spices, and seeds. Cardamom, chilies, cinnamon, cumin, mace, nutmeg, saffron, sesame seeds, and turmeric are among those often used. These spice mixtures are used in many recipes. Indian cuisine varies widely, reflecting the country's different climates, crops, and cultures.

Festivals With its wealth of ancient traditions, India has many holidays and festivals. Each religion has its special days, and other holidays mark political events. Many festivals are annual; some occur every few years.

Holi is observed mainly in northern and central India and celebrates spring and the triumph of good over evil. It is one of the most lively festivals. Everyone takes to the streets with colored water, powdered dyes, and water balloons. They smear, sprinkle, or splatter each other with color. Many other activities may be part of Holi, from wrestling tournaments to giving clothes to new brides. Pongal is a joyful three-day harvest festival of southern India. It features a parade of cattle decorated with beads, bells, and flowers.

✓ **READING CHECK:** *Places and Regions* Why does India have so many languages and such a wide range of customs? because it includes many different ethnic groups

INTERPRETING THE VISUAL RECORD

A dancer performs a peacock dance during a religious festival in Kerala in southern India. Religious festivals are celebrated throughout India and are an important part of many people's lives. All of India's major religions celebrate festivals during the year. Some festivals celebrate the seasons, phases of the Moon, or harvests. Others celebrate important dates in the lives of religious leaders. **How does this festival compare to festivals that you have seen?**

Section 2 Review

go.hrw.com Homework Practice Online
Keyword: SW3 HP25

Identify
Sanskrit, Dalits, Sikhism, Jainism

Define
pantheon, sepoys, boycott, partition, reincarnation, dharma, karma, caste

Working with Sketch Maps
On the map you created in Section 1, label Delhi. Which conqueror destroyed Delhi?

Reading for the Main Idea
1. *Human Systems* What are the origins of Hindi and Hinduism in India?
2. *Human Systems* What were some results of the partition of India?
3. *Human Systems* What are the two most widely used languages in India? How many minor languages and dialects are spoken there?

Critical Thinking
4. **Identifying Cause and Effect** How did growing global trade beginning in the 1400s affect India over time?

Organizing What You Know
5. Create a graphic organizer like the one below. Use it to identify and compare key features of India's main religions.

Hinduism	Islam	Sikhism	Jainism

 **LET'S GET STARTED**

Copy this question onto the chalkboard: *Are revolutions ever peaceful? What do you think a Green Revolution might be?* Discuss responses. Point out that a revolution can be a dramatic change in the way people do things. India's Green Revolution was a modernization of the country's agricultural practices. It changed lives as completely as a political revolution might. Tell students that in Section 3 they will learn more about what is changing—and what is staying the same—in India today.

Building Vocabulary

Write **cottage industries** and **jute** on the chalkboard. Ask students to define *cottage.* *(Possible answer: a small house)* Point out that cottage industries are those conducted in a home or small workshop, not in a large factory. Have students suggest types of economic activities that would work well as cottage industries. *(Possible answer: those that require skilled handwork instead of large machines)* Call on volunteers to read the definitions of both terms.

Section 3 RESOURCES

REPRODUCIBLE

▶ Guided Reading Strategy 25.3
▶ Readings in World Geography, History, and Culture 75 and 76

TECHNOLOGY

▶ One-Stop Planner CD–ROM, Lesson 25.3
▶ Homework Practice Online
▶ HRW Go site

REINFORCEMENT, REVIEW, AND ASSESSMENT

▶ Main Idea Activity 25.3
▶ English Audio Summary 25.3
▶ Spanish Audio Summary 25.3
▶ Section 3 Review, p. 579
▶ Daily Quiz 25.3

Section 3 India Today

READ TO DISCOVER

1. What are the main features of India's economy?
2. How does life in India's villages compare to life in its cities?
3. What challenges does India face today?

WHY IT MATTERS

Rapid population growth, such as India is experiencing, affects many countries. Use **CNNfyi.com** or other **current events** sources to learn about the problems caused by growing populations.

DEFINE

cottage industries
jute

LOCATE

Kolkata (Calcutta)
Bangalore
Mumbai (Bombay)
Chennai (Madras)
Varanasi

India's Economy

India's economy is extremely varied and is expanding rapidly. It includes many different ways of making a living, from subsistence farming to the most advanced computer technology.

Agriculture Farming is the basis of India's economy. It contributes 25 percent of the country's gross domestic product (GDP). Farms cover about half of India's total area. Some of the major crops are rice, wheat, tea, sugarcane, and sorghum. Half of the world's mangoes come from India. No country grows more peanuts, sesame seeds, or tea than India. Only China grows more rice.

Half of India's farms are less than 2.5 acres (1 ha). A majority of farmers own their land. However, as land is divided among a family's sons, the farms become smaller and thus less profitable.

In recent decades, India has become almost self-sufficient in food-grain crops. Yet if the summer monsoon rains are late, inadequate, or too heavy, some people may still go hungry. The government constantly explores ways to make the country's agriculture less dependent on the summer monsoon.

INTERPRETING THE VISUAL RECORD

A pepper harvest in northern India creates a colorful scene. Native to the Americas, peppers were first introduced to Europe by the Spanish and later spread throughout Asia. Spicy peppers are now an important part of Indian cuisine. **How did the diffusion of peppers from the Americas affect the landscape in this photo?**

ALL LEVELS: Copy the following graphic organizer onto the chalkboard, omitting the italicized answers. Have students complete it with characteristics and products produced by each sector of the Indian economy. Ask volunteers to fill in the organizer on the chalkboard. Then ask students to describe how India might expand its economy. *(Possible answers: increase power supply, strengthen infrastructure)*

Agriculture
- *basis of economy*
- *25 percent of GDP*
- *50 percent of land area*
- *rice, wheat, tea, sugarcane, sorghum, peanuts, sesame seeds, mangoes*
- *small farms*
- *depend on monsoon*
- *Green Revolution*

Cottage industries
- *work in the home*
- *employ millions of people, particularly women*
- *silk fabrics, wooden statues, silver and gold lace, other handicrafts*

India's Economy

High-tech businesses
- *concentrated in Bangalore*
- *workers part of growing middle class*

Commercial manufacturing
- *textiles leading export*
- *jute*
- *steel mills*
- *durable goods such as diesel engines and cars*

FOCUS ON SCIENCE, TECHNOLOGY, AND SOCIETY

The Green Revolution The government has made many efforts to increase food production for India's rapidly growing population. The need to grow more food was highlighted in 1943 when a disastrous famine struck eastern India. An estimated 4 million people starved to death. Although food distribution problems played a role in the famine, there was also a clear shortage of food. Government officials knew they had to increase the country's agricultural production. Their efforts, and the partial success achieved between 1967 and 1978, became known as the Green Revolution.

The Green Revolution consisted of three main elements. These were increasing the amount of cultivated land, harvesting two crops per year from existing farmland, and increasing yields with genetically improved seeds. Expanding the amount of farmland had been going on for centuries. Growing two crops each year was more difficult to accomplish. Since the monsoon rains ordinarily made only one crop possible, officials decided to use irrigation to create an artificial second rainy season. They built large and small irrigation projects to trap rainwater from the natural monsoon. Scientists also developed new strains of corn, millet, rice, and wheat that produced more grain per acre. In the areas where they were used, the new types increased yields dramatically.

Within a few decades the Green Revolution changed India from a country plagued by starvation to one that could usually feed its people. However, the Green Revolution had a downside. The irrigation projects, which required huge dams, displaced people and disrupted the environment. The new grain varieties used more dangerous pesticides and fertilizers. Many farmers who could not afford irrigation or chemicals got lower yields than they had from older varieties. Moreover, the new techniques did not work in areas where water was scarce or where farmers had no money for investment.

✓ **READING CHECK:** *Environment and Society* How did the Green Revolution affect food production and the environment in India?

Crop yields increased; need for fertilizer and pesticides increased; irrigation projects displaced people and disrupted the environment.

INTERPRETING THE VISUAL RECORD

About two thirds of all Indians are farmers. While many do not have access to advanced farming technologies, others have benefited greatly from the innovations of the Green Revolution. For example, the use of high-yielding varieties of seeds, particularly for wheat, has led to dramatic increases in some areas. **How is this farmer using animals? How might the use of more advanced technologies change the way this person farms?**

VISUAL RECORD ANSWER

in traditional manner; could allow the work to be done faster and produce higher yields

LEVEL 1: Ask students to say what the phrases *city life* and *village life* mean to them. Write responses on the chalkboard. (*Possible answers: city life—traffic, noise, lots to do, many businesses and shops, homeless people, better medical care, fast pace; village life—open spaces, dirt roads, farming, simple life, quiet, everyone knows everyone, livestock*) Then lead a discussion about how India's cities and villages fit these characteristics. **ENGLISH LANGUAGE LEARNERS**

LEVELS 2 AND 3: Have students review the Level 1 activity. Next, have students write an entry for a biography based on daily life in India. Within each pair, ask one student to imagine that he or she is a city dweller. Ask the other student to assume the role of a villager. When students have completed their biography entries, have them share their work with their partners. Ask for volunteers to read their entries to the class. **COOPERATIVE LEARNING**

Daily Life

Ikat Weaving Ikat (EE-kaht) is a traditional weaving process and an important cottage industry in India. The term, from the Malay *mengikat,* which means "to tie or bind," also refers to the silk or cotton fabric created by the process.

Ikat requires careful preparation because the threads are dyed according to a plan *before* they are placed on the loom. Most patterned fabrics are created by stamping the design on the finished woven material. However, for ikat, the weaver first ties the warp, or lengthwise threads, with strips of bicycle inner tubes, leaves, or plastic. Then the weaver dips the thread in dye. Areas that had been tied do not absorb the color. When the warp threads are dried and woven with plain weft, or crosswise threads, a pattern emerges. The pattern has a blurred effect because the colored areas do not line up perfectly.

ACTIVITY: Have students use cotton string and fabric dyes to experiment with the ikat technique. They might create simple looms to weave their "fabric."

VISUAL RECORD ANSWER

The income helps them buy food, clothing, and other necessities.

INTERPRETING THE VISUAL RECORD

Many Indian women work in cottage industries. Here, women in Rajasthan dye fabrics that they will sell in local markets. **How do you think cottage industries help people meet their basic needs?**

Industry Although India is largely agricultural, its industrial production ranks about tenth among the world's countries. When its GDP is distributed among its billion-plus population, however, the per capita GDP is very low. As a result, most geographers and economists describe India as a developing country.

Millions of Indians make a living by working at home in small-scale industries called **cottage industries**. Many of these skilled workers are women. They weave silk fabrics, carve wooden statues, and make silver and gold lace, as well as many other beautiful handicrafts.

India also has large-scale commercial manufacturing. The country's many cotton and woolen mills produce textiles, the leading export. Factories in Kolkata (Calcutta) process **jute**, a plant fiber used to make burlap and rope. Factories use the steel from India's own steel mills to make durable goods such as diesel engines and cars.

Industry requires power. To increase its industrial production, India must increase its power supply. The country already imports large amounts of oil. India's rivers could supply additional hydropower, but dams can create new problems. To attract more foreign investment, India must strengthen its infrastructure. Airlines, communications systems, railroads, ports, and roads must be improved. These systems are needed to operate factories, move materials and products, and help people do their jobs efficiently. In India most of these systems lag behind what is available in other countries.

The government has been successful in attracting high-tech businesses. For example, Bangalore, in southern India, has so many computer companies that some have compared it to the Silicon Valley of the United States. (See Cities & Settlements: Bangalore.) Many high-tech and service industry workers belong to a growing middle class.

READING CHECK: *Human Systems* What are some ways in which India's market economy is improving? increased agricultural and industrial production, more high-tech businesses

Cities and Villages

India has more than half a million villages, where more than 70 percent of the people live. Most of these villages have fewer than 1,000 residents. At the same time, India has enormous cities. Two of them—Mumbai (Bombay) and Kolkata—rank fifth and eighth among the world's largest cities. More Indians are moving into the cities to look for work, but finding none. In fact, India's big cities are growing about twice as fast as its small towns.

City Life Crowding, noise, smog, and traffic are part of daily life in India's cities. In general, city dwellers wear European-style clothing and read

Teach Objective 3

🌐 **LEVEL 1:** Copy the following statements onto the chalkboard: *India's greatest challenge is curbing population growth and reducing poverty. India's greatest challenge is halting water pollution, deforestation, overgrazing, and soil erosion. India's greatest challenge is reducing conflict with Pakistan and other neighbors, as well as reducing tensions among ethnic and religious groups.* Have students work with a partner to decide which statement they feel is most accurate. Ask volunteers to share their observations and justifications for them. **COOPERATIVE LEARNING**

🌐 **LEVEL 2:** Have students work in small groups of two to three students to create reports for the nightly news documenting one of India's major challenges. Students should use information from the text but may also supplement the information with additional resources. Encourage students to give their report as an actual news feature by videotaping the report and showing it to the class. **COOPERATIVE LEARNING**

English-language newspapers. Most work in factories and offices. Higher education is available in the cities, and women work alongside men in professional and factory jobs.

A small number of rich businesspeople and landowners live in the cities. The working middle class makes up a larger part of urban society. Most live in small apartments and can afford many manufactured goods. However, most of the urban population lives in poverty. Many are homeless and live in the streets. Many others live in shacks built from scraps of cardboard, metal, and wood. They have no clean water or sanitation. Each day they must look for work.

Mumbai has about 18 million residents. It is the center of India's huge movie industry. So many films are produced there that the city has been nicknamed Bollywood. Kolkata has more than 12 million residents and is one of the world's busiest seaports. Although it has a large homeless population, Kolkata is a vibrant place that promotes itself as the City of Joy. Delhi is the hub of northern India. The city's old section contains beautiful Mughal buildings. The new part of the vast city—New Delhi—is India's capital. The East India Company founded Chennai (Madras) on the southeast coast. It is now a major rail and industrial center. Varanasi, on the Ganges River, is a holy city for Hindus.

Village Life For millions of India's villagers, life goes on as it has for generations before them. Only recently have paved roads, electricity, or telephone services reached many villages. Some villages can only be reached on foot. Sanitation is still poor, and many people do not have clean drinking water. Medical services are very limited. More than half of the rural population cannot read or write.

Farm families tend to be large because children have to help in the fields. In addition, both Hindu and Muslim cultures value large families. In general, sons are valued more highly than daughters. They will also take care of their parents when they are old. Therefore, many rural couples continue to have children until they have at least two boys. Men dominate village society. When a woman marries, she becomes a member of her husband's family.

✓ **READING CHECK:** (*Human Systems*) Why do many people leave the villages for the cities? What do they find there? *come to find work, but many unsuccessful; find poverty and homelessness instead*

Challenges

India remains a country of contrasts. While millions struggle with extreme poverty and inequality, a few have wealth and privilege. Life for many Indians has improved in recent decades, but the country faces several issues that may cloud its future.

Population and Poverty Many of India's problems are connected to rapid population growth.

Our Amazing Planet

In Mumbai (Bombay), some 20,000 people known as *dhobis* make a living washing clothes. The *dhobi* identifies the laundry's owner with marks most people cannot see. Sometimes police identify bodies by tracing *dhobi* marks on the clothing.

Kolkata (Calcutta), former capital of British India from 1773 to 1912, is one of India's largest cities. It has a very high population density of about 85,500 persons per square mile, and overcrowding is a major problem. Yet, Kolkata continues to attract new immigrants, particularly from nearby rural areas of eastern India.

Essential Element 4

▶ **Human Systems** ◀

Doing Lunch in India
Religious, ethnic, and/or caste identity are very important to most Indians. Membership in a group may be expressed in many ways, including dietary laws. The different groups may prohibit eating certain meats, spices, vegetables, or other foods. They may also obey various rules for food preparation.

To ensure that their lunches meet all the requirements, many Mumbai workers hire *dabbawalas,* men who pick up just-cooked lunches from the employees' homes and deliver them in metal pails—*dabas*—to the workplace. The lunches get to downtown Mumbai by bicycle, train, and pushcart. After lunch, the *dabbawalas* collect the pails and return them to the employees' homes. In this way, some 3,000 deliverymen and 2,000 supervisors provide more than 100,000 tasty, familiar, and ritually pure lunches every day.

DISCUSSION: Lead a discussion about rituals related to food that students observe. Ask students to reflect on the origins of those rituals—religion, family traditions, popular culture, or other origins.

🔲 **internet** connect

go hrw .com

GO TO: go.hrw.com
KEYWORD: SW3 CH25
FOR: Web sites about daily life in India

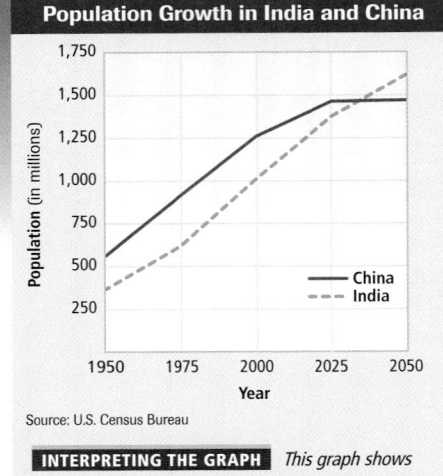

Population Growth in India and China

Source: U.S. Census Bureau

INTERPRETING THE GRAPH *This graph shows past population growth and future projections for the world's two most populous countries.* **Which country has a larger population today? Which country is projected to have a larger population in 2050?**

Because of improved sanitation and health care, the death rate has dropped much faster than the birthrate. The result is a high population growth rate. In addition, more than a third of the people are 14 or younger. Therefore, as this group enters the child-bearing years the population will continue to grow even if the birthrate drops. Government programs encouraging smaller families have had limited success.

The large population often overwhelms services, from communication to transportation. Millions of Indians in both rural and urban areas are still dreadfully poor. Many people never attend school, and this lack of education limits their opportunities.

Environmental Concerns A major challenge for India is to clean up and protect its environment. As much as 70 percent of India's surface water may be polluted. Poorly regulated industries and inadequate sewer systems are partly to blame.

India's large population has strained the country's land resources. Deforestation, overgrazing, and soil erosion have caused serious environmental damage. In fact, half of India's forests have been cut down since the country became independent. The government does, however, support conservation. It has created 80 national parks, 440 wildlife sanctuaries, and 23 tiger reserves. Once on the brink of extinction, India's tigers now number in the thousands. Their future is not secure, however. Illegal trade in tiger skins and bones continues.

INTERPRETING THE MAP *India's political geography has changed greatly since independence in 1947. Over time, India's states have been reorganized to better reflect the country's major language and culture regions. For example, in 1960 the state of Bombay was divided into Gujarat and Maharashtra based mainly on language differences. In 1961 Nagaland became a state for the Naga people after an armed struggle with the national government. Three new states were created in 2000, and India's internal borders seem likely to change further in the future.* **How have forces of conflict and cooperation affected India's political geography?**

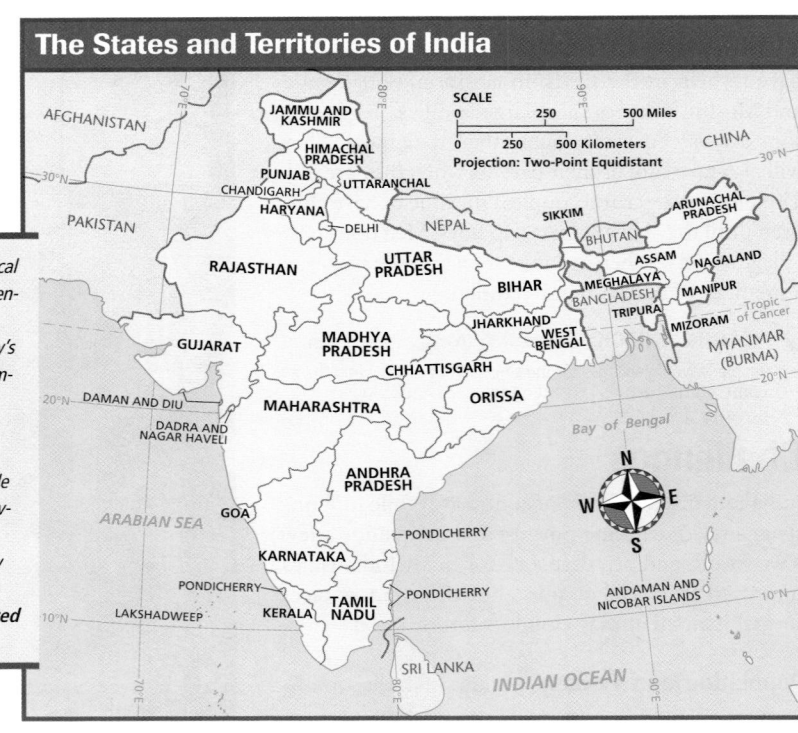

The States and Territories of India

Political Issues India faces challenges from neighboring countries and its own citizens. Ever since the partition, India and Pakistan have clashed over the region of Kashmir in the far northwest. At the time of partition, most people in Kashmir were Muslims. However, unrest in the region led Kashmir's ruler to seek a union with India. This decision led to an armed struggle between India and Pakistan over control of the disputed region. Today Pakistan controls the northern and western areas of Kashmir. India controls southern and southeastern Kashmir as part of the state of Jammu and Kashmir. A line of control separates the two areas of Kashmir. This line was agreed to in 1972 after several clashes between Indian and Pakistani troops. Many Muslim Kashmiris want the entire region to be part of Pakistan. Their movement and the government's efforts to crush it have resulted in the deaths of about 10,000 civilians since 1989. Because both India and Pakistan have nuclear weapons, this tension creates concern around the world. The two countries also disagree over how much water India should take from the upper Indus River. India has border disputes with China as well. China had never accepted the boundaries of northeastern Kashmir. In the 1950s Chinese forces entered Ladakh, a region in Kashmir. This led to border clashes between India and China. Today China controls the northeastern part of Ladakh. However, India does not recognize Chinese claims in the region.

Conflicts between India's Hindus and Muslims sometimes erupt into violence. Leaders from each group have accused the government of favoring the other. Sikhs have also demanded more political power in Punjab. In the early 1980s some Sikh extremists began to call for an independent country of their own called Khalistan, or "Land of the Pure." They began a campaign of terrorism against India's government and people. Ethnic groups in other parts of India have also demanded more autonomy, with violence the frequent result.

✓ **READING CHECK:** *Environment and Society* What issue underlies several of India's challenges? rapid population growth

A Sikh elder prays at the Golden Temple in Amritsar, India. The Golden Temple is the spiritual center for the world's Sikhs and has figured prominently in their struggles with India's government. In 1984, Indian troops attacked Sikh extremists who were using the temple as a fortress and refuge.

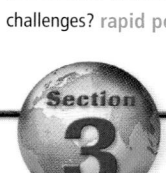

Section 3 Review

go. hrw .com
Homework Practice Online
Keyword: SW3 HP25

Define cottage industries, jute

Working with Sketch Maps
On the map you created in Section 2, label Kolkata (Calcutta), Bangalore, Mumbai (Bombay), Chennai (Madras), and Varanasi. Why is Mumbai sometimes called Bollywood?

Reading for the Main Idea
1. *Environment and Society* In what ways was the Green Revolution unsuccessful?

2. *Human Systems* What types of goods are produced in India's cottage industries? In its commercial industries?

3. *Human Systems* Why do geographers consider India a developing country even though the country's industrial production is among the highest in the world?

Critical Thinking
4. **Analyzing Information** Why is control of Kashmir disputed? Why do you think this conflict has lasted for more than 50 years?

Organizing What You Know
5. Copy the graphic organizer shown below. Fill in each oval with one of the major challenges facing India. Then draw arrows among the ovals to show how the issues are related. Write sentences or phrases near the arrows to describe the connections.

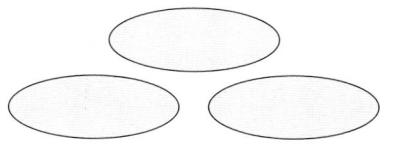

Sequencing

Have students read Cities & Settlements: Bangalore. Emphasize that the city has undergone dramatic changes over the centuries of its existence. Tell students that Bangalore was founded in 1537 by a chieftain, who received the land from an emperor. According to a legend, the city's name comes from *Benda kaalu*, which means "boiled beans." Supposedly, the emperor was lost in the area, and he received a humble meal of boiled beans from a local woman.

From these shadowy beginnings Bangalore became a thriving regional city. From 1831 to 1881 it was British India's military and civil administration station. Now Bangalore has high-tech industries, colleges, universities, and research institutions.

Call on a volunteer to draw a series of boxes connected by arrows on the chalkboard. This sequence diagram is to represent key periods and events in Bangalore's history. Ask remaining class members to determine an appropriate number of boxes, the periods they represent, and the labels for the various events.

Essential Element 5

► Environment ◄ and Society

Snakes in Bangalore

Rapid growth in Bangalore has resulted in the construction of buildings where gardens once flourished. As a result, snakes have been driven from their natural surroundings into areas inhabited by people. Though most of the snakes are not poisonous, they still concern Bangalore residents. Animal welfare groups try to educate children about these reptiles. Measures have also been taken to rescue snakes from the city and release them at Bannerghatta National Park.

MAP ANSWER

Cubbon Park, Botanical Gardens, lakes

Applying What You Know Answers

1. Answers will vary. Students should make connections between the various changes.

2. Possible answers: offer tax incentives to businesses that return to the city, convince apartment owners to offer lower rents to people who work in the city, upgrade utilities, provide better mass transit, and so on

This Cities & Settlements feature addresses National Geography Standards 11 and 12.

CITIES & SETTLEMENTS

Bangalore

Places and Regions Located on the Deccan Peninsula, Bangalore is India's sixth-largest city. With nearly 5 million people, it is one of the fastest-growing cities in Asia. This tremendous growth has brought much change to the city. Bangalore's development illustrates some of the forces that shape a city and its culture. It also shows how a rapidly changing urban environment can affect people's lives.

Garden City to Silicon City

Bangalore may have the most pleasant climate of any Indian city. Humidity is low, and summer temperatures average in the 80s. Even in winter it seldom gets colder than 60°F (16°C). Rainfall is also moderate, averaging less than 40 inches (102 cm) per year. These conditions inspired Bangalore's former Indian and British rulers to create many gardens, lakes, and parks throughout the city. Trees were planted along many streets to shade and beautify them. Because it had such a lovely environment, Bangalore was often called the Garden City. It became a popular vacation and retirement spot for wealthy Indians. It was also a headquarters for British officials during part of the colonial era. Many of these wealthy Indians and British officials built grand mansions in the city.

In the 1950s and 1960s Bangalore's pleasant environment helped to make it India's scientific research center. Indian leaders believed the city would provide a relaxed atmosphere in which scientists could be creative. To support scientific research in Bangalore,

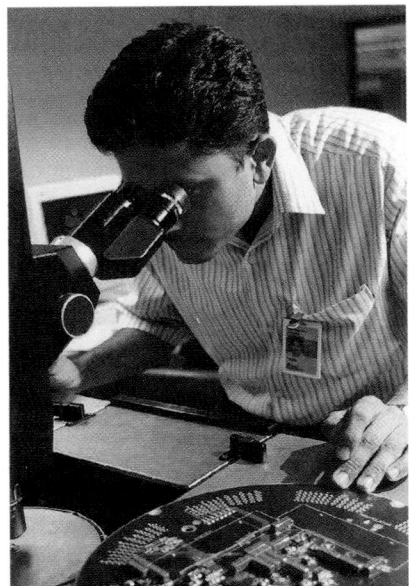

A chip designer inspects a computer chip in Bangalore. The city is home to hundreds of technology companies employing an estimated 75,000 people.

India's government invested heavily in the city. As a result, Bangalore now has three universities, 14 engineering colleges, 47 technical schools, and many research institutes. These include India's most advanced military and space research centers and several private industrial training institutes. These developments brought a large number of highly educated people to Bangalore. Eventually, they helped transform the Garden City into an Indian version of Silicon Valley in the United States.

Booming Bangalore

Because a large percentage of India's population does not use computers, the country is generally not thought of as a high-tech leader. For this reason, you may be surprised to know that India is second only to the United States in computer software exports. Most of this software comes from Bangalore. At first, the city's major export was not software. It was people. In the 1980s American companies began to computerize their operations on a wide scale. Yet there were not enough American software engineers and programmers to fill all the jobs in those fields. As a result, many of Bangalore's computer science graduates left to take high-paying jobs in the United States. However, two developments soon turned Bangalore into a boomtown. American companies found that computer science salaries in India were one fourth the size of those elsewhere. In addition, improved telecommunications made it easier for American companies to contract with Indian software firms.

Going Further: Thinking Critically

Organize the class into groups and have each group select a major world city. Examples might include Buenos Aires, London, Delhi, New York, Paris, Rio de Janeiro, Rome, or Shanghai. Ask students to conduct research on that city's history and create sequence diagrams on butcher paper of the city's major eras and events. Then have each group collect tourist information for the chosen city. They might acquire brochures and other material from travel agents, along with information from the Internet and library resources. Then have students examine the tourist information for bridges, churches, forts, gardens, palaces, streets, and other places that figured prominently in the city's major events. Ask each group to affix pictures of these places and phrases describing them in the proper order on the sequence diagram. Display the sequence diagrams. Then lead a discussion about what the cities have in common and how they differ. **COOPERATIVE LEARNING**

Bangalore's booming high-tech economy has created an increase in traffic congestion and air pollution.

They could also set up their own computer operations in India. Soon more than 100 Indian, American, and multinational computer firms sprang up in Bangalore.

A Changing City

Bangalore changed quickly. Some of the grand old homes were torn down to make room for new high-rise office buildings. Others were gutted and filled with cubicles to become offices. Still others were divided into apartments for the flood of people who came to fill new software jobs. Thousands of cars and motorcycles soon clogged once quiet streets where horse-drawn carriages had traveled as late as the 1970s. The air became more polluted. Bangalore's explosive growth put a strain on its other systems as well. An overloaded electrical system began causing power outages, sometimes several times a day. In addition, the demand for housing and office space caused living and business costs to skyrocket.

Some of the high-tech companies have left Bangalore. Many are now located in so-called technology parks that have sprung up in nearby suburbs. These parks have their own water, power, and communication systems. Some also provide housing, schools, stores, and recreational facilities. Company employees thus have little need to go into the city. In fact, many buildings within Bangalore stand empty. In some ways, the Garden City's rapid growth has wilted. The city government is looking at ways to revive the city's glory.

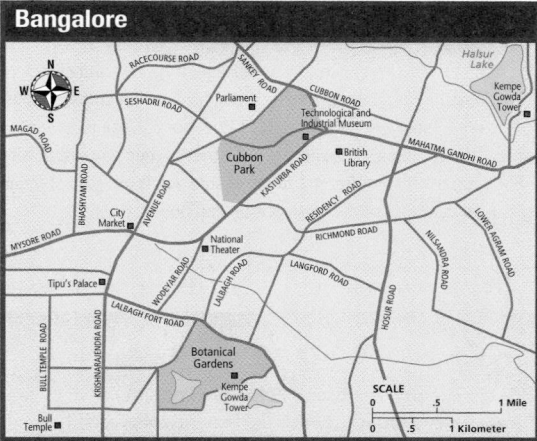

Bangalore

INTERPRETING THE MAP *The heart of Bangalore includes important governmental and cultural buildings, as well as parks, gardens, and lakes. **What features on the map do you see that might have inspired people to call Bangalore the Garden City?***

Applying What You Know

1. **Comparing** How does Bangalore's experience compare to cities or communities in your region that have grown rapidly? How have the political, economic, social, and environmental relationships in those places changed?

2. **Problem Solving** What actions might Bangalore's government take to revive the city?

TECHNOLOGY

▶ Chapter 25 Test Generator (on the One-Stop Planner)
▶ Global Skill Builder CD–ROM
▶ HRW Go site

REINFORCEMENT, REVIEW, AND ASSESSMENT

▶ Chapter 25 Review, pp. 582–83
▶ Chapter 25 Tutorial for Students, Parents, Mentors, and Peers
▶ Chapter 25 Test (form A or B)
▶ Alternative Assessment Handbook

▶ Chapter 25 Test for English Language Learners and Special-Needs Students
▶ Unit 8 Test
▶ Unit 8 Test for English Language Learners and Special-Needs Students

Assess

Have students complete a Chapter 25 Test.

Reteach

Create slips of paper with various topics from the chapter written on them. Examples include monsoons, Mughals, agriculture, and environmental concerns. Have students draw a slip from a hat and create a web page on the topic for an Internet site called everythingaboutindia.com. Display and discuss the pages. **ENGLISH LANGUAGE LEARNERS**

CHAPTER 25 Review Answers

Thinking Critically

1. Answers may vary. Tilt allows more of the Sun's energy to hit Earth at a particular time. (NGS 7)

2. Possible answers: cities—offer a range of goods and services; villages—have few modern conveniences, limited medical services (NGS 12)

3. More of the country's resources might need to be diverted to sustain the population rather than for export or further development. (NGS 11)

Using the Geographer's Tools

1. large—Maharashtra, Madhya Pradesh, Rajasthan, Andhra Pradesh; small—Punjab, Himachal Pradesh, Kerala, Mizoram, Tripura, Manipur, Nagaland; that distribution of power and resources may not be equal

2. more than 2 billion

3. highest—Brahmins; lowest—Sudras; Dalits

Writing

Students should discuss India's economy and should include what types of goods are produced in the industries, who works in the industries, and how these industries differ. Use Rubric 9, Comparing and Contrasting, to evaluate student work.

CHAPTER 25 Review

Building Vocabulary

On a separate sheet of paper, explain the following terms by using them correctly in sentences.

subcontinent	partition	caste
pantheon	reincarnation	cottage industries
sepoys	dharma	jute
boycott	karma	

Locating Key Places

On a separate sheet of paper, match the letters on the map with their correct labels.

Himalayas	Western Ghats	Brahmaputra River
Ganges River	Eastern Ghats	Thar Desert
Gangetic Plain	Narmada River	Kolkata (Calcutta)

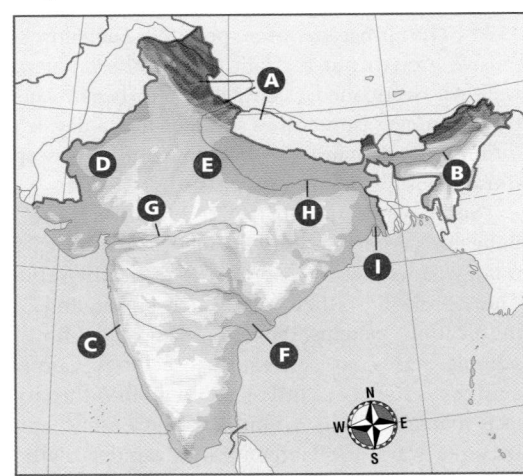

Understanding the Main Ideas

Section 1

1. (*Places and Regions*) What geographic features, such as river systems and climates, characterize India's three major landform regions?

2. (*Physical Systems*) At what time of the year do monsoons bring rainfall to India, and where are the rains the heaviest?

Section 2

3. (*Human Systems*) In what ways have invasions, trade, and outside control affected Indian society and culture?

4. (*Human Systems*) What is India's main religion? Which other religions are found there?

Section 3

5. (*Places and Regions*) What conflicts does India have with Pakistan?

Thinking Critically

1. **Drawing Inferences and Conclusions** How does the timing of the summer monsoons relate to the tilt of Earth on its axis?

2. **Comparing and Contrasting** How does Indian village life compare to city life?

3. **Making Generalizations and Predictions** How might rapid population growth make it hard for India to develop its economy?

Using the Geographer's Tools

1. **Analyzing Maps** Study the map of India's states and territories in Section 3. What are some of India's largest states? What are some of the smallest? What might this indicate about the distribution of political power in India?

2. **Analyzing Graphs** Study the graph of population growth in India and China in Section 3. What was the combined population of India and China in 2000?

3. **Preparing Graphs** Create a pyramid graph of India's four major castes, using the information in Section 2. At the top of the pyramid, list India's highest caste. At the bottom, list the lowest caste. Which group is below the lowest caste?

Writing about Geography

Write an article comparing India's cottage industries and commercial industries. What types of goods are produced in cottage industries? Who makes up the workforce? How are cottage industries different from India's commercial industries? When you are finished with your article, proofread it to make sure you have used standard grammar, spelling, sentence structure, and punctuation.

S K I L L   B U I L D I N G

Geography for Life

Skills Activity: Analyzing Political Instability

(*Places and Regions*) Review the discussion of political issues in Section 3. Which areas in India are politically unstable? How is religion a factor in the country's internal and external conflicts? How does political instability in India compare to political instability in other areas of the world that you have studied? Generally, what are some of the forces that affect political stability?

Portfolio Activity

Have students investigate the role of women in Indian history and society. You might have them concentrate on particular women, such as the following: Lakshmi Bai, the Rani of Jhansi, who led her own rebel troops during the Indian Mutiny; Indira Gandhi, prime minister from 1966 to 1977 and from 1980 to 1984; and Phoolan Devi, who became infamous as a bandit leader but then became a member of India's parliament. Ask students to examine advances in women's rights and the challenges Indian women still face. Then have them create a mural to show their findings. Photograph the mural and place photos in students' portfolios.

Food Festival

India's flavorful, highly varied cuisine reflects the country's many ethnic groups, climates, and vegetation regions. There is no single food that represents India well. Have students conduct research on Indian food and work in groups to prepare sample recipes. Some dishes they might try are tandoori chicken, vegetable or meat curries, rice casseroles called *biryanis,* flat bread called naan, and sweet-and-sour sauces called chutneys. There are many versions of all these foods. Most of the essential ingredients, such as basmati rice, which was originally from India, are available in American supermarkets.

Building Social Studies Skills

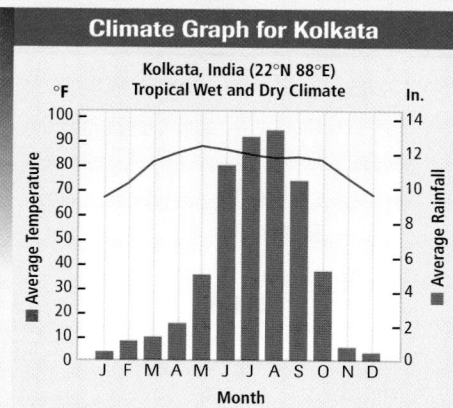

Climate Graph for Kolkata

Kolkata, India (22°N 88°E)
Tropical Wet and Dry Climate

Interpreting Graphs

Use the climate graph to answer the following questions.

1. During which months is the average rainfall above 12 inches?
   a. June, July
   b. April, May
   c. August, September
   d. July, August

2. Which three months would be the least favorable to explore Kolkata by car? Why?

Analyzing Primary Sources

Read the following excerpt from an article that Geoffrey C. Ward wrote after visiting India during the 50-year anniversary of its independence from Britain. Then answer the following questions.

> "For all its newfound modernism India remains steeped in religion. The pious cacophony [discordant sounds] I hear from my window each dawn attests to that. First comes chanting from a Sikh gurdwara [temple], which is soon partly drowned out by the sound of temple bells and the voice of a priest offering prayers from a Hindu temple dedicated to Siva. Then, louder than the rest, comes the wobbly tenor of a Muslim muezzin [crier], proclaiming the greatness of Allah from a mosque."

3. Which religion is not mentioned in the excerpt?
   a. Islam
   b. Judaism
   c. Hinduism
   d. Sikhism

4. In what ways does the writer imply that the past and traditions are still important to modern India?

Geography for Life

Answers should discuss the politically unstable areas of India, how this instability compares to other areas in the world experiencing political instability, and how geography can affect political stability. Students should support their opinions. Use Rubric 37, Writing Assignments, to evaluate student work.

Social Studies Skills

1. d

2. June, July, August; months with greatest rainfall, making travel difficult

3. b

4. Possible answer: Modern India remains concerned about religion.

Alternative Assessment

PORTFOLIO ACTIVITY

Learning about Your Local Geography

Individual Project: Research
Plan, organize, and complete a research project on population growth in your community. In your report answer the following questions: What is the history of population growth in your community during the last 50 years? How many people live in your community today? Is the population increasing or decreasing? What are the reasons for the trend? What do you think will happen in the next 20 years? How has population change affected life in your community?

☑ internet connect

Internet Activity: go.hrw.com
KEYWORD: SW3 GT25

Choose an activity on India to:
- explore the regions of India.
- take the India GeoMap Challenge!
- learn about economic and social issues in India today.

PORTFOLIO ACTIVITY

Students' reports will vary according to sources used but should include the community's population history, growth, and projection. Use Rubric 30, Research, to evaluate student work.

CHAPTER RESOURCE MANAGER

Objectives	Pacing Guide	Reproducible Resources
SECTION 1 **Natural Environments** (pp. 585–88) • Describe the main physical features of the Indian Perimeter. • Identify the types of climates, plants, and animals found in the region. • Identify the natural resources of the region's countries.	**Regular** 1 day **Block Scheduling** .5 day *Block Scheduling Handbook, Chapter 26*	**RS** Guided Reading Strategy 26.1 **SM** Geography for Life Activity 26: Irrigation's Impacts on Pakistan **IC** Lab Activity for Geography and Earth Science, Demonstration 10: Rising Mountains
SECTION 2 **History and Culture** (pp. 590–93) • Describe the history of the countries of the Indian Perimeter. • Analyze some characteristics of the region's cultures.	**Regular** 1.5 days **Block Scheduling** 1 day *Block Scheduling Handbook, Chapter 26*	**RS** Guided Reading Strategy 26.2 **E** Creative Strategies for Teaching World Geography, Lesson 20 **E** Cultures of the World Activity: Region 8 **SM** Critical Thinking Activity 26: A Difficult Birth: The Creation of Pakistan and Bangladesh
SECTION 3 **The Region Today** (pp. 594–97) • Identify the basis of the economies of the Indian Perimeter. • Analyze settlement patterns in the region. • Explain the challenges these countries will face in the future.	**Regular** 1 day **Block Scheduling** .5 day *Block Scheduling Handbook, Chapter 26*	**RS** Guided Reading Strategy 26.3 **RS** Graphic Organizer Activity 26 **PS** Readings in World Geography, History, and Culture 77 and 78 **SM** Map Activity 26: Trekking in Nepal

Chapter Resource Key

PS Primary Sources

RS Reading Support

IC Interdisciplinary Connections

E Enrichment

SM Skills Mastery

A Assessment

REV Review

ELL Reinforcement and English Language Learners

 Transparencies

 CD–ROM

 Video

 Internet

 Holt Presentation Maker Using Microsoft® PowerPoint®

  One-Stop Planner CD–ROM

See the *One-Stop Planner* for a complete list of additional resources for students and teachers.

 One-Stop Planner CD–ROM

It's easy to plan lessons, select resources, and print out materials for your students when you use the **One-Stop Planner CD–ROM with Test Generator**.

internet connect

HRW ONLINE RESOURCES

GO TO: go.hrw.com
Then type in a keyword.

TEACHER HOME PAGE
KEYWORD: SW3 Teacher

CHAPTER INTERNET ACTIVITIES
KEYWORD: SW3 GT26
Choose a topic on the Indian Perimeter to:
- publish a poster on the landscape and culture of the Himalayas.
- learn the history of Sri Lanka.
- examine Pakistani history and culture.

CHAPTER ENRICHMENT LINKS
KEYWORD: SW3 CH26

CHAPTER MAPS
KEYWORD: SW3 MAPS26

ONLINE ASSESSMENT
Homework Practice
KEYWORD: SW3 HP26
Standardized Test Prep
KEYWORD: SW3 STP26
Rubrics
KEYWORD: SS Rubrics

COUNTRY INFORMATION
KEYWORD: SW3 Almanac

CONTENT UPDATES
KEYWORD: SS Content Updates

HOLT PRESENTATION MAKER
KEYWORD: SW3 PPT26

ONLINE READING SUPPORT
KEYWORD: SS Strategies

CURRENT EVENTS
KEYWORD: S3 Current Events

Technology Resources

- One-Stop Planner CD–ROM, Lesson 26.1
- Geography and Cultures Visual Resources 55–59
- Homework Practice Online
- HRW Go site

- One-Stop Planner CD–ROM, Lesson 26.2
- **CNN** Presents Geography: Yesterday and Today, Segment 33: The Tengboche Monastery
- Homework Practice Online
- HRW Go site

- One-Stop Planner CD–ROM, Lesson 26.3
- *ARGWorld* CD–ROM
- Homework Practice Online
- HRW Go site

Reinforcement, Review, and Assessment

- **ELL** Main Idea Activity 26.1
- **ELL** English Audio Summary 26.1
- **ELL** Spanish Audio Summary 26.1
- **REV** Section 1 Review, p. 588
- **A** Daily Quiz 26.1

- **ELL** Main Idea Activity 26.2
- **ELL** English Audio Summary 26.2
- **ELL** Spanish Audio Summary 26.2
- **REV** Section 2 Review, p. 593
- **A** Daily Quiz 26.2

- **ELL** Main Idea Activity 26.3
- **ELL** English Audio Summary 26.3
- **ELL** Spanish Audio Summary 26.3
- **REV** Section 3 Review, p. 597
- **A** Daily Quiz 26.3

Meeting Individual Needs

Ability Levels

Level 1 Basic-level activities designed for all students encountering new material

Level 2 Intermediate-level activities designed for average students

Level 3 Challenging activities designed for honors and gifted-and-talented students

English Language Learners Activities that address the needs of students with Limited English Proficiency

Chapter Review and Assessment

- Chapter 26 Test Generator (on the One-Stop Planner)
- Global Skill Builder CD–ROM
- HRW Go site
- **REV** Chapter 26 Review, pp. 600–01
- **REV** Chapter 26 Tutorial for Students, Parents, Mentors, and Peers
- **A** Chapter 26 Test (form A or B)
- **A** Alternative Assessment Handbook
- **A** Chapter 26 Test for English Language Learners and Special-Needs Students

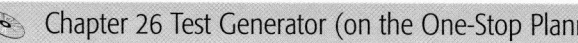

Launch into Learning

Ask students what they already know about Mount Everest, the tallest mountain in the world. *(Students may mention cold temperatures, avalanches, climbing expeditions, location in the Himalayas, or other connections.)* Point out that in 1996, a fierce storm on Everest contributed to the deaths of eight climbers, the worst human disaster to occur on the mountain. Tell students that in this chapter they will learn more about the Himalayas, other environments of the Indian Perimeter, and how those environments affect people's lives.

Using the Physical-Political Map

Have students examine the map on the opposite page. Discuss what characteristics the mainland Indian Perimeter countries of Pakistan, Nepal, Bhutan, and Bangladesh may share and how the island countries of Sri Lanka and the Maldives may be similar.

Why We Should Know More

You may wish to point out to students that there are many reasons why we should know more about the countries of the Indian Perimeter, these among them:

▶ The countries of the Indian Perimeter contain sites of global historical and cultural interest. Spectacular landscapes have drawn adventurers and tourists to the region.

▶ Pakistan and India have nuclear weapons. Political conflicts between the countries could gain global significance.

▶ Some countries of the region are very poor and have rapidly growing populations. People in other parts of the world may benefit also if these countries develop their resources and limit population growth.

CHAPTER 26
The Indian Perimeter

The countries of the Indian Perimeter have very different landscapes and cultures. In this chapter, you will learn about each country's natural environments, history and culture, and situation today.

Detail of a rug from Bhutan

Bragpa man in Bhutan

A Saleem Ale Hem (God's peace be upon you)! I am Rehan. I am an only child and live with my parents in Karachi, a big sprawling city like Los Angeles. On one side is the sea, and on the other is the desert. I love Pakistan. It is a poor country, but it is very beautiful.

I really like to watch Indian movies. Even though relations between India and Pakistan are tense, Indian movies are very popular. Movies open at the same time in India and Pakistan. We do not make movies here.

I attend a boys' private school styled after the British public school system. The school is 157 years old. After school I play tennis or cricket before going home. Our house is connected to my grandparents' house, and we usually eat our meals all together. Dinner is almost always rice with goat meat or beef, along with lentils or vegetables. We never eat pork because we are Muslim and it is forbidden.

Next year, I am going to go to America with my mother. My parents want me to go to a world-class university. Also, Pakistan is not safe. Even in big cities like Karachi, the crime that upper-class people fear most is kidnapping. My father is a doctor. He will have to stay in Pakistan to take care of his parents and keep up his medical practice. My father has to care for his parents because both his brothers are living in the United States.

584

Section 1

OBJECTIVES

1. Describe the main physical features of the Indian Perimeter.

2. Identify the types of climates, plants, and animals found in the region.

3. Identify the natural resources of the region's countries.

LET'S GET STARTED

Copy the following passage onto the chalkboard: *In 1998 Bangladesh suffered from flooding so severe that two thirds of the land was under water. Use the maps in this unit's atlas to list factors that may have contributed to the flooding. Discuss responses. (Students might mention low elevation, heavy rains, the monsoons, or the large number of rivers.)* Tell students that in Section 1 they will learn about how the region's physical geography affects the environment and life in the region.

Building Vocabulary

Write **storm surge** on the chalkboard. Call on a volunteer to read the definition from the text or glossary. Ask students if they have heard the term before and in what context. *(Students may have heard storm surge warnings associated with hurricanes on local or national weather forecasts.)* You may want to ask if any students who have witnessed a storm surge would like to describe the experience.

Section 1

Natural Environments

READ TO DISCOVER

1. What are the main physical features of the Indian Perimeter?
2. What types of climates, plants, and animals are found there?
3. What natural resources do countries in the region have?

WHY IT MATTERS

Bangladesh is often hit by devastating tropical storms, which can kill people and destroy homes and crops. Use **CNNfyi.com** or other **current events** sources to learn about the effects of such storms.

DEFINE

storm surge

LOCATE

Indus River	Tarai
Baluchistan	Palk Strait
Thar Desert	Chittagong Hill Tracts
Mount Everest	

📶 **internet** connect

GO TO: go.hrw.com
KEYWORD: SW3 CH26
FOR: Web sites about the Indian Perimeter

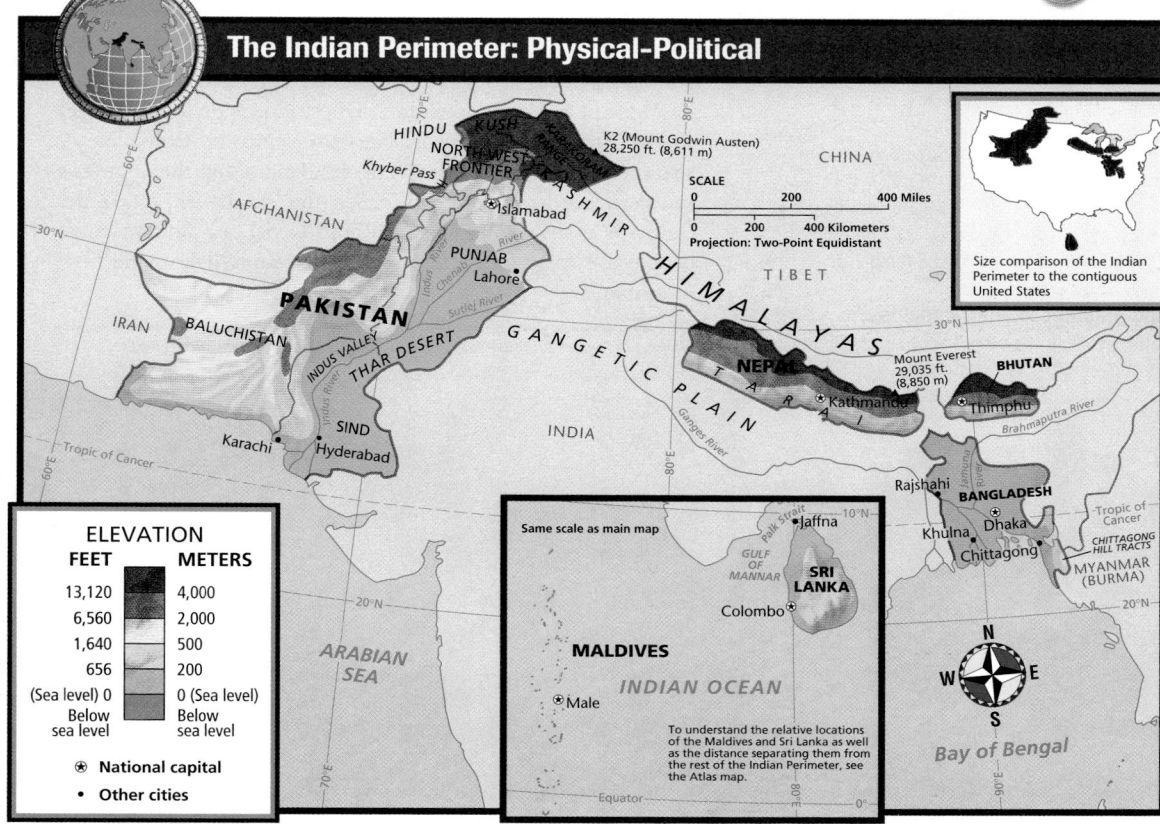

The Indian Perimeter: Physical-Political

ELEVATION

FEET	METERS
13,120	4,000
6,560	2,000
1,640	500
656	200
(Sea level) 0	0 (Sea level)
Below sea level	Below sea level

⊛ National capital

• Other cities

K2 (Mount Godwin Austen) 28,250 ft. (8,611 m)

CHINA

SCALE
0   200   400 Miles
0   200   400 Kilometers
Projection: Two-Point Equidistant

Size comparison of the Indian Perimeter to the contiguous United States

HINDU KUSH

NORTH-WEST FRONTIER

Khyber Pass

KARAKORAM

KASHMIR

AFGHANISTAN

Islamabad

PUNJAB

Lahore

IRAN

BALUCHISTAN

PAKISTAN

INDUS VALLEY

THAR DESERT

Sutlej River

GANGETIC PLAIN

HIMALAYAS

TIBET

NEPAL

Kathmandu

TARAI

Mount Everest 29,035 ft. (8,850 m)

BHUTAN

Thimphu

Brahmaputra River

Karachi

SIND

Hyderabad

INDIA

Ganges River

Tropic of Cancer

Rajshahi

BANGLADESH

Dhaka

Tropic of Cancer

Palk Strait

Jaffna

Khulna

Chittagong

CHITTAGONG HILL TRACTS

MYANMAR (BURMA)

Same scale as main map

GULF OF MANNAR

SRI LANKA

Colombo

ARABIAN SEA

MALDIVES

INDIAN OCEAN

Male

To understand the relative locations of the Maldives and Sri Lanka as well as the distance separating them from the rest of the Indian Perimeter, see the Atlas map.

Bay of Bengal

Equator

Teach Objective 1

 ALL LEVELS: Assign each student one of the Indian Perimeter countries. Each country will be used more than once. Tell students to imagine that they are exchange students assigned to the country. Ask students to write a letter back home to the United States describing the physical landscape of the adopted country. Encourage creativity by having students write about interactions with nature. *(Examples may include mountain climbing, helping to prepare for a cyclone, or describing a flooded village.)* Ask volunteers to read their letters aloud.

Teach Objective 2

LEVEL 1: Copy the following graphic organizer onto the chalkboard, omitting the italicized answers. Ask students to use the unit atlas and Section 1 to complete it. Discuss the organizer. ENGLISH LANGUAGE LEARNERS

Climates, Plants, and Animals of the Indian Perimeter

Lowlands
tropical humid, humid subtropical, arid, tropical wet and dry climates; mango, bamboo, coconut, date palm, rain forests; Bengal tigers, elephants

Highlands
highland arid, and semiarid climates; forests; bears, wild goats, Indian rhinoceroses, snow leopards

Global Perspectives

A Glacial Battlefield

The Siachen Glacier in the Karakorum Range bears the unfortunate distinction of being the world's highest battlefield. Beginning in 1984, Indian and Pakistani troops have fought for control of the strip of inhospitable land 46 miles (74 km) long. Each country is afraid that the other will gain a foothold in the ongoing conflict over Jammu and Kashmir.

So far, very few soldiers on either side have been killed by hostile fire. Temperatures on the glacier can drop to –40°F (–40°C), and some troops are stationed as high as 22,000 feet (6,706 m) above sea level. The punishing altitude, weather, and terrain have caused most of the casualties.

Ironically, the word *Siachen* means "the place of roses." Wild roses grow near the glacier's end point.

ACTIVITY: Have students examine a detailed map of Jammu and Kashmir. Then have them sketch maps showing the relationship of the Siachen Glacier to disputed territory in the region.

CONNECTING TO TECHNOLOGY ANSWER

Possible answers: measuring movement along faults, growth of volcanic cones, erosion of beaches, and so on

Connecting to TECHNOLOGY

Measuring Mount Everest

Technology often changes what we know about Earth. For example, in 1954 surveyors measured the elevation of Mount Everest at 29,028 feet (8,848 m) above sea level. Then, in 1999, researchers funded by the National Geographic Society used Global Positioning System (GPS) technology to measure Everest. Special GPS equipment on Earth's surface sends information to orbiting satellites. Researchers can then use the data that is generated to find the exact location and elevation of any place on Earth's surface.

Getting GPS equipment to the top of Everest would have been impossible only a few years before. An early GPS unit might have weighed 50 pounds (23 kg). These early units were too heavy to carry in the thin air and dangerous conditions of Everest. By 1999 a GPS unit was about the size of a sandwich. The researchers also used a special radar device that could "see" through snow because the summit of Everest is generally snow-covered.

After the equipment was in place, the researchers studied the new GPS data to determine the exact elevation of Mount Everest. They learned that Mount Everest's elevation is actually seven feet higher than previously believed—29,035 feet (8,850 m). The 1999 survey also showed that Mount Everest is moving northeastward at a rate of between 0.12 and 0.24 inches (3 and 6 mm) per year. This movement is caused by the collision of the tectonic plates that created the Himalayas.

Drawing Inferences and Conclusions In what other ways do you think Global Positioning Systems might be helpful in studying Earth?

Physical Features

As the name of this region suggests, the countries of the Indian Perimeter border India. They are Pakistan, Nepal, Bhutan, Bangladesh, Sri Lanka, and the Maldives. These countries have greatly varying physical features. Landforms range from Earth's highest mountain range to islands that barely rise above the sea. Follow along on the chapter map as we look more closely at the natural environments of these countries.

Pakistan and the Himalayan Countries To the northwest of India is Pakistan. About twice the area of California, Pakistan is the largest country of the Indian Perimeter. The towering mountains of the Himalayas, Karakoram Range, and Hindu Kush cover the northern part of Pakistan. Four mountain peaks in Pakistan are higher than 26,000 feet (7,925 m).

South of these mountains lies a region of hills and plateaus, some of which are good for farming. The country's main farming region, however, is the Indus Valley. This large fertile plain was formed from massive amounts of sediment deposited by the Indus River. Although the area receives little rainfall, it is made productive with irrigation. West of the Indus Valley lies Baluchistan, an arid and lightly populated plateau. To the east of the Indus, the Thar Desert stretches into India.

Nepal and Bhutan are landlocked countries in the Himalayas north of India. Both countries are mostly mountainous. Mount Everest, the world's highest mountain at 29,035 feet (8,850 m), lies on the border between Nepal and China. Nepal also includes a strip of the Gangetic Plain along the country's southern border. This plain, called the Tarai (tuh-RY), is the country's main farming area. In both Nepal and Bhutan, most people live in valleys and plains. Few people live in the high mountains.

Pilgrims in Nepal's Khumbu Valley stack rocks as thanks for their sacred journeys and for the good fortune of later travelers.

LEVELS 2 AND 3: Have students use the Level 1 graphic organizer for the basis of an essay on which of the two areas—lowlands or highlands—they would choose to visit first and why. You may have students use additional library or Internet resources to expand the graphic organizer and add details to their essays.

Teach Objective 3

ALL LEVELS: Have students examine the land use and resources map in the unit atlas. Ask students which countries seem rich in resources (*Pakistan and Sri Lanka—most mineral resources*) and which countries seem to be poor in resources (*Bangladesh, Bhutan, Nepal*). Ask them what resources seem plentiful (*natural gas, hydroelectric power*) and which seem scarce (*those with only one or two icons*). Lead a discussion on how distribution of these resources affects the countries of the Indian Perimeter. **ENGLISH LANGUAGE LEARNERS**

Bangladesh and the Island Countries   South of Bhutan and across a narrow extension of India lies Bangladesh. Most of this country is a broad flat alluvial plain. The Ganges and Brahmaputra Rivers join and flow southward through a huge river delta before draining into the Bay of Bengal. This delta is crisscrossed by many smaller rivers and waterways. The rivers overflow their banks every year. These floods deposit new layers of silt on the land, ensuring its fertility. Yet depending on their timing, floods often damage crops. Floods caused by tropical storms can also destroy villages and kill people and livestock.

Floods are also constantly reshaping the topography of Bangladesh. Floods can change the courses of rivers, sweep away years' worth of soil deposits, and form new islands. For this reason, no map of Bangladesh stays accurate for long.

South of India is Sri Lanka, a beautiful tropical island country in the Indian Ocean. The Palk Strait separates Sri Lanka from India. A coastal plain surrounds the mountainous center of the country.

The Maldives lie to the southwest of India in the Indian Ocean. This tropical country is made up of a chain of about 1,200 small coral islands. They have a total land area of just 115 square miles (298 sq km). None of the islands rises more than 8 feet (2.4 m) above sea level.

✓ **READING CHECK:**   *Physical Systems*   What role have rivers played in creating good farmland in the region? **Rivers in Pakistan and Bangladesh have deposited sediment in fertile river plains.**

Climates, Plants, and Animals

The climates of the Indian Perimeter range from arid and semiarid to tropical humid and highland. The peaks of the Himalayas are among the coldest places on Earth. Yet Sri Lanka and the Maldives, which are much lower and lie close to the equator, have warm tropical climates.

As in India, monsoons greatly affect the weather and vegetation of the Indian Perimeter countries. (See the climate graph.) The wet summer monsoon brings rain to most of the subcontinent, including lowland areas of Bhutan and Nepal. The rains support lush tropical rain forests from Bangladesh through the Himalayan lowlands. In fact, much of Bangladesh is covered with lush vegetation. Mango, bamboo, coconut, and date palm are common. About 15 percent of the country is forested. These areas support a variety of wildlife, including the Bengal tiger. Herds of wild elephants live in the Chittagong Hill Tracts in southeastern Bangladesh.

Powerful tropical cyclones often occur in the Bay of Bengal. These storms are called hurricanes in North America and typhoons in the western Pacific. They bring heavy rains and winds that can reach 150 miles (241 km) per hour. The strong winds can cause water to "pile up" in the Bay of Bengal. In what is called a **storm surge**, waters then wash ashore like a very high tide. Waves as high as 20 feet (6 m) may crash onshore. Because coastal areas are flat and low, these storm surges can move far inland and cause severe flooding. Since the early 1700s, when weather records were first kept, more than 1 million people have died because of these storms.

Our Amazing Planet

Mount Everest is named after Sir George Everest, a British surveyor general of India. In Nepal, Mount Everest is called Sagarmatha, which means "Goddess of the Sky." In Tibet it is called Chomolungma, or "Mother Goddess of the Universe."

Climate Graph for Colombo

Colombo, Sri Lanka (7°N 80°E)
Tropical Humid Climate

°F / In.
Average Temperature / Average Rainfall
Month: J F M A M J J A S O N D

INTERPRETING THE GRAPH   *The average temperature in Colombo, Sri Lanka, stays at about 80°F (27°C) year-round. However, rainfall varies from month to month.* **Which months receive the most rainfall? What weather phenomenon might help explain Colombo's rainfall pattern?**

Eye on Earth

Clouded Leopards
The Bengal tiger is not the only magnificent big cat that lives in the region. The clouded leopard can be found in isolated areas of Nepal. Large cloud-shaped spots edged with black mark the pelt. The beautiful fur provides excellent camouflage. However, poachers also prize the pelt. As a result, the clouded leopard has become one of the most endangered cats in the world.

The clouded leopard has several unusual characteristics that help it live much of its life in the trees. The cat has a long heavy tail about the same length as its body that helps it balance on tree branches. Specialized ankle bones allow the cat to climb down a tree headfirst. Clouded leopards can even hang from branches by their hind legs. The cats may leap from trees onto prey below. They make short work of their unlucky targets. Because they have the largest canine teeth in proportion to body size of all cats, clouded leopards have been compared to the saber-toothed tigers of prehistory.

DISCUSSION: Lead a discussion about how other animals with which students are familiar have adapted to a certain niche within their habitats.

GRAPH ANSWER

April, May, October, November; the monsoon

on each of the Section 1 subsections, for a total of nine questions. Have students answer each other's questions. **ENGLISH LANGUAGE LEARNERS**

Close

Have students discuss whether the rewards of living in a flood plain like the one in Bangladesh outweigh the risks.

Review and Assess

Have students complete the **Section Review**. Then have students complete **Daily Quiz 26.1**.

Reteach

Have students complete **Main Idea Activity for English Language Learners and Special-Needs Students 26.1**. Then have each student write three short-answer questions based

Extend

The floods that plague Bangladesh threaten the country's water supply. Disease-causing microbes are spread during floods and live in many shallow wells and ponds. Digging deep wells past the polluted surface water poses other problems. Have interested students investigate the incidence of naturally occurring arsenic in the deep wells, many of which were dug through a UNICEF program. **BLOCK SCHEDULING**

Section 1 Review Answers

Define For definition, see: storm surge, p. 587

Working with Sketch Maps Maps will vary, but listed places should be labeled in their approximate locations. The northeastern area is the most mountainous.

Reading for the Main Idea
1. Rivers frequently change their courses, and floods can wash away or deposit sediment.

2. low elevation, tropical locations of Maldives and Sri Lanka; higher elevation of Nepal and Bhutan in Himalayas

Critical Thinking
3. Floods deposit new layers of silt, helping soil fertility. However, floods sometimes damage crops, sweep away villages, and kill people. (NGS 15)

4. from the south; because wet monsoon winds flow from Indian Ocean (NGS 15)

Organizing What You Know
5. Bangladesh—alluvial plain; Bhutan—mountains; Maldives—none listed; Nepal—mountains, alluvial plain; Pakistan—mountains, alluvial plain, desert; Sri Lanka—mountains

VISUAL RECORD ANSWER

low level of technology; basic level of development

INTERPRETING THE VISUAL RECORD *Members of the Aga Khan Rural Support Program build an irrigation channel from the Batura Glacier to new orchards and fields outside a village in northern Pakistan. Irrigation has long been important in Pakistan. **How would you describe the level of technology being used to build this irrigation channel? What might this indicate about Pakistan's level of development?***

In contrast to Bangladesh, most of Pakistan has dry climates. Much of southern and central Pakistan has an arid climate. Other areas have semiarid climates with cold winters. In the north, the foothills of the Himalayas have a humid subtropical climate. Because of the rain-shadow effect, south-facing slopes of hills are often wooded, while north-facing slopes are often bare and dry.

Mountainous parts of Nepal, Bhutan, and northern Pakistan have highland climates. Forests once covered large areas of these mountains. However, parts of these forests have been cleared. Bears, deer, snow leopards, wild goats, and many other animals live where forests remain. A valley in south-central Nepal is one of the last homes of the Indian rhinoceros, which is threatened by poaching.

✓ **READING CHECK:** *Physical Systems* What factor greatly influences the region's climates and weather? monsoons

Natural Resources

Overall, the countries of the Indian Perimeter are not rich in natural resources. Pakistan and Sri Lanka have the most significant mineral deposits. Pakistan's minerals include iron and copper. Limestone for cement production is also mined there. The country has only small amounts of oil, but it has rich deposits of natural gas. Sri Lanka has deposits of several minerals, such as gemstones, iron, and salt.

Bangladesh has few minerals. However, it has small deposits of oil as well as some coal and natural gas. Rivers and good soil are the country's most important resources. Bhutan and Nepal also have few mineral deposits. However, both of these mountain countries have rich forests, and both have rivers that have potential for generating hydroelectric power.

✓ **READING CHECK:** *Physical Systems* What are the most important natural resources of Bangladesh? rivers and good soil

---

Section 1 Review

go.hrw.com Homework Practice Online
Keyword: SW3 HP26

Define
storm surge

Working with Sketch Maps On a map that you draw or that your teacher provides, label the Indus River, Baluchistan, Thar Desert, Mount Everest, Tarai, Palk Strait, and Chittagong Hill Tracts. Which part of Pakistan is the most mountainous?

Reading for the Main Idea
1. *Physical Systems* Why is the topography of Bangladesh subject to frequent change?
2. *Places and Regions* Why do the Maldives and Sri Lanka have warmer climates than Nepal and Bhutan?

Critical Thinking
3. *Analyzing Information* In what ways is periodic flooding in the region beneficial? How can it be a problem?
4. *Analyzing Information* From which direction do you think Pakistan's moisture-bearing winds come? Why?

Organizing What You Know
5. Create a chart like the one below. Place an X in the box if the landform can be found in that country.

	Mountains	Alluvial plain	Desert
Bangladesh			
Bhutan			
Maldives			
Nepal			
Pakistan			
Sri Lanka			

Mathematics: The Cost of Global Warming

Have students read Geography for Life: Environmental Change and the Maldives. Remind students that some industrial nations think it will cost too much to reduce greenhouse gases that may cause increased global warming. However, the cost of global warming is very real for the Maldives. Organize students into pairs to produce a report detailing the costs of seawall construction on three square islands that each have an area of 1 kilometer. Tell students that their reports should show the cost of building a seawall along all 644 kilometers of the Maldivian coastline as well as a second option of only building seawall for the three small square islands. The students should use $4,700 per foot in figuring their costs. *(Students will first need to determine the number of feet in a kilometer: 1 kilometer is 3280.8 feet)* Students may use an almanac or other resource to convert meters to feet. They may need calculators to determine costs. *(A seawall 644 kilometers long will cost $9,930,325,440 to build. The small square islands will have 4 kilometers of seawall each, or 12 kilometers total. The cost for these seawalls will be $61,679,040 per island or $185,037,120 for all three.)*

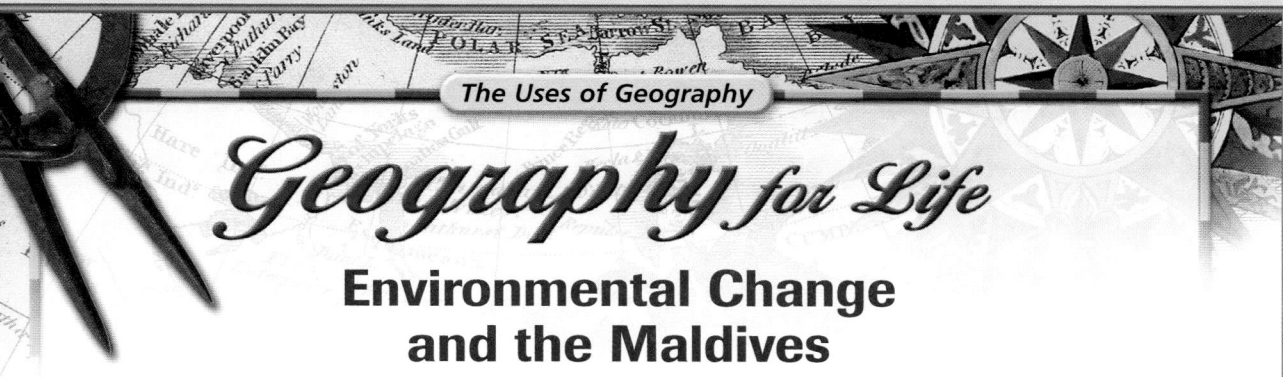

The Uses of Geography

Geography for Life

Environmental Change and the Maldives

Global warming and environmental change present serious threats to the Maldives. The islands are very low in elevation. As a result, just a slight melting of the polar ice caps—resulting in rising sea levels—could spell disaster. In fact, if the sea level of the Indian Ocean rose just 20 inches (51 cm), it could cover roughly 80 percent of the Maldives. Some scientists predict that in the next 100 years, water will completely cover the Maldives.

Recent changes in ocean levels and weather patterns point to other problems. For example, some residents believe the number and intensity of storms in the region has increased. In the past 15 years alone, two powerful storms have caused major damage to the islands. In addition, fishers have complained of a decline in the area's marine life. Residents say that the rising temperature of the ocean waters has killed sea life and damaged coral reefs.

Because of the seriousness of the problems facing the Maldives, its citizens are trying to prepare for the future. For example, the Maldives and other small island countries are pushing for worldwide reductions in greenhouse gases. Many scientists believe that these gases cause global warming. At an early age, students in the Maldives learn about global warming. Television and radio stations also teach the public about the greenhouse effect. With help from Japan, the Maldives built a 6-foot-high (1.8 m) concrete wall around its capital, Male. The island country's government hopes this seawall will protect the city from future storms. However, at $4,700 per foot, this type of wall is too expensive to build on each of the country's 200 inhabited islands. Some government officials have proposed another idea. They have suggested gathering the people of the smaller islands onto the three largest ones and defending them with seawalls.

Applying What You Know

1. **Summarizing** How might global warming and rising sea levels affect the environment of the Maldives?

2. **Problem Solving** What do you think should be done to protect the Maldives from the threat of rising sea levels?

Applying What You Know Answers

1. Global warming and resultant rising sea levels threaten to flood most of the country.
2. Possible answers: strengthen anti-pollution laws to slow global warming, move the population elsewhere, raise buildings and streets

This Geography for Life feature addresses National Geography Standards 4, 14, 15, and 18.

Male • Maldives

OBJECTIVES

1. Describe the history of the countries of the Indian Perimeter.

2. Analyze some characteristics of the region's cultures.

Section 2 RESOURCES

REPRODUCIBLE

▶ Guided Reading Strategy 26.2

▶ Creative Strategies for Teaching World Geography, Lesson 20

▶ Cultures of the World Activity: Region 8

▶ Critical Thinking Activity 26: A Difficult Birth: The Creation of Pakistan and Bangladesh

TECHNOLOGY

▶ One-Stop Planner CD–ROM, Lesson 26.2

▶ CNN Presents Geography: Yesterday and Today, Segment 33: The Tengboche Monastery

▶ Homework Practice Online

▶ HRW Go site

REINFORCEMENT, REVIEW, AND ASSESSMENT

▶ Main Idea Activity 26.2

▶ English Audio Summary 26.2

▶ Spanish Audio Summary 26.2

▶ Section 2 Review, p. 593

▶ Daily Quiz 26.2

🔊 LET'S GET STARTED

Copy the following passage onto the chalkboard: *One of the Indian Perimeter countries has been called Teardrop of India, Resplendent Land, and Pearl of the Orient. Which country is it? What do these phrases suggest about the country? Discuss responses. (Possible answers: Sri Lanka; that the country is very beautiful and has a colorful romantic history)* Tell students they will learn more about the region's turbulent history and fascinating cultures in Section 2.

Building Vocabulary

Write **protectorate** on the chalkboard and underline the base word *protect*. Call on a volunteer to provide a definition for the base word. Point out that the suffix *-ate* means "having." So, a protectorate is a place that has a certain amount of protection. Then have a student read the definition. *(A protectorate gives up certain decision-making powers in exchange for protection by a stronger country.)*

Section 2 History and Culture

READ TO DISCOVER

1. What is the history of the countries of the Indian Perimeter?
2. What are some characteristics of the region's cultures?

WHY IT MATTERS

India, Pakistan, and Bangladesh were once united as one British colony. Use **CNN fyi.com** or other **current events** sources to find out about other countries that have split to form new countries in recent years.

DEFINE

protectorate

LOCATE

Indus Valley

History

The Harappan civilization grew along the Indus Valley in ancient times. This society built well-planned cities such as Mohenjo Daro. By about 1500 B.C. Aryans from central Asia had moved into what is now Pakistan. In time, they spread across northern India. Their culture mixed with the Dravidian culture already present in the area.

Empires—Ancient and Modern Other outsiders eventually took over the northwestern part of the subcontinent. These included the Persians and the armies of Alexander the Great. Two Indian empires, the Mauryan and the Gupta, pushed into parts of what is today the Indian Perimeter. Starting about A.D. 1000, Turkic Muslims from Afghanistan came to the subcontinent, bringing their religion. In time, Islam became

Mohenjo Daro was the largest city of the Harappan civilization, which flourished in the Indus Valley from about 2500 to 1500 B.C. This civilization was the first to use fire-hardened bricks extensively in construction, possibly because of a lack of building stone in the region. Although archaeologists have unearthed many examples of Harappan writing, scholars have been unable to decipher their meaning.

ALL LEVELS: First, lead a discussion about how the following terms relate to the region's early history: *Harappan, Mohenjo Daro, Dravidian, Persians, Alexander the Great, Mauryan, Gupta, Turkic Muslims, Mughal,* and *Europeans.* Then have students work in pairs to write phrases that describe each country's history in the modern period. (*Examples: Pakistan—British colony, independence as West Pakistan 1947, renamed Pakistan 1971; Bangladesh—British colony, independence as East Pakistan 1947, renamed Bangladesh 1971; Sri Lanka—British colony of Ceylon from 1802, independence as Sri Lanka 1948; Maldives—British protectorate, independence 1965; Bhutan—British protectorate,* *independence 1949; Nepal—began reform in 1951, constitutional monarchy today*) Ask for volunteers to describe common traits in the histories (*British colonies, dates of independence*) and note differences (*Nepal never a colony, West-East Pakistan conflict*). **COOPERATIVE LEARNING**

Teacher to Teacher

Melissa Counihan of Rockdale, Texas, suggests the following activity to help students describe the countries of the Indian Perimeter. Have students create colorful and informative travel brochures for the region's countries. Display the brochures.

---

the main religion in the areas that are now Pakistan, Bangladesh, and the Maldives. The Muslim Mughal Empire flourished from the 1500s to the 1700s. About 1500, Europeans began sailing into the Indian Ocean, first to trade and later to set up colonies. Over time, the British came to control most of the subcontinent.

The Modern Period The British granted their Indian colony independence in 1947. India was divided into two countries, India and Pakistan. India was mostly Hindu. Pakistan was mostly Muslim and included present-day Pakistan and Bangladesh. These two parts of the country, separated by hundreds of miles, were known as West Pakistan and East Pakistan. While both were mostly Muslim, West and East Pakistan had important cultural differences. For example, their main languages were different. The new government was centered in West Pakistan, and many people in East Pakistan felt they had no real power. Then, in a war in 1971, East Pakistan broke away from West Pakistan and became the independent country of Bangladesh.

Sri Lanka became a colony of the British in 1802. They called their island colony Ceylon. In 1948 Sri Lanka became independent. The Maldives were a British **protectorate**. A protectorate gives up certain decision-making powers in exchange for protection by a stronger country. The Maldives gained full independence in 1965. Bhutan also was once a British protectorate. It became fully independent in 1949. However, India guides Bhutan's foreign policy.

Nepal was ruled by a series of dynasties until reforms were begun in 1951. Today the country is a constitutional monarchy. However, ethnic troubles, illiteracy, and poverty still make it hard for Nepal's people to build a strong democracy.

✔ **READING CHECK:** *Places and Regions* Why did East and West Pakistan split into modern Pakistan and Bangladesh? There were cultural differences, and many people in East Pakistan felt they had no political power.

Crowds in Bangladesh celebrate Victory Day on December 16, the date on which Bangladesh achieved independence from Pakistan in 1971.

Durbar Square in Patan, Nepal, is home to many Buddhist monuments and Hindu temples. Patan is famous for its artisans and metalworkers, whose work can be seen in the town's many bronze gateways, guardian statues, and other carvings.

Linking Past to Present

Anuradhapura The ancestors of Sri Lanka's Sinhalese people probably came to Sri Lanka from northern India during the 400s B.C. Many settled on the island's dry northern plains, where farming required irrigation. Sinhalese engineers devised a complex system of canals, dams, floodgates, and tanks to water the fields. Remnants of the irrigation works now draw tourists to Anuradhapura, the Sinhalese capital for more than 1,000 years.

Anuradhapura's most revered object is the Sacred Bo Tree. The gnarled tree was planted more than 2,000 years ago, supposedly from a twig from the tree under which the Buddha received enlightenment.

In Sri Lanka, New Year's Day is celebrated in the spring, with elephant races, coconut games, and pillow fights. The date and time for the celebration are determined by the position of the Sun and stars. In 2001 the date was April 13, and the time was 9:27 P.M. Festivities last for many days and include a grand feast with rice from the new crop.

LEVEL 1: Copy the following graphic organizer onto the chalkboard, omitting the italicized answers. Have students use Section 2 to complete it.

People of the Indian Perimeter			
	Ethnic Groups	Languages	Religions
Pakistan		*Urdu*	*Muslim*
Bangladesh	*Bengalis*	*Bengali*	*Muslim*
Bhutan	*Bhote, Nepalese, tribal peoples*	*Dzongkha, English*	*Buddhist*
Nepal	*Indian Aryan, Tibetan*	*Nepali, Sino-Tibetan*	*Hindu*
Sri Lanka	*Sinhalese, Tamils*		*Buddhist, Hindu*
Maldives			*Muslim*

Use the completed chart to lead a discussion of the similarities *(religion, some ethnic groups)* and differences *(many languages, ethnic groups)* that make unity among these countries easy or difficult. **ENGLISH LANGUAGE LEARNERS**

LEVELS 2 AND 3: Tell students to imagine they are social scientists who are writing a book on the cultures of the Indian Perimeter. Then have each student write an outline for the book. *(See Level 1 activity for main culture characteristics.)* Ask for volunteers to share their outlines.

HOMEWORK: On an outline map of the Indian Perimeter, have students shade each country according to its major religions.

INTERPRETING THE VISUAL RECORD

A wide variety of foods are produced and sold in Bangladesh. In addition to staples like rice and fish, Bangladeshi markets may include beans, eggplant, lemons, plantains, red onions, and a wide variety of spices. **Based on these photos, what type of economic system do you think Bangladesh has?**

Culture

All the countries of the Indian Perimeter are multiethnic. Three major religions dominate the region, and many languages are spoken.

People and Languages Pakistan's traditional regions—Baluchistan, the North-West Frontier Province, Punjab, and Sind—are each culturally distinct. However, there are also some common features, such as the importance of Islam. Urdu is Pakistan's official language and is taught in schools along with regional languages. An Indo-European language, Urdu is similar to Hindi, which is widely spoken in India. However, Urdu is written in a form of Persian script and, unlike Hindi and English, is read from right to left. Some Pakistanis also speak English.

In southern Nepal, most ethnic groups are of Indian Aryan ancestry. They speak Indo-European languages, such as Nepali, the country's official language. Others in the north are related to the people of Tibet (now part of China). They speak Sino-Tibetan languages.

Bhutan's population includes three main ethnic groups. The largest group, the Bhote, came from Tibet starting in the A.D. 800s. The second-largest group is made up of more recent immigrants from Nepal. Discrimination against the Nepalese is an important problem. The third group is made up of tribal peoples in eastern Bhutan. They are related to the peoples just across the border in India. Bhutan's official language, Dzongkha, is a Sino-Tibetan language. However, English is widely used in schools.

Most of the people in Bangladesh are Bengalis. The Bengalis are a mix of the region's early settlers with Turks and other Southwest Asians who came as merchants as early as the 1200s. Most speak Bengali.

Nearly 75 percent of Sri Lanka's population are Sinhalese. The Tamils, originally from southern India, make up most of the rest of the population. Tamils generally live in the northern and eastern parts of the country. In recent decades there have been bloody conflicts between the two ethnic groups.

✓ **READING CHECK:** (**Human Systems**) How does Urdu differ from Hindi? Where are these languages widely spoken? Urdu is written in a form of Persian script and is read from right to left. Pakistan, India

Close

Have students call out names of ethnic groups, holidays, or languages from Section 2 and compete to identify them.

Review and Assess

Have students complete the **Section Review**. Then have students complete **Daily Quiz 26.2**.

Reteach

Have students complete **Main Idea Activity for English Language Learners and Special-Needs Students 26.2**. Then have students create covers for a book entitled The History and

Culture of the Indian Perimeter. Display and discuss the book covers. **ENGLISH LANGUAGE LEARNERS**

Extend

Have interested students conduct research on either Mohenjo Daro or Harappa, two important sites of the Indus civilization. Have students focus on the street plans or construction techniques of either city. Students should also compare the ancient cities to Pakistani cities of today. Ask them to try to determine if Indus building practices influenced later construction practices in any way. Have students present their findings to the class. **BLOCK SCHEDULING**

Education Education levels and literacy rates in the Indian Perimeter are generally low. In recent decades these countries have tried to increase literacy. However, there are too few schools and teachers. In general, women are less likely than men to be able to read. This is largely because of cultural attitudes that emphasize women's role in the home.

Religion The region's main religions are Buddhism, Hinduism, and Islam. The religious makeup of each country reflects past events and historical migration patterns.

Pakistan, Bangladesh, and the Maldives are mostly Muslim. Small numbers of Hindus and Christians also live in these countries. At the time of India's partition, the new borders placed districts with Muslim majorities into East and West Pakistan. The number of Hindus in these areas fell sharply when millions fled to India. At the same time, millions of Muslims left India and settled in East and West Pakistan.

In Nepal and Bhutan, most Indian Aryan peoples are Hindu, while peoples of Tibetan origin are generally Buddhist. Bhutan is about 70 percent Buddhist, 25 percent Hindu, and 5 percent Muslim. The kingdom of Nepal is the world's only officially Hindu state. In Sri Lanka most Sinhalese are Buddhist and most Tamils are Hindu.

The region's religions are reflected in its traditions and customs. For example, Islamic influences shape life throughout Pakistan. People stop to pray several times a day, and Muslim holidays such as Ramadan are important. Midday prayers on Friday draw large numbers of people to the country's many mosques. Also in Pakistan, women often wear veils in public. In Sri Lanka, Buddhist festivals are held throughout the year. For example, the traditional Poya Days mark the phases of the Moon. The most important festival in Sri Lanka is held in the mountain city of Kandy each August. According to tradition, a temple there holds the country's most sacred relic, a tooth of the Buddha. During the annual August celebration, a beautifully decorated elephant is paraded through the streets accompanied by dancers and acrobats.

✓ **READING CHECK:** *Human Systems* What are the three main religions of the Indian Perimeter? Buddhism, Hinduism, and Islam

INTERPRETING THE VISUAL RECORD

A gateway leads to Badshahi Mosque in Lahore, Pakistan. This large mosque symbolizes the importance of Islam in Pakistan, a country where some 97 percent of the population is Muslim. Islam is the main link among the many different cultural groups that make up Pakistan. **How might Islam help connect different cultural groups in Pakistan?**

Review

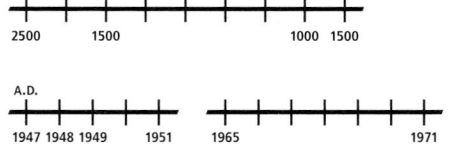
Homework Practice Online
Keyword: SW3 HP26

Define protectorate

Working with Sketch Maps
On the map you created in Section 1, label the Indus Valley. Why do you think the region's first great civilizations grew there?

Reading for the Main Idea

1. *Human Systems* What people brought Islam to the region? What Muslim empire ruled the region?

2. *Human Systems* How did the partition of India affect the countries of this region?

3. *Human Systems* What are some examples of the ways religion is reflected in the region's customs?

Critical Thinking

4. **Analyzing Information** Which country of the region is the most ethnically and linguistically unified?

Organizing What You Know

5. Copy the time lines below. Then identify important events that occurred on or about the dates listed on the time lines.

B.C. ——————— A.D.
2500 1500 1000 1500

A.D.
1947 1948 1949 1951 1965 1971

OBJECTIVES

1. Identify the basis of the economies of the Indian Perimeter.

2. Analyze settlement patterns in the region.

3. Explain the challenges these countries will face in the future.

LET'S GET STARTED

Copy the following instructions onto the chalkboard: *Look at the physical map of the Indian Perimeter in the unit atlas. Recall what you have learned about global warming. What environmental hazard can you predict for the region?* Discuss responses. *(Possible answers: Low-lying areas, such as the Maldives and much of Bangladesh, could be threatened by rising sea levels.)* Tell students they will learn more about these challenges and other aspects of human systems in the Indian Perimeter in Section 3.

Building Vocabulary

Write **graphite** on the chalkboard. Ask students to identify words that might be related to the term. *(Possible answers: graph, graphic, graphology)* All these words have to do with writing; they are based on a Greek word *graphein*, which means "to write." Graphite is the form of carbon used in pencils. It is also used in lubricants, paints, coatings, electrodes, and other products.

Section 3 The Region Today

READ TO DISCOVER

1. What is the basis of the economies of the Indian Perimeter?
2. What are settlement patterns like in the region?
3. What challenges will these countries face in the future?

WHY IT MATTERS

India and Pakistan both have nuclear weapons. Use **CNN fyi.com** or other **current events** sources to learn about the relationship between these two countries today.

DEFINE

graphite

LOCATE

Karachi	Thimphu
Lahore	Dhaka
Islamabad	Colombo
Kathmandu	Male

Economy

The level of economic development in the Indian Perimeter is generally low. The countries depend heavily on agriculture. Most have tried to build new industries. However, these efforts have been slowed by a lack of natural resources. Despite these similarities, there is quite a range between the richest and poorest countries of the region. For example, Nepal and Bhutan are among the poorest countries in the world. Yet Pakistan, while still a poor country, has experienced growth in GDP since independence.

Pakistan, Bangladesh, and Sri Lanka Pakistan has used its significant mineral resources to help develop its manufacturing industries. However, nearly half the labor force still works in agriculture. Although Pakistan's economy has grown, so has its population. This population growth strains the country's ability to provide basic services to its people.

Rice is an important crop in Sri Lanka. Since independence in 1948, the country has steadily increased the amount of land devoted to rice and has adopted new technologies to increase yields.

Teach Objective 1

ALL LEVELS: Copy the following graphic organizer onto the chalkboard, omitting the italicized answers. Have the students complete it. Lead a discussion about which of these products may be imported into the United States and available in local stores. **ENGLISH LANGUAGE LEARNERS**

Products of the Indian Perimeter			
Agricultural	**Minerals**	**Manufacturing**	**Services**
jute, rice, tea, rubber, coconut	*graphite, gems*	*textiles, electricity, processed food*	*tourism*

Teach Objective 2

ALL LEVELS: Ask students to compare the population and settlement patterns of Bangladesh and Bhutan and to draw conclusions about what the population and settlement patterns indicate about the two countries' climates and land use. *(Possible answers: Bangladesh—high population, rural, dense population, particularly in and around Dhaka; Bhutan—lower population density, most population in lowland areas, no cities until 1960s)* Ask for volunteers to share their conclusions. Then lead a discussion about how these factors are reflected in the region's other countries.

Bangladesh is overwhelmingly agricultural. More than half of the people work in farming. Jute, rice, and tea are the most important crops. Farming depends on the monsoon. Variations in the timing and intensity of the monsoon rains make the difference between a good harvest or a poor one. In recent times people have built a number of irrigation projects to control floods and conserve water for the dry months. These projects, the increased use of fertilizers, and new crops have increased farm output.

The government of Bangladesh has tried to industrialize the country. Textile and clothing makers are among the country's largest employers. However, a lack of mineral resources has made other development difficult. A dam in the Chittagong Hill Tracts produces hydroelectric power. Fishing and logging are also important. One common type of bamboo forms the basis of the country's paper industry. Construction is also a growing industry—a result of the country's fast-growing population.

Farming remains important to Sri Lanka's economy. Tea, rubber, and coconut are the main export crops. The country mines a variety of minerals, including gemstones and **graphite**. Graphite is a form of carbon that is used in pencil leads. Manufacturing is growing and now rivals agriculture in importance. Processed foods and textiles are leading manufactured products.

Nepal and Bhutan Nepal and Bhutan are very poor and are still mainly agricultural. Timber is an important resource, with most being exported to India. Tourism is a growing part of the economies of the mountain countries. Many visitors come to hike in the Himalayas. However, some people fear the negative effects of tourism on the environment and cultures of Nepal and Bhutan. Their isolated locations have left many of their ecosystems and traditional ways of life largely intact. Bhutan, in particular, has taken a careful approach to tourism. The country was almost closed to outsiders until the 1970s. Today tourists must pay a fee and follow certain restrictions to limit their effect on the country.

Much of the development in Nepal and Bhutan has been helped by aid from other countries and the United Nations. Both these mountain countries have rivers that can generate hydroelectric power. However, they lack the resources needed to build dams. Aid from India did help Bhutan build one such project. Bhutan now sells extra electricity to India.

✓ **READING CHECK:** *Environment and Society* How have Bangladeshis dealt with the effects of variations in monsoon rainfall on farming? **irrigation projects to control flooding, conserving water for dry months**

Cities, Settlement, and Land Use

Pakistan and Bangladesh each have more than 125 million people. In fact, they are 2 of the 10 most populous countries in the world. The other countries of the region have far smaller numbers of people. The population throughout the region is mostly rural. This pattern reflects the importance of agriculture there.

About a third of Pakistan's population lives in cities. Karachi is the largest city and main seaport. Lahore is the second-largest city. Islamabad, far to the north, is the capital. Pakistan's population is concentrated in the Indus Valley. Most other areas have

Bhutan's rulers banned foreigners from the country until well into the 1900s. This historical isolation, along with Bhutan's inaccessible location in the Himalayas, makes economic development difficult today.

Kathmandu, Nepal, is a city of contrasts. Urban growth, tourism, and increasing connections with the outside world are causing rapid changes throughout the city. Kathmandu's old city is home to many old temples and shrines, but automobile exhaust, construction, and growing shantytowns are putting stress on the city's environment.

Essential Element 5

► **Environment** ◄
and Society

Cleaning Mount Everest
Mountaineering on Mount Everest has come at a cost to the environment. As much as 50 tons of trash has been scattered across the world's highest mountain. This litter includes old tents, ropes, oxygen bottles, stoves, food containers, and human waste.

Cleanup efforts have begun. One group has removed 17,000 pounds of garbage since 1994. To prevent the future buildup of garbage, the group set up a system to pay Sherpas to bring used supplies back down.

Another expedition cleaned up a heavily used part of the mountain. Retrieved gas canisters and used batteries were carried by yak to a nearby city, transported by plane to Kathmandu, shipped to Thailand, and then taken to a hazardous waste site in California.

DISCUSSION: Lead a discussion about ways students can limit the effect they have on the environment when visiting parks or other natural areas.

internet connect

GO TO: go.hrw.com
KEYWORD: SW3 CH26
FOR: Web sites about Mount Everest

595

ALL LEVELS: Call on volunteers to read the subsection on economic and environmental challenges aloud. Then ask students to identify the challenges. *(Possible answers: poverty, health problems, high population growth rates, environmental problems, political problems, rural to urban migration, civil war)* Organize students into pairs and have them rank the challenges in order of the severity with which they affect people's lives. Ask volunteers to share and defend their rankings. **COOPERATIVE LEARNING**

Using National Geography Standard 9:
Human Systems: The Characteristics, Distribution, and Migration of Human Populations on Earth's Surface Have students analyze the factors that keep the population growth rate of Bangladesh high. These factors may include religious beliefs, the role of women in society, the need for farm labor, and a desire for security in old age. Then have students develop ad campaigns that the Bangladeshi government might use to further its policy of encouraging couples to limit family size. Point out that to be successful the ads must take the factors identified earlier into account. Display the ads in the classroom.

Essential Element 5

► **Environment** ◄
and Society

Mosquitoes and the Tarai
Among the ethnic groups of Nepal are the Rana Tharu people. They live in the Tarai of eastern Nepal.

Not long ago, the Rana Tharu lands and way of life were effectively protected by malaria-carrying mosquitoes that thrived in the damp climate. Although the Rana Tharu were not immune, they developed some resistance to malaria. Fear of the deadly disease kept others away. However, during the 1950s DDT use diminished the malaria risk. Soon outsiders arrived to cut down the trees. Deforestation remains a threat to the Rana Tharu way of life.

ACTIVITY: Have students conduct research on various ethnic groups of Nepal and how they deal with the environmental issues they face.

Nepal's mountainous landscapes are a major destination for trekkers and mountain climbers from other countries. However, these international tourists have added to the country's pollution problems. Hikers in Nepal leave behind an estimated 110,000 pounds (50,000 kg) of garbage each year. Here, hikers at Everest Base Camp crush aluminum cans, which are taken back to Kathmandu and recycled.

few people. The rural population lives mostly in small villages. In northwestern Pakistan, villages are sometimes laid out in a ring. The outer walls have no doors or windows, giving the village the appearance of a fortress.

In both Nepal and Bhutan, population density is low overall. However, the average is much higher in the lowlands and valleys where farming is possible. Nepal's most crowded area is the Kathmandu Valley, near the center of the country. The capital and largest city, Kathmandu, is located there. Bhutan had no cities at all until the 1960s. Even today Thimphu, the capital, is a town of only about 30,000 people.

Bangladesh and Sri Lanka are mostly rural and are densely populated. Bangladesh has about 2,324 people per square mile. This is a very high population density by world standards. For comparison, the United States has about 76 people per square mile. In and around Dhaka, in the country's most fertile region, there are more than 2,800 people per square mile. Sri Lanka's population density is much lower, but there are still some 759 people per square mile. The country's population is heavily concentrated along the fertile coastal plain.

In both countries most people live in villages. In good farming areas these villages are located close together. One village merges with the next in many areas. Bangladesh has two major cities. Dhaka is the country's capital, and Chittagong is the major port. Sri Lanka's largest city is Colombo, the capital and leading industrial area.

✓ **READING CHECK:** *Human Systems* How are settlement patterns similar in Bangladesh and Sri Lanka? Both countries are mostly rural, but with closely spaced villages in the fertile farming areas.

Economic and Environmental Challenges

As in many other developing countries, the greatest challenge for the countries of the Indian Perimeter is poverty. Poor sanitation, disease, and poor nutrition cause major health problems. High population growth rates have made these problems worse. The largest cities, such as Karachi and Dhaka, are growing rapidly as people migrate from rural areas. Many people in these cities live in homemade shacks with no fresh water or electricity.

The region's countries also face environmental challenges. For example, deforestation is a serious problem in Nepal. Nepal and Bhutan are trying to limit tourism's destructive effects on their environment. In addition, flooding in Bangladesh and the Maldives, with their low flat terrain, might worsen if global warming causes ocean levels to rise. The threat of rising ocean levels has already affected the Maldives' capital of Male. (See Geography for Life: Environmental Change and the Maldives.)

In addition to problems between Pakistan and India over Kashmir, many countries in the region are struggling with political problems. For example, in Pakistan military leaders have overthrown the elected government three times. Rivalries between richer and poorer parts of society have also led to violence. Achieving democratic government and sharing economic benefits throughout society will be key challenges in the future.

Sri Lanka faces an ethnic conflict between the Hindu Tamil minority and the Buddhist Sinhalese majority. Tensions between the two groups have increased since the late 1970s. Some Tamils demand that they be allowed to form their own country. Fighting and violence continue today.

Close

Tell students to imagine that they are pursuing a political career as prime minister in an Indian Perimeter country. Ask students which country they would prefer to lead and the reasons for their choice. Lead a discussion about campaign promises they would make, problems they would try to solve once in office, and how they would put their plans into action.

Reteach

Have students complete **Main Idea Activity for English Language Learners and Special-Needs Students 26.3.** Then organize the class into groups to present the major points related to the countries' economies, cities and settlement patterns, and challenges to the class. **ENGLISH LANGUAGE LEARNERS, COOPERATIVE LEARNING**

Review and Assess

Have students complete the **Section Review.** Then have students complete **Daily Quiz 26.3.**

Extend

Have interested students conduct research on refugee issues related to the civil war in Sri Lanka. **BLOCK SCHEDULING**

In Sri Lanka high quality "Ceylon" tea is grown on large plantations in the central highlands. Young tender leaves are picked by hand once a week and then quickly processed, producing teas of the highest quality. Women harvesting tea leaves hang large baskets from their heads so that both hands are free to pick leaves.

FOCUS ON GEOGRAPHY

Migration, Tea, and Ethnic Conflict in Sri Lanka   The origins of multiethnic Sri Lanka date back more than 2,000 years. About that time, Hindu Tamils from southern India began trading in Sri Lanka. Those who settled in northern Sri Lanka became known as the Ceylon Tamil. Their descendants are usually considered natives of Sri Lanka, along with the Buddhist Sinhalese.

Another group, the Indian Tamil, have a different history. During the mid-1800s Tamil workers from southern India came to Sri Lanka's central plantation region to harvest the coffee crop. Then they would return to India. When a leaf disease destroyed the coffee business, farmers switched to tea. Tea is an evergreen plant that is harvested year-round. Therefore, the Tamil laborers settled in Sri Lanka permanently. Over time, they formed a poorly paid, overworked underclass.

Conflicts between Hindu Tamils and Buddhist Sinhalese have broken out from time to time over hundreds of years. Fighting in recent years has killed at least 55,000 people. Troubles between Indian Tamils and Ceylon Tamils have added to Sri Lanka's political crisis.

✓ **READING CHECK:**   *Human Systems*   What role did the switch from coffee to tea play in Sri Lanka's modern ethnic conflicts?

Indian Tamil tea pickers, who worked year-round, created a second Tamil group that is considered different from the original Tamil settlers.

Section 3 Review

Homework Practice Online
Keyword: SW3 HP26

Define graphite

Working with Sketch Maps On the map you created in Section 2, label Karachi, Lahore, Islamabad, Kathmandu, Thimphu, Dhaka, and Colombo. Which of these are port cities?

Reading for the Main Idea

1. *Environment and Society*   What is the general level of economic development in the region? What are some indicators of this?

2. *Human Systems*   In which Indian Perimeter country is a civil war currently being fought?

Critical Thinking

3. **Analyzing Information**   How are settlement patterns in Bangladesh related to the way most Bangladeshis make a living?

4. **Identifying Cause and Effect**   What demographic trend has affected Pakistan's economic growth?

Organizing What You Know

5. Copy the following graphic organizer. Use it to identify the leading economic activities of each country.

	Economic activities
Bangladesh	
Bhutan	
Nepal	
Pakistan	
Sri Lanka	

Section 3 Review Answers

Define   For definition, see: graphite, p. 595

Working with Sketch Maps Maps will vary, but listed places should be located in their approximate locations. Karachi and Colombo are port cities.

Reading for the Main Idea
1. generally low level of development; most countries heavily dependent on agriculture, slow industrial growth

2. Sri Lanka

Critical Thinking
3. The population is highly rural because most people are farmers. The fertility of the soil supports a very dense population. (NGS 12)

4. rapid population growth (NGS 9)

Organizing What You Know
5. Bangladesh—agriculture, fishing, logging, paper manufacturing, construction; Bhutan—agriculture, logging, tourism, hydroelectric power; Nepal—agriculture, logging, tourism; Pakistan—agriculture, manufacturing, mining; Sri Lanka—agriculture, manufacturing

Setting the Scene

Bangladesh is often described as one of the world's poorest, most densely populated, and least developed countries. In contrast to the United States, where about 12.7 percent of the population is below the poverty line, almost 50 percent of Bangladesh's people live in poverty. More than 35 percent of Bangladeshis are unemployed. Moreover, agriculture, which employs a majority of the workers, is regularly threatened by floods and cyclones. The country has few natural resources besides its fertile soil. Illiteracy is widespread. How can Bangladeshis pull themselves out of poverty? Thousands of the country's people are relying on their own ingenuity to improve their lives.

Building a Case

Have students read Case Study: Banking on Bangladesh's Female Entrepreneurs. Review the reasons that women have traditionally been unable to help increase their families' income *(forbidden to travel, little schooling, and limited to jobs as maids and beggars)*. Ask how women's use of the small loans differs from men's. *(Women often made long range plans with the money, while men were more likely to spend money on themselves)*. Have students list examples of how Bangladeshi women have used their loans *(cellular phones, making ice cream, repairing radios, weaving floor mats, and processing mustard seeds for oil)*. Review also the positive effects researchers have observed in Bangladesh *(improved nutrition and education among children in households of female borrowers)*.

CASE STUDY

Banking on Bangladesh's Female Entrepreneurs

Human Systems Bangladesh is one of the world's poorest countries. It is a young country that has experienced terrible cyclones and a costly struggle for independence, which have disrupted its economy. Economic development has not reached many small villages. As a result, Bangladesh's rural areas have a very low standard of living. For women in Bangladesh, raising the family's standard of living is a difficult task.

For several reasons, poverty often poses greater challenges for women than for men. In rural Bangladesh, begging and working as maids are the only jobs outside the home available to women. Culture traits can also keep women from working. Women in Bangladesh have traditionally had few contacts with anyone outside their families. Some are forbidden to travel far away from home. In addition, rural Bangladeshi women have little schooling. As a result of all these factors, women are limited in their ability to increase their families' income.

Rural development projects in Bangladesh are helping many women find jobs outside the home, which is raising their standard of living.

However, a creative lending program is showing that women may hold the key to economic development in rural Bangladesh.

Small Loans, Big Results

In the early 1980s the Grameen Bank was founded to offer small loans to rural poor people in Bangladesh. Known as microcredit, these loans typically range from about $100 to $300. Borrowers use the money to start small businesses in their villages. These businesses then provide a steady source of income. Soon after the bank started lending to poor families, officials noticed that the women were often better financial managers than the men. While men were more likely to spend the money on themselves, women used it to make long-range plans. Today about 95 percent of Grameen's borrowers are women.

One such woman is Mosammat Anowara Begum from the village of Chamurkhan. Anowara is a widow with several children. Even though she cannot read or write, the bank offered her a loan. She borrowed $390 from Grameen and used it to buy a cellular phone—the only telephone in her village. Anowara now sells telephone calls for 10 cents a minute. She hopes to pay back the loan in three years. Then Anowara will bring home a steady income of about $2 a day. While that amount may not seem like much, it is almost three times what the average Bangladeshi earns. Anowara's growing income has enabled her daughter to go to college.

Like Mosammat Anowara Begum, Noorjehan Begum borrowed about $320 from the Grameen Bank and bought a cellular phone, which she charges other villagers to use. With the money she earned, Noorjehan paid off her loan and bought a small piece of land and a house.

Drawing Conclusions

Have students use the Internet to find articles in Bangladeshi publications on business deals. *(Students should be able to find information on deals involving millions of taka, the local currency.)* Tell students that in 2000 one U.S. dollar was equal to 51,000 taka (Tk). Then organize the class into groups to represent villages. Issue $1.50, equal to about 76,500 Tk, in play money to each student. The money represents a daily wage in Bangladesh. Ask students if they can participate in the businesses they read about with the money they earn. Then give one person in each "village" a loan of $200 for a business he or she designates. Let the entrepreneur run the business, charging 10¢ per transaction. Ask students for their conclusions. *(Small loans make a difference, while large ones are not feasible.)*

Going Further: Thinking Critically

Microcredit projects also operate in the United States. Have students work in pairs to suggest other legitimate businesses that could be started with a $1,000 loan. (The figure is higher than the Grameen loans to reflect higher costs in the United States.) Next, tell each pair to rank its top 10 ideas and determine which ideas would work the best. Finally, have students pick one idea and write a prospectus for it. A prospectus is a plan for a proposed business. It includes a description of the business and necessary expenditures. Remind students that a banker would require such a document before he or she would make the loan. Ask volunteers to share their prospectuses.

COOPERATIVE LEARNING

Women wait to apply for loans at a Grameen Bank weekly meeting. More than 2.3 million people from some 40,000 villages have borrowed money from the bank.

Anowara's story is just one example of how small loans are changing rural Bangladesh. Many other people have borrowed money from Grameen. They have started more than 400 different kinds of businesses. These include making ice cream, processing mustard seeds for oil, repairing radios, and weaving floor mats. The loans seem to be working. Some 125,000 families with Grameen loans pull themselves out of poverty each year. Researchers have noted that children's nutrition and education have significantly improved in households with female borrowers.

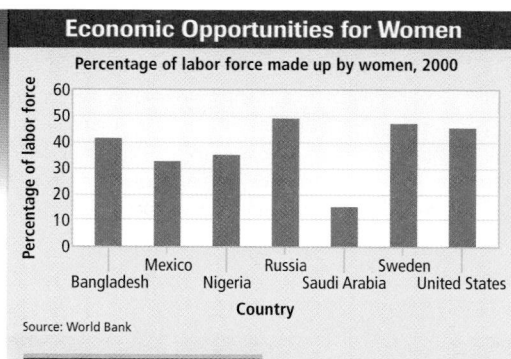

Economic Opportunities for Women

Percentage of labor force made up by women, 2000

Source: World Bank

INTERPRETING THE GRAPH *The percentages of women in the labor forces of Bangladesh and the United States are nearly the same. However, most Bangladeshi women work in agriculture. Still, the Grameen Bank is helping many find other ways to make a living. Officials in other countries, such as Mexico, want to set up similar programs. **What might be some factors that explain differences in economic opportunities for women in these countries?***

A New Approach

Microcredit projects have become so popular that many other countries, including the United States, have experimented with the idea. For example, the Women's Self-Employment Project in Chicago has loaned almost $1 million to a total of about 5,000 women. In all, more than 40 countries have applied the Grameen model, and it has reached nearly 22 million people.

Microcredit projects provide an alternative to conventional methods of helping the world's poor. Pouring large sums of money into industry has been the most common strategy for developing economies. However, these jobs then attract large numbers of people from rural areas to urban areas. By investing money in the rural areas one household at a time, development can be distributed throughout the country. With increasing opportunities for women to earn money, more and more households can lift themselves out of poverty.

Applying What You Know

1. Summarizing How does microcredit create economic opportunities that are not otherwise available to women in Bangladesh?

2. Comparing and Contrasting How is microcredit different from conventional forms of development aid?

CHAPTER
26 Review Answers

Building Vocabulary For definitions, see: storm surge, p. 587; protectorate, p. 591; graphite, p. 595

Locating Key Places

A. Indus River
B. Dhaka
C. Baluchistan
D. Mount Everest
E. Thar Desert
F. Male
G. Colombo
H. Tarai

Understanding the Main Ideas

1. Himalayas—Pakistan, Nepal, Bhutan; plain of the Ganges River—Nepal, Bangladesh

2. The Bhote came from Tibet starting in the A.D. 800s.

3. They divided it into mostly Hindu India and mostly Muslim Pakistan.

4. textiles and clothing, fishing, logging, paper manufacturing, construction

5. hydroelectric power

Applying What You Know Answers

1. loans small amounts of money for small-scale businesses

2. allows people to stay in rural areas, instead of drawing them to industries in cities

TECHNOLOGY
▶ Chapter 26 Test Generator (on the One-Stop Planner)
▶ Global Skill Builder CD–ROM
▶ HRW Go site

REINFORCEMENT, REVIEW, AND ASSESSMENT
▶ Chapter 26 Review, pp. 600–01
▶ Chapter 26 Tutorial for Students, Parents, Mentors, and Peers
▶ Chapter 26 Test (form A or B)

▶ Alternative Assessment Handbook
▶ Chapter 26 Test for English Language Learners and Special-Needs Students
▶ Unit 8 Test
▶ Unit 8 Test for English Language Learners and Special-Needs Students

Assess
Have students complete a Chapter 26 Test.

Reteach
Organize students into six groups—one for each of the region's countries. Have each group prepare an annotated time line of main events in the country's history. Ask students to relate the natural environment and present-day situation to the history. Display and discuss the time lines. **ENGLISH LANGUAGE LEARNERS, COOPERATIVE LEARNING**

CHAPTER 26 Review Answers

Thinking Critically

1. frequency of cyclones, restricted space of the Bay of Bengal, the low flat terrain of Bangladesh (NGS 13)

2. Students might note that unstable governments in Pakistan and civil war in Sri Lanka could undermine efforts toward development. (NGS 11)

3. mountainous northern Pakistan; the Indus Valley (NGS 14)

Using the Geographer's Tools

1. Possible answers: great distance between the two territories and the differences in their physical environments as well as cultural and political differences

2. Rankings may vary, but students should explain their rankings.

3. drawbacks—environmental damage; benefits—increased income and employment

Writing

Students might note connections between Roman Catholics in Northern Ireland and Hindu Tamil in Sri Lanka. Use Rubric 9, Comparing and Contrasting, to evaluate student work.

CHAPTER 26 Review

Building Vocabulary
On a separate sheet of paper, explain the following terms by using them correctly in sentences.

storm surge   protectorate   graphite

Locating Key Places
On a separate sheet of paper, match the letters on the map with their correct labels.

Indus River   Mount Everest   Colombo
Baluchistan   Tarai   Male
Thar Desert   Dhaka

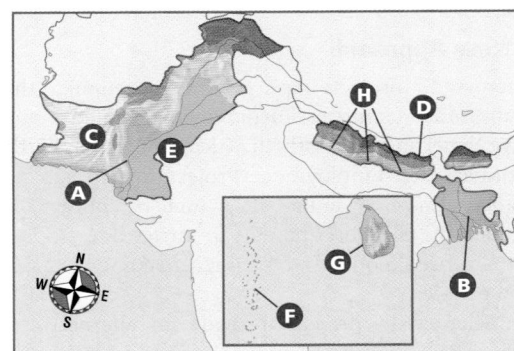

Understanding the Main Ideas
Section 1
1. (*Physical Systems*) Which countries of the Indian Perimeter include parts of the Himalayas? Which include parts of the plain of the Ganges River?

Section 2
2. (*Human Systems*) Where did Bhutan's largest ethnic group, the Bhote, come from?

3. (*Human Systems*) On what basis did the British divide their colony of India upon independence in 1947?

Section 3
4. (*Human Systems*) What are the major industries of Bangladesh?

5. (*Environment and Society*) What is the greatest potential energy resource of Nepal and Bhutan?

Thinking Critically

1. Analyzing Information What combination of factors make Bangladesh so vulnerable to storm surges?

2. Identifying Cause and Effect What political factors might make further development difficult in some countries of the region?

3. Comparing What part of Pakistan bears a resemblance to the main environment of Bhutan? What part might resemble Bangladesh?

Using the Geographer's Tools

1. Analyzing Maps Study the physical and political map of the Indian Perimeter. Why do you think it would be difficult to govern modern-day Pakistan and Bangladesh as one country?

2. Analyzing Tables Study the information from the unit Fast Facts and Comparing Standard of Living tables. Then rank the countries according to their levels of economic development and standard of living. Explain why you have ranked the countries as you have.

3. Preparing Charts Create a two-column chart listing the benefits and drawbacks of tourism for Nepal and Bhutan. What economic benefits does tourism offer for these countries?

Writing about Geography

Use classroom and library materials to research the ethnic and religious conflict in Sri Lanka. Write a news article that compares the Sri Lankan conflict to similar problems in other places, such as Northern Ireland. What role have economic opportunities for ethnic and religious minorities played in conflicts there? What other issues are important? What efforts have been made to end those conflicts? When you are finished with your news report, proofread it to make sure you have used standard grammar, spelling, sentence structure, and punctuation.

SKILL BUILDING

Geography for Life
Applying Geographic Models
(*Physical Systems*) Study the courses of the Indus, Ganges, and Brahmaputra Rivers. Consider the connection between erosion in the Himalayas and deposition of sediment downstream. Create a diagram showing how sediment is eroded and deposited by this process. Consider the interrelationships of climate, slope, and soil.

Portfolio Activity

Stupas, which are Buddhist monuments, are common in the countries of the Indian Perimeter where Buddhism is widely practiced. They usually house objects related to the Buddha or other people who are considered saintly. Pair students and have pairs conduct research on stupas in Bhutan, Nepal, or Sri Lanka. Each pair should build a scale model of a particular example. Students might use papier-mâché or salt-flour dough over a crumpled paper base. Ask students to write a brief report on the stupa to accompany the model. The report should include details on the stupa's construction, history, and role in the religious life of the community. Photograph scale models. Place photos and reports in portfolios.

Food Festival

Mangoes are among the many tropical fruits grown in Bangladesh. Here is a simple recipe from Bangladesh for mango *lassi,* a refreshing drink. Cut a large mango in half, remove the seed, and scoop out the flesh. Process in a blender until smooth. There should be about 1 c. of mango pulp. Mix the pulp with 1 c. plain yogurt, 3 c. milk, 3 c. water, 1 c. sugar, and about 1 tsp. of ground cardamom. Chill well before serving. Pakistanis enjoy mango ice cream. Students might want to try making it with condensed milk, whipping cream, and mango pulp.

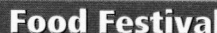

Building Social Studies Skills

Birthrates and Death Rates in Indian Perimeter Countries

Country	Births (per 1,000 people)	Deaths (per 1,000 people)
Bangladesh	29.9	8.6
Bhutan	34.8	13.5
Maldives	36.7	7.7
Nepal	32.5	9.8
Pakistan	29.6	8.8
Sri Lanka	16.1	6.5

Source: *The World Almanac and Book of Facts 2004*

Interpreting Charts

Study the chart above. Then use the information from the chart to help you answer the following questions.

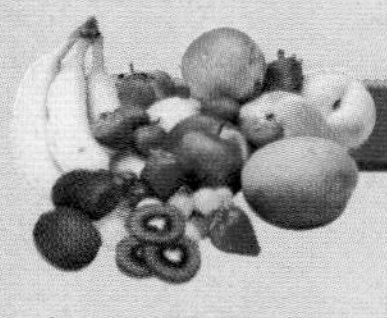

1. Which country has the lowest birthrate?
- **a.** Pakistan
- **b.** Nepal
- **c.** Maldives
- **d.** Sri Lanka

2. Rank the six countries' birthrates and death rates from least to greatest. Which country has the highest birthrate but the second-lowest death rate?

Using Language

The following passage contains mistakes in grammar, punctuation, and usage. Read the passage and then answer the following questions.

(1) Family life in Bangladesh is different from Pakistan. (2) For example, many Bangladeshi women keep close ties to their own families after marriage. (3) However, as in other South Asian countries, marriages arranged by parents. (4) Most couples typically do not know each other prior to there wedding.

3. Which sentence contains a spelling error?
- **a.** 1
- **b.** 2
- **c.** 3
- **d.** 4

4. Which line is a fragment?
- **a.** 1
- **b.** 2
- **c.** 3
- **d.** All are complete sentences.

5. Rewrite sentence (1) to correct the mistake in comparison.

Geography for Life

Diagrams should make clear the connection between erosion and downstream sedimentation. Students should support the information in their drawings with facts from the textbook. Use Rubric 6, Cause and Effect, to evaluate student work.

Social Studies Skills

1. d

2. birthrates—Sri Lanka, Pakistan, Bangladesh, Nepal, Bhutan, Maldives; deathrates—Sri Lanka, Maldives, Bangladesh, Pakistan, Nepal, Bhutan; Maldives

3. d

4. c

5. Family life in Bangladesh is different from family life in Pakistan.

Alternative Assessment

PORTFOLIO ACTIVITY

Learning about Your Local Geography

Group Project: Presenting Information

Conduct research on natural hazards that can strike your area. Examples might be earthquakes, floods, forest fires, hurricanes, mudslides, or tornadoes. Analyze the way people prepare for and react to these hazardous events. Study some examples of how humans have used technology to adapt to the dangers of natural hazards. Compare and contrast the effects of local natural hazards with the effects of tropical cyclones and flooding in Bangladesh. How important is preparing for natural hazards in the daily lives of people in both regions? Use posters or another kind of display to present your findings to the class.

internet connect

go.hrw.com

Internet Activity: go.hrw.com
KEYWORD: SW3 GT26

Choose a topic on the Indian Perimeter to:
- publish a poster on the landscapes and cultures of the Himalayas.
- learn the history of Sri Lanka.
- examine Pakistani history and culture.

PORTFOLIO ACTIVITY

Reports will vary according to the natural hazard on which research is conducted. Students may conclude that people in the United States receive more information, which helps them prepare for natural disasters. Use Rubrics 14, Group Activity, and 29, Presentations, to evaluate student work.

Workshop 1
Going Further: Thinking Critically

Tell students that conducting successful World Wide Web searches is a skill that can be improved with practice, just like athletic or artistic skills. With experience, students can learn to use the proper combination of search engines and terms to locate the information they need quickly and easily. Have students who are familiar with the Internet compile a list of World Wide Web search tips for new computer users. Tell students that they should include information on both broad searches—those that are intended to gather information about a general subject—and narrow searches, which are focused on a more specific topic. Have students write suggestions for particular Web sites and key terms that researchers might use in each type of search. Also have students write paragraphs in which they summarize the importance of the terms used in a search. Students should explain how the choice of words used in a search can affect its results and provide suggestions for selecting the proper search terms. Have students add to their lists other advice that might help novice Internet browsers in conducting online research. Call on volunteers to share their tips with the class.

Geography
Skill-Building Workshop

WORKSHOP 1

Using Media Services: The Internet

The Internet, or the Net, is made up of millions of computers linked worldwide through a telecommunications network. The Net provides access to a great deal of information in many formats and from many sources.

Developing the Skill   An Internet service provider, or ISP, provides access to the Internet. A computer links to the ISP through a modem. Once you are connected, you can "surf" the Internet. Surfing the Internet lets you link to informational sites or send messages called e-mail. Many informational sites are locations on the World Wide Web, sometimes referred to as the WWW or simply the Web. You access the Web by using software called a Web browser. You can call up specific Web sites by entering a special address called a uniform resource locator, or URL. Such addresses typically begin with http:// and end with suffixes such as .com, .edu, .gov, or .org.

Most Web sites have a home, or main, page that will "link" you to additional information on other pages or sites through hyperlinks. These hyperlinks usually appear as underlined terms or phrases. For example, databases like the World Factbook from the U.S. Central Intelligence Agency can be found on the Web. The World Factbook includes hyperlinks to specific categories of information and countries.

You can also search the Web for sites that include information about a particular topic by using a search engine. Search engines will seek Web sites that include key words you have entered. You will need to experiment to find search engines that are most useful to you. Follow instructions on each search engine.

Users must be able to determine whether a Web page contains valuable information. Ask yourself these questions:

• Is a well-respected organization sponsoring the site? Whose views are represented?
• Is the information clear and easy-to-read? What has the designer done to attract your attention?

• Is the Web site easy to navigate, and are most of the page links active?
• Is the information current?
• Is the information biased? What information has been left out of the site? How might information there be interpreted in different ways by different people?

The Internet contains a huge amount of resources, making it an excellent tool for research. However, there is a lot of inappropriate material on the Web. Many well-respected sites have developed safe sites for children and teens. These sites are known as child-safe zones. Even so, you must be careful.

• Do not chat with or meet strangers.
• Do not give out any personal information.
• If you get into an unsecured area, get out immediately and tell your parent or teacher.
• Stay within well-respected search engines.

Practicing the Skill

1. Find Web sites operated by each of the governments of South Asia's countries. List the URL for what appears to be the main Web site for each government.
2. Locate a Web site that offers news from India or Pakistan. Identify the operator of the site. Is it a newspaper, television network, or other news provider?
3. Locate Web sites that provide statistics about countries in South Asia. What kind of data do you find on each site?

Workshop 2
Going Further: Thinking Critically

Have students use the problem-solving process to identify solutions to challenges facing your community. Have students work in groups to identify geographical problems in your area that require attention. *(Possible problems include urban sprawl or rapid depopulation.)* Then have students consult local resources like newspapers, local government publications, or Web sites to gather information about the problem and its causes. Students may also wish to talk to community officials in their searches for information. Then have each group create a list of possible solutions to the problem. Point out to students that these solutions should be reasonable ones that would be viable in your community. They should not call for resources or funding that are unavailable. Have students consider the advantages and disadvantages of each option. You may wish to have students create charts that demonstrate these advantages and disadvantages. Finally have students choose one solution they would recommend. Call on volunteers from each group to present these ideas to the class. Lead a class discussion about whether each proposed solution is rational and predict each plan's effectiveness in resolving local issues.

WORKSHOP 2

Solving Problems

Having effective problem-solving skills is important in geography. For example, some farmers must learn how to transform harsh environments into productive farmlands. Demographers work to find ways to collect more accurate information about populations. Governments study possible solutions to problems like rapid population growth, inadequate health care systems, and pollution.

Developing the Skill The problems listed above are very different from each other. However, solving them can involve similar processes. Once you have identified a problem, these steps can help you find a solution:

- **Gather information** about the issue or problem. Libraries and reliable sites on the World Wide Web are common sources of information about geographic problems. The information you gather will help you understand a problem more clearly and identify possible solutions.
- **List and consider the options** you have. Brainstorm possible options and list them in a chart or idea web. List all the reasonable options that come to mind—you can narrow down your choices later.
- **Consider advantages and disadvantages** of each of your options. For complex problems, you might create a cost-benefits balance sheet

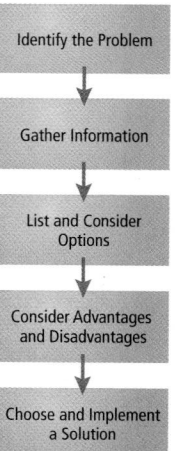

The Problem-Solving Process

Identify the Problem

↓

Gather Information

↓

List and Consider Options

↓

Consider Advantages and Disadvantages

↓

Choose and Implement a Solution

for each option. List the disadvantages or additional problems associated with a particular option on the "cost" side of the balance sheet. On the "benefits" side list the advantages associated with that option. Follow the same process for all of your options. Then decide whether the advantages of each option outweigh the disadvantages.

- **Choose and implement the solution** you think is best from your list of options. The best solution may be clear once you have analyzed your cost-benefits balance sheets. If the best solution is not clear, you may need more information to help you decide. Then create an action plan for implementing your solution.
- **Evaluate the effectiveness** of your solution. What you learn from this evaluation can help you solve other problems in the future. It can also help you refine your own problem-solving process. Review your cost-benefits balance sheet for the option you chose. Note whether you anticipated all of the possible disadvantages and advantages associated with your solution.

Practicing the Skill

1. India and Pakistan have gone to war against each other twice since the late 1940s. Today both countries have nuclear weapons, making the possibility of war between them even more frightening. Imagine you are a member of a United Nations commission trying to forge a lasting peace between the two countries. Identify problems that divide the countries and then identify possible solutions.
2. Mount Everest and other mountains of the Himalayas attract many climbers and other tourists each year. Unfortunately, those visitors leave their garbage behind. How might the government of Nepal work to protect the natural environment from the problems visitors cause? How might the government implement your proposed solution?

Using the Illustrations

Focus students' attention on the photo of Borobudur, one of the world's greatest Buddhist temples. Tell students that Borobudur was built on the island of Java between about A.D. 775 and 850 with almost 2 million cubic feet (56,600 cubic meters) of stone. Five rectangular terraces and three concentric circular terraces form a stepped pyramid that reaches a height of about 103 feet (31.4 m). Hundreds of carved panels decorate the walls. Each of these panels illustrates a step in the Buddhist path to enlightenment. The panels at the lowest level depict scenes from daily life. Each panel in the middle level shows a scene from the life of the Buddha or an element of Buddhist philosophy. Those at the highest levels represent detachment from the physical world. In addition, 72 bell-shaped stupas, or Buddhist shrines, line Borobudur's terraces. Inside many of the stupas are statues of the Buddha, visible through perforations carved into the stone shrines.

Ask students to use what they already know about the Pacific region to predict what natural processes may threaten the structure. *(Volcanoes may bury the site in ash, heavy rain could break through weak walls, and rainforest vegetation could contribute to erosion of the stonework.)* In fact, volcanic ash and vegetation covered the temple so thoroughly it lay hidden for centuries. It was rediscovered in 1814. Repair and reconstruction are ongoing.

Unit Objectives

1. Describe how tectonic activity, climate, and precipitation affect the people of East and Southeast Asia.

2. Analyze how past events have shaped the region's human geography.

3. Explain the economic significance of East and Southeast Asia as a world region.

4. Explore connections among physical geography, historical geography, and cultural geography in East and Southeast Asia.

5. Interpret and construct diagrams to identify connections among geographic data.

6. Use spreadsheets and software to organize and examine geographic information.

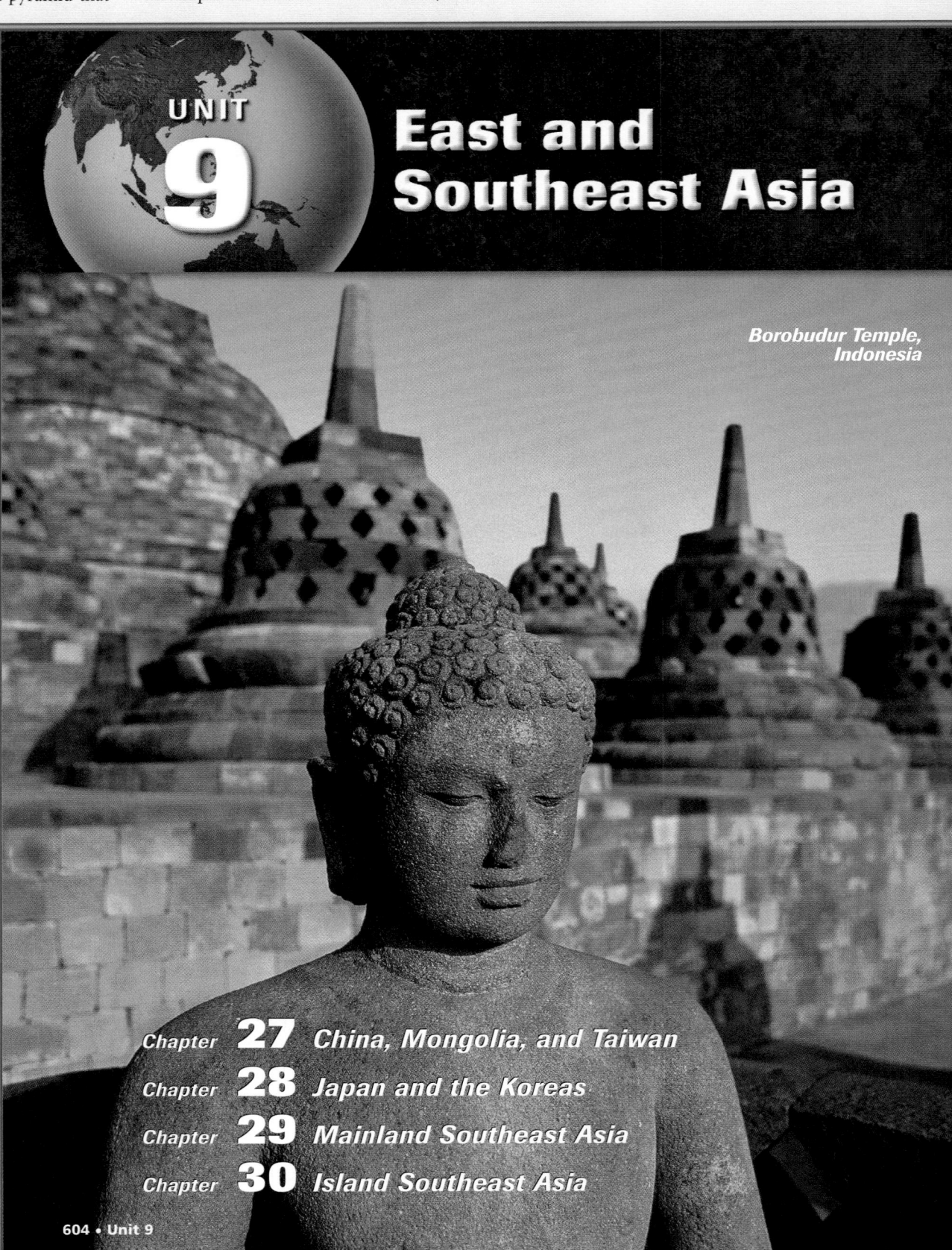

UNIT 9
East and Southeast Asia

Borobudur Temple, Indonesia

CONNECTING TO

Literature

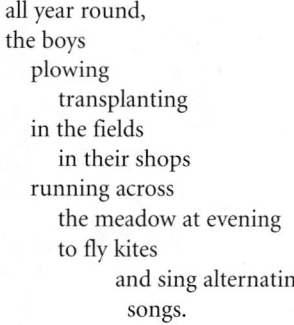

"THOUGHTS OF HANOI" by Nguyen Thi Vinh

Nguyen (nie-EN) **Thi Vinh**
(1924–) was born in northern Vietnam's Red River delta in Southeast Asia. A novelist and poet, she fled to South Vietnam after Communists took over the north in the 1950s. She remained in South Vietnam for a while after it fell to the Communists in 1975. In 1983 she moved to Norway. Her poem "Thoughts of Hanoi" reveals some of her memories of the land of her birth.

The night is deep and chill
as in early autumn. Pitchblack,
it thickens after each lightning
 flash.
I dream of Hanoi:
Co-ngu Road
Ten years of separation
the way back sliced by a
 frontier of hatred.
I want to bury the past
to burn the future
still I yearn
still I fear
those endless nights
waiting for dawn.

Brother,
how is Hang Dao now?
How is Ngoc Son temple?
Do the trains still run
each day from Hanoi
to the neighboring towns?
To Bac-ninh, Cam-giang,
 Yen-bai,

the small villages, islands
of brown thatch in a lush
 green sea?

The girls
 bright eyes
  ruddy cheeks
 four-piece dresses
  raven-bill scarves
 sowing harvesting
  spinning weaving

all year round,
the boys
 plowing
  transplanting
in the fields
 in their shops
running across
 the meadow at evening
 to fly kites
  and sing alternating
   songs.

Stainless blue sky,
 jubilant voices of children
stumbling through the alphabet,
 village graybeards strolling to
  the temple,
grandmothers basking in
 twilight sun,
  chewing betel* leaves
while the children run—

---
* **betel** (BEE-tuhl) **leaves:** *the leaves of a climbing pepper vine; chewed with betel nuts*

Analyzing the Primary Source

1. **Summarizing** What images does the writer remember from her homeland?

2. **Drawing Inferences** What do you suppose the writer means when she writes of her way back to Vietnam being "sliced by a frontier of hatred"?

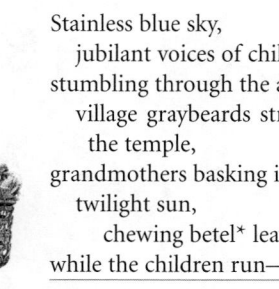

A Closer Connection

The Streets of Hanoi In her poem, Nguyen Thi Vinh mentions Hang Dao. From the context, the reader may conclude that Hang Dao is a place. In fact, it is a main street through Hanoi's Hoan Kiem, or Old Quarter, whose history may go back 2,000 years.

During the 1400s the old city was known as the 36 Streets because the narrow, winding lanes were named for the craft unions located there. For example, bamboo raft makers lived and worked in Hang Be Street. One could find stringed musical instruments in Chan Cam. Cho Gao was a rice market. Pickled fish were the specialty of Hang Mam's workers. Hang Dao was the silk merchant's street. *Dao* referred to the color of apricot blossoms, a bright pink shade that was the specialty of the street's silk dyers. Many of the street names begin with *Hang*, which means "merchandise" or "shop."

Analyzing the Primary Source

Answers

1. children working at various tasks, flying kites, singing songs, reciting the alphabet, running; old people; blue sky; others

2. Possible answer: She would have a hard time returning to Vietnam because of the hatred she feels for elements in the past.

This unit introduces students to East and Southeast Asia. The diversity of the region's landscapes is equalled by the diversity of the ways its people live. For example, in the area's big cities, such as Kuala Lumpur, Malaysia, and Tokyo, Japan, one can visit the world's tallest skyscrapers and catch a ride on the world's fastest trains. In the same region, the remote rain forests of Vietnam shelter a deer species that Western scientists learned about only recently. Economic growth has earned some countries, such as South Korea, Taiwan, and Malaysia, the nickname "the tigers," while farming with traditional methods still sustains millions of poor farmers in other places, such as Laos. City dwellers of Japan, Brunei, and Singapore have a very high standard of living while people surviving in urban slums struggle to feed themselves. Several of the countries have weathered brutal wars, rebellions, and coups, and now look to the future with optimism. In other areas, such as Myanmar, repressive governments continue to limit personal freedoms. In Korea, two countries that share a common heritage are separated by political divisions. The region's big island countries, Indonesia and the Philippines, work to create a sense of nationhood—a task sometimes made difficult by their geography.

Your Classroom Time Line

These are the major dates and time periods for this unit. Have students enter them on the time line you created earlier. You may want to watch for these dates as students progress through the unit.

c.* 6000 B.C. Agriculture is developed around the Chang River in China.

c. 3000 B.C. Farming villages are established along the Huang River.

c. 2000 B.C. The Malay people migrate into what is now Malaysia.

1700s B.C.–1100s B.C. China is ruled by the Shang Dynasty.

551 B.C. Confucius is born.

500s B.C. Taoism develops in China.

500s B.C. The Shwedagon Pagoda is built.

c. 500 B.C. The Chinese begin migrating southward into mainland Southeast Asia.

300 B.C. The Ainu are driven into the northern reaches of Japan.

200s B.C. Chinese farmers use an ox-drawn plow, allowing increased agricultural production.

*c. stands for *circa* and means "about."

POLITICAL MAP ANSWERS

1. Laos, Mongolia
2. Hokkaidō, Honshū, Shikoku, Kyūshū

CRITICAL THINKING ANSWER

3. Possible answer: It might make it harder to control territory and build transportation and communications systems.

UNIT 9 ATLAS

The World in Spatial Terms

East and Southeast Asia:
Political

1. (**Places and Regions**) Which countries in the region are landlocked?
2. (**Places and Regions**) Compare this map to the physical map. Which four major islands make up Japan?

Critical Thinking

3. **Making Generalizations** How might Indonesia's island geography cause problems for its government?

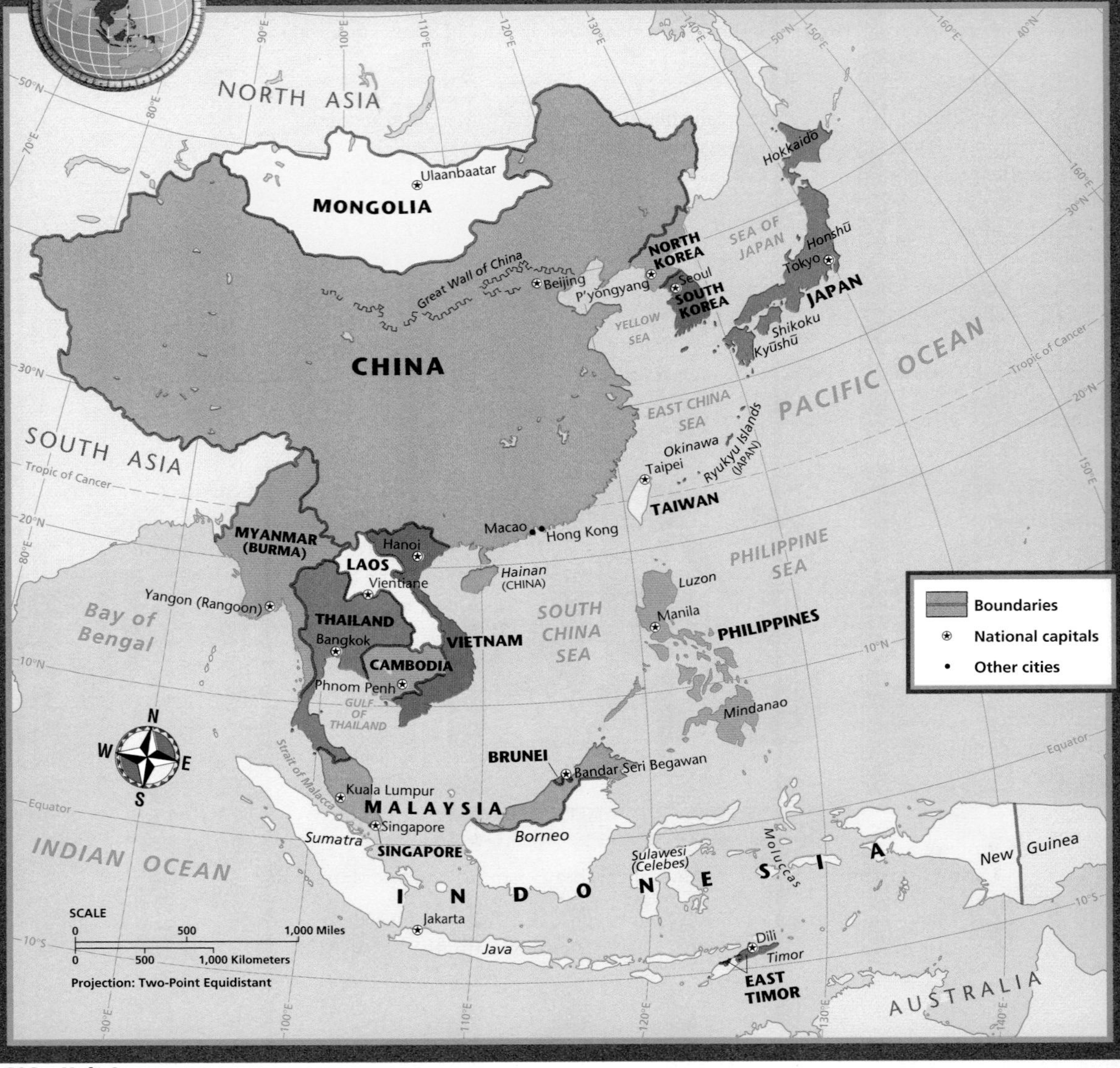

NORTH ASIA

Ulaanbaatar

MONGOLIA

Great Wall of China

Beijing

NORTH KOREA

P'yŏngyang

SEA OF JAPAN

Hokkaido

Honshū

Tokyo

JAPAN

SOUTH KOREA

Seoul

YELLOW SEA

Shikoku

Kyūshū

CHINA

EAST CHINA SEA

PACIFIC OCEAN

Okinawa

Tropic of Cancer

SOUTH ASIA

Taipei

Ryukyu Islands (JAPAN)

Tropic of Cancer

TAIWAN

MYANMAR (BURMA)

Hanoi

Macao

Hong Kong

PHILIPPINE SEA

LAOS

Vientiane

Hainan (CHINA)

Luzon

Yangon (Rangoon)

THAILAND

Bangkok

VIETNAM

SOUTH CHINA SEA

Manila

PHILIPPINES

Bay of Bengal

CAMBODIA

Phnom Penh

GULF OF THAILAND

Mindanao

BRUNEI

Bandar Seri Begawan

Strait of Malacca

Kuala Lumpur

MALAYSIA

Singapore

SINGAPORE

Sumatra

Borneo

Sulawesi (Celebes)

Moluccas

New Guinea

Equator

INDIAN OCEAN

Equator

I N D O N E S I A

Jakarta

Java

Dili

East Timor

EAST TIMOR

AUSTRALIA

Boundaries

⊛ National capitals

• Other cities

SCALE

0 — 500 — 1,000 Miles

0 — 500 — 1,000 Kilometers

Projection: Two-Point Equidistant

Using the Political Map

While students focus on the **political map** on the opposite page, ask them to identify the largest country in the region (*China*) and the smallest (*Singapore*). Then ask them to name the region's island countries (*Japan, Philippines, Indonesia, Taiwan, Singapore, Brunei, part of Malaysia*). Ask students which of the region's countries has two names (*Myanmar [Burma]*) and what they infer from this fact. (*Possible answer: There is disagreement over what the country should be called.*) Tell students that although the country's military regime calls it Myanmar, the United States government does not recognize the regime as legal. Similarly, the regime calls the capital Yangon; we still call it Rangoon.

Using the Physical Map

Focus students' attention on the **physical map**. Have students compare this map to the **political map** to name countries that lie on peninsulas (*Malaysia, Thailand, Cambodia, Vietnam, Laos, part of Myanmar, North Korea, South Korea*). Ask students to identify the region's major river deltas (*Irrawaddy, Mekong, Hong [Red]*). Then remind students of the volcanic activity that occurs along the Pacific Ring of Fire. Ask students to predict how volcanic activity may affect the region's many coastlines and mountainous interiors. (*Possible answers: Tsunamis may flood coastlines. Volcanoes and earthquakes may cause landslides in the mountains.*)

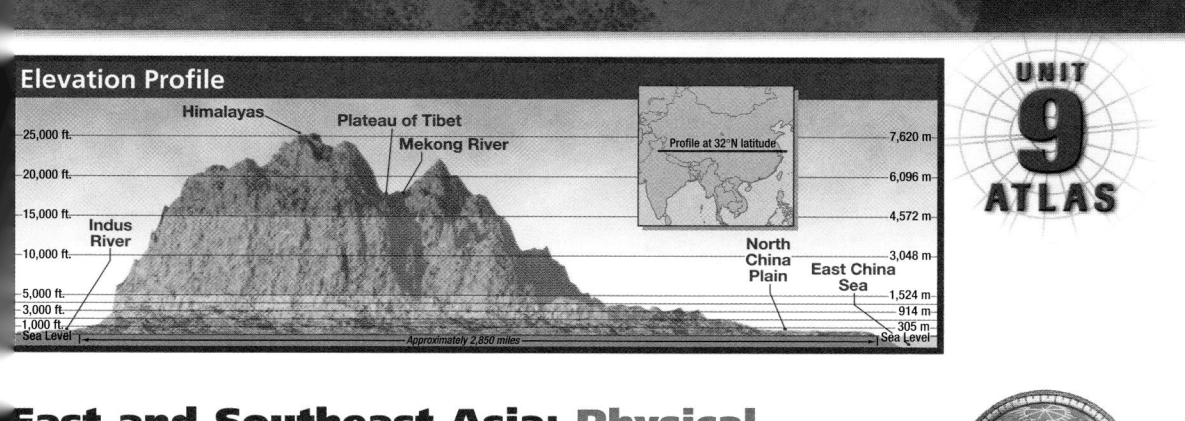

East and Southeast Asia: Physical

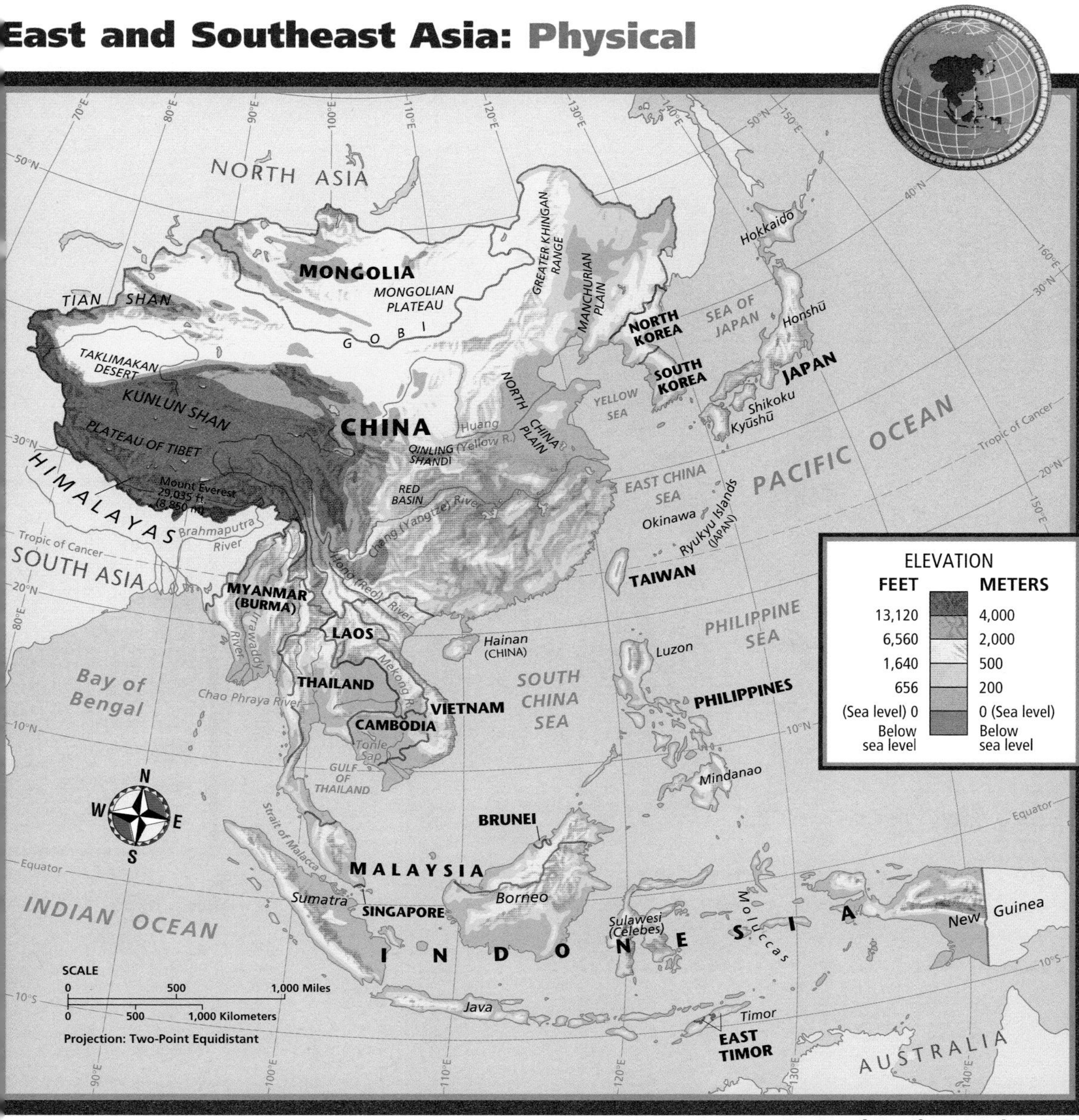

UNIT 9 ATLAS

Your Classroom Time Line, (continued)

200s B.C. Cheng conquers eastern China to become the first emperor of all China. Construction begins on what will become the Great Wall.

202 B.C.–A.D. 220 The Han dynasty spreads Chinese influence to the south.

108 B.C. China invades Korea.

c. A.D. 100 Buddhism spreads to China from India.

500s Buddhism spreads to Japan from China.

c. 700 Hindu and Buddhist kingdoms flourish in Java and Sumatra.

700s Japan develops an imperial government.

800s The Khmer take control of much of Southeast Asia.

c. 850 The Borobudur Temple is completed in Java.

960–1279 The Sung Dynasty rules China.

1100s Shoguns begin to control Japanese politics.

1200s The Thai people migrate into Southeast Asia from China.

1200s Mongols under Genghis Khan conquer the Chinese.

Late 1200s Marco Polo visits China.

Late 1200s The kingdom of Majapahit controls many coastal areas of island Southeast Asia.

1281 The Mongols under Kublai Khan try to invade Japan for the second time.

1300s Arab merchants conduct trade with the peoples of island Southeast Asia.

1521 Ferdinand Magellan lands in the Philippines.

1557 Portuguese traders establish Macao in China.

1565 The first Spanish settlement is established in the Philippines.

607

Using the Climate Map

As students examine the **climate map** on this page, point out that the region stretches from 50° north of the equator to 10° south. Ask students what effect this latitude range probably has on the region *(a wide range of climate types)*. Have them discuss the extremes of the region's climates, which range from subarctic to humid tropical. Have students identify other countries that lie as far north as Mongolia *(such as Canada)* and as far south as Indonesia *(such as Brazil)*.

CLIMATE MAP ANSWERS

1. arid, semiarid, subarctic, humid continental, highland, humid subtropical
2. tropical humid

CRITICAL THINKING ANSWER

3. the Himalayas

UNIT 9 ATLAS

East and Southeast Asia:
Climate

1. (*Places and Regions*) Compare this map to the political map. Which climate types are found in China?
2. (*Places and Regions*) Compare this map to the political map. Which climate type dominates Indonesia, Malaysia, and the Philippines?

Critical Thinking

3. **Analyzing Information** Compare this map to the physical map. Which physical feature likely blocks the wet monsoon airflow from reaching the arid and semiarid regions of China?

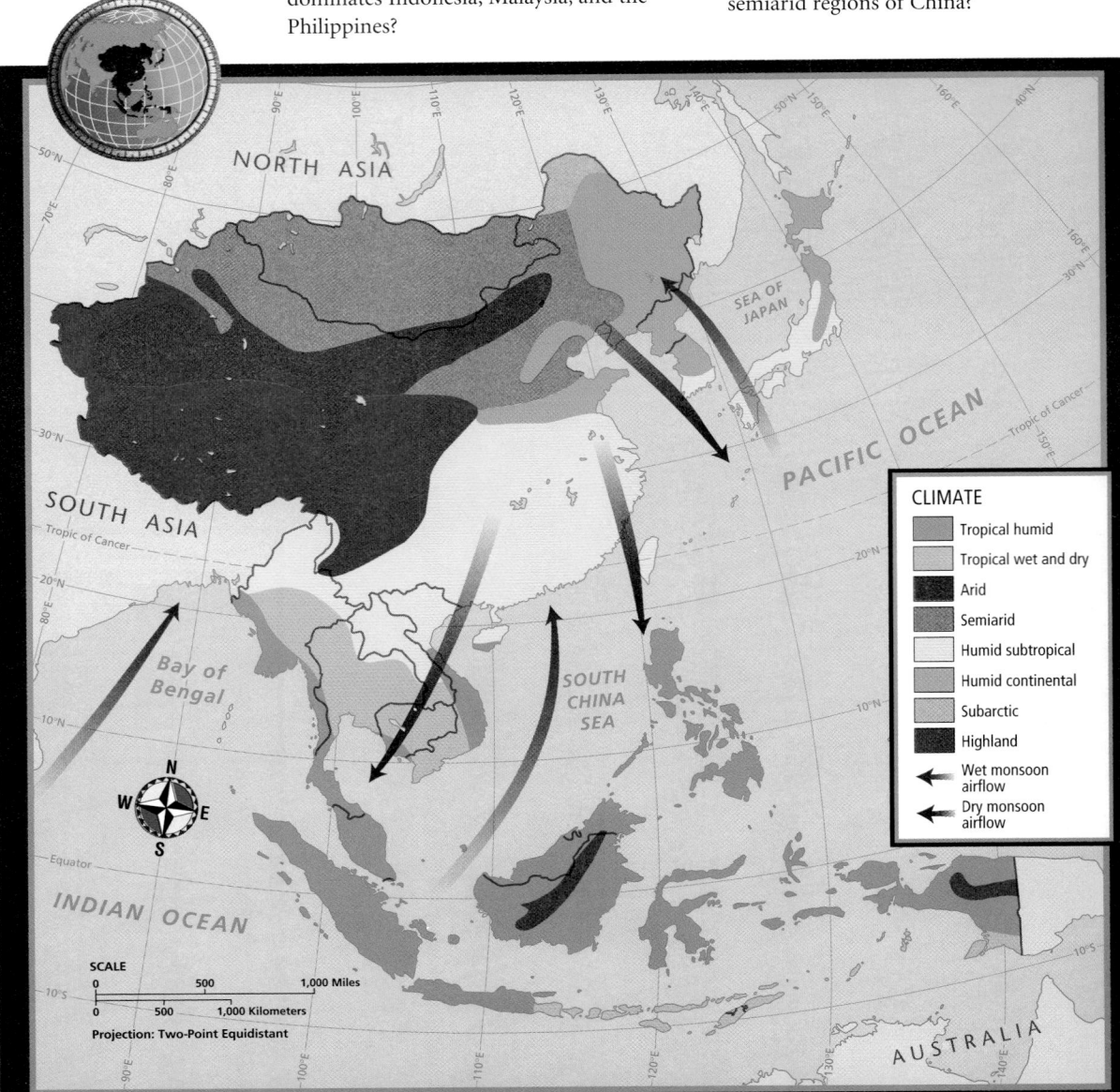

NORTH ASIA
SOUTH ASIA
SEA OF JAPAN
PACIFIC OCEAN
Tropic of Cancer
Bay of Bengal
SOUTH CHINA SEA
INDIAN OCEAN
Equator
AUSTRALIA

CLIMATE
- Tropical humid
- Tropical wet and dry
- Arid
- Semiarid
- Humid subtropical
- Humid continental
- Subarctic
- Highland
- ← Wet monsoon airflow
- ← Dry monsoon airflow

SCALE
0   500   1,000 Miles
0   500   1,000 Kilometers
Projection: Two-Point Equidistant

Have students compare the **precipitation map** on the previous page with the **climate map** below and the physical–political map for Chapter 30. Ask them what type of weather one might experience in central Borneo and central Irian Jaya. *(Possible answer: cool, very rainy)* Then ask how the monsoons appear to affect precipitation. *(The summer monsoon brings heavy rainfall to the region.)* Then ask students to name other factors that might contribute to the region's high precipitation levels *(tropical location, large bodies of water surrounding countries).*

East and Southeast Asia:
Precipitation

UNIT 9 ATLAS

1. (**Places and Regions**) Compare this map to the population map. Which large metropolitan areas are located in areas that receive more than 80 inches (200 cm) of precipitation?

2. (**Places and Regions**) How much precipitation does most of Mongolia receive?

Critical Thinking

3. Making Generalizations How do you think the distribution of precipitation in East and Southeast Asia is related to latitude?

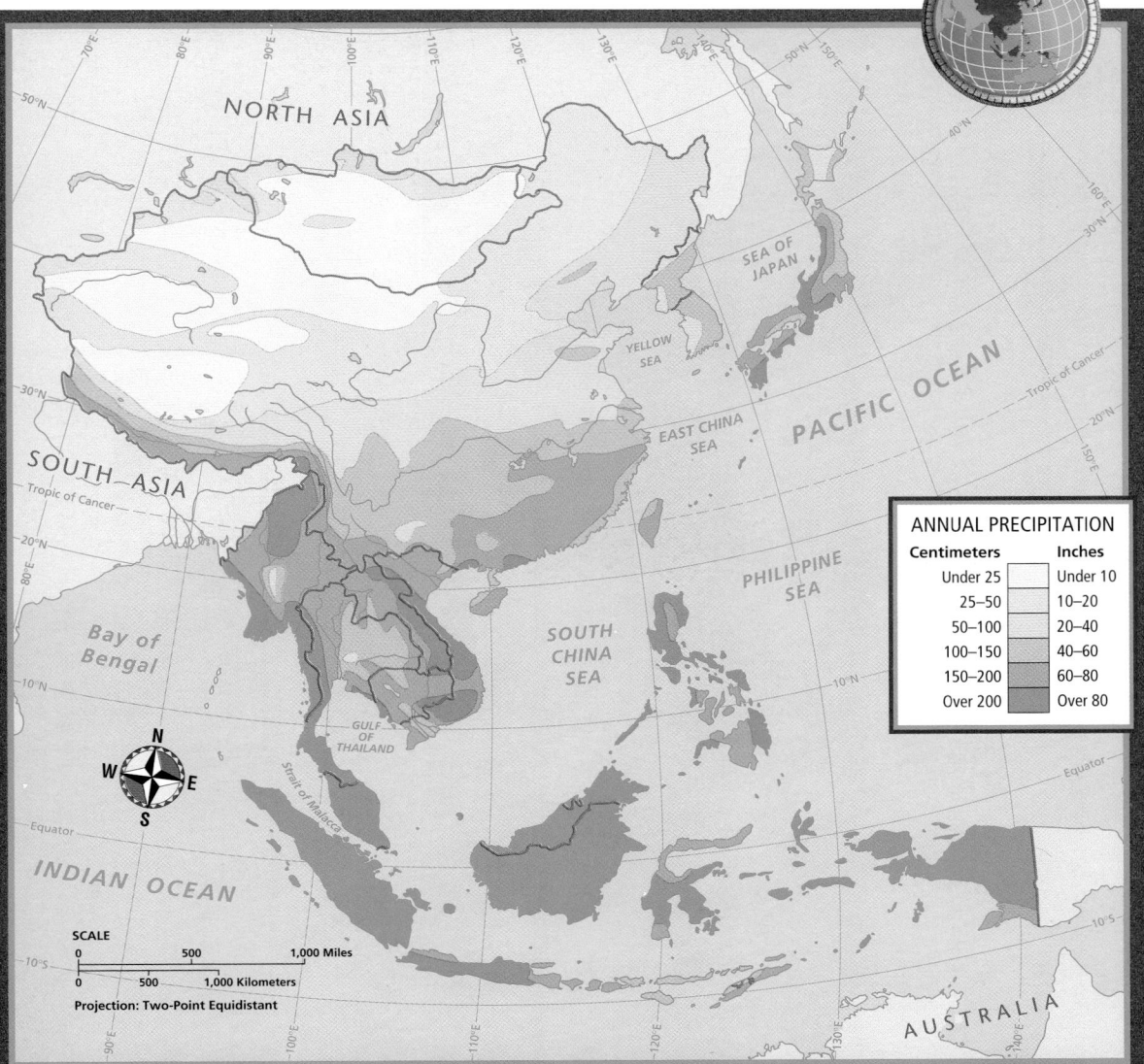

ANNUAL PRECIPITATION

Centimeters	Inches
Under 25	Under 10
25–50	10–20
50–100	20–40
100–150	40–60
150–200	60–80
Over 200	Over 80

SCALE
0   500   1,000 Miles
0   500   1,000 Kilometers
Projection: Two-Point Equidistant

Your Classroom Time Line, (continued)

1941 Japan attacks Pearl Harbor, and the United States declares war on Japan.

1945 The United States bombs Hiroshima and Nagasaki. Japan surrenders to the Allies. World War II ends.

1945 Nationalists and Communists fight for control of China.

1946 The Philippines gains its independence from the United States.

1947 Japan becomes a democracy.

1948 Korea splits into two countries. Kim Il Sung takes control of North Korea.

1948 Burma becomes independent.

1949 Mao Zedong establishes the communist People's Republic of China. Some 2 million Chinese Nationalists flee to Taiwan.

1949 Indonesia wins its independence from the Netherlands.

1950s Japan's economy begins to expand rapidly.

1950 The Korean War begins.

1950 China occupies Tibet.

1953 The Korean War ends.

1954 Vietnam gains independence from France and is divided into two parts.

PRECIPITATION MAP ANSWERS

1. Bandung, Manila, Ōsaka, Singapore, Surabaya, Yangon

2. under 10 inches (25 cm)

CRITICAL THINKING ANSWER

3. Possible answer: Areas in lower latitudes receive more precipitation than areas in higher latitudes.

Using the Population Map

As students examine the **population map** on this page, have them identify the countries that have no metropolitan areas with more than 2 million inhabitants (*Brunei, Cambodia, East Timor, Laos, Malaysia, Mongolia*). Ask which country has a low population density over its entire area (*Mongolia*). Then have students compare this map to other maps in the unit atlas. Point out the big difference in the population densities of Java and Sumatra. Ask students to speculate about reasons for the difference in densities. (*Possible answer: The physical geography of the two islands is similar—mountainous interiors, mainly humid tropical climates, and heavy rainfall. Sumatra, however, has more forest land than Java, which has more land devoted to commercial farming than other uses. This is the only evidence one can find in the maps for why Sumatra is less densely populated.*)

Your Classroom Time Line,
(continued)

1958 Mao Zedong launches the Great Leap Forward in China.

1959 The Dalai Lama leaves Tibet in exile.

1962 A military government takes over Burma.

1963 The Federation of Malaysia is created.

1964–73 U.S. troops fight in Vietnam.

1964 Japan's first *Shinkansen* line connects Tokyo to Ōsaka.

1965 Singapore becomes independent.

1966–76 Mao Zedong implements the Cultural Revolution in China.

1967 The Association of Southeast Asian Nations (ASEAN) is founded.

1975–79 Cambodia is ruled by the Khmer Rouge.

1975 Vietnam is reunified under a communist government.

1975 Indonesia invades East Timor.

1976 Mao Zedong dies, and Deng Xiaoping takes over control of China's government.

1984 Brunei gains its independence from Britain.

1989 Burma is renamed Myanmar.

1989 Thailand outlaws logging for export.

POPULATION MAP ANSWERS

1. China
2. North China Plain

CRITICAL THINKING ANSWER

3. Possible answer: Sumatra, Borneo, New Guinea

UNIT 9 ATLAS

East and Southeast Asia: Population

1. (**Places and Regions**) Compare this map to the political map. Which country in the region has by far the largest number of metropolitan areas with more than 1 million inhabitants?

2. (**Places and Regions**) Compare this map to the physical map. Which large lowland region in northeastern China is densely populated?

Critical Thinking

3. **Analyzing Information** Compare this map to the physical map. If Indonesia's government wanted to lower the population density on Java, to which other islands in the country might they encourage people to move?

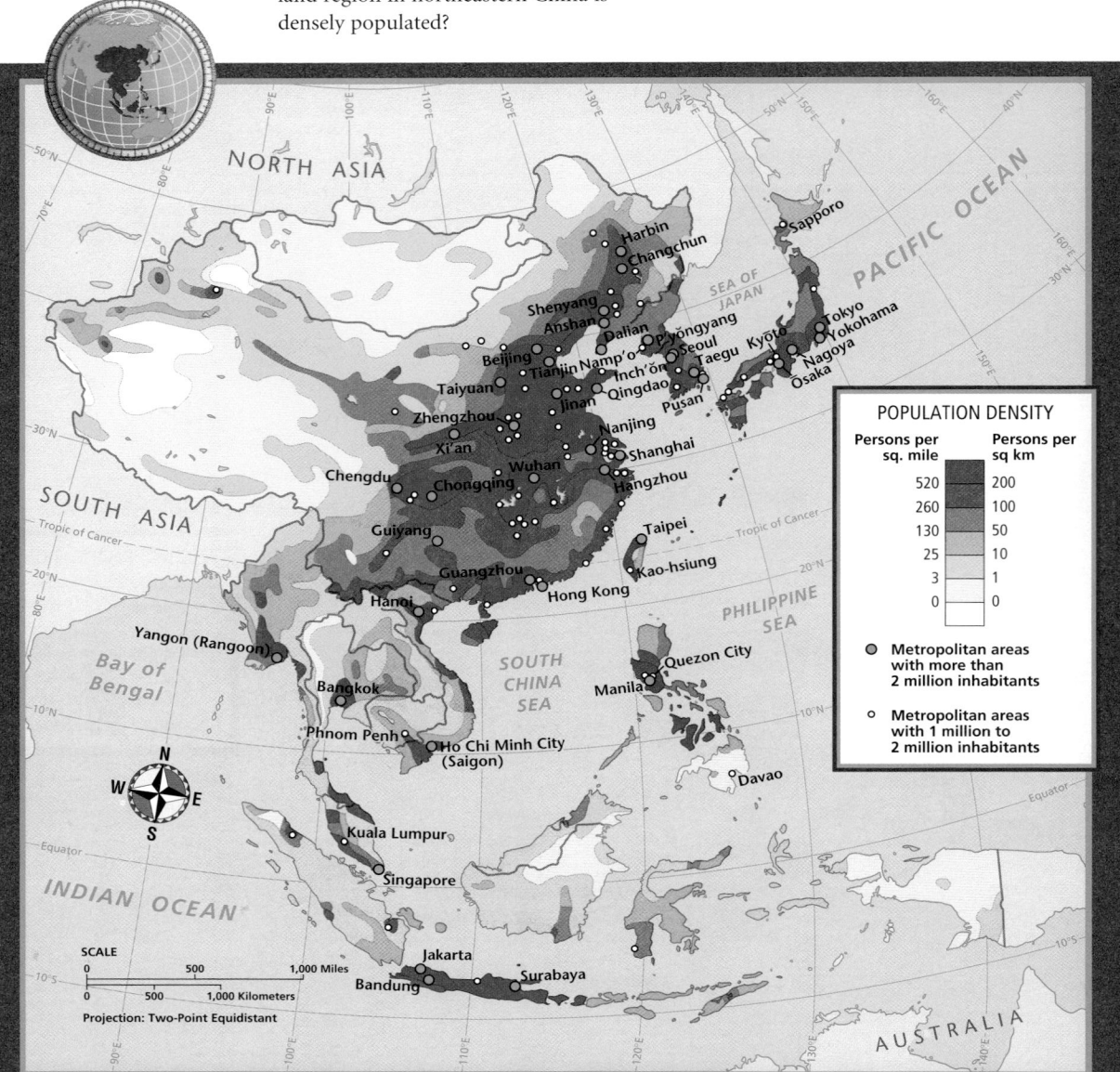

POPULATION DENSITY

Persons per sq. mile	Persons per sq km
520	200
260	100
130	50
25	10
3	1
0	0

● Metropolitan areas with more than 2 million inhabitants

○ Metropolitan areas with 1 million to 2 million inhabitants

SCALE
0 — 500 — 1,000 Miles
0 — 500 — 1,000 Kilometers
Projection: Two-Point Equidistant

Using the Land Use and Resources Map

Focus students' attention on the **land use and resources map** on this page. Ask which countries have developed hydroelectric power in their mountainous regions (*China, Laos, Thailand, and Vietnam*). Then ask students to refer to the **physical map** and **precipitation map**. Ask which island appears to be the best candidate for developing hydroelectric power because it has plentiful rainfall and areas of high elevation (*New Guinea*). Have students identify the country that produces the most nuclear power (*Japan*). Have students examine other maps in the unit atlas and speculate why Japan relies on nuclear power. (*Japan has few rivers large enough to support hydroelectric plants and lacks large fossil fuel deposits.*)

East and Southeast Asia:
Land Use and Resources

UNIT 9 ATLAS

1. **Places and Regions** What type of power is commonly generated in Japan?
2. **Environment and Society** Based on the map, which countries in the region do you think have the most oil?

Critical Thinking

3. **Analyzing Information** Compare this map to the population and political maps. How do you think Japan's location is advantageous for trade with other countries in the region?

Your Classroom Time Line, (continued)

1989 Pro-democracy protesters are killed in Beijing's Tiananmen Square.

1990s Shanghai creates the Pudong Development Zone.

1990 Myanmar's military government refuses to recognize elected democratic leaders.

1990 Mongolia holds its first free elections.

1991 North Korea's economy is devastated by the collapse of the Soviet Union.

1991 Mount Pinatubo erupts in the Philippines.

1993 Construction of China's Three Gorges Dam begins.

1995 The Great Hanshin Earthquake strikes Kōbe, Japan.

1997–98 Fires sweep through Indonesia.

1997 Britain returns control of Hong Kong to China.

1999 East Timor declares independence.

1999 Portugal returns Macao to China.

2000 Leaders of North and South Korea meet to discuss the possibility of reunification.

2002 East Timor is recognized as an independent state and joins the UN.

2003 China becomes the third nation to launch a man into space. Astronaut Yang Liwei orbited Earth 14 times.

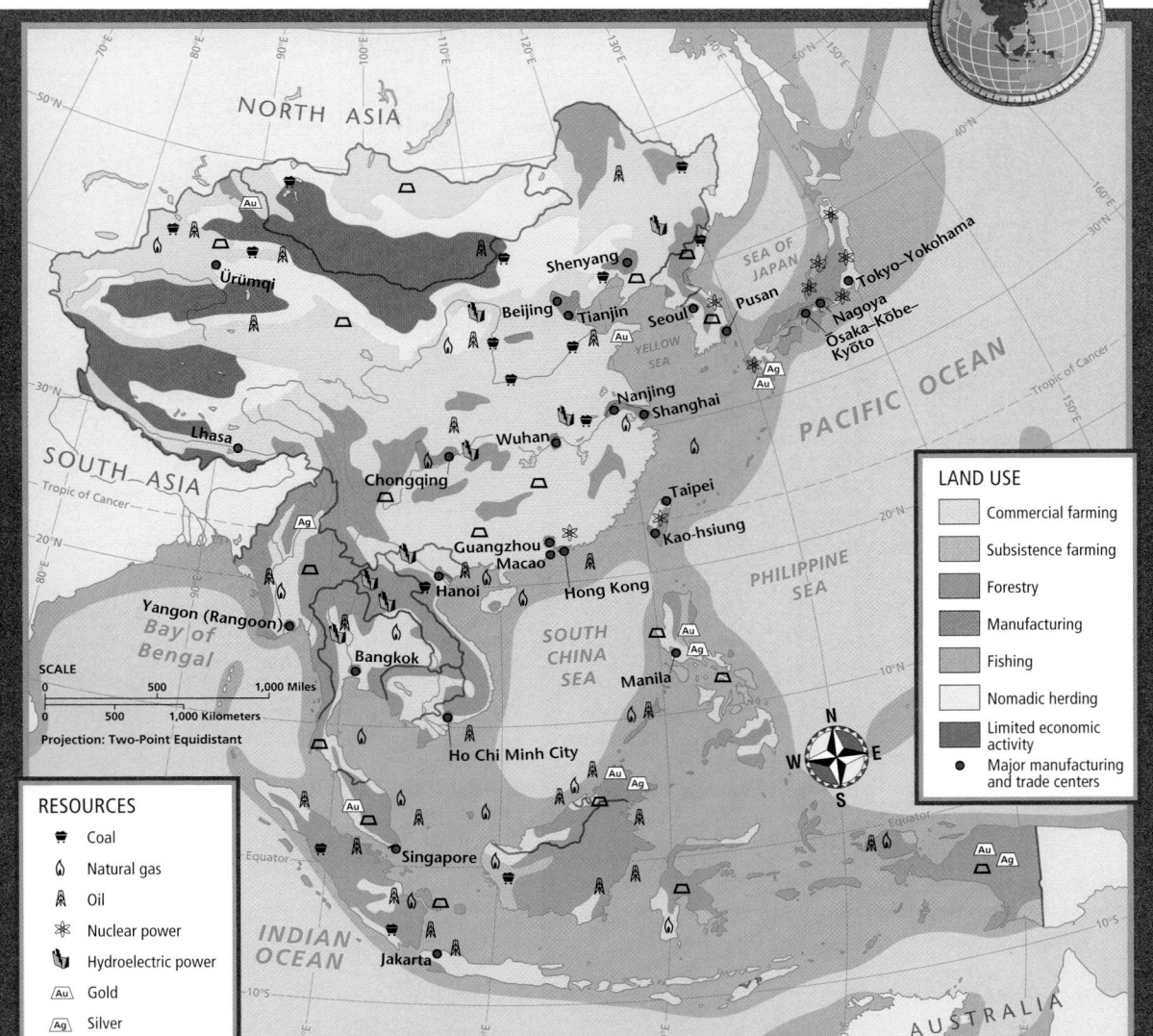

LAND USE
- Commercial farming
- Subsistence farming
- Forestry
- Manufacturing
- Fishing
- Nomadic herding
- Limited economic activity
- ● Major manufacturing and trade centers

RESOURCES
- Coal
- Natural gas
- Oil
- Nuclear power
- Hydroelectric power
- Au Gold
- Ag Silver
- Other minerals

LAND USE AND RESOURCES MAP ANSWERS

1. nuclear power
2. Indonesia, Malaysia

CRITICAL THINKING ANSWER

3. Possible answer: Japan is located close to large populations in China, the Koreas, and Taiwan.

611

Copy the following chart onto the chalkboard, omitting the italicized figures. The figures in the center column represent the total number of miles that all residents of the country traveled by air in one year. Have students use the population figures in the Fast Facts chart to calculate the average number of miles traveled by each resident of each country. (Divide the number of passenger miles by the population figure.) As volunteers provide those figures fill in the third column of the chart on the chalkboard.

COUNTRY	PASSENGER MILES	MILES PER PERSON
Brunei	1.6 billion	*4,000*
Cambodia	not available	*not available*
China	77.4 billion	*59.3*
Indonesia	9.0 billion	*41.1*
Japan	101.2 billion	*796.9*
Laos	48.5 million	*8.5*
Malaysia	21.0 billion	*860.6*
Mongolia	270.9 million	*104.1*
Myanmar (Burma)	220.6 million	*4.5*
North Korea	110.6 million	*4.9*
Philippines	6.4 billion	*80.0*
Singapore	40.7 billion	*9,465.1*
South Korea	34.9 billion	*731.7*
Taiwan	22.8 billion	*1,008.9*
Thailand	23.8 billion	*379.0*
Vietnam	2.4 billion	*29.5*
United States	646.0 billion	*2,197.3*

Source: *The World Almanac and Book of Facts, 2004*

Historical Geography

Chinese Walls Travel guides often describe the Great Wall in northern China, built to protect the country from northern invaders, as the only human-made structure visible from outer space. This is not true, however. Although the wall can be seen from a low orbit, so can many other structures, including highways, railroads, and cities. Farther from the surface, no human-made structures can be seen at all. Furthermore, the Great Wall is more correctly described as a series of many walls rather than a single structure.

For centuries Chinese rulers had walls built to keep out armies from the northern steppe. Records exist that date stretches of wall to the Qin (221–206 B.C.), Han (202 B.C.–A.D. 220), Northern Qi (550-77), Sui (581-618), and Ming (1368-1644) Dynasties. The rulers of these dynasties did not necessarily have the same purposes for building walls, for each had his own goals and his own enemies. The Great Wall's current form is largely the result of work during the Ming dynasty.

CRITICAL THINKING:
Why did many Chinese rulers build walls as a defense against invasion? *(Northern China lacks any physical barriers that could hinder armies moving into the country.)*

UNIT 9 ATLAS

Time Line: East and Southeast Asia

108 B.C. Chinese invade Korea.

C. A.D. 500 Hindu and Buddhist states are established in Java and Sumatra.

1853 U.S. commodore Matthew C. Perry visits Japan.

1949 The Communists win China's civil war. Indonesia wins independence from the Netherlands.

1975 North Vietnamese and other Communist forces take over South Vietnam, ending a long war.

200 B.C.	A.D. 500	1200	1500	1800	1900

202 B.C. Han dynasty rises to power in China and spreads Chinese culture into southern China.

1279 Mongols rule all of China.

1521 Ferdinand Magellan claims the Philippines for Spain.

1912 China becomes a republic.

1945 Japan is defeated after conquering much of East and Southeast Asia in World War II.

1950 North Korea invades South Korea. The Korean War ends in 1953.

1999 East Timor declares independence from Indonesia.

The United States and East and Southeast Asia

Comparing Sizes

internet connect

GO TO: go.hrw.com
KEYWORD: SW3 Almanac
FOR: Additional information and reference sources

Comparing Standard of Living

COUNTRY	LIFE EXPECTANCY (in years)	INFANT MORTALITY (per 1,000 live births)	LITERACY RATE	DAILY CALORIC INTAKE (per person)
Cambodia	56, male 61, female	76	70%	2,078
China	70, male 74, female	25	86%	2,972
Indonesia	67, male 72, female	38	89%	2,850
Japan	78, male 84, female	3	100%	2,874
North Korea	68, male 74, female	26	99%	1,899
Philippines	66, male 72, female	25	96%	2,280
Singapore	76, male 84, female	4	93%	Not available
South Korea	72, male 79, female	7	98%	3,069
Thailand	69, male 74, female	22	96%	2,462
Vietnam	68, male 73, female	31	94%	2,422
United States	74, male 80, female	7	97%	3,757

Sources: *World Almanac and Book of Facts, 2004; Britannica Book of the Year, 2002*

Have students use the completed chart, the Fast Facts per capita GNP figures, and the unit atlas to answer the following questions: What might be some reasons for Laos, Myanmar (Burma), North Korea, and Vietnam having low miles-per-person figures? *(They are relatively small mainland countries, so people do not have as far to go as they do in large countries. Also, they are poor countries, so few people have enough money to fly.)* Why do Singapore and Brunei have such high mileage figures, although they are tiny countries? *(They are wealthy countries, and people can afford to travel.)* How does Malaysia's physical geography seem to affect its miles-per-person figure? *(Because the country is spread across the mainland and islands, distances within the country are great. Travelers must fly or sail to cross the sea.)* Challenge students to use the information provided to write other questions. Call on volunteers to ask their questions and, if possible, provide the answers.

Fast Facts: East and Southeast Asia

UNIT 9 ATLAS

FLAG	COUNTRY / Capital	POPULATION (in millions) / POP. DENSITY	AREA	PER CAPITA GDP (in US $)	WORKFORCE STRUCTURE (largest categories)	ELECTRICITY CONSUMPTION (kilowatt hours per person)	TELEPHONE LINES (per person)
	Brunei Bandar Seri Begawan	0.4 / 176/sq. mi.	2,228 sq. mi. / 5,770 sq km	$ 18,600	48% government / 42% oil, gas prod., services, const.	6,486 kWh	0.25
	Cambodia Phnom Penh	14.1 / 208/sq. mi.	69,900 sq. mi. / 181,040 sq km	$ 1,500	80% agriculture	8 kWh	0.002
	China Beijing	1,304 / 362/sq. mi.	3,705,404 sq. mi. / 9,596,952 sq km	$ 4,400	50% agriculture / 28% services	767 kWh	0.16
	Indonesia Jakarta	219.9 / 314/sq. mi.	741,099 sq. mi. / 1,919,438 sq km	$ 3,100	45% agriculture / 39% services	405 kWh	0.03
	Japan Tokyo	127.7 / 838/sq. mi.	145,883 sq. mi. / 377,835 sq km	$ 28,000	70% services / 25% industry	7,553 kWh	0.58
	Laos Vientiane	5.7 / 63/sq. mi.	91,429 sq. mi. / 236,800 sq km	$ 1,700	80% agriculture	146 kWh	0.01
	Malaysia Kuala Lumpur	24.4 / 193/sq. mi.	127,317 sq. mi. / 329,750 sq km	$ 9,300	28% local trade, tourism / 27% manufacturing	2,599 kWh	0.19
	Mongolia Ulaanbaatar	2.6 / 4/sq. mi.	604,249 sq. mi. / 1,564,998 sq km	$ 1,840	primarily herding, agricultural	846 kWh	0.05
	Myanmar (Burma) Yangon	49.5 / 195/sq. mi.	261,970 sq. mi. / 678,499 sq km	$ 1,660	70% agriculture / 23% services	115 kWh	0.006
	North Korea P'yŏngyang	22.7 / 488/sq. mi.	46,541 sq. mi. / 120,541 sq km	$ 1,000	64% nonagricultural / 36% agriculture	1,231 kWh	0.05
	Philippines Manila	80.0 / 695/sq. mi.	115,831 sq. mi. / 300,001 sq km	$ 4,200	40% agriculture / 19% gov., services	526 kWh	0.04
	Singapore Singapore	4.3 / 17,653/sq. mi.	267 sq. mi. / 692 sq km	$ 24,000	35% fin., bus., serv. / 21% manufacturing	6,666 kWh	0.45
	South Korea Seoul	47.7 / 1,258/sq. mi.	38,023 sq. mi. / 98,479 sq km	$ 19,400	69% services / 22% industry	5,667 kWh	0.55
	Taiwan Taipei	22.6 / 1,815/sq. mi.	13,892 sq. mi. / 35,980 sq km	$ 18,000	58% services / 35% industry	6,216 kWh	0.58
	Thailand Bangkok	62.8 / 318/sq. mi.	198,456 sq. mi. / 513,999 sq km	$ 6,900	54% agriculture / 31% services	1,447 kWh	0.10
	Vietnam Hanoi	81.4 / 648/sq. mi.	127,244 sq. mi. / 329,560 sq km	$ 2,250	63% agriculture / 37% ind., services	341 kWh	0.07
	United States Washington, D.C.	294.0 / 83/sq. mi.	3,717,810 sq. mi. / 9,629,084 sq km	$ 37,600	31% manage., prof. / 29% tech., sales, admin.	12,250 kWh	0.65

Sources: Central Intelligence Agency, *World Factbook 2003; The World Almanac and Book of Facts, 2004.* The CIA calculates per capita GDP in terms of purchasing power parity. This formula equalizes the purchasing power of each country's currency. Data for East Timor was not available.

UNIT 9 Assessment Resources

▶ Unit 9 Test

▶ Unit 9 Test for English Language Learners and Special-Needs Students

internet connect

GO TO: go.hrw.com
KEYWORD: SW3 U9
FOR: Web sites about country statistics

Highlights of Country Statistics
Links to online country statistics for East and Southeast Asia include:
• *CIA World Factbook*
• Library of Congress Country Studies
• Flags of the World

CHAPTER RESOURCE MANAGER

Objectives	Pacing Guide	Reproducible Resources
SECTION 1 **Natural Environments** (pp. 615–20)	• Identify the major landforms and rivers in China, Mongolia, and Taiwan. • Describe the climate types found in the region. • List China's main natural resources.	**Regular** 1 day **Block Scheduling** .5 day *Block Scheduling Handbook, Chapter 27*	**RS** Guided Reading Strategy 27.1 **RS** Graphic Organizer Activity 27 **PS** Readings in World Geography, History, and Culture 68
SECTION 2 **History and Culture** (pp. 621–26)	• Examine some important events in the early history of the region. • Identify the major political events that have affected the region's modern history. • Describe some features of Chinese culture.	**Regular** 1 day **Block Scheduling** .5 day *Block Scheduling Handbook, Chapter 27*	**RS** Guided Reading Strategy 27.2 **PS** Readings in World Geography, History, and Culture 67 **E** Cultures of the World Activity: Region 7 **SM** Critical Thinking Activity 27: Ancient China **SM** Geography for Life Activity 27 **SM** Map Activity 27
SECTION 3 **The Region Today** (pp. 627–31)	• Identify China's major regions. • Describe Mongolia. • Discuss Taiwan's relationship with China.	**Regular** 1 day **Block Scheduling** .5 day *Block Scheduling Handbook, Chapter 27*	**RS** Guided Reading Strategy 27.3 **PS** Readings in World Geography, History, and Culture 69

Chapter Resource Key

PS Primary Sources   **A** Assessment

RS Reading Support   **REV** Review

IC Interdisciplinary Connections

E Enrichment

SM Skills Mastery

ELL Reinforcement and English Language Learners

 Transparencies

 CD-ROM

Video

 Internet

Holt Presentation Maker Using Microsoft® PowerPoint®

One-Stop Planner CD–ROM

See the *One-Stop Planner* for a complete list of additional resources for students and teachers.

 **One-Stop** Planner CD–ROM

It's easy to plan lessons, select resources, and print out materials for your students when you use the *One-Stop Planner CD–ROM with Test Generator*.

 **internet** connect

HRW ONLINE RESOURCES

GO TO: go.hrw.com
Then type in a keyword.

TEACHER HOME PAGE
KEYWORD: **SW3 Teacher**

CHAPTER INTERNET ACTIVITIES
KEYWORD: **SW3 GT27**
Choose an activity to:
• learn about Chinese art.
• create a brochure on the Great Wall of China.
• write a report on diplomatic and trade issues between the United States and China.

CHAPTER ENRICHMENT LINKS
KEYWORD: **SW3 CH27**

CHAPTER MAPS
KEYWORD: **SW3 MAPS27**

ONLINE ASSESSMENT
Homework Practice
KEYWORD: **SW3 HP27**
Standardized Test Prep
KEYWORD: **SW3 STP27**
Rubrics
KEYWORD: **SS Rubrics**

COUNTRY INFORMATION
KEYWORD: **SW3 Almanac**

CONTENT UPDATES
KEYWORD: **SS Content Updates**

HOLT PRESENTATION MAKER
KEYWORD: **SW3 PPT27**

ONLINE READING SUPPORT
KEYWORD: **SS Strategies**

CURRENT EVENTS
KEYWORD: **S3 Current Events**

Technology Resources

- One-Stop Planner CD–ROM, Lesson 27.1
- Geography and Cultures Visual Resources 49–53
- CNN. Presents Geography: Yesterday and Today, Segment 27: The Yangtze (Chang) River Dam
- Homework Practice Online
- HRW Go site

- One-Stop Planner CD–ROM, Lesson 27.2
- Geography and Cultures Visual Resources 54
- CNN. Presents World Cultures: Yesterday and Today, Segment 4: Treasures from China
- CNN. Presents World Cultures: Yesterday and Today, Segment 12: Buddhism in Mongolia
- Homework Practice Online
- HRW Go site

- One-Stop Planner CD–ROM, Lesson 27.3
- *ARGWorld* CD–ROM
- CNN. Presents World Cultures: Yesterday and Today, Segment 19: Hong Kong: Past and Present
- Homework Practice Online
- HRW Go site

Reinforcement, Review, and Assessment

ELL Main Idea Activity 27.1
ELL English Audio Summary 27.1
ELL Spanish Audio Summary 27.1
REV Section 1 Review, p. 620
A Daily Quiz 27.1

ELL Main Idea Activity 27.2
ELL English Audio Summary 27.2
ELL Spanish Audio Summary 27.2
REV Section 2 Review, p. 626
A Daily Quiz 27.2

ELL Main Idea Activity 27.3
ELL English Audio Summary 27.3
ELL Spanish Audio Summary 27.3
REV Section 3 Review, p. 631
A Daily Quiz 27.3

Meeting Individual Needs

Ability Levels

Level 1 Basic-level activities designed for all students encountering new material

Level 2 Intermediate-level activities designed for average students

Level 3 Challenging activities designed for honors and gifted-and-talented students

English Language Learners Activities that address the needs of students with Limited English Proficiency

Chapter Review and Assessment

- Chapter 27 Test Generator (on the One-Stop Planner)
- Global Skill Builder CD–ROM
- HRW Go site
REV Chapter 27 Review, pp. 634–35
REV Chapter 27 Tutorial for Students, Parents, Mentors, and Peers
A Chapter 27 Test (form A or B)
A Alternative Assessment Handbook
A Chapter 27 Test for English Language Learners and Special-Needs Students

Launch into Learning

Display a porcelain object, a dollar bill, and a sheet of your local newspaper. Have students write a few sentences on the importance of porcelain, paper money, and printing to human civilization. *(Possible answers: Porcelain objects serve both functional and decorative purposes, paper money stimulates trade and commerce, and printing has enabled people to record and circulate their ideas.)* Point out that each item is based on a Chinese innovation. Tell students that in this chapter they will learn about other contributions the people of China, Mongolia, and Taiwan have made to world culture.

Using the Physical-Political Map

Have students examine the map on the opposite page. Call on volunteers to locate the region's mountain ranges, plains, plateaus, and rivers. Then ask students to identify the physical features they would encounter if they traveled overland from Lhasa to Harbin and from Guangzhou to Ulaanbaatar.

Why We Should Know More

There are many reasons that American students should know more about China, Mongolia, and Taiwan. Here are a few of them:

▶ China is the most populous country in the world and the third-largest country in area.

▶ Americans buy many goods that were made in China and Taiwan. These countries are important trading partners of the United States.

▶ China's size, economic clout, and status as a nuclear power influence the foreign policy of the United States.

▶ Because the United States has friendly but unofficial diplomatic relations with Taiwan, our country could become involved in conflicts between China and Taiwan.

▶ Mongolia and China have shaped world history. China has made important cultural and technological contributions to the world.

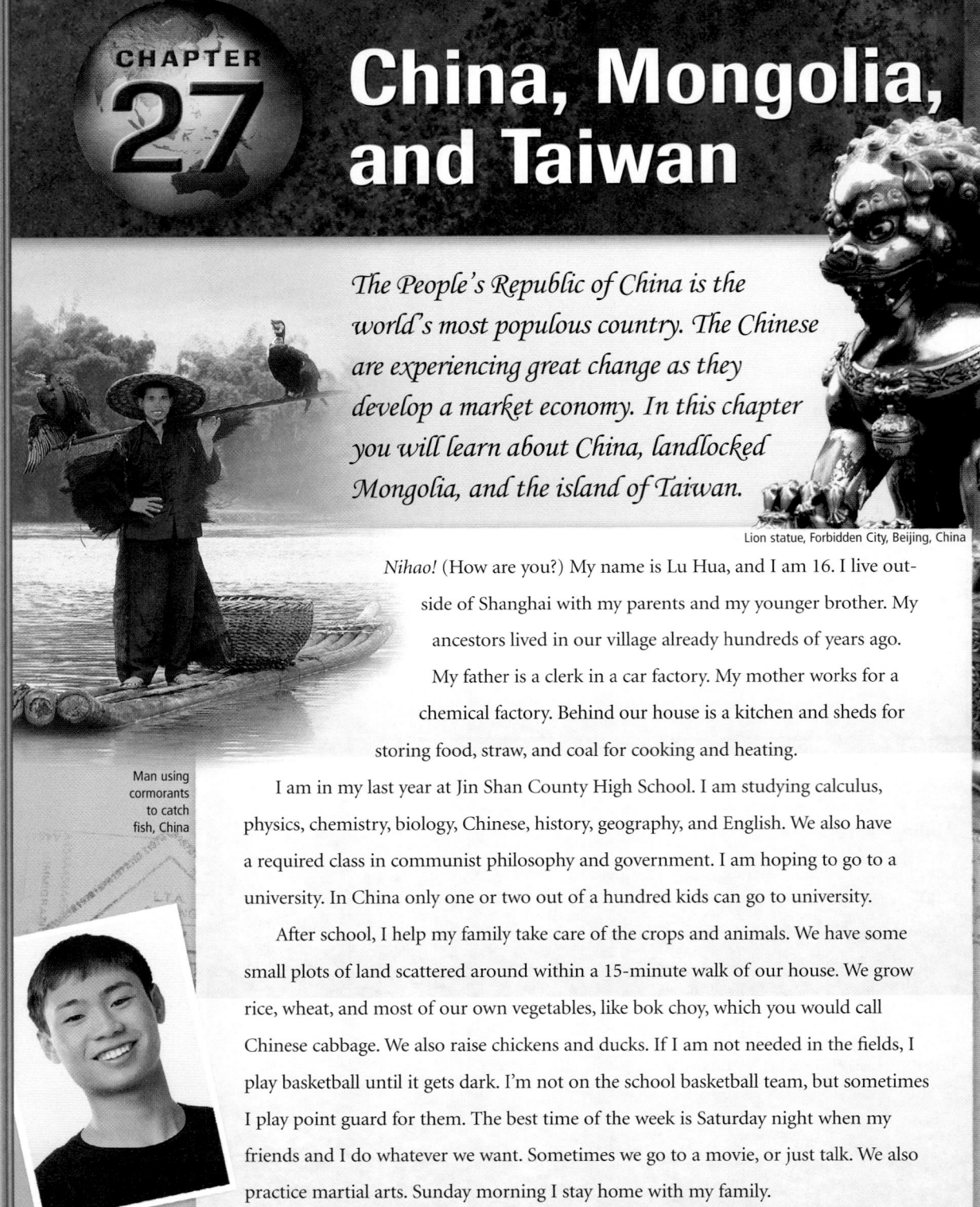

CHAPTER
27

China, Mongolia, and Taiwan

The People's Republic of China is the world's most populous country. The Chinese are experiencing great change as they develop a market economy. In this chapter you will learn about China, landlocked Mongolia, and the island of Taiwan.

Lion statue, Forbidden City, Beijing, China

Man using cormorants to catch fish, China

Nihao! (How are you?) My name is Lu Hua, and I am 16. I live outside of Shanghai with my parents and my younger brother. My ancestors lived in our village already hundreds of years ago. My father is a clerk in a car factory. My mother works for a chemical factory. Behind our house is a kitchen and sheds for storing food, straw, and coal for cooking and heating.

I am in my last year at Jin Shan County High School. I am studying calculus, physics, chemistry, biology, Chinese, history, geography, and English. We also have a required class in communist philosophy and government. I am hoping to go to a university. In China only one or two out of a hundred kids can go to university.

After school, I help my family take care of the crops and animals. We have some small plots of land scattered around within a 15-minute walk of our house. We grow rice, wheat, and most of our own vegetables, like bok choy, which you would call Chinese cabbage. We also raise chickens and ducks. If I am not needed in the fields, I play basketball until it gets dark. I'm not on the school basketball team, but sometimes I play point guard for them. The best time of the week is Saturday night when my friends and I do whatever we want. Sometimes we go to a movie, or just talk. We also practice martial arts. Sunday morning I stay home with my family.

Section 1

OBJECTIVES

1. Identify the major landforms and rivers in China, Mongolia, and Taiwan.

2. Describe the climate types found in the region.

3. List China's main natural resources.

 **LET'S GET STARTED**

Copy the following question onto the chalkboard: *How could a bird help you catch fish?* Discuss responses. Then refer students to the photo of the man using cormorants to catch fish. Explain that the cormorants are trained to dive for fish and give them up at the surface. A cord around the bird's neck keeps it from swallowing its catch. This is only one of the ways that the Chinese and Mongolians use the various elements of their natural environments. Tell students they will learn more about these environments in Section 1.

Building Vocabulary

Write the key terms on the chalkboard and call on a volunteer to read the definitions aloud. Ask students how **paddy**, **intensive agriculture**, and **double cropping** are related. *(Paddy agriculture is a form of intensive agriculture. On paddies, farmers can often harvest two crops a year, an example of double cropping.)* Point out the similarity between the words **aquaculture** and agriculture. Ask students how this similarity is reflected in their meanings. *(In aquaculture, people raise fish in ponds, much as farmers raise crops in fields.)*

Natural Environments

READ TO DISCOVER

1. What are the major landforms and rivers in China, Mongolia, and Taiwan?

2. Which climate types are found in the region?

3. What are China's main natural resources?

WHY IT MATTERS

The construction of China's Three Gorges Dam, set for completion in 2009, is controversial both in China and in other countries around the world. Use  or other **current events** sources to find the latest news about this dam and its effects on the area's people and environment.

DEFINE

paddy
intensive agriculture
double cropping
aquaculture

LOCATE

Himalayas
Kunlun Shan
Tian Shan
Altay Shan
Greater Khingan
  Range

Locate, continued

Plateau of Tibet
Tarim Basin
Turpan Depression
Huang (Yellow) River
Chang (Yangtze) River
Xi River
Red (Sichuan) Basin
North China Plain
Manchurian Plain
Mongolian Plateau
Gobi
Taiwan

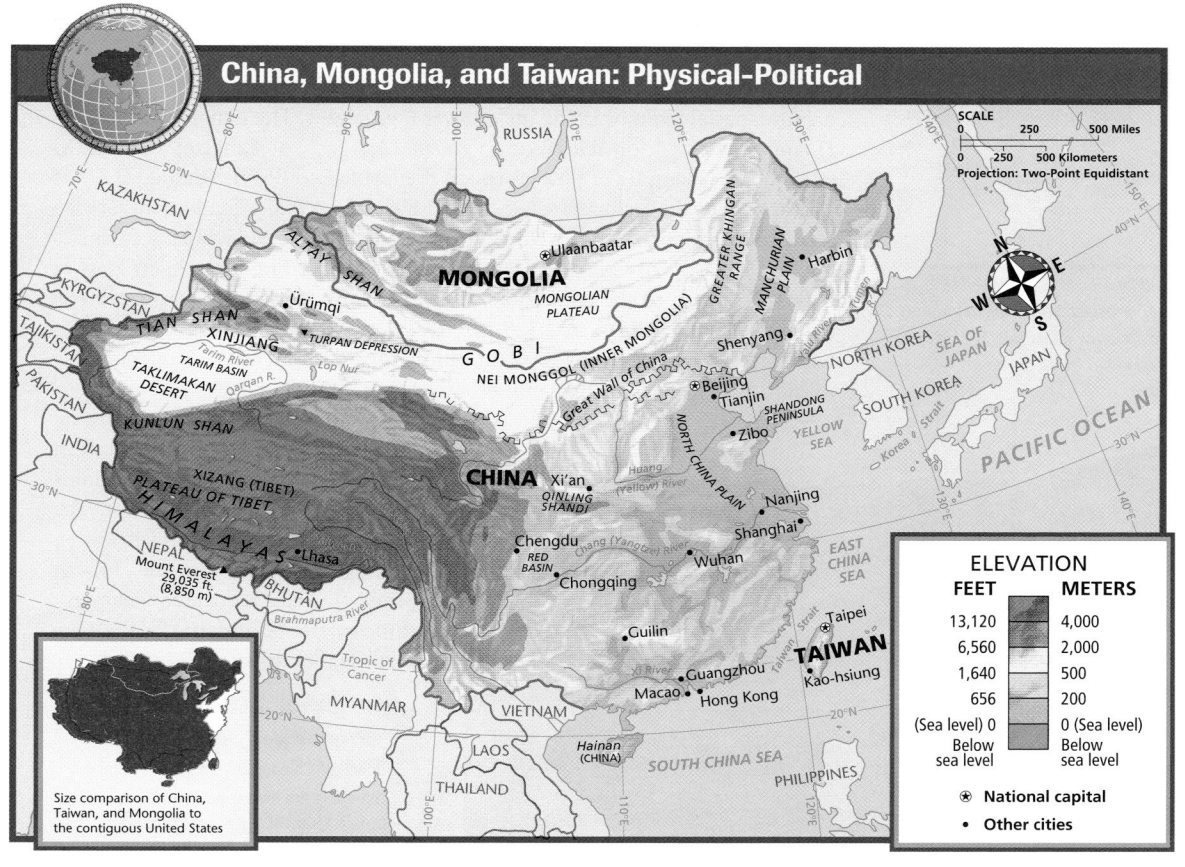

China, Mongolia, and Taiwan: Physical-Political

SCALE
0   250   500 Miles
0   250   500 Kilometers
Projection: Two-Point Equidistant

ELEVATION

FEET	METERS
13,120	4,000
6,560	2,000
1,640	500
656	200
(Sea level) 0	0 (Sea level)
Below sea level	Below sea level

✴ National capital
• Other cities

Size comparison of China, Taiwan, and Mongolia to the contiguous United States

Teach Objective 1

LEVEL 1: Draw a large map of China, Mongolia, and Taiwan on a sheet of butcher paper or posterboard. Include but do not label the region's landforms and rivers. Display the map in the classroom. Write the names of major landforms and rivers on slips of paper, fold the slips, and place them in a large container. Then call on volunteers to take slips of paper from the container and place them on the map next to the features they name. Repeat this process until all slips have been placed. **ENGLISH LANGUAGE LEARNERS**

LEVEL 2: Organize students into several groups and provide each group with a large piece of cardboard, construction paper, scissors, modeling clay, and markers. Have groups use these materials to create three-dimensional maps of China, Mongolia, and Taiwan that include major landforms and rivers. When groups have completed their maps, have them create small flaglike labels for the landforms and rivers. Instruct groups to display and discuss their models with the rest of the class. **COOPERATIVE LEARNING**

Across the Curriculum

► Science ◄

Life on the Plateau of Tibet People who travel to Tibet from low-lying areas can suffer or even die from altitude sickness unless they take time to acclimate. Since the plateau's elevation averages about 16,000 feet (4,880 m), the air above its surface is much thinner than air at sea level.

Recent studies suggest that Tibetans' metabolism can better deal with the lower amounts of breathable oxygen found at these high elevations. It appears that the hearts and brains of the mountain people use glucose and oxygen differently. Also, they have enzymes that process oxygen more efficiently. Some scientists who have compared the Quechua of the Andes to the Sherpa of Tibet found that the Sherpa had adapted more fully to high elevations. Part of the reason for the better adaptation of the Sherpa to high elevations may have to do with the fact that the Sherpa have been indigenous to the Himalayas for longer than the Quechua have been indigenous to the Andes.

☐ internet connect

GO TO: go.hrw.com
KEYWORD: SW3 CH27
FOR: Web sites about Tibet

☐ internet connect

GO TO: go.hrw.com
KEYWORD: SW3 CH27
FOR: Web sites about China, Mongolia, and Taiwan

Our Amazing Planet

Built-up silt deposits on the bottom of the Huang River are so thick that in some areas the riverbed is more than 30 feet (10 m) above the surrounding plain. Levees keep the river from flooding more than 9,000 square miles (23,310 sq km) of the North China Plain.

Nearly all of Mongolia is covered by desert or grassland. As this picture shows, the two are often found side by side. Overgrazing has led to the desert's expansion.

Landforms and Rivers

In area, China is the world's third-largest country after Russia and Canada. China is slightly larger than the United States. From east to west, China and the 48 contiguous U.S. states cover about the same distance. However, from north to south, China extends farther in each direction. (See the size comparison on the chapter map.) China's natural environments include some of the world's driest deserts, highest mountains, and longest rivers.

Mountains cover more than 40 percent of China's land area. The Himalayas, the world's highest mountain range, are located along the southwestern border with Nepal and Bhutan. Mount Everest, the world's highest summit at 29,035 feet (8,850 m), is on the China-Nepal border. To the north are several other major mountain ranges. These include the Kunlun Shan (KOON-LOON SHAHN), Tian Shan, and Altay (al-TY) Shan. In the far northeast is the Greater Khingan (KING-AHN) Range. Much of southeastern China has a rugged terrain of low mountains and steep-sided river valleys.

Between the high mountain ranges of western China are large plateaus and basins. The Plateau of Tibet lies between the Himalayas and the Kunlun Shan. With an average elevation of about 16,000 feet (4,880 m), it is the world's highest plateau. To the north, between the Kunlun and Tian Shan, lies a lower area called the Tarim (DAH-REEM) Basin. The Turpan (toohr-PAHN) Depression, located in the northeastern part of the basin, is the lowest point in China. The Turpan Depression lies 426 feet (130 m) below sea level.

The plains and river valleys of eastern China are the most densely populated areas of the country. Most of the flattest land is located along either the narrow coastal plain or in the major river floodplains. The major rivers are the Huang (HWAHNG), or Yellow; the Chang (CHAHNG), or Yangtze; and the Xi (SHEE). The Chang River flows through the fertile Red Basin. The Red Basin is also known as the Sichuan (SEE-CHWAHN) Basin. Two major plains areas are located in eastern China—the North China Plain and the Manchurian Plain.

To the north of China lies landlocked Mongolia. The Mongolian Plateau makes up much of the country's area. This plateau ranges in elevation from 3,000 to 5,000 feet (915 to 1,525 m). A desert called the Gobi (GOH-bee) extends from north-central China into Mongolia. *Gobi* means "waterless place" in Mongolian. Much of the Gobi is bare rock or gravel.

Southeast of China lies the island of Taiwan (TY-WAHN). Eastern Taiwan has high, steep mountains that rise from the Pacific coast. The western part of the island is flatter, sloping gently toward the Taiwan Strait. Taiwan is located near tectonic plate boundaries, so severe earthquakes threaten the island.

✓ **READING CHECK:**

Places and Regions About how much of China is covered by mountains? Where is the lowest point in China?
more than 40 percent; Turpan Depression

LEVEL 3: Have students locate poems by Chinese poets such as Li Po or Tu Fu that describe the landscapes of their country. Call on volunteers to share their favorite poems with the class. Then have students emulate the style of these poets and compose their own poems about the landscapes of China. Encourage students to use the photographs in this chapter or other sources for inspiration. Ask volunteers to read their poems aloud to the class.

Using National Geography Standard 2:
The World in Spatial Terms: How to Use Mental Maps to Organize Information about People, Places, and Environments in a Spatial Context

Call on volunteers to draw maps of China from memory on the chalkboard. Have other students offer hints and instructions. Inform students that the maps should show major deserts, mountain ranges, plains, rivers, and other physical features. Then tell students that some old maps of China were drawn with north and south reversed—in other words, with north at the bottom of the map. Call on other volunteers to come to the chalkboard to draw maps of China in this configuration.

Climates, Plants, and Animals

The Asian monsoon system influences climates throughout most of China. (See the unit climate and precipitation maps.) Dry winter winds blow from the Asian interior toward the coast. In northern China these winds can be bitterly cold and dusty. Warm humid air from the ocean brings rain and sometimes typhoons during the summer. In general, the eastern third of the country receives the most rain. Rainfall in China tends to decrease to the north and west.

Southeastern China and Taiwan have a mild humid subtropical climate. Much of northeastern China has a humid continental climate. This area has warm humid summers and cold drier winters.

Dry highland climates dominate western China. The Plateau of Tibet has very cold and dry conditions. The high mountains surrounding the plateau have some of the severest weather conditions in the world. The sandy Taklimakan (tah-kluh-muh-KAHN) Desert occupies most of the Tarim Basin in northwestern China. The surrounding mountains create a rain shadow. These mountains block moist ocean air from bringing rain to the desolate basin. North-central China and Mongolia have semiarid and arid climates with the driest areas in the Gobi. Mongolia experiences extreme temperatures. Among the reasons for these extremes are the country's high elevation and its location in the Asian interior at high latitudes. Winter temperatures may drop to −50°F (−46°C), and severe blizzards occasionally hit the country. High temperatures during the dry summers set the stage for raging brush fires.

Because China is so large and has such diverse landforms and climates, the country has a huge number of plant and animal species. In fact, nearly all the major types of Northern Hemisphere plants grow in China. The main exceptions to this are polar tundra plants. China's plant life can be divided into two major regions—that of the dry northwest and the humid southeast. The dry conditions of the northwest produce large areas of steppe grasslands and scattered plants that are resistant to drought. China's humid regions include mangrove swamps along the South China Sea and tropical rain forests on the island of Hainan. China's forests include about 2,500 species of trees.

China's animal life is equally diverse, particularly in the isolated mountains and valleys of Tibet and Sichuan. Wildlife ranges from wild horses and bears to camels and wolves. Some animals that have become extinct in the rest of the world still survive in China. These include the giant panda, Chinese alligator, and Chinese paddlefish, which can grow up to 21 feet (6 m) in length. Many animals are abundant, such as antelope, pheasants, and carp.

✓ **READING CHECK:** *Physical Systems* What factors influence Mongolia's extreme climate conditions? high elevation, interior location, high latitudes

The color of Buddhist monastery ruins blends in with the dry soil and bare rock of Tibet. The river in the background gets its water from Himalayan snow and ice.

INTERPRETING THE VISUAL RECORD

Fewer than 1,000 giant pandas remain in China's bamboo forests, which provide the animals' main source of food. Zoos have worked to increase this number, but pandas do not breed easily in captivity. **How might the panda's limited diet have led to its decreased population?**

Chinese Paddlefish
The Chinese paddlefish is considered by some to be a missing link between bony fish and sharks. The paddlefish shares anatomical characteristics with bony fishes and sharks. Like a shark, the Chinese paddlefish has a skeleton made of cartilage and an upturned tail fin. Unlike a shark, however, the Chinese paddlefish is covered with armorlike scales. Chinese paddlefish can weigh up to 230 pounds (104 kg). They are members of an order that dates back some 100 million years.

Only one other species of paddlefish exists—the North American spoonbill or American paddlefish. The small number and wide separation of the Chinese and American species suggests that the paddlefish family is dying off.

DISCUSSION: Lead a class discussion about geographical factors that could cause two related species to be separated by so great a distance.

VISUAL RECORD ANSWER

Possible answer: Pandas cannot adapt to other environments.

LEVEL 1: Copy the following graphic organizer onto the chalkboard, omitting the italicized answers. Have students use Section 1 and the unit atlas maps to complete the organizer. **ENGLISH LANGUAGE LEARNERS**

LEVELS 2 AND 3: Have students imagine they are helping to develop travel guides for China, Mongolia, and Taiwan. Ask them to write brief entries about typical summers and winters for each of the four regions listed in the Level 1 graphic organizer. *(Students may wish to consult earlier chapters for more details about each climate type.)* Encourage students to create illustrations to accompany their text. Call on volunteers to present and discuss their climate entries.

Climate Types in China, Mongolia, and Taiwan	
Region	**Climate Type**
Southeastern China and Taiwan	*humid subtropical*
Northeastern China	*humid continental*
Western China	*highland*
North-central China and Mongolia	*semiarid and arid*

Global Perspectives

China's Steel Industry
During the first half of the 1900s, the United States led the world in steel production. However, China has since surpassed the United States in this industry. One reason for China's success is its vast supply of mineral resources.

China leads the world in producing many of the metals that are mixed with iron to create steel. For example, China is the world's top producer of manganese. Manganese-based steel is extremely wear-resistant and is used for machines that require high durability and strength. China also leads the world in the production of tungsten, an element used in light bulbs and electronic equipment. Because it has the highest melting point of any known metal, tungsten allows steel to be used in expensive tools that require extreme hardness and resistance to overheating.

ACTIVITY: Have students create maps of China showing the sources of various minerals used in steel production.

INTERPRETING THE VISUAL RECORD

A coal miner drags his load from a tunnel. In some areas miners can use hand tools to remove coal from deposits near the surface. Small private mines like this one generally do not follow safety or environmental guidelines. **What are some ways that government policies may have changed coal mining in China?**

VISUAL RECORD ANSWER

The government is developing alternative energy resources, which may lessen dependence on coal.

Natural Resources

China has huge amounts of energy and mineral resources. It is the world's leading producer of coal, lead, tin, and tungsten. Other mineral resources include iron ore, bauxite, gold, and a wide range of metals. Coal deposits are found throughout the country, but the most important reserves are located in the north and northeast. The widespread use of coal has caused serious air pollution. Oil and natural gas are also found in many areas. China's government is working with Western oil companies to develop both offshore and interior areas for oil and gas production.

Hydropower is a major energy resource in China. Developing this resource further would help lessen dependence on coal and, in turn, reduce pollution. China is currently building the world's largest dam, which could provide vast amounts of hydroelectricity. The dam is located on the Chang River and is known as the Three Gorges Dam.

FOCUS ON GEOGRAPHY

Three Gorges Dam   Detailed planning for the Three Gorges Dam began in the 1950s. After nearly 40 years of delays, dam construction finally began in 1993. It is scheduled to be completed in 2009. When it is finished, the Three Gorges Dam will generate as much power as 15 coal-burning power plants and create a reservoir about 370 miles (595 km) long. However, that reservoir will flood hundreds of towns and cities, forcing between 1 and 2 million people to move to higher ground.

One of the main reasons for building the dam is to control dangerous flooding along the Chang River. The Chinese also want to increase river traffic and trade and generate hydropower. However, the size and scale of the project have many people worried about its environmental impact. The dam will disrupt ecosystems along the river, affecting plants and animals, such as a rare freshwater dolphin species. As much as 240,000 acres (97,200 ha) of farmland will be lost, and more than 1,200 historical sites will be flooded. Researchers are racing to save as much as they can from these sites. Critics also worry about the cost. Originally estimated at $11 billion, the dam could cost up to $50 billion. Many critics argue that several smaller and cheaper dams could be built instead. They

Construction on the Three Gorges Dam in China began in 1993, and completion is expected in 2009. Trials of the dam's system of locks began in 2003. The dam is considered to be the world's biggest hydroelectric project, and over 1 million farmers were relocated to make room for its huge, 370-mile long reservoir.

LEVEL 1: Provide each student with an outline map of China, Mongolia, and Taiwan. Have students use Section 1 and the unit atlas maps to mark the locations of the region's major resources on their maps. Remind students that they will need to include keys that explain the symbols used on the maps. Have students complete their maps by adding titles and brief captions that make general statements about China's major resources. Have students share their maps with the class. **ENGLISH LANGUAGE LEARNERS**

LEVELS 2 AND 3: Organize students into several groups, and ask each group to create a four-page visual essay on China's agricultural resources. Point out that the essays may include charts, graphs, maps, diagrams, sketches, newspaper clippings, pictures from magazines, and photographs downloaded from the Internet. Instruct groups to include an explanatory caption with each visual element included. Have groups display, compare, and discuss their visual essays. **COOPERATIVE LEARNING**

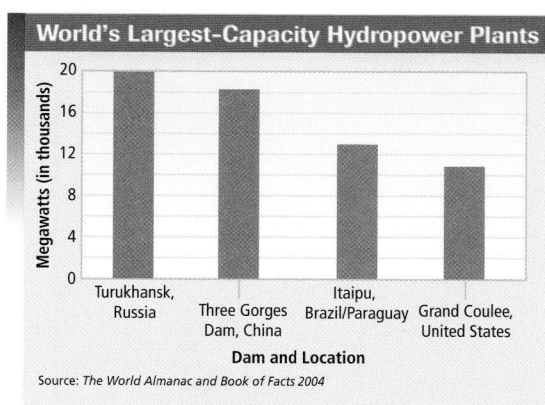

World's Largest-Capacity Hydropower Plants

Megawatts (in thousands)

Turukhansk, Russia · Three Gorges Dam, China · Itaipu, Brazil/Paraguay · Grand Coulee, United States

Dam and Location

Source: *The World Almanac and Book of Facts 2004*

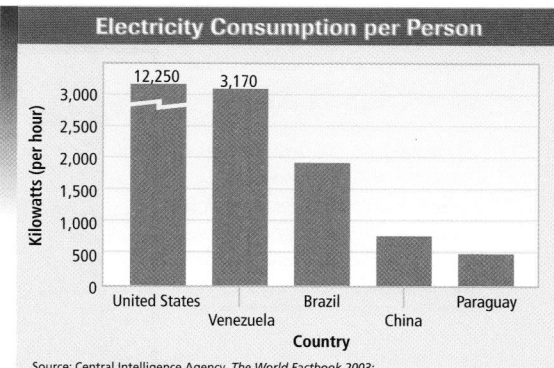

Electricity Consumption per Person

Kilowatts (per hour)

12,250 · 3,170

United States · Venezuela · Brazil · China · Paraguay

Country

Source: Central Intelligence Agency, *The World Factbook 2003; The World Almanac and Book of Facts 2004*

INTERPRETING THE GRAPHS *When fully operational, the Three Gorges Dam is expected to generate more electricity than any other dam now in existence. China's population is more than four times the size of the U.S. population. Electricity consumption per person differs significantly, however.* **How does China's electricity consumption compare to that of the United States? What factors may account for the difference?**

argue that these dams would accomplish the same goals and affect the natural environment less.

✓ **READING CHECK:** *Human Systems* How will the Three Gorges Dam affect settlement patterns and population distribution?

flood hundreds of towns and cities, forcing between 1 and 2 million people to move to higher ground

Agricultural Resources Because of its diverse climates and landform regions, China has a wide variety of soils. However, only about 10 percent of the country is arable. One important farming region is the fertile area along the Chang River. **Paddy** fields—wet lands where rice is grown—are common there. (See the map on the next page.) Because paddy agriculture uses so much human labor, it is a form of **intensive agriculture**. In many paddy areas, two crops are harvested each year from the same plot of land. This practice is known as **double cropping**.

Loess deposits in northern China are very fertile. However, they are easily eroded. The Huang River and its tributaries have eroded a huge loess plateau in the area. This erosion has created a distinctive landscape of steep-sided gullies. The Huang River often overflows its banks during the summer. Heavy rain swells the river, and the thick silt settles to the bottom, raising the level of the riverbed. The yellowish loess soil also gives the river its name, which means "yellow" in Chinese.

To feed its large population, China's farmers grow many different crops. Intensive farming methods boost production to high levels. In fact, China is a world leader in a wide range of farm products, such as ducks, peanuts, and rice. In general, southern China produces crops that need warmer temperatures, such as rice, citrus fruits, tea, and sugarcane. Northern China produces wheat, sorghum, millet, and soybeans. In the west, nomadic herding is common. (See the unit land use and resources map.)

China also has rich fishing resources. The Chinese have a long tradition of both ocean and freshwater fishing. Raising and harvesting fish and marine life in ponds or other bodies of water, known as **aquaculture**, is also important. This practice helps protect against taking too many fish from rivers and seas. It also provides valuable exports such as shrimp.

Like aquaculture, China's silk industry depends on raising small animals under controlled conditions. Silkworms feed on mulberry leaves. Their

INTERPRETING THE VISUAL RECORD

By the 200s B.C., Chinese farmers had developed an ox-drawn iron plow similar to the one used by the farmer shown. This innovation allowed them to increase production. **How might this new technology have contributed to the growth of Chinese society?**

Eye on Earth

Soils of the North China Plain The rich soils of the North China Plain have their origins in the Gobi. This rocky desert is the source for the deep loess deposits that have made the plain a key farming region.

The loess was originally created by Gobi winds that pulverized minerals and organic materials into a fine silt. Ancient dust storms then blew the loess particles to the plain, where loess reaches depths of 300 to 500 feet (90 to 150 m). This natural cycle of loess production and deposition is a continual process. Even though the Gobi has little soil of its own, the people of the North China Plain are dependent upon the desert for their survival.

ACTIVITY: Have students create a diagram showing soil-building processes in the North China Plain.

GRAPH ANSWER

less than one tenth of U.S. consumption; greater industrialization in the United States

VISUAL RECORD ANSWER

Increased production of food allowed the population to grow and economic activities to diversify.

Close

Ask students to identify a physical feature, plant, animal, or resource that they think best symbolizes China, Mongolia, and Taiwan. Have them make drawings of their selected symbols. Call on volunteers to display and explain their symbols of China, Mongolia, and Taiwan.

Review and Assess

Have students complete the **Section Review**. Then have students complete **Daily Quiz 27.1**.

Reteach

Have students complete **Main Idea Activity for English Language Learners and Special-Needs Students 27.1**. Then have students write sentences expressing what they think are the most important aspects of the region's physical features, climates, and natural resources. Have students share their sentences with the class. **ENGLISH LANGUAGE LEARNERS**

Extend

Have interested students conduct research on the typhoons that strike China's coast. Students should explain how typhoons form and how they affect China. **BLOCK SCHEDULING**

Section 1 Review Answers

Define For definitions, see: paddy, p. 619; intensive agriculture, p. 619; double cropping, p. 619; aquaculture, p. 619

Working with Sketch Maps
Maps will vary, but listed places should be labeled in their approximate locations. The Plateau of Tibet is the world's highest plateau.

Reading for the Main Idea
1. They create a rain shadow by blocking moist air from reaching the Taklimakan Desert of the Tarim Basin.

2. The Huang River and its tributaries have eroded a huge loess plateau, creating a distinctive landscape with steep-sided gullies.

3. caused serious air pollution problems

Critical Thinking
4. The collision of two tectonic plates—the Indo-Australian plate and the Eurasian plate—created the Himalayas, the world's highest mountains. **(NGS 7)**

Organizing What You Know
5. Mountains—high mountains, rugged terrain, severe weather; Plateaus and basins—high elevations, dry climates; Plains and river valleys—flat land, many rivers, fertile soils

MAP ANSWER

Possible answer: China could be divided into rich paddy farming regions and non-paddy regions.

620

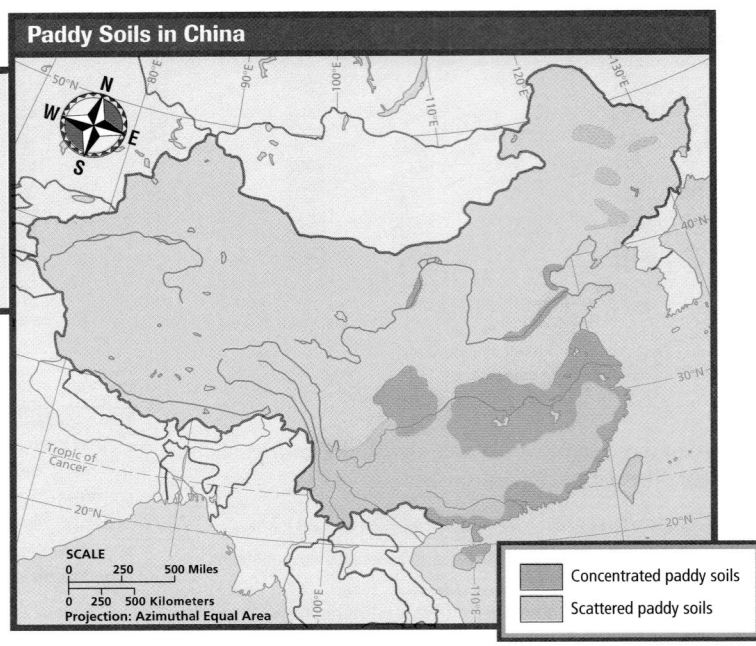

Paddy Soils in China

INTERPRETING THE MAP *Chinese farmers have used paddy soils to grow rice for centuries. Over time, this rice farming has changed the chemical and physical properties of the paddies. How could the distribution of paddy soils in China be used to divide the country into regions?*

SCALE
0   250   500 Miles
0   250   500 Kilometers
Projection: Azimuthal Equal Area

Concentrated paddy soils
Scattered paddy soils

cocoons are unwound for the fiber. China produces more raw silk than any other country.

Population growth and the expansion of agriculture and industries are placing a strain on China's water resources. Droughts and mismanagement have resulted in water shortages across much of the country. Many geographers are concerned that China's water resources will not meet growing demands.

✓ **READING CHECK:** *Human Systems* What factor has helped make China the world's leading producer of many farm products? *intensive farming methods*

Section 1 Review

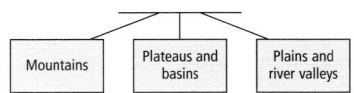

go.hrw.com Homework Practice Online
Keyword: SW3 HP27

Define paddy, intensive agriculture, double cropping, aquaculture

Working with Sketch Maps On a map of China, Mongolia, and Taiwan that you draw or that your teacher provides, label the Himalayas, Kunlun Shan, Tian Shan, Altay Shan, Greater Khingan Range, Plateau of Tibet, Tarim Basin, Turpan Depression, Huang (Yellow) River, Chang (Yangtze) River, Xi River, Red (Sichuan) Basin, North China Plain, Manchurian Plain, Mongolian Plateau, Gobi, and Taiwan. Which plateau is the world's highest?

Reading for the Main Idea
1. *Physical Systems* How do mountain barriers influence the climate of northwestern China?

2. *Physical Systems* How has erosion in northern China created a distinctive landscape along the Huang River?

3. *Environment and Society* How has the use of coal affected China's environment?

Critical Thinking
4. *Analyzing Information* Recall what you learned about tectonic forces in Chapter 4. How do you think tectonic forces have helped shape the landforms of southwestern China?

Organizing What You Know
5. Copy the chart below. Use it to describe the natural environments of the region's mountains, plateaus and basins, and plains and river valleys.

Mountains | Plateaus and basins | Plains and river valleys

Section 2

OBJECTIVES

1. Examine some important events in the early history of the region.

2. Identify the major political events that have affected the region's modern history.

3. Describe some features of Chinese culture.

 **LET'S GET STARTED**

Copy the following instructions onto the chalkboard: *Read the first sentence of Section 2. What effect do you think China's long history has had on its culture? Write down your answer.* Discuss student responses. Point out that many elements of Chinese culture are the products of traditions that date back thousands of years. Tell students that in Section 2 they will learn more about the history and culture of China.

Building Vocabulary

Write the terms **puppet government** and **communes** on the chalkboard. Call on a volunteer to read the definitions aloud. Ask students how they think the term *puppet government* arose. *(Possible answer: An outside force controls a region's government like a puppeteer controls puppets.)* Note the similarity between the words *communes* and *community.* Point out that many communes function like small communities. Then call on volunteers to read the definitions for the other key terms.

Section 2
History and Culture

READ TO DISCOVER

1. What are some important events in the region's early history?
2. What major political events have affected the region's modern history?
3. What are some features of Chinese culture?

WHY IT MATTERS

China's government is working hard to develop the country's economy. However, many political freedoms are still not allowed. Use **CNNfyi.com** or other **current events** sources to read about life in China today.

IDENTIFY

Taoism
Confucianism

DEFINE

puppet government
communes
pictograms
calligraphy

LOCATE

Great Wall
Macao
Hong Kong
Beijing

Section 2 RESOURCES

REPRODUCIBLE

▶ Guided Reading Strategy 27.2
▶ Readings in World Geography, History, and Culture 67
▶ Cultures of the World Activity: Region 7
▶ Critical Thinking Activity 27: Ancient China
▶ Geography for Life Activity 27
▶ Map Activity 27

TECHNOLOGY

▶ One-Stop Planner CD–ROM, Lesson 27.2
▶ Geography and Cultures Visual Resources 54
▶ CNN Presents World Cultures: Yesterday and Today, Segment 4: Treasures from China
▶ CNN Presents World Cultures: Yesterday and Today, Segment 12: Buddhism in Mongolia
▶ Homework Practice Online
▶ HRW Go site

REINFORCEMENT, REVIEW, AND ASSESSMENT

▶ Main Idea Activity 27.2
▶ English Audio Summary 27.2
▶ Spanish Audio Summary 27.2
▶ Section 2 Review, p. 626
▶ Daily Quiz 27.2

Early History

The Chinese have the longest known continuous history of any culture. As early as 6000 B.C. people began farming along the Chang River. In the 3000s B.C. an organized society lived in small farming villages along the Huang River. Evidence of the first dynasty, the Shang, comes from there also. During the Shang period (1700s to 1100s B.C.), the water buffalo was tamed and irrigation systems were dug. Bronze casting reached a remarkable level. Grand tombs, shell money, a fairly accurate calendar, chopsticks, and musical instruments are other Shang-era achievements.

During the 200s B.C. China's first emperor conquered much of eastern China. He ordered the connection of ancient walls as protection against invaders from the north. This would become the Great Wall. The Qin (CHIN), or Ch'in, was the first Chinese imperial dynasty. The term *China* comes from the Qin dynasty. During the Han dynasty (202 B.C.–A.D. 220), China's military power grew stronger, and Chinese culture spread into southern China. Cities grew rapidly, and Chinese art and architecture flourished. Today the Chinese call themselves the Han after this great era.

After the decline of the Han dynasty, several warlords divided and ruled China. Unity was not restored until about A.D. 618. During the later T'ang and Sung dynasties, Chinese culture thrived. China was the most advanced country in the world during these dynasties.

Mongol invaders led by Genghis Khan overthrew the Sung dynasty in the 1200s and ruled all of China by 1279. The Chinese eventually rebelled and regained control of China. Chinese culture again advanced rapidly. However, China's rulers forbade foreign trade and limited outside influences.

Cheng, also known as Shih Huang Ti, the first emperor of China, died in the late 200s B.C. He was buried near the modern city of Xi'an in a tomb that remains unexcavated. Near the tomb stand more than 6,000 warriors and horses made of clay. Each of the statues is unique. The site was discovered in 1974.

✓ **READING CHECK:** **Human Systems** Why did China's first emperor order the connection of ancient walls? as protection against invaders from the north

LEVEL 1: Organize students into groups and have each group illustrate on large note cards some of the major events in early Chinese history. Then direct groups to use their note cards to construct time lines that illustrate China's history on sheets of butcher paper. Encourage groups to add explanatory annotations for each event on their time lines. When groups have completed the task, display their time lines around the classroom. **ENGLISH LANGUAGE LEARNERS, COOPERATIVE LEARNING**

LEVEL 2: Organize students into pairs and assign each pair a particular period in early Chinese history. Have pairs assume the role of newspaper reporters. Direct them to write headlines for articles that pertain to three or four major events in their assigned time periods. Then have each pair write three or four sentences to summarize the article that would accompany each headline. Call on volunteers to read their headlines and summaries to the class. **COOPERATIVE LEARNING**

Linking Past to Present

The Mummies of Ürümqi

Beginning in the late 1970s, archaeologists working in the Tarim Basin have discovered dozens of mummies whose origins can be traced to about 2000 B.C. Bodies have been unearthed in three sites near Ürümqi in far western China. Many of these incredibly well-preserved mummies are still dressed in the colorful clothing and woolen hats they wore in life.

Archaeologists are curious about the origins of these people. Genetically, they seem to resemble Europeans more than Chinese groups. Although little evidence survives of their culture, many scholars believe these people spoke Tocharian, a language related to Germanic and Celtic tongues rather than Asian ones. Even the patterns of their clothing suggest a connection with Europe. Some mummies wear a plaid pattern similar to one worn by ancient Celts. Still, in the absence of concrete evidence their exact origins remain a mystery.

internet connect

GO TO: go.hrw.com
KEYWORD: SW3 CH27
FOR: Web sites about ancient China

VISUAL RECORD ANSWER

Western-style dress

The Great Wall was not built as a single project but evolved over many centuries. The modern appearance of the wall dates to the Ming dynasty (1368–1644).

INTERPRETING THE VISUAL RECORD

Sun Yat-sen poses with his wife, Soong Ch'ing-ling. **How does this picture suggest that Sun wanted China to move away from some traditional ways of life?**

Modern History

For centuries China had contact with the outside world mostly through overland trade routes. The best-known route was the Silk Road. It extended from China through Central Asia to the Black Sea. It was not until the early 1500s, however, that European influences reached China by sea. The first to arrive were the Portuguese, who set up a trading colony at Macao (muh-KOW) in 1557. Over time, Macao became a major trading port. Other European powers followed the Portuguese. Chinese leaders, confident that their society was superior, largely ignored these visitors and their ideas.

China had considered foreign trade as tribute to the Chinese emperor. However, the Europeans rejected the tribute system. They wanted to dominate trade in China. The Chinese were unprepared to deal with these aggressive foreign demands. Over the course of the 1800s, China lost several wars and failed to keep out the European powers, Japan, and the United States. For example, in 1842 China lost Hong Kong to the British. In 1895 Japan captured Taiwan.

Rebels under the leadership of Sun Yat-sen (SOOHN YAHT-SUHN) overthrew the last Chinese dynasty and in 1912 formed the Republic of China. After Sun Yat-sen's death, Chiang Kai-shek (jee-AHNG KY-SHEK) became China's leader. He was a member of China's Nationalist Party. To establish a strong central government, he set out to defeat the warlords who controlled parts of China. Chiang Kai-shek also faced opposition from the Chinese Communist Party.

Taking advantage of a weak and divided China, Japan took over Manchuria in the early 1930s. Manchuria is a resource-rich region in northeastern China. Japan set up a **puppet government**—a government controlled by outside forces—in the region. Then, in 1937, Japan invaded eastern China. The Japanese controlled much of eastern China until the end of World War II.

After Japan's defeat in 1945, the Nationalists and the Communists fought for control of China. The Communists won and set up the People's Republic of China (PRC) under Mao Zedong (MOWD ZUH-DOOHNG) in 1949. The Nationalists retreated to the island of Taiwan and set up a rival government. They called Taiwan the Republic of China, the name it still bears.

✓ **READING CHECK:** **Human Systems** How did the British come to control Hong Kong?
The Chinese lost several wars and failed to keep out foreign powers. The British took over Hong Kong in 1842.

China Under Mao The Communists' first challenge was to feed China's people. In the past, farmers and their families worked their own fields. However, the plots were divided among various heirs when landowners died. Over time, individual plots became smaller and smaller. The Communist leaders believed that larger, government-controlled farms could be worked more

LEVEL 3: Based on one of the headlines that his or her pair wrote in the previous activity, have each student write a complete news story. Then tell each student to create the front page of a newspaper to present the news story. Remind students that their articles should be set in columns and should include appropriate visual materials—maps, sketches, diagrams, and so on. Also point out that each front page should include the name of the newspaper and a box that refers readers to the other story headlines written for the previous activity. Call on volunteers to display and discuss their finished newspaper front pages.

Teach Objective 2

LEVEL 1: Organize students into groups. Give each group three sheets of paper labeled *China in the 1800s and early 1900s, China under Mao Zedong,* and *China after Mao Zedong.* Then have group members work in round-robin style to create fact sheets about each period. As each group member receives a sheet, he or she should write information on the sheet about the appropriate era and then pass it to the next group member. Call on group representatives to share their fact sheets with the rest of the class. Then have students work individually to write generalizations about each of the three time periods covered in the activity. **ENGLISH LANGUAGE LEARNERS, COOPERATIVE LEARNING**

efficiently. They seized all private land and organized farmers and farmland into large, jointly run collective farms. The Communists forcibly relocated families that had owned the land. These families lost their personal property and individual freedoms.

Communism also changed China's traditional family structure. In the past the oldest male had been the dominant family member. Under communism, women had equal status and worked in the fields along with the men. Later the government tried to slow the growth of its huge population by limiting families to one child. Despite opposition, the government still enforces this one-child policy in towns and cities.

The government also controlled the economy, setting fixed goals for agricultural and industrial production. For example, to improve China's economy, Mao launched the Great Leap Forward in 1958. The purpose of this program was to speed up industrialization. However, the program actually delayed the country's economic progress. The government organized workers to build backyard blast furnaces to make steel. Work groups built dams and dikes. The government also organized collective farms into larger **communes**. These agricultural communities organized farming and planned local public services.

Mao's program was a disaster. Poor planning and other problems led to the creation of large inefficient industries. In addition, an overemphasis on industry hurt agriculture and led to a famine that caused millions of deaths. The program was an environmental disaster as well. Forests were cut down to make charcoal to burn in the blast furnaces. No attempts at reforestation were made. The result was massive soil erosion.

The Cultural Revolution (1966–76) also delayed China's economic development and brought chaos to the country. During this period, Mao's followers tried to rid China of people they considered his enemies and critics. Anyone with an education was suspect, particularly intellectuals and scientists. Schools and universities were closed. Many old people and scholars were attacked, sent off to work in the countryside, or even killed.

A New China   After Mao's death in 1976, Deng Xiaoping (DUHNG SHOW-PING) came to power. Deng realized that the Great Leap Forward and the Cultural Revolution had been mistakes. He pushed new policies to modernize China's agriculture, industry, and technology. He also worked to move the country toward a market economy. Agricultural production increased quickly. China's leaders continued Deng's market reforms after his death.

Today many farmers grow and sell their own crops and build their own homes. They do not own the land but lease it from the local government for long terms. Agricultural productivity has increased, and China is self-sufficient in food. A greater variety of fruits, vegetables, and meats has improved the diet of many Chinese. In addition, millions of town village enterprises (TVEs) produce textiles and other export goods. Industry has attracted investment from Chinese in

INTERPRETING THE VISUAL RECORD

In the 1930s Mao Zedong and other Communists walked some 6,000 miles (9,650 km) across China, fighting Nationalist forces along the way. This event, known as the Long March, is shown in the painting above. The Long March inspired many young people to join the Communist Party. **How are Mao and his followers portrayed in this painting?**

Teenagers take in the sights on Wang Fu Jing Street, one of Beijing's main shopping districts.

Essential Element 4

▶ **Human Systems** ◀
China's Population
Home to the largest population in the world, China has long been concerned about overpopulation. For decades, the country has used governmental and social pressure to limit the number of births in each family. Success in rural areas has been limited because economic survival often depends on having enough workers in the family.

In 1979 China enacted a "one child" policy to more effectively limit population growth. Couples with only one child are given cash bonuses, better childcare, and preferential housing assignments. These economic benefits quickly lowered birthrates in urban areas but have been less successful in rural areas. However, families with more than two children are no longer common. Overall, population growth in China slowed in the 1990s.

ACTIVITY: Organize the class into groups. Have one group represent government officials and the other, rural Chinese farmers. Stage a class debate on China's population policy.

VISUAL RECORD ANSWER

as heroic and dynamic

623

Scholars have studied how the Chinese written language reflects the country's geography. Here, you can see that mountain peaks are suggested in the character for mountain. *Similarly, the character for* tree *resembles the trunk and bare limbs of a tree. Many characters first recorded 3,000 years ago are still in use.*

Confucius, or K'ung Ch'iu as he is known in Chinese, is revered for his teachings and the sayings recorded by his disciples.

Singapore and Taiwan. However, about half of China's industrial workers still work in inefficient state-owned industries.

Although China remains a poor country by Western standards, daily life has improved for many people. For example, more consumer goods such as televisions and refrigerators are now available. Nevertheless, challenges remain. One of the most serious is the worsening pollution of China's air and water. In fact, of the 10 cities with the worst air pollution in the world, 9 are in China. Also, the government still has considerable control over the lives of the Chinese people. While the government seems to be in favor of increasing economic freedoms, it does not want political reforms. The government continues to defend actions it took in 1989 to end pro-democracy demonstrations in Beijing (BAY-JING), China's capital. Army troops and tanks crushed peaceful protests, killing or injuring hundreds of people in Tiananmen Square.

✓ **READING CHECK:** *Places and Regions* How has China's command economy changed since the death of Mao? **Market reforms now allow farmers to grow and sell their own crops, build their own private homes, and work in TVEs.**

Culture

Today China has nearly 1.3 billion people, more than four times the population of the United States. The vast majority of the country's people are Han Chinese. The Han speak many Chinese dialects, but Mandarin is the official language. About 70 percent of the country's people speak it. Written Chinese uses symbols called characters. Some of these characters are **pictograms**, or simple pictures of the ideas they represent. (See the illustration.) In all, Chinese writing uses more than 50,000 characters. As you can imagine, learning to read and write Mandarin is a lengthy process. **Calligraphy**—artistic handwriting or lettering—developed along with the writing system. Japanese and Korean calligraphy evolved from Chinese styles.

Buddhism and **Taoism** are the major religions in China. Taoism originated in China in the 500s B.C. Its followers believe there is a natural order to the universe, called the Tao (DOW). A basic idea of Taoism is to live a simple life in harmony with nature. Many Chinese also follow the philosophy of **Confucianism**. Confucianism is based on the teachings of Confucius (551–479 B.C.), a Chinese philosopher. It is more a code of ethics than an organized religion. The Confucian code centers around family loyalty, duty, and education. Confucianism also spread to other Asian countries, particularly Japan, Korea, and Vietnam.

China has some 55 minority groups. Most members of these groups live along China's borders and in the western part of the country. Some of the largest minority groups in the north and northwest are the Mongols, Tibetans, Uighurs (WEE-goohrz), Kazakhs, and Kyrgyz. Ethnic groups in the southeast include the Yao, Zhuang, and Miao.

China's minority groups have distinct cultures and follow different religions. For example, Tibetans and Mongols are Buddhist, while the Uighurs, Kazakhs, and Kyrgyz are Muslim. Several million Han Chinese in the north are also Muslim. China's Communist government discourages religious practice. Nonetheless, it has recognized the religious freedom for some groups, such as the country's Muslims. However, this has not been the case with some other religious groups,

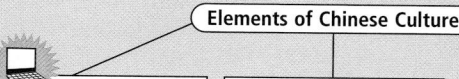
LEVEL 1: Copy the following graphic organizer onto the chalkboard, omitting the italicized answers. Have students complete it. **ENGLISH LANGUAGE LEARNERS**

Elements of Chinese Culture

Language
Mandarin is the official language, but many dialects are spoken. The written language uses some characters derived from pictograms.

Major religions and belief systems
Buddhism
Taoism emphasizes living a simple life in harmony with nature.
Confucianism centers on family loyalty, duty, and education.

Artistic traditions
Poetry; calligraphy; landscape painting; pottery, especially porcelain; wooden houses built on stone foundations, tiled roofs that curve upward

LEVELS 2 AND 3: Ask students to imagine they are exchange students living in China. Have students write letters to friends or family members in the United States describing various aspects of Chinese culture. Letters should mention such topics as language, religions and belief systems, artistic traditions, regional foods, and so on. Encourage students to illustrate their letters with appropriate sketches. Call on volunteers to read their letters to the rest of the class.

These monks are students of the Tibetan form of Buddhism called Gelukpa, or the Yellow Hat sect. Before the Chinese conquest of the region, some 25 percent of Tibetan Buddhists belonged to religious orders.

including some Christian groups and, particularly, China's Tibetan Buddhists. The government has repressed Tibetan culture and religion to prevent any movement toward Tibetan independence.

✓ **READING CHECK:** *Places and Regions* What is China's official and most common language? Mandarin Chinese

Settlement and Traditions About two thirds of China's huge population lives in the eastern half of the country. The coast and major river valleys are densely populated. Most of the population in the east is concentrated in two areas. These are the North China Plain and the lower and upper basins of the Chang River. Other densely settled areas include the Xi and Zhu (JOO) River basins in southeastern China.

Today China is experiencing rapid migration from rural to urban areas. As a result, many cities are expanding rapidly. An estimated 85 million rural migrants live in China's cities. Many of these people are women from China's interior provinces who travel to work in factories along the coast. The money they send back home not only helps their families survive, but also earns respect for the women.

China's many artistic traditions include architecture, literature, music, painting, and pottery. Artists developed a style of landscape painting featuring towering mountains, clouds, and trees. Many Chinese landscapes include descriptive text written in calligraphy. Poetry flowered during the T'ang dynasty. The Chinese have long been innovators in pottery. They developed porcelain more than a thousand years ago. Traders around the world have long desired Chinese porcelain for its fine quality and beauty. Traditional Chinese architecture features wooden buildings on stone foundations. Large tiled roofs curve upward at the corners. These artistic traditions have influenced cultures throughout Asia.

Rice, noodles, and bread are basic Chinese foods. Tofu (soybean curd) and a wide range of vegetables, such as cabbage, are also common. Pork, poultry, and duck are popular, as are fish and other seafoods. Chinese cooking varies from region to region. Food from Sichuan is spicy with chilies, while northern China is famous for roasted duck. Tea is the most popular drink. Chinese food has diffused widely. For example, Chinese restaurants can be found in cities around the world.

✓ **READING CHECK:** *Human Systems* What distinctive Chinese cultural patterns have spread to other parts of the world? artistic traditions such as art, architecture, and pottery and Chinese food

---

Connecting to
THE ARTS

Chinese Jade

Jade is a hard, tough gemstone. It occurs naturally in many colors, but the most valuable jade is usually a brilliant emerald green. Since ancient times, the Chinese have valued jade much like Westerners have valued gold. In Chinese culture jade symbolizes ideas such as purity and indestructibility. Chinese artisans have long carved jade into jewelry, vases, weapons, and other objects. These objects are prized around the world for their beauty, strength, polished surface, and intricate details.

Drawing Inferences and Conclusions Why do you think jade symbolizes purity and indestructibility in Chinese culture?

Across the Curriculum

▶ **Art** ◀

Stupa to Pagoda The stupa, a traditional monument symbolizing the Buddha's death, is a massive, solid dome surrounded by a stone railing and topped by a balcony. As Buddhism spread from India to China, Chinese Buddhists constructed monuments to the Buddha, but they radically modified their structures to fit Chinese architectural styles. The result was the pagoda.

Chinese pagodas are multi-storied masonry towers. They are hollow, not solid. The ground floor of a pagoda usually contains a shrine. Visitors may use the upper stories for viewing the countryside. The most distinctive feature, the graceful up-turned roof, is what gives the pagoda a markedly Chinese style. The pagoda form eventually spread to Japan, where the masonry was replaced by wood, resulting in a lighter, more delicate-looking tower.

ACTIVITY: Have students locate pictures of Chinese religious architecture. Ask them how these structures reflect the description of Chinese architecture in the text.

CONNECTING TO THE ARTS ANSWER

Possible answer: Jade's physical properties make it strong and durable, and it has beautiful colors.

625

Places and Regions

Geography for Life

China's Karst Towers

Known for its massive karst towers, the area around Guilin (GWEE-LIN), China, has one of the most dramatic landscapes on Earth. The word *karst* refers to a limestone landscape with many caves, underground streams, and steep hills or mountains. There are karst landscapes all over the world, but few are as dramatic as the region near the Chinese city of Guilin.

The erosion of limestone by moving water creates karst landscapes. This process is particularly common in tropical areas with heavy precipitation and dense vegetation. Plants release mild acids and carbon dioxide that are absorbed and carried by water. The chemicals and gases in the water then slowly dissolve and carry away rock, sculpting hills and other features. When karst hills reach mountainous size—as they do in Guilin—they are referred to as tower karst.

China's karst towers have inspired the Chinese imagination for more than 1,000 years. According to a Chinese saying, Guilin's scenery is the "first under heaven." A Chinese scholar during the Sung dynasty (A.D. 960–1279) wrote, "I often sent pictures of the hills of Guilin which I painted to friends back home, but few believed what they saw." Many Chinese landscape paintings feature Guilin's karst towers covered with pine trees and shrouded in mist. In recent years the towers have attracted a growing number of tourists.

Karst towers covered with vegetation loom over the Li River, as farmers and their water buffalo cross rice paddies near Guilin.

Applying What You Know

1. **Summarizing** What physical processes have created the distinctive karst towers near Guilin?

2. **Comparing** How do you think the karst towers compare to physical features that have inspired photographers, writers, and artists in the United States?

Section 2 Review

go.hrw.com Homework Practice Online
Keyword: SW3 HP27

Identify Taoism, Confucianism

Define puppet government, communes, pictograms, calligraphy

Working with Sketch Maps
On the map you created in Section 1, label China, Mongolia, the Great Wall, Macao, Hong Kong, and Beijing. What European powers once controlled Macao and Hong Kong?

Reading for the Main Idea

1. **Human Systems** After which dynasty is China's main ethnic group named?

2. **The Uses of Geography** To where did China's Nationalists flee after they were defeated by the Communists in 1949?

3. **Places and Regions** How do China's religions make it a distinctive region? How does the government treat religion?

Critical Thinking

4. **Comparing** Why do you suppose China's Communist government has allowed more economic reform and freedom but continues to limit political rights and freedom?

Organizing What You Know

5. Copy the graphic organizer below. Use it to describe China under the leadership of Mao Zedong and China after Mao. In the center space, note features that remained the same.

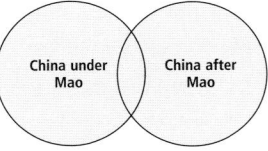

China under Mao / China after Mao

Section 3

The Region Today

READ TO DISCOVER

1. What are China's major regions?
2. What is Mongolia like today?
3. What is Taiwan's relationship with China?

WHY IT MATTERS

International relations between China and Taiwan affect many countries in East Asia and U.S. foreign policy. Use **CNNfyi.com** or other **current events** sources to learn about recent developments between these two countries.

IDENTIFY

Special Economic Zones (SEZs)

DEFINE

martial law

LOCATE

Shanghai
Nanjing
Wuhan

Locate, continued

Chongqing
Guangzhou
Shenyang
Tibet (Xizang)
Xinjiang
Lhasa
Ürümqi
Ulaanbaatar
Taipei
Kao-hsiung

China

Today one out of five of the world's people lives in China. China is a huge country in both population and area. To study it more easily, we will look at China's four major geographic regions.

Southern China   Southern China is bordered by the Qinling (CHIN-LING) Shandi range in the north and the Plateau of Tibet in the west. The East China and South China Seas lie off the eastern coast. Vietnam, Laos, and Myanmar (Burma) are located to the south.

Southern China is the country's most productive region economically. A large percentage of China's population lives there. Despite continuous cultivation for more than 4,000 years, the soil of southern China is still fertile. Alluvial deposits left by flooding rivers and careful farming renew the soils. The region is often called China's rice bowl.

The Chang Delta is a particularly important rice-growing area. Farmers there are able to grow two rice crops a year, plus a vegetable crop. Upstream

Farmers use traditional methods to work terraced vegetable gardens near Guangzhou. China is the world's leading producer of rice, ducks, hogs, chickens, and eggs, and a major producer of fish, tea, and wheat. Products you may not associate with China factor into its agricultural success. The country grows more than 20 percent of the world's corn, cotton, and peanuts, and more than 15 percent of its potatoes. It is the world's leading grower of apples.

LEVEL 1: Use chalk to draw boundaries between China's four major geographic regions on a wall map. Then point to a region and call on a volunteer to provide a fact about the geography of the indicated region. Have students write down the fact. Continue until the students have developed geographic portraits of the regions in their notebooks. Use these descriptions to lead a discussion about the factors that distinguish each region from the others. **ENGLISH LANGUAGE LEARNERS**

LEVELS 2 AND 3: Organize students into several groups and assign each group one of the four geographic regions of China. Have students imagine that they are members of governing bodies in their assigned regions. Have each group compile a list of features that would attract industry or tourism to the region. Instruct groups to use their lists to create brochures titled "Why Come to . . . ?" Suggest that they illustrate their brochures with maps, diagrams, pictures, and other suitable visual materials. Call on groups to display and discuss their brochures. **COOPERATIVE LEARNING**

Essential Element 1

► **The World** ◄
in Spatial Terms

Crowding in Hong Kong

Hong Kong is a small territory of about 422 square miles (1,092 sq km)—about the size of Dallas, Texas. However, more than five times as many people live in Hong Kong than live in Dallas. By comparing the population density of Hong Kong to large U.S. cities, you can see that Hong Kong is extremely crowded.

According to 1999 estimates, Hong Kong is one and a half times as crowded as Chicago, twice as crowded as Los Angeles, and more than six times as crowded as Phoenix. Among U.S. cities, only New York City is more densely populated. Most citizens of Hong Kong live in tightly packed high-rise apartments. Several poor families might share a single apartment of only one or two rooms.

DISCUSSION: Have students use recent population statistics to determine the population densities of other Chinese cities. How do these compare to the densities of U.S. or European cities of comparable sizes?

VISUAL RECORD ANSWER

China saw them as Chinese territories, but residents wanted to keep a capitalist system. The two sides were able to compromise.

INTERPRETING THE VISUAL RECORD

Hong Kong (shown at right) and Macao are classified as special administrative regions of China. They enjoy a high degree of autonomy in all areas except foreign policy and defense. Both have maintained capitalist economies. **How may different points of view have influenced the relationship between China and these territories?**

Tiananmen Square, in the center of Beijing, is one of the largest public squares in the world. At the north end of the square stand the gates to the Forbidden City (inset). The square has been the scene of both communist and anticommunist demonstrations.

---

from the delta, the Chang River valley is the country's major cotton-growing area. Some farmers there use double cropping to grow cotton and a food crop. Cotton is the main raw material for China's textile industry.

The Chang Delta is also China's most populated and industrialized area. The country's largest city, Shanghai, is located there. Shanghai has some 13 million people. It is a huge industrial center and a major sea and river port. New skyscrapers are rising as the city rapidly spreads outward into farming areas. (See Cities & Settlements: Shanghai.) Two big industrial cities, Nanjing and Wuhan (WOO-HAHN), lie upriver. Industrialization there has been based on local iron ore and coal deposits. The city of Chongqing, farther inland along the Chang River, lies in the Red (Sichuan) Basin. That area has productive farms, coal, and minerals.

Farther south, at the mouth of the Xi River, is the famous trading center of Guangzhou (GWAHNG-JOH), once known as Canton. Guangzhou is the largest city south of Shanghai and has also attracted industrial and commercial development. Located on the coast south of Guangzhou is the former British colony of Hong Kong. Hong Kong's territory is about a third the size of Rhode Island. However, Hong Kong has about 7 million people, making it one of the world's most crowded places. Today it is China's major seaport and its banking and international trade center. It has a variety of industries, including textiles, shipbuilding, and manufacturing. Britain returned Hong Kong to China in 1997, and the Chinese government has granted the former colony a special status. This status includes local autonomy and the continuation of a capitalist economy.

Along the coast of southern China are **Special Economic Zones (SEZs)**. These zones are designed to attract foreign companies and investment to the country. The SEZs close to Hong Kong have seen rapid economic growth. Areas that were rice fields a few years ago now bustle with factories, high-rise buildings, and new freeways. Foreign goods and money circulate freely in these seaports.

✓ **READING CHECK:** *Places and Regions* What has aided the growth of southern China's cities? agriculture, river and coastal locations near iron ore and coal, development as trade centers with access to other cities

Northern China Northern China lies north of the city of Nanjing and the Qinling Shandi range. This region includes the Huang River valley and the fertile North

LEVEL 1: Copy the graphic organizer at right onto the chalkboard, omitting the italicized answers. Have students complete the organizer by adding information about Mongolia. **ENGLISH LANGUAGE LEARNERS**

HOMEWORK: Tell students to imagine that for six months they will live with Mongolian nomads. Their hosts herd livestock from horseback and live in *gers*. (Refer students to the photo on the next page.) Each student's luggage may weigh no more than 30 pounds (13.6 kg). Ask each student to create a realistic list of items, along with each item's approximate weight, that he or she would take on the

trip. Encourage students to take Mongolia's climate and culture into account. Then have students discuss their lists and agree on the most important items.

Mongolia: Geographic Facts	
Area	*more than twice as big as Texas*
Population	*2.6 million*
Capital	*Ulaanbaatar*
Main religion	*Tibetan Buddhism*
Major resources	*coal, copper, oil*
Major economic activity	*herding livestock—people outnumbered by livestock*
Industries	*processed foods, clothing, footwear, paper*
Major challenges	*Mongolia faces food shortages and a shortage of water resources. Its isolated and landlocked location make it difficult to attract investment and economic aid.*

China Plain. Chinese culture first developed in this area. It is a very densely populated region.

The Huang River begins in the Plateau of Tibet and flows across northern China to the Yellow Sea. It has changed course and flooded surrounding areas many times. Because so many people have died in floods along the Huang River, it has been called "China's sorrow." Dams along the river store water for irrigation and generating hydroelectricity.

Beijing is China's second-largest city, its cultural center, and its capital. In fact, *Beijing* means "northern capital." The city's history goes back about 2,700 years. The ancient walled part of Beijing is divided into two sections, the Outer City and Inner City. Within the Inner City lies the Imperial City, from which China's emperors once ruled. Within the Imperial City is the Forbidden City, an immense palace complex where the emperors lived. Beijing grew beyond its original walls and spread out across the plains. Today it is a modern city with industry, subways, department stores, hotels, and wide streets.

Northeastern China Northeastern China includes three provinces once known as Manchuria. Oil, coal, iron ore, and other mineral resources are plentiful there. The oil fields have helped make China nearly self-sufficient in energy. The region also contains some of China's remaining forest resources. Important industries include iron and steel, chemicals, paper, textiles, and food processing. Shenyang is the region's largest city.

West of the Greater Khingan Range lies Nei Monggol, or Inner Mongolia. Mongols originally populated this area. Today the Han Chinese far outnumber the Mongols. Nomadic herding is common in this dry region at the edge of the Gobi. Irrigation allows some farming.

Western China Two large autonomous regions make up most of western China, Xizang (SHEE-DZAHNG) and Xinjiang (SHIN-JYAHNG). The land is generally too dry, high, and cold to support a large population. Most people there are either herders or irrigation farmers. These regions have some local government and retain their own culture and languages. Both areas were originally populated by people who were not Han Chinese. However, the number of Han Chinese colonists is rising.

The Forbidden City contains some 800 buildings, which in turn have about 9,000 rooms. Yet the Great Hall of the People, built by the Communist government, has more floor space— 6,060,000 square feet (563,000 sq m)— than all the Forbidden City's palaces put together.

The Potala Palace in Lhasa, Xizang (Tibet), is the traditional residence of the Dalai Lama and the seat of Tibet's government. It has more than 1,000 rooms. The current Dalai Lama, the fourteenth of the line, has been in exile since 1959, when the Tibetans unsuccessfully rebelled against the Chinese. Since that time he has led a peaceful campaign for Tibetan independence.

LEVELS 2 AND 3: Organize students into several groups and have each group create a collage that represents elements of Mongolia's geography. Instruct groups to accompany their collages with explanatory captions. Ask groups to compare and discuss their completed collages. Display the collages in the classroom. **ENGLISH LANGUAGE LEARNERS, COOPERATIVE LEARNING**

Teach Objective 3

ALL LEVELS: Organize students into pairs. Have one member of each pair pretend to represent the Chinese government and the other to represent the Taiwanese government. Have each student consider how the person he or she is impersonating might view the other representative. Then have each pair write a dialogue that discusses the relationship between China and Taiwan and prospects for the future reunion of the two countries. Call on pairs to perform their dialogues for the rest of the class.

Daily Life

Mongolian Foods

Because Mongolia has a short growing season and a shortage of farmland, less than 1 percent of the country is used to grow crops. However, extensive grasslands cover Mongolia and livestock-raising dominates agricultural production. As a result, meat and dairy products have traditionally been the bases of the Mongolian diet.

For breakfast and lunch, Mongolians mainly eat dairy products like cheese, yogurt, or *airag*, which is fermented milk. For dinner, they usually eat meat from the animals they raise—sheep, cattle, camels, goats, and horses. Fatty meat is preferred, and chunks of fat are often used in stews. Mongolians also eat porridge, bread, and noodles, which provide carbohydrates.

CRITICAL THINKING: How might modern technology and transportation have altered the Mongolian diet? *(People can now import grains and canned vegetables and therefore are no longer as dependent on meat.)*

Xizang is the official Chinese name for the region of Tibet. The Chinese have occupied Tibet, once an independent kingdom, since 1950. Tibet is one of the highest and most barren regions on Earth. In fact, some people call it "the roof of the world." Most Tibetans practice a form of Buddhism called Lamaism. The huge Buddhist palace of the Dalai Lama (DAH-ly LAH-muh) towers above Lhasa (LAH-suh), Tibet's capital. The Dalai Lama is the spiritual leader of most Tibetans. In 1959 the Dalai Lama fled to India after an unsuccessful revolt against Chinese control. Tibetans have watched economic development draw many Han Chinese immigrants to the region. In fact, Tibetans fear that they will become an oppressed minority in their own land. Some hope that Tibet will one day again be independent.

North of Tibet, Xinjiang is populated mostly by Muslim Turkic people, particularly the Uighurs and Kazakhs. This dry region has coal, iron ore, copper, and oil. Ürümqi (ooh-ROOHM-CHEE) is the capital city and a manufacturing center. Some Muslims there have sought independence for the region.

✓ **READING CHECK:** (*Places and Regions*) What religion do most Tibetans practice? Lamaism, a form of Buddhism

Mongolia

Mongolia is more than twice as big as Texas but has a population of less than 2.7 million. In fact, Mongolia is the least densely populated country in the world, and livestock outnumber people many times over. A large part of the population herd livestock to earn a living.

Mongolia's natural resources include coal, copper, and oil. Farming is limited, and the country faces food shortages. Water is also in short supply. The capital and only large city is Ulaanbaatar (oo-lahn-BAH-tawr). It has become more modern in recent years as the number of cars, cellular phones, and restaurants has increased. Industrial production includes processed foods, clothing and footwear, paper, and other products.

Mongolia held its first free elections in 1990. Before then, the country had been under the influence of the Soviet Union. The Communist government had suppressed the practice of Tibetan Buddhism, Mongolia's main religion. Mongolia has recently opened its economy to foreign aid and investment. However, its isolated and landlocked location limits Mongolia's economic opportunities. The growth of a market economy may mean a better future for the country.

✓ **READING CHECK:** (*Places and Regions*) How have Mongolia's political and economic characteristics changed since 1990? Mongolia held its first free elections in 1990. The country has opened its economy to foreign aid and investment.

Left: In centuries past, Mongol warriors like the ones shown were known as superb horsemen and fighters. Right: Some Mongolians maintain their ancestors' way of life, following their herds and living in moveable homes called gers.

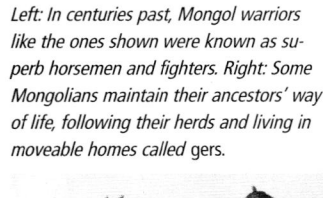

Taiwan

Taiwan is one of Asia's richest and most industrialized countries. It exports computers, scientific instruments, and sports equipment. Taiwan's major trading partners are the United States, Japan, and China. The per capita GDP of Taiwan is nearly five times that of China. (See the unit Fast Facts table.) Taipei (TY-PAY) is Taiwan's largest city, financial center, and capital. On the south end of the island is Kao-hsiung (KOW-SHYOOHNG), the country's second-largest city. The southern city is a center of heavy industry and the main seaport.

Some 2 million Chinese Nationalists who escaped from mainland China after the Communists took over in 1949 settled in Taiwan. The Chinese Nationalist Party controlled Taiwan under **martial law**, or military rule, for 38 years. Only recently have democratic rights been expanded.

China's Communist government claims that Taiwan is really a province of China. On the other hand, Taiwan's government claims to be the legitimate government of China. This disagreement has caused tension throughout the region. It is possible that Taiwan and China may one day be reunited under one government. The two countries already have economic links. For example, the Taiwanese do invest in China's industrializing coastal regions. This increasing economic interdependence is, in some ways, drawing the two countries closer together. However, wide political and economic differences remain. In addition, some Taiwanese political parties oppose reuniting with China.

✓ **READING CHECK:** *Places and Regions* How does Taiwan's per capita GDP compare to China's? more than four times larger

Taipei, the capital of Taiwan, is a bustling city of more than 2 million people. Between 1947 and 1967, the city more than quadrupled in size. High-rise buildings have replaced many more traditional structures.

Section 3 Review

Homework Practice Online go.hrw.com Keyword: SW3 HP27

Identify Special Economic Zones (SEZs)

Define martial law

Working with Sketch Maps
On the map you created in Section 2, label Shanghai, Nanjing, Wuhan, Chongqing, Guangzhou, Shenyang, Tibet (Xizang), Xinjiang, Lhasa, Ürümqi, Ulaanbaatar, Taipei, and Kao-hsiung. Which city is the largest in China?

Reading for the Main Idea

1. *Places and Regions* What is the most common economic activity in Mongolia? What are two problems facing the country?

2. *Places and Regions* What economic ties does Taiwan have with mainland China?

Critical Thinking

3. **Making Generalizations** In what ways could you describe Tibet as both a formal region and a perceptual region?

4. **Drawing Inferences and Conclusions** How do you think Taiwan's physical geography has allowed it to remain free of Communist control from mainland China?

Organizing What You Know

5. Copy the chart below. Use it to describe the major political, economic, social, and cultural features that characterize China's four main regions, Mongolia, and Taiwan.

Region/Country	Characteristics
Southern China	
Northern China	
Northeastern China	
Western China	
Mongolia	
Taiwan	

Analyzing Information

Have students read Cities & Settlements: Shanghai. Ask students to summarize some of the changes that have taken place in Shanghai in recent years. *(Possible answers: new construction, increased foreign investment, the creation of the Pudong Development Zone, and migration to the city from rural areas)* Point out that in 1990 there were practically no foreign-funded business ventures operating in the Shanghai area. By 1997, however, there were nearly 17,500. Ask students how these figures might indicate significant changes in the economy and culture of Shanghai. *(Possible answers: New businesses would attract many workers from elsewhere in China, and this would lead to many new construction projects to meet the housing and other* needs of the newcomers. Foreign investors might also introduce elements of different cultures into the city.) Point out that during the 1990s a building boom occurred in Shanghai as hundreds of thousands of apartments were constructed. Ask students how this also reflects changes in the city *(increased population)*. In addition to residences, many new hotels were built in Shanghai in the 1990s. What does this suggest about the city's attitude toward foreign influence? *(Desire for tourism suggests openness to interaction with other cultures.)* Furthermore, the average income of Shanghai residents has increased markedly. Ownership of consumer goods has grown. For example, practically every family in Shanghai has a color television and a refrigerator. How does this reflect upon the city's recent economic changes? *(It suggests that the city has prospered from the resurgence of foreign investment.)*

Cultural Kaleidoscope

The Bund Remnants of the American and European presence can still be seen in Shanghai, particularly in an area along the Huangpu River called the Bund. Historically the economic and cultural center of American Shanghai, the Bund's buildings reflect a strong Western influence. Mansions, estates, and country clubs remind visitors of New York in the 1930s. Cathedrals and a synagogue also reflect the city's Western ties.

Applying What You Know Answers

1. return of foreign culture, economic influence, and nightlife
2. similarities—cosmopolitan culture, nightlife, much economic opportunity and consumer choice, economic and class differences, all probably centers of popular culture; differences—movement into Shanghai regulated by government, freedom of movement in and out of New York and Paris; more emphasis on politics in Paris, since it is a national capital

MAP ANSWER

larger than Shanghai City; on expressway and on both Chang (Yangtze) and Huangpu Rivers

This Cities & Settlements feature addresses National Geography Standards 4, 12, and 17.

CITIES & SETTLEMENTS

Shanghai

Places and Regions Located where the Huangpu River flows into the Chang River delta, Shanghai is China's busiest port and largest city. Shanghai began as a quiet fishing village and grew to be a busy agricultural center. For a while it was the playground for Europeans in China. Today Shanghai is changing once again. The city is leading China's efforts to reform its communist economy. In the process the people of Shanghai are living through some of the most exciting times their city has ever seen.

History Repeats Itself

Shanghai remained isolated for some 500 years until commercial agriculture developed in the region about A.D. 1000. Cotton production was the basis of Shanghai's economy until the 1840s, when China lost a war with Great Britain. Outside powers then forced China to open Shanghai to foreign development. Zones of the city came under the control of American, British, French, and Japanese companies. Each zone took on the culture of the country that controlled it. For example, the French zone became famous for its cafés and lively nightlife. As a result, in the 1920s and 1930s some people called Shanghai the Paris of the East. The city's international era ended in 1949. Communist forces took control of China that year and drove out all foreigners. For the next 40 years, Shanghai remained closed to outsiders.

Today many of the places that remind residents of their city's foreign past are disappearing in a frenzy of new construction. At the same time, new overseas influences have appeared. China's Communist government has reopened Shanghai to foreign businesses and investment. In the early 1990s the government created the Pudong Development Zone. This zone lies across the Huangpu River on the city's eastern edge.

Shanghai has been called the Pearl of the Orient. The Pearl of the Orient TV Tower contains 11 spheres of varying sizes, representing pearls. The tower houses exhibition space, a hotel, an observation deck, restaurants, and shops.

On what was recently farmland, a "mini-city" called the Pudong New Area has arisen. Within its 200 square miles are high-rise apartment houses, a financial district, and an industrial park. The government has also squeezed in an international airport.

Chinese leaders are using Pudong's high-tech facilities, tax breaks, and skilled workforce to lure foreign companies to Shanghai. More than 1,000 such companies began operations there in the 1990s. They include major corporations from Britain, Germany, Japan, South Korea, and the United States.

The Chinese completed construction of the country's tallest building, the 88-story Jin Mao Tower, in Pudong in 1999. The other notable feature on Pudong's skyline is the Pearl of the Orient TV Tower. The 1,535-foot (468 m) structure has been described as looking like a rocket ship poised for takeoff. The image seems to match the city's rise. By 1997 Pudong already had a GDP larger than that of many small countries. Its economy was growing at the rate of 15 percent per year.

Going Further: Thinking Critically

In 1990 the Shanghai Municipal People's Conference approved a new emblem for the city. Various elements of this official emblem symbolize the city's past, its present growth, and its hopes for the future. In the background is a propeller, chosen as a sign of the city's constant growth and advancement. In front of it is a junk, an ancient Chinese ship, which symbolizes Shanghai's long history as a port. Blossoming from the junk is a white magnolia, the official flower of Shanghai. Because magnolia blossoms always open toward the sky, it symbolizes the pioneering and enterprising spirit of Shanghai's people.

Organize students into small groups and have each group design an official emblem for your community. Tell students that their designs should reflect a sense of your town's history and its future. Encourage students to be creative with the symbols they choose to represent aspects of the community. Have each group draw its design on a posterboard along with an explanation of the symbols used in the design. Display the proposed emblems in your classroom.

INTERPRETING THE MAP *How does the Pudong New Area compare in size to Shanghai City, west of the Huangpu River? How might the Pudong New Area's location aid its development?*

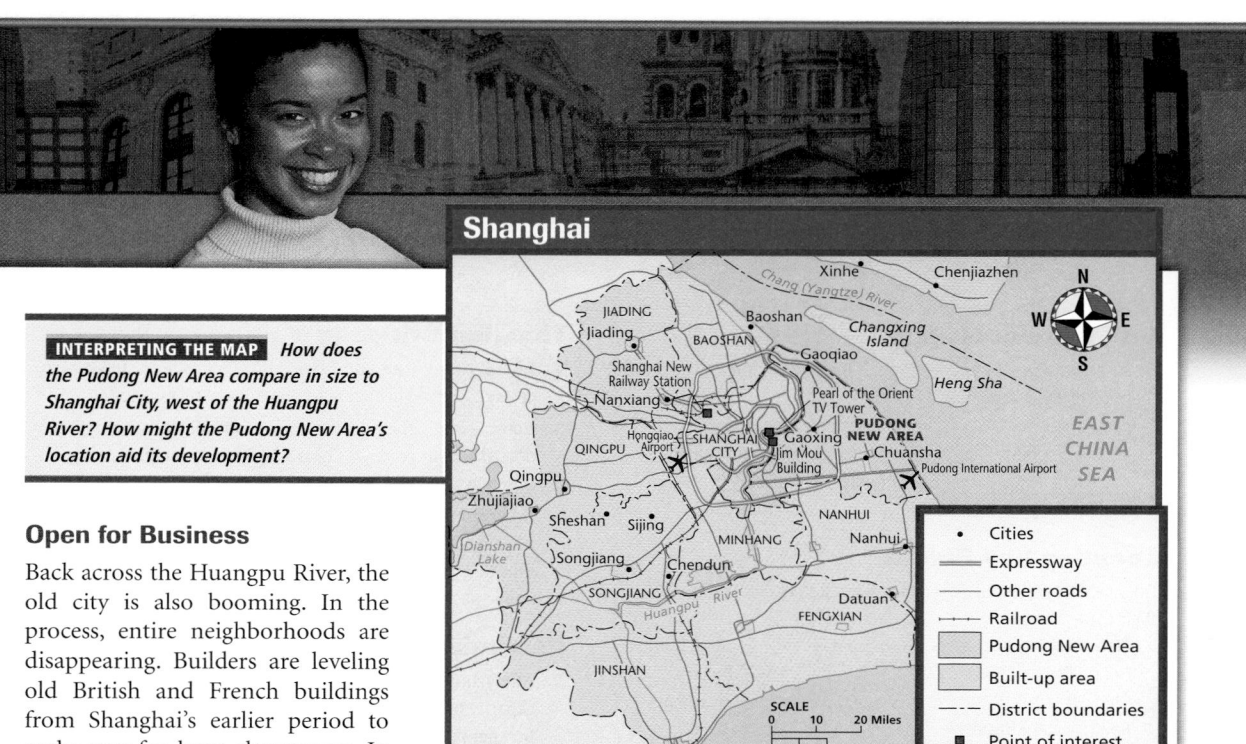

Open for Business

Back across the Huangpu River, the old city is also booming. In the process, entire neighborhoods are disappearing. Builders are leveling old British and French buildings from Shanghai's earlier period to make way for huge skyscrapers. In fact, more than 1,500 high-rise apartments and office buildings have been built in the city since 1990. New elevated freeways also criss-cross the growing urban area. Much of this work has been done by the rural Chinese who flood Shanghai in search of jobs.

These migrants make up almost 25 percent of the city's nearly 14 million people. They are only temporary residents, however. China's government is loosening its control of the country's economy, but it continues to keep tight control over the movement of China's citizens. One must get a permit to stay in Shanghai for longer than three days. For a two-year permit, a newcomer has to prove that he or she has a job and a place to live. Jobs are easy to find, but housing is in short supply. Nearly three fourths of the city's people live in downtown areas, which are particularly crowded. As a result, staying in Shanghai beyond two years is hard. A person must buy a house or apartment, be sponsored by an employer, or marry a resident. None of these alternatives is likely for most rural migrants. They tend to be poorly paid, poorly educated, and looked down upon by the city's permanent residents. This situation deepens the economic and social divisions that the boom has opened in the city.

Despite these problems, Shanghai's boom has radically changed daily life. Shanghai's residents can now enjoy French perfume and Swiss chocolates. They shop in Japanese department stores and luxury shops from New York's Fifth Avenue. Japanese cafés compete with American fast food and Tex-Mex restaurants. Huge German supermarkets provide groceries for meals at home. After dark, Shanghai now offers a scene that rivals the old days. Among the hot spots is Park 97, a nightclub in an old French mansion. Here, young urban professionals in suits or short skirts eat gourmet food and listen to French rap music. Today, however, the crowd is not European. The customers are nearly all Chinese.

Applying What You Know

1. **Summarizing** In what ways is history repeating itself in Shanghai?

2. **Comparing** How do you think life in Shanghai is like life in New York or Paris? What political, economic, and social similarities and differences would you expect to see between the cities? How might environmental change in those cities be similar?

CHAPTER 27 Review Answers

Building Vocabulary For definitions, see: paddy, p. 619; double cropping, p. 619; aquaculture, p. 619; puppet government, p. 622; communes, p. 623; pictograms, p. 624; calligraphy, p. 624; Taoism, p. 624; Confucianism, p. 624; Special Economic Zones (SEZs), p. 628; martial law, p. 631

Locating Key Places

A. Shanghai
B. Plateau of Tibet
C. Beijing
D. Huang (Yellow) River
E. Chang (Yangtze) River
F. Red (Sichuan) Basin
G. North China Plain
H. Taiwan
I. Hong Kong
J. Gobi

Understanding the Main Ideas

1. They include some of the world's driest deserts, highest mountains, and longest rivers.

2. because population growth and the expansion of agriculture and industries require more and more water

3. Shang dynasty

4. Forests were cut to make charcoal, but no attempts at reforestation were made. This caused massive soil erosion.

5. southern China

633

CHAPTER 27 Review and Assessment Resources

TECHNOLOGY

▶ Chapter 27 Test Generator (on the One-Stop Planner)
▶ Global Skill Builder CD–ROM
▶ HRW Go site

REINFORCEMENT, REVIEW, AND ASSESSMENT

▶ Chapter 27 Review, pp. 634–35
▶ Chapter 27 Tutorial for Students, Parents, Mentors, and Peers
▶ Chapter 27 Test (form A or B)
▶ Alternative Assessment Handbook

▶ Chapter 27 Test for English Language Learners and Special-Needs Students
▶ Unit 9 Test
▶ Unit 9 Test for English Language Learners and Special-Needs Students

CHAPTER 27 Review Answers

Thinking Critically

1. Possible answer: Loess deposits are easily eroded and lead to a large amount of silt being carried by the Huang. This silt can raise the river's water level and cause flooding. (NGS 7)

2. Possible answer: Both countries are most densely populated in the east. (NGS 12)

3. Each sees itself as the legitimate government for both China and Taiwan. Closer economic ties could ease disagreements and lead to reunification. (NGS 6)

Using the Geographer's Tools

1. Possible answer: China has low electricity consumption per person, and the huge Three Gorges Dam will provide a huge amount of hydropower.

2. Time lines will vary, but should include major events in China's history. Students should note cultural advances made during various dynasties and times during which China came into conflict with other countries, such as the conflict with Japan from the early 1930s to 1945.

3. Maps will vary but should reflect source material accurately.

CHAPTER 27 Review

Building Vocabulary

On a separate sheet of paper, explain the following terms by using them correctly in sentences.

paddy	communes	Confucianism
double cropping	pictograms	Special Economic
aquaculture	calligraphy	Zones (SEZs)
puppet government	Taoism	martial law

Locating Key Places

On a separate sheet of paper, match the letters on the map with their correct labels.

Plateau of Tibet	North China Plain	Hong Kong
Huang (Yellow) River	Gobi	Beijing
Chang (Yangtze) River	Taiwan	Shanghai
Red (Sichuan) Basin		

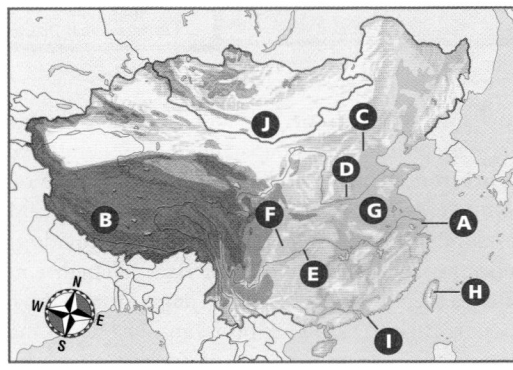

Understanding the Main Ideas

Section 1

1. *Places and Regions* How diverse are China's natural environments? Explain your answer.

2. *Environment and Society* Why is there a growing concern in China over a lack of water resources?

Section 2

3. *Human Systems* During which dynasty did distinctive Chinese culture traits such as the use of chopsticks and the early use of money develop?

4. *Environment and Society* How did Mao's Great Leap Forward cause an environmental disaster?

Section 3

5. *Places and Regions* Which region of China is known as China's rice bowl?

Thinking Critically

1. Identifying Cause and Effect How does the existence of a certain soil type disrupt lives hundreds of miles away?

2. Comparing Compare population maps for Units 2 and 9. How does the distribution of China's population compare to the distribution of population in the United States?

3. Comparing How do you think the differing points of view held by China's and Taiwan's governments make unification difficult? Why do you suppose the two countries have increased economic ties despite these differing points of view?

Using the Geographer's Tools

1. Analyzing Graphs Review the graphs showing the generating capacity of the world's largest dams and electricity consumption per person. How might the data help explain why building the Three Gorges Dam is important to China's leaders?

2. Interpreting Time Lines Create a time line of major events and eras in China's history, using this chapter and the Unit 9 time line for dates. When did major cultural developments occur? When did China come into conflict with other countries?

3. Preparing Maps Work with a partner to research the Great Wall. Prepare a map showing dates of construction of various sections and their condition of preservation. Add information on sections of the wall recently discovered by remote imaging and why estimates of the wall's length vary so widely.

Writing about Geography

Imagine that you are a researcher for an international magazine reporting on the effects of the Three Gorges Dam. How much power will the dam produce? How will it affect the physical environment? How will the distribution of resources and economic conditions in the region be affected? How will population distribution and patterns of settlement be changed? Summarize the answers to these questions in a benefits-cost balance sheet.

SKILL BUILDING

Geography for Life

Using Research Skills

Human Systems Conduct research on an aspect of Chinese culture that has diffused throughout Asia or the world. Examples include acupuncture, paper money, porcelain, or tea. Describe how the item spread beyond China and how its use has changed.

Portfolio Activity

Have students conduct research on the arts and crafts of China. Direct them to use their findings to create a poster that depicts samples of Chinese arts and crafts. Suggest that they include pictures of such arts and crafts as calligraphy, poetry, landscape painting, and lantern making. Point out that each entry on the poster should include a brief description of the art or craft and an example. Ambitious students may wish to create examples of Chinese crafts to accompany their posters. Photograph the posters and examples for inclusion in student portfolios.

Food Festival

Each region of China has its own distinctive cuisine. For example, cooks in the Sichuan region like to use hot peppers. The food of Guangzhou, in comparison, is much milder. Practically all meals in southern China are served with rice. In the north, however, wheat and millet are the most important grains. Have students work in pairs to locate recipes from China's regions. Have each pair prepare a sample of the selected regional cuisine for the class. Serve the meal with hot Chinese tea. If possible, provide chopsticks and instructions on how to use them.

Building Social Studies Skills

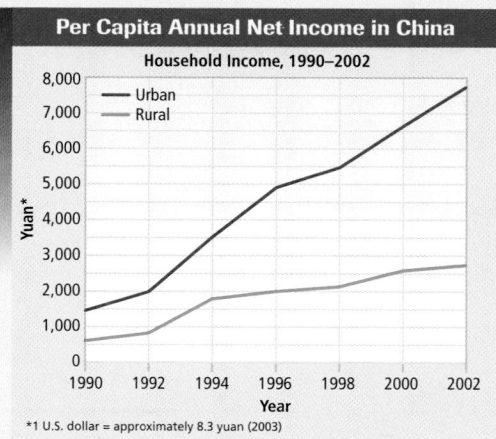

Per Capita Annual Net Income in China

*1 U.S. dollar = approximately 8.3 yuan (2003)

Interpreting Line Graphs

Study the line graph above. Then use the information from the graph to help you answer the questions that follow. Mark your answers on a separate sheet of paper.

1. The gap between urban and rural income was
  a. smallest in 1994.
  b. greatest in 1992.
  c. greatest in 2002.
  d. smallest in 1998.

2. How would you describe urban and rural income growth during the period shown on the graph?

Using Language

The following passage contains mistakes in grammar, punctuation, and usage. Read the passage and then answer the following questions. Mark your answers on a separate sheet of paper.

> **(1) The Huang River rises on the eastern edge of the Plateau of tibet. (2) It flows eastward through the North China Plain and empties into the Yellow Sea. (3) Meaning "yellow river," the yellowish mud carried by the Huang gives it its name."**

3. Which sentence contains an error in capitalization?
  a. 1
  b. 2
  c. 3
  d. all are correct

4. Rewrite sentence 3 to correct the dangling modifier.

Alternative Assessment

PORTFOLIO ACTIVITY

Learning about Your Local Geography

Group Project: Research

China is home to many endangered plant and animal species. Are there some endangered plants and animals in your state or region? Plan, organize, and complete a group research project about endangered species in your area. Contact your state department of wildlife to find out which plants and animals are considered endangered and where they are found. How many endangered species live in your state or area? What laws have been passed to protect them? What threatens their survival? What is being done to protect them? Is loss of habitat part of the reason they are threatened with extinction?

🔲 internet connect

Internet Activity: go.hrw.com
KEYWORD: SW3 GT27

Choose a topic on China, Mongolia, and Taiwan to:
- learn about Chinese art.
- create a brochure on the Great Wall.
- write a report on diplomatic and trade issues between the United States and China.

CHAPTER RESOURCE MANAGER

Objectives	Pacing Guide	Reproducible Resources
SECTION 1 **Natural** **Environments** (pp. 637–40)	• Examine the major landforms of Japan and the Koreas. • Describe the climates found in the region. • Identify some important resources in Japan and the Koreas.	**Regular** 1 day **Block Scheduling** .5 day *Block Scheduling Handbook,* *Chapter 28*	**RS** Guided Reading Strategy 28.1 **RS** Graphic Organizer Activity 28 **IC** Lab Activity for Geography and Earth Science 5
SECTION 2 **History and** **Culture** (pp. 642–48)	• Identify the cultures that influenced the early history of Japan and the Koreas. • Describe some major events in the modern history of the region. • Explain some major features of Japanese and Korean culture.	**Regular** 1 day **Block Scheduling** .5 day *Block Scheduling Handbook,* *Chapter 28*	**RS** Guided Reading Strategy 28.2 **PS** Readings in World Geography, History, and Culture 70 **E** Creative Strategies for Teaching World Geography, Lesson 19 **E** Cultures of the World Activity: Region 7
SECTION 3 **The Region** **Today** (pp. 649–53)	• Identify some characteristics of Japan today. • Contrast life in North and South Korea.	**Regular** 1 day **Block Scheduling** .5 day *Block Scheduling Handbook,* *Chapter 28*	**RS** Guided Reading Strategy 28.3 **PS** Readings in World Geography, History, and Culture 71 **SM** Critical Thinking Activity 28: Is It Still Who You Know? **SM** Geography for Life Activity 28: A Map of Tokyo, A Map of Your Town **SM** Map Activity 28: Japan's Economy

Chapter Resource Key

PS Primary Sources   **A** Assessment   CD–ROM

RS Reading Support   **REV** Review   Video

IC Interdisciplinary Connections   **ELL** Reinforcement and English Language Learners   Internet

E Enrichment   Holt Presentation Maker Using Microsoft® PowerPoint®

SM Skills Mastery   Transparencies

One-Stop Planner CD–ROM

See the *One-Stop Planner* for a complete list of additional resources for students and teachers.

One-Stop Planner CD–ROM

It's easy to plan lessons, select resources, and print out materials for your students when you use the **One-Stop Planner CD–ROM with Test Generator**.

Technology Resources	Reinforcement, Review, and Assessment
One-Stop Planner CD–ROM, Lesson 28.1	**ELL** Main Idea Activity 28.1
Geography and Cultures Visual Resources 49–53	**ELL** English Audio Summary 28.1
Homework Practice Online	**ELL** Spanish Audio Summary 28.1
HRW Go site	**REV** Section 1 Review, p. 640
	A Daily Quiz 28.1
One-Stop Planner CD–ROM, Lesson 28.2	**ELL** Main Idea Activity 28.2
CNN Presents World Cultures: Yesterday and Today, Segment 28: Japan's Royal Family	**ELL** English Audio Summary 28.2
Homework Practice Online	**ELL** Spanish Audio Summary 28.2
HRW Go site	**REV** Section 2 Review, p. 648
	A Daily Quiz 28.2
One-Stop Planner CD–ROM, Lesson 28.3	**ELL** Main Idea Activity 28.3
ARGWorld CD–ROM	**ELL** English Audio Summary 28.3
CNN Presents Geography: Yesterday and Today, Segment 29: The Tokyo Grave Crisis	**ELL** Spanish Audio Summary 28.3
Homework Practice Online	**REV** Section 3 Review, p. 653
HRW Go site	**A** Daily Quiz 28.3

⧉ internet connect

HRW ONLINE RESOURCES

GO TO: go.hrw.com
Then type in a keyword.

TEACHER HOME PAGE
 KEYWORD: SW3 Teacher

CHAPTER INTERNET ACTIVITIES
 KEYWORD: SW3 GT28
 Choose a topic on Japan and the Koreas to:
 • create a newspaper on Japan and the Koreas.
 • research active volcanoes along the Ring of Fire.
 • learn about a typical school day in Japan.

CHAPTER ENRICHMENT LINKS
 KEYWORD: SW3 CH28

CHAPTER MAPS
 KEYWORD: SW3 MAPS28

ONLINE ASSESSMENT
 Homework Practice
 KEYWORD: SW3 HP28
 Standardized Test Prep
 KEYWORD: SW3 STP28
 Rubrics
 KEYWORD: SS Rubrics

COUNTRY INFORMATION
 KEYWORD: SW3 Almanac

CONTENT UPDATES
 KEYWORD: SS Content Updates

HOLT PRESENTATION MAKER
 KEYWORD: SW3 PPT28

ONLINE READING SUPPORT
 KEYWORD: SS Strategies

CURRENT EVENTS
 KEYWORD: S3 Current Events

Meeting Individual Needs

Ability Levels

Level 1 Basic-level activities designed for all students encountering new material

Level 2 Intermediate-level activities designed for average students

Level 3 Challenging activities designed for honors and gifted-and-talented students

English Language Learners Activities that address the needs of students with Limited English Proficiency

Chapter Review and Assessment

Chapter 28 Test Generator (on the One-Stop Planner)

Global Skill Builder CD–ROM

HRW Go site

REV Chapter 28 Review, pp. 656–57

REV Chapter 28 Tutorial for Students, Parents, Mentors, and Peers

A Chapter 28 Test (form A or B)

A Alternative Assessment Handbook

A Chapter 28 Test for English Language Learners and Special-Needs Students

Launch into Learning

Call students' attention to a map of Japan and the Koreas. Point out that the Koreas are located on a peninsula and that Japan is an island country. Ask students to speculate how its location may have affected each country's history. *(Possible answer: Because the Koreas shared the mainland with China, they were probably influenced by Chinese culture. Japan's island location may have shielded it from foreign influences.)* Tell students that Japan once isolated itself by law. In 1635 it was decreed that no Japanese ships could sail for foreign countries, no Japanese could go abroad, and no Japanese could return from living abroad. Tell students that they will learn more about Japan and the Koreas in this chapter.

Using the Physical-Political Map

Have students examine the map on the opposite page. Then ask them to compare the latitude of Japan and the Koreas to the East Coast of the United States as a means of predicting the Asian countries' climates. Ask students to speculate on how the Chishima and Japan Currents affect Japan's climate. *(The icy Chishima Current cools the north; the tropical Japan Current warms and brings moisture to the south.)*

CHAPTER 28

Japan and the Koreas

Japan lies off the Pacific coast of Asia. The Japanese call their country the land of the rising Sun. To the west is the Asian mainland. Stretching from the mainland is a rugged peninsula occupied by North and South Korea.

Geisha in Kyōto, Japan

White-naped cranes

Ohaiyo (Hi!). I'm Akiko and I live in Japan. My dad works for Toyota, and my mom stays home. Every morning except Sunday I put on my school uniform and eat rice soup and pickles. I leave for school on the subway at 6:30 A.M. The subway is so crowded that I can't move. Special workers are hired to push more people into each car. At school, I study reading, math, English, science, and writing. I am still learning to write Japanese. I know 1,800 characters, but I need to know about 3,000 to pass the high school exams. For lunch, I eat rice and cold fish my mom packed for me. Before we can go home, we clean the school floors, desks, and windows. My dad usually isn't home until very late at night, so my mom helps me with my homework in our "big" room, which is 8 feet by 8 feet. In the evenings, I go to a special *juku* school to help me study harder for the important school exams. If I do not do well on these exams, I will not go to a good high school, and my whole family will be ashamed. On Sundays I sometimes go with my parents to visit my grandparents, who are rice farmers. We went to a baseball game once. I like rock music a lot, especially U2.

Section 1

OBJECTIVES

1. Examine the major landforms of Japan and the Koreas.
2. Describe the climates found in the region.
3. Identify some important resources in Japan and the Koreas.

LET'S GET STARTED

Copy the following instructions onto the chalkboard: *Look at the photo of Fuji in Section 1. What do you think the mountain's shape can tell us about how it was formed? Write down your ideas.* Allow students time to write down their answers. Discuss responses. Tell students that Fuji's conical shape indicates that it is a volcano and that volcanoes and earthquakes have long affected life in Japan. Tectonic events are rare in the Koreas, however. Tell students that in Section 1 they will learn more about the region's physical geography.

Building Vocabulary

Write the terms **tsunamis** and **flyway** on the chalkboard. Call on volunteers to locate the definitions in Section 1 and read them aloud. Point out that *tsunami* is a Japanese word that has been incorporated into English. You may want to discuss other Japanese words now common in American speech such as *karate, origami, sumo, sushi,* and *tofu*.

Natural Environments

READ TO DISCOVER

1. What are the major landforms of Japan and the Koreas?
2. Which climates are found in the region?
3. What are some important resources in Japan and the Koreas?

WHY IT MATTERS

Fishing fleets in the Pacific and around the world are facing major challenges from competition, overfishing, and pollution. Use **CNNfyi.com** or other **current events** sources to read about current issues in the fishing industry.

IDENTIFY

Chishima Current (Oyashio)

Japan Current (Kuroshio)

DEFINE

tsunamis

flyway

LOCATE

Hokkaidō

Honshū

Locate, continued

Shikoku

Kyūshū

Inland Sea

Ryukyu Islands

Okinawa

Kuril Islands

Japanese Alps

Fuji

Sea of Japan

Korea Peninsula

Section 1 RESOURCES

REPRODUCIBLE

▶ Guided Reading Strategy 28.1
▶ Graphic Organizer Activity 28
▶ Lab Activity for Geography and Earth Science 5

TECHNOLOGY

▶ One-Stop Planner CD–ROM, Lesson 28.1
▶ Geography and Cultures Visual Resources 49–53
▶ Homework Practice Online
▶ HRW Go site

REINFORCEMENT, REVIEW, AND ASSESSMENT

▶ Main Idea Activity 28.1
▶ English Audio Summary 28.1
▶ Spanish Audio Summary 28.1
▶ Section 1 Review, p. 640
▶ Daily Quiz 28.1

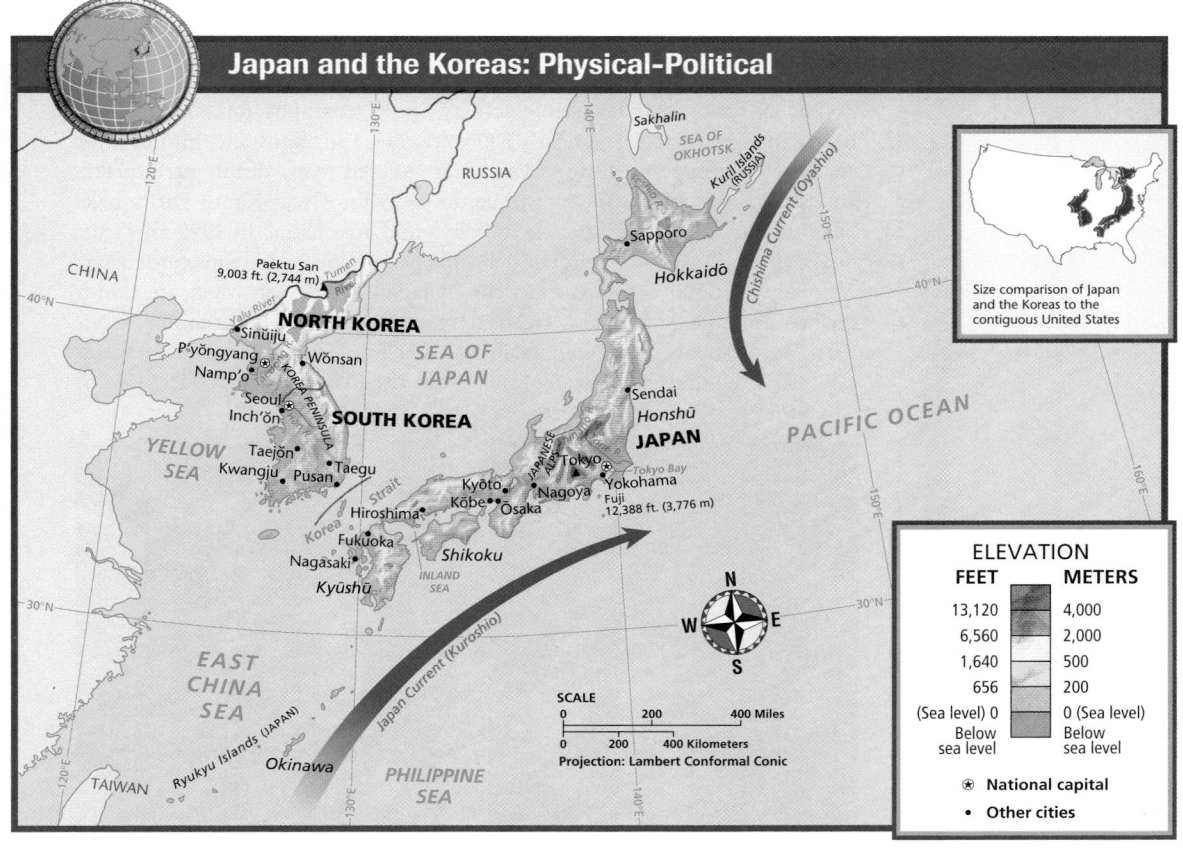

Japan and the Koreas: Physical-Political

Size comparison of Japan and the Koreas to the contiguous United States

Paektu San 9,003 ft. (2,744 m)

Fuji 12,388 ft. (3,776 m)

ELEVATION

FEET	METERS
13,120	4,000
6,560	2,000
1,640	500
656	200
(Sea level) 0	0 (Sea level)
Below sea level	Below sea level

⊛ National capital
• Other cities

SCALE
0   200   400 Miles
0   200   400 Kilometers
Projection: Lambert Conformal Conic

637

Teach Objective 1

ALL LEVELS: Copy the following graphic organizer onto the chalkboard, omitting the italicized answers. Call on students to provide words and phrases to create lists of the major landforms of Japan and the Koreas. In the center, have students list features common to both areas. Then lead a class discussion about the physical processes that have shaped the region. **ENGLISH LANGUAGE LEARNERS**

Physical Features of Japan and the Koreas

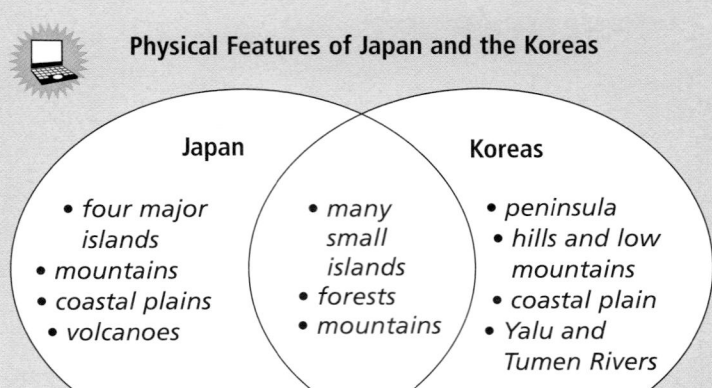

Japan
- *four major islands*
- *mountains*
- *coastal plains*
- *volcanoes*

- *many small islands*
- *forests*
- *mountains*

Koreas
- *peninsula*
- *hills and low mountains*
- *coastal plain*
- *Yalu and Tumen Rivers*

Fuji's volcanic cone is one of Japan's most recognized landmarks. Although Fuji has not erupted since 1707, geologists consider it an active volcano.

Landforms

Four main islands and thousands of smaller ones make up Japan. From north to south, the four main islands are Hokkaidō (hoh-KY-doh), Honshū (HAWN-shoo), Shikoku (shee-KOH-koo), and Kyūshū (KYOO-shoo). Honshū is the largest and most populous island. The Inland Sea separates the three main southern islands. The smaller Ryukyu (ree-YOO-kyoo) Islands to the south are also part of Japan. Okinawa is the largest of these. Japan also claims the Kuril (KYOOHR-eel) Islands to the north. The Soviet Union occupied the Kurils at the end of World War II. Russia now controls them.

More than 70 percent of Japan is mountainous. Japan's longest mountain range is the Japanese Alps. This range forms a rugged volcanic spine on the island of Honshū. Japan's highest peak, Fuji, rises to 12,388 feet (3,776 m) in central Honshū. Fuji's snow-capped volcanic cone has long been a symbol of Japan. The rest of the land—less than 30 percent—is made up of plains. Most are located on the Pacific coast of Honshū. The three most important plains include the Kanto Plain near Tokyo and the Nobi (NOH-bee) Plain near Nagoya. The third, in the Kansai region, lies near the cities of Kōbe (KOH-bay) and Ōsaka (oh-SAH-kah). All three coastal plains are densely populated.

Japan lies along a subduction zone. The Pacific plate dives under the Eurasian and Philippine plates in this zone. This location makes the islands a hotbed of tectonic and volcanic activity. (See Geography for Life: Tectonic Forces in Japan.) Japan has nearly 200 volcanoes, and about one third of these are active. Throughout history, Japan has suffered many deadly earthquakes and volcanic eruptions. For example, in 1923 the Great Kanto Earthquake killed more than 140,000 people in Tokyo and Yokohama. In 1995 an earthquake struck Kōbe, killing more than 6,400 people and causing major damage. As many as 1,500 earthquakes occur in Japan every year. Most are minor quakes. Tectonic activity also creates large sea waves called **tsunamis** (sooh-NAH-mees). The word *tsunami* means "harbor wave" in Japanese. These waves can travel hundreds of miles per hour. Some rise to more than 100 feet (30 m) when they reach shore.

GO TO: go.hrw.com
KEYWORD: SW3 CH28
FOR: Web sites about Fuji

The rugged Korea Peninsula is made up of rocks such as granite and gneiss and has a rocky coastline.

Teach Objective 2

🌐 **ALL LEVELS:** Organize the class into five groups. Assign each group one of these locations: northern Hokkaidō, land along the Yalu River, land along the Han River, Kyūshū, and Tokyo. Have each group write weather reports for a typical summer day and a typical winter day for its assigned locale. When groups have completed their reports, call on volunteers to present them to the class. **ENGLISH LANGUAGE LEARNERS, COOPERATIVE LEARNING**

Teach Objective 3

🌐 **LEVEL 1:** Have each student create an illustration for the cover of a book about the region's natural resources. Instruct students to display what they know about the resources in their illustrations. Display the finished book covers in the classroom. **ENGLISH LANGUAGE LEARNERS**

🌐 **LEVELS 2 AND 3:** Have students write titles and subtitles for the book covers they created in the Level 1 activity. Then have them write summary paragraphs that might appear in advertisements for the books.

West of Japan, across the Sea of Japan and Korea Strait, lies the Korea Peninsula. This 600-mile-long (965-km) peninsula is about the same size as the state of Utah. To the west is the Yellow Sea. Two countries, North Korea and South Korea, occupy the Korea Peninsula. North Korea borders China along the Yalu and Tumen Rivers. It also shares a short border with Russia in the far northeast.

The peninsula's landforms are mostly hills and low mountains. Unlike Japan, the Koreas do not have active volcanoes. The most mountainous region is in the northeast. Several peaks there rise higher than 8,000 feet (2,438 m). Steep mountains plunge into the sea along the east coast. A coastal plain is located on the west coast. This crowded plain contains the peninsula's best farmland and most of its population.

✓ **READING CHECK:** (*Physical Systems*) How do tectonic forces affect Japan?
cause earthquakes, volcanic eruptions, and tsunamis

Climates, Plants, and Animals

The Asian monsoon system influences the climates of Japan and the Korea Peninsula. During the summer, moist Pacific winds sweep across the region from the south. During the winter, dry northwest winds blow from Siberia in eastern Russia. Precipitation is greatest in the summer months. Most areas get between 40–60 inches (100–150 cm) of annual precipitation. (See the precipitation map in the unit atlas.) Typhoons from the tropical Pacific Ocean occasionally strike the Japanese and Korean coasts in the late summer and fall months.

Japan spans almost the same latitudes as the east coast of the United States. Therefore, the climates found in each place are similar. Hokkaidō and northern Honshū have a humid continental climate similar to that of the New England states. The cold **Chishima Current (Oyashio)** from the north cools summers. Cold winds blowing from the Asian mainland bring severe winters to northern Japan. Heavy snows fall in the mountains, particularly on Japan's western slopes. Southern Japan, including Kyūshū, Shikoku, and southern Honshū, has a humid subtropical climate similar to that of the southeastern United States. The warm **Japan Current (Kuroshio)** brings moist marine air to these areas. Summers are quite warm and humid, and winters are mild.

The Korea Peninsula has a similar climate pattern. The north has a humid continental climate, and the south has a humid subtropical climate. North Korea experiences very cold and snowy winters. South Korea's winters are milder. Summers are warm and humid across most of the peninsula.

Temperate and middle-latitude forests cover much of Japan and the Koreas. These forests contain camphor, oak, and pine trees. Maples are also common in Japan and are greatly admired for their beautiful fall colors. Some forests in the Koreas have been cleared. Deforestation and population growth have greatly limited the habitats of many large mammals there. Even in remote areas, bears, leopards, and tigers have almost disappeared. In Japan, however, many mammals are still common in forested mountain areas. These animals include antelope, bears, deer, and foxes. Both Japan and the Koreas are

Our Amazing Planet

Japanese macaques (muh-KAKS), or snow monkeys, live on the northern tip of Honshū at 41° N latitude. This latitude marks the northern limit of monkey habitation in the world. Macaques survive 4 to 5 months of winter snow cover each year by eating tree bark.

Essential Element 5

▶ **Environment** ◀
and Society

Cherry Blossom Festival
In mid-March or April the blooming of cherry trees symbolically begins the Japanese year. At this time schools begin the new academic year and recently hired employees join their companies.

Many Japanese organize flower-viewing parties, called *hanami*. When the cherry blossoms in Tokyo's Ueno Park are in full bloom, as many as 250,000 people come to look each day. They have picnics, sing songs, and photograph the blossoms. About three days after the blossoms are at their best, the petals begin to fall. At this time a spring breeze can cause a *hana fubuki*—a storm of flower petals—that blankets the pavement in white.

ACTIVITY: Ask students what activities or celebrations related to plants occur in your community. (*Possible answers: celebrations related to wildflowers, autumn foliage, harvests, or gardens*) Have them gather information about how the event began, who participates, and what takes place during the observance. Ask students to design a mural to show their findings.

📶 **internet** connect

GO TO: go.hrw.com
KEYWORD: SW3 CH28

FOR: Web sites about Japan and the Koreas

639

Close

Call on volunteers to select random locations on a wall map of Japan and the Koreas. Ask students to suggest ways that the selected area's physical features, vegetation, climate, and resources might affect their daily lives if they lived there.

Review and Assess

Have students complete the **Section Review**. Then have students complete **Daily Quiz 28.1.**

Reteach

Have students complete **Main Idea Activity for English Language Learners and Special-Needs Students 28.1.** Then have each student write a short-answer question for each of the subsections in Section 1. Collect the questions and use them to quiz the class orally. **ENGLISH LANGUAGE LEARNERS**

Extend

Have interested students conduct research on the physical geography of the region and how its landscapes, plants, and animals are portrayed in traditional arts such as pottery, painting, and embroidery. Ask students to match design elements with the species or places they portray. **BLOCK SCHEDULING**

Section 1 Review Answers

Identify For identifications, see: Chishima Current (Oyashio), p. 639; Japan Current (Kuroshio), p. 639

Define For definitions, see: tsunamis, p. 638; flyway, p. 640

Working with Sketch Maps Maps will vary, but listed places should be labeled in their approximate locations. The Kuril Islands are controlled by Russia but claimed by Japan.

Reading for the Main Idea

1. about 70 percent

2. northern areas—humid continental climate; southern areas—humid subtropical climate

3. soil erosion limited; plant and animal species protected; timber imported from Canada, Southeast Asia, United States

Critical Thinking

4. Possible answer: The Japanese feel closely tied to the sea and its resources. Other countries might criticize or condemn Japan for continued whaling. (NGS 10)

Organizing What You Know

5. Japan—islands, mountainous, tectonic activity, humid continental and humid subtropical climates, forested; Koreas—peninsula, hills and low mountains, no volcanoes, humid continental and humid subtropical climates

VISUAL RECORD ANSWER

more fish caught, but fewer fish left

INTERPRETING THE VISUAL RECORD

Tokyo's Tsukiji Market is one of the largest fish markets in the world. Every day, fresh fish are shipped in from around the world. However, even though Japan's fishing fleet is one of the most advanced and successful in the world, the country still imports seafood to meet its huge demand. **How might improved technology have both positive and negative effects on Japan's fishing industry?**

also on a major **flyway**—a migration route for birds. Hundreds of species of migratory birds pass through Japan and the Koreas on their journeys north and south.

✓ **READING CHECK: Physical Systems** How do the locations of warm and cold ocean currents affect the region's climates? The cold Chishima Current (Oyashio) cools summers in northern Japan. The warm Japan Current (Kuroshio) brings moist marine air to southern Japan.

Natural Resources

Mineral and energy resources are quite limited in Japan. As a result, the Japanese rely heavily on oil and coal imports. Japan's industrial economy also depends on imports for many industrial metals and minerals. These imports include iron and aluminum. Because of its expensive oil imports, Japan has made efforts to conserve energy. Nuclear and hydropower plants have helped lower the country's dependence on imported oil. The Koreas are also oil-dependent and use nuclear power and hydropower. North Korea has deposits of iron ore, copper, lead, and coal.

More than 65 percent of Japan is forested—a very high percentage. In lowland areas, deciduous forests are common. Mountain regions have more evergreen trees. Commercial forestry in Japan is carefully controlled to limit soil erosion and protect plant and animal species. To further protect its forest resources, Japan imports much of its timber from Canada, Southeast Asia, and the United States.

Japan's island geography has helped create a culture that depends on the sea for much of its protein. Japanese waters are rich in marine life, and Japan has one of the world's largest fishing fleets. South Korea also has a large fishing fleet. Despite worldwide protests against whaling, the Japanese continue to hunt whales in international waters. In addition, aquaculture supplies fish, shellfish, seaweed, and pearls (from oysters) in Japan. Sometimes this practice is called sea farming.

✓ **READING CHECK: Places and Regions** From where do the Japanese get much of the protein in their diet? seafood

Section 1 Review

Identify Chishima Current (Oyashio), Japan Current (Kuroshio)

Define tsunamis, flyway

Working with Sketch Maps
On a map of Japan and the Koreas that you draw or that your teacher provides, label Japan, North Korea, South Korea, Hokkaidō, Honshū, Shikoku, Kyūshū, Inland Sea, Ryukyu Islands, Okinawa, Kuril Islands, Japanese Alps, Fuji, Sea of Japan, and the Korea Peninsula. Which islands are controlled by Russia but claimed by Japan?

Reading for the Main Idea

1. **Places and Regions** About how much of Japan is mountainous?

2. **Physical Systems** How are the climates of northern Japan and the Koreas different from climates in southern Japan?

3. **Environment and Society** What are some of the geographic and economic effects of Japan's forestry policies?

Critical Thinking

4. *Analyzing* How do you think Japan's dependence on ocean resources influences its culture? How might Japan's whaling policies affect its relations with other countries?

go.hrw.com Homework Practice Online
Keyword: SW3 HP28

Organizing What You Know

5. Create a chart like the one shown below. Use it to compare the natural environments of Japan and the Koreas.

Japan	The Koreas

Religion: Volcanoes in Japanese Religion

Many mountains occupy a key place in Japanese religion and culture. One such mountain is Fuji. Fuji's name means both "everlasting life" and "deity of fire" in the Ainu language, and Ainu belief holds that the peak is home to a powerful fire god. A shrine built in the A.D. 800s was intended to placate the mountain and keep it from erupting. When the Ainu were forced into northern Japan, their Shinto successors retained the shrine and dedicated it to the goddess of flowering trees. Between the 1300s and 1500s a belief arose that the goddess would protect the people from volcanic eruptions. Later, Buddhists who had carried their religion to Japan from the Asian mainland adopted Fuji into their beliefs. They thought the mountain's purity of form and height were an excellent symbol of meditation. They named the summit *zenjo*, a Buddhist term for the state of perfect concentration. Buddhists believe that Fuji is a favorite sanctuary for the deities.

Ask students why they think a culture might incorporate elements of its natural surroundings into religious beliefs and practices. *(Students might suggest that doing so can help people understand and explain their world.)* Throughout history, many religious beliefs and practices have been inspired by the natural world. Have students find other examples of connections between nature and religious beliefs or practices from religions of their choice. Then have them write paragraphs explaining the connections.

Physical Systems

Geography for Life

Tectonic Forces in Japan

Japan is located in one of the most tectonically active areas in the world. Stretching along the Pacific Ring of Fire, the country sits at the intersection of four major crustal plates. Tectonic forces along these plate boundaries have long shaped Japan's physical geography.

The Japanese islands are formed by the upper part of a large mountain chain that rises from the ocean floor. Volcanic processes created this mountain chain. These mountains are the result of the subduction of the Pacific and Philippine plates under the Eurasian and North American plates. This subduction created the Japan Trench as the heavier oceanic plates were pulled down below the continental plates. Material from the oceanic plates was then slowly heated and transformed into magma. Eventually, the magma rose back to the surface. There it formed the mountains and volcanoes that are now Japan.

Tectonic forces also created patterns in Japan's physical geography. Steep geologically young mountains are common. There has not been enough time for most of the land to be worn down by erosion. Also, Japan's many hot springs and most of its lakes are volcanic in origin.

Tectonic forces in Japan pose numerous hazards. Active volcanoes, earthquakes, and tsunamis are serious threats. For example, the Great Kanto Earthquake of 1923 wrecked Tokyo and Yokohoma. The quake struck just before noon. At that time, Japanese in thousands of homes and restaurants were using gas and wood-burning stoves to cook lunch. As wooden buildings collapsed and gas mains broke, fires raged out of control. Broken water mains made it impossible for firefighters to battle the blazes. The Great Hanshin Earthquake that struck Kōbe in 1995 was also a major disaster. It caused about $100 billion in damage and left thousands homeless.

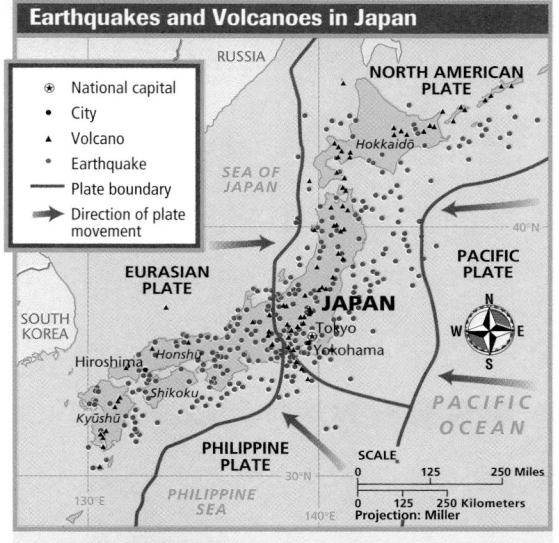

Earthquakes and Volcanoes in Japan

- ⊛ National capital
- • City
- ▲ Volcano
- • Earthquake
- — Plate boundary
- → Direction of plate movement

RUSSIA
NORTH AMERICAN PLATE
Hokkaidō
SEA OF JAPAN
EURASIAN PLATE
PACIFIC PLATE
SOUTH KOREA
JAPAN
Tokyo
Yokohama
Hiroshima
Honshū
Shikoku
Kyūshū
PACIFIC OCEAN
PHILIPPINE PLATE
PHILIPPINE SEA
40°N
30°N
130°E
140°E

SCALE
0   125   250 Miles
0   125   250 Kilometers
Projection: Miller

INTERPRETING THE MAP *Over the years, Japan has spent hundreds of millions of dollars on earthquake prediction research.* **How might advanced warning of earthquakes help prevent disasters?**

Dealing with and preparing for tectonic hazards has been a major concern in Japan. In fact, the Great Kanto Earthquake influenced engineering practices designed to make buildings earthquake-safe. For example, engineers began to focus on building structures flexible enough to withstand the violent shaking caused by earthquakes. Also, local governments made many changes to building codes. These changes included new rules limiting the height of buildings. Still, the 1995 damage in Kōbe shows that tectonic hazards remain a serious threat in Japan.

Kōbe's Nagata Ward was devastated by the 1995 earthquake, which sparked raging fires.

Applying What You Know

1. **Summarizing** Why is Japan so tectonically active?

2. **Cause and Effect** How have tectonic hazards affected engineering practices in Japan? How might new engineering techniques help the Japanese adapt to their environment?

Japanese Hot Springs
Water heated by proximity to magma rises to the surface to form hot springs, called *onsen* in Japanese. The Japanese use these *onsen* as spas where they soak in hot mineral water. There are more than 2,000 outdoor *onsen* in Japan.

The search for alternative energy sources may lead to new uses for *onsen*. The Japanese government is investigating the potential of the springs as locations for geothermal generators. Government officials hope to capture and use the heat to provide needed energy.

DISCUSSION: Lead a discussion about the possible benefits and risks of tapping a new energy source in an area of concentrated tectonic activity.

Applying What You Know Answers

1. located at intersection of four major crustal plates
2. flexible construction ideas, changes in building codes; may allow people to live more safely in dangerous areas

MAP ANSWER

provide more time for evacuation of people, improved fire prevention measures

This Geography for Life feature addresses National Geography Standards 4 and 15.

641

OBJECTIVES

1. **Identify the cultures that influenced the early history of Japan and the Koreas.**

2. **Describe some major events in the modern history of the region.**

3. **Explain some major features of Japanese and Korean culture.**

LET'S GET STARTED

Copy the following question onto the chalkboard: *How might living on an island influence the history and culture of a region?* Discuss responses. *(Possible answers: The island might be safer from conquest, as it has a natural boundary. A culture might be more easily preserved in such an isolated location.)* Tell students that in Section 2 they will learn how the locations of Japan and the Koreas affected the countries' histories and cultures.

Building Vocabulary

Write the terms **samurai** and **shogun** on the chalkboard. Call on volunteers to locate the terms and read the definitions aloud. Ask students in which contexts they may have encountered these words before. *(Many students may have seen the terms in books, movies, or television programs.)* Point out that the portrayal of samurai and shoguns in popular culture is often misleading, inaccurate, or romanticized. Then have volunteers locate and read the definitions of the remaining terms.

History and Culture

READ TO DISCOVER

1. What cultures influenced the early history of Japan and the Koreas?
2. What were some major events in the modern history of the region?
3. What are some notable features of Japanese and Korean culture?

WHY IT MATTERS

For roughly 50 years the boundary between North and South Korea has been one of the world's most tense borders. Use **CNNfyi.com** or other **current events** sources to read about recent events on the Korea Peninsula.

IDENTIFY

Meiji Restoration
Diet

DEFINE

shogun   armistice
samurai   demilitarized zone (DMZ)
annex

LOCATE

Kyōto   Tokyo
Nagasaki   Hiroshima

In this woodcut, a samurai warrior orders farmers to kneel before their lord. During the rule of the shogun, Japanese society was divided into four main classes: artisans, farmers, merchants, and warriors.

Early History

Japan's early inhabitants were the Ainu (I-noo). They may be the descendants of people who migrated into Japan from northern Asia several thousand years ago. In about 300 B.C. invaders from Asia drove the Ainu into northern Japan. The new immigrants introduced rice farming to the islands. Within a few hundred years they had settled all the major Japanese islands. Today the Ainu number only about 20,000. Most live in northern Hokkaidō, where they fish and farm.

Korea's early people came from northern and central Asia. In 108 B.C. the Chinese invaded Korea. This event marked the beginning of a long period of Chinese influence on Korean culture. Eventually, the Koreans recaptured most of the peninsula. Korean culture flourished and became known in Asia for its architecture, ceramics, and painting.

China and Korea greatly influenced Japan's early culture. Over time, however, a distinct Japanese culture emerged. For example, Shintoism became Japan's main religion. Shintoism centers around *kami*. The *kami* are the spirits of natural places, sacred animals, and ancestors.

In the A.D. 700s Japan began to develop a unique political system. The Japanese established a capital at Kyōto (KYOH-toh) in central Honshū. A Japanese emperor was officially in control of Japan's political system. However, by the late 1100s real power rested with a powerful warlord called a **shogun**. Over time the power of the shogun grew. Eventually the

LEVELS 1 AND 2: Copy the following graphic organizer onto the chalkboard, omitting the italicized answers. Call on students to provide phrases to identify the cultures that influenced Japan and the Koreas during their early histories. When students have finished filling out the organizer, lead a class discussion on the various ways in which the countries influenced each other as well as the ways in which outside countries influenced the region. **ENGLISH LANGUAGE LEARNERS**

Early influences on Japan and Korea

Japan
- *the Ainu*
- *invaders from Asia*
- *China and Korea*
- *Portuguese traders*
- *Spanish and Dutch merchants*

The Koreas
- *northern and central Asians*
- *China*

shogun ruled over other wealthy landlords called *daimyo*. The *daimyo* controlled their own local territories. They also had professional warriors called **samurai** who protected them.

The Japanese political system was similar to the feudal system of medieval Europe. There was constant strife as local lords tried to invade each other's territories. However, when the Mongols tried to invade in 1274 and 1281, the Japanese put aside their rivalries. They united to defeat the invaders.

Portuguese traders arrived in Japan in the 1500s. Spanish and Dutch merchants soon followed. These Europeans introduced Christianity to Japan. However, the Japanese drove out European traders and missionaries in the 1600s. They allowed only a few foreigners, mostly Dutch traders, to remain in Japan. They restricted the Dutch traders to an island near the port of Nagasaki. Japanese leaders feared that foreign ideas might cause instability in Japanese society. Japan remained largely cut off from the world from the 1600s to the mid-1800s.

✔ **READING CHECK:** *Human Systems* How did migration and the diffusion of ideas influence cultural change in the region? Asian migrants drove the Ainu into northern Japan and introduced rice farming. The Chinese invasion of Korea influenced Korean culture. Europeans introduced Christianity to Japan.

Modern History

In 1853 U.S. Navy ships under Commodore Matthew C. Perry arrived in what is now called Tokyo Bay. This contact helped open Japan to foreign influences and trade. Over time these influences sparked change in the country. In 1868 a group of samurai demanding reforms overthrew the last shogun. They restored the emperor's power. This political revolution became known as the **Meiji Restoration**. *Meiji* means "enlightened rule." The Japanese then moved the capital from Kyōto to Tokyo. The emperor began to modernize Japan. Many ideas for modernization came from Europe and the United States. Over time the emperor pushed to reform the country's education system, government, industry, and laws. By 1890 Japan had a constitution and parliamentary system of government.

Japan soon became a world industrial and military power. To meet its growing need for natural resources and increase Japanese influence in Asia, Japan expanded its borders. In 1895 the Japanese took the island of Taiwan from China. In 1905 Japan defeated Russia in the Russo-Japanese War. The Japanese then gained control of the southern half of Sakhalin from Russia. In 1910 Japan annexed Korea. To **annex** an area means to formally join it to a country. Later the Japanese gained control of Manchuria, China's mineral-rich northeastern region. By 1920 Japan's empire also included many islands in the North Pacific. Then in 1937 Japan invaded the rest of China. Japan's growing empire provided the Japanese with valuable natural resources and military bases.

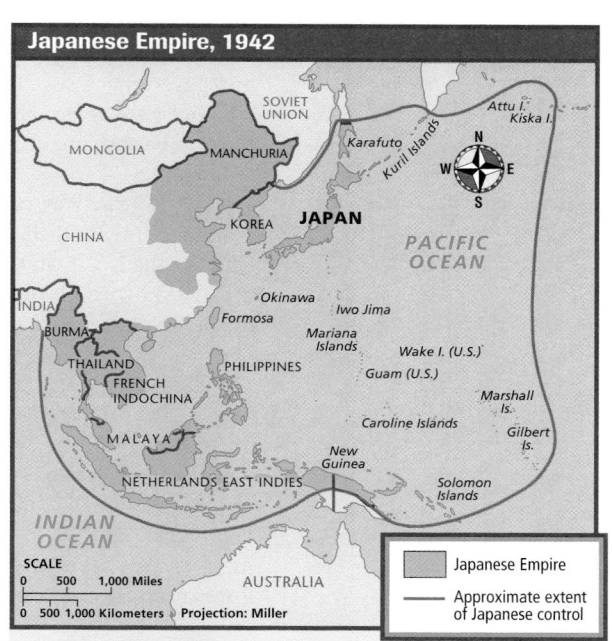

Japanese Empire, 1942

SCALE
0   500   1,000 Miles
0   500   1,000 Kilometers   Projection: Miller

Japanese Empire

Approximate extent of Japanese control

INTERPRETING THE MAP *By 1942 Japan had conquered Korea and large parts of China and Southeast Asia. The Japanese also established fortifications on many small islands in the Pacific.* **What geographic factors led Japan to expand its territory?**

Commodore Matthew Perry Although the United States and Japan are major trade partners today, this was not always true. Until the mid-1800s, Japan allowed only a few merchants, most of them Dutch, to trade in the country.

In 1852 President Millard Fillmore sent Commodore Matthew Perry of the U.S. Navy to Japan to open the country to American trade. Perry presented papers from President Fillmore requesting Japanese protection for shipwrecked U.S. sailors, the right to buy coal, and the opening of one or more ports to trade. Perry left Japan but returned less than a year later with a much larger fleet and gifts intended to impress Japan's rulers. As a result of Perry's resoluteness and display of U.S. naval power and technology, the Japanese decided that they must become as politically strong as the Americans and the Europeans. This decision introduced a period of modernization and opened Japan to foreign trade.

CRITICAL THINKING: How did Perry's voyage affect global trade patterns?

MAP ANSWER

lack of resources, vulnerable to attack from the east

643

Linking Past to Present

The Division of Korea
The division of Korea into two countries after World War II was largely the result of negotiations between U.S. and world leaders. At the Cairo Conference of 1943, the leaders of Great Britain, China, and the United States agreed that Korea should be granted independence from Japan. At a later conference in Tehran, Iran, these countries and the Soviet Union agreed that Korea's government would be decided in an election sponsored by the United Nations. At the end of the war, Korea was temporarily divided at the 38th parallel, with U.S. troops garrisoned south of the line and Soviet forces north of it. The division was to end after elections were held. However, the Soviet Union blocked all efforts by the UN to hold elections, and so the division of Korea was formally established in 1948.

DISCUSSION: Lead a class discussion about partition as a method of conflict resolution. Point out other places in the world where similar systems have developed, like Berlin.

Peace Memorial Park in Hiroshima marks the spot where the first atomic bomb was dropped on August 6, 1945. The blast killed more than 70,000 people and destroyed most of Hiroshima.

Japan signed an alliance with Germany and Italy in 1940. The next year Japan entered World War II by attacking the U.S. naval base at Pearl Harbor in Hawaii. Soon Japan controlled most of Southeast Asia and many Pacific islands. Later in the war, U.S. and Allied forces pushed the Japanese back. The United States dropped atomic bombs on the Japanese cities of Hiroshima and Nagasaki in August 1945. Japan then surrendered.

After World War II Japan set up a democratic government. (See Connecting to Government: Japan's Constitution.) That government includes an elected law-making body called the **Diet** (DEE-uht) and a prime minister. Japan's emperor is still the symbolic leader of the country but has no political power. With U.S. financial aid, Japan began to rebuild its economy and infrastructure.

Japan lost Korea at the end of World War II. The United States and Soviet Union then divided Korea along the 38th parallel (38° N latitude). The Soviets occupied the northern part of the peninsula, and the Americans occupied the south. In North Korea the Soviets set up a communist government called the Democratic People's Republic of Korea. In the south the United Nations supervised an election. South Korea then became the Republic of Korea. In 1949 the Soviet and U.S. occupation forces withdrew.

In 1950 North Korea invaded South Korea, sparking the Korean War. The UN sent troops to defend South Korea. Most were U.S. troops. UN forces drove the North Koreans back and nearly ended the war. However, China's communist government sent troops to support North Korea and again drove the UN forces southward. Eventually, the UN troops were able to push the North Koreans back again. In 1953 the two sides signed an **armistice**, or truce, to end the fighting.

The Korean War caused great damage across the peninsula. More than 1 million Koreans died. At the end of the war, the two sides set up a truce line between the North and South Korean forces. The strip of land along this line became known as the **demilitarized zone (DMZ)**. It stretches east-to-west near the old boundary at the 38th parallel.

FOCUS ON HISTORY

The DMZ Neither North nor South Korea governs the DMZ, which is about 150 miles long (240 km) and 2.5 miles (4 km) wide. At the end of the Korean War, North and South Korean troops withdrew from the DMZ. This movement created a buffer zone between the two countries' armies. No military forces from either side may enter the area.

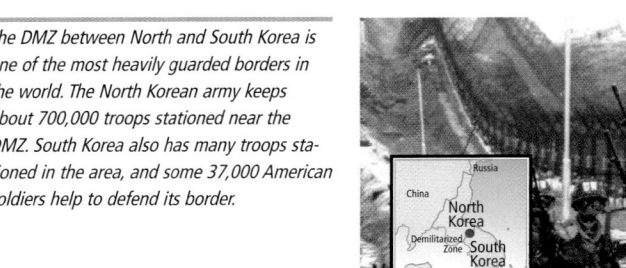

The DMZ between North and South Korea is one of the most heavily guarded borders in the world. The North Korean army keeps about 700,000 troops stationed near the DMZ. South Korea also has many troops stationed in the area, and some 37,000 American soldiers help to defend its border.

LEVEL 1: Call on volunteers to identify a major event in the modern history of Japan or the Koreas. As a student lists an event, write it on a large sheet of paper. When all key events have been listed, tape the sheets onto the chalkboard in random order. Then call on volunteers to name the event that happened first. Ask one student to describe the significance of the event. Then select another volunteer to identify the next event in chronological order. Continue until all events have been properly sequenced and described. **ENGLISH LANGUAGE LEARNERS**

LEVELS 2 AND 3: Organize the class into two groups. Have one group chronicle key events in modern Japanese history and the other do the same for modern Korean history. Provide supplemental resource materials for use in the classroom. Tell students that they have been assigned to write a series of newspaper articles covering key events in each country's history. The articles should present the events as though they had just occurred. Encourage students to include ideas for photographs, maps, or diagrams to accompany their articles. When students have finished their articles, call on volunteers to share them with the class.

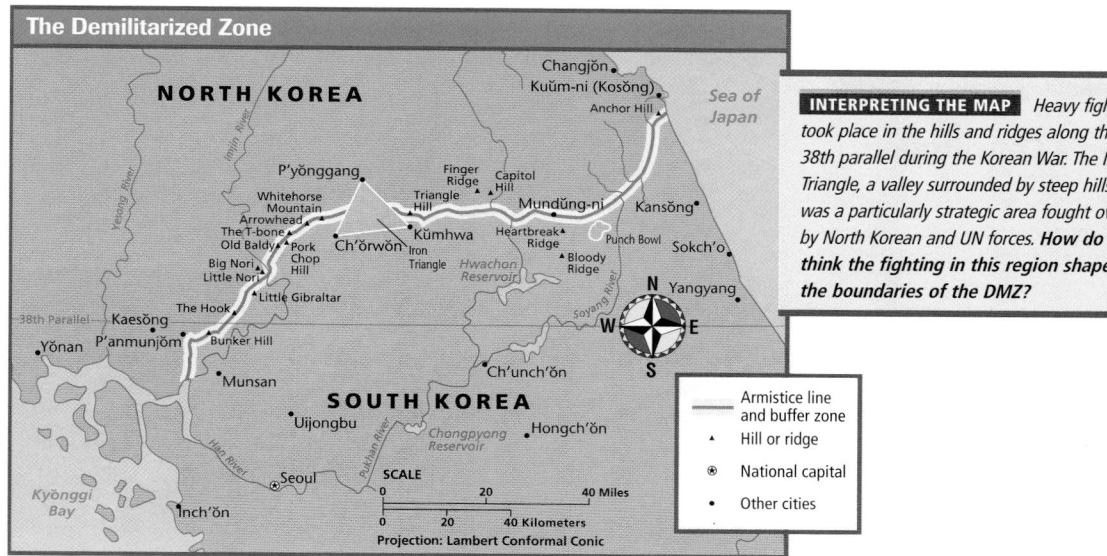

The Demilitarized Zone

INTERPRETING THE MAP *Heavy fighting took place in the hills and ridges along the 38th parallel during the Korean War. The Iron Triangle, a valley surrounded by steep hills, was a particularly strategic area fought over by North Korean and UN forces.* **How do you think the fighting in this region shaped the boundaries of the DMZ?**

The DMZ has become known as one of the most fortified and tense borders in the world. North Korean troops face South Korean and U.S. troops across the divide. The landscape along the edges of the DMZ features barbed wire, concrete walls, guard towers, mines, and tank traps. However, the area between the edges is largely free of humans.

The almost total lack of human activity in the DMZ over the last 50 years has allowed a unique ecosystem to develop. Many rare and endangered animals have found refuge in the DMZ. These include Siberian tigers and Amur leopards. The red-crowned crane and the white-naped crane, two of the world's most endangered birds, also live there. Deforestation and urban growth have reduced the habitats of these animals in other parts of the peninsula.

A group is now working to turn the region into a system of nature reserves and protected areas. This group, called the DMZ Forum, hopes to create a future "peace park" out of the area. Conservation issues will likely remain unresolved for a while, however. First, the Koreas must resolve the political and military differences that separate them.

✔ **READING CHECK:** *Human Systems* What processes created the unique ecosystem between North and South Korea? The creation of the DMZ left an area that has little human activity, allowing endangered animal species to survive there.

Culture

Many elements of Japanese and Korean culture originated in China. For example, Chinese ideas and practices were the basis for Japan's early political systems. Chinese customs, food, and architecture have also greatly influenced the region.

People and Languages Japan and the Koreas are each dominated by a single major ethnic group with a common language. In Japan some 99 percent of the population are ethnic Japanese. The small minority groups are Koreans, Chinese, and Ainu. The Japanese language is spoken throughout the country,

Japanese and Korean Names Traditional naming systems are common in both Japan and Korea. In each system, the family name comes first and is followed by a given name. This is the reverse of European and American systems, in which the family name appears last.

In Japan, girls' given names frequently end in *ko* meaning "child." For example, the name Akiko means "autumn child." Boys are often named for their place in a family using the suffix *ro*, meaning "son." The common name Ichiro, for example, means "first son."

In Korea, the traditional naming system is called *Jokbo* and has been used for more than 600 years. Again, the family name comes first, but it is followed by two personal names. The middle name indicates one's generation and position in the family tree. It is therefore shared by cousins. The given name appears last.

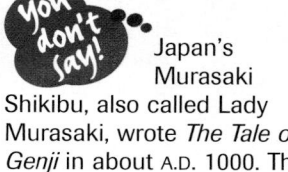 Japan's Murasaki Shikibu, also called Lady Murasaki, wrote *The Tale of Genji* in about A.D. 1000. This work is widely considered to be the world's first novel.

MAP ANSWER

The boundary runs through areas over which the two sides fought, such as the Iron Triangle.

LEVEL 1: Provide students with art supplies for creating collages. Have each student make a collage that depicts various aspects of Japanese and Korean culture. Encourage students to include images and words that demonstrate the manner in which the region's history has affected various elements of its culture. Call on volunteers to present their collages to the class. Display the collages in the classroom. **ENGLISH LANGUAGE LEARNERS**

LEVEL 2: Organize the class into three groups and assign each group one of the region's countries. Have the students in each group list text and images they would include in a geography textbook covering the culture of their assigned country. (*Example: A section on Japan might include Buddhism, Shintoism, and emphasis on education.*) Call on volunteers from each group to read their lists and ask the class to identify the country they are describing. **COOPERATIVE LEARNING**

Across the Curriculum

► Art ◄

Japanese Influence
In the late 1700s Chinese art was very popular with Europeans. As a result, people were prepared to appreciate new forms of Asian art when Japanese wood-block prints started arriving in Europe. These colorful pictures of outdoor scenes and city life became quite popular. Hokusai (1760–1849) and Hiroshige (1797–1858) were two of the most successful printmakers.

The clean lines, balanced design, and subject matter of Japanese wood-block prints appealed to many European artists in the late 1800s. Many tried to copy the Japanese style or included aspects of it in their work. Vincent van Gogh and James A. McNeill Whistler are two notable examples.

ACTIVITY: Have students conduct research on Japanese printmakers and their work. Ask them to look for other examples of Western art that seem to be inspired by Japanese prints. Then have them compare the two groups.

CONNECTING TO GOVERNMENT ANSWER

Culture traits such as democratic ideas were introduced by the Allied powers and became the legal bases for Japan's government, making Japan more like the Western powers.

Connecting to GOVERNMENT

Japan's Constitution

Allied occupation forces directed the creation of Japan's current constitution after World War II. This new democratic constitution took effect on May 3, 1947. It completely changed Japan's form of government.

Japan's earlier Meiji constitution of 1889 had granted supreme power to Japan's emperor. However, the 1947 constitution gave power to the Japanese people through their elected legislature, the Diet. The emperor remained important in government only as "the symbol of the State and the unity of the people." Japan's new constitution also guaranteed individual freedoms, such as freedom of speech, religion, and the press. These individual rights had not been guaranteed under Japan's earlier constitution. Japan's new constitution also granted women the right to vote.

Analyzing Information How is the adoption of the 1947 constitution an example of the spread of culture traits and cultural convergence?

The Daibutsu, *or Great Buddha, in Kamakura, Japan, attracts pilgrims and tourists from all over the country. The hollow bronze statue is 37 feet (11 m) tall. Visitors can enter it through a door in the back.*

and several regional dialects exist. Japanese may be related to Korean. Written Japanese uses a combination of Chinese characters and Japanese symbols called kana.

Nearly the entire population of North and South Korea are ethnic Koreans. Chinese are the main minority group. The Korean language is spoken throughout the peninsula, and there are several major dialects. About half of all Korean words come from Chinese. However, the grammar of Korean is similar to Japanese. Written Korean uses a 24-letter alphabet called hangul.

Religion Buddhism was introduced into Japan from China in the A.D. 500s. Today a combination of Shintoism and Buddhism dominates religion in Japan. Shintoism centers around the worship of natural spirits and ancestors. Buddhism stresses the unimportance of material goods. Both religions have greatly influenced Japan's traditional culture. For example, the same family might hold Shinto marriage ceremonies and Buddhist funerals.

Korea's major religion has historically been a blend of Buddhism and Confucianism. However, Christianity is now a major religion in South Korea. This is true even though Korea was never a European colony. About half of South Korea's population is Christian. North Korea's communist government officially allows freedom of religion. The state controls most religious activity, however.

Settlement and Land Use Japan is about the same size as California. Yet the country is home to more than 126 million people—nearly four times as many people as there are in California. In fact, Japan is one of the world's most densely populated countries. On average there are 867 people per square mile in Japan. Most people live along Japan's narrow coastal plains, which are very crowded. The largest and most densely settled plains are in eastern Honshū. Japan's largest cities are also found there. (See the population map in the unit atlas.)

LEVEL 3: Organize the class into five groups. Have each group create a proposal for a documentary to be titled Cultural Shifts in Japan and the Koreas: Then and Now. Proposals should include summaries of key historical influences on the region's culture and ways in which these events have shaped modern culture. Proposals should also list supplies or props that would be needed for the documentary. Encourage students to create storyboards to accompany their group's proposals. When groups have completed their proposals, call on volunteers to present them to the class. **COOPERATIVE LEARNING**

Teacher to Teacher

William Fisher of Bryan, Texas, suggests the following activity to help students understand different aspects of Japanese culture. Have each student write a haiku—an unrhymed Japanese poem with three lines. The first line has a total of five syllables, the second line seven, and the third line five. You may wish to provide samples of haiku for students to emulate. Students' poems should address aspects of Japanese culture. Have students write their haiku on large sheets of paper and draw illustrations around the poems. Display the illustrated haiku around the classroom.

INTERPRETING THE VISUAL RECORD

Houses are packed together on this densely settled coastal plain near Tokyo. In areas close to cities, population densities can reach several thousand people per square mile. **What human processes contribute to Japan's tremendous population density?**

The Koreas are also densely populated. The most crowded areas lie along the west coast. South Korea's urban areas have grown tremendously in the last decades. About 81 percent of South Koreans now live in urban areas. In North Korea, only about 60 percent of the population is urban.

READING CHECK: *Environment and Society* Where do most Japanese live, and where are the country's largest cities? **along narrow coastal plains, eastern Honshū**

Food Japan's main food is rice, which the Japanese eat at most meals. Fish are the major source of protein. Sushi (SOOH-shee)—vinegared rice and vegetables or raw fish—and sashimi (SAH-shee-mee)—thin slices of raw fish—are popular. Other common foods include cooked vegetables, tofu, and various types of noodles. Tea is the most popular drink. The diet of many Japanese has been changing since World War II. Fast food is more common, and people are eating more meat and dairy products.

Korea's staple food is also rice. Grilled meats and a wide range of vegetables are common. Barley, potatoes, and wheat are grown and used in many foods. One of the most popular dishes in Korea is kimchi (KIM-chee). This spicy dish is a mixture of Chinese cabbage, garlic, ginger, and other vegetables.

Education Japan has an excellent education system and one of the highest literacy rates in the world. This system has helped Japan become an industrial and economic power. Almost all children attend elementary and junior high schools, which are free. Students study many different subjects, such as art, mathematics, science, and social studies. Much time is spent learning how to read and write the Japanese language. There is intense competition among students to get into Japan's best universities. Many high school students attend special "cram" schools to prepare for difficult university entrance exams.

South Korea's education system is similar to Japan's. Most children attend free elementary, middle, and high schools. About one third of high school graduates go on to university. Admission is based on grueling entrance exams. In North Korea, teaching communist ideology is a major focus of the education system. The government also requires students to work as part of their

Essential Element 4

► Human Systems ◄

South Korean Festivals
Visitors to South Korea can enjoy a wide range of festivals. Among the many activities featured at various festivals are snowboarding, ice sculpting, bull riding, papermaking, catching fireflies, flying kites, designing jewelry, and guessing the weight of raw fish.

In October the city of Kwangju hosts the annual Kimchi Festival. Festival organizers hope to attract tourists and educate the world about kimchi. The event begins with an opening ceremony and a parade. At the display area, visitors can taste, buy, or just look at all kinds of kimchi dishes. There are several competitions, with a special category for foreigners.

DISCUSSION: Ask students what food festivals are held in their area and how they reflect the region's culture. What foods are typical of your region?

VISUAL RECORD ANSWER
urbanization, population growth, long life expectancies

Review Answers

Identify For identifications, see: Meiji Restoration, p. 643; Diet, p. 644

Close

Traditional Japanese music features instruments such as the *taiko* (drum), *shoko* (gong), *oteki* (a bamboo flute), *hichiriki* (a double-reeded bamboo instrument), and *koto* (a stringed instrument). Play recordings of Japanese music for the class. Ask students how the melodies might reflect what they know about Japanese culture.

Review and Assess

Have students complete the **Section Review**. Then have students complete **Daily Quiz 28.2**.

Reteach

Have students complete **Main Idea Activity for English Language Learners and Special-Needs Students 28.2**. Then have students use information from the text and class notes to create an annotated time line depicting the major events in the region's history. ENGLISH LANGUAGE LEARNERS

Extend

Have interested students conduct research on the traditional dress, or *hanbok*, of Korea. You may wish to have them create a poster using fabric scraps. More ambitious students may want to construct doll-sized examples. BLOCK SCHEDULING

Section 2 Review Answers

Define For definitions, see: shogun, p. 642; samurai, p. 643; annex, p. 643; armistice, p. 644; demilitarized zone (DMZ), p. 644

Working with Sketch Maps Maps will vary, but listed places should be labeled in their approximate locations. Tokyo became Japan's capital.

Reading for the Main Idea
1. U.S. and Soviet occupation, Korean War

2. caused a group of samurai to restore the emperor's power, which led to modernization

Critical Thinking
3. island geography, mountains, location in eastern Asia; encouraged isolationist foreign policy (NGS 15)

4. Possible answer: European traders—as a source of trade and riches; Japanese leaders—as a society that should be kept free of corrupting foreign influences (NGS 4)

Organizing What You Know
5. 300 B.C.—Asian invaders drive Ainu into northern Japan; 108 B.C.—Chinese invade Korea; A.D. 700s—Japanese develop political system headed by emperor; 1500s—Portuguese traders reach Japan; 1853—Perry reaches Tokyo Bay; 1868—Meiji Restoration returns emperor to power; 1945—United States bombs Hiroshima and Nagasaki; 1950—North Korea invades South Korea, starting Korean War; 1953—Korean War ends.

Korean women perform a traditional fan dance. The dancers' dresses are like those worn centuries ago by the Korean nobility. As they dance, the women make patterns with their fans.

excellent education systems with highly competitive university entrance exams; North Korea: uses education to teach communist ideology, students also work, higher education limited

schooling. Higher education in North Korea is much more limited than in South Korea and Japan. However, adult education is widespread. Many adults attend technical schools at night.

Traditions and Customs Western influences have greatly altered Japanese culture. Still, many traditions are important. Family ties are strong, and respect for elders is important. Traditional dress for both men and women is the kimono—a long robe with wide sleeves. Kimonos are mostly worn for special occasions. Removing one's shoes before entering a house is a common custom. Japanese wear slippers or socks instead of shoes inside the home.

The Japanese have a rich heritage in the arts. Traditional music is played on native instruments such as drums, flutes, and gongs. Japanese theater called Kabuki (kuh-BOO-kee) uses colorful costumes and makeup to portray historical events. In literature several types of Japanese poetry are widely popular. Painting, sculpture, and many other decorative arts also have a long history in Japan.

In the Koreas the central importance of family life has declined somewhat as the peninsula has developed and urbanized. However, many traditions and customs survive. For example, traditional clothing styles are still common, particularly in North Korea. Korean dance is popular in both North and South Korea. Dancers wear traditional clothes called *hanbok*.

✓ **READING CHECK:** *Places and Regions* How is education in Japan and South Korea different from education in North Korea?

Section 2 Review

go. hrw .com **Homework Practice Online**
Keyword: SW3 HP28

Identify
Meiji Restoration
Diet

Define
shogun, samurai, annex, armistice, demilitarized zone (DMZ)

Working with Sketch Maps On the map you created in Section 1, label Kyōto, Nagasaki, Tokyo, and Hiroshima. Which city became Japan's capital after the Meiji Restoration?

Reading for the Main Idea

1. *The Uses of Geography* What events have created the boundary between North Korea and South Korea?

2. *The Uses of Geography* How did trade and the diffusion of ideas lead to cultural change in Japan in the last half of the 1800s?

Critical Thinking

3. Drawing Inferences and Conclusions What geographic factors may have influenced Japan's ability to control its territory and keep European traders out before the mid-1800s? How did this affect Japan's foreign policy at the time?

4. Identifying Points of View How do you think European traders and Japanese leaders viewed Japan differently?

Organizing What You Know

5. Copy the time line below. Use it to describe major events in the region's history.

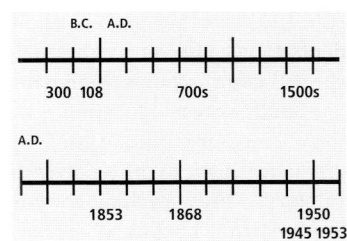

 **LET'S GET STARTED**

Copy the following instructions onto the chalkboard: *Look at the photographs in Section 3 and read the captions. Select one photo and write down two observations you can make about life in the region today by examining it.* Allow students time to complete this activity. Discuss student observations. Tell students that in Section 3 they will learn more about life in modern Japan and the Koreas.

Building Vocabulary

Write **work ethic**, **export economy**, and **trade surplus** on the chalkboard. Point out that all of these terms are used to describe business or the economy in Japan. Also note that they are all compound words. Have students speculate on the meanings of these terms by putting together their knowledge of one or both words in each compound. Then call on volunteers to locate and read the definitions of all the key terms aloud.

Section 3

The Region Today

READ TO DISCOVER

1. What are some characteristics of Japan today?
2. How is life in North Korea different from life in South Korea?

WHY IT MATTERS

Japan is one of the world's leading economic powers. Use **CNN fyi.com** or other **current events** sources to read about current issues facing the Japanese economy.

DEFINE

subsidies
work ethic
export economy
trade surplus
urban agglomeration

LOCATE

Yokohama   Kōbe
Nagoya   P'yŏngyang
Ōsaka   Seoul

Section 3 RESOURCES

REPRODUCIBLE

▶ Guided Reading Strategy 28.3
▶ Readings in World Geography, History, and Culture 71
▶ Geography for Life Activity 28
▶ Critical Thinking Activity 28
▶ Map Activity 28

TECHNOLOGY

▶ One-Stop Planner CD–ROM, Lesson 28.3
▶ CNN Presents Geography: Yesterday and Today, Segment 29: The Tokyo Grave Crisis
▶ Homework Practice Online
▶ HRW Go site

REINFORCEMENT, REVIEW, AND ASSESSMENT

▶ Main Idea Activity 28.3
▶ English Audio Summary 28.3
▶ Spanish Audio Summary 28.3
▶ Section 3 Review, p. 653
▶ Daily Quiz 28.3

Modern Japan

Despite modern influences, traditional Japanese culture remains an important part of Japanese society. Many Japanese still follow traditional ways of life, particularly in rural areas. A land of contrasts, Japan blends traditional and modern and East and West.

Agriculture About 11 percent of Japan is arable, and about 5 percent of Japanese workers are farmers. However, these farmers supply about 70 percent of the country's food needs. They have succeeded by using terraced cultivation and modern farming methods. Government **subsidies**, or financial support, protect Japanese farmers from foreign competition.

Farmers use more than half of Japan's farmland to grow rice. Tea, soybeans, fruits and vegetables, and mulberry trees (for silkworms) come from the south. Northern Japan is colder and has a shorter growing season. Farmers grow wheat, other grains, and vegetables there. Hokkaidō has a successful dairy industry.

Most Japanese farmers live in small villages. The average farm is only 2.5 acres (1 ha). However, expanding cities are taking over more and more good farmland.

Industry Japan began its rapid economic growth in the 1950s. Soon the country became a model for economic success. One of the reasons for this success was Japan's culture. The Japanese have a strong **work ethic**—a belief that work itself has moral value. Long workdays and six-day workweeks are common. Also, most workers feel loyal to

INTERPRETING THE VISUAL RECORD

A farmer in central Japan harvests his rice crop. Farm machinery used in Japanese agriculture is often small and efficient. **Why do you think many Japanese farmers prefer smaller farm machinery?**

VISUAL RECORD ANSWER

Small farms and rugged topography make large machines impractical.

Teach Objective 1

LEVEL 1: Organize the class into five groups. Assign each group one of following topics on Japan: agriculture, industry, international economy, urban geography, or modern ways of life. Have each group list words and phrases that summarize the main points regarding its assigned topic. When groups have completed their summaries, call on volunteers to present their work to the class. **COOPERATIVE LEARNING**

LEVELS 2 AND 3: Ask students to imagine that they have recently moved to Japan. Have them write letters to friends in the United States about their new homes. Ask students to describe characteristics of daily life in Japan, including the economy, urban geography, agriculture, and industry. In particular, encourage students to explain how daily life in Japan is similar to and different from life in the United States. Call on volunteers to read their finished letters to the class.

Global Perspectives

NUMMI In the 1970s an oil shortage in the United States caused many people to buy Japanese cars because they were smaller and more fuel-efficient than American cars. American auto manufacturers feared losing their market to Japanese competition. In 1981 Japan agreed to limit the number of vehicles the country would export to the United States. However, Japanese companies still sought ways to expand their business in the United States.

One result of this search was the creation of New United Motor Manufacturing, Inc. (NUMMI), a joint venture between Toyota and General Motors for building cars in U.S. factories. Under this banner, Japanese management systems and work environments would be introduced into factories staffed by Americans. The first NUMMI plant opened in 1984 and produces vehicles under both the Toyota and Chevrolet—a division of General Motors—names.

VISUAL RECORD ANSWER

Competition hurts some American automakers; others engineer better cars.

their companies. In turn, many companies have offered employees lifetime employment, although this has recently been changing. Japanese industry also benefited from the *keiretsu* system. This system brings together banking, big business, and government to set common goals. The combined efforts of these three groups helped Japanese industry grow.

In the 1990s, however, Japan faced stiff competition from the newly industrialized countries of East and Southeast Asia. The *keiretsu* system could not adapt quickly enough in the rapidly changing world economy. Many observers believe Japan will need to make some changes to regain its economic leadership in Asia. For example, many argue that Japan must allow more foreign competition and foreign investment into the country.

✓ **READING CHECK:** *Human Systems* What is the *keiretsu* system? A system that brings together banking, big business, and government to set common goals.

International Economy Japan has been very successful at selling and manufacturing its products overseas. For example, Japanese companies have auto-manufacturing plants in the United States, which is a major market for Japanese cars. Japanese companies also invest in many other countries.

Japan's main economic competition comes from other Asian countries like China and South Korea. This competition has hurt some Japanese industries, such as shipbuilding and steel production. For example, South Korea is now the world's leading shipbuilder. Japan once dominated this industry. In the future, China may become the leader in shipbuilding because of its lower labor costs. However, Japan remains a world leader in quality manufactured products like automobiles, cameras, and electronics. It is also a leader in robotics and biotechnology.

Japan has an **export economy**. In this type of economy, goods are produced mainly for export rather than for domestic use. Japan imports raw materials and energy and exports high-quality manufactured goods. Japan's export economy is so strong that Japan has built up a huge **trade surplus**. A trade surplus exists when a country exports more than it imports. For example, Japan has a large trade surplus with the United States. This has become an important issue in U.S.-Japanese trade relations. The U.S. government wants Japan to open its markets to more American goods.

INTERPRETING THE VISUAL RECORD

Japan exports between 4 and 5 million automobiles each year and is the world's second-largest automobile producer after the United States. Consumers around the world buy Japanese cars for their high quality, excellent design, and competitive prices. **How do you think the export of Japanese cars to the United States affects automobile production in this country?**

North Korea	South Korea
• *isolated, communist dictatorship*	• *democratically elected government*
• *large military*	• *many rights and freedoms enjoyed by citizens*
• *command economy*	
• *citizens' lives tightly controlled*	• *rapid economic growth*
• *outside the capital, most North Koreans living in poverty*	• *large urban middle class*
	• *export economy*

Urban Geography Japan has many industrial regions. However, the country's industrial-urban core is located on Honshū's eastern coastal plains. Three large functional regions can be found there. The Keihin region includes Tokyo and Yokohama. To the southwest lies the Chukyo region, which includes the city of Nagoya. The cities of Ōsaka, Kōbe, and Kyōto are part of the Keihanshin region even farther west. A dense network of high-speed trains, highways, and air services connects these regions. (See Case Study: High-Speed Rail.)

Tokyo-Yokohama is the largest **urban agglomeration** in the world. An urban agglomeration is a densely populated region surrounding a central city. About 26 million people are jammed into the urban region of Tokyo-Yokohama. This region is Japan's center of commerce, education, entertainment, government, and trade. Because little land is available, real estate prices are among the highest in the world. Yokohama, just to the south on Tokyo Bay, is Japan's busiest seaport.

The Chukyo region around Nagoya is important in the production of textiles, ceramics, and motor vehicles. Kōbe and Ōsaka—Japan's second-largest city—are major seaports. Kyōto is Japan's ancient capital.

Modern Ways of Life Most Japanese are middle-class. Families typically live in suburban areas. Parents often spend several hours each day commuting to and from work in the cities. Most Japanese families are small, with one or two children. Homes are also small and expensive. Sliding wooden screens separate sparsely furnished rooms. Families are used to tight spaces, and children often share bedrooms.

Japanese society has been greatly influenced by Western culture. For example, Western clothing, foods, and music are very popular. Baseball, golf, skiing, and soccer are common forms of recreation. At the same time, the Japanese have also influenced Western culture. Sushi bars and Japanese video games, landscaping, and cartoons can be found around the world.

Japan has one of the lowest birthrates in the world. In addition, life expectancy rates—84 years for women and 78 for men—are among the world's highest. As fewer people are born and people live longer, the average age of Japan's population is increasing. This "graying" population puts increased demands on health care and social services.

The role of women in Japanese society has also changed in recent decades. Women now receive more education than before and are an important part of the workforce.

INTERPRETING THE VISUAL RECORD *Tokyo is one of the largest cities in the world and is the focus of Japan's transportation network.* **What can you see in this photo that shows Tokyo is an important functional region?**

INTERPRETING THE VISUAL RECORD

A family in Japan gets by with very little space when compared to a typical American family. **What might be some reasons that Japanese houses are smaller on average than American houses?**

Capsule Hotels Urban crowding in Japan has led to creativity in the design of homes and businesses. One example is the capsule hotel. This type of lodging, for men only, features a lounge, an exercise room with a pool, a work area, and a restaurant.

The difference between capsule hotels and more traditional hotels lies in the sleeping capsules. Each capsule is just large enough to accommodate a single person. A thin mattress, a small wall-mounted television, a shelf for toiletries, an alarm clock, and a reading light complete the capsules. Capsules are stacked two high and are entered through one end. A plastic shade can be pulled down to cover the opening.

CRITICAL THINKING: Why might capsule hotels be popular among Japanese businessmen? *(Possible answers: Efficient businessmen might use them to avoid long commutes or to save limited travel funds.)*

VISUAL RECORD ANSWER
transportation network

VISUAL RECORD ANSWER
expensive land, lack of level land

LEVEL 3: Organize the class into two groups. Have the students in one group imagine that they are veterans of the Korean War living in North Korea. Have students in the other group imagine they are veterans living in South Korea. Have each side prepare for a debate concerning the benefits and drawbacks of living in the assigned country. Students should prepare opening statements regarding the conditions in their country since the Korean War. Debaters should also be prepared to respond to drawbacks proposed by the opposing side. **COOPERATIVE LEARNING**

Using National Geography Standard 14:

Environment and Society: How Human Actions Modify the Physical Environment Have students investigate how deforestation worsened the famine in North Korea. To demonstrate the effects of the deforestation, have students mound soil in two pans and sprinkle mixed birdseed liberally over one "hill," leaving the other bare. Instruct them to keep the birdseed moist. After the seeds have sprouted, students should water both mounds heavily with a plant mister and observe the difference in runoff. Ask students to draw a conclusion about the effects of runoff on agriculture.

Global Perspectives

Korean Reunification?

In 2000, 50 years after the beginning of the Korean War, the leaders of North and South Korea met for a historic three-day summit at P'yŏngyang. The leaders, pictured on the opposite page, agreed to reunite families separated by political boundaries since the Korean War and to promote economic development in both countries. They also pledged to work toward reunification through increased cultural, athletic, medical, and environmental cooperation and exchanges.

Part of the agreement went into effect immediately. Two hundred families were brought together later that year. In addition, members of the North and South Korean Olympic teams marched together at the 2000 Olympics.

ACTIVITY: Have students use the Internet, almanacs, and other current events sources to find recent information about reunification efforts.

Section 3 Review Answers

Define For definitions, see: subsidies, p. 649; work ethic, p. 649; export economy, p. 650; trade surplus, p. 650; urban agglomeration, p. 651

Working with Sketch Maps Maps will vary, but listed places should be labeled in their approximate locations.

North Koreans honor a giant statue of Kim Il Sung, leader of North Korea from 1948 until his death in 1994. Kim set up a communist government in North Korea with close ties to the Soviet Union and China.

Seoul's Namdaemun Market features a wide range of everyday items and consumer goods.

More than 50 percent of Japanese women are employed. As economic opportunities for women have grown, so have other opportunities. For example, women have become more involved in politics and government.

✓ **READING CHECK:** *Human Systems* How have economic opportunities for women in Japan changed in recent decades? Women receive more education and are an important part of the workforce.

The Two Koreas

The two Koreas share a common culture, but their level of political and economic development is very different. (See the unit Fast Facts table.) For example, the per capita GDP of South Korea is more than 13 times that of North Korea.

North Korea North Korea is an isolated country under a strict communist dictatorship. The country's leader, Kim Jong Il, came to power when his father died in 1994. North Korea has a large military and a command economy. The state controls most aspects of its citizens' lives. For example, the government limits and controls travel within the country. In general, government officials and high-ranking military officers live more comfortably than the average North Korean.

Outside the capital of P'yŏngyang (pyuhng-YANG), most North Koreans live in poverty. The country's farming system is based on inefficient communist state farms. The government lets some farmers have small garden plots for themselves. However, farmers cannot grow enough to feed themselves. Droughts and floods have caused severe food shortages. When the Soviet Union collapsed in the early 1990s, Soviet aid ended. The North Korean economy then collapsed. Food shipments from the United Nations have helped the country feed itself in recent years, but major problems still exist. Thousands of North Koreans have fled to China to seek a better life.

South Korea For many years South Korea's military dominated the government. However, today the country has a democratically elected government. South Koreans enjoy many rights and freedoms not found in the north. In addition, South Korea used U.S. aid to rebuild its industries after the Korean War. It developed from a poor farming country into an industrial "Asian Tiger" in only decades. "Asian Tiger" is a term used to describe countries in East and Southeast Asia that have experienced rapid economic growth in recent decades. Like Japan, South Korea has an export economy. Its major industries are shipbuilding, steel, automobiles, textiles, and electronics.

South Korea's business model is similar to the *keiretsu* model in Japan. Huge family-owned conglomerates called *chaebol* dominate South Korea's economy and political system. Four *chaebol* control most of South Korea's manufacturing and exports. This situation has prevented competition and led to corruption and debt. The government has passed recent reforms

Close

Challenge students to imagine that they are members of a committee whose purpose is moving North and South Korea closer to reunification. Call on volunteers to brainstorm suggestions to ease the process.

Review and Assess

Have students complete the **Section Review**. Then have students complete **Daily Quiz 28.3**.

Reteach

Have students complete **Main Idea Activity for English Language Learners and Special-Needs Students 28.3**. Then have students write three sentences summarizing the main points of each of the section's topics. **ENGLISH LANGUAGE LEARNERS**

Extend

Have interested students conduct research on the trade relationship between the United States and Japan. Students might focus on diplomacy and trade relations or specialized products that each country offers the other. Have students create posters that represent some aspect of their research. Display posters in the classroom. **BLOCK SCHEDULING**

INTERPRETING THE VISUAL RECORD

In June 2000 Presidents Kim Jong II of North Korea (left) and Kim Dae Jung of South Korea (right) met. The two leaders pledged to improve their countries' relations and to work toward the reunification of the peninsula. **How do you think these meetings might affect relations between North and South Korea?**

to try to improve South Korea's economy. They include opening the market to foreign investment and competition.

South Korea has a large urban middle class with access to consumer goods. The country's capital and largest city is Seoul (SOHL). Seoul is a huge city and a growing industrial and cultural center. South Korea's small farming villages are disappearing as the country continues to urbanize.

Korean Reunification   For decades Koreans have hoped to reunify their two countries. In recent years the two Koreas have had better relations. For example, South Korean tourists have been able to visit a few selected areas of North Korea. Talks and official visits continue, but the countries remain worlds apart. Many South Koreans fear that North Korea might develop and possibly use nuclear weapons. Another concern is whether South Korea could afford the cost of reuniting with one of the world's poorest countries.

✓ **READING CHECK:**   *Places and Regions*   How are the economies of North Korea and South Korea different? North Korea has a poor command economy. South Korea has a productive export economy with a per capita GDP more than 12 times higher.

Our Amazing Planet

North Korea has almost no privately owned cars. The government owns nearly all cars and sets them aside for official use only.

Section 3 Review

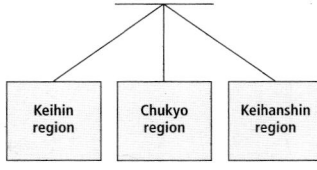

Homework Practice Online
Keyword: SW3 HP28

Define
subsidies
work ethic
export economy
trade surplus
urban agglomeration

Working with Sketch Maps
On the map you created in Section 2, label Yokohama, Nagoya, Ōsaka, Kōbe, P'yŏngyang, and Seoul. Which city is South Korea's capital and economic center?

Reading for the Main Idea
1. **Human Systems**   How did Japan's culture help the country develop economically?

2. **Places and Regions**   What are some important characteristics of the Japanese and South Korean market economies?

3. **Places and Regions**   What are some obstacles to Korean reunification?

Critical Thinking

4. **Comparing** How are politics, economics, and society in North Korea related? How does this situation differ in South Korea?

Organizing What You Know

5. Copy the graphic organizer below. Use it to describe the three functional regions that are the industrial-urban core of Japan.

Keihin region	Chukyo region	Keihanshin region

Section 3 Review Answers

Seoul is South Korea's capital and economic center.

Reading for the Main Idea

1. strong work ethic, company loyalty, *keiretsu* system

2. export economies based on high-quality manufactured goods, Japan's *keiretsu* system similar to South Korea's *chaebol*

3. fears of a North Korean invasion in the south; the cost of reunifying the two countries; the threat of nuclear weapons

Critical Thinking

4. North Korea—Communist dictatorship controls the economy, schools, and travel. South Korea—An elected government opened the country to foreign investment and products and many rights and freedoms not found in North Korea. (NGS 10)

Organizing What You Know

5. Keihin region—center of Japanese commerce, education, entertainment, government, and trade; largest urban agglomeration in the world; high real-estate prices; includes Tokyo and Yokohama; Chukyo region—produces textiles, ceramics, and motor vehicles; includes Nagoya; Keihanshin region—major seaports; includes Ōsaka, Kōbe, and Kyōto

VISUAL RECORD ANSWER

improve communication, reduce conflict, possibly lead to reunification

Setting the Scene

Tokyo is one of the largest and busiest cities in the world. Millions of people live and work within the city limits. An extensive network of roads connects various parts of the city. These roads can be quite confusing to navigate, even for natives of the city. Many roads are very narrow—some wide enough for only a single car to pass. Many twist and turn or zigzag along their routes. This pattern probably dates back to a time before automobiles were common and pedestrians were the only traffic on the streets. Furthermore, few of Tokyo's streets are named. Only the busiest streets carry identifying names. Addresses in Tokyo are usually given as a series of numbers.

Building a Case

Have students read Case Study: Japan's High-Speed Rail. Point out that many Tokyo residents also prefer to take trains within the city rather than drive. Have students brainstorm or conduct research to complete this graphic organizer listing the benefits of driving and taking the subway.

Travel In Tokyo

Driving
- not tied to fixed route or schedule
- air-conditioned comfort

Subway
- extensive network of trains
- regular schedules
- avoids expenses of fuel, toll roads, and parking
- avoids heavy traffic on roads

CASE STUDY

Japan's High-Speed Rail

Human Systems Japanese transportation engineers have developed one of the most sophisticated, efficient, and profitable rail networks in the world. For many Japanese travelers, railways are the preferred mode of transportation. A number of geographic factors have made train travel more attractive than road or air transportation.

Linking Major Urban Centers

The most advanced part of the Japanese rail network is the *Shinkansen*. This high-speed train links major urban centers such as Tokyo and Ōsaka. Only 320 miles (515 km) apart, Tokyo and Ōsaka lie on the southern coast of Japan's largest island, Honshū. These cities have easy access to deepwater ports. This access has helped them develop into the urban and industrial core of Japan. In 1964 the first *Shinkansen* line connected Tokyo and Ōsaka. It was named the "New Tokaido Line" after an ancient highway that once linked Kyōto and Tokyo. By 1975 the Japanese had extended the main rail line to Fukuoka. That city lies at the western edge of Japan's manufacturing belt on the island of Kyūshū. The Japanese have since added other lines radiating northward from Tokyo.

The most impressive aspect of the Japanese *Shinkansen* system is the incredible speeds that trains move. The fastest trains from Tokyo to Ōsaka travel at 160 miles per hour (257 kph). The trains can cover the 664-mile (1,070-km) journey from Tokyo to Hakata in less than seven hours and carry 1,000 passengers. In addition to being fast, the trains run often and on time. These qualities are very important for business passengers. For example, during the morning and evening rush hours, trains run every 7.5 minutes. On average, each train misses its schedule by only 36 seconds!

Factors Behind Success

Why has rail travel been so successful in Japan? Japan is a country where space is limited and land values are very high. As a result, trains have a distinct advantage over road and air travel. Why is this so? Road travel gobbles up larger chunks of real estate than rail networks. For example, cars and trucks need highways, access ramps, gas stations, and parking lots. In turn, increasing traffic constantly creates pressure to build even more roads. These roads then eat up even more space. High fuel costs can also make car and truck travel expensive. In fact, Japanese fuel prices are often about three times higher

Japan's High-Speed Rail System

— Tokaido Shinkansen
— Sanyo Shinkansen
— Tohoku Shinkansen
— Joetsu Shinkansen
— Yamagata Shinkansen
— Akita Shinkansen
— Hokuriku Shinkansen
— Under construction
····· Planned

SCALE
0   200   400 Miles
0   200   400 Kilometers
Projection: Lambert Conformal Conic

INTERPRETING THE MAP Begun in the 1960s, the Shinkansen system now connects most of Japan's major cities, with even more routes planned or under construction. **Which areas on the map have lines planned or under construction? Why do you think these routes have not yet been built?**

Drawing Conclusions

Have students locate and study road maps and subway maps of Tokyo. Ask them to imagine that they are city residents who want to travel from the vicinity of the Imperial Palace (near Tokyo Station) to the Shinjuku district (near Shinjuku Station). Ask them if it would be easier to drive or to take the subway. Discuss student responses. *(Most students will probably say that the subway would be easier. Both stations lie on the Yamanote subway line. Automobile traffic would involve navigating several streets in the heart of downtown Tokyo.)* Repeat the activity with various starting points and destinations.

Going Further: Thinking Critically

Organize students into groups. Have each group choose a U.S. city with an established subway system and repeat the investigation they performed in the last activity. Groups should locate road maps and subway maps for their chosen cities and investigate alternate routes between locations to determine which means of transportation is more efficient. Encourage students to research statistics that reveal the popularity of each travel method in their cities. Does this research support the conclusions they drew from examining the maps?

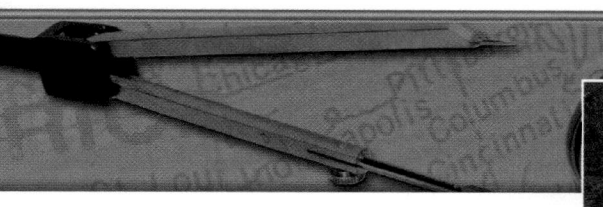

All Shinkansen *trains are monitored by computers in Tokyo. The computers control the trains' speeds and the distances between them.*

than those in the United States. It should be no surprise, then, that many Japanese do not want to take cars on long trips.

Airports present similar disadvantages for mass travel in a country like Japan. For one thing, they also require large areas of land. The most desirable airport locations are close to large urban areas. These locations are often in places where land prices are highest. In addition, the availability of land there is very limited. In Tokyo planners decided to locate the city's main airport an hour away—by train. Workers in Ōsaka built a new airport in the harbor on a human-made island. When you factor in the time it takes to get to and from these airports, trains can be faster than airplanes. Japanese travelers seem to agree. Even with expansion of their airports, Ōsaka and Tokyo are linked by six times as many trains as airline flights. In fact, train travel accounts for 84 percent of all trips between the two cities.

Spreading Success Around

Japan is not the only place where high-speed train travel makes sense. In 1998, 12 countries had passenger trains that average more than 125 miles per hour (200 kph). Sixteen other high-speed railways were in

Originally designed to carry people over long distances, Shinkansen *trains have become popular among Japanese commuters.*

development in places such as South Korea and Sweden.

The same factors that led to the success of the *Shinkansen* are found where other such networks are in development. Many large cities are close together. In addition, road and air travel are stretched to the limit. As a result, high-speed rail is becoming more appealing. Even large countries have found high-speed rail attractive. In 1996, for example, Chinese leaders set a goal to launch a high-speed link between Beijing and Shanghai. Officials hope to cut the travel time between the cities from 17 hours to six or seven hours. The United States recently began operating high-speed train service between New York and Washington, D.C. Projects in Texas and California have also been proposed. Supporters of projects like these believe high-speed rail travel is likely to become an increasingly popular way to travel in densely populated regions. The Japanese *Shinkansen* set the standard for these future projects.

Applying What You Know

1. **Summarizing** What factors have made the *Shinkansen* a successful mode of transportation in Japan?

2. **Making Generalizations** What might be some other solutions to travel problems in countries like Japan?

CHAPTER 28 Review Answers

Building Vocabulary For definitions, see: tsunamis, p. 638; flyway, p. 640; shogun, p. 642; samurai, p. 643; Meiji Restoration, p. 643; annex, p. 643; Diet, p. 644; armistice, p. 644; demilitarized zone (DMZ), p. 644; subsidies, p. 649; work ethic, p. 649; export economy, p. 650; trade surplus, p. 650; urban agglomeration, p. 651

Locating Key Places

A. Seoul
B. Hokkaidō
C. P'yŏngyang
D. Honshū
E. Shikoku
F. Tokyo
G. Sea of Japan
H. Kyūshū

Understanding the Main Ideas

1. Hokkaidō, Honshū, Shikoku, and Kyūshū; Honshū
2. The region is on a major flyway for migratory birds.
3. 1910–1945
4. Japanese uses a combination of Chinese characters and Japanese symbols called kana.

655

TECHNOLOGY

- Chapter 28 Test Generator (on the One-Stop Planner)
- Global Skill Builder CD–ROM
- HRW Go site

REINFORCEMENT, REVIEW, AND ASSESSMENT

- Chapter 28 Review, pp. 656–57
- Chapter 28 Tutorial for Students, Parents, Mentors, and Peers
- Chapter 28 Test (form A or B)
- Alternative Assessment Handbook

- Chapter 28 Test for English Language Learners and Special-Needs Students
- Unit 9 Test
- Unit 9 Test for English Language Learners and Special-Needs Students

CHAPTER 28 Review Answers

Korean uses a 24-letter alphabet called hangul.

5. based on inefficient state farms, some farmers allowed to have small garden plots

Thinking Critically

1. Isolationism limited access to innovations from outside and slowed the diffusion of ideas into the country. **(NGS 4)**

2. Possible answer: The huge bureaucracy created by the government and private parts of the system likely slowed decision-making processes and efforts to implement needed changes on a large scale. **(NGS 10)**

3. to prevent North Korea from invading South Korea, which would disrupt the entire region **(NGS 13)**

Using the Geographer's Tools

1. southward almost to Australia and eastward to the Gilbert Islands

2. Charts will vary but should accurately represent the information presented in the text.

3. They are functional regions because their urban areas and industries are interconnected.

CHAPTER 28 Review

Building Vocabulary

On a separate sheet of paper, explain the following terms by using them correctly in sentences.

tsunamis	armistice
flyway	demilitarized zone (DMZ)
shogun	subsidies
samurai	work ethic
Meiji Restoration	export economy
annex	trade surplus
Diet	urban agglomeration

Locating Key Places

On a separate sheet of paper, match the letters on the map with their correct labels.

Hokkaidō	Kyūshū	P'yŏngyang
Honshū	Sea of Japan	Seoul
Shikoku	Tokyo	

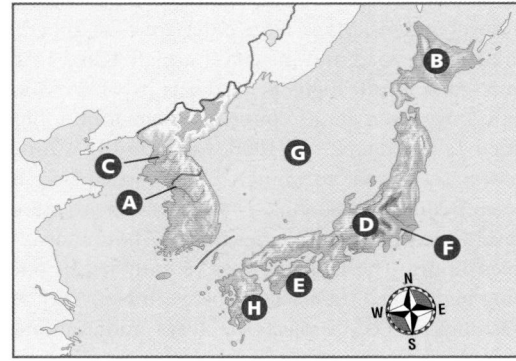

Understanding the Main Ideas

Section 1

1. *Places and Regions* What are Japan's four main islands? Which is the largest and most populated?

2. *Physical Systems* Why might you find so many different species of birds in Japan and the Koreas?

Section 2

3. *Places and Regions* During what period did Japan control the Korea Peninsula?

4. *Human Systems* How are the Japanese and Korean written languages different?

Section 3

5. *Environment and Society* What is agriculture like in North Korea?

Thinking Critically

1. Drawing Inferences and Conclusions How did cultural patterns and attitudes in Japan influence innovation and the diffusion of ideas before the mid-1800s?

2. Analyzing Information Why do you suppose Japan's *keiretsu* system could not adapt quickly enough to the rapidly changing world economy of the 1990s?

3. Evaluating Why do you think the United States maintains a large military presence in South Korea today?

Using the Geographer's Tools

1. Analyzing Maps Study the map of the Japanese Empire in Section 2. How far did Japan expand its empire in the Pacific?

2. Creating Charts Create a chart describing and comparing the languages, religions, land use patterns, systems of education, and customs of Japan, North Korea, and South Korea.

3. Preparing Maps Use the information in Section 3 to prepare a map of Japan's three major urban-industrial regions. Do you think these are formal, functional, or perceptual regions? Why?

Writing about Geography

Imagine you are a high school student in Japan. Write a letter to a friend in the United States in which you describe Japan's educational system. How is it similar to the education system in the United States? How is it different? When you are finished with your letter, proofread it to make sure you have used standard grammar, spelling, sentence structure, and punctuation.

SKILL BUILDING

Geography for Life

Field Work

Human Systems Conduct field work to find some Japanese products that affect your daily life. Start by creating a list of the products in your home that were made in Japan. What types of products are they? How much did they cost? Then interview friends and neighbors in your community to find out what Japanese-made products they find useful. What can these products tell us about Japan's level of development and economy? Why do you think Japanese products are so successful in the United States?

Portfolio Activity

1. Comic books are very popular among Japanese adults and children. Have students create comic books that describe each of the chapter's sections. Encourage students to be creative while maintaining accuracy. Place the comic books in student portfolios.

2. Have students make models or dioramas of typical Japanese or Korean homes, including features of everyday life such as furniture. Take photographs of the dioramas for inclusion in student portfolios.

Food Festival

Sushi is popular in Japan and the United States. Have two or three students search the Internet for sushi rice recipes and prepare the rice at home. Ask other students to bring the remaining ingredients for vegetarian sushi, such as avocados, carrots, green onions, tofu, dried seaweed wrappers called nori, and Japanese horseradish called wasabi. Steer students away from using raw fish, or sashimi, in their sushi because of the expense and health issues. Call on a few volunteers to locate the small bamboo mats used to roll sushi in Asian markets or cookware supply shops and practice using them. Assemble all ingredients and supplies and have a sushi-making party.

Building Social Studies Skills

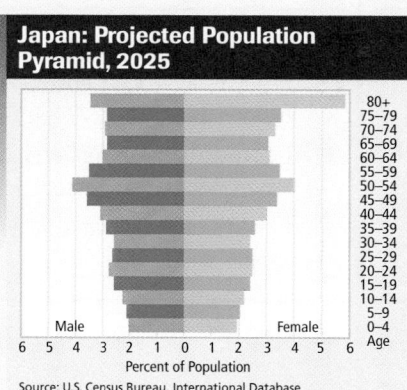

Japan: Projected Population Pyramid, 2025

Percent of Population
Source: U.S. Census Bureau, International Database

Interpreting Graphs

Study the projected population pyramid for Japan in 2025. Then answer the following questions.

1. According to the population pyramid, in 2025 Japan is projected to have
   a. many more younger people than older people.
   b. more people under the age of 4 than between 50–54.
   c. more people over the age of 40 than below that age.
   d. more men than women over the age of 80.

2. What factors might account for the way Japan's population pyramid is projected to look in 2025?

Analyzing Secondary Sources

Read the following passage and answer the questions that follow.

"Japan's earlier Meiji constitution of 1889 had granted supreme power to Japan's emperor. However, the 1947 constitution gave power to the Japanese people through their elected legislature, the Diet. The emperor remained important in government only as 'the symbol of the State and the unity of the people.' Japan's new constitution also guaranteed individual freedoms, such as freedom of speech, religion, and the press. These individual rights had not been guaranteed under Japan's earlier constitution."

3. Which of the following things did Japan's 1947 constitution *not* do?
   a. guarantee individual freedoms
   b. make the emperor more powerful
   c. allow freedom of the press
   d. take power from the emperor and give it to the people

4. What role did the emperor play in the government set up by the new constitution?

Writing

Student letters should demonstrate an understanding of the educational system in Japan. Use Rubric 42, Writing to Inform, to evaluate student work.

Geography for Life

Student work will vary based on the products available to students. Conclusions about the Japanese economy should be reasonable and supported by the products found. Use Rubric 12, Drawing Conclusions, to evaluate student work.

Social Studies Skills

1. c

2. Possible answer: a low growth rate and a longer life expectancy

3. b

4. He remained a symbol of the State and the unity of the people.

Alternative Assessment

PORTFOLIO ACTIVITY

Learning about Your Local Geography

Individual Project: Research
Like Japan and the Koreas, the United States is on a major flyway for migrating birds. How does this affect the wildlife in your area? Use your local library to learn about migratory birds that pass through your area each year. What are some of these types of birds? Where are they traveling from, and where are they going? At what time of the year do they travel? Write a short report answering these questions. You might want to create a map of the United States that notes important migratory bird routes.

internet connect

Internet Activity: go.hrw.com
KEYWORD: SW3 GT28

Choose a topic on Japan and the Koreas to:
- create a newspaper on Japan and the Koreas.
- research active volcanoes along the Ring of Fire.
- learn about a typical school day in Japan.

PORTFOLIO ACTIVITY

Student projects will vary based on the community and the birds that can be found there. Reports and maps should accurately reflect student research. Use Rubric 30, Research, to evaluate student work.

CHAPTER RESOURCE MANAGER

	Objectives	Pacing Guide	Reproducible Resources
SECTION 1 **Natural Environments** (pp. 659–62)	• Identify the major landforms and rivers of mainland Southeast Asia. • Describe the region's climates, vegetation types, and animals. • Name some of mainland Southeast Asia's main resources.	**Regular** .5 day **Block Scheduling** .5 day *Block Scheduling Handbook, Chapter 29*	**RS** Guided Reading Strategy 29.1 **RS** Graphic Organizer Activity 29
SECTION 2 **History and Culture** (pp. 663–67)	• Explain some of the major events in mainland Southeast Asia's history. • Describe the main cultural features of this region.	**Regular** .5 day **Block Scheduling** .5 day *Block Scheduling Handbook, Chapter 29*	**RS** Guided Reading Strategy 29.2 **PS** Readings in World Geography, History, and Culture 72 and 74 **E** Creative Strategies for Teaching World Geography, Lesson 18 **E** Cultures of the World Activity: Region 7
SECTION 3 **The Region Today** (pp. 669–73)	• Describe the economies and politics of mainland Southeast Asia. • Analyze the types of agriculture practiced in the region. • Examine the issues and challenges faced by the region's countries.	**Regular** 1 day **Block Scheduling** .5 day *Block Scheduling Handbook, Chapter 29*	**RS** Guided Reading Strategy 29.3 **SM** Critical Thinking Activity 29: Myanmar and the Tatmadaw **SM** Geography for Life Activity 29: The Shadowed Ground of Cambodia **SM** Map Activity 29: Mainland Southeast Asia: Exports and Urbanization

Chapter Resource Key

PS Primary Sources	**A** Assessment	CD–ROM	**One-Stop** Planner CD–ROM
RS Reading Support	**REV** Review	Video	
IC Interdisciplinary Connections	**ELL** Reinforcement and English Language Learners	Internet	See the *One-Stop Planner* for a complete list of additional resources for students and teachers.
E Enrichment	Transparencies	Holt Presentation Maker Using Microsoft® PowerPoint®	
SM Skills Mastery			

 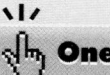 **One-Stop** Planner CD–ROM

It's easy to plan lessons, select resources, and print out materials for your students when you use the **One-Stop Planner CD–ROM with Test Generator**.

 **internet** connect

HRW ONLINE RESOURCES

GO TO: go.hrw.com
Then type in a keyword.

TEACHER HOME PAGE
  KEYWORD: SW3 Teacher

CHAPTER INTERNET ACTIVITIES
  KEYWORD: SW3 GT29
  Choose an activity to:
  • learn about the architecture of mainland Southeast Asia.
  • create a brochure on Laos and Vietnam since the Vietnam War.
  • understand the causes and effects of deforestation in mainland Southeast Asia.

CHAPTER ENRICHMENT LINKS
  KEYWORD: SW3 CH29

CHAPTER MAPS
  KEYWORD: SW3 MAPS29

ONLINE ASSESSMENT
  Homework Practice
  KEYWORD: SW3 HP29
  Standardized Test Prep
  KEYWORD: SW3 STP29
  Rubrics
  KEYWORD: SS Rubrics

COUNTRY INFORMATION
  KEYWORD: SW3 Almanac

CONTENT UPDATES
  KEYWORD: SS Content Updates

HOLT PRESENTATION MAKER
  KEYWORD: SW3 PPT29

ONLINE READING SUPPORT
  KEYWORD: SS Strategies

CURRENT EVENTS
  KEYWORD: S3 Current Events

Technology Resources

- One-Stop Planner CD–ROM, Lesson 29.1
- Geography and Cultures Visual Resources 49–53
- Homework Practice Online
- HRW Go site

- One-Stop Planner CD–ROM, Lesson 29.2
- Homework Practice Online
- HRW Go site

- One-Stop Planner CD–ROM, Lesson 29.3
- *ARGWorld* CD–ROM
- **CNN** Presents Geography: Yesterday and Today, Segment 30: The Highway to Prosperity
- Homework Practice Online
- HRW Go site

Reinforcement, Review, and Assessment

- **ELL** Main Idea Activity 29.1
- **ELL** English Audio Summary 29.1
- **ELL** Spanish Audio Summary 29.1
- **REV** Section 1 Review, p. 662
- **A** Daily Quiz 29.1

- **ELL** Main Idea Activity 29.2
- **ELL** English Audio Summary 29.2
- **ELL** Spanish Audio Summary 29.2
- **REV** Section 2 Review, p. 667
- **A** Daily Quiz 29.2

- **ELL** Main Idea Activity 29.3
- **ELL** English Audio Summary 29.3
- **ELL** Spanish Audio Summary 29.3
- **REV** Section 3 Review, p. 673
- **A** Daily Quiz 29.3

Meeting Individual Needs

Ability Levels

Level 1 Basic-level activities designed for all students encountering new material

Level 2 Intermediate-level activities designed for average students

Level 3 Challenging activities designed for honors and gifted-and-talented students

English Language Learners Activities that address the needs of students with Limited English Proficiency

Chapter Review and Assessment

- Chapter 29 Test Generator (on the One-Stop Planner)
- Global Skill Builder CD–ROM
- HRW Go site
- **REV** Chapter 29 Review, pp. 676–77
- **REV** Chapter 29 Tutorial for Students, Parents, Mentors, and Peers
- **A** Chapter 29 Test (form A or B)
- **A** Alternative Assessment Handbook
- **A** Chapter 29 Test for English Language Learners and Special-Needs Students

Launch into Learning

Ask students to describe Vietnam in phrases. Many students will probably mention the Vietnam War. Then tell students that in 1997 a deer species previously unknown to Western scientists was found in Vietnam's dense forests. Ask students how an entire species could escape detection throughout the years when the media were focused on Vietnam. *(Possible answers: Perhaps the deer moved into the area after the war, or the forest region was relatively unaffected by the fighting.)* Point out that, although many features of mainland Southeast Asia are known to Westerners, political turmoil and rough terrain have kept some hidden. Tell students they will learn more about familiar and unfamiliar aspects of mainland Southeast Asia in this chapter.

Using the Physical-Political Map

Have students examine the map on the opposite page. Ask volunteers to identify a peninsula *(Indochina or Malay Peninsula)* and delta *(Irrawaddy, Chao Phraya, Mekong, or Hong)*. Point out the Tonle Sap in Cambodia. Tell students that during the rainy season the lake is more than twice the size it is during the dry season.

CHAPTER
29
Mainland Southeast Asia

Mainland Southeast Asia includes five countries: Cambodia, Laos, Myanmar (Burma), Thailand, and Vietnam. The region's mountains, rain forests, and river valleys have nurtured distinct cultures. The entire region has tropical environments, with most of the population concentrated along major river valleys.

Banana seller, Vietnam

Sawaddee! (May you have good fortune!) I am Chosita, and I am 14 years old. I live in Bangkok with my parents and older sister. My mother is a colonel in the Royal Thai Air Force and my father is a government economist.

My sister and I get up very early for school because traffic in Bangkok is really bad. By 6:00 A.M. we are on the road. If traffic isn't too bad, we get to school in about 30 minutes. However, if traffic is heavy, it takes more than an hour. We go to school from June to September and from November to February. Our big "summer" vacation is from March to May.

Our school is near Jatujak Park, where the city government has relocated all the street vendors in Bangkok to help ease downtown traffic. However, people come to Jatujak to buy everything from food to shoes, clothing, and souvenirs, so traffic is still a big problem.

In my free time, I like to listen to music, watch TV, or go to the beach. The sea in the south is beautiful, and the beaches are not very crowded. I'm not sure what I want to be when I grow up—maybe an economist like my father or an accountant for a big firm. I still have quite a few more years to think about it.

Detail of temple interior, Vietnam

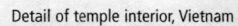

Section 1

OBJECTIVES

1. Identify the major landforms and rivers of mainland Southeast Asia.

2. Describe the region's climates, vegetation types, and animals.

3. Name some of mainland Southeast Asia's main resources.

 **LET'S GET STARTED**

Copy the following instructions onto the chalkboard: *Look at the pictures in Section 1 of this chapter. What do they suggest about the climate and vegetation of mainland Southeast Asia?* Discuss responses. *(Possible answer: Lush forests and flooding suggest that the region receives heavy rains and probably has a wide variety of plant life.)* Point out that much of the region is covered by thick rain forests. Tell students they will learn more about the climates, biomes, and landforms of this tropical region in Section 1.

Building Vocabulary

Write **arboreal** on the chalkboard and underline the root word *arbor*. Below it write other words that start with the same root, like *arboretum* and *Arbor Day*. Ask students what these terms mean. *(Possible answers: An arboretum is a place where trees are cultivated. Arbor Day is a holiday celebrated with the planting of trees.)* Ask students to conclude from these examples what arbor means *(tree)*. Point out that arboreal animals are species that live in trees. Call on volunteers to suggest examples of these creatures.

Natural Environments

READ TO DISCOVER

1. What are mainland Southeast Asia's major landforms and rivers?

2. Which climate and vegetation types are found in the region, and what animals live there?

3. What are some of mainland Southeast Asia's main resources?

WHY IT MATTERS

Biologists searching Earth's tropical regions, including those in mainland Southeast Asia, are still finding new plant and animal species. Use **CNN fyi.com** or other **current events** sources to read about some recent discoveries.

DEFINE

arboreal

LOCATE

Bay of Bengal	Khorat Plateau
Andaman Sea	Irrawaddy River
Malay Peninsula	Chao Phraya River
Gulf of Thailand	Mekong River
South China Sea	Hong (Red) River
Gulf of Tonkin	Tonle Sap
Indochina Peninsula	

Section 1 RESOURCES

REPRODUCIBLE
▶ Guided Reading Strategy 29.1
▶ Graphic Organizer Activity 29

TECHNOLOGY
▶ One-Stop Planner CD–ROM, Lesson 29.1
▶ Geography and Cultures Visual Resources 49–53
▶ Homework Practice Online
▶ HRW Go site

REINFORCEMENT, REVIEW, AND ASSESSMENT
▶ Main Idea Activity 29.1
▶ English Audio Summary 29.1
▶ Spanish Audio Summary 29.1
▶ Section 1 Review, p. 662
▶ Daily Quiz 29.1

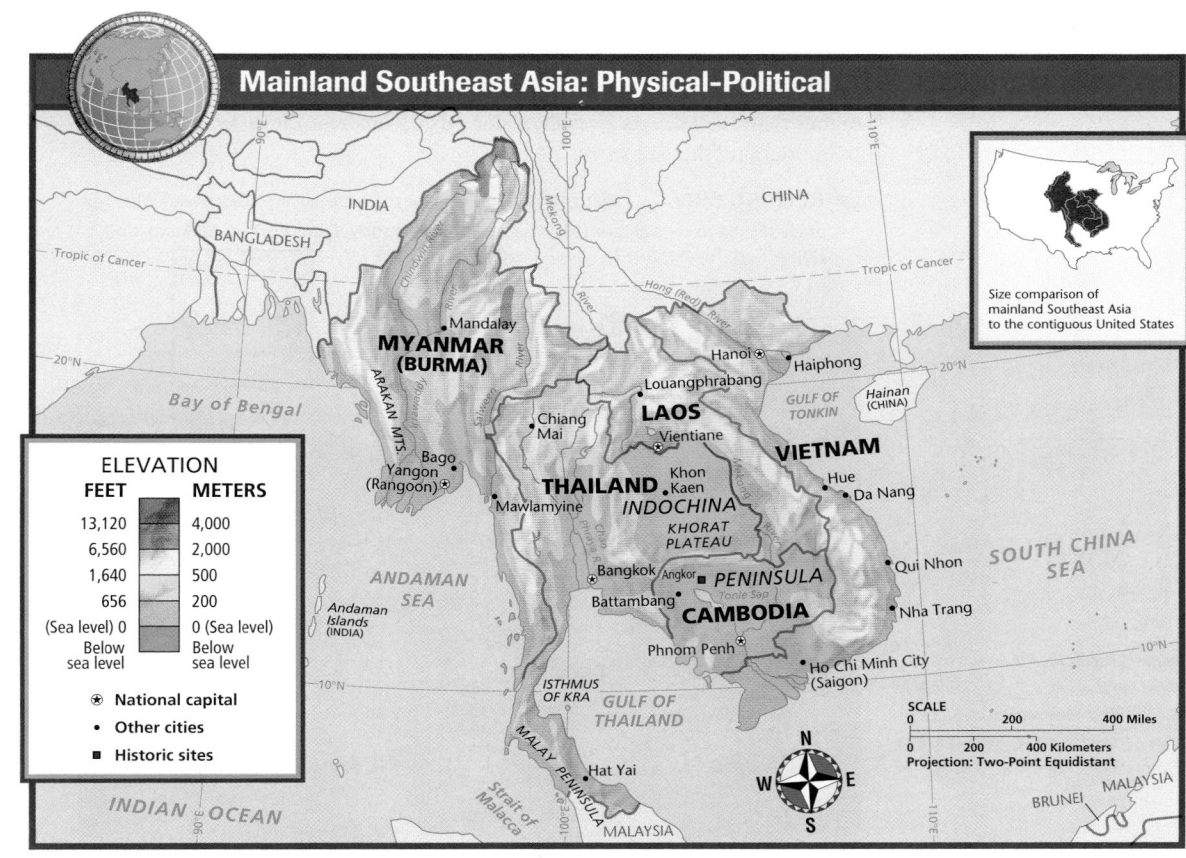

Mainland Southeast Asia: Physical-Political

Size comparison of mainland Southeast Asia to the contiguous United States

ELEVATION

FEET	METERS
13,120	4,000
6,560	2,000
1,640	500
656	200
(Sea level) 0	0 (Sea level)
Below sea level	Below sea level

⊛ National capital
• Other cities
■ Historic sites

SCALE
0   200   400 Miles
0   200   400 Kilometers
Projection: Two-Point Equidistant

ALL LEVELS: Copy the following graphic organizer onto the chalkboard, omitting the italicized answers. Then have students complete the chart with details about the three landform regions of mainland Southeast Asia. Ask students if these constitute formal, functional, or perceptual regions *(formal regions)*. **ENGLISH LANGUAGE LEARNERS**

Landform Regions of Mainland Southeast Asia

Mountains	Plains and plateaus	River valleys and deltas
• *located in northern Southeast Asia* • *rugged landscapes* • *fan out from Himalayas and Plateau of Tibet through Myanmar, Thailand, Laos, Vietnam*	• *found in central Southeast Asia* • *cover much of Thailand and Cambodia* • *includes Thailand's Khorat Plateau*	• *Irrawaddy, Choa Phraya, Hong (Red), and Mekong Rivers* • *fertile alluvial soils, heavy farming* • *dense populations, most of the region's cities* • *rivers used for local transportation*

Using Illustrations

Ko Phi Phi Direct students' attention to the photograph of the Phi Phi Islands on this page. Ask students to describe the islands' landforms, as seen in this picture. *(Possible answers: mountains; steep cliffs; karst towers; narrow sandy beach)* Have students recall what they have learned about karst in previous chapters *(limestone formations carved by moving water).* Then ask students what the presence of a karst landscape suggests about the climate of southwestern Thailand. *(Possible answers: It receives heavy precipitation.)* What other elements of the photograph suggest that the area receives heavy rainfall? *(lush tropical vegetation)* Ask students what they think might be the most important economic activity in the islands. What evidence in the picture suggests this? *(tourism; passenger boat)* You may also wish to point out that several American movies have been filmed in the Phi Phi Islands.

internet connect

GO TO: go.hrw.com
KEYWORD: SW3 CH29
FOR: Web sites about tourism in mainland Southeast Asia

internet connect

GO TO: go.hrw.com
KEYWORD: SW3 CH29
FOR: Web sites about mainland Southeast Asia

Landforms and Rivers

Mainland Southeast Asia stretches southward from the Asian landmass. To the west lie the Bay of Bengal and Andaman Sea. The narrow Malay Peninsula, which includes part of Thailand, extends to the south. In the east, the Gulf of Thailand, South China Sea, and Gulf of Tonkin surround the Indochina Peninsula. Laos is the region's only landlocked country.

The area has three main landform regions. In the north, rugged mountain ranges fan out from the Himalayas and the Plateau of Tibet. They stretch into Myanmar (MYAHN-mar), Thailand, Laos, and Vietnam. A central region of plains and low plateaus lies to the south in Thailand and Cambodia. Thailand's Khorat (koh-RAHT) Plateau is there. River valleys and deltas make up the third main landform region.

Four major rivers flow southward from Asia's mountainous interior. The Irrawaddy (ir-ah-WAH-dee) empties into the Bay of Bengal. The Chao Phraya (chow PRY-uh) flows into the Gulf of Thailand. The Mekong (MAY-KAWNG), the region's longest river, borders Thailand. It flows through Laos, Cambodia, and Vietnam to the South China Sea. The Hong (Red) River flows across northern Vietnam into the Gulf of Tonkin. The valleys and deltas of these large rivers have fertile alluvial soils, which support intensive farming and dense populations. The rivers have also been useful for local transportation. In addition, they have influenced the locations and growth of the region's cities.

Southeast Asia's largest freshwater lake is Tonle Sap (tohn-LAY SAP) in Cambodia. During the dry season, a river flows south from the lake into the Mekong River. However, during the wet season, the Mekong's water level is higher than the lake. As a result, water pushes upstream into Tonle Sap, more than doubling the lake's size.

✔ **READING CHECK:** *Places and Regions* What are the region's major rivers? What type of soils would you find in their valleys and deltas? *Irrawaddy, Chao Phraya, Mekong, and Hong (Red) Rivers; fertile alluvial soils*

Southern Thailand's Phi Phi Islands, or Ko Phi Phi in the Thai language, are famous for spectacular karst landscapes. Limestone cliffs have eroded to create these towering karst formations.

Teach Objective 2

ALL LEVELS: Organize the class into three groups and assign each group one of the following topics: climate, vegetation, or animal life. Tell the groups that they are writing travel features for a local magazine about mainland Southeast Asia. Each group should prepare an article about the assigned topic for readers unfamiliar with the region but who might be interested in visiting. Point out that the articles should explain the relationships among landforms, climates, and biomes, and that students will need to define any geographic terms they use. Remind groups to use standard grammar, spelling, sentence structure, and punctuation in their articles. **COOPERATIVE LEARNING**

Teach Objective 3

LEVEL 1: Call on volunteers to name some of the natural resources of mainland Southeast Asia. Have students use this list and the information in the text to design a cover for a book about Southeast Asia's natural wealth. The cover should include the book's title and drawings or pictures reflecting the region's main resources. **ENGLISH LANGUAGE LEARNERS**

LEVELS 2 AND 3: Have students write dust jacket blurbs for the covers they designed in the Level 1 activity. Blurbs should address the relationship between the region's resources and human activities there.

INTERPRETING THE VISUAL RECORD

A boat approaches a house isolated by the flooded Mekong River. Floodwaters can rise to 22 feet (7 m) along the Mekong during the rainy season. **How does this photo suggest that Cambodians have adapted to living in an area prone to floods?**

Climates, Plants, and Animals

All of mainland Southeast Asia has tropical or subtropical climates. Precipitation is high throughout the region. Most coastal areas have a tropical humid climate with heavy rainfall. Interior plains and plateaus have a tropical wet and dry climate. The driest region is the Khorat Plateau in eastern Thailand. A tropical wet and dry climate extends across central Myanmar, Thailand, Cambodia, and southern Vietnam. A humid subtropical climate region is found in northern Myanmar, Laos, and northern Vietnam.

Mainland Southeast Asia is greatly affected by the monsoon wind system. This system causes a rainfall pattern with distinct wet and dry seasons. Summer winds from the southwest bring heavy rainfall. The mainland gets most of its rainfall during this season. Winter brings dry winds that blow from the northeast.

The monsoons create serious natural hazards. During the wet monsoon, severe flooding is common. On the other hand, fires can be a major problem in the dry season. Between monsoons in the fall, Vietnam is sometimes hit by powerful typhoons. Myanmar lies in the path of typhoons that come in from the Bay of Bengal.

The region's tropical and subtropical climates support a great number of plants and animals. While much of the environment has been greatly modified by people, large areas remain almost untouched. Rare and endangered plants and animals live in these areas. Some thickly forested mountain areas are so rugged and inaccessible that scientists are still discovering new species there. To protect the environment, some of the region's countries have created large national parks and reserves.

The thick tropical rain forests of many lowland and coastal areas support **arboreal**, or tree-dwelling, animals such as monkeys. A wide range of birds also live in the rain forests. Many fish species and other marine life live in coastal mangroves. These mangroves are particularly common in places where mud is deposited along the shore. Drier inland areas have monsoon forests. These forests of broadleaf trees are not as dense as tropical rain forests.

✓ **READING CHECK:** *Places and Regions* What climates does the region have? Where is the driest region? tropical humid, tropical wet and dry, humid subtropical; Khorat Plateau in eastern Thailand

Our Amazing Planet

In the 1990s biologists working in the remote mountains of Laos and Vietnam discovered at least five new mammal species. These species included a striped rabbit, a deer that barks like a dog, and a 200-pound wild ox.

Leopard cats are native to the forests of Southeast Asia. Sometimes hunted for their pelts, these arboreal cats are now protected by law.

Global Perspectives

Damming the Mekong River As many as a dozen dams are being built on the 2,600-mile Mekong River. All of the dams currently under construction are in China, but the governments of Cambodia, Laos, and Vietnam have also expressed interest in building dams. These new dams would allow farmers to bring water into arid areas and to diversify their harvests. They would also help to control the Mekong's annual flooding and generate electricity. However, the environmental damage the dams might cause has thus far prevented their construction.

Already, the Chinese dams cause problems downstream. They lower the river's level, deplete fish stocks, and sharply reduce silt deposits. This silt enriches the soil in which half of Vietnam's rice crop grows. Additional dams along the river's course could increase environmental damage.

ACTIVITY: Organize the class into two groups. Have the groups debate whether more dams should be built on the Mekong. You may wish to have groups conduct further research to prepare for the debate.

VISUAL RECORD ANSWER

house built on stilts

Close

Ask students to summarize the relationships among Southeast Asia's location, climate, and vegetation.

Review and Assess

Have students complete the **Section Review**. Then have students complete **Daily Quiz 29.1.**

Reteach

Have students complete **Main Idea Activity for English Language Learners and Special-Needs Students 29.1.** Then have students design text and graphics for web pages about mainland Southeast Asia's landforms, climates, biomes, and resources. **ENGLISH LANGUAGE LEARNERS**

Extend

Laterite, a soil type commonly found in tropical rain forests, is not particularly fertile. To demonstrate this, have interested students plant seeds in two pots—one filled with laterite, available from aquarium supply stores, and the other with regular potting soil. Have students water the plants evenly and report their observations about plant growth to the class. **BLOCK SCHEDULING**

Section 1 Review Answers

Define For definition, see: arboreal, p. 661

Working with Sketch Maps Maps will vary, but listed places should be labeled in their approximate locations. The Mekong is the region's longest river.

Reading for the Main Idea
1. rugged mountains in the north, central plains and low plateaus, major river valleys

2. creates distinct wet seasons (summers) and dry seasons (winters) along with serious natural hazards such as flooding and fires

3. interfere with natural soil-forming processes of the river valleys

Critical Thinking
4. found in coastal areas where mud is deposited (NGS 7)

Organizing What You Know
5. The Irrawaddy flows through Myanmar to the Bay of Bengal. The Chao Phraya flows through Thailand to the Gulf of Thailand. The Mekong flows through Laos, Cambodia, and Vietnam to the South China Sea. The Hong (Red) flows through northern Vietnam to the Gulf of Tonkin.

Natural Resources

Southeast Asia's tropical rain forests hold valuable hardwoods such as mahogany, teak, and ebony. Large areas have been logged for these woods or cleared by growing populations. Until recently there has been little concern for conservation or for replanting trees.

The continued clearing of tropical rain forests has damaged the habitat of endangered wildlife such as tigers and elephants. Removing trees and other plants has also led to flooding and soil erosion. In addition, a soil type common in the area, called laterite, hardens and becomes useless when the forest cover is removed. Some countries have tried to slow the deforestation. For example, Thailand has set up national parks to protect key areas and banned logging for export. However, in most countries logging and habitat loss continue at a rapid pace. (See Case Study: Studying Deforestation.)

Mainland Southeast Asia also has valuable minerals and fossil fuels. Iron, manganese, tin, and tungsten are found there. The region exports gems, including sapphires and rubies. In addition, Thailand has natural gas, Myanmar has oil, and Vietnam has coal. The large rivers have potential for the development of hydropower. Laos plans to build dams for hydropower in the near future. However, building dams may interfere with the natural soil-forming processes of the river valleys.

Although logging for export has been outlawed in Thailand since 1989, illegal operators continue to smuggle teak out of the country. Often elephants move the heavy logs. However, harsh treatment threatens the working elephants, and habitat destruction endangers the survival of those still living in the wild.

✓ READING CHECK:
Environment and Society How has clearing the region's rain forests affected the environment? loss of habitat for wildlife, flooding, soil erosion, damage to laterite

Section 1 Review

 Homework Practice Online Keyword: SW3 HP29

Define arboreal

Working with Sketch Maps
On a map of mainland Southeast Asia that you draw or that your teacher provides, label the Bay of Bengal, Andaman Sea, Malay Peninsula, Gulf of Thailand, South China Sea, Gulf of Tonkin, Indochina Peninsula, Khorat Plateau, Irrawaddy River, Chao Phraya River, Mekong River, Hong (Red) River, and Tonle Sap. Which is mainland Southeast Asia's longest river?

Reading for the Main Idea
1. *Places and Regions* What are the major landform regions of mainland Southeast Asia?
2. *Physical Systems* How does the monsoon system influence the region's climates?
3. *Environment and Society* How may building dams affect the environment?

Critical Thinking
4. *Analyzing Information* How is the distribution of mangroves in the region related to soils?

Organizing What You Know
5. Copy the chart below. Use it to describe mainland Southeast Asia's four major rivers.

River	Flows through	Empties into
Irrawaddy		
Chao Phraya		
Mekong		
Hong (Red)		

Section 2

OBJECTIVES

1. Explain some of the major events in mainland Southeast Asia's history.

2. Describe the main cultural features of this region.

 **LET'S GET STARTED**

Copy the following instructions onto the chalkboard: *Examine the chapter and unit maps. What countries do you think may have had the strongest influences on the history and culture of mainland Southeast Asia? Why?* Discuss responses. Point out that although powerful Asian countries like China and India have been influential in the region, European powers have also been major players in mainland Southeast Asia's history. Tell students that they will learn more about the history and cultures of the region in Section 2.

Building Vocabulary

Write the term **domino theory** on the chalkboard. Ask students what happens when the first in a row of closely placed dominoes standing on end is knocked over. *(The rest fall down too.)* You may wish to bring dominoes to class to illustrate this. Ask students how a similar idea might relate to international politics. *(The domino theory holds that if one country falls under the sway of communism, the same event will occur in nearby countries.)* Call on a volunteer to read the definition from the text or glossary.

History and Culture

READ TO DISCOVER

1. What are some major events in mainland Southeast Asia's history?
2. What are the main cultural features of the region?

WHY IT MATTERS

Geographers, archaeologists, and other scientists study sites where plants and animals have been domesticated in mainland Southeast Asia and around the world. Use CNNfyi.com or other **current events** sources to find recent news on domestication.

DEFINE

domino theory

LOCATE

Angkor Wat

History

Southeast Asia has been inhabited for a long time. The earliest ancestors of the region's peoples arrived at least 1.5 million years ago. Modern humans (*Homo sapiens*) have lived in Southeast Asia for at least 40,000 years.

Southeast Asia was an important center of plant domestication. Early peoples grew rice, citrus fruits, and bananas. By about 3000 B.C. rice farming had been established, and people had domesticated buffalo, pigs, and cattle. Evidence indicates that people from southern China began to migrate through the region at least 2,500 years ago. Over time, settlements headed by chiefs grew, and trade developed with China and India. Merchants from India probably introduced Sanskrit writing and Hinduism during this period.

Angkor Wat in Cambodia, built in the 1100s by King Suryavarman II as a Hindu temple, is the largest religious complex ever constructed. Inset: Statues line the road into the Bayon Temple near Angkor Wat.

LEVEL 1: Organize students into pairs and have each pair outline the history of mainland Southeast Asia. Have students illustrate their outlines with sketches that recall some aspect of each period in the region's history. **ENGLISH LANGUAGE LEARNERS, COOPERATIVE LEARNING**

LEVEL 2: Have students conduct further research on one period of history from the outline they created for the Level 1 activity to learn about key events, groups that entered the region, and important figures that influenced the region's history during that period. Have students create time lines based on their research and share their findings with the class.

LEVEL 3: Have each student choose a country of mainland Southeast Asia and conduct research to learn about its history during and after the European colonial period. Ask students to identify which European power or powers—if any—established colonies in their chosen countries and to explore the reactions of the local peoples to the Europeans' arrival. Then have students examine trade patterns established by the Europeans during the colonial period. Ask students if these patterns have changed since the countries of mainland Southeast Asia gained their independence. Finally, ask students what indications of the colonial past remain in their chosen countries today. **BLOCK SCHEDULING**

Linking Past to Present

Angkor In addition to housing Angkor Wat, the largest religious complex in the world, the city of Angkor was the capital and crowning jewel of the vast Khmer empire. The artistic and architectural accomplishments at Angkor have been compared to those of ancient Egypt, Greece, or Rome. Particularly famous are the stone carvings that decorated Angkor's buildings, but some of the city's other features were equally impressive. An advanced irrigation system, for example, allowed the city to feed a population that may have reached one million people.

The city of Angkor was abandoned in 1431. However, the temple at Angkor Wat remained a center for Buddhist worship for several centuries. Today the ruins are a favorite destination for pilgrims, tourists, and filmmakers. Unfortunately, countless artifacts and sculptures have been damaged or stolen. Images cut from facades are sold for as much as $300,000. Tombs have also been raided for their treasures. In fact, almost 40 percent of the city's original sculpture work has disappeared.

The Shwedagon Pagoda in Yangon, Myanmar, is a shrine sacred to the country's Buddhists. Tradition dates the gold-covered pagoda to the 500s B.C. The structure seen here, though, was rebuilt in the 1770s.

Our Amazing Planet

In 1994 the U.S. space shuttle *Endeavour* scanned the Angkor area with radar. This and other remotely sensed images revealed the presence of previously unknown buildings.

Early Cultures and Settlement Over the centuries various peoples moved into the region, particularly from China. The largest highly developed culture group in the region was the Khmer (kuh-MER). The Khmer dominated what is now Cambodia beginning in the A.D. 800s. By the end of the 1100s, their empire included most of mainland Southeast Asia. Angkor Wat, a huge temple complex built by the Khmer, reflects their advanced civilization and Hindu religion. In the 1200s the Thais (TYZ) migrated from southern China into the Khmer regions. Buddhism, introduced earlier from India and Sri Lanka, spread across mainland Southeast Asia, replacing Hinduism.

✓ **READING CHECK:** *Human Systems* From where did Hinduism and Buddhism diffuse into the region? India and Sri Lanka

Colonialism and Independence Europeans came to Southeast Asia in the early 1500s. Portuguese and Dutch sailors set up posts for trading in spices and other goods. Traders from Great Britain and France followed. Eventually, most of the region was colonized. Burma (now Myanmar) was a British colony. The French controlled Cambodia, Laos, and Vietnam, all of which they called Indochine, or French Indochina. Only Siam (now Thailand) was never colonized by Europeans.

During the 1800s the British and French set up plantations for growing export crops. They also built roads and railroads. They set up English- and French-language schools and introduced Christianity. Many Chinese and Indians migrated into the region during this period. They came to work on French and British plantations, mines, and railroads.

The Japanese invaded Southeast Asia during World War II. Only parts of Burma remained free of Japanese control. After the war, nationalist groups in the region tried to end colonialism. Over time four newly independent countries emerged. French Indochina was split into three countries—Vietnam, Cambodia, and Laos. Burma became independent from Britain. In 1989 Burma changed its name to Myanmar.

The transition to independence was not easy. In Vietnam, internal conflict led to a civil war involving the United States. (See Connecting to History: The Vietnam War.) The United States was involved because it wanted to stop the spread of communism in Asia. U.S. policy was based on the **domino theory**. This was the idea that if one country fell to communism, neighboring countries would follow like falling dominoes. The war in Vietnam also disrupted the neighboring countries of Laos and Cambodia.

✓ **READING CHECK:** *Human Systems* Why did migrants from India and China come to the region during the European colonial era? to work on French and British plantations, mines, ports, and railroads

Culture

Each country in the region has one dominant cultural group. These are the Burmans, Thais, Khmer, Lao, and Vietnamese. However, there are differences within each group. For example, China has influenced culture in northern Vietnam, while Khmer influences are more pronounced in southern Vietnam. Each of the region's countries is also home to many minority groups. These groups have their own cultures and languages.

People, Languages, and Religion Mainland Southeast Asia has three main language families. They are spoken by the largest ethnic groups. In the west the Burmans speak a Sino-Tibetan language related to Chinese. Languages of the Tai family are spoken in Thailand and Laos. These languages—Thai and Laotian—may have originated in southwestern China. In the east the Vietnamese and Khmer peoples speak languages from the Austro-Asiatic family.

Many of the region's smaller ethnic groups live in mountain areas. In fact, more than 50 ethnic groups live in the Vietnamese highlands. They include the Cham, Yao, Hmong (MONG), and Muong. Many of these people have maintained their traditional ways of life. (See Geography for Life: Keeping Traditional Ways of Life in Myanmar.) For example, they practice animism, wear unique clothing and jewelry, and remain cut off from mainstream cultures.

Most of the region's major cities have large Chinese populations. About 14 percent of Thailand's population is ethnic Chinese. In addition, the colonial languages

Connecting to HISTORY

The Vietnam War

Vietnam's struggle for independence from France was long and hard. After years of fighting following World War II, the Vietnamese defeated the French in 1954. A peace settlement then divided Vietnam into two parts. North Vietnam had a Communist government, with its capital at Hanoi. In South Vietnam a pro-Western government was set up with its capital at Saigon. The two Vietnams could not come to an agreement about reunification, or rejoining as one country.

Communist groups supported by North Vietnam fought to control South Vietnam. The United States supported the South Vietnamese forces, entering directly into the conflict in late 1964. A long and bitter war followed. Nearly 2 million Vietnamese and more than 58,000 Americans lost their lives during the Vietnam War. The war even spread across the border into Laos and Cambodia. Most U.S. forces left South Vietnam in 1973. In 1975, North Vietnamese forces occupied Saigon, which they renamed Ho Chi Minh City in honor of their late leader.

More than 1 million South Vietnamese tried to escape the Communist takeover. Many became refugees in nearby countries. Some eventually settled in Western countries, particularly the United States, Australia, and France.

Drawing Inferences and Conclusions How did the Vietnam War affect the cultural geography of the United States?

In this 1975 photo, refugees flee to Saigon from cities taken by Communist forces. Local people offer food and drink.

ALL LEVELS: Copy the graphic organizer at right onto the chalkboard, omitting the italicized answers. Call on students to suggest words or phrases to describe cultural elements of each country in mainland Southeast Asia. Then lead a class discussion about physical and human factors that have influenced the region's cultures. **ENGLISH LANGUAGE LEARNERS**

HOMEWORK: Have students write letters to imaginary pen pals in mainland Southeast Asia. Encourage students to mention specific events in the region's history and aspects of the region's culture. Have students exchange letters with partners and write responses to the letters they receive.

Mainland Southeast Asia's Cultures

Myanmar
• *Sino-Tibetan language (Burmese)*
• *many ethnic groups*
• *mostly Buddhists, some animists*

Vietnam
• *Austro-Asiatic language (Vietnamese)*
• *many ethnic groups*
• *mix of Confucianism and Buddhism, some animists*
• *varied food tradition*

Cambodia
• *Austro-Asiatic language (Khmer)*
• *mostly Buddhist*

Thailand
• *Tai language (Thai)*
• *mostly Buddhist, but Hindu Brahmins lead ceremonies*
• *curries common in diet*

Laos
• *Tai language (Laotian)*
• *some animists*

Daily Life

Betel-Chewing Customs

Visitors to Vietnam may notice that some residents appear to have bloodred lips. This is a result of chewing betel nuts, a common custom in Vietnam.

People who follow this custom actually chew a mixture of nuts from the areca palm and leaves from the betel pepper plant along with powdered lime. This blend is said to relieve pain, ease digestion, and sweeten the breath. It is also a common gift among the Vietnamese and sometimes appears at weddings. The juice that results from chewing the mixture is red. This color is a traditional Vietnamese symbol of good luck and happiness.

ACTIVITY: Have students research other customs of mainland Southeast Asia. Encourage them to share their findings with the class.

You don't say! The Vietnamese mid-autumn festival of Têt-Trung-Thu is also called the Children's Festival. According to Vietnamese folklore, parents worked so hard harvesting crops in the fall that they use this festival to make up for lost time with their children.

VISUAL RECORD ANSWER

provide sense of common identity, historical tradition

INTERPRETING THE VISUAL RECORD

Thai Buddhists bring offerings of food and incense to a Bangkok temple. Theravada Buddhism has been Thailand's main religion since the 1300s. **How might Buddhist traditions help unify the Thai people?**

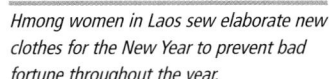
Hmong women in Laos sew elaborate new clothes for the New Year to prevent bad fortune throughout the year.

---

of French and English are often spoken in the region. For example, English has become the language of international business in Southeast Asia. Some people in Vietnam, Cambodia, and Laos still speak French.

The region's dominant religion is Buddhism. Thailand, Cambodia, and Myanmar all have large Buddhist majorities. Most of the region's Buddhists practice a form of Buddhism called Theravada. They claim that Theravada, one of 18 major branches of Buddhism, is closest to the Buddha's original teachings. Buddhism and Hinduism coexist in a unique way in Thailand. There Hindu Brahmins lead most of the royal or official ceremonies.

Most Vietnamese practice a mix of Confucianism, which came from China, and Buddhism. Christianity and Islam are also present in the region. Animist religions are common in the highlands of Laos, Vietnam, and Myanmar.

✓ **READING CHECK:** **Human Systems** What European languages are still spoken in mainland Southeast Asia? What religions are practiced there? English, French; Buddhism mostly, also Confucianism, Islam, Christianity

Food, Traditions, and Customs In the region's big cities, American fast food restaurants draw many new customers. Still, the most important food throughout mainland Southeast Asia is rice. In most places, rice is part of every meal. Other typical foods are fish and vegetables. Native tropical fruits like bananas, citrus, and durian are also available. Durian is a fruit known for its sweet flavor and unpleasant smell. The durian's odor is so strong, in fact, that some hotels and buses post signs forbidding the fruit. Spices such as ginger and chili peppers add flavor to regional dishes. Lime juice, lemon grass, and coriander add tang. Ground peanuts and coconut milk are other flavorings. Fish sauces are popular throughout the region. These sauces, which have different names in different countries, are based on the liquid poured off of salted, fermented fish. Sometimes chilies or sugar flavor the fish sauce. Spicy sauces with Indian origins called curries are popular in Thailand. Curries usually flavor rice or vegetables. Thais often wash down their spicy food with sugarcane juice. Vietnamese food is particularly varied, with nearly 500 traditional dishes. Some use exotic meats, such as bat or cobra. Plain white rice with various sauces is now typical fare, however.

Buddhism shapes people's lives in mainland Southeast Asia. For example, Thai men often spend time working and serving in monasteries. In Laos, all

Close

Ask students to name various foreign powers that have influenced the history of mainland Southeast Asia. (*Possible answers: China, France, Great Britain, India, Japan, Netherlands, Portugal*) As each country is named, call on volunteers to suggest elements of Southeast Asia's culture that reflect that country's influence in the region.

Review and Assess

Have students complete the **Section Review**. Then have students complete **Daily Quiz 29.2**.

Reteach

Have students complete **Main Idea Activity for English Language Learners and Special-Needs Students 29.2.** Then have students write headlines for newspaper articles about the region's history or culture. **ENGLISH LANGUAGE LEARNERS**

Extend

Have interested students conduct research on King Rama IV, also called King Mongkut of Siam, who kept his country from being colonized and whose life was the inspiration for the musical *The King and I.* Have students compare what they learn to any of the books, movies, or plays about the king. **BLOCK SCHEDULING**

Buddhist men have traditionally been expected to become monks for a while. One of Myanmar's main holidays is the Festival of Light, which marks an event in the Buddha's life. Paper lanterns light up the streets, and families visit the local shrine. Also in Myanmar, a water festival called Thingyan marks the country's Buddhist New Year. The festival celebrates the cleansing of the soul and washing away of the old year. During this mid-April festival, people in the streets soak each other with water. Since the event falls during the hottest season, the buckets of cold water may be welcome! Other countries celebrate the new lunar year in similar ways.

Throughout the region, cultures of urban and rural areas may differ widely. Many rural people follow the same practices generation after generation. For example, village religious festivals may celebrate local animist beliefs. Country people are more likely to wear garments such as the *panung* in Thailand and the *longyi* in Myanmar. A *panung* is a colorful cotton or silk cloth wrapped tightly around the body. A *longyi* is a long skirt. Western clothing is common in urban areas.

✔ **READING CHECK:** *Places and Regions* How is Buddhism a unifying element of the region's culture? Even though traditions and customs vary throughout the region, in many places Buddhism shapes people's lives.

INTERPRETING THE VISUAL RECORD

Two different education systems are pictured in these photographs. At left, Thai boys study at the Wat Po monastery in Bangkok. At right, Vietnamese girls attend a state school in Ho Chi Minh City. **What are some ways in which these schools might differ?**

Section 2 Review

Define domino theory

Working with Sketch Maps On the map of mainland Southeast Asia that you created in Section 1, label the countries of mainland Southeast Asia and Angkor Wat. Shade in the countries that made up French Indochina. Which culture group built Angkor Wat?

Reading for the Main Idea

1. *Places and Regions* What were some of the first crops grown in mainland Southeast Asia?

2. *Places and Regions* Which two European countries had large colonies in mainland Southeast Asia? How did the U.S. military become involved in the region?

3. *Human Systems* What is the dominant religion throughout mainland Southeast Asia? What is unique about religious practices in Thailand?

Critical Thinking

4. **Analyzing** How has mainland Southeast Asia's location between India and China affected its culture?

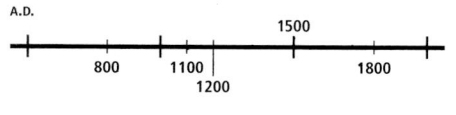
go.hrw.com **Homework Practice Online**
Keyword: SW3 HP29

Organizing What You Know

5. Create two time lines like the ones shown below. Use the time lines to trace the region's history through important periods and events.

```
A.D.                    1500
  |————|———|——————|——————|
      800  1100         1800
            1200
```

```
A.D.
1900                              1975
  |——————————————————|——|——|——|
                    1945 | 1964  1973
                       1954
```

Section 2 Review Answers

Define For definition, see: domino theory, p. 664

Working with Sketch Maps Maps will vary, but listed places should be labeled in their approximate locations. Cambodia, Laos, and Vietnam made up French Indochina. The Khmer built Angkor Wat.

Reading for the Main Idea
1. rice, citrus fruits, bananas

2. Great Britain and France; to support South Vietnam and stop spread of communism in Asia

3. Buddhism; most royal or official ceremonies led by Hindu Brahmins

Critical Thinking
4. Ethnic groups from China have migrated into the region, and some languages are related to Chinese. Hinduism and Buddhism were introduced from India and Sri Lanka. (NGS 9)

Organizing What You Know
5. 800s–1100s—Khmer Empire; 1200s—Thais migrating into Khmer region; 1500s—Europeans coming to region; 1800s—European colonies in region; 1954—Vietnam independence, division; 1964—U.S. involvement in Vietnam War; 1973—end of U.S. involvement; 1975—Communist takeover of South Vietnam

VISUAL RECORD ANSWER

subject matter, teaching methods, admissions requirements

Economics: Cultural Heritage Tourism

For years, Thailand has depended on its rich cultural heritage to attract visitors, and now its neighbors want to follow its lead. Faced with increased competition for tourist dollars, Thai travel groups have begun to emphasize the rich traditions of the various ethnic groups who live in Thailand to spur tourists' interest in rural areas and thereby increase revenues. In some cases, the travel agencies have even encouraged or paid the people in these areas to maintain their traditions.

Thailand is not alone in encouraging its people to maintain traditional ways of life to attract tourists. Tourism agencies in Laos stress the traditional cultures of the country's residents.

In Vietnam, young people are encouraged to dress in traditional styles, especially in tourist destinations like Hue. Visitors to the city can also dress in Vietnamese clothing and dine in one of the country's historic imperial villas.

Call on students to look through their textbooks to find examples of other countries that draw upon their cultural heritage to promote tourism. Organize students into small groups and have each group conduct research on one of these countries. Have students examine various ways in which their chosen countries may have shaped their cultures to encourage tourism. Have each group share its findings with the class. **COOPERATIVE LEARNING, BLOCK SCHEDULING**

Kayan Women Myanmar's Kayan, or Padaung, women of Myanmar have traditionally worn metal rings around their necks. With time, the rings' weight deforms the collarbone and ribs, creating the illusion of a long neck.

Several hundred Kayan have fled the turmoil in Myanmar and settled in northern Thailand. They are allowed to stay because the so-called giraffe women attract tourists to rural villages. Tour operators may pay a woman $60 or so per month for her exotic look. The practice of fitting young girls with the metal rings had been dying out. Now that long necks are moneymakers, the custom has been revived.

DISCUSSION: Are Kayan women being exploited or supporting themselves? Should travelers support or avoid Kayan villages?

Applying What You Know Answers

1. resisted attempts to assimilate them, demanded political freedom and basic human rights, taken up arms against the government
2. Possible answer: less likely to maintain traditional ways because they are more often exposed to new ideas and technology

This Geography for Life feature addresses National Geography Standards 4, 6, 9, and 10.

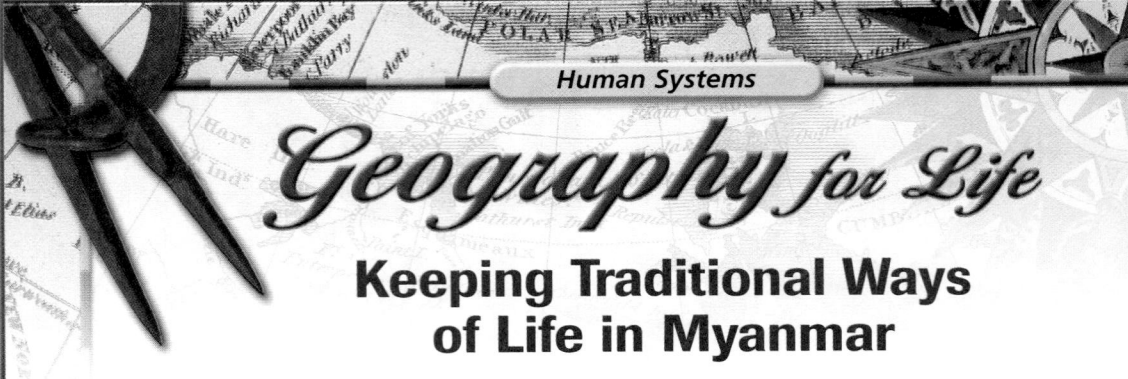

Human Systems

Geography for Life

Keeping Traditional Ways of Life in Myanmar

Traditions—established ways of thinking, acting, and behaving—are valued for many reasons. Traditional activities help people connect with their ancestors, land, and environment. In a world of rapid change, traditions offer a sense of stability. On the other hand, some traditions are controversial. For example, the shifting agriculture practiced by some peoples can damage the environment. Other customs may sharply limit the choices that women can make.

Indigenous cultures are particularly rooted in tradition. However, these cultures are under increasing pressure from the modern world. Modernization and economic development make it harder and harder for some to maintain their ways of life. These peoples are often viewed as backward by the majority culture in control of a country's politics, economic growth, and land use policies. The majority may dominate indigenous minority groups and force them to assimilate into modern society.

One example of the challenges facing traditional cultures is in Myanmar. Myanmar's largest ethnic group—the Burmans—make up about 68 percent of the country's population. The Burmans occupy the lowlands, river valleys, and large cities of Myanmar. As the country's largest ethnic group, they dominate Myanmar's government. However, Myanmar is also home to many smaller ethnic groups with different languages and cultures. Some of these groups include the Shan,

Karen, Rakhine, Mon, Chin, and Kachin. In general, Myanmar's minority groups live in the hills and uplands near the country's borders. In fact, they are often referred to collectively as the hill peoples. Most of these peoples live in small villages and practice animism.

Since Myanmar became independent from Great Britain in 1948, many of the country's minority groups have struggled to maintain their identity and culture. Rapid economic and social change has altered old ways of life, particularly among younger generations. Minority groups have resisted government attempts at cultural assimilation. They have also demanded political freedom and basic human rights. Some have even taken up arms against the government. For example, the Kachin Independence Army (KIA) and Karen National Union (KNU) are fighting for independent states. Such conflicts are not uncommon in other parts of the world.

Applying What You Know

1. **Summarizing** How have the peoples who live in the hills and uplands of Myanmar struggled to maintain their traditional ways of life?

2. **Supporting a Point of View** Do you think younger generations are more or less likely to maintain traditional ways of life? Why?

Left: Heavy neck rings are common among the Kayan (Padaung). Right: Fishers in east-central Myanmar use traditional fishing nets. Many people in the area belong to the Mon ethnic group.

Section 3

OBJECTIVES

1. Describe the economies and politics of mainland Southeast Asia.

2. Analyze the types of agriculture practiced in the region.

3. Examine the issues and challenges faced by the region's countries.

 **LET'S GET STARTED**

Copy the following question and instruction onto the chalkboard: *Would you rather live in a rural area or in the city? Write down your answer and a few sentences explaining your choice.* Discuss responses. Explain that most mainland Southeast Asians live in rural areas. Tell students that in Section 3 they will learn more about where and how the people of mainland Southeast Asia live.

Building Vocabulary

Write the terms *wats* and *klongs* on the chalkboard. Call on volunteers to read the definitions from the text or glossary. Point out that both terms are Thai words. Ask students to think of English words that could be used to describe the same things. *(Possible answers: wat—temple or monastery; klong—canal, waterway, aqueduct, or sewer tunnel)*

Section 3 The Region Today

READ TO DISCOVER

1. What are the economies and politics of mainland Southeast Asia like?

2. What types of agriculture are practiced in the region?

3. What issues and challenges do the region's countries face?

IDENTIFY

ASEAN

DEFINE

wats
klongs

LOCATE

Bangkok
Ho Chi Minh City
Hanoi
Yangon (Rangoon)
Phnom Penh
Vientiane

WHY IT MATTERS

Some countries in mainland Southeast Asia still bear environmental scars left by years of warfare. Use  or other **current events** sources to learn how these countries are working to repair the damage.

Section 3 RESOURCES

REPRODUCIBLE

▶ Guided Reading Strategy 29.3
▶ Critical Thinking Activity 29: Myanmar and the Tatmadaw
▶ Geography for Life Activity 29: The Shadowed Ground of Cambodia
▶ Map Activity 29: Mainland Southeast Asia: Exports and Urbanization

TECHNOLOGY

▶ One-Stop Planner CD–ROM, Lesson 29.3
▶ CNN Presents Geography: Yesterday and Today, Segment 30: The Highway to Prosperity
▶ Homework Practice Online
▶ HRW Go site

REINFORCEMENT, REVIEW, AND ASSESSMENT

▶ Main Idea Activity 29.3
▶ English Audio Summary 29.3
▶ Spanish Audio Summary 29.3
▶ Section 3 Review, p. 673
▶ Daily Quiz 29.3

Economic and Political Development

The countries of mainland Southeast Asia have different levels of development. Some of them, such as Laos and Cambodia, are very poor and undeveloped. In contrast, Thailand has experienced rapid industrialization. In fact, it is one of the "Asian Tigers." This group of East and Southeast Asian countries includes Malaysia, Singapore, South Korea, and Taiwan. All have experienced rapid economic expansion since the 1980s.

Myanmar is rich in natural resources. However, the country's military government has kept its economy and people isolated. The Burmese have little freedom and remain poor. Both Laos and Vietnam have Communist governments. They have been trying to improve their economies by slowly accepting some features of capitalism. Vietnam has dynamic private businesses, but government policies still limit their growth. Still, Thailand is the region's leading economic power. It is a constitutional monarchy. Industry, agriculture, fishing, mining, and tourism drive Thailand's economy.

A Lao farming family loads hay into a cart.

INTERPRETING THE VISUAL RECORD

These shrimp were harvested from one of Thailand's seafood farms. Many farms release wastewater and fertilizer into coastal mangrove forests. This destroys the trees. **How might damage to mangrove forests affect Thailand's coastal environment?**

VISUAL RECORD ANSWER

disappearance of mangrove-dwelling species, increased coastal erosion

Teach Objective 1

ALL LEVELS: Organize the class into five groups. Assign each group a country of mainland Southeast Asia and provide it with a large rectangular piece of poster board. Have group members refer to the pictures in the unit Fast Facts table to re-create their countries' flags on the paper. Then have each group write down facts about the country's economy and politics on the image of the flag and to conclude whether the country has a developed economy or a developing one. Have each group present its poster to the class. Then lead a discussion about differences in the levels of development and standards of living among the region's countries. **ENGLISH LANGUAGE LEARNERS, COOPERATIVE LEARNING**

Teach Objective 2

LEVEL 1: Copy the following graphic organizer onto the chalkboard, omitting the italicized answers. Have students complete it by identifying the crops produced in each agricultural system. Then lead a discussion about the differences between market-oriented and subsistence agriculture in mainland Southeast Asia. **ENGLISH LANGUAGE LEARNERS**

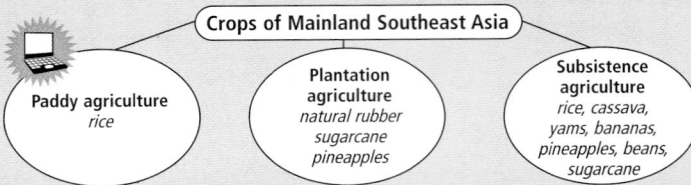

Crops of Mainland Southeast Asia

Paddy agriculture
rice

Plantation agriculture
*natural rubber
sugarcane
pineapples*

Subsistence agriculture
rice, cassava, yams, bananas, pineapples, beans, sugarcane

Linking Past to Present

Hue The city of Hue was the capital of the Nguyen dynasty, the last royal family in Vietnam's history. Its grand architecture stands testament to the city's former role. The Palace of Supreme Harmony—Dien Thai Hoa in Vietnamese—was the home of the imperial court. Access to the palace was controlled by the Ngo Mon, or Midday Gate. The Heavenly Lady Pagoda, Chua Thien Mu, was a national center for the practice of Buddhism. The imperial tombs near Hue were designed to create a feeling of harmony with each other and with their surroundings.

Today Hue is a tourist destination and reminder of Vietnam's past. Local girls still wear traditional Vietnamese clothing, including the tunic and trousers outfit called *ao dai* and conical straw hats called *non bai tho*. Inside the brim of each hat is a poem that can only be read when the hat is held up to the light.

ACTIVITY: Have students locate pictures and information about historic monuments in Hue or other cities in the region to share with the class.

GRAPH ANSWER

subsistence agriculture

Percentage of Population Living in Rural Areas

Cambodia 82%

Thailand 80%

Laos 80%

Vietnam 75%

Myanmar 72%

Rural
Urban

Source: *The World Almanac and Book of Facts 2004*

INTERPRETING THE GRAPH

Each country in Southeast Asia has only one or two large cities, like Hanoi, Vietnam, shown above. However, most people live in small villages. **How do most rural residents make a living?**

All the countries of Southeast Asia are members of the Association of Southeast Asian Nations, or **ASEAN**. ASEAN was founded in 1967. It promotes economic development as well as social and cultural cooperation among its members. The ASEAN countries have treaties of cooperation with each other. In the 1990s they moved to form a free-trade region.

READING CHECK: *Human Systems* Why was ASEAN formed? to promote economic development and cooperation in the region

Settlement and Cities Most people in mainland Southeast Asia live in rural areas and are subsistence farmers. In general, coastal and lowland areas are more densely populated, while highlands are less crowded. The most crowded regions are along the rivers. These regions have both the main rice-growing areas and the largest cities. Because of their location along riverbanks, however, the big cities are subject to occasional flooding.

The region's cities are growing rapidly and display many contrasts. For example, motorcycles speed past pedicabs. Pedicabs are three-wheeled carriages with a separate seat for a driver who pedals. In some cities skyscrapers tower over huts, and street vendors compete with shopping malls.

With more than 7 million people, Bangkok is by far the largest city of mainland Southeast Asia. Thailand's capital is one of the most dynamic cities in Asia, with its modern technology blending with older ways of life. For example, Bangkok has more than 400 *wats*, which are Buddhist temples that also serve as monasteries. The *wats* are among the many attractions that draw thousands of tourists to the city each year.

Ho Chi Minh City, formerly called Saigon, is Vietnam's biggest city. Located in southern Vietnam, it is a thriving industrial center. It is also home to a busy social scene. Almost any type of entertainment, from disco to ancient forms of Vietnamese music, can be found in its nightspots. In the north, Hanoi is Vietnam's capital. Much of the country's history is evident in Hanoi's temples, monuments, and faded French colonial mansions. Hue has often been called the most beautiful city in Vietnam. Its historical attractions include a citadel surrounded by a wall 7 miles (11 km) long. Not far from Hue are tombs of the last Vietnamese royal family to rule the country.

The largest city in Myanmar is Yangon, the capital, also called Rangoon. It is a very green city, with so many trees that some travelers have compared it to a jungle. The city's skyline is dominated by the Shwedagon Pagoda, an enormous golden shrine which is said to hold eight of the Buddha's hairs. Gardens, museums, and a zoo also attract visitors.

The other national capitals are not as large as these cities. Phnom Penh (puh-NAWM PEN), Cambodia, saw most of its people driven out by the country's Communist government in the 1970s. The city's population is slowly increasing. Vientiane (vyen-TYAHN), also known as Viangchan, on the Mekong River, is the most important port for Laos.

READING CHECK: *Human Systems* How are ancient and modern times evident in the cities of mainland Southeast Asia? different forms of transportation, skyscrapers contrasted to *wats*, various types of entertainment, historical sites contrasted to industry

LEVEL 2: Organize the class into small groups. Have each group conduct research to find crops that the plantations of mainland Southeast Asia export to the United States. You may wish to have students bring samples of these products to class. Ask students to recall the products European plantations in the region produced for export. Compare the products exported now with those produced then. Ask students how plantation agriculture has changed since the countries became independent. Then lead a discussion about factors that have led to these changes. **COOPERATIVE LEARNING**

LEVEL 3: Have each student select a crop produced in mainland Southeast Asia and identify both traditional and modern methods of cultivating it. (Some crops will be selected by more than one student.) Then have students create charts comparing the two methods. Tell students that their comparisons should include information about each method's costs, equipment requirements, yields, and environmental effects. Ask students to note where in the region each method of cultivation is more prevalent and why.

Agriculture

Agriculture plays a central role in the economies of all the countries in mainland Southeast Asia. Farmers in the river lowlands grow rice on fertile slopes along the riverbanks. The wet tropical climate lets farmers grow two or three crops each year. Workers move the seedlings by hand to flooded paddy fields. Harvesting is also done by hand. Because paddy agriculture uses so much human labor, it is a form of intensive agriculture.

Paddy farming is particularly successful in areas with good water resources. In these areas farming supports dense populations. For example, the delta regions of the Mekong and Hong (Red) Rivers are very fertile, productive, and crowded. This is true also in the Irrawaddy and Chao Phraya valleys of Myanmar and Thailand. Farmers in mainland Southeast Asia produce enough rice to feed the local population and to export. In fact, Thailand and Vietnam are two of the world's leading exporters of rice. (See the graph.) Many farmers grow vegetables or fruit to sell or trade in local markets. Farmers also raise ducks, water buffalo, pigs, and fish.

The region's many subsistence farmers grow a variety of crops besides rice. These crops include cassava, yams, bananas, pineapples, beans, and sugarcane. In the rugged and forested mountains, poor soils make farming difficult. Farmers must leave their fields after only a few years when soil fertility decreases. Then they move to a new area. This shifting agriculture is common in Laos, northern Thailand, and northern Myanmar. Governments today discourage this practice because it can damage rain forests.

While many people are subsistence farmers, plantation agriculture is also important. For example, Thailand is the world's leading producer of natural rubber. Other plantation crops include sugarcane and pineapples.

Some people grow opium in the rugged mountains of Myanmar, Thailand, and Laos. Opium is a major crop for some poor mountain peoples. This illegal crop is processed to make the deadly and addictive drug heroin. Opium dealing produces riches and risks for corrupt officials, smugglers, and drug dealers. Opium is smuggled out of the region and into the United States, Europe, and other parts of Asia.

✔ **READING CHECK:** *Environment and Society* Where is paddy farming most successful in the region? areas with good water resources, delta regions

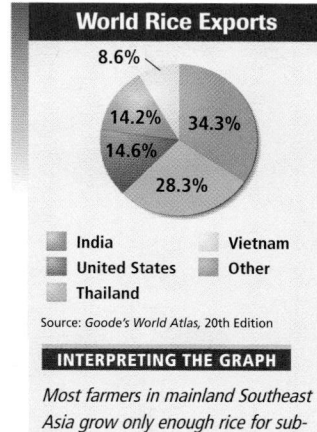

World Rice Exports

8.6%
14.2%   34.3%
14.6%
28.3%

■ India   ■ Vietnam
■ United States ■ Other
■ Thailand

Source: *Goode's World Atlas*, 20th Edition

INTERPRETING THE GRAPH

Most farmers in mainland Southeast Asia grow only enough rice for subsistence, but rice exports have increased in recent years. Most of the region's cultivated land is used to grow this single crop. **Why do you think so much of Southeast Asia's farmland is dedicated to growing rice?**

INTERPRETING THE VISUAL RECORD

Irrigation in this Vietnam rice field is a labor-intensive project. **How does this irrigation system work?**

GRAPH ANSWER

It is both a staple food crop and a profitable cash crop.

VISUAL RECORD ANSWER

The bucket is used to lift water from the pond over the ridge and to pour it onto the crop.

LEVELS 1 AND 2: Before class, prepare large paper cut-outs of Cambodia, Laos, Myanmar, Thailand, and Vietnam. Do not label them. Organize the class into five groups and give each group one of the cutouts. Have each group identify its country and list on the cutout three political or economic issues or challenges it faces. Then have groups assemble their cutouts on the chalkboard to configure a map of mainland Southeast Asia. Each group should report its challenges to the class while placing its cutout. **ENGLISH LANGUAGE LEARNERS, COOPERATIVE LEARNING**

LEVEL 3: Have students bring to class newspaper or magazine articles that discuss an issue or challenge facing mainland Southeast Asia. The articles can discuss the issues presented in this section or may introduce challenges not listed here. Have each student exchange his or her article with another student. Then ask each student to identify the challenge presented in the article he or she received. Review the characteristics of traditional, command, and market economies and ask students to evaluate the roles of these economic systems in the challenges described. Students should also note any solutions that have been proposed to resolve the challenges. Have students write short paragraphs summarizing the possible effects of these proposals on the region's politics and economies. **BLOCK SCHEDULING**

Cooperative Learning

U.S. Relations with Southeast Asia Military rule and Communist governments in mainland Southeast Asia have affected the countries' relationships with the rest of the world. These foreign ties can have a profound influence on a country's political and economic landscapes.

Organize the class into five groups. Assign each group one of the countries covered by this chapter. Have each group conduct research on the history of diplomatic relations between the assigned country and the United States within the last 20 years. Direct the class to address the following issues: Does the United States actively support or oppose the current government? Has the U.S. government taken steps to intervene in the country's domestic or international political situations? Are the two countries trade partners? How have relations between the countries changed in recent years? Have students prepare written reports with their findings and share them with the class.

Vendors at Bangkok's floating markets sell fish, fruit, meat, rice, and vegetables to locals and tourists.

FOCUS ON GEOGRAPHY

Bangkok's *Klongs* The farm and the city meet on the canals of Bangkok, called **klongs**. The people of Bangkok use this network of canals to travel around the city and to transport and sell goods. Some travelers have compared Bangkok to Venice, Italy, another city with a canal system.

In the 1900s Bangkok grew tremendously. New forms of transportation such as the automobile spread. Many *klongs* were filled in and paved to create space for cars and trucks. Today most of the canals have been converted to roads, and cars are more common than boats.

Despite these changes, *klongs* are still a part of life in Bangkok. Boats and river taxis ply the waters of Bangkok's remaining canals, transporting people around the city. As the city's traffic has grown, more people have been using the canals for local travel. Special commuter boats take people to and from work. The canals are also vital for moving fresh fish and produce around the city. In some places *klongs* are still the center of neighborhood life. Traditional houses built on stilts line the water. Families bathe on their front doorsteps. Boats loaded with mangoes, durians, and other produce or goods join together to form floating markets.

✓ **READING CHECK:** *Environment and Society* What role have *klongs* played in the development of Bangkok? used for travel and transport; many filled in and paved over as the city grew and modernized; still used for transport, local travel, center of some neighborhood life

Issues and Challenges

The major challenges in mainland Southeast Asia involve both political and economic issues. Corruption and dictatorial governments have caused serious problems for most of the region's countries. These political problems have in turn led to economic difficulties.

Cambodia has experienced terrible problems since independence. From 1975 to 1979, the country fell under the brutal rule of a Communist group called the Khmer Rouge ("Red Khmer"). Its leaders wanted to change Cambodia into a rural peasant society. To further their plan, the Khmer Rouge forced all citizens to work as field laborers. They emptied the cities, separated families, and targeted educated people for execution. The country's intellectual and skilled worker classes were practically wiped out. Perhaps 1 million Cambodians were killed. Starvation, disease, forced marches, and other hardships killed many more people. An invasion by Vietnamese forces finally ended the terror. However, millions of land mines are left over from decades of fighting. They maim or kill farmers and other civilians who accidentally disturb the hidden explosives. Today Cambodia has a stable elected government, but it remains a poor farming country with a troubled heritage.

A military government took control of Myanmar in 1962. Pro-democracy groups have tried to regain political freedoms but so far have failed. Their leaders and supporters have been jailed and harassed. Minority and rebel groups in the northern mountains have also battled Myanmar's government. The government controls many parts of the economy, such as the rice trade and heavy industry. Because of the political situation, many foreign investors have stayed away. The country remains poor, and living standards have not improved.

Tourism might be a source of income for Myanmar. However, leaders of the democratic movement there favor a boycott on tourism. They say that most of the money spent by tourists goes directly to the military government.

Aung San Suu Kyi is the best known opponent of military rule in Myanmar. Her party, the National League for Democracy (NLD), won control of the country's parliament in 1990, but the military refused to give up power. The government placed Aung San Suu Kyi and other NLD members under house arrest. For her efforts to bring democracy to Myanmar, she received the Nobel Peace Prize in 1991. Although some political prisoners were released from house arrest in 2003, the United States and some Asian countries continue to push for the reform of Myanmar's government.

Both Vietnam and Laos have Communist governments. They have been slowly moving toward market economies, but many challenges remain. Both countries are poor, with most people working in subsistence agriculture. For years Laos was almost cut off from the Western world. This lack of outside influence now draws tourists who want to see a glimpse of traditional Southeast Asian life. In the late 1990s Vietnam and the United States began forming closer relations. American businesses, products, and tourists are now increasing in Vietnam.

Thailand's economy grew by almost 9 percent a year between 1985 and 1995. It was the highest growth rate in the world during that time period. Investment from Japan helped Thailand develop textile, electronics, and automobile assembly plants. However, economic growth slowed in the late 1990s.

Aung San Suu Kyi stands in front of the National League for Democracy flag.

✓ READING CHECK: (*Human Systems*) How has Myanmar's political situation hurt economic development there? The government controls many parts of the economy. Because of the political situation, many foreign investors and tourists have stayed away.

Section 3 Review

Homework Practice Online
Keyword: SW3 HP29

Identify ASEAN

Define wats, klongs

Working with Sketch Maps
On the map of mainland Southeast Asia that you created in Section 2, label Bangkok, Ho Chi Minh City, Hanoi, Yangon (Rangoon), Phnom Penh, and Vientiane. Which city has been compared to Venice, Italy? Why?

Reading for the Main Idea

1. (*Places and Regions*) How is Thailand's economy unique in mainland Southeast Asia?

2. (*Human Systems*) What makes paddy farming a form of intensive agriculture?

Critical Thinking

3. **Analyzing Information** How do rivers influence the location and patterns of movement of products and people in the region? How do you think the rivers might be used to attract business investment?

4. **Making Generalizations** Why do you think the region's countries have worked to form closer economic and trading ties with each other? What might such ties mean for the economic and political futures of the region's countries?

Organizing What You Know

5. Copy the chart below. Use it to describe the major issues and challenges facing the countries of mainland Southeast Asia.

Country	Issues and challenges
Myanmar	
Thailand	
Cambodia	
Laos	
Vietnam	

Setting the Scene

To examine the role of underlying causes in geographical issues, first have students read Case Study: Studying Deforestation. Then review the three main reasons why the rain forests of Southeast Asia are being cut down (*for fuel, to clear land for farming, and for commercial logging*). Ask students to identify the three main results of deforestation (*extinction of plants and animals, flooding and soil erosion, and smoke from burning forests*). Point out that for centuries the people of Southeast Asia have cut trees for fuel and cleared forests for farmland. However, commercial logging is a more recent development.

Building a Case

Lead a discussion about why these ancient activities now threaten the very survival of the forests. (*Possible answer: When the population was small, few trees were being cut, so the forests could recover from the damage done to them. With more people cutting down trees, the forests do not have time to grow back fully before the trees are cut down again. A large and growing population, therefore, is an underlying cause of deforestation.*) Ask students what cause underlies the deforestation that results from logging. (*Possible answers: need for income or desire for profit*) Call on volunteers to suggest ways this underlying cause could be eliminated. (*Possible answer: by developing alternate industries that do not damage the rain forests*)

CASE STUDY

Studying Deforestation

Environment and Society Environmental geographers around the world are concerned about deforestation in Southeast Asia. The region has about one fourth of the world's remaining tropical rain forests. These forests are being destroyed at an alarming rate. Some observers believe that most of the region's tropical rain forests could be gone in just 20 to 30 years.

Geographers bring a spatial and environmental perspective to studying this problem. They begin by asking geographic questions from a spatial perspective. Where is deforestation occurring? What are the patterns and causes of deforestation? How is deforestation in Southeast Asia related to economic development and world trade? Geographers also ask questions from an environmental perspective. How does deforestation affect the environment? How does it affect people?

Patterns and Causes

Geographers use satellite images and other tools to study the problem. They have found that Southeast Asia has one of the highest deforestation rates in the world. Between 1990 and 1995, about 56,255 square miles (145,700 sq km) of rain forest were lost. This figure represents an area about the size of Florida. The countries with the highest deforestation rates included the Philippines, Thailand, and Malaysia. (See the map.)

INTERPRETING THE MAP *Which country in Southeast Asia had the highest rate of deforestation? Which country has the bulk of the region's remaining rain forests?*

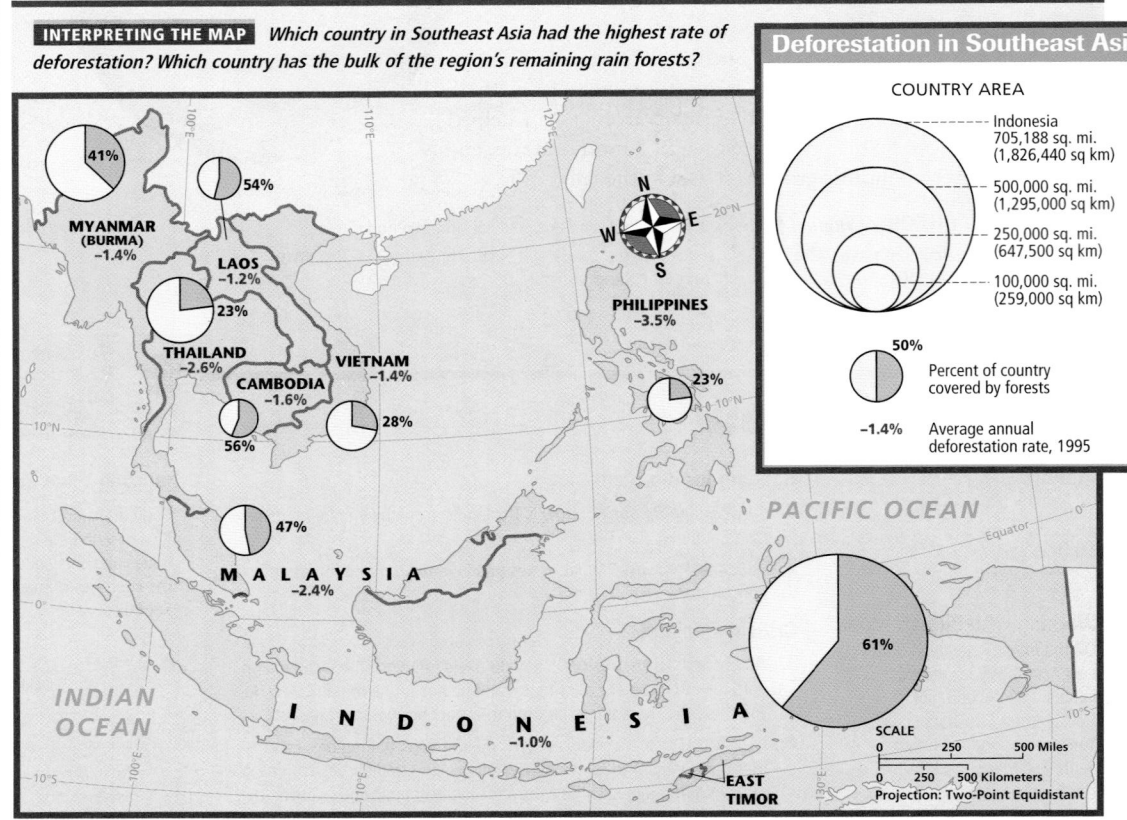

Drawing Conclusions

Tell students about the transmigrant program sponsored by Indonesia's government. This program sends families from over-crowded Java to less densely populated islands, including Sumatra, Borneo, Sulawesi, and New Guinea. The program has been criticized because the transmigrants often try to carve farms out of the rain forest. They also displace indigenous peoples. Ask students if Indonesia should continue this program or find other ways to relieve overcrowding. Call on a volunteer to summarize the discussion. *(Possible answer: A government policy can have wide-ranging effects on environmental issues.)*

Going Further: Thinking Critically

Organize the class into small groups to conduct their own case studies. Have each group research a region, either in the United States or in another country, experiencing deforestation or some other type of environmental damage. Ask students to concentrate on these questions:

• What are the causes of the environmental damage?
• What reasons underlie the more direct causes?
• Are there government policies that make the problem worse?
• Do other government programs work to solve the problem?
• How might the underlying causes of the problem be eased?

Have each group present its case study to the class.

Left: In 1997, airplanes dropped water on the Indonesian fires. Right: A man looks out at the burning rain forest. For many years his people had used this part of Indonesia for hunting and for gathering medicinal plants.

In Southeast Asia there are many causes of deforestation. For example, people in many areas cut trees and use the wood for fuel. In addition, trees are cut and burned to clear fields for farming. As populations in Southeast Asia continue to grow rapidly, more forested areas are being used for farming. After nutrients in the soil are used up, farmers move on to other areas to grow crops. This pattern of shifting agriculture clears large forest areas.

Logging is another cause of deforestation. Valuable hardwood and wood-based products are important exports for countries like Malaysia, Indonesia, Myanmar, Cambodia, and Laos. Most logging is done by large corporations. Developed countries such as Japan import the timber. To limit deforestation, some Southeast Asian countries have enacted total or partial bans on logging. However, in many other countries, income from timber exports is used to develop national economies.

Environmental Effects of Deforestation

Geographers also study the effect of deforestation on people and environments. Some believe the most serious result of deforestation is the extinction of plant and animal species. A great diversity of plant and animal life thrives in the rain forests of Southeast Asia. For example, Indonesia may have about 11 percent of the world's plant species. About 12 percent of all mammal species and 17 percent of all bird species may also live there. Deforestation destroys the habitats of many of these rare plants and animals.

Deforestation also causes flooding and soil erosion, which can more directly affect humans. When areas are stripped of their forest cover, the soil is exposed directly to water's erosive power. As a result, the clearing of forests has led to erosion and flooding in many Southeast Asian countries, such as Thailand.

Smoke created by burning forests has been another problem. In 1997 and 1998, for example, a severe drought allowed fires to burn out of control, particularly in Indonesia. Some of the fires had been set to clear land. Smoke from these huge forest fires endangered people's health, shut down airports and schools, and disrupted the tourist industry.

Future Outlook

As you can see, deforestation in Southeast Asia is a very complicated problem. It is tied to the region's economies, world trade patterns, and people's need to use the resources around them to survive. To solve this problem, geographers and other scientists must learn more about the causes, rates, patterns, and environmental effects of deforestation.

Applying What You Know

1. Summarizing How is deforestation linked to economic development, population growth, and environmental change in Southeast Asia? What have some governments done to slow deforestation?

2. Problem Solving What do you think should be done to halt the destruction of Southeast Asia's tropical rain forests? Work with a partner to develop a proposed solution and present it to your class in a report.

TECHNOLOGY

▶ Chapter 29 Test Generator (on the One-Stop Planner)
▶ Global Skill Builder CD–ROM
▶ HRW Go site

REINFORCEMENT, REVIEW, AND ASSESSMENT

▶ Chapter 29 Review, pp. 676–77
▶ Chapter 29 Tutorial for Students, Parents, Mentors, and Peers
▶ Chapter 29 Test (form A or B)
▶ Alternative Assessment Handbook

▶ Chapter 29 Test for English Language Learners and Special-Needs Students
▶ Unit 9 Test
▶ Unit 9 Test for English Language Learners and Special-Needs Students

Assess

Have students complete a Chapter 29 Test.

Reteach

Organize the class into small groups. Have each group design and complete a series of graphic organizers that answer the Read to Discover questions for the three sections. Then have groups present their graphic organizers to the class. **ENGLISH LANGUAGE LEARNERS, COOPERATIVE LEARNING**

CHAPTER 29 Review Answers

Thinking Critically

1. Possible answer: The region's rugged, isolated interior is still largely unknown by outsiders. **(NGS 4)**

2. on the banks of major rivers or in river deltas; water and soil resources for agriculture and transportation **(NGS 12)**

3. Plantations grow these and other crops for sale, generally for export. Subsistence farmers grow these and other crops for their own consumption. Subsistence farmers are probably poorer and consume what they produce. They usually live in rural areas or villages. Plantation workers are probably also rural, but they probably buy the food they need for personal consumption. **(NGS 9)**

Using the Geographer's Tools

1. Students should put Thailand at the top of their rankings and note that it is politically stable. Countries with Communist or military governments are generally poor.

2. Maps will vary, but listed features should be labeled in their approximate locations. Areas with fertile soils generally have large populations.

3. Possible answer: Its location as a buffer between French and British colonies may have helped it remain independent.

CHAPTER 29 Review

Building Vocabulary

On a separate sheet of paper, explain the following terms by using them correctly in sentences.

arboreal · *wats*
domino theory · *klongs*
ASEAN

Locating Key Places

On a separate sheet of paper, match the letters on the map with their correct labels.

Malay Peninsula · Irrawaddy River · Mekong River
Indochina Peninsula · Chao Phraya River · Hong (Red) River
Khorat Plateau

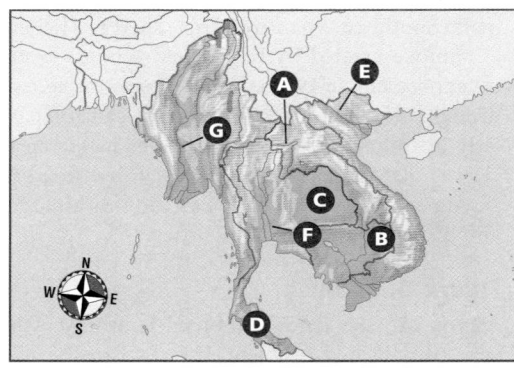

Understanding the Main Ideas

Section 1

1. (Places and Regions) What are the major climates in mainland Southeast Asia, and where are they found?

2. (Physical Systems) What type of soils have made the region's river valleys rich farming areas?

Section 2

3. (Places and Regions) Which of the region's countries was never a European colony? What two European countries had colonies in the region?

4. (Places and Regions) What three major language families are found in the region? Who are some of the primary speakers of the languages in each group?

Section 3

5. (Human Systems) In what areas do most of the region's people live? What are the large cities of the region like?

Thinking Critically

1. **Drawing Inferences** Why do you think previously unknown animal species are still being discovered in the region?

2. **Analyzing Information** Where are the region's major cities located? What advantages might these locations offer?

3. **Comparing** Some crops, such as sugarcane and pineapples, are grown by both plantation owners and subsistence farmers. How are these two types of farming different in the region? How might the lives of subsistence farmers and plantation workers differ?

Using the Geographer's Tools

1. **Analyzing Statistics** Use the unit Fast Facts and Comparing Standard of Living tables to rank the region's countries in order, from most to least developed. Then note the political situation in the region's countries. How does the political geography affect the countries' economic geography?

2. **Creating Maps** Create a map of the region that shows rivers, deltas, mountains, plateaus, and plains. Then shade areas that have fertile soils and dense populations. Write a short paragraph noting the connections between the two.

3. **Creating Maps** Prepare a map showing past European colonies in the region. How might Thailand's (Siam's) relative location have helped it to remain independent?

Writing about Geography

Write a newspaper article about life in Myanmar today. Be sure to discuss the country's recent political, economic, social, and environmental changes. How are these areas related? How do they affect life in Myanmar? Compare the situation in Myanmar to the United States. How are they similar? How are they different?

SKILL BUILDING

Geography for Life

Using Research Skills

(Environment and Society) Plan a research project on deforestation in mainland Southeast Asia. First, review the Case Study: Studying Deforestation. Then create three questions that you would like to research. For example, you might want to know more about how deforestation has affected a particular country. Use newspaper articles, documentaries, or statistics to research your questions. Then write a short report that answers your three questions.

1. Have students conduct research on Buddhist myths and traditional Thai crafts. Then have each student make a Thai mask to represent a character from Buddhist mythology. Have students explain their masks to the class. Photograph student work for portfolios.

2. Have students design a series of commemorative postage stamps for one of the mainland Southeast Asian countries. The stamps might depict wildlife, plants, natural resources, famous people, historical events, or customs. Encourage students to use library or Internet resources as necessary for additional information. Place the stamp designs in student portfolios.

Food Festival

Students are probably familiar with soy sauce, but few may know about the fish sauces of mainland Southeast Asia. These pungent strong-flavored condiments are made from the liquid poured off of salted fermented fish. The ancient Romans enjoyed a similar preparation, called *garum*. In Thailand, fish sauce is called *nam pla*. In Vietnam it is *nuoc mam*. Asian markets carry a wide variety of fish sauces. Bring a bottle of one of these products to class, along with boiled rice. If students like the fish sauce, you may want to have them prepare recipes using fish sauces and to bring samples to share.

Building Social Studies Skills

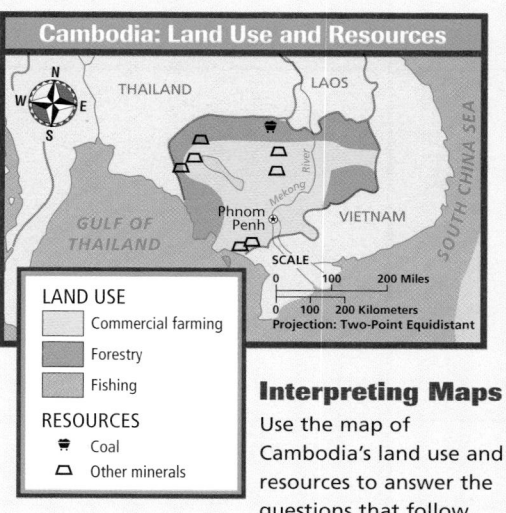

Cambodia: Land Use and Resources

LAND USE
- Commercial farming
- Forestry
- Fishing

RESOURCES
- Coal
- Other minerals

Interpreting Maps

Use the map of Cambodia's land use and resources to answer the questions that follow.

1. Which of the following can be concluded from the information shown on the map?
   a. The country has no major manufacturing and trade centers.
   b. Coal is found only along Cambodia's southern coast.
   c. Commercial farming is common in central areas of the country.
   d. Few forested areas are found in Cambodia.

2. Why do you think commercial farming is located where it is?

Analyzing Primary Sources

Karin Muller spent seven months hitchhiking through Southeast Asia. Read her observations about Vietnam and answer the questions.

> "Rice in Vietnam . . . is not just that fluffy white stuff in a box that you dig out once a month as an accompaniment to your favorite lemon chicken. Rice is life. Almost no meal goes by without it . . . The poor mountain Vietnamese say "there is no money" and "there is no rice" interchangeably because if they had money, they would use it to buy rice. In many areas rice is money. It is the traditional currency."

3. Based on what you have read, which of the following is true?
   a. Rice is Vietnam's least important food.
   b. Most of Vietnam's rice comes from mountain areas.
   c. Vietnamese eat rice mostly at special occasions.
   d. Rice is a very important part of the Vietnamese diet.

4. Why would a poor mountain Vietnamese person equate rice to money?

Alternative Assessment

PORTFOLIO ACTIVITY

Learning about Your Local Geography

Group Project: Research

Plan, organize, and complete a group research project on native arboreal animals in your area or state. First, use your local library to learn about some arboreal animals common near your community. Then have each individual in your group select one animal and learn more about it. In which tree species do these animals live? How are they adapted to life in the trees? What is their range of habitat in the United States and the rest of the world? To which other animals are they related? You might want to create maps and diagrams to illustrate the answers you find to these research questions. When you are done, share your results with the other group members.

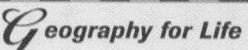

internet connect

Internet Activity: go.hrw.com
KEYWORD: SW3 GT29

Choose a topic on mainland Southeast Asia to:
- learn about the architecture of ancient and modern mainland Southeast Asia.
- create a brochure on Laos and Vietnam since the Vietnam War.
- understand the causes and effects of deforestation in mainland Southeast Asia.

Writing

Student articles should describe specific changes that have resulted from the internal conflict in Myanmar. Comparisons with the United States should note many differences between the two countries. Use Rubric 42, Writing to Inform, to evaluate student work.

*G*eography for Life

Student questions should be designed to gather meaningful information. Reports should reflect the results of student research. Use Rubric 1, Acquiring Information, to evaluate student work.

Social Studies Skills

1. c

2. Possible answer: Commercial farming is located along the Mekong River and its tributaries. The river and tributaries provide the water needed for large-scale farming.

3. d

4. Because rice is so important, he or she would use whatever money is available to buy it.

PORTFOLIO ACTIVITY

Student projects will vary based on the arboreal animals found in or near your community. Reports and associated materials should accurately reflect student research. Use Rubric 14, Group Activity, to evaluate student work.

CHAPTER 30 Island Southeast Asia

CHAPTER RESOURCE MANAGER

	Objectives	Pacing Guide	Reproducible Resources
SECTION 1 **Natural Environments** (pp. 679–82)	• Describe the landforms of island Southeast Asia and some physical processes that shaped them. • Identify the region's climates, biomes, and natural resources.	**Regular** 1 day **Block Scheduling** .5 day *Block Scheduling Handbook, Chapter 30*	**RS** Guided Reading Strategy 30.1 **RS** Graphic Organizer 30 **IC** Lab Activity for Geography and Earth Science, Demonstration 7: How's Your Lava Life? **SM** Critical Thinking Activity 30 **SM** Map Activity 30: Diversity in the Philippines
SECTION 2 **History and Culture** (pp. 684–89)	• Identify the early peoples who migrated to the region. • Analyze how colonialism affected island Southeast Asia's history. • Describe what the region's people and culture are like today.	**Regular** 1.5 days **Block Scheduling** 1 day *Block Scheduling Handbook, Chapter 30*	**RS** Guided Reading Strategy 30.2 **E** Cultures of the World Activity, Region 7 **SM** Geography for Life Activity 30: Sacred Buildings
SECTION 3 **The Region Today** (pp. 690–93)	• Analyze how island Southeast Asia's economy has changed in recent decades. • Describe the region's cities. • Discuss the challenges the people of the region face.	**Regular** 1 day **Block Scheduling** .5 day *Block Scheduling Handbook, Chapter 30*	**RS** Guided Reading Strategy 30.3 **PS** Readings in World Geography, History, and Culture 73

Chapter Resource Key

PS Primary Sources
RS Reading Support
IC Interdisciplinary Connections
E Enrichment
SM Skills Mastery

A Assessment
REV Review
ELL Reinforcement and English Language Learners
Transparencies

CD-ROM
Video
 Internet
 Holt Presentation Maker Using Microsoft® PowerPoint®

One-Stop Planner CD–ROM

See the *One-Stop Planner* for a complete list of additional resources for students and teachers.

One-Stop Planner CD–ROM

It's easy to plan lessons, select resources, and print out materials for your students when you use the **One-Stop Planner CD–ROM with Test Generator**.

Technology Resources

- One-Stop Planner CD–ROM, Lesson 30.1
- Geography and Cultures Visual Resources 49–53
- Homework Practice Online
- HRW Go site

- One-Stop Planner CD–ROM, Lesson 30.2
- **CNN** Presents Geography: Yesterday and Today, Segment 31: The Borobodur Temple of Indonesia
- Homework Practice Online
- HRW Go site

- One-Stop Planner CD–ROM, Lesson 30.3
- *ARGWorld* CD–ROM
- Homework Practice Online
- HRW Go site

Reinforcement, Review, and Assessment

ELL Main Idea Activity 30.1
ELL English Audio Summary 30.1
ELL Spanish Audio Summary 30.1
REV Section 1 Review, p. 682
A Daily Quiz 30.1

ELL Main Idea Activity 30.2
ELL English Audio Summary 30.2
ELL Spanish Audio Summary 30.2
REV Section 2 Review, p. 689
A Daily Quiz 30.2

ELL Main Idea Activity 30.3
ELL English Audio Summary 30.3
ELL Spanish Audio Summary 30.3
REV Section 3 Review, p. 693
A Daily Quiz 30.3

internet connect

HRW ONLINE RESOURCES

GO TO: go.hrw.com
Then type in a keyword.

TEACHER HOME PAGE
　KEYWORD: SW3 Teacher

CHAPTER INTERNET ACTIVITIES
　KEYWORD: SW3 GT30
　Choose an activity to:
　• test knowledge with an interactive map.
　• create a poster about rain forests.
　• analyze tectonic forces that cause volcanoes and earthquakes.

CHAPTER ENRICHMENT LINKS
　KEYWORD: SW3 CH30

CHAPTER MAPS
　KEYWORD: SW3 MAPS30

ONLINE ASSESSMENT
　Homework Practice
　KEYWORD: SW3 HP30
　Standardized Test Prep
　KEYWORD: SW3 STP30
　Rubrics
　KEYWORD: SS Rubrics

COUNTRY INFORMATION
　KEYWORD: SW3 Almanac

CONTENT UPDATES
　KEYWORD: SS Content Updates

HOLT PRESENTATION MAKER
　KEYWORD: SW3 PPT30

ONLINE READING SUPPORT
　KEYWORD: SS Strategies

CURRENT EVENTS
　KEYWORD: S3 Current Events

Meeting Individual Needs

Ability Levels

Level 1 Basic-level activities designed for all students encountering new material

Level 2 Intermediate-level activities designed for average students

Level 3 Challenging activities designed for honors and gifted-and-talented students

English Language Learners Activities that address the needs of students with Limited English Proficiency

Chapter Review and Assessment

 Chapter 30 Test Generator (on the One-Stop Planner)

 Global Skill Builder CD–ROM

 HRW Go site

REV Chapter 30 Review, pp. 696–97

REV Chapter 30 Tutorial for Students, Parents, Mentors, and Peers

A Chapter 30 Test (form A or B)

A Alternative Assessment Handbook

A Chapter 30 Test for English Language Learners and Special-Needs Students

Launch into Learning

In advance, fill small bags or jars with black pepper, cloves, mace, and nutmeg. Have students pass around and smell the spices. Ask students what foods come to mind with the various scents. (*Example: Mace is commonly used to flavor doughnuts.*) Tell students that Europeans were first attracted to island Southeast Asia by the spices they could buy there. Then tell students that in this chapter they will learn more about how the region's natural environments have influenced its development.

Using the Physical-Political Map

Have students examine the map on the opposite page. Ask them to identify physical features that are common to several of the countries. (*Examples: mountains and islands*) Have students note unusual characteristics of the physical geography and the size of these countries. (*Possible answers: Malaysia is split between a peninsula and part of an island. Singapore, Brunei and East Timor are very small, while Indonesia is spread over a large area.*)

Why We Should Know More

You may want to reinforce students' interest in island Southeast Asia by pointing out the following concepts:

▶ Some of the region's countries have close trade links to the United States.

▶ Many U.S. citizens have immigrated from island Southeast Asia.

▶ Environmental issues in the region, such as widespread fires that ravaged Indonesia in 1997, can affect countries around the world.

▶ Ethnic and political unrest in the region can affect people elsewhere. For example, disruption of major shipping lanes would have a global impact.

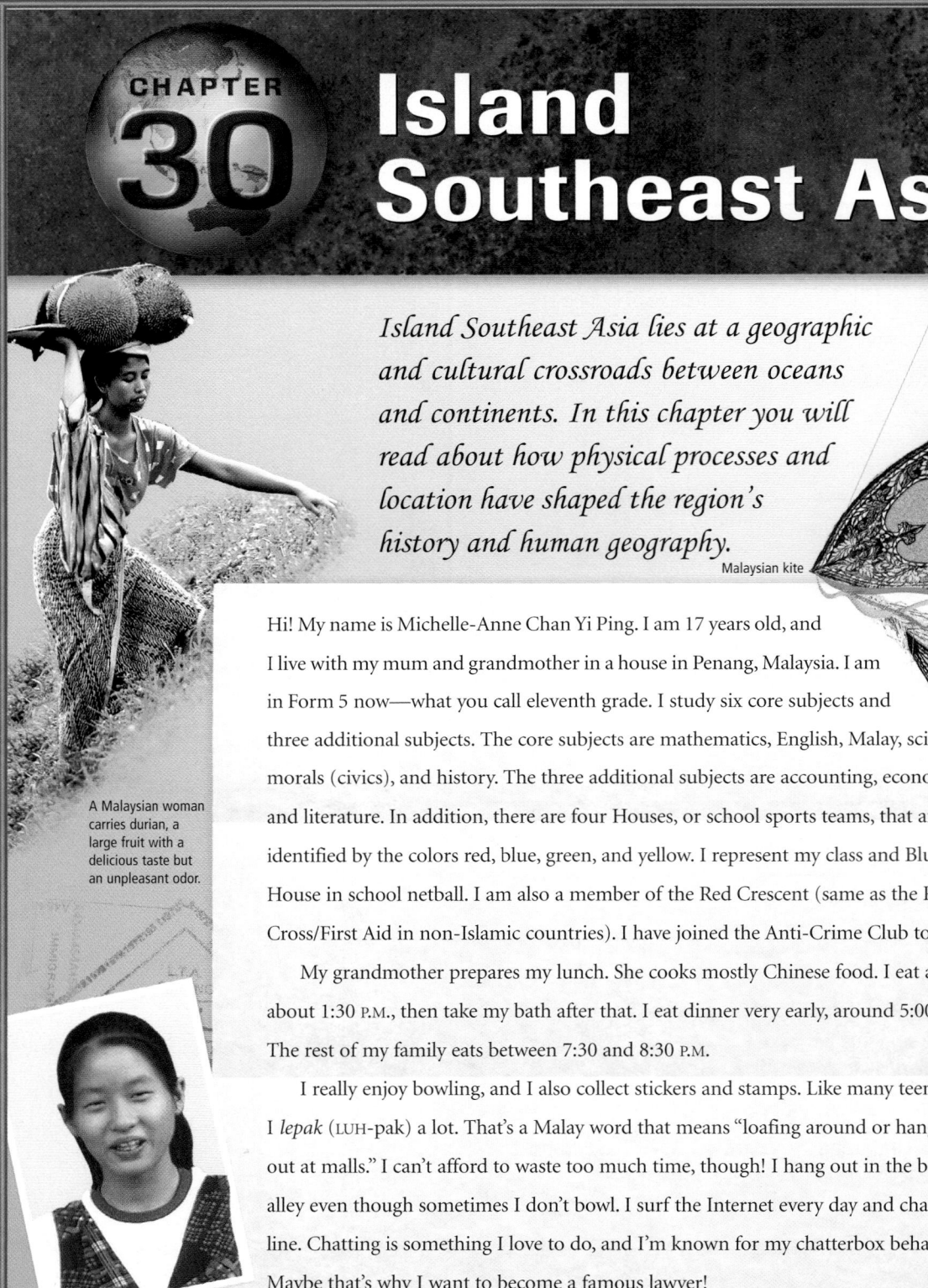

CHAPTER 30 Island Southeast Asia

Island Southeast Asia lies at a geographic and cultural crossroads between oceans and continents. In this chapter you will read about how physical processes and location have shaped the region's history and human geography.

Malaysian kite

A Malaysian woman carries durian, a large fruit with a delicious taste but an unpleasant odor.

Hi! My name is Michelle-Anne Chan Yi Ping. I am 17 years old, and I live with my mum and grandmother in a house in Penang, Malaysia. I am in Form 5 now—what you call eleventh grade. I study six core subjects and three additional subjects. The core subjects are mathematics, English, Malay, science, morals (civics), and history. The three additional subjects are accounting, economics, and literature. In addition, there are four Houses, or school sports teams, that are identified by the colors red, blue, green, and yellow. I represent my class and Blue House in school netball. I am also a member of the Red Crescent (same as the Red Cross/First Aid in non-Islamic countries). I have joined the Anti-Crime Club too.

My grandmother prepares my lunch. She cooks mostly Chinese food. I eat at about 1:30 P.M., then take my bath after that. I eat dinner very early, around 5:00 P.M. The rest of my family eats between 7:30 and 8:30 P.M.

I really enjoy bowling, and I also collect stickers and stamps. Like many teenagers, I *lepak* (LUH-pak) a lot. That's a Malay word that means "loafing around or hanging out at malls." I can't afford to waste too much time, though! I hang out in the bowling alley even though sometimes I don't bowl. I surf the Internet every day and chat online. Chatting is something I love to do, and I'm known for my chatterbox behavior. Maybe that's why I want to become a famous lawyer!

 **LET'S GET STARTED**

Copy the following instructions onto the chalkboard: *Use the physical-political map to estimate the region's size. Where would you draw lines to measure the greatest distances?* Discuss responses. *(Possible answer: It is about 3,500 miles from the northern tip of Sumatra to southeastern Irian Jaya).* Point out how these great distances compare to the contiguous United States. Tell students that in Section 1 they will learn more about the physical geography of the islands within this vast area.

Building Vocabulary

Write the key terms on the chalkboard and call on volunteers to read the definitions from the text or glossary. Ask which terms relate to landforms and climates (**archipelago**, **lahars**) and which relates to plants and animals (**endemic species**). Point out that the word *lahars* is from a Javanese word. *Archipelago* is based on a Greek word, *pelagos*, which means "sea." The root can be found in other words that relate to the sea, such as *pelagic*, which is used to describe birds that spend much of their lives on the open ocean.

Section 1

Natural Environments

READ TO DISCOVER

1. What landforms are found in island Southeast Asia, and what are some physical processes that have shaped them?

2. What climates, biomes, and natural resources does the region have?

WHY IT MATTERS

Earthquakes and volcanoes often shake the countries of island Southeast Asia. Heavy rains after a volcanic eruption can increase the amount of damage. Use **CNNfyi.com** or other **current events** sources to learn about the damage that these events can cause.

DEFINE

archipelago
lahars
endemic species

LOCATE

Malay Peninsula	Java Sea
Malay Archipelago	South China Sea
Sumatra	Timor Sea
Irian Jaya	Strait of Malacca
New Guinea	Java
Borneo	Spratly Islands

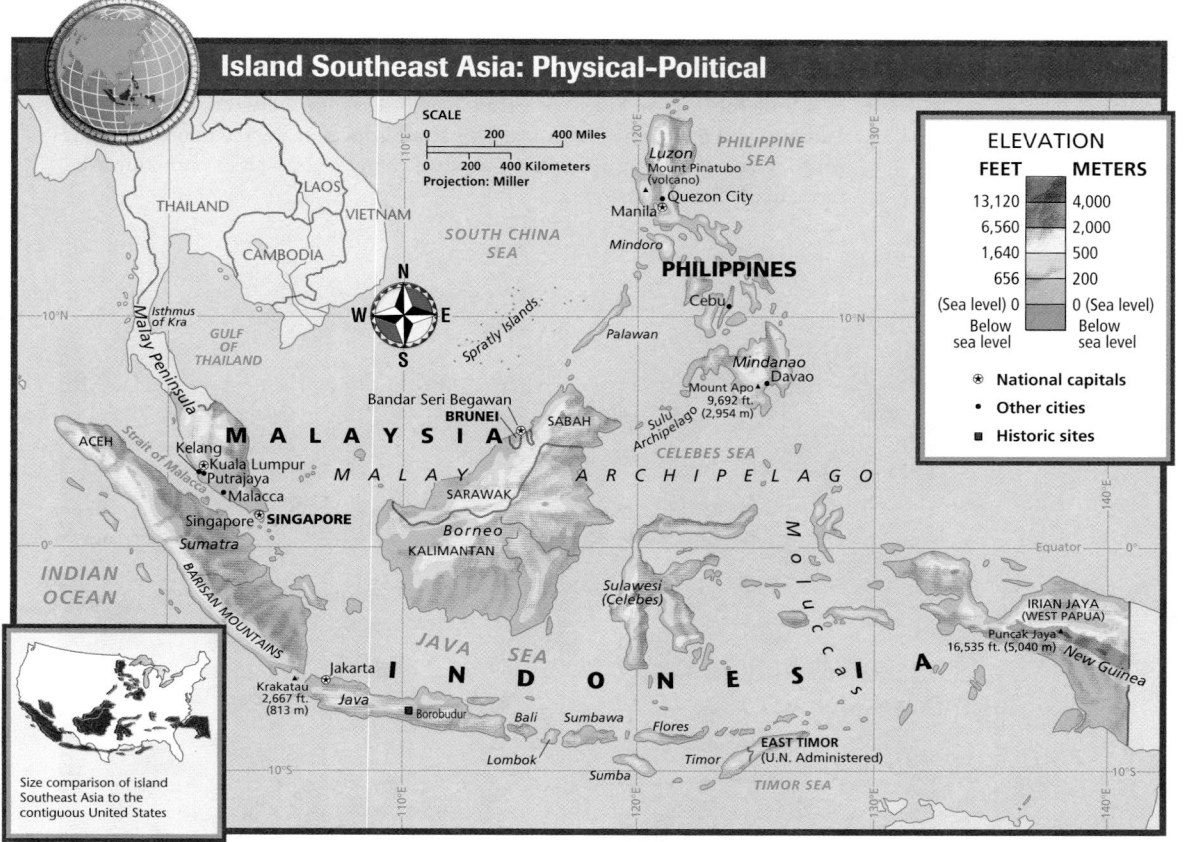

Size comparison of island Southeast Asia to the contiguous United States

Teach Objective 1

ALL LEVELS: Pair students and have pairs use Section 1 and the map of tectonic plates in Chapter 4 to identify the areas of island Southeast Asia included in the Indo-Australian Plate, Eurasian Plate, and Philippines Plate. Ask them to describe areas that have frequent volcanic activity and provide an example. *(Possible answer: areas along the edges of the plates; Philippines)* Ask them also to describe those areas that are relatively stable and provide an example. *(Possible answer: areas far from the plate edges; Borneo)* Then have each pair write a paragraph to describe how tectonic activity has affected the region's landforms. **COOPERATIVE LEARNING**

Teach Objective 2

LEVEL 1: Copy the following graphic organizer onto the chalkboard, omitting the italicized answers. Have students complete it. Then lead a discussion about the relationships among the region's climates, vegetation, wildlife, and natural resources. **ENGLISH LANGUAGE LEARNERS**

internet connect

GO TO: go.hrw.com
KEYWORD: SW3 CH30
FOR: Web sites about island Southeast Asia

INTERPRETING THE VISUAL RECORD

Mount Pinatubo, a volcano on the island of Luzon in the Philippines, erupted violently in June, 1991. It was the largest eruption anywhere in more than 50 years. Because scientists had predicted the event, rescuers were able to save thousands of lives. This photo was taken between eruptions. **What evidence of erosion do you see in the photo?**

Landforms

The Malay Peninsula and the Malay Archipelago (ahr-kuh-PEH-luh-goh) stretch from the Southeast Asian mainland almost to Australia. An **archipelago** is a large group of islands. We call this region island Southeast Asia. Look at the chapter map. The region sits between the Pacific and Indian Oceans. To the north and west lies the Asian mainland. Australia and the Pacific Ocean lie to the south and east.

The region's countries include Brunei, East Timor, Indonesia, Malaysia, the Philippines, and Singapore. East Timor, a former Portuguese and Indonesian territory, declared independence in 1999.

The Islands Island Southeast Asia contains more than 20,000 islands. These islands make up the Malay Archipelago. They extend along the equator from Sumatra to Irian Jaya (also called West Papua) on the island of New Guinea. Other major islands include Borneo and the islands of the Philippines. Some of the region's islands are among the largest in the world. Only the North Atlantic island of Greenland is larger than New Guinea. Borneo is the world's third-largest island. Malaysia's Malay Peninsula is almost an island. The Isthmus of Kra connects it to the Asian mainland.

The region's larger islands have high mountains. The highest peaks are in Irian Jaya on New Guinea. Peaks there rise to more than 16,000 feet (5,000 m). Some mountains are tall enough to have glaciers and snowfields.

Many seas and narrow straits separate the islands and the Malay Peninsula from one another. Some of these bodies of water are the Java Sea, South China Sea, and the Timor Sea. The Strait of Malacca lies between the island of Sumatra and the Malay Peninsula. This strait is located along a major shipping route. Trade along this route has benefited many nearby cities and countries, particularly Singapore.

Climates of Island Southeast Asia			
	Tropical humid	Tropical wet and dry	Highland
Countries or regions	Brunei, Malaysia, Philippines, Singapore, most of Indonesia	eastern islands of Indonesia, southern Irian Jaya	central Borneo, central Irian Jaya
Factors that influence the climate	tropical location	tropical location, monsoons	high elevation

LEVELS 2 AND 3: Have students use the chart in the previous activity and other information in Section 1 to write weather forecasts for wildlife conservation specialists planning expeditions into various areas of island Southeast Asia. You may want to have them include what the scientists expect to find on their expeditions. Then have students use Section 1 and the Land Use and Resources Map in the unit atlas to list the major natural resources for each of the five countries in island Southeast Asia. (*Brunei—oil, natural gas; Indonesia—rubber, oil, natural gas, gold, coal, other minerals, seafood; Malaysia—rubber, oil, natural gas, gold, other minerals, seafood; Philippines—oil, natural gas, gold, silver, other minerals. Singapore does not have significant natural resources.*) Then call on individual students to answer the following questions: What are some of the other minerals that are commonly found in the region? (*tin, copper*) Considering their island location, what resource can we assume all the countries use to some extent? (*seafood*)

Tectonic Processes

Tectonic activity is one of the physical processes that have shaped island Southeast Asia. The region is located in one of the world's most geologically active areas. It lies along and sits between several tectonic plate boundaries. Earthquakes and volcanic eruptions are common in areas near those boundaries. Volcanic islands have formed along deep ocean trenches at the edges of the Indian and Pacific Oceans. The Indonesian island of Java, for example, has about 13 active volcanoes.

Dangerous volcanic mudflows, called **lahars** (LAH-hahrz), sometimes rush down steep volcanic slopes. Lahars can bury river valleys and towns. However, volcanoes also provide Indonesia and the Philippines with rich soils. Undersea earthquakes and volcanoes make tsunamis an occasional threat to coastal areas in Indonesia and the Philippines.

A more stable area of Earth's crust lies under the shallow sea between the Malay Peninsula and Borneo. No plate boundaries exist in this area. As a result, Borneo, the Malay Peninsula, and eastern Sumatra do not have active volcanoes.

✓ **READING CHECK:** (**Physical Systems**) What physical processes created some of the region's islands? tectonic processes, particularly volcanic activity

Climates, Biomes, and Natural Resources

All of island Southeast Asia is located in the tropics. This tropical location strongly influences the islands' climates and biomes.

Climates Much of the region has a tropical humid climate. The weather is hot and damp all year. Rainfall is heavy most months—80 to 100 inches (200 to 250 cm) per year are normal. Some mountain areas receive even more rain. Afternoon thunderstorms are common.

The Philippines lie in the path of typhoons that sweep in from the Pacific Ocean. These huge storms can cause terrible destruction. Typhoons bring heavy rain and powerful winds. They also cause sea levels to reach dangerous heights. Most typhoons strike between August and October when ocean temperatures are warmest. Sometimes the storms are so severe that they cause long-lasting damage to the country's farmland and economy.

A few of the region's eastern islands have a tropical wet and dry climate. This climate region stretches from eastern Java to southern Irian Jaya. These areas have both very rainy and dry seasons. This wet-and-dry pattern results from the influence of the monsoon flow. Because this area is south of the equator, from November through March humid monsoon winds blow from the Indian and Pacific Oceans. The wet season occurs at this time. The dry season falls between June and September. At that time, monsoon winds blow from the dry interior of Australia.

Where would you expect to find the coolest climates in this tropical region? The only areas without tropical climates are found in the higher elevations of the region's mountains. Inland Borneo and Irian Jaya have highland climates. Resorts in the highlands offer an escape from the constant heat of the lower elevations.

Our Amazing Planet

The volcano Krakatau (kra-kuh-TOW) is located between Sumatra and Java. Tsunamis caused by Krakatau's eruption in 1883 rose as high as 100 feet (30 m) and killed about 36,000 people.

Climate Graph for Singapore

Paya Lebar, Singapore (1°N, 104°E)
Tropical Humid Climate

(bar graph showing Average Temperature in °F and Average Rainfall in In. by month J F M A M J J A S O N D)

INTERPRETING THE GRAPH *As you see in this climate graph, the wettest months in Singapore are November, December, and January. However, temperatures stay warm throughout the year.* **What is the widest variance in the monthly rainfall totals? How does Singapore's location affect temperature?**

GRAPH ANSWER

about five and one half inches; warm temperatures created by location in the tropics

681

Plants and Animals   The region's tropical climates support ancient tropical rain forests. Indonesia alone has about 10 percent of the world's remaining tropical rain forests. Thick mangrove forests grow in coastal areas. Mangroves are trees or shrubs that have exposed supporting roots. Birds, fish, and other small marine animals live in these tidal areas.

Today most of the forests and animals are at risk. Rain forests are being cut down at a fast rate. Loggers and developers are harvesting valuable hardwoods and clearing land for buildings. Farmers clear land for crops. Many indigenous peoples in Borneo, Irian Jaya, and the Philippines may also lose their forest homelands and traditional ways of life. Some countries have banned logging or created national parks to protect their tropical rain forests.

Tropical rain forests have the greatest variety of plant and animal life on Earth. In fact, island Southeast Asia has one of Earth's highest levels of biodiversity. There are many **endemic species** in the region. Endemic species are those native to a certain area. Southeast Asia's endemic species include the Komodo dragon, Javan rhinoceros, and orangutan. As people change the animals' habitats—by farming, logging, or setting fires—many of these animals are becoming endangered.

INTERPRETING THE VISUAL RECORD

With a length of about 10 feet (3 m) and weighing some 300 pounds (135 kg), the rare Komodo dragon is the biggest lizard on Earth. The Komodo tastes the air with its foot-long forked yellow tongue to locate prey. Komodos live only on a few small Indonesian islands. **What factors might threaten the Komodo dragon's survival?**

Natural Resources   Tropical rain forests are among island Southeast Asia's many valuable natural resources. The region also has rich fisheries and volcanic soils that are good for farming. Rubber tree plantations are important in some countries, particularly Malaysia and Indonesia. The region produces many minerals, including copper, gold, and tin. Some countries, particularly Brunei and Indonesia, produce oil and natural gas. The tiny country of Brunei sits on top of a major oil-producing field.

Scientists believe that large oil and natural gas deposits lie near the Spratly Islands, southwest of the Philippines. China, Malaysia, the Philippines, Taiwan, and Vietnam all claim these uninhabited islands because they have valuable natural resources.

✓ **READING CHECK:**   *Physical Systems*   What climates and biomes are distributed throughout the region? tropical humid climate, some tropical wet and dry and highland climate areas; tropical rain forest and mangrove forests

Section 1 Review

go.hrw.com
Homework Practice Online
Keyword: SW3 HP30

Define
archipelago, lahars, endemic species

Working with Sketch Maps
On a map of island Southeast Asia that you draw or that your teacher provides, label the Malay Peninsula, Malay Archipelago, Sumatra, Irian Jaya, New Guinea, Borneo, Java Sea, South China Sea, Timor Sea, Strait of Malacca, Java, and the Spratly Islands. Which island has about 13 active volcanoes?

Reading for the Main Idea
1. *Physical Systems*   What area of island Southeast Asia has the least tectonic activity? Why?

2. *Places and Regions*   How does the region's location affect its climate? What are two other factors that influence climates there?

3. *Physical Systems*   What do the Komodo dragon, Javan rhinoceros, and orangutan have in common?

Critical Thinking
4. Analyzing Information   Why might people consider the Strait of Malacca another important natural resource?

Organizing What You Know
5. Create a chart like the one shown below. Use it to identify the region's major natural hazards and their effects on the environment.

Natural hazards	Effects

Biology: Mobility and Species Survival

Emphasize that rising sea levels have isolated many species found on Southeast Asia's islands. Some species can adapt more easily to a limited range than others. Ask students what determines where animals can live and reproduce. (*Possible answers: appropriate climate, reliable food sources, suitable living or nesting places*) Point out that for a species to survive, individual animals may have to cross geographical barriers to find food and mates. The ability of individual animals to expand their home range may determine a species' survival.

Ask students what types of animals could expand their range more easily than others (*birds, fish, flying insects*). To determine how the ranges of different types of animals compare, have students choose a North American amphibian, bird, mammal, and reptile species. Have pairs of students use field guides and other resources to create a map of North America with the ranges of their chosen species shaded in different colors. Have students formulate a hypothesis to explain the relationship between an animal's mobility and its range. (*Birds tend to have the largest ranges, while some amphibians have the smallest, with reptiles and mammals somewhere in betweeen. Birds can fly great distances. Although they may live all of their adult lives on dry land, many amphibians must lay eggs in freshwater, so they cannot successfully migrate through arid lands.*)

Physical Systems

Geography for Life

Wallace's Line

Indonesia's great biodiversity results in part from its location. The country lies in a transition zone between ecosystems. Asian animal and plant species live in its western islands. Species related to those of Australia and New Guinea live in the eastern islands.

English naturalist and biogeographer Alfred Russel Wallace (1823–1913) studied the reasons for this distribution of species. Wallace spent eight years in the region. He noted the difference between the birds on the island of Bali and those just 20 miles east on Lombok. The birds on Bali appeared to be related to those on Java, mainland Malaysia, and Sumatra. The Lombok birds resembled those of New Guinea and Australia. Based on what he learned, Wallace drew a line across a map of the region. This line separated areas with Asian plants and animals from areas with Australian species. It eventually became known as Wallace's line. (See the map.)

What created this curious distribution of ecosystems? Recall what you learned about the theory of continental drift. Long ago, all of Earth's land surface was part of one huge landmass called Pangaea. Then Pangaea broke into what are today's continents and islands. As the pieces of Pangaea drifted apart, Australia, New Guinea, and nearby islands became isolated from other continents. Even during Earth's ice ages, lower sea levels failed to relink these regions. Thus, plant and animal life that evolved in Australia is very different from that in much of island Southeast Asia.

Many biogeographers today do not see the boundary between the two ecosystems as a simple line. They think of the island region between Java and New Guinea as an ecotone—a transition zone between adjoining ecosystems. Geographers refer to this transition zone as Wallacea, in honor of Alfred Wallace.

Applying What You Know

1. **Summarizing** What past physical processes account for Indonesia's biodiversity today?

2. **Analyzing Information** Where do you think you might find other regions with high levels of biodiversity? Why?

Philippines Wildlife
Like Indonesia, the Philippines may lose rare species to poaching and habitat destruction.

Among the most spectacular of the endangered creatures is a national symbol, the monkey-eating eagle. It has a six-foot (2 m) wingspan and a squared-off tail that allows it to rise almost vertically through the trees. Other animals at risk include dugongs, similar to manatees; small water buffaloes called *tamaraws;* flying lemurs, which are not true lemurs and glide rather than fly; Palawan bear cats, also known as *binturongs;* and tarsiers, which have huge eyes and appealing faces. Many fruit- and nectar-eating bats are also endangered.

ACTIVITY: Have students conduct research on these endangered species and propose strategies for their conservation.

Applying What You Know Answers

1. As the pieces of Pangaea drifted apart, life in the two areas developed separately.
2. other areas of tropical rain forest and areas where past climate patterns and physical features allowed species to develop separately

This Geography for Life feature addresses National Geography Standards 1, 7, 8, and 17.

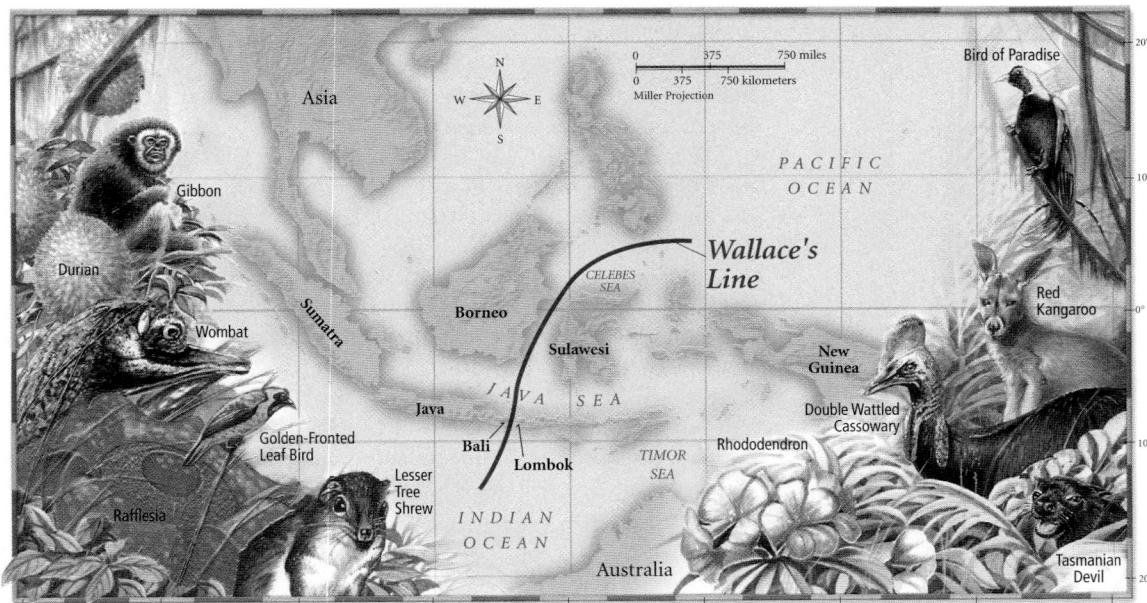

OBJECTIVES

1. **Identify the early peoples who migrated to the region.**

2. **Analyze how colonialism affected island Southeast Asia's history.**

3. **Describe what the region's people and culture are like today.**

 **LET'S GET STARTED**

Copy the following passage onto the chalkboard: *Look at the large photo in the unit opener. What other statues in the textbook look similar? What can you conclude about culture in island Southeast Asia?* Discuss responses. *(Similar statues from China or India indicate that the region shares religious traditions with those countries.)* Then point out a photo of a mosque and tell students that more Muslims live in Indonesia than in any other country. In Section 2 students will learn more about the region's history and culture.

Building Vocabulary

Write the key terms on the chalkboard. Call on a student to read the definition of the word *heterogeneous* from a dictionary and contrast it with the definition of **homogeneous** as it appears in the Section 2 text. Ask another student to propose a meaning for **slash-and-burn** agriculture based on the term's parts. Finally, have volunteers read the definitions aloud from the glossary.

VISUAL RECORD ANSWER

Hinduism

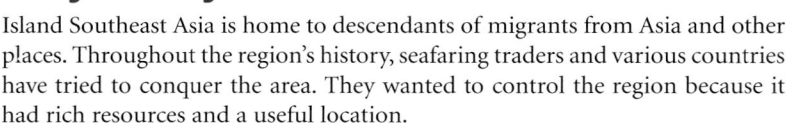**Section 2 History and Culture**

READ TO DISCOVER

1. What early peoples migrated to island Southeast Asia?
2. How did colonialism affect the region's history?
3. What are the people and culture of island Southeast Asia like today?

WHY IT MATTERS

The Philippines was once the largest overseas possession of the United States. Use **CNNfyi.com** or other **current events** sources to learn about the relationship between the two countries today.

DEFINE

homogeneous
slash-and-burn agriculture

LOCATE

Borobudur
Bali
Manila
Jakarta

Early History

Island Southeast Asia is home to descendants of migrants from Asia and other places. Throughout the region's history, seafaring traders and various countries have tried to conquer the area. They wanted to control the region because it had rich resources and a useful location.

Human remains in the Philippines date back more than 30,000 years. The first Malay people from Asia probably migrated into the region about 2000 B.C. They mixed with the peoples who had long lived in the region. Over time, many other peoples came to island Southeast Asia.

Hindus from what is now India influenced the area early in its history. By about A.D. 700, Hindu and Buddhist kingdoms were well established in Java and Sumatra. These kingdoms controlled trade and built huge monuments, including Borobudur in central Java. This temple was completed about A.D. 850. Some of these monuments have been restored.

Majapahit (mah-jah-PAH-hit), the largest early kingdom, existed from the late 1200s to about 1500. Majapahit was centered on the islands of Java and Bali, but controlled many coastal areas. Trade and cities grew throughout the kingdom and the area.

Chinese merchants also sailed their ships long distances to trade among the islands. Some of the merchants then began to settle in coastal cities. Today ethnic Chinese are an important minority in the populations of many island Southeast Asian countries.

By the 1300s Arabs from Southwest Asia were also trading in the region. The Arabs introduced Islam to coastal peoples there, and the religion gradually spread. Areas in northern Sumatra and on the Malay Peninsula became early centers of Islam. Today Islam is island Southeast Asia's main religion.

INTERPRETING THE VISUAL RECORD

Indian influences on Indonesia appear in the movements of this Balinese dancer, who probably began training when she was four or five years old. **Which religion came to Bali from India?**

✔ **READING CHECK:** *Human Systems* What people began migrating to the region about 4,000 years ago? first Malay people from Asia

LEVEL 1: Copy the following graphic organizer onto the chalkboard, omitting the italicized answers. Have students fill in details. **ENGLISH LANGUAGE LEARNERS**

LEVELS 2 AND 3: Have students use the graphic organizer in the Level 1 activity as a basis for this one. Tell students to imagine that they are among the early peoples who migrated to island Southeast Asia. Ask them to write a diary entry for the day their boat lands in the region, a day on the Borobudur construction site, time spent trading with Chinese merchants, or another episode in the region's early history. Have students use the illustrations in this chapter and the unit opener as inspiration for detail. You may want students to consult other resources also.

First Humans	Chinese	Hindus and Buddhists
in Philippines more than 30,000 years ago	*traders; settled in coastal cities*	*kingdoms established in Java and Sumatra by A.D. 700; controlled trade, built monuments; Majapahit largest kingdom*

Early Peoples of Island Southeast Asia

Malays	Arabs
migrated from Asia about 2000 B.C.	*trading in the region by 1300s; introduced Islam*

Colonial Era and Independence

As in many other parts of the world, island Southeast Asia came under the control of European colonial powers. (See Connecting to History: A Colonial History.) The Portuguese, who came in the 1500s, were the first Europeans to arrive. They were searching for spices such as cloves, nutmeg, and pepper and therefore called the area the Spice Islands. In the 1600s and 1700s the Dutch drove out the Portuguese. Portugal lost control of all its lands in the region except the island of Timor.

European Influence The explorer Ferdinand Magellan reached the Philippines in 1521 and claimed the islands for Spain. The Spaniards who followed wanted to Christianize and colonize the islands. Roman Catholicism, the religion brought by the Spaniards, is the main faith of the Philippines today. In addition, Manila became a major port for trade with China and the Spanish colonies in the Americas. In 1898, after the Spanish-American War, the United States took over the Philippines. These islands were the first large overseas U.S. territory.

The Dutch were much less interested than the Spaniards in converting the region's peoples to Christianity. In contrast, the Dutch went to the Spice Islands for commerce. They controlled the spice and tea trade of what became known as the Dutch East Indies. Today these islands make up Indonesia. The Dutch ruled from Batavia—now called Jakarta—their main port on Java.

The British set up colonies in Malaya on the Malay Peninsula and along the northern coast of Borneo. In 1819 they founded Singapore, which became a

The Philippines were named for Philip II of Spain, pictured above. The first permanent Spanish settlement in the Philippines was established in 1565. The islands remained part of the Spanish Empire for more than 300 years.

INTERPRETING THE VISUAL RECORD

This design recalls the Portuguese arrival in the islands of Southeast Asia. The detail is from a piece of batik, a fabric from Indonesia. Batik is made by applying wax to fabric and then dyeing it with various colors. Areas covered by wax resist the dye. **What elements of European culture are visible in the design?**

internet connect

GO TO: go.hrw.com
KEYWORD: SW3 CH30
FOR: Web sites about Philippine festivals

VISUAL RECORD ANSWER

Possible answers: clothing, flags, masted ships

ALL LEVELS: Write the names of the six countries that affected island Southeast Asia's colonial history on the chalkboard: Portugal, Spain, Netherlands, Britain, United States, and Japan. Call on a student to identify one way that one of these countries related to or affected the region. Then have the first student call on another to add another detail. Have a volunteer write down the students' suggestions. Continue until all the major facts have been noted and most students have contributed information. *(Example: Netherlands—during 1600s and 1700s drove Portuguese from*

the region, except from Timor; more interested in trade than converting people to Christianity; controlled spice and tea trade in the Dutch East Indies [now Indonesia]; ruled from Batavia [now Jakarta]) Have all students copy the chart into their notebooks. **ENGLISH LANGUAGE LEARNERS**

HOMEWORK: Provide each student with an outline map of island Southeast Asia. Have students shade each country according to which colonizing power affected it. Ask students to create a coloring or marking scheme to accommodate more than one foreign power playing a role in the country's history.

The Sultan of Brunei

Paduka Seri Baginda Sultan Haji Hassanal Bolkiah Mu'izzadin Waddaulah has been sultan of Brunei since 1967. He holds power over his subjects in many ways. The sultan is also the country's prime minister, minister of finance, and minister of home affairs. He appoints commissioners to the Supreme Court. The sultan names the council members in charge of local government. Members of his family hold high positions in the government. There is only one legal political party.

Brunei's petroleum wealth has made the sultan extremely rich. In 1999 his account totaled about $30 billion. Oil revenues flowing into government coffers make health care and education free for all citizens. About half the population works for the government. However, Brunei's oil will not last forever. Brunei needs to diversify its economy before the wells run dry.

DISCUSSION: Lead a class discussion about how Brunei might prepare for the future.

CONNECTING TO HISTORY ANSWER

Possible answers: religion, such as Roman Catholicism from Spain, and language, such as English from the United States

---

Connecting to
HISTORY

A Colonial History

By the early 1900s foreign powers ruled all of island Southeast Asia. The United Kingdom, the Netherlands, Portugal, and the United States each controlled parts of the region. France ruled parts of nearby mainland Southeast Asia.

In some of the colonies, rebels fought for independence. In the Philippines, for example, rebels first fought against Spanish troops. When Filipino rebels declared independence in 1898, they fought against U.S. troops. The fighting was bitter and bloody. In 1946 the Philippines gained independence from the United States. Still, some Filipinos wanted fewer links to the United States. As a result, the United States closed its last military base in the Philippines in 1992. The Philippines remains a U.S. ally.

Making Generalizations What are some European and American cultural influences that you might expect to find in the region today?

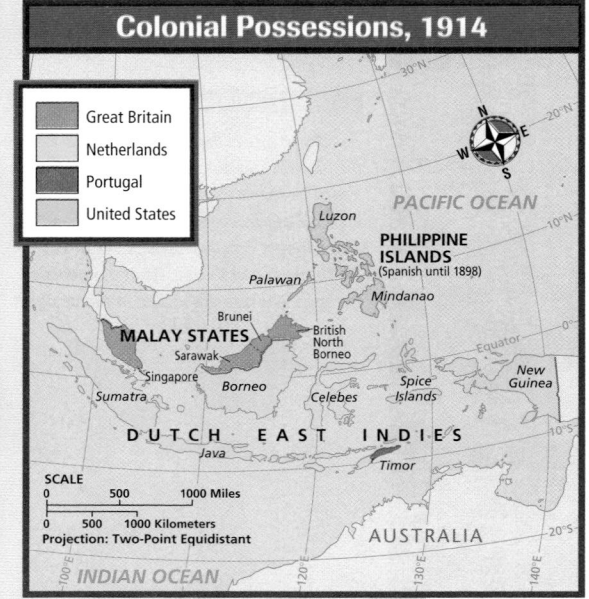

Colonial Possessions, 1914

Great Britain
Netherlands
Portugal
United States

PACIFIC OCEAN

Luzon
PHILIPPINE ISLANDS
(Spanish until 1898)
Palawan
Mindanao
Brunei
MALAY STATES
Sarawak
British North Borneo
Singapore
Sumatra
Borneo
Celebes
Spice Islands
New Guinea
Java
DUTCH EAST INDIES
Timor

SCALE
0   500   1000 Miles
0   500   1000 Kilometers
Projection: Two-Point Equidistant

AUSTRALIA

INDIAN OCEAN

---

major port for British and Chinese trade. The British used many local workers to build roads, plantations, and schools. Workers from China and India also came to the plantations. Today Chinese and South Asians make up large ethnic groups in the region.

War and Independence In 1941 and 1942 Japan invaded European and U.S. territories in island Southeast Asia. During World War II the Japanese wanted the region's natural resources, particularly oil. They occupied much of the region until they surrendered at the end of World War II in 1945.

Soon after the war, the Philippines gained independence. The colonial system began to crumble throughout the area. The Dutch tried to reestablish their rule after the war, but the Indonesians resisted. As a result, the Dutch gave up the colony in 1949. Malaya won independence in 1957. Then in 1963 Malaya joined former British territories in Singapore and northern Borneo to form the Federation of Malaysia. Singapore later broke away from Malaysia in 1965. Brunei, a British colony, gained independence in 1984. This tiny country is ruled by a sultan. A sultan is the ruler of a Muslim country.

✓ **READING CHECK: Human Systems** What European countries colonized the region? Britain, Netherlands, Portugal, Spain

Culture

How do you think island Southeast Asia's history has influenced its peoples and cultures? In what ways might you see these influences today?

People and Languages A history of migration and colonization has created a diverse population in island Southeast Asia. People from many different ethnic groups live in the region. In Indonesia, for example, no one ethnic group makes up a clear majority of the population. Nearly half of Indonesians are Javanese. Malays and others make up large minority groups.

Malaysia's population is somewhat less diverse. Nearly 60 percent of Malaysians are ethnic Malay. Still, this country also has large minority groups, like the descendants of Chinese migrants. In fact, ethnic Chinese and South Asians dominate much of Malaysia's economy.

Chinese live throughout the region, particularly in large cities. In Singapore, Chinese make up a majority—more than 75 percent of the population. Tensions between the city's mostly Chinese population and the Malays in Malaysia led Singapore to seek independence in 1965.

Each country in the region has one or more official languages. For example, Singapore has four official

Organize the class into six groups—one for each of the following topics: people and languages, settlement patterns, religion, food and rice farming, education, and traditions and customs. Have each group create a poster about its assigned topic. Ask students to use both words and graphics on their posters. Then have each group present its poster to the class. Display posters around the classroom. **ENGLISH LANGUAGE LEARNERS, COOPERATIVE LEARNING**

LEVELS 2 AND 3: Organize the class into groups. Have each group compile several questions which would be useful in a discussion of this topic: Of the many cultures that have played a role in the history of island Southeast Asia, which one has had the most lasting effect on the people and culture today? (*Examples: What beliefs and practices combine elements from more than one culture? Are many customs of daily life based on religious belief and, if so, how closely? How have music, literature, and other arts been influenced by these cultures? What monuments remain from the various cultures, and how highly are they valued by the people who live near them?*) Have each group report on its questions. You may want students to conduct research on some of them. **COOPERATIVE LEARNING**

---

languages, including English and Malay. Chinese dialects are spoken in many large cities. In addition, indigenous peoples throughout the region speak local languages.

The Philippines is the region's most **homogeneous** (hoh-muh-JEE-nee-uhs) country. The word *homogeneous* means "of the same kind." More than 90 percent of the country's people are ethnic Malays. Pilipino, which is based on a native language called Tagalog, is one of the Philippines' official languages. English is also an official language in this former U.S. territory.

Settlement and Land Use   Island Southeast Asia's population is not evenly distributed. For example, Java has more than half of Indonesia's population of about 225 million. The Javanese live on an island smaller than New York State, which has fewer than 20 million people. The Indonesian government encourages citizens to move to less-populated islands. Between 1969 and 1994, some 8.5 million Indonesians were relocated. This policy has not been popular with the residents of those islands, however.

The country of Singapore, which occupies a small island, is almost completely urban. About 70 percent of tiny Brunei's people also live in cities. The larger countries are more rural. Many people are farmers. About a third of Indonesians and half of Malaysians and Filipinos live in cities. Still, many people are moving from rural areas to cities in search of work. Two of the most populous cities are Jakarta and Manila.

Religion   As you have read, Buddhism, Hinduism, and Islam have long been practiced in the region. Indonesia is the world's most populous Islamic country. Nearly 90 percent of its people are Muslims. Hinduism is practiced in some areas, such as on the Indonesian island of Bali. Buddhism is most common in Malaysia and Singapore, where many Chinese live. Europeans brought Christianity, and today Christians live throughout the region. In the Philippines Christians make up more than 90 percent of the population.

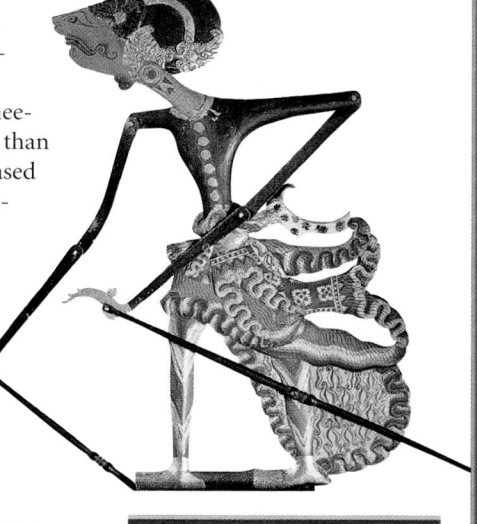

INTERPRETING THE VISUAL RECORD

Traditional theater performances help preserve the Hindu-Buddhist heritage of Java. For example, wayang puppets relate stories from Hindu epics. Wayang means "shadow." Elaborately painted leather puppets are placed behind a rice paper screen, with a candle or oil lamp as a light source behind them. The audience watches the shadows projected by the puppets. A wayang performance may last an entire night. **How might such a performance help unify a community?**

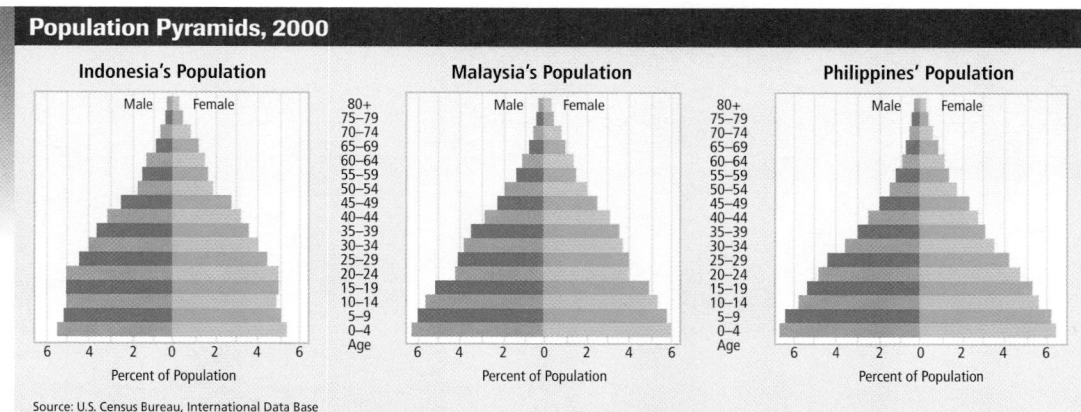

Population Pyramids, 2000

Indonesia's Population / Malaysia's Population / Philippines' Population

Age groups: 80+, 75–79, 70–74, 65–69, 60–64, 55–59, 50–54, 45–49, 40–44, 35–39, 30–34, 25–29, 20–24, 15–19, 10–14, 5–9, 0–4

Male / Female — Percent of Population (6 4 2 0 2 4 6)

Source: U.S. Census Bureau, International Data Base

INTERPRETING THE GRAPH   *Study the population pyramids. In which country has the population growth rate remained the most constant? In which country does the growth rate appear to have slowed somewhat? What do the graphs indicate about future population growth in all three of these countries?*

VISUAL RECORD ANSWER

Possible answers: reminds community members of their past; provides opportunity for socializing; organizes people to put on the show

GRAPH ANSWER

Philippines; Indonesia; that they will continue to grow at a fairly rapid rate

Daily Life

Rice Farming Wet-rice cultivation requires back-breaking work. It is feasible only where labor is plentiful.

The rice seeds are first planted in seedbeds away from the paddy. When the seedlings have sprouted they are brought to the flooded fields. Then the farmer pushes the seedlings into the mud by hand. A skilled worker can plant more than one plug per second, while maintaining neat rows.

This cultivation method gives the rice plants a head start, before weeds can take over. If the rice were planted as seeds, farmers would have to contend with more weeds, spend more money on weed-killing chemicals, and lose valuable time. Instead, they get three harvests per year from the same fields.

You don't say! Villagers in Bali shape intricate sculptures weighing more than 200 pounds (90 kg) from brightly-colored rice paste. They take these creations to a temple as offerings for their gods.

VISUAL RECORD ANSWER

by building terraces

Food Farmers grow many kinds of foods in the region. However, rice is the main food crop, or staple, for most of the people. Rice is served with many other foods and spices, such as curries and chili peppers.

FOCUS ON GEOGRAPHY

Growing Rice Farmers in the region grow rice in three ways. Wet-rice, or paddy, cultivation is the most productive and common method. Rice paddies are constructed with dikes in lowland areas or with mud terraces in hilly areas. Water flow down steep slopes is controlled, and erosion is limited. This, along with the area's warm and wet climate, allows farmers to grow more than one rice crop each year.

Paddy cultivation is a form of intensive agriculture. Many workers are needed to harvest a particular crop. This type of rice farming supports large concentrated populations in island and mainland Southeast Asia. In addition, farmers can raise ducks, fish, and shrimp in the paddies. These food sources add protein to the diet. Paddy workers may use the water buffalo, a domesticated animal, for plowing and heavy farm duties.

In the tropical wet-and-dry climate areas, farmers use dry-rice cultivation. Farmers plow fields and plant rice seeds. Rivers may flood the area during the wet season. People practice a third type of rice cultivation in forested areas. This method is known as **slash-and-burn agriculture**, a form of shifting cultivation. Farmers clear or slash small areas of forest and burn the fallen trees. After a few years the soil's nutrients have been used up, and farmers move on to a new area. Dry farming and slash-and-burn agriculture are also used throughout the region to grow many other crops.

✓ **READING CHECK:** *Environment and Society*
How do farmers in the region grow rice? wet-rice, or paddy; dry-rice; slash-and-burn

INTERPRETING THE VISUAL RECORD

Rice paddies like these in Indonesia are common throughout island Southeast Asia. **How have farmers limited erosion in the rice paddies pictured below?**

Trees provide both fruit and firewood for the local people.

The waters of the paddy field are full of nutrients provided by the small plants and animals that grow there.

Insects and other parasites that live on the rice plants are eaten by the frogs and fish that live in the rice paddy.

Close

Refer students to the population pyramids in Section 2. Call on volunteers to create one-sentence summaries of what the diagrams depict or what can be inferred from them. *(Example: In 2000, all three countries had more children under the age of five than in any other age category.)*

Review and Assess

Have students complete the **Section Review**. Then have students complete **Daily Quiz 30.2**.

Reteach

Have students complete **Main Idea Activity for English Language Learners and Special-Needs Students 30.2.** Then have each student write 10 questions based on Section 2. Ask students to exchange papers, answer the questions, and return them for checking. Discuss questions that caused difficulty. **ENGLISH LANGUAGE LEARNERS**

Extend

Have interested students conduct research on the Philippines' revolt against American authority (1899–1901) and the role that the revolt played in Filipinos' perceptions of national unity. **BLOCK SCHEDULING**

Bruneian students read in a religion class. The girls wear a traditional head covering called a tudong. The boys wear the songkok. Men who have gone on a pilgrimage to Mecca (Makkah) wear a white songkok.

Other Traditions and Education   Many traditional clothing styles are still worn partly because they are ideal for the region's hot humid climates. For example, for business and other important occasions, Filipino men wear the *barong tagalog.* This light shirt is made from cotton or from fibers of the banana or pineapple plant. Malaysian men and women often wear sarongs. They wrap these long strips of cloth around their bodies.

People in the region use special methods to make some traditional clothing. For example, Indonesians create colorful fabrics called batiks (buh-TEEKS). Coating areas of the cloth with wax creates patterns on a batik. Uncoated areas can then be dyed with bright colors.

Education is a key to the region's future prosperity. To aid economic development, governments have tried to improve educational opportunities for all people. Schools have also been used to create a sense of national identity in the region's multiethnic countries. For example, in the Philippines the Pilipino language is used more and more in schools. Indonesian schools emphasize what are called the Pancasila, or "Five Principles." Indonesia's early leaders believed their new country should be based on those principles. One of the principles is the importance of national unity among the country's many ethnic groups. The others are belief in one God, a just and civilized humanity, democracy, and social justice.

✓ **READING CHECK:** *Environment and Society* How does clothing help people adapt to the region's hot humid climate? wear light clothing, such as the *barong tagalog* in the Philippines, often made from local fibers

Review

Homework Practice Online
Keyword: SW3 HP30

Define
homogeneous, slash-and-burn agriculture

Working with Sketch Maps
On the map you created in Section 1, label the countries of island Southeast Asia and Borobudur, Bali, Manila, and Jakarta. What was Jakarta called during the colonial era?

Reading for the Main Idea

1. **Human Systems** From where and when did Malays migrate to the region? What drew Europeans to the region during the colonial era?

2. **Environment and Society** How do farmers modify the region's environment to grow rice?

Critical Thinking

3. **Comparing and Contrasting** How are the people and culture of the Philippines unique in the region?

4. **Drawing Inferences and Conclusions** Why do you think the Indonesian government wants people to move to less-populated islands?

Organizing What You Know

5. Create a word web like the one shown below. Use the word web to describe the peoples and cultures of the region.

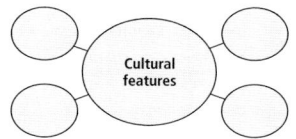

Cultural features

Section 2 Review Answers

Define For definitions, see: homogeneous, p. 687; slash-and-burn agriculture, p. 688

Working with Sketch Maps Maps will vary, but listed places should be labeled in their approximate locations. Jakarta was called Batavia.

Reading for the Main Idea
1. from Asia probably in 2000 B.C.; desire to Christianize and colonize, to exploit region's spices and other resources, and to expand commerce and trade

2. create paddies and terraces to hold water; plow fields; cut and burn forest areas

Critical Thinking
3. most homogenous, mostly ethnic Malay and Christian (NGS 4)

4. to relieve crowding on Java and other densely populated islands, promote economic development in other areas, ease some environmental problems (NGS 9)

Organizing What You Know
5. Possible answers: diverse people and languages, many Malay and Chinese; population not evenly distributed, Java most densely populated in Indonesia; many live in rural areas; many Muslims, Christians, also Buddhists, Hindus; rice is staple food crop; farmers use wet-rice, dry-rice, and slash-and-burn farming; schools used to foster sense of national identity; special methods for making clothes; light clothes for hot, humid climate

Section 3

OBJECTIVES

1. Analyze how island Southeast Asia's economy has changed in recent decades.

2. Describe the region's cities.

3. Discuss the challenges the people of the region face.

LET'S GET STARTED

Copy the following instructions onto the chalkboard: *Indonesia's national motto is "Unity in diversity." How might this motto relate the country's physical geography to its history and culture?* Discuss responses. *(Possible answer: Indonesia promotes a unified national culture although its population includes many ethnic groups and the country is made up of thousands of islands.)* Tell students they will learn more about tensions in the region between national identity and ethnic identity in Section 3.

Building Vocabulary

Write **kampongs** on the chalkboard and call on a volunteer to read the definition from the text or glossary. Point out that the word comes from the Malay language. A kampong was originally a village made up of houses built on stilts along a shoreline. Now the term also refers to slums that grow up around the region's big cities.

Section 3 RESOURCES

REPRODUCIBLE

▶ Guided Reading Strategy 30.3
▶ Readings in World Geography, History, and Culture 73

TECHNOLOGY

▶ One-Stop Planner CD–ROM, Lesson 30.3
▶ Homework Practice Online
▶ HRW Go site

REINFORCEMENT, REVIEW, AND ASSESSMENT

▶ Main Idea Activity 30.3
▶ English Audio Summary 30.3
▶ Spanish Audio Summary 30.3
▶ Section 3 Review, p. 693
▶ Daily Quiz 30.3

Section 3 The Region Today

READ TO DISCOVER

1. How has island Southeast Asia's economy changed in recent decades?
2. What are the cities of the region like?
3. What challenges do the people of the region face?

WHY IT MATTERS

Sometimes ethnic and religious conflicts have caused problems in island Southeast Asia. Use **CNNfyi.com** or other **current events** sources to identify international efforts to ease such conflicts around the world.

DEFINE

kampongs

LOCATE

Kuala Lumpur
East Timor
Aceh
Moluccas
Mindanao
Sulu Archipelago

Economic Development

The countries of island Southeast Asia have mostly free-market economies. Governments have encouraged private industry and business. In addition, businesses in other countries have invested in the region.

Cooperation among the region's countries has helped economic development. All of the countries are members of the Association of Southeast Asian Nations (ASEAN). The region's leaders are also working to improve economic and trade ties to countries outside the region.

A swordfish adorns this Singaporean coin. Some coins display flowers to enhance Singapore's garden city image.

These boats bring teak and other types of lumber to Jakarta for distribution and export. Teak has long been valued for its beauty and durability. Although teak is grown on plantations, some trees in the rain forests are cut down illegally. This unauthorized logging is one cause of deforestation.

Economic Growth Rapid industrialization during the 1980s and 1990s brought great economic growth. Some countries became known as the Tigers of the Pacific Rim because of their spectacular growth. These newly industrialized countries included Singapore and Malaysia. Singapore was one of the earliest success stories. The city is located on the Strait of Malacca, a major shipping route. This location helped the city and its economy grow. Today Singapore is a major trade and industrial center. Many financial and high-technology companies have also opened offices in the country. This development has helped make Singapore's per capita GDP among the highest in the world.

Brunei also has a high GDP because of its oil and natural gas reserves. In fact, these reserves and the country's refineries account for more than half of the country's GDP. Brunei's government has used some of the oil income to benefit the country's citizens. For example, medical care is free. The government also helps pay food and housing costs.

LEVEL 1: Organize the class into groups. Assign one of the following topics to each group: ASEAN, industrialization, agriculture, fossil fuels, and tourism. Have each group use travel brochures, maps, drawings, and other materials to create a collage about the assigned topic. Then ask all groups to work together to combine their collages into one mural. Have them insert arrows, phrases, or sentences between the collages to indicate how the various aspects of the region's economy are related. Display the mural in the classroom. (*Example: This sentence might connect the industrialization and* agriculture collages: *Although the region is industrializing, agriculture is still very important.*) ENGLISH LANGUAGE LEARNERS, COOPERATIVE LEARNING

LEVELS 2 AND 3: Tell students to imagine that they work for a public relations firm hired by ASEAN to stimulate investment in the countries of island Southeast Asia. The firm's campaign will include a television commercial. Have each student write a paragraph of text for the commercial. Call on volunteers to read their paragraphs to the class.

A growing economy has also improved life for many Malaysians. Malaysia has tried to maintain its economic strength by selling many government-owned industries to private citizens. However, the Philippines and Indonesia have not done as well, because of their huge populations and recent political problems. Still, overall the region's future could be bright. Rich natural resources and a large and skilled labor force could fuel continued economic growth.

Agriculture Even with recent industrialization, agriculture has remained vital to the region. Wet-rice cultivation is the most common form of agriculture. The region's countries also produce and export coffee, fruit, spices, sugarcane, and tea. Rubber trees, which came to the region from South America, are also valuable. Malaysia and Indonesia, along with Thailand, are now the world's largest producers of natural rubber.

Fisheries provide this island region with seafood, the major source of protein. Traditional fishers have sailed nearby waters for thousands of years. Today, however, their small boats must compete with large commercial ships. Overfishing now poses a threat to local fisheries.

Tourism Tourism is also a big industry. For example, the Indonesian island of Bali is popular with tourists from around the world. The island's mixed Hindu-Buddhist culture and beautiful rice-paddy scenery attract thousands of visitors each year. Bali's skilled artists, dancers, and weavers help increase the island's popularity.

✓ **READING CHECK:** (*Human Systems*) Which countries have experienced rapid industrialization? Singapore, Malaysia

Island Southeast Asia	Major Exports
Brunei	crude oil
Indonesia	textiles/garments, wood products, electronics, footwear
Malaysia	electronic equipment, palm oil
Philippines	electronics and telecommunications, machinery and transport
Singapore	computer equipment, rubber

Source: *National Geographic Atlas of the World, Seventh Edition*

INTERPRETING THE CHART *Most countries of island Southeast Asia export electronics and products relating to technology.* **What does this chart indicate about the level of economic development in these countries?**

Daily Life

A Balinese Ritual
Colorful ceremonies are among the many reasons why Bali is a popular tourist destination. The Balinese mark each of life's stages with a ritual. The cremation ceremony may be the most spectacular. Preparations can be so extensive—and expensive—that the body is buried temporarily while the family settles details. Hindus who are members of the Brahmin caste are cremated immediately after death.

On a day that has been judged favorable, the deceased is carried from the burial ground, or the home, in a high tower decorated with bamboo, paper, tinsel, cloth, mirrors, and flowers. Several men carry the tower on their shoulders. The tower of a very important person may be so big that hundreds of men are required to carry it. To prevent the dead person's spirit finding its way back home, the bearers shake the tower, spin it, run it in circles, and throw water at it.

After being placed in a sarcophagus shaped like a bull, winged lion, or elephant-fish, the body is burned, along with the tower and all its decorations.

CHART ANSWER

that economic development is fairly advanced

VISUAL RECORD ANSWER

compares to structures in China and Japan

INTERPRETING THE VISUAL RECORD

Lake Bratan and the Ulun Danu Temple on the Indonesian island of Bali typify the region's beautiful landscapes and architectural features. The temple, which includes Buddhist and Hindu elements, was built in 1633 and is dedicated to a Hindu water goddess. **How does the temple's architecture compare to styles represented in other regions?**

Teach Objective 2

ALL LEVELS: Copy the following graphic organizer onto the chalkboard, omitting the italicized answers. Have students fill in features the region's cities share and ways they are different. **ENGLISH LANGUAGE LEARNERS**

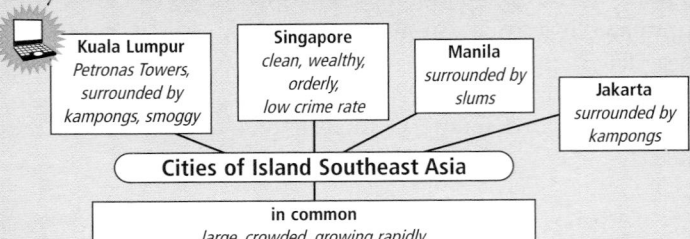

Teach Objective 3

LEVEL 1: Tell students to imagine that they will attend a political rally in an island Southeast Asian country. The theme of the rally is "Our Future is Now." Have students create banners for the rally, using phrases and illustrations that suggest solutions to the challenges facing a particular country in the region. Display and discuss the banners.

LEVELS 2 AND 3: Have students write speeches for political leaders to deliver at the "Our Future is Now" rally. Call on volunteers to read their speeches to the class.

Essential Element 4

▶ **Human Systems** ◀

Race Relations in Malaysia Since 1971 the Malaysian government has enacted race-relations policies that favor native Malays. Malays are given preference for government jobs and in college admissions. The intent is to raise the Malays' economic status to equal that of the more economically successful Chinese minority.

ACTIVITY: Have students conduct research on Malaysia's racial policies, as stipulated in the New Economic Policy, and debate their fairness and effectiveness.

Kuala Lumpur gets its name from a Malay phrase meaning "muddy estuary." The city lies near the mouth of a river on the Malay Peninsula.

Kampong architecture reflects the region's physical environment. Stilts keep the houses safe from high tides or floods. The roofs are made of tiles woven from native plants. The Toraja people built these traditional houses in Sulawesi, one of Indonesia's larger islands.

Urban Environments

The region's economic development has fueled the rapid growth of its cities. As you have read, many people are moving to cities to search for work. The largest cities are the capitals of the major countries. These include Jakarta in Indonesia, Kuala Lumpur in Malaysia, Manila in the Philippines, and Singapore.

All four cities have modern urban centers and government buildings. However, major differences exist among them. Singapore is orderly, wealthy, and very clean. Crime rates are low there. On the other hand, Kuala Lumpur, Manila, and Jakarta have smog and traffic problems. Manila and Jakarta have large slums.

Kampongs, which are villages built on stilts, make up the traditional Malay housing style. Today the term also refers to the crowded slums around Jakarta and other large cities.

Why is Singapore so different from its neighbors? The country's government has worked hard to clean up slums and provide better housing. (See Cities & Settlements: Singapore.) In addition, Singapore has strict laws against even minor offenses, such as littering. The government even outlaws chewing gum and certain kinds of music and movies. Some people believe that such strict limits on personal freedoms are a good trade-off for less crime, a clean city, and a strong economy. However, others believe Singapore could be just as successful with fewer rules.

✓ **READING CHECK:** (*Human Systems*) How has economic growth affected the region's cities? fueled rapid urban growth

Issues and Challenges

The region's countries face a number of issues. As you have read, the rapid growth of cities presents a major challenge. Other tasks include overcoming political problems and ethnic and religious tensions.

Political Challenges Protecting political and personal freedoms represents one of the region's challenges. The Philippines and Indonesia, for example, have held truly free elections only in recent years.

Indonesia faces other challenges as well. Terrorism, including the 2002 bombing in Bali, has severely hurt Indonesia's tourism and economy. Ethnic and religious groups continue to push for independence. In 1999 East Timor voted for independence. Following the 1999 election, violence broke out between independence supporters and their opponents. The United Nations sent in troops to stop the fighting. East Timor's new government took power in 2002.

Other hot spots simmer. People in Aceh, on the northern tip of Sumatra, and Irian Jaya (also called West Papua) also want independence. At times, violence breaks out between independence supporters and government forces. Christian and Muslim residents of the Moluccas (moh-LUH-kuhs), another area of Indonesia, have also fought.

Religious differences have also led to conflict in the Philippines. The southern part of this mostly Roman Catholic country is home to many Muslims, some of whom want independence. The United States has aided the Philippines by training Filipino troops to fight terrorism.

Other Challenges Environmental problems like deforestation, loss of wildlife diversity, overfishing, and air and water pollution present difficult challenges. In addition, with the exception of Singapore, many people in the region are still poor. A few business families, military officers, and politicians control most of the power and money. Some corrupt officials have managed to stay in power for decades. Poor workers from Indonesia have moved to richer Malaysia and Singapore to look for jobs. Raising the standard of living for all the region's people will be difficult.

READING CHECK: *Human Systems* What challenges face the region today? *protecting freedoms, cultural conflicts, independence movements, deforestation, protecting biodiversity, overfishing, pollution, corruption*

Soaring 1,483 feet (452 m) into the air, the Petronas Towers in Kuala Lumpur ranked as the tallest buildings in the world when they were completed in 1998. These twin skyscrapers of glass, steel, and concrete are a visual symbol of Malaysia's commitment to being modern.

Section 3 Review

Homework Practice Online
Keyword: SW3 HP30

Define
kampongs

Working with Sketch Maps On the map you created in Section 2, label Kuala Lumpur, East Timor, Aceh, Moluccas, Mindanao, and Sulu Archipelago. Which island region declared independence in 1999?

Reading for the Main Idea

1. *Human Systems* Why are some of the countries in island Southeast Asia known as the Tigers of the Pacific Rim?

2. *Environment and Society* What factor helped make Singapore a major trade and industrial center?

Critical Thinking

3. **Comparing and Contrasting** What are some differences between the city of Singapore and other big cities in the region? What are some reasons for these differences?

4. **Making Generalizations** Why do you suppose Indonesia's various ethnic groups each view their country differently?

Organizing What You Know

5. Copy the table below. Use it to identify economic, environmental, and political challenges the region's countries face.

Challenges		
Economic	Environmental	Political

Comparing Places

Have students read Cities & Settlements: Singapore. Point out that governments are involved in city functions to varying degrees. Explain that Singapore's government is closely involved in the details of running the city. People must follow the rules or risk punishment. Some punishable offenses include eating in the subway and chewing gum in public. People caught littering must undergo counseling and pay a fine. Because Singapore's government operates much like an well-run corporation, the country has been referred to as a nation-corporation. However, for those willing to follow the rules, Singapore offers a wide range of attractions, from lively festivals and parades to a nighttime cable-car ride down a mountainside. The city's low crime rate makes tourists feel at ease.

Island Southeast Asia has several other big cites. Both Jakarta and Manila have millions of inhabitants. Jakarta's name means "City of Great Victory." From its beginning as a small harbor town, it has grown to become a center of government, business, and industry. Manila's name comes from a white flowered mangrove plant known as the *nilad*. Manila probably began as a port town during the 1100s. During the 1500s the Spanish arrived and continued trade patterns begun by the Chinese and Arabs. Spanish rule ended in 1898, when the United States acquired the Philippines.

CITIES & SETTLEMENTS

Singapore

Human Systems Singapore has grown from a small port in the 1800s into a densely populated high-tech city. Efficient use of land, particularly for housing, has been crucial in planning the city's growth. Through housing policies, Singapore's government has tried to improve living conditions, encourage economic growth, and use land efficiently. It has also used housing policies to promote good relations among different ethnic groups.

Development and Growth

Changes have been particularly dramatic in recent decades. In the early 1950s about 75 percent of Singaporeans lived in crowded slums on the island's southern shore. The other 25 percent lived in the more rural northern part of the island in traditional kampongs. In the 1960s providing low-cost housing became a major goal of Singapore's government. The newly created Housing and Development Board began to build high-rise high-density apartments for public housing. The government gave the board the power to clear slums and forcibly resettle residents.

The program was very successful. Thousands of people moved into the new apartments. Government loan programs helped even low-income families buy their apartments. By 1988, 2.3 million people—88 percent of Singapore's population—lived in apartments provided by the board. Most tenants owned their apartments. The government saw apartment ownership as giving people a "stake in Singapore" and encouraging political stability. At the same time, housing construction became a major industry, helping to fuel economic growth.

Singapore, also known as the Garden City because of its parks and tree-lined streets, began its modern history in the early 1800s. A representative of the British East India Company established the area as a trading site and gained possession of the harbor for the United Kingdom. Singapore's location on major shipping lanes has helped it become one of the world's busiest ports.

The government also built new roads and highways, along with a mass-transit rail system. Housing complexes rose where highways and the mass-transit railroad met. These new towns of up to 200,000 residents were meant to redirect urban growth away from the old city center. The government hoped these new towns would provide employment for most of their residents. In reality, many residents commuted to jobs either in the central business district or in Jurong, a huge industrial area west of the city center.

Housing and Ethnicity

Singapore's government has also tried to shape the role of ethnicity in society with its housing policies. Singapore has three major ethnic groups—Chinese, Indian, and Malay. In most cases, people of a single ethnic group had occupied Singapore's old slums and kampongs. The government used the resettlement program to break up the pattern of single-ethnic-group

Going Further: Thinking Critically

Organize the class into small groups. Allow each group to choose one of island Southeast Asia's cities. Besides those discussed elsewhere in this chapter, possibilities include Bandar Seri Begawan, Brunei; Davao and Quezon City, Philippines; Bandung and Surabaya, Indonesia; and Malacca, Malaysia. Have students conduct additional research on their city, concentrating on the relationships involved in political, economic, social, and environmental changes. Ask students to pay particular attention to the role that governments play in the city's functions and how the city compares to Singapore. Then have each group present its findings to the class. While one group is presenting, have the other students take notes about key features of each city. When the reports are complete, lead a discussion about the traits or relationships the cities have in common and the ways in which they are different.

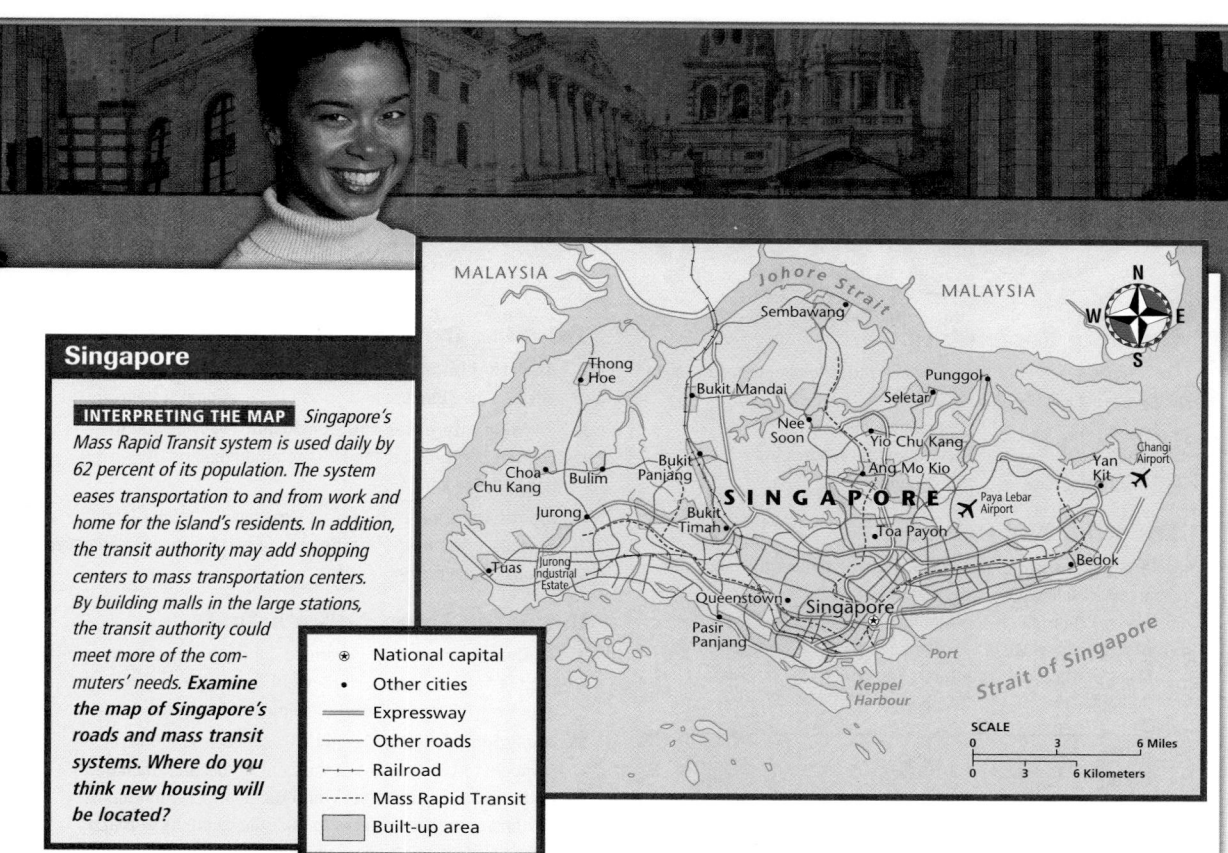

Singapore

INTERPRETING THE MAP *Singapore's Mass Rapid Transit system is used daily by 62 percent of its population. The system eases transportation to and from work and home for the island's residents. In addition, the transit authority may add shopping centers to mass transportation centers. By building malls in the large stations, the transit authority could meet more of the commuters' needs. **Examine the map of Singapore's roads and mass transit systems. Where do you think new housing will be located?***

neighborhoods. Singapore's leaders hoped that integrating the new apartment complexes would encourage tolerance of ethnic differences.

To accomplish this, a mix of Chinese, Indians, and Malays were assigned to every apartment block. The government also sponsored residents' committees and education and recreation programs for each apartment complex. These were designed to create a sense of community among the apartment residents.

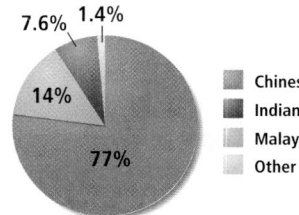

Singapore's Ethnic Groups

- 7.6%
- 1.4%
- 14%
- 77%

- Chinese
- Indian
- Malay
- Other

However, despite the hopes of the Housing and Development Board, few tenants thought of their apartment block as a real community. Most people maintained their ties to relatives and friends of the same ethnic group. Residents began selling their apartments to members of other ethnic groups so they could be near friends and family. The result was a trend toward apartment blocks of one ethnic group. In 1989 the government tried to halt this trend. It passed new rules to prevent people from selling their apartments or turning them over to members of other ethnic groups.

A Thriving City-State

Singapore's government has used housing policy to do more than simply alter settlement patterns. It has also tried to change attitudes. Not all of these goals have been met. However, the country has prospered in spite of its crowding and ethnic issues.

Applying What You Know

1. **Summarizing** What did Singapore's government hope its housing policies would accomplish?

2. **Analyzing Information** In what ways were Singapore's efforts to create new local communities successful? In what ways were they unsuccessful? Why?

CHAPTER 30 Review Answers

Building Vocabulary For definitions, see: archipelago, p. 680; lahars, p. 681; endemic species, p. 682; homogeneous, p. 687; slash-and-burn agriculture, p. 688; kampongs, p. 692

Locating Key Places

A. Strait of Malacca
B. Singapore
C. Philippines
D. Jakarta
E. Manila
F. Malay Peninsula
G. Kuala Lumpur
H. Malay Archipelago

Understanding the Main Ideas

1. location along tectonic plate boundaries

2. Location in tropics brings hot, wet climate; monsoons bring alternating wet-and-dry seasons.

3. Chinese; urban areas

4. Islam; Christianity, Buddhism, Hinduism

5. raising standard of living for all people; resolving political and cultural conflicts; slowing deforestation and loss of wildlife diversity, ending overfishing, cleaning up air and water; decreasing corruption, protecting freedoms

TECHNOLOGY
▶ Chapter 30 Test Generator (on the One-Stop Planner)
▶ Global Skill Builder CD–ROM
▶ HRW Go site

REINFORCEMENT, REVIEW, AND ASSESSMENT
▶ Chapter 30 Review, pp. 696–97
▶ Chapter 30 Tutorial for Students, Parents, Mentors, and Peers
▶ Chapter 30 Test (form A or B)
▶ Alternative Assessment Handbook

▶ Chapter 30 Test for English Language Learners and Special-Needs Students
▶ Unit 9 Test
▶ Unit 9 Test for English Language Learners and Special-Needs Students

Assess
Have students complete a Chapter 30 Test.

Reteach
Organize the class into three groups—one for each section. Have groups outline assigned sections. Then have groups rotate papers. Each group should add detail to the outline it receives. Rotate papers two more times to complete the process. Call on students to read the outlines. **ENGLISH LANGUAGE LEARNERS, COOPERATIVE LEARNING**

CHAPTER 30 Review Answers

Thinking Critically

1. Countries there use schools to foster a sense of national identify. **(NGS 10)**

2. takes advantage of region's wet climate, produces the largest amount of rice **(NGS 15)**

3. Students might say that European languages and religions probably would not be common in the region. Also, without its history as the colony of mostly one country, a large country like Indonesia might be broken into many countries. **(NGS 17)**

Using the Geographer's Tools

1. Similar growth rates indicate similar standards of living.

2. Students should recognize that Singapore has the highest standard of living. Students might say that the Philippines has the lowest. Students should demonstrate that they understand the concept of standard of living as explained in Chapter 6.

3. Maps should identify conflicts in East Timor, Aceh, Irian Jaya, Moluccas, Mindanao, and Sulu Archipelago.

CHAPTER 30 Review

Building Vocabulary
On a separate sheet of paper, explain the following terms by using them correctly in sentences.

archipelago
lahars
endemic species
homogeneous
slash-and-burn agriculture
kampongs

Locating Key Places
On a separate sheet of paper, match the letters on the map with their correct labels.

Malay Peninsula
Malay Archipelago
Strait of Malacca
Philippines
Manila
Jakarta
Kuala Lumpur
Singapore

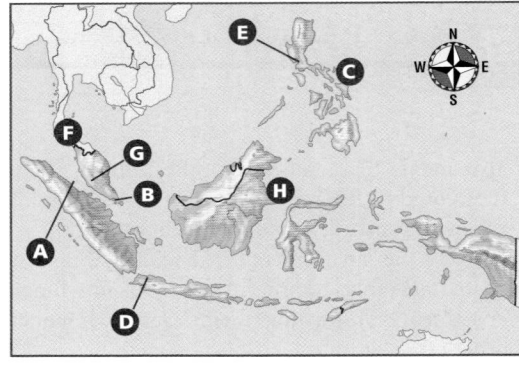

Understanding the Main Ideas
Section 1
1. _Physical Systems_ Why is the land of island Southeast Asia so unstable in some areas?

2. _Places and Regions_ How do the region's location and monsoons influence its climates?

Section 2
3. _Human Systems_ Which ethnic group makes up the majority of Singapore's population? Do members of that ethnic group tend to live in urban or rural areas throughout island Southeast Asia?

4. _Places and Regions_ What is the dominant religion of island Southeast Asia? What other religions are practiced there?

Section 3
5. _Places and Regions_ What economic, environmental, and political challenges face the region today?

Thinking Critically
1. **Analyzing Information** How has the region's cultural diversity affected the structure and goals of educational systems there?

2. **Drawing Inferences and Conclusions** Why do you think wet-rice cultivation is the most common method of growing rice in the region?

3. **Identifying Cause and Effect** How do you think island Southeast Asia might be different today if Europeans had never colonized the region?

Using the Geographer's Tools
1. **Analyzing Population Pyramids** Study the population pyramids in Section 2. What clues might these pyramids give us about the relative standard of living in the countries?

2. **Analyzing Statistics** Review the Fast Facts and Comparing Standard of Living tables at the front of this unit. Then go to **go.hrw.com** to access the World Factbook. Use information you find from these sources to rank the countries of island Southeast Asia by standard of living. Write a paragraph explaining your ranking.

3. **Preparing Maps** Draw an outline map of island Southeast Asia and identify areas that are experiencing ethnic and religious conflict.

Writing about Geography
Review the various ways farmers in island Southeast Asia modify their physical environment to grow rice. Compare those methods with the ways farmers in dry climate regions of the United States modify their environment to grow wheat. You may want to use the Internet or consult library resources for more information. Write a short report about your findings. When you are finished with your report, proofread it to make sure you have used standard grammar, spelling, sentence structure, and punctuation.

SKILL BUILDING

Geography for Life
Creating a Benefits-Cost Balance Sheet
Environment and Society As you have read, the tropical rain forests of island Southeast Asia are being cleared rapidly. What might be some advantages and disadvantages of clearing the forests for farmland, timber, or other purposes? Identify and evaluate those benefits and costs in the form of a two-column balance sheet.

Have students conduct research on the *wayang* shadow puppets of Java. These puppets are usually made from sheets of leather pierced with decorative holes. Have students create their own puppets and write a shadow play about events or issues in island Southeast Asia. Heavy paper can be substituted for leather. Students can mount their puppets on bamboo skewers and use brads to make the limbs flexible. The performance takes place behind a screen—perhaps an old white sheet—and in front of a light source. After students have performed their plays, place puppets and notes in portfolios.

Food Festival

During the time the Dutch occupied Indonesia they adopted a style of dining called a rijsttafel (RYS-tah-fuhl), which means "rice table" in Dutch. A rijsttafel consists of rice accompanied by many small, highly-seasoned dishes, including fruits, meats, seafoods, vegetables, and sauces. Have students locate and prepare recipes for the side dishes. Provide a big bowl of rice, and let students help themselves. Note that peanut sauce is a popular condiment in Indonesia. Some students are severely allergic to peanuts. You may want to eliminate peanut sauce or peanuts from the menu completely.

Building Social Studies Skills

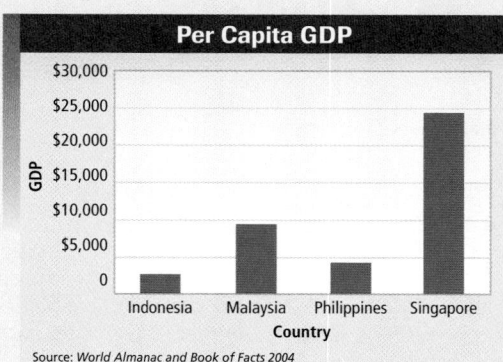

Per Capita GDP

Source: *World Almanac and Book of Facts 2004*

Interpreting Graphs

Study the bar graph above. Then use the information in the graph to help you answer the questions that follow. Mark your answers on a separate sheet of paper.

1. Singapore's per capita GDP is
   a. less than $10,000.
   b. more than five times the per capita GDP of Malaysia.
   c. more than $20,000.
   d. the lowest of the four countries.

2. Rank the countries from first to last by the value of each country's per capita GDP.

Analyzing Secondary Sources

Read the following passage and answer the questions. Mark your answers on a separate sheet of paper.

> "Farmers in the region grow rice in three ways. Wet-rice, or paddy, cultivation is the most productive and common method. Rice paddies are constructed with dikes in lowland areas or with mud terraces in hilly areas. Water flow down steep slopes is controlled, and erosion is limited. This, along with the area's warm and wet climate, allows farmers to grow more than one rice crop each year."

3. In the selection the word *terraces* refers to
   a. beautiful structures on the region's old colonial farmhouses.
   b. scenic spots along rivers in Malaysia.
   c. city gardens that have been transformed into rice farms.
   d. flat areas carved into hillsides so that rice can be grown there.

4. Why do you think being able to grow more than one rice crop a year is important in island Southeast Asia? Give specific reasons for your answer.

Writing

Students should discuss various irrigation and soil conservation methods used by U.S. wheat farmers and the impact of those methods on the environment. Use Rubric 38, Writing to Classify, to evaluate student work.

Geography for Life

Costs might include loss of wildlife habitat, loss of biodiversity, smoke from land-clearance fires, diminished tourist revenue, and increased flooding from deforested slopes. Benefits might include increased farmland and increased revenue for governments and corporations. Students should support their opinions. Use Rubric 7, Charts, to evaluate student work.

Social Studies Skills

1. c

2. Singapore, Malaysia, Philippines, Indonesia

3. d

4. Possible answer: Growing large amounts of rice is essential to feeding the region's huge and growing population.

Alternative Assessment

PORTFOLIO ACTIVITY

Learning about Your Local Geography

Group Project: Field Work

Plan, organize, and complete a research project with a partner about the plant and animal life of your state. Identify important geographic questions about the species you identify in the area. What environmental conditions, such as climate, are common there? What species have adapted to that environment? What effect has interaction with humans had on the local environment and the species that live there? Organize your questions and answers in a table and identify connections between your local environment and that of island Southeast Asia. Finally, prepare a short report comparing the natural environments of your state and island Southeast Asia.

internet connect

Internet Activity: go.hrw.com
KEYWORD: SW3 GT30

Choose a topic about island Southeast Asia to:
• test your knowledge of island Southeast Asia with an interactive map.
• research the rain forests of island Southeast Asia and make a poster.
• analyze tectonic forces that cause volcanoes and earthquakes.

PORTFOLIO ACTIVITY

Students' reports will vary according to environmental conditions, the amount of biodiversity, and extent of human interaction in the region. Use Rubric 14, Group Activity, and Rubric 30, Research, to evaluate student work.

Workshop 1
Going Further: Thinking Critically

Before class, prepare slips of paper that name everyday skills that younger children may not yet have mastered but at which your students are adept. Possibilities include starting a car, programming a VCR, loading a dishwasher, baking a cake, or scoring a soccer goal. Then organize the class into small groups—one group for each slip of paper you prepared. Ask each group to draw a skill from a hat. Allow groups to trade skills if they wish. Then have each group prepare a diagram that demonstrates how to complete the task. Tell students they should not include titles on their diagrams at this point.

Then have the groups exchange their diagrams and challenge each group to identify the skill that is being illustrated in the diagrams it receives. If groups are able to identify the skill, ask them to write suggestions for revisions that might make the diagrams clearer. If a group cannot identify the skill presented on the diagram, tell group members what the diagram is intended to depict and ask them how the process can be better illustrated. Return the diagrams to the groups that produced them and have groups revise their diagrams, incorporating the suggested improvements. Then ask a volunteer from each group to present the group's diagram to the class. Lead a class discussion about the importance of clarity in making diagrams.

Geography
Skill-Building Workshop

Creating a Diagram

Determine the Purpose → Research → Gather Materials → Illustrate the Pieces → Review and Revise

WORKSHOP 1

Using Diagrams

Some people are visual learners. They learn best when information is displayed in a visual format, such as a photograph, map, or diagram. Diagrams are illustrations that show us how to do something or how parts of a whole are related. You are probably familiar with many kinds of diagrams. For example, you may have used a diagram to put together a model airplane or ship. Car owner's manuals also have diagrams. They may show an owner how to accomplish important tasks, such as changing a tire. Public buildings often have diagrams that direct people to exits. Many kinds of diagrams are useful in geography. For example, the diagram on this page shows the kind of tectonic activity that makes parts of East Asia so prone to earthquakes.

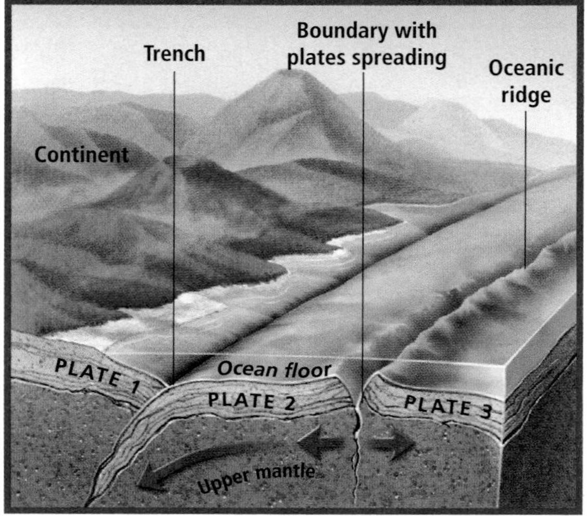

Labels, arrows, and different colors help explain what happens when one tectonic plate slides under another.

Developing the Skill Another useful diagram is a flowchart. A flowchart is a graphic representation of a series of related events or the steps in a process. Note the steps, or activities, in the flowchart at the top of this page. Arrows, or connectors, show the progress of activities. We can use this flowchart to help us plan steps for preparing other diagrams:

- Determine the purpose of your diagram. Are you showing a process, special features, a distribution, relationships, or something else?
- Conduct research to find the information you need. Research may also help you decide what kind of diagram best presents your information.
- Gather the materials you will need, such as paper, pencils, other art tools, or special software.
- Illustrate the pieces of the diagram, including labels that will help the reader follow along.
- Review and revise your diagram and labels for accuracy.

Movement at Plate Boundaries

Trench
Boundary with plates spreading
Oceanic ridge
Continent
PLATE 1
Ocean floor
PLATE 2
PLATE 3
Upper mantle

Practicing the Skill

1. Create a diagram that illustrates features in your school. For example, illustrate an evacuation plan for emergencies. You might also illustrate the suggested arrangement of a classroom or auditorium for a special event.
2. Research the process involved in producing rice. Then create a flowchart illustrating this process. Major steps will include planting, harvesting, and processing rice. You might want to create illustrations for each major step in the harvesting process.

Workshop 2
Going Further: Thinking Critically

Point out to students that a popular feature of many spreadsheet programs is the ability to convert numerical data into graph form. Coupled with the ability to perform calculations on large amounts of data, this feature makes spreadsheet software a popular tool in organizing information for presentations.

Have students examine the spreadsheets about economic growth in China that they created in the activity below. Ask them to list the steps that would be necessary to create a line graph based on the data. *(Possible answers: drawing a grid, assigning values along each index, plotting points along the indices, drawing the line)* Tell them that because the information has been entered into a spreadsheet program, creating a graph takes only a few keystrokes. Have students experiment with the various types of graphs your spreadsheet software can create. Ask students to suggest the advantages and disadvantages of each type.

Organize the class into small groups and have each group use information in this unit's Fast Facts tables to create a spreadsheet. Students might, for example, create formulas to determine how many people in each country are engaged in particular economic activities. Then have students use the spreadsheet program to create graphs that reflect their findings. Remind students to choose the type of graph most appropriate to the information they wish to convey.

Sample Population Data: Country A

E6 = 10.06

Workbook1

A	B	C	D	E	F	G	H	I
	2000	2001	2002	2003	2004	2005		
Total Population (×1,000)	1,198,500	1,211,210	1,223,890	1,236,260	1,248,100	1,259,090		
Birthrate (per 1,000)	17.70	17.12	16.98	16.57	16.03	15.23		
Death Rate (per 1,000)	6.49	6.57	6.56	6.51	6.50	6.46		
Natural Growth Rate (%)	11.21	10.55	10.42	10.06	9.53	8.77		
Total Number of Births (×1,000)	21,040	20,630	20,670	20,380	19,910	19,090		

Sheet1 / Sheet2 / Sheet3

WORKSHOP 2

Using Spreadsheets and Software

Sometimes you need to work with a large amount of numerical information in a simple and efficient manner. People use electronic spreadsheets, or worksheets, to do just that. You can use a spreadsheet for just text. However, a spreadsheet's real advantage is its ability to calculate values from preset formulas. It can even recalculate data automatically when entries change.

Developing the Skill All spreadsheets, regardless of the software, will follow the same design. Vertical columns are assigned letters in alphabetical order. Horizontal rows are arranged numerically. A cell is the intersection of one column and one row. Identifying a cell is similar to finding a location on a map using lines of latitude and longitude. The cell is named for its corresponding column and row. For example, the cell at the intersection of Column B and Row 2 is labeled B2. (See the spreadsheet above.)

To reach a specific cell you have several choices. You may use the mouse, arrow keys, or tab key. The computer will tell you which cell you are in by outlining the four sides of the cell. The cell will also be shown in a separate window at the top of the screen. In the example above you can see that cell E6 has a value of 10.06%. This is determined by studying the cell itself or by viewing the *status* or *formula line* found at the top of the worksheet.

Instead of calculating a value on a calculator, you can put a formula, such as C2 + C5, in a spreadsheet cell. Spreadsheets use standard formulas to create a simple mathematical equation. To create a formula, select the cell where you want to display the formula results. Type an equal sign (=) and then the rest of the formula. You may add, subtract, multiply, and divide.

For example, the formula = *A1* + *A2* would add the value in cells A1 and A2. The formula = *C5* − *C4* would subtract the value in cell C4 from the value in cell C5. A slash (/) is used for division and an asterisk (*) for multiplication. Adding is the most common function of many spreadsheets. As a result, most spreadsheet software uses an *AutoSum* key (Σ) that you can place in a cell to add values automatically. For your convenience, many spreadsheet programs include a number of predefined formulas.

Another key feature of spreadsheets is the ability to calculate data from multiple worksheets. Suppose you are doing a report on precipitation around the world over a certain period of time. You could have separate worksheets for the continents, with precipitation amounts for individual countries on each continent. In addition, you could have a worksheet to show world precipitation changes over the same number of years. As you enter new country information, the numbers on the world worksheet change as well.

Practicing the Skill

Practice using a spreadsheet.
1. Open a new spreadsheet file.
2. Type the information as shown in the example on this page.
3. Estimate data you might expect to find for Country A in 2000 and place that data in Column H. Label this column "2000."
4. Label column I "Total." In cell I7 calculate the total births in Country A from 1994 to 2000. Use the formula = *B7* + *C7* + *D7* + *E7* + *F7* + *G7* + *H7* or use the *AutoSum* key (Σ).
5. Use a new spreadsheet to analyze data from a real country. For example, track annual growth in GDP, exports, or imports to analyze economic development in China in the 1990s.

PRACTICING THE SKILL

Students' spreadsheets should accurately reflect information about countries of their choice. Data contained in the spreadsheet should be relevant to the subject studied.

Using the Illustration

Direct students' attention to the photograph of the Great Barrier Reef on this page. Point out that the reef covers some 135,000 square miles (349,650 sq km) off the northeastern coast of Australia. Not actually a single reef, the Great Barrier Reef is composed of about 2,500 individual reefs and 800–900 islands. The entire structure was built by coral—tiny marine organisms whose skeletons make up reefs— and so the Great Barrier Reef has been called the largest structure in the world built by living creatures. Scientists have determined that the building of the Great Barrier Reef began more than 5 million years ago.

Coral reefs provide habitats for thousands of oceanic species. The Great Barrier Reef is also home to many endangered species. Humpback whales and dugongs live there, as do six of the world's seven species of sea turtle. Turtles like the one seen here make their way to the reef each year to breed. To protect the reef's fragile ecosystem, Australia allows tourists to visit only about 5 percent of its total area. Government regulations designed to protect the reef have also been established. Ask students to suggest some threats that tourism or other economic activities pose to the Great Barrier Reef. *(Possible answers: pollution from litter or fuel, destruction of habitat caused by mining or other activities, reduction of species by fishing)*

Unit Objectives

1. Describe the landforms and climates of the Pacific world.

2. Identify the culture groups of the region and relate their experiences to historical events.

3. Explain the relationship of the region's resources to the economic development of Australia, New Zealand, and the Pacific Islands.

4. Identify the environmental challenges that confront the Pacific world.

5. Use media services to present geographic information.

6. Understand the decision-making process and how it can be used to resolve geographic issues.

UNIT 10 The Pacific World

Green turtle,
Great Barrier Reef

Chapter **31** *Australia and New Zealand*

Chapter **32** *The Pacific Islands*

CHAPTER 31 Australia and New Zealand

describes the countries' landforms, climates, water and mineral resources, unique plant and animal life, history, agriculture, industry, and population distribution.

CHAPTER 32 The Pacific Islands

introduces the distinctive characteristics of the islands and island groups of the Pacific region.

CONNECTING TO Literature

A SECRET COUNTRY *by John Pilger*

John Pilger
(1939–) was born and educated in Sydney, Australia. Pilger is a journalist, filmmaker, and playwright. In this selection from his book *A Secret Country*, Pilger describes the beach in the Bondi neighborhood of Sydney, where he grew up. His reflections provide valuable insights into the human geography of his homeland.

By December, when the king tides have arrived from across the south Pacific, the salt spray blows up from the beach. It stiffens the air, covers windows with a stocky mist, corrodes paint on cars and mortar between bricks, and tastes like Bondi and summer. . . .

In Bondi, even the crankiest streets have a glimpse of the Pacific, if not of the beach itself. Whatever the state of life in the streets, the great sheet of dazzling blue-green is always there, framed between chimneys and dunnies [outhouses]. On weekdays my friend Pete and I would 'scale' a Bondi Beach tram if we spotted an old 'jumping jack' type, which had a peculiarly high suspension and could be bounced off the rails by a swarm of eleven-year-olds synchronising their efforts. And of course Len's [the tram conductor] apoplexy [great anger] was part of the fun. . . .

All principal beaches in Australia are public places. This is not so in the United States and Europe, where the private possession of land and sea is rightly regarded by visiting Australians as a seriously uncivilised practice. Although private property is revered by many Australians, there are no proprietorial [ownership] rights on an Australian beach. Instead, there is a shared assumption of tolerance for each other, and a spirit of equality which begins at the promenade steps. . . .

Australia, a society with a deeply racist past, has absorbed dozens of diverse cultures peacefully. The beach and the way of life it represents are central to this. A spectacle at Bondi in the 1950s . . . was the arrival on the beach of the first post-war immigrants. . . . Bolting lemming-like into a deceptively light surf, they would be duly rescued by lifesavers with a large trawling net. The ritual was repeated as each national group arrived. . . .

For most Australians, who live in congested coastal cities, the foreshore, the beach, is the one link with our ancient continent. . . . We see and understand little beyond the last of the urban red-tiled roofs, but many of us understand well the rhythm of water on sand, of wind on current. A Bondi child will know the feel of a westerly, a nor'easterly and a 'southerly buster.' There is a grace about this life.

Analyzing the Primary Source

1. **Comparing** How do beaches in Australia differ from beaches in the United States?

2. **Analyzing Information** According to Pilger, how have Australia's beaches shaped the country's society?

In this unit, students will learn about the people and geography of Australia, New Zealand, and the Pacific Islands.

Australia's unique ecosystems are the result of the continent's isolation. Some of the country's endemic species, once protected by isolation, are now threatened by imported species. Geological activity makes New Zealand's landscape highly varied. Both countries have majority populations of British heritage and minority indigenous populations. The Aborigines of Australia and the Maori of New Zealand have distinct vibrant cultures.

Australia and New Zealand have democratic governments. Most of the people enjoy a high standard of living. The economies combine agriculture, manufacturing, and services.

The Pacific Islands' beauty disguises some economic difficulties. Many of the islanders depend on aid from other countries.

Your Classroom Time Line

These are the major dates and time periods for this unit. Have students enter them on the time line you created earlier. You may want to watch for these dates as students progress through the unit.

c.* 38,000 B.C. Aborigines move into Australia.

c. 1300s B.C. Fiji and Tonga may have been settled at this time.

c. A.D. 1000 The Maori move into New Zealand.

1500s Europeans explore the Pacific Islands.

1520–21 Ferdinand Magellan sails across the Pacific Ocean.

1642 Abel Tasman reaches New Zealand.

1769 James Cook visits New Zealand.

1788 Great Britain establishes its first penal colony in Australia.

1820s Van Diemen's Land (Tasmania) and Western Australia are founded.

1840 Britain signs a treaty with the Maori and takes control of New Zealand.

1845–72 The Maori Wars are fought in New Zealand.

* c. stands for *circa* which means "about."

POLITICAL MAP ANSWERS

1. Wellington
2. United States, France, United Kingdom

CRITICAL THINKING ANSWER

3. Possible answer: Australia, because it is by far the largest country in the region

UNIT 10 ATLAS

The World in Spatial Terms

The Pacific World:
Political

1. **Places and Regions** Which of the region's capital cities is located farthest south?
2. **Human Systems** Which outside countries control islands in the Pacific region?

Critical Thinking

3. **Analyzing** Based on information on the map, which country would you expect to have the largest economy? Why?

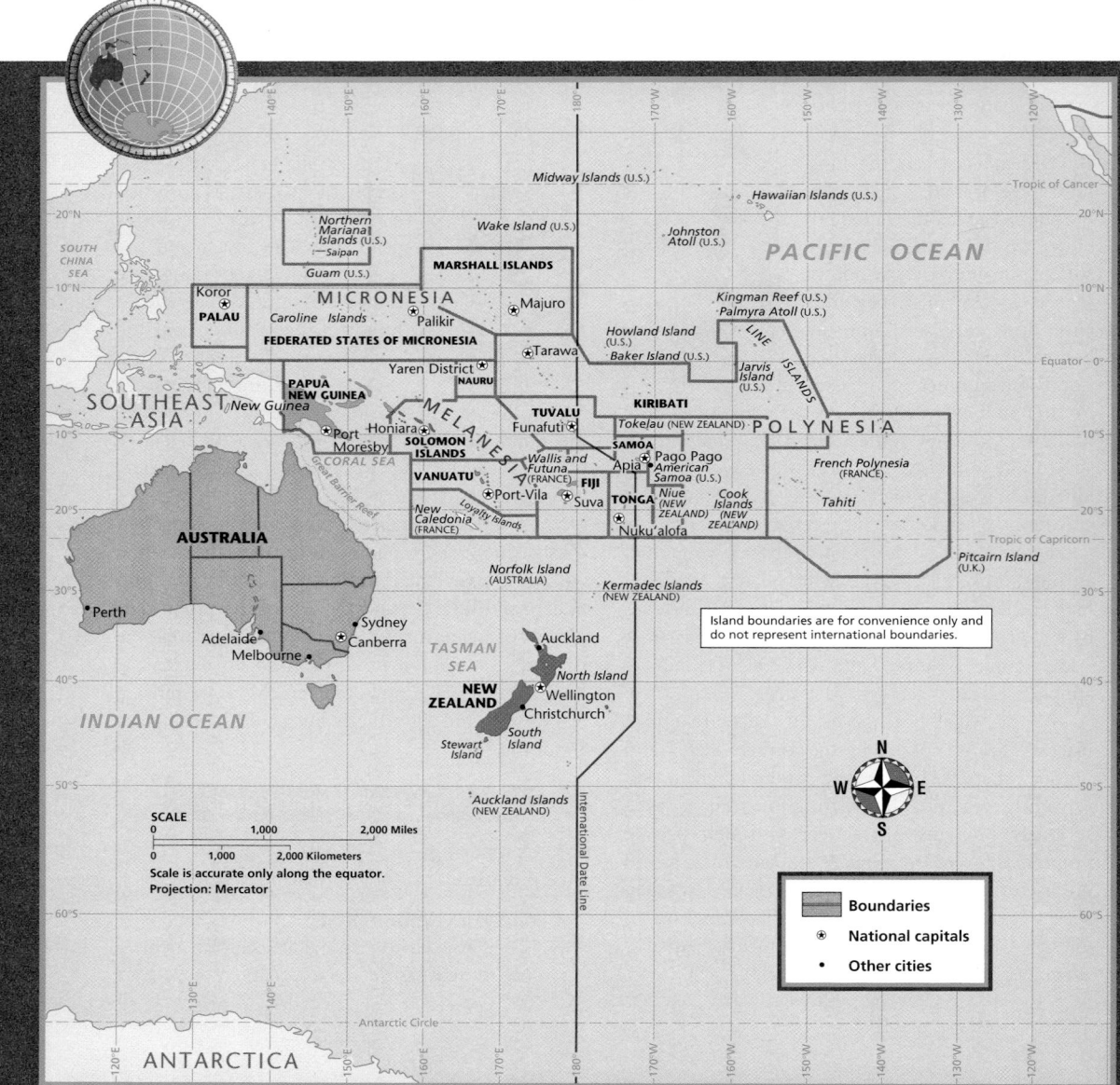

Using the Political Map

Focus students' attention on the **political map** on the opposite page. Point out that the islands of this region are often grouped for convenience into three clusters. Ask students to name the groups *(Melanesia, Micronesia, Polynesia)*. Tell the class that the prefixes in these terms describe the dark skin of many island residents *(mela-)*, the small size of the islands *(micro-)*, and the large number of islands *(poly-)*. Call on volunteers to identify some of the islands within the three groups and their capital cities. Point out that the three terms do not relate to political boundaries. Then ask students to identify the capitals of Australia *(Canberra)* and New Zealand *(Wellington)*.

Using the Physical Map

Direct students' attention to the **physical map** on this page. Have students identify the major river system of Australia *(the Darling-Murray system)*. Ask students to locate the physical feature in which the source of this river system lies *(Great Dividing Range)*. Then ask students to identify and compare the highest points in Australia and New Zealand. *(Mount Cook, New Zealand's highest peak, is almost twice as tall as Mount Kosciusko, the highest peak in Australia.)*

Your Classroom Time Line, (continued)

1851 Gold is discovered in Australia.

Mid–1800s More British colonies are established in Australia.

1898 The United States takes Guam from Spain after the Spanish-American War.

1901 Six colonies unite to form the independent Commonwealth of Australia.

1907 New Zealand gains its independence.

1913 Canberra is founded.

1940s–60s France, Britain, and the United States begin nuclear testing in their Pacific territories.

1941–45 World War II is fought in the Pacific.

1967 Aborigines are granted Australian citizenship.

1970s–90s Papua New Guinea fights to prevent Bougainville from gaining independence.

1970s Tensions between ethnic groups in Fiji escalate.

1970s Many Asians migrate into Australia.

Late 1980s Aborigines begin to protest mining on sacred lands in Australia.

1987 Maori is named an official language of New Zealand.

Late 1990s France ends underground nuclear testing in its Pacific territories.

2000 Rebels try to overthrow the governments of Fiji and the Solomon Islands but are unsuccessful.

The Pacific World: Physical

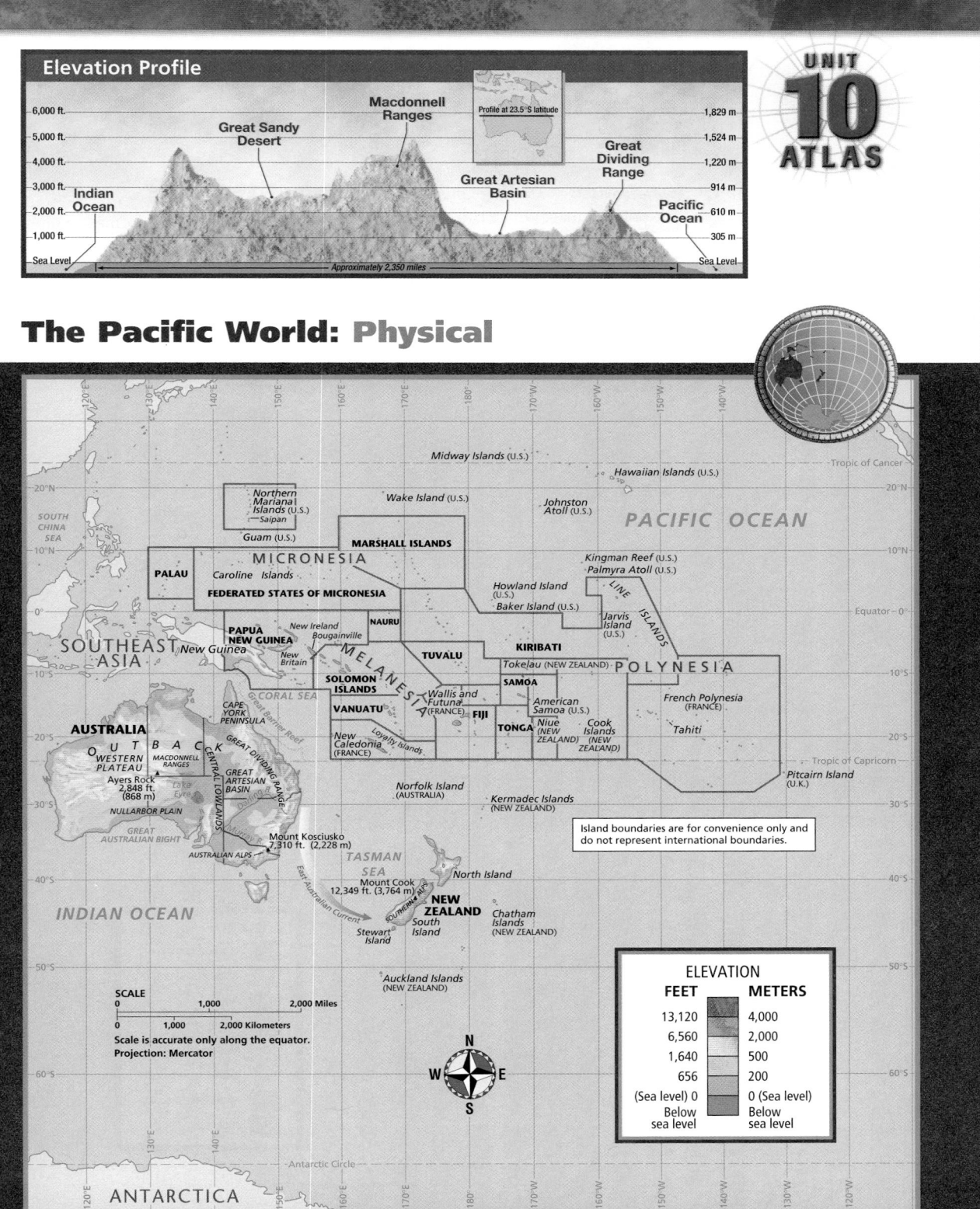

Elevation Profile

UNIT 10 ATLAS

Using the Climate Map

Have students examine the **climate map** on this page. Point out that nearly all of the world's climate types are represented in this region. Have students compare this map to the **physical map** of the region. For each climate type, call on a volunteer to describe the relative location of a place with that climate.

(Examples: tropical humid—northeast coast of Papua New Guinea; tropical wet and dry—north coast of Australia; arid— central Australia) Ask which climate appears to be most common in the Pacific Islands *(tropical humid)*.

CLIMATE MAP ANSWERS

1. arid and semiarid

2. It probably brings rainy weather from the west to New Zealand.

CRITICAL THINKING ANSWER

3. Possible answer: along the coast in areas with mild climates

UNIT
10
ATLAS

The Pacific World:
Climate

1. (*Physical Systems*) Compare this map to the political map. Which two climates cover most of Australia?

2. (*Physical Systems*) Compare this map to the physical map. How might the East Australian Current affect New Zealand's weather?

Critical Thinking

3. **Analyzing** Based on information on the map, where would you expect most people in Australia to live?

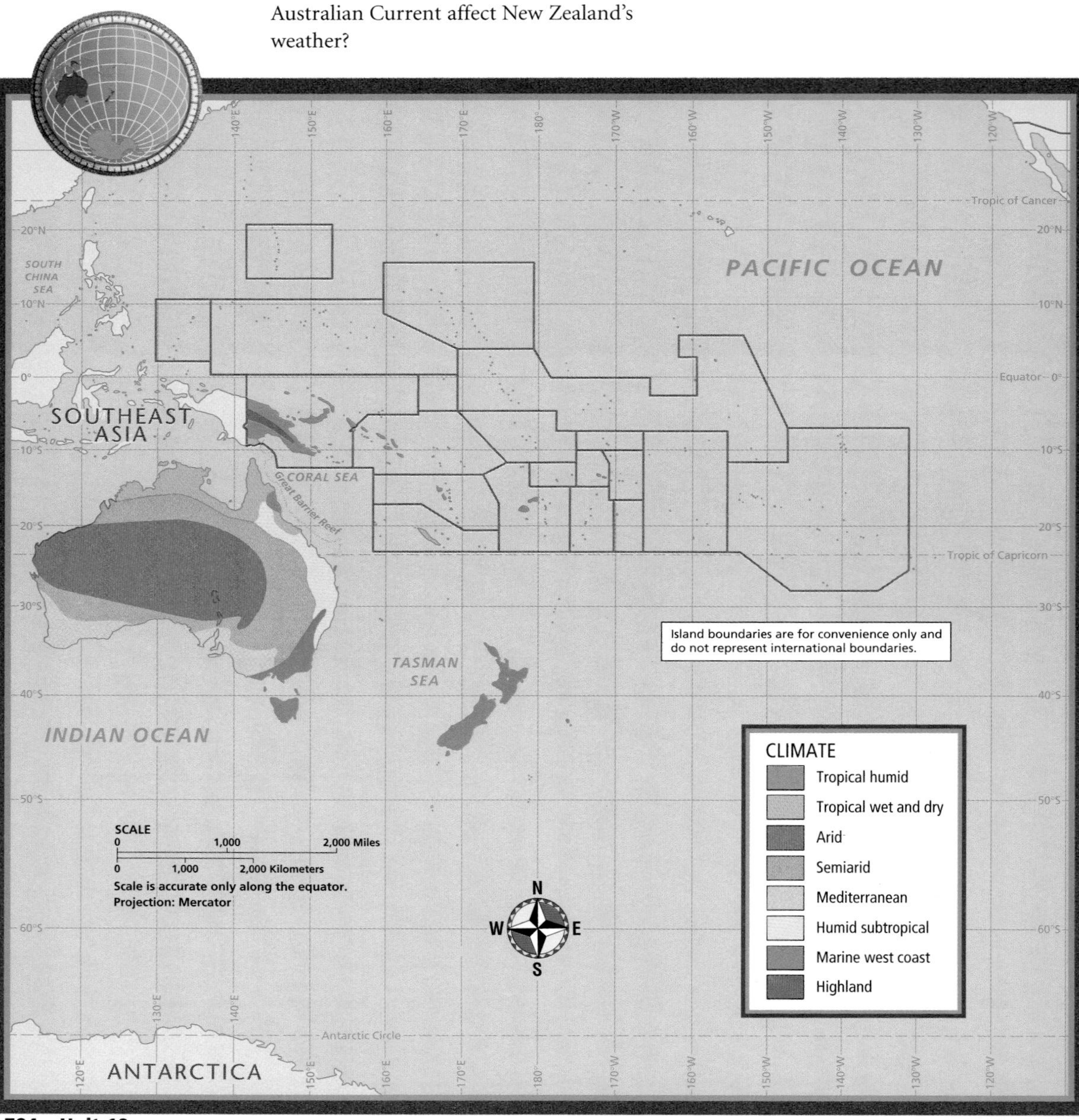

SOUTH CHINA SEA

PACIFIC OCEAN

Tropic of Cancer

20°N

10°N

SOUTHEAST ASIA

0° Equator

10°S

CORAL SEA

Great Barrier Reef

20°S

Tropic of Capricorn

Island boundaries are for convenience only and do not represent international boundaries.

30°S

TASMAN SEA

40°S

INDIAN OCEAN

50°S

SCALE

0 1,000 2,000 Miles

0 1,000 2,000 Kilometers

Scale is accurate only along the equator.
Projection: Mercator

60°S

N
W E
S

CLIMATE

- Tropical humid
- Tropical wet and dry
- Arid
- Semiarid
- Mediterranean
- Humid subtropical
- Marine west coast
- Highland

Antarctic Circle

ANTARCTICA

Focus the students' attention on the **precipitation map** on this page. Call on a volunteer to describe the amount of precipitation that falls over most of the Pacific Islands *(more than 80 inches each year)*. Ask students why they think these islands receive so much precipitation. *(They are located in the middle of a warm ocean in the low latitudes.)* Then ask students to identify which coast of Australia receives the most precipitation *(east coast)*. Have students examine the **physical map** to suggest why this coast receives so much rain. *(Possible answer: A mountain barrier blocks moisture from the Pacific and creates rain on the coast.)*

The Pacific World:
Precipitation

UNIT
10
ATLAS

1. **Physical Systems** Compare this map to the political map. Which part of Australia receives the most precipitation?

2. **Places and Regions** Compare this map to the political and physical maps. How do the Southern Alps affect the distribution of precipitation in New Zealand?

Critical Thinking

3. **Making Generalizations** Compare this map to the political map. How do you think latitude affects the amount of precipitation in central Australia?

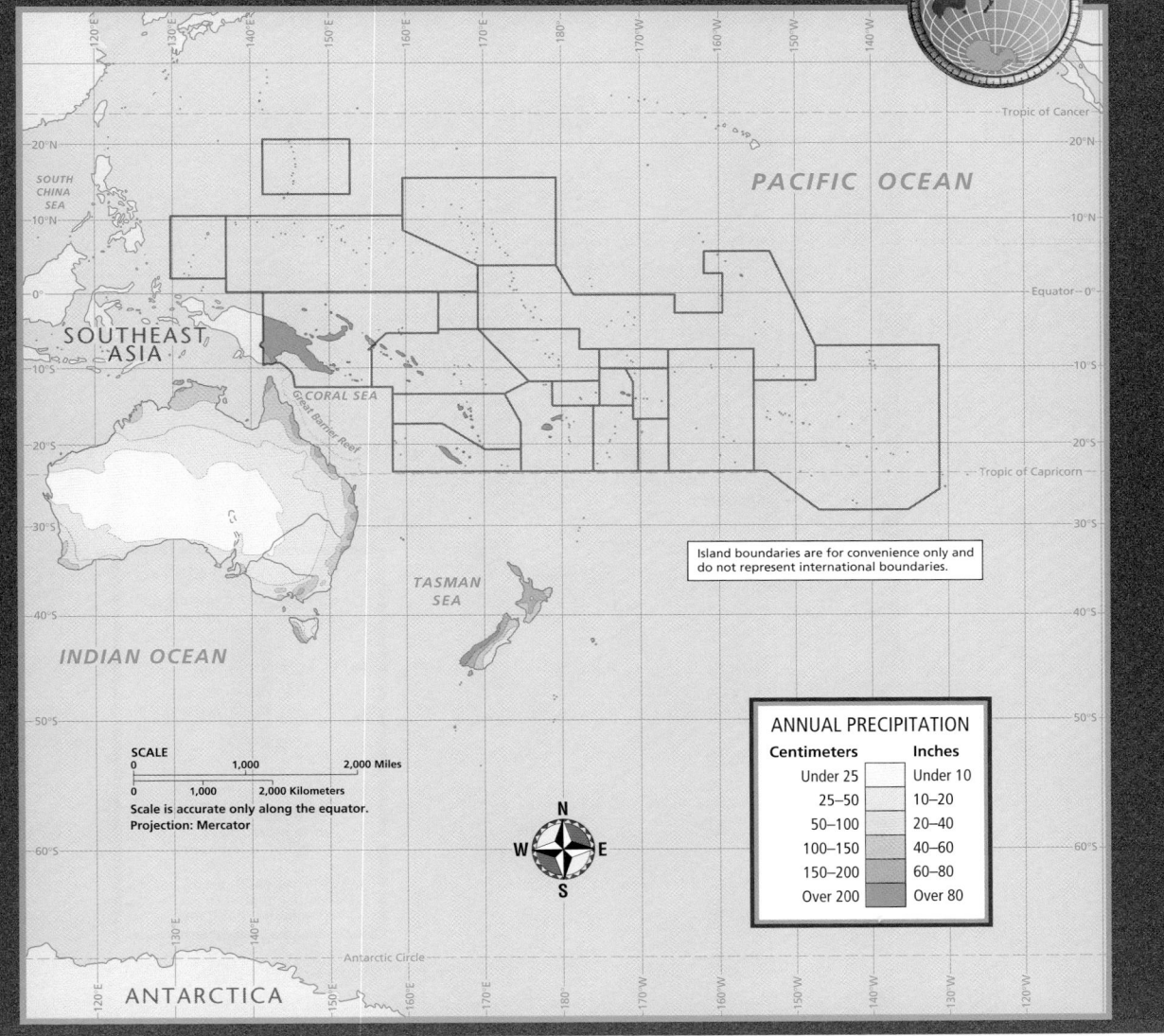

ANNUAL PRECIPITATION

Centimeters	Inches
Under 25	Under 10
25–50	10–20
50–100	20–40
100–150	40–60
150–200	60–80
Over 200	Over 80

SCALE
0 — 1,000 — 2,000 Miles
0 — 1,000 — 2,000 Kilometers
Scale is accurate only along the equator.
Projection: Mercator

Island boundaries are for convenience only and do not represent international boundaries.

705

Using the Population Map

Focus students' attention on the **population map** on this page. Ask them to write statements to summarize the region's population density. *(Examples: Population density is low in the region and concentrated in a few cities. Population density is heaviest along the southeast coast of Australia.)* Call on volunteers to read their sentences.

Ask students which of New Zealand's islands appears to have more people *(North Island)*. Have students compare this map to the **physical** and **climate maps**. Ask: Why does Papua New Guinea have an east-west strip of sparse population density? *(It has a mountainous area with a highland climate.)*

POPULATION MAP ANSWERS

1. the southeast
2. Lowland areas along the coast have higher population densities than mountainous areas.

CRITICAL THINKING ANSWER

3. Possible answer: It is probably much smaller because it includes many small islands, has few large cities, and most of Australia has a low population density.

UNIT 10 ATLAS

The Pacific World:
Population

1. (**Places and Regions**) Compare this map to the political map. Which part of Australia has the highest population density?

2. (**Environment and Society**) Compare this map to the physical map. How do New Zealand's landforms appear to influence the distribution of its population?

Critical Thinking

3. **Comparing** How do you think the total population of the Pacific region compares to other major world regions? Why?

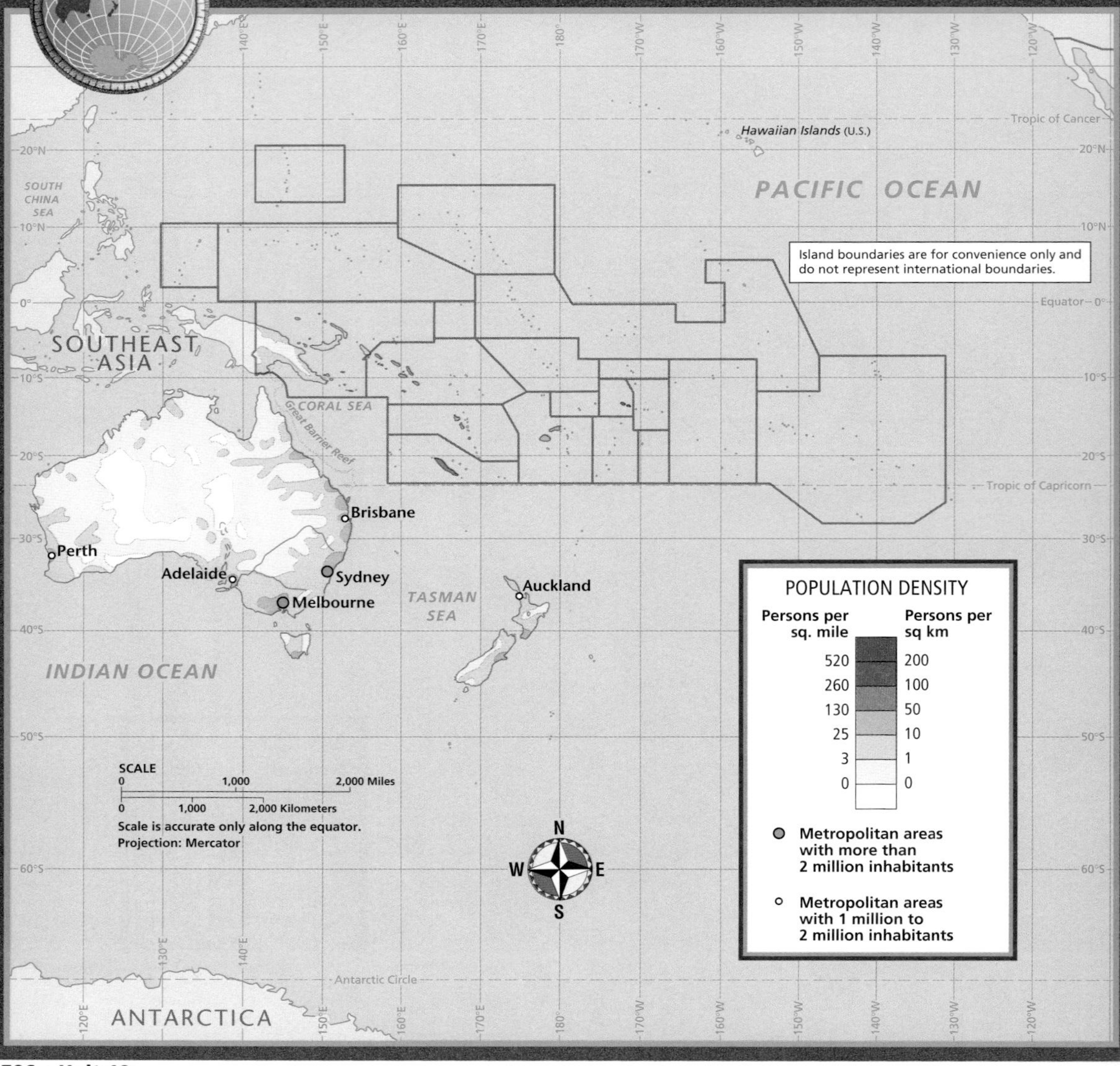

Island boundaries are for convenience only and do not represent international boundaries.

Hawaiian Islands (U.S.)

Tropic of Cancer

PACIFIC OCEAN

SOUTH CHINA SEA

Equator—0°

SOUTHEAST ASIA

CORAL SEA

Great Barrier Reef

Tropic of Capricorn

Brisbane

Perth

Adelaide
Sydney
Melbourne

TASMAN SEA

Auckland

INDIAN OCEAN

SCALE
0   1,000   2,000 Miles
0   1,000   2,000 Kilometers
Scale is accurate only along the equator.
Projection: Mercator

Antarctic Circle

ANTARCTICA

POPULATION DENSITY

Persons per sq. mile	Persons per sq km
520	200
260	100
130	50
25	10
3	1
0	0

● Metropolitan areas with more than 2 million inhabitants

○ Metropolitan areas with 1 million to 2 million inhabitants

Using the Land Use and Resources Map

As students examine the **land use and resources map**, explain that the interior of Australia has been described as a storehouse of mineral wealth. Ask students what they think this statement means. (*Possible answer: The region has rich deposits of coal, gold, diamonds, uranium, and other minerals.*) Then ask what minerals are found in New Zealand (*coal, gold, and other minerals*). Have students compare this map to the **political map**.

Ask students what Pacific Island country has gold deposits (*Papua New Guinea*).

Ask students what industry is dominant across most of Australia and New Zealand (*livestock raising*). Then ask which country in the region has the largest forested area (*Papua New Guinea*).

The Pacific World:
Land Use and Resources Map

UNIT 10 ATLAS

1. (*Places and Regions*) Compare this map to the political map. Which natural resources are found in New Zealand?
2. (*Places and Regions*) Which resources are found in western Australia?

Critical Thinking

3. **Analyzing** Compare this map to the political map. How do you think the location of resources in interior Australia has affected the movement of products, capital, and people?

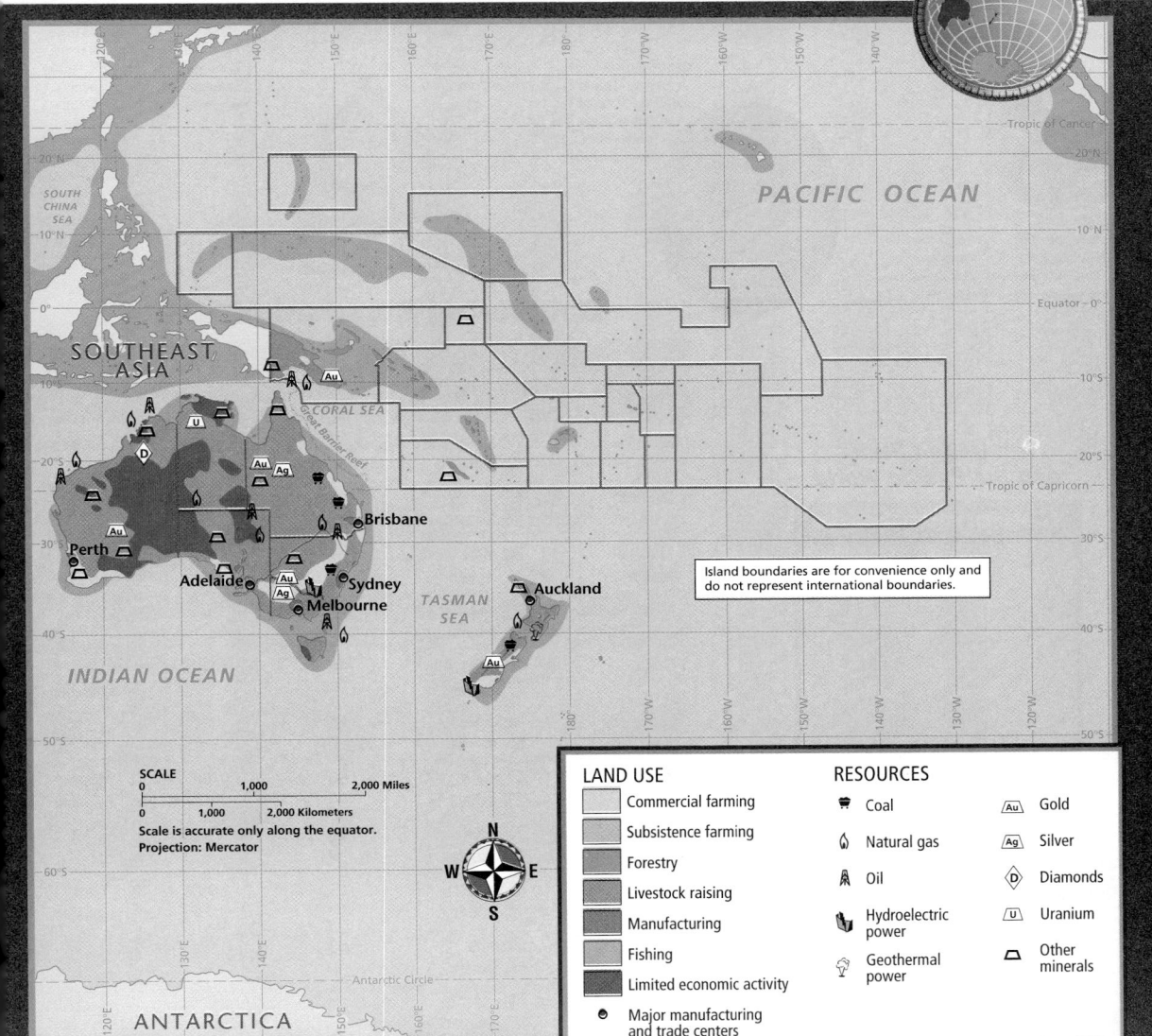

LAND USE
- Commercial farming
- Subsistence farming
- Forestry
- Livestock raising
- Manufacturing
- Fishing
- Limited economic activity
- Major manufacturing and trade centers

RESOURCES
- Coal
- Natural gas
- Oil
- Hydroelectric power
- Geothermal power
- Au Gold
- Ag Silver
- Diamonds
- U Uranium
- Other minerals

Island boundaries are for convenience only and do not represent international boundaries.

SCALE
0 — 1,000 — 2,000 Miles
0 — 1,000 — 2,000 Kilometers
Scale is accurate only along the equator.
Projection: Mercator

Fast Facts Activity

Direct student's attention to the Australian flag pictured in the Fast Facts table. Ask students if they recognize the design in the upper left corner of the flag. (*It is the flag of the United Kingdom.*) Then ask students which other countries in the region have incorporated the British flag into theirs (*Fiji, New Zealand, Tuvalu*). Call on a volunteer to speculate about why these countries use a British emblem in their flags. (*They were once British colonies.*)

Point out the arrangement of five stars on the right side of Australia's flag. Tell students that it represents the constellation known as the Southern Cross, which cannot be seen north of the equator. Ask students why they think Australians chose to include this symbol on their flag. (*Possible answer: Australia is the only inhabited continent located entirely in the Southern Hemisphere. Australians are proud of their distinction as the "Land Down Under.*") Have students identify other countries that have the Southern Cross depicted on their flags (*New Zealand, Papua New Guinea, Samoa*). Tell students that the stars on other countries' flags also have meanings. The nine stars on Tuvalu's flag represent the country's nine islands, and the 12-pointed star on Nauru's flag honor the island's 12 ethnic groups. Its position below the yellow stripe on the flag represents Nauru's position below the equator.

Stars are not the only celestial bodies represented on the flags of the Pacific World. Ask students which countries depict the sun (*Kiribati*) and the moon (*Palau*) on their flags.

Historical Geography

Island-hopping During World War II, Allied forces fighting the Japanese in the Pacific adopted a strategy of island-hopping, attacking and taking strategic islands controlled by Japanese forces. They believed this would leave Japanese garrisons on other islands cut off from supplies and reinforcements as Allied troops got closer and closer to Japan.

The American offensive was launched from Kiribati's Gilbert Islands. From there, troops moved to Butaritari (then called Makin Island) and Tarawa (also in Kiribati), the Marshall Islands, and the Caroline Islands of Micronesia. In 1944, American troops took the Northern Mariana Islands and Guam. From airstrips on these islands, American planes were able to launch raids on Japan itself. That same year, American, Australian, and New Zealand troops retook New Guinea, giving them a base from which to attack the Japanese in the Philippines.

CRITICAL THINKING:
How does the political geography of the Pacific reflect American activities in World War II? (*Many islands are owned by or tied politically to the United States.*)

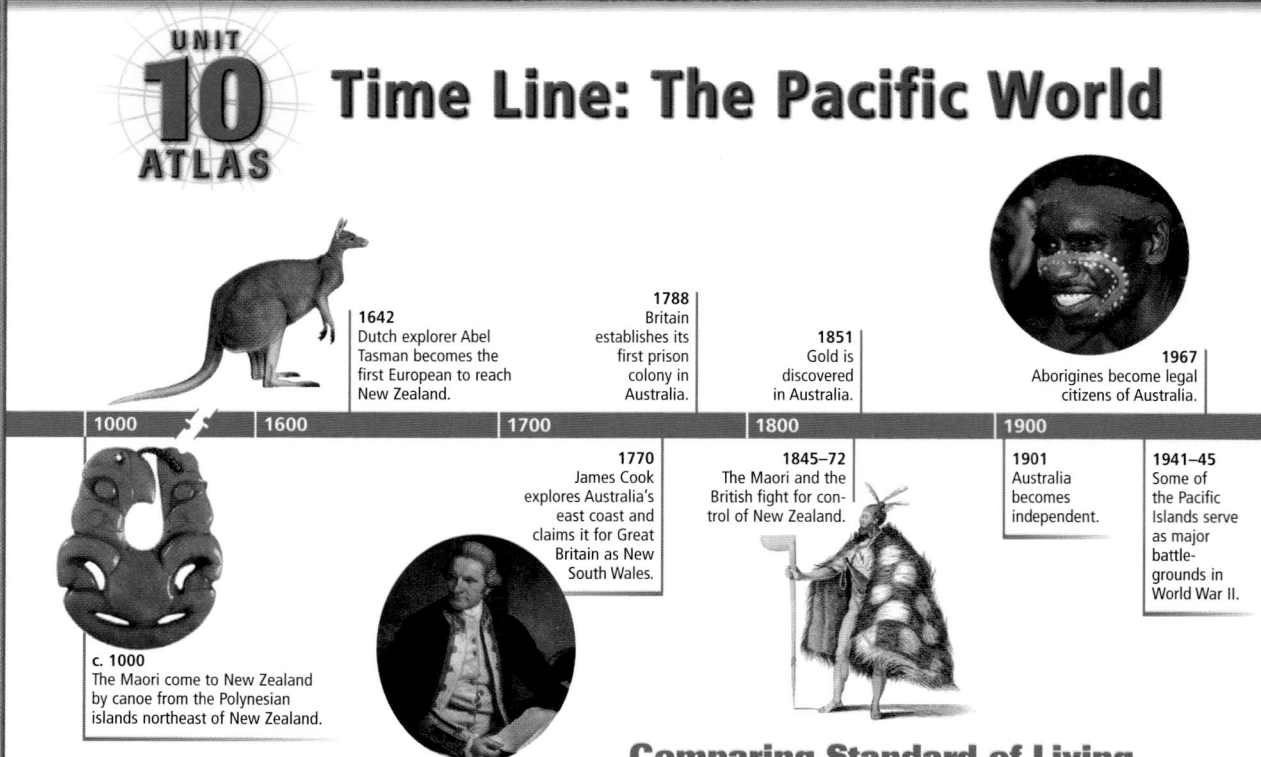

UNIT 10 ATLAS

Time Line: The Pacific World

1642 Dutch explorer Abel Tasman becomes the first European to reach New Zealand.

1788 Britain establishes its first prison colony in Australia.

1851 Gold is discovered in Australia.

1967 Aborigines become legal citizens of Australia.

| 1000 | 1600 | 1700 | 1800 | 1900 |

1770 James Cook explores Australia's east coast and claims it for Great Britain as New South Wales.

1845–72 The Maori and the British fight for control of New Zealand.

1901 Australia becomes independent.

1941–45 Some of the Pacific Islands serve as major battlegrounds in World War II.

c. 1000 The Maori come to New Zealand by canoe from the Polynesian islands northeast of New Zealand.

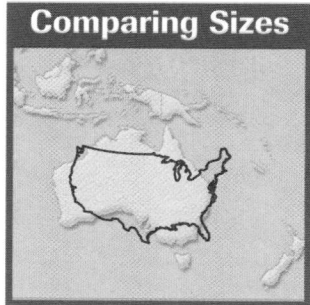

The United States and the Pacific World

Comparing Sizes

Comparing Standard of Living

COUNTRY	LIFE EXPECTANCY (in years)	INFANT MORTALITY (per 1,000 live births)	LITERACY RATE	DAILY CALORIC INTAKE (per person)
Australia	77, male 83, female	5	100%	3,190
Kiribati	58, male 64, female	51	90%	2,977
Marshall Islands	68, male 71, female	32	94%	Not available
New Zealand	75, male 81, female	6	100%	3,315
Palau	66, male 73, female	16	98%	Not available
Papua New Guinea	62, male 66, female	55	66%	2,168
Samoa	67, male 73, female	30	100%	Not available
Solomon Islands	70, male 75, female	23	54%	2,130
Tonga	66, male 71, female	13	99%	Not available
Vanuatu	60, male 63, female	58	53%	2,737
United States	74, male 80, female	7	97%	3,757

internet connect

GO TO: go.hrw.com
KEYWORD: SW3 Almanac
FOR: Additional information and reference sources

Sources: *World Almanac and Book of Facts, 2004; Britannica Book of the Year, 2002*

Tell students that stars may be an important symbol on the Pacific world's flags, but in some parts of the region people can no longer see the stars as clearly as they once could. Scientists blame this on light pollution—excess artificial light that prevents the night sky from becoming completely dark. In some areas, this light keeps the sky bright enough to render the stars of the Milky Way invisible at night.

Copy the following table onto the chalkboard. Then have students examine the Fast Facts tables to make connections between the information in the two tables. *(Possible answer: Countries with high populations and electricity consumption levels have more light pollution than those with small populations that*

use less electricity.) Then have students examine the population map in the unit atlas to draw conclusions about the causes of Australia and New Zealand's high light-pollution levels. *(Most people in those countries live in large cities where more light is produced, so fewer people can view the stars.)*

COUNTRY	POPULATION THAT EXPERIENCES LIGHT POLLUTION	POPULATION THAT CANNOT VIEW THE MILKY WAY
Australia	71 percent	48 percent
Fiji	19 percent	less than 1 percent
New Zealand	87 percent	45 percent
Papua New Guinea	13 percent	less than 1 percent
Vanuatu	8 percent	2 percent
United States	99 percent	71 percent

Source: Light Pollution Science and Technology Institute

Fast Facts: The Pacific World

UNIT 10 ATLAS

FLAG	COUNTRY / Capital	POPULATION (in millions) / POP. DENSITY	AREA	PER CAPITA GDP (in US $)	WORKFORCE STRUCTURE (largest categories)	ELECTRICITY CONSUMPTION (kilowatt hours per person)	TELEPHONE LINES (per person)
	Australia / Canberra	19.7 / 7/sq. mi.	2,967,907 sq. mi. / 7,686,844 sq km	$ 27,000	73% services / 22% industry	9,346 kWh	0.54
	Fiji / Suva	0.8 / 119/sq. mi.	7,054 sq. mi. / 18,270 sq km	$ 5,500	70% subsistence agriculture	577 kWh	0.11
	Kiribati / Tarawa	0.1 / 356/sq. mi.	313 sq. mi. / 811 sq km	$ 840	71% agriculture	66 kWh	0.04
	Marshall Islands / Majuro	0.06 / 807/sq. mi.	70 sq. mi. / 181.3 sq km	$ 1,600	58% services / 21% agriculture	Not available	0.08
	Micronesia, Fed. States of / Palikir	0.1 / 141/sq. mi.	271 sq. mi. / 702 sq km	$ 2,000	67% government	Not available	0.09
	Nauru / Yaren District	0.01 / 1,550/sq. mi.	8 sq. mi. / 21 sq km	$ 5,000	Not available	2,146 kWh	Not available
	New Zealand / Wellington	3.9 / 37/sq. mi.	103,738 sq. mi. / 268,680 sq km	$ 20,200	65% services / 25% industry	9,001 kWh	0.46
	Palau / Koror	0.02 / 111/sq. mi.	177 sq. mi. / 458 sq km	$ 9,000	26% services / 19% trade	Not available	Not available
	Papua New Guinea / Port Moresby	5.7 / 33/sq. mi.	178,703 sq. mi. / 462,839 sq km	$ 2,300	85% agriculture	244 kWh	0.01
	Samoa / Apia	0.2 / 162/sq. mi.	1,137 sq. mi. / 2,945 sq km	$ 5,600	65% agriculture / 30% services	549 kWh	0.06
	Solomon Islands / Honiara	0.5 / 45/sq. mi.	10,985 sq. mi. / 28,451 sq km	$ 1,700	75% agriculture / 20% services	62 kWh	0.01
	Tonga / Nuku'alofa	0.1 / 376/sq. mi.	289 sq. mi. / 749 sq km	$ 2,200	65% agriculture	244 kWh	0.11
	Tuvalu / Funafuti	0.01 / 1,126/sq. mi.	10 sq. mi. / 26 sq km	$ 1,100	68% agric., fish. / 22% services	Not available	Not available
	Vanuatu / Port-Vila	0.2 / 37/sq. mi.	4,710 sq. mi. / 12,199 sq km	$ 2,900	65% agriculture / 30% services	191 kWh	0.03
	United States / Washington, D.C.	294.0 / 83/sq. mi.	3,717,810 sq. mi. / 9,629,084 sq km	$ 37,600	31% manage., prof. / 29% tech., sales, admin.	12,250 kWh	0.65

Sources: Central Intelligence Agency, *World Factbook 2003; The World Almanac and Book of Facts, 2004*
The CIA calculates per capita GDP in terms of purchasing power parity. This formula equalizes the purchasing power of each country's currency.

UNIT 10 Assessment Resources

► Unit 10 Test

► Unit 10 Test for English Language Learners and Special-Needs Students

internet connect

GO TO: go.hrw.com
KEYWORD: SW3 U10
FOR: Web sites about country statistics

Highlights of Country Statistics
Links to online country statistics for the Pacific world include:
• *CIA World Factbook*
• Library of Congress Country Studies
• Flags of the World

Australia and New Zealand

CHAPTER RESOURCE MANAGER

Objectives	Pacing Guide	Reproducible Resources
SECTION 1 **Australia** (pp. 711–19) • Identify the main features of Australia's natural environments. • Describe Australia's history and culture. • Examine some important features of Australia's human systems.	**Regular** 2 days **Block Scheduling** 1.5 days *Block Scheduling Handbook,* *Chapter 31*	**RS** Guided Reading Strategy 31.1 **PS** Readings in World Geography, History, and Culture 79 **E** Creative Strategies for Teaching World Geography, Lessons 22 and 23 **E** Cultures of the World Activity: Region 9 **SM** Map Activity 31: Exotic Animals **SM** Critical Thinking Activity 31
SECTION 2 **New Zealand** (pp. 722–27) • Identify some important features of New Zealand's natural environment. • Describe New Zealand's history and culture. • Examine New Zealand's economy and identify the major economic challenge the country faces.	**Regular** 1.5 days **Block Scheduling** 1 day *Block Scheduling Handbook,* *Chapter 31*	**RS** Guided Reading Strategy 31.2 **PS** Readings in World Geography, History, and Culture 80 **E** Cultures of the World Activity: Region 9 **SM** Geography for Life Activity 31: Where Should New Zealand's Urban Areas Be?

Chapter Resource Key

PS Primary Sources

RS Reading Support

IC Interdisciplinary Connections

E Enrichment

SM Skills Mastery

A Assessment

REV Review

ELL Reinforcement and English Language Learners

Transparencies

CD-ROM

Video

Internet

Holt Presentation Maker Using Microsoft® PowerPoint®

One-Stop Planner CD-ROM

See the *One-Stop Planner* for a complete list of additional resources for students and teachers.

 **One-Stop** Planner CD–ROM

It's easy to plan lessons, select resources, and print out materials for your students when you use the ***One-Stop Planner CD–ROM with Test Generator***.

Technology Resources

- One-Stop Planner CD–ROM, Lesson 31.1
- *ARGWorld* CD–ROM
- Geography and Cultures Visual Resources 61–65
- **CNN**. Presents Geography: Yesterday and Today, Segment 34: The Great Barrier Reef
- Homework Practice Online
- HRW Go site

- One-Stop Planner CD–ROM, Lesson 31.2
- *ARGWorld* CD–ROM
- Homework Practice Online
- HRW Go site

Reinforcement, Review, and Assessment

ELL Main Idea Activity 31.1
ELL English Audio Summary 31.1
ELL Spanish Audio Summary 31.1
REV Section 1 Review, p. 719
A Daily Quiz 31.1

ELL Main Idea Activity 31.2
ELL English Audio Summary 31.2
ELL Spanish Audio Summary 31.2
REV Section 2 Review, p. 727
A Daily Quiz 31.2

⨍ internet connect

HRW ONLINE RESOURCES

GO TO: go.hrw.com
Then type in a keyword.

TEACHER HOME PAGE
KEYWORD: SW3 Teacher

CHAPTER INTERNET ACTIVITIES
KEYWORD: SW3 GT31
Choose a topic on Australia and New Zealand to:
- explore Australia's Great Barrier Reef.
- meet the Aborigines of Australia.
- learn about the tectonic processes in New Zealand.

CHAPTER ENRICHMENT LINKS
KEYWORD: SW3 CH31

CHAPTER MAPS
KEYWORD: SW3 MAPS31

ONLINE ASSESSMENT
Homework Practice
KEYWORD: SW3 HP31
Standardized Test Prep
KEYWORD: SW3 STP31
Rubrics
KEYWORD: SS Rubrics

COUNTRY INFORMATION
KEYWORD: SW3 Almanac

CONTENT UPDATES
KEYWORD: SS Content Updates

HOLT PRESENTATION MAKER
KEYWORD: SW3 PPT31

ONLINE READING SUPPORT
KEYWORD: SS Strategies

CURRENT EVENTS
KEYWORD: S3 Current Events

Meeting Individual Needs

Ability Levels

Level 1 Basic-level activities designed for all students encountering new material

Level 2 Intermediate-level activities designed for average students

Level 3 Challenging activities designed for honors and gifted-and-talented students

English Language Learners Activities that address the needs of students with Limited English Proficiency

Chapter Review and Assessment

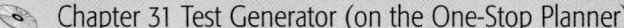

- Chapter 31 Test Generator (on the One-Stop Planner)
- Global Skill Builder CD–ROM
- HRW Go site
- **REV** Chapter 31 Review, pp. 728–29
- **REV** Chapter 31 Tutorial for Students, Parents, Mentors, and Peers
- **A** Chapter 31 Test (form A or B)
- **A** Alternative Assessment Handbook
- **A** Chapter 31 Test for English Language Learners and Special-Needs Students

Launch into Learning

Read the following passage aloud: *"I drew aside a curtain to peep at the weather, and found the glass of the window panes so heated that I was glad quickly to withdraw my hand....The wind was still rising as the day wore on, moaning and whistling through the streets and hollows in fiery suffocating blasts, that seemed to issue from a furnace heated seven times."* Ask students what they think is being described. Tell them the passage is from an account by a settler who witnessed a huge wildfire that devastated southeastern Australia in 1851. Wildfires are common in Australia under certain climatic conditions. Tell students that in this chapter they will learn more about this austere land, its more temperate neighbor, New Zealand, and the people of both countries.

Using the Physical-Political Map

Have students examine the map on the opposite page. Then have students compare Australia's physical features with those of the United States. *(Australia has more desert areas, fewer rivers, and lower mountains.)* Ask students how New Zealand's physical features compare to Australia's. *(New Zealand has higher mountains and few plains areas.)*

Why We Should Know More

You may want to reinforce interest in Australia and New Zealand by pointing out the following facts:

▶ Australia and New Zealand are both home to many rare and endangered species of plants and animals.

▶ Australia's Great Barrier Reef is the world's largest group of coral reefs. It hosts a complex ecosystem that is threatened by many aspects of modern life. We can learn from Australia's efforts to save it.

▶ Most of both countries' residents share cultural traits such as clothing, food, and entertainment with U.S. citizens.

▶ Both Australia and New Zealand are trying to redress injustices done to native peoples and are working to preserve minority traditions.

CHAPTER 31 Australia and New Zealand

Australia and New Zealand are wealthy democratic countries located in the Southern Hemisphere. In this chapter you will learn about the distinctive and fascinating landscapes of these two countries.

Aboriginal art, *The Big Feast*

Koalas live in and feed on eucalyptus trees.

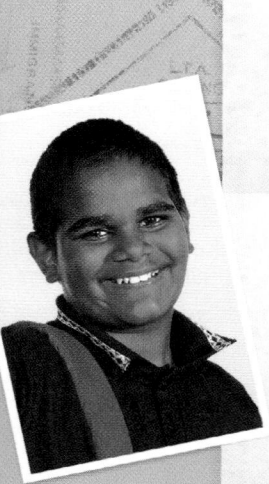

Hi! My name is Jared. My twin sister's name is Ashleigh. We live in Cooroy, Australia, about two hours north of Brisbane. We live with our mother. Our father doesn't live with us. He lives up north on the traditional lands of our people, the Djabugayndgi, at the edge of the Cape York Peninsula. Our house in Cooroy is on seven acres and has a big creek with a dam on it. The dam forms a big pond, or what we call a billabong. We can jump into the billabong from the trees along the edge. One tree used to have a swing, but it broke.

In the morning we shower and have orange juice, cereal, toast, and eggs before we catch the bus to school. We are in the ninth grade at Noosa High School. Roll call at school is at 8:30 A.M. Ashleigh and I are both studying math, science, English, history, and physical education. We also get to choose three electives. I'm taking art, speech, and woodworking. Ashleigh is taking art, music, and dance.

At 3:00 P.M. we catch the bus home. On Fridays we often stop at a friend's house and watch American and Australian TV shows. In my free time, I skateboard and play video games. On Saturdays I play sports. I have been invited to the state trials in rugby, softball, and three-on-three basketball. When I grow up, I'd like to be a rugby player. Ashleigh likes to play guitar and keyboard. She listens to rap music, pop, and reggae. She also designs houses. She wants to be an architect. We both like to go to the beach.

LET'S GET STARTED

Copy the following instructions onto the chalkboard: *What comes to mind when you think of Australia? Write down as many words or phrases as you can.* Discuss responses. *(Students may mention Aborigines, crocodiles, kangaroos, koalas, or other images from mass media.)* Point out that the country's portrayal in popular culture often contrasts with the fact that most Australians reside in cities and live like urban Americans. Tell students that in Section 1 they will learn more about Australians' daily lives and other topics.

Building Vocabulary

Write the key terms on the chalkboard in a random array. Call on volunteers to read the terms' definitions aloud from the text or glossary. Then ask students to draw arrows connecting two or more terms and to explain how these words may be related. *(Possible answer: **Artesian wells** may provide enough water for **extensive agriculture**.)* Point out that **Aborigines** comes from the Latin words *ab*, meaning "from," and *origine*, meaning "the beginning." The word can be applied to the first inhabitants of any region.

Section 1

Australia

READ TO DISCOVER

1. What are the main features of Australia's natural environments?
2. What are Australia's history and culture like?
3. What are some important features of Australia's human systems?

WHY IT MATTERS

Protecting the environment from nonnative species is a challenge in Australia and many other countries. Use **CNN fyi.com** or other **current events** sources to find an issue involving nonnative species and environmental protection.

IDENTIFY

Aborigines

DEFINE

artesian wells
outback
marsupials
extensive agriculture
exotic species

LOCATE

Great Dividing Range
Central Lowlands
Western Plateau
Great Barrier Reef

Locate, continued

Cape York Peninsula
Tasmania
Murray River
Darling River
Great Artesian Basin
Lake Eyre
Sydney
Melbourne
Brisbane
Adelaide
Perth
Canberra

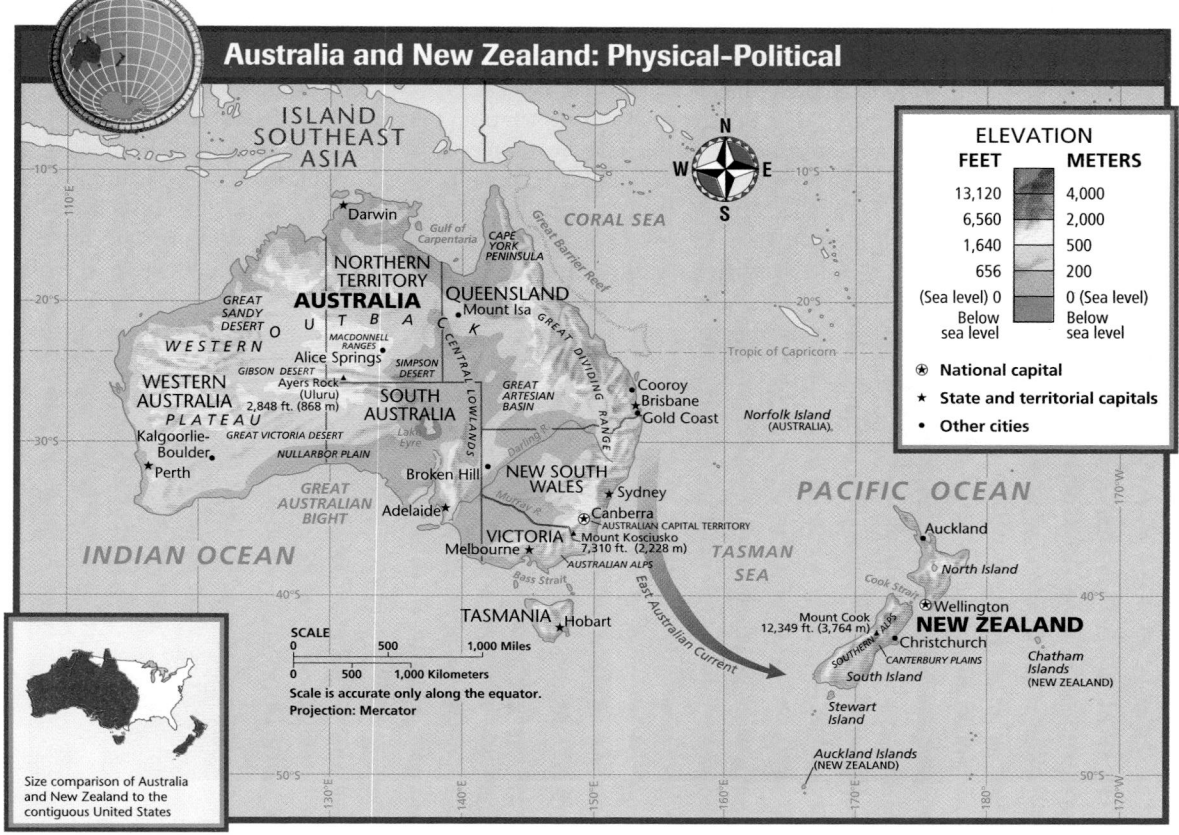

Australia and New Zealand: Physical-Political

ISLAND SOUTHEAST ASIA

Darwin

Gulf of Carpentaria

CORAL SEA

CAPE YORK PENINSULA

NORTHERN TERRITORY

GREAT SANDY DESERT

AUSTRALIA

QUEENSLAND

Mount Isa

MACDONNELL RANGES

WESTERN

Alice Springs

GIBSON DESERT

SIMPSON DESERT

WESTERN AUSTRALIA

Ayers Rock (Uluru) 2,848 ft. (868 m)

SOUTH AUSTRALIA

PLATEAU

Kalgoorlie-Boulder

GREAT VICTORIA DESERT

NULLARBOR PLAIN

Lake Eyre

Perth

GREAT AUSTRALIAN BIGHT

Broken Hill

NEW SOUTH WALES

Adelaide

Sydney

VICTORIA

Canberra
AUSTRALIAN CAPITAL TERRITORY

Melbourne

Mount Kosciusko 7,310 ft. (2,228 m)

AUSTRALIAN ALPS

Bass Strait

TASMANIA

Hobart

INDIAN OCEAN

Tropic of Capricorn

GREAT ARTESIAN BASIN

CENTRAL LOWLANDS

GREAT DIVIDING RANGE

Cooroy
Brisbane
Gold Coast

Norfolk Island (AUSTRALIA)

PACIFIC OCEAN

TASMAN SEA

East Australian Current

Mount Cook 12,349 ft. (3,764 m)

Auckland

North Island

Wellington

NEW ZEALAND

Christchurch

SOUTHERN ALPS

CANTERBURY PLAINS

South Island

Cook Strait

Chatham Islands (NEW ZEALAND)

Stewart Island

Auckland Islands (NEW ZEALAND)

ELEVATION

FEET		METERS
13,120		4,000
6,560		2,000
1,640		500
656		200
(Sea level) 0		0 (Sea level)
Below sea level		Below sea level

⊛ National capital
★ State and territorial capitals
• Other cities

SCALE
0 — 500 — 1,000 Miles
0 — 500 — 1,000 Kilometers
Scale is accurate only along the equator.
Projection: Mercator

Size comparison of Australia and New Zealand to the contiguous United States

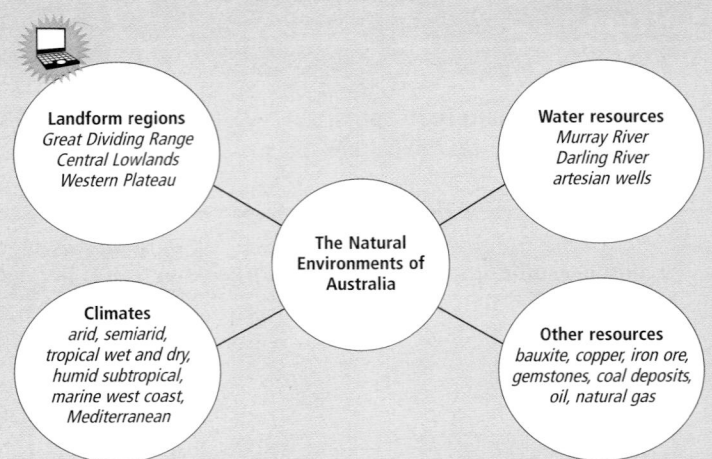

LEVEL 1: Copy the graphic organizer at right onto the chalkboard, omitting the italicized answers. Use it to help students describe Australia's natural environments. Call on students to provide words and phrases that describe the landform regions, water resources, climates, and other resources of the continent. Then lead a class discussion on the forces that have shaped Australia's landforms and the factors that influence the distribution of its climates, plants, and animals. **ENGLISH LANGUAGE LEARNERS**

▶ Physical Systems ◀

Australia's "Lost Cities"
The winds that have swept across the landscape of northern Australia for millions of years have carved some interesting rock formations. Among the most dramatic are the "lost cities" of the Caranbirini Conservation Reserve and the nearby Abner Range. Dozens of sandstone columns grouped together like skyscrapers rise more than 80 feet (25 m) above the surrounding terrain. Although the columns themselves are devoid of vegetation, their bases are surrounded by open woodlands that shelter a number of plants and animals. Because these stone "cities" lie in a transition zone between Australia's tropical northern shores and its desert interior, many of the creatures that live there are found nowhere else.

CRITICAL THINKING: How did the physical processes that shaped Australia differ from those that created continents in the Northern Hemisphere? *(Possible answer: Australia has never been covered by the glaciers that carved northern continents. Wind has been the most important force in shaping its landscapes.)*

---

internet connect

GO TO: go.hrw.com
KEYWORD: SW3 CH31
FOR: Web sites about Australia and New Zealand

Australia has many unusual landforms. This gorge at Alice Springs is near the country's center. Far to the south, near Adelaide, lies Kangaroo Island. Almost one third of the island is within national and conservation parks.

Natural Environments

Australia is known as the Land Down Under because of its location south of, or "under" the equator. The name *Australia* comes from a Latin word that means "southern." The country is located between the Indian and Pacific Oceans. Almost the same size as the contiguous United States, Australia is the only country that is also a continent. It is also the smallest, flattest, and second-driest continent. As you will learn, flat topography and dry climates are major features of Australia.

Landforms and Rivers Geographers divide Australia into three main landform regions. In the east lies a highland region called the Great Dividing Range. The other two landform regions are the Central Lowlands and Western Plateau. In addition, the Great Barrier Reef lies off the northeast coast. This group of coral reefs is 1,250 miles (2,010 km) long. It is famous for its size and varied tropical sea life.

The Great Dividing Range stretches from the Cape York Peninsula to Tasmania. These highlands are the eroded remains of an old mountain range. They divide the flow of Australia's rivers. Those that flow down the eastern slopes empty into the Pacific Ocean. Those that flow west drain into the Central Lowlands. The Murray River and Darling River—Australia's major river system—flow west from this range.

The low Great Dividing Range is also Australia's main mountain system. Mount Kosciusko (kah-zee-UHS-koh), at only 7,310 feet (2,228 m), is the highest elevation on the continent. It is in the highest part of the Great Dividing Range—the Australian Alps.

The Central Lowlands stretch from the Gulf of Carpentaria to the Indian Ocean. Beneath this low area is the Great Artesian Basin, which has huge amounts of groundwater. **Artesian wells**—wells in which water flows naturally to the surface—are common here. In the Central Lowlands lies Lake Eyre (AYR), Australia's lowest point at 52 feet (16 m) below sea level. Lake Eyre is a salt lake that in many years is completely dry. North of Lake Eyre is the

Alice Springs, Northern Territory

Kangaroo Island, South Australia

LEVEL 2: Organize the class into three groups and assign each group one of the following landform regions: Great Dividing Range, Central Lowlands, or Western Plateau. Trace a large map of Australia onto butcher paper and draw the three regions onto it. Then have each group draw the main physical features of its assigned region, such as mountains and water resources, on the map. Have students add pictures or drawings also of plants and animals found in their assigned regions to the map. You may want to have students consult other resources for more information about where various species live. Finally, ask students to create symbols to represent the climates and mineral, energy, and agricultural resources of their regions and to add these symbols to the map. Point out how some environmental features overlap regional boundaries. For example, both the Central Lowlands and Western Plateau have arid and semiarid climates.

LEVEL 3: Have each student write a short poem to describe a physical feature of Australia. *(Possible subjects: major rivers, Great Barrier Reef, outback, Great Dividing Range)* Ask volunteers to recite their poems to the class.

Simpson Desert. This desert has sand dunes that are from 70 to 120 feet (21 to 37 m) high. Winds move and shape these dunes.

The Western Plateau covers about two thirds of Australia. It has the oldest rocks on the continent. For millions of years, these rocks have been eroded. Deserts cover the central part of the area. In the south is the Nullarbor Plain—a dry flat limestone plateau.

✓ **READING CHECK:** *Places and Regions* What are the three formal landform regions of Australia? Great Dividing Range, Central Lowlands, Western Plateau

Climates Australia is a desert continent with green edges. About two thirds of Australia has an arid or semiarid climate. (See the unit climate map.) Most of the heart of Australia is extremely dry. Around these desert areas are ribbons of semiarid climate. Rainfall in these areas is often unreliable. Long droughts may be followed by short powerful storms and floods. During droughts, wildfires often sweep across the land. The dry interior of Australia is called the **outback**.

There are several reasons why Australia has such dry climates. Much of the country is between about 20° and 30° south latitude. These areas are often warm subtropical high-pressure zones with dry air. Another factor that contributes to Australia's dry climates is its generally low elevation. Only the Great Dividing Range is high enough to cause air to rise and cool, which creates rain. However, the mountains keep moisture from reaching much of the interior. West of the mountains is a rain shadow.

Temperatures in Australia are generally warm. Average January (summer) temperatures are higher than 85°F (29°C) in most of the interior and much of the north and northwest. Average July (winter) temperatures are warmest in the north and cool in the south. High elevations in the southeastern Great Dividing Range are the only areas that have cold weather. Winter skiing is possible there.

Much of northern Australia has a tropical wet and dry climate. This area has summer monsoons that bring heavy rainfall. Some parts of the Cape York Peninsula get more than 150 inches (381 cm) of rainfall per year. Winters are still warm, but may have long droughts. Along the east coast are humid subtropical and marine west coast climates. In the southeast, westerly winds bring winter storms and rain. Unlike most parts of Australia, the southeast is generally well watered. The mild climate and rainfall generate small streams and rivers. Two parts of the southern edge of Australia have a Mediterranean climate. Mild rainy winters and warm dry summers are normal there.

✓ **READING CHECK:** *Physical Systems* How does the Great Dividing Range affect Australia's climates? prevents moisture from reaching much of interior; high elevations have cold weather and snow in winter

Plants and Animals Australia is known for its strange plants and animals. Most are endemic species. Because of past movements of Earth's tectonic plates, Australia has been separated from the other continents for about 35 million years. During this time, its plants and animals developed in isolation. This condition has caused a unique biogeography. For example, cats are native to every continent except Australia and Antarctica. Also, hoofed animals like deer and cattle are not native to Australia. Most of Australia's mammals are **marsupials**—mammals that have pouches to carry their young. These include the kangaroo, koala, and wallaby (WAH-luh-bee).

INTERPRETING THE VISUAL RECORD

These 300-foot cliffs stretch for nearly a hundred miles along the edge of the Nullarbor Plain. Nullarbor comes from a Latin phrase that means "no tree." **What force is eroding these cliffs?**

Eye on Earth

Kakadu's Climates and Animals The tropical wet and dry climate region of northern Australia is characterized by wide variations in weather. In fact, the Aborigines who live in Kakadu National Park east of Darwin divide the year into six distinct seasons. For example, January and February are in *Gudjewg,* the season of violent thunderstorms and floods. *Wurrgeng,* in June and July, is the time of low humidity and cool nights.

Kakadu's weather variations are reflected in the wide range of animal species that live in the park. Among its most remarkable residents is the saltwater crocodile, one of the most dangerous predators in the world. Salties, as they are called, kill several people every year. Because they live in salt water, they can swim huge distances across the ocean looking for new homes. Salties have been spotted more than 600 miles (960 km) from shore, encrusted with barnacles.

ACTIVITY: Have students conduct research on Kakadu National Park and write scripts for rangers' guided tours of the park.

VISUAL RECORD ANSWER

wave action

LEVEL 1: Organize students into four groups. Have each group create a time line that includes information on the cultures that developed in Australia before or since 1700. Direct students to use the textbook to find information for their time lines. Also have students draw pictures to illustrate some of the various cultures on the time lines. Ask students how the cultures mentioned on their time lines have contributed to Australia today. Have students present their time lines to the class. **ENGLISH LANGUAGE LEARNERS, COOPERATIVE LEARNING**

LEVELS 2 AND 3: Have students imagine that they are participating in an archaeological dig at a site in Australia at which artifacts of an early Aborigine culture have been found. Have students conduct research about the Aborigines and write reports describing the artifacts that have been found. Reports should include information identifying the civilization that created the artifacts and descriptions of the sites where the artifacts were found.

Essential Element 5

► **Environment** ◄
and Society

Duck-Billed Platypus
One of Australia's unusual creatures is the duck-billed platypus—a small, furry, egg-laying mammal. The platypus lives in the lakes and streams of eastern Australia and Tasmania and spends nearly half of every day eating fish, frogs, earthworms, and other small organisms. Researchers have found that special receptors in the platypus' snout detect electrical fields produced by its prey.

Because many platypuses live within a few miles of big cities, they are vulnerable to human carelessness. In a recent study, about 10 percent of the platypuses found in urban areas had been cut by or become tangled in fishing line.

Australia's compass termites build huge mounds that always point north. These mounds can be as large as 13 feet (4 m) high, 8 feet (2.5 m) long, and 3 feet (1 m) thick.

GO TO: go.hrw.com
KEYWORD: SW3 CH31
FOR: Web sites about Australia's unique animals

Rain forest, New South Wales

Border Ranges National Park, pictured above left, lies on the rim of a huge extinct volcano. More than 170 species of birds live in the park. In the western Macdonnell Ranges, above right, careful observers may see the rock wallabies that live there.

Semiarid landscape, Northern Territory

Australia's biomes mirror its climates. The interior is grassland and desert. Grassland areas have acacia (e-KAY-shuh) shrubs, bunch grasses, and scattered eucalyptus (yoo-kuh-LIP-tuhs) trees. About 500 kinds of eucalyptus are found all across Australia. Monsoonal northern Australia has large areas of savanna. Plants and animals there depend on seasonal rains. The Cape York Peninsula has tropical rain forests. Large trees and dense vegetation are common there. Animals include the tree kangaroo and many native birds, such as parrots and cockatoos. The south and southwest have Mediterranean scrub forests. The southeast and east coast have temperate forests.

✓ **READING CHECK:** *Places and Regions* What is unique about Australia's plant and animal life? It developed in isolation; endemic marsupials are the dominant mammals.

Natural Resources Australia is rich in mineral and energy resources. As you have learned, water resources are scarce in many areas, particularly the arid interior.

Mineral resources include bauxite, copper, iron ore, and many other valuable minerals. (See the map of land use and resources in the unit atlas.) Many of these minerals are found in the dry interior. Mining centers such as Broken Hill and Mount Isa have been operating for many years. The Broken Hill mines in southeast Australia produce lead, silver, and zinc. Other mines yield valuable gems, such as diamonds, opals, and sapphires. Energy resources include coal, oil, and natural gas. Along Australia's east coast are large coal deposits. Most of the oil and natural gas comes from offshore fields. The main fields are in the Bass Strait near Tasmania and off the coast of western Australia.

The country's farming resources are more limited. Most areas have poor soils, and there is not much water. Only about 6 percent of the land is good for farming. The best farming areas are in the southeast. Because of the limited amount of good farmland, many areas are used for grazing. The many artesian wells and groundwater sources make this possible. Much of the water is too salty for people, but can be used for sheep.

✓ **READING CHECK:** *Environment and Society* What factors affect the location of different types of economic activities in Australia? oil and minerals extracted at their sources; animals graze on marginal lands

Cane toads have very poisonous skin, which kills animals that attack or eat them. Brought to Queensland in the 1930s to eat pests, cane toads have disrupted native wildlife in much of northern Australia.

Teach Objective 3

🌐 **LEVEL 1:** Call on students to list some of Australia's natural resources. Then have students create charts to identify the country's primary, secondary, tertiary, and quaternary economic activities. Lead a class discussion on the relationship between Australia's resources and the economic activities that are conducted there. Ask students to speculate about factors that influence the locations of economic activities within the country.

🌐 **LEVELS 2 AND 3:** Organize the class into four groups and have each group choose a natural resource found in Australia. Then have each group create a fictitious company and establish a plan to make use of its chosen natural resource. Remind students they will need to decide where their companies will be located, explain why they selected these sites, and describe how the company will use the natural resources. Have groups share their plans. As a class discuss the possible consequences of each group's decisions.

History and Culture

Australia's native peoples have one of the world's oldest continuous cultures. However, Australian culture has been shaped by its history as a British colony. Although it is far from Europe, Australia's dominant culture is European.

Early History and Settlement Australia's first peoples were the **Aborigines** (a-buh-RIJ-uh-nees). They came to Australia from Southeast Asia at least 40,000 years ago. Early Aborigines lived a nomadic way of life. They hunted with spears, nets, and boomerangs—curved throwing sticks. There were many groups, each with a different name, speaking hundreds of different languages. Although estimates vary, at least 300,000 Aborigines probably lived in Australia when European settlers arrived in the late 1700s.

The British settled Australia as a prison colony. The first settlement, set up in 1788, later became the city of Sydney. By 1830, nearly 60,000 prisoners had been sent to Australia. Other people came to farm or raise sheep. In 1851 gold was discovered, attracting even more people. Many settlers forced Aborigines off their land. Aborigines had no resistance to the diseases brought by Europeans, and many of them died. In Tasmania, the Aborigines were completely wiped out.

In the mid-1800s more towns and colonies were founded. Eventually, six large colonies developed. (See Connecting to History: Australia's States and Territories.) In 1901 these six colonies joined to form the Commonwealth of Australia. The new country was a close ally of Great Britain. During World Wars I and II many Australians fought alongside British troops.

✓ **READING CHECK:** *Human Systems* About how many Aborigines lived in Australia when European settlers arrived? probably at least 300,000

Aborigine elders from Melville Island, off the north coast of Australia, display their spears.

Connecting to

HISTORY

Australia's States and Territories

Australia's first colony, New South Wales, was founded in 1788. In the 1820s Van Diemen's Land (later renamed Tasmania) and Western Australia were added. Later, colonies were created with land taken from New South Wales. These colonies included South Australia, Victoria, and Queensland. In 1901 Australia became independent from Great Britain. Soon after, the country's political geography looked much like it does today. The six colonies became states, and two territories—Northern Territory and the Australian Capital Territory—were created. (See the chapter map.)

Drawing Inferences and Conclusions Look at the chapter map. Which major cities do you think developed in each colony?

Australia in 1829	Australia in 1851	Australia by 1911
INDIAN OCEAN / CORAL SEA / WESTERN AUSTRALIA 1829 / NEW SOUTH WALES 1788 / INDIAN OCEAN / VAN DIEMEN'S LAND 1825	INDIAN OCEAN / CORAL SEA / WESTERN AUSTRALIA / NEW SOUTH WALES / SOUTH AUSTRALIA 1836 / VICTORIA 1851 / INDIAN OCEAN / VAN DIEMEN'S LAND	INDIAN OCEAN / CORAL SEA / NORTHERN TERRITORY 1911 / QUEENSLAND 1859 / WESTERN AUSTRALIA / SOUTH AUSTRALIA / NEW SOUTH WALES / VICTORIA / AUS. CAPITAL TERR. 1911 / INDIAN OCEAN / TASMANIA 1911

Cooperative Learning

Prison Colony Ballads
Organize the class into groups to conduct research about Australia's early history during the 1700s and 1800s. Have students concentrate on the prison colony era. Direct the groups to write traditional ballads about these early settlers' lives.

Specific subjects for the ballads might include Irish political prisoners sent to Australia for revolting against English rule, bushrangers (escaped convicts who survived in the wilderness), women or children prisoners, convicts who won their freedom and stayed in Australia to become prosperous merchants, or those who suffered terribly in conditions similar to slavery. Call on volunteers to perform their ballads for the class.

CONNECTING TO HISTORY ANSWER

Canberra and Sydney in New South Wales; Hobart in Tasmania; Perth in Western Australia; Adelaide in South Australia; Melbourne in Victoria; Brisbane in Queensland

715

ALL LEVELS: Have each student write an informative synopsis about Australia for a book describing the countries of the world. Each synopsis should include details of the country's natural environment, history and culture, and economic activities. Students might wish to model their entries after the brief descriptions of countries found in geographical dictionaries or encyclopedias. Encourage students to write about how various physical and human processes have interacted to shape modern Australia.

LEVELS 2 AND 3: Organize the class into groups. Have each group create a few pages for an Australian newspaper. Ask students to include information about the country's physical features and history as well as issues facing the country today. Students should include articles, cartoons, editorials, and other typical newspaper features. Then have a spokesperson from each group present his or her group's newspaper pages to the class. **COOPERATIVE LEARNING**

The Didgeridoo The didgeridoo is probably the oldest musical instrument in the world still in use. Aborigines may have been playing the didgeridoo when they first came to Australia some 40,000 years ago.

To make a didgeridoo, Aborigines find a branch that has been eaten out by insects. The branch is cut to a length of about 4.2 feet (1.3 m), hollowed, and cleaned. To play the instrument, the musician blows across one end to create a droning sound. Sometimes Aborigine musicians try to mimic sounds found in nature, such as animal calls or thunder, by modulating the drone.

Different events require different kinds of didgeridoos. An initiation ceremony requires a low-pitched instrument, while a higher-pitched didgeridoo is used to accompany dancing. Many Aborigines believe that playing the didgeridoo is one way to recall the mysteries of the Dreamtime.

ACTIVITY: Play recordings of didgeridoo music to students. Have them identify any sounds from nature that they hear in the music.

VISUAL RECORD ANSWER

well-kept historical buildings

716

Some Common Australian Words	
Word	**Definition**
biscuit	cookie
clicks	kilometers (or miles) per hour
dunny	restroom with just a toilet
esky	ice chest
g'day	hello
mozzies	mosquitoes
Oz	Australia
roo	kangaroo
saltie	saltwater crocodile
shark biscuit	inexperienced surfer

INTERPRETING THE VISUAL RECORD

Visitors can find fashionable galleries, hotels, and other establishments on Collins Street in Melbourne, Australia. **How can you tell from the photograph that Melbourne has had a relatively long history of prosperity?**

People and Languages Australia's colonial history shaped its society. About 92 percent of Australia's 19 million people are of British or other European ancestry. English is the official language. However, it is spoken with a distinct "Aussie" accent and many special Australian words. (See the table.) Asians make up about 7 percent of the population. Many Asians began moving to Australia in the 1970s. They have added to the country's growing cultural diversity. Aborigines are a small but important group. About 200,000 Aborigines now live in Australia. Many of them have mixed European and Aboriginal ancestry.

Settlement and Land Use Most people live in cities along the southeastern coast. These include Sydney, Melbourne, Brisbane, and Adelaide. Perth is a large city in western Australia. All together, about 85 percent of Australians live in cities.

Australia's settlement pattern is tied to the country's colonial history. Major cities grew as ports during colonial times. Even today, each state and territory has just one major city. (See the chapter map.) The only large city not on the coast is the capital, Canberra. It was not founded until 1913. Inland settlements, such as Kalgoorlie-Boulder, are generally mining or agricultural towns. Alice Springs, the outback's major town, is important for transportation and tourism. Where people live is also tied to Australia's natural environment. Most people settled in the southeast. This area has pleasant climates and reliable rainfall. In the dry interior, a lack of water makes settlement and farming risky. Even today, very few people live there.

Religion and Education About 90 percent of Australians are Christian. Asian immigration has brought many Buddhists and Muslims. Aborigines who follow traditional ways emphasize spiritual ties to the land. They believe that their ancestors created the world during what the Aborigines call the Dreamtime. These ancestors became part of nature. Such beliefs help Aborigines feel close to their ancestral lands.

Australia has a good education system. State and territorial governments run schools with help from the national government. In the outback, some students get lessons by radio, e-mail, video, and even satellite connections. One important national goal is to improve the situation of Aborigines. On average, they lag far behind other Australians in education. Historically, Aborigines have not had access to a good education in Australia. In fact, Aborigines did not even become legal citizens until 1967.

Traditions, Customs, and Food Swimming, surfing, and going to the beach are popular in Australia, as are organized sports. Most of these sports, such as rugby and cricket, are originally British games. Bruising Australian Rules football is very popular in the south. Australian artists, filmmakers, musicians, and writers have produced works known around the world. Aboriginal art is also popular. These paintings on tree bark or rocks feature human and animal figures.

Foods in Australia often mix Mediterranean and Asian styles. Common foods like beef and lamb are often grilled or roasted and served with potatoes and other vegetables. Italian and Greek foods are popular. Many immigrants to

Music: Australian Popular Music

Trade involves not only the exchange of goods but also the sharing of thoughts and ideas. This exchange of ideas can lead to the diffusion of elements of popular culture. For example, increased contact between the United States and Australia in recent decades has introduced residents of each country to music popular in the other. Features of each country's musical tradition have been adopted and modified by artists in the other country. As a result, many artists have achieved fame in both countries.

Obtain recordings of songs by several popular Australian musicians spanning several years. Play samples of these recordings for the class. Have students note similarities and differences between Australian popular music and American music of the same year. *(You may wish to play samples of American music to facilitate student comparisons.)* After you have played several samples, ask how Australian music styles seem to have changed over time. Call on volunteers to speculate about influences that might have led to these changes. *(Possible answer: Students might note similarities to American or British music and conclude that Australian artists were influenced by artists from those countries.)* Then ask students how they think Australian musicians might have influenced artists from other parts of the world. Show students examples of Australian music videos, films, commercials, or other elements of popular culture. Lead a discussion on how they depict the landscapes or people of Australia. Ask students if they see any changes in these representations over time.

Human Systems

Geography for Life

Australia's Changing Trade Patterns

Most of the world's developed countries have become rich through the production and export of manufactured goods. However, Australia's economy has long been based on the export of raw materials. The country's leading exports today include coal, gold, meat, wool, iron ore, and wheat. Australia ships these materials mainly to Japan, European Union (EU) countries, ASEAN (Association of Southeast Asian Nations) countries, and the United States. Factories in those countries make products using those raw materials.

While Australia exports mostly raw materials, it imports mostly manufactured goods. These imports include computers, machinery, office machines, and transportation equipment. Most of these imports come from the United States, Japan, ASEAN countries, and EU countries like the United Kingdom and Germany.

Since World War II Australia's government has tried to diversify the country's economy. Its goal has been to make and export more manufactured goods. Doing so would help the country become less dependent on the export of raw materials. Many factories now specialize in light manufacturing. Many also process farm products or minerals for sale overseas. In addition, Australia produces household appliances, paper products, processed foods, and textiles. The country also has a growing wine industry.

Just as Australia's economy has changed since World War II, so have its trading partners. Australia has long had strong trade relations with Britain. In fact, Britain was Australia's main trading partner for many years. However,

Wool from different breeds of sheep becomes products made by various segments of the wool industry, such as carpets, fine fabric, and upholstery. The fleece from the largest sheep may weigh 35 pounds (16 kg).

these ties are not nearly as strong today. This change began after World War II as U.S. influence in the Pacific region began to grow. Trade and other ties between the United States and Australia strengthened. Trade with Japan and China also began to grow. Then in 1972 Britain joined what is now the EU and ended policies that favored trade with Australia and its other former colonies.

Today EU countries are significant importers of Australian goods. However, Japan and the United States are the largest single buyers of Australia's main exports. Japan, for example, imports Australian raw materials and natural resources that it lacks at home. In trade, as in other ways, Australia is increasingly looking toward Asia for its future.

Australia's Major Trading Partners

Exports

10%
11%
14%
20%
45%

Imports

12%
14%
28%
22%
24%

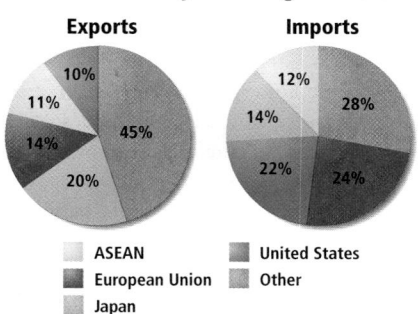

ASEAN
European Union
Japan
United States
Other

Source: Central Intelligence Agency, *The World Factbook 2000*

Applying What You Know

1. **Summarizing** How and why have Australia's economy and trade patterns changed since World War II?

2. **Drawing Inferences and Conclusions** How might changing trade patterns increase cultural and other ties between Australia and Asia?

Global Perspectives

The New Silk Road Just as the ancient Silk Road allowed Chinese goods to reach western markets, the Australian government is seeking global routes for its products. It encourages Australian business owners to establish enterprises online. They call the Internet the New Silk Road and they want to adapt the country's trade patterns to fully incorporate new technology.

Conducting business online would allow Australian exporters to reach new markets and would therefore increase the country's export levels. It would also reduce the costs of many business ventures—a prospect that is particularly important to small- or medium-sized firms.

CRITICAL THINKING: Do you think the name *New Silk Road* is applicable to the Internet? Why or why not?

Applying What You Know Answers

1. Australia's government has tried to diversify the country's economy and export more manufactured goods.

2. Exposure to countries other than Britain might help foster cultural and economic exchanges.

This Geography for Life feature addresses National Geography Standards 8, 9, and 16.

717

Teacher to Teacher

Frank Thomas Jr., of Austin, Texas, suggests the following activity to help students relate to Australian culture. Have each student consult the table of Australian slang words in this section and write a short story that includes at least six of the listed words. Have students read their stories to the class. You might also wish to have students translate their classmates' stories back into standard American English.

HOMEWORK: Tell students to imagine that they are explorers who are leading expeditions across Australia during the 1860s. Have students choose one of the following starting points for their journeys: Perth, Melbourne, Sydney, or Brisbane. Then have students plot courses across the continent for their expeditions to follow. Ask each student to write a series of journal entries in which he or she records information about the landforms, bodies of water, plant and animal life, and other aspects of the natural environment encountered during the expedition. The entries should also describe any physical hardships or dangers caused by the environment. Call on volunteers to read their journal entries to the class.

Global Perspectives

Aboriginal Rights For years, Aboriginal art and designs have been reproduced on common objects, ranging from carpets to T-shirts. Now Aborigine artists are seeking copyright protection for their ancient artwork, much of which was originally created as part of religious experiences.

Some Aborigines are also seeking payment for scientific discoveries related to their knowledge of the environment. For example, the Aborigines use the smokebush plant to treat many ailments. A pharmaceutical company recently received a license to develop an anti-AIDS drug from the plant. The Aborigines are seeking payment for their knowledge, but their case is controversial. Some company officials claim that they did not know the Aborigines used the smokebush for medicinal purposes.

ACTIVITY: Have students conduct research on intellectual and cultural property laws. Then stage a class debate about the Aborigines' rights to copyright symbols of their culture.

VISUAL RECORD ANSWER

Possible answer: It has a Mediterranean climate.

CHART ANSWER

minerals, agricultural products

INTERPRETING THE VISUAL RECORD

Workers pick grapes near Adelaide, South Australia. Why do you think the southern coastal region is a good one for growing grapes?

Australia came from these countries. Asian foods have also become popular. Large cities now have many Indonesian, Vietnamese, and Chinese restaurants.

✓ **READING CHECK:** *Places and Regions* What features of Australia's history and culture make it a distinctive region? Aboriginal culture; history as a prison colony; foods with a blend of Mediterranean and Asian styles

Australia Today

Australia is a developed country with a market economy. It has good transportation systems and health care and a tradition of stable democratic government. The per capita GDP is about $23,200—very high by world standards. Also, life expectancy and literacy rates are higher than in the United States.

Australia exports mostly raw materials and imports mostly manufactured goods. (See the chart.) Most people work in service industries like education, government, or tourism. Since World War II, Australia's major trading partners have changed. Australia used to trade mainly with Britain and other European countries. Today, however, the country trades mainly with Asia and the United States. (See Geography for Life: Australia's Changing Trade Patterns.)

Mining, Agriculture, and Tourism Mining is an important part of Australia's economy. In fact, Australia is the world's leading producer of bauxite, diamonds, opals, and lead. It is also a major producer of coal, copper, iron ore, silver, and many other minerals. Most minerals are exported to Japan or other countries in East Asia. Many of Australia's mineral resources are found in the dry interior.

Only 6 percent of Australia's land is good for farming. However, the country's large size and modern technology make it a major exporter of farm goods. The main products are wool, meat, and wheat. Wool supports the economy. The country has about 150 million sheep, or about 15 percent of all the sheep in the world. As a result, Australia is the world's leading producer of wool, supplying about 30 percent of the world's total. Most sheep are raised on the western slopes of the Great Dividing Range and around Perth. Cattle are raised in the north. Both sheep and cattle are raised on huge ranches called stations. (See Cities & Settlements: Life in the Outback.) These ranches are examples of **extensive agriculture**. This kind of agriculture uses much land but small inputs of capital and labor per unit area. Wheat is Australia's most important crop. It is concentrated in the southeast. Tropical crops like bananas, pineapples, and sugarcane grow along Queensland's wet coast. Australia produces many other fruits and vegetables also.

Tourism is a large and growing industry. Millions of people come to Australia each year to enjoy beaches and beautiful landscapes. Major vacation areas include the Great Barrier Reef, Queensland's Gold Coast, Ayers Rock (Uluru), and major cities like Sydney. Many people from other countries visit Australia even though they face long travel times to get there. The country is about a 15-hour flight from Los Angeles and a 24-hour flight from London.

Australia's Major Exports and Imports

Exports	Imports
dairy products, meat, fish, wool, forestry products, manufactures	machinery and equipment, vehicles and aircraft, petroleum, consumer goods, plastics

Source: Central Intelligence Agency, *The World Factbook 2003*

INTERPRETING THE CHART *Alumina is a component of bauxite, the main aluminum ore. It has other uses too, such as in ceramics and pigments. Because so many mineral deposits are far from population centers, mining these minerals can be expensive. Foreign investment is crucial. What types of products does Australia export?*

Close

Point out that Australia fascinates many Americans and attracts thousands of tourists from this country every year. Ask students if they would like to visit Australia. Have them give reasons to support their answers.

Review and Assess

Have students complete the **Section Review**. Then have students complete **Daily Quiz 31.1**.

Reteach

Have students complete **Main Idea Activity for English Language Learners and Special-Needs Students 31.1**. Then have each student write a list of 10 facts he or she has learned about Australia's natural environments, history and culture, and current challenges. **ENGLISH LANGUAGE LEARNERS**

Extend

Have interested students conduct research on the Tasmanian wolf (or Tasmanian tiger, *Thylacinus cynocephalus*), a marsupial that is believed to have become extinct in the 1930s. Ask students to present their findings to the class in the form of a storyboard for an educational video. **BLOCK SCHEDULING**

Australia
Ayers Rock

Issues and Challenges Important challenges include addressing Aboriginal claims to land and protecting the environment. In the late 1980s Aborigines began protesting mining on sacred lands. They also claimed that these lands belonged to them. This dispute has become a national issue. Judges have said that Aborigines have a right to claim traditional lands. As a result, many farmers, mining companies, and ranchers worry that they might lose control of the lands they use to earn a living.

Another challenge is protecting the environment. Australia's environment has changed greatly during the last 200 years. For example, more than a third of the country's woodlands have been cleared or altered. People usually cleared land to create areas for farming and raising animals. Hardest hit were the country's rain forests. About 75 percent of these forests were destroyed. This has reduced habitat for native wildlife.

Europeans brought many new plants and animals to Australia. New types of plants and animals that people introduce to an area are called **exotic species**. In Australia they include camels, cane toads, prickly pear cacti, and rabbits. Many have spread across the country and become problems. These new animals usually have no natural predators. For example, the British brought rabbits for sport hunting. However, the number of rabbits rose quickly. They damaged grazing lands, causing erosion. They became such a problem that people purposely introduced a disease among them. This succeeded in lowering their numbers. However, rabbits and other exotic species still cause problems.

✓ **READING CHECK:** *Human Systems* What activities are important to Australia's market economy? mining, agriculture, manufacturing, international trade, tourism

INTERPRETING THE VISUAL RECORD

Because of the rock's mineral content, Uluru changes color as the sun rises and sets. The Aborigines who own the site, the Anangu, believe that only two special men from the local group may climb Uluru. **How can you tell from the photo that these tourists and the Aborigines view Uluru differently?**

Section

1 Review

go.hrw.com
Homework Practice Online
Keyword: SW3 HP31

Identify
Aborigines

Define artesian wells, outback, marsupials, extensive agriculture, exotic species

Working with Sketch Maps On a map of Australia that you draw or that your teacher provides, label the Great Dividing Range, Central Lowlands, Western Plateau, Great Barrier Reef, Cape York Peninsula, Tasmania, Murray River, Darling River, Great Artesian Basin, Lake Eyre, Sydney, Melbourne, Brisbane, Adelaide, Perth, and Canberra. Which interior area has large amounts of groundwater?

Reading for the Main Idea

1. *Physical Systems* How do latitude and elevation influence Australia's climate regions?

2. *Human Systems* How have the human characteristics of Australia changed since the late 1700s?

3. *Human Systems* What political, economic, social, and demographic data give clues to Australia's level of development?

Critical Thinking

4. **Drawing Inferences and Conclusions** How is the distribution of plants and animals in Australia related to climate?

Organizing What You Know

5. Create a chart like the one shown below. Use it to describe Australia's three major landform regions.

Great Dividing Range	Central Lowlands	Western Plateau

Section 1 Review Answers

Identify For identification, see: Aborigines, p. 715

Define For definitions, see: artesian wells, p. 712; outback, p. 713; marsupials, p. 713; extensive agriculture, p. 718; exotic species, p. 719

Working with Sketch Maps Maps will vary, but listed places should be labeled in their approximate locations. The Great Artesian Basin has groundwater.

Reading for the Main Idea
1. Subtropical latitudes contain high-pressure zones that block moisture. The Great Dividing Range creates a rain shadow in the interior.

2. Aborigines have declined in numbers. People with British ancestors are now the largest group. Asian immigration is increasing.

3. stable government, high per capita GDP, good transportation and health care, high life expectancy and literacy rates

Critical Thinking
4. Its biomes roughly mirror its climates. (NGS 7)

Organizing What You Know
5. Great Dividing Range—eroded mountain range; divides flow of rivers; Central Lowlands—low elevations; artesian wells, deserts; Western Plateau—oldest rocks on continent; deserts

VISUAL RECORD ANSWER

The tourists are climbing Uluru, which most Aborigines would not do.

719

Identifying Differences among Regions

Have students read Cities & Settlements: Life in the Outback. Then ask students to recall the images they suggested for the Let's Get Started activity at the beginning of this section—Aborigines, crocodiles, kangaroos, koalas, and so on. Which of these images relate to the Australian outback? What new images of the outback does the feature suggest? *(Possible answers: flies, dust, people living underground)* How do these images help define the outback as a region? *(Possible answer: They help describe the climate, which is a unifying factor for the region.)*

Review the definitions of formal, functional, and perceptual regions. Ask students to relate the definitions of each type of region to the feature's description of the outback. *(Possible answers: formal region—the huge dry interior of the continent; functional—towns and cattle stations along the Stuart Highway, with Alice Springs as the region's node; perceptual—areas linked by radio, friends' homes even if they are far apart)* Lead a discussion about how the physical geography of the outback affects the concept of community. Do outback residents define community differently than the students do?

This Cities & Settlements feature addresses National Geography Standards 4, 9, 12, 15, and 16.

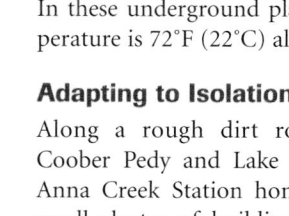

CITIES & SETTLEMENTS

Life in the Outback

Places and Regions West of Australia's major mountains and rivers, a vast arid region extends across the country to the Indian Ocean. Some Australians call this desolate expanse the "back of the beyond," but it is better known as the outback. Temperatures here can reach 120°F (48°C) in the shade. Rainfall averages under 10 inches (25 cm) per year. The outback's riverbeds hold only rock and sand, except after heavy downpours. Dust storms are more common than rain! When it fills with water, Lake Eyre, at the edge of the Great Victoria Desert, is the country's largest lake. However, the lake fills completely only about twice in a century. Instead, most of the time the "lake" is a huge salt flat.

Adapting to the Environment

The outback covers about 75 percent of Australia. It is the ancestral home of Australia's first people, the Aborigines. Today only about 10 percent of Australians live on this desolate unforgiving land. Population density averages fewer than two people per square mile, and large areas are completely uninhabited. Many outback dwellers live 300 miles or more from the nearest store. Doctors and schools are even farther away. Such conditions have created a rugged and independent people. Yet the region defeats those who believe they can truly conquer it. The people of the outback survive only by adapting to the arid environment.

Coober Pedy, a town west of Lake Eyre in South Australia, provides a clear example of this adaptation. This isolated community is one of the few stops on the only paved road that crosses the entire outback. Called the Stuart Highway, this route connects the northern and southern coasts to Alice Springs, in the Northern Territory. With just 27,000 people, "the Alice" is one of the outback's largest towns.

Fewer than one percent of South Australia's 1.5 million people live in the state's interior. A visit to Coober Pedy quickly reveals why. The people of Coober Pedy have had to take extraordinary measures to live in such a harsh environment. In the Aborigine language, *Coober Pedy* means "White Man's Burrow." The name fits. Coober Pedy exists only because of the opal mines nearby. The town is the world's largest producer of these precious stones. In addition, much of Coober Pedy is itself underground. To escape the blistering heat, businesses, churches, and many homes are below the surface. These establishments, and about half the residents, occupy old mine shafts or specially dug homes called dugouts. In these underground places the temperature is 72°F (22°C) all year.

Adapting to Isolation

Along a rough dirt road between Coober Pedy and Lake Eyre lies the Anna Creek Station homestead. This small cluster of buildings is the station's headquarters. Covering some

People shop for opals in one of the many belowground stores in Coober Pedy.

Going Further: Thinking Critically

Organize students into groups. Have each group create lists of the formal, functional, and perceptual regions to which they belong. Then ask one student from each group to read the group's list. Lead a discussion about how the groups' lists are similar and different and how the various culture traits that students bring to their groups help determine the regions to which they feel they belong. Discuss also how modern communications may have enlarged the size of perceptual regions. **COOPERATIVE LEARNING**

Anna Creek family members and jackeroos pose for an informal portrait. The children's best friends may live hundreds of kilometers away. Still, some residents compare their far-flung community's closeness to a small town's.

18,600 square miles (30,000 sq km), Anna Creek is the largest cattle station in Australia. However, only 15 people live here. Besides the station manager and his family, the population includes a cook, pilot, and teacher. Several ranch hands—or "jackeroos" as cowboys are known in Australia—complete the group.

For those who are not used to it, station life can be uncomfortable. The ground is hot and dry. Snakes sun themselves on low sand dunes behind the buildings. Flies seem to be everywhere. The wind carries a fine red dust that covers everyone and everything. For some of the station's residents, dealing with the loneliness is difficult. After all, Coober Pedy, 100 miles to the west, is the closest town. Yet other station residents could not imagine living in a city again.

Few of the station's teachers have lasted more than a year before returning to the coast. Even when the station has an on-site teacher, basic education comes from the School of the Air. Using a high-frequency radio, the children take part in classes broadcast from the coastal city of Port Augusta. Every grade level has a half-hour class each day. Each student also receives an

A worker watches a small part of the Anna Creek herd, which had been reduced by drought. Although horses are seldom used for roundups, many outback families keep them for other purposes.

individual 10-minute radio session with his or her radio teacher once a week. Other lessons arrive on videotape. Students mail in their assignments. The teacher grades them and returns them by mail. Students work five or six hours a day on these assignments. A parent or a hired teacher, as at Anna Creek Station, supervises. Use of the Internet, e-mail, and video conferencing is increasingly important for instruction.

March through November is the busiest time at Anna Creek. The cattle must be gathered before the heat and the flies make life miserable for people and animals. Calves must be tagged, horns cut, and some livestock shipped to market. The roundups are challenging. Some 16,000 head of cattle may be spread across an area larger than the state of Maryland. The jackeroos go out for weeks at a time to various parts of the station. Like education in the outback, this job also depends on technology. Jackeroos use motorcycles instead of horses on these roundups, and workers in airplanes spot cattle from the sky.

Applying What You Know

1. **Analyzing Information** How have technological innovations allowed people to live and work in the outback?

2. **Comparing** How is education in the outback similar to education in the United States? How are the two systems different?

Cultural Kaleidoscope

Central Australia's Boat Race Each year the locals around Alice Springs hold a boat race known as the Henley-on-Todd. What is unusual about this boat race is that it does not need water. The Todd River is normally a dry riverbed near Alice Springs. To encourage fun and entertain visitors, contestants build boats from various materials and leave the bottom open to allow room for their legs to stick out. When the race begins, the boat crews run as fast as they can, holding their boats around them as they race to the finish. In 1997 the residents of Alice Springs thought they would have to cancel the race because there was a flood and the Todd River actually had water in it.

Applying What You Know Answers

1. Radio, video, and computer technology such as e-mail allow children to be educated. Motorcycles and airplanes assist with ranching tasks.

2. similarities—instruction by grade level, both individual and group learning, the use of computers in learning, out-of-class assignments (homework); major difference—outback instruction by two-way radio instead of in a classroom

OBJECTIVES

1. **Identify some important features of New Zealand's natural enviroment.**

2. **Describe New Zealand's history and culture.**

3. **Examine New Zealand's economy and identify the major economic challenge the country faces.**

 **LET'S GET STARTED**

Copy these instructions onto the chalkboard: *When your team is preparing for a game, how do you get motivated? How do you get inspired to win?* Discuss responses. (*Possible answers: pep rallies, pep talks from the coach, team rituals, or other practices*) Tell students that before games, New Zealand's national rugby team performs a dance. Focus students' attention on the photo in this section and read the caption about the *haka*. Tell students they will learn more about New Zealanders and their land and cultures in Section 2.

Building Vocabulary

Call on a volunteer to read the definition of **economy of scale**. Ask students to suggest industries that probably develop such an economy easily and some that do not. (*Examples: easily—manufacturers of tennis shoes, computer chips; with difficulty—makers of specialized medical equipment, unicycles*) Point out that in a large country with extensive trade links, like the United States, even these industries can find buyers. They would have a difficult time succeeding in a small, isolated country such as New Zealand, however.

 **RESOURCES**

REPRODUCIBLE

▶ Guided Reading Strategy 31.2
▶ Readings in World Geography, History, and Culture 80
▶ Cultures of the World Activity: Region 9
▶ Geography for Life Activity 31: Where Should New Zealand's Urban Areas Be?

TECHNOLOGY

▶ One-Stop Planner CD–ROM, Lesson 31.2
▶ Homework Practice Online
▶ HRW Go site

REINFORCEMENT, REVIEW, AND ASSESSMENT

▶ Main Idea Activity 31.2
▶ English Audio Summary 31.2
▶ Spanish Audio Summary 31.2
▶ Section 2 Review, p. 727
▶ Daily Quiz 31.2

 **Section 2**

New Zealand

READ TO DISCOVER

1. What are some important features of New Zealand's natural environment?
2. What are New Zealand's history and culture like?
3. On what is New Zealand's economy based, and what economic challenge does the country face?

WHY IT MATTERS

Depending heavily on world trade may cause problems for New Zealand. Use **CNNfyi.com** or other **current events** sources to investigate dependence on world trade in New Zealand or other countries.

IDENTIFY

Maori

DEFINE

economy of scale

LOCATE

North Island
South Island
Cook Strait
Southern Alps
Canterbury Plains
Auckland
Wellington

Natural Environments

The South Pacific island country of New Zealand is about 1,000 miles (1,609 km) southeast of Australia. Like Australia, New Zealand has a British colonial history and a high standard of living. However, New Zealand is also very different. It is much smaller, more mountainous, and has a wetter and milder climate. The country has two major islands—North Island and South Island. (See the chapter map.) They are separated by Cook Strait. Some smaller islands in the Pacific are also part of New Zealand.

North Island In the north are peninsulas with forests and fertile lowlands. New Zealand is located on the Pacific Ring of Fire, so it is a tectonically active country. The central and western parts of North Island have active volcanoes,

Tree ferns grow in Waitaanga Forest, North Island. These plants, which can be more than 30 feet (9 m) tall, have been called living fossils because the species has survived for some 150 million years. A fern leaf is one of New Zealand's national symbols.

ALL LEVELS: Copy the following graphic organizer onto the chalkboard, omitting the italicized answers. Call on students to provide words and phrases that describe the physical geography of North Island and South Island to complete the diagram. In the center, have students list features the islands have in common. Then lead a class discussion comparing and contrasting the physical features of the two islands. Ask students to name some physical processes that have helped to shape the islands. *(Possible answers: tectonic activity, glaciation)* **ENGLISH LANGUAGE LEARNERS**

North Island
forests,
volcanoes,
geysers,
hot springs, hills

Both Islands
lowlands

South Island
mountains,
glaciers,
lakes, plains

geysers, and hot springs. These are created by the collision of the Pacific and Indo-Australian Plates. Subduction along this plate boundary causes earthquakes as well. The eastern part of North Island has rugged hills and small coastal lowlands.

South Island South Island is larger and has higher elevations than North Island. A steep mountain range called the Southern Alps runs along the west coast. The Southern Alps include Mount Cook, New Zealand's highest peak, which rises to 12,349 feet (3,764 m). The Southern Alps are famous for their beautiful scenery, with many glaciers and mountain lakes. Below are thick green forests. Along the east coast of South Island lie the Canterbury Plains. Different grain crops and livestock feeds are grown in this fertile lowland area.

✓ **READING CHECK:** *Physical Systems* What physical processes have created volcanoes, geysers, and hot springs on North Island? tectonic processes

Climates and Biomes All of New Zealand has a mild marine west coast climate. However, temperatures and precipitation vary across the country. In general, North Island is warmer than South Island and rarely receives snow. Westerly winds bring moisture. When these winds hit the Southern Alps, they drop rain and snow. As a result, western South Island receives much more precipitation than the east, which is in a rain shadow. Most of the west coast averages more than 100 inches (254 cm) of rainfall each year. Some eastern areas receive less than 20 inches (51 cm).

Most of New Zealand has a temperate forest biome. Forests cover about 30 percent of the country. These forests have mostly evergreen trees. Plant and animal life includes many endemic species. The most well known of these are flightless birds like kiwis (KEE-weez), which live in forests. Kiwis sleep during the day and look for food at night. Moas, a much larger kind of flightless bird, became extinct several hundred years ago. Unlike Australia, New Zealand has no endemic mammals except bats. New Zealand also has many exotic species. European settlers brought cats, cattle, deer, and sheep.

✓ **READING CHECK:** *Places and Regions* Which climate and biome are found throughout New Zealand? marine west coast climate, temperate forest biome

Our Amazing Planet

The tuatara (TOO-uh-TAHR-uh), a small reptile, is found only on a few islands of New Zealand. Tuataras are the last living members of an ancient group of reptiles related to the dinosaurs. They appeared on Earth more than 200 million years ago. Some tuataras may live to be 100 years old.

Eye on Earth

Kiwis under Attack One of New Zealand's national symbols, the flightless kiwi bird, is threatened with extinction. Conservationists warn that if rates of decline do not change, the kiwi could be gone in only five to ten years. In the 1920s there were about 5 million kiwis in New Zealand; now there are about 70,000.

Kiwis are under attack at every stage of their life cycle. Possums and ermines eat kiwi eggs, cats and ermines prey on chicks—only five percent survive to adulthood—and dogs and ferrets kill mature birds.

Predators in an island environment can devastate a bird population because many island birds have never developed any defense mechanisms. Domestic cats alone are blamed for the extinction of more than 40 bird species from New Zealand.

CRITICAL THINKING: Why are island bird populations particularly vulnerable to predation? *(Many island species evolved isolated from predators, so they have no natural defenses. Limited land area also prohibits escape to other areas.)*

Lake Matheson lies at the foot of Mount Cook and Mount Tasman on New Zealand's South Island. This tiny lake is protected from the wind, so its still surface reflects the mountains clearly.

LEVEL 1: Organize the class into five groups and assign each group one of the following topics about New Zealand: the Maori, the British in New Zealand, cities and settlement patterns, religion and education, or traditions and food. Have each group create a poster with information about its assigned topic. Encourage students to use the unit atlas and illustrations within the chapter for reference. **ENGLISH LANGUAGE LEARNERS, COOPERATIVE LEARNING**

LEVEL 2: Tell students to imagine that they are New Zealand citizens who recently made contact with U.S. teenagers in an online chat room. Have each student write an e-mail letter to a new American friend, telling the friend why New Zealand's history and culture make it such an interesting country.

LEVEL 3: Pair students and have the students in each pair exchange the letters they wrote for the Level 2 activity. Then ask each student to write a response to his or her classmate's letter, relating how the history and culture of New Zealand differ from the history and culture of the United States. Ask students also to note any similarities. Finally, have them give their letters to their partners to read. Instruct students to discuss among themselves any misconceptions either writer may have included in his or her descriptions. **COOPERATIVE LEARNING**

This example of Maori folk art was created in 1888 for a visit by Te Kooti Rikirangi, a Maori guerrilla and religious leader. Maori artisans were highly skilled in wood carving and stone carving. This painted work, therefore, may show the influence of European art forms.

Natural Resources   New Zealand's main resource is good land for farming. More than 50 percent of the land supports crops or livestock. Forests are also important. Pulp and paper are produced from the fast-growing *radiata* pine tree, an exotic species brought from California. Much hydroelectric power is produced from New Zealand's rivers. This kind of energy supplies 65 percent of the country's electricity. Geothermal power is also produced. New Zealand does not have many large mineral deposits. The most important are coal, gold, iron ore, and natural gas.

✓ **READING CHECK:**   *Places and Regions*   What is New Zealand's main resource?
good farmland

History and Culture

Like Australia, New Zealand was colonized by British settlers. What effects do you think this had on New Zealand?

Early History and Settlement   The first people in New Zealand were the **Maori** (MOWR-ee). They came from Pacific islands to the north about 1,000 years ago. Anthropologists call the early Maori moa hunters because moas were their main prey. Other groups of Maori came later. Most settled on North Island. Their culture was based on farming, fishing, and hunting.

In 1642 Dutch explorer Abel Tasman became the first European to reach New Zealand. However, the Dutch did not return to settle the islands. In 1769 British explorer James Cook landed on North Island and made contact with the Maori. He explored both North Island and South Island.

The first European settlers in New Zealand came from the British colony in Australia. They were missionaries, traders, and whalers. In 1840, British settlers and Maori leaders signed a treaty that gave the British control of the islands. In exchange, the British agreed to protect Maori rights. The British set up several settlements on North Island in the 1840s. However, troubles between the Maori and British soon arose. Some British settlers began taking

ALL LEVELS: Have each student write five questions about New Zealand's economy. Call on volunteers to read their questions to the class. Call on other students to answer them. Continue calling on students to ask questions until all major points from the section have been covered. Then lead a discussion about New Zealand's need to diversify its economy and how the country might go about doing so. Ask how the country's location and resources have limited diversification in the past. **ENGLISH LANGUAGE LEARNERS**

LEVELS 2 AND 3: Have students name examples of the power sources, agricultural products, and industrial products found in New Zealand. Write each example on the chalkboard as it is named. Then work with students to create a poster about the economic geography of New Zealand. Have students design the poster, draw illustrations to represent the item they listed on the chalkboard, and write captions for each illustration. Lead a discussion on how changes in technology and transportation have affected New Zealand's economy.

Maori lands. Also, diseases introduced by Europeans killed many Maori. These problems led to the Maori Wars of 1845–72, which the Maori lost.

In 1907 New Zealand became an independent country within the British Empire. It continued to develop its farming economy. New refrigeration methods allowed farmers to ship meat and dairy products to Britain. Like Australia, New Zealand sent troops to fight alongside the British in World Wars I and II.

✔ **READING CHECK:** *Human Systems* How did British settlement affect New Zealand's human geography? changed dominant cultural group from Maori to British

People and Languages Most New Zealanders have British ancestors and speak English. Asians and Pacific Islanders are both small but growing minorities. New Zealand's largest minority group, the Maori, makes up nearly 10 percent of the population.

FOCUS ON CULTURE

The Maori The Maori are related culturally to other peoples of the Pacific Islands. According to Maori legend, they came to New Zealand in seven canoes. Then a Maori hero named Maui created North Island by fishing it from the sea. Maori society was made up of different tribes *(iwi)* ruled by chiefs *(ariki)*. Tribes lived in villages. Land belonged to smaller groups *(hapuu)* within each tribe. Maori artists decorated canoes and houses with beautiful designs. Tattooing was common. Chiefs and warriors had facial tattoos *(moko)* that symbolized their high place in society.

Maori culture has changed greatly since Europeans came. Today most Maori live in cities. Many have mixed Maori and European heritage. While Maori have adapted to Western society, many are behind other New Zealanders in education and employment.

INTERPRETING THE VISUAL RECORD

Left: Maori train for a celebration. Their war canoes are in the background. Right: New Zealand's national rugby team, the All Blacks, performs a Maori dance and chant called a haka *before every international match. A New Zealand team performed a* haka *overseas for the first time in 1888, in Great Britain.* **By what process would the** haka **enter popular culture?**

ALL LEVELS: Organize the class into groups. Have each group develop a page or series of pages for a New Zealand newspaper. Ask students to include information about the country's natural environments, history, culture, and economy using articles, cartoons, editorials, illustrations, weather reports, and other typical newspaper features. Then have a spokesperson from each group present his or her group's newspaper to the class. Display the newspapers around the classroom. **ENGLISH LANGUAGE LEARNERS, COOPERATIVE LEARNING**

Using National Geography Standard 11:
Human Systems: The Patterns and Networks of Economic Interdependence on Earth's Surface

Many of New Zealand's major agricultural products, such as butter, cheese, lamb, and mutton, must be refrigerated to be transported over long distances. The country's island location precluded export to distant consumers until the 1880s, when refrigerated ships began taking cargoes to Great Britain. Have students conduct research on New Zealand's economic history and then create idea webs or flowcharts that show the effect of refrigerated ships not only on farming but also on secondary, tertiary, and quaternary industries.

Section 2 Review Answers

Identify For identification, see: Maori, p. 724

Define For definition, see: economy of scale, p. 727

Working with Sketch Maps
Maps will vary, but listed places should be labeled in their approximate locations. Auckland is the country's primate city.

Reading for the Main Idea
1. generate heavy precipitation in western South Island and create a rain shadow in eastern South Island

2. New refrigeration methods allowed farmers to ship meat and dairy products to Britain.

3. Many industries are based on agriculture, such as the production of processed foods like butter and cheese.

Critical Thinking
4. Possible answer: availability of good farmland (NGS 12)

Organizing What You Know
5. North Island—peninsulas with forests and fertile lowlands in north; volcanoes, hot springs, and geysers in west; hills and lowlands along east coast; marine west coast climate; South Island—Southern Alps along west coast; glaciers and mountain lakes above thick forests near Alps; fertile Canterbury Plains along east coast; marine west coast climate

VISUAL RECORD ANSWER

Possible answers: can move more people with fewer vehicles, can move uphill easily

726

Many Maori have maintained traditional ways of life. For example, Maori often greet each other by pressing their noses together *(hongi)*. Dances called action songs are also common, as are traditional carved meeting houses. Although nearly all Maori speak English, some prefer to use the Maori language in their homes. In 1987 Maori became an official language of New Zealand.

✓ **READING CHECK:** **Human Systems** How have the Maori maintained traditional ways since the arrival of Europeans? kept traditional customs; formal Maori gatherings; use of Maori language

Settlement Most of New Zealand's 3.8 million people live in lowland areas along the coast. About 75 percent live on North Island. More than 80 percent live in cities. New Zealand's primate city is Auckland. More than 30 percent of all New Zealanders live in or around Auckland. Other major cities include Wellington, which is the capital, and Christchurch.

Traditions, Customs, and Food Like in Australia, outdoor activities are popular in New Zealand. A mild climate makes camping, hiking, and sailing possible year-round. Skiing is popular at resorts in the Southern Alps. Organized sports include rugby and cricket, both of which are played throughout the country. New Zealand also competes in international yachting races. In fact, practically every sport one can imagine is available. New Zealanders even enjoy creating new outdoor activities.

In New Zealand sheep outnumber people 13 to 1, so workers who can shear sheep quickly and well are valued. Recordholders can clip the wool from a sheep in less than a minute. The best shearers in the world come to New Zealand to compete against each other. Champions earn prize money and local fame.

Favorite foods in New Zealand include clam soup, lamb, and sweet potatoes. A meringue, fresh fruit, and cream dish called a pavlova is the national dessert. Tea is popular, reflecting the country's British heritage.

✓ **READING CHECK:** **Places and Regions** Where do most New Zealanders live? lowland areas along coast; 75 percent on North Island; more than 80 percent in cities; more than 30 percent in or around Auckland

Economy and Issues

New Zealand's economy is a combination of farming, manufacturing, and tourism. Major exports include wool, meat, fish, and dairy products. Historically, the country's economy has been based on the export of farm goods. However, manufacturing and services have become much more important in the past 20 years.

New Zealand's pastures support millions of sheep and cattle. Major crops include wheat, barley, fruits, potatoes, and vegetables. The country is the

INTERPRETING THE VISUAL RECORD *Cable cars provide transportation for residents and visitors in Wellington, New Zealand.* **Why do you think that cable cars are an efficient form of transportation in this city?**

Close

Challenge students to guess the meanings of these New Zealand slang words: benzine *(gasoline)*, candy floss *(cotton candy)*, chocker *(completely full)*, The Ditch *(Tasman Sea)*, fizzy *(soft drink)*, rellies *(relatives)*, and togs *(swimsuit)*. Point out that many Maori words have entered common New Zealand speech. Examples include *kapai*, which means "very good," *kia ora*, which means "hello," and *hoihoi*, which means "be quiet."

Review and Assess

Have students complete the **Section Review**. Then have students complete **Daily Quiz 31.2.**

Reteach

Have students complete **Main Idea Activity for English Language Learners and Special-Needs Students 31.2.** Then have students compile fact sheets on New Zealand's physical features, history, cities, and economy. Students' fact sheets should cover the main ideas of the section. Call on volunteers to read their fact sheets to the class. **ENGLISH LANGUAGE LEARNERS**

Extend

Have interested students work in pairs to conduct research on the history of Maori political activism. Have each pair write and present a mock interview between a news reporter and a Maori leader. **COOPERATIVE LEARNING, BLOCK SCHEDULING**

world's largest producer of kiwi fruit. Many industries are closely tied to agriculture. For example, factories make processed foods like butter and cheese. Most of these are exported. Other industries include wood and paper production, textiles, and machinery. The main industrial center is Auckland. New Zealand has a growing film industry. Many movies and television programs are filmed there. One reason is the wide range of settings the varied landscape provides. Tourism is also important to the economy.

One challenge with developing industries in New Zealand is the country's small population. It makes New Zealand a small market. Therefore, it is harder for industries to develop an **economy of scale**—a large production of goods that reduces the production cost of each item. For example, if a company produced machines but was only able to sell 10 per year, it might not make a profit. However, if the company sold 10,000 machines per year, it would lower the cost of producing each machine and make a large profit.

Another challenge facing New Zealand is diversifying its economy. The country depends heavily on exports. Therefore, changes in world markets can have important effects on its economy. For example, like Australia, Great Britain used to be New Zealand's main trading partner. In 1973 Britain joined what is now the European Union (EU). Britain then had to raise its tariffs on goods from non-EU trade partners. As a result, prices on products from New Zealand rose in Britain. This increase caused the number of goods New Zealand sold to Britain to fall. Since then, New Zealand has diversified its exports. It has also found new trading partners. New Zealand now trades with Australia, the United States, and Japan. However, global trade is still vital to the country.

✔ **READING CHECK:** *Human Systems* How have changes in world trade patterns affected New Zealand's economy? Great Britain joining the EU led to a decline in trade with New Zealand and to a change in New Zealand's major trading partners

INTERPRETING THE VISUAL RECORD *Sheep graze in a pasture on South Island.* **How does the environment of this sheep station appear to differ from the one you read about in Cities & Settlements: Life in the Outback?**

Homework Practice Online
Keyword: SW3 HP31

Section 2 Review

Identify
Maori

Define
economy of scale

Working with Sketch Maps
On a map of New Zealand that you draw or that your teacher provides, label North Island, South Island, Cook Strait, Southern Alps, Canterbury Plains, Auckland, and Wellington. What is New Zealand's primate city?

Reading for the Main Idea

1. *Physical Systems* How do the Southern Alps affect New Zealand's precipitation patterns?

2. *Environment and Society* How did technological changes in the early 1900s affect New Zealand's export economy?

3. *The Uses of Geography* In New Zealand's market economy, how are major commercial industries tied to its agricultural products?

Critical Thinking

4. Analyzing Information What do you think might have attracted early settlers to New Zealand?

Organizing What You Know

5. Create a chart like the one shown below. Use it to describe the natural environments of North Island and South Island.

North Island	South Island

TECHNOLOGY

▶ Chapter 31 Test Generator (on the One-Stop Planner)
▶ Global Skill Builder CD–ROM
▶ HRW Go site

REINFORCEMENT, REVIEW, AND ASSESSMENT

▶ Chapter 31 Review, pp. 728–29
▶ Chapter 31 Tutorial for Students, Parents, Mentors, and Peers
▶ Chapter 31 Test (form A or B)
▶ Alternative Assessment Handbook

▶ Chapter 31 Test for English Language Learners and Special-Needs Students
▶ Unit 10 Test
▶ Unit 10 Test for English Language Learners and Special-Needs Students

Assess

Have students complete a Chapter 31 Test.

Reteach

Organize students into two groups. Assign Australia to one group and New Zealand to the other. Have each group write a basic plan or treatment for a documentary film about travel in its assigned country. Have each group share its treatment with the class. ENGLISH LANGUAGE LEARNERS, COOPERATIVE LEARNING

CHAPTER 31 Review Answers

5. Both have a British colonial history, indigenous minority groups, developed economies, and Southern Hemisphere locations. English is the main language in both. Australia is much bigger and has a larger population. It is drier, hotter, and less mountainous and does not have as high a percentage of arable land as New Zealand but has more abundant mineral resources.

Thinking Critically

1. Possible answer: The British settled in areas with mild climates, adequate rainfall, and good farmland. (NGS 12)
2. Possible answers: Introduced species, agriculture, and mining have disrupted natural wildlife. (NGS 14)
3. Possible answers: mostly urban; English main language; high standard of living; sports and outdoor activities popular; mostly Christian (NGS 10)

Using the Geographer's Tools

1. Possible answer: Settlers wanted to control their own affairs, rather than be subject to other groups.
2. primary goods; most land used for animal grazing
3. Answers will vary.

CHAPTER 31 Review

Building Vocabulary

On a separate sheet of paper, explain the following terms by using them correctly in sentences.

artesian wells   extensive agriculture
outback   exotic species
marsupials   Maori
Aborigines   economy of scale

Locating Key Places

On a separate sheet of paper, match the letters on the map with their correct labels.

Western Plateau   Southern Alps
Tasmania   Sydney
North Island   Central Lowlands
Great Dividing Range   Great Barrier Reef

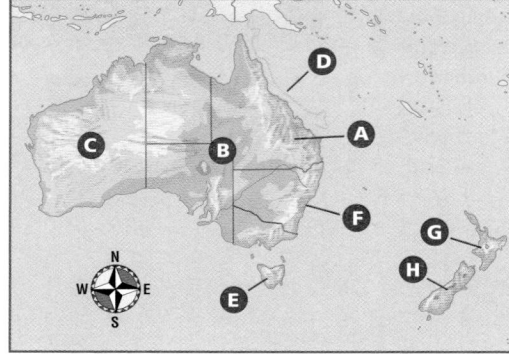

Understanding the Main Ideas
Section 1

1. (Physical Systems) What are some of the reasons Australia is dominated by arid and semiarid climates?
2. (Human Systems) What is the pattern of the distribution of major cities in Australia? Why do few people live in the continent's interior?

Section 2

3. (The Uses of Geography) How have the physical characteristics of New Zealand changed during the last 1,000 years?
4. (Places and Regions) What are some important characteristics of New Zealand's market economy?
5. (Places and Regions) How are the physical and human geography of Australia and New Zealand similar? How are they different?

Thinking Critically

1. Drawing Inferences and Conclusions How do you think environmental conditions in Australia influenced where British colonists settled?
2. Identifying Cause and Effect How do you think the development of Australia's economy has affected the environment?
3. Comparing How are life and culture in New Zealand similar to life and culture in the United States?

Using the Geographer's Tools

1. Analyzing Maps Study the map of Australia's states and territories in Section 1. Why do you think the colony of New South Wales had land taken away to create other colonies?
2. Analyzing Charts Study the chart of Australia's major exports and imports in Section 1. What types of goods does Australia export? How do these exports relate to land use in Australia?
3. Preparing Graphs Use the information in Section 1 to create a pie graph of Australia's different ethnic groups. How do you think Australia's population will change during the next 20 years?

Writing about Geography

Imagine you are a government consultant in Australia. You have been asked to write a report comparing different points of view on mining in lands the Aborigines hold sacred. You may want to use Internet or library resources to find more information. Write a short report. First, describe how Aborigines feel about mining operations in their sacred lands. Then describe a miner's point of view. Finally, suggest some ways that this difficult issue could be resolved. When you are finished with your report, proofread it to make sure you have used standard grammar, spelling, sentence structure, and punctuation.

SKILL BUILDING

Geography for Life

Drawing Sketch Maps

(Places and Regions) Review the unit political map. Then draw a sketch map of Australia showing its six states, two territories, five major cities, and lines of latitude and longitude. On what are the country's state and territorial boundaries based? Study the names of Australia's states and territories. How are they geographical names? What clues do they provide about Australia's culture and history?

Portfolio Activity

1. Have students conduct research on Australia's Great Barrier Reef to learn about the effects of tourism on the reef and efforts to protect its inhabitants. Ask students to create posters displaying their findings. Photograph the posters for student portfolios.

2. Have students create dioramas showing the past, present, and future of New Zealand's kauri forests. Kauri trees, once common on North Island, can grow more than 150 feet (45 m) tall, but many have been cut down and made into lumber. Place photographs of the dioramas in student portfolios.

Food Festival

A pungent, salty, savory spread marketed under the name Vegemite® is a favorite food in Australia and New Zealand. It is made from brewer's yeast extract and flavorings. Buy a jar in a specialty market or from an online supplier. To sample this regional treat, spread a very thin layer on buttered bread. One jar will probably serve the entire class. Students may also want to locate recipes using Vegemite® and prepare them for sampling. Note that this spread probably does not compare to any other food with which students are familiar. It is an acquired taste.

Building Social Studies Skills

Interpreting Graphs

Study the pie graph below. Then use the information from the graph to help you answer the questions that follow.

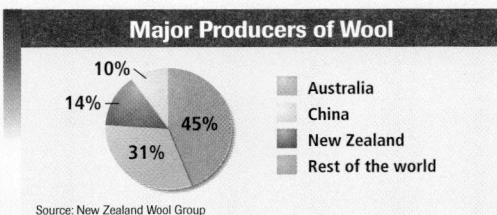

Major Producers of Wool

- 45% Australia
- 31% China
- 14% New Zealand
- 10% Rest of the world

Source: New Zealand Wool Group

1. What answer most accurately reflects the information from the graph?
   a. New Zealand produces more than half of the world's wool.
   b. China is the world's largest wool producer.
   c. Australia produces more of the world's wool than New Zealand and China combined.
   d. Australia produces more than four times the wool produced in China.

2. How do the countries shown in the graph rank, from the biggest to the smallest producers of the world's wool?

Analyzing Primary Sources

Read the following excerpt from an article by Bill Bryson. Then answer the questions.

> **"I first realized that I was going to like the outback when I read that the Simpson Desert, an area bigger than some European countries, was named in 1929 for a manufacturer of washing machines. . . . It wasn't so much the pleasingly unheroic nature of the name as the realization that an expanse of land of more than 50,000 square miles didn't even have a name until 70 years ago.**
>
> **But then that's the thing about the outback—it's so vast and forbidding that much of it has yet to be charted. . . ."**

3. Which of the following is *not* a reason that the writer likes the outback?
   a. Part of it was named for someone who was not a heroic explorer.
   b. The Simpson Desert is in Europe.
   c. Much of it has not been charted.
   d. Part of it was unnamed for a long time.

4. Which might the writer enjoy more—the Mojave Desert or Los Angeles? Why?

Alternative Assessment

PORTFOLIO ACTIVITY

Learning about Your Local Geography

Individual Project: Research
Plan, organize, and complete a research project on exotic plants and animals in your area. In your report answer the following questions: What are some of the plants and animals that have been introduced into your area? Where did they come from? Why were they introduced? Have any of these exotic species damaged the environment? If so, how? How have they affected endemic species? Prepare a report on your project and present it to your class.

internet connect

Internet Activity: go.hrw.com
KEYWORD: SW3 GT31

Choose a topic on Australia and New Zealand to:
- explore Australia's Great Barrier Reef.
- meet the Aborigines of Australia.
- learn about tectonic processes in New Zealand.

Writing

Student reports should contrast Aborigines' concerns about the sacredness of the land with miners' interests in economic development. Suggestions for resolving the conflict should be logical and well supported. Use Rubric 9, Comparing and Contrasting, to evaluate student work.

Geography for Life

Maps will vary, but should recognizably depict the listed places. The boundaries are based mostly on lines of latitude and longitude and on rivers. Many names, such as Western Australia, state relative location. Some names, such as New South Wales, refer to elements of British heritage. Use Rubric 20, Map Creation, to evaluate student work.

Social Studies Skills

1. c
2. Australia, New Zealand, China
3. b
4. Mojave Desert; uninhabited expanse of land

PORTFOLIO ACTIVITY

Students' reports should identify species introduced to the area, their origins, the purposes for the introductions, and any damages they may have caused to endemic species. Use Rubric 30, Research, to evaluate student work.

CHAPTER RESOURCE MANAGER

	Objectives	Pacing Guide	Reproducible Resources
SECTION 1 **Natural Environments** (pp. 731–34)	• Identify the main physical features of the Pacific Islands region and the physical processes that affect them. • Explain how the Pacific Islands region is divided into subregions. • Describe the climates, biomes, and resources found in the region.	**Regular** 1 day **Block Scheduling** .5 day *Block Scheduling Handbook, Chapter 32*	**RS** Guided Reading Strategy 32.1 **RS** Graphic Organizer Activity 32 **SM** Geography for Life Activity 32: Biogeography of the Pacific Islands
SECTION 2 **History and Culture** (pp. 735–39)	• Identify some important events in the region's history. • Describe the traditions and culture of the region.	**Regular** 1.5 days **Block Scheduling** 1 day *Block Scheduling Handbook, Chapter 32*	**RS** Guided Reading Strategy 32.2 **E** Cultures of the World Activity: Region 9 **SM** Critical Thinking Activity 32: World War II in the Pacific and the Battle of Midway
SECTION 3 **The Region Today** (pp. 741–43)	• Describe the economies of the Pacific Islands. • Identify some demographic characteristics of the region. • Examine some challenges that the people of the region face.	**Regular** 1 day **Block Scheduling** .5 day *Block Scheduling Handbook, Chapter 32*	**RS** Guided Reading Strategy 32.3 **SM** Map Activity 32: Samoa **PS** Readings in World Geography, History, and Culture 81

Chapter Resource Key

PS Primary Sources   **A** Assessment   💿 CD-ROM

RS Reading Support   **REV** Review   📼 Video

IC Interdisciplinary Connections   **ELL** Reinforcement and English Language Learners   🌐 Internet

E Enrichment   💻 Holt Presentation Maker Using Microsoft® PowerPoint®

SM Skills Mastery   Transparencies

 One-Stop Planner CD–ROM

See the *One-Stop Planner* for a complete list of additional resources for students and teachers.

 **One-Stop** Planner CD–ROM

It's easy to plan lessons, select resources, and print out materials for your students when you use the ***One-Stop Planner CD–ROM with Test Generator***.

 **internet** connect

HRW ONLINE RESOURCES

GO TO: go.hrw.com
Then type in a keyword.

TEACHER HOME PAGE
KEYWORD: SW3 Teacher

CHAPTER INTERNET ACTIVITIES
KEYWORD: SW3 GT32
Choose an activity to:
• tour the South Pacific.
• learn the traditions of the Pacific Islands.
• explore the diversity of the region.

CHAPTER ENRICHMENT LINKS
KEYWORD: SW3 CH32

CHAPTER MAPS
KEYWORD: SW3 MAPS32

ONLINE ASSESSMENT
Homework Practice
KEYWORD: SW3 HP32
Standardized Test Prep
KEYWORD: SW3 STP32
Rubrics
KEYWORD: SS Rubrics

COUNTRY INFORMATION
KEYWORD: SW3 Almanac

CONTENT UPDATES
KEYWORD: SS Content Updates

HOLT PRESENTATION MAKER
KEYWORD: SW3 PPT32

ONLINE READING SUPPORT
KEYWORD: SS Strategies

CURRENT EVENTS
KEYWORD: S3 Current Events

Technology Resources

	Reinforcement, Review, and Assessment
One-Stop Planner CD–ROM, Lesson 32.1	**ELL** Main Idea Activity 32.1
Geography and Cultures Visual Resources 61–65	**ELL** English Audio Summary 32.1
Homework Practice Online	**ELL** Spanish Audio Summary 32.1
HRW Go site	**REV** Section 1 Review, p. 734
	A Daily Quiz 32.1
One-Stop Planner CD–ROM, Lesson 32.2	**ELL** Main Idea Activity 32.2
Geography and Cultures Visual Resources 66	**ELL** English Audio Summary 32.2
Homework Practice Online	**ELL** Spanish Audio Summary 32.2
HRW Go site	**REV** Section 2 Review, p. 739
	A Daily Quiz 32.2
One-Stop Planner CD–ROM, Lesson 32.3	**ELL** Main Idea Activity 32.3
ARGWorld CD–ROM	**ELL** English Audio Summary 32.3
Homework Practice Online	**ELL** Spanish Audio Summary 32.3
HRW Go site	**REV** Section 3 Review, p. 743
	A Daily Quiz 32.3

Meeting Individual Needs

Ability Levels

Level 1 Basic-level activities designed for all students encountering new material

Level 2 Intermediate-level activities designed for average students

Level 3 Challenging activities designed for honors and gifted-and-talented students

English Language Learners Activities that address the needs of students with Limited English Proficiency

Chapter Review and Assessment

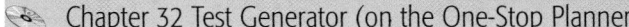

Chapter 32 Test Generator (on the One-Stop Planner)

Global Skill Builder CD–ROM

HRW Go site

REV Chapter 32 Review, pp. 746–47

REV Chapter 32 Tutorial for Students, Parents, Mentors, and Peers

A Chapter 32 Test (form A or B)

A Alternative Assessment Handbook

A Chapter 32 Test for English Language Learners and Special-Needs Students

Launch into Learning

Have students examine the pictures of the Pacific Islands in this chapter. Then call on volunteers to suggest words or phrases to describe what they see. *(Possible answers: tropical climate, palm trees, sandy beaches, clear waters, traditional cultures, volcanoes)* Point out that although these islands have great natural beauty, they are not trouble free. Many have seen their cultures and ecosystems damaged by international activities. Tell students that they will learn more about the landscapes, history, and cultures of the Pacific Islands in this chapter.

Using the Physical-Political Map

Have students examine the map on the opposite page. Ask them to note some physical characteristics of this region. *(Possible answers: small islands clustered in isolated groups, surrounded by the Pacific Ocean)* Have students identify the subregions of the Pacific Islands *(Melanesia, Micronesia, and Polynesia)*. Then draw students' attention to a wall map. Call on a volunteer to locate Easter Island, or Rapa Nui, which lies far east of the other Pacific Islands.

Why We Should Know More

You may wish to point out that there are many reasons why we should study the Pacific Islands:

▶ Many coral reefs and other fragile ecosystems in the Pacific region are threatened.

▶ The U.S. armed forces have been involved in two wars in the Pacific—the Spanish–American War in 1898 and World War II from 1941 to 1945.

▶ Rising sea levels could threaten Pacific Islands with low elevations.

▶ The natural beauty and pleasant climates of the Pacific Islands attract many tourists. Students may wish to visit the region someday.

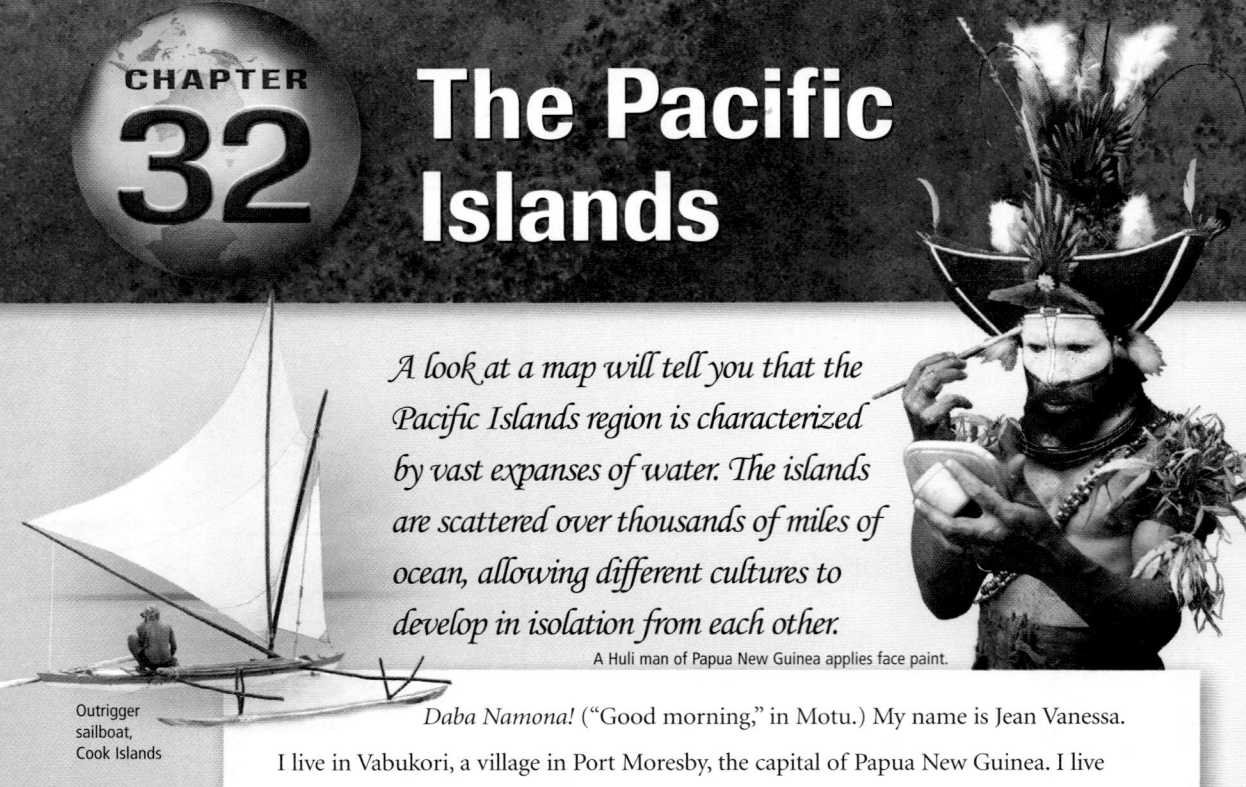

CHAPTER 32

The Pacific Islands

A look at a map will tell you that the Pacific Islands region is characterized by vast expanses of water. The islands are scattered over thousands of miles of ocean, allowing different cultures to develop in isolation from each other.

A Huli man of Papua New Guinea applies face paint.

Outrigger sailboat, Cook Islands

Daba Namona! ("Good morning," in Motu.) My name is Jean Vanessa. I live in Vabukori, a village in Port Moresby, the capital of Papua New Guinea. I live with my mother, grandmother, four younger brothers, and my younger sister. My mother is a journalist, and my grandmother bakes bread for the family in a drum oven. Sometimes we sell the bread. My grandfather, whom I loved very much, passed away last year. He was Motu, but my grandmother's language is Susu.

Our house is big, built on stilts, and painted blue. It is a few yards away from the village square where we have meetings and church gatherings and play sports. Our house has one big bedroom. My mum, my sister, my three little brothers, and I sleep in this room. The others sleep in the main living room area.

For breakfast we usually have tea and bread with fillings such as Vegemite,® jam, or peanut butter, and sometimes spaghetti or baked beans. Cereal is expensive, so it's a real treat. On weekends, we often have fried flour (pancakes) for breakfast. After breakfast, I take a bus to school. All my classes are taught in English.

I really like Christmas and New Year's. In Vabukori, we celebrate with much feasting and singing. The villagers are divided into two groups. On Christmas Day one group cooks for the other one. The other group sings songs about Bible stories from dawn to dusk. On New Year's Day the groups exchange roles.

730

Section 1

OBJECTIVES

1. Identify the main physical features of the Pacific Islands region and the physical processes that affect them.

2. Explain how the Pacific Islands region is divided into subregions.

3. Describe the climates, biomes, and resources found in the region.

LET'S GET STARTED

Copy the following passage onto the chalkboard: *Compare the physical map in this unit's atlas to the map of tectonic plates in Unit 1. How do you think many of the Pacific Islands were formed?* Discuss responses. *(Possible answer: undersea volcanoes that reached the ocean's surface)* Tell students they will learn more about the formation of islands and other aspects of the Pacific's natural environments in Section 1.

Building Vocabulary

Write **atoll** on the chalkboard. Call on a volunteer to locate and read the term's definition from the glossary or text. Then have students sketch how an atoll might look, based on this definition. Show students pictures of actual atolls to which they can compare their drawings.

Section 1

Natural Environments

READ TO DISCOVER

1. What are the main physical features of the Pacific Islands, and what physical processes affect them?

2. How is the Pacific Islands region divided into subregions?

3. What climates, biomes, and resources does the region have?

WHY IT MATTERS

Before air travel became common, travel across the Pacific Islands region was slow and difficult. Use **CNNfyi.com** or other **current events** sources to learn about how air travel has changed people's lives.

IDENTIFY

Intertropical Convergence Zone (ITCZ)

DEFINE

atoll

LOCATE

Tahiti
Melanesia
Micronesia
Polynesia

Locate, continued

Papua New Guinea
Solomon Islands
Vanuatu
New Caledonia
Fiji
Kiribati
Northern Mariana Islands
Marshall Islands
Guam
Marquesas Islands
Samoa

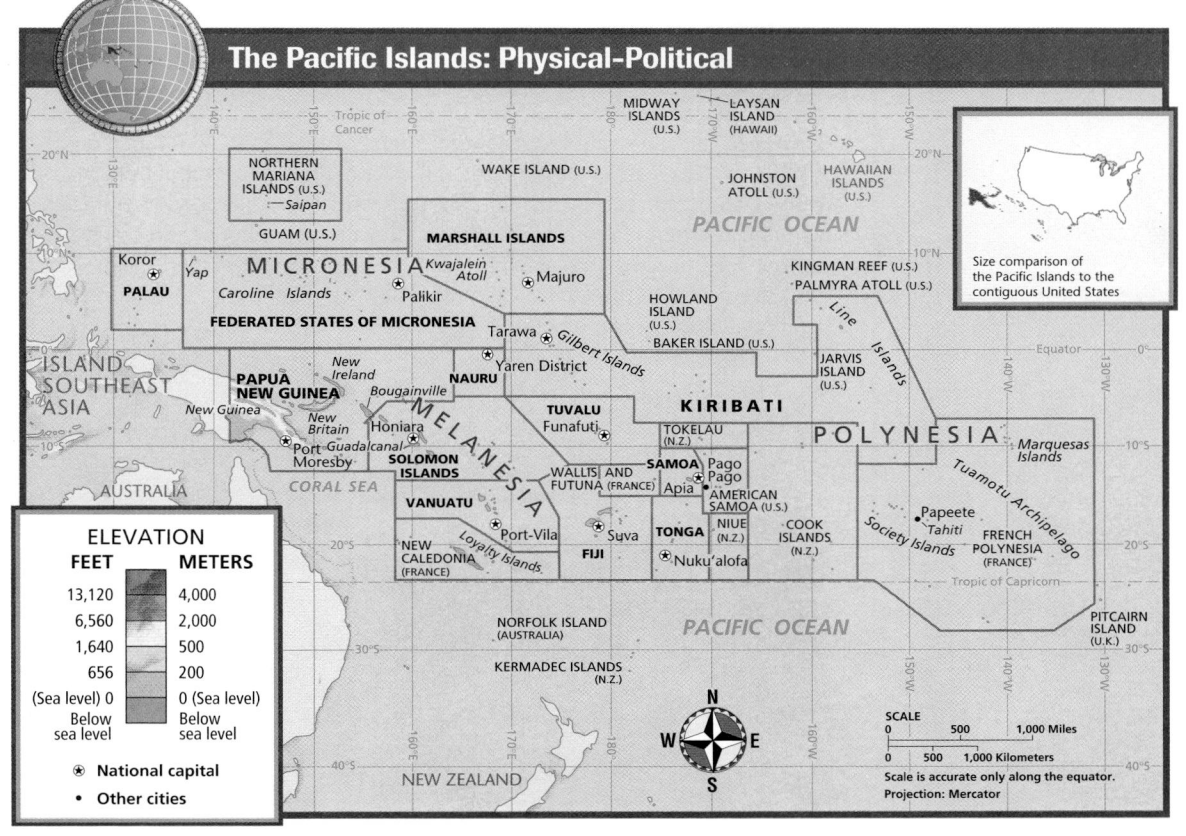

The Pacific Islands: Physical-Political

Size comparison of the Pacific Islands to the contiguous United States

ELEVATION

FEET	METERS
13,120	4,000
6,560	2,000
1,640	500
656	200
(Sea level) 0	0 (Sea level)
Below sea level	Below sea level

⊛ National capital
• Other cities

ALL LEVELS: Copy the following graphic organizer onto the chalkboard, omitting the italicized answers. Call on students to provide words and phrases to describe similarities and differences between the Pacific's high islands and low islands. **ENGLISH LANGUAGE LEARNERS**

HIGH ISLANDS   **LOW ISLANDS**

- *formed by volcanoes or continental rock*
- *mountainous and rocky*
- *freshwater*
- *rich, volcanic soils*

- *subject to natural phenomena such as tsunamis and volcanoes*
- *generally found in groups*

- *made of coral*
- *small and flat*
- *often ring-shaped*
- *little freshwater*
- *thin soil*

Global Perspectives

Global Warming and the Pacific Rising ocean levels caused by melting polar ice threaten many of the Pacific Islands. In 1999, two uninhabited islands in the South Pacific—Tebua Tarawa in Kiribati and Tepuka Savilili in Tuvalu—simply disappeared beneath the waves. On other islands, salt water has contaminated drinking water and killed crops.

Pacific countries have created regional environmental groups to address the threat of global warming. These groups have attended international meetings to discuss their concerns with industrial nations like the United States that generate most of the greenhouse gases scientists think may be responsible for global warming.

ACTIVITY: Tell students to imagine that they represent a Pacific Island country at an international environmental conference. Ask each student to write a speech that outlines the environmental concerns of his or her chosen country.

The Formation of an Atoll

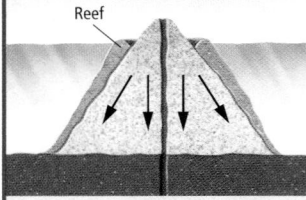

A. A coral reef forms along the edges of a volcanic island.

Reef

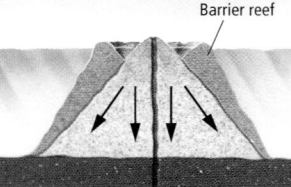

B. As the island sinks into the ocean floor, the coral reef grows upward and forms an offshore barrier reef.

Barrier reef

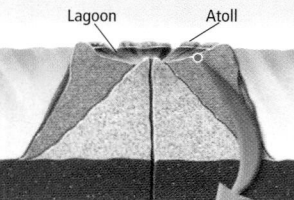

C. When the island is submerged, the reef forms an atoll. A ring of coral islands surrounds a shallow lagoon.

Lagoon   Atoll

A coral polyp builds a limestone skeleton that bonds it to a reef. The skeletons of many dead coral polyps form a reef. The coral reef pictured shelters a wide variety of marine life and attracts many fish.

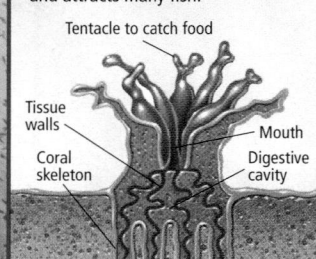

Tentacle to catch food

Tissue walls

Coral skeleton

Mouth

Digestive cavity

An Ocean Realm

The Pacific Ocean is the largest natural feature on Earth. It covers about a third of the world's surface. More than 10,000 islands lie within this region. Yet the total land area of the islands is very small. Most of them are tiny coral islands where no people live.

High and Low Islands Geographers classify the islands of the Pacific as either high islands or low islands. High islands can be further divided into continental and oceanic islands. Both types of high islands tend to be mountainous and rocky, and both may have volcanoes. Continental islands are formed from continental rock and lie on a shallow continental shelf such as that of Australia. New Guinea, the world's second-largest island, is a continental island. Oceanic islands such as Tahiti are simply volcanic mountains that have grown from the ocean floor to its surface.

Low islands form from coral. They are usually small and flat. Low islands also tend to have a characteristic ring shape. A ring-shaped coral island, or a ring of several islands linked by underwater coral reefs, is called an **atoll**. Within this ring lies a body of water called a lagoon. Kwajalein (KWAH-juh-luhn) Atoll in the Marshall Islands is about 80 miles (130 km) long and 20 miles (30 km) wide. However, most atolls are much smaller.

 **FOCUS ON GEOGRAPHY**

Coral Reefs Coral is formed by colonies of tiny marine animals. Millions of these tiny creatures' skeletons build up into reefs on a volcanic base. Coral reefs have been called the rain forests of the ocean because of their biological diversity. Reefs cover less than 0.2 percent of the ocean floor. However, about 25 percent of all ocean species inhabit coral reefs, which provide food and shelter. Coral reefs around the world face natural threats like typhoons and crown-of-thorns starfish, which eat coral. Reefs can recover from these assaults. However, they are now also threatened by human activities. Pollution and destructive fishing practices are among the dangers.

✓ **READING CHECK:** *Physical Systems* How might the destruction of the world's coral reefs affect the biological diversity of the oceans?

probably cause devastation, because 25 percent of all ocean species inhabit the coral reefs

Physical Processes High islands and low islands tend to have very different environments. The high islands' volcanic soils are usually rich. Many high islands support rain forests. Variations in elevation, rainfall, and soil lead to differences in plant life within one island as well as between islands. Low islands, on the other hand, are usually much less fertile. They often have no sources of freshwater.

The Pacific is the site of active tectonic processes. Many islands have active volcanoes, and many more volcanoes lie deep beneath the ocean's surface. Earthquakes are common in some areas. Tsunamis can present a danger as well to coastal areas.

Tectonic processes have also shaped the Pacific Ocean's floor. The Pacific's average depth is 14,000 feet (4,300 m). However, the deepest point, in the Mariana Trench, is more than 36,000 feet (11,000 m) below sea level. Oceanic trenches are the result of subduction. Volcanic ridges, visible on the map as chains of islands, often run parallel to the trenches. The Northern Mariana Islands and Tonga Islands are examples of volcanic island chains alongside deep trenches.

✓ **READING CHECK:** *Physical Systems* Which type of Pacific island can often support rain forests, and why? high islands, because their volcanic soil is usually rich

Three Island Groups

Geographers divide the Pacific Islands into three large subregions—Melanesia, Micronesia, and Polynesia. These groupings are based on the cultural variations and spatial arrangement of the islands.

Melanesia lies closest to Australia. It includes the eastern half of New Guinea, which makes up most of Papua (PA-pyooh-uh) New Guinea. From there Melanesia stretches east to include the Solomon Islands, Vanuatu (van-wah-TOO), New Caledonia, and Fiji. Most of the Melanesian islands have mountains and volcanoes.

Micronesia lies east of the Philippines, mostly north of the equator. Micronesia includes the Caroline Islands, the Gilbert Islands of Kiribati, the Northern Mariana Islands, and the Marshall Islands. These are a mix of high islands and low islands. The Northern Mariana Islands are a chain of volcanoes. Guam, the largest, is made up of a limestone plateau in the north and volcanic hills in the south. In contrast, the Gilbert Islands are all coral atolls.

Polynesia is the largest of the three subregions. It covers a huge triangle with its corners at Easter Island, the Hawaiian Islands, and New Zealand. See the unit atlas for Easter Island's isolated location about 2,000 miles (3,200 km) west of South America. Polynesia also includes the Cook, Marquesas, Samoa, Society, and Tonga Islands, and the Tuamotu Archipelago.

✓ **READING CHECK:** *Places and Regions* What are the three subregions of the Pacific Islands? Melanesia, Micronesia, Polynesia

GO TO: go.hrw.com
KEYWORD: SW3 CH32
FOR: Web sites about the Pacific Islands

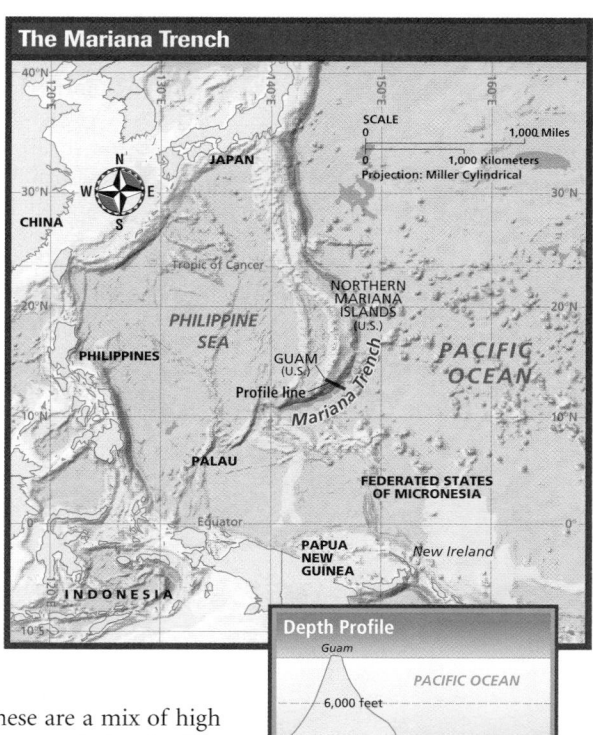

The Mariana Trench

SCALE
0   1,000 Miles
0   1,000 Kilometers
Projection: Miller Cylindrical

JAPAN

CHINA

Tropic of Cancer

PHILIPPINE SEA

PHILIPPINES

GUAM (U.S.)

NORTHERN MARIANA ISLANDS (U.S.)

PACIFIC OCEAN

Profile line

Mariana Trench

PALAU

FEDERATED STATES OF MICRONESIA

Equator

PAPUA NEW GUINEA

New Ireland

INDONESIA

Depth Profile
Guam

PACIFIC OCEAN

6,000 feet

12,000 feet

18,000 feet

Mariana Trench

24,000 feet

30,000 feet

36,000 feet

The Mariana Trench is the deepest known place on Earth's surface. The trench is more than 1,580 miles (2,550 km) long.

The Fly River is one of the major rivers of New Guinea. Named for the ship HMS *Fly*, whose captain discovered its mouth in 1842, the river flows about 650 miles (1,045 km) from the Star Mountains to the Gulf of Papua.

Section 1 Review Answers

Identify For identification, see: Intertropical Convergence Zone (ITCZ), p. 734

Define For definition, see: atoll, p. 732

Working with Sketch Maps Maps will vary, but listed places should be labeled in their approximate locations. Melanesia and Polynesia lie mostly south of the equator; Micronesia lies mostly north of the equator.

Reading for the Main Idea
1. tectonic processes, buildup of coral; volcanoes, earthquakes, tsunamis, typhoons

2. cultural variation and the spatial arrangement of the islands

Critical Thinking
3. mountains in island's interior; latitude, monsoons, trade winds, ITCZ, elevations (NGS 4)

4. Because they are formed from coral, low islands are not likely to contain useful minerals such as metals. (NGS 7)

Organizing What You Know
5. Surface landforms and elevation—usually mountainous; usually low and flat; Soil-building process—volcanic; formed from coral; Soil fertility—usually fertile; less fertile; Vegetation—can include rain forests; none listed

VISUAL RECORD ANSWER

north; Northern Hemisphere; because the ITCZ is north of the equator, indicating that the surface is warmer there

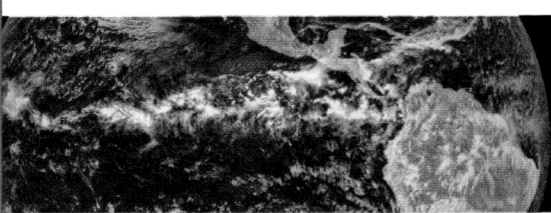

INTERPRETING THE VISUAL RECORD

*The band of clouds in the photo marks the Intertropical Convergence Zone (ITCZ), which moves north or south of the equator throughout the year to the area with the warmest surface temperature. **Does this photo show the ITCZ north or south of the equator? Which hemisphere was experiencing summer when this photo was taken? How can you tell?***

Climates, Biomes, and Resources

Most of the region's islands lie between the Tropic of Cancer and the Tropic of Capricorn. As you might expect, climates there are generally hot with high rainfall. Only Papua New Guinea, with its mountainous interior, has areas with highland climates. Here mountain peaks have tundra and alpine grasslands despite their location in low latitudes. Some parts of the region have distinct wet and dry seasons. For example, in the far western Pacific, wet and dry seasons are influenced by monsoons. In Papua New Guinea and some other areas, heavy rainfall supports thick tropical rain forests.

Trade winds play an important role in the climates of the islands. Northeast trade winds flow from the Tropic of Cancer. Southeast trade winds flow from the Tropic of Capricorn. The area where these prevailing winds meet near the equator is called the **Intertropical Convergence Zone (ITCZ)**. This area has generally calm humid air at low elevations. However, the warm, moist air of the ITCZ condenses and cools as it rises, causing heavy rainfall. Typhoons most often occur in the western Pacific, while the southeastern Pacific has very few. These swirling oceanic storms can be very destructive.

Fish and shellfish are important resources for the region. Lobsters, octopuses, sharks, shrimp, and tuna are just a few of the sea creatures people catch. Cultured pearls are harvested from oysters in French Polynesia and the Cook Islands. Other resources are less plentiful. Papua New Guinea and some other islands of Melanesia export timber. Only the large continental high islands have useful metal deposits. Papua New Guinea has major gold and copper deposits. New Caledonia is rich in nickel.

✓ **READING CHECK:** *Places and Regions* What are the main natural resources of the Pacific? fish, shellfish, pearls, timber, gold, copper, and nickel

 ## Section 1 Review

go. hrw .com Homework Practice Online
Keyword: SW3 HP32

Identify
Intertropical Convergence Zone (ITCZ)

Define
atoll

Working with Sketch Maps
On a map of the Pacific Islands region that you draw or that your teacher provides, label Tahiti, Papua New Guinea, Solomon Islands, Vanuatu, New Caledonia, Fiji, Kiribati, Northern Mariana Islands, Marshall Islands, Guam, Marquesas Islands, and the Samoa Islands. Outline Melanesia, Micronesia, and Polynesia. Which of the three subdivisions lie mostly south of the equator? Which lies mostly north of the equator?

Reading for the Main Idea
1. *Places and Regions* What physical processes have shaped the region's islands? What environmental hazards affect the region?

2. *Physical Systems* What factors form the basis for dividing the Pacific Islands into three subregions?

Critical Thinking
3. *Analyzing Information* Why does New Guinea have highland climates? What factors affect the region's climates?

4. *Drawing Inferences and Conclusions* How do you think the distribution of mineral resources in the region relates to how different islands are formed?

Organizing What You Know
5. Create a chart like the one shown below. Use it to make generalizations about the two types of islands found in the region.

	High islands	Low islands
Surface landforms and elevation		
Soil-building process		
Soil fertility		
Vegetation		

Section 2

OBJECTIVES

1. Identify some important events in the region's history.

2. Describe the traditions and culture of the region.

LET'S GET STARTED

Copy the following passage onto the chalkboard: *Fijian women have traditionally greeted each other by complimenting weight gain. In 1995 Fiji started receiving American TV shows. A few years later, a researcher found that many Fijian teenage girls were dieting.* Ask students what they can conclude about Fiji from this series of events. Discuss their responses. Tell students they will learn more about the history and culture of the Pacific Islands in Section 2.

Building Vocabulary

Write **trust territories**, **pidgin languages**, and **matrilineal** on the chalkboard. Have students look up the definitions in the section and read them aloud. Point out that the word *territories* comes from the Latin root *terra*, meaning "land," and that the word *pidgin* may come from the Chinese pronunciation for the word *business*. Tell students also that the word *matrilineal* can be traced to the Latin word *mater*, which means "mother." Ask students to suggest other words derived from *mater*. (*Examples: maternal, matrimony*)

Section 2 — History and Culture

READ TO DISCOVER

1. What are some important events in the history of the region?
2. What are the traditions and culture of the region like?

WHY IT MATTERS

Early contacts with Europeans introduced new diseases into the Pacific Islands region. Use CNNfyi.com or other **current events** sources to learn about modern-day outbreaks of diseases.

DEFINE

trust territories
pidgin languages
matrilineal

LOCATE

Wake Island
American Samoa

History

Researchers have used archaeology and other evidence to learn that waves of people from Southeast Asia settled the region. Much later, Europeans and others arrived and brought many changes.

Migration Patterns   Humans may have lived on New Guinea at least 33,000 years ago. Human migration into the Pacific may have begun even earlier than that. (See Geography for Life: Migration into the Pacific.) Over thousands of years, different peoples spread to different island groups. They sometimes mixed with earlier settlers. Over time, the peoples of Micronesia and Polynesia developed some distinct cultural features and became different in appearance. While the peoples of the two regions are different from each other, they share many cultural features. On the other hand the third region, Melanesia, is more distinct.

Polynesians created these stone figures, the tallest of which stands 37 feet (11 m) high, on Easter Island, or Rapa Nui (RAH-pah NOO-ee), as it is also known. After a period of peace and prosperity that lasted from about A.D. 1000 to 1500, rapid population growth and deforestation created an environmental crisis. Hunger and fighting over scarce resources devastated the island's people.

Section 2 RESOURCES

REPRODUCIBLE

▶ Guided Reading Strategy 32.2
▶ Cultures of the World Activity: Region 9
▶ Critical Thinking Activity 32: World War II in the Pacific and the Battle of Midway

TECHNOLOGY

▶ One-Stop Planner CD–ROM, Lesson 32.2
▶ Geography and Cultures Visual Resources 66
▶ Homework Practice Online
▶ HRW Go site

REINFORCEMENT, REVIEW, AND ASSESSMENT

▶ Main Idea Activity 32.2
▶ English Audio Summary 32.2
▶ Spanish Audio Summary 32.2
▶ Section 2 Review, p. 739
▶ Daily Quiz 32.2

Teach Objective 1

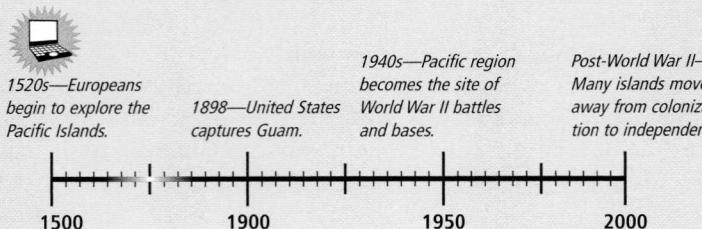

LEVEL 1: Copy the following graphic organizer onto the chalkboard, omitting the italicized answers. Have students complete it to examine foreign influences in the history of the Pacific Islands region. **ENGLISH LANGUAGE LEARNERS**

LEVELS 2 AND 3: Have students complete the Level 1 activity. Then organize the class into pairs. Provide each pair with an outline map of the Pacific Islands region. Have pairs mark locations where significant events in the region's history took place. Students should note the dates of the events and write a sentence or two describing each event in the margin. Call on volunteers to share their maps with the class. **ENGLISH LANGUAGE LEARNERS, COOPERATIVE LEARNING**

1520s—Europeans begin to explore the Pacific Islands.

1898—United States captures Guam.

1940s—Pacific region becomes the site of World War II battles and bases.

Post-World War II— Many islands move away from colonization to independence.

| 1500 | 1900 | 1950 | 2000 |

Our Amazing Planet

The French—like the British in Australia—used their colony in New Caledonia as a prison. They sent more than 22,000 French prisoners there between 1864 and 1897.

INTERPRETING THE VISUAL RECORD

Captain James Cook, one of Europe's greatest explorers, led three voyages to the Pacific Ocean. William Hodges was the official artist on Cook's second trip. Hodges painted this view of Matavai Bay, in Tahiti, in the 1770s. **How might this painting have influenced Europeans' perceptions of the Pacific Islands?**

Cultural patterns, languages, and physical traits there all differ from those in Micronesia and Polynesia. Many Melanesians seem to be genetically linked to the Aborigines of Australia. However, peoples and cultures within Melanesia vary a great deal. Movement and mixing of peoples may have gone on for thousands of years. The picture is not complete, and more research may provide new details.

✓ **READING CHECK:** **Human Systems** Where did the first settlers of the Pacific Islands come from? Southeast Asia

European Arrival Europeans began to explore the Pacific Islands in the 1500s. Ferdinand Magellan, a Portuguese navigator working for Spain, sailed across the Pacific in 1520–21. Other explorers followed. Spanish, Dutch, English, and French sailors came to explore, trade, spread Christianity, and claim territory. Later, Germany, Japan, and the United States entered the race for colonies in the area. The United States captured Guam and the Philippines from Spain during the Spanish-American War in 1898. By the end of the 1800s, foreign powers controlled nearly the whole region.

At first, colonial control was limited. American and European whale hunters sailed into the Pacific, sometimes setting up small outposts. Before the discovery of petroleum, whale oil was very valuable for lighting and industrial uses. With little or no regulation by the colonial countries, the whalers and traders badly exploited the people of the islands. This disrupted traditional cultures in many areas. Some Pacific islanders were enslaved. Unknowingly, the Europeans also spread deadly diseases, including measles and influenza.

Over time, colonial rule became more organized. Colonial powers set up plantations and military bases. The British brought thousands of workers from India to work on sugar plantations in Fiji. Despite the colonial presence, the Pacific Islands remained uninvolved in global politics until the 1940s. World War II then brought sudden changes to the Pacific region. Many

VISUAL RECORD ANSWER

It might have influenced them to think of the region as a peaceful paradise.

LEVELS 1 AND 2: Assign each student a subregion of the Pacific Islands. Have students imagine that they live on an island in their assigned area and that they are writing to a pen pal in the United States. Ask students to describe in their letters daily life and traditions on their island. ENGLISH LANGUAGE LEARNERS

LEVEL 3: Organize the class into five groups and have each group act as a team of anthropology students researching a culture group of the Pacific Islands. Have each group conduct research and create a report that includes a description of the studied group's behavior and theories about the historical development of its culture. Call on volunteers to present their reports to the class. COOPERATIVE LEARNING

HOMEWORK: Have students create charts, drawings, maps, or other graphics to illustrate aspects of Pacific culture. For example, charts may indicate the languages spoken in each island group, drawings may depict food, and text may include recipes.

---

islands became battlefields between 1941 and 1945. Others were used as bases. Armies, planes, and ships moved through the region. Japan conquered many islands early in the war. Over time, the United States and its allies pushed back the Japanese and defeated them. At the end of the war, the United Nations made some islands **trust territories**. These were areas placed under the control of another country while they prepared for independence.

Independence Since the end of World War II, the islands of the Pacific have moved away from colonialism. Some have become fully independent. A few other islands are still colonies of or are otherwise associated with an outside country. Guam, Wake Island, and American Samoa are U.S. territories. The Northern Mariana Islands form a commonwealth with the United States, similar to Puerto Rico's situation. The Federated States of Micronesia are in free association with the United States. This status allows their citizens to work in this country. In return, the United States can keep military bases on the islands. Australia, Chile, France, New Zealand, and the United Kingdom also have territories in the Pacific.

✓ **READING CHECK:** *Human Systems* Why did Europeans and other outsiders come to the Pacific Islands? to explore, trade, spread Christianity, claim territory, hunt whales, set up plantations and military bases

Traditions and Culture

Each of the societies of the Pacific developed its own culture. Despite the many variations, it is possible to make some generalizations about cultural features. Groups on the same island or within an island chain often shared cultural characteristics. Some similarities extended through a subregion or even through the entire region.

Today the Pacific Islands are home to a great number of different ethnic groups and languages. How did this happen? Huge stretches of ocean between islands allowed different cultures and languages to develop independently from one another. On New Guinea, thick rain forests and rugged mountains separated different groups of people in a similar way. Today the peoples of Papua New Guinea speak more than 700 different languages. Some of them are spoken by only a few hundred people.

Reflecting the lasting influence of colonialism, English and French are used in government and education in many parts of the region. To communicate within island groups where languages differ, some Pacific peoples have developed simplified languages based on English. These are called **pidgin languages**.

Education Almost all children in Polynesia and Micronesia now receive education through the high school level. Schooling is not yet available to everyone in Melanesia. There are several universities in the Pacific Islands. Most teach mainly in English, but a French-language university has locations in both New Caledonia and Tahiti.

A piece of traditional stone money leans against the wall of this men's meeting house on the island of Yap. Although the U.S. dollar is the common currency in Yap, the stone disks are used for some major transactions, such as land purchases.

Daily Life

Traditional Polynesian Religion Everything and everyone in traditional Polynesian culture is thought to have a special spiritual power. Polynesians called this sacred power *mana*. The amount of *mana* in a person or object can be increased or decreased by human actions. To maintain *mana* the islanders developed an elaborate set of rules. While many Polynesians practice Christianity today, some of the rules for protecting *mana* are still observed. It is considered in poor taste, for example, to pass one's hand over another's head or stand on higher ground than someone of greater rank. These actions are believed to drain a person's *mana*.

CRITICAL THINKING: Which of our customs might be compared to maintaining *mana* among Polynesians? *(Possible answers: using proper terms of address, standing when another person enters the room, offering a chair to an older person)*

internet connect

GO TO: go.hrw.com
KEYWORD: **SW3 CH32**
FOR: Web sites about Polynesian customs

Linking Past to Present

The Voyage of the *Hokule'a*

In the past, Pacific Islanders designed huge canoes that could carry enough supplies for long voyages. These traditional vessels were up to 72 feet (22 m) long and had double hulls that looked like two canoes attached side by side. They were propelled by sails and steered by an oar in the stern.

In 1999 the *Hokule'a*, a reconstruction of a Polynesian canoe, sailed from Hawaii to Easter Island, perhaps the world's most-remote inhabited island. One mission of the crew, which has been sailing the *Hokule'a* for more than 25 years, is to raise public awareness of great Polynesian seafaring achievements of the past.

ACTIVITY: Have students conduct research on the routes taken by the original settlers of Polynesia and plot them on a map.

Connecting to TECHNOLOGY

Navigation Skills of the Pacific Islanders

To the untrained eye, the open ocean looks empty and featureless. How would you find your way from one tiny island to another? What if you were in a wooden sailboat without a compass, radio, or other modern instruments?

The sailors of the Pacific Islands used many skills to navigate the ocean. At the start of a trip, they would take their bearings from landmarks on shore to plot their direction. Then they would note the rising points of certain stars and steer by "star paths." Winds and currents were also familiar to the sailors. The flights of birds and the movements of clouds provided clues. The sailors noted the reflections of lagoons on the bottoms of clouds. They could even tell the way waves were deflected off of distant islands. Understanding all these signs helped these sailors reach their destinations, sometimes across thousands of miles of ocean. People of the Pacific Islands still use many of these skills.

Expert navigators of the Pacific Islands teach others how to use a compass based on the locations of certain stars.

Religion Early traditional religions of the Pacific had some similarities. People commonly worshipped several gods and goddesses. Each supernatural being was linked to specific areas of human life or natural phenomena. Carved statues, costumes, masks, and dance were usually part of religious rituals. The spirits of ancestors also were important to some peoples.

Christianity is the main religion in the Pacific Islands today. Roman Catholic and Protestant missionaries spread their faiths through the area during the colonial period. Churches are important to community life in many parts of the region. However, people still practice local religions in parts of Melanesia, including Papua New Guinea.

✓ **READING CHECK:** *Human Systems* How did French, English, and Christianity spread to the Pacific Islands? spread by European colonists and missionaries during the colonial period

Food Before Europeans came, three root crops—sweet potatoes, taro, and yams—were among the key foods in the Pacific Islands. Three tree crops—bananas, breadfruit, and coconut—were also important to the diet. The coconut palm was particularly useful. Pacific Islanders ate the soft flesh of the coconut and drank the milk. They used the shells as containers and got fibers from the coconut's outer husk. The people also made roofs and baskets from coconut palm tree fronds. Rice was the only grain grown in the region. It was grown only in parts of the Mariana Islands.

Around the Pacific, fish are the main source of protein. Before European contact, the domesticated animals of the Pacific Islands were limited to chickens, dogs, and pigs. People on some islands ate fruit bats. Not all these animals were found on all islands. For example, no dogs lived in the Marquesas Islands. Easter Island had only chickens. Many islands had no domesticated animals at all. In general, people served meat—especially pigs—only at special celebrations. In Samoa, for example, an occasion's importance could be measured by the number of pigs served.

Today imported foods such as canned meat, boxed cake mixes, and soft drinks are a growing part of people's diet. The need for money to buy imported goods has pushed many farmers to switch to growing cash crops such as sugarcane for export.

Close

Call on students to provide key facts about the history and culture of the subregions of the Pacific Islands. List the facts on the chalkboard as students share them with you. Then have students vote on which aspects of Pacific culture they find most interesting.

Review and Assess

Have students complete the **Section Review**. Then have students complete **Daily Quiz 32.2**.

Reteach

Have students complete **Main Idea Activity for English Language Learners and Special-Needs Students 32.2**. Then have students work in pairs to summarize the history and culture of the Pacific Islands region. Call on volunteers to present their summaries to the class. **ENGLISH LANGUAGE LEARNERS, COOPERATIVE LEARNING**

Extend

Have interested students conduct research on Nan Madol, a ceremonial center built on 92 artificial islets within a lagoon on Pohnpei in Micronesia. Ask students to examine the site's history and its importance for tourism. **BLOCK SCHEDULING**

Traditional Societies Historically, people in the region tended to be organized into clans or tribes. They lived in places that ranged from small villages to communities with thousands of people. Patterns of social interactions within these groups had some features in common.

Polynesian groups often had complex rules of behavior and social ranks. On some islands, like Tahiti and Tonga, political status was kept within certain families. Individuals of high rank had great power over the common people. Chiefs distributed land and organized work such as the digging of irrigation systems. Competition for land, resources, and status often led to war.

Elsewhere in the Pacific, people placed less emphasis on inherited rank. People could gain status by giving feasts or presenting valuable gifts. They could also gain status by organizing trade with other groups or other islands. Many Melanesian peoples selected a leader through a kind of competition. A leader won supporters through his abilities or his wealth. Marriage or descent from an important family could also help determine leadership. The amount of power a leader held varied from group to group.

One interesting feature of Micronesia was that local groups were often **matrilineal**. That is, people traced kinship through the mother. When a marriage took place, the husband became part of the wife's clan, rather than the other way around. In some societies, women held high status. For example, Tongan women outranked men in several situations.

Art in the Pacific Islands was often connected to religion. Wood carvings, for example, usually showed gods and ancestors. Today island artists create similar carvings and sell them to tourists.

This Fijian chief's home contains traditional handicrafts. Fiji has a Great Council of Chiefs that elects the country's president.

✓ **READING CHECK:** *Human Systems* How did a Melanesian leader win his position? by demonstrating his wealth and abilities; through marriage into or descent from an important family

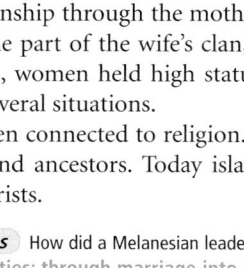

Section 2 Review

go.hrw.com **Homework Practice Online**
Keyword: SW3 HP32

Define
trust territories, pidgin languages, matrilineal

Working with Sketch Maps
On the map you created in Section 1, label Wake Island and American Samoa. In which of the Pacific Islands' three subregions is American Samoa located?

Reading for the Main Idea
1. *Environment and Society* What food crops did Pacific Islanders grow before contact with Europeans?
2. *Human Systems* What six foreign countries have territories in the Pacific?

Critical Thinking
3. *Analyzing Information* How did physical features affect migration and the distribution of culture groups in the region?
4. *Analyzing Information* How is the colonial history of the Pacific Islands reflected in the region's culture today? How has importing food changed the region?

Organizing What You Know
5. Create word webs like the one shown below to describe the people, languages, religions, educational systems, food, and traditional customs of the region.

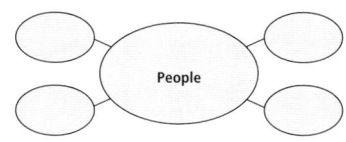
People

Section 2 Review Answers

Define For definitions, see: trust territories, p. 737; pidgin languages, p. 737; matrilineal, p. 739

Working with Sketch Maps Maps will vary, but listed places should be labeled in their approximate locations. It is part of Polynesia.

Reading for the Main Idea
1. sweet potatoes, taro, yams, bananas, breadfruit, coconut, rice
2. Australia, Chile, France, New Zealand, United Kingdom, United States

Critical Thinking
3. Ocean and mountain barriers allowed cultural groups to develop independently of one another. (NGS 9)
4. English and French languages, Christianity, and English- and French-language universities common; growing part of people's diet, farmers switching to cash crops for money to buy imported food (NGS 10)

Organizing What You Know
5. people—many ethnic groups; languages—mostly French and English, pidgin, hundreds of different languages in Papua New Guinea; religions—mostly Christian, some local religions; educational systems—high school in Polynesia and Micronesia, not available to everyone in Melanesia; some universities; food—traditional crops, fish, various meats, imported foods; traditional customs—matrilineal society

Archaeology: Lapita Pottery

Archaeologists use the term *Lapita* to refer to a number of related elements of historic Pacific culture. The term has been used to describe the early cultures of many of the Pacific Islands and to name a distinct people who may have originated this culture. However, it was first used to describe a type of pottery found throughout Melanesia that has shaped archaeologists' theories about the origins of human settlement in the Pacific.

Lapita pottery is not a single style or pattern but an extensive series of ceramic vessels whose shapes and designs varied over time. Interestingly, the earliest examples found of this pottery are the most elaborately decorated. These early pieces have complex shapes and bear geometric patterns of stamped and appliquéd designs. Later pieces follow much simpler designs and are less elaborately decorated. These changes allow archaeologists to trace the spread of Lapita pottery—and possibly human settlement—through the Pacific because they can match newly discovered fragments or shards to the pieces they have already found and dated.

Organize the class into groups and have each group conduct research to find pictures of Lapita pottery from various time periods. Lead a discussion about changes in the pottery and how they can be used to trace its diffusion through the region. **COOPERATIVE LEARNING**

Human Systems

Geography for Life

Migration into the Pacific

To learn about the past, geographers study many clues. One great historical mystery is how the remote and numerous Pacific Islands were settled. Geographers have drawn from different areas of study to map the migration and settlement of the Pacific Islands.

One idea developed after a geographer drifted on a raft in the ocean for some time. He proposed that South Americans could have drifted west across the Pacific to reach Polynesia. Evidence from the field of botany may support this theory. Polynesians raise sweet potatoes. However, the plant is native to South America.

Geographers also study what cultures leave behind. For example, a type of pottery called Lapita ware is found throughout the western Pacific region. Lapita ware is named for a certain site in New Caledonia. These ceramic vessels are of many types and are highly decorated. The people who made this pottery had excellent navigation and canoe-building skills. They were also farmers. Archaeologists have compared samples of pottery from the region's different islands. They concluded that the Lapita culture spread from Fiji to Tonga and Samoa. Fiji and Tonga may have been settled by 1300 B.C.

Physical geography can also offer hints about the islands' settlement. For example, when ocean levels were lower during the ice ages some Pacific Islanders may have traveled across land bridges. At that time, New Guinea and Australia were joined by dry land.

Today most Polynesians believe their people came from places to the west—like Indonesia and the Philippines. The islands of Southeast Asia are large and close together. It is likely that mainland people who needed new fishing areas and land set off for the nearby islands. Overpopulation of their homeland was probably a major factor. After these travelers perfected their navigation skills, their voyages took them farther east. Eventually, they migrated through the Melanesian islands into Polynesia.

People from other parts of the world, such as South America, may have visited the islands. However, the people who actually settled the islands were most likely from Southeast Asia. Geographers think Polynesian migration and settlements east and north eventually formed a huge triangle in the Pacific Ocean. (See the map.) The cultures that evolved in central Polynesia had their beginnings in Tonga and Samoa. From there, some people moved eastward to the Marquesas Islands. During the next 200 years, one group went south to Easter Island and another to the Hawaiian Islands. By about A.D. 500, Polynesians had spread throughout the region, with the exception of New Zealand, which they reached in about A.D. 1000.

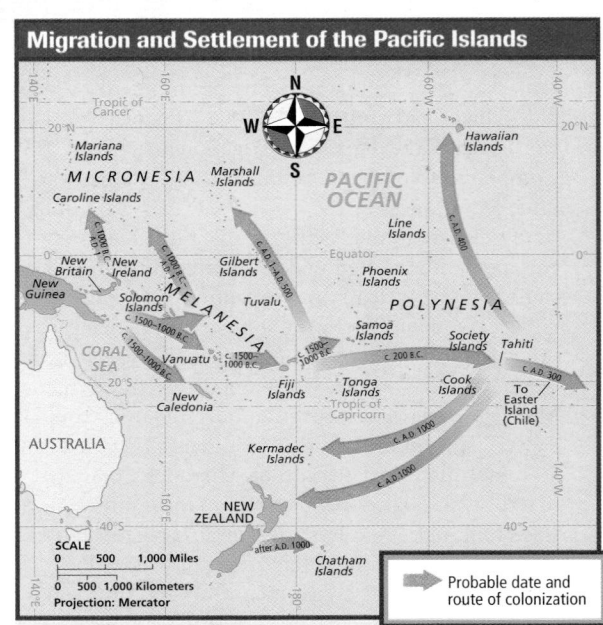

Migration and Settlement of the Pacific Islands

SCALE
0   500   1,000 Miles
0   500   1,000 Kilometers
Projection: Mercator

Probable date and route of colonization

INTERPRETING THE MAP   *Which of the three subdivisions of the Pacific Islands was settled first? How might the winds in this area have affected early migration patterns?*

Applying What You Know

1. **Summarizing**   What are two theories about how the Pacific Islands were settled?

2. **Evaluating**   Based on what you know, what geographic tools would you use to learn about how your state was settled?

Section 3

OBJECTIVES

1. **Describe the economies of the Pacific Islands.**

2. **Identify some demographic characteristics of the region.**

3. **Examine some challenges that the people of the region face.**

 **LET'S GET STARTED**

Copy the following statement and questions onto the chalkboard: *"The Pacific Islands are the closest thing we have to Paradise on Earth." Do you think this statement is true? Why or why not?* Discuss responses. Point out that although the islands have beautiful scenery and a mild climate, the people who live there have endured hardships and face uncertain futures. Tell students they will learn more about what the future holds for the Pacific Islands in Section 3.

Building Vocabulary

Write **copra** and **phosphates** on the chalkboard. Call on a volunteer to read the definitions aloud. Tell students that the word *copra* was introduced to English through Portuguese from Malayalam, a language of southern India. Ask students to speculate on how the word may have traveled to the Pacific Islands. The word *phosphate* is derived from a Greek root, *phōs,* meaning "light." Phosphates are compounds containing phosphorus, which glows in the dark and ignites when exposed to air.

Section 3
The Region Today

READ TO DISCOVER

1. What are the economies of the Pacific Islands region like?
2. What are some demographic characteristics of the region?
3. What challenges do the people of the region face?

WHY IT MATTERS

Like those in many other parts of the world, the cities of the Pacific Islands are growing rapidly. Use **CNN fyi.com** or other **current events** sources to learn more about why so many people are moving to cities and how this shift affects different countries.

IDENTIFY

Exclusive Economic Zone (EEZ)

DEFINE

copra
phosphates

LOCATE

Yap   Suva
Nauru   Papeete
Port Moresby

Section 3 RESOURCES

REPRODUCIBLE
► Guided Reading Strategy 32.3
► Map Activity 32: Samoa
► Readings in World Geography, History, and Culture 81

TECHNOLOGY
► One-Stop Planner CD–ROM, Lesson 32.3
► Homework Practice Online
► HRW Go site

REINFORCEMENT, REVIEW, AND ASSESSMENT
► Main Idea Activity 32.3
► English Audio Summary 32.3
► Spanish Audio Summary 32.3
► Section 3 Review, p. 743
► Daily Quiz 32.3

Economy

The region's economies have changed as they have become linked with the global economy. However, regional trade networks linked the islands long before European contact. The uneven distribution of resources, even on the larger islands, made trade essential. Trade networks not only stretched throughout island chains but also operated across wide areas of the Pacific. Traders carried goods such as feathers, food, mats, shells, spices, and wood in their canoes. Trading sometimes took on symbolic or political meanings. For example, many people in the Carolines offered yearly gifts to the chiefs of villages on the island of Yap. In exchange, the Yapese chiefs allowed people from the Carolines to continue to use land that the Yapese had claimed.

More recently, development in the region has been slow. Industries here face a number of hurdles. Local markets are small, and raw materials are limited. The need to import raw materials and export finished products adds to the cost of goods made in the region. For these reasons, the Pacific Islands region overall is poor. On many islands, people still rely on fishing and subsistence farming for food. Many of the main crops are those grown long ago. Coconut oil and **copra**—dried coconut meat—have been major exports. Some plantations grow introduced crops like cacao, pineapples, and vanilla.

Each country of the Pacific controls an **Exclusive Economic Zone** (**EEZ**). These zones stretch 200 nautical miles (370 km) from each country's shores. The countries can charge fees for economic activities within their EEZs. Fees paid by foreign businesses, particularly big tuna-fishing companies, provide much-needed income.

Mining has become important in Papua New Guinea, New Caledonia, and Nauru. The people of the Pacific may someday profit from the mining of metal deposits from the ocean floor. The technology needed to mine these minerals may not be practical for many years, however.

Workers process pineapple in a Cook Islands factory. Growing pineapples has caused extensive soil erosion on some of these islands.

741

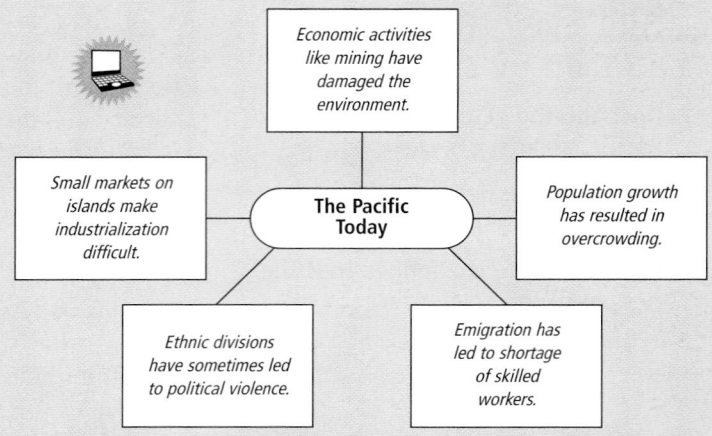

ALL LEVELS: Copy the following graphic organizer onto the chalkboard, omitting the italicized answers. Call on students to provide words and phrases that link the Pacific Islands' economies, demography, and challenges. Lead a discussion about how changes in population can affect a country's economic development and lead to challenges for its people. Ask students to provide specific examples from the Pacific Islands.

Some islands have tried to move toward a manufacturing economy. The Northern Mariana Islands, Cook Islands, Fiji, and Tonga export textiles and clothing. Tourism provides more income than manufacturing, however. Clear blue water, white sand, and island culture draw visitors from around the world, particularly Japan and the United States. Tourism is a vital industry for some places, such as Tahiti and Tonga. Other islands have attracted few tourists.

✓ **READING CHECK:** *Places and Regions* What role might technology play in the development of the region's resources? might help in the development of mineral resources on the ocean floor

Population and Migration Today

Overall, the population of the Pacific Islands is low. Even Papua New Guinea, which is larger than California, has only about 5 million people. However, some of the smaller island countries are very crowded.

There are few big cities in the region. Port Moresby in Papua New Guinea stands out as the largest city. It has a population of about 200,000. Suva, the capital of Fiji, is home to about 170,000 people. Papeete, a city on the island of Tahiti, has a population of about 24,000. It is the capital of French Polynesia and a regional center for tourism, transportation, and trade.

Although the cities of the Pacific are not large, the region saw rapid urban growth in the late 1900s. Population movement from island to island has also increased. People tend to move to cities and the more populated islands. At the same time, there is a great deal of emigration or movement out of the region. Many Pacific Islanders move to Australia, New Zealand, and the United States. The reasons for all three of these patterns are similar. The populations of most islands in the region have very high rates of natural increase. In fact, some of those rates are among the highest in the world. Birthrates in the Pacific have remained high while death rates have fallen. Populations have grown rapidly, straining resources. This is particularly true in the smaller islands. The search for jobs, education, and a better standard of living pushes people to move to other islands or to other regions.

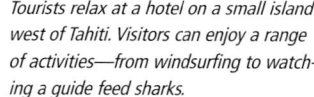

Tourists relax at a hotel on a small island west of Tahiti. Visitors can enjoy a range of activities—from windsurfing to watching a guide feed sharks.

While emigration may keep a population from growing too quickly, it can also cause problems. Young productive workers are often the ones who move away. As a result, some islands have a shortage of skilled labor.

✓ **READING CHECK:** *Human Systems* What are the causes of migration within and out of the region? the search for jobs and a better standard of living

Facing Challenges

In addition to rapid population growth, the region faces other challenges. These include concerns about how economic development will affect the environment. Political problems and issues of nuclear testing and climate also cloud the region's future.

With their small land area, many islands are particularly vulnerable to environmental destruction. For example, cutting forests in Melanesia could lead to rapid soil erosion. Also, mining has polluted some streams in Papua New Guinea. In addition, overfishing may reduce future catches.

The tiny island country of Nauru—less than 8 square miles (21 sq km) in area—displays an extreme case of environmental exploitation. Mining **phosphates**—chemicals used to make fertilizer—brings essential income to Nauru. The government shares some of the profits with Nauru's people. It has invested the rest of the money. Those investments may provide income after the supply of phosphates runs out. However, strip mining has steadily ruined most of the island's surface.

Another concern involves past nuclear weapons testing. France, Great Britain, and the United States used their Pacific territories as nuclear testing grounds. They exploded bombs from the 1940s to the 1960s. France continued underground testing until the late 1990s, even though French Polynesians and other countries protested. Some people fear the radiation from the tests may cause health problems in the future.

Finally, the possibility of global warming is a special worry for the people of the Pacific. As you read earlier, many researchers believe worldwide temperatures are rising. As a result, melting polar ice might raise ocean levels. If the ocean rises even slightly, many low-lying islands will be submerged or become more vulnerable to storms.

Some islands have also suffered from political violence in recent years. Beginning in the 1970s, Papua New Guinea's military fought against a group demanding independence for the island of Bougainville. This fighting continued through the 1990s. Ethnic divisions have led to another conflict in Fiji. Indo-Fijians make up more than half of that country's population. They are descended from workers the British brought from India. Beginning in the 1970s, some ethnic Fijians became concerned about losing political control to Indo-Fijians. Tensions between the two groups have led to violence.

✓ **READING CHECK:** *Environment and Society* How have humans modified the environment in parts of the region? mining in Papua New Guinea, Nauru; past nuclear testing; possibly global warming

Phosphate mining on Nauru has left about 90 percent of the island a wasteland. In the photo above, ancient coral remains after the phosphate material has been removed.

go.hrw.com
Homework Practice Online
Keyword: SW3 HP32

Section 3 Review

Identify Exclusive Economic Zone (EEZ)

Define copra, phosphates

Working with Sketch Maps On the map you created in Section 2, label Yap, Nauru, Port Moresby, Suva, and Papeete. Which is the region's largest city?

Reading for the Main Idea

1. *Human Systems* What two roles did traditional trade networks play in the Pacific?

2. *Places and Regions* What are five sources of income for the Pacific Islands region?

3. *Environment and Society* Why is the possibility of global warming particularly worrisome to the peoples of the Pacific?

Critical Thinking

4. **Evaluating** How have policies about resource development affected environments in Melanesia and Nauru? What economic impact have these policies had?

Organizing What You Know

5. Create a cause-effect diagram like the one below. Use it to identify the patterns of human movement in the Pacific Islands today.

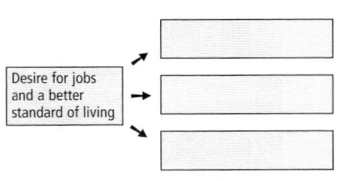

Desire for jobs and a better standard of living →

Case Study

Setting the Scene

Have students read Case Study: Exotic Invaders. Call on students to review some of the threats exotic animals pose to the endemic species of a region. *(Possible answers: New species could be predators. The region may not have sufficient resources to support both old and new species.)* Ask why island species are more vulnerable to these threats than their mainland counterparts. *(The islands' small size and isolation leave no place to go when exotics take over.)* Point out to students that animals are not the only exotic invaders that wreak havoc on ecosystems. New plants can also alter or destroy a region's natural environment.

Building a Case

In the early 1900s the banana poka, a South American variety of the passion fruit, was introduced into Hawaii. Birds and feral pigs that ate its fruit spread banana poka seeds across the islands. The fast-growing plant thrived in its new home, spreading through the native koa forests. Its bright flowers and luxuriant vines draped over tree branches are beautiful, but the banana poka is deadly to many native Hawaiian species. The vines choke the koa trees that support them and have driven three bird species—the Hawaii Creeper, the Hawaii Akepa, and the Akiapolaau—almost to extinction.

CASE STUDY

Exotic Invaders

Environment and Society The Pacific Islands are home to some of the most unusual creatures on Earth. Unfortunately, the introduction of exotic plants and animals threatens or endangers many of these species. When new plants or animals are brought to an island, disaster can result. In the most extreme cases, the original plants or wildlife die off and are replaced by the invaders.

The Brown Tree Snake

The brown tree snake provides a dramatic example of the effect an exotic invading animal can have on island wildlife. This snake was once found only in Australia, eastern Indonesia, New Guinea, and the Solomon Islands. During the 1940s and 1950s, the snakes probably stowed away in ships' cargoes and came to Guam accidentally. Soon they began to multiply and spread throughout the island. Today, in some of the island's forest areas, as many as 12,000 snakes live within just one square mile. In the 1960s local residents began to notice that there were fewer and fewer birds. Since then, native birds like the Guam flycatcher have practically disappeared from

the island. Today native forest birds can only be found on the island's northernmost tip. Guam could become the first place on Earth to lose all its native birds.

A Detective Story

At first, researchers were not sure why the island's birds were disappearing. Possible reasons for the decline included disease, hunting, loss of habitat, pesticides, and predators. An introduced animal could also be responsible. Brown tree snakes, cats, dogs, and rats were all potential suspects.

Researchers used a geographic approach to solve the puzzle. First, they noticed that the snake was the only suspect not found on three nearby islands. These islands did not experience the same decline in bird populations as Guam. The scientists also noted that birds and bird eggs are important sources of food for the snake. The brown tree snake then became the number one suspect. Next, researchers gathered data about the date, location, and number of bird and snake sightings on Guam. Poultry owners helped by reporting when they first saw the snakes. Government biologists provided information about the number of birds on different parts of the island over previous years. Scientists then plotted this data on a map. They saw that as the snake's range moved north the range of native birds grew smaller and smaller. This evidence proved that the snake was the main culprit in the decline of native birds.

Drawing Conclusions

Ask students to suggest ways in which the banana poka might threaten endemic plant and animal species. *(Its vines cover koa trees and block the sunlight; its weighty vines break the limbs off trees; or it takes over or infests the habitat of various bird species.)* Discuss the environmental changes that could result from unchecked growth of the plant. Ask students to think of ways humans might try to control the spread of the banana poka. What consequences could result from each method? *(Possible answers: herbicides—could be harmful to native plant and animal species and humans; introduction of species that eat the poka—could lead to similar infestations)*

Going Further: Thinking Critically

Organize students into groups and assign each group a plant species that has been introduced into an area of the United States where it is not native. *(Possible species: kudzu in the Southeast, purple loosestrife in the Northeast, salt cedar in the Southwest, or leafy spurge in the Northwest)* Have groups conduct research and write reports that answer the following questions:

- When and why was the new species introduced to the region?
- What threat, if any, does the exotic species pose to the region's native plants and animals?
- What measures have the region's people taken to control the spread of the plant?
- Have local residents found new uses for the plant?

The Rabbits of Laysan

Laysan Island is another example of the disastrous effects introducing exotic animals can have on island habitats. About 1903, workers brought rabbits to this small sandy strip of land north of Hawaii. A year later, the workers left. In 1923, visitors returned and found Laysan Island a barren wasteland with only a few stunted trees. At some time the rabbits had multiplied to more than 5,000 and eaten most of the plants. Several types of native birds had also disappeared. Even the rabbits were dying out. With most of the plants gone, the rabbit population had shrunk to about 200. The remaining rabbits were removed. Within 10 years, plants had taken root again and several kinds of birds had returned to the island. Scientists now consider Laysan a success story.

The small photo opposite shows a brown tree snake. Goats, such as the feral animal in the inset photo below, are an exotic species in Hawaii. They have eaten practically all the native plants in some areas. In the background you can see what part of Kauai's coast looked like before (left) and after (right) the goats stripped the area's vegetation.

Islands of Trouble

Islands' small size and their isolation from other areas of land make them particularly vulnerable to invading species. Unlike plants and animals living on continents, those on islands quickly run out of new places to go when an exotic invader takes over. Islands are like fragile little rafts. They have limited space and a limited food supply. When one creature takes too much space and food, the other creatures often find themselves with nowhere to go.

Applying What You Know

1. **Summarizing** What steps did researchers take to solve the mystery of the disappearing birds?

2. **Identifying Cause and Effect** What role did people play in bringing the brown tree snake to Guam and rabbits to Laysan? What were the results of each event?

TECHNOLOGY

▶ Chapter 32 Test Generator (on the One-Stop Planner)
▶ Global Skill Builder CD–ROM
▶ HRW Go site

REINFORCEMENT, REVIEW, AND ASSESSMENT

▶ Chapter 32 Review, pp. 746–47
▶ Chapter 32 Tutorial for Students, Parents, Mentors, and Peers
▶ Chapter 32 Test (form A or B)
▶ Alternative Assessment Handbook

▶ Chapter 32 Test for English Language Learners and Special-Needs Students
▶ Unit 10 Test
▶ Unit 10 Test for English Language Learners and Special-Needs Students

Assess

Have students complete a Chapter 32 Test.

Reteach

Assign each student a section of Chapter 32 and have him or her write sentences about the section's main ideas. Have students read their sentences aloud for you to write them on the chalkboard so as to create a chapter summary. Then have students work in pairs to review the summary.
ENGLISH LANGUAGE LEARNERS, COOPERATIVE LEARNING

CHAPTER 32 Review Answers

Thinking Critically

1. Environmental hazards include typhoons, earthquakes, tsunamis, and volcanic eruptions; flooding caused by rising ocean levels threatens low-lying islands and the people living there; typhoons, tsunamis, earthquakes, and volcanoes also threaten the lives of people in the region. (NGS 15)

2. Many islanders rely on fishing and subsistence agriculture for food. However, income from plantation crops, such as cacao or pineapples, pays for increased food imports. (NGS 4)

3. They have introduced new ethnic groups, languages, and religions. Emigration has meant an outflow of skilled workers; war has brought foreign control and a strong military presence that still exists; trade has brought goods, such as imported foods; trade has also led farmers to grow cash crops, such as cacao or pineapples, for export. (NGS 9)

Using the Geographer's Tools

1. Answers will vary but should reflect the information in the source material.

2. Answers will vary, but students should note important clues, such as the Mariana Trench and landform ridges.

CHAPTER 32 Review

Building Vocabulary

On a separate sheet of paper, explain the following terms by using them correctly in sentences.

atoll
Intertropical Convergence Zone (ITCZ)
trust territories
pidgin languages

matrilineal
copra
Exclusive Economic Zone (EEZ)
phosphates

Locating Key Places

On a separate sheet of paper, match the letters on the map with their correct labels.

Tahiti
Papua New Guinea
Solomon Islands
Fiji

Northern Mariana Islands
Wake Island
Nauru

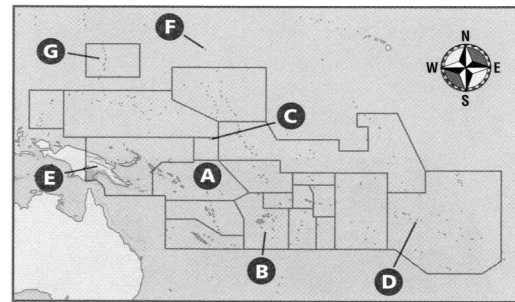

Understanding the Main Ideas

Section 1

1. (Places and Regions) What are the two types of islands in the Pacific? What are their origins?

Section 2

2. (Human Systems) What geographic factors helped a large number of languages to develop in the region?

3. (Environment and Society) What food crops have traditionally been important in the region? What uses did Pacific Islanders find for the coconut palm?

Section 3

4. (Human Systems) Why is manufacturing not a major factor in the region's economy?

5. (Human Systems) What are the three patterns of human movement in the Pacific Islands today?

Thinking Critically

1. Summarizing What environmental hazards affect the Pacific Islands region? What impact might these hazards have on the land and people living there?

2. Comparing How do fishing, subsistence farming, and plantation agriculture play different roles in the region?

3. Analyzing Information How have migration, war, and trade affected the people and culture of the region?

Using the Geographer's Tools

1. Interpreting Charts Review the unit Fast Facts chart, the unit Comparing Standards of Living chart, and information from this chapter. Then use that information and the World Factbook (go.hrw.com) to rank the countries of this region by level of development. Write a short paragraph explaining why you ranked the countries as you have.

2. Evaluating Maps Look at the map of the ocean floor in Section 1. What does this map tell you about the region's geography? What are some clues that this is a tectonically active region?

3. Creating and Interpreting Maps Construct a map of the Pacific Islands region. Label the subregions of the Pacific Islands. Then use arrows to show the origin and direction of migration of people into the region over time. Label each arrow with the origin of migration. How would you expect these migrations to have influenced the region's cultures?

Writing about Geography

Write a speech from the point of view of a resident of French Polynesia calling for greater political independence from France. Include specific reasons why you feel this way. When you are finished with your speech, proofread it to make sure you have used standard grammar, spelling, sentence structure, and punctuation.

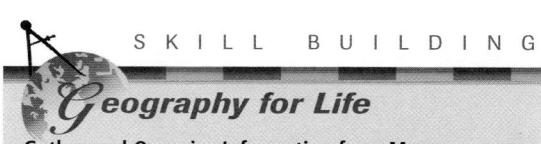

S K I L L   B U I L D I N G

Geography for Life

Gather and Organize Information from Maps

The World in Spatial Terms Using the physical-political map of the region, pick five Pacific islands. For each island, measure its approximate distance from Honolulu, Hawaii; Port Moresby, Papua New Guinea; Sydney, Australia; Suva, Fiji; and Papeete, Tahiti. Finally, create a chart to organize this information.

1. Have students conduct research on a traditional religion of the Pacific Islands. Ask them to focus on one religious ceremony and to write a brief description of it. Encourage students to draw a scene from the ceremony to accompany their descriptions. Place both items in student portfolios.

2. Organize the class into pairs. Have each pair conduct additional research on nuclear testing in the Pacific from the 1940s to the 1990s. Then ask them to write correspondence between the leaders of an island and the governments of the countries that conducted these tests about the tests' effects on Pacific environments. Place correspondence in portfolios.

Food Festival

The luau, featuring a roast pig, is a traditional feast. It would be hard to duplicate a luau, but students can enjoy another Polynesian pork favorite—Spam®! During World War II, U.S. soldiers fighting in the Pacific received the canned meat product in their rations. They shared it with the local people. Spam® is still extremely popular among Pacific Islanders. Have students find and prepare Asian recipes using Spam® and bring samples to share.

Building Social Studies Skills

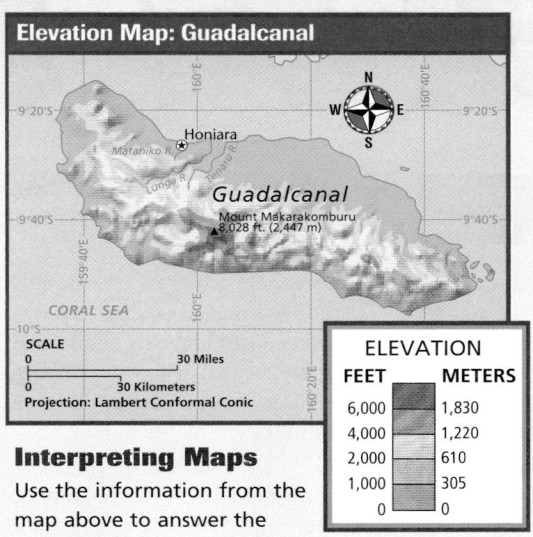

Elevation Map: Guadalcanal

Honiara

Matanibo R.

Guadalcanal

Mount Makarakomburu
8,028 ft. (2,447 m)

CORAL SEA

SCALE
0 — 30 Miles
0 — 30 Kilometers
Projection: Lambert Conformal Conic

ELEVATION

FEET	METERS
6,000	1,830
4,000	1,220
2,000	610
1,000	305
0	0

Interpreting Maps

Use the information from the map above to answer the questions that follow.

1. What is the elevation of Guadalcanal's highest area?
   a. less than 1,000 feet (305 m)
   b. between 1,000 and 2,000 feet (305 m and 610 m)
   c. above 9,000 feet (2,740 m)
   d. above 6,000 feet (1,830 m)

2. How does the elevation of the island change from north to south?

Analyzing Secondary Sources

Read the following passage and answer the questions.

"Today the Pacific Islands are home to a great number of different ethnic groups and languages. How did this happen? Huge stretches of ocean between islands allowed different cultures and languages to develop independently from one another. On New Guinea, thick rain forests and rugged mountains separated different groups of people in a similar way. Today the peoples of Papua New Guinea speak more than 700 different languages. Some of them are spoken by only a few hundred people."

3. Which feature was *not* a major factor in the development of the region's diverse cultural geography?
   a. ice cap
   b. rain forest
   c. mountain range
   d. ocean

4. What do the region's oceans, forests, and mountains have in common?

3. Maps should show arrows from Southeast and East Asia, and from Europe. The migration created cultural diversity and change over time, including the development of various ethnic groups, languages, religions.

Writing

Student speeches should state and support an opinion. Support for student positions should be rooted in fact. Use Rubric 43, Writing to Persuade, to evaluate student work.

Geography for Life

Student charts should accurately reflect distances between listed locations. Use Rubric 7, Charts, to evaluate student work.

Social Studies Skills

1. d
2. Northern parts of the island are generally lower, becoming higher in the south.
3. a
4. Possible answer: They all separate different ethnic groups and languages.

Alternative Assessment

PORTFOLIO ACTIVITY

Learning about Your Local Geography

Individual Project: Navigating with Landmarks
Just as Pacific Islanders navigated across vast ocean distances, you navigate from place to place every day. Like those sailors, you probably do not use a map or compass to find your way. Write detailed descriptions of some of your daily journeys, such as from home to school or from school to a job. Include how you get from place to place and what landmarks you use to find your destination.

internet connect

Internet Activity: go.hrw.com
KEYWORD: SW3 GT32

Choose a topic about the Pacific Islands to:
- tour the South Pacific.
- learn the traditions of the Pacific Islands.
- explore the diversity of the region.

PORTFOLIO ACTIVITY

Student descriptions should describe in detail student travels. They should also discuss specific landmarks along the traveled route. Use Rubric 40, Writing to Describe, to evaluate student work.

Workshop 1
Going Further: Thinking Critically

Point out to students that many software programs can present information in a manner similar to overhead transparencies. Using these programs, presenters can create visually striking series of images that incorporate text, graphics, and even animation. Many businesses regularly use this type of software during meetings, conventions, and sales presentations. In some classrooms, too, computer-generated presentations have replaced overhead transparencies.

Organize the class into several groups and have each group compare the advantages and disadvantages of presentation software and overhead transparencies. *(Possible answers: for software: advantages—visually pleasing presentations, can incorporate graphics or animations, can be saved and reused; disadvantages—must be prepared in advance, cannot be changed easily during presentation, more expensive; for overhead transparencies: advantages—easily modified and adapted, easy to prepare; disadvantages—smeared ink, difficult to store transparencies for later use)* Then tell students to imagine that a local school district must decide whether it should purchase presentation software or overhead projectors for a new school. Have each group write a brief proposal in which it argues for the purchase of one or the other. Ask each group to share its ideas with the class.

PRACTICING THE SKILL

1. Students' presentations should discuss the effects of exotic species in the Pacific World. Information included on overhead transparencies should be clear and well-organized.
2. Students' transparencies should clearly outline the history of Australia and New Zealand. Text should be clear and concise.
3. Students' proposals should demonstrate the results of their research. Transparencies should be appropriate to the content of the proposals.

Geography
Skill-Building Workshop

WORKSHOP 1

Using Media Services: Overhead Transparencies

As you read in an earlier workshop, media services include a variety of tools that can be used to communicate information. These tools include sites on the World Wide Web, CD–ROMs, laser discs, DVDs, and a variety of software programs. Other tools also include VCRs, music CDs, and slides.

You have likely seen a teacher use a common media tool called an overhead projector. You can use an overhead projector to display information from transparencies. Transparencies allow you to quickly reorder your presentation or to repeat and emphasize something. All you need for an overhead presentation are the projector, transparency film, some transparency pens, a screen, and an electrical outlet.

Developing the Skill Putting together overhead presentations is not difficult. The following hints will help:

- Choose the main points and keep the number of words to a minimum. List just a few points on each transparency. You might list just one point on a transparency if it is particularly important or needs extended discussion.
- Either create your transparencies as you go, as if you were using a chalkboard, or use a computer to make professional-looking transparencies. If you use a computer, you will need to photocopy the original image onto the transparency.
- Make sure the lettering is big enough for people to see. If you choose to make handwritten transparencies, take your time. You want the writing to be clear and neat.

Using Transparencies

1. Choose your main points. Keep it simple!
2. Create your transparencies by hand or computer.
3. Make it neat and easy to read.
4. Use color.
5. Include graphics and pictures.

- Keep writing away from the edges of a transparency so that the text is visible on the screen.
- Colored text can draw attention to certain words. Use a pen of a different color to highlight, underline, or circle the text or to draw arrows, checks, or stars for emphasis. Be sure to use a water-soluble pen if you need to reuse the transparency.
- You might want to include graphics to keep your presentation interesting.

Before you begin a presentation, practice laying down the transparencies so they appear right-side up on the screen. You might want to number the transparencies in case you drop them. Also, darken the room slightly so your audience can see and read properly. Stand next to the screen after you place a transparency. Your audience will not have to look back and forth between you and the screen. Point to certain messages for emphasis by using a pen or pointer. You can also use a sheet of paper to reveal a portion of the transparency at a time. Doing so may allow you to use fewer transparencies. Finally, remember to talk to the audience, not to the screen.

Practicing the Skill

1. Create a transparency presentation about exotic species that have been introduced into the Pacific Islands region.
2. Use transparencies to summarize key points about the history of Australia and New Zealand.
3. Conduct research to learn how former Olympic cities have reused their Olympic sites. Then develop a proposal for reusing Sydney's 2000 Olympic site. Create transparencies to illustrate your proposal.

Organize the class into groups to apply the decision-making process to issues facing your community. Ask students to imagine that they have been asked to work with local charitable organization to resolve a problem in your area. Have each group identify an organization that it would like to help. *(Possible organizations: operators of food banks, homeless shelters, community childcare facilities, and so on)* Then have each group decide how it could best contribute to the organization. First have each group conduct research to learn about its chosen organization's activities in your area. Encourage students to consult newspapers and local media to find information or to contact the organization itself for further information. Then have students identify ways in which they could contribute to the organization's efforts. For example, to assist a food bank students could collect canned goods at school or in their neighborhoods, make a cash donation, volunteer to work at a food collection or distribution center, and so on. Then ask students to examine the likely results of each contribution they have listed. Finally, have students decide which method of contribution they believe is most helpful to the community. Call on volunteers from each group to explain their decisions.

Parliament House in Canberra, Australia

WORKSHOP 2

Making Decisions

You make decisions every day. Some decisions are easy to make and take little time. More difficult decisions can take much longer to make. Perhaps you are deciding whether you should join an organization, buy a new CD player, or take a certain class. Such decisions often require that you gather information, identify options, predict consequences, and take an appropriate action to implement a decision.

A similar process can be used to make decisions in geography. Suppose, for example, that a government in the Pacific Islands is debating whether to adopt new laws that would protect important natural resources and local ecosystems. Government officials would need to learn about the area, including local ecosystems and how human activities affect those ecosystems. Officials would also have to consider the kinds of protections that might be useful. In addition, they would need to predict how those laws might affect the local environment, economic development, and the ways of life for people there. Finally, the government would have to decide what laws to pass and then do so.

Developing the Skill   As you can see, four key steps make up the decision-making process. You might want to copy the flowchart on this page and make notes about each step.

Gather Information   To learn more about the issue, gather the necessary facts. Look for possible causes and solutions. Also, prioritize your needs and wants before you move to the next step of identifying your options. You might use a variety of databases, such as sites on the

The Decision-Making Process

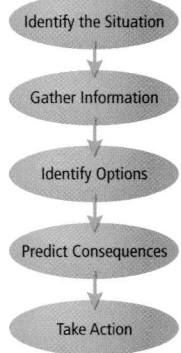

- Identify the Situation
- Gather Information
- Identify Options
- Predict Consequences
- Take Action

World Wide Web, to find needed information. You can also use newspapers, other periodicals, one or more questionnaires, and field work to gather information.

Identify Options   Draw conclusions from the gathered information and identify your options. Then weigh the advantages and disadvantages of each. You might organize your options in a list or chart.

Predict Consequences   It is important to understand the consequences of each option you have. Who would be affected by each option? What might those effects be? Which consequences are positive? Which are negative? You might rank the consequences from best to worst.

Take Action   Follow through with your decision. Choose and implement the best option. You might want to monitor the effects of your actions. The information you learn could be useful in making decisions on future issues you might face.

Practicing the Skill

Consider a long-running debate in Australia about whether the country should become a republic or maintain ties to the British monarchy. In 1999 Australian voters rejected a proposal to make their country a republic. However, some Australians continue to work to replace the British monarch with a president or other head of state from Australia itself. Imagine that you have been asked to work on a new commission. That commission has been charged with recommending whether or not Australia should become a republic. Working with a small group of four or five students, use the decision-making process to develop a proposal for Australia's future political status.

Gazetteer

Phonetic Respelling and Pronunciation Guide

Many of the place-names in this textbook have been respelled to help you pronounce them. The letter combinations used in the respelling throughout the narrative are explained in the following phonetic respelling and pronunciation guide. The guide is adapted from *Merriam-Webster's Collegiate Dictionary, Merriam-Webster's Geographical Dictionary,* and *Merriam-Webster's Biographical Dictionary.*

MARK	AS IN	RESPELLING	EXAMPLE
a	alphabet	a	*AL-fuh-bet
ā	Asia	ay	AY-zhuh
ä	cart, top	ah	KAHRT, TAHP
e	let, ten	e	LET, TEN
ē	even, leaf	ee	EE-vuhn, LEEF
i	it, tip, British	i	IT, TIP, BRIT-ish
ī	site, buy, Ohio	y	SYT, BY, oh-HY-oh
	iris	eye	EYE-ris
k	card	k	KAHRD
ō	over, rainbow	oh	OH-vuhr, RAYN-boh
ů	book, wood	ooh	BOOHK, WOOHD
ȯ	all, orchid	aw	AWL, AWR-kid
ȯi	foil, coin	oy	FOYL, KOYN
aů	out	ow	OWT
ə	cup, butter	uh	KUHP, BUHT-uhr
ü	rule, food	oo	ROOL, FOOD
yü	few	yoo	FYOO
zh	vision	zh	VIZH-uhn

*A syllable printed in small capital letters receives heavier emphasis than the other syllable(s) in a word.

Abu Dhabi (24°N 54°E) capital of the United Arab Emirates, **433**

Abuja (ah-BOO-jah) (9°N 7°E) capital of Nigeria, **499**

Acapulco (17°N 100°W) city on the southwestern coast of Mexico, **221**

Accra (6°N 0°) capital of Ghana, **499**

Addis Ababa (9°N 39°E) capital of Ethiopia, **517**

Adriatic Sea sea between Italy and the Balkan Peninsula, **349**

Aegean (ee-JEE-uhn) **Sea** sea between Greece and Turkey, **349**

Afghanistan landlocked country in Southwest Asia, **433**

Africa second-largest continent; surrounded by the Atlantic Ocean, Indian Ocean, and Mediterranean Sea, **S37–S38**

Ahaggar Mountains mountain range in southern Algeria, **483**

Albania country in the western Balkan region of Europe on the Adriatic Sea, **349**

Alberta province in western Canada, **189**

Aleutian Islands volcanic island chain extending from Alaska into the Pacific Ocean, **165**

Alexandria (31°N 30°E) city in northern Egypt, **483**

Algeria country in North Africa located between Morocco and Libya, **483**

Algiers (37°N 3°E) capital of Algeria, **483**

Alps major mountain system in south-central Europe, **349**

Altiplano broad, high plateau in Peru and Bolivia, **257**

Amazon River major river in South America, **257**

Amman (32°N 36°E) capital of Jordan, **449**

Amsterdam (52°N 5°E) capital of the Netherlands, **305**

Amu Dar'ya (uh-MOO duhr-YAH) river in Central Asia that drains into the Aral Sea, **403**

Amur (ah-MOOHR) **River** river in northeast Asia forming part of the border between Russia and China, **381**

Andes (AN-deez) great mountain range in South America, **257**

Andorra European microstate in the Pyrenees mountains, **S34**

Andorra la Vella (43°N 2°E) capital of Andorra, **S34**

Angkor ancient capital of the Khmer Empire in Cambodia, **659**

Angola country in southern Africa, **533**

Ankara (40°N 33°E) capital of Turkey, **449**

Antananarivo (19°S 48°E) capital of Madagascar, **533**

Antarctica continent around the South Pole, **S41**

Antarctic Circle line of latitude located at 66 1/2° south of the equator; parallel beyond which no sunlight shines on the June solstice (first day of winter in the Southern Hemisphere), **S21–S22**

Antarctic Peninsula peninsula stretching toward South America from Antarctica, **S41**

Antigua and Barbuda island country in the Caribbean, **239**

Antwerp (51°N 4°E) major port city in Belgium, **305**

Apennines (A-puh-nynz) mountain range in Italy, **349**

Apia (14°S 172°W) capital of Western Samoa, **731**

Appalachian Mountains mountain system in eastern North America, **165**

Arabian Peninsula peninsula in Southwest Asia between the Red Sea and Persian Gulf, **433**

Arabian Sea sea between India and the Arabian Peninsula, **433**

Aral (AR-uhl) **Sea** inland sea between Kazakhstan and Uzbekistan, **403**

Arctic Circle line of latitude located at 66 1/2° north of the equator; the parallel beyond which no sunlight shines on the December solstice (first day of winter in the Northern Hemisphere), **S21–S22**

Arctic Ocean ocean north of the Arctic Circle; world's fourth-largest ocean, **S21–S22**

Argentina second-largest country in South America, **257**

Armenia country in the Caucasus region of Asia; former Soviet republic, **381**

Ashgabat (formerly Ashkhabad) (38°N 58°E) capital of Turkmenistan, **403**

Asia world's largest continent; located between Europe and the Pacific Ocean, **S35–S36**

Asmara (15°N 39°E) capital of Eritrea, **517**

Astana (51°N 71°E) capital of Kazakhstan, **403**

Astrakhan (46°N 48°E) old port city on the Volga River in Russia, **381**

Asunción (25°S 58°W) capital of Paraguay, **257**

Atacama Desert desert in northern Chile, **257**

Athens (38°N 24°E) capital and largest city in Greece, **349**

Atlanta (34°N 84°W) capital and largest city in the U.S. state of Georgia, **165**

Atlantic Ocean ocean between the continents of North and South America and the continents of Europe and Africa; world's second-largest ocean, **S21–S22**

Atlas Mountains African mountain range north of the Sahara, **483**

Auckland (37°S 175°E) New Zealand's largest city and main seaport, **711**

Augrabies (oh-KRAH-bees) **Falls** waterfalls on the Orange River in South Africa, **533**

Australia only country occupying an entire continent (also called Australia); located between the Indian Ocean and the Pacific Ocean, **S22**

Austria country in central Europe south of Germany, **327**

Azerbaijan country in the Caucasus region of Asia; former Soviet republic, **381**

Bab al-Mandab narrow strait that connects the Red Sea with the Gulf of Aden and Indian Ocean beyond, **433**

Baghdad (33°N 44°E) capital of Iraq, **433**

Bahamas island country in the Atlantic Ocean southeast of Florida, **239**

Bahrain country on the Persian Gulf in Southwest Asia, **433**

Baja California peninsula in northwestern Mexico, **221**

Baku (40°N 50°E) capital of Azerbaijan, **381**

Bali island in Indonesia east of Java, **679**

Balkan Mountains mountain range that rises in Bulgaria, **349**

Baltic Sea body of water east of the North Sea and Scandinavia, **305**

Baltimore (39°N 77°W) city in Maryland on the western shore of Chesapeake Bay, **165**

Bamako (13°N 8°W) capital of Mali, **499**

Bandar Seri Begawan (5°N 115°E) capital of Brunei, **679**

Bangkok (14°N 100°E) capital and largest city of Thailand, **659**

Bangladesh country in South Asia, **585**

Bangui (4°N 19°E) capital of the Central African Republic, **499**

Banjul (13°N 17°W) capital of Gambia, **499**

Barbados island country in the Caribbean, **239**

Barcelona (41°N 2°E) Mediterranean port city and Spain's second-largest city, **349**

Basel (48°N 8°E) city in northern Switzerland on the Rhine River, **327**

Basseterre (17°N 63°W) capital of St. Kitts–Nevis, **239**

Bay of Bengal body of water between India and the western coasts of Myanmar (Burma) and the Malay Peninsula, **563**

Bay of Biscay body of water off the western coast of France and the northern coast of Spain, **305**

Beijing (40°N 116°E) capital of China, **615**

Beirut (34°N 36°E) capital of Lebanon, **449**

Belarus country located north of Ukraine; former Soviet republic, **381**

Belém (1°S 48°W) port city in northern Brazil, **S32**

Belfast (55°N 6°W) capital and largest city of Northern Ireland, **305**

Belgium country between France and Germany in western Europe, **305**

Belgrade (45°N 21°E) capital of Serbia and Montenegro on the Danube River, **349**

Belize country in Central America bordering Mexico and Guatemala, **239**

Belmopan (17°N 89°W) capital of Belize, **239**

Benghazi (32°N 20°E) major coastal city in Libya, **483**

Benin (buh-NEEN) country in West Africa between Togo and Nigeria, **499**

Bergen (60°N 5°E) seaport city in southwestern Norway, **305**

Berlin (53°N 13°E) capital of Germany, **327**

Bern (47°N 7°E) capital of Switzerland, **327**

Bhutan South Asian country in the Himalayas located north of India and Bangladesh, **585**

Birmingham (52°N 2°W) major manufacturing center of central Great Britain, **305**

Bishkek (43°N 75°E) capital of Kyrgyzstan, **403**

Bissau (12°N 16°W) capital of Guinea-Bissau, **499**

Black Sea sea between Europe and Asia, **349**

Blue Nile East African river that flows into the Nile River in Sudan, **517**

Bogotá (5°N 74°W) capital and largest city of Colombia, **257**

Bolivia landlocked South American country, **257**

Bombay *See* Mumbai.

Bonn (51°N 7°E) city in western Germany; replaced by Berlin as the capital of reunified Germany, **327**

Borneo island in the Malay Archipelago in Southeast Asia, **679**

Bosnia and Herzegovina country in the western Balkans region of Europe between Serbia and Croatia, **349**

Bosporus a narrow strait separating European and Asian Turkey, **449**

Boston (42°N 71°W) capital and largest city of Massachusetts, **165**

Botswana country in southern Africa, **533**

Brahmaputra River major river of South Asia that begins in the Himalayas of Tibet and merges with the Ganges River in Bangladesh, **563**

Brasília (16°S 48°W) capital of Brazil, **257**

Bratislava (48°N 17°E) capital of Slovakia, **327**

Brazil largest country in South America, **257**

Brazilian Highlands region of old, eroded mountains in southeastern Brazil, **257**

Brazzaville (4°S 15°E) capital of the Republic of the Congo, **499**

Bridgetown (13°N 59°W) capital of Barbados, **239**

Brisbane (28°S 153°E) seaport and capital of Queensland, Australia, **711**

British Columbia province on the Pacific coast of Canada, **189**

British Isles island group consisting of Great Britain and Ireland, **305**

Brittany region in northwestern France, **305**

Brunei (brooh-NY) country on the northern coast of Borneo in Southeast Asia, **679**

Gazetteer

Brussels (51°N 4°E) capital of Belgium, **305**
Bucharest (44°N 26°E) capital of Romania, **349**
Budapest (48°N 19°E) capital of Hungary, **327**
Buenos Aires (34°S 59°W) capital of Argentina, **257**
Bujumbura (3°S 29°E) capital of Burundi, **517**
Bulgaria country on the Balkan Peninsula in Europe, **349**
Burkina Faso (boor-KEE-nuh FAH-soh) landlocked country in West Africa, **499**
Burma *See* Myanmar.
Burundi landlocked country in East Africa, **517**

Cairo (30°N 31°E) capital of Egypt, **483**
Calcutta *See* Kolkata.
Calgary (51°N 114°W) city in the western Canadian province of Alberta, **189**
Callao (kah-YAH-oh) (12°S 77°W) port city in Peru, **S32**
Cambodia country in Southeast Asia west of Vietnam, **659**
Cameroon country in Central Africa, **499**
Campeche (20°N 91°W) city in Mexico on the west coast of the Yucatán Peninsula, **221**
Canada country occupying most of northern North America, **189**
Canadian Shield major landform region in central Canada along Hudson Bay, **189**
Canberra (35°S 149°E) capital of Australia, **711**
Cancún (21°N 87°W) resort city in Mexico on the Yucatán Peninsula, **221**
Cantabrian (kan-TAY-bree-uhn) **Mountains** mountains in north-western Spain, **349**
Cape Horn (56°S 67°W) cape in southern Chile; southernmost point of South America, **257**
Cape of Good Hope cape on the southwest coast of South Africa, **533**
Cape Town (34°S 18°E) major seaport city and legislative capital of South Africa, **533**
Cape Verde island country in the Atlantic Ocean off the coast of West Africa, **499**
Caracas (kuh-RAHK-uhs) (11°N 67°W) capital of Venezuela, **257**
Cardiff (52°N 3°W) capital and largest city of Wales, **305**
Caribbean Sea arm of the Atlantic Ocean between North America and South America, **239**
Carpathian Mountains mountain system in Eastern Europe, **327**
Casablanca (34°N 8°W) seaport city on the western coast of Morocco, **483**
Cascade Range mountain range in the northwestern United States, **165**
Caspian Sea large inland salt lake between Europe and Asia, **403**
Castries (14°N 61°W) capital of St. Lucia, **239**
Cauca River river in western Colombia, **257**
Caucasus Mountains mountain range between the Black Sea and the Caspian Sea, **381**
Cayenne (5°N 52°W) capital of French Guiana, **257**
Central African Republic landlocked country in Central Africa located south of Chad, **499**
Central America narrow southern portion of the North American continent, **239**

Central Lowlands area of Australia between the Western Plateau and the Great Dividing Range, **711**
Central Siberian Plateau upland plains and valleys between the Yenisey and Lena Rivers in Russia, **381**
Chad landlocked country in northern Africa, **499**
Chang (Yangtze) **River** major river in central China, **615**
Chao Phraya (chow PRY-uh) **River** major river in Thailand, **659**
Chelyabinsk (chel-YAH-buhnsk) (55°N 61°E) manufacturing city in the Urals region of Russia, **381**
Chernobyl (51°N 30°E) city in north-central Ukraine; site of a major nuclear accident in 1986, **381**
Chicago (42°N 88°W) major city on Lake Michigan in northern Illinois, **165**
Chile country on the west coast of South America, **257**
China country in East Asia; most populous country in the world, **615**
Chişinău (formerly Kishinev) (47°N 29°E) capital of Moldova, **349**
Chongqing (30°N 108°E) city in southern China along the Chang River, **615**
Christchurch (44°S 173°E) city on the eastern coast of South Island, New Zealand, **711**
Ciudad Juárez (32°N 106°W) city in northern Mexico opposite El Paso, **221**
Cologne (51°N 7°E) manufacturing and commercial city along the Rhine River in Germany, **327**
Colombia country in northern South America, **257**
Colombo (7°N 80°E) capital city and important seaport of Sri Lanka, **585**
Colorado Plateau uplifted area of horizontal rock layers in the western United States, **165**
Columbia River river that drains the Columbia Basin in the northwestern United States, **165**
Comoros island country in the Indian Ocean off the coast of Africa, **533**
Conakry (10°N 14°W) capital of Guinea, **499**
Congo Basin region in Central Africa, **499**
Congo, Democratic Republic of the largest and most populous country in Central Africa, **499**
Congo, Republic of the Central African country located along the Congo River, **499**
Congo River major navigable river in Central Africa that flows into the Atlantic Ocean, **499**
Copenhagen (56°N 12°E) seaport and capital of Denmark, **305**
Córdoba (31°S 64°W) large city in Argentina northwest of Buenos Aires, **S32**
Cork (52°N 8°W) seaport city in southern Ireland, **305**
Costa Rica country in Central America, **239**
Côte d'Ivoire (KOHT dee-VWAHR) (Ivory Coast) country in West Africa, **499**
Crimean Peninsula peninsula in Ukraine that juts southward into the Black Sea, **381**
Croatia country and former Yugoslav republic in the western Balkans region of Europe, **349**
Cuba country and largest island in the Caribbean, **239**
Cuzco (14°S 72°W) city southeast of Lima, Peru; former capital of the Inca Empire, **257**
Cyprus island republic in the eastern Mediterranean Sea, **449**
Czech Republic Central European country and the western part of the former country of Czechoslovakia, **327**

Dakar (15°N 17°W) capital of Senegal, **499**
Dallas (33°N 97°W) city in northern Texas, **165**

Damascus (34°N 36°E) capital of Syria and one of the world's oldest cities, **449**

Danube River major river in Europe that flows into the Black Sea in Romania, **349**

Dardanelles narrow strait separating European and Asian Turkey, **449**

Dar es Salaam (7°S 39°E) capital and major seaport of Tanzania, **517**

Dead Sea salt lake on the boundary between Israel and Jordan in southwestern Asia, **449**

Deccan Plateau the southern part of the Indian subcontinent, **563**

Delhi (29°N 77°E) city in India, **563**

Denmark country in northern Europe, **305**

Detroit (42°N 83°W) major industrial city in Michigan, **165**

Devil's Island (5°N 53°W) French island off the coast of French Guiana in South America, **S31**

Dhaka (24°N 90°E) capital and largest city of Bangladesh, **585**

Dinaric Alps mountains extending inland from the Adriatic coast to the Balkan Peninsula, **349**

Djibouti country located in the Horn of Africa, **517**

Djibouti (12°N 43°E) capital of Djibouti, **517**

Dnieper River major river in Ukraine, **381**

Doha (25°N 51°E) capital of Qatar, **433**

Dominica Caribbean island country, **239**

Dominican Republic country occupying the eastern part of Hispaniola in the Caribbean, **239**

Donets Basin industrial region in eastern Ukraine, **373**

Douro River river on the Iberian Peninsula that flows into the Atlantic Ocean in Portugal, **349**

Drakensberg mountain range in southern Africa, **533**

Dublin (53°N 6°W) capital of the republic of Ireland, **305**

Durban (30°S 31°E) port city in South Africa, **533**

Dushanbe (39°N 69°E) capital of Tajikistan, **403**

Eastern Ghats mountains on the eastern side of the Deccan Plateau in southern India, **563**

Ebro River river in Spain that flows into the Mediterranean Sea, **349**

Ecuador country in western South America, **257**

Edmonton (54°N 113°W) provincial capital of Alberta, Canada, **189**

Egypt country in North Africa located east of Libya, **483**

Elburz Mountains mountain range in northern Iran, **433**

El Salvador country on the Pacific side of Central America, **239**

England southern part of Great Britain and part of the United Kingdom in northern Europe, **305**

English Channel channel separating Great Britain from the European continent, **305**

equator the imaginary line of latitude that lies halfway between the North and South Poles and circles the globe, **S21–S22**

Equatorial Guinea Central African country, **499**

Eritrea (er-uh-TREE-uh) East African country located north of Ethiopia, **517**

Essen (51°N 7°E) industrial city in western Germany, **327**

Estonia country located on the Baltic Sea; former Soviet republic, **327**

Ethiopia East African country in the Horn of Africa, **517**

Euphrates River major river in Iraq in southwestern Asia, **433**

Europe continent between the Ural Mountains and the Atlantic Ocean, **S33–S34**

Fergana Valley fertile valley in Uzbekistan, Kyrgyzstan, and Tajikistan, **403**

Fès (34°N 5°W) city in north-central Morocco, **483**

Fiji South Pacific island country; part of Melanesia, **731**

Finland country in northern Europe located between Sweden, Norway, and Russia, **305**

Florence (44°N 11°E) city on the Arno River in central Italy, **349**

France country in west-central Europe, **305**

Frankfurt (50°N 9°E) main city of Germany's Rhineland region, **327**

Freetown (9°N 13°W) capital of Sierra Leone, **499**

French Guiana French territory in northern South America, **257**

Funafuti (9°S 179°E) capital of Tuvalu, **731**

Gabon country in Central Africa located between Cameroon and the Republic of the Congo, **499**

Gaborone (24°S 26°E) capital of Botswana, **533**

Galway (53°N 9°W) city in western Ireland, **305**

Gambia country along the Gambia River in West Africa, **499**

Ganges River major river in India flowing from the Himalayas southeastward to the Bay of Bengal, **563**

Gangetic (gan-JE-tik) **Plain** vast plain in northern India, **563**

Gao (GOW) (16°N 0°) city in Mali on the Niger River, **499**

Gaza Strip area occupied by Israel from 1967 to 1994; partly under Palestinian self-rule since 1994, **449**

Geneva (46°N 6°E) city in southwestern Switzerland, **327**

Genoa (44°N 10°E) seaport city in northwestern Italy, **349**

Georgetown (8°N 58°W) capital of Guyana, **257**

Georgia (Eurasia) country in the Caucasus region; former Soviet republic, **381**

Germany country in central Europe located between Poland and France, **327**

Ghana country in West Africa, **499**

Glasgow (56°N 4°W) city in Scotland, United Kingdom, **305**

Gobi desert that makes up part of the Mongolian plateau in East Asia, **615**

Golan Heights hilly region in southwestern Syria occupied by Israel, **449**

Göteborg (58°N 12°E) seaport city in southwestern Sweden, **305**

Gran Chaco (grahn CHAH-koh) dry plains region in Paraguay, Bolivia, and northern Argentina, **257**

Great Artesian Basin Australia's largest source of groundwater; located in interior Queensland, **711**

Great Barrier Reef world's largest coral reef; located off the northeastern coast of Australia, **711**

Great Basin dry region in the western United States, **165**

Great Bear Lake lake in the Northwest Territories of Canada, **189**

Great Britain name for island and country of northern and western Europe, **291**

Great Dividing Range mountain range of eastern Australia, **711**

Gazetteer

Greater Antilles larger islands of the West Indies in the Caribbean Sea, including Cuba, Hispaniola, Jamaica, and Puerto Rico, **239**

Great Lakes largest freshwater lake system in the world; located in North America, **175**

Great Plains plains region in the central United States, **165**

Great Rift Valley valley system extending from eastern Africa to Southwest Asia, **517**

Great Slave Lake lake in the Northwest Territories of Canada, **189**

Greece country in southern Europe located at the southern end of the Balkan Peninsula, **349**

Greenland self-governing province of Denmark between the North Atlantic and Arctic Oceans, **305**

Grenada Caribbean island country, **239**

Guadalajara (21°N 103°W) city in west-central Mexico, **221**

Guadalquivir (gwah-dahl-kee-VEER) **River** river in southern Spain, **349**

Guam (14°N 143°E) South Pacific island and U.S. territory in Micronesia, **731**

Guatemala most populous country in Central America, **239**

Guatemala City (15°N 91°W) capital of Guatemala, **239**

Guayaquil (gwy-ah-KEEL) (2°S 80°W) port city in Ecuador, **214**

Guiana Highlands elevated region in northeastern South America, **257**

Guinea country in West Africa, **499**

Guinea-Bissau (GI-nee bi-SOW) country in West Africa, **499**

Gulf-Atlantic Coastal Plain North American landform region stretching along the Atlantic Ocean and Gulf of Mexico, **149**

Gulf of Bothnia part of the Baltic Sea west of Finland, **305**

Gulf of California part of the Pacific Ocean east of Baja California, Mexico, **221**

Gulf of Finland part of the Baltic Sea south of Finland, **S33**

Gulf of Guinea (GI-nee) gulf of the Atlantic Ocean south of western Africa, **499**

Gulf of Mexico gulf of the Atlantic Ocean between Florida, Texas, and Mexico, **165**

Gulf of St. Lawrence gulf between New Brunswick and Newfoundland Island in North America, **189**

Guyana (gy-AH-nuh) country in South America, **257**

Haiti country occupying the western third of the Caribbean island of Hispaniola, **239**

Halifax (45°N 64°W) provincial capital of Nova Scotia, Canada, **189**

Hamburg (54°N 10°E) seaport on the Elbe River in northern Germany, **327**

Hanoi (ha-NOY) (21°N 106°E) capital of Vietnam, **659**

Harare (18°S 31°E) capital of Zimbabwe, **533**

Havana (23°N 82°W) capital of Cuba, **239**

Helsinki (60°N 25°E) capital of Finland, **305**

Himalayas mountain system in Asia; world's highest mountains, **585**

Hindu Kush high mountain range in northern Afghanistan, **433**

Hispaniola large Caribbean island divided into the countries of Haiti and the Dominican Republic, **239**

Ho Chi Minh City (formerly Saigon) (11°N 107°E) major city in southern Vietnam; former capital of South Vietnam, **659**

Hokkaidō (hoh-KY-doh) major island in northern Japan, **637**

Honduras country in Central America, **239**

Hong Kong (22°N 115°E) former British colony in East Asia; now part of China, **615**

Hong (Red) River major river that flows into the Gulf of Tonkin in Vietnam, **659**

Honiara (9°S 160°E) capital of the Solomon Islands, **731**

Honshū (HAWN-shoo) largest of the four major islands of Japan, **637**

Houston (30°N 95°W) major port and largest city in Texas, **165**

Huang (Yellow) River one of the world's longest rivers; located in northern China, **615**

Hudson Bay large bay in Canada, **189**

Hungary country in central Europe between Romania and Austria, **327**

Iberian Peninsula peninsula in southwestern Europe occupied by Spain and Portugal, **349**

Iceland island country between the North Atlantic and Arctic Oceans, **305**

India country in South Asia, **563**

Indian Ocean world's third-largest ocean; located east of Africa, south of Asia, west of Australia, and north of Antarctica, **S22**

Indochina Peninsula peninsula in Southeast Asia that includes the region from Myanmar (Burma) to Vietnam, **659**

Indonesia largest country in Southeast Asia; made up of more than 17,000 islands, **679**

Indus River major river in Pakistan, **585**

Inland Sea body of water in southern Japan between Honshū, Shikoku, and Kyūshū, **637**

Interior Plains vast area between the Appalachians and Rocky Mountains in North America, **165**

Iran country in southwestern Asia; formerly called Persia, **433**

Iraq (i-RAHK) country located between Iran and Saudi Arabia, **433**

Ireland country west of Great Britain in the British Isles, **305**

Irian Jaya western part of the island of New Guinea that is part of Indonesia, **679**

Irish Sea sea between Great Britain and Ireland, **305**

Irrawaddy River important river in Myanmar (Burma), **659**

Islamabad (34°N 73°E) capital of Pakistan, **585**

Israel country in southwestern Asia, **449**

İstanbul (formerly Constantinople) (41°N 29°E) largest city and leading seaport in Turkey, **449**

Italy country in southern Europe, **349**

Jakarta (6°S 107°E) capital of Indonesia, **679**

Jamaica island country in the Caribbean Sea, **239**

Japan country in East Asia consisting of four major islands and more than 3,000 smaller islands, **637**

Java major island in Indonesia, **679**

Jerusalem (32°N 35°E) capital of Israel, **449**

Johannesburg (26°S 28°E) city in South Africa, **533**

Jordan Southwest Asian country stretching east from the Dead Sea and Jordan River into the Arabian Desert, **449**

Jordan River river in southwestern Asia that separates Israel from Syria and Jordan, **449**

Jutland Peninsula peninsula in northern Europe made up of Denmark and part of northern Germany, **305**

Kabul (35°N 69°E) capital and largest city of Afghanistan, **433**

Kalahari Desert dry plateau region in southern Africa, **533**

Kamchatka Peninsula peninsula along Russia's northeastern coast, **381**

Kampala (0° 32°E) capital of Uganda, **517**

Kao-hsiung (23°N 120°E) Taiwan's second-largest city and major seaport, **615**

Karachi (25°N 69°E) Pakistan's largest city and major seaport, **585**

Karakoram Range high mountain range in northern India and Pakistan, **585**

Kara-kum (kahr-uh-KOOM) desert region in Turkmenistan, **403**

Kashmir mountainous region in northern India and Pakistan, **563**

Kathmandu (kat-man-DOO) (28°N 85°E) capital of Nepal, **585**

Kazakhstan country in Central Asia; former Soviet republic, **403**

Kenya country in East Africa south of Ethiopia, **517**

Khabarovsk (kuh-BAHR-uhfsk) (49°N 135°E) city in southeastern Russia on the Amur River, **381**

Khartoum (16°N 33°E) capital of Sudan, **517**

Khyber Pass major mountain pass between Afghanistan and Pakistan, **585**

Kiev (50°N 31°E) capital of Ukraine, **381**

Kigali (2°S 30°E) capital of Rwanda, **517**

Kilimanjaro (3°S 37°E) (ki-luh-muhn-JAHR-oh) highest point in Africa (19,341 ft.; 5,895 m); located in northeast Tanzania near the Kenya border, **517**

Kingston (18°N 77°W) capital of Jamaica, **239**

Kingstown (13°N 61°W) capital of St. Vincent and the Grenadines, **239**

Kinshasa (4°S 15°E) capital of the Democratic Republic of the Congo, **499**

Kiribati South Pacific country in Micronesia and Polynesia, **731**

Kjølen (CHUHL-uhn) **Mountains** mountain range on the Scandinavian Peninsula, **305**

Kōbe (KOH-bay) (35°N 135°E) major port city in Japan, **637**

Kolkata (23°N 88°E) giant industrial and seaport city in eastern India, **563**

Korea Peninsula peninsula on the east coast of Asia, **637**

Koror (7°N 134°E) capital of Palau, **731**

Kosovo province in southern Serbia, **349**

Kuala Lumpur (3°N 102°E) capital of Malaysia, **679**

Kuril (KYOOHR-eel) **Islands** Russian islands northeast of the island of Hokkaidō, Japan, **381**

Kuwait country on the Persian Gulf in southwestern Asia, **433**

Kuwait City (29°N 48°E) capital of Kuwait, **433**

Kuznetsk Basin (Kuzbas) industrial region in central Russia, **381**

Kyōto (KYOH-toh) (35°N 136°E) city on the island of Honshū and the ancient capital of Japan, **637**

Kyrgyzstan (kir-gi-STAN) country in Central Asia; former Soviet republic, **403**

Kyūshū (KYOO-shoo) southernmost of Japan's main islands, **637**

Kyzyl Kum (ki-zil KOOM) desert region in Uzbekistan and Kazakhstan, **403**

Labrador mainland region in the territory of Newfoundland and Labrador, Canada, **189**

Lagos (LAY-gahs) (6°N 3°E) former capital of Nigeria and the country's largest city, **499**

Lahore (32°N 74°E) industrial city in northeastern Pakistan, **585**

Lake Baikal (by-KAHL) world's deepest freshwater lake; located north of the Gobi in Russia, **381**

Lake Chad shallow lake between Nigeria and Chad in western Africa, **499**

Lake Malawi (also called Lake Nyasa) lake in southeastern Africa, **533**

Lake Maracaibo (mah-rah-KY-buh) extension of the Gulf of Venezuela in South America, **257**

Lake Nasser artificial lake in southern Egypt created in the 1960s by the construction of the Aswān High Dam, **483**

Lake Nicaragua lake in southern Nicaragua, **239**

Lake Poopó (poh-oh-POH) lake in western Bolivia, **257**

Lake Tanganyika deep lake in the Great Rift Valley in Africa, **517**

Lake Titicaca lake between Bolivia and Peru at an elevation of 12,500 feet (3,810 m), **257**

Lake Victoria large lake in East Africa surrounded by Uganda, Kenya, and Tanzania, **517**

Lake Volta large artificial lake in Ghana, **499**

Laos landlocked country in Southeast Asia, **659**

La Paz (17°S 68°W) administrative capital and principal industrial city of Bolivia at an elevation of 12,001 feet (3,658 m); highest capital in the world, **257**

Lapland region extending across northern Finland, Sweden, and Norway, **305**

Las Vegas (36°N 115°W) city in southern Nevada, **165**

Latvia country on the Baltic Sea; former Soviet republic, **327**

Lebanon country in Southwest Asia, **449**

Lesotho country completely surrounded by South Africa, **533**

Lesser Antilles chain of volcanic islands in the eastern Caribbean Sea, **239**

Liberia country in West Africa, **499**

Libreville (0° 9°E) capital of Gabon, **499**

Libya country in North Africa located between Egypt and Algeria, **483**

Liechtenstein microstate in west-central Europe located between Switzerland and Austria, **327**

Lilongwe (14°S 34°E) capital of Malawi, **533**

Lima (12°S 77°W) capital of Peru, **257**

Limpopo River river in southern Africa forming the border between South Africa and Zimbabwe, **533**

Lisbon (39°N 9°W) capital and largest city of Portugal, **349**

Lithuania European country on the Baltic Sea; former Soviet republic, **327**

Ljubljana (lee-oo-blee-AH-nuh) (46°N 14°E) capital of Slovenia, **349**

Lomé (6°N 1°E) capital of Togo, **499**

London (52°N 0°) capital of the United Kingdom, **305**

Luanda (9°S 13°E) capital of Angola, **533**

Lubumbashi (loo-boom-BAH-shee) (12°S 27°E) industrial city in the Democratic Republic of the Congo, **477**

Gazetteer

Luxembourg small European country bordered by France, Germany, and Belgium, **305**

Luxembourg (50°N 7°E) capital of Luxembourg, **305**

Luzon chief island of the Philippines, **679**

Macao (22°N 113°E) former Portuguese territory in East Asia; now part of China, **615**

Macedonia Balkan country; former Yugoslav republic, **349**

Madagascar largest of the island countries off the eastern coast of Africa, **533**

Madrid (40°N 4°W) capital of Spain, **349**

Magdalena River river in Colombia that flows into the Caribbean Sea, **257**

Magnitogorsk (53°N 59°E) manufacturing city in the Urals region of Russia, **381**

Majuro (7°N 171°E) capital of the Marshall Islands, **731**

Malabo (4°N 9°E) capital of Equatorial Guinea, **499**

Malawi (muh-LAH-wee) landlocked country in Central Africa, **533**

Malay Archipelago (ahr-kuh-PE-luh-goh) large island group off the southeastern coast of Asia including New Guinea and the islands of Malaysia, Indonesia, and the Philippines, **679**

Malay Peninsula peninsula in Southeast Asia, **679**

Malaysia country in Southeast Asia, **679**

Maldives island country in the Indian Ocean south of India, **585**

Male (5°N 72°E) capital of the Maldives, **585**

Mali country in West Africa along the Niger River, **499**

Malta island country in southern Europe located in the Mediterranean Sea between Sicily and North Africa, **349**

Managua (12°N 86°W) capital of Nicaragua, **239**

Manama (26°N 51°E) capital of Bahrain, **433**

Manaus (3°S 60°W) city in Brazil on the Amazon River, **257**

Manchester (53°N 2°W) major commercial city in west-central Great Britain, **305**

Manila (15°N 121°E) capital of the Philippines, **679**

Manitoba prairie province in central Canada, **189**

Maputo (27°S 33°E) capital of Mozambique, **533**

Marseille (43°N 5°E) seaport in France on the Mediterranean Sea, **305**

Marshall Islands Pacific island country in Micronesia, **731**

Maseru (29°S 27°E) capital of Lesotho, **533**

Masqat (Muscat) (23°N 59°E) capital of Oman, **433**

Mato Grosso Plateau highland region in southwestern Brazil, **S31**

Mauritania African country stretching east from the Atlantic coast into the Sahara, **499**

Mauritius island country located off the coast of Africa in the Indian Ocean, **533**

Mazatlán (23°N 106°W) seaport city in western Mexico, **221**

Mbabane (26°S 31°E) capital of Swaziland, **533**

Mecca (21°N 40°E) important Islamic city in western Saudi Arabia, **433**

Mediterranean Sea sea surrounded by Europe, Asia, and Africa, **S22**

Mekong River important river in Southeast Asia, **659**

Melanesia island region in the South Pacific that stretches from New Guinea to Fiji, **731**

Melbourne (38°S 145°E) capital of Victoria, Australia, **711**

Mexican Plateau large, high plateau in central Mexico, **221**

Mexico country in North America, **221**

Mexico City (19°N 99°W) capital of Mexico, **221**

Miami (26°N 80°W) city in southern Florida, **165**

Micronesia island region in the South Pacific that includes the Mariana, Caroline, Marshall, and Gilbert island groups, **731**

Micronesia, Federated States of island country in the western Pacific, **731**

Milan (45°N 9°E) city in northern Italy, **349**

Minsk (54°N 28°E) capital of Belarus, **381**

Mississippi River major river in the central United States, **165**

Mogadishu (2°N 45°E) capital and port city of Somalia, **517**

Moldova European country located between Romania and Ukraine; former Soviet republic, **349**

Monaco (44°N 8°E) European microstate bordered by France, **305**

Mongolia landlocked country in East Asia, **615**

Monrovia (6°N 11°W) capital of Liberia, **499**

Montenegro See Serbia and Montenegro.

Monterrey (26°N 100°W) major industrial center in northeastern Mexico, **221**

Montevideo (mawn-tay-bee-THAY-oh) (35°S 56°W) capital of Uruguay, **257**

Montreal (46°N 74°W) financial and industrial city in Quebec, Canada, **189**

Morocco country in North Africa south of Spain, **483**

Moroni (12°S 43°E) capital of Comoros, **533**

Moscow (56°N 38°E) capital of Russia, **381**

Mount Elbrus (43°N 42°E) highest European peak (18,510 ft.; 5,642 m); located in the Caucasus Mountains, **381**

Mount Everest (28°N 87°E) world's highest peak (29,035 ft.; 8,850 m); located in the Himalayas, **585**

Mozambique (moh-zahm-BEEK) country in southern Africa, **533**

Mumbai (Bombay) (19°N 73°E) India's largest city, **563**

Munich (MYOO-nik) (48°N 12°E) major city and manufacturing center in southern Germany, **327**

Murray-Darling Rivers major river system in southeastern Australia, **711**

Myanmar (MYAHN-mahr) (Burma) country in Southeast Asia between India, China, and Thailand, **659**

Nairobi (1°S 37°E) capital of Kenya, **517**

Namib Desert Atlantic coast desert in southern Africa, **533**

Namibia (nuh-MI-bee-uh) country on the Atlantic coast in southern Africa, **533**

Nanjing (32°N 119°E) city along the upper Chang River in China, **615**

Naples (41°N 14°E) major seaport in southern Italy, **349**

Nassau (25°N 77°W) capital of the Bahamas, **239**

Nauru South Pacific island country in Micronesia, **731**

N'Djamena (12°N 15°E) capital of Chad, **499**

Negev desert region in southern Israel, **449**

Nepal South Asian country located in the Himalayas, **585**

Netherlands country in west-central Europe, **305**

New Brunswick province in eastern Canada, **189**

New Caledonia French territory in the South Pacific Ocean east of Queensland, Australia, **731**

New Delhi (29°N 77°E) capital of India, **563**

Newfoundland and Labrador province in eastern Canada including Labrador and the island of Newfoundland, **189**

New Guinea large island in the South Pacific Ocean north of Australia, **731**

New Orleans (30°N 90°W) major port city in Louisiana located on the Mississippi River, **165**

New York Middle Atlantic state in the northeastern United States, **165**

New York most populous city in the United States, **165**

New Zealand island country located southeast of Australia, **711**

Niamey (14°N 2°E) capital of Niger, **499**

Nicaragua country in Central America, **239**

Nice (44°N 7°E) city on the southeastern coast of France, **305**

Nicosia (35°N 33°E) capital of Cyprus, **449**

Niger (NY-juhr) country in West Africa, **499**

Nigeria country in West Africa, **499**

Niger River river in West Africa, **499**

Nile Delta region in northern Egypt where the Nile River flows into the Mediterranean Sea, **483**

Nile River world's longest river (4,160 miles; 6,693 km); flows into the Mediterranean Sea in Egypt, **483**

Nizhniy Novgorod (Gorki), Russia (56°N 44°E) city on the Volga River east of Moscow, **381**

North America continent including Canada, the United States, Mexico, Central America, and the Caribbean islands, **S21**

North China Plain region of northeastern China, **615**

Northern European Plain broad coastal plain from the Atlantic coast of France into Russia, **S33**

Northern Ireland the six northern counties of Ireland that remain part of the United Kingdom; also called Ulster, **305**

Northern Mariana Islands U.S. commonwealth in the South Pacific, **731**

North Island one of the two main islands of New Zealand, **711**

North Korea country on the northern part of the Korea Peninsula in East Asia, **637**

North Pole the northern point of Earth's axis, **S41**

North Sea major sea between Great Britain, Denmark, and the Scandinavian Peninsula, **305**

Northwest Highlands region of rugged hills and low mountains in Europe, including parts of the British Isles, northwestern France, the Iberian Peninsula, and the Scandinavian Peninsula, **291**

Northwest Territories division of a northern region of Canada, **189**

Norway European country located on the Scandinavian Peninsula, **305**

Nouakchott (nooh-AHK-shaht) (18°N 16°W) capital of Mauritania, **499**

Nova Scotia province in eastern Canada, **189**

Novosibirsk (55°N 83°E) industrial center in Siberia, Russia, **381**

Nuku'alofa capital of Tonga, **731**

Nunavut territory of northern Canada, **189**

Nuuk (Godthåb) (64°N 52°W) capital of Greenland, **305**

Ob River large river that drains Russia and Siberia, **381**

Oman country on the Arabian Peninsula; formerly known as Masqat (Muscat), **433**

Ontario province in central Canada, **189**

Orange River river in southern Africa, **533**

Orinoco River river in South America, **257**

Orizaba (19°N 97°W) volcanic mountain (18,700 ft.; 5,700 m) southeast of Mexico City; highest point in Mexico, **221**

Ōsaka (oh-SAH-kuh) (35°N 135°E) major industrial center on Japan's southwestern Honshū island, **637**

Gazetteer

Oslo (60°N 11°E) capital of Norway, **305**

Ottawa (45°N 76°W) capital of Canada; located in Ontario, **189**

Ouagadougou (wah-gah-DOO-GOO) (12°N 2°W) capital of Burkina Faso, **499**

Pacific Ocean Earth's largest ocean; located west of North and South America and east of Asia and Australia, **S21–S22**

Pakistan South Asian country located northwest of India, **585**

Palau South Pacific island country in Micronesia, **731**

Palikir capital of the Federated States of Micronesia, **731**

Pamirs mountain area mainly in Tajikistan in Central Asia, **403**

Panama country in Central America, **239**

Panama Canal canal connecting the Pacific Ocean and Caribbean Sea; located in central Panama, **239**

Panama City (9°N 80°W) capital of Panama, **239**

Papua New Guinea country on the eastern half of the island of New Guinea, **731**

Paraguay landlocked country in South America, **257**

Paraguay River river that divides Paraguay into two separate regions, **257**

Paramaribo (6°N 55°W) capital of Suriname in South America, **257**

Paraná River large river in southeastern South America, **257**

Paris (49°N 2°E) capital of France, **305**

Patagonia arid region of plains and windswept plateaus in southern Argentina, **257**

Peloponnese (PE-luh-puh-neez) peninsula forming the southern part of the mainland of Greece, **349**

Persian Gulf body of water between Iran and the Arabian Peninsula, **433**

Perth (32°S 116°E) capital of Western Australia, **711**

Peru country in South America, **257**

Philadelphia (40°N 75°W) important port and industrial center in Pennsylvania in the northeastern United States, **165**

Philippines Southeast Asian island country located north of Indonesia, **679**

Phnom Penh (12°N 105°E) capital of Cambodia, **659**

Phoenix (34°N 112°W) capital of Arizona, **165**

Plateau of Brazil area of upland plains in southern Brazil, **257**

Plateau of Tibet high plateau in western China, **615**

Poland country in central Europe located east of Germany, **327**

Polynesia island region of the South Pacific Ocean that includes the Hawaiian and Line Island groups, Samoa, French Polynesia, and Easter Island, **731**

Po River river in northern Italy, **349**

Port-au-Prince (pohr-toh-PRINS) (19°N 72°W) capital of Haiti, **239**

Port Elizabeth (34°S 26°E) seaport in South Africa, **533**

Portland (46°N 123°W) seaport and largest city in Oregon, **165**

Port Louis (20°S 58°E) capital of Mauritius, **533**

Port Moresby (10°S 147°E) seaport and capital of Papua New Guinea, **731**

Port-of-Spain (11°N 61°W) capital of Trinidad and Tobago, **239**

Porto-Novo (6°N 3°E) capital of Benin, **499**

Portugal country in southern Europe located on the Iberian Peninsula, **349**

Port-Vila (18°S 169°E) capital of Vanuatu, **731**

Gazetteer

Prague (50°N 14°E) capital of the Czech Republic, **327**
Praia (PRIE-uh) (15°N 24°W) capital of Cape Verde, **499**
Pretoria (26°S 28°E) administrative capital of South Africa, **533**
Prince Edward Island province in eastern Canada, **189**
Pripet Marshes (PRI-pet) marshlands in southern Belarus and northwest Ukraine, **381**
Puerto Rico U.S. commonwealth in the Greater Antilles in the Caribbean Sea, **239**
Pusan (35°N 129°E) major seaport city in southeastern South Korea, **637**
P'yŏngyang (pyuhng-YANG) (39°N 126°E) capital of North Korea, **637**
Pyrenees (PIR-uh-neez) mountain range along the border of France and Spain, **349**

Qatar Persian Gulf country located on the Arabian Peninsula, **433**
Qattara Depression lowland region in northern Egypt, **483**
Quebec province in eastern Canada, **189**
Quebec City (47°N 71°W) provincial capital of Quebec, Canada, **189**
Quito (0° 79°W) capital of Ecuador, **257**

Rabat (34°N 7°W) capital of Morocco, **483**
Rangoon *See* Yangon.
Red Sea sea between the Arabian Peninsula and northeastern Africa, **433**
Reykjavik (RAYK-yuh-veek) (64°N 22°W) capital of Iceland, **305**
Rhine River major river in west-central Europe, **327**
Riga (57°N 24°E) capital of Latvia, **327**
Río Bravo Mexican name for the river between Texas and Mexico, **221**
Rio de Janeiro (23°S 43°W) city in southeastern Brazil, **257**
Río de la Plata estuary between Argentina and Uruguay in South America, **257**
Riyadh (25°N 47°E) capital of Saudi Arabia, **433**
Rocky Mountains major mountain range in North America, **165**
Romania country in the eastern Balkans region of Europe, **349**
Rome (42°N 13°E) capital of Italy, **349**
Rosario (roh-SAHR-ee-oh) (33°S 61°W) city in eastern Argentina, **S32**
Roseau (15°N 61°W) capital of Dominica in the Caribbean, **239**
Ross Ice Shelf ice shelf in Antarctica, **S41**
Rub' al-Khali (Empty Quarter) uninhabited desert area in southeastern Saudi Arabia, **433**
Russia world's largest country, stretching from Europe and the Baltic Sea to eastern Asia and the coast of the Bering Sea, **381**
Rwanda country in East Africa, **517**

Sahara desert region in northern Africa; world's largest desert, **483**
St. George's (12°N 62°W) capital of Grenada in the Caribbean Sea, **239**
St. John's (17°N 62°W) capital of Antigua and Barbuda in the Caribbean Sea, **239**
St. Kitts-Nevis Caribbean country in the Lesser Antilles, **239**
St. Lawrence River major river linking the Great Lakes with the Gulf of St. Lawrence and the Atlantic Ocean in southeastern Canada, **189**
St. Lucia Caribbean country in the Lesser Antilles, **239**
St. Petersburg (formerly Leningrad) (60°N 30°E) Russia's second-largest city and former capital, **381**
St. Vincent and the Grenadines Caribbean country in the Lesser Antilles, **239**
Sakhalin Island Russian island north of Japan, **381**
Salvador (13°S 38°W) seaport city of eastern Brazil, **257**
Salzburg city in Austria, **327**
Samarqand (40°N 67°E) city in southeastern Uzbekistan, **403**
Samoa South Pacific island country in Polynesia, **731**
Sanaa (15°N 44°E) capital of Yemen, **433**
San Diego (33°N 117°W) seaport and city in California, **165**
San Francisco (38°N 122°W) seaport and city in California, **165**
San José (10°N 84°W) capital of Costa Rica, **239**
San Juan (19°N 66°W) capital of Puerto Rico, **239**
San Marino microstate in southern Europe surrounded by Italy, **349**
San Marino (45°N 12°E) capital of San Marino, **S34**
San Salvador (14°N 89°W) capital of El Salvador, **239**
Santiago (33°S 71°W) capital of Chile, **257**
Santo Domingo (19°N 70°W) capital of the Dominican Republic, **239**
São Francisco River river in eastern Brazil, **257**
São Paulo (24°S 47°W) Brazil's largest city, **257**
São Tomé (1°N 6°E) capital of São Tomé and Príncipe, **499**
São Tomé and Príncipe island country located off the Atlantic coast of Central Africa, **499**
Sarajevo (sar-uh-YAY-voh) (44°N 18°E) capital of Bosnia and Herzegovina, **349**
Saskatchewan province in central Canada, **189**
Saudi Arabia country occupying much of the Arabian Peninsula in southwestern Asia, **433**
Scandinavia peninsula of northern Europe occupied by Norway and Sweden, **305**
Scotland northern part of the island of Great Britain, **305**
Sea of Azov sea in Ukraine connected to and north of the Black Sea, **S33**
Sea of Japan body of water separating Japan from mainland Asia, **637**
Sea of Marmara sea separating European and Asian Turkey, **449**
Sea of Okhotsk inlet of the Pacific Ocean on the eastern coast of Russia, **381**
Seattle (48°N 122°W) largest city in the U.S. Pacific Northwest; located in Washington, **165**
Seine River river that flows through Paris in northern France, **305**
Senegal country in West Africa, **499**
Senegal River river in West Africa, **499**
Seoul (38°N 127°E) capital of South Korea, **637**
Serbia and Montenegro formerly Yugoslavia, **349**
Seychelles island country located east of Africa in the Indian Ocean, **533**
Shanghai (31°N 121°E) major seaport city in eastern China, **615**

Shannon River river in Ireland; longest river in the British Isles, **305**

Shikoku (shee-KOH-koo) smallest of the four main islands of Japan, **637**

Siberia vast region of Russia extending from the Ural Mountains to the Pacific Ocean, **381**

Sierra Madre Occidental mountain range in western Mexico, **221**

Sierra Madre Oriental mountain range in eastern Mexico, **221**

Sierra Nevada one of the longest and highest mountain ranges in the United States; located in eastern California, **165**

Sinai (SY-ny) **Peninsula** peninsula in northeastern Egypt, **483**

Singapore small island country located at the tip of the Malay Peninsula in Southeast Asia, **679**

Skopje (SKAW-pye) (42°N 21°E) capital of Macedonia, **349**

Slovakia country in central Europe; formerly the eastern part of Czechoslovakia, **327**

Slovenia country in the western Balkans region of Europe; former Yugoslav republic, **349**

Sofia (43°N 23°E) capital of Bulgaria, **349**

Solomon Islands South Pacific island country in Melanesia, **731**

Somalia East African country located in the Horn of Africa, **517**

South Africa country in southern Africa, **533**

Southern Alps mountain range in South Island, New Zealand, **711**

South Island one of the two main islands of New Zealand, **711**

South Korea country occupying the southern half of the Korea Peninsula, **637**

South Pole the southern point of Earth's axis, **S41**

Spain country in southern Europe occupying most of the Iberian Peninsula, **349**

Sri Lanka island country located south of India; formerly known as Ceylon, **585**

Stockholm (59°N 18°E) capital of Sweden, **305**

Strait of Gibraltar (juh-BRAWL-tuhr) strait between the Iberian Peninsula and North Africa that links the Mediterranean Sea to the Atlantic Ocean, **349**

Strait of Magellan strait in South America connecting the South Atlantic and South Pacific Oceans, **S31**

Sucre (19°S 65°W) constitutional capital of Bolivia, **257**

Sudan East African country; largest country in Africa, **517**

Suez Canal canal linking the Red Sea to the Mediterranean Sea in northeastern Egypt, **483**

Sumatra large island in Indonesia, **679**

Suriname (soohr-uh-NAH-muh) country in northern South America, **257**

Suva (19°S 178°E) capital of Fiji, **731**

Swaziland country in southern Africa, **533**

Sweden country in northern Europe, **305**

Switzerland country in west-central Europe located between Germany, France, Austria, and Italy, **327**

Sydney (34°S 151°E) largest urban area and leading seaport in Australia, **711**

Syr Dar'ya (sir duhr-YAH) river draining the Pamirs in Central Asia, **403**

Syria Southwest Asian country located between the Mediterranean Sea and Iraq, **449**

Syrian Desert desert region covering parts of Syria, Jordan, Iraq, and northern Saudi Arabia, **449**

Tagus River longest river on the Iberian Peninsula in southern Europe, **349**

Tahiti French South Pacific island in Polynesia, **731**

Taipei (25°N 122°E) capital of Taiwan, **615**

Taiwan (TY-WAHN) island off the southeastern coast of China, **615**

Tajikistan (tah-ji-ki-STAN) country in Central Asia; former Soviet republic, **403**

Taklimakan Desert desert region in western China, **615**

Tallinn (59°N 25°E) capital of Estonia, **327**

Tampico (22°N 98°W) Gulf of Mexico seaport in central eastern Mexico, **221**

Tanzania East African country located south of Kenya, **517**

Tarai (tuh-RY) region in Nepal along the border with India, **585**

Tarawa capital of Kiribati, **731**

Tarim Basin arid region in western China, **615**

Tashkent (41°N 69°E) capital of Uzbekistan, **403**

Tasmania island state of Australia, **711**

Tasman Sea part of South Pacific Ocean between Australia and New Zealand, **711**

Tbilisi (42°N 45°E) capital of Georgia in the Caucasus region, **381**

Tegucigalpa (14°N 87°W) capital of Honduras, **239**

Tehran (36°N 52°E) capital of Iran, **433**

Tel Aviv (tehl uh-VEEV) (32°N 35°E) largest city in Israel, **449**

Thailand (TY-land) country in Southeast Asia, **659**

Thar (TAHR) **Desert** sandy desert of northwestern India and eastern Pakistan; also called the Great Indian Desert, **563**

Thessaloníki (41°N 23°E) city in Greece, **349**

Thimphu (28°N 90°E) capital of Bhutan, **585**

Tian Shan (TIEN SHAHN) high mountain range separating northwestern China from Russia and some Central Asian countries, **403**

Tiber River river that flows through Rome in central Italy, **349**

Tibesti Mountains mountain group in northwest Chad, **499**

Tierra del Fuego group of islands at the southern tip of South America, **257**

Tigris River major river in southwestern Asia, **433**

Tijuana (33°N 117°W) city in northwestern Mexico, **221**

Timor island of the Malay Archipelago north of Australia, **679**

Tiranë (ti-RAH-nuh) (42°N 20°E) capital of Albania, **349**

Togo West African country located between Ghana and Benin, **499**

Tokyo (36°N 140°E) capital of Japan, **637**

Tombouctou (Timbuktu) (17°N 3°W) city in Mali and an ancient trading center in West Africa, **499**

Tonga South Pacific island country in Polynesia, **731**

Toronto (44°N 79°W) capital of the province of Ontario, Canada, **189**

Transantarctic Mountains major mountain range that divides Antarctica into East and West, **58**

Trinidad and Tobago Caribbean country in the Lesser Antilles, **239**

Tripoli (33°N 13°E) capital of Libya, **483**

Tropic of Cancer parallel 23^1/$_2$° north of the equator; parallel on the globe at which the Sun's most direct rays strike Earth during the June solstice (first day of summer in the Northern Hemisphere), **S21–S22**

Tropic of Capricorn parallel 23^1/$_2$° south of the equator; parallel on the globe at which the Sun's most direct rays strike Earth during the December solstice (first day of summer in the Southern Hemisphere), **S21–S22**

Tunis (37°N 10°E) capital of Tunisia, **483**

Tunisia country in North Africa located on the Mediterranean coast between Algeria and Libya, **483**

Turin (45°N 8°E) city in northern Italy, **349**

Gazetteer

Turkey country of the eastern Mediterranean occupying Anatolia and a corner of southeastern Europe, **449**

Turkmenistan country in Central Asia; former Soviet republic, **403**

Tuvalu South Pacific island country in Polynesia, **731**

Uganda country in East Africa, **517**

Ukraine country located between Russia and Eastern European; former Soviet republic, **381**

Ulaanbaatar (oo-lahn-BAH-tawr) (48°N 107°E) capital of Mongolia, **615**

United Arab Emirates country located on the Arabian Peninsula, **433**

United Kingdom country in Europe occupying most of the British Isles; Great Britain and Northern Ireland, **305**

United States North American country located between Canada and Mexico, **165**

Ural Mountains mountain range in west-central Russia that divides Asia from Europe, **381**

Uruguay South American country on the northern side of the Río de la Plata between Brazil and Argentina, **257**

Uzbekistan country in Central Asia; former Soviet republic, **403**

Vaduz (47°N 10°E) capital of Liechtenstein, **S34**

Valletta (36°N 14°E) capital of Malta, **349**

Valparaíso (33°S 72°W) Pacific port for the national capital of Santiago, Chile, **S32**

Vancouver (49°N 123°W) Pacific port in Canada, **189**

Vanuatu South Pacific island country in Melanesia, **731**

Vatican City (42°N 12°E) European microstate surrounded by Rome, Italy, **349**

Venezuela country in northern South America, **257**

Victoria (4°S 55°E) capital of the Seychelles, **533**

Vienna (48°N 16°E) capital of Austria, **327**

Vientiane (18°N 103°E) capital of Laos, **659**

Vietnam country in Southeast Asia, **659**

Vilnius (54°N 25°E) capital of Lithuania, **327**

Vinson Massif (78°S 87°W) highest mountain (16,066 ft.; 4,897 m) in Antarctica, **S41**

Virgin Islands island chain lying just east of Puerto Rico in the Caribbean Sea, **239**

Vistula River river flowing through Warsaw, Poland, to the Baltic Sea, **327**

Vladivostok (43°N 132°E) chief seaport of the Russian Far East, **381**

Volga River Europe's longest river; located in west-central Russia, **381**

Wake Island (19°N 167°E) U.S. South Pacific island territory north of the Marshall Islands, **731**

Wales part of the United Kingdom occupying the western portion of Great Britain, **305**

Warsaw (52°N 21°E) capital of Poland, **327**

Washington, D.C. (39°N 77°W) U.S. capital; located between Virginia and Maryland on the Potomac River, **165**

Wellington (41°S 175°E) capital of New Zealand, **711**

West Bank area of Palestine west of the Jordan River; occupied by Israel in 1967; political status is in transition, **449**

Western Ghats (GAWTS) steep, rugged hills facing the Arabian Sea on the western side of the Deccan Plateau in India, **563**

Western Plateau large plain covering more than half of Australia, **711**

Western Sahara disputed territory in northwestern Africa; claimed by Morocco, **483**

West Siberian Plain region with many marshes east of the Urals in Russia, **381**

White Nile part of the Nile River system in eastern Africa, **517**

Windhoek (22°S 17°E) capital of Namibia, **533**

Windsor (42°N 83°W) industrial city across from Detroit, Michigan, in the Canadian province of Ontario, **189**

Winnipeg (50°N 97°W) provincial capital of Manitoba in central Canada, **189**

Witwatersrand (WIT-wawt-uhrz-rahnd) a range of low hills in north central South Africa, **533**

Wuhan (31°N 114°E) city in south central China, **615**

Xi River river in southeastern China, **615**

Yamoussoukro (7°N 5°W) capital of Côte d'Ivoire, **499**

Yangon (Rangoon) (17°N 96°E) capital of Myanmar (Burma), **659**

Yaoundé (4°N 12°E) capital of Cameroon, **499**

Yekaterinburg (formerly Sverdlovsk) (57°N 61°E) city in the Urals region in Russia, **381**

Yemen country located in the southwestern corner of the Arabian Peninsula, **433**

Yenisey (yi-ni-SAY) major river in central Russia, **381**

Yerevan (40°N 45°E) capital of Armenia, **381**

Yucatán Peninsula peninsula in southeastern Mexico, **221**

Yugoslavia former country of six republics on Europe's Balkan Peninsula, **349**

Yukon Territory Canadian territory bordering Alaska, **189**

Zagreb (46°N 16°E) capital of Croatia, **349**

Zagros Mountains mountain range in southwestern Iran, **433**

Zambezi (zam-BEE-zee) **River** major river in central and southern Africa, **533**

Zambia country in southern Africa, **533**

Zimbabwe (zim-BAH-bway) country in southern Africa, **533**

Zürich (47°N 9°E) Switzerland's largest city, **327**

This glossary contains terms you need to understand as you study world geography. A brief definition or explanation of the meaning of the term as it is used in *World Geography Today* follows each term. The page number refers to the page on which the term is introduced in the textbook.

Phonetic Respelling and Pronunciation Guide

Many of the key terms in this textbook have been respelled to help you pronounce them. The letter combinations used in the respelling throughout the narrative are explained in the following phonetic respelling and pronunciation guide. The guide is adapted from *Merriam-Webster's Collegiate Dictionary*, *Merriam-Webster's Geographical Dictionary*, and *Merriam-Webster's Biographical Dictionary*.

MARK	AS IN	RESPELLING	EXAMPLE
a	alphabet	a	*AL-fuh-bet
ā	Asia	ay	AY-zhuh
ä	cart, top	ah	KAHRT, TAHP
e	let, ten	e	LET, TEN
ē	even, leaf	ee	EE-vuhn, LEEF
i	it, tip, British	i	IT, TIP, BRIT-ish
ī	site, buy, Ohio	y	SYT, BY, oh-HY-oh
	iris	eye	EYE-ris
k	card	k	KAHRD
ō	over, rainbow	oh	OH-vuhr, RAYN-boh
ù	book, wood	ooh	BOOHK, WOOHD
ȯ	all, orchid	aw	AWL, AWR-kid
ȯi	foil, coin	oy	FOYL, KOYN
aù	out	ow	OWT
ə	cup, butter	uh	KUHP, BUHT-uhr
ü	rule, food	oo	ROOL, FOOD
yü	few	yoo	FYOO
zh	vision	zh	VIZH-uhn

*A syllable printed in small capital letters receives heavier emphasis than the other syllable(s) in a word.

abdicate Resign, **387**

Aborigines (a-buh-RIJ-uh-nees) Australia's first peoples, **715**

abyssal plains Areas of the ocean floor where rocks gradually sink because they have no supporting heat below them; the world's flattest and smoothest regions, **65**

acculturation Process in which an individual or group adopts some of the traits of another culture, **96**

acid rain Polluted rain that can damage trees and kill fish in lakes, **77**

African National Congress (ANC) A political organization that pushed for an end to apartheid in South Africa, **539**

Afrikaners (a-fri-KAH-nuhrz) Dutch, French, and German settlers and their descendants in South Africa, **538**

agribusiness The operation of specialized commercial farms for more efficiency and profits, **126**

alliances Agreements between countries to support one another against enemies, **328**

alluvial fan Fan-shaped deposit of mud and gravel often found along the bases of mountains, **69**

alluvial soils Soils deposited by streams or rivers, **158**

animist religions Religions in which people believe in the presence of the spirits and forces of nature, **102**

annex To formally join an area to a country, **643**

Antarctic Circle The parallel 66 1/2° south of the equator, **30**

apartheid (uh-PAHR-tayt) Official policies that forced black South Africans to live in separate areas and use separate facilities from white South Africans, **539**

aquaculture Raising and harvesting fish and marine life in ponds or other bodies of water, **619**

aqueducts Artificial channels for transporting water, **78**

aquifers Rock layers where groundwater is plentiful, **78**

arable Fit for growing crops, **175**

arboreal Tree-dwelling, **661**

archipelago (ahr-kuh-PEH-luh-goh) Large group of islands, **680**

Arctic Circle A parallel 66 1/2° degrees north of the equator, **31**

arid Dry, **54**

armistice Truce, **644**

artesian wells Wells in which water flows naturally to the surface, **712**

ASEAN Association of Southeast Asian Nations, an organization founded to promote economic development as well as political cooperation among its members, **670**

atlas A collection of maps in one book, **11**

atmosphere The envelope of gases that surrounds a planet like Earth, **35**

atoll A ring-shaped coral island or ring of several islands linked by underwater coral reefs, **732**

autarky (AW-tahr-kee) A system in which a country tries to produce all the goods that it needs, **388**

autonomy Self-government, **351**

ayatollahs Religious leaders of the highest authority among Shia Muslims, **443**

Glossary

balance of power Condition existing when countries or alliances have such equal levels of strength that war is prevented, **328**

barrier islands Coastal islands created from sand deposited by ocean waves and currents in shallow water, **150**

basins Low areas of land, often surrounded by mountains, **151**

bauxite Ore from which aluminum is made, **241**

Bedouins Nomadic herders in Southwest Asia, **441**

Berbers Cultural group that has lived in North Africa since long before waves of Arab armies crossed the continent, **489**

bilingual Able to speak two languages, **169**

biodiversity The level of variety of plants and animals, **534**

biosphere The part of Earth that includes all life forms, **35**

birthrate The number of live births each year for every 1,000 people living in a place, **89**

Bolsheviks A communist group that overthrew the government during the Russian Revolution in 1917, **387**

boycott Refusal to buy certain goods from a country or business, **570**

buffer state A small country between two larger, more powerful countries, **265**

cacao Type of tree from which we get cocoa beans, **246**

calligraphy Artistic handwriting or lettering, **624**

cantons Largely self-governing states within a country such as Switzerland, **336**

capitalism An economic system in which businesses, industries, and resources are privately owned, **115**

caravans Groups of people traveling together for protection, **408**

Caricom Caribbean Community and Common Market, **252**

cartography The study of maps and mapmaking, **5**

cash crops Crops grown for sale in a market, **230**

caste Group of people who are born into a certain position in society, as in Hinduism, that restricts the occupation and associations of their members, **571**

central business district (CBD) City center dominated by large stores, offices, and buildings, **123**

Chibcha Early people of the Colombian Andes who developed gold-working skills, **263**

Chishima Current (Oyashio) Cold ocean current that cools northern Japan in the summer, **639**

city-states Self-governing cities and their surrounding areas, as in ancient Greece, **360**

climate Weather conditions in a region over a long time, **41**

climate graphs Graphs showing average temperature and precipitation in a place, **18**

colonies Territories controlled by people from a foreign land, **167**

command economy An economic system in which the government decides what to produce, where to make it, and what price to charge, **115**

commonwealth Self-governing territory associated with another country, **250**

communes Collective farms grouped together to organize farming and plan public services, **623**

communism An economic and political system in which the government owns or controls almost all the means of production, **115**

compass rose A map element with arrows pointing in all four principal directions, **14**

complementary region A region formed by the combination of two areas with different activities or strengths, each of which benefits the other, **342**

condensation The process by which water vapor changes from a gas into liquid droplets, **46**

confederation A group of states joined together for a common purpose, **336**

Confucianism A philosophy based on the teachings of Confucius, an ancient Chinese philosopher, **624**

coniferous forests Forests of cone-bearing evergreen trees, **55**

conquistadores (kahn-kees-tuh-DAWR-ez) Spanish conquerors of foreign lands during the colonial era, **225**

consensus General agreement, **197**

constitutional monarchy A type of government with a king or queen as head of state and a parliament as the lawmaking branch, **307**

contiguous Connecting or bordering, **14**

Continental Divide The crest dividing North America's major river systems into those flowing eastward and those flowing westward, **153**

continental drift The process by which Earth's plates slowly move across the upper mantle, **63**

continental shelves Areas where continental surfaces extend under the shallow ocean water around the continents, **66**

continents Large landmasses on Earth's surface, **10**

contour plowing Plowing fields across a hill, rather than up and down the hill, **75**

copra Dried coconut meat, **741**

core Earth's center, where pressures and temperatures are very high, **63**

cork Bark that is stripped from the trunks of cork oaks, **353**

Corn Belt A region of the U.S. Midwest that specializes in growing corn, **175**

cosmopolitan Having many foreign influences, **316**

Cossacks People from the southern steppe frontiers of the Russian Empire who played an important role in that empire's expansion, **387**

cottage industries Small-scale industries based in the home, **576**

coup (KOO) A change of government caused by a group taking control by force, **266**

creole A language blending African, European, or indigenous Caribbean languages, **251**

crop rotation The practice of planting different crops in a field in alternating years, **76**

cultural boundaries Boundaries that are based on culture traits, **129**

culture All the features of a people's way of life, **94**

culture region An area in which people have many shared culture traits, **95**

culture traits Learned activities and behaviors that people often take part in, **94**

cyclones Winds around centers of low atmospheric pressure, **42**

czar (ZAHR) Title of the emperor of Russia before the Russian Revolution, **387**

Dairy Belt A region of the U.S. Midwest that specializes in dairy products, **175**

Dalits Hindus in India who are not part of any caste, **571**

death rate Total number of deaths each year for every 1,000 people in a place, **89**

deciduous forests Forests made up of trees that lose their leaves during part of the year, **55**

deforestation Destruction or loss of forests, **76**

degrees Units used to measure distances between parallels and between meridians, **10**

delta Accumulation of sediment at the mouth of a river, **69**

demilitarized zone (DMZ) A buffer zone into which opposing armies may not enter, **644**

democracy A system of government in which the people decide who will govern, **130**

demography Statistical study of human populations, **87**

depressions Large low areas, **484**

desalinization Process of removing salt from ocean water, **70**

desertification Spreading of desert conditions, **501**

developed countries Countries with high levels of industrialization and high standards of living, **117**

developing countries Countries with less productive economies than developed countries and low standards of living, **117**

dharma For Hindus, the importance of doing one's duty according to one's station in life, **571**

dialect A regional variety of a language, **100**

dictator A leader who rules with almost absolute authority, **226**

Diet (DEE-uht) Japan's elected lawmaking body, **644**

diffusion A process occurring when an idea or innovation spreads from one person or group to another and is adopted, **97**

dikes Walls built to hold back water, **293**

doldrums Areas with no prevailing winds, **44**

domestication The process in which people grow plants and tame animals for their own use, **120**

domino theory The idea that if one country fell to communism, neighboring countries would follow like falling dominoes, **664**

double cropping Harvesting two crops in the same plot each year, **619**

drainage basin A region drained by a river and its tributaries, **71**

dryland farming Growing crops that can rely on limited rainfall rather than irrigation, **412**

dual economies Economies in which some goods are produced for export while other goods are produced for local consumption, **509**

dynasty Line of hereditary rulers, **438**

economy of scale Large production of goods that reduces the production cost of each item, **727**

ecosystems Communities of plants and animals, **50**

ecotourism Type of tourism focusing on guided travel through natural areas, **248**

edge cities Clusters of large buildings away from the central business district of a city, **124**

El Niño (ehl NEEN-yoh) Ocean and weather pattern in which the southeastern Pacific Ocean is warmer than usual, affecting regional climates, **260**

emigrants People who leave a country to live somewhere else, **89**

enclaves Areas that are completely surrounded by another region, **362**

endemic species Plants and animals native to a certain place, **682**

environment Combination of Earth's four spheres, including all of the biological, chemical, and physical conditions that affect life, **35**

equator An imaginary line that circles the globe halfway between the North Pole and South Pole, **10**

equinox The time when both of Earth's poles are at a 90 degree angle from the Sun and the Sun's direct rays strike the equator, **31**

erg A sea of sand created by high, shifting sand dunes, **484**

erosion Movement of surface material from one location to another by water, wind, and ice, **67**

escarpment Steep face at the edge of a plateau or other raised area, **534**

estuaries Semi-enclosed bodies of water, seawater, and freshwater formed where a river meets an inlet of the sea, **71**

ethnic groups Human populations that share a common culture or ancestry, **95**

ethnic religions Religions found among people of one ethnic group and that generally have not spread into other cultures, **102**

Eurasia The landmass of Europe and Asia combined, **382**

European Union Organization of European countries featuring close cooperation on trade, economic, political, and social issues, **315**

evaporation Process by which liquid changes to gas, **46**

exclave An area separated from the rest of a country by the territory of other countries, **338**

Exclusive Economic Zone (EEZ) Zone extending 200 nautical miles (370 km) from a country's shore that includes resources controlled by that country, **741**

exotic rivers Rivers that begin in humid regions and then flow across dry areas, **434**

exotic species New types of plants and animals that people introduce to an area, **719**

export economy A type of economy in which goods are produced mainly for export rather than for domestic use, **650**

extensive agriculture A kind of agriculture using much land but small inputs of capital and labor per unit area, **718**

fall line A natural boundary between two landform regions with different elevations marked by rapids and waterfalls, **150**

Glossary

famine A widespread shortage of food, **307**

faults Places where rock masses have been broken apart and are moving away from each other, **66**

favelas (fah-VE-lahs) Large slums around Brazilian cities, **270**

fellahin Peasant farmers of North Africa, **492**

fjords (fee-AWRDZ) Narrow deep inlets of the sea set between high rocky cliffs, **292**

flyway Migration route for birds, **640**

folds Places where rocks have been compressed into bends by colliding plates, **66**

formal region Region with one or more common features that make it different from surrounding areas, **6**

fossil fuels Energy resources—formed from the remains of ancient plants and animals—including coal, natural gas, and petroleum, **79**

fossil water Groundwater that is not replenished by rain, **78**

free enterprise System that lets competition among businesses determine the price of products, **115**

free port Port where almost no taxes are placed on the goods unloaded from other places, **492**

front Zone of contact between two air masses of widely different temperatures or moisture levels, **44**

functional region Region made up of different places that are linked and function as a unit, **6**

fundamentalism Movement in which people believe in strictly following certain established principles or teachings, **98**

genocide The intentional destruction of a people, **527**

gentrification Process of buying run-down homes in older areas of a community and restoring them, **181**

geography The study of the physical, biological, and cultural features of Earth's surface, **3**

geometric boundaries Boundaries that follow geometric patterns, **129**

geothermal energy Heat of Earth's interior, **81**

geysers Hot springs that shoot water into the air, **320**

ghetto Section of a city where a minority group is forced to live, **339**

glaciers Thick masses of ice, including great ice sheets and bodies of ice that flow down mountains like slow rivers, **68**

globalization The process in which connections around the world increase and cultures become more alike, **98**

global warming The process in which Earth grows warmer over a period of time, **41**

Gran Chaco (grahn CHAH-koh) Very flat plains between the Andes and Brazilian Highlands of South America, **258**

graphite A form of carbon that is used in pencil leads, **595**

great-circle route Shortest route between any two places on the planet, **12**

greenhouse effect The process in which Earth's atmosphere traps heat energy, **41**

grid Lines crisscrossing each other in uniform intervals to create a pattern of squares, **9**

gross domestic product (GDP) The total value of goods and services created within a country, **116**

gross national product (GNP) The total value of goods and services that a country produces in a year, **116**

groundwater Water found below ground, **72**

gulag Network of prison labor camps in the Soviet Union, **388**

Gulf Stream Warm ocean current that moves warm tropical water northward along eastern North America, **45**

gum arabic Sap of acacia trees; a sticky substance that binds the ingredients in many candies and medicines, **525**

haciendas (hah-see-EN-duhs) Large Spanish colonial estates usually owned by wealthy families but worked by many peasants, **225**

hajj Religious journey to Mecca required of Muslims, **102**

headwaters First and smallest streams formed from the runoff of a mountain, eventually forming rivers, **71**

heavy industry Industry focusing on manufacturing, usually based on metals, **393**

hemispheres Halves of the globe divided by the equator and prime meridian, **10**

hieroglyphs Pictures and symbols used in the ancient Egyptian writing system, **487**

hinterland A region that lies far away from major population centers, **192**

Holocaust The murder of millions of Jews during World War II, **456**

homogeneous (hoh-muh-JEE-nee-uhs) Of the same kind, such as a population made up mostly of the same ethnic group, **687**

hot spot Place where magma wells up to the surface from Earth's mantle, **152**

humidity The amount of water vapor in the air, **46**

humus Broken-down plant and animal matter in soil, **75**

hurricanes The most powerful and destructive tropical cyclones, **48**

hydroelectric power Electricity produced by moving water, **81**

hydrologic cycle Movement of water through the hydrosphere, **71**

hydrosphere Earth's water in all of its forms, **35**

icebreakers Ships that can break paths through ice, **384**

imams Muslim religious leaders, **440**

immigrants People who come to a new country to live, **90**

Inca Founders of South America's greatest early civilization, **263**

indigenous Native to an area, **244**

industrialization The process by which manufacturing based on machine power becomes widespread in an area, **116**

informal sector Sector of an economy made up of people who do not work for formal businesses, including many self-employed people and small family-owned businesses, **543**

infrastructure System of roads, ports, and other facilities needed by a modern economy, **117**

innovation New idea that a culture accepts, **96**

intensive agriculture A form of agriculture using a great amount of human labor and/or capital to farm small amounts of land, **619**

Intertropical Convergence Zone (ITCZ) An area where northeast and southeast trade winds meet near the equator, creating humid, often rainy conditions, **734**

irrigation A process in which water is artificially supplied to the land, **76**

isthmus Narrow strip of land with water on either side connecting two larger land areas, **222**

ivory Cream-colored material that makes up the tusk of an elephant, **522**

Jainism Religion with origins in India that emphasizes a strict moral code based on preserving life, **572**

Japan Current (Kuroshio) Warm ocean current bringing moist marine air to Japan, **639**

jute Plant fiber used to make products like burlap and rope, **576**

kampongs Traditional villages usually built on stilts in parts of Southeast Asia, **692**

karma For Hindus, Buddhists, and Jains, the positive or negative force caused by a person's actions, **571**

klongs Canals of Thailand, **672**

Kurds A minority group in Turkey and neighboring countries, **457**

lahars (LAH-hahrz) Dangerous volcanic mudflows, **681**

landlocked Having no coast on the ocean, **271**

landscapes Scenery of places, including physical, human, and cultural features, **4**

La Niña An ocean and weather pattern that causes the eastern Pacific Ocean to be colder than normal, **260**

latifundia (lah-ti-FOOHN-dee-uh) Large estates in South America, **265**

latitude Lines on a globe drawn in an east-west direction and measured in degrees, **9**

leaching Downward movement of minerals and humus in soils, **75**

legend Map key, **14**

lichens Small plants that consist of algae and fungi, **157**

light industry Industry using low-weight materials and focusing on the production of consumer goods, **393**

lingua franca A language of trade and communication used by people who speak different languages, **101**

literacy rate Percentage of people who can read and write, **117**

lithosphere Solid crust of Earth, **35**

Llanos (YAH-nohs) Large plains area in northeastern Colombia and western Venezuela, **258**

loess Fine-grained, windblown soil that is very fertile, **299**

longitude Lines on a globe drawn in a north-south direction and measured in degrees, **9**

Glossary

magma Liquid rock within Earth, **63**

magnesium Light metal that is valuable in certain industries, such as aerospace, **451**

mandates Territories temporarily placed under another country's control, **455**

mangrove Type of tree with roots that grow in saltwater; found in tropical coastal areas around the world, **241**

manioc Tropical plant with starchy roots, **267**

mantle The section of Earth's interior that lies above the outer core and has the most mass, **63**

Maori (MOWR-ee) First peoples of New Zealand, **724**

map projections Different ways of representing a round Earth on a flat map, **11**

maquiladoras (mah-kee-lah-DOHR-ahs) Special factories in Mexico owned mainly by American companies, **232**

market economy An economic system in which people choose freely what to buy and sell, **115**

market-oriented agriculture Agricultural system in which farmers grow products to sell to consumers, **126**

marsupials Mammals that have pouches to carry their young, **713**

martial law Military rule, **631**

matrilineal A social system characterized by tracing kinship only through the mother, **739**

Megalopolis A group of cities in the northeastern United States that has grown into one large, built-up area; also used to describe similar areas around the world, **173**

Meiji Restoration Japanese political revolution that overthrew the shogun and restored the power of the emperor, **643**

Mercosur South American organization whose purpose is to expand trade, improve transportation, and reduce tariffs among member countries, **270**

meridians Lines of longitude, **10**

mestizos People of mixed European and American Indian ancestry, **226**

meteorology The study of weather, **5**

metropolitan area A city and its surrounding built-up areas, **174**

microstates Very small countries, such as Andorra, Monaco, and San Marino, **357**

migration The process of moving from one place to live in another, **89**

millet Type of drought-resistant grain, **507**

minifundia (mi-ni-FOOHN-dee-uh) Small South American farms created when large colonial estates were broken up, **268**

missionaries People whose goal it is to spread their religion, **102**

monoculture Cultivation of a single crop, **409**

monotheism The belief in one god, **102**

monsoon A wind system in which winds reverse direction and cause seasons of wet and dry weather, **51**

moons Smaller objects that orbit a planet, **26**

mosques Islamic houses of worship, **102**

Glossary

mulattoes People with both African and European ancestry, **245**

multilingual Able to speak two or more languages, **336**

nationalism A feeling of pride and loyalty for one's country or culture group, **130**

nationalized The process by which organizations or businesses become owned and operated by the government, **308**

natural boundaries Boundaries that follow a feature of the landscape, **129**

natural hazards Events in the physical environment that can destroy human life and property, **155**

navigable Able to be used for shipping, **294**

neutral Characterized by not taking sides in conflicts, **336**

newsprint Inexpensive paper used mainly for newspapers, **158**

nomads People who move often from place to place, **409**

North American Free Trade Agreement (NAFTA) An agreement between Canada, Mexico, and the United States to eliminate tariffs on many products flowing between these three countries, **183**

North Atlantic Drift Warm ocean current that moderates climates in northwestern Europe, **297**

oasis A place where a spring bubbles to the surface in desert areas, **436**

OPEC Organization of Petroleum Exporting Countries, which influences oil prices by controlling supply, **443**

ore Mineral-bearing rock, **79**

orographic effect Cooling effect that occurs when air is forced to rise over a mountain, resulting in a wetter windward side and a drier leeward side, **47**

outback Dry interior of Australia, **713**

paddy Wet land where rice is grown, **619**

Pampas Wide fertile grasslands in southern South America, **258**

pantheon All the gods of a religion, **568**

parallels Lines of latitude, **10**

parliament A legislature lead by a prime minister or premier, **196**

partition Division of a place, as in the 1947 partition of India, **570**

pastoralism A type of subsistence agriculture involving herding animals, **125**

perception Our awareness and understanding of the environment around us, **6**

perceptual regions Regions that reflect human feelings and attitudes, **6**

permafrost Permanently frozen soil below the ground's surface, **56**

perspective The way a person looks at something, **4**

petrochemicals Certain products made from oil, **80**

pharaohs Ancient Egyptian monarchs, **487**

phosphates Chemicals used to make fertilizer, **743**

pictograms Simple pictures used to represent ideas, **624**

pidgin languages Simplified languages used to help people with different languages communicate, **737**

piedmont An area at or near the foot of a mountain region, **150**

planets Major bodies that orbit a star, **25**

plantations Large farms that produce one major crop often for export, **168**

plateau An elevated flatland that rises sharply above nearby land on at least one side, **69**

plate tectonics The theory that Earth's crust is divided into rigid plates that slowly move across the upper mantle, **63**

plaza Open space in towns in Mexico and other Spanish-speaking countries, often in front of a church, **225**

polar regions High-latitude, cold regions around the North and South Poles, **29**

polders Lands reclaimed from the sea in the Netherlands, **293**

polytheism The belief in many gods, **102**

population density The average number of people living in an area; usually expressed as persons per square mile or square kilometer, **88**

population pyramids Graphs showing the percentages of males and females by age group in a country's population, **18**

potash A mineral used to process wool and to make fertilizers, glass, and soft soaps, **451**

precipitation Condensed droplets of water that fall as rain, snow, sleet, or hail, **16**

prevailing winds Winds that blow from the same direction most of the time, **44**

primate city A city that ranks first and dominates a country in terms of population and economy, **312**

prime meridian Imaginary line drawn from the North Pole through Greenwich, England, to the South Pole, **10**

protectorate Country that forfeits certain decision-making powers in exchange for protection by a stronger country, **591**

provinces Government districts similar to states, **190**

pull factors Factors attracting people to a new location, **90**

puppet government Government controlled by outside forces, such as another country's government, **622**

push factors Factors causing people to leave a location, **90**

quotas In trade matters, limits on the amount of a product that can be imported, **130**

rain shadow A drier area on the leeward side of a mountain range, **47**

reforestation The replanting of a forest, **76**

refugees People who have been forced to leave an area, **90**

reg A gravel-covered plain in a desert area, **484**

region An area with one or more common features that make it different from surrounding areas, **6**

regionalism A feeling of strong political and emotional loyalty to one's own region, **200**

reincarnation For Hindus, the repeated rebirth of one's soul in different forms, **571**

Renaissance (re-nuh-SAHNS) A period from about the 1300s to the 1500s marked by a renewed interest in learning in Europe, **356**

revolution One elliptical orbit of Earth around the Sun, **27**

rift valleys Places on Earth's surface where the crust stretches until it breaks, **65**

rotation One complete spin of Earth on its axis, **27**

Rus (ROOS) Scandinavian traders who were some of Kiev's early leaders; also the word from which Russia gets its name, **386**

Sahel (sah-HEL) A region of semiarid climate extending from Senegal and Mauritania to Sudan in Africa, **501**

samurai Professional Japanese warrior employed by a shogun for protection, **643**

sanctions Penalties intended to force a country to change its policies, **539**

Sanskrit Ancient language from which Hindi developed, **568**

Santeria Originating in Cuba, a religion that blends African traditions and Christian beliefs, **251**

satellite A body that orbits a larger body, **26**

savannas Areas of tropical grasslands, scattered trees, and shrubs, **51**

secular Nonreligious, **463**

sediment Small particles of weathered rock, **67**

separatism The belief that certain parts of a country should be independent, **200**

sepoys Indian troops under the command of British officers during the colonial era in India, **569**

sequent occupance A process of settlement by successive groups of people in which each group creates a distinctive cultural landscape, **306**

serfs Poor peasant farmers in Russia who worked for lords and were bound to the land, **387**

shatter belt A zone of frequent boundary changes and conflicts, often located between major powers, **389**

shifting cultivation A process in which farmers clear trees or brush for short-term planting before moving on to clear another area, **125**

Shi'ism One of the two main branches of Islam, **440**

shogun A powerful warlord in early Japan, **642**

Sikhism A religion with origins in India that combines the Muslim belief in one God with the Hindu belief in reincarnation, **572**

Silicon Valley An area south of San Francisco, California, that became the leading center of computer technology in the United States, **178**

silt Fertile, finely ground soil deposited by flowing water, **487**

sinkholes Steep-sided depressions that form when the roof of a cave collapses, **222**

sisal (SY-suhl) Strong, durable plant fiber used to make rope and twine, **522**

slash-and-burn agriculture A form of shifting cultivation in which farmers slash small areas of forest and burn the fallen trees, **688**

Slavs The main people to settle in what are now Russia, Ukraine, Belarus, and eastern parts of central and southern Europe, **386**

smelters Factories that process metal ores, **394**

smog Air pollution resulting from chemical reactions involving sunlight and automobile and industrial exhaust, **179**

socialism An economic system in which the government owns and controls the means of producing goods, **321**

soil exhaustion A condition in which soil has lost nutrients and becomes nearly useless for farming, **76**

soil salinization Salt buildup in the soil, **76**

solar energy Energy from the Sun, which reaches Earth as light and heat, **27**

solar system The Sun and the group of bodies that revolve around it, **25**

solstice The time that Earth's poles point at their greatest angle toward or away from the Sun, **30**

sorghum A type of drought-resistant grain, **507**

souks Open-air markets in the Arab world, **461**

soviets Local governing bodies of the Soviet Union, **388**

Special Economic Zones (SEZs) Zones designed to attract foreign companies and investment to areas of China, **628**

staple The main food of a region, **507**

storm surge Waters washed ashore by large storms, **587**

subcontinent Very large landmass that is smaller than a continent, such as the Indian subcontinent, **564**

subsidies Financial support from the government, **649**

subsistence agriculture Farming in which food is produced by a family just for its own needs, **125**

sultans Ottoman rulers, **455**

Sunni One of the two main branches of Islam, **440**

superpower A huge powerful country, **183**

Swahili (swah-HEE-lee) Language in East Africa with roots in the languages of the African coast with added Arabic words, **522**

taiga A forest of mainly evergreen trees that covers half of Russia, **384**

Taoism A religion that teaches there is a natural order to the universe, called the Tao, **624**

tariffs Taxes on imports and exports, **130**

tar sands Rock or sand layers that contain oil, **261**

temperature Measurement of heat, **41**

tepuís (tay-PWEEZ) A chain of high plateaus in the Guiana Highlands, **258**

terrorism The use of fear and violence as a political force, **271**

textiles Cloth products, **173**

theocracy A country governed by religious law, **443**

topography Elevation, layout, and shape of the land, **17**

tornadoes Violent twisting spirals of air, **48**

totalitarian government A government in which one person or a small group of people rule a country and in which the people have no say, **130**

Glossary

trade deficit A situation in which the value of a country's exports is less than the value of the country's imports, **183**

trade surplus A condition existing when a country exports more than it imports, **650**

traditionalism Following longtime practices and opposing many modern technologies and ideas, **98**

transhumance The practice of moving herds from mountain pastures in the summer to lowland pastures in the winter, **409**

tree line The line of elevation above which trees do not grow, **259**

trench A deep valley marking a collision of plates, where one plate slides under another, **66**

tributary Any smaller stream or river that flows into a larger stream or river, **71**

Tropic of Cancer The parallel 23 1/2° north of the equator, **31**

Tropic of Capricorn The parallel 23 1/2° south of the equator, **30**

tropics Warm, low-latitude areas near the equator, **29**

trust territories Areas placed under the control of another country while the territories prepare for independence, **737**

tsetse fly African fly carrying a human disease called sleeping sickness, **519**

tsunamis (sooh-NAH-mees) Large sea waves created by undersea tectonic activity like earthquakes, **638**

typhoons The term for hurricanes in the western Pacific Ocean, **48**

uninhabitable Unable to support human life and settlements, **319**

universalizing religions Religions that seek followers all over the world, **102**

urban agglomeration Densely populated region surrounding a central city, **651**

urbanization Growth in the proportion of people living in towns and cities, **120**

veld (VELT) Grassland regions of South Africa, **535**

voodoo Haitian version of traditional African religious beliefs that are blended with elements of Christianity, **251**

wadis Dry streambeds that fill with water only after rainfall, **484**

watershed The entire region drained by a river and its tributaries, **71**

water table The groundwater level at which all the cracks and spaces in rock are filled with water, **72**

wats Buddhist temples that also serve as monasteries, **670**

weather The condition of the atmosphere at a given time and place, **41**

weathering The process by which rocks break and decay over time, **67**

wetlands Landscapes that are covered with water for at least part of the year, **72**

Wheat Belt The region of the U.S. interior west specializing in growing wheat, **178**

work ethic The belief that work itself has moral value, **649**

world cities Huge urban areas that are the most important centers of economic power and wealth, **122**

yurts Movable, round houses of wool felt mats in Central Asia, **410**

zinc An element important in metal processing and other industries, **406**

Zionism The movement that called for Jews to set up their own country or homeland in Palestine, **456**

abdicate/abdicar Abandonar un cargo de jefe de gobierno, **387**

Aborigines/aborígenes Primeros habitantes de Australia, **715**

abyssal plains/suelos abisales Áreas del fondo del océano en que las rocas se hunden gradualmente debido a que no existen fuentes de calor que las mantengan en su sitio; suelen ser las regiones más llanas del mundo, **65**

acculturation/aculturación Proceso en el cual un individuo o un grupo adopta algunas características de otra cultura, **96**

acid rain/lluvia ácida Lluvia contaminada que puede dañar a los árboles y matar a los peces de los lagos, **77**

African National Congress/Congreso Nacional Africano (**ANC**, por sus siglas en inglés; **CNA**, en español) Organización política que luchó por poner fin a la separación racial en Sudáfrica, **539**

Afrikaners/afrikaners Colonizadores holandeses, franceses y alemanes y sus descendientes establecidos en Sudáfrica, **538**

agribusiness/agroeconomía Granjas especializadas que operan con mayor eficacia y ganancias económicas, **126**

alliances/alianzas Acuerdos entre diferentes países para respaldarse y defenderse de sus enemigos, **328**

alluvial fan/abanico pluvial Depósito de lodo y grava en forma de abanico que con frecuencia se acumula en las faldas de las montañas, **69**

alluvial soils/suelos pluviales Depósitos de tierra arrastrada por la corriente de los ríos, **158**

animist religions/religiones animistas Religiones cuyos integrantes creen en la presencia de espíritus y en las fuerzas de la naturaleza, **102**

annex/anexar Unir formalmente un territorio a un país, **643**

Antarctic Circle/Círculo Polar Antártico Paralelo 66 1/2° al sur del ecuador, **30**

apartheid/apartheid Política oficial que obligaba a las personas de raza negra en Sudáfrica a vivir en áreas separadas y a usar instalaciones diferentes de las que usaban los de raza blanca, **539**

aquaculture/acuacultura Cultivo y recolección de peces y vida marina en estanques y otros depósitos de agua, **619**

aqueducts/acueductos Canales artificiales usados para transportar agua, **78**

aquifers/acuíferos Capas de roca en las que abunda el agua, **78**

arable/cultivable Apropiado para el cultivo, **175**

arboreal/arborícola Que vive en los árboles, **661**

archipelago/archipiélago Grupo grande de islas, **680**

Arctic Circle/Círculo Polar Ártico Paralelo 66 1/2° al norte del ecuador, **31**

arid/árido Seco, **54**

armistice/armisticio Tregua, **644**

artesian wells/pozos artesianos Pozos en los que el agua fluye de manera natural a la superficie, **712**

ASEAN/ASEAN (por sus siglas en inglés) Asociación de Naciones del Sureste Asiático, organización creada para promover el desarrollo económico y los acuerdos políticos de las naciones que la integran, **670**

atlas/atlas Colección de mapas reunidos en un solo libro, **11**

atmosphere/atmósfera Capa de gases que envuelve a un planeta como la Tierra, **35**

atoll/atolón Isla coralina en forma de anillo o anillo de varias islas menores unidas por un arrecife de coral submarino, **732**

autarky/autarquía Sistema en el que un país trata de producir todos los bienes que necesita, **388**

autonomy/autonomía Autogobierno, **351**

ayatollahs/ayatolas Líderes religiosos de la máxima autoridad entre los shiíes musulmanes, **443**

balance of power/equilibrio de poder Condición que surge cuando varios países o alianzas mantienen niveles tan similares de poder evitar guerras, **328**

barrier islands/islas de barrera Islas costeras formadas por depósitos de arena arrastrada por las mareas y las corrientes de aguas poco profundas, **150**

basins/cuencas Zonas de tierras bajas, con frecuencia rodeadas por montañas, **151**

bauxite/bauxita Mineral con el que se elabora el aluminio, **241**

Bedouins/beduinos Pastores nómadas del suroeste de Asia, **441**

Berbers/berberes Grupo cultural que habita en el norte de África desde mucho antes que las oleadas de ejércitos árabes llegaran al continente, **489**

bilingual/bilingüe Persona que habla dos idiomas, **169**

biodiversity/biodiversidad Grado de variedad de plantas y animales, **534**

biosphere/biosfera Parte de la Tierra donde habitan todas las formas de vida que hay en ella, **35**

birthrate/tasa de natalidad Número de nacimientos anuales por cada 1,000 habitantes de un lugar, **89**

Bolsheviks/bolcheviques Grupo comunista que derrocó al gobierno durante la Revolución Rusa de 1917, **387**

boycott/boicot Negativa a comprar bienes provenientes de un país o de una empresa, **570**

buffer state/estado tapón País pequeño que se ubica entre dos países más grandes y poderosos, **265**

cacao/cacao Árbol del que se obtienen las semillas de cacao, **246**

calligraphy/caligrafía Escritura o rotulación artística, **624**

Spanish Glossary

cantons/cantones Grandes estados con autogobierno dentro de un país, como Suiza, **336**

capitalism/capitalismo Sistema económico en el que los negocios, las industrias y los recursos son de propiedad privada, **115**

caravans/caravanas Grupos de personas que viajan juntas para protegerse, **408**

Caricom/Caricom (por su abreviatura en inglés) Comunidad del Caribe y Mercado Común, **252**

cartography/cartografía Estudio y elaboración de mapas, **5**

cash crops/cultivo para la venta Productos cosechados para su venta en el mercado, **230**

caste/casta Grupo de personas nacidas con cierta posición social, como en el hinduismo, que restringe la ocupación y asociación de sus miembros, **571**

central business district/distrito comercial (**CBD**, por sus siglas en inglés) Zona de una ciudad dominada por grandes comercios, oficinas y edificios, **123**

Chibcha/Chibcha Antigua civilización de los Andes colombianos que desarrolló grandes habilidades orfebres, **263**

Chishima Current/corriente de Chishima (Oyashio) Corriente oceánica fría que disminuye la temperatura del norte de Japón en el verano, **639**

city-states/ciudades estado Ciudades autogobernadas y sus territorios aledaños, como en la antigua Grecia, **360**

climate/clima Condiciones meteorológicas de una región durante un periodo largo, **41**

climate graphs/gráficas del clima Gráficas que muestran la temperatura promedio y la precipitación pluvial de una región, **18**

colonies/colonias Territorios controlados por personas de otro país, **167**

command economy/economía dirigida Sistema económico en el que el gobierno decide lo que produce, dónde lo produce y el precio de venta de esos productos, **115**

commonwealth/mancomunidad Territorio autogobernado asociado con otro país, **250**

communes/comunas Granjas colectivas agrupadas para organizar las cosechas y la prestación de servicios, **623**

communism/comunismo Sistema económico y político en que el gobierno posee y controla casi todos los medios de producción, **115**

compass rose/rosa de los vientos Elemento de un mapa que señala con una flecha la posición de los cuatro puntos cardinales, **14**

complementary region/región complementaria Región formada por la combinación de dos zonas con diferentes actividades y niveles de poder para el beneficio de ambas, **342**

condensation/condensación Proceso por el cual el vapor pasa del estado gaseoso al estado líquido y forma pequeñas gotas, **46**

confederation/confederación Grupo de estados unidos con un propósito común, **336**

Confucianism/confucianismo Filosofía basada en las enseñanzas de Confucio, un antiguo filósofo chino, **624**

coniferous forests/bosques de coníferas Bosques poblados por árboles de semillas de cono y hojas siempre verdes, **55**

conquistadores/conquistadores Españoles que conquistaron territorios extranjeros durante la época colonial, **225**

consensus/consenso Acuerdo general, **197**

constitutional monarchy/monarquía constitucional Tipo de gobierno con un rey o una reina como jefe de estado y un parlamento legislador, **307**

contiguous/contiguo Que conecta o limita, **14**

Continental Divide/división continental Cresta que divide los principales ríos de Estados Unidos en dos partes, los que fluyen al este y los que fluyen al oeste, **153**

continental drift/desplazamiento continental Movimiento lento de las placas de la Tierra sobre el manto superior, **63**

continental shelves/arrecifes continentales Áreas en que la superficie continental se extiende a las aguas poco profundas que rodean a los continentes, **66**

continents/continentes Grandes masas de tierra sobre la superficie terrestre, **10**

contour plowing/arado de contorno Arado de una colina realizado en forma horizontal y no en forma vertical, **75**

copra/copra Carne de coco deshidratada, **741**

core/núcleo Centro de la Tierra donde la temperatura y la presión son muy altas, **63**

cork/corcho Material extraído de la corteza de los árboles de corcho, **353**

Corn Belt/región maicera Región del medio oeste de Estados Unidos especializada en el cultivo de maíz, **175**

cosmopolitan/cosmopolita Que recibe influencia de muchas culturas extranjeras, **316**

Cossacks/Cosacos Pueblo de la frontera sur del imperio ruso que jugó un importante papel en la expansión del imperio, **387**

cottage industries/industrias caseras Industrias de pequeña escala establecidas en el hogar, **576**

coup/golpe de estado Cambio de gobierno que ocurre cuando otro grupo toma el poder por la fuerza, **266**

creole/criollo Idioma que combina varios idiomas de África, Europa y el Caribe, **251**

crop rotation/rotación de cultivos Práctica que consiste en sembrar diferentes cultivos en años alternados, **76**

cultural boundaries/fronteras culturales Fronteras basadas en las tendencias culturales de diferentes grupos, **129**

culture/cultura Todas las características de vida de un grupo de personas, **94**

culture region/región cultural Área cuyos habitantes comparten varias tendencias culturales, **95**

culture traits/tendencias culturales Actividades y conductas aprendidas en las que participan las personas, **94**

cyclones/ciclones Vientos que se desplazan alrededor de centros de baja presión atmosférica, **42**

czar/zar Título del emperador de Rusia antes de la Revolución Rusa, **387**

Dairy Belt/región lechera Región del medio oeste de Estados Unidos especializada en la elaboración de productos lácteos, **175**

Dalits/dalits Habitantes de clase social baja en la India que no pertenecen a ninguna casta, **571**

death rate/tasa de mortalidad Número de muertes anuales por cada 1,000 habitantes de un lugar, **89**

deciduous forests/bosques de hoja caediza Bosques poblados por árboles que pierden sus hojas en cierto periodo del año, **55**

deforestation/deforestación Destrucción o pérdida de los bosques, **76**

degrees/grados Unidades usadas para medir la distancia entre meridianos y entre paralelos, **10**

delta/delta Acumulación de sedimentos en la desembocadura de un río, **69**

demilitarized zone/zona desmilitarizada (**DMZ**, por sus siglas en inglés) Zona de protección en la que los ejércitos enemigos no pueden entrar, **644**

democracy/democracia Sistema de gobierno en el que los ciudadanos deciden quién los gobernará, **130**

demography/demografía Estudio estadístico de los asentamientos humanos, **87**

depressions/depresiones geográficas Grandes áreas de terreno localizadas a muy baja altitud, **484**

desalinization/desalinización Proceso de remoción de la sal del agua del océano, **70**

desertification/desertificación Propagación de condiciones desérticas, **501**

developed countries/países desarrollados Países con altos niveles de industrialización y elevados estándares de vida, **117**

developing countries/países en desarrollo Países con economías productivas menores que las de los países desarrollados y bajos estándares de vida, **117**

dharma/dharma Según las prácticas hinduistas, importancia de cumplir con los deberes personales de acuerdo con la etapa en que vive cada quien, **571**

dialect/dialecto Variación regional de un idioma, **100**

dictator/dictador Líder que gobierna con autoridad absoluta, **226**

Diet/Dieta Órgano legislador de Japón que opera por elección, **644**

diffusion/difusión Proceso que ocurre cuando una idea innovadora se extiende y es adoptada por otras personas o grupos, **97**

dikes/dique Enorme muro usado para almacenar agua, **293**

doldrums/depresiones barométricas Áreas en las que no existen vientos predominantes, **44**

domestication/domesticación Proceso mediante el cual las personas cultivan la tierra y doman animales para su beneficio, **120**

domino theory/teoría del dominó Idea que explica que si un país cae víctima del comunismo, los países que lo rodean también caerán como si fueran fichas de dominó, **664**

double cropping/doble cosecha Cosecha de dos cultivos diferentes en la misma tierra en un solo año, **619**

drainage basin/cuenca de drenado Región drenada por un río y sus tributarios, **71**

dryland farming/cultivo de sequía Cultivos que pueden desarrollarse con lluvia limitada en vez de riego, **412**

dual economies/economías duales Economías en las que ciertos bienes son producidos exclusivamente para su venta y otros para el consumo local, **509**

dynasty/dinastía Línea de gobernantes por herencia, **438**

Spanish Glossary

economy of scale/economía de escala Producción masiva que reduce el costo de producción de cada artículo, **727**

ecosystems/ecosistemas Comunidades de plantas y animales, **50**

ecotourism/ecoturismo Tipo de turismo enfocado en visitas guiadas a regiones de interés natural, **248**

edge cities/ciudades satélite Grupos de grandes edificaciones lejanas de los distritos comerciales de una ciudad, **124**

El Niño/El Niño Patrón climatológico y de corrientes marinas que origina un aumento en la temperatura de las corrientes del sureste del océano Pacífico y afecta las condiciones climatológicas de varios países, **260**

emigrants/emigrantes Personas que abandonan su país para irse a vivir a otro lugar, **89**

enclaves/enclaves Áreas rodeadas completamente por otra región, **362**

endemic species/especies endémicas Plantas y animales nativos de cierto lugar, **682**

environment/medio ambiente Combinación de las cuatro esferas de la Tierra y que incluyen todas las condiciones biológicas, químicas y físicas necesarias para la vida, **35**

equator/ecuador Línea imaginaria que rodea al globo terrestre justo a la mitad de la distancia entre los polos Norte y Sur, **10**

equinox/equinoccio Momento en que los polos de la Tierra forman un ángulo de 90 grados en relación con el Sol, lo cual ocasiona que los rayos solares caigan directamente sobre el ecuador, **31**

erg/erg Mar de arena creado por el desplazamiento de altas dunas de arena, **484**

erosion/erosión Desplazamiento de fragmentos de una superficie ocasionado por la acción del agua, el viento o el hielo, **67**

escarpment/acantilado Cara sumamente empinada de una placa o área elevada de la superficie terrestre, **534**

estuaries/estuarios Extensiones semicerradas de agua, agua salada de mar y agua dulce, que se forman donde un río desemboca en una caleta del mar, **71**

ethnic groups/grupos étnicos Poblaciones humanas que comparten una cultura o herencia ancestral, **95**

ethnic religions/religiones étnicas Religiones creadas por personas de un grupo étnico que por lo general no se extienden a otras culturas, **102**

Eurasia/Eurasia Masa continental formada por Europa y Asia juntas, **382**

European Union/Unión Europea Organización de países europeos que muestra gran, cooperación en asuntos comerciales, económicos, políticos y sociales, **315**

evaporation/evaporación Proceso mediante el cual un líquido pasa al estado gaseoso, **46**

Spanish Glossary

exclave/exclave Área separada del resto de un país por la presencia geográfica de otro país, **338**

Exclusive Economic Zone/zona de exclusividad económica (**EEZ**, por sus siglas en inglés) Zona de 200 millas náuticas (370 km) a partir de la costa que contiene los recursos naturales propiedad de ese país, **741**

exotic rivers/ríos exóticos Ríos que nacen en regiones húmedas y fluyen a regiones más secas, **434**

exotic species/especies exóticas Plantas y animales que los humanos llevan a una región donde antes no existían, **719**

export economy/economía de exportación Tipo de economía cuyos bienes se producen casi exclusivamente para su exportación y no para su consumo local, **650**

extensive agriculture/agricultura extensiva Procesos agrícolas que cuentan con grandes extensiones de tierra pero pocos recursos económicos y mano de obra, **718**

fall line/línea de caída Frontera natural entre dos regiones con diferente elevación marcada por rápidos y caídas de agua, **150**

famine/hambruna Condición provocada por una escasez extrema de alimentos, **307**

faults/fallas Lugares donde las masas rocosas se fragmentan y se separan unas de otras, **66**

favelas/favelas Grandes asentamientos humanos localizados en las afueras de ciudades brasileñas, **270**

fellahin/fellahin Nombre dado a los agricultores del norte de África, **492**

fjords/fiordos Estrechos y profundos brazos de mar que penetran en los acantilados, **292**

flyway/ruta de vuelo Camino que siguen las aves en su paso migratorio, **640**

folds/pliegues Lugares donde el desplazamiento de las placas comprime las rocas y forma dobleces en la superficie, **66**

formal region/región formal Región con una o más características comunes que la hacen diferente de las zonas que la rodean, **6**

fossil fuels/combustibles fósiles Recursos energéticos formados por los restos de plantas y animales a lo largo de la historia; entre ellos se incluyen carbón mineral, gas natural y petróleo, **79**

fossil water/agua fosilizada Agua subterránea no renovada por las lluvias, **78**

free enterprise/libre empresa Sistema en el que la competencia entre empresas determina el precio de los productos, **115**

free port/zona libre Zona en la que los productos importados están casi exentos de impuestos, **492**

front/frente Zona de contacto entre dos masas de aire con diferente temperatura o nivel de humedad, **44**

functional region/región funcional Región conformada por diferentes lugares que operan en conjunto, **6**

fundamentalism/fundamentalismo Corriente cuyos seguidores obedecen estrictamente ciertas enseñanzas o principios establecidos, **98**

genocide/genocidio Aniquilamiento intencional de un pueblo, **527**

gentrification/aburguesamiento Adquisición de hogares en ruinas, en las zonas viejas de una comunidad, con la finalidad de restaurarlos, **181**

geography/geografía Estudio de las características físicas, biológicas y culturales de la superficie de la Tierra, **3**

geometric boundaries/fronteras geométricas Fronteras que siguen patrones geométricos, **129**

geothermal energy/energía geotérmica Calor proveniente del interior de la Tierra, **81**

geysers/géiseres Manantiales que lanzan chorros de agua caliente al aire, **320**

ghetto/ghetto Sector de una ciudad donde se ven obligados a vivir los grupos minoritarios, **339**

glaciers/glaciares Gruesas masas de hielo conformadas por una placa superficial y un gran bloque inferior que fluyen lentamente en los ríos congelados, **68**

globalization/globalización Proceso mediante el que las comunicaciones alrededor del mundo se han incrementado haciendo a las culturas más parecidas, **98**

global warming/calentamiento global Aumento de la temperatura de la Tierra en cierto espacio de tiempo, **41**

Gran Chaco/Gran Chaco Gran planicie localizada entre los Andes y las mesetas de Brasil en América del Sur, **258**

graphite/grafito Tipo de carbón usado para fabricar la punta de los lápices, **595**

great-circle route/ruta del gran círculo Ruta más corta entre dos lugares cualesquiera de la tierra, **12**

greenhouse effect/efecto invernadero Proceso mediante el cual la atmósfera de la Tierra atrapa energía calorífica, **41**

grid/enrejado Líneas verticales y horizontales que se cruzan a intervalos regulares para formar un patrón cuadriculado, **9**

gross domestic product/producto nacional bruto (**GDP**, por sus siglas en inglés; **PNB**, en español) Valor total de los bienes y servicios creados en un país, **116**

gross national product/producto interno bruto (**GNP**, por sus siglas en inglés; **PIB**, en español) Valor total de los bienes y servicios producidos por un país en un año, **116**

groundwater/aguas subterráneas Mantos de agua que se localizan debajo de la tierra, **72**

gulag/gulag Red de campos de trabajos forzados de las prisiones de la antigua Unión Soviética, **388**

Gulf Stream/corriente del Golfo Corriente oceánica tibia que transporta las aguas tropicales hacia el norte a lo largo de la costa este de América del Norte, **155**

gum arabic/goma arábiga Sustancia espesa extraída de los árboles de acacia, usada para combinar los ingredientes de diversos dulces y medicamentos, **525**

haciendas/haciendas Grandes fincas en la colonia española pertenecientes a familias adineradas en cuyas tierras trabajaban los campesinos de la región, **225**

hajj/hajj Peregrinación religiosa a la Meca que realizan de manera obligatoria todos los musulmanes, **102**

headwaters/cabeceras Corrientes de poco caudal que nacen en las montañas y que conforme descienden se convierten en ríos, **71**

heavy industry/industria pesada Industria enfocada en la fabricación de productos, especialmente aquellos hechos con metales, **393**

hemispheres/hemisferios Cada una de las mitades en que se divide la Tierra a partir del ecuador y el primer meridiano, **10**

hieroglyphs/jeroglíficos Dibujos y símbolos usados en el sistema de escritura del antiguo Egipto, **487**

hinterland/interior Región muy alejada de los grandes centros de población, **192**

Holocaust/holocausto Asesinato de millones de judíos durante la Segunda Guerra Mundial, **456**

homogeneous/homogéneo Que posee elementos del mismo tipo, como una población conformada por habitantes de un solo grupo étnico, **687**

hot spot/zona caliente Lugar en el que el magma sale del interior de la Tierra a la superficie, **152**

humidity/humedad Cantidad de vapor de agua que hay en el aire, **46**

humus/humus Restos orgánicos de plantas y animales acumulados en el suelo, **75**

hurricanes/huracanes Los ciclones tropicales de mayor poder destructivo, **48**

hydroelectric power/energía hidroeléctrica Electricidad producida con la fuerza del agua, **81**

hydrologic cycle/ciclo hidrológico Movimiento del agua en la hidrosfera, **71**

hydrosphere/hidrosfera El agua que existe en la Tierra, en todas sus formas, **35**

icebreakers/rompehielos Barcos equipados con un casco metálico que les permite abrirse camino en el hielo, **384**

imams/imanes Líderes religiosos musulmanes, **440**

immigrants/inmigrantes Personas que llegan a vivir a un nuevo país, **90**

Inca/inca Fundadores de la civilización antigua más importante de América del Sur, **263**

indigenous/indígena Persona nacida en una región específica, **244**

industrialization/industrialización Proceso por el cual la fabricación de productos basada en el uso de máquinas se extiende en toda una región, **116**

informal sector/sector informal Sector de la economía conformado por personas que no tienen empleos formales, incluidos quienes generan empleos por su cuenta y los pequeños comercios familiares, **543**

infrastructure/infraestructura Sistema de carreteras, puertos y otras instalaciones necesarias en una economía moderna, **117**

innovation/innovación Idea nueva aceptada por toda una cultura, **96**

intensive agriculture/agricultura intensiva Forma de agricultura que usa una gran fuerza laboral y capital en pequeñas extensiones de terreno, **619**

Intertropical Convergence Zone/zona de convergencia intertropical (**ITCZ**, por sus siglas en inglés) Área cercana al ecuador en la que se encuentran los vientos provenientes del noreste y del sureste, generando condiciones de humedad y a menudo de lluvia, **734**

irrigation/riego Proceso mediante el cual el agua se hace llegar a los cultivos de manera artificial, **76**

Spanish Glossary

isthmus/istmo Franja estrecha de tierra rodeada de agua por ambos lados que une dos regiones mayores, **222**

ivory/marfil Material de color blanco del que están hechos los colmillos de los elefantes, **522**

Jainism/jainismo Religión originada en India que enfatiza un código moral estricto basado en la preservación de la vida, **572**

Japan Current/Corriente de Japón (Kuroshio) Corriente de aguas tibias que transporta la mayor parte de la humedad de origen marino que llega a Japón, **639**

jute/yute Planta fibrosa que sirve para elaborar diversos productos como redes y cuerdas, **576**

kampongs/kampongs Aldeas tradicionales de algunas partes del sudeste de Asia construidas generalmente sobre pilotes, **692**

karma/karma Para los hinduistas, budistas y jainistas, es la fuerza positiva o negativa que surge como consecuencia de las acciones de las personas, **571**

klongs/klongs Canales de Tailandia, **672**

Kurds/kurdos Grupo minoritario que habita en Turquía y en países vecinos, **457**

lahars/lahars Deslizamientos peligrosos de lodo en los volcanes, **681**

landlocked/lugar bloqueado Sitio que no tiene litorales, **271**

landscapes/paisajes Panorama de un lugar que abarca características físicas, humanas y culturales, **4**

La Niña/La Niña Patrón climatológico y oceánico que hizo disminuir la temperatura de las aguas del este del océano Pacífico, **260**

latifundia/latifundio En América del Sur, gran extensión de terreno, **265**

latitude/latitud Líneas de un mapamundi dibujadas en dirección este a oeste y que se miden en grados, **9**

leaching/lixiviado Desplazamiento de minerales y humus hacia la parte inferior del suelo, **75**

legend/leyenda Clave de un mapa, **14**

lichens/líquenes Plantas pequeñas formadas de algas y hongos, **157**

light industry/industria ligera Industria que usa materiales ligeros en la fabricación de diversos bienes de consumo, **393**

Spanish Glossary

lingua franca/lengua franca Lenguaje de intercambio y comunicación comercial que usan personas que hablan distintos idiomas, **101**

literacy rate/tasa de analfabetismo Porcentaje de la población que no sabe leer ni escribir, **117**

lithosphere/litosfera Parte sólida de la corteza terrestre, **35**

Llanos/llanos Grandes planicies del noreste de Colombia y el oeste de Venezuela, **258**

loess/limo Suelo de grano fino que arrastra el viento y es muy fértil, **299**

longitude/longitud Líneas de un mapamundi dibujadas en dirección de norte a sur y que se miden en grados, **9**

magma/magma Rocas en estado líquido del interior de la Tierra, **63**

magnesium/magnesio Metal ligero muy valioso en ciertas industrias, como la aeroespacial, **451**

mandates/mandatos Territorios de un país que permanecen temporalmente en control de otros países, **455**

mangrove/mangle Tipo de árbol con raíces que crecen en agua salada; suelen encontrarse en las costas de las regiones tropicales de todo el mundo, **241**

manioc/mandioca Planta tropical de raíces almidonadas, **267**

mantle/manto Sección del interior de la Tierra localizada entre el centro y la mayor parte de la masa terrestre, **63**

Maori/maoríes Primeros pobladores de Nueva Zelanda, **724**

map projections/proyecciones de mapa Diferentes maneras de representar la curvatura de la Tierra con dibujos planos, **11**

maquiladoras/maquiladoras Fábricas mexicanas cuyos propietarios son principalmente empresas estadounidenses, **232**

market economy/economía de mercado Sistema en el que las personas eligen libremente dónde comprar y dónde vender sus bienes, **115**

market-oriented agriculture/agricultura orientada al mercado Sistema agrícola en el que los agricultores siembran un producto para venderlo a los consumidores, **126**

marsupials/marsupiales Mamíferos que tienen un pequeño saco en su vientre para transportar a sus crías, **713**

martial law/ley marcial Ley militar, **631**

matrilineal/matriarcado Sistema social en el que la transferencia del reino se realiza sólo por medio de la madre, **739**

Megalopolis/megalópolis Grupo de ciudades del noreste de Estados Unidos que se ha desarrollado abarcando una enorme área común; este término también describe grupos similares en otras partes del mundo, **173**

Meiji Restoration/restauración Meiji Revolución política japonesa que derrocó al shogún y restauró en el poder al emperador, **643**

Mercosur/Mercosur Organización de América del Sur creada con el propósito de extender el comercio, mejorar las vías de transporte y reducir los aranceles entre los países que lo integran, **270**

meridians/meridianos Líneas de longitud del globo terrestre, **10**

mestizos/mestizos Raza creada con la mezcla de europeos y nativos de América, **226**

meteorology/meteorología Estudio del clima, **5**

metropolitan area/área metropolitana Ciudad muy grande y zonas urbanas que la rodean, **174**

microstates/miniestados Países muy pequeños como Andorra, Mónaco y San Marino, **357**

migration/migración Proceso en el que un grupo de personas abandonan su lugar de origen para irse a vivir a otro, **89**

millet/mijo Tipo de grano resistente a la sequía, **507**

minifundia/minifundios Pequeñas granjas de América del Sur que surgieron con la separación de las colonias, **268**

missionaries/misioneros Personas que se dedican a difundir su religión, **102**

monoculture/monocultivo Cultivo de un solo producto, **409**

monotheism/monoteísmo Creencia que considera la existencia de un solo dios, **102**

monsoon/monzón Sistema en el que los vientos cambian de dirección y crean temporadas de clima muy húmedo y muy seco, **51**

moons/lunas Pequeños cuerpos que giran alrededor de un planeta, **26**

mosques/mezquitas Centros islámicos de adoración, **102**

mulattoes/mulatos Personas con ancestros europeos y africanos, **245**

multilingual/políglota Persona que habla varios idiomas, **336**

nationalism/nacionalismo Sentimiento de orgullo y lealtad por el país o grupo cultural propio, **130**

nationalized/nacionalizar Proceso por el cual organizaciones o comercios dejan de ser operados por particulares y pasan a manos del gobierno, **308**

natural boundaries/fronteras naturales Límites creados por las características de un paisaje, **129**

natural hazards/peligros naturales Sucesos de un medio ambiente natural que pueden destruir la vida y las propiedades de los humanos, **155**

navigable/navegable Que puede usarse como vía de transporte acuático, **294**

neutral/neutral Que no toma partido en ningún conflicto, **336**

newsprint/papel periódico Papel de bajo precio que se usa para imprimir los diarios, **158**

nomads/nómadas Personas que frecuentemente cambian de lugar de residencia, **409**

North American Free Trade Agreement/Tratado de Libre Comercio de América del Norte (**NAFTA**, por sus siglas en inglés; **TLC**, en español) Acuerdo entre Estados Unidos, México y Canadá que elimina los aranceles de muchos productos que circulan entre estos tres países, **183**

North Atlantic Drift/corriente del Atlántico Norte Corriente de aguas tibias que regula el clima de la parte noroeste de Europa, **297**

oasis/oasis Lugar donde un manantial brota a la superficie en un área desértica, **436**

OPEC/OPEC (por sus siglas en inglés; **OPEP** en español) Organización de Países Exportadores de Petróleo, creada con la finalidad de regular el precio del petróleo mediante el control de suministros, **443**

ore/mena Roca con minerales, **79**

orographic effect/efecto orográfico Efecto de enfriamiento que ocurre cuando una corriente de aire sube por la cuesta de una montaña produciendo un clima húmedo en un lado de la montaña y seco en el otro, **47**

outback/llanura desértica Región árida del centro de Australia, **713**

paddy/arrozal Tierra húmeda en la que se cultiva el arroz, **619**

Pampas/las pampas Zona de pastizales muy fértiles localizada al sudeste de América del Sur, **258**

pantheon/panteón Todos los dioses de una religión, **568**

parallels/paralelas Líneas de latitud, **10**

parliament/Parlamento Legislatura encabezada por un Primer Ministro, **196**

partition/división División de una región, como la de India, en 1947, **570**

pastoralism/pastoral Tipo de agricultura de subsistencia complementada por el pastoreo, **125**

perception/percepción Visualización y comprensión del ambiente que nos rodea, **6**

perceptual regions/regiones perceptuales Regiones que reflejan el comportamiento y sentimiento humanos, **6**

permafrost/permafrost Suelo que permanece congelado bajo la superficie, **56**

perspective/perspectiva Punto de vista de una persona sobre cualquier tema, **4**

petrochemicals/petroquímicos Productos hechos con petróleo como materia prima, **80**

pharaohs/faraones Monarcas del antiguo Egipto, **487**

phosphates/fosfatos Sustancias químicas que sirven para elaborar fertilizantes, **743**

pictograms/pictogramas Imágenes simples que representan ideas, **624**

pidgin languages/lenguaje franco Versión simplificada del lenguaje creada para facilitar la comunicación entre personas que hablan diferentes idiomas, **737**

piedmont/tierras bajas Área localizada al pie de una región montañosa, **150**

planets/planetas Cuerpos que giran alrededor de una estrella, **25**

plantations/plantaciones Grandes extensiones en las que se produce un cultivo principal para su exportación, **168**

plateau/meseta Región plana que tiene una elevación pronunciada al menos en uno de sus lados, **69**

plate tectonics/tectónica de placas Teoría que explica que la corteza de la Tierra está dividida en placas rígidas que se desplazan lentamente sobre el manto superior, **63**

plaza/plaza En ciudades de México y otros países de habla hispana, espacio abierto, con frecuencia localizado frente a una iglesia, **225**

polar regions/regiones polares Zonas frías de mayor latitud localizadas en las cercanías de los polos Norte y Sur, **29**

polders/pólderes Tierras ganadas al mar en los Países Bajos, **293**

polytheism/politeísmo Creencia que considera la existencia de varios dioses, **102**

population density/densidad de población Promedio de habitantes de una región; por lo general se expresa como un número de habitantes por milla o kilómetro cuadrado, **88**

population pyramids/pirámide de población Gráfica que muestra el porcentaje de hombres y mujeres por edad en la población de un país, **18**

potash/potasio Mineral que se usa para procesar la lana y en la fabricación de fertilizantes, vidrio y jabones suaves, **451**

precipitation/precipitación Gotas condensadas de agua que caen en forma de lluvia, nieve, aguanieve o granizo, **16**

prevailing winds/vientos dominantes Vientos que soplan en la misma dirección la mayor parte del tiempo, **44**

primate city/ciudad primaria La ciudad más importante de un país en términos de población y economía, **312**

prime meridian/primer meridiano Línea imaginaria entre el Polo Norte y el Polo Sur que pasa por el meridiano de Greenwich, Inglaterra, **10**

protectorate/protectorado País que cede la responsabilidad de la creación de leyes a un país más poderoso a cambio de protección, **591**

provinces/provincias Distritos de gobierno similares a los estados, **190**

pull factors/factores de atracción Características propicias que hacen que un grupo de personas emigre a otra región, **90**

puppet government/gobierno marioneta Gobierno controlado por fuerzas externas, **622**

push factors/factores de abandono Características que hacen que un grupo de personas abandone una región, **90**

quotas/cuotas En términos comerciales es el límite en la cantidad que puede importarse de un producto determinado, **130**

rain shadow/sombra de lluvia Área seca opuesta al extremo húmedo en cadena montañosa, **47**

reforestation/reforestación Plantación de nuevos árboles donde se han cortado otros, **76**

refugees/refugiados Personas obligadas a abandonar la región donde viven, **90**

reg/reg Planicie cubierta de grava en un área desértica, **484**

region/región Área con una o más características comunes que la hacen diferente de otras zonas cercanas, **6**

regionalism/regionalismo Sentimiento político de lealtad a la región donde se vive, **200**

reincarnation/reencarnación Según los hinduistas es la serie de nacimientos repetidos en los que el alma adopta una forma distinta cada vez, **571**

Spanish Glossary

Renaissance/Renacimiento Periodo entre los años 1300 y 1500, aproximadamente, marcado por un renovado interés de la humanidad en el aprendizaje, **356**

revolution/revolución Una vuelta a la órbita elíptica de la Tierra alrededor del Sol, **27**

rift valleys/valles de fisura Puntos de la superficie de la Tierra en los que la corteza se estira hasta romperse, **65**

rotation/rotación Giro completo de la Tierra sobre su eje, **27**

Rus/rus Comerciantes escandinavos que actuaron como los primeros líderes de Kiev; esta palabra es el origen del nombre de Rusia, **386**

Sahel/sahel Región de clima semiárido que se extiende de Senegal y Mauritania hasta Sudán en África, **501**

samurai/samurai Guerrero profesional japonés empleado por el shogún para su protección, **643**

sanctions/sanciones Castigos impuestos a un país para obligarlo a cambiar sus políticas, **539**

Sanskrit/sánscrito Antigua lengua que dio origen al Hindi, **568**

Santeria/santería Religión originada en Cuba que combina tradiciones africanas con creencias cristianas, **251**

satellite/satélite Cuerpo que gira alrededor de otro cuerpo de mayor tamaño, **26**

savannas/sabanas Áreas de pastizales tropicales, árboles dispersos y matorrales, **51**

secular/seglar Que no está ligado a la religión, **463**

sediment/sedimento Pequeñas partículas de rocas fragmentadas, **67**

separatism/separatismo Creencia de que ciertas partes de un país deben ser independientes, **200**

sepoys/cipayos Tropas indias comandadas por el ejército británico en India durante la época de la colonia, **569**

sequent occupance/ocupación sucesiva Establecimiento de grupos sucesivos en el que cada grupo crea un paisaje cultural que lo identifica, **306**

serfs/siervos Campesinos pobres en Rusia que trabajaban la tierra para los señores feudales sin tener otra alternativa, **387**

shatter belt/zona de conflictos Zona de cambios y conflictos frecuentes, por lo general localizada entre grandes potencias, **389**

shifting cultivation/cultivo de reemplazo Técnica agrícola en la que se cortan árboles o se podan para sembrar cultivos de temporada en una zona a la vez, **125**

Shi'ism/chiitas Una de las dos ramas principales del Islam, **440**

shogun/shogún Señor feudal poderoso en el antiguo Japón, **642**

Sikhism/sikismo Religión creada en India que combina la creencia musulmán en la existencia de un solo dios y el punto de vista hinduista sobre la reencarnación, **572**

Silicon Valley/Silicon Valley (Valle del silicón) Área localizada al sur de San Francisco, California, que se convirtió en el principal centro de tecnología de computadoras de Estados Unidos, **178**

silt/limo Suelo fértil cubierto de arena de grano fino, por lo general depositada por el flujo de agua corriente, **487**

sinkholes/hoyos de hundimiento Depresiones de paredes empinadas que se forman cuando la bóveda de una cueva se derrumba, **222**

sisal/sisal Fibra fuerte y duradera obtenida de una planta que se usa para fabricar cuerdas trenzadas y sogas, **522**

slash-and-burn agriculture/agricultura de corte y quema Forma del cultivo de reemplazo en el que se talan pequeñas áreas de bosque y se queman los troncos caídos antes de sembrar, **688**

Slavs/eslavos Los grupos más importantes de habitantes establecidos en las regiones actualmente conocidas como Rusia, Ucrania, Bielorrusia y otras regiones del centro y el sur de Europa, **386**

smelters/fundidoras Fábricas de procesamiento de minerales, **394**

smog/smog Contaminación del aire producida por la reacción química de la luz solar y los contaminantes emitidos por los automóviles y las fábricas, **179**

socialism/socialismo Sistema económico en el que el gobierno posee y controla los medios de producción, **321**

soil exhaustion/agotamiento del suelo Condición en la que el suelo pierde nutrientes y se convierte en tierra no apta para el cultivo, **76**

soil salinization/salinización del suelo Acumulación de sales en el suelo, **76**

solar energy/energía solar Energía que proviene del Sol y llega a la Tierra en forma de luz y calor, **27**

solar system/sistema solar El Sol y el grupo de cuerpos que giran a su alrededor, **25**

solstice/solsticio Momento en que los polos de la Tierra forman el mayor ángulo posible en relación con el Sol, **30**

sorghum/sorgo Un tipo de grano resistente a la sequía, **507**

souks/souks Mercados árabes al aire libre, **461**

soviets/soviéticos Cuerpos locales de gobierno de la antigua Unión Soviética, **388**

Special Economic Zones/zonas de economía especial (SEZs, por sus siglas en inglés**)** Zonas diseñadas para atraer la inversión de compañías extranjeras en ciertas áreas de China, **628**

staple/cultivo básico Cultivo principal de una región, **507**

storm surge/marejada de tormenta Aguas agitadas en la costa debido a la acción de una gran tormenta, **587**

subcontinent/subcontinente Grandes masas de tierra menores que un continente, como el subcontinente de la India, **564**

subsidies/subsidios Apoyo financiero del gobierno, **649**

subsistence agriculture/agricultura de subsistencia Tipo de agricultura en el que una familia cultiva un producto sólo para su propio consumo, **125**

sultans/sultanes Gobernantes otomanos, **455**

Sunni/sunitas Una de las dos ramas más importantes del Islam, **440**

superpower/superpotencia País muy poderoso, **183**

Swahili/suahili Lengua utilizada en África Occidental que tiene raíces en las lenguas de la costa africana y combina algunas palabras de origen árabe, **522**

taiga/taiga Bosque poblado por árboles siempre verdes que abarca la mitad de Rusia, **384**

Taoism/taoísmo Religión que explica que existe un orden natural llamado Tao en el universo, **624**

tariffs/aranceles Impuestos sobre importaciones y exportaciones, **130**

tar sands/arena alquitranada Capas de arena y roca que contienen aceite, **261**

temperature/temperatura Sistema de medición de calor, **41**

tepuís/tepuís Cadena de mesetas en las altiplanicies de las Guayanas, **258**

terrorism/terrorismo Uso del temor y la violencia como fuerza política, **271**

textiles/textiles Productos elaborados con telas, **173**

theocracy/teocracia País gobernado por una ley religiosa, **443**

topography/topografía Elevación, características y forma del terreno, **17**

tornadoes/tornados Espirales de aire que giran violentamente, **48**

totalitarian government/gobierno totalitario Gobierno en el que una persona o pequeño grupo de personas gobiernan un país sin dar a los habitantes voz ni voto, **130**

trade deficit/déficit comercial Situación en la que el valor de las exportaciones de un país es menor que el valor de sus importaciones, **183**

trade surplus/excedente comercial Condición que resulta cuando un país exporta más productos de los que importa, **650**

traditionalism/tradicionalismo Seguimiento de prácticas antiguas y opuestas a ideas y tecnologías modernas, **98**

transhumance/trashumar Práctica que consiste en llevar los rebaños de los pastizales de las montañas donde se alimentaron en el verano a tierras más bajas durante el invierno, **409**

tree line/límite de los árboles Altura a partir de la cual ya no crecen los árboles, **259**

trench/trinchera Valle profundo formado por un choque de placas en el que una de ellas se desliza debajo de la otra, **66**

tributary/tributario Corriente menor que alimenta un río más caudaloso, **71**

Tropic of Cancer/Trópico de Cáncer Paralelo 23 ¹/₂° al norte del ecuador, **31**

Tropic of Capricorn/Trópico de Capricornio Paralelo 23 ¹/₂° al sur del ecuador, **30**

tropics/trópico Área de clima cálido y baja latitud, cercanas al ecuador, **29**

trust territories/territorios bajo administración fiduciaria Áreas cuyo control fue cedido a otro país después de la Segunda Guerra Mundial como preparativo para su independencia, **737**

tsetse fly/mosca tse tse Mosca africana portadora de una enfermedad llamada mal del sueño, **519**

tsunamis/tsunamis Enormes olas creadas por movimientos tectónicos como los terremotos submarinos, **638**

typhoons/tifones Término usado al oeste del océano Pacífico para nombrar a los huracanes, **48**

uninhabitable/inhabitable Que no presenta condiciones adecuadas para la vida humana, **319**

universalizing religions/religiones universales Religiones que buscan seguidores en todo el mundo, **102**

urban agglomeration/aglomeración urbana Región densamente poblada en los alrededores de una ciudad central, **651**

urbanization/urbanización Crecimiento en la proporción de los habitantes de poblados y ciudades, **120**

veld/veld Regiones de pastizales localizadas en África del Sur, **535**

voodoo/vudú Versión haitiana de ciertas creencias tradicionales religiosas africanas combinadas con elementos cristianos, **251**

wadis/vados Cuencas vacías de ríos que sólo se llenan con la lluvia, **484**

watershed/cuenca Lecho que antes ocupaba un río y sus tributarios, **71**

water table/mesa de agua Nivel al que llegan las aguas subterráneas cuando llenan las grietas de las rocas, **72**

wats/wats Templos budistas que también sirven como monasterios, **670**

weather/tiempo Condiciones atmosféricas de un lugar y momento determinados, **41**

weathering/desgaste Proceso de descomposición de las rocas con el paso del tiempo, **67**

wetlands/terreno pantanoso Paisaje cubierto de agua al menos una parte del año, **72**

Wheat Belt/región triguera Región central de Estados Unidos especializada en el cultivo del trigo, **178**

work ethic/ética laboral Creencia de que el trabajo en sí tiene valor moral, **649**

world cities/urbes del mundo Grandes áreas urbanas en las que se encuentran los centros de poder económico más importantes, **122**

yurts/yurtas Casas móviles de lana con forma circular comunes en la parte central de Asia, **410**

zinc/zinc Elemento muy importante en el procesamiento de metales y otras industrias, **406**

Zionism/sionismo Movimiento que exhortaba a los judíos a formar su propia nación en Palestina, **456**

Index

Index

Index

Index

Index

Lake Sarez, 404
Lake Superior, 194
Lake Tanganyika, 518
Lake Turkana, 518
Lake Victoria, 518, 519, 527
Lake Volta, 546
landfills, 79, 80
landforms: 68–69; of Australia, 712–13; of Canada, 150; of Caribbean islands, 11; of Central America, 11; of China, 617, 619; of Mexico, 222; and natural environment, 4; of North Africa, 484; of Persian Gulf, 434; of Russia, 382; shape of, 69, *d69;* and soil variation, 75; of South Africa, 534; of South America, 258; and tectonic processes, 63–67, 712–13; types of, 62; of United States, 150
landlocked, (def.) 271
land reclamation, 316
landscape, (def.) 4
land use, *m145, m215, m285, m377, m429, m477, m559, m611, m707*
languages: of Africa, 522, *c540;* in Central Asia region, 410; in Eastern Mediterranean, 457; of Europe, 537; families of, 100, *m101;* and geography, 100–01; in Indian Perimeter, 592; Indo-European, 592; in Persian Gulf region, 439; principal world, *c100;* Sino-Tibetan, 592; in Southeast Asia, 665; special characteristics of, 100. *See also* cultures; specific languages, regions, and countries
La Niña, 260
L'Anse aux Meadows (Our Amazing Planet), 190
Laos, *t613,* 627, 658, 660–61, 664–66, 673, 675, *m606, m607*
Lapita, 740
latifundia, (def.) 265
Latin alphabet, 402, 410
latitude, (def.) 9–12, *d9,* 29, 41, 44–46, 56, 83, 129, 154, 383, 519, 713
Latvia, *t288,* 338, 341
Laurasia, 64
Laurentian Mountains, 194
lava, 53, 65
Laysan Island, 745
leaching, (def.) 75
lead, 406, 618, 714
Lebanon, *t431,* 440, 450, *m451,* 454, 455, 456, 457, 460–63
leeward, (def.) 47, 57
legends, (def.) 14, 16
Lenin, Vladimir, *p387,* 388, 398, *p409*
Leningrad. *See* St. Petersburg
Lesotho, *t480,* 532, 535
Lesser Antilles, 240
Liberia, *t480*
Libya, *t478, t480,* 482, 485, 486, 512
lichens, (def.) 157
Liechtenstein, *t288, m327*
life expectancy, 92, 117–18, *t146, t216, t286, t378,* 411, *t430, t478, t560, t612, t708*
light industry, (def.) 393
Lima, Peru, 270
limestone plateau, 713
Limpopo. *See* Crocodile River
lingua franca, (def.) 101
Lisbon, Portugal, 353
literacy rates, *t116,* (def.) 117, *t146, t216, t286, t378,* 411, *t430, t478, t560,* 593, *t612, t708*
literature, 438, 626, 648
lithosphere, (def.) 35, 37
Lithuania, *t288,* 338, *m339,* 341
Llanos, 258
loess, (def.) 299, 340, 619
logging, 76, 158, 384, 595, 662, *p662,* 675, 682
London, England, 32, 121, 122, 310, 322, 718
longitude, (def.) 9, *d9,* 32, 129, 154, 155
"Loo-Wit," 139
Los Angeles, California, 32, 67, 77, 97, *p124,* 181, 584, 718
Lost Pines, 83
Louisiana, 153, 160
low islands, 732
Luxembourg, 315, 317
Lyon, France, 312

Macao (China), 622
Macdonnell Ranges, *p714*
Macedonia, *p90, t288,* 361, 362, 363
Mackenzie Delta, 188
Mackenzie River, 153
Madagascar, *t480,* 532, 534, 536, *m536,* 542, *m542, p545*
Madeira, 353
Madras (India), 577
Magellan, Ferdinand, 685, 736
magma, (def.) 63, 152
magnesium, (def.) 451
Mahabharata, **"Hundred Questions,"** 553
Mahfouz, Naguib, 489
Mainz, Germany, *p329*
Majapahit, 684
malaria, 99, 546–47, *m546*
Malawi, *t480,* 532
Malay Archipelago, 680
Malay Peninsula, 660, 680, 681, 684
Malaysia, 118, *m607, t612, t613,* 674–75, *m679,* 680, 683, 686, 691–93
Maldives, *m554, t560, t561,* 586, 587, 589, 591, 593, 596
Male, Maldives, 596
Mali, *t480*
Malta, *t288*
Manchester, England, 323
Manchuria (China), 629
Manchurian Plain, 616
Mandarin (Chinese), 99, 624
mandates, (def.) 455
Mandela, Nelson, 539, *p539*
mangoes, 672
mangrove, (def.) 241
Manila, Philippines, 687
manioc, (def.) 267
Manitoba (Canada), 190, 192, *p192,* 196
mantle, (def.) 63
manufacturing, 116
Maori: 724–26; art of, *p724,* 725; wars, 725
Mao Zedong, 622, *p623*
mapmaking. *See* cartography
map projections, (def.) 11–12
maps: elements of, 13–15; insert, 2, 3, 6, 7, 17, 36, 41, 42, 61, 72, 73, 76, 78, 114, 127, 397, 451, 463, 485, 590, 592, 719; interpreting, 38; special-purpose, 15–16; study of, 5, 11; types of, 16–20; weather, 47
maquiladoras, (def.) 232
Mariana Islands, 737, 738, 742
Mariana Trench, *m733*
marine ecosystem, 57
Marine West Coast Climate, 55
market economy, (def.) 115
market-oriented agriculture, (def.) 126
Marquesas Islands, 738
Mars, 25, 35
Marseille, France, 312
Marshall Islands, *t708, t709, m731,* 732, 733
marshes, 72, 484
marsupials, (def.) 713
martial law, (def.) 631
Martinique, 240

Index

Index

Index

Index

Index

For permission to reprint copyrighted material, grateful acknowledgment is made to the following sources:

The Asia Society: "Thoughts of Hanoi" by Nguyen Thi Vinh from *A Thousand Years of Vietnamese Poetry,* translated by Nguyen Ngoc Bich. Copyright © 1974 by The Asia Society.

Luba Brezhneva c/o Tanner Propp, LLP: From *The World I Left Behind: Pieces of a Past* by Luba Brezhneva, translated by Geoffrey Polk. Copyright © 1995 by Luba Brezhneva.

Jed Mattes Inc, New York: From "Australian Outback" by Bill Bryson from *National Geographic Online Magazine,* accessed May 10, 2001, at http://www.nationalgeographic.com/traveler/outback2/html. Copyright © 1999 by Bill Bryson.

John R. Mullin: Quote by John R. Mullin from "Architects, Planners and Residents Wonder How to Fill the Hole in the City" by Kirk Johnson and Charles V. Bagli from *The New York Times,* September 26, 2001, p. B9. Copyright © 2001 by John R. Mullin.

National Council for Geographic Education: From "Preferred Locations in North America: Canadians, Clues, and Conjectures" by Herbert A. Whitney from *Journal of Geography,* vol # 83, no. 5, September-October 1984, p. 222. Copyright © 1984 by National Council for Geographic Education.

W.W. Norton & Company, Inc.: From *Nature's Metropolis: Chicago and the Great West* by William Cronon. Copyright © 1991 by William Cronon.

Harold Ober Associates Incorporated: From "On the Beach" from *A Secret Country: The Hidden Australia* by John Pilger. Copyright © 1989 by John Pilger.

Penguin Books Ltd.: From *The Vinland Sagas,* translated by Magnus Magnusson and Hermann Pálsson (Harmondsworth: Penguin Classics 1965). Copyright © 1965 by Magnus Magnusson and Hermann Pálsson. From *The Epic of Gilgamesh,* translated by N. K. Sandars (Penguin Classics 1960, Third Edition, 1972). Copyright © 1960, 1964, 1972 by N. K. Sandars.

The Estate of Richard Rive: From *African Songs* by Richard Rive. Copyright © 1963 by Richard Rive. Published by Seven Seas Publishers, Berlin.

Wendy Rose: From "Loo-Wit" from *Bone Dance: New and Selected Poems 1967–1992* by Wendy Rose (University of Arizona Press, 1994). Copyright © 1985 by Wendy Rose. First appeared in *The Halfbreed Chronicles & Other Poems* by Wendy Rose (West End Press 1985/1993).

Viking Penguin, a division of Penguin Putnam Inc.: From *The Incas: The Royal Commentaries of the Inca* by Garcilaso de la Vega, translated by Maria Jolas. Copyright © 1961 by The Orion Press, Inc.

Villard Books, a division of Random House, Inc.: From *Fifty Years of Europe* by Jan Morris. Copyright © 1997 by Jan Morris.

Wallace Literary Agency, Inc.: "Hundred Questions" from *The Mahabharata* by R. K. Narayan. Copyright © 1978 by R. K. Narayan. Published by Viking Penguin.

Sources Cited:

From "Vancouver: Open for Business" by Lindsay Elliott from *BC Business,* vol. 26, no. 3, March 1998, page 73. Published by Canada Wide Magazine Ltd.

From "Mali's Nomads Coaxed to Adopt Modern Life; After a five-year conflict, government aims to integrate wandering Tuaregs into society" by Jennifer Ludden from *The Christian Science Monitor,* March 6, 1996. Published by The Christian Science Publishing Society, Boston, MA.

From "Prince Edward Island" by Ian Darragh from *National Geographic,* May 1998. Published by The National Geographic Society, Washington, D.C.

From "Blue Nile" by Virginia Morell from *National Geographic,* vol.198, no. 6, December 2000. Published by The National Geographic Society, Washington, D.C.

From "India" by Geoffrey C. Ward from *National Geographic,* May 1997. Published by The National Geographic Society, Washington, D.C.

From "London" by Simon Worrall from *National Geographic,* June 2000. Published by The National Geographic Society, Washington, D.C.

From "A Spare and Separate Walk of Life; Nomads Survive the Sahara Amid Dwindling Resources" by Stephen Buckley from *The Washington Post,* December 8, 1996.

Abbreviations used: (t) top, (c) center, (b) bottom, (l) left, (r) right, (bkgd) background

ILLUSTRATIONS

All work, unless otherwise noted, contributed by Holt, Rinehart & Winston.

Skills Handbook: Page S13 (c), MapQuest.com, Inc.; (bc), Leslie Kell.

Atlas Maps: MapQuest.com, Inc.

Chapter One: Page 3 (cl), 4 (tl), 6 (cl), 7 (tl), (tr), 9 (bl), (bc), 11 (br), 12 (tc), (c), (bc), 13 (bl), 16 (tl), (tr), 17 (tl), (tr), MapQuest.com, Inc.; 18 (tc), Ortelius Design; 18 (cl), (cr), MapQuest.com, Inc.; 19 (t), (bl), Leslie Kell; 20 (cr), MapQuest.com, Inc.; 21 (b), Ortelius Design; 22 (cl), 23 (tl), MapQuest.com, Inc.

Chapter Two: Page 26 (tl), Don Dixon; 26 (br), Uhl Studios, Inc.; 29 (cr), 30 (b), Argosy; 32-33 (bc), 36 (tl), 37 (tr), 38 (cl), MapQuest.com, Inc.; 39 (tl), Leslie Kell.

Chapter Three: Page 41 (cr), 42 (bl), MapQuest.com, Inc.; 43 (bl), Uhl Studios, Inc.; 44 (tl), 47 (b), (tl), 50 (bl), MapQuest.com, Inc.; 51 (tl), John White/The Neis Group; 52-53, 54 (tl), 55 (tc), (bl), (bc), 58 (b), 60 (cl), 61 (tl), MapQuest.com, Inc.

Chapter Four: Page 63 (br), Uhl Studios, Inc.; 64 (t), (b), MapQuest.com, Inc.; 65 (t), 66 (tl), Mark Heine; 66 (br), 67 (cr), MapQuest.com, Inc.; 69 (tr), 70 (br), Mark Heine; 72 (bl), 73 (tl), MapQuest.com, Inc.; 75 (tr), Mark Heine; 76 (tc), 78 (cl), 79 (tr), 80 (b), 82 (b), MapQuest.com, Inc.; 83 (b), Leslie Kell; 84 (cl), Mark Heine; 85 (tl), Leslie Kell.

Chapter Five: Page 87 (bc), MapQuest.com, Inc.; 88 (tl), Leslie Kell; 88 (b), 89 (c), MapQuest.com, Inc.; 91 (br), 92 (tl), 93 (tr), Leslie Kell; 96 (tl), 98 (cl), MapQuest.com, Inc.; 100 (bl), Leslie Kell; 101 (t), 103 (t), 110 (cl), MapQuest.com, Inc.; 111 (cl), Leslie Kell.

Chapter Six: Page 113 (b), Leslie Kell; 114 (cl), 115 (cr), MapQuest.com, Inc.; 116 (bl), 117 (b), Leslie Kell; 118 (tl), (cl), 119 (b), MapQuest.com, Inc.; 121 (t), Mark Heine; 122 (bl), David Chapman; 123 (bc), MapQuest.com, Inc.; 124 (t), 125 (br), 127 (tr), David Chapman; 127 (c), 130 (br), 131 (tc), MapQuest.com, Inc.; 134 (cl), David Chapman; 135 (tl), Leslie Kell.

Unit Two: Page 140 (c), 141 (c), MapQuest.com, Inc.; 141 (t), Ortelius Design; 142 (c), 143 (c), 144 (c), 145 (c), MapQuest.com, Inc.; 146 (cl), Ortelius Design; 147 (tl), MapQuest.com, Inc.

Chapter Seven: Page 149 (b), 150 (cl), 153 (cr), 154 (cr), 156 (tc), (c), 159 (bl), 160 (tl), 161 (bl), 162 (cl), MapQuest.com, Inc.; 163 (tl), Leslie Kell.

Chapter Eight: Page 165 (b), MapQuest.com, Inc.; 165 (cr), Craig Attebery/Jeff Lavaty Artist Agent; 166 (bl), 168 (tl), 169 (br), 170 (bl), MapQuest.com, Inc.; 172 (tr), Mark Heine; 173 (bl), 175 (tl), MapQuest.com, Inc.; 175 (bl), Leslie Kell; 175 (tr), Nenad Jakesevic; 176 (tl), 178 (tl), 179 (tc), 180 (cl), 181 (br), 182 (tr), 185 (tr), 186 (cl), MapQuest.com, Inc.; 187 (tl), Leslie Kell.

Chapter Nine: Page 189 (b), MapQuest.com, Inc.; 191 (cr), (br), Leslie Kell; 193 (cl), 194 (bc), 196 (tl), 198 (b), 200 (cl), 201 (c), 202 (cl), MapQuest.com, Inc.; 203 (tl), Leslie Kell.

Unit Three: Page 210 (c), 211 (c), MapQuest.com, Inc.; 211 (t), Ortelius Design; 212 (c), 213 (c), 214 (c), 215 (c), MapQuest.com, Inc.; 216 (cl), Ortelius Design; 217 (tl), (cl), (bl), 218 (tl), (cl), (tl), (cl), MapQuest.com, Inc.

Chapter Ten: Page 221 (b), 222 (tc), 223 (tr), 224 (bc), 227 (tr), 229 (bl), MapQuest.com, Inc.; 230 (tl), Leslie Kell; 230 (cr), 231 (bc), 232 (b), (tr), 233 (tr), 236 (cl), MapQuest.com, Inc.; 237 (tl), Leslie Kell.

Chapter Eleven: Page 239 (b), 240 (cl), MapQuest.com, Inc.; 241 (tr), Leslie Kell; 243 (bl), (tr), 244 (cl), MapQuest.com, Inc.; 245 (br), Leslie Kell; 249 (cr), 252 (cl), 254 (cl), 255 (tl), MapQuest.com, Inc.

Chapter Twelve: Page 257 (b), 258 (tr), MapQuest.com, Inc.; 259 (b), Mark Heine; 260 (tc), 261 (cl), MapQuest.com, Inc.; 262 (b), Mark Heine; 263 (bc), 264 (tr), 267 (tr), 270 (tl), 272 (bl), 273 (t), 274 (cl), MapQuest.com, Inc.; 275 (tl), Leslie Kell.

Unit Four: Page 280 (c), 281 (c), MapQuest.com, Inc.; 281 (t), Ortelius Design; 282 (c), 283 (c), 284 (c), 285 (c), MapQuest.com, Inc.; 286 (cl), Ortelius Design; 287 (tl), (c), (cl), (bl), 288 (tl), (cl), (bl), 289 (tl), (cl), MapQuest.com, Inc.

Chapter Thirteen: Page 291 (b), 292 (c), MapQuest.com, Inc.; 293 (b), Ortelius Design; 296 (c), 297 (tr), 298 (c), 299 (cr), 300 (tl), 301 (tr), 302 (cl), MapQuest.com, Inc.; 303 (tl), Leslie Kell.

Chapter Fourteen: Page 305 (b), 306 (b), (tl), 307 (tr), 310 (bl), 312 (tl), MapQuest.com, Inc.; 313 (cr), (br), Leslie Kell; 316 (tl), 318 (bl), 319 (bl), 322 (cl), 323 (t), 324 (cl), 325 (tl), MapQuest.com, Inc.

Chapter Fifteen: Page 327 (b), 328 (tl), MapQuest.com, Inc.; 330 (bl), Leslie Kell; 330 (cl), 333 (tr), MapQuest.com, Inc.; 334 (tr), Leslie Kell; 335 (cr), 336 (bl), 339 (tr), MapQuest.com, Inc.; 340 (bl), Leslie Kell; 341 (tc), 342 (bc), 343 (t), (cr), 344 (bl), (tr), 346 (cl), MapQuest.com, Inc.; 347 (tl), Leslie Kell.

Chapter Sixteen: Page 349 (b), 350 (bc), 351 (tc), (c), MapQuest.com, Inc.; 352 (tl), Leslie Kell; 353 (tr), Mark Heine; 354 (cr), 355 (bc), 356 (bl), 358 (cl), MapQuest.com, Inc.; 359 (tr), Leslie Kell; 360 (bc), 361 (tr), 362 (b), 363 (tc), 365 (cr), 366 (cl), MapQuest.com, Inc.; 367 (c), Leslie Kell.

Unit Five: Page 372 (c), 373 (c), MapQuest.com, Inc.; 373 (t), Ortelius Design; 374 (c), 375 (c), 376 (c), 377 (c), MapQuest.com, Inc.; 378 (cl), Ortelius Design; 379 (tl), (cl), MapQuest.com, Inc.

Chapter Seventeen: Page 381 (b), 385 (cl), 388 (cl), MapQuest.com, Inc.; 389 (cr), Leslie Kell; 389 (tl), MapQuest.com, Inc.; 390 (bl), Leslie Kell; 394 (tc), 395 (bl), 396 (cl), 397 (tc), 400 (cl), MapQuest.com, Inc.; 401 (tl), Leslie Kell.

Chapter Eighteen: Page 403 (b), 407 (br), MapQuest.com, Inc.; 407 (bl), Leslie Kell; 408 (cl), MapQuest.com, Inc.; 410 (bl), Leslie Kell; 411 (cr), 413 (bc), 416 (bl), 418 (cl), 419 (tl), MapQuest.com, Inc.

Unit Six: Page 424 (c), 425 (c), MapQuest.com, Inc.; 425 (t), Ortelius Design; 426 (c), 427 (c), 428 (c), 429 (c), MapQuest.com, Inc.; 430 (cl), Ortelius Design; 431 (tl), (cl), (bl), MapQuest.com, Inc.

Chapter Nineteen: Page 433 (b), 435 (cr), 436 (c), 437 (c), 440 (tl), 442 (cl), MapQuest.com, Inc.; 443 (b), Leslie Kell; 446 (cl), MapQuest.com, Inc.; 447 (tl), Leslie Kell.

Chapter Twenty: Page 449 (b), 450 (bc), 451 (tc), 453 (tl), 454 (bc), 455 (c), 456 (bl), 459 (bl), 461 (cr), 463 (c), 465 (tl), 465 (tr), 466 (cl), MapQuest.com, Inc.; 467 (tl), Leslie Kell.

Unit Seven: Page 472 (b), 473 (c), MapQuest.com, Inc.; 473 (t), Ortelius Design; 474 (c), 475 (c), 476 (c), 477 (c), MapQuest.com, Inc.; 478 (cl), Ortelius Design; 479 (tl), (cl), (bl), 480 (tl), (cl), (bl), 481 (tl), (cl), (bl), MapQuest.com, Inc.

Acknowledgments

Chapter Twenty-One: Page 483 (b), 485 (tr), MapQuest.com, Inc.; 486 (tl), Leslie Kell; 490 (bl), 492 (bl), MapQuest.com, Inc.; 493 (tr), Leslie Kell; 495 (bl), (tr), 496 (cl), MapQuest.com, Inc.

Chapter Twenty-Two: Page 499 (b), 500 (tc), 503 (bl), 504 (tr), 505 (br), 508 (tr), MapQuest.com, Inc.; 508 (cl), 510 (tl), Leslie Kell; 510 (b), Mark Heine; 511 (tr), 514 (cl), 515 (tl), MapQuest.com, Inc.

Chapter Twenty-Three: Page 517 (b), 518 (cr), 521 (bc), 522 (tr), 523 (tl), MapQuest.com, Inc.; 524 (tl), Leslie Kell; 526 (c), 528 (tc), MapQuest.com, Inc.; 529 (bl), Leslie Kell; 530 (cl), 531 (tl), MapQuest.com, Inc.

Chapter Twenty-Four: Page 533 (b), 534 (br), 536 (cl), 537 (bc), MapQuest.com, Inc.; 540 (bc), Leslie Kell; 542 (c), MapQuest.com, Inc.; 543 (tr), Leslie Kell; 544 (tl), (br), 545 (cr), 546 (bl), MapQuest.com, Inc.; 547 (tl), Leslie Kell; 548 (cl), 549 (tl), MapQuest.com, Inc.

Unit Eight: Page 554 (c), 555 (c), MapQuest.com, Inc.; 555 (t), Ortelius Design; 556 (c), 557 (c), 558 (c), 559 (c), MapQuest.com, Inc.; 560 (cl), Ortelius Design; 561 (tl), 561 (cl), MapQuest.com, Inc.

Chapter Twenty-Five: Page 563 (b), 569 (r), 570 (bl), 572 (bl), 575 (cr), 577 (c), MapQuest.com, Inc.; 578 (tl), Leslie Kell; 578 (b), 579 (cr), 581 (bl), 582 (cl), MapQuest.com, Inc.; 583 (tl), Leslie Kell.

Chapter Twenty-Six: Page 585 (b), MapQuest.com, Inc.; 587 (br), Leslie Kell; 589 (bl), 590 (cl), 592 (cr), 593 (tr), 594 (cl), 595 (cr), MapQuest.com, Inc.; 599 (bl), Leslie Kell; 600 (cl), MapQuest.com, Inc.; 601 (tl), Leslie Kell.

Chapter Twenty-Seven: Page 615 (b), 618 (cr), MapQuest.com, Inc.; 619 (tr), (tl), Leslie Kell; 620 (tc), 621 (cr), MapQuest.com, Inc.; 624 (cl), The John Edwards Group; 627 (bl), 628 (c), 633 (tr), 634 (cl), MapQuest.com, Inc.; 635 (tl), Leslie Kell.

Chapter Twenty-Eight: Page 637 (b), 638 (tl), 640 (tl), 641 (tr), 643 (cr), 644 (cl), (bc), 645 (tl), 651 (c), 652 (tl), (c), 654 (bl), 656 (cl), MapQuest.com, Inc.; 657 (tl), Leslie Kell.

Chapter Twenty-Nine: Page 659 (b), 660 (cr), 661 (tc), 663 (bl), 666 (tl), MapQuest.com, Inc.; 670 (bl), Leslie Kell; 670 (tl), MapQuest.com, Inc.; 671 (cr), Leslie Kell; 674 (b), 676 (cl), 677 (tl), MapQuest.com, Inc.

Chapter Thirty: Page 679 (b), 680 (cl), MapQuest.com, Inc.; 681 (br), Leslie Kell; 682 (cl), MapQuest.com, Inc.; 683 (b), Ortelius Design; 683 (b), Mark Heine; 686 (bl), MapQuest.com, Inc.; 687 (b), Leslie Kell; 688 (cl), Nenad Jakesevic; 691 (tr), Leslie Kell; 691 (c), 693 (cr), MapQuest.com, Inc.; 695 (bl), Leslie Kell; 695 (t), 696 (cl), MapQuest.com, Inc.; 697 (tl), Leslie Kell.

Unit Ten: Page 702 (c), 703 (c), MapQuest.com, Inc.; 703 (t), Ortelius Design; 704 (c), 705 (c), 706 (c), 707 (c), MapQuest.com, Inc.; 708 (cl), Ortelius Design; 709 (tl), (cl), (bl), MapQuest.com, Inc.

Chapter Thirty-One: Page 711 (b), 713 (cr), 715 (br), MapQuest.com, Inc.; 716 (tl), 717 (bl), 718 (bl), Leslie Kell; 718 (tl), 719 (tl), 723 (bl), 726 (bl), 728 (cl), MapQuest.com, Inc.; 729 (tl), Leslie Kell.

Chapter Thirty-Two: Page 731 (b), MapQuest.com, Inc.; 732 (l), Mark Heine; 733(cr), 737 (tr), 739 (c), 740 (tr), 741 (cr), 742 (c), 743 (tr), 746 (cl), 747 (tl), MapQuest.com, Inc.

PHOTO CREDITS

Features, Icons and Backgrounds: *Unit Opener Backgrounds:* (texture) CORBIS Images; (globes) Mountain High Maps® Copyright ©2003; Digital Wisdom, Inc. *Chapter Opener and Section Backgrounds:* (passport) Sam Dudgeon/HRW Photo; (textures) CORBIS Images; (globes) Mountain High Maps® Copyright ©2003 Digital Wisdom, Inc.; *Geography for Life Backgrounds:* (map) Map drawn by J. Asherton, engraved by J. Shury, Oregon Historical Society; (compass rose) CORBIS Images; (compass) Digital Imagery® copyright 2003 PhotoDisc. Inc. *Case Study Icons:* (notebook, compass, magnifying glass) Digital Imagery® copyright 2003 PhotoDisc, Inc.; (map, pencil) HRW Library Photo; (computer monitor) Image Copyright ©2003 PhotoDisc, Inc.; *Cities & Settlements Icons:* (Kurdish woman) Phillis Picardi/International Stock Photography; (Young cowboy) Paul Chesley/National Geographic Society Image Collection; (Woman in turtle neck) Peter Griffith/Masterfile; (Buildings collage) R. Ian Lloyd/Masterfile, J. A. Kraulis/Masterfile, Ron Stroud/Masterfile, Miles Ertman/Masterfile, CORBIS Images. *Internet**connect** Mouse Icon:* Digital Imagery® copyright 2003 PhotoDisc, Inc. *Our Amazing Planet Icon:* Mountain High Maps® Copyright ©2003 Digital Wisdom, Inc.

Front Matter: iv (tl), VCG/FPG International; iv (tr), Carr Clifton/Minden Pictures; iv (br), ©Yvette Cardozo/Getty Images/Stone; v (b), ©Shaun Egan/Getty Images/Stone; vi (t), George Hunter/ H. Armstrong Roberts; vi (bl), Greg Ryan/Sally Beyer/Positive Relections; vii (b), Richard Turpin/Aspect Picture Library Ltd./Corbis Stock Market; viii (t), Wayne Eastep; viii (br), Robert Frerck/Woodfin Camp & Associates; ix (tr), Albert Normandin/Masterfile; ix (cr), Frank Fournier/ Woodfin Camp & Associates; ix (b), ©Kevin Schafer/kevinschafer.com; x (t), Miwako Ball/International Stock Photography; x (br), Blaine Harrington; xi (tl), Telegraph Colour Library/FPG International; xi (b), Leo de Wys/eStock Photo; xii (t), ©Chad Ehlers/Getty Images/Stone; xiii (t), W.P. Fleming/Viesti Collection, Inc.; xiv (t), ©Bob Krist/Getty Images/Stone; xiv (tc), CORBIS/Roger Ressmeyer; xv (t), Photo ©Stefan Schott/Panoramic Images, Chicago 1998; xvi (t), Corbis Images; xvii (t), Carr Clifton/Minden Pictures; xxii (tl), CORBIS/ Tiziana and Gianni Baldizzone; xix (br), Digital imagery® copyright 2003 PhotoDisc, Inc.; xxii (tr), Art Wolfe, Inc.; xxiii (tl), Tom Stewart/Corbis Stock Market; xxiii (br), Daniel J. Schaefer; S1 (cl), John Langford/HRW Photo; S1 (bl), John Langford/HRW Photo; S2 (b), ©Nik Wheeler/CORBIS; S2 (br), Digital imagery® copyright 2003 PhotoDisc, Inc.; S3 (tl), Jeffrey Aaronson/Network Aspen; S3 (tc), Index Stock Imagery, Inc.; S3 (bl), Amit Bhargava/Newsmakers/Getty Images; S3 (c), Norman Owen Tomalin/Bruce Coleman, Inc.; S3 (cr), ©Fritz Polking/Peter Arnold, Inc.; S4 (br), Sam Dudgeon/HRW Photo; S4 (bl), Larry Kolvoord Photography; S4 (c), Runk/Schoenberger/Grant Heilman Photography; S4 (tr), Wolfgang Kaehler Photography; S5 (bl), Sam Dudgeon/HRW; S7 (tr), Sam Dudgeon/HRW Photo; S8 (br), Sam Dudgeon/HRW; S9 (l), Sam Dudgeon/HRW Photo; S10 (c), Sam Dudgeon/HRW; S10 (tr), John Langford/HRW; S12 (tr), Digital imagery® copyright 2003 PhotoDisc, Inc.; S12 (cl), David Phillips/HRW; S12 (bl), Kyodo News Service; S13 (t), Digital imagery® copyright 2003 PhotoDisc, Inc.; S14 (cl), Sam Dudgeon/HRW Photo; S15 (l), Sam Dudgeon/HRW Photo; S16 (br), Sam Dudgeon/HRW Photo; S17 (bl), Sam Dudgeon/HRW; S20 (br), Digital imagery® copyright 2003 PhotoDisc, Inc.

Unit One: 1, CORBIS/Keren Su. **Chapter One:** 2 (bl), Victor Boswell/National Geographic Society Image Collection; 2 (r), Corbis Images; 2 (cl), ©Digital Vision/HRW; 3 (bl), ©Yvette Cardozo/Getty Images/Stone; 4 (tl), Boyd Norton/Evergreen Photo Alliance; 4 (br), Carr Clifton/Minden Pictures; 5 (br), CORBIS/Roger Ressmeyer; 6 (cl), David Muench Photography; 7 (tl), ©Earth Imaging/Getty Images/Stone; 7 (tr), Don Couch/HRW Photo—Postcard reprinted by permission from Jeffery Stanton; 8 (tr), William Campbell/Peter Arnold, Inc.; 11 (tl), Erich Lessing/Art Resource, NY; 15 (tl), The Art Archive/Dagli Orti (A); 15 (br), The Art Archive/Dagli Orti (A). **Chapter Two:** 24 (r), ©Earth Imaging/Getty Images/Stone; 24 (bl), CORBIS/Roger Ressmeyer; 25 (cr), Digital imagery® copyright 2003 PhotoDisc, Inc.; 27 (tr), NASA Image eXchange/www.nix.nasa.gov; 28 (bl), Space Telescope Science Institute/NASA/Science Photo Library/Photo Researchers, Inc.; 28 (bc), Jack Zehrt 1993/FPG International; 31 (tr), NASA's Global Hydrology and Climate Center, Huntsville, AL.; 34 (br), ©Douglas Peebles; 35 (atmosphere), VCG/FPG International; 35 (lithosphere), ©Douglas Peebles; 35 (hydrosphere), H. Richard Johnston/FPG International; 35 (biosphere), Nawrocki Stock Photo; 36 (tl), Calvin Larsen/Photo Researchers, Inc. **Chapter Three:** 40 (cl), ©Alan R. Moller/Getty Images/Stone; 40 (bl), Sepp Seitz /Woodfin Camp & Associates; 40 (r), Greg Gawlowski/Lonely Planet Images; 41 (br), Larry Kolvoord Photography; 42 (br), ©Gillette Photography/ImageState; 44 (tr), ©2000 Kent Wood, All Rights Reserved; 46 (bl), Wayne Scarberry/The Image Works; 47 (tl), David Madison/Bruce Coleman, Inc.; 47 (tr), ©John Elk III; 48 (br), Reuters/Juan Carlos Ulate/Hulton Archive; 49 (tl), NOAA (National Oceanic & Atmospheric Administration); 49 (tc), NOAA (National Oceanic & Atmospheric Administration); 49 (tr), NOAA (National Oceanic & Atmospheric Administration); 50 (b), ©James Martin/Getty Images/Stone; 54 (tr), Kal Muller/Woodfin Camp & Associates; 55 (tr), ©Dimaggio/Kalish/Peter Arnold, Inc.; 55 (bl), Bruce Coleman, Inc.; 55 (bc), Jane Van Eps/SuperStock; 56 (b), Calhoun/Bruce Coleman, Inc.; 57 (tl), ©David Ryan/Lonely Planet Images; 59 (tr), ©2003 Galen Rowell/Mountain Light Photography; 59 (tl), ©2003 Galen Rowell/Mountain Light Photography. **Chapter Four:** 62 (r), Carr Clifton/Minden Pictures; 62 (cl), Corbis Images/HRW; 62 (bl), Dorling Kindersley; 65 (br), "World Ocean Map", Bruce C. Heezen and Marie Tharp, 1977, ©Marie Tharp 1977. Reproduced by permission of Marie Tharp, 1 Washington Avenue, South Nyack, NY 10960; 66 (bl), CORBIS/Tom Bean; 66 (br), CORBIS/Dave Bartruff; 67 (br), Carr Clifton/Minden Pictures; 67 (cr), Wendell Metzen/Bruce Coleman, Inc.; 68 (b), Bill Hatcher/National Geographic Society Image Collection; 68 (tl), Andrew Miles; 68 (inset), Carr Clifton/Minden Pictures; 71 (tr), Joseph Hutchins Colton, Johnson's New Illustrated Family Atlas with Physical Geography (New York, 1864), pp. 10–11.; 72 (b), Frans Lanting/Minden Pictures; 73 (tl), Digital image ©2003 CORBIS; Original image courtesy of NASA/CORBIS; 74 (br), Walter H. Hodge/Peter Arnold, Inc.; 76 (tr), D. Wiggett/Natural Selection; 77 (br), David R. Frazier Photolibrary; 77 (t), www.visibleearth.nasa; 78 (tr), CORBIS/Ecoscene/Andrew Brown; 79 (tr), CORBIS/AFP Photo/Vanderlei Almeida; 81 (tr), Photo ©Stefan Schott/Panoramic Images, Chicago 1998. **Chapter Five:** 86 (r), ©Chad Ehlers/Getty Images/Stone; 86 (bl), Michael & Barbara Reed/SuperStock; 87 (bl), Erica Lansner/Black Star; 87 (br), Jeffrey Aaronson/Network Aspen; 89 (tl), ©Sally Mayman/Getty Images/Stone; 90 (b), Peter Turnley/Black Star; 94 (b), Bill Bachmann/Network Aspen; 95 (tr), Eric Sanford/International Stock Photography; 96 (t), Wolfgang Kaehler Photography; 97 (bl), Laura J. Hannam/Pictor Uniphoto; 98 (tl), ©Yann Layma/Getty Images/Stone; 99 (bl), Galen Rowell/Mountain Light Photography; 102 (bl), ©Ulrike Welsch; 104 (tc), Jewish Museum, London/Bridgeman Art Library; 104 (c), Richard Nowitz/Words & Pictures/Picture Quest; 104 (b), CORBIS/Annie Griffiths Belt; 104 (tr), Scala/Art Resource, NY; 104 (tl), Lisa Quinones/Black Star; 105 (tr), Mark Downey/Viesti Collection, Inc.; 105 (c), CORBIS/Lindsay Hebberd; 105 (b), CORBIS/Gian Berto Vanni; 105 (tl), ©Dinodia/TRIP Photo Library; 105 (tc), Milind A. Ketkar/Dinodia Picture Agency; 106 (tl), CORBIS/AFP; 106 (tc), CORBIS/Luca I. Tettoni; 106 (tr), Josef Beck/FPG International; 106 (cl), Dinodia/SOA; 106 (b), ©ML Sinibaldi/Corbis Stock Market; 107 (tc), CORBIS/Bohemian Nomad Picturemakers; 107 (tr), Courtesy of Information Division, Taipei Economic and Cultural Office in Chicago; 107 (c), ©Alan Thornton/Getty Images/Stone; 107 (tl), Private Collection/Bridgeman Art Library; 107 (b), Courtesy of Information Division, Taipei Economic and Cultural Office in Chicago; 108 (tr), Erich Lessing/Art Resource, NY; 108 (bc), Image Copyright ©2003 PhotoDisc, Inc.; 108 (c), CORBIS/Elio Ciol; 108 (tl), Scala/Art Resource, NY; 108 (b), CORBIS/Mark Thiessen; 109 (tl), Ronald Sheriidan/Ancient Art & Architecture Collection, Ltd.; 109 (tr), Bonhams, London/Bridgeman Art Library, New York/London; 109 (tc), Werner Forman/Art Resource, NY; 109 (cr), CORBIS/AFP; 109 (b), CORBIS/Reuters NewMedia, Inc. **Chapter Six:** 112 (cl), ©Shaun Egan/Getty Images/Stone; 112 (r), ©Glen Allison/Getty Images/Stone; 112 (bl), Runk/Schoenberger/Grant Heilman Photography; 114 (br), Carolyn Schaefer/SCHAE/ Bruce Coleman, Inc.; 115 (br), ©Cosmo Condina/Getty Images/Stone; 115 (cr), Hulton Archive; 118 (tl), ©Glen Allison/Getty Images/Stone; 118 (cl), John Elk III/Bruce Coleman, Inc.; 120 (br), ©Werner Forman/CORBIS; 123 (bl), David R. Frazier Photolibrary; 124 (bl), ©Pete Seaward/Getty Images/Stone; 125 (tr), ©Doris DeWitt/Getty Images/Stone; 126 (tr), Arthur C. Smith III/Grant Heilman Photography; 127 (bl), ©Paula Bronstein/Getty Images/Stone; 128 (br), Popperfoto/Hulton Archive; 129 (br), ©Tom Bean/Getty Images/Stone; 130 (br), Amit Bhargava/Newsmakers/Getty Images; 131 (tl), Christine Osborne Pictures/MEP; 132 (br), Agencia

Estado/AP/Wide World Photos; 133 (bl), E. R. Degginger/Bruce Coleman, Inc.; 133 (t), www.visibleearth.nasa.gov; 137 (t), CORBIS/Lowell Georgia.

Unit Two: 138-139 (all), CORBIS/Michael S. Yamashita; 139 (tr), Science Photo Library/Photo Researchers, Inc.; 139 (c), Ray Manley/SuperStock; 146 (tl), Collection of The New-York Historical Society; 146 (soldier), 4th Ind. Company Maryland State Troops. plate 11 by Charles Lefferts. Collection of The New York Historical Society; 146 (Washington), White House Collection, courtesy White House Historical Association; 146 (settlers), ©Shelburne Museum, Shelburne, Vermont, detail of the painting "Conestoga Wagon" by Thomas Birch; 146 (prospector), CORBIS/Bettmann; 146 (uniform), The Granger Collection, New York; 146 (Uncle Sam), The Granger Collection, New York; 146 (Bush), AP/Wide World Photos; 146 (br), CORBIS/David Muench; 147 (b), CORBIS/Michael T. Sedam. **Chapter Seven:** 148 (cr), Raymond Gehman/CORBIS; 148 (bl), Michio Hoshino/Minden Pictures; 148 (cl), Chuck Place Photography; 151 (br), Chase Swift/CORBIS; 152 (bl), Ansel Adams Publishing Rights Trust/CORBIS; 152 (br), Roger Ressmeyer/CORBIS; 153 (tr), Digital image ©2003 CORBIS Original image courtesy of NASA/CORBIS; 154 (b), CORBIS/Robert Estall; 155 (tr), CORBIS/Jim Zuckerman; 156 (tr), Carr Clifton/Minden Pictures; 156 (bl), Carr Clifton/Minden Pictures; 157 (tr), Philip Rosenberg Photography; 158 (bl), ©Bruce Forster/Getty Images/Stone; 159 (tr), CORBIS/Jeffrey L. Rotman; 160 (tr), CORBIS/Bill Ross. **Chapter Eight:** 164 (bl), ©Walter Hodges/Getty Images/Stone; 164 (tr), CORBIS/Buddy Mays; 164 (cl), Will Crocker/Getty Images/The Image Bank; 166 (t), Chuck Place Photography; 168 (bl), CORBIS; 169 (br), Mark Gibson Photography; 170 (tl), ©Christian Heeb/Gnass Photo Images; 171 (tr), CORBIS/Bettmann; 173 (b), CORBIS/Michael S. Yamashita; 174 (bl), CORBIS/Kelly-Mooney Photography; 176 (tl), David R. Frazier Photolibrary; 176 (bl), CORBIS/Robert Holmes; 177 (tr), ©Bruce Hands/Getty Images/Stone; 177 (b), ©Ric Ergenbright; 178 (tl), Tim Fitzharris/Minden Pictures; 179 (tr), Carr Clifton/Minden Pictures; 180 (cl), ©Ric Ergenbright; 181 (br), David R. Frazier Photolibrary; 181 (tr), CORBIS/Roger Ressmeyer; 183 (tl), CORBIS/Jim Sugar Photography; 184 (tr), CORBIS/Bettmann. **Chapter Nine:** 188 (bl), Eastcott/Momatiuk/Woodfin Camp & Associates; 188 (tl), Jim Brandenburg/Minden Pictures; 188 (tr), The Lowe Art Museum, The University of Miami/SuperStock; 190 (t), The Huntington Library, Art Collections and Botanical Gardens, San Marino, CA/SuperStock; 192 (cl), George Hunter/SuperStock; 192 (tl), Stephan Poulin/SuperStock; 193 (bl), George Hunter/H. Armstrong Roberts; 194 (br), Mark Antman/The Image Works; 195 (br), Mauritius/SuperStock; 196 (tl), George Hunter/H. Armstrong Roberts; 197 (cr), John Eastcott/YVA Momatiuk/DRK Photo; 199 (br), Michael S. Yamashita/CORBIS; 200 (b), Nik Wheeler; 201 (tr), Walter Bibikow/Viesti Collection, Inc.; 204 (tr), Enrique Shore/Reuters/TimePix; 204 (cr), Mike Segar/Reuters/TimePix; 205 (bl), Jeff Christensen/Reuters/TimePix; 205 (tr), ©Monika Graff/The Image Works.

Unit Three: 208-209 (all), CORBIS/Kevin Schafer; 209 (tr), CORBIS/Wolfgang Kaehler; 209 (bc), SuperStock; 216 (cl), Museo Civico, Como, Italy/Art Resource, NY; 216 (c), Museo Nacional de Historia, Castillo de Chapultepec, Mexico City, Mexico/Art Resource, NY; 216 (tc), CORBIS/Lake County Museum; 216 (cr), CORBIS/Bettmann; 216 (tr), CORBIS/Dave G. Houser; 218 (b), ©Renee Lynn/Getty Images/Stone; 219 (b), Steve Vidler/SuperStock. **Chapter Ten:** 220 (bl), Steve Ewert Photography/HRW Photo; 220 (tl), Stewart Aitchison/D. Donne Bryant Photography; 220 (tr), Scala/Art Resource, NY; 222 (tl), ©Robert Frerck/Odyssey/Chicago; 223 (tr), George Grall/National Geographic Society Image Collection; 223 (tc), James Hanken/Bruce Coleman, Inc.; 224 (br), D. Donne Bryant/D. Donne Bryant Photography; 225 (bl), Giraudon/Art Resource, NY; 226 (tl), J. Sarapochiello/Bruce Coleman, Inc.; 226 (bl), Nik Wheeler; 228 (tr), Eleanor S. Morris; 228 (bl), Alon Reininger/Woodfin Camp & Associates; 229 (cr), Sam Dudgeon/HRW Photo; 230 (br), Steve Vidler/SuperStock; 231 (tr), CORBIS/Jorge Uzon/AFP; 231 (b), ©Robert Frerck/Odyssey/Chicago; 232 (tr), Bob Daemmrich/The Image Works; 233 (tr), Joel Sartore/National Geographic Society Image Collection; 234 (t), ©Robert Frerck/Odyssey/Chicago; 234 (bl), ©Robert Frerck/Odyssey/Chicago; 235 (t), taken from Travel Guide Mexico ©Dorling Kindersley. **Chapter Eleven:** 238 (bl), Steve Ewert Photography/HRW Photo; 238 (tl), Karl Kummels/SuperStock; 238 (tr), CORBIS/Kevin Schafer; 240 (tl), Steve Vidler/SuperStock; 241 (br), ©Bob Krist/Getty Images/Stone; 242 (tr), ©Kevin Schafer/kevinschafer.com; 243 (tr), Rob Huibers/Panos Pictures; 244 (b), ©David Hiser/Getty Images/Stone; 245 (tr), SuperStock; 246 (b), ©2003 Greg Johnston; 247 (tr), August Upitis/SuperStock; 247 (br), ©Kevin Schafer/kevinschafer.com; 248 (tr), ©Kevin Schafer/kevinschafer.com; 249 (b), Wolfgang Kaehler Photography; 250 (tr), ©Nick Dolding/Getty Images/Stone; 250 (bl), Catherine Karnow/Woodfin Camp & Associates; 250 (tl), Wolfgang Kaehler Photography; 251 (cr), Michael K. Nichols/National Geographic Society Image Collection; 252 (tl), ©2003 Greg Johnston; 252 (b), Steve Vidler/SuperStock; 253 (t), Steve Vidler/SuperStock. **Chapter Twelve:** 256 (bl), Steve Ewert Photography/HRW Photo; 256 (tl), ©Kevin Schafer/kevinschafer.com; 256 (tr), ©Gianni Dagli Orti/CORBIS; 258 (tr), ©Kevin Schafer/kevinschafer.com; 258 (bl), Angelo Cavalli/SuperStock; 260 (inset), ©Kevin Schafer/kevinschafer.com; 260 (tl), Norman Owen Tomalin/Bruce Coleman, Inc.; 261 (cl), D. Donne Bryant Photography; 261 (cr), D. Donne Bryant Photography; 262 (tc), NASA/Science Photo Library/Photo Researchers, Inc.; 263 (bl), ©Robert Frerck/Odyssey/Chicago; 263 (cr), The Art Archive/Museum of Mankind London/Eileen Tweedy; 264 (cl), The Art Archive/Simon Bolivar Amphitheatre/Dagli Orti (A) ; 265 (br), RAGA/Corbis Stock Market; 266 (bl), Robert Perron; 267 (tr), Robert Frerck/Odyssey/Chicago; 268 (bl), ©Peter Lang/The PhotoWorks/D. Donne Bryant Photography; 268 (b), Pedro Raota/SuperStock; 269 (br), ©M. Edwards/Still Pictures/Peter Arnold, Inc.; 269 (cr), Juca Martins/f4/D. Donne Bryant Photography; 270 (tl), ©Mauricio Simonetti/f4/D. Donne Bryant Photography; 270 (br), Auscape (Parer-Cook)/Peter Arnold, Inc.; 271 (tr), ©Peter Wilson/CORBIS; 272 (b), Jeremy Homer/CORBIS; 273 (br), CORBIS/Yann Arthus-Bertrand.

Unit Four: 278-279 (all), ©Bob Krist/Getty Images/Stone; 279 (tr), CORBIS/Werner Forman; 279 (bc), AKG, Berlin/SuperStock; 286 (cl), The Art Archive; 286 (tl), Alinari/Art Resource, NY; 286 (tc), The Art Archive/Bibliotheque des Arts Decoratifs Paris/Dagli Orti (A); 286 (cr), CORBIS/Hulton-Deutsch Collection Limited; 286 (tr), AP/Wide World Photos; 289 (b), SuperStock. **Chapter Thirteen:** 290 (tl), Miles Ertman/Masterfile; 290 (bl), CORBIS/Macduff Everton; 290 (tr), CORBIS/David Lees; 292 (bl), Corbis Images; 293 (c), Benjamin Rondel/Corbis Stock Market; 294 (tl), Pierre Boulat/Woodfin Camp & Associates; 295 (t), Bill Brooks/Masterfile; 295 (b), ©George Grigoriou/Getty Images/Stone; 296 (b), Ron Sanford/Tony Stone Images; 297 (br),

Acknowledgments

CORBIS/Frank Lane Picture Agency/Fritz Polking; 298 (tl), Len Kaufman; 299 (br), Peter Weimann/Animals Animals/Earth Scenes; 300 (t), Richard Turpin/Aspect Picture Library Ltd./Corbis Stock Market; 301 (tl), Gaetano Barone/SuperStock. **Chapter Fourteen:** 304 (bl), Steve Ewert Photography/HRW Photo; 304 (tl), Steve Vidler/eStock Photo; 304 (tr), London, British Museum/AKG Photo, London; 306 (tl), Telegraph Colour Library/FPG International; 308 (bl), AKG Photo, London; 309 (tr), David Lomax/Robert Harding; 310 (b), Gerry Johansson/eStock Photo; 311 (bl), Musée d'Orsay/AKG Photo, London; 312 (tl), Picture Finders Ltd./eStock Photo; 312 (br), Jean Paul Nacivet/eStock Photo; 313 (tl), Michael Jenner/Robert Harding; 314 (tl), Natalie B. Fobes/Tony Stone Images; 315 (cr), Steve Vidler/eStock Photo; 317 (tr), Gavin Hellier/Robert Harding; 318 (br), Everett C. Johnson/eStock Photo; 319 (tr), Vladpans/eStock Photo; 319 (b), Tom Stewart/Corbis Stock Market; 320 (br), Photographers Library LTD/eStock Photo; 320 (tl), CORBIS/Hubert Stadler; 321 (tr), Jan Kofod Winther; 322 (bl), Telegraph Coulour Library/FPG International. **Chapter Fifteen:** 326 (bl), Steve Ewert Photography/HRW Photo; 326 (cl), Dankwart Von Knobloch/SuperStock; 326 (tr), Kunsthistorisches Museum, Vienna, Austria/Erich Lessing/Art Resource, NY; 327 (tr), Wolfgang Kaehler Photography; 328 (t), Steve Vidler/SuperStock; 329 (bl), Frank Rumpenhorst/AP/Wide World Photos; 330 (tl), Wolfgang Kaehler Photography; 331 (tr), Argus Fotoarchiv/Peter Arnold, Inc.; 332 (bl), SuperStock; 333 (tl), CORBIS/ Reuters NewMedia Inc.; 335 (br), Wolfgang Kaehler Photography; 336 (tl), Robin Smith/SuperStock; 337 (tl), SuperStock; 338 (cl), Fitzwilliam Museum, University of Cambridge, UK/Bridgeman Art Library; 339 (br), Horst Schafer/Peter Arnold, Inc.; 340 (tr), Kurt Scholz/SuperStock; 341 (tr), Fred Bruemmer/Peter Arnold, Inc.; 342 (bl), Hubertus Kanus/SuperStock; 342 (tr), ©J. Icachsen/Art Directors & TRIP Photo Library; 343 (br), Wolfgang Kaehler Photography; 345 (cr), Sovfoto/Eastfoto. **Chapter Sixteen:** 348 (bl), MercuryPress.com; 348 (tl), Grant V. Faint/Getty Images/The Image Bank; 348 (tr), Erich Lessing/Art Resource, NY; 350 (br), Nathan Benn/Woodfin Camp & Associates; 351 (br), ©Ulrike Welsch; 351 (tl), ©Daniel Aubry/Odyssey/Chicago; 355 (bl), ©Louis Goldman/FPG International LLC; 357 (bl), Dorling Kindersley; 360 (br), ©Scala/Art Resource; 362 (tl), Giraudon/Art Resource, NY; 363 (tl), Mario Corvetto/Evergreen Photo Alliance; 364 (br), Peter Schmid/SuperStock; 365 (t), ©Figaro Magazine/Robert Tixador/Liaison Agency.

Unit Five: 370-371 (spread), ©David Sutherland/Getty Images/Stone; 371 (tr), CORBIS/Marc Garanger; 371 (c), SuperStock Inc. Collection, Jacksonville/SuperStock; 378 (cl), Giraudon/Art Resource, NY; 378 (c), Victoria and Albert Museum, London/Art Resource, NY; 378 (tr), Giraudon/Art Resource, NY; 378 (cr), CORBIS/Bettmann; 379 (b), Vladimir Jirnov/Bruce Coleman, Inc. **Chapter Seventeen:** 380 (bl), Steve Ewert Photography/HRW Photo; 380 (tr), SCALA/Art Resource, NY; 380 (tl), Mark Wadlow/Russia and Eastern Images; 382 (b), Randall Hyman; 383 (br), Sovfoto/Eastfoto; 384 (bl), Sovfoto/Eastfoto; 385 (tr), Galen Rowell/ Peter Arnold, Inc.; 385 (t), Sovfoto/Eastfoto; 386 (bl), The Art Archive/Hermitage Museum Saint Petersburg/Dagli Orti(A); 387 (b), Hulton Archive/Getty Images; 387 (t), CORBIS/ Bettmann; 388 (b), Deborah Harse/Peter Arnold, Inc.; 388 (t), CORBIS/UPI/Bettmann; 389 (tr), Sovfoto/Eastfoto; 391 (t), Marka/eStock Photo; 393 (b), ©Paul Chesley/Getty Images/Stone; 394 (t), CORBIS/Steve Raymer; 394 (b), CORBIS/Fabian/Sygma; 395 (tr), AP/Wide World Photos; 395 (bl), ©Jerry Alexander/Getty Images/Stone; 396 (bl), CORBIS/Peter Turnley; 397 (tl), Sovfoto/Eastfoto; 398 (tc), Helga Lade/Peter Arnold, Inc.; 399 (tr), The Art Archive/Tretyakov Gallery Moscow/Dagli Orti (A); 399 (tl), IFA/Peter Arnold, Inc.; 399 (tc), Jeff Greenberg/Peter Arnold, Inc. **Chapter Eighteen:** 402 (bl), ©2003 Brian A. Vikander; 402 (tr), Wayne Eastep; 402 (tl), CORBIS/David Turnley; 404 (t), Wayne Eastep; 405 (tr), Digital Image ©2003 CORBIS; Original image courtesy of NASA/CORBIS. ; 405 (br), ©Wolfgang Kaehler/CORBIS; 406 (tl), Hope Ryden/National Geographic Society Image Collection; 408 (bl), Wolfgang Kaehler Photography; 409 (tr), ©2003 BRIAN A. VIKANDER; 409 (tr), ©David Samuel Robbins/ CORBIS; 410 (t), Caroline Penn/Panos Pictures; 410 (cl), Nevada Wier; 411 (tr), Chris Tordai/Panos Pictures; 411 (cr), David Samuel Robbins; 412 (bl), Nevada Wier; 413 (tr), CORBIS/Bettmann; 413 (br), ©2003 BRIAN A. VIKANDER; 414 (tr), Marcus Rose/Panos Pictures; 414 (tl), Gregory Wrona/Panos Pictures; 415 (tr), Brian Goddard/Panos Pictures; 416 (tr), @ I. Burgandinov/Art Directors & TRIP Photo Library; 417 (tl), ©P. Bucknall/Art Directors & TRIP Photo Library.

Unit Six: 422–423 (all), CORBIS/Yann Arthus-Bertrand; 423 (c), Louvre, Paris, France/Giraudon/Art Resource, NY; 423 (tr), Nik Wheeler/Black Star Publishing/Picture Quest; 430 (cl), CORBIS/Bettmann; 430 (tc), CORBIS/Araldo de Luca; 430 (c), CORBIS/Burstein Collection; 430 (cr), CORBIS/Bettmann; 430 (tr), CORBIS/Francoise de Mulder. **Chapter Nineteen:** 432 (bl), Steve Ewert Photography/HRW Photo; 432 (tl), Adam Woolfitt/CORBIS; 432 (tr), Archivo Iconografico, S.A./CORBIS; 436 (tl), Wolfgang Kaehler Photography; 437 (bl), CORBIS; 437 (br), CORBIS; 438 (bl), University Library Istanbul/The Art Archive; 439 (br), AFP/CORBIS; 440 (tl), ©Abbas/Magnum; 441 (br), Nik Wheeler; 442 (tl), ©Alexandra Avakian /Contact Press Images; 444 (tl), Syed Jan Sabawoon/EPA/AP; 445 (bl), Burnett Moody/Bruce Coleman, Inc.; 445 (bc), Bill Foley/Bruce Coleman, Inc.; 445 (br), Bill Foley/Bruce Coleman, Inc. **Chapter Twenty:** 448 (cl), Richard T. Nowitz; 448 (tr), The Art Archive/British Museum; 448 (bl), ©2003 Zafer KIZILKAYA; 450 (bl), Steve Vidler/SuperStock; 450 (cl), ©Look GMBH/eStock Photography/Picture Quest; 451 (tl), ©1998 Peter Armenia; 452 (tr), David Wells/The Image Works; 452 (bl), @J. Wakelin/Art Directors & TRIP Photo Library; 454 (br), Robert Frerck/Woodfin Camp & Associates; 455 (tr), ©Robert Frerck/Odyssey/Chicago; 456 (br), ©1998 Peter Armenia; 457 (cr), Rich Pomerantz; 457 (br), ©Robert Frerck/Getty Images/Stone; 458 (tl), Christine Osborne Pictures/MEP; 459 (br), CORBIS/Richard List; 460 (cl), Nik Wheeler; 461 (tr), Nik Wheeler; 461 (b), ©Alexandra Avakian/Contact Press Images; 462 (b), Wolfgang Kaehler

Acknowledgments

Photography; 462 (cl), Rich Pomerantz; 463 (tl), Nik Wheeler; 464 (tc), Guy Marche/FPG International; 469 (tl), http://edcwww.cr.usgs.gov/earthshots.

Unit Seven: 470-471 (spread), Steve Vidler/SuperStock; 471 (c), Picture Finders Ltd./eStock Photography/Picture Quest; 471 (tr), Richard S. Durrance/National Geographic Society Image Collection; 478 (tc), CORBIS/Anthony Bannister/Gallo Images; 478 (c), The Art Archive/British Museum/Harper Collins Publishers; 478 (diamond), CORBIS/ D. Boone; 478 (tr), The Art Archive/London Museum/Eileen Tweedy; 478 (cr), CORBIS/Walter Dhladhla/AFP; 478 (cl), The Metropolitan Museum of Art, The Michael C. Rockefeller Memorial Collection, Gift of Nelson A. Rockefeller, 1972. (1978.412.323) Photograph by Malcolm Varon. Photography ©1986 The Metropolitan Museum of Art. **Chapter Twenty-One:** 482 (bl), Steve Ewert Photography/HRW Photo; 482 (tl), ©Sylvain Grandadam/Getty Images/Stone; 482 (tr), The Art Archive/Musee des Arts Africains et Oceaniens/Dagli Orti; 484 (br), ©Victor Englebert; 484 (bl), SuperStock; 485 (tl), ©Victor Englebert; 487 (br), ©Planet Art; 488 (t), Mimmo Jodice/CORBIS; 488 (br), Giorgio Ricatto/SuperStock; 488 (t), Thomas J. Abercrombie/National Geographic Society Image Collection; 489 (cl), TRIP/Viesti Collection, Inc.; 489 (tr), AP/Wide World Photos; 490 (br), SuperStock; 491 (tr), ©Ben Edwards/Getty Images/Stone; 492 (tl), Stone/Michael Shopenn; 494 (tr), ©John Lamb/Getty Images/Stone; 495 (tr), AP/Wide World Photos; 497 (cl), GEOPIC/Earth Satellite Corporation. **Chapter Twenty-Two:** 498 (bl), Steve Ewert Photography/HRW Photo; 498 (cl), Lineair/R. Giling/Peter Arnold, Inc.; 498 (cr), M & E Bernheim/Woodfin Camp & Associates; 500 (tl), Wolfgang Kaehler Photography; 501 (tl), ©BIOS/Peter Arnold, Inc.; 502 (inset), ©Still Pictures/M. Edwards/Peter Arnold, Inc.; 502 (tr), ©Still Pictures/M. Edwards/Peter Arnold, Inc.; 503 (cr), Frank Fournier/Woodfin Camp & Associates; 504 (tl), Frank Fournier/Woodfin Camp & Associates; 504 (tr), Martha Cooper/Peter Arnold, Inc.; 506 (b), ©Victor Englebert; 507 (tl), Mark Edwards/Still Pictures /Peter Arnold, Inc.; 509 (cr), ©Lineair (R. Giling)/Peter Arnold, Inc.; 509 (br), ©Wendy Stone/Odyssey/Chicago; 511 (tr), AP/Wide World Photos/David Guttenfelder; 512 (tl), Bob Burch/Bruce Coleman, Inc.; 512 (br), M. & E. Bernheim/Woodfin Camp & Associates; 513 (t), Wolfgang Kaehler Photography. **Chapter Twenty-Three:** 516 (bl), Betty Press/Picture Group/Panos Pictures; 516 (tl), Daniel Westergren/National Geographic Society Image Collection; 516 (tr), The Art Archive; 518 (b), CORBIS/ Sharna Balfour/Gallo Images; 519 (tr), Boyd Norton; 520 (inset), Jason Lauré/Lauré Communications; 520 (c), Jason Lauré/Lauré Communications; 521 (bl), Dominic Harcourt-Webster/Panos Pictures; 521 (tr), CORBIS/Morton Beebe, S.F.; 522 (cl), Jason Lauré/Lauré Communications; 522 (tr), Boyd Norton; 523 (t), Betty Press/ Woodfin Camp/Picture Quest; 524 (cl), CORBIS/Jim Sugar Photography; 525 (br), Gerald Cubitt; 526 (t), B. Brander/Photo Researchers, Inc.; 527 (br), Boyd Norton; 528 (tr), Crispin Hughes/Panos Pictures; 528 (cl), ©Charles Henneghien/ Bruce Coleman, Inc.; 529 (br), ©2000 Alison M. Jones. **Chapter Twenty-Four:** 532 (cl), Friedrich Von Horsten /Animals Animals/Earth Scenes; 532 (bl), ©Jan Butchofsky/ Houserstock; 532 (tr), Clem Haagner/Bruce Coleman, Inc.; 534 (t), John Elk III/Bruce Coleman, Inc.; 534 (b), Gerald Cubitt; 535 (br), Burnett Moody/Bruce Coleman, Inc.; 535 (tr), Albert Normandin/Masterfile; 536 (tr), Patti Murray/Animals Animals/Earth Scenes; 537 (br), S. Trevor/D.B./Bruce Coleman, Inc.; 538 (tr), Martin Gostelos/Bruce Coleman, Inc.; 538 (cl), Hulton Getty Collection/Hulton Archive; 539 (br), A. Ramey/ Woodfin Camp & Associates; 540 (tl), Piero Guerrini/Woodfin Camp & Associates; 541 (tl), Wolfgang Kaehler Photography; 541 (cr), Volkmar Wentzel/National Geographic Society Image Collection; 542 (bl), Gerald Cubitt; 543 (br), William R. Curtsinger/ National Geographic Society Image Collection; 544 (tr), Gerald Cubitt; 545 (tr), ©Victor Englebert; 551 (tr) Jason Lauré/Lauré Communications.

Unit Eight: 552-553 (all), ©Earl Bronssteen/Panoramic Images, Chicago 2003; 553 (tr), SuperStock; 553 (c), CORBIS; 560 (cl), Borromeo/Art Resource, NY; 560 (c), The Art Archive/Victoria and Albert Museum/Eileen Tweedy; 560 (tr), Hulton Archive/Getty Images; 560 (cr), @ H. Rogers/Art Directors & TRIP Photo Library; 561 (b), CORBIS/Tom Brakefield. **Chapter Twenty-Five:** 562 (bl), Steve Ewert Photography/HRW Photo; 562 (tr), Floyd Holdman/International Stock Photography; 562 (cl), Christie's Images/SuperStock; 564 (b), Steve McCurry/Magnum Photos; 565 (tr), Earth Satellite Corporation/Science Photo Library/Photo Researchers, Inc.; 566 (inset), CORBIS/Jayanta Shaw/Reuters; 566 (tl), SuperStock; 567 (bl), ©Joel Simon/ Getty Images/Stone; 567 (tr), CORBIS/©Jeremy Horner; 570 (bl), SuperStock; 570 (tl), ©Stone/Hulton Getty; 571 (tr), SuperStock; 572 (br), Gerald Brimacombe/International Stock Photography; 573 (tr), Miwako Ball/International Stock Photography; 574 (br), Ric Ergenbright; 575 (b), SuperStock; 576 (tl), SuperStock; 578 (br), James Strachan/ Robert Harding; 580 (tr), Jeremy Bright/ Robert Harding; 580 (c), CORBIS/Savita Kirloskar; 581 (t), Chris Stowers/Panos Pictures. **Chapter Twenty-Six:** 584 (bl), Steve Ewert Photography/HRW Photo; 584 (tr), Holton Collection/SuperStock; 584 (tl), CORBIS/Tom Owen Edmunds; 586 (b), Hugh Burden/SuperStock; 586 (bl), Charles Corfield; 588 (tl), CORBIS/Jonathan Blair; 589 (b), ©geyer@malediven.at; 590 (b), ©R. Graham/Art Directors & TRIP Photo Library; 591 (tr), CORBIS/Agence France Presse/Jewel Samad; 591 (bl), William McCloskey; 592 (tl), Baldev Kapoor/ SuperStock; 592 (tl), SuperStock; 593 (tr), Peter Symasko/International Stock Photography; 594 (b), Miwako Ball/International Stock Photography; 595 (br), Karl Kummels/SuperStock; 596 (cl), Binod Joshi/AP/Wide World Photos; 597 (tr), CORBIS/ Jeremy Horner; 597 (tl), Steve Vidler/SuperStock; 598 (bl), Shehzad Noorani/Woodfin

Camp & Associates; 598 (tc), Fred Hoogervorst/Panos Pictures; 599 (t), Zed Nelson/Panos Pictures; 602 (tr), Jeff Greenberg/Photo Researchers, Inc.

Unit Nine: 604-605 (spread), ©David Ball/Getty Images/Stone; 605 (c), Wolfgang Kaehler Photography; 605 (tr), Ed Gifford/Masterfile; 612 (tl), CORBIS/©Jack Fields; 612 (tc), CORBIS/Bettmann; 612 (tr), ©Magaret Gowan/Getty Images/Stone; 612 (cl), Reunion des Musees Nationaux/Art Resource, NY; 612 (c), CORBIS/Gianni Dagli Orti (A). **Chapter Twenty-Seven:** 614 (bl), Steve Ewert Photography/HRW Photo; 614 (tl), Keren Su/China Span; 614 (tr), ©2003 Brian A. Vikander; 616 (bl), CORBIS/Dean Conger; 617 (tr), Thomas Laird/Peter Arnold, Inc.; 617 (br), Lynn M. Stone/Bruce Coleman, Inc.; 618 (b), AFP/Getty Images; 618 (tl), Keren Su/China Span; 619 (br), ©Yann Layma/Getty Images/Stone; 621 (cr), Stone/Nigel Hicks; 621 (inset), Keren Su/China Span; 622 (t), James Montgomery/Bruce Coleman, Inc.; 622 (bl), China Stock; 622 (inset), Joe McNally/Getty Images/The Image Bank; 623 (tr), ©2003 Brian A. Vikander; 623 (br), Paul Chesley/Network Aspen; 624 (bl), ©Keren Su/FPG International LLC; 625 (t), Xue Jun Yuan/Getty Images/The Image Bank; 625 (br), CORBIS/Asian Art & Archaeology, Inc.; 626 (tr), Dennis Cox/China Stock; 627 (bl), Cameramann International, Ltd.; 627 (inset), Cameramann International, Ltd.; 628 (tr), Jeffrey Aaronson/Network Aspen; 628 (bl), Dennis Cox/ China Stock; 628 (inset), ©2003 Brian A. Vikander; 629 (b), Suolang Luobu/Sovfoto/ Eastfoto; 630 (br), Mahaux Photo/Getty Images/The Image Bank; 630 (bl), Mongolian Eight flags soldiers from Ching's military forces, engraved by R.Rancati (colour engraving)/Private Collection/ Bridgeman Art Library, London/New York; 631 (tr), Jeffrey Aaronson/ Network Aspen; 632 (tr), Wolfgang Kaehler Photography. **Chapter Twenty-Eight:** 636 (bl), Steve Ewert Photography/HRW Photo; 636 (tr), Wolfgang Kaehler Photography; 636 (tl), Wolfgang Kaehler Photography; 638 (tl), CORBIS/John & Dallas Heaton; 638 (tr), H. Edward Kim/National Geographic Society Image Collection; 639 (cr), Stephen Walker/Peter Arnold, Inc.; 640 (tl), Paul Chesley/Network Aspen; 641 (bl), Michael Yamashita; 642 (bl), CORBIS/Asian Art & Archaeology, Inc.; 644 (tl), Bruno P. Zehnder/Peter Arnold, Inc.; 644 (br), Kyodo News Service; 646 (tr), Kyodo News Service; 646 (bl), CORBIS/ John Dakers; Eye Ubiquitous; 647 (cr), Dr. Darlyne A. Murawski/National Geographic Society Image Collection; 647 (tl), Kyodo News Service; 648 (tr), Blaine Harrington; 649 (br), Jeffrey Aaronson/Network Aspen; 650 (b), Kyodo News Service; 651 (tr), Michael Yamashita; 651 (br), David A. Harvey/National Geographic Society Image Collection; 652 (bl), Catherine Karnow/Woodfin Camp & Associates; 652 (tl), Jeffrey Aaronson/ Network Aspen; 653 (tl), CORBIS/AFP; 655 (tr), Kyodo News Service; 655 (bl), CORBIS/Reuters NewMedia Inc. **Chapter Twenty-Nine:** 658 (bl), Steve Ewert Photography/HRW Photo; 658 (r), Dennie Cody/FPG International; 658 (cl), Radhika Chalasani/Network Aspen; 660 (b), ©Connie Coleman/Getty Images/Stone; 661 (br), Michael Dick/Animals Animals/ Earth Scenes; 661 (tl), Mike Yamashita/Woodfin Camp & Associates; 662 (cl), Robert Frerck/Woodfin Camp & Associates; 663 (cr), Harvey Lloyd/FPG International; 663 (b), Jeffrey Aaronson/Network Aspen; 664 (tl), Jeffrey Aaronson/Network Aspen; 665 (b), CORBIS/Bettmann Archive; 666 (bl), Jeffrey Aaronson/Network Aspen; 666 (tl), Paul Chesley/Network Aspen; 667 (tl), Wolfgang Kaehler Photography; 667 (tr), ©Owen Franken/Getty Images/Stone; 668 (br), ©James Strachan/Getty Images/Stone; 668 (bl), Liverani-UNEP/Peter Arnold, Inc.; 669 (bl), ©Brecelj & Hodalic/Still Pictures/Peter Arnold, Inc.; 669 (cr), Fritz Prenzel/Peter Arnold, Inc.; 670 (tl), ©Paul Chesley/Getty Images/Stone; 671 (bl), Dennie Cody/FPG International; 672 (tl), Telegraph Colour Library/FPG International; 673 (cr), Jeffrey Aaronson/Network Aspen; 675 (tr), Mike Yamashita/Woodfin Camp & Associates; 675 (tr), Mark Edwards/Still Pictures/Peter Arnold, Inc. **Chapter Thirty:** 678 (tl), Mike Yamashita/ Woodfin Camp & Associates; 678 (bl), @ D. Saunders/Art Directors & TRIP Photo Library; 678 (tr), Don Couch/HRW Photo - Mustapha Mazurki's Wau Bulan kite from www.SKY-DANCER.com; 680 (b), CORBIS/Les Stone/Sygma; 682 (cl), Picture Finders Ltd./eStock Photo; 684 (cl), Wolfgang Kaehler Photography; 685 (tl), SuperStock; 685 (bl), Werner Forman Archive Private Collection, Prague/Art Resource, NY; 687 (tr), Werner Forman Archive/Private Collection/Art Resource, NY; 688 (b), Stone/Denis Waugh; 689 (tl), CORBIS/Michael S. Yamashita; 690 (cr), SuperStock; 690 (bl), John Zoiner; 691 (bl), ©Jerry Alexander/Getty Images/ Stone; 692 (b), Wolfgang Kaehler Photography; 693 (tr), Cesar Pelli & Associates/ Photo by Jeff Goldberg/Esto; 694 (tr), ©David Saunders/Getty Images/Stone.

Unit Ten: 700-701 (all), Jeff Hunter/Getty Images/The Image Bank; 701 (tr), ©Auscape/Peter Arnold, Inc.; 701 (c), David Burnett/Contact Press Images/Picture Quest; 708 (c), The Art Archive/ Harper Collins Publishers; 708 (cr), CORBIS/Historical Picture Archive; 708 (cl), CORBIS/Werner Forman; 708 (tr), CORBIS/Charles & Josette Lenars; 708 (tl), CORBIS/Michael Maslan Historic Photographs. **Chapter Thirty-One:** 710 (bl), Penny Tweedie/HRW Photo; 710 (tl), Joseph Van Os/Getty Images/The Image Bank; 710 (tr), Leo de Wys/eStock Photo; 712 (bl), Gisela Damm/eStock Photo; 712 (br), R. Ian Lloyd/Masterfile; 713 (tr), ©David Doubilet; 714 (tl), Brian Sytnyk/ Masterfile; 714 (tr), R. Ian Lloyd/ Masterfile; 715 (tr), ©Bill Bachman; 716 (bl), Wes Thompson/Corbis Stock Market; 717 (tr), Robin Smith/SuperStock; 718 (tl), Jean-Paul Ferrero/Auscape International; 719 (cr), Steve Vidler/eStock Photo; 719 (t), Corbis Images/HRW; 722 (br), Brian Sytnyk/ Masterfile; 723 (bl), ZEFA GmbH/eStock Photo; 724 (tr), Macduff Everton; 725 (bl), Andris Apse/APSEA/Bruce Coleman, Inc.; 725 (br), Michael Steele/Allsport; 726 (bl), CORBIS/Paul A. Souders; 727 (tr), Picture Finders Ltd./eStock Photo. **Chapter Thirty-Two:** 730 (bl), ©2000 Zafer A. Kizilkaya; 730 (tl), CORBIS/Dennis Marsico; 730 (tr), CORBIS/Wolfgang Kaehler; 732 (b), CORBIS/ Australian Picture Library; 734 (tl), www.visibleearth.nasa.gov; 735 (b), Peter Hendrie/ Getty Images/The Image Bank; 735 (inset), Sylvaine Achernar/Getty Images/The Image Bank; 736 (b), The Art Archive/Eileen Tweedy; 737 (tr), Steve Vidler/eStock Photo; 738 (bl), Steve Thomas The Last Navigator; 739 (tr), CORBIS/ Wolfgang Kaehler; 741 (r), William Albert Allard/National Geographic Society Image Collection; 742 (bl), Steve Vidler/eStock Photo; 743 (tr), Carl N. McDaniel; 744 (inset), Mike Tinsley/Auscape International Pty Ltd; 744 (b), David Alan Harvey/National Geographic Society Image Collection; 745 (inset), Chris Johns/National Geographic Society Image Collection; 745 (b), Chris Johns/National Geographic Society Image Collection; 749 (tr), SuperStock.